# Human Factors Design Handbook

Information and Guidelines
for the Design of Systems,
Facilities, Equipment,
and Products for Human Use

## OTHER MCGRAW-HILL HANDBOOKS OF INTEREST

*American Institute of Physics* · American Institute of Physics Handbook
*American Society of Mechanical Engineers* · ASME Handbooks:
    Engineering Tables        Metals Engineering—Processes
    Metals Engineering—Design    Metals Properties
*Beeman* · Industrial Power Systems Handbook
*Brady and Clauser* · Materials Handbook
*Brater and King* · Handbook of Hydraulics
*Burington* · Handbook of Mathematical Tables and Formulas
*Burington and May* · Handbook of Probability and Statistics with Tables
*Callender* · Time-Saver Standards for Architectural Design Data
*Carrier Air Conditioning Company* · Handbook of Air Conditioning System Design
*Considine* · Chemical and Process Technology Encyclopedia
*Considine* · Energy Technology Handbook
*Considine* · Process Instruments and Controls Handbook
*Crocker and King* · Piping Handbook
*Crouse and Anglin* · Automotive Technician's Handbook
*Davis* · Handbook of Applied Hydraulics
*DeChiara and Callender* · Time-Saver Standards for Building Types
*Dudley* · Gear Handbook
*Fink* · Electronics Engineers' Handbook
*Fink and Beaty* · Standard Handbook for Electrical Engineers
*Gaylord and Gaylord* · Structural Engineering Handbook
*Harris* · Dictionary of Architecture and Construction
*Harris* · Historic Architecture Sourcebook
*Harris* · Handbook of Noise Control
*Harris and Crede* · Shock and Vibration Handbook
*Hicks* · Standard Handbook of Engineering Calculations
*Higgins and Morrow* · Maintenance Engineering Handbook
*Hopf* · Handbook of Building Security Planning and Design
*Juran* · Quality Control Handbook
*LeGrand* · The New American Machinists' Handbook
*Machol* · System Engineering Handbook
*Maynard* · Industrial Engineering Handbook
*Merritt* · Standard Handbook for Civil Engineers
*Optical Society of America* · Handbook of Optics
*Parmley* · Standard Handbook of Fastening and Joining
*Peckner* · Handbook of Stainless Steels
*Perry* · Chemical Engineers' Handbook
*Perry* · Engineering Manual
*Rohsenow and Hartnett* · Handbook of Heat Transfer
*Rothbart* · Mechanical Design and Systems Handbook
*Smeaton* · Motor Application and Maintenance Handbook
*Smeaton* · Switchgear and Control Handbook
*Society of Manufacturing Engineers:*
    Die Design Handbook        Handbook of Fixture Design
    Manufacturing Planning and    Tool and Manufacturing Engineers
      Estimating Handbook        Handbook
*Staniar* · Plant Engineering Handbook
*Woodson* · Human Factors Design Handbook

# Human Factors Design Handbook

■

Information and Guidelines
for the Design of Systems,
Facilities, Equipment,
and Products for Human Use

## Wesley E. Woodson
President, Man Factors, Inc.

**McGraw-Hill Book Company**

New York   St. Louis   San Francisco   Auckland   Bogotá   Hamburg
Johannesburg   London   Madrid   Mexico   Montreal   New Delhi   Panama
São Paulo   Singapore   Sydney   Tokyo   Toronto

**Library of Congress Cataloging in Publication Data**

Woodson, Wesley E
    Human factors design handbook.
    Bibliography: p.
    Includes index.
    1. Human engineering—Handbooks, manuals, etc.
I. Title.
TA166.W57  620.8′2  80-13299

ISBN 0-07-071765-6

4567890 KPKP 898765432

The editors for this book were Harold B. Crawford and Geraldine Fahey, the designer was Edward J. Fox, and the production supervisor was Teresa F. Leaden. It was set in Helvetica by Haddon Craftsmen.

Printed and bound by The Kingsport Press.

# Contents

*Note:* A more complete table of contents is provided at the beginning of each chapter.

Index follows Chapter 5

# Preface

Advances in technology have made it possible to make tremendous improvements in our daily living, working, and recreational situations and conditions, providing us with more efficient and comfortable places in which to work and live; improvements in the products, equipment, and tools we use so that less energy is required; and faster and more efficient transportation systems so that we can get to where we want to go in a minimum amount of time and with more assurance of getting there safely.

However, in spite of these potential advantages, many of the facilities, products, and devices that we use are often frustrating because of their complexities, inconveniences, and stress-producing demands and because of the difficulty of maintaining them at a level that will allow them to provide the advantages they were intended to provide.

Many of these problems occur because, in his or her enthusiasm to provide a new product and its proposed benefits, the designer has failed to appreciate the importance of the direct interface features of the product in terms of their compatibility with human capabilities and limitations. In some cases it would seem that the designer almost purposely made the user's task difficult—in a few cases appearing to invite the user to make errors. Although the latter is probably not the case (the designer undoubtedly felt that other matters had higher priority), the fault often is still one of design failure—because of either inadvertent oversight or purposeful compromise for technical reasons.

It has often been noted that poorly designed product-user interfaces result from lack of knowledge by the designer of human factors. This does not mean that designers are completely negative toward human factors, but rather that they evaluate and resolve human factors questions on the basis of personal feelings or experience. Unfortunately, this is seldom sufficient, and in some instances it may introduce personal biases that are completely contrary to fact. In addition, without a thorough appreciation of what human factors is all about, designers often attempt to define human factors problems only as they see them, and they may not recognize the real needs of the eventual product user in terms of making the product completely compatible with the user's basic sensory, motor, mental, and physical characteristics as a critical element in the final product-user system.

Human factors covers such a wide variety of special knowledge that no single human factors specialist knows all he or she needs to know about the human. It is therefore little wonder that designers cannot be expected to be completely informed on all human factors areas. Therefore, numerous attempts have been made to collate as much human factors information as possible into textbooks, guidebooks, and other human factors data references so that both the human factors specialist and engineers and designers can refer to these during their design efforts.

This handbook is yet another compilation of human factors information designed to be used by the designer as a ready reference guide. It is somewhat different

**Preface**

from many other human factors references in that it is directed specifically toward the engineer or designer, as opposed to being directed primarily toward the human factors specialist. In this respect, it addresses human factors problems as the designer typically views them. For example, the handbook is divided into chapters that are more or less related to the steps according to which any design program usually progresses, i.e., conceptual development at the overall system level, specification at the subsystem level, and detailed design at the component level. This was done because the human factors questions that should be addressed at each of these levels are different.

The last two chapters, however, are somewhat independent, since it is recognized that we cannot anticipate all the types of design situations a designer will face. Therefore, Chapter 4 provides an entirely different type of information, more related to the way in which a design problem should be approached, investigated, evaluated, or tested. This chapter will be useful in the event that the previous chapters have not provided the exact design guidance or background technical information that the designer may need.

Ultimately, everything that one designs impacts on the human in one way or another. Someone will have to fabricate the product, package it, distribute it, unpack it and prepare it for use, operate or use it, service and maintain it, and finally dispose of it. For this reason, designers should be constantly alert to the human factors implications of their proposed design for the various people who may come in contact with it.

This handbook obviously cannot cover all possible use situations of every possible or probable design. However, it can provide a wide variety of information that, it is hoped, will be useful in identifying and solving most of the human factors problems for most design programs. It will not, however, substitute for the use of common sense and judgment by the designer. By "common sense," we mean careful consideration of all the factors that may impinge on product effectiveness in the final operational sense. But in considering these factors, keep in mind that the ultimate success of a product depends on how well the user performs the tasks associated with it.

No single reference (not even this one) could cover all the questions a designer may have regarding the human factors associated with every design program. Therefore, there are frequent references to other material that will supplement the information presented in this volume. Readers are urged, therefore, not to rely completely on what is presented here, but also to avail themselves of the information provided by these other important sources.

It is hoped that this handbook will provide a basic reference source for a broad sample of engineers and designers in the various fields of architectural, equipment, consumer product, furnishing, instrument, tool, vehicle, and environmental systems and component design and that it will help them become more aware of the importance of human factors in product design and also assist them in solving specific product-user interface design problems.

*Wesley E. Woodson*

# Acknowledgments

One could not hope to acknowledge individually all the individuals who directly or indirectly have contributed to the fund of knowledge we normally attribute to human factors engineering. The information in this handbook, for instance, is the combined product of many professional disciplines, including engineering, design, medicine, physiology, anthropology, biology, personal hygiene, and psychology. The information is representative of many years of research and design practice. Therefore, as the author, I can pay tribute only to the dedicated concern and perseverance of the multitude of professional and technical specialists who, individually and collectively, have generated the bits and pieces of information that make up the content of this handbook. In addition, I would also like to pay tribute to the various special organizations that examine and generate guidance information related to the human-use aspects of design.

# Introduction

PURPOSE OF THIS HANDBOOK

The main purpose of this handbook is to provide a general reference to key human factors questions and human-product interface design suggestions in a form that engineers and designers can utilize with a minimum of searching or study. The handbook is not intended as a compendium of human factors data to be used by human factors or human engineering specialists (the latter have their own special references, which are noted from time to time throughout this volume).

INTENT OF HUMAN FACTORS ENGINEERING

A single, simple definition of human factors engineering is difficult (if not impossible) to come by, for human factors engineering implies and involves a variety of interests, intentions, and pursuits, depending on whom one talks with. For example, if you talk with a human factors researcher, human factors engineering generally is defined as the investigation of the input-output aspects of human-machine systems, with the primary objective of developing quantifiable measures of human performance under discretely specified conditions of the system environment. The researcher is interested in developing basic data that can be applied to many design situations. More applications-oriented human factors engineering specialists typically address themselves to interpreting the basic human factors research data as they apply to specific human-product interface situations. Unfortunately, In many cases this results in a specialist who is familiar with only a limited variety of system categories, such as aircraft flight control, communications systems display and control, or vehicular safety.

In spite of these seeming restrictions resulting from specialization, the intent of human factors engineering on the whole is to focus on and resolve human-product interface problems and solutions wherever or whatever they are. Philosophically, then, human factors engineering looks at a design from the standpoint of user efficiency, or total human-product output effectiveness. Inherent in this philosophy are the following objectives:

1. To make the user's contribution to product output as efficient as possible so that the basic product output is not compromised by human failures.
2. To make the combined user-product involvement as safe as possible so that neither human nor product failures will compromise the user's health or damage the hardware. Inherent in this objective also is the avoidance of injury to others and of damage to adjacent hardware.
3. To minimize the stress that a product imposes on the user as he or she uses, operates, services, or maintains it. This includes such stresses as an undue energy demand, frustration in trying to deal with the product at any point in the human-product interaction, and worry about whether one is using the product properly.
4. To maximize the acceptability of the product, not only in terms of its attrac-

## Introduction

tiveness (to encourage its purchase or maintain its satisfactoriness throughout ownership), but also in terms of giving users the feeling that the product allows them to use it efficiently and keep it in good working order with a minimum of effort.

A definition then might be as follows: *Human factors engineering is the practice of designing products so that the user can perform required use, operation, service, and supportive tasks with a minimum of stress and a maximum of efficiency.* To accomplish this, the designer must understand and acknowledge the needs, characteristics, capabilities, and limitations of the intended user and design "from the human—out." In other words, the designer should make the design "fit the user," as opposed to trying to make the user "fit the design."

### ERGONOMICS VERSUS HUMAN FACTORS ENGINEERING

It will be noted that the term "ergonomics" can be used as a subtitle for this handbook because "human factors engineering" and "ergonomics" have been used interchangeably in discussing the general topic of product design as it relates to user efficiency. In spite of some observers' attempts to define the two terms differently, in actual fact those who work under the guise of "human factors engineers" and those who call themselves "ergonomists" actually use the same basic information and perform the same kinds of work with respect to design. Their expertise and purposes are for all intents and purposes identical. The only difference one might perceive is that in the United States the term "human factors engineering" is used more widely, while in other countries the term "ergonomics" has been more predominant. It has often been noted, however, that the term "human factors engineering" is closely associated with U.S. military design programs, whereas in other countries "ergonomics" has been closely allied with industrial human factors work.

### HOW TO USE THIS HANDBOOK

Although it is expected that, after familiarizing themselves with the organization and content of this handbook, readers will develop their own personal methods of use, the material was organized purposely on the basis of the following usage philosophy:

1. Project managers and their immediate concept development staffs will utilize Chapter 1 as the primary entry point. That is, this chapter is oriented toward human factors questions relating to top-level design concept planning, in which it is extremely important to start out right in terms of appropriate decisions regarding how the human is to be used in the system. If these decisions are appropriate, many of the later human engineering requirements are more easily met. If the wrong decisions are made during these early stages, many constraints will occur that will make it necessary to compromise the human factors engineering solutions.

2. Subsystem designers are the target group in Chapter 2. In many instances, following top-level conceptual decisions, the various subsystem design problems are handed over to individual engineers or designers who expand the concept to a second, third, or lower level of detail, where the end result will be actual design specification. Once again, it is important that appropriate consideration be given to the human factors in order that further constraints will not be instituted to prevent detailed design from being optimized

as far as the operator, maintainer, or user is concerned. It will also be noted that the subsystems discussed in Chapter 2 often occur in a variety of general systems, although they obviously may be the primary system too. Thus, for instance, the subsystems dealing with the general problems of personal hygiene would be similar in the case of a bathroom for a home, hotel, or motel; a public rest room; and a lavatory for a commercial airliner, from the standpoint of a subsystem in one of these larger systems. On the other hand, manufacturers of modular bathrooms or lavatories would consider these a system in terms of their particular interest.

3. After system and/or subsystem decisions are made, someone has to be responsible for the detailing of hardware to the component level. Chapter 3 addresses human engineering requirements, criteria, and standards for this level of detail. Thus the information relates to components that may occur in several different systems and/or subsystems. For instance, a visual display may be an important component in command and control systems, in a kitchen or galley, or in an appliance. And seats, for example, occur in architectural settings, in vehicular settings, and in industrial production operations.

4. Although a great many topics are covered in Chapters 1 to 3, it can be expected that unique and unforeseen design situations will arise in which direct application of ready-made suggestions is not possible. Chapter 4 provides a number of nonspecific design-related data concerning human characteristics, capabilities, and limitations. It is hoped that readers can use these to help them resolve their more unusual and unique design problems. And, finally, it is highly probable that readers may not find anything in Chapters 1 to 4 that applies directly or that helps answer unique questions. Chapter 5 has been designed to provide readers with ideas on how they can generate certain kinds of human factors data themselves and thus develop a rationale for designing one way or another.

# Human Factors Design Handbook

Information and Guidelines
for the Design of Systems,
Facilities, Equipment,
and Products for Human Use

# 1

# Systems Conceptualization

## GENERAL SYSTEMS PRECEPTS

### Systems Concept

A "system" as used in this chapter (as opposed to "subsystem" or "component" as used in later chapters) refers to larger, more complex, mission-oriented groupings of elements into an integrated, functional whole. The system typically includes a physical facility, equipment, furnishings, and fixtures and involves a variety of people who use, operate, or maintain it.

### System Development

Although not all systems develop from scratch (i.e., some are merely updates or expansions of previous systems), the same general steps are usually taken:

1. Concept formation
2. Preliminary design and evaluation
3. Detailed design
4. Prototype development and test
5. Design modification (if required)
6. Production and delivery (may include preliminary field evaluation before full production)

### Human Factors in System Development

Human factors should be considered at the concept formation stage and at all the succeeding development stages. However, at the systems concept level, the principal considerations should include the following:

1. Deciding what roles humans are to be assigned in system operation, i.e., administrative management, operation, maintenance, and general use
2. Deciding where, when, and how humans will interact with subsystems and components, directly or indirectly
3. Deciding what has to be done to provide humans satisfactory living and working environments to ensure not only their safety but also their efficient performance and comfort
4. Determining what human constraints impact on the system design and eventual system performance and deciding how to ensure that humans will not become the weak link in the system

### Principal Objectives

The principal objectives should be to design a system that:

1. Is "adapted" to the human, as opposed to creating a system in which the human has to do all the adapting
2. Provides the human in the system the wherewithal to perform in the best manner of which he or she is capable
3. Does not subject the human to extreme physical or mental stress or to possible injury or death as a result of either some equipment malfunction or unpremeditated operator error
4. Provides personal satisfaction for the user in terms of both successful operation and pride of ownership

### Chapter Organization

The following design sheets provide general guidelines appropriate to all systems. They deal with such areas as defining the scope of the human factors problems and efforts, defining system mission requirements, determining where to use humans within the system, and organizing to accomplish human factors research and application tasks.

Following this general introductory material is specific information organized into four gross systems categories:

1. Architectural systems
2. Transportation systems
3. Military and space systems
4. Industrial systems

This is followed by three other brief sections dealing with agricultural, communications, and consumer products systems.

Within each of the four major systems categories are specific lesser systems categories typical of the most common problem areas in system design. Although these might appear to the reader to be subsystems within a larger system (e.g., a bathroom, under the heading of "architectural systems"), they are considered here to be of sufficient size and complexity to warrant the system connotation, and each requires slightly different considerations from the broad concept planning point of view. It is hoped that these gross systems categories will provide a familiar "entry heading" for those planners and designers who most often have the responsibility for overall conceptual design formulation and development. At least they will define the primary human factors areas of concern and encourage the planner or designer to plan ahead and avoid creating constraints that may be difficult to overcome at the subsystem and component design level.

## SCOPE

### Determining the Scope of Human Factors Concern in Systems Conceptualization

Avoid the temptation to limit attention to those human factors that pertain to the key operators or maintainers of new equipment. Although an emphasis on primary users may be important during the conceptual phase of system development, the ultimate success of the system operation invariably depends on the effective performance of all subsystems and components, new or old, including all the people who may be involved, directly or indirectly.

### 1. Potential People Categories That Should Be Considered

Managers and supervisors
Administrators and assistants
Equipment operators
Technical and maintenance personnel
Housekeepers and laborers
Customers and clients
Production personnel
Service personnel
Delivery personnel
Weapons operators
Communications personnel
Visitors and VIPs
First-aid and hygiene personnel
Counselors and training personnel
Security personnel
Disaster control personnel

### 2. Potential User Characteristics That Should Be Considered

Age
Sex
Cultural background
Training and experience
Size
Mobility
Dexterity
Coordination
Reaction time
Sensory response
Cognitive response
Motor response
Health and handicaps
Physiological tolerance
Motivation
Fatigue limits
Strength
Metabolic requirements
Adaptive limits
Equilibrium
Intelligence level

### 3. System Features That Should Be Considered

Procedures, instructions, manning, and training for the job, task, or activity
Information input (display)
Information output (control and speech)
Body control and support
Reach and clearance
Demands: speed, accuracy, duration, and strength
Environmental stresses and hazards: temperature, noise, and radiation
Task stresses: complexity and inconvenience
Operational stresses: combat, space, crash, and fire
Consumer acceptance: appearance

*Note:* The system user is concerned with operational effectiveness, not with the designer's problems of creating the system. In many cases the user does not have the time or inclination to compensate for design deficiencies; the designer does.

## ANALYSIS PROCEDURE

### General Procedure for Analyzing and Developing the Human Factors Requirements and Solutions in Major Systems Development

#### 1. Mission Analysis
All major systems concepts are developed around some stated mission. The proposed mission should be analyzed in terms of clarifying its purpose and objectives. These will be the underlying basis for all succeeding decisions regarding both the projected hardware and the facility and personnel requirements.

#### 2. Definition of Operational Requirements
Once the general mission purpose and objectives are firmed up and everyone agrees on

them, reasonably detailed operating requirements should be defined in order to clarify the demands that will be made on the eventual elements of the system. In the initial stages, these requirements may have to be of a fairly qualitative nature. But in the final stages, requirements should be refined to the point where quantitative values are established for each requirement. These requirements will be the basis for defining functions that have to be performed by physical elements, such as hardware, facilities, or software, and/or by operators, technicians, maintainers, or managers.

### 3. Definition of Operational Constraints

The general mission should be further studied, and the initial operating requirements should be reviewed with respect to the conditions under which operations must take place; the rules and regulations that the system must adhere to; and the physical, political, social, and economic factors to which the system must be responsive. These constraints, along with the basic operating requirements, provide the basis for defining the functions the system must include and for analyzing alternative means to accomplish the functions.

### 4. Function Identification

Functions (or processes) should be defined to fulfill each of the requirements identified above. They should be defined in such a way that there is minimal bias regarding how a function will be accomplished. That is, functions should not be described as either "machine" or "human"; rather, there should be independent process descriptions that allow one to consider alternative methods for accomplishing a given process. These descriptions will serve as the basis for examining alternative techniques for performing functions using objective evaluative criteria rather than subjective opinions (which are subject to argument).

### 5. Function Allocation

Once the various functions are clearly stated and agreed to by all those involved in system concept development and approval, alternative approaches to function accomplishment should be examined. As indicated above, objective evaluation criteria must be established against which to compare alternative function accomplishment methods, modes, or techniques. An important aspect of function allocation is the decision making regarding whether certain functions should be performed by humans rather than by some physical feature, electromechanical device, tool, or software procedure. While it is important during this process to have experts in electromechanical design, structures, materials, and other physical disciplines, it is equally important to have the participation of specialists in the physiological, biological, and psychological disciplines, since these people know more about the capabilities, limitations, and behavioral characteristics of humans as they may be considered to perform particular functions.

### 6. Definition and Analysis of the Human-System Interface

Once an initial decision has been made regarding which functions humans may be called upon to perform in system operation and use, a list of interface features should be developed in order to analyze the nature of the probable task the humans will be called upon to perform. That is, interface features imply human sensory, cognitive, and motor activity, and it is important to analyze these activities in terms of basic human capacity, temporal and accuracy limits and therefore establish a basis for defining interface design characteristics. All possible interface intersections should be listed, regardless of immediately apparent importance; i.e., not only should the typical control-display interfaces be identified, but also the host of other interfacing features such as furnishings, vehicles, tools, job aids, and life support and habitability elements required to complete the total human operating, living, and survival environment should be identified as well.

The final list developed by this analysis provides the basis for preparing preliminary system, subsystem, and component specifications. However, prior to finalizing these specifications, an investigation should be made of alternative state-of-the-art hardware (as well as other interface elements) to determine *(a)* what may already be available to perform the function (and meet human interface requirements), *(b)* what new devices will have to be developed and designed and/or what modifications of off-the-shelf hardware choices may be required, and *(c)* the requirements for interfacing new and old elements to satisfactorily complete the total system.

### 7. Preliminary Task Analysis

This is a step that, although typical in military systems development, is often neglected in domestic systems conceptualization. It is important, however, in that it provides the basis for adding appropriate human factors information and/or requirements in the primary hardware specification. Basically, task analysis should focus on creating preliminary descriptions of what humans will do in the system, how they will do it, and what the critical input-output characteristics are between human-machine and operating environment. These descriptions should be encoded in such a way that the earlier functional definitions show a one-to-one relationship with the hardware being used to perform a task. The descriptions should further identify at least the following:

*a.* The location of the task activity
*b.* The physical elements associated with the task (e.g., equipment, displays and controls, tools, documents, furnishings, communication devices, internal and external visual and auditory implications, and environmental factors, including ambient illumination, noise, vibration, and acceleration); and the people elements associated with the task (e.g., other operators)
*c.* Input-output requirements (the type and amount of information)
*d.* Time and accuracy requirements
*e.* Potential failure modes and effects (including effects on operator performance and potential hazards)
*f.* Implications regarding operator skill requirements

Various analysts have devised a number of analytic worksheets for accomplishing task analyses, and a few of these are provided as examples. However, the important thing is to develop one which "fits" the particular system problem and which provides maximum visibility of all the interrelationships that are foreseen and of all the possible problem areas that designers need to address when they go about selecting and/or developing hardware, facilities, and environmental control systems. In other words, the form is not so important as the assurance that all the related factors are being considered.

### 8. Preliminary Human-System Concept Evaluation

Before a system specification is finalized, it is often desirable to perform a limited concept evaluation. Although this might be done merely by reviewing all the documentation developed during the above steps, some more "visual evaluation" is sometimes helpful.

Typically this is done by most designers when they create artists' sketches, three-dimensional models, and/or preliminary full-scale mockups. Unfortunately, the objective of many of these devices is to "sell." Although this is evaluative in one sense, such devices are frequently not utilized for the more objective analysis of how well the proposed system will work. The principal purpose of the three-dimensional scale model or mockup should be to reexamine certain human-machine-facility interactions in terms of spatial association, activity flow, convenience, ingress and egress, and other factors associated with effective human performance, e.g., visual sight lines, possible noise problems, housekeeping or maintenance problems, and emergency evacuation problems.

Although the model is obviously a useful sales tool, this should not prevent use of the device as a definitive design evaluation tool by the engineering and design staff. Although this may appear obvious, many such devices (originally created by the design staff) have been taken away from the designer by marketing people before they have been fully exploited to evaluate the proposed system concept.

For certain systems, it may be desirable to create full-scale models to test geometric features prior to finalizing a particular subsystem element, e.g., an operator console, a driver station, or a jetliner buffet or lavatory. Such mockups provide an excellent chance to use the proposed configuration, with "live subjects" performing the expected tasks. Although more of this should be done later, during the predesign phase of system development, it may also be important even at the conceptual stage.

## SYSTEM MISSION

### Mission Requirements Identification Checklist

When and how will the mission begin and end?
Where will the mission take place?
How will the mission interact with other system missions?

| | PREPARATION | OPERATION | PARK/STORE | SERVICE/MAINTAIN |
|---|---|---|---|---|
| **EXAMPLE FUNCTIONS** | • PARCEL LOAD<br>• OCCUPANT INGRESS<br>• START ENG<br>• BACK OUT | • STEER, ACCELERATE, SHIFT, BRAKE, ADJ SYSTEMS<br>• DAY/NITE OPNS<br>• NEGOTIATE URBAN/ RURAL ROADWAYS/ TRAFFIC<br>• ADJ TO WEATHER/ TEMP | • STEER/BRAKE<br>• UNLOAD/EXIT<br>• LOCK | • REFUEL/OIL REPL<br>• LUBE<br>• TIRE CHANGE<br>• PARTS REMOVE/REPLACE<br>• REPAIR BODY/GASS<br>• CLEAN |
| **EXAMPLE CONSIDERATIONS** | DOOR CLEARANCE<br>SEAT ADJ<br>RESTRAINT CONVENIENCE<br>REAR VISION<br>PARCEL POSITIONING | DRIVER STATION CON-VENIENCE<br>EXTERIOR VIEW (FRONT, SIDE, REAR)<br>INTERIOR DELETHALIZED<br>STEERING QUALITY/BRAKE, CLUTCH, ACCEL FORCES<br>RIDE QUALITY VS VEH CONTROLLABILITY<br>CRASH SURVIVAL/RESCUE<br>INTERIOR HABITABILITY (VENT, TEMP, ILLUM) | EXT VEH BOUNDRY VISIBILITY<br>EGRESS/LOADING CONVENIENCE<br>PARKING BRAKE ACCESSIBILITY<br>LOCKS/HANDLES<br>IGN LOCK/SYSTEMS SHUT DOWN<br>TRUNK LID CONTACT HAZARDS | SERVICE ACCESS/HAZARD CONTROL<br>HOOD RELEASE ACCESS<br>JACKING CONVENIENCE/ SAFETY<br>DIAGNOSTIC SIMPLICITY<br>EXT SHEET METAL CONTACT HAZARDS |

Personal vehicle system mission profile.

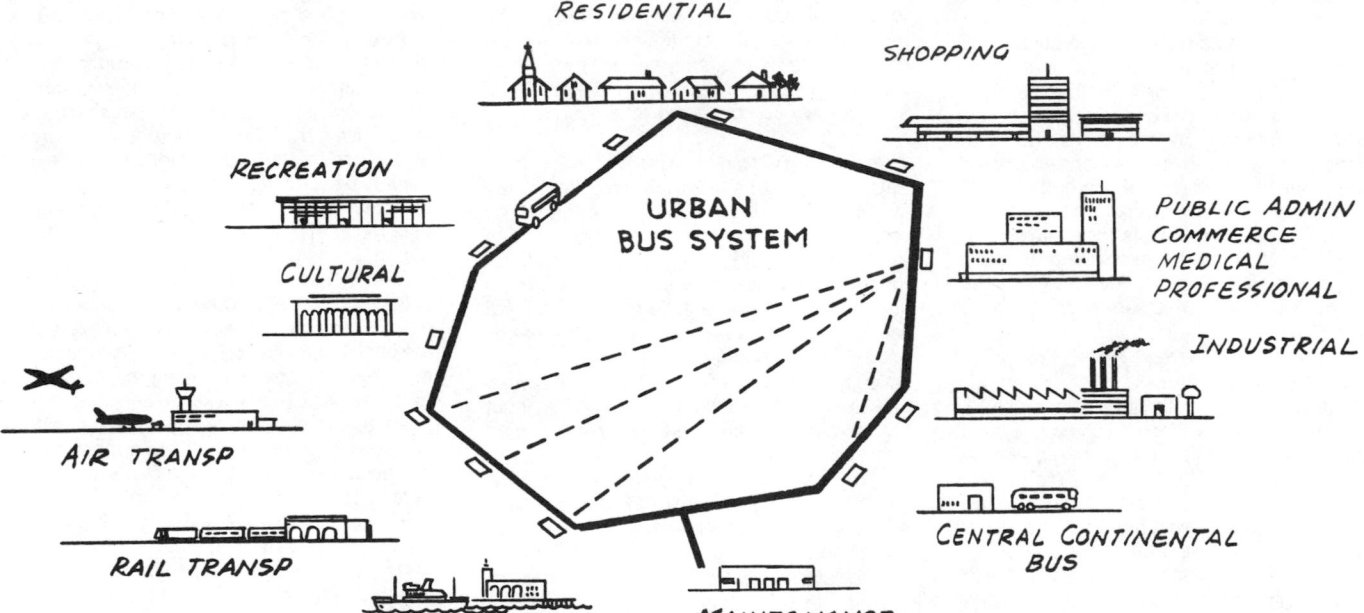

Pictorialization of an urban transit bus system mission.

What are the time constraints for the mission?

What are the environmental conditions under which the mission must operate?

What are the sociopolitical conditions or terms under which the mission must operate?

What are the cost constraints within which the system mission must be conceived and operate?

What are the personnel availability constraints within which the system or mission must operate?

What are the ecological constraints within which the system must operate?

What signals must be sensed, detected, and processed and what information must be transferred and stored to meet mission objectives?

What natural and man-made hazards may occur during the mission (e.g., extraterrestrial hazards, undersea hazards, earthquakes, floods, acceleration, explosions, fire, and shock)?

What physical movements of people, equipment, or materials will occur during the mission?

What probable human roles are directly or indirectly associated with the mission (e.g., management, operation, maintenance, and support)?

What current systems or subsystems, if any, are to be used as part of the proposed mission or system?[1]

---

[1]Answers to the above and other pertinent questions should be quantified wherever possible. Although some will require fairly blue-sky estimates at the beginning, reiterative review as other information is generated soon provides a solid basis for system specification. From the human factors viewpoint, these answers establish the basis for identifying critical and noncritical human-system interfaces and the need to perform a special human factors study, particularly if human failure and/or injury may occur during the mission.

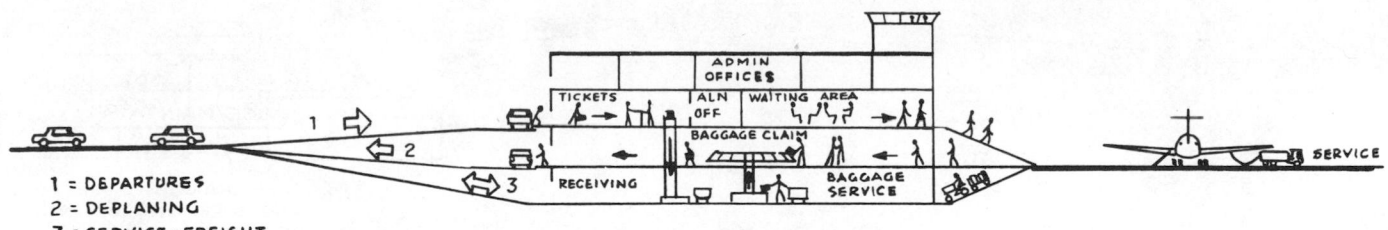

1 = DEPARTURES
2 = DEPLANING
3 = SERVICE · FREIGHT

Mission pictorialization: passenger, baggage, and cargo handling.

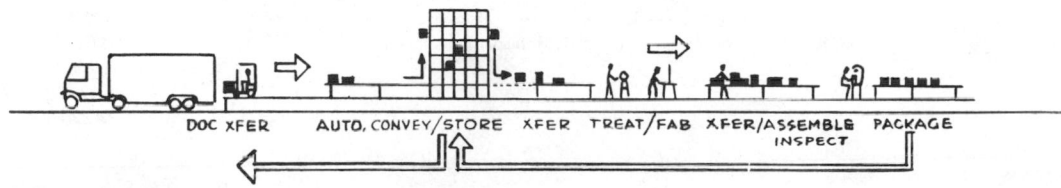

Mission pictorialization: manufacturing process.

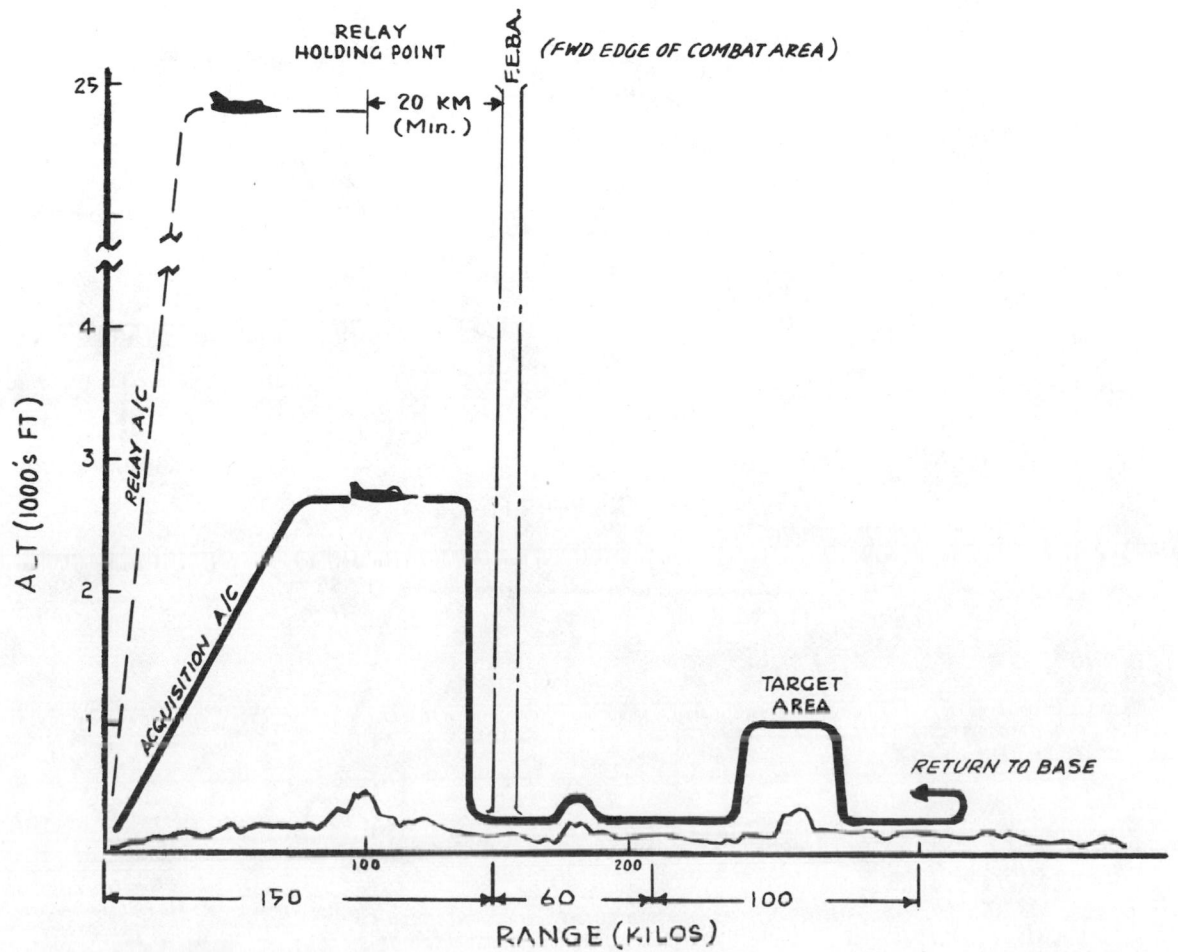

Low-altitude air reconnaisance mission profile.

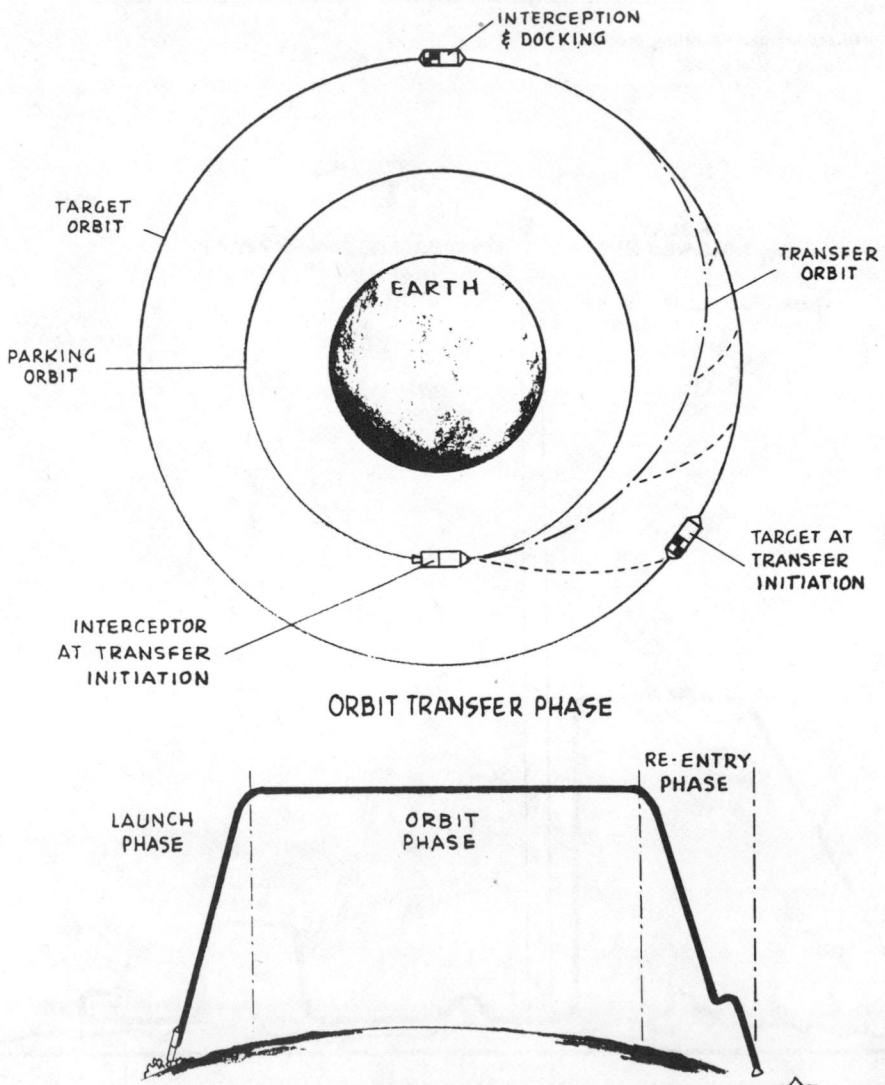

SURFACE
STATION

TPI 101-C
OR
ADSAF

|  | Sensor Aircraft | Relay Aircraft |
| --- | --- | --- |
| Mission radius | 20 to 150 IVM from Feba | 15 to 60 IVM from Base |
| Data taking time | 30 min maximum<br>10 min minimum | 30 min maximum<br>10 min minimum |
| Altitude | 500 to 15,000 ft | 5,000 to 40,000 ft |
| Velocity | Subsonic and supersonic | Subsonic |

**Air-ground data transmission system mission concept.**

INTERCEPTION
& DOCKING

TARGET
ORBIT

EARTH

TRANSFER
ORBIT

PARKING
ORBIT

TARGET AT
TRANSFER
INITIATION

INTERCEPTOR
AT TRANSFER
INITIATION

ORBIT TRANSFER PHASE

RE-ENTRY
PHASE

LAUNCH
PHASE

ORBIT
PHASE

| 10 MIN | 30 DAYS | 1 HOUR | 5 MIN |

MISSION PICTORIAL

**Manned space system concept.**

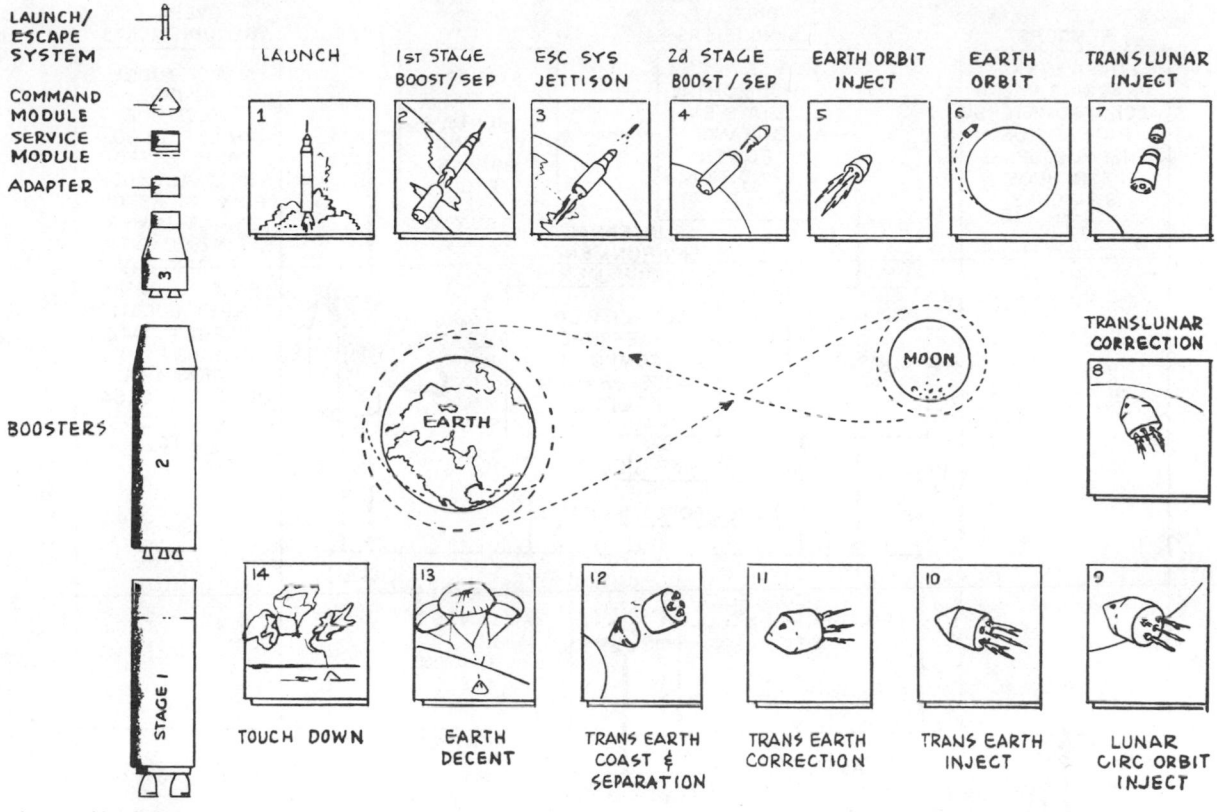

Lunar-orbit mission.

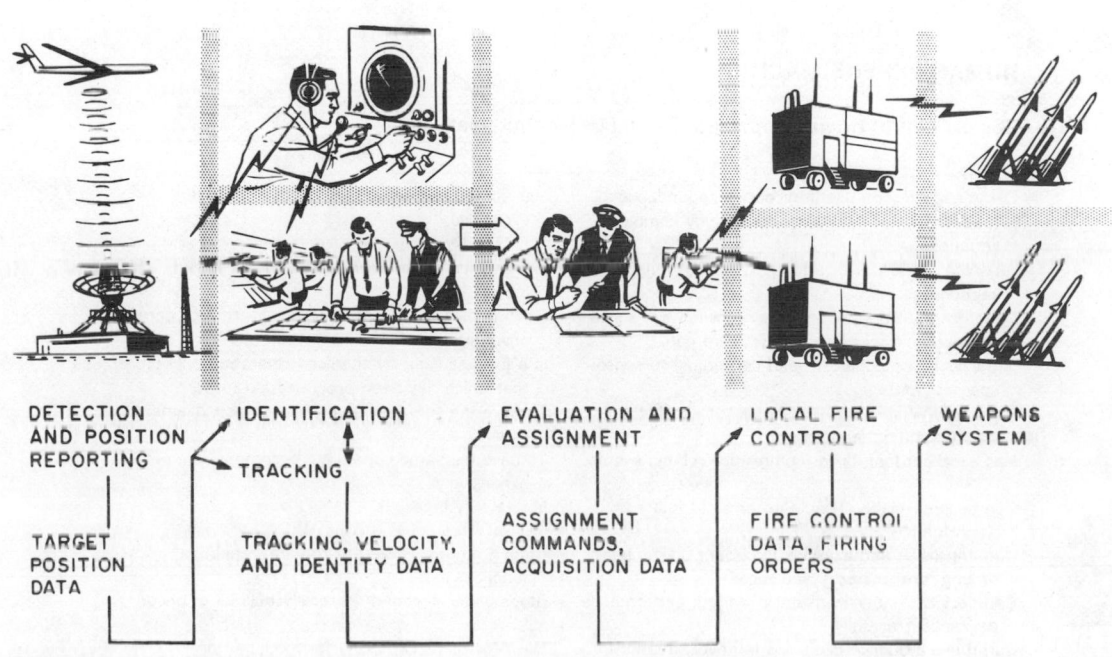

Military interdiction concept. (From Morgan et al., *Human Engineering to Equipment Design*, McGraw-Hill, p. 6.

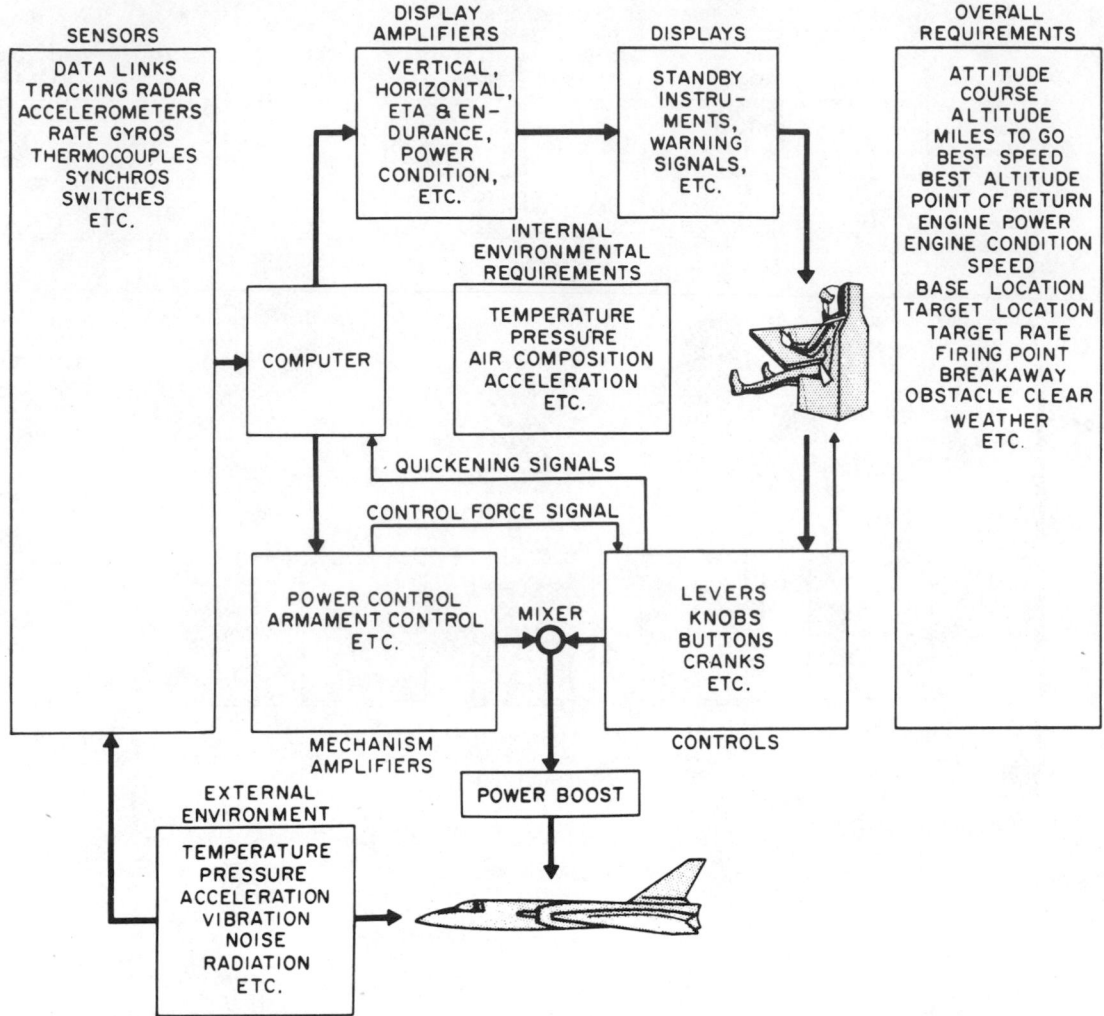

Aircraft concept. (From Morgan et al., *Human Engineering to Equipment Design*, McGraw-Hill, p. 4.)

## HUMAN VERSUS MACHINE

### Comparison of Human Capabilities with Machine Alternatives

| Human | Machine |
|---|---|
| Can recognize and use information redundance (pattern) in the real world to simplify complex situations. | Has limited perceptual constancy and is very expensive. |
| Has high tolerance for ambiguity, uncertainty, and vagueness. | Is highly limited by ambiguity and uncertainty in input. |
| Can interpret an input signal even when subject to distraction, high noise, or message gap. | Performs well only in a generally clean, noise-free environment. |
| Is a selecting mechanism and can adjust to sense specific inputs. | Is a fixed sensing mechanism, operating only on that which has been programmed for it. |
| Has very low absolute thresholds for sensing (e.g., vision, audition, and touch). | To have the same capability, becomes extremely expensive. |
| Has excellent long-term memory for related events. | To have the same capability, becomes extremely expensive. |
| Can become highly flexible in terms of task performance. | Is relatively inflexible. |
| Can improvise and exercise judgment on the basis of long-term memory and recall. | Cannot do this; is best at routine, repetitive functions. |
| Can perform under transient overload; performance degrades gracefully. | Stops under overload; generally fails all at once. |
| Can make inductive decisions in novel situations; can generalize. | Has little or no capability for induction or generalization. |

*(continued)*

## Comparison of Human Capabilities with Machine Alternatives *(continued)*

| Human | Machine |
|---|---|
| Can modify performance as a function of experience; can "learn to learn." | Is not characterized by trial-and-error behavior. |
| Can override own actions, should the need arise. | Can do only what it is built to do. |
| Is reasonably reliable; can add reliability to system performance by selection of alternatives. | Is reliable only at the expense of increased complexity and cost, and then only for routine functions. |
| Complements the machine, i.e., can use it in spite of design failures, can use it for a different task, or can use it more efficiently than it was designed to be used. | Has no such capability. |
| Complements the machine by aiding in sensing, extrapolating, decision making, goal setting, monitoring, and evaluating. | Has no capacity for performance different from what was originally designed. |
| Can acquire and report information incidental to the primary mission. | Cannot do this. |
| Can perform time-contingency analyses and predict events in unusual situations. | Does very poorly at this. |
| Is relatively inexpensive for corresponding complexity and is generally in good supply, but must be trained. | Is more limited in terms of complexity and supply by cost and time. |
| Is light in weight and small in size for function achieved for most situations. | To have functional equivalence of the human, requires more weight, power, and cooling facilities. |
| Is relatively easy to maintain; demands a minimum of "in-task" extras. | Maintenance problems become disproportionately serious as complexity increases. |

## Comparison of Human Limitations with Machine Alternatives

| Human | Machine |
|---|---|
| Is a poor monitor of infrequent events or of events which occur frequently over a long period of time. | Can be constructed to reliably detect infrequent events or events which occur frequently over a long period of time. |
| Has a limited channel capacity. | May have as much channel capacity as can be afforded. |
| Is subject to coriolis effects, motion sickness, disorientation, etc. | Is not subject to these effects. |
| Has extremely limited short-term memory for factual material. | May have as much short-term (buffer) memory as can be afforded. |
| Is not well suited to data coding, amplification, or transformation tasks. | Is well suited to these tasks. |
| Performance is degraded by fatigue and boredom. | Performance is degraded only by wearing out or by lack of calibration. |
| Performance is degraded by long duty periods, repetitive tasks, and cramped or unchanged positions. | Is less affected by long duty periods and performs repetitive tasks well; some may be restricted by position. |
| Becomes saturated quickly in terms of the number of things that can be done and the duration of the effort. | Can do one thing at a time so fast that it seems to do many things at once, for a long period of time. |
| May introduce errors by misidentification, reintegration, or closure. | Utilizes these processes. |
| Expectation or cognitive set may lead an operator to see what he or she expects or wants to see. | Does not exercise these processes. |
| Much human mobility is predicated and based on gravity relationships. | May be built to perform independently of gravity. |
| Is adversely affected by high $g$ forces. | Is unaffected by $g$ forces. |
| Can generate only relatively small forces and cannot exert large forces for very long or very smoothly. | Can generate and exert forces as needed. |
| Generally requires a review or rehearsal period before making decisions based on items in memory. | Goes directly to stored information for a decision. |
| When performing a tracking task, requires frequent reprogramming; does best when changes are under 3 rad/s. | Has no such limitations. |
| Has a built-in response latency of about 200 $\mu$s in a go–no-go situation. | Has no response latency. |

*(continued)*

**Comparison of Human Limitations with Machine Alternatives** *(continued)*

| Human | Machine |
|---|---|
| Is not well adapted to high-speed, accurate search of large volumes of information. | Computers are designed to do just this. |
| Does not always follow an optimum strategy. | Will always follow the strategy designed into it. |
| Has physiological, psychological, and ecological needs. | Has only ecological needs. |
| Is subject to anxiety, which may affect performance efficiency. | Is not subject to this factor. |
| Is dependent upon the social environment, both present and remembered. | Has no social environment. |
| Diurnal cycle imposes cyclic degradation of behavior. | The machine cycle may be whatever is desired. |
| Is subject to stress as a result of interpersonal problems. | Is not affected by such problems. |
| Great differences exist among unselected individuals. | There are no unselected machines. |

## Typical Functions Assigned Either to Man or Machine[2]

### Some Comparisons of the Limitations of Machine and Human Functions, in Relative Terms

| Conditions for function | Limitations | |
|---|---|---|
| | Human | Machine |
| Sensing Display | Limited to certain ranges of energy change affecting human senses | Range extends far beyond human senses (X-rays, infrared, etc.) |
| | Sensitivity: very good | Sensitivity: excellent |
| Filtering | Easy to reprogram | Difficult to reprogram |
| Identifying Display | Can be varied over relatively wide range of physical dimensions | Can be varied only in very narrow range of physical dimensions |
| | Channel capacity: small | Channel capacity: large |
| Filtering | Easy to reprogram | Difficult to reprogram |
| Memory | Limits to complexity of models probably fairly high, but not precisely known | Potential limits of complexity very high |
| | Limits to length of sequential routines fairly high, but time-consuming to train | Potential limits of length of routines very high |
| Interpreting Display | Same as Identifying | Same as Identifying |
| Filtering | Easy to reprogram | Difficult to reprogram |
| | Highly flexible, i.e., adaptable | Relatively inflexible |
| | May be reprogrammed by self-instruction following input changes contingent on previous response (dynamic decision making) | |
| Shunting | Can be readily reprogrammed to lower levels of functioning | Difficult to reprogram |
| Memory | Limitation to rule storage not known | Limits of rule storage ("logic") quite high |
| | Speed of reinstatement of rule sequences relatively low (as in computing) | Speed of using rule sequences high (e.g., computing) |
| | Use of novel rules possible (inventing) | Limited in use of novel rules |

[2]Gagne, R. M., *Psychological Principles in System Development,* Holt, Rinehart & Winston, New York, 1962.

## TASK DEFINITION

### Suggested Steps for Developing Preliminary Task Descriptions during System Conceptualization

Although no two systems will be exactly alike and although the steps may not always be the same, the following list of steps may be useful in generating task descriptions to the level necessary to evaluate conceptual design alternatives:

Analyze initial mission descriptions, purpose, goals, and operational requirements to try to isolate obvious potential human activities. Start by listing mission events in order of their occurrence, from the time the mission begins to the time it is concluded. Although primary interest should be in those activities dealing with the main operating events, consider also all the supporting events and activities.

Create a graphic representation of the sequence of events showing key human and hardware elements, flow of information, materials, mobile components, and people. Sometimes this is referred to as a "pictogram" or "storyboard" method of activity portrayal. At any rate, the purpose should be to create a pictorial scenario of what is expected to happen during the proposed system operation. The objective should be to make it clear (without verbal explanation) what the system does and how it does this via hardware elements, human elements, and even "unseen elements," e.g., telecommunications, paper transfer, and direct verbal communications.

Analyze each of the principal activity clusters and define the tasks involved in each activity, e.g., monitoring a display, operating a control, and writing or inputting an instruction.

Create a graphic representation of each activity or task event, first to show information flow and then to show operator and equipment tasks. Information flow charts usually begin with a simple graphic showing operation and decision points—how they probably are linked sequentially and with feedback loops. Later these should be refined to differentiate between operations and decisions that are expected to be performed by humans and those which are to be performed by machines. Reference to human versus machine capabilities should be made as these decisions are being formulated.

Next, create an operational sequence diagram that will show the probable mechanical or manual links between human and hardware elements, the visual and auditory links between human and hardware elements and human and human elements, and any other electrical or physical links between hardware elements.

Prepare a general task description and equipment requirement statement for each activity cluster and create a preliminary operator position identification, e.g., supervisor, radar operator, and vehicle driver.

### Preliminary Information-Flow Diagraming Example

The scenario for information flow should first be developed independently of whether the decision and operation points should be via human or machine.

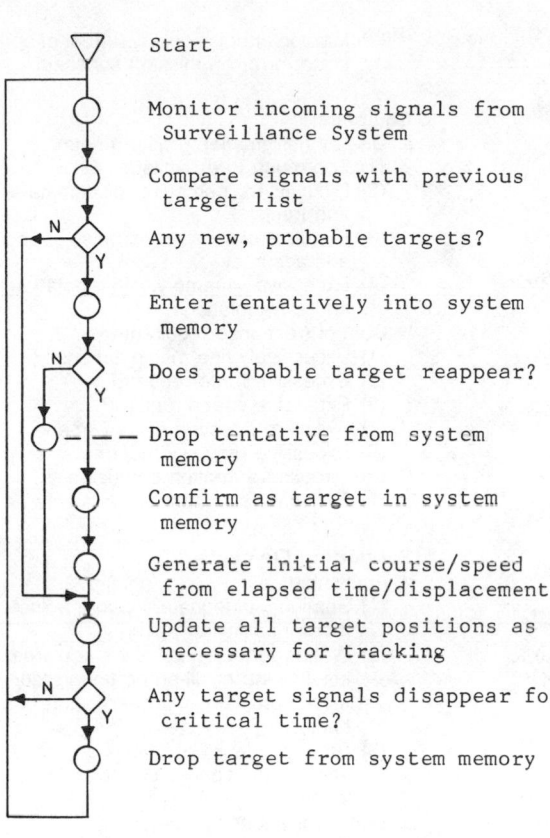

Start

Monitor incoming signals from
 Surveillance System

Compare signals with previous
 target list

Any new, probable targets?

Enter tentatively into system
 memory

Does probable target reappear?

Drop tentative from system
 memory

Confirm as target in system
 memory

Generate initial course/speed
 from elapsed time/displacement

Update all target positions as
 necessary for tracking

Any target signals disappear for
 critical time?

Drop target from system memory

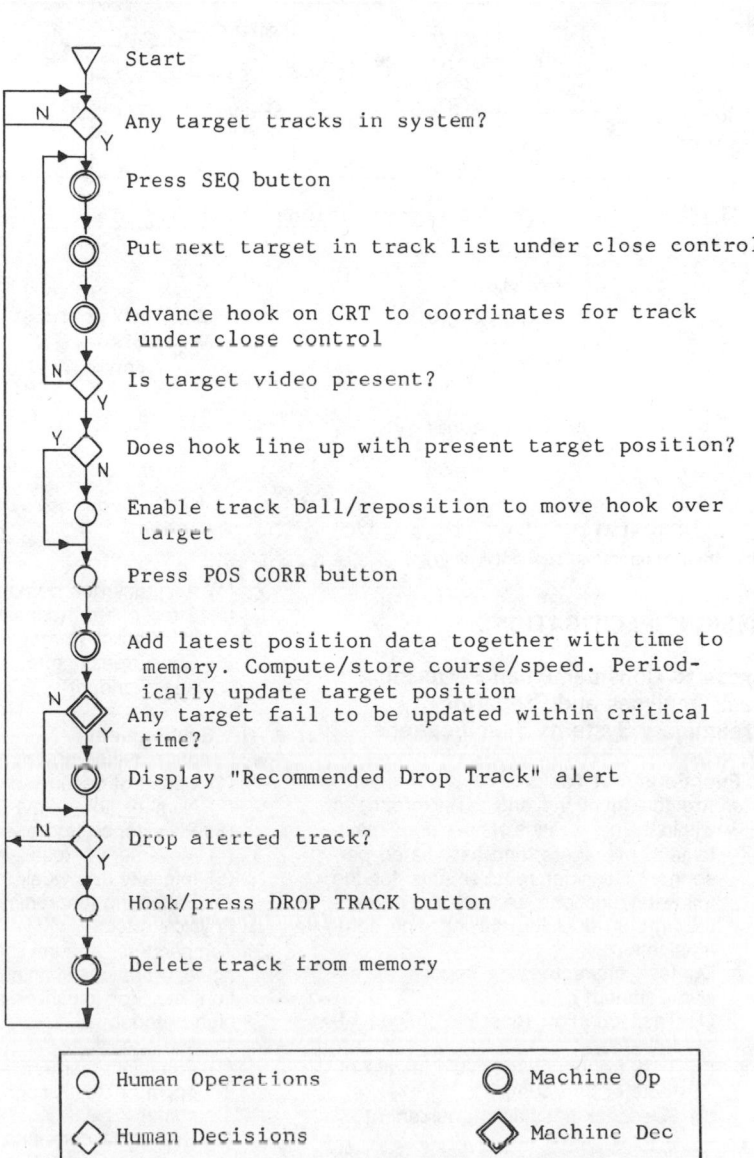

Start

Any target tracks in system?

Press SEQ button

Put next target in track list under close control

Advance hook on CRT to coordinates for track
 under close control

Is target video present?

Does hook line up with present target position?

Enable track ball/reposition to move hook over
 target

Press POS CORR button

Add latest position data together with time to
 memory. Compute/store course/speed. Periodically update target position

Any target fail to be updated within critical
 time?

Display "Recommended Drop Track" alert

Drop alerted track?

Hook/press DROP TRACK button

Delete track from memory

| ⭘ Human Operations | ◎ Machine Op |
|---|---|
| ⬦ Human Decisions | ⬥ Machine Dec |

Refined information flow after human and machine assignments are made.

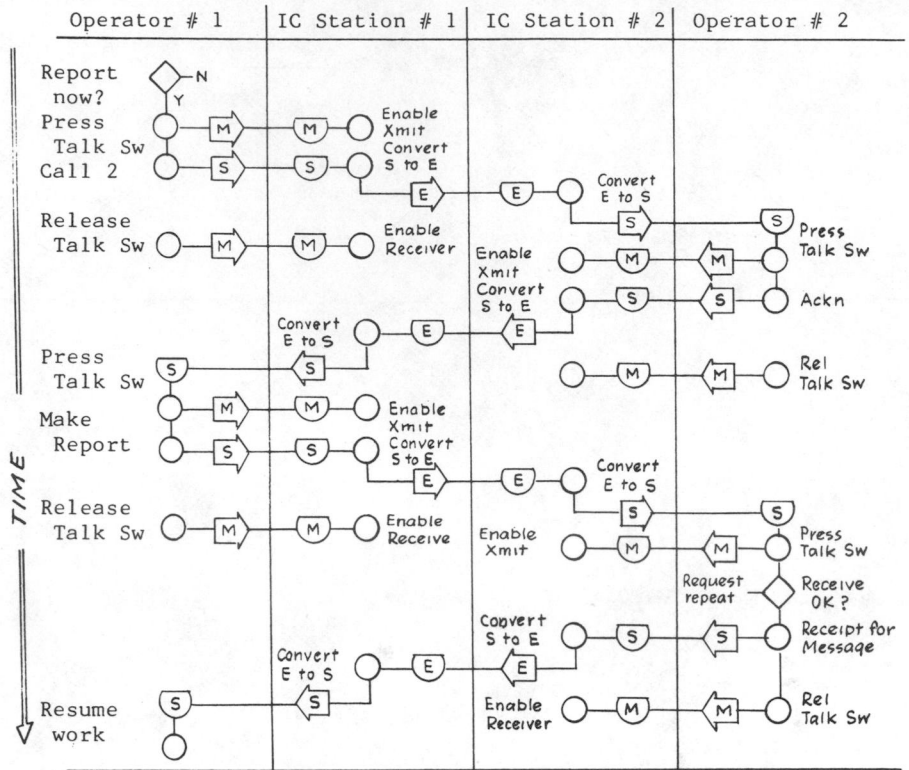

| Operator # 1 | IC Station # 1 | IC Station # 2 | Operator # 2 |

Report now?
Press Talk Sw Call 2
Release Talk Sw
Press Talk Sw
Make Report
Release Talk Sw
Resume work

Enable Xmit Convert S to E
Enable Receiver
Convert E to S
Enable Xmit Convert S to E
Enable Receive
Convert E to S

Convert E to S
Enable Xmit Convert S to E
Enable Xmit
Convert S to E
Enable Receiver

Convert E to S
Press Talk Sw
Ackn
Rel Talk Sw
Press Talk Sw
Request repeat
Receive OK?
Receipt for Message
Rel Talk Sw

TIME

**Symbols**

◇ Decision
○ Operation
⬠ Transmission
⬡ Receipt
⬭ Delay
▢ Inspect/Monitor
▽ Store

**Notes on Operational Sequence Diagram**

**Links**

M   Mechanical or Manual
E   Electrical
V   Visual
S   Sound

Station or subsystems are shown by columns. Sequential time progresses down the page.

**Hypothetical operation sequence diagram.**

## DESIGN SPECIFICATIONS

### Areas to Consider When Performing Task Analyses and Preparing Preliminary Systems Specifications[3]

**1. Functions and Tasks**
a. Are the functions and tasks proposed within the capabilities of operators, maintainers, managers, and associated personnel? Consider requirements for the following functions: sensory and perceptual, motor, decision making, and communication.
b. Do task characteristics impose excessive demands?
   (1) Task duration (possible fatigue effects)
   (2) Task frequency (possible fatigue effects or boredom)
   (3) Feedback (insufficient guidance)

[3]Adapted from David Meister, *Human Factors: Theory and Practice*, Interscience Publishers, a division of John Wiley & Sons, Inc., New York, 1971.

   (4) Accuracy (too demanding)
   (5) Effect of error (consequence criticality or ability to recover)
   (6) Concurrency effects (interference and overload)

**2. The Environment**
a. Events requiring operator response
   (1) Speed of occurrence (too fast)
   (2) Number (too many)
   (3) Persistence (too brief)
   (4) Movement (excessive)
   (5) Intensity (too weak to perceive)
   (6) Patterning (unpredictable)
b. Physical effects
   Temperature, humidity, ventilation, noise, vibration, illumination, acceleration, etc. (too much or too little or too high or too low)
c. Mission conditions
   (1) Potential emergencies (ability of the operator to recognize and overcome)
   (2) Response demands (ability of the operator to meet speed and accuracy demands)

   (3) Mission effect criticality (effect of operator error on mission success)

**3. Equipment**
a. Display (information requirements)
   (1) Too much to assimilate
   (2) Difficult to perceive, discriminate, and track
   (3) Requires excessive response speed and accuracy
   (4) Excessive memory, interpolation, and extrapolation
b. Control (response requirements)
   (1) Excessively fine manipulations
   (2) Excessive force required
   (3) Excessive speed required
   (4) Awkward direction of movement
   (5) Excessive range of motion
   (6) Direction-of-motion confusion
   (7) Too many simultaneous movements

**4. Supporting Elements**
a. Furnishings
   (1) Seating (inadequate support, adjustment, mobility, or security)
   (2) Writing surface (lack of surface area, insufficient or ill-proportioned surface area, or improper angle or height)
   (3) Storage (lack of storage space, inadequate space, or inconvenient location)
b. Communications
   (1) Telephone (inadequate number of internal and external connections, telephone sets, and terminal controls)
   (2) Intercom (inappropriate use in terms of contribution to general noise level or because of it)
   (3) Public address (feedback problems and inappropriate use in terms of contribution to general noise level or because of it)
   (4) Radio (inadequate fidelity and volume-level control)
   (5) Signs and signals (inadequate conspicuousness, visibility, or legibility)
   (6) Message delivery (inadequate collection facilities or distribution system)
c. General illumination, heating, and air conditioning
   (1) General illumination (day and night)
   (2) Supplementary illumination (at specific work stations)
   (3) Emergency lights (in the event of power failure)
   (4) Natural ventilation (air circulation)
   (5) Mechanical heating, cooling, and ventilation (control within comfort range)
d. Personnel safety
   (1) Restraint systems
   (2) Safety guards
   (3) Personal protective equipment and garments
   (4) Fire extinguishers (general and local)
   (5) First aid (stations and special fixtures)
   (6) Materials and substances (combustion and toxicity)
   (7) Survival equipment

## SYSTEM CLASSIFICATIONS[4]

### The Designing of Man-Machine Systems

| Kind of System and Its Mode of Operation | Components | Couplings between Components | Examples |
|---|---|---|---|
| 1. Manual system—Operator directed, flexible | hand tools or aids | one human operator | cook plus utensils, craftsman plus tools, singer plus amplifying equipment |
| 2. Mechanized system—System directed, rigid | powered mechanical subsystems | on-line human operators, tracks, conduits, etc. | railway system, assembly line |
| 3. Automatic system—Preset, programmed, or adaptive | powered mechanical subsystems | cables, pipes, conduits, levers, etc., forming a control circuit | clock, process plant, telephone exchange, digital computer |
| 4. Collaborative human machine system—Exploratory and flexible | one or more human operators, one or more complete automatic systems | complex displays and controls | multiple-access computers, *Sketchpad* |
| 5. Mechanical subsystem—Operator controlled and inflexible | highly interdependent physical parts forming indistinguishable components and couplings | | engine, automobile, machine tool |
| 6. Administrative system—Goal directed and hierarchical | human operatives with tools or aids | rules, messages, human administrators, and informal contacts | army of foot soldiers, a business, a school |
| 7. Voluntary system—Self-rewarding and collaborative | any number of persons each of whom is also a biological system and some of whom also act as administrative subsystems | affection, shared aims, laws, customs, managers, physical presence, mutual aid, common language, ancestry, etc. | family, religious order, club, society, (university) |
| 8. Environmental system—Permissive of a range of human activities and contacts: prohibitive of others | inhabitants and facilities within an environment, the outside world | spaces and the barriers between and around the components | occupied building, city, or region |
| 9. Biological system—Homeostatic, adaptive, evolutionary, growing, differentiating, and self-reproducing | cells, organs, subsystems, all of which are also physical systems | nerves, glands, chromosomes, etc., past experience and environment | cells, plants, animals, "human operators" |
| 10. Physical system—Dynamically stable but subject to eventual decay | elementary particles, planets, seas, land, etc. | gravitation, electrical forces, radiation, physical motions, and forces | solar system, molecule, crystal, cloud, strut, tie, beam, shell |
| 11. Symbol system—Semantic, analogous, ambiguous, or precise | words, signs, symbols, numbers, etc. | syntactical rules | languages, mathematics, codes, etc. |

Note: Systems classified according to mode of operation and the physical nature of their components and couplings.

[4] J. C. Jones, "The Designing of Man-Machine Systems," in Singleton et al.

## PROGRAM ORGANIZATION

### Organizing for Human Factors Participation in System Development[5]

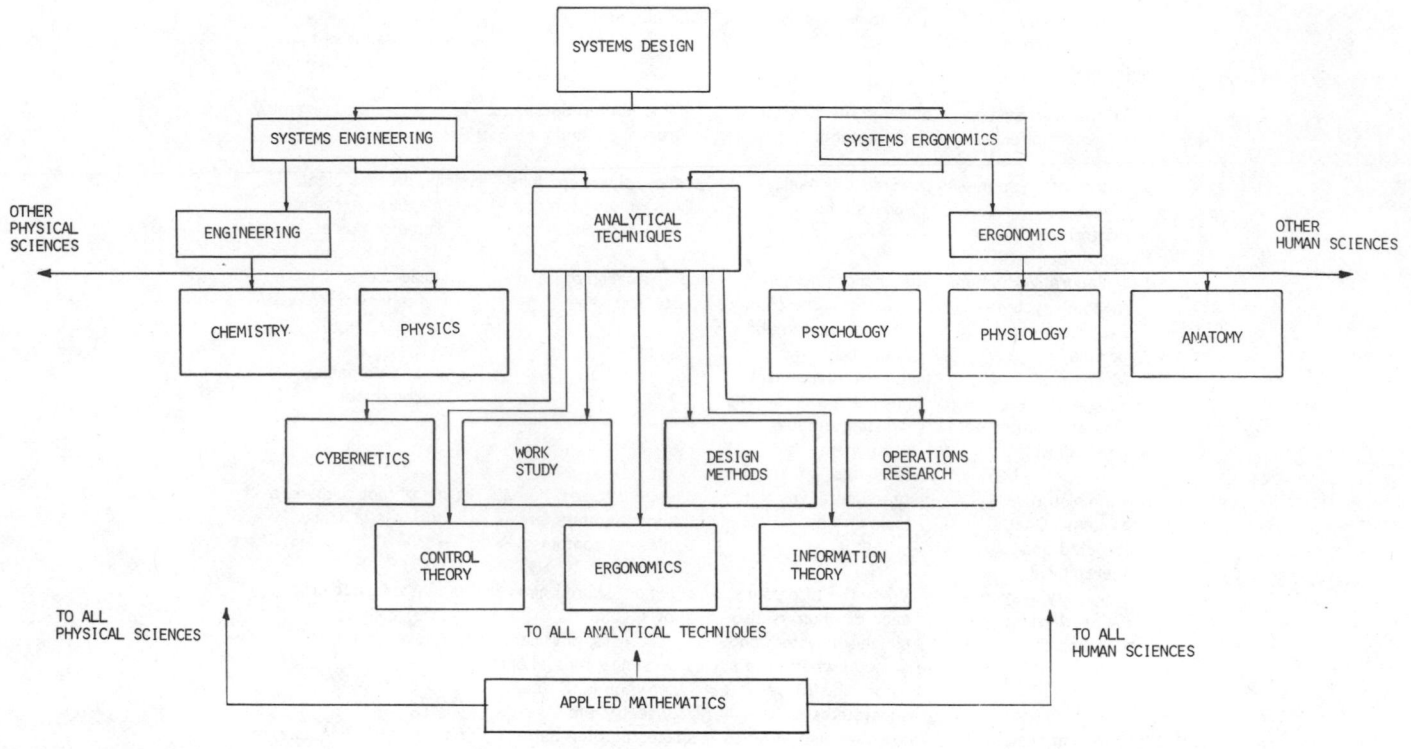

*"Ergonomics," a term used by Europeans and others, is synonomous with "human factors" as used in the United States.

## PROJECT STAFFING

### Expertise Required for Human Factors Effort during System Conceptualization, Design, Development, and Test

Most system development projects proceed systematically through the following phases: concept formulation, preliminary design, detailed design and development, production, and test and evaluation. The following areas of expertise are typically required during various system development phases.

### 1. Basic System Experience
A variety of people specialize in human factors research and applied human engineering. They have various educational backgrounds, e.g., anthropometry, medicine, physiology, and psychology. Some are more research-oriented, and others are more design-oriented. For total system development, it is wise to pick someone who is more or less a "generalist," i.e., someone who is not biased toward one discipline (e.g., anthropometry or medicine) or toward one aspect of the system (e.g., life support, controls and displays, or training). Finally,

[5]W. T. Singleton et al. (eds.), *The Human Operator in Complex Systems*, Taylor & Francis Ltd., London, 1967.

pick someone who has had total system planning and development experience, preferably with systems akin to the one being planned. This type of person understands both the human factors requirements and the problems of planning, designing, and producing systems within the typical industrial constraints of time and cost.

### 2. Specialty Experience
Add to the human factors staff and/or employ consultants when it can be determined that the proposed system may have some unique human factors problems. For example:

Space projects require special medical and physiological expertise.
Complex information-processing systems projects require perceptual-motor psychologists.
Systems that must "fit" a wide range of users of different sizes, with different biomechanical capabilities, etc., require anthropometric expertise.

### 3. Research Experience
Add to the human factors staff and/or hire a consultant or human factors subcontractor when it can be determined that the proposed system development may require special scientific study or experimentation. For example:

A behavioral scientist for sensory, motor, and cognitive studies
An anthropometrist for dimensional surveys of the human body
A training psychologist for skill development studies

## STANDARDS

It is generally recognized that the humans in any system can perform better when the interfaces between them and the system are somewhat consistent (or standardized) relative to other, similar systems in which the same individuals may become involved. Although human factors standards have been specially created for certain systems (e.g., the military), there has not yet been similar standardization in other types of systems. The planner or designer can approach the objective of complete standardization to some extent by using various professional engineering and government regulatory standards (although it must be recognized that there will be frequent discrepancies between these in the human-system interface design area). A few of the more useful standards that should be reviewed during system concept formulation are the following:

American National Standards Institute standards: Industry standards developed by the American National Standards Institute, 1430 Broadway, New York, N.Y. 10018. These are often referred to as "ANSI standards."

SAE recommended practices: Society of Automotive Engineers, Inc., 400 Commonwealth Drive, Warrendale, Pa. 15096. These standards relate to automotive, industrial, and similar machine systems.

OSHA standards: Health and safety standards for industry developed by the U.S. Department of Labor, Occupational Safety and Health Administration, Washington, D.C. 20001.

FMVSS standards: Safety standards for transportation systems developed by the U.S. Department of Transportation.

ISO standards: Various standards developed by the International Organization for Standardization are available from the American National Standards Institute (see above).

Other suggested standards are noted under various topical headings in this chapter.

## GENERAL

This section deals with human factors considerations that should be addressed during the initial conceptualization of any architectural system, including the community plan.

The section begins with general guidelines that are applicable to all systems. These are followed by discussions of more specific systems: a new community, individual family residences, office buildings, and so on.

Although, for practical reasons, not all the possible architectural systems have been addressed, it is hoped that sufficient related information is included so that readers can extrapolate to some special system they have responsibility for conceptualizing.

### User-Oriented Conceptual Planning

Start with the user. Recognize his or her characteristics and constraints. Determine the user's needs. Create a place for the user to perform whatever tasks he or she expects to do.

This approach differs from some traditional ones which, from a human factors point of view, should be avoided; i.e., according to these approaches, the designer should start by creating:

A design that is interesting to look at
A design that reflects his or her ingenuity with respect to the use of special materials and techniques
The most profitable design

Although these factors are, of course, important, do not allow them to get in the way of properly planning to meet user needs. The following steps are suggested in conceptualizing an architectural system "from the human—out":

Step 1: Define and examine the needs of the total user population; i.e., do not concentrate only on the primary resident, for example, but look at the needs of his or her visitors or clients and the people who will serve the primary resident in the proposed facility.
Step 2: Examine and define the various tasks that each of the above users has to perform. Determine what these tasks imply in terms of space, environmental control, supporting furnishings, and utilities.
Step 3: Explore the interactive as well as the isolative needs of the various users and their furnishings and equipment. Examine alternative arrangements to determine the most convenient organization of people, furnishings, spaces, buildings, etc.
Step 4: Create an enclosure for the most effective alternative defined in Step 3 and add appropriate partitioning to provide desired environmental control, privacy, and security.
Step 5: Select an appropriate site that will accommodate the building or buildings defined in Step 4 and locate, position, and arrange the building or buildings with respect to appropriate site and building access.

Now you are ready to examine the concept in terms of aesthetic features, including architectural style, special material effects, and landscaping. These features are generally

### User Population Characteristics

| Characteristic | Architectural Implication |
|---|---|
| Cultural factors | Considerable variation exists among people with respect to their cultural background, including social mores, religious attitudes, intellectual development, skill development, attitudes toward others, and where and how they live in terms of spatial features and modern technological amenities. Language differences create an important barrier to communications in many system operational settings. |
| Body size | People of different nationalities, as well as individuals of the same nationality, vary considerably in terms of size. There are also differences in size between children and adults, between men and women, and between members of special user populations. Differences in size impact on architectural space, including clearances and reach distances. |
| Mobility | The agility of various individuals varies considerably (e.g., between the young and the old and between handicapped and nonhandicapped persons), and mobility may be restricted by the garments people wear. The impact of restricted mobility on human-architectural interfaces may be critical to the operational utility of a system concept. |
| Strength | Very young and very old people have considerably less strength than those in the middle range, women are generally weaker than men, and handicapped persons may have virtually no strength. Architectural features that require lifting, pushing, pulling, or twisting must be tailored to the weakest member of the expected user population. |
| Sensory factors | Principal sensory factors associated with architectural systems relate to vision, hearing, and touch. Although only persons with so-called normal capacities may make up the expected user populations of special systems (because of operator selection restrictions), most general system concepts require consideration of the more limited capacities of elderly and handicapped individuals, especially the visually and aurally handicapped. |
| Motor skills | A limited number of people have superior motor skill capabilities as a result of either innate capability or training. Others are limited both innately and by lack of training. Still others are even more limited by physical handicaps. |
| Cognitive skills | Variation in cognitive skill occurs because of age differences, differences in education and/or technological opportunity, and innate mental handicaps. Understanding the operational aspects of the proposed architectural concept is critical to its effective use. |

cosmetic, and although they are important in terms of making the system pleasant to look at, they can for the most part wait until the five steps outlined above have been completed.

### Defining the Needs, Characteristics, and Limitations of the Expected User Population

A clear definition of the needs, characteristics, and limitations of the expected user population should be sought before any architectural concept is frozen. Otherwise, the system may not be completely compatible with user expectations or abilities to use the system effectively. Because people are different, it is a mistake to assume that a system can be designed for the so-called average person. Understanding these differences and accommodating the proposed concept to them are vital to the eventual operation of the system. The accompanying table lists the important characteristics that should be addressed in defining the user population.[6]

[6]It is important to note that, because of the differences noted above, a given architectural concept may have to be modified to fit each user population. This is particularly important where, as a result of lack of understanding, the eventual users of the system could misuse it and cause it to fail, possibly resulting in injury to themselves or others. For further information regarding population differences, see the reference cited below. A. Chapanis (ed.), *Ethnic Varibles in Human Factors Engineering*, The Johns Hopkins Press, Baltimore 1975.

### User Efficiency

It is often said that "user efficiency does not sell products—appearance does." From a human factors point of view, however, efficiency is of prime importance to the eventual effectiveness of any system. The table on page 19 provides suggestions regarding user efficiency variables that should be considered carefully during the conceptual phase of any architectural system development.

### Handicapped Users

Special consideration should be given to the needs of the handicapped when it is obvious that they too can be expected to utilize a proposed architectural system. The following description of typical handicaps provides a general idea of how the architectural system will have to be modified so that the handicapped can use it effectively.

1. The blind and partially blind: Such individuals must depend on sound and/or touch in order to locate, identify, and interpret their physical surroundings. Those who have been blind from birth must depend on touch since they have never seen alphanumeric or other symbols. The partially blind require that visual patterns and brightnesses be maximized.
2. The deaf and partially deaf: Such in-

dividuals have less difficulty within the architectural context because they receive most of their information visually. However, they cannot hear warning sounds, which may be the only signal of impending danger.

3. The orthopedically handicapped: Most current standards for the handicapped emphasize special designs related to the use of wheelchairs because of the constraints that such devices place on the mobility of their users. However, there are many other considerations that too often are neglected by the designer. Among these is the fact that many orthopedically handicapped individuals have hand deformities that prevent them from grasping, from applying force, from reaching, etc. In addition, consideration should be given to the person who may be required to use one or more crutches —an entirely different set of constraints.

*Special note:* Many aged persons have many of the above handicaps, which highlights the importance of not accepting the "average person" approach to architectural design. Also, it is highly probable that many facilities may have to be used by individuals who are only temporarily handicapped. Many of the special features required by the handicapped may also be helpful to the average user, e.g., more maneuvering room, better handrails, and better handles. On the other hand, care must be taken to avoid creating aids for the handicapped that may cause difficulties for nonhandicapped users.

## Sociocultural Variables

It is important to recognize that people with different cultural backgrounds view architectural features differently. Some of the more important differences to evaluate are:

### 1. Attitudes toward Privacy
Some cultural groups demand complete privacy, others appreciate privacy, and still others rely merely on subtle cues that signal a desire for privacy.

### 2. Family Structure
In certain cultural groups, several generations live together within a single dwelling; in others, several families live in separate but joined dwellings; and in still others, the family members function as individuals partially separated within the single dwelling.

### 3. The Role of Women
In some cultural groups, women are isolated; in others, the distinction relates entirely to functional factors (e.g., the mother-child relationship); and in still others, completely non-segregated attitudes prevail (e.g., the working wife whose husband shares the household tasks equally).

### 4 Recreational Patterns
Some families are oriented toward more formal and sedentary recreational pursuits; others are more physically oriented, usually toward outdoor activities; and still others are travel-oriented and treat facilities merely as a temporary base of operations.

## User Efficiency

| Parameter | Variables |
|---|---|
| Vision | What a person sees clearly establishes the basic input to that person. His or her use response depends on how well the architectural concept implies what the designer intends the user to do with it. The critical variables include the following:<br>1. Visibility: Are critical features in sight, or are they obscured by intervening elements, glare, or shadows?<br>2. Legibility: Are critical features clear, or are they distorted by lack of contrast, parallax, exaggerated embellishment, or illusory geometrics?<br>3. Conspicuousness: Are features that are important to detecting, recognizing, and understanding lost in the background?<br>4. Recognizability: Are features natural, familiar, and/or similar to the observer's expectations, or are they distorted or purposely made to look like what they are not? |
| Hearing | What people hear not only affects their ability to communicate but may also affect their general capacity to perform other tasks. The critical variables include the following:<br>1. Audibility: If certain sounds must be heard, the acoustic environment must be designed to carry the sounds and not block them.<br>2. Intelligibility: The acoustic environment must be designed so that it will not distort the sounds intended for the listener.<br>3. Signal-to-noise ratio: The combined communications and acoustic system must be designed to maximize the probability that extraneous noises will not obscure the desired sound signal.<br>4. Noise annoyance: Adequate noise attenuation must be provided to minimize the possible deleterious effects that an annoying noise can have on individual task performance. |
| Stability | How well a person performs ambulation or biomechanical or other manipulative tasks depends on the stability-aiding elements of the architectural system and/or the possible impediments designed into the system. In addition, there are critical visual interactions that may add to the instability of the user. Among the typical features to examine are the slope of floors, walkways, stair treads, handrails, and door thresholds. Structural vibration also impacts on user stability. |
| Mobility | How well people perform dynamic tasks in which they must move their bodies and limbs depends both on the clearances provided around their task envelope and on the supporting area provided to maintain stability. |
| Convenience | How well people perform various tasks depends to a great extent on how conveniently they can move from one place to another. This requires careful consideration of functional relationships, the sequence of events, time constraints, and emergency demands in order to create a logical and energy-saving arrangement of spaces and activities within spaces. Lack of convenience not only reduces immediate user efficiencies but also may add to fatigue and possible operator failures. |

### 5. Shopping Habits
Some families shop on the basis of day-to-day replenishment, and others shop infrequently but store for the long run.

### 6. Job Patterns
To some, the job is a means to an end (in which case, ease of getting back home is extremely important); others are job-oriented and would just as soon the separation between home and job remain clear and distinct; and still others prefer to work at home and thus desire a single living and working setting.

### 7. Technological Experience
Although technology continues to advance even to the most primitive areas, care must be exercised not to assume that all cultures either want or will appreciate many of the amenities offered by the more technologically advanced societies, especially if these destroy certain living patterns held sacred for ethnic or religious reasons.

*Note:* This brief review makes it obvious that it may be impossible to create architectural concepts that will match the tremendous variety of sociocultural demands of any given user group. It is also apparent that the rapid changes occurring in even the most disadvantaged countries are continually modifying many of the above observations. Therefore, although planning and conceptual design should reflect sensitivity to these sociocultural factors, the most obvious payoff will come from ensuring that the very basic human-architectural interface design is correct, i.e., by providing such things as adequate space so that individuals and their furnishings or equipment can operate effectively; a reasonably normal visual, acoustic, and thermal environment; and a sensible organization of architectural elements so that people are not continually confused. Direct response to potentially significant sociocultural needs, then, should be examined on the basis of each specific

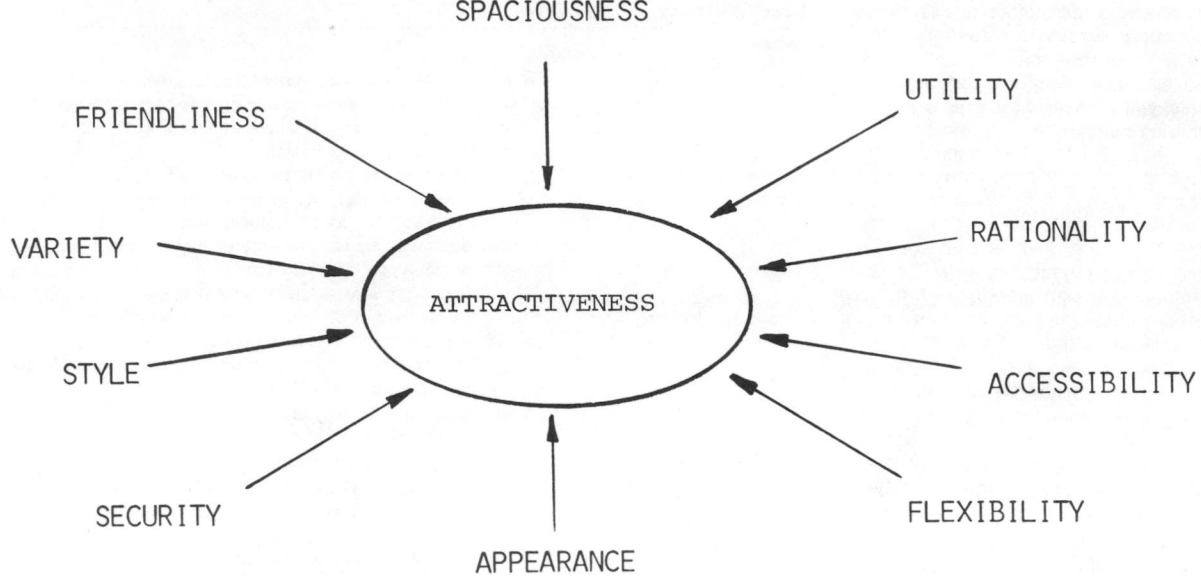

client's time frame and not on the basis of generalizations. For example, just because generalizations about a certain cultural characteristic of the past indicate unique architectural responses, the designer should not assume that changes have not occurred and that the particular client does not desire something more akin to modern concepts. Planners and designers are encouraged to explore carefully each client's feelings about his or her past in order to determine the most effective direction to go with regard to a particular concept.

## Psychological Considerations

The architect is usually concerned about whether the user will be attracted by the design of the community, home, building, or other structural edifice—not only when it is first observed, but also as it is occupied over a longer period of time. The model shown above may be useful in considering the more important ingredients that play a part in determining user reaction.

The accompanying table expands on these ingredients, listing common responses given by typical research test subjects who were asked to verbalize their reactions to architectural environmental features.

In general, one should seek a more or less middle ground in trying to create a balanced combination of the above factors. However, the adjustment between descriptors will be different for different types of architectural systems. For example, the objectives for a satisfactory home environment are not necessarily the same as those for a satisfactory office or factory environment. By the same token, similar adjustment is required for subsystems within the home, office, or factory; e.g., the psychological needs in the bedroom are different from those in the bathroom, and the needs of the production department are different from those of the company library.

**Semantic Descriptors for Assessing Observable Physical Features in Architecture**

| Factor Categories | Descriptor Scaling Examples | |
|---|---|---|
| Spaciousness | Generous | Cramped |
| | Ample | Limited |
| | Empty | Crowded |
| Friendliness | Warm | Cold |
| | Intimate | Detached |
| | Relaxed | Stiff |
| Variety | Stimulating | Boring |
| | Dynamic | Static |
| | Diverse | Monotonous |
| Utility | Purposeful | Unnecessary |
| | Efficient | Confusing |
| | Practical | Frivolous |
| Rationality | Organized | Uncoordinated |
| | Logical | Confusing |
| | Simple | Complex |
| Flexibility | Adjustable | Constrained |
| | Mobile | Fixed |
| | Expandable | Contained |
| Accessibility | Open | Closed |
| | Direct | Indirect |
| | Formal | Casual |
| Security | Familiar | Unknown |
| | Safe | Uncertain |
| | Protected | Exposed |
| Appearance | Graceful | Awkward |
| | Contemporary | Obsolete |
| | Meaningful | Obscure |

## Personal Space

Individuals perceive their relationships with others in terms of the distance between themselves and the people they can see. At least four distinct territorial categories have been defined by various researchers:

1. Public: Those areas where the individual has freedom of access, but not necessarily of action.
2. Home: Those areas where the regular participants have regular freedom of behavior and a sense of control over the area.
3. Interactional: Those areas where social gatherings may occur. An invisible

boundary and a territorial claim are implicit, though not officially promulgated by the people present.

4. Body: The area immediately surrounding the individual's body. This area is most private and inviolate to the individual.

Although absolute distance criteria are probably not pertinent, the following approximations are useful for architectural space consideration:

1. Intimate—0 to 18 in (46 cm): People desire involvement with each other, both physically and emotionally.
2. Casual-personal—30 to 48 in (76 to 122 cm): People will accept minimal contact (a handshake), but wish to retain freedom from domination.
3. Social-consultative—7 to 12 ft (2.1 to 3.7 m): People desire to maintain a certain formality and the freedom to break away when desired.
4. Public—30 to 1500 ft (9 to 450 m): People want to be able to remain unnoticed and to flee or be independent from the crowd.

Many factors are related to the individual's need for personal space:

1. The desire to converse privately in a subdued voice
2. The desire to interact intimately with a loved one
3. The desire to avoid physical contact with another person or the offensive odor of another person
4. The desire to see the eyes of another person clearly
5. The desire to view another person completely at a single glance
6. The desire to be an observer, but not an active participant

As these factors imply, architectural features can bear importantly on user reaction to space.

## Perceptual Quality of the Designed Environment

The following principles, although not easily quantified, provide a further, subjective scaling of design characteristics that may reduce the probability of negative response to the perceived environment:

1. Order: Most people are impelled to seek order and understanding, but they also need sufficient variety to be stimulated by what they see.
2. Outline: The outline of the "whole" should represent grace and balance, not awkward angularity, overpowering massiveness, or unintentional asymmetry.
3. Identifiable references: Environmental references—i.e., paths, edges, districts, nodes, landmarks, runs, mergings, portals, areas, volumes, and acoustic divisions—should be clear.
4. Functional form: Space should appear "positive" rather than "negative"; i.e., it should seem to have been purposefully designed, rather than left to chance.
5. The whole versus a sequential experience: Perceptual confidence comes from an understanding of the whole, as opposed to a sequential experience, which leads to continuing assumptions and doubts.
6. Familiarity: An impression of security based on the repetition of familiar patterns should be created, but without incurring boredom or monotony.
7. Reliability: Visual illusions that could lead to incorrect assumptions and loss of confidence on the part of the observer should be avoided.
8. Cultural identity: Cultural differences reflect individual needs to identify with the traditional, as opposed to challenges to keep up with what is fashionable.
9. Aesthetic objective: The aesthetic objective should be relevant to human needs rather than architectural monuments; i.e., it should provide psychosocial values with which to identify, it should express the user's individuality (not the designer's), and it should provide general perceptual enrichment.

## Definition of Critical Human-Architectural Interfaces

Although the specific human-architectural interfaces and the level of criticality of each interface may vary from one system to another, those listed in the table above should be considered during the conceptualization of any new system.

## Considering Ease of Maintenance during the Planning Stage of Design

If it is considered at all during the planning stage, facility maintenance traditionally becomes a question of: How can this facility be maintained as we have designed it? In other words, too often the question of ease of maintenance does not come up when key architectural configuration decisions are being made. In spite of this traditional attitude, all facilities have to be maintained, and by human beings. The following key maintenance functions should be part of any design concept

## Human-Architectural Interfaces

| Interface | Consideration |
|---|---|
| Site location | Distance to related facilities: residential, commercial, and industrial areas; airport and bus terminals; etc. Accessibility (or nonaccessibility) to highway, roadway, street, rail, and waterway systems: vehicular or pedestrian, etc. Environmental factors: noise, air pollution, possible natural disasters, traffic problems, and aesthetic factors |
| Site amenities | Parking, public transit access, walk-in customer exposure, landscape, illumination, security, and emergency access |
| Building or buildings, external | Number, size, and location; entrances (number, location, and type); identification and illumination; special amenities (utilities, sidewalks, and stairs); external maintenance requirements; etc. |
| Building or buildings, internal | Compartmentalization (number, size, and arrangement); doors and windows (type, number, size, and location); stairs, ramps, elevators, and escalators; heating, air conditioning, ventilation, and illumination; built-in features (cabinets, plumbing fixtures, and electrical and pneumatic systems); floor covering; safety equipment; communications systems; acoustics; and internal maintenance requirements |
| Safety | Fire protection, emergency escape, first aid, and disaster control |

trade-off analysis:

1. Daily housekeeping: cleaning floors, walkways, windows, walls, ceilings, etc.
2. Periodic inspection and repair: inspecting and repairing windows, roofs, walls and woodwork, hot water heaters, plumbing, etc.
3. Periodic refurbishment: repainting exterior and interior surfaces, replacing roofs, replacing plumbing fixtures, etc.
4. Landscape maintenance: watering lawns and shrubs, removing trash, etc.

## Common Human Factors Problems Associated with Maintenance-Related Design

1. One cannot get to the spot that requires inspection, adjustment, cleaning, removal, replacement, or refurbishment.
2. There is insufficient space to do the job once a person has reached or located the maintenance problem.
3. There is insufficient illumination to see what needs to be seen.
4. There is a lack of appropriate service connections to enable use of the necessary tools at the work site.
5. The device to be repaired or replaced is buried into the structure, requiring major destruction and eventual repair.
6. Main service shutoffs (e.g., water, electricity, or gas) are variously distributed, hidden, and/or inaccessible, requiring an inordinate amount of time to find them.
7. The composite land-site–structural relationship precludes the normal and safe use of common maintenance aids such as ladders or scaffolds.

## Architectural Safety

An objective during conceptual planning should be to create an environment in which the user can be as safe as possible. Although this is a tall order, many of the accidents that frequently occur in homes, offices, schools, factories, and elsewhere are due as much to the facility design as they are to user errors. The following typical safety considerations are applicable to all architectural systems:

1. Use nonflammable, nontoxic materials.
2. Eliminate sharp edges, corners, etc., that could cause injury.
3. Create properly designed stairs, ramps, and walkways.
4. Do not use large ceiling-to-floor glass windows or doors without appropriate barriers to prevent people from walking through them when they are closed.
5. Ground all electrical controls, cover outlets, and otherwise prevent people from receiving electric shocks.
6. Provide adequate illumination so that people can see where they are going and avoid tripping over a walkway obstruction or step.
7. Use nonskid materials on floors, walkways, and stairs, especially if there is a possibility of their becoming wet.
8. Provide appropriate handrails around balconies and alongside stairs and use railing designs that children cannot fall through or get their heads caught in.
9. Cover moving parts of machines to prevent people from getting their hands or clothing caught.
10. Avoid locating heaters where they can be touched inadvertently or where pilot lights could ignite the structure or adjacent materials or cause an explosion as a result of gas fumes from a nearby vehicle.
11. Provide adequate emergency escape routes that can be used in the event that normal passageways and exits are impassable.
12. Consider the problems of window washing and of house or building repair in terms of typical unsafe practices associated with ladders and scaffolds.
13. Provide appropriate fencing around special facilities from which children should be barred (e.g., swimming pools, high-voltage wires, and heaters).
14. Provide fire sprinkler and alarm systems.

## Design for the Handicapped

Practically all aspects of architectural facility concepts, including the overall community planning, should be examined from the standpoint of use by elderly and handicapped persons. Over half of the population of the United States soon will fall into this general category, and since a great many of these people are taxpayers (with a right to access to public facilities) and/or consumers with considerable purchasing power, they have the right to expect facilities to be designed for their use as well as for the use of the so-called normal population.

### 1. Considerations for the Blind

Blind or partially sighted individuals (many of whom are elderly) get about by depending on sound signals and tactile cues. They require the following special features:

a. Well-defined, rectilinear walkways, street corners, and curbs which the blind person can touch with a cane
b. Pathway obstructions that go all the way to the floor or ground so that the blind person's cane does not pass beneath the object and thus allow the person to run into the object
c. Nothing at head height, such as signs, guy wires that support telephone poles, and trees with low branches
d. Special braille signs for key public locations for identifying a building name and number, a street corner, or a bus stop
e. Sound signals so that the blind person will know when a DON'T WALK signal is on or whether an elevator is going up or down
f. Guardrails and/or special tactile identification of pathways to keep the blind person from veering into the street

### 2. Considerations for the Deaf

Whenever an audio signaling device (warning) is used for the general population because of the chance that people may not be looking in the direction of a hazard, an accompanying visual and/or tactile (vibration) signal should be devised for deaf persons to draw their attention to the hazard also.

### 3. Considerations for the Orthopedically Handicapped

Architectural mobility for the orthopedically handicapped can be increased by the following:

a. Adequate clearance; smooth ground and floor surfaces, especially at thresholds of doorways, curbings, and ramps for change of elevation; and reachable heights for such items as drinking fountains, telephones, and built-in worktable tops and shelves
b. Limited force application requirements for opening doors
c. Door handles and cabinet handles that can be pushed rather than grasped or squeezed and turned
d. Stairs that are not steep and railings that can be grasped and held firmly in the arthritic hand

### 4. Considerations for People with Different Handicaps

Extreme care should be exercised in developing mobility aids or concepts for people with one specific handicap, since the same facility may have to be used by people with other handicaps (e.g., a smooth, ramped intersection corner designed to aid the wheelchair user may remove the very tactile cues that tell the blind person where the street begins).

In addition, care should be exercised in terms of how some aids to the handicapped may affect the use of the facility or device by nonhandicapped persons. In many cases, however, the aid may help both the handicapped and the nonhandicapped person. For example, larger, clearer street signs are needed in most cities today both for the normally sighted motorist and for the partially sighted person, who might be able to use these signs if they were not so small.

THE COMMUNITY

## COMMUNITY PLANNING

In planning a new community or developing modifications to an older one, the needs of the community residents and their visitors should be uppermost in the mind of the planner. Although this may seem an obvious and unnecessary statement, the average planner often does not think of the user of his or her community except in rather gross and vague terms, e.g., "The residents will be proud of their community because it is attractive."

The human needs addressed in this section pertain to more down-to-earth factors, such as finding one's way around, identifying buildings and streets, and not having to travel a long distance to the market or having to go to several parts of town to complete a simple business transaction involving local government offices.

Unfortunately, the needs of each individual vary, and so do the needs of different individuals as they play out the special roles of housewife, businessman or businesswoman, factory worker, police officer, plumber, and clergyman, for example. In order to sort out these varied needs and thus be better prepared to understand how they can be accommodated best for the benefit of all concerned, planners should develop written or verbal descriptions of the behavioral circuits for all the expected community citizenry. These are, in a sense, scenarios of activity as each individual carries out his or her own personal mission. The scenarios need not be long or detailed, but they should be complete enough to give some idea of the importance of events, probable frequency, time implications, and so on.

With such information at hand, the planner is better prepared to consider alternative ways in which common elements of the community can be put together to serve all users in an approximately balanced manner.

The design sheets which follow deal with a number of planning and design principles and guidelines that should be useful to the community planner and architect.

## Community Planning Model

The key community activity elements shown here must be arranged to provide balanced interaction, isolation, and accessibility. If a priority is given, it should be to the primary community resident.

## Community Facilities User Identification

Avoid the tendency to arrange community facilities purely on the basis of general facility relationships or, even more importantly, on the basis of the primary facility user or tenant. Although the facility tenant is, of course, a prime consideration in planning the location of the facility with respect to other parts of the community, many other users should also be considered. The table on page 24 lists potential

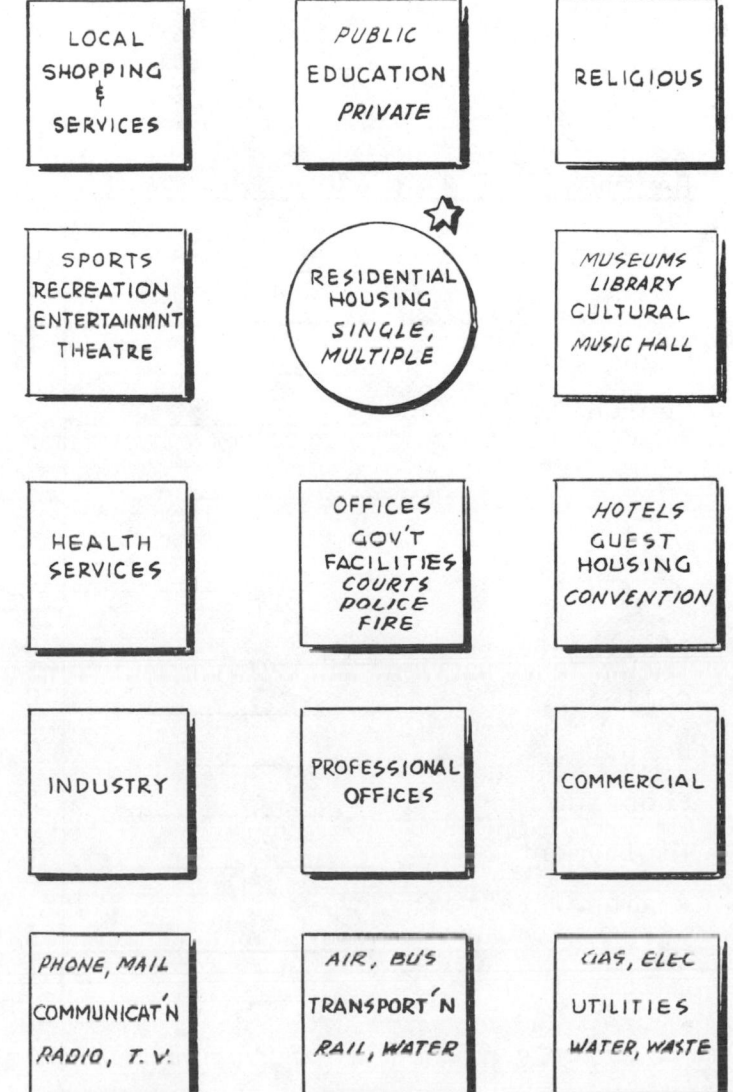

| User | Special Needs |
|------|---------------|
| Primary tenants (residential) | Convenience to, but isolation from, heavy traffic arteries |
| Primary tenants (hotel and motel) | Convenience to business, public transportation, and special entertainment facilities |
| Primary tenants (others, e.g., employers and employees) | Convenience to key traffic arteries, related business, and public transit |
| Facility visitors (clients and customers) | Convenience to key traffic arteries and public transit |
| Facility service personnel (mail and package delivery) | Convenience to related service facilities |
| Facility repair personnel (structural, utility, etc.) | Convenience to related service facilities |
| Disaster service personnel (police, fire, and ambulance) | Convenience to both key traffic arteries and local streets and to related service facilities |

users and their special needs.

Convenience should be thought of from the following points of view:

1. Distance
2. Directness of access
3. Mode of access (by foot, private vehicle, public transit, or special vehicle, e.g., heavy fire or construction equipment or helicopter)

Isolation should be thought of from the following points of view:

1. Reduction of noise, air pollution, and odors
2. Traffic congestion
3. Privacy and security (without creating convenience difficulties for critical service personnel)

## Distance and Time between Home and Community Facilities

An important goal in community planning is to provide enough facilities of each type that are reasonably accessible to all the community's citizens. The accompanying rough guidelines provide an approximate goal for the planner in terms of the most important community facilities.

| | |
|---|---|
| Elementary School _____ | 0.5 mile |
| Jr. High School _____ | 0.5 mile |
| High School _____ | 1.0 mile |
| Church _____ | 1.0 mile |

| | |
|---|---|
| Playing Field _____ | 0.5 hour |
| College _____ | 0.75 hours |
| Hospital _____ | 0.75 hours |
| Shopping _____ | 0.75 hours |
| Employment _____ | 1.0 hour |
| Regional Recreation _____ | 1.0 hour |
| Cultural _____ | 1.5 hours |

## An Urban Activity Model[7]

*Urban Activity Categories*

1.00 Consumption-related activities
2.00 Celebration and congregation-related activities
3.00 Work-related activities
4.00 Service-related activities
5.00 Recreation and relaxation-related activities
6.00 Communication and learning-related activities

*Category 1.00: Consumption-Related Activities*

1.01 Children are able to perform errands for adults at nearby shops and stores.
1.02 Children are able to set up makeshift roadside stands to sell lemonade, cookies, etc. to passers-by.
1.03 People are able to reach grocery stores by public transportation.
1.04 People are able to drive to a variety of shopping areas to do their grocery shopping.
1.05 People are able to shop for groceries on their way home from work (or have someone else do this for them).
1.06 People are able to walk to necessary grocery stores.
1.07 People are able to window shop.
1.08 People are able to reach a variety of larger department stores by public transportation.
1.09 People are able to reach a variety of larger department stores by driving to them.
1.10 People are able to find parking when driving to shops and stores.
1.11 People are able to eat in outdoor restaurants during the warmer seasons of the year.
1.12 People are able to dine in a variety of restaurants in the larger area of the community.
1.13 Local establishments provide catering/take-out services.

*Category 2.00: Celebration and Congregation-Related Activities*

2.01 People are able to partake in a variety of programs which bring many people together in the larger area of the community (e.g., concerts, theater, athletic events, etc.)
2.02 People are able to partake with other community members in a seasonal schedule of activities (e.g., art shows, caroling, picnics, festivals, etc.)
2.03 People are able to discuss concerns about the community with friends at meeting places provided for this purpose.
2.04 Children and adolescents are able to partake in the activities of community youth groups.
2.05 People are able to partake in community activities with older people.
2.06 People are able to partake in community activities with younger people.

2.07 People are able to partake in community activities with people of the same age.

*Category 3.00: Work-Related Activities*

3.01 People are able to walk to and from work.
3.02 People are able to use public transportation facilities to go to and from work.
3.03 People are able to comfortably drive to and from work.
3.04 Children are able to attend day-care facilities.
3.05 Children are able to find small jobs in the community.
3.06 Older people are able to find small jobs in the community.

*Category 4.00: Service-Related Activities*

4.01 Garbage collection occurs frequently and sufficiently without major disruption of traffic or sleep.
4.02 All sidewalks and roads are well illuminated at night.
4.03 All sidewalks are regularly cleaned by the people in the community.
4.04 All roads are regularly cleaned by the municipality.
4.05 All sidewalks are regularly repaired and maintained by the municipality.
4.06 All roads are regularly repaired and maintained by the municipality.
4.07 People are able to get action on complaints and requests through political channels in the community.
4.08 Ambulances respond quickly and reach their destinations without delays.
4.09 Fire equipment responds quickly and is able to reach an emergency without delays.
4.10 All laws and ordinances are satisfactorily enforced.
4.11 The community's sidewalks are adequate to handle all pedestrian traffic.
4.12 The community provides local outpatient health.
4.13 The community provides legal-aid services.
4.14 The community has an adequate number of practicing physicians to handle all patients.
4.15 The available hospital facilities adequately serve the needs of the community.
4.16 The public schools provide satisfactory education for the children of the community.
4.17 The community provides adequate automobile service facilities.
4.18 The community provides adequate laundry and dry-cleaning facilities.
4.19 The community provides adequate off-street parking for all vehicles.
4.20 Hotels, motels, and other overnight accommodations are placed conveniently from the residences.

*Category 5.00: Recreation and Relaxation-Related Activities*

5.01 People are able to watch traffic and other people from benches or from similar observation places in the community.
5.02 People are able to partake in a variety of indoor and outdoor recreational activities throughout the year within the larger

area of the community.
5.03 The community provides for quiet relaxation, especially on weekends.
5.04 The community provides for boisterous, outgoing recreational activities.
5.05 Playgrounds and other recreational areas for children allow for play which is without interference from children of other age groups.
5.06 Children are able to safely ride bicycles in the community.
5.07 People are able to enjoy safe walks in the community.
5.08 People are able to enjoy walks which are visually stimulating.
5.09 The community provides adequately for the keeping of pets without their becoming a nuisance.

*Category 6.00: Communication and Learning-Related Activities*

6.01 Elementary schools are located within walking distance from the community residences.
6.02 Children are able to walk to school safely.
6.03 The community is free from persistent foul or annoying odors.
6.04 The community contains a number of characteristic, identifiable smells.
6.05 The community contains places of lively activity.
6.06 The community is free from dominant visual clutter and ugliness.
6.07 People are able to orient themselves easily with the help of signs and symbols when looking for a particular place or places.
6.08 People are able to view exhibits in museums, galleries, etc. in the larger area of the community.
6.09 The community provides adequate library facilities.

## User Activities and Needs

The community layout should be responsive to the needs of individuals in terms of the activities they pursue, primarily in going from facility to facility. Individuals should be able to identify where they are and where they want to go first. Then they should be able to get there by the most direct route possible with a minimum of interference. The following suggestions are made for various facility activities, starting with the key user category, the local citizen.

**PRIVATE RESIDENTIAL FACILITIES** The illustration on page 27 shows a general prioritization of facility-need relationships for the private resident, whether he or she lives in a single-family residence or a multifamily residence.

**EDUCATIONAL FACILITIES** Public educational facilities include both the precollege institution and the college or university. Although planners must also consider other private institutions, they may have less control over the location of these facilities. The key issue for locating public schools is access between home and school for students; access for faculty members and administrators is an important but less critical consideration. The following is a general checklist for determining the location of public schools:

[7]T. R. Martineau, "The Urban Activity Model," in W. J. Mitchell (ed.), *Environmental Design: Research and Practice*, Proceedings of the EDRA Conference, University of California, Los Angeles, January 1972.

1. Grade schools should be located in terms of minimized walking distance between the school and the related residential neighborhood. It is important to arrange school-residence relationships to minimize the necessity for small children to cross heavily traveled highways and streets. Grade schools should be considered primarily a single neighborhood facility.

2. Junior high schools should be conveniently located with respect to related residential neighborhoods and should be arranged to minimize the necessity to cross heavily traveled highways and streets, although this need not be given the same high priority as in the case of grade schools. Since more real estate is usually required for these facilities, consider a more central location with respect to several neighborhoods.

3. High schools should be centrally located with respect to several neighborhoods, but also should be arranged to minimize unusually heavy traffic-crossing requirements. In most communities, many high school students drive their own automobiles to and from school. Thus, special consideration should be given not only to the vehicular routing between residential and school areas but also to the effect this may have on other types of traffic, such as general "going-to-work traffic," commercial traffic, and transient traffic.

4. Colleges and universities should be planned and located more or less as independent student communities, and both resident and off-campus students must be taken into account. Depending on the nature of the institution, consideration should be given to the plan for a total college-residential arrangement that will provide satisfactory local convenience with a minimum of interaction with the rest of the community.

**Common Factors**
1. Public transit access
2. Off-street parking
3. Security services
4. Special community usage (local community sharing of the auditorium and sports facilities)
5. Exclusions (isolation from certain types of businesses such as liquor stores, bars, and adult entertainment centers)

**RELIGIOUS FACILITIES** Although local governmental bodies do not always exercise control over the location of churches and other religious facilities, there will be fewer problems in the future if there is a logical plan for including suggested locations for these facilities and, preferably, for holding these land sites for potential church use. The following is a general checklist for determining the location of religious facilities:

1. Set aside future church property in, or adjacent to, residential neighborhoods, rather than commercial, industrial, or recreational neighborhoods.
2. Set aside sufficient property for off-street parking.
3. Provide reasonable separation between church property and public schools (to minimize potential vandalism) and commercial property of an antireligious character.
4. Provide reasonable separation between church property and noise-producing environments (heavily traveled roads, airports, railroad rights-of-way, etc.).
5. Locate religious facilities near planned public transit routes.

**HEALTH FACILITIES** Advanced planning for the location of hospitals, convalescent homes, and similar health-oriented facilities is important for several reasons:

1. Because emergencies may happen in the home, at work, or on the highway, there should be rapid access to health facilities from all parts of the community. Although centralization is often suggested, this may create considerable conflict with downtown traffic.
2. Health facilities should be isolated from typical noise-generating aspects of the community, e.g., traffic, schoolchildren, factory equipment, aircraft, and railroads. Although modern techniques make it possible to lower noise levels within a facility, an improper location adds immeasurably to the cost.
3. Large hospitals, especially, generate tremendous parking needs; therefore, sufficiently large land sites must be identified early in the planning stage.
4. Health facilities should be located near planned public transit routes.
5. Health facilities should not be located in disaster-prone areas (near earthquake faults, below watersheds or dams, etc.) because of the difficulty of moving patients.

The following is a general checklist for determining the location of various health facilities:

1. Major medical centers should be located somewhat centrally but peripherally to primary residential areas, with immediate access from main thoroughfares and highways.
2. Convalescent homes should be located in suburban areas, preferably where adequate open space can be retained around the convalescent home. Convenient, but not necessarily major, street access should be considered so that the facility will be isolated from traffic and noise, which tend to build as the community is further developed.
3. Retirement homes should be located similarly to convalescent homes, except that convenient public transit availability is perhaps more critical. Land-site considerations should include the possible need for outdoor recreational facilities such as a golf course, swimming pool, or picnic area.

**RECREATIONAL FACILITIES** The increasing demand for recreational facilities of all types places a considerable burden on the community planner, principally because of the amount of land that such facilities require. The following is a checklist for determining the location of various types of recreational facilities:

**1. Public Parks**
a. Neighborhood parks: These are a combination of playground and picnic ground within walking distance of a defined residential neighborhood. They should not require crossing heavily traveled streets.
b. General community parks: These consist of one or more larger park areas that might include, in addition to playgrounds and picnic grounds, an open-air theatre, equestrian paths, zoological facilities, etc. They require easy access by vehicle and local bus and also adequate parking, especially for holiday occasions.
c. Downtown parks: One or more small downtown parks can provide noontime relief for business and professional workers. They should be accessible, with minimum requirement to cross heavily traveled downtown streets.

**2. Amusement Parks**
a. Public and/or commercial educational parks (including botanical gardens, wild-animal parks, and sea-life parks): These should be located outside the main periphery of the community, but with good access by private and public conveyance.
b. Carnival and concession parks: These should be located outside the main periphery of the community, but with good access by private and public conveyance.

**3. Sports Parks**
a. Amateur sports parks (including golf, tennis, baseball, etc.): These should be located on the periphery of the community, with repetition to provide convenience to all residential areas of the community. Good access is required for vehicular, bike, and public transit traffic.
b. Professional sports parks (including golf, tennis, baseball, football, basketball, hockey, etc.): These should be located on the periphery of the community or just outside.

*Note:* All recreational facilities should be planned with adequate parking, and special consideration should also be given to the peak traffic problems associated with special holidays and sporting events.

**ENTERTAINMENT FACILITIES**

**Theatres**
a. Local motion picture theatres: These should be located strategically within or near typical residential neighborhoods that are large enough to support such facilities. Public parking is an important factor in land-site selection. Such facilities should also be convenient for pedestrian and public transit traffic.
b. Downtown theatres (including motion picture theatres, music halls, opera houses, etc.): These should be located downtown and should be accessible to both residents and visitors. Adequate off-street parking, space for dispensing and picking up private and public transit patrons, and access for delivery of entertainers' equipment are important factors to consider.

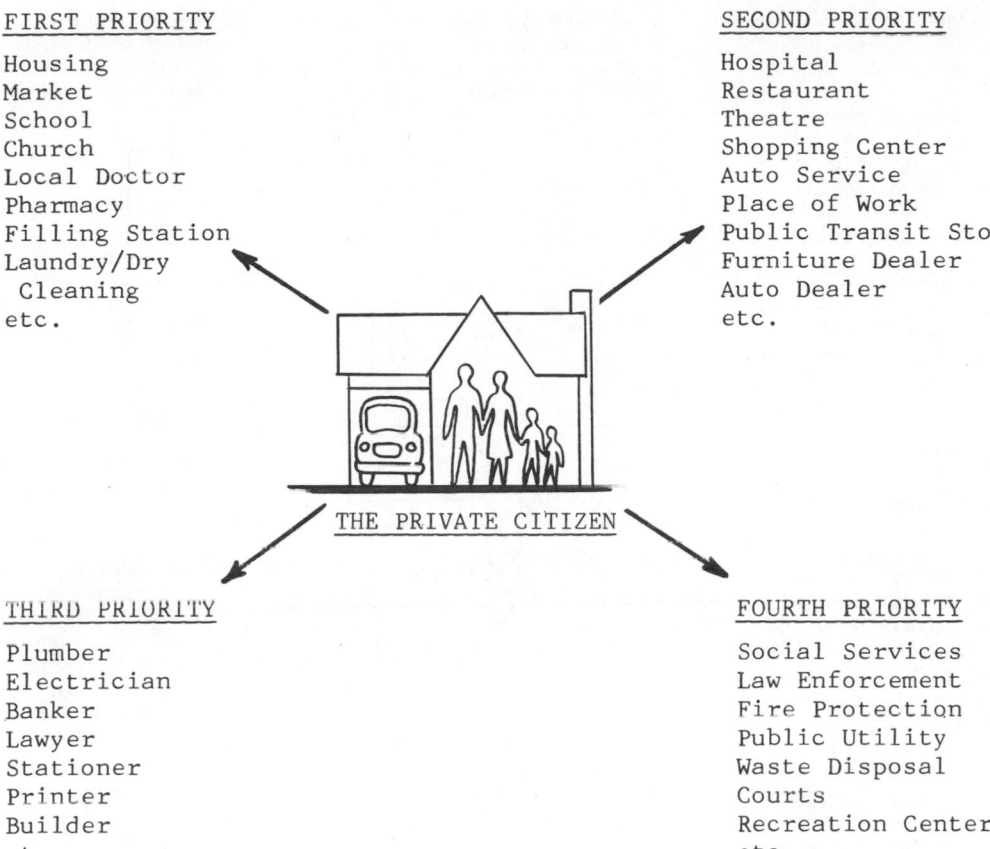

FIRST PRIORITY

Housing
Market
School
Church
Local Doctor
Pharmacy
Filling Station
Laundry/Dry
  Cleaning
etc.

SECOND PRIORITY

Hospital
Restaurant
Theatre
Shopping Center
Auto Service
Place of Work
Public Transit Stop
Furniture Dealer
Auto Dealer
etc.

THE PRIVATE CITIZEN

THIRD PRIORITY

Plumber
Electrician
Banker
Lawyer
Stationer
Printer
Builder
etc.

FOURTH PRIORITY

Social Services
Law Enforcement
Fire Protection
Public Utility
Waste Disposal
Courts
Recreation Center
etc.

---

c. Dinner theatres: These should be located outside the periphery of the community, but with good access and parking for private conveyance.

**CONVENTION FACILITIES** Publicly supported convention facilities should be located downtown, wherever feasible, in order to provide good access between these facilities and local hotels. However, an equally important factor to consider is the accessibility of such facilities to key outside transportation (e.g., the airport), major expressways or freeways, etc. Similar consideration should be given to convenience between convention facilities and entertainment facilities.

**COMMERCIAL FACILITIES**

**1. Neighborhood Shopping and Services**
Proper zoning for commercial shopping and services within various residential neighborhoods is needed, not only to provide convenience for residents, but also to preclude the proliferation of sometimes unsightly commercial buildings in an otherwise purely residential atmosphere. As a general rule, smaller shopping areas should be limited to those residential areas in which people typically walk to stores. On the other hand, larger shopping complexes should be located centrally, but peripherally to several residential neighborhoods, assuming that the general pattern would be to shop by private vehicle—in which case parking becomes a major consideration.

**2. Commercial Services**
Services required by commercial establishments should be located for maximum convenience to these establishments in terms of delivery, but there should not be direct individual interface. This suggests a more or less central location away from residential neighborhoods. However, there is a considerable advantage to confining different types of commercial services to specific areas within the community so that individuals, when necessary, can find a particular kind of service in one location, rather than spread out in different parts of the city. Most commercial services require easy delivery access, both incoming and outgoing, which means that there must be good access to major external transportation terminals and to key arterial streets and highways both for delivering within the city and for shipping inter-city.

**INDUSTRIAL FACILITIES** Industrial facilities, whether light or heavy, should be located away from both residential and light commercial areas because of the unique nature of their traffic and possible environmental pollution. Heavy industry should be located the farthest out, but it should be convenient to major highways for trucking access, to railroads, to waterways, and to air terminals. Consideration should be given to locating heavy industrial facilities where any expelled fumes, smoke, or dust will not be blown across the city by prevailing winds. However, industry must also be located so that workers do not spend considerable time getting to and from their homes.

**COMMUNITY ADMINISTRATIVE FACILITIES**
Whereas, in the past, the "central courthouse" concept was extremely popular for the small town, modern community planners are warned not to consider the location of their jobs in the completed facility the central issue in community planning. The convenience of the local citizen should still be the key issue, and the convenience of the community administrative facility should be attuned to the needs of the typical citizen who wants to make contact with public officials. Public administrative facilities should be located so that citizens can find them easily; consideration should also be given to how easily public servants, such as police and fire fighters, can service the public. The following are guidelines to locating public offices:

1. Facilities with a direct public interface: Facilities to which the individual citizen must go (for building permits, licenses, etc.) should be located not in crowded downtown areas but in peripheral areas where public and private access and parking are available.
2. Facilities from which special services emanate to various parts of the community (police and fire fighters): These facilities should be distributed variously to increase the efficiency of response, i.e., to control crime, put out fires, etc.
3. Facilities without a direct public interface (mayor's and city council's offices and technical departments): These facilities

should be located together primarily for effective administrative interaction and secondarily for convenience of public interaction. The actual site is perhaps of less importance in terms of general accessibility to various parts of the community than in terms of the almost inevitable expansion that will come with community growth. Other special administrative and service facilities should be located on the periphery of the community, according to the shape and size of the community area. These include such services as road and street maintenance and maintenance of public transit system vehicles, for which storage requirements continually expand as the community expands. Detention facilities such as jails or prisons should be isolated from the rest of the community wherever feasible. Although such facilities are often considered a part of typical administrative facilities, the average citizen fears the possible escape of a prisoner and reacts negatively to such facilities' being near his or her residence, schools, or church.

4. Public disaster command post: Although city planners often think of their regular administrative facility as the primary command post for control of all activities within the community, consideration should be given to the location and development of a separate emergency command post to which key disaster control personnel could go to oversee emergency procedures. The site for such a facility should be chosen on the basis of the invulnerability of communications and personnel to interference by man-made or natural causes, rather than on the basis of convenience, other than rapid access by key personnel.

### Street and Roadway Geometry

The layout of the street and roadway system is important to both local users and visitors, since it determines how easily they can locate themselves and decide where they want to go. The important considerations are noted below.

### 1. Use a Rectilinear North-South–East-West Plan

a. Most people relate where they are going to a north-south reference. They use this reference as they follow a projected path on a road map. They expect to turn left

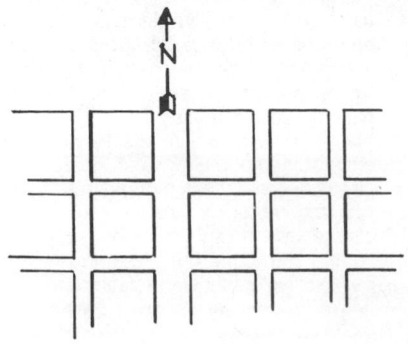

or right as they progress north or south or east or west; therefore, they are more positive when the roadways are at right angles to each other.

b. People become confused as they traverse curved or angled streets. Not only do they lose track of what is north and south or east and west, but all intersecting cues become distorted—nothing is "equal or square anymore."

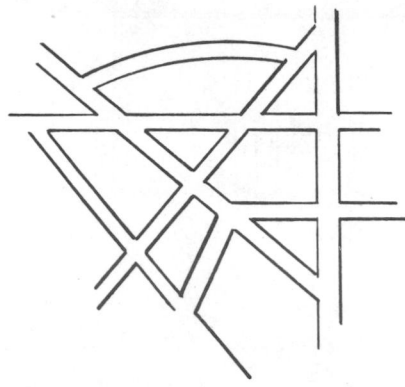

c. Multiple-entry intersections (five or more streets) make it difficult for people to follow their course across the intersection.

d. Traffic circles or "town squares" interrupt people's perception of where a street continues on the other side, making it difficult to find the continuation.

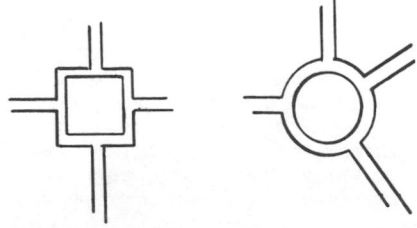

e. Street jogs confuse people who are trying to follow through across an intersection street. Similarly, a person using a cross street is confused by a cross street, which appears sooner on one side than on the other.

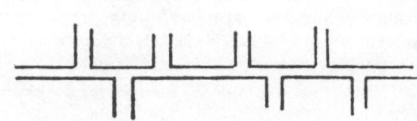

### 2. Develop and Maintain Constant Roadway Width

a. People soon recognize the importance and nature of a given street or roadway

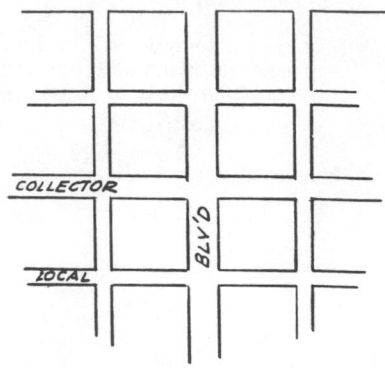

by its size (and sometimes by its surface characteristics). They soon associate wider streets with through traffic, and narrower streets with cross traffic and typically interrupted traffic (i.e., frequent stops). A plan should be developed in which street widths are standardized for various traffic and operating categories; otherwise, people become confused with respect to street identification and often fail to note stop signs or signals.

b. Changing width of a continuing roadway is not only confusing but also extremely hazardous. If street paving cannot be continued beyond a certain point be-

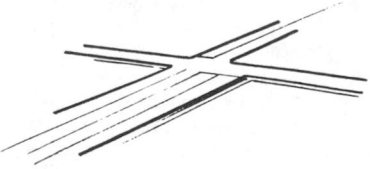

cause of the cost involved, the roadway width should continue, even though only the center portion of the street is paved for the time being.

c. Multilaned freeways, boulevards, etc., should continue the same number of lanes (although expansion is less serious

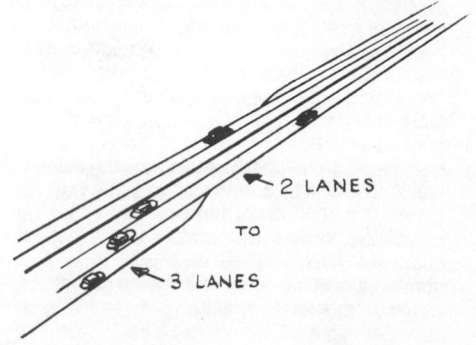

than contraction). Unless such changes are properly signed way in advance, motorists become trapped when their particular lane runs out.

d. Bike lanes should be "added," rather than taken away from the regular street or highway width. They should also be isolated by a berm or raised surface to keep motorists off the bike route and cyclists out of the street.

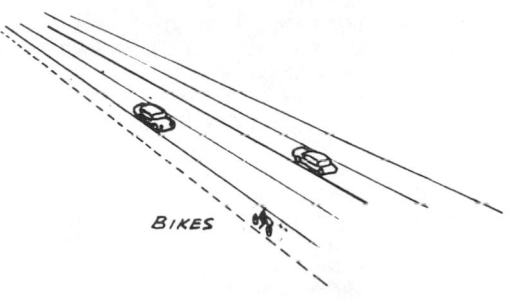

tion (preferably north and south) and numbers for streets running in the other direction (i.e., "First," "Second," etc., for streets running east and west). (See the illustration on page 30.) Below are some factors to consider in creating a viable and nonconfusing system:

1. Avoid using a community name for a freeway; observers become confused if they do not intend to take the freeway to the community of the same name. That is, if they want to go to another area, they may believe that the freeway will not take them in the correct direction. For example, if the community at one end of the freeway is called "Santa Ana," do not call the freeway "the Santa Ana Freeway." Although this may not confuse local residents, it may cause people who want to go in the opposite direction *not*

to take that particular freeway, even though they should. People also may not differentiate between roadway and community designations when they are the same.

2. Keep street names short and use names that are easy to pronounce. Motorists are easily confused when they have to remember long, hard-to-pronounce street names as they try to follow a path through a new community.

3. Do not use a geographically related name for a street that is not coincident with the name; e.g., "North Ventura" should not be used if the street does not run generally in a northerly direction.

4. Avoid using street names that sound the same but are spelled differently, e.g., "Sleighter" and "Slater." A motorist may have obtained directions over the phone

## Sidewalk Configurational Policy

Regardless of local standards, several user-related considerations are important regarding the design of community sidewalks. Some of these are discussed below.

### 1. Sidewalk Setback

For at least two important reasons, all sidewalks should be set back from the street. First, the sidewalk should be kept level as it crosses typical driveways, which must be sloped to the street. Second, pedestrians should be kept away from the curb. Children are especially vulnerable because, as they play, they may inadvertently step off the sidewalk and into the street.

### 2. Ramps for Wheelchairs

Most states require special wheelchair ramps so that handicapped persons can negotiate between sidewalk and street with a minimum chance of tipping over their wheelchairs. The threshold of the ramp should be serrated so that blind persons can feel the ramp threshold and thus not inadvertently step into the street in front of a passing vehicle.

### 3. Downtown Sidewalks

Wide sidewalks are preferred in order to "carry" the larger volume of foot traffic. The width is especially important at corners where people collect to wait for traffic signals to change. A nonsidewalk setback similar to that noted in item 1 above should be provided to keep people from being accidentally pushed into the street.

## Street and Roadway Identification and Address System

Although the planner has considerable latitude in devising a name and numbering system for streets and addresses, certain systems are more easily understood by people who are trying to find their way around the community. One of the best systems is to use proper names for streets running in one direc-

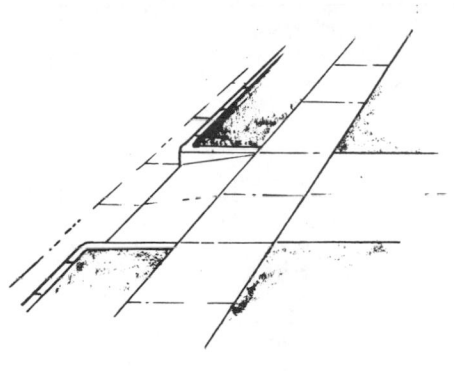

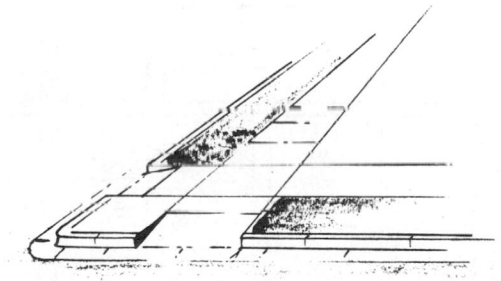

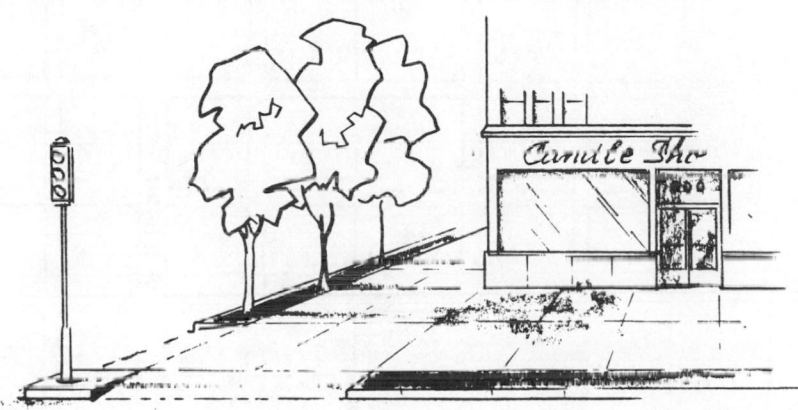

and not realize that there are two different streets with the same-sounding name.

5. Do not use the same proper name for two different streets, one of which is called "Street," and the other "Avenue." Many people give their address using only the proper name and fail to let the other person know that there is more than one street with the same proper name.

6. Do not select an encoding system that uses letters or proper names arranged alphabetically if it can be anticipated that the system will run out as a result of future expansion. In other words, there are only 26 letters in the alphabet.

7. Try to maintain consistency in address numbering; i.e., numbers should progress in the same direction from a common reference point, and odd numbers should always appear on the same side of the street, with even numbers on the other.

8. Avoid changing the number sequence in the middle of a block, even though there may be an intervening street breaking into the block on just one side. Starting from a reference point, the first block should be numbered from 1000, the second block from 2000, and so on. This helps observers keep track of where they are without having to look at every number in the block.

9. Develop a consistent policy with regard to how streets and roadways will be "signed." That is, the size and shape and the location of identification signs and address numbers should be consistent throughout the community; people should not have to hunt for the sign or address number.

*Note:* Although there are many standards for signing, these are not always appropriate in terms of size and other legibility factors. The alphanumeric font and size should be compatible with the needs of observers relative to how far away they must be able to read the sign.

## Environmental Control

Habitability within the proposed community will be influenced by how well the overall plan provides control over both natural and man-made pollution. The most serious possibilities include traffic noise; odors, fumes, or smoke from factories; and pollution of adjacent waterways. These factors should be considered as early as possible because their possible impact can be mitigated much more easily before the community is laid out. Although corrective devices can be added to control air and water pollution, it is difficult to control noise and odors except by properly routing traffic away from residential areas or by placing odor-producing factories in positions where the odor is blown away from, rather than toward, the community.

Effective noise control, as well as a general improvement in the quality of life, can be obtained by planned green belts, parks, and earthen barriers between heavily traveled roadways and nearby dwellings or facilities within which noise is undesirable. The sketch on page 31 illustrates how a major freeway can be lowered with land barriers and foliage on each side to block and/or muffle the traffic noise. Obviously, the proper foliage treatment can also provide desirable parklike surroundings that people can use for outdoor recreation, picnicking, or resting. To have sufficient

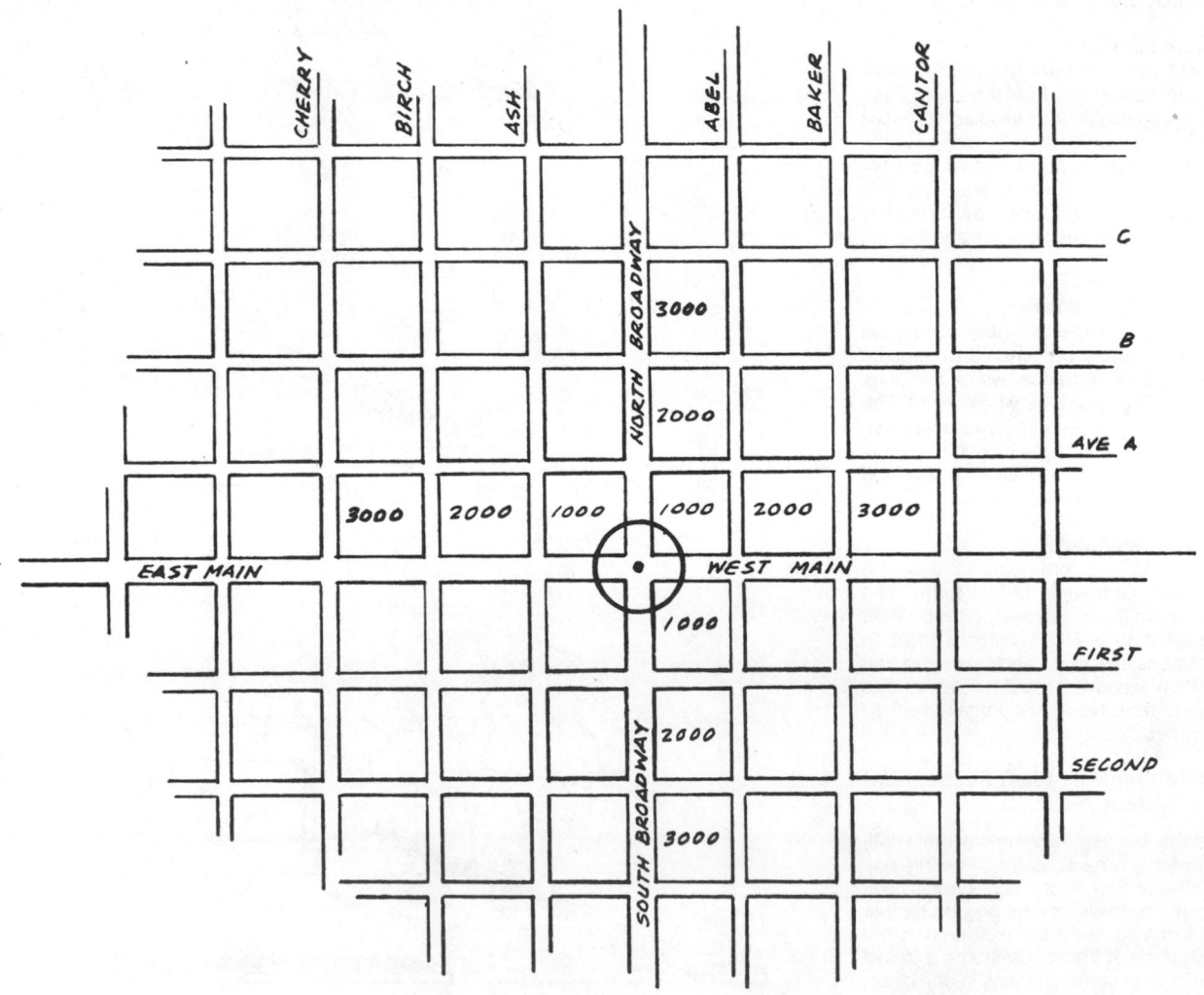

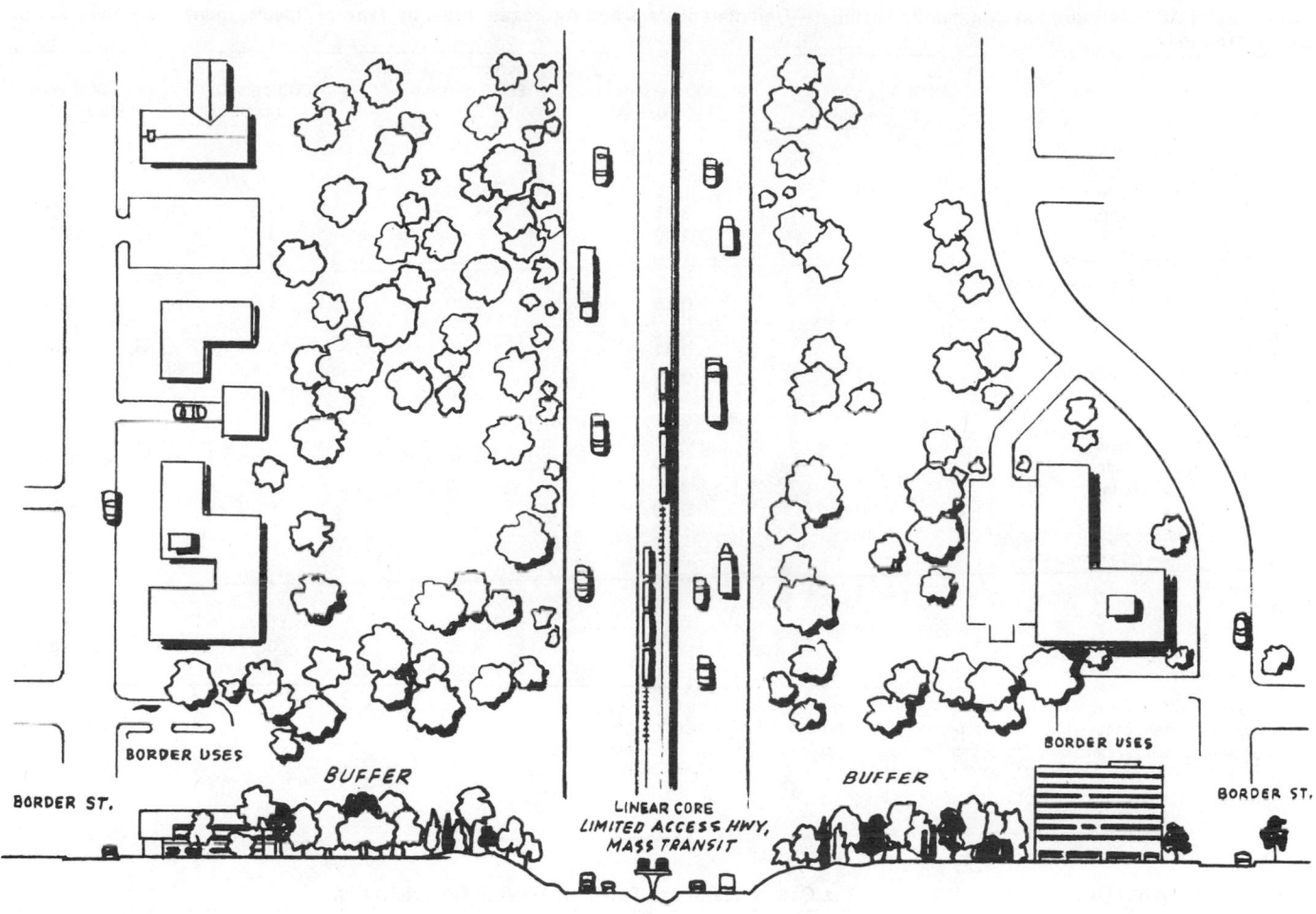

BORDER USES

BUFFER

BORDER ST.

LINEAR CORE
LIMITED ACCESS HWY,
MASS TRANSIT

BUFFER

BORDER USES

BORDER ST.

land space, such planning has to be considered as early as possible. This is even more critical with respect to locating airports.

### Neighborhood Community Facilities as a Whole[8]

**GROUPING** The facilities should, if possible, be grouped together in the direction of the major traffic flow from the development area to the outside, accessible by direct pedestrian and automobile routes. Such grouping will encourage the use of all facilities.

The existence of a physical center of the neighborhood stimulates the growth of community relationships and the acceptance of

[8]J. De Chiara and L. Koppelman, *Planning Design Criteria*, Van Nostrand Reinhold Co., New York, 1969.

community responsibilities by the residents.

As most community facilities require comparatively flat land, topography will, to some extent, govern their grouping and location. Special situations may occur in which such grouping will not be advisable, especially where existing facilities must be taken into account.

Within the group, the various community facilities should be physically separated from each other to prevent conflict of circulation. It is especially important that pedestrian access to the school be separate from all vehicular access to other facilities.

**TOTAL LAND REQUIREMENTS** For many neighborhood planning purposes, it is desirable to know community facility land requirements as a whole. The table on page 35 is a summation of these requirements. It should be

noted that this table combines recommended and assumed areas; the values given are therefore not to be considered mandatory standards.

**USE OF EXISTING FACILITIES** Before final decisions are made in regard to the provision of neighborhood community facilities, the area should be examined for available existing facilities. Special care must be taken to check the capacity as well as the location of such facilities. There may be citywide or district facilities that can also be used by the neighborhood and that will in reality be so used if they provide good service and are readily accessible.

The possibility of using these facilities should be investigated if any of them exist within acceptable distance from the development.

**Land Area of All Neighborhood Community Facilities—Component Uses and Aggregate Area, by Type of Development and Population of Neighborhood[a]**

| Type of Development | Neighborhood Population | | | | |
|---|---|---|---|---|---|
| | 1000 persons 275 families | 2000 persons 550 families | 3000 persons 825 families | 4000 persons 1100 families | 5000 persons 1375 families |
| One- or two-family development[b] | | | | | |
| *Area in Component Uses* | | | | | |
| 1. Acres in school site | 1.20 | 1.20 | 1.50 | 1.80 | 2.20 |
| 2. Acres in playground | 2.75 | 3.25 | 4.00 | 5.00 | 6.00 |
| 3. Acres in park | 1.50 | 2.00 | 2.50 | 3.00 | 3.50 |
| 4. Acres in shopping center | 0.80 | 1.20 | 2.20 | 2.60 | 3.00 |
| 5. Acres in general community facilities[c] | 0.38 | 0.76 | 1.20 | 1.50 | 1.90 |
| *Aggregate Area* | | | | | |
| 6. Acres: total | 6.63 | 8.41 | 11.40 | 13.90 | 16.60 |
| 7. Acres per 1000 persons | 6.63 | 4.20 | 3.80 | 3.47 | 3.32 |
| 8. Square feet per family | 1050 | 670 | 600 | 550 | 530 |
| Multifamily development[d] | | | | | |
| *Area in Component Uses* | | | | | |
| 1. Acres in school site | 1.20 | 1.20 | 1.50 | 1.80 | 2.20 |
| 2. Acres in playground | 2.75 | 3.25 | 4.00 | 5.00 | 6.00 |
| 3. Acres in park | 2.00 | 3.00 | 4.00 | 5.00 | 6.00 |
| 4. Acres in shopping center | 0.80 | 1.20 | 2.20 | 2.60 | 3.00 |
| 5. Acres in general community facilities[c] | 0.38 | 0.76 | 1.20 | 1.50 | 1.90 |
| *Aggregate Area* | | | | | |
| 6. Acres: total | 7.13 | 9.41 | 12.90 | 15.90 | 19.10 |
| 7. Acres per 1000 persons | 7.13 | 4.70 | 4.30 | 3.97 | 3.82 |
| 8. Square feet per family | 1130 | 745 | 680 | 630 | 610 |

[a]This table combines the recommended or assumed values.
[b]With private lot area of less than ¼ acre per family (for private lots of ¼ acre or more park area may be omitted).
[c]Allowance for indoor social and cultural facilities (church, assembly hall, etc.) or separate health center, nursery school, etc.
[d]Or other development predominantly without private yards.

## GENERAL STRUCTURE

The standard system for coding land use activity is comprised of 9 one-digit categories (2 of which have been assigned to "manufacturing"), 67 two-digit categories, 294 three-digit categories, and 772 four-digit categories. The categories at the four-digit level identify land use activity in the greatest detail, and as the system is aggregated to the three-, two-, and one digit levels the categories become more generalized. The structure of this classification system, therefore, permits an agency to select the level of detail considered most appropriate for analysis and presentation of its data.

## Highway Classification

Highway systems are grouped into a number of different classifications for administrative, planning, and design purposes. The Federal Aid financing system, state-county-city's administrative systems, and commercial-industrial-residential-recreational systems are examples of the variety of highway classifications.

In the most basic classification system for design work, highways and streets are grouped into: (1) interstate, primary (excluding interstate), secondary, and tertiary road classes in rural areas, and (2) expressway, arterial, collector, and local road classes in urban areas. These classifications usually carry with them a set of suggested minimum design standards which are in keeping with the

### The Categories at the Two-Digit Level of Generalization

| Code | Category | Code | Category |
|---|---|---|---|
| 1 | Residential | 7 | Cultural, entertainment, and recreational |
| 2 and 3 | Manufacturing | | |
| 4 | Transportation, communication, and utilities | 8 | Resource production and extraction |
| 5 | Trade | 9 | Undeveloped land and water areas |
| 6 | Services | | |

### The Categories at the Two-Digit Level of Generalization

| Code | Category | Code | Category |
|---|---|---|---|
| 1 | Residential | 11 | Household units |
| | | 12 | Group quarters |
| | | 13 | Residential hotels |
| | | 14 | Mobile home parks or courts |
| | | 15 | Transient lodgings |
| | | 19 | Other residential |
| 2 | Manufacturing | 21 | Food and kindered products—manufacturing |
| | | 22 | Textile mill products—manufacturing |
| | | 23 | Apparel and other finished products made from fabrics, leather, and similar materials—manufacturing |
| | | 24 | Lumber and wood products (except furniture)—manufacturing |
| | | 25 | Furniture and fixtures—manufacturing |
| | | 26 | Paper and allied products—manufacturing |
| | | 27 | Printing, publishing, and allied industries |
| | | 28 | Chemicals and allied products—manufacturing |
| | | 29 | Petroleum refining and related industries |

importance of the system and are governed by the specific transportation services the system is to perform. The principal consideration for designating roads into systems are the travel desires of the public, land-access requirements based on existing and future land use, and continuity of the system. Four basic purposes of urban street systems have been suggested:

1. Expressway system (including freeways and parkways)—providing for expeditious movement or large volumes of through traffic between areas and across the city, and not intended to provide land-access service.

2. Major arterial system—providing for the through traffic movement between areas and across the city, and direct access to abutting property; subject to necessary control of entrances, exists, and curb use.

3. Collect or street system—providing for traffic movement between major arterials and local streets, and direct access to abutting property.

4. Local street system—providing for direct access to abutting land, and for local traffic movements.

These basic purposes of city street systems are similar to those of rural interstate, primary, secondary, and tertiary highways, respectively, so far as the various degrees of accommodation of through traffic and land access is concerned. However, regional as well as national highway transportation requirements must be met by rural highways. The tables compare the overall criteria of urban street and rural highway classifications.

The principles and elements of geometric design for both urban and rural facilities are generally the same. However, to meet urban and rural traffic demands, design details are often varied because speeds, traffic composition, lengths and purposes of trips, etc., are not the same.

| 3 | Manufacturing (continued) | 31 | Rubber and miscellaneous plastic products—manufacturing |
| | | 32 | Stone, clay, and glass products—manufacturing |
| | | 33 | Primary metal industries |
| | | 34 | Fabricated metal products—manufacturing |
| | | 35 | Professional, scientific, and controlling instruments: photographic and optical goods; watches and clocks—manufacturing |
| | | 39 | Miscellaneous manufacturing NEC |
| 4 | Transportation, communication, and utilities | 41 | Railroad, rapid rail transit, and street railway transportation |
| | | 42 | Motor vehicle transportation |
| | | 43 | Aircraft transportation |
| | | 44 | Marine craft transportation |
| | | 45 | Highway and street right-of-way |
| | | 46 | Automobile parking |
| | | 47 | Communication |
| | | 48 | Utilities |
| | | 49 | Other transportation, communication, and utilities, NEC |
| 5 | Trade | 51 | Wholesale trade |
| | | 52 | Retail trade—building materials, hardware, and farm equipment |
| | | 53 | Retail trade—general merchandise |
| | | 54 | Retail trade—food |
| | | 55 | Retail trade—automotive, marine craft, aircraft, and accessories |
| | | 56 | Retail trade—apparel and accessories |
| | | 57 | Retail trade—furniture, home furnishings, and equipment |
| | | 58 | Retail trade—eating and drinking |
| | | 59 | Other retail trade, NEC |
| 6 | Services | 61 | Finance, insurance, and real estate services |
| | | 62 | Personal services |
| | | 63 | Business services |
| | | 64 | Repair services |
| | | 65 | Professional services |
| | | 66 | Contract construction services |
| | | 67 | Governmental services |
| | | 68 | Educational services |
| | | 69 | Miscellaneous |
| 7 | Cultural, entertainment, and recreational | 71 | Cultural activities and nature exhibitions |
| | | 72 | Public assembly |
| | | 73 | Amusements |
| | | 74 | Recreational activities |
| | | 75 | Resorts and group camps |
| | | 76 | Parks |
| | | 79 | Other cultural, entertainment, and recreational, NEC |
| 8 | Resource production and extraction | 81 | Agriculture |
| | | 82 | Agriculture related activities |
| | | 83 | Forestry activities and related services |
| | | 84 | Fishing activities and related services |
| | | 85 | Mining activities and related services |
| | | 89 | Other resource production and extraction, NEC |
| 9 | Undeveloped land and water areas | 91 | Undeveloped and unused land area (excluding noncommercial forest development) |
| | | 92 | Noncommercial forest development |
| | | 93 | Water areas |
| | | 94 | Vacant floor area |
| | | 95 | Under construction |
| | | 99 | Other undeveloped land and water areas, NEC |

Source: *Standards for Street Facilities and Services*, Procedure Manual 7A, National Committee on Urban Transportation, Public Administration Service, Chicago, 1958, p. 11.

### Urban Street Classification Criteria

| Element | System | | | |
|---|---|---|---|---|
| | Expressway | Major Arterial | Collector | Local |
| **Service function** | | | | |
| Movement | Primary | Primary | Equal | Secondary |
| Access | None | Secondary | Equal | Primary |
| Principal trip length | Over 3 miles | Over 1 mile | Under 1 mile | Under ½ mile |
| Use by transit | Express | Regular | Regular | None, except CBD |
| **Linkage** | | | | |
| Land uses | Major generators and CBD | Secondary generators and CBD | Local areas | Individual sites |
| Rural highways | Interstate and state primary | State primary and secondary | County roads | None |
| Spacing | 1–3 miles | 1 mile | ½ mile | |
| Percentage of system | 0.8 | 20–35 | | 65–80 |

### Rural Road Classification Criteria*

| Element | System | | | |
|---|---|---|---|---|
| | Interstate | Primary | Secondary | Tertiary |
| **Service function** | | | | |
| Movement | Primary | Primary | Equal | Secondary |
| Access | Controlled | Secondary | Equal | Primary |
| **Linkage to:** | | | | |
| Geographic | Major cities | Smaller cities | Smaller cities and regions | Farm-to-market |
| Urban streets | Expressways | Expressways and major arterials | Major arterials and collectors | Collectors and local |
| Percentage of system | 2 | 17 | 10 | 71 |

*Includes surfaced roads only.

Houses should always front on a minor street. If the subdivision is properly designed, there will be a minimum of traffic on the minor street providing a maximum of privacy and safety.

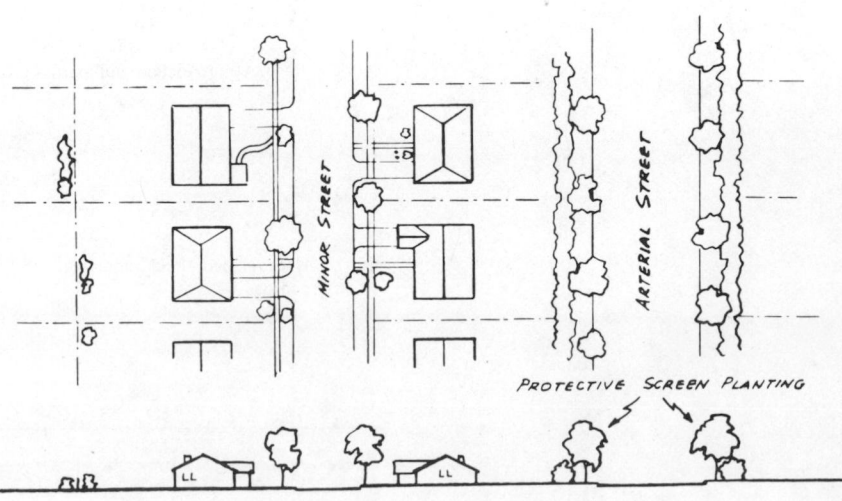

## General Highway and Street Standards

| Type of Facility | Function and Design Features | Spacing | Right-of-way | Widths Pavement | Desirable Maximum Grades | Speed | Other Features |
|---|---|---|---|---|---|---|---|
| Freeways | Provide regional and metropolitan continuity and unity. Limited access; no grade crossings; no traffic stops. | Variable; related to regional pattern of population and industrial centers | 200–300 ft | Varies; 12 ft per lane; 8- to 10-ft shoulders both sides of each roadway; 8- to 60-ft median strip. | 3% | 60 mi/h | Depressed, at grade, or elevated. Preferably depressed, through urban areas. Require intensive landscaping, service roads, or adequate rear lot building set-back lines (75 ft) where service roads are not provided. |
| Expressways | Provide metropolitan and city continuity and unity. Limited access; some channelized grade crossings and signals at major intersections. Parking prohibited. | Variable; generally radial or circumferential | 200–250 ft | Varies 12 ft per lane; 8- to 10-ft shoulders; 8- to 30-ft median strip. | 4% | 50 mi/h | Generally at grade. Requires landscaping and service roads or adequate rear lot building set-back lines (75 ft) where service roads are not provided. |
| Major Roads (Major Arterials) | Provide unity throughout contiguous urban area. Usually form boundaries for neighborhoods. Minor access control; channelized intersections; parking generally prohibited. | 1½ to 2 miles | 120–150 ft | 84 ft maximum for four lanes, parking and median strip. | 4% | 35 to 45 mi/h | Require 5 ft wide detached sidewalks in urban areas, planting strips (5 to 10 ft wide or more) and adequate building set-back lines (30 ft) for buildings fronting on street; 60 ft for buildings backing on street. |
| Secondary Roads (Minor Arterials) | Main feeder streets. Signals where needed; stop signs on side streets. Occasionally form boundaries for neighborhoods. | ¾ to 1 mile | 80 ft | 60 ft | 5% | 35 to 40 mi/h | Require 5 ft wide detached sidewalks, planting strips between sidewalks and curb 5 to 10 ft or more, and adequate building set-back lines (30 ft). |
| Collector Streets | Main interior streets. Stop signs on side streets. | ¼ to ½ mile | 64 ft | 44 ft (2- to 12-ft traffic lanes; 2- to 10-ft parking lanes) | 5% | 30 mi/h | Require at least 4 ft wide detached sidewalks; vertical curbs; planting strips are desirable; building setback lines 30 ft from right of way. |
| Local Streets | Local service streets. Nonconducive to through traffic. | at blocks | 50 ft | 36 ft where street parking is permitted. | 6% | 25 mi/h | Sidewalks at least 4 ft in width for densities greater than 1 d.u./acre, and curbs and gutters. |
| Cul-de-sac | Street open at only one end, with provision for a turn-around at the other. | only wherever practical | 50 ft | (90 ft dia. turn-around) 30 to 36 ft (75-ft turn-around) | 5% | | Should not have a length greater than 500 ft. |

SOURCE: *George Nez, Standards for New Urban Development—The Denver Background, Reprinted by Permission of Urban Land, vol. 20, no. 5, Urban Land Institute, 1200 18th Street, N. W., Washington, D. C.*

When houses must front on an arterial street, a marginal access street should be provided for these houses. This will eliminate any conflict between through and local circulation. *Note:* Provide an alley for rear lot access. (From J. DeChiara and L. Koppelman, *Planning Design Criteria,* Van Nostrand-Reinhold Co., New York, 1969.)

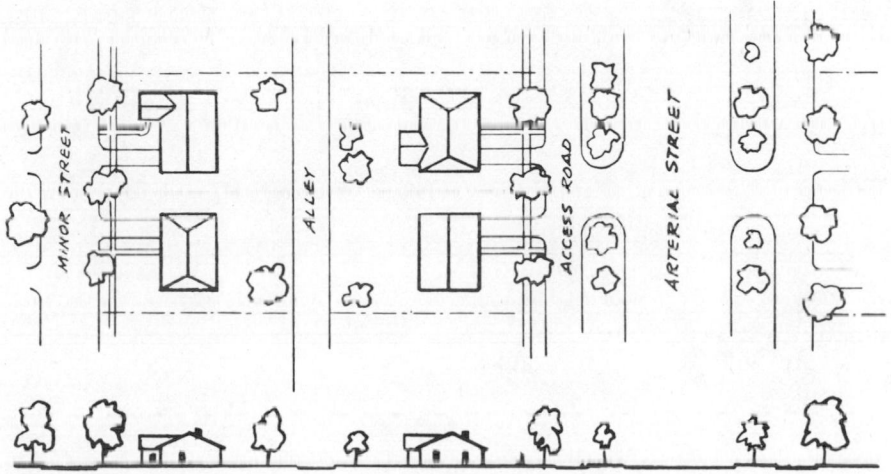

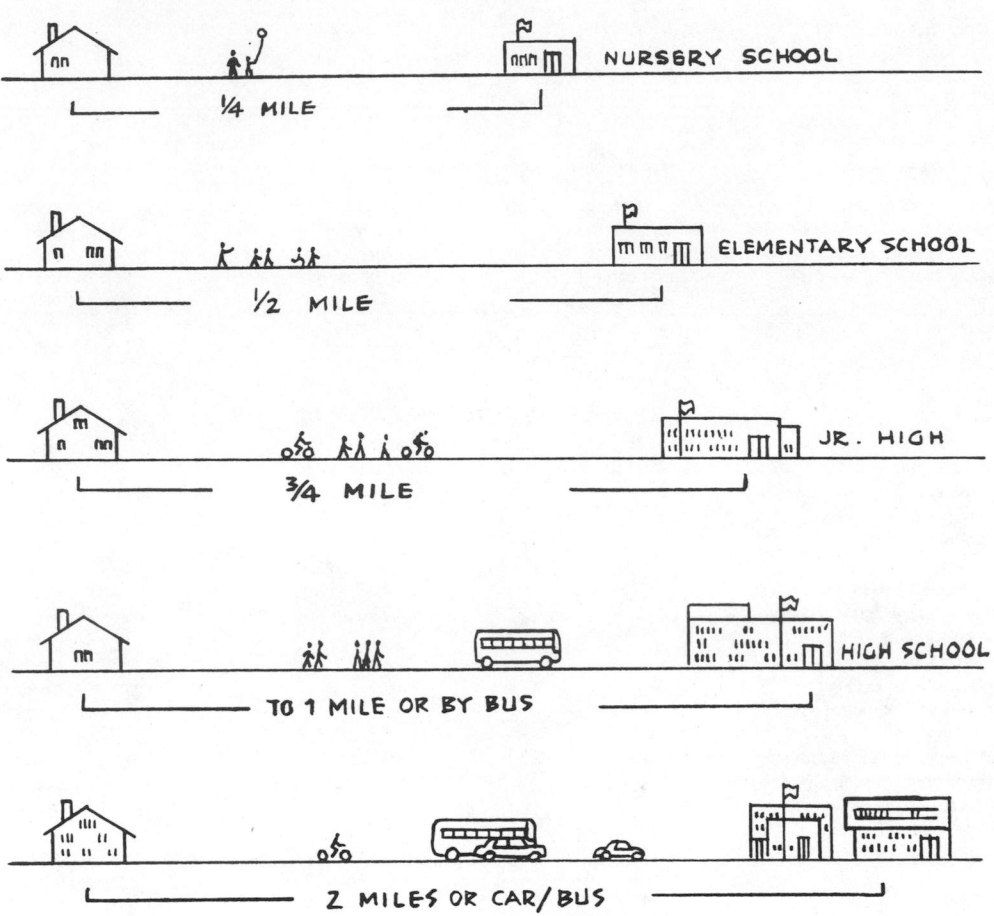

NURSERY SCHOOL

¼ MILE

ELEMENTARY SCHOOL

½ MILE

JR. HIGH

¾ MILE

HIGH SCHOOL

TO 1 MILE OR BY BUS

2 MILES OR CAR/BUS

**Maximum walking distances for students.**

All distances given are considered to be maximum.

In high density, urban areas most schools are located within the maximum recommended walking distances.

In low density, rural areas many schools are located beyond maximum recommended walking distances. They must have bus service.

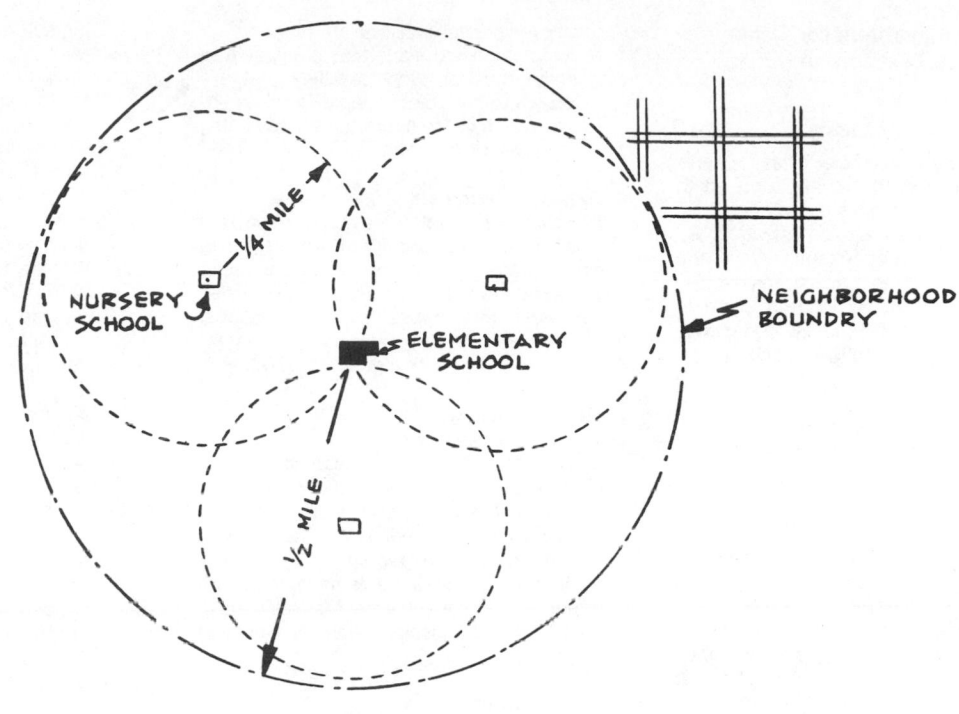

Typical neighborhood organization.

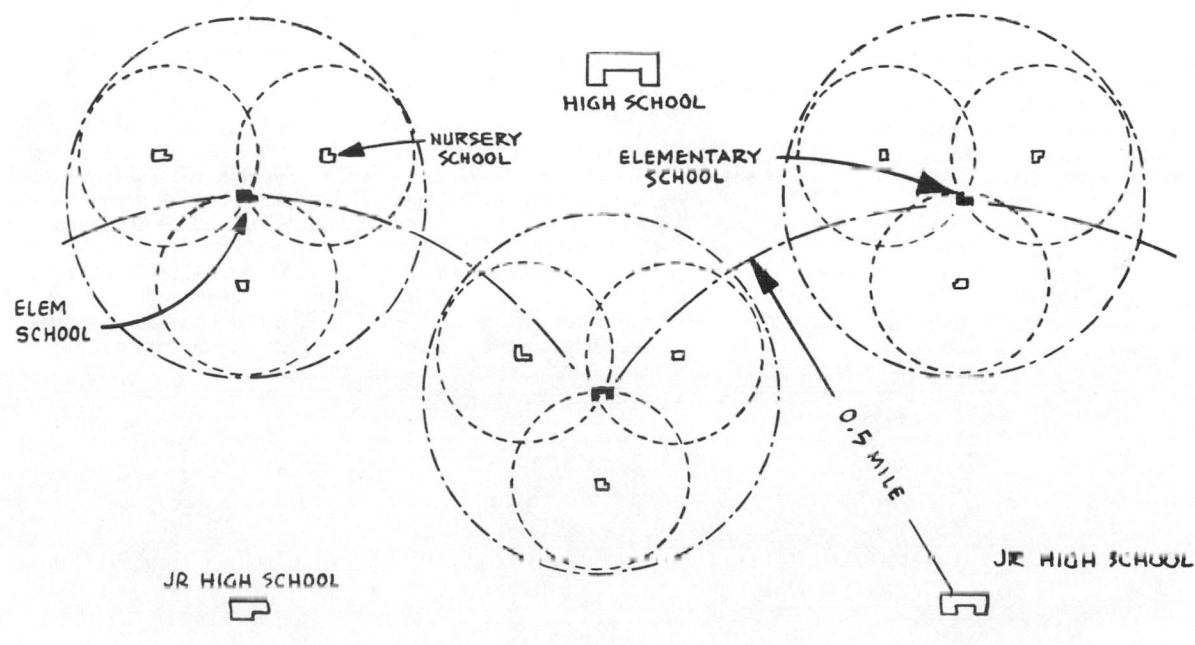

Typical district organization.

## Location Considerations for Other Key Community Features and Functions

### 1. Airports
a. There should be sufficient flat topography so that aircraft are not required to take off or land across obstructing hills.
b. Noise isolation should be provided. Although adequate access roadways and reasonable distances are important, the airport should not be located close to residential or other special community functions, e.g., hospitals, schools, recreational facilities, or churches. Early zoning policies are necessary to keep these latter functions and/or residential development from surrounding the planned airport site.

### 2. Government Offices
a. Centralize government offices with respect to principal user functions, i.e., business, legal services, law enforcement services, etc.
b. Local citizens should be able to reach government offices by means of public transit services.

### 3. Disaster Services
a. Decentralize fire stations and ambulance and hospital services; i.e., equalize travel requirements for all principal community areas.
b. Provide access to main thoroughfares; i.e., minimize the time required to pass congested business or residential streets and potential traffic barriers.

### 4. Shopping Centers
a. Locate regional shopping centers so that they serve large residential areas.
b. Provide local shopping-center sites to serve specific areas such as residential neighborhoods, center-city areas, and separate commercial areas.

### 5. Churches
a. Set aside specific sites for a range of religious denominations for each defined community area.
b. Set aside specific sites within each planned downtown area where center-city residential facilities are planned.
c. Avoid choosing a site that is adjacent to businesses or other facilities that are known to be objectionable to parishoners, e.g., bars, restaurants, supper clubs, theatres, or the local jail.

### 6. Railroad Terminals
Although the site for a passenger terminal should be generally centralized with respect to access from all parts of the planned community, rail lines should be as far from residential, school, church, hospital, and recreational functions as practicable because of the potential noise and air pollution generated by trains.

### 7. Heavy Industry
Like railroads, heavy industry often produces undesirable noise and air pollution. In addition, certain hazards may be associated with specific industrial facilities and operations, e.g., fires, explosions, and chemical contamination. Set aside sites for expected heavy industry that will minimize these problems for the entire community, but especially for residential, school, church, hospital, and recreational areas.

### 8. Detention Facilities
Detention facilities should not be located adjacent to, and preferably should be considerably distant from, residential, school, church, hospital, and recreational areas. Although temporary detention facilities, such as local jails, are necessary within the principal boundaries of the community, most people are innately opposed to having any major penal facility located near or within the community because of the real or imagined hazards sometimes associated with detention facility operations.

## Planning Checklist for Minimizing Architectural Barriers to the Mobility of Handicapped Citizens

1. Provide wheelchair-use ramps at all intersections, footbridges, parking-lot-to-sidewalk interfaces, public buildings, etc.
2. Provide wheelchair-use clearance for all public-use doorways, elevator doors, rest-room entries, and interior stall entrances and hallways.
3. Provide smooth thresholds and crossing surfaces at all public-use doorways, railway crossings, etc.
4. Provide optimized identification signs for the partially sighted at all intersections used by these handicapped individuals (i.e., position signs at eye level so that users can get close to the signs and provide optimized lettering in terms of font, stroke width, letter height, and figure-ground contrast).
5. Provide tactile identification devices at all intersections used by the totally blind and/or audio signaling devices that both identify the intersection and give information about when to cross safely.
6. Establish a general policy regarding placing obstructions within normal pathways (i.e., mailboxes, trash containers, telephone booths, utility poles, and guy wires within the sidewalk perimeter) and/or provide suitable guard devices around these obstructions so that the blind can identify and avoid them.
7. Position any controls used by the handicapped so that they are accessible both to the handicapped person in a wheelchair and to the nonhandicapped or handicapped user who may be standing.
8. Establish a policy regarding the positioning of public-use devices such as drinking fountains and telephones so that they are accessible to both the handicapped person in a wheelchair and the nonhandicapped or handicapped person who normally stands to use these devices.
9. Make public sidewalks wide enough so that handicapped persons are not forced to walk or maneuver their wheelchairs too close to the street.
10. Provide surface guide markers that can be easily identified by the blind or partially blind by touching these markers with a cane.
11. Consider the effect that inclement weather conditions may have on particular barrier-removal systems, e.g., snow and ice that could cover surface markers.
12. Do not use or allow to be used any gratings in sidewalk or street surfaces into which the wheel of a wheelchair could drop or become caught.

RESIDENTIAL

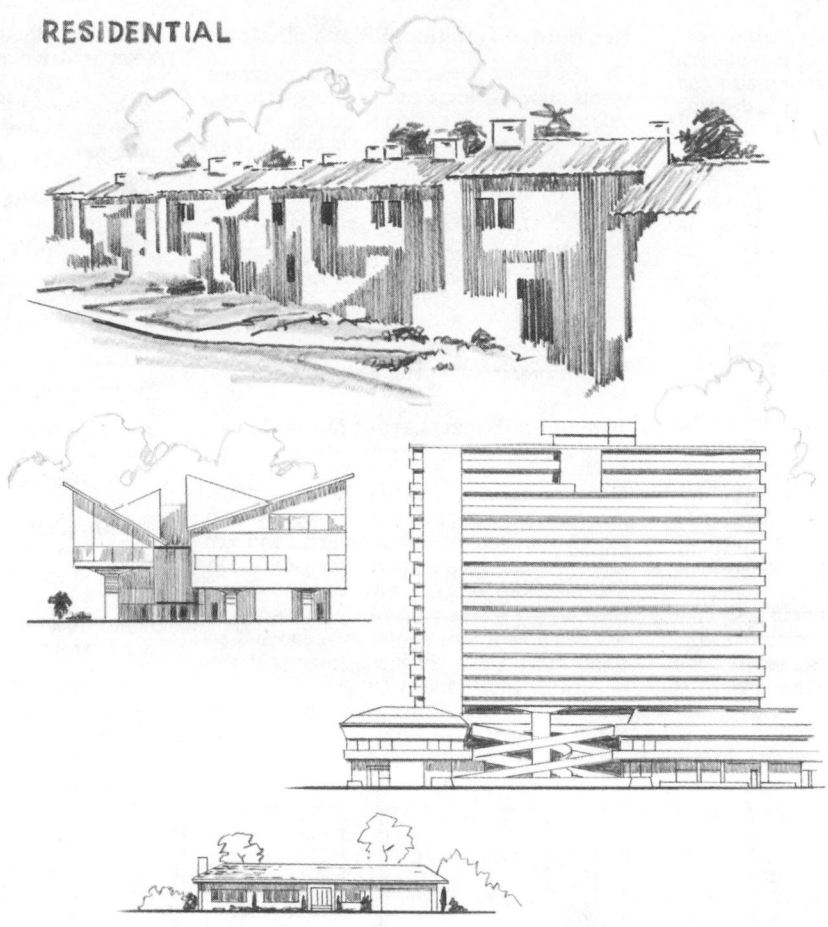

## RESIDENTIAL

### Residential Systems

Whether planning a single-family or a multifamily dwelling, a separate house, or an apartment building, planners should approach residential systems from the standpoint of optimizing dynamic activity relationships. The human is, of course, the focal point of all such activities, and the objective should be to satisfy human needs for all expected activities, such as the following:

1. Sleep and rest
2. Personal hygiene and dressing and undressing
3. Preparing and eating food
4. Performing household chores (e.g., laundry, sewing, and cleaning)
5. Performing exterior maintenance and gardening
6. Studying and record keeping
7. Entertaining guests
8. Child care
9. Performing hobby tasks
10. Performing shopping chores

In addition to these activities of residents, attention should also be given to the following supportive activities of nonresidents:

1. Deliveries (mail, milk, newspapers, furnishings and appliances, packages, etc.)
2. Special services (plumbing, electrical work, building repair, etc.)
3. Parking (for both residents and guests)

The principal objectives in creating a residential system plan should be:

1. Convenience (i.e., arrangement of both external and internal activity areas so that unnecessary steps, traffic interferences, and communication difficulties between household members are minimized)
2. Privacy (i.e., appropriate visual and sound barriers to provide individuals with the necessary isolation for specific tasks such as resting, sleeping, and studying; separation of adults' and children's activities; and specific areas for personal activities such as bathing)
3. Personal interaction (i.e., for group and family activities)
4. Security (i.e., protection from vandalism and from natural or man-induced disasters such as high winds, inclement weather, floods, earthquakes, and fire)

Finally, considerable attention should be given to the storage requirements associated with all activities, including the following:

1. Clothing
2. Foodstuffs
3. Books and records
4. Housekeeping equipment and materials
5. Exterior maintenance and gardening equipment and materials
6. Linens
7. Eating and cooking utensils
8. Hobby, entertainment, and play equipment
9. Family automobiles and recreational vehicles and equipment
10. Waste and garbage
11. Other special storage requirements (for drugs and first-aid supplies, weapons, suitcases, empty boxes, etc.)

Whether the residence plan calls for a fully partitioned interior or a more "open space" configuration, all the above activities and requirements should be addressed, starting with the individual and his or her activities and working "outward." That is, planning should begin with a thorough analysis of the individual and the tasks that he or she must perform; then the planner must determine what materials and equipment or furnishings the individual needs to perform each task. This analysis establishes the space requirements for each activity center. Activity centers should then be analyzed to determine the most effective arrangement of activity spaces. This determines the size and shape of the total dwelling. A final analysis should then be made of external activities; this analysis determines the size and shape of the land site upon which the dwelling is to be placed. This is referred to as "designing from the human—out" and is quite different from typical planning approaches. That is, most architectural plans are developed from the opposite point of view. Land sites are selected first, and then a building size and form are chosen that will fit the land site; this is followed by an attempt to divide the building into workable activity modules. Whether the human and his or her activities are properly accommodated is often a matter of sheer coincidence or luck.

Obviously, some compromise between the two approaches is usually required since available land, costs, and other factors make the ideal "human—out" approach somewhat hypothetical. However, unless this approach is considered, designers seldom know what they are compromising or how much they may have compromised the human-activity objective until it is too late to do much about it. Experience indicates that a better compromise is made when the design is approached from the user-activity reference point than when it is approached from the other end of the reference chain.

## Residential Functional Planning Model

The key family, guest, and service activity elements shown below must be arranged to provide balanced interaction, isolation, and accessibility. If priority is given, it should be given to residents.

Among the more important layout considerations are the following:

1. Ease of ingress and egress
2. Ease of traffic flow
3. Privacy
4. Security
5. Emergency disaster control
6. Environmental control

## Functional Space Layout Model for a Single-Family Dwelling

Certain adjacency implications are illustrated in the model on page 41 (e.g., the family sleeping quarters are associated with a bathroom, and the kitchen is associated with both indoor and outdoor dining), certain separations are illustrated (e.g., the guest areas and entrance are separated from the family areas and entrances), and differentiation is made between guest and family parking. Other implications to draw from the model are that, although some things should be located adjacent to each other and although some should purposely be separated, a number of typical functions have no strong relationships that require special attention during the layout process; i.e., many alternatives are available to match the client's preferences.

## RESIDENTIAL—SINGLE FAMILY

### Single-Family Residential Planning

Although it is not a necessary requirement that activity areas of the residence be separated into commonly defined compartments, such as living room, bedroom, or bathroom, such identifications provide a convenient classification for discussing such areas relative to examining alternate arrangements. The following guidelines are useful in identifying arrangement objectives.

#### 1. Entrances

There should always be one entry that is primarily "guest-oriented" and one or more others that are for family use only and/or for infrequent use by service people. The following considerations are important:

a. The guest entrance should be readily visible and identifiable from the normal approach by both social guests and others

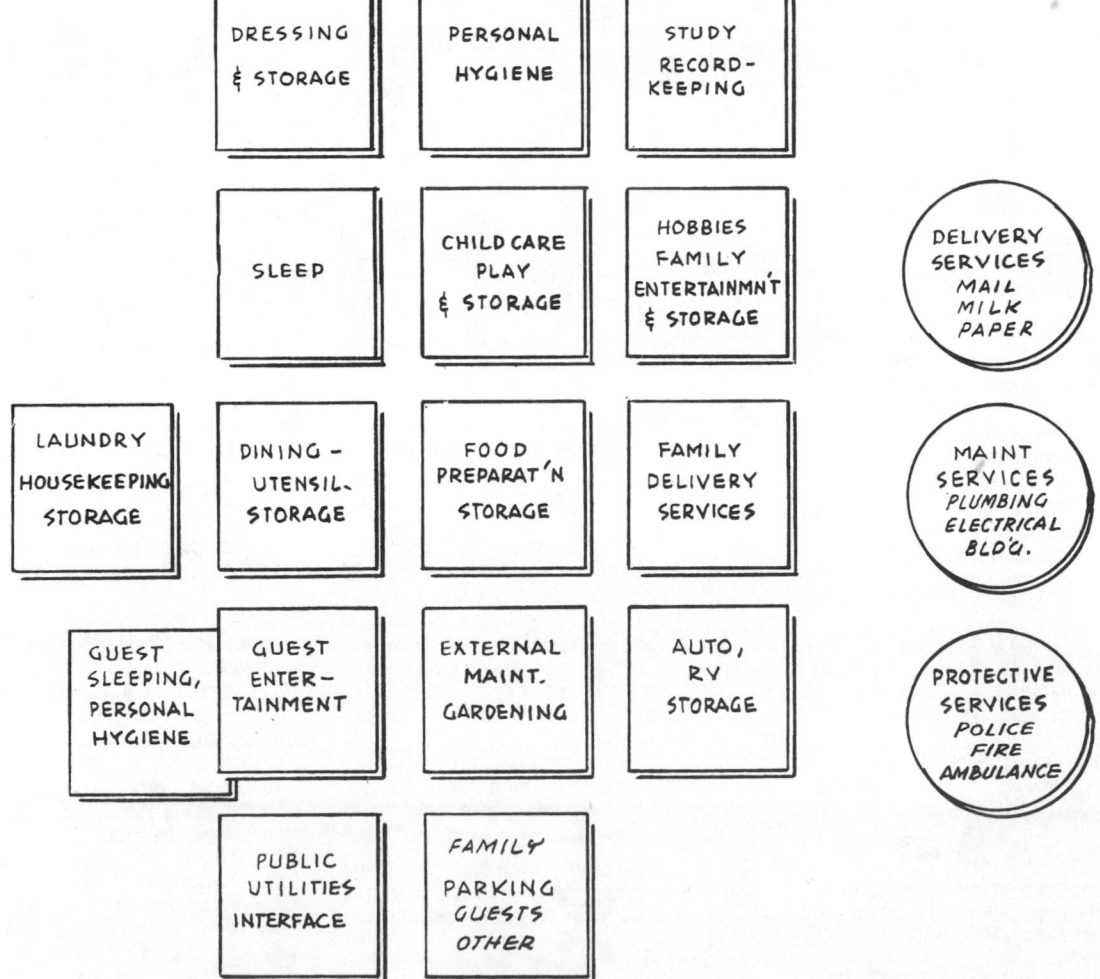

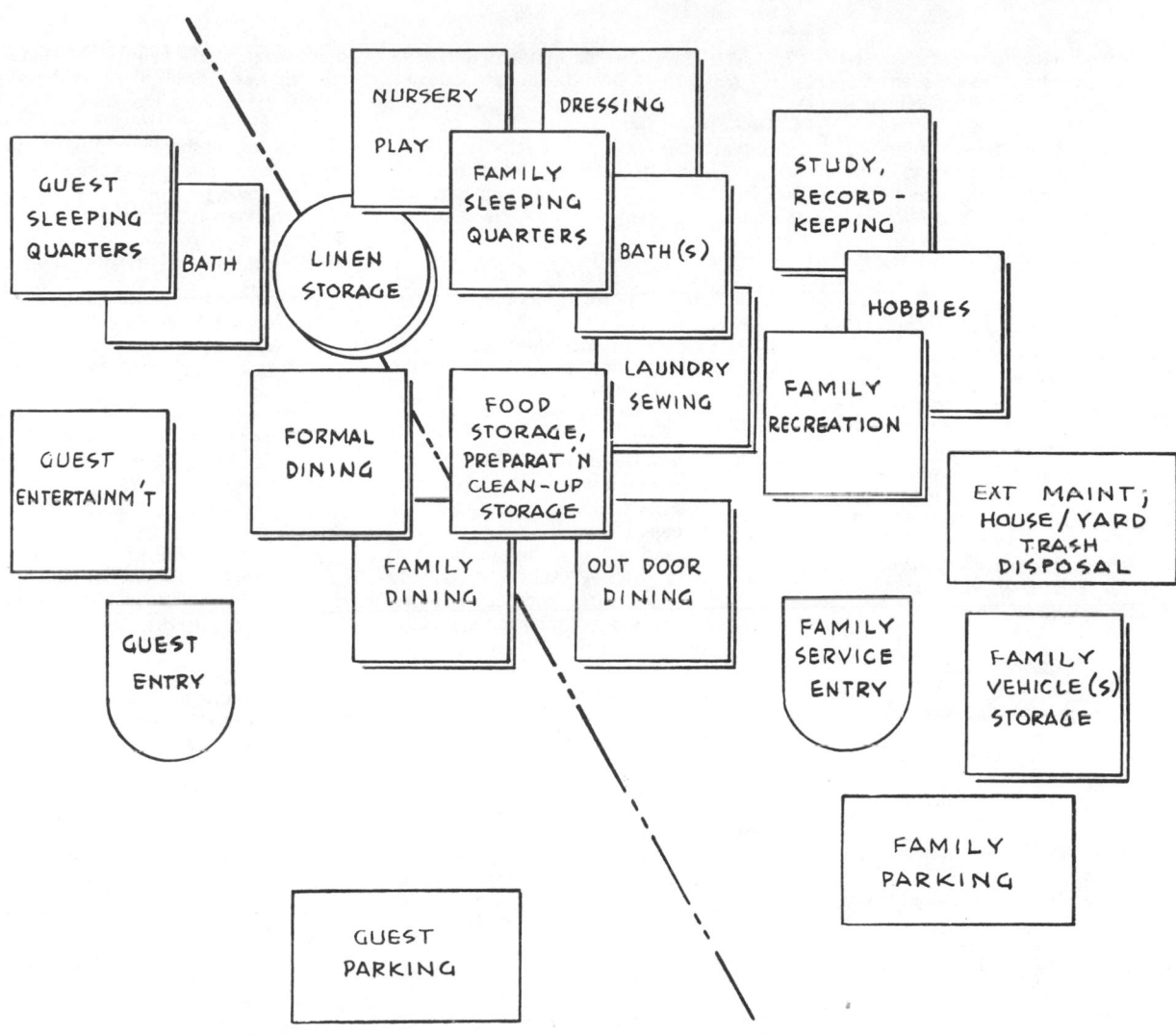

such as the mail carrier and the paper carrier.

b. Family and service entrances need not be readily visible or identifiable by guests, but they should be located for maximum convenience by family members (including access between the family car and the kitchen area for grocery transfer, access to and from rear yard facilities and internal facilities such as the kitchen and bathroom, etc.).

c. All entrances should have an adequate threshold area so that the user is not inconvenienced as he or she opens the door. Such external thresholds should have adequate cover to protect the user from rain, and there should be proper lighting so that the user can find a lock, a doorknob, etc. The threshold to the guest entrance should be large enough so that several guests can gather on the threshold, under cover, without being knocked over when the door is opened. Lighting should be provided so that the guests can be seen and recognized. (Many porch lights are positioned so that guests' faces are in shadow or so that their shadow makes it difficult for them to see where they are stepping.)

d. All entry doors should be provided with vandal-proof locks, and any windows in doors should be located so that the glass cannot be broken for easy access to an inside locking bolt or door handle.

e. A sufficient number of doors should be provided and located to ensure that emergency escape is possible from any part of the house in case of fire. If windows are designated as alternative escape routes, they should be accessible from inside and should open to a plausible and safe escape route on the outside.

f. At least one entrance should be accessible for use by a person who is confined to a wheelchair; i.e., the threshold between the walkway and the porch should be properly ramped.

g. Internal entryway or foyer space should be provided so that there is adequate room for guests to collect inside and remove their outer clothing without being jostled by the closing of the door. A cloak closet should be located in this area so that wet garments will not drip onto the floor of a living-room area. Avoid combining the entryway with a living room, thus requiring people to track through the living room to get to other parts of the house.

## 2. Bathrooms

The number, type, and location of bathrooms should be determined on the basis of the following considerations:

a. The size of the family should be the primary determinant, although the size of the dwelling also has a bearing on the number of bathrooms that should be provided. Although in the past it was typical practice to provide only one bathroom, most people now expect at least two (sometimes a bath and a half). Since the purpose of having two or more bathrooms is to allow people to use these facilities simultaneously, all bathrooms should be accessible to anyone (contradictory to many current approaches, in which one bathroom is secreted within a master bedroom suite).

b. The convenience of bathrooms to specific activity areas should determine their location. For example, at least one bathroom should be convenient to sleeping areas, one to the kitchen and family recreational area, and perhaps one to special guest areas. An important consideration should be the minimizing of bathroom noises that could be heard by guests and the seclusion of the bath room from direct view from typical guest-occupied areas. When special overnight guest facilities are included in the house plan, it is always desirable to provide a separate, convenient guest bathroom. Finally, it is desirable to have one bathroom (usually a half bath) accessible from outside the house so that children, bathers from a pool, etc., can reach this facility without having to go through the rest of the house.

c. All bathrooms should be large enough to accommodate local storage of linens, paper, and cleaning supplies.

d. At least one bathroom should be large enough to accommodate a wheelchair user. Although the homeowner may not be handicapped, there is always the possibility that he or she may have to use a wheelchair temporarily or may have a handicapped guest. It is practically impossible to modify a bathroom that is too small to accommodate a wheelchair user at some later date without extensive structural modification.

### 3. Bedrooms and Dressing Spaces

a. Location: Since sleep and rest are of prime importance, bedroom areas should be isolated from typical noise sources, both internal and external. The following should be considered in locating bedrooms and related dressing spaces:

(1) Bedrooms should be located away from noisy streets.

(2) Bedrooms should be located away from typically noise-producing areas of the house, e.g., recreation rooms and family rooms that may contain a hi-fi or TV set.

(3) Bedrooms should be convenient to bathroom facilities.

(4) Bedrooms should be convenient to primary linen storage and laundry facilities.

(5) Family bedrooms should be reasonably isolated from the guest bedroom and its accompanying bathroom.

b. Size: The size of bedrooms should be based on use dynamics rather than on "static furniture accommodation"; i.e., consider the bedroom not only as a place to sleep but also as a place to store clothing, to dress and undress, and to perform certain grooming functions. If a particular home plan is meant for a family with children, it is important to consider the children's needs for play space, including storage of toys.

c. Emergency escape: Emergency escape from bedrooms in case of fire is an especially important consideration in terms of both bedroom location and window location and configuration. Sleepers who are suddenly confronted with a fire emergency often do not have time to escape by normal exits or routes. Special attention should be given to the size and placement of windows relative to occupants' being able to get them open and crawl out of them and also to the exterior escape mode after they have crawled

out of a window; this interacts with exterior house configurational decisions.

### 4. Kitchen and Dining Areas

Food-preparation and dining activities should be addressed simultaneously since the complex interaction between these functions and the related facilities demands special attention to the following:

a. Location: The following should be considered in locating kitchen and dining areas:

(1) Delivery of food supplies to the kitchen and removal of kitchen waste to the outside

(2) Access between the kitchen and dining areas

(3) Access between the kitchen and family-exclusive areas, e.g., bedrooms, bathrooms, and the family recreation room

(4) Isolation of the kitchen (in terms of odors, visibility, etc.) from primary guest areas (i.e., nonfamily guests)[9]

(5) Access between the kitchen and the patio area (i.e., without traversing other parts of the house)

(6) Access between the formal dining area and the primary guest entertainment or visitation area, e.g., the living room

b. Size: The size of the kitchen and/or dining areas is more a matter of client lifestyle; e.g., an eventual owner who likes to use the kitchen and dining areas as a focal point may desire rather spacious accommodations. On the other hand, another client may desire more of an "efficiency arrangement" wherein a fairly minimal space is required for the kitchen or dining area. Regardless of orientation, certain minimum space requirements dictate consideration of the following:

(1) Specific spaces for typical kitchen appliances (refrigerator, range, etc.).

(2) Specific storage and counter space for sink, dishwasher, trash compactor, dishes and flatware, cooking utensils, crystal, dry food staples, canned goods, kitchen and dining linens, kitchen housekeeping supplies, trash, miscellaneous paper products, etc. Counter space should be provided for specific food-preparation activities (cleaning, cutting, mixing, setting out serving dishes and cooking utensils, etc.).

(3) Specific storage within the formal dining area, for either built-in or purchased cabinets, which would contain dishes and flatware, crystal, and table linens.

### 5. Laundry, Ironing, and Sewing Areas

The frequent and regular tasks of gathering,

processing, and returning clothing, linens, etc., between and among bedrooms, bathrooms, the kitchen, the dining room, etc., requires careful consideration not only of the arrangement of these but also of the primary area or areas in which the actual laundry, ironing, or sewing is done. Perhaps the only recommendation should be that, whatever the proposed arrangement, the following objectives should be sought:

a. Minimize travel distance.

b. Organize to make single-trip patterns feasible (i.e., passing several pickup areas to and from the laundry facility, ironing and mending adjacent to the laundry, etc.).

c. Provide adequate interim storage for soiled items adjacent to washing facilities and the collection point and for laundry soaps, bleaches, etc., in the laundry facility.

d. Provide clearance around key laundry equipment for ease of loading and unloading and for possible use of laundry carts or baskets.

*Note:* Although some designers tend to assume that modern homemakers do not desire outdoor drying facilities, there are still many people who prefer to hang laundry out in the fresh air. Unless the designer can clear this issue with a particular client, outdoor drying convenience should be given serious consideration. Similarly, basement drying should also be considered as a second alternative in the event that a basement is being planned for the proposed home. Again, distance and directness of clothing transfer are important factors in locating the laundry facility.

### 6. Other Areas

Although numerous other areas are typically considered in the design of homes (e.g., living room, family room, study, and nursery), these generally do not impose quite the same demands relative to convenience. However, all such spaces should still be examined on the basis of activity versus traffic convenience, dynamic clearance for humans and furnishings, etc. Many of the guidelines noted above also imply reciprocal needs such as convenience to the bathroom, food service, and housekeeping.

In addition, however, during the conceptual stage of design, special attention should be given to the exterior amenities associated with the home, including location and layout of sidewalks, driveways, the garage, and other possible outbuildings. It is especially important not to leave the exterior plan until the main house configuration is fixed. Often this creates constraints on where these amenities can be located, resulting in unsatisfactory solutions from the user's standpoint. The following areas are especially significant in terms of their interaction with various activities and parts of the house:

a. Swimming pool: Provide direct access from the pool to a bathroom so that bathers can get from the pool to a dressing area without dripping water through the house.

---

[9]Many clients desire a more informal approach to entertaining and therefore do not require that the kitchen area be completely isolated from the view of guests—and, in fact, some encourage guest participation in food-preparation activities. However, it is suggested that when this approach is taken, the arrangement of kitchen and guest areas be designed so that guests are not forced to look directly into the kitchen, although this may be more a matter of furniture arrangement.

b. Patio: Provide direct access to and from the kitchen.

c. Driveway: Provide direct access to and from both guest and service (to kitchen) entrances.

d. Rear yard: Provide direct access (preferably from the rear of the property) for delivering and removing larger, heavier equipment or materials, such as would be needed for gardening activities. Rear access is also desirable for utility service.

e. Guest parking: This important feature is often ignored until it is too late, especially when a property has minimum frontage.

## 7. Other Factors

Numerous other factors impinge directly on decisions being made during the conceptual stage of home planning and design. These include the following:

a. Environmental control: The environmental control system or systems should be considered at the same time that structural systems and space organization are being determined, not after these aspects have already been decided upon. Otherwise, it often is difficult, if not impossible, to optimize air circulation, minimize noise and vibration transmission, or provide adequate access to environmental control equipment (air conditioner, furnace, etc.) that must be inspected or serviced.

b. Exterior drainage: Although most communities impose strict standards regarding property drainage, too often a designer fails to study the external building plan with regard to dispersing runoff from rooftops, and in many cases no attempt is made to provide proper rain gutters and/or drainage pipes as part of the initial home building package. In some cases, this lack of forethought forces a new homeowner to modify the roof in order to install gutters.

c. Exposed utility poles or boxes: Consideration should be given to minimizing the unsightly utility pole or box at the front of a property and/or the utility lines that have to run between poles and the house. In addition, few people appreciate having their property overrun when utility service people have to service electrical or telephone lines.

d. Trash storage and pickup: Although a seemingly minor point, too little attention is given to the trash-collection chores of the homeowner and collector. An inconspicuous place for collecting trash and a convenient path to the expected collection point should be provided.

e. Exterior lighting: Adequate exterior lighting should be a basic part of initial home conceptualization, not an add-on. Two primary considerations are visibility for regular, safe outside activities (walking about, night yard recreation, viewing guests, etc.) and protection against vandalism and burglary. In addition, adequate exterior convenience outlets should be planned on all sides of the house so that the homeowner will have convenient power sources for additional decorative illumination.

f. Other exterior service conveniences: At least one hose bib should be provided on each side of the house so that the homeowner will have a water source convenient to any yard area (without resorting to hundreds of feet of hose). Similar electric power convenience is desirable since many homeowners use electric yard tools.

g. General safety: Local ordinances and standards generally regulate most of the safety needs to be considered in planning a new home. The following are typical problem areas:

(1) Unsafe stairs, ramps, sidewalks, railings, balconies, etc.

(2) Use of flammable materials.

(3) Electrical systems that are not shockproof.

(4) Circuit-breaker box located where it is inaccessible in the dark.

(5) Gas shutoff where it cannot be found quickly in the dark.

(6) Driveway and street configuration that forces the driver to back into traffic that he or she cannot see.

(7) Special booby traps, e.g., low-hanging roof projections, windows that fold out across sidewalks, planter boxes that project too near typical pathways, and floor-to-ceiling glass doors and windows that people can accidentally walk through. (See the section on subsystem and component safety recommendations.)

## Typical Storage Requirement Considerations

| | Clothes | Linens | Towels and washcloths | Dishes, crystal, and flatware | Cooking utensils | Foodstuffs | Toiletries | Soaps and cleansers | Cleaning tools | Shop tools | Sewing materials | Books | Office supplies | Household records | Soiled clothes and linen hamper | Toys and games | Audio tapes and records | Luggage | Trash and garbage | First-aid and medical supplies | Tricycles, bicycles, etc. | Garden tools and supplies | Spare boxes, newspapers, etc. |
|---|---|---|---|---|---|---|---|---|---|---|---|---|---|---|---|---|---|---|---|---|---|---|---|
| Main entryway | I | | | | | | | | | | | | | | | | | | | | | | |
| Sleeping area | A | C | | | | | | | C | | | | | I | | | | | | | | | |
| Dressing area | I | | | | | | I | | C | | | | | | C | | | I | | | | | |
| Personal hygiene area | A | | I | | | | I | I | C | | | | | | | | | | | I | I | | |
| Recreation area | | | | C | | C | | | C | | | I | | | | I | I | | | | | | I |
| Kitchen | | | I | I | I | I | | I | I | | | I | | | | I | | | I | I | | | |
| Dining area | | I | | I/A | A | A | | | C | | | | | | | | | | | | | | |
| Nursery | I | I | | | | | I | C | C | | | | | | I | I | | | | I | | | |
| Play area | | | | | | | | | C | | | I | | | | I | I | | | I | | | |
| Study area | | | | | | | | | C | | | I | I | I | | | | | | I | | | |
| Laundry | C | C | C | | | | | I | I | | | | | | C | | | | | | | | I |
| Shop | | | | | | | | | I | I | I | | | | | | | | I | | | I | |
| Sewing center | | | | | | | | | C | I | I | | | | | | | | | | | | |
| General storage area | | | | | | | | | | | | | | | | | | | | I | | | I |
| Garage | | | | | | | | | I | I | | | | | | | | | | | | I | I |

I—required in immediate area.

A—required in adjacent area.

C—required in convenient area, not necessarily in the immediate or adjacent areas.

## Specification Writing

The culmination of the conceptual stage is preparation of a design and/or production specification. It should include specifics that relate to optimizing human user interfaces. The following checklist is a useful guide in preparing the specification:

*Architectural Human Engineering Specification Checklist*

1. Street side
    a. Pedestrian sidewalks, surfaces, grade, and illumination
    b. Driveways, grade, passenger exit, surface, and illumination
    c. Address and illumination
2. Approach
    a. Pedestrian sidewalks, surfaces, grade, and illumination
    b. Gates, handles, and locks
    c. Steps, ramp, porch, railing, surfaces, grade, tread or riser, illumination, and protection from weather
    d. Door size, door weight, door handles, door locks, view ports, and mail slot
    e. Doorbell, communications, and mailbox
3. Entrance
    a. Interior door clearance area
    b. Cloak closet
    c. Illumination and light switches
4. Interior
    a. Door handles, door locks, window handles, and window locks
    b. Light switches
    c. Electric outlets
    d. Circuit-breaker box or boxes
    e. Water heater
    f. Heating and ventilating vents and handles
    g. Built-ins, e.g., fireplace, storage cabinets, appliances, and control handles and indicators
    h. Illumination
    i. Floor covering
    j. Stairs, landing, balconies, and railings
    k. Plumbing fixtures, controls, and indicators
    l. Thermal and acoustic insulation
5. Exterior
    a. Vehicle storage: door controls and illumination
    b. Electric outlets
    c. Hose bibs
    d. Yard illumination
    e. Rain gutters
    f. Fencing, gates, handles, and locks
    g. Patio and surfaces
    h. Trash collection and pickup
6. Other
    a. Maintenance: window cleaning, installing and removing storm windows, gas main cutoff, water main cutoff, electrical main cutoff, structural refurbishment, painting, etc.
    b. General safety: nonflammable materials, accessibility of water supply and of fire-fighting apparatus to dwelling, adequate electrical grounding and circuit breakers, adequate number of circuits, proper drainage for flood disasters, protection from landslide, appropriate barrier to street vehicle penetration, and access for

visual security monitoring
    c. Service accessibility: mail, milk, paper, and other deliveries; furniture and appliance delivery and removal; and gardening equipment and trash cleanup and pickup.[10]

## RESIDENTIAL—MULTIFAMILY

### Multifamily Residential Planning

The key family, guest, and service activity elements of the single-family residence, discussed above, are similar to those of multifamily residential facilities. In addition, however, the following special requirements should be addressed:

#### 1. Common-Use Interior Activity Areas
    a. Reception area and lobby, central mailbox, call box, and paper delivery area
    b. Central laundry facility
    c. Central recreational and hobby facilities
    d. Resident parking
    e. Central trash and garbage collection
    f. Stairs and elevator

In addition, consideration must be given to noise attenuation between neighbors, building management, and maintenance (for larger complexes); building security; fire protection; and signing and marking.

#### 2. Common-Use External Activity Areas
    a. Guest and delivery parking

[10]More detailed information and guidelines may be found in other chapters dealing with specific subsystems and components.

    b. Guest and delivery ingress and egress
    c. Children's play area
    d. External maintenance (electric and water outlets and tool and equipment storage)

In addition, consideration must be given to external lighting, signing and marking, vehicular approach, and accessibility for emergency activities (police, fire, ambulance, etc.). In the case of fire fighting, consider the availability of water supply and the ease of ladder use (for multistory units).

**COMMON PLANNING ERRORS**  Some of the most common planning errors for multifamily dwelling facilities are the following:

1. Lack of adequate guest parking
2. Lack of adequate noise control
3. Lack of dual emergency escape routes
4. Lack of convenient delivery features (for delivering heavy, large furnishings)
5. Lack of sufficient "dead storage" space per tenant
6. Maintenance inconvenience (e.g., window cleaning)
7. Barriers for handicapped users
8. Hidden identification and address signs
9. Camouflage of main entrance
10. Lack of adequate, general trash-collection space (including the maneuvering space for trash trucks)
11. Lack of play space for tenant children, which increases their chances of being exposed to street or internal area traffic, swimming pool hazards, etc.

### High-Density, Multifamily Residential Planning Model

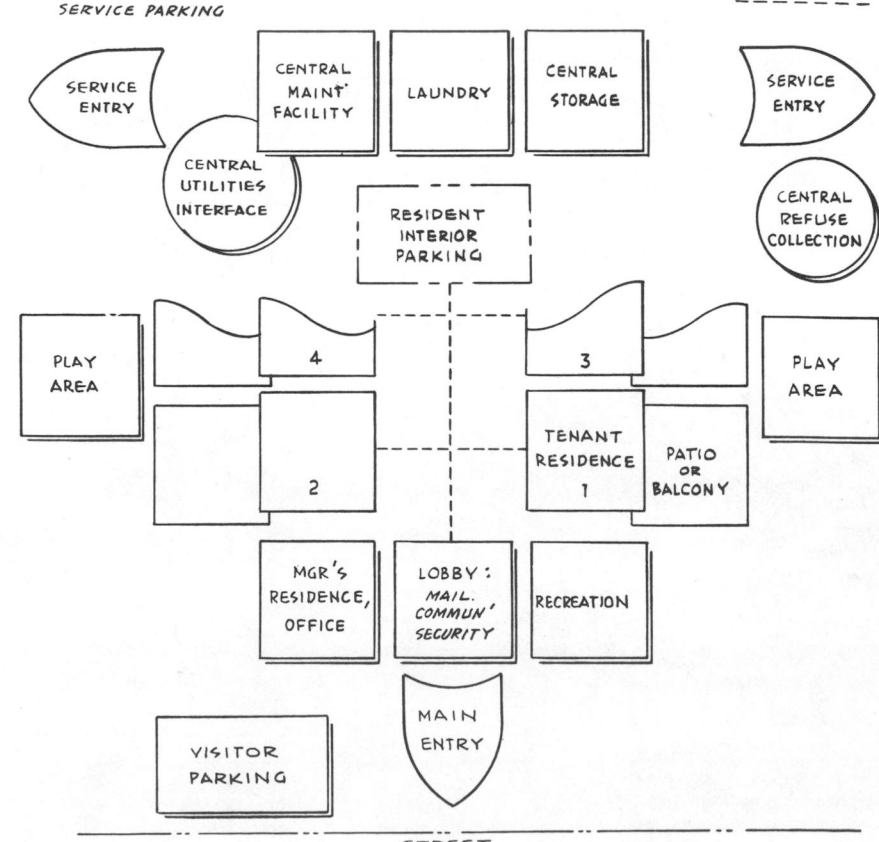

Among the more important layout considerations are the following:

1. Multiple emergency escape routes
2. General security
3. Building services and maintenance accessibility
4. Acoustic isolation between residences

## Information Requirements Checklist for Apartment Complex Planning and Development[11]

*Summary Site Analysis*

1. *Marketability:*
   a. Demand for multifamily housing:
      (1) At what rents
      (2) Distribution (0-BR, 1-BR, 2-BR, 3-BR)
      (3) Size of rooms
   b. Existing population and potential growth
   c. Type of existing tenants living in apartments:
      (1) With children and how many
      (2) Without children
      (3) Elderly
      (4) Single occupancy
      (5) Income brackets
      (6) Age brackets
   d. Industries in the area and their future plans
2. *Pertinent Information of the Surrounding Area:*
   a. Existing street layout and how it may affect the parcel in question
   b. Proposed street changes:
      (1) Widening
      (2) Elimination of streets
      (3) Map changes
   c. Location of main arteries (parkways, freeways, and highways)
   d. Mobility from site (in all directions)
   e. Zoning and proposed changes
   f. Kind of buildings:
      (1) Single-family
      (2) Multifamily
      (3) Commercial and industrial
   g. Appearance and general character:
      (1) Design of the exteriors
      (2) Condition of buildings, grounds, and streets
   h. Off-street parking:
      (1) Is the surrounding area provided with adequate off-street parking?
      (2) Are the existing streets wide enough for street parking and easy access for cars and service vehicles?
   i. Proximity of parks, public playgrounds, other recreation areas, and waterways, if any, to site (see 7, *Community Facilities*)
   j. Hazards:
      (1) Noise
      (2) Proximity of airports, railroads, and trucking highways
      (3) Smoke and fumes
      (4) High-tension wires
      (5) Ravines
   k. General trend:
      (1) Stability of area

[11]Samuel Paul, *Apartments: Their Design and Development*, Reinhold Book Corporation, New York, 1967.

      (2) Building expansion
      (3) Deterioration
3. *Transportation Available:*
   a. Other than automobile:
      (1) Rapid transit
      (2) Bus
      (3) Railroad
      (4) Taxis and other vehicles, such as helicopters, hydrofoils, ferryboats, airplanes
   b. Time of travel to center city and to job location
   c. Automobile travel:
      (1) To center city
      (2) To job location
   d. Cost of daily traveling:
      (1) Daily fares
      (2) Gas and oil, parking charges, if by car
   e. Schedule of transport services
4. *Zoning of Site:*
   a. Density, coverage, height, yard requirements, and parking
   b. Proposed changes, if any
5. *Planning Boards:*
   a. Rules and regulations that control land development other than zoning
6. *Deed Restrictions*
7. *Community Facilities (Distance from Site and Methods of Getting There):*
   a. Schools:
      (1) Public, parochial, or preschool
      (2) Grammar, junior high, high school
   b. Shopping:
      (1) Necessities
      (2) All others
   c. Religious buildings:
      (1) Denominations
   d. Recreation:
      (1) Theaters
      (2) Playgrounds, beaches, swimming pools, bowling alleys, and others
   e. Hospitals, medical centers, and clinics
   f. Cultural:
      (1) Libraries, art galleries, museums, and others
8. *Community Services:*
   a. Garbage and refuse collection
   b. Police and fire protection
   c. Snow removal
   d. Street cleaning
   e. Street maintenance
   f. Street lighting
9. *Size and Shape:*
   a. If irregular, can plot be utilized efficiently?
   b. If small, can economical project be built?
10. *Topography:*
    a. Rugged terrain, gently sloping, or flat
    b. Rock exposure or filled-in land
    c. Type of surface soil
    d. Surface drainage and ground water
    e. Natural features:
       (1) Trees, streams, lakes, adjoining parks, and rock outcroppings
11. *Subsurface Conditions (Information Usually Received from Borings):*
    a. Composition of soil
    b. Evidence of rock or filled-in land
    c. Soil bearing capacity (necessity of piling)

    d. Underground streams
    e. Water level
    f. Percolation of soil
12. *Utilities:*
    a. Storm and sanitary:
       (1) Combined or separate
       (2) Nearby body of water or drainage ditch
       (3) Depth of sewers
       (4) Adequacy of sizes and pitch for additional loads anticipated
       (5) Public or private
    b. Water supply:
       (1) Pressure
       (2) Reservoir, well, or other
       (3) Rates
       (4) Who pays for installation (from what point to what point)?
    c. Gas:
       (1) High or low pressure
       (2) Natural or manufactured
       (3) Rates
       (4) Who pays for installation (from what point to what point)?
    d. Electricity:
       (1) Overhead or underground
       (2) Current available
       (3) Rates
       (4) Who pays for installation (from what point to what point)?
    e. Telephone service
13. *Features:*
    a. Views
    b. Trees, streams, lakes, and parks
14. *Cost of Site:*
    a. Potential yield (number of families):
       (1) Land cost per family
       (2) Land cost per room
    b. Rent limitations for area:
       (1) Rent per room per month
       (2) Relation of total rent to cost of site
    c. Cost of abnormal site conditions:
       (1) Excessive fill and grading
       (2) Piling
       (3) Rock excavation
       (4) Possible retaining walls
       (5) Cost of bringing utilities to site

## Unique Layout Considerations

Although it is recognized that considerable cost savings accrue by arranging multitenant facilities in a somewhat regimented fashion, this often leads to interior arrangement of each tenant's residence in a less than efficient configuration. The guidelines for a single home should be applied wherever practical. In addition, the following unique layout considerations relating to the total complex should be addressed:

### 1. Visitor Parking and Main Entrance

Both the visitor parking area and the main building entrance should be conspicuous from typical street approaches, which implies an obvious adjacency between these two elements. Guests should not be required to walk great distances from their parked vehicle to the entrance. The parking and walkway both should have adequate night illumination.

### 2. Main Entrance and Lobby, Resident Manager's Quarters, and Guest Lounge or Recreation Area

All these facilities are important to infrequent

guests as well as to guests whom a tenant may want to entertain outside his or her apartment or condominium; thus these facilities need to be fairly close to one another. Such facilities should be located generally outside a main security perimeter to prevent unwanted visitors from penetrating the private tenant area.

### 3. Indoor Resident Parking
When indoor parking is provided, it should have an entrance that is not generally exposed to the public, and if practical it should have an automatic gate controlled from the resident's car.

### 4. Central Maintenance, Laundry, and Storage Facilities
These facilities should be readily accessible only to building maintenance personnel and residents via special stairs and/or elevators, and they should be convenient to service entries that are not available to guests. Similarly, a central utilities facility should be directly accessible from one of the service entries.

### 5. Play Areas
If the multifamily complex is designed for use by families with children, special enclosed play areas should be provided with convenient interior access from tenant residences. Other special facilities such as a pool, sauna, hobby shops, etc., should be centrally located as a general rule, and accessible only to residents.

### 6. Stairs and Elevators
More than one stair and elevator system is recommended for complexes of more than two stories (although each community generally sets forth standards for such installations). The location of the dual facilities should be selected to create the best balance between travel distances from each tenant residence, the principal concern being to provide adequate emergency escape in the event of fire. Avoid adjacent stairs and elevators, which can cause traffic jams.

### 7. Exterior Privacy
Most multistoried residential facilities provide some sort of balcony for each tenant. The need for privacy as well as freedom from other annoyances (people throwing debris onto a lower balcony, smoke drifting into a balcony from a neighboring tenant's barbecue, etc.) should be considered in initial planning. Considerable privacy can be obtained by means of proper overhangs, directional orientation, and/or special barriers designed into the original structure.

Avoid configurations in which balconies on opposite sides of a courtyard face directly toward each other.

### 8. Internal Privacy
In addition to the obvious need to insulate between residences for acoustic reasons, consider methods for providing reasonable visual privacy for individuals who may stand momentarily outside a friend's apartment. One method is to provide an alcove or setback at the entrance to each apartment so that visitors do not have to stand directly in the hallway.

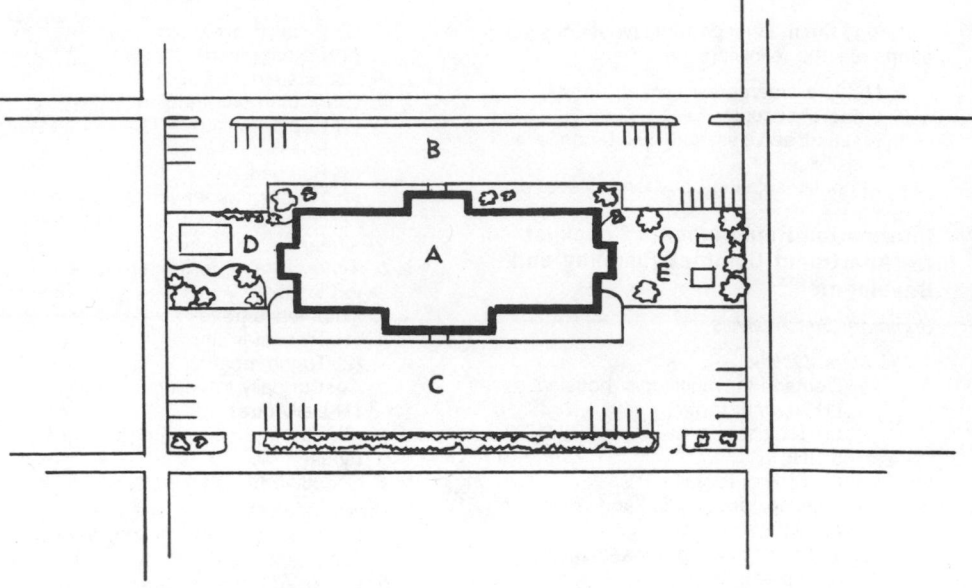

Idealized site layout. (A—Apartment building; B—Tenant and service parking; C—Guest parking; D—Mixed recreational area; and E—Children's play area.)

### 9. Special Safety Considerations
The following are unique objectives for the typical high-density residential complex:

a. Each apartment should have more than one exit. The common practice of considering a balcony an emergency exit is not an acceptable compromise. Even though an agile tenant may be able to use an emergency rope device, the elderly, the handicapped, and especially small children cannot do so safely.

b. Do not arrange kitchens so that they are immediately next to primary hallways (prime escape routes). Because the majority of internally generated fires may occur in the kitchen, such fires typically can block the primary exit.

c. Provide emergency devices. Fire codes usually require the location of fire extinguishers at strategic points both internally and externally. Of equal importance is provision of emergency lighting and emergency warning communications.

d. Do not use flammable materials. Although it is difficult to control the materials used by tenants, the materials used in common areas of the multifamily complex should be a nonflammable variety.

e. Provide for building maintenance and security monitoring. Systems for monitoring general maintenance features within the building and general security should be considered for the large complex. A careful study is, of course, necessary to devise a system that does not encroach upon the privacy of tenants.

### Land-Site and Facilities Layout Objectives

The sketch above shows an idealized (although perhaps impractical) layout of an apartment building and grounds and illustrates a number of important user-oriented planning objectives:

1. Identifiable main and guest entrance and convenient parking.

2. Semi-isolated tenant parking and multiple entrances—also for service parking and entrance.

3. General separation of adult- and child-oriented outdoor recreation areas (although children would also associate with parents at times in area D).

4. Multiple street and site approaches to reduce traffic conflicts.

5. Parking and building relationships readily apparent even before the guest motorist enters the site.

6. Regularity of layout to simplify visitors' immediate recognition of where they should go to enter the site. (Note that it is easier to sign and mark sites that are rectilinear.)

7. Building accessible to emergency vehicles and equipment (firefighters, police, etc.) from more than one entrance and side.

8. Block plan easily approached from typical public transportation.

### Multibuilding Layout

The example at the top of page 47 is an actual plan of an apartment complex and site layout; it illustrates some important points:

1. It may be desirable to separate buildings in order to accommodate more than one life-style, e.g., families with no children and families with children.
2. A separate garage facility is provided.
3. There is lower apparent density because the buildings are separate.

Note that many of the objectives identified for the idealized site layout are still maintained, i.e., separate visitor and guest entrances and parking, separation of play areas for small children and recreation areas used either by adults and older children or by small children under adult supervision, multiple access for emergency vehicles and equipment, and convenient service parking and entry.

## Multibuilding Estate Layout

Very large apartment or condominium complexes often are laid out to create an almost estate or parklike atmosphere. Although some of the advantages of the more regular or rectilinear patterns noted in the previous examples are lost, there is still an opportunity to retain reasonable visitor orientation, as shown in the architectural example below. Some of the special features that emphasize user-related objectives are as follows:

1. The cluster concept creates a certain neighborhood identification for residents and also minimizes the general high-density, "row-on-row" effect that might develop when so many units are aligned in regular patterns.
2. The cluster concept provides a simpler means for separating childless family groups from those with children.
3. Parking areas can be much smaller and, because they are separated, do not give the impression of large masses of paving and a sea of vehicles.
4. Although internal roadways are slightly curved, a reasonably clear pattern is maintained so that visitors can easily find their way to the area they are looking for. Signing and marking should not be difficult.

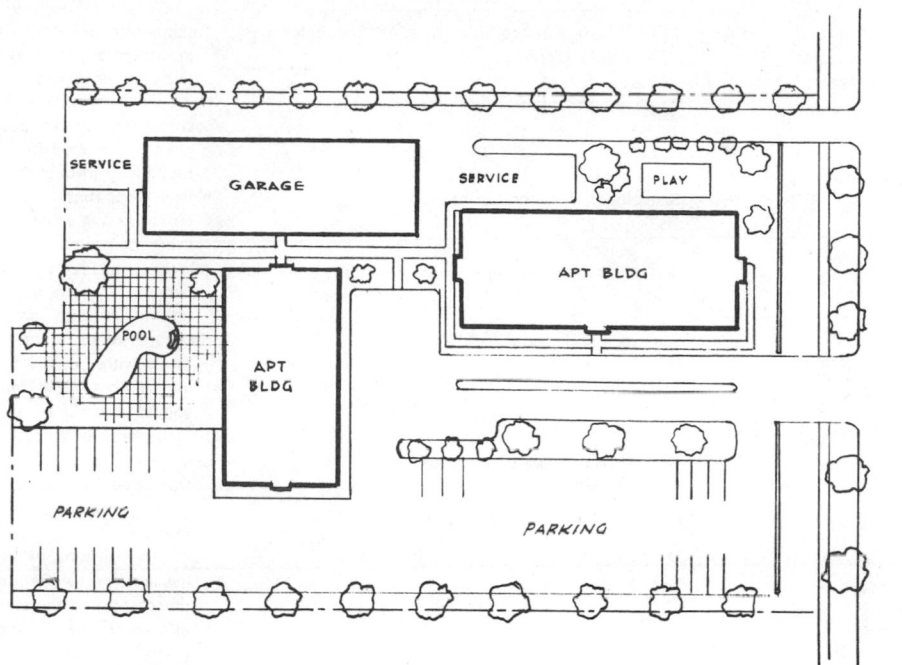

Multibuilding layout example.

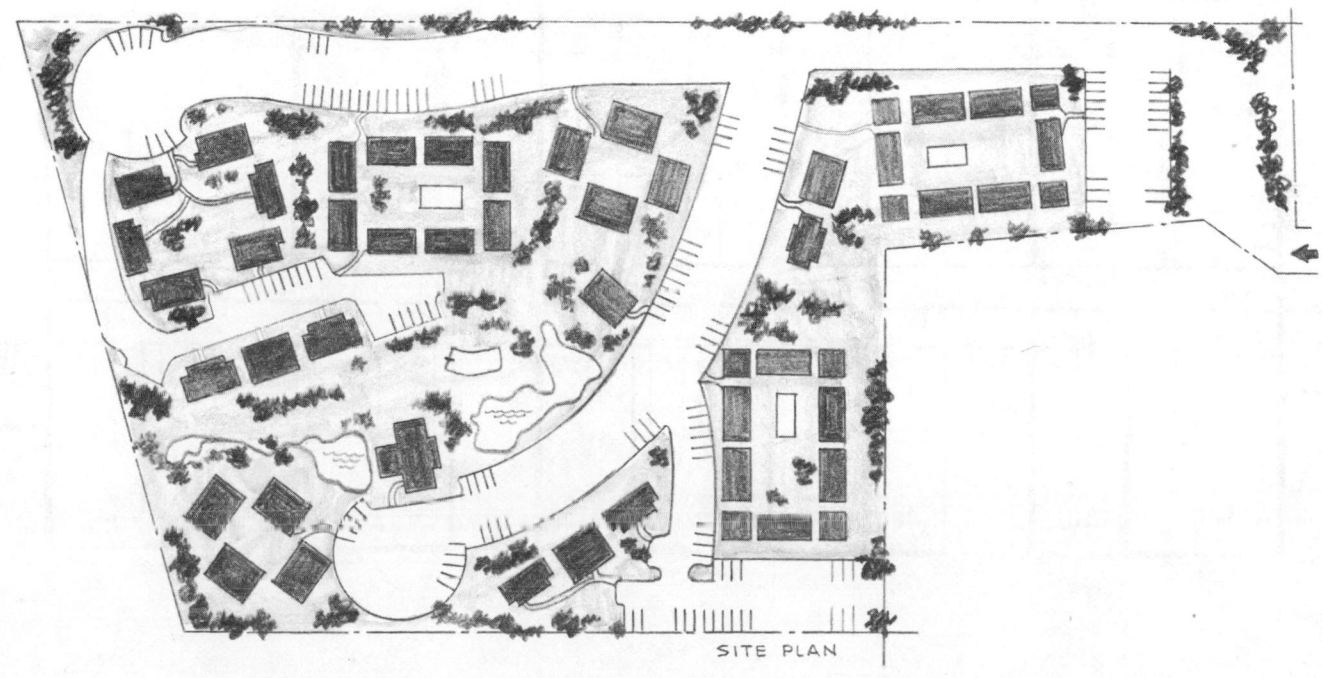

SITE PLAN

Multibuilding, estate layout example.

## Individual Apartment Layout Objectives

Because of the many physical constraints associated with multifamily facility planning, it is obviously difficult to create so-called optimized layouts for each of the apartments within the facility. However, it is helpful to have a set of objectives in order to avoid overlooking certain desirable features. The table on page 48 lists some of the more important objectives to be considered in analyzing various layout alternatives.

| Objective | Considerations |
|---|---|
| 1. Minimum distance between apartment entrance and the kitchen | Tenants often have their arms full of heavy groceries, or they may have heavy trash or garbage containers to dispose of. |
| 2. Minimum traffic through the living room | Not only is the wear and tear on the living-room carpet undesirable, but also most people prefer not to have other family members annoy guests by walking through the living room. |
| 3. Isolation of bathroom noises | Noises emanating from the bathroom are often embarrassing when guests are present in the living-room area. |
| 4. Natural light and view of the exterior | All rooms (including the kitchen) should have an exterior view and the opportunity for introducing natural light into spaces. |
| 5. Patios and balconies | Each apartment owner generally enjoys his or her own private outdoor facility, which is visually inaccessible to the immediate neighbors. These should normally be accessible from a living or dining area or both. |
| 6. Bedrooms isolated from noise | Generally, people prefer bedrooms to be isolated from areas where the noise of entertaining could interrupt the sleep of another member of the family. |
| 7. Two exits | For safety reasons, at least two exits should be available in the event of an emergency. Since most internally initiated fires start in the kitchen, at least one exit should bypass this potential barrier. |

## Individual Apartment Layout Examples

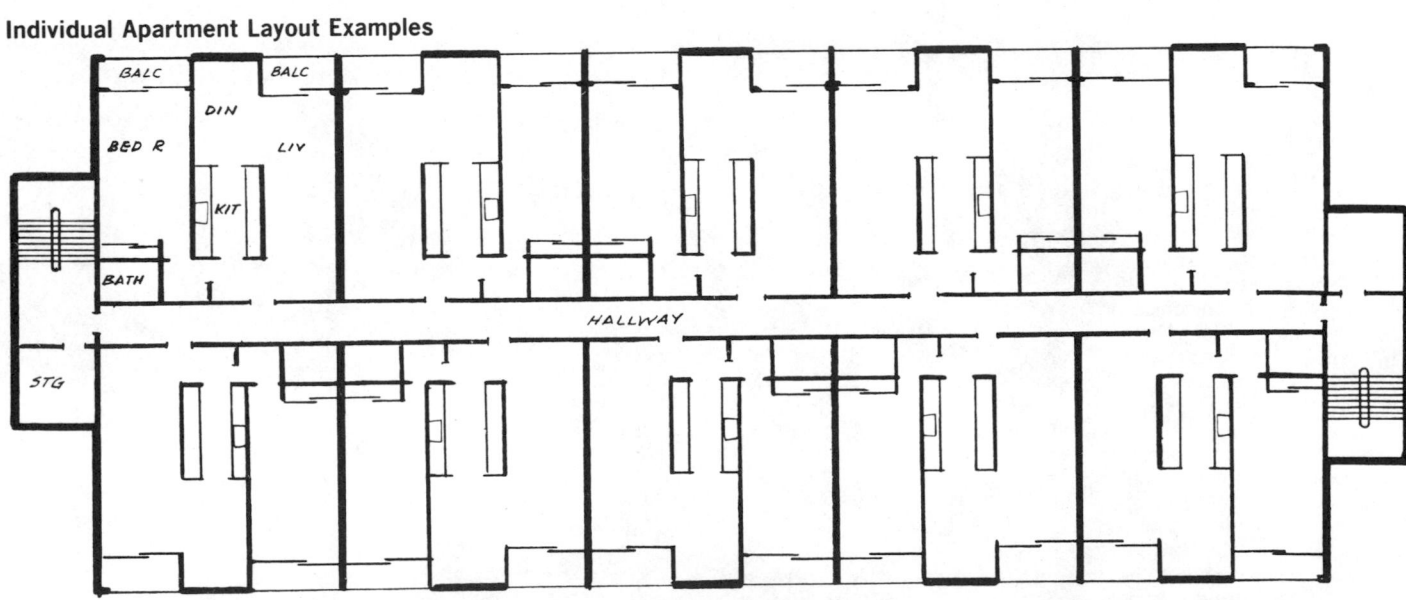

TYPICAL FLOOR PLAN

Example 1.

Although few, if any, ideal examples of apartment layouts could be presented, the accompanying sketches are reasonably good compromises illustrating desirable and undesirable layout features.

In Example 1, the following points should be noted:

1. Multiple access and escape routes via two staircases and two elevators are a desirable feature. By distributing these facilities, the architect has maximized convenience and also provided the extra safety feature of minimizing traffic jams.

2. The architect has provided a reasonably good solution to the problems of access between apartment entrance and kitchen, minimal traffic through the living room, and convenience between kitchen and dining areas.

3. Balconies have been inset rather than extended, providing maximum individual privacy for each tenant.

4. Note that, in some apartments, guests have access to a bathroom only by passing through family bedrooms. Although not completely objectionable, this is not as desirable as a less isolated location.

In this example, the architect has provided many of the same desirable features found in Example 1, e.g., multiple access and escape routes via stairs and elevators and almost equal distance between these and individual apartments (considerably shorter distances).

One significantly new improvement is provided in this plan; i.e., there are more spaces with exterior viewing capability, especially in the case of kitchens in the corner apartments.

Note also that bathrooms are located for convenience to all other spaces, as opposed to some of the inconvenience observed in Example 1.

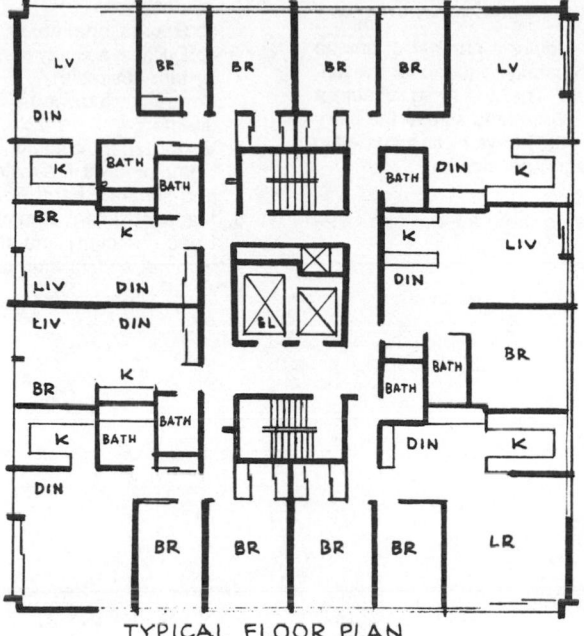

TYPICAL FLOOR PLAN

Example 2.

In this plan the architect has provided unusually good isolation of guest and family areas, excellent convenience between entry and kitchen and between kitchen and dining area, and very desirable bathroom centralization. In addition, balcony privacy is assured, although perhaps not as well as in Example 1; i.e., extending the balconies may make visual isolation less satisfactory. In addition, this plan provides exterior views and natural light for every space except bathrooms and kitchens in two of the apartments.

The questionable feature of this plan is the general concentration of stairway traffic; i.e., although there would be little problem for individual tenants on one floor (because there would be so few people involved), if the building contained a considerable number of floors, there would be the potential for traffic jams at the bottom-floor level.

It is hoped that these three examples will provide useful ideas to readers in evaluating their own plans. It is possible, for example, to utilize the good features of the several layouts and, by an appropriate rearrangement of general building shape, to overcome some of the less desirable features that have been noted.

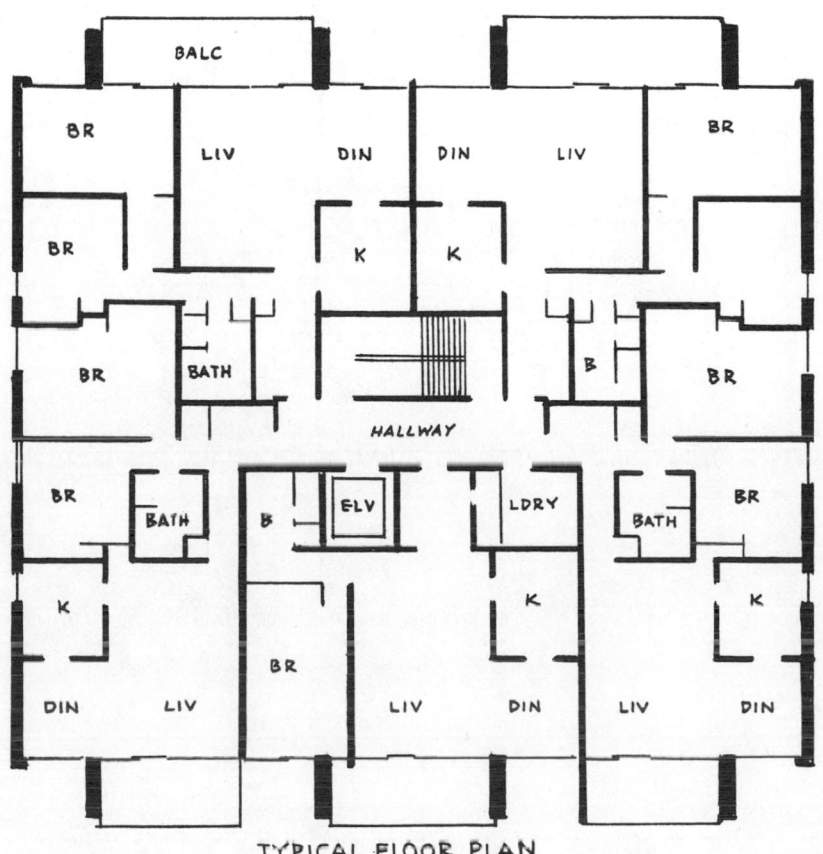

TYPICAL FLOOR PLAN

Example 3.

## Special Amenities Considerations

Multiunit facilities require a number of unique amenities not necessarily required for the single-family dwelling. The following checklist identifies these requirements so that the planner can make sure they have been included in planning and design specifications:

1. Exterior lighting
   a. Parking area, both interior and exterior
   b. Walkways
   c. Entrances
   d. Recreational areas
   e. Building security (automatic timer system desirable)
2. Central mailbox or mailboxes and tenant directory
3. Visitor-tenant communication system
4. Common-area fire-extinguishing system and separate hand extinguishers
5. Manager-tenant communication system
6. Special security and key system
7. General and/or individual dead storage facility
8. Building identification signs:
   a. Individual buildings
   b. Individual apartments within each building
   c. Emergency exit signs
   d. Parking signs
   e. Other safety signs, e.g., for play and pool areas
9. Fire hydrants

*Note:* Consult local ordinances for the latest community building codes and standards.

# PUBLIC ACCOMMODATION

## PUBLIC ACCOMMODATIONS

### Public Accommodations Planning Model

The following are among the more important layout considerations (see model on page 52):

1. Initial guest convenience to main entrance, check-in area, and guest quarters
2. Secondary guest convenience to dining, recreation, and parking

3. Guest service efficiency
4. Hotel administrative efficiency
5. Acoustic isolation of guest quarters

### Planning Information Development

Certain background information should be gathered prior to making important decisions regarding the nature, size, or location of a proposed public accommodation facility. The following checklist may be used as a point of departure in looking for such information.

### 1. Nature of the Proposed Facility

a. The principal user age group or groups to which the facility will be oriented
b. The ethnic distribution of the expected guest population
c. The general socioeconomic distribution of the expected guest population
d. Orientation with respect to leisure, business, conventions, and exhibits
e. Orientation with respect to predominant guest occupancy, i.e., daily transient, by

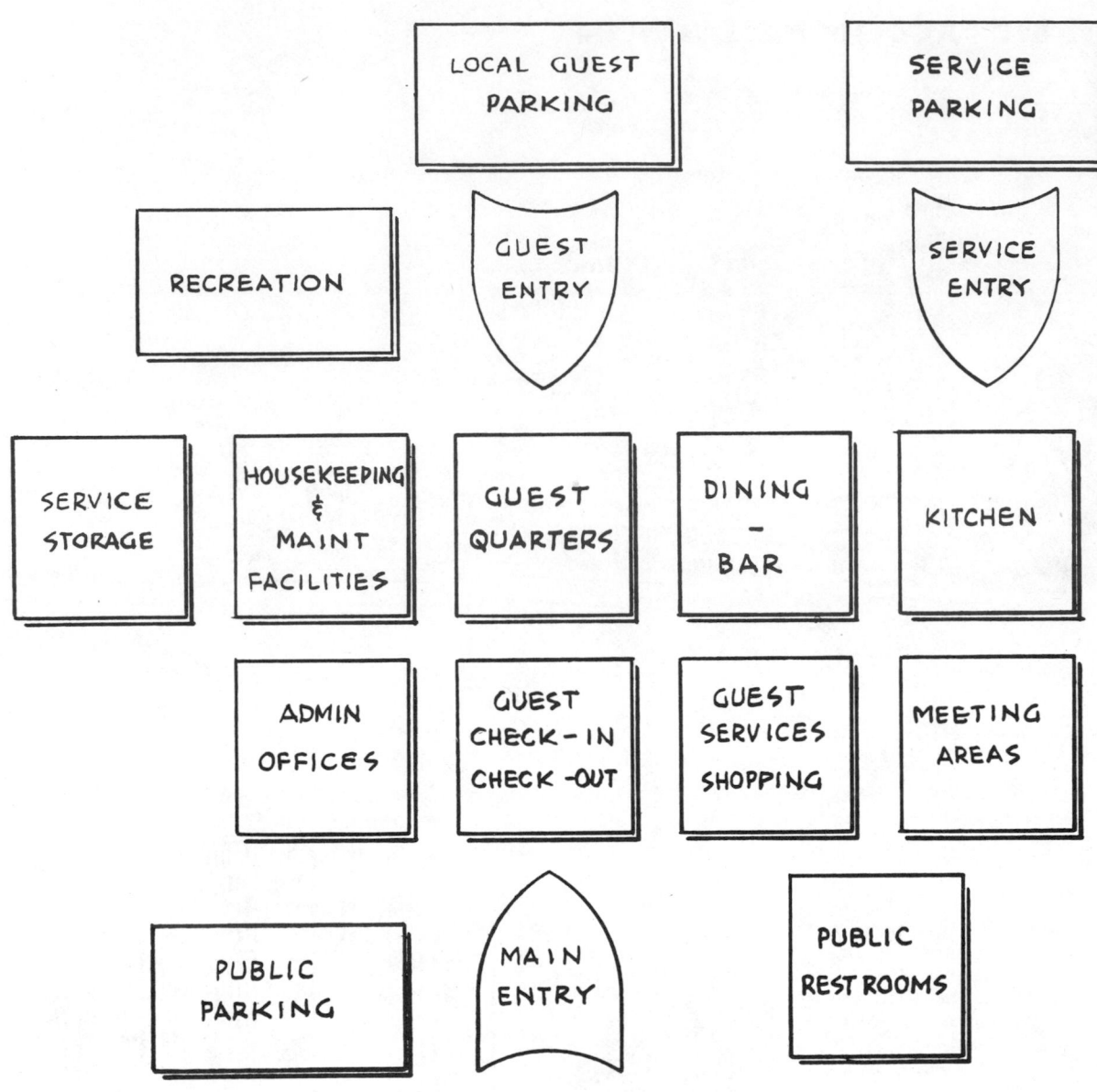

the week, by the month, or extended residency
f. Local versus transient guest emphasis
g. Local versus cosmopolitan architectural theme
h. Downtown versus residential or rural character
i. Year-round versus seasonal use
j. Principal transportation modes

**2. Size of the Proposed Facility**
In addition to the obvious economic determinants, consider:

a. Size of convention and/or exhibit potential
b. Local conference-use potential
c. Projected guest registration
d. Nature and "scale" of adjacent community facilities
e. Projected guest parking requirements (both resident and nonresident guests)

**3. Site Selection Considerations**
a. Accessibility relationships (location of principal transportation facilities and systems, business establishments, leisure-time features, nearby historical sites, and/or other community business potential)
b. Topographic characteristics as they relate to user-oriented facility layout, views from the facility, and views of the facility from key thoroughfares and highways
c. Weather and other environmental impact considerations (wind, noise, odors, etc.)
d. Potential for future expansion in an orderly, user-oriented manner
e. Relation of the site to future changes in the character of the immediate community surround

**Public Accommodations Systems**

The principal functional considerations for public accommodations systems relate to the overnight guest and include the following:

1. Sleeping
2. Personal hygiene
3. Dressing and temporary apparel and baggage storage
4. Communications (telephone, etc.)

In addition, depending on the nature of the proposed system, the following functions should be considered:

1. Dining
2. Vehicle parking
3. Recreational facilities (pool, sauna, tennis courts, etc.)
4. Special group meeting facilities (business, conventions, industry exhibits, etc.)

In support of the above guest-oriented facilities, many other functions must be considered, including:

1. Administrative (guest register, record keeping, etc.)
2. Maid service and laundry
3. Facility maintenance and security
4. Guest assistance (baggage handling, room service, transportation, etc.)
5. Other (cleaning, drugs and notions, gifts

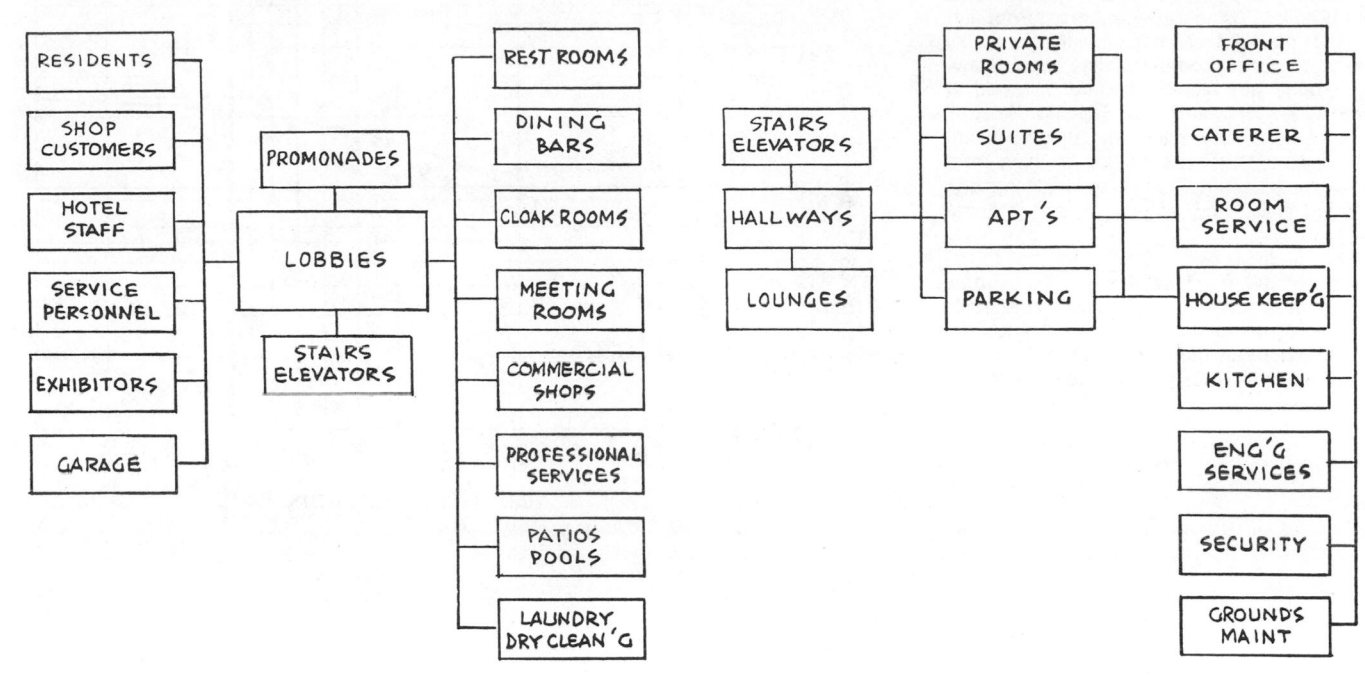

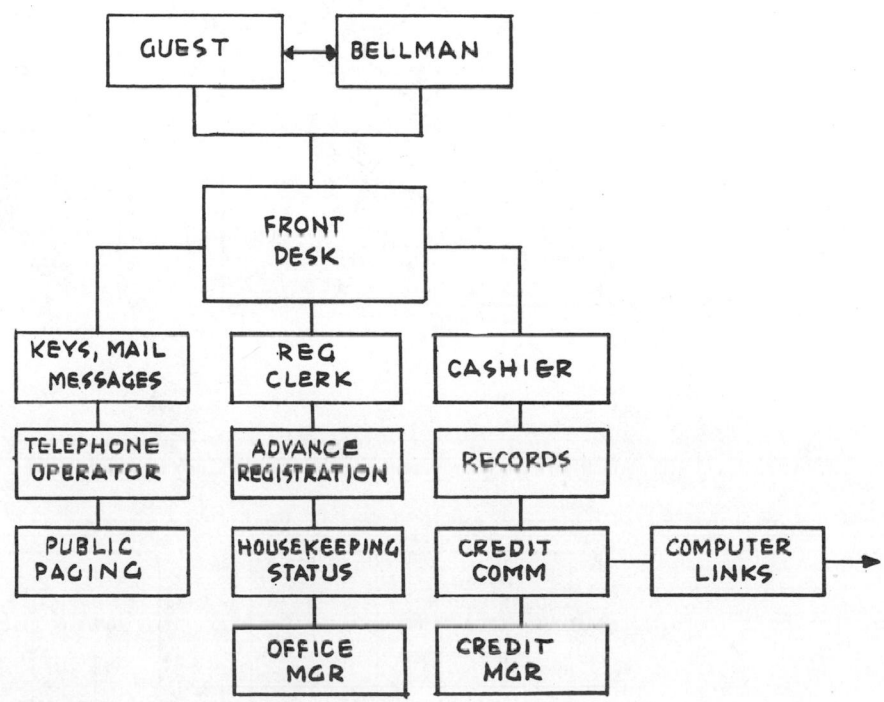

and flowers, liquor and other shops such as a barber shop and beauty salon, travel agency, tour agency, etc.)

In conceptualizing the public accommodations facility, whether all the above functions are contained and regulated as a whole or are subleased and possibly detached but adjacent facilities, each user's needs and conveniences should be analyzed before launching into some preconceived facility layout. In general, the needs of guests have priority since it is their patronage that creates a successful operation. It is the overnight guests that should be considered first, starting with their arrival, proceeding through the series of typical activities they and their party will participate in, and ending finally with their departure.

Obviously, however, unless other participants or users associated with serving guests are also efficient (i.e., unless all their functional needs are satisfied reasonably well), guests still will not be happy with the system and thus probably will not patronize the complex again. The following checklists, guidelines, and other information should help the planner develop the best overall plan to meet the needs of all participants in the public accommodations system.

The size and complexity of public accommodations facilities obviously will vary from one concept to another; however, because of the typically large number of people and activities, special attention should be given to each of the more significant activity-space interface interactions. The accompanying diagram illustrates a typical array of interacting activity and space elements.

Typical activity interactions with the front desk.

## Estimating Hotel Space Requirements

Preliminary estimates of space requirements can be made using data generated from surveys of hotels that seem to provide reasonably adequate accommodations. The accompanying graphs are based on data provided in *Time-Saver Standards for Building Types*.[12] These graphs allow initial estimating for a variety of typical facilities including lobby, front office, public corridors, lounges, dining and kitchen area, and employee and storage areas.

The reader is warned, however, not to assume that these estimates are to be used as final figures for planning, since the specific mission of the proposed new facility and its new and perhaps more comprehensive equipment and furnishings may dictate more or less space.

[12] J. De Chiara and J. H. Callender, *Time-Saver Standards for Building Types*, McGraw-Hill Book Company, New York, 1973.

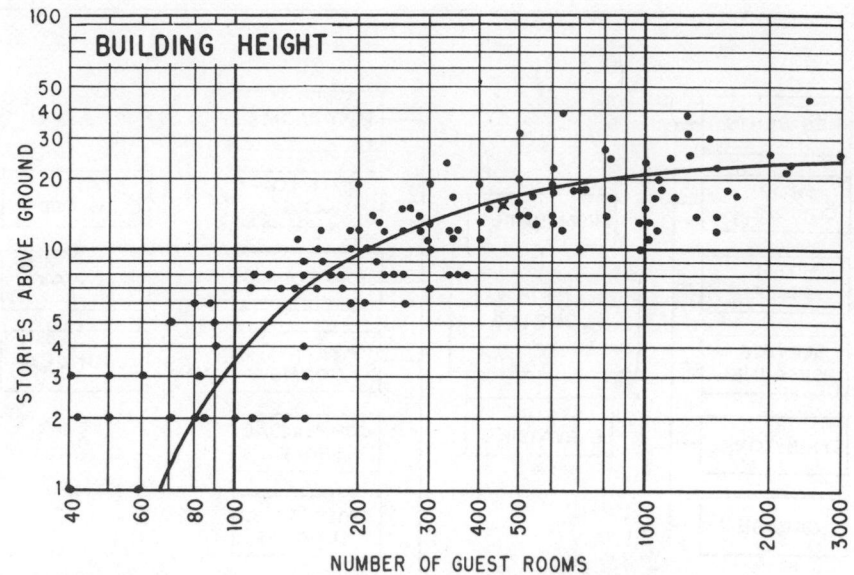

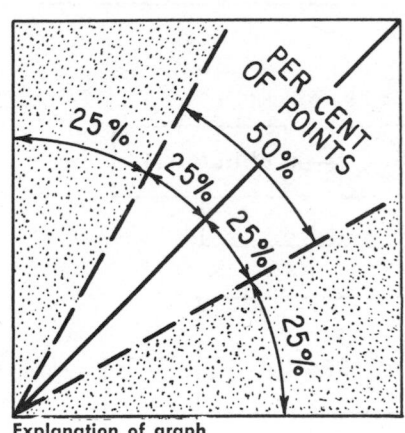

Explanation of graph

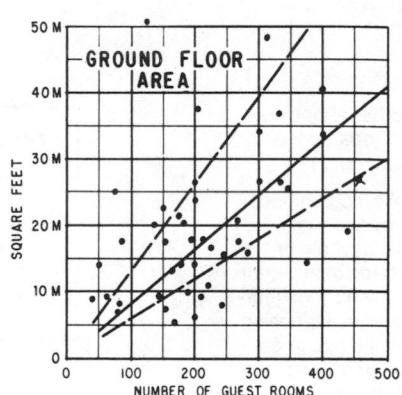

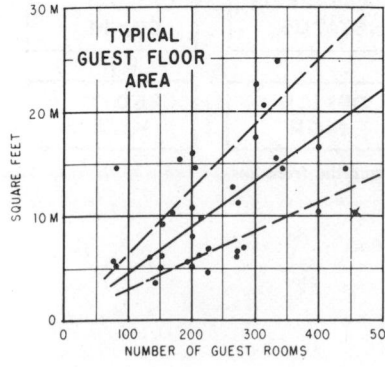

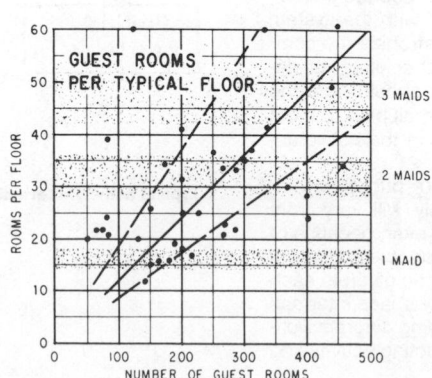

## General Human Factors Principles to Consider During Conceptual Planning and Design

### 1. Guest Accommodation

First priority should be given to making all guest accommodations comfortable, convenient, and safe. Depending on the type of facility, a variety of guests should be considered: overnight guests; nonovernight guests who may utilize dining, meeting, or convention facilities; and interim guests of house guests. The more important considerations are:

a. Ease of locating and identifying the facility in terms of its location (i.e., with respect to highways and streets), signing, and approach system geometry.

b. Ease of locating the entry to the site, parking, and identification of, and convenience to, the main facility entrance (i.e., not only should these elements be easily recognizable in terms of their location, arrangement, and physical appearance, but also proper signing should leave no doubt in the minds of guests as to where they should go to gain entrance into the facility). Appropriate external illumination must be provided in order for the above to occur. The distance between parking and buildings should be minimal.

c. Ease of locating the check-in area once the guest is inside. The check-in desk should be conspicuous from the point at which the guest first enters the facility. Appropriate, visible, and legible signing is an important aspect of the confidence that a guest requires upon entering.

d. Ease and convenience of the front desk to the guest quarters. This should be considered a prime prerequisite. Avoid arrangements that are confusing or that require considerable walking by guests. Although most hotels provide baggage service (which includes guidance to the guest quarters), a design objective should be to create a system that is simple enough so that guests can find their own rooms easily by themselves.

e. Convenience between the guest quarters and typical hotel facilities that guests will use during their stay (dining facilities, a bar, a barber or beauty shop, dry-cleaning facilities, a drugstore, etc.) and particularly the recreational facilities provided (swimming pool, game rooms, sporting facilities, etc.). Consider also special localized facilities that should be especially convenient to guests' rooms, e.g., soft-drink and ice dispensing machines.

f. Adequate guest-room size, configuration, furnishing, lighting, security, air conditioning, and sound isolation. Treat the guest quarters as a "home away from home." Provide adequate internal and external communication and control of the thermal environment.

g. For motel or motor hotel systems, provide adequate and convenient local parking for each guest room. For all systems, provide adequate and convenient visitor parking, typically separate from overnight guest parking.

### 2. Accommodation of Administrative and Maintenance Personnel

The myriad of interactions between the hotel or motel management and the service personnel should be analyzed carefully in order to provide a basis for identifying, specifying, and arranging special as well as general accommodations for their respective tasks. Individual performance efficiency with minimum interference of guests should be of prime concern, and time is perhaps the critical factor. Consider the following:

a. General management: The location of the facility manager should be chosen both to maximize his or her convenience to guest contacts and to facilitate immediate supervision of the administrative staff (desk clerks, the cashier, the business agent, etc.).

b. Front-desk activity: The size and shape of the front-desk facilities should be based on a combination of numbers of clerical persons required (for peak loads) and their special furnishings and equipment. As indicated above, this particular facility must also have proper interfacing characteristics with guests; i.e., it should be easily identified, not prone to traffic jams at peak arrival or departure times, etc.

c. Housekeeping activity: First priority should be given to the time- and energy-saving convenience of those responsible for the service of the guest quarters. This activity should be addressed not only in terms of the overall facility layout but also in terms of the layout of individual guest quarters. In addition, certain architectural features, such as clearance beneath fixtures and cabinets, coved baseboards, and pullman counters, have been found to facilitate cleaning tasks.

Although generally of secondary priority, housekeeping convenience of general and common facilities is equally important in terms of maintaining an acceptable appearance. Especially important to housekeeping effectiveness are the provision and proper location of housekeeping supplies and equipment in order to reduce the time spent by housekeepers and the number of steps they must take.

d. Maintenance activities: Both internal and external maintenance considerations should be taken into account in terms of the general layout of the facility, special work-area and storage-area requirements, and, last but not least, the type and amount of landscaping that may be planned. In addition, special consideration should be given to the accessibility of the service area or areas with respect to the use of large vehicles for deliveries or debris disposal.

e. Food service activities: The dining room, coffee shop, bar, or other food service systems associated with hotels or motels should be designed according to good restaurant guidelines. Whether they are housed within the general hotel structural complex or developed as separate units, consider the guest convenience to and from the main hotel complex, especially during inclement weather.

f. Special entertainment, convention, and exhibit facility activity: Such facilities should normally be considered adjuncts to the main hostelry mission of the hotel or motel; i.e., their placement and traffic patterns should be such as to minimize interference with guests' activities (many guests may not participate in events occurring within the special facilities and may become annoyed by heavy traffic through regular guest areas or by the excessive noise that such facilities tend to generate). Special attention should be given to the food service problems associated with these special facilities and to the unique problems associated with exhibitors and entertainers and their equipment. Added storage is a major consideration because these special activities require a considerable amount of portable furnishings, including chairs, tables, and stages. A special loading dock may also be required for delivery and removal of exhibits, musical instruments, and the like.

g. Other factors: The following considerations relate to the above:

(1) Fire safety (exits and emergency lighting)
(2) Security (including vandalism and theft)
(3) Accessibility to various delivery functions (mail, vendors, etc.)
(4) Accessibility to emergency crews and equipment (fire fighters, police, etc.)
(5) Environmental control and illumination for common-use areas
(6) Signing and marking
(7) General communications (administrative and public address)
(8) Public rest rooms (associated with the lobby, restaurant and bar, convention area, and swimming pool and with such activities as golf or tennis, which may require consideration of a locker room)
(9) Audio-visual aids (power, screens, and storage)
(10) Access to critical areas by the handicapped

## Guest Quarters

Although many alternative designs will provide satisfactory guest-room accommodation, the accompanying example illustrates a number of desirable features to consider in planning such facilities. Among the more important points that this example illustrates are the following:[13]

1. Double doors associated with the entry vestibule provide extra noise protection from activities that go on in the corridor.
2. Certain minimum clearances are observed, which ensures adequate passage through doorways and between furnishings.
3. Adequate clearance in front of the commode prevents the person who is opening the door from being caught between the door and the commode.
4. Clearance around the beds facilitates makeup.

[13]A number of quarters should be specially designed to accommodate handicapped guests. These will require additional clearance around beds and clearance and handrails in the bathroom.

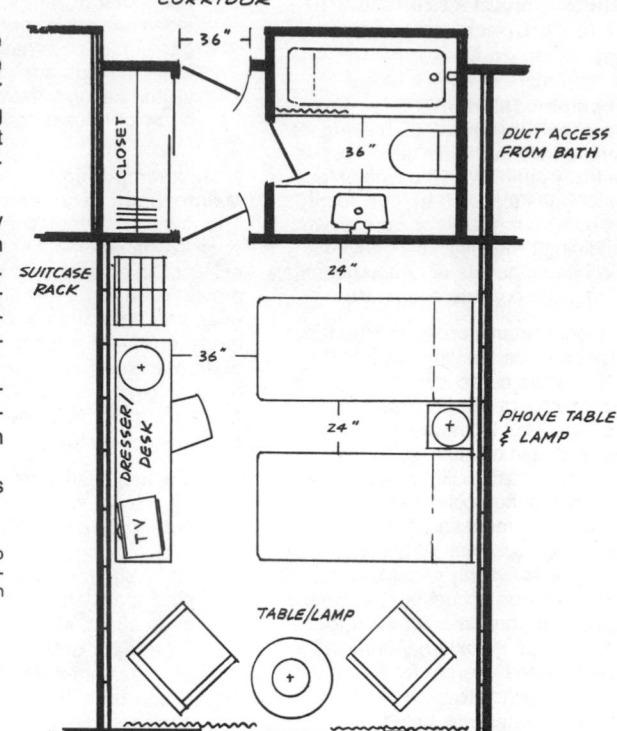

When floor space has to be limited for one reason or another, care should be taken to seek out special furniture arrangements in order not to crowd guests, especially during the daytime. Typical approaches have been to create special furniture that can be moved from one position to another, as in the accompanying example. This example also illustrates a special technique for providing linen and bedding storage within the headboard unit. Such a device further provides the alternative daytime davenport for visiting with other guests.

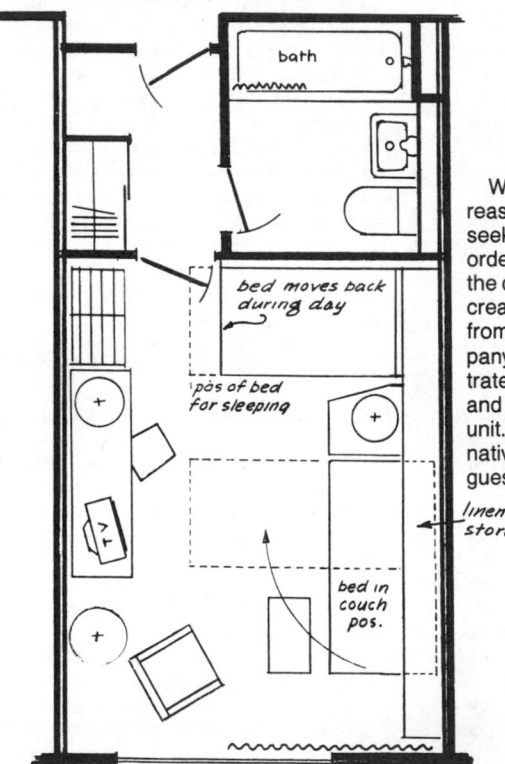

More expanded quarters can take on any form, of course. Refer to the guidelines for apartments in the section on multifamily residential systems. The accompanying example illustrates a logical and convenient arrangement that includes a kitchenette and separate sleeping area, making such quarters complete enough for longer-duration occupancy.

*Note:* The bathroom area in the accompanying example was enlarged, reflecting the desirability of more convenience for the guest who is staying longer.

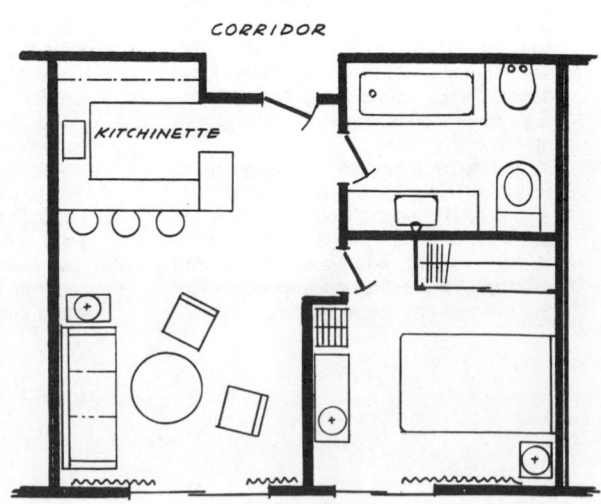

## Environmental Considerations

### 1. Noise and Vibration Control

Because of the nature of hotel and motel activity and the unique requirements for special air-conditioning and other noise-producing equipment, attention must be given to the problem of noise and vibration control. The following points provide a rationale for this concern and indicate some of the planning features that are significant in providing proper acoustic control:

a. Traffic noise: Although traffic may subside to some extent in some areas where public accommodations are located, traffic noise should be an important consideration in terms of site planning and facility organization. Guest sleeping areas obviously should be isolated as much as possible from traffic.

b. General operating noise: Primary noise-producing equipment should be isolated not only from guest sleeping areas but also from certain conference areas. It is easier to minimize noise problems at the source than to create barriers between the source and an area that requires quiet.

c. Guest-produced noise: An inevitable problem associated with hotel and motel operation is that some guests try to sleep while others come and go, entertain, or otherwise create commotion and noise. By means of proper arrangement of hallways, elevators, stairs, etc., the basic guest accommodations plan should provide as much separation as possible between these and sleeping quarters.

d. Entertainment noise: A common error is to place an entertainment facility right in the midst of the complex, with the result that the noise produced by a band or other entertainment tends to permeate the entire facility—obviously an undesirable approach to planning.

e. Systems vibration: Vibration (and often an attendant noise) is particularly annoying to guests who wish to sleep or rest. In planning, be especially cognizant of the possible production of vibration within and through the facility by passing traffic, low-flying aircraft, air-conditioning systems, etc.

### 2. Heating, Cooling, and Ventilating

Both natural ventilation and artificial air conditioning should be provided for all but the few uniquely blessed public accommodations because of the round-the-clock, year-round operating conditions typically required by these facilities. The following special considerations are germane at the planning stage:

a. Provide enough independent control so that all areas can adjust the immediate area temperature and ventilation to match the individual activity need. Guests require different temperatures and airflow for active versus rest cycles, public areas require adjustment to match small versus large crowds, and so on.

b. Provide humidity control as well as temperature and ventilation and consider the use of electronic dust-collecting systems to aid in preventing the spread of contagious diseases and in minimizing cleaning and refurbishment requirements.

c. Avoid the semiclosed air-conditioned concept, wherein fresh air cannot be introduced; this is especially important in the event of a breakdown in the air-conditioning system.

### 3. Illumination

Although decorative illumination is a desirable feature of the public accommodation facility, special care should be taken during planning not to let aesthetic lighting schemes interfere with ordinary seeing tasks. This includes consideration of both natural and artificial lighting features and systems. The principal areas that are commonly poorly executed in terms of illumination are the following:

a. Parking areas
b. Entryways
c. Restaurants and dining areas
d. Stairways and balconies
e. Hallways in guest-room areas

Especially important is the consideration of emergency lights that illuminate in the event of a power failure and the sudden extinguishing of normal lighting.

## Hotel and Motel Safety Considerations[14]

For the following reasons, safety is particularly critical for hotel and motel operations:

1. The majority of the occupants are transient, are unfamiliar with the facility, and in many instances are too tired and/or too preoccupied with other concerns to exercise normal safety-conscious behavior.
2. Large numbers of people who may be sleeping in many separate rooms require separate warning and a variety of evacuation solutions.
3. Large groups of people congregated together in public areas create severe crowd control and evacuation problems.
4. It is difficult to devise a single disaster plan for mixtures of adults, children, the disabled, and the aged.

[14]*Life Safety Code,* NFPA no. 101, National Fire Protection Association, Boston, Mass.; and United Kingdom Home Office, *Guide to Fire Precautions Act, 1971,* H. M. Stationery Office, London, 1971.

5. The fire loading of the typical hotel or motel is often substantial because of the large amount of furnishings that such a facility contains.

During conceptual planning, consider at least the following safety features:

1. Materials flammability
2. Fire warning systems
3. Water supplies for fire fighting
4. Emergency exit doors
5. Sprinkler systems
6. Escape routes and distances
7. Fire extinguishers
8. Access for fire-fighting vehicles and equipment
9. Barriers to control the spread of fire (including building separation)
10. Warning and exit signs, markers, and instructions

In addition, consider the following specific design features:

1. Nonslip walkways and stair treads
2. Appropriate handrails, balcony and staircase guards, and pool guards
3. Electrical system grounding and outlet child-proofing
4. Elimination of single-step changes in walkways and room thresholds
5. Adequate illumination, including emergency lights
6. Elimination of tripping hazards, sharp corners, and edges or protrusions on structural or built-in cabinetry, railings, etc.
7. Exclusion of guests from hazardous operating machinery
8. Avoidance of floor-to-ceiling glass windows or doors through which individuals might walk
9. Minimization of escape-route travel distance and maximization of the number and identifiability of exits; escape routes that are direct, i.e., that do not require the user to make many changes of direction, traverse many stairs, or pass through many doors; and avoidance of dead ends

| Maximum travel distances | Circumstances | NFPA m | NFPA ft | United Kingdom m | United Kingdom ft |
|---|---|---|---|---|---|
| | | Within rooms | | | |
| Places of assembly, convention rooms, restaurants, ballrooms | 1 exit only | | | 9 | 30† |
| | 2 or more exits* | 45 | 150 | 18 | 60† |
| | with automatic sprinklers | 60 | 200 | | |
| | 3 or more exits | | | 30 | 100† |
| Areas of high fire risk (kitchens, boilers, houses, etc.) | 1 exit only | | | 6 | 20 |
| | 2 or more exits | | 75 | | |
| Apartments | | 15 | 50 | | |
| | | Corridors | | | |
| Hotel guest rooms to safe exit (protected stairs, etc.) | 1 direction (dead end) | 10.6 | 35 | 7.5 | 25 |
| | 2 alternate directions | 30 | 100 | 18 | 60 |
| | with automatic sprinklers | 45 | 150 | | |
| (also apartments) | external access ways | 60 | 200 | | |

*At an angle of 45° or more from any point.
†Disposition of room contents must not increase the distance more than 1½ times.

# OFFICE BUILDINGS

## OFFICE BUILDINGS

### Public and Private Office Buildings

One of this country's foremost designers of office buildings has noted that "the human factor must be predominant in offices and that although factories are built around the machines they house, an office building should be built primarily around [the people] and the functions it houses." Although an infinite variety of office building types could be addressed, this section deals principally with the large office building system. However, many of the planning guidelines presented here are equally applicable to smaller office complexes.

The following guidelines outline some of the more important human factors considerations that should be addressed during initial office building planning.

### 1. Building-Site Configuration

Since the typical large office building causes a concentration of people and vehicles within a fairly restricted area and since it often produces fairly high-density traffic loads at regular intervals, it is critical that approaches to, and departures from, the site be laid out to minimize traffic conflicts; that adequate and convenient parking be provided to maximize the ease of transfer from vehicles to and from the building or buildings; and that the approaches, parking, sidewalks, and building entrances be not only arranged but also signed for ease of understanding by the people who will be using the building, especially the transient visitor who is arriving for the first time.

### 2. Interior Layout of Office Spaces

Unlike many other large architectural systems, the office building generally is subject to almost continual rearrangement as tenant replacements occur. This requires careful consideration not only of the shape, size, and placement of the fixed features of the building's interior (to provide adequate flexibility for future refurbishment) but also of the basic geometric concept as it may provide an apparent logic for tenants and visitors.

### 3. The Office Complex as Part of a Larger Architectural System

Often the office complex may be an integral part of a larger system, e.g., a factory, a laboratory, or a governmental complex. In this case, the office portion should not be allowed to become "lost" within the overall architectural plan; i.e., its exterior appearance should provide a clear indication to the visitor of what it is (e.g., an office building, as opposed to a manufacturing facility), how to get to it, where to park, etc.

### 4. The Windowless Office Building

Although there are obvious advantages for the designer in utilizing a windowless building concept for office space (i.e., it simplifies many environmental control solutions), most office workers need visual contact with the outside world. Even though most modern office buildings are air conditioned, the value of visual contact with what is going on outside cannot be overemphasized. On the other hand, "solid glass" concepts, popular with some designers, are equally inhuman in that they create many visual, environmental, and physical arrangement problems for the office workers and their clients and particularly for the individuals who must maintain the facility.

### 5. Key Subsystems Absolutely Necessary for All Office Buildings

a. Rest rooms for tenants, visitors, building management and maintenance personnel, and special tenants who may supply unique services such as security, food service, or other catering

b. Building management and maintenance areas

c. Building engineering and utilities areas
d. Service and delivery docks
e. Corridors, stairs, and elevators
f. Entrances and exits for tenants, visitors, services, and emergency escape
g. Parking for tenants, visitors, and services

**6. Special Subsystems That May Be Desirable for Common Usage**
a. Cafeteria and kitchen
b. Inside parking
c. Mail room

d. Tenant recreation (lounges, gymnasium, locker rooms, etc.)
e. Limited catering and shopping facilities (barber and beauty shop, drugstore, pastry shop, etc.)
f. Branch bank

**7. Other Features Critical to the Initial Planning**
a. Safety features (sprinkler systems, fire alarm system, fire hoses, hand extinguishers, emergency exits, and fire escapes)
b. Occupant conveniences (drinking fountains, public telephones, vending machines, etc.)
c. Environmental control (illumination, temperature, ventilation, humidity, noise, etc.)
d. Services (electric power, trash collection, building and office housekeeping and maintenance, security, etc.)

**Public Office Building Planning Model**

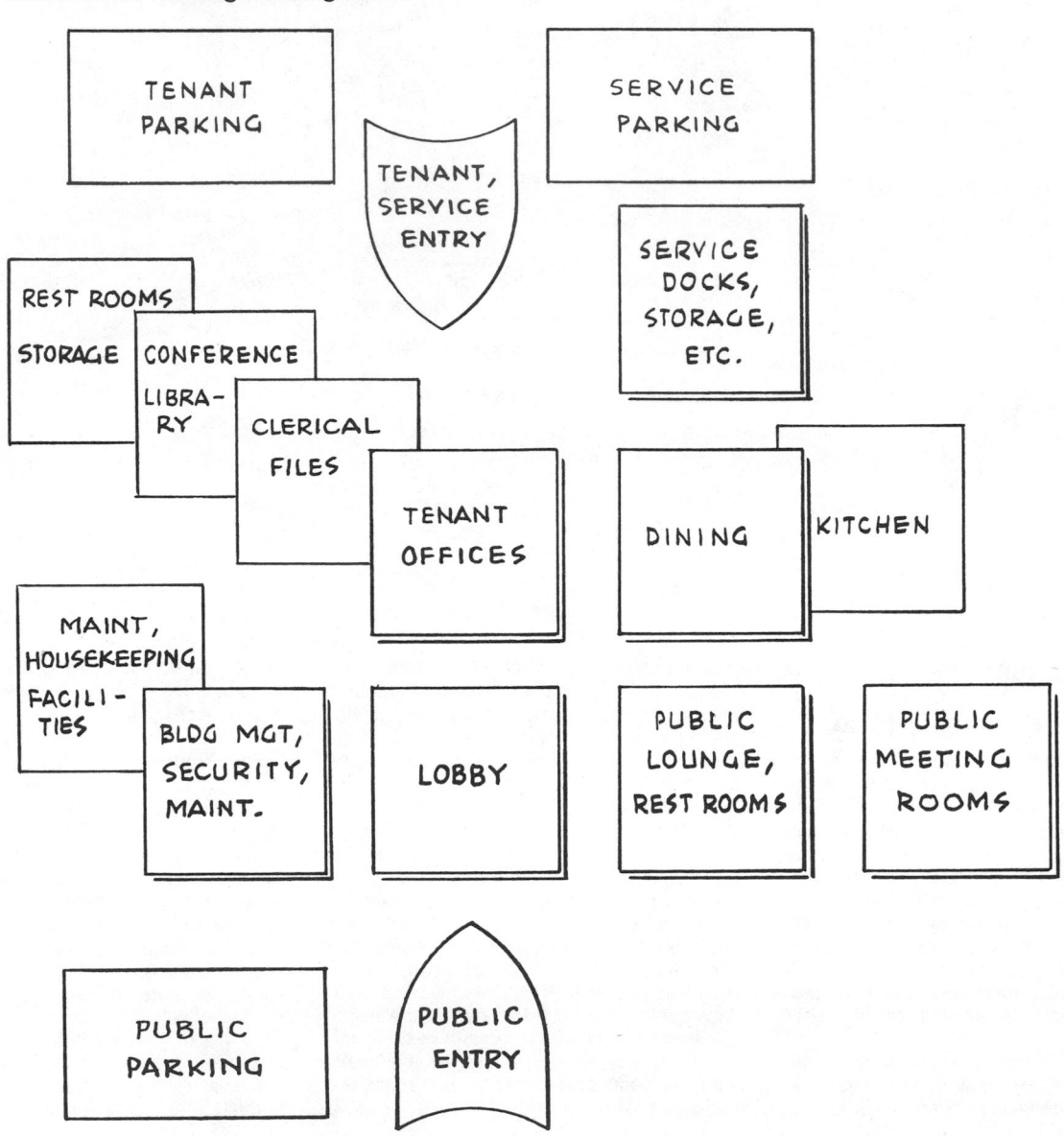

The following are among the more important layout considerations:

1. Initial visitor and client convenience to the main entrance, to the office of the person the visitor or client wishes to visit, to public rest rooms, and possibly to public meeting rooms
2. Tenant ease of ingress and egress to the building and to the tenant's office and convenience of his or her office layout for each task to be performed
3. Ease and convenience of building management and maintenance

**Information Requirements Checklist
for Office Building Planning and
Development
Marketing Factors**

| Activity | Example |
|---|---|
| Accounting | Private and group account processing, client counseling, bookkeeping, and financial planning |
| Banking | Central or branch bank, teller service, accounts and record keeping, safety deposit, loans, and financial advisory services |
| Commercial | Wholesale distributor sales center, private detective agency, employment agency, travel agency, secretarial services agency, mailing services, real estate agency, etc. |
| Governmental | Public records; offices of city, county, state, and federal elected official administrative, clerical, and engineering functions; issuance of legal permits and licenses; employment and welfare counseling and disbursement; courts, detention facilities; etc. |
| Legal | Attorneys, paralegal assistants, private counseling, research, and record keeping |
| Medical and dental | Doctor, nurse, and patient reception; diagnostic services; minor treatment and recovery area; laboratory testing; pharmacy; etc. |

The above checklist is, of course, only representative and is intended to imply that the nature of expected client populations requires considerable attention to differences in needs for office space. Of particular concern should be the type of client each tenant will have and the frequency and number of special considerations of these clients (e.g., the handicapped).

**Site Selection Factors**

| Features | Typical Details |
|---|---|
| Location | Area, frontage, topography, access from highways and street, and convenience to related facilities (courts, hospitals, and other office buildings) |
| Aspect | Views, benefit of height and visibility from roadways, and possible hazards to traffic entering or leaving the site |
| Land use | Present use, restrictions to demolition or clearance, development potential and controls, plot ratio allowed, and expansion potential |
| Restrictions | Easements, rights and covenants on site, tenancy restrictions to occupation, height, etc. |
| Noise | Road traffic flows, night traffic, and rail and aircraft noise levels |
| Transportation | Availability and frequency of public transportation and availability of airport terminals and rail terminals |
| Neighborhood | Character and condition of adjacent and nearby properties |
| Economic activity | Prominent commercial and manufacturing premises, shopping centers, markets, restaurants, etc. |
| Services | Locations and capacities of utility services (electricity, gas, telephone, water, sewage disposal, police and fire protection, etc.) |

**Exterior Configurational Implications
Relative to Entrance Recognition**

It is desirable to select a building configuration that provides an immediate indication of where the main entrance is located. Although tenants soon learn to find less conspicuous entrances, the visitor can easily become confused by certain exterior features of the office building. Some of the configurational aspects of entrance recognition are shown in the accompanying sketches.

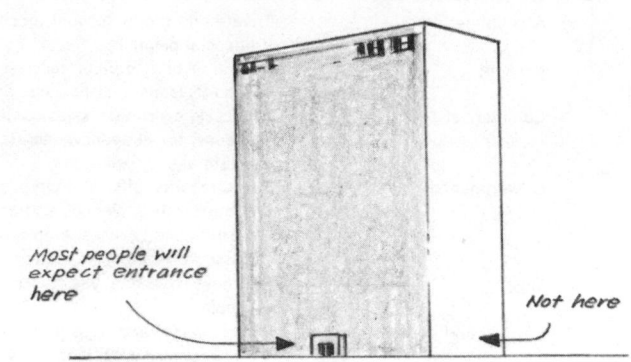

1. The accompanying sketch shows that people generally expect a main entrance to be on the broad side of a building, as opposed to being on one end.

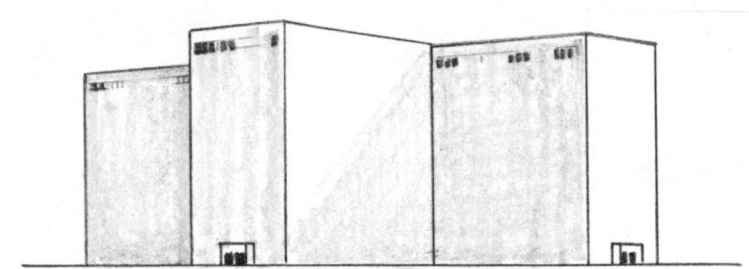

2. Symmetrical configurations such as the one shown in the accompanying sketch create confusion because there are essentially equal choices. If one entrance is to be used as the main entrance, it should have a special size, shape, or other emphasis, and it should be clearly marked.

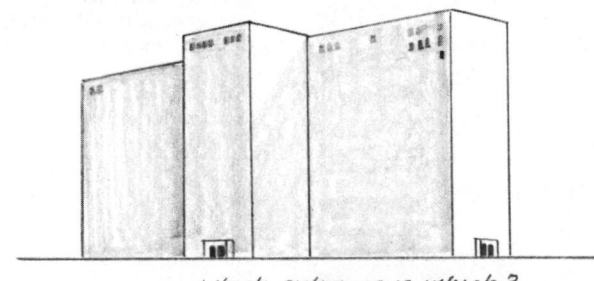

3. A slight geometric modification, as shown in the accompanying sketch, helps the observer discriminate among possible choices; i.e., the reduced size of one wing provides a cue that it is different and is probably the most likely entry point.

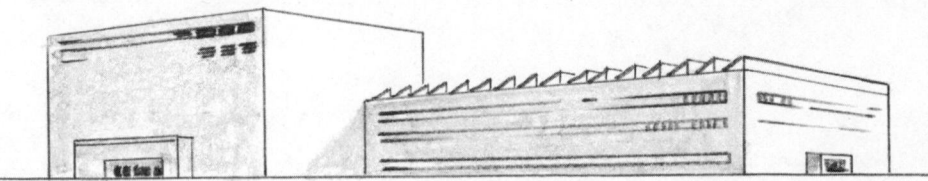

4. When an office section of a larger complex is provided, it is important that the exterior appearance make it obvious to the observer which part of the overall complex is the office section, rather than some other section, such as the apparent factory section shown in the accompanying sketch.

## Exterior Building Configuration and Identification

Although business identification is an impor- tant aspect of all architectural planning, iden- tification of the large office building is perhaps even more critical because of the nature of its transient user needs. Many visitors may visit the building only once or infrequently, and when they do, they often are in a hurry. The accompanying illustrations emphasize several important aspects of building identification:

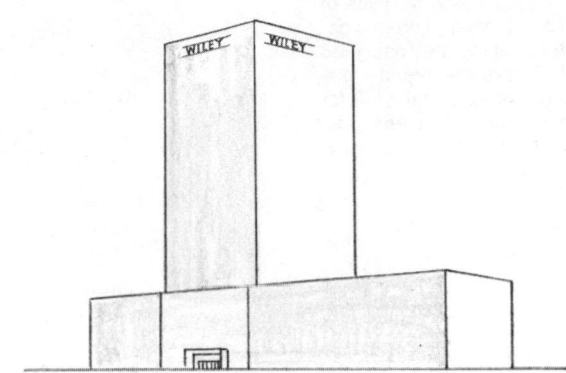

1. Identification from a distance helps the person plan ahead and become pre- pared to look for specific streets, num- bers, and/or special directional signs to parking and driveways.

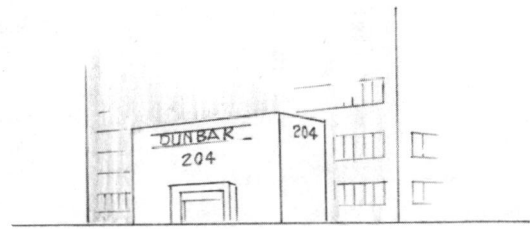

2. Identification near the main building site is important, especially when there are a number of similar office buildings located side by side. Thus, as the accompanying sketch illustrates, the identification (e.g., the building name and the street num- ber) must be placed lower, or nearer the observer's normal line of sight. Note that the size of identification signs and/or numbers must be commensurate with the viewing distance considered appro- priate for the observer's planning needs.

3. Identification of parking entrances is es- pecially critical when the office building is located in a high-density area. The loca- tion and size of the sign must be such that it can be seen in time for the person who is driving to plan any lane changes, and it should not be blocked from view by other cars or pedestrians that may be located between the driver and the sign.

Avoid the temptation to leave identification problems until later. In many cases the shape of the proposed structure and/or special ideas about decorative symbolism make it impossi- ble to add an adequate identification system.

**Exterior Building Configuration,
Location, and Planning for Parking**

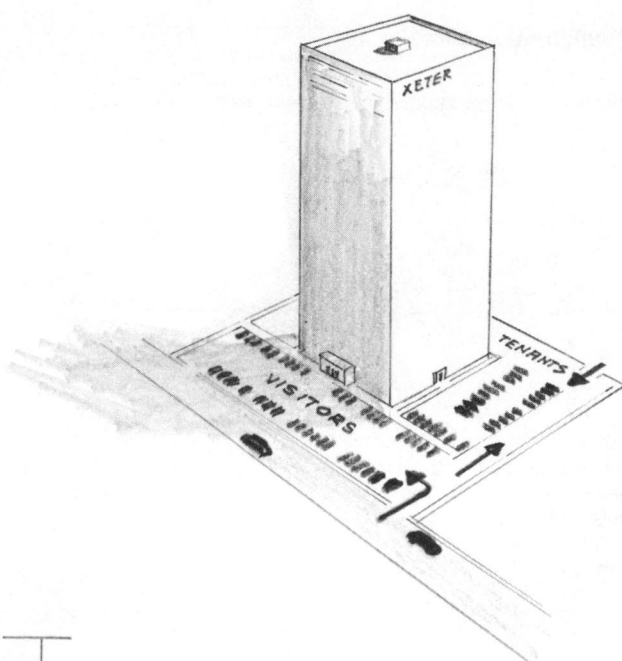

There is typically a continuing exchange of vehicles in and out of the parking facilities of the average large office complex. Tenants object fiercely to outsiders' taking their assigned parking places, and visitors are equally dismayed when there is no parking space left for them—especially when time is of the essence. The accompanying illustrations point up several important considerations.

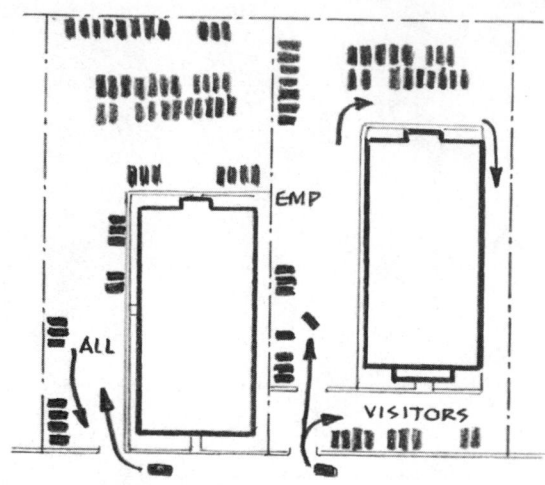

POOR    BETTER

Separate the tenant and visitor parking areas.
Locate the building so that the parking area provided for visitors is adjacent to the main entrance they will use.
Provide separate entrances and exits.
Consider indoor parking areas; these can be within the building and/or within an adjacent parking structure.

Plan the parking-entrance-building configuration so that service entrances and their associated vehicle parking will not interfere with either tenant or visitor parking areas.
In planning all parking, consider the infrequent problems of removing a disabled vehicle, e.g., maneuvering space for the tow truck and clearing away snow.

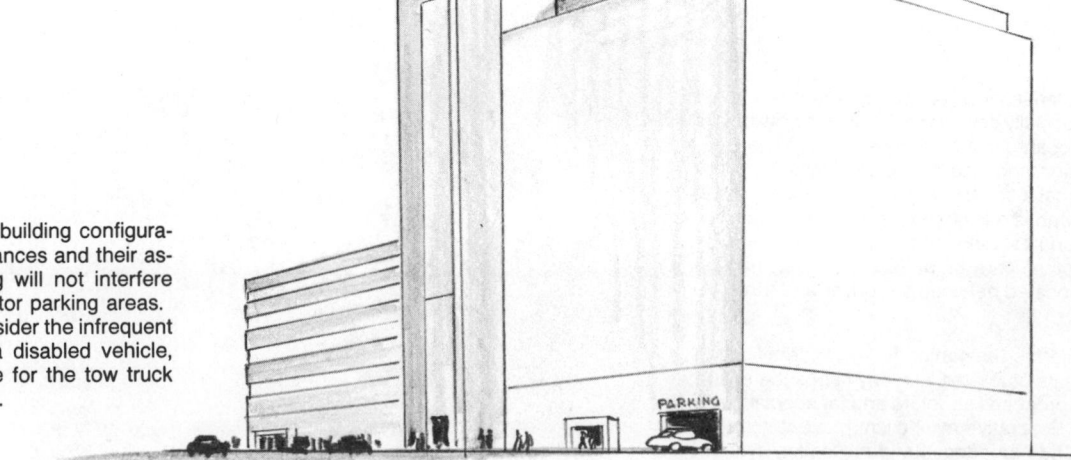

## Building Configuration and Expected Internal Traffic Flow

First-time visitors to a new building depend on a mental picture of the relationship (directional) between the entrance and the general floor plan (routes) to help them find their way to individual offices and back again. The accompanying illustrations emphasize some of the desirable relationships:

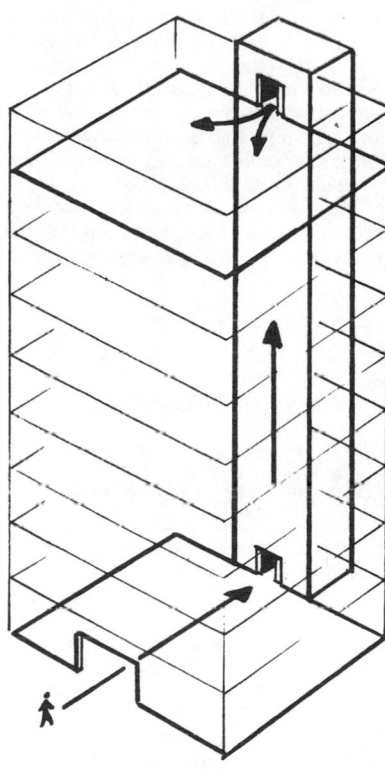

1. The horizontal pattern should emphasize rectilinear, straight-ahead, to-the-left, and to-the-right pathways. Jogs and curvilinear or angular pathways disturb the "logic" of the average visitor.

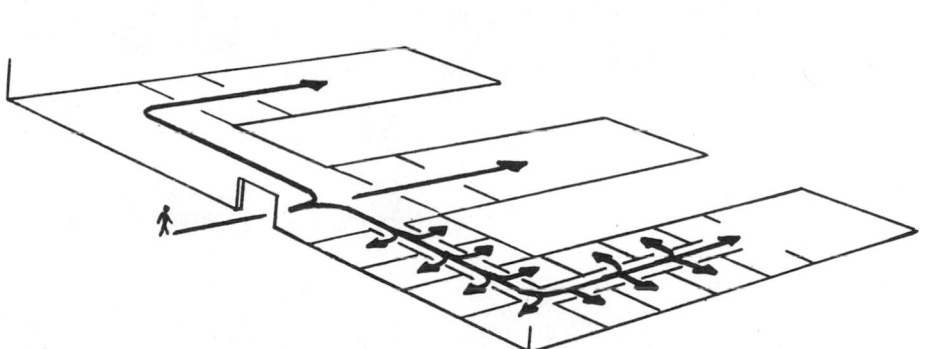

2. Vertical transfer via either the peripheral service-well system or the center service-well system, as shown in the right-hand sketches, requires "mental translation of directions"; the pattern shown in the lower sketch does not.

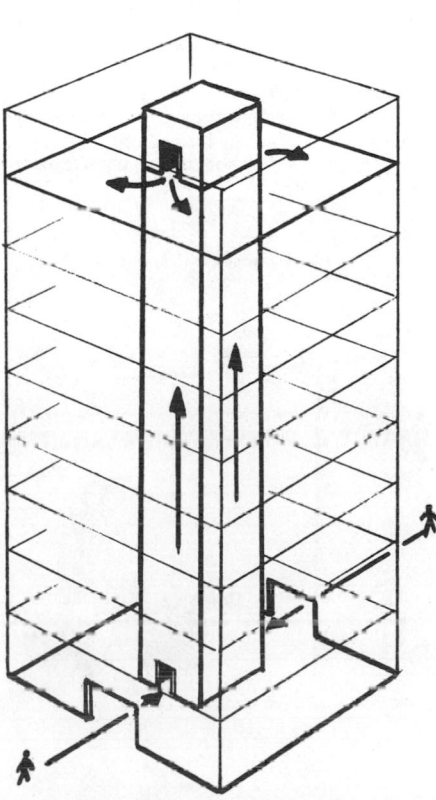

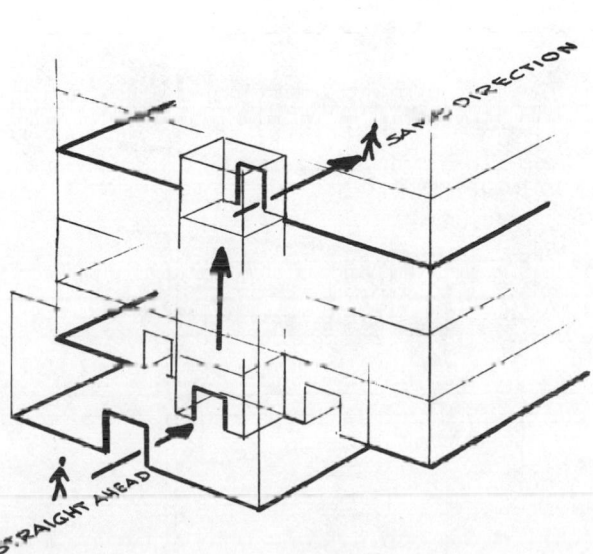

### Implications of Building Configuration and Window Patterns

The primary purpose of windows in an office building is to provide outside viewing for occupants. In spite of this obvious purpose, there is often a tendency to consider windows and window patterns as a key aesthetic feature of the building during initial conceptualization. In their enthusiasm for this latter objective, planners may forget, or at least may downplay, the primary visual function, failing to take problems of glare and maintenance into account. The accompanying illustrations emphasize several important human factors considerations.

### 1. Window Location
Windows should be located so that they will provide some benefit to the occupants, not merely symmetrical exterior geometric patterns.

### 2. Window Size
Windows should be large enough to allow for viewing and natural illumination, but not so large as to increase glare and make cleaning difficult and time-consuming.

### 3. Glare
Although interior techniques can be used, consider glare control techniques in the initial structure configuration.

### 4. Cleaning
Window cleaning from the outside is hazardous at best. Consider ways to minimize the necessity for suspending window cleaners outside the building (balconies, windows that can be cleaned from the inside, etc.).

### 5. Interior Hazards
Avoid floor-to-ceiling windows; people fall into them and furniture may be pushed into them. (See Chapter 3 for specific window design recommendations.)

Glare, cleaning problems

Reduced glare, expanded view!

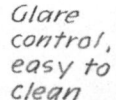

Hazardous, hard to control glare, big job to clean!

Glare control, easy to clean

## Interior Office Building Planning Considerations

An important planning question is: What functions do I locate on which floor of the multistory office complex? The following checklist provides a number of guidelines.

### 1. Ground Floor (Street Level)
a. Lobby—information directory or personnel station
b. Stairs and elevators
c. Building management offices
d. Security office or offices
e. Special tenant facilities (i.e., a bank, a restaurant, a travel agency, and commercial shops that require walk-in traffic exposure and convenience)

### 2. Upper Floors
a. Foyer (adjacent to elevators and stairs)
b. Rest rooms
c. Professional tenant offices
d. Emergency stairs

### 3. Special Areas
a. Underground parking level or levels, with stair, elevator, and ramp access.
b. Tenant cafeteria (preferred at midlevel for essentially equal convenience to all upper floors). This requires a special service elevator.
c. Sun deck (preferred at the upper level for tenant use only). The location should ensure adequate sun exposure during the times it would normally be used.
d. Combined public and tenant patio (preferred at ground level, probably associated with a ground-floor restaurant establishment).
e. Shopping arcade (preferred at ground level for easy access for walk-in traffic).

The next planning question is: How do I plan the interior space on specific floors to provide the maximum utility for various tenants? In the case of a special long-term tenant for whom the floor space is to be customized, a detailed analysis of that customer's needs and unique desires is, of course, necessary. For general planning purposes, however, the suggestions on the following pages provide useful guidance. However, these suggestions are intended primarily to ensure adequate consideration of the human factors.[15]

[15]For other important design requirements (i.e., of a structural, materials, or utilities nature), see J. De Chiara and J. H. Callender, *Time Saver Standards for Building Types*, McGraw-Hill Book Company, New York, 1973; and Ramsey and Sleeper, *Architectural Graphic Standards*, John Wiley & Sons, Inc., New York.

## General Interior Planning Guidelines

Although there are considerable differences between various building floor plans, key human factors considerations should be kept in mind. The two accompanying illustrations emphasize some of these factors.

### Travel Distances
1. Locate stairs and elevators so as to reduce the distances tenants must travel and to make more than one emergency escape pattern possible.
2. Locate rest rooms so as to reduce the distances tenants must travel; also provide visitor facilities off the lobby near the entrance (i.e., avoid hiding rest rooms from visitors).[16]
3. Locate maintenance closets so that the building maintenance personnel do not have to carry equipment long distances.
4. Locate utility rooms or closets so that technicians do not have to travel considerable distances between these and office spaces.

[16]The rest room is the most important facility within any public building. The past practice of trying to make such facilities inconspicuous and/or private is not consistent with human need.

5. Provide a service elevator which is independent of the normal passenger elevators and which connects with a service entrance rather than with normal tenant or visitor entrances.
6. Plan emergency escape facilities that minimize the distances that the person in a wheelchair must travel to an elevator or to an exit. Many such special facilities are placed where the person in a wheelchair has to travel further than anyone else.

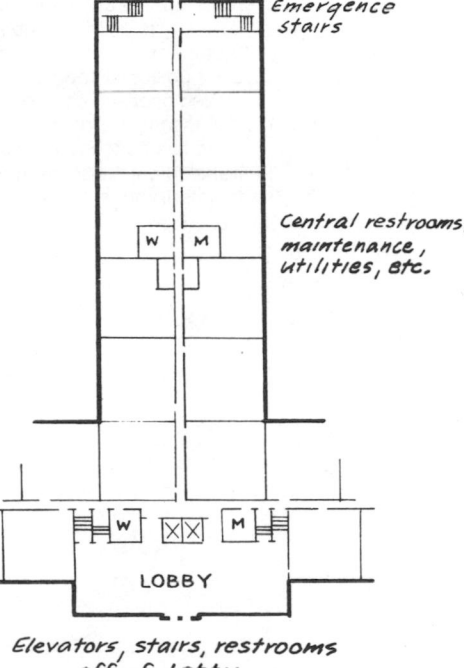

Emergence stairs

Central restrooms, maintenance, utilities, etc.

LOBBY

*Elevators, stairs, restrooms off of Lobby*

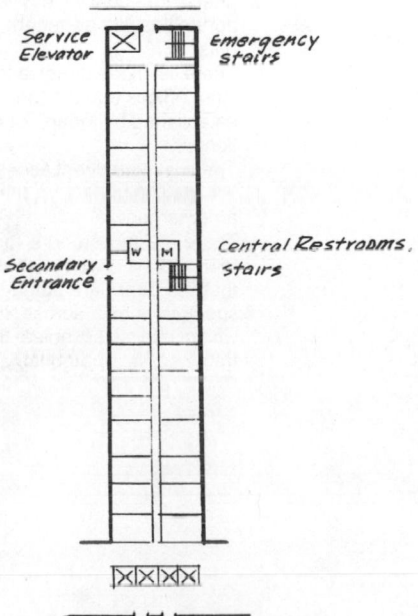

Service Elevator

Emergency stairs

Secondary Entrance

Central Restrooms, stairs

*Main Entrance*

**General Permanent Space
Compartmentalization**

When permanent walls between general tenant office spaces are planned, consider the following guidelines:

1. Avoid deep, narrow spaces. Not only executives, but also all staff members, need to have visual access to the outside.
2. The equal-sided space is generally more easily utilized by the individual tenant for subpartitioning and/or "open-space planning." In addition, it offers more opportunity to provide window and light opportunities for more people.
3. Door positions should be staggered, not placed opposite each other, as shown in the accompanying illustrations. Not only does this concept introduce more noise transfer, but it also creates potential traffic interference.

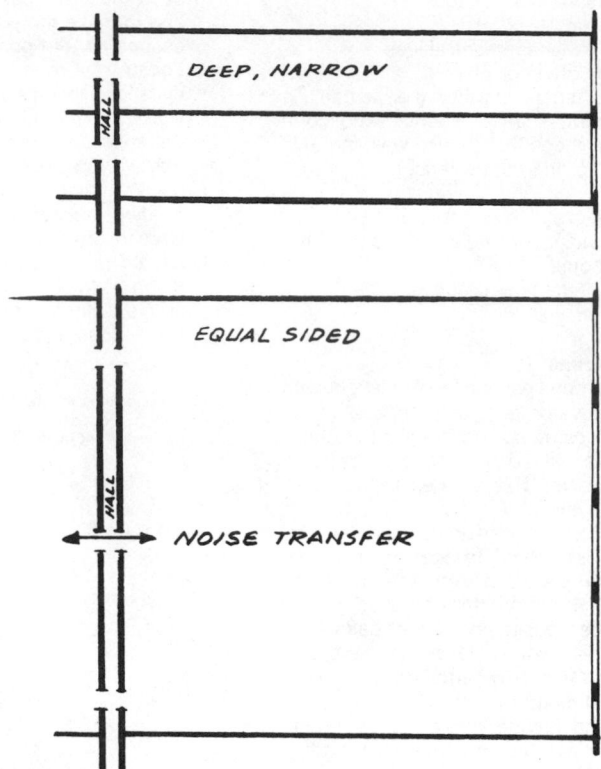

4. Where feasible, the space configuration shown in the accompanying sketch is more desirable, for several reasons:
   a. There are equal window and light opportunities for all tenant staff members.
   b. There is more direct access to any inner offices that the tenant may have established by means of partial partitions.
   c. There is more direct access by all tenant staff members to the corridor in an emergency.
   d. There are more doors for the tenant to use as he or she wishes. Note also that the doors are staggered with respect to those across the corridor, which minimizes noise transfer and traffic conflict potential.

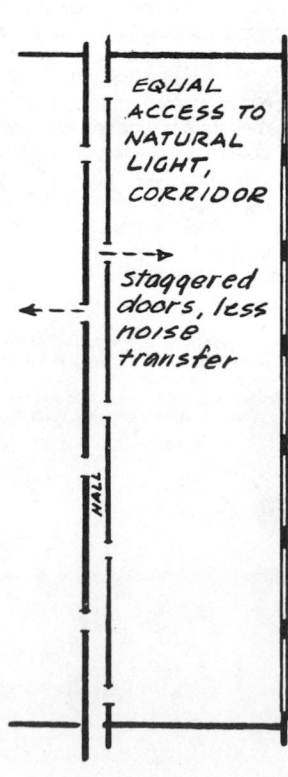

### Corridor- and Stair-Width Guidelines

Although people can pass by each other in fairly cramped quarters if necessary, they do not like to, and in certain cases it is hazardous to crowd people, such as during an emergency exit situation. The accompanying illustrations and dimensional guidelines are representative of the desirable widths for the following office building situations.

**1. Main Corridors for Ground-Floor Level**
People often walk in pairs, and generally one couple dislikes touching another couple as they pass.

**2. Primary Corridors for Upper Levels**
Typically, a couple needs to pass at least a single individual going in the opposite direction.

**3. Secondary Corridors, Any Level**
People generally are not walking in pairs, and/or traffic is light and local.

**4. Within-Office Areas**
The aisle is typically bordered by half partitions (at least on one side).

**5. Elevator Foyer**
Although a larger space is desirable at ground levels for very large office buildings, the dimensions shown in the accompanying sketch should be about minimum for any floor level; 25 percent less is generally adequate when elevators are on only one side of the corridor.

**6. Stair Width**
The width between dual handrails shown in the accompanying sketches below should be the minimum for any stairwell.[17]

[17]See Chap. 2 for additional criteria for corridors, stairs, etc. Also consult local ordinances regarding standards for corridors and stairs.

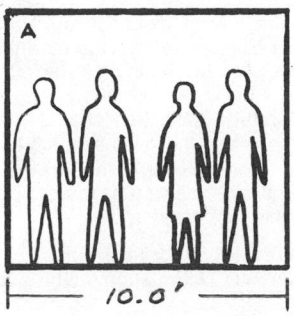

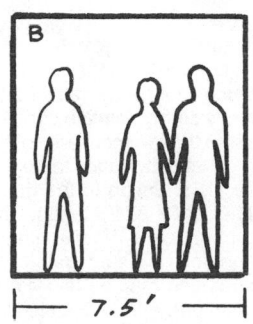

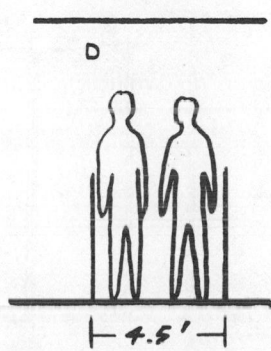

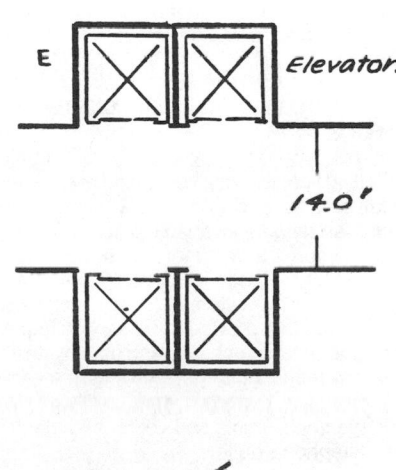

## Individual Tenant Space Requirements

Wherever possible, plan the basic internal floor-space requirements per tenant on the basis of the tenant's needs (as opposed to creating the external walls and letting the tenant use what space is left inside as best he or she can). Some of the typical tenant needs that should be considered in developing floor-space envelopes are the following.

### 1. Reception Area

Although some tenants may be able to manage without a client or visitor reception area, this is not desirable. The reception area should provide enough space for a receptionist and his or her desk and chair, possibly a switchboard and file, and suitable waiting-room furnishings, e.g., a davenport, chairs, a coffee table, and an end table and magazine rack. Although the reception area can be made larger or smaller according to tenant needs, it should be considered from the standpoint of its location with respect to how the rest of the space might be partitioned. That is, the reception space acts as a lobby from whence visitors will proceed into the inner office complex.

### 2. Storage

Avoid the temptation to consider the internal space requirements of the tenant only in terms of his or her potential furnishings, e.g., desks, chairs, and file cabinets. Generally, most tenants will have to add portable storage cabinets or build in such cabinets, thus reducing the originally planned personnel and furnishings space. This factor alone suggests that the so-called minimum floor-space criteria found in many architectural handbooks should not be used; i.e., plan on the generous side wherever possible.

### 3. Lighting, Ventilation, and Utilities Service

Avoid the general approach of permanently installing light fixtures, ventilating outlets, electrical outlets, and/or telephone outlets. Because all tenants invariably place their equipment where they want it, there are no successful methods for providing acceptable locations for these items. Use modern, flexible techniques that provide basic service capacity in ceiling or floor recesses and employ flexible, modular techniques for removing and replacing interface materials and hardware, e.g., ceiling panels, ducts and vents, and electrical and telephone outlets.

### 4. Built-in Cabinetry and Modular Partitions

Develop modular components (or select off-the-shelf components) that allow for easy change and/or addition of landlord-provided cabinetry or partitions. That is, either provide no such amenities as part of the tenant package, or plan a system that allows these elements to be added, changed, relocated, or removed with a minimum requirement for refurbishment (i.e., because of damage to the basic structure).

*Special note:* Consider the needs for control flexibility when planning lighting and air conditioning. It should be possible for tenants to adjust temperature and ventilation at various points within their particular space, not at some single point that may not represent the specific conditions in some other part of the space (note that when tenants partition their space, they may destroy well-intended, general ventilation-flow patterns that the architect devised). Similarly, tenants should have the capability of selectively turning various light fixtures on or off within their space. However, it is also desirable to consider an added, overall system switch so that tenants can turn off all lights (and possibly electric power) as they leave, without having to check out various parts of their own space to make sure that a light has not been left on in an inner office.

### 5. Other Tenant Space Services

Although these depend on the basic objectives of the specific office builder and/or the client, several other service needs of special tenants should be considered:

   *a.* Private rest room or rest rooms
   *b.* Kitchenette (with appropriate water, power, sink, etc.)
   *c.* Provisions for a wet bar or bars
   *d.* Private shower or showers
   *e.* Private elevator

### 6. Individual Tenant Safety

Avoid the tendency to consider safety only from the point of view of overall tenant safety. The crux of any safety solution should always be to make sure that all occupants of the building have an equal opportunity to be safe. This requires specific attention to the hazard possibilities within each basic tenant space. Of particular importance is the planning of alternative escape possibilities, regardless of how an individual space may be laid out or partitioned by its tenant. Although a significant aspect of this problem may reside in the building management's rules and policies, even the best rules are of little help if the original structural plan does not provide adequate alternatives.

Other significant factors are the selection of nonflammable materials, proper execution of the utility system layout and installation, and adequate accessory items such as sprinkler systems, fire and smoke detection equipment, communications systems, and emergency power and illumination. Although planners cannot prevent tenants from introducing hazards (e.g., flammable materials), they can make sure that floor coverings, drapes, and other permanent or management-supplied items are not dangerous.

## Planning Considerations for Specific Architectural Features

Certain basic features, such as window systems, doors, staircases, and elevators or escalators, may be selected quite early in architectural planning. Although, in most cases, certain local ordinances must be observed in making these selection decisions, the following key human factors considerations are important to remember.

### 1. Door Concepts

Recent door concepts include the floor-to-ceiling configuration illustrated in the accompanying sketch. These concepts are not recommended from the human factors standpoint for two important reasons:

   *a.* Such doors are heavier and are particularly hard for women, older persons, and handicapped individuals to handle.
   *b.* These doors have an undesirable psychological impact; i.e., they are out of scale in terms of their relation to human body size. Although some people accept them eventually, others may feel dwarfed and threatened by them.

### 2. Doorway Dimensions

Clearance for the transfer of furniture is as important as clearance for the passage of people. It is undesirable to make movers stand a typical desk on end, twist it around the door jamb, or remove the door in order to get it in or out of an office.

### 3. Door and Window Hardware

Avoid leaving the choice of opening and closing hardware to someone else. The size and shape of handles, the forces required, the direction of motion, etc., should be compatible with human expectations and capabilities, especially those of the infirm and handicapped. Note the L-shaped handles in the accompanying sketch. (See also Chapter 3.)

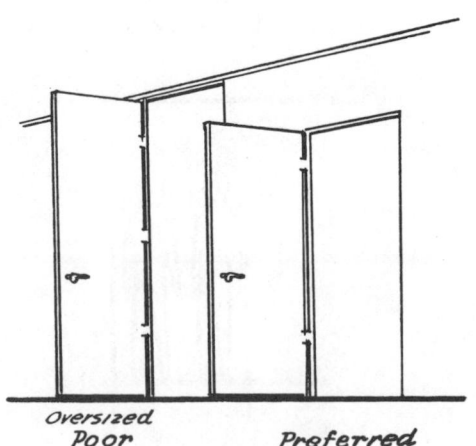

*Oversized Poor*     *Preferred*

## Hints for Estimating Floor-Space Requirements

To estimate floor-space requirements in offices, consider both the area occupied by furnishings and that occupied by office personnel under all anticipated dynamic movement patterns (i.e., moving a chair away from a desk, swinging around to a file cabinet, reaching for the wastebasket, getting out of a chair, and passing by furniture and/or other people). The accompanying sketches provide general dimensional suggestions for typical situations.

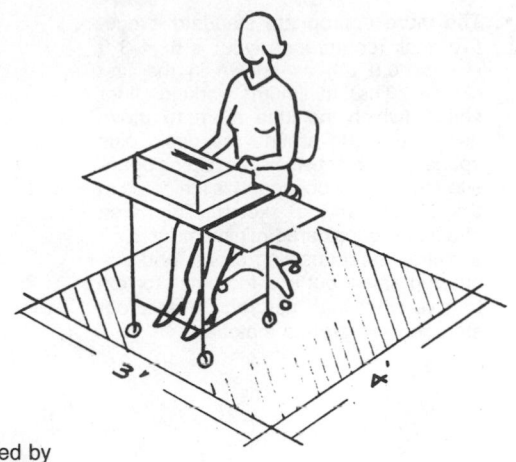

1. Perhaps the smallest area is needed by a typist working at a machine, as shown in the accompanying sketch. No furnishings or wastebasket add to the required floor area.

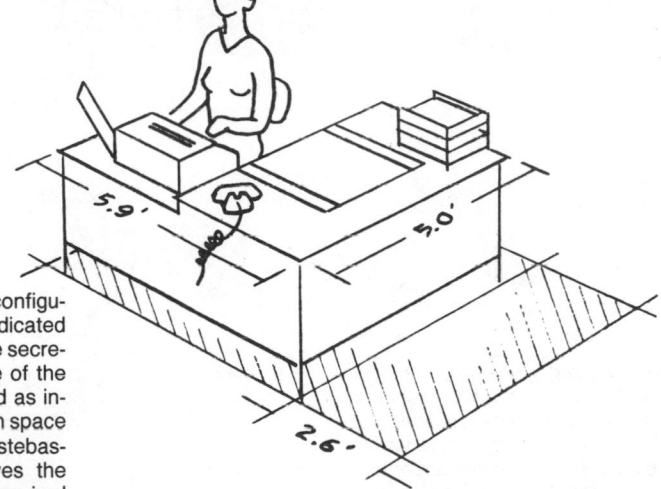

2. The typical L-shaped secretarial configuration requires a floor area as indicated in the accompanying sketch. If the secretary greets visitors on either side of the desk, additional space is required as indicated. Generally there is enough space beneath the desk for a small wastebasket. The secretary usually leaves the desk toward the rear; the space required for this is included within the total area shown.

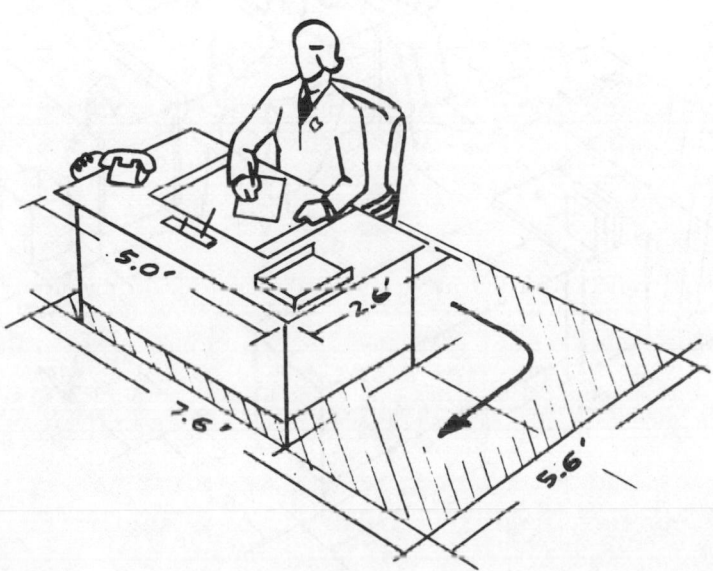

3. The minimum office desk size that is acceptable for any kind of office work is shown in the accompanying sketch, along with the additional floor area required so that the occupant can move away from and around his or her desk. This assumes that such a worker does not ordinarily greet or communicate with others at the desk, and thus no added space has been indicated for visitor floor space.

4. The more appropriate standard executive desk (conference type) is 6 × 3 ft (1.8 × 0.9 m), as shown in the first sketch. The minimum working floor space (which includes room to move away from and around the desk, plus space for one or two visitors sitting opposite the primary occupant) is indicated in the accompanying sketch. This area should be considered minimum for such activities as counseling clients and discussing sales, but it is too small for extensive work that may require adjacent storage of reference materials.

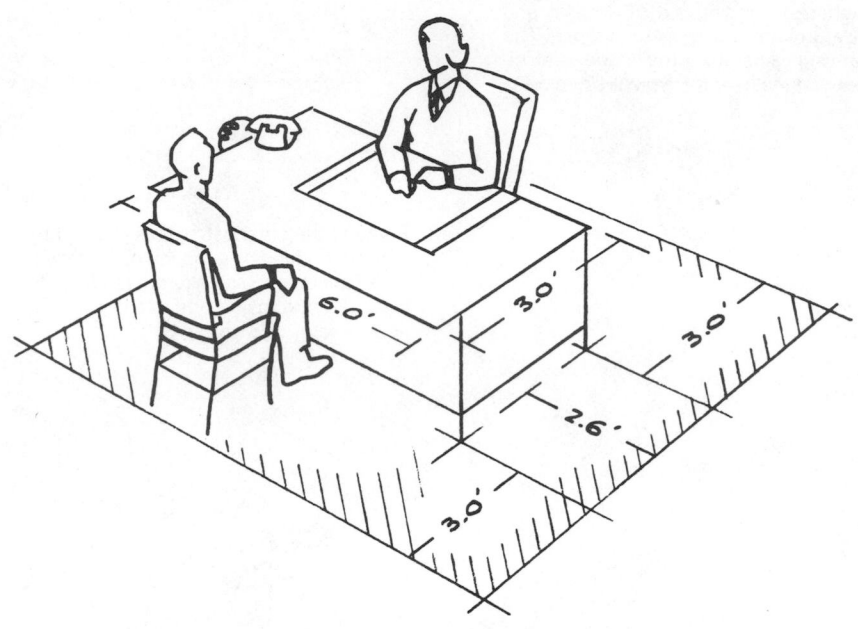

5. A preferred working space for the executive who not only meets a few clients or counselees but also does a considerable amount of work requiring additional file storage space is shown in the accompanying sketch. Note the additional clearance for walking past a visitor (i.e., it should be possible to walk past a visitor while he or she sits in the side chair undisturbed).

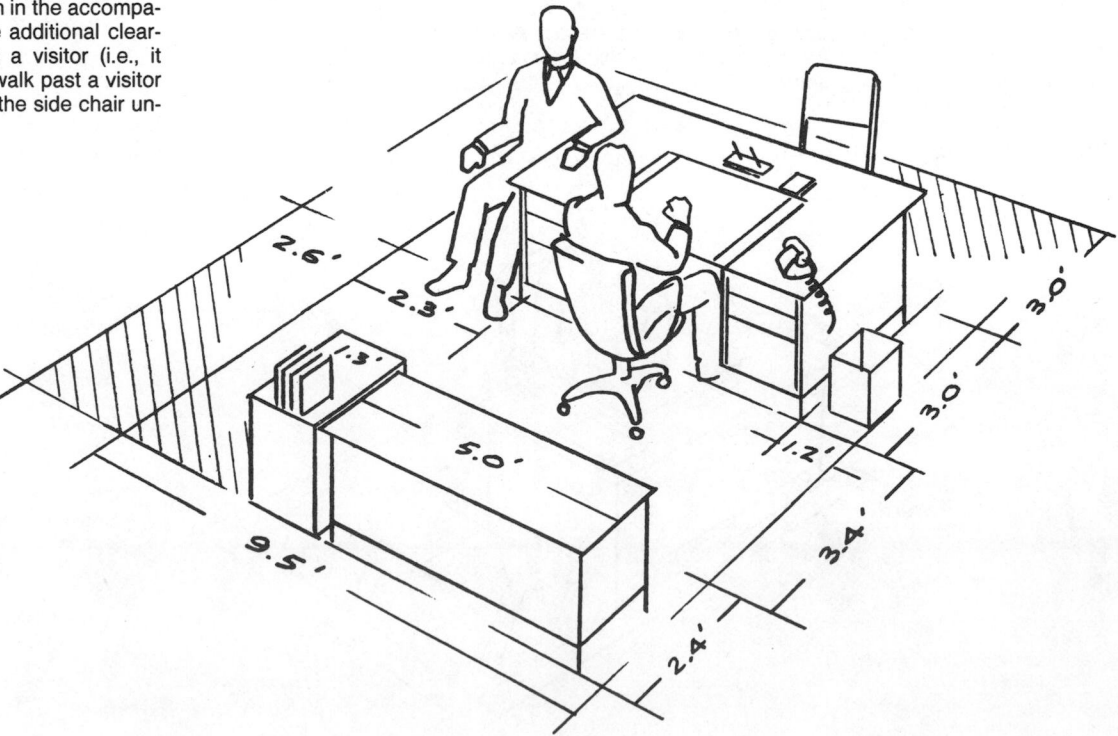

6. Minimum clearance in front of the standard metal storage cabinet is shown in the accompanying sketch. If, however, it is expected that other people may have to pass by while the secretary or other office person is working at the cabinet, an additional 24 in (61 cm) of clearance is necessary to avoid conflicts. Even more clearance is required if the individual passing is pushing a portable cart or carrying heavy packages. Therefore, a good rule of thumb is always to provide at least 30 in (76 cm) behind anyone who is working from a standing (dynamic) position.

7. Minimum clearance in front of the typical metal file cabinet is shown in the accompanying sketch. Once again, these dimensions assume that no one is passing behind the person. If this is expected, use the 30-in (76-cm) clearance rule of thumb.

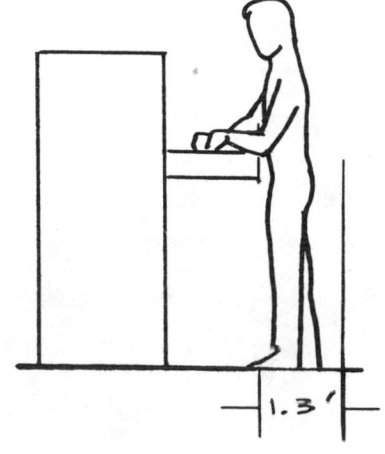

8. Various other working situations call for consideration of the specific dimensions of the equipment being used. As shown in the accompanying sketch, certain basic clearances are required just so that the operator (in this case, a draftsman) can move the stool around to more convenient positions. In addition, note the suggested clearance behind the board to allow large drawings to drape across the board without being creased or torn. (See also the section on special workplace layout in Chapter 2.)

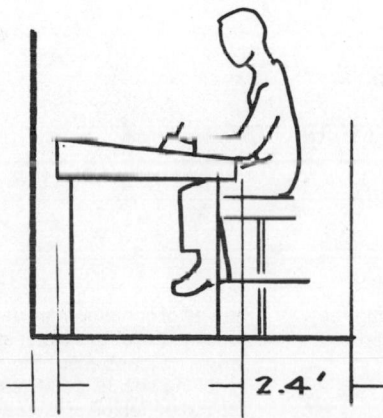

## Typical Dimensional Characteristics of Office Furniture

For estimating floor space, it is recommended that the following guidelines be used:

Desks: Secretary—60 × 30 in (152 × 76 cm); executive secretary—72 × 36 in (183 × 91 cm); supervisory personnel—72 × 36 in (183 × 91 cm); other executives and managers—72 × 36 in (183 × 91 cm) or larger

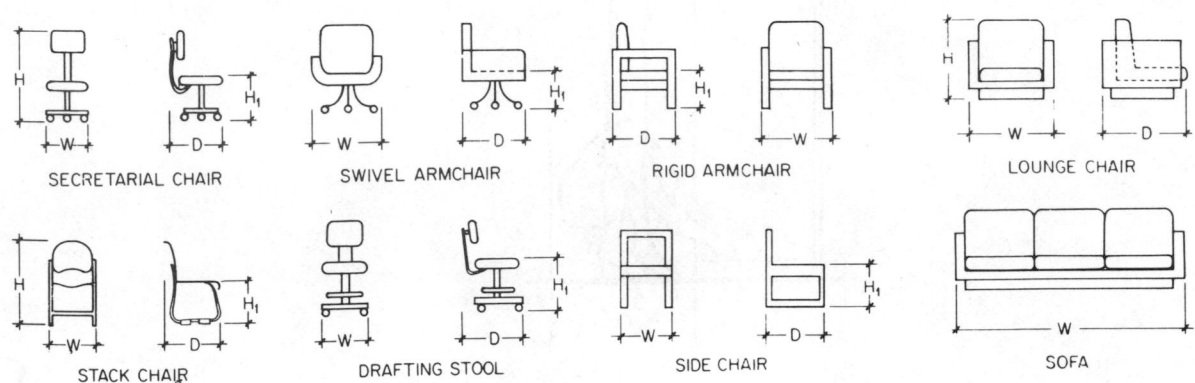

SECRETARIAL CHAIR SWIVEL ARMCHAIR RIGID ARMCHAIR LOUNGE CHAIR

STACK CHAIR DRAFTING STOOL SIDE CHAIR SOFA

Other furniture:

CHAIR DIMENSIONS

| | SECRETARIAL | | SWIVEL ARMCHAIR | | RIGID ARMCHAIR | | STACK CHAIR | | RIGID AND ADJUSTABLE DRAFTING STOOL | | SIDE CHAIR | |
|---|---|---|---|---|---|---|---|---|---|---|---|---|
| | STD. | RANGE | STD. | RANGE | STD. | RANGE | STD. | RANGE | STD | RANGE | STD. | RANGE |
| W | 1'-5" | 1'-4"–1'-8" | 2'-4" | 1'-8"–2'-6" | 1'-10" | 1'-6"–2'-3" | 1'-9" | 1'-6"–1'-11" | 1'-6" | 1'-5"–2'-0" | 1'-8" | 1'-4"–2'-0" |
| D | 1'-7½" | 1'-6–2'-0" | 2'-3" | 1'-8"–2'-6" | 1'-10" | 1'-7"–2'-8" | 1'-9" | 1'-7"–1'-10" | 1'-8" | 1'-6"–2'-0" | 1'-10" | 1'-6"–2'-8" |
| H | 2'-6" | 2'-5"–2'-10" | 2'-9" | 2'-6"–3'-0" | 2'-6" | 2'-4"–2'-10" | 2'-6" | 2'-4"–2'-9" | 3'-0" | 2'-11"–3'-6" | 2'-6" | 2'-4"–2'-10" |
| H₁ | 1'-5" | 1'-4"–1'-8" | 1'-5" | 1'-4"–1'-10" | 1'-6" | 1'-4"–1'-7" | 1'-5" | 1'-5"–1'-6" | 2'-4" | 1'-5"–2'-10" | 1'-6" | 1'-5"–1'-7" |

LOUNGE CHAIR AND SOFA DIMENSIONS

| | LOUNGE CHAIR | | SOFA |
|---|---|---|---|
| | STD. | RANGE | |
| W | 2'-6" | 2'-6"–3'-4" | D, H AND H₁ SIMILAR |
| D | 2'-7" | 2'-2"–3'-4" | 2 SEATS-5'-0"–6'-7" |
| H | 2'-6" | 2'-1"–3'-4" | 3 SEATS-6'-0"–7'-6" 4 SEATS-7'-8"–9'-0" |
| H₁ | 1'-3" | 1'-0"–1'-6" | |

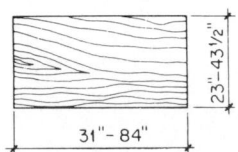

ARTIST AND DRAFTING DESKS OR TABLES

*Note:* Because of the constant change in office furniture design, it is suggested that a survey of contemporary designs be made prior to using the above dimensions. This is particularly important, also, for professional groups requiring special equipment, e.g., medical examination tables and dentists' chairs. In addition, it is wise to anticipate the possible use of the so-called space systems furnishings, which integrate work and storage functions within adjustable, modular configurations.

From J. De Chiara and J. H. Callender, *Time-Saver Standards for Building Types,* McGraw-Hill Book Company, 1973.

## Conventional Office Layouts

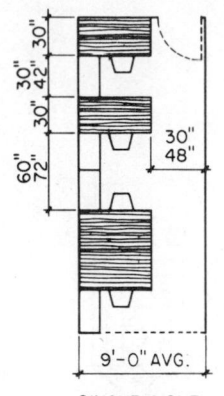

9'-0" AVG.

SINGLE AISLE

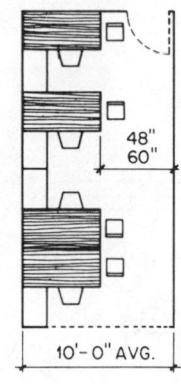

10'-0" AVG.

NOTES:

DIMENSIONS SHOWN ARE BASED ON 2'-6"x 5'-0" DESKS

FOR PLANNING PURPOSES SECRETARIAL AND CLERICAL AREAS REQUIRE 45 TO 60 SQ. FT. PER PERSON INCLUDING AISLES, ADD 10 TO 15 SQ. FT. FOR SIDE CHAIRS

"BACK TO BACK" AND "FACE TO FACE" PLACEMENT OF DESKS CAN SAVE SPACE BUT SHOULD BE AVOIDED IF POSSIBLE.

MULTIPLE-PERSON OFFICES ALL INFORMATION CONTAINED ON THIS PAGE CAN APPLY TO MULTIPLE PERSON OFFICES.

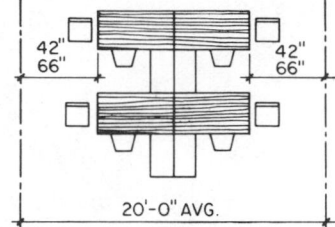

20'-0" AVG.

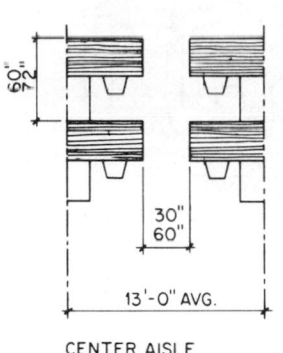

13'-0" AVG.

CENTER AISLE

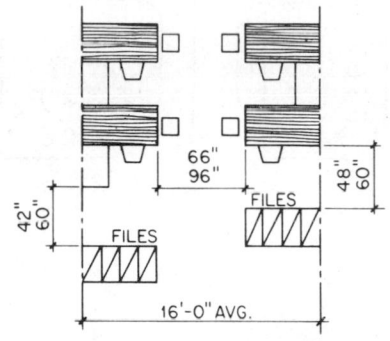

16'-0" AVG.

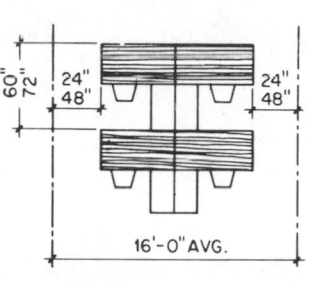

16'-0" AVG.

SIDE AISLE

## General Clerical Work Spaces

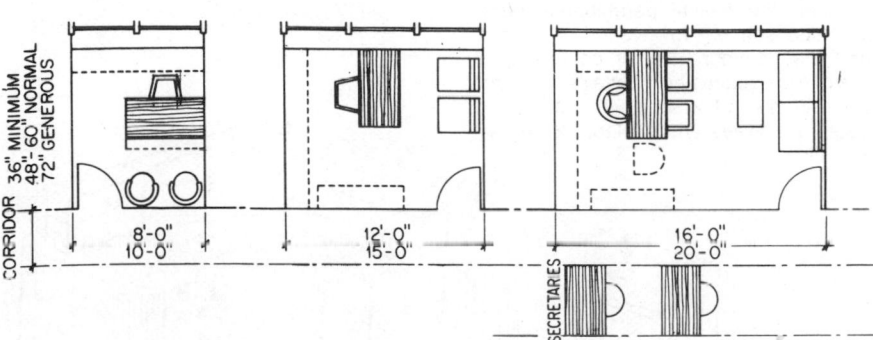

## Private Offices

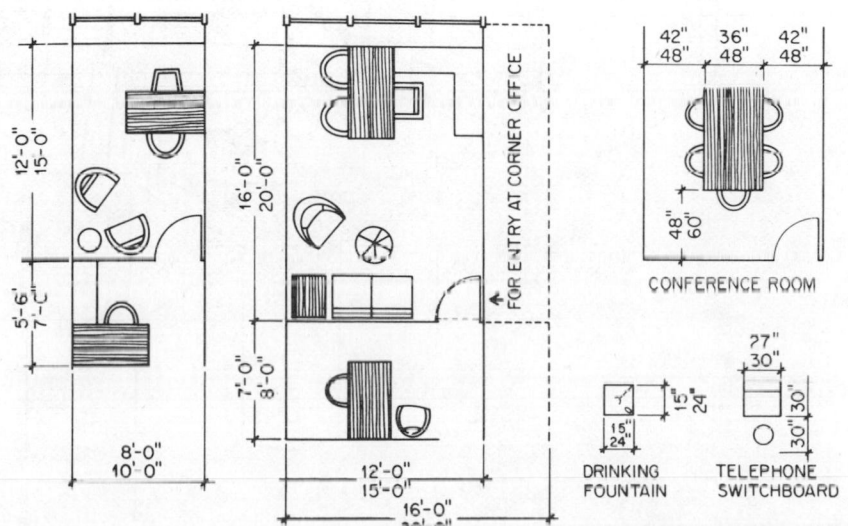

CONFERENCE ROOM

DRINKING FOUNTAIN

TELEPHONE SWITCHBOARD

From J. De Chiara and J. H. Callender, *Time-Saver Standards for Building Types*, McGraw-Hill Book Company, 1973.

**Arrangement Flexibility through Space Systems and Modular Furnishings[18]**

The accompanying sketch shows how one arrangement (A) is easily modified to another arrangement (B) in a matter of a few hours by using some of the modular furnishings available today.

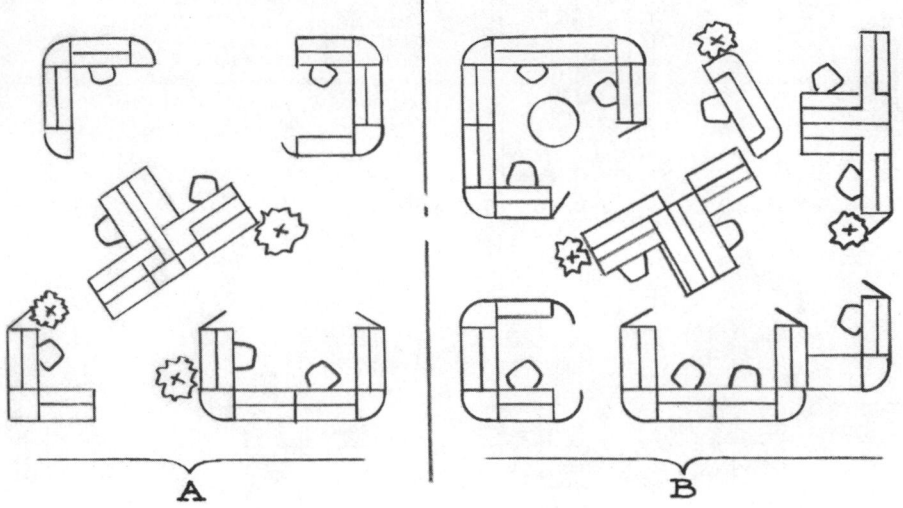

A                                    B

*Note:* Considerable interest has been shown recently in the "office landscape" approach. This generally provides a more open feeling than the typical permanently partitioned office space. However, it is recommended that some rectilinear patterning of clusters of work spaces be maintained in order that (1) visitors can find their way around and (2) orderly pathways are available for emergency exiting.

B

[18]*Action Office*, Herman Miller Inc., Zeeland, Mich.

SCHOOLS

## SCHOOLS

Philosophies concerning the creation of an optimum learning environment are constantly changing, and this has resulted in considerable experimentation in educational architecture. This section deals not with these philosophical aspects but rather with the fairly constant physical interface needs of students, faculty members, administrators, and others as they utilize facility subsystems in their everyday work, play, or recreation on the school site. Although many of the same types of building elements found in other systems are required on the school campus, the school presents numerous rather unique facility requirements because of both the nature of the educational processes and the variety of users involved. The following are among the general, yet unique, considerations to be addressed during the conceptual stage of planning.

### 1. Site Location
School sites should be selected so that they are as convenient to the potential student population as practical and yet isolated from such disruptive influences as heavy traffic, high-density commercial and industrial areas, and establishments such as liquor stores.

### 2. Size of Site
To avoid the necessity of splitting and separating facilities and thus causing special problems for both students and faculty, sufficient property should be acquired to provide for expected future expansion. Special consideration should be given to the parking requirements of secondary and college systems, since the foreseeable trend is for more and more students to have their own cars and for educational institutions to sponsor an increased number of public-oriented sporting events, thus adding to the parking needs of the campus.

### 3. Accessibility for the Handicapped
The requirement for equal educational opportunity for the handicapped has forced all school planners to give special attention to designing out architectural barriers. This makes it important to consider these problems before some aesthetic concept precludes an effective solution to the access problem of the handicapped student and other handicapped persons.

### 4. Security and Resistance of Facilities to Vandalism
Presently, and for the foreseeable future, schools seem to be prime targets for deviant behavior, not only by a few of the students within the particular institution, but also by outsiders bent on disruptive objectives. This has raised many questions regarding general architectural concepts that have fostered a more open and friendly campus atmosphere. Although the facilities obviously should not

give the appearance of a jail, they should minimize the opportunity for theft and destruction and should make it possible to provide proper on-campus disciplinary control.

## School Facilities Planning Model

Among the more important planning considerations are the following:

1. Students: optimum learning environment and safety
2. Faculty: optimum teaching environment and access to students
3. Administration: optimum work environment and communications
4. Maintenance: ease of maintenance and freedom from vandalism

A primary objective in planning school facilities, regardless of the level or type, should be to seek a reasonable balance in meeting the needs of each of the above user groups. If priority is given, it should be in the order presented; i.e., the needs of students should be considered first; those of the faculty, second; and so on.[19]

## Special Planning Emphasis by School Type

### 1. Elementary Schools

a. The physical compatibility of facilities for use by children and adults is an obvious requirement that architects try to meet. However, the unique furnishings requirements of children often make the tasks of the adult faculty extremely difficult. This is true not only of the classroom furnishings but also of stairs, bathroom facilities, and many other jointly used features. By the same token, children's tasks may be made difficult when administrative facilities are designed primarily for adult staff members.

b. Play facilities are a unique requirement of the lower grades, and special attention should be given not only to the various types of play activities and devices provided for the children but also to the problems of faculty monitoring and/or participation.

c. The convenience and safety of student arrival and departure are a unique consideration of the elementary school facility plan. Students may walk to and from school, or they may travel by bicycle, bus, or private automobile. This requires consideration of site layout in terms of public transportation, special school busing systems, and parking for vehicles and bicycles.

### 2. High Schools

a. The school library, study halls, assembly halls, etc., are an important feature of the high school facility system, requiring anticipation of the special space and convenience needs of students at this level.

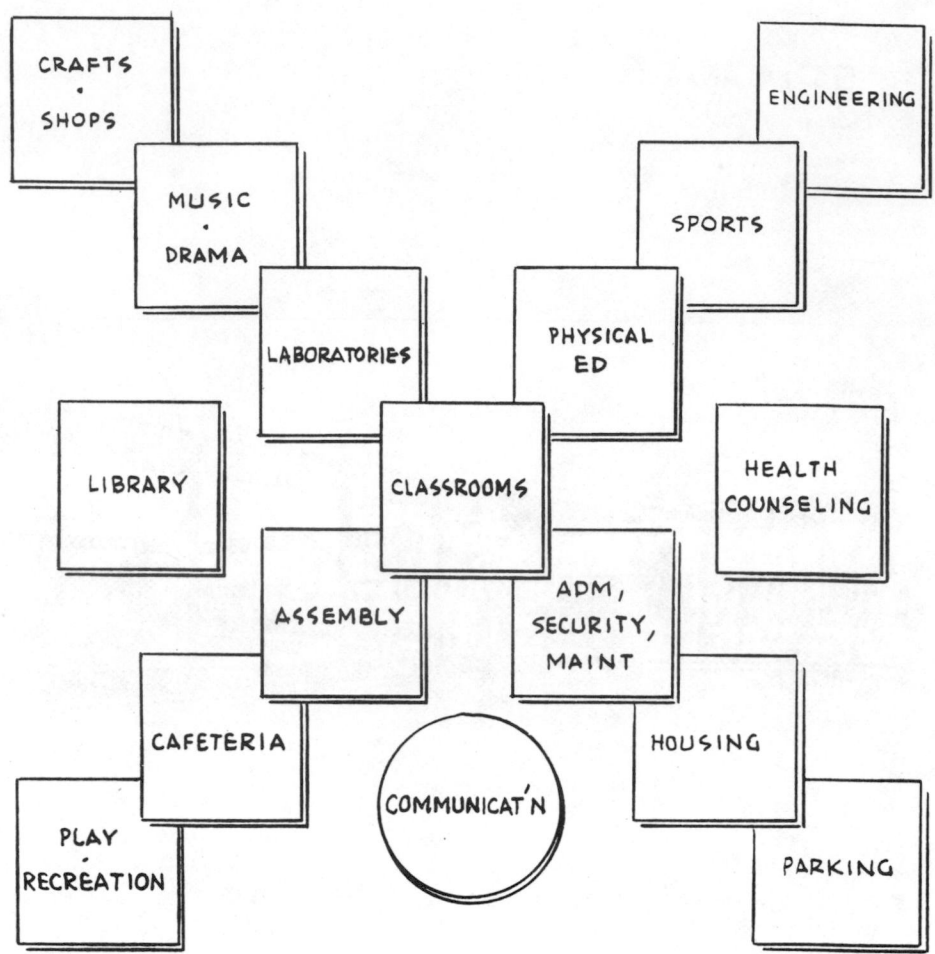

[19]Most architectural handbooks contain extensive sections dealing with each type of educational facility. In addition, the reader is encouraged to consult. G. Coates, *Alternative Learning Environments,* Dowden, Hutchinson & Ross, Inc., Stroudsburg, Pa., 1974; and A. P. Taylor and G. Vlastos, *School Zone: Learning Environments for Children,* Van Nostrand Reinhold Co., New York, 1975.

b. The gymnasium, swimming pool, track and field, tennis courts, etc., are an important part of high school planning. Sporting events require consideration of additional public parking and unique equipment storage.

c. Shops, laboratories, and craft facilities are typically introduced at the high school level, requiring consideration of the special facilities for storage, display, experimentation, etc., associated with these activities and also consideration of the unique safety requirements associated with use of toxic, combustible, and flammable substances.

d. Culturally oriented activities require consideration of special practice as well as storage and audience-participation facilities. Special note should be made of the potential theft and vandalism problems associated with musical instruments.

### 3. Colleges and Universities

a. On-campus housing is a unique consideration of college campus planning, particularly if the site is somewhat isolated and/or if administrative policies require first-year students to live on campus. The modern college campus often must accommodate both married and single students, and the majority of students will have private cars, which create obvious parking needs.

b. The field house, stadium, athletic fields, etc., are an important part of the modern college campus, from both the educational and the recreational points of view. In many cases, indoor as well as outdoor sporting events draw exceptionally large public attendance, requiring that facilities be oriented not only to student needs but also to public needs (including especially the consideration of parking and traffic accommodation and control).

c. Student unions are an accepted part of college facility requirements and should be considered during initial planning; they should not be thought of as something that can be added later, when the economic situation is more favorable. The student union has become the center not only of student relaxation but also of student self-government and sociopolitical activity.

d. Technically oriented facilities (computers, nuclear reactors, engineering test laboratories and field facilities, radio and TV stations, etc.) have become an important part of advanced educational philosophy. Because of the space and safety implications of these facilities, they must be addressed during the initial planning stage.

e. Public-oriented cultural and academic interface requires the careful consideration of both student and public use of such community-related facilities as theatres, auditoriums, and libraries.

f. Other special facilities should also be considered during the planning stage at the college level. Some of these might include (1) educational training schools that are open to parents who would like their children to participate in advanced or experimental educational programs and (2) special research institutes created to study specific scientific or engineering areas for government or industry. In such cases, the planning should consider the physical advantages and disadvantages of combined use of the university facility versus separate facilities, though they may be located on the same site.

## Precollege Facilities Differentiation

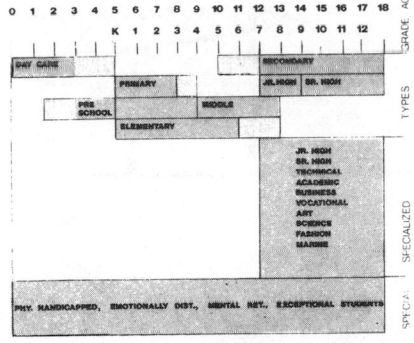

Because of the overlapping of names, grades, levels, and age groups, it is difficult to be specific about the architectural approach to planning a proper school facility. In addition, because of changing philosophies regarding educational methods and techniques, it is difficult to maintain a consistent concept for each type of facility. The accompanying chart, however, should help clarify some of the relationships between common age, grade, and facility types.

Regardless of the nomenclature applied to the facility and/or the educational process that is chosen for a particular program, the student, his or her teacher, and others associated with the administration and maintenance of the final facility must all interface effectively with its basic physical features and the thermal, acoustic, and atmospheric environment that is designed into it.

As can be imagined from the above chart, differences in size, mental and physical differences, and experiential differences are ex-

tremely great across the total age–facility-type span indicated. There are also considerable differences between the students (especially the younger ones) and the adults who teach in the facility, administer it, and maintain it. These differences are what make school architecture uniquely difficult.

## School-Site and Facilities Layout Considerations

### 1. Single, Large Structures versus Smaller, Independent Structures
Smaller, independent structures are more desirable because they make it possible to provide natural window light from opposite sides of a room. However, certain types of facilities benefit more from being relatively windowless and combined into a larger structure, e.g., a music building with practice rooms of various sizes or a gymnasium combining one large playing area for basketball, etc., with separate locker rooms, a swimming pool, and other physical education areas.

### 2. Indoor versus Outdoor Corridors
In milder climates, outdoor corridors are desirable because they give students an opportunity to get out of doors periodically as they go from one classroom to another. Outdoor corridors should have covers to protect students from rain.

### 3. Facilities Adjacency
In general, the following campus facilities should be located adjacent to each other:

a. Facilities for administrators and visitors
b. Facilities for faculty and students
c. Facilities for students and the primary assembly facility
d. The physical education facility and outdoor sports facilities
e. Classrooms and the bus loading area
f. The public parking area and the outdoor sports area and auditorium

### 4. Sound Isolation
Certain facilities generate loud sounds and thus should be spatially isolated from other facilities; e.g., music practice facilities and physical education facilities should be isolated from classrooms and libraries.

### 5. Centralization
Certain facilities should be centralized with respect to the overall campus layout in order to maximize the convenience of these facilities to all potential users:

a. The cafeteria

b. Rest rooms
c. The student assembly or auditorium

### 6. Parking
Single, mass parking areas are least desirable. Several smaller and strategically located parking areas not only provide shorter travel distances but also help segregate traffic categories, i.e., vehicles of students, faculty and administrators, visitors, and service personnel.

### 7. Terrain Factors
Although some aesthetic benefits are obtained by selecting a hilly site and positioning certain buildings so as to provide inspiring views, important practical benefits accrue from the more or less level site. Perhaps the most important benefit is to the handicapped, for whom vertical movement is considerably stressful and time-consuming. When the terrain is not reasonably flat, care should be taken to design and arrange buildings so the separate facilities to be used by given groups of users are on approximately the same level.

### 8. Maintenance
The layout and organization of buildings and other facilities on the campus are extremely important to those who must perform janitorial and landscape maintenance. In spite of the desirability of providing adequate open space between buildings and/or of placing buildings close together for economic reasons, either of these approaches can work to the disadvantage of maintainers. In the case of janitorial personnel, great travel distance increases the time required to perform duties; in the case of those who perform landscape maintenance, crowded conditions make it difficult to manipulate service vehicles among the buildings.

### 9. Visitor Orientation
An organized (more formal) rectilinear layout of campus facilities provides visitors a better chance to maintain their bearings and to follow directions for getting from one place to another. Curvilinear roads and pathways (although aesthetically pleasing and perhaps less boring) create confusion. Equally important is the planning of roadways on the large campus with a minimum of dead ends.

### 10. Security
Early consideration of security problems during campus layout planning can often eliminate later problems, some of which can be solved only by employing large security staffs and/or building distasteful fencing with locked gates. Neither of these solutions sets well with students or visitors.

## Structural Configuration Considerations

A wide variety of building configurations is, of course, acceptable for most school facilities. However, several important guidelines are noted within the accompanying sketches:

1. Cross ventilation: Although air conditioning may be planned, it is desirable to introduce a proper flow of natural air when temperatures warrant.
2. Even light distribution: The structural configuration should provide evenly distributed natural light wherever possible, even though a satisfactory artificial illumination system is being proposed. However, special attention should be given to control of glare.

Separated classrooms provide natural light and ventilation from two sides. The outdoor corridor allows students to enjoy a moment of fresh air as they go from one classroom to another.

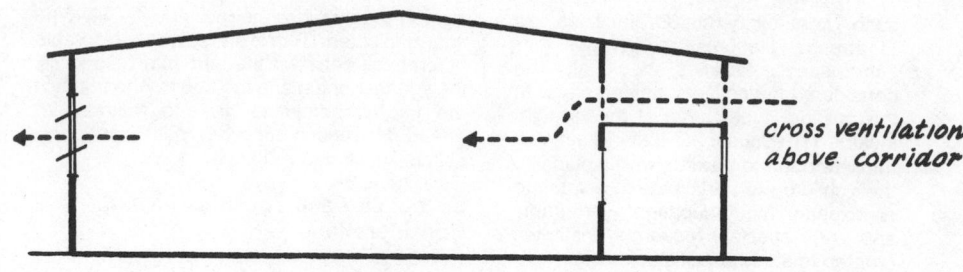

cross ventilation above corridor

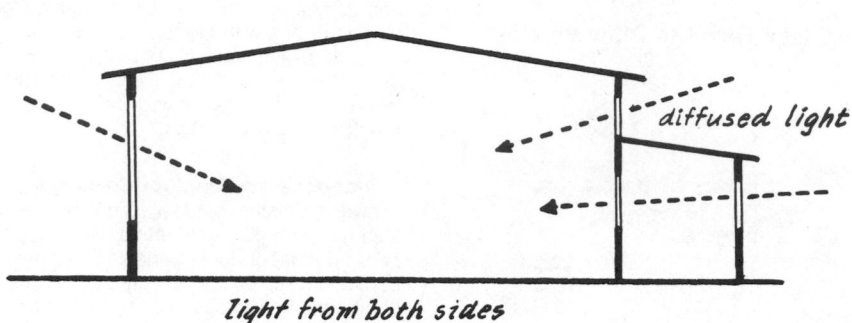

diffused light

light from both sides

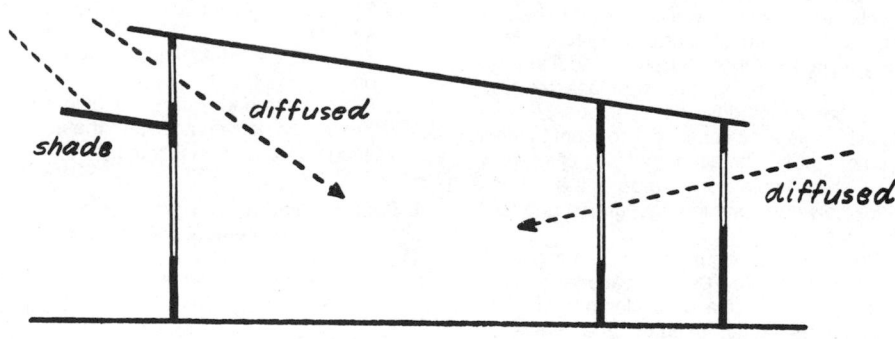

shade

diffused

diffused

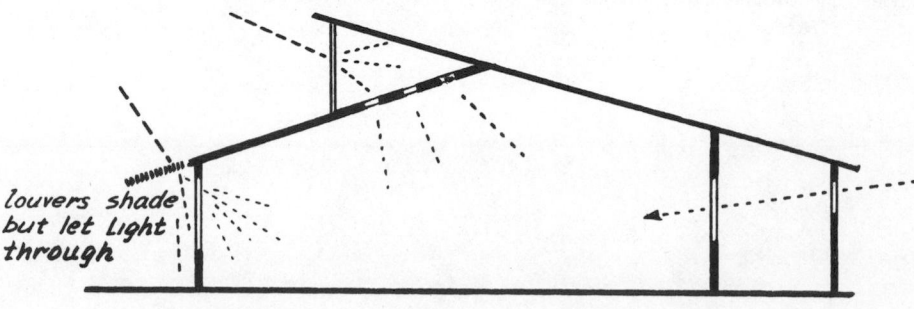

louvers shade but let light through

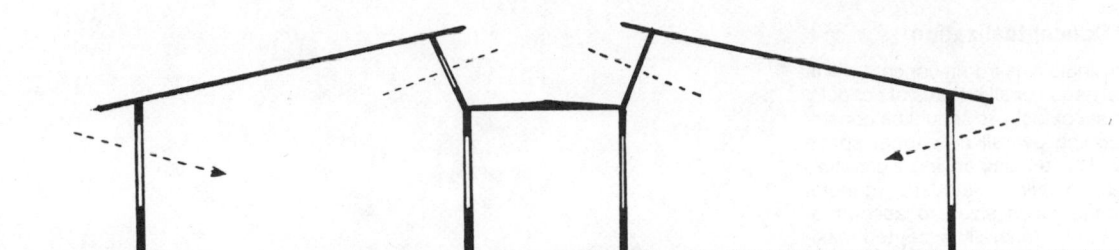

*central corridor-all ventilation and
illumination on both sides*

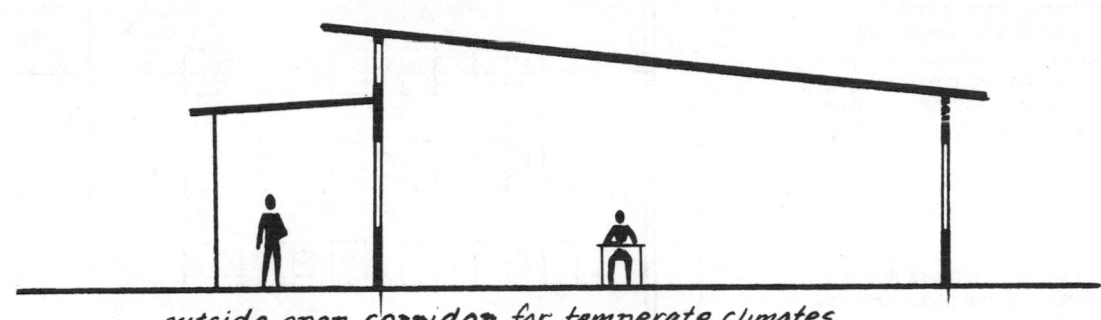

*outside open corridor for temperate climates*

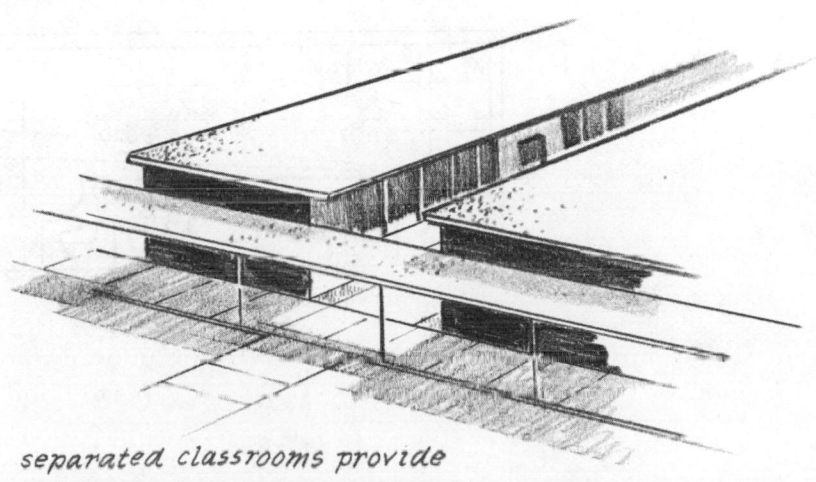

*separated classrooms provide
natural light and ventilation
from two sides - outdoor
corridor provides fresh air
between classes*

## Classroom Conceptualization

Obviously, no single classroom concept will fit all educational needs or all classes of schools. However, certain objectives should be considered in developing overall classroom space requirements. The accompanying illustration represents a so-called standard, general classroom configuration sized to accommodate 30 students (a generally accepted maximum for most classroom teaching). Note the following with regard to this example:

1. Horizontal and vertical sight lines are important for both the students and the teacher.
2. Natural light enters from the students' left side.
3. Entry is at the forward part of the classroom, not because this is best from the standpoint of class disturbance, but because it is best from the standpoint of observation by the teacher and/or by a monitor from outside the classroom.
4. Blackout drapes at the windows are provided for use during film presentations.
5. The offset position of the teacher's desk provides all students a clear view of the blackboard.

The dimensions shown are presented only as a minimum guideline. Other considerations will suggest the need to add space for a cloakroom, storage for materials and visual aids equipment, or special furnishings.[20]

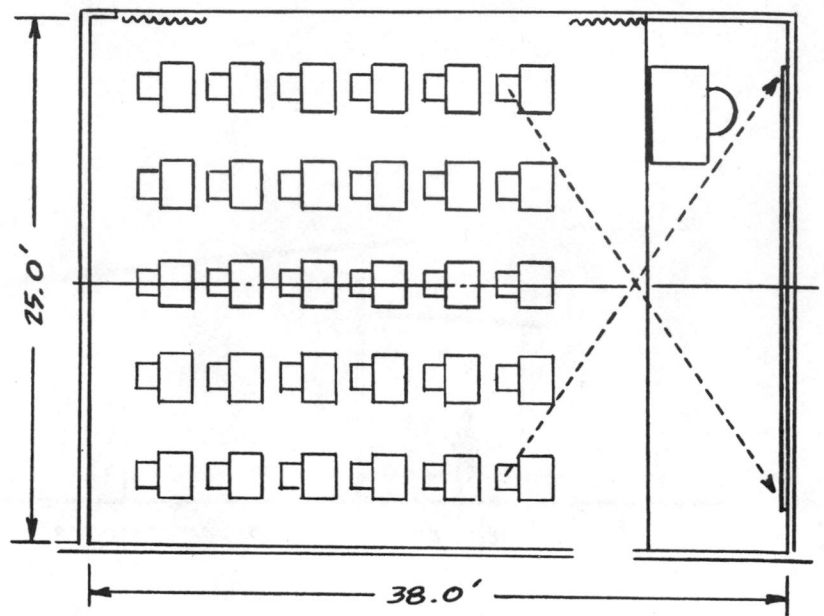

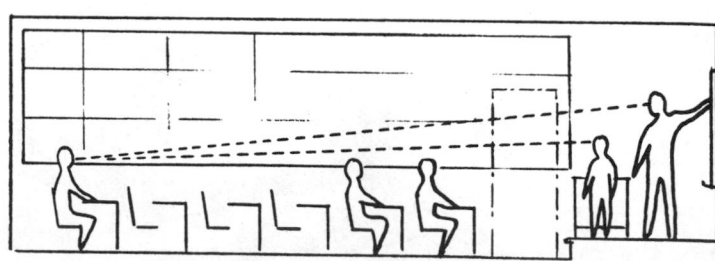

[20]For additional guidelines for a variety of classroom types and activities, the reader is referred to J. De Chiara and J. H. Callender, *Time-Saver Standards for Building Types*, McGraw-Hill Book Company, New York, 1973. Some of these are shown on the following pages.

**Elementary and Secondary School
Classroom Examples**

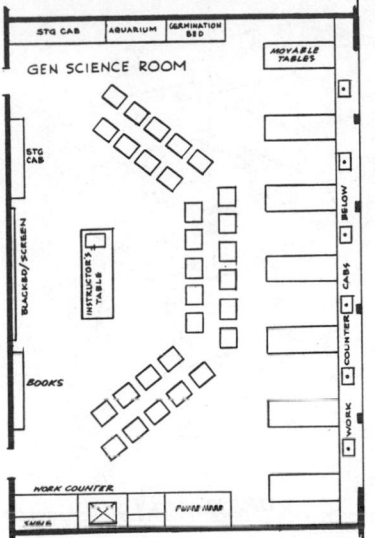

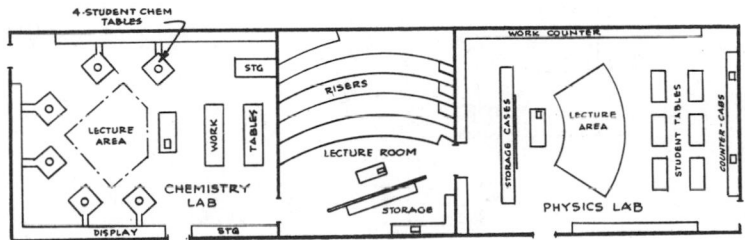

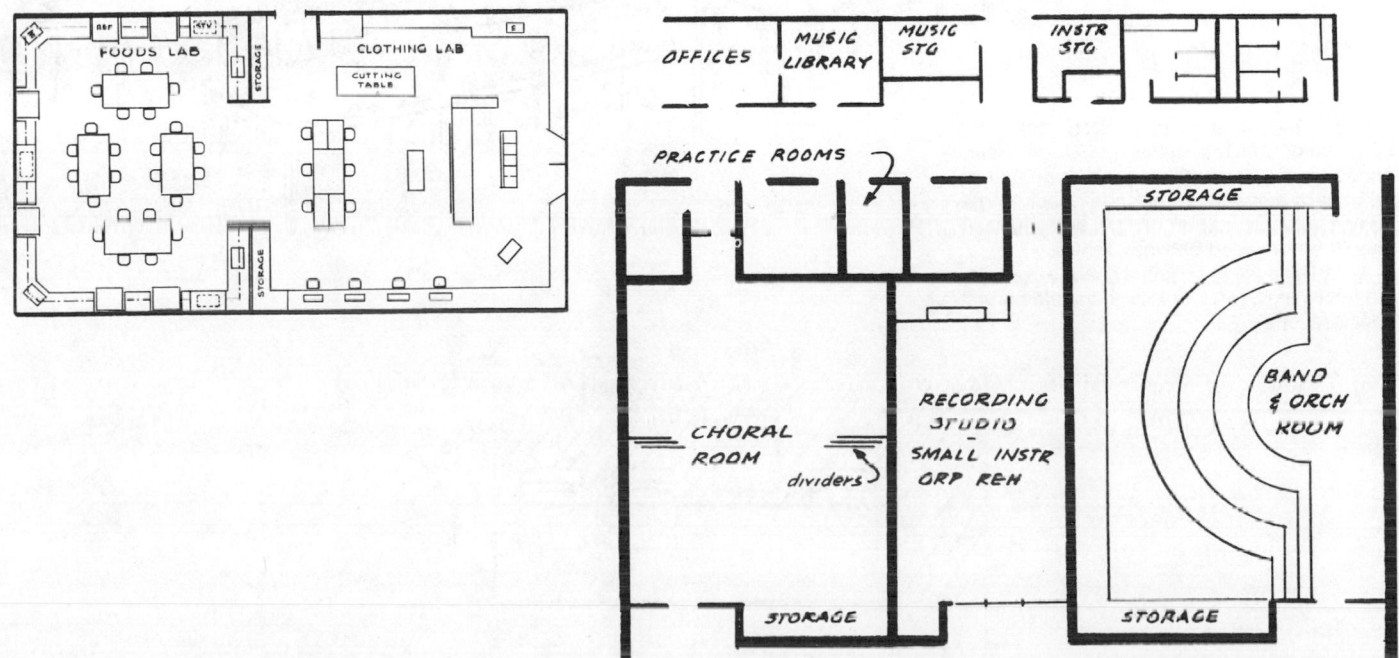

## Typical Classroom Furnishings and Space Guidelines

In planning the size and shape of a classroom, consider the clearances required around furnishings so that students can move their chairs to get out from behind desks, tables, or counters and so that students and the teacher can pass among desks or other furnishings. The critical clearance dimensions relate to the perimeter of furnishings; i.e., avoid the arbitrary centerline-to-centerline spacing standards that are typically suggested.

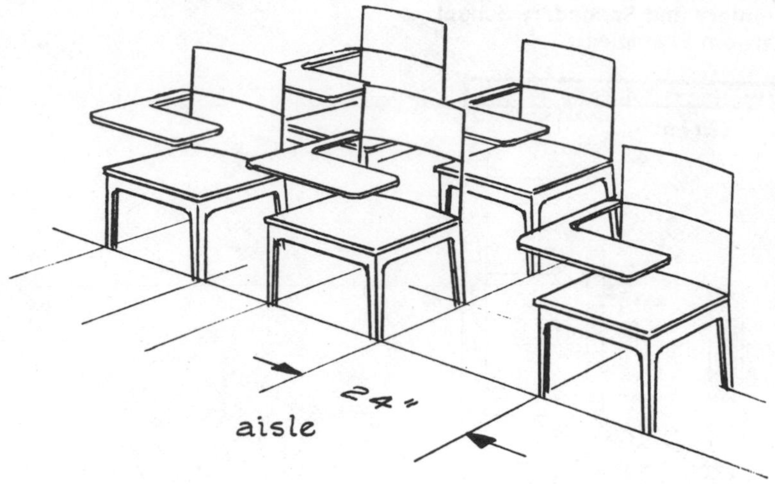

24"
aisle

(minimums)

The space required to pass behind the desk and chair and the space required to sit down at the desk and get up from it determine the floor area for each desk and, eventually, the floor area of the classroom.

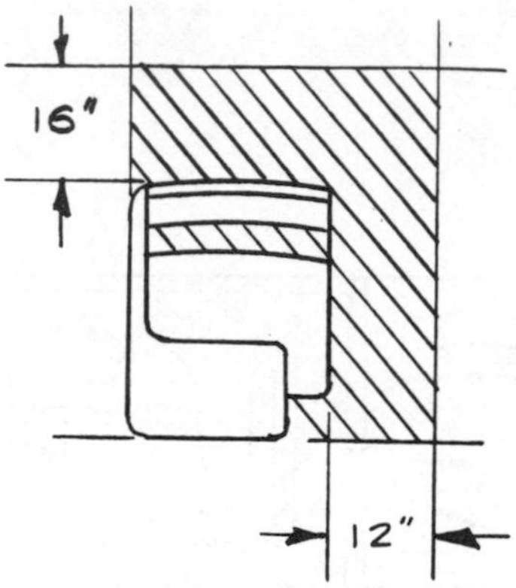

16"

12"

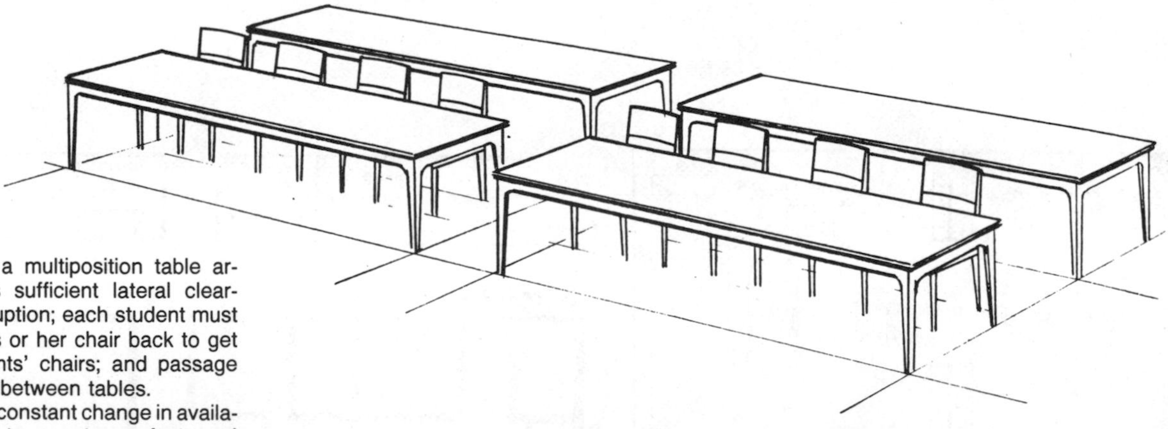

Each student at a multiposition table arrangement requires sufficient lateral clearance to avoid interruption; each student must be able to move his or her chair back to get behind other students' chairs; and passage should be provided between tables.

*Note:* Because of constant change in available furnishings, refer to recent manufacturers' specifications.

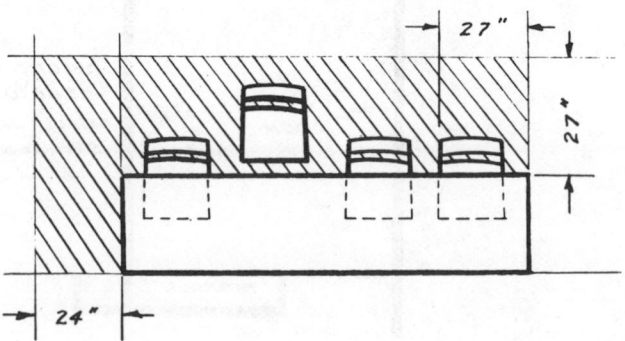

27"

27"

24"

## Furnishings Guidelines for Elementary and Secondary Schools

### Working Heights in Inches for Elementary and Secondary School Children

| Item | Elementary | | | | | | | | | Junior high | | | Senior high | | |
| | Kindergarten | | | Grades 1–3 | | | Grades 4–6 | | | Grades 7–9 | | | Grades 10–12 | | |
| | Min. | Opti-mum | Max. | Min. | Opti-mum | Max. | Min. | Opti-mum | Max. | Min. | Opti-mum | Max. | Min. | Opti-mum | Max. |
|---|---|---|---|---|---|---|---|---|---|---|---|---|---|---|---|
| Cabinet, display (top) | | 54 | | | 56 | | | 66 | | | 74 | | | 77 | |
| Cabinet, display (bottom) | | 26 | | | 29 | | | 34 | | | 38 | | | 39 | |
| Cabinet, pupil use (top) | | | 50 | | | 56 | | | 65 | | | 74 | | | 79 |
| Chairs and bench | 10 | 11 | 11 | 10 | 12 | 13 | 12 | 14 | 16 | 13 | 15 | 17 | 14 | 16 | 18 |
| Chalkboard (top) | 68 | 70 | 73 | 72 | 73 | 74 | 76 | 77 | 78 | 79 | 80 | 82 | 80 | 82 | 84 |
| Chalkboard (bottom and chalkrail) | 20 | 22 | 25 | 24 | 25 | 26 | 28 | 29 | 30 | 31 | 32 | 34 | 32 | 34 | 36 |
| Counter, cafeteria | 21 | 27 | 32 | 25 | 31 | 34 | 29 | 36 | 39 | 32 | 40 | 45 | 33 | 42 | 48 |
| Counter, classroom work (standing) | 20 | 24 | 26 | 24 | 26 | 29 | 28 | 30 | 34 | 31 | 34 | 38 | 32 | 36 | 39 |
| Counter, general office | 20 | 27 | 32 | 24 | 31 | 34 | 28 | 36 | 39 | 31 | 40 | 45 | 32 | 42 | 49 |
| Desk and table, classroom | 17 | 18 | 19 | 18 | 20 | 22 | 21 | 23 | 25 | 23 | 26 | 28 | 24 | 27 | 29 |
| Desk, typing | | | | | | | | | | | 26 | | | 26 | |
| Door knob | 19 | 27 | 32 | 24 | 31 | 35 | 28 | 36 | 40 | 30 | 40 | 46 | 31 | 42 | 49 |
| Drinking fountain | 20 | 24 | 27 | 24 | 27 | 29 | 28 | 32 | 34 | 32 | 36 | 40 | 32 | 40 | 44 |
| Fire extinguisher (tank)* | | | | | | | | | | | | | | | |
| Hook, coat | 32 | 36 | 48 | 38 | 41 | 51 | 47 | 48 | 58 | 53 | 54 | 64 | 54 | 55 | 68 |
| Lavatory and sink | 20 | 23 | 25 | 24 | 26 | 27 | 28 | 29 | 31 | 32 | 33 | 35 | 32 | 35 | 38 |
| Light switch | 27 | 27 | 46 | 31 | 35 | 49 | 36 | 40 | 56 | 40 | 46 | 64 | 42 | 50 | 68 |
| Mirror, lower edge | | | 35 | | | 38 | | | 43 | | | 48 | | | 52 |
| Mirror, upper edge | 46 | | | 56 | | | 65 | | | 71 | | | 71 | | |
| Panic bar | 21 | 27 | 32 | 25 | 31 | 34 | 29 | 36 | 39 | 32 | 40 | 45 | 33 | 42 | 48 |
| Pencil sharpener | 20 | 27 | 33 | 25 | 31 | 35 | 28 | 36 | 40 | 32 | 40 | 46 | 32 | 42 | 49 |
| Rail, hand and directional | 20 | 21 | 32 | 24 | 24 | 34 | 28 | 29 | 39 | 31 | 32 | 45 | 32 | 33 | 48 |
| Shelf, hat and books | | 41 | 48 | | 46 | 51 | | 54 | 58 | | 60 | 64 | | 62 | 68 |
| Soap dispenser | 20 | 27 | 33 | 25 | 31 | 35 | 28 | 36 | 40 | 32 | 40 | 46 | 32 | 42 | 49 |
| Stool, drawing | | 19 | | | 21 | | | 26 | | | 28 | | | 29 | |
| Table, drawing | | 26 | | | 29 | | | 34 | | | 38 | | | 39 | |
| Table and bench, work (standing) | 25 | 26 | 28 | 26 | 29 | 32 | 30 | 34 | 38 | 36 | 38 | 41 | 37 | 39 | 42 |
| Tackboard (top) | 72 | 84 | | 72 | 84 | | 72 | 84 | | 72 | 84 | | 72 | 84 | |
| Tackboard (bottom) | 20 | 22 | 25 | 24 | 25 | 26 | 28 | 29 | 30 | 31 | 32 | 34 | 32 | 34 | 36 |
| Telephone, wall mounted | | | 35 | | | 37 | | | 43 | | | 48 | | | 52 |
| Toilet stall, top of partition | 44 | 44 | | 52 | 52 | | 61 | 61 | | 67 | 67 | | 69 | 69 | |
| Towel dispenser | 23 | 27 | 46 | 28 | 31 | 49 | 33 | 36 | 56 | 37 | 40 | 64 | 37 | 42 | 68 |
| Urinal (bottom) | | | | 3 | 3–15 | 17 | 3 | 3–17 | 20 | 4 | 4–18 | 22 | 4 | 4–19 | 24 |
| Wainscotting | 54 | 54 | 54 | 54 | 54 | 54 | 54 | 54 | 54 | 60 | 60 | 60 | 60 | 60 | 60 |
| Water closet (seat) | 10 | 10½ | 12 | 11 | 11½ | 12 | 13 | 13½ | 14 | 14 | 14½ | 15 | 14½ | 15 | 15 |
| Window ledge | | | 29 | | | 30 | | | 34 | | | 38 | | | 41 |

*Recessed at baseboard height.*

Note: Since there is considerable size variation among children in the first three grades (within a single class), it is recommended that adjustable-height furniture be considered.

From J. De Chiara and J. H. Callender, *Time-Saver Standards for Building Types*, McGraw-Hill Book Company, 1973.

## Special School Facilities

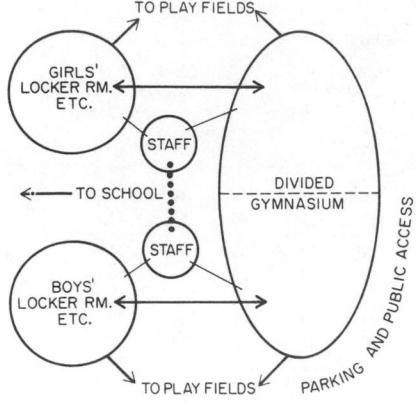

**Recommended Dimensions in feet for Gymnasiums**

| School | W | L | W₁* | L₁* | Seats |
|---|---|---|---|---|---|
| Small elementary | 36 | 52 | | | |
| Large elementary | 52 | 72 | | | |
| Junior high school* | 65 | 86 | 42 | 74 | 400 |
| Small senior high school† | 79 | 96 | 50 | 84 | 700 |
| Large senior high school† | 100 | 104 | 50 | 84 | 1,500 |

\* $W_1$ and $L_1$ are dimensions of basketball court.
† Use folding partition.

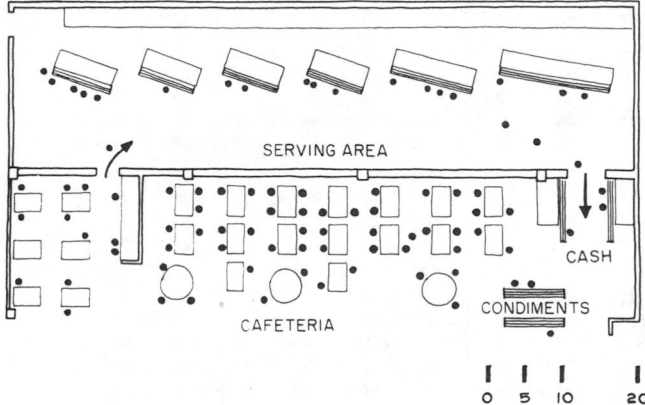

**DINING FACILITIES** A "scramble system" is often best for school cafeterias because it provides fast service on a daily basis. Duplication of service modules minimizes cross-traffic interference. (See also the section on kitchen and restaurant subsystem design in Chapter 3.)

**Food Service Space Requirement Chart for School, College, and Commercial Lunch Programs**

| Planned enrollment or patrons | Seats required | Area designation, sq ft | | Number of counters† |
|---|---|---|---|---|
| | | Kitchen* | Serving | |
| 400 | 170 | 1,500 | 700 | 1 |
| 500 | 210 | 1,650 | 800 | |
| 600 | 250 | 1,800 | 1,540 | 2 |
| 700 | 290 | 1,950 | 1,540 | |
| 800 | 335 | 2,100 | 1,920 | |
| 900 | 375 | 2,250 | 1,920 | |
| 1,000 | 420 | 2,400 | 2,310 | 3 |
| 1,100 | 460 | 2,550 | 2,310 | |
| 1,200 | 500 | 2,700 | 2,690 | |
| 1,300 | 540 | 2,850 | 2,690 | |
| 1,400 | 585 | 3,000 | 2,690 | |
| 1,500 | 625 | 3,150 | 3,080 | 4 |
| 1,600 | 670 | 3,300 | 3,080 | |
| 1,700 | 710 | 3,450 | 3,460 | |
| 1,800 | 750 | 3,600 | 3,460 | |
| 1,900 | 790 | 3,750 | 3,460 | |
| 2,000 | 835 | 3,900 | 3,850 | 5 |
| 2,100 | 875 | 4,050 | 3,850 | |
| 2,200 | 920 | 4,200 | 4,230 | |
| 2,300 | 960 | 4,350 | 4,230 | |
| 2,400 | 1,000 | 4,500 | 4,620 | 6 |
| 2,500 | 1,040 | 4,650 | 4,620 | |
| 2,600 | 1,085 | 4,800 | 5,000 | |
| 2,700 | 1,125 | 4,950 | 5,000 | |
| 2,800 | 1,170 | 5,100 | 5,000 | |
| 2,900 | 1,210 | 5,250 | 5,390 | 7 |
| 3,000 | 1,250 | 5,400 | 5,390 | |
| 3,100 | 1,290 | 5,550 | 5,770 | |
| 3,200 | 1,335 | 5,700 | 5,770 | |
| 3,300 | 1,375 | 5,850 | 5,770 | |
| 3,400 | 1,420 | 6,000 | 6,160 | 8 |
| 3,500 | 1,460 | 6,150 | 6,160 | |
| 3,600 | 1,500 | 6,300 | 6,540 | |

\* Kitchen space:
    150–650 students = 3 and 4 sq ft per student
    650–2,000 students = 2 and 2¼ sq ft per student
    2,000–6,000 students = 1½ and 1¾ sq ft per student
† Counter = 35 to 40 lineal feet of serving equipment.

From J. De Chiara and J. H. Callender, *Time-Saver Standards for Building Types*, McGraw-Hill Book Company, 1973.

**Physical Education Facilities**

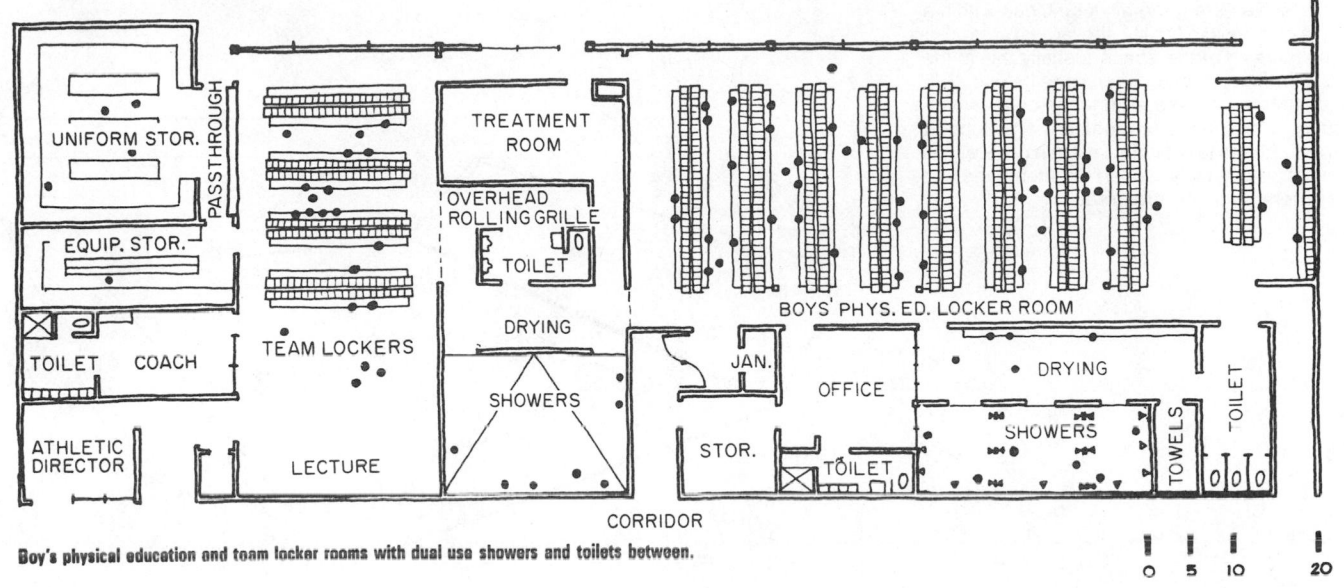

Boy's physical education and team locker rooms with dual use showers and toilets between.

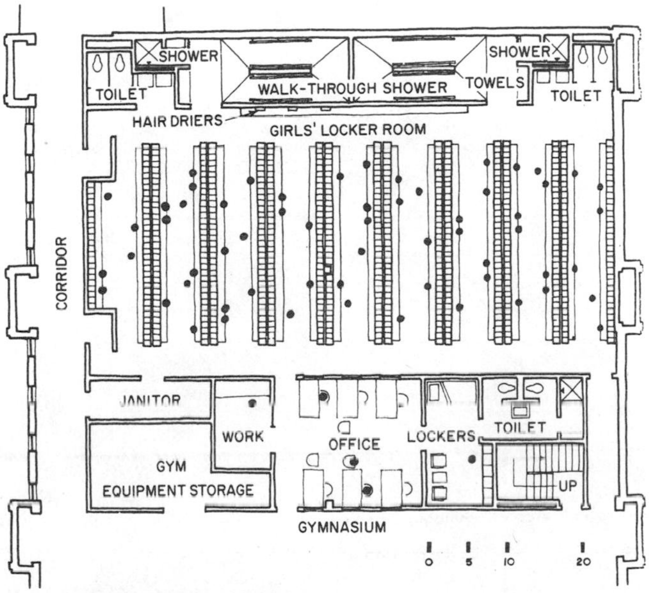

High school girl's locker room (1500).

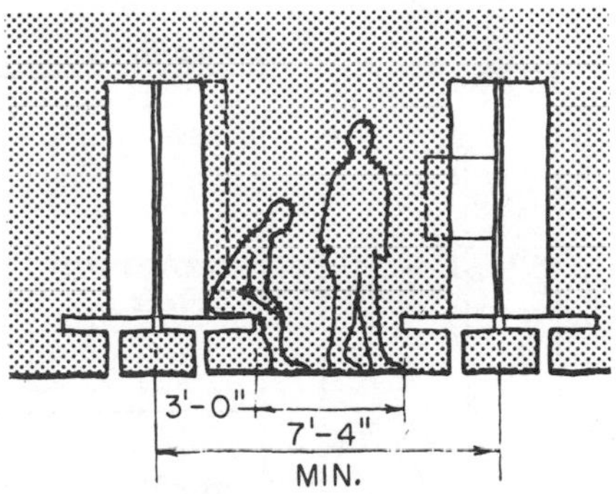

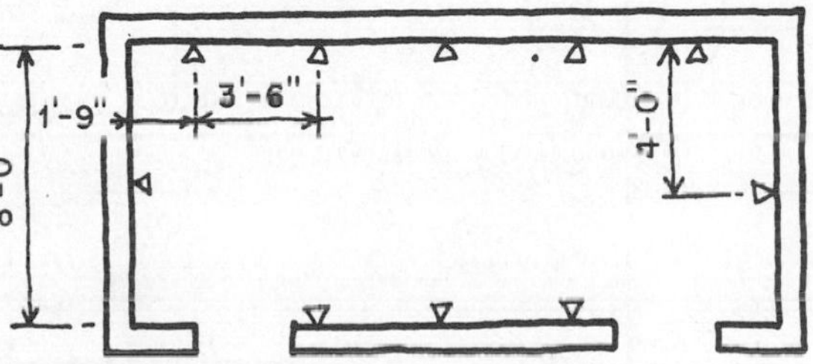

Conventional shower (in-wall piping); 14 sq ft per head.

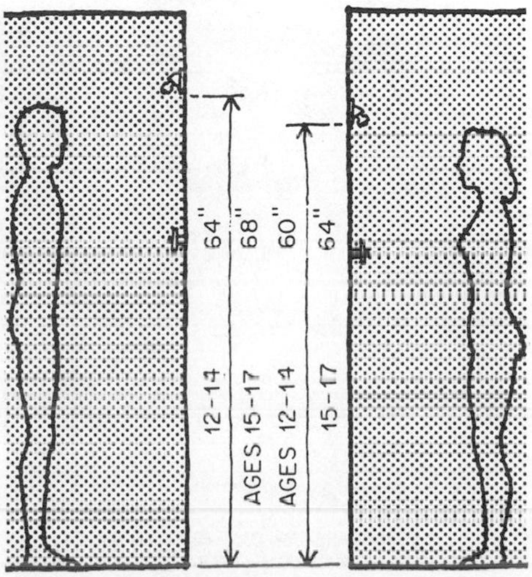

From J. De Chiara and J. H. Callender, *Time-Saver Standards for Building Types*, McGraw-Hill Book Company, 1973.

3'-0" MAX.
2'-6" MIN.

12" X 12"
X60" . . . . 12" X 12" X 30"
9" X 12" X 30"
X72" . . . . 12" X 12" X 36"
9" X 12" X 36"
• RATIO: 6 TO 1

• GOOD VENTILATION—
SINCE HANGING OF
GYM SUITS POSSIBLE

3'-0"
7'-4"
MIN.

• COMBINED WIDER AISLE AND
DRESSING AREA
• SEAT FOR EACH LOCKER BANK—
LESS CONGESTION
• SEAT PROVIDES AISLE PROTECTION
FROM DOOR SWING
• POSITIVE THROUGH—LOCKER
VENTILATION (SEE DETAIL SKETCH)
• ECONOMICAL USE OF FLOOR AREA

**Gymnasium Facilities**

Although various athletic and physical education activities are generally combined within a single gymnasium facility, the largest court dictates the size of the space (usually this is the basketball court; see dimensions below). Sufficient perimeter area around the court is necessary to accommodate officials and support personnel and to provide a margin of safety as players go beyond the court perimeter and into bleachers or other structures.

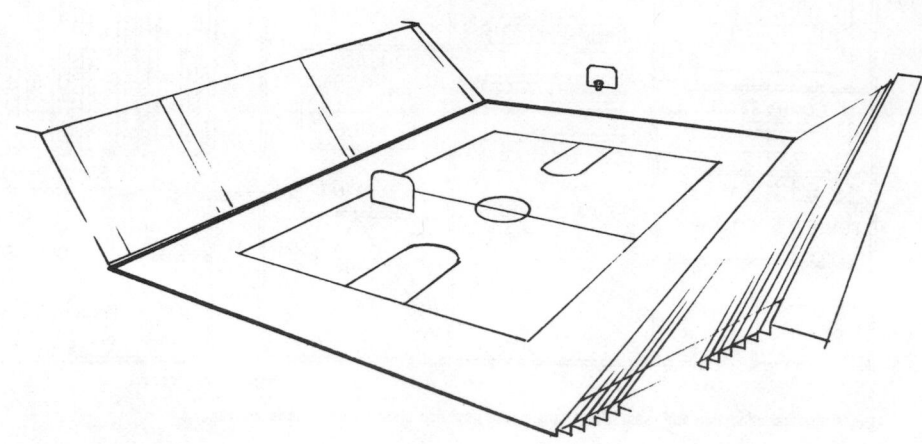

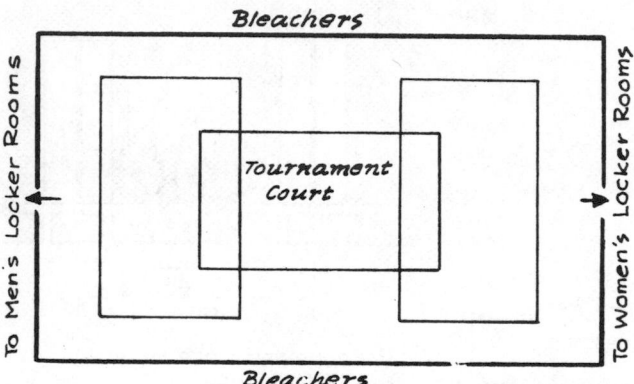

Combination practice and tournament court.

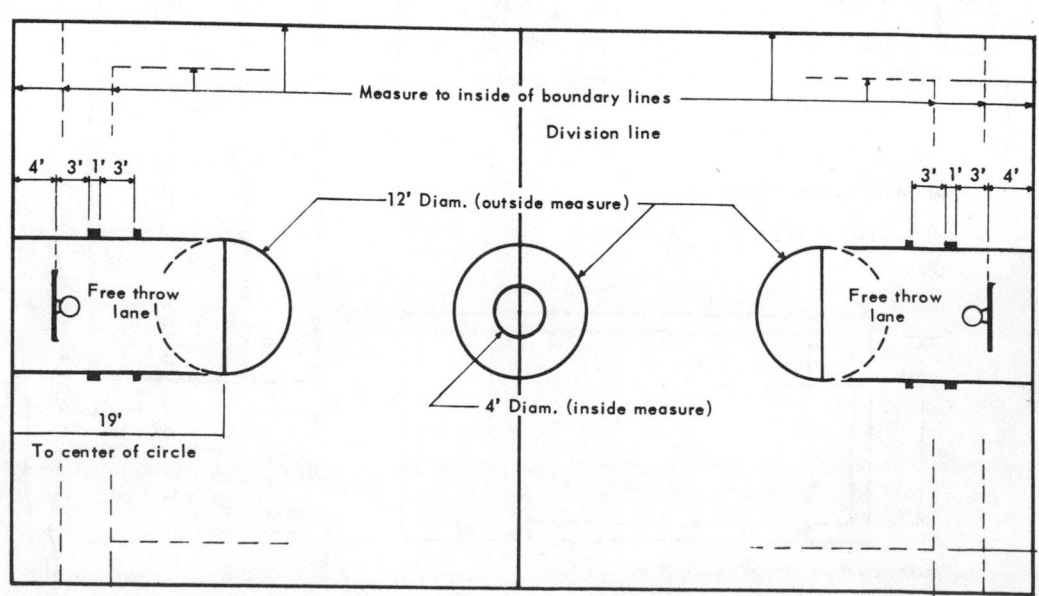

From J. De Chiara and J. H. Callender, *Time-Saver Standards for Building Types,* McGraw-Hill Book Company, 1973.

BASKETBALL COURT

**Swimming Pool Facilities**

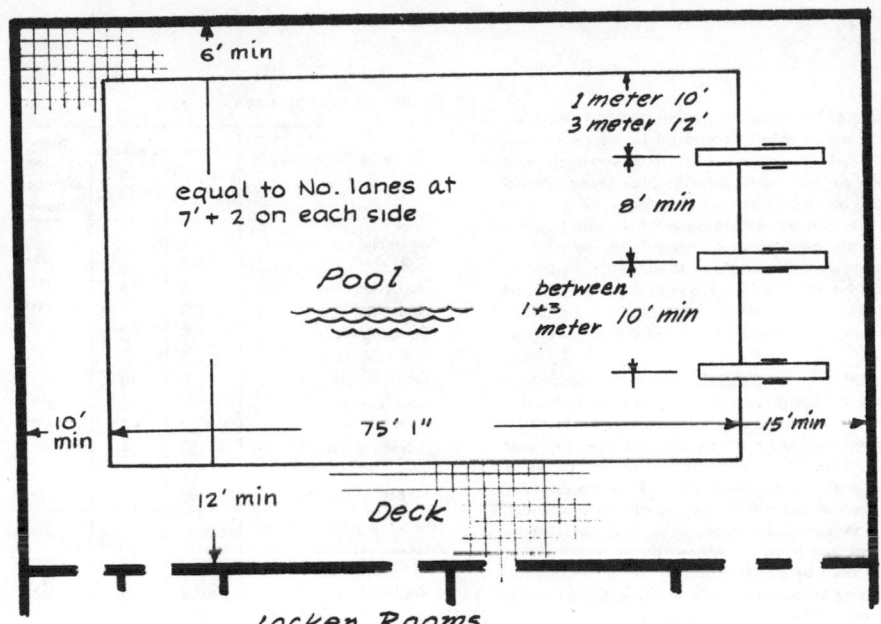

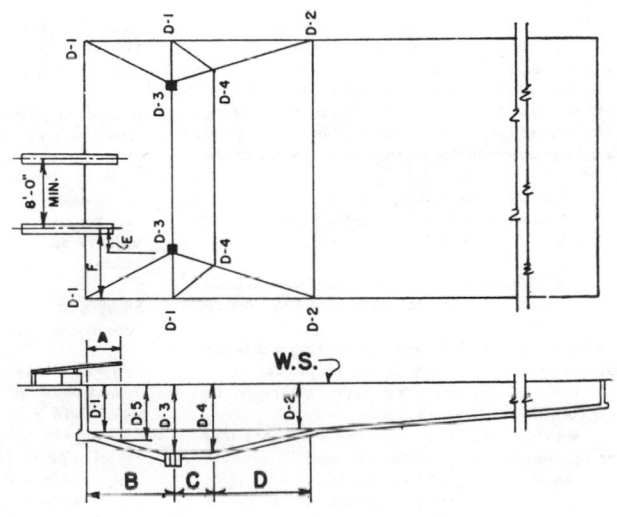

| STANDS & BOARDS | | Depth—Feet & Inches | | | | | Length of Section—Feet & Inches | | | | | |
|---|---|---|---|---|---|---|---|---|---|---|---|---|
| | | D-1 | D-2 | D-3 | D-4 | D-5 | A | B | C | D | E | F |
| 3-Meter Board | Min. | 5-0 | 4-6 | 10-0 | 9-9 | 8-6 | 5-0 | *6-0 | *9-0 | 20-0 | 1-0 | 8-0 |
| | Max. | | 5-6 | | | | 6-0 | 10-0 | | | | |
| 1-Meter Board | Min. | 5-0 | 4-6 | 8-6 | 8-3 | 7-6 | 5-0 | *6-0 | *9-0 | 15-0 | 1-0 | 8-0 |
| | Max. | | 5-6 | | | | 6-0 | 10-0 | | | | |
| Deck Level Board | Min. | 5-0 | 4-6 | 8-0 | 7-6 | | 2-6 | †6-0 | †6-0 | 12-0 | 1-0 | 8-0 |
| | Max. | | 5-6 | | | | 4-0 | 10-0 | | | | |

*As D-2 varies between min. and max., D may vary, but slope of D may not exceed 1 ft vert. to 4 ft horiz. D-1 shall be at end wall of diving area, or not more than 12 in. from it*

\* B & C May vary to attain 15'—0" Min.
† B & C May vary to attain total 12'—0" Min.

From J. De Chiara and J. H. Callender, *Time-Saver Standards for Building Types*, McGraw-Hill Book Company, 1973.

## Playground Planning Considerations

### General Equipment Selection Factors

The following general factors should be considered in selecting equipment for playlots and playgrounds.

**Developmental and Recreational Values** All equipment should contribute to the healthy growth and recreational enjoyment of the child, so that he learns to coordinate, cooperate, compete, create, enjoy, and acquire confidence. Play equipment should:

1. Develop strength, agility, coordination, balance, and courage.

2. Stimulate the child to learn social skills of sharing and playing with others, and to compete in a spirit of fair play.

3. Encourage each child to be creative and have play experiences which are meaningful to him.

4. Permit the child to have fun and a sense of complete enjoyment.

5. Assist the child in making the transition from playlot to playground.

**Child Preference and Capacity** Play equipment, to be selected with due regard to the child's changing preference, maturity, and capacity, should:

1. Be scaled and proportioned to meet the child's physical and emotional capacities at different age levele.

2. Permit the child to do some things alone without direct adult supervision or assistance.

3. Provide a wide variety of play opportunities to accommodate changing interests of the child.

4. Free the child's imagination.

5. Meet a variety of interests, abilities, and aptitudes.

**Safety of Participants** All play equipment should be designed and built for safety of the participants, and:

1. Be free of all sharp protruding surfaces caused by welds, rivets, bolts, or joints.

2. Have sufficient structural strength to withstand the expected loads.

3. Be designed to discourage incorrect use and to minimize accidents; examples are seats that discourage children from standing in swings, slides that require children to sit down before sliding, and steps or ladders that discourage more than one participant at a time.

4. Have hand or safety rails on all steps and ladders, and nonskid treads on all steps.

5. Be installed in accordance with the specific directions of the manufacturer.

6. Be placed over suitable surfaces that will reduce the danger of injury or abrasions in the event a child falls from the climbing, moving or sliding equipment. (A safe landing surface should be provided at the end of a slide chute.)

**Durability of Equipment** Equipment that is durable should be selected. It should be made of materials which are of sufficient strength and quality to withstand normal play wear. Wood should be used only where metal or plastics have serious disadvantages. All metal parts should be galvanized or manufactured of corrosion-resistant metals. All movable bearings should be of an oilless type. Equipment should be designed as vandal-resistant as possible (for example, wire-reinforced seats for swings).

**Equipment with Eye Appeal** All play equipment should be designed and selected for function, for visual appeal to stimulate the child's imagination, with pleasing proportions and with colors in harmonious contrast to each other and the surroundings. Play equipment may have a central theme, to reflect historical significance of the area, a storybook land, a nautical motif or a space flight motif. The theme may be carried out by constructing retaining or separation walls to resemble a corral, ship, or airplane, and by appropriate design of such elements as paving, benches, and trash cans.

**Ease of Maintenance** Equipment should be selected which requires a minimum of maintenance. Purchased equipment should be products of established manufacturers who can provide a standard parts list. Equipment parts which are subject to wear should be replaceable. Color should be impregnated into the material, if feasible, to avoid repainting. Sand areas should be surrounded by a retaining wall and be maintained regularly to remove foreign objects and loosen the sand as a suitable play medium.

**Supervision** Equipment should be selected that requires a minimum of direct supervision.

**Playground Equipment for Elementary School Children** The following table indicates types, quantities, and minimum play space requirements totaling about 6,600 sq ft; this area, plus additional space for circulation, miscellaneous elements, and buffer zones, will accommodate a full range of playground equipment serving approximately 50 children at one time.

| Equipment | Number of pieces | Play space requirements, ft |
|---|---|---|
| Balance beam............. | 1 | 15 x 30 |
| Climbers................. | 3 | 21 x 50 |
| Climbing poles .......... | 3 | 10 x 20 |
| Horizontal bars .......... | 3 | 15 x 30 |
| Horizontal ladder.......... | 1 | 15 x 30 |
| Merry-go-round........... | 1 | 40 x 40 |
| Parallel bars.............. | 1 | 15 x 30 |
| Senior swing set (6 swings) .... | 1 | 30 x 45 |
| Slide ................. | 1 | 12 x 35 |

**Children's playgrounds**

| Type of Equipment or Area | Area per Unit (Sq. Ft.) | Capacity in Children | Suggested Number Included |
|---|---|---|---|
| **Apparatus** | | | |
| Slide | 450 | 6 | 1[b] |
| Horizontal Bars | 180 | 4 | 3[b] |
| Horizontal Ladders | 375 | 8 | 2[b] |
| Traveling Rings | 625 | 6 | 1 |
| Giant Stride | 1,225 | 6 | 1 |
| Small Junglegym | 180 | 10 | 1 |
| Low Swing | 150 | 1 | 4[a] |
| High Swing | 250 | 1 | 6[a] |
| Balance Beam | 100 | 4 | 1 |
| See-saw | 100 | 2 | 4 |
| Medium Junglegym | 500 | 20 | 1 |
| **Misc. Equip't & Areas** | | | |
| Open Space for Games (Ages 6-10) | 10,000 | 80 | 1[a] |
| Wading Pool | 3,000 | 40 | 1[a] |
| Handcraft, Quiet Games | 1,600 | 30 | 1[a] |
| Outdoor Theater | 2,000 | 30 | 1 |
| Sand Box | 300 | 15 | 2 |
| Shelter House | 2,500 | 30 | 1[c] |
| **Special Sports Areas** | | | |
| Soccer Field | 36,000 | 22 | 1 |
| Playground Baseball | 20,000 | 20 | 2 |
| Volley Ball Court | 2,800 | 20 | 1 |
| Basketball Court | 3,750 | 16 | 1 |
| Jumping Pits | 1,200 | 12 | 1 |
| Paddle Tennis Courts | 1,800 | 4 | 2[d] |
| Handball Courts | 1,050 | 4 | 2 |
| Tether Tennis Courts | 400 | 2 | 2[d] |
| Horseshoe Courts | 600 | 4 | 2 |
| Tennis Courts | 7,200 | 4 | 2[d] |
| Straightaway Track | 7,200 | 10 | 1[d] |
| Landscaping | [a]6,000 | | |
| Paths, Circulation, etc. | [a]7,000 | | |

(a) Minimum desirable.

(b) One or all of these units may be omitted if playground is not used in conjunction with a school.

(c) May be omitted if sanitary facilities are supplied elsewhere.

(d) May be omitted if space is limited.

From J. De Chiara and J. H. Callender, *Time-Saver Standards for Building Types*, McGraw-Hill Book Company, 1973.

**Playground Equipment Examples**
Playground equipment for the handicapped has to be designed to accommodate the constraints created by wheelchairs; this should be an important consideration in the conceptual planning stage of schools for younger children.

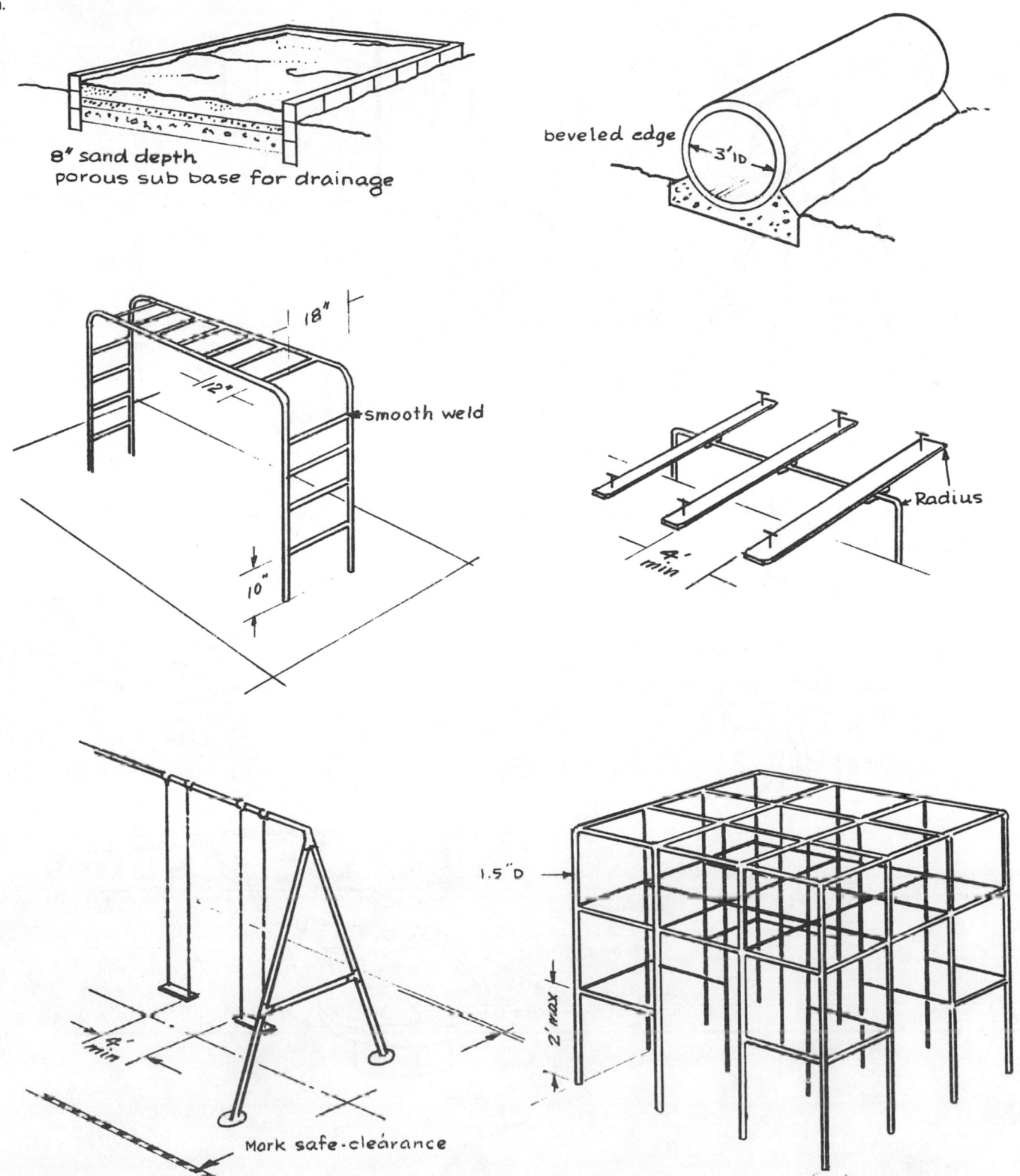

8" sand depth
porous sub base for drainage

beveled edge    3' ID

18"
12"
smooth weld
10"

4' min

Radius

4' min

Mark safe-clearance

1.5" D

2' max

**Playground Equipment Examples**

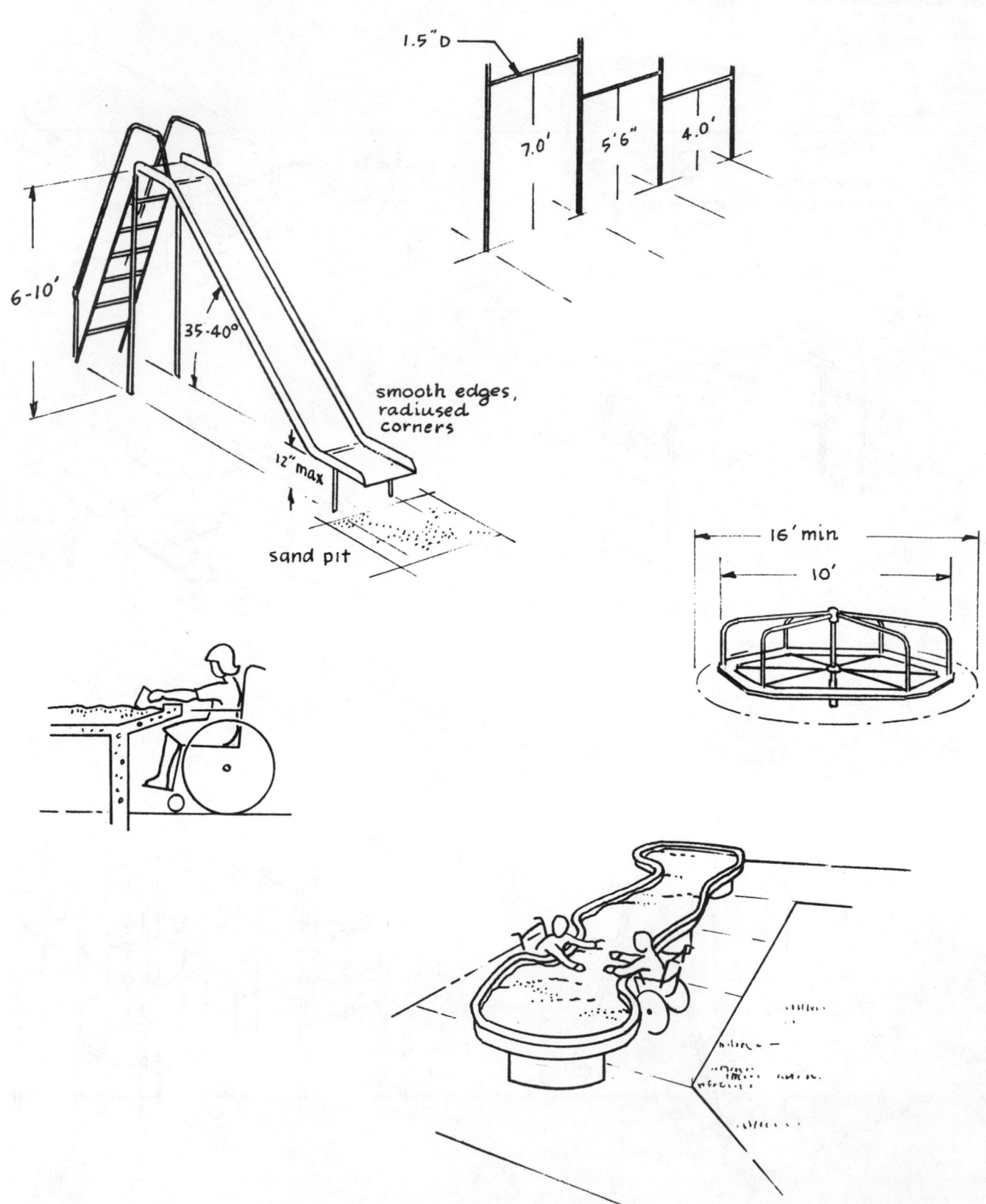

1.5" D

7.0'    5'6"    4.0'

6-10'

35-40°

smooth edges,
radiused
corners

12" max

sand pit

16' min

10'

## College Dormitory Planning

According to a study on university residential planning, the single dormitory room must accommodate at least a bed, a desk, a wardrobe, a bookcase, a soft chair, and the clearance space required by the entry door, *plus* the additional "use space" required around each element so that it can be used appropriately.

The accompanying sketches provide general element-use space requirements for each of the identified furniture elements.

It is suggested, however, that several other furnishing elements also be considered in the development of room sizes and shapes because these are equally necessary to meet the needs of each student:

1. A wastebasket
2. A hamper for dirty clothes
3. A place to hang outer garments that may be wet
4. A bulletin board on which to mount memos, pictures, schedules, etc.
5. A stand or shelf for an alarm clock, radio, or personal mementos
6. A cabinet for storage of personal hygiene and grooming aids

Obviously, a considerable variety of room layouts could be developed, depending on the size and shape of the spaces being considered. Usually, planners are of necessity interested in the most cost-effective layout. General objectives in developing any layout are the following:

1. To provide a well-defined resting, dressing, and grooming area
2. To provide a well-defined study area
3. To provide an adequate visiting area that does not require a visitor or visitors to occupy the host's sleeping area

When two students are to occupy the same general space, it should be arranged for minimum interference by either student of the other's critical activities, i.e., sleeping and studying. Consider at this level a common visiting area that can be used by the students and their friends. A few examples are shown in the accompanying sketches (which are not drawn to scale).

From *User Requirements*, URBS Publication No. 5, University Residential Building Systems, University of California, Berkeley, 1969.

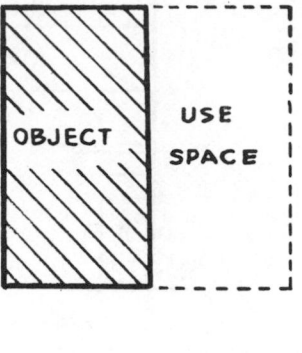

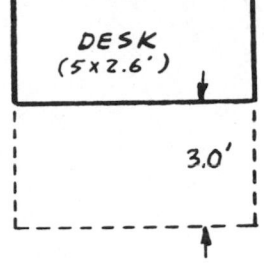

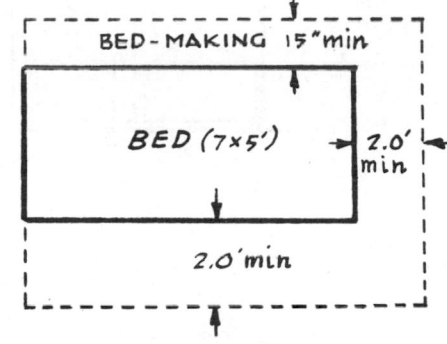

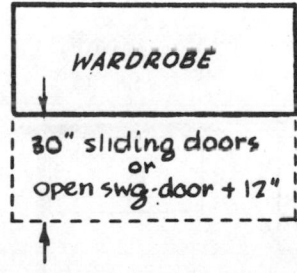

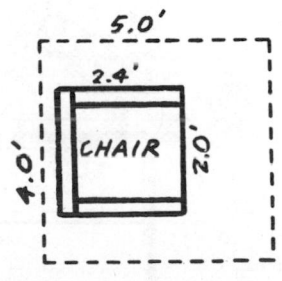

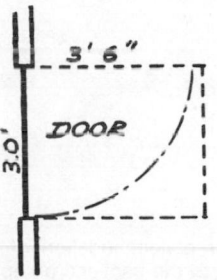

**SINGLE-OCCUPANCY ROOMS** Although it is possible to squeeze the student into less space, approximately 120 ft² (10.8 m²) is recommended as a minimum.

Note that in both examples, the bed has been positioned as far from the corridor doorway as possible to minimize interference with sleep.

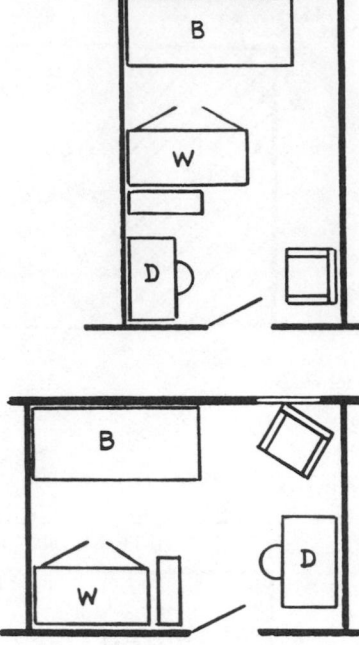

**DOUBLE-OCCUPANCY ROOMS** It is important in planning double-occupancy layouts to maintain as much equality as possible between the two spaces, not only in terms of actual floor space, but also in terms of the apparent convenience of each space.

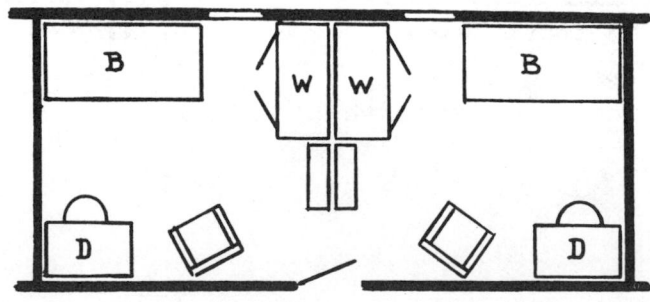

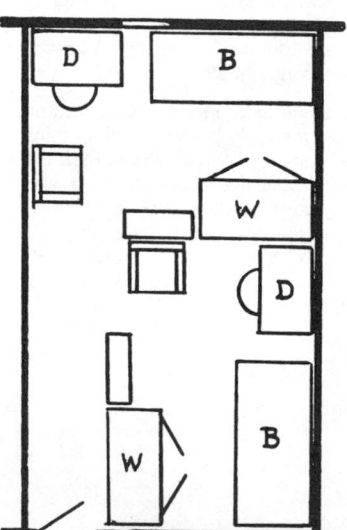

For accommodation of married students, refer to the sections on apartment planning.

Other alternatives are important to consider when planning dormitory space, including the suite concept. A minimal example is illustrated in the accompanying sketch, where a common social area has been added to the double-occupancy space. Also, this example illustrates an important addition of extra storage space adjacent to the student's quarters (as opposed to central storage at some distant point).

For planning information relative to other dormitory facilities—e.g., rest rooms, lounges, cafeterias, kitchens, and laundries—refer to Chapter 2 for the appropriate subsystem design.

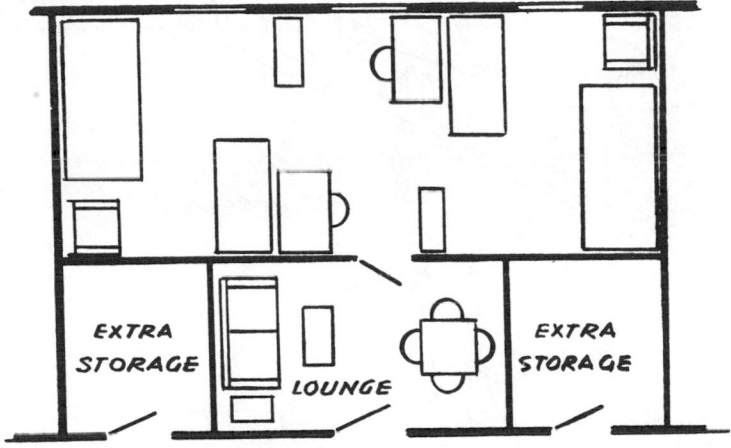

## Environmental Control

As in the case of all multiple-occupancy situations, control of the thermal and noise environment of the dormitory room becomes especially important, and usually equally difficult. Consideration should be given to the use of false partitions to help create a reasonably well-controlled environment for dual-occu-

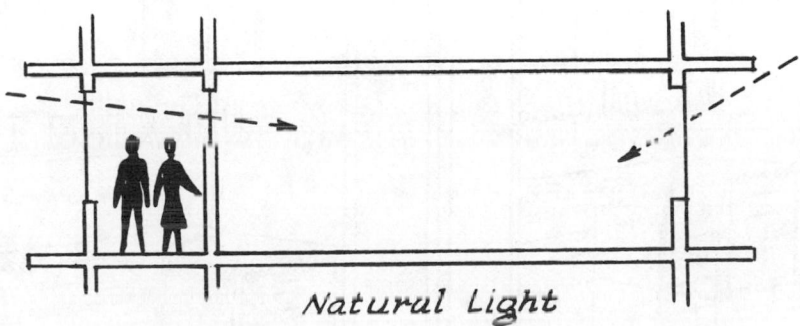

pancy spaces. Situations requiring such control include one student sleeping while the other studies or comes in late at night.

The accompanying sketch illustrates an important environmental consideration for the dormitory room, i.e., natural light from two sides, as opposed to the traditional method of introducing light from only one end of the room. Such techniques help maintain equality between the two spaces, mentioned earlier.

## Other Amenities

Electric outlets, telephones, and even private lavatories should be considered during the planning stage, for all these elements improve the quality of the student's environment.

## Safety

Refer to general housing safety recommendations, as well as local ordinances and other recognized standards.

**Customized Furnishings**

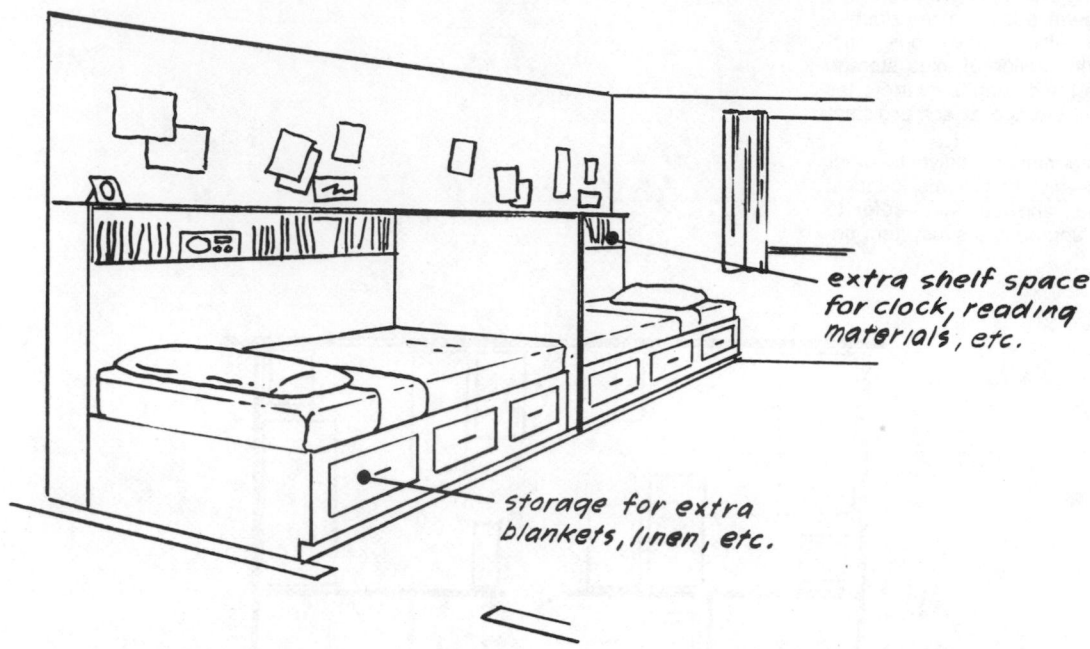

extra shelf space
for clock, reading
materials, etc.

storage for extra
blankets, linen, etc.

By customizing the furnishings of the dormitory room (rather than using standard furniture), considerable space saving may be accomplished. More importantly, good custom designs usually provide the student with considerably more useful storage space, as illustrated in the accompanying two sketches.

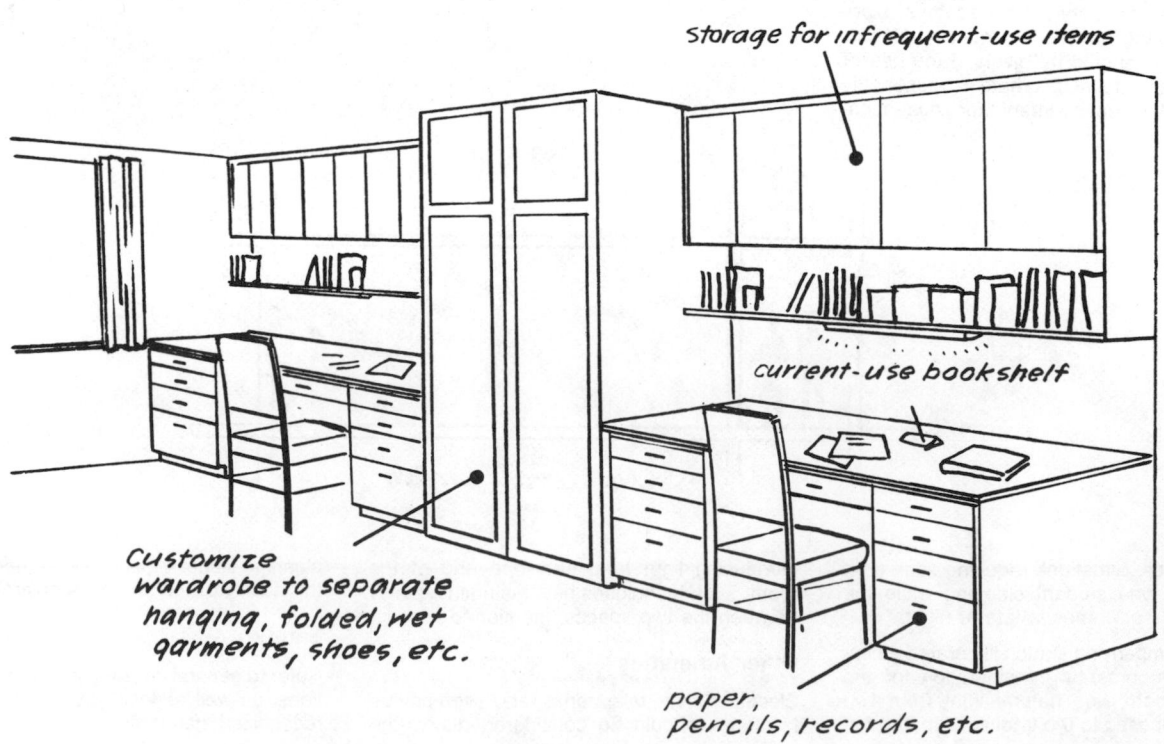

storage for infrequent-use items

current-use bookshelf

Customize
wardrobe to separate
hanging, folded, wet
garments, shoes, etc.

paper,
pencils, records, etc.

## Checklist for Other College and University Facilities[21]

The following checklist identifies some of the facilities that should be considered during college or university campus planning:

1. Offices (administration, faculty, maintenance, and security)
2. Eating facilities (student, faculty, and guest cafeterias, dining rooms, coffee shop, and snack bars)
3. Rest rooms and lounges (students, administration, faculty, guests, and maintenance personnel)
4. Post office and mail collection

[21]For specialized school facilities (e.g.; medical and dental facilities), refer to J. De Chiara and J. H. Callender, *Time-Saver Standards for Building Types,* McGraw-Hill Book Company, New York, 1973.

5. Communications (telephone, intercom, public address system, radio, and TV)
6. Publications (campus news, yearbook, and technical publications)
7. Book and stationery store
8. Student union and alumni offices and facilities
9. Chapel and social services center
10. Visitors' information center
11. Outdoor theatres
12. Transportation center and terminals
13. Housing (guests, administration, and faculty)
14. Fire station
15. Ground maintenance, landscape equipment, and nursery
16. Branch bank
17. Furnishings warehouse and shipping and receiving center
18. Agricultural experiment areas
19. Visual aids development facility
20. Campus radio and TV station

21. Health clinic and first-aid center or centers

*Note:* The removal of architectural barriers to the physically handicapped has become a public issue in recent years, especially as it applies to providing equal educational opportunity for all students. Refer to the appropriate guidelines for each subsystem as well as to the general principles of designing for the handicapped.

The campus walkway and roadway system should be an integral part of planning the above as well as planning the standard educational facilities, for the effectiveness with which this system interconnects the various facilities for both campus and itinerant pedestrians, bicyclists, and motor vehicle operators is of utmost concern to the campus administration. A key factor to be addressed is: How can we effect an optimum compromise between maintaining a semiprivate system and providing for public access?

PUBLIC LIBRARIES

# PUBLIC LIBRARIES

## Public Library Concept Development

The public library generally is no longer simply a place where books are stored and read or where materials are checked out to be taken home and read. The modern library has become a much broader system; in addition to performing the typical document storage and handling functions, it is a place in which to exhibit, provide entertainment, educate, and develop craft skills. Library specialists sometimes differ in their personal approach to the public library; some feel that they are primarily guardians of the materials they have been given (and thus are more concerned with keeping the public's hands off their treasured volumes), while others firmly believe that books should be in use as constantly as possible (and thus are more concerned with making it easy for the public to obtain materials). The obviously desirable architectural objective should, of course, be to provide a balance between effective safeguard of the library contents, and maximum exposure of these contents to the public. From a purely human factors point of view, the public should probably be given first priority; i.e., the most important considerations should be ease of getting into the facility, of finding one's way around the facility, of locating the book one desires and finding a good place to use it while in the facility, and of checking out materials to take home.

Many of the library-user interfaces are similar to those found in other public office buildings, and the reader should refer to the sections dealing with office building planning. In addition, however, the following unique planning considerations relate particularly to library management and use functions.

## LIBRARY STAFF FUNCTIONS
1. Administrative
2. Materials acquisition
3. Materials cataloging and filing and record keeping
4. Materials distribution
5. Overdue follow-up
6. Bookbinding and book repair
7. Research
8. Display and exhibit development
9. Materials check-out, return, and redistribution
10. Building maintenance

## PATRON FUNCTIONS
1. Materials identification
2. Browsing
3. Reading and study
4. Copying
5. Listening
6. Check-out and return

## ANCILLARY FUNCTIONS
1. Lectures
2. Recitals and drama
3. Special materials exhibits
4. Education and crafts

## COMMON FUNCTIONS
1. Parking (staff, patrons, and delivery people)
2. Personal hygiene

## TYPICAL LIBRARY SPACES
1. Book stacks (reference and lending)
2. Periodical, newspaper, and map racks
3. Check-out desk area
4. Card catalog file area
5. Reading areas
6. Audio-visual studies areas
7. Recital and lecture halls
8. Meeting rooms
9. Rest rooms
10. Administrative offices
11. Workrooms (staff and public)
12. Archival rooms and vaults
13. Receiving and shipping
14. Storage (materials and equipment, maintenance supplies, and special vehicles, e.g., a bookmobile)

## ENVIRONMENTAL FACTORS
1. Nominal thermal regulation for habitability of staff and patrons
2. Low noise levels
3. High-quality illumination
4. Special environments for certain materials (archival books and other documents, film, etc.) and control of temperature, humidity, dust, etc.
5. Special acoustic qualities for recital and lecture spaces
6. Special lighting effects for exhibits, drama, etc.

**SECURITY** As noted earlier, philosophies differ with regard to public access to library materials and the importance of the loss of books and other materials. Special consideration, however, should be given to materials whose loss could be serious. These include such items as archival, one-of-a-kind volumes and special art objects which would be extremely expensive to replace or which could never be replaced. It is general practice to place such items in special areas, the access to which is controlled by the library staff. The planning should anticipate this need, but consider techniques that minimize potential loss and yet provide reasonable access—without which, the public is not being served properly.

**SAFETY** The library, like other public buildings, is subject to hazards requiring careful consideration of traffic flow for emergency escape in the event of fire, properly designed stairways, etc. One special hazard exists where films are stored, requiring consideration of techniques to control such possibilities as fire in the film vault.

## General Layout Considerations

Although ease of finding one's way around most public buildings suggests a plan form that is rectilinear, there are advantages to other plans for the library. One such plan is illustrated in the accompanying sketch. The following are some of the advantages of such a plan:

1. Bookshelves are distributed more or less equally, which shortens the distance to stacks.
2. The shape lends itself to ease of monitoring activity in the stack area.
3. The arrangement provides convenience for study in the several alcoves at the ends of the stacks.
4. Increased open shelving is provided on the periphery.
5. All traffic entering or leaving the library passes by the front charge desk.
6. The central reading room can be used for discussion group activity as well as for individual reading.
7. The circular traffic pattern tends to reduce traffic movement and to minimize conflicts at traffic intersections.

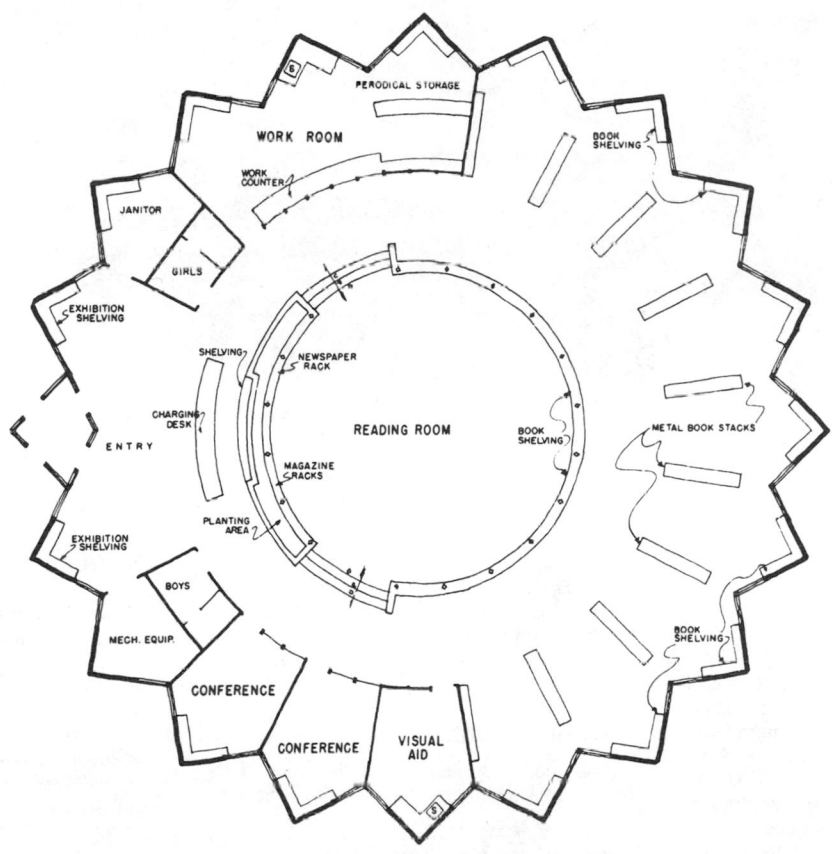

From J. H. Callender, *Time-Saver Standards for Architectural Design Data*, McGraw-Hill Book Company.

# CHURCHES

## CHURCHES

### Church Facilities Concept Development

Since there is a variety of philosophies regarding church facility design, it is vital that designers explore with their clients both these philosophical concepts and the basic functional needs typical of all building design. This section deals only with the latter aspects of church planning. Among the functions to be addressed for any church project are the following:

1. Worship service procedure as it pertains to the tasks of ministerial and lay persons, special furnishings, music, lighting and communication, and, finally, the congregation itself
2. Group instructional activities for both children and adults
3. Social activities involving large and small groups, within buildings and/or outside, and dining and kitchen facilities
4. Administrative functions including clerical and supervisory tasks, record keeping, publication, and accounting,
5. Ministerial activities including sermon preparation, counseling, and staff conferences
6. Lay program work, committee meetings, activity planning, program implementation, etc.
7. Music program activities including rehearsal, concerts, and special events
8. Housekeeping and ground mainte-

nance, storage, shops, etc.
9. Special events (weddings, funerals, etc.)
10. Communications
11. Playgrounds, nursery, and sports facilities
12. Parking
13. Rest rooms

The major facilities generally required for the complete church project are the following:

· Sanctuary
· Administration building or buildings
· Education building or buildings
· Maintenance facilities
· Social hall
· Ministerial housing (if on site)

### Church Facilities Planning Model

As illustrated on page 101, the important planning considerations are the following:

Congregation: Effective worship, education, and social facilities
Staff: Efficient work space and convenience among facilities

Special attention should be given to the problem of concurrent use of facilities, i.e., the ability to expand, isolate, and control. The latter is of special concern to smaller organizations, which may be unable to hire regular custodians. Above all, designers should not become so enthusiastic about building a monument that they create barriers to effective use and maintenance.[22]

## SANCTUARY PLANNING CONSIDERATIONS

### 1. Entrances and Exits
Convenient entrances and exits must be considered for a variety of users (the ministers, the choir, the choir director, the organist, the ushers, and the congregation). Both exterior and interior entrances must be provided as necessary to allow people to move from one end of the sanctuary to the other without being seen by the congregation or exposed to inclement weather.

### 2. Seating
Seating should be provided not only for the congregation but also for the ministers and their assistants, the choir and its director, the ushers, and the audio-visual equipment operators.

### 3. Lighting
Both general and supplementary lighting must be provided both for adequate seeing tasks (e.g., reading and walking safely) and for special effects (illuminating specific decorative items, stage lighting effects, etc.).

### 4. Communications
Depending on the size of the sanctuary, sound-amplification systems should be considered. In addition, consideration should be

[22]For more detailed design recommendations, see J. De Chiara and J. H. Callender, *Time-Saver Standards for Building Types,* McGraw-Hill Book Company, New York, 1973; and Rudolf Herz (ed.), *Neufert Architects' Data,* Archon Books, The Shoe String Press, Inc., Hamden, Conn.,

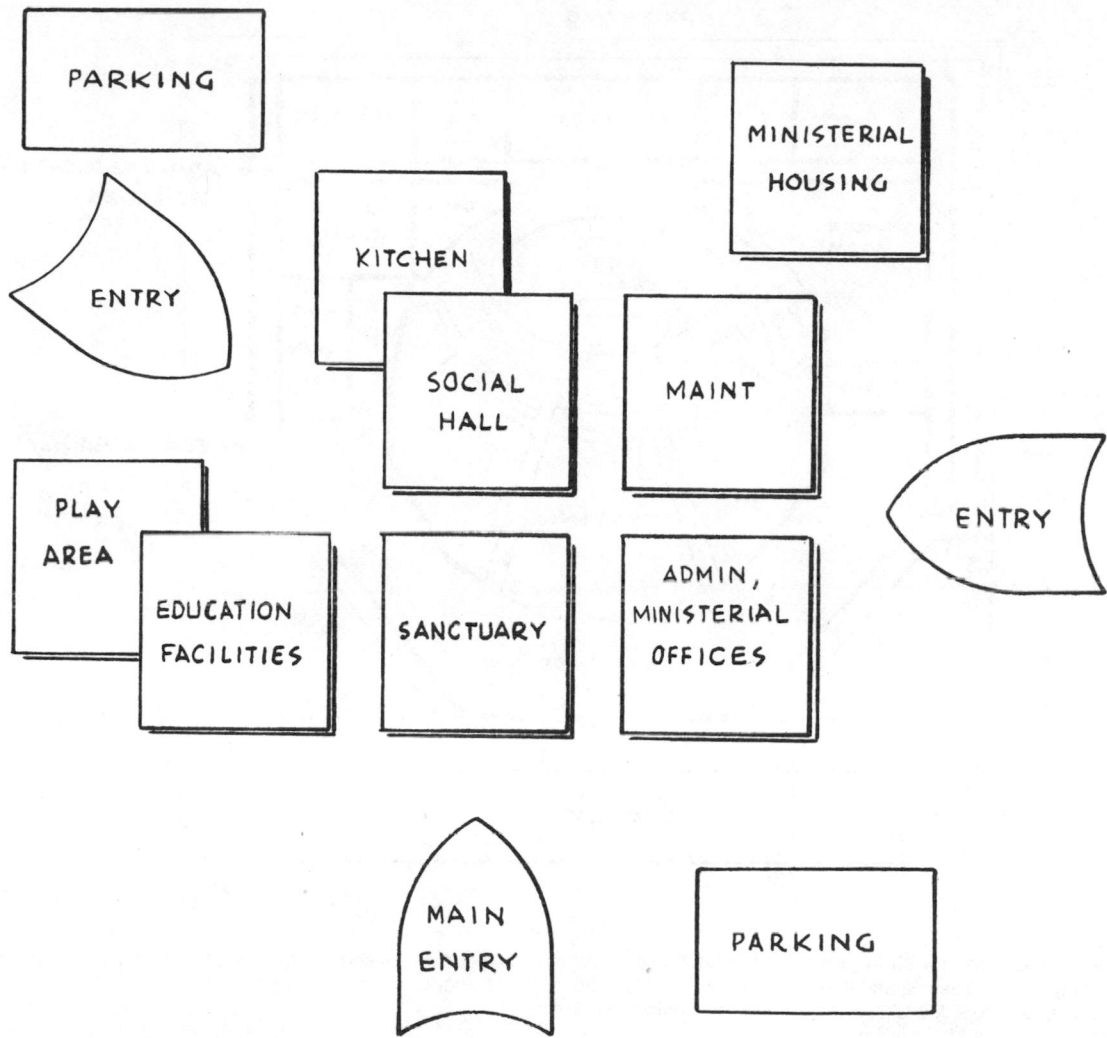

given to other aids such as special equipment for the hard of hearing, a recording system, and audio signaling between key participants, e.g., the minister, the organist, and the ushers.

## 5. Acoustics

In order to preclude interruption of a service, it is important to consider techniques for isolating the main sanctuary (nave) from the adjacent areas where people tend to generate noise, i.e., the entry vestibule, anterooms where the choir may gather, and areas where parents with small babies sit so that they can leave if a child becomes noisy and restless.

## 6. Special Areas

Several special areas often included as part of the sanctuary building require consideration during the planning stage in order to optimize their functional relationship with the main nave and chancel areas. These include:

Sacristy for communion preparation
Minister's study
Choir assembly area
Organ chambers
Choir and organ loft

Baptismal chamber
Audio-visual system monitoring booth
Bride's room
Ushers' materials facility (should include air-conditioning system controls, lighting controls, etc.)

## 7. Line of Sight

In spite of certain traditions in sanctuary layout, it is important that certain participants see each other. The more important visual links to consider are:

a. The minister should be able to see the ushers, the organist, and the choir director.

b. The organist should be able to see the minister, the choir director, and the ushers.

c. The ushers should be able to see the minister and the organist.

d. The congregation should be able to see the minister, the choir, and special performers when the sanctuary is used for special events.

e. The audio-visual monitors should be able to see the general house lights, special lighting effects, and the exit lights.

## Accommodations for the Elderly and the Handicapped

More than any other public facility, the church should be accessible to the elderly and the handicapped, since these people are typically among the most regular attenders. The key features to be considered are:

1. Wheelchair ramps and a minimal requirement to negotiate stairs
2. Better-than-average illumination
3. Special aids for the hard of hearing
4. Rest rooms specially equipped for the handicapped
5. Parking for the handicapped

## Special Safety Considerations

In addition to the typical safety considerations for any public building, special attention should be given to exterior safety problems that are somewhat unique to the church situation, especially the problem of the safety of children with respect to vehicular traffic around and within the church site. A typical problem occurs when church buildings are built directly adjacent to sidewalks and streets

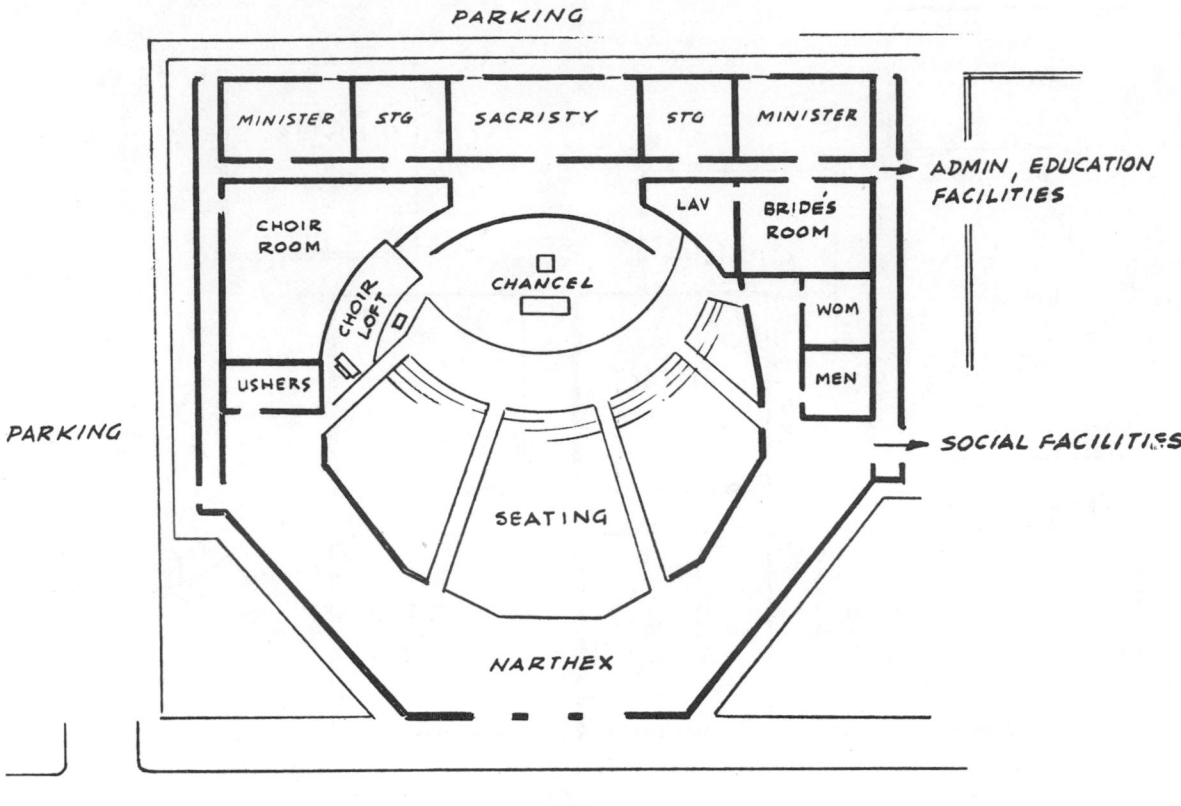

Children often are unattended for short periods between Sunday school classes, and they tend to roam and play on the sidewalks, often inadvertently running into the street in front of cars which are picking up passengers.

*Special note:* Church buildings are a prime target for experimenting with unusual design features. Planners and designers are cautioned against letting their enthusiasm for aesthetic experimentation introduce unnecessary problems for the eventual user, including difficulties in seeing, hearing, and negotiating stairs and pathways; uncomfortable seating; and, most important, difficulties in maintaining the facility.

### Sanctuary Planning: Problem-Area Checklist

#### 1. Seating Arrangement
No parishioner should be required to sit further away from the pulpit than about 75 ft (23 m) or forced to turn his or her head to see what is going on; i.e., the lateral viewing angle should not exceed about 40°.

#### 2. Choir Orientation
The sound of the choir should be directed at the congregation, and the congregation should be able to see the performers (although many people believe they do not need to see the performers, it is natural to look in the direction from which the sound comes). The choir should be able to see the choir director directly (not via mirrors), and the direc-

tor and the organist should be able to see each other (i.e., the organist should not be placed behind the director).

#### 3. Organ-Chamber Orientation
Although tradition has dictated separate pipe chambers located at various points within the sanctuary for special effects, the average listener is often confused and/or disturbed by such arrangements, and organists seldom can control sound volume properly, especially when one chamber is closer to them than another.

#### 4. Instrumental Accommodations
Many sanctuary designs do not provide sufficient seating space for larger instrumental groups that may participate in special programs. A particular problem occurs because of irregular floor levels created for other uses.

#### 5. Noise Isolation
There should be adequate noise isolation between the main sanctuary (nave) and other facilities in the same building (i.e., isolation from noise emanating from the narthex, from nearby rest rooms, from Sunday school areas, etc.). A special problem also is created by the selection of light fixtures that vibrate as a result of certain organ sounds. Full carpeting is recommended to minimize the noise of foot traffic.

#### 6. Lighting
Pulpits should be properly illuminated so that

the minister can read his notes. The sanctuary lighting should be adequate to allow the congregation to read songbooks or programs, and the choir loft should be properly illuminated so that choir members, the organist, and the choir director can read their music.

#### 7. Air Conditioning
Systems should be designed to accommodate the different needs of the choir and the minister, who may be wearing robes, as opposed to the congregation, which is garbed in street clothes.

#### 8. Exterior Visual Distraction
Although windows are desirable in the sanctuary, many are misplaced in terms of the potential distraction caused by people passing by outside. In some modern designs, considerable glare problems are created by large window expanses placed directly behind the pulpit, which make it difficult to see the minister.

#### 9. Maintenance Problems
A major maintenance problem is often created when the design includes a great number of light fixtures mounted extremely high above the sanctuary floor (a floor which may not be level or which may be complicated by pews that make it difficult to place ladders for relamping the light fixtures).

Many decorative features are created that naturally collect dust, and because they are in view, they must be cleaned on a regular basis. Although the roof is typically an important identifying feature that planners and designers use as a vehicle for expressing some special religious aesthetic, many such roof designs leak when it rains—and, worst of all, many are practically impossible to inspect or repair because of steep slopes, numerous water traps, etc.

*Special note:* When a pipe organ installation is being considered, consult both the organ builder and an acoustic engineering specialist regarding the size, shape, and location of organ chambers. Some of the special problems to be addressed are the size and shape of the pipe chamber or chambers in order to fit equipment and pipes into the chamber, proper construction and materials to ensure good sound reflection and prevent undesirable vibration (including vibrations that may be induced in adjacent structures or vent piping), and ease of access both for the installation and for later maintenance. Anticipate the possible future expansion of the organ system in order to avoid makeshift arrangements at a later time.

It is important to consult both the organ builder and an acoustic engineering specialist in order to guard against typical bias; i.e., builders of organs want to show off their equipment as a concert instrument. Although this is important, other considerations are equally important to the church's director of music.

## Sanctuary Planning Example

The layout sketch on page 102, although not intended to represent any particular facility, provides several important ideas to be considered in the planning of any new sanctuary. Note the following:[23]

[23]For other church facilities (e.g., churches, auditoriums, offices, and schools), see J. De Chiara and J. H. Callender, *Time-Saver Standards for Building Types*, McGraw-Hill Book Company, New York, 1973.

1. Convenience of the main entrance to the street
2. Parking for the general congregation and the staff
3. Large narthex (which is used as a gathering place, in spite of the best intentions of the ushers)
4. Sound barrier between the main sanctuary and any outside walls
5. Reasonable viewing distance between any seat in the sanctuary and the pulpit
6. A choir position that allows sound to focus fairly well on all parts of the congregational seating and also allows choir members to view the pulpit
7. Complete inside access around the main sanctuary
8. Good sight lines between the minister, choir, organist, and ushers
9. Good staff access to other church facilities and good congregational access to social gathering areas
10. Rest rooms for both staff and others

HOSPITALS

## HOSPITALS

### Hospital System Planning Model

The prime planning objective in hospital design should be to seek a balanced compromise between maximizing hospital staff efficiency and maintaining adequate respect for patients' personal needs, both physical and psychological. Foremost among patients' needs (other than medical and therapeutic care) are the minimization of fear, boredom, and the feeling that they have lost complete control over their lives.

### Hospital Facilities Concept Development

In conceptualizing and planning either the general or the specialized hospital facility, much depends on the nature and scope of the services required by the public or private client. However, all hospital and convalescent facilities have many common human interface needs and design requirements, which is the subject of this section. It is perhaps important in the beginning, however, to remind the planner and designer that the primary purpose of any hospital facility is to serve the patient, a fact that is often overlooked in the rush to make the jobs of the hospital staff and the doctors easier. High priority should therefore be given to making the patient's task of getting to the hospital, checking into the hospital, receiving a proper diagnosis and treatment, having an acceptable recovery environment, and

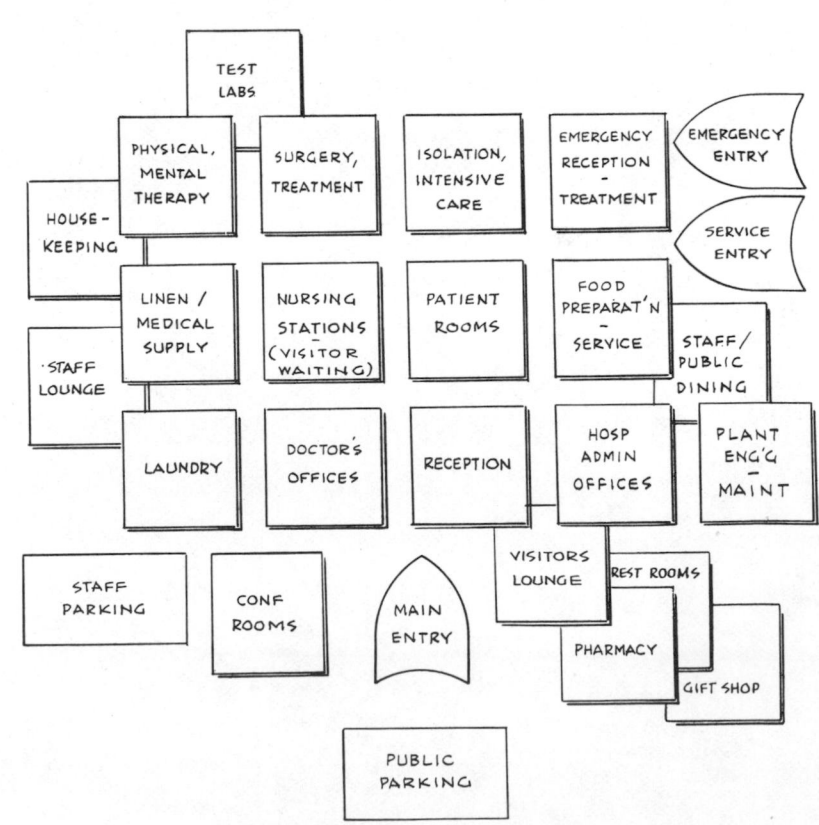

finally getting discharged with a minimum of strain as pleasant and convenient as possible. Obviously, it should be possible for the hospital staff and the doctors to perform their tasks as efficiently as possible, but not at the expense of the patient. Some of the observed deficiencies noted below should serve to emphasize the need for careful consideration of priorities.

## PATIENTS

1. Patients cannot find a parking place.
2. Patients cannot find the entrance to the hospital building.
3. Patients cannot locate the check-in or information desk.
4. Patients cannot find a place to sit while waiting to be admitted or to see a doctor.
5. The rooms are institutional-looking and often frightening.
6. The rooms lack privacy and freedom from sounds made by other patients or their visitors.
7. The room facilities are not designed for ease of use from the bed.
8. There is lack of space to receive visitors.

## NURSES AND AIDES

1. The space layout requires considerable walking.
2. The equipment is awkward to operate.
3. Work spaces are not conveniently located or laid out for maximum efficiency.
4. Relaxation facilities are inadequate.

## DOCTORS

1. The space layout requires considerable walking and is time-consuming.
2. Work spaces (operating rooms, etc.) are not laid out for maximum efficiency.
3. Relaxation facilities are inadequate.

## VISITORS

1. Confusing floor plans make it hard to find one's way around.
2. There is not enough parking.

## Hospital Facilities Shopping List

It is doubtful that any list of possible facilities would include all possible considerations. The accompanying table, however, should provide the planner with a good start.

## General Conceptual Considerations

The following user-related considerations should be addressed early in the planning stage of any type of hospital facility.

### 1. Hospital Location

Whenever possible, locate the hospital facility in a quiet neighborhood where it is away from heavily traveled, noisy highways or streets and yet adequately accessible to and from such routes. The access route should be direct and well marked so that emergency vehicle operators do not waste time trying to locate the facility.

### 2. Site Availability to Public Transit

The site should be chosen with the potentiality of public transportation access in mind, since both staff members and visitors often need to use public transportation to get to the hospital.

| Administrative Personnel | Spaces Required |
|---|---|
| Hospital director and assistants | Offices |
| Business manager and assistants | Conference rooms |
| Secretarial and clerical personnel | Admissions and check-out center or centers |
| Bookkeeper | Staff lounges and rest rooms |
| Engineering staff | Kitchen and dining facilities |
| Housekeeping staff | Mechanical and maintenance spaces |
| Maintenance staff | Laundry |
| Ambulance staff | Pharmacy |
| General pharmacist | Storage (records, housekeeping |
| Kitchen and dining-room staff | materials and equipment, linens |
| | and supplies, drugs, furniture, |
| | etc.) |
| | Temporary morgue |
| | Vault |
| | Parking |

| Medical Personnel | Spaces Required |
|---|---|
| Medical director and assistants | Offices and conference rooms |
| Staff doctors | Nursing stations |
| Itinerant doctors | Examination rooms |
| Director of nursing staff | Operating and treatment rooms |
| Nurses and aides | Laboratories |
| Dieticians | Nursery |
| Food preparers and aides | Emergency receiving rooms |
| Laboratory technicians | Kitchen and dining facilities |
| Physical therapists | Staff lounges, rest rooms, lockers, and dressing spaces |
| Pharmacists | Parking |

| Patients | Spaces Required |
|---|---|
| Emergency patients | Private and multipatient rooms |
| Patients requiring minor first aid | Lavatories and bathrooms |
| Patients requiring intensive care | Physical and mental therapy spaces |
| Patients requiring major surgery | Isolation wards and intensive care areas |
| Maternity patients | Nursery and play areas |
| Patients with communicable diseases | Day rooms and solariums |
| Psychiatric patients | Personal storage space |
| Children and infants | Emergency parking |
| Cancer patients | |
| Others | |

| Visitors and Others | Spaces Required |
|---|---|
| Patients' visitors | Lobby, waiting rooms, and public |
| Social workers | rest rooms |
| Vendors | Information center |
| Delivery personnel | Public eating facilities (kitchen) |
| Security personnel | Shop selling gifts, newspapers, and flowers |
| Funeral parlor personnel | Mall room |
| | Private sleeping rooms for |
| | overnight visitors |
| | Parking |

### 3. Planning for Emergency Evacuation Possibilities

Although one hopes that a disaster requiring emergency evacuation of hospital patients and staff members never occurs, this is a critical consideration in terms of both site selection and site and building layout.

### 4. Site Layout

Hospital grounds should be laid out to maximize the logic of the plan for visitors, especially those trying to bring an injured or very sick patient to the hospital for the first time. Although an attractive site is important, a confusing one may mean the difference between life and death.

### 5. Building Layout

The layout of buildings and building interiors should also evidence a logic that visitors can readily understand from their first-time, limited viewpoint. Although the staff may eventually learn to navigate from one place to another, the first-time visitor is easily confused by ir-

## Flow Charting for Preliminary Facility Planning[24]

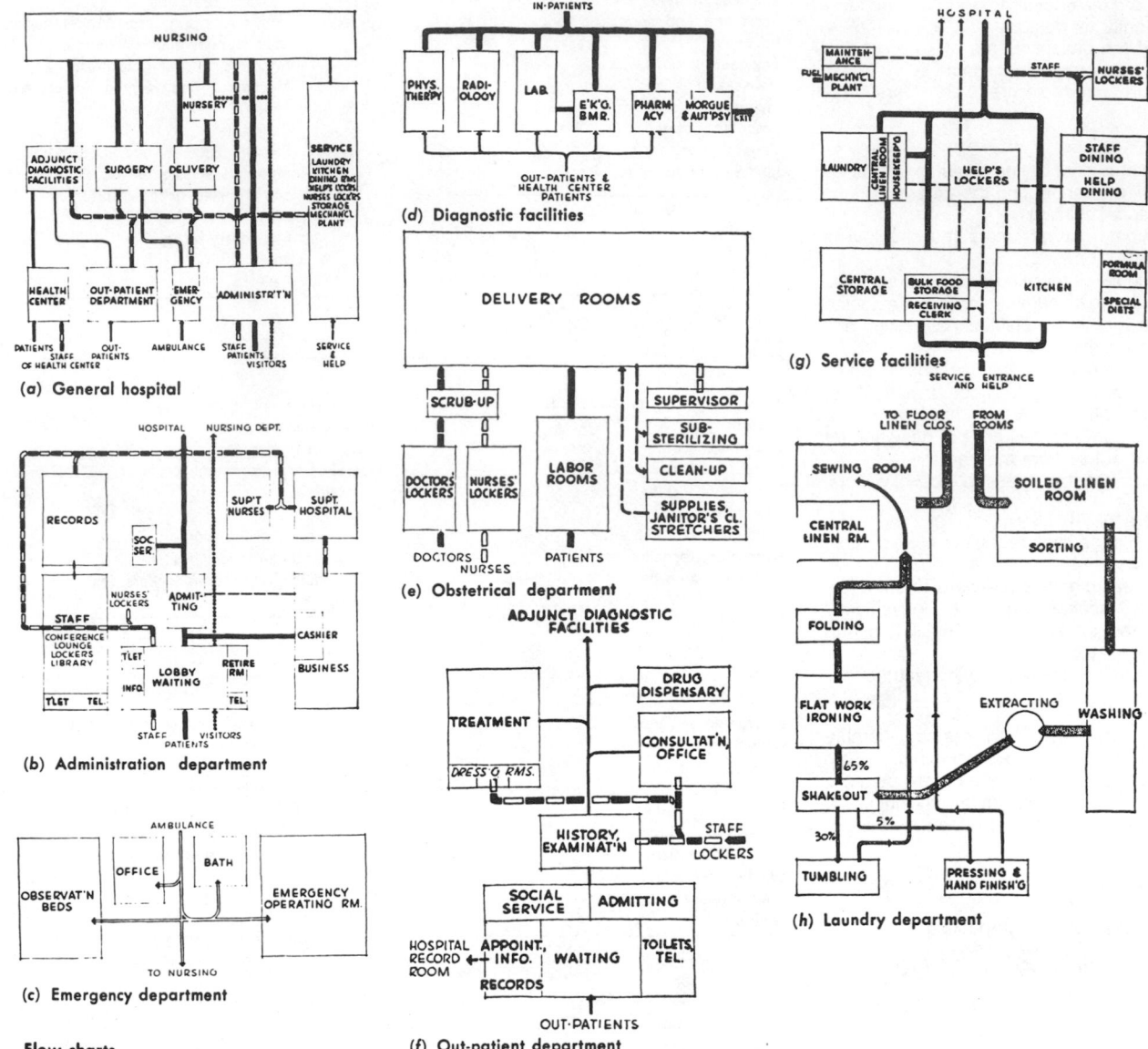

(a) General hospital

(b) Administration department

(c) Emergency department

**Flow charts**

(d) Diagnostic facilities

(e) Obstetrical department

(f) Out-patient department

(g) Service facilities

(h) Laundry department

*From Design and Construction of General Hospitals by Public Health Service, U.S. Department of Health, Education and Welfare (1953).*

regular pathways, curved or angular geometries, irregular elevations, etc.

### 6. Identification and Instructional Signing
Signing requirements should be addressed at the same time that grounds and building layouts are under consideration. Do not expect a sign to make up for a poor layout.

### 7. Removal of Architectural Barriers
The hospital, more than any other facility, is subject to constant use by those who are handicapped. Do not assume that all handicapped persons are being assisted by a nonhandicapped person; many are not. Plan for the handicapped in the basic design so that

[24]From J. De Chiara and J. H. Callender, *Time-Saver Standards for Building Types*, McGraw-Hill Book Company, New York, 1973.

modifications or add-ons are not required; e.g., a basic ramp entry should be the primary entry, not an auxiliary one.

### 8. Protection from Inclement Weather
Staff members, patients, and visitors should never be required to walk in the rain from building to building, be buffeted by wind on an outdoor balcony or roof solarium, or be unable to utilize special therapy facilities outside. These conditions should be considered in the initial concept.

*Note:* The planner or designer who is not acquainted with hospital design should obtain additional information from such sources as the Bacon Library of the American Hospital Association in Chicago and the U.S. Public Health Service in Washington, D.C.

### Site Planning

The sketch on page 107 illustrates several important site planning considerations:

1. Central core (includes elevators, information center, admissions and checkout, administration, etc.) equally accessible from both the main and the emergency entrances
2. Ample parking for both the public and staff members arranged so that no one has to cross busy streets or roadways
3. Auxiliary lifts and stairs at each end of the building
4. Covered, drive-through entrances
5. Short distance from entries to the information point

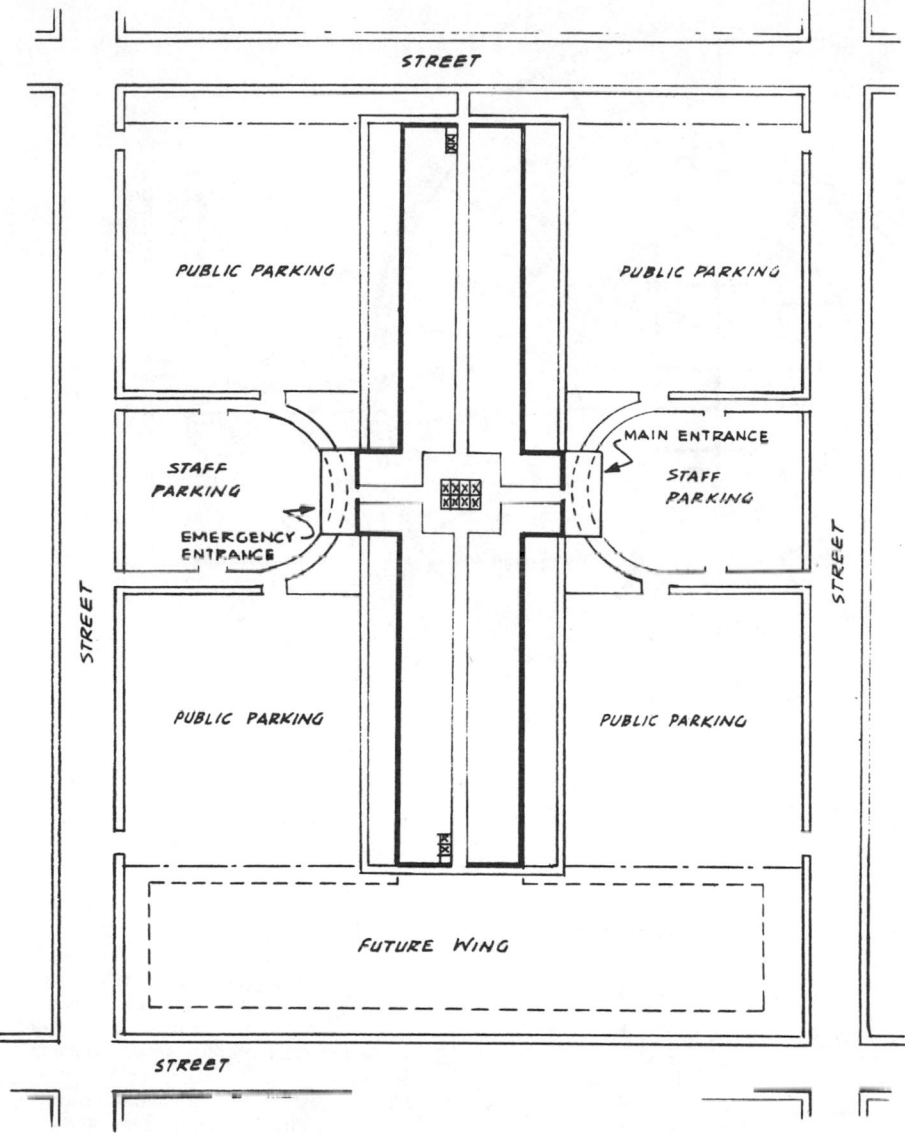

## Floor Planning

The sketch below illustrates several important floor planning considerations:

1. By using the central-core (elevators) concept, two mirror-image wings are created. These could be identified as "North Wing" and "South Wing," or they could be identified by a color code (the color of the walls, the carpeting, etc.).
2. The nurses' station, services spaces, and visitor lounges are located adjacent to the elevator foyer so that visitors see information points as soon as they get off the elevator.

Other alternatives are, of course, possible. The sketch at the top of page 108 illustrates one unique floor plan that provides efficient traffic flow for the nursing staff and provides especially good patient monitoring in terms of direct view of individual rooms.

## Planning Patients' Rooms

Considerable attention typically is given to the efficiency of a room layout from the nurse's point of view and, of course, from the point of view of making do with the least amount of floor area. The sketch at the bottom of page 108 provides some important patient-oriented planning considerations:

1. Although patients can have privacy by closing the intervening folding partition, opening the partition enables them to see each other and visit.
2. Each patient can see out of his or her individual window, which is important to those who are able to sit up and enjoy this feature (the windowsill, of course, must be low enough).
3. The shared lavatory-shower facility is uniquely accessible and yet not exposed when passersby happen to look into the room. Washbasins are provided outside the lavatory so that they can be used while another person is showering or using the commode.

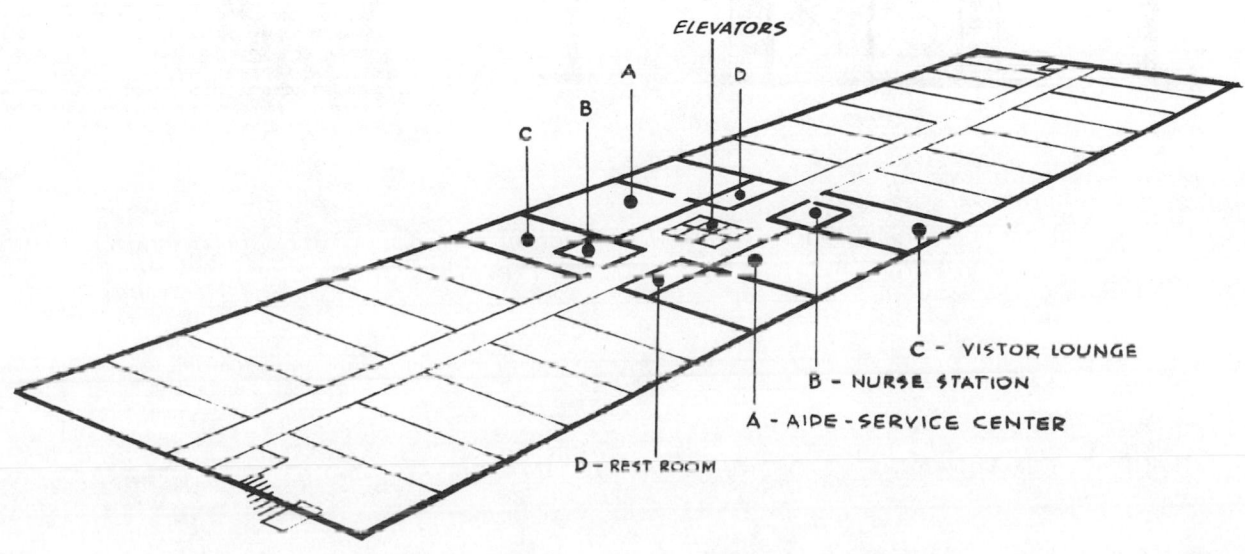

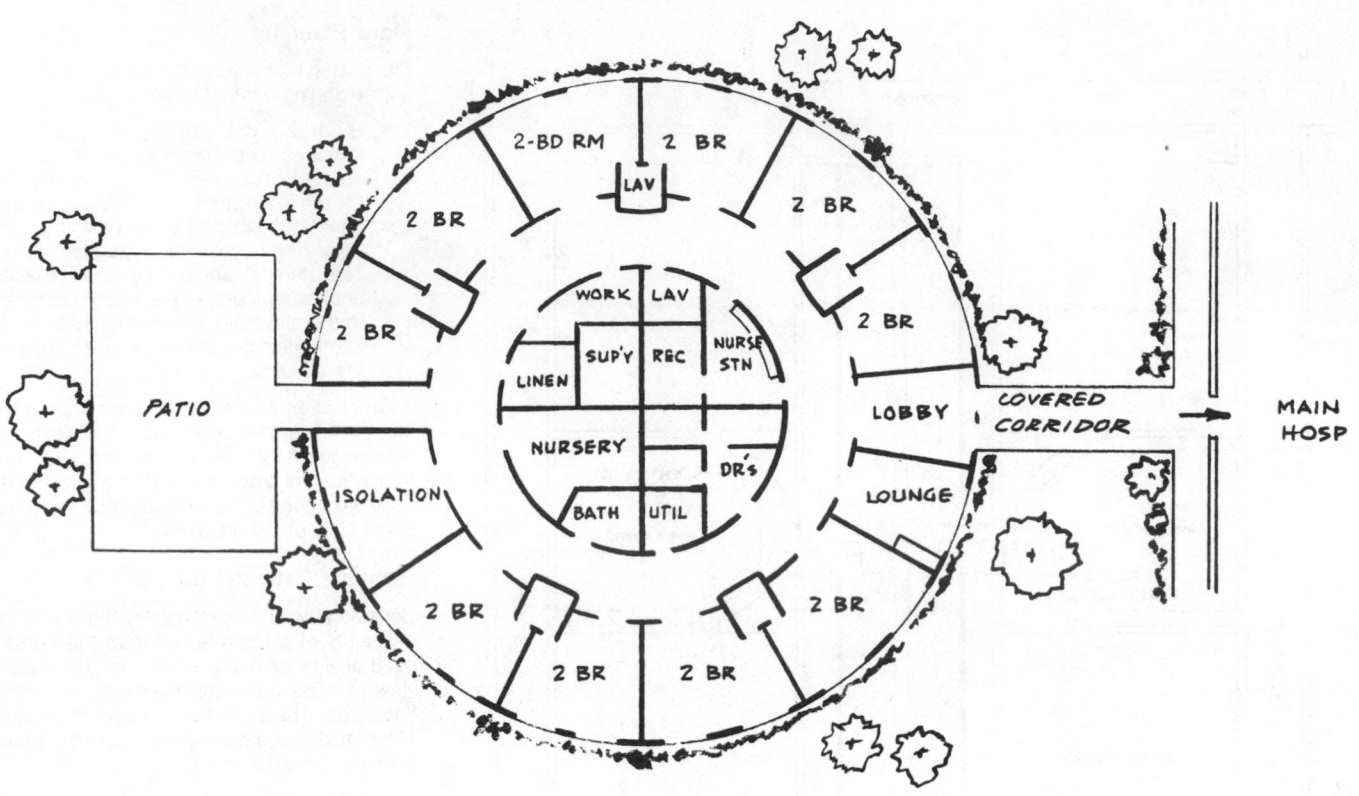

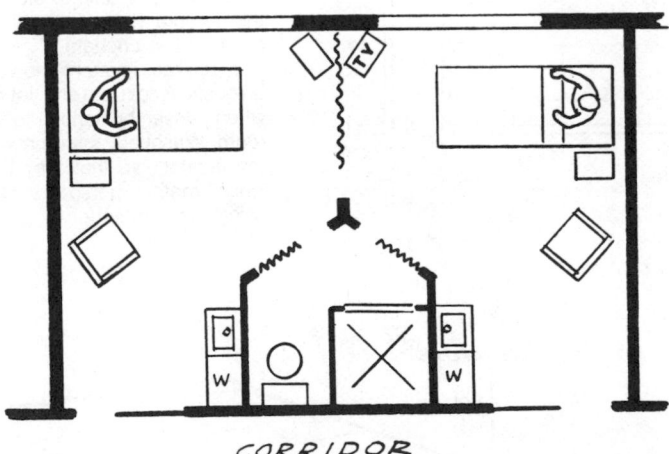

CORRIDOR

4. The sliding doors to each portion of the room are much easier to manage from a wheelchair. Windows in these doors permit monitoring by people outside and also allow the recuperating patient to keep tabs on the activity outside the room.

5. The particular arrangement shown provides ample visiting space and equally good access for replacing furnishings from time to time (e.g., special beds may replace standard ones for certain patients).[25]

[25]For other special administrative and medical staff workplace layout suggestions, see J. De Chiara and J. H. Callender, *Time-Saver Standards for Building Types*, McGraw-Hill Book Company, New York, 1973; Rudolf Herz (ed.), *Neufert Architects' Data*, Archon Books, The Shoe String Press, Inc., Hamden, Conn.; and Allen and Karolyi, *Hospital Planning Handbook*, John Wiley & Sons, Inc., New York.

# THEATRES

## THEATRES

### Theatre Concept Development

Whether a proposed theatre is restricted to a single type of theatrical performance or is to accommodate a variety of performances (sometimes simultaneously), similar user activity categories should be addressed in order to ensure that each user group is properly accommodated. The principal user categories include the following:

1. Patrons
2. Performers
3. Performance support personnel
4. Patron support personnel
5. Theatre management
6. Facilities operation, housekeeping, and maintenance

The typical subsystems required in most theatres include the following:

1. Parking (for each of the above categories)
2. Entrances and exits (separate customer, performer, service, and other entrances and exits)
3. Basic patron facilities (lobby, rest rooms, cloakrooms, refreshment booths, stairs, elevators, and escalators)
4. Basic theatre management facilities (offices, rest rooms, cash vault, plant engineering and equipment spaces, housekeeping and maintenance shops

and storage, employee lounge, etc.)
5. Auditorium (including seating, lighting, air conditioning, special projection, and special-effect light and sound booths)
6. Main stage and backstage (including stage props management, revolving and elevating stage elements, stage lighting, and performer preparation spaces, e.g., dressing, instrument storage, and rehearsal)
7. Other performance support facilities (shops and storage for stage props, lighting and sound equipment, costumes, furniture storage, etc.)

*Note:* Although several user categories could obviously utilize a common subsystem, (e.g., rest rooms and storage), care should be taken during initial concept development not to create undesirable constraints on certain users or to create interference patterns that will affect persons who are under considerable time duress; e.g., performers and stage support personnel need isolation from theatre customers.

Each theatre user group's key task and activity requirements should determine where common facilities can be used.

### Key Human Factors Considerations by User Activity Requirements

#### 1. Visibility

a. Patrons need to see in order to get to

and from their seats, often in semidarkness.
b. All patrons should be provided approximately equal visual access to all activity onstage (but not be exposed to what is going on offstage).
c. Ushers must guide patrons to their seats, often in semidarkness.
d. Performers and stagehands must move quickly and safely both backstage and onstage, often in the dark.
e. Stage lighting operators must be able to monitor the effect of their light control manipulations (preferably from the patrons' viewpoint).
f. There should be intervening, preparatory areas that allow patrons and others to adapt visually to a darkened auditorium or a bright external sunlit environment.

#### 2. Audibility

a. Patrons should have equal auditory access to the important sounds emanating from a performance, but they should be isolated from unwanted sounds coming either from within or from outside the auditorium.
b. The acoustic characteristics onstage should fit performers' requirements to hear one another.
c. The acoustic characteristics within the auditorium should be designed to minimize distortion of desired sounds from the stage and to minimize possible patron-generated noise.

### 3. Thermal Comfort
Proper temperature, humidity, and ventilation control should be provided (particularly within the auditorium) to ensure patron comfort and freedom from drafts, undesirable odors, etc.

### 4. General Safety
The overall facility concept should be examined carefully from the standpoint of the safety of all theatre user categories, but especially in terms of the following:

   a. Emergency escape
   b. Nonflammable materials
   c. Emergency lighting
   d. Number, location, and operation of emergency exits
   e. Disaster control (fire-extinguishing systems and equipment)

### 5. Other Considerations
Special attention should be given to the layout of the theatre in terms of the following:

   a. Patron queuing both outside and inside the theatre to ease the flow of both entering and exiting customers
   b. Isolation of patrons from backstage activities to avoid unnecessary interference with performers and/or others who must perform time-constrained tasks and to preclude the possibility of damage to stage equipment, instruments, etc.
   c. General layout clarity and convenience, especially in terms of patrons' needs to locate their seats, find rest rooms, etc.
   d. External considerations, including isolation of patrons', theatre operators' and performers' parking and entrances; covered walkways to protect patrons who may have to wait outside in inclement weather; and sufficient space outside for people to make an emergency exit without having to run directly into an adjacent street, where they may be exposed to heavy traffic

## Conventional Community Theatre

Although details may vary, the illustration on page 111 identifies the major theatre subsystems that impact on human users, either as individual categories or in combinations of categories.

   User categories include the following:

Patrons
Performers
Performance support personnel
Administrative and managerial personnel

   Subsystems include the following:

Entrance and foyer
Audience seating
Stage
Stage support and performer preparation area

## Special Considerations for Each Theatre Subsystem

### 1. Entrance
A sufficient number of entrances should be provided for the anticipated exchange of patrons entering and leaving the theatre. An entry vestibule large enough to accommodate purchasers of tickets (out of the rain) should be provided.[26]

### 2. Foyer
A large foyer should be provided (for each floor) because this must accommodate patrons both before the performance and during intermissions. Rest rooms, lounges, and administrative and management offices should be provided off the foyer. Refreshment centers are also appropriate here, in addition to ticket booths and a cloakroom.

[26]The entrance and foyer area should be so arranged that patrons' eyes have time to adapt to bright sunlight following exposure to the typically low light levels of the theatre. Similar visibility problems occur for persons working backstage, where scenery must be changed in semidarkness and where performers must move from unlit to brightly lit areas, and vice versa.

### 3. Audience Seating Area or Areas
The audience seating area should be designed for optimum viewing and listening and should have both comfortable seating and convenient and safe access to seats. A sloping floor with staggered seating can provide optimum viewing.

### 4. Orchestra Pit
A large orchestra pit should be provided when this facility is planned. Instrumentalists must have room to maneuver and must be able to see the conductor. Access to and from the pit (without exposure to the audience) is necessary.

### 5. Stage
A generous stage should be considered. It is simple to contract the stage size, but it is extremely difficult to enlarge it for special performances (i.e., ones for which the chorus and orchestra may be positioned on the stage together).

### 6. Backstage
The behind-the-scenes activity of typical stage productions requires considerable space, including access from one side to the other for moving scenery and performers. Included among the backstage facilities are scenery storage and preparation, performers' dressing rooms and rest rooms, costume storage, curtain and light controls, and rehearsal studios. A shipping dock should be provided.

### 7. Special Features
Among the special features to be considered are a projection booth for motion pictures and sound control, mounting and control of special lighting effects, elevators for patrons and for staging, acoustic control features (sound reflection and absorption), and illumination for housekeeping and maintenance. A basement is desirable for many purposes, among which are storage, maintenance shops, and art studios for creating advertising materials.

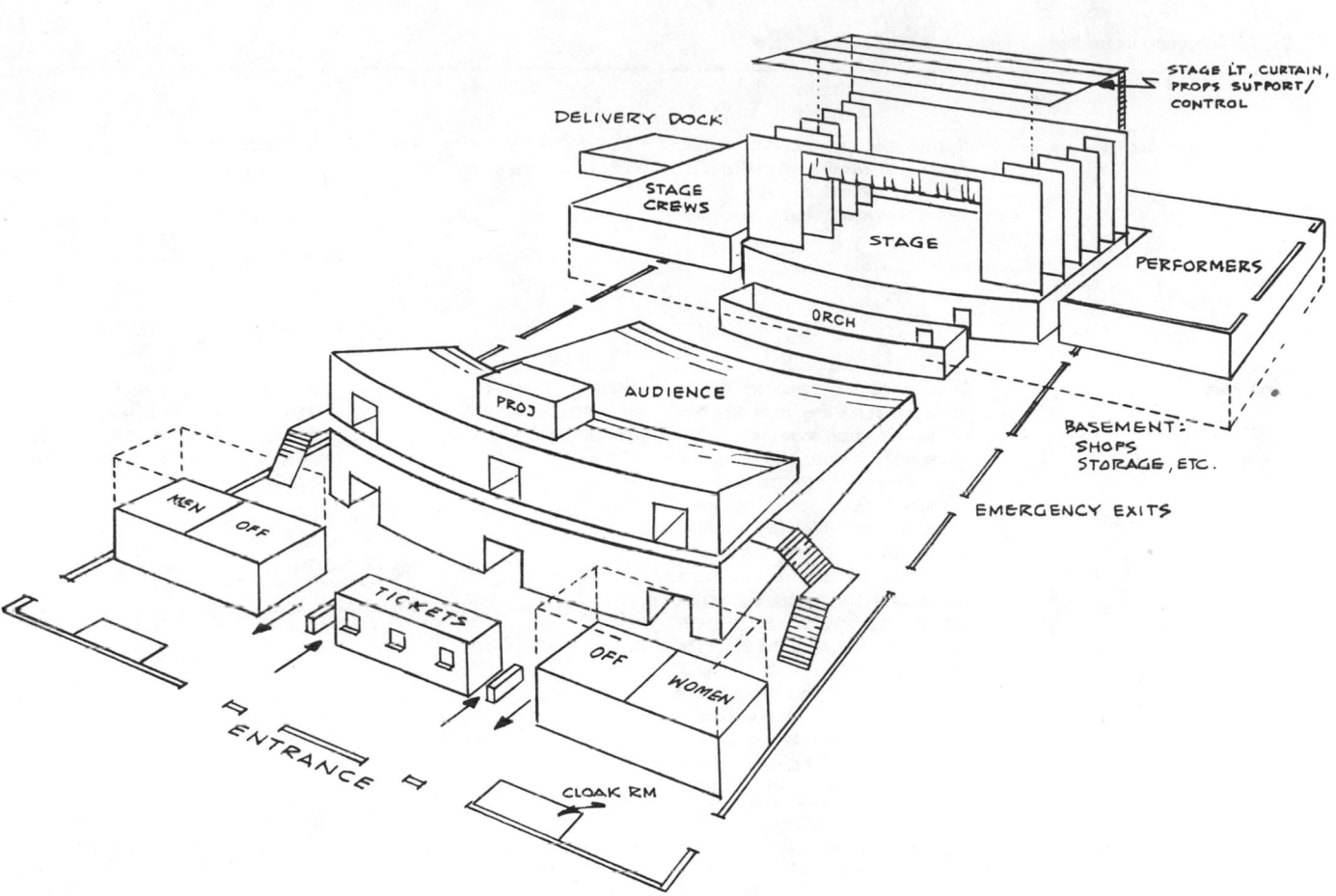

STAGE LT, CURTAIN, PROPS SUPPORT/ CONTROL

DELIVERY DOCK

STAGE CREWS

STAGE

PERFORMERS

ORCH

AUDIENCE

PROJ

BASEMENT: SHOPS STORAGE, ETC.

EMERGENCY EXITS

MEN

OFF

TICKETS

OFF

WOMEN

ENTRANCE

CLOAK RM

**Spatial Requirements for Various Types of Theatrical Productions**

| | General characteristics | Acting area size | Shape |
|---|---|---|---|
| Pageant and symphonic drama . . . . . . . | Dramatic episodes, processions, marches, dances, and crowd scenes. Masses of performers engaged in simple but expansive movements before very large audiences. | From 2,000 to 5,000 sq ft, depending on the scale of the pageant. | Rectangular with aspect ratio between 1 to 3 and 2 to 3. |
| Grand opera. . . . . . . . . . . . . . . . | Large numbers of performers on the acting area at one time; often more than one hundred in big scenes and finales. Movement is martial processions and group dances and the costumes are elaborate. Soloists perform downstage center, close to the footlights but within the bounds of the conventional proscenium, principals play twosome and group scenes in the area near the audience, and choruses and supernumeraries require space upstage. The ballet and the chorus of soldiers, pilgrims, peasants, or what not, sometimes fill the entire acting area. The performance is viewed objectively by the audience and does not benefit by intimate contact between performance and audience. | Minimum: 1,000 sq ft<br>Usual: About 2,500 sq ft<br>Reasonable maximum: 4,000 sq ft | Quadrilateral with an aspect ratio between 1 to 2 and 2 to 3. Sides converge toward the back of the stage, following the sight lines from the extreme lateral positions. |
| Vaudeville, revue. . . . . . . . . . . . . . | Vaudeville and revue emphasize the human scale. Although the vaudevillian keys his performance for the last row in the gallery, the form is characterized by intimate direct relationship between performer and audience: monologues straight to the front, confidential asides to the front row, and audience participation in illusions. Other acts (acrobatics, etc.) are played across the line of audience vision for maximum effect. | Minimum: 350 sq ft<br>Usual: About 450 sq ft<br>Reasonable maximum: 700 sq ft | Rhomboid with aspect ratio about 1 to 3. Sides converge toward back of stage following the sight lines from the extreme lateral seats. |
| Dance . . . . . . . . . . . . . . . . . . . | Graceful and expressive movements of human figures in designed patterns, chiefly in two dimensions but with the third dimension introduced by leaps and carries. Occasional elevation of parts of the stage floor. Singles, duets, trios, quartets, groups. The movement demands maximal clear stage space. | Anything under 700 sq ft is constricting.<br>Reasonable maximum: 1,200 sq ft | Rhomboid with aspect ratio about 3 to 4. May project into and be surrounded by audience (open stage or arena) since frontal aspect of performers has minimal and space-filling quality has maximal significance. |
| Musical: folk opera, operetta, musical comedy, musical drama | These forms embody on a smaller scale the production elements of grand opera, plus a certain freedom and a quest for novelty which encourage the development of new performance devices. Close audience contact of soloists and specialists is borrowed from vaudeville and revue. Big scenes involve many dancers, singers, and showgirls, often with space-filling costume and movement. Fifty people on stage at one time is not unusual. | Minimum: 600 sq ft<br>Usual: About 1,200 sq ft<br>Reasonable maximum: 1,800 sq ft | Proscenium:<br><br>Rhomboid with aspect ratio between 1 to 2 and 2 to 3. Sides converge toward the back of the stage following the sight lines from the extreme lateral seats.<br><br>Arena:<br><br>Circle, square, or rectangle (3 by 4 aspect ratio) or ellipse (3 by 4 aspect ratio). |

**From** J. De Chiara and J. H. Callender, *Time-Saver Standards for Building Types*, McGraw-Hill Book Company, New York, 1973, pp. 298, 304, 307.

| Arrangement | Proscenium | Orchestra | Comment |
|---|---|---|---|
| Long dimension of acting area perpendicular to general sight line. Audience entirely on one side, elevated to perceive two-dimensional movement. Large openings at ends and in side opposite audience for processions, group entrances, and exits. Some elevation of portion of acting area opposite audience, purely for compositional reasons. | Either no proscenium with performers entering the "pageant field" from beyond the lateral sight lines, or structural or natural barriers to delineate the side limits of the acting area and conceal backstage apparatus and activity. "Curtains" of sliding panels, lights or fountains for concealing the acting area; often the concealment is by blackout only. | Space for 100 musicians between audience and acting area. Conductor must see performance. | Primarily an outdoor form, it is often staged in makeshift or adapted theatres, utilizing athletic fields and stands or natural amphitheatres. A few permanent pageant theatres have been built. |
| Long dimension perpendicular to the general sight line. Audience elevated to perceive two-dimensional movement. | Width equal to the long dimension of the acting area. | Pit for 60 to 80 musicians. Conductor must have good view of action. | Movement in two dimensions in acting area is a significant visual component, predicating elevation of the seating area to make this movement visible. |
| Long axis of the acting area perpendicular to the optimum sight line. Audience grouped as close as possible to the optimum sight line. The forestage is an essential part of the acting area; steps, ramps, and runways into the house are useful. | Width equal to the long dimension of the acting area. Flexibility is to some advantage in revue but of little value in vaudeville. | Music and music cues closely integrated with both vaudeville and revue performances. Pit space for from 15 to 30 musicians. Conductor and percussionist must have good view of the action. | Most of the visual components of vaudeville and revue are such that they are perceived best in the conventional audience-performance relationship. The comic monologist who must confront his audience is defeated by the open stage and arena arrangements. |
| Nearly square acting area so that dance patterns may be arranged in depth and movement may be in many directions including along the diagonals. Many dance figures require circular movement. Many entrances desirable, especially from the sides of the acting area. | Proscenium not really necessary; though useful as concealment for lighting instruments and dancers awaiting entrances, other devices such as pylons, movable panels, and curtains may be substituted. | Music almost always accompanies dancers. For dance as part of opera or musical show, orchestra is in pit. For dance as specific performance, as in ballet, orchestra may be in remote location and music piped in. Maximum orchestra for dance: 60 musicians in pit for classical ballet. Minimum: one drummer. | Dance in its various manifestations is the performance form best suited to the open stage or arena since it possesses the least amount of facial-expression significance and the greatest amount of movement and pattern in two or three dimensions. Elevation of the audience to perceive best the patterns of dance is desirable. |
| Proscenium: Long axis of acting area perpendicular to the optimum sight line. Mechanized mobility of structural parts to produce changes in acting area arrangement are desirable. Forestages, sidestages, acting area elevators. Arena: Numerous wide entrances for actors and stage hands via the aisles or through tunnels under the seating banks. Ramps preferable to stairs or steps. Experimentation possible in rendering stage flexible by lifts, and in development of flying systems over the acting area. | Usually as wide as the acting area, but should be adaptable to changes in the arrangement of the acting area described in the preceding column. Arena: None | Music an integral auditory component, sometimes integral visually. Elevating orchestra pit to accommodate from 20 to 40 musicians. Arena: Orchestra pit beside the acting area parallel to long axis and opposite principal entrance. This unavoidably imparts a performer orientation toward the orchestra and favors the seats in that general direction. | The assumption by ballet of a greater share in the performance of musical comedy indicates the need for a high general sight line from the audience. A phenomenon of the last 20 summers has been the growth of the musical theatre arena under canvas by which huge audiences have been enabled to see revivals of standard and Broadway musicals at similar prices though with general reduction of scenic investiture to that which is possible in the arena form. The movement has been economically feasible and generally profitable. |

**Spatial Requirements for Various Types of Theatrical Productions (Continued)**

Legitimate drama. . . . . . . . . . . . . . . Of all production types, legitimate drama places the greatest emphasis upon the scale of the human actor. The importance of the individual actor requires that stage space and scenery do not dwarf him. Dominance of plot, locale, and characterization requires verisimilitude in the size and relationship of scenic objects. Too small an acting area crowds actors and furniture, hampers stage action, and detracts from the dramatic effect which is the sole aim of the performance. Too large an acting area diminishes the actor in scale and renders his performance ineffective by weakening the effect of his gestures and movement.

Minimum: 240 sq ft (12 by 20 ft)
Usual: About 525 sq ft (15 by 35 ft)
Reasonable maximum: 1,000 sq ft (25 by 40 ft)

Proscenium:

Quadrilateral with an aspect ratio about 1 to 2. Sides converge toward the back of the stage following the sight line from the extreme lateral seats.

Open stage:

Semicircle, quadrilateral, or polygon projecting from a proscenium or from an architectural facade.

Arena:

Circle, square, rectangle, polygon, or ellipse with about 3 by 4 aspect ratio. Entrances from diagonal corners and in middle of one or both long sides.

## Community Theaters

### Typical space requirements

| Spaces | Areas* (sq ft) | Remarks |
|---|---|---|
| Vestibule and gallery. . . . . . | 1,200 | Less area would hamper use of space as gallery and meeting place. Area may be increased in proportion as auditorium capacity exceeds 800. Good lighting is necessary. |
| Checkroom . . . . . . . . . . . | 240 | Minimum unless checkroom does not serve auditorium or unless patrons do not check overcoats. |
| Lobby . . . . . . . . . . . . . | 1,000 | See Vestibule; mechanical ventilation needed here. |
| Ticket office . . . . . . . . . . | 50 | Minimum; for larger houses additional administration office (50–80 sq ft) is required. Ticket windows (2) and wall space (approx. 4 by 8 ft) are necessary. |
| Lounge-rehearsal room | 750 | Minimum size, equal to acting area of stage; mech. vent. needed. |
| Administrative . . . . . . . . . | 350 | Minimum; area varies. Outside light and air needed. |
| Men's toilets. . . . . . . . . . | 250 | Consult codes; areas ample for 800 capacity; either mech. vent. or outside light and air needed. |
| Women's toilets . . . . . . . . | 250 | |
| Auditorium . . . . . . . . . . . | 5,600 | Minimum for conventional seating; may increase to 7,000–8,000 sq ft for aisleless seating. Area includes forestage (removable seats). Outside light undesirable. |
| Radio studio . . . . . . . . . . | 300 | Can be reduced to 200 sq ft; no outside light; mech. vent. needed. |
| Control room . . . . . . . . . | 70 | Minimum; mech. vent. needed. |
| Director's room . . . . . . . . | 20 | Minimum, but adequate. |
| Quiet room . . . . . . . . . | 30 | Acts as sound insulation between circulation and radio unit. |
| Projection room . . . . . . . . | 200 | Ample, includes toilet and lavatory; consult code requirements. |
| Spotlight booth . . . . . . . . | 400 | Area may be divided into three booths: one on center with stage, one at each side of auditorium. |

### Typical Space Requirements

| Spaces | Areas* (sq ft) | Remarks |
|---|---|---|
| Stage . . . . . . . . . . . . . | 3,500 | Ample; 2,800 sq ft minimum; 3,500 usual avg. except for encircling stage. Air conditioning in conjunction with auditorium desirable; no outside light; top of stage house louvered (consult codes); if conventional stage, minimum height, floor to grid, is 70 ft. |
| Stage workshop . . . . . . . . | 1,500 | Sometimes reduced to 1,200 sq ft. Outside light, if clear glass, preferably from north; if obscure, orientation unimportant. |
| Scene storage . . . . . . . . . | 1,000 | Minimum; larger if possible. |
| Costume workshop . . . . . . | 420 | May reduce to 300 sq ft; north light desirable. |
| Costume storage . . . . . . . | 210 | Minimum; no outside light; preferably ventilated; must be dry. |
| Costume dyeing . . . . . . . . | 80 | Minimum; no outside light required; unless outside air provided, must be mechanically ventilated. |
| Six dressing rooms† . . . . . . | 680 | Each room requires access to two lavatories; size not changed with size of building; stars' dressing rooms each need private toilet and shower; all preferably air-conditioned. |
| Makeup room† . . . . . . . . | 130 | Minimum; used also for dressing, requires two lavatories; preferably air-conditioned. |
| Two chorus rooms† . . . . . . | 440 | Reasonable minimum; three lavatories needed in each; preferably air-conditioned. |
| Two bathrooms . . . . . . . . | 300 | Reasonable minimum. |
| Stage manager . . . . . . . . | 150 | Minimum. |
| Discussion room . . . . . . . | 750 | Can be used for rehearsal; area determined by acting area. |

* Based on auditorium capacity of 800.
† Dressing, chorus, make-up rooms require mirrors, preferably 3-sided type, movable; and overhead lighting, mirror-lighting equipment.

| Arrangement | Proscenium | Orchestra | Comment |
|---|---|---|---|
| The realistic style of dramatic production confines the performance to an acting area entirely inside the proscenium. The apron is not used. Most historic styles and much modern dramatic theory demand more freedom of audience-performance relationship than the realistic style and call for the projection of the performance toward, into, and around the audience. For this projecting aprons, forestages, sidestages, runways, steps and ramps into the aisles are all to some degree useful. To meet the demands of different styles and stylists, the acting area for drama must be capable of assuming many shapes. To confine it within the proscenium opening is adequate for the realistic style but inadequate for the others; to project it toward, into, or around the audience in any rigidly unalterable form is likewise adequate for one style but inadequate for others. | Width equal to long dimension of the acting area. Moving panels to vary width, openings in proscenium splay to form side-stages, movable pylons or columns by which opening may be subdivided are all desirable. Flexibility and mobility are increasingly desirable. The application of motive power under remote control to the movement of structural parts to produce different arrangements appears desirable but is costly. Manually alterable parts, particularly forestage proscenium panels and sections of the stage floor, if not unwieldy, are reasonable substitutes. | Orchestral music is sometimes an integral visual part of the performance, but most generally it is a purely auditory component. It is not generally necessary for the orchestra to be seen by the audience, but because cueing of music is so exacting, the conductor must see the action. It is reasonable to provide a pit for from 15 to 30 musicians, but the flexibility cited at the left must be provided, either by portable pit covers, steps, and platforms or by mechanized orchestra lifts. There is opportunity for originality of arrangement. | The various forms of theatre used by legitimate drama are discussed fully earlier in this chapter. |

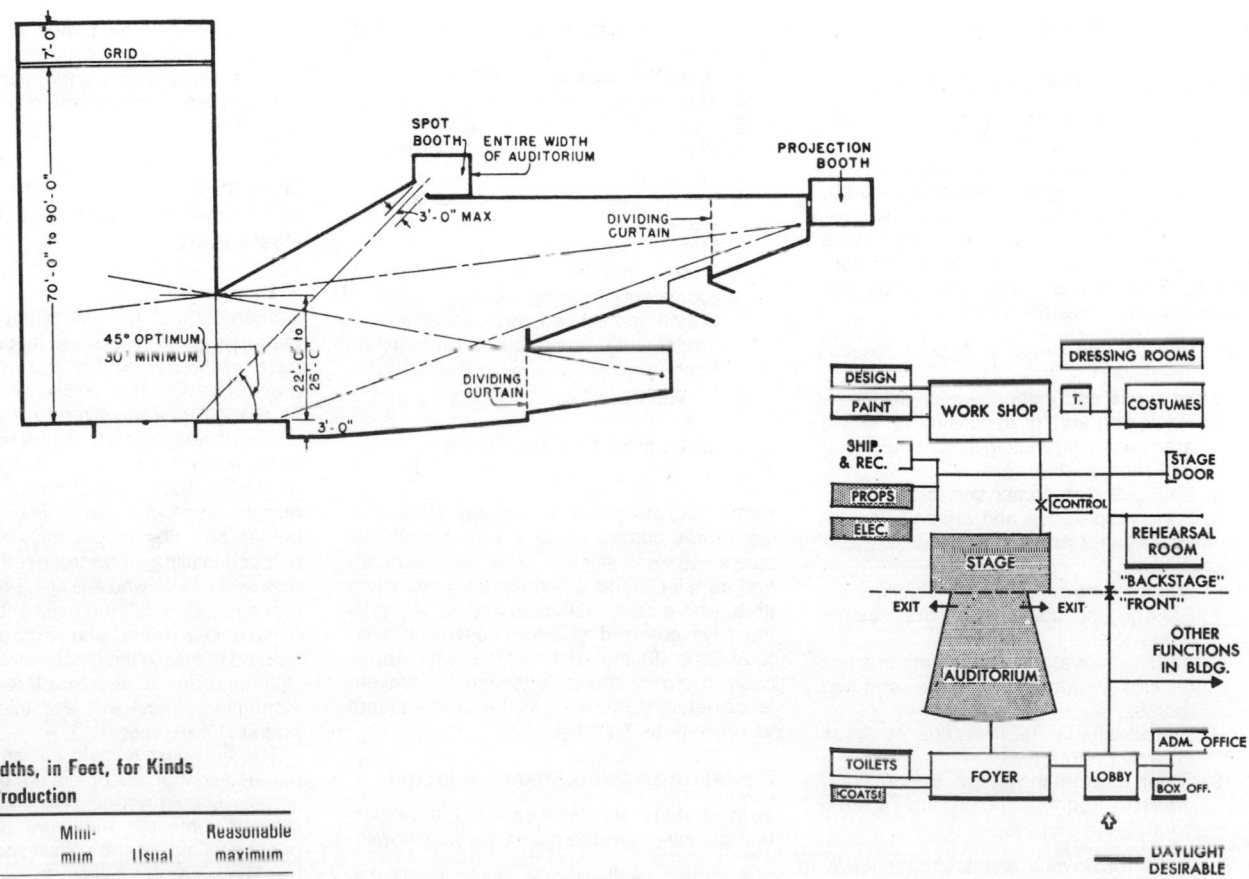

### Proscenium Widths, in Feet, for Kinds of Theatrical Production

| | Minimum | Usual | Reasonable maximum |
|---|---|---|---|
| Drama | 26 | 30 to 35 | 40 |
| Vaudeville, revue | 30 | 35 | 45 |
| Musical comedy, operetta | 30 | 40 | 50 |
| Presentation, opera | 40 | 60 | 00 |

From J. De Chiara and J. H. Callender, *Time-Saver Standards for Building Types*, McGraw-Hill Book Company, 1973.

# RESTAURANTS

## RESTAURANTS AND BARS

### Restaurant Concept Development

Although restaurant concepts will vary depending on the size and extent of service planned, certain basic human interface features remain relatively constant, and the problems that patrons or staff members will have depend on how well these interfaces have been designed to fit human needs and constraints. The principal functions to be addressed include the following:

1. Parking and public transportation accessibility
2. Entrance and exit convenience
3. Reception and/or access to the dining area (including waiting area, cloakroom, rest rooms, etc.)
4. Dining and customer service
5. Food preparation and storage (including delivery and trash and waste disposal)
6. Dishwashing and garbage disposal
7. Storage for dishes, flatware, linens, utensils, etc.
8. Administrative and management areas (including staff lounge, locker, and rest rooms)
9. Housekeeping and maintenance areas (including storage)
10. Other (performing stage, dance floor, dressing rooms, audio-visual system control and storage, etc.)

Each of the following architectural subsystems requires special individual as well as collective consideration:

1. Parking area or areas both for patrons' and staff members' vehicles and for service vehicles

2. Foyer, waiting room, lounge and rest rooms, and cloakroom
3. Special lounge and bar (storage)
4. Main dining room or rooms
5. Intermediate service centers
6. Kitchen (food, utensil, and equipment storage)
7. Service staff lounge, rest rooms, lockers, etc.
8. Manager's office
9. Special performer areas (instrument rooms and dressing rooms)
10. Sound and lighting control booths
11. Engineering spaces (environmental control)
12. Emergency exits

### Special Human Factors Problem Areas

Human factors problems typically occur in at least three primary areas: (1) the transitional space, where customers enter the restaurant and transfer to the dining area; (2) the dining area, where both customers and service staff must be provided uniquely convenient features; and (3) the kitchen area, which obviously requires special attention to convenience details to make the kitchen staff's output as efficient as possible.

### Restaurant Facility Planning Model

As illustrated at the top of page 117, the important planning considerations are as follows:

1. Ample, well-laid-out, easily identifiable parking that will not cause vehicles to queue up in the street
2. An ample reception and waiting area with minimum conflict between incoming and departing guests

3. An ample, quiet, and pleasant dining area that provides reasonable privacy for individual diners and groups of diners
4. Efficient kitchen and service arrangements

### General Considerations in Planning the Transitional Space of the Restaurant

#### 1. Entrance

Remember that the main entrance to the restaurant is also the main exit for customers who have finished dining. Since people tend to approach the door (from either inside or outside) as though they were the only ones using it, avoid the temptation to minimize the entryway width (both inside and outside). In addition, remember that people often are in the process of putting on outer garments or adjusting umbrellas, and they are generally busy with small talk and jostling, unaware that they may interfere with others who are going or coming. It is recommended that the main entrance consist of separate entry and exit doors, clearly marked to advise the customer which is which. (Although not all people will read and follow instructions, most will, and this reduces the potential traffic conflict.)

The area just outside the entry should be planned so that customers are protected; that is, they should not have to step directly into the rain, onto snow-covered sidewalks, into bright sunlight, or into passing pedestrian traffic. The area just inside the entry should be large enough so that customers can step aside to reconnoiter, fold umbrellas, remove outer garments, and adjust their eyes (if the interior illumination is considerably darker than the out-of-door entry illumination).

## 2. Foyer

A fairly spacious foyer should be considered. This should be arranged so that new patrons can immediately survey and plan their next step. That is, patrons will be looking for a place to put their hats and coats, perhaps the rest rooms, a maître d' or hostess or the pathway to the dining area, or a place to sit down and wait. Again, remember that the foyer will also be used by exiting patrons, and the layout should provide a clear delineation of the preferred paths for incoming as opposed to exiting patrons.

Depending on the type of restaurant, there may be the requirement for a cashier's stand in, or adjacent to, the foyer. Often this can be used as a natural divider between incoming and exiting patrons.

## 3. Dining-Area Entry

The transitional point between the foyer and the dining area should be clearly delineated so that it is obvious to incoming patrons. Avoid single step-downs or step-ups from the foyer to the dining area (a pet architectural device that leads to many accidents). This transition point should be well illuminated, not treated as a secret of the management. If the levels between the foyer and the dining area are to be different, create at least two steps and also provide a ramp for customers who cannot safely negotiate steps.

It is suggested that consideration be given to the problem of noise that may emanate from the foyer, which often is disturbing to diners. Obviously, certain acoustic treatments are available. However, a change in direction can also help ameliorate the problem.

A typical plan incorporating some of the features discussed above is shown in the accompanying illustration.

## General Considerations in Planning the Dining Area

### 1. From the Diner's Point of View

Upon entering the dining area, diners need to have a good preview of the total area so that (in case they are not being ushered by a maître d') they can choose the area and/or table that they believe meets their particular requirements and decide which is the best way to get there. Although there may be something intriguing about "hiding" various tables and creating devious pathways and nooks and crannies (which also are of some value from the standpoint of providing privacy and freedom from noise), care should be taken not to confuse patrons completely.

Fairly direct, clear, and generous pathways should be provided for patrons going to and from tables and/or dining areas. Not only is it annoying to bump into waiters and other patrons, but also some safety hazards are created when the layout forces crowding.

Reasonable privacy is desired by most diners. Plans that are based on "hub-to-hub" tables and seats do not provide the privacy most people want. In addition, most diners want to be free from physical disturbance, and therefore the plan should provide adequate space around tables and chairs so that waiters and waitresses can have free access

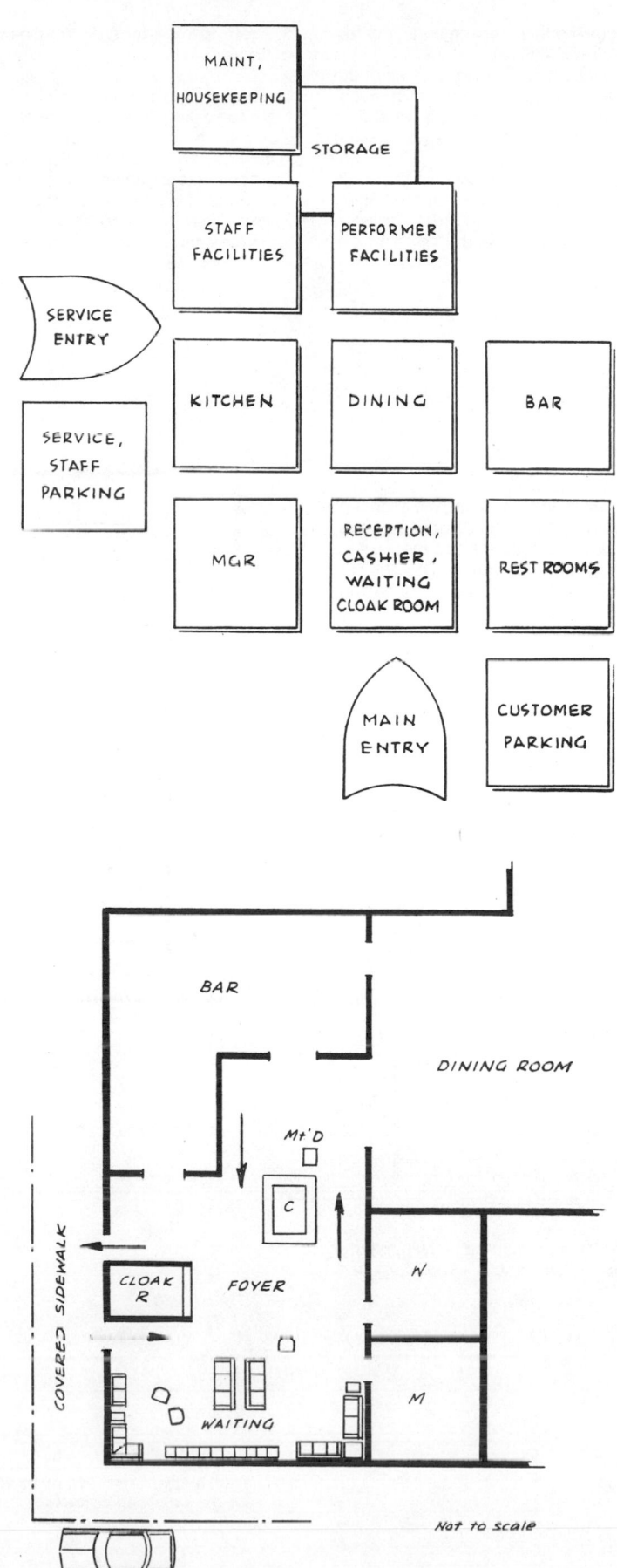

without bumping their serving carts into the backs of diners' chairs.

The plan should provide pleasant views for the diners of the outdoors or of attractive decorative elements in the room. Diners should not have to look at the kitchen or serving activities.

Special attention should be given to the ventilation and acoustic aspects of the dining room. No one likes to sit in a draft or listen to the clatter of dishes in the kitchen.

### 2. From the Waiter's or Waitress's Point of View

Convenience to and from the kitchen and/or strategically placed service centers is, of course, of prime concern to waiters, waitresses, and busboys. Pathways to and among diners' tables are equally important in terms of clearance, directness, and sufficient illumination so that restaurant employees can see where they are going.

Particularly important to waiters and waitresses is elimination of "blind spots," i.e., possible traffic conflicts at doorways, partition thresholds, stairs, etc.

*Note:* Avoid the temptation to create visibility problems for both patrons and employees, i.e., extremely low-level lighting that makes it impossible for a patron to read the menu and for a waiter or waitress to write out an intelligible order.

### Dining-Room Examples

1. Initial view of possible booths and tables; good clearances and organization enhance traffic flow; service center or centers reduce number of steps that waiters or waitresses must take; and booths provide reasonable privacy and reduce general noise level.

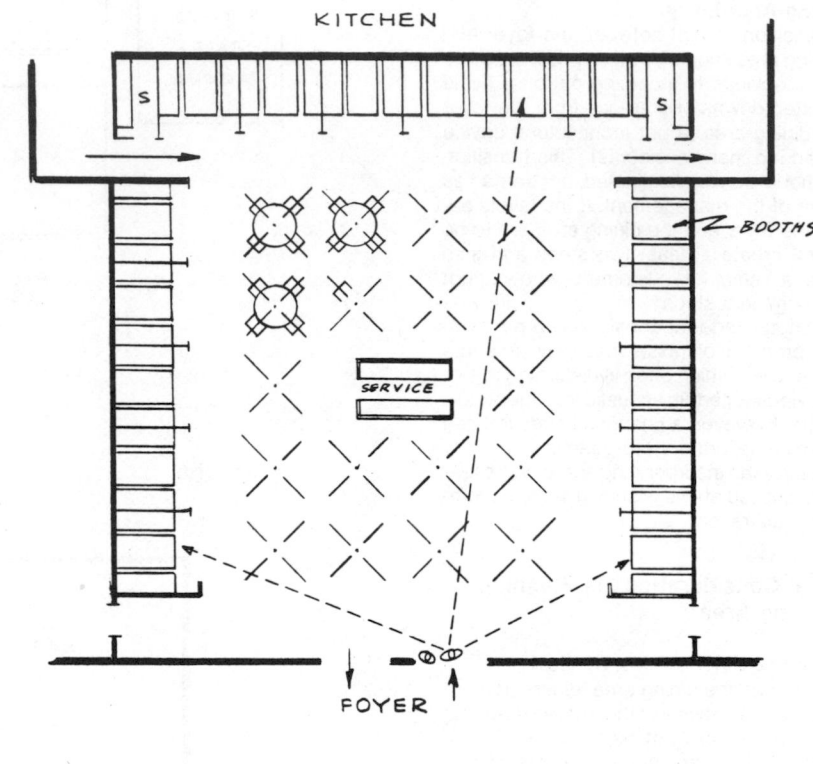

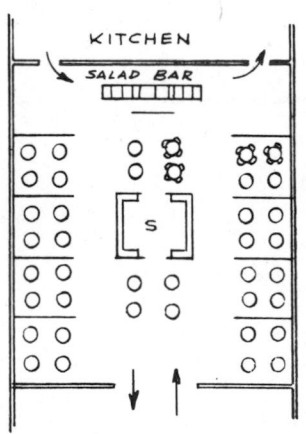

2. Clusters of tables separated by partitions minimize noise transfer even more.

3. Wide angle of view can be provided even for the fast-food cafeteria.

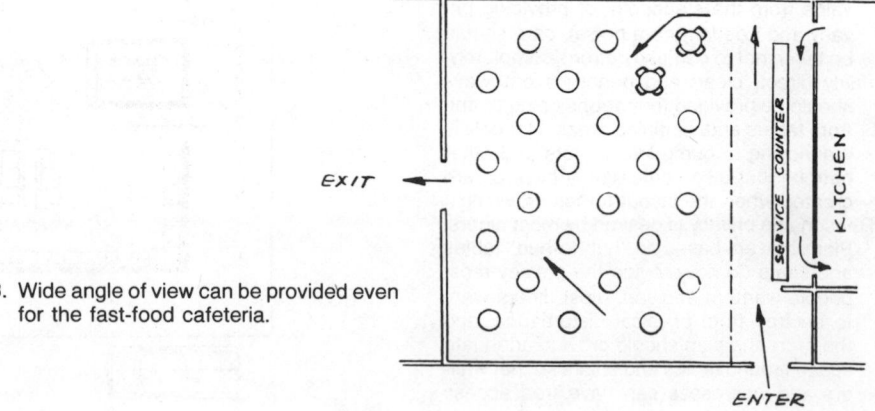

## Calculating Dining-Area Requirements

From the patron's point of view, clearance is perhaps the most important factor. Although the particular furnishings selected also must be considered in arriving at the final clearance and area dimensions, consider especially the points illustrated in the accompanying sketches:

1. In order for people to sit comfortably at a table, there must be clearance for their feet. This suggests a minimum 40-in (102-cm) diameter for a table to seat four people.

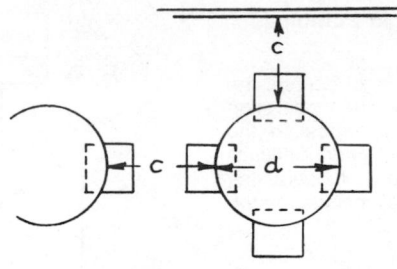

*Clearance between tables*

2. In order for people sitting back to back to extricate themselves from the table, there must be at least 47 in (119 cm) between tables. This can be reduced by about 5 in (13 cm) if the chairs are arranged so that they are not back to back; however, this creates a clearance problem for the waiter or waitress who is trying to pass between chairs.

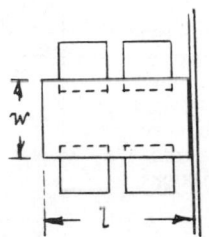

*Side by side clearance*

3. So that people sitting side by side do not have to touch each other, provide at least 24 in (61 cm) between seat centers. The table should be at least 30 in (76 cm) wide to provide space for dishes and clearance for feet.

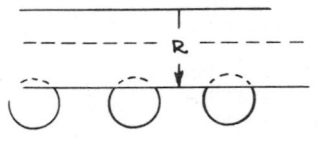

*Across-counter reach*

4. A counter top must obviously be deep enough to accommodate the patron's dishes, but not so deep that the waiter or waitress cannot reach across it to clean off the counter. Patrons at a counter must be seated further apart so that they can get in and out (the centerline separation should be at least 27 in, 69 cm).

*Desirable viewing angle*

5. Eye heights between patron and waiter behind the counter should be relatively equal, but people (total strangers) sitting across the counter from each other should not have to be stared at. Separate them by at least 15 ft (4.5 m) or more.

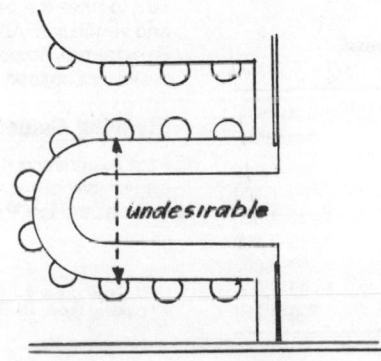

*undesirable*

## General Considerations in Planning the Restaurant Kitchen

Individual worker efficiency is, of course, the most important aspect of kitchen operation and therefore should be uppermost in the mind of the planner and designer. Difficulty arises because of the variety of operations that all must be efficient. They include the following:

1. Receiving (all supplies)
2. Storage and retrieval (dry foodstuffs, produce, meats, refrigerated and frozen products, etc.)
3. Food preparation
   a. Precooking and dispensing stage (cutting, trimming, etc.)
   b. Cooking, baking, frying, etc. (involves acquisition of utensils and use of ranges, ovens, fryers, and other equipment)
   c. Dispensing (involves acquisition of dishes, flatware, crystal, trays, etc., and critical interaction between cooks and waiters and waitresses)
4. Cleanup, disposal, and storage of cleaned elements
   a. Garbage disposal
   b. Preparation of dirty dishes for washing and disposal of soiled linens
   c. Dishwashing and storage of clean dishes, utensils, flatware, cutlery, crystal, and special food-preparation equipment
5. Housekeeping (maintaining the kitchen facilities, i.e., cleaning counters, worktables, griddles, ovens, ranges, floors, venting hoods, etc.)

## Safety

Consult local ordinances regarding the required safety features for commercial kitchens. In addition, however, consider particularly the following hazards, since they imply special planning considerations:

1. Splatter of hot grease and potential ignition of same
2. Breakage of crockery and glass and attendant cuts
3. Falls due to liquid spilled on floors
4. Cuts suffered while using special meat-cutting equipment
5. Staff collision injuries due to crowding, swinging doors, and blind intersections

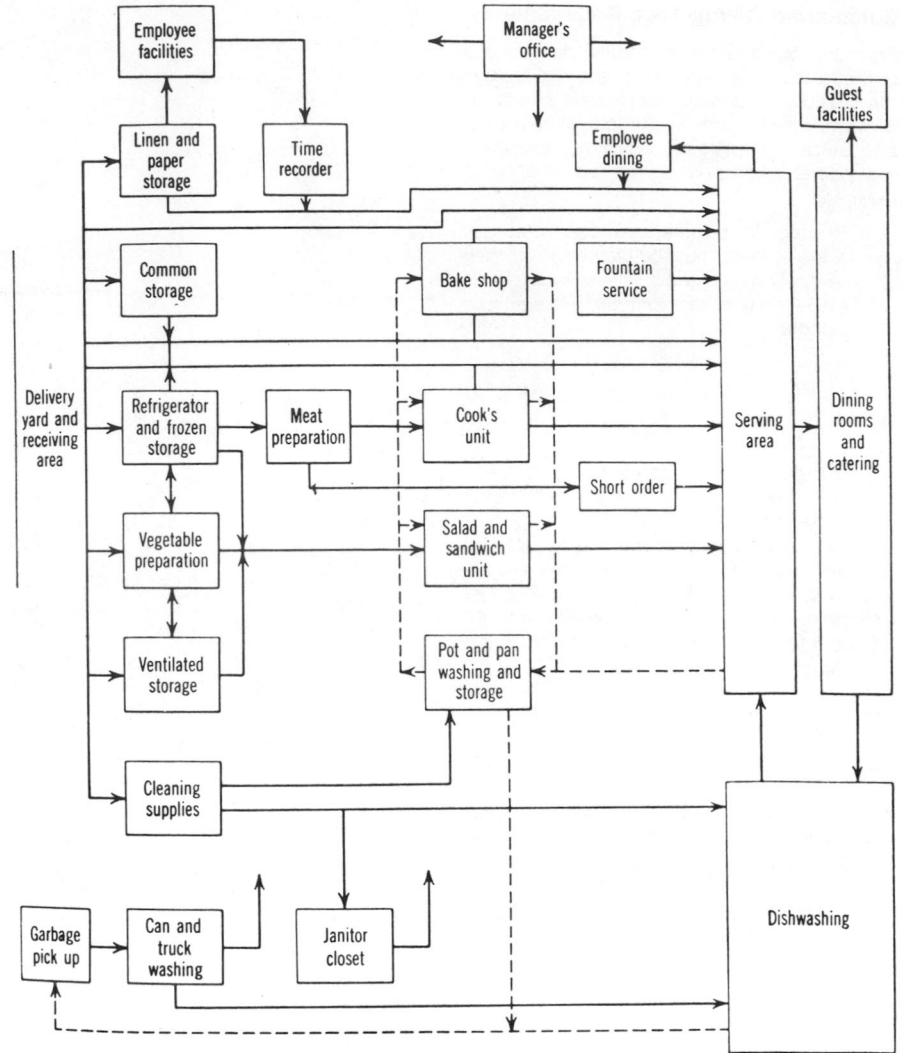

### Habitability

Not only should the health of kitchen workers be given high priority in planning the environmental control systems, but also the importance of worker efficiency without undue stress must be considered an important factor in reducing accidents. Because of the heat typically generated in the kitchen, it is important to have the best possible air conditioning and ventilation. Above-average lighting quality is extremely important because of the time pressures imposed on kitchen staff members.

### Planning Guidelines

Past experience dictates space needs based on number of meals served, as shown in the table at the left.[27]

### Variation in Space Needs in Relation to Numbers Served

| Meal load | Square feet per meal | Variation in square feet |
|---|---|---|
| 100–200 | 5.00 | 500–1,000 |
| 200–400 | 4.00 | 800–1,600 |
| 400–800 | 3.50 | 1,400–2,800 |
| 800–1,300 | 3.00 | 2,400–3,900 |
| 1,300–2,000 | 2.50 | 3,250–5,000 |
| 2,000–3,000 | 2.00 | 4,000–6,000 |
| 3,000–5,000 | 1.85 | 5,500–9,250 |

[27]J. De Chiara and J. H. Callender, *Time-Saver Standards for Building Types*, McGraw-Hill Book Company, New York, 1973.

**Kitchen Work-Flow Layout Example[28]**

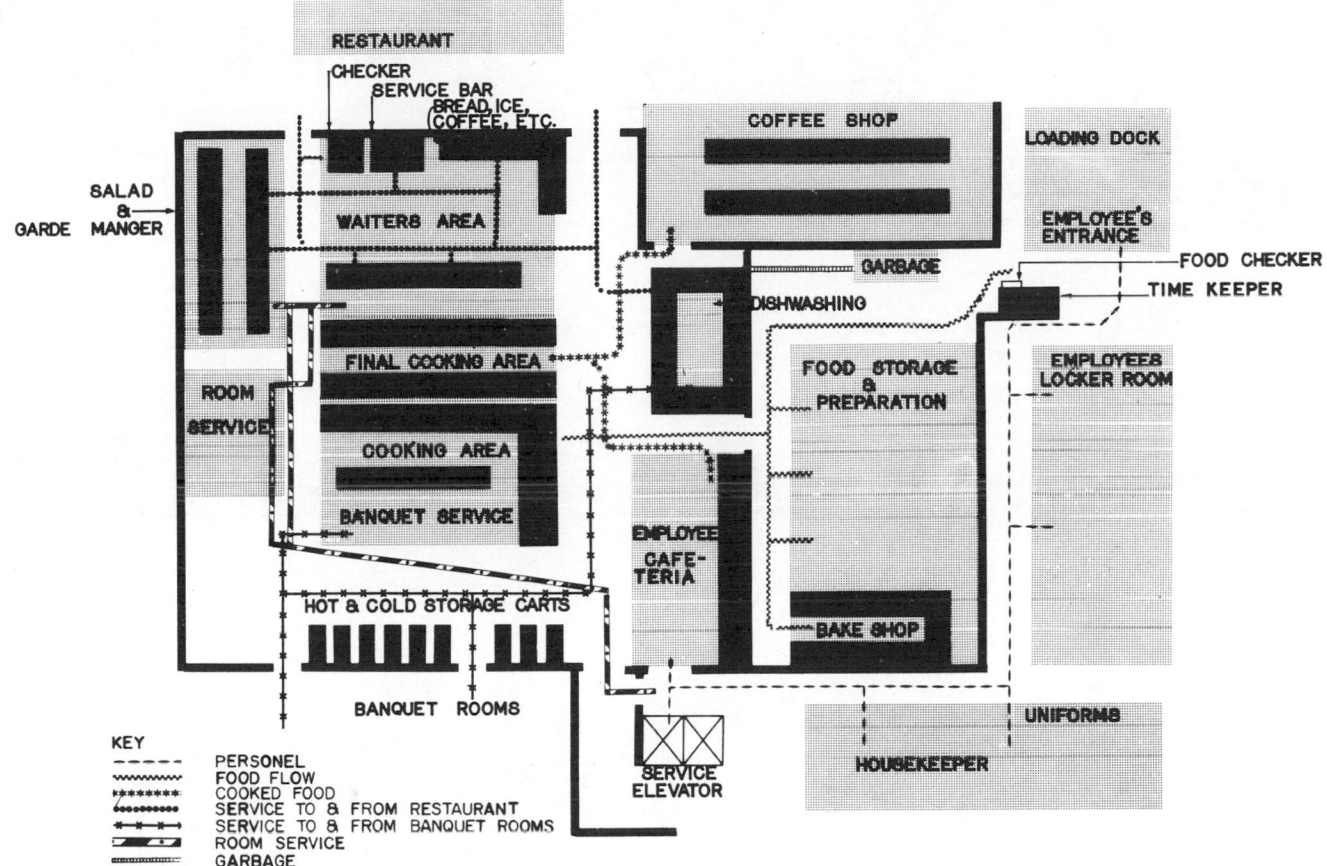

KEY

- – – – – PERSONEL
- ~~~~~ FOOD FLOW
- ✱✱✱✱✱ COOKED FOOD
- ●●●●●● SERVICE TO & FROM RESTAURANT
- ×+×+× SERVICE TO & FROM BANQUET ROOMS
- ▰▰▰ ROOM SERVICE
- ∿∿∿∿ GARBAGE

**Flow diagram of service areas.**

## Additional Support Facilities for Restaurant and Kitchen Employees

1. Rest rooms
2. Locker and lounge areas
3. Showers
4. Staff dining area
5. Staff parking
6. Manager's office

*Note:* The kitchen and employee areas should be considered off limits as far as customers are concerned. Therefore, the initial plan should provide suitable physical controls to prevent customers from purposely or inadvertently walking into these areas without special invitation.

[28]J. De Chiara and J. H. Callender, *Time-Saver Standards for Building Types,* McGraw Hill Book Company, New York, 1973.

# FACTORIES

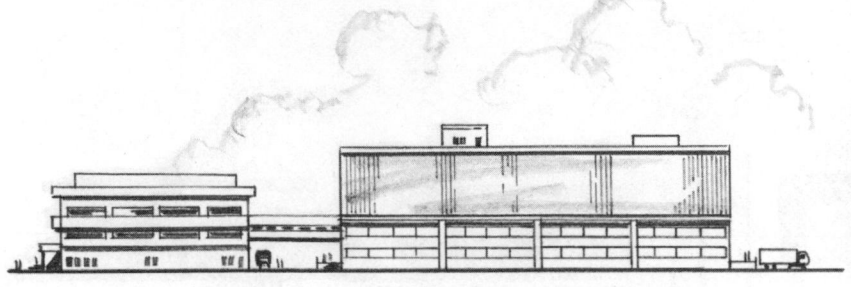

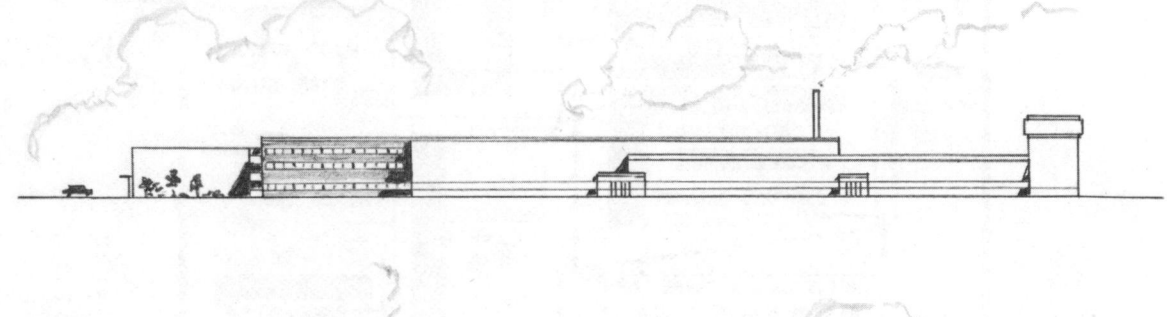

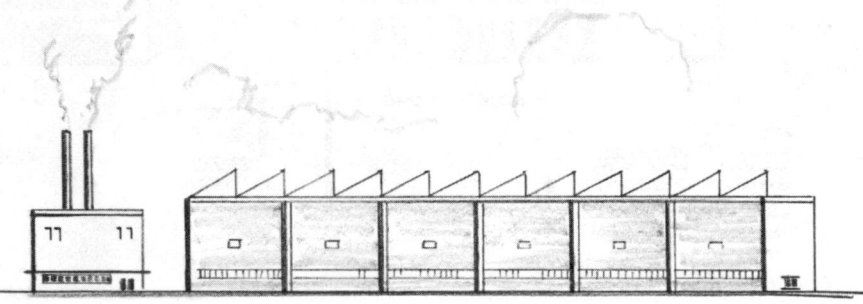

## FACTORIES AND WAREHOUSES

### Factory and Warehousing Facilities Planning Model

Among the more important planning considerations are the following (see model on page 123):

1. Flexibility for reorganizing spaces to accommodate product manufacturing shifts and expansion of plant facilities
2. Hazard monitoring and control
3. Efficient work flow
4. A nonstressful working environment for workers
5. Security
6. Minimal pollution of the surrounding environment

The primary objective in planning factory and warehousing facilities should be to expedite the flow of materials into, through, and out of the facility. However, people are also involved, and they either enhance or degrade this objective, depending on how well their welfare and task interface designs are formulated in the

initial planning stage. The following general functional requirements are common to almost all factories and should be addressed both individually and collectively during the planning stage:

1. Materials, equipment, and components receiving
2. Checking, sorting, and preliminary storage
3. Materials preparation
4. Manufacturing (cutting, forming, treating, drilling, etc.)
5. Assembly (electrical, mechanical, welding, soldering, etc.)
6. Finishing (anodizing, painting, etc.)
7. Inspection and testing
8. Packing
9. Predelivery storage
10. Delivery
11. Waste disposal
12. Management and supervision
13. Inventory control

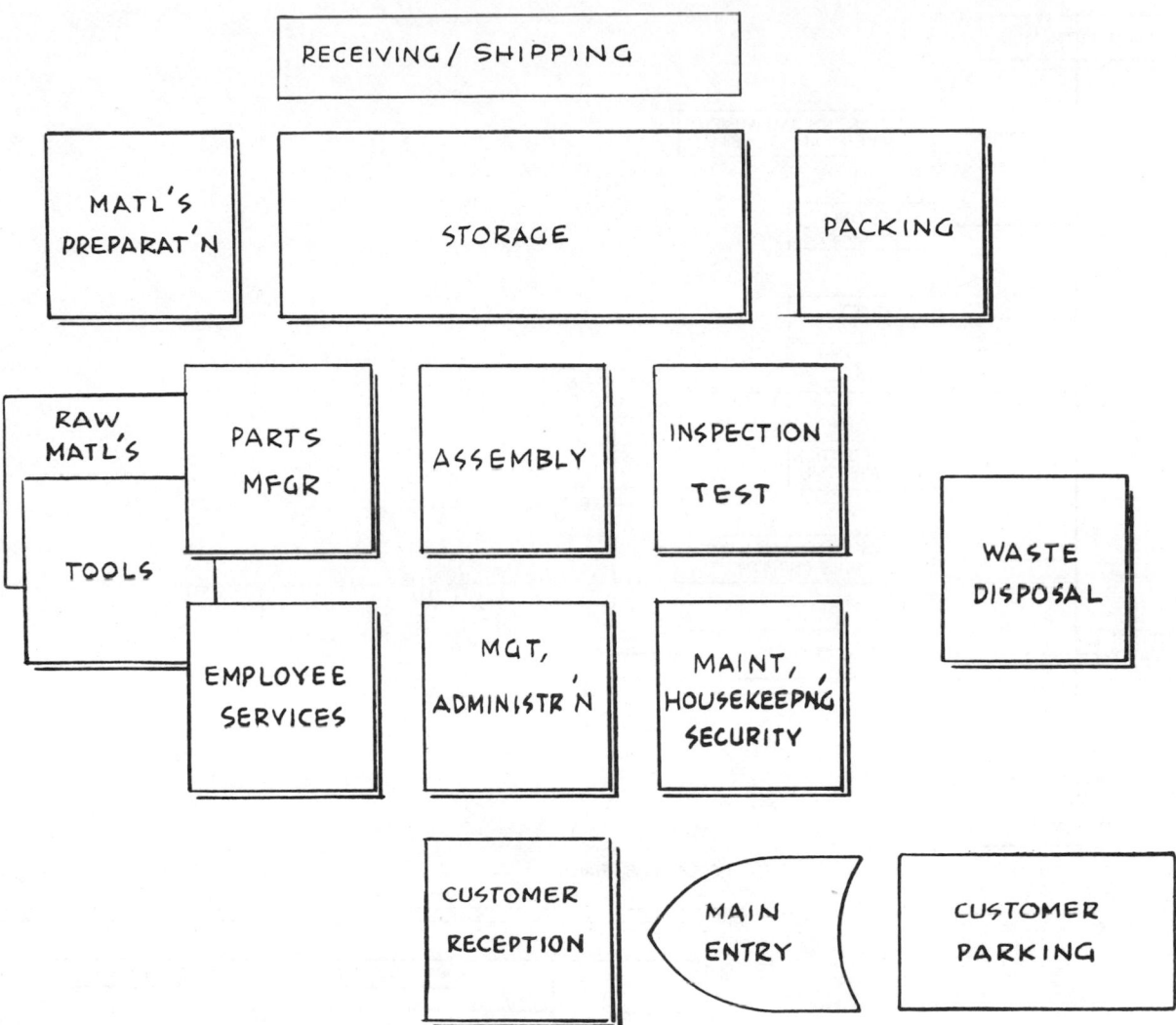

14. Communications
15. Maintenance and housekeeping
16. Employee services
17. Security

## Special Human Factors Problem Areas

As indicated by the development of the Occupational Safety and Health Administration of the U.S. government (OSHA), personal health and safety are a prime concern. Whether the factory manufactures heavy equipment, hazardous materials, or toys, the planner should anticipate as many of the hidden as well as the human-precipitated hazards as possible and should plan the facility to prevent these hazards from occurring or to mitigate their possible effects.

Equally important, however, is the planning of an efficient working environment so that the workers can do the best job of which they are capable.

## Checklist for Identifying Typical Factory Hazards[29]

1. Toxic fumes, dusts, and other substances
2. Chemical contaminants
3. Explosives and explosive material
4. Scalding liquids and steam
5. Falling objects, flying particles, and sparks
6. Moving machinery, vehicles and parts and swinging cables
7. Cutting tools, machines, and sharp materials and parts
8. Presses and crushers
9. High-voltage electrical equipment and lines
10. Radiation
11. Irregular floors and walkways
12. High noise levels
13. Falls from scaffolds, ladders, and lifts

[29]See also the section on safety in Chapter 3.

## Checklist for Identifying Worker Efficiency Considerations

1. Visual: Provide proper illumination for each task.
2. Auditory: Provide protection from ear damage, communication interference, and undue annoyance and psychological stress.
3. Biomechanical: Ensure that human-facility interface features are designed to be used within human mobility and strength limits.[30]
4. Cognitive: Ensure that the physical features of the facility are compatible with typical human expectancies and mental capacities for comprehension.
5. Energy consumption: Create a plan that conserves human energy.
6. Working environment: Provide an environment that will allow workers to operate within their most productive range.

[30]Special attention should be given to the removal of architectural barriers for the handicapped worker during the initial planning stage.

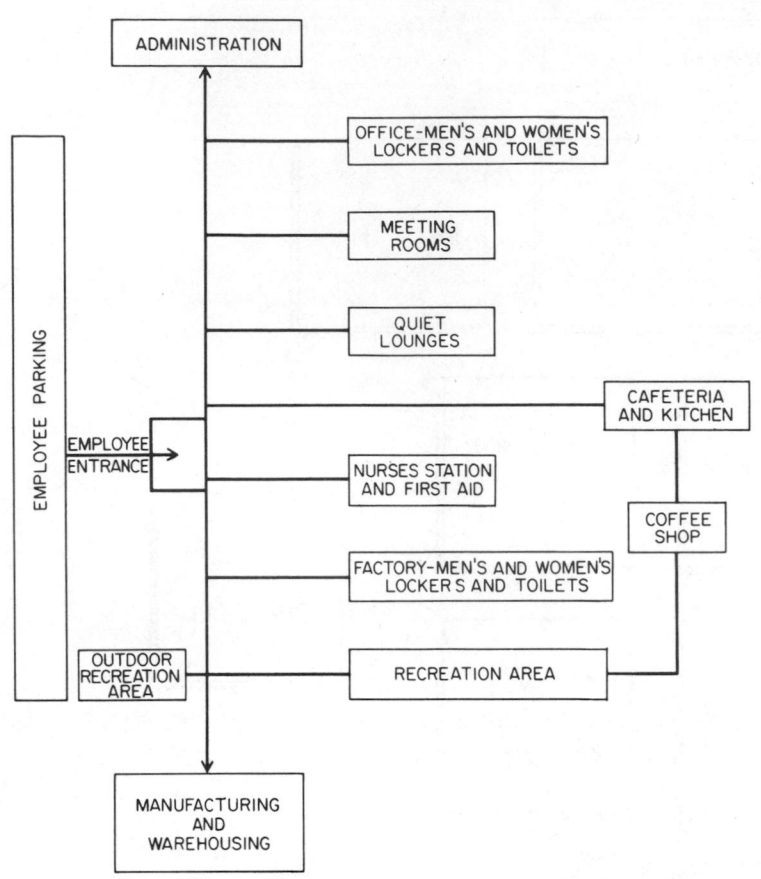

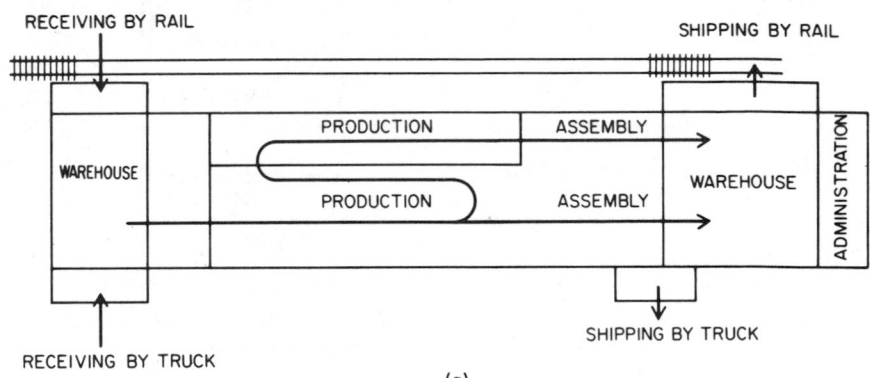

(a)

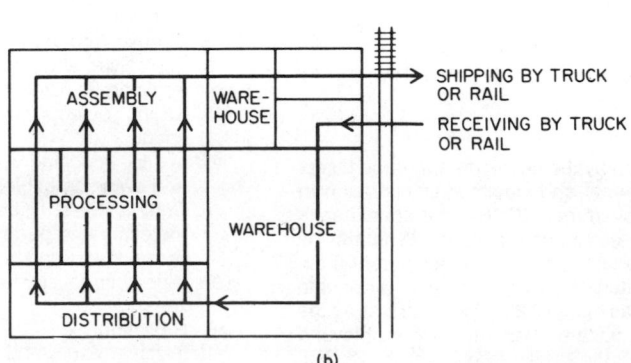

(b)

**(a) Layout by product. (b) Layout by process.**

From J. De Chiara and J. H. Callender, *Time-Saver Standards for Building Types,* McGraw-Hill Book Company, 1973.

## Minimum Toilet Fixture Requirements (New York State Labor Code)

| No. of MEN | Water Closets | Urinals | No. of WOMEN | Water Closets | No. MEN or WOMEN | Wash Basins |
|---|---|---|---|---|---|---|
| 1–9 | 1 | 0 | 1–15 | 1 | 1 20 | 1 |
| 10–15 | 1 | 1 | 16–35 | 2 | 21 40 | 2 |
| 16–40 | 2 | 1 | 36–55 | 3 | 41 60 | 3 |
| 41–55 | 2 | 2 | 56–80 | 4 | 61 80 | 4 |
| 56–80 | 3 | 2 | 81–110 | 5 | 81 100 | 5 |
| 81–100 | 4 | 2 | 111–150 | 6 | 101 125 | 6 |
| 101–150 | 4 | 3 | 151–190 | 7 | 126 150 | 7 |
| 151–160 | 5 | 3 | 191–240 | 8 | 151–175 | 8 |
| 161–190 | 5 | 4 | 241–270 | 9 | 176 200 | 9 |
| 191–220 | 6 | 4 | 271–300 | 10 | 201 225 | 10 |
| 221–270 | 6 | 5 | 301–330 | 11 | 226 250 | 11 |
| 271–280 | 7 | 5 | 331–360 | 12 | 251 275 | 12 |
| 281–300 | 7 | 6 | 361–390 | 13 | 276 300 | 13 |
| 301–340 | 8 | 6 | 391–420 | 14 | 301 325 | 14 |
| 341 360 | 8 | 7 | 421–450 | 15 | 326 350 | 15 |
| 361 390 | 9 | 7 | 451 480 | 16 | 351 375 | 16 |
| 391 400 | 10 | 7 | 481 510 | 17 | 376 400 | 17 |
| 401 450 | 10 | 8 | 511 540 | 18 | 401 425 | 18 |
| 451 460 | 11 | 8 | 541 570 | 19 | 426 450 | 19 |
| 461 480 | 11 | 9 | 571 600 | 20 | 451 475 | 20 |
| 481–520 | 12 | 9 | 601 630 | 21 | 476 500 | 21 |
| 521–540 | 12 | 10 | 631–660 | 22 | 501 525 | 22 |
| 541–570 | 13 | 10 | 661–690 | 23 | 526 550 | 23 |
| 571–580 | 14 | 10 | 691–720 | 24 | 551 575 | 24 |
| 581 630 | 14 | 11 | 721 750 | 25 | 576 600 | 25 |
| 631 640 | 15 | 11 | 751 780 | 26 | 601 625 | 26 |
| 641–660 | 15 | 12 | 781 810 | 27 | 626 650 | 27 |
| 661–700 | 16 | 12 | 811–840 | 28 | 651 675 | 28 |
| 701–720 | 16 | 13 | 841 870 | 29 | 676 700 | 29 |
| 721–750 | 17 | 13 | 871 900 | 30 | 701 725 | 30 |
| 751 760 | 18 | 13 | 901 930 | 31 | 726 750 | 31 |
| 761 810 | 18 | 14 | 931 960 | 32 | 751 775 | 32 |
| 811 820 | 19 | 14 | 961 990 | 33 | 776 800 | 33 |
| 821 840 | 19 | 15 | 991 1020 | 34 | 801 825 | 34 |
| 841 880 | 20 | 15 | | | 826 850 | 35 |
| 881–900 | 20 | 16 | | | 851 875 | 36 |
| 901–930 | 21 | 16 | | | 876 900 | 37 |
| 931–940 | 22 | 16 | | | 901 925 | 38 |
| 941 990 | 22 | 17 | | | 926 950 | 39 |
| 991–1000 | 23 | 17 | | | 951 975 | 40 |
| | | | | | 976 1000 | 41 |

## WASH FOUNTAINS REQUIRED

| Number of Fixtures | Persons Accommodated By: | | | |
|---|---|---|---|---|
| | 54" CIRCULAR (8 each) | 54" SEMI-CIRCULAR (4 each) | 36" CIRCULAR (5 each) | 36" SEMI-CIRCULAR (3 each) |
| 1 | 1 175 | 1 80 | 1 100 | 1 60 |
| 2 | 176–375 | 81 175 | 101 225 | 61 125 |
| 3 | 376 575 | 176–275 | 226–350 | 126 200 |
| 4 | 576–775 | 276–375 | 351 475 | 201 275 |
| 5 | 776–975 | 376–475 | 476 600 | 276 350 |
| 6 | 976 1175 | 476 575 | 601 725 | 351 425 |
| 7 | | 576 675 | 726 850 | 426 500 |
| 8 | | 676 775 | 851 975 | 501 575 |
| 9 | | 776 875 | 976 1100 | 576 650 |
| 10 | | 876 975 | | 651 725 |
| 11 | | 976–1075 | | 726 800 |
| 12 | | | | 801 875 |
| 13 | | | | 876 950 |
| 14 | | | | 951–1025 |

## Employee Facilities

Both the quantity and the quality of the product depend not only on the sequence, precision, and efficiency of the factories, tools, and machines but on the proficiency, pride, and fitness—both mental and physical—of the personnel. The development of factory design in recent years has become more and more concerned with creature comforts for the employees.

The facilities should be near the work space, so that no time is lost getting back and forth, but they should be sufficiently insulated from the sights and sounds of the work area itself so that a real change of scene is provided. If a pleasant outside view is available, it should obviously be used.

A clear distinction should be made between quiet lounging places and recreation and cafeteria areas. The problems are interesting, the solutions may be various, but the reigning criteria seem to be constant—cheerfulness, comfort, and durability.

The areas in this category include the following:

Cafeteria and kitchen
Coffee lounges
Recreation areas (indoor and outdoor)
Quiet lounges
Factory men's and women's lockers and toilets
Office men's and women's lockers and toilets
Meeting rooms
First Aid and nurses station

From J. De Chiara and J. H. Callender, *Time-Saver Standards for Building Types*, McGraw-Hill Book Company, 1973.

## Offices

The office facilities may include any or all
of the following:

General office
Message center
Billing office
Cashier
Telephone room
Foreman's office
Office manager
Terminal manager
Operations manager
Salesmen's room
Record room
Heater room
Central checking
Drivers' locker room
Transportation department
Dormitory
Cafeteria
Drivers' ready room

## Other Facilities

Maintenance shop
Fueling area (near shop)
Weighing area
Truck and trailer parking area (two parking
  spaces per dock stall recommended)
Employee and visitor parking area

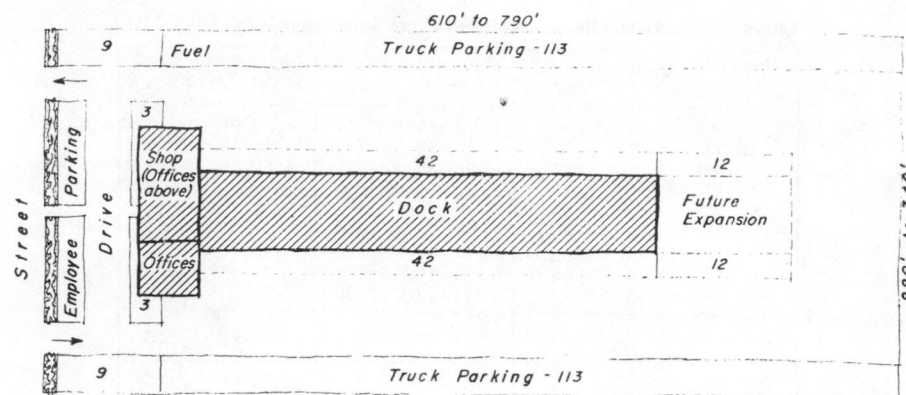

Truck terminal with short side to street.

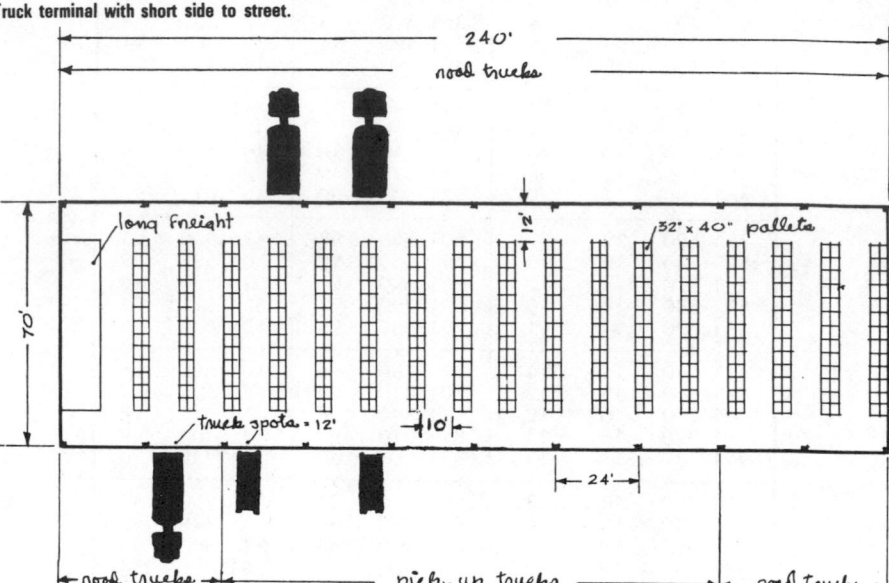

Truck terminal dock plan for fork-lift truck and pallet storage.

## Minimal Parking Space or Dock
## Approach Length and Width (Width — 12 ft)

| Overall length of tractor-trailer, feet | Apron length, feet | Dock approach length, feet |
|---|---|---|
| 40 | 43 | 83 |
| 45 | 49 | 94 |
| 50 | 57 | 107 |
| 55 | 62 | 117 |
| 60 | 69 | 129 |

## Percent of Grade for Material
## Handling Equipment

| Type of equipment | Allowable percent of grade* |
|---|---|
| Powered handtrucks . . . . . . . . . . | 3 |
| Powered platform trucks. . . . . . . . | 7 |
| Low-lift pallet or skid trucks . . . . . . | 10 |
| Electric fork trucks . . . . . . . . . . . | 10 |
| Gas fork trucks . . . . . . . . . . . . . | 15 |

* Contact manufacturer and check manufacturer's specifications
before operating beyond allowable percent of grade.

Most standard truck dockboard lengths
range from 6 to 10 ft. For most applications,
dockboards should be 6 ft wide. Use 7-ft wide
dockboard for loading or unloading unit loads
with fork truck.

8. *Provide area for access to trucks.* A
minimum area measured inside the plant from
the edge of the dock should be kept clear and
unobstructed for the movement of freight and
materials handling equipment. The depth of the
area must allow for maneuverability of materials handling equipment in and out of vehicles
and for two-way cross traffic behind the dock.

From J. De Chiara and J. H. Callender, *Time-Saver Standards for Building Types,* McGraw-Hill Book Company, 1973.

## Calculation of Storage Space (Area Utilization)

Gross
warehouse area = inside total square footage
of warehouse

   Net
storage area = actual area occupied by
inventory, not including
aisles plus space for empty
pallets plus shipping and
receiving areas plus
allowance for "honey-
combing" plus special
inventory (inspection, etc.)

  Interference = irregularities due to
columns, odd corners, etc.

Miscellaneous = space for offices,
equipment, etc.

Gross
warehouse area = net storage area plus aisles
plus interference plus
miscellaneous

Rule of thumb (for general package warehousing):

$$\frac{\text{gross warehouse area}}{\text{net storage area}} = \frac{3}{2}$$

or Net storage area + (50% net storage area)
= gross warehouse area

This rule is accurate for average warehouses,
but actual analysis of the layout is recommended.

"Honeycombing" is a warehouse term used
when space is not fully occupied because of
partial withdrawal of inventory. Maximum
honeycombing factors are in the range of 75
to 90 percent of maximum capacity, depending
upon the activity, number, and quantity of
items stored.

**Straight-Line Flow or Assembly-Line Principle** Straight-
line flow is inherently efficient and usually is
adopted in warehouses adjacent to production
areas. Conveyors and pallet systems illustrate
typical methods of efficient handling.

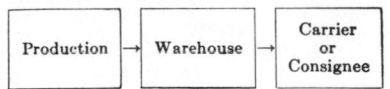

Order pick lines in a grocery warehouse are
characterized by the assembly-line principle.

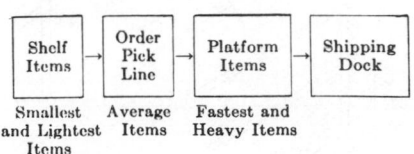

**Shipping and Receiving Areas** The receiving area
of a warehouse should be located adjacent to
incoming rail or truck facilities and as convenient as possible to the storage area. The
receiving dock is usually separated from the
shipping area if possible to minimize cross
traffic and possible confusion. The number of
unloading positions required is dependent
upon the volume of receipts or the maximum
number of cars or trucks spotted at the same
time. The light weight of portable aluminum
or magnesium dockboards is desirable when
power equipment is not available for positioning units of conventional steel construction.

Weather protection at the unloading positions permits continuous handling operations.
Loading platforms located outside the warehouse building can be designed for one-way or
two-way traffic where required.

The proper control, checking, and sorting of
inbound materials is important for the prompt
and efficient delivery of outbound shipments.
The size of the receiving area is determined
by the analysis of the temporary storage lag
needed to perform the necessary inbound
handling and inventory control operations.

A shipping area (or dock) receives materials
for outbound shipment after selection and
transfer from storage. The preassembly of
orders according to plan requires sufficient
room to perform packing, packaging, or preparation operations prior to shipment.

The size of the shipping area is dependent
on the makeup time of filling orders and the
quantity of simultaneous loading operations
during peak periods.

Utilizing centrally located short-lot and bin-storage operation, coordinated by use of dragline conveyors

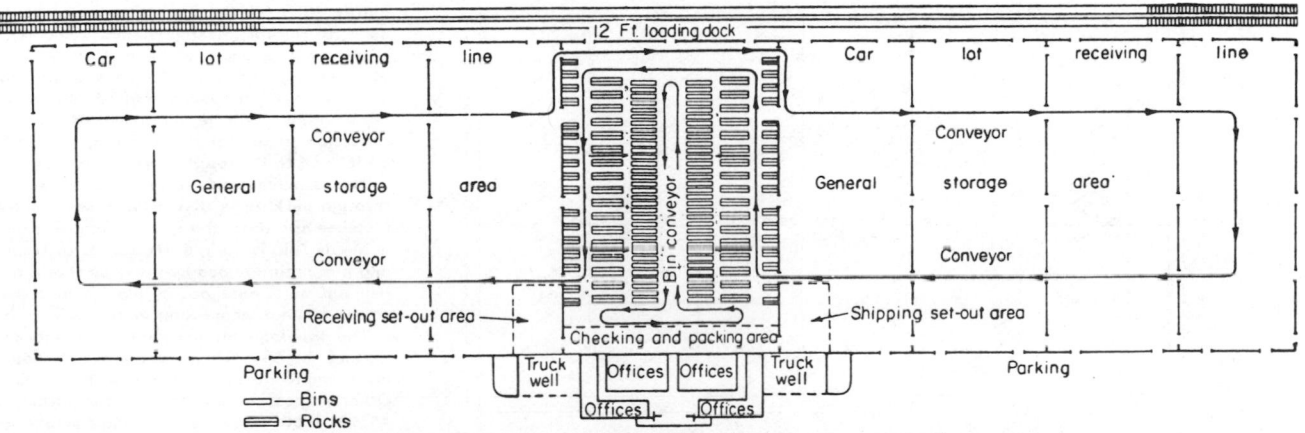

This bin layout emphasizes a direct-flow replenishment and stock selection operation

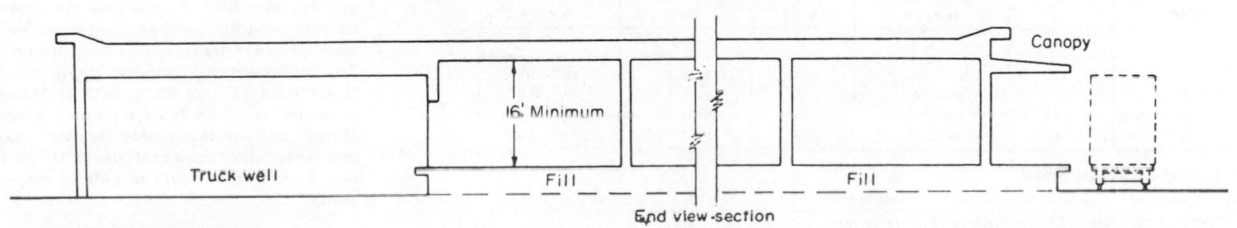

**Typical large warehouse layout.**

From J. De Chiara and J. H. Callender, *Time-Saver Standards for Building Types*, McGraw-Hill Book Company, 1973.

### Hypothetical Relationship of Parking Area Requirements to Location

| Location | No. of employees at peak shift overlap | Percent as drivers or auto psgrs. | Number of autos to be parked* | Approx. site, sq ft | Area, acres |
|---|---|---|---|---|---|
| Urban | 1,600 | 60 | 740 | 222 | 5.0 |
| Suburban | 1,600 | 80 | 990 | 297 | 6.8 |
| Rural | 1,600 | 95 | 1,180 | 354 | 8.0 |

\* Assuming car occupancy of 1.3 persons per car.

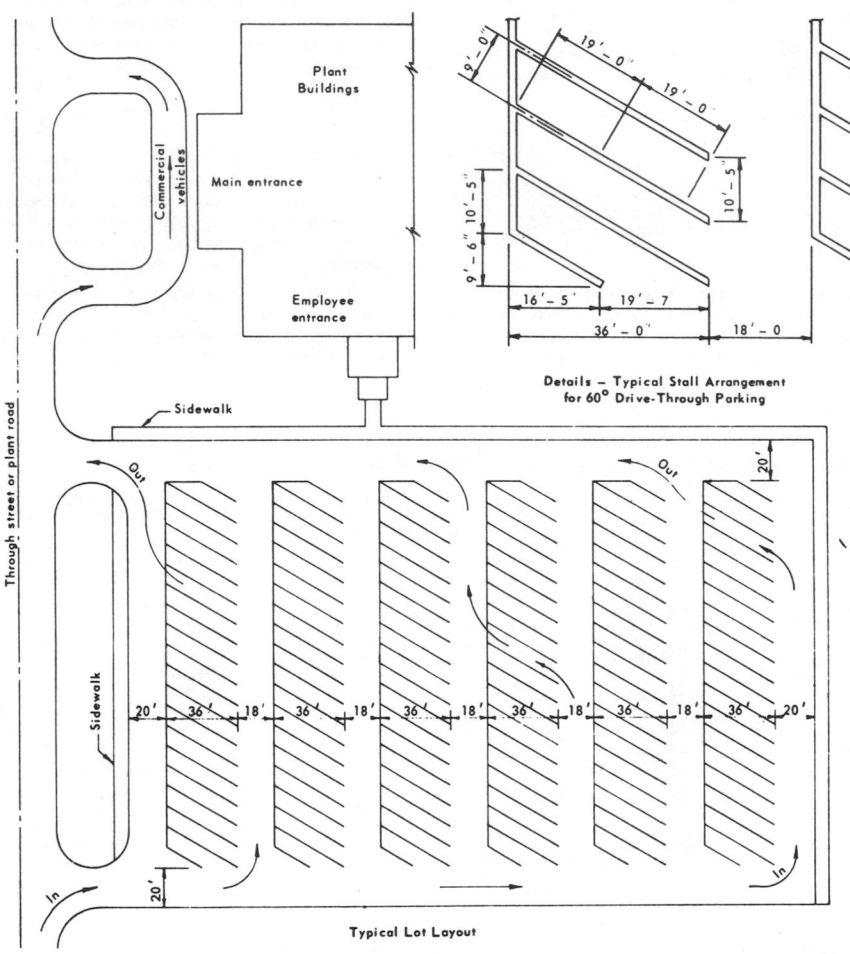

Details – Typical Stall Arrangement for 60° Drive-Through Parking

Typical Lot Layout

### DESIRABLE STALL AND AISLE DIMENSIONS FOR DRIVE-THROUGH PARKING

| Angle of Parking | Width of Stall | Depth of Stall Perpendicular to Aisle | Width of Aisle | Unit Parking Depth | Width of Stall Parallel to Aisle |
|---|---|---|---|---|---|
| α | W | L | A | UPD | W |
| 90 | 10' | 38' – 0" | 24' | 62' – 0" | 10' – 0' |
| 90 | 9' | 38' – 0" | 26' | 64' – 0" | 9' – 0' |
| 60 | 9' | 36' – 0" | 18' | 54' – 0" | 10' – 5' |
| 53 | 9' | 35' – 10" | 18' | 53' – 10' | 11' – 3' |

**Fig.1. Drive-through lot layout.**

*Parking Facilities for Industrial Plants,* Institure of Traffic Engineers, Washington, D.C., 1969.

From J. De Chiara and J. H. Callender, *Time-Saver Standards for Building Types,* McGraw-Hill Book Company, 1973.

**Stall Arrangements** Decisions about the choice of angle and the layout of aisles must be based on individual site conditions. The placement and number of entrances and exits, and the site shape and contour are the major controls. At large plants, blocks of parking by groups of three to five hundred cars are preferable to larger aggregations. Pedestrian-vehicle conflicts can be reduced, and assigned parking for different shifts and employment groups can be better controlled, through the use of such relatively small blocks.

The following general practices are desirable: use natural grades to facilitate drainage; provide for counterclockwise traffic aisle flow, since left turns are easier than right turns for drivers; have parked vehicles face downhill rather than uphill, to allow for stalled vehicles or winter weather conditions.

This report includes layout details for only one type of parking. Figure 1 illustrates a stall arrangement and an aisle design that have not been widely published—the drive-through double stall pattern, usable in either 90° or acute angle parking layouts.

In general, angle parking is preferred for large industrial parking facilities. First, properly designed angle parking can employ space as effectively as right-angle parking. Second, it virtually forces one-way movements, thereby simplifying control, reducing conflicts, and ensuring that daily parking practices conform to the established design. Third, it provides for easier turning movements into and out of stalls.

Drive-through angle parking design offers the further advantages of minimizing backing out of stalls and directing all aisle travel in the same direction. It conserves space more effectively than other angle parking designs. Typically, the angled drive-through layout requires 36 ft for the double stall and an 18-ft aisle (to permit passing stalled vehicles), for a unit parking depth of 54 ft. Compared with 90° parking, the space loss along the length of the bay—eight spaces in 500 ft according to the example—will be compensated for by the reduction in unit parking depth, from 62 or 64 ft to 54 ft, if enough bays can be used.

The disadvantage of this design of drive-through parking is that it increases the travel distance and time of a search pattern if the lot is nearly full. It also is imperative to keep the end circulation aisles two-way so that a driver will not be forced out of the lot in order to return to another parking aisle.

The drawing also gives dimensions for angle parking at 53° 8', an angle which has the layout convenience of being a 3-4-5 triangle. Other angles commonly used for parking are 45, 55, or 60°. However, any angle smaller than the 3-4-5 configuration tends to be wasteful of space, without offering any significant advantage.

Where two-way aisle flow may be desirable, as in visitor parking lots, 90° parking is more appropriate. Site dimensions sometimes may be such that 90° unit parking depths are most appropriate regardless of other circumstances. The minimum 90° parking depth reported to Committee 6T was 61 ft, with preferences expressed for 62-64 ft as desirable dimensions. When unit parking depths are less than desirable, shortened stall lines (10–15 ft long) may encourage drivers to pull all the way into stalls.

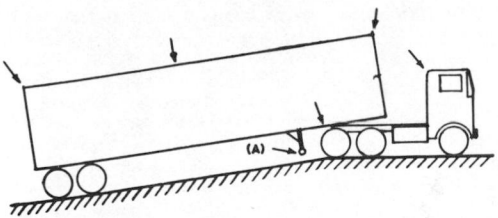

When designing for ramps, dips, or crowns in the terminal area, special care must be taken to provide clearance at the points indicated in the diagram. Actual dimensions must be obtained. Cab clearances are more critical when the combination is jackknifed. Landing gear height (A) may be as low as 10 in.

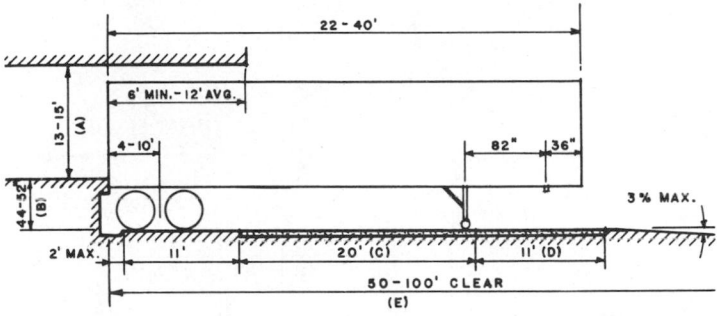

NOTES: (A) Should be at least 6 in. over legal height for level area, more for slope.

(B) Dock height, 48 to 52 in. for road trailers, 44 in. for city trucks.

(C) Concrete apron of the dimensions shown will accommodate trailers from 22 to 40 ft long.

(D) Additional slab length recommended to support tractor wheels.

(E) General rule for distance required: total length of tractor-trailer times 2.
Trailer width—8 ft
Trailer stall width—10 ft minimum, 12 ft recommended.

Recommended dimensions and clearances for truck loading docks.

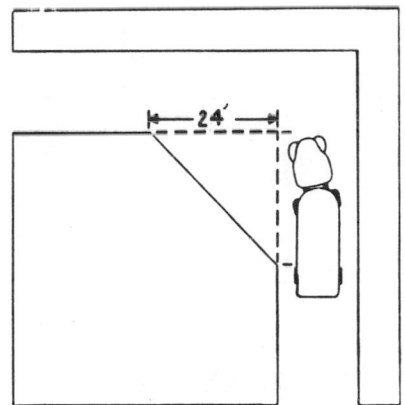

Turning clearance for driveway.

## Site

*Location:* In selecting a site, consider the following factors:

1. Proximity to pickups, deliveries, and connecting carriers

2. Accessibility to main traffic arteries

3. Obstructions such as bridges, underpasses, and railroad crossings

4. Zoning

5. Urban and regional plans; future growth pattern of city

6. Transportation facilities for employees

7. Utilities

*Grade:* Site should be approximately level: maximum slope 3 percent; minimum slope for drainage, 1 percent. Storm drains recommended 60 to 75 ft on centers, 100 ft maximum.

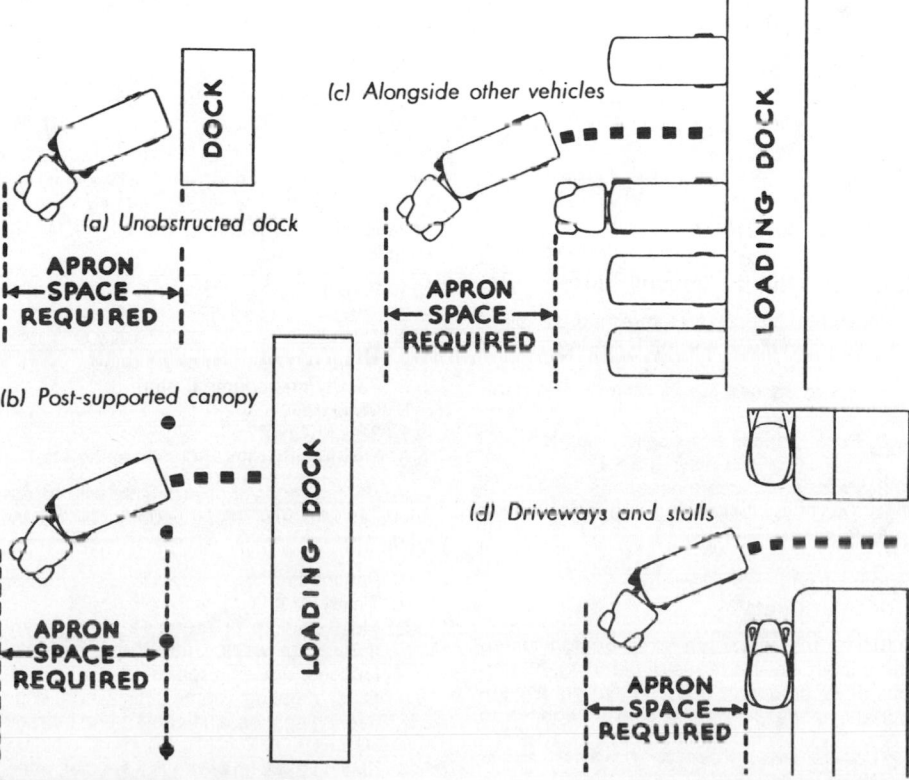

Apron space required for various conditions.

From J. De Chiara and J. H. Callender, *Time-Saver Standards for Building Types,* McGraw-Hill Book Company, 1973.

SPORTS

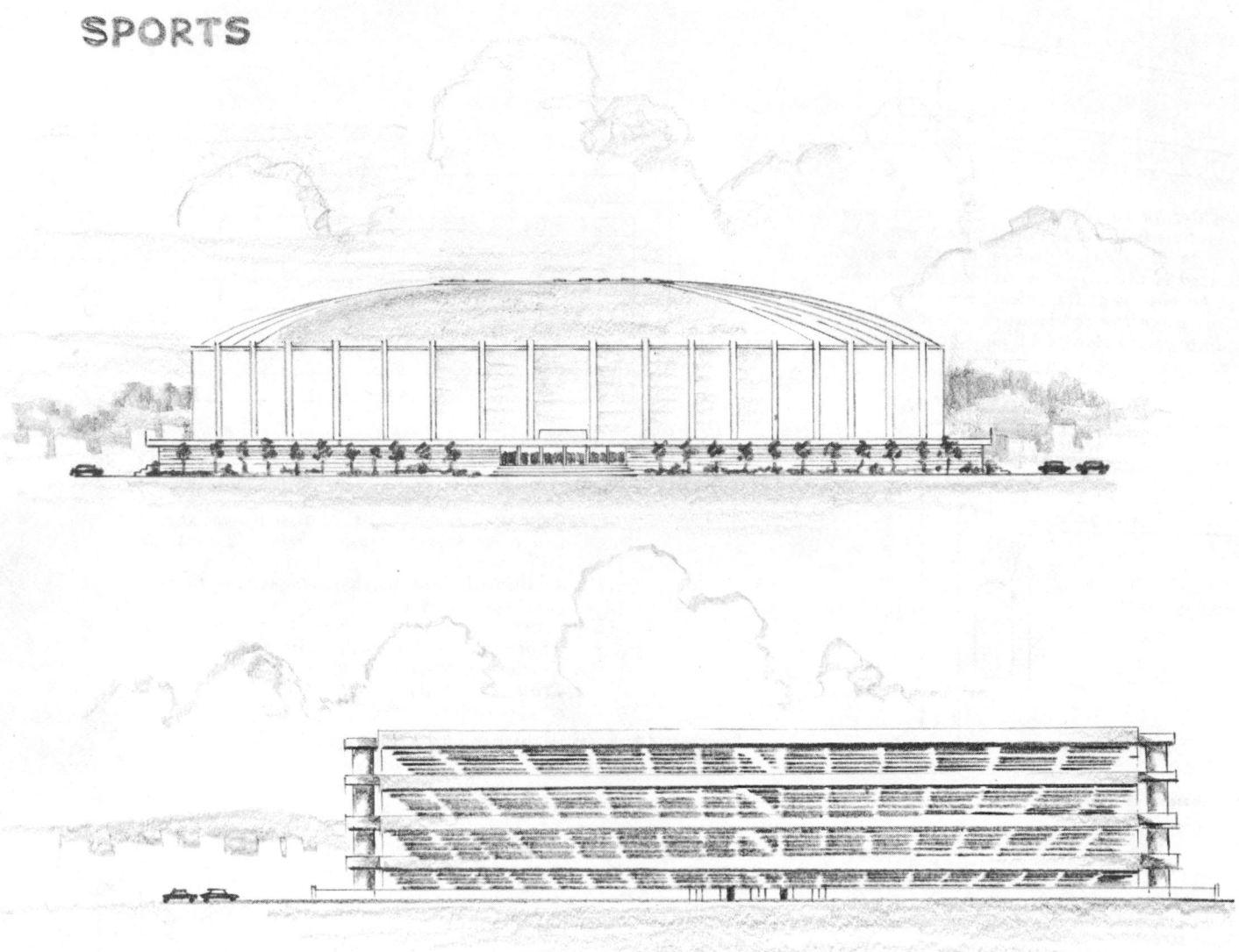

## SPORTS FACILITIES

### Sports Facilities Planning Model

As illustrated on page 131, the important planning considerations are the following:

1. Public access, traffic control, and parking
2. Public seating accessibility, comfort, and visibility of the playing area
3. Participant accommodations
4. Facility housekeeping and maintenance
5. Security

### Sports Facilities Planning Considerations[31]

Although the sports fan might be considered the primary user in terms of planning, numerous other groups of people have equally important needs. The major user categories to

[31]J. De Chiara and J. H. Callender, *Time-Saver Standards for Building Types,* McGraw-Hill Book Company, New York, 1973.

be addressed in planning the sports facility include:

1. Fans
2. Special guests (VIPs)
3. Officials
4. Athletes, coaches, and staffs
5. Facility management staff
6. Vendors
7. Security staff
8. Maintenance and housekeeping staff

Key subsystems that ordinarily will be associated with any major sporting facility include:

1. Parking lot
2. Ticket booths
3. Main stadium or arena structure, corridors, stairways, lifts, rest rooms, lounges, dressing rooms, press box, offices, meeting rooms, utility and engineering spaces, and vendor and storage spaces
4. Playing area (may require special, alternative facilities for different events, e.g.,

basketball, hockey, tennis, football, swimming, or track)

### Special Human Factors Problem Areas

1. Access to and from the facility parking area often is a major problem, not only for the sports fans, but also for the planner, who may not have control over related community thoroughfares and property. Failure to consider this problem early in the planning stage will invariably result in either considerable cost to modify existing roadways, or both expense and disgust on the part of fans when a poorly planned access creates long and time-consuming traffic jams. An important objective should be to create multiple approaches to reduce the traffic load at any one entry and to lay out the site-roadway configuration to prevent traffic conflicts near the site.
2. Parking sites for major facilities generally have to be quite large to accommodate

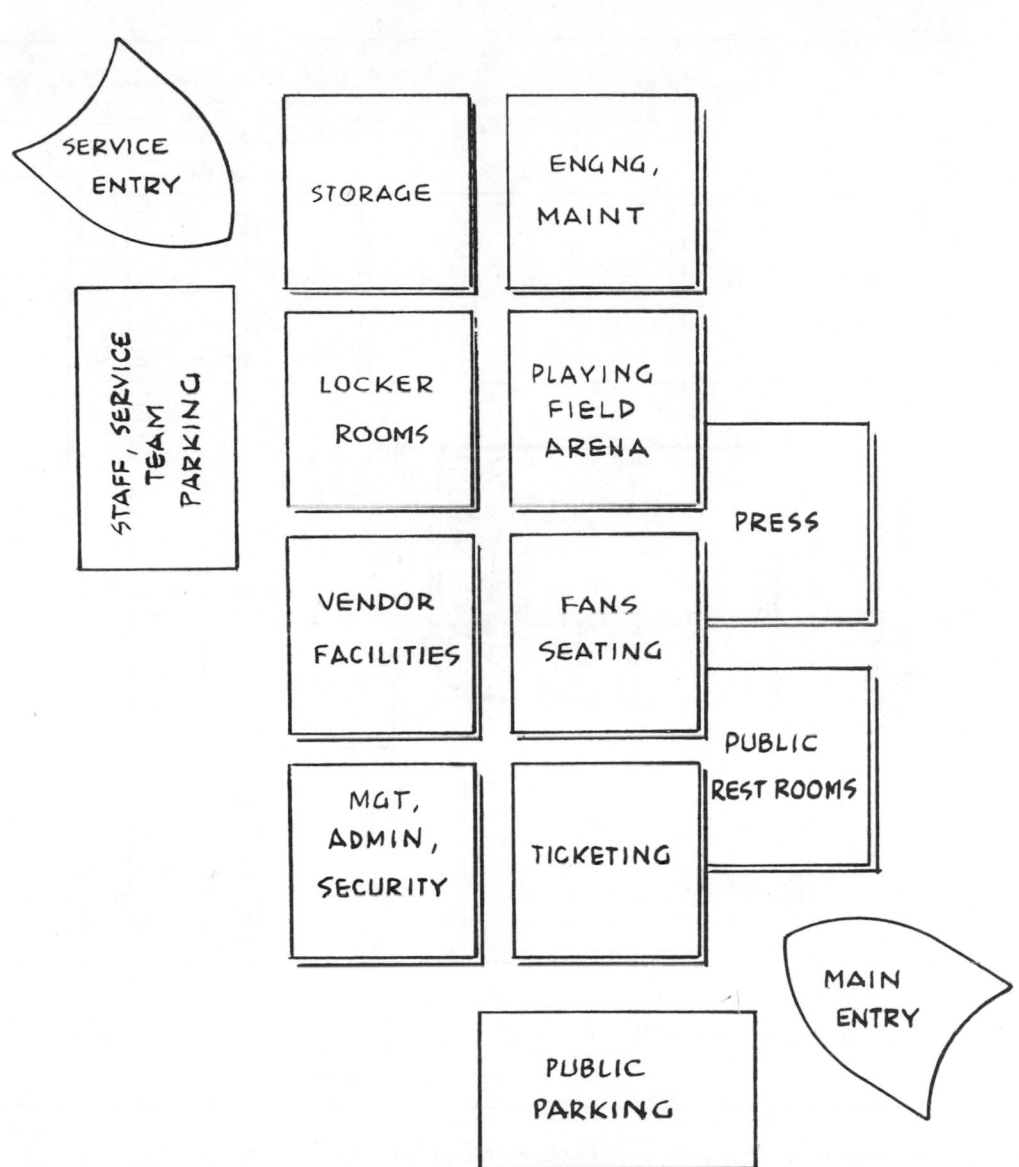

the typical vehicle load. If the parking pattern is not clear to drivers, considerable congestion can result, and many drivers will become lost out on the periphery. An objective should be to create a plan in which traffic flows logically and cannot be circumvented by the motorist who is anxious to get parked first.

3. Access from the parking area to the stadium or arena entrance and ticket windows is often extremely confusing to the new fan, who may have difficulty seeing the entrance from the point where his or her car is parked. An obvious objective, of course, should be to make these distances as short as practical. However, an equally important consideration is to provide obvious pedestrian pathways to the entrance (a convenience missing from most plans). Finally, the stadium, parking area, and pedestrian pathways should be well signed, not only so that

fans can find the entrance, but, equally as important, so that they can find their cars again. A good addition is to slope the parking area upward toward the stadium, which gives people who are out on the periphery a clear view of the facility and of any instructional signs that have been placed at various locations for their benefit.

4. Seat location can become confusing and difficult for the fan since the tendency is to make all corridors, ramps, stairs, and doorways look alike. Consider the problem of clear signing during the planning stage, since this forces one to see the relationship between building configuration and how to make it clear to fans what they are supposed to look for and do to find their proper location. Color coding is a useful technique to consider.

5. A good view of the playing area is, of course, very important to the fan. Early

consideration of viewpoints can eliminate the problem of the unfortunate fan who has to sit behind a pillar or who cannot see over the head of the person in front. Seating comfort can be provided by following prescribed criteria for good seat design. Too many planners fail to consider seat design and purchase seats purely on the basis of low cost and ease of maintenance (also important issues). In seeking effective seat-view configurations, avoid creating steep stairways; these can become hazardous, especially for the excited fan who insists on running up and down them from time to time.

*Note:* Special attention should be given to the handicapped fan who cannot use stairways or sit in general seating areas. Provide elevators, ramps, special wheelchair areas, and, of course, rest rooms designed for the handicapped.

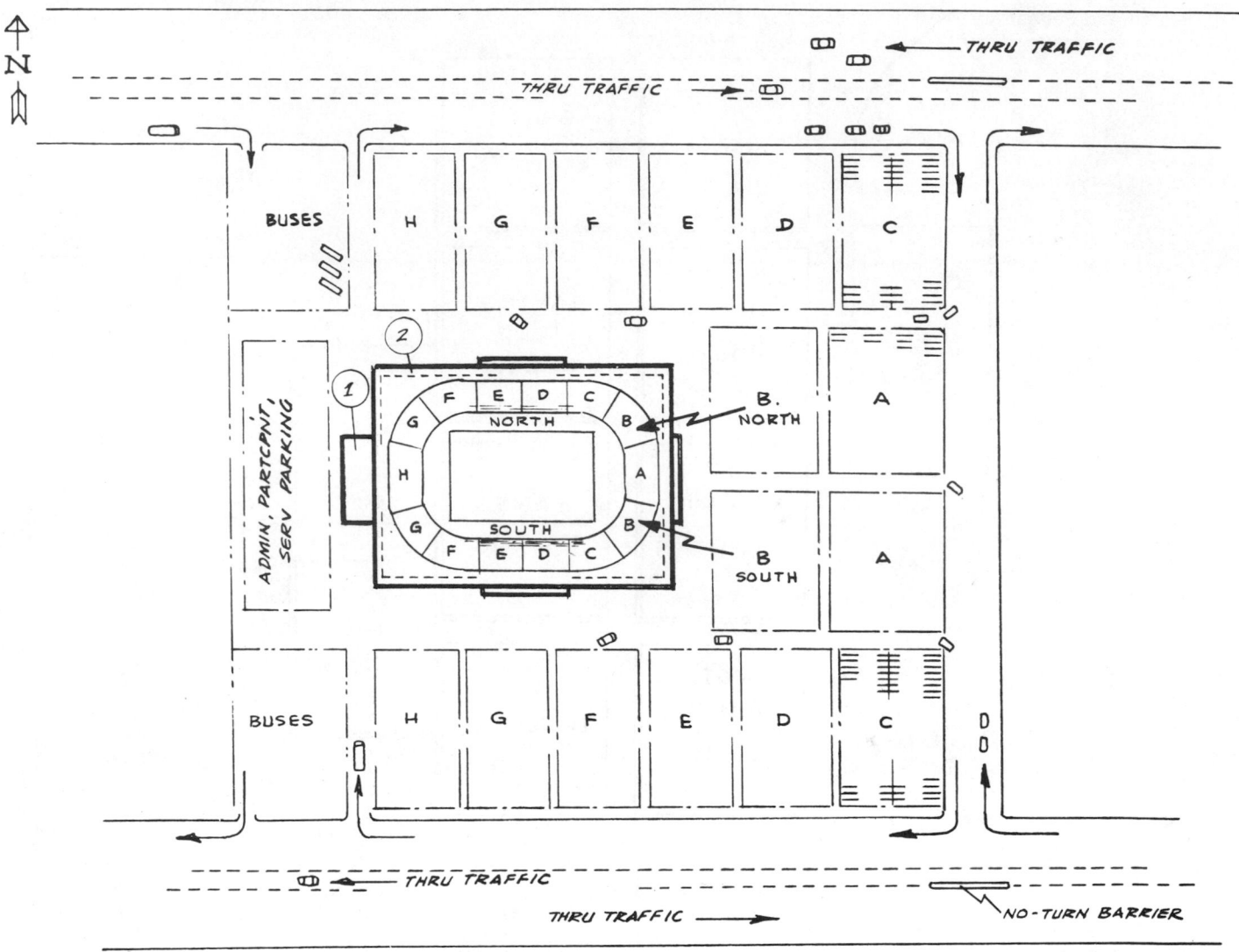

**Theoretical sports facility concept to illustrating traffic, parking, and seating identification principles.**

1—Administrative offices, service delivery, team quarters, etc.
2—Perimeter corridor follows outer rectilinear format to maintain user's north-south and left-right references.

Street traffic does not tie up because of crossovers.

Parking and aisle system lends itself to easy identification and marking.

Customer parking area designations are compatible with seating sections.

Basic stadium and arena shape is rectilinear and oriented with north-south reference; useful for seating identification system.

A-sections at main entrance keep fans oriented.

Parking for fans' private vehicles is separate from parking for buses, parking for administrative personnel, and other special parking.

## Vehicular and Pedestrian Flow Planning Example

The accompanying illustration shows how vehicular traffic can be controlled in terms of being directed to specific areas with minimum loss of control. In addition, specific pedestrian pathways (sidewalks) are provided to minimize the hazards created by fans who walk among parked and moving vehicles. Note that the vehicular-pedestrian geometry provides a natural flow to and from the stadium.

Temporary barriers can be placed at the points marked with an asterisk to prevent drivers from straying beyond the intended parking area being filled at a particular moment. Special traffic, such as press, participants, and vendors, can be routed on through these points at the discretion of parking-lot attendants.

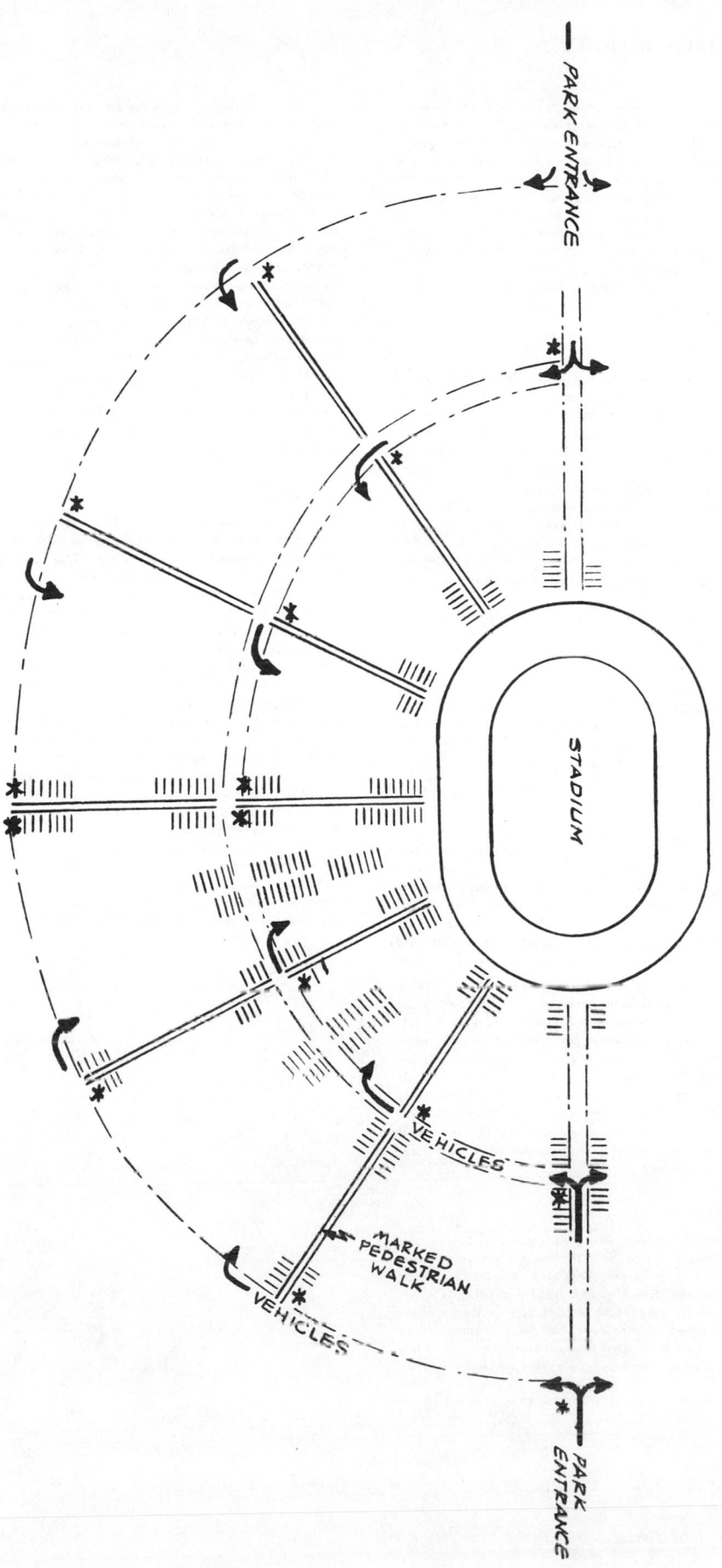

## SPECTATOR REQUIREMENTS

### Seating

Most arenas now being built are employing theater-type upholstered seats. The minimum recommended width is 19 in., and they should vary up to 23 in. center to center for the prime areas. In the wedge-shaped sections in curved rows, a mix of widths is usually used to make the ends of each row come out as flush as possible. Where risers are 5 in. or over, seat stanchions should be riser-mounted to facilitate cleaning. Seats should be self-rising with perforated acoustical treatment on the seat bottoms.

### Concessions

Concession stands for food, beverages, souvenirs, and coat checking should be provided at convenient locations in the promenade areas. Counter areas should be as long as practical and if possible recessed in alcoves to prevent backup of patrons into circulation spaces. Storage space should be provided immediately adjacent to each counter area so that food items can be restocked to the sales area during a game or performance without a trip to the central supply point. If it is contemplated by the arena management that an outside concession firm be brought in to run the operation, it should be selected as early as possible and participate in the planning process. If the firm is an experienced national operation, it will have strong points of view on counter locations, size, visibility, and utility requirements among others. If not built into the original building, the concessionaire's desires will likely prevail at a later date and unsightly and costly additions result.

### Toilet Rooms

Sets of men's and women's rest rooms should be provided at one or more locations on each public level. Their layout must provide for peak loads during the 15- to 20-minute intermission periods when hundreds of patrons will pass through each room. It is ideal if a one-way traffic flow can be developed with an in and out doorway separated by some distance. Within, the space should be divided with the water closets and urinals located near the entrance and the lavatories in a space near the exit. Also it is desirable if possible to design each toilet room so that half of the space can be closed off by some means during events of small attendance. This will save a good deal of operating cost for cleaning. Plumbing-line capacity should be studied carefully for peak use and generous pipe spaces with good access provided.

### Graphics

A good graphics and signing control program is important not only for an attractive appearance, but for controlling and expediting crowd movement. Signing can help establish a clear pattern of movement which can easily be comprehended by the patrons. Seat colors in the arena can be keyed to ticket colors to identify the various areas or categories of seating. This can be done on a horizontal basis with rings of seats changing color as they change from one price category to another. Or the arena can be divided into quadrants each with its own color key. In cases where the arena sits within a large parking field, this color system may even extend to the exterior and guide patrons to the proper entrance as they park and approach the building. Within the seating area, signs designating sections, rows, seats, etc., should be large, clear, and located in easily read places. Signs for rest rooms, concessions, telephones, etc., should also be of good size and clear and consistent in style. In the lobby ticketing area, space must be provided for coming attraction signs, current-event pricing, and seating plans for various event setups. It has proved successful also to have a scale model of the arena seating including colors and section identification within the ticket sales area to assist patrons with their ticket purchases.

### Scoreboard

Two basic types of scoreboards are in common use. The center-hung 4-sided type is one, wall-mounted single-faced the other. The central type is usually on a drop cable system which allows it to be lowered to the floor for maintenance. The central speaker cluster can also be combined with this type scoreboard, but it should be checked early whether the same suspension height is appropriate for both scoreboard visibility and sound distribution. When the wall-mounted type is used, at least two units will be required so that all spectators will have a proper view. Very often the building management will arrange for advertising display to be incorporated into the scoreboard design as a revenue-producing device. If so, the decision should come as early as possible, as it will have obvious effect on size and detailing. The boards, of whichever type, must have provisions for the major sports that are likely to use the arena and have a portable control console that can operate from several positions depending on the sport involved.

From J. De Chiara and J. H. Callender, *Time-Saver Standards for Building Types*, McGraw-Hill Book Company, 1973.

## OPERATING REQUIREMENTS

### Administrative Offices

Areas for the building manager, accounting, personnel, booking, publicity, and engineer are generally provided within the building. In addition, office space may be required for the various teams who use the building, whether they are only tenants or are owned by the arena owner. Additionally, office space should be available for use by shows booked into the arena for an extended period (circus, ice shows, etc.). Also, the owner of the arena, if it is a private venture, will usually require a suite of rooms including his office, private bath, and a conference/meeting room suitable for entertaining dignitaries. Food may be catered to this area from the central club kitchen; thus it should be within easy access. It is possible in some instances that a portion of the offices mentioned could be located in other space remote from the arena. This decision and a full program of office requirements should be developed at an early stage of the design/planning process.

### Ticketing Facilities

This area will vary depending upon the intended scope of events to be booked. However, in most situations, ticket booths will be required in the lobby area or an outer lobby. They should be accessible to the public during non-event periods without losing security to the remainder of the building. Madison Square Garden has 25 booths, the Forum, 20. Immediately to the rear of the booths should be a large ticket room for storage and sorting advance sale tickets. Also required will be a money room with vault, group sales office, ticket manager's office, and a work area for storing event posters and making up ticket pricing boards.

### Storage

Large bulk storage areas will be needed for a variety of uses. The temporary seating setups for the arena floor will require space to store both chairs and riser platforms. These are usually stacked on metal pipe racks as high as ceilings will permit and handled with forklift trucks. Space for storing the hockey dasher boards and glass, basketball floor and goals, and indoor track must also be provided. All of these should be so located relative to the arena floor as to minimize time and cost for the setting up of each event.

### Locker and Dressing Rooms

If the arena is the permanent home of two professional teams (hockey and basketball, for example), a pair of separate home team dressing rooms will be required (Fig. 12). As illustrated, the teams can share toilets, shower room, a training area, and the trainer's office. The hockey dressing area should be somewhat larger than that for the basketball dressing area because of larger team size and more cumbersome equipment. A pair of rooms for visiting teams somewhat smaller than the rooms for home teams, can be located adjacent to or nearby with home team rooms as shown. Several smaller dressing and interview rooms should be planned in this area. Some can be for individual use, others for four to six people, and each with appropriate toilet facilities. All these spaces should be located at arena floor level with convenient vomitory access to the playing floor. Public exiting traffic should be routed away from the dressing area corridors.

### Press Facilities

A press workroom with adjacent Teletype room and toilet should be located near the lower seating area. It is also desirable to include a lounge in this group with facilities to set up a small bar and food service from the main concession kitchen. A small photographer's work area and darkroom should also be provided at the arena floor level. Location of the press seating varies widely. Many arenas which have been built with elaborate press booths high above the floor have discovered them unused, reporters preferring to sit at courtside near the action. Radio and TV announcers, however, usually prefer to sit high for an overall view of the action. Booths for this purpose can be located over vomitory openings or suspended from the ceiling or balcony structure.

### Concession/Vendors Storage

Large bulk storage areas will be required for the concessionaires supplies of dry food goods, beverages, meat, general supplies, souvenirs and programs. This may include walk-in refrigerator space and cold rooms as specified by the operator. Also needed will be a concession manager's office, a security area for counting money and a vault. Ample vendors' stations will be needed at several points around the arena. They must be located within easy reach of the seating and be laid out to allow fast refill of the seat vendor's stock. Separated in-out doors are helpful.

### Employee Toilets/Lockers

As seen from the following space allocation summary, several categories of employee spaces will be necessary. General cleaning and maintenance help, ushers, and concession employees each need separate toilet/locker facilities. As local conditions might warrant, space may also be needed for security guards and parking lot employees. Definitive space needs for each group will depend upon a management analysis of the numbers of staff required.

## SPACE ALLOCATIONS

The following space allocations for the Forum in Inglewood, California, an arena of 18,42 seats, can serve as a planning guide and check list of required facilities:

| | Area, sq |
|---|---:|
| 1. Play floor surface | 26,900 |
| 2. Lobby promenade | 20,000 |
| 3. Concession stands | 2,500 |
| 4. Public toilets | 4,800 |
| 5. Home team lockers/toilets | 4,300 |
| 6. Visiting team lockers/toilets | 2,100 |
| 7. Dressing/interview rooms | 1,200 |
| 8. Press work area | 600 |
| 9. Darkroom | 150 |
| 10. Men employee toilets/lockers | 1,200 |
| 11. Women employee toilets/lockers | 700 |
| 12. Men ushers' toilets/lockers | 400 |
| 13. Women usherettes' toilets/lockers | 600 |
| 14. Men concession toilets/lockers | 350 |
| 15. Women concession toilets/lockers | 500 |
| 16. Truck dock | 4,500 |
| 17. Receiving area | 1,300 |
| 18. Storage — bulk | 8,800 |
| 19. Storage — concessions/vendors | 6,000 |
| 20. Storage — temporary seating | 6,600 |
| 21. Storage — dasher glass | 250 |
| 22. Ice machine | 250 |
| 23. Administrative offices | 9,000 |
| 24. Ticketing facilities | 7,000 |
| 25. Private club dining and kitchen | |
| 26. Pay telephones — 22 booths | |

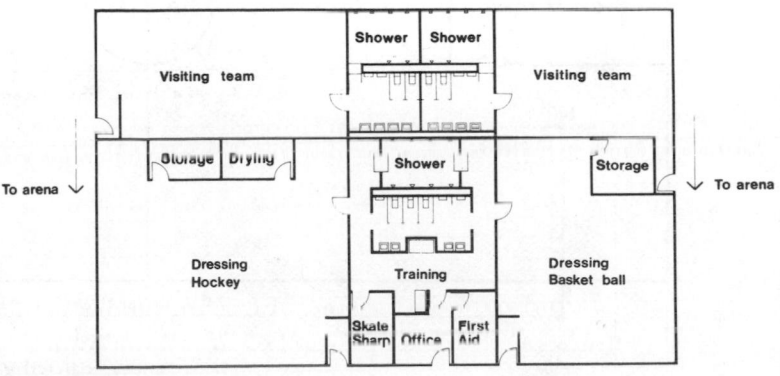

Dressing rooms.

From J. De Chiara and J. H. Callender, *Time-Saver Standards for Building Types,* McGraw Hill Book Company, 1070.

**Basic Playing Area Guidelines**

**BASKETBALL**

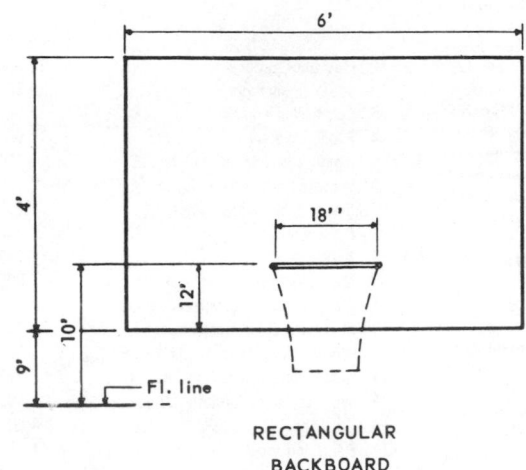

RECTANGULAR
BACKBOARD
AND GOAL

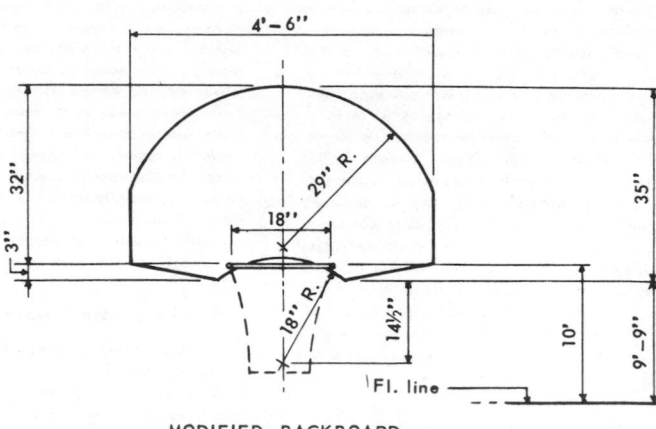

MODIFIED BACKBOARD

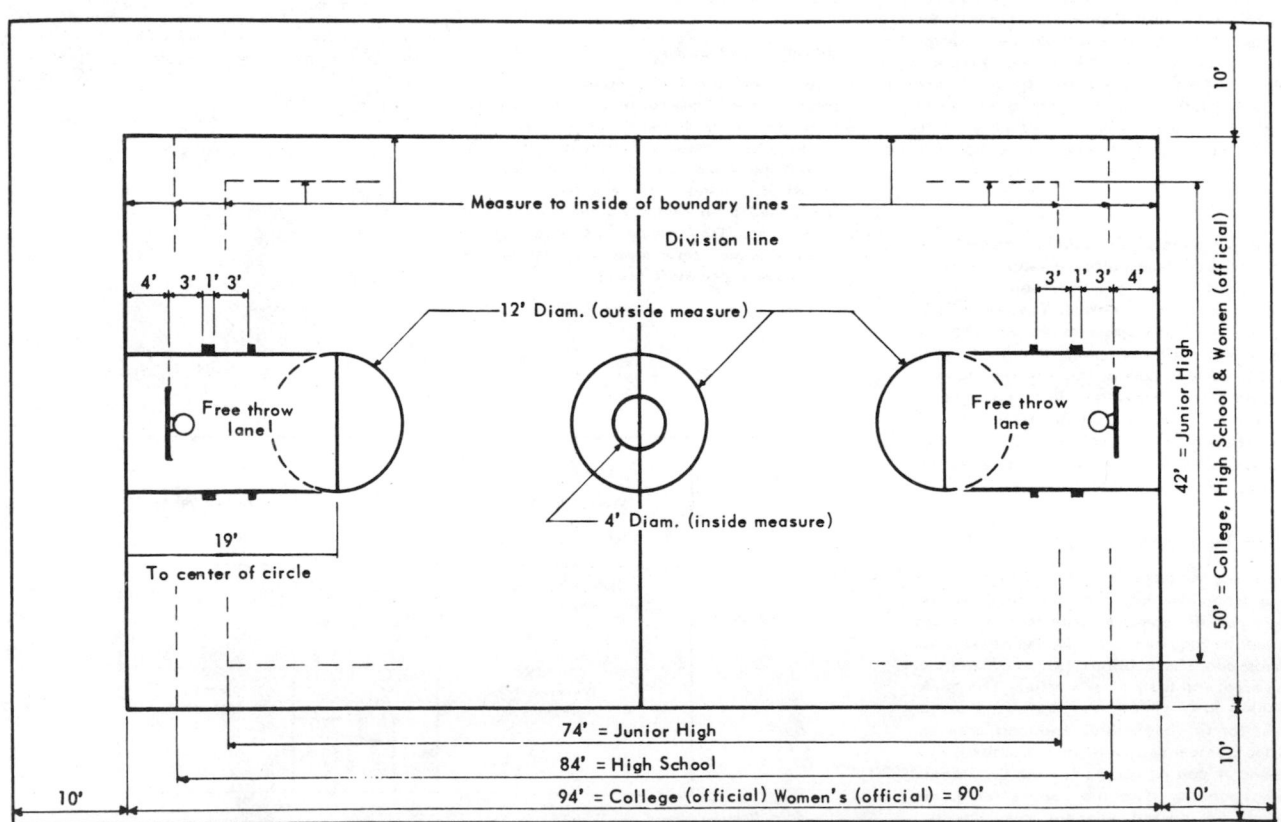

BASKETBALL COURT          All Lines 2'' wide — unless otherwise note

J. De Chiara and J. H. Callendar, *Time-Saver Standards for
Building Types*, McGraw-Hill Book Co., New York, 1973.

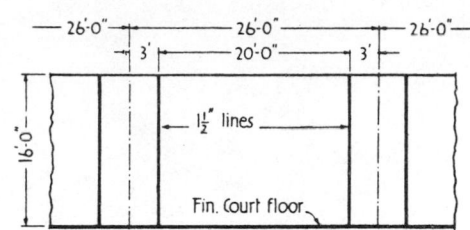

ELEVATION OF BACKSTOP

## HANDBALL

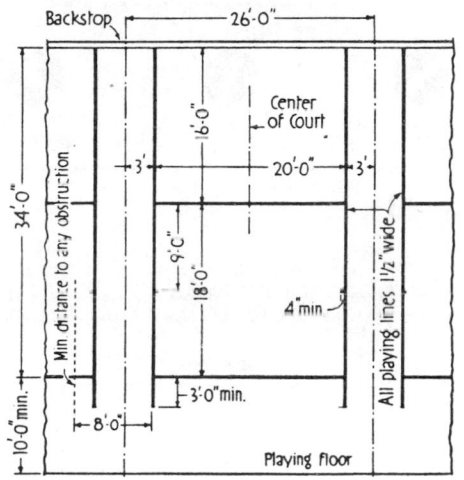

Playing floor

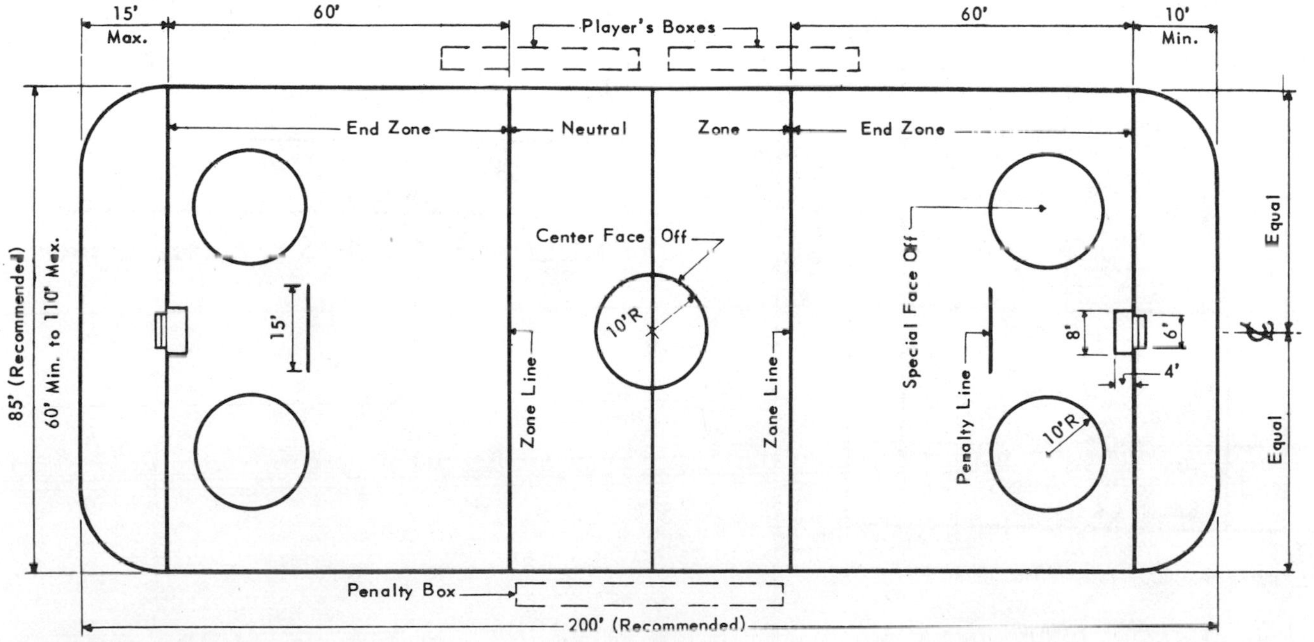

(a) ICE HOCKEY RINK

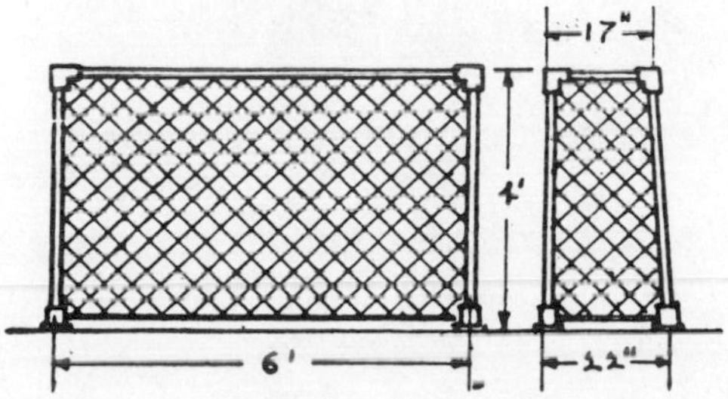

(b) ICE HOCKEY GOAL

J. De Chiara and J. H. Callendar, *Time-Saver Standards for Building Types*, McGraw-Hill Book Co., New York, 1973.

**TENNIS**

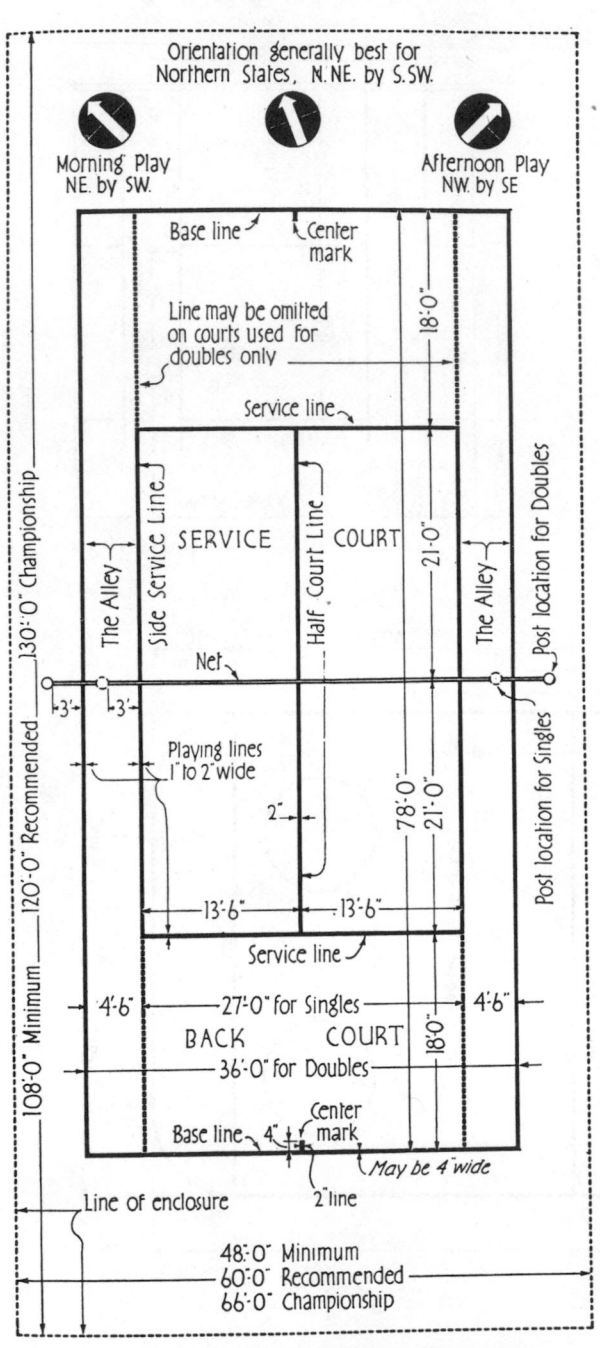

## DOUBLE or SINGLE COURT

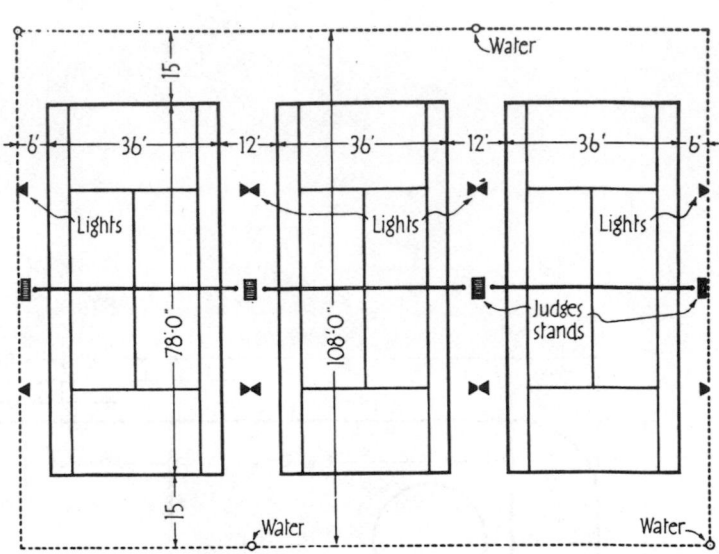

## MULTIPLE COURTS   *(Minimum dimensions)*

J. De Chiara and J. H. Callendar, *Time-Saver Standards for Building Types*, McGraw-Hill Book Co., New York, 1973.

**FOOTBALL**

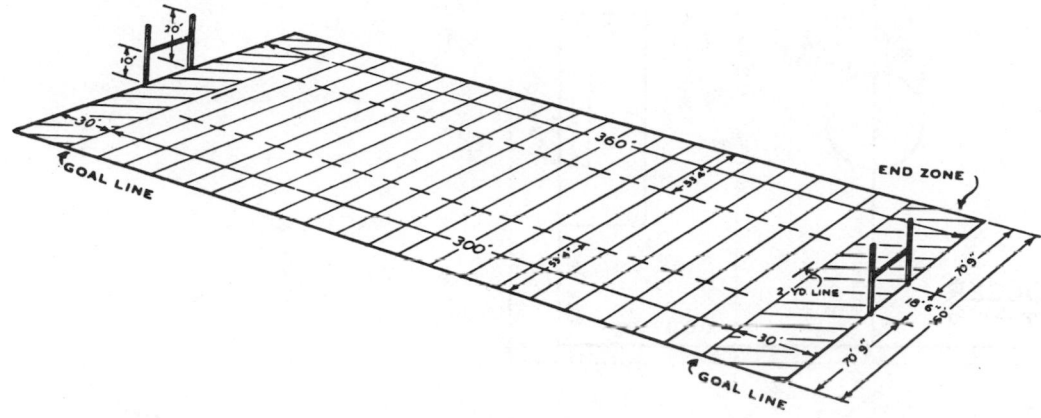

GOAL LINE
END ZONE
2 YD LINE
GOAL LINE

**COMBINED FOOTBALL-TRACK AND FIELD**

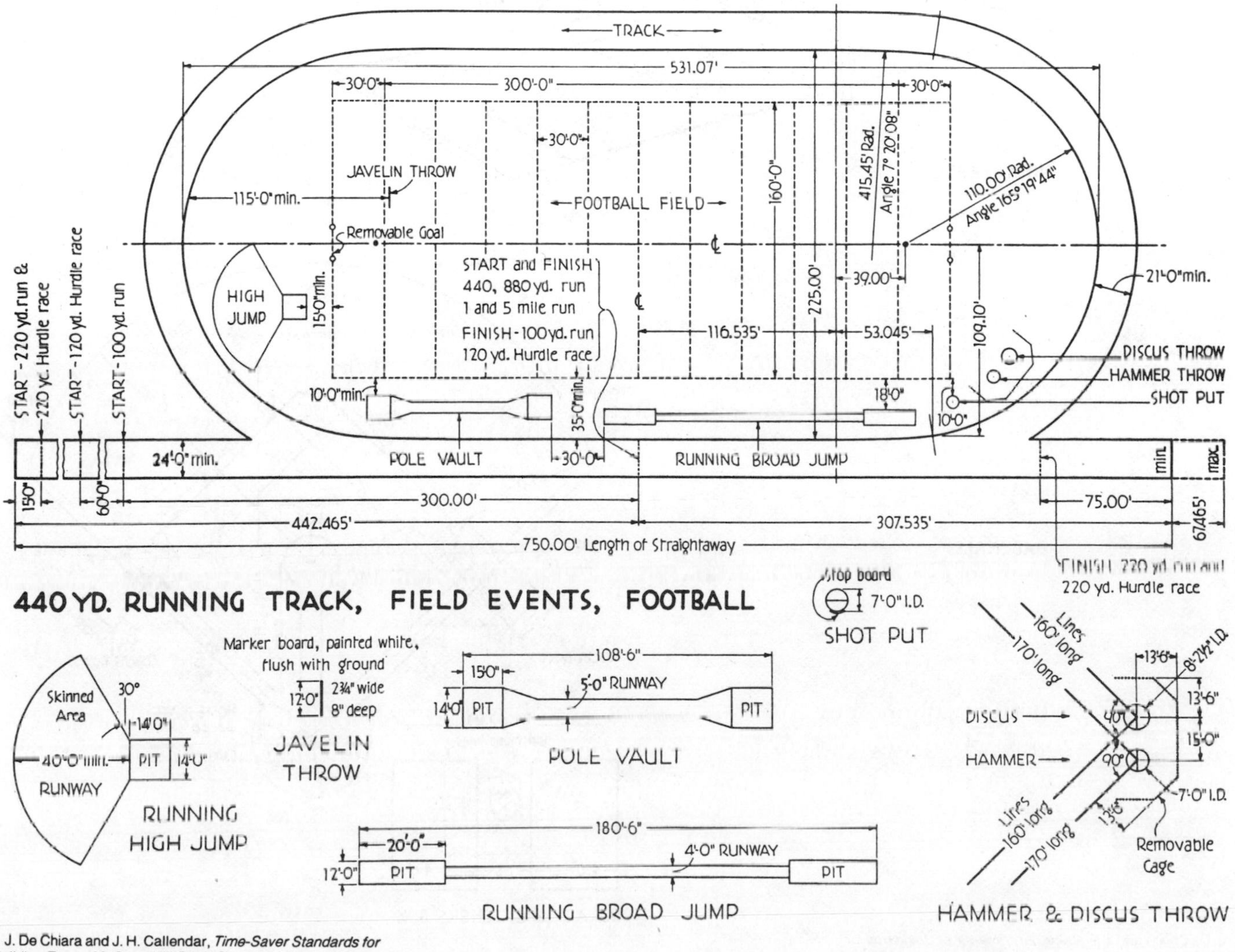

←— TRACK —→

531.07'

30'-0" | 300'-0" | 30'-0"

30'-0"

JAVELIN THROW

←— FOOTBALL FIELD —→

115'-0" min.

Removable Goal

160'-0"

415.45' Rad.
Angle 7° 20' 08"

110.00' Rad.
Angle 165° 19' 44"

HIGH JUMP

15'-0" min.

START and FINISH
440, 880 yd. run
1 and 5 mile run
FINISH-100 yd. run
120 yd. Hurdle race

225.00'

39.00'

214'-0" min.

STAR⁻ 220 yd. run & 220 yd. Hurdle race
STAR⁻ 120 yd. Hurdle race
START-100 yd. run

116.535'

53.045'

109.10'

DISCUS THROW
HAMMER THROW
SHOT PUT

10'-0" min.

35'-0" min.

18'-0"

10'-0"

POLE VAULT

←30'-0"

RUNNING BROAD JUMP

24'-0" min.

15'-0"

6'-0"

300.00'

75.00'

min.

max.

6'.465'

442.465'

307.535'

750.00' Length of Straightaway

FINISH 220 yd. run and
220 yd. Hurdle race

# 440 YD. RUNNING TRACK, FIELD EVENTS, FOOTBALL

Stop board

7'-0" I.D.

**SHOT PUT**

Marker board, painted white,
flush with ground
2¾" wide
8" deep

12'-0"

**JAVELIN THROW**

108'-6"

15'-0"

5'-0" RUNWAY

14'-0" PIT

PIT

**POLE VAULT**

Skinned Area

30°

14'-0"

40'-0" min.

PIT

14'-0"

RUNWAY

**RUNNING HIGH JUMP**

180'-6"

12'-0"

20'-0"

PIT

4'-0" RUNWAY

PIT

**RUNNING BROAD JUMP**

Lines 160' long
170' long
13'-6"
8'-7½" I.D.

DISCUS →
HAMMER →

40°

90°

13'-6"

15'-0"

7'-0" I.D.

Lines 160' long
170' long

13'-6"

Removable Cage

**HAMMER & DISCUS THROW**

J. De Chiara and J. H. Callendar, *Time-Saver Standards for
Building Types*, McGraw-Hill Book Co., New York, 1973

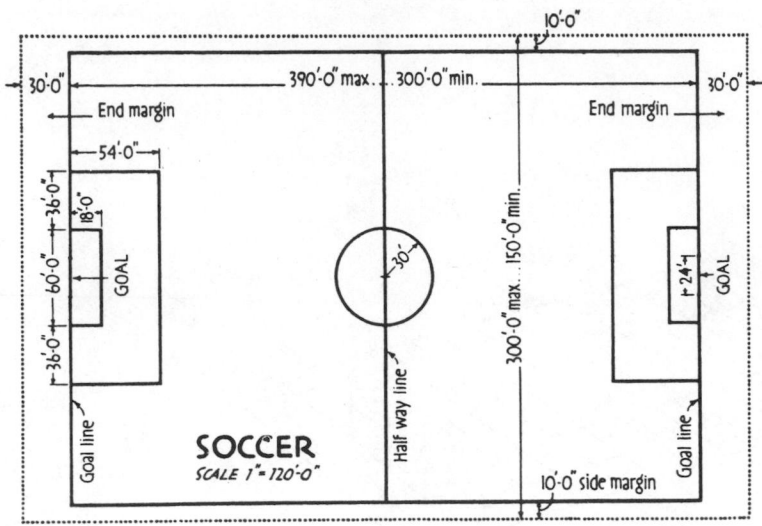

**SOCCER**
*SCALE 1" = 120'-0"*

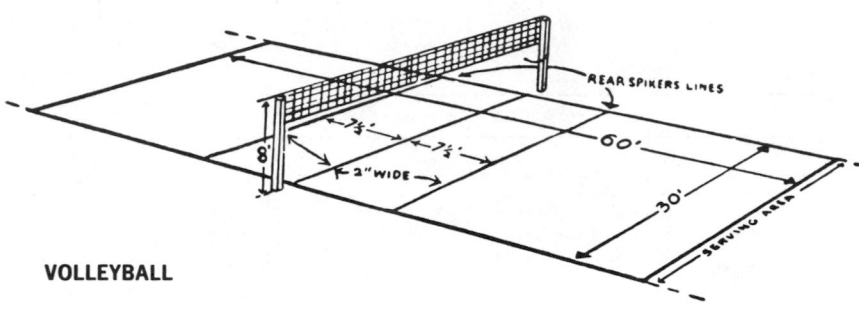

**VOLLEYBALL**

**BASEBALL**

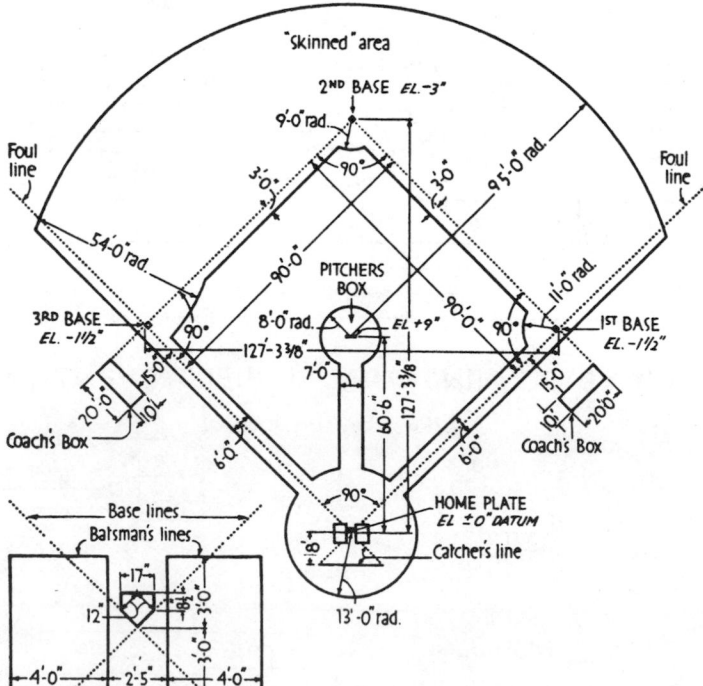

**DETAIL of CATCHER'S BOX**

J. De Chiara and J. H. Callendar, *Time-Saver Standards for Building Types*, McGraw-Hill Book Co., New York, 1973.

## Special Storage Considerations for the Sports Facility

Since many sports facilities must be designed for multiple uses (sports, conventions, circuses, musical shows, etc.), it is important to consider the storage requirements of all these events during the planning stage. The following are some of the items that must be not only stored but also handled readily during performances and other events:

1. Personal clothing and articles of participants
2. Athletic clothing and equipment
3. Special costumes
4. Musical instruments and music stands
5. Scenery, curtains, and modular booth accessories
6. Extra seats and chairs and special furniture
7. Special wheeled equipment (floor cleaners and ice sweepers)
0. Special field and court equipment (basketball goals, field goals, and equestrian equipment)
9. Special sound equipment, scoreboards, TV equipment, etc.
10. Records and money
11. Vendors' supplies
12. Advertising materials and equipment
13. Electrical accessories and lighting equipment
14. Internal and external maintenance materials, tools, and equipment
15. Special parking-lot accessories (signs, markers, and stanchions)
16. Shop materials (lumber and other materials used to make special stage props, field and court barriers, etc.)

## Lighting

Consult appropriate lighting handbooks and/or architectural guides for lighting requirements for the various elements of the sports facility listed below:

1. Parking
2. Building exteriors
3. Stairs and corridors
4. Locker rooms and rest rooms
5. Sporting events

Visual task and lighting principles are also discussed in more detail in Chapter 4.

## Safety

General safety considerations for all categories of sports facilities include poorly designed and/or improperly lighted corridors, stairways, ramps, and escalators, which may cause people to trip or fall; lack of proper separation of pedestrian and vehicular traffic, which may lead to pedestrians' being run down by a vehicle; and crowding, which may lead to personal entrapment and/or injury during a panic exit due to the layout of escape routes, exit doors, etc.

Special consideration should, however, be given to the following possible safety hazards as they apply to specific sports facility users.

### 1. Spectators
a. Steeply angled bleachers
b. Railings with sharp corners and/or lack of handrails
c. Lack of protective barriers to keep players and other objects from entering the bleacher area
d. Narrow aisles between seats, which make it difficult to walk past someone who is sitting along the row of seats
e. Lack of adequate illumination on bleacher stairways

### 2. Sports Participants
a. Sharp or unpadded field or court features into which players may run or fall
b. Lack of barriers between players and sideline participants or onlookers
c. Improper court or field illumination, e.g., bright light that may cause a player to run into another player or a hazardous object
d. Slippery locker-room flooring

### 3. All Parking-Lot Drivers and Pedestrians
a. Lack of properly defined vehicle and pedestrian separation in terms of fences, walls, curbs, etc.
b. Guy wires across the path of a pedestrian who may be blinded by stray headlights or field lights
c. Lack of adequate parking-area illumination, which may cause pedestrians to trip over irregular surfaces, berms, etc.
d. Lack of illuminated signs to guide both pedestrians and drivers so that they are less apt to cross each other's path

### 4. First-Aid and Medical Facilities
All public sports facilities should have an appropriate number of strategically located first-aid stations to provide assistance to both spectators and sports participants.

## AMUSEMENT CENTERS

### Amusement Center Planning Considerations

Although the variety of amusement center concepts is almost limitless, certain critical user functions are fairly common and should be addressed as seriously as the primary theme of the center. The following are among the most important considerations:

1. Site access for both private and public transportation
2. Adequate and well-planned parking for both patrons and center operations personnel
3. Entrance, ticketing, and information to accommodate maximum patron load
4. A site layout that provides sufficient organizational clarity to ease patrons' problem of keeping themselves oriented throughout their stay at the center
5. An adequate number of easily located patron support services such as rest rooms, dining facilities, personal storage facilities, drinking fountains, rest seating (with shade), telephones, a lost-and-found center, and informational signs
6. Adequate security for both patrons and center personnel (including their facilities, equipment, etc.) so that patrons cannot "wander behind the scenes"
7. Accessibility for handicapped patrons
8. Safety (especially where special exhibits, rides, or animals could create physical hazards)
9. Damage-resistant as well as wear-resistant design throughout
10. Site organization and structural and materials design that will minimize the time necessary to maintain the facility
11. Examination of queuing strategies for various anticipated exhibits, rides, entertainment, etc., during the initial planning stages rather than later, when spaces left over may create constraints that will not allow the best queuing methods to be used
12. Consideration of possible natural or man-produced disasters and the requirements for rapid egress of both patrons and center personnel, the use of disaster control equipment, and the problem of crowd control
13. Consideration of the effect the center may have on adjoining communities, i.e., noise, glare from lights, traffic, etc.
14. Consideration of potential expansion possibilities so that none of the above factors become distorted by the add-ons

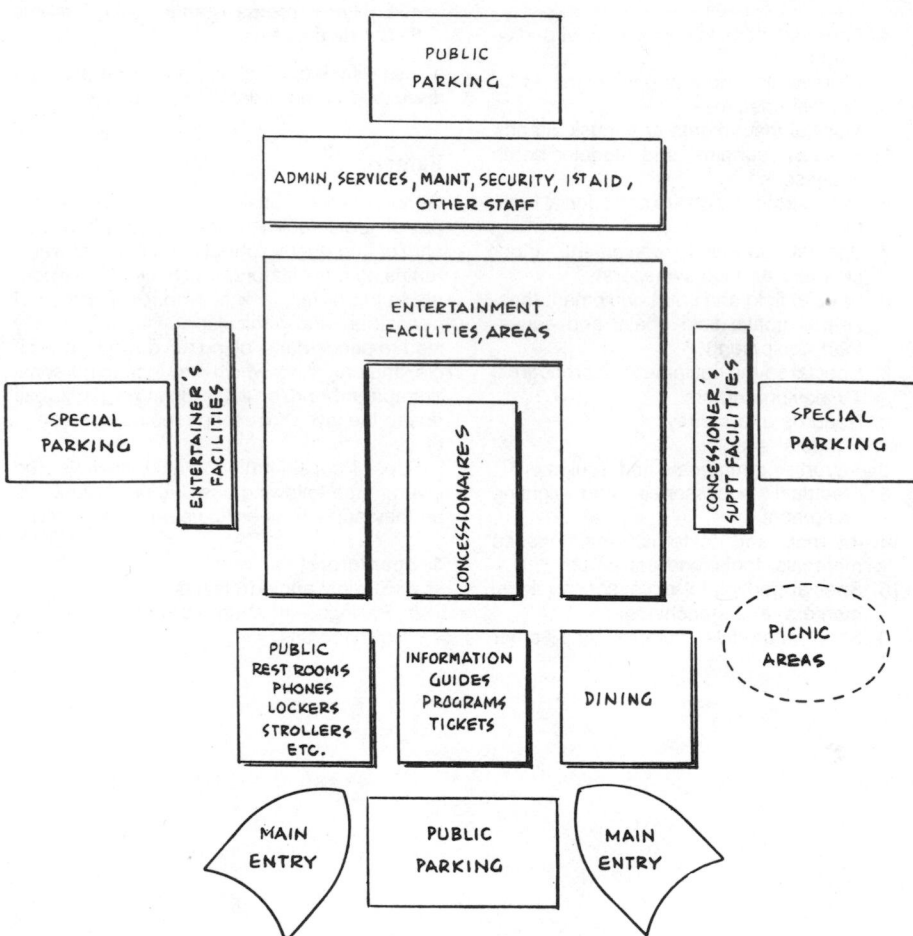

### Amusement Center Planning Model

The more important considerations include the following:

1. Site access and parking
2. A logical layout to ease personal location, crowd control, housekeeping, and maintenance
3. Personal security and safety
4. Crowd control

## Design for Safety

Although architectural planners may not have control over all the special amusement center equipment, such as special rides and exhibits, they should have control over, and therefore planning responsibility for, permanent features that relate to the hazards that may be associated with some of these special equipment features. Consider the following:

1. Wherever animal exhibits are part of the center plan, provide appropriate barriers to keep customers away from the animals and, obviously, to keep the animals from escaping.
2. Whenever special rides such as a merry-go-round, a Ferris wheel, or similar devices are expected, provide appropriate barriers to keep children from getting too near the moving elements and possibly being struck by the device.
3. Plan for permanent, safe electrical power transfer for rides, concessions, and other center exhibits so that concessionaires do not lay loose cables where patrons can trip over them and so that children cannot come in contact with high-voltage boxes or connections.
4. Provide smooth walking surfaces throughout the center. These should be permanent (i.e., paved), nonskid, and well drained.
5. Provide separate pathways for center maintenance and delivery vehicles so that they do not mingle with the customer traffic, either on the main grounds or on approaches to the grounds.
6. Provide adequate illumination in all areas, for both customers and operating personnel, so that they do not run into barriers, guy wires, or other objects in their path. Since most youngsters tend to stray from the normal pathways, other areas should be made inaccessible or should be illuminated.
7. If water features are planned (e.g., ponds, lakes, or waterfalls), lay these out so that small children (who are apt to stray from their parents) cannot fall into the water.
8. Provide strategically located and easily identifiable first-aid facilities for emergency treatment of both customers and operating personnel.
9. When laying out the center, consider the possibility of an emergency disaster requiring the grounds to be evacuated quickly and safely. This may require consideration not only of customers but also of other personnel, animals, etc.

## PARKS

### Public Park Facilities Planning Considerations

The public park, of course, varies from the small urban or residential community recreational space to the large, semiwild or wilderness space where hunting and camping can be accommodated. This section treats some of the human factor considerations common to all out-of-door recreational facilities:

1. Entrances and exits
2. Roadways and parking areas
3. Public rest-room facilities
4. Lighting
5. Signing

**ENTRANCES AND EXITS** Typically, public park users would like complete freedom to enter or leave the park whenever and wherever they please. Unfortunately, the planner must seek a compromise between letting users have as much of this freedom as possible and exercising sufficient control to minimize such problems as protecting park facilities and natural features from destruction or damage, protecting park users against traffic hazards, and minimizing difficulties associated with park surveillance and maintenance. Thus, it is desirable to define park boundaries either by the natural barriers available or by means of man-made barriers and appropriately designed entrances and exits. Even the small neighborhood park benefits from a decorative fence which will prevent children from inadvertently running into the street and which will keep stray animals from injuring children inside the park. Larger parks should have guarded entrances and gates in order to control who and what goes into the park and what is taken out. The very large wilderness park is, of course, another matter. Although primary guarded entrances are still appropriate, their benefits depend entirely on the capacity of the natural barriers to encourage use of the designated entrances.

**ROADWAYS AND PARKING AREAS** The two major problems regarding most park roadways relate to the adequacy of roadways to allow for passage of vehicles, especially large recreational vehicles, and to maintenance of safe speeds on these roadways (a major problem in recent years with the advent of dune buggies, motor bikes, etc., which, the riders or drivers believe, are meant only for racing). In the latter case, consideration must be given to the use of speed barriers (berms placed across the roadways at periodic intervals and/or regular police patrol). It is also suggested that, in planning the entrances and exits and

the inner park roadways, the geometric relationships between the exterior roads or streets and the immediate inner roadway be such that people will not be tempted to use the park as a thoroughfare, i.e., to travel through it only to get from one side to the other, with no intention of enjoying the park as a facility.

**REST ROOMS** With the exception of the wilderness park, most park facilities should be provided with an appropriate number of properly designed rest-room facilities. Such facilities should be designed to require a minimum of maintenance and should be constructed and equipped to minimize the possibilities of misuse and vandalism. Additionally, rest rooms should be accessible to the handicapped; i.e., they should have smooth approaches and large doorways, and there should be appropriate fixtures and handrails.

**LIGHTING** In spite of the desire to leave parks as natural as possible, they are used by people who need to see. In addition, the urban park benefits from good lighting in terms of reduced vandalism and criminal activity. Of course, it is also important that the method for lighting itself be as vandal-proof as possible.

**SIGNING** The question of signing should be addressed during the early stages of park planning so that the signing system can successfully explain or identify areas, roadways, etc., for the individual who may not know how the park is laid out or what facilities are available and where they are. Although it is desirable to design signs that are compatible with the rustic environment, this should not preclude their being legible and understandable to the observer. Consider the entire signing system at the beginning, including instructions, identification, and guidance information.

### Park Service and Maintenance

The following list of possible park services should be reviewed during the planning stage in order to approach the accommodation of such service in the most cost-effective manner.

#### 1. Park Patron Services
a. Picnic tables
b. Fire pits and barbecue pits or stoves (including gas)
c. Trash receptacles
d. Water outlets
e. Emergency communications and first-aid stations
f. General provisions store
g. Park benches (neighborhood park)
h. Prepared hiking or walking paths, bridges, stairways, etc.

i. Shelters for use during inclement weather
j. Prepared campsites (including flat areas for tent, vehicle, cooking equipment, and dining table and seats)

For large parks where considerable camping is planned, it is recommended that a special area be set aside for the handicapped user. This area should be as flat as possible, and/or appropriate ramps and pads should be provided so that persons using wheelchairs can get about easily within each campsite and can go to and from nearby rest-room facilities.

#### 2. Park Management Services
A variety of management services should be considered for the large park or recreational area, including the following:

a. Park ranger residence
b. Park ranger entrance and exit station
c. Service, fire, and inspection roads (not generally used by the public)
d. Park housekeeping (weed cutting, tree trimming and removal, trash pickup and disposal, etc.)
e. Facilities maintenance (road repair, fence repair, utilities repair, etc.)
f. Security (traffic control, crime prevention, and animal control)
g. Fire control, including provision of appropriate fire-fighting vehicles, equipment, and storage
h. First-aid station, equipment, and provisions
i. Landscape maintenance (watering, trimming, etc.)

Special attention should be given to the design of park facilities in terms of the ability of the materials selected to withstand frequent and continued use with minimum wear, damage due to misuse or vandalism, weather effects, and, of course, flammability. In addition, there is no reason why park facilities should not also be functionally convenient and comfortable. Avoid the tendency to believe that a rustic design cannot be convenient or comfortable. Also do not assume that most people want to preserve the beauty of your park and therefore will respect the natural and man-made facilities provided. In fact, most users will be inconsiderate, if not disrespectful, and careless at one time or another. It is these latter considerations that must establish the basic design goals even though they may compromise the more desirable functional goals of providing the most natural and enjoyable environment for those park patrons who will normally use the park as it was intended to be used.

# TRANSPORTATION

## GENERAL

This section deals with human factors considerations that should be addressed during the initial conceptual phases of any transportation system development. The key elements addressed include the primary transporter or vehicle, the roadway upon which the transporter moves, and the special terminal interfaces where various people and transporter systems interact.

The section begins with a series of general guidelines applicable to all elements of the transportation system; these are followed by separate sections dealing with the more specific questions to be addressed for each of the system elements.

## Transportation Systems Planning Model

Although a particular transportation system may be of primary interest, it is generally the case that it must be made compatible with any or all of the total system elements indicated in the accompanying illustration. The key to this compatibility lies in a satisfactory solution to users' problems of interfacing with any or all of these elements; i.e., users usually do not confine their travel to a single element, even though that may be the major interaction for most of the trip.

## Considerations for International Travel

Because of the opportunities that modern transportation system technology offers for international travel, it is important to address human factors on a global basis. That is, although a specific type of system may be proposed for fairly local use, it is probable that the system will at some time be used by individuals with widely differing backgrounds in terms of experience, language, and understanding. Thus a local urban transit system might be used by individuals from many parts of the world as they travel on business or for personal pleasure.

| TRUCKS | PRIVATE AUTOS | BICYCLES MOTORCYCLES |
| BUSES | TRAVELER PACKAGE SHIPMENT | SUBWAYS |
| AIRPLANES | | SHIPS |
| | RAIL | |
| STREETS, HIGHWAYS, ROAD BEDS | TERMINALS | TRAFFIC CONTROL, SERVICES, MAINT |

Among the typical system features to be considered from this international point of view are the following:

1. Standardization of the driver controls in automobiles that may be rented by individuals of various nationalities so that the vehicles can be operated effectively and safely
2. Standardization of roadway patterns, marking, signing, and signaling devices so that drivers can "mesh" effectively and safely with the local traffic
3. Standardization of special terminal facilities so that travelers do not become confused or lost or cause special problems for terminal staff or vehicular systems operators
4. Standardization of facilities, equipment, and communications used by the operators of vehicular systems that may be acquired from a variety of equipment manufacturers (e.g., aircraft purchased by a carrier in one country from a manufacturer in another country and communication protocol and language used by various air, land, or sea traffic control organizations) so that the mix of foreign carriers does not become snarled at terminal areas, etc.
5. Standardization of equipment and facilities so that maintenance can be accomplished by a variety of technicians with varying backgrounds

Special attention should also be given to world travelers who have special problems, e.g., the elderly, the handicapped, children traveling alone, and mothers traveling with small children.

The user population to be considered for transportation systems includes the following key individuals:

1. Vehicle operators
2. Passengers
3. Vehicle maintenance technicians
4. Terminal operators (management, ticketing and baggage-handling personnel, housekeeping and maintenance personnel, concessionaires, carrier service attendants, security personnel, etc.)
5. Visitors
6. Taxi drivers
7. Public transit drivers
8. Traffic controllers
9. Fire and disaster crews
10. Other supporting personnel (travelers' aids, medical staff, porters, immigration staff, etc.)

## Characteristics of the User Population

The user population for transportation systems is perhaps broader and more complex than the user population for any other system; this makes it especially important to consider the individual differences of each user group not only during initial planning but also at each stage of each subsystem design and development. Among the most important characteristics to consider are the following.

### 1. Differences in Human Body Size
Any transportation system must accommodate persons of all sizes, from tiny babies to the largest adult. Although there are certain conventions (e.g., designing to fit a range from the 5th-percentile female to the 95th-percentile male), the traveler who does not fall within this arbitrary design range still must use the system.

### 2. Male-Female Differences
Recognition of certain male and female characteristics is required for public transportation, especially with respect to separate facilities.

### 3. Ethnic Differences
Differences in terms of language, experience, custom, etc., should be considered wherever the convenience and safety of the traveler are critical.

### 4. Age Differences
Although the middle-aged group typically establishes the norm with respect to the planning and design of a transportation system, consideration must be given to the unique requirements of children in terms of their limited understanding and to the needs of the elderly in terms of their lack of mobility, their slower responses, and their visual, aural, and other limitations.

### 5. The Handicapped
More and more, the physically handicapped expect to be able to utilize public transportation. Therefore, special attention should be given to their problems during the initial system planning phase. Corrections are difficult to make once basic barriers are established by the initial concept.

### 6. Behavioral Attitudes
Users of transportation systems represent a unique blend of the broadest possible range of attitudes. That is, users' attitudes will range from considerable respect for equipment and facilities to complete disrespect for either these elements or their fellow travelers. Anticipation of this is particularly important in order to preclude both operational and service problems.

### 7. System Use Patterns
A significant factor in system planning is rate of use of a proposed system so that the most cost-effective solution to the ebb and flow of travel patterns can be arrived at. This applies not only to the overall system, where roadway capacity and terminal capacity are of concern, but also to the individual vehicle size in terms of numbers of passengers and amount of baggage.

**General Classification of Human Considerations for Public Transportation**

| Area of Consideration | Principal Human Considerations |
|---|---|
| Vehicle and terminal | Space and seating. Ingress and egress (entrances and exits) Environmental stress (acceleration, deceleration, noise, vibration, temperature, atmospheric pressure, ventilation, etc.) |
| Convenience and mobility | Time: total trip time, waiting, walking, transfer, terminal organization, and special equipment operation. Operations: scheduling, frequency, and reliability of service. Intelligibility of information. Accessibility for elderly and handicapped persons |
| Safety and security | Emergency provisions: alarms, restraints, procedures, and first aid. Security provisions: crime and vandalism control. Accident prevention |
| Social factors | Personal privacy. Crowding. Freedom to choose traveling companions |
| Psychological factors | Comfort. Freedom from anxiety and fear. Confidence with respect to what to do and where to go. Pleasurable features, i.e., features that are aesthetically pleasing, interesting, and nonmonotonous |
| System environment | Minimal congestion and delay. Minimal noise, air, and visual pollution and minimal physical interference |
| Maintenance | Reliability. Ease of maintenance. Resistance to wear, abuse, and vandalism |

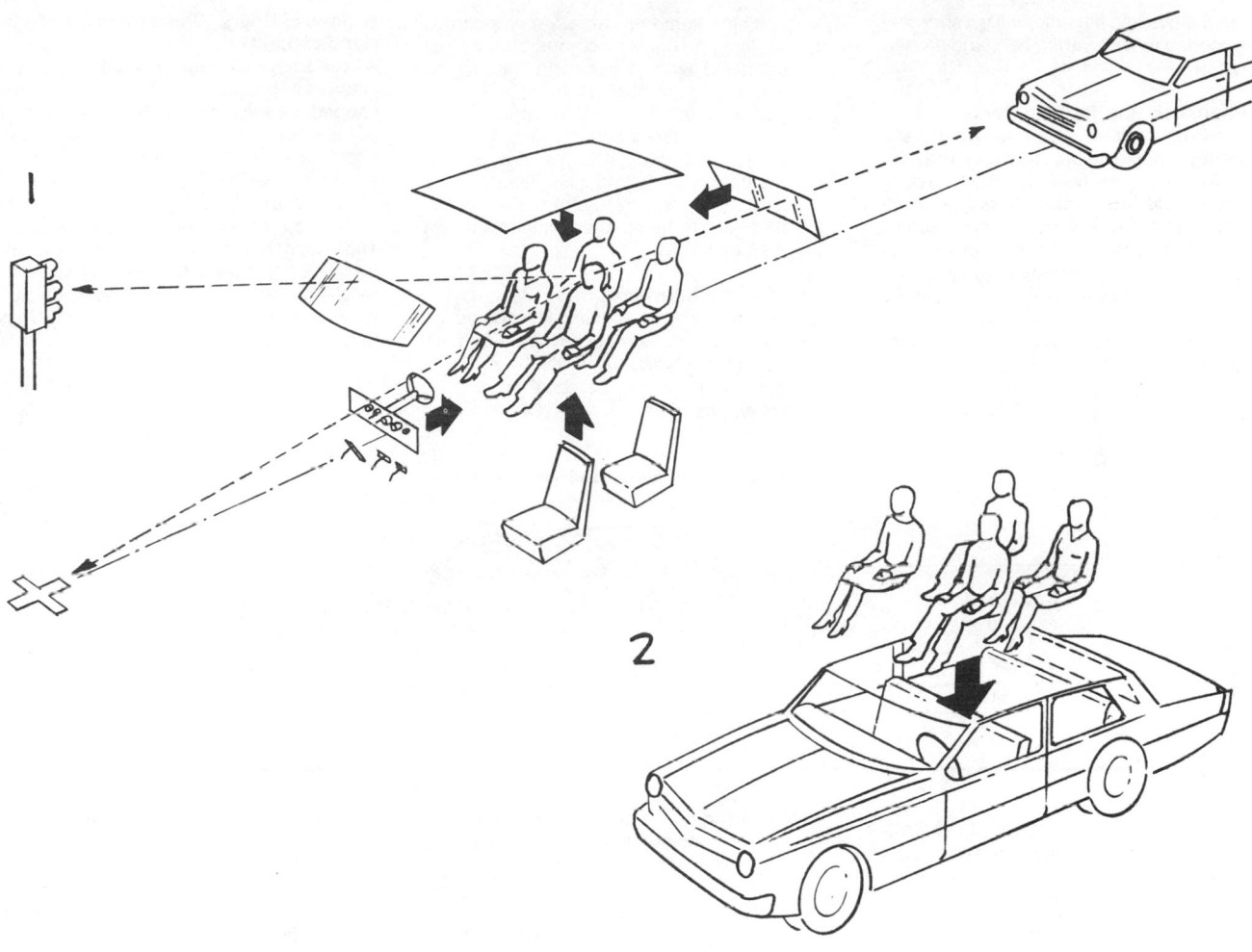

## ROAD VEHICLES

### General

1. The human factors approach: Start with the occupants (especially the driver) and create the vehicle around the spatial and sensorimotor needs of the occupants. This is the preferred approach.

2. The traditional approach. Create a "package concept" and then manipulate the occupants to make them fit into the package. This approach invariably compromises the occupants' efficiency, convenience, and comfort.

### HUMAN FACTORS DESIGN OBJECTIVES BY VEHICLE USER CATEGORY

#### 1. All Occupants

a. Ease and safety of entry and exit (especially emergency escape) require a properly sized and shaped door opening and an optimum physical relationship between the opening and the seats, plus elimination of intervening impediments (especially snag-producing elements). Special care should also be taken to design doors and door-manipulating hardware so that doors are secure when closed and also are not confusing or difficult to lock, unlock, or open. The adjustable windows should be equally easy to operate (sometimes windows are the only escape route in an emergency).

b. Body, limb, and head clearance should be sufficient to minimize crowding and inadvertent bumping when the vehicle sways or goes over a bump.

c. All interior surfaces should be made as noninjurious as possible (e.g., there should be no sharp protrusions, edges, or corners, and appropriate padding should be provided) so that if an occupant strikes a surface when the car is moving, when getting in or out of the car, or during a crash, he or she will not be cut, bruised, pinched, or otherwise injured.

d. All seating should be designed to fully support occupants in a nonstressful position so that unnecessary fatigue is minimized during long duration occupancies.

e. All interior materials used in the vehicle should be nonflammable.

f. Shatterproof glazing should be used in all windows.

g. Equally effective crash-restraint systems should be provided for all passengers.

#### 2. Drivers

a. The driver station should be laid out so that all the driver's visual and manipulative tasks can be performed as easily and efficiently as possible (i.e., as correctly and quickly as necessary to provide safe response to the driving environment).

b. The controls and displays provided for driving should be designed so that they will provide the necessary input and output for proper vehicle control and also will not be confusing and cause the driver to make errors or take his or her attention away from the external information needed to cope with traffic.

c. The vehicle dynamics and vehicle control systems should be designed so that they allow not only an expert driver but also an inexperienced driver to cope with all driving conditions (especially emergency conditions associated with skids, rough surfaces, etc.) and are compatible with all drivers' sensorimotor response capabilities and limitations (i.e., strength, response time, coordination, and dexterity).

### 3. Maintenance and Service Personnel

a. Servicing features (fuel, oil, water, and battery) should be placed where they are easily accessible (with minimum hazard) and should have easily manipulated intervening features (caps, covers, etc.).

b. Typical service and maintenance features (battery replacement, engine timing, fan-belt adjustment or replacement, etc.) should not be designed only for convenience and ease of operation; special consideration should also be given to the related hazards created by the design and/or location of adjacent, inter-

vening elements (e.g., components within an engine compartment that someone who is servicing the engine might touch or get caught in).

c. Consider the potential hazards of tire changing, lubricating the vehicle on a grease rack, etc., in terms of designing the basic vehicle and its components as well as the special supportive devices that have to be used in conjunction with these services.

### 4. Pedestrian Interfaces

All external surfaces of vehicles should be made as noninjurious as possible so that if a pedestrian is inadvertently struck by a vehicle or walks or runs into it, personal injury is as minimal as possible.

### 5. Special Users: Children and the Handicapped

All the above objectives should be evaluated in terms of possible use of a vehicle by handicapped persons. Although this is most critical in the case of handicapped passengers, one should also consider the possibility of service and maintenance personnel with certain types of handicaps.

Children require special attention as passengers, not only because of their size, but also because they may create safety hazards (e.g., small children may try to operate vehicle controls, windows, doors, etc., when they should not).

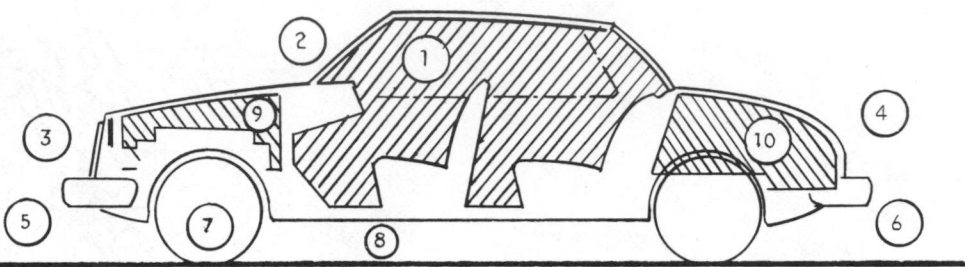

### The Automobile

**PLANNING CONSIDERATIONS FOR AUTOMOBILES** Although many other factors also must be considered in conceptualizing a new automobile, human factors should be addressed first, in the following general order.

1. *Passenger Space,* i.e., a space that is large enough and geometrically appropriate for the number of passengers (large adults) the automobile is planned to accommodate. Do not "short-change" passengers who are expected to be riding in the vehicle only once in awhile.
   *Passenger Seating,* i.e., properly designed seats and restraints for *all* passengers. Do not force some passengers to sit in less-than-adequate seats or be less protected because these are used infrequently.
   *Driver Control, Display, and Visibility,* i.e., create an optimum driving configuration so that the driver has the best possible chance of operating the vehicle safely and efficiently in all expected environments.
   *Environmental Control,* i.e., adequate heating, cooling, ventilation, and noise and vibration control to minimize stress on passengers and especially on the driver.
   *Passenger Compartment Structural Integrity,* i.e., the compartment envelope should be able to withstand expected crash loads, possible penetrations, etc.
   *Interior Safety,* i.e., adequate passenger restraint and freedom from structural contact injury should be provided to minimize the possibility of injury or death during a crash.
   *Ingress and Egress,* i.e., ease of getting

into and out of the vehicle should be a prime consideration, under both normal and emergency circumstances.

2. *Visibility,* i.e., provide the necessary visibility for the driver not only in terms of angle of view but also in terms of keeping that visibility under typical environmental disruptions, e.g., rain, snow, ice, or fog.
   *Crash Protection,* i.e., provide glazing that will not splinter or shatter and possibly cut passengers during a crash.

3. *Road Illumination,* i.e., provide a properly controllable headlight subsystem to ensure that the driver can see where he or she is going and can observe other vehicles, pedestrians, or obstructions.
   *Identifying Lights,* i.e., provide auxiliary light subsystems that indicate when the driver plans to turn, which is the front or rear of the vehicle, and when the driver is going to stop.
   *Headlight Beam Control,* i.e., provide control of the headlight beam direction so that the driver can reduce the potential visual disturbance that the headlights may cause for an oncoming motorist.

4. *Rear-Collision Protection,* i.e., provide proper energy absorption and isolation of the fuel tank so that, in the event of a crash, the fuel tank is protected and passengers are not injured.

5. *Minimization of Injuries to Pedestrians,* i.e., provide a front-end configuration that is minimally injurious in the event the vehicle runs into a pedestrian.
   *Crash Energy Absorption,* i.e., provide a front-end design that will reduce the amount of energy transferred to the passenger compartment and will limit

the possibility of penetration of the compartment.

6. *Rear-Impact Energy Absorption,* i.e., provide a design that will reduce structural distortion and the possibility of penetration of the passenger compartment.

7. *Ease and Safety of Tire Replacement,* i.e., adequate consideration should be given to the problems of replacing a flat or blown tire, including the safety aspects of jacking up the vehicle.

8. *Side-Impact Energy Absorption,* i.e., provide adequate internal padding and a design that will reduce structural distortion and the possibility of penetration of the passenger compartment.

9. *Ease and Safety of Maintenance and Service,* i.e., provide proper access to frequently serviced components; ease of refueling and of refurbishing, removing, and replacing parts; and protection of the service technician and the owner from maintenance hazards, such as electric shock, hot surfaces, moving parts, and fire.

10. *Baggage,* i.e., provide reasonable storage capacity and ease of loading and unloading, removing and replacing the spare tire, etc.

*Note:* The typical priorities used by many automobile manufacturers are somewhat in the opposite of the above order; i.e., a package size is generally established on the basis of engine, baggage, and other requirements. Passengers are then "squeezed into" whatever space is left over. Although it is recognized that compromises often are necessary, passengers will fare much better when the above order of considerations is followed.

**OTHER HUMAN FACTORS CONSIDERATIONS** Other key human factors considerations should be addressed during the initial concept development stage. These include the following.

### 1. Riding and Handling Quality

Obviously, the automobile should not possess qualities that would make it difficult to control, especially under unusual conditions, i.e., on rough or slick roads, when it is heavily loaded, or when unusual maneuvers are necessary, such as making a sudden turn to avoid another vehicle or a pedestrian or object in the road. Equally important is a ride quality that will not create unreasonable stresses on the passengers. This latter consideration is secondary to the former since people vary in their interpretations of what is a satisfactory ride; e.g., a sports-car enthusiast wants to "feel the road," while a Wall Street executive probably does not.

### 2. Reliability

Obviously, most consumers appreciate a vehicle that is reliable and does not require continual repair. Most important, however, is a design that will not let the driver down in a pinch, i.e., an automobile whose brakes, steering, motor, tires, etc., will not fail.

### 3. Ease of Maintenance

Much of the cost of maintenance is related to the difficulty and time required to determine where a malfunction is and to the difficulty of getting to a part, removing it, and replacing it. Ease of maintenance can be increased primarily during the conceptual stage, before difficulties have been created by placing things in awkward places.

### 4. Design-Induced Driver Problems

A great many of the driver's problems are often created because a designer has chosen to use the driver interface as a prime target for artistic or aesthetic experimentation. That is, instead of recognizing that, to make the driver's tasks easier, controls and displays should be designed, located, and labeled in a certain way, the designer chooses to experiment with various methods and techniques for mechanizing these displays and controls, arranging them to create a special effect or adding decorative chrome that reflects in the windshield or into the driver's eyes, for example. In addition, designers often ignore good seating design criteria, thus creating difficulties in postural control or leg manipulation for the driver (e.g., soft cushions may allow the heavy driver to sink too low to see out or may cause clothing to bunch, and incorrect cushion dimensions may create pressures on the back of the driver's legs).

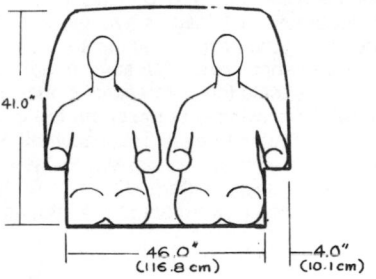

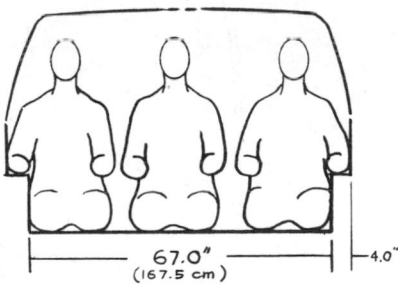

**PASSENGER SPACE REQUIREMENTS** An honest appraisal of automobile passenger space requirements requires that the defined space be based on the largest expected passenger's dimensions. The typical practice of referring to two-abreast seating when the space accommodates only a passenger and a half is capricious and arbitrary. The accompanying dimensional criteria provide proper accommodation for the typical complement of full-size American passengers.

Fore and aft crowding not only is uncomfortable for the rear-seat passenger but also may be the cause of unnecessary injury during a crash. The recommended clearance is for the case in which the front seat is adjusted to its aftmost position.

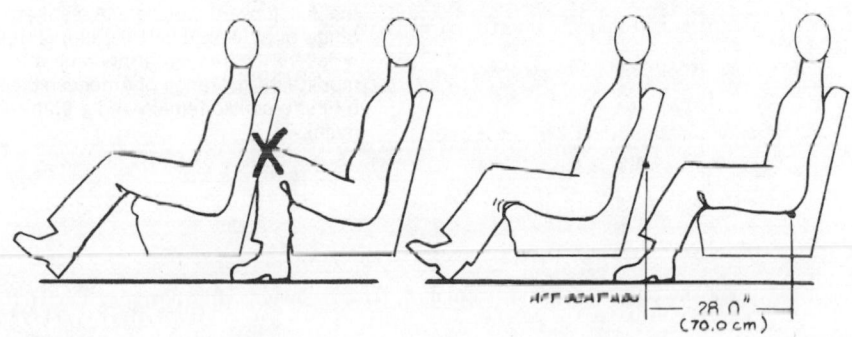

**INGRESS AND EGRESS CLEARANCE** As automobiles have become lower and lower, principally for styling reasons, the difficulties for the larger individual with respect to entry and exit have become considerably greater, to the point where the probability of striking one's head is quite high. Although many injuries are undoubtedly sustained as a result of such styling, the consumer continues to accept the inconvenience (and the sore head). Good practice should follow the dimensional criteria shown here, which are based on the average clearance needed by a reasonably tall American driver who still would tend to perform a normal stooping pattern, either to enter the vehicle facing forward or to back into the seat.

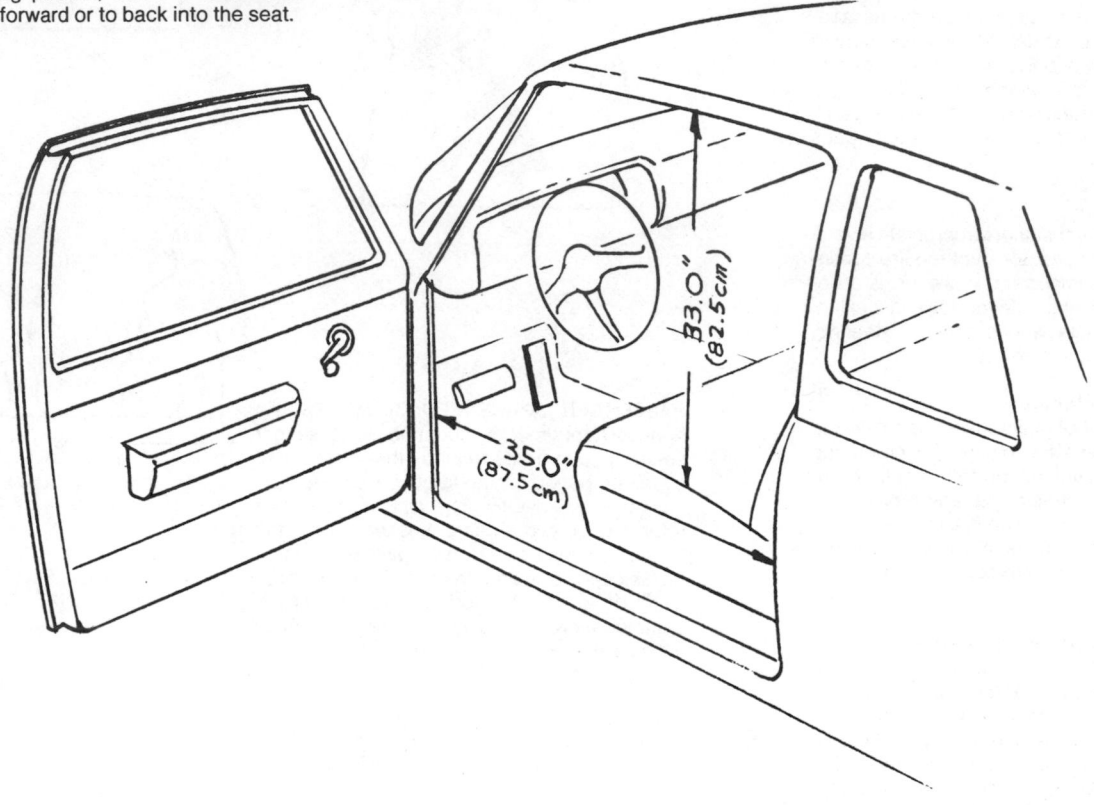

In addition to the doorway opening clearance, adequate space between the rim of the steering wheel and the seat should be provided. This minimum clearance (e.g., 7 in, 18 cm) should be measured when the seat is in the full forward position. A seat-adjustment range of at least 8.0 in (20 cm) is required in order for the seating arrangement to accommodate the full range of American drivers (i.e., a 5th-percentile female and a 95th-percentile male).

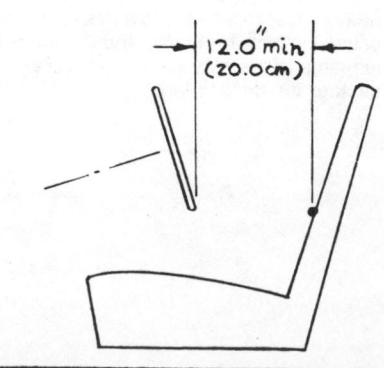

**GENERAL CHECKLIST FOR INITIAL CONCEPT DEVELOPMENT**

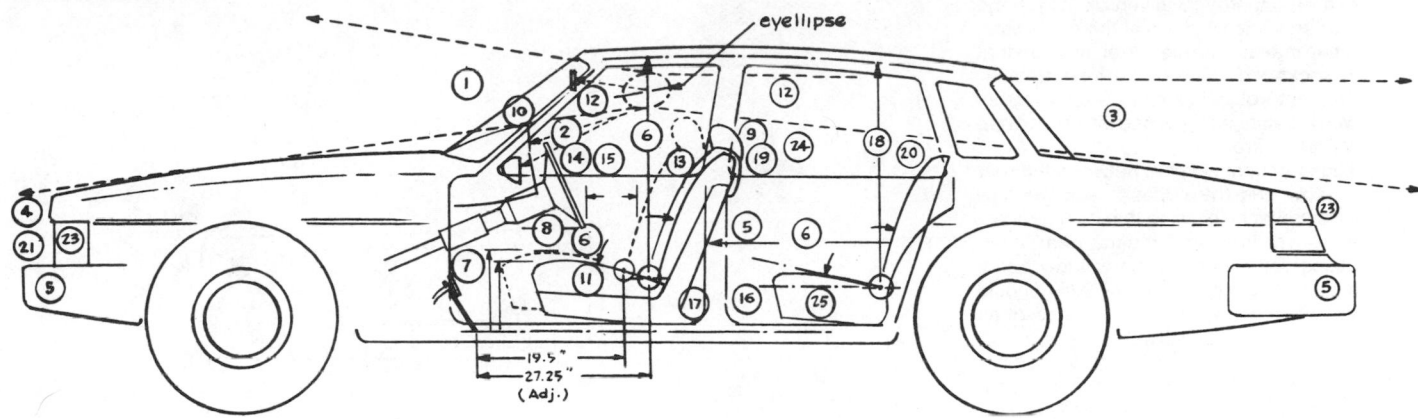

1. Exterior visibility (signs and signals, over the hood and dashboard, and vehicles in front and to the side)
2. Interior visibility (dash displays and hand controls)
3. Exterior visibility (vehicles in the rear and to the side)
4. Noninjurious exterior pedestrian interface
5. Energy-absorbing front, rear, and side; cabin envelope integrity in crash
6. Driver position (reach controls and clear head)
7. Ease of ingress and egress
8. Clearance for legs and knees around dashboard and steering wheel
9. Headrest—whiplash protection
10. Energy-absorbing surfaces, and buried or noninjurious controls
11. Posture-control seat (adjustable fore and aft and up and down)
12. Glare and reflection control
13. Crash restraint system
14. Display and control illumination
15. Labels, instructions, and warnings
16. Ease of ingress and egress
17. Knee and foot clearance
18. Headroom
19. Noninjurious front-seat back
20. Crash restraint system
21. Roadway illumination
22. Interior illumination
23. Running and signal lights
24. Environmental control (temperature, vibration, ventilation, and noise)
25. Fireproof interior materials

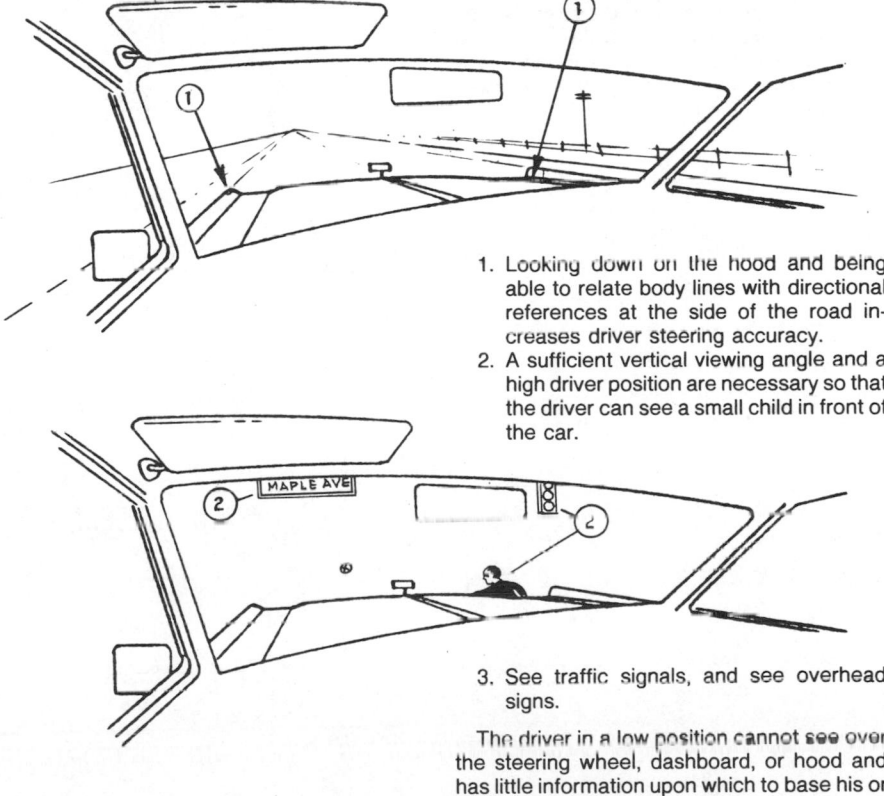

1. Looking down on the hood and being able to relate body lines with directional references at the side of the road increases driver steering accuracy.
2. A sufficient vertical viewing angle and a high driver position are necessary so that the driver can see a small child in front of the car.

3. See traffic signals, and see overhead signs.

The driver in a low position cannot see over the steering wheel, dashboard, or hood and has little information upon which to base his or her steering or speed.

**DRIVER POSITIONING FOR SAFE EXTERNAL VIEWING** For several reasons, listed below, it is most critical to seat the driver of an automobile (and of any other highway vehicle, for that matter) relatively high so that he or she can see the road and the hood of the vehicle. It is equally important to minimize any serious obstructions to lateral viewing and/or viewing out the rear.

1. Rearview mirrors should be large enough so that the driver can see parts of a rear-approaching vehicle in both the rear and side windows at the same time. If this makes a single mirror so large that it blocks too much of the forward view, the same objective can be approached with a combination of one inside and one outside mirror.

2. The structural element between the rear window and the opposite side window must never be so wide that it will prevent the driver from seeing parts of an overtaking car in each of the window openings. In fact, this separating pillar should not even obstruct the driver's view of an overtaking motorcyclist.

3. The same obstruction-removal objectives should be sought for the side next to the driver, except that now the B pillar (the one next to the driver) becomes equally important.

4. Although the driver's ability to see where he or she is backing is not considered so important by safety experts, it *is* important to the person on the street. A key consideration, obviously, is to let the driver see that all-important part of the car that may strike a post or doorway first: the right rear fender.

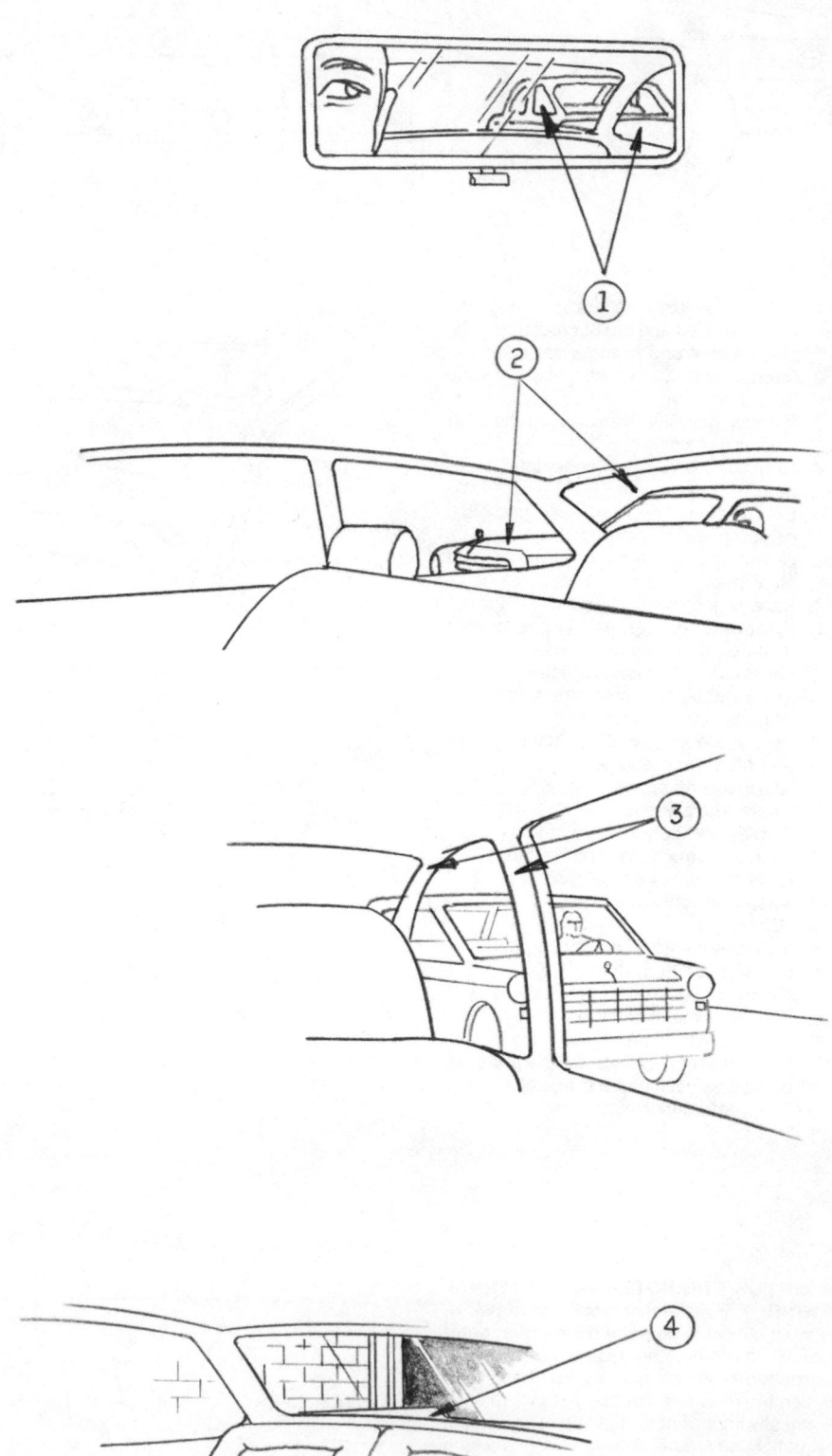

## OPTIMIZED DRIVER STATION PROFILE[32]

These specifications provide optimum positioning for the range of American drivers from the 5th-percentile female through the 95th-percentile male to ensure the best posture, external and internal visibility, and access to all controls without requiring the driver to lean or reach and thus risk losing visual contact with the road. Approximation of these dimensions will ensure the maximum efficiency and comfort for all drivers.

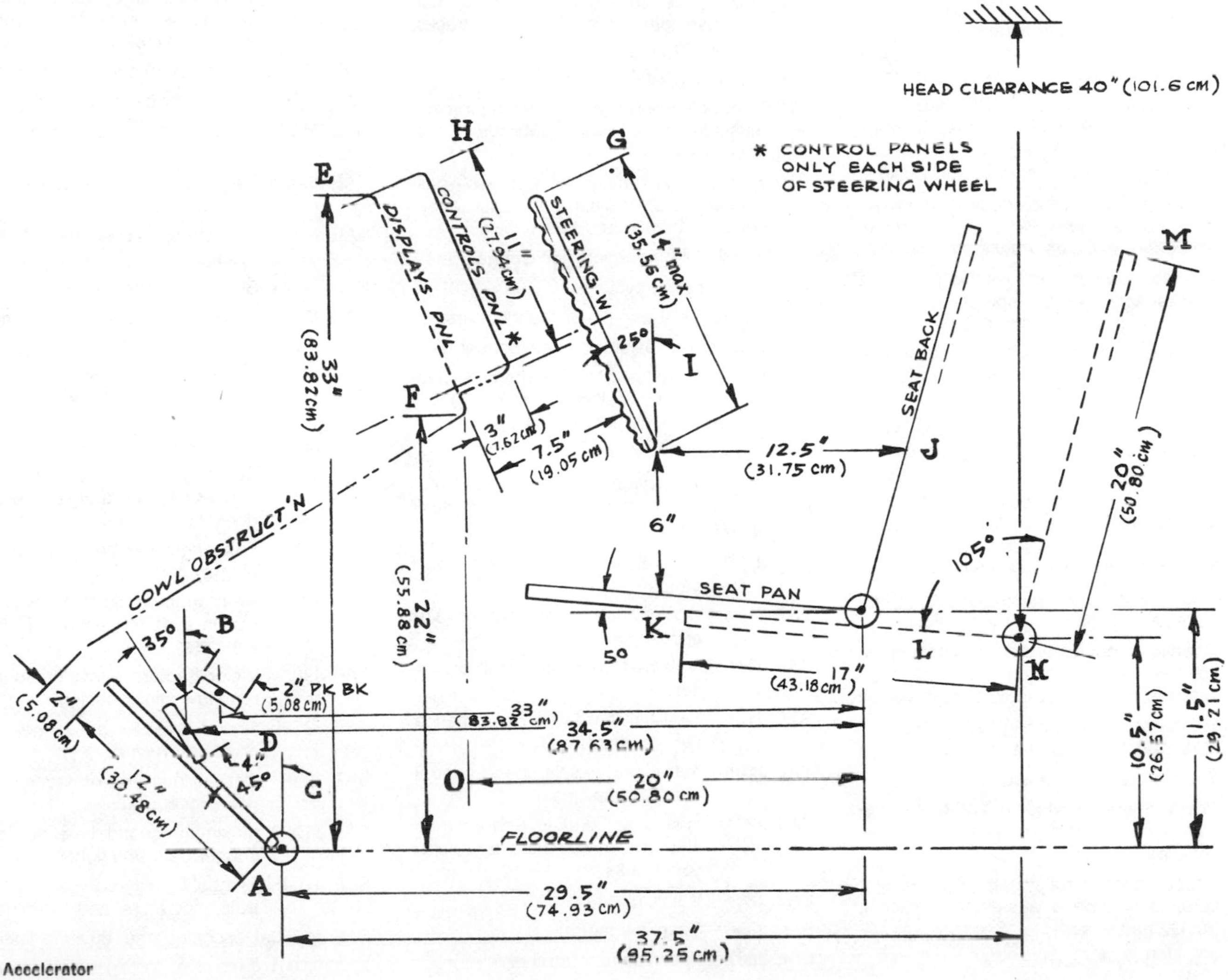

HEAD CLEARANCE 40″ (101.6 CM)

* CONTROL PANELS ONLY EACH SIDE OF STEERING WHEEL

A—Accelerator
B—Service brake
C—Accelerator slope
D—Pedal dimension
E—Top of dashboard
F—Bottom of dashboard
G—Wheel diameter
H—Control panel
I—Wheel slope
J—Minimum clearance
K—Minimum clearance
L—Seat angles
M—Backrest height
N—Seat-pan length
O—Knee clearance

[32] W. E. Woodson, et al., *Instrument and Control Location, Accessibility and Identification*, Report MFI 60 106, U.S. Department of Transportation, NHTSA Contract FH-11-6907, 1969

**RELATED GOVERNMENT AND INDUSTRY STANDARDS** The following government and industry standards relating to the human factors aspects of automobile design should be consulted during conceptual development. Although some of the information may not necessarily coincide with some human engineering design principles, the objectives are common and therefore provide a sound basis for conceptual analysis.

*Federal Motor Vehicle Safety Standards (FMVSS)*

101 Control location, identification, and illumination—passenger cars, multipurpose passenger vehicles, trucks, and buses

102 Transmission shift lever sequence, starter interlock, and transmission braking effect—passenger cars, multipurpose passenger vehicles, trucks, and buses

103 Windshield defrosting and defogging systems—passenger cars, multipurpose passenger vehicles, trucks, and buses

104 Windshield wiping and washing systems—passenger cars, multipurpose passenger vehicles, trucks, and buses

105 Hydraulic brake system—passenger cars, multipurpose passenger vehicles, trucks, and buses

107 Reflecting surfaces—passenger cars, multipurpose passenger vehicles, trucks, and buses

108 Lamps, reflective devices, and associated equipment—passenger cars, multipurpose passenger vehicles, trucks, trailers, buses, and motorcycles

111 Rearview mirrors—passenger cars and multipurpose passenger vehicles

112 Headlight concealment devices—passenger cars, multipurpose passenger vehicles, trucks, buses, and motorcycles

113 Hood latch systems—passenger vehicles, trucks, and buses

114 Theft protection—passenger cars

118 Power-operated window systems—passenger cars and multipurpose passenger vehicles

124 Accelerator control systems—passenger cars, multipurpose passenger vehicles, trucks, and buses

125 Warning devices

201 Occupant protection in interior impact—passenger cars

202 Head restraints—passenger cars

203 Impact protection for the driver from the steering control system—passenger cars

204 Steering control rearward displacement—passenger cars

205 Glazing materials—passenger cars, multipurpose passenger vehicles, motorcycles, trucks, and buses

206 Door locks and door retention components—passenger cars, multipurpose passenger vehicles, and trucks

207 Seating systems—passenger cars, multipurpose passenger vehicles, trucks, and buses

208 Occupant crash protection

209 Seat-belt assemblies—passenger cars, multipurpose passenger vehicles, trucks, and buses

210 Seat-belt assembly anchorages—passenger cars, multipurpose passenger vehicles, trucks, and buses

212 Windshield mounting—passenger cars

213 Child seating systems

214 Side-door strength—passenger cars

216 Roof crush resistance—passenger cars

302 Flammability of interior materials—passenger cars, multipurpose passenger vehicles, trucks, and buses

*Society of Automotive Engineers Standards, Recommended Practices (ARP's), and Information Reports*

SAE J100 Passenger car glazing shade bands

SAE J128 Occupant restraint system evaluation—passenger cars

SAE J140a Seat-belt hardware test procedure

SAE J141 Seat-belt hardware performance requirements

SAE J195 Automatic vehicle speed control—motor vehicles

SAE J201 In-service brake performance test—passenger cars and light-duty trucks

SAE J264 Vision glossary

SAE J268a Rearview mirrors

SAE J287 Driver hand control reach

SAE J367 Passenger car door system crush test procedure

SAE J369a Flammability of automotive interior materials—horizontal test method

SAE J374a Passenger car roof crush test procedure

SAE J383 Motor vehicle seat-belt anchorages—design recommendations

SAE J578c Color specification for electric signal lighting devices

SAE J639 Safety practices for mechanical vapor compression refrigeration equipment or systems used to cool passenger compartments of motor vehicles

SAE J673 Automotive vehicle glazing

SAE J834a Passenger car rear vision

SAE J839b Passenger car side-door latch systems

SAE J902b Passenger car windshield defrosting systems

SAE J903c Passenger car windshield wiper systems

SAE J914a Side-turn signal lamps

SAE J915 Automatic transmissions—manual control sequence

SAE J921b Motor vehicle instrument panel laboratory impact test procedure—head area

SAE J941e Motor vehicle driver's eye range

SAE J944a Steering control system—passenger car laboratory test procedure

SAE J953 Passenger car back-light defogging system

SAE J963 Anthropomorphic test device for use in dynamic testing of motor vehicles

SAE J985 Vision factors considerations in rearview mirror design

SAE J986b Sound level for passenger cars and light trucks

SAE J989 Carbon monoxide concentration test procedure

SAE J1030 Maximum sound level for passenger cars and light trucks

SAE J1048 Symbols for motor vehicle controls, indicators, and tell-tales

SAE J1050a Describing the driver's field of view

SAE J1052 Motor vehicle driver and passenger head position

## Trucks

**PLANNING CONSIDERATIONS FOR TRUCK DESIGN CONCEPTUALIZATION**  Depending on the purpose of the truck, the following functional considerations should be addressed:

1. Driver and assistant work station
2. Sleeping accommodation
3. Cab ingress and egress
4. Cargo loading and unloading provisions and package tie-down
5. Tractor-trailer connect and disconnect and trailer parking
6. Recreational vehicle (RV) camper module loading and unloading
7. Access for maintenance (cab over engine)
8. Dump-truck operation
9. Special cargo compartment environmental control (refrigeration, liquid filling and dispensing, etc.)

**DRIVER WORK STATION**  Because of the necessity to place the cab in a relatively high position and because of the obvious rear-vision obstruction created by the cargo component, large truck configurations, especially, create several unique human factors problems over and above those associated with the automobile. Refer to the section of the automobile for basic driver dimensional configurations. Although some truck configurations will require additional driver controls, it is important to recognize that truck drivers are also automobile drivers, and thus the general arrangement of controls and displays common to both types of vehicles should be similar; otherwise, truck drivers may make errors when they transfer from one type of vehicle to the other. The special truck driver work-station considerations to address are the following:

POOR

PREFERRED

### 1. Cab Ingress and Egress
When the cab is placed high above the ground, special steps and handrails, accessible door handles and locks, etc., must be examined in terms of both the mounting and the dismounting processes; i.e., it is not sufficient to place these devices wherever there is a convenient attachment point. Consider the order in which they will be used and the best location to put them in order to provide the driver or passenger the best advantage for supporting and controlling body weight and movement. And make sure that the steps are not so far from the ground (or so far apart) that the act of mounting creates too much strain on the operator or offers the operator an opportunity to take a wrong step or lose his or her grip on a handhold or rail.

### 2. Frontal Visibility
When the cab is unusually wide and high, consider the problem of the driver's ability to see the front corners of the cab in order to clear adjacent vehicles or objects. This suggests a seating arrangement that places the driver close to the windshield and high enough to provide a good down angle of view. Consider the possibility of additional viewing windows placed in the lower front portions of the cab.

### 3. Rear Visibility
Because the typical cargo compartment occludes the rear viewing potential, appropriately designed and positioned exterior mirror systems are usually necessary. Refer to the guidelines on the following pages.

### 4. Cab Environment
Trucks that are expected to perform long-duration (long-haul) operations should always provide an acceptable driver environment (in terms of temperature, ventilation, noise, and vibration) in order to ensure minimal fatigue and optimum driver performance.

### 5. Service and Maintenance
Examine the overall service and maintenance requirements of the proposed truck system as early in the conceptual phase as possible in order to preclude awkward working arrangements that often are the result of "jury-rigging" to fit the constraints of a preconceived cab configuration. This is especially critical for very large rigs, in which the height of components from the ground makes it difficult to reach these elements without using special aids. Remember that maintenance often occurs out on the road, where such aids are not available.

### 6. Cargo Loading and Unloading
For large and/or heavy cargo-handling systems, consider self-contained loading and unloading subsystems that are suitable for operation by one person. In approaching the cargo-handling problem, examine all potential loading-site situations, including special docking interfaces. The analysis of the loading and unloading sequence should include the following as a minimum: *(a)* ground-up cargo manipulation, *(b)* dock-truck exchange, and *(c)* on-board cargo movement and tie-down.

Special loading and unloading systems are required for the transfer of liquid products. Consider these in terms of spillage and especially in terms of safety aspects when chemicals or other toxic substances are involved.

### 7. Special Operating Problems
Control of a vehicle both when it is loaded and when it is unloaded is an especially critical consideration in developing the truck system concept. Although truck drivers are usually well trained, it is still important to create a system that does not perform considerably differently when it is loaded from the way it performs when it is unloaded. Be especially cognizant of the hazards created by smaller trucks when they are empty (the rear is much lighter than the front) because these types of vehicles often are driven by less skilled drivers who are unaware of the drastic change in vehicle controllability.

**DRIVER VIEWING POSITION AND CAB CONCEPT**  As the illustrations above and on the bottom of page 156 show, the cab concept has considerable influence on the viewing position of the driver. Whether the truck is small or large, the cab-over-engine configuration offers increased opportunity to place the driver where he or she can see more readily what may be directly in front of the vehicle.

As can be seen, the primary problem is the large, bulky hood, which obstructs the driver's view of the area around the front of the vehicle.

In order to provide improved viewing, however, controls and instrument panels must
also be located so as to maximize down-angle
view, as indicated in the accompanying
sketch.

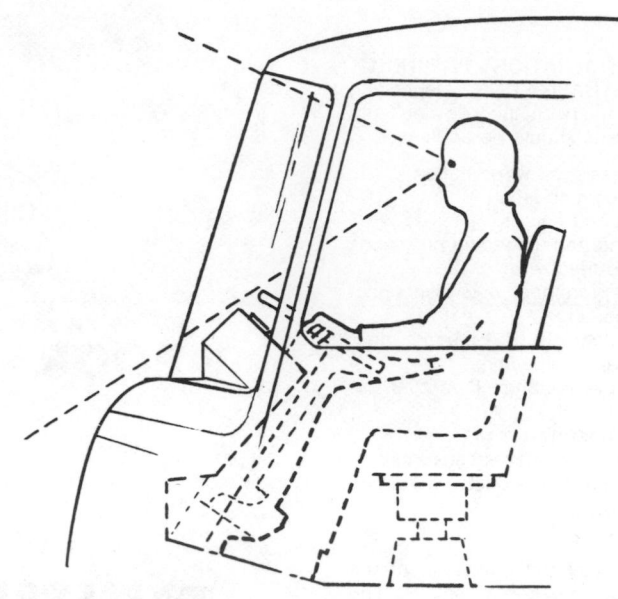

POOR

PREFERRED

Exterior mirrors located on both sides of the cab should be remotely controllable by the driver. It is unsatisfactory to depend on manual positioning because most operators forget from time to time to position these mirrors until after they are on the road.

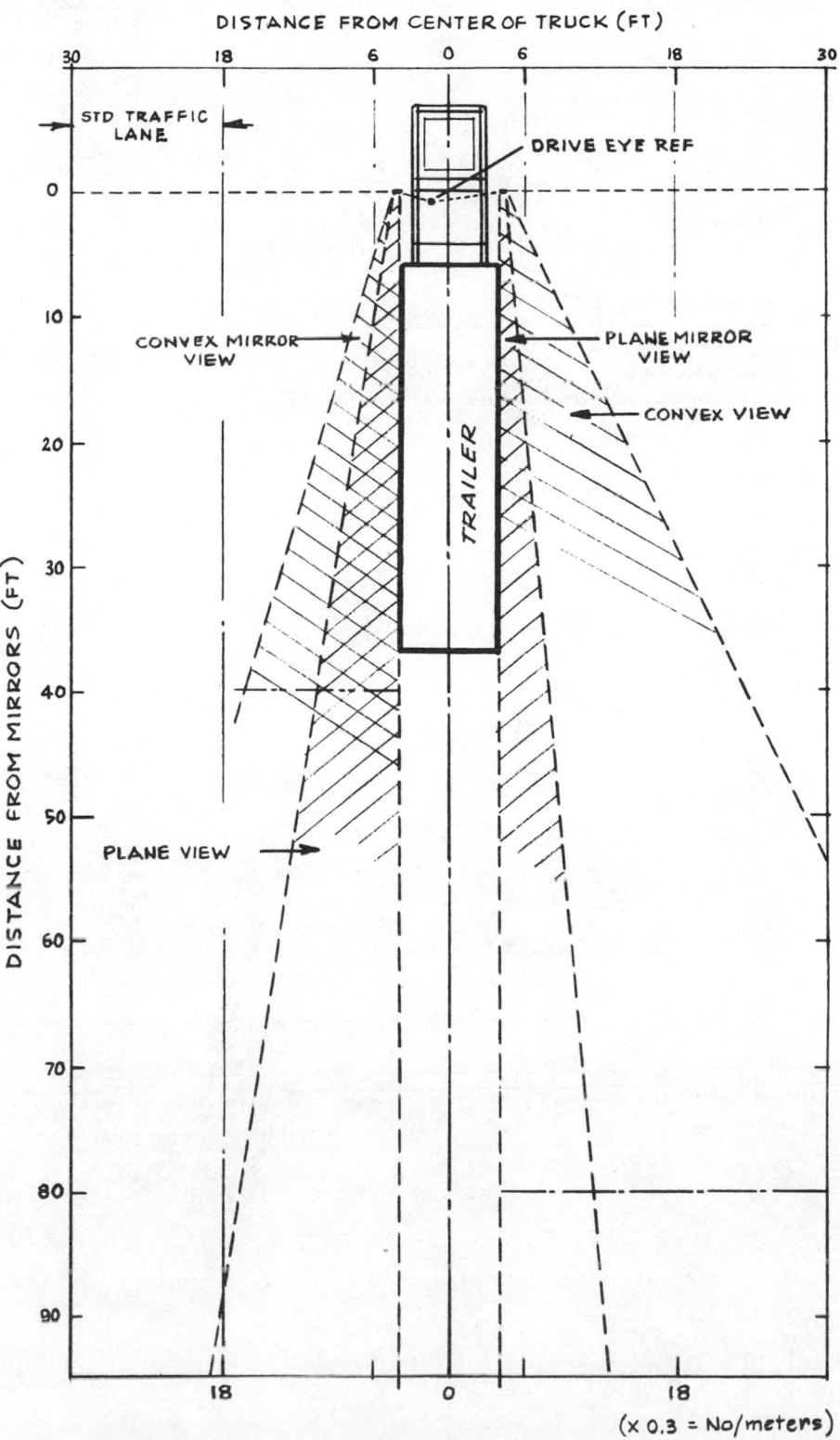

Recommended horizontal fields of view to be provided by rearviewing mirrors on tractor-trailer trucks (e.g., combination of plane and convex mirrors mounted on forward part of truck cab).

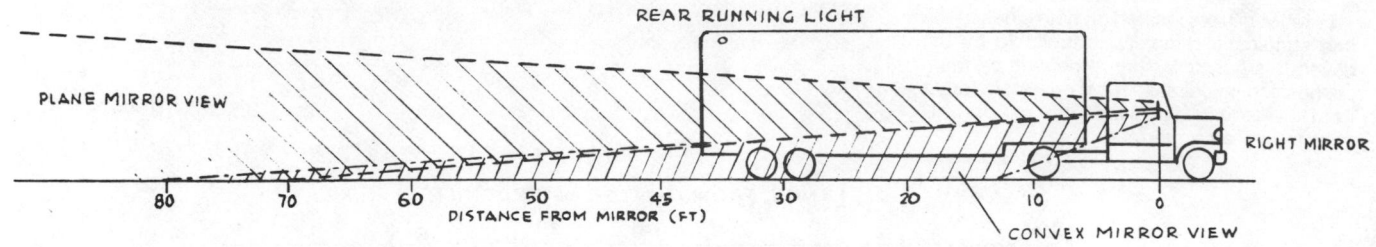

Note bottoms of fields overlap

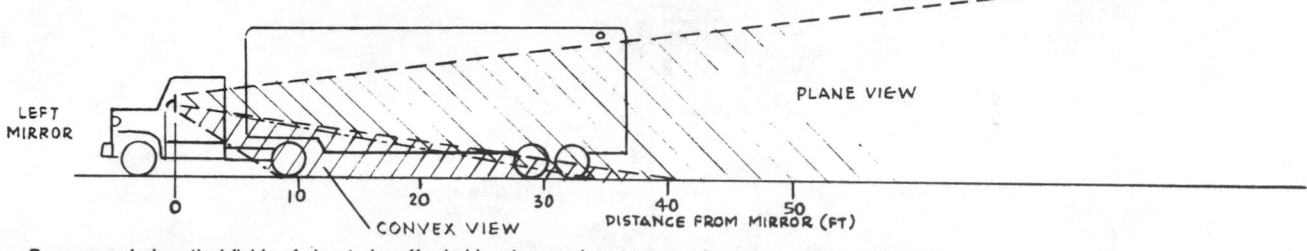

Recommended vertical fields of view to be afforded by plane and convex rearview mirror system mounted on tractor-trailer.

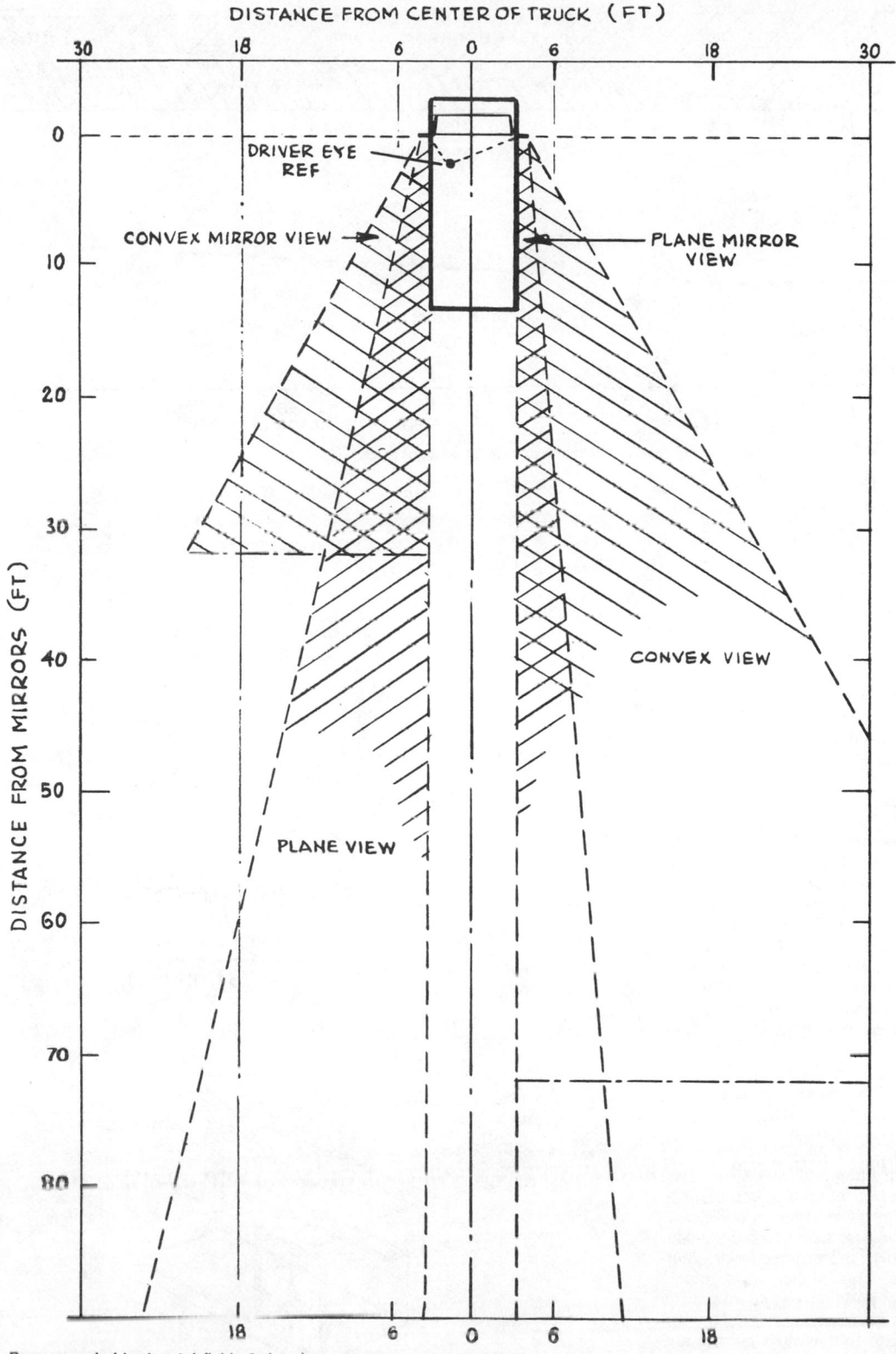

Recommended horizontal field of view for straight truck mirror system.

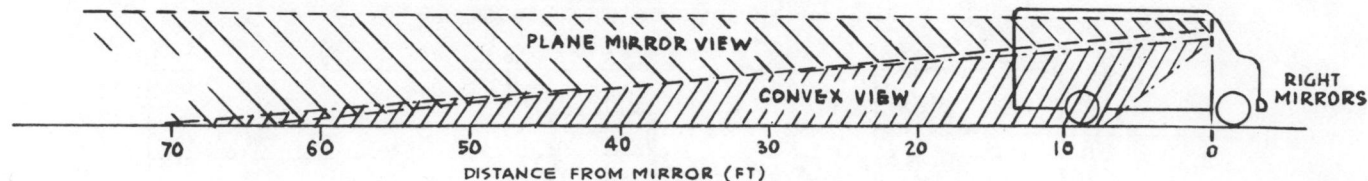

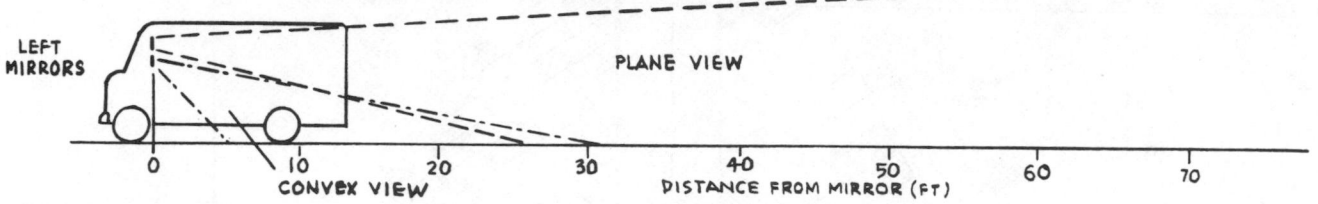

Recommended vertical field of view afforded by rearview mirror systems for straight truck.

### TRACTOR-TRAILER CAB CONSIDERATIONS

Long-haul tractor trailers impose considerable stress on drivers, and consideration should be given to minimizing these stresses, especially in terms of avoiding cramped spaces, providing ease of ingress and egress, and controlling noise and vibration.

#### 1. Access

The accompanying sketch illustrates some of the critical dimensional considerations for cab design. Special attention is directed to the boarding ladder, steps, and handrails. Too often these are located wherever there appears to be a suitable structural connection or surface. The step-handrail relationship should be such that the user is not thrown off balance as he or she mounts the steps and/or proceeds through successive steps to the cab. Note the fairly large threshold step, or platform, provided at the point where the user may have to change his or her body orientation for final cab entry and/or egress at the beginning of dismounting.

#### 2. Sleeper Cabs

Sleeper cabs are commonly used in cross-country rigs, in which case it is important to provide a sleeping cubicle that is sufficiently large to ensure the user's rest. The accompanying sketch shows preferred interior clearance dimensions. The interior walls should have some type of padding to prevent bruises if the occupant is rolled about. Acoustic treatment and adequate temperature control and ventilation should be provided. Note also the suggestion for storage of personal items, e.g., things that the user might want to take out of his or her pockets before sleeping.

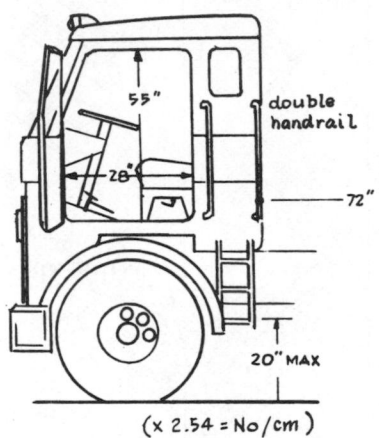

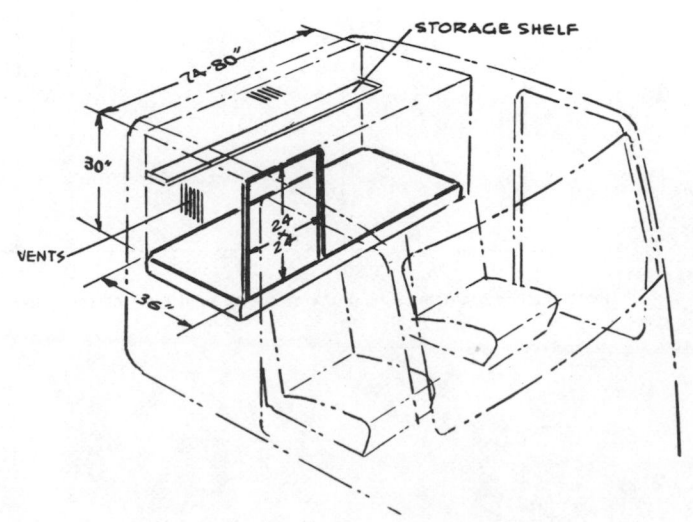

**DELIVERY VEHICLES**  Delivery trucks which are used for frequent stops and deliveries are generally designed for "stand operation"; the driver has to get in and out of the vehicle many times during a tour of duty, and standing is considered less fatiguing than getting up and down from a seat. The difficulty arises in locating foot controls where they can be used as effectively as from a seated position.

Although a reasonably satisfactory foot-pedal arrangement can be developed via the mockup approach, it is suggested that other alternatives also be considered in conceptualizing this type of vehicle. One approach would be to eliminate foot controls in favor of all hand controls. This would make it easier for the operator to place his or her feet in the best position to maintain the proper balance.

Another approach is to consider the alternative sit-stand configuration, as shown in the accompanying sketch. The seat is positioned so that the operator's relative visual and control relationships are the same, whether he or she is standing or sitting on the seat. This approach provides some important relief for the driver who has to remain on the job for considerable periods of time. Foot-control placement is more difficult with this approach, however, and the all-hand-control approach is one way of avoiding the foot-control problem.

**RECREATIONAL VEHICLES AND MOTOR-HOME VEHICLES** Although intended for a distinctly different purpose, recreational vehicles and motor homes often begin with a truck chassis and therefore are considered special-purpose trucks here. Since the variety of possible accommodations for such vehicles is almost limitless, the conceptual considerations below are limited to the most common human factors problem areas.

**1. Modular Camper Concepts**

   *a.* Ease of camper shell installation or removal is particularly important to the owner who uses the pickup for other purposes. Provide a simple, safe method by which the owner can back the truck accurately underneath the camper unit and then lower the unit for securing it to the truck.

   *b.* Provide external access to replenishment and disposal components and to external power, gas, and water hookups.

   *c.* Provide internal storage systems to prevent articles from being dislodged during travel.

   *d.* Provide at least two exits so that occupants can escape if the primary one is blocked by fire. Use only nonflammable materials in construction.

**2. Motor-Home Concepts**

Since motor-home operation is quite similar to camper operation, all the above suggestions apply.

   The major additional consideration relates to the fact that the rear of the motor home may be occupied while the vehicle is under way. Adequate occupant restraints should be provided for seats that may be used, and all interior surfaces should be made as safe as possible; i.e., rounded corners and edges, padding, and appropriately positioned handrails should be used to keep passengers from falling in the event they are walking about while the vehicle is moving.

**GENERAL ACCOMMODATIONS CHECKLIST**

   1. Toilet, sink, and shower

   2. Range, oven, refrigerator, sink, and garbage disposal

   3. Storage for food, dining and kitchen equipment and supplies, clothing, linens and housekeeping supplies and equipment, water (both hot and cold), and cooking gas

   4. Seats, tables, and beds

   5. Entertainment (radio, TV, record player, and games)

   6. Illumination and air conditioning

   7. Emergency equipment (fire extinguishers, jacks and tire repair equipment, extra gasoline and oil, road flares, first-aid kit, etc.)

**OTHER OPERATING CONSIDERATIONS**

**1. Vehicle Leveling**

In order for certain systems to function properly, it may be necessary to level the vehicle.

**2. Vehicle Stability**

Recreational vehicles are characteristically affected by wind both when they are moving and when they are parked. Appropriate schemes should be provided either to minimize the possibility that the vehicle will become unstable and uncontrollable while traveling in a high wind, or to reduce the effect of winds on the vehicle when it is parked and occupied as a temporary home.

**3. Resource Conservation**

A method for monitoring the use of certain resources is important so that stoves are not left on, supplies are not drained unnecessarily, etc.

**4. Security**

Owners of recreational vehicles often park in remote areas, which may expose them to the dangers of burglary, vandalism, or bodily harm. Windows and doors should be designed to minimize the possibility of wrongful entry.

## Public Transit Buses

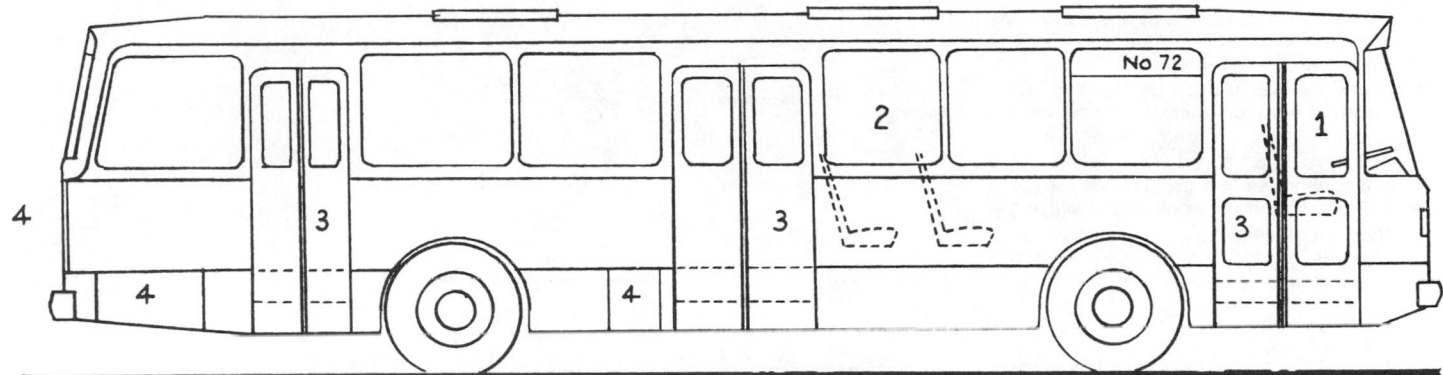

### PRIMARY HUMAN FACTORS CONSIDERATIONS FOR PUBLIC TRANSIT BUSES

1. Driver: Exterior visibility, control and display visibility and accessibility, visual monitoring of passengers, lack of reflections at night, and comfortable seat for long-duration occupancy
2. On-board passengers: Level floor, wide aisle, accessible handrails, good seats with knee and elbow clearance, visibility to look for stops (either seated or standing), illumination to read and move about safely, air conditioning, minimal noise, and reasonable ride quality
3. Boarding passengers: Adequate bus identification that can be seen at a distance and nearby and convenient entry with handrails
4. Service personnel: Convenient access to all components that frequently require servicing or maintenance

Any bus that will be used by the public should be configured so that handicapped persons and the elderly are not excluded from its use and/or put under stress because of the difficulties imposed. The principal features to consider are special wheelchair ramps or elevators, appropriately positioned handrails to aid the unsteady senior citizen, good illumination for the partially sighted, and an appropriate technique for securing wheelchairs aboard.

Nonflammable materials should be used throughout the public transit bus, and suitable window designs should be provided to allow passengers to exit from windows in an emergency.

Doorways should be wide enough for easy exit or entry. Note that the accompanying example provides two two-way doors as well as one one-way door in the rear for exit only. This makes for extra convenience and safety.

### VEHICLE CONCEPTUALIZATION FROM THE DRIVER'S POINT OF VIEW
In addition to the primary task of controlling the vehicle, the bus driver is required to perform a wide variety of other tasks. Collectively, these establish the functional requirements for many bus features. The following functional requirements are important to consider.

#### 1. Driving
a. External visibility: City buses frequently operate under crowded traffic conditions; therefore, it is important that the driver be able to see the relationship between the bus exterior boundaries and adjacent obstacles. The driver must be able to see potential passengers waiting at the curbside, so as not to run into them, and to see a small child who may have wandered in front of the bus. In order for the driver to see clearly at night, steps must be taken to minimize the reflections that interior lighting might cause on the windshield.
b. Internal visibility: The driver, of course, must be able to see the controls and instruments while driving. In addition, the driver needs to observe passengers who are standing or walking in the aisle, passengers who are creating a problem for other passengers, etc.

#### 2. Passenger Service
a. Making change: Although many system operators have gone to an exact-change or token operation because of frequent en-route robberies, the designer should be conscious of the driver change-making task, wherein the convenience of the coin machine, the need to see the fare being deposited, etc., are important design considerations.
b. Observing boarding and disembarking passengers: The good driver needs to see that passengers are boarding safely so that he or she does not start the bus up suddenly and throw them off their feet. In addition, the good driver needs to see disembarking passengers, especially at rear doors, so that he or she knows the passengers are clear before

closing a door and moving the bus.
c. Disembarking signal monitoring: The bus driver must be able to hear a disembarking passenger's signal so that he or she can stop the bus at the appropriate stop.
d. Controlling the passenger environment: The driver is responsible for providing appropriate illumination and temperature within the bus. This requires an appropriately selected and located set of controls so that the driver can find and operate them while driving.

### 3. Driver Work-Station Geometry

The accompanying illustrations show the key geometric and dimensional features required to provide a work station that is compatible with the range of adult drivers from a 5th-percentile female driver to a 95th-percentile male driver. The criteria represent a more or less ideal design that would not be restricted because of other factors such as structural thickness and curvature. Because one cannot attain these ideal geometries, these criteria should be looked upon as a guide; i.e., one should develop an adjustable mockup of the proposed work station using these data, but adjusting them as necessary to fit within the existing constraints.

When and how to compromise is often a moot question, but consider these factors:

1. Continuous, extending reaching increases driver fatigue.
2. A pedal may not be pressed as hard, or far, if it is at the limits of the driver's leg-foot reach.
3. Holding one's foot at an awkward angle (as on the accelerator) tires the driver overall, not just his or her foot or ankle.
4. A steering wheel which is too big or which is positioned in such a way that the driver has to go through drastic torso and arm motions to turn corners not only tires the driver but may also encourage "sloppy" steering.
5. An optimally designed seat, even one that appears to be too upright or too firm, may in the long run minimize fatigue with long usage periods. A seat that is too soft often adds to fatigue because it fails to provide support and because of the "bunching" it causes. Proper seat-pan angle and seat-pan-to-backrest angle are the most important contributors to fatigue minimization. A straight, rather than contoured, seat pan and back cushion are preferred. The backrest should, however, have a slight concavity in order to counteract the effects of sidesway.

It is recommended that a lap belt be provided for the driver and that the belt be an-

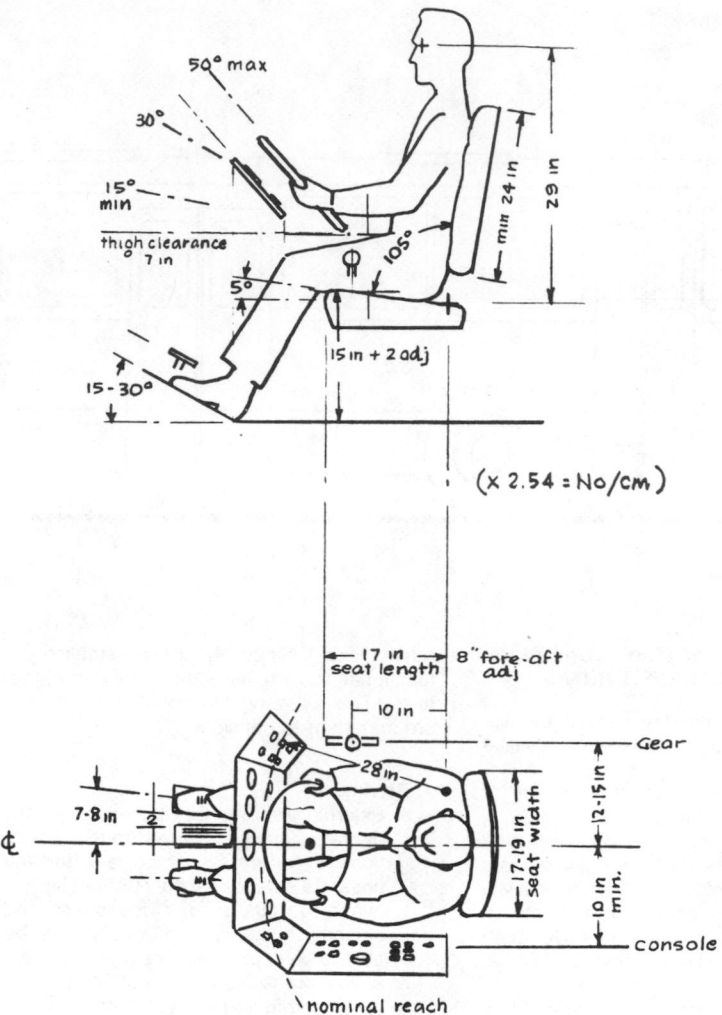

(x 2.54 = No/cm)

chored to the seat in such a way that the belt geometry will stay constant regardless of where a particular driver adjusts his or her seat.

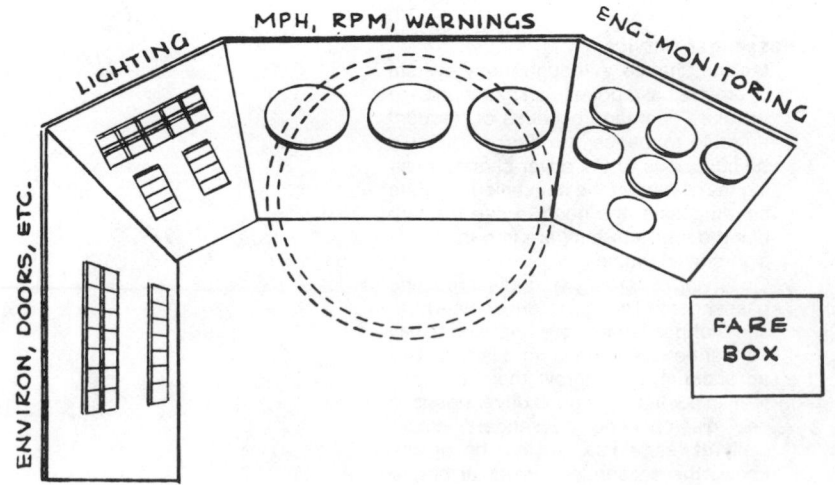

### 4. Control and Display Layout

Although various bus configurations vary in terms of the subsystems that have to be monitored and controlled by the driver, all layouts can and should be organized by system function. As noted in the accompanying illustration, electrical and environmental systems should be located on the left of the driver because these are the first systems he or she activates and checks before moving the bus; i.e., since one normally proceeds sequentially from left to right, the first elements to be used are placed on the left. The central panel contains those instruments (and related controls) that are monitored most often as the driver is maneuvering the bus; i.e., although they are not the second step in precheckout, they occupy a preferred position relative to the nominal in-motion driving task. The other subsystems, including engine, brakes, fuel, etc., are located on the right-hand panel. Finally, the fare box is placed at the most obvious driver-customer interface point.

### 5. Reflections in the Windshield

One of the serious problems in bus design (particularly the design of city buses) involves the reflections the driver sees in the windshield at night when the interior lights are on in the passenger compartment. A similar problem, but one that is more easily solved, is the reflection of the driver's instrument lights.

Although it might be possible to calculate reflection patterns, the simplest method for examining these problems and developing a reasonable solution to them is via the mockup. As the evaluation proceeds, it will become apparent that all lighting fixtures in the passenger compartment must be oriented and hooded in such a way that they will not reflect in the windshield. In addition, it will become apparent that all interior surfaces should be of a dark color. Finally, it may become necessary to provide some type of light barrier directly behind the driver to block out serious reflections, i.e., those that might cause visual confusion as the driver attempts to watch for traffic and signs out front.

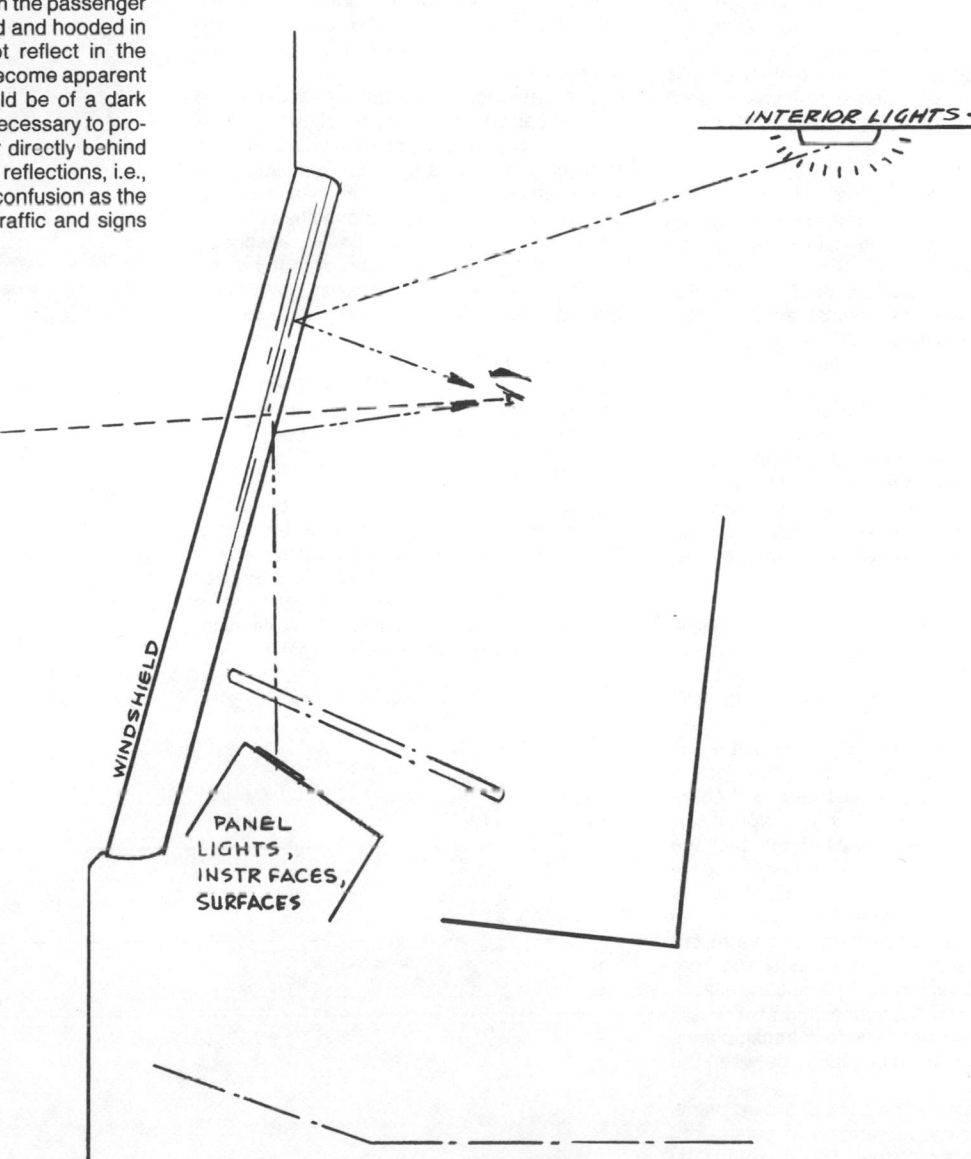

watch for reflections in the windshield

Compound curvatures in the windshield itself should be avoided whenever possible, since these create considerably more difficulty in isolating reflections.

Avoid shiny metal or plastic surfaces throughout the interior. Use dark matte finishes and/or textured coverings.

## VEHICLE CONCEPTUALIZATION FROM THE PASSENGER'S POINT OF VIEW

### 1. Entry Threshold

The initial step onto the bus should be reasonably convenient; i.e., the passenger should not be made to stretch to gain the first step up from the ground or curb. It is recommended that a ramp concept be considered for the main entrance in order to accommodate wheelchair users. There should be sufficient floor space directly adjacent to the driver and the fare box to accommodate the wheelchair user as he or she stops to make change. Appropriate handrails should be provided on both sides of the entry ramp or steps, and the steps should be well lighted at night. Nonskid treads should be provided so that passengers whose shoes are wet will not slip and fall.

### 2. Aisles

Aisles should be level so that they will not add to the possibility that standing or walking passengers may lose their footing. Aisles should be wide enough so that the standing or walking passenger does not bump seated passengers. Overhead as well as vertical and/or seat-mounted handholds should be provided so that standing passengers can steady themselves while the bus is moving. Aisles should be lighted at night (this is more appropriately accomplished when the fixtures are close to the floor, since overhead illumination may not reach the floor when the bus is crowded with standing passengers). Avoid single step-ups and step-downs; a single-step change is usually not observed, and people will trip and fall.

### 3. Seating

Fancy seats are not necessary on the typical public transit bus since the duration of occupancy is seldom very long. Seats should provide comfortable posture, however, and each seat should be wide enough to minimize inter-passenger crowding. Equally important is the fore and aft spacing; room for both knees is important to the large passenger, and some space is needed to accommodate typical carry-on packages (unless an overhead package space is to be provided, which is not usually the case for public transit buses). Armrests should be provided along the aisle, not so much to serve as a place to rest one's arm, but to provide some passenger security and freedom from encroachment by someone standing beside the seat. No more than two seats should be placed side by side in transit buses, and no armrest should be placed between the two seats.

A handrail across the back of a seat pair is suggested; it aids passengers in getting out (especially the passenger who is seated by the window), and it acts as a steadying device in the event of sudden deceleration of the bus.

Some space and special securing facilities should be provided aboard the public transit bus for accommodating wheelchair users. Although some individuals could possibly transfer from a wheelchair to a regular seat, there is still the problem of the wheelchair to be dealt with.

### 4. Other Passenger Features

Passenger reading lights should be provided. These should be designed and positioned so that they serve their purpose (i.e., merely providing such lights does not guarantee that they will illuminate the passenger's reading material properly). Care should also be taken to minimize the possible reflection of these lights in the windshield, which makes the driver's external viewing task more difficult.

Passenger facilities for notifying the driver of a desire to get off at the next stop should be located so that they can be easily reached, by both seated and standing passengers.

Exit-door systems should be designed to preclude injury to the passenger; i.e., they should be designed so that a folding door will not strike a passenger who is standing too close, pinch a passenger's hand if it is in the wrong place, etc.

The passenger environment should be as pleasant as possible. Consideration should be given to providing adequate soundproofing from engine and traffic noise and adequate temperature, humidity, and ventilation control to combat unusual weather conditions. In addition, the ride quality should be reasonably nonannoying; i.e., there should not be too much jolting, vibration, sideway, or forward pitch during turning or stopping sequences.

### 5. Vandalism

Special consideration should be given to the problem of vandalism, especially in terms of materials selected for use in the bus.

### 6. External Features

The principal external feature of concern to the passenger is bus identification. Signs should be located both on the front and on the sides of the bus. The front sign should be positioned so that sun reflection will not obscure the information. All signs should have illumination so that they can be read at night.

## PUBLIC TRANSIT BUS SEATING CRITERIA

1. The fore and aft spacing between seats should be a minimum of 14 (36 cm), and preferably 16 in (41 cm) or more.
2. The height of the leading edge of the seat pan should not exceed 14 in (36 cm) or be less than 13 in (33 cm).
3. A seat-pan angle of 5° with an included angle between the seat pan and seat back of 105° is recommended for proper posture control.

    Avoid "scoop" designs, as shown in the accompanying sketch, for they cause the occupant to slip down and forward.
4. The seat backrest should be at least 24 in (61 cm) high. The backrest should be straight up and down, not contoured. The backrest should, however, have a slight lateral concavity in order to minimize the effects of sidesway when the bus is negotiating a corner turn.
5. Handrails across the upper edge of the seat or handholds placed at the aisle corner of the seat should be provided to assist passengers in walking up and down the aisle and/or in getting out of the seat.
6. An armrest should be provided on the aisle side of the seat to help prevent standing passengers from encroaching on the seat space.

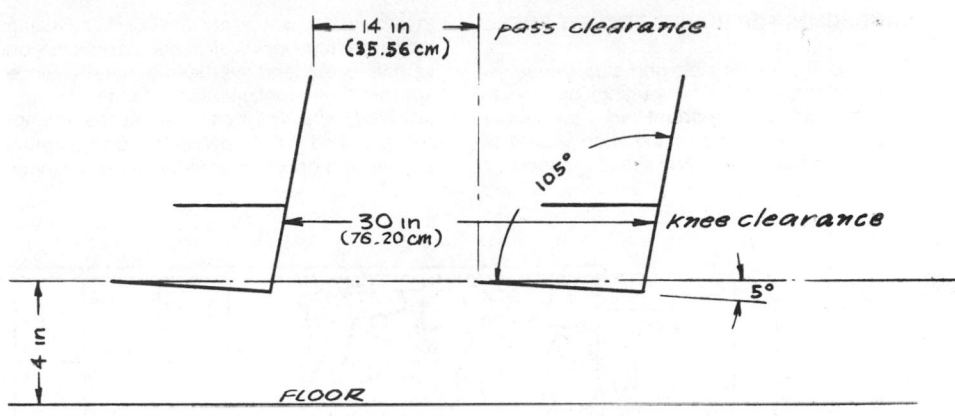

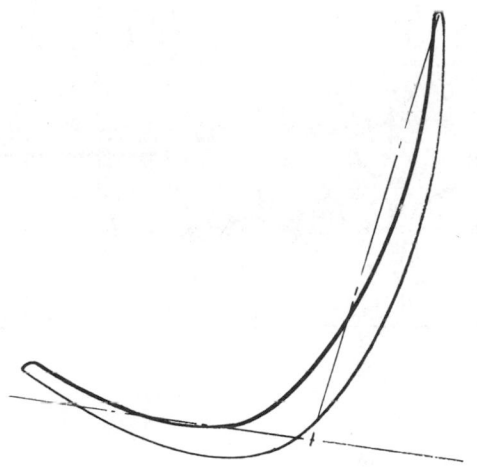

DO NOT SELECT STYLIZED CONTOUR
CONCEPT WITH BUILT-IN "SLUMP"

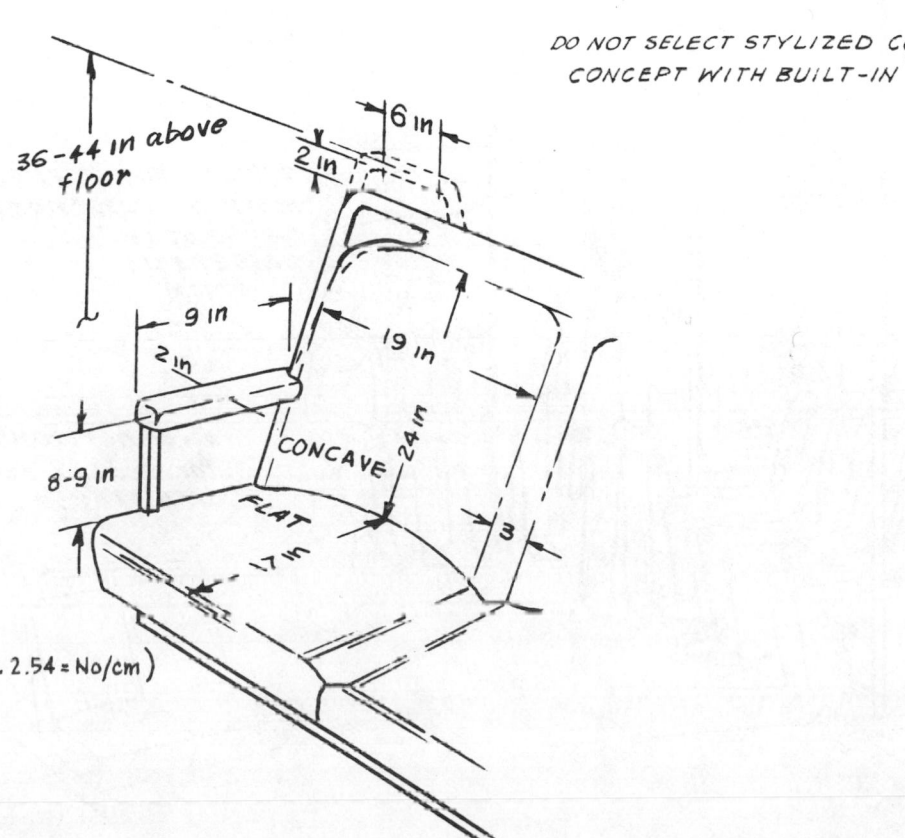

(x 2.54 = No/cm)

**HANDHOLDS FOR PUBLIC TRANSIT BUSES**

Because city buses start and stop frequently (and sometimes with little warning), handholds are particularly important for passenger safety. Also, because of the wide variety of passenger sizes and infirmities, it is important to provide a variety of handholds, including those on the backs of seats, handholds on vertical poles, and overhead handrails. Since specific bus configurations cannot be anticipated, specific recommendations cannot be provided. The particular configuration should be mocked up in full scale, and various handholds and rails should be placed experimentally to determine the type and placement that will provide maximum passenger access without interference.

Handrail diameter should be approximately 1½ in (3.8 cm).

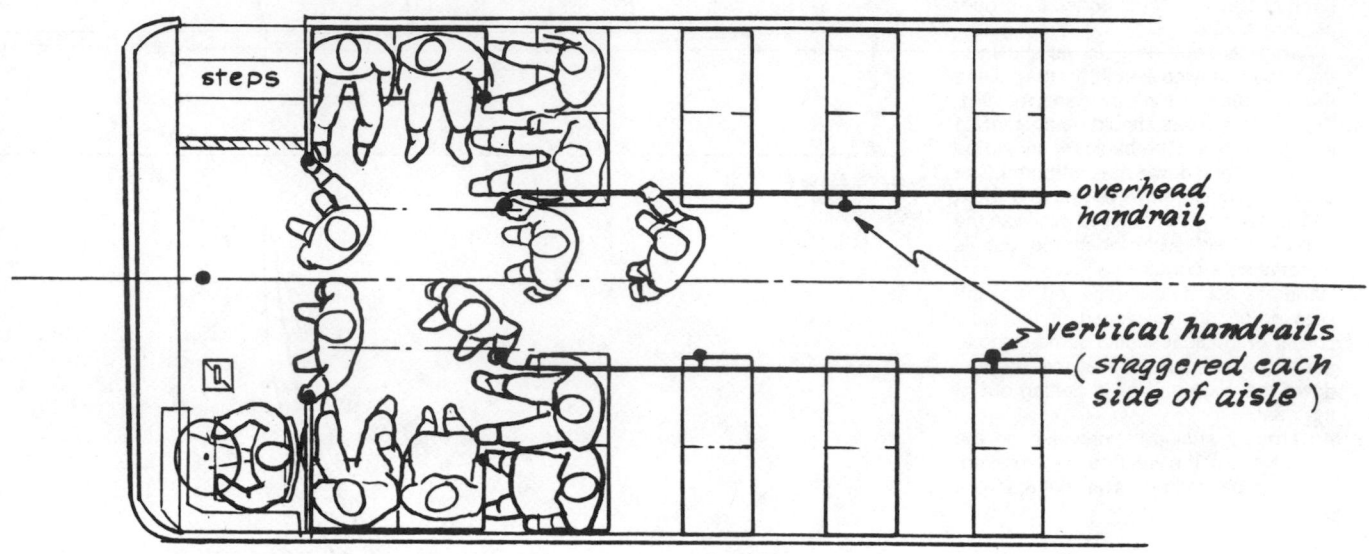

steps

*overhead handrail*

*vertical handrails (staggered each side of aisle)*

*vertical stanchions for moving passengers, overhead rails for standing passengers who may not be near stanchion,*

*lower horizontal rails for seated passenger's assistance.*

**CONSIDERATIONS FOR HANDICAPPED USERS**[33] Special provisions must be made to accommodate handicapped users who are confined to wheelchairs. The principal considerations relate to boarding and leaving the bus, maneuvering the wheelchair within the bus, securing the wheelchair against sudden stops and/or lateral forces occurring during turns, and providing special spaces for the wheelchair, both when the occupant remains in it and when he or she transfers from the wheelchair to a bus seat and keeps the folded wheelchair close by.

The accompanying sketches depict some of the possible boarding solutions that have been proposed by various bus manufacturers: a ramp entry for a full-size bus at a standard curb, a "kneeling" bus design in which a ramp provides for minor variations in curb- or street-height irregularities, and an integral elevator configuration.

Note that in all these approaches, the floor section of the main aisle must be coincident with the ramp threshold level.

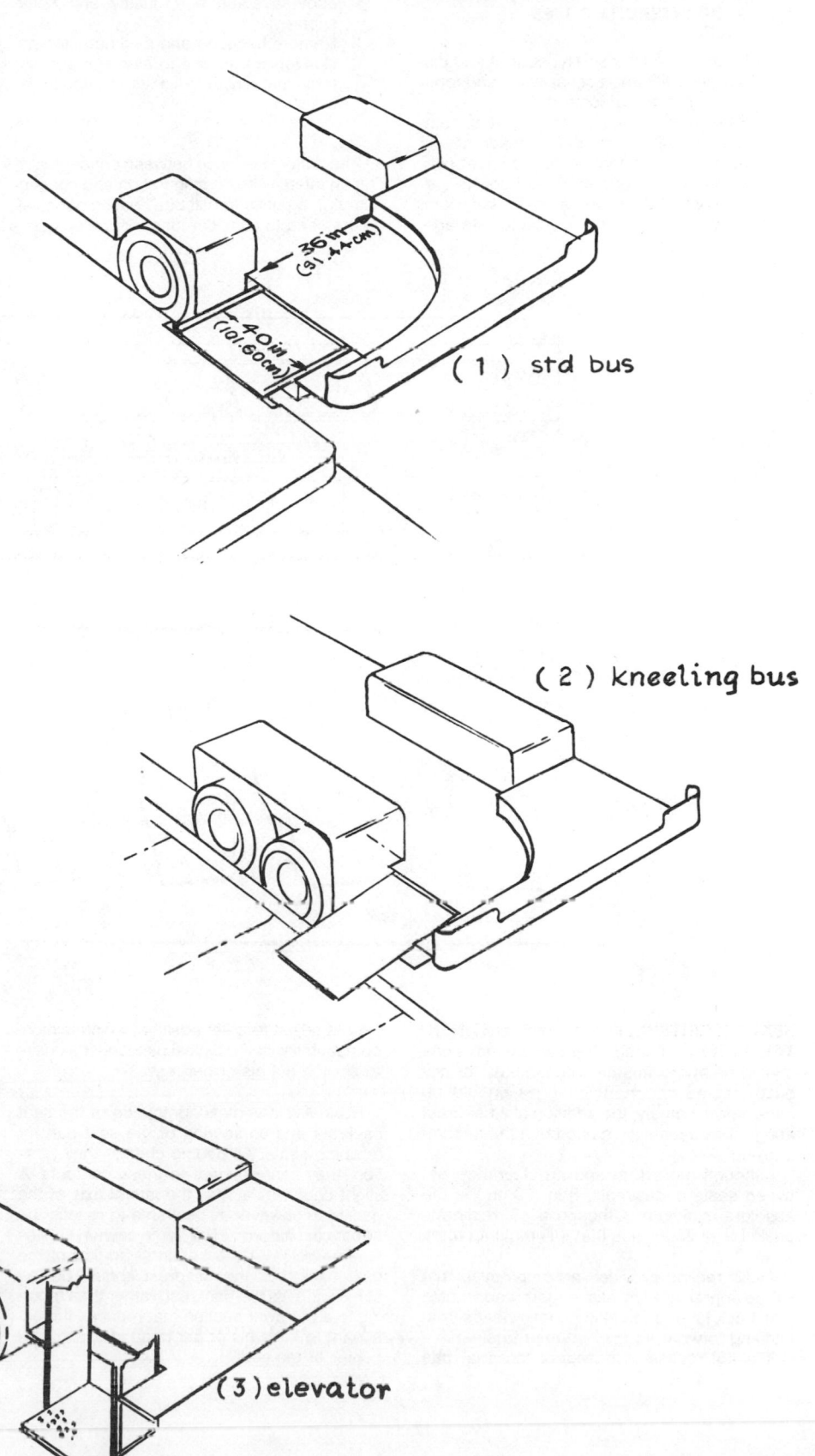

(1) std bus

(2) kneeling bus

(3) elevator

[33]T. L. Black and J. A. Mateyka, *Bus Design for the Elderly and the Handicapped*, SAE Paper 760082, Feb. 23, 1976.

## Intercity Buses

### PRIMARY HUMAN FACTORS CONSIDERATIONS FOR INTERCITY BUSES

1. Driver: Exterior visibility, control and display visibility and accessibility, and long-duration seating comfort
2. Passengers: Long-duration seating comfort, including appropriate headrest and back recline, adequate leg room, above-average environmental control (especially ventilation), an on-board rest room, local garment and package storage, reading lights, adequate aisle space to bring carry-on baggage aboard, and above-average ride quality and noise suppression
3. Remote baggage and package storage
4. Convenient access to all major subsystems that require quick turnaround maintenance

The major difference between conceptualizing an intercity bus configuration and conceptualizing a public transit bus configuration lies in the need to consider longer-duration occupancy and the special features required to reduce the effects of driver and passenger fatigue. It is also important to recognize that en-route speeds will be considerably higher than for the public transit bus system and that special precautions are therefore required with respect to possible emergencies, including the possibility of accidents. The interior should be made as noninjurious as possible to protect the driver and the passengers in the event of a crash, and provisions should be made for emergency escape and for ease of rescue by fire and accident workers in the event of a major crash.

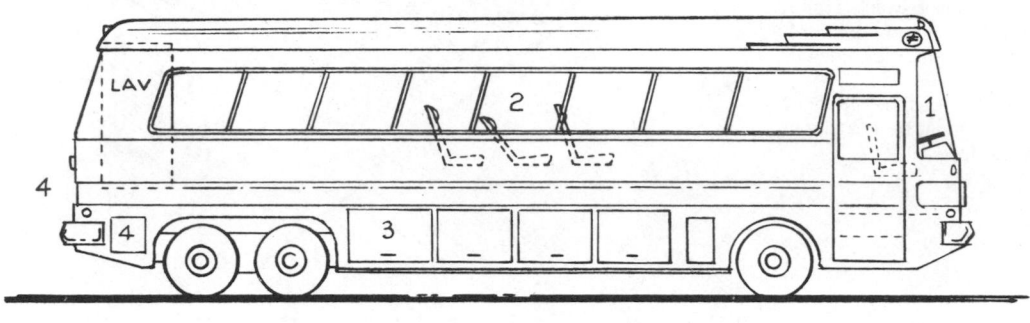

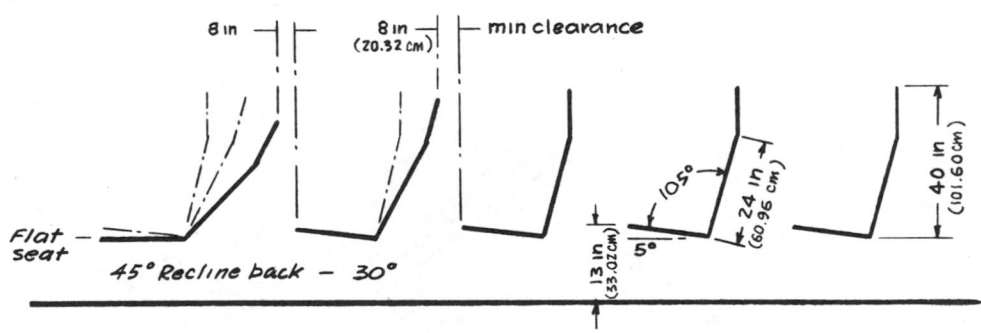

### SEATING CRITERIA FOR TRANSCONTINENTAL BUSES

Although the basic dimensions should be approximately the same as for city buses, some important changes should be considered, namely, the addition of a headrest and recline capability, as noted in the accompanying sketch.

Although more fore and aft clearance between seats is desirable, 8 in (20 cm) is the absolute minimum. Although a 45° recline is desirable, it is obvious that this requires more space.

A 30° recline provides a compromise that will be appreciated by passengers who want to lean back far enough to keep their heads from pitching forward as they attempt to sleep.

If a 45° recline is provided, the seat pan should adjust to a flat position, which is more compatible with the typical passenger's desire to stretch out his or her legs.

*Note:* Avoid vertical contouring of the seat backrest and contouring of the seat pan for reclining seats. As people change their position, they cannot adapt to these contours. A slight concavity across the lateral axis of the backrest, however, is desirable to reduce the effects of sideway. The same caution applies to the design of the headrest portion of the seat back; i.e., the headrest should be the same width as the backrest, rather than tapering to a narrower section that requires the occupant to keep his or her head in exactly the center of the seat.

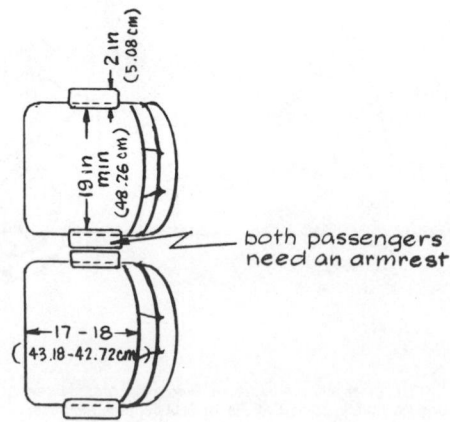

## OTHER AMENITIES AND CONSIDERATIONS

As shown in the accompanying sketch, it is sometimes desirable to place seats the equivalent of one step higher than the aisle (in order to keep the overhead lower). When such a configuration is used, provide aisle lighting so that passengers do not forget the step-down and thus stumble as they try to get out of their seats.

At least the following interior lighting should be provided:

1. Overhead general lighting for boarding
2. Aisle lighting
3. Reading lights that are selected and positioned so that one individual's light does not bother a fellow passenger who is trying to sleep

Overhead (captive) baggage compartments are desirable so that passengers can place garments and small luggage or parcels above their seats. Although open bins are common, these are not as effective as closed bins; the latter not only are easier to use (since open bin configurations must have some type of strap or elastic barrier to keep packages in place) but also prevent loose baggage from becoming a hazard in the event of a sudden stop or crash.

If smoking is to be allowed, ashtrays should be located in the armrest of each seat (to keep one passenger from dropping ashes on another as he or she tries to reach a wall-mounted tray). Mounting an ashtray on the back of a folding seat is not satisfactory because when the seat back is down, the ashtray is usually inaccessible.

Dimensions for overhead bins should be determined through mockup evaluations because the primary problem of head clearance when a passenger is entering or leaving the seat will depend on several factors that can be evaluated only in the mockup.

Seat identification should be in a conspicuous place, i.e., one that can be viewed from the standing position, and should be located so that it is obvious which seat is being referenced.

Avoid vehicle configurations in which the aisle floor is not level; not only is a change in angle difficult for certain passengers (e.g., the aged) to negotiate, but also most people will not notice the change in angle and thus may fall.

The same caution should be exercised with respect to the aisle overhead; i.e., avoid any ceiling patterns that require a passenger to lower his or her head in order to avoid contact.

It is desirable to provide air conditioning for all buses, but especially transcontinental buses. Not only does air conditioning regulate temperature and ventilation, but properly designed systems also minimize the effects of exterior noise, generated either by the bus itself or by other traffic.

For ventilation purposes, the fresh-air supply should be 18 to 20 ft³/min (8500 to 9400 cm³/min) per passenger, or a minimum of 25 percent of the total airflow required to heat or cool the vehicle. The system should be quiet in operation (i.e., it should not exceed about 70 dB).

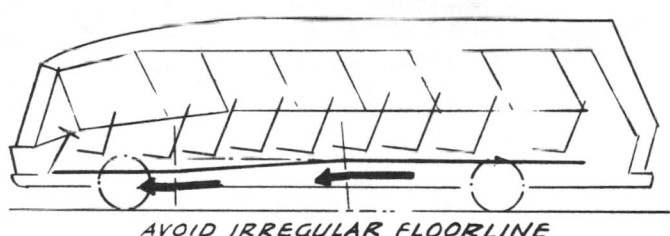

AVOID IRREGULAR FLOORLINE

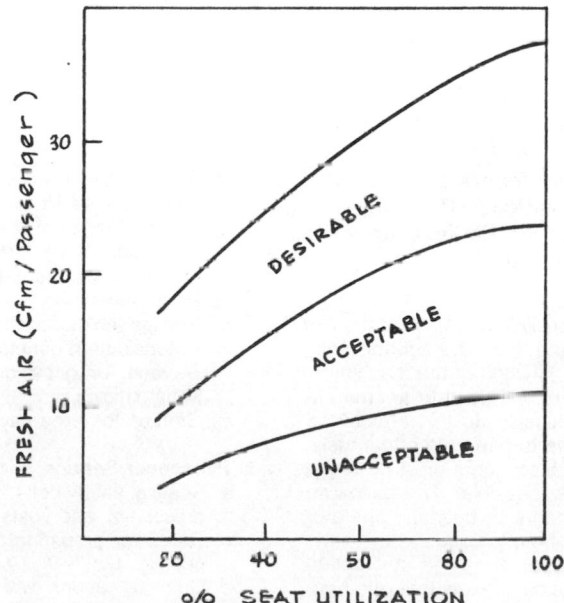

Fresh air supply necessary to suppress body odors.

It is usually not recommended that modern buses have passenger windows that open; not only does this tend to disturb some passengers, but it also creates problems for the air conditioning system and for the driver, who must close windows from time to time. Push-out safety windows for escape are, however, important.

## COMMERCIAL AIRCRAFT

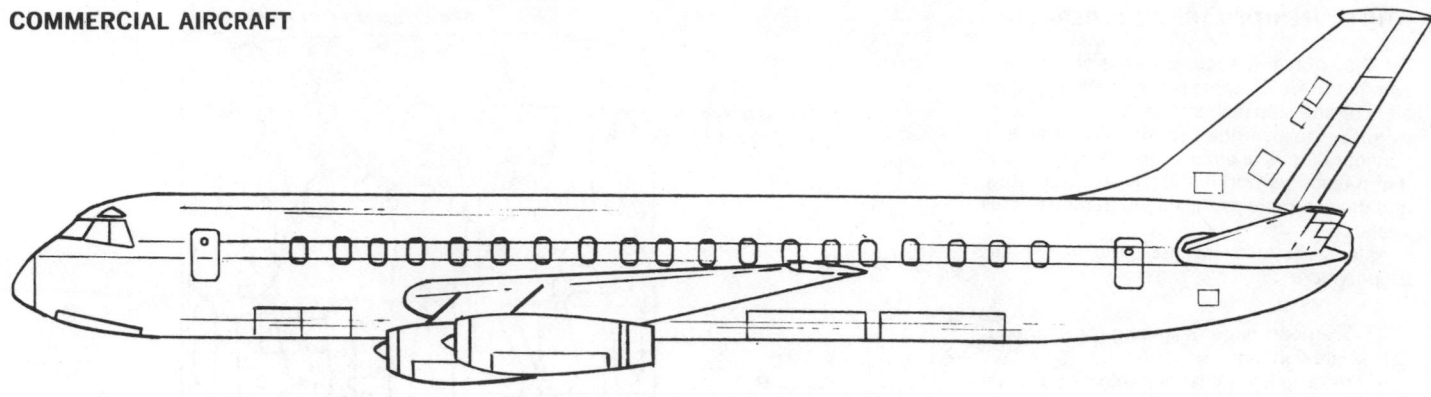

**Typical aircraft system human-machine interface areas.**

Flight Crew Work Station:
Visibility
Display and control layout
Seating restraint
Communications
Cockpit lighting

Flight Attendant Stations:
Entrance door
Galleys
Garment storage
Communications
Seating restraints
Food service equipment

Passenger Accommodations:
Seating restraints
Outside view
Eating and entertainment
Reading illumination
Parcel storage
Environmental controls
Oxygen masks

Rest rooms
Special lounges

Emergency Escape, Survival, and Rescue:
Exits, chutes, and rope
Life rafts and flotation cushions
Survival kits and first-aid supplies
Radio beacons

Exterior Lights:
Navigation, landing, and anticollision
In-flight inspection

Ground Support:
Refueling
Ground inspection
Emergency repair
Ramp towing
Baggage and cargo handling
Lavatory service
Galley service
Overhaul and repair

### Functional Requirements and Human Factors Considerations That Should Be Addressed at the Earliest Stages of Conceptual Design

Although it is obviously important that the aircraft fly properly and that it be economically feasible, the prime function of the machine is to carry people from one point to another; in order for the machine to do this effectively, various people have to perform certain tasks correctly, and the passengers must be made as comfortable as possible. The following functional requirements, by user and operator categories, should be addressed at the earliest stages of conceptual design in order to minimize the typical constraints that other engineering considerations may impose.

#### 1. Passenger Needs
a. Convenient ingress to the aircraft
b. Easy access to, and identification of, seats
c. Comfortable seats
d. Windows for external viewing
e. Safety belts

f. Reading lights, general cabin illumination, and aisle illumination
g. Flight attendant call buttons
h. Emergency oxygen masks, life preservers, cushions, and vests
i. Rest rooms
j. Serving tables for eating
k. Entertainment outlets and controls
l. Storage for outer garments and personal articles
m. Storage for large baggage

#### 2. Passenger Service Personnel Needs
a. Seating, safety belts, oxygen masks, life preservers, and vests
b. Galley for preparing food and carts for serving
c. Communications with flight crew and passengers
d. Storage for personal articles
e. Controls for operating doors, emergency exit chutes, and entertainment equipment

#### 3. Flight Crew Needs
a. Seats, safety belts, and oxygen masks (command pilot, copilot, flight engineer, and government inspector)

b. External viewing capability for taxiing, takeoff, in-flight observation, and landing
c. Controls and displays for aircraft operation, internal and external communication, and cabin environment control
d. Storage for personal articles
e. Instrument panel and general cockpit illumination

In addition, the following must be provided for all persons aboard the aircraft:

a. Cabin pressurization
b. Temperature, humidity, and ventilation control
c. Life rafts for overseas configurations

#### 4. Ground Crew Needs
a. Access for baggage and cargo, catering service, and housekeeping equipment loading and unloading
b. Access for refueling, engine inspection, and servicing
c. Access for overhaul and repair
d. Ground support equipment and communication connections
e. Means for towing

### 5. Other Needs
  a. External lights (navigation, identification, landing, anticollision, inspection, etc.)
  b. External marking and instructions for rescue operations
  c. Emergency exits, markings, and instructions
  d. On-board fire-fighting equipment and emergency lighting

## Passenger Accommodation Considerations

The following are the most common areas of complaint by airline passengers and are due to improper planning in the earlier stages of system conceptualization.

### 1. Door-to-Seat Distance
The longest distance a passenger should have to walk to get to his or her seat should not exceed about 50 ft (15 m). It should be possible for the passenger visually to survey the total array of seats from the entrance so that he or she can determine where there is an empty seat.

### 2. Seat Identification
Seat identification should be located so that it is readily seen, even when the aisle is crowded with passengers, and the identification label should provide an immediate and obvious indication of which seat is which (i.e., a window seat or an aisle seat).

### 3. Seat-to-Lavatory Distance
The longest distance a passenger should have to walk to a lavatory should not exceed about 50 ft (15 m). It should be possible for passengers to see the lavatory identification and OCCUPIED and UNOCCUPIED signs from a seated position.

### 4. Aisle Width
It should be possible for the largest expected passenger to walk down the aisle without bumping into the arms or shoulders of seated passengers.

### 5. Seat Width and Interpassenger Clearance
The distance between armrests should accommodate the largest expected passenger's hip width without the passenger's hips fitting tightly between the armrests. Armrests should be wide enough so that adjacent passengers can both rest their arms on the armrest (i.e., they should not have to compete for a place to put their arms).

### 6. Fore and Aft Seat Separation
There should be sufficient clearance between seats so that when a person reclines a seat back, the person behind still has plenty of knee clearance, room to put a good-sized magazine on his or her lap, and space to enjoy a meal without staring at the top of the front passenger's head.

### 7. Window-Seat Relationship
Window size and placement should be compatible with window use; i.e., the passenger directly adjacent to the window should not cover it up for his or her fellow passenger or have to "scrunch down" in order to see out.

### 8. Individual Passenger Service Modules
Overhead service modules should be located so that they can be reached easily by the smallest expected adult passenger. The service controls should be positioned so that each passenger can reach his or her own particular set of controls. Reading lights should be positioned and designed so that they illuminate only the specific passenger's reading material or dinner tray, not that of an adjacent passenger.

### 9. Lavatory Size and Layout
Adjacent lavatories should be arranged so that a person who is about to enter one lavatory is not struck by the door being opened by someone who is leaving the other lavatory. Lavatory doors should not face the general array of seated passengers (this is an embarrassment to many people). The interior of each lavatory should be large enough and/or arranged so that the largest expected adult passenger can maneuver inside it, including closing and opening the door. The space should also be compatible with the requirements of a mother who is trying to attend to a small child. At least one of the lavatories should be large enough (and properly equipped) for wheelchair users.

### 10. Seat-Belt and No-Smoking Signs
Such signs should be readily visible and readable from any passenger seat. The signs should also be visible and readable even when they are not illuminated so that passengers can identify them and know where to look when the captain or flight attendant indicates that they are in effect (i.e., avoid camouflaging important signs).

### 11. Seat-Reclining Controls
Place seat-reclining and other controls where the passenger can see them (positions inside the armrest or on the front, out of sight, are confusing).

### 12. Seat Design
Select seat designs that do not create special comfort problems for certain people, e.g., contours that fit only a few people's backs; seat-pan cushions that are too long for short-legged people; headrests that are too low for tall people and hit them in the shoulders, causing them to be uncomfortable; and cushions that are too soft, causing clothing to "bunch" and bind. Make sure that the upright position of the seat does not push the passenger forward into an extremely uncomfortable position. Seat arrangements should not be used that require an "inside" passenger to crawl across more than one person to get to the aisle and back again.

## Flight Crew Accommodation Considerations

The following are the most common areas of complaint by airline flight crews and are due to improper planning in the earlier stages of system conceptualization.

### 1. Flight Compartment Overhead Clearance
Flight crew members should not have to bend over or watch out for low overhead features when they enter the flight deck compartment or try to get into their seats.

### 2. Pilot Seat Ingress and Egress
There should be sufficient clearance between any center console and each of the pilot's seats so that entry or exit is convenient; it should not be necessary to take special precautions to start with the proper foot or lift one's leg over the seat.

### 3. Joy-Stick (Wheel) Clearance
The main control device should be designed and/or mounted so that the pilot does not have to go through a series of contortions in order to get past the control column when getting into or out of his or her seat. The wheel should be designed so that the pilot can easily view any displays located directly forward of the wheel. No panel controls should be placed in such a way that the pilot has to reach through or around the control wheel to operate them.

### 4. Centrally Located Shared Controls and Displays
Either pilot should be able to reach shared controls without having to lean an appreciable distance. Instruments should be designed and/or positioned so that there are no parallax problems for either pilot (e.g., numbers, scales, or labels that are covered by bezels or other controls and pointer-scale relationships that are distorted because of the angle of view).

### 5. Instrument-Panel Illumination
All displays and labels should be equally readable during the day and at night, and the illumination system should be of even quality and brightness so that, if a pilot turns down the illumination level, one or more of the indicia do not disappear.

### 6. Window Reflections
By means of proper selection, positioning, etc., colors and lights used in the cockpit should be prevented from reflecting in the windshield or side windows and preventing the pilots from seeing out. The faces of instruments should be tilted so that the cover glass does not reflect the pilot's bright shirt front.

### 7. Crew Seat Design
All crew seats should be designed to minimize possible fatigue and stress due to long periods of use. Good posture control is vital; i.e., avoid soft cushions, sling backs, etc.

### 8. Adjustable Foot Pedals
Foot pedals should be adjustable so that the short-legged pilot can bring them within reach without having to move the seat forward; remote adjustment is desirable.

## Pilot Visibility

Flight safety requires that the pilot be provided maximum opportunity to see outside the aircraft in all directions. However, for practical reasons it is usually necessary to settle for something less. The external viewing requirements are illustrated by the accompanying sketches.

### 1. Standard Commercial Aircraft
  a. Normal in-flight operation requires a general view over the instrument panel and about 30 to 40° upward vision.
  b. A minimum of 15° down vision is required for landing, taxiing, and visual communication with the ground crew.
  c. Aft vision is required to inspect engines during flight.

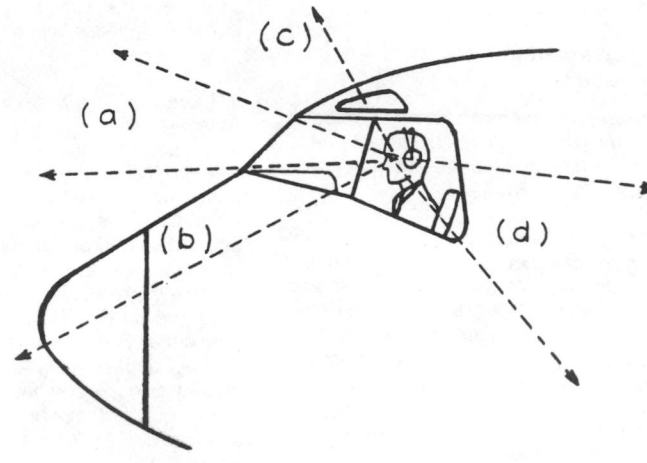

### 2. New High-Performance Aircraft
Aircraft such as the new supersonic type land at such extreme angles that special arrangements must be made to lower the nose section, as shown.

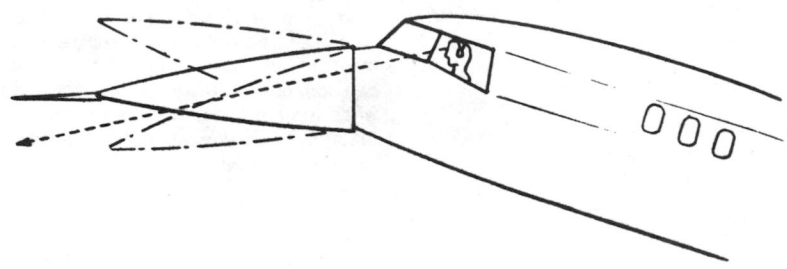

### 3. Small Commercial and Private Aircraft
Maximum opportunity usually exists for increasing pilot visibility all around, as shown in the accompanying sketch. If a high wing concept is proposed, there should be generous overhead visibility since operations within the typical small commercial field require constant surveillance to avoid accidents.

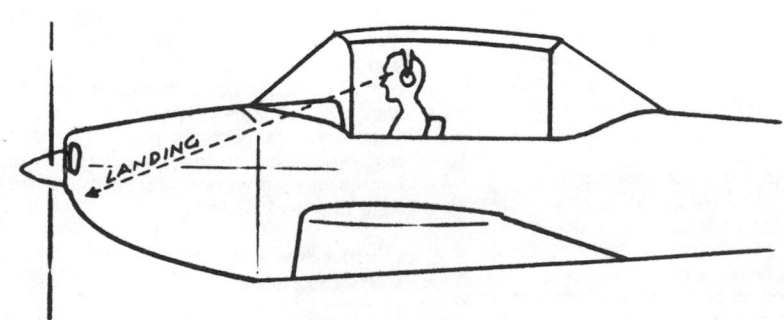

*Note:* Because of the considerable number of instruments that must be located in front of the pilot, panels tend to extend vertically and obstruct forward viewing. However, pilots do not necessarily use direct forward view for the most critical phase of flying; rather, they normally look about 30 to 40° to the side so that they can judge both forward and downward progress simultaneously.

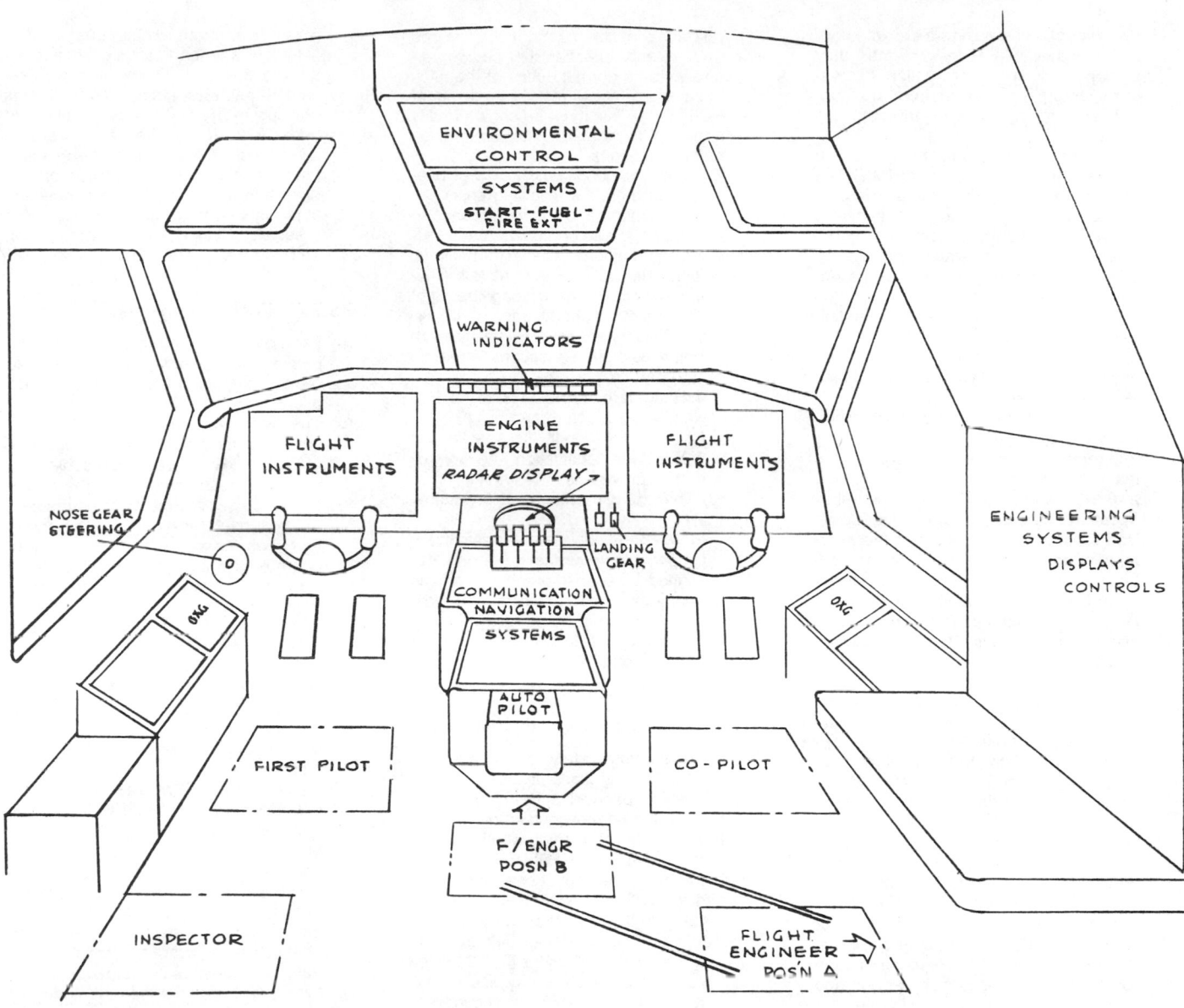

**Typical commercial flight deck layout convention.**

### Pertinent Federal Aviation Administration Regulations[34]

Before beginning the design of any commercial aircraft, the designer should review FAA requirements pertaining to crew and passenger accommodation, such as the following.

*Criteria for determining minimum flight crew.* The following are considered by the FAA in determining the minimum flight crew under § 25.1523:

a. *Basic workload functions.* The following basic workload functions are considered:
   (1) Flight path control.
   (2) Collision avoidance.
   (3) Navigation.
   (4) Communications.

[34]U.S. Department of Transportation. Federal Aviation Administration, *Federal Aviation Regulations*, vol. III.

   (5) Operation and monitoring of aircraft engines and systems.
   (6) Command decisions.

b. *Workload factors.* The following work load factors are considered significant when analyzing and demonstrating workload for minimum flight crew determination:

   (1) The accessibility, ease and simplicity of operation of all necessary flight, power, and equipment controls, including emergency fuel shutoff valves, electrical controls, electronic controls, pressurization system controls, and engine controls.
   (2) The accessibility and conspicuity of all necessary instruments and failure warning devices such as fire warning, electrical system malfunction, and other failure or caution indicators. The extent to which such instruments or devices direct the proper corrective action is also considered.
   (3) The number, urgency, and complexity of operating procedures with particular consideration given to the specific fuel management schedule imposed by center of gravity, structural or other considerations of an air-worthiness nature, and to the ability of each engine to operate at all times from a single tank or source which is automatically replenished if fuel is also stored in other tanks.
   (4) The degree and duration of concentrated mental and physical effort involved in normal operation and in diagnosing and coping with malfunctions and emergencies.
   (5) The extent of required monitoring of the fuel, hydraulic, pressurization, electrical, electronic, deicing, and other systems while en route.

(6) The actions requiring a crewmember to be unavailable at his assigned duty station, including: observation of systems, emergency operation of any control, and emergencies in any compartment.

(7) The degree of automation provided in the aircraft systems to afford (after failures or malfunctions) automatic crossover or isolation of difficulties to minimize the need for flight crew action to guard against loss of hydraulic or electric power to flight controls or to other essential systems.

(8) The communications and navigation workload.

(9) The possibility of increased workload associated with any emergency that may lead to other emergencies.

(10) Incapacitation of a flight crewmember whenever the applicable operating rule requires a minimum flight crew of at least two pilots.

c. *Kind of operation authorized.* The determination of the kind of operation authorized requires consideration of the operating rules under which the airplane will be operated. Unless an applicant desires approval for a more limited kind of operation, it is assumed that each airplane certificated under this Part will operate under IFR conditions.

## § 25.771 Pilot compartment.

(a) Each pilot compartment and its equipment must allow the minimum flight crew (established under § 25.1523) to perform their duties without unreasonable concentration or fatigue.

(b) The primary controls listed in § 25.779(a), excluding cables and control rods, must be located with respect to the propellers so that no member of the minimum flight crew (established under § 25.1523), or part of the controls, lies in the region between the plane of rotation of any inboard propeller and the surface generated by a line passing through the center of the propeller hub making an angle of five degrees forward or aft of the plane of rotation of the propeller.

(c) If provision is made for a second pilot, the airplane must be controllable with equal safety from either pilot seat.

(d) The pilot compartment must be constructed so that, when flying in rain or snow, it will not leak in a manner that will distract the crew or harm the structure.

(e) Vibration and noise characteristics of cockpit equipment may not interfere with safe operation of the airplane.

## § 25.773 Pilot compartment view.

(a) *Nonprecipitation conditions.* For nonprecipitation conditions, the following apply:

(1) Each pilot compartment must be arranged to give the pilots a sufficiently extensive, clear, and undistorted view, to enable them to safely perform any maneuvers within the operating limitations of the airplane, including taxiing, takeoff, approach, and landing.

(2) Each pilot compartment must be free of glare and reflection that could interfere with the normal duties of the minimum flight crew (established under § 25.1523). This must be shown in day and night flight tests under nonprecipitation conditions.

(b) *Precipitation conditions.* For precipitation conditions, the following apply:

(1) The airplane must have a means to maintain a clear portion of the windshield, during precipitation conditions, sufficient for both pilots to have a sufficiently extensive view along the flight path in normal flight attitudes of the airplane. This means must be designed to function, without continuous attention on the part of the crew, in—

(i) Heavy rain at speeds up to 1.6 $Vs_1$, with flaps retracted; and

(ii) The icing conditions specified in § 25.1419 if certification with ice protection provisions is requested.

(2) The first pilot must have a window that—

(i) When the cabin is not pressurized, is openable under the conditions prescribed in subparagraph (1) of this paragraph and provides the view specified in that paragraph; and

(ii) Gives sufficient protection from the elements against impairment of the pilot's vision.

## § 25.775 Windshields and windows.

(a) Nonsplintering safety glass must be used in internal glass panes.

(b) Windshield panes directly in front of the pilots in the normal conduct of their duties, and the supporting structures for these panes, must withstand, without penetration, the impact of a four-pound bird when the velocity of the airplane (relative to the bird along the airplane's flight path) is equal to the value of $Vc$, at sea level, selected under § 25.335(a).

(c) Unless it can be shown by analysis or tests that the probability of occurrence of a critical windshield fragmentation condition is of a low order, the airplane must have a means to minimize the danger to the pilots from flying windshield fragments due to bird impact. This must be shown for each transparent pane in the cockpit that—

(1) Appears in the front view of the airplane;

(2) Is inclined 15 degrees or more to the longitudinal axis of the airplane; and

(3) Has any part of the pane located where its fragmentation will constitute a hazard to the pilots.

(d) The design of windshields and windows in pressurized airplanes must be based on factors peculiar to high altitude operation, including the effects of continuous and cyclic pressurization loadings, the inherent characteristics of the material used, and the effects of temperatures and temperature differentials. The windshield and window panels must be strong enough to withstand the maximum cabin pressure differential loads combined with critical aerodynamic pressure and temperature effects, after failure of any load-carrying element of the windshield or window. It may be assumed that, after a single failure that is obvious to the flight crew (established under § 25.1523), the cabin pressure differential is reduced from the maximum, in accordance with appropriate operating limitations, to allow continued safe flight of the airplane with a cabin pressure altitude of not more than 15,000 feet.

## § 25.777 Cockpit controls.

(a) Each cockpit control must be located to provide convenient operation and to prevent confusion and inadvertent operation.

(b) The direction of movement of cockpit controls must meet the requirements of § 25.779. Wherever practicable, the sense of motion involved in the operation of other controls must correspond to the sense of the effect of the operation upon the airplane or upon the part operated. Controls of a variable nature using a rotary motion must move clockwise from the off position, through an increasing range, to the full on position.

(c) The controls must be located and arranged, with respect to the pilots' seats, so that there is full and unrestricted movement of each control without interference from the cockpit structure or the clothing of the minimum flight crew (established under § 25.1523) when any member of this flight crew, from 5'2" to 6'0" in height, is seated with the seat belt fastened.

(d) Identical powerplant controls for each engine must be located to prevent confusion as to the engines they control.

(e) Wing flap controls and other auxiliary lift device controls must be located on top of the pedestal, aft of the throttles, centrally or to the right of the pedestal centerline, and not less than 10 inches aft of the landing gear control.

(f) The landing gear control must be located forward of the throttles and must be operable by each pilot when seated with seat belts fastened.

(g) Control knobs must be shaped in accordance with § 25.781. In addition, the knobs must be of the same color, and this color must contrast with the color of control knobs for other purposes and the surrounding cockpit.

(h) If a flight engineer is required as part of the minimum flight crew (established under § 25.1523), the airplane must have a flight engineer station located and arranged so that the flight crewmembers can perform their functions efficiently and without interfering with each other.

## § 25.779 Motion and effect of cockpit controls.

Cockpit controls must be designed so that they operate in accordance with the following movement and actuation:

(a) Aerodynamic controls:
   (1) *Primary.*

| Controls | Motion and effect |
|---|---|
| Aileron | Right (clockwise) for right wing down. |
| Elevator | Rearward for nose up. |
| Rudder | Right pedal forward for nose right. |

   (2) *Secondary.*

| Controls | Motion and effect |
|---|---|
| Flaps (or auxilliary lift devices). | Forward for flaps up: rearward for flaps down. |
| Trim tabs (or equivalent). | Rotate to produce similar rotation of the airplane about an axis parallel to the axis of the control. |

(b) Powerplant and auxiliary controls:
   (1) *Powerplant.*

| Controls | Motion and effect |
|---|---|
| Throttles | Forward to increase forward thrust and rearward to increase rearward thrust. |
| Propellers | Forward to increased rpm. |
| Mixture | Forward or upward for rich. |
| Carburetor air heat. | Forward or upward for cold. |
| Super-charger. | Forward or upward for low blower. For turbosuper-chargers, forward, upward, or clockwise, to increase pressure. |

   (2) *Auxiliary.*

| Controls | Motion and effect |
|---|---|
| Landing gear | Down to extend. |

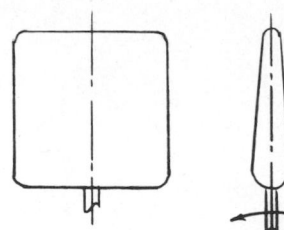 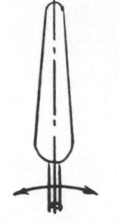

**FLAP CONTROL KNOB**

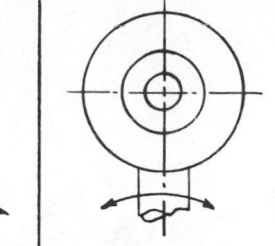 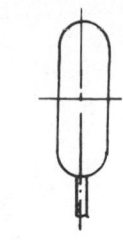

**LANDING GEAR CONTROL KNOB**

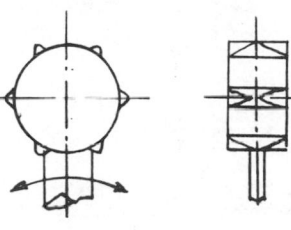

**MIXTURE CONTROL KNOB**

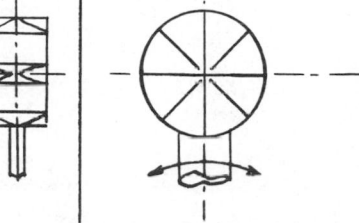

**SUPERCHARGER CONTROL KNOB**

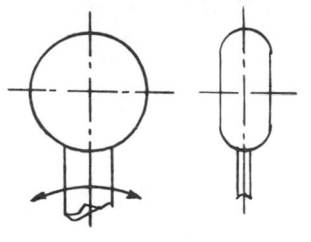

**THROTTLE CONTROL KNOB**

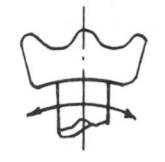 

**RPM CONTROL KNOB**

## § 25.781 Cockpit control knob shape.

Cockpit control knobs must conform to the general shapes (but not necessarily the exact sizes or specific proportions) in the accompanying figure:

## § 25.1411 General.

(a) *Accessibility.* Required safety equipment to be used by the crew in an emergency such as automatic liferaft releases, must be readily accessible.

(b) *Stowage provisions.* Stowage provisions for required emergency equipment must be furnished and must—
(1) Be arranged so that the equipment is directly accessible and its location is obvious; and
(2) Protect the safety equipment from inadvertent damage.

(c) *Emergency exit descent device.* The stowage provisions for the emergency exit descent device required by § 25.807(c)(4) must be at the exits for which they are intended.

(d) *Liferafts.* The stowage provisions for the liferafts described in § 25.1415 must accommodate enough rafts for the maximum number of occupants for which certification for ditching is requested. Liferafts must be stowed near exits through which the rafts can be launched during an unplanned ditching. Rafts automatically or remotely released outside the airplane must be attached to the airplane by means of the static line prescribed in § 25.1415.

(e) *Long-range signaling device.* The stowage provisions for the long-range signaling device required by § 25.1415 must be near an exit available during an unplanned ditching.

(f) *Life preserver stowage provisions.* The stowage provisions for life preservers described in §25.1415 must accommodate one life preserver for each occupant for which certification for ditching is requested. Each life preserver must be within easy reach of each seated occupant.

(g) *Life line stowage provisions.* If certification for ditching under § 25.801 is requested, there must be provisions to store life lines. These provisions must—
(1) Allow one life line to be attached to each side of the fuselage; and
(2) Be arranged to allow the life lines to be used to enable the occupants to stay on the wing after ditching.

## § 25.1415 Ditching equipment.

(a) Ditching equipment used in airplanes to be certificated for ditching under § 25.801, and operated under the operating rules of this chapter, must meet the requirements of this section.

(b) Each liferaft and each life preserver must be approved. In addition—
(1) Unless excess rafts of enough capacity are provided, the buoyancy and seating capacity beyond the rated capacity of the rafts must accommodate all occupants of the airplane in the event of a loss of one raft of the largest rated capacity; and
(2) Each raft must have a trailing line, and must have a static line designed to hold the raft near the airplane but to release it if the airplane becomes totally submerged.

(c) Approved survival equipment must be attached to each liferaft.

(d) There must be an approved long-range signaling device for use in one liferaft.

(e) For airplanes not certificated for ditching under § 25.801 and not having approved life preservers, there must be an approved flotation means for each occupant. This means must be within easy reach of each seated occupant and must be readily removable from the airplane.

## § 25.783 Doors.

(a) Each cabin must have at least one easily accessible external door.

(b) There must be a means to lock and safeguard each external door against

opening in flight (either inadvertently by persons or as a result of mechanical failure). Each external door must be openable from both the inside and the outside, even though persons may be crowded against the door on the inside of the airplane. Inward opening doors may be used if there are means to prevent occupants from crowding against the door to an extent that would interfere with the opening of the door. The means of opening must be simple and obvious and must be arranged and marked so that it can be readily located and operated, even in darkness. Auxiliary locking devices may be used.

(c) Each external door must be reasonably free from jamming as a result of fuselage deformation in a minor crash.

(d) Each external door must be located where persons using them will not be endangered by the propellers when appropriate operating procedures are used.

(e) There must be a provision for direct visual inspection of the locking mechanism by crewmembers to determine whether external doors, for which the initial opening movement is outward (including passenger, crew, service, and cargo doors), are fully locked. In addition, there must be a visual means to signal to appropriate crewmembers when normally used external doors are closed and fully locked.

(f) Cargo and service doors not suitable for use as an exit in an emergency need only meet paragraph (e) of this section and be safeguarded against opening in flight as a result of mechanical failure.

(g) Each passenger entry door in the side of the fuselage must qualify as a Type A, Type I, or Type II passenger emergency exit and must meet the requirements of §§ 25.807 through 25.813 that apply to that type of passenger emergency exit. If an integral stair is installed at such a

**Comparison of Average Emergency Escape Times for Various Descent Devices**

| Descent device | Average time (sec) | Number of civilian subjects | Average time (sec) | Number military subjects |
|---|---|---|---|---|
| Slide | 5.18 | 31 | 1.75 | 14 |
| Ladder | 10.61 | 31 | 3.45 | 14 |
| Rope | 9.36 | 22 | 1.69 | 14 |
| Jacob's ladder | 11.52 | 25 | 5.38 | 14 |

passenger entry door, the stair must be designed so that when subjected to the inertia forces specified in § 25.561, and following the collapse of one or more legs of the landing gear, it will not interfere to an extent that will reduce the effectiveness of emergency egress through the passenger entry door.

## § 25.805 Flight crew emergency exits.

Except for airplanes with a passenger capacity of 20 or less in which the proximity of passenger emergency exits to the flight crew area offers a convenient and readily accessible means of evacuation for the flight crew, the following apply:

(a) There must be either one exit on each side of the airplane or a top hatch, in the flight crew area.

(b) Each exit must be of sufficient size and must be located so as to allow rapid evacuation of the crew. An exit size and shape of other than at least 19 by 20 inches unobstructed rectangular opening may be used only if exit utility is satisfactorily shown, by a typical flight crewmember, to the Administrator.

## § 25.807 Passenger emergency exits.

(a) *Type and location.* For the purpose of this Part, the types and locations of exits are as follows:

(1) *Type I.* This type must have a rectangular opening of not less than 24 inches wide by 48 inches high, with corner radii not greater than one-third the width of the exit. Type I exits must be floor level exits.

(2) *Type II.* This type must have a rectangular opening of not less than 20 inches wide by 44 inches high, with corner radii not greater than one-third the width of the exit. Type II exits must be floor level exits unless located over the wing, in which case they may not have a step-up inside the airplane of more than 10 inches nor a stepdown outside the airplane of more than 17 inches.

(3) *Type III.* This type must have a rectangular opening of not less than 20 inches wide by 36 inches high, with corner radii not greater than one-third the width of the exit, located over the wing, with a step-up inside the airplane of not more than 20 inches and a stepdown outside the airplane of not more than 27 inches.

(4) *Type IV.* This type must have a rectangular opening of not less than 19 inches wide by 26 inches high, with corner radii not greater than one-third the width of the exit, located over the wing, with a step-up inside the airplane of not more than 29 inches and a stepdown outside the airplane of not more than 36 inches.

(5) *Ventral.* This type is an exit from the passenger compartment through the pressure shell and the bottom fuselage skin. The dimensions and physical configuration of this type of exit must allow at least the same rate of egress as a Type I with the airplane in the normal ground attitude, with landing gear extended.

(6) *Tail cone.* This type is an aft exit from the passenger compartment through the pressure shell and through an openable cone of the fuselage aft of the pressure shell. The means of opening the tail cone must be simple and obvious, and must employ a single operation.

(7) *Type A.* An emergency exit may be designated as a Type A exit if the following criteria are met:

(i) There must be a rectangular opening not less than 42 inches wide by 72 inches high, with corner radii not greater than one-sixth of the width of the exit.

(ii) It must be a floor level exit.

(iii) Unless there are two or more main (fore and aft) aisles, the exit must be located so that there is passenger flow along the main aisle to that exit from both the forward and aft direction.

(iv) There must be an unobstructed passageway at least 36 inches wide leading from each exit to the nearest main aisle.

(v) If two or more main aisles are provided, there must be unobstructed cross aisles at least 20 inches wide between main aisles. There must be a cross aisle leading directly to each passageway between the exit and the nearest main aisle.

(vi) There must be at least one seat

adjacent to each such exit that could be occupied by a flight attendant.

(vii) Adequate assist space next to each Type A exit must be provided at each side of the passageway, to allow the crewmember(s) to assist in the evacuation of passengers without reducing the unobstructed width of the passageway below that required by subdivision (iv) of this subparagraph.

(viii) At each non-over-wing exit there must be installed a slide capable of carrying simultaneously two parallel lines of evacuees.

(ix) Each overwing exit having a step-down must have an assist means unless the exit without an assist means can be shown to have a rate of passenger egress at least equal to that of the same type of non-over-wing exit. If an assist means is required it must be automatically deployed, and automatically erected, concurrent with the opening of the exit and self-supporting within 10 seconds.

Stepdown distance as used in this section means the actual distance between the bottom of the required opening and a usable foothold, extending out from the fuselage, that is large enough to be effective without searching by sight or feel.

(b) *Accessibility.* Each required passenger emergency exit must be accessible to the passengers and located where it will afford the most effective means of passenger evacuation. Openings larger than those specified in this section, whether or not of rectangular shape, may be used if—

(1) The specified rectangular opening can be inscribed within the opening and

(2) The base of the inscribed rectangular opening meets the specified step-up and step-down heights.

(c) *Passenger emergency exits.* The prescribed exits need not be diametrically opposite each other nor identical in size and location on both sides. They must be distributed as uniformly as practicable taking into account passenger distribution. The first floor level exit on each side of the fuselage must be in the rearward part of the passenger compartment unless another location affords a more effective means of passenger evacuation. Where more than one floor level exit per side is prescribed, at least one floor level exit per side must be located near each end of the cabin, except that this provision does not apply to combination cargo/passenger configurations. Exits must be provided as follows:

(1) Except as provided in subparagraphs

(2) through (8) of this paragraph, the number and type of passenger emergency exits must be in accordance with the following table:

(2) Two Type IV exits may be installed instead of each Type III exit prescribed in subparagraph (1) of this paragraph.

(3) If slides meeting the requirements of § 25.809(f)(1) are installed at floor level

| Passenger seating capacity (cabin attendants not included) | Emergency exits for each side of the fuselage | | | |
|---|---|---|---|---|
| | Type I | Type II | Type III | Type IV |
| 1 through 10 | | | | 1 |
| 11 through 19 | | | 1 | |
| 20 through 39 | | 1 | | 1 |
| 40 through 59 | 1 | | | 1 |
| 60 through 79 | 1 | | 1 | |
| 80 through 109 | 1 | | 1 | 1 |
| 110 through 139 | 2 | | 1 | |
| 140 through 179 | 2 | | 2 | |

exits (other than overwing exits), the passenger emergency exit relationship specified in subparagraph (1) of this paragraph may be increased by—

(i) Not more than five passengers on airplanes with at least two of these exits; and

(ii) Not more than 10 passengers on airplanes with at least four of these exits. However, no increase in passenger seating capacity is allowed under this subparagraph if an increase in passenger seating capacity is obtained under subparagraph (4) of this paragraph.

(4) An increase in passenger seating capacity above the maximum permitted under subparagraph (1) of this paragraph but not to exceed a total of 299 may be allowed in accordance with the following table for each additional pair of emergency exits in excess of the minimum number prescribed in subparagraph (1) of this paragraph for 179 passengers:

| Additional emergency exits (each side of fuselage) | Increase in passenger seating capacity allowed |
|---|---|
| Type A | 100 |
| Type I | 45 |
| Type II | 40 |
| Type III | 35 |

(5) For passenger capacities in excess of 299, each emergency exit in the side of the fuselage must be either a Type A or a Type I. A passenger seating capacity of 100 is allowed for each pair of Type A exits and a passenger seating capacity of 45 is allowed for each pair of Type I exits.

(6) If a passenger ventral or tail cone exit is installed and can be shown to allow a rate of egress at least equivalent to that of Type III exit with the airplane in the most adverse exit opening condition because of the collapse of one or more legs of the landing gear, an increase in passenger seating capacity beyond the limits specified in subparagraph (1), (4), or (5) of this paragraph may be allowed as follows:

(i) For a ventral exit, 12 additional passengers.

(ii) For a tail cone exit incorporating a floor level opening of not less than 20

inches wide by 60 inches high, with corner radii not greater than one-third the width of the exit, in the pressure shell and incorporating an approved assist means in accordance with § 25.809(f)(1), 25 additional passengers; or

(iii) For a tail cone exit incorporating an opening in the pressure shell which is at least equivalent to a Type III emergency exit with respect to dimensions, step-up and stepdown distance, and with the top of the opening not less than 56 inches from the passenger compartment floor, 15 additional passengers.

(7) For airplanes on which the vertical location of the wing does not allow the installation of overwing exits, an exit of at least the dimensions of a Type III must be installed instead of each Type III and each Type IV exit required by subparagraph (1) of this paragraph.

(8) Each emergency exit in the passenger compartment in excess of the minimum number of required emergency exits must meet the applicable requirements of §§ 25.809 through 25.812, and must be readily accessible.

(d) *Ditching emergency exits for passengers.* If the emergency exits required by paragraph (c) of this section do not meet subparagraphs (1) and (2) of this paragraph, exits must be added to meet them:

(1) A Type IV exit on each side of the airplane, both above the waterline, with a passenger seating capacity of 10 or less.

(2) A Type III exit for airplanes with a passenger seating capacity of 11 or more, with at least one emergency exit above the waterline for each unit (or part of a unit) of 35 passengers, but no less than two such exits, with one on each side of the airplane. However, where it has been shown through analysis, ditching demonstrations, or any other tests found necessary by the Administrator, that the evacuation capability of the airplane during ditching is improved by the use of larger exits or by other means, the passenger exit ratio may be increased.

(3) If side exits cannot be above the waterline, the side exits must be replaced by an equal number of readily accessible overhead hatches of not less than the dimensions of a Type III exit except that, for airplanes with a passenger capacity of 35 or less, the two required Type III side exits need be replaced by only one overhead hatch.

(4) Two Type IV exits may be installed instead of each required Type III exit.

## § 25.809 Emergency exit arrangement.

(a) Each emergency exit, including a flight crew emergency exit, must be a movable door or hatch in the external walls of the fuselage, allowing unobstructed opening to the outside.

(b) Each emergency exit must be openable

from the inside and the outside except that sliding window emergency exits in the flight crew area need not be openable from the outside if other approved exits are convenient and readily accessible to the flight crew area.

(c) The means of opening emergency exits must be simple and obvious and may not require exceptional effort. Internal exit-opening means involving sequence operations (such as operation of two handles or latches or the release of safety catches) may be used for flight crew emergency exits if it can be reasonably established that these means are simple and obvious to crew members trained in their use.

(d) There must be a means to lock each emergency exit and to safeguard against its opening in flight, either inadvertently by persons or as a result of mechanical failure. In addition, there must be a means for direct visual inspection of the locking mechanism by crewmembers to determine that each emergency exit, for which the initial opening movement is outward, is fully locked.

(e) There must be provisions to minimize the probability of jamming of the emergency exits resulting from fuselage deformation in a minor crash landing.

(f) Each landplane emergency exit (other than exits located over the wing) more than 6 feet from the ground with the airplane on the ground and the landing gear extended must have an approved means to assist the occupants in descending to the ground as follows:

(1) The assisting means for each passenger emergency exit must be a self-supporting slide or equivalent, and must be designed so that it is—

(i) Automatically deployed, and automatically erected, concurrent with the opening of the exit except that the assisting means may be erected in a different manner when installed at service doors that qualify as emergency exits, and at passenger doors; and

(ii) Erectable within 10 seconds and of such length that the lower end is self-supporting on the ground after collapse of any one or more landing gear legs.

(2) The assisting means for flight crew emergency exits may be a rope or any other means demonstrated to be suitable for the purpose. If the assisting means is a rope, or an approved device equivalent to a rope, it must be—

(i) Attached to the fuselage structure at or above the top of the emergency exit opening, or, for a device at a pilot's emergency exit window, at another approved location if the stowed device, or its attachment, would reduce the pilot's view in flight;

(ii) Able (with its attachment) to withstand a 400-pound static load.

(g) The proper functioning of each emergency exit must be shown by tests.

(h) If the trailing edge of the flaps in the landing position is more than 6 feet above the ground with the airplane on the ground and the landing gear extended, or if the wing is more than 6 feet above the ground with the landing gear extended and the flaps are unsuitable as a slide, means must be provided to assist evacuees (who have used the overwing exits) to reach the ground.

### § 25.811 Emergency exit marking.

(a) Each passenger emergency exit, its means of access, and its means of opening must be conspicuously marked.

(b) The identity and location of each passenger emergency exit must be recognizable from a distance equal to the width of the cabin.

(c) Means must be provided to assist the occupants in locating the exits in conditions of dense smoke.

(d) The location of each passenger emergency exit must be indicated by a sign visible to occupants approaching along the main passenger aisle. There must be a locating sign—

(1) Above the aisle near each over-the-wing passenger emergency exit, or at another ceiling location if it is more practical because of low headroom;

(2) Next to each floor level passenger emergency exit, except that one sign may serve two such exits if they both can be seen readily from the sign; and

(3) On each bulkhead or divider that prevents fore and aft vision along the passenger cabin, to indicate emergency exits beyond and obscured by it, except that if this is not possible the sign may be placed at another appropriate location.

(e) The location of the operating handle and instructions for opening must be shown—

(1) For each passenger emergency exit, by a marking on or near the exit that is readable from a distance of 30 inches; and

(2) For each Type I or Type II passenger emergency exit with a locking mechanism released by rotary motion of the handle, by—

(i) A red arrow, with a shaft at least three-fourths inch wide and a head twice the width of the shaft, extending along at least 70° of arc at a radius approximately equal to three-fourths of the handle length; and

(ii) The word "open" in red letters 1 inch high, placed horizontally near the head of the arrow.

(f) Each emergency exit that is required to be openable from the outside, and its means of opening, must be marked on the outside of the airplane. In addition, the following apply:

(1) The outside marking for each passenger emergency exit in the side of the fuselage must include a 2-inch colored band outlining the exit.

(2) Each outside marking including the band, must have color contrast to be readily distinguishable from the surrounding fuselage surface. The contrast must be such that if the reflectance of the darker color is 15 percent or less, the reflectance of the lighter color must be at least 45 percent. "Reflectance" is the ratio of the luminous flux reflected by a body to the luminous flux it receives. When the reflectance of the darker color is greater than 15 percent, at least a 30-percent difference between its reflectance and the reflectance of the lighter color must be provided.

(3) In the case of exits other than those in the side of the fuselage, such as ventral or tail cone exits, the external means of opening, including instructions if applicable, must be conspicuously marked in red, or bright chrome yellow if the background color is such that red is inconspicuous. When the opening means is located on only one side of the fuselage, a conspicuous marking to that effect must be provided on the other side.

(g) Emergency exits need only be marked with the word "Exit."

### § 25.812 Emergency lighting.

(a) An emergency lighting system, independent of the main lighting system, must be installed which includes:

(1) Illuminated emergency exit marking and locating signs, sources of general cabin illumination, and interior lighting in emergency exit areas.

(2) Exterior emergency lighting.

(b) Each passenger exit sign and each exit locating sign must have white letters at least 1 inch high on a red background at least 2 inches high. These signs may be internally electrically illuminated, or self-illuminated by other than electrical means, with an initial brightness of at least 160 microlamberts. The colors may be reversed in the case of internally electrically illuminated signs if this will increase the illumination of the exit.

(c) General illumination in the passenger cabin must be provided so that when measured along the centerline of main passenger aisles at seat armrest height and at 40-inch intervals, the average illumination is not less than 0.05 foot-candle. A main passenger aisle is considered to extend along the fuselage from the most forward passenger emergency exit or cabin occupant seat, whichever is farther forward, to the most rearward passenger emergency exit or cabin occupant seat, whichever is farther aft.

(d) The floor of the passageway leading to each floor-level passenger emergency exit, between the main aisles and the exit openings, must be provided with illumination.

(e) The emergency lighting system must be designed as follows:

(1) The lights must be operable manually from the flight crew station and (if required by the operating rules of this chapter) from a point in the passenger compartment that is readily accessible

to a normal flight attendant seat. Means must be provided to safeguard against inadvertent operation of the manual controls.

(2) When armed or turned on, the lights must remain lighted or become lighted upon interruption (except an interruption caused by a vertical separation of the fuselage during crash landing) of the airplane's normal electric power.

(f) Exterior emergency lighting must be provided at each overwing exit so that the illumination is—

(1) Not less than 0.02 foot-candle (measured on a plane parallel to the surface) on a 2-square-foot area where an evacuee is likely to make his first step outside the cabin;

(2) Not less than 0.05 foot-candle (measured normal to the direction of the incident light) for a minimum width of 2 feet along the 30 percent of the slip-resistant escape route required in § 25.803(e) that is farthest from the exit; and

(3) Not less than 0.02 foot-candle on the ground surface with the landing gear extended (measured on a horizontal plane) where an evacuee using the established escape route would normally make first contact with the ground.

(g) The means required in § 25.809(f)(1) and (h) to assist the occupants in descending to the ground must be illuminated so that the deployed assist means is visible from the airplane.

(1) If the assist means is illuminated by exterior emergency lighting, it must provide—

(i) Illumination at each overwing emergency exit of not less than 0.02 foot-candle on the ground surface with the landing gear extended (measured in a horizontal plane) where an evacuee using the established escape route would normally make first contact with the ground; and

(ii) Illumination at each non-over-wing emergency exit, of not less than 0.03 foot-candle (measured normal to the direction of the incident light) at the ground end of the assist means and, for each non-over-wing exit in the side of the fuselage, over a spherical surface 10° to either side of the center of the assist means and from 30° above to 5° below the 16° position of the assist means.

(2) If the assist means is self-illuminated, the lighting provisions—

(i) May not be adversely affected by stowage; and

(ii) Must provide sufficient ground surface illumination so that obstacles at the end of the assist means are clearly visible to evacuees.

(h) The energy supply to each emergency lighting unit must provide the required level of illumination for at least 10 minutes at the critical ambient conditions after emergency landing.

(i) If storage batteries are used as the energy supply for the emergency lighting system, they may be recharged from the airplane's main electric power system:

*Provided*, That, the charging circuit is designed to preclude inadvertent battery discharge into charging circuit faults.

(j) Components of the emergency lighting system, including batteries, wiring relays, lamps, and switches must be capable of normal operation after having been subjected to the inertia forces listed in § 25.561(b).

(k) The emergency lighting system must be designed so that after any single vertical separation of the fuselage during crash landing—

(1) Not more than 25 percent of all electrically illuminated emergency lights required by this section are rendered inoperative, in addition to the lights that are directly damaged by the separation;

(2) Each electrically illuminated exit sign required under § 25.811(d)(2) remains operative exclusive of those that are directly damaged by the separation; and

(3) At least one required exterior emergency exit light for each side of the airplane remains operative exclusive of those that are directly damaged by the separation.

## § 25.813 Emergency exit access.

(a) There must be a passageway between individual passenger areas, and leading from each aisle to each Type I and Type II emergency exit. These passageways must be unobstructed and at least 20 inches wide.

(b) For each passenger emergency exit covered by § 25.809(f), there must be enough space next to the exit to allow a crewmember to assist in the evacuation of passengers without reducing the unobstructed width of the passageway below that required for the exit.

(c) There must be access from each aisle to each Type III or Type IV exit. The access must not be obstructed by seats, berths, or other protrusions which would reduce the effectiveness of the exit. However, for airplanes having a maximum passenger seating capacity not exceeding 19, there may be minor obstructions if there are compensatory factors to maintain the effectiveness of the exit. For airplanes having a maximum seating capacity of 20 or more, the projected opening of the exit provided must not be obstructed by a seatback in any position at the outboard seat locations. However, if the lateral distance between an outboard seat and the exit is not less than the width of the narrowest passenger seat installed on the airplane, that seat need not meet the seatback obstruction provision of this paragraph.

(d) If it is necessary to pass through a passageway between passenger compartments to reach any required emergency exit from any seat in the passenger cabin, the passageway must be unobstructed. However, curtains may be used if they allow free entry through the passageway.

(e) No door may be installed in any partition between passenger compartments.

(f) If it is necessary to pass through a doorway separating the passenger cabin from other areas to reach any required emergency exit from any passenger seat, the door must have a means to latch it in open position. The latching means must be able to withstand the loads imposed upon it when the door is subjected to the ultimate inertia forces, relative to the surrounding structure, listed in § 25.561(b).

## § 25.815 Width of aisle.

The passenger aisle width at any point between seats must equal or exceed the values in the following table:

| Passenger seating capacity | Minimum passenger aisle width (inches) | |
| --- | --- | --- |
| | Less than 25 inches from floor | 25 inches and more from floor |
| 10 or less | 12 | 15 |
| 11 through 19 | 12 | 20 |
| 20 or more | 15 | 20 |

## § 25.817 Maximum number of seats abreast.

On airplanes having only one passenger aisle, no more than 3 seats abreast may be placed on each side of the aisle in any one row.

## § 25.785 Seats, berths, safety belts, and harnesses.

(a) Each seat, berth, safety belt, harness, and adjacent part of the airplane at each station designated as occupiable during takeoff and landing must be designed so that a person making proper use of these facilities will not suffer serious injury in an emergency landing as a result of the inertia forces specified in § 25.561.

(b) Each seat and berth must be approved.

(c) Each occupant of a seat that makes more than an 18 degree angle with the vertical plane containing the airplane centerline, must be protected from head injury by a safety belt and an energy absorbing rest that will support the arms, shoulders, head, and spine, or by a safety belt and shoulder harness that will prevent the head from contacting any injurious object. Each occupant of any other seat must be protected from head injury by—

(1) A safety belt and shoulder harness that will prevent the head from contacting any injurious object;

(2) A safety belt plus the elimination of any injurious object within striking radius of the head; or

(3) A safety belt and an energy absorbing rest that will support the arms, shoulders, head, and spine.

(d) If the seat backs do not have a firm hand hold, there must be a hand grip or rail along each aisle to enable occupants to steady themselves while using the aisles in moderately rough air.

(e) Each projecting object that would injure persons seated or moving about the airplane in normal flight must be padded.

(f) Each berth must be designed so that the forward part has a padded end board, canvas diaphragm, or equivalent means, that can withstand the static load reaction of the occupant when subjected to the forward inertia force specified in § 25.561. Berths must be free from corners and protuberances likely to cause serious injury to a person occupying the berth during emergency conditions.

(g) Each crewmember seat at flight deck stations must have provisions for a shoulder harness. These seats must meet the strength requirements of paragraph (i) of this section.

(h) Cabin attendant seats must be in the passenger compartment near approved floor level emergency exits.

(i) Each seat berth, and its supporting structure, must be designed for an occupant weight of 170 pounds, considering the maximum load factors, inertia forces, and reactions between the occupant, seat, and safety belt or harness, at each relevant flight and ground load condition (including the emergency landing conditions prescribed in § 25. 561). For berths, the forward inertia force must be considered in accordance with paragraph (f) of this section and need not be considered with respect to the safety belt. In addition—

(1) The structural analysis and testing of the seats, berths, and their supporting structures may be determined by—

(i) Assuming that the critical load in the forward, sideward, downward, and rearward directions (as determined from the prescribed flight, ground, and emergency landing conditions) acts separately: and

(ii) Using selected combinations of loads if the required strength in each specified direction is substantiated:

(2) Each pilot seat must be designed for the reactions resulting from the application of the pilot forces prescribed in § 25.395; and

(3) The inertia forces specified in § 25. 561 must be multiplied by a factor of 1.33 (instead of the fitting factor prescribed in § 25.625) in determining the strength of the attachment of—

(i) Each seat to the structure; and

(ii) Each belt or harness to the seat or structure.

## VENTILATION AND HEATING

### § 25.831 Ventilation.

(a) Each passenger and crew compartment must be ventilated, and each crew compartment must have enough fresh air (but not less than 10 cu. ft. per minute per crewmember) to enable crewmembers to perform their duties without undue discomfort or fatigue.

(b) Crew and passenger compartment air must be free from harmful or hazardous concentrations of gases or vapors. In meeting this requirement, the following apply:

(1) Carbon monoxide concentrations in excess of one part in 20,000 parts of air are considered hazardous. For test purposes, any acceptable carbon monoxide detection method may be used.

(2) Carbon dioxide in excess of three percent by volume (sea level equivalent) is considered hazardous in the case of crewmembers. Higher concentrations of carbon dioxide may be allowed in crew compartments if appropriate protective breathing equipment is available.

(c) There must be provisions made to ensure that the conditions prescribed in paragraph (b) of this section are met after reasonably probable failures or malfunctioning of the ventilating, heating, pressurization, or other systems and equipment.

(d) If accumulation of hazardous quantities of smoke in the cockpit area is reasonably probable, smoke evacuation must be readily accomplished, starting with full pressurization and without depressurizing beyond safe limits.

(e) There must be a means to enable the crew to control the temperature and quantity of ventilating air supplied to the crew compartment, independently of the temperature and quantity of ventilating air supplied to other compartments.

## PRESSURIZATION

### § 25.841 Pressurized cabins.

(a) Pressurized cabins and compartments to be occupied must be equipped to provide a cabin pressure altitude of not more than 8,000 feet at the maximum operating altitude of the airplane under normal operating conditions. If certification for operation over 25,000 feet is requested, the airplane must be able to maintain a cabin pressure altitude of not more than 15,000 feet in the event of any reasonably probable failure or malfunction in the pressurization system.

(b) Pressurized cabins must have at least the following valves, controls, and indicators for controlling cabin pressure:

(1) Two pressure relief valves (at least one of which is the normal regulating valve) to automatically limit the positive pressure differential to a predetermined value at the maximum rate of flow delivered by the pressure source. The combined capacity of the relief valves must be large enough so that the failure of any one valve would not cause an appreciable rise in the pressure differential. The pressure differential is positive when the internal pressure is greater than the external.

(2) Two reverse pressure differential relief valves (or their equivalents) to automatically prevent a negative pressure differential that would damage the structure. One valve is enough, however, if it is of a design that reasonably precludes its malfunctioning.

(3) A means by which the pressure differential can be rapidly equalized.

(4) An automatic or manual regulator for controlling the intake or exhaust airflow, or both, for maintaining the required internal pressures and airflow rates.

(5) Instruments at the pilot or flight engineer station to show the pressure differential, the absolute pressure in the cabin, and the rate of change of the absolute pressure.

(6) Warning indication at the pilot or flight engineer station to indicate when the safe or preset pressure differential and absolute cabin pressure limits are exceeded. Appropriate warning markings on the cabin pressure differential indicator meet the warning requirement for pressure differential limits and an aural or visual signal (in addition to cabin altitude indicating means) meets the warning requirement for absolute cabin pressure limits if it warns the flight crew when the cabin absolute pressure is below that equivalent to 10,000 feet.

(7) A warning placard at the pilot or flight engineer station if the structure is not designed for pressure differentials up to the maximum relief valve setting in combination with landing loads.

### § 25.1445 Equipment standards for the oxygen distributing system.

(a) When oxygen is supplied to both crew and passengers, the distribution system must be designed for either—

(1) A source of supply for the flight crew on duty and a separate source for the passengers and other crewmembers; or

(2) A common source of supply with means to separately reserve the minimum supply required by the flight crew on duty.

(b) Portable walk-around oxygen units of the continuous flow, diluter-demand, and straight demand kinds may be used to meet the crew or passenger breathing requirements.

### § 25.1447 Equipment standards for oxygen dispensing units.

If oxygen dispensing units are installed, the following apply:

(a) There must be an individual dispensing unit for each occupant for whom supplemental oxygen is to be supplied. Units must be designed to cover the nose and mouth and must be equipped with a suitable means to retain the unit in position on the face. Flight crew masks for supplemental oxygen must have provisions for the use of communication equipment.

(b) If certification for operation up to and including 25,000 feet is requested, an oxygen supply terminal and unit of oxygen dispensing equipment for the immediate use of oxygen by each crewmember must be within easy reach of that crewmember. For any other occupants, the supply terminals and dispensing equipment must be located to allow the use of oxygen as required by the operating rules in this chapter.

(c) If certification for operation above 25,000 feet is requested, there must be oxygen dispensing equipment meeting the following requirements:

(1) There must be an oxygen dispensing unit connected to oxygen supply terminals immediately available to each occupant, wherever seated. If certification for operation above 30,000 feet is requested, the dispensing units providing the required oxygen flow rate must be automatically presented to the occupants. The total number of dispensing units and outlets must exceed the number of seats by at least 10 percent. The extra units must be as uniformly distributed throughout the cabin as practicable.

(2) Crewmembers on flight deck duty must be provided with demand equipment. In addition, there must be an oxygen dispensing unit, connected to an oxygen supply terminal, immediately available to each flight crewmember when seated at his station.

(3) There must be at least two outlets and units of dispensing equipment of a type similar to that required by subparagraph (1) of this paragraph in—

(i) Each washroom; and
(ii) Each lavatory, if separate from the washroom.

(4) Portable oxygen equipment must be immediately available for each cabin attendant.

## § 25.1449 Means for determining use of oxygen.

There must be a means to allow the crew to determine whether oxygen is being delivered to the dispensing equipment.

## § 25.1443 Minimum mass flow of supplemental oxygen.

(a) If continuous flow equipment is installed for use by flight crewmembers, the minimum mass flow of supplemental oxygen required for each crewmember may not be less than the flow required to maintain, during inspiration, a mean tracheal oxygen partial pressure of 149 mm. Hg when breathing 15 liters per minute, BTPS, and with a maximum tidal volume of 700 cc. with a constant time interval between respirations.

(b) If demand equipment is installed for use by flight crewmembers, the minimum mass flow of supplemental oxygen required for each crewmember may not be less than the flow required to maintain, during inspiration, a mean tracheal oxygen partial pressure of 122 mm. Hg, up to and including a cabin pressure altitude of 35,000 feet, and 95 percent oxygen between cabin pressure altitudes of 35,000 and 40,000 feet, when breathing 20 liters per minute BTPS. In addition, there must be means to allow the crew

to use undiluted oxygen at their discretion.

(c) For passengers and cabin attendants, the minimum mass flow of supplemental oxygen required for each person at various cabin pressure altitudes may not be less than the flow required to maintain, during inspiration and while using the oxygen equipment (including masks) provided, the following mean tracheal oxygen partial pressures:

(1) At cabin pressure altitudes above 10,000 feet up to and including 18,500 feet, a mean tracheal oxygen partial pressure of 100 mm. Hg when breathing 15 liters per minute, BTPS, and with a tidal volume of 700 cc. with a constant time interval between respirations.

(2) At cabin pressure altitudes above 18,500 feet up to and including 40,000 feet, a mean tracheal oxygen partial pressure of 83.8 mm. Hg when breathing 30 liters per minute, BTPS, and with a tidal volume of 1,100 cc. with a constant time interval between respirations.

(d) If first-aid oxygen equipment is installed, the minimum mass flow of oxygen to each user may not be less than four liters per minute, STPD. However, there may be a means to decrease this flow to not less than two liters per minute, STPD, at any cabin altitude. The quantity of oxygen required is based upon an average flow rate of 3 liters per minute per person for whom first-aid oxygen is required.

(e) If portable oxygen equipment is installed for use by crewmembers, the minimum mass flow of supplemental oxygen is the same as specified in paragraph (a) or (b) of this section, whichever is applicable.

## LIGHTS

### § 25.1381 Instrument lights.

(a) The instrument lights must—
(1) Make each instrument, switch, and other device for which they are provided easily readable; and
(2) Be installed so that—
(i) Their direct rays are shielded from the pilot's eyes; and
(ii) No objectionable reflections are visible to the pilot.

(b) Unless undimmed instrument lights are satisfactory under each expected flight condition, there must be a means to control the intensity of illumination.

### § 25.1383 Landing lights.

(a) Each landing light must be approved, and must be installed so that—
(1) No objectionable glare is visible to the pilot;
(2) The pilot is not adversely affected by halation; and
(3) It provides enough light for night landing.

(b) Except when one switch is used for the lights of a multiple light installation at

one location, there must be a separate switch for each light.

(c) There must be a means to indicate to the pilots when the landing lights are extended.

## § 25.1385 Position light system installation.

(a) *General.* Each part of each position light system must meet the applicable requirements of this section and each system as a whole must meet the requirements of §§ 25.1387 through 25.1397.

(b) *Forward position lights.* Forward position lights must consist of a red and a green light spaced laterally as far apart as practicable and installed forward on the airplane so that, with the airplane in the normal flying position, the red light is on the left side, and the green light is on the right side. Each light must be approved.

(c) *Rear position light.* The rear position light must be a white light mounted as far aft as practicable, and must be approved.

(d) *Light covers and color filters.* Each light cover or color filter must be at least flame resistant and may not change color or shape or lose any appreciable light transmission during normal use.

(e) *Passing light.* If an additional steady red light (commonly known as a passing light) is installed, it must be—
(1) Within the left landing light unit:
(2) On the centerline of the airplane nose: or
(3) In the leading edge of the left wing, outboard of the propeller disc.

## § 25.1387 Position light system dihedral angles.

(a) Each forward and rear position light must, as installed, show unbroken light within the dihedral angles described in this section.

(b) Dihedral angle L (left) is formed by two intersecting vertical planes, the first parallel to the longitudinal axis of the airplane, and the other at 110° to the left of the first, as viewed when looking forward along the longitudinal axis.

(c) Dihedral angle R (right) is formed by two intersecting vertical planes, the first parallel to the longitudinal axis of the airplane, and the other at 110° to the right of the first, as viewed when looking forward along the longitudinal axis.

(d) Dihedral angle A (aft) is formed by two intersecting vertical planes making angles of 70° to the right and to the left, respectively, to a vertical plane passing through the longitudinal axis, as viewed when looking aft along the longitudinal axis.

## § 25.1389 Position light distribution and intensities.

(a) *General.* The intensities prescribed in this section must be provided by new equipment with light covers and color filters in place. Intensities must be determined with the light source operating at a steady value equal to the average luminous output of the source at the normal operating voltage of the airplane. The light distribution and intensity of each position light must meet the requirements of paragraph (b) of this section.

(b) *Forward and rear position lights.* The light distribution and intensities of forward and rear position lights must be expressed in terms of minimum intensities in the horizontal plane, minimum intensities in any vertical plane, and maximum intensities in overlapping beams, within dihedral angles *L, R,* and *A,* and must meet the following requirements:

(1) *Intensities in the horizontal plane.* Each intensity in the horizontal plane (the plane containing the longitudinal axis of the airplane and perpendicular to the plane of symmetry of the airplane) must equal or exceed the values in § 25.1391.

(2) *Intensities in any vertical plane.* Each intensity in any vertical plane (the plane perpendicular to the horizontal plane) must equal or exceed the appropriate value in § 25.1393, where *l* is the minimum intensity prescribed in § 25.1391 for the corresponding angles in the horizontal plane.

(3) *Intensities in overlaps between adjacent signals.* No intensity in any overlap between adjacent signals may exceed the values given in § 25.1395, except that higher intensities in overlaps may be used with main beam intensities substantially greater than the minima specified in §§ 25.1391 and 25.1393 if the overlap intensities in relation to the main beam intensities do not adversely affect signal clarity. When the peak intensity of the forward position lights is more than 100 candles, the maximum overlap intensities between them may exceed the values given in § 25.1395 if the overlap intensity in Area *A* is not more than 10 percent of peak position light intensity and the overlap intensity in Area *B* is not greater than 2.5 percent of peak position light intensity.

## § 25.1391 Minimum intensities in the horizontal plane of forward and rear position lights.

Each position light intensity must equal or exceed the applicable values in the following table:

| Dihedral angle (light included) | Angle from right or left of longitudinal axis, measured from dead ahead | Intensity (candles) |
|---|---|---|
| *L* and *R* (forward red and green) | 0° to 10° | 40 |
| | 10° to 20° | 30 |
| | 20° to 110° | 5 |
| *A* (rear white) | 110° to 180° | 20 |

## § 25.1393 Minimum intensities in any vertical plane of forward and rear position lights.

Each position light intensity must equal or exceed the applicable values in the following table:

| Angle above or below the horizontal plane: | Intensity |
|---|---|
| 0° | 1.00 *l.* |
| 0° to 5° | 0.90 *l.* |
| 5° to 10° | 0.80 *l.* |
| 10° to 15° | 0.70 *l.* |
| 15° to 20° | 0.50 *l.* |
| 20° to 30° | 0.30 *l.* |
| 30° to 40° | 0.10 *l.* |
| 40° to 90° | 0.05 *l.* |

## § 25.1395 Maximum intensities in overlapping beams of forward and rear position lights.

No position light intensity may exceed the applicable values in the following table, except as provided in § 25.1389(b)(3).

| | Maximum intensity | |
|---|---|---|
| Overlaps | Area *A* (candles) | Area *B* (candles) |
| Green in dihedral angle *L* | 10 | 1 |
| Red in dihedral angle *R* | 10 | 1 |
| Green in dihedral angle *A* | 5 | 1 |
| Red in dihedral angle *A* | 5 | 1 |
| Rear white in dihedral angle *L* | 5 | 1 |
| Rear white in dihedral angle *R* | 5 | 1 |

Where—

(a) Area A includes all directions in the adjacent dihedral angle that pass through the light source and intersect the common boundary plane at more than 10° but less than 20°; and

(b) Area B includes all directions in the adjacent dihedral angle that pass through the light source and intersect the common boundary plane at more than 20°.

## § 25.1397 Color specifications.

Each position light color must have the applicable International Commission on Illumination chromaticity coordinates as follows:

(a) *Aviation red*—
"*y*" is not greater than 0.335; and
"*z*" is not greater than 0.002.
(b) *Aviation green*—
"*x*" is not greater than 0.440–0.320L "*y*";
"*x*" is not greater than *y*-0.170; and
"*y*" is not less than 0.390–0.170*x*.
(c) *Aviation white*—
"*x*" is not less than 0.350;
"*x*" is not greater than 0.540; and
"*y* $y_0$" is not numerically greater than 0.01. "$y_0$" is being the *y* coordinate of the Plancklan radiator for which $x_0=x$.

## § 25.1401 Anticollision light system.

(a) *General.* The airplane must have an anticollision light system that—

(1) Consists of one or more approved anticollision lights located so that their light will not impair the crew's vision or detract from the conspicuity of the position lights: and

(2) Meets the requirements of paragraphs (b) through (f) of this section.

(b) *Field of coverage.* The system must consist of enough lights to illuminate the vital areas around the airplane considering the physical configuration and flight characteristics of the airplane. The field of coverage must extend in each direction within at least 30° above and 30° below the horizontal plane of the airplane, except that a solid angle or angles of obstructed visibility totaling not more than 0.03 steradians is allowable within a solid angle equal to 0.15 steradians centered about the longitudinal axis in the rearward direction.

(c) *Flashing characteristics.* The arrangement of the system, that is, the number of light sources, beam width, speed of rotation, and other characteristics, must give an effective flash frequency of not less than 40, nor more than 100, cycles per minute. The effective flash frequency is the frequency at which the airplane's complete anticollision light system is observed from a distance, and applies to each sector of light including any overlaps that exist when the system consists of more than one light source. In overlaps, flash frequencies may exceed 100, but not 180, cycles per minute.

(d) *Color.* Each anticollision light must be aviation red and must meet the requirements of § 25.1397(a).

(e) *Light intensity.* The minimum light intensities in all vertical planes, measured with the red filter and expressed in terms of "effective" intensities, must meet the requirements of paragraph (f) of this section. The following relation must be assumed:

$$I_e = \frac{\int_{t_1}^{t_2} I(t)dt}{0.2+(t_2-t_1)}$$

where:

$I(e)=$ effective intensity (candles).

$I(t)=$ instantaneous intensity as a function of time.

$t_2-t_1=$ flash time interval (seconds).

Normally, the maximum value of effective intensity is obtained when $t_2$ and $t_1$ are chosen so that the effective intensity is equal to the instantaneous intensity at $t_2$ and $t_1$.

(f) *Minimum effective intensities for anticollision lights.* Each anticollision light effective intensity must equal or exceed the applicable values in the following table.

| Angle above or below the horizontal plane: | Effective intensity (candles) |
|---|---|
| 0° to 5° | 100 |
| 5° to 10° | 60 |
| 10° to 20° | 20 |
| 20° to 30° | 10 |

(e) There must be at least one hand fire extinguisher for use by the flight crewmembers; and

(f) There must be at least the following number of hand fire extinguishers conveniently located in passenger compartments:

| Passenger capacity: | Minimum number of hand fire extinguishers |
|---|---|
| 7 through 30 | 1 |
| 31 through 60 | 2 |
| 61 or more | 3 |

## § 25.857 Cargo compartment classification.

(a) *Class A.* A Class *A* cargo or baggage compartment is one in which—

(1) The presence of a fire would be easily discovered by a crewmember while at his station; and

(2) Each part of the compartment is easily accessible in flight.

(b) *Class B.* A Class *B* cargo or baggage compartment is one in which—

(1) There is sufficient access in flight to enable a crew member to effectively reach any part of the compartment with the contents of a hand fire extinguisher;

(2) When the access provisions are being used, no hazardous quantity of smoke, flames, or extinguishing agent, will enter any compartment occupied by the crew or passengers:

## Service System

It is particularly critical that delays not be incurred because of aircraft servicing or because of passenger and baggage exchange. The obvious design objective should be to create a system in which all these activities can be accomplished simultaneously. As indicated by the accompanying sketches, both the prime vehicle and the interfacing equipment must be considered during conceptual development so that the numerous attachments and exchanges can take place without interference.

In this example, the system concept provides multiple service points spaced to accommodate simultaneous vehicle and equipment positioning. Passenger exchange is also kept apart from service activities that might annoy deplaning or enplaning passengers.

It is also important to the particular airline that ground support equipment already in use by that particular company be considered prior to launching an entirely new concept for ground support equipment. In addition to the added cost, there are important human factors considerations in terms of the additional training required.

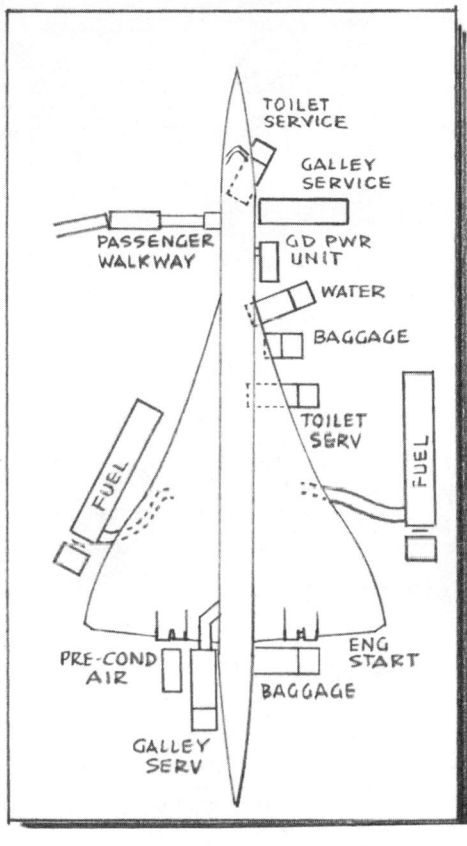

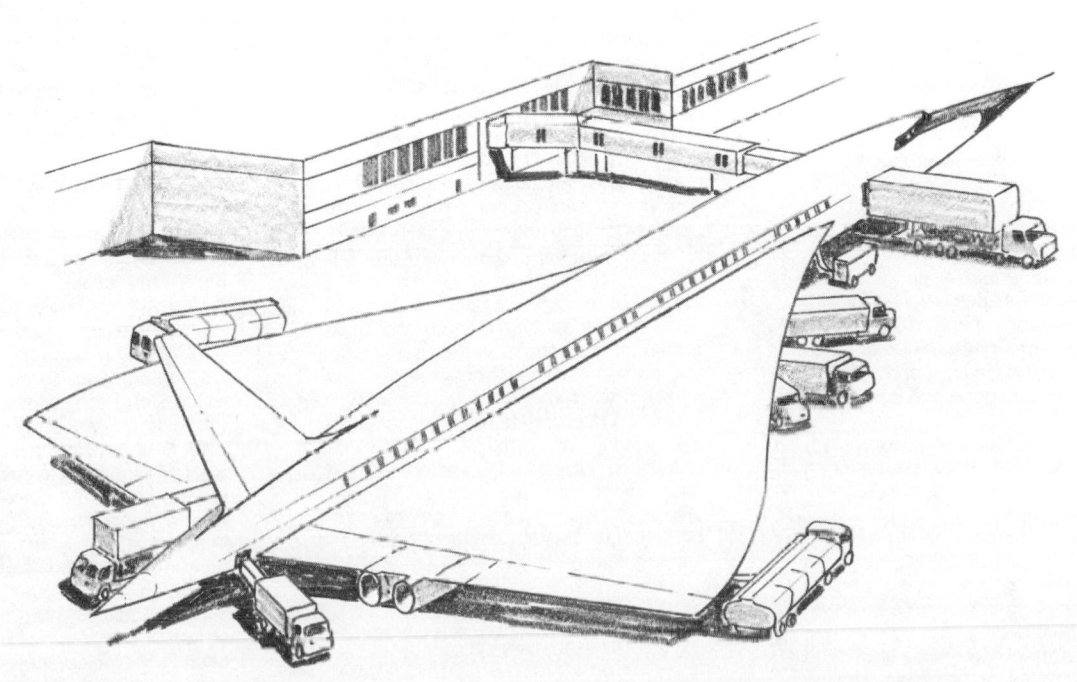

## RAILWAY VEHICLES

1. Train Crew Locomotive Accommodation:
   Exterior inspection
   Cab ingress and egress
   Control stations
   System compartment access
   Environmental control
   Exterior visibility
   Forward roadbed illumination and external signal lights
   Audio-visual warning systems

2. Passenger Railcar Accommodation:
   Exterior car identification
   Ingress and egress
   Intercar access
   Aisleways
   Seating
   Personal article and baggage storage
   Rest rooms
   Snack bar and dining car
   Environmental control
   Public address system

3. Servicing Personnel Maintenance Accessibility:
   Systems, equipment, and component access
   Maintenance hazard protection
   Housekeeping
   Baggage handling
   Coupling and yard handling

Typical train system–human interface considerations.

## Primary Human Factors Considerations for Railway Vehicles

Although there is a wide range of variations in rail system concepts (intercontinental transportation, urban transit, etc.), certain common interfaces between the train crew, the passengers, the maintenance and service personnel, and the vehicle features remain constant for all systems. The following should be considered in the early stages of system development in order to ensure the efficiency, comfort, and safety of train crews, support personnel, and passengers.

### 1. On-Board, En-Route Train Crew Operations

a. Ingress and egress to the locomotive cab as well as to the principal power system compartment are critical in terms of both nominal and emergency operating scenarios. There should be direct access between the locomotive and other rail cars as well as between the typical external entry modes.

b. The layout of the engineer and assistant work stations should be designed for efficient monitoring of the external roadway, individual systems instrumentation, manipulation of appropriate controls, and the general convenience and comfort of operators.

c. The layout of the various on-board systems and equipment should be designed for ease of inspection, troubleshooting, servicing, and repair.

d. Adequate environmental control should be provided to protect crew members from the typical thermal, noise, and vibration stress produced by on-board systems and equipment.

e. Appropriate support amenities should be provided the crew (a lavatory, drinking water, storage for personal articles, etc.).

f. Appropriate communication systems should be provided both for on-board intercrew communications and for communication with external transportation system facilities.

g. Special consideration should be given to emergency- and safety-related scenarios, e.g., derailment, collision, loss of power for environmental systems, and engineer incapacitation.

### 2. Passenger Accommodation[35]

a. Exterior car identification should be clear and understandable, and signs and marking should be placed where boarding passengers can readily find them. Instructions should be designed to minimize the possible confusion created by a long series of cars when a passenger (limited by his or her observing position) must quickly locate a particular car. A typical problem occurs because signs cannot be read until the passenger is alongside the car or because signs are obscured by other passengers standing in front of them.

b. Entrances should be located and positioned for the ease of passengers carrying suitcases. They should provide a level entry for wheelchair use, and there should be appropriate handrails to assist the elderly and children. Entrances should be properly illuminated so that passengers can see where they are stepping. When steps are required to accommodate stations where there is no loading platform, these should be designed to minimize long stretches to reach the first step from the ground, and stair treads should be nonslip so that passengers who may have had to walk in rain or snow will not slip and fall climbing the steps. No unusual structural projections or sharp corners should protrude into the entryway; these could be a hazard for entering passengers. Handrails should be designed to accommodate both entry and exit (i.e., the mobility procedure is different for each situation).

c. Entry vestibules should be large enough to accommodate the passenger who is carrying suitcases, packages, or a small child. Vestibules should be well lighted, and appropriate signs should be provided so that the passenger knows which direction to go when reaching the vestibule. Doors leading from vestibules to the main car compartment should operate easily (i.e., they should not require considerable force to open, and they should not operate in a fashion that requires awkward manipulation of packages or handles).

d. Aisles within the passenger car should be wide enough to allow passengers to negotiate the baggage or parcels they may be carrying. There should be handholds along the pathway, especially for the elderly, who may be unsteady on their feet. Aisles should be well lighted.

e. Seating should be designed for maximum comfort and convenience; i.e., the seat dimensions and the seat-pan and seat-back angles should provide good posture control. Seats used in trains that travel longer distances should have recline capability. There should be plenty of foot and leg space between seats (including the clearance behind a reclined seat). Independent reading lights should

---

[35]Entryways, aisles, seating and/or special wheelchair tie-down spaces, rest rooms, etc., should be given special attention in terms of the problems of passengers who may be in wheelchairs.

be provided for each passenger, and these should be positioned so that one passenger's light does not annoy an adjacent passenger who may want to sleep. Two-abreast seating should be maintained so that a passenger who is seated by a window is not required to step around more than one person to get into the aisle. Fixed armrests should be provided on the aisle and window sides of seats, and a folding armrest should be provided between seats. This latter armrest should be wide enough so that both passengers can rest their arms at the same time. Seat-reclining controls should be provided on the fixed armrests. In addition, controls for individual reading lights, attendant call buttons, etc., should also be provided on these armrests. (Airlines have offered these special amenities for some time, and, in order for rail travel to be attractive, similar conveniences should be provided rail travelers.) Consider the provision of removable trays at each seat so that food service or other activity is possible at the individual passenger's seat.

f. Whenever possible, rest rooms should be provided at both ends of a passenger car so that some passengers are not required to negotiate the entire length of the car to reach these facilities. Rest rooms should be properly signed so that passengers know where they are and when they are available.

g. Storage for personal baggage, garments, and parcels should be provided at each seat section. Such storage should be readily accessible for the shortest expected adult passenger. Consider parcel compartment designs that can be closed, which will keep the packages from being dislodged in the event the train suddenly decelerates.

h. Each passenger car should have the following special amenities:

(1) Drinking water
(2) Magazine and newspaper rack
(3) Trash receptacle
(4) Fire extinguisher or extinguishers
(5) Emergency lights (inertial actuated, self-powered)
(6) Emergency ventilation system (in the event the power for the main air conditioning or heating fails)
(7) First-aid kit

i. The entire interior of each passenger car should be made safe in terms of removal of sharp projections, corners, edges, etc.

j. Seat identification should be prominently and clearly displayed so that passengers can easily locate their seats.

k. Provide a public address system so that stations as well as other information or instructions can be announced to all the passengers simultaneously.

l. Provide an appropriate car environmental control panel so that the car attendant can adjust lighting, heat, cooling, and ventilation for each car. Apply appropriate techniques for the control of noise, and design car suspension to minimize the effects of vibration and sidesway.

## 3. Equipment Maintenance Features

Rail vehicles, like other public transportation system equipment, require daily servicing and maintenance which is time-critical. The ease with which support personnel can accomplish their various tasks is directly related to how much attention is given to this aspect of system operation in the initial stages of conceptual design—to the physical arrangement of equipment and components, the type of enclosures that may be used around equipment, the types of materials used, etc. Special note should be taken of the problems that the activities of one crew may create for another, especially the following:

a. If maintenance technicians have to carry heavy, dirty, or oily equipment through passenger cars, there is a likelihood that this will add to the difficulties of housekeeping crews.

b. If equipment compartments do not have easily operated covers, there is a good chance that the interior trim and materials will be soiled or damaged and thus will have to be replaced to keep them looking decent.

c. If the interior decor, materials, etc., are not carefully selected and appropriately attached, typical cleaning and servicing procedures will soon result in a shabby-looking interior. The bulkhead, ceiling, and floor covering should be easily replaceable so that the basic car can be maintained in a relatively new-looking state. One of the most common complaints of rail travelers is that rail systems (unlike airplanes) are frequently dirty and unattractive.

d. The following critical exterior maintenance considerations should be addressed during initial conceptual planning:
(1) Car coupling and decoupling hazards
(2) Electrical grounding hazards
(3) Lack of illumination at key inspection and repair points
(4) Lack of appropriate safety interlocks to ensure that all maintenance personnel are aware of maintenance in progress (to prevent inadvertent car movement, power activation, etc.)
(5) Adequate provisions for replacement (with ease) of major equipment components (large power units, tanks, wiring cables, etc.)
(6) Adequately designed and positioned worker aids for climbing to and from, and for walking on top of, the locomotive and other rail cars

## Typical Engineer and Assistant Operating Tasks to Be Considered in Developing a Locomotive Cab Concept

Ingress and egress
Starting and stopping the train
Controlling speed
Negotiating turnouts and crossovers and moving to a side track
Negotiating grades
Forming a train and/or locomotive consist
Detaching a locomotive consist

Passing trains and equipment
Obtaining clearance and receiving and transmitting messages
Managing auxiliary systems
Responding to malfunctions

## Typical Locomotive Systems

Train brake system
Drive system
Propulsion system
Engine system
Pneumatic system
Dynamic brake system
Auxiliary systems
Internal environment systems
External environment systems
Locomotive brake system and track and train situation signal systems
Communications system or systems

## Other Cab Features

Exterior visibility (windows, windshield wiper and washer, defogger, and defroster)
Lavatory
Drinking water and refrigerator
Fire extinguisher
Engineer and brakeman seats
Crashworthiness features

Note: Typically, two principal operator positions are required: one for the engineer and one for an assistant. The engineer's position is generally placed on the right side of the cab near a window so that the engineer can turn to see to the rear. Where practical, the main headlight should be placed below the windshield so that light will not reflect on it, especially in foggy weather. Display control consoles should be laid out so that operators can easily see ahead of the locomotive at the same time they are monitoring and/or manipulating principal control devices. The displays should be positioned and/or hooded, however, so that they will not reflect in the windshield and obscure or confuse frontal external visibility.

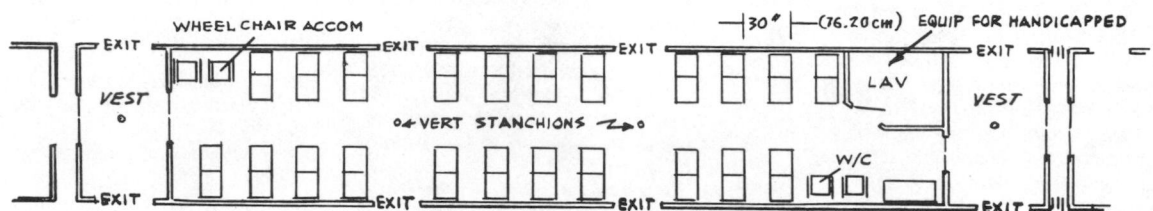

1. Transcontinental coach: seat spacing for reclining, double lavatories, sound-isolated vestibules, overhead storage, drinking-water dispensers, newspaper racks, individual reading lights, and attendant service and entertainment outlets

2. Commuter Rapid-Transit Coach: added entrances and exits, standing room with handrails, closer seat spacing, wheelchair provisions (including lavatory), and nonindividualized illumination

Passenger coach conceptual suggestions. (The above are for illustrative purposes only and are not drawn to scale.)

## SPECIAL TRANSIT VEHICLES

### Elevated and Suspended Rail System Vehicles

Passenger cars used for roadbeds that are elevated or suspended present several unique problems, in addition to the interior problems and the problems of ingress and egress of typical rail systems.

### 1. Emergency Egress and Rescue
In the event such vehicles become stranded high above the ground because of a system malfunction, there are obvious hazards to passengers who might try to leave the vehicle by the normal exit doors. Conceptually, it is more desirable to confine the passengers until the car can be rescued with them safe inside.

### 2. Crash Protection
Typically, elevated or suspended cars need to be constructed of extremely light materials to save weight. This requires consideration of special concepts of energy absorption and passenger restraint to protect passengers in the event the car collides with another car.

### 3. Ventilation and Air Conditioning
Since passengers cannot get off at will in the event of a malfunction and since they must wait in the car for some time before rescue, emergency power must be provided to keep the car ventilated and cooled or heated.

### 4. Emergency Communication
People who are stranded temporarily may become frightened because of the close confinement high above the ground, and therefore communication with system authorities should be provided so that passengers will know that someone is aware of their plight and also so that they will know when they can expect to be rescued.

# DESIGN FOR THE HANDICAPPED

## Vehicular Design to Accommodate the Handicapped

### 1. Wheelchair Users

Accommodation of the wheelchair user, whether he or she is a driver or a passenger, involves primarily allowance for movement of the wheelchair in entering or leaving the vehicle, maneuvering within certain public conveyances, and securing the wheelchair once it and its occupant are aboard. Wheelchair dimensions alone are not sufficient to define the clearance needs at doorways; generally, additional space is required for turning, folding, and parking the wheelchair, and special attention must be given to threshold and floor characteristics (how level, whether there are gaps between the vehicle threshold and the loading surface, etc.) in order to minimize the possibility that the chair will be upset. A unique and critical requirement for vehicular systems (e.g., public transit systems) is the necessity of securing the wheelchair against dynamic forces when the vehicle is under way and, more importantly, in the event of a crash. (Note that wheelchairs themselves are not designed to be crashworthy.)

In addition to providing the extra clearances for the wheelchair itself, special attention must be given to the accessibility of special vehicular controls and other features that the handicapped person may have to reach and operate from the wheelchair and/or operate as a driver with less-than-normal use of limbs, hands, or fingers.

### 2. The Blind and Partially Sighted

Attention should be given to critical visual interfaces that the partially blind have to see or read, such as displays, signs, and markings. This includes not only the legibility of words and symbols, but also the selection of colors that can be recognized by the color-deficient, illumination levels and contrasts that make up for the reduced visual response sensitivity of the elderly, and/or elimination of extreme glare, which adversely affects even the reduced visual capacity of certain visually handicapped persons.

### 3. The Deaf and Partially Deaf

Care should be taken in developing certain critical communication systems to recognize that many people may not hear warning signals or advisory communications by transit personnel or be able to understand verbal messages that are minimally satisfactory (i.e., those with poor transmission quality).

### 4. Persons with Other Handicaps

Some transit system users, although ambulatory, may lack full use of their hands, may be too weak to apply normal force to door handles or controls, or may not be able to push heavy doors. Similarly, many people cannot reach overhead handrails, grip and turn doorknobs, or bend sufficiently to keep from striking their heads on low overheads.

## Design Criteria for the Handicapped

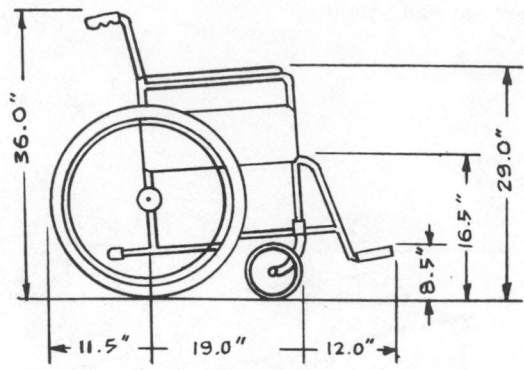

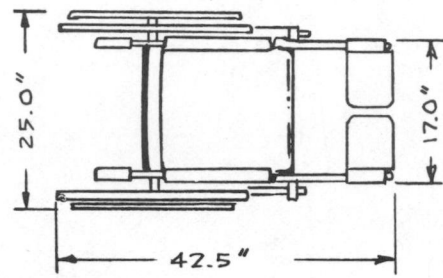

*Collapsible Wheelchair Dimensions*

Length    42 in (107 cm)
Width (open)    25 in (64 cm)
Seat height    19.5 in (49.5 cm)
Armrest and floor height    29 in (74 cm)
Push-handle height    36 in (91 cm)
Width (folded)    11 in (28 cm)
Fixed turn radius    18 in (46 cm) (wheel to wheel)
Fixed turn radius    31.5 in (80 cm) (front to rear)

Desirable turn space    63 × 56 in (160 × 142 cm)
Passing (two wheelchairs)    60 in (152 cm)

*Adult Occupant Limits*

Vertical reach (unilateral)    60 in (152 cm)
Average horizontal reach (table height)    30 in (76 cm)
Average diagonal to wall (height above floor)    48 in (122 cm)

# TRANSPORTATION TERMINALS

## TERMINAL FACILITIES

### Transportation System Terminal Facilities Model

The following are among the most critical considerations:

1. Efficient movement of passengers, baggage, and cargo
2. Rapid and efficient turnaround for the primary vehicular system
3. Effective 24-hr terminal facilities maintenance

### Terminal Facilities Concept Development

Although each type of transportation requires slightly different facilities and site conditions in order to accommodate the particular prime vehicular system (airplanes, buses, trains, ships, etc.), certain basic requirements should be considered for all terminal concepts.

### 1. For Passengers and Their Guests
   a. Site access (by foot, private automobile, taxi, or bus)
   b. Public parking for private vehicles (both short- and long-term)
   c. Vehicle-to-terminal transfer (from the curbside or from a parking lot)
   d. Ticketing and baggage check

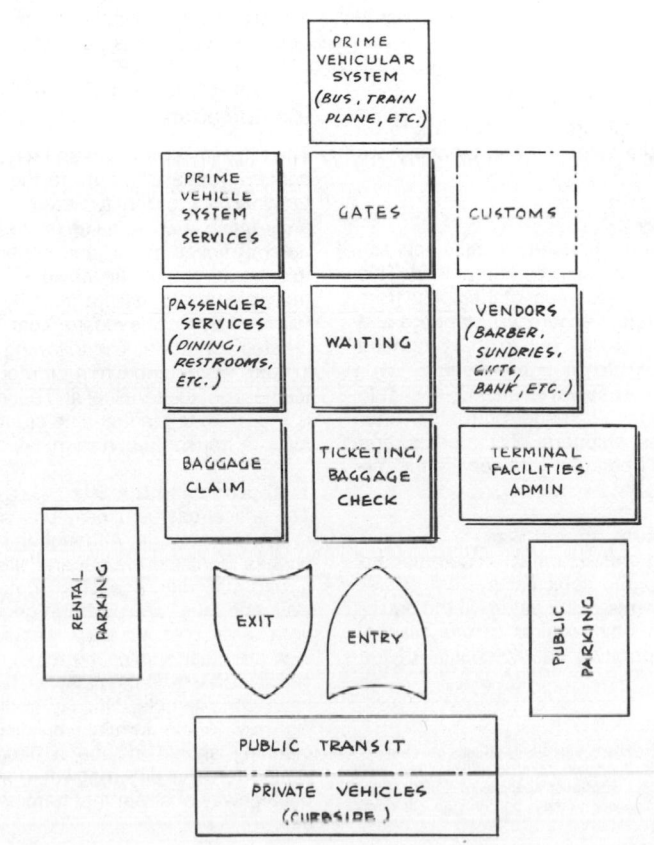

e. Waiting (i.e., seating, temporary storage for personal articles, rest rooms, dining or snacking facilities, information centers, and arrival and departure communications)

f. Transfer from terminal to prime transportation vehicular system (gates, ramps, mobile lounge, etc.)

g. Passenger assistance (porters, baggage carts, strollers, wheelchairs and carts, telephones, and public address system)

### 2. For Terminal Operations Personnel

a. Site access

b. Private employee parking and terminal access

c. Lounges and storage for personal articles

d. Administrative offices

e. Housekeeping and maintenance offices, shops, storage, and delivery facilities and parking

### 3. For Terminal Tenant Operations Personnel[36]

a. Site access (service)

b. Special parking (personnel and service vehicles)

c. Primary public interface facility (internal and external)

d. Offices, storage, delivery facility, kitchen, etc.

e. Special prime vehicle services (docking, refueling, baggage handling, maintenance and repair, equipment storage, etc.)

f. Personnel facilities (lounges, locker rooms, toilet, conference rooms, etc.)

g. Dispatch and shipping facilities

h. Systems control (air traffic control tower and center, weather office, pilots' ready rooms, etc.)

## Site Planning

The key human factors considerations to be addressed in site selection and planning are the following.

### 1. Accessibility

The site should be as nearly equally accessible to all expected users as is practical. This requires careful assessment of the points from which potential passengers will emanate and how they will probably get to the proposed site, i.e., via private or public vehicle, over what streets or roadways, etc. One should consider not only passengers but also other users, such as shippers of materials and equipment and operators who service the terminal.

### 2. Site Configuration and Size

The site shape and size must accommodate the particular prime vehicular system operator's needs in terms of the nature of the equipment operation. This requires careful assessment of the operating characteristics of the

operator's equipment; e.g., airplanes need runways, docking ramps, etc.; railroads need tracks and switching and docking facilities; and bus and truck operators need parking, docking, and maneuvering facilities. However, in addition to the obvious concern for the prime vehicular system operator's needs, careful attention must also be given to the other side of the system: passengers, shipping clients, and terminal operators. For example, one should be careful not to allow concern for an airline operator's problems of maneuvering and docking airplanes create impossibly long and difficult transfer and walking problems for passengers.

### 3. Community Interaction

The location of a terminal site should be compatible with the general environmental needs of the community in terms of traffic congestion, noise, safety, and air pollution. Consideration should also be given to the interactions between the proposed terminal and other terminals in terms of roadway interconnections, local public transportation, and the effects that time and distance have on the interchange of passengers between terminals.

### 4. Terminal Facilities Layout

The proposed site should create the least amount of constraint on the layout of terminal facilities in terms of spacing buildings and orienting roadways, track, or runways. A common problem of many large terminal sites is that the odd shape of the original site creates a confusing maze of pathways and building arrangements. A great many of the eventual site users will be unfamiliar with the site and may have time constraints, and there may be environmental conditions that will further add to their possible misorientation when site layouts are not obvious and logical to them.

## Terminal User Facility Orientation and Identification

The most important human factors in terminal concept formation relate to the ability of the terminal user to identify elements of the facility and remain oriented as he or she approaches, enters, moves about, and departs. Although this would appear an obviously easy set of requirements to fulfill, a majority of terminals fail miserably in this regard. That is, the design creates confusion. The following scenarios illustrate where and when the need to identify elements of the terminal and become oriented is important to the first-time or infrequent visitor to a transportation terminal.

### 1. Approaching the Site

The critical user is usually one who is driving to the terminal site. Although this user probably has identified the general site location on a road map with respect to some major highway, and thus can reach the general terminal area easily, unless the street or highway markings are clear and/or the terminal building itself is visible from the highway, he or she may have considerable difficulty getting from the highway to the terminal building area. Site planning should include a direct and well-signed terminal-site roadway from the principal highway or street that terminal visitors will use.

### 2. Going from the Terminal-Site Entrance to the Terminal Building and Other Facilities

Again the critical user is driving his or her own car or a rental car to the terminal. The route from the terminal-site entrance to the terminal building, a public parking area, or a rental-car return area should be as direct and uncluttered by intervening interruptions (intersections, other buildings, other signs, other parking areas, etc.) as possible. In other words, the layout and signing for this portion of the user's terminal approach should leave as little doubt in the user's mind as possible as to whether he or she is on the right track. It is desirable to create a site layout which makes it possible for individuals to keep the terminal building in sight throughout the approach and in which the roadway does not gyrate through many curves or turns.

### 3. Going from the Parking Facility to the Terminal Building

The critical user is one who is pressed for time. Thus, it should be possible for this user to keep track of which way to go, after leaving his or her car in a parking lot. Either plan a parking transportation system that will take the user directly to the terminal building entrance, or locate the parking facility so that the user can keep the terminal building in sight. The distance to the terminal building is obviously important if the user is to walk (carrying luggage). If an enclosed parking garage concept is to be used, careful attention should be given to the marking and signing so that the user can easily locate the signs. In deciding on a parking facility concept (i.e., whether it is to be a lot arranged in a certain way or a parking structure that has various levels and ramps), consider carefully the method by which the facility is to be encoded and marked. Avoid the use of abstract alphanumerics, the logic of which generally is not readily apparent to the visiting observer.

### 4. Becoming Oriented to the Terminal Building

Terminal buildings, especially large airport terminals, have a way of becoming very confusing to the casual user. This is often due to the fact that, in laying out the building, the designer gave priority to the space demands of parked aircraft. When aesthetic whims are added, such buildings become almost impossible to perceive as a logical and systematic geometric pattern. The key orientation and identification considerations for building conceptualization should be the following:

a. Entrances should be readily identifiable, not only by the signing, but also by the geometric features. All pathways should lead directly to these entrances; users should not have to take paths that turn, jog, circle, or curve.

b. Entrances should open directly to the things the user expects to see; i.e., as the user enters the building, he or she should have a direct view of the ticket counters, the information desk, indications of where the gates are, the rest rooms, etc.

c. There should also be obvious placement of those interior features which a board-

---

[36]Prime vehicular system operators, on-site government operations, rental-car agencies, travel agencies, social service agencies, public transportation agencies (e.g., a local bus or taxi), and concessionaires (a bank, barber shop, gift shop, restaurant, etc.).

ing passenter will need if he or she has to wait for a flight, i.e., the waiting-room area or areas, dining room, bar, snack bar, newsstand, etc. Concessionaire stands can be located in a variety of places since passengers who have a longer wait can look for these at their leisure.

d. Disembarking passengers have the same needs as those of boarding passengers, except in reverse. The most criti-

cal scenario is that of disembarking passengers who must make a quick connection between the plane they left and the one they have to catch. The arrangement concept should minimize the problems of getting out of the gate area of one airline and into the gate area of the other—preferably without having to go all the way back to the originating ticket-counter area of the terminal. An especially important layout consideration is

the separation of areas for incoming passengers and reception areas for people who are meeting debarking passengers. Traffic jams occur when the design allows these people to block the path of passengers who are in a hurry to get to the next departure gate.

## Terminal-Site Planning (Air Terminal)

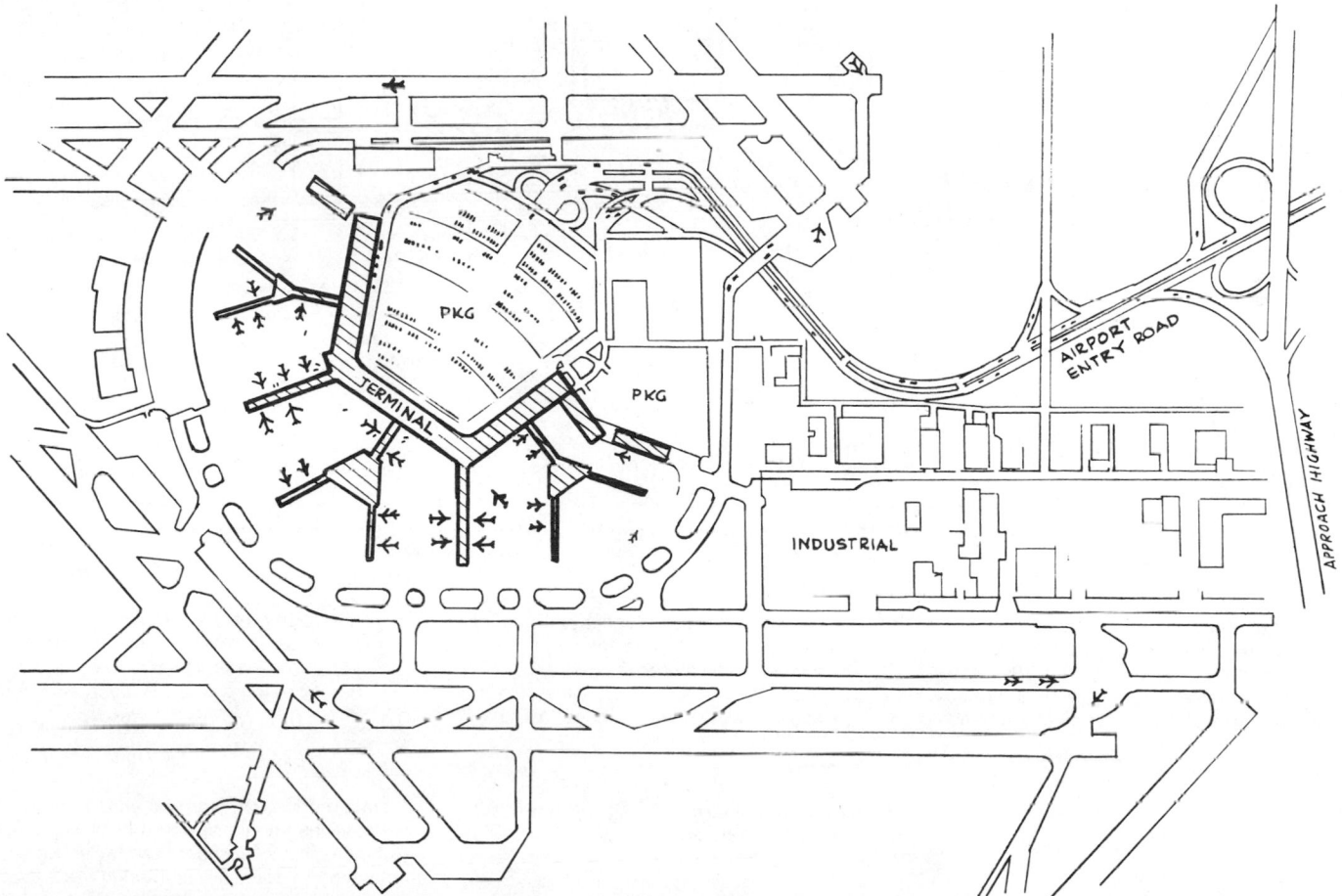

Example of poor terminal-site planning. (This terminal-site layout creates utter confusion for the traveler who must approach or leave the area for the first time. Note especially the lack of north-south geographic reference, the frequent turns from the highway to the terminal area, and the confusing industrial area on the way into the terminal area.)

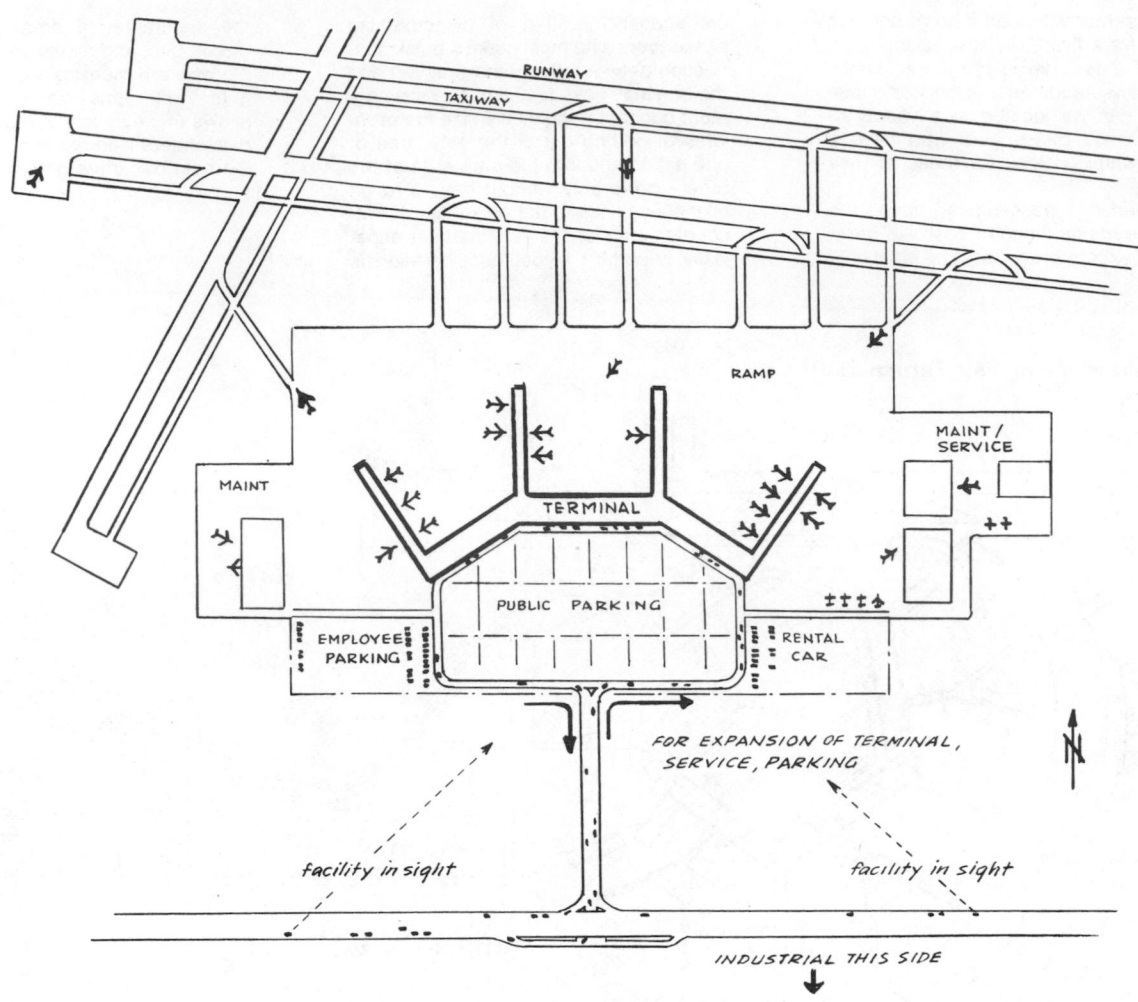

1. Regardless of required runway axis (due to prevailing winds), plan the layout of terminal buildings and the approach to be "squared" with a principal highway.
2. Roadways into the terminal area should be organized in rectilinear patterns, and incoming traffic and departing traffic should not be forced to cross each other.
3. Plan for possible expansion. Do not allow industrial or business facilities to encroach on the approach areas in such a way as to preclude later terminal expansion or obstruct the view of the terminal for patrons who are approaching it.

## Pedestrian Movement by Special Transit Subsystems

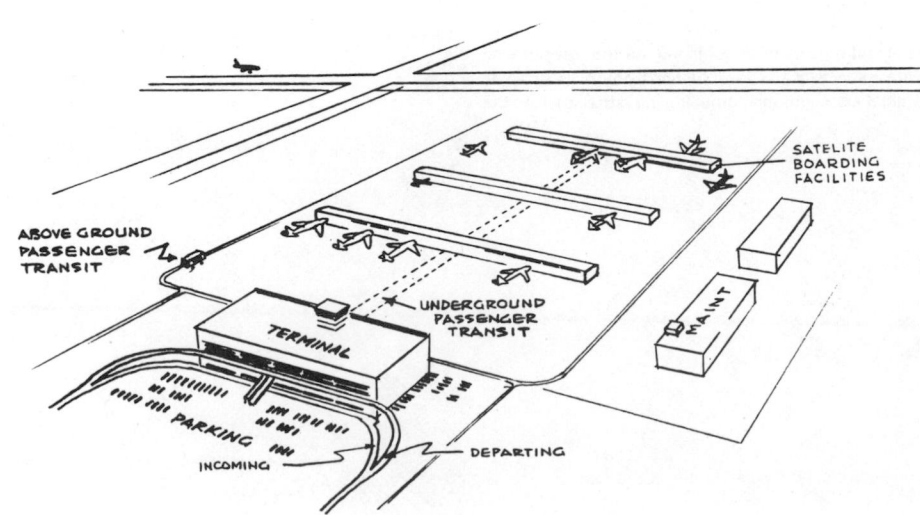

A major concern in planning very large terminal facilities should be the minimizing of the distances that passengers and facility personnel have to walk. As illustrated in the accompanying sketch, various mobile transit configurations should be considered to reduce the typically long distances between the point at which passengers first arrive at, or depart from, the principal terminal building and the various gates where passengers board or deplane. As indicated, interfacility transit systems can be used above ground, as in the case of the large mobile lounge (a typical example may be found at the Dulles International Airport outside Washington, D.C.), or smaller underground trains of passenger vehicles, generally rail, can be used.

In addition to the advantages of faster and less fatiguing transfer of passengers, there are fewer possible hazards and less confusion for passengers since such systems remove the conflicts between various types of single-passenger transporters (e.g., electric carts used to aid the elderly or the disabled), service and baggage vehicles, and pedestrian traffic.

### Terminal-Site Planning (Bus and Rail)

Unlike air terminal sites, the bus or rail terminal is usually located within the city and is often constrained by adjacent street and building patterns. However, the following basic considerations are still important to the ease with which passengers and visitors can find their way around:

1. Provide sufficient terminal building set-back so that site visitors are able to get off the street while they reconnoiter to determine where to deposit or pick up passengers at the curbside and/or where to locate the public parking lot.
2. The site drive should be laid out so that entering and departing vehicles do not conflict with one another.
3. Separate busline vehicle entry and departure drives should be provided so that these vehicles do not conflict with patron or visitor vehicles, city buses, or cabs.
4. Provide for taxi waiting lines.
5. Provide separate roadways and parking for service and freight vehicles so that these vehicles do not intermix with patrons' vehicles more than necessary. Whenever possible, do not locate freight facilities so that service trucks enter the site from the same side as vehicles proceeding to the passenger terminal building.

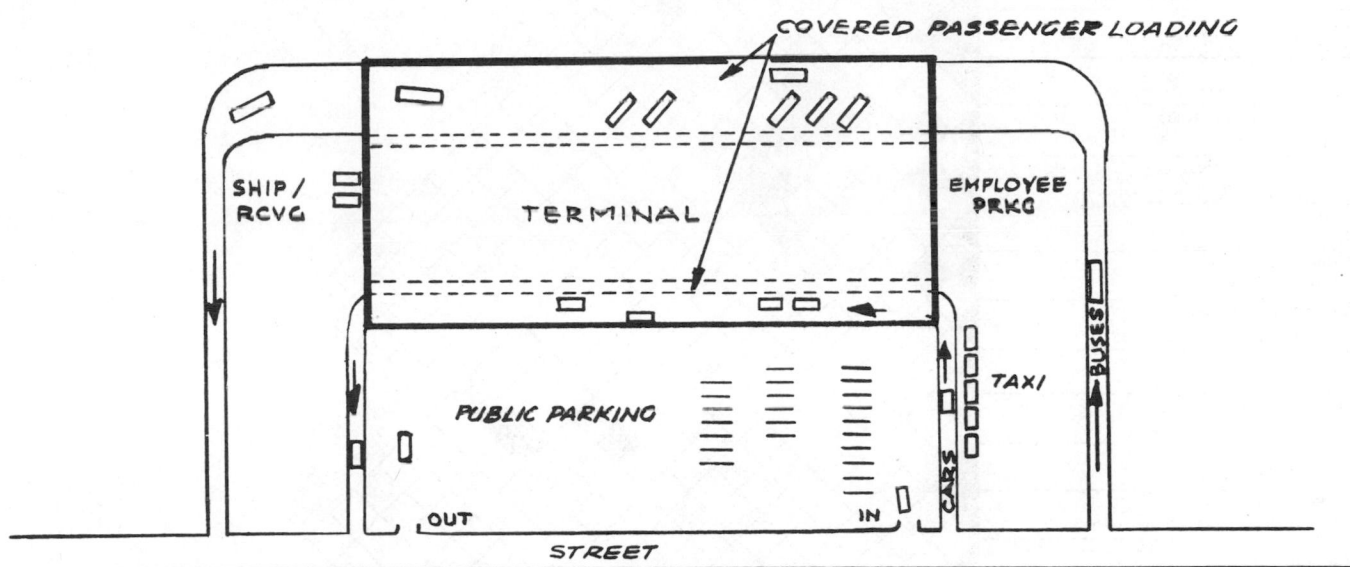

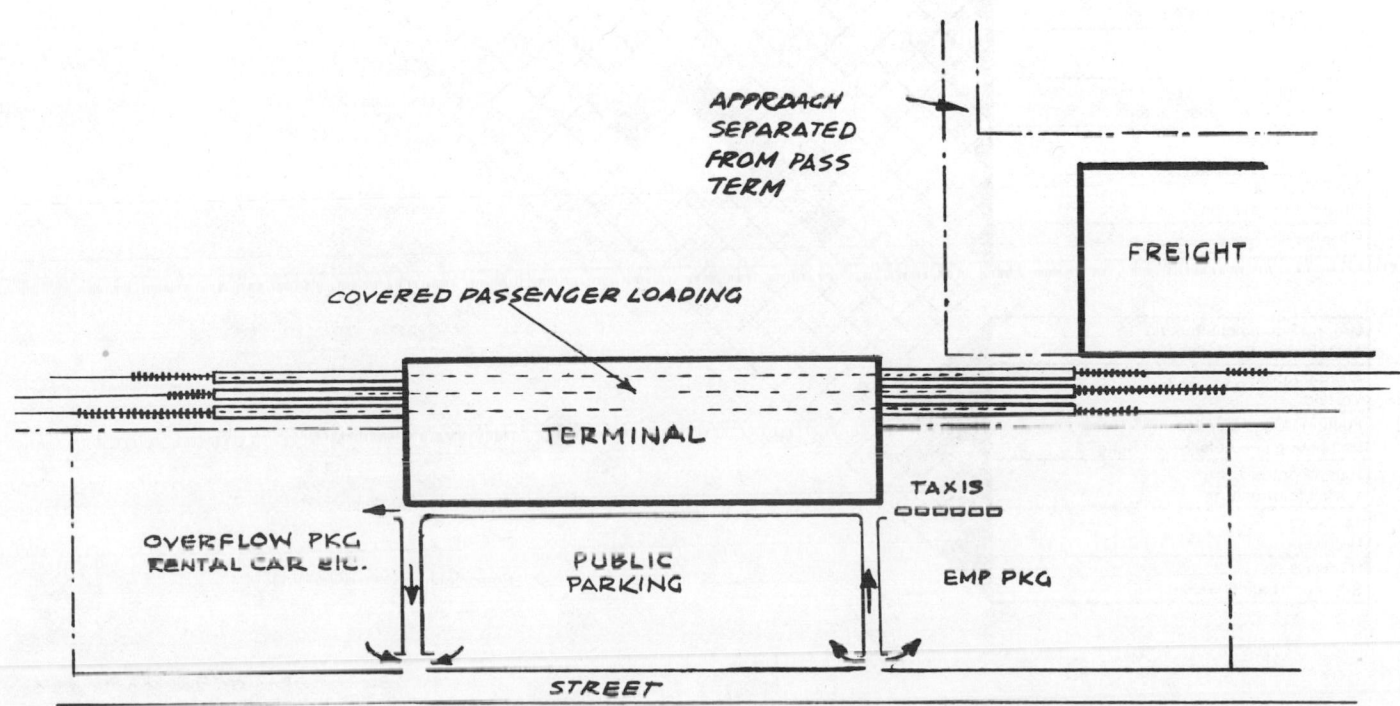

**Air Terminal Functional Interface Requirements[37]**

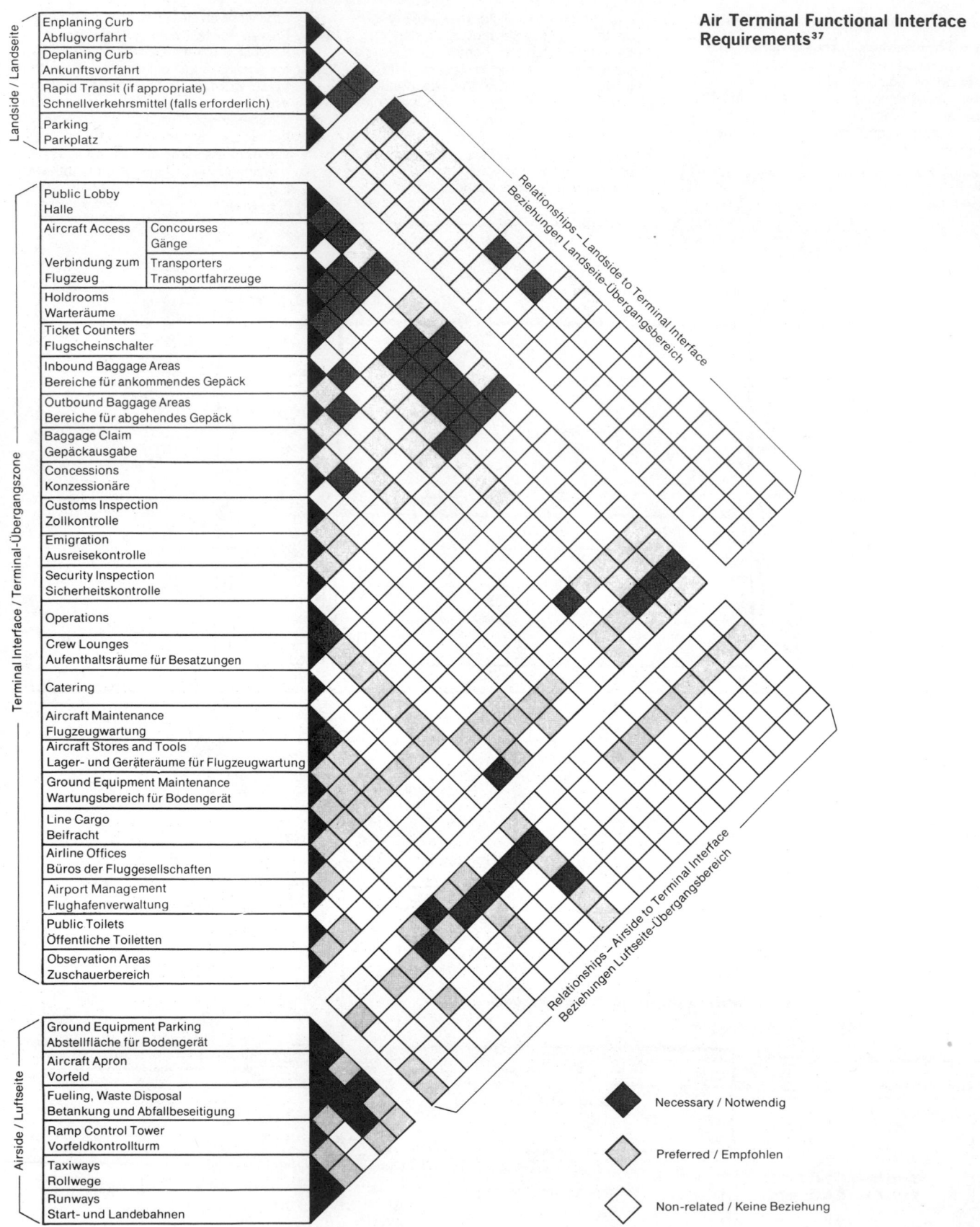

Landside / Landseite

Enplaning Curb / Abflugvorfahrt
Deplaning Curb / Ankunftsvorfahrt
Rapid Transit (if appropriate) / Schnellverkehrsmittel (falls erforderlich)
Parking / Parkplatz

Terminal Interface / Terminal-Übergangszone

Public Lobby / Halle
Aircraft Access / Verbindung zum Flugzeug — Concourses / Gänge — Transporters / Transportfahrzeuge
Holdrooms / Warteräume
Ticket Counters / Flugscheinschalter
Inbound Baggage Areas / Bereiche für ankommendes Gepäck
Outbound Baggage Areas / Bereiche für abgehendes Gepäck
Baggage Claim / Gepäckausgabe
Concessions / Konzessionäre
Customs Inspection / Zollkontrolle
Emigration / Ausreisekontrolle
Security Inspection / Sicherheitskontrolle
Operations
Crew Lounges / Aufenthaltsräume für Besatzungen
Catering
Aircraft Maintenance / Flugzeugwartung
Aircraft Stores and Tools / Lager- und Geräteräume für Flugzeugwartung
Ground Equipment Maintenance / Wartungsbereich für Bodengerät
Line Cargo / Beifracht
Airline Offices / Büros der Fluggesellschaften
Airport Management / Flughafenverwaltung
Public Toilets / Öffentliche Toiletten
Observation Areas / Zuschauerbereich

Airside / Luftseite

Ground Equipment Parking / Abstellfläche für Bodengerät
Aircraft Apron / Vorfeld
Fueling, Waste Disposal / Betankung und Abfallbeseitigung
Ramp Control Tower / Vorfeldkontrollturm
Taxiways / Rollwege
Runways / Start- und Landebahnen

Relationships – Landside to Terminal Interface / Beziehungen Landseite-Übergangsbereich

Relationships – Airside to Terminal Interface / Beziehungen Luftseite-Übergangsbereich

◆ Necessary / Notwendig
◇ Preferred / Empfohlen
◇ Non-related / Keine Beziehung

[37]E. G. Blankenship, *The Airport*, Frederick A. Praeger, Inc., New York, 1974, p. 30.

**Air Terminal Activity Circulation**
**Model[38]**

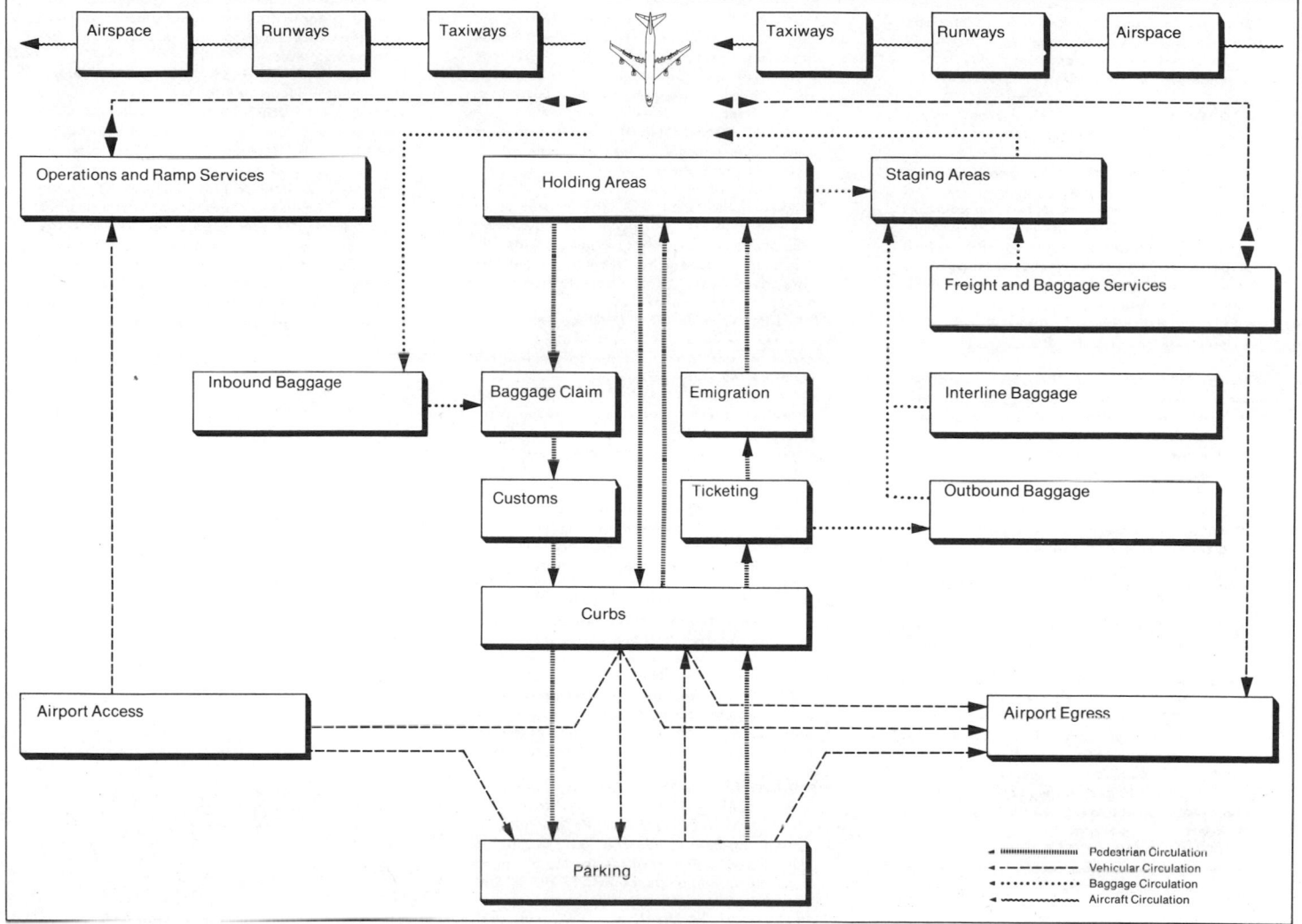

[38]E. G. Blankenship, *The Airport*, Frederick A. Praeger, Inc.,
New York, 1974, p. 41.

## Glossary (Common Usage in Terminal Design)[39]

**AMENITIES:** That part of a terminal building housing convenimnce, service, and diversion facilities for the passengers, tenants, and public.

**AVERAGE PEAK HOUR:** The peak hour of the average peak day. The peak hour is the one-hour period of any peak day during which the highest percentage of the day's traffic is experienced. The average peak day is the average of the top 37 days (10 percent) of a year in terms of traffic volume.

**BAGGAGE DIVERTER:** A mechanical device for transferring baggage from a moving conveyor belt to a baggage claim counter in such manner that the baggage is evenly distributed along the baggage counter.

**BOARDING CONTROL POINT:** The point at which a passenger's credentials are inspected to assure that he is authorized to board a particular flight. Normally, this point is located in the vicinity of the gate from which the flight will depart.

**BOARDING PASSENGER:** Any originating or connecting passenger authorized to board a flight.

**CONNECTING PASSENGER:** A passenger who arrives on one flight only for the purpose of transferring to another flight to reach his destination. These passengers are broken down into two categories: intraline and interline passengers.

**CUSTOMS:** This is an area under federal jurisdiction through which passengers arriving from foreign countries are required by law to pass in order to make a declaration related to baggage which is accompanying them upon entry to the United States. This area is used for receipt of a declaration and/or examination of baggage. If duty is required, the customs agent will receive same in the customs area. Special attention must be paid to the design of this area because of changing techniques of operation.

**DEPARTURE ROOM:** An assembly area, including the boarding control point, located at a gate position(s) for passaengers pending availability of aircraft for boarding.

**DEPLANING:** Any passenger, cargo, baggage, visitor, etc., which is related to the unloading from an arriving flight.

**DOMESTIC PASSENGERS:** All passengers traveling in the United States or its territories are considered as domestic. Foreign nationals within the confines and territory require no special checking and operate as domestics.

**ENPLANING:** Any passenger, cargo, baggage, visitor, etc., which is related to the boarding of a departing flight.

**FIS:** FIS is an abbreviation for Federal Inspection Services. It is utilized as an all-inclusive term for the U.S. Public Health, Immigration, and Naturalization Service, the Department of Agriculture, and U.S. Customs.

**GATE:** A location to which aircraft are brought for the purpose of discharging and loading passengers and their baggage.

**GATE CONCOURSE:** An extension from the main terminal building primarily intended to provide protected access for passengers between the main terminal building and the gates. In addition to the passenger corridor, the concourse may include airline functional areas and minimum consumer services.

**GROUND TRANSPORTATION:** The independently operated transportation vehicles scheduled for passengers' use between airports and the areas served thereby is called ground transportation.

**IMMIGRATION:** This area is devoted to the examination of passports of United States nationals and aliens seeking to enter the United States. Consideration for design and function of this area must be correlated with federal authorities.

**INTERLINE CONNECTING(ION):** A term used to describe passengers and baggage which arrive on the flight of one airline and depart on a flight of another.

**INTOWN TERMINAL:** A facility located apart from the airport, usually in the downtown area of the city, at which passengers may be processed, baggage is checked to passengers' destinations, and from which ground transportation is provided.

**INTRALINE CONNECTING(ION):** A term used to describe passengers and baggage which arrive on one flight and depart on another flight of the same airline.

**IN-TRANSIT PASSENGER:** If an internationally bound aircraft stops at an airport for refueling or discharge of passengers and a remaining number of passengers are to be detained in the aircraft for another destination, the convenience of providing a totally segregated lounge facility may be warranted for the continuing passengers. This facility is referred to as an in-transit area. No FIS inspection is, required, but security of the area is important.

**LONG-HAUL** A term used to define flights or traffic which travel over a relatively long distance as opposed to those which travel over a shorter distance. Normally, long-haul passengers arrive at their originating airport earlier than short-haul passengers, carry more baggage than short-haul passengers, and are accompanied to or are met at the airport by more persons than short-haul passengers.

**ORIGINATING PASSENGER:** A passenger who is starting his trip.

**OUTBOUND BAGGAGE ROOM:** The area to which checked baggage of originating passengers is delivered for sorting by flights prior to its being dispatched to the aircraft for loading.

**PUBLIC HEALTH SERVICE:** The function of the Public Health Service is to determine whether an arriving passenger will present a health hazard to the general population. This may require inoculation, special examination, and possibly quarantine. Design requires correlation with federal authorities.

**READY ROOM:** An area adjacent to the normal work areas in which personnel whose duties are performed out-of-doors may assemble, be protected, and from which they may receive their work assignments. These rooms should be concealed from public view.

**SELF-CLAIM BAGGAGE:** A method under which passengers have direct access to terminating baggage in a controlled area. As passengers leave the area, an attendant retrieves baggage claim checks and matches them to strap checks to assure that passengers have selected only baggage to which they are entitled.

**SHORT-HAUL:** A term used to define flights or traffic which travel over a relatively short distance as opposed to those which travel over a long distance. Normally, short-haul passengers arrive at the airport of origin later than long-haul passengers, carry less baggage than long-haul passengers, and are accompanied to or met at the airport by fewer persons than long-haul passengers.

**STANDBY PASSENGER:** A passenger not holding confirmed space but who is on hand at departure time for space that might become available.

**TERMINATING PASSENGER:** A passenger who has arrived at his destination.

**THROUGH PASSENGER:** A passenger who arrives and departs on the same flight.

**TRANSFER BAGGAGE ROOM:** The area to which checked baggage of connecting passengers is delivered for sorting by flights prior to its being dispatched to the aircraft for loading. This may be combined with the outbound baggage room at some locations.

**UNIT TERMINAL:** One of several functionally complete terminal areas (which may be in the same or several buildings) each of which houses the activities of one or more airlines.

[39]J. De Chiara and J. H. Callender, *Time-Saver Standards for Building Types,* McGraw-Hill Book Company, New York, 1973, pp. 772–775.

## Selected User Interface Considerations for Initial Terminal Facility Conceptualization[40]

### 1. Entrances, Exits, Corridors, and Risers

a. A sufficiently broad sidewalk is required outside main entrances in order to accommodate incoming passengers, baggage, baggage carts, passengers who are waiting for transportation, and people who are walking from one entrance to another or to a waiting vehicle. A minimum of 15 ft (4.6 m) should be provided. An overhanging roof should protect passengers from rain as they transfer from a vehicle to the sidewalk.

b. Extra-wide double doors should be provided so that arriving and departing passengers do not conflict with one another. An automatic tread-actuated door system that opens both doors at once is recommended, with a clear opening of at least 10 ft (3.1 m). It is recommended that more than one entrance be provided to help distribute passengers among the various ticket booths inside the terminal building.

c. When a two-level structural concept is practical, consider separating arriving and departing passengers so that one group transfers on one level, and the other group on the other level.

d. Entrance configuration should be obvious in terms of general appearance (i.e., avoid the temptation to camouflage an entrance for aesthetic reasons, such as by using an uninterrupted glass facade). Provide identifying signs to indicate which airline, bus, or railroad system is nearest a particular entrance. Orient the signs so that they are readable from the viewpoint of vehicle drivers and passengers (i.e., not as though drivers and passengers were facing the building), and make sure that the sign lettering is readable from expected viewing distances.

e. Provide flat entry pathways and thresholds so that people in wheelchairs can negotiate them safely and easily.

f. A double entry with an intervening vestibule should be considered for terminal buildings located in cold climates; this helps keep cold winds from entering the main building.

### 2. Main Ticketing and Baggage Check-in Area

a. The size and shape of this area should be based on the anticipated number of airline, bus, or train ticketing booths; the expected peak passenger queuing loads; and the obvious milling around created by nonpassenger as well as passenger patrons.

b. Information signs should be centrally located so that they are immediately visible to the incoming or transferring passenger and/or to the person who wants to know what plane, bus, or train is expected at what time and at which gate. If

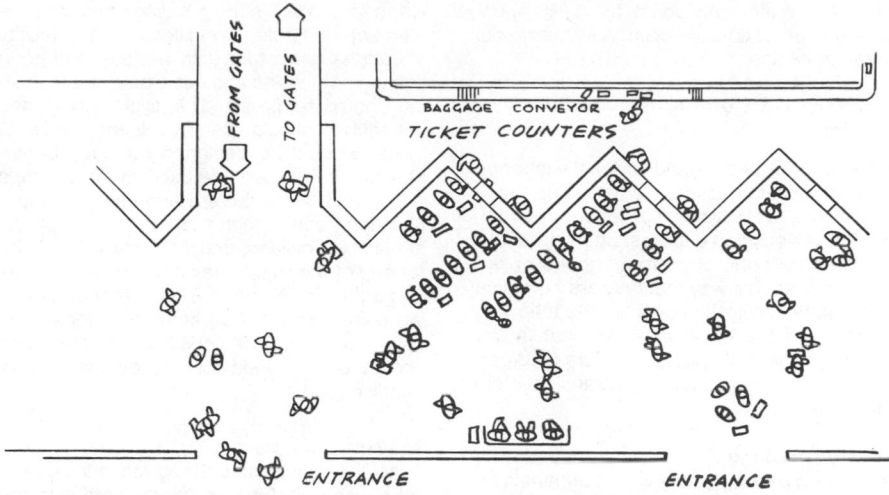

A diagonal ticket-counter concept helps minimize interference between passenger queues and other persons who are walking, standing, or milling about in the ticketing area.

the area is large, consider several such installations, spaced appropriately along the ticketing-area corridor. Signs should be oriented so that they are readable from the typical viewing angle of the observer (i.e., observers should not be required to walk to a position in front of the sign before they can read it).

c. Several special facilities should be located within, or directly adjacent to, the main ticketing area:
   (1) Information desk
   (2) Traveler's aid desk
   (3) Rest rooms
   (4) Telephones
   (5) Seating

### 3. Passenger Waiting and Services Area or Areas

One or more passenger waiting and services areas should be provided for the terminal in which considerable waiting may occur between services. The following should be considered:

a. Rest rooms, shoeshine stand, and parcel lockers

b. Dining, snack, and cocktail concessions and drinking fountains

c. Bank and money exchange facility

d. Drugstore, gift shop, flower shop, duty-free shops, liquor store, haberdasher, women's clothing store, newsstand, etc.

e. Insurance vendor's desk

f. Special airline lounges

g. Flight, bus, and train schedule board

h. Traveler's aids (baggage carts, strollers, etc.)

Note: When separate immigration facilities are required, any or all of the above should be considered within the restricted areas used by passengers awaiting departures to foreign countries.

### 4. Terminal Administrative Facilities

The following are among the important terminal facilities that will be needed both by the terminal management and by other nonpassenger service persons (e.g., government transportation agents):

a. General administrative and clerical offices

b. Employee lounges, rest rooms, locker rooms, dressing rooms, private dining rooms, etc.

c. Communications centers (for security, public address systems, transportation systems control, etc.)

d. Maintenance and housekeeping facilities, including shops, equipment storage, and service docks and parking

e. First-aid station

### 5. Baggage-Claim and Local Transportation Area

Most modern transportation terminal concepts separate the baggage-claim and local transportation functions from the previously outlined functions. Among the more typical facility requirements are the following:

a. Incoming baggage collectors and dispensers (diverters, carousels, racetrack conveyors, etc.)

b. Temporary parcel storage (lockers and/or room)

c. Rental-car concessionaire booths

d. Telephones

e. Taxi, limousine, and city bus stands

f. Rest rooms

### 6. Other Amenities

a. In-terminal hotel

b. Theatre

c. Public stenographic service

d. Stairs, ramps, escalators, and elevators

e. Environmental control (illumination, temperature, etc.)

f. Observation deck (airport terminal)

g. Nursery, play rooms, TV areas, etc.

Note: All transportation terminal facilities should be accessible to the handicapped traveler.

Air terminal buildings must be planned to ameliorate the detrimental effects of high

[40]For a more detailed discussion, see J. De Chiara and J. H. Callender, *Time-Saver Standards for Building Types*, McGraw-Hill Book Company, New York, 1973.

level jet aircraft noise, both for passengers and for all personnel associated with terminal operation or use.

## Rapid-Transit Commuter System Terminals

Terminal facilities provided for local exchange of transit system passengers (including those using buses, rail vehicles, suspended rail vehicles, and air-cushion vehicles) often either are neglected in terms of patrons' needs or are designed in such a way that they are attractive but not necessarily serviceable. The following are some of the basic local terminal facility considerations that should be addressed during the planning stages of any transit system development.

### 1. Patron Parking
Fringe-area parking at key transit terminals on the perimeters of a city not only should be sufficiently large to accommodate expected user vehicles, but also should provide reasonable security for the vehicles left temporarily in these facilities.

### 2. Waiting Area and Shelter
A sheltered waiting structure should be provided to protect waiting passengers from inclement weather conditions. The facility should have heat for cold weather and good lighting both inside and out. Comfortable seating should be provided. A system map and schedules should be prominently posted. These should be designed for easy understanding by the average user (avoid art maps and complex street maps and use clear identification code techniques). A telephone should be provided, and if the station is at the far end of the line, there should be rest rooms and a drinking fountain. The shelter and loading platform should be accessible for wheelchair users, and the schedule and map data should be designed for use by the visually handicapped.

### 3. Safety
Special attention should be given to the potential hazards that may be associated with the terminal facility and the equipment and area around the facility, especially with regard to people walking or falling on the tracks.

### 4. Vending Machines
When self-service ticket and/or change vending machines are provided, these should be easy to use. The machines should also be vandal-proof.

## Downtown Bus Stops

Downtown bus stops, although not strictly terminals in the sense of the facilities previously discussed, also have critical human factors considerations. Among the most important are the following.

### 1. Safety
Although midstreet island stops are appropriate in a few cases, the curbside stop is preferable because it prevents people from walking across traffic to get to the island. It is suggested that a guardrail be placed between waiting passengers and the street. It should be designed to force patrons to queue up behind it.

### 2. Signing and Illumination
Bus-stop identification, schedules, and other pertinent information should be legible not only to those with normal eyesight but also to the visually handicapped.

HIGHWAYS

## STREETS, ROADS, AND HIGHWAYS

### Street, Road, and Highway Systems
### Concept Development

The primary objective of any street, road, or highway system is, of course, to provide vehicular and pedestrian links between geographic points. Since the main purpose is to make it possible for people and their vehicles to get from one point to another, any proposed system should be people-oriented. From a user's point of view, any system should provide the following.

### 1. Convenience

There should be the opportunity for as rapid and direct movement as possible from where a person is to where that person wants to go.

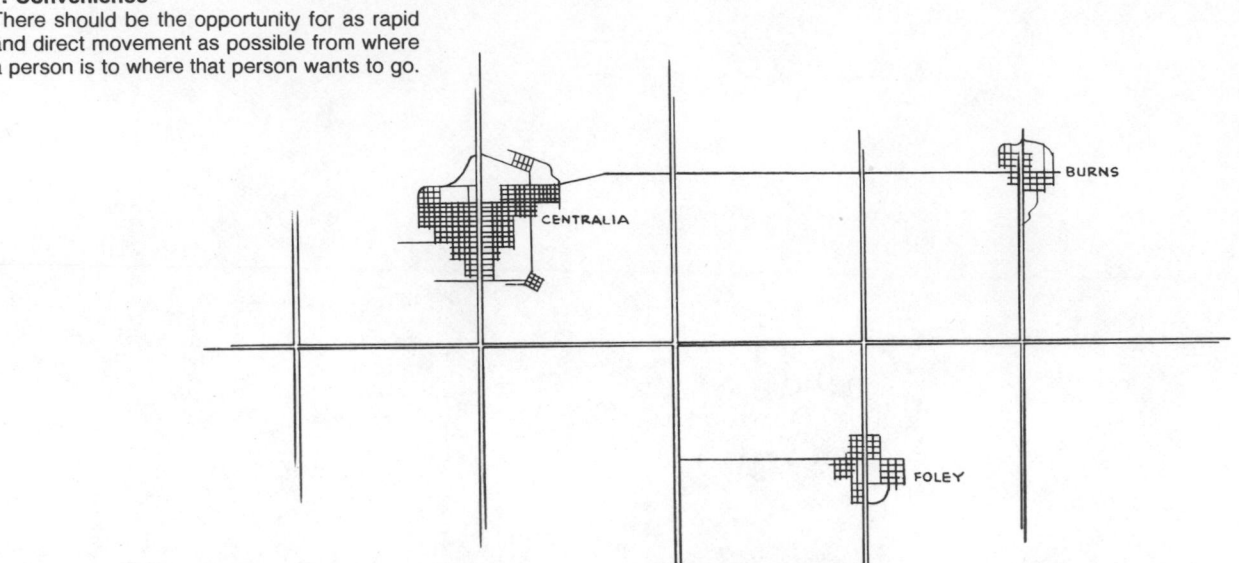

FROM COMMUNITY TO COMMUNITY

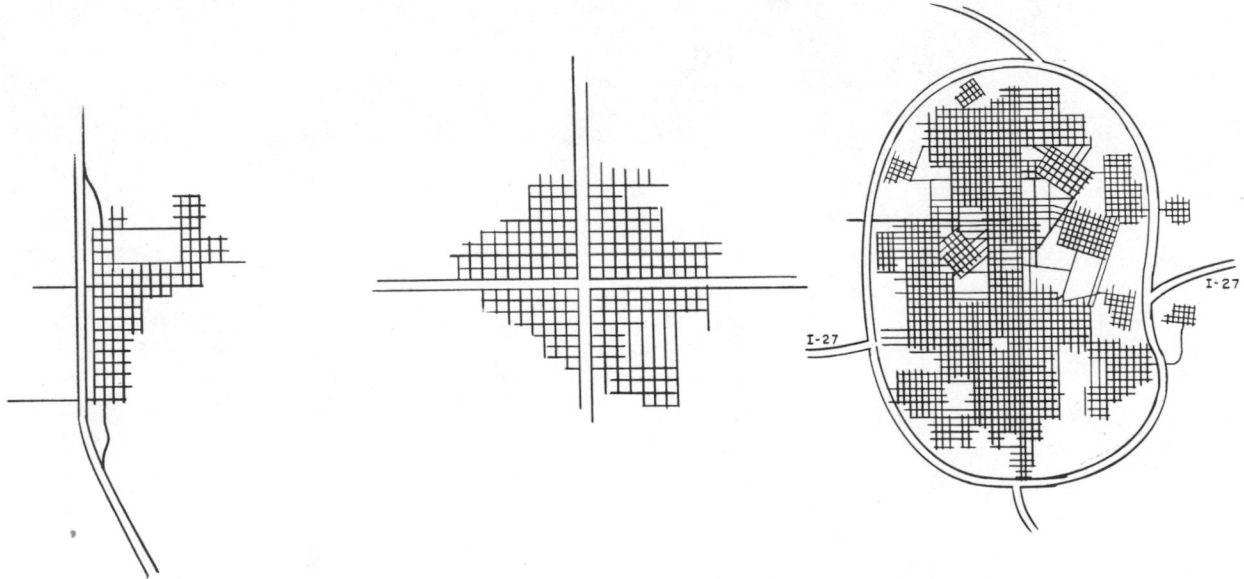

THROUGH AND AROUND A COMMUNITY

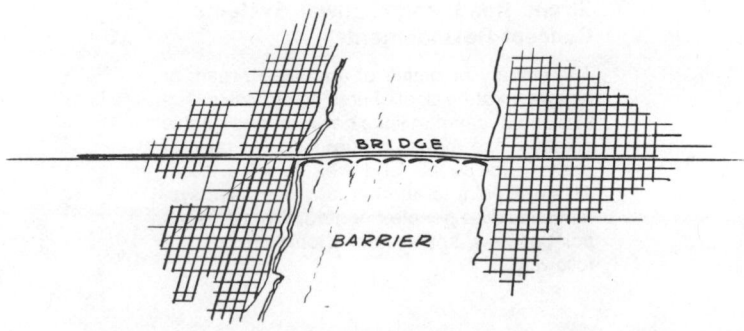

WITHIN THE COMMUNITY

User convenience differs. For example, pedestrians need sidewalks, street crosswalks, and pedestrian overpasses and underpasses to cross high-speed expressways and freeways; local motorists need arterial streets between major shopping areas, restricted but connecting residential streets, and means for crossing or by-passing major expressways and freeways; and transient motorists need noninterrupted pathways around or across communities, methods for changing from one major highway to another, and periodic access to local services.

## 2. Interference
There should be as few interruptions, impediments, and delays as possible, commensurate with the safety requirements of the street and roadway system.

TUNNEL

PEDESTRIAN BRIDGE

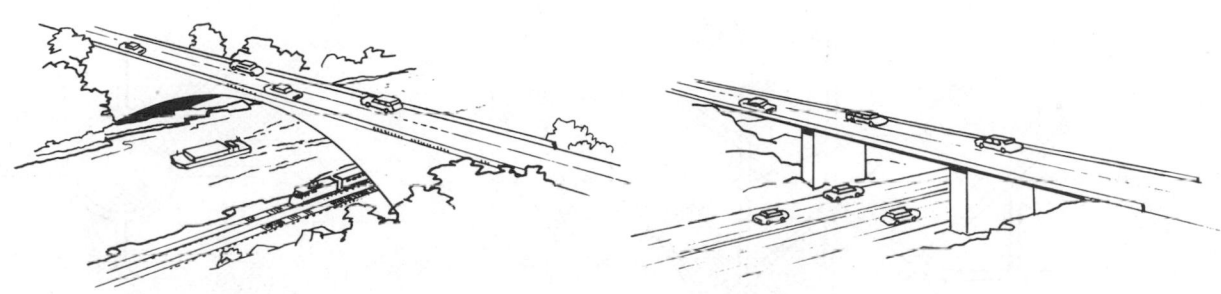

BRIDGE OVER NATURAL BARRIER,    OVER TRAFFIC

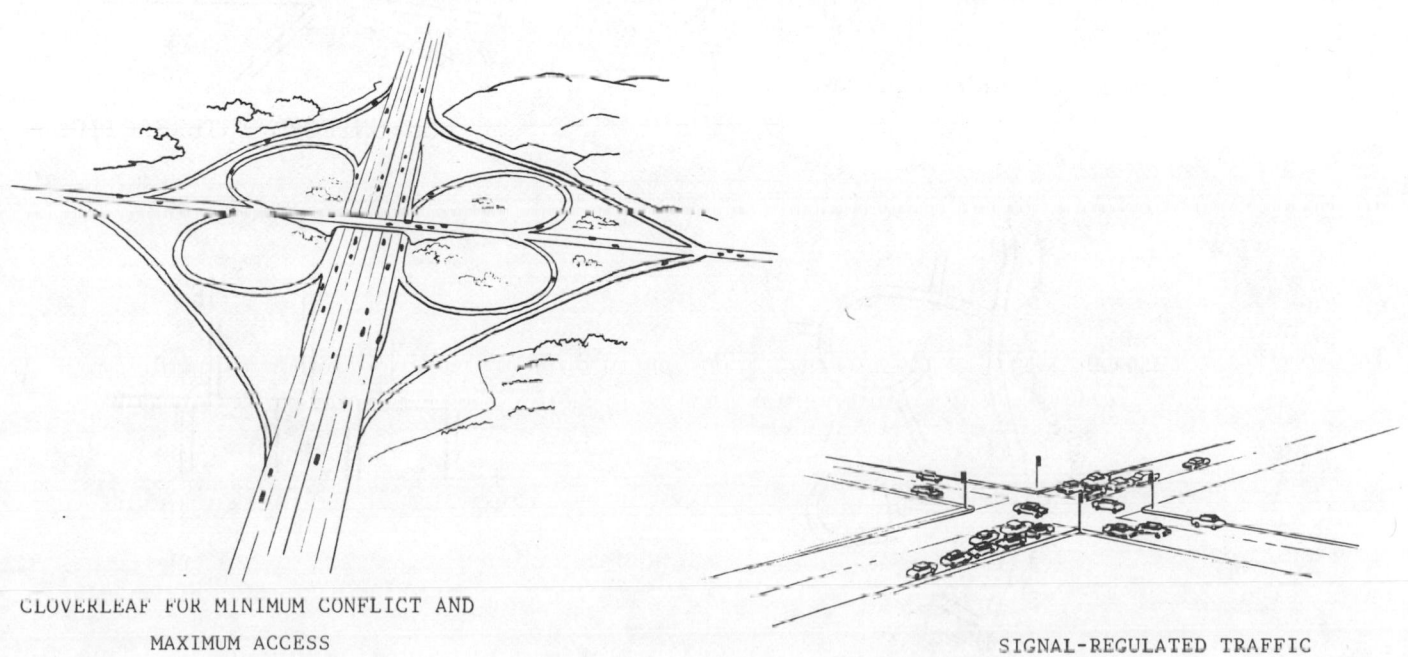

CLOVERLEAF FOR MINIMUM CONFLICT AND
MAXIMUM ACCESS

SIGNAL-REGULATED TRAFFIC

### 3. Orientation

The system organization, layout, and identification features should make it easy for motorists and pedestrians to know where they are, to know what path they should follow to get where they want to go, and to know they are in the right place when they get there.

A uniform street grid oriented to a north-south and east-west reference is easy to understand and follow. Odd and nonuniform patterns are hard to understand and retain in one's mind; angular, rather than perpendicular, intersections cause people to lose their direction; and traffic circles cause people to forget where they are. Streets that do not continue across an intersection leave one in doubt as to where the continuing path is.

### 4. Safety

The system should not include any built-in booby traps for the motorist or pedestrian, and any features that could be struck should be designed so that injury potential will be as minimal as possible.

Typical built-in booby traps to watch for include the following:

*For Motorists*

a. Blind corners, curves, and hilltops
b. A sudden stop at the end of a sharp, blind curve

c. Sudden narrowing of the road and disappearance of a lane
d. Short preparation distance between an advisory sign and the decision and maneuver point
e. Sharp, unbanked curves
f. Varying freeway interchange concepts (right versus left, under versus immediate approach to left overpass route, etc.)
g. Nonuniform curvature (radius) of on- and off-ramp turn
h. A common approach road used for both on and off ramps to and from a freeway
i. Ill-defined road edges (lack of contrast)
j. Lack of lane marking on multilaned street, frontage, or off-ramp roadways
k. Advisory or warning signs not located on the side of the road on which motorists normally look for them
l. Signs positioned so that other vehicles block motorists' view of them

m. Overhead advisory lane arrows that do not coincide with lanes from the driver's viewing angle and lane designations that shift back and forth on succeeding signs
n. Highway names that imply a location rather than a route (e.g., the name "Riverside" implies that the road goes to that city, even though motorists need to

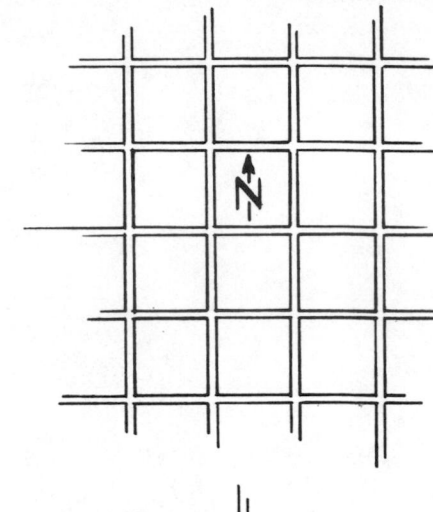

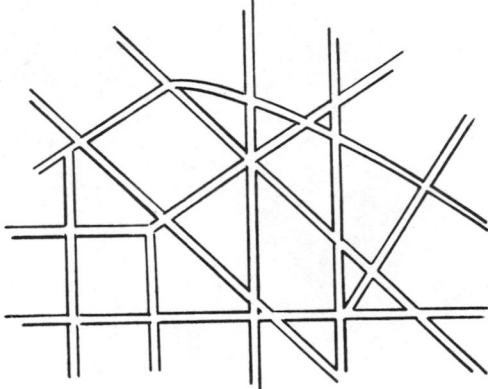

HAPHAZARD PATTERNS

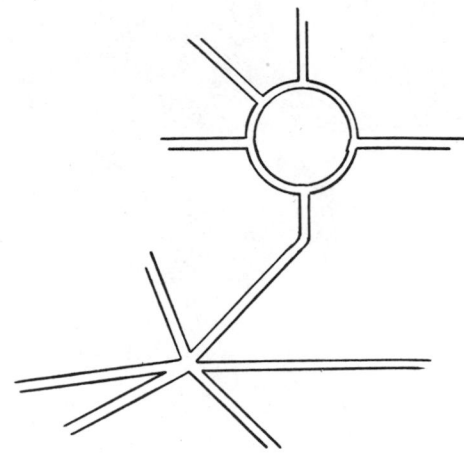

NON-UNIFORM INTERSECTIONS

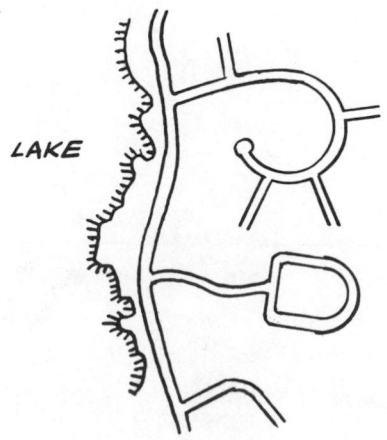

LAKE

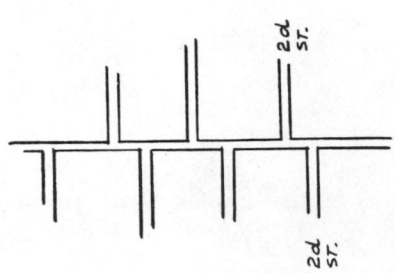

take the highway in the opposite direction)

o. Marked bike lanes that are also used for curbside parking
p. Deep ditches alongside the road without appropriate guardrails
q. Abrupt changes or irregularities in the road surface
r. Lack of a "slow truck" lane on steep hills

*For Pedestrians*

a. Crosswalks at blind curves
b. Automobile stop lines not set back from the crosswalk
c. Lack of sidewalk along the roadway
d. No place to walk along the freeway or cross the center divider in an emergency
e. No sidewalk on bridges
f. A sloping driveway that crosses a sidewalk
g. Lack of a wheelchair ramp at an intersection
h. Lack of intersection illumination
i. Light-signal duration that is too short to allow pedestrians to cross fully, and lack of a safety platform in the center
j. Lack of crosswalk safety marking and no tactile cues for the blind to use in following the crosswalk

*For Bicycle and Tricycle Riders*

a. No separate bike lanes and lanes that disappear at corners and on bridges
b. Curbside automobile parking allowed in bike lanes
c. Sidewalks that are directly adjacent to residential streets, making it possible for children to ride off the curb and into the street

Other safety considerations in planning street, road, and highway systems include the following:

a. Perpendicular drive and road entries to main highways or thoroughfares
b. Energy-absorbing barriers at bridge abutments and roadway divisions
c. Crash barriers at the ends of pedestrian center-island platforms
d. Guardrails between center-island pedestrian waiting platforms and adjacent vehicle lanes
e. Freeway emergency pull-off areas
f. Oncoming headlight glare barriers
g. Nonskid road surfaces
h. Rumble strips and raised lane markers for the nighttime warning of motorists
i. Illumination at all intersections, on bridges, and in tunnels
j. Following the *Manual on Uniform Traffic Control Devices* principles and standards for implementation of appropriate marking and signing [41]

[41] U.S. Department of Commerce, Bureau of Public Roads, *Manual on Uniform Traffic Control Devices for Streets and Highways*, June 1961. For additional information relative to street and highway planning and design, see D. E. Capelle et al. (eds.), *An Introduction to Highway Transportation Engineering*, Institute of Traffic Engineers, 1968; and R. F. Baker et al. (eds.), *Handbook of Highway Engineering*, Van Nostrand Reinhold Co., New York, 1975.

## Typical Roadway Classifications and General Characteristics

| Class of Roadway | Application | Characteristics |
|---|---|---|
| Street | Urban and suburban areas | Usually two lanes, two ways, but may be one way; occurs in residential and commercial districts. |
| Arterial road | Urban, suburban, and rural areas | Usually four lanes, two ways; typically a numbered route occurs in residential and commercial as well as through undeveloped areas. |
| Expressway | Urban, suburban, and rural areas | Four or more lanes, two ways, with median divider; typically numbered route; partial access control; usually skirts residential and commercial areas. |
| Freeway | Urban, suburban, and rural areas | Four or more lanes, two ways, with median divider and full access control; short interchange spacing in urban and suburban areas and long interchange spacing in rural areas. |
| Service road | Urban, suburban, and rural areas | Usually two lanes, one way, leading into freeway. |
| Collector and distributor road | Urban, suburban, and rural areas | Connects two freeways or freeway and arterial road; two or more lanes, usually one way. |
| Parallel arterial road | Urban, suburban, and rural areas | Same as arterial road except parallels freeway; usually one way each side; usually in commercial areas only. |
| Dual-dual | Urban, suburban, and rural areas | Two parallel freeways; one may be reserved for noncommercial traffic. |

## Other Roadway Feature Classifications and General Characteristics

| Feature Classification | Application | Characteristics |
|---|---|---|
| Bridge | Any area (may include toll plaza) | Varies with bridge configuration; usually two or more lanes, two ways; no turnoffs; may be two levels (e.g., one for motor vehicles and the other for rail vehicles). |
| Multilevel overpass | Interchange separation | Two or more lanes, one way; may have turnoffs; may include cloverleaf for alternative traffic directional distribution. |
| Tunnel | Any area (may include toll plaza) | Varies with tunnel configuration; usually two lanes, two ways; no turnoffs; usually lighted. |
| Traffic circle | Any area | Level interchange with 360° movement of traffic; roads from all directions, usually with traffic controls. |
| Driveway and alleyway | Any area | Usually two lanes, two ways; occurs in residential and commercial districts, traffic controls. |
| Loading zone | Any area | Off-street loading recommended, although curbside loading typical. |
| Parking zone | Any area | Varies with configurations; off-street parking recommended, especially in commercial districts, business areas, and shopping centers; usually with traffic controls. |
| Cul-de-sac | Areas where traffic restriction is needed | Turning circle required to reverse traffic. |

## Recommended Design Values for Sight Distance[42]

Minimum sight distances to be provided at all points of a road are shown in the accompanying table.

Crest vertical curves on two-way roads should generally be designed on the basis of minimum sight distance to minimize the total length of road that is subject to restricted sight distance.

Where it is desired to provide sufficient sight distance to allow safe overtaking, use the values given in the accompanying table.

The accompanying figure illustrates the use of these criteria for specifying no-passing lane marking based on the premise that a vehicle must have returned to its lane before the start of a barrier line.

### Minimum Sight Distance, in meters, to Be Applied in Design

| Speed, km/h | Vertical Curves | | Horizontal Curves | |
|---|---|---|---|---|
| | 5-s Travel Time Critical | Stopping Distance Critical | 5-s Travel Time Critical | Stopping Distance Critical |
| 60 | 85 | — | 85 | |
| 70 | 100 | — | 100 | 100 |
| 80 | 110 | — | — | 120 |
| 90 | 125 | — | — | 140 |
| 100 | 140 | — | — | 170 |
| 110 | — | 210 | — | 210 |
| 120 | — | 250 | — | 250 |
| 130 | — | 300 | — | 300 |
| Eye Height, m | 1.15 | 1.15 | 1.15 | 1.15 |
| Object Height, m | 0.1 | 0.6 | 0.1 | 0.5 |

### Sight Distance for Overtaking Conditions

| Speed, km/h | Proposed Design SD for Overtaking Zone $D_c$, m | Minimum Length of Overtaking Zone $D_o$, m | Distance for Completion of Overtaking $d$, m |
|---|---|---|---|
| 60 | 215 | 170 | 96 |
| 70 | 280 | 215 | 120 |
| 80 | 345 | 270 | 150 |
| 90 | 410 | 325 | 170 |
| 100 | 475 | 390 | 200 |
| 110 | 540 | 440 | 225 |
| 120 | 605 | 500 | 250 |
| 130 | 670 | 560 | 285 |

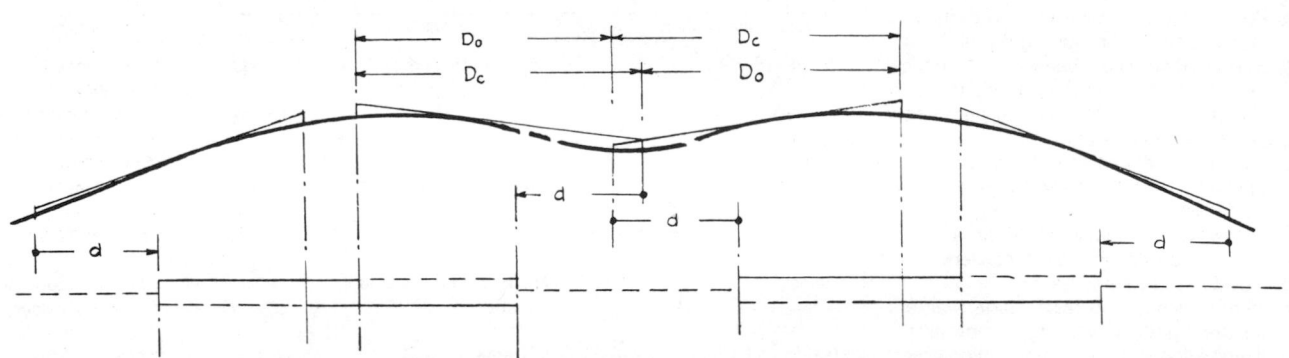

EXAMPLE OF REQUIRED LINE MARKING FOR
RESTRICTED VISIBILITY

$D_c$ = Design sight distance, overtaking lane

$D_o$ = Minimum length of overtaking zone

$d$ = Length to complete overtaking maneuver from critical point (excludes distance travelled by opposing vehicle.)

[42]L. J. Louis, "Sight Distance Requirements of Rural Roads —A Review," *Australian Road Research,* vol. 7, no. 2, pp. 32–44, June 1977.

## Freeway Lane-Drop Concepts

Under certain circumstances it is necessary to reduce the number of lanes on a freeway. In many current designs, considerable confusion is created because the lane drops are not physically well defined; they are difficult to see because they tend to blend into the background, or they are hidden by topographic features, such as the crest of a grade. In addition, advance warning signs or traffic control devices are frequently misleading or obscure. The following guidelines will help eliminate some of the obvious confusion that motorists may experience:

1. The lane drop should be placed where the surface of the roadway remains continuously visible for a significant distance.
2. The lane drop should not be located near "attention-dividing" conditions, e.g., at ramps or where there is complicated directional signaling or signing.
3. The lane-drop taper should provide conspicuous visual clues which inform the driver that the lane is ending and which allow a smooth rather than an abrupt lane transition in the taper area.
4. The lane drop should be located on the side of the freeway that is best with respect to given traffic and geometric considerations.
5. The lane should "appear" to end on the same side of the freeway as the operational lane drop.
6. When a lane drops at an exit ramp, an escape area of appropriate dimensions should be provided to allow for smooth transition into through lanes.
7. When a lane is added at an on ramp and dropped at a nearby off ramp, entering drivers should be warned well in advance that the lane in which they are traveling is not a continuous lane for through traffic.
8. Traffic control devices used in conjunction with lane-drop configurations should be consistent and designed so that they can be "read" sufficiently ahead of decision-making points.

## Uniform Traffic Control for U.S. Highways[43]

### DEFINITIONS RELATING TO SIGNS

1. *Guide Sign:* A sign used to direct traffic along a route or toward a destination or to give information concerning places or points of interest.
2. *Lane-Use Sign:* A sign indicating regulations governing use of specific lanes.
3. *Legend:* Any message on a sign, whether expressed in words or symbols.
4. *Public Parking Area (or Facility).* A parking facility available for use by the general public, with or without payment of a fee.
5. *Regulatory Sign:* A sign used to indicate regulations governing use of the highway.

6. *Traffic Sign:* A device mounted on a fixed or portable support whereby a specific message is conveyed by means of words or symbols, officially erected for the purpose of regulating, warnings, or guiding traffic.
7. *Warning Sign:* A sign used to indicate actual or potential hazards to highway users.

### DEFINITIONS RELATING TO MARKINGS

1. *Barrier Line:* A line which, when placed parallel to a center or lane line, or to another barrier line, indicates that all traffic must keep to the right thereof.
2. *Center Line:* A line indicating the division of the roadway between traffic moving in opposite directions.
3. *Channelizing Line:* A line which directs traffic and indicates that traffic should not cross but may proceed on either side.
4. *Delineator:* A light-reflecting device mounted at the side of the roadway, in series with others, to indicate the alinement of the roadway.
5. *Edge Line:* A line which indicates the edge of the roadway.
6. *Lane Line:* A line separating two lanes for traffic moving in the same direction.
7. *Stop Line (or Limit Line):* A line which indicates where vehicles should stop when directed by a traffic officer or traffic control device.
8. *Traffic Markings:* All lines, patterns, words, colors, or other devices, except signs, set into the surface of, applied upon, or attached to the pavement or curbing or to objects within or adjacent to the roadway, officially placed for the purpose of regulating, warning, or guiding traffic.

## Traffic Sign Categories[44]

Regulatory signs. Provide notice of traffic laws and regulations, which usually are unenforceable in the absence of such signs.

Warning signs. Call attention to conditions that are potentially hazardous to traffic operations and that otherwise may not be immediately apparent.

Guide signs. Show route designations, destinations, directions, distances, services, points of interest, and other geographical or cultural information.

## Sign Shape Codes

Octagon is reserved exclusively for the stop sign.

Equilateral triangle with one point downward is reserved exclusively for the yield sign.

Round is reserved for the advance warning of a railroad crossing and for the civil defense evacuation route marker.

Pennant, an isosceles triangle with its longest axis horizontal, is used to warn of "no passing" zones.

Diamond is used only to warn of existing or possible hazards on the roadway or adjacent thereto.

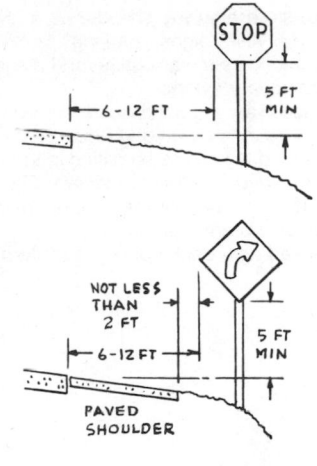

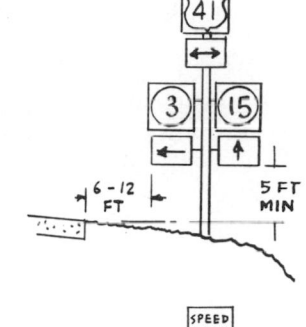

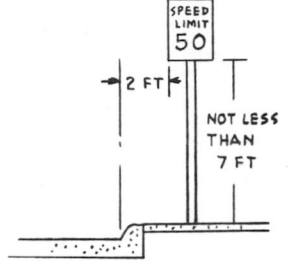

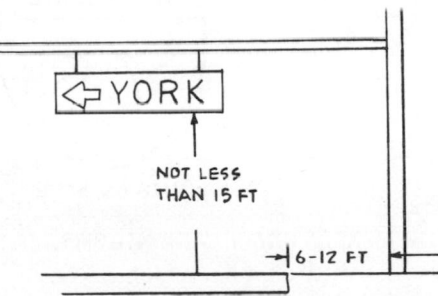

Rectangle, ordinarily with the longer dimensions vertical, is used for regulatory signs—with the exception of certain route markers and recreational-area guide signs.

Trapezoid shape may be used for recreational-area guide signs.

Pentagon, with the point up, is used for school advance warning and school crossing signs.

## Sign Color Codes

Red—used as a background color for *stop* signs, multiway stop supplemental signs, *Do Not Enter* messages, *Wrong Way* signs, and

[43]U.S. Department of Commerce, Bureau of Public Roads. *Manual on Uniform Traffic Control Devices for Streets and Highways.* June 1961.

[44]U.S. Department of Commerce, Bureau of Public Roads. *Manual on Uniform Traffic Control Devices for Streets and Highways.* June 1961.

on interstate highway shields; as a legend color for yield signs, parking prohibition signs, and the circular-outline and diagonal-bar prohibitory symbol.

Black—used as a background on one-way signs, certain truck weigh-station signs, and night speed-limit signs specified in the *Manual on Uniform Traffic Control Devices.* Black is also used for messages on white, yellow, and orange signs.

White—used as a background on route markers, guide signs, and regulatory signs (except stop signs) and for the legend on brown, green, blue, black, and red signs.

Orange—used exclusively as a background color on highway construction and maintenance signs.

Yellow—used as a background color for warning signs.

Brown—used as a background color for guide and information signs related to points of recreational or cultural interest.

Green—used as a background color for guide signs (except those using brown or white) and mileposts, and as a legend color with background for permissive parking regulations.

Blue—used as a background color for information signs related to motorist services and the evacuation route marker.

## Sign Standards Examples

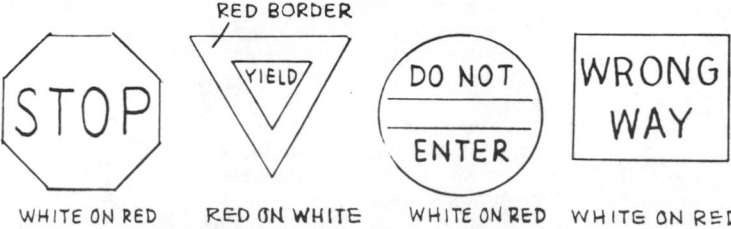

RED BORDER

WHITE ON RED          RED ON WHITE          WHITE ON RED          WHITE ON RED

WHITE ON BLACK          BLACK ON WHITE

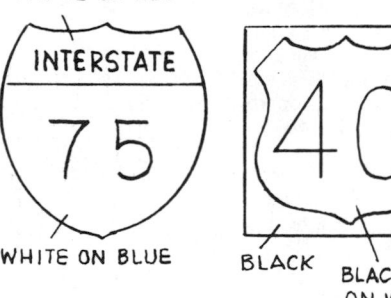

WHITE ON RED

WHITE ON BLUE

BLACK     BLACK ON WHITE

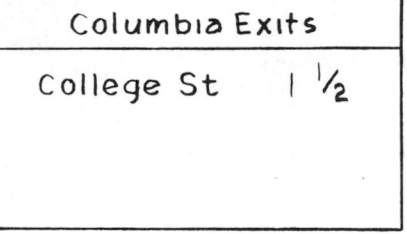

WHITE ON GREEN

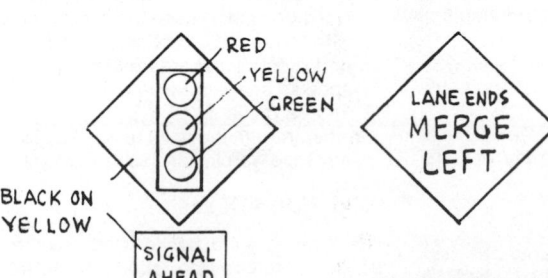

RED
YELLOW
GREEN

BLACK ON YELLOW

BLACK ON YELLOW

BLACK ON YELLOW

BLACK ON YELLOW

BLACK ON YELLOW

## Roadway Sign Location

Street and road signs must be located where the motorist expects to find them, and they should be positioned far enough in advance of the motorist's proposed change of path so that he or she has time to prepare for the change.

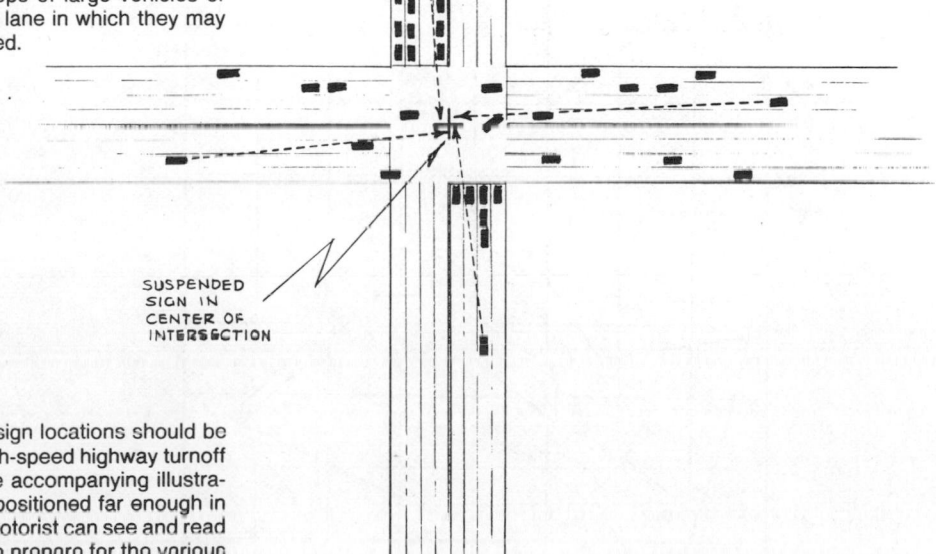

Ordinarily, a motorist expects a street sign to be located on his or her side of the street, as the accompanying illustration shows. How-ever, as indicated, a motorist also needs to verify that he or she is still traveling on the right street. This type of location for street signs is applicable to all but the very widest streets.

Signs used for city streets should be as large as necessary to be seen far enough in advance (i.e., lettering heights should be based on the average speed range of the street system) so that motorists do not have to slow down at every intersection to read the signs.

When very wide streets and broad intersec-tions are involved, a large central sign in the center of the intersection is desirable because motorists may be too far from a curbside sign to see it above the tops of large vehicles or from the most distant lane in which they may be traveling or stopped.

A minimum of two sign locations should be considered for the high-speed highway turnoff situation shown in the accompanying illustra-tion. Sign A must be positioned far enough in advance so that the motorist can see and read it and still have time to prepare for the various maneuvers that he or she must make. Sign B also should be provided to reconfirm that the off ramp is the one the motorist desired.

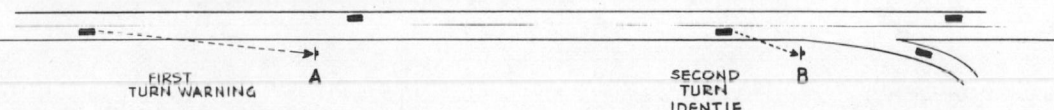

**Signal Lights**[45]

Early planning should include the location of appropriate signal lights to ensure the most effective flow of traffic in both urban and rural areas. A three-color system should be used: red for "stop," yellow for "caution," and green for "go ahead." Because there are many color-blind drivers, the colors selected for these basic hues should fall within the boundaries shown in the accompanying CIE chromaticity diagram.

All traffic signal lights should be located consistently throughout a given system; i.e., they should be located on the corners nearest the lane of traffic being regulated. When very wide freeway configurations are subject to signal lights, it is desirable to place a second set of lights in the center of the roadway.

Additional early-warning signal lights in advance of an impending signaled intersection are desirable for high-speed roadways in order to give the motorist time to slow down for a red light that is about to be sequenced.

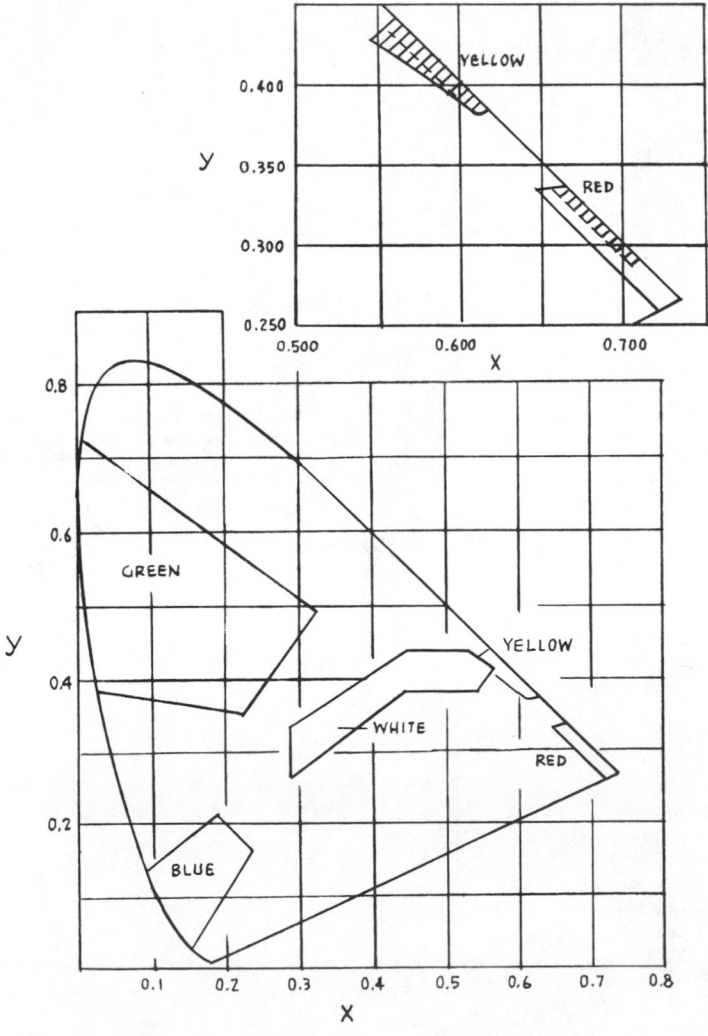

[45]U.S. Department of Commerce, Bureau of Public Roads, *Manual on Uniform Traffic Control Devices for Streets and Highways*, June 1961.

## Arrows and Signs Painted on Roadway Surfaces

In order for the driver to recognize markings on the roadway surface, letters, arrows, etc., must be modified to accommodate the distorted viewpoint of the driver, as shown in the accompanying illustration.

The horizontal subtense of the base or the arrow is given by

$$\alpha = \frac{W}{L} \text{ rad}$$

The vertical substense of the arrow is the angle, $\gamma$, where

$$\tan \phi = \frac{L}{a}$$

$$\tan (\phi + \gamma) = \frac{L + h}{a}$$

Using the identity $\tan A + \tan B = \tan (A + B) (1 - \tan A \tan B)$,

$$\frac{L}{a} + \tan \gamma - \frac{L + h}{a} \left(1 - \frac{L \tan \gamma}{a}\right)$$

$$\tan \gamma \left(1 + \frac{L(L + h)}{a^2}\right) = \frac{L + h}{a} - \frac{L}{a}$$

i.e.,

$$\tan \gamma = \frac{ah}{L^2 + hL + a^2}$$

Hence, for small $\gamma$,

$$\gamma = \frac{ah}{L^2 + hL + a^2} \text{ rad}$$

If $L > a$ or $> h$,

$$\gamma = \frac{ah}{L^2} \text{ rad approximately}$$

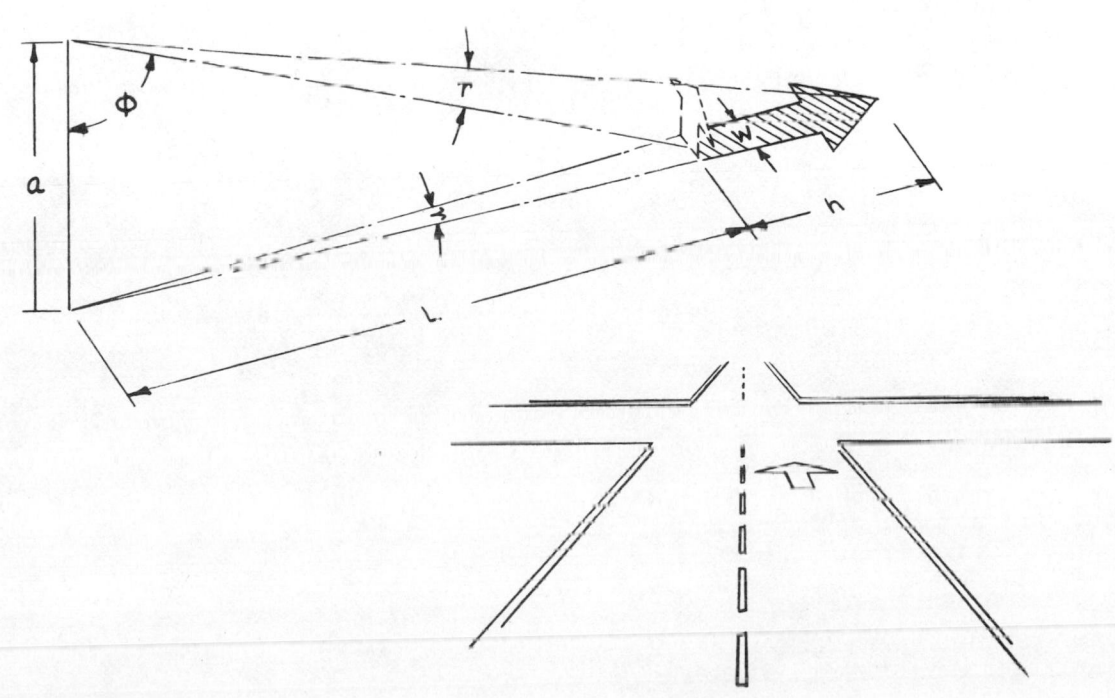

### Other Considerations Pertinent to Conceptual Planning

#### 1. Traffic-Noise Attenuation

The problem of potential traffic noise, especially in certain areas of a city, should be addressed very early in the planning stage, since there are various ways to avoid creating the problem in the first place. These include routing heavily traveled thoroughfares away from residential areas, hospitals, schools, etc., and taking advantage of natural noise barriers, such as hills and groves of trees, and of prevailing winds that will carry the sounds away from critical areas.

#### 2. Landscaping

Advanced landscape planning is important not only from the standpoint of aesthetic objectives but also in terms of precluding visual sight problems for motorists and of providing a reasonable system for maintenance that will not introduce traffic bottlenecks, e.g., the interference that is often caused by highway landscape crews and equipment when they have to park along the highway or by sprinkler systems that spray water across a roadway, thus creating hazards such as wet roadways and spray on the motorist's windshield.

#### 3. Other Built-in Hazards

  a. Bridge rail designs that block the motorist's view of merging traffic because the designer was not aware of how important it is to create a railing pattern that the motorist can see through or past.
  b. No pull-off space on the centerline side of a multilane freeway. In many instances a motorist who is in trouble cannot get across the several lanes to the typical emergency shoulder.
  c. No pull-off space on very long bridges or in tunnels.
  d. Lack of provisions for reducing the blinding effects of oncoming headlight glare, especially at curves.
  e. Introduction of sharp dips or humps in the roadway, which can cause a motorist who is driving too fast to lose control.
  f. Use of alternating concrete-macadam (light and dark) merging. Although lane patterns are marked, this may confuse the motorist (especially at night), who tends to follow the color of the surface rather than the lane marking.

### Street and Roadway Systems Planning Relative to Mobility of Handicapped Pedestrians

In most urban areas there is a need to increase the mobility capability of certain handicapped persons. The principal user of concern is the person who is blind or partially sighted. Although it is important to provide sufficient smooth shoulder along unimproved residential roadways even for the pedestrian who is not handicapped, it is extremely important to plan for such pedestrian paths for the blind. Those who are totally blind or almost blind usually must find their way by using a cane. The pathways must provide a high contrast that can be detected by the partially blind who may be able to see light and shadow, but not detail. These pedestrian pathways should be designated and controlled in such a way that obstructions such as fire hydrants and telephone poles are set outside the path.

The urban signing system should be standardized as much as possible. That is, only certain kinds of posts or poles should be allowed at intersections, and these should be placed in the same relative positions so that blind persons can easily determine where they are at an intersection and so that they can know which poles have tactile identification devices on them.

In fully developed areas where sidewalks are built, these should be large enough for two pedestrians in wheelchairs to pass each other; they should also be wide enough to accommodate a fire hydrant or other utility within the sidewalk area without taking up the space needed by handicapped pedestrians. No poles should be supported by guy wires that could be run into by a blind person who is feeling along the surface with a cane. All standard street, bus, and other roadway signs must be placed above the heads of pedestrians.

# MILITARY SYSTEMS

## GENERAL

### General Considerations in Development of Military Systems

Military missions demand maximum performance efficiency from both the hardware and the personnel elements of their systems. Thus the military establishment places more strict requirements on the designer and manufacturer in terms of making sure that people can operate equipment with minimum error. In addition, the military establishment is particularly sensitive to the problem of creating equipment that is too sophisticated for the types of people it has available to operate its equipment, since this not only induces system failures but also imposes lengthy and costly training requirements on the services.

In initiating any new military system concept, the designer and/or planner should be thoroughly familiar with at least two important documents:

1. MIL-H-46855 *Human Engineering Requirements for Military Systems Equipment and Facilities:* This document outlines the responsibilities of both the procuring agency and the system manufacturer for creating a system in which system operators are assigned appropriate roles in equipment operation and maintenance, as well as a system designed to minimize operator and maintainer failures, excessive training demands, and unnecessary hazards when the equipment is delivered and put in operation. The document specifies the analysis, documentation, design review, and testing required to ensure that the final system will meet the needs of the procuring agency and the operating forces.

2. MIL-STD-1472 *Human Engineering Design Criteria for Military Systems, Equipment and Facilities:* This document provides an array of good human engineering design principals, criteria, and guidelines that are based on research and operating experience. Development contractors are expected to follow these guidelines in developing their system concepts and their final design of human-machine interface hardware. When they think they cannot, they are required to explain this in writing.

These two documents apply to all U.S. military development. In addition, various services have their own special human engineering requirements. Although similar to the documents described above, these special documents are written to address specific and unique problems of each particular service. Examples of these include the following:

1. U.S. Army
   a. MIL-HDBK-759 *Military Standardization Handbook: Human Factors Engineering Design for Army Materiel*
   b. AMCP 706-134 *Engineering Design Handbook—Maintainability Guide for Design*
2. U.S. Air Force
   a. AFSC DH 1-3 Human Factors Engineering Series 1-0
   b. AFSC DH 1-5 Environmental Engineering Series 1-0
   c. AFSC DH 1-6 System Safety Series 1-0
   d. AFSC DH 1-9 Maintainability
   e. AFSC DH 2-2 Crew Stations and Passenger Accommodations Series 2-0
   f. AFSC DH 2-8 Life Support Series 2-0
   g. AFSC DH 3-3 Ground Equipment and Facilities Series 3-0

Two supplementary references are also especially relevant to military systems development. Although not official references, these were prepared by and for the government and government contractors:

*Human Engineering Guide to Equipment Design* (edited by H. P. VanCott and R. G. Kinkade): This volume, sponsored by the Joint Army–Navy–Air Force Steering Committee, is published by, and available through, the U.S. Government Printing Office, Washington, D.C.

*Bioastronautics Data Book* (edited by J. F. Parker and V. R. West): This volume was sponsored by the Scientific and Technical Information Office of the National Aeronautics and Space Administration and is also available through the U.S. Government Printing Office, Washington, D.C.

The reader's attention is called to the fact that the military requirements noted at the beginning of this discussion apply to total systems conceptualization and design. This includes not only the hardware but also the software, the logistics, and the personnel system that utilizes and maintains the completed system. In addition, this "total concept" implies that attention has to be given to all parts of a system; i.e., the fact that a particular manufacturer is required to produce only a subsystem, part, or component does not mean that this manufacturer can ignore the interaction of this subsystem, part, or component with all other subsystems, parts, or components. Thus, the planner must examine, for example, a total ship in terms of power plant, weapons, transported systems (e.g., aircraft), and personnel accommodations (berthing, mess, personal hygiene, etc.).

In general, military system users will fall within a well-defined age group, the range of their body sizes will be fairly well defined because of selection procedures, and their backgrounds, in terms of education and experience, will be fairly well defined by service entrance requirements. Note that this is considerably different from the situation that exists with regard to nonmilitary systems, which must be designed to fit a much broader range of users who are often ill defined because of lack of research on the general population. Many of the known facts about anthropometry, physiology, psychology, human-machine system interaction, etc., have been generated from military-sponsored studies. These data are available in the references noted above, which provide a ready source of information that can be used very early in the system concept development stage. The following are among the key human factors considerations which will invariably be addressed in system conceptualization and concerning which the planner and designer can find suitable information available:

Military personnel anthropometric data
Military personnel biomechanical performance data
Military personnel information-processing performance data
Military personnel control response performance data
Military personnel physiological and psychological data
Environmental reaction data

Appropriate examination and utilization of these data can provide the planner of a new system concept with sound bases for determining when and how to use people in the proposed system (i.e., when a human should be required to perform a system function or when the function should be automated), the best type of interface for the particular manual function in order to ensure maximum human response effectiveness, and the kind of living or working environment that is required to preclude injury or maintain effective human performance.

## Typical Functional Considerations for Military Systems

**1. Hand-Held Weapons**
   a. Unpacking, assembly, and emplacement
   b. Adjustment and calibration
   c. Carrying and stowage
   d. Loading, aiming, and firing
   e. Disassembly and cleaning
   f. Service and repair
   g. Personnel hazards
   h. Training and training devices

**2. Large Nonhand-Held Weapons**
   a. Transport
   b. Uncovering, erection, and emplacement
   c. Adjustment and calibration
   d. Loading, aiming, and firing
   e. Disassembly and cleaning
   f. Service and repair
   g. Personnel hazards
   h. Training and training devices

**3. Vehicular Systems: Personnel and Weapons Carriers**
   a. Ingress and egress
   b. System check-out
   c. System operation
   d. Personnel accommodations
   e. Service, maintenance, and repair
   f. Stowage
   g. Personnel hazards
   h. Training and training devices

**4. Communications Command and Control**
   a. Installation, activation, adjustment, and calibration
   b. Operation
   c. Maintenance
   d. Dismantling, loading, and transport
   e. Personnel accommodation
   f. Personnel hazards and working environment

**5. Ground-Based Housing**
   a. Space accommodation and arrangement
   b. Living and working environment
   c. Security and safety
   d. Housekeeping and maintenance
   e. Assembly and transport (mobile)

**6. Personnel System**
   a. Manning requirements
   b. Selection requirements
   c. Training requirements

**7. Logistics**
   a. Service and maintenance plan
   b. Parts requirements
   c. Inventory control
   d. Support (stowage, transportation, and communication)

**AREAS OF ANALYSIS** For all military systems, the following areas of analysis should be addressed from the very beginning and refined throughout the system development:

1. System mission, operational requirements, and function definition: This information provides the basis on which all other concept and design decisions are made and justified.
2. Human versus machine function allocation: This is a trade-off necessary to ensure that humans are assigned tasks that they are best capable of doing.
3. Operator and maintainer hardware interface selection and design: This information ensures that the devices the human must use are appropriate to his or her physical and mental capabilities and do not make demands upon the human that may cause failure to perform as planned.
4. Human-environment compatibility: This information determines how to preclude and/or minimize health hazards and also defines habitability control limits that will ensure the maximum performance efficiency of operators, maintainers, and personnel in general.
5. Job and task descriptions: This information helps determine the kinds of skills that are required and the number of people that will be needed to operate and maintain the system. It also provides the basis for defining training requirements necessary to bring personnel skills to the level required to ensure satisfactory system operation.
6. Training concepts and plan: This information is necessary in order to determine training content and method, training aids, devices and/or simulation hardware support, and the scheduling re-

quired to design, develop, fabricate, and install all elements noted, not the least important of which is scheduling of personnel, instructors, and students.
7. Training equipment and facilities: This is a critical area in many instances because it involves fairly extensive hardware and/or facilities design and construction. It must be started early enough to ensure that classrooms, simulators, and other complex systems required for training are ready prior to the delivery of the basic system.
8. Operating and training publications: Operations, maintenance, and other technical publications are required to support both the training operations and the regular operations of the system. The length of time between the date on which sufficient information is available and the date on which the documents are needed generally is very short, and thus sensitive communication between the design and publications staffs is required on a continuing basis throughout system development.
9. Test and evaluation: This is one of the most important requirements imposed by military system agencies, since it implies the need to "prove the system" (both human and machine) prior to delivering it for field use. Human engineering test and evaluation, however, begins early in any program and continues at successive points to the last and final field test. Plans for this program must dovetail with other engineering test plans in order to preclude duplication or omission.

Military agencies typically require documentation of the above activities. These documen-

tation end items often are identified as follows:

1. Personnel and equipment data: A general file of all data generated during system development and relating to the human factors.
2. Human engineering: Specific human-machine interface design results.
3. Quantitative and qualitative personnel requirements information: Data on tasks, skill levels, job descriptions, and procedures suitable for defining training objectives, personnel selection, etc.
4. Training concept and training plans: A detailed plan and schedule for preparing the training cadre and the students in time to operate and maintain the delivered system.
5. Training equipment development plan: Hardware and facilities procurement specification and schedule.
6. Technical orders and manuals development plan: Preparation of all publications required to support training and the operation of the system.
7. Personnel subsystem test and evaluation plan: A plan integrated with general engineering test and evaluation to ensure proper involvement of human factors specialists in predesign, detailed design, engineering testing, mockup evaluations, system simulations, supporting human factors research, and predelivery and postdelivery field tests in which both people and hardware or software components are evaluated in real-time operation of the system.
8. Final human engineering reports: A separate human factors report generally must accompany the final engineering report delivered by a contractor to the procuring agency.

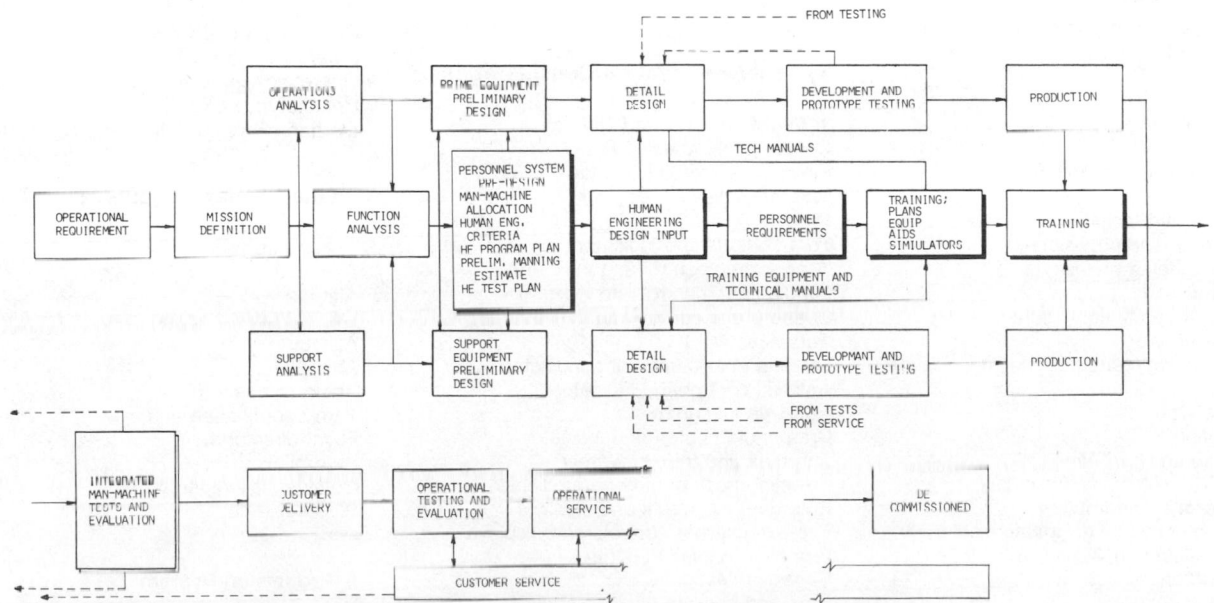

A typical military system development progression illustrating how human factors activities should progress parallel with prime and supporting hardware development. Note that all three emanate from the same initial decisions concerning operational requirements, mission definition, and functional requirements and that all interact throughout the development so that the final human machine system can provide a total compatibility among all elements. (From D.W. Conover and W.E. Woodson, *Human Engineering Guide for Equipment Designers,* University of California Press, 1964.)

## AIRCRAFT

### Human Factors Considerations in Military Aircraft Conceptual Design

Although many of the human factors considerations are the same in the case of both military aircraft and private or commercial aircraft, certain military operational requirements create unique problems that must be resolved in the design of military aircraft systems. The following are among these special problem areas.

#### 1. High Altitudes

Certain types of military aircraft operate at extremely high altitudes, where the hazards of loss of pressurization require use of special pilot protective garments and consideration of unique emergency ejection systems.

#### 2. High Performance

Unusual dynamic stress on the pilot during critical maneuvers requires giving special consideration to the pilot's protection, while at the same time ensuring that the pilot's performance as a controller remains intact.

#### 3. Combat Hazards

Armor may be required to protect the pilot from gunfire or shrapnel, glare shield may be required to protect the pilot's eyes from nuclear flash blindness, and/or the pilot must be protected for emergency ditching, i.e., when the pilot is required to eject at high or low altitudes, when the aircraft is in difficult positions, or when the pilot has to survive in rough or frigid seas.

### Typical Aircraft Design Areas Requiring Consideration of Human Factors

#### 1. Airframe Design

Facilities for passage and movement of crew
Flooring
Headroom
Passageways
General illumination
Sharp protrusions
Interior doors and hatches
Exterior doors
Door checks and locks
Emergency escape provisions
Emergency landing provisions
Ditching structure
Installation of heavy equipment
Turnover structure
Openings for maintenance, loading, airdrop, etc.
Body pressurizing
Soundproofing
Upholstering and carpeting

#### 2. Transparent Assemblies

Materials selection (e.g., impact, color, and shatter-resistance)
Visual distortion
Angle of incidence
Degree of curvature
Light transmissibility and reflection
Sighting area
Scanning area
Sliding canopies and locks
Bullet-proofing and bird-proofing
Windshield clearing
Thermal barrier

#### 3. Landing Gear Retracting System

Manually operated
Power-operated
Gear position indicator
Landing gear controls
Emergency control
Malfunction warning
Ground safety locks

#### 4. Personnel Safety

Walkways
Steps, handgrips, and rails
Armor plate

#### 5. Other

External maintenance access openings and covers
Cargo doors and opening and closing systems
Weapons attachment
Soundproofing
Watertightness
Pressure-tightness

#### 6. Aircrew Stations

Workplace arrangement
Vision (internal and external)
Seating (nominal and ejection)
Separable crew compartment
Flight controls and displays
Engineering controls and displays
Special operations controls and displays
Internal and external communication controls and displays
Weapons systems controls and displays
Personal protective systems connections, controls, and displays
Map stowage
Internal and external lighting systems controls and displays
Internal environmental systems controls and displays

### Typical Crew Station Considerations

#### 1. Flight Control and Display

Control stick/wheel
Elevator, aileron, and rudder
Rudder pedals
Brakes
Trim controls and indicators
Wing sweep control and indicator
Landing flap control and indicator
Speed brake control and indicator
Automatic flight
Landing gear control and indicator
Antiskid control and indicator
Nose wheel control
Drag chute control and indicator
Throttles and thrust reverser
Arresting hook control and indicator
Emergency controls and displays
Fire-extinguisher controls and displays
Propeller feathering control
Canopy release
Seat and capsule ejection control
Bailout alert display
Power plant controls and indicators
Fuel tank selector and indicator
Fuel dump switch
Afterburner control and indicator
Armament controls and displays
Gunsight and gunswitch
Bombsight, display and switch
Rocket selector and indicator
Radio and navigation system controls and indicators
Missile guidance control and display
Cabin pressure controls and displays
Air-conditioning control and display
Defogging and deicing controls and indicators
Canopy controls
Personal equipment controls, displays, and disconnects
Seat and pedal adjustment controls
Checklists, map and data case, and procedures charts

#### 2. Navigator and bombardier control and display

Chart board
Forward and downward vision
Side and upward sighting vision
Window shades
Systems and devices controls and indicators
Map and special instruments stowage
Oxygen display control and connections
Special lighting and control

#### 3. Systems Engineer Control and Display

Systems controls and indicators
Worktable and work light
Operations manuals stowage
Oxygen display, control, and connections
Stray light curtain

#### 4. Radio Operator Control and Display

Systems controls and indicators
Worktable, transmit key, and work light
Radio log and manuals stowage
Emergency repair kit stowage
Headset and handset stowage
Oxygen display, control, and connections

#### 5. Radar Operator Control and Display

Radar controls, indicators, and scopes
Chart board and work light
Emergency repair kit stowage
Operations manuals stowage
Headset stowage
Oxygen display, control, and connections

### Visual Display Requirements Checklist

#### 1. For Flight

Airspeed
Altitude
Vertical rate of speed
Azimuth
Attitude
Angle of attack
Turn coordination and rate
Flight director(s)
Auto pilot
Outside air temperature
Weather radar
etc.

#### 2. Propulsion System

RPM
Temperature (oil, exhaust, jet pipe, etc.)
Pressure (oil, fuel, torque, ratio, etc.)
Quantity (oil and fuel)
Fuel flow
Exhaust nozzle position
etc.

### 3. Navigation
Clock
CDI/OBS (VOR and ILS)
RMI (VOR and ADF)
HSI/PDI (VOR, ILS and inertial)
Marker beacons
etc.

### 4. Environmental System
Cabin pressure (altitude, rate, and differential)
Cabin temperature
Cabin ventilation
Oxygen (quantity, pressure, and flow)

### 5. Other Systems
Communications (UHF, VHF, HF, and Crypto)
Lights (internal)
Lights (external)
Fuel
Landing gear
Electrical
Hydraulic
Pneumatic
Fire warning
Control
Weapons
etc.

*Definitions:*

ADF = Automatic Direction Finder
CDI = Course Deviation Indicator
HSI = Horizontal Situation Indicator
ILS = Instrument Landing System
PDI = Pictorial Deviation Indicator
OBS = Omni Bearing Selector
VOR = VHF Omni Range
UHF = Ultra High Frequency
VHF = Very High Frequency
HF = High Frequency

## Crew Protection and Safety Requirements Checklist

### 1. On Board
Seats and restraints
Oxygen masks/systems and walk-around units
Emergency lighting
Bailout signals and communications
Radiation and glare protection
Protective armor
Auxiliary power
Fire extinguishers
Cargo tie-downs
Absence of sharp structure
Labels, signs, and markings
etc.

### 2. Escape Provisions
Emergency exits (doors, hatches, windows, and fuselage cuts; canopy ejection, etc.)
Ejection seats and ejection capsules
Parachutes
Anti-*g* suits
Anti-exposure garments
Portable oxygen equipment
Ground exit equipment (ropes, poles, ladders, chutes, etc.)

### 3. Survival Equipment
Life vests
Life rafts and weather shelters
Tethers
Anti-exposure garments
Communications (radio, beacons, flares, auxiliary power units, etc.)
Food and water, associated preparation aids
First aid
Ground rescue marker equipment
Weapons
Flashlight
Knife
Shelter
etc.

## Unique Considerations Related to Combat Operations

### 1. Vision
Military pilots have unique visual problems that must be addressed during the conceptual stage of aircraft system development. Among the more significant are the following:

a. Formation flying: Formation flying, especially at night, requires consideration of the tasks of both internal instrument viewing and external adjacent aircraft viewing. In combat, it is generally required that all telltale lights be extinguished that might give the aircraft's position away to the enemy. In order to see their associates' aircraft in silhouette, pilots must be completely dark-adapted.

b. Aerial combat: It is vital that the pilot be able to see possible enemy aircraft approaching from any angle, which requires an excellent viewing angle from the side, top, and rear as well as in front of the aircraft. In addition, rearview mirrors are desirable so that the pilot can keep tabs on what is behind while visually attending to other forward viewing tasks.

c. Ground reconnaissance: In many types of military missions, the pilot and/or another flight crew member is required to inspect the ground for targets. When this is done on an almost continuing basis, the aircraft must be designed to provide wide down-angle vision.

### 2. Control
In high-performance aircraft operations, not only the speed but also the dynamics of certain maneuvers create observational and control problems for the pilot and the crew members. In addition, combat injuries should be anticipated in terms of how a pilot will be able to control the aircraft for return to base. Typical considerations are the following:

a. High *g* forces: High *g* forces may preclude a pilot from operating critical controls properly; thus control designs should be made as insensitive to such a possibility as is practical.

b. Disabling wound: The pilot is subject to loss of an arm or leg because of enemy gunfire; however, if control concepts provide a reasonable opportunity to use another limb, the pilot should be able to return to base.

c. Control damage: Certain critical parts of the aircraft determine whether the machine is flyable. Whenever possible, control damage should not preclude a return to base.

## Cockpit Standardization

Military agencies are particularly adamant about cockpit standardization in order to minimize pilot confusion when transferring from one aircraft to another. Following are selected examples of U.S. Air Force standardization guidelines.[46]

1. The pedal "brake off" angle, with respect to the vertical shall remain constant throughout the entire rudder pedal travel.
2. Three inches of rudder pedal adjustment is acceptable when seats provided with $+2.500$ in of vertical and $1.500$ in of fore and aft adjustments are utilized.
3. Rudder pedal adjustment shall be in increments of 1 in or less.

4. The stick and throttle reference point is defined as the point at which the pilot's second finger is in contact with the forward face of the control.
5. Reference spec MIL-B-6584 for brake pedal angles and dimensions.
6. All measurements are based upon the seat reference point at the centerline of the seat in the neutral position.

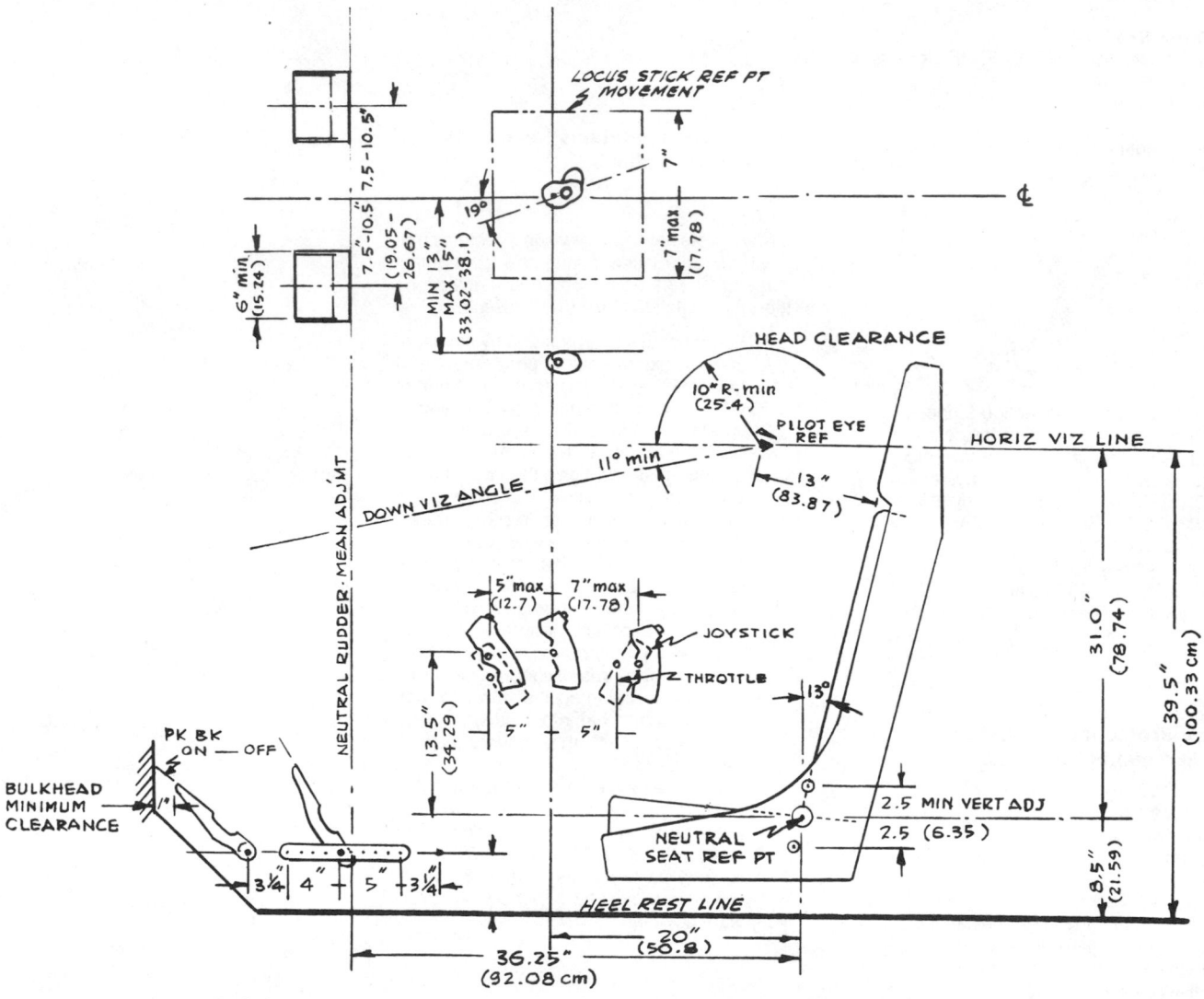

Cockpit-basic dimensions, fixed wing.

[46]Crew Stations and Passenger Accommodations, AFSC DH 2-2, Design Handbook Series 2-0, Aeronautical Systems Headquarters, Air Force Systems Command, Wright-Patterson Air Force Base, Ohio.

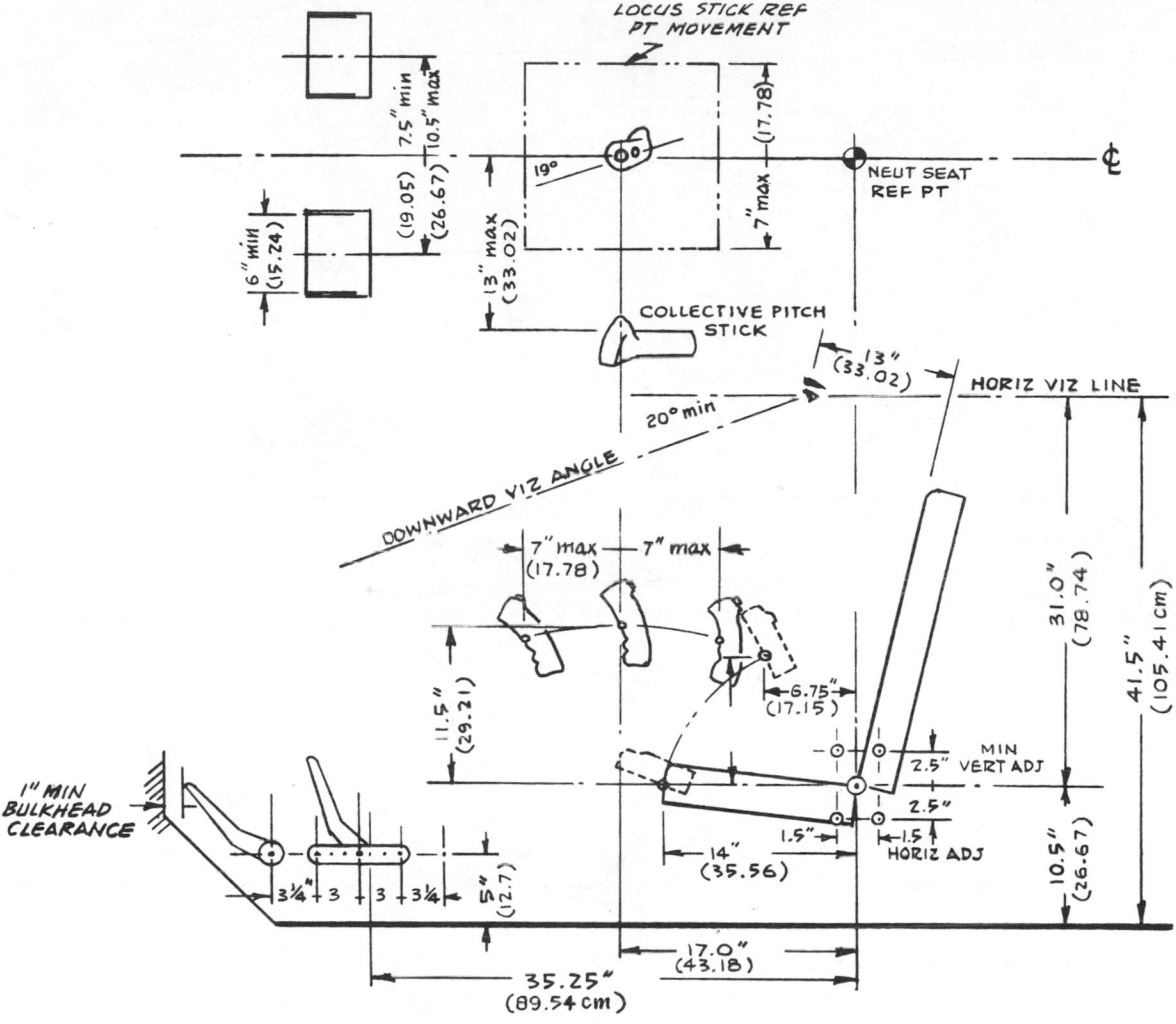

Cockpit-basic dimensions, helicopter.

1. Adequate clearance shall be provided to allow unrestricted operation of the rudder pedal throughout the full range of travel.
2. The pedal "brake off" angle with respect to the vertical shall remain constant throughout the entire rudder pedal travel.
3. The collective pitch stick travel shall not exceed 12 in from pull-down position of the grip.
4. The seat shall be provided with fore and aft and vertical adjustment independent fore and aft and vertical adjustment is desirable. However, diagonal adjustment is acceptable.
5. Rudder pedal adjustment shall be in increments of 1 in or less.
6. The stick reference point is defined as the point at which the pilot's second finger is in contact with the forward face of the stick grip.
7. Reference spec MIL-B-6584 for brake pedal angles and dimensions.
8. All measurements are based upon the seat reference point at the centerline of the seat in the neutral position.

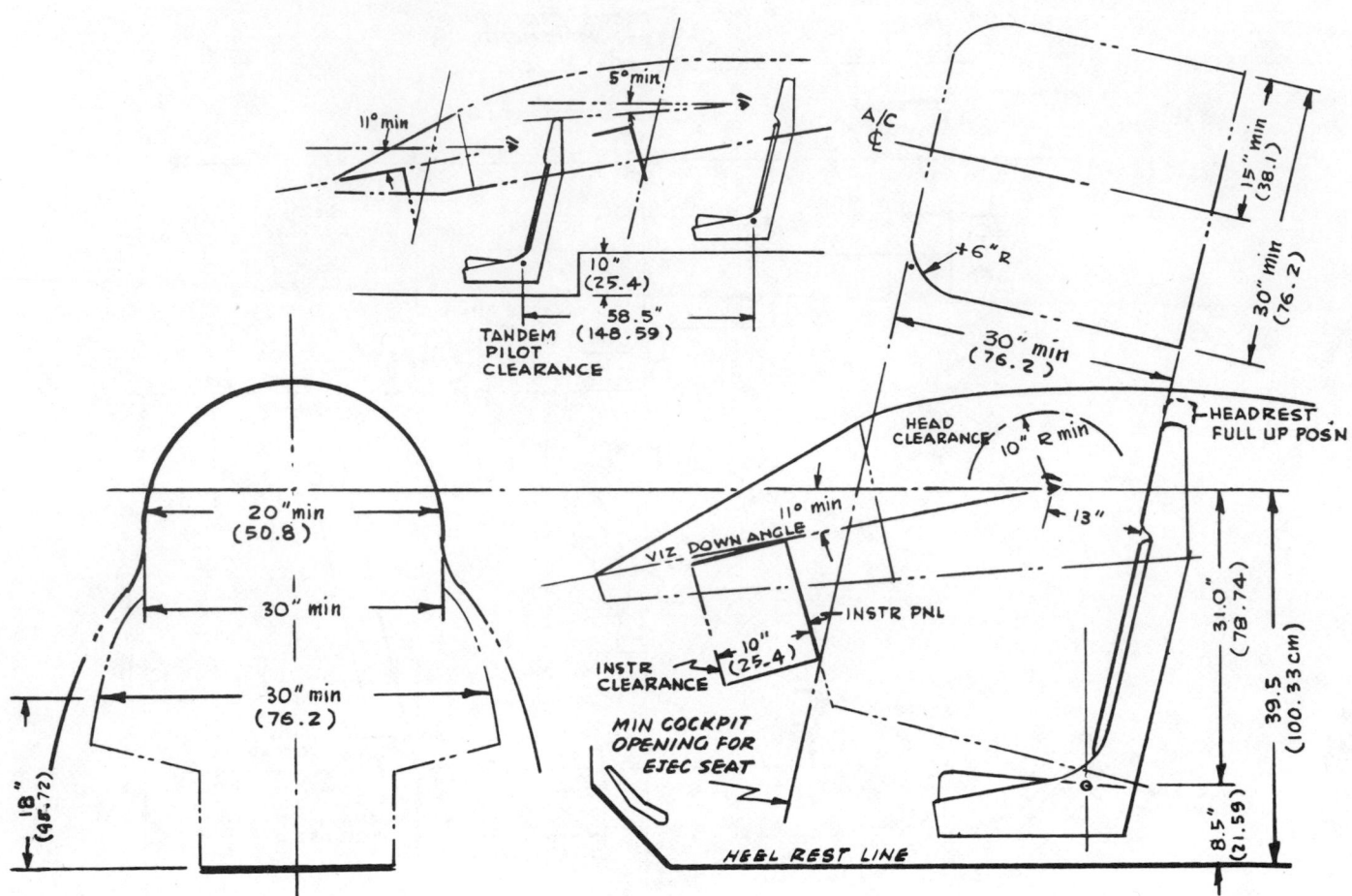

Cockpit-clearance dimensions, ejection seat.

1. There shall be no projections such as the throttle, landing gear control, instrument panel, etc., into the ejection seat envelope that would interfere with seat ejection.
2. All measurements are based upon the seat reference point at the centerline of the seat in the neutral position.
3. The 30-in minimum ejection clearance line (parallel to the ejection path and measured perpendicularly to the plane of the seat back) shall be provided from the seat reference point. For airplanes not requiring ejection seats the minimum cockpit opening shall be 24 by 24 in.
4. The seat shall be provided with vertical adjustment as shown.

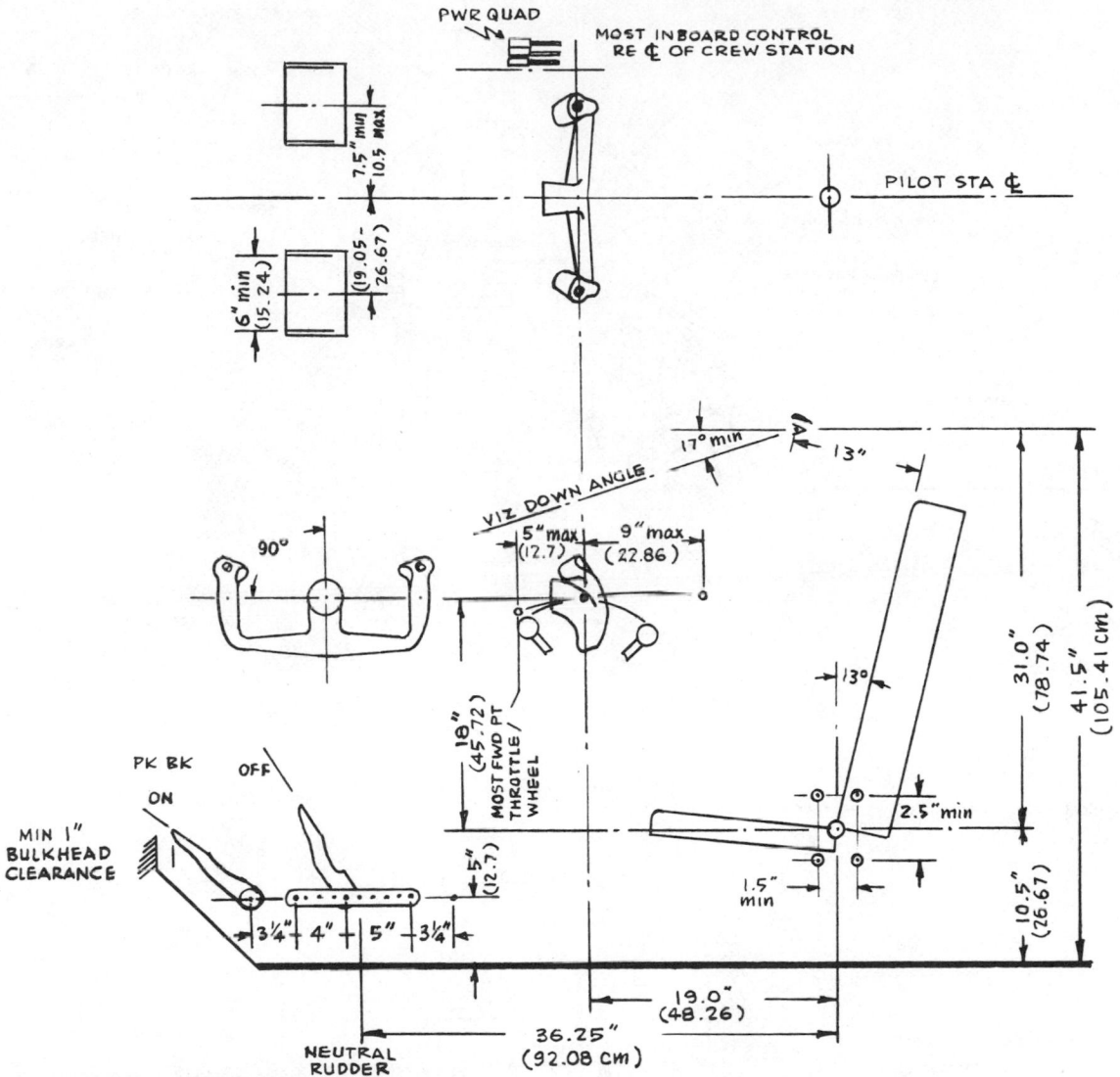

Cockpit-basic dimensions, cargo and bomber.

1. When ejection seats are required, the vertical seat adjustment shall be parallel to the ejection rails.
2. The rudder pedal "brake off" angle with respect to the vertical shall remain constant throughout the entire rudder pedal travel.
3. Rudder pedal adjustment shall be in increments of 1 in or less.
4. The wheel reference point is defined as the point at which the pilot's second finger is in contact with the forward face of the control wheel.
5. The seat shall be provided with fore and aft and vertical adjustment. Independent fore and aft and vertical adjustment is desirable. However, diagonal adjustment, as shown, is acceptable in lieu of vertical adjustment.
6. Reference spec MIL-B-6584 for brake pedal angles and dimensions.
7. All measurements are based upon the seat reference point at the centerline of the seat in the neutral position.

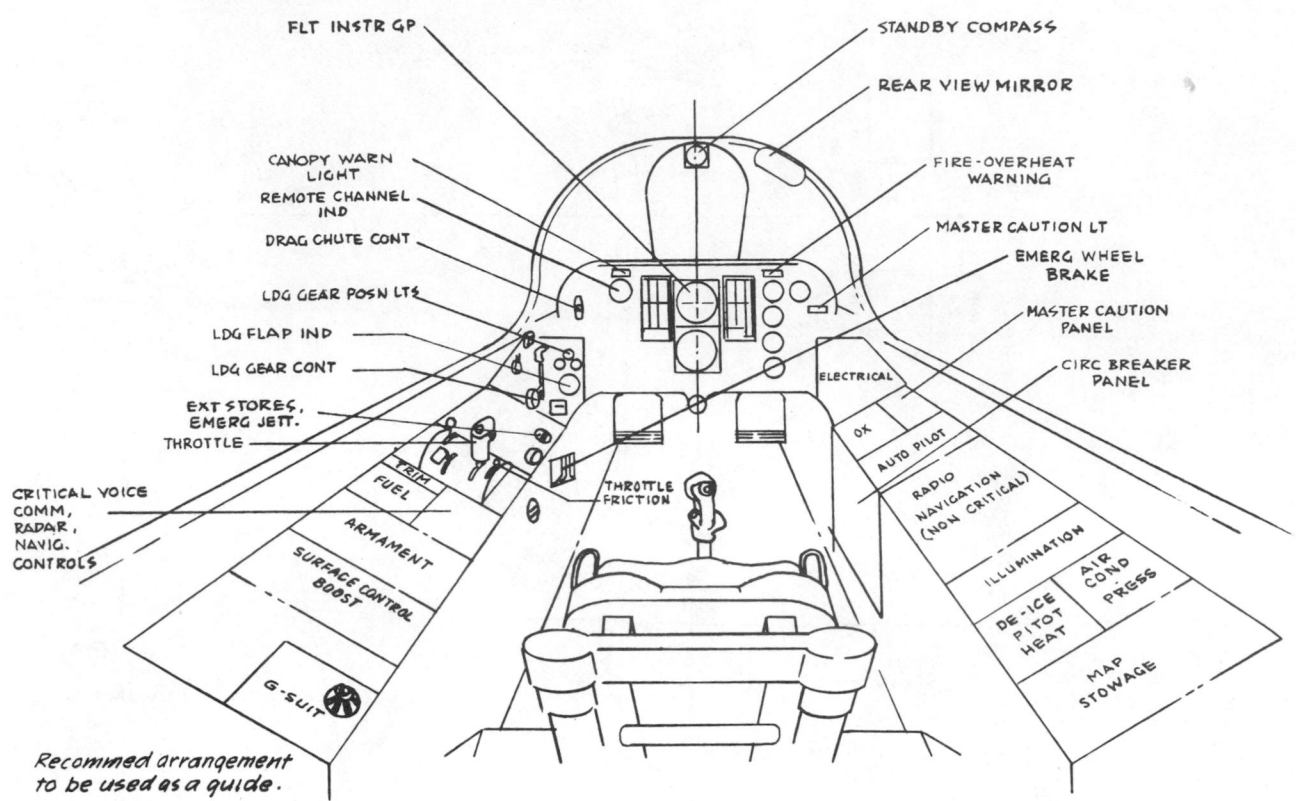

*Recommend arrangement
to be used as a guide.*

Cockpit-typical arrangement, side-by-side, four engine.

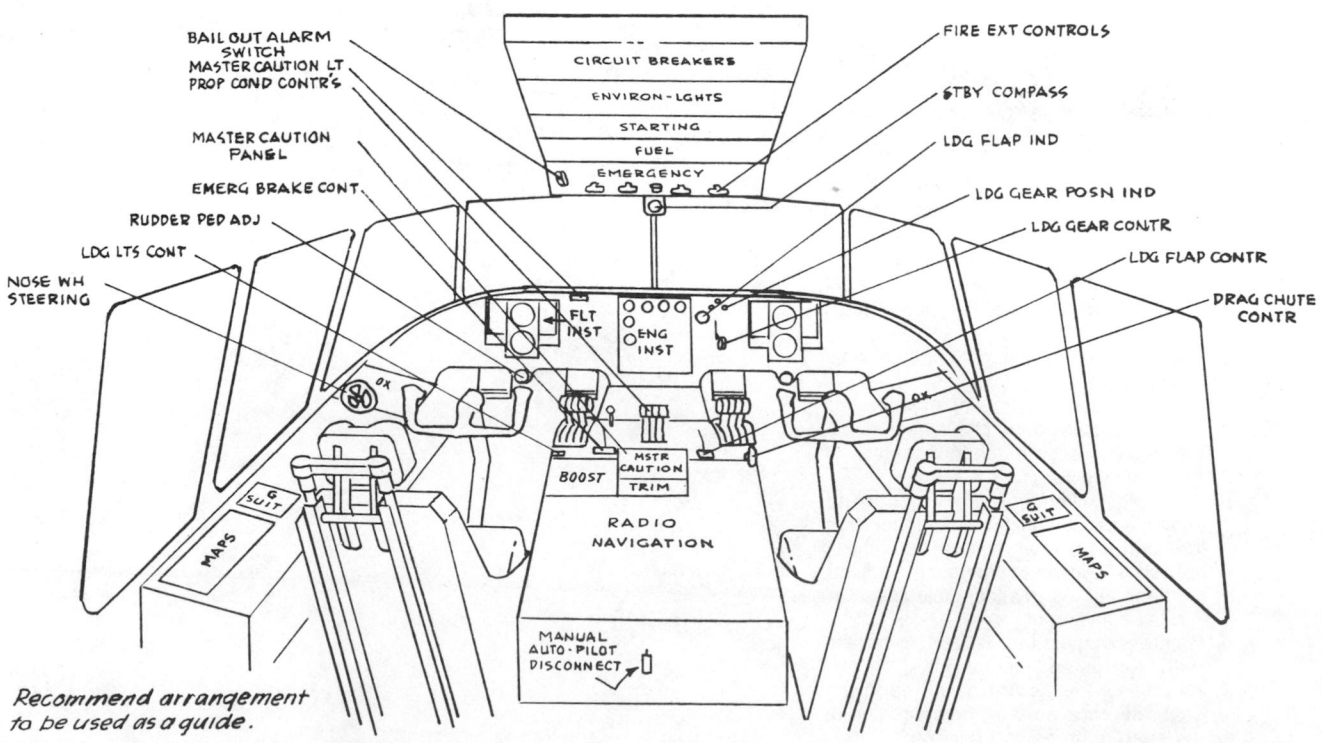

*Recommend arrangement
to be used as a guide.*

Cockpit-typical arrangement, fighter type aircraft.

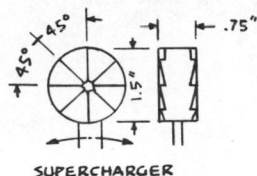

SUPERCHARGER

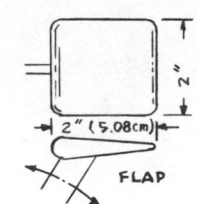

FLAP

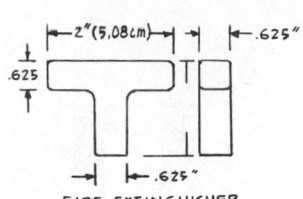

FIRE EXTINGUISHER

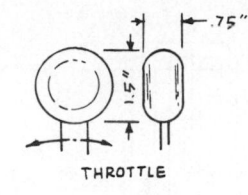

THROTTLE

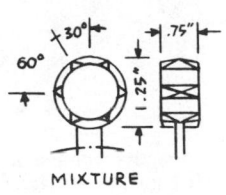

MIXTURE

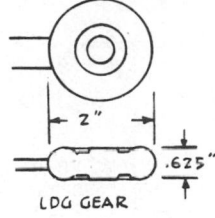

LDG GEAR

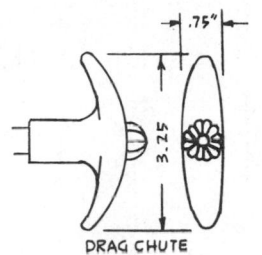

DRAG CHUTE

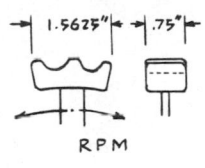

RPM

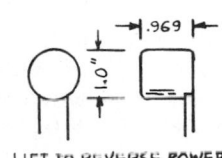

LIFT TO REVERSE POWER

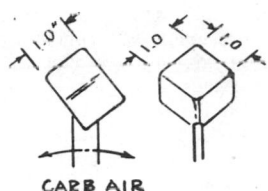

CARB AIR

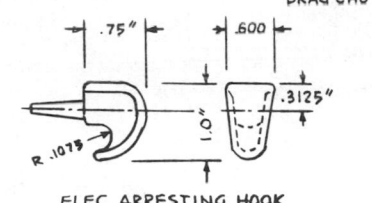

ELEC ARRESTING HOOK

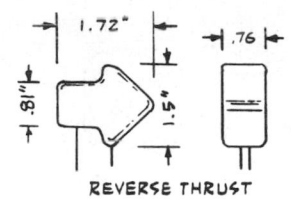

REVERSE THRUST

( 1″ = 2.54 cm )

Standard knob shapes.

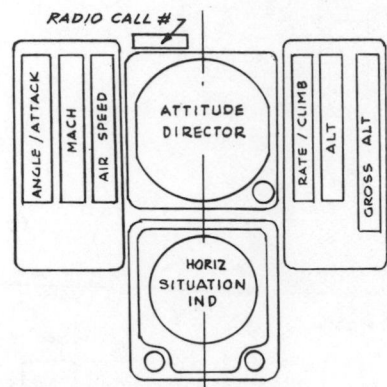

Standard basic advanced flight instrument
arrangement (Fig. 1).

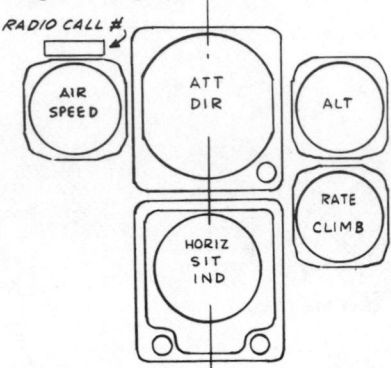

Standard basic advanced flight instruments
with conventional indicators (Fig. 2).

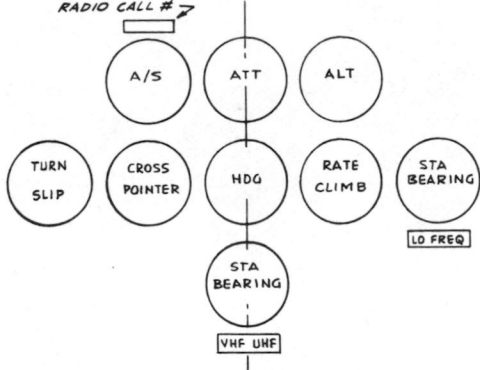

Standard basic flight instrument group with
conventional instruments (Fig. 4).

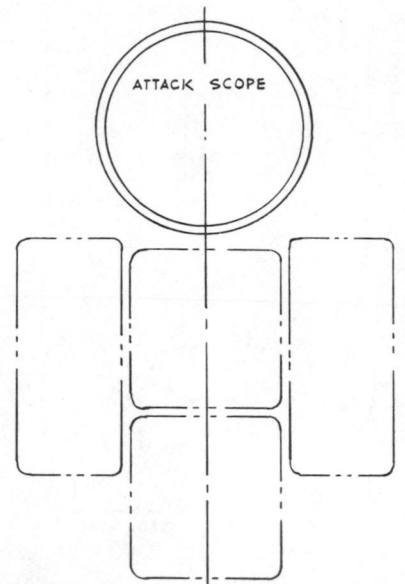

Attack or radar scope location (Fig. 3).

Instrument arrangements.

1. Vertical line through the directional indicator dial indicates centerline of pilot's head.
2. Other instruments not shown here, which are required, shall be located in the most convenient place to suit the particular design of each type airplane, and shall have engineering approval by the procuring activity.
3. Any deviation from the proposed instrument arrangement as shown on this drawing shall have engineering approval by the procuring activity.
4. The panels containing instruments, other than self-contained gyro instruments, may be inclined subject to approval of the procuring activity.
5. When advanced instrument Figs. 1 and 2 are provided only for the pilot, conventional instruments shall be arranged in accordance with Fig. 4.

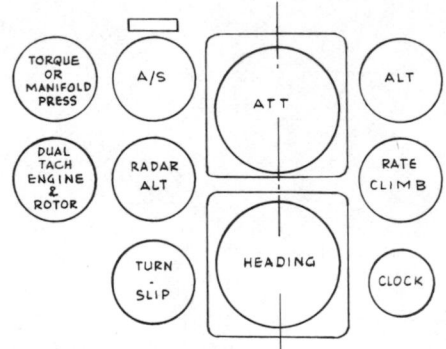

Standard basic flight instrument arrangement
for rotary wing aircraft (Fig. 5).

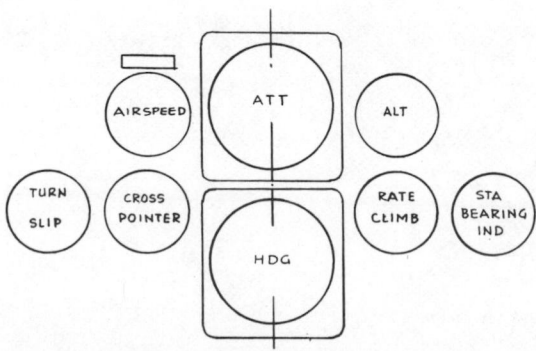

Standard basic flight instrument group with large
heading and attitude indicator (Fig. 6).

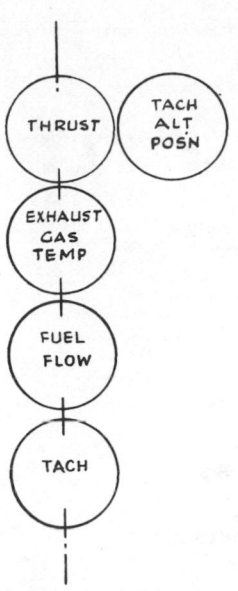

Standard engine instrument arrangement—jet aircraft (Fig. 7).

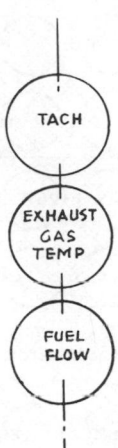

Standard engine instrument arrangement—jet aircraft—without thrust indicator (Fig. 8).

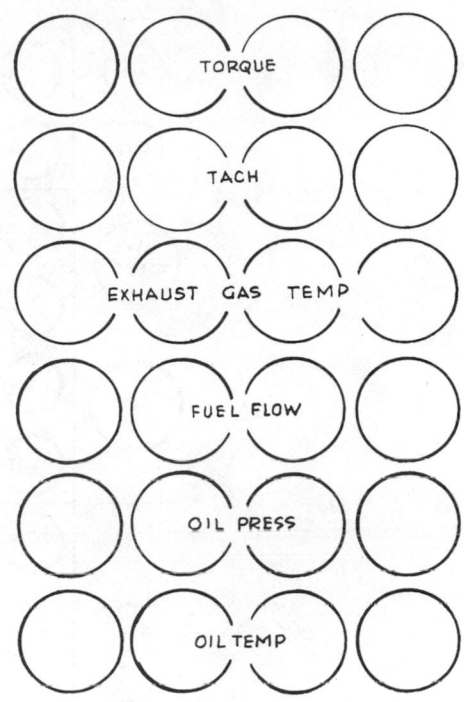

Standard vertical tape engine instrument arrangement—jet aircraft (Fig. 10).

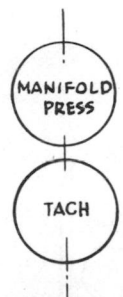

Standard turboprop engine instrument arrangements (Fig. 11).

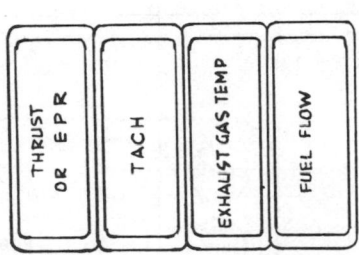

Standard engine instrument arrangement—reciprocating engine aircraft (Fig. 9).

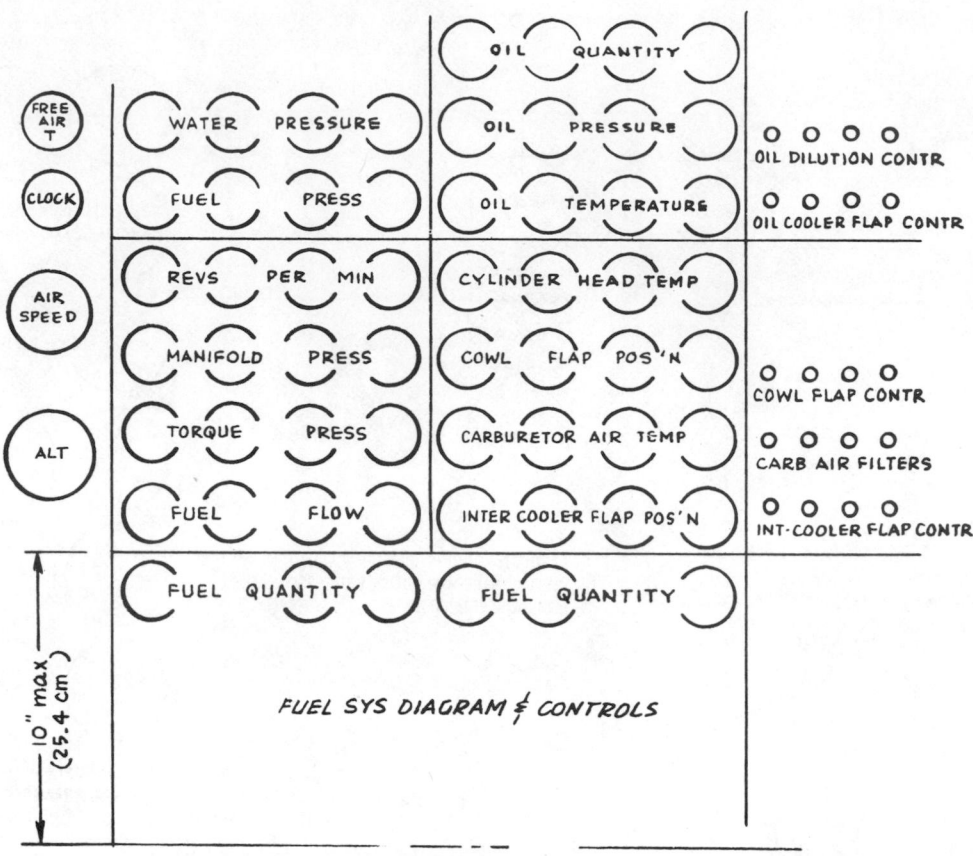

Flight engineer's panel—heavy bombardment and cargo
aircraft—reciprocating engines (Fig. 12).

1. Dividing lines shall be 0.062+0.016
   wide. (The above markings shall be finished with luminescent fluorescent material conforming to MIL-L-25142).
2. Main borders shall be 0.156+0.031
   wide.
3. Numerals designating engine numbers
   shall be 0.250 minimum height.

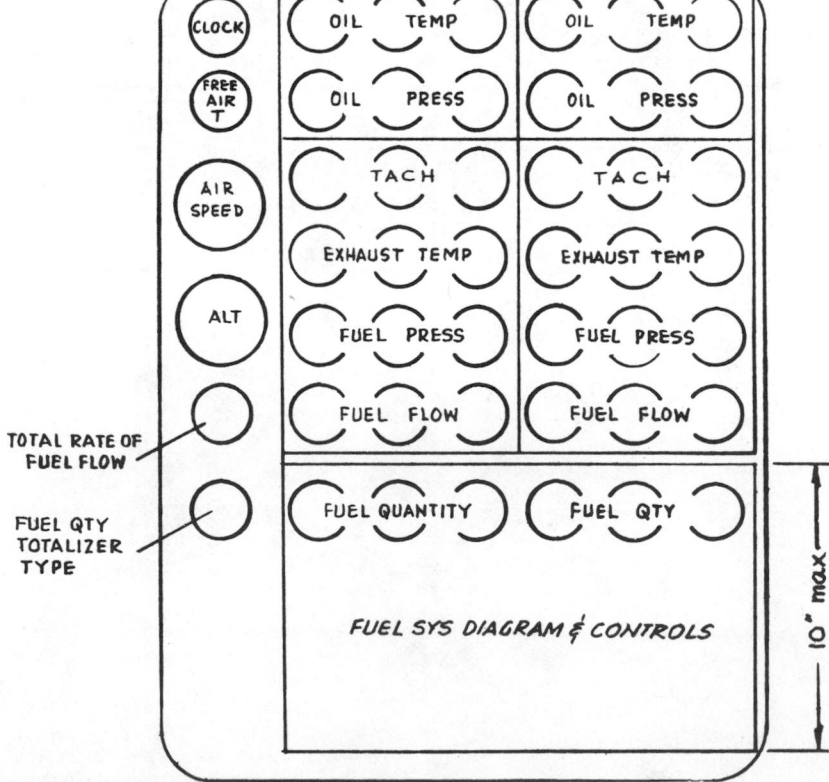

Flight engineer's panel—multi-engine jet (Fig. 13). When direct reading manifold gages are used, the purge
valves shall be located in a position readily accessible to the flight engineer.

## MARINECRAFT

### Human Factors Considerations in Marinecraft Concept Development

#### 1. Environment

Both surface and undersea operational environments create numerous unique human factors problems that should be addressed early in the conceptual development of any seaborne vessel and the various pieces of equipment that may be placed aboard for special mission accomplishment. The principal factors to consider are the following:

a. Ship motion (e.g., pitch, roll, heave, and buffet caused by rapid crossing of wave crests): These motions impact on personnel not only in terms of potential sickness but also in terms of task interference and general safety (i.e., loss overboard).

b. Equipment noise and vibration: Because of the necessity to conserve space and minimize vessel size, personnel and equipment must necessarily be housed in close proximity. Adequate noise and vibration control is required in order to maintain crew performance at acceptable levels. Control must be exercised not only in working spaces but also in living spaces.

c. Weapons systems noise and hazards: Guns, missiles, and aircraft operating aboard various vessels each create special noise and hazard conditions. Although priority is necessarily given to maximizing the effectiveness of each of these systems, it is possible to degrade such performance if insufficient consideration has been given to reducing the effects of weapon noise and certain failure modes on the crews that must control the weapons. Early consideration of the hazards to the crew usually provides more opportunity to minimize potential crew performance degradations or hazards to personnel in the event of system failures (explosion, fire, aircraft crashes, etc.).

d. Atmospheric control: Military vessels (especially submersible vessels) present unusual atmospheric control requirements in order to maintain the crew in a healthy and efficient state. The idea that unfavorable environments are just part of the job is no longer valid within the context of the more sophisticated and demanding systems of today, in which human failure can no longer be tolerated. Air conditioning for crew living spaces is as important as it is for electronic equipment if the total human-machine system effectiveness is to be maintained. It is difficult to add atmospheric control if insufficient space and power were provided by the basic concept.

#### 2. Space

Although on-board space is generally at a premium, lack of space is often the result of a failure to appreciate the way in which space impacts on crew performance. Traditional ship architecture tends to follow outmoded trends toward designing crew spaces according to "minimums," as opposed to creating spaces that will maximize crew effectiveness. The principal space-related features of vessels that make a considerable difference in crew performance effectiveness are the following:

a. Head clearance: Traditionally, low overheads are created because between-deck dimensions are based on minimum head clearances; in many cases it is forgotten that pipes and ducts, which will consume part of the space, will be installed. The worst problems are created when multiple berthing is imposed and the berths have so little vertical separation that crew members cannot sit up in bed. This is more than a matter of convenience, as lack of head clearance often leads to unnecessary head injuries and/or to inefficiencies such as slowing the movement of personnel through passageways.

b. Cramped living and working conditions: Lack of adequate planning in the initial conceptual phase creates spaces that interfere with job efficiency because of restrictions on mobility. More prevalent however, is the restriction imposed on the general habitability of crew living quarters where there is insufficient space for storing personal belongings, for relaxing, or for performing personal hygiene tasks, such as toilet, showering, and dressing.

c. Passageway clearance: Not only is it important to create passageways and hatches that are large enough for crew transfer, but it is also mandatory that such passageway clearance accommodate passage of crew members and the equipment they may have to transfer from one part of the ship to another.

d. Space organization: Organization of ship space requires obvious compromises. However, human requirements for basic living and work convenience are the key to effective crew performance and eventual mission success. The crew-occupied space must be compatible not only with environmental needs (e.g., isolation from excess ship motion, vibration, and noise) but also with typical personnel traffic patterns as these relate to crew fatigue and time lost en route from one space to another. Traffic patterns should be thoroughly analyzed from both a nominal and an emergency standpoint.

#### 3. Special Features

The following special features aboard ship are particularly important to crew efficiency and safety:

a. Lighting: Although electric power is always at a premium, good lighting both for general purposes and for special tasks is important to crew performance efficiency and safety. Adequate lighting does not necessarily imply tremendous power demands if the lighting plan is properly conceived and executed. Many lighting problems are created because too little attention was given in the initial stage of design conceptualization to the types of lighting fixtures or to the light reflection factors in selecting overhead, bulkhead, decking, furnishing, or equipment colors.

b. Ventilation: Even the most sophisticated air-conditioning system often fails to provide a cost-effective living and working environment if initial planning for proper ventilation is not taken into account. Not only personal efficiency but also health is affected when ventilation is not adequate to provide sufficient air exchange within closed compartments to maintain the proper oxygen content, remove toxic elements from the air, minimize undesirable odors, and so on.

c. Communications: Effective communication among the ship's personnel is obviously critical for efficient ship control and systems operation. Traditionally, ship communications systems have been almost universally poor in terms of intelligibility of voice transmissions. Intelligibility standards should be stressed during initial specifications preparation, and system design and/or selection should be monitored closely. Communications failures due to misinterpretation of low-fidelity voice transmission should not occur under today's technological state of the art.

d. Safe design: Numerous features that traditionally present hazards to personnel are present throughout most vessels. These should be reviewed carefully during concept planning so that they do not recur as a result of default, i.e., previous design practices which are assumed to be acceptable. Among the key issues are sharp corners and edges; improperly designed ladders, stairs, railings, or handholds; slippery deck or stair surfaces; low overheads created by various equipment installations (booms, ducts, platforms, weapons, etc.); and special equipment or devices that create hazards in the event of some failure during operation (cargo or fuel line transfer systems, cable failures in deck cargo-handling systems, carrier aircraft barrier systems, etc.).

e. Escape, survival, and rescue: Although considerable guidance is available in regard to this important aspect of ship design, each new configuration often presents unique problems that should be addressed as early as possible in conceptual development. Some of the typical emergencies that should be reviewed in detail are those related to entrapment below deck in the event of hull penetration and flooding, explosion and fire, nuclear reactor failure, and shifting of cargo.

f. Crew protection in combat: Protection of deck crews from typical enemy action during combat, although a familiar requirement in ship design, must be given special consideration with each new vessel. This is particularly true as new deck-mounted weapons are introduced, since the proposed protective device (e.g., armor) must provide not only protection but also access for the operating crew and reasonable assurance that other crew members will not be exposed to unusual hazards created either by the weapon's operation or by the fact that it may be the target of enemy fire.

g. Special equipment-produced hazards: Special hazards may be produced by radar, laser, or other exotic systems planned for a new ship system. Each must be examined in detail in order to create a proper preventive concept to mitigate hazards to the crew.

### Crew Living-Quarter Concepts

In light of the trend toward an all-volunteer Navy, most governments have come to stress the importance of providing the best possible living quarters commensurate with practicality and cost. Although a great deal of the emphasis has often been placed on the "cosmetic aspects" of these spaces, the planner should recognize that there are significant gains to be made by planning ahead in terms of the amount of space allocated and the manner in which this space is used or arranged. Some of the pertinent considerations are presented below.

#### 1. Amount of Space
The amount of space allowed for each person aboard ship should be based not on rank but on need. Every man or woman requires approximately the same amount of space for sleeping, for storing clothing, and for dressing, showering, etc. Where needs differ in terms of space requirements, decisions should be based on the special tasks that various members of the crew have to perform. For example, top-level officers require private conference space, middle-management officers and noncoms require special space for bookwork, and others require special space for record keeping.

#### 2. Space Configuration
Provision of private spaces should be based not on rank but on specific needs to separate persons or activities. The lowest-ranking crew member has the same basic psychological needs for privacy as the highest-ranking officer. However, privacy can be provided in different ways. The officer may need a suite of spaces in order to keep berthing, personal hygiene articles, and personal bookwork close to a conferencing space, but these require physical partitioning to avoid certain conflicts, e.g., visual and audio attenuation. However, it should be remembered that enlisted men and women also have needs that suggest reasonable separation or adjacency of spaces. For example, a sleeping crew member should not have to put up with the leisure-time noise or other disturbances of crew members who are on a different shift. There is a considerable rationale for creating shipboard living-quarter arrangements that would be similar to the domestic shore-based apartment that may be shared by several persons; i.e., the suite would provide separate sleeping quarters, a study area, and adjacent leisure-time and lavatory spaces.

#### 3. Personal Storage
Every man or woman aboard ship has the same basic need to have adequate space to store his or her personal belongings in a secure and convenient manner. There can be no defensible rationale for forcing an enlisted man or woman to stuff garments and personal articles into a bag or under a mattress. Persons need (and are required) to look as neat as their commanding officers, and they should have the opportunity to do so. Similarly, there is little defense for allowing commanding officers to bring trinkets, pictures, or other personal items aboard and for providing special shelves or cabinets for display of these articles, while directing enlisted persons to forgo these important amenities, which add to satisfaction with one's personal space.

#### 4. Appearance of Quarters
The argument that officers, as a general rule, are more used to attractive surroundings than many enlisted personnel because of their previous civilian experience is less valid today than in the past. Today, in schools, homes, hotels, restaurants, etc., practically everyone is exposed to the very finest in architecture, interior decor and furnishings, and other environmental features. Differences in cost are also often given as a reason for not providing attractive crew quarters. Modern technology, however, has eliminated most of these differences. Thus, the systems planner should recognize the importance of appearance in planning not only the officers' quarters but also the crew quarters. Some of the major features to consider in terms of appearance include the use of good lighting; attractive color schemes, carpeting, and deck, bulkhead, and overhead sheathing; and furnishings that do not look "institutional."

Two important considerations are pertinent to the question of appearance, however. First, materials must meet fireproofing requirements because of the seriousness of the fire problem aboard ship. Second, interior decor and furnishings must be appropriately sturdy, wear-resistant, and noninterfering in terms of task activities that will be performed; i.e., exotic colors and patterns, dark colors, poorly distributed illumination, etc., should not be allowed just to satisfy a single designer's or skipper's personal decorating whim.

#### 5. Special Facilities
Although they need not be directly adjacent to crew quarters, several important amenities are worth considering during initial concept development because they require additional space. Among these are special lounges, snack bars, game rooms, and library and study areas. In considering these facilities, one other tradition must be confronted, i.e., the separation of spaces for officers, noncoms, and enlisted persons. Modern Navy planners are just beginning to recognize the fallacy of "off-limits" traditions. The modern naval crew should be considered a team. Although the members' specific responsibilities differ, the sum total of their efforts must result in the success of an operation. Thus from a human factors point of view, most facilities aboard ship (other than private quarters) should be open to the entire crew. Not only does this increase crew morale, but it also provides an important variety of alternatives to reduce boredom during long sea-duty periods.

### MISSILES

#### Human Factors Considerations in Missile Systems Concept Development

The primary human factors considerations in missile system development relate to (1) handling, (2) checkout, (3) launch, and (4) tracking. Each of these areas of concern varies depending on the particular system; e.g., missiles may be carried by hand or transported by ground vehicles, rail vehicles, aircraft, or marinecraft. Missiles may be launched from hand-held launchers, from vehicles, or from underground or aboveground launch pads. Missiles may also be tracked from any combination of ground, air, and seaborne stations. Finally, missile checkout may occur in a special facility far from the scene of combat, or it may have to be performed in the field within the range of enemy fire.

#### 1. Hand-Held Missiles and Launcher Systems
The following are some of the principal human factors considerations in developing hand-held systems:

a. Proper size, weight, and balance to ensure the efficient and safe handling of the weapon as the operator picks it up, carries it, places it into the aiming and firing position, and returns it to the carrying or storage position

b. Aiming and firing features that make it easy to hold the weapon steady, match the eye to the aiming system, and actuate the firing mechanism

c. Hazard control features to prevent injury to the operator and/or adjacent personnel (flash blindness, flash burn, toxic fume inhalation, etc.)

*Note:* It may be necessary to package the system in a modular fashion in order to meet the first requirement, including the possibility of transport of separate elements by more than one person. When this is done, assembly and disassembly become another important human factors consideration in terms of ease and accuracy of performing these tasks.

#### 2. Missiles and Rockets Launched from a Mobile Platform
The following are some of the principal human factors considerations in developing these systems:

a. Loading the missile, i.e., the ease and safety aspects of picking up the missile, transferring it, and placing it on the launcher. The obvious hazards associated with dropping the missile during any stage of handling must be analyzed, and a suitable design must be defined that will minimize typical human errors of tripping, falling, losing hold, or inadvertently striking the missile against structural elements of the prime carrier or launcher device.

b. Control station configuration, i.e., the layout of the specific workplace or workplaces in which the principal operator (and/or the principal operator's assistants) will perform appropriate checkout, arming, aiming, firing, and other tasks. The safety implications of the system for other personnel and/or equipment that might ordinarily be in the vicinity of the system should also be of concern.

c. Control of special hazards such as possible burns or flash blinding from missile plume, inhalation of toxic gases, and injuries incurred when a missile explodes on the launcher.

### 3. Ground Launch Systems for Large Missiles

Because of the size of certain very large missile systems, the principal handling has to be performed by equally large, automated transporters, cranes, and erecting devices, and the actual checkout, countdown, firing, and control of the missile must occur within the protective confines of specially prepared, blast-proof facilities. Therefore, human factors considerations should be addressed in each of the following areas:

a. Transporter operation
b. Launch-pad facility operation
c. Launch control facility operation
d. Other support facilities (i.e., ground, sea, and/or airborne tracking stations)

A significant aspect of the larger and more complex missile systems is the coordination of large numbers of people, all of whom must contribute on a timely basis to the sometimes long and arduous series of steps leading to the final firing of the missile. Since many such weapons may never be fired until a very strategic moment, they may remain in a "ready" state for months or even years. In order to maintain such a state of readiness, system checkout and simulated firing exercises must be conducted on a regular basis. Thus, there are essentially two systems designed to work independently and/or together through certain stages of checkout and firing. Both the hardware and the personnel subsystems must be monitored and tested regularly. Since most such weapons are of a strategic nature and since inadvertent launching could contribute to a major confrontation between world powers, a key aspect of system design is to create preventive features and procedures to preclude unauthorized weapon firing.

The power and energy of the larger missile systems present a major safety problem, because of both the missile itself and the warhead, especially if it is nuclear. It is particularly important that definitive safety analyses be performed early in the conceptual stages, including scenarios taking into account the human failure that typically occurs when there is a large group of people. Special attention also should be given to the design of various launch-area and control-center facilities in terms of such special hazards as the following:

a. Falling from high platforms at the launch-pad site
b. Escape and rescue of personnel working at the higher levels of a launching platform or tower in the event of a premature explosion
c. Handling of, and/or protection from, high-pressure hose breaks
d. Launch-pad monitoring to make sure that personnel are "clear" prior to a missile launch

The launch control center is a major area for human factors analyses during concept formulation, whether it is land-based or sea-based. Particular attention should be given to the selection and conceptualization of the various individual and large-screen, team visual displays necessary to maintain continuing communication among the many operators, technicians, and commanders.

Habitability within the blast-proof control center should be of prime concern during the initial planning stage since such facilities often are extremely complex and costly and do not lend themselves to easy modification later on. Particularly important are such considerations as emergency power for air conditioning, illumination, and special filtering systems required in the event of nuclear attack and/or local disaster.

Finally, because of the high injury potential of many large systems, early consideration should be given to emergency treatment facilities within the launch and control complex, including the special equipment and procedures that should be provided for rescue and treatment of the injured in the event of a major disaster on the launch pad.

Although they are all of a slightly different nature, each of the various types of launch and control facilities requires approximately the same consideration of human-machine operations, housing, and maintenance; i.e., although an underground launch site is somewhat different in execution from an aboveground site, the basic living and working requirements are similar. Likewise, the seaborne launch and control (primarily submarine) system, although usually somewhat smaller, requires consideration of the total range of human-machine interactions and housing and protection requirements.

## ARMORED VEHICLES

### Special Human Factors Considerations for Armored Vehicle Conceptualization

#### 1. Ingress and Egress

Although there are many constraints to contend with in designing an armored vehicle such as a tank (e.g., small and perhaps poorly located hatches), rapid and safe ingress and egress is often a key operational consideration, particularly in combat. Hatches should also be located to minimize the exposure of personnel who are either entering or departing because the typical top hatch places the user in a prime position for the enemy sniper's shot. Emergency escape from more than one hatch also is an important consideration in the event the tank crew is forced to leave and can pick the safe side from which to depart.

#### 2. Interior Crew Positioning

In spite of the desirability of a low tank profile from the military operations point of view, crew members can perform better and withstand the sometimes tortuous dynamics of the tank's progress over rough terrain longer if they are positioned well, i.e., if a reasonably normal sitting posture is provided.

#### 3. Exterior Visibility

A major problem to be resolved in tank design is that of seeing out when the tank is "buttoned up" for combat. Although hard to obtain, 360° visibility is obviously desirable. Equally important, however, are the vertical viewing-angle requirements necessary for some severe types of terrain operations. Optical systems are obvious methods for obtaining these objectives, but to be effective, they must be defined before the basic external package has precluded their potential effectiveness.

#### 4. Internal Crew Environment

The typical armored vehicle concept creates tremendously severe environmental problems for the crew. Special attention early in the concept formation may, however, allow minimization of the effects of noise, vibration, pitch and roll, and temperature, humidity, and ventilation (especially the latter, since the "buttoned-up" condition implies a need for a semiclosed life support system). Safe interior design to prevent contact injuries requires special consideration for removing sharp corners and projections and for providing suitable restraints and handholds and comfortable, shock-absorbing seating.

#### 5. Field Maintenance and Repair

Although the typical armored vehicle cannot be pushed to the nearest garage by the crew, simple field service and maintenance can be designed into the basic system if considered early in the planning stage. Such maintenance should not require special tools. The vehicle should run on readily available fuels. Replaceable parts should be readily accessible.

## FIELD ARTILLERY

### Special Human Factors Considerations in Field Artillery Weapon System Conceptualization

#### 1. Operations

a. Disengagement of the weapon from the prime mover
b. Manhandling the weapon into position
c. Site survey and final weapon positioning and securing
d. Prefiring weapon preparation
e. Loading
f. Aiming
g. Firing
h. Expended shell expulsion and disposal
i. Communications
j. Cleaning and field service
k. Securing the weapon
l. Reengagement of the weapon to the prime mover

#### 2. Environmental Factors

a. Climate (cold, rain, snow, ice, etc.)
b. Darkness
c. Hazards associated with firing (noise, recoil, flash blindness and burns, flying particles, noxious and toxic gases and fumes, etc.)
d. Topographic deterrents (rough terrain, interfering vegetation, unstable surfaces, etc.)

### 3. Personal Equipment Constraints

a. Cumbersome and constraining winter garments, gloves, mittens, helmet, face mask, etc.
b. Special equipment interferences (backpack, belt-hung articles, small arms, etc.)

**TYPICAL PROBLEM AREAS** Most field artillery weapons are heavy and awkward to handle, and they present serious personnel hazards both during their initial maneuvering and during firing sequences. As indicated by the above operating, environmental, and operator considerations, the basic packaging design must reckon with numerous, often conflicting objectives. Although it is difficult to anticipate and suggest specific design features for all future field artillery weapons concepts, several typical problem areas have been identified with most past and current systems:

1. Too little attention is paid to creating suitable handholds so that crew members can maintain a secure grip on the fieldpiece as it is being manipulated.
2. Too little attention is paid to creating an efficient operator position so that a gunner can assume the proper posture for observing various displays, sighting, and manipulating various firing and sighting controls.
3. Too little attention is paid to creating an efficient ammunition loading system that will accommodate the sometimes heavy and awkward shells and guide them into the breech easily and safely without requiring unnecessarily great precision.
4. Too little attention is paid to designing system elements so that they will not require extra care when being cleaned, serviced, or replaced and to minimizing the need for special tools and other aids for dismantling and reassembling various parts of the fieldpiece.
5. Too little attention is paid to designing weapons that either cannot be jammed or can be unjammed simply and safely.

## SMALL ARMS

### Typical Human Factors Considerations for Conceptualization of Hand-Carried Small-Arms Weapons

#### 1. Weight
A shoulder-fired weapon should not weigh more than 7 lb (3.2 kg) (unless it rests on top of the operator's shoulder, as in a bazooka-type launcher). A hand-held weapon (pistol) should weigh no more than 3.5 lb (1.6 kg).

#### 2. Center of Gravity
Hand-held weapons should be designed so that when the operator holds the weapon in the normal firing position, its center of gravity is within 20 in (51 cm) of the operator's body. If the center of gravity must be farther away, provide special support in the form of a tripod or similar device.

#### 3. Size and Shape
Avoid very long weapon configurations, particularly because of the difficulties they create in manipulating the weapon, i.e., picking it up, carrying it without interference, maneuvering it for aiming in close quarters, etc. Critical dimensions relate to gripping (both hands), reaching a trigger, aligning the eye to sighting devices, and reaching specific elements for cocking, loading, adjusting sights, etc.

#### 4. Manipulation and Carrying Aids
Consider the potential difficulties of rapid acquisition of the weapon from a holster, from a back-carry sling position, from the ground, or from another weapon holder. The prime objective should be a design that allows the operator to swing the weapon rapidly and directly into a firing position; designs that require the weapon to be repositioned, rotated, etc., should be avoided.

#### 5. Weapon Operation Interference
Design the weapon package, including grips, sights, etc., so that it can be used by an operator who may be encumbered by typical military

clothing, a helmet, a gas mask, gloves, or mittens.

#### 6. Safety Devices
Safety lock systems should be devised which will not seriously increase the time required to fire the weapon but which will provide protection against inadvertent operation while the weapon is being carried, cleaned, adjusted, or stored.

#### 7. Firing Hazards
The general weapon design should minimize the potential for operator injury in terms of violent recoil effects (i.e., injury from the sight or other elements that might strike the face), flash burns, pinching by sliding parts, skin puncture by sharp protrusions, and cartridge expulsion after firing.

#### 8. Sighting Features
The sighting system should be designed so that all components are positively retained, so that the rear aperture will not flutter during recoil, so that windage and elevation controls will not be disturbed by recoil, etc. The main sighting concept should make it possible for gunners to see both the target and the necessary reference marks, the background environment, etc. Settings should be immediately obvious so that operators are not required to move their bodies or the weapon. Appropriate index-mark illumination for night firing should be provided.

### Dimensional Guidelines for Rifles and Machine Guns

1. Weapons must be designed so suitably clothed and equipped user personnel, with applicable body dimensions between the 5th and 95th percentiles, can perform all required tasks—both field operations and maintenance—easily and efficiently in daylight and at night, and in either the standing or prone position.
2. When the user must reassemble parts under field conditions, their mating sur-

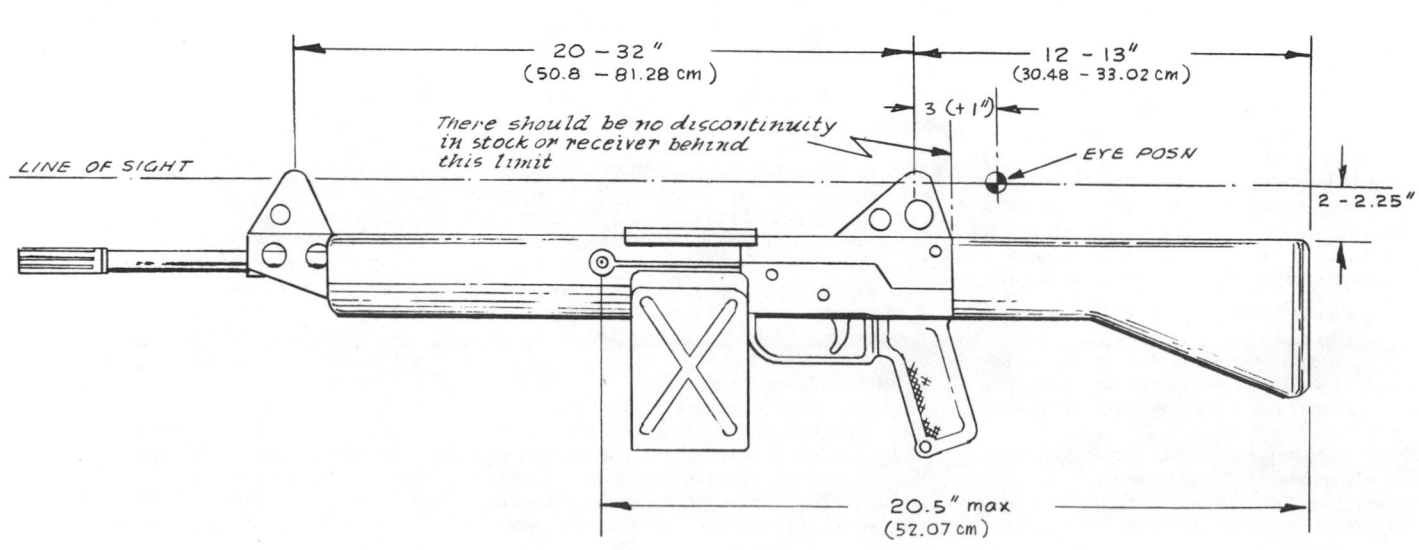

faces should be beveled to simplify assembly.

3. Surface of the weapon (or its attachments) that normally contact the user's body or clothing during firing or maneuver should be smooth, without sharp edges or discontinuities.

4. Any part of the weapon (or attachments) that contacts the user's skin should have thermal insulation.

*Note:* MIL-HDBK-759 *(Military Standarization Handbook: Human Factors Engineering Design for Army Material)*

## PORTABLE PACKAGING

### Human Factors Considerations in the Conceptualization of Various Portable Equipment and Container Packages

To come under the heading "portable," any equipment or container package must be compatible with the limitations of the smallest and weakest military person who may at some time or other have to pick it up and carry it. When a package is too large or heavy to be safely and effectively carried by two people, it should be considered transportable by some other means. The key factors to consider for portable packaging conceptualization include the following.

### 1. Shape and Size
The shape and size of any portable package must be compatible not only with the user's anatomic characteristics but also with the specific operational intentions, i.e., whether the package is to be lifted and moved only a short distance or whether it is to be carried for a considerable distance, over a flat surface, up or down stairs or ladders, etc.

### 2. Weight and Balance
The weight of any portable package must, of course, be compatible with the user's anatomic and strength capabilities (not his or her maximum capacity, but some value suitable to the operational requirements). Weight distribution is often a critical variable since a poorly distributed weight may cause the user to lose his or her balance and/or to be unable to manipulate the package for lifting and depositing without dropping it.

### 3. Grasping and Holding the Package
Consider whether it is desirable to lift and carry the package with one or two hands, to carry it in a "bear-hug" fashion, or to use appropriately positioned handles.

### 4. Backpack Packages
Design of backpack concepts requires consideration not only of the "fit" of the package as it is being carried (i.e., the proper weight distribution so as not to interfere with walking) but also of the method for picking it up and transferring it to and from the carry position.

### 5. Contact Hazards
Not only should portable packages be designed so that there are no sharp protrusions, corners, or edges that may cut into hands, but there should also be no external elements that can snag on clothing or catch on things, thus interfering with the safe and orderly manipulation of the package. In addition, there must be sufficient finger clearance along the bottom of the package to preclude mashing fingers when the package is set down.

### 6. Handholds and Handles
Avoid creating packages without handholds or with handles that are not appropriately placed to aid in properly balancing and manipulating the package. Handles should be designed so that they will not cut into the user's hand or fingers because of the heavy weight of the package. Handle positions may have to be provided for independent handling functions; e.g., the right position for carrying may not be satisfactory for lifting the package or depositing it in a truck or on a shelf.

### 7. Container Removal
Containers should be designed so that the user removes the container or cover from the equipment or contents it protects, as opposed to removing the equipment or contents from the container.

### 8. Labels and Instructions
Provide conspicuous and legible labels to avoid handling misuse.

# SPACE SYSTEMS

## GENERAL

### General Human Factors Considerations for Space Systems Concept Development

Human factors considerations permeate every aspect of space system operation. Although many of the subsystems may include much automation, the human remains a key element in space systems and equipment checkout, inspection, servicing, maintenance, and operation. The following are key areas within the total space complex which require a detailed examination of the human's potential roles and the design of suitable interfaces to allow these roles to be efficiently performed.

#### 1. At the Launch Site
a. Booster and payload assembly, test, transfer from the assembly facility to the launch pad, inspection, adjustment, and fueling on the launch pad
b. Booster and payload launch control, emergency abort, in-flight communica-

tions and control, on-board data receipt, malfunction analysis, and in-flight maintenance instruction
c. Personnel hazards control at the launch site (noise, explosion, fire, electric shock, toxic contaminants, personnel falling from high platforms, objects being dropped from platforms on personnel below, etc.)
d. Astronaut holding facility, mission simulators, emergency medical facilities, special mission laboratories, and shop facilities

#### 2. Within Manned Payload Vehicles
a. Ingress and egress (for both normal and emergency escape prior to launch and for in-space extravehicular and/or intervehicular transfer)
b. View ports
c. Crew accommodations and furnishings
d. Life support systems (food, water, and waste management)
e. Crew protective systems (space garments, radiation shielding, meteor

shielding, reentry heat shield, and internal environmental control)
f. Crew monitoring (physiological and psychological)
g. Special performance aids (space tools, tethers and restraints, mobility aids, and remote manipulators)
h. Other (special foods and containers, in-space communicating and visual signaling devices, and astronaut extravehicular self-propulsion systems)

## CREW ACCOMMODATIONS

### Special Considerations in Conceptualizing Space Crew Accommodations

#### 1. Reduced Gravity
A major consideration in conceptualizing crew accommodations for orbital or outer-space operation is the influence that the absence of gravity has on crew performance and health.

The lack of gravity requires that everything within the space capsule, including the astronauts themselves, be tied down, since everything that is not secured to the space capsule structure will float free. Although the absence of gravity is sometimes useful in that astronauts can move more easily from one point to another once they have pushed off from a solid object, they may have serious problems, such as the following:

a. They cannot slow themselves down and thus may strike another object much harder than was anticipated.

b. They cannot remain attached to a floor surface, as they would if gravity held them there. Thus, they must have some way of attaching themselves to surfaces if they wish to.

c. If astronauts are not firmly attached to an object they are working on and if they try to apply force or torque to a device mounted to the object, the chances are that they will be propelled away from the device or be rotated about the device. Thus, astronauts must be secured to an object and/or must have a tool that will accomplish the task without imparting reciprocal force to them.

d. The earth's gravity causes a person's muscles and cardiovascular system to be exercised on a regular basis. In space, these systems tend to atrophy (deteriorate or waste away). Thus, we must provide special exercising devices and/or procedures to help crew members maintain an appropriate state of health.

e. Since small particles also float in the reduced-gravity environment, special systems have to be designed to keep food particles from choking the crew members, cut whiskers from getting into their eyes, and pencils, small parts, paper, etc., from getting away from the person who is trying to work with them.

f. Without appropriate tethering, handholds, or footholds and/or without appropriate spacing of adjacent surfaces, it is possible for a crew member in the zero-gravity environment to become suspended in midspace and not be able to get back to any surface. Likewise, a crew member who has been caused to rotate while out of touch with a solid surface cannot readily stop spinning, which can lead to serious disorientation and nausea.

## 2. Deep Space Vision

Lack of atmospheric particles in space eliminates the light-scattering phenomena that we find useful in visual recognition tasks. In space, an object is extremely dark on its shadow side and extremely bright on the side facing direct sunlight. Thus it becomes difficult to recognize objects that are viewed outside the manned space capsule. A round ball viewed at a distance will appear not as a ball but as half a ball. Special viewing surrogates and/or optical systems are required for visual recognition of objects outside the space capsule.

By the same token, if astronauts are expected to work outside, special illumination will be required so that they can see objects that are in shadow.

Because of the above phenomena, it is almost impossible for an astronaut to judge relationships between several objects outside the space capsule. Although relative motion still can be recognized under certain circumstances, an object moving toward the observer may not be recognized until too late. For example, in the mating of two space vehicles, it is extremely unlikely that an astronaut could judge the rate of closure; thus special techniques must be devised either to mate the vehicles automatically or to display closing rates and paths in a form that will allow the astronaut to make the appropriate corrections.

## 3. Temperature Extremes

Astronauts are subject to intolerably high temperatures wherever the sunlight strikes them and to intolerably cold temperatures on their shadow side. Thus, if astronauts are actually working outside the controlled space vehicle environment, they must have space suits that provide heating and cooling to compensate for these extreme temperature differences.

## 4. Benefits of the Unusual Space Environment

As noted earlier, astronauts can propel themselves easily without the typical restrictions of the earth's gravity. It is also not necessary to create typical furnishings for them to sit or lie down on; all that is necessary in terms of providing a place to rest and relieving the stress involved in standing is to secure astronauts so that they do not float away. Astronauts literally can be "hung on a hook."

Although during early manned space experiments it was thought that astronauts would become disoriented and possibly sick in the reduced-gravity environment, experience has shown that humans not only adapt quickly to this environment but also find it exhilarating. Perhaps the most beneficial aspect of the reduced-gravity environment is the fact that astronauts can move much larger packages than they could on earth. As long as the mass differential between the human and the package and/or a solid surface is provided from which to initiate package movement, there are no limits to the size of packages that could be moved—except, of course, for the problem of seeing around a package to guide its path.

*Note:* Although there may be obvious reasons for automating certain features of an in-space operation, there are few reasons why humans cannot be a very effective component in any space operation. Since much of future space exploration will involve making decisions only after the space hardware has been confronted with unknown factors, humans should and can be used effectively to discover, recognize, decide on, and execute appropriate responses to their discoveries.

## Special Dynamic Environment Considerations[47]

Tolerance to *g* forces can be increased to a measurable extent by placing the subject so that acceleration forces are transverse to the long axis of his body. The illustration at the left indicates the optimum position for maximum tolerance.

[47]W. E. Woodson and D. W. Conover, *Human Engineering Guide for Equipment Designers*, University of California Press, Berkeley, 1964.

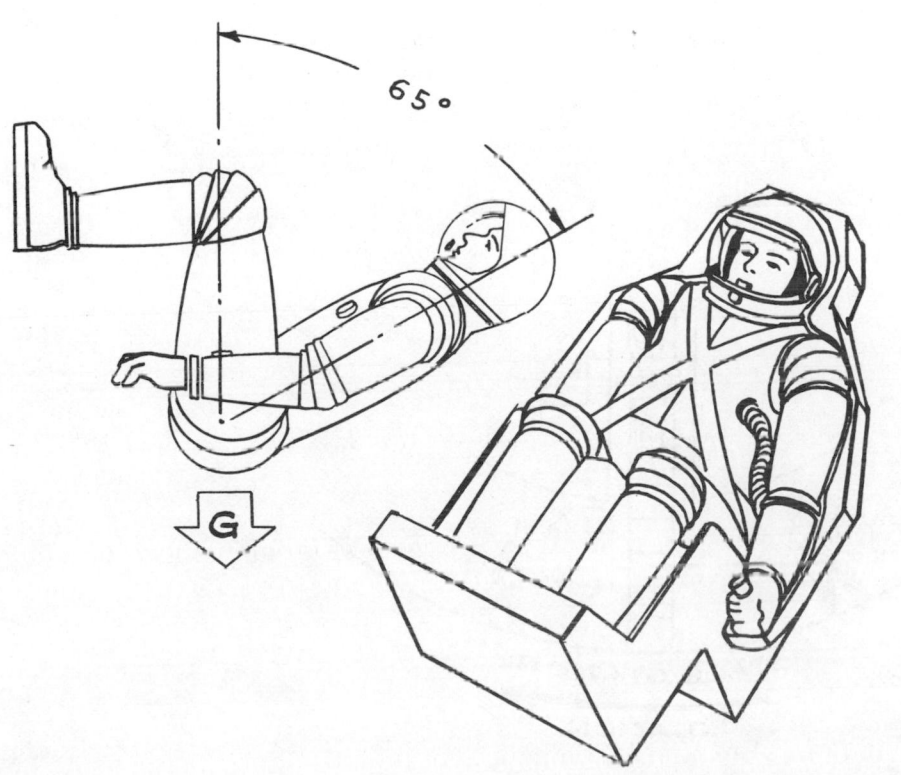

65°

G

CONTOURED ASTRONAUT COUCH

In addition, it is important to distribute the force as evenly over the body as possible—but at the same time to contain the fleshy portions so that they cannot be deformed by the force. Recent studies have also shown that increased tolerance will be obtained by the use of positive pressure-breathing.

Experimental use of water immersion techniques have also demonstrated remarkable increases in $g$ tolerance—as yet, however, not for practical purposes.

Where critical $g$ levels are anticipated, use auditory rather than visual signals. Reaction time to auditory signals is shorter, and they can often be heard even after the subject has begun to black out.

As one would expect, $g$ forces restrict the ease with which a person can move his arms, legs, head, etc. For example, he cannot move his head away from the headrest if more than 4 $g$ are acting against his head, or he cannot raise an arm off the armrest if more than 8 $g$ are acting against the arm. He cannot raise a leg against more than about 3 $g$, although he can still perform limited wrist motion even against 25 $g$. Armrests and controls should be placed very close to the operating position if extreme forces are expected.

In addition to the typical $g$ forces just discussed, another force problem is faced which creates physical difficulties for the human operator. This is called Coriolis, which is the component of force present whenever movement takes place within a rotating vehicle. A typical case in point might occur in a rotating space station, the rotation being provided to create artificial gravity.

In such a case, the Coriolis force will affect the occupants of the space station differently, depending upon the direction in which they are moving, and also their distance from the hub of the spin axis.

If a vehicle had a 20-ft radius and was spinning to simulate a 1.0-$g$ field, a 150-lb man climbing a ladder from the periphery of the station to the hub at 3.5 ft/s would experience a 41-lb force, which is in the same direction as that of the rotation.

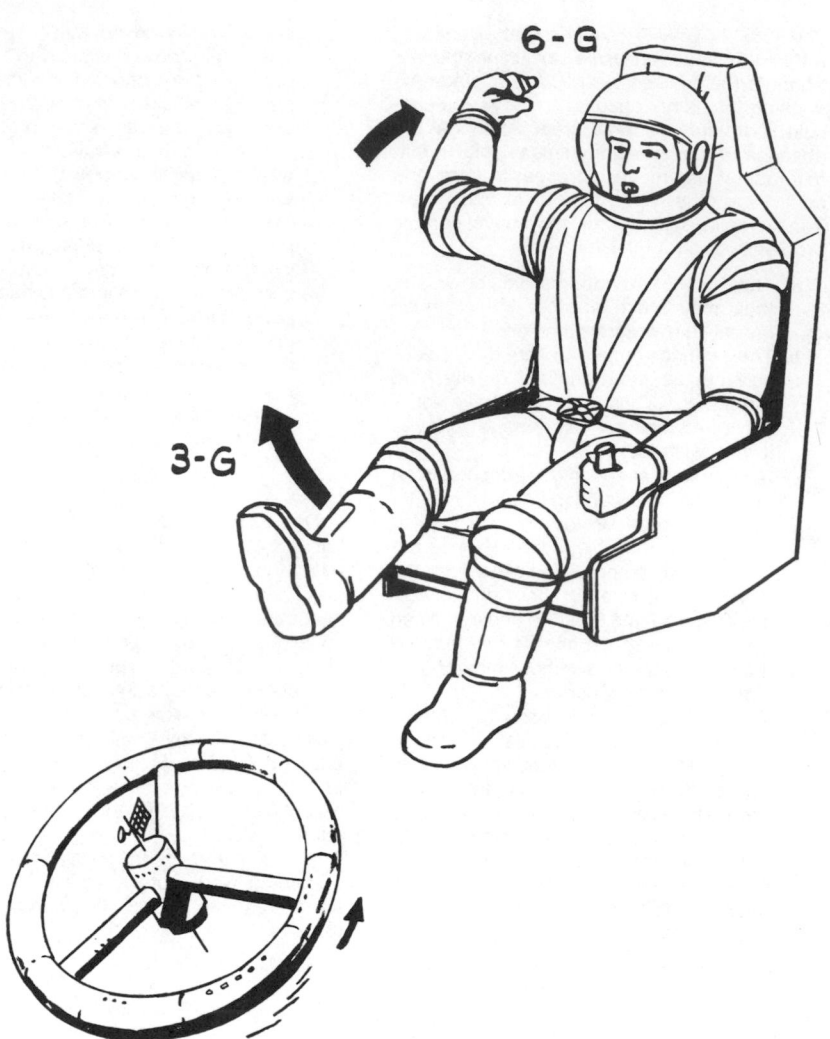

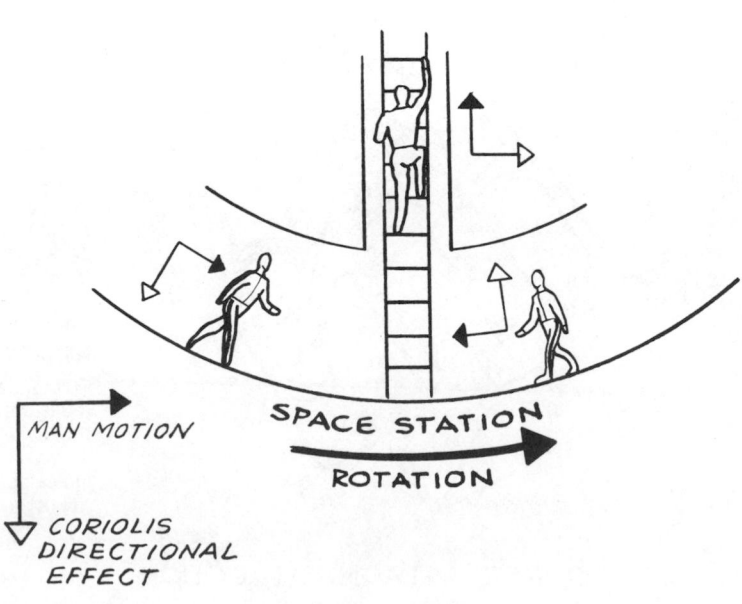

In the accompanying illustrations are shown typical design considerations to overcome deleterious effects of Coriolis.

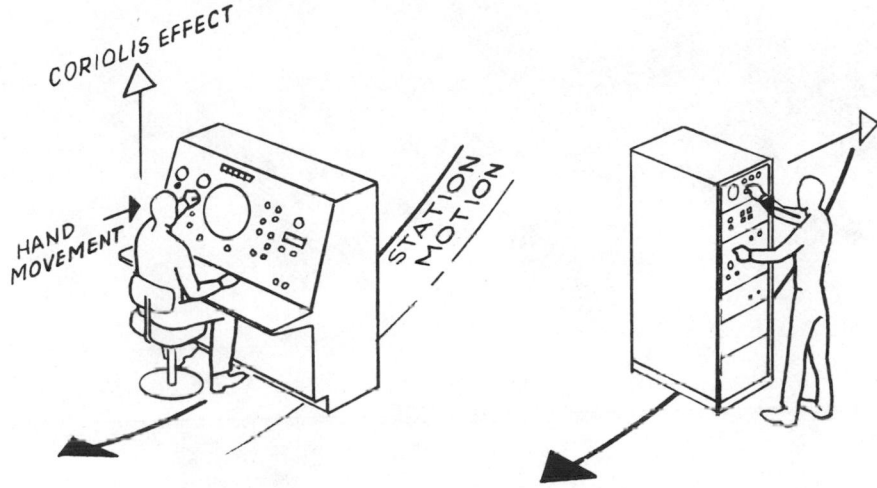

LOCATE CONSOLE
ATHWARTSHIP SO THAT
TYPICAL LATERAL
HAND MOVEMENT
ISN'T DISTURBED

LOCATE TALL RACKS
SO THAT ACTION OF
PULLING OUT CHASSIS
ISN'T INFLUENCED
BY CORIOLIS

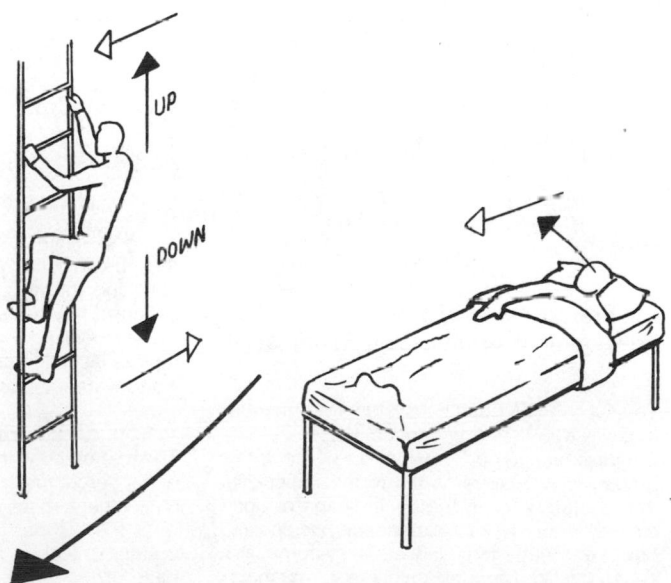

LOCATE PLANE OF
LADDER SO THAT
CORIOLIS WON'T
THROW MAN BACK-
WARD

IF BUNK IS ORIENTED
WITH FOOT TOWARD
DIRECTION OF ROTATION
OCCUPANT CAN RAISE
UP OUT OF BED EASIER

# INDUSTRIAL SYSTEMS

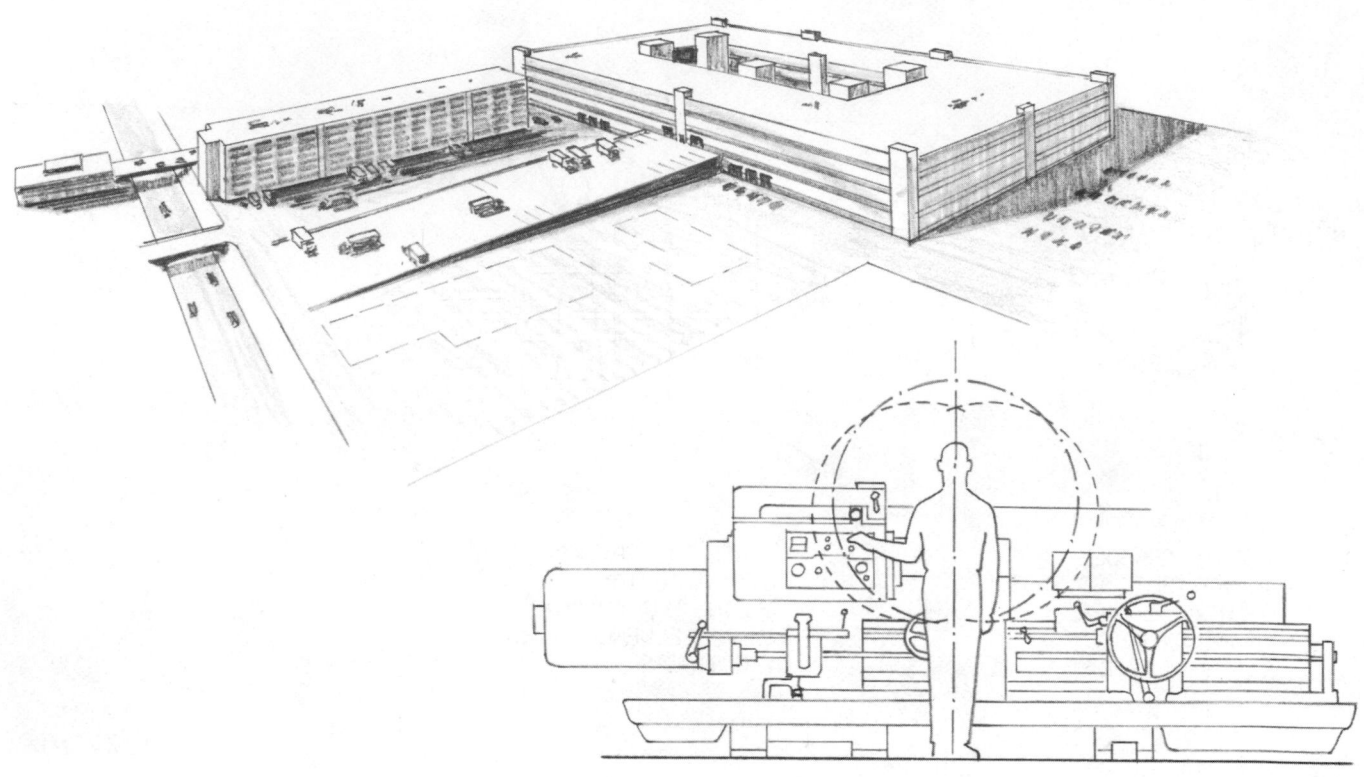

## GENERAL

### General Considerations in Development of Industrial Systems

The success of industrial systems is generally measured in terms of productivity or output. Effective human performance is a key factor in producing a desired output in the factory, in the laboratory, or in the construction or agricultural industry. Past experience has shown, however, that many industrial systems are conceptualized around machinery, materials, and architectural features, with too little thought given to what workers should be doing, how they should be doing it, and where they should be doing it. As a result, these systems have tended to be designed more to accommodate the machines and materials than to make sure that workers can contribute maximally to the eventual industrial output. Fortunately, this bias toward the machine rather than the worker in the industrial system is changing through the efforts of ergonomics, or the human factors approach to the design of work systems.

Ergonomists consider that the work system must be designed to perform a particular work task and that it is constituted of the human supported by his or her work equipment and the surrounding environment. The key human factors issues in conceptualizing an effective work system include the following.

### 1. Role of the Worker
The role of the worker should be based on his or her perceptual, cognitive, physical, and psychological capabilities and limitations, not on what the technicalities of preconceived electromechanical or other purely physical features of the technologically oriented inventor create for the worker to do. Just as industrial system managers would not want a computer to make their decisions, industrial machines should not direct the work activities of the industrial workers. Each member of the industrial team should have a meaningful job to do.

### 2. Role of the Industrial Machine
The role of the industrial machine should be to support the worker in accomplishing his or her assigned task. The machine should be assigned functions to complement the worker's capabilities, relieve certain stresses, and per-

form activities which are needed but which are beyond human strength, speed, or perceptual capacities.

### 3. Role of the Industrial Environment
The role of the industrial environment (i.e., the architectural facility, temperature, noise, illumination, etc.) should be to provide working conditions that will ensure that both the worker and the machine can function at maximum capacity. It is often noted that planners are more concerned about machines than they are about workers; e.g., air conditioning may be provided for computers, while factory workers "sweat" on the assembly line.

### 4. Worker Safety
The design of the work process should be based not only on the objective of maximizing output but also on that of safeguarding the worker's well-being and health. Safety results only when both machine and human failures are anticipated and designed out of the system. Although safety training and supervisional procedures are important, hazards are created at the beginning by lack of proper consideration of the modes of machine, environmental. and human failure.

## Human Factors Principles to Consider during the Conceptual Phase

The principal areas in which human factors principles are important considerations during the conceptual phase of industrial systems development are the following:

### 1. Design of the Workplace
a. Ingress and egress and internal traffic flow
b. Working position, seating, and layout of materials and tools
c. Communications
d. Illumination
e. Temperature, ventilation, and humidity control
f. Noise and vibration control
g. Clearance and protection from hazards (materials and vehicular movement, electric shock, toxic contaminants, dust, smoke, fumes, fire, etc.)

### 2. Design of Machines, Tools, and Equipment
a. Visual displays, instructions, labels, and signs
b. Controls and handles
c. Protective features (guards, handrails, fail-safe switches, locks, warning devices, blast and noise barriers, etc.)
d. Ease of maintenance and service (i.e., inspection, adjustment, calibration, troubleshooting, removal and replacement, and portability)
e. Supplementary illumination for critical seeing tasks
f. Safety marking (i.e., the appropriate use and location of standard safety symbols, colors, marking, and signs)

### 3. Design of Supportive Implements and Devices
a. Vehicular equipment
b. Scaffolds, platforms, ladders, and stairs
c. Lifts, loaders, cranes, and materials pallets
d. Automated materials-handling devices (conveyor belts, elevators, remote manipulators, etc.)
e. Process-monitoring consoles
f. Cleaning and maintenance (vats, chambers, furnaces, etc.)

## Specific, Critical Human Factors Considerations in Industrial System Workplace Design

### 1. Worker Posture
Whether workers are standing or seated at a workbench or machine or are riding on some type of industrial vehicle, dolly, or crane, their working posture is extremely important. That is, if the hardware forces workers to remain in an awkward position for a long time, they obviously will become fatigued and will be more apt to make mistakes or incur some type of physical disability over a period of time.

### 2. Worker Mobility and Balance
Tasks that involve considerable muscular activity and/or the continual maintenance of balance place considerable stress on workers, especially when the tasks continue for extended durations. Tasks should be designed in such a way that workers do not have to continually remind themselves of potential loss of balance, possible muscle overstrain, the potential for rupture, etc. The workplace within which tasks are performed should not be restrictive in the sense that workers cannot assume a normal posture and balance; use symmetrical, ballistic movements; or distribute the strain by using first one hand and then the other.

### 3. Weight, Force, and Precision Requirements
Tasks should be created that do not place excessive (and especially continuous) demands on workers in terms of lifting weights, applying force, and making precise movements.

### 4. Visibility
Task environments and the design of the various task elements should be based on the fact that people need to see what they are doing. They need to see where they are walking and where the perimeters of a package that they are carrying are with reference to the pathway obstructions. They need to see clearly the spatial and positional relationships between a tool and the material they are about to bend, cut, drill, or treat, and they need to see clearly the markings on tools, machine adjustment controls, and monitoring or inspection displays.

### 5. Environmental Conditions
Although the healthy worker can adapt quite readily to a considerable variation in temperature, humidity, or ventilation, the stress of compensating is sometimes cumulative and can lead to sudden collapse in the event of a sudden task stress (not to mention the potential for irritations and interemployee distress). Most serious, however, are the effects on worker performance and morale when noise, vibration, and atmospheric contaminants are not properly controlled. Uncontrolled noise not only leads to temporary and/or permanent hearing losses but also may interfere with crucial communications (e.g., hearing warning signals). Uncontrolled vibration may interfere with specific task performance, cause annoyance stress to the point of incurring employee aggression or absenteeism, and affect the employee's health after a time. Respiratory disabilities are perhaps the most insidious of the uncontrolled hazards, second only to the hazards associated with toxic substances where workers are exposed to toxic fumes, dusts, chemicals, etc. It is important to consider all the potential environmental hazards in terms of system design control, as opposed to discovering that such hazards exist after the system is designed. The latter condition is too often used as a guiding principle during system concept development; it is based on the assumption that it is easy to "fix" a problem if it occurs.

### 6. Human Energy and Efficiency and Motion Economy
Time and motion specialists espouse the desirability of designing work systems and procedures to economize on worker motion and thus improve human output efficiency and reduce fatigue potential. Attention should be given to motion economy during the conceptual stage of industrial system design because it is here that modifications in hardware, procedure, and worker utilization are most easily made. As indicated earlier, once the machine is built and the architectural features of the factory are fixed, the human must do all the adapting. Various industrial handbooks present the principles of motion economy, but perhaps the best known is the one by Barnes.[48]

Human energy consumption, on the other hand, is a factor that is less well understood in terms of its overall effect on human performance; i.e., although excessive energy demands for long durations obviously lead to physical fatigue, mental fatigue may be the worst enemy of the worker in terms of how it contributes to serious task failures. A basic objective of work design should be to conserve the worker's energy, but not to the extent that boredom is introduced. Actually, boredom may contribute to the energy cost.

Ultimate worker efficiency is necessarily a combination of many variables. Although the design cannot be expected to compensate for all the human variability (much of which may have little to do with the work or the job), the job or task design should provide workers with the best possible chance to perform their assigned tasks as effectively as they can. The application of good human engineering and time and motion principles early in the conceptual stages of system development generally leads to energy conservation and output efficiency.

## MOTION ECONOMY

### Principles of Worker Efficiency and Motion Economy

#### 1. Plant Layout
a. Organize plant facilities so that workers can easily learn where things are located with respect to one another, i.e., so that they can recognize how materials, people, and products flow through the physical plant system.
b. Provide recognizable visual references so that workers know where they are and where they are going. This includes not only signs and other codes but also visually distinctive architectural features.
c. Provide convenience between internal and external features (employee parking and work stations, delivery vehicle access and shipping and receiving storage, customer parking and administrative stations, etc.).
d. Provide convenience between separate buildings and special testing and/or storage facilities; i.e., minimize travel distances between facilities that interact frequently.
e. Provide appropriate separation between facilities where serious noise or safety hazards may compromise worker efficiency or safety.
f. Arrange for internal functions that are related and interact frequently to be located close together.
g. Provide a sufficient number of conven-

[48]H. M. Barnes, *Motion and Time Study*, John Wiley & Sons, Inc., New York, 1968.

## Industrial Plant Planning[49]

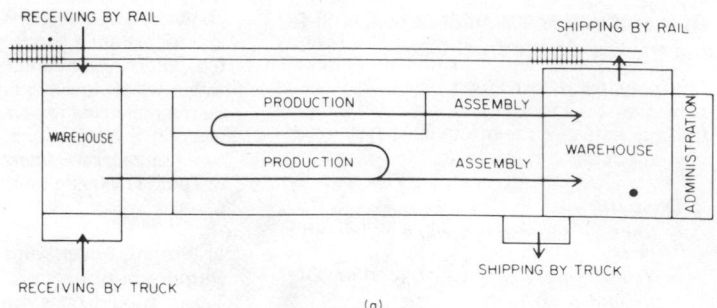

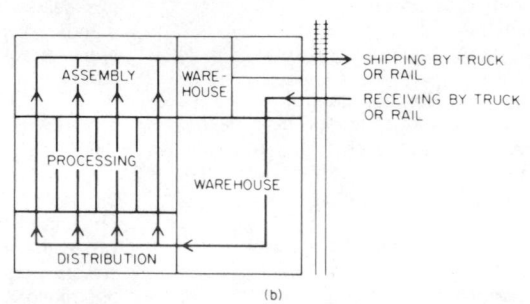

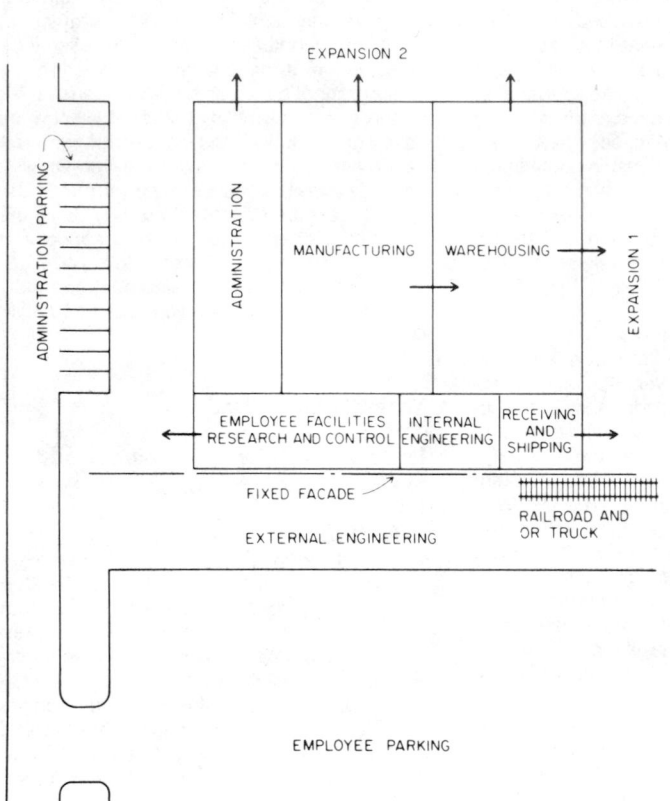

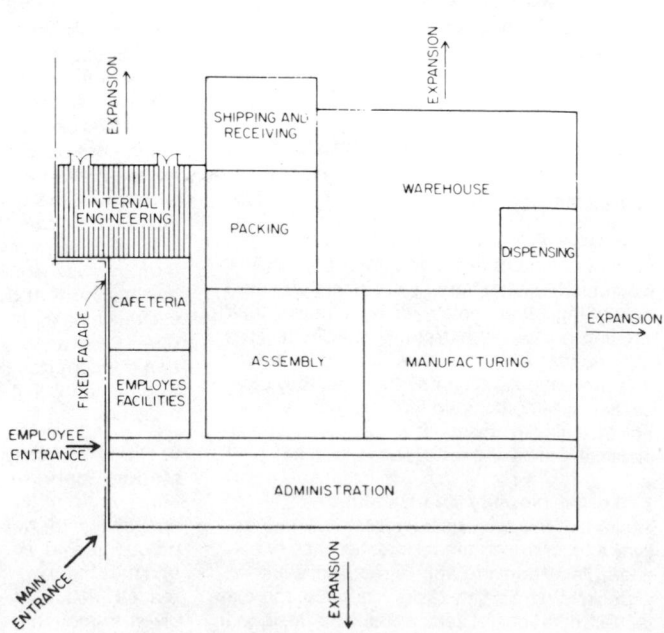

[49]J. De Chiara and J. H. Callender, *Time-Saver Standards for Building Types,* McGraw-Hill Book Company, New York, 1973.

iently located employee support services to minimize travel time (rest rooms, cafeterias, tool cribs, employee personnel services, etc.).

## 2. Individual Work-Station Planning

a. Provide a definite, fixed place for all tools and materials.

b. Locate tools, materials, and machine controls and displays close to the point of use.

c. Arrange for delivery of materials to a point close to the work site and the point of use. The worker should not have to go for parts or materials.

d. Design the height of the workplace and the seat so that alternate sitting and standing are possible.

e. Configure the workplace and seat arrangement to allow the operator to maintain the least fatiguing posture.

f. Design the work process so that both the worker's hands begin and complete their motions at approximately the same time.

g. Design the process so that both the worker's hands are not idle at the same time.

h. Lay out the work so that the worker's arm motions are generally in opposite and symmetrical directions and are approximately simultaneous.

i. Lay out the work so that momentum can be used to advantage—but not if it must be overcome by muscular effort.

j. Lay out the work so that smooth, continuous, curved motions of the arms and hands occur, as opposed to straight, irregular, suddenly changing motions.

k. Lay out the work so that ballistic motions are used, as opposed to restricted, precisely controlled movements.

l. Create work-flow patterns that are natural, rhythmic, and easy.

m. Lay out the work station so that as few eye fixations, as close together as possible, are required.

n. Wherever a specific movement requires repetitive or precise control, assign the motion to the hand or digit that provides the required degree of strength or dexterity, but avoid overloading a particular hand or finger unnecessarily.

o. Provide jigs, fixtures, and foot-operated aids to relieve the hands of unnecessary and restrictive tasks.

p. Combine two or more tools wherever practicable.

q. Pre-position tools and materials as much as possible.

r. Provide an auditory signal to indicate when long-duration machine operations are completed so that the operator is free to perform other tasks.

s. Place controls and displays where they require the least excess movement of torso, head, limbs, or eyes.

t. Provide lighting that is appropriate to the specific tasks, as opposed to general illumination for an entire area.

## 3. Interaction between Work Stations

a. Arrange machines, workbenches, etc., so that work can be passed directly and naturally from one to the other, i.e., not across aisles, around corners, or past an intervening work station.

b. Arrange a production line so that products finished at one station can be moved to an interim waiting point adjacent to the next operation; individual stations will then not be held up in accepting the next work package.

## MATERIALS HANDLING

### Human Factors Considerations in Materials-Handling Systems Conceptualization

Although a great deal of the handling of materials coming into, or being shipped from, the industrial complex, as well as the actual handling within the facility, may be accomplished by machines of one kind or another, humans interface with this process either directly or indirectly via machine controls and displays. The primary human factors considerations fall into two major categories, as described below.

### 1. Prevention of Materials Damage

Waste of materials due to damage is of considerable economic consequence to the manufacturer. Therefore, it is extremely important to examine the materials-handling concept in terms of the potential errors the human element could make with respect to movement of all materials in the receipt, manufacturing, and shipping phases of the proposed operation. This examination should include consideration not only of raw materials and substances but also of packaged items. The potential for human error often exists at the following points during materials handling:

a. Dock transfer to and from a transportation system (rail, truck, ship, etc.)
b. Shipping and storage-area transfer (i.e., to and from the loading dock and areas designated for temporary storage of either the preprocessed materials or the finished and packaged products)
c. Transfer within the plant (i.e., between processing, assembly, test, and packaging stations)

### 2. Prevention of Personal Injuries

Personal injuries also have a considerable economic impact on an industrial operation in terms of lost time and possible legal liability. More important from a human factors point of view, however, is the direct consequence to the worker and his or her family. This latter consideration should guide the planner because it assumes that most workers do not want to be injured and that it is therefore incumbent on the systems designer to create as safe a place to work as is technologically possible. The typical areas in materials handling where personal injury often occurs include the following:

a. Direct interface with raw materials that are heavy, sharp, rough, toxic, breakable, or explosive
b. Storage and stacking, where materials fall on people
c. Automatic material movement, where workers may be struck by the material or machine or may fall into or be caught by machine elements
d. Toxic liquids or chemicals, which may spill onto workers during transfer

### TRANSPORT AND LIFTING APPLIANCES

Narrow-gauge railway superseded by overhead trolleys or trackless electric trolleys with trailers. Long distance transport by lorries. Internal transport by traveling cranes of different types, fork lift trucks, etc.

**CRANES** Simplest lifting appliance for vertical elevation is the hoist (also electric) having a 0.5- to 5.0-ton capacity. Additional horizontal movement by wheel carriage (1) and (2). Wheel carriages controlled from shop floor or from driver's seat (3) or from remote control cabin (6).

Bracket jib cranes make load lifting above any part of an area possible. Exact positioning of load difficult, however, due to sloping extension device.

Combination of several wheel carriages possible for transport of heavy pieces (4).

Bracket cranes (8) are for transport along the length of the building. Jib cranes (7) are for transport of production parts to neighboring bay. Connecting platform (9) is for changeover of wheel carriage.

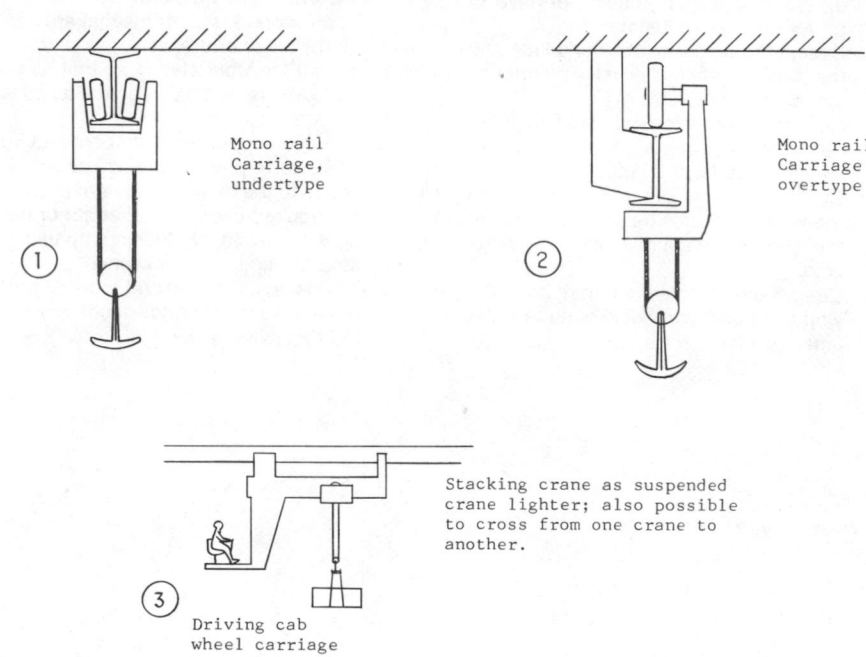

① Mono rail Carriage, undertype

② Mono rail Carriage overtype

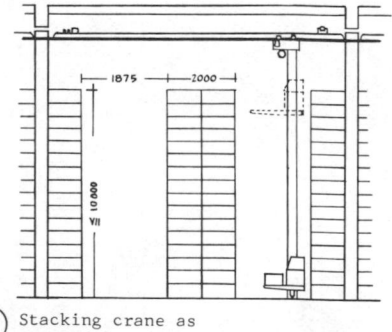

Stacking crane as suspended crane lighter; also possible to cross from one crane to another.

③ Driving cab wheel carriage

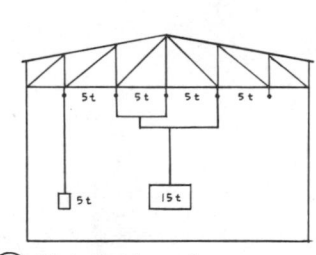

④ Distribution of a large weight

⑤ Stacking crane as suspended crane

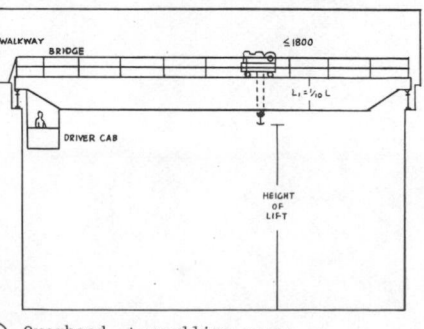

⑥ Overhead, travelling crane remote-controlled

Stacking crane (5) is a combination fork-lift truck and crane. Utilization of unused roof space and thus better use of storage space. Height of stack up to 10 m (33 ft).

Outside cranes require enclosed or rain-protected motors.

Design of structure depends on height of lift and crane capacities.

For mechanical workshops a lifting capacity of 4.50 to 6.00 m (15 to 20 ft) is sufficient, thus room heights are from 6 to 10 m (20 to 33 ft). From top edge of crane bridge to bottom edge of roof girder > 1.80 m (6 ft).

The height of wheel carriage depends on size of products and not necessarily on height of lift.

Standard traveling speed of wheel carriage is approx 30 m/min (100 ft/min).

## OTHER TYPES OF HANDLING PLANT

Other types include roller, curved roller, spiral roller, screw or worm, screw ribbon, wire mesh and belt, slat, portable belt conveyors, bucket conveyors, etc.

Fork-lift truck (11) loads from either side or is a lifting device on lorries which makes loading ramps unnecessary. (From E. Neufert, *Architects' Data,* Archon Books, 1970.)

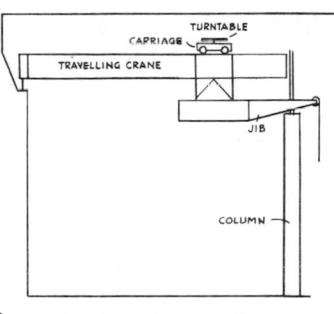

⑦ Overhead, underhung jib crane will serve adjoining bay

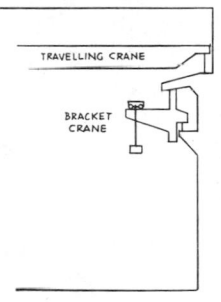

⑧ Bracket crane below travelling crane

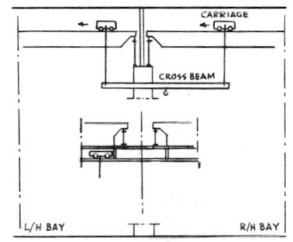

⑨ Crane arrangement, multi-bay factories. Detail of connecting device through crane track girder.

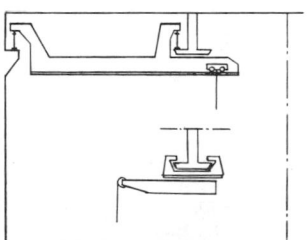

⑩ Travelling crane with jib. Inset, fixed radius jib crane in center of building.

| Capacity | Height | Length | Width |
|----------|--------|--------|-------|
| 1000 kg  | 2200   | 1725   | 955   |
| 2000 kg  | 2200   | 2190   | 1400  |
| 3000 kg  | 2200   | 2885   | 1600  |

Length of fork lift: 0.80–1.4 m

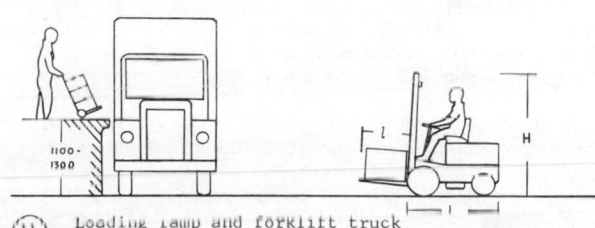

⑪ Loading ramp and forklift truck

## Semiautomatic Warehouse Systems

**CONTINUOUS TROLLEY CONCEPTS** Tripping and collision hazards must be considered. Design of the trolley track should be such that personnel cannot step into the track slot and turn an ankle. Forklifts should be painted with appropriate hazard markings, and audio alarms should make it possible for operators to warn other workers.

**PARTIALLY COMPUTER-CONTROLLED TROLLEY, ELEVATOR, AND CONVEYOR CONCEPTS** Inadvertent activation while performing maintenance tasks and limited visibility in high-density stack areas are common hazards. Manual operation during system down periods requires special consideration of access into areas and positions that may not have been planned for human occupation, and audio alarms should be provided to warn workers that conveyors are moving.

## Human Factors Considerations in Conceptual Design of Materials-Handling Vehicles

Special vehicles designed for the express purpose of handling materials within the industrial plant should be approached from the operator's point of view, i.e., in terms of what the operator has to see and operate in order to manipulate the vehicle and its load in an efficient and safe manner. Some of the key issues are the following.

### 1. Visual Clearance of Vehicle and Load
The vehicle operator should be positioned so that he or she can see the perimeters of the vehicle and its load with reference to pathway obstructions.

### 2. Visual Observation of Load Transfer
The vehicle operator should be positioned so that he or she can maintain continuous observation of the load as it is raised and lowered and otherwise moved into position, i.e., onto a pallet, into a storage area, into a truck, etc.

### 3. Vehicle Controls and Displays
Controls and displays required to operate the vehicle should be located and positioned so that the operator can reach or see them, while at the same time visually monitoring vehicle and load clearance and load manipulation.

### 4. Operator Support and Restraint
Select the most efficient posture for the vehicle operator in terms of his or her ability to control the vehicle and load, considering alternate standing and seated modes. Consider the possible need to provide restraints to keep the operator from falling from the vehicle.

### 5. Operator Protection
Provide appropriate guards and/or screens to prevent materials or components being transported from toppling onto the operator. Consider weight and balance dynamics in terms of tipping over the vehicle. Provide a warning

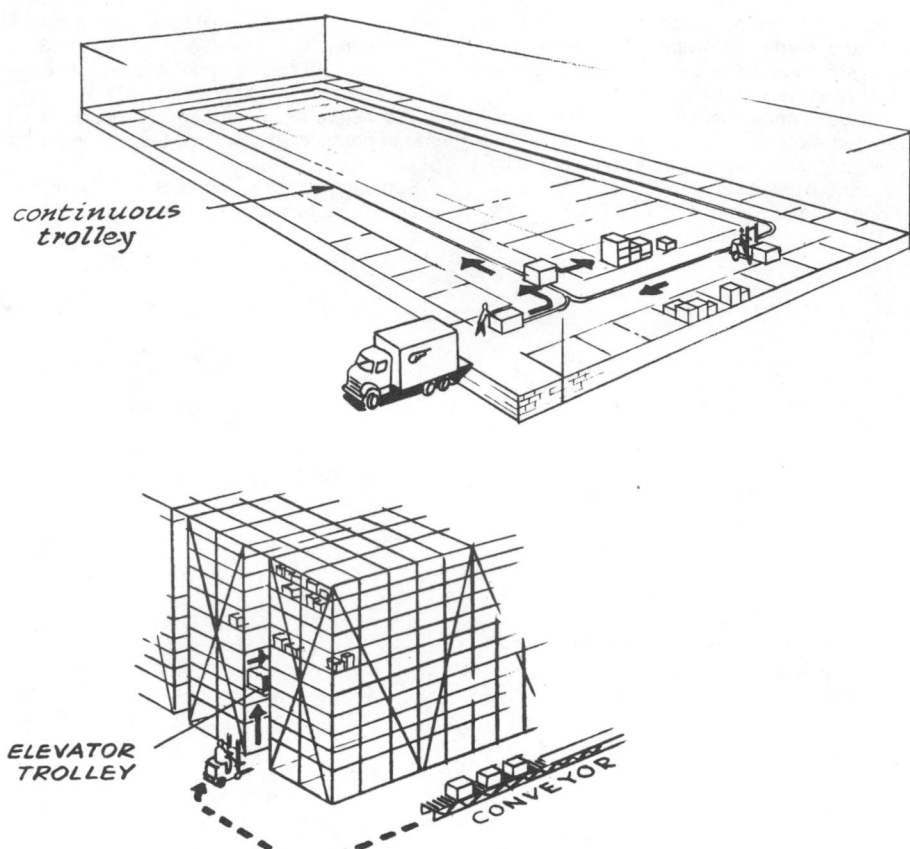

*continuous trolley*

*ELEVATOR TROLLEY*

*CONVEYOR*

system to indicate when too heavy a load is being manipulated or when an impending load shift is approaching the imbalance point.

### 6. Fail-Safe Features
Consider the typical misuse behaviors of the average driver and provide fail-safe features to prevent the driver from leaving the vehicle insecured on an incline, leaving the vehicle in gear while he or she is away from the work position, driving too fast, etc.

## MANUFACTURING

### Human Factors Considerations in Product Manufacturing Systems Conceptualization

Although the nature of product manufacturing varies considerably, depending on the particular product, and although many of the operations involve automatic manipulation of materials, humans interface with this process at many points, either directly or indirectly through machine controls and displays. The primary human factors considerations fall into the following main categories.

### 1. Prevention of Materials Waste
Waste of materials and damage of components and parts used in manufacturing the product are of considerable economic consequence. Therefore, it is extremely important to examine the entire manufacturing process concept in terms of potential human errors that could be made by the worker at the machine, at the assembly bench, in the test lab, or at the packing and shipping center. Human error often occurs at the following points during product manufacture:

   *a.* Cutting and forming raw materials
   *b.* Fastening materials together
   *c.* Treating materials
   *d.* Testing materials
   *e.* Assembling components and parts
   *f.* Testing assemblies
   *g.* Packaging assemblies

### 2. Prevention of Tool and Machine Damage
Although many of the fabricating machines and tools are standard items procured from another manufacturer, these should be selected in part on the basis of the fact that they are minimally prone to misuse by the average factory worker. Tooling that is new should definitely be designed with this in mind. Some of the typical misuse potentials are:

   *a.* Putting the wrong kind of material into a machine
   *b.* Putting material into a machine in the wrong way
   *c.* Improperly securing material in a machine
   *d.* Overtorquing material fasteners
   *e.* Scarring critical sliding surfaces with hand tools
   *f.* Burning out a motor by leaving it on too long
   *g.* Failing to monitor a machine operation and allowing the machine to damage itself when the material in process fails.

h. Inadvertently putting or dropping a hand tool into exposed gears or into a chain or belt system

i. Damaging a machine while moving it from one place to another

### 3. Prevention of Personal Injuries

Personal injuries on the job have considerable economic impact on an industrial operation, in terms of both the lost time and the liability potential. Again, however, the impact is greatest in terms of the potential effect on the employee and his or her family. Manufacturing processes should be conceived on the assumption that no employee wants to be injured and that it is therefore incumbent upon the planner to create a system that is as safe as is technologically possible. The following are typical of the injuries that can occur in the manufacturing process:

a. Entrapment in exposed moving machinery elements (gears, chains, belts, sliding devices, wheels, cranks, etc.)

b. Being hit by flying particles, broken materials, and sparks

c. Being struck by moving beams or materials or by machine components

d. Being crushed by stamps and dies

e. Being burned by hot machine surfaces, by hot materials being produced from a machine, or by a blast furnace

f. Being blinded by operations such as welding

g. Being electrocuted by electrical shorts or exposed high-voltage contact points

h. Being inadvertently exposed to electromagnetic or nuclear radiation

A critical incident analysis of personal injury should be performed early in the manufacturing concept formulation. This analysis should be done not only in terms of the manufacturing process personnel but also in terms of factory maintenance people, supervisory personnel, and visitors, who may be less familiar with the hazard potentials and safety rules of the installation.

In conjunction with this analysis, consider also the requirements for responding to the potential injuries that could occur in spite of a planner's best intentions and/or precautionary design. Consider alarms, special facilities such as contamination-removal showers, first-aid equipment, and a nearby first-aid station. Last, but not least, include in the basic layout plan for any fabrication facility easy access for outside emergency help and/or movement of emergency medical equipment to injured persons.

## HEAVY CONSTRUCTION

### Human Factors Considerations in Heavy Construction Systems Concept Development

Large earthmoving machines, construction cranes, heavy materials-loading devices, and the like generally dwarf the human operator because of their necessarily large size. Yet few of these systems can be made sufficiently automatic to eliminate either the machine operators or the support personnel who must work closely with the machine system. Most

such systems present similar human factors problems which should be addressed as early as possible in the concept formation stage; otherwise, human tasks become extremely difficult, and the potential for personal injury is great. The following key human factors problem areas should be examined and resolved along with the basic machine design questions.

### 1. Operator Position and Visual Surveillance of Machine Operation

Because of the size of many large construction machines, it is difficult to find an appropriate position from which an operator can see all that is going on, i.e., the perimeters of the machine for maneuvering clearance and avoiding irregularities that could cause the machine to be upset, the juxtaposition of other machine features that the operator may be maneuvering to lift or dump materials, and/or personnel who may be working within the area near the machine.

### 2. Machine Upset

A serious problem in heavy construction machine operation is the potential for upsetting the machine as a result of either its movement across irregular terrain or inadvertent imbalancing due to improper loading or load mismanagement.

### 3. Inadvertent Contact with High-Voltage Wires, Obstructions, and/or Personnel

Incidents involving contact with high-voltage wires are reported frequently, indicating that too little thought is given to this hazard. Although operator carelessness may explain many of these incidents, it is apparent that operators are often overloaded in terms of what they are required to watch and do. Similarly, many accidents in which machines have been run into obstructions and/or personnel have been due to the fact that the operators could not see where they were going or were too busy with other coincidental operating tasks.

### 4. Injuries Associated with Mounting and Dismounting and/or Being Thrown from a Machine

Personnel often are required to work at locations high above the ground without suitable guards, ladders, platforms, or tethers. The machines typically used for these operations are designed without due regard for sharp protrusions, corners, or edges, which personnel bump into or fall against. Ladder and step rungs and/or platform surfaces may be slippery (especially when the worker's shoes are covered with mud or snow) and thus contribute to falls.

### 5. Loss of Control of Cables and Loads

Machines that utilize cable-lifting systems often are sensitive to high wind disturbance; the hoisted load may become unmanageable and thus be allowed to drop or swing into adjacent structures or the crane itself, or even the total system may become unbalanced and upset.

### 6. Inappropriate Movement of Machine Elements

Occasionally the machine operator inadvert-

ently moves the crane member, the hoisting cable, the scoop, or another device in the opposite direction from the one intended. Since even the most skilled operators do this from time to time, it is apparent that the basic control system is incompatible with the natural direction-of-motion expectancies of the operator.

### 7. Working Environment

Heavy construction operations are often subject to fairly stressful and inconveniencing environments, i.e., cold, heat, snow, rain, ice, high wind, dust, noise, and vibration. Too often, these conditions are accepted as hazards of the trade, and little concerted effort is made to design the system to minimize the effects of such environmental stress on operator performance.

### 8. Communications

The combination of distance between operating personnel and the typical noise generated by machinery creates a serious problem in terms of communications between operating personnel. In addition to warning systems, special voice communication systems should be considered an integral part of the total system package.

### 9. System Maintenance

Most heavy construction equipment must be serviced out of doors, in many instances at the actual construction site. Considerations for the ease and safety of field maintenance should be part of the early concept design formulation, as opposed to figuring out how to maintain the system after certain constraints have been built in. In addition to the important objective of accessibility, special consideration must be given to the typical problems of handling extremely large and heavy components. Much of this work must also be accomplished by personnel who are burdened by heavy winter garments.

### Occupational Safety and Health Administration Standards

Occupational Safety and Health Administration safety standards should be reviewed as a major consideration during initial industrial systems concept development. Some of the key standards for review include the following.

*Part 1910*

Subpart D: Walking—working surfaces

Subpart E: Means of egress

Subpart F: Powered platforms, personnel lifts, and vehicle-mounted work platforms

Subpart G: Occupational health and environmental control

Subpart H: Hazardous materials

Subpart I: Personal protective equipment

Subpart J: General environmental controls

Subpart K: Medical and first aid

Subpart M: Compressed gas and compressed air equipment

Subpart N: Materials handling and storage

Subpart O: Machinery and machine guarding

*Note:* For a more complete listing and description, refer to the *Federal Register* of October 18, 1972.

## GENERAL

### General Considerations for Agricultural Systems Concept Development[50]

In conceptualizing the typical farm facility, human factors considerations should be addressed for the following:

1. Main and secondary site access
2. Living center
3. Service yard
4. Public parking
5. Family garden
6. Poultry center
7. Dairy center
8. Grain and feed center
9. Other animal centers
10. Machinery center
11. Bulk feed and grain storage
12. Utility systems
13. Security and lighting

Since the typical farm is often isolated, it requires considerable self-sufficiency, and many of the human factors considerations become more critical than they would be if the system were more accessible, in terms of time and distance, to typical urban amenities such as supplies, fire and police protection, doctors, and hospitals.

The farm system should be approached in much the same manner as any other system in terms of laying out facilities for convenience, functional relatedness, work flow or sequence of operations, and, most important, ease of handling emergencies.

Topographic and soil characteristics, of course, influence the placement of crops; thus the location of the principal farm structures must represent a suitable compromise between access to the main highways and access to the principal crop fields.

Finally, natural environmental considerations necessarily influence not only the location but also the orientation of the principal farm structures. Principal environmental considerations include sun path and prevailing winds. The most suitable compromise between environmental conditions involves other natural features which are important to a farm system that is both effective and economical, i.e., natural foliage that may serve as a windbreak, natural drainage, and natural water resources (see the illustrations on page 245).

Fires are a particularly critical hazard for isolated farmers and their families because of the difficulty of controlling them once they have started. Appropriate spacing of farm structures and positioning with respect to the possible spread of fire by prevailing winds are both important considerations.

Buildings and roads should be designed for maximum utility in terms of the kinds of machinery that will traverse or utilize these facilities. Consider "drive-through" structures and turnarounds so that the machine operator is not required to do a lot of backing.

Build up roadways, service areas, and build-

[50]J. De Chiara and J. H. Callender, *Time-Saver Standards for Building Types*, McGraw-Hill Book Company, New York, 1973.

ings so that water drains away. Orient buildings and roadway relationships so that snow will not build up across roads or entrances. Provide gates and/or appropriate cattle guards so that the farmer can control the animals and yet traverse the farm without having to mount and dismount from the vehicle any more than is absolutely necessary. The entire farm should be fenced.

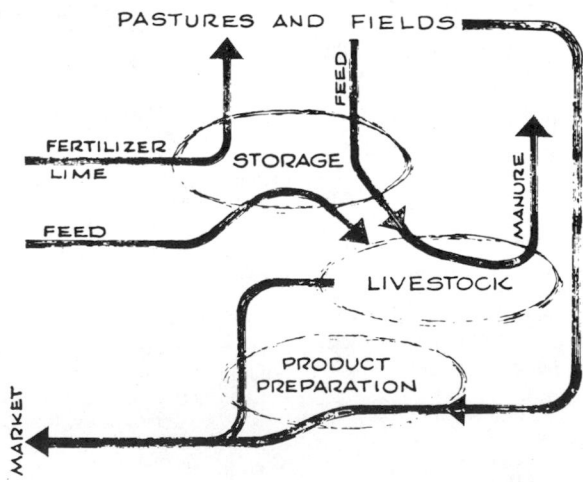

## FARM MACHINES

### Conceptual Development of Farm Machines

Perhaps the most significant human factors consideration in developing farm machinery is the fact that these systems will often be operated by persons with minimum skill and/or

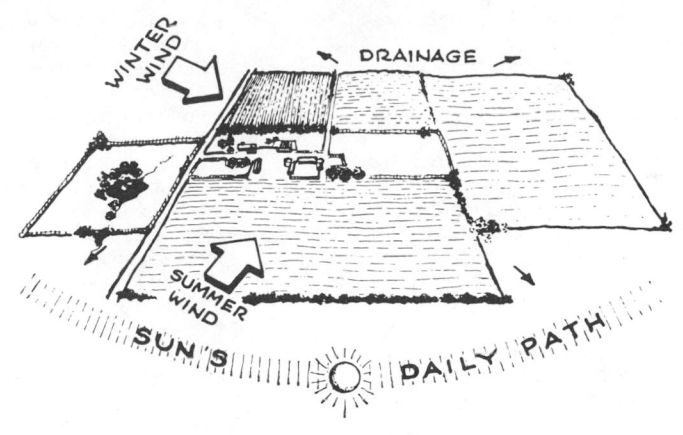

understanding of mechanical systems; i.e., not only will the farmer himself operate much of the machinery, but also his wife and children often will assist him both around the immediate farm and in the fields. This requires that farm machines be made simple to operate and as free of hazards as possible.

In addition, since the farmer will expect to perform the major share of the machinery servicing and maintenance, machines should be designed so that they can be maintained with a minimum of training and special skill and without the need for special tools and test equipment.

The following are some of the more important considerations for the design of farm machinery.

### 1. Mobile Machines (Tractors, Combines, Soil Conditioners, Planters, etc.)

a. They should be easy to hook together.
b. They should be easy to operate (start, steer, stop, back, adjust, monitor, etc.).
c. They should be easy to mount and dismount.
d. They should be difficult to operate improperly, turn over, or become dislodged from.
e. They should be easy to service and inspect, and it should be possible to remove and replace parts using only common tools.
f. They should have special features such as comfortable seats and closed, heated, cooled, and well-illuminated cabs.

### 2. Fixed Machines (Huskers, Conveyors, Pumps, Auxiliary Power Systems, Milkers, etc.)

a. They should be simple to operate.
b. They should be easy to load and unload.
c. They should be simple and safe to unjam.
d. They should be easy to service and maintain with common tools.
e. They should have appropriate safety features, guards, automatic power cutoffs, proper electrical grounding, etc.

## INDUSTRIAL AND AGRICULTURAL SYSTEMS STANDARDS

### Industrial and Agricultural System and Equipment Design Standards and Recommended Practices

The Society of Automotive Engineers has generated the following recommended standard practices relative to the design of heavy industrial and agricultural machine systems:

SAE J38 — Lift arm safety device for loaders
SAE J67 — Shovel dipper, clam bucket, and dragline bucket rating
SAE J94 — Combination taillamp and floodlamp for industrial equipment
SAE J95 — Headlamps for industrial equipment
SAE J96 — Flashing warning lamp for industrial equipment
SAE J98 — Safety for industrial wheeled equipment
SAE J99 — Lighting and marking of industrial equipment on highways
SAE J115 — Safety signs for construction and industrial equipment
SAE J137c — Lighting and marking of agricultural equipment on highways
SAE J153 — Safety considerations for the operator
SAE J154 — Operator enclosures (cabs)—human factors design considerations
SAE J167a — Overhead protection for agricultural tractors—test procedures and performance requirements
SAE J168a — Protective enclosures for agricultural tractors—test procedures and performance requirements
SAE J169 — Design guidelines for air-conditioning systems for construction and industrial equipment cabs
SAE J185 — Access systems for construction and industrial equipment
SAE J208c — Safety for agricultural equipment
SAE J209 — Instrument face design and location for construction and industrial equipment
SAE J220 — Crane boom stop

SAE J232 — Industrial rotary mowers
SAE J284a — Safety alert symbol for agricultural construction and industrial equipment
SAE J297 — Operator controls on industrial equipment
SAE J298 — Universal symbols for operator controls on industrial equipment
SAE J389a — Universal symbols for operator controls on agricultural equipment
SAE J742b — Front-end loader bucket rating
SAE J765a — Crane load stability test code
SAE J774c — Emergency warning device
SAE J833a — U.S. male and female physical dimensions for construction and industrial equipment design
SAE J841e — Operator controls on agricultural equipment
SAE J898a — Control locations for construction and industrial equipment design
SAE J899 — Operator's seat dimensions—construction and industrial equipment design
SAE J909b — Attachment of implements to agricultural wheeled tractors equipped with quick-attaching coupler for three-point free link hitch
SAE J919b — Operator sound-level measurement procedure for powered mobile construction machines—singular-type test
SAE J925 — Minimum access dimensions for construction and industrial machinery
SAE J943a — Slow-moving vehicle identification emblem
SAE J956 — Remote and automatic control systems for construction and industrial machinery

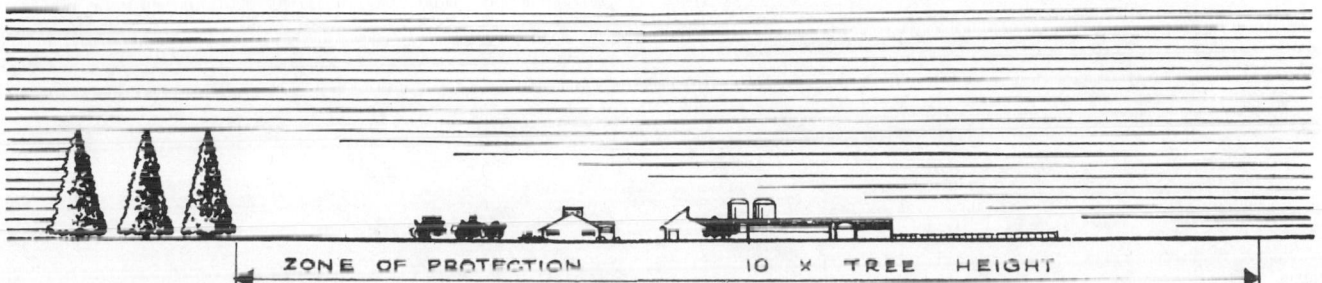

ZONE OF PROTECTION          10 x TREE HEIGHT

SAE J974 Flashing warning lamp for agricultural equipment

SAE J975 Headlamps for agricultural equipment

SAE J976 Combination taillamp and floodlamp for agricultural equipment

SAE J983 Crane and cable excavator basic operating control arrangement

SAE J994b Performance, test, and application criteria for electrically operated backup alarm devices

SAE J1001 Safety criteria for industrial flail mowers

SAE J1006 Performance test for air-conditioned agricultural equipment

SAE J1012 Agricultural equipment enclosure pressurization system test procedure

SAE J1013 Measurement of whole body vibration of the seated operator of agricultural equipment

SAE J1029 Lighting and marking of construction and industrial machinery

SAE J1040 Performance criteria for roll-over protective structures (ROPS) for construction, earth-moving, forestry, and mining machines

SAE J1051 Force-deflection measurements of seat and back cushions for agricultural, construction, and industrial equipment

SAE J1071 Operator controls for motor graders

SAE J1084 Operator protective structure performance criteria for certain forestry equipment

SAE J1105 Performance, test, and application criteria of electrically operated forward warning horn for mobile construction machinery

SAE J1129 Operator cab environment for heated, ventilated, and air-conditioned construction and industrial equipment

SAE J1163 Method for determining operator seat location on agricultural and construction machines

SAE J1166 Operator station sound-level measurement procedure for powered mobile earth-moving machinery—work cycle test

SAE J1194 Roll-over protective structures (ROPS) for wheeled agricultural tractors

# COMMUNICATIONS

## GENERAL

### Communications Systems

In conceptualizing a communications system, whether it is large and complex (such as a worldwide satellite system) or relatively simple and limited (such as an internal closed-circuit television system in a factory), the common denominator should always be the human need. That is, communications between people should be analyzed and planned on the basis of human task needs and thus should be designed to meet these needs in the most effective manner. This requires special consideration of human input-output characteristics, or, if you will, human channel capacities and sensory channel sensitivity. In developing a communications system concept, consider at least the following:

1. Communications mode (i.e., visual, auditory, tactile, and/or some combination)
2. Communicating environmental constraints (i.e., noise, environmental interference, and the need for security or privacy)
3. The time factor (i.e., how much informa-

tion must be transmitted within some prescribed time frame)
4. Reliability requirements (i.e., whether you can afford to repeat or lose some of the information without degrading system performance objectives)
5. Convenience (i.e., whether immediate accessibility is critical)
6. Multiuser requirements (i.e., whether the same communication is important to several users simultaneously)
7. Personal innuendo (i.e., the value of face-to-face expression)
8. International language (i.e., whether it is critical that people with differing language backgrounds understand the communication)
9. Freedom from interruption (i.e., whether it is critical that a given communication not be interrupted prior to its complete transmission)
10. System cost, both the initial cost and the cost of maintenance
11. The effectiveness of the human-machine interface design (i.e., ease of use of handsets, microphones, earphones, controls, etc), considering the environmental and other constraints involved in use of the interface (temperature, noise, vibration, clothing restrictions, etc.)

### General Considerations for Development of Communications System Concepts

The following discussion is limited primarily to visual and auditory communications between people within the working community, as opposed to communications for the purpose of recreation or entertainment. A primary objective should be to create a system that supports other human task activities, e.g., the exchange of information among individuals and the transmission of instructions, commands, warnings, orders, etc. Such systems should be based on the capabilities and limitations of the human as well as on the information transfer needs of the users. Consider the following:

1. Type of information (spoken messages, pictures, coded messages, digital data, etc.)
2. The time factor (i.e., how rapidly the information must be transmitted)
3. Security (i.e., whether the information is to be accessible to only one person or to anyone who might be within hearing or seeing distance of the receiving device)
4. Number of intended recipients (i.e., whether the message is for general broadcast for the benefit of a large

group of recipients)

5. The receiving environment (i.e., whether certain interferences at the receiving station may mask the incoming communication)
6. Reliability requirements (i.e., whether the situation permits repetition of the message if it is not understood the first time)
7. Quality (i.e., whether the information requires high fidelity in order to be fully understood)
8. Language (i.e., whether it is critical that people of differing language backgrounds understand the communication)
9. Personal factors (i.e., how much value there is to seeing facial expressions or hearing voice inflections)
10. Directionality (i.e., how important it is to associate the transmission with a direction or source)
11. Message capture (i.e., whether it is necessary to have a record of the transmission)
12. Condition of the addressee (i.e., whether he or she is busy, is using his or her hands or eyes, or is under special physical stress, such as high $g$ forces)

Communications hardware technology has reached the state at which the planner can prescribe almost any level of fidelity or reliability and almost any mode. Therefore, the choice is limited primarily by cost. However, since the primary purpose is to allow people to communicate with one another, systems concepts should generate from human needs and from basic human sensorimotor characteristics and limitations. In defining the characteristics of any proposed system, consider the following:

*For Auditory Communications*

1. Loudness (intensity)
2. Pitch (frequency)
3. Timbre (signal and voice profile)
4. Signal-to-noise ratio
5. Signal pattern (duration, intermittency, and frequency shift)
6. Sound location
7. Public as opposed to private

*For Visual Communications*

1. Brightness
2. Color
3. Contrast
4. Duration
5. Pattern and shape
6. Alphanumeric legibility

The following are some of the important human factors objectives for communications systems:

1. Detectability: The intended receiver can sense the signal.
2. Recognizability: The intended receiver can tell what the signal is.
3. Intelligibility: The intended receiver can tell what the signal means.
4. Conspicuousness: The signal is attention-getting.

Although communications systems designers obviously are concerned with the basic transmission medium (i.e., they must select radio, radar, sonar, or another basic scheme for the system), the human factors relate primarily to the immediate interface between the system and the sender, listener, or visual observer. Although tactile communication interfaces may have some application in isolated instances, the primary questions are whether the message originator should transmit by voice, encoded aural signal, or written or graphic means; whether the message will be received by aural or visual means; and whether there is increased value (e.g., reliability) if the message is transmitted using both visual and auditory modes.

**1. When to Use Auditory Communication**

a. When the message is simple
b. When the message is short
c. When the message does not necessarily have to be referred to later
d. When the message deals with events in time
e. When the message calls for immediate action
f. When the visual system is already overloaded
g. When the intended receiver cannot look at a visual display or may have to move about continually
h. When the receiving environment is incompatible with seeing, (e.g., when it is too bright or when it is necessary to avoid visual signals that could interfere with the intended receiver's adaptation to the dark)
i. When the intended receiver's visual capacity is impaired as a result of a dynamic stress (e.g., high $g$ forces)
j. When the intended receiver's eyes may be closed (e.g., when he or she is asleep)
k. When a visual signal strength is borderline and the addition of a coincident auditory signal would provide redundant, reinforcing confirmation

**2. When to Use Visual Communication**

a. When the message is very complex
b. When the message is long
c. When the message needs to be referred to later
d. When the information deals with location in space
e. When the message does not necessarily call for immediate action (e.g., when the receiver needs to finish some other task)
f. When the auditory system of the receiver is overburdened
g. When the receiving environment is too noisy to ensure reliable receipt of an aural message
h. When the receiver will remain in a position where he or she can continue to watch the visual display
i. When the information characteristics cannot be reliably described by words
j. When face-to-face communication may provide added understanding between communicators

**3. When to Use a Private Rather than a Public Interface**

An important decision in developing any communication system concerns whether, and under what circumstances, to use individual rather than group visual or auditory displays. Although there are no hard-and-fast rules, the following should be considered:

a. Use group display interfaces when:
   (1) Several people should receive the same information simultaneously.
   (2) Several people should all receive the information in the same format.
   (3) Some people require general information at the same time that they are busy sending or receiving individual communications.
   (4) The transmitter may not know who the persons are who should receive the information.
   (5) The individual channel that the transmitter wishes to use is temporarily unavailable.
   (6) The receiving individual is temporarily away from his or her individual communication interface.
   (7) Input to a general status display is being generated by several individuals on a continuing basis.
   (8) It is inconvenient for the required number of listeners or viewers to gather around an individual display device.
b. Use individual display interfaces when:
   (1) The transmission involves only two persons.
   (2) The information need not be shared with any other persons.
   (3) The transmission is or should be private.
   (4) The information is being modified by the individual for his or her own benefit.
   (5) The individual needs to record the information for future reference.

**4. When to Use Handsets, Intercoms, or Public Address Systems**

Although there are obvious overlaps in the manner in which handsets, intercoms, and public address systems can or should be used, the following are guidelines for making this decision:

a. Handsets should be used primarily when privacy is desired.
b. Intercoms should be used primarily when the individual is too busy to look up a number and dial it.
c. Public address systems should be used primarily for making public announcements.

**5. Other Communicating Techniques and Rationales for Their Use**

a. Colored and/or flashing light signals, horns, bells, claxons, sirens, etc., are particularly useful for issuing warnings and conditional communication because they can be made conspicuous without materially interfering with other communications.
b. Digitized communications can be used to reduce the amount of paper transfer.
c. Encoded signals can be used to maintain communication security.
d. Flag, light, and hand signals can be used to minimize the cost of more expensive systems.

## ALARMS

### Guidelines for Selecting Auditory Alarm Systems[51]

### Types of Alarms, Their Characteristics and Special Features

| Alarm | Intensity | Frequency | Attention-getting ability | Noise-penetration ability | Special features |
|---|---|---|---|---|---|
| Diaphone (foghorn) | Very high | Very low | Good | Poor in low-frequency noise Good in high-frequency noise | |
| Horn | High | Low to high | Good | Good | Can be designed to beam sound directionally Can be rotated to get wide coverage |
| Whistle | High | Low to high | Good if intermittent | Good if frequency is properly chosen | Can be made directional by reflectors |
| Siren | High | Low to high | Very good if pitch rises and falls | Very good with rising and falling frequency | Can be coupled to horn for directional transmission |
| Bell | Medium | Medium to high | Good | Good in low-frequency noise | Can be provided with manual shutoff to insure alarm until action is taken |
| Buzzer | Low to medium | Low to medium | Good | Fair if spectrum is suited to background noise | Can be provided with manual shutoff to insure alarm until action is taken |
| Chimes and gong | Low to medium | Low to medium | Fair | Fair if spectrum is suited to background noise | |
| Oscillator | Low to high | Medium to high | Good if intermittent | Good if frequency is properly chosen | Can be presented over intercom system |

### Summary of Design Recommendations for Auditory Alarm and Warning Devices

| Conditions | Design recommendations |
|---|---|
| 1. If distance to listener is great | 1. Use high intensities and avoid high frequencies |
| 2. If sound must bend around obstacles and pass through partitions | 2. Use low frequencies |
| 3. If background noise is present | 3. Select alarm frequency in region where noise masking is minimal |
| 4. To demand attention | 4. Modulate signal to give intermittent "beeps" or modulate frequency to make pitch rise and fall at rate of about 1–3 cps |
| 5. To acknowledge warning | 5. Provide signal with manual shutoff so that it sounds continuously until action is taken |

[51]C. T. Morgan, J. S. Cook, A. Chapanis, and M. W. Lund (eds.), *Human Engineering Guide to Equipment Design*, McGraw-Hill Book Company (*1st Ed*), New York, 1964.

## ARCHITECTURAL FACTORS

### Architectural Planning for Optimum Communication

Communications systems requirements should be considered at the beginning of architectural concept development because the effectiveness of the final communication depends on the inherent characteristics of the architectural configuration; that is, avoid the tendency to try to add a communications system to an architectural plan that is frozen. Few, if any, communications systems are completely satisfactory unless the architectural and communications systems are optimized as an integrated concept.

Key factors in planning the architectural and communications systems include the following:

1. Plan specifically for the total array of expected communications, i.e., direct person-to-person speech (without benefit of intervening aids such as the telephone), private telephone conversations, public address systems, emergency alarms, special intercom connections, mobile communications (e.g., by means of walkie-talkies), closed-circuit television, signal lights, special visual displays (e.g., movies and other audio-visual aids), and directional and instructional signs.
2. Establish requirements concerning privacy and prevention of annoyance.

3. Determine both inside and outside communication interference possibilities and identify the exact nature of these in terms of loudness, frequency distribution, etc.
4. Examine and define the potential physical interactions between each type of communicating mode and the proposed architectural features that may produce undesirable effects on the communication link (reverberation, echos, sound time lag, sound shadow, visual barriers, effect on sound localization, visual glare, seeing distance, illumination level, etc.)
5. Organize the facility in terms of ease of developing a logical and meaningful signing system.

Key architectural factors that should be considered relative to auditory communications systems include the following:

1. The shape and size of interior spaces and other occupied areas (e.g., hallways and stairwells)
2. The surface texture and shape of floors, walls, and ceiling and of furnishings that are expected to be relatively permanent
3. The physical positions of speakers (e.g., people) and listeners
4. The position of communications system hardware elements (e.g., loudspeakers, intercoms, and microphones)
5. The number of people who may occupy a space and thus will tend to absorb sound

Key architectural factors that should be considered relative to visual communications systems include the following:

1. Viewing distances
2. Ambient lighting effects
3. Artificial lighting effects (e.g., decorative lighting)
4. Day versus night ambient and artificial lighting effects and their influence on internal seeing conditions
5. Individual display and light control needs relative to interference among different facility users

Maintenance of both the communications system and the related architectural features of the planned facility should also be considered early and simultaneously. Like plumbing, many communications systems become inaccessible for maintenance because of the tendency to hide wiring and hardware. Although the aesthetic objective is laudable, all visual and audio communications systems hardware that can be expected to need service or replacement should be made accessible, especially so that wear and tear on architectural features (e.g., ceiling and wall panels) is minimized.

Finally, if a proposed communications system (for a public auditorium, church, campus, factory, etc.) cannot be fully implemented at the beginning or if there is any possibility of future system expansion, plan the architectural facility for ease of adding new wiring, controls, connections, etc.

## CONSUMER PRODUCTS

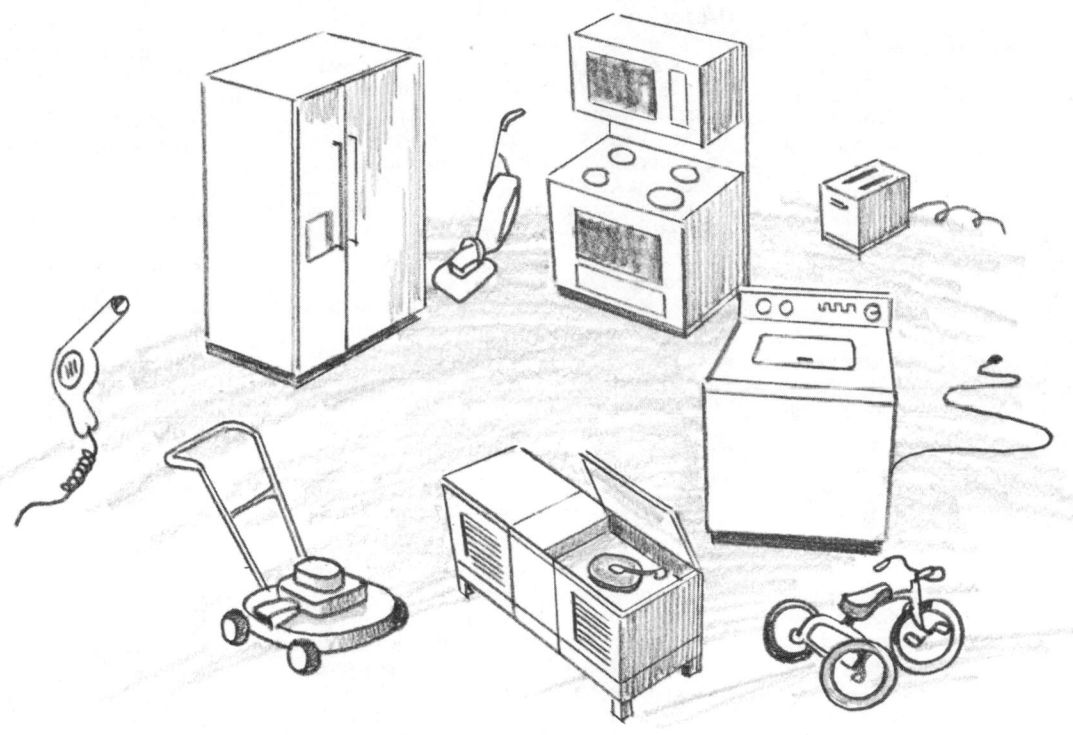

### GENERAL

#### Human Factors in Consumer Product Conceptualization and Design

There are at least three distinct, but not necessarily mutually exclusive, areas in which human factors should be considered relative to the conceptualization and design of consumer products such as appliances, entertainment and recreational equipment, tools, and toys. These are discussed briefly below.

#### 1. Safety
Products should not be designed so that they could fail and cause injury to the user or so that they could easily be misused in a manner that could lead to injury to the user, nor should they be made out of materials that might be toxic, poisonous, or otherwise hazardous to the health of the user. Lack of a proper safety analysis during the conceptual design stage typically results in the following:

a. Products in which toxic or flammable materials are used
b. Products which are made of breakable materials or which have sharp protrusions, corners, or edges
c. Products in which exposed moving parts cause pinches, cuts, or amputations
d. Products that can be used improperly to

strike another person or to shatter glass
e. Products that can be swallowed
f. Products that can cause electric shock because of improper insulation or grounding
g. Products that people can easily fall from
h. Products that can produce injurious noise, extreme heat, or flying particles that could puncture a person's skin or eye
i. Products that could cause rupture or strain when lifted or injury to a foot if dropped
j. Products that can cause burns

#### 2. Operability and Maintainability
In order for a product to be truly acceptable, it has to be easy to operate and maintain. Most consumers fail to appreciate this aspect of a product fully until after they have purchased it. There is considerable evidence that the typical consumer will not purchase another product of the same make once he or she has discovered that it is difficult to operate or maintain. Typical features that show up in many consumer product designs are the following:

a. The product is hard to pick up, hold onto, manipulate, or maneuver.
b. The product is hard to open and close, and the latches and handles are difficult to manipulate.
c. The controls and displays are poorly la-

beled, hard to read, difficult to understand, and awkward to adjust or operate.
d. The product is difficult to load and unload and requires the user to stretch, reach, bend, etc.
e. The product requires too much force or too much precision to operate properly, resulting in frequent errors in use.
f. The product is packaged in such a way that it is difficult to get to parts that need adjustment or replacement.
g. A presumably movable product is not provided with appropriate handholds or casters so that it is easy to move to a new location.
h. The product contains hard-to-find parts, with the result that acquiring replacement parts later is impossible or time-consuming.
i. It is practically impossible to keep the product properly adjusted long enough for it to do its job.
j. The product wears the operator out because of its weight, the awkward way it has to be held, the noise it makes, or the way it vibrates.

#### 3. Attractiveness
Although most manufacturers and designers are fully aware of the importance of product attractiveness to salability, and thus pay more attention to this aspect of product design than

to any other, attractiveness is an elusive variable when one tries to define it. A critical comment is in order also: Although attractiveness is an admirable and desirable trait in product conceptualization, it is extremely important that, in an effort to make something attractive, the designer be careful not to compromise product safety, operability, or ease of maintenance.

Since attractiveness is truly a human factors consideration, it is important to recognize that it may be a transient phenomenon; i.e., what one considers attractive this year is not necessarily considered attractive several years from now. This point is made to refute the argument that the aesthetic function is the most important consideration in consumer product conceptualization; i.e., a product that works well can continue to do so over many years and through many different packaging treatments, but only if the initial conceptual objective was to make the product work well. It should also be emphasized that a product can be made attractive after the objectives of operability and maintainability have been developed. Some of the hazards of starting with the aesthetic objective are the following:

a. Aesthetic features may interfere with product recognition. For example, there is often a tendency to camouflage a product, i.e., to make it look like something that it is not.
b. Aesthetic features may make visually displayed information impossible to read, as when white labels are printed on bright gold metallic backgrounds.
c. Exotic-looking control knobs may interfere with the user's ability to adjust the control easily and precisely, or the fancy surface treatment or shape may actually hurt the user's hand or fingers.
d. A streamlined cover or case may make it difficult to pick up the device or may cause a person to drop the object because there is no way to get a good grip on it.
e. An attempt to completely hide all visible means of fastening a cover on a product may cause the owner to damage the cover while trying to get it off for maintenance or replacement of a part.
f. Ornaments placed on a product to make it attractive often turn out to be hazards, or at least annoyances, because the user is constantly catching his or her clothing or hands on the ornamentation.
g. Color used for aesthetic reasons often becomes "visual noise" around an important warning display, making the display insufficiently conspicuous to attract the user's attention.
h. Fancy logos are often much more conspicuous, readable, and intelligible than more important user instructions. In some cases, important labeling is actually left off because the designer felt the label would disturb the overall aesthetic appearance that he or she was trying to create.
i. Furniture and related products are often less than useful, operable, comfortable, or movable because of the interferences created by aesthetic objectives. For example, deep, puffy cushions do not provide needed posture control, and they

make it impossible for the short-legged user to sit comfortably.
j. Products in which sharp edges and corners are typical styling features often injure the user.

## Misuse Behavior as a Key Safety Consideration

Liability should be an important consideration in conceptualizing any consumer product; i.e., the possibility of consumer injury and a potential lawsuit should stimulate any planner to investigate thoroughly the possible failure modes of a proposed product design, including especially the misuse behaviors of expected consumers.

Avoid the excuse that "I can't worry about every moron who may buy or use my product." These words have come home to haunt many manufacturers and, more recently, many designers. Worrying about potential misuse of a product can become an advantage in that it generates product features that will make the product more salable, e.g., easier to operate without making mistakes.

The following are some of the typical misuse behaviors that consumer product planners should consider.

### 1. Children
a. Children are extremely curious. As a result, they investigate and examine or try to play with products or product elements which they should not touch. If there is a potential hazard, design the product so that children cannot reach or operate the device.
b. Small children investigate by touching things, putting their fingers or hands into openings, putting things into their mouths, or at least trying to put their tongues on things. Make sure that objects that are small enough to swallow cannot be removed, that holes are either too small to allow passage of fingers or arms or large enough so that fingers and arms can be pulled back out, and that surface finishes are not toxic and cannot flake off.
c. Children like to climb or reach for something on top of a surface which they cannot see. Package products so that there are no climbing aids. Design hazardous products, like stoves, so that children cannot reach the controls or burners.
d. In spite of a designer's best intentions, most children will use a toy in an entirely different manner from the way it was intended to be used; e.g., they will use it to pound with, stick it in their own or another child's eyes, or try to pull it apart to see what makes it work. Make sure that toys are sturdy enough so that they will not break or come apart; soft enough so that they will not break something else or injure another child; devoid of sharp edges, corners, and projections; and made of materials that are not toxic.
e. Children are unsteady on their feet, and they often run so fast that they are unable to maintain their balance or control their bodies. As a result, they often fall or run into objects. Products around which children may play should be designed with the idea that a child might fall or run

into them; i.e., there should be no sharp edges or corners, and they should be made of unbreakable materials.

### 2. Adults
a. Adults often attempt to operate a product without looking at the instructions first. As a result, they may do something to create a hazard for themselves and/or for others around them. Products should be designed as nearly as possible to be understandable without requiring the user to read special instructions; i.e., the critical instructions and labels should tell how to start, adjust, and turn off the product.
b. Adults are often absentminded or easily distracted, and they often forget to be careful. Products that have a hazard potential should have safety lock-out features and/or added protective features that will minimize the hazard in the event the user forgets and turns a wrong knob, for example.
c. Adults may be in a hurry, may bump into things in their haste, or may not look to see where they are putting their hands or feet. Products should be designed so that even if the user does tend to hurry, injury will not result.
d. Adults tend to believe that they are smart enough to take shortcuts (even though there may be a warning to the contrary). Often these shortcuts end in tragedy. Products that have such hazard potential should be designed so that shortcuts are out of the question.
e. Many adults are willing to take risks, even though they know there is a chance of making a mistake or a miscalculation, either because they think they can get away with it or because they truly believe their time is more important. Products should be designed so that they cannot be operated in any way other than the right way; i.e., there should be no opportunity for risk taking.

### 3. The Handicapped
Many handicapped consumers are forced to try to operate products that were not designed with their infirmities in mind. As a result, they are subject to hazards that other adults are not exposed to. Since the number of handicapped persons is increasing and since they are becoming a significant portion of the consumer product market, products should be designed so that people with handicaps can use them efficiently.

## Design Conceptualization for an International Consumer Market

Designing a product that may be marketed around the world requires knowing something about the differences between consumers that are critical to the product's safe and efficient use and to its acceptability. The following are some of the human factors that are important to an understanding of a product's operation and acceptability in different cultures.

### 1. Anthropometric Differences
People of various nationalities differ in terms of their body dimensions. Some are taller, heavier, and stronger than others. Some have

larger hands, longer reaches, and more strength than others. Products designed for American consumers might be too large or too high or have reach requirements that are too great for Oriental consumers. By the same token, certain products designed for the smaller Oriental might not provide sufficient clearance for the average American.

### 2. Technological Experience

An ethnic population that is not used to many of the modern products that a more technically oriented population is used to will neither appreciate the technical complexities that may be built into a product nor understand how to operate or maintain the product. Thus the complexity of a product should be geared to the expected user population. Even though a product may not seem as sophisticated and efficient as it could be, its final effectiveness and acceptance might be much higher if it were less complex.

### 3. Language Barriers

A major problem for designers of equipment that may be used in many parts of the world is the difficulty of labeling a product and/or creating appropriate instructions for each potential user. Translating instruction manuals is fairly simple; however, labeling to provide universal understanding presents a more difficult problem. As will be noted elsewhere in this handbook, using international symbols in labeling is one way to ensure that people who speak different languages can still understand and operate a product or device.

### 4. Ethnic Differences in Terms of Styling

Although of less importance than in past years, various styling features have considerable significance to members of some religious or ethnic groups. It should be noted that much of this can be treated in a cosmetic manner; i.e., after a basic product package has been developed for the total market, cosmetic features appropriate to particular user markets can be added to increase user acceptance.

## PACKAGING

### Specific Human Factors Considerations in Consumer Product Packaging Conceptualization

1. The package should be easily movable and transportable.
2. The package should be designed so that it will "stand on its own"; i.e., avoid a design that has to be propped up or held in place.
3. If the device has to be assembled before it can be used, design the components so that the parts are easily identifiable, so that it is obvious how they fit together, and so that they will go together without forcing and without using special tools.
4. If the device has to be disassembled between uses, design the parts so that special tools are not required to take it apart.
5. Use captive fasteners so that they will not be lost.
6. If assembly or disassembly instructions or other warnings or instructions are

critical to the preparation, operation, or maintenance of the product, place these instructions or warnings on the device at points where the user normally would see them as he or she begins assembly or operation.
7. The package should not have any spring-loaded devices that could pop out when the product is disassembled.
8. The package should not have any parts which are sharp or pointed or which are so flimsy that they are easily bent or broken. If sharp features are necessary to operation of the product, provide protective covers so that the user is not accidentally stabbed or cut.
9. The product should be packaged in such a way that it is easy to clean; i.e., avoid narrow slots or crevices in which dirt could become lodged.
10. The package should have smooth surfaces that will not snag hands, clothing, cleaning cloths, etc. Use exterior materials and/or finishes that are minimally subject to wear and damage.
11. The package should be shaped and/or have appropriate handholds so that the user can manipulate or hold the package in place. Provide adequate finger clearance so that the package can be set down without pinching the user's fingers.
12. If necessary, the package should be designed so that it can be stacked and/or easily placed in a carrying case or shipping carton.
13. The package should be designed so that internal parts cannot be shaken loose while the product is being moved.
14. The package should be designed so that liquids, powders, or similar materials are not easily spilled when the product is moved.

### Human Factors Considerations in Packaging for Attractive Appearance

There is an almost infinite variety of possibilities with regard to making products attractive, which is an important human factor in encouraging the consumer to purchase a particular product. However, it is extremely important that the appearance factor not interfere with the operability, maintainability, or safety of the product.

The principal areas to consider when adding to the attractiveness of a product are the following:

1. Color
2. Pattern
3. Surface texture
4. Shape features

The following guidelines concern pitfalls that should be avoided when adding to the attractiveness of a product:

1. The product should look like what it is. Avoid the temptation to create a concept that camouflages the real identity of the product. Although sometimes there is an aesthetic value to making a product blend with a general decorative scheme, this usually can be done without destroying key product identifying features.

2. The general appearance of the product package should provide some clues as to what the observer is supposed to do with it. Once again, avoid the temptation to camouflage the features that are obvious indicators of how the product is to be approached and used. Typical errors include a camouflaged latch, handle, control device, door, or drawer; the purposeful omission of a label; the minimization of visual contrast between labels and backgrounds; and the stylization of a label to the point where the observer cannot tell whether it is a label or merely a decorative feature.
3. The product's appearance should imply simplicity, not complexity. The typical consumer has come to expect apparent organization among the various exterior product features, the absence of extraneous features that he or she obviously will not use, and a general product appearance that is uncluttered by bizarre patterns that tend to hide the key user interface elements.
4. The product should imply efficiency. The first impression should dispel any feeling that the product is crude, cumbersome, out of balance, etc. For example, the typical consumer has been conditioned to associate solid structure with sturdiness and strength, symmetry with balance, and smooth lines with efficiency. Although an asymmetrical package is quite acceptable as long as the observer recognizes that the asymmetry is purposeful, he or she will be disturbed if the lack of symmetry looks like an oversight. Do not confuse "streamlining" with efficiency (a typical error in the design of certain products such as chairs, tools, and hand-held appliances).
5. The product should appear coherent and complete to the observer. Avoid the use of features that create the impression that the product was thrown together, that parts were merely "tacked on," or that something might be missing. Avoid features that may be difficult to assemble correctly; otherwise, it may appear as though the various elements of the product do not fit properly. Many packages have failed to reach their appearance objective because there was too much opportunity for the factory assembler to attach parts in such a way that they looked misaligned.
6. The product's proportions should relate to the human scale. The product should not overwhelm the user psychologically, and the parts should not be so large or so small that the user has difficulty getting hold of or manipulating them.
7. The shape of the direct interface devices (handles, knobs, controls, etc.) should be compatible with what the user has to do with them. Avoid the temptation to overdecorate such interfacing devices so that they are hard to get hold of or so that they may hurt the hand when they are gripped.
8. The visual contrast required for specific visual display visibility should not be

compromised by colors or patterns used to decorate the exterior of the package.

9. Appearance constancy should be addressed in terms of making sure that the product is recognizable and attractive from all angles from which it will be viewed. Avoid the tendency to concentrate on only one side, which gives the impression that the designer did not care about the other sides.

10. Avoid package features that create odd illusions, e.g., a pattern of lines or other geometric features that look like a grinning face or tilted stripes that make the package look as if it is leaning.

11. Choose colors which provide "life" and visual interest but which are not so intense that they are overwhelming or conflict with the color environment within which they will appear. Do not use colors for aesthetic purposes that will conflict with specific operational colors; i.e., if color codes are used, make sure that these codes are not rendered ineffective by the colors used as a background.

It is easy to create undesirable illusions when implementing certain appearance objectives. Beware of the following:

1. A package that appears larger at the top than at the bottom gives the impression of instability or top-heaviness.
2. A package that appears too tall and slender gives the impression that it may fall over.
3. Evenly spaced concentric circles or spirals appear to alternately advance and recede, thus creating visual pulsation.
4. An extended straight line or edge appears to sag in the middle.
5. Crossing lines appear to bend near the point of crossing.

On the other hand, desirable illusions can be created, such as the following:

1. A positively sloping line can give the illusion of forward movement, and a negatively sloping line can give the illusion of slowing down; e.g., vertical patterns that slope toward the direction in which a vehicle would ordinarily move can make the vehicle look as though it is moving forward, even though it is standing still.
2. A long horizontal line along the side of a short automobile will make the car look longer.
3. A very tall, narrow package or wall panel will be less disproportionate if a horizontal line crosses the expanse about midway.
4. A border around the edge of a package form provides the observer with a firmer basis for judging the package shape and size.
5. The overall size of a package is more pleasing to the observer if it is compatible with the sizes of other packages with which it will be viewed. This suggests that, in some cases, a product should be packaged in a larger container than is actually required to house the internal components.
6. Consumers frequently associate bulk

(e.g., an overstuffed chair) with comfort. Even though the bulk may not be necessary, the illusion it creates might be necessary in order to make the user feel that the product possesses what he or she considers to be desirable features.

## PRODUCT SAFETY

### Safety Concepts by Product Category

Although the following safety considerations by no means represent all the potential safety problems to be addressed for each of the product categories, they provide a point of departure for planning product design concepts.

**1. Electrical Appliances**
   a. Electrical grounding
   b. Overload fuses
   c. Nonmetallic user interface hardware
   d. Automatic power cutoff with cover removal
   e. Adequate insulation
   f. Warning and caution labels and instructions for operation and maintenance

**2. High-Temperature-Producing Products**
   a. No direct contact between the internal heat source and the case
   b. Insulation to attenuate radiant heat transfer to the exterior
   c. Low-heat-transfer hardware to user interfaces (e.g., handles and controls)
   d. Nonflammable materials and finishes

**3. Mechanical Products with Moving Parts**
   a. Covers over moving gears, chains, belts, and activating levers
   b. Captive spring-loaded parts
   c. Fracture-, chip-, and distortion-resistant metal, composition, or plastic parts

**4. Products with Explosive Components**
   a. Nontoxic substances
   b. Minimal energy release commensurate with purpose
   c. Structural integrity of the container
   d. A replacement package and a method of replacement that preclude mishandling and other forms of misuse

**5. Toys for Small Children**
   a. Nonflammable materials
   b. Nontoxic materials
   c. Unbreakable materials
   d. Sufficient size to preclude swallowing the toy or putting it far enough into the mouth to strangle and no small parts that can be removed and put into the mouth
   e. No sharp corners, edges, or protrusions
   f. No cords or wires that could be swallowed or wrapped around the user or a playmate
   g. No metal parts that could be inserted into an electric outlet
   h. No articulating element in which the child could be pinched

**6. Tools and Cutlery**
   a. Appropriate electrical and mechanical component grounding and guards
   b. Guards to prevent contact by flying particles or sparks

   c. Safety catches and covers
   d. Heat insulation around heat-producing parts (nonflammable)
   e. No cords, if possible; otherwise, quick release connectors
   f. Appropriate warning labels and instructions on the product
   g. Safety features associated with replacement of parts and cleaning or servicing

**7. Furniture**
   a. Structural integrity for expected loads
   b. Antitip construction
   c. Nonflammable and nontoxic materials
   d. No sharp corners or edges
   e. Lockable casters (when used)
   f. A foldable design that will not cut or pinch the user or collapse accidentally during use and a design that lets the user know when the product is fully locked in place

**8. Other Products**
   a. Lockable medicine cabinet doors
   b. Refrigerator doors that can be opened from inside in the event a child gets shut in accidentally
   c. Stove burners and controls arranged so that a child cannot get to them
   d. Space heaters designed so that children cannot be burned by hot exteriors
   e. Ladders which will not collapse and which the user can be sure are locked in place
   f. Stair rails designed so that small children cannot get their heads caught between the rail supports or fall between the supports
   g. Baby furnishings designed to prevent the child from falling between the mattress and the rails, from lowering a rail or getting caught between the rail supports, and from being cut or pinched by rail-lowering hardware; nontoxic materials and finishes on parts of the product that a child might touch with his or her mouth

In conceptualizing any consumer product, consider "Murphy's law": If anything can possibly go wrong, it will.

SELECTED READING LIST

DeGreene, Kenyon B.: *Systems Psychology,* McGraw-Hill Book Company, New York, 1970.

Gagne, Robert M. (ed.): *Psychological Principles in System Development,* Holt, Rinehart and Winston, Inc., New York, 1962.

Hammer, Willie: *Handbook of System and Product Safety,* Prentice-Hall, Inc., Englewood Cliffs, N.J., 1972.

Harrigan, John E., and Janet R. Harrigan: *Human Factors Program for Architects, Interior Designers and Clients,* Blake Printing and Publishing, 1976.

Meister, David: *Behavioral Foundations of System Development,* John Wiley & Sons, Inc., New York, 1976.

Meister, David, and Gerald F. Rabideau: *Human Factors Evaluation in System Development,* John Wiley & Sons, Inc., New York, 1965.

# Subsystems Design

## INTRODUCTION

Although many of the subsystems discussed in this chapter are sufficiently complex to be referred to as "systems," they are included here because they generally occur as part of a much larger system. When they are addressed during the overall *system* conceptual planning phase, they often are merely referred to, and no specific recommendations are made concerning the way in which they should be designed.

Thus, a particular designer may be given the task of designing a subsystem without receiving much direction from the original overall system planners, which means that this designer is left with the job of analyzing the functional requirements and actually "conceiving" as well as designing the subsystem.

In this chapter, subsystem considerations are discussed in more detail than the systems conceptualization considerations in Chapter 1. It will also be noted that in many instances, considerable attention is directed to the typical problem of selecting from among alternative already designed and produced systems and components. Typically, architects, interior designers, and industrial designers should not create new designs, nor do they usually have the time to do so; rather, they must search out the system and component products that most effectively meet their immediate needs. Their job, then, is to select the right systems and components and put them together in the most user-effective manner.

## INITIAL PLANNING

The following planning steps should be taken before designing any subsystem:

1. Determine the constraints placed on the subsystem by the overall system concept. These include both operational and cost constraints.
2. Make a detailed operational and functional requirements analysis for the subsystem itself, even though a preliminary analysis has already been made and a general specification has already been prepared.
3. Determine which functions should be performed by some type of equipment and which should be performed manually by a human and establish specifications for both types of functions.
4. Identify the equipment, displays and controls, furnishings, and other support items that are generally available off the shelf; i.e., avoid reinventing the wheel.
5. Develop grouping and traffic analyses to describe how the elements in the subsystem will probably be used and prepare a rough scenario of the various operations and/or the tasks that will be performed.
6. Establish criteria for searching out and evaluating off-the-shelf elements for the subsystem and perform a trade-off analysis to determine the best selection.
7. Begin putting elements together in alternative combinations and arrangements, utilizing appropriate analytic methods, such as time-line analysis, link analysis, and mockup evaluations.
8. Prepare a detailed specification for the subsystem, including performance requirements for each element in the system. Include salient human engineering requirements as well as engineering requirements.

## DESIGN

The design process, whether new or off-the-shelf products or components are utilized, should always start with the user; i.e., start with the user at a task level and work outward, step by step. The following are typical steps:

1. Determine who the user or users will be: a child, a housewife, an office worker, an engineer, a technician, a client, an equipment operator, a deliveryman, a housekeeper, a maintenance worker, and so on. Then determine what functions the user is to perform, individually or in conjunction with others.
2. Determine how each user will perform the functions he or she is responsible for and what tools, equipment, component interfaces, and furnishing support each will probably have to use.
3. Define the immediate "use space" required by the users and their related devices and tools and then define the interrelationships among several use spaces so that necessary communications and traffic flow are optimized.
4. Maximize interface details both for each immediate subsystem user and for multiuser interactive features; i.e., design and/or select displays, controls, consoles, seats, illumination, etc., which are well engineered from the human factors point of view and which will help assure that each user has the maximum opportunity to perform whatever task he or she is assigned as efficiently as possible.
5. Maximize the environmental conditions surrounding user spaces so as to minimize the possibility that noise, vibration, extreme temperature, or poor ventilation may degrade the user's performance, cause annoyance to the user, or affect his or her health and safety.

Throughout each of these steps, continue to evaluate, test, and reiterate in order to develop a balanced and successful trade-off between human and hardware priorities. Above all, keep in mind that the ultimate success of the system depends in most cases on how well the user performs, which is a combination of the user's appreciation, skill, and physical and psychological interaction with the system hardware.

## TEST AND EVALUATION

It is particularly important in subsystem design to verify concept details at each step in the design process. Avoid the temptation to assume that because certain subsystems have traditionally been designed in a certain way, they have been successful in the past. Remain sensitive to the possibilities of improvement in user performance. In many cases this requires merely a more thorough search for improved products or components, i.e., ones that evidence good human engineering in their design.

One of the most important steps in evaluating a proposed subsystem concept is to construct mockups of alternative designs so that the concept can be evaluated from the point of view of both physical space and procedures. Avoid the mistake of believing that a three-dimensional task environment can be evaluated by looking at drawings alone. People work, live, and play in three-dimensional environments, and therefore it is vital that we confirm that a proposed space and human-equipment interface allows the user to perform his or her various tasks as efficiently as possible. Remember that a given design should adapt to the user, rather than the other way around. Although humans are obviously quite adaptable, they often are forced to pay a considerable penalty in terms of impaired performance and physical or psychological stress.

Finally, learn to evaluate the system from the standpoint of its performance rather than its appearance. Many seemingly attractive systems or products are nothing more than that. They are nice to look at, but you would not want to use them!

## GENERAL PRINCIPLES FOR ARCHITECTURAL ENTRYWAY PLANNING AND DESIGN

1. Provide two or more entries for all facilities that are to be occupied by people (preferably located at opposite ends of the facility) so that there will always be an alternative emergency escape route and so that the various user functions will not interfere with one another more than necessary.
2. Locate entryways so that internal and external functional relationships and associated activities are as convenient as possible; e.g., the travel distances should be as short as practicable between the family car or delivery vehicle and the kitchen, between the kitchen and the outdoor patio, between the swimming pool and a bathroom, and between the laundry room and the outdoor clothesline.
3. Do not hide or camouflage main entries; i.e., place them where they can be seen from the typical visitor's approach view. A hidden entry provides ideal camouflage for a burglar who is trying to force open the door.
4. Create an easily negotiated approach to the entry, especially when there may be a significant grade change; i.e., provide properly designed ramp, stairs, handrails, path surfacing, and illumination.
5. Create an appropriate docking interface for service entrances that must provide special vehicle interfacing. Keep in mind the special pedestrian relationships, especially where safety may be concerned.

6. Create an appropriate external threshold (a porch, platform, deck, balcony, etc.) that will safely accommodate expected pedestrians and provide sufficient clearance to manipulate packages, furniture, equipment, wheelchairs, etc.; to clear swinging doors; and to avoid conflicts between entering and exiting traffic.
7. Create a similarly appropriate internal threshold (e.g., a hall, lobby, or reception area), keeping in mind the potential conflicts created by putting on and removing outer clothing, clearing swinging doors, and manipulating furniture or equipment to other parts of a facility.
8. Provide proper illumination for both the external and the internal entry thresholds so that people can see where they are going, what they are doing, and with whom they may be conversing. If identification displays are required, make sure that these also are illuminated.
9. Provide an appropriate communications system, such as a buzzer, bell, intercom, or visual identification.
10. Provide protection from inclement weather.

## ENTRY AND EXIT LOCATION

1. The main entry should be convenient for visitors from their point of arrival.
2. There must be more than one exit in case of fire. Exits from opposite sides of the house are desirable in case one is blocked. Exit through a garage is unacceptable because of the high possibility that a fire may be associated with an automobile.
3. Other convenient entries should be located relative to external and internal positional relationships.

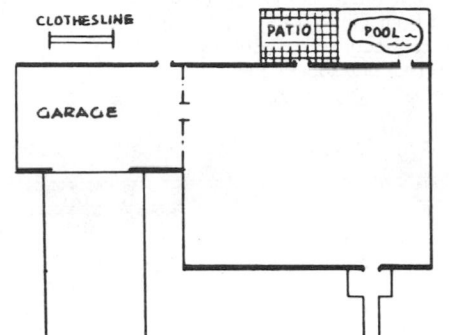

## MULTIFAMILY DWELLINGS

Tradition has allowed apartments, motels, and hotels to be designed with only one entrance and exit. From a safety standpoint this is unacceptable. Equally unacceptable is the assumption that occupants can escape through windows or from balconies. Children, the elderly, and handicapped persons are usually unable to use the latter methods of exiting without serious injury.

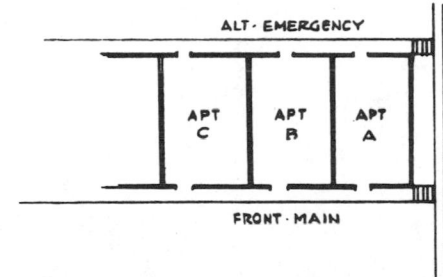

## COMMERCIAL AND INDUSTRIAL BUILDINGS[1]

1. Main entries should be readily obvious to visitors, clients, and other infrequent users, and such entries should be convenient to their point of arrival and/or planned parking area.
2. Separation of entries by user function is desirable to minimize traffic-flow conflicts and bottlenecks due to internal activities such as loading and unloading and to provide privacy and security.
3. Provide multiple entries for building occupants when there is a considerable influx or departure problem.

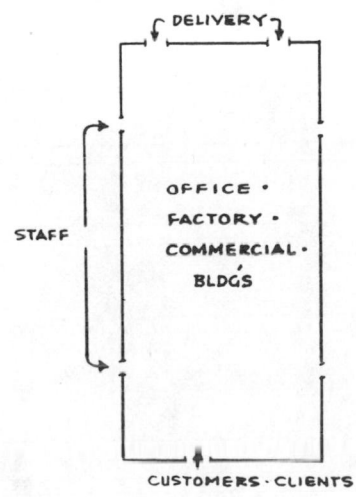

[1]Emergency exits, including alternative escape routes, should be accessible to blind persons and to persons in wheelchairs.

## HIDDEN OR CAMOUFLAGED ENTRYWAYS

Avoid the tendency to hide or camouflage main or primary entryways with aesthetic features. Although the designer and/or a customer may feel that such features are important in terms of appearance, they are confusing, time-consuming, and frustrating to the visitor.

The route or routes to the entryway should also be readily obvious and should be made as direct as possible.

When special entrances are provided (an emergency entrance for hospital patients, a delivery entrance for package delivery, etc.), design and mark these entryways (and the routes to and from them) so that people will not waste time or become confused as they try to find the correct entrance.

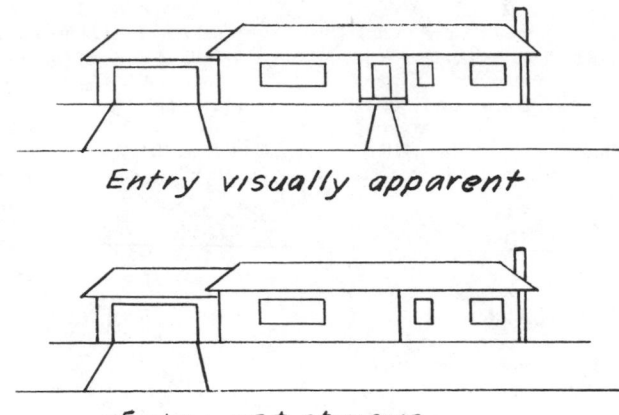

Entry visually apparent

Entry not obvious

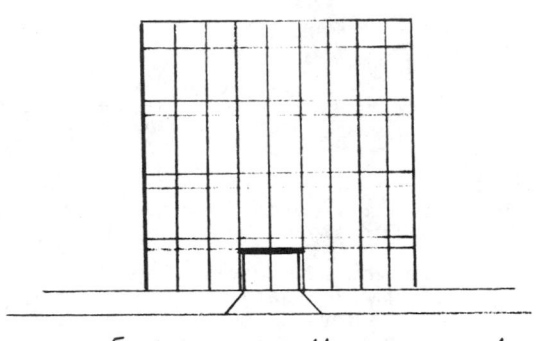

Entry visually apparent

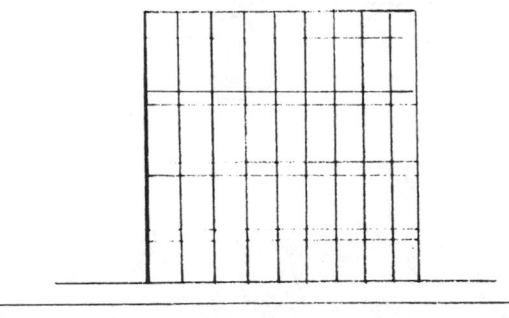

Entry not obvious

## MAJOR GRADE-APPROACH CONSIDERATIONS

Whenever feasible, the approach to all facilities should be as level as possible in order to ensure minimum inconvenience and maximum accessibility for the elderly and the handicapped. The accompanying illustrations emphasize some of the most common problems created by lack of attention to the approach to entryways:

1. Avoid long, unbroken stair patterns. When the natural grade precludes the use of ramps or other methods for providing level entry approaches, provide rest platforms. If the building is for public use, provide a series of ramps or elevators or escalators. The typical architectural pattern shown in the accompanying illustration may look interesting and impressive, but it is tiring, frustrating, and extremely hazardous to use.

*Stairway with "rest break"*

*Hazardous and fatigueing*

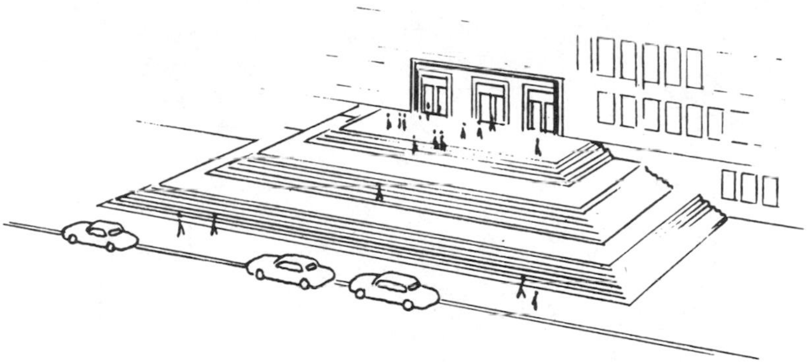

2. A long (and apparently unbroken) approach to an entry is undesirable. Not only does such an approach discourage visitors as they anticipate the long walk, but it also poses the problems of lengthy exposure to inclement weather conditions and of snow and ice on the walkway (not to mention the problem of clearing the pathway). More significant, however, is the single step shown in the accompanying illustration. Typically, the observer will perceive the path as continuously smooth and uninterrupted and may trip over the step.

*Avoid single step*

*Several steps more conspicuous*

PREFERRED METHODS FOR GRADUAL APPROACH GRADE

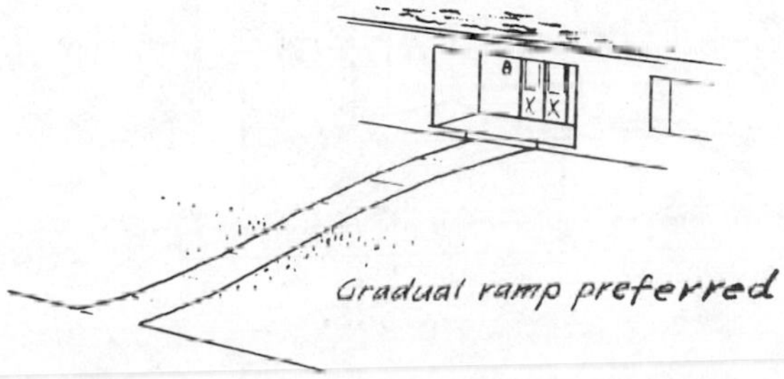

*Gradual ramp preferred*

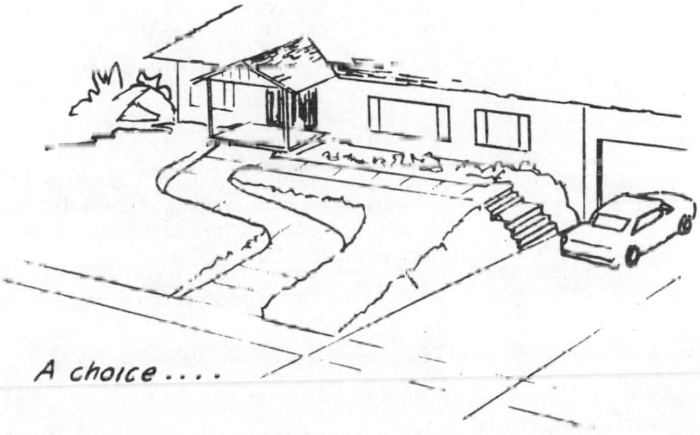

*A choice....*

## ENTRYWAY DESIGN TO PROVIDE PROTECTION FROM INCLEMENT WEATHER

Provide cover at all entryway thresholds, not only to reduce the discomfort of users caused by rain, snow, wind, etc., but also to minimize interference or hazards caused when snow, ice, or rain collects on the entry threshold.

Avoid styling features, such as the one shown in the accompanying sketch, which obviously neither provide protection nor prevent rain, snow, or ice from collecting on the threshold platform.

*No eaves — no protection*

The accompanying sketches illustrate various approaches to providing cover at the entrance:

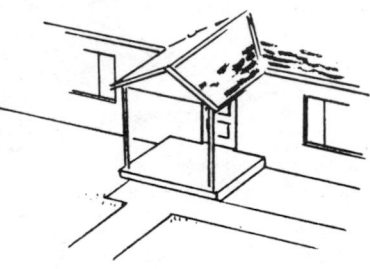

1. The inset type provides protection not only from rain but also from wind.
2. The covered entryway is suitable for locations where rain or wind is less frequent or severe.

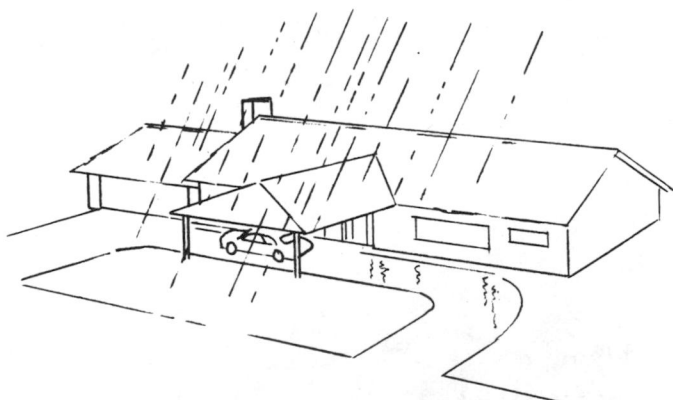

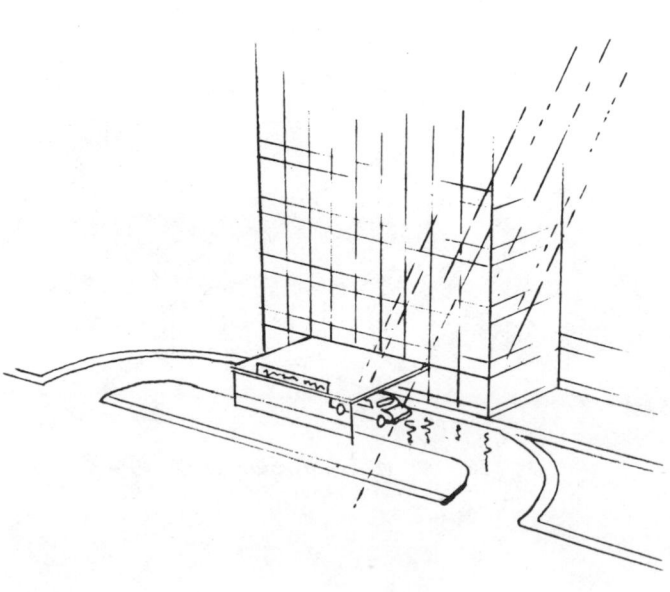

3. Complete protection from the vehicle to the residence or building is desirable in order to eliminate the need to run or use an umbrella.

## RAMP GUIDELINES

Practically all facilities may be used at one time or another by elderly or handicapped persons. Therefore, all facilities should be laid out in such a manner as to ease the problems of entry and exit for these users.

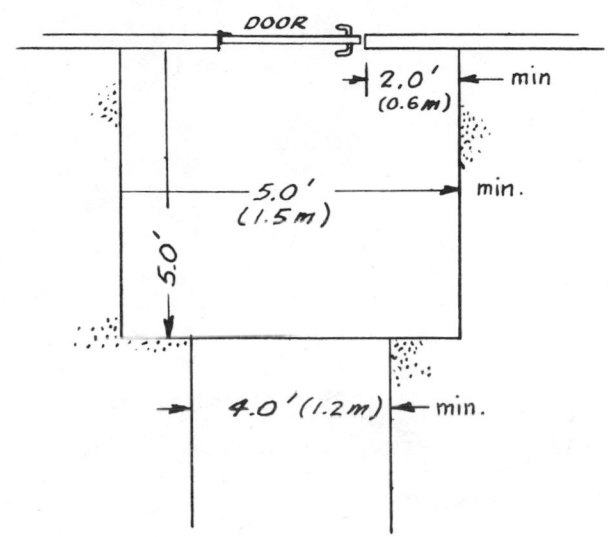

### 1. Large Threshold at Entrance
The dimensions shown in the accompanying sketch make it possible for a person in a wheelchair to maneuver the wheelchair in conjunction with opening the door. This is also the minimum space suitable for others to collect and wait for admission to the facility.

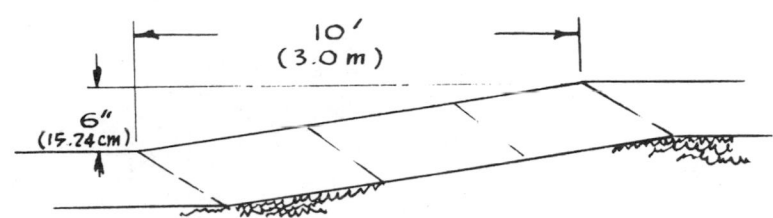

5 PERCENT GRADE MAX ACCEPTABLE

### 2. Ramp Grade
Steep ramps are hard for persons in wheelchairs to negotiate and also are difficult for the elderly to walk on without stubbing a toe.

### 3. Ramp Length
Avoid creating long, continuous ramps unless a resting platform is provided now and then.

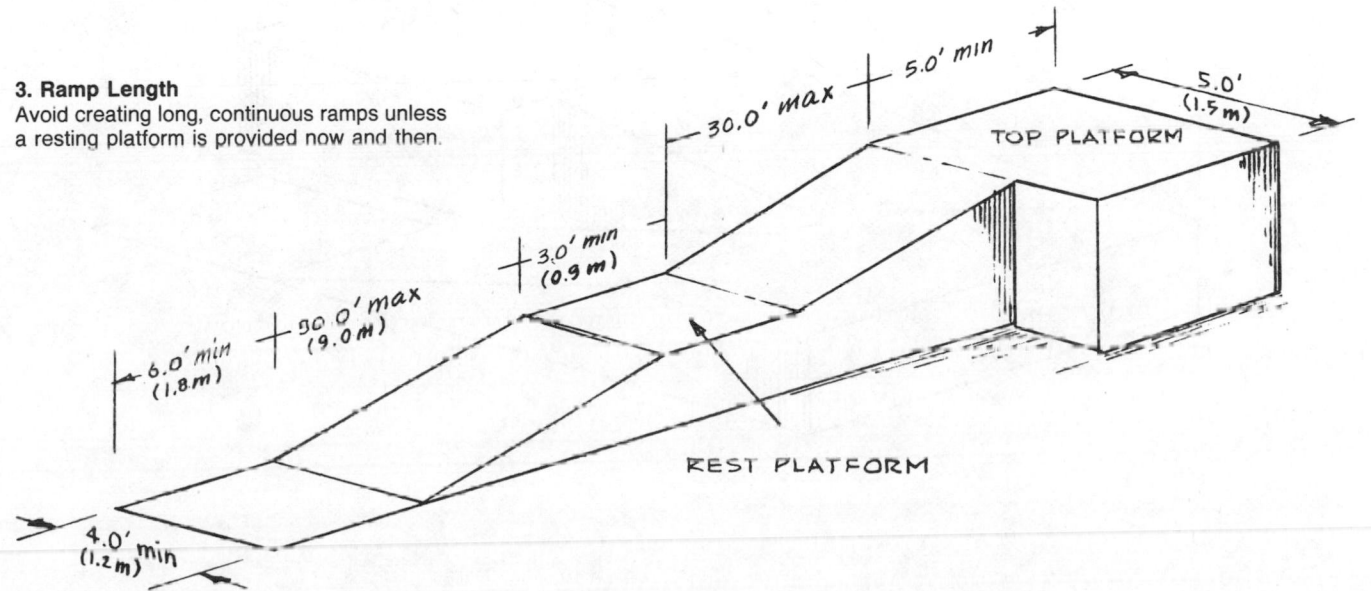

TOP PLATFORM

REST PLATFORM

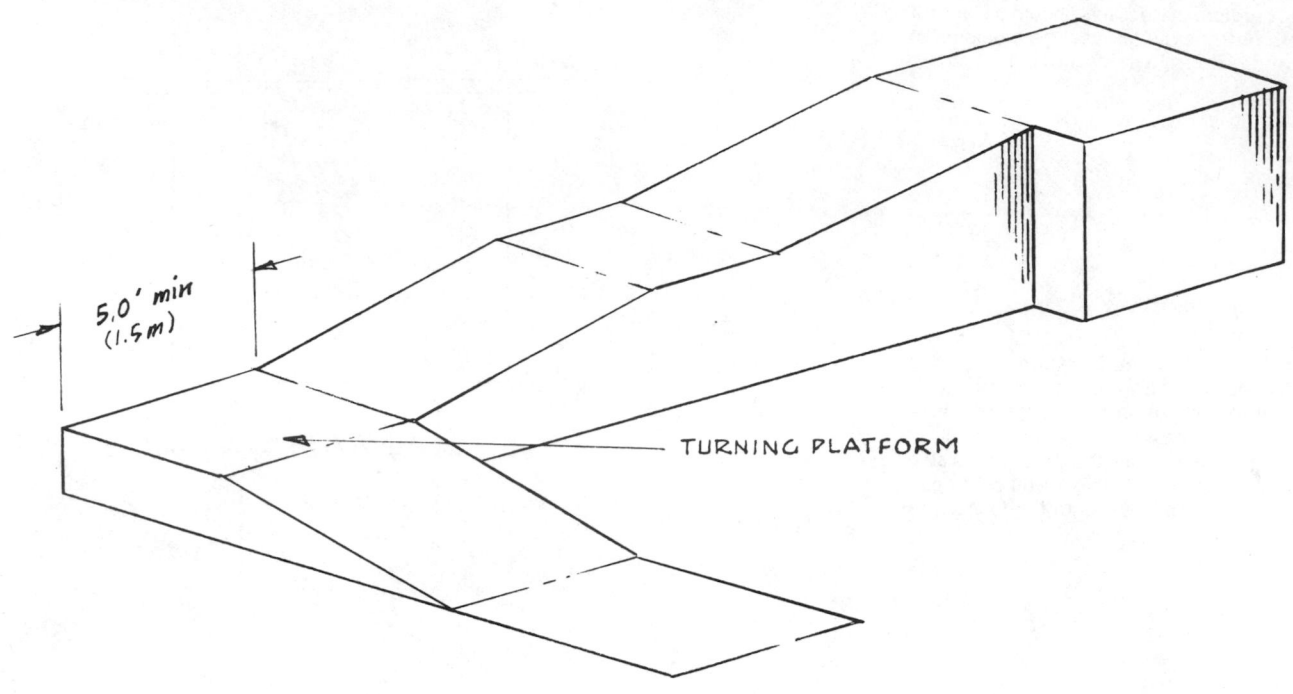

5.0' min
(1.5m)

TURNING PLATFORM

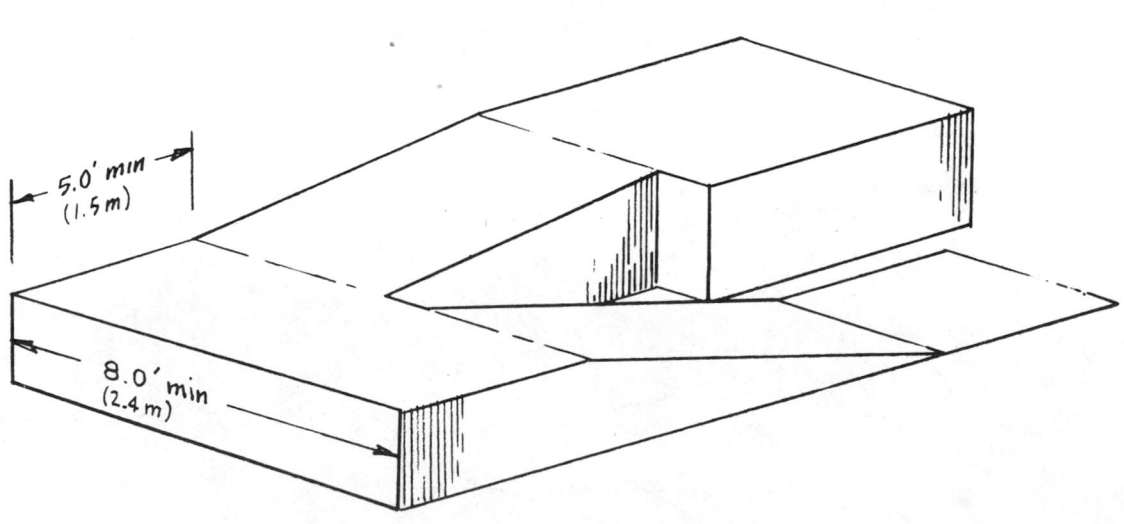

5.0' min
(1.5m)

8.0' min
(2.4m)

### 4. Handrails

Handrails should be provided on all ramps or platforms that are more than 2 ft (0.6 m) above grade.

Note the preparatory horizontal element at both the bottom and the top of the ramp handrail shown in the accompanying sketch. This helps the user adjust to the change from a horizontal to a ramped surface.

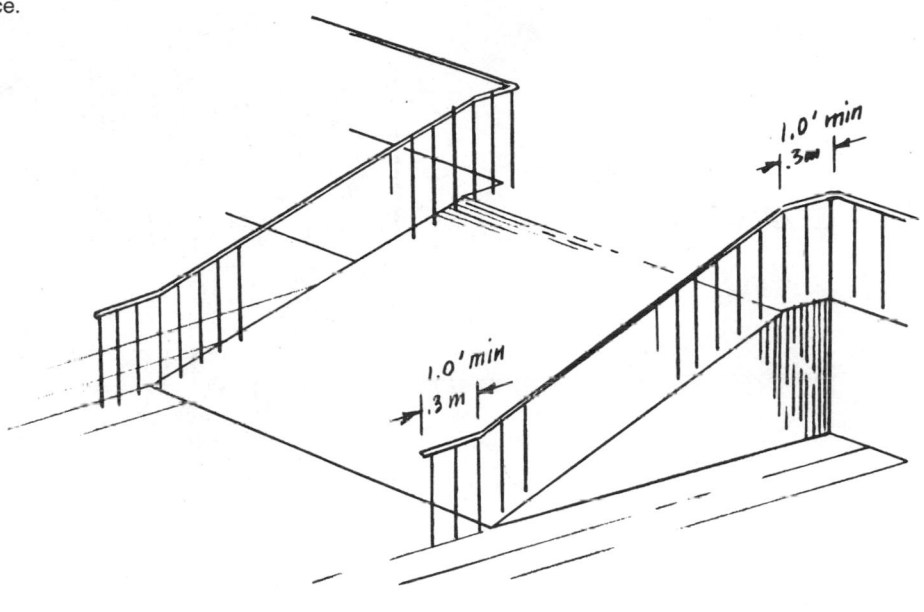

### 5. Stairs

Since stairs may at some time be used by both elderly persons and handicapped persons not confined to wheelchairs, stairs should be designed so that they are negotiable without stubbing a toe. Elderly and handicapped persons often are not aware that they are not picking up their feet high enough to clear the next tread.

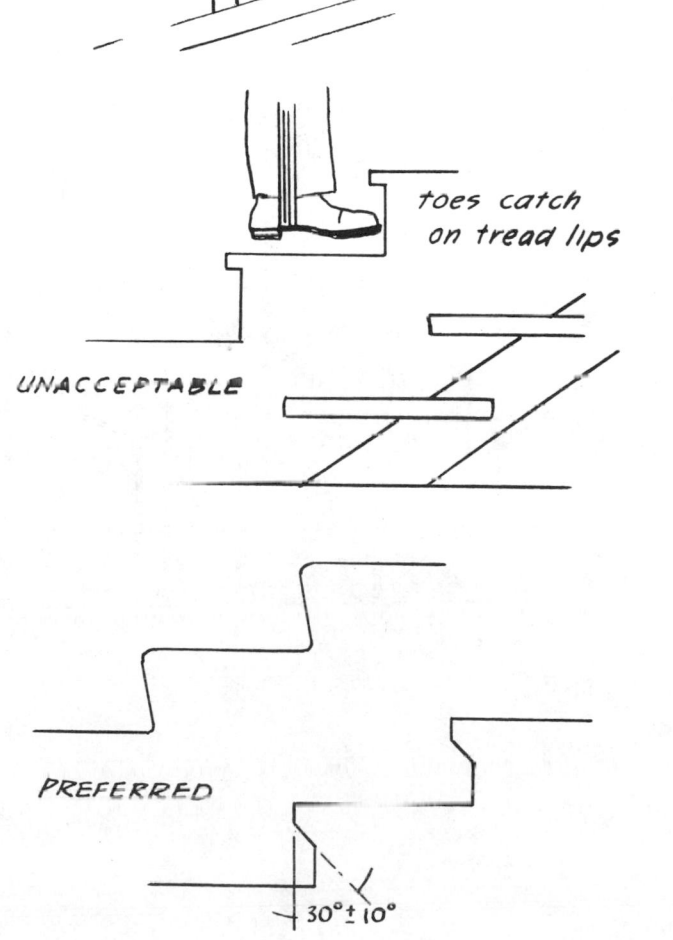

## INSIDE THRESHOLDS

The two accompanying illustrations provide
guidelines for ensuring adequate clearance for
the typical threshold activities.

## VEHICULAR ENTRYWAYS

### 1. Separate Vehicular and Pedestrian Entryways

Avoid combining pedestrian and vehicular entryways, especially in industrial, public, and similar facilities. A separate pedestrian entryway, such as shown in the accompanying illustration, is desirable to avoid hazardous conflict between pedestrians and vehicles. Separate entryways alone will not suffice, however. Appropriate marking on surfaces both inside and outside should also be provided.

Separation of the two entryways also helps prevent pedestrians from using the vehicular entryway and from wandering into the path of a vehicle, once they are outside.

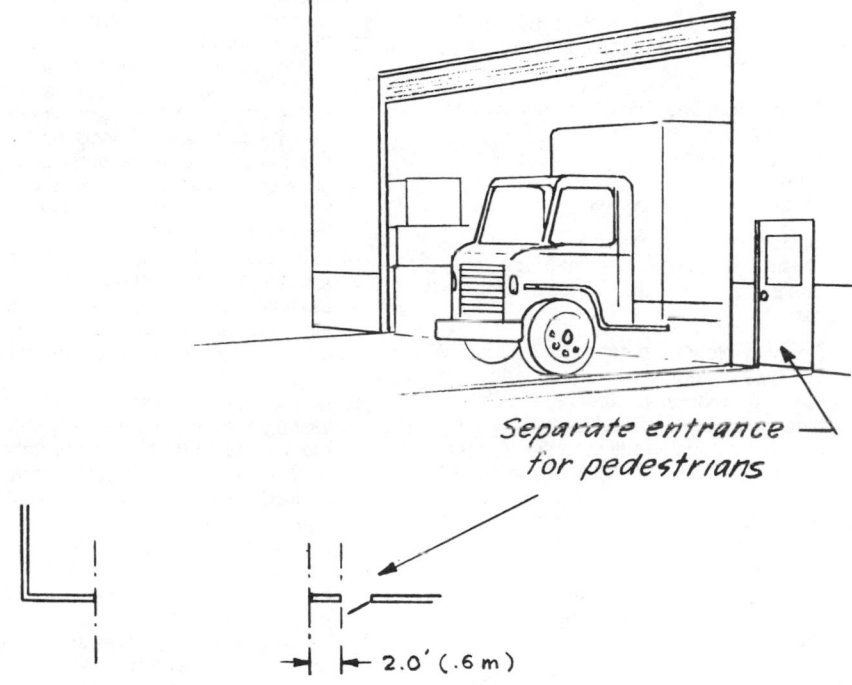

*Separate entrance for pedestrians*

2.0' (.6 m)

### 2. Entry Approach Inclines

Too steep an approach to the vehicular entryway prevents the driver from seeing required clearance needs.

Cars should be level just prior to entering.

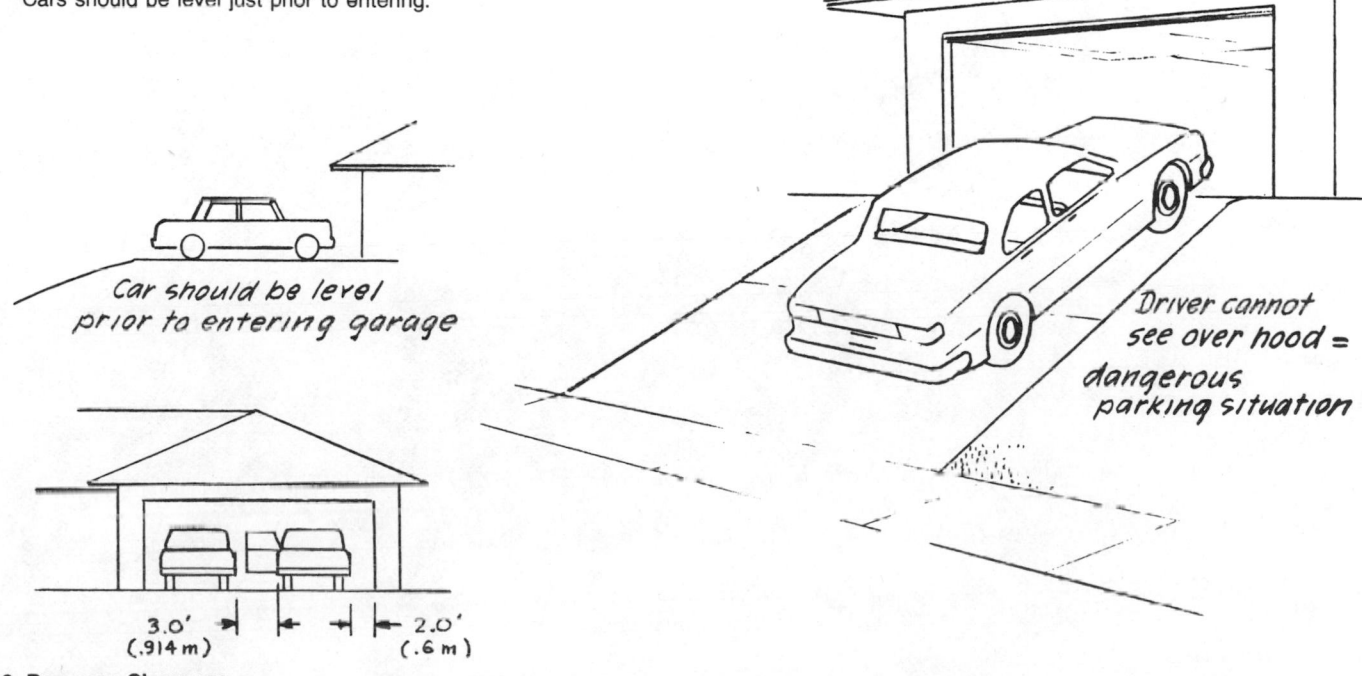

*Car should be level prior to entering garage*

*Driver cannot see over hood = dangerous parking situation*

3.0' (.914 m)    2.0' (.6 m)

### 3. Doorway Clearance

## GENERAL PRINCIPLES

The primary functions provided by doors, hatches, and equipment closures include the following:

1. Access from one side of an enclosure to another
2. A security barrier
3. A privacy shield
4. An environmental barrier

In considering doors for a given project, one needs to address several user interface variables. The following is a useful checklist for the designer:

1. Doorway clearance needs
2. The type of door or doors (material; door opening geometry; door manipulating hardware, latching, and/or locking system; see-through requirements; door position retention; the type of threshold; etc.)
3. Automatic versus manual door articulation
4. Identification
5. Sealing against drafts, extreme pressure differentials, water leakage, dirt and dust, etc.

From a user's point of view, several dos and don'ts should be noted:

1. Do not use revolving doors or all-glass frameless doors; these are apt to be misused and cause personal injury.
2. Avoid the use of extremely large and/or heavy doors if they are to be manipulated by hand.
3. Provide view ports on free-swinging doors.
4. Do not combine vehicle and pedestrian doors; i.e., provide separate entryways for pedestrians and vehicles.
5. Provide ample doorway and door clearances; i.e., avoid crowding a person who is approaching or passing through a doorway or past an opening door.
6. Avoid using fancy "decorator" door handles that are hard to grasp or manipulate without hurting one's hand.
7. Place door handles where they provide the best advantage for gripping and manipulating the door; there should also be sufficient clearance between the handle and the adjacent structure to minimize the possibility that a person will scrape or pinch his or her hand or fingers.
8. When required, select door latching or locking systems that are easy to operate and locate them where visual guidance for a key is reasonably near the operator's eye level and/or viewing perspective.
9. Select door hardware that does not have sharp corners or edges that could cause injury if a person fell or bumped against the device.
10. All metal hatches and equipment doors or covers should be designed with radiused corners and edges so that users will not be cut by the sharp features. This applies to doors on vehicular equipment as well as to equipment closures.
11. Use materials for doors that are not affected by changing weather conditions so that the doors will not stick when wet or rattle when dry.
12. Select and/or design doorway and door thresholds that are relatively smooth (i.e., level with adjoining floor surfaces) so that pedestrians will not trip, wheelchairs will not be tipped over, and vehicles or equipment cars will not be suddenly jolted by the abruptness of the threshold.
13. Passenger doors for automobiles should be reinforced to minimize possible penetration by side impact from another vehicle, and doorframe and door latching systems should be designed to preclude inadvertent unlatching of the door in the event of a crash.
14. Door handles on automobiles (both inside and outside) should be inset so that neither a pedestrian (outside) nor a passenger (inside) can be injured by contact with an exposed handle.
15. Place door or other access handles on vehicles where they can be seen and easily manipulated; i.e., avoid "hiding" such devices.
16. Design opening and closing hardware for airline hatch doors so that it can be easily manipulated by flight attendants as well as by the crew.
17. Design hardware for airline emergency door releases so that the average passenger can easily locate and operate the device without assistance from a crew member (who may be incapacitated) and make sure that the instructions for operating the device are easy to follow.

## ARCHITECTURAL DOORS

The dimensions indicated in the accompanying sketch are recommended for most pedestrian doors. This clearance envelope and tread design will admit all adults and a person in a wheelchair. Smaller openings not only preclude the wheelchair user but also limit the passage of large pieces of furniture. (Note that the width includes any encroachment of the door and door hardware.) Avoid larger doors than shown; they become difficult for children and wheelchair users to manipulate, and when the door is of the swinging type, considerable clearance is required for the door swing.

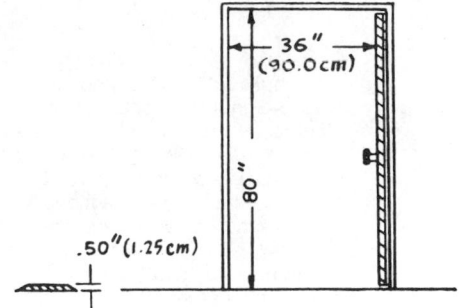

Sliding doors have advantages over swinging doors in terms of the swing clearance noted above; in addition, an open swinging door takes up valuable wall space that could otherwise be used for furniture or equipment. When a sliding door is used, however, it is vital that the door handle remain easily accessible and that the door be designed so that fingers cannot be crushed accidentally when a person opens the door.

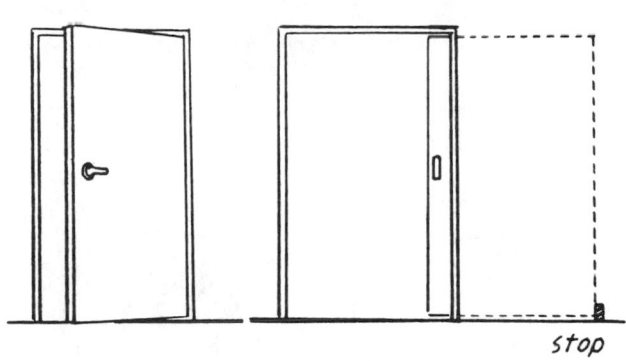

Sliding glass doors provide good natural light and a pleasant view of the outside. However, the glass should always be framed. The lower portion of the frame should have a "bumper" at least 12 in (30 cm) above the floor to prevent children's tricycles or wheelchair footrests from striking the glass.

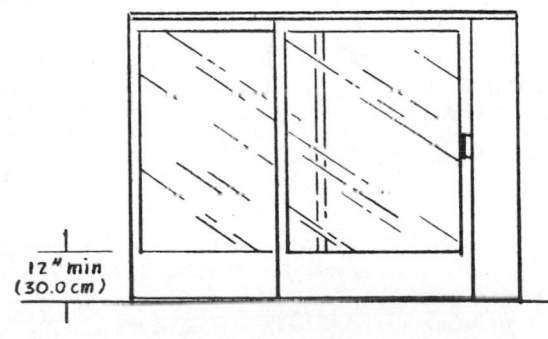

Double doors are useful in many applications because only one door can be used to admit people, thus preventing wind, rain, or snow from entering. When it is necessary to move extra-large furnishings or equipment through the doorway, both doors provide this convenience with clearance to spare. Exterior main entry doors should have a visual access of some kind so that the host can determine who is outside. Note the security grille in the accompanying sketch.

### Characteristics of Swinging Doors

Generally, exterior swinging doors should open outward so that people are minimally hindered in the event they are making an emergency escape. However, if a screen door is added, as shown in the accompanying illustration, it is not practical for both doors to swing in the same direction. Typically accepted practice is to have the screen door open out, and the primary door open in. It is recommended that screen doors have a sturdy panel at the bottom, plus some kind of grille protector to prevent small children from pushing on the screen to get out.

Screen out   Door in

As shown in the accompanying illustration, door hardware should indicate to the user whether to push, pull, or turn a handle to open a door. Avoid the temptation to camouflage these features.

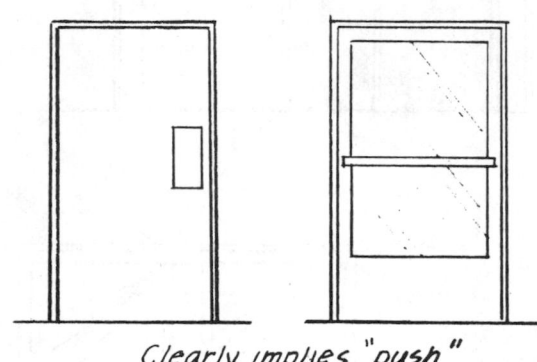

Clearly implies "push"

Implies a latch and "turning"

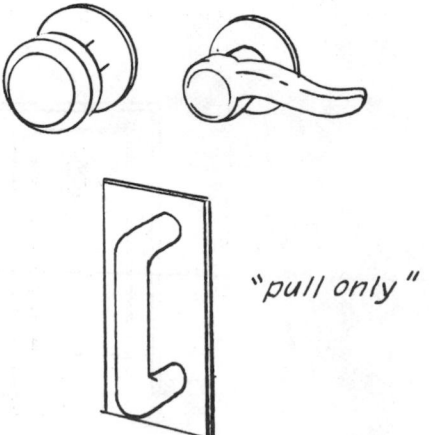

"pull only"

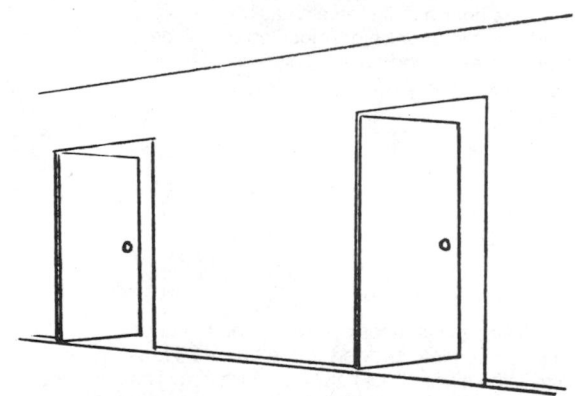

Doors should not open into hallways because of the possibility that someone might be struck by a door that is being opened from the other side.

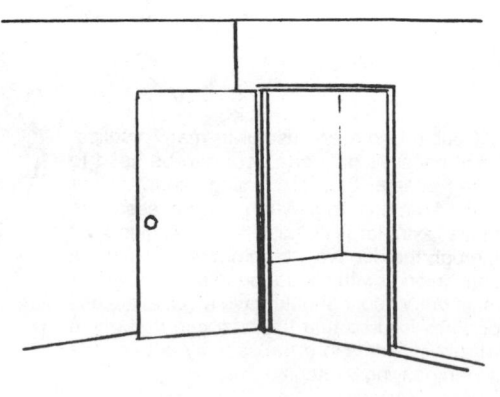

Doors located at the corner of a room should open toward the corner.

Free-swinging doors are useful in many industrial and commercial applications because people whose hands are full can push them open with their feet. There should be a spacer between double free-swinging doors, however, because people tend not to notice when the doors are not aligned and thus may be struck by the adjacent door if it is pushed toward them. All free-swinging doors should have view ports so that a door is not accidentally pushed into the face of an unsuspecting person on the other side.

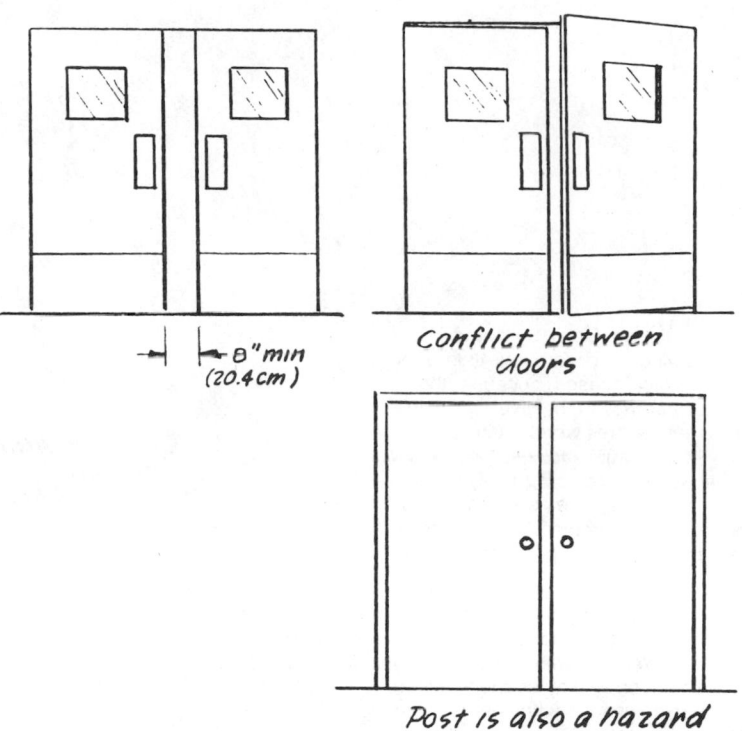

8"min
(20.4cm)

Conflict between doors

Avoid double door configurations that have a narrow divider between them, as shown in the accompanying illustration. Experience has demonstrated that people tend not to observe the divider and to run into it.

Post is also a hazard

Decorative doors are generally acceptable unless they have a built-in hazard, as in the case of the Dutch door, shown in the accompanying illustration. People tend not to give themselves adequate clearance around the upper half when it is swinging clear, as shown. Children often bump their heads on the upper half. Such doors are also hard to control in a sudden wind gust when the two halves are not fastened together.

Avoid !

Watch for pinching hazards

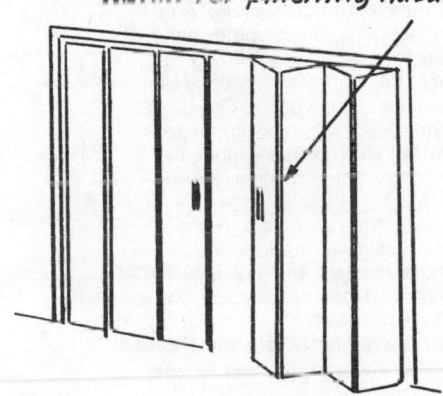

Accordion doors provide some useful advantages such as less weight, less loss of wall space, and little, if any, encroachment into the room area. However, consider the specific accordion design in terms of potential injury, namely, pinching of fingers that are carelessly placed near the folds.

## Vehicular Storage and Industrial Doors

Large, heavy doors are generally difficult for people to handle, and they may be hazardous if they are not properly designed to prevent human errors.

*Swinging door hazard*

### 1. Residential Garage Door Systems

An overhead, swinging, or tambour-type door is desirable in most cases because these doors provide clearance at the level where the homeowner often locates workbenches, laundry facilities, or storage shelves. Unless an automatic garage door opener (highly recommended) is provided, the swinging door causes the user more problems and tends to be more hazardous. Typical problems include the following:

1. The user has to go through more fatiguing gyrations to open and close the door.
2. Wind can interfere with opening and closing the door or may suddenly close the door on an unsuspecting person who is walking beneath it.
3. The tension springs may break, making the door a lethal missile.

Few of these problems are associated with the tambour door.

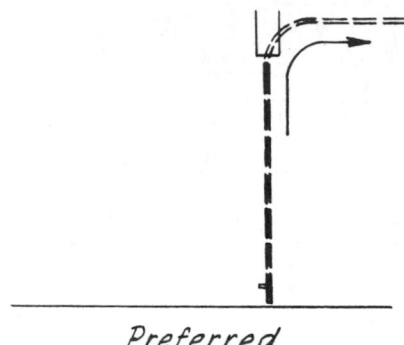

*Preferred*

### 2. Large Industrial Door Systems

In general, most large industrial doors should not be manually operated unless they are made from extremely light materials. Automated, powered systems, on the other hand, must include suitable safety features so that persons cannot be suddenly trapped or crushed between two closing doors. Roll-up or sectional, horizontal sliding door systems generally are preferred (few swinging-door systems are acceptable for industrial applications).

It is recommended that both garage and industrial doors be designed to allow the passage of natural exterior light and thus improve interior illumination. Transparency is, however, not usually a necessity.

Strive to provide a nonhazardous threshold design and a design that makes it easy to keep door tracks clean.

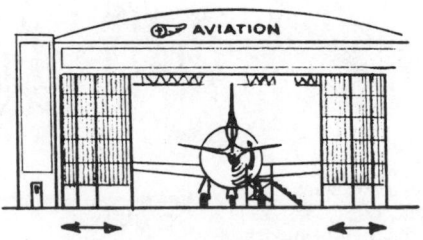

**Selection and Location of Door
Manipulation Devices**

### 1. Shape and Size of Door Handles

Although most adults can manipulate a door fairly well using a variety of handles, children and certain handicapped users have difficulty with some types. The best type for all users is the L-shaped handle, shown in the accompanying illustration. This type of handle does not have to be squeezed, which is an extremely difficult task for some users, and its appearance gives an immediate indication of what one is to do with it. Avoid oddly shaped handles that are hard to get hold of and fancy decorator handles that have sharp patterns that cut into the user's hand or fingers.

Care should be taken in relating door handle selection and the planned installation. The accompanying sketch illustrates a typical problem: not enough finger clearance remains once the door jamb is installed.

### 2. Door Locks

Door lock systems should be selected and located so that users can see what they are doing when they attempt to find the keyhole. Combination handles and locks, as shown in the accompanying illustration, are not recommended because when the handle is placed for easy use, the lock is too low. A separate lock placed near eye level is preferred.

All external architectural doors should be provided with dead-bolt type burglar resistant locks.

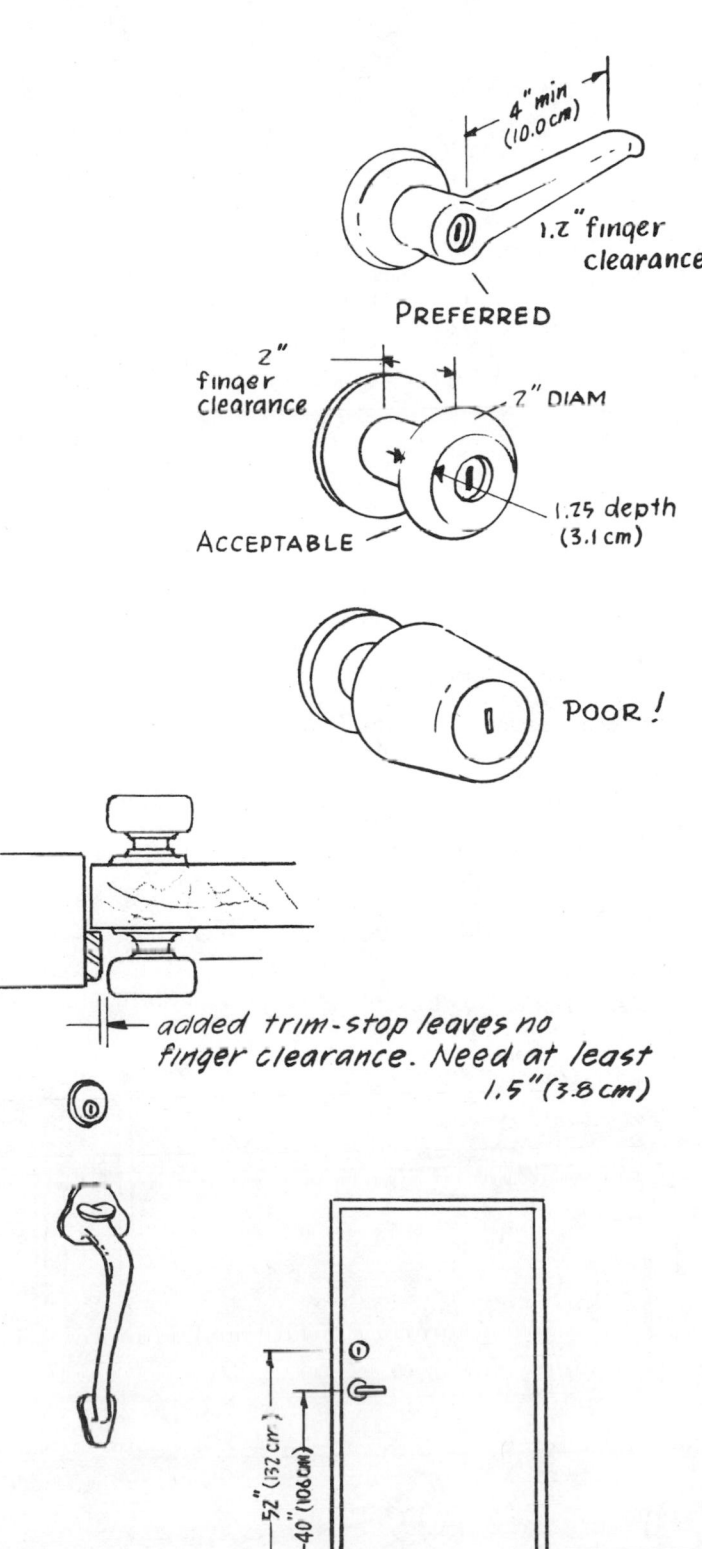

4" min (10.0 cm)

1.2" finger clearance

PREFERRED

2" finger clearance

2" DIAM

1.25 depth (3.1 cm)

ACCEPTABLE

POOR!

added trim-stop leaves no finger clearance. Need at least 1.5" (3.8 cm)

52" (132 cm)

40" (108 cm)

### 3. Finger Latches and Door Pulls

The principal considerations in the design and/or selection of door latch and pull devices include finger clearance and force requirements either to unlatch the device or to pull the door open.

Although many different latch configurations are available or possible, the accompanying sketches show the clearance needs of the average user. Note that more finger clearance is needed for an outdoor application because the user may be wearing gloves.

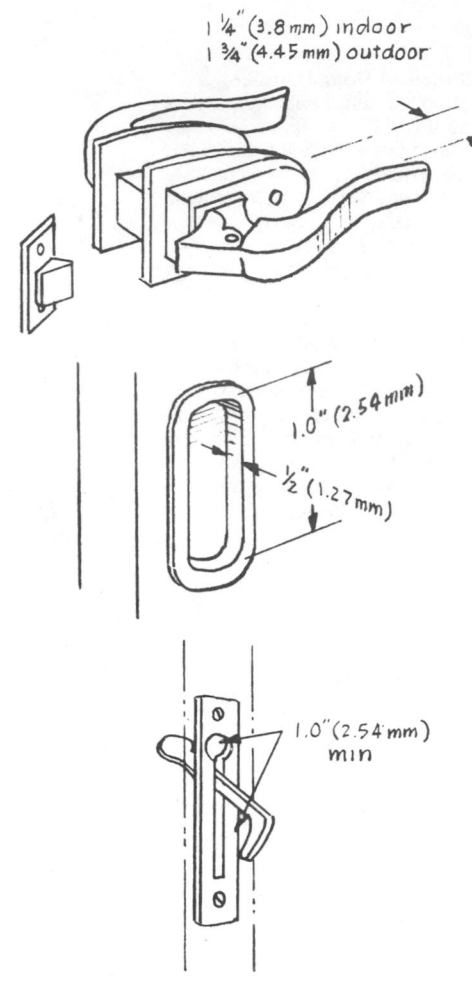

1 ¼" (3.8 mm) indoor
1 ¾" (4.45 mm) outdoor

1.0" (2.54 mm)

½" (1.27 mm)

1.0" (2.54 mm) min

As indicated in the accompanying sketch, depth of the door pull is especially important because children, elderly persons, and those with orthopedic handicaps do not have sufficient fingertip strength to move a heavy door. Although the minimum depth shown is generally acceptable, a deeper finger stall is desirable; i.e., ¾-in (1.9-cm) depth allows people with weaker fingers to manipulate a door more easily.

### 4. Door Closure and Restraint Devices

Automatic door closure systems should have both adjustment and temporary securing features to make them maximally effective.

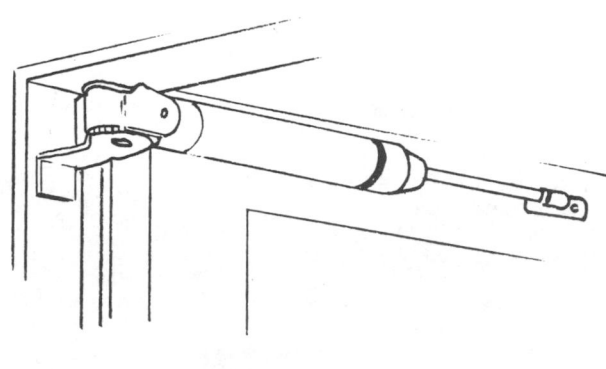

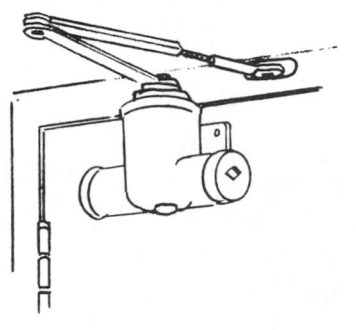

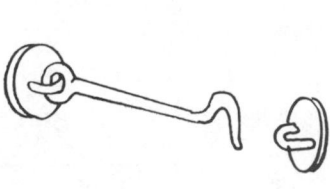

## Safety and Emergency Door Features

A typical "panic" door installation (see the accompanying sketch) should provide emergency exit without requiring the user to manipulate locks or latches. The door must obviously open out, or with the flow of traffic. The force required to unlatch the system should be sufficiently low so that the weakest member of the user population can unlatch the door (the door must also not be so heavy that such persons cannot push it all the way open). A door "catch" should be located outside so that the first person opening the door can secure it in an open position.

The standard height criterion used by most architects is 36 in (91 cm) above the floor level. However, this is too high for a person in a wheelchair, and therefore the lower height of 34 in (86 cm) is recommended.

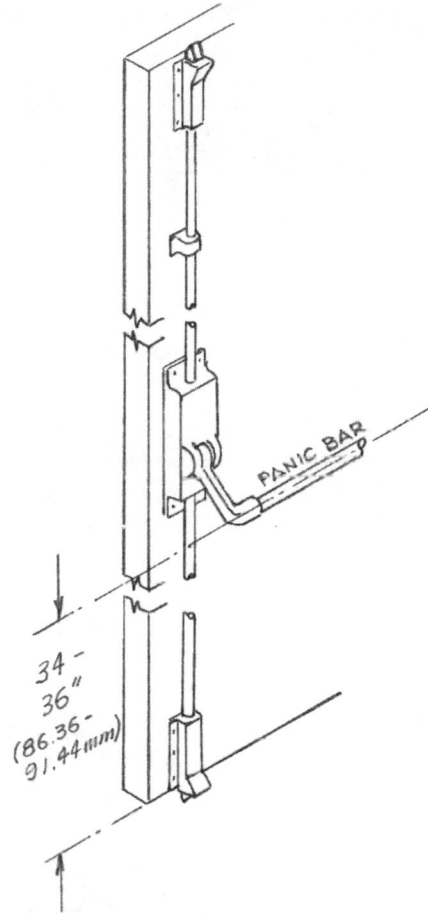

The latch shown in the accompanying sketch should automatically retract when the door opens so that if someone lets the door swing closed, it will not relock.

If such doors are not meant to be used except in an emergency, provide an auditory alarm to indicate when the door has been opened. If the door is a fire door, place a sign on the door identifying it as such.

**OPEN ELEVATOR DOORS** Commercial elevator systems (see the accompanying sketches) should always be provided with a guard gate or bars, as appropriate. There should also be an auditory alarm to indicate either that the sidewalk doors are being opened or that the guard gate is being lowered in preparation for an elevator arrival.

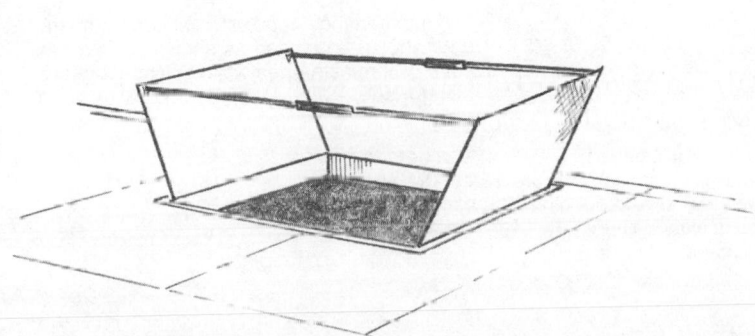

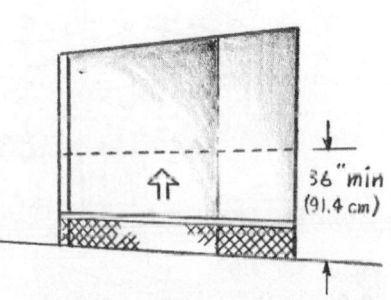

## EQUIPMENT AND CABINET DOORS

Cabinet doors and doors for gaining access to equipment and home appliances should be designed and located to simplify the user's problem of getting the doors open and holding them open while he or she is working or putting something inside, and they should be planned to "fit" the constraints of the expected range of installations. The accompanying sketches illustrate a few of the common problems associated with typical door designs.

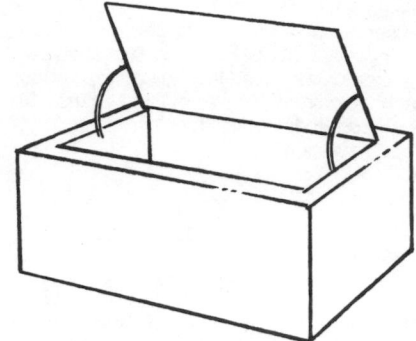

A "lid-type" door should be hinged to prevent the cover from being lost, and guides should be provided so that there is a means for holding the cover open. The same objectives pertain to the design of doors for a typical office storage cabinet.

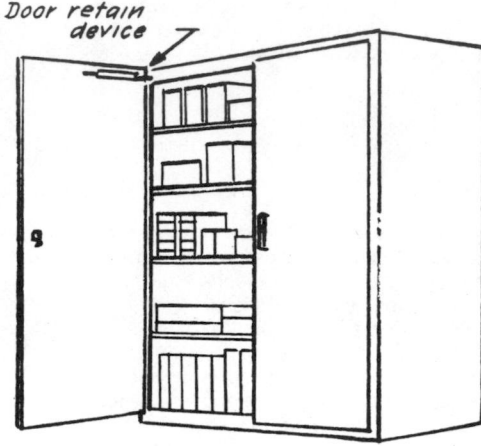

Door retain device

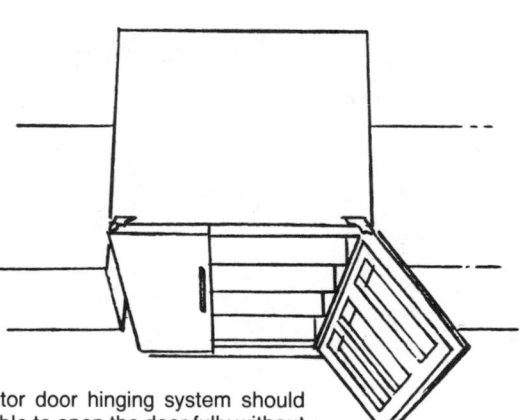

A refrigerator door hinging system should make it possible to open the door fully without interfering with adjacent cabinets, but it should be designed to close automatically (slowly, of course).

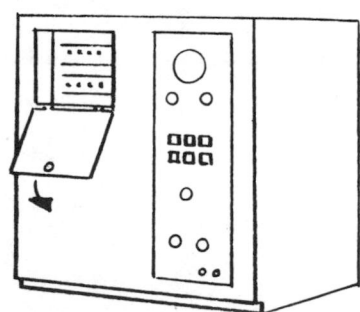

A maintenance or adjustment compartment door should be hinged as shown so that the cover will remain open while adjustments are being made.

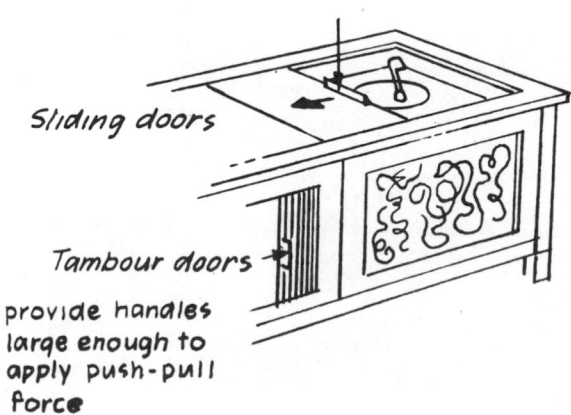

Sliding doors

Tambour doors

provide handles large enough to apply push-pull force

Typical compartment closure schemes for entertainment cabinets are acceptable as long as the doors do not bind. The handles should be designed so that the user's fingers are not pinched.

### Injury-Causing Door Features

The following common errors in design may cause considerable difficulty for the user.

*Poor installation = bruised hand!*

*Correct*

### 1. Incorrect Door Hinge Pattern

Equipment or cabinet doors should not be mounted so that the user's hand might be injured if he or she opens the door too far. In some instances the user may be applying considerable force in order to overcome the restriction of a door clasp; when the clasp suddenly lets go, the user's knuckles may be cracked on an adjacent structure before he or she can reduce the amount of force being applied.

Never locate the handles of adjacent doors so that they could coincide during an opening procedure.

### 2. Sharp Corners

Although the cost of eliminating sharp corners may not seem warranted to the manufacturer, the typical equipment user is not very sympathetic to the so-called cost-effective sharp corner that may just have punctured his or her hand. Be especially alert for sharp corners that people might inadvertently hit with their heads.

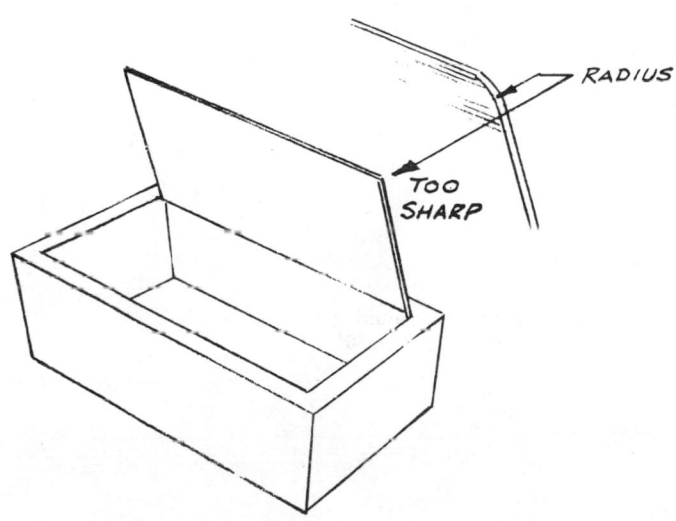

*RADIUS*

*TOO SHARP*

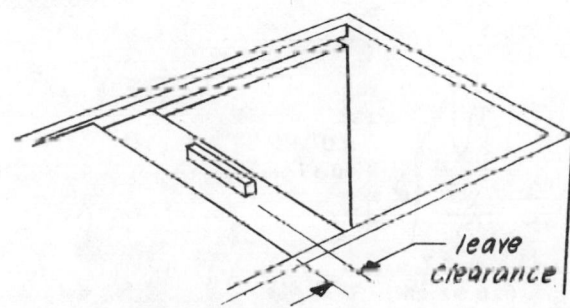

*leave clearance*

### 3. Sliding Doors

Provide stops so that people will not pinch their fingers as they slide a door against another part of the cabinet. At least 1½ in (3.8 cm) of clearance is needed to provide finger protection.

**Equipment Door Size**

Equipment door size obviously relates to the
specific activities that may take place as technicians put their hands and/or an object
through the opening. The door size should
permit the following:

1. Using a common screwdriver with freedom to turn the hand through 180°

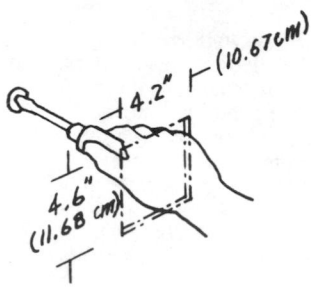

2. Using pliers and similar tools that require
   gripping

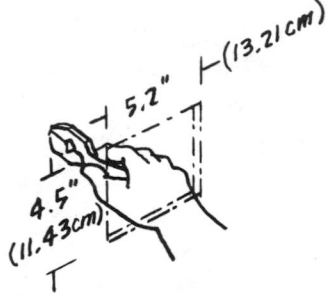

3. Using a T-handle wrench with freedom to
   turn the tool and hand through 180°

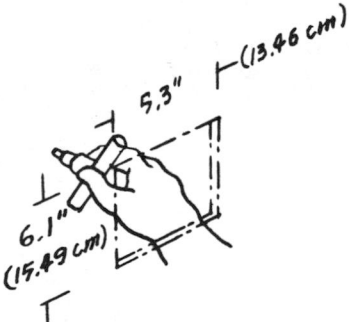

4. Using an open-end or box-end wrench
   with freedom to turn the wrench through
   at least 60°

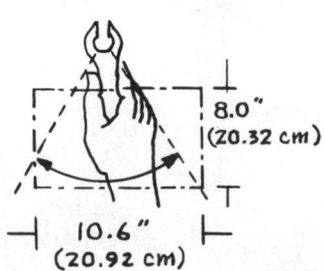

5. Grasping and manipulating small objects (up to 2¼ in, 5.7 cm wide) with one hand

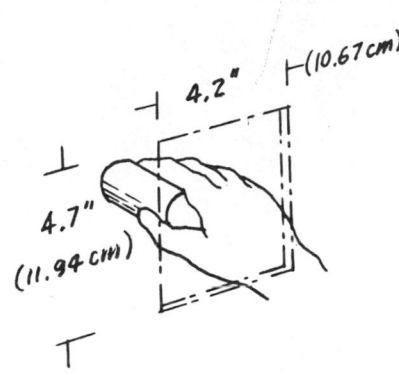

6. Grasping large objects with one hand

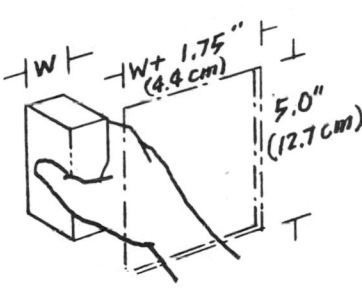

7. Grasping large objects with two hands, with the hands extended through the opening up to the length of the fingers

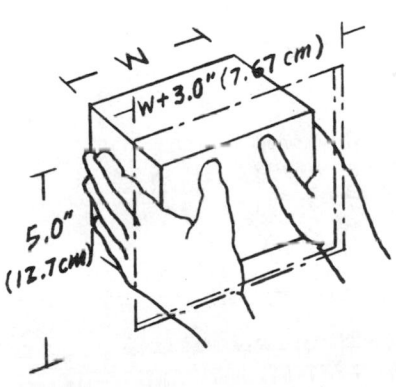

8. Grasping large objects with two hands, with the arms extended through the opening

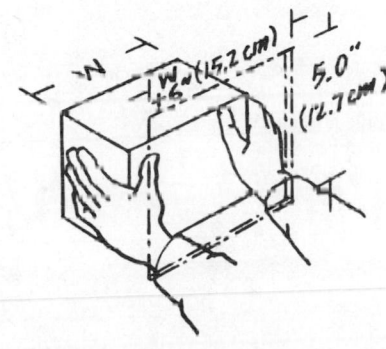

## VEHICULAR DOORS

Actually, door shape, size, and location should always be considered in the initial phase of vehicular profile or concept development. When these overall profiles are initially established on the basis of styling alone, door design options are already so limited that the door's ultimate utility, as far as the user is concerned, is only barely acceptable. Key issues that should be addressed early in the vehicular concept stage are shown in the accompanying illustrations:

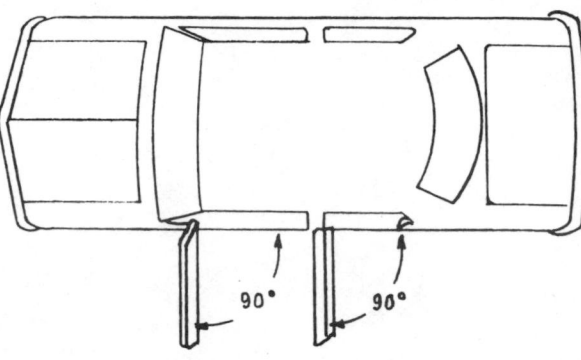

1. Doors should be attached so that they can be opened at least 90°. Intermediate detented positions should be provided so that the door will remain in place.

Note: Doors open forward

2. The door opening dimensions and shape should be compatible with the way in which the user ordinarily would want to enter and sit down.

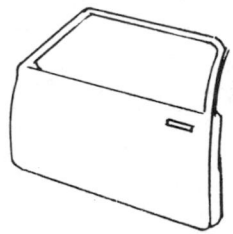
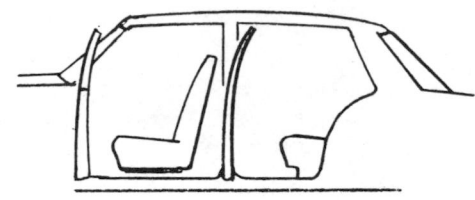

3. All exposed door corners that face the user as he or she gets in or out should be radiused to prevent injury.

4. In spite of their popularity, nonframed glass door windows are a potential hazard and should not be considered.

5. Engine and trunk lids should have a spring-assist and positioning lock and should not swing through the envelope that the mechanic or owner normally occupies while opening or closing the lid.

6. The interior deck of a trunk should be flush with the bumper so that the owner does not have to lift packages over the trunk lip or bumper.

**SPECIAL PROBLEM AREAS** When special features such as the optional window wing (shown in the accompanying sketch) are provided, watch for unusual hazard conditions associated with certain design configurations. In this case, the sharp corner presented by the window wing can cause serious injury to the person as he or she leans forward to duck under the low overhead. This is accentuated if the door is not opened all the way, as in close parking situations.

*Avoid !*

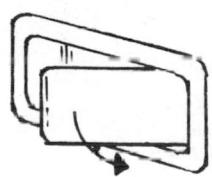

*One-hand, separate unlatching- pull is less confusing*

6.0" (37.5 cm)

2.0" min (5.0 cm)

*Push latch*

The vehicle door handle and lock combination shown in the accompanying illustration is preferred because the user can easily identify the plunger for unlatching and can keep his or her hand comfortably and securely on the handle while pulling open the door or closing it in a high wind.

*awkward motions*

Although acceptable, this design is more confusing in terms of how it is to be operated.

This design is unacceptable — requires awkward hand position.

Avoid the use of very wide doors. As illustrated in the accompanying sketch, in close parking situations the user may not be able to get in or out of the car. The wide door presents a problem in terms of placing a door handle (on the inside) that is accessible to a small person, particularly when the door is opened wide.

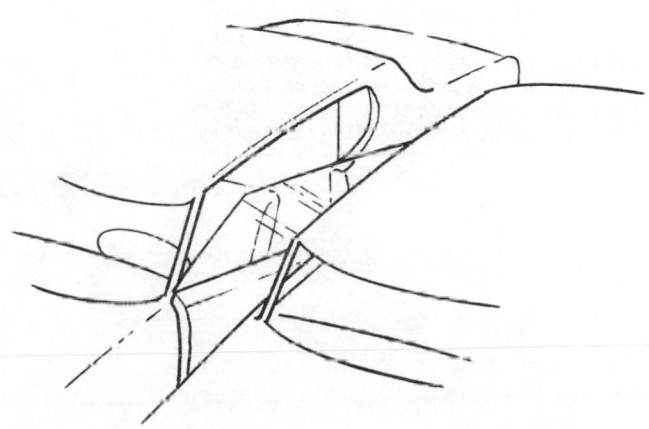

The same precautions noted for the design of exterior door hardware should be taken with respect to interior door hardware.

Because of recent requirements that interior vehicle hardware be made as noninjurious as possible, so that occupants will not be thrown against sharp protrusions during a crash, many designers have tended to hide door handles. However, it is confusing to the user under normal operations to have to hunt for a door handle, and in an emergency there may not be time to do so. Door handle location and operation should be obvious. It is quite feasible to do this and still retain a noninjurious design. The accompanying sketch illustrates a typically preferred approach; a long handle is provided so that the user can grasp it regardless of how far the door is opened (to pull it shut), and the simple plunger latch control is obvious and noninjurious.

Similar precautions apply to the standard mechanical window crank, as shown in the accompanying illustration. The end knob should rotate freely. A fairly thin, large-diameter soft knob provides ease of manipulation plus safety characteristics. Cranks for raising the window should rotate toward the user.

To avoid confusing the user, electrically powered window controls should be arranged in one of the layouts shown in the accompanying sketches. Labels should be provided, however, so that the user will know that these are the window and/or lock controls.

Individual, plunger-type door lock controls are recommended for the following reasons:

They are more common, and therefore most users expect them.
They are accessible to both the occupant and the rescuer; other types usually are not accessible to the rescuer.
They are simple to use, and their method of operation is readily apparent.

However, door lock controls should be positioned as indicated in the accompanying sketch so that the user can reach them. Avoid the error of locating them near the aft doorframe merely because that is where they are typically found; i.e., when the occupant's seat is considerably forward of the rear edge of the door, the plunger should be moved to coincide with the operator's reach.

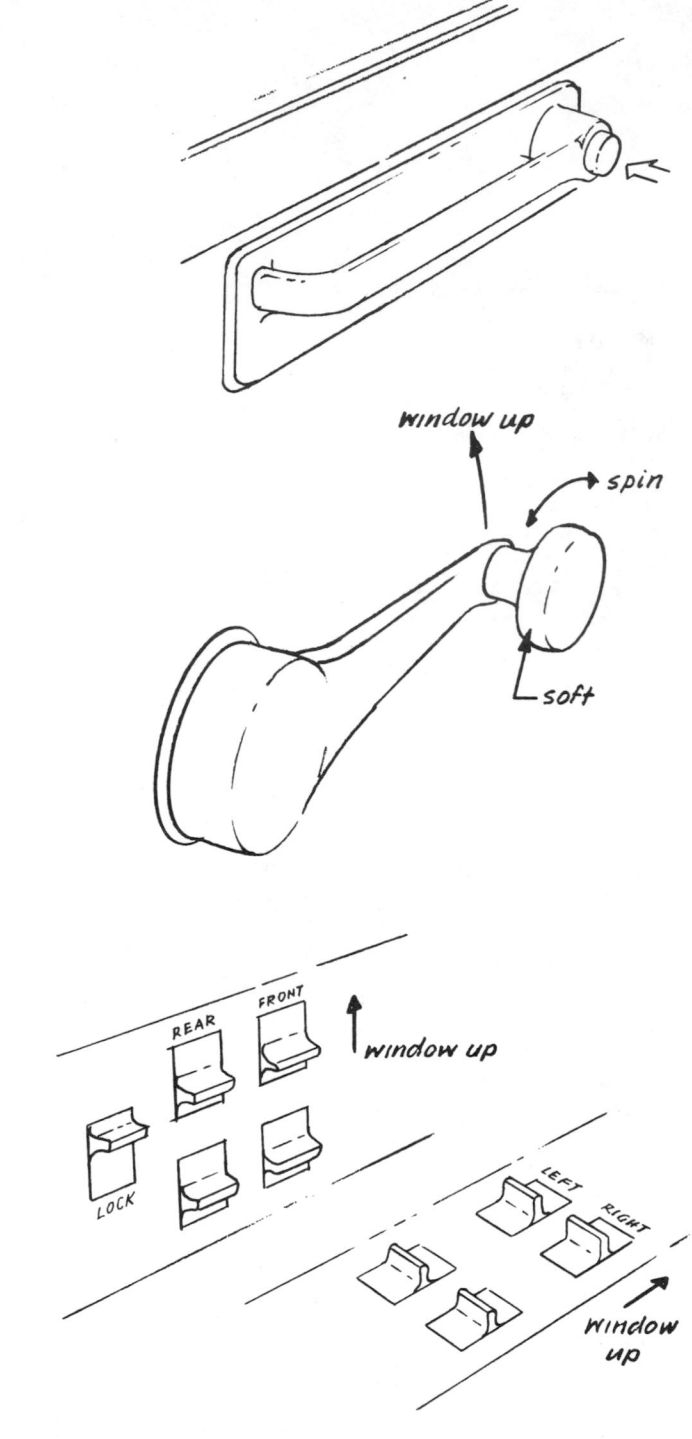

If automatic, electrically powered door lock systems are used (i.e., ones that lock automatically when the vehicle is put into motion), they must be designed to fail in the unlocked position. Although individual plunger locks could be used by the occupants, rescuers could not open the doors in the event passengers were incapacitated.

**MASS-TRANSIT VEHICLE DOORS** Although all door systems used in mass-transit vehicle systems should be openable by both operators and passengers, the primary door opening systems should be designed to be under the control of the on-board operator to prevent passengers from opening doors either intentionally or unintentionally.

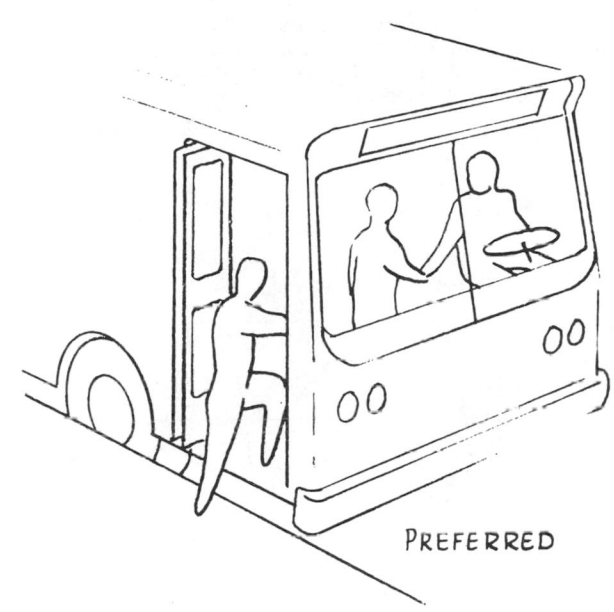

PREFERRED

Select door system configurations that do not pose hazards to a boarding passenger who is standing too close to the vehicle when the door is being opened. As the accompanying sketches indicate, a sliding or folding door system generally provides a slim profile, whereas swinging doors are obviously a hazard.

The most common design failure in passenger entry doors for commercial aircraft is insufficient head clearance for taller passengers. The problem exists primarily in smaller aircraft. Consider conspicuous warning signs when head clearance is inadequate.

AVOID!

Sliding clamshell-type doors are generally preferred for rail cars used in rapid transit, e.g., subways and other urban systems. Emergency bumper stops are vital to prevent passenger entrapment.

## VEHICULAR HATCHES

Since many vehicular hatches have an emergency escape function, they are considered auxiliary doors. Nevertheless, they must be large enough to make escape easy and sure, and they should be free from elements upon which people could become snagged. The dimensions shown in the accompanying illustrations represent the best guidelines for various typical hatch-use functions.

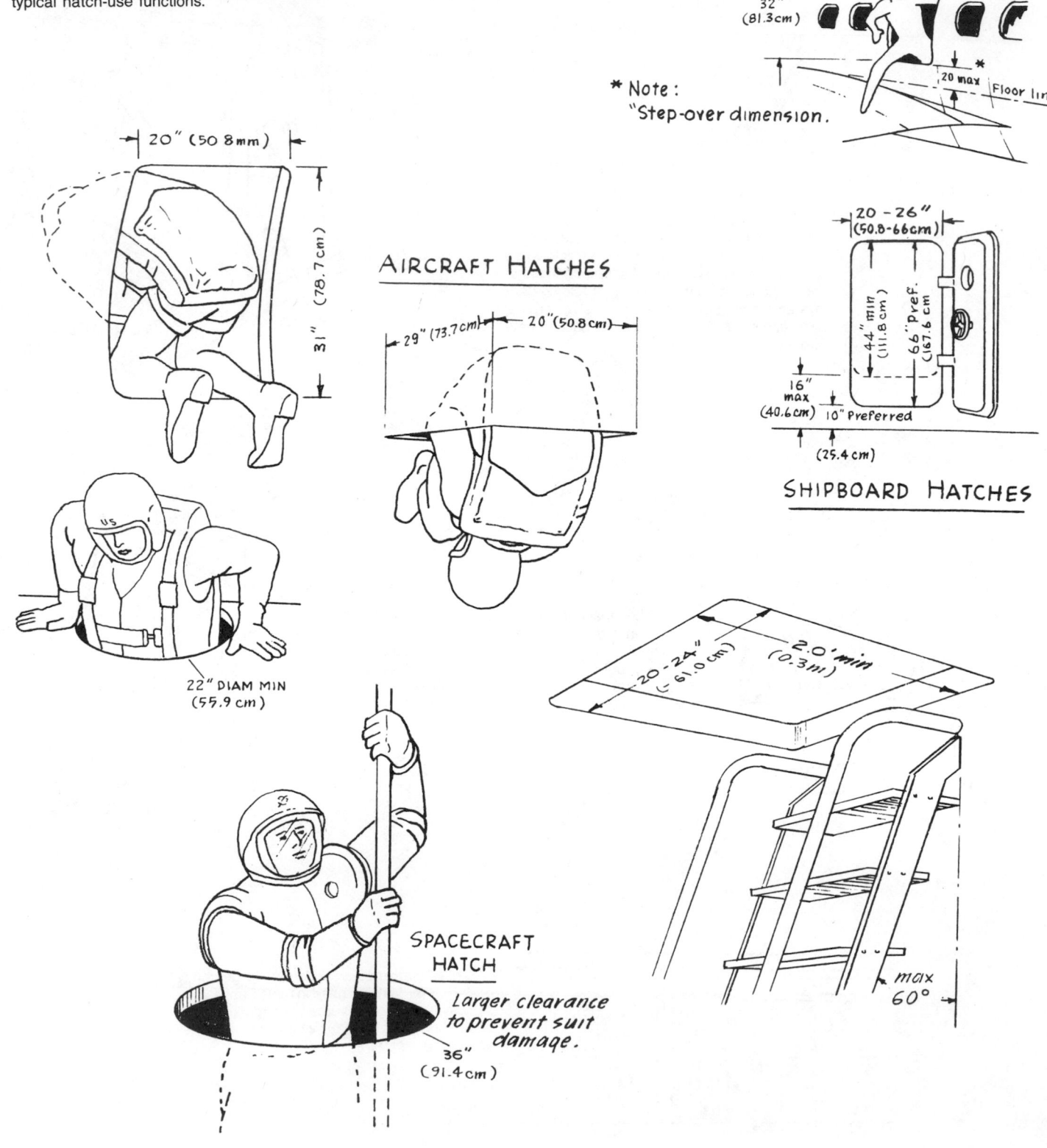

\* Note:
"Step-over dimension."

AIRCRAFT HATCHES

SHIPBOARD HATCHES

22" DIAM MIN
(55.9 cm)

SPACECRAFT
HATCH

Larger clearance
to prevent suit
damage.

## Aircraft Canopy Hatch

Beyond the obvious requirement to design the fighter-type canopy so that it permits the pilot or pilots to have clear, unobstructed ingress and egress to their pilot station, special consideration must be given to rapid and safe removal of the canopy in the event of emergency abandonment of the aircraft. One must address at least the following emergency operational problems in defining the nature of the canopy configuration and removal system:

1. Ditching at zero speed, at ground level, from under water, at high altitude, or at high speed
2. Alternative ditching modes, i.e., by using the ejection seat and/or by going over the side
3. Control of an ejected canopy to prevent its contacting personnel or aircraft before it is clear
4. A manual backup canopy release in addition to automatic canopy ejection, which is required in most emergencies

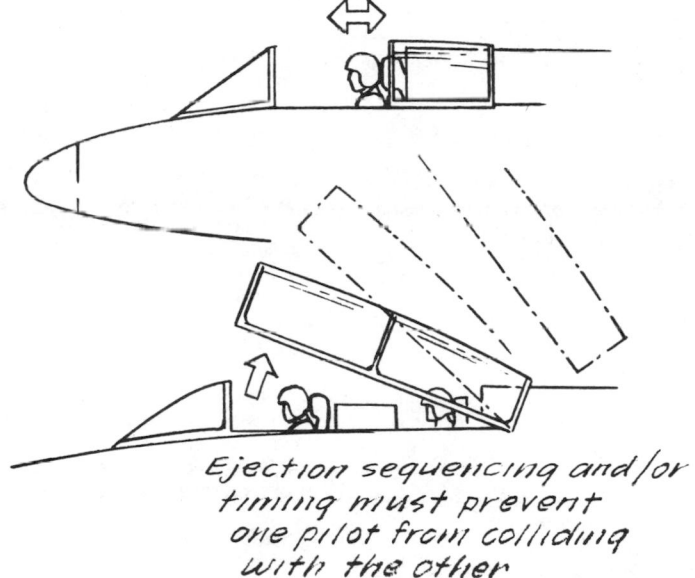

*Ejection sequencing and/or timing must prevent one pilot from colliding with the other*

## Spacecraft Emergency Hatch

Certain types of spacecraft (such as the Apollo-type spacecraft shown in the accompanying illustration) typically have a single hatch, both for normal and for emergency ingress and egress. Such hatches should always permit opening both by crew members from the inside and by support crew members from the outside. If water landings are the expected mode, the location of the hatch must be such as to minimize the potential for shipping water when the hatch is opened while the spacecraft is in the water.

*Note:* Hatch or canopy opening hardware design must be compatible with the typical limitations imposed by the personal protective garments or equipment that pilots and astronauts are expected to wear.

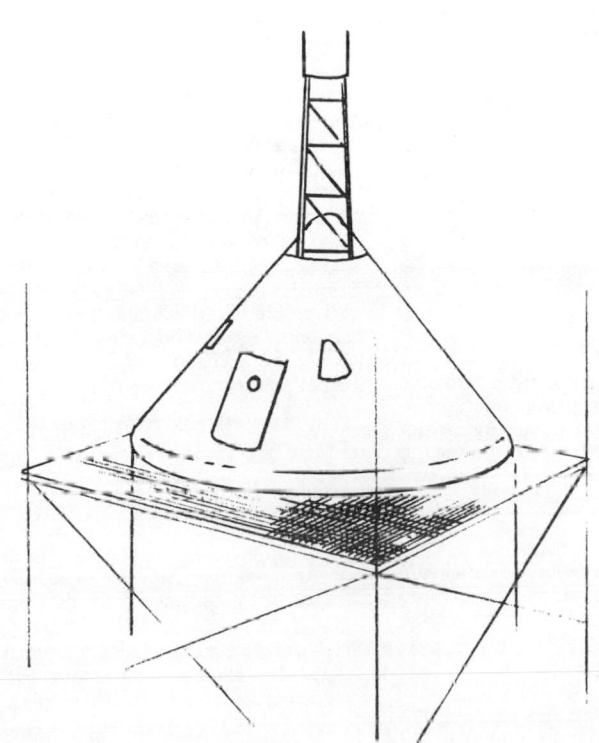

# GENERAL

## Window Functions

The basic function of a window subsystem should be identified before a custom window for any particular application is selected or designed. Note that the aesthetic function should be considered only after the other functional needs are completely satisfied. In many cases, the earlier functional needs can be satisfied, while at the same time making the window design quite attractive. The primary functions of windows are as follows:

1. To provide a view of the external world from inside an enclosed architectural facility, vehicle, or other occupied system
2. To provide natural light for the interior of an occupied facility
3. To provide ventilation
4. To provide an auxiliary emergency escape route
5. To provide a visual communication link between two sides of an otherwise solid wall
6. To provide an aesthetically desirable focal point for an architectural system

The secondary or auxiliary functions of windows may also include the following:

1. To provide natural light filtering
2. To provide a thermal and/or acoustic barrier
3. To provide a moisture and/or dust barrier
4. To provide one-way viewing for covert observational requirements

## Special Window Design and Selection Considerations

1. Location with respect to functional needs
2. Size with respect to ventilation, escape, and maintenance requirements
3. Security configuration (locks, guards, etc.)
4. Access and ease of opening and closing, cleaning, etc.
5. Location and/or materials relative to possible breakage
6. Screening the open window to prevent insect penetration
7. Glare control
8. Supplemental windows (storm windows, draft shields, etc.)

## Typical User Problems

A brief review of the most common problems people have with windows provides further insight into the features that good window design should have. The following are among the most frequent complaints made by homeowners, office workers, and others:

### 1. Ventilation versus Draft
Poor window design and/or location makes it practically impossible to obtain the desired ventilation without introducing undesirable drafts, rain, dirt, or dust.

### 2. Observation
People cannot see out from their typical working or resting position, cannot observe children playing outside, or cannot observe arriving guests or an unknown visitor at the door.

### 3. Visual Task Efficiency
People cannot perform required visual tasks efficiently or without becoming fatigued because of direct glare from the window and/or because of the extreme contrasts created between the bright window and the interior workplace arrangement.

### 4. Window Operating Difficulties
Windows are difficult to open and close because of their location and/or because of the particular opening and closing hardware system selected.

### 5. Lack of Security
With few exceptions, most commonly produced window systems have extremely poor locking systems and/or guard systems or have no such systems.

### 6. Window Installation Methods
Many window systems are installed with little consideration for the additional window-related features that may have to be added; i.e., there may be a lack of wall space and/or a solid surface on which to mount blinds or drapes.

### 7. Window Maintenance
Many window configurations create extremely difficult cleaning problems and replacement problems, such as removal and replacement of broken glass or obsolete hardware.

### 8. Special Hazards
Hazards are created by windows which have no frames, which extend to the floor, which are very large and/or very high, which require the use of a ladder for cleaning, or which are located so that they open out across an adjacent path.

Sometimes a window is the last remaining escape in an emergency, and if it is too small or is located out of reach, this final route is unavailable.

### 9. Expense
Some window designs are extremely costly, not only at the time of purchase, but also on a continuing basis as maintenance and replacement requirements occur. The "all-glass" concept, which comes and goes in terms of popularity, creates severe glare control problems, adds to the cost of heating and cooling, is a continual cleaning problem, and generally results in considerable replacement expense for the owner.

## Windows versus Windowless Architecture

Although there are certain times when a windowless environment is important for the work being carried on within a facility, such as a photographic laboratory or a light and vision experimental laboratory, people tend to fare better psychologically when they have visual access to the outside world. Therefore, most living and working spaces should have windows. The objectives for windows include the following:

1. Specific observation of outside activities, such as when a mother watches her children at play; when a plant supervisor checks a specific operation that is under way; when an air controller observes aircraft landings, taxiing, and takeoff; or when a railroad switchman observes railyard movements.
2. Observation and/or inspection of what is happening inside a facility, such as when a person looks inside a restaurant before entering, when a shopper examines articles in a retail store window display, or when a patrol officer makes a security inspection.
3. Observation of the out-of-doors is very important in many situations, such as schools, factories, hospitals, and convalescent homes. A moment of relaxed observation of the outside world often gives the office or industrial worker a "second wind" during a long, hard work period.
4. Although there are conflicting opinions about the value of a sunlit living and working environment in relation to general health, most experts are of the opinion that people who work or live in windowless environments are physiologically and psychologically less well off than those who have the benefit of sunlit living and working environments.
5. Emergencies will always arise when power failures cause loss of illumination, ventilation, or air conditioning. Therefore, from a cost-effectiveness point of view, the windowless architectural facility may lose its primary advantage of ease of interior environmental control.

# ARCHITECTURAL WINDOWS

## User-Oriented Considerations in the Selection and Use of Windows in Architectural Systems

Decisions concerning the location, number, size, and type of windows to be used in architectural systems should be based upon a user-oriented philosophy. That is, one should approach these decisions from the point of view of those who will be inside the occupied spaces of the home, office, factory, school, hospital, or church, rather than from the point of view of creating architectural effects for the benefit of passersby. Decisions concerning the selection and use of windows should be based on questions such as the following:

### 1. Where Should Windows Be Located?
a. They should be located so that they will maximize desired exterior viewing angles from key visual reference points.
b. They should be located so that they will distribute incoming, natural light effectively for visual tasks and for the enhancement of the internal psychological environment.
c. They should be located so that good cross ventilation can be obtained by opening various windows.
d. They should not be located where they present unnecessary hazards.
e. They should be located so that they are accessible for cleaning and repair.

## 2. How Many Windows Should There Be, and What Size Should They Be?

a. There should be enough windows and/or sufficiently large windows to permit the desired natural light to enter where it is needed. However, the windows should not be so numerous or so large that there is a problem of light control or of cleaning and maintenance.

b. There should not be so many windows that there is not enough wall space to arrange furnishings.

c. There should be enough properly located windows to provide secondary emergency escape.

d. There should not be so many windows or so much glass that a thermal or noise insulation burden is placed on the owner of the facility.

## 3. What Types of Windows Should Be Selected?

a. Select windows that permit the desired exterior viewing.

b. Select windows that permit maximum ventilation control.

c. Select windows that provide desired light control.

d. Select windows that provide both security from burglary and ease of use for emergency escape.

e. Select windows that are easy to clean from the inside.

## 4. What Decorative Features Should Be Selected?

a. Select windows that are compatible with the basic architectural style of the proposed facility.

b. Select windows that convey a desired symbolism.

c. Select windows that will not be impossible to clean, repair, or replace.

An important aspect of window selection is the particular hardware that goes with the window, i.e., devices which will be used to open and close, lock, or adjust the window which will be used to replace broken glass.

Consider the type of glazing that best fits each application, i.e., glass that provides thermal insulation, semireflecting glass, colored glass, or plastic safety material.

Consider the type of window framing material with regard to weathering, rust, corrosion, need for frequent painting, etc.

## 5. Screens

Except in unusual climates (i.e., areas where there is never a problem with insects), window types should be selected that will accept screens. Screen removal and replacement should be simple and such that potential damage to the screens is minimal. The method of attachment should be such that the screens are accessible from inside the building unless the windows are positioned so that the screens can be reached and removed or replaced by a person who is on ground level or on a solid surface.

*Note:* The size, shape, and weight of removable windows and screens must always be compatible with the reach and strength capabilities of the persons who are normally involved in cleaning them; e.g., residential windows, which are typically cleaned by women, are smaller than industrial and commercial windows.

## General Characteristics of Window Types[2]

1. Double-hung: Not apt to sag; easy to install screens or storm windows; only 50% openable; does not protect from rain when open; hard to operate over an obstruction; inflowing air cannot be diverted up or downward; horizontal members obstruct view.

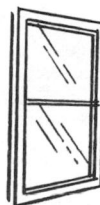

2. Double-hung (reversed): Same as above but easier to clean.

3. Casement (out): 100% vent opening; will deflect drafts; does not protect from rain; a hazard close to walkways; vertical members obstruct view; may sag if not structurally strong; inflowing air cannot be diverted downward; interferes with drapes or furniture.

4. Casement (in): Same in many respects to above but easier to clean.

5. Awning, canopy: Not apt to sag; screens easy to install; 100% opening; will deflect drafts; offers rain protection when partly open; diverts inflowing air upward; vertical members obstruct view; inflowing air cannot be diverted downward; a hazard close to walkways; can be cleaned from inside.

 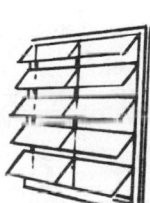

6. Pivoted (vertical): Not apt to sag; 100% opening; easy to clean; will deflect drafts; no protection from rain when open; hazard when located close to walkway; inflowing air cannot be diverted downward; interferes with drapes and furniture; screens or storm windows difficult to provide; fairly easy to clean.

[2]Reprinted from *Selection of Windows*, Technical Reprint no. 5, Building Research Institute.

7. Pivoted (horizontal): Same as above, except offers rain protection when partly open; diverts inflowing air upward; inconvenient operation when over an obstruction; horizontal members obstruct view.

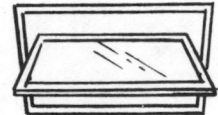

8. Top hinged (out): Not apt to sag; screen and storm windows easy to install; provides 100% opening; offers rain protection when partly open; inconvenient operation when over an obstruction; hazard when close to walkway; inflowing air cannot be diverted downward; sash has to be removed for washing.

9. Bottom hinged (in): Not apt to sag; 100% opening; easy to clean; will deflect drafts; offers rain protection when partly open; diverts inflowing air upward; interferes with drapes and furniture.

10. Fixed sash: Not apt to sag; large size practical; sash has to be removed for cleaning.

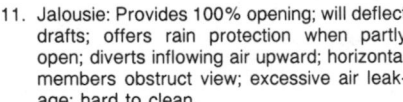

11. Jalousie: Provides 100% opening; will deflect drafts; offers rain protection when partly open; diverts inflowing air upward; horizontal members obstruct view; excessive air leakage; hard to clean.

12. Monitor: Not apt to sag; 100% opening; will deflect drafts; offers rain protection when partly open; diverts inflowing air upward; hazard when located close to walkway; horizontal members obstruct view; sash has to be removed for cleaning.

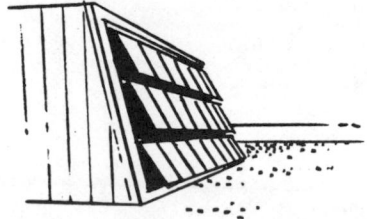

13. Projected: Offers rain protection when partly open; diverts inflowing air upward; hazard when located close to walkways; horizontal members obstruct view; somewhat difficult to clean from inside; not apt to sag.

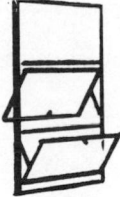 

14. Horizontal sliding: Not apt to sag; screen and storm windows easy to install; large sizes practical; only 50% openable; does not protect from rain when open (although screen helps); inconvenient when located over obstruction; vertical members may obstruct view; inflowing air cannot be directed or diverted; storm windows difficult to provide; sash has to be removed for cleaning (large size windows difficult to handle).

**OTHER ALTERNATIVES**

Combination

Security

Basement - Utility

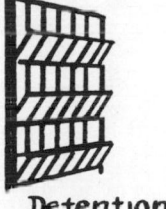

Detention

### Window Type and Hardware Considerations

Key considerations from the user's point of view should be the following:

1. Ease of opening and closing
2. Ease of latching and locking
3. Ease of cleaning

For the following reasons, select sliding window configurations whenever appropriate and practical:

1. They do not encroach upon interior space, nor do they project outside, possibly into the path of someone who is walking past an open window.
2. The opening operation is direct and obvious.
3. Screening is usually simpler.
4. They generally can be cleaned from inside.

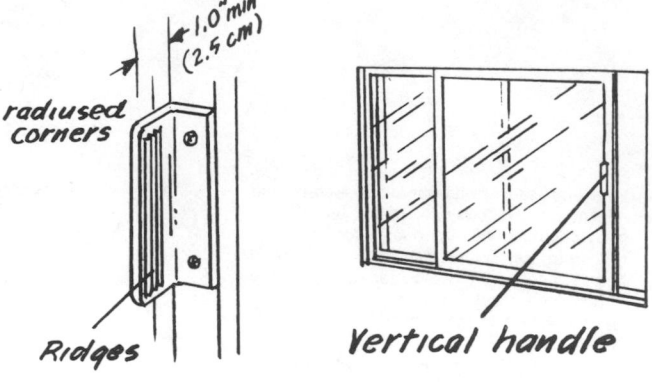

radiused corners

1.0" min (2.5 cm)

Ridges

Vertical handle

The following precautions should be considered:

1. Select windows that are easy to slide, i.e., that do not bind or hang up.
2. Select windows with manipulating hardware that is easy to get hold of.

The next most preferred window types are those which utilize a mechanical hand crank (as opposed to those which are opened or closed without some device to hold them in place). The examples shown in the accompanying sketches are typical. Note the suggested hardware dimensions; these are desirable so that the user can manipulate the crank or lock easily.

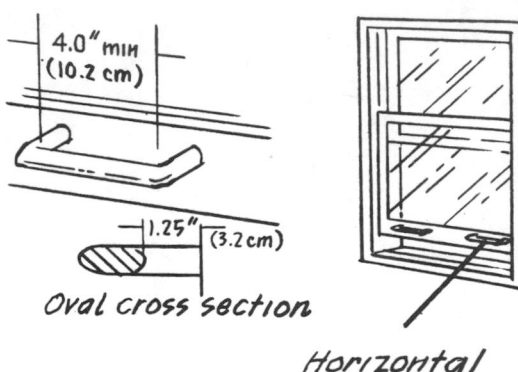

4.0" min (10.2 cm)

1.25" (3.2 cm)

Oval cross section

Horizontal handles

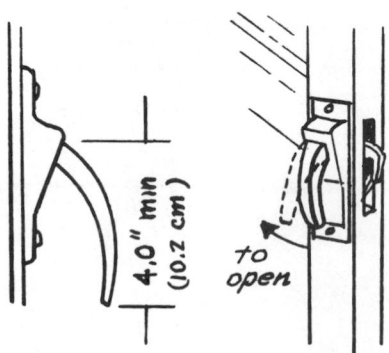

4.0" min (10.2 cm)

to open

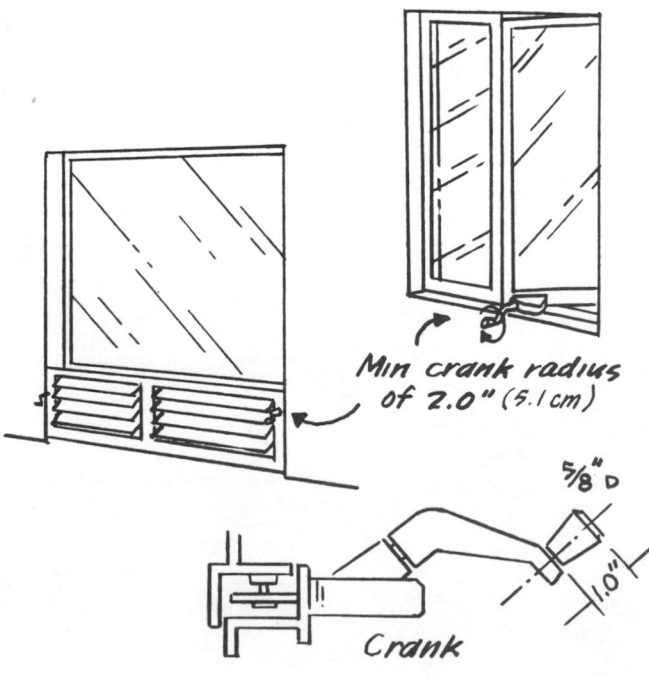

Min crank radius of 2.0" (5.1 cm)

5/8" D

1.0"

Crank

## Avoiding User-Induced Injuries

Window concepts and window placement should be analyzed carefully to determine what possible user-induced injuries might occur. Many hazards that could lead to injury can be minimized by proper design. The accompanying sketches illustrate several common design-related hazards that often occur in the home environment:

1. Avoid large glass expanses that go all the way to the floor unless special safety glazing is used.
2. Sliding glass doors should always have a solid section at the bottom, not only so that people can see when the door is closed, but also so that children cannot run tricycles into the glass.
3. Avoid placing large glass windows where furnishings might inadvertently be pushed into them.
4. Avoid placing large windows at the bottom of narrow stairs.
5. Provide a method for getting to the outside of large glass windows so that they can be safely cleaned.

6. Placing windows where it is difficult to reach them either to open or close them or to clean them is a gross design error. As shown in the accompanying sketches, not only may this cause an injury, but it may also encourage people to take chances that could lead to even more serious injuries.

7. Care in the total planning is also required to avoid placing certain kinds of windows so that they can lead to the kind of human error (and obvious injury) that the accompanying sketch indicates is about to occur.

8. When a window is obviously the only alternative emergency escape route, select a window that a person can get out of quickly and safely and locate the window where it can be reached easily. Consider not only access from inside but also a safe escape route once the person is through the window.

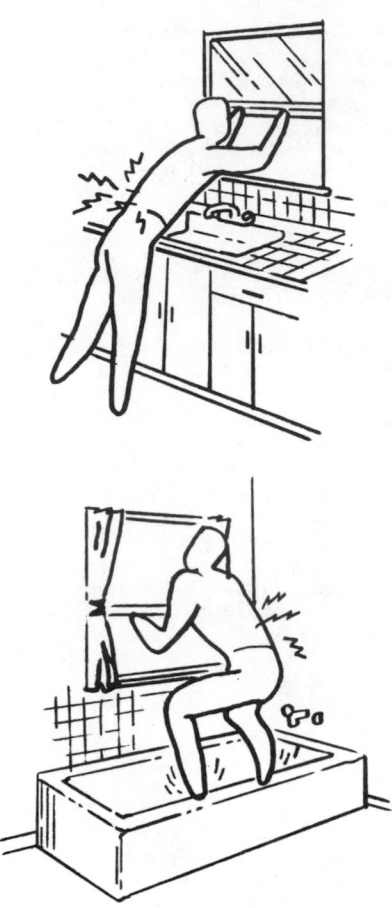

*Poor*

*Correct*

*Don't invite mis-use!*

9. "Jut-out" windows such as those illustrated in the accompanying sketch are extremely hazardous to both sighted and blind persons. As illustrated, sighted individuals tend to turn from their position in front of a store window that is flush with the sidewalk and to walk directly into the protruding window before checking to see whether the coast is clear. This occurs most often if the window that juts out has no frame around it. The blind person who is using a cane to trail the storefront wall often runs directly into the jut-out window before the cane makes contact, especially if the jut-out window ledge is above knee level.

*People move before they look – Avoid creation of booby traps!*

## Window Glazing Guidelines

### FHA Safety Requirements

Safety glass should be used on all exterior doors with large glass areas that do not have a conspicuous visual barrier. The glass must either be tempered or wired.

For single-strength door panes, the short dimension must not exceed 15 in (38 cm).

Glass doors in which the glass goes all the way to the floor must be at least $7/32$ in (0.55 cm) thick, and horizontal muntins or bars (push rails) must be provided 36 in (91 cm) above the floor (otherwise, safety glass is required).

Glazed doors for shower and/or tub enclosures must be made of safety glass.

Glass areas and thicknesses required for various maximum wind loads are as shown in the accompanying table.

The values shown in the table apply to sheet glass not more than 30 ft (9.2 m) above ground level and are based on a length/width ratio of 1:2 or less.[3]

### Acoustic Barrier Effect of Glazing Methods

Ordinary openable windows, closed but not weatherstripped, common window glass up to 20 dB

Single fixed weatherstripped glass (3 to 5 mm thick)    25 dB

Single fixed or openable weatherstripped windows (6- to 10-mm glass)    30 dB

Single fixed or openable weatherstripped windows (10- to 20-mm glass)    35 dB

Double windows with 50- to 100-mm air space (3- to 6-mm glass)    35 dB

Double windows with 100- to 200-mm air space (3- to 6-mm glass)    40 dB

Sealed double windows, separate frames, with 150- to 300-mm air space (4- to 12-mm glass)    45 db

Glass 3 mm thick provides an average sound attenuation of 23 dB; glass 4 mm thick, 24 dB; glass 6 mm thick, 25 dB; and glass 12 mm thick, 27 dB. The effect of the proportion of window area versus a solid brick wall is approximately as shown in the accompanying table.

### Thermal Barrier Effect of Glazing Methods and Size of Window

For specific information regarding the thermal transfer and/or barrier effects of windows and glazing materials, refer to the latest edition of the *ASHRAE Guide and Data Book*. As a general consideration, however, one must address the thermal transfer question as a problem of creating a balance between minimizing the loss of internal heat during the winter months and resisting the penetration of external heat during the summer months. This involves, in addition to examination of the particular window and window glazing plan, examination of sun-angle patterns and consideration of the possible use of special architectural features such as roof eaves and fixed or movable louvers and window designs (single versus double panes, coatings, etc.).

[3]For other configurational requirements, see *Minimum Property Standards*, Federal Housing Administration.

### Windowpane Size and Thickness and Materials to Minimize Accidental Breakage

| Thickness, mm | Maximum Area, m² | |
| --- | --- | --- |
| | Ordinary Glass | Laminated Glass* |
| 3 | 0.2 | Not recommended |
| 4 | 0.3 | Not recommended |
| 5 | 0.4 | 1.0 |
| 6 | 0.7 | 3.0 |
| 10 | 1.5 | 6.0 |
| 12 | 3.0 | 7.0 |

*Laminated glass should contain a plastic interlayer not less than 0.76 mm thick.

| To 80 mph | To 90 mph | To 120 mph |
| --- | --- | --- |
| 16 ft²—$1/8$ in | 11 ft²—$1/8$ in | 7 ft²—$1/8$ in |
| 40 ft²—$3/16$ in | 27 ft²—$3/16$ in | 18 ft²—$3/16$ in |
| 75 ft²—$1/4$ in | 51 ft²—$1/4$ in | 34 ft²—$1/4$ in |
| 160 ft²—$1/2$ in | 113 ft²—$1/2$ in | 73 ft²—$1/2$ in |

| Window Area as Percent of Total Wall Area | Estimated Insulation Value, dB | |
| --- | --- | --- |
| | Single Pane | Double Pane |
| 100 | 20 | 40 |
| 75 | 21 | 41 |
| 50 | 23 | 43 |
| 25 | 26 | 45 |
| 10 | 30 | 47 |
| 0 | 50 | 50 |

## VEHICULAR WINDOWS

### 1. Driver Visibility

From the human factors point of view, the most important consideration relates to the ability of the driver to see everything outside the vehicle that he or she should see. The size, shape, and transmissibility characteristics of the windows should be such that there is minimal obstruction of the driver's exterior view as a result of roof support structure, reflections on either the interior or the exterior surface of the window glass, optical distortions in the glass itself, or the inability to keep windows clear of fog, mist, rain, snow, or ice.

**ANGLE OF VIEW** Essentially, the driver should be able to see typical traffic features (e.g., other vehicles, the roadway, pedestrians, and signs or signals) in all directions, either directly or by means of appropriately positioned rearview mirrors. The driver should also be able to see key referencing elements of the vehicle, i.e., the front and rear corners or fender limits.

**GLARE AND REFLECTION** Windows should be configured (along with other elements of the vehicle, such as instruments and interior reflecting surfaces) to minimize the amount of direct sun glare and/or reflected patterns on the glazing that might interfere with the driver's ability to see clearly.

### 2. Passenger Visibility

Although of lesser priority, it is important that passengers also be able to see out of a vehicle, whether it is an automobile, a bus, a transport aircraft, a train, or another type of vehicular system. Although the things that passengers have to see are perhaps less critical in most cases than those which the driver must see, passengers of public transit vehicles need to be able to observe signs that help them determine where they are or when they should get off. This is particularly important for the passengers of urban transit buses, interurban transit buses, and subway trains. For this reason, passenger windows in these vehicles should be designed so that passengers can accomplish these seeing tasks when either sitting or standing.

### 3. Ventilation

Vehicular windows should be designed for ease in controlling direct ventilation within the vehicle (except for special cases, such as the air-conditioned train or bus).

### 4. Emergency Escape

Windows often may be the only means for emergency escape and/or rescue. Window opening shape and dimensions, as well as opening methods, should be made compatible with the physical and mental capabilities and limitations of the driver or operator and the passengers.

### 5. Crashworthiness

Shatterproof window glazing is required in all vehicles that may be subject to crash conditions. It may also be desirable to consider "pop-out" windshields for private or smaller highway vehicles where it is possible that passengers could be thrown into the windshield.

### 6. Window Clearing

Most modern vehicles are now equipped with reasonably effective windshield wiping, washing, defogging, and/or defrosting systems. Some are being equipped with similar systems to clear the rear and side windows. It is recommended that all vehicles have systems to perform these clearing functions for all windows that a driver may, at one time or another, need to look out of in order to drive as safely as possible.

### 7. Pressurized Windows

All window systems in aircraft that fly above 20,000 ft (6100 m) should be designed to maintain cabin pressure under any condition of flight.

### 8. Aircraft Windshield Penetration

The forward windshield and/or the windows in any high-performance aircraft should be designed to preclude the penetration of the glazing by foreign objects, e.g., birds. Refer to FAA regulations pertaining to windshield "bird-proofing."

### 9. Distortion

Windshield glazing materials should always be of sufficiently high quality and configuration so as not to alter the appearance of objects viewed through the window or seriously distort the color of important traffic control devices such as signal lights, color-coded signs, and runway or taxiway markings. Select glazing materials that will not deteriorate as a result of weather, sun, heat, cleaning compounds, etc.

## Safety Considerations

All glazing in vehicular windows, whether for automobiles, trucks, buses, aircraft, or marine-craft, should be shatterproof.

There should be some means for clearing rain, snow, ice, mud, or collected dirt from outer surfaces and for clearing moisture condensation from inner surfaces.

Objectives for clearing windows should be based on the critical operator viewing needs, i.e., the viewing angles the operator requires in order to see ahead, to the side, and to the rear of the vehicle.

External surface washing capability should be provided for the critical windows, i.e., the windshield and the rear window.

Window clearing controls should be located where they can be reached quickly, and they should operate in a simple, direct manner so that the operator is not delayed while trying to understand complex labels or instructions. These labels must be visible at night as well as during the day.

(Refer to SAE standard practices for a complete description of window design requirements.)

## Window Operating Convenience

Window controls should be located conveniently, and they should operate smoothly and easily. A window crank with a 2- to 3-in (5.1- to 7.6-cm) radius is preferred, with a maximum torque of no more than 40 in-lb (4.5 m·N). No more than two to three turns should be required to raise or lower a window. Electrically powered windows should include an emergency lock-out so that a child will not accidentally crush a hand or choke. The driver should be able to prevent passenger window controls from being operated.

Avoid booby traps that can cause injuries and/or inconveniences for the vehicle user, particularly the following:

1. Sharp corners: As illustrated by the accompanying sketch, sharp corners can become a hazard to the person who is hurrying to get into a car. This applies not only to the upper rear corner of a door-frame or frameless window glass, but also to special features such as the window wings frequently provided to control ventilation.

2. Window opening size and shape: Windows that may have to be used for emergency egress must be of a sufficient size and an appropriate shape so that a person can get his or her body through the opening. Preferably, the opening system should cause the window glass to disappear completely, leaving the opening clear of glazing.

3. Powered system malfunction: Powered systems should not fail with the window closed if the particular window is intended as an auxiliary emergency escape route.

*Shatter-proof - pop-out windshield*

*Wipers front and rear*

*Fast response for complete closure*

*Smooth operating - No binding or cocking!*

Avoid sharp corners or use frameless windows

## AVERAGE AUTOMOBILE DIMENSIONAL GUIDELINES

Very large cars take approximately 26 ft (7.8 m) turning radius. Although careful and skillful drivers can turn much shorter in the average large or small car, it is recommended that the maximum or conservative dimension be used whenever possible.

Although it has been standard practice for some time to specify garage and parking-lot spaces using considerably smaller values than shown in the accompanying illustration, such standards do not account for the difficul-ties that less skilled drivers have in maneuvering a vehicle in and out of tight quarters. In addition, current standards, because of the ever present objective of putting as many cars into as little space as possible, are based on mixes of large and small cars and assume that most people can get in and out of a car without fully opening the doors and/or that all drivers can park parallel to surface markings pro-vided.

Note especially that the spaces in parking areas set aside for persons who must manipu-late a wheelchair in and out of a car must be wider than the other spaces.

|     |                    | Small Cars          | Large Cars          |
|-----|--------------------|---------------------|---------------------|
| L   | Overall length     | 15.0 ft (4.5 m)     | 20.25 ft (6.1 m)    |
| W   | Overall width      | 5.9 ft (1.8 m)      | 6.81 ft (2.0 m)     |
| H   | Overall height     | 5.0 ft (1.5 m)      | 5.4 ft (1.6 m)      |
| WD  | Width plus one door | 8.7 ft (2.6 m)     | 10.49 ft (3.2 m)    |
| W2D | Width plus two doors | 11.48 ft (3.4 m)  | 14.18 ft (4.3 m)    |
| WB  | Wheelbase          | 8.82 ft (2.7 m)     | 12.48 ft (3.7 m)    |
| T   | Tread C/C          | 4.68 ft (1.4 m)     | 5.23 ft (1.6 m)     |
| RC  | Road clearance     | 4.7 in (11.9 mm)    | 6.5 in (16.5 mm)    |
| OHF | Overhang, front    | 2.36 in (7.7 mm)    | 3.38 in (11 mm)     |
| OHR | Overhang, rear     | 3.63 in (11.9 mm)   | 5.7 in (16.6 mm)    |
| AA  | Approach angle     | 14.4°               | 29.71°              |
| AD  | Departure angle    | 9.6°                | 21.21°              |
| AR  | Ramp breakover     | 7.3°                | 16.21°              |

Full open

T    WD   W2D   W

AR

24" (60.4cm)

36"    36" (91.4cm)

12" (30.5cm)

20" (50.8 cm)    20"

H

AA    RC    AD

OHF    WB    OHR

L

Tire bumper stops prevent car bumper overlapping curb

Curb extends behind longest vehicle keeps passing cars from hitting parked cars

60" (152.4cm)    40" (101.6cm)

HANDICAPPED STALL    HANDICAPPED STALL    STD STALL

10' (32.8 m)    12' (39.4 m)    10' (32.8 m)    10'

30' (98.4 m)

WALL OR PARKED VEHICLES

## MULTIVEHICLE PARKING

The preferred arrangement for multivehicle parking is shown in the accompanying sketch. Although perpendicular parking will accommodate more cars in less space, many drivers have considerable difficulty manipulating their cars so that they are not skewed within the marked stalls. Also, it is preferred that the arrangement allow drivers to turn left into the stall because drivers are positioned on the left side of a car and can see what they are doing better than when they have to look across a large expanse of hood to the far corner of a car.

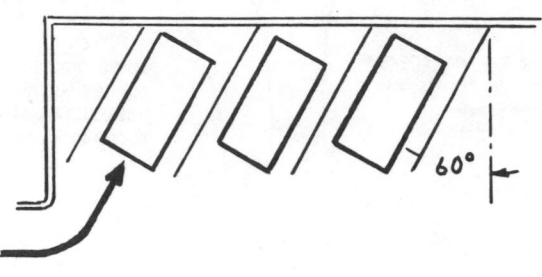

The preferred back-out clearance and circulation pattern for the parking garage or parking lot are shown in the accompanying sketch. This clearance permits the worst combination of two long vehicles, side by side, and the clearance will be sufficient so that even the poorest driver can be sure that his or her right front fender and bumper have cleared the rear fender and bumper of the car to the right. Although it is recognized that this circulation pattern takes more space than current parking pattern standards, it precludes some of the typical rear-end collisions that occur when two people try to back out at the same time, and incoming traffic and departing traffic are kept separated.

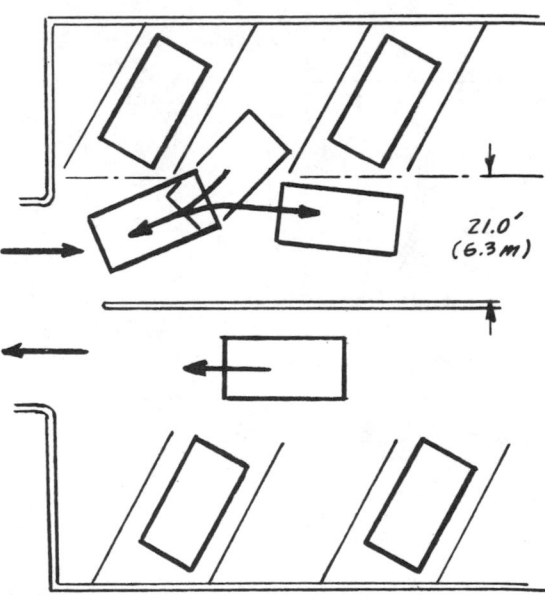

Complete separation of the entrance and exit is even more desirable, as indicated in the accompanying sketch.

The current trend toward smaller cars makes it desirable to consider separate sections for larger or standard-size vehicles and for smaller ones; this provides an opportunity to save considerable space in the multivehicle parking lot or parking garage. When a separate small-car area is set aside, the spacing for each car can be reduced from the recommended 10 ft (3.1 m) centerline to centerline, and the rear backing clearance can be reduced to 21 ft (6.4 m) for perpendicular stalls and 18 ft (5.5 m) for 60° stalls. This reduction is practical not only because there is a basic reduction in the width of the smaller vehicles but also because drivers can see the perimeters of the vehicle better and thus can be more precise in their maneuvering. The spacing for drivers who are in wheelchairs, however, should still be about 12 ft (3.7 m).

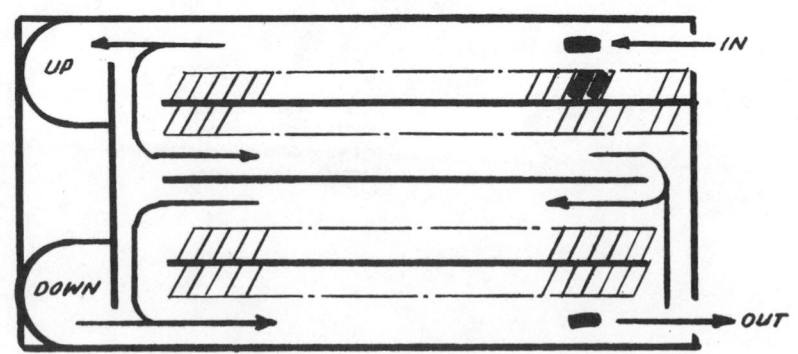

## PEDESTRIAN ACCESS

Especially in multistory parking buildings, one should pay particular attention to the matter of how persons who have parked in the proposed structure are to get to and from their cars. Although part of this question involves the physical aspect, i.e., by elevator and/or by stairs, another and equally important part relates to identification of the owner's vehicle position so that he or she can find it again.

The multilevel parking garage should be laid out so that the relationship between floors and pedestrian departure and return features is coincident; i.e., avoid some of the typical approaches in which, for example, an elevator entrance is placed at one level, and a stairway for that same level is placed at a point that appears to the parking client to be on the next higher or lower floor or level. The preferred arrangement is to place both the elevators and the stairs in the same area or in adjacent areas.

## ILLUMINATION

Often multilevel parking garages are designed with less than desirable illumination. This occurs most often with the "open-sided" designs because designers believe there will be enough natural light for motorists to see. Unfortunately, this often works in reverse; i.e., the glare contrasts against a rather dark interior during the day, making it difficult for motorists to see. Although it is generally not necessary to have as extensive artificial illumination during the day as at night, in the case of the multilevel parking garage the opposite may be true. For the same reason, the interiors of the open-sided parking garage should also be painted a light color, thereby taking advantage of "bounce" light and thus reducing the necessity for extra daytime artificial lighting.

## PARKING-SPACE ALLOWANCES

The number of parking spaces to be provided for various types of buildings will, of course, vary with the type of building and with the availability of public transportation. The following guidelines provide a good starting point for planning purposes:

*Residential*

1. Single-family dwelling (separate house)  3.0
2. Multifamily dwelling (efficiency apartment)  1.0 per dwelling unit
3. One- or two-bedroom apartment  1.5 per dwelling unit
4. Apartment with three or more bedrooms  2.0 per dwelling unit

*Commercial*

1. Offices  3.3 per 1000 ft²
2. General retail  4.0 per 1000 ft²
3. Shopping centers  5.5 per 1000 ft²
4. Restaurant  0.3 per seat
5. Motor hotel  1.0 per guest room  0.5 per staff member

*Industrial Facilities*  0.6 per employee

*Churches*  0.3 per seat

1.0 per staff member

*Hospitals*  1.2 per bed

*Note:* There should be a minimum of three stalls for handicapped persons per 100 cars for any public facility. These special parking places should be located the minimum distance possible from main entrances.

## PARKING-AREA LIGHTING

Needless to say, parking facilities should be well lighted at night. Use of fluorescent-type lighting is preferred in order to ensure maximum diffusion and minimum shadow. Avoid the temptation to use a few fixtures at relatively high positions above the ground in order to reduce the initial cost. Generally this requires higher footcandle levels, and in the long run these fixtures may cost just as much to maintain as more lower-powered fixtures spaced closer together. A minimum of about 3 (maintained) fc measured at ground level is required for people to find their way around in the typical parking lot at night.

## PARKING-LOT SIGNING

A logical signing system should be developed at the same time that the parking arrangement is being considered. All signs and markings should be visible (not obscured by the tops of cars) and readable, i.e., well illuminated at night, large enough for the typical viewing distances, and legible (appropriate contrast and alphanumeric font, etc.).

## OTHER FEATURES

**1. Curbs and Tire Bumpers**
In spite of the higher initial cost, curbs or tire bumpers are desirable to keep order in the average parking lot. In addition, such devices may prevent damage to vehicles and thus possible liability for the parking-lot proprietor.

**2. Speed Control Berms**
Raised ridges in the parking-lot surface at strategic points are useful to keep drivers from driving too fast. These should always be painted to contrast with the primary surface, however; otherwise, pedestrians may trip over them.

**3. Wheelchair Facilities**
There should be special ramps to allow persons in wheelchairs to get from their cars (at ground level) to whatever primary walkways are associated with the particular facility. These should be located close to the special

parking spaces provided, and the route between these spaces and the ramps should not cause handicapped persons to be exposed any more than is absolutely necessary to moving vehicles as they go from their cars to the special ramps.

**4. Marking of Pedestrian Paths**
When very large parking lots are involved, consider the addition of special walkways to various parts of the lot and/or plan to mark safe paths to these areas on the lot surface.

## COVERED GARAGE PARKING: EXITS AND ENTRANCES

Care should be exercised in planning garage exits so that departing motorists can see ahead. The accompanying illustrations depict several critical considerations (lateral view and view over the hood and/or beneath the upper windshield brow):

Provide the motorist a good peripheral view before he or she crosses the pedestrian right-of-way.

Avoid inclines that are so steep that the car's hood prevents the driver from seeing where he or she is going.

Allow the car to travel over a sufficiently large flat space prior to crossing the sidewalk and entering the street.

In residential design, avoid creating short driveways on inclines; i.e., provide a level space between the garage and the sidewalk or street on which a car can sit without being in danger of rolling out of the driveway in the event a parking brake fails.

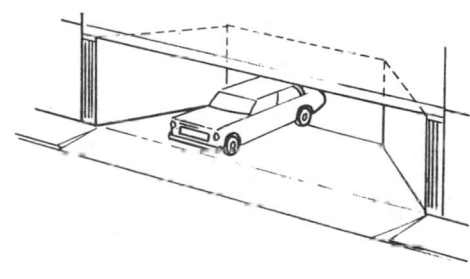

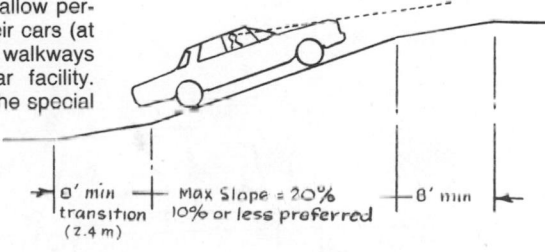

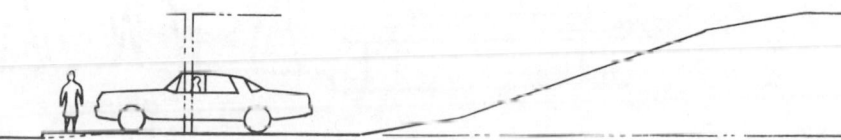

8' min transition (2.4 m)    Max Slope = 20% 10% or less preferred    8' min

## EXTERIOR SIDEWALKS

Exterior sidewalks should be wide enough so that persons who are traveling along the walkway do not feel as if they are walking a tightrope. The guidelines illustrated in the accompanying sketches take into account elderly persons' unsteadiness as well as the need to feel that one can walk rapidly and easily without scraping walls or falling off a curb:

1. The minimum width for a sidewalk that normally would be used by only one person at a time is about 30 in (76 cm) (36 in, 91 cm, is preferred because individuals are usually carrying something).

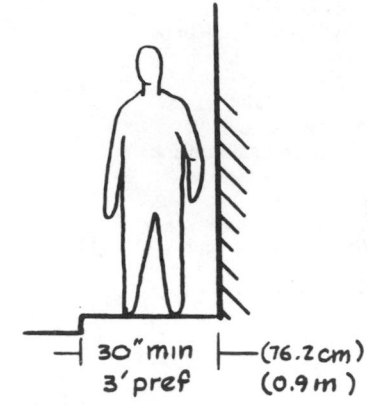

2. Lightly traveled sidewalks should be at least 4 ft (1.2 m) wide so that two adults can pass each other without bumping.

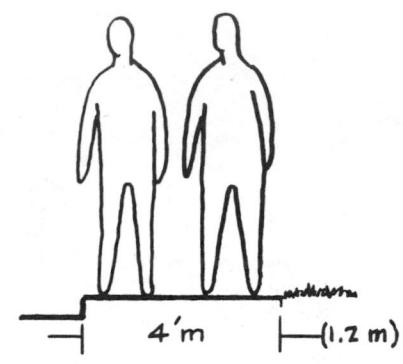

3. Moderately traveled sidewalks in suburban areas should be at least 7 ft (2.1 m) wide in order to accommodate the most common passing situation, in which two people pass a single person going in the opposite direction, without crowding.

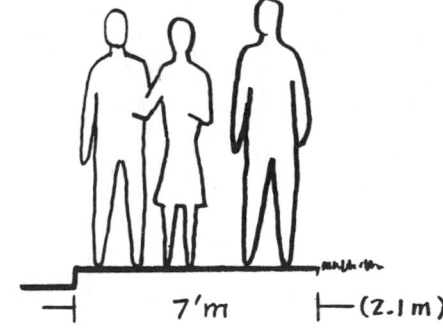

4. The busy sidewalk, especially in commercial urban environments, should be at least 12 ft (3.7 m) wide. The accompanying sketch illustrates some of the needs that dictate ample space for widely varying sidewalk activities.

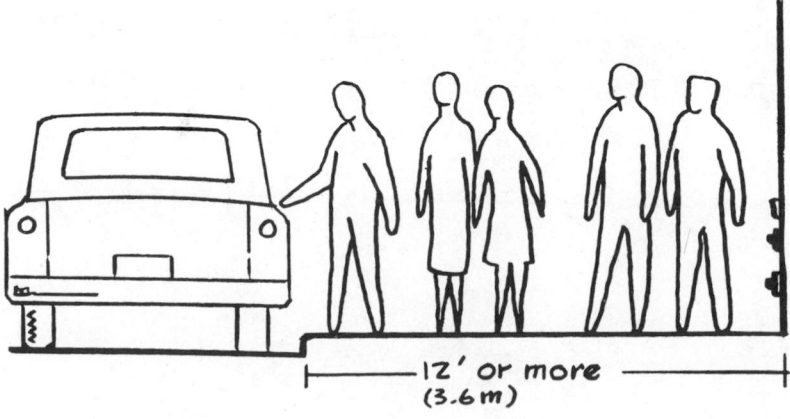

Outdoor sidewalks should always be laid out so that they are relatively level (except for sufficient slope to cause rainwater to drain away).

Sidewalks should be set back from the street. This prevents pedestrians from having to walk precariously close to vehicular traffic and provides sufficient space so that driveway and/or wheelchair ramps do not cross the primary walkway, which is sometimes difficult for elderly persons to negotiate without turning an ankle.

Sidewalks should be at approximately the same level as adjacent lawns so that someone who accidentally steps off the walkway will not turn an ankle.

Curbs should not be more than 10 in (25 cm) high (this is the accepted maximum for a single step).

Whenever a walkway is more than 10 in (25 cm) above the ground, a suitable railing should be provided. The railing should contain some type of intermediate grille or other barrier to prevent small children from falling off the walkway.

Walkways should have nonskid surfaces, especially at ramps, so that pedestrians will not slip if the surface is covered with water or other liquid that might have been spilled on the walkway.

Avoid placing a single step along a walkway; people tend not to notice single steps. If such a configuration seems necessary, mark it conspicuously for daytime warning and make sure that it is properly illuminated at night.

If decorative walkways are planned, examine the layout in terms of possible hazards, such as stepping-stones that are too far apart or wooden planks that are irregular or full of rough splinters that may cause a person to trip.[4]

---

[4]See also the section on design of ramps for the handicapped.

**Design Specification for Wheelchair
Ramps between Sidewalk and Street**

The criteria shown in the accompanying illustration will provide not only optimum passage of the wheelchair but also tactile cues for blind pedestrians and visual cues that will help prevent other pedestrians who do not notice sidewalk irregularities from stumbling.

Each of the special features has been derived from a number of attempts to create the optimum compromise for all users.

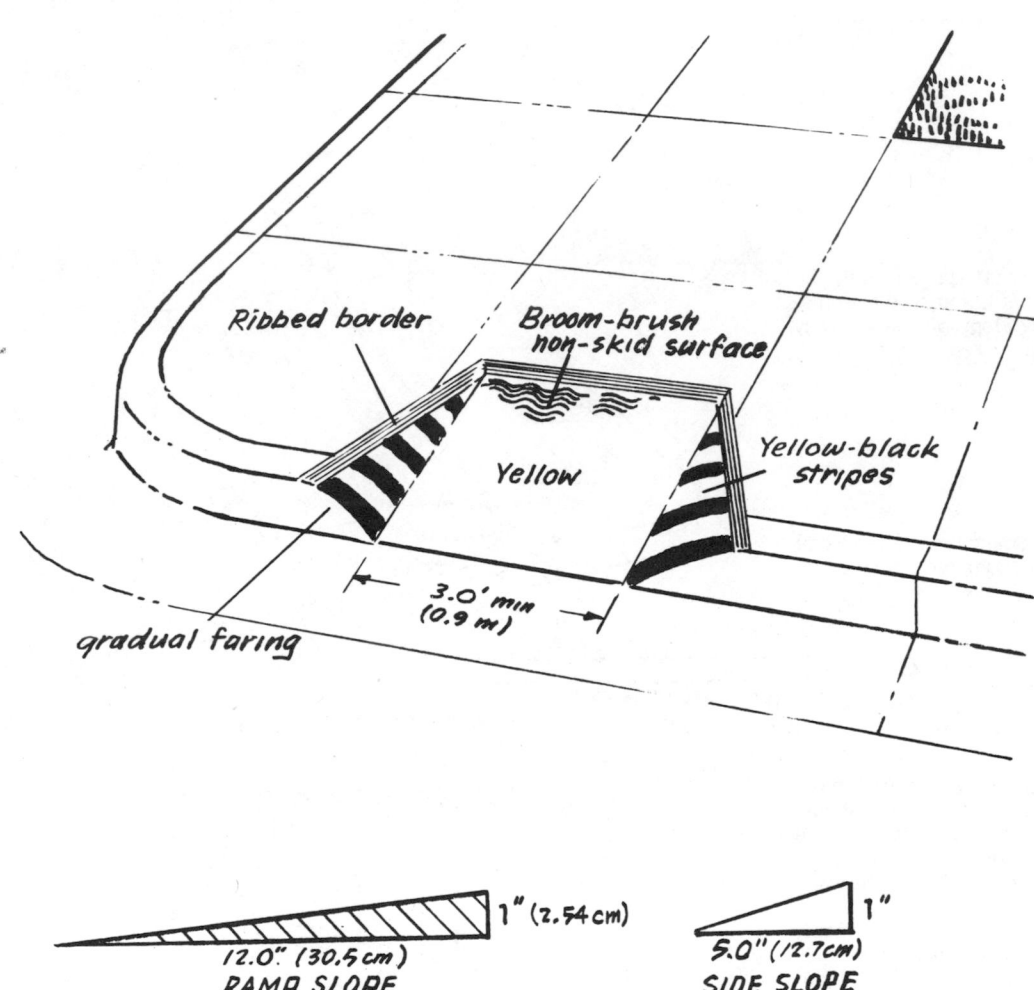

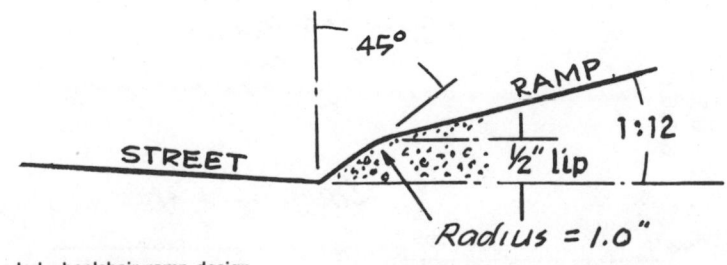

Recommended wheelchair ramp design.

## INTERIOR HALLWAYS AND AISLES

Interior hallways, usually being defined by walls on both sides, should be wide enough so that people do not have to walk carefully in order to keep from bumping into a wall, into other people, or into equipment that is attached to the wall or is being carried or moved on wheels. Although this may sound like an unnecessary caution, most standards for hallway widths do not account for all these factors. For example, a hallway standard is often applied without consideration of the fact that a drinking fountain, a public telephone, or a fire extinguisher will later be hung on one of the walls.

The accompanying sketches reflect preferred dimensions to ensure lack of conflict during passage of persons and/or equipment through typical hallway configurations:

1. Although a minimum of 30 in (76 cm) is wide enough for one person, designers should always plan for the person in a wheelchair situation, which requires more space. The extra width will also accommodate a person on crutches.
2. A person should not have to step aside or crowd against a wall to let another person pass.
3. When mobile equipment can be identified, the additional space required for moving such equipment past people who are using the hallway should be determined and designed for accordingly.
4. Aisleway widths for public transportation systems should not be narrowed to the point where people who are using the aisle bump into and annoy fellow passengers. If it is expected that the transportation vehicle will be accommodating persons in wheelchairs, be sure that the aisle is widened so that wheelchairs can pass freely, or lay out the vehicle so that persons in wheelchairs can park their wheelchairs near the door (e.g., in a public transit bus).

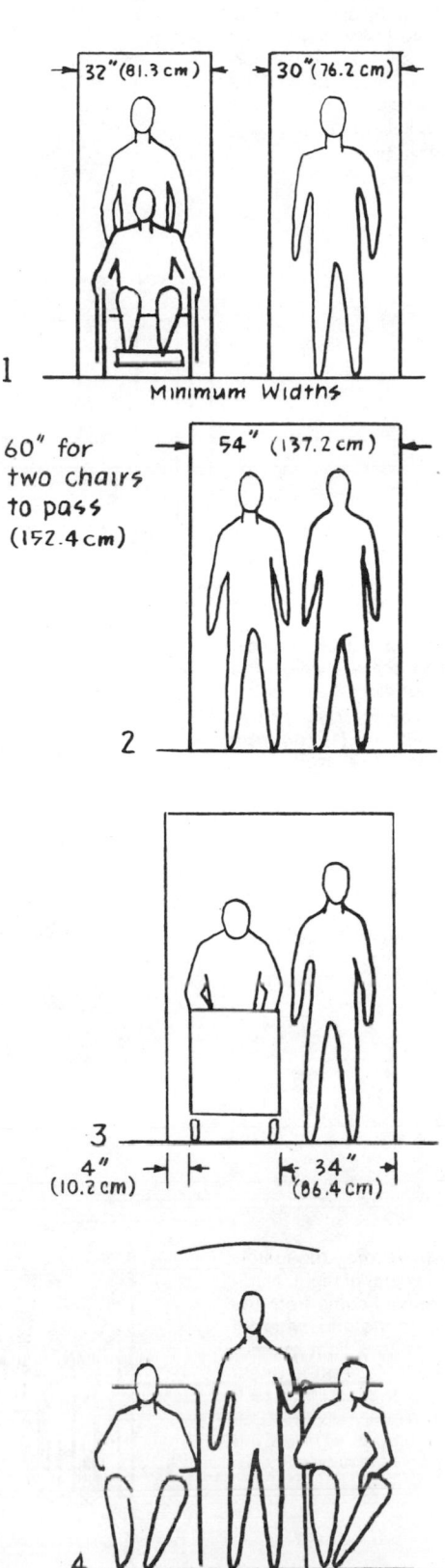

### Hallway Path Definition

A defined pathway limit slightly apart from the walls of a hallway tends to keep people from wandering or absentmindedly cutting the corner at a change of hallway direction. Avoid the use of large-radius cove design at the juncture of the wall and floor; people tend to trip or step on the radiused area and fall.

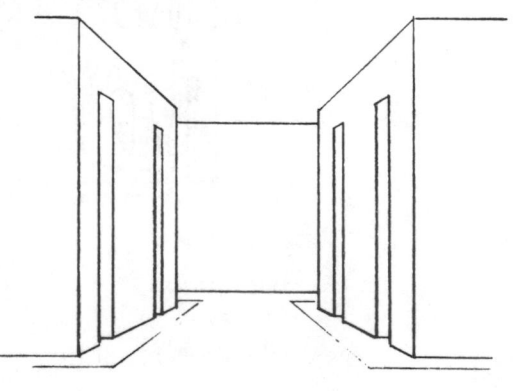

PREFERRED          HAZARDOUS !

In laying out hallways, do not place doorways directly adjacent to corners where the hallway junctures with another hall and/or the single hallway merely turns the corner. The doorway should be at least 5 ft (1.5 m) from the corner; more distance is preferred.

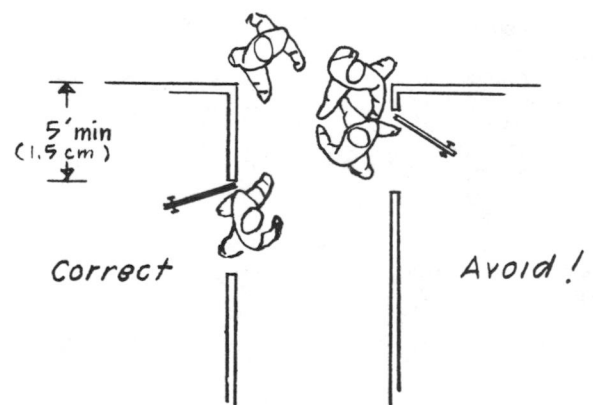

5' min
(1.5 cm)

Correct          Avoid !

Make sure that hallways are adequately lighted both during the day and at night. Natural light during the day should come from the side of a hallway, not from the end, where it tends to contrast with the less light exterior and blind the person who is facing the bright sunlight.

If a building is not ordinarily occupied at night except by security personnel, plan the hallway lighting circuits so that only some of the fixtures will burn.

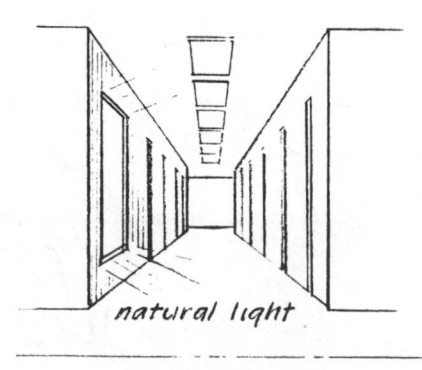

natural light

Correct

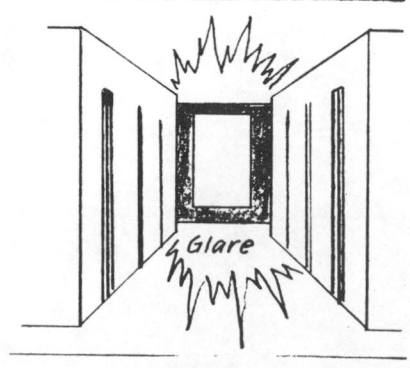

Glare

Poor !

### Design-Generated Confusion

Hallway paths should be laid out in a rectilinear pattern so that people make approximately 90° turns left or right. Although radiused corners often are desirable because they tend to reduce conflicts between people who are approaching the corner at the same time, large curved hallways are not desirable; people tend to get lost and find themselves in doubt as to which direction they are going in.

Hallways should have smooth, level floors (i.e., avoid inclines and single step-ups). If a change of grade is necessary, provide at least two steps, and make sure that the steps are well lighted and conspicuously defined. If the installation serves persons in wheelchairs, add a suitable ramp to one side of the hall. This should have a railing separating it from any steps so that nonhandicapped persons will know which is the ramp aisle and which is the stairway.

Hallway flooring should be made of nonskid materials. If carpet is used, make sure that it is properly laid so that it will not slip or bunch as people shuffle or drag their feet on it.

Floor patterns should be chosen that will not create unusual visual illusions, as shown in the accompanying illustration.

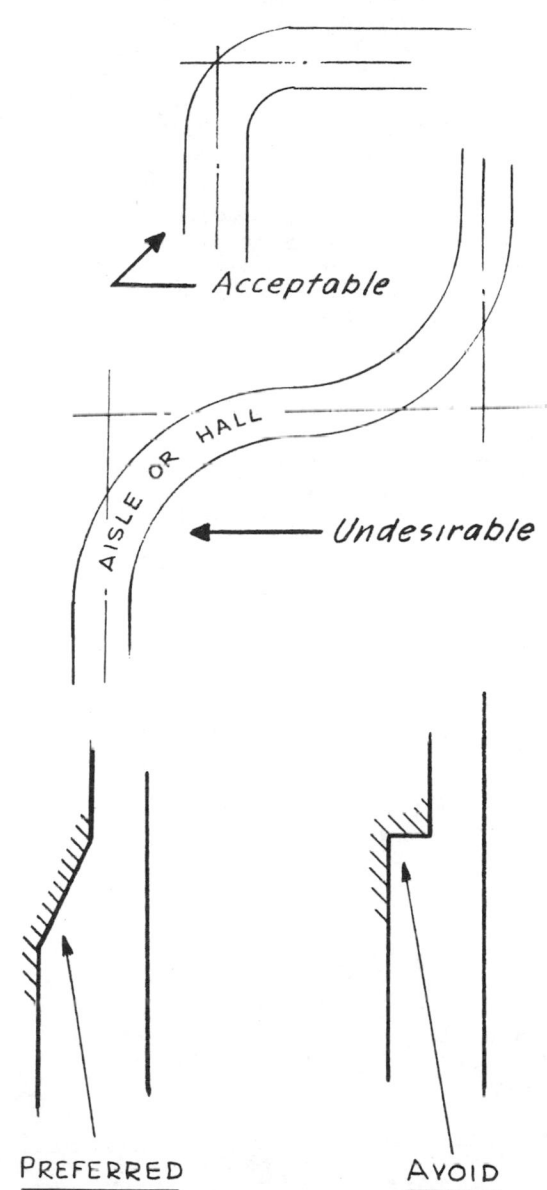

where width of hallway changes

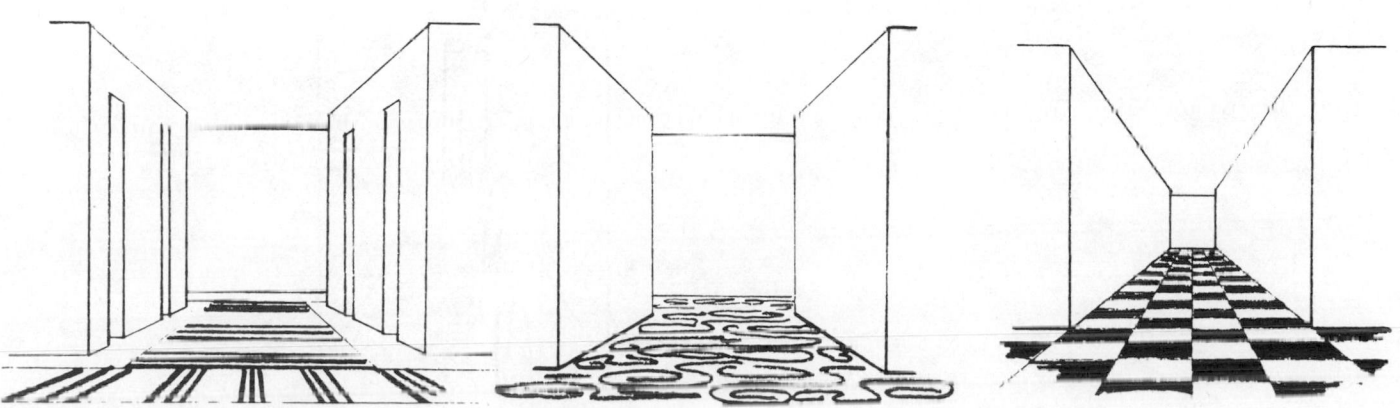

Although there have been no studies of the psychological effects of hallway proportion on user performance, it has been observed that certain kinds of proportional exaggeration (e.g., very long halls with low ceilings and very narrow halls with very high ceilings) tend to produce negative reactions.

In some instances, low ceilings have been used on long halls to accommodate overhead air-conditioning ducting. In other cases, however (especially in home design), very low hallway ceilings have been used purely for effect.

Because the hallway walls tend to create a "confined" appearance, care should be exercised not to use very low ceilings for more than a few feet (6 ft, 1.8 m or less). Otherwise, people may feel that they have to duck to keep from hitting their heads. No ceiling should be less than 7 ft (2.1 m).

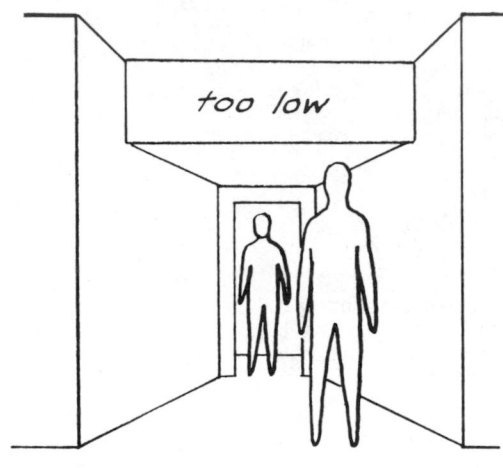

Very high ceilings in a long hallway tend to create the impression of being in the bottom of a chasm.

When low ceilings are used, make sure that the ceiling color is light. When high ceilings are used, the color should be darker and more subdued.

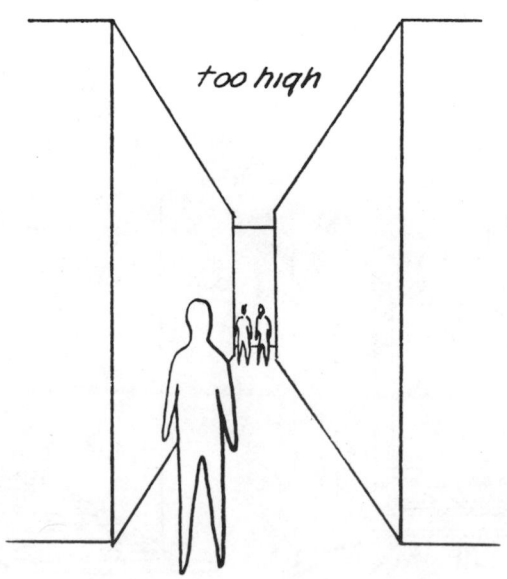

**SPECIAL WALKWAYS AND PASSAGEWAYS**

The accompanying sketches illustrate a few of the special types of walkways and passageways that often are associated with industrial or military applications. The accompanying sketch shows the minimum passage envelope for the adult male, walking at a slow and careful pace. Passageways of this type should be laid out in a straight line whenever possible, with no intervening obstacles which people have to step over, or which might snag their clothing.

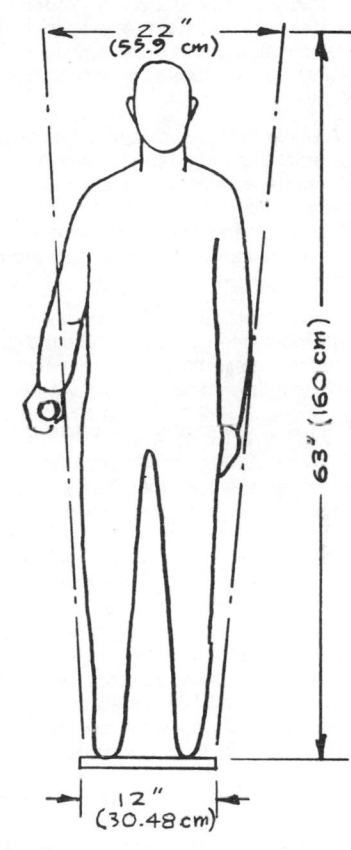

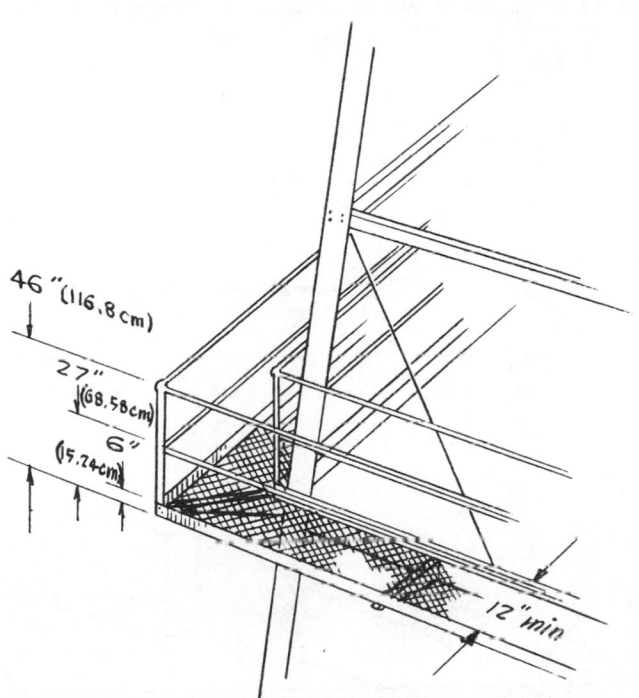

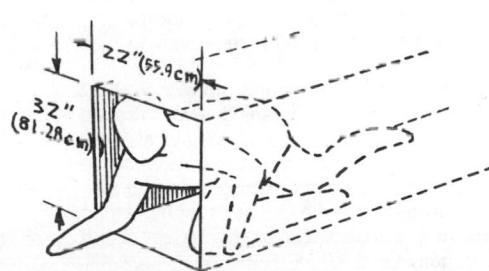

CRAWL - THROUGH

Walkways and catwalks should have non-skid surfaces, especially when they are outdoors and when they are extremely high above the ground-floor level indoors.

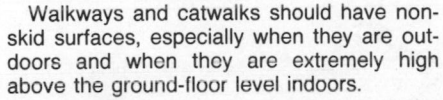

COMMUNICATION TUNNEL WITH MANUAL TROLLEY SYSTEM

Catwalks should have handrails on both sides, a midrail, and a toe board for safety.

Rough edges, bolt ends, and other protrusions must be avoided. The interior surfaces of crawl spaces and tunnels should be painted a light color.

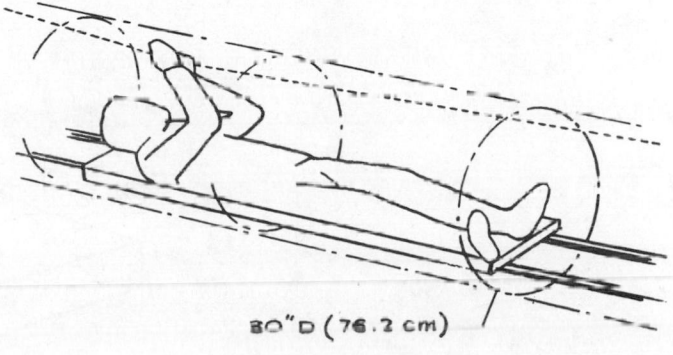

30"D (76.2 cm)

## RAMPS

The primary purpose of a ramp is to provide a smooth change of elevation for wheeled vehicles, mobile equipment, and wheelchairs. Unless a ramp is quite shallow, it is not (as some people think) preferred over a properly designed staircase for many applications. The principal considerations for deciding on the use of a ramp as opposed to a staircase are the following:

1. Wheeled vehicles, mobile equipment, or wheelchairs must be accommodated.
2. There is sufficient space to meet the ramp angle guidelines.
3. The ramp can be positioned where opening windows or doors will not interfere with it.

It should be noted in considering ramps that, although more energy may be required to go up a ramp, the major hazards occur when people are going down a ramp. Although healthy, nonhandicapped individuals can negotiate a fairly steep ramp (up to about 15°), the main criteria for designing most ramps should be based on the needs of the elderly or handicapped user. The guidelines shown in the accompanying sketches are generally accepted by those who design for the handicapped person, who must travel by wheelchair.

A slope of 5 percent is the maximum allowed for a ramp before it is required that a handrail be added. When the slope becomes greater, it is very taxing for most wheelchair users to "pull the hill." Thus this upper limit is a preferred slope guideline for all ramps, whether they are for indoor or outdoor applications.

No ramp should exceed about 8 percent slope under any circumstances.

If a ramp slope is created that is over the 5 percent grade, handrails should be provided to assist individuals who are weaker and/or less steady on their feet. There should be a separate railing for both sides of such ramps.

Nonskid surfaces should be provided on all ramps.

Ramps should not extend more than 30 ft (9.2 m) before an intervening level "resting platform" is provided. There should also be similar flat platforms both at the bottom of the ramp and at any point at which the ramp system changes direction. The accompanying guidelines are best for wheelchair users but are equally valid for all users.

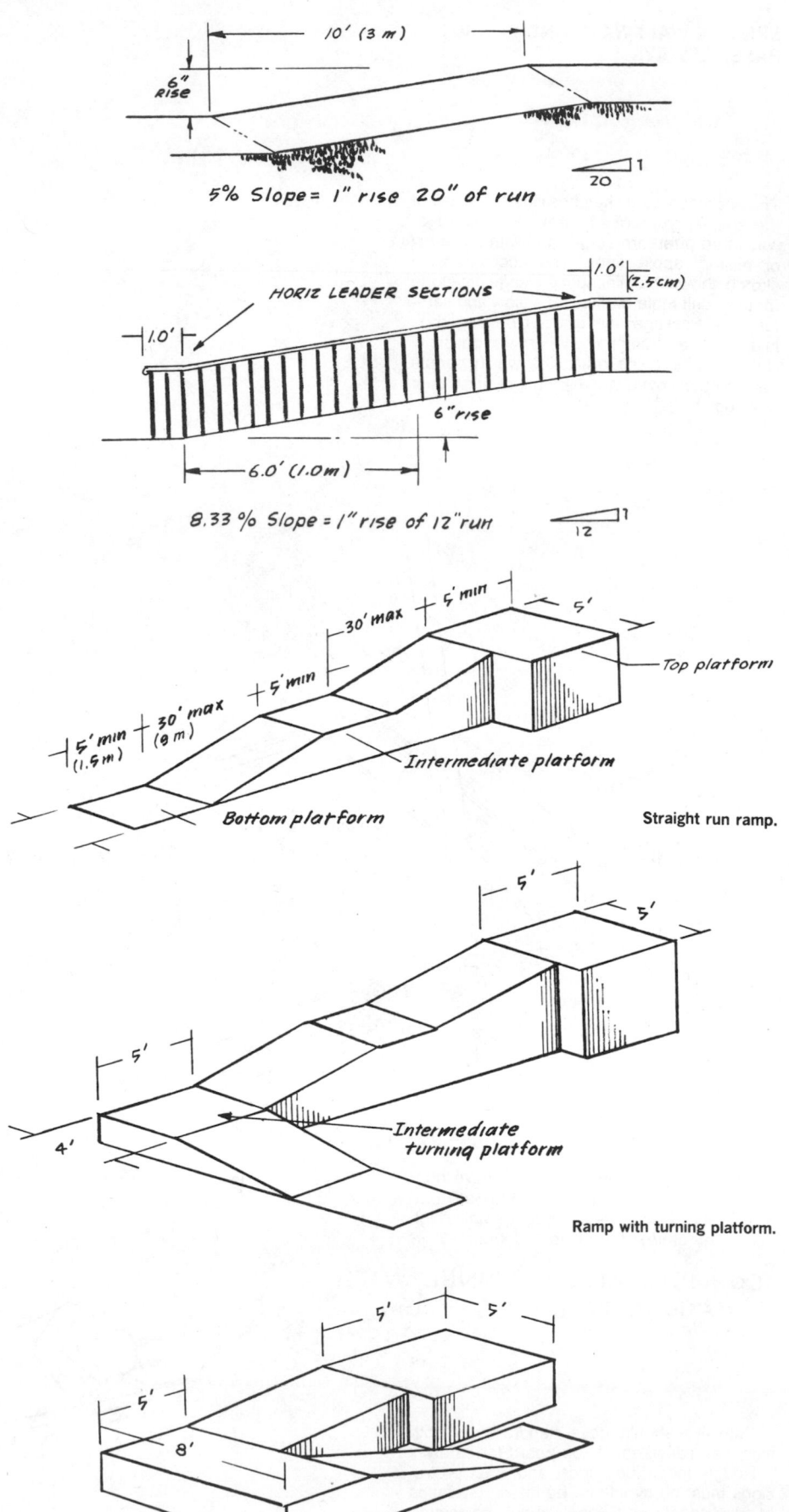

5% Slope = 1" rise 20" of run

8.33 % Slope = 1" rise of 12" run

Straight run ramp.

Ramp with turning platform.

Ramp with intermediate switch-back platform.

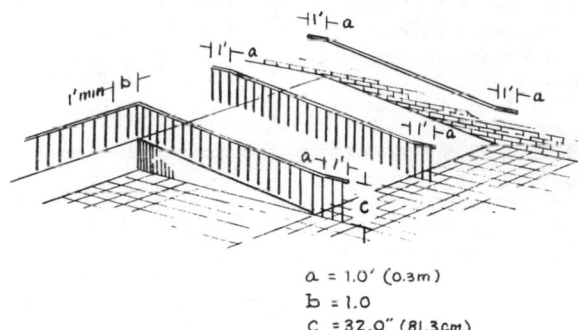

Ramps should level off minimum of 1.0' before meeting floor to allow for 1.0' extension of rail before turn

In order to improve traffic flow in busy public buildings, consider a combination ramp and staircase system, which allows the nonhandicapped to "move through" without waiting for someone who is using the ramp (who is usually slower). This scheme prevents a handicapped person from trying to hurry to get out of the way, which could cause an accident.

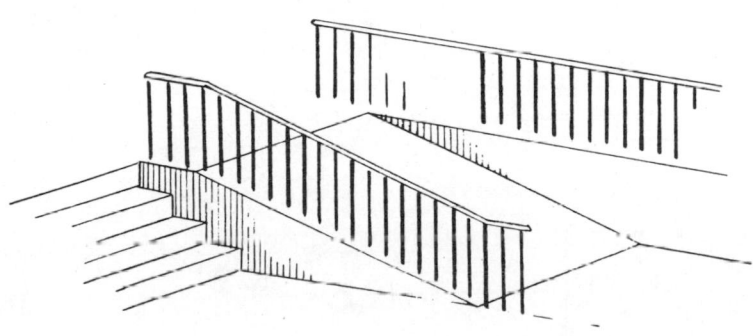

a = 1.0' (0.3m)
b = 1.0
c = 32.0" (81.3cm)

## Special Ramp Situations

### Sloping (Ramp) Floors in Assembly Halls and Theatres

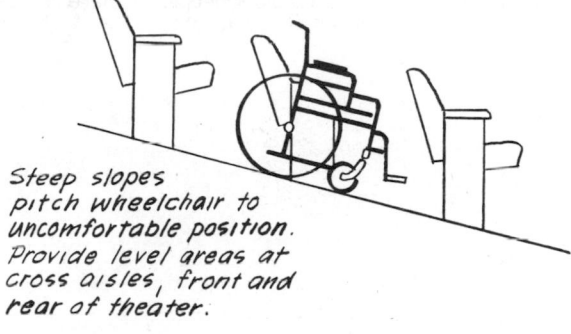

Steep slopes pitch wheelchair to uncomfortable position. Provide level areas at cross aisles, front and rear of theater.

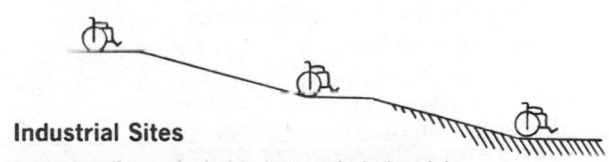

### Industrial Sites

It is sometimes desirable in certain industrial plants to provide a combination pedestrian–mobile equipment ramp. The design shown in the accompanying illustration (although not acceptable for handicapped users) can be used in special situations. Note especially the dimensions for ramp cleat spacing and height of the handrail.

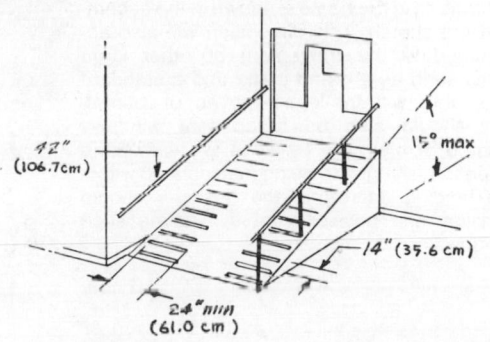

42" (106.7cm)

15° max

14" (35.6 cm)

24" min (61.0 cm)

## STAIRS

### Optimizing Stair Design

Studies have shown that a staircase slope of approximately 30° requires the least expenditure of energy on the part of the average adult. Such a slope requires a tread depth of approximately 11 in (28 cm) and a riser height of about 6.7 in (17 cm).

This, of course, does not adequately take into consideration the two groups who have perhaps more stairway accidents than any other, i.e., children and the elderly. Unfortunately, it is probably impractical to design separate staircases for everyone (except in special cases that might be identified as custom installations). Therefore, it is recommended that all staircases (both indoors and outdoors) be designed with the following characteristics:

Tread     11.0 in (27.9 cm)
Riser      6.5 in (16.5 cm)
Nosing    1 to 1.5 in (2.5 to 3.8 cm)

As indicated in the accompanying sketches, certain stair features are undesirable, principally in regard to the "nosing." The reverse taper configuration is preferred over the "lip" because the latter tends to trip older people and the handicapped, who are not always able

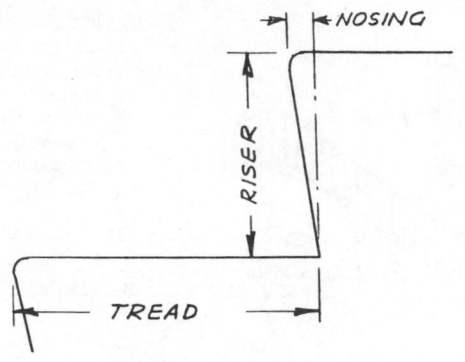

PREFERRED DESIGN

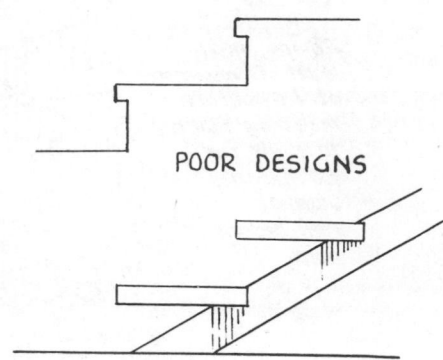

POOR DESIGNS

to lift their feet to clear the nosing. The "open-face" configuration creates a similar problem, and, in addition, some people are confused by the "see-through" staircase.

Other important optimizing features include, obviously, the proper width (a minimum width between handrails of 30 in, 76 cm, for one person at a time and of 52 in, 132 cm, for two persons at a time), the overhead clearance, and the height of the handrail (all stairs should have at least one handrail, and preferably two).

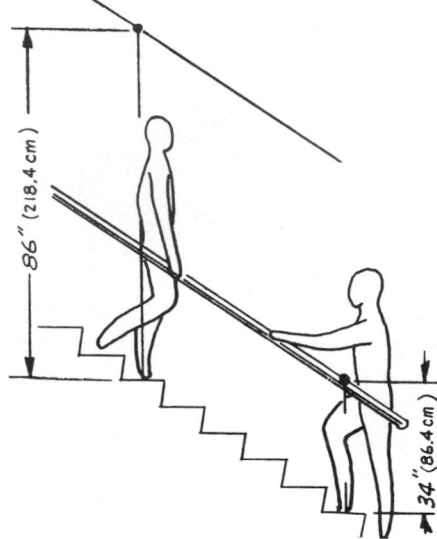

### Stair Design in Relation to Equipment and Furniture Handling

Although equipment and furnishings can often be transferred between the floors of a structure by means of an elevator, sometimes, especially in the home, the only avenue is by the staircase.

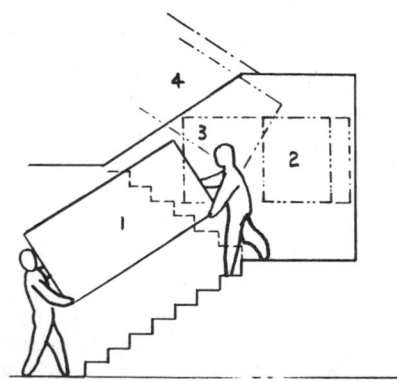

The recommended spacing between stair rails for safely moving the typical large piece of furniture in the home is shown in the accompanying sketch. This dimension will also accommodate the movement of other large items such as a grand piano and a standard pool table (with the legs removed, of course).

It should also be noted that windows located at staircase landings are vulnerable when furnishings are being manipulated within the landing; therefore, the designer should consider the possibility of eliminating such windows.

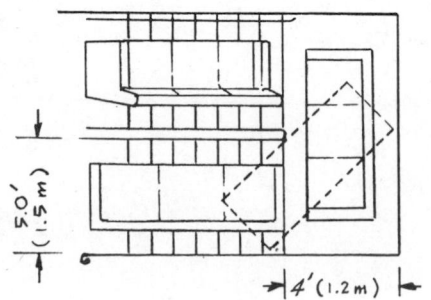

## Special Safety Considerations in Stair Design

Spiral staircases should not be used as a general rule. The varying tread depth (the inside and the outside dimensions differ) often leads to missteps and accidents.

All main stairways within a single structure or dwelling should be alike in terms of tread-riser configuration.

Do not leave the backside of a staircase open, or people (especially the blind) may walk into the stair structure.

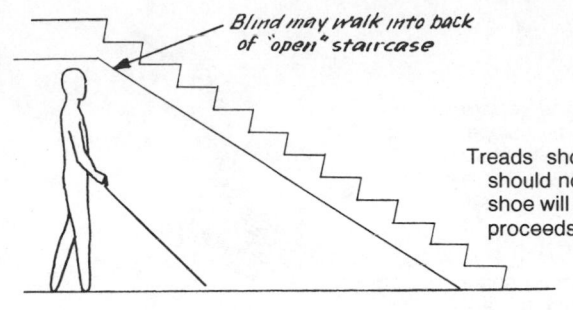

Blind may walk into back of "open" staircase

Treads should be of nonskid material but should not be so rough that the sole of the shoe will not slide slightly, especially as one proceeds up the stairs.

knuckle clearance

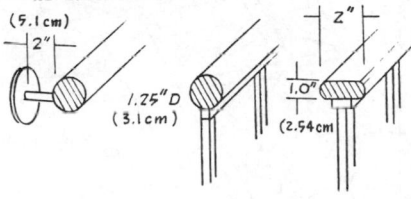

GOOD DESIGN

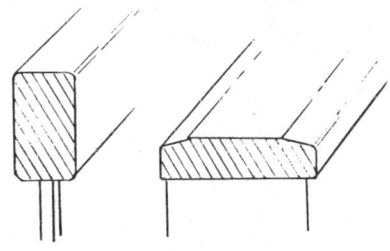

POOR DESIGN

Select handrails that people can grasp easily.

Protruding, "saberlike" rail ends should not be left exposed, and nonslippery handrail material should be used.

Open stair rail designs should not be used because it is too easy for people (especially children) to fall through. Equally important is the use of a closure design that is in itself not hazardous to a child who insists on playing on the staircase.

If vertical post designs are used, the spacing between the posts must be such that a child cannot get his or her head caught between them. Spacing should not exceed about 5 in (13 cm).

Other fancy, filligree-type supports should also be examined before they are used to determine whether they have sharp projections or edges and/or possible entrapment potential.

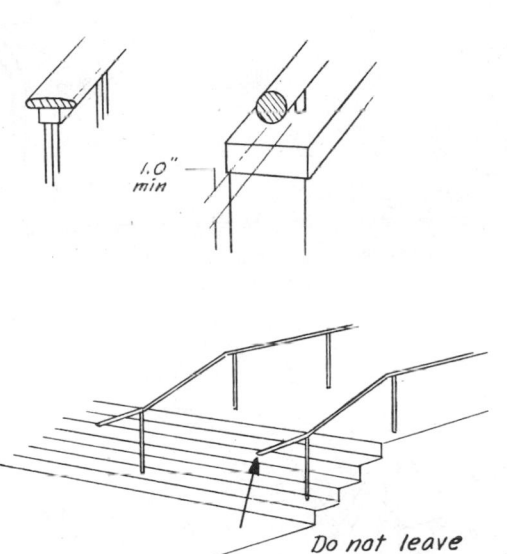

Do not leave protruding, saber-like rail ends exposed — use non-slip railing materials.

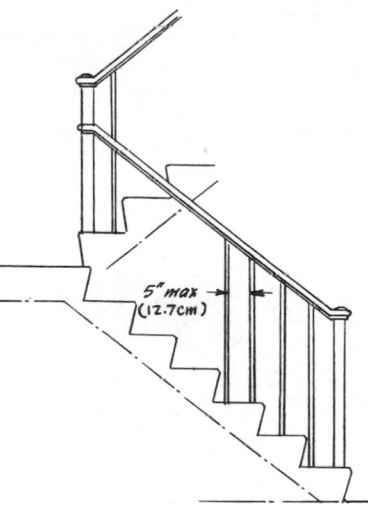

5" max (12.7cm)

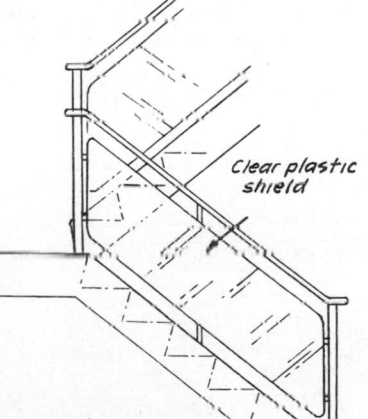

Child's rail

Clear plastic shield

If one wishes to maintain a see-through quality about the design, one good approach is to use a transparent shield, as illustrated in the accompanying sketch.

Children need to have handrails that are not too high for them to grasp easily. Since children tend to fall on stairs more often than anyone else, special pains should be taken to design stairs that they can use as safely as possible.

## Special Industrial Stairs

When specialized stair applications require some compromise of the preferred slope, tread and riser, and other previously noted features, one should still consider the possible hazards that may be associated with stair use.

The accompanying sketches illustrate some of the considerations that should be addressed:

The slope should never be steeper than 50°; otherwise, the user cannot decide whether to walk down head first or back down.

Always provide a safety screen behind open stair designs.

Use similar screening at landings.

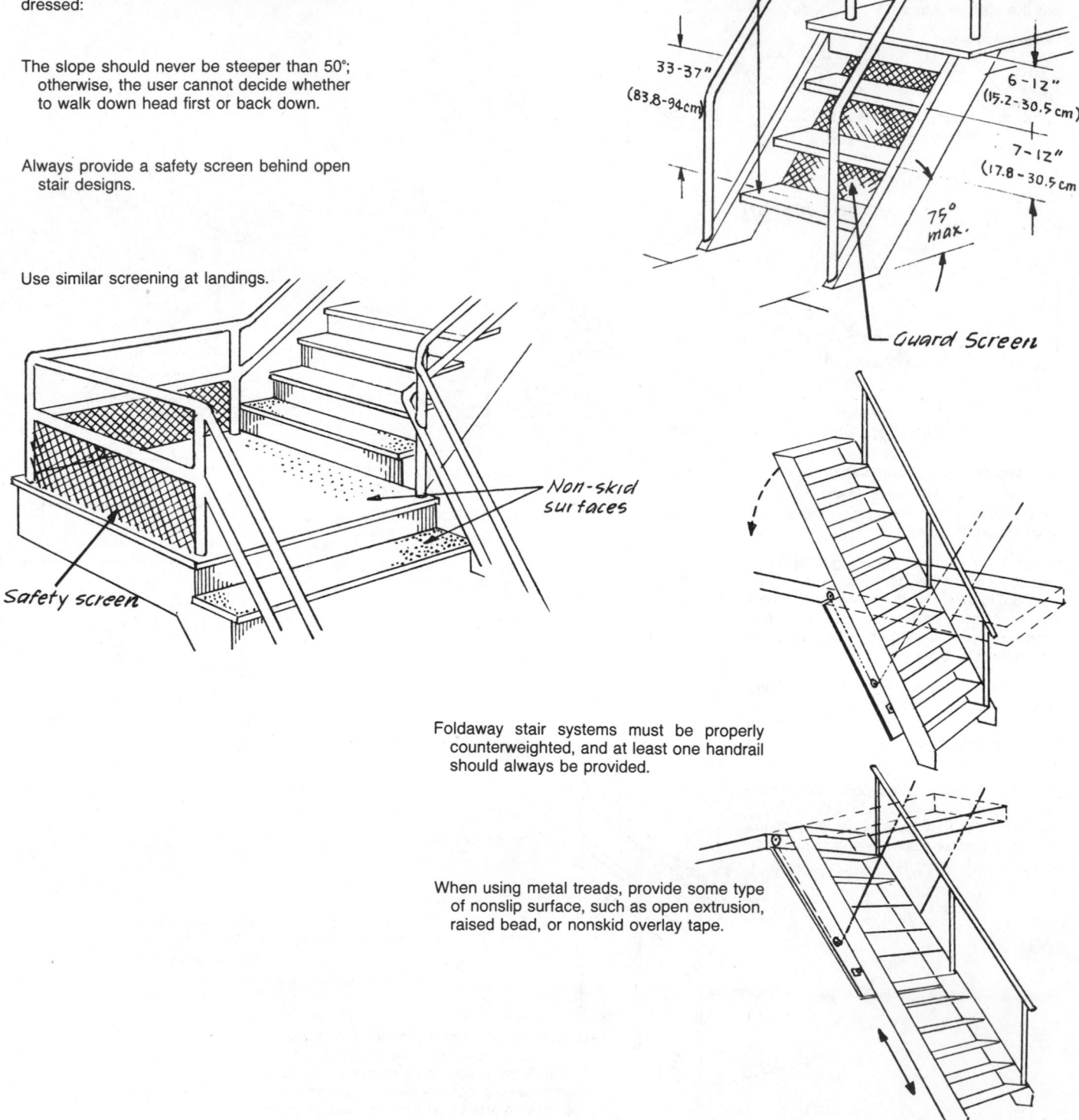

HEAD CLEAR. 7' (2.1 m)

33-37" (83.8-94 cm)

6-12" (15.2-30.5 cm)

7-12" (17.8-30.5 cm)

75° max.

Guard Screen

Non-skid surfaces

Safety screen

Foldaway stair systems must be properly counterweighted, and at least one handrail should always be provided.

When using metal treads, provide some type of nonslip surface, such as open extrusion, raised bead, or nonskid overlay tape.

## LADDERS

Once the incline requirement reaches about 50°, the designer is no longer within acceptable stair concepts; i.e., the device should be conceived as a ladder which should be employed only where it is expected that agile adults will be using the system. Ladder angles between 50° and 75° ordinarily should be designed with flat treads at least 4 in (10.2 cm) deep (with at least 6 in, 15 cm, of toe clearance behind the tread) and with a handrail that starts and ends at a point 36 in (91 cm) vertically above the bottom tread and similarly above the top tread and/or platform.

When a ladder becomes steeper than 75°, a railing should no longer be used, and rungs should replace the flat treads; this is done because at these steeper angles, the user should not be encouraged to come down the ladder frontward.

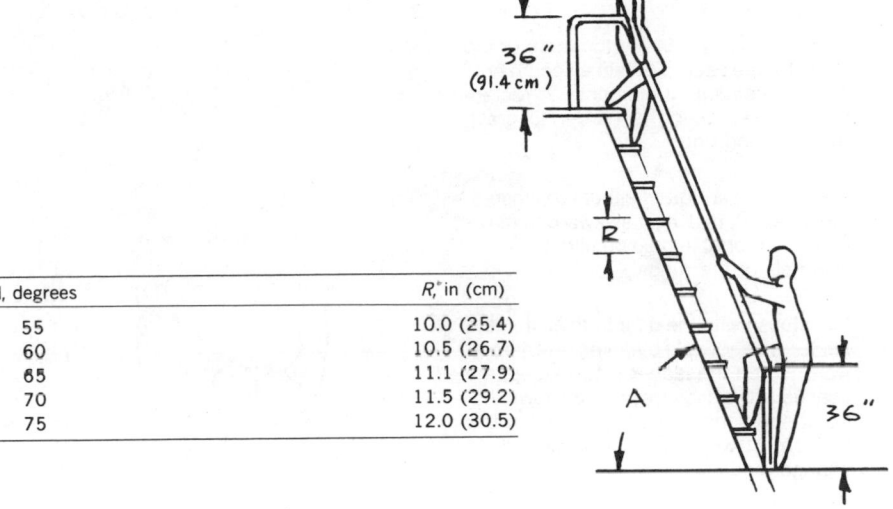

| A, degrees | R, in (cm) |
|---|---|
| 55 | 10.0 (25.4) |
| 60 | 10.5 (26.7) |
| 65 | 11.1 (27.9) |
| 70 | 11.5 (29.2) |
| 75 | 12.0 (30.5) |

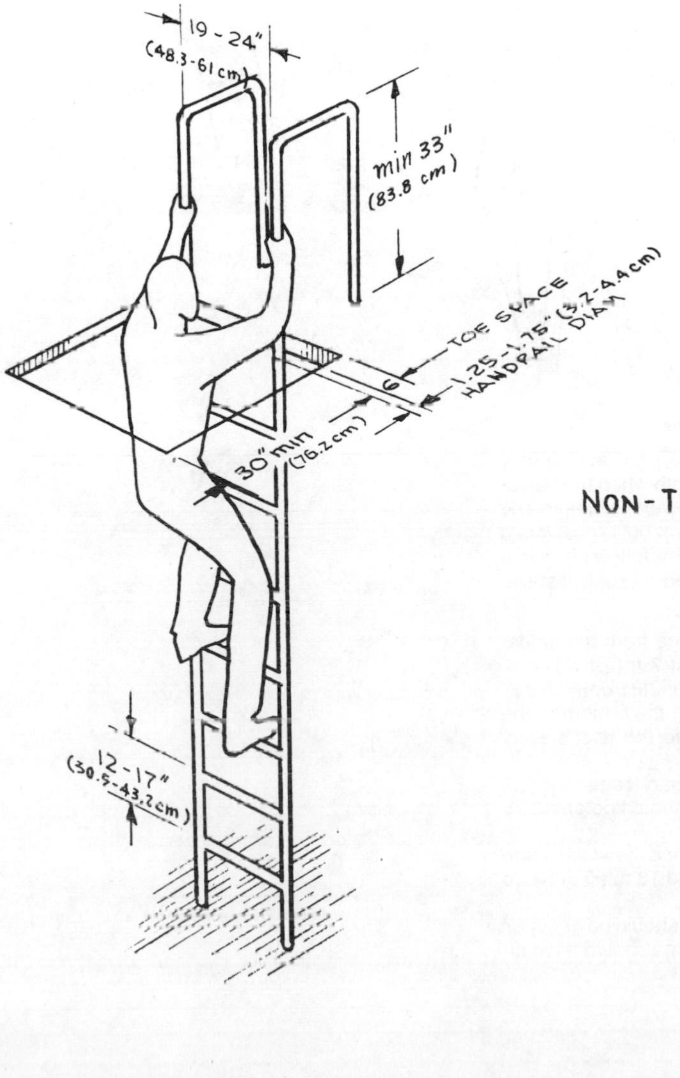

NON-TREAD LADDERS

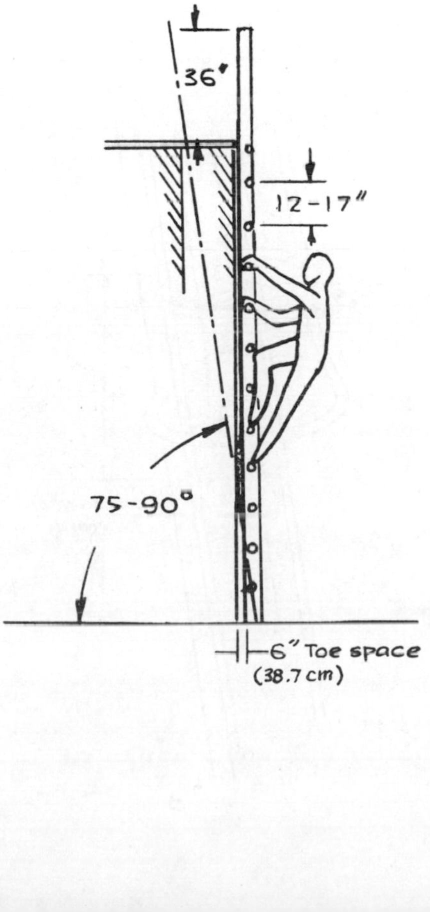

## Common Design and Application Errors with Respect to Ladders

The most common ladder design and application errors occur relative to the final interface between the user's hands and feet. Several of these are shown in the accompanying illustrations:

1. The rail or rung is either too large or too small for the user to obtain a secure hold on these structural elements. A rectangular cross section is harder to grasp than a round one.

2. Vertical rails are often either too close or too far apart, making it awkward to maintain a comfortable hold on either the vertical rails or the rungs.

3. Flat steps invite the user to think that the ladder is negotiable by going down frontward. When the ladder is too steep, the user usually discovers partway down that he or she cannot safely proceed in this fashion, at which point it is difficult to change modes.

4. To be safe, a flat tread must always be horizontal. Often the ladder is set up so that the flat tread is not horizontal.

Ladders that are erected 75° to 90° should always have round vertical and tread features to ensure maximum flexibility, obviousness of use mode, and safety.

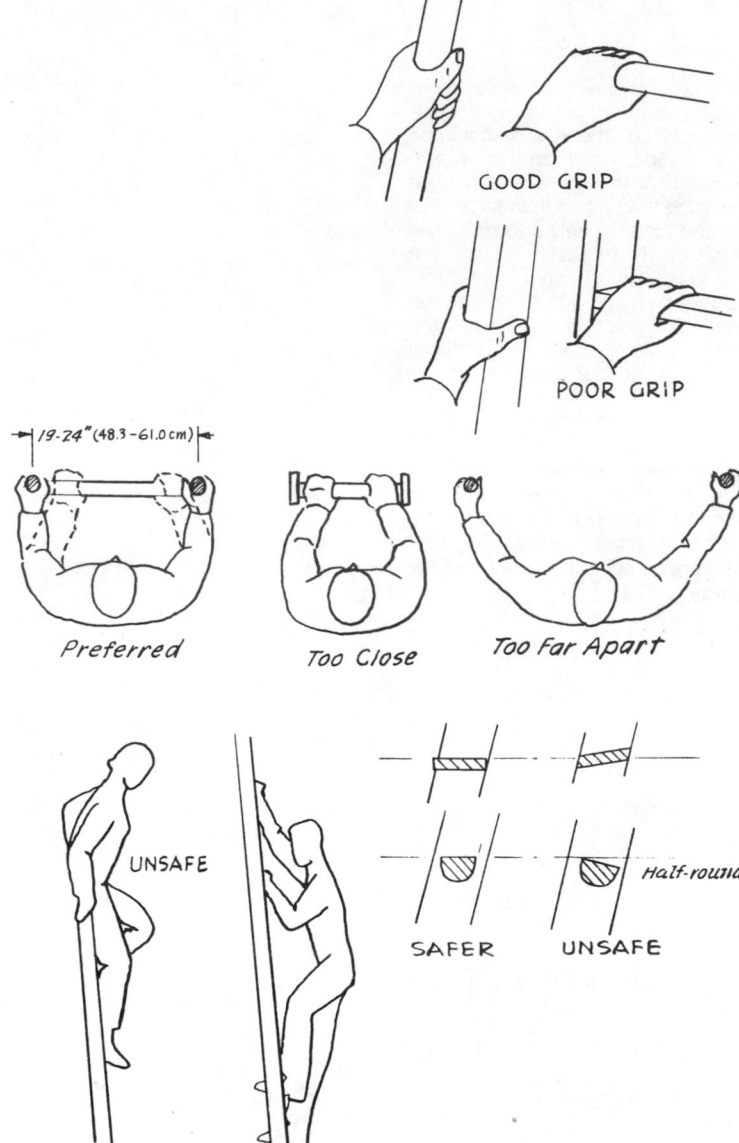

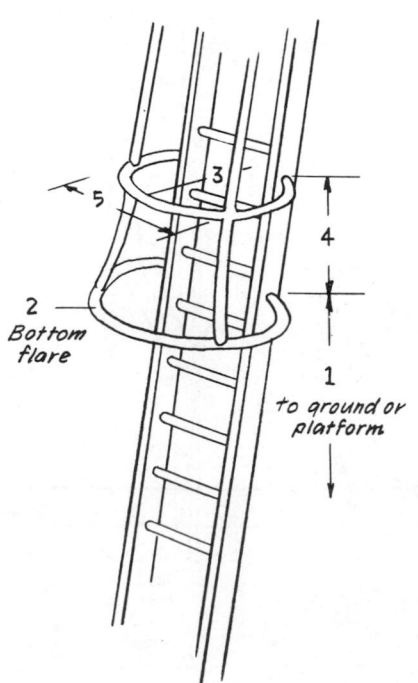

A safety cage should be provided on very high ladder subsystems, both when the ladder extends for a considerable vertical length and when the ladder is fairly short but is positioned high above the ground. The following guidelines should be considered in such installations:

1. The height of the cage from the ladder base should be about 7 ft (2.1 m).
2. Approximately a 32-in (81-cm) "flare" should be provided at the bottom of the cage in order to ease the user's entry and exit.
3. The depth of the primary cage from the center of the ladder should be about 28 in (71 cm).
4. The maximum distance between cage ribs should not exceed 18 to 20 in (46 to 51 cm).
5. The width of the cage should be about 27 in (69 cm); i.e., the cage should have an oval shape.

## Special Ladder Applications

Special step and handrail design requirements are often associated with transportation system vehicles and equipment and with industrial, construction, and agricultural vehicles and equipment. Planning for the ingress and egress or mounting of these special devices should always occur before basic configurations make it impossible to create a suitable ladder, step, and/or handrail subsystem. In most of these cases, the relationship between the ladder or step element and the associated handholds or handrails is particularly critical to the safety of the person using them. The accompanying sketch illustrates an obvious concern by the designer to integrate the ladder handrail into the overall machine concept early enough so that the driver will have a reasonably safe approach to the driver's compartment and back again to the ground. Although not quite so obvious, the short ladder and platform provided at the rear also demonstrate the designer's concern for the person who has to service the engine.

In spite of the apparent logic of planning steps and handrails from the very beginning, it is common for large machinery to be designed and packaged almost completely before any thought is given to where steps and handrails for entering or leaving the machine should be placed. The following principles should be observed in considering the design of these special types of entry devices:

1. The first step from the ground must be reachable by the shortest expected user, and at least two handholds must be accessible to this person while he or she is still on the ground.
2. The steps or rungs must accept the user's shoe, with the shoe being located for a firm step; i.e., the midpoint of the shoe, not just the toe, must rest solidly on the step. Each step or rung should be sufficiently wide so that the person can stop and rest both feet on it.
3. The position of each succeeding step and its associated handhold must be planned so that the user's final entry into the machine or vehicle will be compatible with sitting in the seat. Remember that when a person climbs a ladder, the hand and leg that are making the next move are on opposite sides of the body. If handholds and steps are not planned to conform to this natural "climb pattern," the person will more than likely end up with the wrong foot ready to enter the vehicle; i.e., usually a person cannot hold and step from the same side without swinging. Do not create a system that forces the user to step on a hubcap, tire, or other irregular surface.
4. The specific contour of handrails must follow the gripping, pulling, and supporting pattern associated with both entry and exit; entry and exit are generally quite different and in some cases require more than one handrail geometry.
5. Both ladder rungs or steps and handrails must have nonskid or non-slip surfaces. Handrails should be placed on both sides.

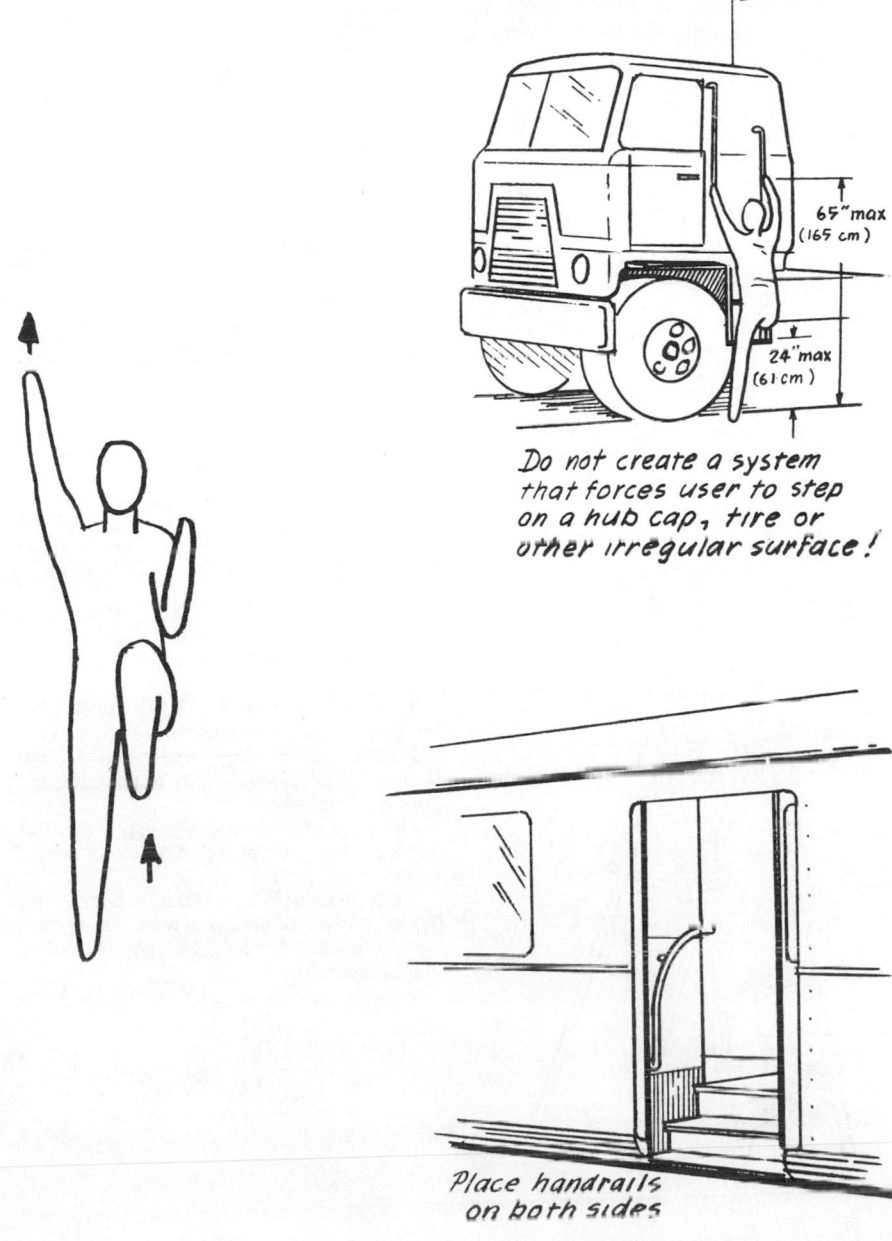

65"max
(165 cm)

24"max
(61 cm)

*Do not create a system that forces user to step on a hub cap, tire or other irregular surface!*

*Place handrails on both sides*

## Portable Stepladders

Portable, collapsible stepladders should be designed to maximize safety. Stepladders more than 12 ft (3.7 m) high are not recommended because it is practically impossible to construct a ladder above that height that will not "weave" when the user is working near the top of it. As the accompanying sketches indicate, certain key features are extremely important in assuring the safe use of all stepladders.

The short household stepladder should have the following characteristics:

The stepladder should not be more than 27 in (69 cm) high.

The riser height should be approximately 9 in (23 cm).

The top step or platform should be approximately 16 × 7 in (41 × 18 cm).

The footprint should be sufficiently large to minimize the possibility that the ladder will be unbalanced when the user is standing on the top platform. The minimum angles for vertical supports are shown.

Non-slip treads

Rubber feet

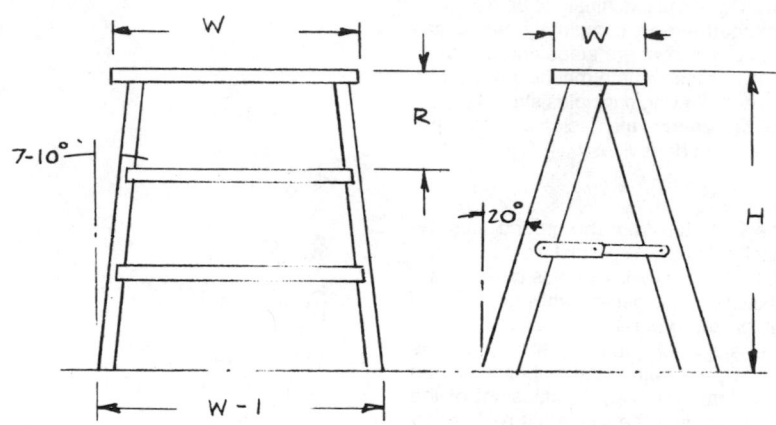

Ladder should safely support a 250 lb person on next to top step

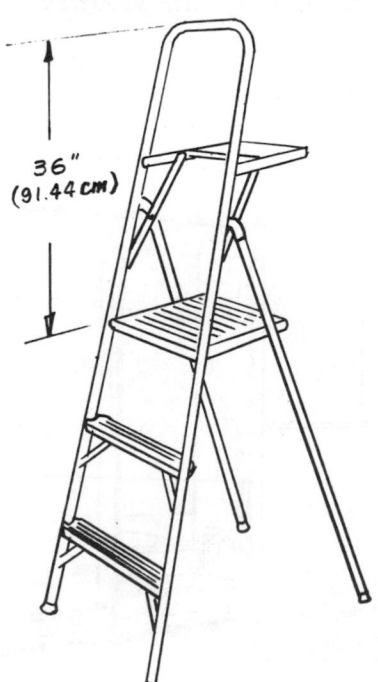

36" (91.44 cm)

Taller stepladders may have special features, such as those shown in the accompanying sketches. Folding braces on ladders should have locking detents to keep the ladders from collapsing.

All treads should be slip-resistant, and rubber feet should be provided on the bottom of vertical legs.

Wooden ladders should not be painted. If a ladder is used for electrical work, its structure should be made of fiber glass or other nonconducting material.

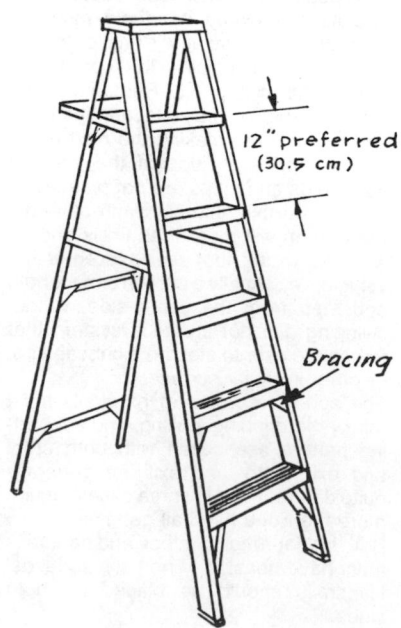

12" preferred (30.5 cm)

Bracing

## ELEVATORS

Critical variables to consider are the following:

1. Capacity
2. Velocity
3. Acceleration and deceleration
4. Normal and emergency ingress and egress
5. External and internal controls
6. Emergency communication
7. Illumination
8. Ventilation
9. Door operating variables (e.g., speed and safety bumper)
10. Entrance and threshold matching
11. Handrails

Elevator standards generally are established by local ordinances, which should be consulted. It is also suggested that the American Standard Safety Code for Elevators (A17.1-1965) be used to establish basic mechanical requirements.

The accompanying sketch provides a number of important human-machine interface guidelines that are often ignored or forgotten. These features are especially important to the elevator user.

In addition, consideration should be given to the problem of passenger escape and rescue in the event of a power failure.

### General Guide for Determining Number and Type of Elevators[5]

Elevators should be provided for any residence facility that is more than one story high because of the possibility of use by persons in wheelchairs.

Elevators should be provided for residences and office buildings that are four or more stories high (public buildings that are frequented by handicapped persons should have elevators if they are more than one story high and if there are no other ramped approaches to the second story). The accompanying table provides guidelines for determining the number of elevators required in a facility.[6]

External and internal control panels should be located and arranged so that orthopedically handicapped persons can locate and reach the controls. There should be an audio signal for all passengers, and, in addition, the tonal signal should provide a cue for blind persons as to which way the elevator is expected to move, i.e., a higher-pitched signal for "up," and a lower-pitched signal for "down."

Visual signs, floor indicator displays, and panel controls should be arranged and labeled so that they are visible, legible, and understandable. Avoid special abbreviations that are known only to the tenants. The floor-selection panel shown in the accompanying sketch illustrates desirable control-button identification; i.e., the vertical arrangement provides vertically distributed numbers that relate to the vertical arrangement of the floors. When there are several subterranean floors, separate the buttons for these floors from upper-floor controls by an intervening space and a general panel label that identifies the lower button array as "basement."

Push buttons, as well as floor-identification indicators, should illuminate when a control button is actuated and/or when a particular floor is reached.

Outside control buttons should be shape-coded as shown in the accompanying sketch, and the translucent button should illuminate in the colors shown.

Interiors of elevators should be properly illuminated so that deep shadows are not cast by passengers. It is recommended that only light colors for walls, ceiling, and floor be used in order to help distribute interior illumination. There should be an emergency light source in the elevator that is not dependent on system power.

*Note:* It is suggested that elevators in facilities frequented by the blind have control buttons that can be tactually identified. Consider also an audio signal so that the blind person can "count" floors as they are passed.

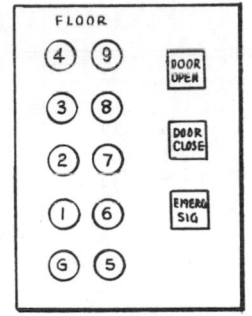

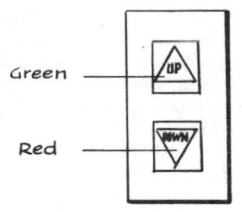

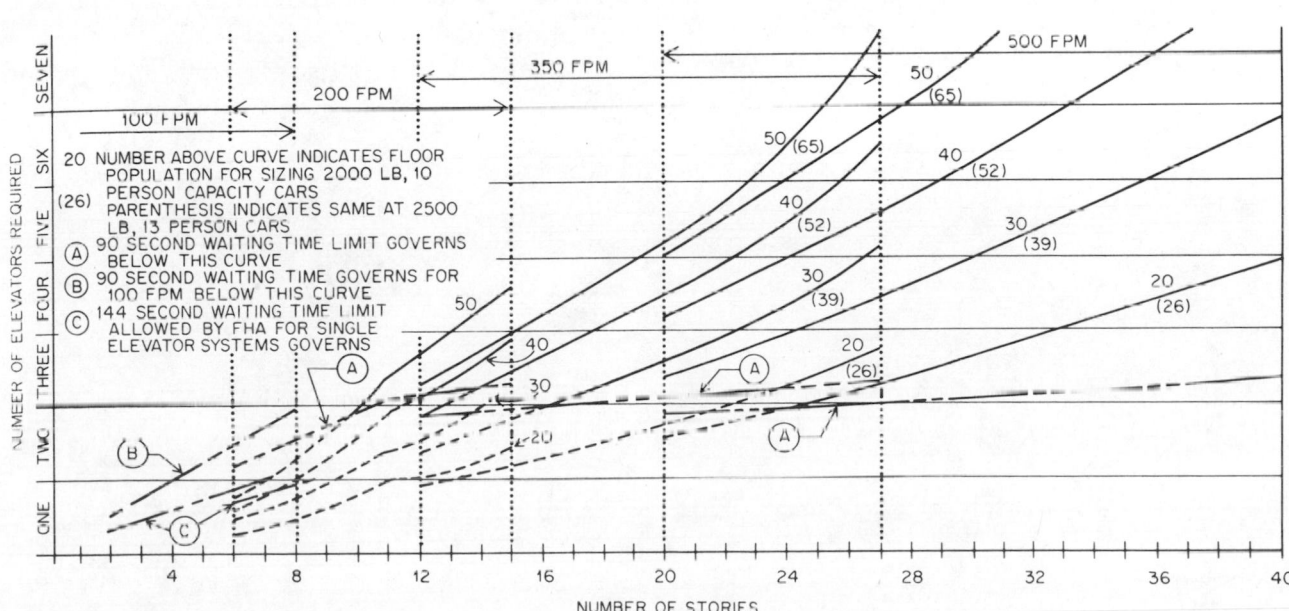

[5]J. De Chiara and J. H. Callender, *Time-Saver Standards for Building Types*, McGraw-Hill Book Company, New York, 1973, p. 83.

[6]J. De Chiara and J. H. Callender, *Time-Saver Standards for Building Types*, McGraw-Hill Book Company, New York, 1973, p. 1039.

**Total number of persons who can be served by one elevator of a given capacity and speed[1]**

| | | Total number of persons[2] | | | | | | | | |
|---|---|---|---|---|---|---|---|---|---|---|
| | | Floor-to-floor stops[3] | | | | Alternate-floor stops[4] | | | | |
| | | 1,200 lb | 2,000 lb | | | | 1,200 lb | 2,000 lb | | |
| No. of stories | No. of stops | 100 fpm | 100 fpm | 200 fpm | 300 fpm | No. of stops | 100 fpm | 100 fpm | 200 fpm | 300 fpm |
| 5 | 5 | 461 | 618 | — | — | 3 | — | — | — | — |
| | | | | | | 3 | 480 | 715 | — | — |
| 6 | 6 | 391 | 521 | — | — | 3 | — | — | — | — |
| | | | | | | 4 | 420 | 595 | — | — |
| 7 | 7 | 336 | 465 | 500 | — | 4 | — | — | — | — |
| | | | | | | 4 | 375 | 550 | 566 | — |
| 8 | 8 | 313 | 431 | 465 | — | 4 | 375 | 550 | 566 | — |
| | | | | | | 5 | 339 | 495 | 535 | — |
| 9 | 9 | 295 | 408 | 431 | 465 | 5 | 339 | 495 | 535 | — |
| | | | | | | 5 | 302 | 434 | 504 | 545 |
| 10 | 10 | 281 | 387 | 408 | 431 | 5 | 302 | 434 | 504 | 545 |
| | | | | | | 6 | 274 | 397 | 468 | 521 |
| 11 | 11 | — | — | 387 | 419 | 6 | — | — | 468 | 521 |
| | | | | | | 6 | — | — | 444 | 483 |
| 12 | 12 | — | — | 368 | 402 | 6 | — | — | 444 | 483 |
| | | | | | | 7 | — | — | 419 | 468 |
| 13 | 13 | — | — | 350 | 387 | 7 | — | — | 419 | 468 |
| | | | | | | 7 | — | — | 400 | 447 |
| 14 | 14 | — | — | 335 | 375 | 7 | — | — | 400 | 447 |
| | | | | | | 8 | — | — | 379 | 428 |
| 15 | 15 | — | — | — | 363 | 8 | — | — | — | 428 |
| | | | | | | 8 | — | — | — | 413 |
| 16 | 16 | — | — | — | 350 | 8 | — | — | — | 413 |
| | | | | | | 9 | — | — | — | 397 |

[1] *In using this table to determine the number of elevators required, the local authority should exclude persons occupying the bottom terminal-stop floor from its count of building occupants who will require service.*

[2] *Where a dash, instead of a number, appears in any box it means that installation of an elevator of that capacity or speed is not recommended for the particular number of stories.*

[3] *"Floor-to-Floor Stops" assumes a single elevator stopping at every floor.*

[4] *"Alternate-Floor Stops" assumes two elevators, each stopping at alternate floors, except that one elevator is assumed for five-, six-, and seven-story buildings. The top number in each box relates to the alternate-stop elevator whose top terminal stop is the floor below the top floor; the bottom number in each box relates to the alternate-stop elevator which includes the top floor among its stops.*

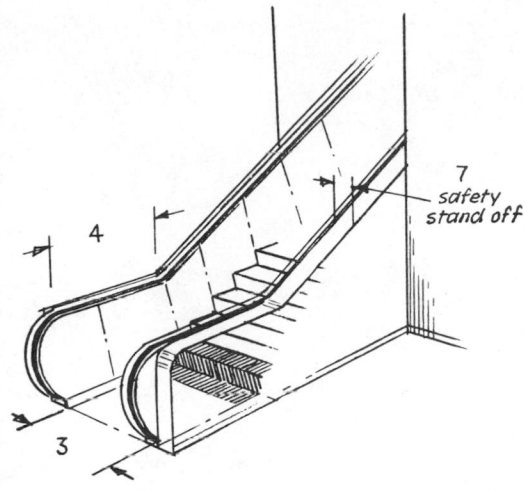

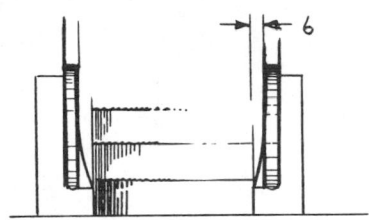

## ESCALATORS

The escalator generally is an efficient method of moving people in certain department store and lower office building interfloor transfer. The escalator industry has developed a fairly consistent set of standards which it follows to ensure the safety of users. In addition, the following special human factors considerations are important in the selection and installation of such facilities:

1. The angle of incline should be approximately 30°.
2. Speed should not exceed about 120 ft/min (37 m/min).
3. The width between handrails should be approximately 32 in (81 cm).
4. Users should be able to grasp the handrail at least 6 ft (1.8 m) prior to the first rise of the moving steps so that they can be prepared for the rate of stair movement.
5. The handrail and step movement should be synchronous so that users do not have to continually change their hand grip.
6. The inside, fixed panel should curve toward the handrail so that users' legs or clothing will not drag.
7. There should be at least 8 in (20 cm) of clearance between the handrail and any wall alongside the railing so that users cannot inadvertently run their hands or arms into any wall or adjacent structure.
8. The handrail should be approximately 3 to 4 in (7.6 to 10.2 cm) wide and have a curved surface that is easy to grasp.
9. The tread depth should not be less than about 13 in (33 cm) so that even the largest shoe can rest solidly on the tread while the escalator is in transit.
10. There should always be good illumination at entry and departure points.
11. The escalator and/or the escalator well installation should be designed so that users cannot catch their hands or arms between the moving railing and the adjacent structure.
12. Steps should be suitably marked so that the user can easily see the edges of the step and therefore avoid stepping on parts of two steps, which could upset the individual as the steps change from a common level to elevated levels.
13. There should be no sharp edges, corners, or gaps that could cause injury if a person fell down on the steps.
14. Appropriate signing should be provided so that people can easily identify where the escalator is, which direction it is going, and what the rules are concerning use (no shopping carts, no children without parental supervision, etc.).
15. All escalators should be clustered together (i.e., up and down assemblies) so that people do not have to hunt for the escalator to the next level.

## MOVING WALKWAYS

The moving walkway may be a useful device for minimizing some of the difficulties that elderly persons and persons carrying heavy baggage experience in airports, large shopping malls, and other public facilities where people must walk long distances. Some of the more important human factors considerations are the following:

1. Speed should never exceed about 100 ft/min (31 m/min), and preferably should not exceed 90 ft/min (28 m/min). An alternative walkway should be provided for those who wish to walk faster, and/or the moving walkway should be made wide enough so that fast walkers can pass those who are holding their position and "riding" the system.
2. The width of the walkway should not be less than 32 in (81 cm) so that persons in wheelchairs can also use the walkway. Widths for more than three persons are undesirable.
3. The moving walkway can be ramped as much as 7° to 10° and still be acceptable to older passengers. Although some safety codes permit up to 15°, this is too steep for older persons who are unsteady on their feet or who have difficulty seeing. Good illumination should be provided not only at the entrance and exit points but also along the walkway if it is wide enough for persons to "walk the ramp."
4. The ramp or walkway surface must be as smooth as practical; i.e., there should be no "bumps" created by the walkway propelling or supporting system. Moving walkway surfaces typically are ribbed to provide good nonslip characteristics. The rib pattern should be clearly visible so that persons who are entering the walkway can observe the approximate speed of the walkway before stepping onto it.
5. Systems should be designed so that, in case of malfunction, the walkway "coasts" to a stop rather than stopping abruptly.

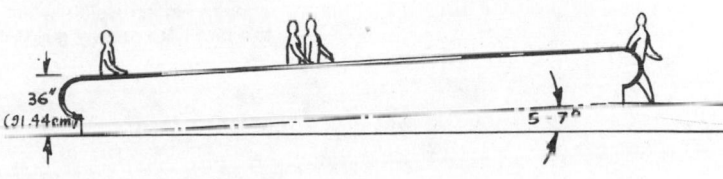

## BATHROOMS

The residential bathroom should be conceived as a series of the following related but independent activity centers.

### 1. Head and Hand Grooming Center

This activity center typically is oriented around a washbasin and mirror, where the user performs one or more of the following tasks:

- a. Washing hands and face
- b. Brushing teeth
- c. Washing, drying, brushing, and combing hair
- d. Shaving
- e. Applying cosmetics
- f. Manicuring
- g. Performing minor first aid

### 2. Bathing Center

This activity center typically involves either partial or full immersion of the body in water, either in a tub or in a shower, and must provide access to soap and other bath items, washcloths, and towels.

### 3. Toilet Center

This activity center typically involves use of a commode and, in some cases, a bidet, and it must provide access to toilet tissue.

### 4. Storage Center or Centers

The typical residential bathroom should provide storage for articles used in each of the above activity centers plus bathroom housekeeping items, including at least the following:

- a. Washcloths, hand towels, and bath towels
- b. Personal grooming equipment and cosmetics
- c. Toilet tissue and feminine hygiene supplies
- d. Medicine and first-aid supplies
- e. Housekeeping supplies and equipment

The selection and arrangement of bathroom fixtures and cabinetry should be based on the premise that the facility will be used by a variety of people, including adults, children, and elderly and handicapped persons, and that in some cases more than one person may use the facility simultaneously. It is especially important to consider bathroom layout from the standpoint of the dynamic movement envelopes of people as they utilize various bathroom features. Avoid the temptation to crowd cabinets and fixtures merely to fit a preconceived space.

## The Half Bath

The half bath, or "powder room," should be considered only an auxiliary facility. Although it obviously requires less space than other types of bathrooms, care must be taken not to make it so small and crowded that the user is inconvenienced, either while utilizing the two primary fixtures or when entering, exiting, or manipulating the door.

As long as there are other major bathroom facilities within the residence or office, many arrangements are possible. The accompanying sketches illustrate two arrangements and show the critical clearance dimensions that must be observed. A common design error occurs when it is assumed that, because smaller fixtures are used, they can be located closer together. Remember that the size of the user does not change just because the fixtures are smaller.

Although one can dispense with some of the normal storage elements of the primary bathroom, the half bath should always include an ample towel-rack fixture, an appropriately accessible toilet-tissue dispenser, a mirror, and space for a wastebasket.

Powder rooms are often located near the front part of a house or apartment so that they are handy for guests. Do not neglect consideration of the acoustic problem; it is not only annoying but also extremely embarrassing when the sounds of the toilet flushing can be heard in nearby guest areas. Similarly, provide good ventilation for such facilities, since it is equally embarrassing when odors drift from the facility to nearby guest-occupied areas.

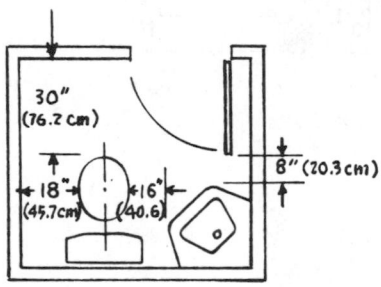

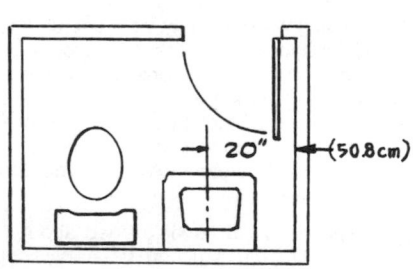

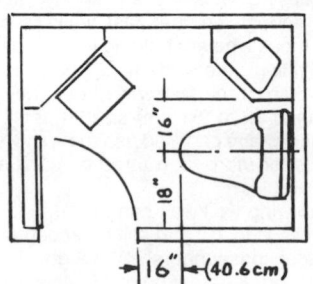

### The Three-Quarter Bathroom

The three-quarter bathroom may be used as either an auxiliary or a primary facility. Preferably, it should not be used as a primary facility except in hotels, motels, and minimal-efficiency apartments, primarily because women, especially, prefer a tub bath to a shower as a regular practice.

Although many arrangements for the three-quarter bath concept are possible, the accompanying sketches illustrate the critical dimensions to be considered.

Whenever the three-quarter bath becomes the primary facility, it should be designed so that it is accessible for persons who are permanently or temporarily confined to a wheelchair. The accompanying sketches provide important dimensional guidelines for making such a facility maximally accessible.

Although most home buyers probably will not want to accept some of the less attractive, special fixtures that have been created for wheelchair users, minimal inconvenience still depends on the ability of the wheelchair user to get as close to the fixtures as possible; thus one must be particularly careful not to design cabinets under which the chair cannot pass. Handrails can be designed so that they are not too unattractive, and they can be helpful even to those who are not handicapped.

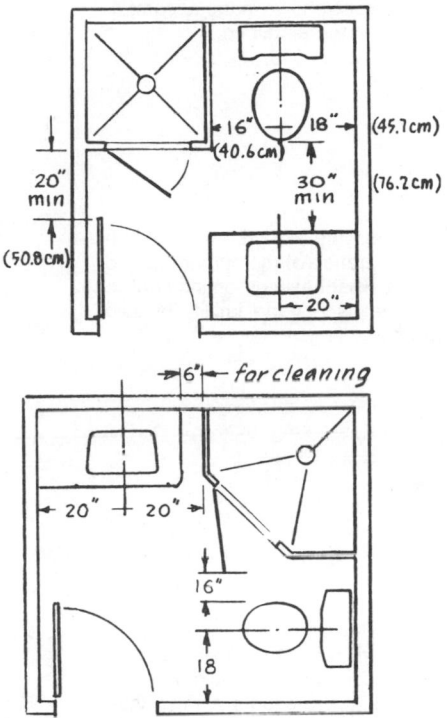

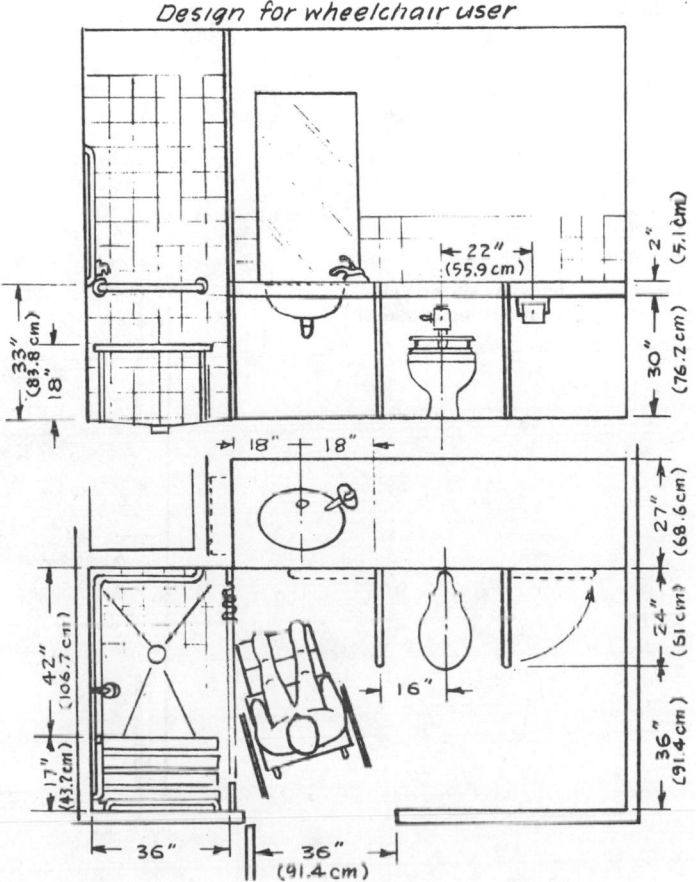

Design for wheelchair user

## The Full Bathroom

The full bathroom contains at least one commode, one washbasin, and a combination tub and shower. Many variations and combinations are acceptable as long as the typical clearance dimensions shown in the accompanying illustrations are maintained.

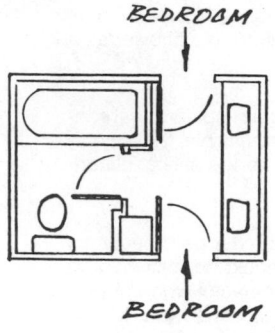

In the accompanying sketch, the architect has provided separation of the commode and bathtub from the washbasin grooming center, thus allowing the use of the facility by two people at the same time.

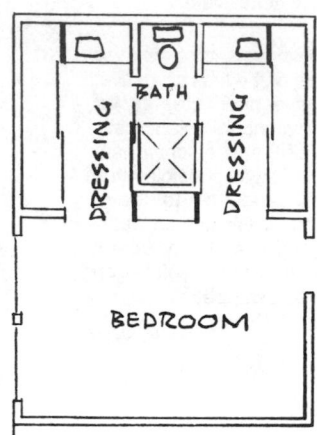

Even more dual-use convenience can be obtained by an arrangement similar to that shown in the accompanying illustration. Although this actually represents a dual-use–three-quarter combination, the basic principle of dual use is demonstrated, including dual dressing capability.

An additional convenience, as shown in the accompanying sketch, is simultaneous use of the commode and bathtub.

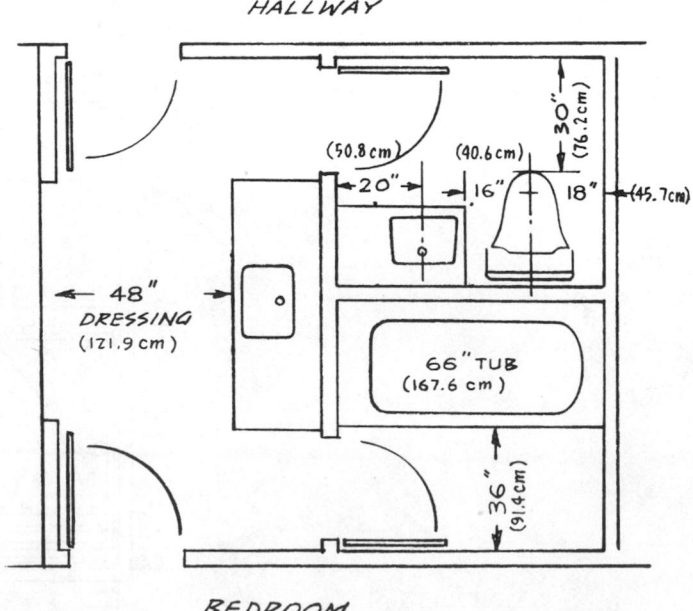

### Guidelines for Selection and Installation of Bathing Facility Fixtures

Although the typical tub and shower configuration is generally created using off-the-shelf fixtures, a certain amount of customizing (actual positioning) is left up to the designer. The following guidelines contain suggestions for making things safer and more convenient for the user:

1. Place the shower and bath controls where they can be reached easily from both the showering and the bathing position and also from outside the tub. Note that the tub position for controls is closer to the normal sitting position than typically seen in the average installation.

2. Provide insets for soap dispensers and for shampoo and ointment containers. Note that the shower control is recessed within the inset so that if the bather should slip, he or she will not accidentally strike the control.

3. Handrails should be provided to aid the bather in getting up and down. The high rail is helpful if the bather loses his or her balance.

4. Select a tub that has a slight lip around the edge; this prevents water from running onto the floor.

5. Select a tub whose edge is not so high that it is difficult to step over. Minimum internal tub dimensions should be 25 in (64 cm) wide by 48 in (122 cm) long. Longer lengths are desirable so that a bather can lie down in the tub with less crowding.

6. Place at least one towel rack close at hand so that the bather can reach a towel without dripping water all over the floor.

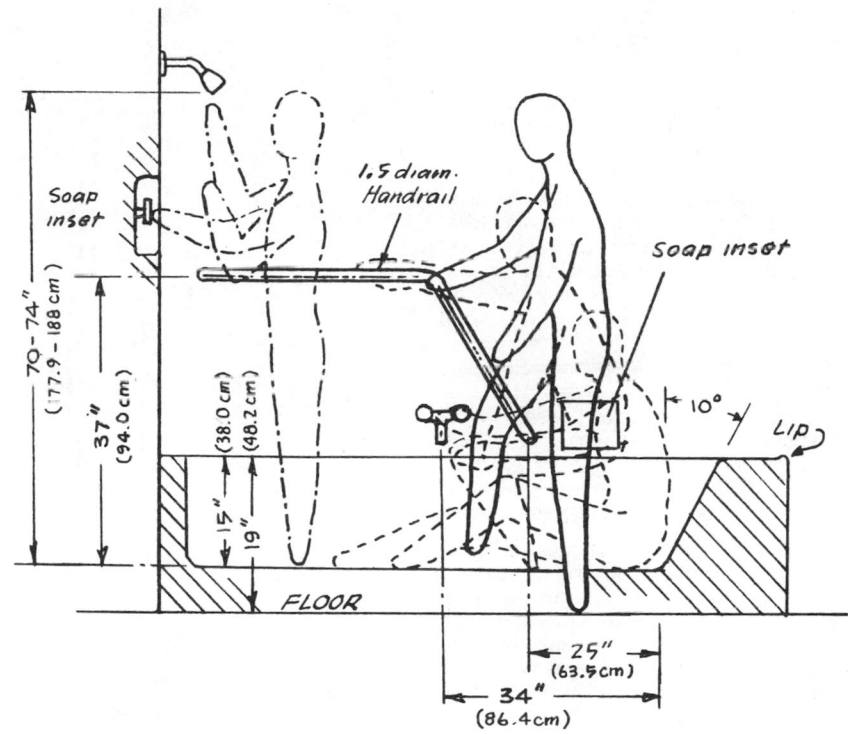

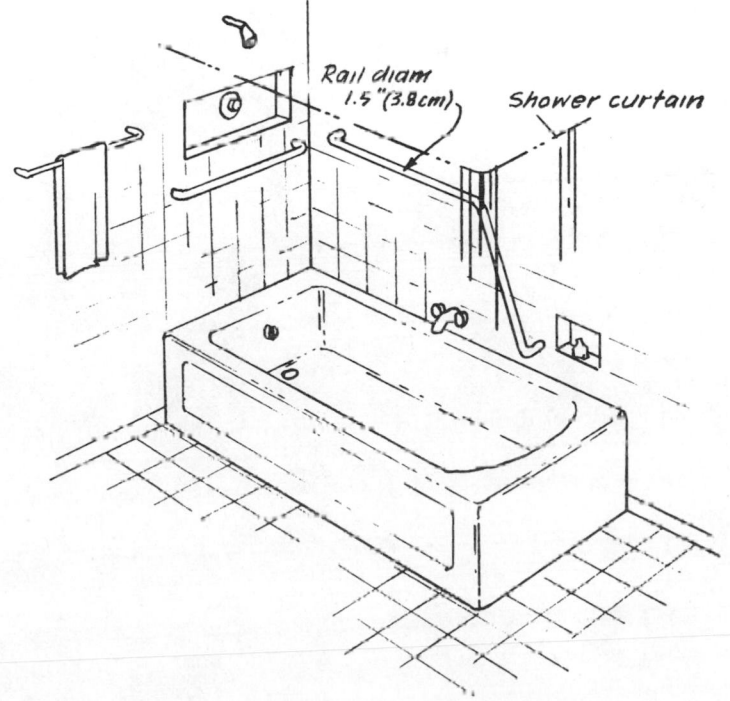

**Tub and Shower Fixture Guidelines
for Facilities Used by the
Orthopedically Handicapped**

There obviously must be enough room for the
wheelchair user to maneuver about within the
bathroom. In addition, a sliding rather than a
hinged door should be provided. Minimum dimensional guidelines are noted in the accompanying sketch.

Other special considerations include the following:

1. A seat placed at the end of the tub and
   designed so that the handicapped person can easily shift back and forth from
   the wheelchair to the tub.
2. A handrail positioned so that the user
   can obtain maximum support. Also look
   for a tub that has a grip on the outside.
   This allows certain individuals to support
   their full body weight as they lower or
   raise themselves to and from the bottom
   of the tub.
3. Look for specially designed shower ex-
   tenders to make manipulation of the
   shower head as easy as possible while
   the user is supporting himself or herself
   from the handrail.
4. Look for faucet handles that are easily
   pushed, rather than gripped and turned.

*Note:* Although the above suggestions provide generally useful aids for the wheelchair
user, the designer should investigate more
fully the range of user problems of a particular
client. These suggestions assume that the
handicapped users still have considerable use
of their arms and hands and sufficient arm
strength to move themselves with both arms.
Some persons, however, may be unable to
use their arms and hands.

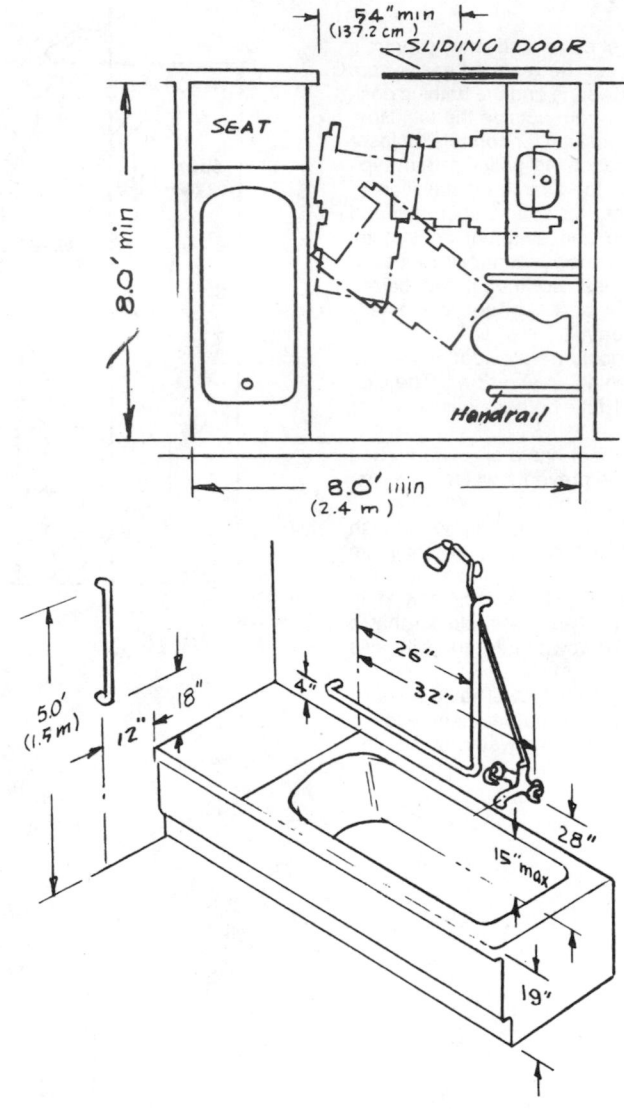

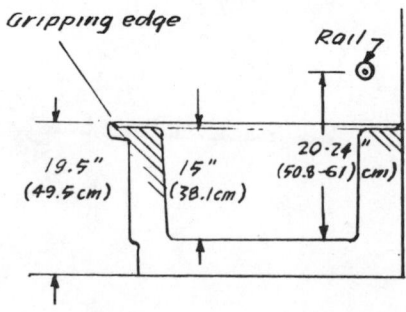

## Shower Stalls

Shower installations that are not intended for use by wheelchair users should be designed taking into account the considerations and dimensional guidelines illustrated in the accompanying sketches.

An idealized shower is illustrated in the accompanying sketch. Note the following:

A seat is included so that the user can sit down to scrub his or her feet. The controls are accessible from outside as well as from inside (and are mounted on a module to minimize injuries).

A portable shower head as well as the typical fixed shower head is provided.

A handrail is provided.

A recessed toiletry and soap shelf is provided. The drain is at the far end of the stall so that the user does not have to stand on it.

A shower partition and door should not be used with a tub unless the minimum interior clearance is at least 36 in (91 mm).

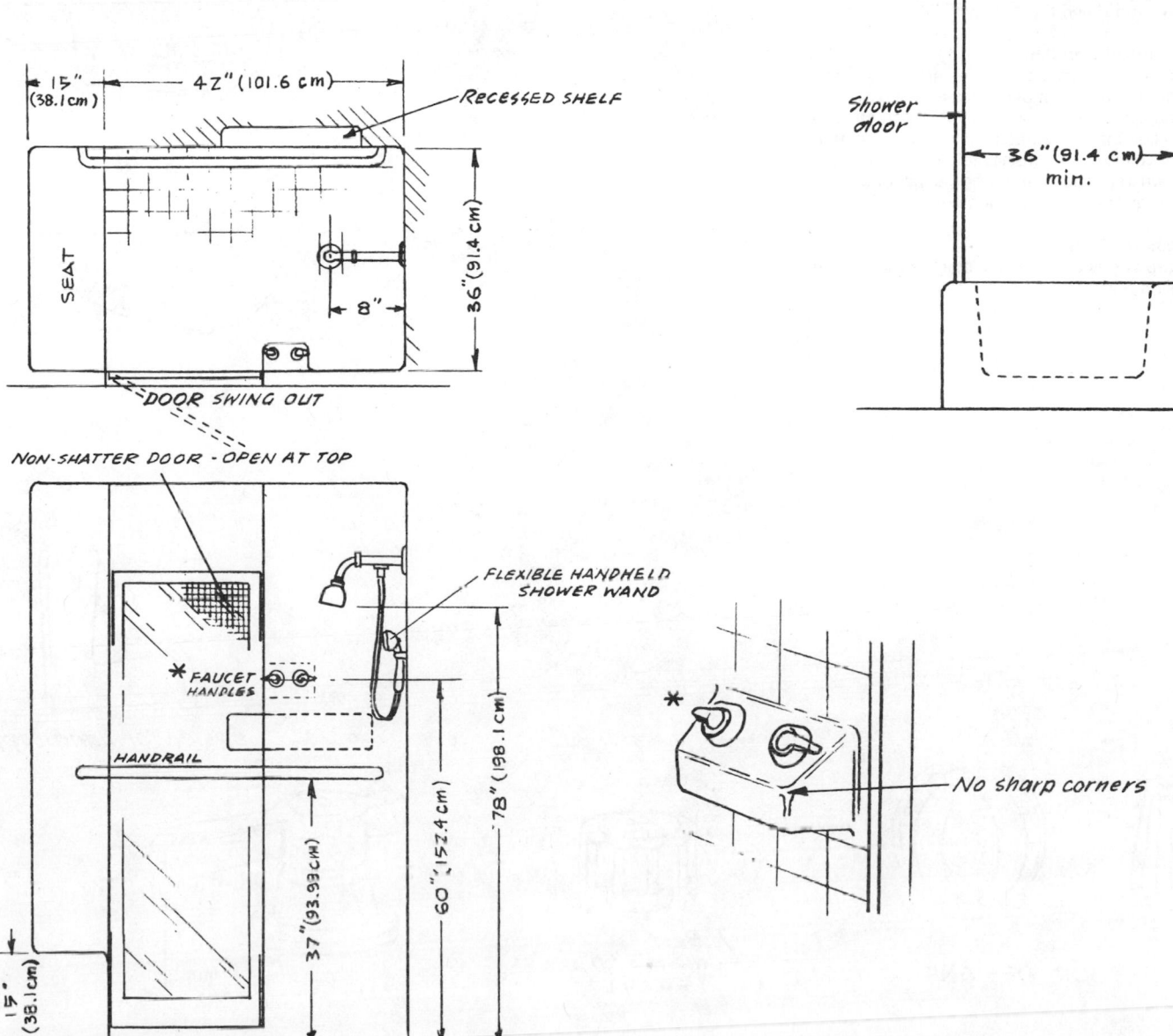

## Washbasin and Fixture Selection

The desire to create unusual aesthetic effects in bathrooms has led to fixture designs that are sometimes poorly engineered from the human factors point of view. A washbasin can, however, be both attractive and convenient and easy to use if one is careful to look for certain features that have a significant impact on the utility of the basin or its hardware, i.e., faucets and control devices. The following guidelines are suggested:

### 1. Basin Shape and Faucet Configuration
The basin shape and faucet configuration should combine to capture water rather than splatter it all over the user and/or the floor.

### 2. Basin Size and Shape
The basin must be large enough so that users can get their hands and faces inside it without bumping their heads on the faucet or having water run down their arms.

### 3. Control Handles
The shape of control handles should be compatible with the most efficient gripping and handle manipulation objectives. Oddly shaped handles may be attractive, but they are often very difficult to manipulate. The handles shown in the accompanying illustrations show poor vs preferred configurations.

### 4. Sharp Corners
Sharp edges or corners should be avoided on basins and hardware.

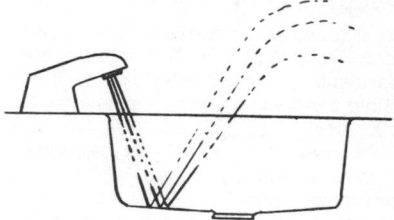

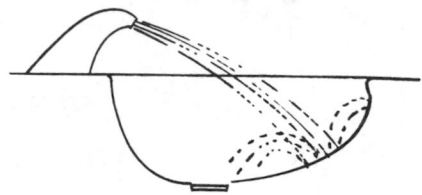

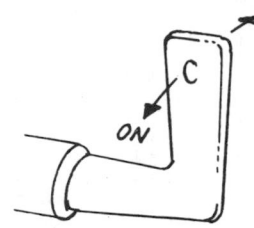

*Easier for handicapped*

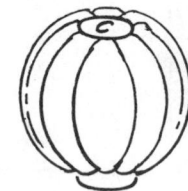

POOR DESIGNS

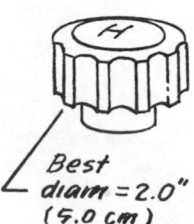

*Best diam = 2.0"
(5.0 cm)*

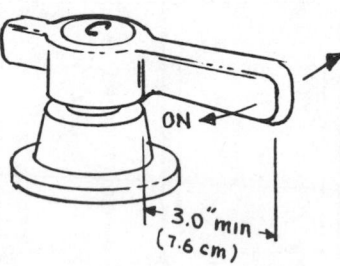

*3.0" min
(7.6 cm)*

BETTER DESIGNS

### 5. Separate Controls versus Single Combined Controls

Separate controls for hot and cold water are preferred in many applications because it is easier to manipulate them more precisely and because there is less confusion about which is which and about how the controls operate.

Hot water should always be on the left, and cold water on the right. The handles should rotate as shown in the accompanying sketches (note that this is somewhat contrary to the operation of single valve controls).

A single control is often desired because it allows the user to use one hand to operate the control and the other hand to test the water temperature as it is being adjusted.

The common type of control shown in the accompanying sketch, although theoretically very handy, is difficult to operate precisely because the pulling action required to open and close the valve is not the best direction of motion as far as human capabilities are concerned.

Better single control configurations are illustrated in the accompanying sketches. Not only is the configuration shown in the upper sketch the more compatible in terms of precise manipulation, but also the handle shape indicates readily whether cold or hot water will probably predominate when the handle is pulled forward to turn the water on.

The handle shown in the lower sketch is also good, and in addition it allows the user to change the position of the spout.

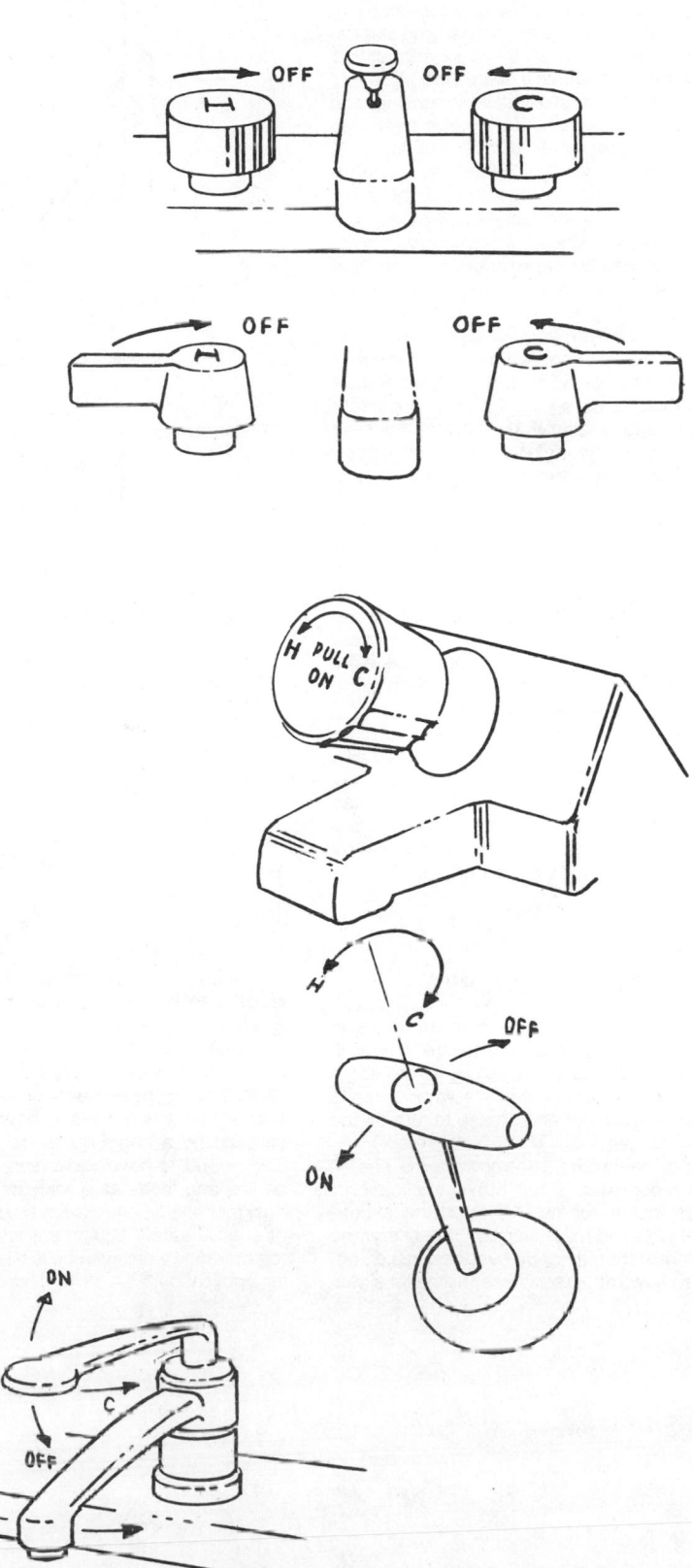

Recent studies regarding bathroom usage by Dr. Alexander Kira[7] have produced some interesting suggestions regarding washbasin shape and size. The accompanying sketches illustrate some of the principal features suggested by Kira.

As can be observed, the general shape of the basin shown in the accompanying illustration allows users to wash their hair without fear of bumping their heads, it places the stream of water where it is most effective for rinsing, and it provides sufficient space so that users' elbows do not drip water all over the floor.

When looking for a suitable basin to propose for a particular architectural client, avoid the tendency to select very small basins just because they are cheaper or because insufficient space has been provided. In order that the necessary space accommodation can be built into the plan, space considerations should not be frozen until an appropriate array of fixtures have been evaluated. Although a fixture as small as 15 in (38 cm) in diameter can be used successfully for, say, a powder room, nothing smaller than about 18 by 20 in (46 by 51 cm) should be considered acceptable for three-quarter or full bathroom concepts.

Although various studies have shown that people of differing stature prefer washbasins at differing heights, the 30-in (76-cm) height shown in the accompanying sketch is required so that an adult can bend and place his or her head in the basin for hair washing (i.e., without having water run off onto the floor). In addition, this is the maximum height for children to be able to reach into the basin.

Of course, it would be preferable to create adjustable-height basins (which, incidentally, is feasible with the advent of flexible piping). This would allow a sink to be lowered for use by small children and raised so that a wheelchair could fit beneath it or so that it would be more convenient for very tall persons.

*(Illustration adapted from Kira, 1976)*

## Special Industrial Washbasins

In many industrial settings, it is desirable to select a special washbasin design which allows several workers to wash up at the same time or which permits workers whose hands are covered with dirt and grease to turn on the water with their feet.

Special basin and control systems should also be considered for other applications, such as locker rooms. Although the locker-room application may seem unnecessary, the advantages should not be overlooked; i.e., not only can several persons use the basin simultaneously, but also the single configuration takes less time to clean than an equivalent number of separate basins.

Industrial facilities in which potentially hazardous materials are used should also have special safety basins and/or showers permitting workers who have been accidentally sprayed by a chemical to be rushed to the device and to have water sprayed over their bodies and faces as quickly as possible. The designer should check with the manufacturers of special safety equipment and/or plumbing to determine which device is best for a particular application.

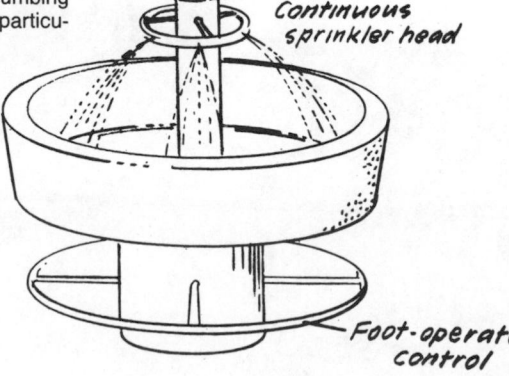

Continuous sprinkler head

Foot-operated control

[7]Alexander Kira, *The Bathroom,* The Viking Press, Inc., New York, 1976.

**Integration of the Head and Hand Grooming Center**

A fully satisfactory head and hand grooming center requires more than a washbasin. The accompanying sketch illustrates a number of the special integrating features that should be considered in developing the residential bathroom.

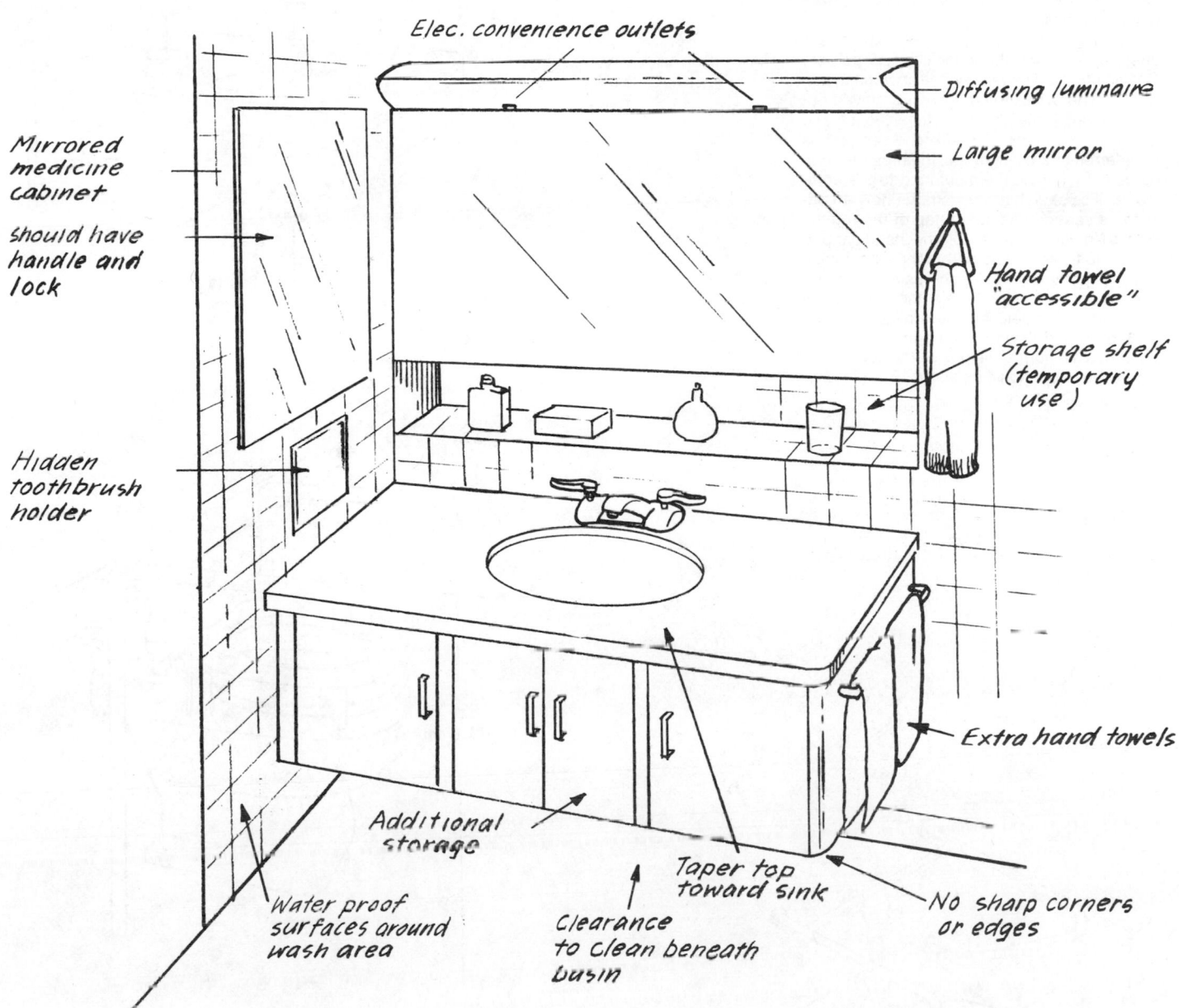

Elec. convenience outlets

Diffusing luminaire

Large mirror

Mirrored medicine cabinet

should have handle and lock

Hand towel "accessible"

Storage shelf (temporary use)

Hidden toothbrush holder

Extra hand towels

Additional storage

Taper top toward sink

No sharp corners or edges

Water proof surfaces around wash area

Clearance to clean beneath basin

*Note:* The cabinetry and pullman top should have radiused corners so that people will not be injured if they accidentally bump into the cabinet.

### Special Bathroom Features for the Handicapped

Although in certain cases a bathroom must be given special treatment in order to accommodate a particular handicapped user, in many instances preplanning can make it easier to adapt these features to the bathroom that already exists.

The accompanying sketch illustrates how a pullman type of washbasin configuration can be made attractive and yet usable by persons in wheelchairs. Some of the special basins that have been designed to accommodate the handicapped are as attractive as the standard residential basins.

Transfer from a wheelchair to a commode is often difficult for the handicapped person. One solution is illustrated in the accompanying sketch. The swinging arms can be moved out of the way while the wheelchair is placed alongside the commode. After the person has transferred to the commode, the handrails can be used by folding them out next to the commode. This sketch also illustrates how an adjustable seat can be set on top of the regular seat to make it easier for the handicapped person to transfer to and from the wheelchair.

The accompanying sketch illustrates a more permanent installation for a person who is not confined to a wheelchair but who has difficulty getting up and down from a commode.

*Note:* The normal height of the toilet seat above the floor should be about 15 in (38 cm).

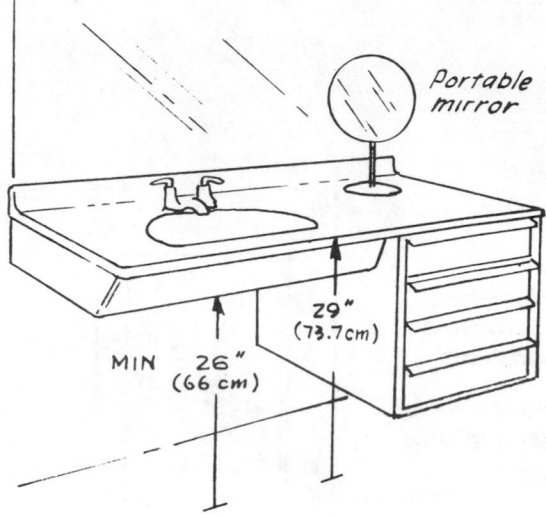

Portable mirror

29" (73.7cm)

MIN 26" (66 cm)

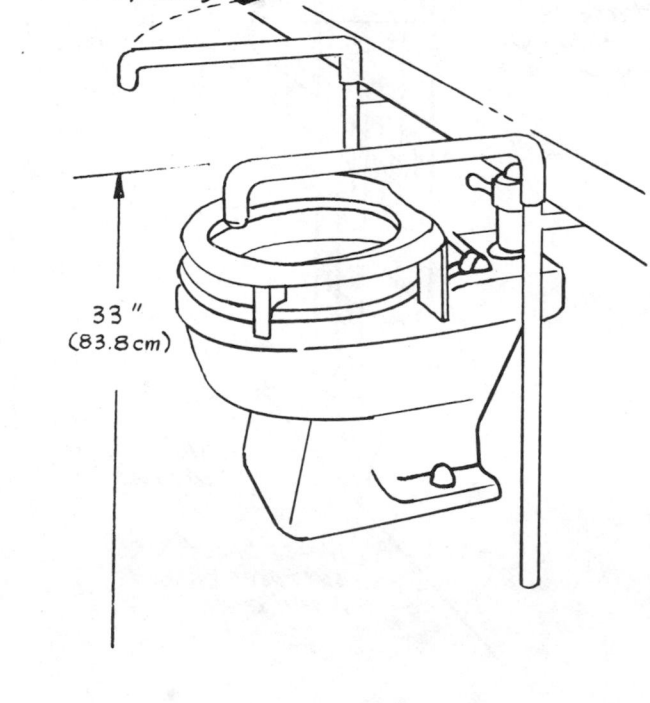

Swing-away

33" (83.8cm)

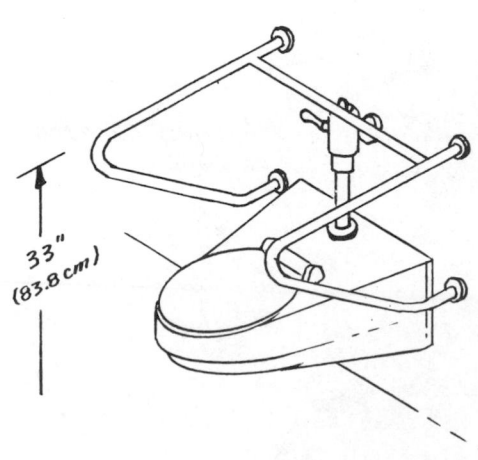

33" (83.8cm)

## REST ROOMS

Except for private or very small offices, separate rest-room facilities should be provided for men and women. Single rest rooms jointly used in the smaller office should be provided with an inside lock to ensure the privacy of the occupant.

It is desirable in most public buildings to locate rest rooms for men and women fairly close together since most people expect them to be in the same area. This is particularly important in public buildings, where the facilities are used primarily by visitors who will not be familiar with the building layout.

Public rest rooms should be located near the entrance to a building. The tradition of "hiding" the rest room because it should not be too conspicuous is contrary to human needs. Rest rooms used primarily by the building staff and its occupants should be centrally located on each floor so that users have approximately equal distances to travel to the facility.

Women's rest rooms should have a lounge adjacent to the basic rest room. Although recent laws forbid different treatment for men and women, there are obvious differences between the sexes, and from a human factors point of view, these differences should be accommodated.

Although the number of rest rooms obviously must vary according to the basic layout of a particular building, the number of fixtures provided should be determined by the number of people who are expected to use them. The accompanying chart provides general guidelines for determining the number of basic fixtures. For up to 10 users, two toilets are required in a women's rest room, and one toilet and one urinal in a men's rest room. One basin is required in both the men's and the women's rest rooms.

LEGEND

TOILETS  (Women) ————————

(Men) —— — ——

Urinals  - - - - - - -

Wash Basins —— - - ——

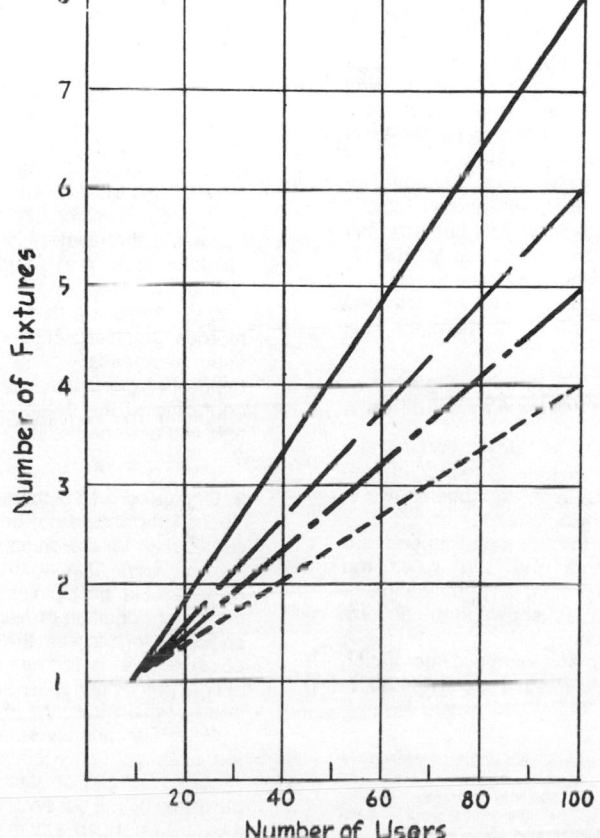

## Special Requirements for Public Rest Rooms

1. Individual stalls should be provided for all toilets. These should have doors that can be secured from the inside.
2. Individual privacy panels should be provided all urinals.
3. The following accessory fixtures should be provided in all rest rooms:
   a. Paper towel dispensers (rolling cloth towel dispensers are not recommended) and/or hot-air hand dryers and soap dispensers
   b. Sanitary napkin dispensers (women's rest rooms) and disposal containers
   c. Mirrors above all washbasins
   d. Shelves above all washbasins.
   e. Toilet paper dispensers (all toilet stalls)
   f. Garment racks or hooks
4. All large rest rooms should have a special maintenance area which includes a special sink for rinsing mops and storage space for cleaning implements and supplies and for paper supplies.
5. Adequate illumination should be provided, including luminaires above each mirror for grooming and several general luminaires for overall mobility and maintenance illumination.
6. Adequate air conditioning and/or natural ventilation should be provided, depending upon the climate and/or seasonal temperature conditions.
7. All rest rooms should be laid out in such a way that there is an intervening antechamber and/or privacy screen so that persons using the facility cannot be seen from outside as someone enters or leaves the rest room. Care must be taken to make this separating space large enough so that people who are coming and going can pass without interference. It is recommended that separate entry and exit doors be provided.
8. There should always be at least one toilet stall, one urinal, and one washbasin designed especially for handicapped persons (see the guidelines for the removal of architectural barriers).
9. Rest-room signs should be provided wherever the public may use the rest rooms.[8]

### 1. Separate Rest Rooms in Public Buildings

Use terms such as "Boys" and "Girls," or "Men" and "Women" and/or international symbols, such as those shown in the accompanying sketch.

If facilities for the handicapped are provided within the men's or the women's rest rooms, also provide the international symbol shown in the accompanying sketch.

Both printed and symbolic information should also be made tactually identifia-

[8]Rest rooms in dormitories, aboard ship, etc., should have additional facilities such as showers, dressing areas with seats, storage facilities for personal belongings, clothing hooks, and racks for towels, the latter being located for maximum convenience to washbasins and showers.

ble; i.e., the print or symbol should be raised, and braille code symbols should be placed below each letter or symbol so that partially sighted and blind persons can also identify the rest-room door.

### 2. The Single Lavatory

When only a single lavatory that may be used by the public is provided, use the words "lavatory," "toilet," or "water closet," plus the international symbol denoting that it is for use by the handicapped. If the lavatory may be used by blind persons, provide raised lettering and/or symbols.

*Note:* Avoid the use of artistic or humorous graphics. Although these may seem aesthetically pleasing or amusing to the designer, they are often merely confusing to the person who needs to find and use the facility.

### 3. Occupied and Unoccupied Signs

Such signs should not be considered a substitution for the primary rest-room or lavatory sign. One or the other of the signs should be illuminated when that particular condition prevails. Such signs should be legible from the position of the observer who is farthest away, e.g., an airliner passenger seated in the most remote position from the lavatory.

*Note:* The primary lavatory sign, like the OCCUPIED or UNOCCUPIED sign in an airplane, rail car, or bus, should always be visible by the observer who is farthest away. Do not use a primary sign that is

not illuminated either internally or by reflected light.

### Rest-Room Dimensional Guidelines

The accompanying sketch illustrates some of the key fixture spacing considerations in planning any public rest-room facility.

There should be at least 36 in (91 cm) of clearance behind any washbasin or urinal fixture so that an individual can pass other people without disturbing them. An additional 15 in (38 cm) should be added where it is necessary for a wheelchair to pass.

Regular toilet stalls should be at least 32 in (81 cm) wide, and there should be rigid monitoring of the installation of toilets and partitions to make sure that the fixtures are centered within the stall envelope. Many poor installations cause the toilet user to be crowded against one of the stall partitions.

The guidelines shown in the accompanying sketches pertain to the special requirements of the wheelchair user. Thus the reader should refer to other sections for appropriate washbasin heights for the nonhandicapped user and/or should place other fixtures according to normal architectural practice.

Fixtures and partitions for rest rooms should be suspended so that it is easy to clean beneath them.

*Note:* Certain dimensional considerations related to privacy are important in mounting toilet fixture stalls and partitions.

For the sake of safety, use a nonskid floor covering. For ease of cleaning, provide a slight slope to a floor drain placed near one corner of the rest room. Use wall material that resists damage and is easy to clean.

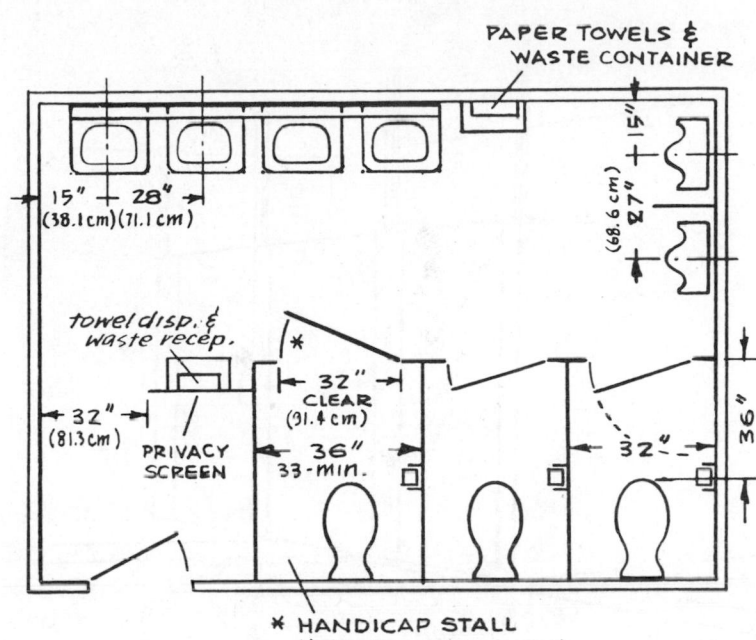

PAPER TOWELS &
WASTE CONTAINER

15" (38.1cm)    28" (71.1cm)

(68.6 cm) 27"    15"

towel disp. &
waste recep.

* 

32" CLEAR (91.4cm)

32" (81.3cm)

PRIVACY
SCREEN

36"
33-min.

36"

32"

* HANDICAP STALL
Note door opens out

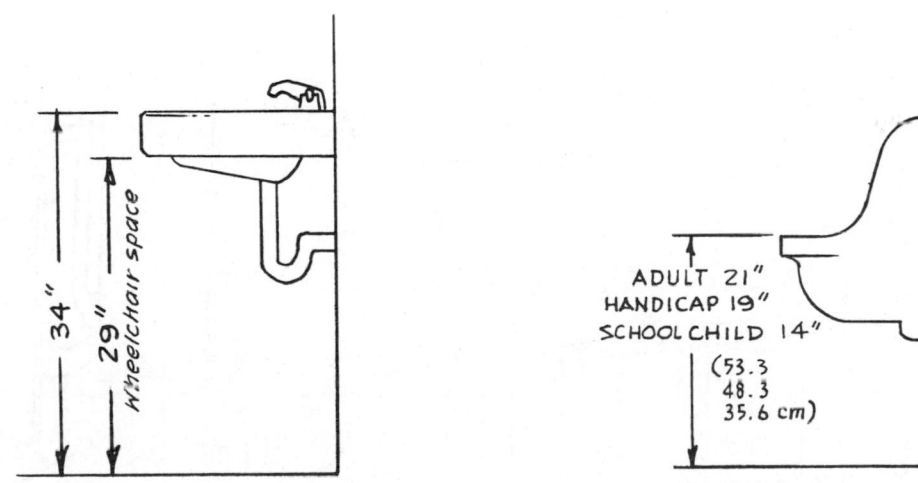

34"    29" wheelchair space

ADULT 21"
HANDICAP 19"
SCHOOL CHILD 14"

(53.3
48.3
35.6 cm)

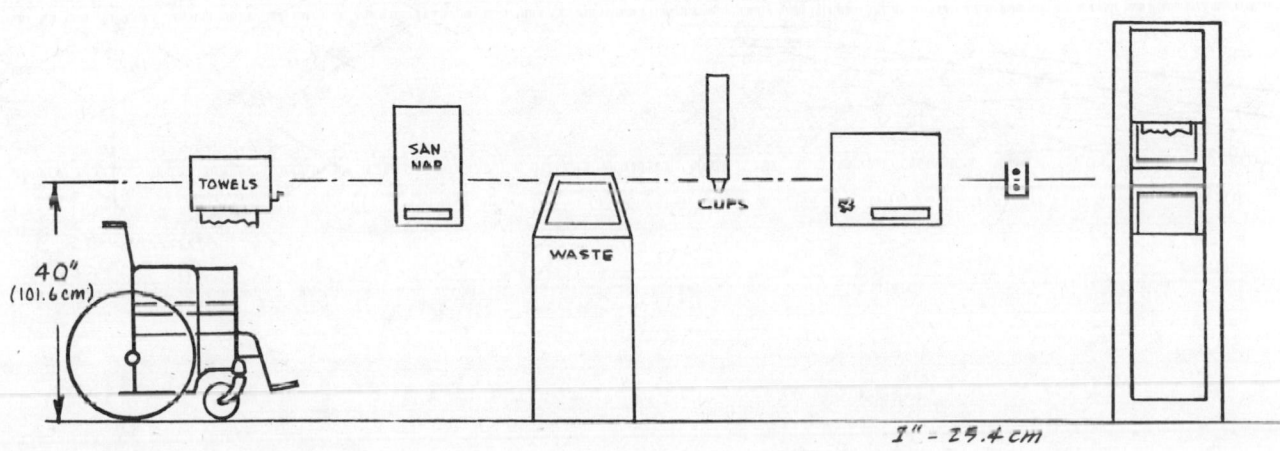

TOWELS    SAN NAP    WASTE    CUPS

40"
(101.6 cm)

1" - 25.4 cm

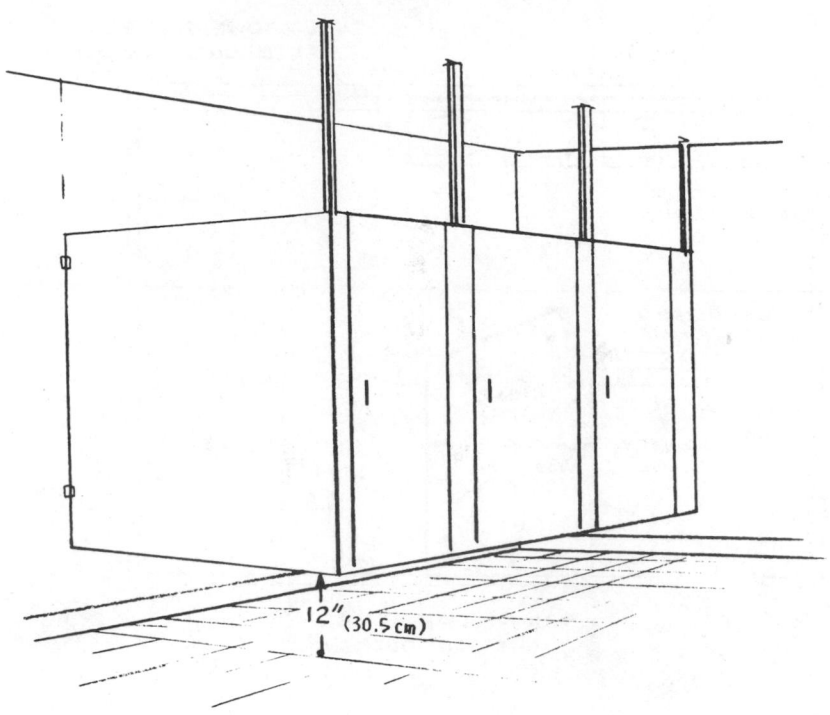

12" (30.5 cm)

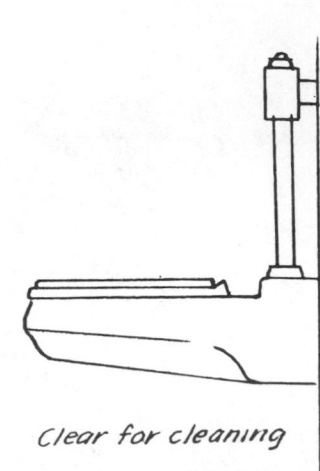

Clear for cleaning

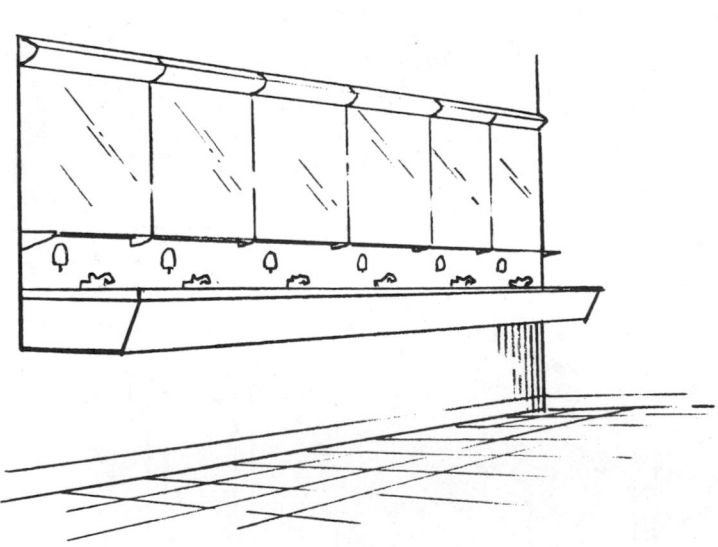

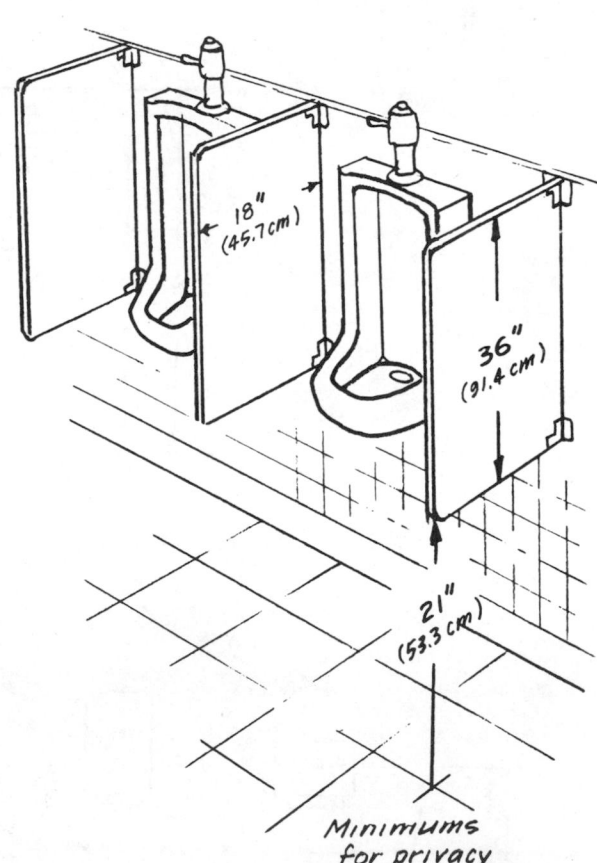

18" (45.7 cm)

36" (91.4 cm)

21" (53.3 cm)

Minimums for privacy

**Shipboard Rest Rooms and Lavatories**

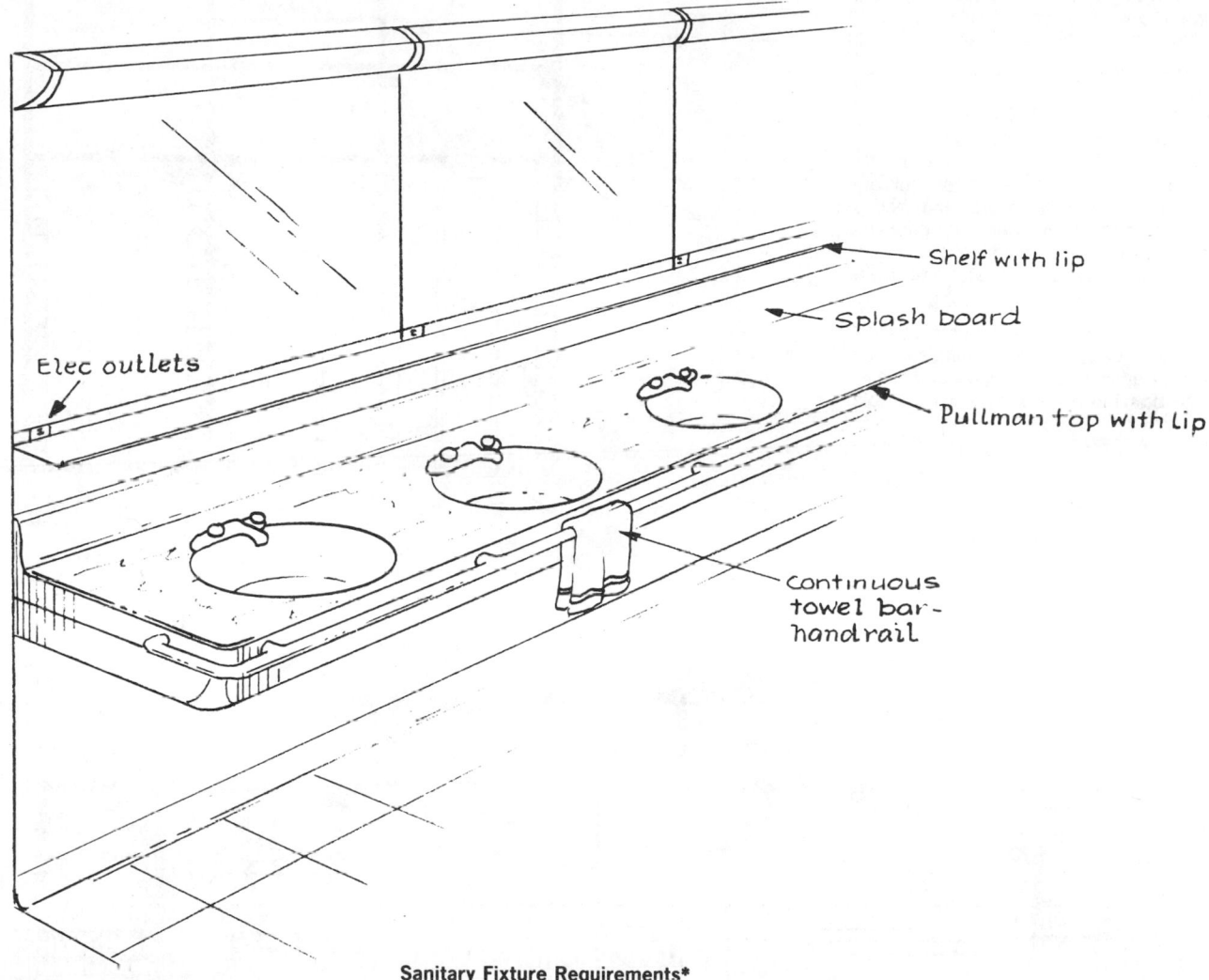

Elec outlets

Shelf with lip

Splash board

Pullman top with lip

Continuous towel bar-handrail

The following special features should be considered for shipboard rest rooms and lavatories, as well as for such facilities in other systems that are subject to considerable dynamic motion:

1. A lip should be provided around the edge of any shelf to minimize the possibility that water or objects will be projected from the surface.
2. Appropriately placed handrails should be provided to prevent the person using the facility from being thrown to the floor or against some object or structure.
3. Floor and deck surfaces should be nonskid even when wet.
4. There should be no sharp edges or corners on cabinets or fixtures where a person could be thrown against them and injured.
5. Toilet stalls should be equipped with privacy curtains rather than doors, which often swing when the vessel pitches or rolls.

**Sanitary Fixture Requirements***

| Accommodation | Washbasins | Toilets | Urinals | Showers |
|---|---|---|---|---|
| Private and semiprivate heads: | | | | |
| 1-man stateroom | 1 | 1 | 0 | 1 |
| 2-man stateroom | 1 | 1 | 0 | 1 |
| 4 men (2 staterooms, shared head) | 2 | 1 | 0 | 1 |
| 4-man bunk room | 2 | 1 | 0 | 1 |
| 4 men (2 bunk rooms, shared head) | 2 | 2 | 0 | 2 |
| 6-man bunk room | 2 | 2 | 1 | 2 |
| General berthing-area heads: | | | | |
| 9 men | 2 | 2 | 1 | 2 |
| 12 men | 3 | 3 | 1 | 3 |
| 15 men | 3 | 3 | 1 | 3 |
| 18 men | 4 | 4 | 1 | 4 |
| 21 men | 4 | 4 | 2 | 4 |
| 24 men | 5 | 4 | 2 | 5 |
| 27 men | 5 | 4 | 2 | 5 |
| 30 men | 5 | 4 | 2 | 5 |
| Nonberthing heads: | | | | |
| Preferred | 1 for 6 men | 1 for 10 men | 1 for 16 men | |
| Minimum | 1 for 10 men | 1 for 12 men | 1 for 24 men | |
| Single lavatory | 1 | 1 | | |

*These guidelines were developed during a habitability study for the U.S. Navy and apply primarily to dormitory-type living environments aboard ship. However, they are also applicable to any dormitory-type environment, for either men or women. In the case of women's quarters, the urinal allotment would obviously be replaced by an equal number of toilets.

## LOCKER-ROOM GUIDELINES

The accompanying illustrations show two examples of locker-room arrangement and critical dimensions. Key considerations should also include the following:

1. Good ventilation in toweling and dressing areas
2. Water control curbs
3. Nonskid floor surfaces throughout
4. Good lighting
5. Ample accessories such as mirrors, clothing and towel racks and hooks, waste receptacles, drinking fountains, and lockable, ventilated lockers
6. A first-aid room directly adjacent to the locker-room facilities

*Note:* In communal shower rooms, adjustable-height, nonclogging shower heads at least 4 ft (1.2 m) apart should be provided. There should be both individual and master controls.

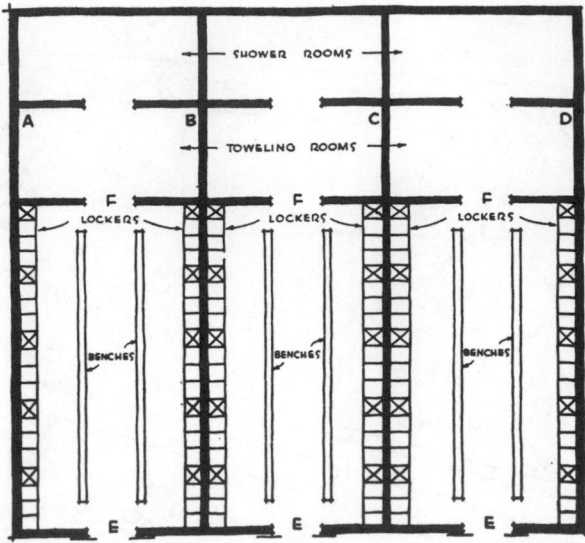

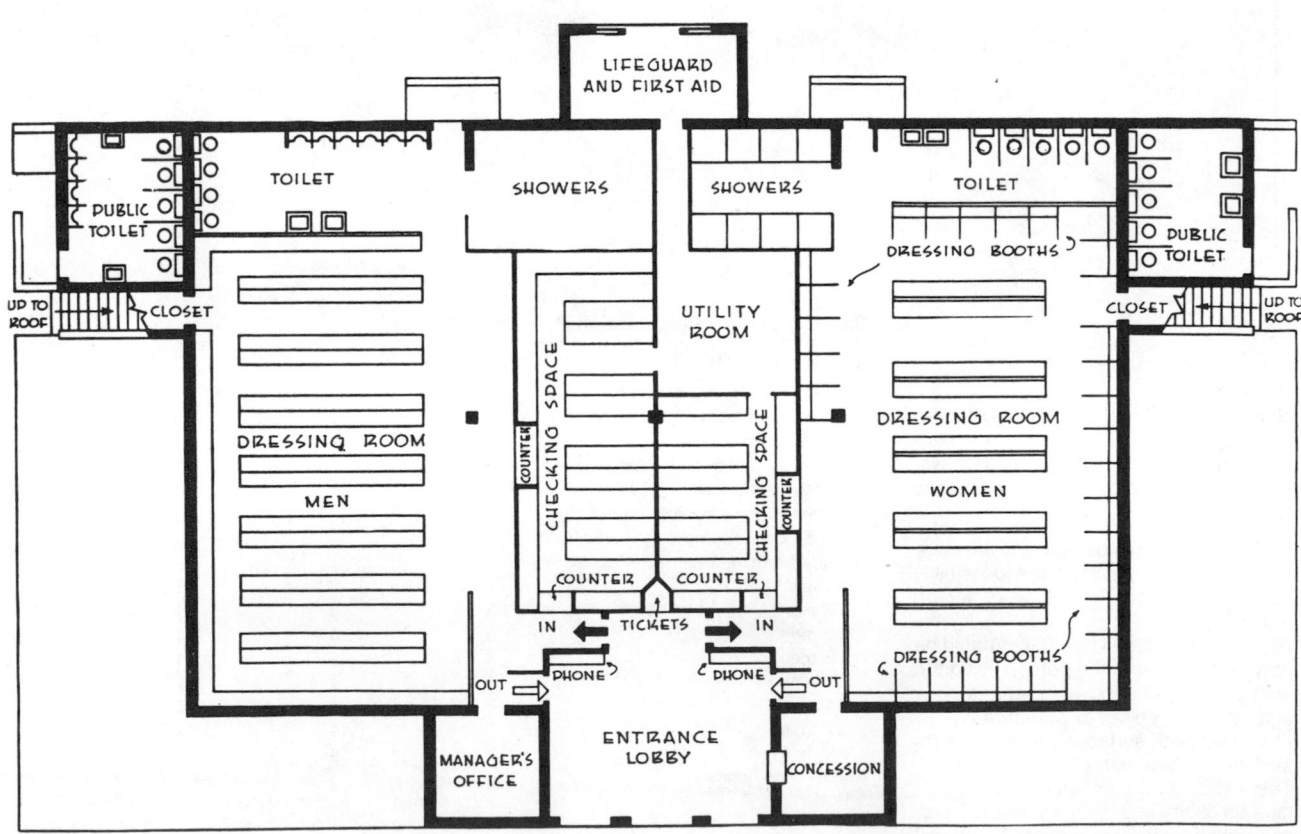

PLAN OF TYPICAL BATHHOUSE  FOR 750 PERSON POOL

SCALE ⅟₁₆" = 1'-0"

From J. De Chiara and J. H. Callender, *Time-Saver Standards for Building Types,* McGraw-Hill Book Company, New York, 1973.

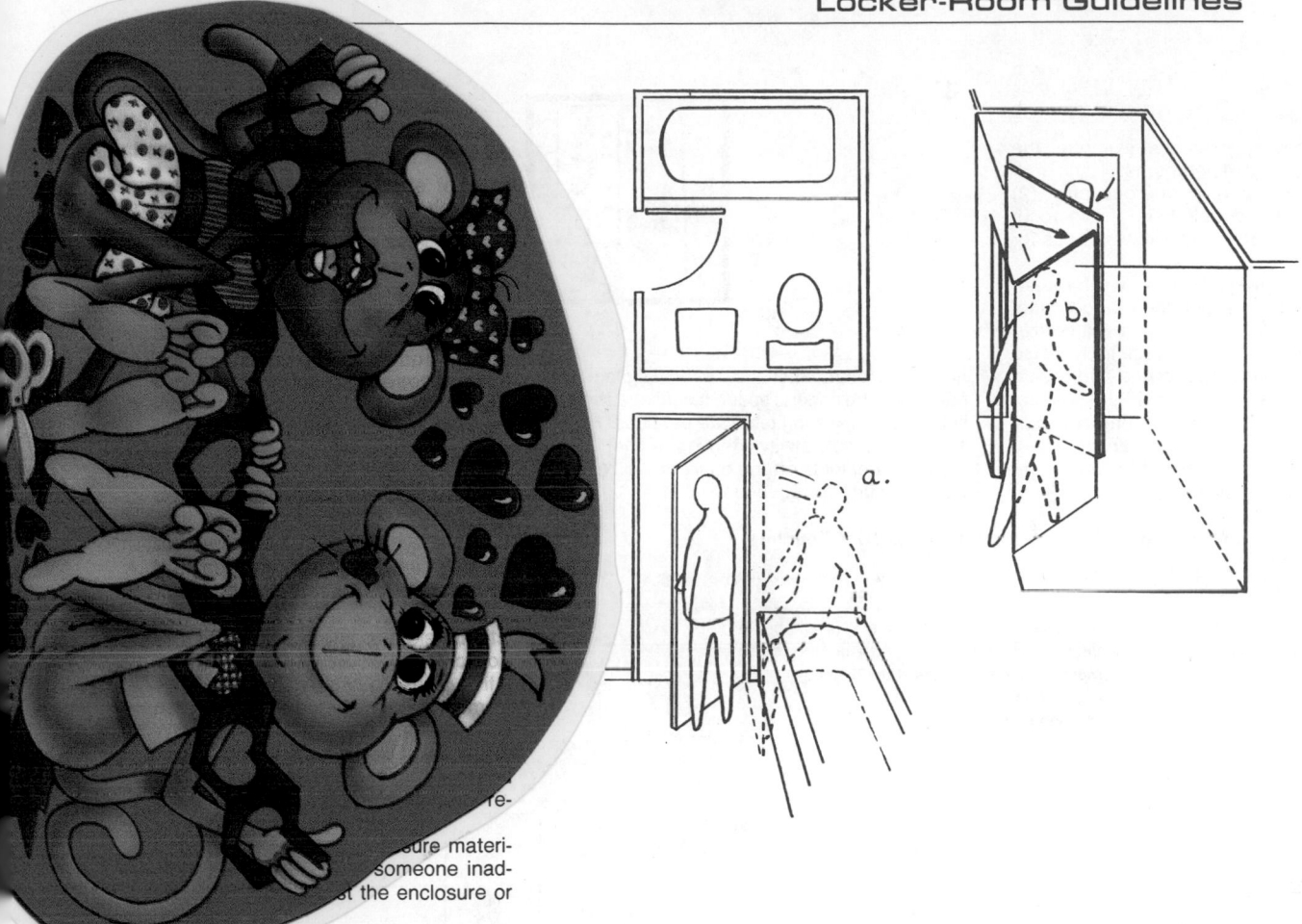

re-
ure materi-
someone inad-
the enclosure or
slippery floor materials in an effort to provide a shiny, attractive surface.

9. Medicine cabinets positioned so that if
people leave the door open while bending over the washbasin, they almost invariably bump their heads on the open
door.

10. Lack of a convenient shelf, with the result that a person may put an electric
appliance down on a wet surface.

## Maintenance and Housekeeping Considerations

1. Seamless joints and surfaces are extremely important in the design and installation of cabinets and fixtures in the
bathroom since such seams trap mixtures of water, dust, soap, and cosmetic
substances.

2. Smooth hardware designs are important;
fancy filigree entraps water, soap, and
dirt and is extremely hard to clean.

3. Suspended cabinets and fixtures are
preferred because floor-mounted features tend to collect dirt at the base, this

is hard to remove by the typical methods
of sweeping or mopping.

4. Plenty of clearance must be provided
around such fixtures as the toilet tank if
it is necessary to remove the tank lid in
order to adjust, service, or repair tank
components.

5. Intersections between the floor and wall
and pullman-type sink tops should be
radiused for ease of cleaning.

6. Standard practice invariably locates important plumbing connections and pipes
within unaccessible walls. Care should
be taken during initial planning to provide
quick access to the points which may
leak or which require frequent replace-

ment of seals or parts.

7. Toilet fixtures should be selected that
provide easy access for replacing toilet-
seat covers (which often become loose
and are difficult to replace because the
fasteners are hidden inside the tank).
These fasteners invariably rust unless
they are made of nonrusting materials,
and it becomes impossible to loosen the
rusted nuts; then they have to be cut off,
which often mars the porcelain.

8. So that they will not stick, doors, windows, and the frames for these should
be made of materials that are not affected by the moisture that is typically
produced within the bathroom.

## BEDROOMS

The accompanying sketches illustrate the three standard types of bedrooms: (1) the bedroom with a single bed for one occupant, (2) the bedroom with a double bed for two occupants, and (3) the bedroom with twin beds for two occupants.

Space should always be provided for standard furniture elements and for the necessary clearance to move around them for normal use and housekeeping, and space should also be provided for garment storage. Note also the desirability of providing for nightstands.

Since many variations are not only possible but also desirable to satisfy varying client desires, the illustrations do not include room dimensions, nor do they necessarily represent any preferred room shape. For initial planning purposes, the general space guidelines shown in the accompanying table are provided. However, one should always derive the final floor-area requirements on the basis of the type of furniture currently in vogue.

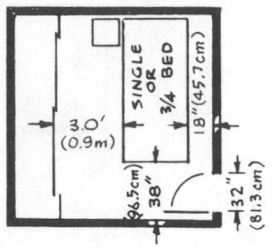
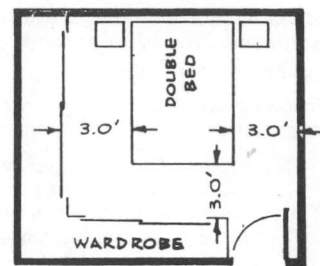
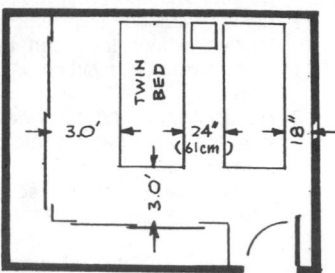

**Typical Minimum Bedroom Square Footage (Not Counting Closets)**

| Dimensions | Application |
|---|---|
| 9 × 10 ft | Single adult; one child; nursery (single bed or crib) |
| 10 × 12 ft | Single adult; one youth (single bed) |
| 12 × 16 ft | Married couple (standard double bed) |
| 16 × 16 ft | Married couple (twin beds) |
| 16 × 18 ft | Married couple (king-size bed) |

## Bedroom Planning

Bedroom size and shape should be planned around the use of standard furnishings that people can purchase at a local furniture store. The guidelines shown in the accompanying illustrations are based on typical bedroom furniture that is generally available in the United States. In addition to the standard bed sizes shown, however, one should also consider some of the newer types and sizes of beds, such as queen- and king-size beds and the more recent water beds, some of which are quite large.

In considering the various furnishings, keep in mind the following arrangement constraints:

1. The head of a bed should be against a clear wall, not against a window or a built-in closet door.
2. Beds should be located in such a way that there are no drafts across the head of the bed.
3. High chests and dressers with mirrors should be backed against a clear wall, not a window.
4. Plenty of clear dressing space must be provided in front of dressers and chests.
5. Remember that furnishings may interfere with window manipulation and curtain operation.
6. Plan double-occupancy rooms and furnishings so that there is an effective separation of the dressing activities of the two occupants—i.e., "his" closet and dresser and "her" closet and dresser—and the associated dressing space required by each person.

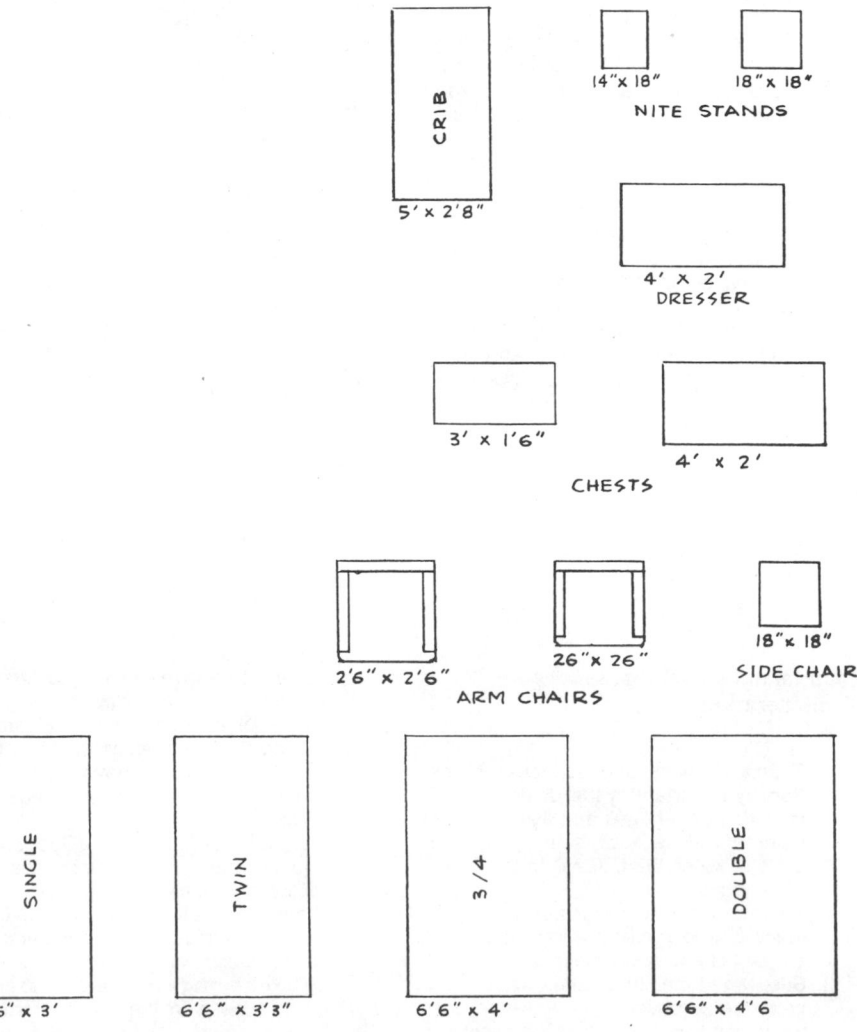

TYPICAL BED DIMENSIONS

## Built-in Closets for Bedrooms

The accompanying illustrations present general guidelines for providing closet space and various built-in conveniences in bedroom-related closets.

Note that there are several basic storage categories:

1. Storage of hanging garments, both long and short
2. Shelf or drawer storage of garments that normally should be folded and should lie flat
3. Shoe storage
4. Hat storage
5. Miscellaneous storage, which includes storage of ties, grooming items, luggage, etc.

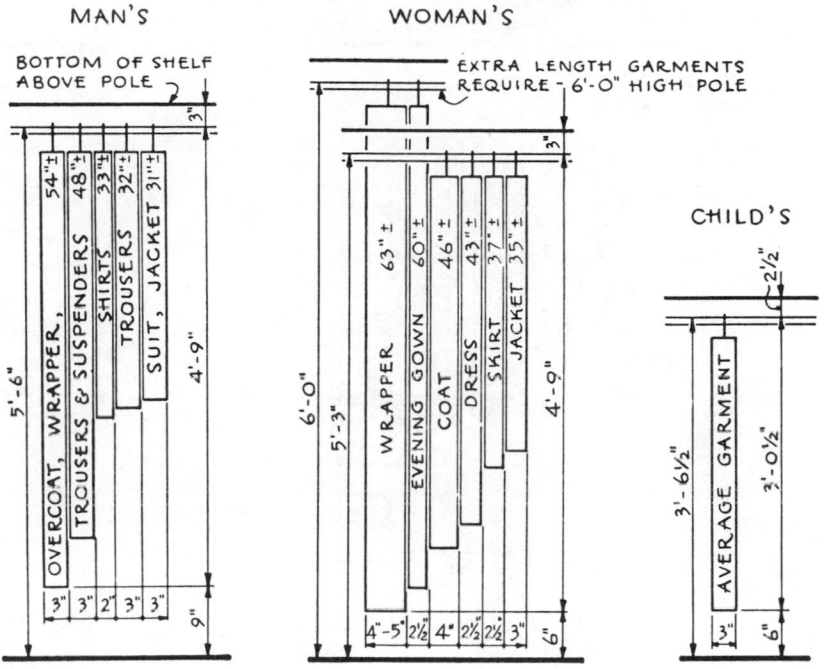

Sizes of clothes hung in closet

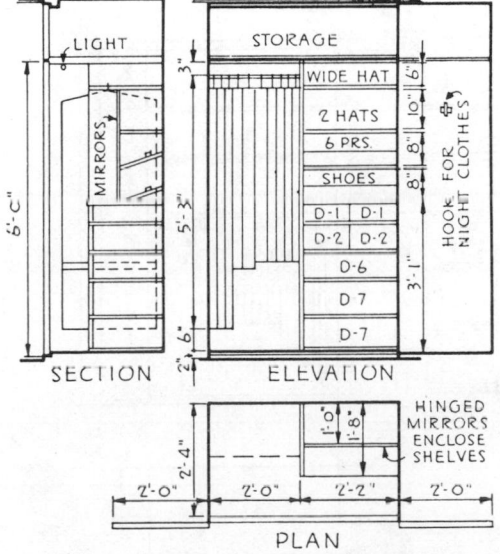

A four-foot closet combining hanging and shelf space with drawers for stockings, underthings, and what-not. Shoes are easily seen and chosen from the almost eye-level cleat rack above the drawers. Hat storage on the shelves.

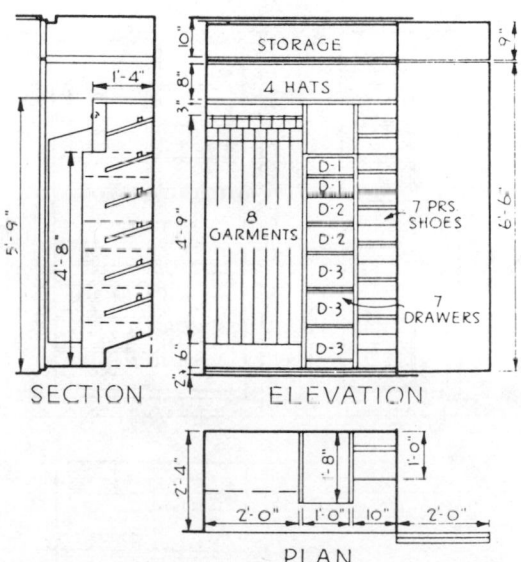

A four-foot closet with seven drawers for shirts, socks, underwear, etc., and a vertical-tier of shoe racks (as above). Night clothes and bathrobe hooks are best on the right hand door, necktie racks flat against the left hand door.

From J. De Chiara and J. H. Callender, *Time-Saver Standards for Building Types*, McGraw-Hill Book Company, New York, 1973.

## Closets for Children

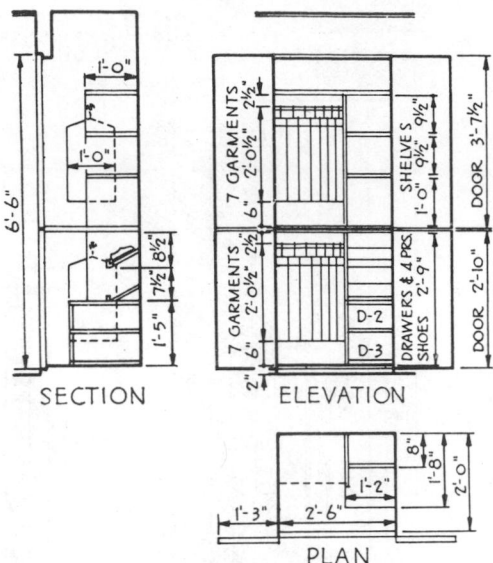

SECTION    ELEVATION

PLAN

Closet for infants up to about 5 years old Low hanging pole shelves and drawers permit habits of care and orderliness to be developed at an early age. Upper part would be used by adults. Note two sets of doors.

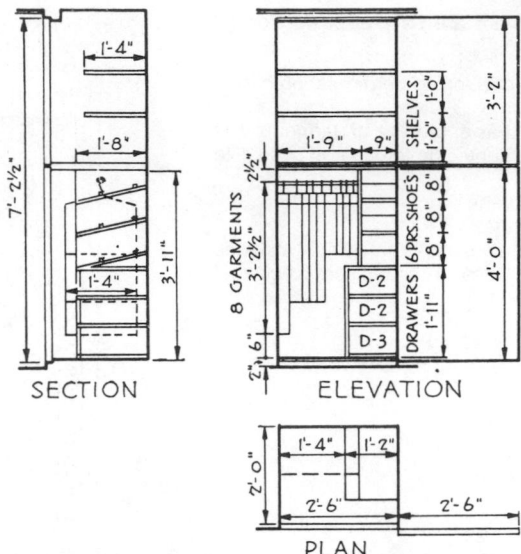

SECTION    ELEVATION

PLAN

Small closet designed for a child of from 6 to 10 years. Pole at higher but easily reached level. Drawers and shoe racks at convenient heights. Ample shelf room provided above for the storage of possessions.

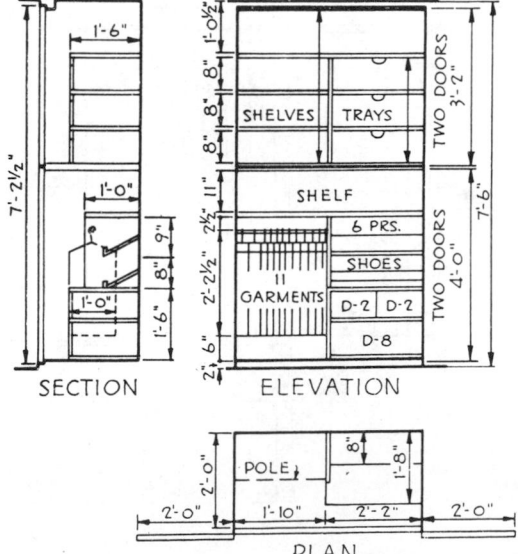

SECTION    ELEVATION

PLAN

Alternate, and larger, closet for an infant up to 5 years of age. Trays or drawers for folded garments at an upper level for adult use. Hanging space, drawers and shelf available to child using the lower doors.

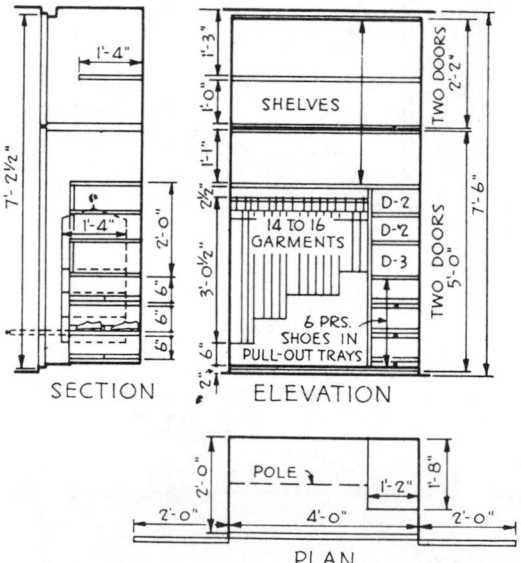

SECTION    ELEVATION

PLAN

Closet for youngster up to 10 years old, providing greater length of hanging pole and different shoe arrangement, trays instead of cleat racks. A large shelf for hats, toys, or "collections" available to child.

From J. De Chiara and J. H. Callender, *Time-Saver Standards for Building Types*, McGraw-Hill Book Company, New York, 1973.

## Walk-in Closets

The chief considerations for walk-in closet design relate to accessibility and clearance. They include the following:

1. A minimum 30-in (76-cm) doorway width (32 in, 81 cm, if the closet is to be used by a person in a wheelchair) should be provided.
2. There should be a minimum of 32 in (81 cm) between garments hung opposite each other.
3. There should be at least 1 in (2.5 cm) of clearance between the garments and the adjacent wall.
4. A garment rack or pole should be provided that is high enough so that clothes will not drag on the floor; about 44 in (112 cm) is required for children's clothes (on hangers), and about 68 in (173 cm) for women's dresses.
5. There should be good lighting inside the closet so that colors are visible.
6. The light switch for the closet should be placed on the wall outside the closet—not inside, where it can be hidden by clothing.
7. Where possible, good ventilation should be provided inside the closet.
8. If an overhead shelf is provided, as shown in the accompanying sketches, it should not extend out so far as to keep the light from shining on the clothing.

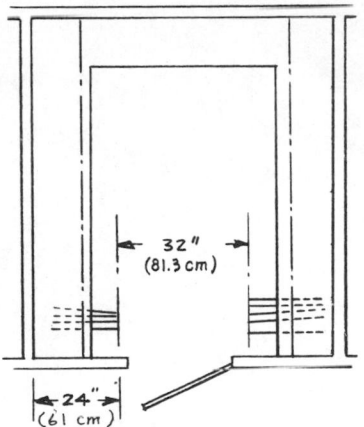

*Place lightswitch outside closet*

## SHIPBOARD BERTHING

Contrary to past practice aboard U.S. Navy ships, berths (beds) should be at least 84 in (213 cm) long and 36 in (91 cm) wide so that they will fit the largest members of the Navy's shipboard crews.

There should be at least 32 in (81 cm) of vertical clearance between stacked bunks so that crew members can get in and out easily, can sit on the edge of the bunk to put on shoes and socks, and can sit up (partially) in bed to read a book.

The crew member's bunks (unlike the situation in an officer's quarters) are their "home away from home." They can call their bunks their own, and it is here that they find some privacy and store their few personal belongings. Therefore, berthing should make it possible to shut out some of the everyday sights and sounds by means of appropriately sound-proofed partitions and privacy curtains. The curtains should be designed so that they can be snapped to the bunk structure and thus keep the crew member from rolling out of bed when the ship rolls.

The accompanying sketch illustrates some important features that were developed in a recent shipboard habitability study.

Berthing for officers where private state rooms are provided should follow good bedroom practice, but with the special requirement that things must be tied down to keep them in place whenever the ship pitches or rolls.

Special attention should be given to the selection of pleasant colors for shipboard berthing areas (as opposed to the standard Navy gray).

Ventilation within stacked bunks should be controllable by the occupant when the curtains are closed.

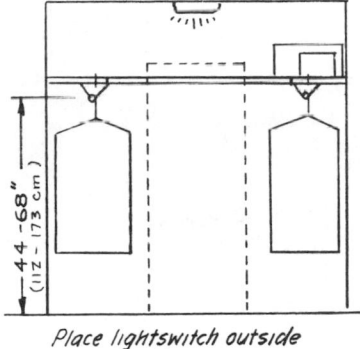

*Upper mattress should not encroach into lower bunk area when upper bunk is occupied.*

## Shipboard Wardrobes

Although certain wardrobes can be designed with objectives similar to those for closets in residential facilities (primarily for key officers aboard ship), more austere facilities must be designed for the major share of a ship's crew. The accompanying general requirements for minimum wardrobe design are based on the personal belongings that each crew member normally has aboard.

An acoustic grommet or bumper is required to prevent the door from making noise when it is being closed. Expanded versions of this design are desirable when space allows.

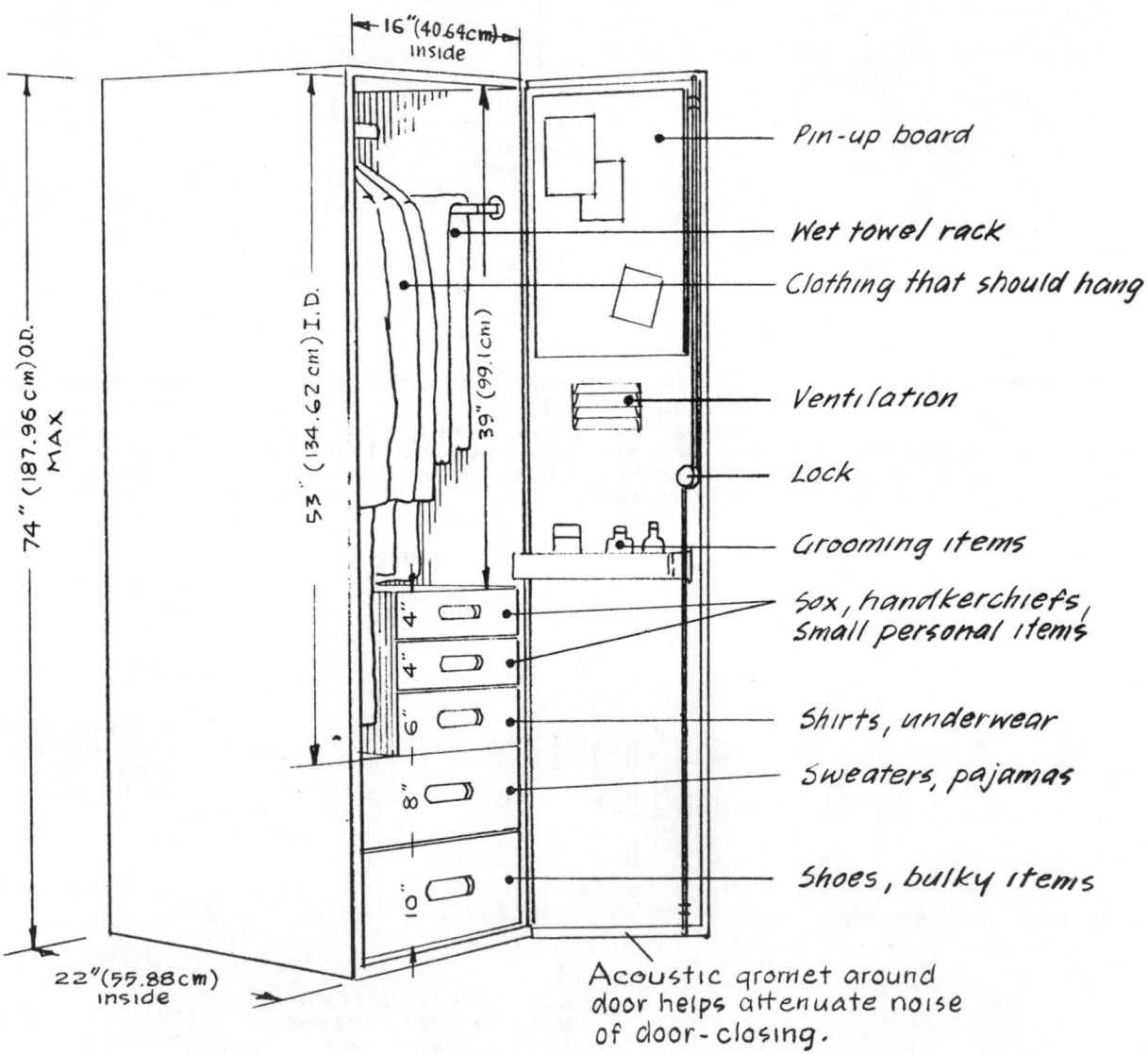

Pin-up board

Wet towel rack

Clothing that should hang

Ventilation

Lock

Grooming items

Sox, handkerchiefs, small personal items

Shirts, underwear

Sweaters, pajamas

Shoes, bulky items

16" (40.64cm) inside

74" (187.96 cm) O.D. MAX

53" (134.62 cm) I.D.

39" (99.1cm)

4"  4"  6"  8"  10"

22" (55.88cm) inside

Acoustic grommet around door helps attenuate noise of door-closing.

## Berthing System Integration

Adapting the berthing subsystem to oddly shaped spaces is often difficult. Avoid the tendency to cramp an individual berth just to "squeeze it in."

An objective should be to arrange related crew-member functional groups in the same area so as to include not only their bunks but also their lockers, sanitary facilities, and any lounging space in an integrated manner. The accompanying sketches show an example of "fitting" one crew's berthing aboard a particular ship.

Key issues addressed here were the semi-isolation of the actual sleeping area from the noisier dressing area. Note that acoustic treatment consisted of carpeting and of individual curtains on each bunk and on doorways to sleeping compartments. (All special materials must be fireproof.)

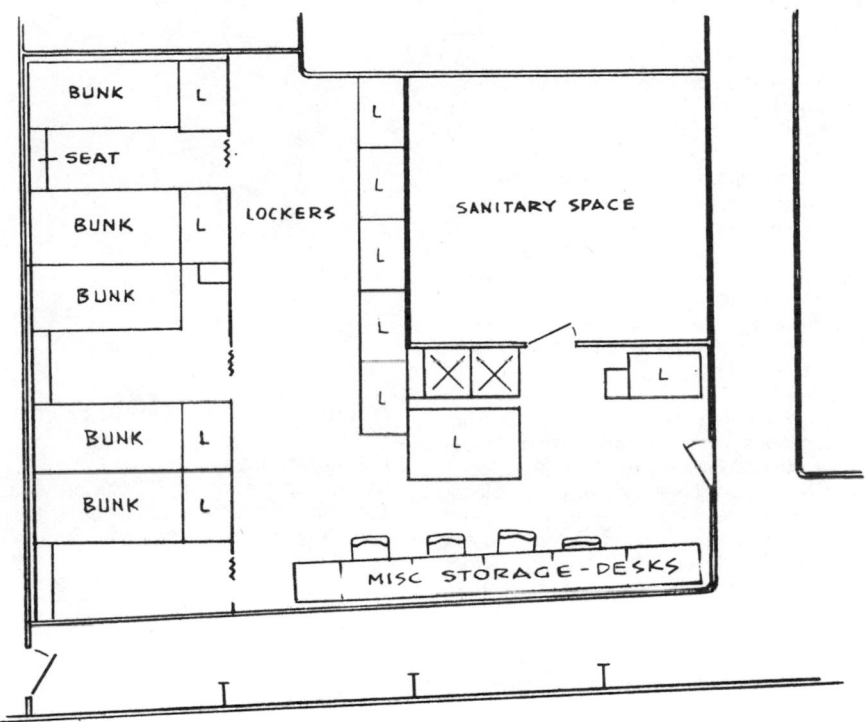

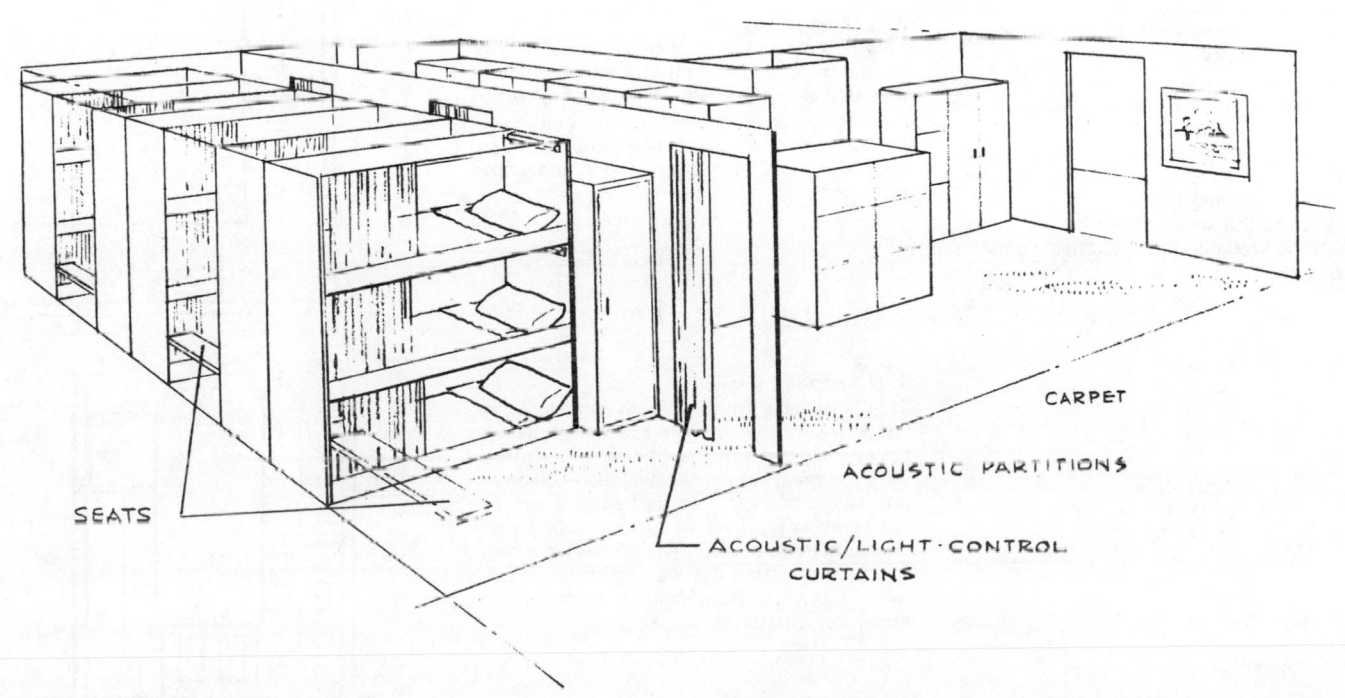

## RESIDENTIAL KITCHEN PLANNING

The residential kitchen should be planned in terms of work centers. The following are typical:

1. Food preparation: One or more locations where raw materials are cleaned, cut, mixed, etc., prior to cooking or serving
2. Cooking and baking: One or more locations where food is cooked or baked, e.g., on a stove, in an oven, or in a special electric cooking vessel or device
3. Food storage: One or more locations where raw or already prepared materials and food are stored before and after meals (refrigerator, freezer, pantry, cupboards, etc.)
4. Culinary utensil storage: One or more locations where pots, pans, cutlery, special electric mixing devices, etc., are stored
5. Dish and flatware storage: One or more locations where serving dishes, glassware, and flatware are stored
6. Washing and cleanup: One or more locations where dishes, pots and pans, and other implements used for cooking are emptied, rinsed, stacked, and either hand-washed or deposited in an automatic dishwasher
7. Miscellaneous storage: One or more locations where soaps, detergents, cleaning materials and devices, and other special serving devices such as trays and cutting boards, are stored

Primary kitchen arrangement objectives should be the following:

1. Locate stored items close to the primary task involving these items.
2. Arrange the work centers so that activity flows logically from one point to another in the order in which tasks usually are performed.
3. Arrange the work centers so that there is adequate clearance to accommodate the task mobility requirements.
4. Arrange the general kitchen so that extraneous traffic will not interfere with the in-kitchen tasks.
5. Arrange the kitchen to accommodate other individuals who may join in preparing and serving food.
6. Arrange the windows so that a person who is working in the kitchen can look outside.

### Traditional Kitchen Layout Examples

#### 1. Single-Wall Layout

This is suitable for small "efficiency" apartment kitchens where space is limited and there is typically a single user.

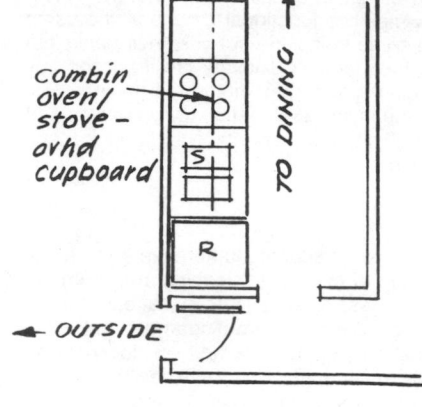

#### 2. Pullman Layout

This is suitable for a smaller residence. A parallel cabinet and appliance arrangement minimizes the necessity to string out the work flow.

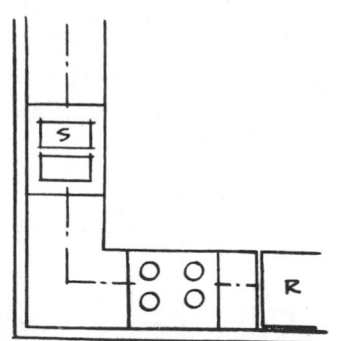

#### 3. L-Shaped Layout

This is suitable for a small to medium-sized residence. It allows combining work-center relationships and permits the addition of extra storage space. Its most important virtue is the reduction of traffic congestion and the ease with which additional kitchen help is accommodated.

#### 4. U-Shaped Layout

This is suitable for a medium-sized to large residence where enough space can be allowed between rows of cabinets. It can be varied in many ways, including the addition of an "island counter," so that several people can work in the kitchen simultaneously with minimum traffic conflict. The island layout also tends to minimize the walking distance between common work centers such as the stove, refrigerator, and sink.

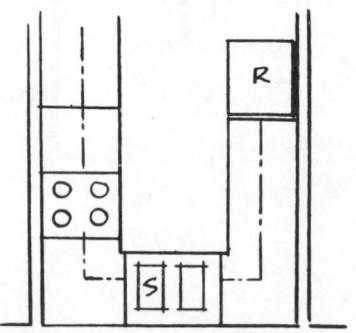

### 5. Island Concept

Perhaps the greatest benefit provided by the island concept is that a primary work center, such as the stove or sink, can be equally convenient to many other work centers. This is important to the person who cooks for many guests.

When an island concept is used, consider the addition of a small utility counter for mixing, cutting, or temporary storage of cooking utensils. Also take advantage of the island to provide additional storage space below the range or sink counter.

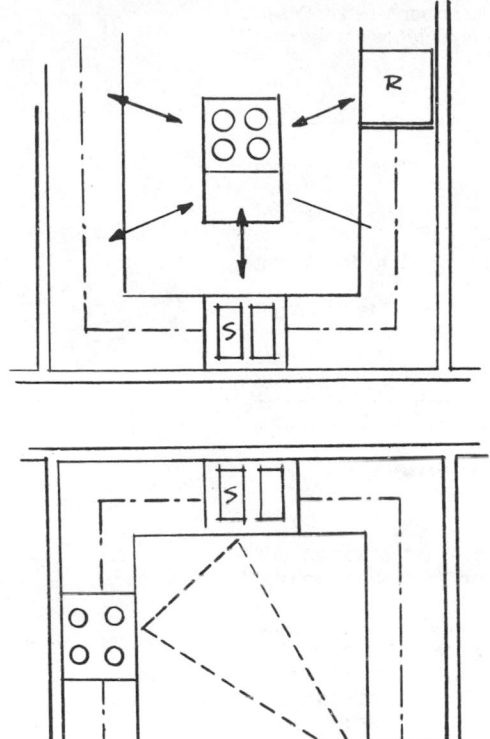

### Work-Triangle Concept

The basis of good kitchen layout has for many years been considered to be the so-called work triangle. This consists of an arrangement that limits the walking distances between the three key work centers to no more than 22 ft (6.6 m). The key centers, as shown in the accompanying illustration, include the sink, range, and refrigerator. Obviously, storage requirements are associated with each work center, i.e., food storage, utensil storage, dish storage, etc.

### Work Flow

Equally important in kitchen arrangement is the work-flow principle. An attempt should be made to arrange elements in such a way that the necessity to randomly retrace steps as one works in the kitchen is minimized. The basic sequence typically is as follows:

1. Acquiring and cleaning raw materials
2. Mixing (often adding water)
3. Cooking or baking
4. Dishing up
5. Serving

After the meal, this sequence generally is carried out in reverse.

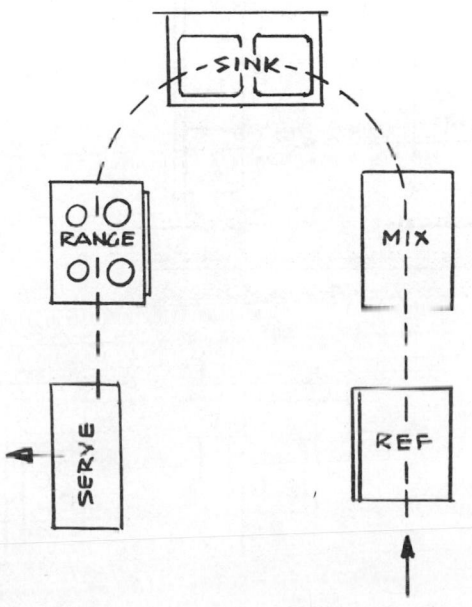

The various tasks have been estimated to take up the following percentages of total time in the kitchen:

1. Tasks performed at the sink — 43 to 50 percent
2. Tasks performed at the range — 14 to 20 percent
3. Tasks performed at mixing counters — 12 to 15 percent
4. Tasks performed at the refrigerator — 7 to 8 percent
5. Going to and from the dining area — 7 to 8 percent
6. Storing dishes — 5 to 8 percent

## General Traffic Considerations

1. Lay out the kitchen so that extraneous traffic does not normally flow through the work triangle.
2. Provide clear space between facing cabinets or appliances of at least 48 in (122 cm). When such fixtures are at right angles to each other and are separated by a passageway, the fixtures should be at least 30 in (76 cm) apart.
3. In L-shaped or U-shaped layouts, the minimum edge distance between an appliance and an adjacent corner should be 9 in (23 cm) for the edge of a sink, 16 in (41 cm) for the refrigerator, and 14 in (36 cm) from the edge of the nearest range burner.

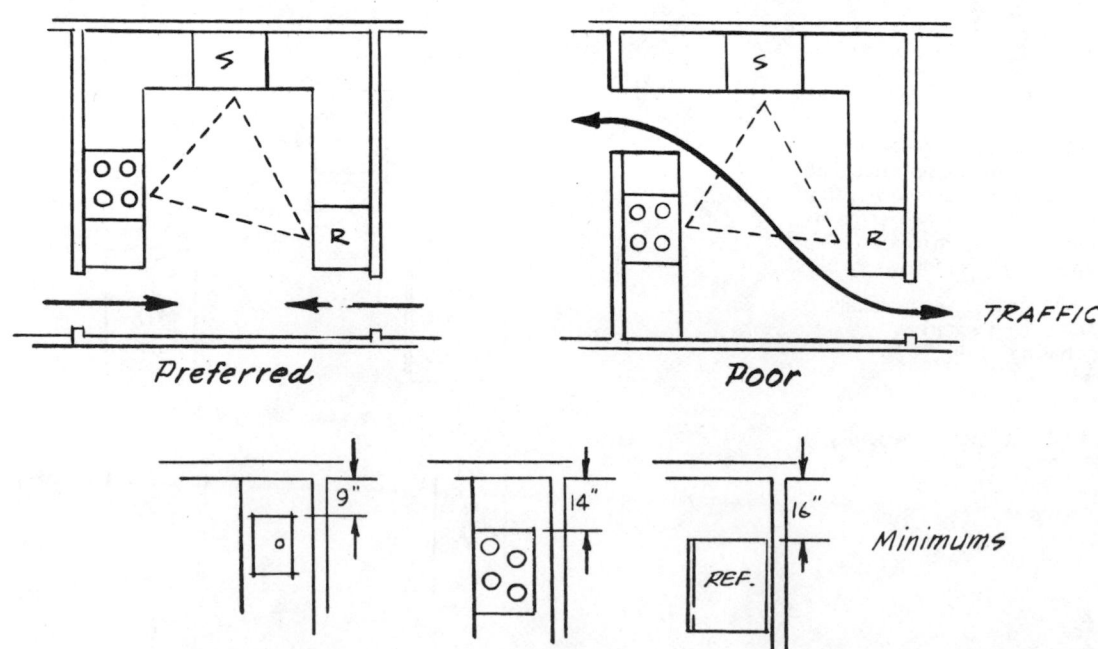

## Key Kitchen Workplace Dimensions

Basic requirements for kitchen cabinets, established by the U.S. Department of Housing and Urban Development (HUD), are shown in the accompanying table and illustration.

Other related specifications are as follows:

1. Work centers may be combined; kitchen multiuse space shall equal or exceed the largest frontage of any one of the work centers being combined, plus 6 in (15 cm).
2. Provide at least 9 in (23 cm) from the edge of the sink or range to any adjacent corner cabinet and 16 in (41 cm) from the latch side of the refrigerator.
3. 24-in (61-cm) sinks are permissible when dishwashers are provided.
4. Counter-space minimums are as follows:[9]
   52 in (132 cm) for one- or two-bedroom units
   60 in (152 cm) for three-bedroom units
   72 in (183 cm) for four-bedroom units
5. Base cabinet width minimums are as follows:
   68 in (173 cm) for one- or two-bedroom units
   72 in (183 cm) for three-bedroom units
   84 in (213 cm) for four-bedroom units
6. Wall cabinet width minimums are as follows:
   68 in (173 cm) for one- or two-bedroom units
   72 in (183 cm) for three-bedroom units
   84 in (213 cm) for four-bedroom units

## Kitchen Cabinet Requirements

| Work centers | Minimum frontage, in | | |
| --- | --- | --- | --- |
| | Two bedrooms | Three bedrooms | Four or more bedrooms |
| Sink | 24 | 32 | 32 |
| Counter and base cab at each side | 20 | 24 | 30 |
| Range | 24 | 30 | 30 |
| Counter and base cab at one side | 20 | 24 | 30 |
| Refrigerator (space) | 36 | 36 | 36 |
| Counter at latch side | 15 | 15 | 18 |
| Mixing (base and wall cabinet) | 36 | 36 | 42 |

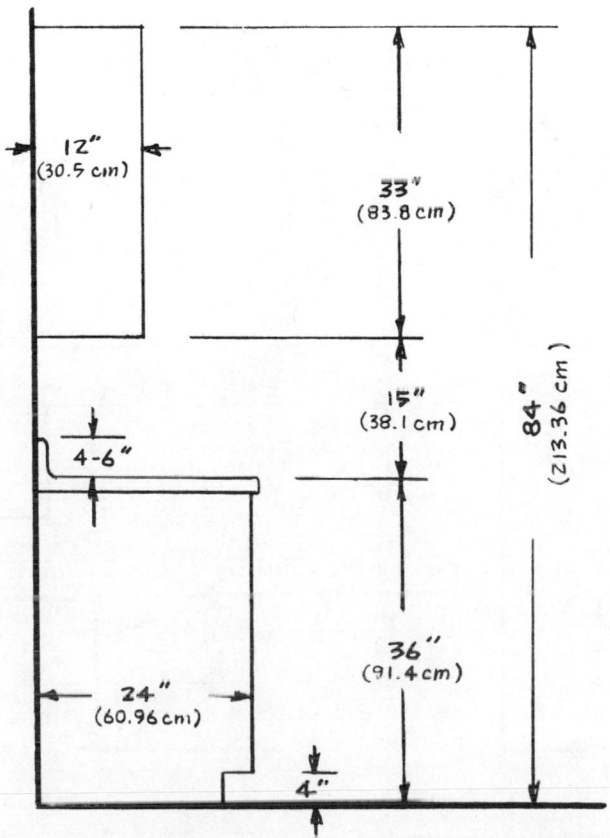

[9]The following are lineal dimensions.

**Typical Architectural Guidelines for
Kitchen Cabinet Planning**[10]

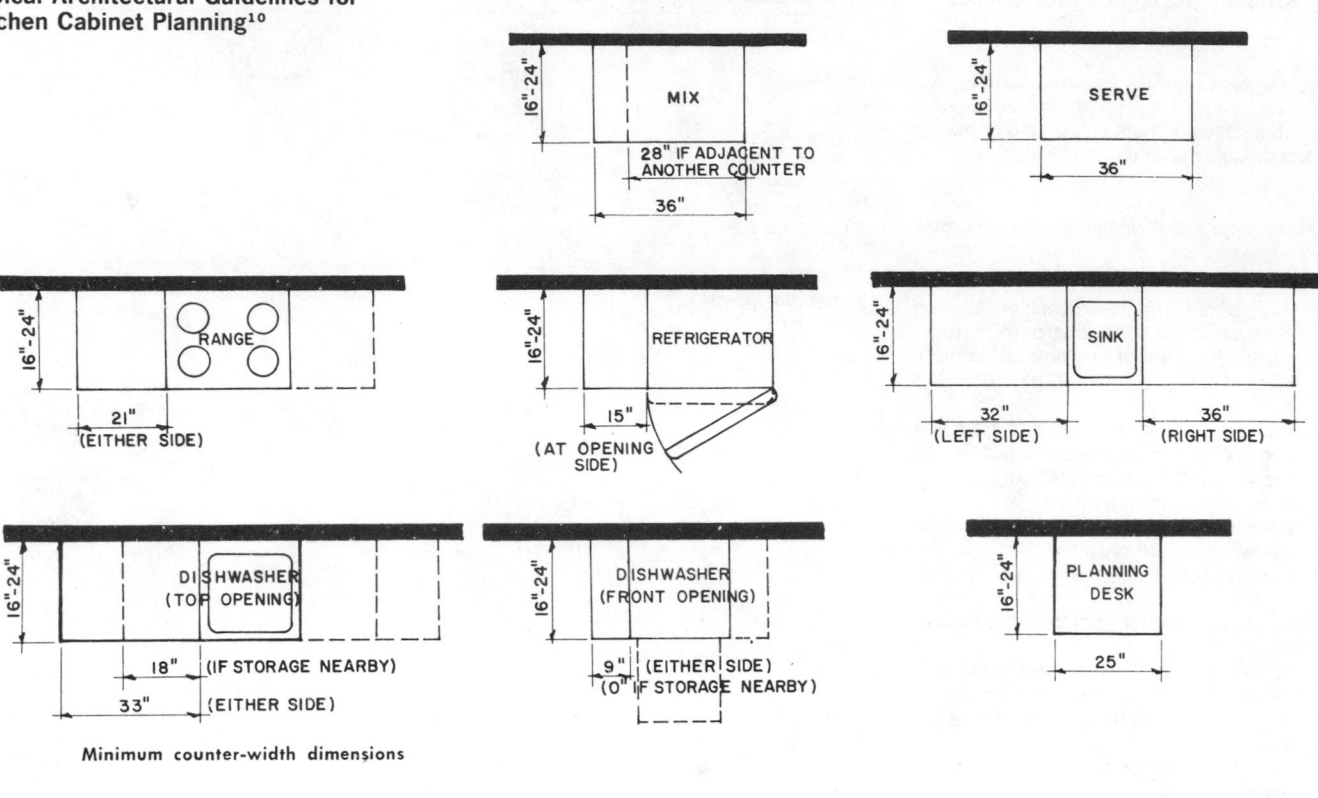

Minimum counter-width dimensions

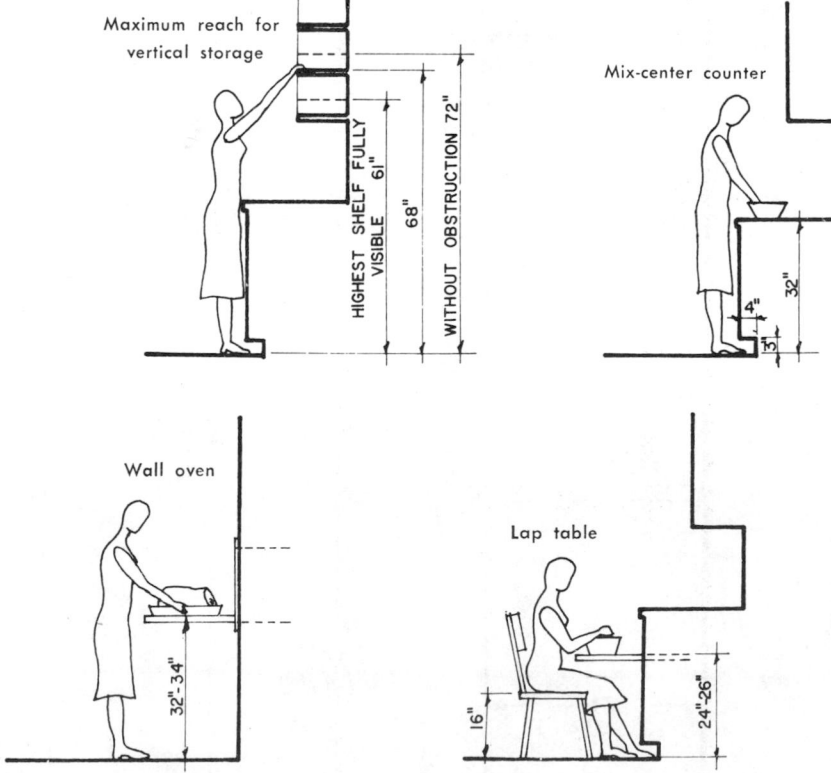

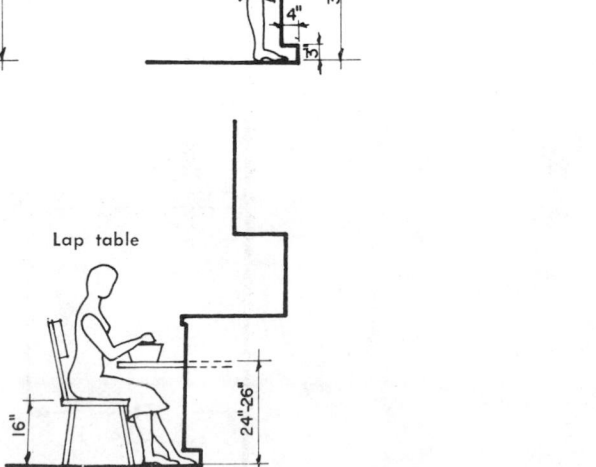

Comfortable working heights

[10]J. De Chiara and J. H. Callender, *Time-Saver Standards
for Building Types,* McGraw-Hill Book Company, New York,
1973.

## Planned Storage

Avoid the tendency to make storage an after-thought. That is, consider the arrangement and size of a kitchen not only in terms of where you plan to place the major fixtures and appliances but also in terms of the supporting storage spaces for each of the work centers. Also, avoid thinking of storage only in terms of a "bin" into which things are thrown. Each type of storage requirement should be addressed in terms of what is to be stored, the shapes and sizes of articles to be stored, how the items are normally positioned and stacked, and, obviously, approximately how many of each item there will be. Certain items are stored best in drawers; others, in bins; and still others, on shelves.

Consider the problems of weight. For instance, a large drawer full of flatware or heavy steel cookware will be sufficiently heavy so that it may require special drawer slides.

Drawers should be used which cannot be pulled all the way out, possibly leading to the dumping of the drawer's contents. Doors on cabinets should swing in the direction most convenient for the person who is trying to get into the cupboard or cabinet. Avoid very large swinging doors because they get in the way when opened. Do not place a door where the user might inadvertently mash his or her hand against an adjacent structure or appliance when opening the door.

The first step in analyzing storage needs should be to determine how much of each item must be stored. The accompanying tables can be used as a guide.

### Equipment and food supplies stored at range center

| Item | Number stored | | Storage space per item, in.* | | |
|---|---|---|---|---|---|
| | Limited | Liberal | Side to side | Front to back | Height |
| *Equipment* | | | | | |
| Potato masher | 1 | 1 | $3\frac{1}{2}$ | 13 | $4\frac{1}{2}$ |
| Knives, forks, spoons | 3 | 3 | $3\frac{1}{2}$ | 13 | 3 |
| Frying pan, $10\frac{1}{2}$-in. | 1 | 1 | 11 | $17\frac{1}{2}$ | $5\frac{1}{2}$ |
| Frying pan, 9-in. | 1 | 2 | $9\frac{1}{2}$ | 16 | 5 |
| Frying pan, 6 in. | 0 | 1 | 6 | 12 | 5 |
| Pot lids | 2 | 4 | $10\frac{1}{2}$ | $10\frac{1}{2}$ | 1 |
| Potholders | 4 | 8 | 7 | 7 | 2† |
| *Food supplies* | | | | | |
| Rice, 1-lb pkg. | 1 | 1 | $2\frac{1}{2}$ | 4 | $6\frac{1}{2}$ |
| Spaghetti, 1-lb pkg. | 1 | 1 | $2\frac{1}{2}$ | $11\frac{1}{2}$ | 6 |
| Coffee, 1 lb can | 1 | 1 | $5\frac{1}{2}$ | $5\frac{1}{2}$ | 4 |
| Oatmeal, 3-lb box | 1 | 1 | 6 | 6 | 11 |
| Macaroni, 1-lb pkg. | 1 | 1 | 2 | $5\frac{1}{2}$ | 9 |
| Tea, 8-oz pkg. | 1 | 1 | $2\frac{1}{2}$ | $4\frac{1}{2}$ | 7 |

*Dimension of the item (including lid, if any) plus clearance for handling.
†Provides for stack of 6 potholders.

From J. De Chiara and J. H. Callender, *Time-Saver Standards for Building Types*, McGraw-Hill Book Company, New York, 1973.

**Storage Requirements[11]**

## Equipment and food supplies stored at sink center

*In addition to the items listed below, allow space for hand tools (such as can opener, small vegetable brush, paring knives, rubber plate scraper), cleaning supplies (such as* soap, soap powder, cleanser, paper towels), garbage and trash containers, and possibly a stool for sitting.

| Item | Number stored | | Storage space per item, in.* | | |
| | Limited | Liberal | Side to side | Front to back | Height |
|---|---|---|---|---|---|
| *Equipment* | | | | | |
| Dishpans, nested | 2 | 2 | 16½ | 18½ | 8 |
| Dishdrainer | 1 | 2 | 14½ | 18½ | 6 |
| Double boiler | 1 | 1 | 7½ | 12 | 10½ |
| Pressure saucepan | 0 | 1 | 9 | 17 | 7½ |
| Saucepan, 6-qt | 0 | 2 | 10½ | 10½ | 9 |
| Saucepan, 4-qt | 1 | 1 | 9 | 11 | 7½ |
| Saucepan, 3-qt | 2 | 2 | 8½ | 15 | 8 |
| Saucepan, 2-qt | 1 | 1 | 7½ | 14 | 7 |
| Saucepan, 1-qt | 1 | 1 | 6½ | 13 | 6 |
| Colander | 1 | 1 | 11½ | 13 | 6 |
| Coffee pot, 6-cup | 1 | 1 | 6½ | 9 | 10 |
| Dishtowels | 8 | 12 | 12 | 11 | 5(8) |
| Handtowels | 8 | 12 | 12 | 10 | 5(8) |
| Aprons | 4 | 6 | 11 | 10 | 5(4) |
| Dishcloths | 6 | 12 | 8 | 8 | 4(6) |
| *Food supplies* | | | | | |
| Potatoes, lb | 10 | 10 | 9 | 11 | 8 |
| Onions, lb | 3 | 3 | 9 | 7 | 8 |
| Fruit, lb | 3 | 3 | 9 | 7½ | 5 |
| Lentils and peas, 2-lb pkg. | 1 | 1 | 3½ | 5 | 9½ |
| Dry beans, 2-lb pkg. | 1 | 1 | 3½ | 5 | 8½ |
| Prunes, 1-lb pkg. | 1 | 1 | 3 | 5 | 8 |
| Canned food, No. 2 can | 6 | 8 | 4 | 4 | 5½ |

*Dimensions include clearance for handling.
†Number in parentheses refers to number of items in stack for which storage space dimension is given.

[11]J. De Chiara and J. H. Callender, *Time-Saver Standards for Building Types*, McGraw-Hill Book Company, New York, 1973.

## Equipment and food supplies stored at serve center

*In addition to the items listed below, provide a drawer for silverware and space for such miscellaneous items as lunch boxes, serving tray, and hot-plate pads.*

| Item | Number stored* | | Storage space per item, in.† | | |
| --- | --- | --- | --- | --- | --- |
| | Limited | Liberal | Side to side | Front to back | Height |
| **Equipment** | | | | | |
| Paper napkins, box | 1 | 2 | 8 | 8 | 3½ |
| Tablecloth, luncheon | 0 | 1 | 10 | 14 | 3§ |
| Tablecloth, dinner | 1 | 2 | 10 | 19 | 2§ |
| Cups | 8(4) | 12(6) | 4½ | 5½ | 6 |
| Cereal dishes | 6(2) | 8(2) | 7½ | 7½ | 5 |
| Dinner plates | 8(1) | 12(1) | 11 | 11 | 4½ |
| Salad or pie plates | 8(1) | 12(1) | 9 | 9 | 4½ |
| Fruit dishes | 6(1) | 12(1) | 5½ | 5½ | 6 |
| Saucers | 8(1) | 12(1) | 7½ | 7½ | 4½ |
| Juice glasses‡ | 6 | 8 | 3 | 3 | 5 |
| Pitchers, large | 1 | 2 | 7½ | 10½ | 10 |
| Pitchers, medium | 1 | 1 | 7 | 8 | 10 |
| Water glasses‡ | 8 | 12 | 3½ | 3½ | 6 |
| Bowls, oval | 2(1) | 3(2) | 13½ | 9½ | 9½ |
| Bowls, round | 2(1) | 4(2) | 9½ | 9½ | 7½ |
| Creamer | 1(1) | 1(1) | 5 | 7 | 5 |
| Gravy boat | 0 | 1(1) | 6 | 10½ | 5½ |
| Jelly-relish dishes | 2(1) | 2(1) | 7½ | 7½ | 2 |
| Platter, large | 1(1) | 1(1) | 16½ | 13 | 2½ |
| Platter, medium | 1(1) | 2(1 or 2) | 14 | 11 | 2½ |
| Platter, small | 0 | 1(1) | 12 | 9 | 2½ |
| Serving plates | 0 | 2(1 or 2) | 11 | 11 | 4½ |
| Sugar | 1(1) | 1(1) | 5½ | 6½ | 5½ |
| Tray, medium | 0 | 1(1) | 15¼ | 11¼ | 3 |
| Refrigerator dishes, set of 4 | 1(1) | 1(1) | 8 | 8¼ | 7 |
| Toaster | 1 | 1 | 6–7 | 9–12 | 7–8 |
| Waffle iron | 0 | 1 | 10–14 | 8–14 | 3–5 |
| **Food supplies** | | | | | |
| Prepared cereals, 11-in.-tall box | 2 | 4 | 3 | 8 | 14 |
| Cookies, 1-lb pkg. | 1 | 2 | 3 | 6½ | 11½ |
| Crackers, 1-lb pkg. | 1 | 2 | 4½ | 10½ | 6½ |
| Peanut butter, 1-lb 4-oz jar | 1 | 1 | 3 | 3 | 6½ |
| Mayonnaise, 1-pt jar | 1 | 1 | 3½ | 3½ | 6 |
| Jam and pickles, 1-pt jar | 1 | 3 | 3½ | 3½ | 2 |
| Bread | 1 | 2 | 5½ | 12 | 6 |
| Cake | 1 | 1 | 9½ | 9½ | 2½ |

*Number in parentheses refers to number of stacks.

†One-half in. added to side-to-side and front-to-back measurement of item or stack and ½ to ¾ in. to height to permit safe handling. For stacked items, clearance is sufficient to remove single item from stack.

‡Glasses placed three rows to a shelf instead of stacking.

§Provides space for two tablecloths.

## Equipment and food supplies stored at mix center

*In addition to equipment and supplies listed below, allow space for such miscellaneous items as cookbooks, wax paper, and certain* *essential hand tools (hammer, pliers, screw driver, and knife sharpener).*

| Item | Number stored | | Storage space per item, in.* | | |
| | Limited | Liberal | Side to side | Front to back | Height |
|---|---|---|---|---|---|
| **Equipment** | | | | | |
| Electric mixer | 1 | 1 | 7½–12 | 10–14 | 10–17 |
| Flour sifter | 1 | 1 | 6½ | 9 | 7 |
| Mixing bowl, 3½-qt | 1 | 1 | 12½ | 12½ | 6 |
| Mixing bowl, 2-qt | 1 | 2 | 9½ | 9½ | 5½ |
| Mixing bowl, 1-qt | 1 | 1 | 7½ | 7½ | 5 |
| Pint measure | 1 | 1 | 4½ | 6½ | 5½ |
| Cup measure, set | 1 | 1 | 4 | 5 | 5 |
| Baking dish, 10½-in. diam | 0 | 1 | 11 | 12½ | 4½ |
| Baking dish, 9½-in. diam | 1 | 1 | 10 | 11½ | 4½ |
| Loaf pan | 1 | 2 | 6 | 10½ | 3½ |
| Biscuit pan | 1 | 1 | 10 | 13½ | 3 |
| Pie pans | 1 | 3 | 10 | 10 | 2½ |
| Cake pans | 2 | 2 | 12 | 12 | 2½ |
| Muffin pan | 1 | 2 | 11 | 14 | 2½ |
| Cookie (baking) sheet | 1 | 2 | 12½ | 16 | 2 |
| Egg beater | 1 | 1 | 4 | 12½ | 4 |
| Cookie cutter | 1 | 1 | 3 | 3 | 3½ |
| Rolling pin | 1 | 1 | 3 | 19 | 3½ |
| Mixing and blending forks | 2 | 6 | 3 | 12½ | 2½ |
| Measuring spoons, 4 sets | 1 | 2 | 3 | 6 | 2½ |
| Egg whisk | 0 | 1 | 4 | 12½ | 2½ |
| Knives and spatulas | 2 | 6 | 3 | 14 | 2 |
| **Food supplies** | | | | | |
| Cornmeal, 5 lb | 1 | 1 | 8 | 8 | 9 |
| Flour, 5 lb | 1 | 1 | 8 | 8 | 9 |
| Sugar, 5 lb | 1 | 1 | 8 | 8 | 9 |
| Pancake flour, 2-lb pkg. | 0 | 1 | 2½ | 6½ | 10½ |
| Cake flour, 2¾-lb pkg. | 1 | 1 | 3 | 7 | 10½ |
| Vinegar, 1-qt bottle | 1 | 1 | 4 | 4 | 10 |
| Powdered sugar, 1-lb pkg. | 1 | 1 | 2½ | 4 | 8½ |
| Brown sugar, 1-lb pkg. | 1 | 1 | 2 | 4 | 8 |
| Coconut, 7-oz pkg. | 1 | 1 | 2 | 4 | 8 |
| Shortening, 3-lb can | 1 | 1 | 5½ | 5½ | 8 |
| Cornstarch, 1-lb pkg. | 1 | 1 | 2½ | 4 | 7½ |
| Cocoa, 1-lb pkg. | 1 | 1 | 3 | 4 | 7½ |
| Raisins, 15-oz pkg. | 1 | 1 | 2½ | 4½ | 7½ |
| Flavorings, 6-in.-tall bottle | 3 | 5 | 1½ | 2½ | 7 |
| Salt, 1-lb 10-oz pkg. | 1 | 2 | 4 | 4 | 7 |
| Baking powder, 1-lb pkg. | 1 | 1 | 3½ | 3½ | 6½ |
| Baking soda, 1-lb pkg. | 1 | 1 | 2½ | 4 | 6½ |
| Package desserts, 3⅜-oz pkg. | 1 | 3 | 2 | 4 | 5½ |
| Spices, 4½-in.-tall can | 2 | 3 | 2½ | 3½ | 5½ |
| Spices, 3-in.-tall can | 4 | 6 | 1½ | 2½ | 4 |

*Dimension of the item (including lid, if any) plus clearance for handling.

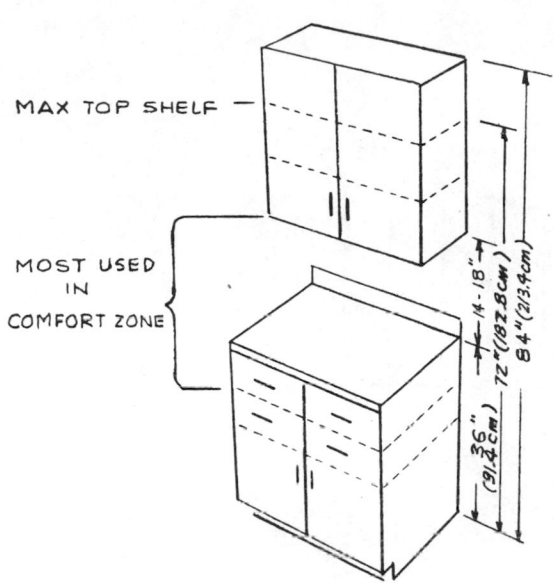

MAX TOP SHELF

MOST USED
IN
COMFORT ZONE

14-18"
72"(182.8cm)
84"(213.4cm)
36" (91.4cm)

## Activity-Storage Relationships

Although few people agree with the designer's decisions concerning where things should be stored in the kitchen, some consideration must be given to likely relationships that will occur so that the shape of the storage (drawer depth, shelf spacing, etc.) can accommodate what is expected at each location. The accompanying generalizations are meant to assist in developing an appreciation for activity-storage relationships.

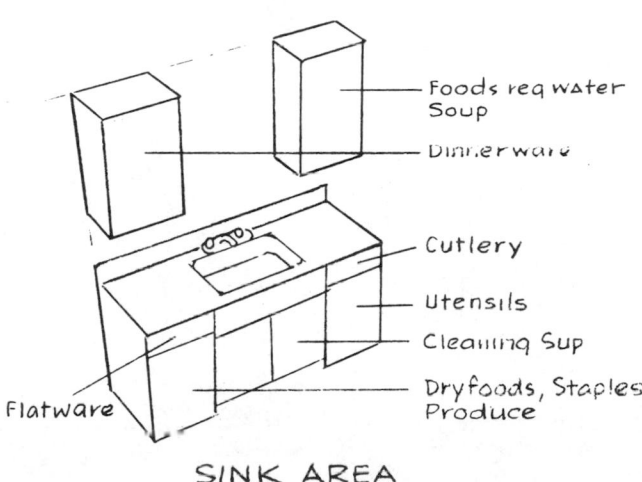

Foods req water
Soup

Dinnerware

Cutlery

Utensils

Cleaning Sup

Dryfoods, Staples
Produce

Flatware

SINK AREA

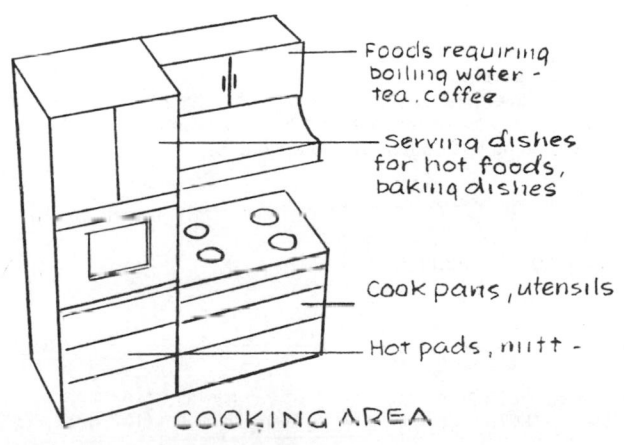

Foods requiring
boiling water -
tea, coffee

Serving dishes
for hot foods,
baking dishes

Cook pans, utensils

Hot pads, mitt -

COOKING AREA

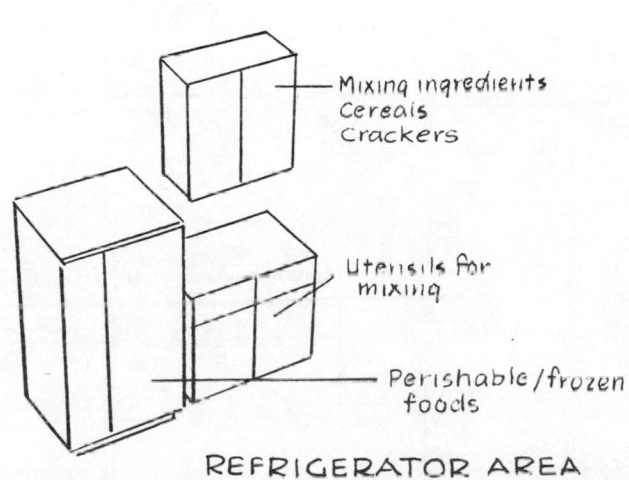

Mixing ingredients
Cereals
Crackers

Utensils for
mixing

Perishable/frozen
foods

REFRIGERATOR AREA

### Design for Health and Safety

#### 1. Circuits and Switches
Use ground fault circuit interrupters (GFCI) for all kitchen circuits and locate switches far enough from the sink so that no one whose hands are in water or who is touching a sink fixture can reach them.

#### 2. Lighting
Good-quality illumination is required throughout the kitchen, not only for inspecting food and washing dishes but also for utilizing cutlery in a safe manner. Key points to note—i.e., the stove top, underneath cabinets, and around the sink—are illustrated in the accompanying sketch.

#### 3. Appliance Selection and Installation
Select cooking appliances which have controls that are out of reach of small children and which do not require the person who is using them to reach across hot burners or utensils.

Mount stove hoods where they are not easily bumped when the person who is using the stove bends over the stove top. The hood should have radiused corners.

Make sure that the stove or oven selected is well insulated so that the exterior surface will not become hot when the appliance is in operation.

#### 4. Gas Leakage
All gas inlet pipes should have shutoff valves and should be securely anchored to prevent their being dislodged by movement of the stove.

#### 5. Hood Location

| A | B |
|---|---|
| Less than 18 in (46 cm) | 56 in minimum (142 cm) |
| 18 in (46 cm) or more | 60 in minimum (152 cm) |

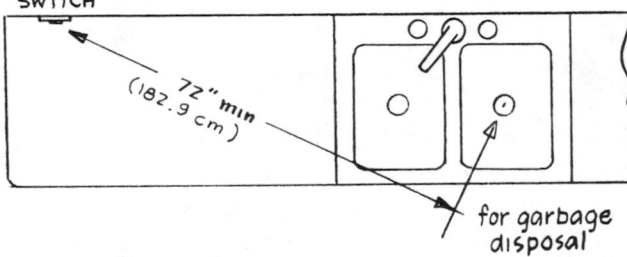

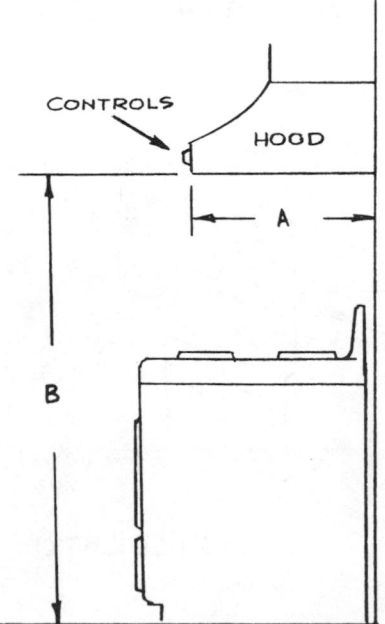

## Special Designs for the Handicapped Homemaker

1. Leave frequently used pots and pans exposed and within reach.
2. Select a range that has controls near the front.
3. Consider a mobile oven rather than the built-in type because such ovens are more accessible.
4. Provide several alternative work shelves positioned at a convenient height for mixing and provide a hole in which the mixing bowl is captured so that it does not have to be held.
5. Provide clearance beneath the sink for wheelchair armrests.
6. Provide shallow, roll-out shelves for storage of small items.
7. Provide clearance for wheelchair footrests.
8. If the handicapped person can stand, provide strategically placed handholds along the edges of the cabinets so that the person can hold on and/or steady himself or herself while working.
9. Locate a telephone in the kitchen.

## Special Considerations for the Blind or Partially Sighted Homemaker

1. Provide high-level (a minimum of 50 fc at the working surface) shadowless lighting throughout the kitchen.
2. Use contrasting colors to help the partially sighted homemaker differentiate dishes and pots from work surfaces, special controls from their immediate background, electric cords and outlets from work surfaces or the splashboard, etc.
3. Provide tactile coding on range and oven controls and labels.
4. Do not include any cabinetry that is suspended in the path that the homemaker is expected to take.
5. Select appliances that have auditory cues to indicate when cooking is completed.
6. Make sure that there are no sharp corners on cabinets or appliances.
7. Provide a lip around counter tops where water is used or where mixing is done, so that liquids or objects will not drain or roll off onto the floor.
8. Consider providing a tactile coding system for identifying storage areas, i.e., a braille or other tactile code on the outside of cabinet doors or drawers to indicate what is stored within.

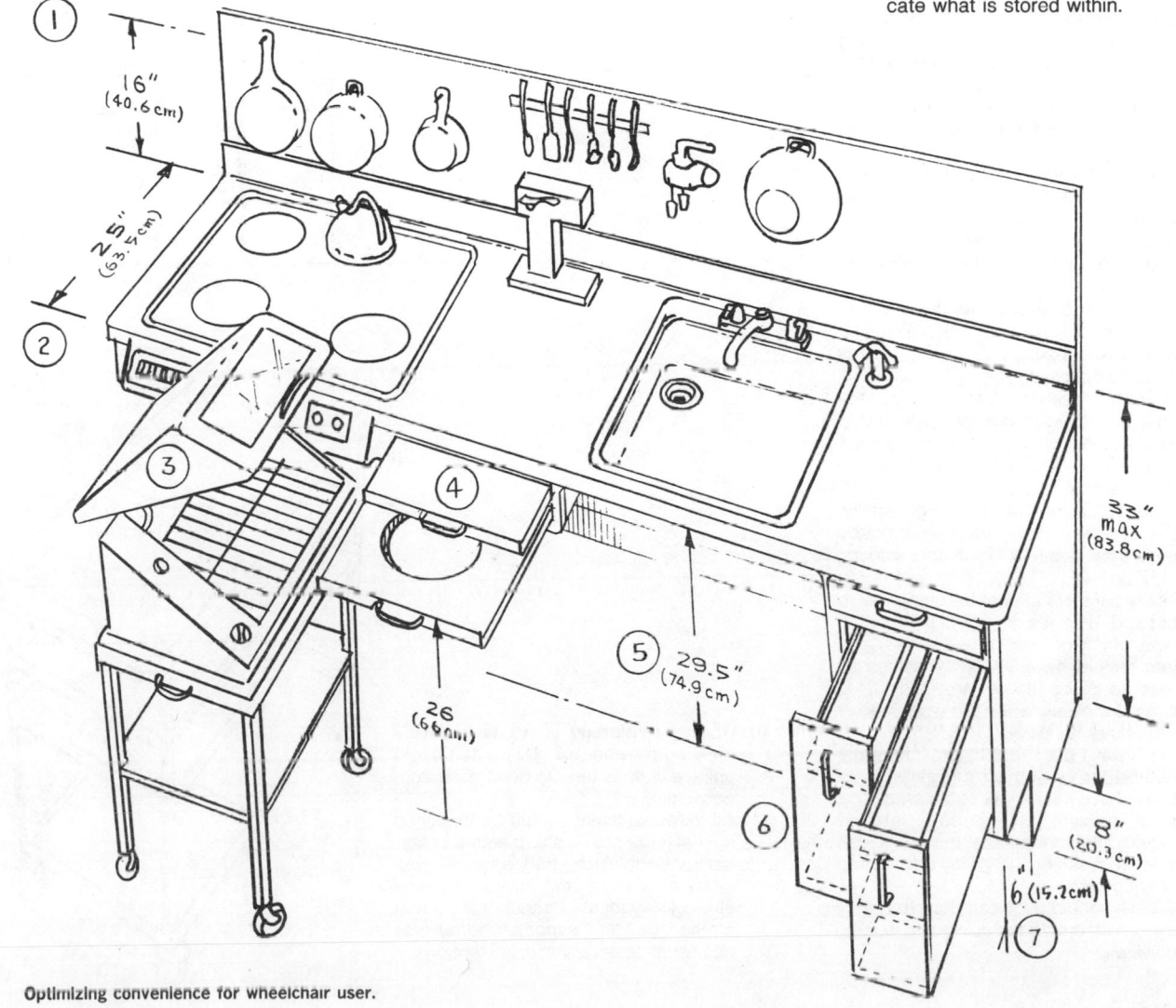

Optimizing convenience for wheelchair user.

## GALLEYS

The galley typically provided aboard commercial aircraft, commercial and private ships and boats, etc., requires considerable ingenuity not only in terms of finding sufficient space for everything and fitting this space to irregular contours but, most important, in terms of creating a "captive design" that will keep everything in place when the motions of the airplane or boat are sufficient to cause objects to shift position, doors to fly open, etc.

In spite of a typical lack of space, one should still attempt to apply the same organizational principles in arranging the galley that were presented in the section on kitchens. Although there is usually less urgency aboard private craft, time and efficiency are extremely critical for commercial systems. An important objective should be to design storage systems so that they serve both pre-serving and post-serving functions. Since each new galley must be customized for the particular vehicle, there is little point in suggesting specific arrangements. However, the following guidelines are applicable to all galley installations:

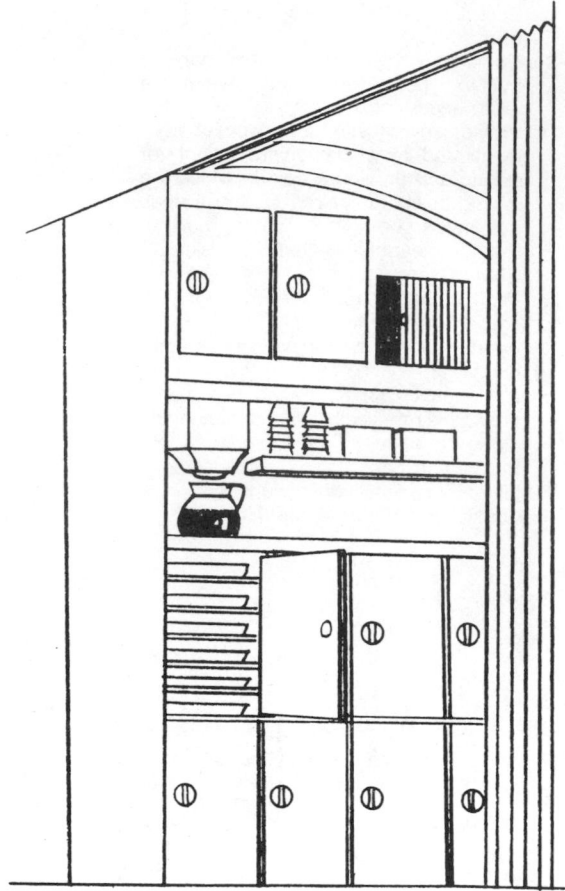

1. Everything should be storable in a captive state, i.e., by means of latching doors, secured trays, shelf rails, or special brackets.
2. Counter tops must have "spill lips" to prevent liquids from spilling over the edge.
3. Devices that must remain in service, such as coffee pots, should be deep-set within retaining burners.
4. Plug-in warmers or cookers should plug in and not require electrical cord lead-ins.
5. Serving carts should be designed to complement and extend the galley subsystem whenever possible. Such carts should also be designed to "capture" all items carried, and, in addition, they must be provided with wheel locks to prevent the cart from being displaced if extensive vehicular movement is experienced.
6. Galley lighting should be of high quality, although the intensity level need not be more than about 30 fc at the working surface.
7. Galley cabinets should be designed so that spilled liquids cannot drip into electrical wiring or control switches. Cabinets should have as few crevices as possible since these are difficult to clean, especially when the work load of attendants is heavy.
8. It is always desirable to provide some method for closing off the galley from the rest of a passenger cabin area. This helps prevent unnecessary interference by passengers and also keeps them from seeing the galley area during cleanup time.
9. Galleys should be located so that they are convenient to servicing and loading activities.

10. Above all, storage of waste materials must be provided for. Discarded packaging materials tend to become space-consuming.
11. All galley cabinets should be designed so that there are no sharp edges or corners against which attendants can suddenly be thrown and injured. Provide strategically located handholds so that attendants can support themselves against excessive vehicular motions.

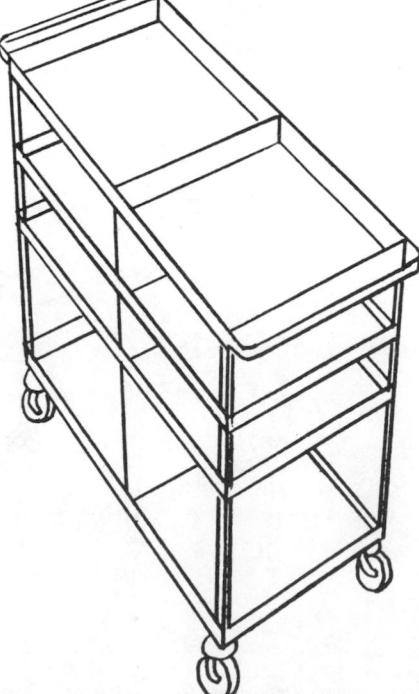

## DINING FACILITIES

A primary design objective should be to make the diner as comfortable as possible by creating an optimum relationship between the chair and the tabletop. The accompanying guidelines are recommended. One should start with a comfortable chair that is about 17 in (43 cm) high. The seat should have a slight fore-aft slope of 5°, and the enclosed angle between the seat and the seat back should be 105°. The length of the seat should be 17 in (43 cm). The vertical height of the tabletop above the seat should be no more than 12 in (30 cm), except in the case of tables that are to be used by persons in wheelchairs (see below).

When a combination serving and eating table or bar is desired, the accompanying geometric guidelines are recommended. The stool or seat height should be about 25 in (64 cm) above the floor. This requires that a footrest be provided, either on the stool or built into the table or cabinet, as shown. It is important to provide at least a 12-in (30-cm) table overhang for adequate knee clearance.

If a fixed seat is used, i.e., one that is affixed permanently to the floor, the seat should rotate for ease in getting in and out. A round seat is preferred to one that is square or rectangular. A round seat should have a diameter no larger than 16 in (41 cm); a square seat should measure about 17 × 17 in (43 × 43 cm). Smaller sizes are not desirable.

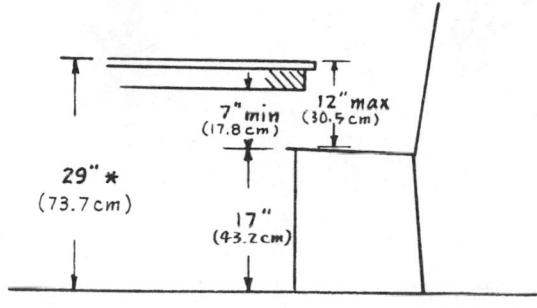

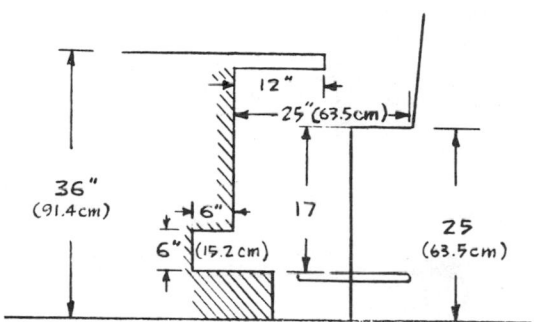

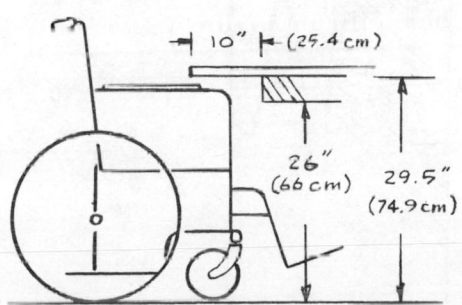

Dimensions for dining table to accommodate wheelchair

When the combination service and dining configuration includes an overhead storage cabinet and/or a stove ventilating hood, care should be taken to ensure that the cabinet does not create a hazard; i.e., it should not be likely that either the person doing the serving or the diner will lean forward and bump his or her head on the cabinet. The approximate dimensions shown in the accompanying sketch can be used as a guideline. Note also that the counter must not be too deep; otherwise, the person doing the serving cannot reach to the far edge to serve diners or wipe off the counter.

The 36-in (91-cm) counter height is recommended when the counter contains a range top because this is a comfortable height for stirring food in cooking utensils on the stove.

Combination situations in the typical restaurant require an additional consideration, i.e., the frequent communications between the customer and the waiter or waitress. An objective should be to establish an approximately horizontal face-to-face relationship. The approximate dimensions shown will provide a good and comfortable compromise between the several geometric relationships:

1. Eye-to-eye level
2. Comfortable dining-table height
3. Comfortable counter-top serving height
4. Reasonable storage access behind the counter

*Note:* Counter depth should be determined on the basis of the space required by the diner, plus the reach limits of the waiter or waitress who must clean the counter.

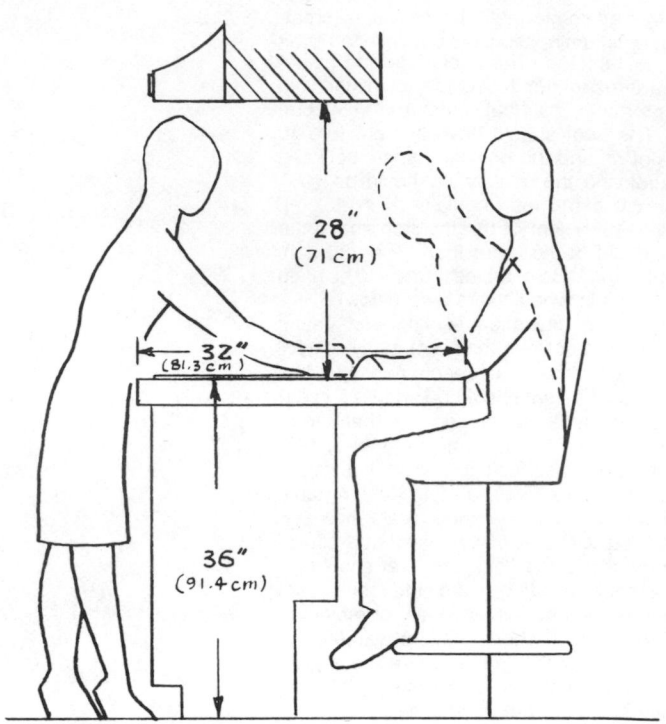

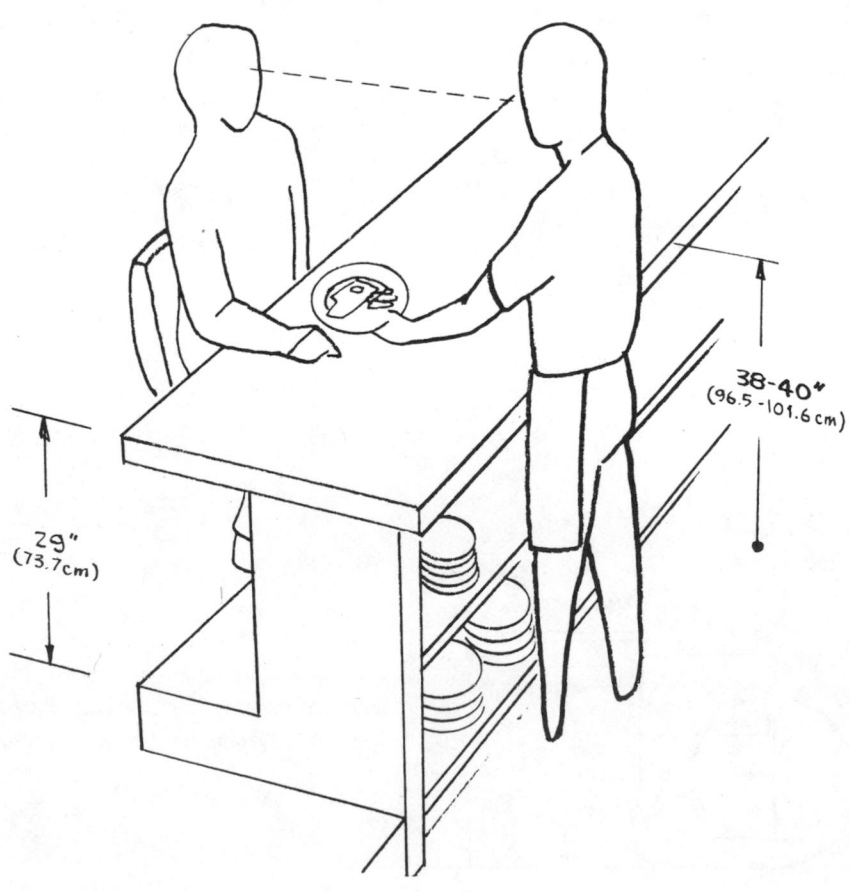

Lateral spacing for comfortable food service and noninterference for the single or multiple diner situation requires proper accommodation of typical dishes, glassware, and flatware. The important considerations include the following:

1. Sufficient space for a full table setting should be provided.
2. Comfortable fore and aft positioning of the diner should be provided.
3. There should be comfortable spacing between adjacent diners to minimize interference while eating.

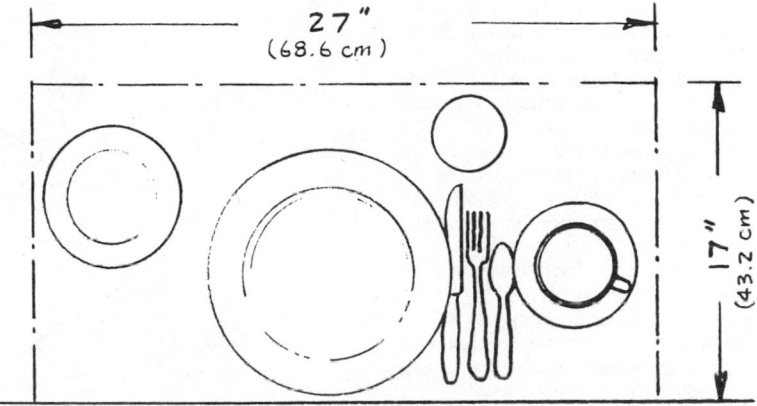

Comfortable table setting envelope

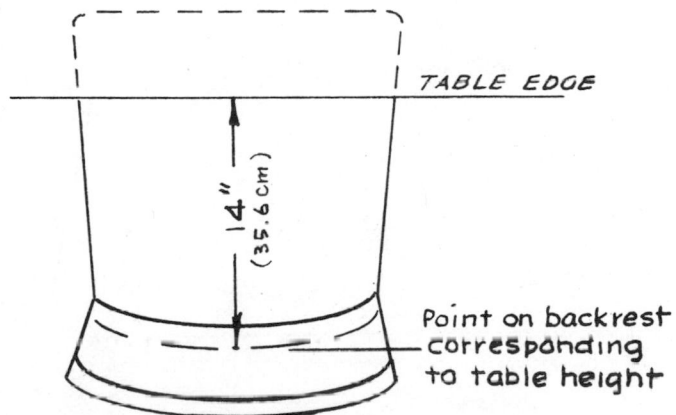

Comfortable eating distance
(Adults)

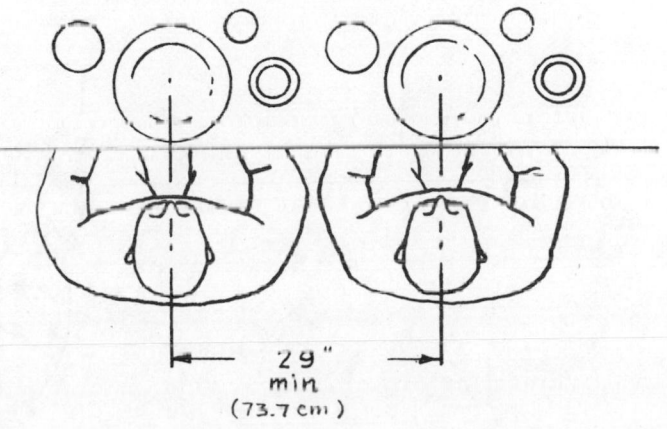

4. In addition to adequate center-to-center spacing of adjacent diners to accommodate them while they are eating, there should be sufficient space between the chairs so that people can get into the chairs with minimum disturbance of the other diners. If fixed seats are involved, it is preferred that the seats be round and that they revolve so that the person getting in and out will have less problem clearing his or her legs between the table and/or the adjacent seat.

5. Foot and knee interference between facing diners should be minimized. Toe-to-toe clearance is particularly important in booth design, where there is no leeway to move seats.

6. Similar knee and foot clearance is needed for diners who are sitting at the corner of a table or counter, i.e., 90° to one another. The minimum clearance for facing or corner-positioned wheelchairs is 21 in (53 cm).

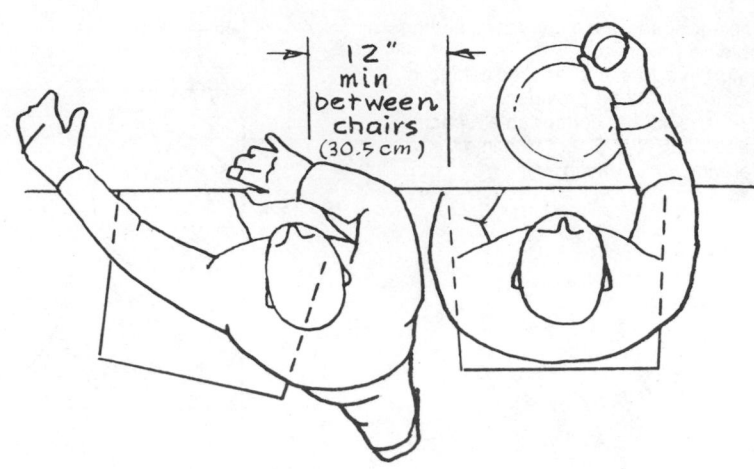

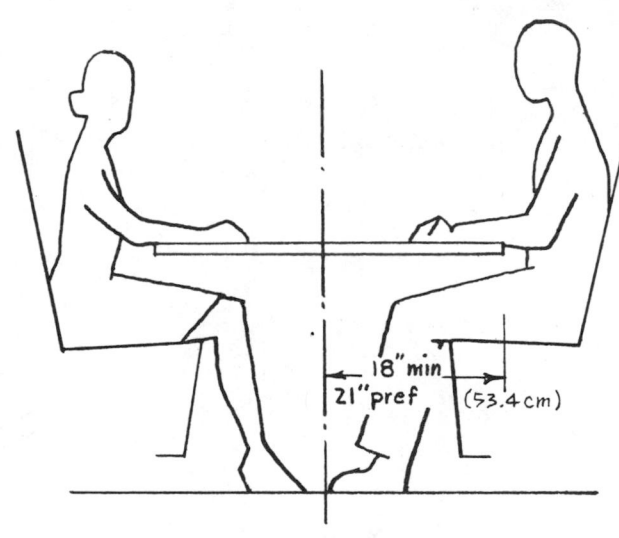

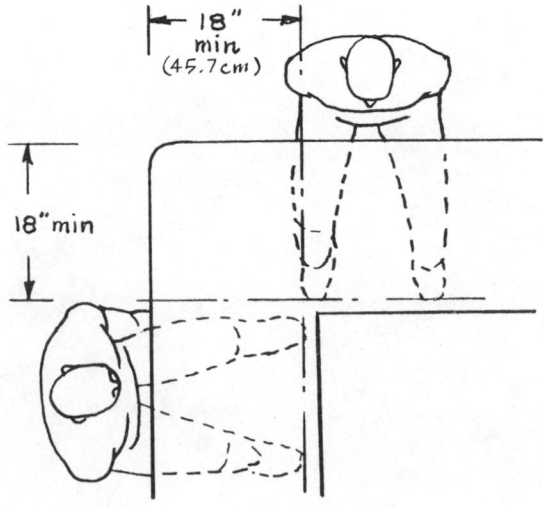

7. Clearance around tables should be adequate to permit other persons to pass by seated diners and to allow a diner to get up suddenly from the table. The minimum for a typical home environment would be as shown in the accompanying sketch; i.e., this would prescribe the envelope within which no other furniture encroaches.

In planning the commercial dining room, the best table-chair orientation is 45° to the general traffic flow, as shown in the accompanying sketch. This set of dimensions should be considered minimum; more clearance is desirable.

The high-efficiency dining-room or cafeteria arrangement shown in the accompanying sketch illustrates suggested clearance dimensions for that type of facility.

The important point is that the space envelope for any dining area should be determined on the basis of the furniture, people, and traffic flow. Avoid trying to crowd furniture into a space that was decided on arbitrarily.

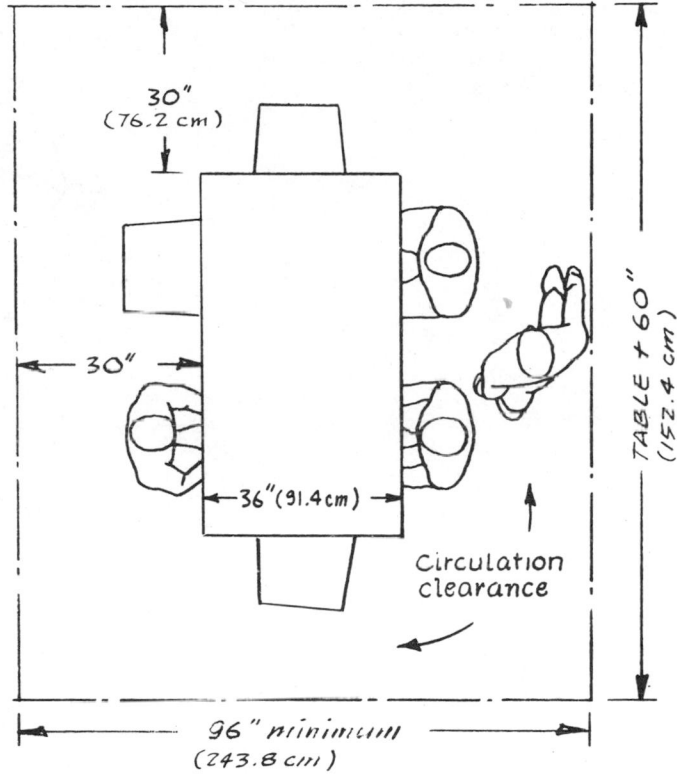

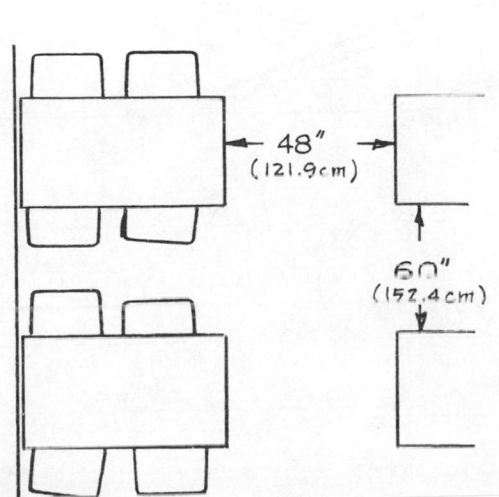

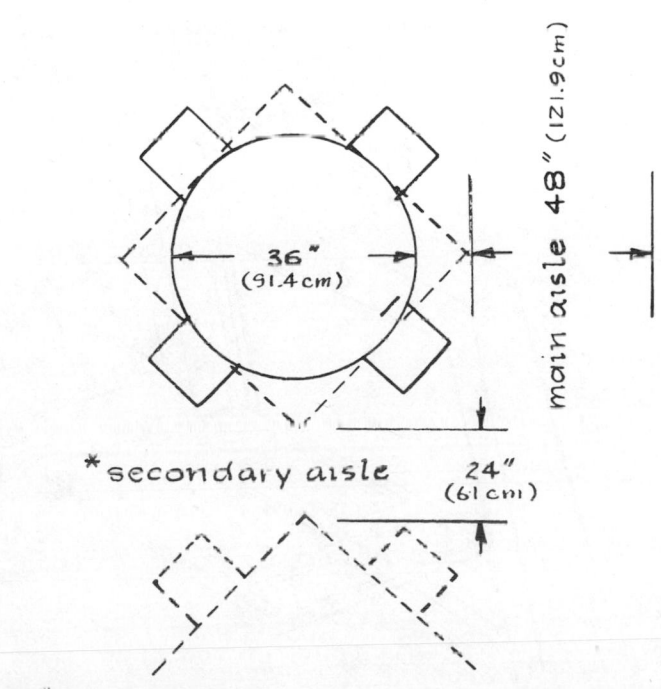

## SPECIAL REQUIREMENTS FOR DINING ABOARD MOVING VEHICLE SYSTEMS

All built-in food service features used aboard ships, aircraft, and other mobile systems must be designed to help retain foodstuffs and utensils while the vehicle is in motion. The accompanying sketches emphasize typical methods and applications.

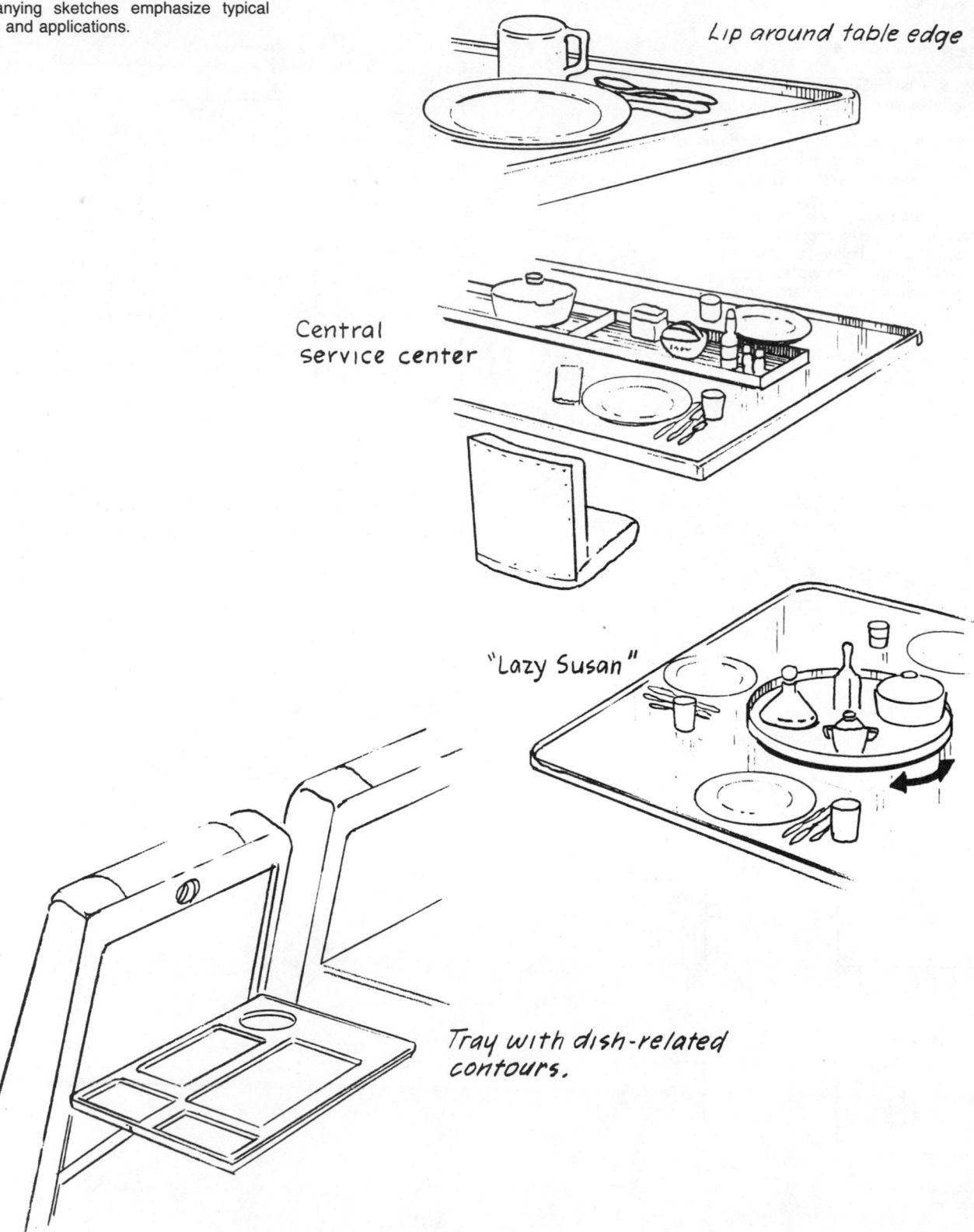

Lip around table edge

Central service center

"Lazy Susan"

Tray with dish-related contours.

## CAFETERIA FOOD SERVICE FIXTURES

A wide variety of standard as well as custom food service fixtures are generally available to the designer. However, some of these are less convenient than others. The following are some of the important considerations in selecting an appropriate fixture:

1. Tray slide racks should be high enough to allow a wheelchair armrest to pass underneath; i.e., there should be at least 30 in (76 cm) of clearance, but the tray height should not exceed 34 in (86 cm).
2. There should be at least 10 in (2.5 cm) of clearance to pass dishes through the glass sanitary shield opening.
3. There should be overhead mirrors so that the patron in a wheelchair can see what food is available.
4. If there is to be a guardrail to control traffic, it should provide an aisle at least 44 in (112 cm) wide.

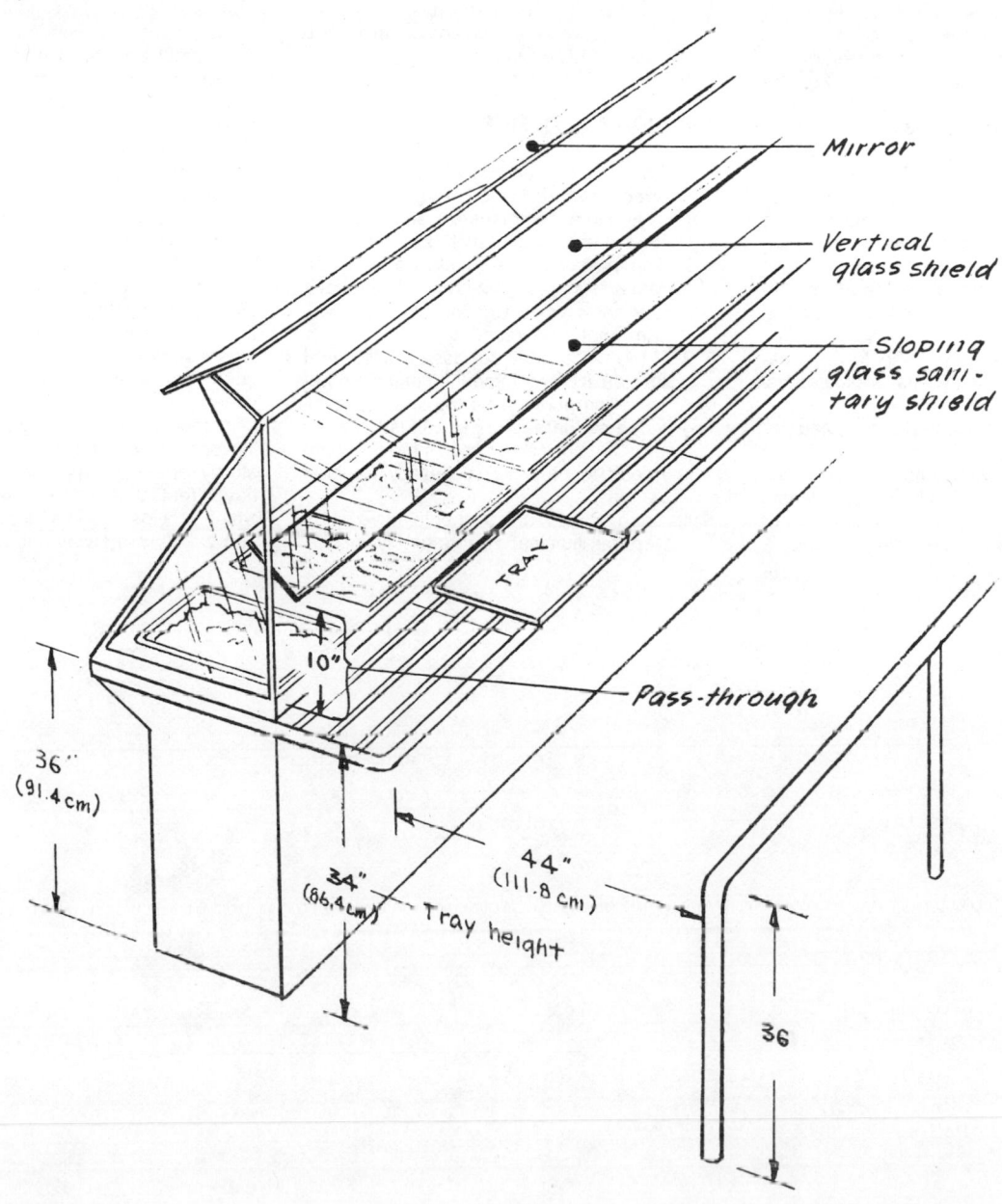

## GENERAL GUIDELINES FOR CREATING AN EFFECTIVE OFFICE

1. Provide a private office (i.e., a floor-to-ceiling partition or walls) only when necessary. A private office is required where professional or supervisory counseling or confidential discussions are a regular occurrence.
2. Provide open but semipartitioned and/or landscaped offices when the conditions warranting a private office are not present. The open office provides for better interoffice communication, less feeling of being crowded, easier traffic flow (because of the lack of doors that must be closed), and better ventilation.
3. Provide transparent but sound-absorbing partitions where supervisory visual access is required; most workers need to have protection from the noise emanating from neighboring offices and also need a sense of "place," which the semipartition concept provides.
4. Do not put more than two persons in one office or partitioned space. Provide general workrooms and/or conference spaces for those times when workers need to get together for a common activity.
5. Do not separate workers (by means of walls or partitions) who must share items of equipment on a continuing basis.
6. Do not place noisy equipment in the same work area with workers who need the equipment only infrequently; i.e., copiers, sorters, etc., should be separated from clerks, typists, bookkeepers, etc.
7. Use modular office partitioning and furnishing concepts wherever frequent changes in work effort or activities occur; permanent walls generally waste space when an activity changes and/or when undesirable crowding occurs.
8. Determine office location on the basis of the convenience of customers and clients, staff members, support personnel, and/or housekeeping and service personnel. Priority should be based on expected frequency of contact.
9. Centralize library, copy service, filing, rest-room, and other commonly used facilities. Convenience should be equalized among all workers. This also applies to office-exit relationships.
10. Do not isolate some workers from window exposure. All workers like to have a view of the outside.
11. Keep frequently interacting functions on the same floor.
12. Separate customer and worker spaces, except where there is a necessary relationship. Consider private avenues between the main entrance and the special customer-related activities requiring confidentiality.
13. Organize the office space so that there are straight-line pathways, i.e., avoid making workers or customers take devious, winding paths.

## THE PRIVATE OFFICE

1. The smallest acceptable private office space is about 7.5 × 7.5 ft (2.3 × 2.3 m). This can accommodate a small table or desk and a maximum of four side chairs.
2. The private office should be designed and arranged so that no one has to face a window while working or conversing with another person.
3. If the office does not have a window of its own, it is recommended that the floor-to-ceiling partition be transparent or divided so that the upper half is transparent. Curtains can be used if the occupant needs to have visual privacy for brief periods.
4. Each private office should have good general illumination (preferably a luminous ceiling system). If the office is less than about 15 ft (4.6 m) square, the ceiling and walls should be painted a light color to provide adequate light distribution.
5. The ventilation system should be such that independent control can be accomplished within the office, even though the door may be shut. Do not place a general temperature-sensing control for a suite of offices in one office.

## THE LANDSCAPED OFFICE

The landscaped office concept can be accomplished either through the use of special furnishings (i.e., the furniture arrangement can be used to separate various workers visually) or by means of partial partitions. In recent years many planners have created rather chaotic situations by misusing these unique and useful partitioning schemes; i.e., they have strayed from any systematic organization of the elements, scattering workers in random fashion throughout a generally large space. The result has been to confuse most of the workers (and all the visitors) by the lack of aisles and landmarks.

Although it is often desirable to get away from the severe, straight halls and aisles of the conventional office by using curved partitions, it is desirable (even when using a curved partition system) to retain a certain amount of pathway continuity. When using these partitioning schemes, it is particularly important not to create unnecessary dead ends. Not only are these confusing to the person who is trying to locate someone, but they may also be a safety consideration in the event of an emergency escape.

In designing or selecting a modular partitioning or furniture system, choose one which is sturdy and self-supporting, which does not have "feet" that can become tripping hazards, which provides good acoustic characteristics, and which wears well and is easily cleaned.

## PARTITIONING FOR SEMIPRIVACY

Many modular partitioning systems are available to the designer. The accompanying sketches illustrate several points to consider in choosing a system:

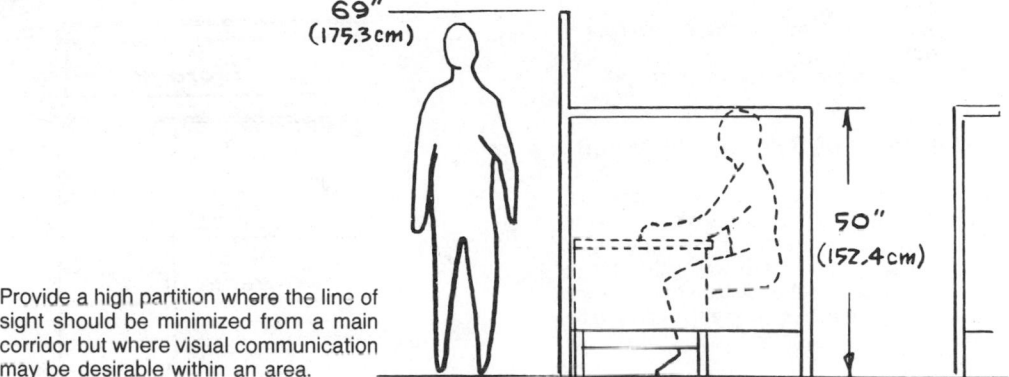

1. Provide a high partition where the line of sight should be minimized from a main corridor but where visual communication may be desirable within an area.

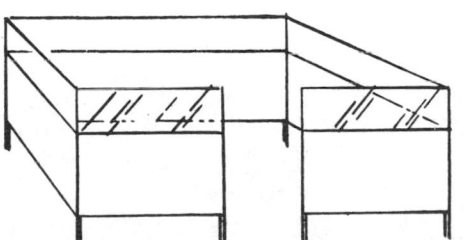

2. By using a transparent section at the top of the partition, it is possible to provide an internal line of sight and still reduce the transfer of unwanted sounds.

3. Combine high and low partitions, but with air circulation at foot level.

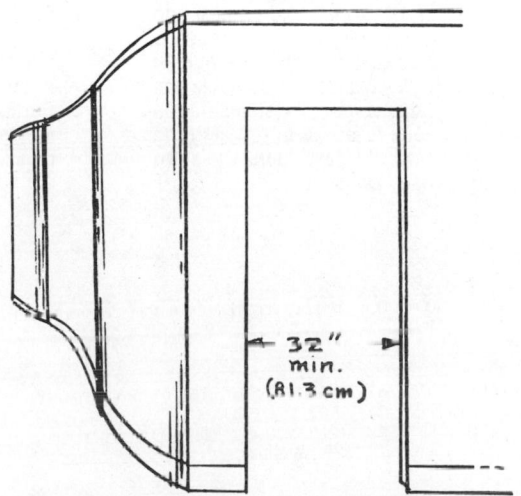

4. Use curved partitions to soften the severely institutional look created by straight lines.

## OFFICE SUPPORT REQUIREMENTS CHECKLIST

1. Electric outlets
2. Telephone outlets
3. General materials storage
4. Drinking fountain
5. Rest rooms
6. Security systems, alarms, fire extinguishers, etc.
7. Housekeeping equipment storage

## FACILITIES FOR THE HANDICAPPED

Most offices will be occupied by either handicapped workers or handicapped clients at one time or another. Therefore, every office should be examined in terms of removing those barriers which make it either difficult or impossible for a handicapped person to enter and use the office. The principal considerations are the following:

1. Wide doorways and hallways
2. Ramps and elevators
3. Tactile cues for the blind
4. Special rest-room fixtures
5. Special furniture to accommodate wheelchairs

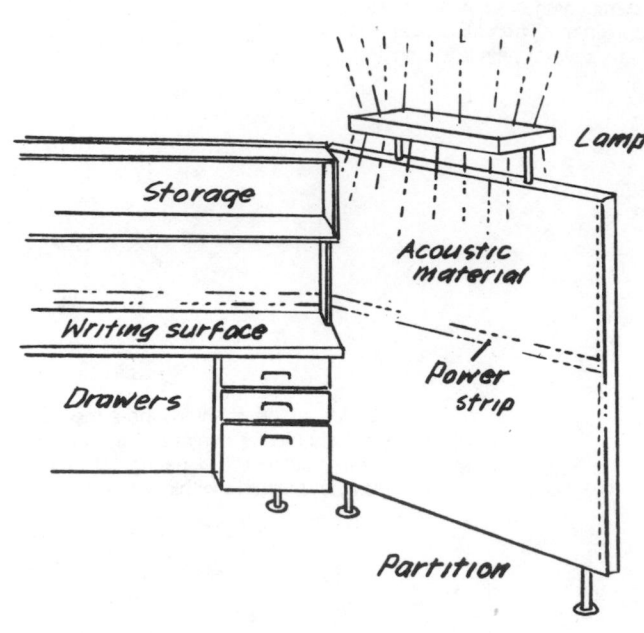

## OFFICE IDENTIFICATION AND SIGNING

From a human factors point of view, proper office identification and signing are one of the most important requirements of office planning. Too often this requirement either is forgotten or is purposely overlooked for aesthetic reasons. *Do not ignore this requirement!* There is no reason to create confusion purposely; in addition, it is possible to make the signing scheme both useful and attractive.

## ENVIRONMENTAL CONTROLS

Good lighting and comfortable air conditioning are prerequisites of efficient office work. Refer to the sections on environmental control and on lighting. These controls, as well as acoustic planning, should be accomplished as part of the initial office facility planning. When this has not been done, utilize special consultants in each of these areas to assist in devising the best systems to overcome any faults the basic building may already possess.

## "OPEN OFFICE," MODULAR SYSTEMS

The popularity of open office systems is due to the flexibility that such systems offer, plus the fact that they considerably reduce the amount of time and labor required to rearrange an office space when a change in workplace layout is required.

In choosing a particular system, the following should be considered:

1. Partitioning modules can be made to stand sturdily without being secured to the primary building structure. Partitions should be designed so that other modules can be attached at any level.
2. Partitions should provide reasonable nonpenetration of sound and should permit the application of sound-absorbing material when necessary.
3. Partitions should contain wiring or cables for desired electric power, plus communication lines and ambient lighting module additions. Cable connections should be accessible for both overhead and floor-level attachment.
4. Partitions should have adjustable legs so that the partition can be either lowered to the floor or raised to allow ventilation beneath the panel.
5. Hung-type modules should include the following:[12]
   Writing and drawing surfaces
   Office machine support surfaces (typewriters, etc.)
   Drawers
   Bookcases
   Cabinets (with sliding, tambour, or folding doors)
   Special supports for telephones
   Supplementary and overhead ambient light fixtures
6. Ease of assembly and disassembly should be considered; i.e., avoid systems which can be safely erected or changed only by factory specialists or which can be wired only by licensed electricians.

[12]Hung-type modules are more desirable from the standpoint of ease of cleaning beneath them.

## AUDITORIUM DESIGN GUIDELINES

### Patron Seating Arrangement

Primary attention should be given to the positions of the worst seats in the house in terms of viewing distances and angles. Consider the following:

1. No patron should be seated so close to the stage that he or she has to look up more than about 15° continuously (a maximum of 30°) in order to observe actors, singers, or other performers on the stage or screen.
2. No patron should have to keep his or her head turned continuously to the right or left more than about 15° in order to watch a performance.
3. Surveys of where people sit indicate that they invariably avoid seats that are outside the angular range illustrated in the accompanying sketch, even though they could probably still see adequately from these seats.
4. Patrons seated in balconies should not have to look down more than about 30° in order to see the action at the leading edge of the stage.
5. Patrons in the last row of balcony seats should be able to see as much of the upper portion of scenery backdrops as required to appreciate its relationship to the stage action.
6. Patrons in front seats on the main floor should be able to see the stage surface.
7. The angle and distance for viewing a screen from the front row should be as shown in the accompanying sketch.
8. The maximum viewing distance to recognize a performer's facial expression is about 75 ft (23 m); gross gestures, about 125 ft (38 m); and gross movements, about 150 ft (46 m).
9. The relative arrangement of seating and stage appurtenances (e.g., scenery and curtains) should be such that patrons see only what they are intended to see.

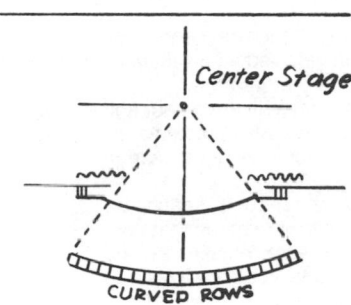

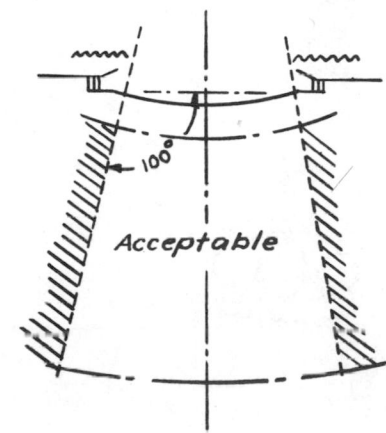

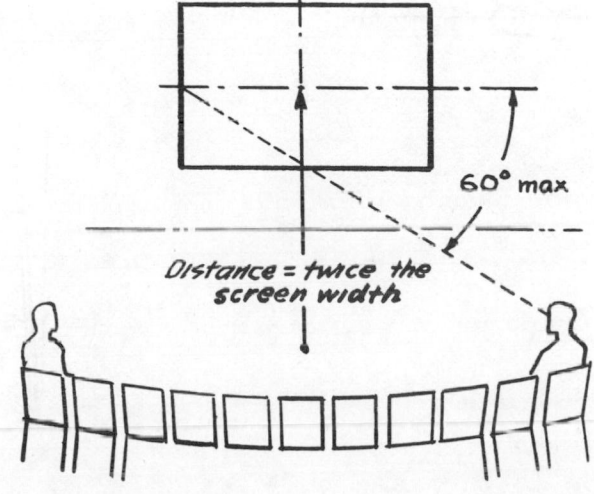

## The Difficult Compromise

Although the following guidelines often create problems of compromise, they are helpful in establishing patron sight-line goals. Remembering first the desire to minimize patron discomfort (i.e., having to keep the head in awkward positions for long periods), consider the suggestions illustrated by the accompanying sketches: arrange seats so that patrons can see past patrons in front of them.

Use a sloping floor rather than a step configuration wherever possible, but do not make it so steep that people fall down or so that persons in wheelchairs cannot negotiate the aisles.

There should be enough space between seats so that patrons can get in and out with minimum annoyance to other patrons.

Tilt a movie screen so that both main-floor and balcony observers have an approximately equal chance to observe a good screen image.

Although a generally "fan-shaped" seating arrangement is desirable, it requires that considerable individual seat adjustment be made in order to create *straight aisles* and not to include hazardous pockets where patrons could become entrapped during an emergency evacuation.

Aisles should be large enough for two people to pass each other comfortably in the darkened theatre.

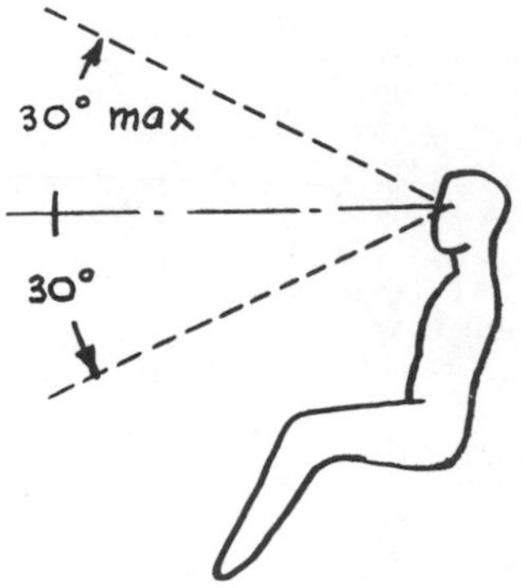

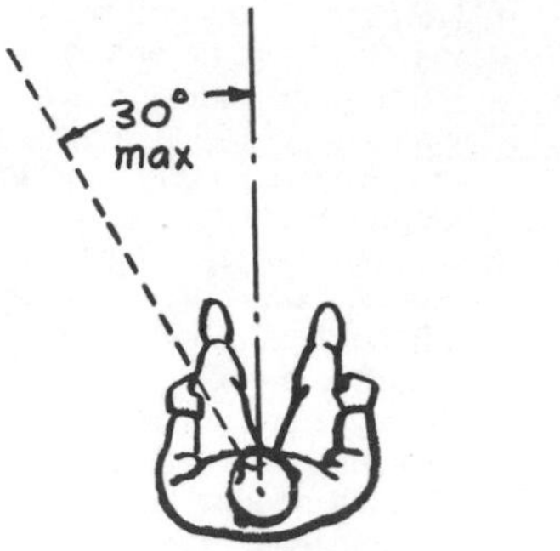

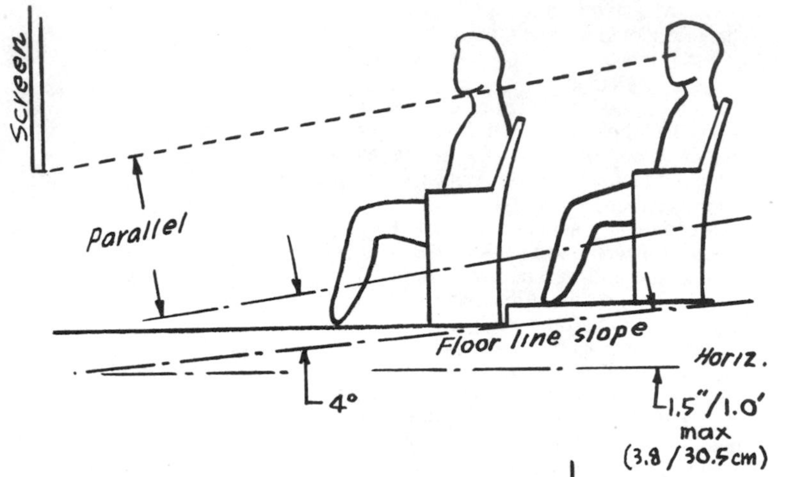

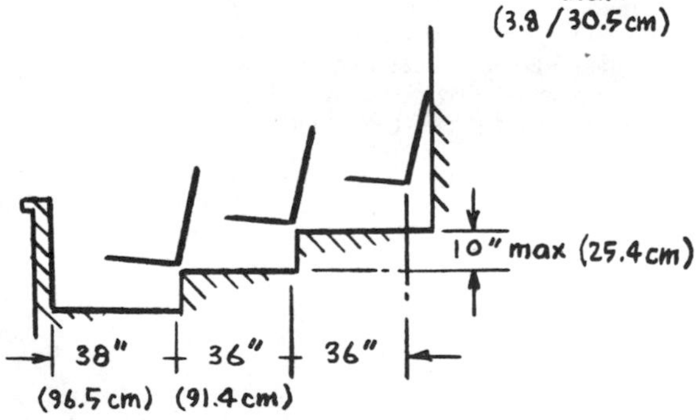

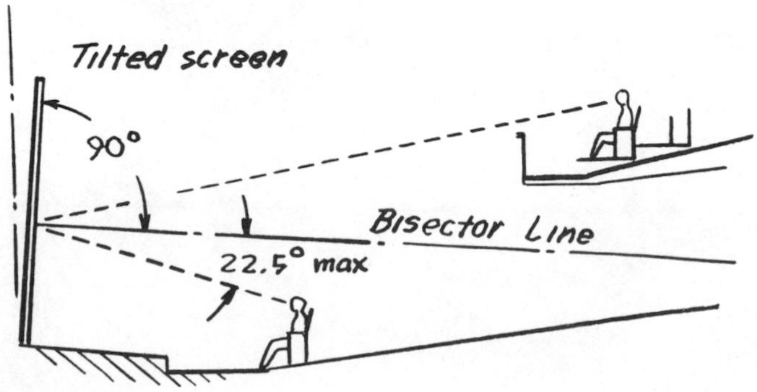

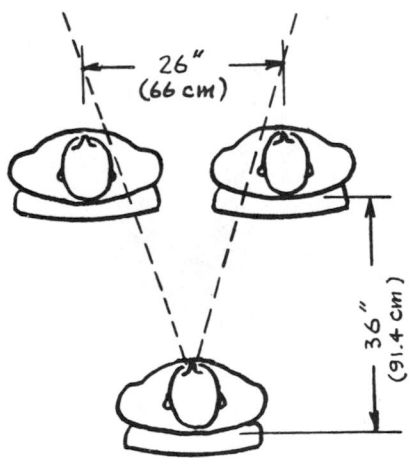

Because the central section of any seating arrangement contains the best viewing positions, it is suggested that a central aisle be avoided and that a configuration such as that shown in the accompanying illustration be used.

Avoid very long rows of seats. It should not be necessary for the patron in the center seat of any row to have to pass over more than about seven seats to get to one aisle or another.

To help the patron find his or her row and seat, provide visible and legible letters and numerals of sufficient size, but located where the patron would expect to find them. Row letters should be illuminated so that they can be seen when the theatre is darkened during a performance.

Provide at least 30 fc of general illumination evenly distributed throughout the seating area so that patrons can read their programs.

## Safety

Refer to local ordinances for safety requirements. In addition, consider at least the following:

1. At least one emergency exit per 100 patrons and a minimum of two per theatre, located at opposite ends of the theatre
2. Emergency lights (self-powered and self-actuated)
3. Fire-retardant carpet and upholstery materials
4. Sprinkler system plus strategically located fire extinguishers

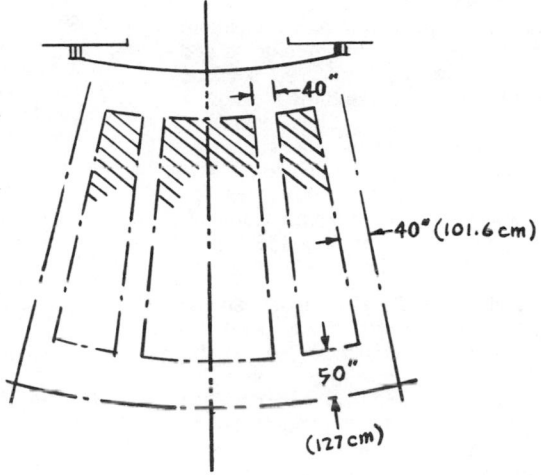

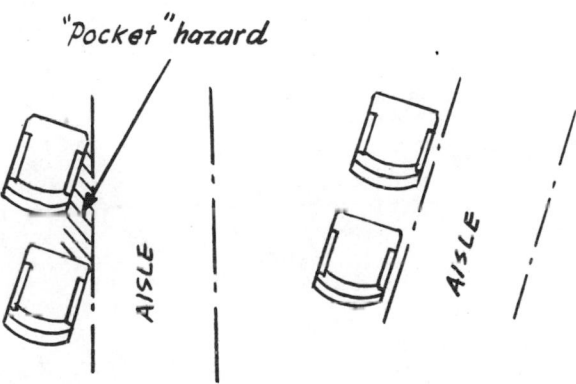

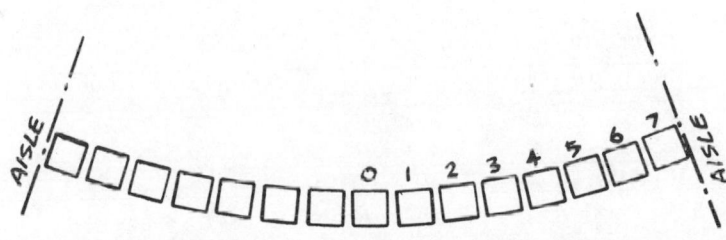

*Max of 7 seats "to" an aisle*

## GENERAL REQUIREMENTS FOR AUDITORIUM SOUND, LIGHT, AND TEMPERATURE CONTROL

### Sound Systems

Sound systems should be designed in conjunction with the architectural plan, rather than added on after the architectural features and constraints are already fixed. The acoustic features of the auditorium and the location of the microphones and loudspeakers must combine to create speech or music that sounds as natural as possible (i.e., there should be minimum distortion in terms of frequency, intensity, and/or identification of sound source direction). Among the important sound system characteristics to consider are the following:

1. Audience: The speech or music should appear to come from the speakers, actors, or musicians when they can be seen by the audience. Microphones and loudspeakers should be positioned so that individual sounds from actors on opposite sides of a stage or sounds from instrumental groups on opposite sides of an orchestra appear to emanate from the portions of the stage where the actors or orchestra members are being observed. A sound control operator should be positioned approximately in the center of the audience area so that he or she can monitor and balance the sound from the audience's point of view.
2. Performers: The sound system should be designed so that a single lecturer or speaker, stage actors, and/or musicians can hear themselves and their fellow performers. Means should be provided for "covert communication" between stage performers, stage or orchestra directors, and sound booth operators.

### Lighting Systems

Separately controllable lighting systems should be provided for the following:

1. General ambient illumination: Evenly distributed, general ambient illumination should be provided for audience areas, the general stage area, and backstage areas to facilitate movement of people and equipment and to allow the audience to read programs.
2. Auditorium aisle illumination: Aisleways should be illuminated when the main house lights are dimmed.
3. Stage lighting: A variety of stage lighting systems should be considered, depending on the types of performances anticipated for the auditorium. The following are among the more important considerations:
   a. Footlights, floodlights, and spotlights for illuminating the performers must be provided.
   b. Overhead stage lights for illuminating major portions of a stage (e.g., illuminating music for an orchestra) should be provided.
   c. Special-effects stage lights should be provided to create various color and brightness conditions to represent day and night changes during a play or opera or to create special psychological effects, as when red is used to indicate terror.
   d. Emergency lights must be provided throughout the auditorium so that, in the event of a power failure, the audience, the performers, and the theatre staff personnel can find their way to safety.

   Independent control of each of the above lighting systems should be provided. Dimmer circuits should be provided for each system. Lighting control stations should be located so that operators have an optimum position for observing the effects of their light system adjustment.
4. Orchestra pit lighting: An appropriate number of conveniently located electric outlets should be provided within an orchestra pit so that musicians' music stand lights can be connected with a minimum of loose extension cables entwined across the pit (an extreme hazard when orchestra members have to find their way to and from their positions in the dark).
5. Speaker lectern light control: A supplementary lectern luminaire should be provided so that lecturers and speakers can see their notes. In addition, it is desirable to provide on the lectern controls for house lights so that the speaker can turn off house lights, turn on projectors, raise and lower a movie or slide screen, etc.

### Temperature Control Systems

It is generally desirable to provide an automatically controlled temperature and ventilation system that adjusts to the conditions of the auditorium. A separate air-conditioning system may be desirable for the stage areas, where heat from special lighting often makes the performance areas almost unbearably hot.

Care should be taken during the design of the auditorium stage areas and during the air-conditioning system installation not to introduce undesirable noises that may interfere with satisfactory hearing conditions, i.e., noises made by equipment or airflow. Similarly, ventilation within the audience area must not introduce drafts that will make patrons uncomfortable. Locate system sensors where the audience sits, not in out-of-the-way places such as hallways, ceilings, or vestibules.

## GENERAL CONCEPTUAL PRINCIPLES

1. Avoid placing materials directly on the floor without a pallet or other support.
2. Assure adequate storage space at the workplace for the expected amount of materials, both before and after materials operations.
3. Use the same container throughout the system; i.e., avoid frequent changes of containers.
4. Provide adequate materials-handling clearances around each workplace.
5. Minimize the number of materials handlings; whenever practical, arrange for productive operations or inspections while materials are being moved.
6. Minimize the number of manual handlings, especially when materials are large or heavy.
7. Arrange for alternative handling methods in case of emergency.
8. Provide appropriate methods for anticipating bottlenecks or breakdowns of automatic materials-moving systems.
9. Plan movement in direct paths; i.e., avoid zigzagging and backtracking.
10. Minimize movements between floors and between buildings.
11. Minimize movement distance, especially for large, bulky materials.
12. Minimize rehandling.
13. Minimize the number of moves per process.
14. Minimize walking.
15. Utilize gravity whenever feasible.
16. Use stacking containers where practicable.
17. Use collapsible containers to reduce the amount of storage space required for empty containers.
18. Provide guards around automatic, moving materials-handling equipment.
19. Provide good lighting.
20. Provide "dead-man" switches for automatic equipment requiring operators.
21. Provide warning signs and signals around automatically operated hazardous equipment.
22. Deliver materials at a proper rate so that they do not pile up or make an operator wait.
23. Provide adequate communication between key operator positions.

## MATERIALS-HANDLING DEVICES

1. Place operators where they can see what they are doing.
2. Provide a control system that allows operators the best method for manipulating or guiding the device.
3. Provide positive control over the motion of the handler, i.e., putting it into motion, regulating its speed, and stopping it quickly.
4. In the case of a large, heavy device, provide the operator with an audio signaling device to warn bystanders.
5. Provide a parking brake

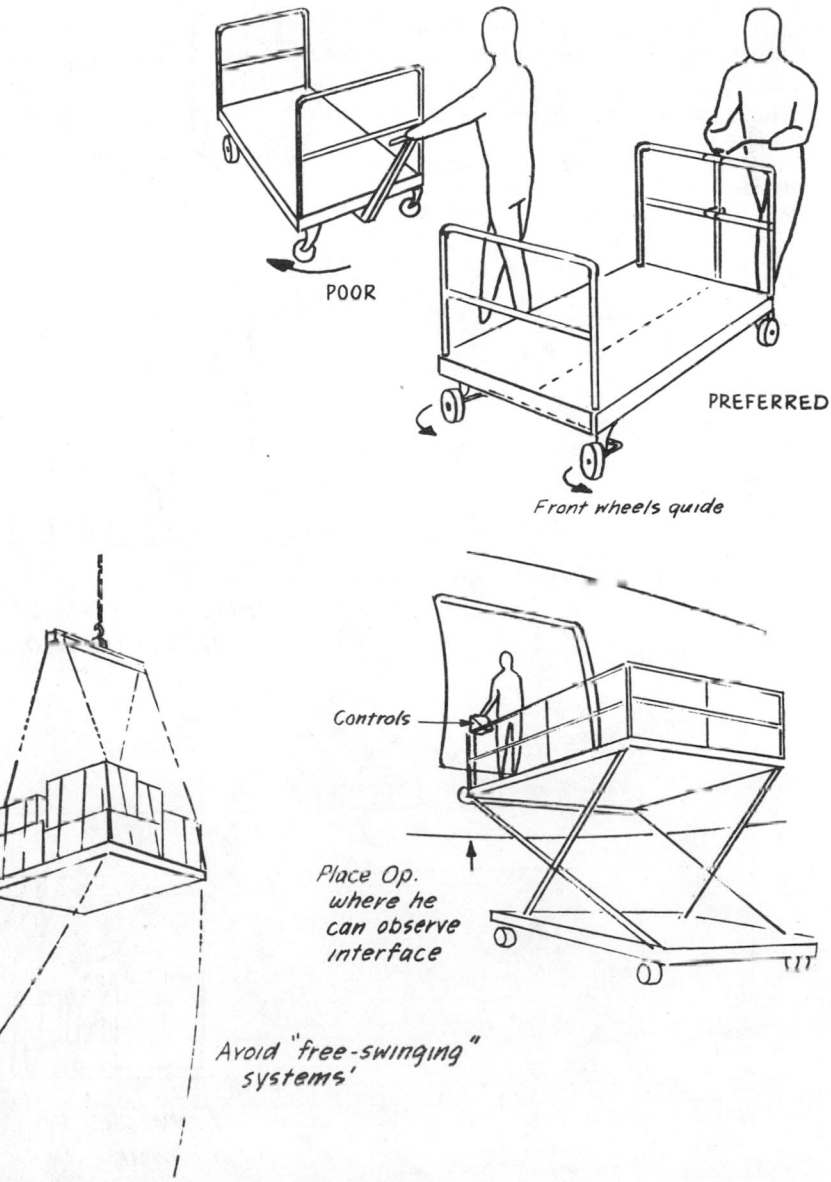

POOR

PREFERRED

Front wheels guide

Controls

Place Op. where he can observe interface

Avoid "free-swinging" systems'

## GENERAL PRINCIPLES FOR CONCEPTUALIZING SPECIAL WORKPLACES

### Working Position

Workers or operators should stand when:

1. The task they must perform requires reaching beyond a point that typically can be reached comfortably while seated.
2. They have to move frequently from one type of task or workplace to another.
3. Certain tasks require lateral reaching beyond a specific task site, i.e., when it is easier to step to another position.
4. They frequently interface with others who are standing.
5. Several persons have to work jointly at a large display area such as a map.
6. They have to reach long distances to touch, transport, or adjust controls.

Workers or operators should be provided an alternative sit-stand configuration when:

1. They normally remain at a task more than 30 min, but others frequently must observe their operation.
2. They need the advantages that seating provides for their major task, but often have to move to another position for a temporary standing task.

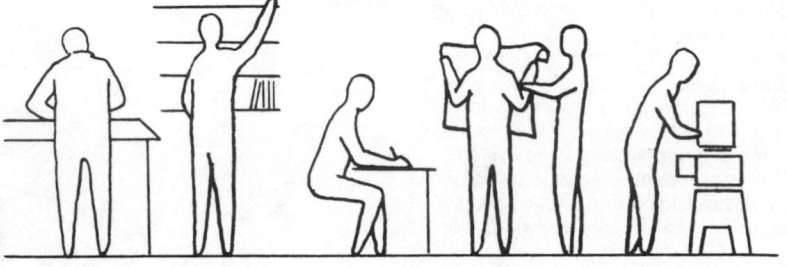

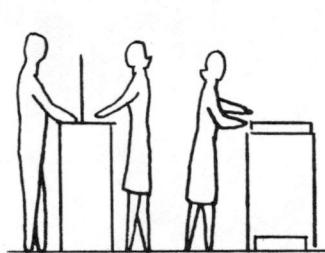

*Have workers stand task require frequent position change and task are short.*

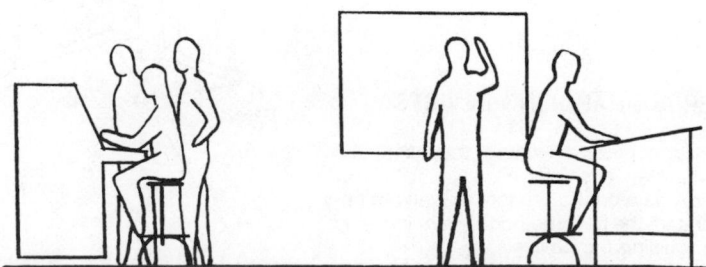

*Provide sit-stand configuration primary task is long, but frequently interspersed with standing operations - getting up and down is fatigueing.*

Workers or operators should sit when:

1. They must work at the same task for an extended time period (30 min or more).
2. Task precision demands steadiness both for them and for the devices being used.
3. They need to be restrained to prevent their displacement by dynamic environmental forces.
4. They have to use their feet to operate controls.
5. They are required to apply maximum force to operate a control.
6. They have to perform extensive writing tasks.

*Note:* Although the above principles apply generally, it is obvious that each decision must be based on a thorough analysis of all the interacting factors, e.g., restraint, steadiness, mobility, interoperator coordination, and potential fatigue. For example, a draftsman usually can perform fairly precise drawing movements either standing or sitting. On the other hand, a vehicle operator generally cannot perform the typical hand and foot control operations as well when standing as when seated.

In addition, the nature of the equipment configuration may require that elements be widely spaced, in which case excessive reaching can be minimized by providing a more mobile operator facility.

Above all, avoid the tendency to repeat traditional workplace configurations without making sure these are compatible with all the task demands.

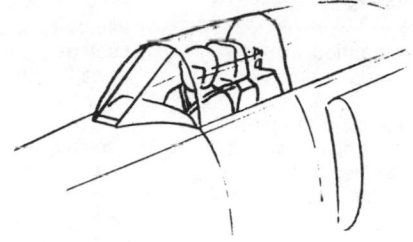

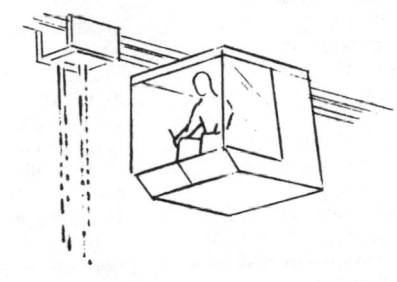

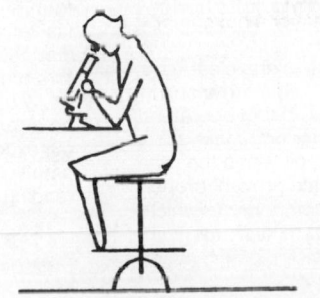

## General Worker and Operator Positioning Considerations

### 1. Avoiding Awkward Positioning

a. Minimize the necessity for operators to lean; i.e., arrange the workplace so that operators can maintain a more or less normal alignment of their bodies, especially when standing.

b. Operators should not have to use their maximum reach capability; i.e., they should be able to shift their bodies to a better position.

c. Operators should not have to sit or stand for long periods with their head, torso, or limb positions skewed; i.e., they should be able to keep their heads, necks, torsos, and limbs in a symmetrical relationship as much as possible.

d. Operators should not be forced to work frequently or for long periods with their hands and arms above normal elbow level.

e. Avoid positioning operators in supine or prone positions.

f. If the limbs must be extended for long periods, provide some kind of limb support.

g. Configure the workplace so that operators can see what they are doing without having to assume an awkward head or body position to see around their hands or the control device.

### 2. Creating Arrangements That Are Compatible with Normal Limb Movement Patterns and Reach Limitations

a. Minimize the need for operators to move their torsos during continuous control operation, except where a ballistic advantage is desirable.

b. Repetitive arm and leg motions should be in the direction that limbs articulate normally about typical shoulder, elbow, wrist, hip, knee, or ankle pivot points.

c. When both hands or both feet are involved simultaneously, create a configuration in which the motions are opposite rather than in the same direction; i.e., the right hand should rotate in the opposite direction to the left, and the right foot should push forward while the left relaxes in the aft direction, as in aircraft rudder pedal movement.

d. Minimize the necessity for operators to shift their position in a seat in order to look at a display or operate a control. In certain situations, however, it may be desirable to provide seat movement, i.e., fore and aft, laterally, or rotating.

e. Avoid workplace arrangements that force operators to stand too close to, or lean across, a hazardous element, such as a moving belt or chain or a hot component.

f. Take into account the extra clearance requirements and/or reduced mobility caused by special apparel that operators may be required to wear and make sure that there are no features within the mobility envelope that could prevent proper limb movement, i.e., components which might be bumped, which limit full limb movement, or upon which clothing might become snagged.

### 3. Force Application

a. Place manipulable objects or controls in positions that are compatible with the best geometric relationship for applying pushing, pulling, or rotating force by operators whenever such forces cannot be minimized.

b. Provide an appropriately positioned support against which operators can minimize counterforce effects, e.g., an armrest, handrest, or backrest.

c. Arrange force-demanding controls in positions where operators can apply the necessary force without disturbing their normal body position, especially when this might interfere with a primary visual or control activity and/or with necessary body-referencing posture.

### 4. Minimizing Fatigue-Producing Workplace Arrangements

a. Provide backrests for seated operators.
b. Provide armrests.
c. Provide handrests when operators are using a continuously operated controller, such as a desk-mounted joy stick or roller-ball controller.
d. Provide footrests.[13]

### 5. Minimizing Potential Safety Hazards

a. The operator's posture should be such that his or her body is in a position that can accommodate dynamic force inputs, such as excessive spinal loads during deceleration, the impact of a sudden landing, or the buffeting of a vehicle as it impacts against rough terrain or ocean waves.

b. Provide body restraints for vehicle operators which will protect them from impacting the interior structure or equipment and which will not allow them to slip into a position in which the restraint itself, such as a seat belt, becomes a hazard.

c. Make the workplace as safe as possible in terms of locating and designing components so that they will not penetrate an operator's skin upon contact, i.e., so that they will spread the energy rather than concentrate it.

## Key Factors to Consider for Each Type of Operator Working Position

### 1. Standing Operators

A smooth, level surface should be provided.

There should be sufficient surface area for operators to establish an adequate spread of their feet, to move when necessary, and to brace their feet when required.

A nonslip surface should be provided if a platform is subject to movement, as in a vehicular environment.

A resilient surface should be provided if operators must stand all day.

Visually observed workplace elements should be arranged so that they can be seen without excessive movement, so that there is no parallax for reading instruments, and so that reading distance and size of display are

---

[13]Refer also to the discussion of problems of awkward positioning.

compatible with readout precision.

Manipulative tasks should be arranged and/or oriented to be compatible with respect to reach convenience, required motion patterns and excursions, force application requirements, precision demands, and response speed. Special attention should be given to the implications of eye-hand coordination.

### 2. Sit-Stand Operators

The eye reference for both seated and standing operators should be approximately the same.

The arrangement of visual displays that have to be monitored by both seated and standing operators should be such that there is the same level of readout accuracy from both positions; i.e., operators should not make reading errors because of parallax.

If both seated and standing operators may have to use the same control device, place this in a position that will minimize interoperator interference.

Provide a footrest for seated operators.

### 3. Seated Operators

Provide a seat that ensures optimum working posture for the tasks being performed.

Arrange visual displays and controls so that an operator's hand will not cover critical displays and so that the displays are normal to the expected viewing line of sight.

If the primary viewing task is directed outside the workplace (vehicle control), position the most important internal displays so that they can be viewed without excessive eye movement from the nominal exterior line of sight.

Position all controls so that, in manipulating them, operators do not appreciably move their nominal eye reference and possibly miss seeing important events occurring outside or on the principal internal display.

### 4. Miscellaneous Considerations

a. Organizational standardization versus individual adjustment: When similar workplaces are to be repeated, a compromise should be sought between the benefits of a standardized workplace (which permits operators to shift from one workplace to another without confusion) and the benefits that accrue from allowing individual operators to modify their workplace to fit each type of activity more conveniently (i.e., to rearrange tools, materials, or other aids so that they are more accessible).

b. Illumination: Do not assume that a generalized type of illumination will satisfy all task-seeing requirements; i.e., the type and location of light fixtures should be determined according to the varying needs of the operator. Especially important is the creation of seeing conditions that minimize glare and reflection problems.

c. Storage: Almost all workplaces require the provision of storage space for books, files, tools, or writing and drawing equipment and materials. Equally important is the provision of storage space for communications devices, discarded items, and personal articles such as glasses or eye shields.

d. Service and maintenance: Careful consideration should be given to ensuring that, in arranging the workplace and/or packaging of equipment, access to critical elements for service or other maintenance is provided.

e. Design for handicapped operators: Whenever a workplace may be used by a person in a wheelchair, a blind person, or a deaf person, special care must be taken not only to dimension the workplace properly but also to provide those special features necessary for the handicapped operator to perform more effectively within his or her limitations (e.g., tactile features for the blind and special visual signals for the deaf).

f. Environmental considerations: Be aware of the possibility that a particular individual workplace may reduce the effectiveness of general ambient thermal, ventilation, and noise control systems, thus requiring additional features to ensure that the operator does not become overheated or is not subjected to respiratory contamination or to interfering or annoying noises that may be produced within his or her particular workplace.

g. Manipulative clearance for large materials: Often large materials must be maneuvered into place before they can be worked on. Provide adequate clearance so that such materials can be maneuvered into place easily.

## BASIC TASKS IN RELATION TO WORKPLACE DIMENSIONING

1. Writing: The writing surface should be at approximately elbow level and should be relatively horizontal (maximum 5° slope); greater slopes make it difficult to keep a pen or pencil from rolling off the surface.
2. Typing: The center of the keyboard should be at about elbow level.
3. Data entry keyboard: The center of the keyboard should be at about elbow level. An associated visual display should be positioned so that the operator's line of sight is perpendicular to the face of the CRT.
4. Drawing: A drawing board used for precision work should slope approximately 3° to 4°; this provides the best compromise between reaching, viewing, and precise instrument manipulation. For less precise drawing, sketching, or artwork, the drawing surface should be adjustable, so that the artist can match the size of the work with the best position for sketching, painting, etc. A sit-stand arrangement is most desirable for average mechanical drafting work since it allows the worker to reach all areas of a large drawing easily.

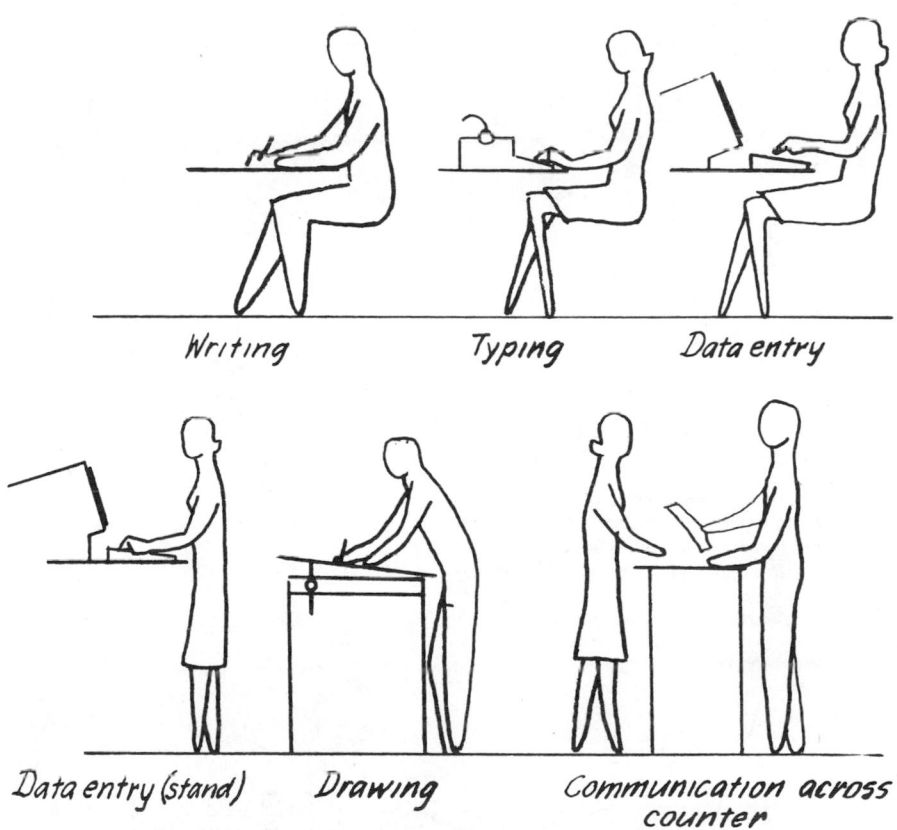

*Writing*      *Typing*      *Data entry*

*Data entry (stand)*      *Drawing*      *Communication across counter*

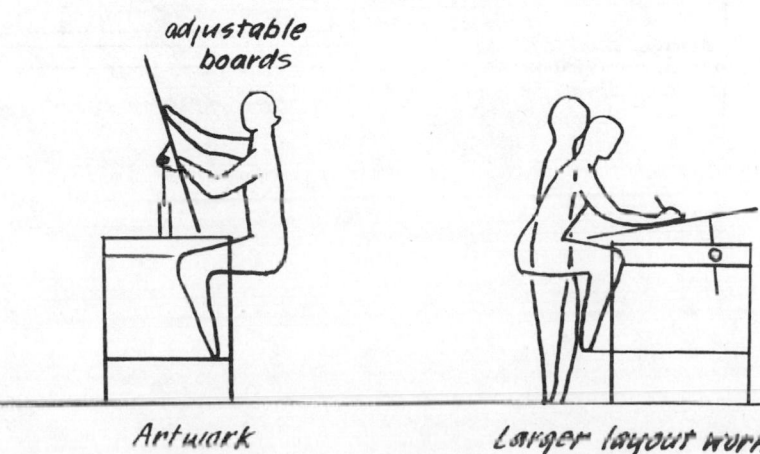

adjustable boards

*Artwork*      *Larger layout work*

5. Storing and retrieving shelves: Design and/or select storage shelving that allows the worker to reach and grasp objects to be stored or retrieved easily and without fear of dropping them. Avoid deep, high shelving since it is easy for small objects to be pushed to the rear of a shelf, where they are neither visible nor manually accessible.

6. Office machine use: Either select machines that have been properly designed (including special stands) or select or build tables that assure proper positioning of the principal work level of the machine. The work level should be at approximately the operator's elbow height.

7. Tool operation: Various tools require that the work level be carefully chosen, especially if there is a requirement for precision and/or the application of controlled force.

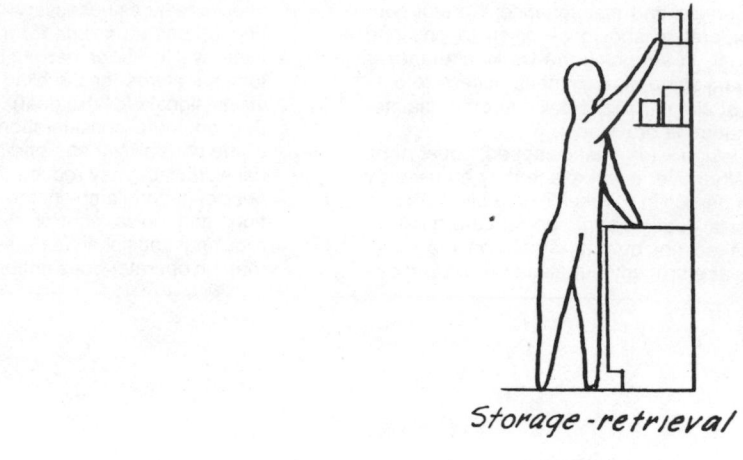

*Storage-retrieval*

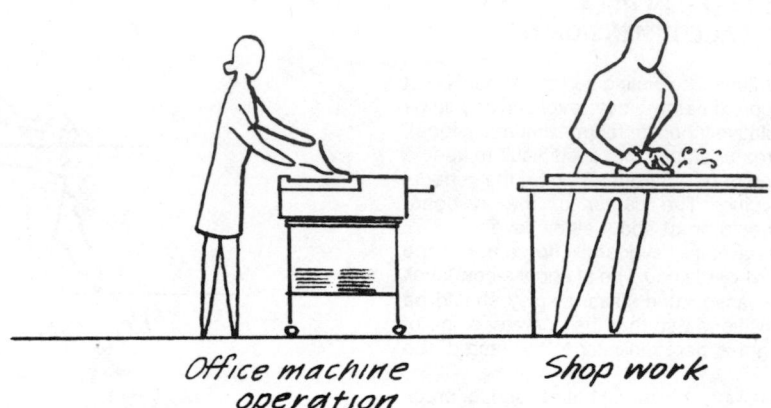

*Office machine operation*          *Shop work*

## Nominal Guideline Dimensions

Although the accompanying general dimensional suggestions are recommended to approximate the optimum, one should be careful not to assume that they necessarily apply exactly to a specific problem. Use of a test mockup to verify these is recommended.

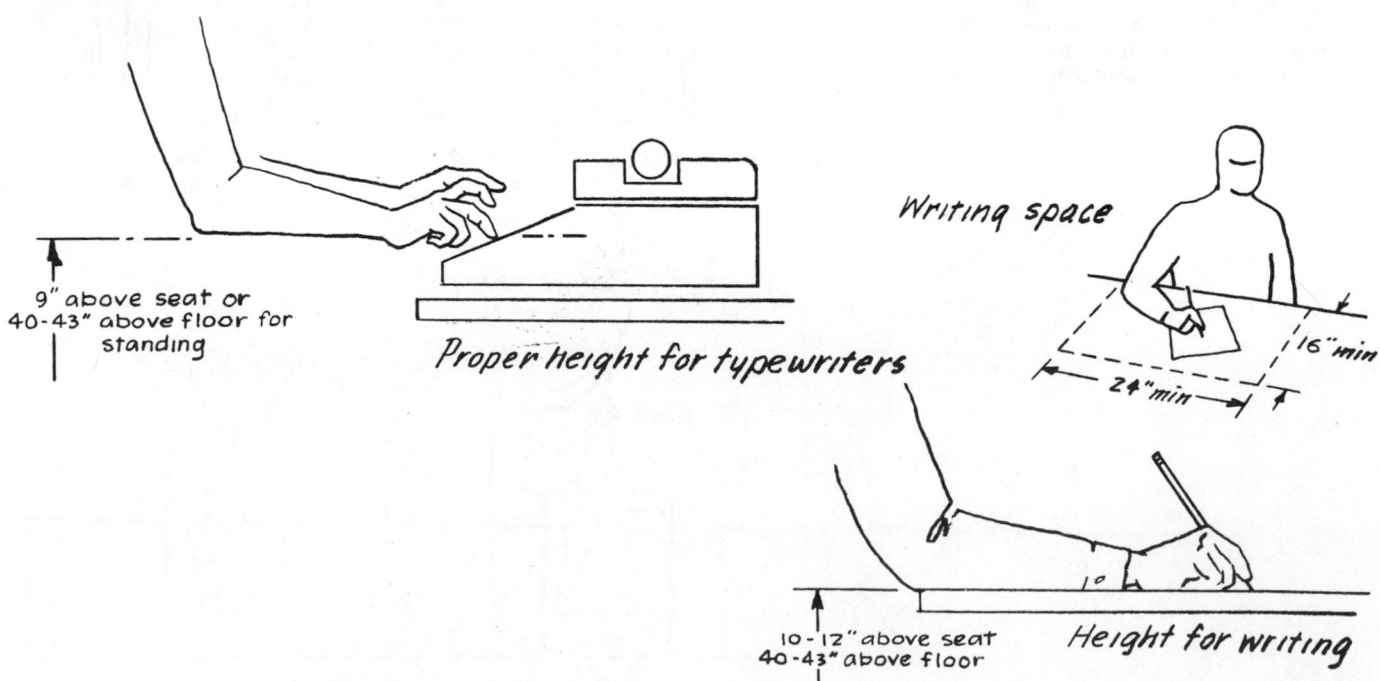

9" above seat or 40-43" above floor for standing

*Proper height for typewriters*

*Writing space*

16" min

24" min

10-12" above seat 40-43" above floor

*Height for writing*

**USE OF OPTICAL INSTRUMENTS** Both eye alignment and control manipulation must be considered in determining the proper working height for optical instruments. Because of the multiple points of articulation in an operator's back and neck, it is difficult to anticipate exactly how operators may have to position their bodies and heads in order to align their eyes with the eyepiece of a particular instrument. The most practical approach is to obtain the actual instrument that is to be used and then mock up a mounting geometry that has sufficient adjustment for operators of various sizes to try out. In some cases it will not be possible to "fix" the work height so that it is acceptable for all operators; an adjustable worktable must then be provided.

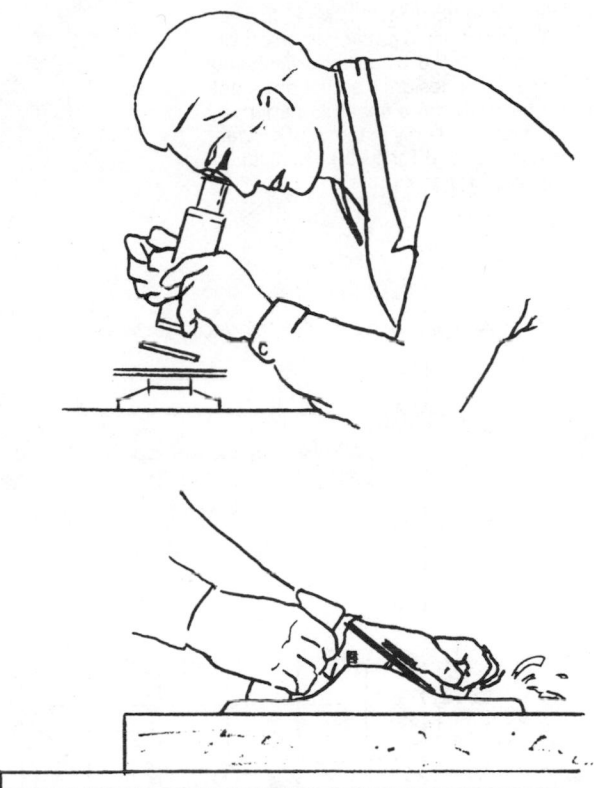

## Nominal Guidelines for Tool Use

Assuming the typically available tool and clamping devices obtainable from the average hardware store, a standard workbench height of 36 in (91 cm) is generally acceptable for most applications.

More difficult is the selection or positioning of machine tools that require precise visual inspection and control adjustment. Especially important is removal or positioning to avoid visual blocks.

Table heights must be lower for ease of manipulating or working on large equipment.

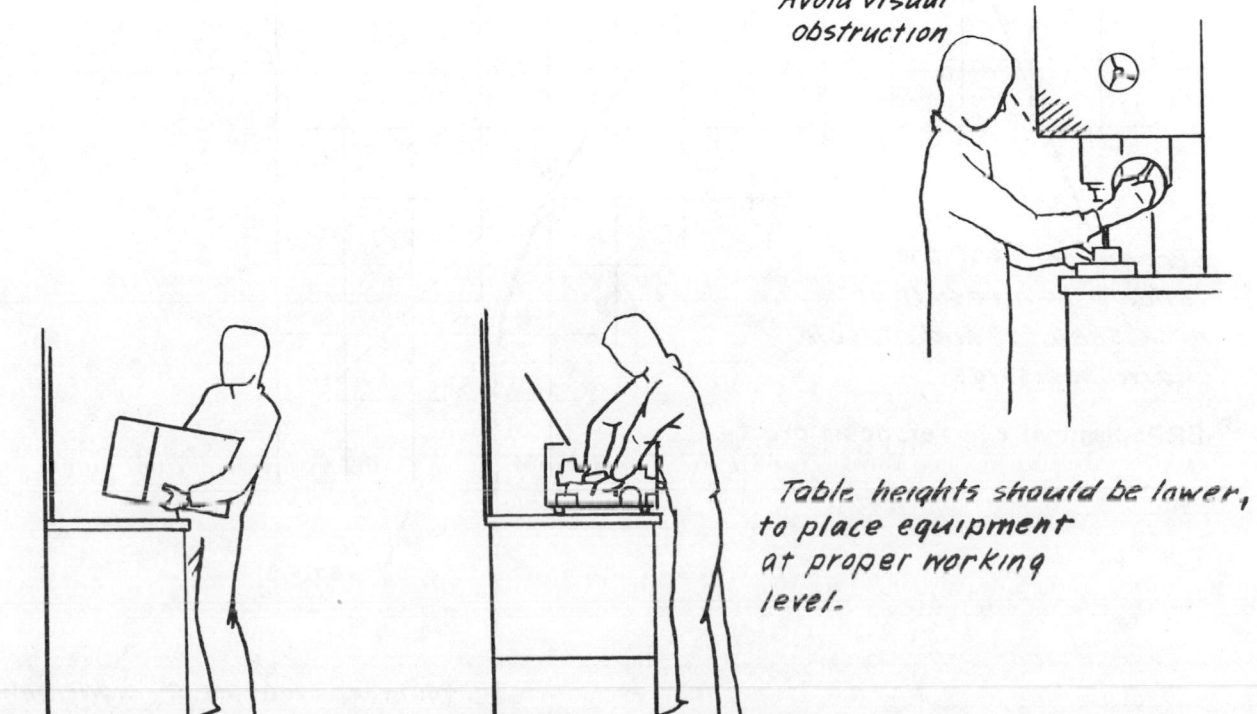

Avoid visual obstruction

Table heights should be lower, to place equipment at proper working level.

**Eye-Positioning Path for Aligning on an Optical Device**[14]

The accompanying diagram illustrates the approximate eye-positioning path of a person sitting in a fixed seat and using an articulating optical device. (The designer is cautioned not to consider this diagram a final recommendation; rather, it is useful only in terms of preparing a mockup test bed for further evaluation with an expected range of users.)

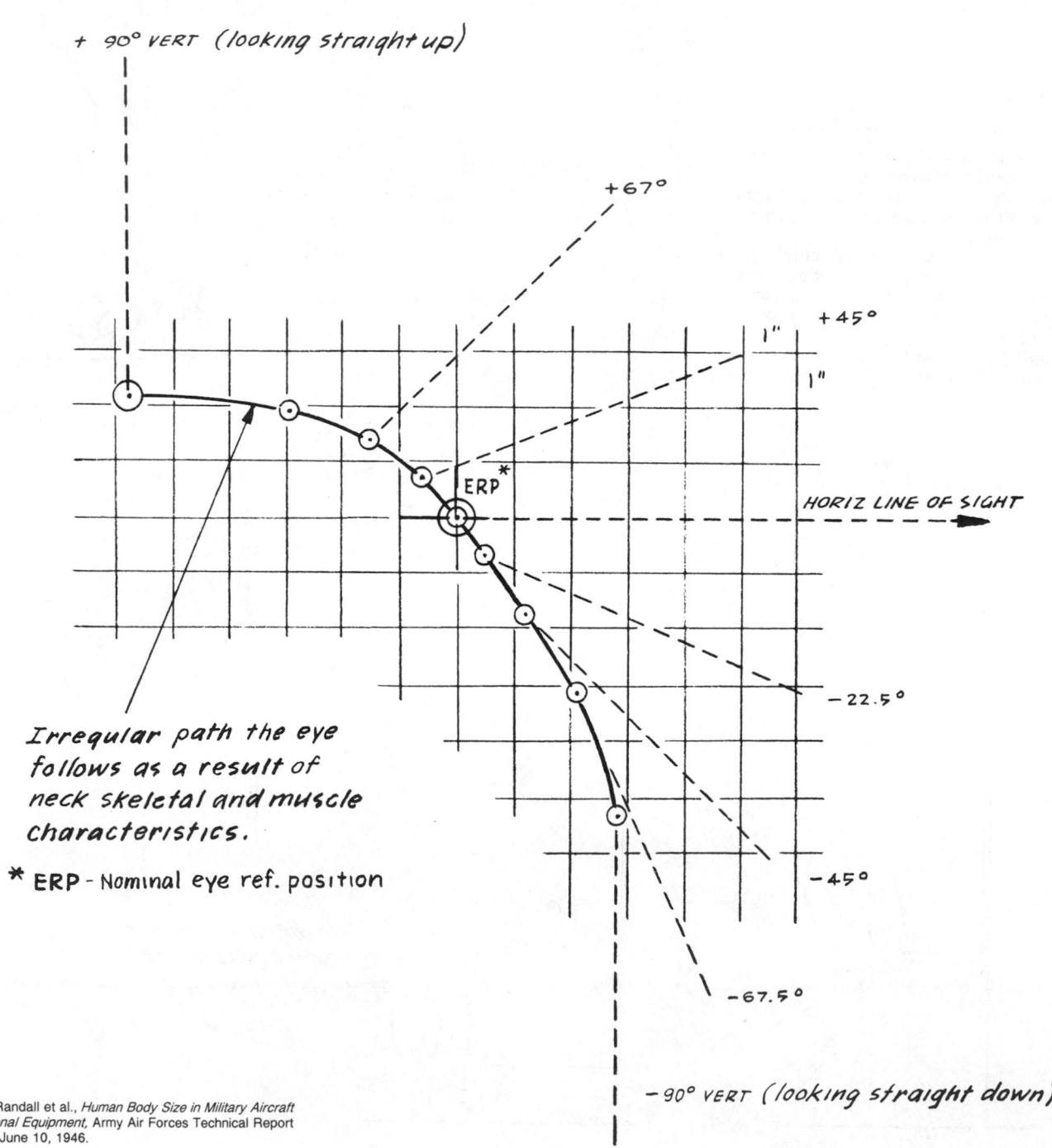

+ 90° VERT (looking straight up)

+67°

+45°

1"

1"

ERP*

HORIZ LINE OF SIGHT

−22.5°

−45°

−67.5°

Irregular path the eye follows as a result of neck skeletal and muscle characteristics.

* ERP − Nominal eye ref. position

− 90° VERT (looking straight down)

[14]F. E. Randall et al., *Human Body Size in Military Aircraft and Personal Equipment,* Army Air Forces Technical Report no. 5501, June 10, 1946.

## ELECTRONIC CONSOLES AND RACK-MOUNTED PANELS

### 1. Seated Operator Consoles

Operators who are seated at an electronic console are necessarily limited in terms of how far they can reach. In addition, their viewing angles may be critical in order that display faces are reasonably perpendicular to their normal lines of sight.

In general, consider the suggestions illustrated by the accompanying sketch:

*a.* Position the primary display—say, a large CRT—slightly below the operator's eye level, but sloped to match his or her normal head slump of about 10°.

*b.* Adjustment controls that are used most frequently should be positioned at approximately elbow level.

*c.* If a finger joy stick or roller-ball controller is required, place it nearer the back edge of the desk, not near the outer edge, where the operator will not have a resting surface for his or her arm.

*d.* Displays or controls that are used infrequently can be located in less convenient positions, as long as they are not "detached" from related functional monitoring areas.

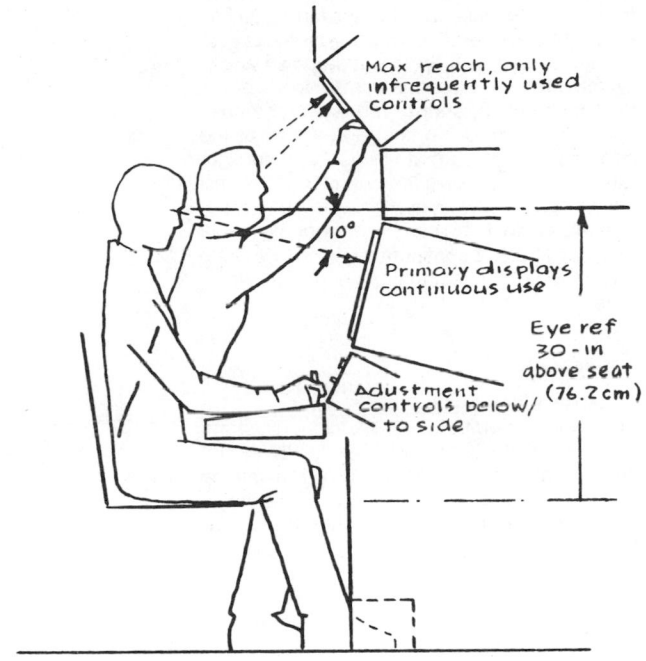

### 2. Standing Operator Cabinets

See the accompanying sketch.

### 3. See-Over Consoles

If the operator needs to see over the console to monitor other displays, reduce the console height so that the operator is not required to stretch his or her torso or neck.

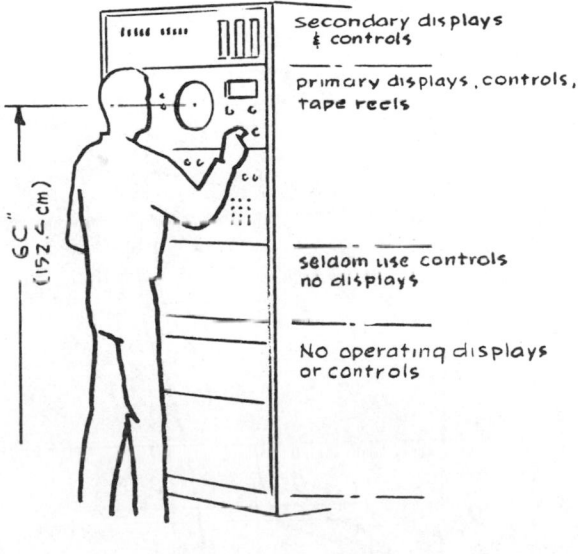

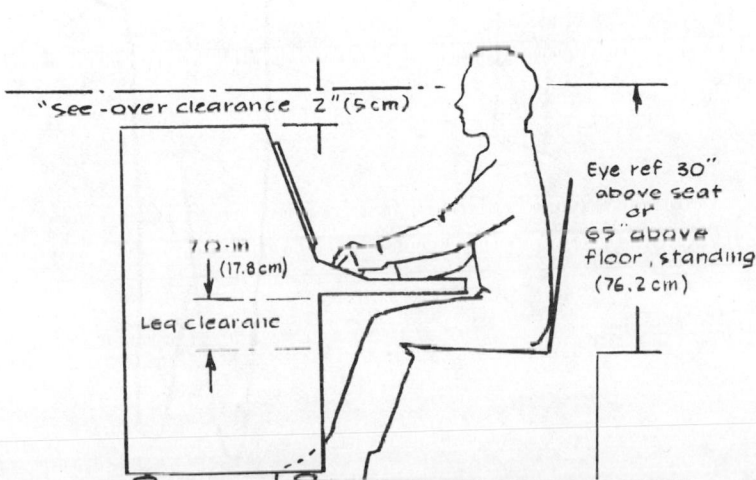

### 4. Extreme See-Over Configurations

When it is necessary for an operator to "see down" (as illustrated in the accompanying sketch), one should examine the total facility architecture in conjunction with the design of the operator's control station. As the illustrative example implies, the operator's position for viewing objects outside the workplace has to reflect consideration for *(a)* what the operator has to see, i.e., the required down angle; *(b)* the operator's position relative to the structural obstruction; and *(c)* the operator's position in relation to the shape and size of the control package so that the package configuration (at least) allows the operator to take full advantage of whatever clear view the structure allows. In some cases it may be desirable to use a sit-stand configuration in order to take advantage of the fact that the operator is freer to lean over the console.

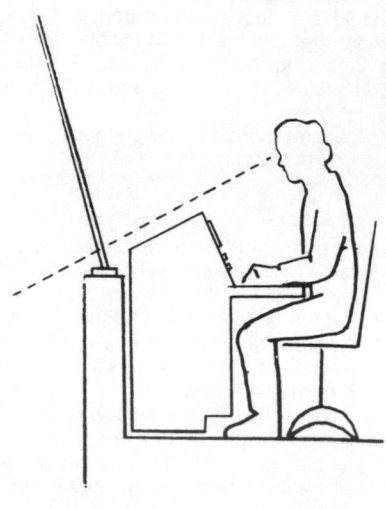

### 5. Large Plot Subsystems

The two accompanying sketches illustrate general dimensional guidelines for large plot subsystems; these are based on *(a)* limitations created by knee clearance if the operator sits and *(b)* convenient use area when the operator stands.

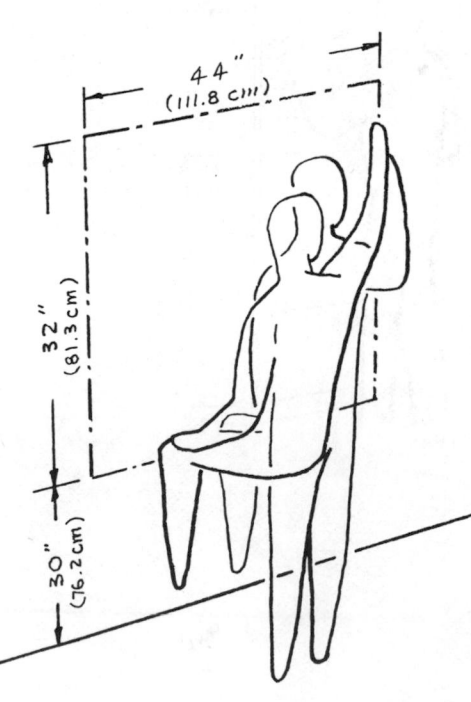

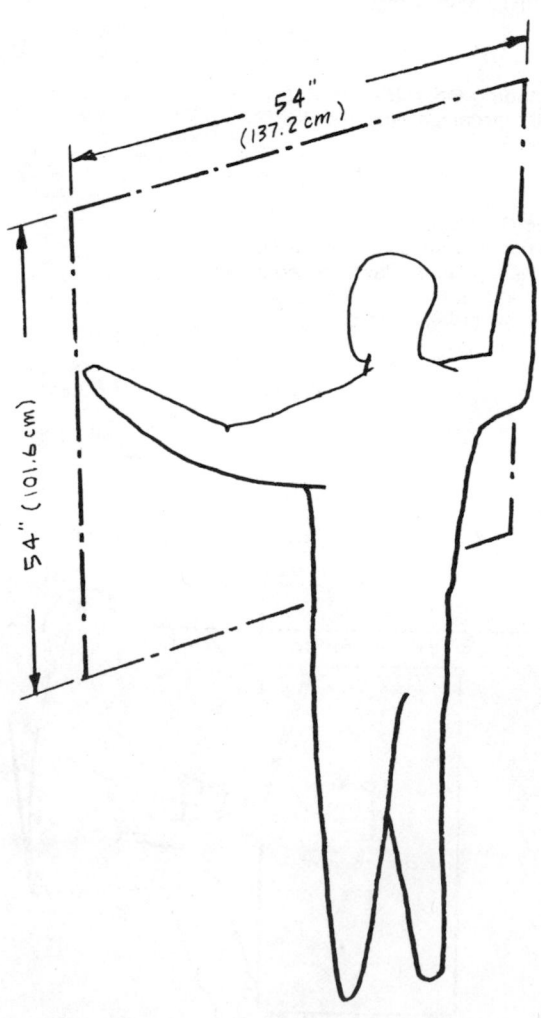

## Other Considerations

As a general rule, panels for either seated or standing operation are easier to use when the displays are laid out for horizontal, sequential scanning. This is due to the fact that an operator's eye motions and head movements are less tiring in the horizontal direction. Studies also have shown that operators can respond slightly faster when they attempt to read instruments or other visual information from horizontally arrayed data than when this is presented in vertical arrays.

On the other hand, standing operators, because they are free to walk backward and to the side, can more easily adjust to control and display arrangements that are organized in vertical patterns.

## Display Positioning Relative to Illumination and Reflection or Glare

Large CRTs and/or glass-covered instruments are subject to interference by ambient lighting conditions. Whenever possible, one should attempt to control both the positioning of displays and the ambient lighting conditions. Although an important objective usually is to position a display so that the face is perpendicular to the operator's normal line of sight, it may be necessary to compromise in order to eliminate undesirable reflections from artificial and/or natural light sources.

When control of ambient lighting is not the responsibility of the designer, other steps must be taken to minimize the effects of possible glare or reflection. The glare shield illustrated in the accompanying sketch is typically used to accomplish this.

Horizontal arrangement for seated operation.

Horizontal and vertical for standing operator.

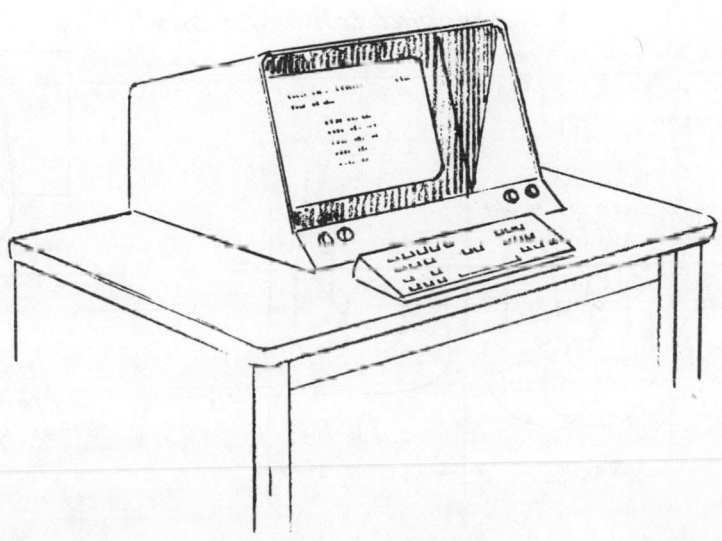

### Arranging for Added Convenience

Although the typical console or equipment operator has considerable reach capability, it is desirable to minimize the necessity for reaching, especially if it may be required on a frequent basis. The so-called wraparound concept is helpful as long as care is taken to ensure that unusual interferences are not introduced.

The wraparound console is especially helpful for the seated operator, as shown in the accompanying illustration. If the seat has to be secured to the floor (as it does aboard ship), a careful analysis of reach limits must be made. If a movable seat is acceptable, the arrangement can be expanded, taking advantage of a chair with casters.

A sit-stand operation (see the accompanying sketch) that includes both worktable and panel rack operations can be accommodated best if the rack can be made movable.

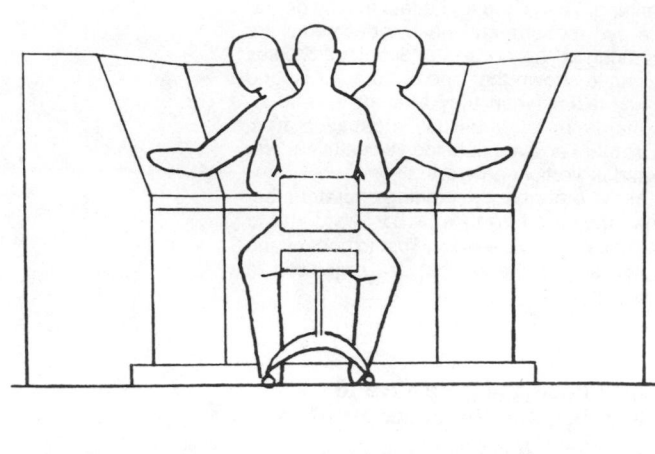

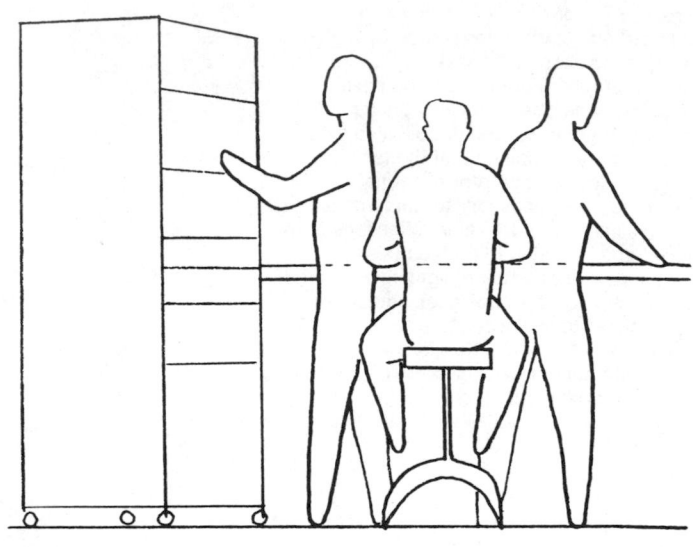

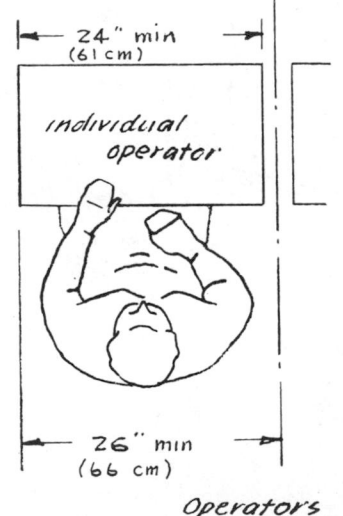

*Start with basic single-operator requirements*

*expand for additional operators*

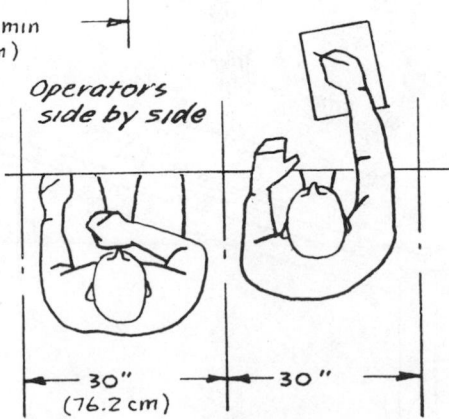

*Operators side by side*

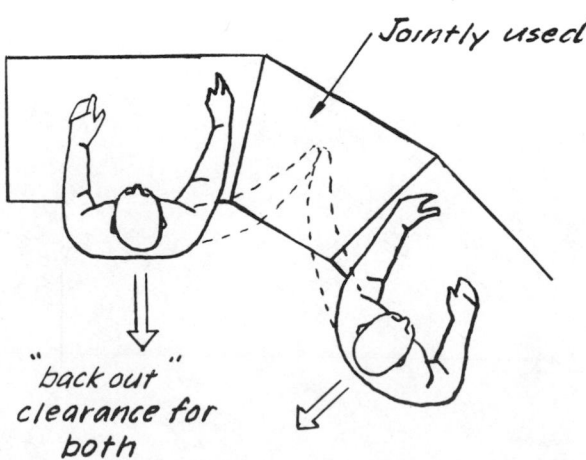

*Jointly used*

*"back out" clearance for both*

## LARGE GROUP DISPLAY AND OBSERVER ARRANGEMENT

When a number of operators must view a large visual display in conjunction with console operation, the large group display should be arranged so that the various observers can see the display from their normal working positions without having to peer around their neighbors or their neighbors' equipment. Many alternative arrangements are, of course, possible, and sometimes a compromise must be made because of facility space constraints.

The accompanying illustrations and guidelines are helpful in making sure that all the sight-line requirements have been evaluated.

As the two accompanying sketches illustrate, angle of view may be improved by seating the operators further apart.

Operators and their consoles can be staggered in order to maintain the recommended screen-observer line-of-sight limit of 60°, as shown in the accompanying illustration.

When the recommended clear line of sight cannot be accomplished by lateral spacing, consider placing the operators in the rear on platforms so that they can see over the heads of those in front.

Possible but awkward

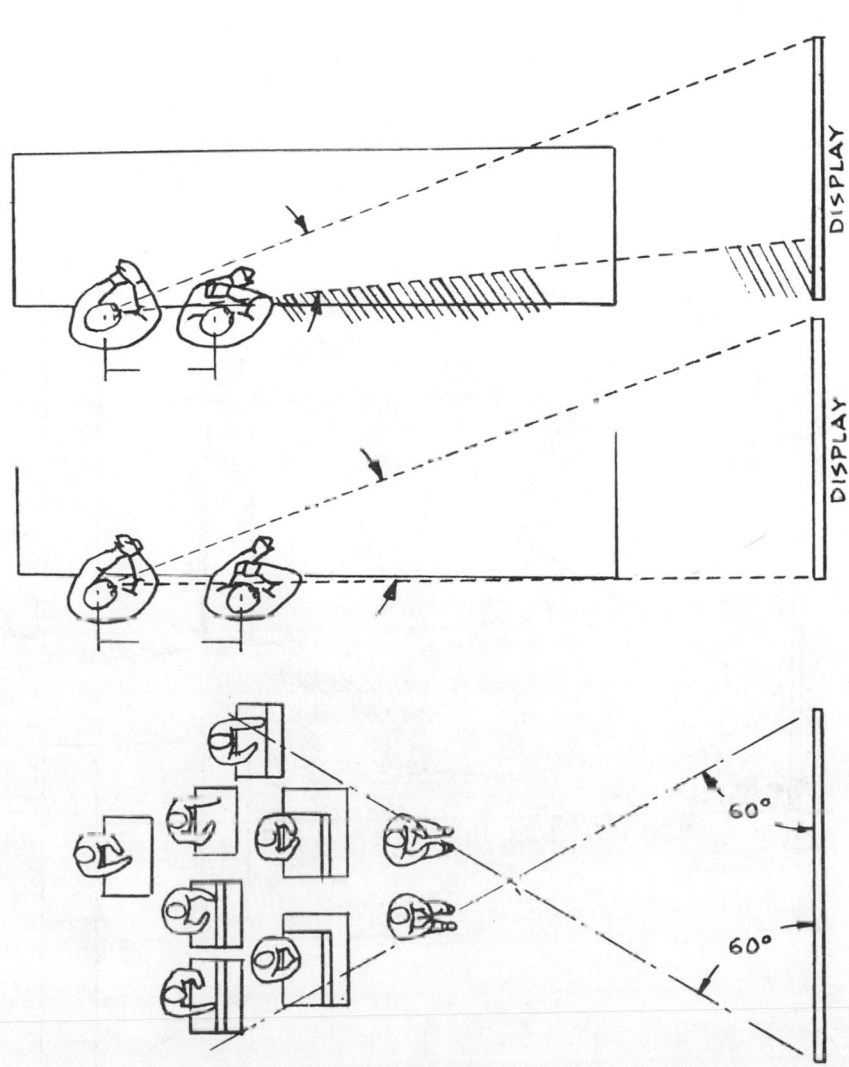

If, because of space constraints, the large display cannot be placed far enough away from a group of observers, consider breaking up the display so that the edges can be rotated for a more suitable angle of viewing.

When the large display is created by a projector, take care to arrange the projector and the observers so that intervening observers do not block the projected image.

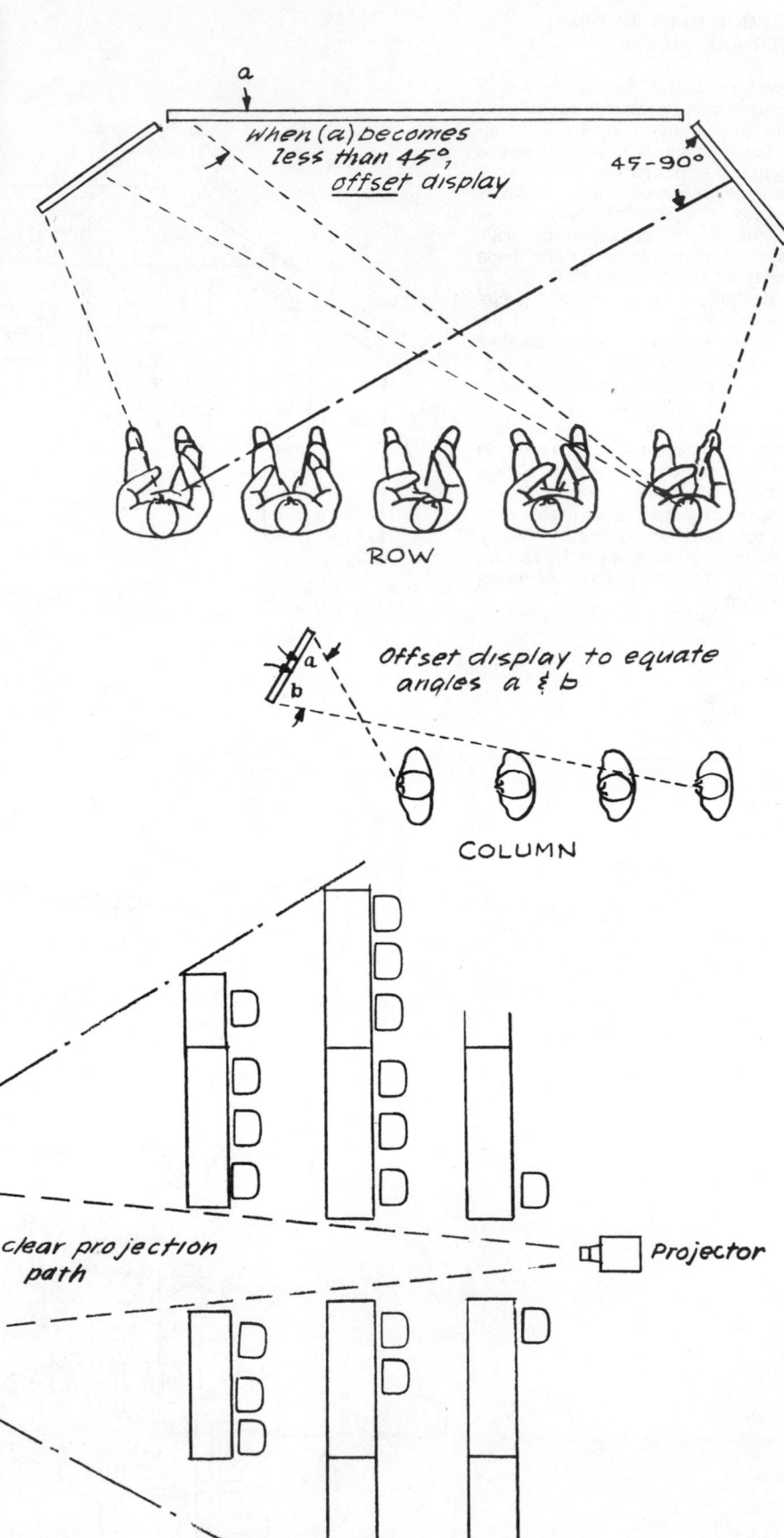

*when (a) becomes less than 45°, offset display*

45-90°

ROW

*Offset display to equate angles a & b*

COLUMN

SCREEN

60°

60°

*Keep clear projection path*

Projector

**Typical Military Operator Dimensional Guidelines[15]**

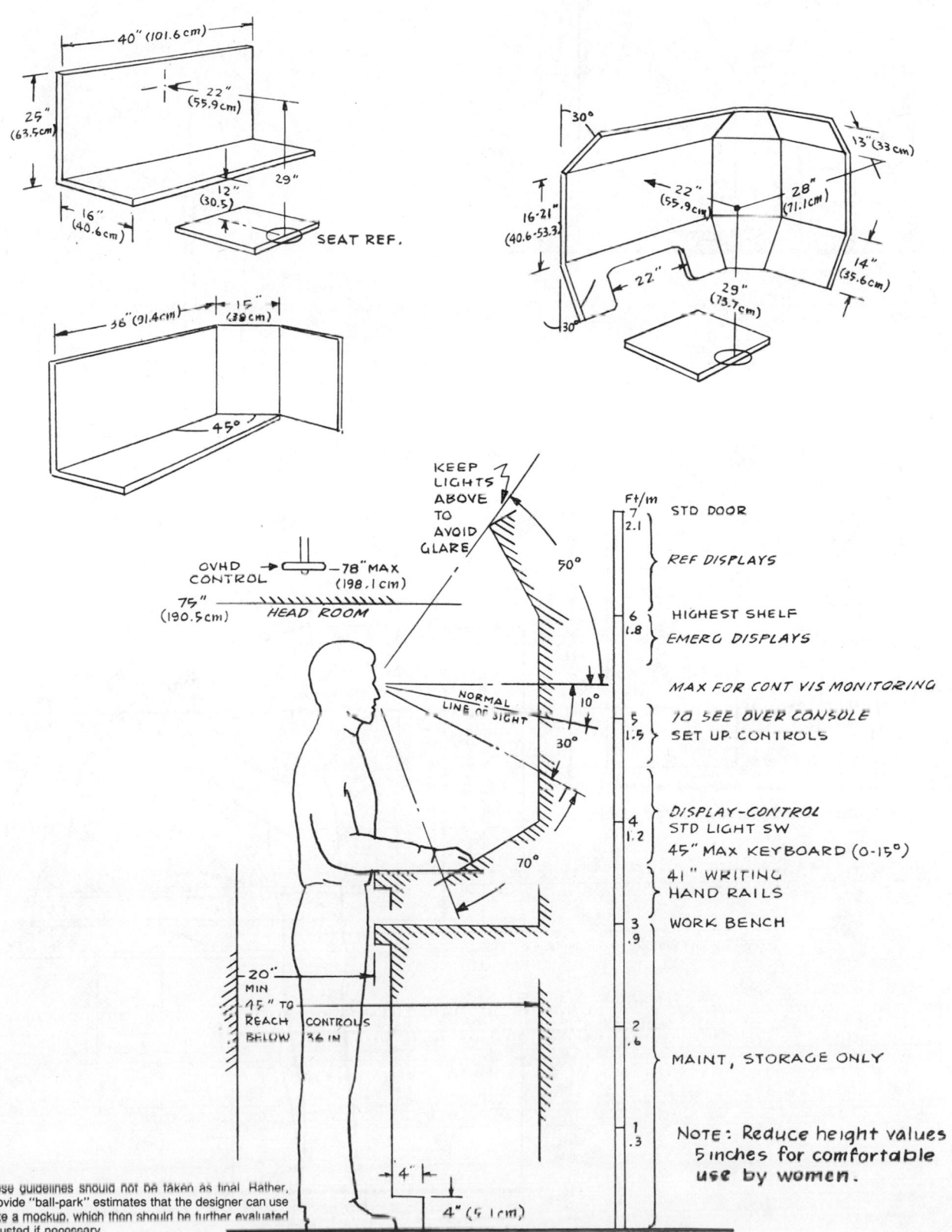

¹⁵These guidelines should not be taken as final. Rather, they provide "ball-park" estimates that the designer can use to create a mockup, which then should be further evaluated and adjusted if necessary.

**GUIDELINES FOR POSITIONING
CRT-TYPE DISPLAYS**

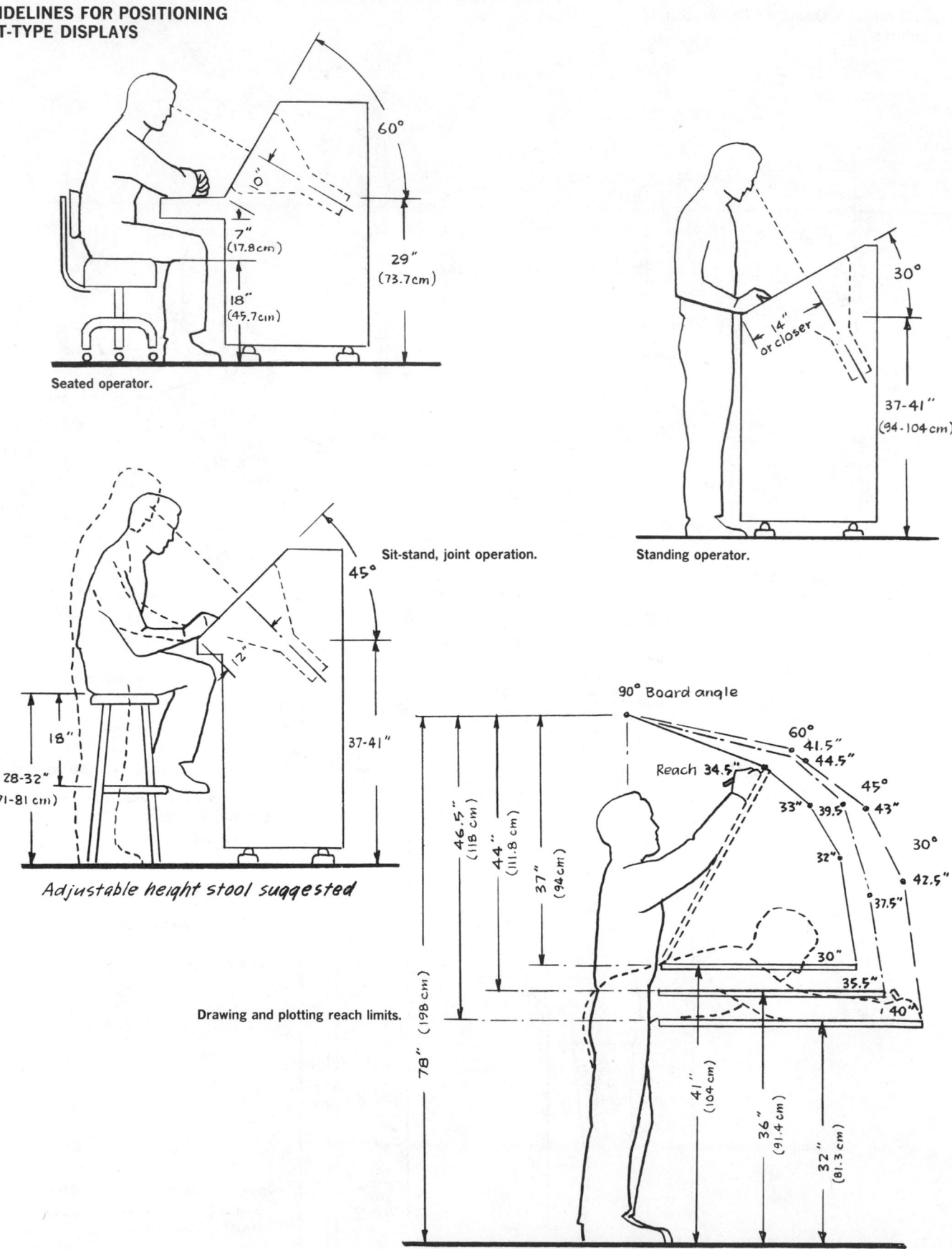

60°

10°

7"
(17.8cm)

29"
(73.7cm)

18"
(45.7cm)

Seated operator.

30°

14"
or closer

37-41"
(94-104 cm)

Standing operator.

45°

12"

Sit-stand, joint operation.

18"

28-32"
(71-81 cm)

37-41"

Adjustable height stool suggested

90° Board angle

60°    41.5"
       44.5"

Reach 34.5"

33"   39.5"   43"
                45°

      32"      30°
                42.5"

            37.5"

       30°

46.5"
(118 cm)

44"
(111.8 cm)

37"
(94 cm)

35.5"

40"

78" (198 cm)

Drawing and plotting reach limits.

41"
(104 cm)

36"
(91.4 cm)

32"
(81.3 cm)

NOTE: Reach criteria should be reduced approx
10-15 o/o for women.

## Miscellaneous Console Considerations

Avoid the tendency to make all elements of a console operator's operation fixed. That is, it is sometimes desirable to modularize portions of the console so that the modules can be rearranged to be more compatible with the particular activity. One typical example is the separate packaging of a keyboard. The operator may perform several different operations at the console, only one of which requires the keyboard. If the keyboard is fixed in a central position, it is obvious that that prime space cannot be used in any other manner.

When a pencil joy stick or track-ball controller is involved in the console use, be sure that the control position is far enough forward so that the operator has a place on the desk for his or her arm.

CRT viewing angle is important in that when the device is used for long periods, the operator should not have to hold his or her head in an awkward position. Although we often try to set the CRT angle in a so-called compromise position, this may still create problems for either the shortest or the tallest operator. One obviously desirable approach is to make the CRT package adjustable, as shown in the accompanying sketch. Not only does this allow operators to find the best vision angle to reduce neck fatigue, but it also allows them to adjust the face of the display to minimize any annoying reflections of themselves or of objects behind or above them in the room.

A convenient and effective workplace can be created using independent modules, as shown in the accompanying illustration.

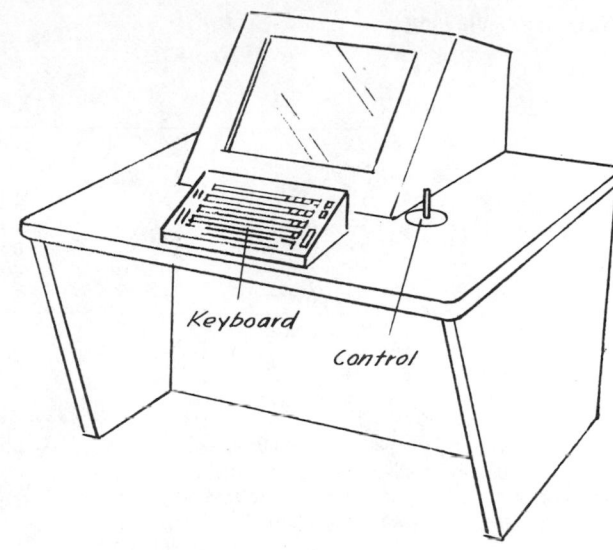

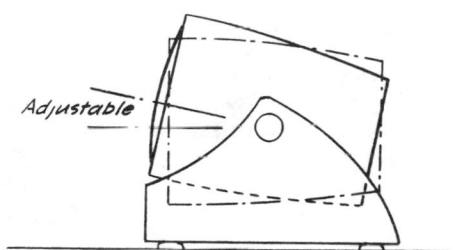

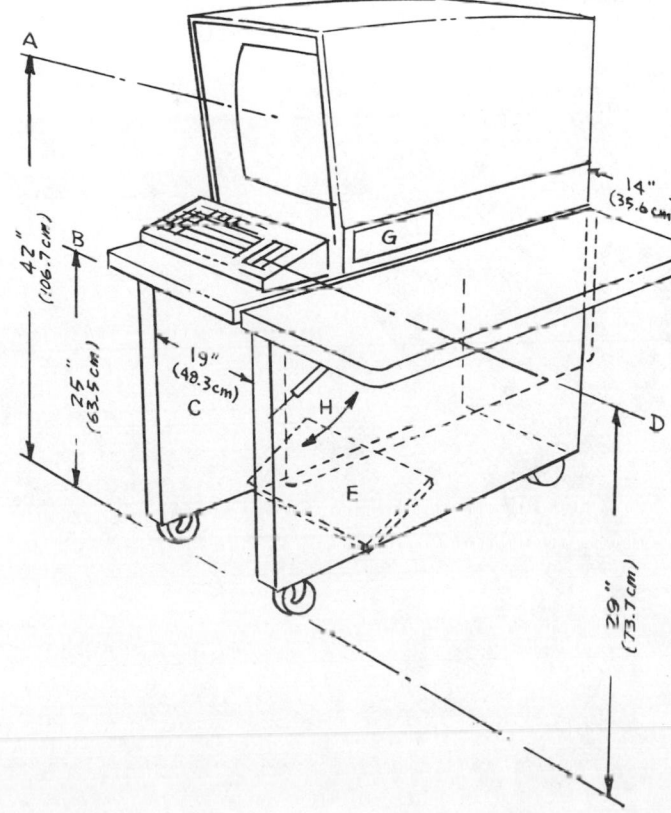

A—The center of a vertically oriented CRT should be approximately as shown. Note that the case extends over the tube to minimize glare from overhead ambient light.
B—Vertical knee clearance minimum.
C—Lateral knee clearance minimum.
D—Center height of keyboard.
E—Footrest.
F—Minimum width for fold-down writing surface.
G—CRT controls (under cover).
H—Supporting bracket for fold-down desk. Casters with locks allow the unit to be moved around to fit the user's needs.

## Television Viewing

| Screen Size* | Minimum Viewing Distance | Maximum Viewing Distance |
|---|---|---|
| 17 in (43 cm) | 4 ft 11 in (1.47 m) | 14 ft 9 in (4.43 m) |
| 19 in (48 cm) | 5 ft 1 in (1.53 m) | 15 ft 2 in (4.58 m) |
| 21 in (53 cm) | 6 ft 4 in (1.89 m) | 19 ft (5.7 m) |
| 23 in (58 cm) | 6 ft 6 in (1.95 m) | 19 ft. 4 in (5.8 m) |
| 25 in (64 cm) | 7 ft 6 in (2.25 m) | 21 ft. 9 in (6.5 m) |

*Diagonal.

For arranging observers in direct TV viewing situations, use the guidelines shown in the accompanying table. For projected television viewing, use the following guidelines:[16]

W = Screen width

Ambient light level should be between 0.1 and 2.0 ftc for comfortable viewing.

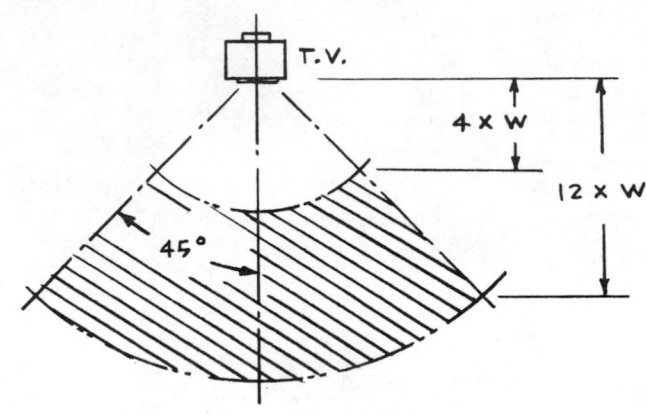

Preferred viewing area for watching T.V.

1. Screen brightness    2 to 20 ft (0.6 to 6.1 m) L
2. Brightness ratio    2:1—excellent
     3:1—very good
     10:1—acceptable
3. Contrast ratio    100:1 for pictorial images
     25.1 for printed characters
     5.1 for white characters on black background

[16] A minimum contrast ratio of 30:1 is for the poorest conditions, e.g., the observer who is furthest away, high ambient room lighting levels, and poor materials.

**Miscelleneous Guidelines for Other
Typical Operator Working Conditions**

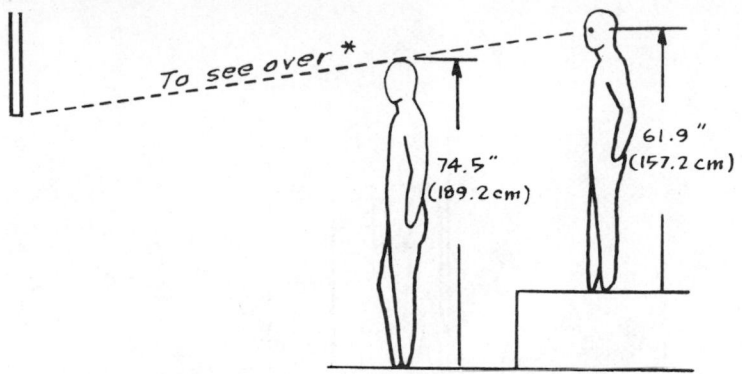

*NOTE: *Values for male observers. For mixed
male/female situations, consider
tall male in front, short female in
rear.*

**Manual accessibility and clearance.**

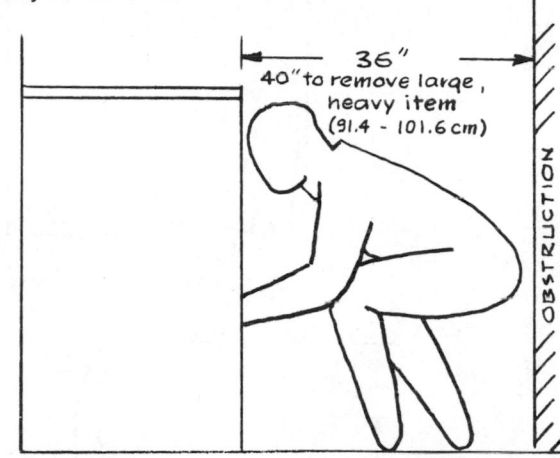

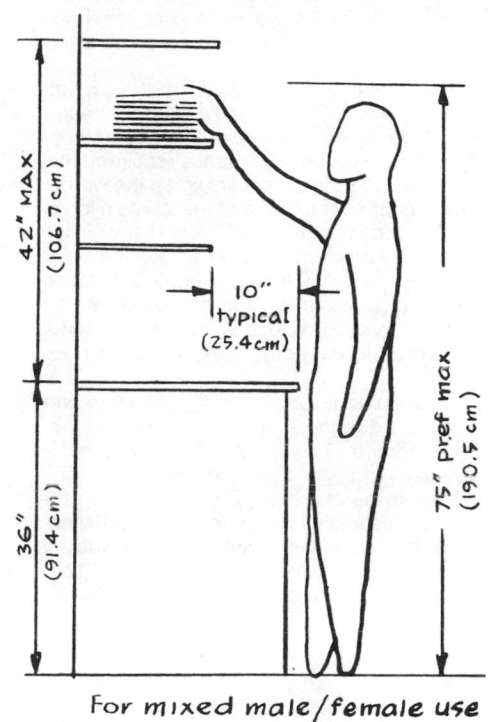

For mixed male/female use

**Visual accessibility.**

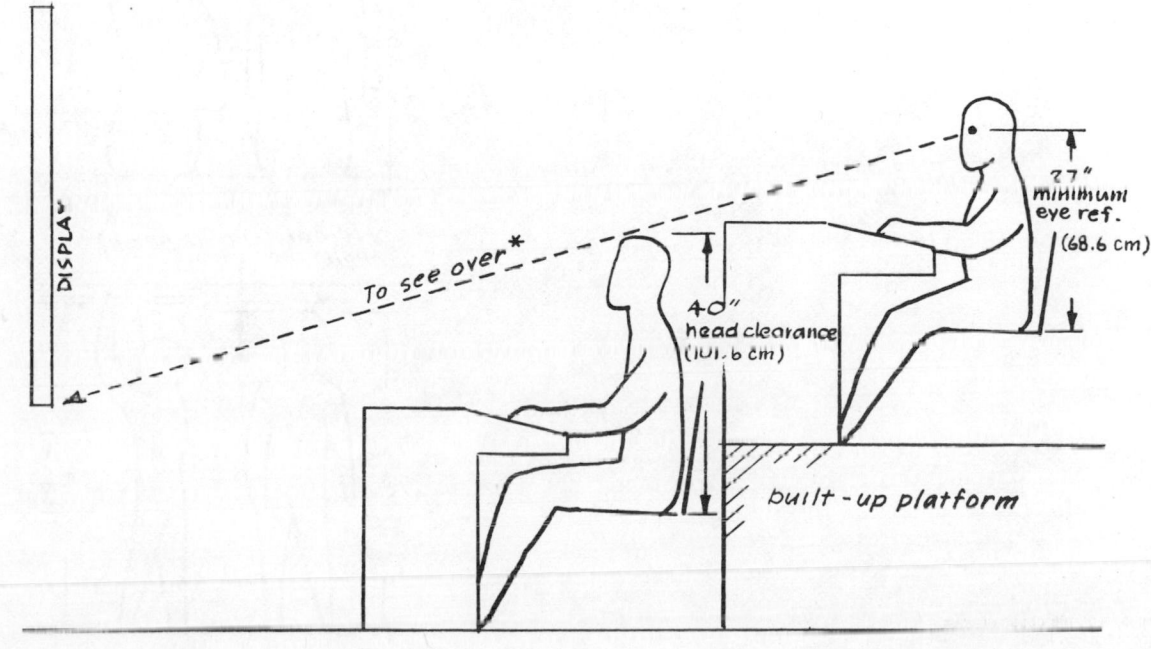

When purchasable workplace furnishings are to be accommodated, their use-dimension characteristics should be examined carefully so that the general workplace geometry that is created (and/or the proposed arrangement in an existing facility) will provide adequate clearance.

The typical examples shown in the accompanying illustrations emphasize the importance of relating drawer-pull distance and operator or worker body clearance requirements.

The worst situation occurs when the worker has to bend over to see and reach into a lower drawer or compartment. Although it is obvious that the worker could step aside and work from one side of the drawer, this position makes it awkward to scan printed or written material, and the worker is forced to assume difficult and tiring body, neck, and head positions.

When several workers may use the workplace, the designer must consider the fact that one worker may be larger or smaller than the others. Although the general dimensions shown in the accompanying illustrations provide a reasonable compromise in most situations, the designer is urged to make a mockup of the proposed workplace and to check it out using several test subjects of different sizes.

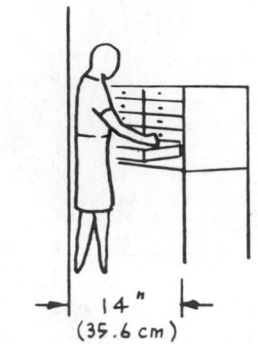

14" (35.6 cm)

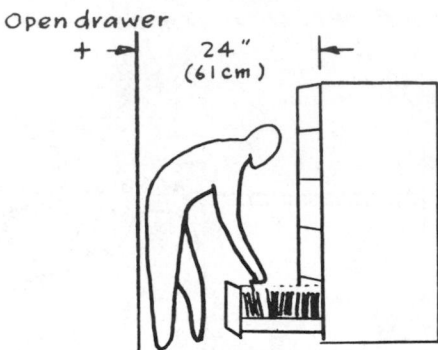

Open drawer

24" (61 cm)

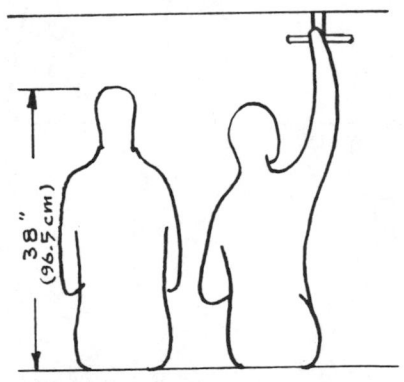

38" (96.5 cm)

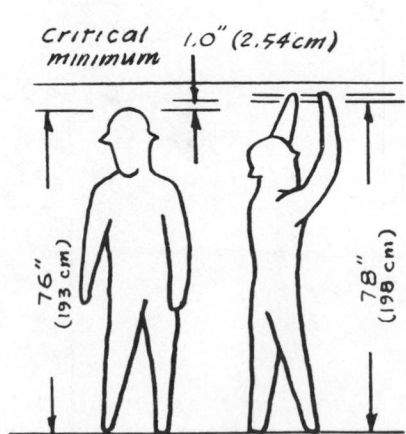

Critical minimum    1.0" (2.54 cm)

76" (193 cm)    78" (198 cm)

**WORKING IN CLOSE QUARTERS** As a general rule, avoid creating situations in which workers have to work in close quarters and/or in the awkward positions illustrated in the accompanying sketches.

When this cannot be avoided, consider the general limiting dimensions provided in these illustrations. Critical factors are the following:

1. There must be sufficient vertical clearance for the worker to enter and leave the environment.
2. There must be sufficient horizontal and lateral clearance for the worker to perform the task (considering the nature, size, and configuration of any tool being used).
3. In the supine position, it is critical that the worker's eyes do not have to be too close to the specific work point; otherwise, the worker's eyes cannot adjust for clear viewing.
4. Openings must be large enough to accommodate both the tool package and the operator's hand, finger, or arms, as shown in the accompanying illustrations.

*Note:* These opening dimensions are for shirt-sleeve conditions. If heavier clothing may be worn, the openings have to be at least 1 in (2.5 cm) larger on all sides (1.5 in, 3.8 cm, is better).

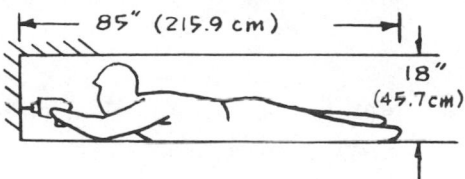

85" (215.9 cm)

18" (45.7 cm)

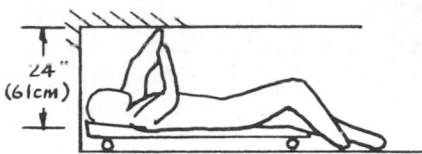

24" (61 cm)

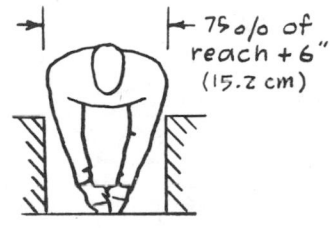

75% of reach + 6" (15.2 cm)

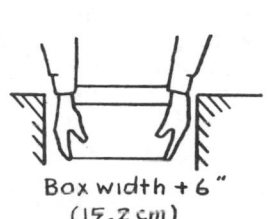

Box width + 6" (15.2 cm)

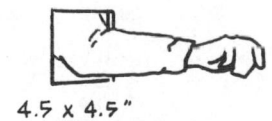

4.5 x 4.5" (11.4 x 11.4 cm)

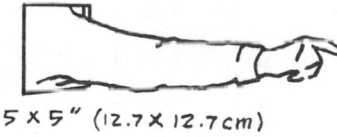

5 x 5" (12.7 x 12.7 cm)

4" (10.2 cm)

2" (5.1)

Flat hand

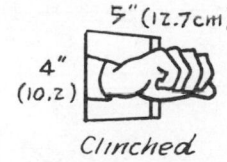

5" (12.7 cm)

4" (10.2)

Clinched

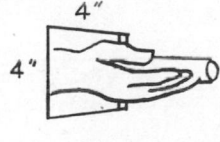

4"

4"

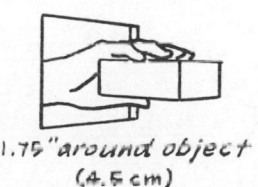

1.75" around object (4.5 cm)

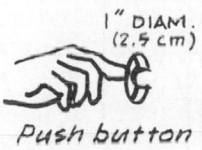

1" DIAM. (2.5 cm)

Push button

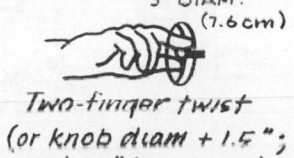

3" DIAM. (7.6 cm)

Two-finger twist (or knob diam + 1.5"; add 1.0" for gloves)

## LIBRARY FACILITIES

Dimensional guidelines for key adult library workplaces are illustrated by the accompanying sketches.

These dimensions should be altered to fit young people as follows:

1. For teenagers:
   Top shelf height      62 in (158 cm)
   Reading table height      27 in (69 cm)
   Counter height      35 in (89 cm)
   Catalog worktable height      35 to 38 in (89 to 97 cm)
   Map and book table height      34 in (86 cm)
2. For grade school children:
   Top shelf height      48 in (122 cm)
   Reading table height      21 in (53 cm)
   Counter height      30 in (76 cm)
   Map and book table height      30 in (76 cm)

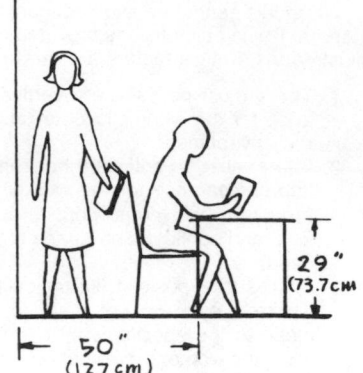

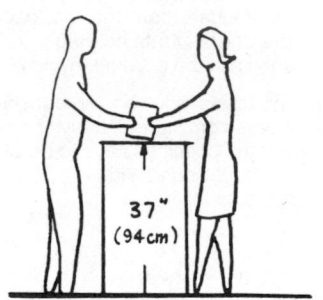

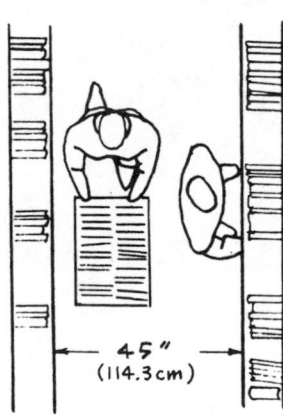

**Estimated Number of Books per Foot of Linear Shelf Space and per Single-Face Bookcase Section for Different Subject Matter**

| Subject | Volumes per foot | Volumes per Single-Face Bookcase Section |
|---|---|---|
| Circulating nonfiction | 8 | 168 |
| Fiction | 8 | 168 |
| Economics | 8 | 168 |
| General literature | 7 | 147 |
| History | 7 | 147 |
| Art | 7 | 147 |
| Technical and scientific books | 6 | 126 |
| Medical books | 5 | 105 |
| Public documents | 5 | 105 |
| Bound periodicals | 5 | 105 |
| Law books | 4 | 84 |

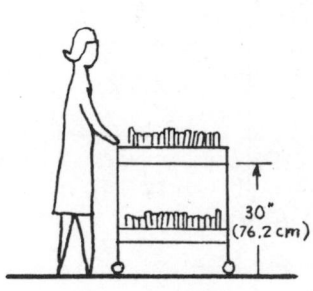

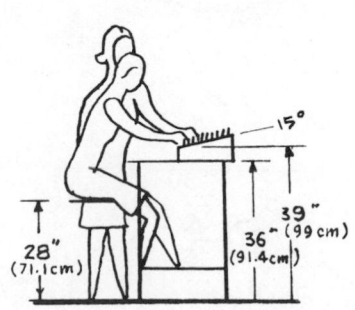

General Guidelines (Female-limiting)

## Reading and Study Carrels

Visual and auditory isolation is often desirable for library patrons. The accompanying sketches illustrate configurations as well as dimensional guidelines for designing and/or selecting the various types of carrels.

Carrels with temporary book storage shelves.

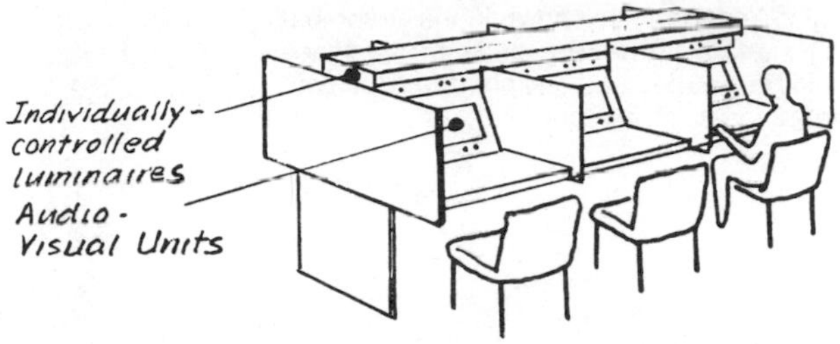

Individually-controlled luminaires

Audio-Visual Units

Carrels with special audio-visual equipment.

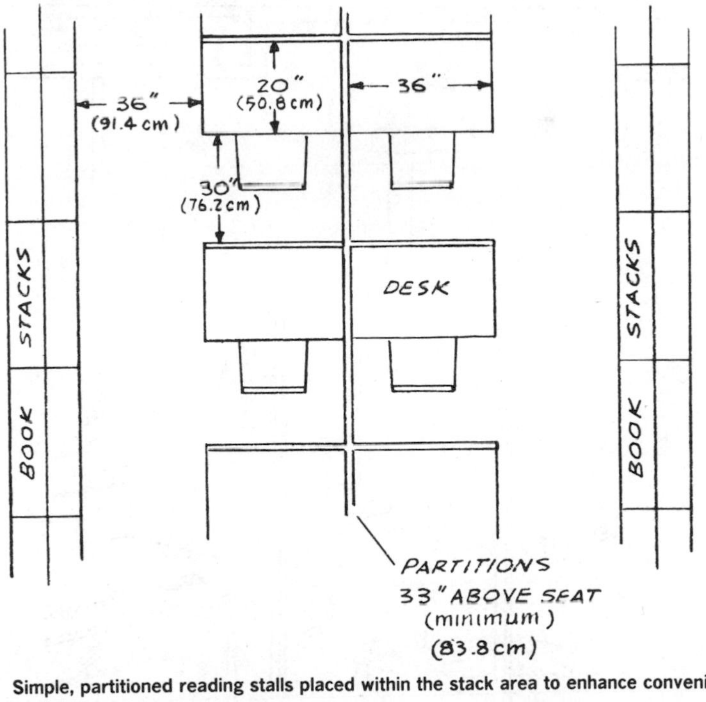

36" (91.4 cm)

20" (50.8 cm)

36"

30" (76.2 cm)

BOOK STACKS

DESK

BOOK STACKS

PARTITIONS 33" ABOVE SEAT (minimum) (83.8 cm)

**Simple, partitioned reading stalls placed within the stack area to enhance convenience to source materials.**

*Note:* Partitions should be high enough (30 in, 76 cm, above the desk) to prevent one patron from seeing another while seated, and they should provide some sound absorption to reduce the overall noise level.

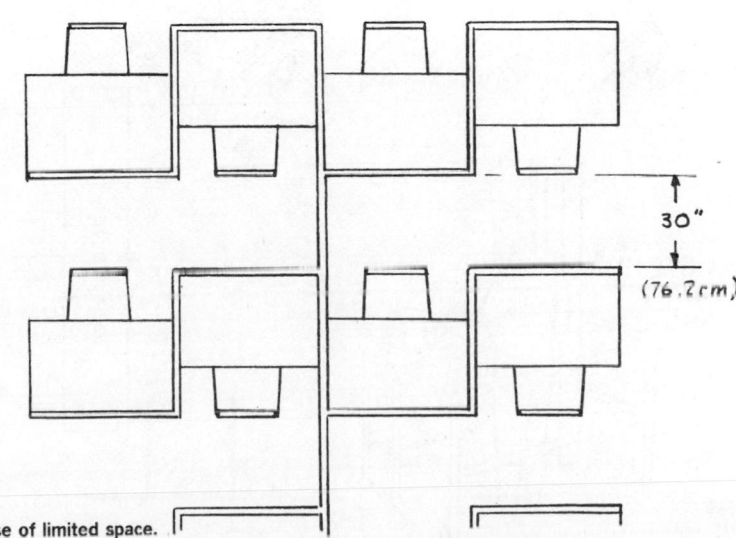

30" (76.2 cm)

**Minimum carrel configuration to make maximally effective use of limited space.**

## FACTORY WORKPLACES

Workplaces in which fabrication and assembly operations are performed should be laid out not only in terms of placing the critical visual and manual interfaces so that they are compatible with efficient manipulation of materials, parts, and tools but also in terms of work flow.

The accompanying examples represent typical layouts that have been used from time to time to increase the overall efficiency of factory workers.

The safety aspects of the factory workplace are also important, as shown in the accompanying sketch. All moving parts of machines should be guarded and equipped with fail-safe emergency cutoff switches.

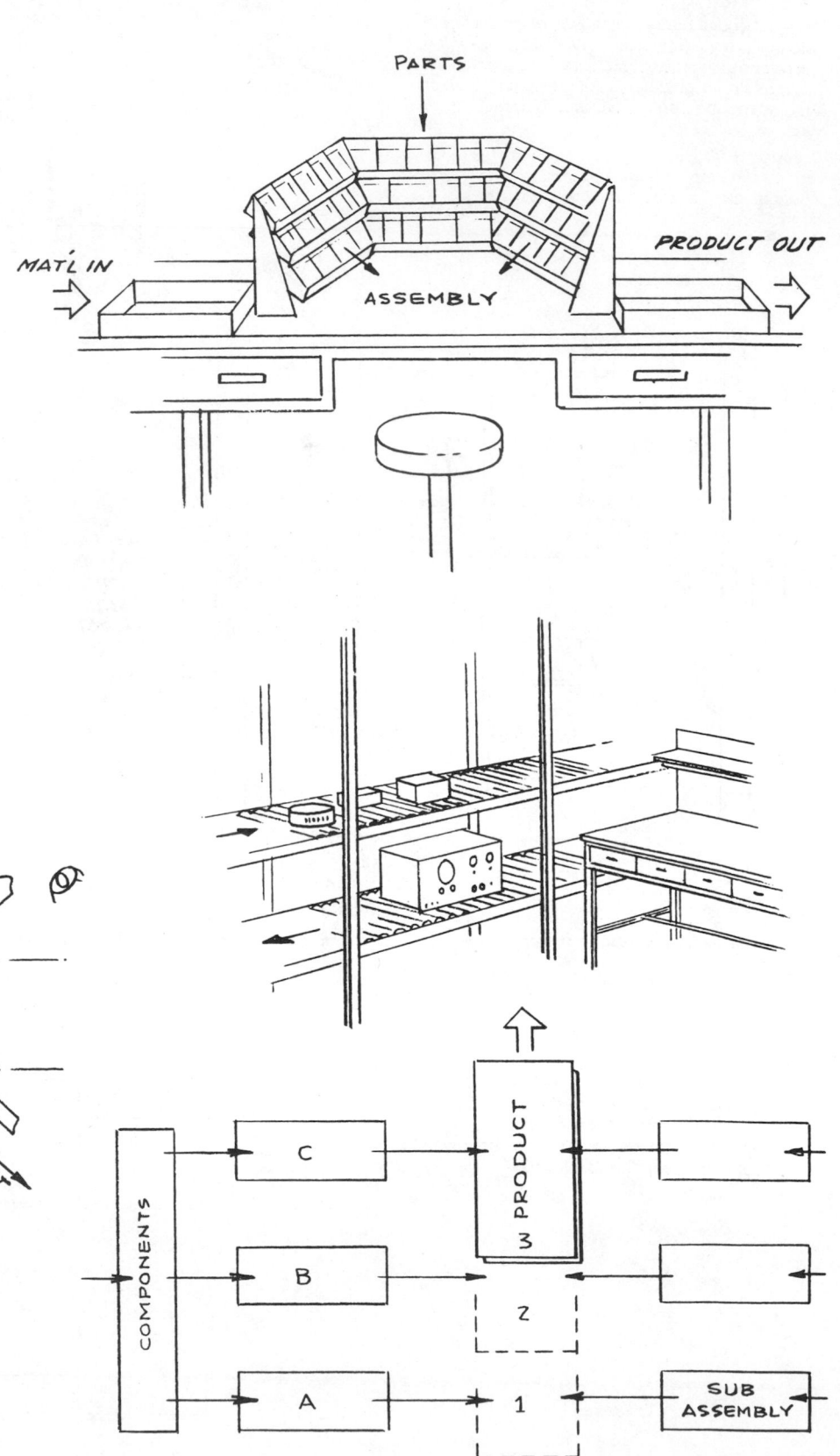

ASSEMBLY LINE WORK FLOW

## MAINTENANCE WORKPLACES

Although equipment maintenance often is not considered as important as the operator's interface, design of the maintenance interface is important to the maintenance technician and therefore should be considered at the same time that operator interfaces are being designed. Several typical examples of maintenance problems, created by lack of proper attention, are illustrated in the accompanying sketches.

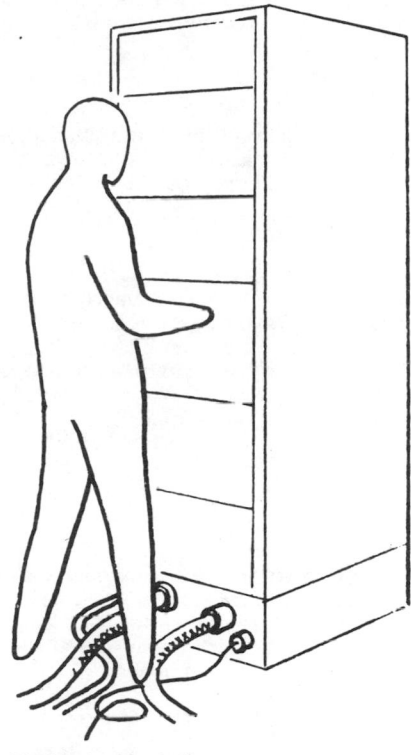

Electrical cables in the way.

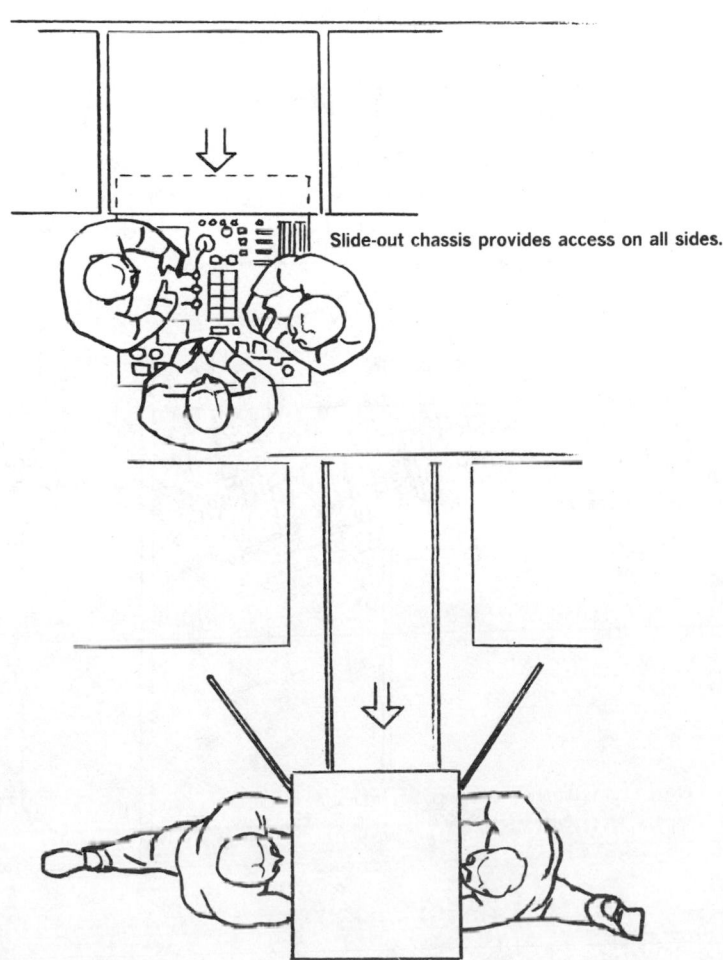

Slide-out chassis provides access on all sides.

Cabinet tracks allow the cabinet to be removed from among a row of cabinets that block access for maintenance.

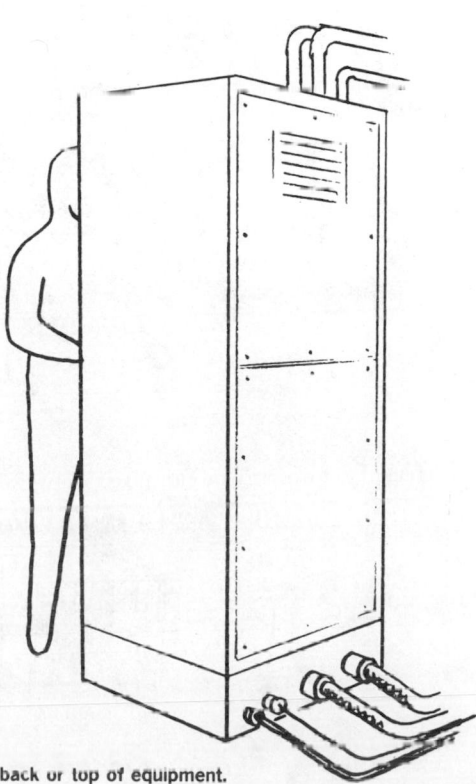

Should exit the back or top of equipment.

Different equipment packaging concepts either aid or hinder the maintenance technician, as illustrated by the accompanying sketches:

Access is better if the case can be lifted off the chassis. More freedom to work is provided, and there is no cover to prop up.

Avoid designs in which the chassis has to be lifted out of the case or in which a cover has to be removed or propped out of the way.

Design covers so that the minimum number of fasteners have to be removed.

Fasteners should be captive so that they do not become mislaid or lost.

When high voltages are involved, removal of the case or cover should automatically shut the power off. A secondary switch or jumper should be used to reactivate the power when maintenance is being performed.

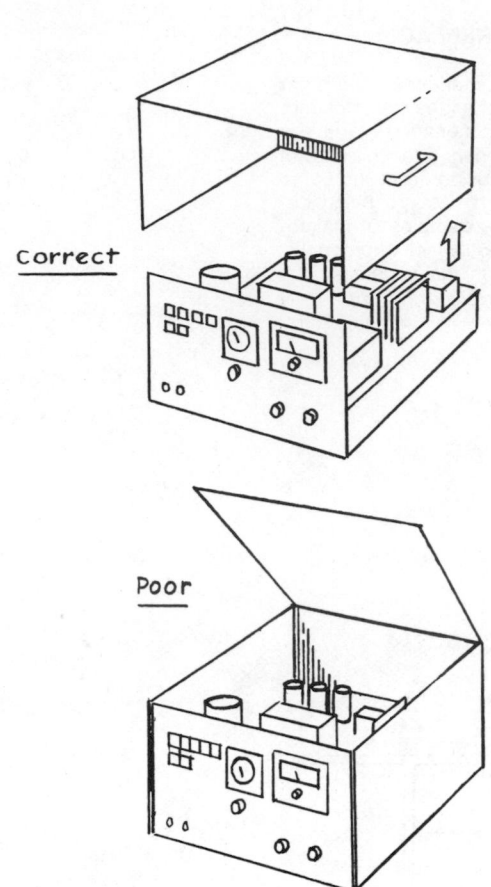

Correct

Poor

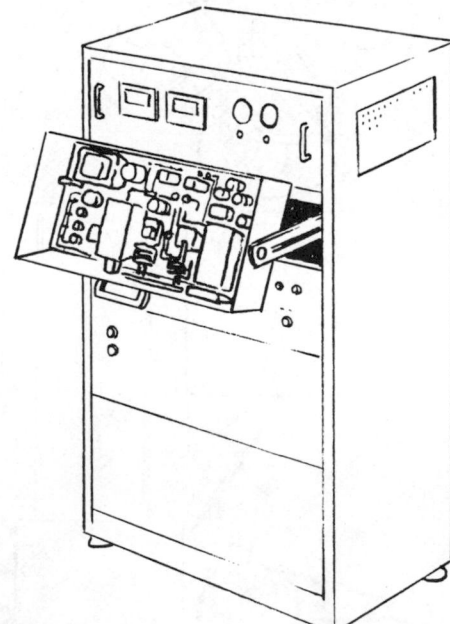

A rotating chassis makes both the top and the bottom accessible.

The fold-out package makes the entire inner assembly accessible.

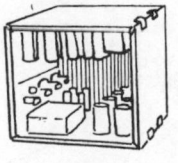

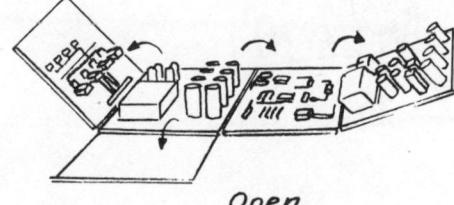

Closed          Open

Although this is not always possible, try not to create maintenance working situations in which the technician or mechanic has to work in awkward positions for extended lengths of time. Often this happens because sufficient planning for maintenance was not done before a design was frozen.

The most serious and most common difficulties are illustrated in the accompanying sketches.

Working overhead with arms outstretched is extremely tiring, and there is the added possibility that debris or parts may fall into the technician's eyes.

Working close to the floor or ground is tiring, and it also may make it difficult to get one's eyes in the proper position to see what needs to be done. Even the common gas water heater used in homes illustrates how designers tend to ignore this problem. To light, adjust, or simply inspect the working parts of the heater, one almost has to stand on one's head. In addition, while a person is peering into the darkness to find a pilot light, his or her eyes and face are in the direct line of an explosion, should one occur.

The accompanying sketch illustrates a common problem of most car owners and mechanics. Too little thought was given to the reach distance for engine components that are frequently serviced or replaced. Most serious is the possibility that the mechanic will reach across hot parts, moving belts, etc.

*Poor working positions!*

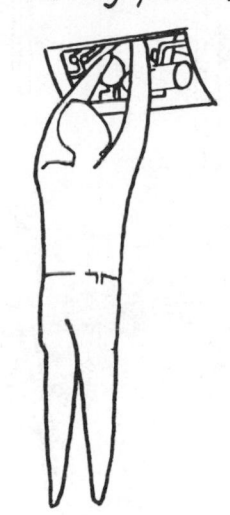

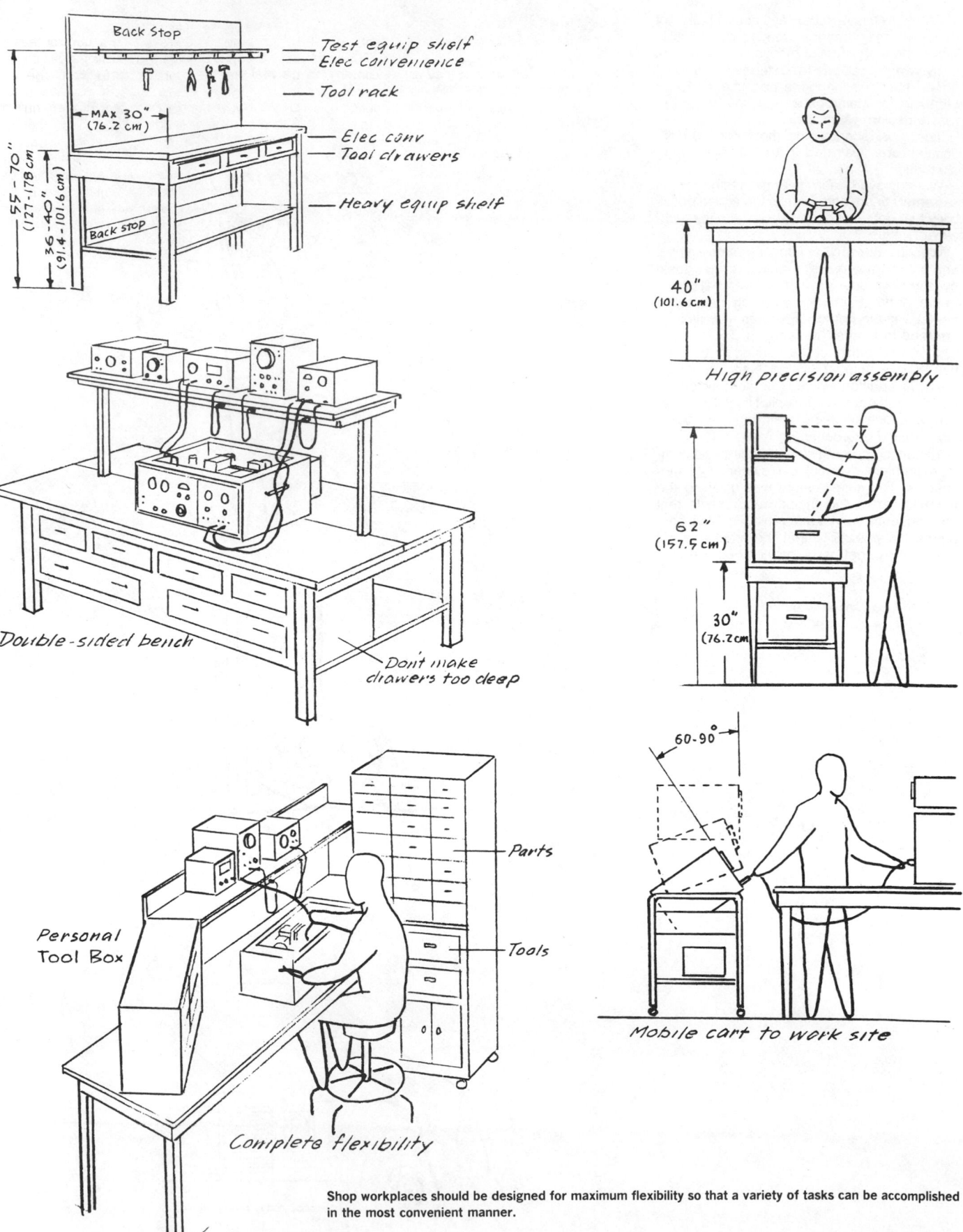

Back Stop

— Test equip shelf
— Elec convenience

— Tool rack

— Elec conv
— Tool drawers

— Heavy equip shelf

MAX 30"
(76.2 cm)

55-70"
(127-178 cm)

36-40"
(91.4-101.6 cm)

Back Stop

Double-sided bench

Don't make drawers too deep

Personal Tool Box

Parts

Tools

Complete flexibility

High precision assembly

40"
(101.6 cm)

62"
(157.5 cm)

30"
(76.2 cm

60-90°

Mobile cart to work site

Shop workplaces should be designed for maximum flexibility so that a variety of tasks can be accomplished in the most convenient manner.

## OPERATOR CONTROL PANEL LAYOUT

### Basic Principles

Four basic principles should be considered in planning the layout of an operator workplace or, more specifically, the equipment control interface (usually a panel containing controls and displays). These are discussed below.

### 1. Sequence of Operations

The order in which operators will normally use controls and displays is important in terms of their being able to find what they want and being able to go through the sequence with a minimum possibility of missing a step and also in terms of reducing unnecessary motion due to "backtracking." Although some operations may not have a single sequential pattern, there are usually one or two elements that proceed according to a fairly standard sequence of steps.

### 2. Functional Grouping

Certain functions within almost any operation are related to each other, and/or certain controls are related to certain displays. Related functions should be arranged together so that their association is readily apparent to the operator. When such functions are not grouped, operators tend to get lost and to wander from one point to another trying to find what they want. Although there may be a few elements within the workplace that cannot be placed together for obvious reasons (i.e., a foot pedal obviously cannot be placed where an associated visual display would be), an attempt should be made to avoid scattering controls and displays anywhere on the panel without regard to functional associations.

### 3. Frequency of Use

Some controls or displays will probably be used more frequently than others. These should be located so that they are maximally convenient; i.e., visual displays should be near the normal sight line, and controls should be near the hand when the arm is in a normal position. For example, a frequently used visual display should not be located where the operator must continually turn his or her head or body in order to see it, and a control should not be placed where the operator continually has to reach a long distance for it.

### 4. Emergency Use

Once in awhile there will be a control that is critical under emergency situations; i.e., it may be vital that the operator get to it and operate it successfully in the shortest possible time. Obviously, if such a control is hidden, out of reach, or otherwise difficult to get to and operate, the operator may not be able to reach it in time.

*Note:* Each of the above principles must be considered in light of a particular operator interface problem; there is no set priority that can be all-prevailing.

### Primary Visual Reference Principle

When it has been determined that an operator is expected to visually monitor a primary display for a great share of the time, that display should be located for maximum comfort, usually directly in front of the operator and approximately at eye level or slightly below. In addition, the things that the operator should be able to see fairly regularly in conjunction with the principal viewing task should be within about a 30° cone around the principal line-of-sight line. This includes not only visual displays and indicators but also critical controls that the operator may have to look at while he or she is adjusting them.

A typical console situation in which the visual reference principle must be applied is the large CRT display requiring the operator to continually watch for and track sonar or radar targets. The center of the CRT should coincide with the comfortable line of sight. If the operator has to share more than one line of sight, seek a compromise reference, as shown in the accompanying illustration.

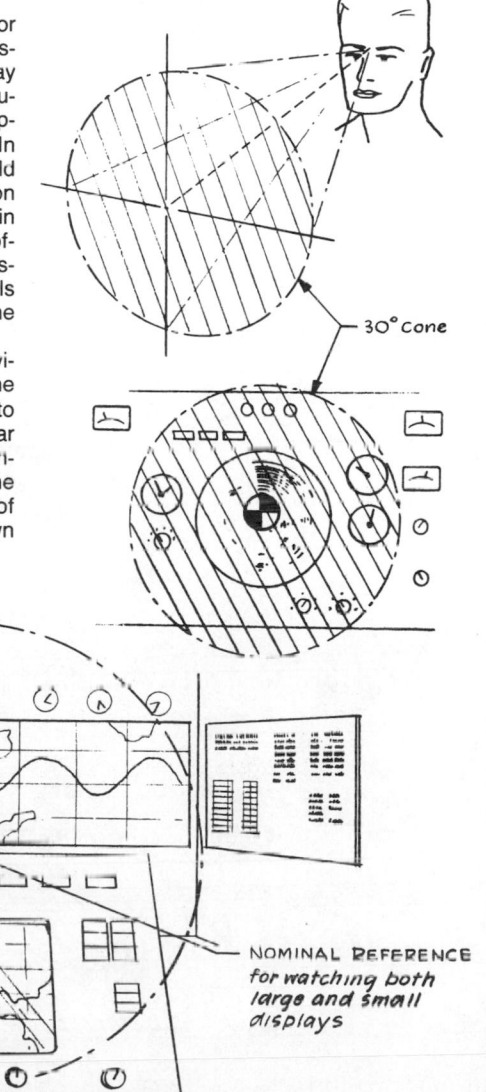

30° cone

NOMINAL REFERENCE
*for watching both
large and small
displays*

## Positioning Panel Elements Relative to Line of Sight

Visual displays, including indicator lights, instruments, panel and control labels, etc., are useless to the operator if he or she cannot see them. This requires that no control (joy stick, steering wheel, etc.) be placed between the observer's eye and other displays. In addition, it requires that care be taken in the placement of control and display labels so that when critical, the operator's hand will not cover up an important label that may be used in adjusting the control.

As a general principle, labels should be located in a consistent manner with respect to being above or below a control or display; i.e., avoid having some labels above a control, while others are below a control—at least on the same panel.

However, the choice of label position should also reflect consideration of the expected operator's eye reference; i.e., if the control is above the eye reference position, the label is more likely to be visible than if it were placed above the control.

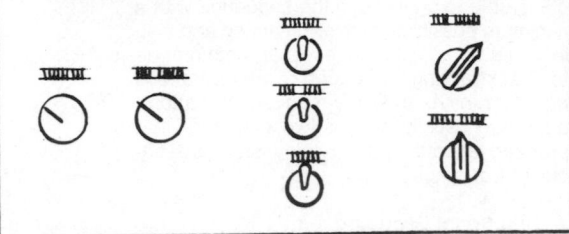

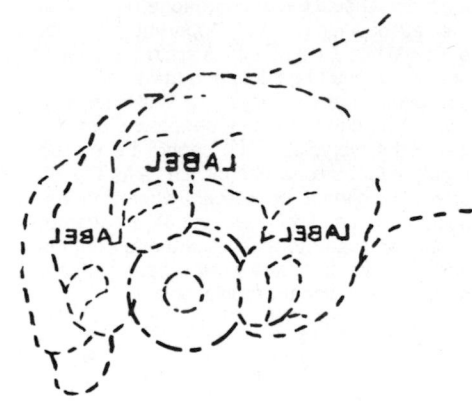

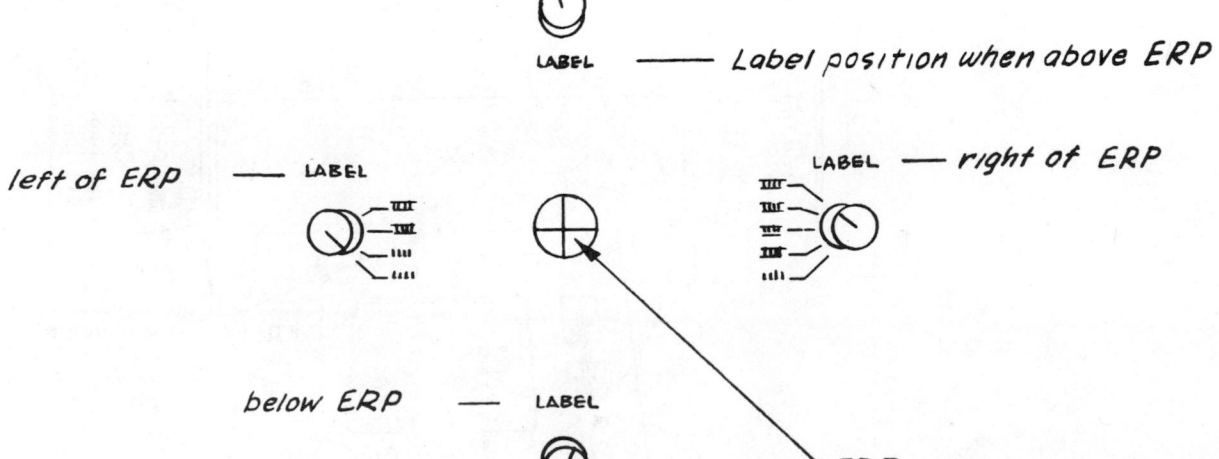

LABEL —— Label position when above ERP

left of ERP —— LABEL

LABEL —— right of ERP

below ERP —— LABEL

ERP

## Display Arrangement

When the circumstance includes an array of displays that must be monitored on a regular basis to determine whether conditions are approximately "normal," the check-reading task is made easier for the operator if the displays are arranged so that their indications (e.g., pointer positions) generally reside in the same relative place on the instrument (see the accompanying sketch). Arrange the normal pointer reading at the twelve o'clock position when the array is vertical and at the nine o'clock or three o'clock position when the array is horizontal.

If displays are ordinarily read in sequence, arrange them in order, either horizontally or vertically. The normal reading pattern is from left to right or from top to bottom.

When a display is adjustable by means of an associated control, the control should be located as close to the display as practical. However, since there may be a directional relationship, as shown in the accompanying sketch, care must be taken to position the control relative to the display so that there is no doubt in the operator's mind as to the direction in which the control should be moved in order to cause the displayed element to move in the right direction. Controls which are placed below their respective displays are less confusing than those which are placed to the right or left of the displays.

When it is impractical to place controls close to the displays they affect, try to arrange both the displays and the controls in similar patterns (see the accompanying sketch) so that there is a similarity between the two.

When a control causes something to move in space outside the console (e.g., an antenna), control-element directions of motion should also be similar, i.e., clockwise, counterclockwise, left to right, up and down, etc.

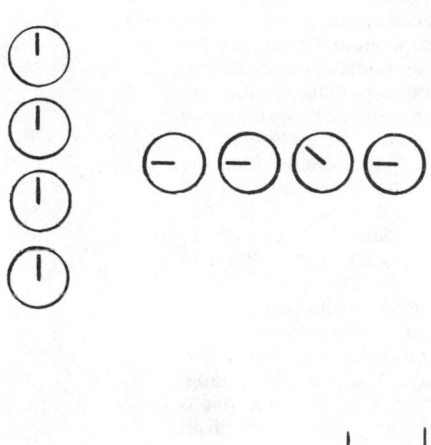

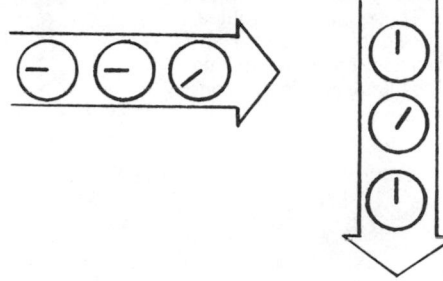

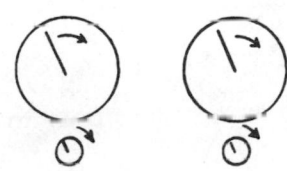

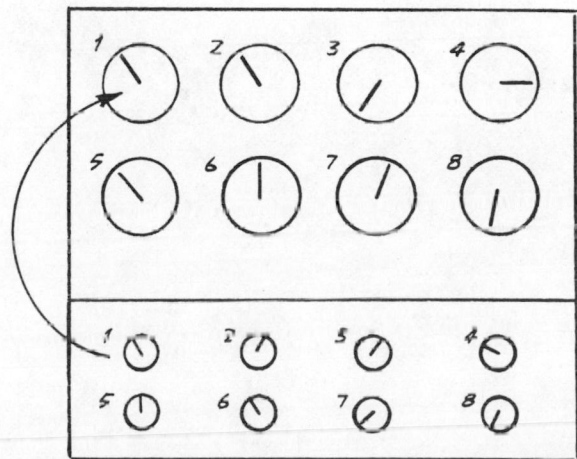

Although not highly recommended (except where panel space is limited), relationships between concentric-shaft knobs and their display counterparts should be defined as shown in the accompanying sketch; i.e., the top knob is associated with the first display on the left, and the bottom knob is associated with the last display on the right.

In certain situations aboard an aircraft (namely, the engineer's or other auxiliary crew member's workplace), relationships between controls on the panel and external counterparts (e.g., the engines) should be as shown in the accompanying sketch. The general principle to be followed is that operators face the panel as though the engines were in front of them, depending on the direction in which they are facing. It is highly desirable, however, to face all operators "forward" with respect to the direction in which the vehicle is traveling.

Movement of controls should generally be forward, upward, and/or clockwise to effect an increase in function. This is illustrated in the accompanying sketch. There is one notable exception, however; pilot controls that relate to increasing the aircraft's speed should always move forward. These include throttles, trim controls that operate to increase the aircraft's speed, and folding wing configurations (even though the wings actually fold aft to increase the aircraft's speed).

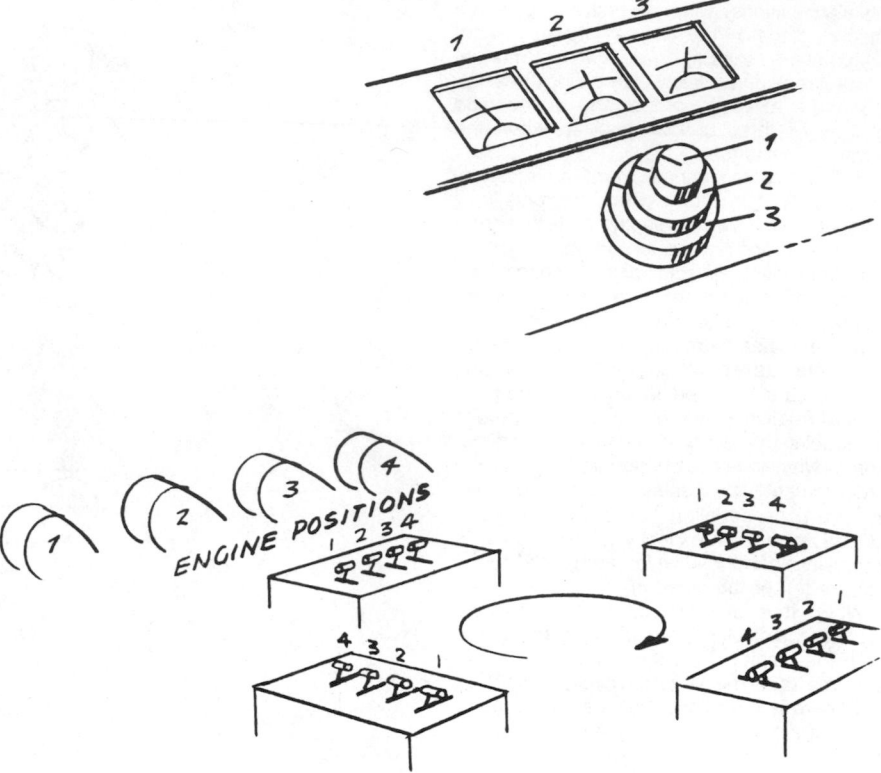

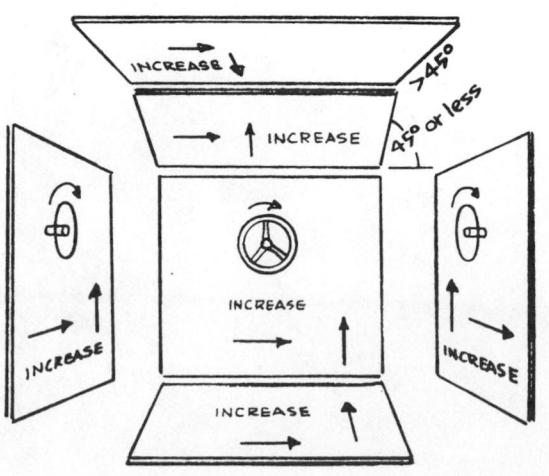

An important objective in panel layout should be to make the functional grouping of controls and displays readily apparent to the operator. There are various effective methods for doing this, some of which are illustrated in the accompanying sketches:

1. Spatial separation is effective when there is sufficient panel area. This is the preferred method.
2. When the panel is necessarily space-limited, one can use lines or borders around the various functions. This is particularly effective if the functional grouping is irregular.
3. If one functional group is especially important for the operator, it can be made more conspicuous by using a different color for the entire background for the group from that which is used for the rest of the panel.
4. Subpanels—either the protruding or the inset type, as shown by the accompanying sketches—are effective for setting off the various functional groupings.
5. In some cases, a separate package can be attached to the main package, especially if the package is small but needs a special physical orientation to make it easier to use.

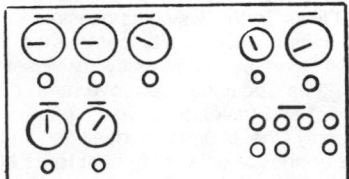

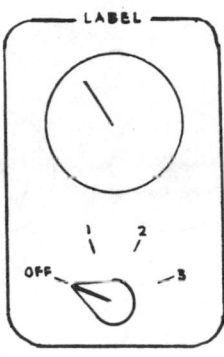

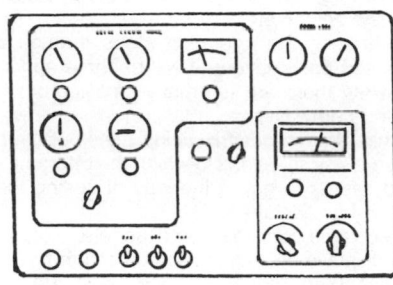

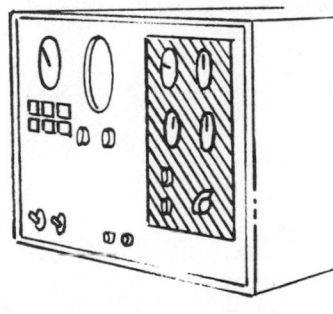

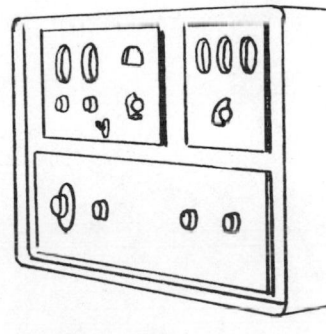

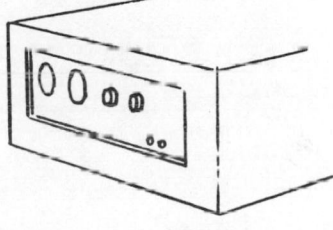

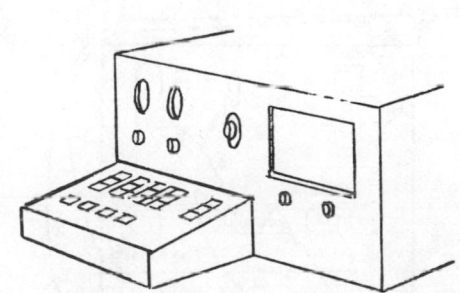

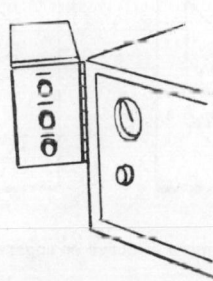

## Panel Arrangement According to Sequence and Frequency of Expected Use

Whenever possible, try to determine the order in which an operator will probably use each display or control on the proposed panel and to estimate the frequency of the various sequential links. In this way, it is possible to lay out the instruments and/or controls so that those which are used frequently are closer together, thus requiring less eye and hand movement and travel distance.

Generally, the equipment designer knows how he or she expects equipment to be operated. Where in doubt, the designer should search for someone who may have experience operating similar equipment. Together they usually can determine the best layout for the new equipment.

By tracing the operating steps through a proposed panel layout, it will become apparent when there are too many crossing or reversing pathways.

While this process is being accomplished, observe any situations in which the operator's hand seems to get in the way of seeing the next visual display.

Also remember that, as a general rule, visual displays should be located at the top of the panel, and controls should be located at the bottom (the natural positional relationship between the eye and the hand).

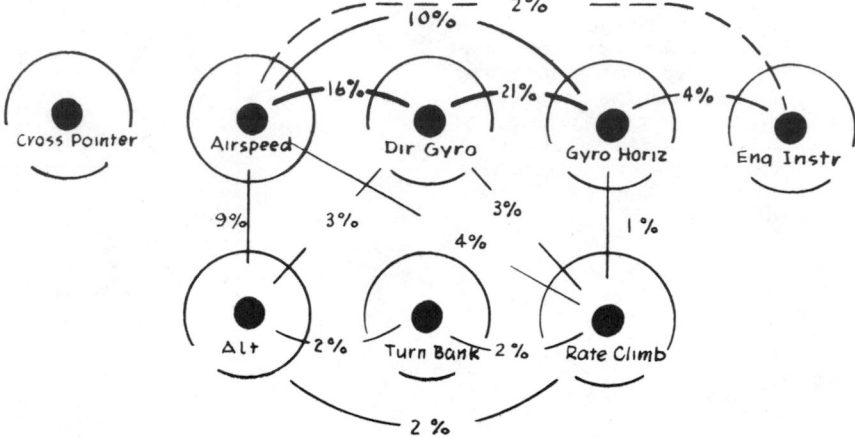

Use link percentages for displays.

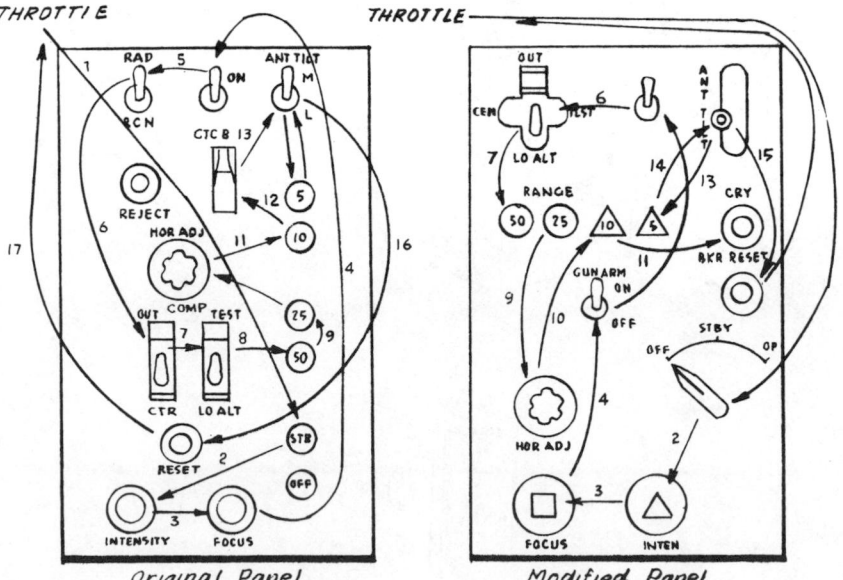

Sequence analysis demonstrates the more effective arrangement.

## VEHICLE OPERATOR WORK STATIONS

As the accompanying illustrations show, the vehicle operator's position relative to external and internal viewing requirements depends on the size and shape of the proposed vehicle:

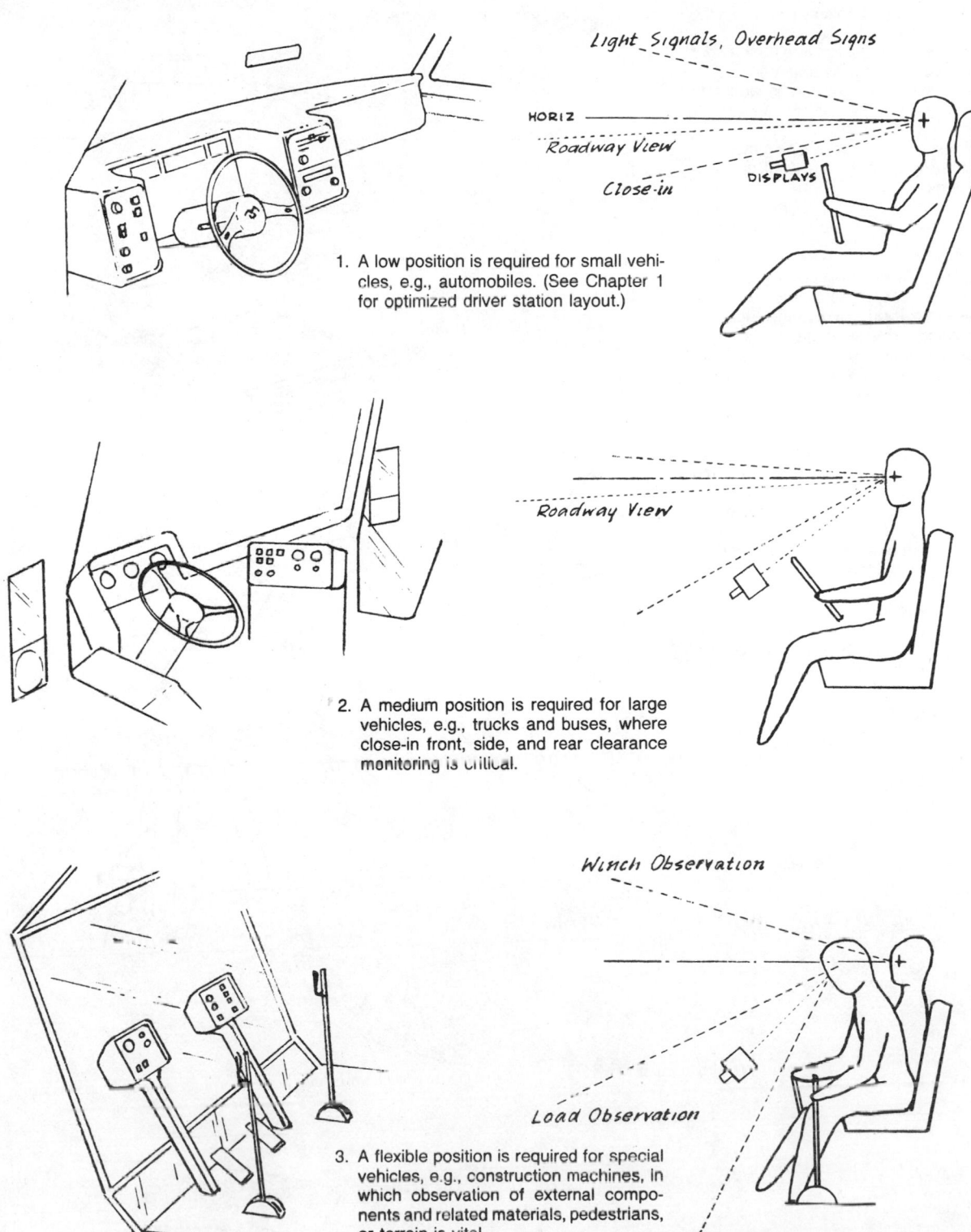

1. A low position is required for small vehicles, e.g., automobiles. (See Chapter 1 for optimized driver station layout.)

2. A medium position is required for large vehicles, e.g., trucks and buses, where close-in front, side, and rear clearance monitoring is critical.

3. A flexible position is required for special vehicles, e.g., construction machines, in which observation of external components and related materials, pedestrians, or terrain is vital.

## External Visibility Objectives

The proper reference eye position, plus the vehicle body and window configuration, should be combined to meet the following requirements:

1. To see the roadway surface to within 10 ft (3.1 m) of the front of the vehicle
2. To see an overhead signal light when the vehicle is stopped at a stop line
3. To see the foremost left and right fender excursions (mobility clearance)
4. To see a pedestrian who is about to cross in front of the vehicle (A-pillar occlusion angle)
5. To observe a rear-approaching motorcyclist in the left or right lane (B-pillar occlusion angle)
6. To observe the aftmost left and right fender excursions (backing mobility clearance)

Although rearview mirrors should also be provided (i.e., an interior mirror plus right and left rearview mirrors), everything possible should be done to arrange the driver's eye position and the vehicle's window and exterior body contours (especially the fender lines) so that direct external viewing requirements can be met.

These viewing needs should be met for all drivers whenever possible (at least 90 percent of the population), considering the expected variations in seat adjustment.

Placement of rearview mirrors should not incur viewing occlusions, nor should seat belts and hardware or headrests be designed or positioned so as to block any of the above viewing requirements.

It should be possible to adjust rearview mirrors when sitting with the eyes in the normal driving position; i.e., the driver should not have to lean to make such adjustments.

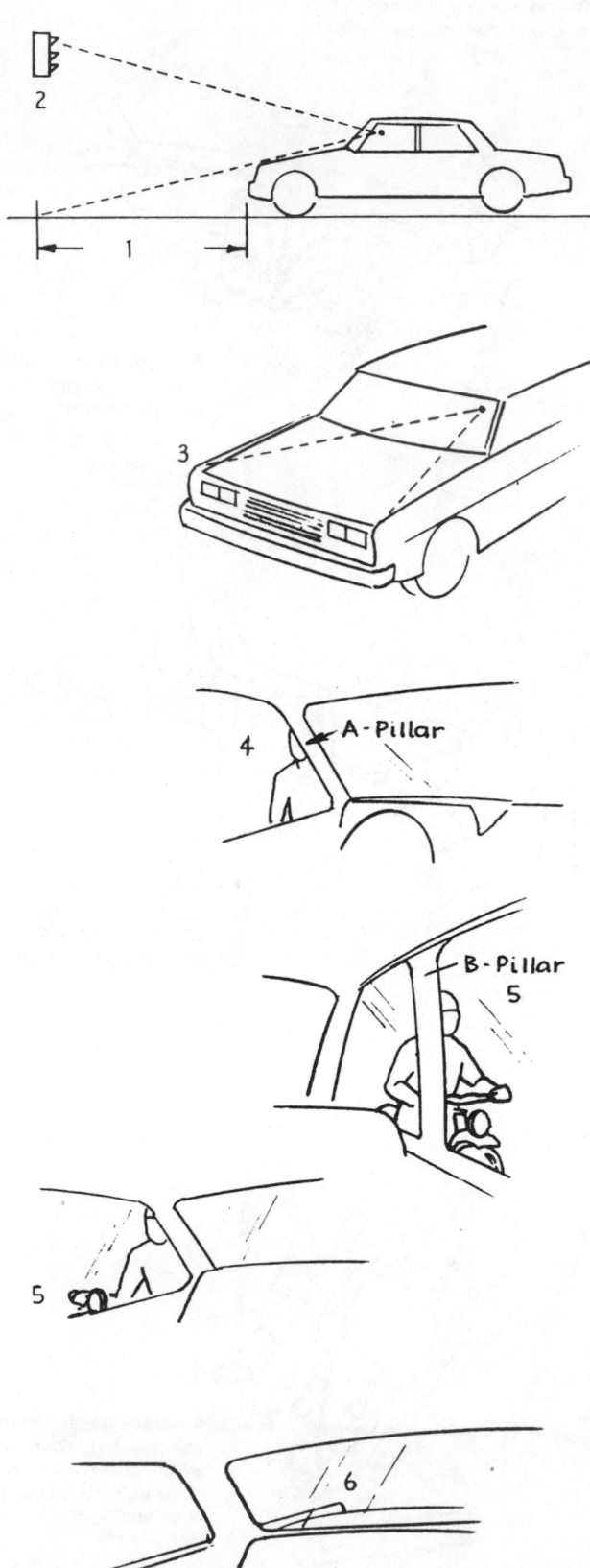

## Layout of Automobile Driving Controls and Displays

The automobile control and display location and arrangement should be standard among all automobiles. This is because a great many drivers transfer among a variety of automobile makes and models. If the driver station layout varies considerably, drivers can become confused and may select the wrong control, look at the wrong display, or become so distracted trying to find the right control or display that they fail to watch the road. The following suggestions concerning standardization are based on a number of studies that have attempted to establish the most common driver expectancies in terms of locating and operating controls and displays in automobiles.

As indicated in the accompanying sketch, primary visual displays often used in conjunction with observing the road ahead are positioned so that they can be seen above the steering wheel rim. Other controls and displays are placed on panels to the side of the wheel, on wheel spokes, or on levers mounted on the steering column so that the controls are within fingertip reach (within 6 in, 15 cm, of the wheel rim). All controls and displays should be properly labeled for both day and night observation.

*Note:* In the layout shown in the accompanying sketch, even the ashtray is located so that the driver is not required to move his or her body or eyes appreciably to find it while driving.

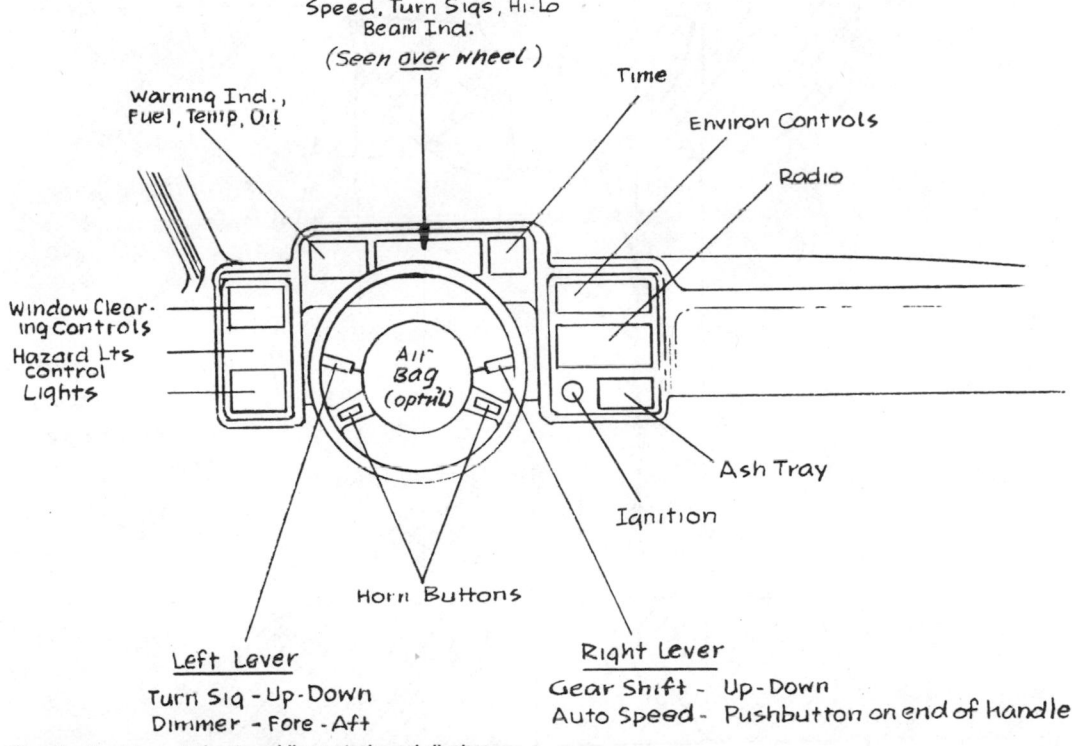

**Standardized layout of automobile controls and displays.**

## Application Guidelines for Location of Automobile Driver Controls

### 1. Basic Functional Categories

a. Vehicle control (steering, accelerating, braking, using the clutch, etc.)
b. Visibility control (internal and external lights, window clearing, and rear viewing)
c. Warning systems (turning signals and horn)
d. Environmental systems control (fresh air, airflow direction, airflow rate, heating, and cooling)
e. Driver information, communications, audio entertainment, etc.
f. Driver seat adjustment
g. Exit systems (door latches and locks)
h. Convenience systems (window lifts, tailgate actuator, etc.)

### 2. Use Priorities

a. Priority 1: Controls that require (or may require) immediate error-free response while the vehicle is under way. Such controls should be located so that the hand travels a minimum distance away from the normal "hand-on-the-wheel" reference position.
b. Priority 2: Controls that, although reach time is not as critical, should be placed where the driver can see them peripherally while watching the road (without excessive head or torso movement).[17]

[17]Foot controls obviously are exceptions to these first two priorities.

c. Priority 3: Controls that are used only when the vehicle is stopped. Although they do not have to be reached quickly or be visible while the driver watches the road, they should not be hidden from view, nor should they be out of reach from the driver's normal position.

Priority 1 controls are often referred to as "fingertip reach controls"; priority 2 controls should be within the 30° vision cone for driving. The sketches on page 406 further illustrate the various priority conditions. The tables that follow define location and configurational suggestions for the design of private automobiles and smaller trucks or recreational vehicles.

**Priority 1, Fingertip Control
Guidelines**

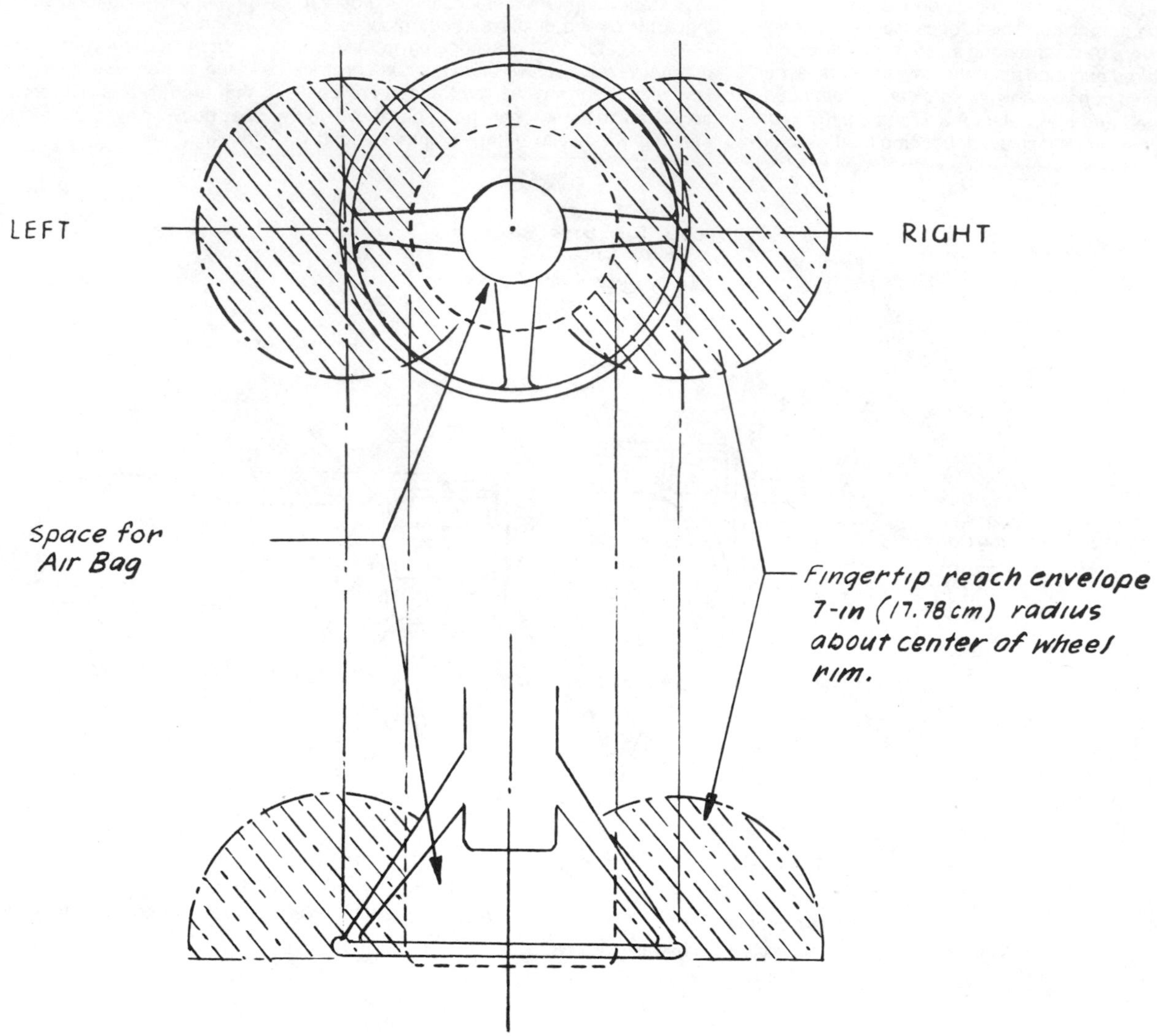

LEFT                    RIGHT

*Space for
Air Bag*

*Fingertip reach envelope
7-in (17.78 cm) radius
about center of wheel
rim.*

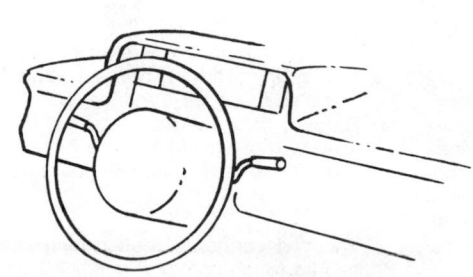

*Column-mounted levers
(called "stalks")*

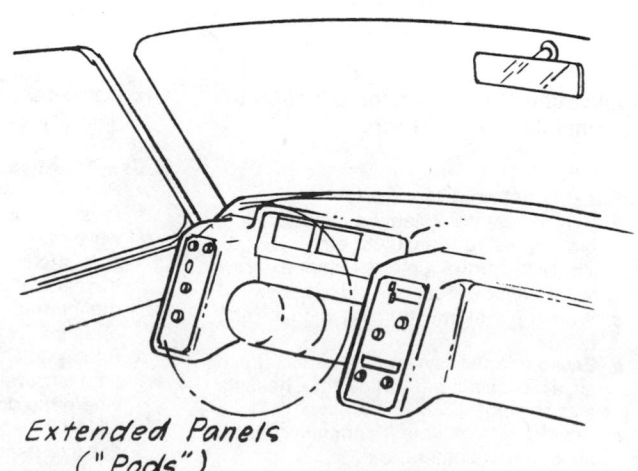

*Extended Panels
("Pods")*

**Alternative configurations.**

**Automobile Driver Hand Control Location Guidelines**

### Control Location Guidelines

| Function (Manual) | Left-Hand Operation | Right-Hand Operation | Either Hand Operation by | Left-Foot Operation | Right-Foot Operation | Operation by Either Foot | Instrument Panel | Stalk | Panel Pod | Steering Column | Center Console | Door Panel | Windowsill | Floor | Remarks |
|---|---|---|---|---|---|---|---|---|---|---|---|---|---|---|---|
| **Priority 1:** | | | | | | | | | | | | | | | |
| Defog, windshield | x | | | | | | x | | | | | | | | |
| Defog, rear window | x | | | | | | x | | | | | | | | |
| Deice, windshield | x | | | | | | x | | | | | | | | |
| Deice, rear window | x | | | | | | x | | | | | | | | |
| Gear select, automatic | | x | | | | | x | | | | | | | | |
| Headlight, high-low beam select | x | | | | | | x | | | | | | | | |
| Horn, audio | | x | | | | | | | | | | | | | Steering wheel spokes |
| Horn, optical | x | x | | | | | | x | | | | | | | Steering wheel spokes* |
| Ignition | | | x | | | | | | | x | x | | | | Combination ON/OFF, lock, start |
| Steering wheel | | | x | | | | | | | | | | | | |
| Turn signal select | x | | | | | | | x | | | | | | | |
| Starter | | x | | | | | | | | x | x | | | | Separate starter |
| Washer, windshield | x | | | | | | x | | | | | | | | |
| Washer, rear window | x | | | | | | x | | | | | | | | |
| Wiper, windshield | x | | | | | | x | | | | | | | | |
| Wiper, rear window | x | | | | | | x | | | | | | | | |
| Wiper, windshield, intermittent | x | | | | | | x | | | | | | | | |
| **Priority 2:** | | | | | | | | | | | | | | | |
| Automatic speed set (cruise) control | | | x | | | | | x | | | | | | | End push button |
| Clock set | | | x | | | | | | | | | | | | Adjacent to clock |
| Choke, hand | | | x | | | | x | | | | | | | | |
| Door lock, mechanical | | | x | | | | | | | | | x | | | |
| Door lock, master-powered | x | | | | | | | | | | | x | | | |
| Environmental control group: | | | x | | | | x | | | | | | | | |
|   Fresh air | | | | | | | | | | | | | | | |
|   Airflow direction | | | | | | | | | | | | | | | |
|   Airflow rate | | | | | | | | | | | | | | | |
|   Heating | | | | | | | | | | | | | | | |
|   Refrigeration | | | | | | | | | | | | | | | |
| Lights: | x | | | | | | | | | | | | | | |
|   Headlights, ON/OFF | | | | | | | | | | | | | | | |
|   Parking lights | | | | | | | | | | | | | | | |
|   Side markers, tail, license | | | | | | | | | | | | | | | Activate automatically when parking lights or headlights are on |
|   Panel illumination and dimmer | | | | | | | | | | | | | | | |
|   Fog light (front) | | | | | | | | | | | | | | | |
|   Fog light (rear) | | | | | | | | | | | | | | | |
|   Headlight automatic dimmer sensitivity adjust | | | | | | | | | | | | | | | |
|   Headlight cover | | | | | | | | | | | | | | | |
|   Headlight cleaner | | | | | | | | | | | | | | | |
|   Dome light | | | | | | | | | | | | | | | Switch at driver's position |
|   Dome light | | | x | | | | x | | | | | | | | Switch and/or ON fixture |
|   Map light | | | x | | | | x | | | | | | | | Switch and/or ON fixture |

*The three foot controls shown below also are priority 1 items. However, they obviously are not within "fingertip reach."

| Function (Manual) | Left-Hand Operation | Right-Hand Operation | Either Hand Operation by | Left-Foot Operation | Right-Foot Operation | Operation by Either Foot | Instrument Panel | Stalk | Panel Pod | Steering Column | Center Console | Door Panel | Windowsill | Floor | Remarks |
|---|---|---|---|---|---|---|---|---|---|---|---|---|---|---|---|
| Accelerator | | | | | x | | | | | | | | | x | |
| Brake, service | | | | | | x | | | | | | | | x | |
| Clutch | | | | x | | | | | | | | | | x | |

(continued)

**Control Location Guidelines (continued)**

| Function (Manual) | Left-Hand Operation | Right-Hand Operation | Either Hand Operation by | Left-Foot Operation | Right-Foot Operation | Operation by Either Foot | Instrument Panel | Stalk | Panel Pod | Steering Column | Center Console | Door Panel | Windowsill | Floor | Remarks |
|---|---|---|---|---|---|---|---|---|---|---|---|---|---|---|---|
| Cigar lighter | x | | | | | | x | | | | | | | | |
| Mirror, exterior, remote adjust: | | | | | | | | | | | | | | | |
| Left | x | | | | | | | | | x | | | | | |
| Right | | x | | | | | x | | | | | | | | |
| Odometer set | | | x | | | | | | | | | | | | Adjacent to odometer |
| Parking brake release (foot-operated parking brake) | x | | | | | | | | | | | | | | Below instrument panel |
| Parking brake release (between seats) | | x | | | | | | | | | | | | | |
| Radio and tape communications group | | x | | | | | x | | | | | | | | |
| ON/OFF, volume | | | | | | | | | | | | | | | |
| Channel select | | | | | | | | | | | | | | | |
| Frequency adjust | | | | | | | | | | | | | | | |
| Tone adjust | | | | | | | | | | | | | | | |
| Speaker select | | | | | | | | | | | | | | | |
| Antenna extension adjust | | | | | | | | | | | | | | | |
| Tape insert | | | | | | | | | | | | | | | |
| Track select | | | | | | | | | | | | | | | |
| Mike mount | | | | | | | | | | | | | | | |
| Mike ON/OFF | | | x | | | | | | | | | | | | On mike |
| Seat adjust, mechanical | x | | | | | | | | | | | | | | Left side of seat |
| Seat adjust, powered | x | | | | | | | | | | | | | | Left side of seat |
| Steering wheel position adjust | x | | | | | | | | | x | | | | | Only non-air-bag option |
| Throttle, hand | | | x | | | | x | | | | | | | | |
| Top roof hatch open and close, mechanical | | x | | | | | | | | | | | | | At hatch position |
| Top roof hatch open and close, electric | | x | | | | | x | | | | | | | | Or at hatch position |
| Window open and close, mechanical, side | x | | | | | | | | | | | x | | | |
| Window open and close, powered, side | x | | | | | | | | | | | x | | | Or armrest |
| Window open and close, powered, rear | x | | | | | | | | | | | x | | | With other electric window controls |
| *Priority 3:* | | | | | | | | | | | | | | | |
| Door handle | x | | | | | | | | | | | x | | | |
| Hood release | | x | | | | | | | | | | | | | Beneath instrument panel |
| Road hazard warning lights | x | | | | | | x | | | | | | | | |
| Trunk lock release | | x | | | | | | | | | | | | | Locate in lockable glove compartment |
| Brake, parking | x | | | | | | | | | | | | | x | |

### Priority 2, 30° Cone Reference Criteria

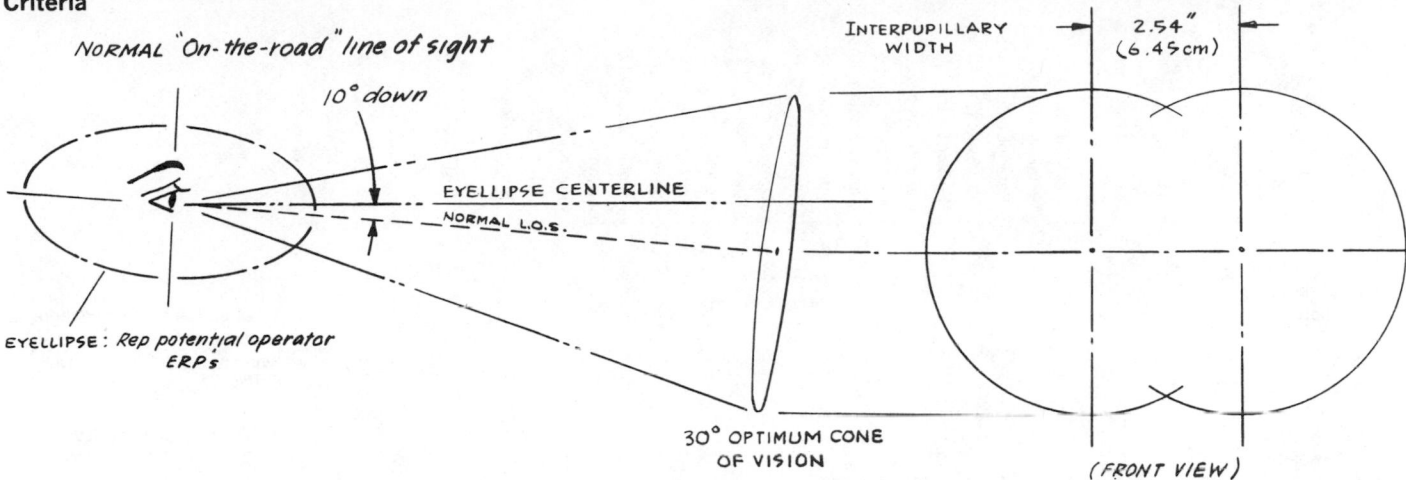

It is desirable to locate all priority 2 controls within or near the visual envelope shown in the accompanying illustration. Otherwise, the driver's eyes—and, more important, his or her attention—may wander from the roadway and traffic in front of the vehicle. A split-second lapse of attention may be sufficient for the driver to miss seeing an important situation leading to a potential collision.

### Guidelines for Selecting Control Type and Operation*

| Function | Push Button | Rotary Detent Knob | Rotary Control Knob | Tab Switch | Rocker Switch | Thumb Wheel | Pull Knob | Pull and Rotate Knob | Pull T-Handle | Slide Lever | Concentric Knobs | Level Action | Rotary Handle | End Push Button | Handle Slide Switch | Pull Knob (column) | Tab switch | Between-Seats Lever | Push ON | Push OFF | Pull ON | Pull OFF | Right Forward, Increase; Up, Increase | Remarks |
|---|---|---|---|---|---|---|---|---|---|---|---|---|---|---|---|---|---|---|---|---|---|---|---|---|
| Ignition: lock/on/start | x | | | | | | | | | | | | | | | | | | | | | | x | Mechanically detented, rotary, key-actuated switch. Restart should not require return of switch to OFF position. |
| Automatic gear shift | | | | | | | | | | | | | | | | | | x | | | | | | See FMVSS 102. |
| Hand choke or throttle | | | | | | | x | | | | | | | | | | | | | | x | | | |
| Parking brake release (foot-operated system) | | | | | | | | | x | | | | | | | | | | | | x | | | |
| Automatic speed set | | | | | | | | | | | | | x | | | | | | x | | | | | Press service brake to release. |
| Windshield wiper | | x | | | | | | | | | | | | | | | | | | | x | | | |
| Rear window wiper | | x | | | | | | | | | | | | | | | | | | | x | | | |
| Windshield washer | x | | | | | | | | | | | | | | | | | | x | x | | | | Combine in center of wiper knob. |
| Rear window washer | x | | | | | | | | | | | | | | | | | | x | x | | | | |
| Headlights ON/OFF parking lights ON/OFF† | x | | | | | | | | | | | | | | | | | | | | | | x | Actuation automatically activates side marker, tail, and license lights. |
| Panel illumination | | x | | | | | | | | | | | | | | | | | | | | | x | |
| Combination panel and dome illumination | x | x | | | | | | | | | | | | | | | | | | | | | x | Clockwise rotation increases panel illumination; further clockwise rotation past detent actuates dome light. |
| Fog lights | x | | | | | | | | | | | | | | | | | | x | x | | | | |
| Hazard warning | x | | | | | | | | | | | | | | | | | | x | x | | | | |
| Headlight, high and low beam | | | | | | | | | | | | x | | x | | | | | | | | | x | Push for high beam; push again for low beam. |

*Note:* Although the above guidelines are meant especially for the noncommercial driving situation and the private vehicle, it should be pointed out that similar control locations, types, and motions should also be considered for commercial vehicles. This is so because the commercial driver often gets out of a truck or bus and immediately operates a private vehicle. If the two vehicles are considerably different, the driver is very apt to make mistakes.

*Drivers will expect certain types of controls and control movement to be associated with certain functions. When a different type of control is used for a given function or if the control must be moved in a different way from what is expected, the driver may become confused, have to take time to look at the control, try it several times, etc. This could be a severe safety hazard if the driver does not see a situation that could lead to a collision while he or she is investigating the control.

†Although a combined pull-rotate headlight switch is acceptable, it is not recommended.

*(continued)*

## Guidelines for Selecting Control Type and Operation* (continued)

| Function | Panel-mounted: Push Button | Rotary Detent Knob | Rotary Control Knob | Tab Switch | Rocker Switch | Thumb Wheel | Pull Knob | Pull and Rotate Knob | Pull T-Handle | Slide Lever | Concentric Knobs | Stalks: Level Action | Rotary Handle | End Push Button | Handle Slide Switch | Column: Pull Knob | Tab switch | Between-Seats Lever | Motion: Push ON | Push OFF | Pull ON | Pull OFF | Right Forward, Increase; Up, Increase | Remarks |
|---|---|---|---|---|---|---|---|---|---|---|---|---|---|---|---|---|---|---|---|---|---|---|---|---|
| Automatic head light dimming sensitivity adjust | | | x | | | | | | | | | | | | | | | | | | | x | | |
| Turn signals | | | | | | | | | | | | x | | | | | | | | | | | | |
| Audio horn | | | | | | | | | | | | | | | | | | | | | | | x | Push button on each steering wheel spoke, facing driver. |
| Optical horn | | | | | | | | | | | | x | | | | | | | | | | | x | Pull toward driver to activate. |
| Interior environment control | | | | | | | | | | | | | | | | | | | | | | | | |
| Fresh air | | | | | | | | | | x | | | | | | | | | | | x | | | |
| Airflow direction | | | | | | | | | | x | | | | | | | | | | | | | | Move in direction of flow. |
| Airflow rate | | x | | | | | | | | x | | | | | | | | | | | x | x | | |
| Heat | | | x | | | | | | | | | | | | | | | | | | x | | | |
|  | | | | | | | | | | x | | | | | | | | | | | x | x | | |
| Refrigeration | | | x | | | | | | | | | | | | | | | | | | x | | | |
|  | | | | | | | | | | x | | | | | | | | | | | x | x | | |
| Emergency defog and/or defrost | x | | x | | | | | | | | | | | | | | | | x | x | | x | | |
| Radio and tape controls: | | | | | | | | | | | | | | | | | | | | | | | | |
| ON/OFF and volume adjust | | | x | | | | | | | | | | | | | | | | | | x | | | |
| Manual frequency adjust | | | x | | | | | | | | | | | | | | | | | | x | | | |
| Treble and bass adjust | | | x | | | | | | | | | | | | | | | | | | | | | Clockwise—more treble. |
| Speaker selection | | | x | | | | | | | | | | | | | | | | | | | | | Clockwise—Right rear speaker increase. |
|  | | | | | | | | | | | x | | | | | | | | | | | | | Ganged rotary controls. acceptable, i.e., volume and tone, tune speakers. |
| Antenna, raise and lower | | | | x | | | | | | | | | | | | | | | | | | x | | |
| Tape channel select | x | | x | | | | | | | | | | | | | | | | x | x | | x | | |
| Driver window crank | | | | | | | | | | | | | | | | | | | | | | | | Counterclockwise for raising the driver's window. |
| Electric window lift(s) | | | | x | | | | | | | | | | | | | | | | | | | | Forward or up for raising window. |
| Parking brake and hood release | | | | | | | | | x | | | | | | | | | | | | x | | | |

*Note:* Although the above guidelines are meant especially for the noncommercial driving situation and the private vehicle, it should be pointed out that similar control locations, types, and motions should also be considered for commercial vehicles. This is so because the commercial driver often gets out of a truck or bus and immediately operates a private vehicle. If the two vehicles are considerably different, the driver is very apt to make mistakes.

*Drivers will expect certain types of controls and control movement to be associated with certain functions. When a different type of control is used for a given function or if the control must be moved in a different way from what is expected, the driver may become confused, have to take time to look at the control, try it several times, etc. This could be a severe safety hazard if the driver does not see a situation that could lead to a collision while he or she is investigating the control.

†Although a combined pull-rotate headlight switch is acceptable, it is not recommended.

## Miscellaneous Driver Interface Considerations Relative to Anticipated Driver Problems

Because one can anticipate certain driver behaviors, it is necessary to consider means for designing so that drivers do not get themselves in trouble.

There are times when the driver will have to look at the shift control position indicator. If it is located on a console between the seats, the driver's eyes will most likely be "off the road" for a considerable time, even though he or she knows this may be hazardous. The control position display should be located up close to the driver's normal on-the-road viewing axis.

Drivers are forced to "look at" certain types of panel controls in order to see where a controller is positioned and/or to position the controller where they want it to be. In order for drivers to align a pointer with an index mark or label, they have to see both elements. We often forget to illuminate the control element for night use. Thus, either drivers have to "feel" for the right position (which usually requires several tries), or they take their eyes off the road and try to squint in the dark to see where the pointer is resting, a distraction that could be hazardous.

Too often little regard is given to the importance of the precise relationship of a pointer display to the scale mark or label to which it relates. A good case in point is the automatic gear shift display. Many of the configurations are extremely sloppy; i.e., the pointer often can be positioned between indexing characters. The situation is worse at night if the illumination system allows two letters to light up simultaneously. Typically, the pointer is never illuminated. The accompanying sketch shows a preferred scheme in which a "window" illuminates both the letter and the pointer, as long as the control is within the window area. Detents should provide tactile feedback, also.

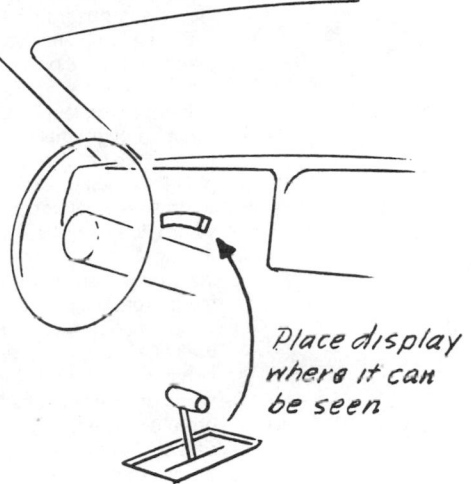

*Place display where it can be seen*

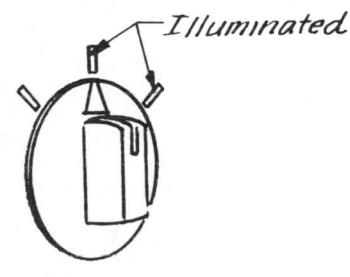

*Illuminated*

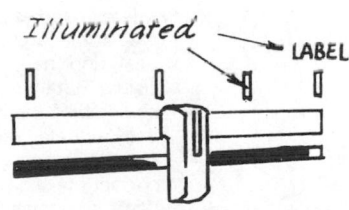

*Illuminated* → *LABEL*

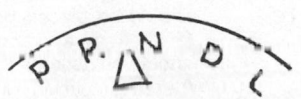

*quadrant illuminates*

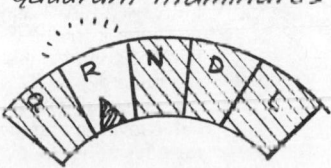

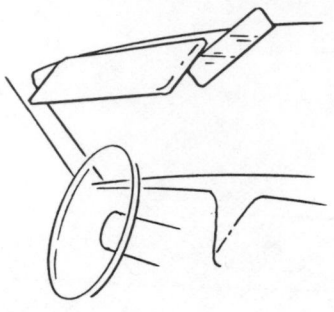

Poor planning often results in the accident illustrated in the accompanying sketch; i.e., moving the visor has displaced the mirror, which could present a considerable hazard. Although one can assume that the driver will be careful, experience indicates that many drivers are not careful. The time and attention required to readjust the mirror may be just enough to cause the driver not to notice an impending hazard on the roadway.

Similar problems occur relative to steering when levers are placed too close to the wheel rim.

It is poor layout when drivers are forced to reach (and get their eye reference out of position) to adjust a rearview mirror; i.e., drivers should be able to adjust the mirror from their normal eye reference position since that is where they will be viewing the mirror while driving. It is unforgivable, then, to provide an inside remote mirror control which also requires reaching.

Automatic window controls and door locks are usually reserved for the affluent buyer who considers them a luxury option. Use of automatic options such as these should be based on their effect on driver performance and safety. Drivers frequently find that a window on the opposite side of the car needs to be closed while they are underway. Instead of stopping, many drivers will try to reach across the car to operate the control. If the car is fairly narrow, this is probably all right. But if the car is wide, all kinds of problems can occur; e.g., the driver's attention is diverted too long from the road, or there may be false inputs to the steering wheel or accelerator pedal. Automatic controls should be used for a performance reason.

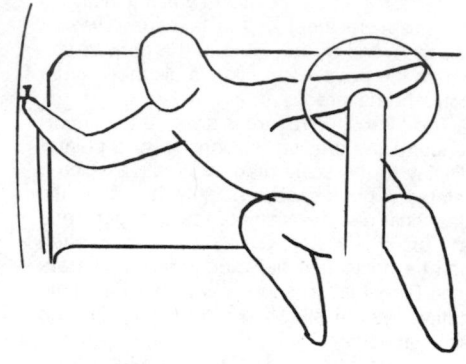

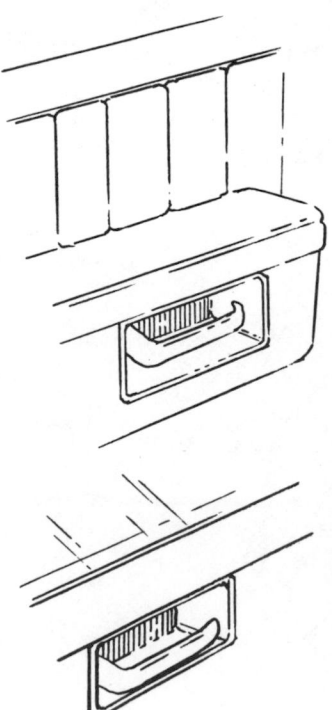

Door handles should be located where they are visible and quickly accessible. In years past, door handles were very accessible and visible. But they were also a hazard since one could strike them and incur injury or inadvertently actuate them and open a door while the car was under way. Overzealous attempts to correct these deficiencies often result in "hidden" and confusing door handle configurations, such as those illustrated in the accompanying sketch. Looking down at the armrest, the observer seldom can see the handle. Although the vehicle owner may get used to this, a guest may not be able to find the handle in time to get out in an emergency. Handles should be placed nearer the observer's eye level; from a practical point of view, they should be near the windowsill, as illustrated in the accompanying sketch.

Handles should be shaped so that the hand slips normally into position for operating the handle. Many handles are made so that users must twist their hands or wrists into unusual positions in order to get hold of them. The handles shown in the accompanying sketches are generally best.

Foot pedals should be at approximately the same level so that the driver does not have to raise his or her foot very high in order to keep from catching a toe on the adjacent pedal.

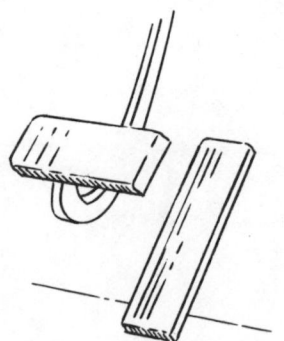

## Illumination and Glare

Vehicle drivers are confronted with a unique visibility problem in that they must operate their vehicles under many different ambient illumination conditions. Observe the following while planning the new vehicle and its driver control and display environment:

1. Avoid using any highly reflective material in the general forward viewing area; i.e., no glossy paints or chrome trim should be used on the instrument panel, window frames, mirrors, hood, or fenders.

2. Artificial lighting should be provided for all critical control labels so that they can be read during the day and at night.

3. Artificial lighting should be provided for all critical visual displays so that they can be read during the day and at night.[18]

4. Provide adjustable glare shields mounted on the header above the windshield.

5. Provide an adjustable headlight glare control for the interior rearview mirror.

6. Provide a slightly tinted (neutral color) windshield glazing (see the appropriate federal motor vehicle safety standard).

7. Provide a glare shield for the instrument control panel and/or an inset design to prevent external ambient light from obscuring instruments and label detail and/or to prevent these details from reflecting on the windshield at night. Use dark colors and a matte finish on any interior surface (windowsills, steering wheel or other control devices, upholstery, etc.) that might reflect on the windshield during the day or at night.

8. Slope the cover glass on instruments to minimize reflections of the sun and the driver on the glass.

## Making the Driver Station and Passenger Compartment Safe

Because of the potential hazards associated with moving vehicles (i.e., sudden changes of direction and deceleration), the interiors of all vehicles should be examined closely to make sure that they contain no special "contact hazards." The accompanying sketches illustrate some of the typical areas in automobiles requiring special consideration:

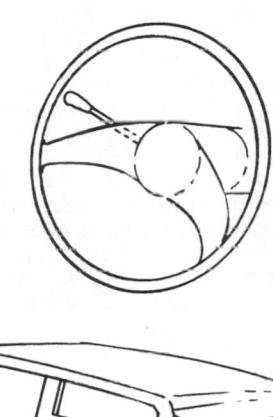

1. Steering wheel: The hub should be broad and set back from the rim, and levers should not protrude in the driver's direction. A collapsible steering column is desirable.

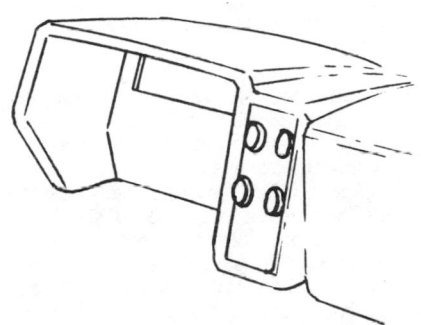

2. Frontal areas: Energy-absorbing padding should be provided on all dash surfaces, and all controls should be inset and collapsible and/or should present a large contact surface.

3. Glare shields: These should be made of frangible materials, and the hardware that holds them should not protrude into the expected head movement envelope.

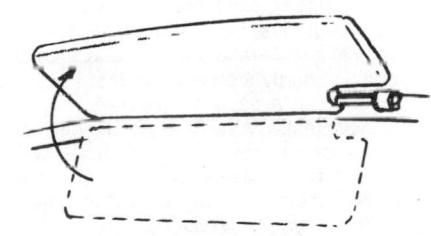

4. Interior rearview mirror: The mirror should have a padded frame and be mounted so that it is easily broken away in the event the driver's head is thrown into the mirror during a crash.

5. Door and window handles: All door-mounted hardware should be designed and positioned to minimize the possibility of an occupant's being impaled on the device during a crash. On the other hand, the door handle must not be hidden or made so obscure that the occupant cannot find it quickly and operate it easily in an emergency.

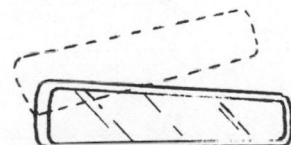

6. Seat security: All seats (and adjusting systems) must be fail-safe; i.e., the seat and/or adjusting system must not suddenly let go and allow the occupant to be thrown forward during a crash.

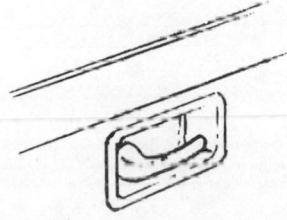

[18] Dimming capability should be provided for instrument control lighting so that the driver can adjust the levels for the best seeing conditions.

## RESTRAINT SYSTEMS

1. Some type of restraint against crash forces should be provided for all automobile passengers.

2. Restraint systems should be made and installed in such a manner that they do not interfere with the driver's tasks while he or she is driving.

3. If a seat-belt system is used, the seat belts must not incur injury during a crash, especially in terms of cutting across the occupant's neck or abdomen.

4. The seat-belt geometry should be comfortable; i.e., the belts should not rub on the occupant's neck or shoulder, cross or press uncomfortably on a female driver's breasts, or make it difficult for the driver to lean forward (i.e., against the retractor forces of a self-adjusting belt system).

5. The seat-belt webbing should retract into a position that makes it easy for the occupant to get in and out of the car, and it should not fall loosely so that it can be trapped by the closing door and/or fall behind the seat so that the next user cannot locate the latch plate that he or she expects to use in fastening the belt.[19]

6. Passive restraint systems are generally of two types: passive belt systems and air bags. Passive belt systems should be designed with the above suggestions in mind. Air-bag systems should be designed so that they effectively restrain the occupant's upper torso and head and also prevent the buttocks from sliding forward. This requires carefully planning the location of the stowed air bag in relation to the position and varying sizes of occupants. The air bag should inflate upon the sudden deceleration of the vehicle under typical frontal crash conditions (but must not inflate under normal vehicle decelerations). The bag should deflate shortly thereafter so that the occupant can escape if necessary.

7. Special restraint systems should be provided for children since the above systems are effective only for adult occupants (see current accepted child restraint system products).

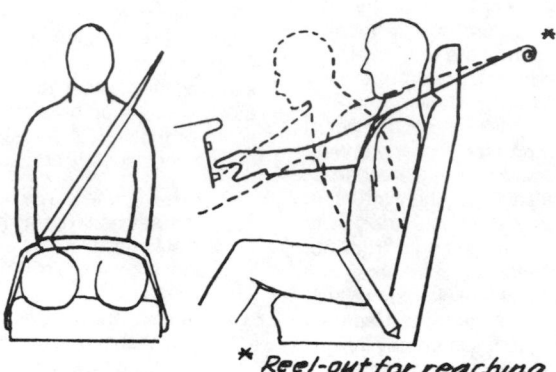

*\* Reel-out for reaching*

*Air bag inflated*

## Seat-Belt Design Guidelines

Although a lap belt alone provides some measure of protection for the vehicle passenger, a combination lap belt and shoulder harness is more effective because the shoulder harness minimizes the possibility that the occupant's head will be thrown against a structural barrier. The combination integrated shoulder harness and lap belt is considered an active three-point system; i.e., it is secured at three places, and the occupant manually dons the system. The important considerations are as follows:

1. The latch plate should be positioned so that it is easily accessible; i.e., it should not be hidden behind the seat or be inaccessible because of an interfering door-mounted armrest. It is preferred that the latch plate and webbing be anchored to a vehicle-sensitive emergency-locking retractor so that the webbing can be pulled smoothly and easily across the lap.

2. The buckle should extend on a semiflexible stem from the opposite side of the occupant so that it, too, is easily accessible; i.e., it should not be buried between seat cushions, where it is hard to get hold of.

3. The shoulder harness and lap-belt webbing elements should be sewn together so that the two elements are both donned with the simple movement of only one element, i.e., the latch plate. The lower juncture between shoulder and lap webbing should occur approximately on top of the occupant's pelvic arch. The upper departure of the shoulder webbing should be slightly above and behind the shoulder of the tallest occupant to ensure that the webbing will cross the shoulder midway between the neck and shoulder tip. If these upper and lower departure points are properly maintained, the webbing will cross the occupant's chest at approximately the sternum point, which is the most comfortable wearing position.

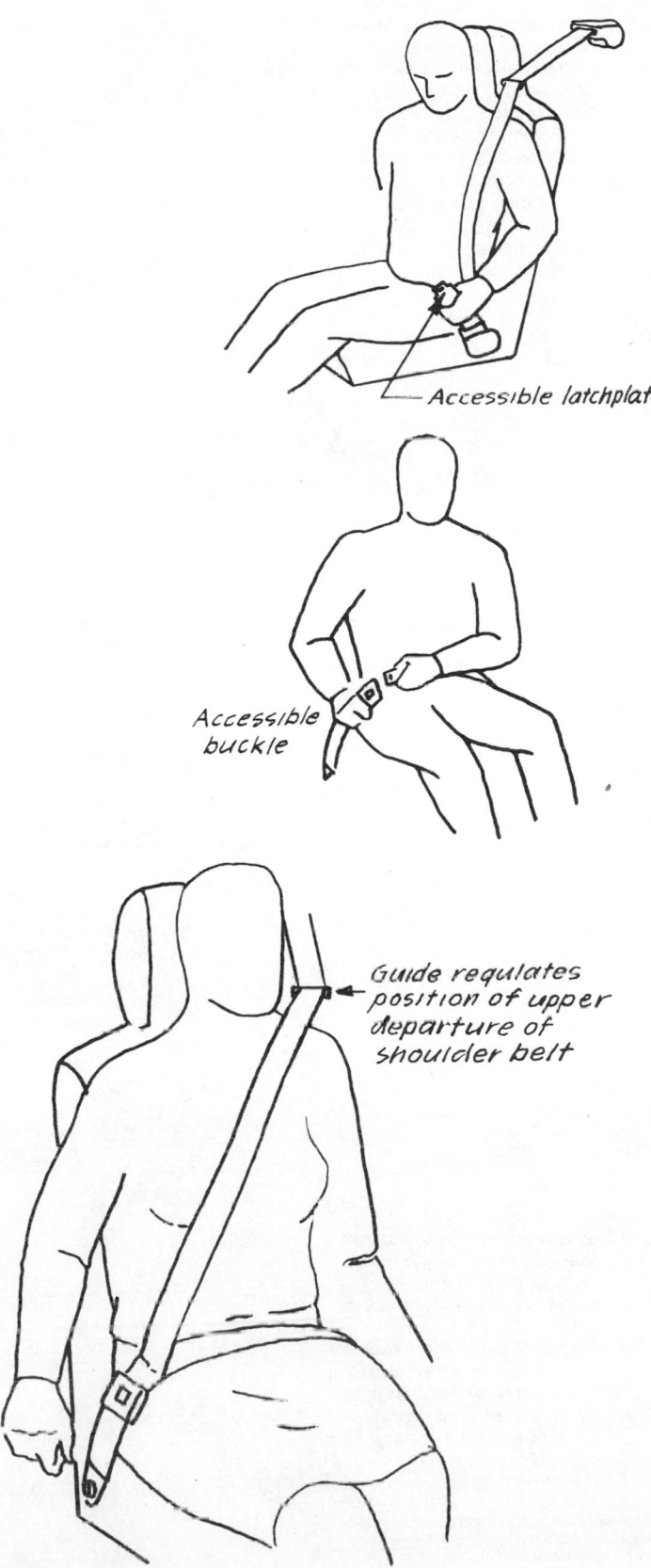

Accessible latchplate

Accessible buckle

Guide regulates position of upper departure of shoulder belt

## Seat-Belt Geometry for Maximum Comfort

People are more apt to wear seat belts if they are comfortable. To obtain this comfort, use the guidelines provided in the accompanying sketch. The key issues are positioning the departure points for the upper and lower ends of the torso belt so that the belt crosses the wearer's chest approximately at the sternum (midway between the breasts), does not rub against the neck, and does not "pull down" on the shoulder. The dimensions shown provide this desired geometry for at least 90 percent of the driver population as long as the departure points ($\pm$ 1.0 in, 2.5 cm, at the lower point; $\pm$ ½ in, 1.3 cm, at the upper point) are maintained.

The lap-belt approach angle shown is necessary to cause the belt to stay on the wearer's lap and not creep up on his or her abdomen, which could result in injury during a crash.

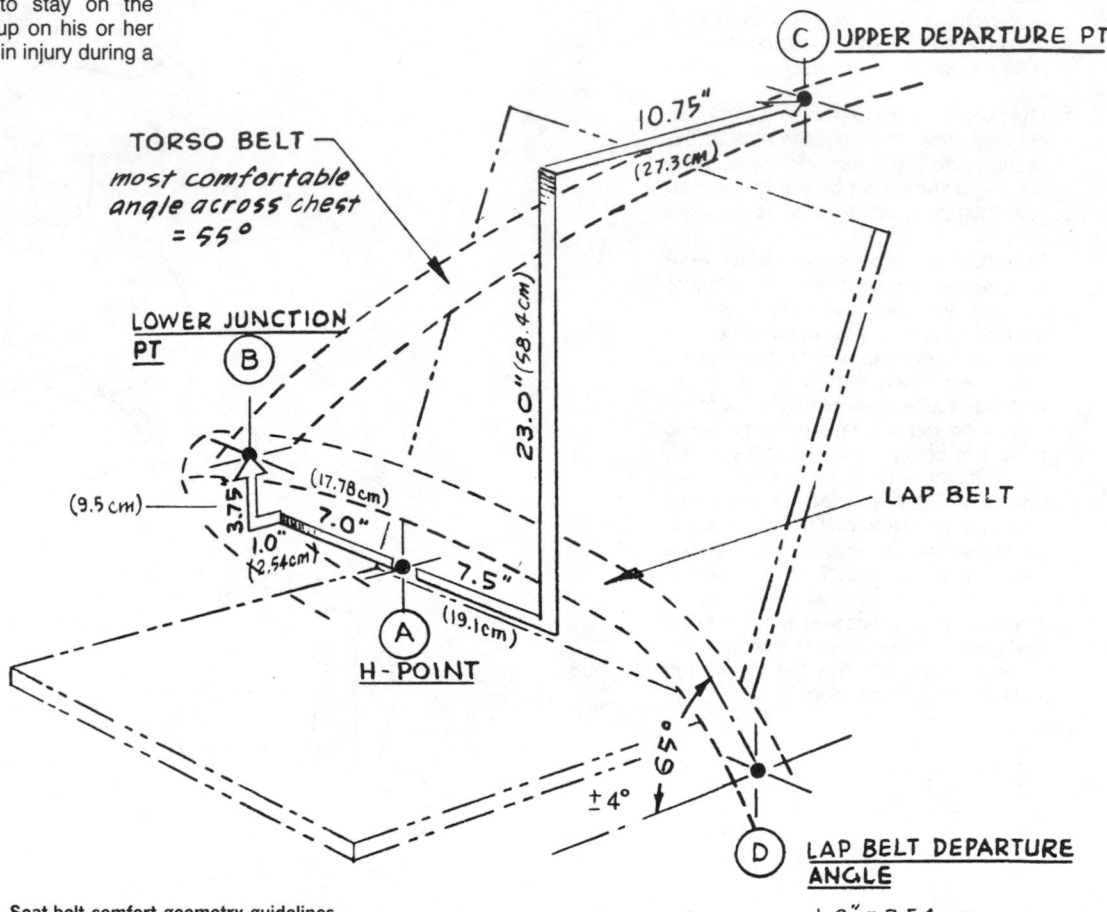

Seat-belt comfort geometry guidelines.

*Note:* Points B and C are referenced to the standard automotive industry reference, e.g., the H-point (A). For a definition of the H-point, see SAE J826b of the Society of Automotive Engineers.[20]

It should be noted that the recommended geometry must remain constant while seat position is adjusted, preferably by attaching the webbing retractor belt guide to the seat itself and/or by adding seat-mounted guides to maintain the lap-belt anchor points.

[20]*Devices for Use in Defining and Measuring Vehicle Seating Accommodation*, SAE J826b, part 2, Society of Automotive Engineers, Inc., Warrendale, Pa.

### Other Active Seat-Belt Considerations

In order for an active seat belt to be donned easily, care must be taken in the design of the various hardware components, as shown in the accompanying illustrations:

1. The latch-plate element must be designed so that it is possible to grasp it firmly and yet leave the end of the plate exposed sufficiently to insert it into the buckle.
2. Although the buckle end of the belt does not have to be pulled out, it is necessary to hold it in a position that makes latch-plate insertion possible. As in the case of the latch plate, there must be sufficient gripping area behind the buckle so that the buckle will be exposed.

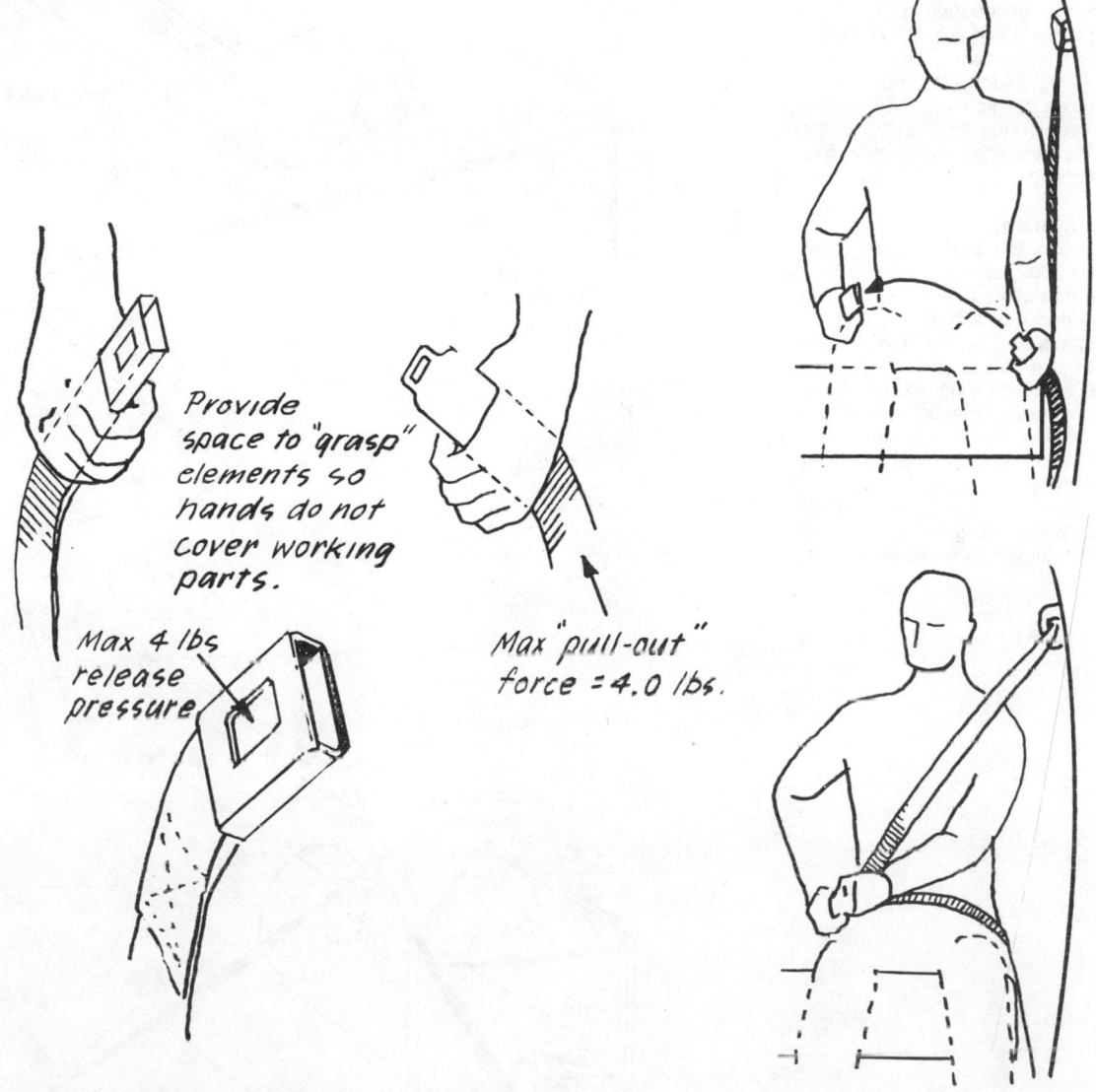

Provide space to "grasp" elements so hands do not cover working parts.

Max 4 lbs release pressure

Max "pull-out" force = 4.0 lbs.

3. The torso webbing pressure on the occupant's shoulder should not exceed 4 lb (28 kN). If it becomes necessary to increase this amount to cause proper belt retraction, some type of tension-relieving system should be added.
4. The latch plate should always return to a common, accessible position, either beside the seat or near the shoulder.

## Passive Seat Belts

The passive seat-belt concept provides automatic donning and doffing; i.e., when the door is opened (and/or when some motorized device is provided to pull the webbing away from the occupant), the seat belt is pulled away from the occupant's body so that he or she can get out of the car without manually manipulating the buckles or webbing. When the person gets back into the car and closes the door, the webbing once again envelopes his or her body and provides restraint.

### 1. Two-Point System

Typically this consists of a single shoulder belt plus a knee bolster on the lower edge of the dash. This bolster prevents the occupant from sliding forward during a crash, and the torso webbing keeps the occupant's upper torso from slamming forward into the dashboard. A typical configuration is shown in the accompanying sketch. Although the retractor can be placed in the inside position, it is preferred in the outside position so that the webbing is let out as the driver or passenger leans forward to reach the controls.

### 2. Three-Point System

This includes both a lap and an upper torso belt. The system operates much like the two-point system in that it is actuated by opening and closing the door (or with motorized pullers). Added ingress and egress may be obtained by means of a special hook located as shown in the accompanying sketch. This hook, however, must automatically dump the webbing before the vehicle can be driven off. Otherwise, the hook could be used to defeat the purpose of the system.

*Note:* Both systems should be laid out as nearly as possible according to the seat-belt comfort geometry guidelines noted previously.

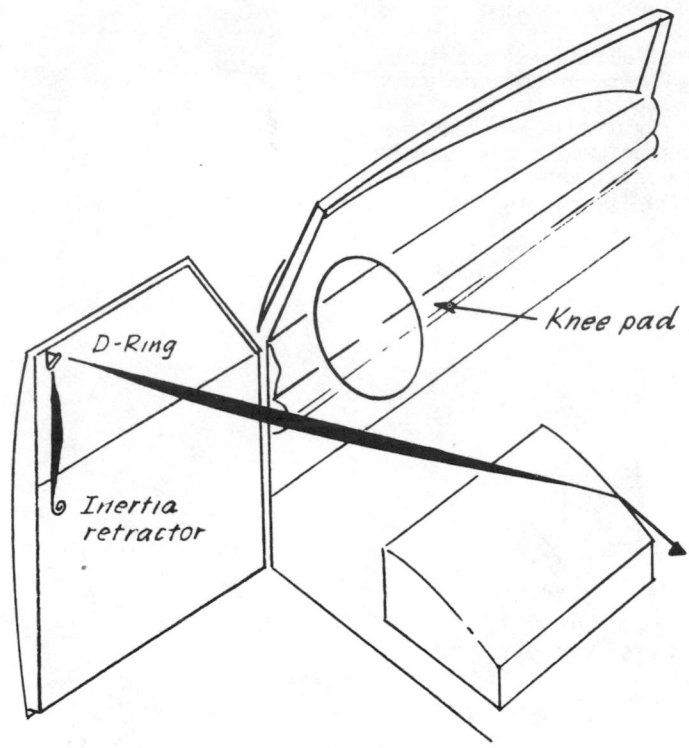

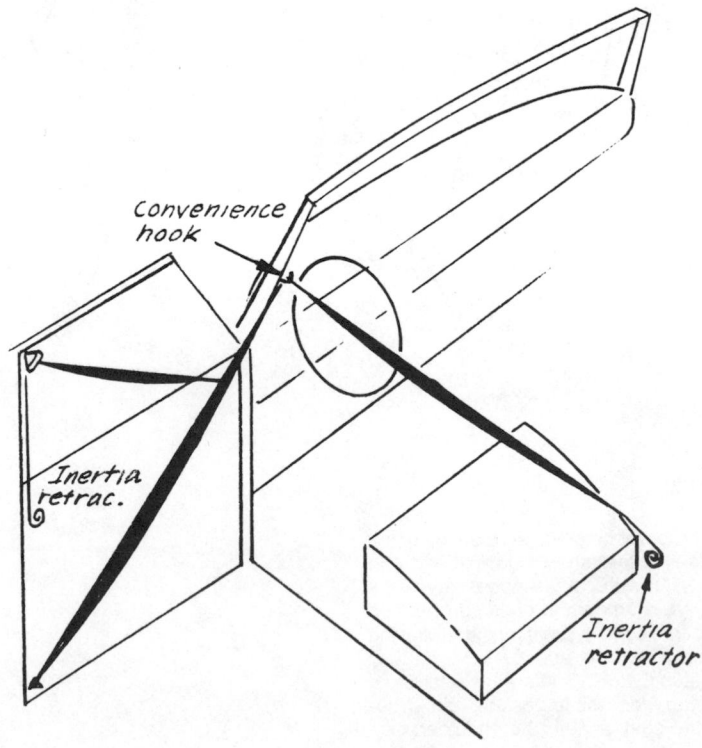

### Other Passive Seat-Belt System Considerations

**MANUALLY OPERATED CONVENIENCE HOOK** Care should be taken to place the hook device where it is within easy reach but where the occupant's head will not strike it in the event of an accident. The hook should not have sharp edges or corners, but it should hold the belt securely (so that it does not slip off easily) until the ignition is actuated, which "dumps" the webbing so that it will fall around the occupant.

**INTERFERENCE OF WEBBING WITH DOOR MANIPULATION** When special motorized track systems are installed, they should not cause the in-transit webbing to "capture" the occupant's arms as he or she is opening or closing the door. For instance, avoid placing the track above a door handle because that is where the person's hand will be when the belt begins to articulate.

**EMERGENCY RELEASE** It is vital to provide a means for belt release in an emergency. If a release buckle system is used, place it in the inside position because that is where the occupant will expect to find it. Also use a design that makes the release button visible from the occupant's viewpoint (see the accompanying illustration). Making the button red also helps identify it as an emergency item.

**MOTORIZED RETRACTION RATES** If the belt retracts too slowly, people become impatient; if it retracts too fast, they may be startled. A good compromise is to have the cycle fall between 1.5 and 3 s.

**BENCH-SEAT APPLICATION** Passive belt systems can be used with bench seats as long as it is recognized that there is no way to accommodate a middle-position occupant; i.e., it is practically impossible for a person to get past a system that is positioned on the outside in order to get to the middle position of a bench seat.

It is equally important to recognize that it is not practical to utilize a two-point passive system in a bench-seat configuration—at least on the front passenger's side of the vehicle—because if a tall driver moves the bench all the way aft for comfort, a small front passenger will probably be too far away from the protective knee pad for it to be effective. Therefore, if a two-point passive system is used with a bench seat on the driver's side of the car, a three-point passive system should be provided for the front passenger.

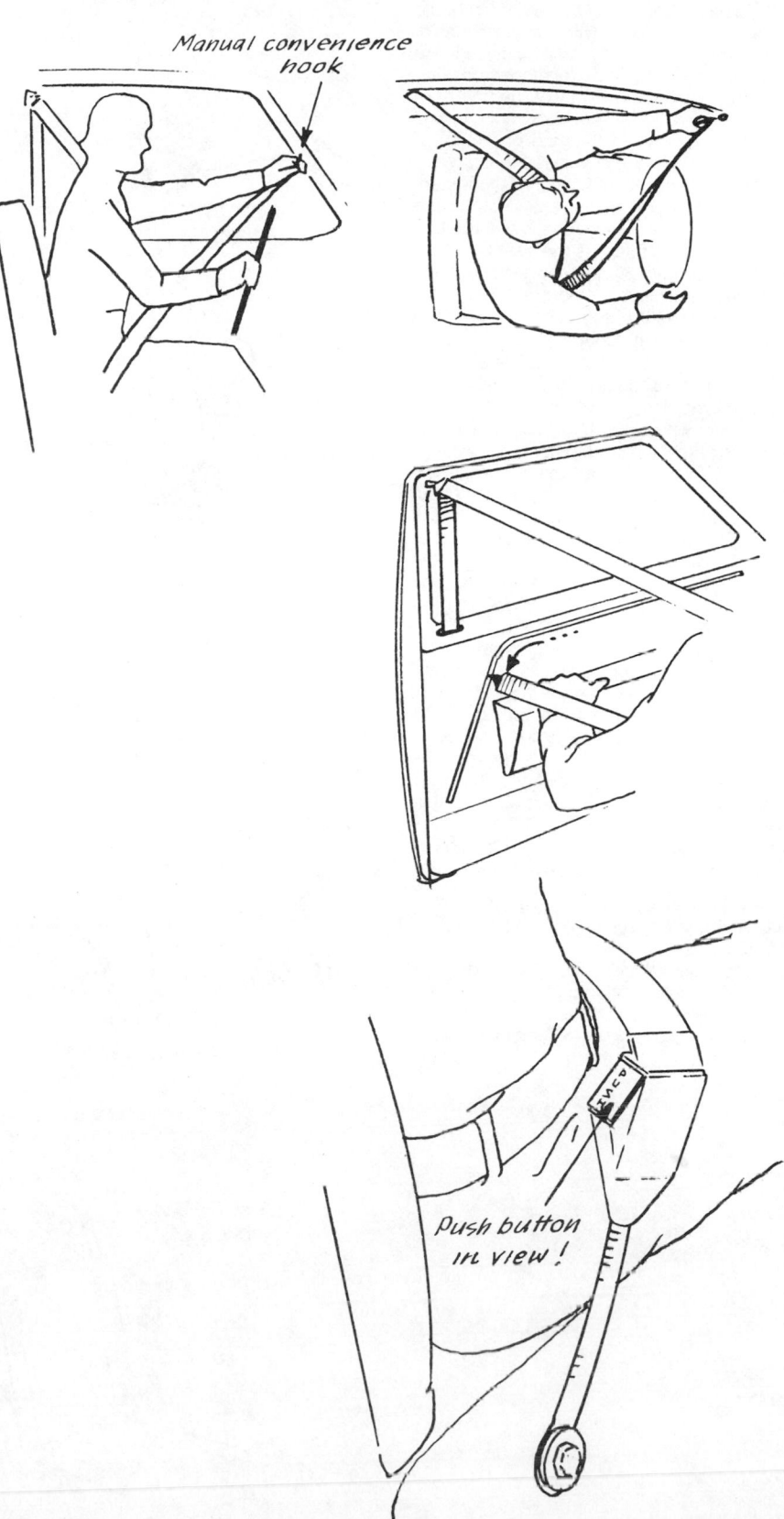

Manual convenience hook

Push button in view!

### Special Considerations for Seat-Belt System Design and Installation Relative to Occupant Safety

Ideally, design of the seat-belt system should take place concurrently with development of the seat and vehicle body designs since there are significant physical relationships that should be maintained. First, the geometry of the seat belt and torso harness should remain constant regardless of the size or weight of the occupant. If very soft adjustable seats are used, it is difficult to maintain this constant geometric configuration; i.e., the angle of the seat belt will become more shallow (if the belt is anchored to the floor) as the seat is moved forward to accommodate a shorter occupant. When the seat cushion is soft and this angle becomes too shallow, the occupant tends to depress the cushion during a crash deceleration, his or her pelvis rotates, and the seat belt rides up on the abdomen, often resulting in severe internal injuries. If the shoulder harness pulls down on the shoulder because of the positioning of the torso belt anchoring, this pelvic rotation is increased even more.

Although some forward motion of the occupant's upper torso is acceptable, the buttocks should remain as securely against the seat back as possible. The two-point system, using a knee bolster, as shown in the accompanying sketch, is one method for minimizing the forward motion of the buttocks.

It is obvious that the proper geometric conditions could be maintained more effectively if seat cushions were relatively firm and the seat was not adjustable so that the belts could be anchored to the vehicle structure. This could be accomplished for all seats except the driver's seat, which drivers of various sizes must be able to adjust to fit their control interfaces.

Whatever the system, the following loads on the occupant should not be exceeded:

| | |
|---|---|
| Forward head deceleration | 80 $g$/3 ms |
| Rotational head deceleration | 7000 to 10,000 rad/s² |
| Horizontal chest deceleration | 60 $g$/3 ms |
| Horizontal pelvis deceleration | 50 to 80 $g$ |

It may be necessary to consider special load-limiting devices in order to keep loads within these limits.

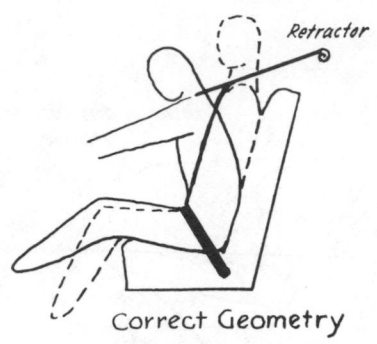

Correct Geometry

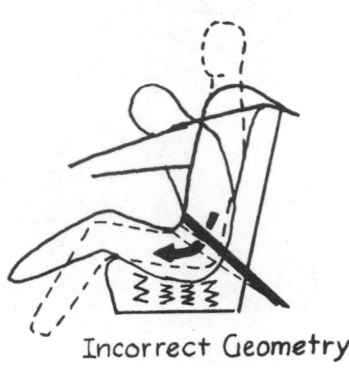

Incorrect Geometry

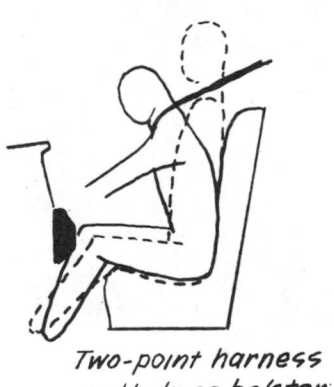

Two-point harness with knee bolster

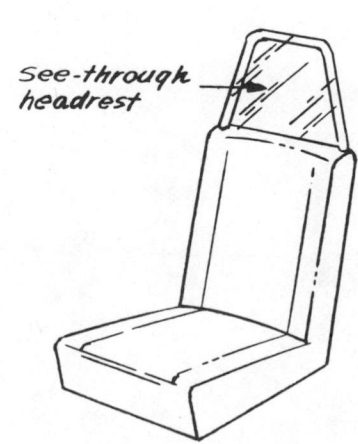

See-through headrest

### Whiplash Prevention

When a vehicle is impacted from the rear, the occupants may be subject to severe whiplash injuries because their heads have no support. Headrests are one solution. However, an adjustable headrest does not ensure injury prevention because many people fail to adjust a headrest properly. A permanent headrest is preferred. It should be designed to provide minimum clearance between the headrest and the occupant's head during normal operation, but not so much that the head and neck can rotate too far (a maximum clearance of about 4 in, 10.2 cm, is suggested).

Since a permanent headrest may create a visual obstruction for the driver, it is suggested that the headrest be made from a transparent material so that the driver can see through it. A successful design is shown in the accompanying sketch.

### Webbing Retractors

The "inertial reel" type of retractor is recommended, as opposed to others, such as the common belt-sensitive systems. The latter systems tend to "lock" as the user attempts to buckle up.

Any retractor should be of the emergency locking type, however; i.e., the system should be free under gradual movements of the wearer, but under specified decelerative loads it should immediately lock out the belts from further webbing extension.

*Note:* Although the air-bag restraint is considered by many to be the most desirable because it creates no interferences of any kind and does not require any operator action, such a system does not usually provide total protection; i.e., it protects the occupant only when the crash deceleration is head on. It is suggested that when air bags are provided, an additional seat belt also be available to the occupant.

*A seat restraint system, either belt or air bag, will not be effective if a reclining seatback option is provided.*

**Miscellaneous Dynamic Task and Design Considerations Associated with Vehicular Systems**

### 1. Steering

A wide variety of vehicles, of course, require steering wheels. Whenever possible, avoid using wheels with very large diameters. As indicated in the accompanying sketches, the large, flat wheel requires frequent awkward (and tiring) body motions. A smaller wheel designed to minimize operator reach problems could be used in vehicles that have power steering. Unfortunately, the traditional large wheel is often used because it is believed necessary in order for the driver to apply enough force.

### 2. Lever Operations

Avoid the tendency to place levers so that they are awkward for the operator to reach. Typically, awkward location is dictated by the mechanical linkage. When this is the case, the linkage should be rearranged.

### 3. Severe Motion

When it is anticipated that the vehicle system will be subject to severe motion conditions, provide a means for the operator to "hang on."

### 4. Visibility

Visibility for backing or observing other machines, workers, or objects should be planned.

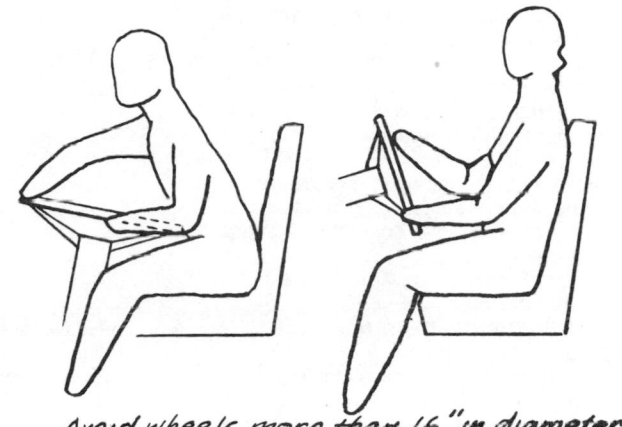

Avoid wheels more than 16" in diameter

Avoid awkward reach requirements

Provide handrail

**PRINCIPAL HUMAN FACTORS PROBLEM AREAS IN AGRICULTURAL AND CONSTRUCTION MACHINE DESIGN**

1. Ingress and egress (steps, handrails, doors, and door opening devices)
2. Ingress and egress to operator position (clearance)
3. Arrangement of seat controls and displays (visibility and reach)
4. Exterior viewing (critical operating areas and components)
5. Lighting (interior and exterior for safe operation and identification)
6. Operator habitability (control of temperature, ventilation, vibration, and noise)
7. Maintenance (inspection, servicing, troubleshooting, and removal and replacement of parts)
8. Safety (impact, roll-over, electric shock, burns, toxic materials and substances, falls, and accidental contact with sharp elements)

**PRIORITIES DURING DESIGN CONCEPTUALIZATION**

1. Locate the operator position for maximum task surveillance.
2. Arrange task controls and displays for maximum monitoring and manipulative effectiveness and minimum strain and fatigue.
3. Provide adequate control over potentially degrading environmental factors to ensure operator performance efficiency.
4. Provide necessary safety features to preclude injury to other persons who may be working in the vicinity.
5. Provide adequate features to facilitate the servicing and maintenance functions required to keep the machine operating effectively.

1. Position the operator for the best view for steering and monitoring key machine operations.
2. Provide good seating to maintain posture and security during a rough ride.
3. Arrange controls and displays for easy monitoring and reach.
4. Provide appropriate exterior lights for both steering and equipment monitoring at night.
5. Provide a safety cab to protect the operator from noise, dust, and weather.
6. Provide good ingress and egress, including handrails and platforms.
7. Eliminate all sharp corners, protrusions, and edges.
8. Provide adequate access for inspection, servicing, and repairs.

**Operator Controls for Industrial Equipment[21]**

The brake pedal should be actuated by the right foot. When traction control is not foot actuated, brakes may be operated by either foot.

When separate brake pedals are provided for independent right-hand and left-hand braking, it should be possible to obtain combined and/or equalized control.

When a foot pedal traction control is used it should be positioned for left-foot operation.

If the control is hand operated, the control should move aft for disengagement.

When a combination clutch and brake control is used, it should be positioned for left-foot operation; forward and/or downward for clutch disengagement and brake engagement.

When a hand-operated clutch control is used, its movement should be rearward and/or downward for disengagement.

An engine speed control should move generally forward or upward to increase speed.

When the speed control is foot operated, it should move forward and/or downward for speed increase.

When a hand-operated forward-reverse lever is used, forward motion should coincide with forward vehicle movement, and vice-versa.

Lift control levers should be located for right-hand operation.

If a foot pedal is used for lift control, downward motion of the toe should lower the lift and vice versa.

If a single lever (other than tiller type) is used for steering, right control motion should turn the vehicle to the right, and vice-versa.

[21]Reprinted from *Operator Controls on Industrial Equipment—SAE Recommended Practice*, SAE J297, Society of Automotive Engineers, Inc., Warrendale, Pa.

When a tiller control is used, a right lateral movement should result in a left vehicle movement.

When two levers are provided for steering, the right lever should control the right-hand element, the left lever the left element. Forward lever movements should cause that particular element being controlled to move forward and increase in speed, and vice versa.

*Note:* It should be noted that the several direction-of-motion relationships listed above and in other machine guidelines prepared by the SAE are drawn, for the most part, from traditional practice, although some parallel results of human engineering research (see the section on control design in Chapter 3). It is recommended that, when a new control function is required, the specific type of control and its movement patterns be based on the human engineering recommendations presented in Chapter 3. For guidelines for motor grader controls, see SAE J1071.

## Construction and Industrial Equipment Instrument Guidelines[22]

The accompany illustrations present general guidelines for the design of instrument faces and their arrangement in construction and industrial equipment, as recommended by the Society of Automotive Engineers.

In general, instrument faces should be designed according to good human engineering practices, as discussed in Chapter 3 in the section on visual displays. Since the accompanying illustration represents only a few of the possible instrument requirements for many types of equipment and machines, it behooves the designer to establish sound functional requirements before attempting to select a particular visual display. Then the designer should create a display that provides the necessary information in the simplest, clearest, and most legible format possible. The panel arrangement should be based on an iterative layout study in which the several criteria are exercised for the best trade-off between the following:

Sequential operation
Frequency of use
Functional grouping
Emergency accessibility

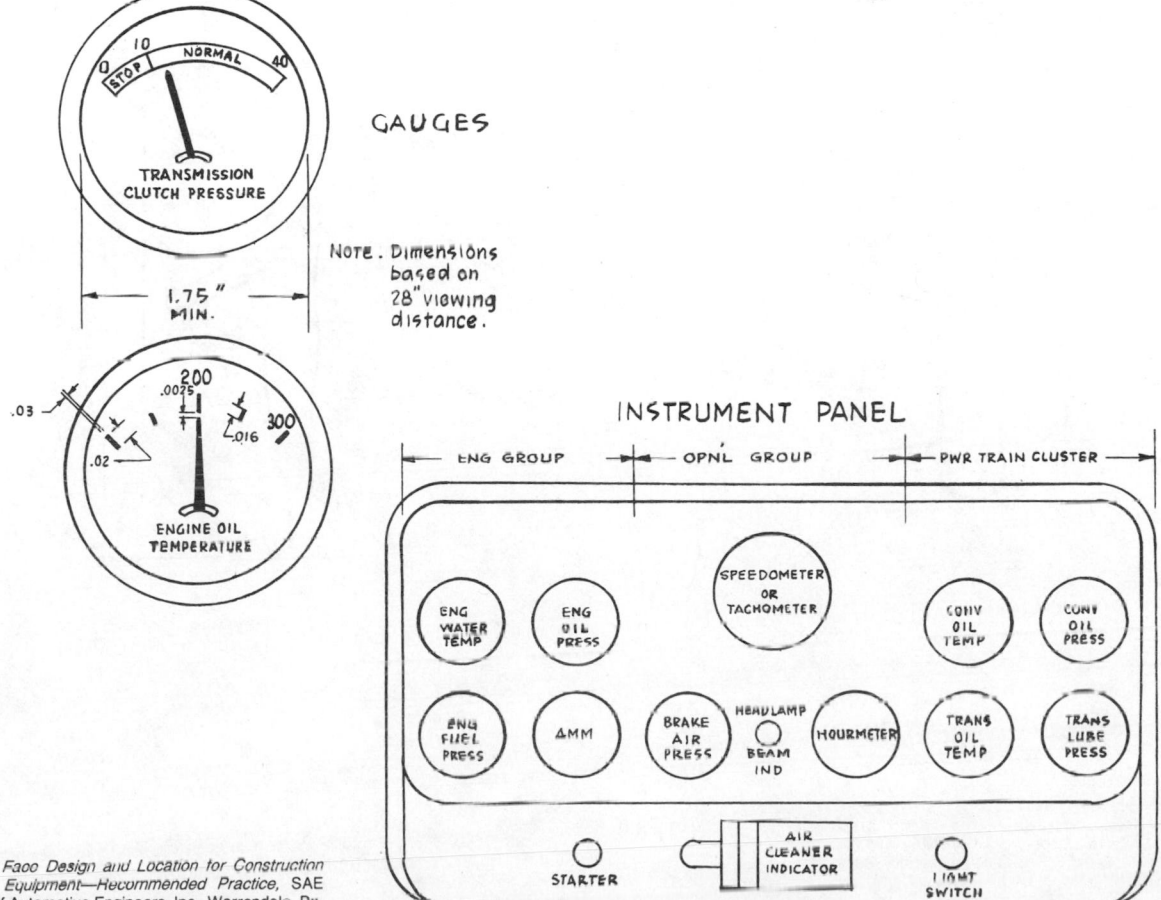

[22]*Instrument Face Design and Location for Construction and Industrial Equipment—Recommended Practice,* SAE J209, Society of Automotive Engineers, Inc., Warrendale, Pa.

## Industrial and Agricultural Machine Enclosure Guidelines[23]

The accompanying operator cab clearance guidelines were generated by the automotive and machinery industry to guide designers. The space envelope occupied by an operator of such equipment must be sufficiently generous to minimize the possibility that the operator will strike the structure or other elements of the cab, especially when the vehicle moves over rough terrain. These guidelines should be used only as the first step, however. A full-scale mockup should be created to provide a more specific evaluation of each specific operator work station, including all the controls and other elements required in the cab enclosure, each of which can become a contact hazard.

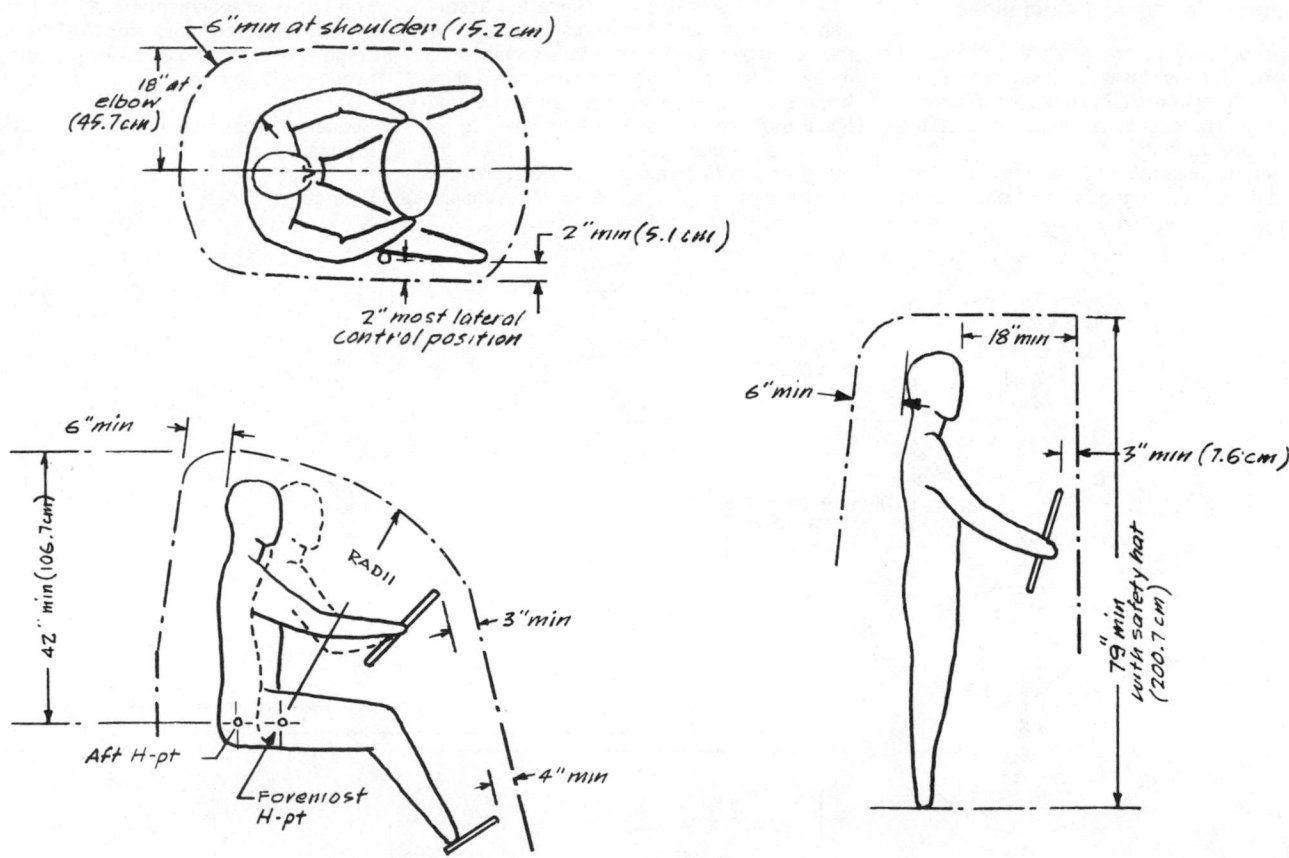

[23]*Operator Enclosures (Cabs)—Human Factor Design Considerations*, SAE J154, Society of Automotive Engineers, Inc., Warrendale, Pa. See also SAE J833 for operator dimensional reference.

## Location of Operator Mounting Steps and Handrails or Handholds

In planning the location of steps and handholds to help the operator mount or dismount, consider the interaction between the operator's hands and feet at each discrete pause position and also the intervening maneuvers, which are dynamic. The operator requires not only a sure footing that does not exceed the limits of comfortable leg and foot manipulation but also strategic handhold locations that are "natural." As shown in the accompanying illustration, continuous handrails are preferred to individual handholds because they make it possible for the operator's hands to follow through without having to "search." It is also important that dual handrails be provided when the "path" is skewed, as shown, because the operator's body progresses through a series of axial and center-of-gravity shifts that could cause the operator to lose his or her grip or footing with only one hand for support. Note also that the suggested handrail configuration provides both "pulling" and static gripping capability.

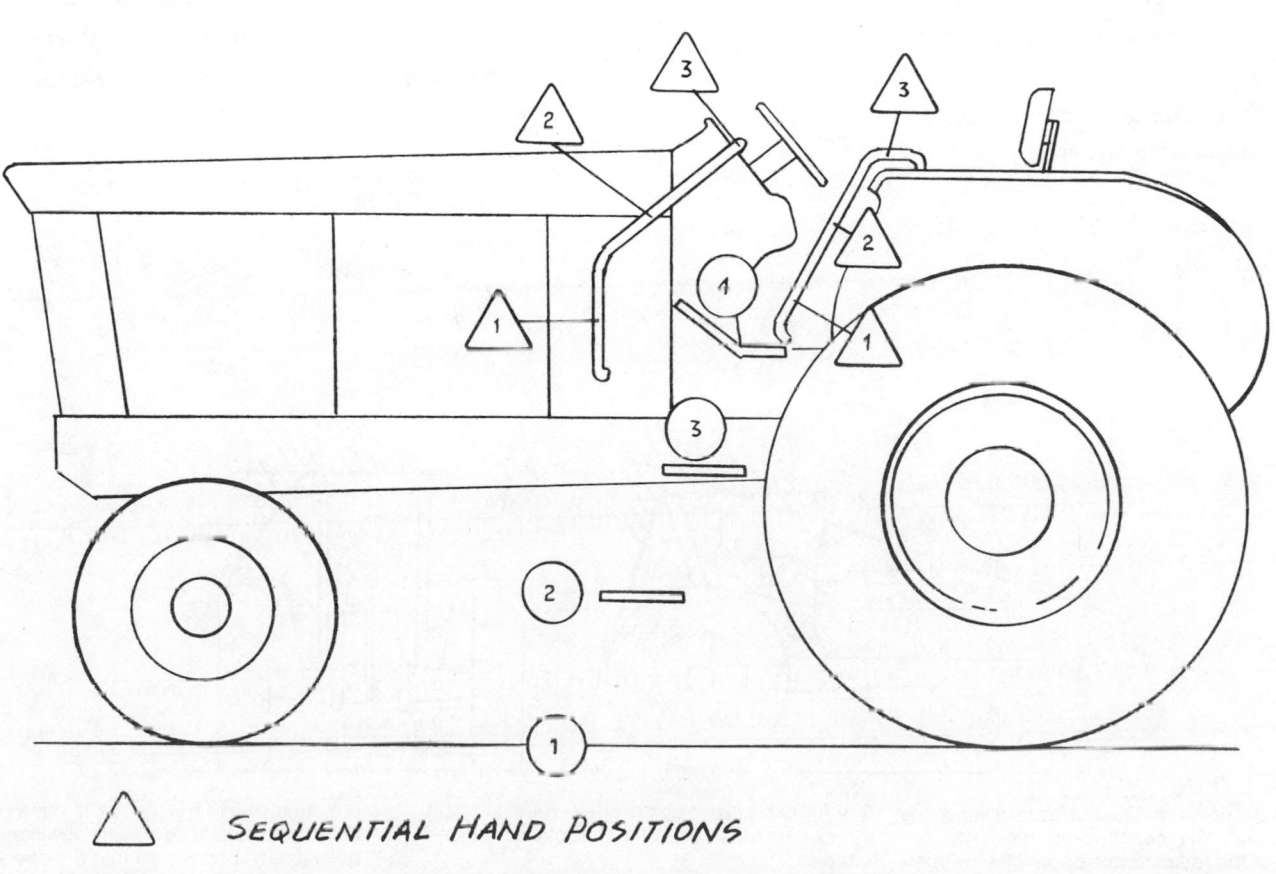

SEQUENTIAL HAND POSITIONS

SEQUENTIAL FOOT POSITIONS

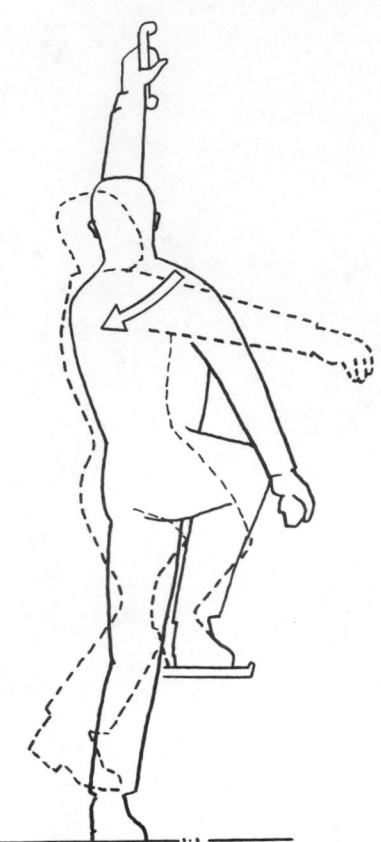

A single, off-center handhold causes the body to swing around gripping point.

**Plan handhold positions to fit ingress pathway dynamics.**

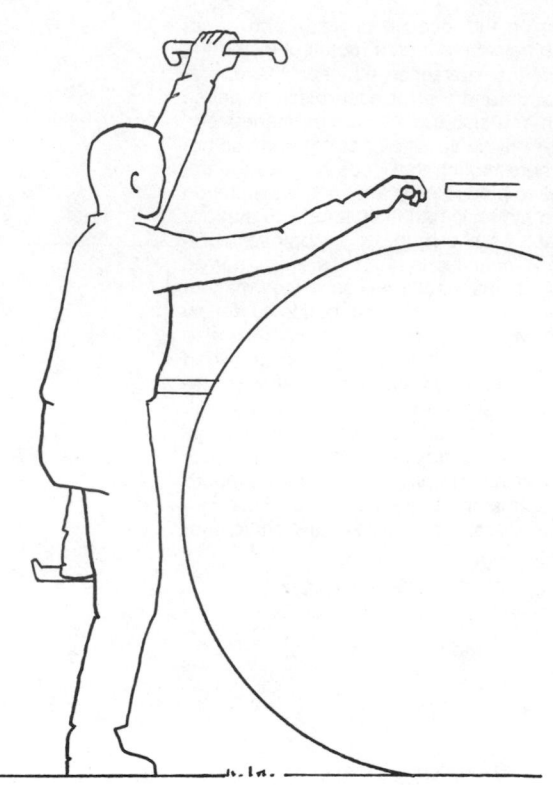

Two-hand assist, located midway in ingress path provide easier, safer method.

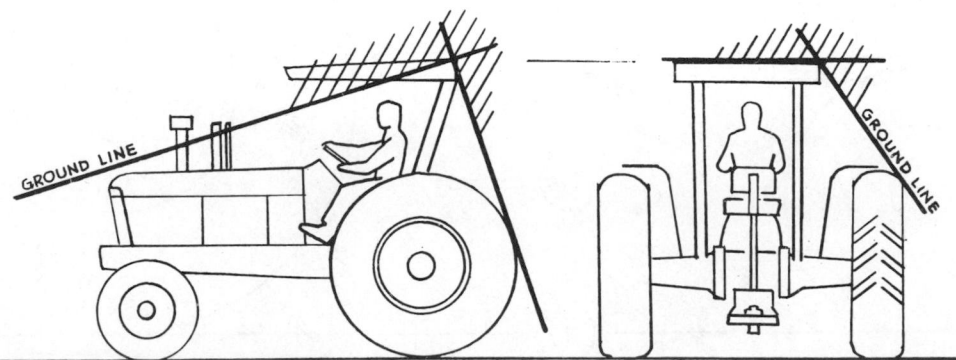

Vehicle upset is a major hazard for many agricultural and construction machines because of the rough terrain on which they operate. Roll-over guards such as those shown in the accompanying sketches should be considered mandatory.

In addition to roll-over protection, the partial or full cab cover also provides protection from rocks or other debris that could fall on the driver.

Seat belts are suggested for such machines not only to protect the operator from being thrown from the vehicle but also to help control the operator's position and thus provide a more steady base from which to observe and manipulate. Such belts should, however, have a quick-release buckle so that the operator is not detained in the event the vehicle must be abandoned quickly.

## U.S. Federal Motor Vehicle Safety Standards

Std. No.

101 Control location, identification, and illumination—passenger cars, multipurpose passenger vehicles, trucks, and buses

102 Transmission shift lever sequence, starter interlock, and transmission braking effect—passenger cars, multipurpose passenger vehicles, trucks, and buses

104 Windshield wiping and washing systems—passenger cars, multipurpose passenger vehicles, trucks, and buses

107 Reflecting surfaces—passenger cars, multipurpose passenger vehicles, trucks, and buses

108 Lamps, reflective devices, and associated equipment—passenger cars, multipurpose passenger vehicles, trucks, buses, trailers, and motorcycles

111 Rearview mirrors—passenger cars and multipurpose passenger vehicles.

118 Power-operated window systems—passenger cars and multipurpose passenger vehicles

123 Motorcycle controls and displays

124 Accelerator control systems

125 Warning devices

201 Occupant protection in interior impact—passenger cars

202 Head restraints—passenger cars

203 Impact protection for the driver from the steering control system—passenger cars

204 Steering control rearward displacement—passenger cars

205 Glazing materials

206 Door locks and door retention components—passenger cars, multipurpose vehicles, and trucks

207 Seating systems—passenger cars, multipurpose passenger vehicles, trucks, and buses

208 Occupant crash protection—passenger cars, multipurpose passenger vehicles, trucks, and buses

209 Seat-belt assemblies—passenger cars, multipurpose passenger vehicles, trucks, and buses

210 Seat-belt assembly anchorages—passenger cars, multipurpose passenger vehicles, trucks, and buses

213 Child seating systems—passenger cars

214 Side-door strength—passenger cars

215 Exterior protection—passenger cars

216 Roof crush resistance—passenger cars

217 Bus window retention and release

219 Windshield zone intrusion

301 Fuel tanks, fuel tank filler pipes, and fuel tank connections—passenger cars

301–375 Fuel system integrity

302 Flammability of interior materials—passenger cars, multipurpose vehicles, trucks, and buses

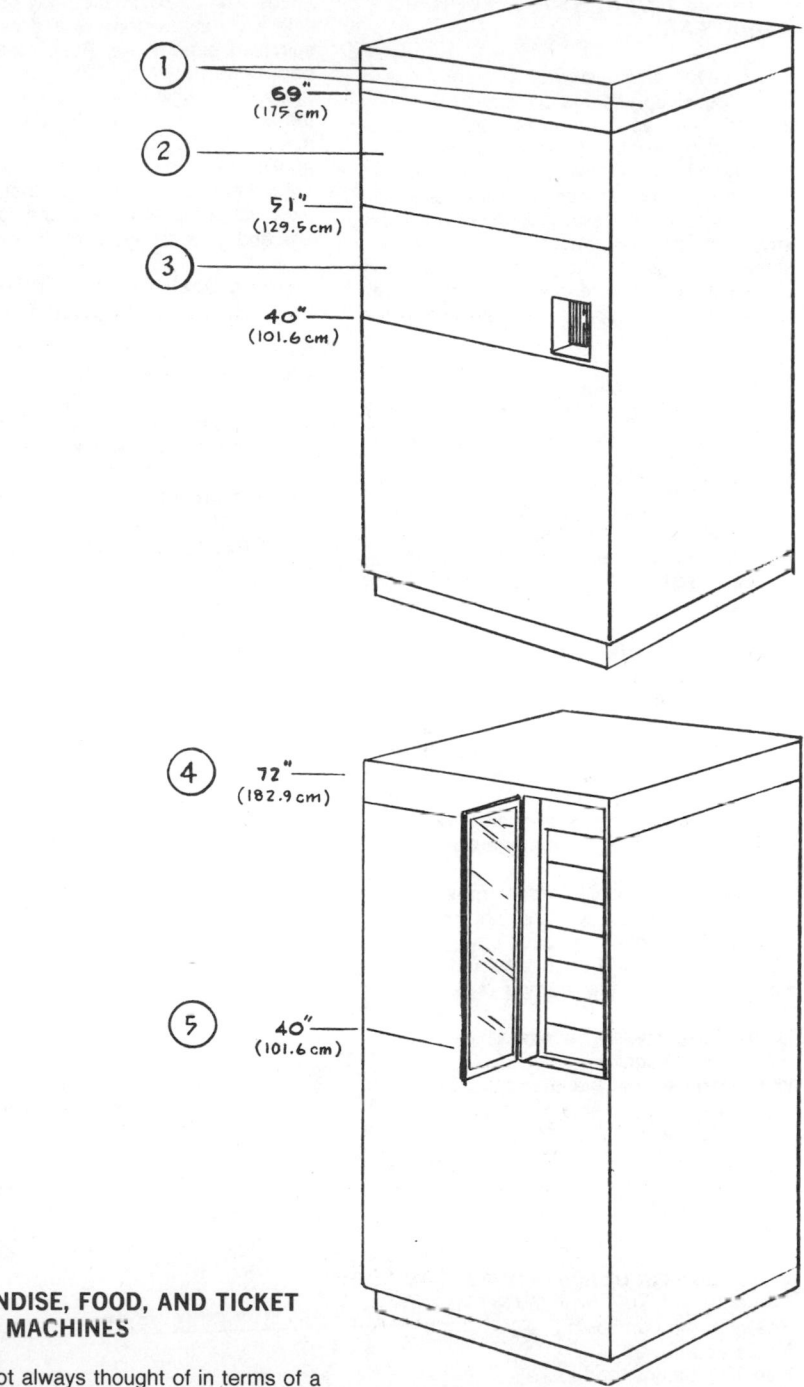

## MERCHANDISE, FOOD, AND TICKET VENDING MACHINES

Although not always thought of in terms of a workplace, merchandise vending machines often create considerable confusion and work for the user if they are not properly designed. Consider the following:

1. What the machine contains should be prominently displayed both on the front and on the sides.
2. Operating instructions should be located near eye level. The instructions should indicate sequential steps for inserting money or tokens, operating the controls, and finding the dispensed product.
3. The controls should not be lower than waist height, nor should the patron have to bend over to obtain the product being dispensed.

4. If a vertical food dispenser is used, it should not require a reach higher than that of the person with the shortest arm, and the product should be no higher than necessary so that the patron can see to select an item.
5. The 40 in (102 cm) dispensing height is about the lowest level that will not cause the patron to have to lean over to obtain the dispensed item.
6. Any access doors should close automatically after the item has been withdrawn.
7. All vending machines should be designed so that they are untippable.

## STORAGE SPACE AS A SPECIAL WORKPLACE

The location and design of a storage space should receive the same systematic appraisal and attention as other work stations, especially when the storage space is utilized frequently. Above all, avoid the temptation to treat the problem of storage space as something to be taken care of after all the more important operator work-station problems are solved. Design of the storage subsystem should consist of more than "making do" with whatever leftover space one can find in the facility or vehicle.

### 1. General Principles
Provide storage space where it is needed, not at some distant, out-of-the-way, or inaccessible location.
Design the space to "fit" the items to be stored; do not design just a generalized volume of space or shelf area.
Provide enough space not only for the predicted use but also for unexpected or future expansion.

### 2. Content Considerations
Size, shape, and weight
Special storing conditions, e.g., package position and protective and/or safety precautions to prevent deterioration and damage
Label exposure, i.e., the necessity to have identification labels immediately accessible as the storage container or storeroom door is opened and the contents are viewed

### 3. Special Feature Considerations
Content separation, securing, and preserving
Content adjustment as needs change
Ease of ingress and of storage and retrieval
Ease of cleaning the storage area (dirt and spills)
Content security, i.e., inaccessibility to unauthorized persons
Illumination so that the user can move about safely and/or can see to identify and manipulate stored items, make space adjustments, or clean
Environmental control, i.e., control of heating, cooling, ventilation, humidity, shock and vibration, etc.

### 4. Key Element Design Considerations
Location
Shelving and partitioning
Doors and covers
Handles, latches, and locks
Content securing

### 5. Storage Type and Location in Relation to Order and Frequency of Use
In many situations storage cabinets are built with little thought given to what type of storage is needed at what locations, e.g., a kitchen in the typical home. Although this is not always the case, kitchen cabinets may be created merely to fill in the spaces between appli-

ances. The cabinetmaker may be told to provide a variety of drawers and cabinets to fill specified dimensions. As a result, the user often finds that things cannot be stored next to the area in which they are needed because the storage space provided does not accommodate them. Examine the use sequence and/or the frequency for each given storage requirement and make sure that the storage plan has some direct relationship to the storage and retrieval needs of the eventual user.

### Storage Shelf Depth in Relation to Shelf Height and Package Weight

Avoid the temptation to provide general shelf storage space without regard to human limitations in terms of reaching and lifting. Obviously, people cannot reach very far into a shelf that is very high or very low. In addition, people cannot support or manipulate as heavy a package when a shelf is either very high or very low as they can when the shelf is somewhere between waist and shoulder height.

The accompanying sketch provides some general guidelines with respect to package weight and shelf height. Obviously, these are not absolute values, since the size, shape, and other geometric features of specific packages tend to modify the capabilities of the individual who is storing or retrieving them. Note that, because of a balancing problem, a worker in the kneeling position cannot support a very heavy or large package.

The accompanying sketches illustrate the difficulties created by extremely deep shelving. The following are general guidelines for specifying shelf depth:

1. Below waist height, the maximum shelf depth should not exceed about 18 in (46 cm).
2. Above shoulder height, the maximum shelf depth should not exceed about 12 in (31 cm).
3. At waist to shoulder height, the maximum shelf depth should be about 24 in (61 cm).

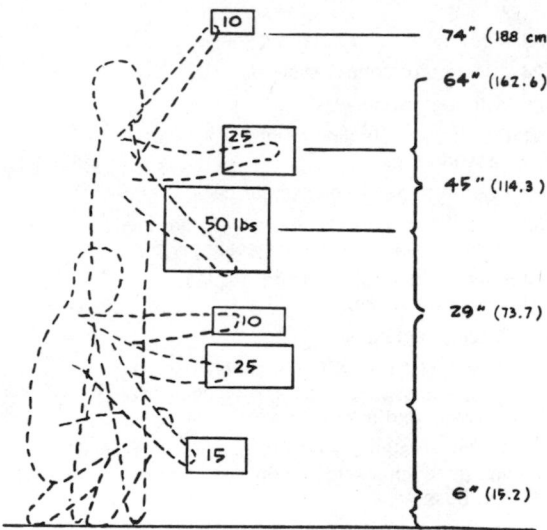

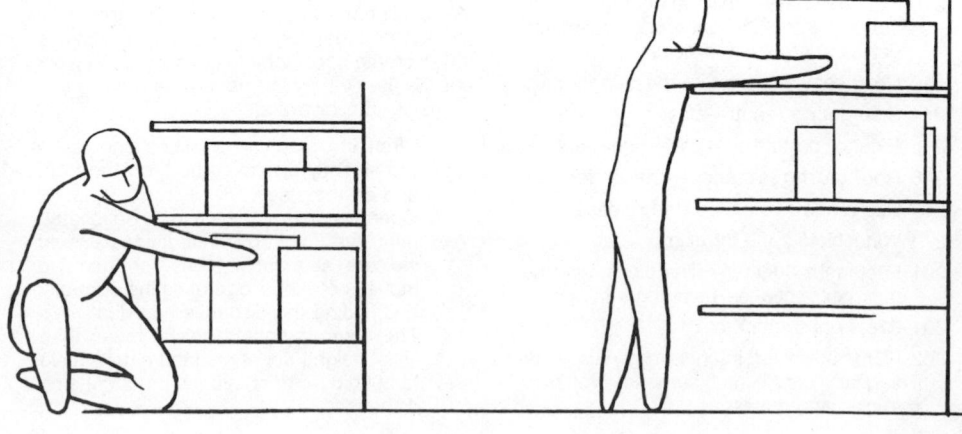

## Shelving and Partitioning

In planning and/or selecting modular shelving, it is desirable to consider some of the added conveniences that could be provided for the user, as shown in the accompanying illustrations.

Special storage units designed for separation of such materials as stationery can be extremely useful. However, they should be designed so that materials cannot be shoved back so far that they are difficult to retrieve.

Using cutouts is one obvious way to make material retrieval easier when the spaces between shelves are too small for a person to get his or her entire hand into the opening.

Vertical storage spaces like the one shown in the accompanying sketch are often handy beside the worker or operator. However, they also collect dirt and other debris. Clean-out holes should be provided so that the storage bin can be cleaned out once in awhile.

Movable shelves are generally more useful because they can be adjusted to meet changing storage needs (see the accompanying illustrations). Movable partitions are equally beneficial.

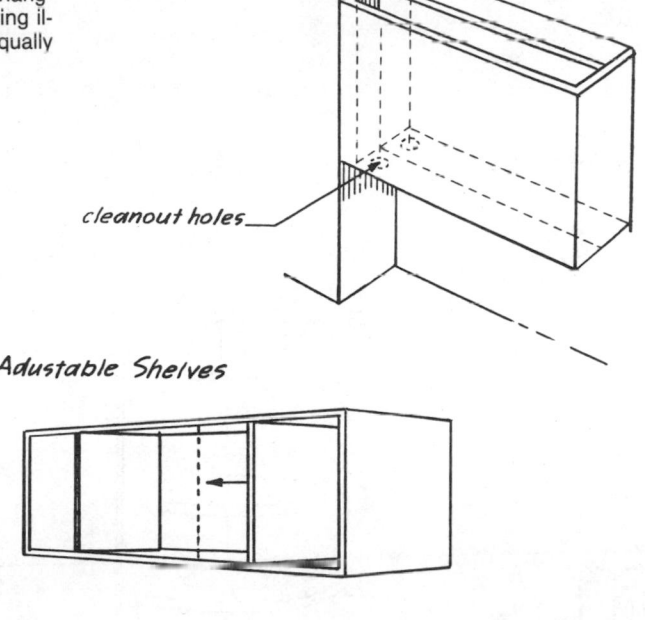

cleanout holes

Adjustable Shelves

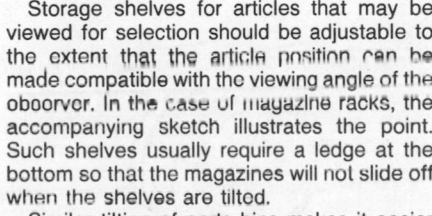

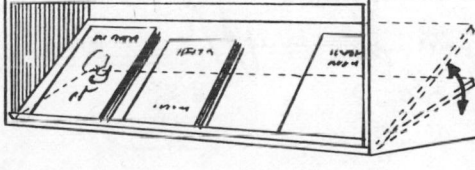

Storage shelves for articles that may be viewed for selection should be adjustable to the extent that the article position can be made compatible with the viewing angle of the observer. In the case of magazine racks, the accompanying sketch illustrates the point. Such shelves usually require a ledge at the bottom so that the magazines will not slide off when the shelves are tilted.

Similar tilting of parts bins makes it easier for workers to look into the bin to see which part they want.

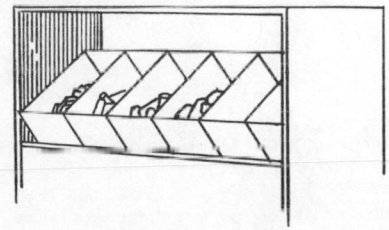

A desirable design objective should be to
"bring the stored item to the user." That is,
instead of designing fixed shelves, bins, etc.,
consider the possibility of making these elements slide out, rotate, or move into more convenient positions for the user to store or retrieve articles. The accompanying sketches
illustrate some of the schemes that various
designers have used successfully.

In executing such designs as shown here, it
is extremely important that the mobility devices operate smoothly and easily and that
there be proper means for securing shelves or
drawers so that they stay where the user
wants them. Equally important is ensuring that
a drawer, for instance, cannot inadvertently be
pulled all the way out, causing its contents to
be spilled on the floor.

Always provide handles that are visible and
easy to grasp on all such mobility systems.

Heavy-duty, smoothly functioning slide or
motion systems should be provided, especially if the shelf or drawer supports heavy
articles.

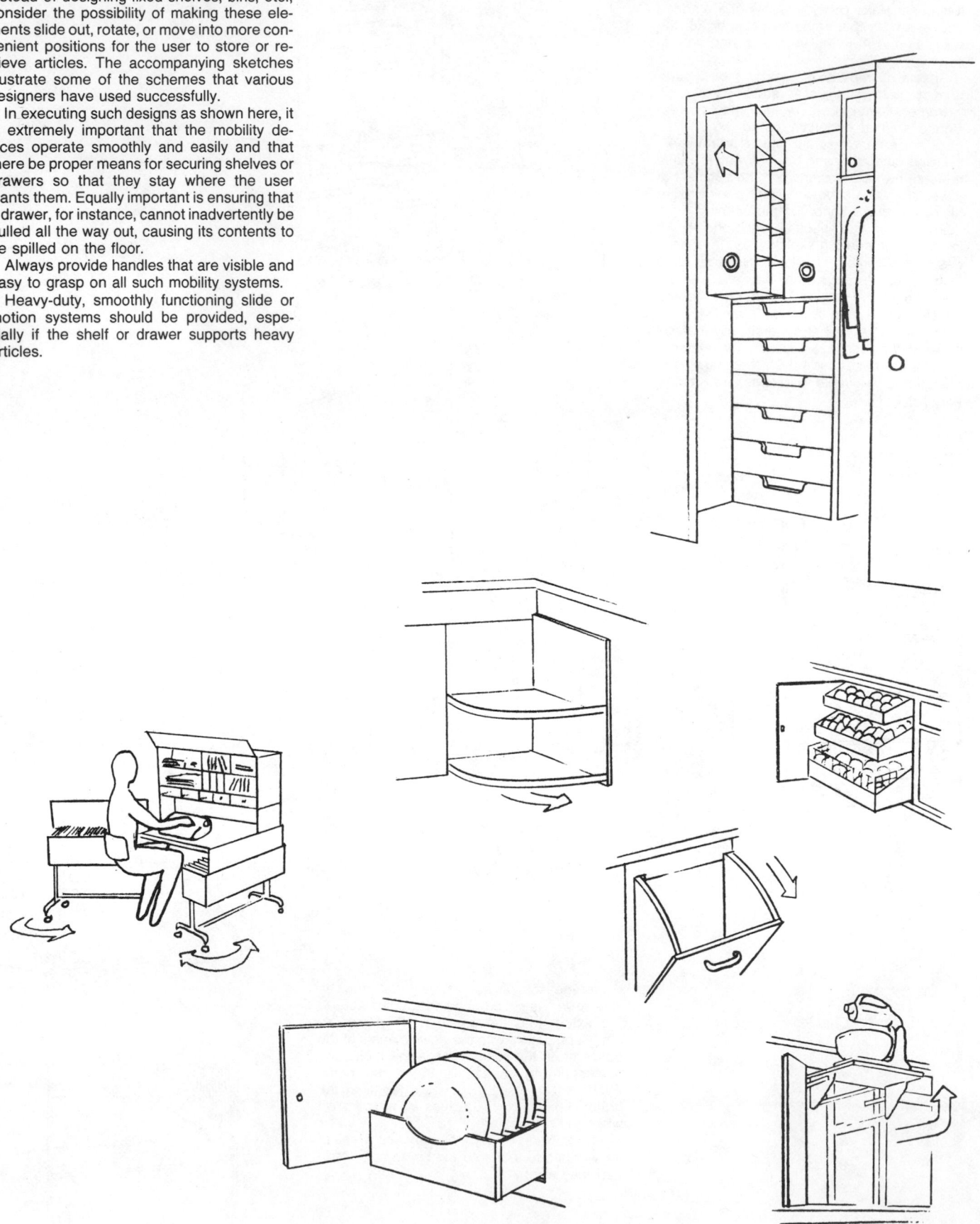

## Storage Cabinet Drawer and Door Design

Although there are some aesthetic as well as constructional advantages to certain types of cabinets, these sometimes create problems for the user. The accompanying sketches provide some guidelines for designers, whether they are selecting ready-made units or designing new cabinets:

1. Lap-front drawers or doors are not preferred because they make it easier to catch and pinch one's fingers and because any liquids that are spilled down the front of the cabinet tend to enter the drawer or cabinet.
2. Handles should be on the front of drawers and doors so that the user knows where they are.
3. Corner adjacent drawers should be avoided, and similar door arrangements should be planned so that when two adjacent doors swing open, their handles do not coincide.

Magnet or spring-return types of door closures are preferred over latching types (except for those cases where dynamic forces might cause the door to fly open, as in vehicular applications).

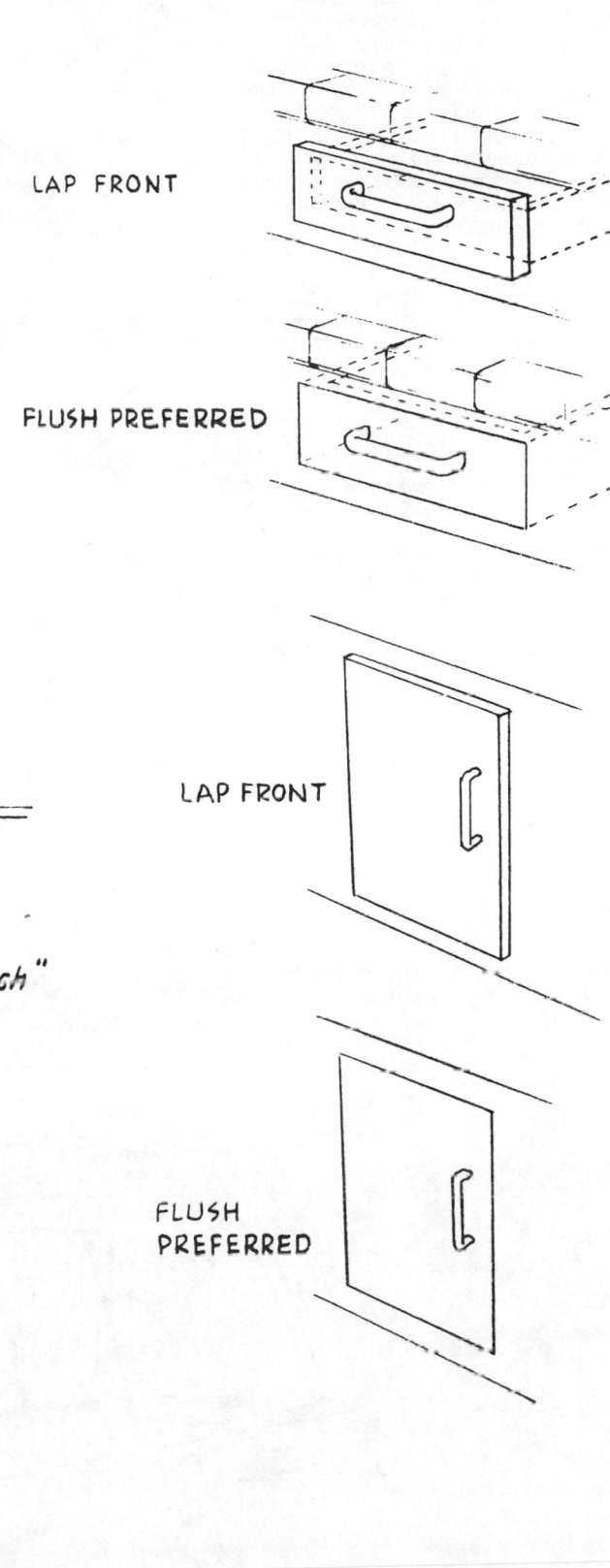

LAP FRONT

FLUSH PREFERRED

LAP FRONT

FLUSH PREFERRED

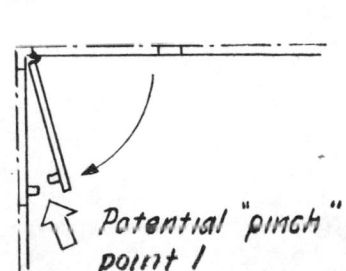

*Potential "pinch" point !*

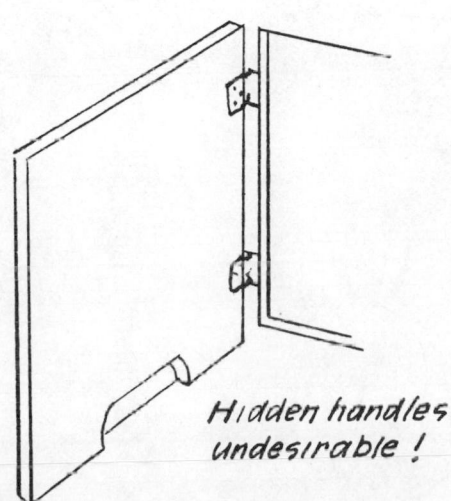

*Hidden handles undesirable !*

## Storage Requirements in Moving Vehicles

In addition to the basic requirements discussed above regarding planning for the more stable, fixed storage place, when planning storage aboard various vehicles (whether for private or commercial use), the designer must consider the added difficulties caused by the dynamic environment. The critical issue involves making sure that cargo stays put! Although most designers are aware of this need in commercial applications and/or are forced to comply with specifications that define cargo tie-down requirements, a similar concern is often not displayed with respect to private vehicles.

The accompanying sketches, although not representative of all possible situations, provide at least three basic approaches for cargo storage capture: (1) Place stored items inside a closed compartment; (2) provide railings or insets to minimize lateral shifting; and/or (3) provide tie-down features so that cargo can be strapped in place.

The family station wagon shown in the accompanying sketch illustrates a problem that is typically ignored. Many a passenger has become a statistic because some loose cargo became a missile during a crash. Something as small and light as a tape cassette has been known to kill a passenger because it was left unsecured on the back window ledge of the family automobile.

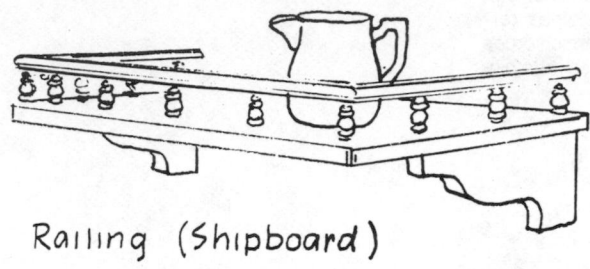

Overhead, closed compartment
(Aircraft - Bus)

Railing (Shipboard)

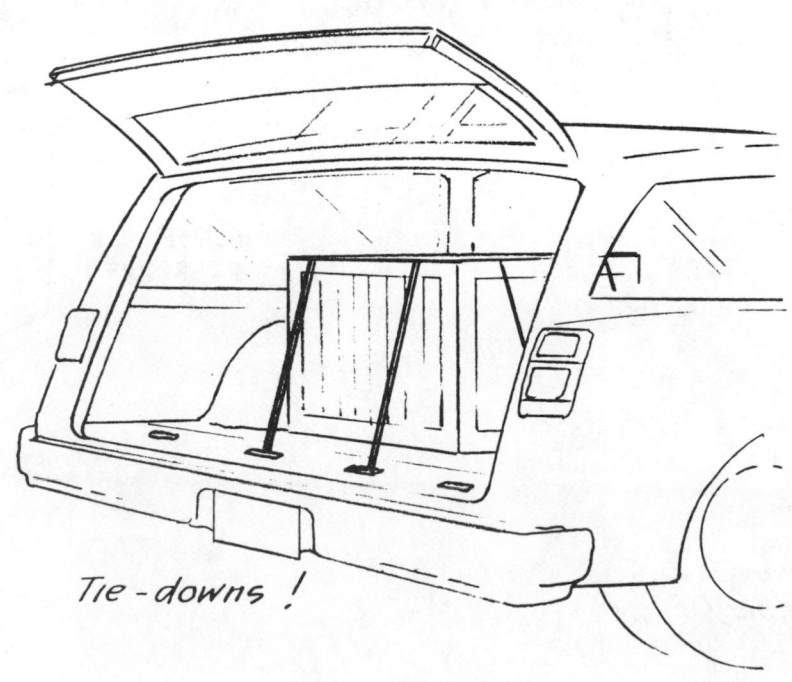

Tie-downs !

# LIGHTING SYSTEM DESIGN

## 1. Purpose

Lighting is typically provided for two main purposes: (a) to enable people to see what they need to see in order to perform various tasks and (b) to create decorative effects. It is important to differentiate between these purposes because the objectives of each are different. In some cases decorative lighting schemes may interfere with task illumination unless there is a constant awareness that the observer, to see well, must have enough of the proper type of light, placed where it does the most good, and that there should be no extraneous light or reflections that might interfere with the observer's seeing process.

Seeing tasks are not always the same; thus the amount and type of lighting and the method for providing illumination should be tailored to fit the particular seeing task. A few examples will serve to illustrate how different tasks and lighting system requirements can be:

a. Reading a book or newspaper
b. Reading instruments or cathode ray tube displays in a darkened room or cockpit
c. Performing surgery or sewing on dark material
d. Inspecting for flaws in material or metal surfaces
e. Matching colors for material selection
f. Developing film in a darkroom

## 2. Types of Lighting

a. Natural light: The proper manipulation and use of natural light is very desirable for many applications, especially because it tends to make things we look at seem more "natural."
b. Artificial light: Artificial light is used when there is not enough natural light available for the desired seeing task and/or when it may be necessary to control the color of the light for a special seeing task.

## 3. Lighting Methods

The two basic types of illumination either for illuminating seeing tasks or for creating decorative lighting effects are (a) general illumination and (b) supplementary illumination

a. General illumination: General illumination should be used when the seeing requirements are broad and varied, including seeing to move about safely, observing other people or general features within the surrounding area, performing housekeeping chores, moving materials or equipment, and performing general maintenance tasks.
b. Supplementary illumination: Supplementary illumination is required when general illumination is not available and/or when the nature of the task is such that the general illumination is insufficient, is not appropriately directed, or is of such a quality or color as to make specific seeing tasks difficult to perform. Supplementary illumination also is generally used for decorative effects since it can be precisely controlled in terms of color, direction, and amount.

## 4. Lighting Techniques

a. Floodlighting: As the term implies, this type of lighting is usually external to the object or objects that are to be illuminated, literally flooding them with light. Floodlighting is generally used to illuminate large areas, either indoors or outdoors, but it can also be used in more concentrated, localized situations such as for illuminating a control panel, a doorway or platform, a theatre stage, a parking lot, or an airport apron. Although some control can be exercised over the flood pattern, it should be noted that the illuminated area covered by a floodlighting system will vary from the beam center to its periphery. Care must also be taken in using the floodlighting technique to consider the problems that might be associated with shadows, which are a by-product of the floodlighting technique.
b. Internal lighting: A wide variety of internally generated lighting techniques are available to make objects or visual displays "self-luminous." Typical applications for which internal lighting is appropriate include signs, panel labels, vehicle or operator console instruments and displays, shadow boards, and marker and signal lights. Internal lighting can be created in many ways, including the use of luminous phosphors, back-lighted screens, edge-lighted plastics, and end-lighted plastic "light pipes." The primary advantage of internal lighting is that more precise control can be obtained over where the light will appear.

## 5. Light Level

Avoid the temptation to believe that merely providing higher and higher illumination levels will result in improved seeing conditions. Although the level or intensity of light is obviously important, light can be too bright as well as too dim for the seeing task. Also remember that as light levels increase, glare problems also increase. There is usually a point below which seeing is difficult and another point above which seeing either does not improve or becomes more difficult for each type of seeing task.

## 6. Color

The color of the light used (i.e., for general illumination, supplementary illumination, or signal lighting) is important from several standpoints:

a. A broad-spectrum light (white light) is desirable to make the colors of objects or people seem natural. Natural light is used as the base-line reference since it is by this reference that we observe most things during everyday living. Several types of artificial light have been created to simulate natural light, but some are better than others for certain purposes. Care must be taken to select the artificial light color that is most appropriate for the task or seeing conditions being created. Incandescent luminaires typically provide more reds and yellows, which impart a warm appearance to viewed objects. Fluorescent lights, on the other hand (depending on the specific luminaire), tend to accentuate the blue. Often, a colored material selected under fluorescent light will not look the same when the material is viewed under natural sunlight.
b. Monochromatic light (i.e., red, yellow, blue, and green) may be desirable for certain types of seeing tasks and/or for decorative purposes. Such colors are not, however, desirable for general seeing conditions; they make surface colors appear abnormal and/or impossible to recognize. Monochromatic light is particularly useful for creating color-coded displays, e.g., traffic signal lights, airport runway and taxiway lights, and warning displays.

## 7. Light Fixture Type and Usage

An extensive array of lighting fixtures, lamps, and materials is available for creating almost any type of lighting that is required. Some precautions need to be taken, however, in selecting commercially available fixtures. Some provide effective task illumination, but others are merely attractive to look at. When looking for appropriate fixtures off the shelf, designers should analyze the seeing task fully so that they can evaluate the merits (or, more often, the demerits) of the fixtures they are contemplating for a particular application.

An important factor to consider is the location of lighting fixtures, for even a good fixture's value can be lessened if it is improperly positioned relative to the seeing task. In addition, it is important to consider the light control requirements in terms of the possible need for dimming and/or selectively illuminating only certain fixtures at certain times during an operation.

### General Illumination Guidelines

| Task Requirements | Light Level, fc | Type of Illumination |
|---|---|---|
| Small detail; low contrast; prolonged viewing; fast, error-free response | 100 | Supplementary lighting fixture located near visual task |
| Small detail, fair contrast, close but short-duration work, speed not essential | 50–100 | Supplementary lighting and/or well-distributed and diffused general lighting |
| Typical desk and office work | 40–60 | General lighting with diffusing fixture directly overhead |
| Sports (e.g., tennis and basketball) or indoor recreational games (e.g., ping pong and billiards) | 30–50 | General lighting with sufficient number of fixtures to provide even court or table illumination |
| Recreational reading and letter writing | 25–45 | Supplementary lighting, positioned over reading so that page glare does not occur |
| General housekeeping, detail not required | 10–25 | General lighting |
| Visibility for moving about, avoiding people and furniture, and negotiating standard stairs | 5–10 | General and/or supplementary lighting (with care taken not to allow supplementary sources to project in the user's eyes) |

*Note:* The above guidelines are only approximations. They are higher than some recommendations, not for seeing, but because these levels provide an additional psychological benefit as well. Levels relate to light levels measured at the primary seeing point, e.g., the desk or table surface on the floor or stair tread level.

Brightness ratios between the seeing task and the immediate surroundings should not exceed 5:1; between the task and the remote surroundings, 20:1; and between the immediate work area and any other remaining visual environment, 80:1.

Natural or white artificial light should be used regardless of the type of illumination; i.e., these levels do not apply to monochromatic light sources.

### Special Situations Where Visual Dark Adaptation Is Required

| Work Situations | Light Level, fc | Type of Illumination |
|---|---|---|
| Darkened control rooms and cockpits at night | 0.01–0.4 | Illumination on work surfaces |
| | 0.03–0.10 | Black and red instrument face detail (scale marks, numerals, and labels) |
| | 0.10–1.00 | Indicator lights (not to exceed 1 in, 2.5 cm square) |
| Ready rooms where crew members are required to maintain some dark adaptation prior to a mission at night | 0.50–2.00 | Illumination at tabletop |
| Movie theatre during film running | 0.50–2.00 | Illumination in aisle and immediate vestibule |

*Note:* Red illumination is required in military workplaces and ready rooms. Low-level white light is satisfactory in theatres.

**ILLUMINATION LEVEL GUIDELINES FOR MAINTAINING PROPER BRIGHTNESS CONTRAST BETWEEN VISUAL TASK AND BACKGROUND**

### Brightness Ratios

| Comparison | Environmental Classification* | | |
|---|---|---|---|
| | A | B | C |
| Between tasks and adjacent darker surroundings | 3:1 | 3:1 | 5:1 |
| Between tasks and adjacent lighter surroundings | 1:3 | 1:3 | 1:5 |
| Between tasks and more remote darker surfaces | 10:1 | 20:1 | † |
| Between tasks and more remote lighter surfaces | 1:10 | 1:20 | † |
| Between luminaires and adjacent surfaces | 20:1 | † | † |
| Between the immediate work area and the rest of the environment | 40:1 | † | † |

*Note:* Direct glare arises from a light source within the visual work field. It should be controlled by:

1. Avoiding bright light sources within 60° of the center of the visual field. Since most visual work is at or below the eye's horizontal position, placing luminaires high above the work area minimizes direct glare.
2. Using indirect lighting.
3. Using more relatively dim light sources, rather than a few very bright ones.
4. Using polarized light, shields, hoods, or visors to block the glare in confined areas.

*A—Interior areas where reflectances of entire space can be controlled for optimum visual conditions. B—Areas where reflectances of immediate work area can be controlled, but there is only limited control over remote surroundings. C—Areas (indoors and outdoors) where it is completely impractical to control reflectances and difficult to alter environmental conditions.

†Brightness-ratio control not practical.

## LIGHT-SOURCE COLOR

The color of the light source should be chosen carefully to fit the application, as is indicated by the accompanying table.

**Effect of Colored Light on Appearance of Object Color**

| Object Color | Red Light | Blue Light | Green Light | Yellow Light |
|---|---|---|---|---|
| White | Light pink | Pale blue | Pale green | Pale yellow |
| Red | Brilliant red | Dark bluish red | Yellowish red | Bright red |
| Light blue | Reddish blue | Bright blue | Greenish blue | Light reddish blue |
| Dark blue | Dark reddish purple | Brilliant blue | Dark greenish blue | Light reddish purple |
| Green | Olive green | Green-blue | Brilliant green | Yellow-green |
| Yellow | Red-orange | Light reddish brown | Light greenish yellow | Brilliant light orange |
| Brown | Brownish red | Bluish brown | Dark olive brown | Brownish orange |

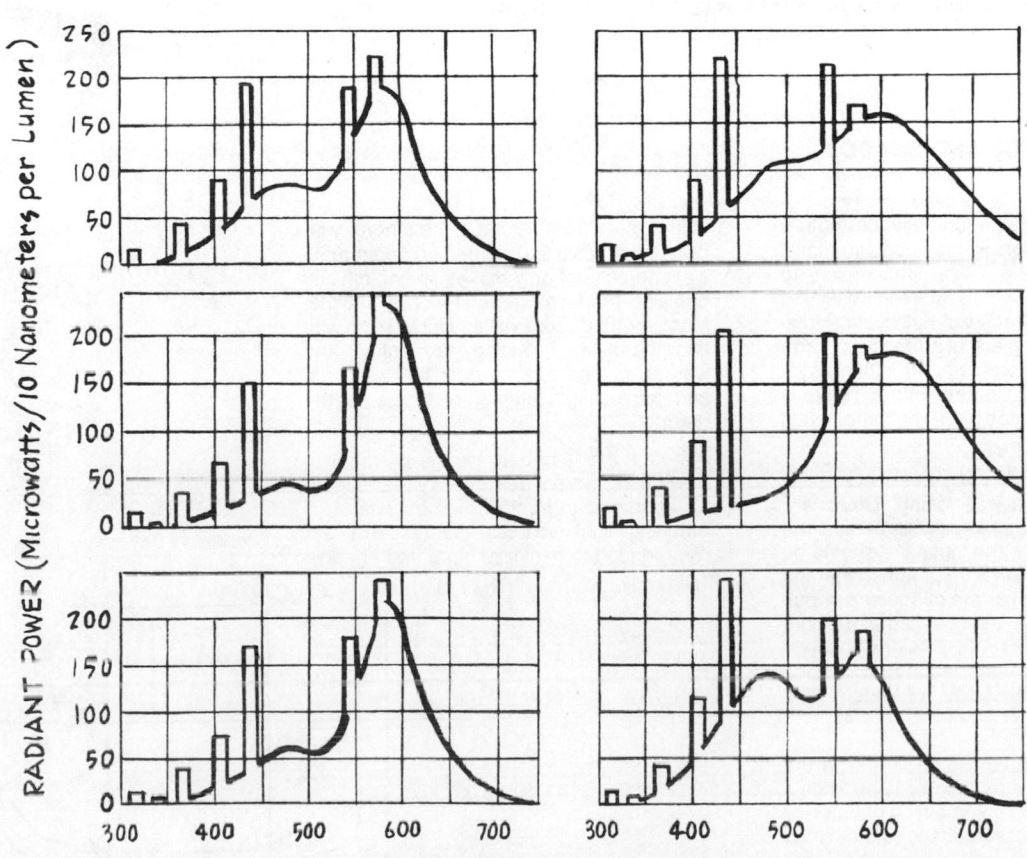

Spectral distributions of a commercial fluorescent lamp.

## APPEARANCE RATINGS OF TYPICAL SURFACE COLORS WHEN VIEWED UNDER VARIOUS ARTIFICIAL LIGHT SOURCES

| Color | Daylight | Fluorescent lamps | | | | Incandescent Lamps |
|---|---|---|---|---|---|---|
| | | Standard Cool White | Deluxe Cool White | Standard Warm White | Deluxe Warm White | |
| Maroon | Dull | Dull | Dull | Dull | Fair | Good |
| Red | Fair | Dull | Dull | Fair | Good | Good |
| Pink | Fair | Fair | Fair | Fair | Good | Good |
| Rust | Dull | Fair | Fair | Fair | Fair | Good |
| Orange | Dull | Dull | Fair | Fair | Fair | Good |
| Brown | Dull | Fair | Good | Good | Fair | Good |
| Tan | Dull | Fair | Good | Good | Fair | Good |
| Golden yellow | Dull | Fair | Fair | Good | Fair | Good |
| Yellow | Dull | Fair | Good | Good | Dull | Fair |
| Olive | Good | Fair | Fair | Fair | Brown | Brown |
| Chartreuse | Good | Good | Good | Good | Yellowed | Yellowed |
| Dark green | Good | Good | Good | Fair | Dull | Dull |
| Light green | Good | Good | Good | Fair | Dull | Dull |
| Peacock blue | Good | Good | Dull | Dull | Dull | Dull |
| Turquoise | Good | Fair | Dull | Dull | Dull | Dull |
| Royal blue | Good | Fair | Dull | Dull | Dull | Dull |
| Light blue | Good | Fair | Dull | Dull | Dull | Dull |
| Purple | Good | Fair | Dull | Dull | Good | Dull |
| Lavender | Good | Good | Dull | Dull | Good | Dull |
| Magenta | Good | Good | Fair | Dull | Good | Dull |
| Gray | Good | Good | Fair | Soft | Soft | Dull |

*Note:* Good—color appears most nearly as it would under an ideal white-light source, such as north skylight. Fair—color appears about as it would under an ideal white-light source, but is less vivid. Dull—color appears less vivid. Brown—color appears to be brown because of small amount of blue light emitted by lamp. Yellowed —color appears yellowed because of small amount of blue light emitted by lamp. Soft—surface takes on a pinkish cast because of red light emitted by lamp.

## GLARE AVOIDANCE AND SHADOW PREVENTION

Because everyone has at one time or another experienced seeing problems due to shadows and/or glare, it would seem unnecessary to advise designers to exercise care in designing so that these conditions will not occur. However, shadows and glare are the most commonly encountered problems in lighting system designs. The three most serious conditions are illustrated by the accompanying sketches:

Shadows are cast by the observer because of improper luminaire positioning. Obviously, the light source should not be behind the observer. Typically, the luminaire should be located slightly to the right or left so that the light comes from over the observer's shoulder. General practice suggests that the light come from over the left shoulder because most people are right-handed. A lighting mockup is recommended for determining the best location.

The light source should not be visible to the observer; i.e., the light bulb or lamp should be screened either by some type of baffle or by a distributing screen. Once again, since layout errors have a way of slipping by, use a lighting mockup to make sure that the observer cannot see the light source directly from his or her various working positions.

Ambient glare from frontal or peripheral windows and/or similarly large and bright artificial sources not only makes seeing difficult but also produces general worker fatigue. Many methods are available for dealing with this problem, including reorienting the worker, putting shades over the windows, and adding light-filtering materials to the glazing.

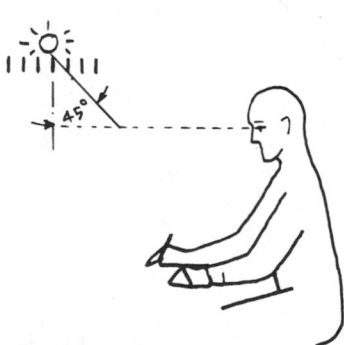

Although this is not always the prerogative of the lighting system designer, try to eliminate glossy surfaces in front of the observer.

Diffusing light sources (fluorescent) are recommended to minimize glare and shadow effects.

## INTEGRAL LUMINOUS CEILING LIGHTING SYSTEMS

A popular concept has been the ceiling which provides a combination of overhead general lighting, acoustic treatment, and heating and ventilation, plus the basic supporting structure (as shown in the accompanying sketch). When used properly, this concept provides excellent seeing conditions for large-area installations. A common failure, however, occurs when the architect or interior designer paints the surfaces of the fixture insets a dark color, which absorbs much of the light from each fixture. This also makes the contrast between lighting fixtures and support structure extreme, which causes the lights to glare rather than create a pleasing atmosphere.

Many other overhead lighting systems fail to provide the desired seeing conditions for the same reason. Always minimize the contrast between the lighted fixture and its surrounding support structure or decorative trim.

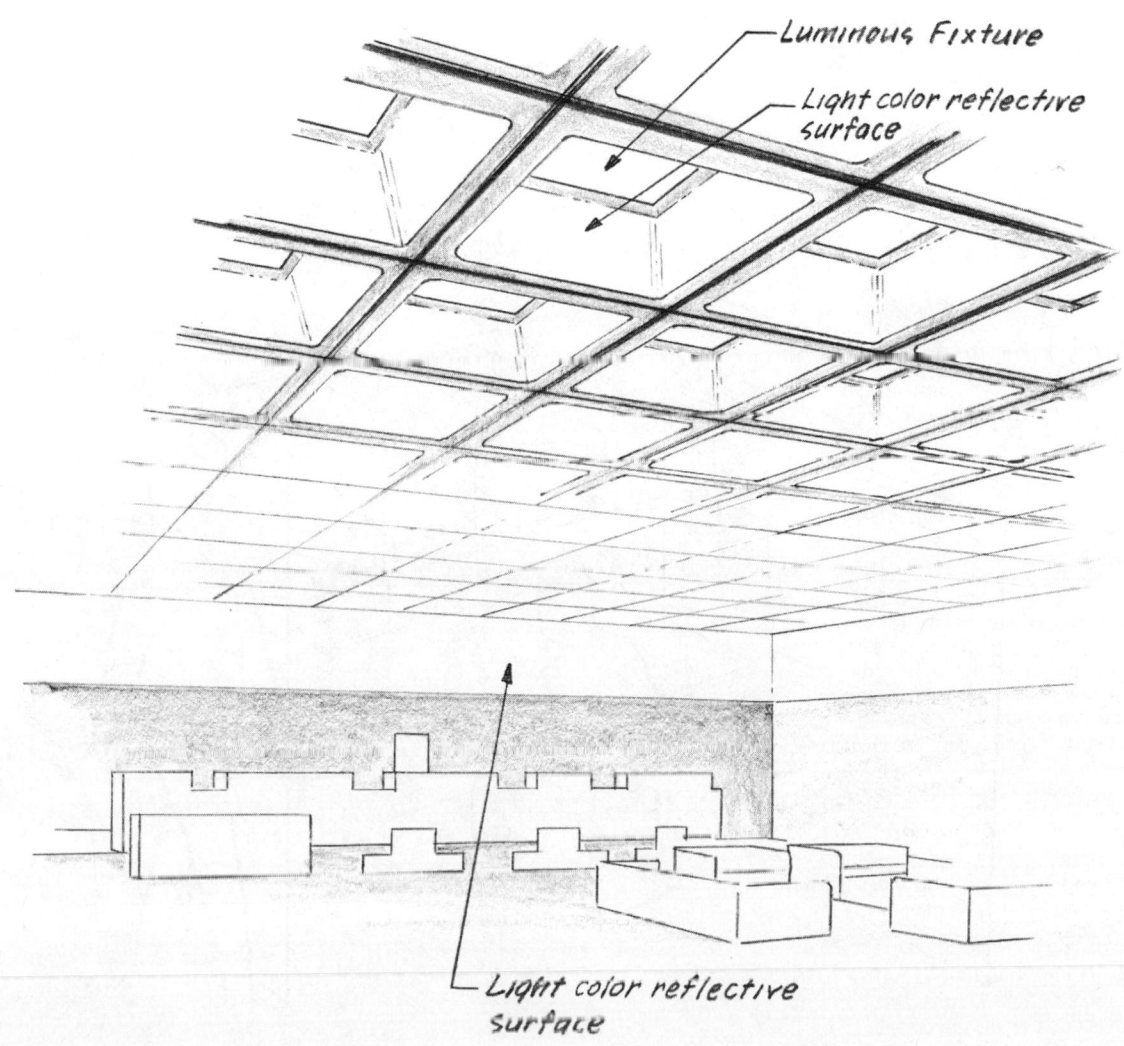

Luminous Fixture

Light color reflective surface

Light color reflective surface

## MODULARIZED, COMBINATION LIGHTING SYSTEMS

Energy conservation often can be obtained in conjunction with good lighting, as illustrated in the accompanying sketch. The objective here is to create a combination ambient and task lighting system that can "move with the workplace." Instead of trying to create a general ceiling lighting system that will be appropriate for all possible seeing situations, the designer should consider the modular approach, which illuminates only the areas that need it; in addition, since the luminaires are tailored to the modular workplace, the lighting will always be right for that workplace. This is important to note because, after a general lighting system has been installed, many seeing problems are created by the placement of furnishings in such a way that they block the intended light rays and the general light distribution.

Systems such as that shown in the accompanying illustration must be designed so that the luminaires on top of the modular cabinets are above the heads of persons passing by. In addition, to be most effective, such overhead lighting requires a normal-height (rather than a double-height) ceiling of light color.

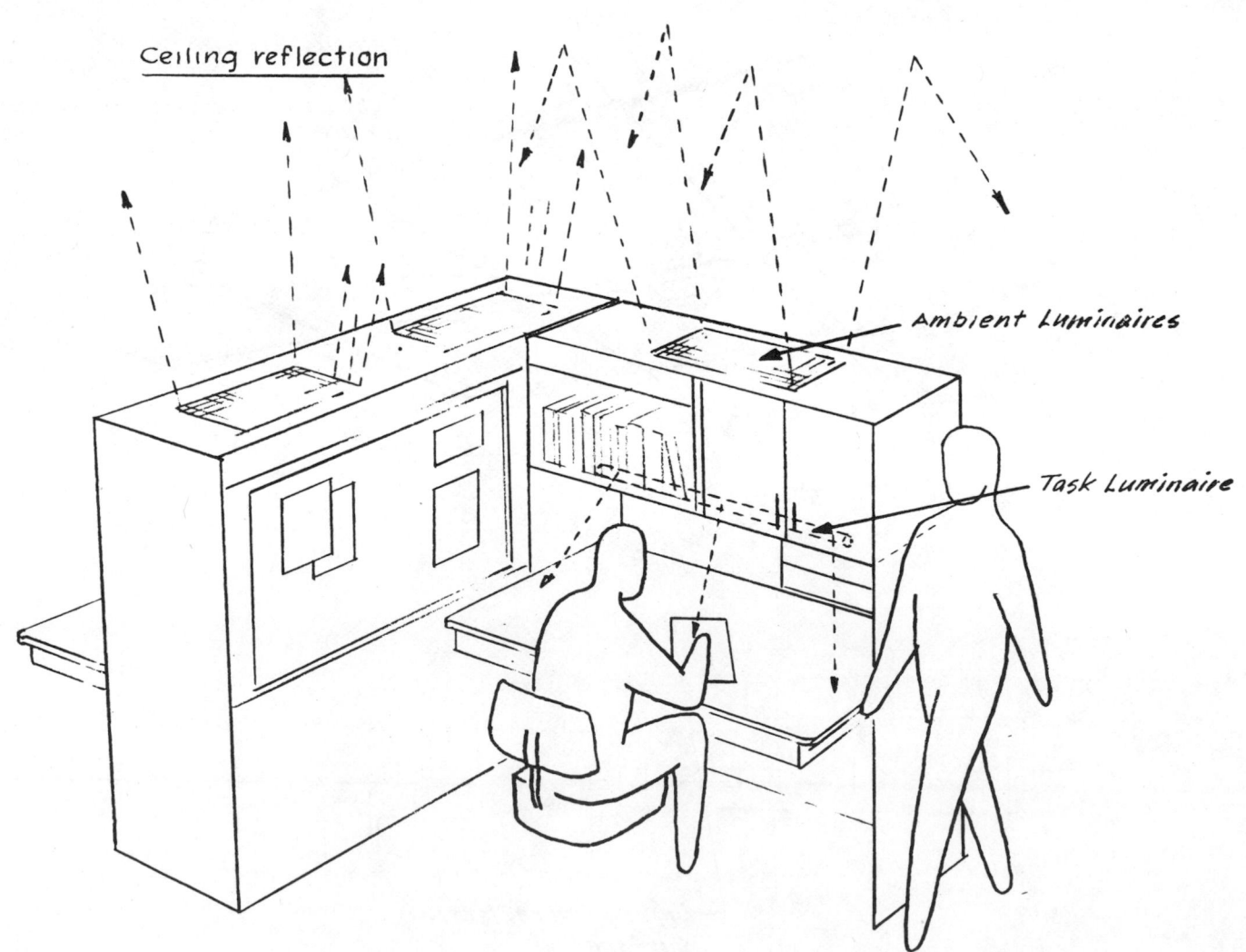

Ceiling reflection

Ambient Luminaires

Task Luminaire

## LIGHT-LEVEL CONTROL

Light-level adjustment is desirable for a wide variety of applications because it enables the operator to optimize his or her seeing conditions. The principal applications are the following:

1. Instruments and control panels: The ability of operators to adjust the light levels of a vehicle or of console instruments and/or the illumination of control panel labels is extremely important when they are confronted with changing ambient lighting conditions and/or operating needs; i.e., operators should be able to balance brightness levels among various visual displays, reduce interfering glare when they must maintain lower visual adaptation levels, and/or optimize special display viewing conditions (e.g., CRT reference scales).
2. Ambient illumination level: It is often desirable for operators to be able to adjust ambient room lighting in order to improve their ability to see special visual displays, e.g., to detect targets on a CRT, to view projected slides or movies, or to view a TV program. In addition, the ability to modify the ambient lighting level of various other architectural spaces may be important for psychological reasons, such as when lower lighting is desired for an intimate dining environment.

*Note:* It is generally desirable that a dimming system be designed so that the light cannot be turned completely off because, if a system has been previously turned off, it may create a problem for the succeeding operator or user, who may come into a dark room or sit at a darkened control panel and not be able to find a light switch or recognize obstacles while trying to locate the switch.

## ENERGY SAVING

Because of the need to save energy, it often becomes necessary to use fewer lights and/or to guarantee that lights are turned off when they are not being used. Consider the following:

1. Use selective circuit control concepts so that it is possible to turn off various luminaires. Care must be taken in creating such system control capability not to destroy the general acceptability of seeing conditions; i.e., the light distribution (although less intense) should continue to be balanced for a given visual operating area.
2. Provide central control for large-area lighting systems so that the person who leaves last can turn off all remaining lights.
3. Consider the use of automatic light control (timing systems) so that certain lights are turned on or off to match operating needs. Similar automatic control of special exterior light control shutters should be considered; i.e., as long as natural light can be used, it is foolish to waste artificial light.
4. Supplementary lighting may provide greater energy saving than general lighting because the user can often obtain all the light needed (both for the immediate task and for general mobility) from the supplementary lighting system.

Using a dimmable luminous ceiling lighting system is a better energy-saving approach than selectively turning out every other luminaire because the latter approach creates light and dark spots, and often a dark spot occurs directly over a desk or other operator position.

The same consideration applies to hallway illumination; i.e., it is better to turn off one lamp within a fixture than to turn off every other luminaire along the hallway.

An automatic "user-presence" sensing system can be used effectively to turn out lights in unoccupied spaces.

**LIGHTING INDUSTRY STANDARD PRACTICE**

| Facility | Light Level, fc |
|---|---|
| Art galleries: | |
|   General | 30 |
|   Paintings | 30–60 |
|   Sculpture | 30–100 |
| Auditoriums: | |
|   Lobbies | 30 |
|   Seating area (nonperformance) | 10–30 |
|   Seating area (during performance, in aisle) | 1–3 |
|   Orchestra (music stands) | 15–20 |
| Banks: | |
|   Lobby | 30 |
|   Officer platform area | 30–50 |
|   Teller line | 50–75 |
|   Rear posting and keypunch area (work surface) | 75–100 |
| Depots and terminals: | |
|   Waiting rooms | 30 |
|   Ticket counter | 50–75 |
|   Baggage room | 50–75 |
|   Platform | 30 |
|   Rest rooms | 30 |
| Hospitals: | |
|   Lobbies and hallways | 50–75 |
|   Patient rooms | 3–50 |
|   Laboratory and examination rooms | 75–100 |
|   Surgery and delivery | To 2500 |
| Hotels: | |
|   Lobbies | 30–75 |
|   Front office | 50–75 |
|   Halls and elevators | 20–30 |
|   Rest rooms and guest rooms | 20–50 plus reading lights to 75 |
|   Kitchen | 100 |
|   Laundry | 50–75 |
| Office buildings: | |
|   Lobbies, halls, and elevators | 20–30 |
|   Offices | 75–100 |
| Residences: | |
|   Entryway and porch | 10 |
|   Hallways and stairs | 10–25 |
|   Kitchen | 50–75 |
|   Laundry | 50 |
|   Library | 50–70 |
|   Sewing room | 50–100 |
|   Bathroom—general | 30 |
|     sink and mirror | 50 |
|   Workshop | 70 |
| Restaurants: | |
|   Quick-service counter | 50 |
|   Intimate dining area | 15–30 |
|   Kitchen | 70 |
|   Cashier area | 50 |
| Schools: | |
|   Study areas and classrooms | 50–100 |
|   Drafting room | 100 |
|   Offices | 50 |
|   Rest rooms | 30 |
| Retail stores: | |
|   Circulation area | 30 |
|   Showcase | 100–150 |
| Sports: | |
|   Archery | 10 |
|   Badminton | 30 |
|   Baseball—seating | 3–10 |
|     field | 100–150 |
|   Basketball | 50 |
|   Gymnasium, general exercise room, and | 30 |
|     locker room | 20 |
|   Hockey | 20 |
|   Boxing | 200 |
|   Football | 100 |
|   Swimming pool, general overhead and | 10 |
|     under water | 100 |
|   Tennis | 30–50 |
|   Volleyball | 20 |

**INDUSTRIAL/MANUFACTURING LIGHTING PRACTICES**

| Facility | Light Level, fc |
|---|---|
| Assembly: | |
|   Rough seeing | 30 |
|   Rough and difficult seeing | 50 |
|   Medium-difficulty seeing | 100 |
|   Fine seeing | 500 |
| Auto body: | |
|   Frame assembly | 50 |
|   Chassis assembly | 100 |
|   Final inspection | 200 |
| Bakery: | |
|   General | 30–50 |
|   Hand decoration | 100 |
| Bookbinding: | |
|   General | 70 |
|   Embossing | 200 |
| Chemical laboratory | 30–50 |
| Cleaning and pressing: | |
|   General inspection and | 50 |
|   spotting | 500 |
|   Hand pressing | 150 |
|   Repair and alteration | 200 |
| Electrical equipment manufacturing: | |
|   General | 50 |
|   Coil winding | 100 |
| Foundries: | |
|   General | 30 |
|   Molding and pouring | 50 |
|   Grinding and chipping | 100 |
|   Fine inspection | 500 |
| Glassworks: | |
|   General | 30–50 |
|   Fine grinding and polishing | 100 |
|   Engraving | 200 |
| Iron and steel manufacturing | 20–30 |
| Machine shops: | |
|   General | 50 |
|   Fine bench work | 100 |
| Paint manufacturing: | |
|   General | 30 |
|   Color matching | 200 |
| Printing industries: | |
|   Presses | 70 |
|   Composing | 100 |
|   Proofing | 150 |
| Sheet metal works: | |
|   General cutting and bending | 50 |
|   Scribing and inspection | 200 |
| Shoe manufacturing | 200–300 |
| Textile mills: | |
|   General | 30 |
|   Carding, spooling, and spinning | 50–150 (depending on color) |
| Inspection and weaving | 100 |
| Welding: | |
|   General | 50 |
|   Precision arc | 1000 |
| Woodworking: | |
|   General | 30 |
|   Finishing | 100 |

*Note:* Where color matching or working on darker colors is involved, higher illumination levels are required. In order to avoid spectral glare, when the work involves close visual inspection of materials with a high gloss finish, light levels should not be too high. Although the above standards appear to be very specific, the reader is cautioned not to assume that the specified illumination levels are either adequate or necessary. Each visual task should be examined in detail to determine seeing difficulty. A lighting mockup is desirable to test the proposed lighting scheme as well as the proposed levels.

## FORMULAS APPLICABLE TO GENERAL LIGHTING REQUIREMENTS

$$\text{No. of fixtures} = \frac{\text{footcandles desired} \times \text{room area per}}{\text{CU} \times \text{MF} \times \text{lamps per fixture} \times \text{lumens per lamp}} = \frac{\text{total lumens}}{\text{lumens per fixture}}$$

$$\text{Average footcandles} = \frac{\text{lamp lumens} \times \text{CU} \times \text{MF}}{\text{room area (ft}^2)} = \frac{\text{lumens per lamp} \times \text{CU} \times \text{MF}}{\text{area per lamp (ft}^2)} =$$

$$\frac{\text{lamp lumens per fixture} \times \text{CU} \times \text{MF}}{\text{area/fixture (ft}^2)}$$

$$\text{Area/lamp (ft}^2) = \frac{\text{lumens per lamp} \times \text{CU} \times \text{MF}}{\text{footcandle level desired}}$$

$$\text{Area/fixture (ft}^2) = \frac{\text{lumens per lamp} \times \text{no. of lamps per fixture} \times \text{CU} \times \text{MF}}{\text{footcandle level desired}}$$

$$\text{Total watts/ft}^2 = \frac{\text{footcandle level desired}}{\text{overall lumens per watt} \times \text{CU} \times \text{MF}}$$

where CU is the coefficient of utilization and MF is the maintenance factor.

## GUIDELINES FOR LIGHTING FIXTURE INSTALLATION: GENERAL ILLUMINATION

To provide the best seeing conditions as well as the most pleasant seeing conditions (i.e., minimum annoyance), consider the *luminous ceiling*. The intensity of the light will not have to be as high, and the distribution will be even and free from extreme brightness contrasts.

The next best system consists of *translucent fixtures,* which reflect light from the ceiling and also direct light through the fixture screen. The egg-crate configuration has the advantage of not collecting dirt or insects, which eventually cut down the amount of light obtained.

*Translucent globe fixtures* provide excellent general light distribution and minimize the collection of dirt, which can reduce the lighting effectiveness of the fixture.

*Note:* Suspended fixtures require sufficient ceiling height in order to prevent direct glare from the fixture.

Indirect lighting systems rely on "bounce light"; i.e., the light is never direct and thus annoying to the observer. This type of lighting is useful when there is no requirement for a high degree of brightness on the visual task.

The following are suggested applications:

Luminous ceilings—professional offices
Translucent fixtures—work spaces with low ceilings
Translucent globe fixtures—areas with higher ceilings
Indirect lighting—areas where reading or other critical tasks are not required

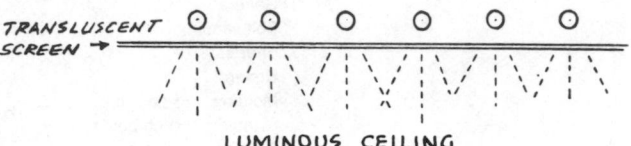

LUMINOUS CEILING

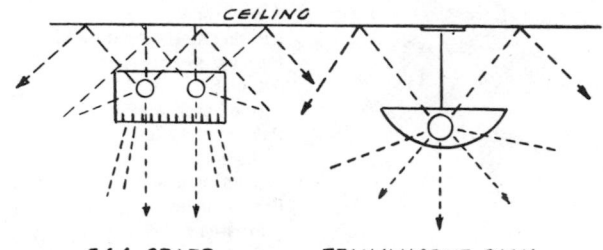

EGG CRATE          TRANSLUCENT DISH

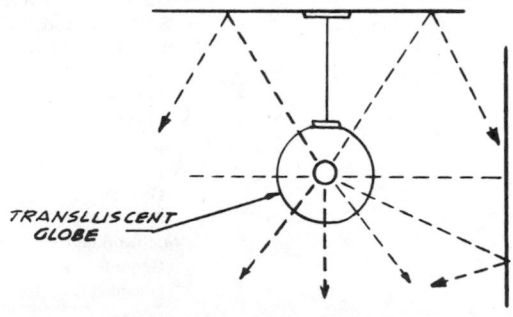

TRANSLUCENT GLOBE

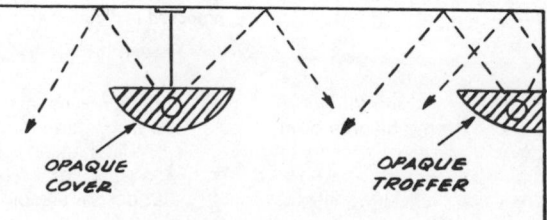

OPAQUE COVER          OPAQUE TROFFER

The *inset screened luminaire* is less desirable than previously discussed fixtures because it does not provide an opportunity for the light to reflect from the ceiling and thus improve general, even distribution. It can be effective, however, for "over-the-task" illumination as long as the fixture and task positional relationships remain constant.

The *hooded fixture* provides high efficiency for high-bay factory or sports applications.

The *cannister fixture,* although popular for decorative lighting, is perhaps the least desirable luminaire system for good seeing. It tends to create a "spotlighting" condition in which extreme shadow is created whenever anyone walks beneath the light. Although it can be used as a general illumination system, an inordinately large number of fixtures are required in order to provide reasonable light distribution.

In selecting any of the above, one should pay special attention to the manner in which bulb or lamp replacement and filter removal and cleaning can be accomplished.

Suggested applications include the following:

Inset screened fixtures: These are appropriate for a small office where one or two fixtures can provide even light distribution over the entire space.

Hooded fixtures: These are recommended for high-bay areas where fairly high-intensity light is desirable, but not for close individual visual work. The fixtures need to be high enough to minimize direct view glare from high-intensity lamps.

Cannister fixtures: These are recommended for decorative purposes and/or where there is a special need for localized illumination of a doorway threshold, etc. Care should be exercised in using such systems where the casting of shadows could create seeing problems.

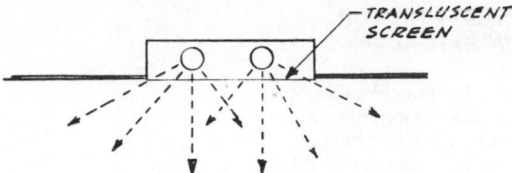

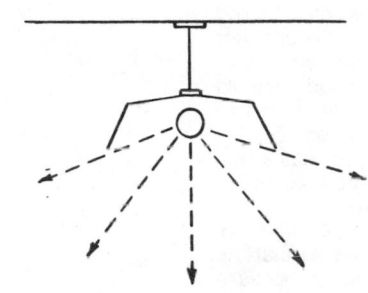

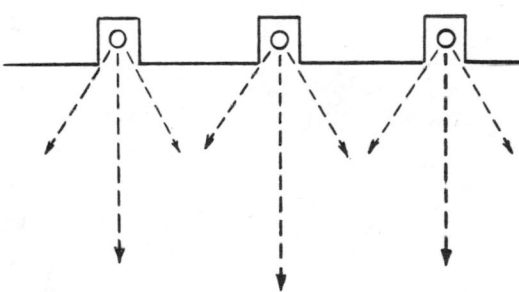

## FIXTURE SERVICING

The ease with which light fixtures can be cleaned and relamped is perhaps as important as the overall lighting system design concept. It has been estimated that a typical lighting system output is reduced by 25 percent six weeks following its installation. In selecting fixtures for any lighting system, consider the following servicing features:

1. The fixture should be readily accessible using only the simplest stepladder. Avoid placing fixtures at extreme heights or in positions where it becomes necessary to employ special devices and/or where there is access only through difficult crawl spaces.

2. It should be easy to remove intervening light control panels without damaging them, and the method of fastening and releasing these devices should be obvious and not require special tools. Many current schemes that are offered by manufacturers or designed on a custom basis seem to purposely camouflage the method for removing a ceiling panel, egg-crate screen, or translucent filter panel. Removable screens or panels should not be so large that they are unwieldy to handle efficiently, and they should not be made of material that is too fragile and easily damaged.

3. Light-filtering or light-distributing panels or covers should be made of materials that do not discolor with age.

4. Fixture housings should be designed so that there is sufficient clearance to get hold of the lamps and manipulate them in and out of sockets.

5. Choose fixtures which either allow dirt or dust to drop through naturally (open egg-crate screen) or which are semisealed against dust and dirt collection.

6. Preferably, use fluorescent fixtures that contain lamps no longer than 40 in (102 cm). Longer tubes are difficult to handle safely.

*Note:* Do not assume that by leaving relamping to a maintenance crew, damage to fixtures will be avoided. In spite of their supposed experience, many maintenance workers are more careless than regular staff personnel in the office.

## SUPPLEMENTARY LIGHTING FOR THE INDIVIDUAL WORKPLACE

Floodlighting techniques for individual work-place illumination should be designed so that the operator can select or adjust the best position for his or her seeing task; sometimes this changes.

The key consideration is to place the luminaire so that it does not cause light rays to reflect from the work surface into the operator's eyes and so that the light source (lamp) is not directly visible to the operator.

In general, the best lamp position is behind and to the side of the operator. If the operator is right-handed, the lamp should be to his or her left; if the operator is left-handed, it should be to his or her right. Thus it is necessary to provide a flexible positioning system.

The accompanying sketch illustrates a typical flexible-arm luminaire that provides a wide range of positioning possibilities.

For some special tasks, such as engraving, the light source must be placed so that the light rays actually cast shadows because these provide important guidance cues for the worker.

A low position for the light source, as shown in the accompanying sketch, is typical of the special situations in which shadow casting is advantageous.

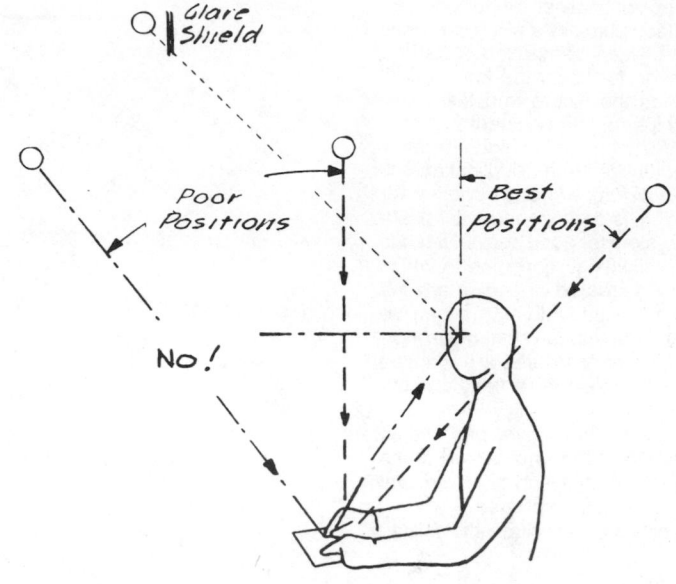

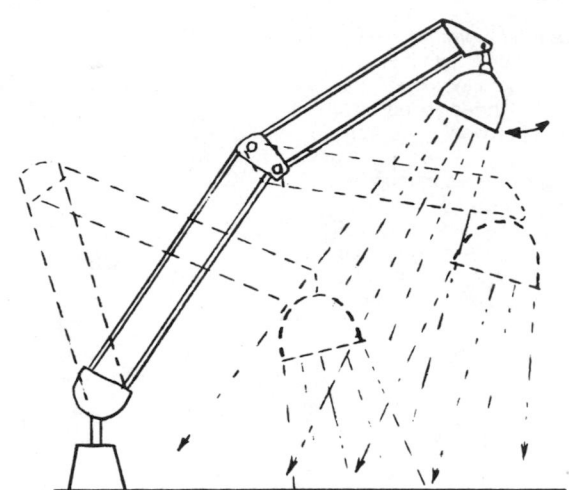

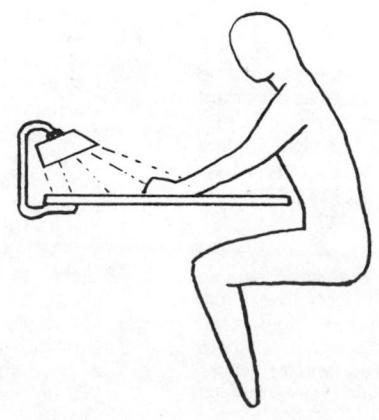

Illumination for personal grooming tasks requires careful fixture selection and placement. The lamp sources and adjacent surface colors used should provide light which is ample and properly distributed and which is flattering to the complexion.

The accompanying sketch illustrates an arrangement of fixtures in a pattern that illuminates both sides of the user's face, plus his or her hair. The illumination level at the person's face should not be less than 50 fc. Note that fluorescent light is suggested since it provides minimal shadow effects.

The wall lamps shown in the accompanying sketch are more typical of a decorative approach and tend not to provide as good illumination as one would like. When this type of lighting is used, it is extremely important that the lamps be placed at approximately "face height," or about 28 in (71 cm) above the seat. Lampshades should be as transparent as possible, and adjacent wall areas should be light in color. Warm colors are required since these are flattering to the complexion.

Flexible, pull-down fixtures are effective when shades are transparent.

Lampshades for all other lamps should also be transparent.

Various lamps are available that permit adjusting the position of the light for the best illumination.

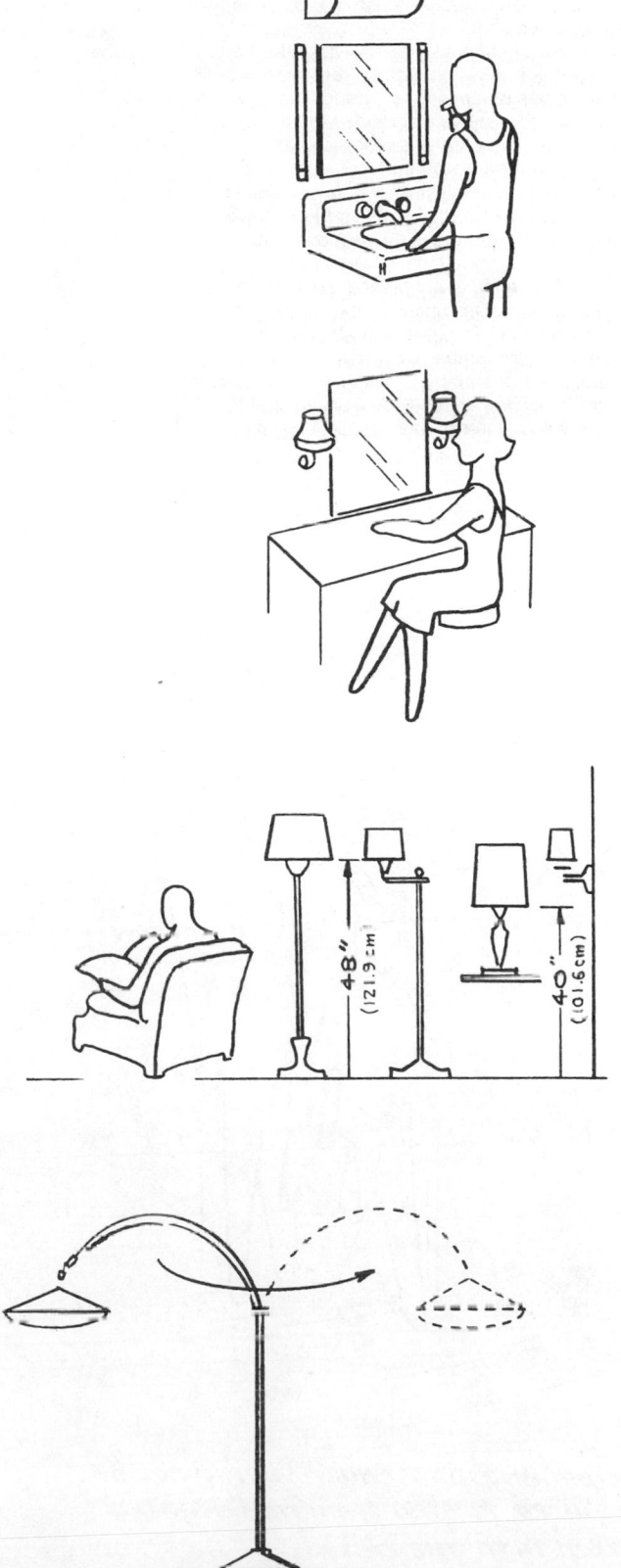

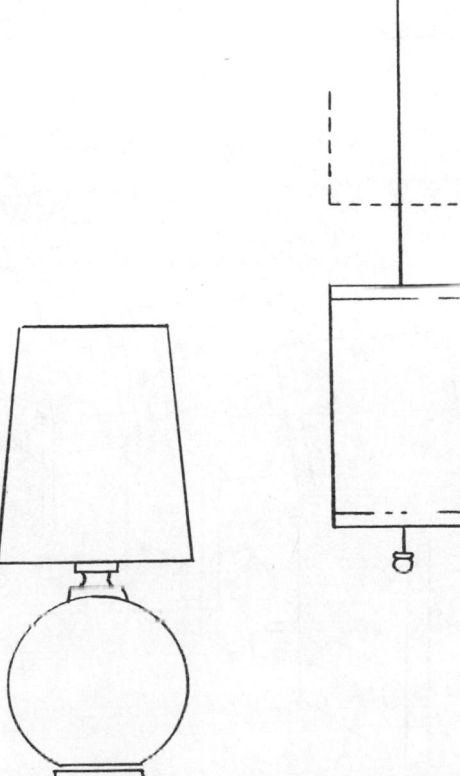

## ILLUMINATING FLOORS, WALKWAYS, AND STAIRS

An overhead floodlighting system is not adequate for illuminating floor, walkway, or stair surfaces at night or in darkened rooms. The light is blocked by the pedestrian's body, causing dangerous shadows.

An entryway such as that shown in the accompanying illustration should be provided with two types of luminaires: one to illuminate the faces of the guests standing at the threshold of the doorway and the other to illuminate the steps and the porch surface.

Although standard practice is to provide a minimum of about 3 fc, it is desirable to have at least 10 fc (for those whose eyesight is poor).

Luminaires should be placed so that the light patterns overlap whenever possible. If for some reason this is not possible, it is critical that good illumination be provided at the threshold of the stairs.

Aircraft passengers should not have to step out of a well-lit interior into the dark night.

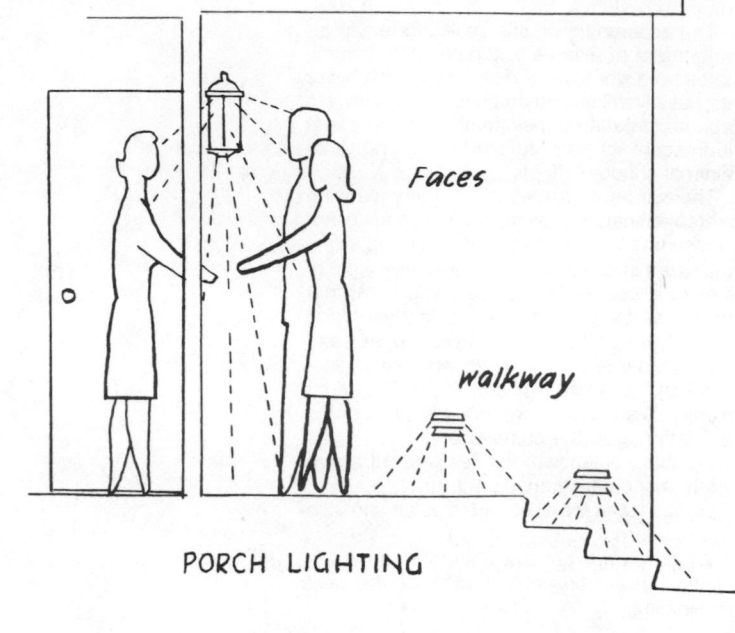

PORCH LIGHTING

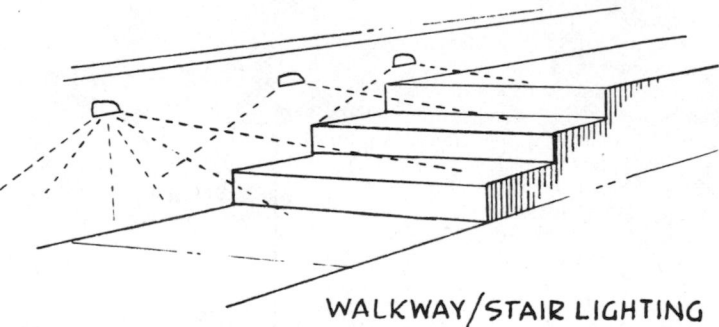

WALKWAY/STAIR LIGHTING

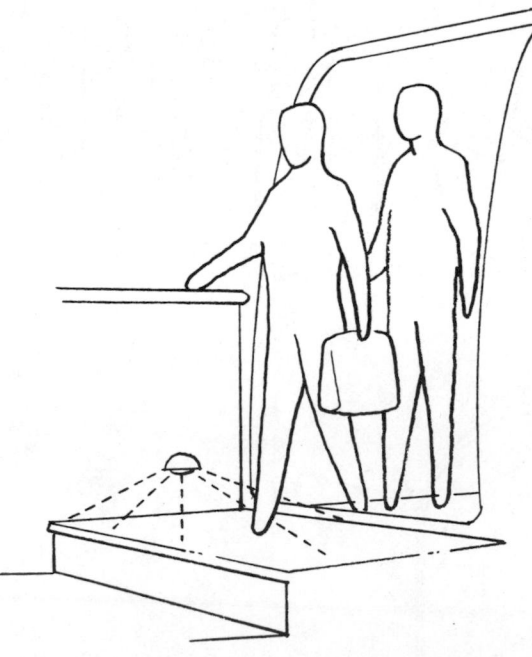

de-planing passenger should not be required to step from well-lighted cabin into the dark

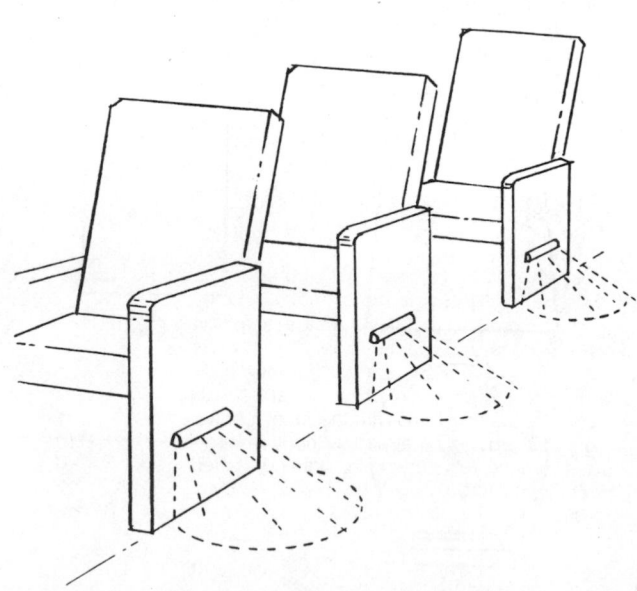

THEATRE FLOOR LIGHTING

## OUTDOOR FLOODLIGHTING

The principal objectives of outdoor floodlighting include the following:

1. Illumination of roadways, walkways, parking areas, runways, and taxiways
2. Illumination of obstacles, including people, vehicles, obstructions, and stairways
3. Illumination of signs and pathway markers
4. Decorative illumination of buildings, sculpture, etc.
5. Illumination of external work and storage areas
6. Illumination for the purpose of crime prevention

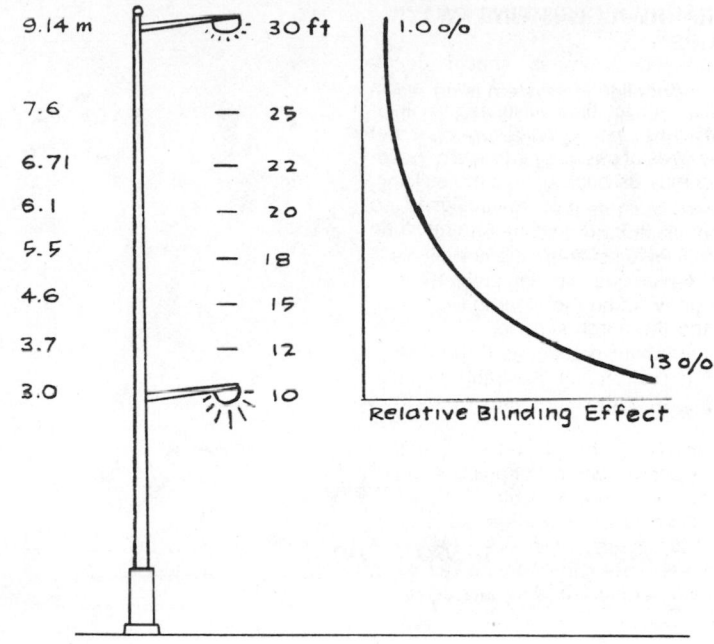

## ROADWAY LIGHTING[24]

Traffic safety requires proper illumination of the roadway, vehicles, and pedestrians. In addition, consideration should be given to unusual conditions that make visibility important, such as road work that is in progress.

Although automobile headlights theoretically provide a major share of roadway illumination, fixed roadway lighting is important, especially at intersections, on particularly hazardous curves, on bridges, in tunnels, and at railroad crossings. In addition to the obvious requirement to provide sufficient, relatively evenly distributed floodlighting, consideration must be given to the potential interference factors that floodlighting may create. A primary objective of good floodlighting practice should always be to put the light on the object to be seen, not in the eyes of the observer.

As indicated in the accompanying sketch, roadway floodlights should be high enough to minimize glare when the fixture (light source) is visible to the motorist. The position of the streetlight should be such that it will maximally illuminate the pedestrian who is crossing the roadway.

Light sources used to illuminate roadway signs must be designed and/or positioned so that the light does not shine into the eyes of motorists. Roadway lights must not create confusing shadows across the roadway.

## PARKING-LOT ILLUMINATION

Evenly distributed, nonglaring light should be provided for parking lots. Many good lighting systems are available to the designer, but some are more decorative than practical from the standpoint of effective seeing.

In certain situations it is critical that floodlights not interfere with airborne traffic; pilots have enough problems with the many confusing lights around an airport without having a poorly conceived parking-lot lighting system direct light into their eyes while they are trying to land.

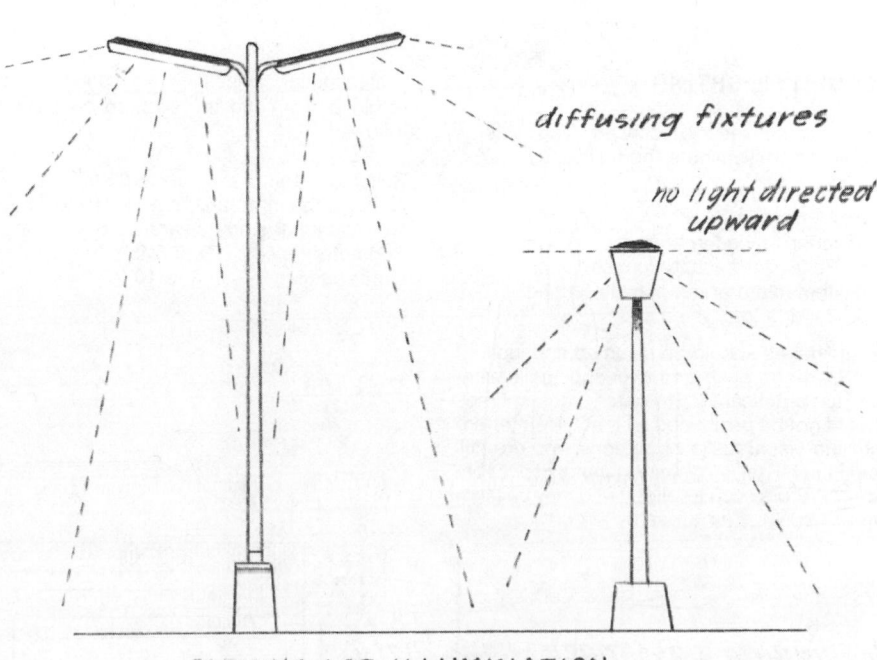

PARKING LOT ILLUMINATION

[24]For roadway lighting standards, see *IES Lighting Handbook*, Illuminating Engineering Society, New York.

## URBAN PARKWAY LIGHTING

Heavily traveled parkways should be illuminated throughout their peak load areas and especially at multiroad exchange points.

*Decorative lighting of buildings* must not shine in the eyes of passing motorists or pedestrians who may be approaching the building.

*Illuminated signs* should be designed so that the letters are self-luminous rather than being floodlit; deep shadows make such signs difficult to read.

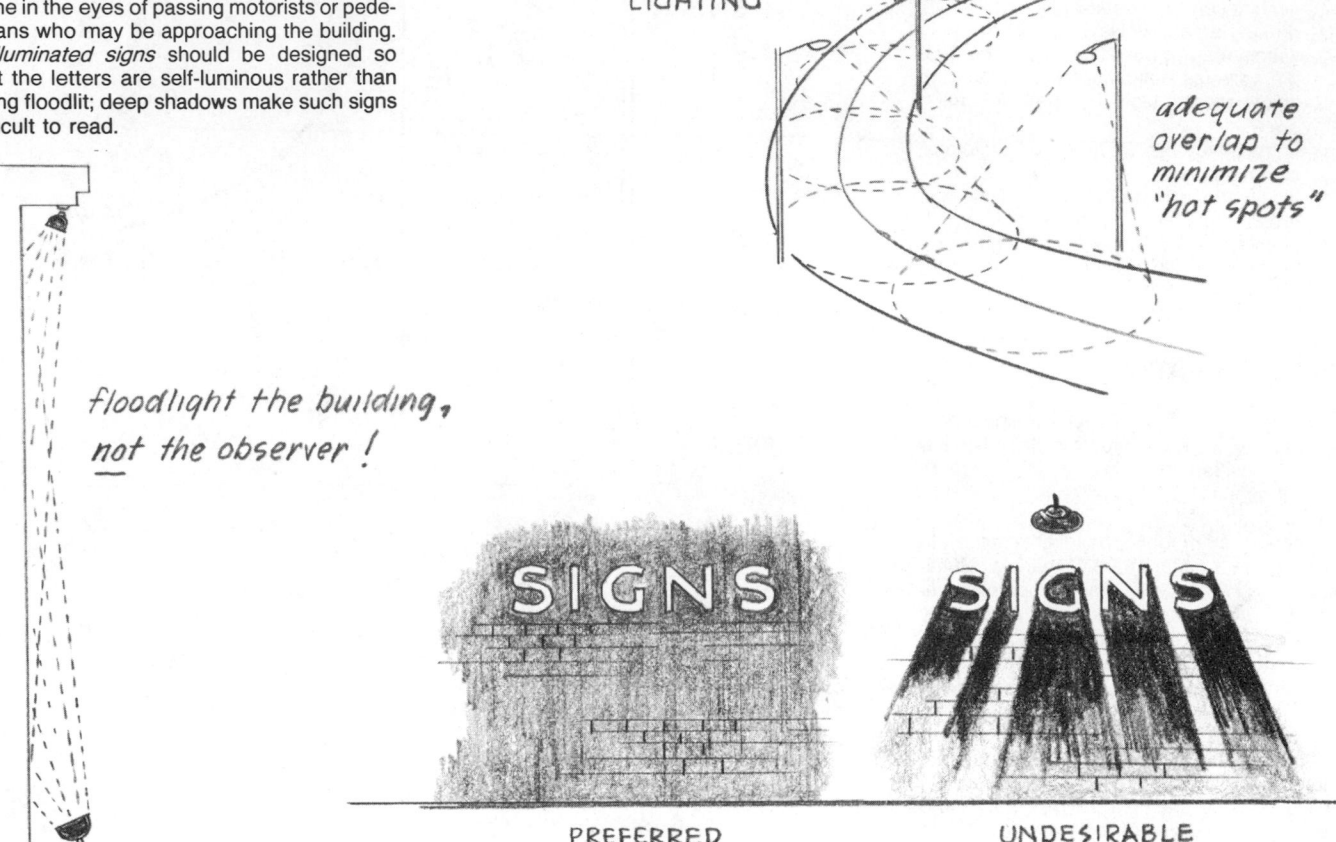

PARKWAY LIGHTING

adequate overlap to minimize "hot spots"

floodlight the building, not the observer!

SIGNS

PREFERRED

SIGNS

UNDESIRABLE

## SECURITY LIGHTING

Typical objectives for security lighting systems should be to illuminate the following:

1. All entrances to buildings
2. All gates
3. Inner fenced-in areas
4. Perimeter security fences
5. Immediate areas outside fences
6. Parking lots

Luminaires should be selected and located to provide as even and overlapping illumination as practicable. However, light sources should not be positioned so that they interfere with the visual tasks of persons who are still working at night or of passing motorists or special operators such as pilots, who may be landing or taking off at a nearby airport.

Illumination guidelines for specific points within and around the secured area are as follows:

Building entrances        10 to 30 fc
Areas adjacent to buildings        10 to 15 fc
General inner storage yard        5 to 10 fc
Perimeter fence        3 to 5 fc
Outer parking lot        3 to 10 fc

The accompanying sketch illustrates various positions that should be considered for security luminaires (the specific spacing and number of luminaires depend on the type of luminaire selected). Although white light provides the best illumination for normal atmospheric conditions, amber lights should be considered for locales where fog may be a more or less regular condition.

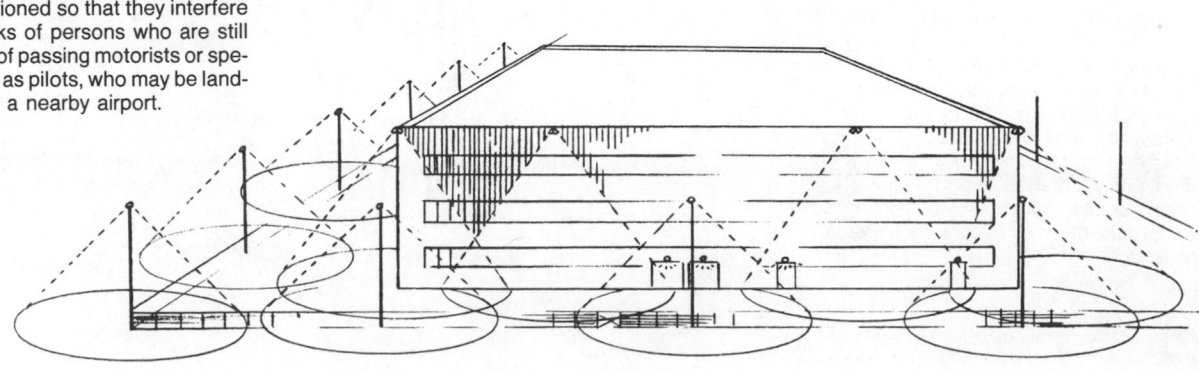

## SPECIAL CONSIDERATIONS FOR INTERIOR VEHICLE LIGHTING: GENERAL

Vehicles that are operated at night present special problems of reflections on windshields emanating from the lighted instruments, the illuminated structure, the seats, and the passengers. The accompanying illustrations point up some of these problems.

Although each vehicle configuration presents a special case, the best way to control these undesirable reflections is to use the following techniques:

1. Keep illumination levels no higher than is required to provide adequate seeing.
2. Locate luminaires so that they cannot be seen in the windshield.
3. Use glare shields over instrument panels.
4. Recess instruments that are to be illuminated.
5. Provide curtains between the passenger and driver compartments.
6. Control the slope of the windshield so that reflections (if they cannot be prevented) do not occur on the parts of the windshield which are critical for driving.
7. Use low-reflectance surfaces on structures that could be reflected in the windshield.
8. Provide continuously controllable illumination at the command of the driver.

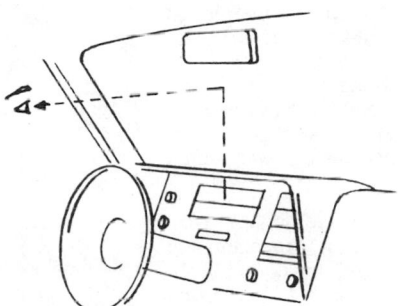

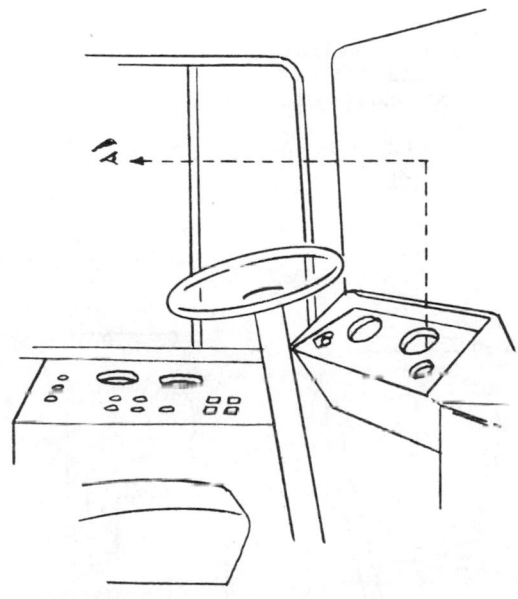

### Interior Vehicular Lighting

Although the accompanying illustrations are of aircraft, the principles apply to all vehicles that are used at night.

**FOR PASSENGERS** Provide both general overhead ambient light for moving in aisles and performing tasks in rest rooms. At seats, provide reading lights that are positioned so that passengers can control their own light, i.e., so that one passenger can see to read while his or her fellow passenger sleeps. Provide floor illumination so that passengers can find their way around the cabin even though the overhead lights or reading lights are extinguished. Provide emergency lights which illuminate automatically in the event of a crash (and/or can be turned on by a crew member for noncrash situations) and which are not dependent on the primary electric power source of the vehicle.

**FOR FLIGHT DECKS** Provide general overhead lights for nonflight maintenance, internal instrument and panel lights for normal flight operation, panel floodlights for use during thunderstorms or at other times when the panel lighting is not sufficient, and floor or deck illumination for moving about the compartment. Instrument lighting should be red to maintain dark adaptation. Both red and white instrument panel lights should be continuously controllable by the crew, and a sufficient number of circuits should be provided to allow different parts of the panel to be controlled separately.

Instrument lights must not reflect on any critical windscreen or window surface.

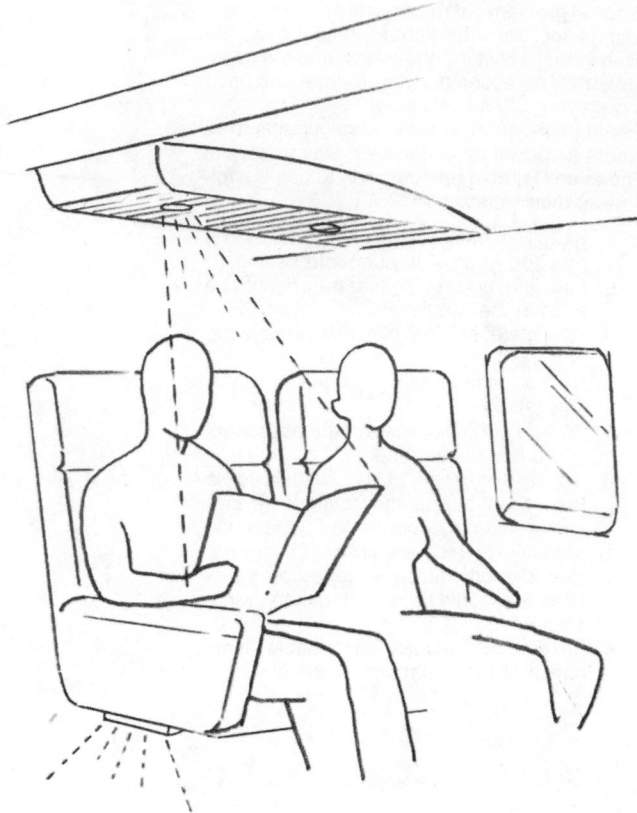

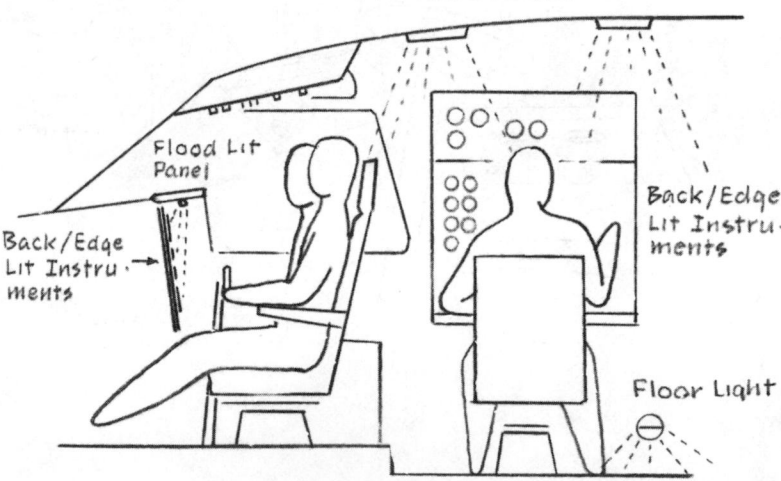

Overhead Floods

Flood Lit Panel

Back/Edge Lit Instruments

Back/Edge Lit Instruments

Floor Light

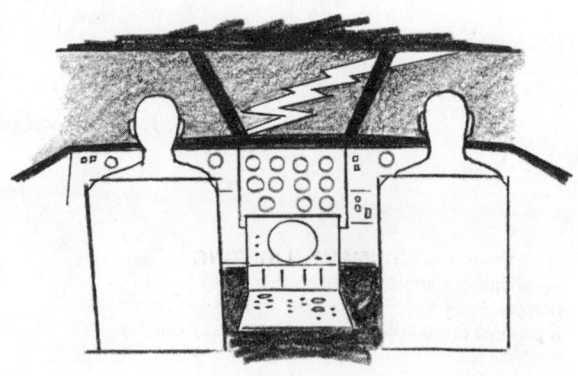

## Console Panel and Instrument Illumination

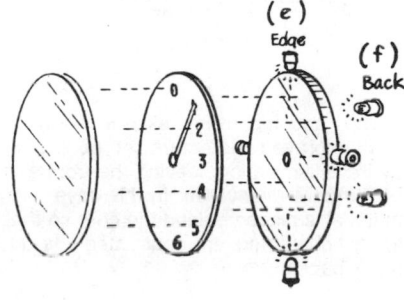

### Illumination Guidelines for Special Military Applications

| Task and Area | Illumination Level, fL | Lighting Equipment Features |
|---|---|---|
| Dark adaptation required | 0.02–0.1 | Rheostat control throughout the range; red light* |
| Dim-out and CRT target detection | 0.02–1.0 | Rheostat control throughout the range; red or white light* |
| Dim-out, CRTs, status boards, map boards, and console operation | 1.0–3.0 | Separate rheostat control for each piece of equipment; white light* |
| Hallways, stairways, rest rooms, and storage areas | 10.0 minimum | General fixed lighting |
| Equipment maintenance, tape reel removal, and intermittent record keeping | 25.0 minimum | General and/or supplementary fixed lighting |
| Classroom, drafting, teletype, and telemetry readout | 35.0 minimum | General and/or supplementary fixed lighting |
| Fine detail, optical, machining, and assembly work | 50.0 minimum | Special supplementary lighting |
| Critical visual tasks, e.g., surgery, small instrument repair, prolonged map work, and laboratory work using a microscope | 100.0 minimum | General plus supplementary lighting |

*There should be at least two lamps per instrument and/or panel label for all vehicle applications so that failure of a lamp will not leave the display completely unilluminated.

**FLOODLIGHTING TECHNIQUES** Adjustable-position luminaires are preferred over fixed fixtures because they allow the operator to place the fixture in a position that best suits his or her needs. Especially important is the operator's ability to illuminate panel control and display labels without causing light to fall on special displays such as the CRT.

As shown in the first sketch *(a)*, an egg-crate baffle is useful to control the direction of the light rays. It is also possible to inset a CRT so that it is in shadow, *(b)* as shown in the second sketch. The flexible arm support shown in the third sketch *(c)* allows the operator to minimize shadows that may be caused by control knobs and switches.

Care must be taken to make the contrast between labels and panel background sufficient so that the floodlighting will make the labels legible.

When there are several deep-set instruments on the console panel, it is desirable to create an internal display illumination system. Although this can be done by means of a floodlight within the instrument case *(d)* (in the accompanying sketch, note the baffle, which keeps light sources from being seen by the operator), an edge-lighting or back-lighting technique is preferred.

**INTERNAL INSTRUMENT LIGHTING** The floodlight illuminates both the dial face and the pointer. Edge lighting or back lighting requires a positive or negative dial mask through which light shines.

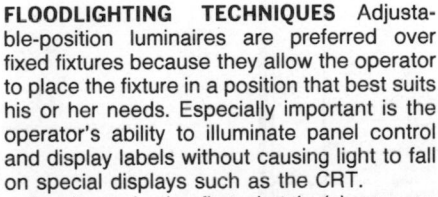

*Flood Lit*

floodlight illuminates both dial face and pointer

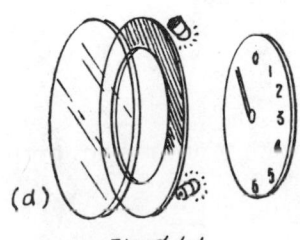

*Edge or Back Lit*

requires positive or negative dial mask through which light shines

**TRANSILLUMINATION TECHNIQUES** A transparent plastic core provides an effective means for transferring light from a lamp to screen print or engraved labels on a panel. The light from the lamp may be introduced either from the back or from the edge of the plastic core. Light rays bounce back and forth inside this core and break out wherever the plastic surface has an engraving or contact with another material.

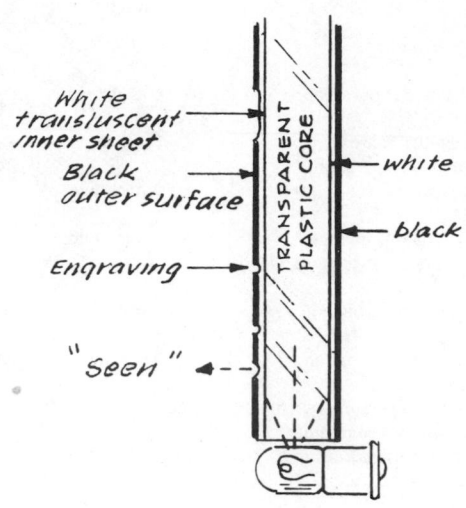

As illustrated in the accompanying sketch, a two-layer sandwich of white, covered with black material, is sealed to the transparent light-transmitting core. Labels are engraved through the black outer material (which is translucent to let the light pass). Wherever these engravings occur, the light will show through, making the character visible both at night, when the panel is lighted, and during the day, as the ambient light reflects from the white character layer.

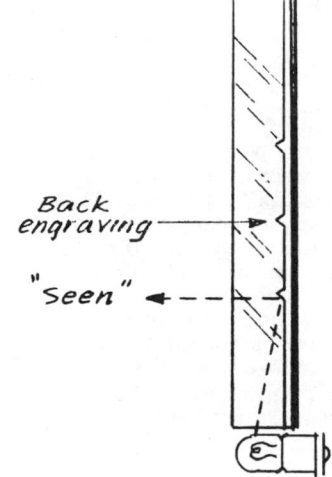

The accompanying sketch shows how back-engraved characters can be illuminated by transillumination, leaving the front surface of the core uncovered. In this case, a black material is placed behind the core so that the lighted characters are contrasted against this black background.

Light piping is also a useful technique in which transillumination is employed (as shown in the accompanying sketch). As long as the clear plastic rod is not in contact with anything, the light will travel from one end to the other and break out at the far end when it is in contact with some other material.

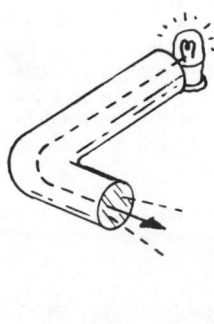

Transillumination techniques can also be used effectively for illuminating control knobs. The first approach shown in the accompanying illustration utilizes a cutout in the knob skirt to expose a back- or edge-lighted panel character.

In the accompanying sketch, a clear plastic knob is used. It collects light from the back- or edge-lighted panel and passes the light out through knob pointers which are engraved in the same manner as described earlier for the edge-lighted panel.

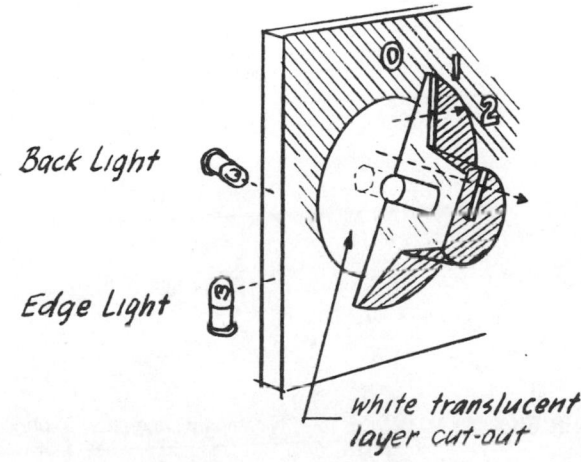

In placing lamps for back or edge lighting, one must be careful to locate them so that intervening holes (i.e., for knob shafts) do not cast shadows and thus prevent a label from being fully illuminated.

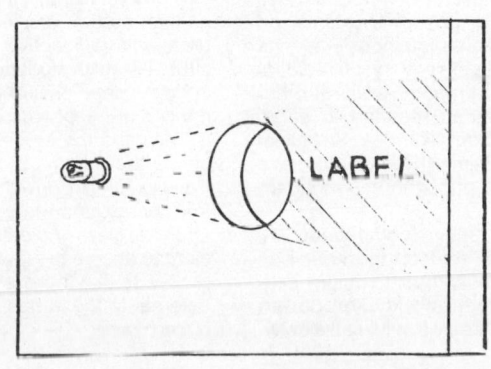

Different colors for labels can be obtained by using colored lamps or placing a colored filter between the lamp and the clear plastic light-transmitting core.

### Electroluminescent Lighting

Electroluminescent (EL) lighting is excellent in terms of even distribution. Two basic types of EL lamps are available: rigid sheets with steel, glass, or thermosetting plastic substrates and flexible EL lamps with metallic foil or thermoplastic substrates. These provide an evenly "lit" surface that can be used behind black-and-white screens or engraved layers such as described in the section on transillumination techniques. In effect, EL lamps provide a "sheet of light." Their greatest drawback is that they do not at present provide a very high light level, and therefore they are applicable only where the ambient light levels in the work space are fairly low (i.e., below about 15 fc). EL lamps are available in several colors, as indicated by the accompanying graph.

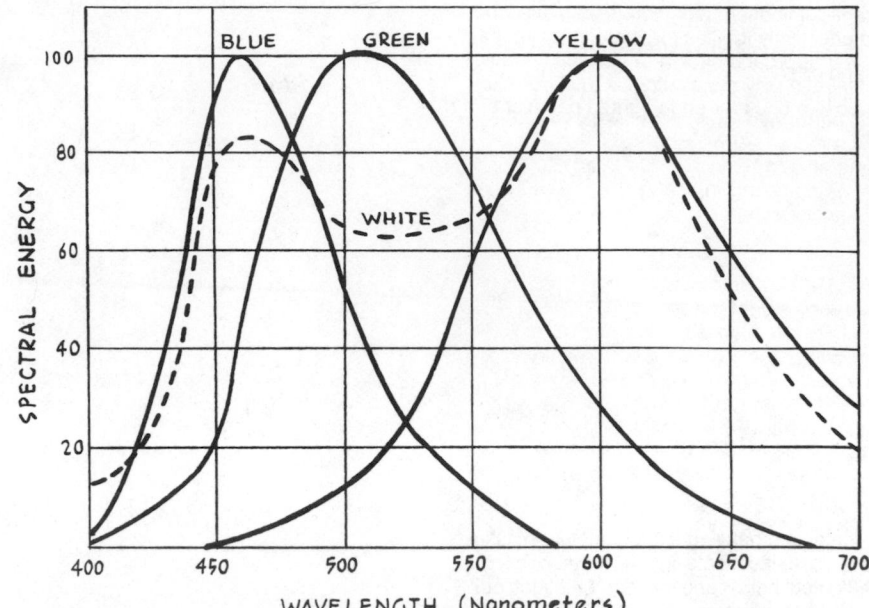

Spectral energy distributions for typical blue, green, and yellow EL phosphors, plus the SED of a "white" phosphor mixture (all peaks normalized at 100 percent).

### Radioactive Light Sources

The invisible emissions from radioactive materials can be used to activate phosphors. Small amounts of radium compound have been incorporated in phosphor paints on watch dials to produce self-luminous markings since shortly after the discovery of radioactivity. In recent years, the availability of other radioactive materials has resulted in safer, more stable self-luminous sources. Strontium 90, krypton 85, and tritium (hydrogen 3) have been used to activate phosphors in most applications.

Strontium 90 and radium compounds are available in powder or solution form and can be combined with phosphors to make self-luminous paints. Tritium and krypton 85 are gases and are used as fills in small sealed lamps having internal phosphor coatings.

The luminance of "radioactive light sources" is dependent on the phosphor used, the amount and kind of radioactive material, and the optical characteristics of paint vehicles and glass or plastics covers. Luminances as high as 5 fL can be attained. In most applications, radioactive sources are applied to surfaces which must be seen by their own light in fairly dark environments only, and luminances are kept well below 1 fL.

Tritium is the safest available material because the energy of its beta emission (electrons which activate the phosphors) is at very low energy, and there is no gamma radiation. Ordinary glass or plastics enclosures are sufficient to absorb the low-energy electrons completely so that there is no radiation of concern external to the lamps. The other isotopes require heavier shielding and have measurable levels of radiation external to the lamps, even in low-luminance versions.

Gaseous krypton 85 and tritium must be hermetically sealed in lamps. Strontium 90 is incorporated in phosphor paints but is sealed for safety and to prevent phosphor degradation.

The life of radioactive sources is determined by the rate of decay of beta emission and by degradation of the phosphor by the emission. The alpha emission of radium degrades phosphors severely, so that, although the half-life of radium 226 is about 1600 years, the half-life of radium paints is generally less than a year. Sealed lamps made with tritium, krypton, or strontium are likely to have half-lives not very much less than the half-lives of their beta emission, which are, approximately, tritium: 12 years; krypton 85: 10 years; and strontium 90: 28 years.

## Illumination for Armored Vehicles

Following exposure to extreme glare (sunlight on sand or snow on a bright day), illumination of a higher order may be required for difficult tasks performed within the armored vehicle compartment. Auxiliary or supplementary lighting should be provided for difficult tasks such as map and instrument panel reading. Use the guidelines shown in the accompanying table.

**ABSOLUTE MINIMUMS**  The illumination required for daylight operation (10.0 fc) is considerably more than is required for night operation. Each crew position should be provided with both a red lighting system for night and a white lighting system for daytime.

Maximum allowable brightness contrasts are as follows:

### Levels of Illumination for Vehicle Fighting Compartments

| Task | Night Operations (Red), fc | Daylight Operations (White), fc |
|---|---|---|
| Map reading | 1.0 | 10.0 |
| Clearing machine gun | 0.4 | 4.0 |
| Controls operation | 0.3 | 4.0 |
| Storage | 0.002 | 0.1 |

5:1 between the task and the adjacent surroundings
7:1 between the dimmest and the brightest instruments
20:1 between the task and remote surfaces
40:1 between the light source (or shy) and the adjacent surface
80:1 between the immediate work area and the remainder of the visual environment

*Note:* Sharp gradients (ratios of 10:1 or greater) should be avoided.

## Space Vehicle Illumination

### Illumination Required for Various Tasks in a Space Vehicle

| | Visual Acuity | Percent Contrast | Threshold Background Luminosity, fL Required Discriminations per second | Speed of Vision | Speed of Vision Factor, Discriminations per second | Acceleration Force g | Acceleration Force Factor | Percent Accuracy Required | Accuracy Factor | Minimum Luminosity Required, fL | Approximate Illumination Required, fc |
|---|---|---|---|---|---|---|---|---|---|---|---|
| | A | B | C | D | E | F | G | H | I | J* | K* |
| Map and chart reading | 0.25 | 90 | 0.004 | 30 | 20.0 | 4 | 3 | 1 | 50 | 12.0 | 20.0 |
| Instrument reading | 0.25 | 150 | 0.003 | 50 | 50.0 | 4 | 3 | 1 | 50 | 22.5 | 30.0 |
| Operating radios | 0.25 | 150 | 0.003 | 50 | 50.0 | 4 | 3 | 1 | 50 | 22.5 | 30.0 |
| Operating flight controls | 0.1 | 20 | 0.004 | 5 | 2.5 | 4 | 3 | 1 | 50 | 15.0 | 30.0 |
| Performing calculations | 0.25 | 50 | 0.01 | 20 | 10.0 | 1 | 1 | 1 | 50 | 5.0 | 10.0 |
| Preparing reports | 0.25 | 50 | 0.01 | 20 | 10.0 | 1 | 1 | 1 | 50 | 5.0 | 10.0 |
| Cleaning compartments | 0.1 | 30 | 0.03 | 5 | 2.5 | 1 | 1 | 1 | 50 | 3.75 | 10.0 |
| Scullery tasks | 0.25 | 20 | 0.1 | 20 | 10.0 | 1 | 1 | 1 | 50 | 50.0 | 60.0 |
| Maneuvering in the area | 0.05 | 30 | 0.0001 | 5 | 2.5 | 1 | 1 | 1 | 50 | 0.0125 | 0.02 |
| Entering and leaving the compartment | 0.05 | 30 | 0.0001 | 5 | 2.5 | 1 | 1 | 1 | 50 | 0.0125 | 0.02 |
| Reading for pleasure | 0.25 | 150 | 0.003 | 50 | 50.0 | 1 | 1 | 1 | 50 | 7.5 | 10.0 |
| Playing recreational games | 0.2 | 50 | 0.006 | 30 | 20.0 | 1 | 1 | 1 | 50 | 6.0 | 10.0 |
| Preparing food | 0.25 | 90 | 0.004 | 30 | 20.0 | 1 | 1 | 1 | 50 | 4.0 | 5.0 |
| Eating food | 0.25 | 30 | 0.035 | 5 | 2.5 | 1 | 1 | 1 | 50 | 4.375 | 10.0 |
| Inventorying stores | 0.25 | 90 | 0.004 | 20 | 10.0 | 1 | 1 | 1 | 50 | 2.0 | 5.0 |
| Self-relief of personnel | 0.1 | 30 | 0.003 | 5 | 2.5 | 1 | 1 | 1 | 50 | 0.375 | 1.0 |
| First aid | 0.5 | 50 | 0.1 | 15 | 7.0 | 1 | 1 | 1 | 50 | 35.0 | 60.0 |
| Personal cleanliness | 0.5 | 50 | 0.1 | 5 | 2.5 | 1 | 1 | 1 | 50 | 12.5 | 20.0 |
| Emergency repair | 0.25 | 30 | 0.035 | 10 | 4.5 | 1 | 1 | 1 | 50 | 6.875 | 10.0 |

$$*J = (C)(E)(G)(I); \quad K = \frac{J}{\% \text{ reflection}}$$

**Aircraft Display Illumination**

**Recommendations for Display Lighting**

| Condition of Use | Lighting Technique | Brightness of Markings, fL | Brightness Adjustment |
|---|---|---|---|
| Indicator reading, dark adaptation necessary | Red flood, indirect, or both, with operator choice | 0.02–0.1 | Continuous throughout range |
| Indicator reading, dark adaptation not necessary but desirable | Red or low-color-temperature white flood, indirect, or both, with operator choice | 0.02–1.0 | Continuous throughout range |
| Indicator reading, dark adaptation not necessary | White flood | 1.0–20.0 | Fixed or continuous |
| Panel monitoring, dark adaptation necessary | Red edge lighting, red or white flood, or both, with operator choice | 0.02–1.0 | Continuous throughout range |
| Panel monitoring, dark adaptation not necessary | White flood | 10.0–20.0 | Fixed or continuous |
| Possible exposure to bright flashes, restricted daylight | White flood | 10.0–20.0 | Fixed |
| Chart reading, dark adaptation necessary | Red or white flood with operator choice | 0.1–1.0 (on white portion of chart) | Continuous throughout range |
| Chart reading, dark adaptation not necessary | White flood | 5.0–20.0 | Fixed or continuous |

## GUIDELINES FOR SELECTION AND UTILIZATION OF SOUND SYSTEMS

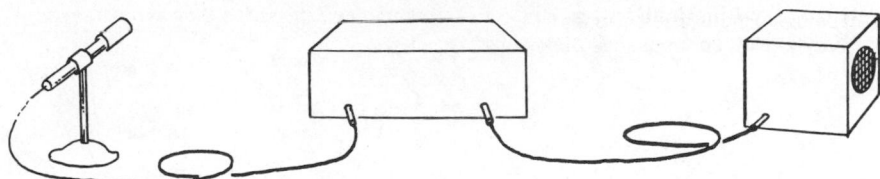

SIGNAL CAPTURE    SIGNAL AMPLIFICATION    SIGNAL DISTRIBUTION

**Signal characteristics.**

The quality of a sound system will be only as good as the quality of the poorest element within the system, e.g., the microphone, amplifier, loudspeaker, or headset. However, the ultimate satisfaction of the listener also depends on the proper handling of the acoustic environment and on the location of the input and output elements of the system.

First, select quality components; the microphone, amplifier, and speakers or headsets should faithfully capture and transmit the speech or music without introducing extraneous noise or distortion.

Second, determine the environmental conditions within which the sound must originate and within which the output must be heard and consider where and how to place and/or use the input and output components. Consider the following.

| Signal Type | Signal Characteristics Frequency Range, Hz | Dynamic Range, dB |
|---|---|---|
| Minimum speech recognition | 75–7,000 | 50 |
| Voice recognition | 50–11,000 | 60 |
| Music and vocalists | 30–12,000 | 80 |
| Orchestral instruments | 15–16,000 | 125 |

### 1. Microphones
   a. Single versus multiple inputs
   b. Mobility of speaker, vocalist, or instrumentalist
   c. Speaker's manual constraint (e.g., to hold the microphone)
   d. Ambient noise
   e. Special speaker constraints (headgear, oxygen mask, etc.)

When special microphone-use situations require that the microphone be contained in a mask, make sure that the mask is designed so that the speech is not distorted by the enclosure; i.e., it should not be muffled or sound as though the speaker is talking in a barrel, and breathing sounds should not be sufficient to mask the speech.

If a console operator's hands are so completely occupied that he or she cannot pick up a microphone, a fixed microphone should be placed as close to the user's mouth as practical. If the ambient noise level is high, consider using a highly directional and/or noise-canceling microphone mounted on a flexible arm so that the operator can adjust the microphone to the most convenient position.

Lectern microphones should not be highly directional because speakers tend to wander or turn away. Either mount the microphone on an adjustable arm or use a lapel microphone so that the speaker can move about without increasing the distance between his or her lips and the microphone.

Professional performers generally learn to handle the independent, highly directional microphone quite well, keeping it close to their lips and aligned for the best sound capture. It is, however, desirable to provide a microphone stand so that the performer can easily retrieve and/or store the microphone between numbers.

When the communications operator has a free hand, the independent hand-held microphone is effective since the user can follow the operator's lip position even though his or her head moves.

Select omnidirectional microphones when it is necessary to pick up the voices of several people.

Omnidirectional microphones are effective for small groups of people who are sitting around a desk or conference table. The average distance of talkers should be calculated, and more than one microphone should be provided, depending on the effective pick-up distance of the microphone selected.

If ambient noise levels are high, however, use directional and/or noise-canceling microphones, selecting ones that are portable so that they can be passed around the table or conference area.

When vocal or instrumental groups need to have a combination of group and/or individual amplification, use a combination of omnidirectional and directional microphones so that the sound control engineer can selectively choose overall and/or individual sounds, i.e., pick up an individual soloist or instrumental choir and yet maintain sufficient accompaniment to provide proper solo-accompaniment balance.

Performances on large stages often require the use of a number of microphones strategically positioned so that the individual voices of actors positioned at different points on the stage can be balanced in terms of average loudness. If the stage is shallow, this usually can be accomplished with a row of microphones arranged across the front of the stage.

When the stage is deep, several microphones usually have to be suspended both laterally and in depth throughout the stage area.

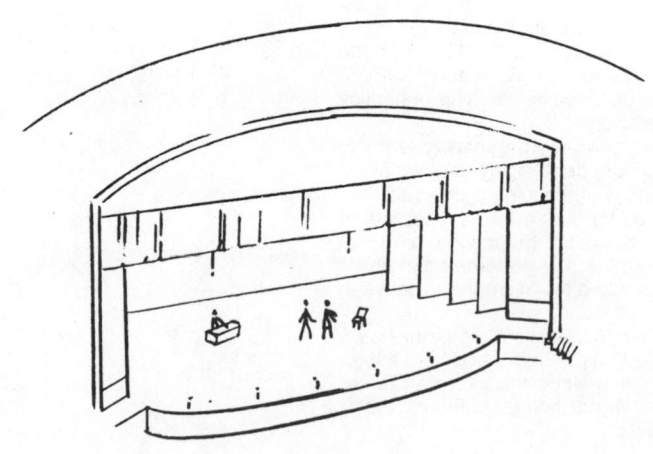

The most desirable approach is to use cordless microphones so that an individual actor will have complete mobility and yet not be able to "walk away" from the microphone.

## Microphone Packaging

Although it may seem somewhat trivial to the audio engineer or sound system designer, the size and shape of the hand-held, head-mounted, or chest-mounted microphone are important to the user. Several use-related considerations are discussed below and illustrated in the accompanying sketches.

**HEAD-MOUNTED MICROPHONES** When choosing a head-mounted microphone assembly, make sure that the microphone can be properly adjusted, i.e., positioned directly in front of, and close to, the speaker's lips. The assembly must not be heavy (no more than about 1 lb, 0.5 kg), and the support must remain secure at the position selected; i.e., it should not be easily pushed away or slip easily out of the adjusted position.

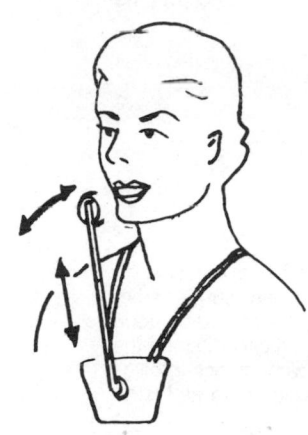

**CHEST-SUPPORTED MICROPHONES** If the selected microphone unit is heavy, choose a chest-supported assembly. This, too, must have sufficient adjustability to allow the user to position the microphone in front of, and close to, his or her lips.

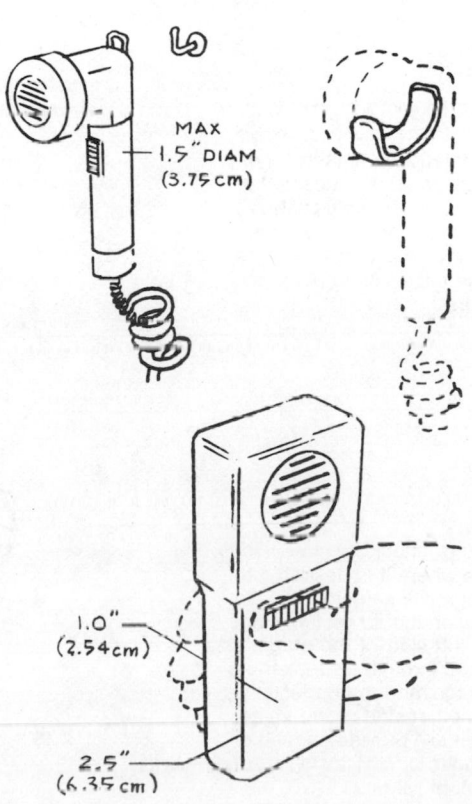

**HAND-HELD MICROPHONES** Select microphone packages that are light in weight and easy to grasp securely.

The configuration shown in the accompany sketch ensures that the operator's hands will not cover the microphone opening, and the switch is centrally located so that the microphone can be operated easily with either hand. Such microphone packages should have some means for conveniently storing them while they are not in use (it is undesirable to lay a microphone on a desk because it invariably falls to the floor). To minimize cord interference, pick a microphone that has a coiled power cord. Walkie-talkie packages often are too large to hold conveniently. The maximum dimensions shown should be sought. Note also the centrally located switch and the setback to reduce the likelihood that the operator's thumb will cover the microphone opening.

In addition, determine the listening environment and decide whether the output should be via earphones (because of noise or the fact that the listener has to be protected by special headgear) or loudspeakers. Loudspeakers are preferred for reasons of comfort and also because they allow the sound to reach more listeners. Consider the following.

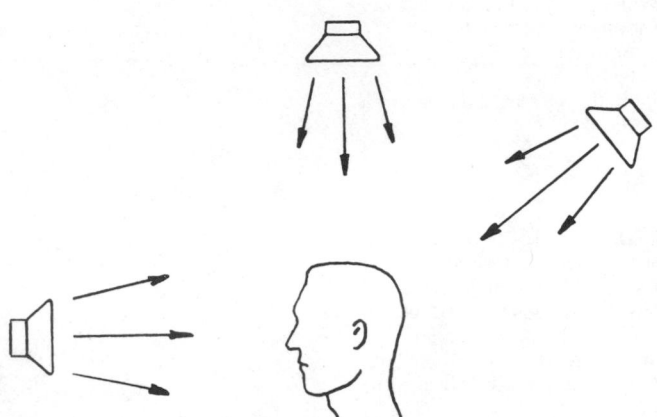

### 2. Loudspeakers

As a general rule, loudspeakers should be positioned so that the sound is directed "at" the listener.

*Direct sound at the listener*

Although speakers can be positioned so that the sound reflects off hard surfaces before reaching the listener, this should be done only when music is being amplified, since the listener is concerned not with speech intelligibility but with blended sounds from an instrumental group.

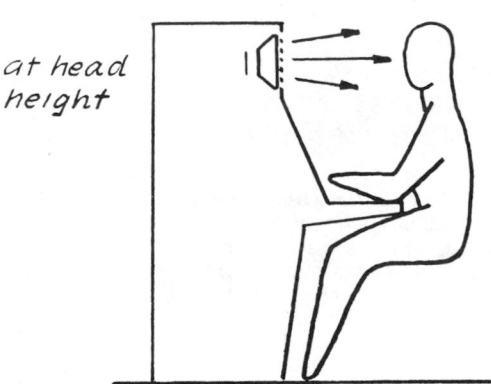

*at head height*

When the sound is intended primarily for an individual listener, place the loudspeaker close to the listener at approximately ear level.

Use several loudspeakers when the sound is meant to reach many people at the same time. This reduces the necessity to overdrive individual loudspeakers.

Several loudspeakers should be used for large stage situations where it is desirable to maintain audio-visual positional integrity with what the listener sees on the stage; i.e., what is happening on the left side of the stage is "miked" and distributed through a speaker on the left side of the stage, and the opposite is accomplished on the other side of the stage. The center speaker should provide enough of both left and right sounds to fulfill the needs of the most laterally distant listeners.

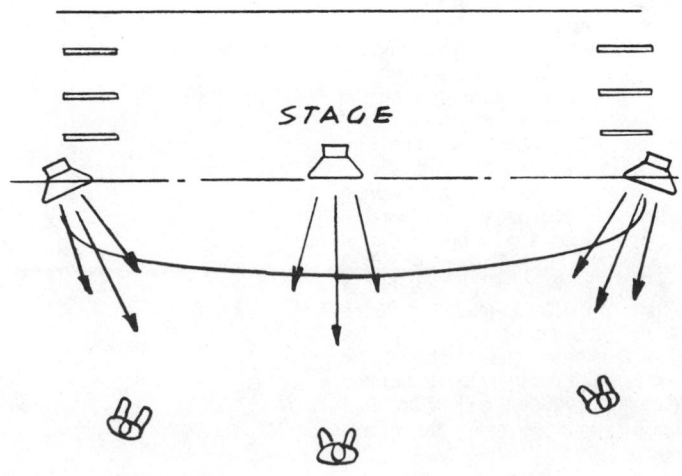

*STAGE*

Even when the speech of a single lecturer is being amplified by a sound system, it is important that the voice of the lecturer seem to be originating from where he or she is standing. The loudspeaker should be located either in the lectern (directed toward the audience) or above the lecturer's head. Avoid locating the loudspeaker too far to one side and never place it behind the audience. If the auditorium is too large or if the ceiling is too low for the voice of a single speaker to reach all listeners in the back of the room, use speakers (in addition to the front speakers) at several points on the ceiling in order to provide adequate sound levels so that the people in the rear can hear.

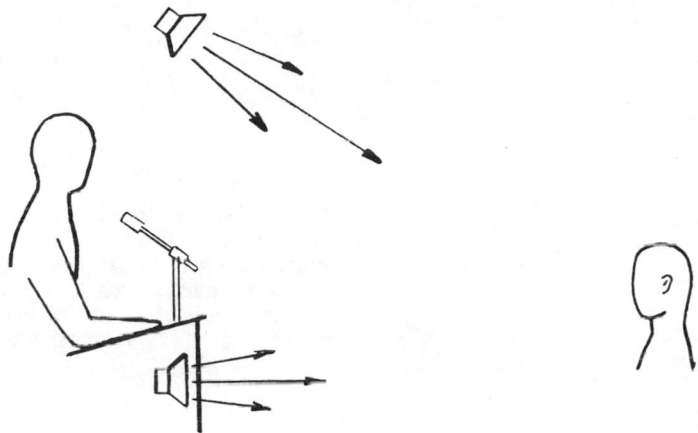

The preferred speaker location in very large auditoriums is shown in the accompanying sketch. Use a number of speakers (some of which should be designed especially for higher frequencies, and others for low frequencies) in order to provide adequate sound levels without introducing distortion.

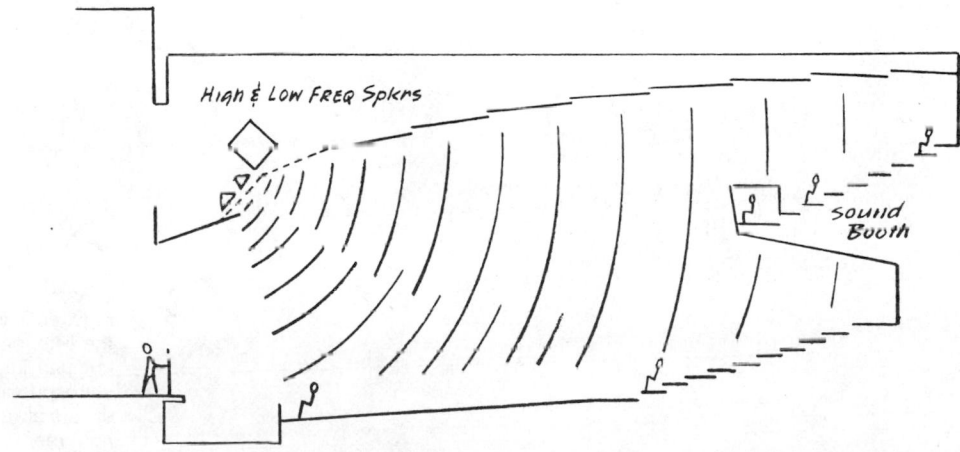

NOTE: Sound system should be located amidst audience, not backstage.

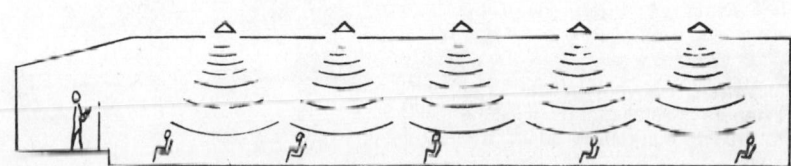

Sound amplification for motion pictures requires speaker location based on the same principles used for live performances; i.e., the sound should seem to come from the visual source (screen area). In the case of smaller screens, for which it is not necessary to create stereophonic relationships between opposite sides of the screen, a single overhead speaker is adequate.

For "wide-screen" movies, it is helpful to provide a number of speakers in order to create the audio-visual spatial impact necessary to provide total audio-visual fidelity. The preferred method is to utilize a sound-transparent screen through which the sound is projected.

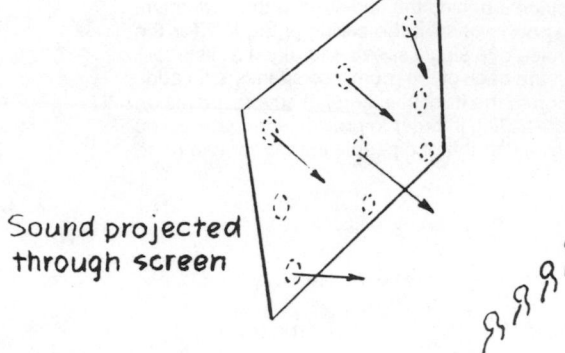

Sound projected through screen

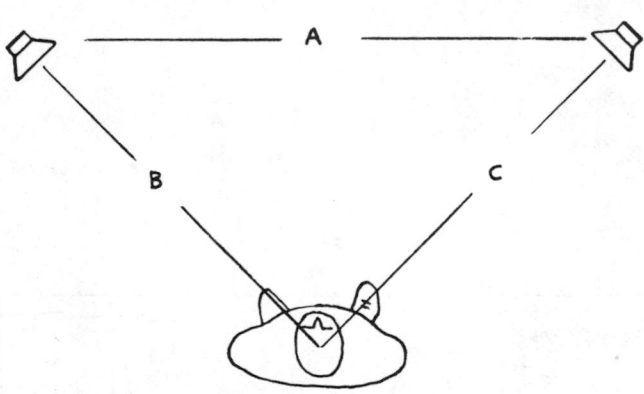

For the most realistic stereophonic response, loudspeakers should be placed so that the distances A, B, and C shown in the accompanying sketch are equal.

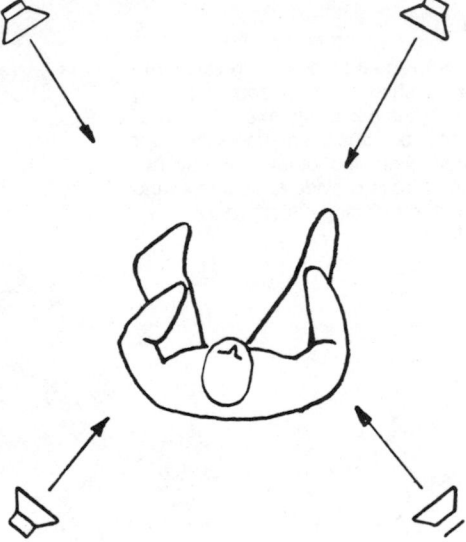

As long as no visual-audio association is required, it is often more dramatic to surround the listener with sound, as shown in the accompanying sketch. If the performance is stereophonic, the left front and left rear speakers are always coupled, and the right front and right rear speakers are coupled.

*Note:* Whenever practical, as with individual listeners or small groups of listeners, it is wise to provide controls to adjust stereo relationships.

Dynamic sound movement can further enhance the listener's enjoyment, especially in the case of motion picture presentations.

As shown in the accompanying sketch, a series of speakers are arranged laterally across the audience's viewing-listening axis, and speakers are successively reinforced from left to right (or fore to aft for nonvisual sound dynamics).

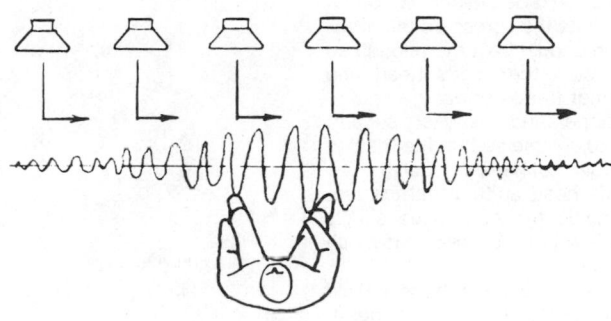

*Note:* Such dynamic motion simulation can also be used with earphones since the process is accomplished electronically.

Multichannel listening via a single loudspeaker is generally less satisfactory than listening by means of several speakers (each channel is piped through its own independent speaker).

To be most effective, speakers (and channels) must be arranged at least 10° apart in order for the listener to differentiate reliably (i.e., to localize the sound from an individual speaker).

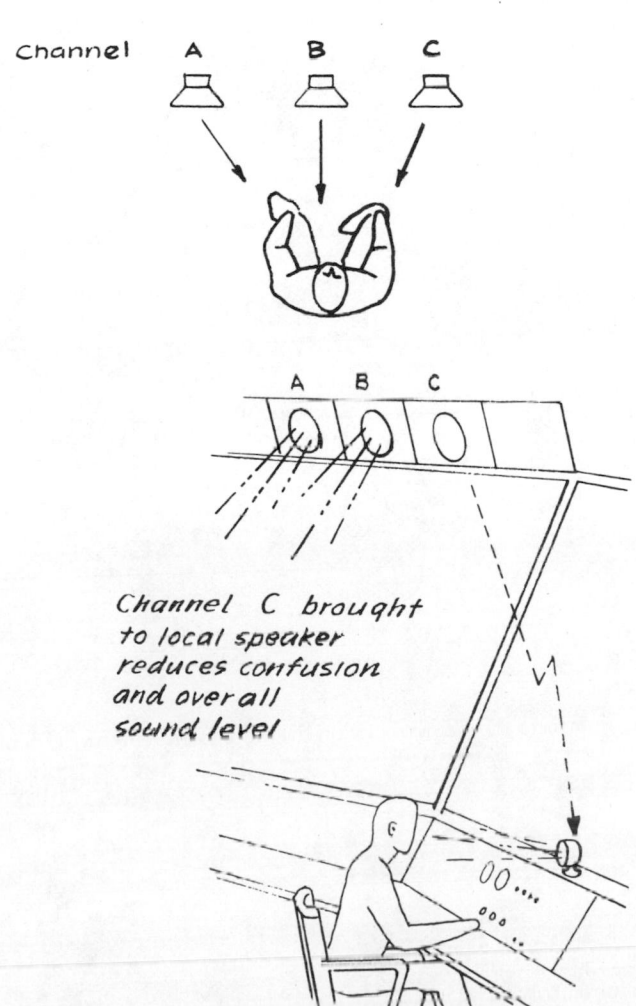

Channel C brought to local speaker reduces confusion and over all sound level

It is desirable to provide the operator with a separate volume control for each speaker and also to give the operator the alternative of selecting an automatic gain control when signal strength varies significantly.

In some situations (e.g., a busy air control tower), it is useful to provide a local speaker at each operator station so that when things become so busy that each operator needs to maintain constant contact on a single channel, each one can "pull down" the channel he or she needs to monitor and thus minimize its contribution to the overall sound confusion within the tower.

**3. Earphones and Headsets**

Selection of the proper headset and/or earphones requires careful consideration of the following:

a. Fit is perhaps the most important consideration as far as the user is concerned. The headset should be light in weight and adjustable.

b. Earphones should be cushioned so that they feel soft as they press on the head. The cushion should be large enough so that it contacts the user's head and cheek, but not the outer ear.

c. The earphone and headset support should be adjustable so that the bracket assembly does not have to touch the top of the user's head and so that the tension that holds the earphones snugly against the head will not create too much pressure.

d. The cavity inside the earphone should not be so large that the sound-generating element is too far away from the ear canal.

e. Headsets and earphones designed for maximum ambient noise occlusion include an ear insert. Preferably, this should be custom molded to fit each ear. If not, the ear insert must be extremely soft and pliable so that it can be adjusted for a tight fit without creating undue pressure. Ear inserts must be made of non-toxic materials.

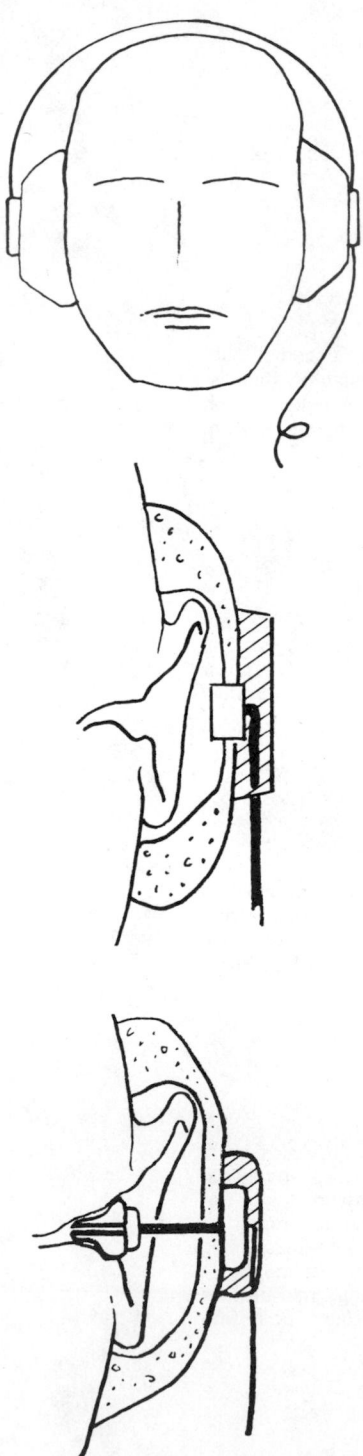

### Sound System Operating Controls

Whether the system is complex or simple, it should have a control station that does not require a sound system engineer to operate it. The following considerations relate to the design of the system operator's interface:

1. Provide a separate ON/OFF switch with a light to indicate when the system is powered.

2. Provide a separate control over the primary system features that the nontechnical person would expect:
   Speaker or channel selection
   Microphone selection[25]
   Volume adjustment
   Bass and treble adjustment

3. Apply good human engineering principles to panel layout (functional grouping, sequence of operation, proper labeling, etc.).

4. Provide easily read and understood visual displays (meters, tele-lights, etc.).

5. Use appropriate control types for selection and continuous adjustment functions.
   Provide feedback to keep the operator continuously advised.
   Use push buttons, toggle switches, or rotary switches, which are easier and less confusing to operate than slide and thumb-wheel controls.

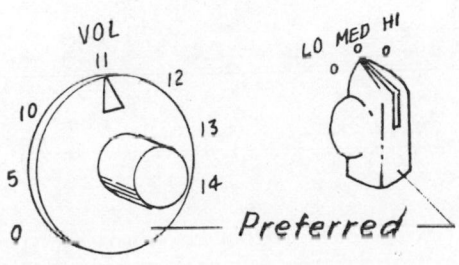

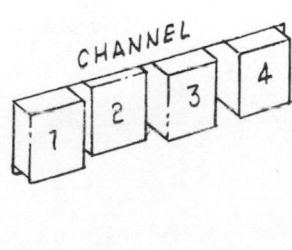

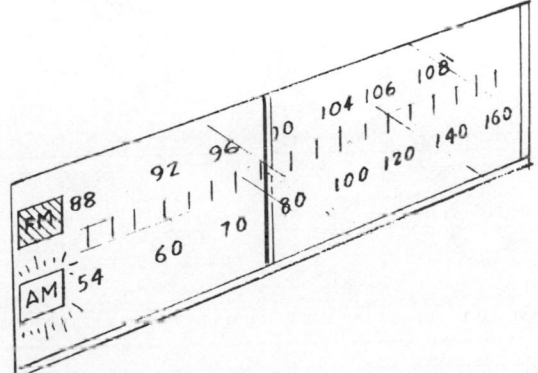

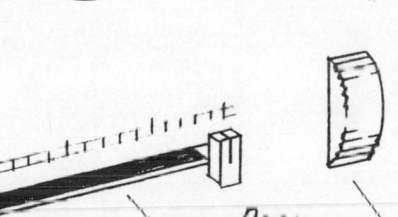

# Component and Product Design

**3**

## INTRODUCTION

This chapter deals with the more immediate interface between a system, subsystem, or product user and the hardware. Although a system or subsystem may be well conceived in terms of what it does and what it requires the user to do, it can fail or become operationally less effective if the immediate interface design contains features that cause the user difficulties in making effective contact with it, either directly in terms of physical contact or indirectly in terms of perceiving through visual or auditory channels.

In many cases, designers do not have the time to design brand-new components for every system with which they are concerned, nor is it always appropriate for them to do so. Therefore, as a first step, designers should look for components and product items that have already been produced and proved effective. Although most designers are well aware of this approach and search the marketplace for components and products that can be incorporated into new systems or subsystems, the tendency is to concentrate on the engineering features and/or on the aesthetic features that products appear to possess. If human factors are considered at all, this area is treated from the point of view of personal assessment, with minimal critical evaluation in terms of specific human capabilities and limitations across the broad spectrum of anthropometrics, psychological or physiological interaction, and eventual user performance effectiveness.

Many people tend to believe that when the general system concept is right, the details at the component level are relatively insignificant. If this were true, "superstars" would not be constantly looking for a better golf club, tennis racket, tool, violin bow, or data keyboard, for example. To experts, details are the key to performing better than their competitors.

Space, of course, does not permit discussion of every possible component or product that a designer might be concerned with. However, it is hoped that the information presented in this chapter, which covers some of the most common components that one typically has to consider in developing a system or subsystem, includes a sufficiently large number of variables and examples of how these impact on user performance at the component level to allow the reader to extrapolate to products that are not covered in this handbook.

In searching the marketplace for off-the-shelf components and products to use in a system, one should allow sufficient time not only to cover as many products as possible but also to evaluate each of those items which offer a reasonably effective solution to a particular problem. A designer who wants to be satisfied with a new automobile, for instance, should not only examine the overall engineering characteristics of the vehicle but also get in and drive it enough to recognize that the driver station is poorly laid out, for example, or that the instruments are difficult to see and read. Too often, consumers have been brainwashed into believing that difficult controls and displays, uncomfortable furniture, and hard-to-operate fasteners or locks are things they have to live with. This is seldom true.

Most poor designs are due to the fact that the designer was not aware of human factors and therefore spent all his or her time making sure that the engineering features were satisfactory, that the component or product was easy to manufacture, or that the product was attractive. As a result, the consumer reacts like a man who gets used to a rock in his shoe; even though the rock makes a sore spot on his foot or causes him to walk in a peculiar way, he tends to go about his business without stopping to remove it.

Sometimes after examining the marketplace, it becomes apparent that there are no components or products available that are well engineered from a human factors point of view. One has to create something new. The information that is provided here is equally applicable, and in many cases it may correct long-standing product errors that should have been corrected long before. Here again, one should try to provide sufficient lead time in a system program to examine as many alternatives as possible, evaluating each one dynamically and thus ensuring maximum effectiveness at the immediate user-system interface point, e.g., the component level.

## DESIGNING FOR THE EXTREMES VERSUS THE AVERAGE

In no phase of system development is it more important to know whether to design for the average user population or for the extremes of the population than at the component level, for this is where effects are often the greatest. Typical examples are designing a work seat for everyone when the requirements of the task require that each worker be positioned exactly right in order to do the task properly and with minimum stress; designing a display for average visual or auditory capabilities when it is obviously possible that users with subnormal perceptual capacities could make costly mistakes; and creating a tool that people are expected to know how to use when a great share of the consumers who will buy the tool have little or no skills relative to tool use.

## BEHAVIORAL EXPECTANCIES

Sometimes a new idea seems to have all the ingredients necessary to improve user efficiency, but because the immediate interface is so radically different from what people are used to or expect, we tend to induce operator error and confusion. There are no magic suggestions for making sure that one has not designed contrary to user expectancies except to test each idea on a sufficiently large and representative sample of the expected user population to demonstrate either that few people seem to have problems or that appropriate indoctrination and training will overcome negative expectancies and that subsequent reversals are not probable.

## EXPECTED PAYOFF

What are the expected payoffs that accrue from being more sensitive to component detail? First, it is generally at the immediate component interface that people make mistakes, many of which may result in product damage or personal injury. Therefore, from the standpoint of product liability, we should try to prevent human error.

Second, it is at the immediate interface that a great many of the confusions or stresses occur which make product users dissatisfied enough with a product not to purchase it ever again. This potential loss of future sales should stimulate the designer to try to provide the most effective interface possible the first time.

Third, in this day of proliferating technology and hardware, competition is great with respect to making any product available to the largest market. If a product requires special user skill to overcome basic component interface deficiencies, that product may end up only in the hands of a small number of consumers because the rest of the potential buyers are looking for something they can use easily and correctly without having to become experts.

## CHAPTER OVERVIEW

The topics and examples selected for inclusion in this chapter represent components and/or product items that most often occur in almost any system or subsystem. For example, there is generally some kind of visual display, auditory display, control device, handle, fastener, tool, or furnishing item required in order for a system or subsystem to be operable. Although these components can be generally called out during concept development and although some of the detailed specifications for the components can be described in sufficient detail to make sure that key operator issues have been properly identified, the specific details at the component level need serious attention to ensure that the final system can take advantage of all the capability of its intended user.

It is not enough to call out the need or requirement for a sign, an instrument, or a label; each of these interfaces has to be designed in a certain way in order that the user can interact with the system in the manner intended by the designer.

It is not enough to call out the need or requirement for a control, a handle, a fastener, etc.; each of these devices has to be designed so that the user can get hold of it, grip it, and manipulate it in the manner intended by the designer.

It is not enough to call out the need or requirement for a work surface or seat; each of these devices has to be designed so that the user can perform the tasks that the device is supposed to support in the easiest, most convenient, and least stressful manner possible.

This chapter provides information dealing with the following component and product elements:

1. Visual displays
2. Auditory displays
3. Controls
4. Fasteners
5. Tools
6. Furniture and appliances

The suggestions and guidelines associated with each of the above generally apply to any type of system or subsystem; i.e., whether one is concerned with a military, commercial, architectural, or vehicular product or whether the system application deals with specialized populations (e.g., engineers) or general populations (i.e., typical consumers), basic human interface requirements generally apply.

It should be noted, however, that many of the suggestions and guidelines *are* general in nature. Some of the dimensional guidelines are merely beginning references. The reader is cautioned to examine each suggestion in light of his or her specific operating need, to use common sense in applying the suggestions, and, most importantly, to test and evaluate each specific proposed design in a simulated operational setting, using a representative number of expected user subjects to confirm that the design is actually compatible under use conditions.

## GENERAL GUIDELINES FOR THE SELECTION AND DESIGN OF VISUAL DISPLAYS

### General Principles

1. *Use the simplest display concept* commensurate with the information transfer needs of the operator or observer. The more complex the display, the more time it takes to read and interpret the information provided by the display, and the more apt the observer or operator is to misinterpret the information or fail to use it correctly.

2. *Use the least precise display format* that is commensurate with the readout accuracy actually required and/or the true accuracy that can be generated by the display-generating equipment. Requiring operators to be more precise than necessary only increases their response time, adds to their fatigue or mental stress, and ultimately causes them to make unnecessary errors.

3. *Use the most natural or expected display format* commensurate with the type of information or interpretive response requirements. Unfamiliar formats require additional time to become accustomed to them, and they encourage errors in reading and interpretation as a result of unfamiliarity and interference with habit patterns. When new and unusual formats seem to be needed, consider experimental tests to determine whether such formats are compatible with basic operator capabilities and limitations and/or whether the new format does in fact result in the required performance level.

4. *Use the most effective display technique* for the expected viewing environment and operator viewing conditions (lighting, acceleration, vibration, operator position, mobility restrictions, etc.). Match the display technique to the operator's constraints; do not make the operator match the display.

5. *Optimize the following display features:*

   a. Visibility: Viewing distance in relation to size, viewing angle, absence of parallax and visual occlusion, visual contrast, minimal interference from glare, and adequate illumination

   b. Conspicuousness: Ability to attract attention and distinguishability from background interference and distraction.

   c. Legibility: Pattern discrimination, color and brightness contrast, size, shape, distortion, and illusory aspects.

   d. Interpretability: Meaningfulness to the intended observer within the viewing environment; requirements for interpretation, extrapolation, special learning, and training; and general reliability in terms of retention of meaning

*Note:* Consideration should be given to possible visual anomalies of the expected user population, i.e., lack of normal visual acuity, color deficiencies, and nearsightedness and farsightedness. These user limitations should be considered not only in terms of selecting the proper display concept but also in terms of executing the specific features that go to make up the display.

Consideration should also be given to creating displays that are easy to read and interpret, as opposed to displays that are merely attractive and/or startling to look at. Avoid the temptation to make displays decorative at the expense of their being readable and interpretable.

When you do not have the choice of creating a display, i.e., when you are required to select ready-made components, carefully study the available products until you find the one that most nearly meets the requirements and guidelines provided in the remaining parts of this section.

Design or select dynamic displays (instruments that have moving elements, warning indicators, etc.) that let the observer know if the display is malfunctioning. If it is not obvious (e.g., a pointer that suddenly drops to zero when it should be indicating some value), consider a special indication of malfunctioning, such as a warning light or flag.

## VISUAL DISPLAY SELECTION

| Display Requirements | Type of Display | Rationale | Example |
|---|---|---|---|
| Go/no-go, start/stop, ON/OFF | Light | Easy to tell when the light is on or off | |
| Identification | Light or lights<br>Symbol or symbols | Easy to detect; can be coded by means of location, number, color, repetition, or on-off duration | |
| Instructional | Labels and printed words<br>Arrows<br>Pictorial symbols | Single word or multiword message clarifies meaning; arrow implies direction; pictorial illustrates feature | |
| Exact quantity | Digital readout | Single number value requires no further interpolation or extrapolation | |
| Approximate quantity or positional relationship | Moving pointer and fixed scale with numbers; moving scale and numbers with fixed pointer | General pointer orientation gives quick cue to scale-pointer relationship; rate of change; reference to scale limits | |
| Data entry | Keyset | Alphanumeric keys provide immediate data selection reference | |
| | Pointer knob | Easy to position pointer adjacent to desired scale mark or number or label | |
| Tracking and hooking targets | CRT with associated data input and pick-off controller (joy stick, track ball, pantograph, etc.); indexing marks introduced on CRT with continuous capability for repositioning marker and adding alphanumerics or reference symbology | Provides convenient means for "pointing" to target and adding orientation reference marks or alphanumerics | |

## VISUAL DISPLAY SELECTION (continued)

| Display Requirements | Type of Display | Rationale | Example |
|---|---|---|---|
| Geographic position | Plan position reference analog: map or electronic map and grid display | Shows direct comparison of operator's position with geographic area and features; can show superimposed targets of interest | |
| Aircraft attitude | Electromechanical display of position of aircraft relative to pitch and roll and angle of attack; may be abstract or pictorial | Shows direct comparison of "own vehicle position" relative to horizon | |
| Command guidance | Electronic or mechanical analog of predicted position, path, and energy consumption; surrogate picture of external view (heads-up display) | Allows observer to anticipate what will occur (e.g., predicts) and provides substitution for external view when conditions limit direct view | |
| Equipment performance analysis | Meters<br><br>CRT<br><br>Pen recorder | Single parameter is easy to interpret<br>Multiparameter relationships are easy to see<br>Permanent record for later analysis | |
| Surface traffic control | Lights, signals, signs, and markers | Multiple observer usage requires displays on or juxtapositioned with roadways, runways, and taxiways | |

## VISUAL DISPLAY SELECTION (continued)

| Display Requirements | Type of Display | Rationale | Example |
|---|---|---|---|
| Large-group monitoring and signboard and scoreboard displays | Static, painted, back-lighted, or multilight signs (lights effective only at night) | Least expensive to install and maintain; permanent | |
| | Slide projection (front and rear; effective only in darkened room) | Static information; temporary exposure of successive pictorial or data items | |
| | Matrix type, moving message, or pictorial sign (multilight) | Effective at night for advertising or sports scoreboards | |
| | Mechanical character-changing message board | Useful for message status boards in transportation terminals | |
| | CRT and TV | Use only indoors for short viewing distance | |
| | Motion pictures | Primary use for movies, although has potential for training simulation | |
| | Edge-lighted plotting displays (darkened rooms only) | Command and control operation centers—plotting by hand on rear side | |

## LIGHTS TO DISPLAY INFORMATION

The following are examples of how a simple light or set of lights can be used to convey information by the fact that the light is on or off, by the color of the light, by the position and/or shape of the light, and/or by whether the light is steady or flashing.

### 1. Power ON Lights

These are used to indicate that electrical power is available to the equipment. Such indicators should be limited to one per piece of equipment or work station since several lights will confuse the operator unless each pilot light is labeled.

Pilot lights should not be brighter than necessary to determine that they are either lit or not lit (depending on the prevailing ambient illumination). Otherwise, they may interfere with other visual tasks since they tend to cause glare if the general ambient light conditions are fairly dim.

### 2. Function Active Lights

These are used to indicate that a particular function is occurring, as in the case of the hand-held recorder shown in the accompanying sketch; i.e., the pilot light advises the operator that the recorder is in the "dictate mode" and that the tape reel is running.

### 3. Special Hazard Condition Lights

Some equipment, products, or tools present potential hazards that the operator may not be aware of unless a warning light is provided. Two examples are illustrated. The soldering iron will be hot when power is fed to the soldering element and the operator forgets and becomes careless about touching the soldering element and/or lays the tool down before it has cooled. The pilot light is helpful in this case to remind the operator to be careful.

In the case of the stove with several burners, a pilot light arranged with each burner and/or control is helpful in reminding the user not to touch a live or hot burner. This hazard has become especially critical with the advent of the newer ceramic-top stoves, which provide no visual cue that a burner is hot; with an older stove, the change in color tells the user that a burner is hot, but no such visual indicator is provided on the ceramic-top stove.

### 4. Operating Mode Lights

A single push-button control switch is often used on some equipment in order to reduce the number of switches. This is practical because the push button can be back-illuminated, not only with a single light, but, as in the case of the copying machine shown on page 476, also with both a white and a red back-lighting source; the red light indicates that the machine is in a warm-up mode, and the white light indicates that the machine is ready for copying.

Many different approaches can be taken in using a simple light to indicate an operating mode. However, it is important to recognize that such advisory lights must be understood by the particular user; i.e., the mere fact that a light comes on does not necessarily mean that information is being conveyed.

In some instances, however, almost everyone associates a certain type of light signaling with a particular function. One example is the simple light on the automobile driving panel that indicates when the headlights are switched to the high-beam condition. The recent use of blue as opposed to red has confused many drivers. Such arbitrary changes should be avoided (see illustration on page 470).

A similar panel light is the turn signal indicator. Although some designers have used a single indicator light for both left- and right-turn indication, this is poor practice. In addition, since a primary concern of the driver is to know whether the control is signaling right or left, the position of the two indicators, as well as their shape, should reinforce the needed identification. These lights, by tradition, should be green.

Entire control panel lit — signifies power ON

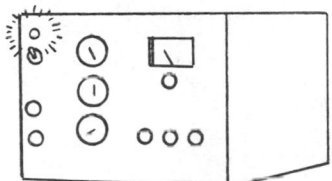

Power ON pilot light

Record mode pilot light

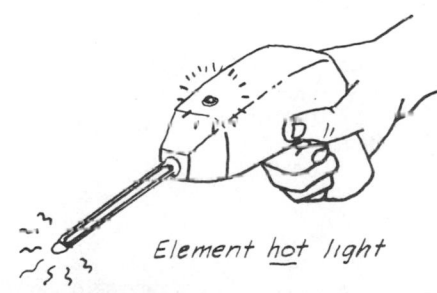

Element hot light

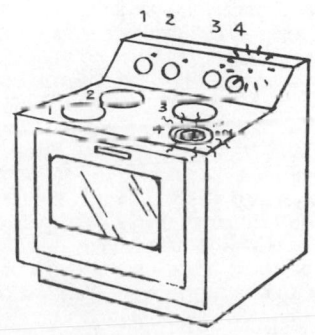

Burner No. 4 ON

### 5. Localizer Lights

Simple pilot lights can be effectively used to help identify one of a series of systems or devices. The first sketch below illustrates how a pilot light could be activated to indicate to the operator which speaker he or she is listening to. The second sketch below shows a similar application relative to identifying, say, a hatch that is still open prior to a submarine dive. The possible applications of this positional light cue are many, and the designer has only to remember that the light indicators and the devices they represent are spatially related.

### 6. Malfunction Lights

A simple light can be used as an initial cue that some malfunction has occurred or is about to occur. If there is more than one such light at the same work station, however, appropriate labeling is required to ensure that the operator does not become confused.

In some situations it may be more appropriate to use only the malfunction light, as opposed to pilot lights that indicate that everything is functioning properly. This approach is particularly relevant if there is a requirement to keep the general ambient light level very low so that operators are able to maintain their dark adaptation.

A particularly good example of the use of a malfunction caution or warning light is shown in the sketch of an automobile gas gauge (see below). The light comes on to tell the driver that the vehicle is almost out of gas. This warning light may prevent the driver who does not watch the gauges on a regular basis from running out of gas.

### 7. Outdoor Signal Lights

Outdoor signal lights are widely used for controlling traffic, warning observers about obstacles, etc. The illustrations on page 477, which show fairly common signal lights, are meant to remind readers of the possibilities for simple light signaling and information transfer.

The single flashing signal light shown in the accompanying sketch not only attracts attention but, because it is amber in color, also implies the need for caution. The most effective flash rate is between 3 and 10 per second.

The standard stop-light system shown in the accompanying illustration was not always as effective as it currently is because the signal lights used to be variously arranged in terms of vertical versus horizontal positioning and/or the specific location of the red, green, or amber light. Not until it was recognized that some observers are color-deficient was it considered necessary to standardize the position of the lights and colors. Since standardization, the color-deficient observer can tell whether to stop or go by the position of the lighted element.

Light signals are also commonly used on police and rescue vehicles. The recent change from a red flashing beacon to a blue one was based on the fact that there are so many red lights now that the red police light failed to elicit the attention it deserved. It should be noted, however, that blue is not necessarily a better signal color than red, except for the fact that it is different.

Standard automobile signal lights are now accepted for turn signals and for port and star-

board running or perimeter lights. However, there are still such differences as the use of red turn signals at the rear in the United States and of amber turn signals in Europe. An important functional consideration is that the vehicle's direction should be recognizable from the color and position of the light, as has been true for aircraft and seacraft. That is, the running lights on the front of an automobile or truck should indicate whether the vehicle is approaching or going away from the observer.

Most people recognize that navigation and positional signals for aircraft and seacraft are standardized in conjunction with "rules of the road" for flying the airways and sailing the seaways. It is important not to create new signal system concepts that may be different from these accepted standards and thus introduce confusion for the many persons familiar with the standards. The accompanying illustration shows how a vessel would appear as it ap-

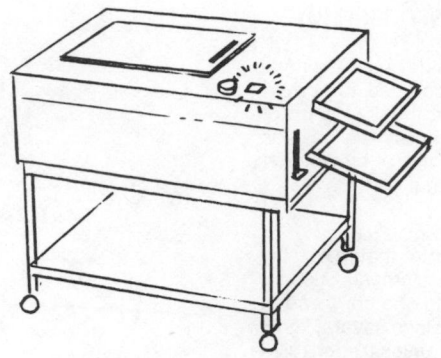

Ready-to-copy = Green-lighted switch cap

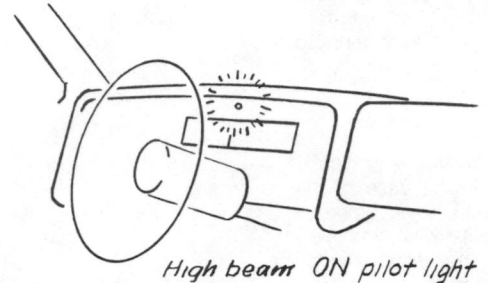

High beam ON pilot light

Right-turn "shaped" light

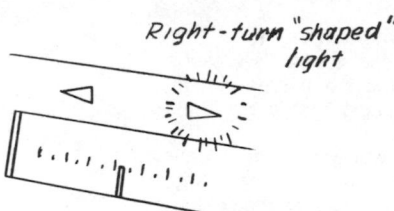

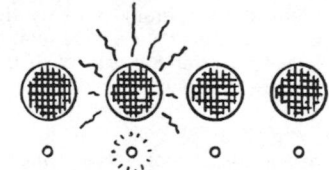

Channel 2 active pilot

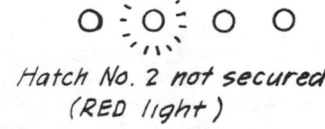

Hatch No. 2 not secured (RED light)

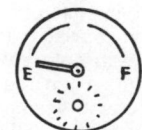

Fuel Low - pilot on or adjacent to display

proaches the observer at night; i.e., the approaching vessel will always have a green light on its starboard side and a red light on its port side. Note how the amber light is then used to relay vessel type to the observer.

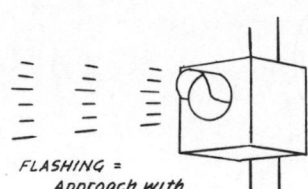

*FLASHING =*
*Approach with*
*caution*

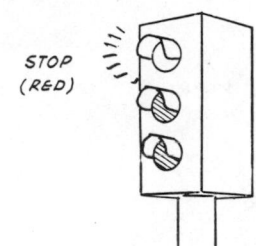

*STOP*
*(RED)*

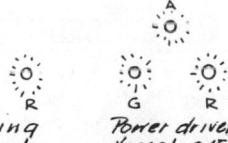

G    R
*Sailing*
*Vessel*

G    R
*Power driven*
*Vessel < 150'*

A

*Power Driven*
*> 150'*

A

*Tug - Tow length*
*> 600'*

G = green
R = red
A = amber

Formation lights are a unique requirement for military aircraft so that they can fly together at night. The key issue in formation of a light configuration is to identify those features of an adjacent aircraft which help determine its position, direction, and distance away. Also, in order to maintain a "covert" relationship in battle, these formation lights should be visible only to members of the formation. Thus, the design of each formation lighting system has to be customized for each particular vehicle model.

## LIGHT DISPLAY VARIABLES

### Viewing Angle

In most situations it is desirable that the light display be visible from the widest viewing angle possible. An extended lens is suggested to increase the effective viewing angle. Except for special cases where it may be desirable to limit the viewing angle so that only certain observers can see the light, avoid burying the light below the surface of an instrument panel or other structural surface.

As shown in the accompanying illustrations, when it is desirable to make the light visible only to certain observers, use louvers and/or turn the display toward the intended observer. The limited view is appropriate for such devices as a stop light which is intended only for one group of motorists and which would be confusing for other motorists.

### Colored Lights

Certain color-coding conventions are fairly well established and should be followed; i.e., unless it can be demonstrated that changing the color code would enhance performance, do not change colors that already have significant meaning to the majority of observers. The following are some of the conventions:

Red lights: Use for danger, warning, fire, taillights for highway vehicles, and portside navigation lights.

Yellow or amber lights: Use for caution, slow, power on, front turn signals for highway vehicles, and marinecraft identification.

Green lights: Use for go, ready, functioning correctly, and interior turn signal indicator for highway vehicles.

Blue lights: Use for high-beam headlight indicator for highway vehicles and taxiway lights for airfields.

White lights: Use for airfield runway lights.

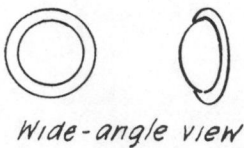

Wide-angle view

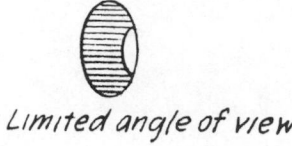

Limited angle of view

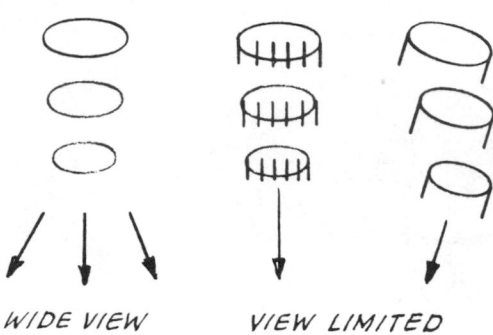

WIDE VIEW          VIEW LIMITED

## SIZE AND COLOR CODING

When a number of light indicators are required within a single operator station or several related operator stations and when the level of importance varies for different functions, a combination of encoding techniques can be effectively applied.

The accompanying table illustrates the use of size and color codes to help differentiate among the several functions and urgency categories. It must be recognized, of course, that the value of such encoding lies in the fact that the devices are repeated in several places, thus allowing operators to learn to recognize the size differences.

### Military Standards for Indicator Colors

Because of their special importance to the military, the colors of indicator lights of various types are specified and are rigorously controlled to ensure that all observers (including those who may have some inherent visual color deficiency) will recognize the color they are observing. For color-coding specifications, refer to MIL-C-25050.

**Coding of Simple Indicator Lights**

| Size and Type | Color | | | |
| --- | --- | --- | --- | --- |
| | Red | Yellow or Amber | Green | White |
| ½-in (12.7-mm) diameter or smaller; steady | Malfunction; failure, stop action, no-go | Marginal condition; possible problem arising | Condition acceptable; OK to go ahead; in tolerance; ready to perform function | Indicates status of function or physical position; some action in progress |
| 1-in (25.4-mm) diameter or larger; steady | Master summation (system or subsystem) | Extreme caution (impending danger) | Master summation (system or subsystem) | |
| 1-in (25.4-mm) diameter or larger; flashing (3 to 5 per second) | Emergency condition (impending personnel or equipment disaster) | | | |

## Color as an Indicator of Operational Condition

Red: Red should be used to alert an operator to the fact that some situation makes the system inoperative, e.g., an error, a failure, or a no-go or life-endangering condition.

Yellow: Yellow should alert an operator to a situation in which caution, recheck, or delay is necessary, i.e., a condition that, if not attended to, could lead to a dangerous situation.

Green: Green should indicate that equipment is operating satisfactorily or that one can proceed. It can be used to indicate the successful completion of steps within a process, thereby establishing a basis for continuing with the next step.

White: White should indicate such items as status, alternative functions and selection modes, a test in progress, or any similar items that imply neither success nor failure of system conditions.

Blue: Preferential use of blue should generally be avoided because it has no standard meaning, except when it has been assigned a special significance within a given operating system; e.g., a blue reference light has been used by some automobile designers to indicate that headlights are in the high-beam condition.

A flashing red light should be used when an extreme danger is present; i.e., the flashing light indicates a situation of more importance than a continuous red light. A flash rate of between three and five flashes per second, with approximately equal on and off times, is recommended.

The flash coding technique is also suggested for other colored lights when it is desirable to increase the conspicuousness of the signal. Typical examples include a flashing amber traffic signal in advance of a regular stop-lighted intersection and an alerting signal on a communications status board.

For additional suggestions regarding the use of color coding for light indicators, refer to U.S. Army MIL-HDBK-759.

## Specific Functional Requirements for Warning, Caution, and Advisory Light Indicators

In military equipment installations, it is required that master warning and/or caution light indicators be placed directly in front of operators to ensure that they will not miss seeing the light when it illuminates. Illumination of such a light indicates to operators that they should immediately look for specific indicators that will tell them what is wrong. These latter indicators are typically located where they can be functionally grouped. Some of the types of warning, caution, and advisory functions that are identifiable by means of annunciator light displays are listed below:

*Warning Indicators (Red Light)*
Bailout
Cabin pressure failure
Canopy unlocked
Emergency fuel failure
Engine overheated
Engine on fire
Fuel system failure
Landing gear lower failure

*Caution Indicators (Yellow Light)*
Aileron trim failure
Anti-skid-off
Automatic pilot failure
Constant speed drive failure

Defroster failure
Door open
Engine icing
Generator # - Inoperative
Oil pressure low

*Note:* Although these examples relate to aviation, the principle of devising separate master and functional indicators to warn, caution, or advise an operator through the use of colored lights has obvious application to a host of situations, including commercial systems. The concept is especially applicable to highway vehicles, communication centers, and such domestic products as farm or construction machines, stoves, and washers and dryers.

*Advisory Indicators (White, Green, or Blue Light)*
Alternate current
De-ice
Direct current
External power on
Marker beacon
Radar on
Trimmed for takeoff

## FLASHING LIGHTS

Flashing lights are recommended wherever it can be anticipated that an observer needs reminding that a situation exists or is about to occur; in certain cases this need arises when there is a high probability that the observer may be visually occupied and needs the benefit of a more conspicuous visual signal. The typical applications and conventions are as follows:

A preparatory flashing amber light in advance of a traffic signal.

An anticollision flashing light on aircraft because of the special alerting qualities of a flashing (and/or rotating) light. Although the basic anticollision light is typically red, high-intensity white flashing lights have been shown to be conspicuous at greater distances.

Flashing lights (both red and blue) for police and emergency vehicle identification and alerting.

A coded flashing light to aid identification of an airfield.

*Note:* The main benefit of the flashing light is that it is more conspicuous. On the other hand, too many flashing lights easily destroy the conspicuousness of any one signal. The same is true of color. Red was at one time the most effective color to use for taillights because they were easily identified among the typical white street lights. On a crowded freeway, however, there are so many red taillights that the motorist often has difficulty knowing what is going on ahead.

## BRIGHTNESS

The desired brightness level of any signal light depends on the ambient lighting conditions, the brightness of other lights in the field of view, and the need to identify a change in situation (i.e., a given signal may be made to glow brighter when a functional change has occurred).

Often it is helpful to increase the conspicuousness of a conditional change by combining a brighter light with flashing. Care should always be exercised, however, not to blind an observer at a critical moment by increasing the brightness of a light. A typical case in point is the change in brightness of the taillight of a car when the brakes are applied. Although the objective is to make the change in brightness sufficient to let the observer in the rear know for sure that the brake has been applied, the observer may be blinded by an extremely bright light and make precarious evasive maneuvers that are not warranted.

## POSITION OF LIGHT SIGNALS

To be effective, of course, a signal or pilot light has to be in a spot where the observer is most likely to be looking. Other positional considerations are also important, such as the following:

A power pilot light should be located in the place where the observer first looks at the equipment (typically on the upper part of the equipment).

A high-beam indicator light should be located at the top of the vehicle instrument panel near the observer's typical line of sight as he or she watches the road. The same applies to turn indicator lights.

Perimeter-defining lights on vehicles should be as near to the edges of the vehicle envelope as is practical; otherwise, the observer may come too close.

Traffic signal lights should be positioned high enough so that they can be seen over the tops of other vehicles, but not so high that when the driver stops, he or she cannot see the signal because of the limits of the vehicle's upper windshield frame.

Warning lights (and/or lighted indicator lights) should be located within about 15° of the principal line of sight of an operator in order to make sure that the operator will notice a warning while concentrating on something else (e.g., watching the road ahead, in the case of a motorist, or watching a CRT display, in the case of a sonar or radar operator).

Stop lights used on the rear of automobiles should be placed at a height above the road approximating the eye level of the driver of the car behind (or slightly above this level). This has been shown to reduce the incidence of rear-end collisions.

## COMBINATIONS OF PILOT LIGHTS, ANNUNCIATORS, AND LIGHTED PUSH BUTTONS

It is important to an operator to be able to tell quickly which lighted indicator is a pilot light, an annunciator, or a telelight and which is a lighted push button. Avoid the temptation to select hardware components that all look alike; e.g., use a round fixture for a pilot light, a rectangular fixture for an annunciator or telelight, and a square fixture for a push button.

The accompanying illustration shows how much easier it is to tell the difference between components when each has a different shape.

It is also desirable to select push-button hardware in which the button extends about 1/8 in (0.32 cm) out from the panel and annunciator hardware that is more or less flush with the panel. The best pilot light is one that also extends from the panel so that an operator can tell from an angle when the light is on.

Some of the new flush control concepts that require the operator merely to touch an outlined section of a panel do not comply with the above suggestion. There are no studies to indicate whether operators are confused by such concepts. However, one can be certain of no confusion if the above suggestions are followed.

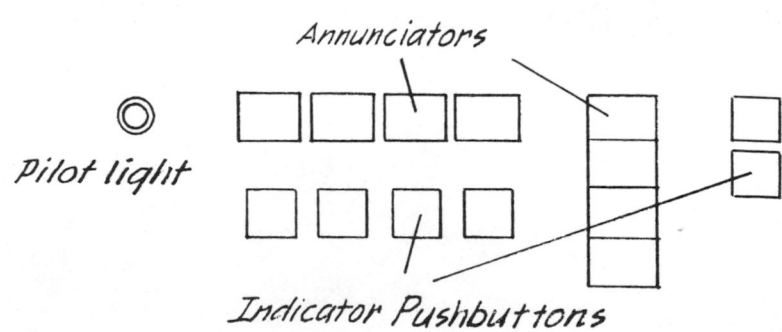

Annunciators

Pilot light

Indicator Pushbuttons

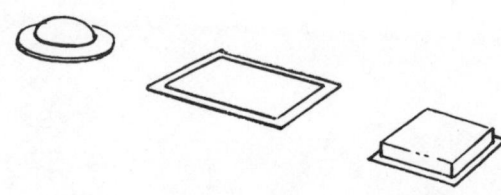

## WORD DISPLAYS: SIGNS AND LABELS

The use of words as a display device adds another dimension for increasing information transfer. However, it may also add complexity and increase the time an observer requires to recognize and understand the meaning of the information one is trying to convey. The most common word display is perhaps the sign, which can be used for identification, for warning or caution, for guidance, and for instruction.

Graphics specialists often are given the task of producing signs. In some instances, they are concerned primarily with the attractiveness of the sign rather than with its readability. Refer your graphics specialist to the guidelines presented in this handbook if you are not sure that he or she is familiar with what is required to make a sign readable.

### Sign Standards

Certain types of signing have been standardized in order that the size, shape, color, and wording of particular signs will be the same wherever they may occur. For example, the U.S. Department of Commerce, Bureau of Public Roads, issues a *Manual on Uniform Traffic Control Devices for Streets and Highways*. Similar standards have been devised for various industrial safety signing. It is important to acquire and use these standards rather than to invent new signs.

Traffic signs fall into several distinct categories: (1) regulatory signs, (2) warning signs, (3) guide signs, and (4) pavement and curb markings. Although the manual also covers street-name signs, the specifications are not necessarily acceptable from a human engineering standpoint because they are not large enough in most cases and because they cannot be used by the blind.

Key issues for signing are the following:

Legibility
Size in relation to viewing distance
Illumination
Clarity of meaning
Location relative to how and when the sign is
    to be used by the observer

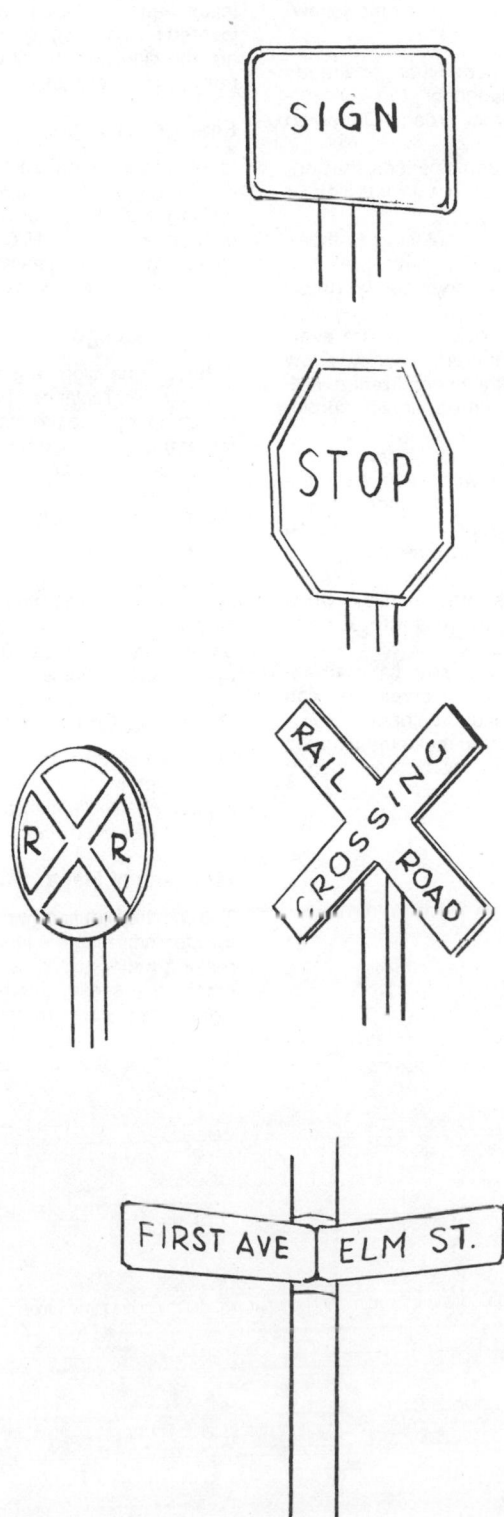

## Sign Colors

Color can be used to encode and/or categorize various signs. For example, highway signs are typically red, yellow, black, or white, with other colors such as green and blue interspersed for less critical information.

The important factors to consider in deciding to use special color codes are the following:

Adhere to certain color codes that are already standardized: red—danger or stop; yellow—caution or be prepared; green—OK or go ahead.

Select colors or color combinations that enhance the legibility of the sign lettering or symbols.

Limit the number of colors selected to those which the observer can differentiate reliably, i.e., those which the observer can verbalize and recognize independently of a comparison with another color close by. The average individual can differentiate only a few colors in terms of faithful verbalization: red, yellow, orange, blue, green, brown, purple, white, and black.

In general, select saturated colors, as opposed to pale colors when you want the colors to have special significance.

Use luminous or fluorescing color to enhance color discrimination under marginal lighting or viewing conditions.

Illuminate colored signs with essentially white light in order to ensure reliability of color identification.

Use aviation red, yellow, or blue lights to ensure that color-deficient observers also can recognize these particular colors.

*Do not* alter critical colors to make them match an aesthetic color scheme.

## Orientation Signs

Designers often fail to recognize the importance of providing signs to help people identify where they are and where they are going. Typical of these are signs that identify an entrance, a parking area, a bus stop, and an elevator. Once again, such signs should be legible and intelligible, and they should be located so that the observer can easily associate the sign with the physical features pertinent to the situation.

## Emergency Signs

Identification of features that may be needed in an emergency is especially important. The typical exit sign is generally prescribed by local ordinance. Other, similar signs include those indicating an emergency escape route, a fire extinguisher, and a first-aid station.

## Company Logos

Although one might argue with a specific logo used by a particular company, it is important to the company (and probably to most of its customers) that the logo be retained in any signing created for that company.

## Caution Signs

In addition to the caution signs one generally associates with large outdoor installations (e.g., traffic control), similar signing is applicable to other situations, such as the sign posted on an automobile gas filler cover indicating that unleaded fuel is to be used.

## Operating Control Labels

The push buttons for elevator call are typical of this type of signing. Usually, the button that is pushed also illuminates to indicate that the request has been made.

## Mechanical Flags

The mechanical flag has been used successfully for indicating the functional condition of a parking meter and/or a control panel instrument. In a sense, this provides a go–no-go signal using words rather than a light.

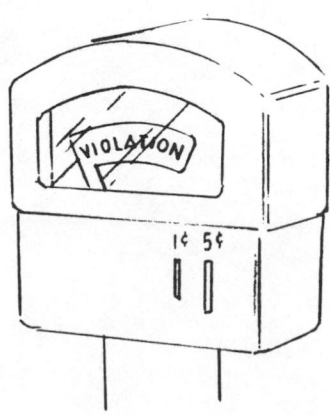

## WARNING SIGNS

Regardless of whether the proposed sign falls within the specific classification of a warning or caution sign or signal or a hazard advisory or instruction, the primary objective should always be to make the device noticeable, recognizable, and understandable. Some of the key factors to consider are the following.

### 1. Conspicuousness

The sign should stand out among adjacent visual distractions because of its size, shape, color, or symbology or because it is moving or flashing. The sign or signal must be located where it is expected that most people will be looking; it should not be hidden behind a post or an operator control or below the heads of other observers or the tops of equipment, parked cars, etc. Above all, do not camouflage the sign or signal to keep it from spoiling the decor planned for the site.

### 2. Emphasis

The most prominent feature of a warning sign should be words or symbols that imply danger. Include such words as "danger," "hazard," "caution," and "warning." Use standardized symbology that by convention immediately indicates the nature of the hazard, e.g., a skull and crossbones for poison, the "fan" symbol for radiation, and the slow vehicle triangle for a roadway hazard. Add unique borders around the sign, such as yellow and black stripes or a checkerboard design.

### 3. Legibility

When words and messages are used, make sure that they are legible by selecting the proper letter style and size and by making the contrast between the letters and the background sufficient for whatever lighting condition may exist. Provide a sufficient border around the lettering so that other background influences, such as other signs, lights, or patterns, will not cause confusion.

### 4. Simplicity

Use as few words as possible; i.e., avoid long, complex advisory information or instructions that take too much time to read. Tell the observer exactly what to do or not to do. Avoid acronyms and abbreviations.

### 5. Intelligibility

Make it clear what the hazard is and what may happen if the warning is ignored. Especially avoid contradictions or implications that imply some doubt as to the actuality or seriousness of the hazard.

### 6. Visibility

Make sure the sign is visible under *all* expected viewing conditions, i.e., during the day and at night, if the artificial light fails, under marginal atmospheric conditions, in bright sunlight, etc.

### 7. Permanence

Choose devices and/or sign materials that are resistant against aging, wear, soil, and deterioration due to sunlight or cleaning methods, and use techniques that are resistant to vandalism.

### 8. Standardization

Use standard signing words, symbols, etc., wherever they already exist, such as those recommended by OSHA, the FHA, the Department of Commerce, and the military. Although some of these standards may not be in complete accord with the recommendations suggested here (i.e., they may not represent the best human engineering), they are usually so well established that changes to comply with human engineering principles might confuse the observer. When no such standard exists, however, apply the above principles and guidelines in creating any new signing.

---

## STREET SIGNING AND MARKING

The practice of placing signs directly on street surfaces has proved useful for instructing and guiding motorists. Special care must be taken in designing such signs so that the letters (which are not viewed from the same angle from which most signs are seen) are readable at the expected approach speeds and visual angles of the motorist.

### Combining Words and Lights

As noted earlier, providing both a light and a word expands the possibilities for signing visual information displays. The addition of the words increases the probability that the meaning of the sign will always be clear, and the addition of the light provides a means for emphasizing the particular sign or signal mode and/or allows one to use the same display unit to present alternate information. One of the most common devices is the lighted pedestrian sign that switches from a "walk" to a "wait" instruction.

Combination light-word signal devices have long been successfully used for vehicle and equipment control monitoring. These are variously referred to as "telelights," "annunciator lights," etc.

One has the choice of using dark printing with a lighted background or the reverse, i.e., illuminated printing with a dark background. The general rule is to reserve the lighted background for emergency or warning displays, since this makes the display much more conspicuous and attention-demanding.

The question is invariably raised as to whether the lettering of an inactive display should be visible. As a rule, it should be visible because it is important for the operator to have previous knowledge of what display comes up, and where.

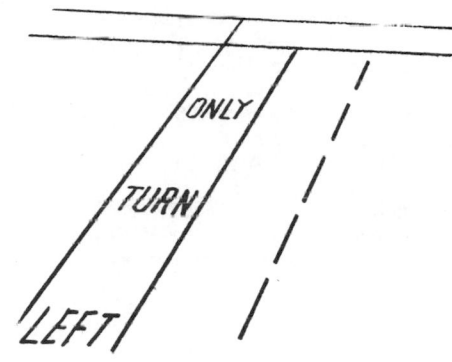

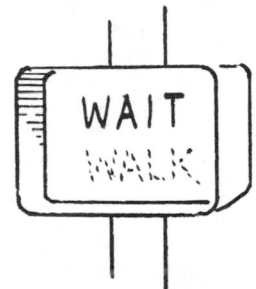

## SIGN SHAPE

Sign shape can be used as a method for identifying categories of information, as in the case of highway signs and industrial safety signs. The most recognizable shapes under all conditions (i.e., size, distance, ambient light, and atmospheric interference) are shown in the accompanying sketches. Other shapes, such as stars, shields, and for example the shape of a state, can be used, but with caution. Some of the factors to consider in using shape are the following:

Will the shape mean anything to the average observer? For example, street and highway sign shapes are categorized for various regulatory, guidance, or information purposes, but very few motorists remember these categories.

Will the shape be compatible with the spacing of words or symbol labels; i.e., will the shape crowd the lettering so that it will not be legible?

Will the shape continue to maintain its recognizability under all expected size, distance, or other viewing conditions, or will the critical features tend to blend and disappear?

Are there too many other signs of the same shape (and possibly color) that would tend to diminish the significance of the intended sign?

*Note:* In the examples shown, it should be noted that the basic categories of shapes are circles, triangles, rectangles, and crosses. They are varied in terms of orientation or expansion and contraction. Pentagons, hexagons, and octagons do not retain their recognizability; i.e., they fade into circles when viewed from great distances.

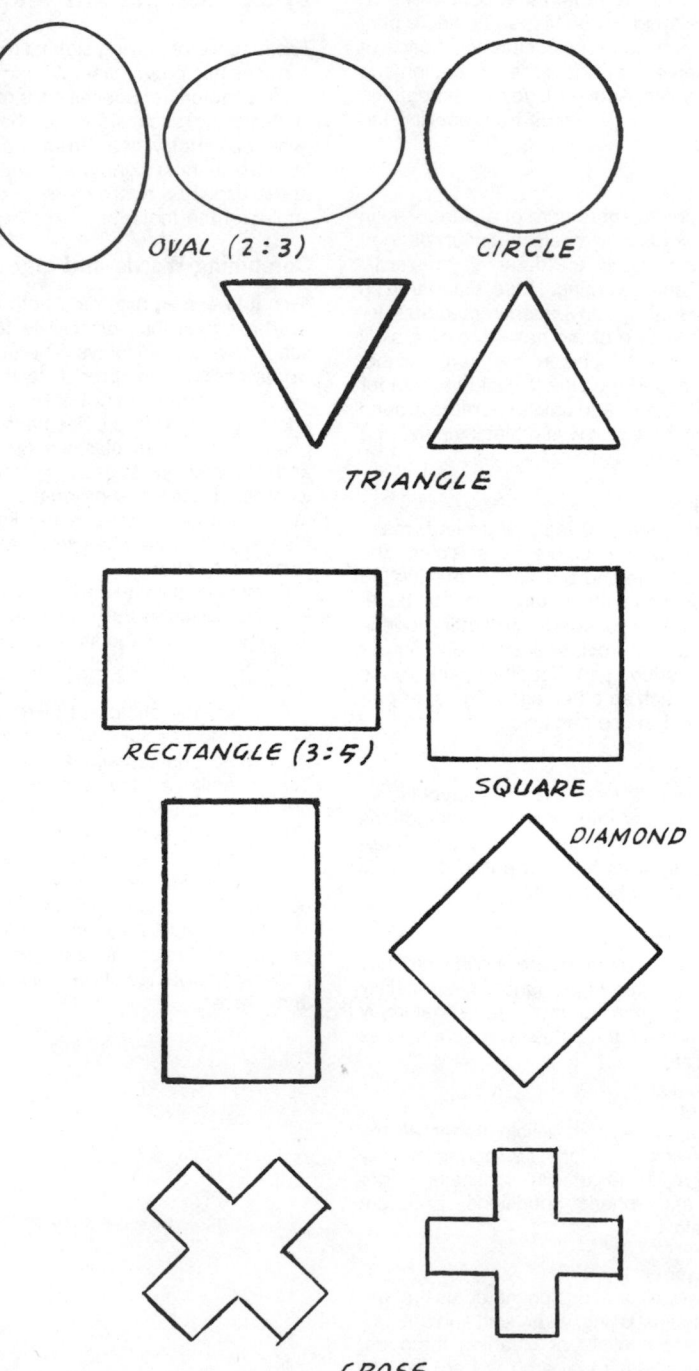

OVAL (2:3)  CIRCLE

TRIANGLE

RECTANGLE (3:5)  SQUARE

DIAMOND

CROSS

## ARROW AND POINTER GRAPHICS

Directional pointers, unlike those for instrument displays, do not have to mate with other markings.

Design or select directional pointers which are recognizable as pointers and which are sufficiently conspicuous to stand out among an assortment of other geometric designs and backgrounds.

 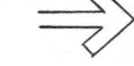 

GOOD          BETTER          BEST

The designs shown in the accompanying illustrations have various deficiencies, although they are used frequently. The first two lack boldness and are often not recognizable under certain illumination conditions; the third is not recognizable under poor illumination conditions or from considerable distances; and the fourth and fifth are easily confused with other kinds of symbols because they lack sufficiently conspicuous geometric cues.

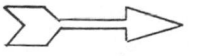   

The idealized pointer proportions shown in the accompanying illustration provide proper geometric emphasis to maximize recognition.

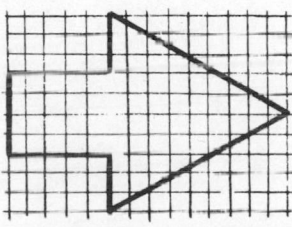

Pointer proportions for directional signs

### The Arrow as a Directional Aid

Although many people are aware of the value of the arrow symbol as a directional aid for highway signs and other architectural guidance signs, not many designers have taken advantage of this special graphic device to help the consumer in other situations where guidance is necessary. The following guidelines may be helpful in deciding when to use the arrow symbol and how to use it effectively:

The arrow tends to attract attention since it implies "Follow this."

The use of several arrows properly aligned above separate aisles or highway lanes provides immediate identification of guidance information related to the specific aisles or lanes.

An arrow with a tail that obviously changes direction provides immediate recognition of a directional change. In a series of arrows, it is immediately recognized when one or more of the arrows point to another direction, as opposed to continuing in the direction the observer might be going.

A single arrow accompanied by simple place names immediately implies to the observer that these places may be found by following in the direction in which the arrow points.

Beware of the use of an arrow pointing upward. Unless observers can plainly associate a staircase or inclining ramp, this type of arrow orientation merely confuses them.

Arrows are very useful on complex machine panels (especially those which may be used by the public, such as a vending machine). Observers generally will follow the arrows, while they may often ignore the best instructional material. The addition of "step numbers" almost assures that observers will follow instructions.

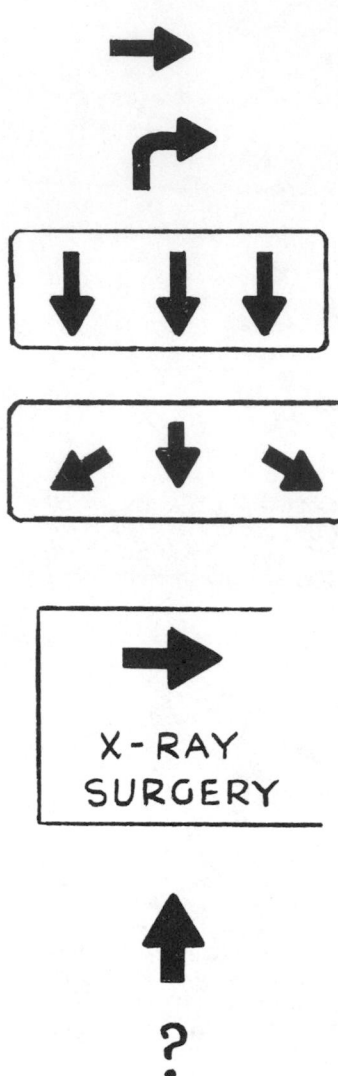

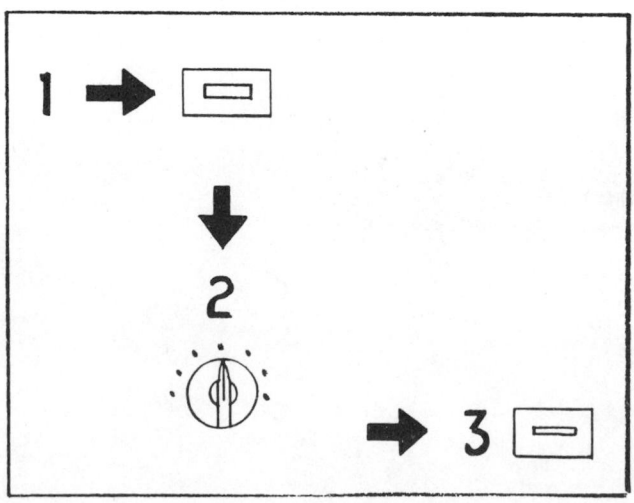

## ALPHANUMERIC CHARACTERS

Each alphanumeric character's recognizability depends on the proper emphasis of specific features such as diagonals, flat tops, and easily discernible loops. Some type fonts tend to obscure these features. Some examples of how certain characters can be distorted to the point of confusion are shown in the accompanying illustration. The following points are important:

The letter A must have a clearly delineated space above the horizontal stroke.

The letter B should have approximately equal loops at the top and bottom.

C's and G's are easily confused with each other or with O's if the break is not clearly discernible, if the horizontal stroke of the G is not long enough, or if the O tends to appear too square.

The E must not have its horizontal strokes too close together.

The M and the W must have a sufficiently long center section.

The P must have a large enough loop.

The S and the 5 are easily confused if the S is too square or if the horizontal top of the 5 is not long enough.

The 1 and the I are easily confused with each other unless they are made to look different. In this instance the serif is perhaps most helpful.

The U and the V are easily confused unless the vertical sides of the U are maintained, as opposed to the slanting sides of the V.

The Y needs a long tail to be discernible from the V. However, the V-shaped top also must be definite to keep the Y from being mistaken for a T.

When the 6 and the 9 have loops that are too large and tails that are too curved, they are not easily recognized. The loops must be readily apparent, but the tails need to be straight.

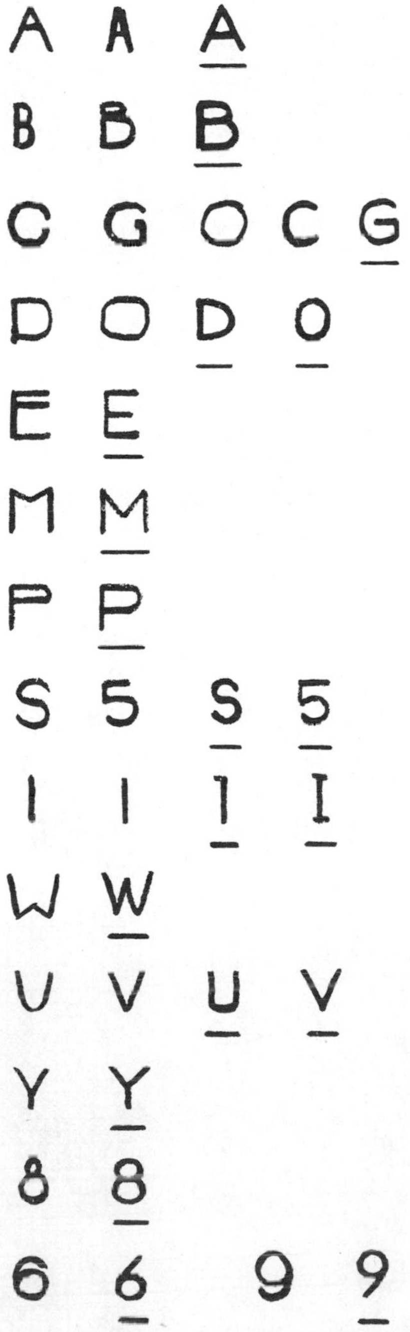

— Underlined are preferred

## GUIDELINES FOR OPTIMIZING LETTERS AND NUMBERS TO BE USED ON SIGNS, CONTROL PANELS, AND INSTRUMENT FACES

In spite of some controversy over the best style of character for use in signing, labeling, and instrument marking, at least the following parameters are agreed upon as important to consider in selecting a given character style for a specific application:

1. Proper height/width ratio, e.g., character proportions
2. Proper stroke width, e.g., width of the lines making up the character
3. Maximum figure-ground contrast
4. Adequate overall size for the viewing distance expected
5. A simple rather than an elaborate style, i.e., generally a block style without serifs
6. Upright rather than slanted (italic) characters
7. Adequate intercharacter spacing
8. All capital letters for short labels and signs (as opposed to capital and lower-case letters for extended, instructional materials)

### Special Type Fonts

Special type fonts such as those shown in the illustrations on this page and on page 488 have been devised for use on military control panels and visual displays. Individually, these characters are considered maximally legible. They are available in Le Roy Lettering Guide templates and are generally preferred over the standard Le Roy characters.

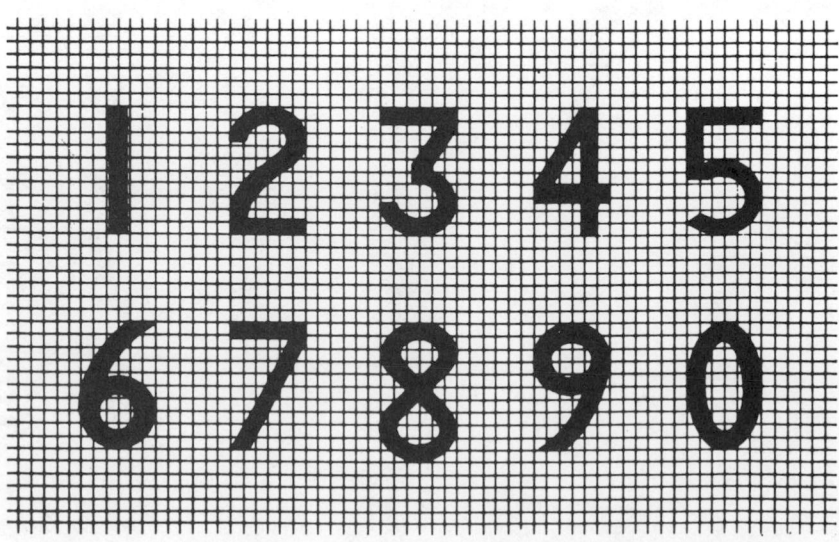

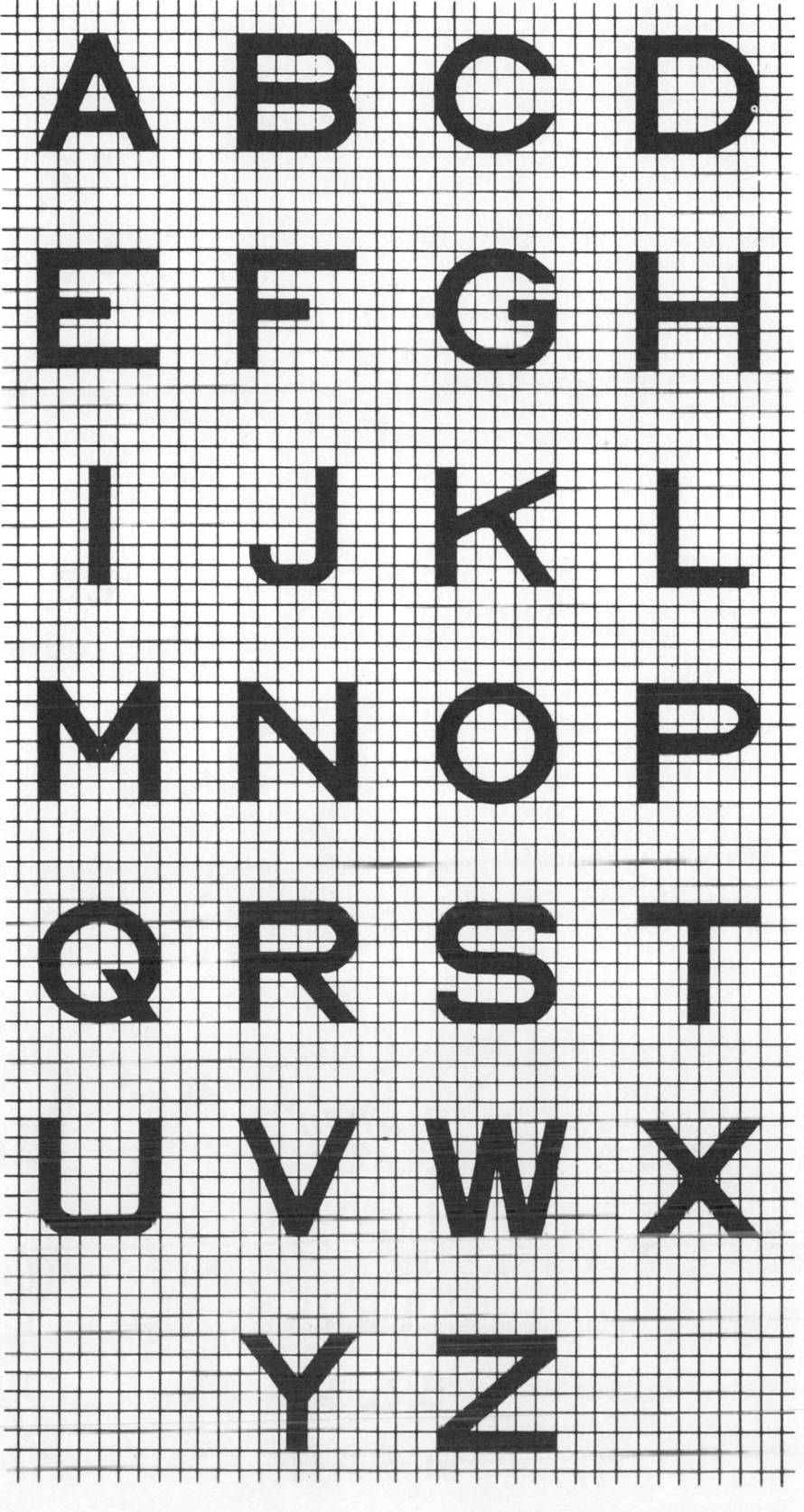

## Choosing Correct Type Styles from Standard Type-Font Catalogs

Typically one must select type styles from the engraver's or printer's available type. In doing so, one should look for certain key letters or numbers that are the most difficult to read. The several comparison characters shown in the accompanying illustration are especially important to examine because they are most easily confused.

Simple block styles should be used for control panel and instrument labeling applications. Some of the standard type styles that are acceptable include Folio Book, News Gothic, Trade Gothic, Futura Medium, and Spartan Medium.

However, since the particular printer or engraver may not have these styles, one should learn to examine the available styles and select the best one for a particular application.

Type faces to be avoided are those which:

Have uneven stroke widths
Have extended serifs
Include internal patterns or stripes
Are italicized
Are stenciled
Appear like handwritten script
Are shadowed or made to look three-dimensional
Look like Old English script
Are distorted to look tall and thin or wide and fat

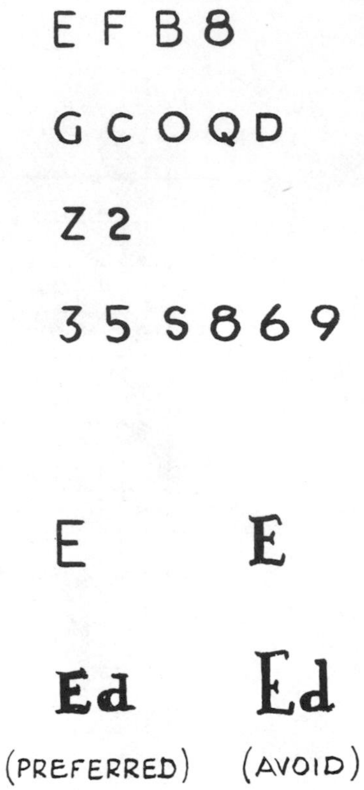

E F B 8

G C O Q D

Z 2

3 5 S 8 6 9

E       E

Ed      Ed

(PREFERRED)   (AVOID)

Examples of typically poor type styles that can be found in standard type catalogs are shown in the accompanying illustrations.

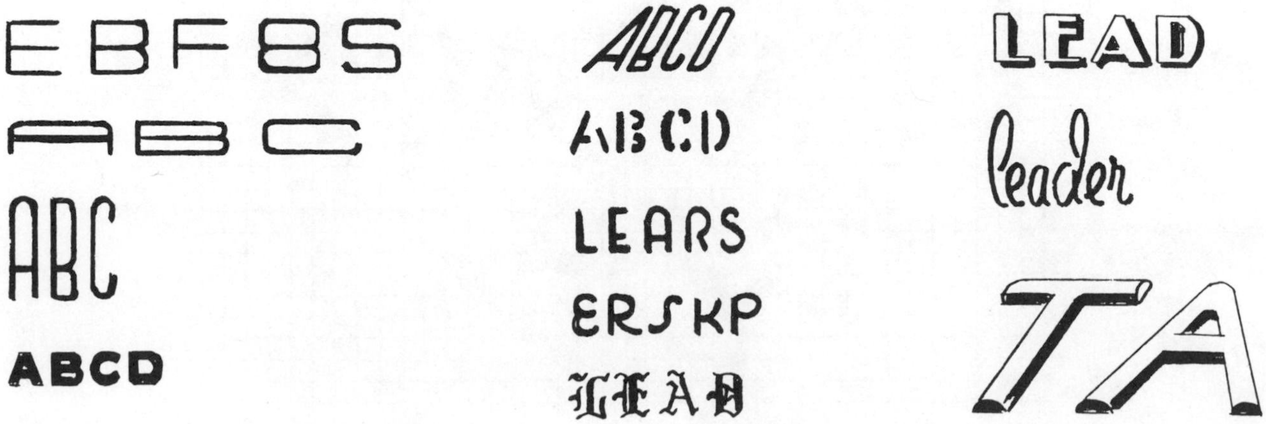

EBF8S
ABC
ABC
ABCD

ABCD
ABCD
LEARS
ERSKP
LEAD

LEAD
leader
TA

**STROKE WIDTH** The stroke width of letters and numerals is important in determining how legible the characters will be. If the stroke width is too narrow, there will not be enough definition to make the character clearly recognizable. If the stroke width is too thick, some of the more critical features of the character may be obscured so that it is unrecognizable. For example, the closed portions of a letter such as a B or an R or of a number such as a 6 or 9 may be obscured by a thick stroke width. Similarly, the open space between the horizontal strokes of an F or an E may not be discernible when the stroke width is too thick.

The preferred stroke width for *dark characters* on a light background is 1:6, i.e., the stroke width versus the character height.

The preferred stroke width for *light characters* on a dark background is 1:7.

The rationale for two different stroke widths is based on the fact that light falling on the lighter surface tends to "spill over" into the darker area. In the case of a lighter character, this tends to make the stroke width broaden out; in the opposite case, the stroke width appears narrower. The above recommendations compensate for this apparent artifact of light reflection.

Above all, avoid selecting extremes, either stroke widths that are extremely narrow or ones that are extremely bold.

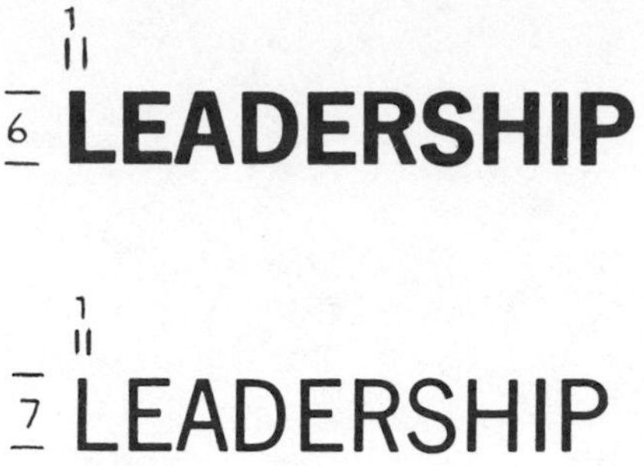

$\frac{1}{6} = \frac{\text{II}}{}$ **LEADERSHIP**

$\frac{1}{7} = \frac{\text{II}}{}$ LEADERSHIP

**LEADERSHIP**

LEADERSHIP

LEGIBILITY

1384

NH ED

AND FOR NOW

Because letters and numbers are most often grouped, it is important to recognize that the criteria for a single character's maximum legibility do not always apply. For example, since grouped numbers or letters become too spread out for the observer to grasp the significance of the total pattern, characters should be selected that are taller than they are wide.

Depending upon the specific typeface available, the most readable height/width ratio should be somewhere between 5:3 and 3:2.

Letters or numbers within a group should be about one stroke width apart ($\pm$ ½ stroke width).

The space between groups of words or number groups should be about three stroke widths ($\pm$ ½ stroke width).

The space between lines of number groups or words should be at least one-third the height of the tallest character, but in no case should the space between the tallest character of a lower line be less than one stroke width from a character above it that projects below the line.

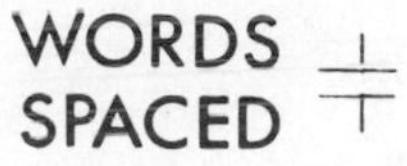

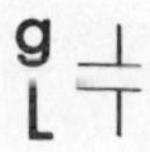

WORDS
SPACED

g
L

The preferred height/width ratio for typical block-type letters and numerals ranges from 5:3 to 3:2, depending on the typeface available.

For emphasis, it is satisfactory to use wider characters with ratios up to 1:1. Avoid selecting characters that are wider than they are tall, however, since such distortion tends to increase recognition time.

LEADERSHIP

LEADERSHIP

When horizontal space is limited, it is permissible to use narrower characters. In some cases, this is even more desirable than trying to crowd the so-called optimum character into a limited space or reducing the space between words.

LEADERSHIP

Avoid selecting extremely narrow characters since they tend to appear blurred and to increase both observation time and errors.

LEADERSHIP

The name of a manufacturer on a product usually is not a critical operator use consideration. A unique typeface here may be an advantage since it clearly separates the name from other operationally critical information.

LEADER

## FIGURE-GROUND CONTRAST

### Brightness and Color

To ensure adequate legibility, there must be adequate visual contrast between alphanumeric characters and the background against which they are viewed. The following should be considered:

Under normal illumination conditions (i.e., where the observer does not have to be dark-adapted), use dark characters against a light background.

Where the observer must maintain a dark-adapted condition, use light characters on a dark background.

The character should be at least twice as light (or dark) as the background. Greater contrast is preferred and is especially important if the expected observer population includes persons with impaired vision, if the viewing conditions will be less than optimal, or if physiological stress factors are anticipated, such as oscillation vibration, motion, or high *g* forces.

Avoid the use of glossy or highly reflective metallic materials for either the lettering or the background (except where the lettering must be illuminated remotely, e.g., by automobile headlights). Spectral unevenness usually causes portions of the letter or numeral to become illegible as a result of the variation in light reflection.

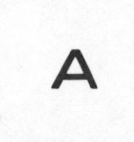

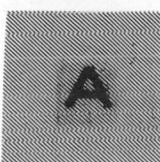

## Suggested Color Contrast Selections in Order of Expected Visual Efficiency

| Conditions | Characters | Background |
|---|---|---|
| Average or higher levels and quality of illumination | Black | White |
| | Black | Yellow |
| | White | Black |
| | Dark blue | White |
| | White | Dark red, green, and brown |
| | Black | Orange |
| | Dark green and red | White |
| | White | Dark gray |
| | Black | Light gray |
| Poor level and quality of illumination | Black | White |
| | White | Black |
| | Black | Yellow |
| | Dark blue | White |
| | Black | Orange |
| | Dark red and green | White |
| Dark adaptation required | White | Black |
| | Yellow | Black |
| | Orange | Black |
| | Red* | Black |
| | Blue and green | Black |

*Low-level red light is required to maintain the lowest dark adaptation level.

*Note:* When lettering is internally illuminated so that the letter is always bright against an essentially black background, almost any color is satisfactory as long as the brightness is above the minimal threshold, i.e., as long as the light is visible. Opaque letters must be illuminated with an essentially white light; otherwise, the colored illumination may decrease the contrast between the character and its background.

## CHARACTER SIZE

All things being generally equal (legibility factors, illumination, atmospheric interference, etc.), the larger letters and numbers are, the better they are seen and recognized—up to the point where the observer would be unable to see an entire character at a single glance.

Many people make a common error, however, in applying the above generalization; i.e., they believe that they should also make the character bolder as it is made larger for more distant viewing. As noted in the discussion of stroke width, to maintain legibility while increasing the size of a character, one must also maintain the proper stroke-width relationship.

The accompanying charts provide guidelines for determining how high a letter or numeral has to be for given viewing distances.

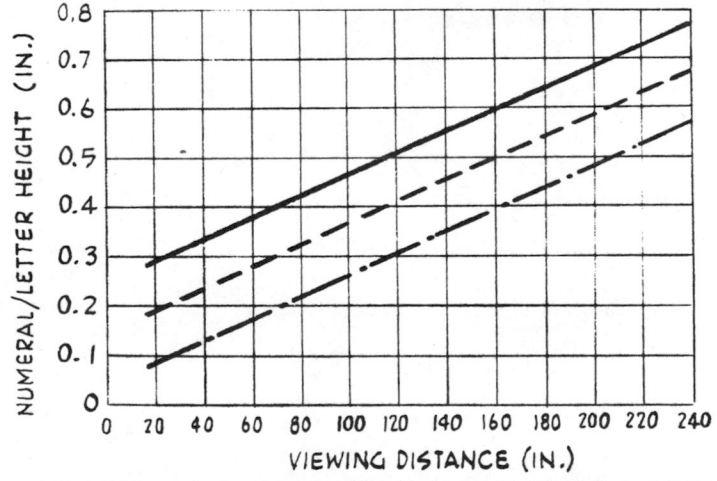
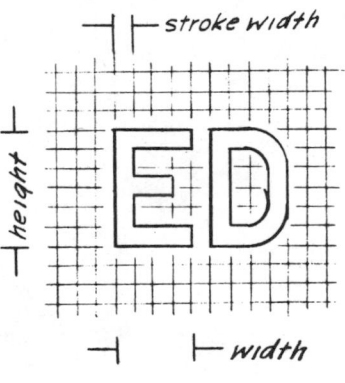

Letter height versus viewing distance and illumination level (minimum space between characters, one stroke width; between words, six stroke widths). (— For instruments where the position of the numerals may vary and the illumination is between 0.03 and 1.0 fL. --- For instruments where the position of the numerals is fixed and the illumination is 0.3–1.0 fL, or where position of the numerals may vary and the illumination exceeds 1.0 fL. —·— For instruments where the position of the numerals is fixed and the illumination is above 1.0 fL.)

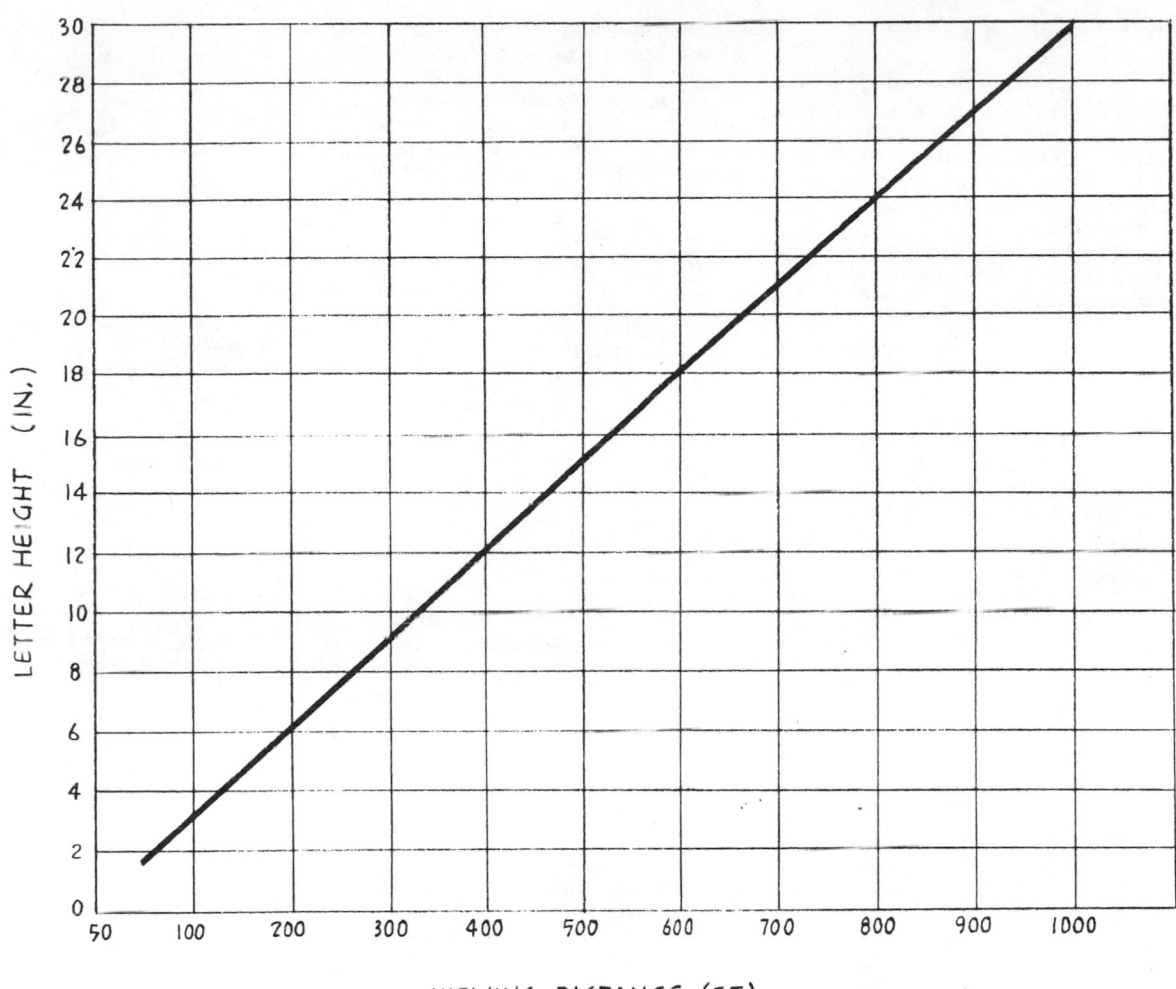

Minimum letter size recommended for highway signs. To calculate character height for greater viewing distances, letter height $= \dfrac{\text{viewing distance (ft)} \times 0.3 \text{ in}}{10}$

*Note.* The above is based on the assumption that the observer is relatively still. If a driver is moving rapidly toward or past a sign, the letter-height values should be increased approximately 10 percent, since the observer must be able to read the sign far enough in advance of a decision point to be prepared for the selected maneuver.

## MESSAGE INTELLIGIBILITY

To be effective, words have to be selected carefully so that their meaning is perfectly clear to the observer. One should first try to determine the background of the expected user population. The lay person, for example, probably will not understand some words that may be clear to an engineer. Wherever possible, use common words and abbreviations so that everyone will understand the meaning intended.

### Word Labels and Instructional Materials

1. Use common terms that originate from typical language usage and/or from standard lists of terms for special fields or groups of users (pilots, military personnel, etc.).

**Examples of General versus Special Usage Terminology**

| General (Preferred) | Special |
|---|---|
| Speed | Velocity |
| Rear window | Back light |
| Rest room | Head |
| Brightness | Intensity |
| Adjust | Calibrate |
| Water analyzer | Ionic chromatograph |

2. Use whole words rather than abbreviations wherever space permits. If abbreviations are required, refer to the list of standard abbreviations created by the appropriate "user" group (military, Department of Transportation, etc.). However, check these lists for duplications; it is undesirable to use the same abbreviation for two different functions.

| Preferred | Avoid |
|---|---|
| Range | RNG |
| Bearing | BNG |
| Ticket | TKT |
| Air conditioner | A/C |
| Transmitter | TM |
| Tachometer | TCM |
| Power amplifier | PA |

Acronyms may be used sparingly if they have been well established over a long period. They are least desirable for consumer products because of the wide variation in user experience.

3. To identify instruments, use terms that indicate what the instrument measures rather than the name of the device.
OHMS not OHMETER
RPM not SPEEDOMETER
HRS/MIN not TIMER

4. Avoid the use of words that may be interpreted one time as a noun and another time as a verb whenever both may occur at the same workplace.
FIRE = a fire has occurred (warning)
FIRE = to fire a weapon (command)

5. Always use capital letters for labels and short instructions because they can be read at a greater distance than capital and lowercase letters. Capital and lowercase letters should be used only for extended sentence messages where it is necessary to provide punctuation.

6. Generally, instruction labels should be as brief as possible within the limits of clarity. If an instruction implies several sequential steps, arrange each step on a separate line, preferably with a dot at the beginning of each line.

| *THIS* | *NOT THIS* |
|---|---|
| · JETTISON CANOPY<br>· FEET IN STIRRUPS<br>· RAISE ARMRESTS<br>· SIT ERECT<br>· SQUEEZE TRIGGER | FOR EMERGENCY EJECTION IN FLIGHT JETTISON CANOPY BY PULLING CANOPY JETTISON HANDLE. HOOK HEELS IN FOOT RESTS. RAISE BOTH ARMRESTS TO HORIZONTAL POSITION. ASSUME ERECT POSTURE, ACTUATE EJECTION TRIGGER FOR SEAT EJECTION |

The original instruction was:

WALK UP ONE FLOOR
WALK DOWN TWO FLOORS
FOR IMPROVED ELEVATOR
SERVICE

What the instruction really meant was: If you intend to go up only one floor or down two floors, walk instead of taking the elevator because this will relieve the use load and therefore provide better service for everyone.

Following are two ways to simplify the message:

| SAVE TIME BY WALKING IF YOU ARE GOING UP ONE OR DOWN TWO FLOORS | TO GO UP ONE FLOOR<br>OR<br>DOWN TWO FLOORS<br>*PLEASE WALK* |
|---|---|

7. Avoid the use of words that may have a negative psychological effect; e.g., the word "speed" itself implies traveling at a fast rate and therefore may encourage a motorist to drive fast.

| *THIS* | *NOT THIS* |
|---|---|
| SLOW TO 10 MPH | SPEED ZONE AHEAD<br>10 MPH |
| SLOW DOWN<br>WORKMEN IN ROADWAY | SLOW MEN WORKING |
| SLOW DOWN<br>ANIMALS IN ROADWAY | CAUTION<br>DEER CROSSING |
| DANGER<br>LETHAL VOLTAGES | DANGER<br>HIGH VOLTAGE |
| SLOW DOWN<br>ROCKS ON ROAD | WATCH FOR<br>FALLING ROCKS |

8. Avoid using the number 1 in combination with letters when encoding such things as airport terminal gates. For example, if you are planning to encode the gates with letter-number combinations, such as K-1, K-2, K-3, etc., do not include the letter I because it is easily confused with the number 1; i.e., I-1 might look like an H from a distance.

9. When special precautionary words are required, select ones that provide a sense of urgency, hazard, or danger. Avoid words that seem to leave it up to the observer to decide whether there is a danger involved, since he or she may not have the necessary experience to make such a judgment.

| Preferred Nomenclature | Poor |
| --- | --- |
| Warning | Note |
| Hazard | Precaution |
| Caution | Slow men working |
| Danger | Slippery |
| Watch out for workmen | Slide area |
| Watch for children | |

10. The lettering for all labels should be oriented so that they read from left to right, not around corners, on the side, or up and down.

11. A manufacturer's label must not overshadow the primary display labels and markings. Place the manufacturer's label in an inconspicuous place, in small print, as shown in the illustrations at the right.

12. Control and display labels should be located consistently, either below or above the panel component, especially on the same panel. The decision to locate the label above or below components should be based on the eye position of the observer; i.e., if the panel labels are above the eye reference level, the controls may extend out so far that they obscure a label placed above the control, or the opposite might be true if the control panel is considerably below the eye reference level. Random placement of labels on the same panels makes them difficult to find and associate with the pertinent function.

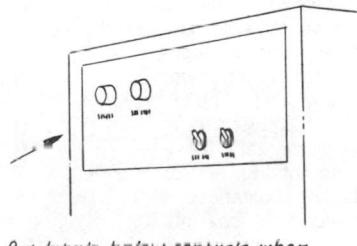

Put labels below controls when viewed from below,

above when viewed from above.

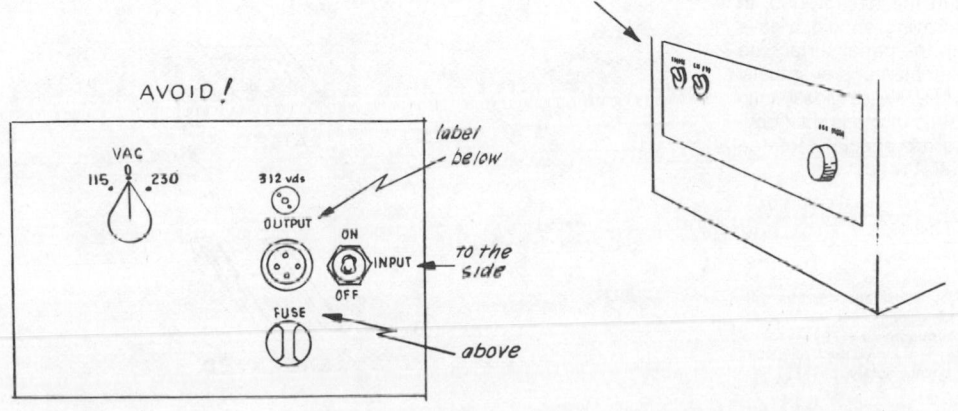

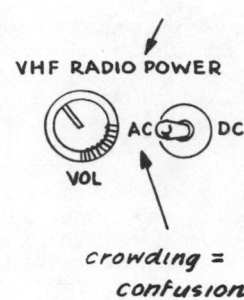

VHF RADIO POWER

*crowding = confusion*

13. Avoid running adjacent labels together. This often happens because control elements were spotted on the panel without considering where the labels would go or how much space they might require.

14. Utilize size and/or color coding to help the observer differentiate between levels of importance on your panel; i.e., the *system* label should be slightly larger than subsystem labels, and these should be larger than component labels. The examples shown in the accompanying sketches should be used as a general guideline.[1]

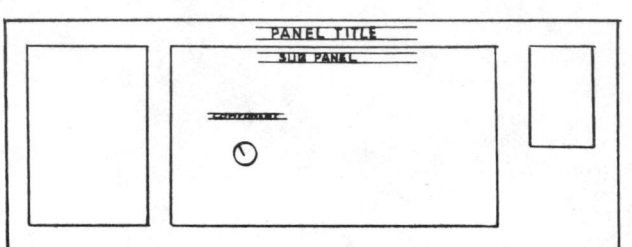

*Primary title about half again larger than intermediate title - this half again larger than smallest title.*

15. As a general rule, all labels should be made as permanent as possible; i.e., they should be wear- and damage-resistant. If labels are to be read under low light levels, they must be illuminated by means of ambient room lighting, special supplementary floodlighting, back or edge lighting, and/or luminescent materials or substances. If raised letters are provided (i.e., cast into a panel or package housing), there must be additional coloring to make the letters contrast with the background. Do not assume that a raised letter that is the same color as the basic background will be readable just because it is raised. Engraved or depressed lettering should also be of a color that contrasts sharply with the background. In addition, the engraving should always be filled level with the panel surface so that it will not collect dirt and eventually become unreadable. Avoid glossy inks or paints and plastic or glass dust covers, unless these are specially treated to minimize spectral reflectivity.

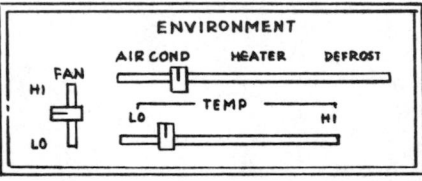

*Typical good practice*

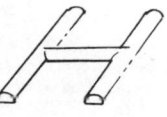

RAISED

ENGRAVED

[1]The character height of the smallest label must be sufficient for the maximum viewing conditions expected. Avoid the tendency to settle for some arbitrary policy set by your engineering department for all equipment turned out by the company; this may be completely unsatisfactory.

## LABEL SPECIFICATIONS

Starting with the smallest lettering size that will be compatible with the typical average viewing distance, establish specifications for each of the labeling categories illustrated by the accompanying sketch.

### Typical Panel Labeling Standards

| Label Designation | Letter Size | Location |
|---|---|---|
| 1. Panel title | 18 pt (0.187 in) | Centered; ¼ in from top edge of panel |
| 2. Panel subsection | 14 pt (0.156 in) | Centered at top of subsection; ¾ in from top edge of panel |
| 3. Subtitle | 12 pt (0.125 in) | ¼ in above components or ⅛ in above labels of individual components |
| 4. Toggle switch | 10 pt (0.093 in) | ¼ in above and below standard switch |
| 5. Single component | 12 pt (0.125 in) | ¼ in above component |
| 6. Rotary switch positions | 10 pt (0.093 in) | ¼ in from apex of pointer, line from pointer to label |

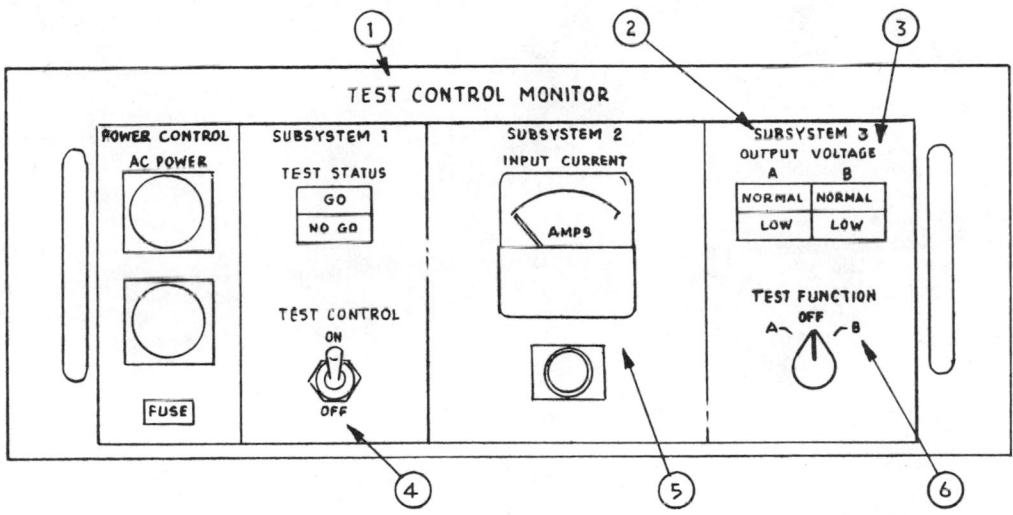

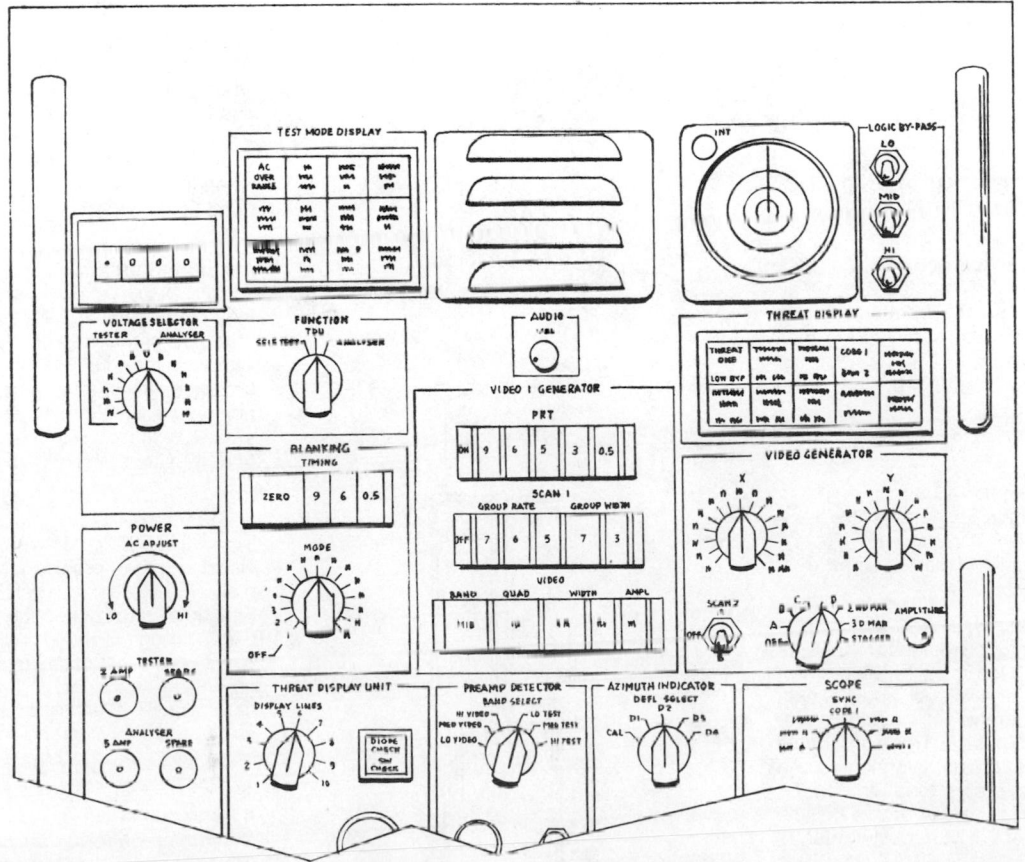

Example of a relatively well-executed labeling program for a fairly complex panel.

## LABELS PLACED ON CONTROLS

### Word Displays in Conjunction with Controls

Although perhaps more appropriately discussed under the general topic of "labeling," words used directly on certain types of control devices increase the utility and efficiency of control identification and operation by combining the identifying variables of control position, light, and words. The examples shown in the accompanying illustration are representative of only a few of the applications that would be appropriate. See the section on controls for a more complete discussion of this application of word and light combinations.

Several precautions are in order relative to placing word labels directly on the control device:

The control must not rotate so that the label may become unreadable.

If the control surface area is too small for a fully legible label, place the label adjacent to the control; i.e., do not feel that you have taken care of your obligation just because a label is there—it has to be readable.

Push buttons should extend out from a control panel, as compared with a telelight (i.e., not a control); otherwise, an operator will be unable to tell which similar-looking word-light displays are really controls and which are merely advisory displays.

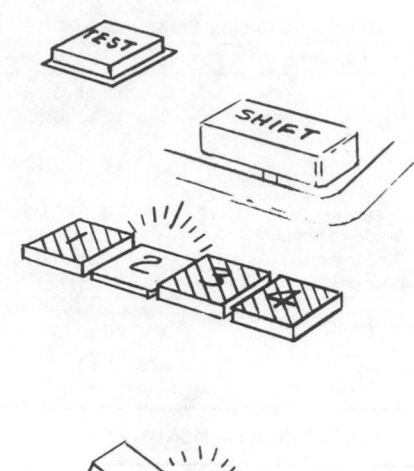

## TACTILE SIGNS FOR THE BLIND AND PARTIALLY SIGHTED

Public signing plans should consider the problems of blind and partially sighted users. The most reliable type of signing for these people is the tactile sign in which raised lettering and/or braille code is used.

Typical signs that should be provided for the blind include:

Street signs
Bus-stop signs
Intersection signal control signs
Exterior building signs (especially for restroom buildings)
Interior office and public facilities signs (rest rooms, elevators, etc.)

The accompanying sketches illustrate a suggested approach to providing tactile street signing. The tactile sign must be placed approximately at head or eye level, and it should face the street so that the person knows with which street to associate it. The sign should have raised letters for the partially sighted and non-braille user, plus braille code for individuals who are trained in the use of this device.

For identification of braille codes, contact a local or regional office of the Braille Institute of America.

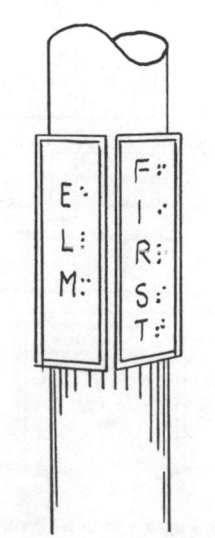

## PICTORIAL SYMBOLS

Pictorial symbols make it possible for persons with different language backgrounds to recognize and understand a single symbol. Unfortunately, it is not always possible to pictorialize all informational conditions since some things would not be recognized by the observer because he or she has never seen the actual object. An example is the choke system used on some vehicles. A symbol that is significant to engineers because they have seen the actual system or device not only is meaningless to average drivers but also may encourage them to believe the symbol implies something entirely different; e.g., the typical hourglass pictorial suggested by the automotive designer could be construed to mean "time" to the lay person. Some principles to consider in deciding whether a pictorial signing method should be used include the following.

### 1. Pictorial Comprehensibility

If a particular situation includes items that cannot be completely pictorialized, i.e., if some items are easy to pictorialize, but others are not, it is better to stick to a word labeling system; observers will only be confused if they find some pictorials and some word labels on an operator panel.

### 2. Combined Pictorial and Word Labeling

Whenever there is any doubt as to whether some observers will be able to understand the meaning of a pictorial, use both pictorial and word labels.

### 3. Size-Distance Compatibility

Some pictorial patterns may be effective only when the viewing distance and lighting conditions are optimum; be sure that a particular pictorial pattern does not lose its identity when it becomes smaller or more distant and/or when the ambient lighting and/or atmospheric conditions are not good.

### 4. Orientation

Do not place pictorials on components that may be reoriented, i.e., turned or moved so that the pictorial may not appear right side up. Although the observer might figure out what the pictorial is after considerable study, a prime purpose of the single pictorial is to elicit a quick response with a single symbol, as opposed to several words in a label.

### 5. Sufficient Detail

In creating a pictorial to represent some actual visual element, provide just enough detail to make the symbol recognizable, and no more. Fine detail often cannot be seen under certain distance and lighting conditions and may serve only to distort the impression the symbol creates. On the other hand, be extremely careful not to overstylize pictorial symbols just to create an artistic rendition. Such stylization often makes all symbols begin to look alike.

### 6. Isolation

Some type of border should always be used around a pictorial; otherwise, it may blend with background images.

STAIRS - UP

STAIRS - DOWN

STAIRS UP & DOWN

ESCALATOR UP

DOWN

ELEVATOR

OR

TOILET
MEN

WOMEN

CLOAK ROOM

MEN'S

WOMEN'S

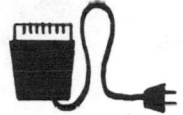
ELEC RAZOR OUTLET

USED BLADE DEPOSIT

USED LINEN

USED TOWELS

USED CUPS

TRASH

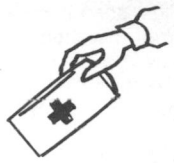
SANITARY NAPKINS

NO FOREIGN MATERIAL IN TOILET

FASTEN SEAT BELT

NO SMOKING

OR

READING LIGHT

FLIGHT ATTENDANT

SICKNESS CONTAINER

OXYGEN

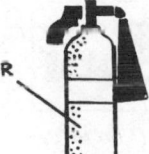
FIRE EXTINGUISHER

BAGGAGE LOCKERS

PORTER

TAXI

CAR RENTAL

PHONE

HANDICAPPED ACCESS

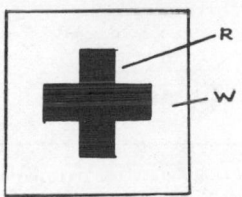
R
W
FIRST AID

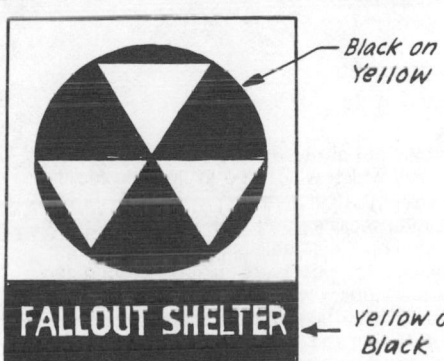
Black on Yellow

FALLOUT SHELTER ← Yellow on Black

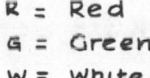
R = Red
G = Green
W = White

**EXAMPLES OF PICTORIAL SIGNS FOR PUBLIC ACCOMMODATIONS**

With careful thought, signs and symbols can be used in ways that will make them more effective.

The aircraft pictorial shown in the accompanying illustration can be used to indicate the location of the airport or the passenger gate merely by "pointing" the symbol in the appropriate direction. This removes the need to add an arrow.

TO AIRPORT OR TO BOARDING GATE

To the right

left

Some pictorials may be used for slightly different purposes, e.g., the baggage symbol. By adding slightly to the pictorial, the observer can be informed as to whether the location is for baggage check-in, inspection, or storage.

BAGGAGE CHECK-IN

Similar clarification for public transportation stops is helpful; in addition to showing a picture of the vehicle, providing a stop sign or platform helps clarify that the location is specific, i.e., that this is where one should stand to wait for a bus or train.

BUS STOP

TRAM (Rail) STOP

Arrows are always helpful in letting observers know which way to go to find the facility identified by a pictorial.

Several facilities and their location can be identified on the same sign, thus making it unnecessary for observers to look around for separate signs.

With a white sign background, outline the field definition and black figure. This is the only condition in which the symbol outline should be delineated.

Individual pictorials should be tested to determine the appropriate size for the expected viewing distance.

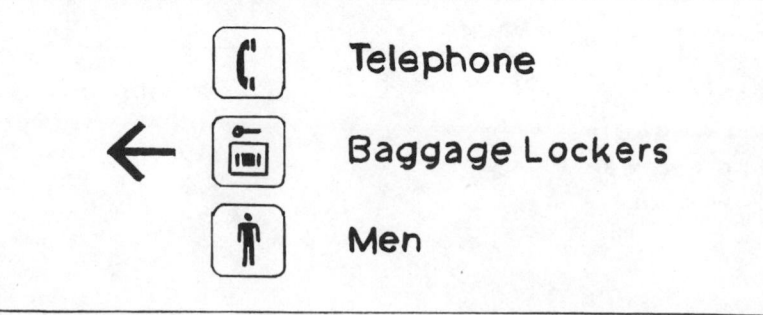

## Legibility Criteria[2]

The following diagram illustrates the results of pragmatic testing of several symbols, (Ticket Purchase, Elevator, and Taxi), and represents a rough guide to size/distance relationships.

For the purpose of this illustration "legibility" was defined as the recognition of the various elements that make the sym-

bol understandable without the aid of wording or preconditioning. "Recognition" of the symbols after they are learned is another matter, and we feel cannot be meaningfully tested at this time except in the case of those few well known symbols such as First Aid, Men, Women, etc. The testing was done in daylight using symbols

with black "figures" on white "symbol fields" displayed on a black sign background.

The illustration shows the result of the testing of the Ticket Purchase symbol. The distances from which the Taxi symbol was legible were 10% greater, and for the Elevator symbol, 30% less.

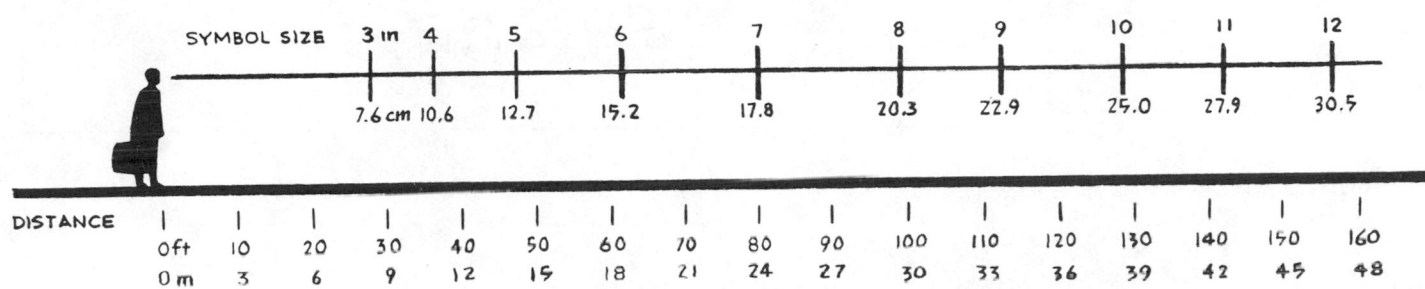

One of the most important aspects of good signing is siting. The closer to one's natural line of vision, the better.

A useful rule of thumb is to avoid exceeding a 10 degree angle from the natural line of vision. This formula has value, primarily with regard to height, except in the case of a roadway or corridor type of condition where "natural line of vision" can be reasonably defined.

If conditions require that the viewing angle exceeds 10 degrees, the size/distance relationship may have to be adjusted, (for example, a sign at 15 feet

above the floor level will probably have to be larger than the same sign at 8 feet to be as effective) or another smaller sign may have to be added for short-distance reading.

It must be pointed out that legibility varies greatly from one symbol to another, or from one type style to another, and that color relationships, lighting, spacing, and viewing angle may also affect legibility. We recommend pragmatic testing of symbols and lettering on-site, or in simulated on-site conditions.

If an attempt is made to equalize

symbols of unequal legibility in a signing system by varying their size, the result would be visually chaotic. We recommend that the legibility characteristics of the least legible symbols determine the size of all the symbols in a given system. This would provide a sense of order and adequate legibility throughout.

The intensity of internal lighting of symbols on translucent background material should be minimal to prevent loss of legibility due to "halation", the spreading of light.

[2]U.S. Department of Transportation, *Symbol Signs: Development of Passenger/Pedestrian Oriented Symbols for Use in Transportation-related Facilities*, DOT-OS-40192 (NTIS PB-239-352), 1974

## Location Signs

The illustrations below show two solutions to typical symbol sign problems. The first is a sign band and the second a hanging sign occupying less horizontal dimension.

*Note:* The size of the lettering should be based on the height of the lowercase letters, not on the height of the capital letters.

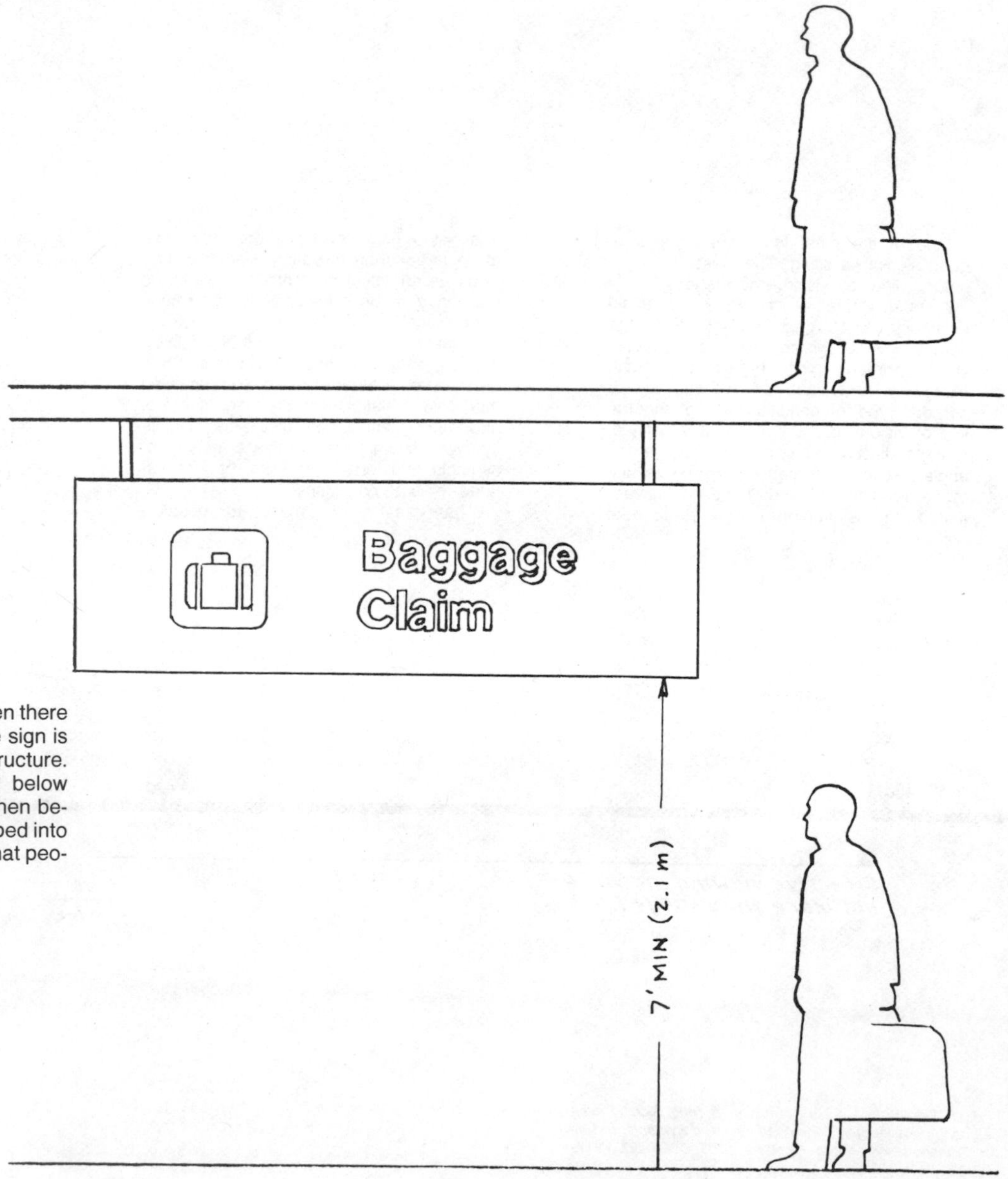

The suspended sign is preferred when there is sufficient ceiling height because the sign is less easily confused with the facility structure. Suspended signs should not extend below about 7 ft (2.1 m), however, for they then become obstructions that are easily bumped into either by very tall people or by things that people may be carrying.

## Multidirectional Signs

Where several directions are indicated on one sign it is best to keep the arrow in a constant position relative to the symbols, regardless of arrow direction.

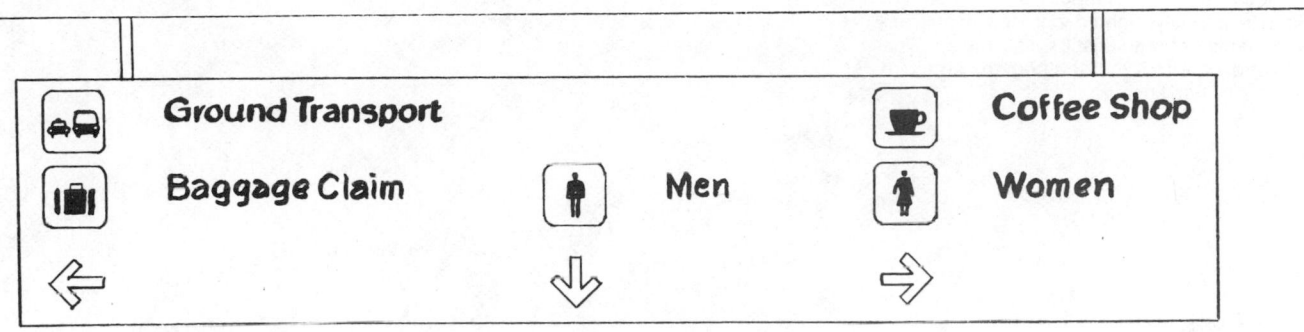

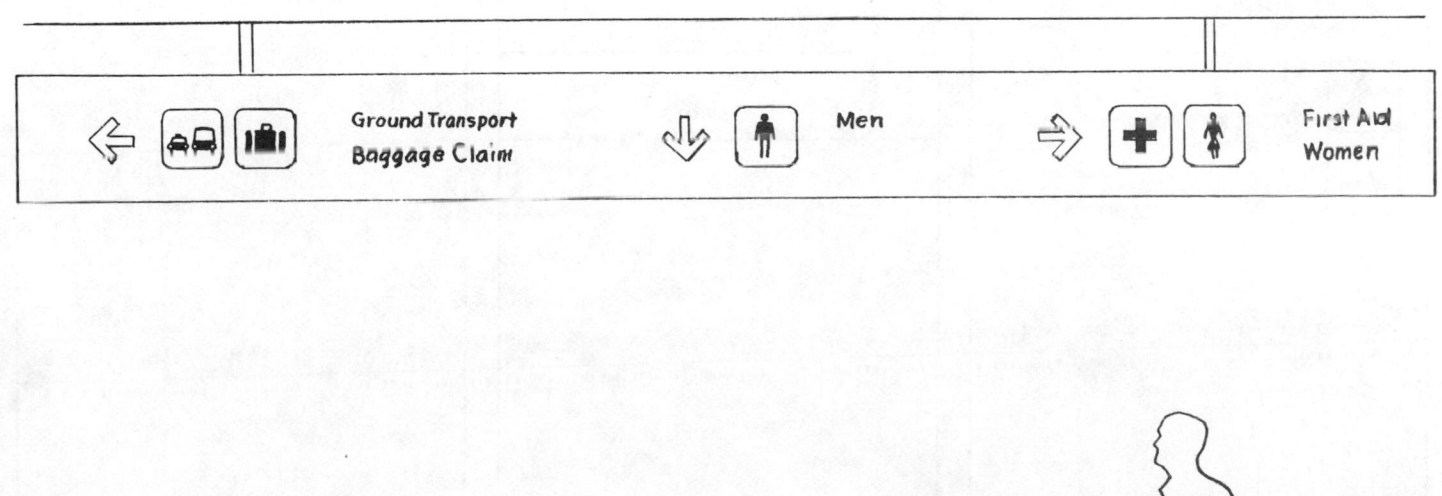

**Door Signs**

The plaque shown below in the example on the left illustrates a desirable application technique for doors or surfaces that require periodic refinishing. This type of sign may also be projected from the wall for better visibility in 'corridor' conditions.

Critical signs that should be available to blind or partially sighted persons should include tactile characteristics, i.e., raised symbols and/or letters that the person can touch with his or her fingers.

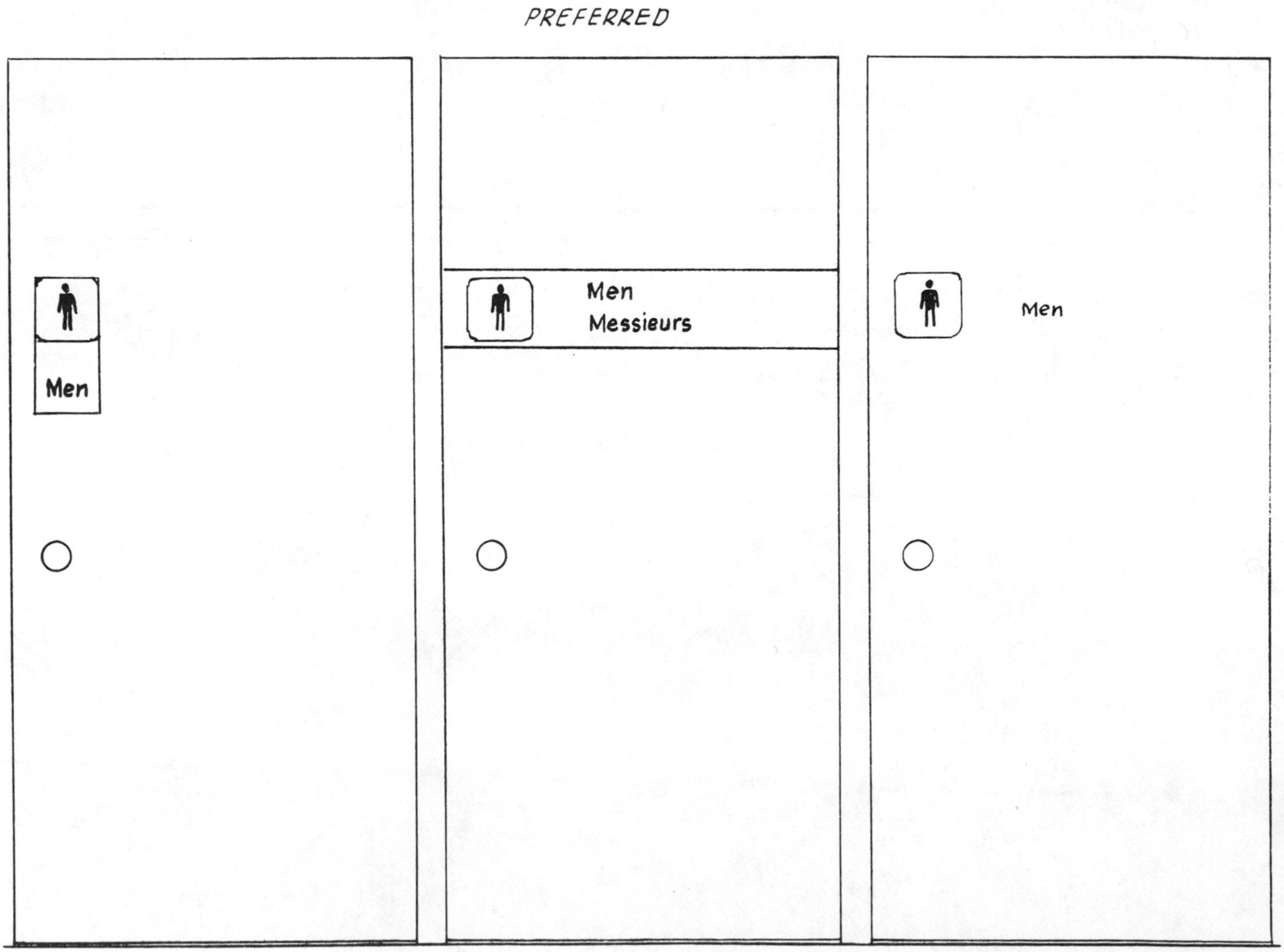

*PREFERRED*

Men

Men
Messieurs

Men

## Freestanding Signs

Curb side signage should be carefully sited to provide good sightlines from cars, buses and trucks. The use of reflective material for the symbol field may be desirable if ambient light is not adequate for night viewing.

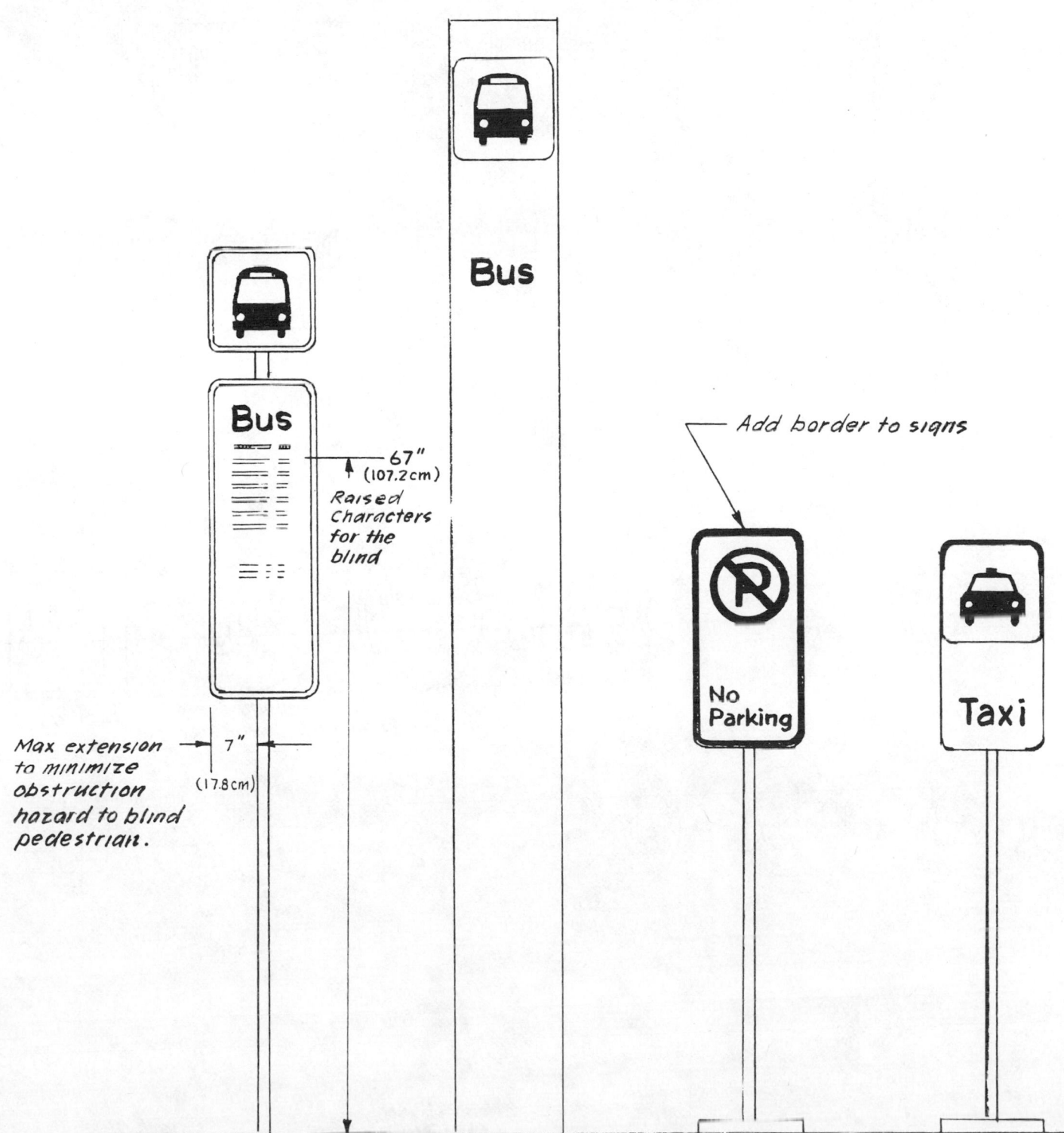

67" (107.2cm)

Raised Characters for the blind

Max extension to minimize obstruction hazard to blind pedestrian.

7" (17.8cm)

Add border to signs

Bus

No Parking

Taxi

## Roadway, Traffic Control Signs

Although roadway signing concepts still vary widely throughout the world, there has been a considerable effort at standardization among countries, especially in the use of pictorial or international signing. The examples shown in the accompanying sketches reflect some of the different types of signing that are used.

A specific warning is indicated by placing any of the accompanying symbols within these shapes. The inner symbol may be either black or dark blue.

A specific restriction or prohibition is indicated by placing any of the following symbols within these shapes. The inner symbol may be either black or blue.

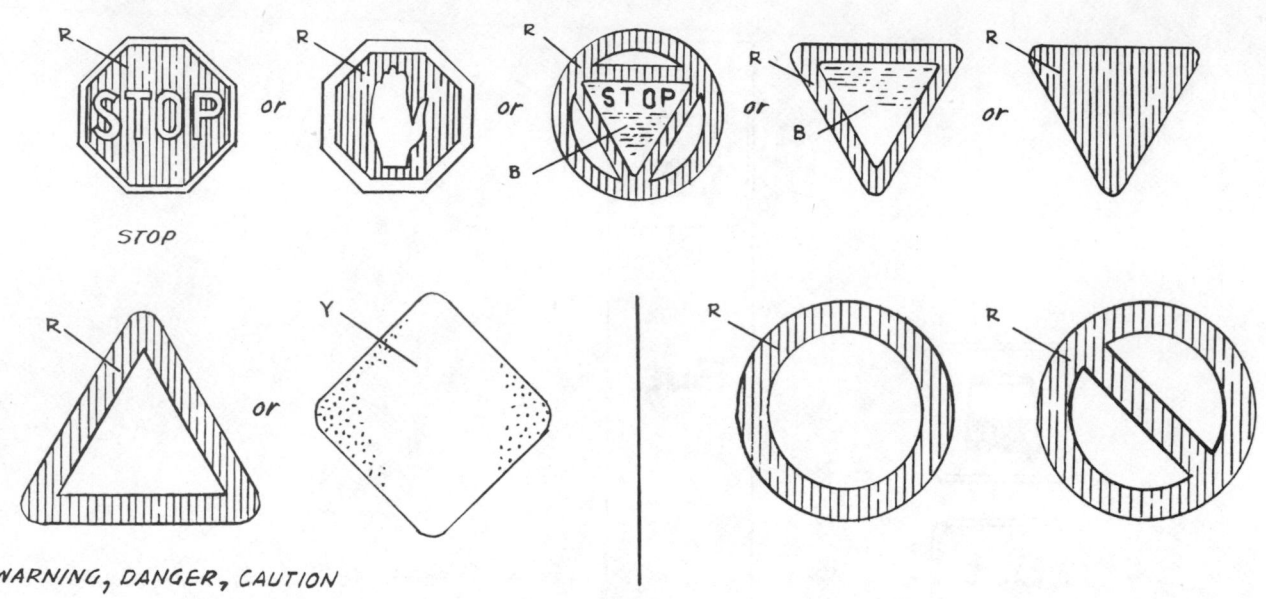

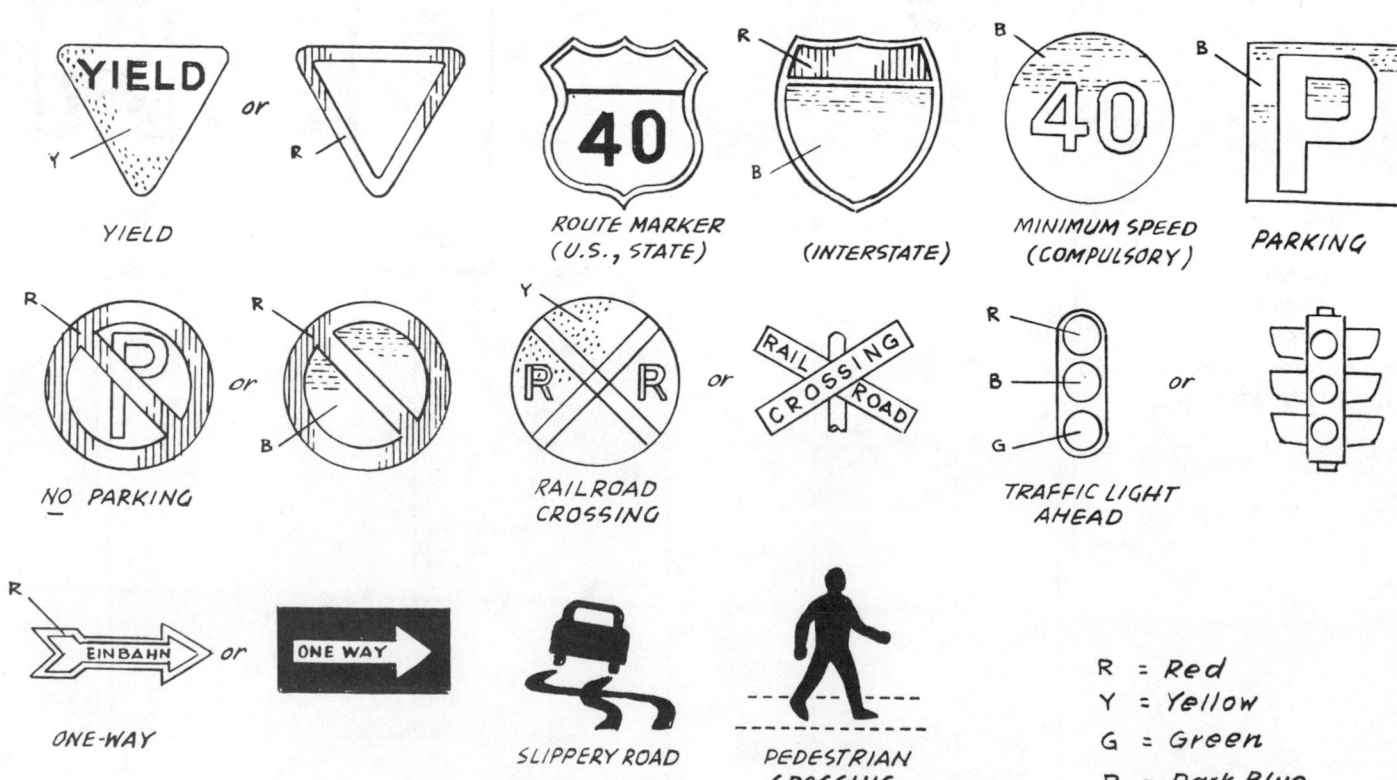

### International (Pictorial) Road Signs

The concept of using pictorial signs has gained wide acceptance around the world because such graphic symbols can generally elicit reliable interpretation by people of different language backgrounds. The examples shown in the accompanying illustrations are representative.

Within the international signing concept, graphics are placed in one of three types of borders, shown in the accompanying illustration. A red triangular border is used for danger signs, a red circular border is used for instructional signs, and a blue border is used for purely informational signs. The various graphics are black silhouettes within a particular border according to the requirement objective.

When an informational sign is meant to denote a restriction such as "no entry," a diagonal slash mark is placed across the graphic.

DANGER

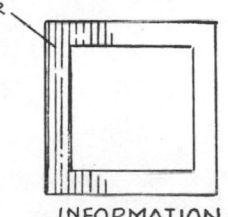

INFORMATION

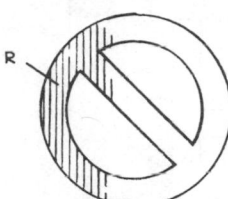

INSTRUCTIONS

STEEP ASCENT

STEEP DESCENT

FALLING ROCK

LOOSE GRAVEL

CHILDREN CROSSING

ANIMAL CROSSING

WILD ANIMAL CROSSING

LOW-FLYING PLANES

TRUCK CROSSING

TRAIN CROSSING, NO GATES

ROAD WORK AHEAD

NO MOTOR VEH. OF ANY KIND

NO BUS ENTRY

NO COMMCL

NO VEH./TRAILER

  or

NO PASSING (COMMCL VEH.)

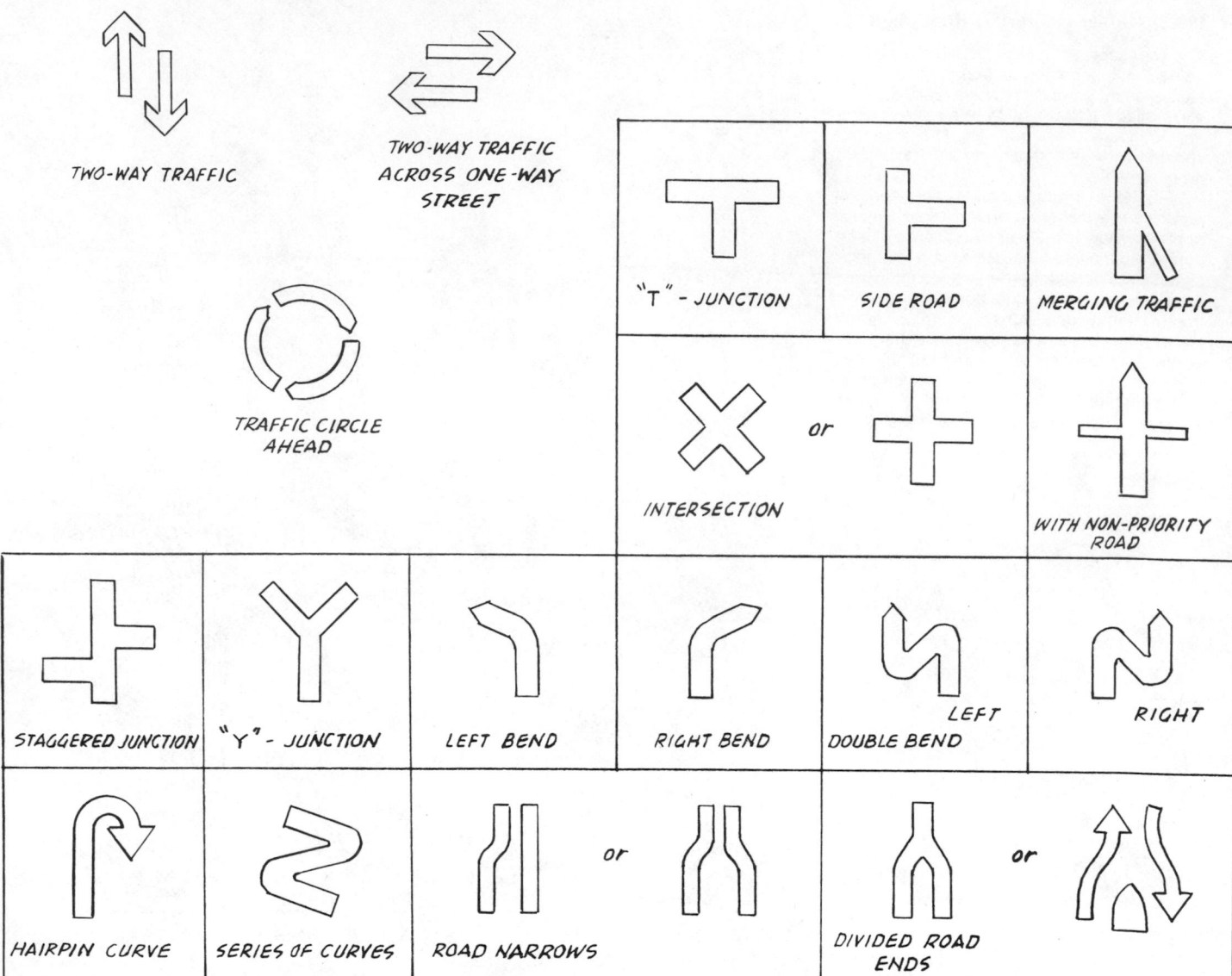

TWO-WAY TRAFFIC

TWO-WAY TRAFFIC ACROSS ONE-WAY STREET

TRAFFIC CIRCLE AHEAD

"T" - JUNCTION

SIDE ROAD

MERGING TRAFFIC

INTERSECTION    or

WITH NON-PRIORITY ROAD

STAGGERED JUNCTION

"Y" - JUNCTION

LEFT BEND

RIGHT BEND

DOUBLE BEND    LEFT

RIGHT

HAIRPIN CURVE

SERIES OF CURVES

ROAD NARROWS    or

DIVIDED ROAD ENDS    or

Quasi-pictorial graphics devised to identify roadway features.

## UNIFORM ROADWAY SIGNING IN THE UNITED STATES

Signs are functionally categorized as follows:

Regulatory signs: These note traffic laws and regulations, disregard of which may subject the offender to a fine or other punishment.
Warning signs: These call attention to conditions that are potentially hazardous.
Guide signs: These give route designations, destinations, directions, distances, and other travel information.

Through its *Manual on Uniform Traffic Control Devices for Streets and Highways,* the U.S. Department of Commerce, Bureau of Public Roads, attempts to standardize all roadway signs in terms of their shape, color, dimensions, symbology, wording, lettering, and illumination or reflectorization. The following is an example of the information found in this manual.

### Shapes

The significance of sign shapes is standardized as follows:

The octagon shall be reserved exclusively for the Stop sign.

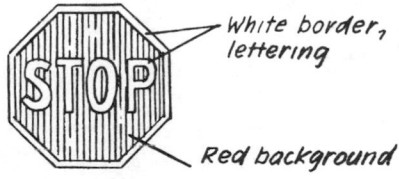

The equilateral triangle, with one point downward, shall be reserved exclusively for the Yield sign.

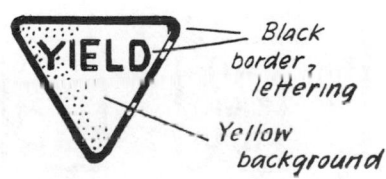

The round shape shall be used for the advance warning of a railroad crossing, and for the civil defense Evacuation Route Marker. It is also used for some State Route Markers.

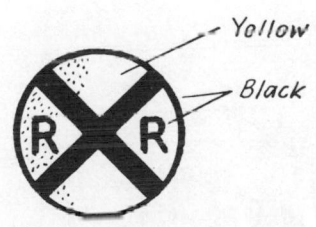

The diamond shape shall be used only to warn of existing or possible hazards either in the roadway or adjacent thereto.
Regulatory signs, with the exception of certain route markers, shall be rectangular, ordinarily with the longer dimension vertical. Guide signs, with the exception of certain route markers, shall be rectangular, ordinarily with the longer dimension horizontal.

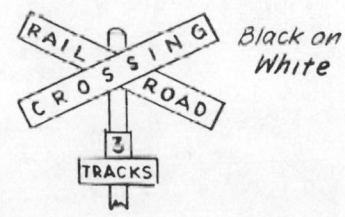

Other shapes are reserved for special purposes; for example, the shield or other characteristic design for route markers on Interstate, U.S., and State highway routes, and the crossbuck for railroad crossings.
The sign shall be used in advance of every railroad crossing, even if protected by crossbucks, signals, gates, or flagmen, except in the following instances:

1. At a minor siding or spur which is infrequently used and which is guarded when in use by a member of the train crew.
2. In the business districts of large cities where the crossings are fully protected and the physical conditions are such that even a partially effective display of the sign is impossible.

## Colors

Warning signs, with the exception of the railroad crossbuck sign, shall have a background of "highway yellow" with black legend.

Regulatory signs, with the exception of the red Stop sign, the yellow Yield sign, and certain urban parking signs (sec. 1B-31), shall have a white background with black legend or, optionally for signs exceeding 30 inches by 36 inches in size, a black background with white legend.

*Interstate.*—Standards for the Interstate System permit only a white background for all black and white regulatory signs.

Guide signs, with the exception of certain route markers and Rest Area and Services signs, shall use only the colors white, black, and green. Destination, Distance, and Information signs, except Rest Area and Services signs for which special distinctive designs are permitted, shall have a black legend on a white background, white legend on a black background, or white legend on a green background. U.S. route markers, and auxiliary markers used in assemblies with them to show junctions, turns, directions, and alternates, shall have a black legend on a white background.

## Dimensions

The sign dimensions prescribed in this Manual shall be standard for application on public highways. The size of a sign must depend primarily on the length of its message and the size and spacing of the letters that form the message, or on the size of any required symbol, when the complete legend is designed for adequate legibility. In this respect some flexibility is desirable to permit the use of a uniform sign plate for any particular series of signs.

The standard dimensions shown herein have been designed to provide clear legibility of the signs during the time and throughout the distance necessary for approaching traffic to read and comprehend their messages under "normal" highway conditions. Increases above these standard sizes are desirable where greater legibility or emphasis is needed, and for expressways special categories of large signs are prescribed. In determining whether the standard size is adequate, consideration should be given to such elements as prevailing speeds and volumes, the width of roadway or number of lanes, the degree of hazard (as appraised by a field survey of sight limitations, intersection complications, etc., or as revealed by accident records), and the competition offered by other signs, lighting, displays, or background. Any doubt should be resolved in favor of a larger sign.

*Note:* All signs have borders of the same color as the lettering or symbols.

Diamond shapes warn of existing hazards (black on yellow).

Vertical, rectangular regulatory signs (black on white).

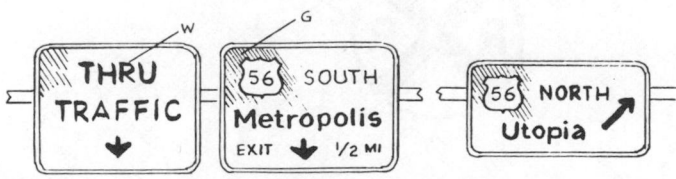

Horizontal, rectangular guide signs (white on green).

## Design of Route Markers

The U.S. Route Marker (M1-1) when used as a Confirming or Reassurance marker shall consist of a shield-shaped plate, of a standard size of 16½ inches by 16 inches, carrying the State name, the letters US and the route number in black legend on a white background. State Route Markers are designed by the individual State highway departments, but should be of approximately the same size as U.S. Route Markers.

When used in Junction assemblies,

Route Turn assemblies, Directional assemblies, and Trailblazer assemblies, U.S. and State Route Markers shall consist of a square plate carrying only the white shield (or other distinctive shape) of the Route Marker with the route number in black. The area of the plate outside the Route Marker outline shall be black, without a border. The standard Assembly Route Marker plate shall be 24 inches square, with a small size 18 inches square for use on minor roads and secondary streets.

So that markings placed on roadway surfaces will remain legible while the motorist is moving toward the markings, such elements as the letters, numerals, and arrows must be elongated as indicated in the accompanying illustrations.

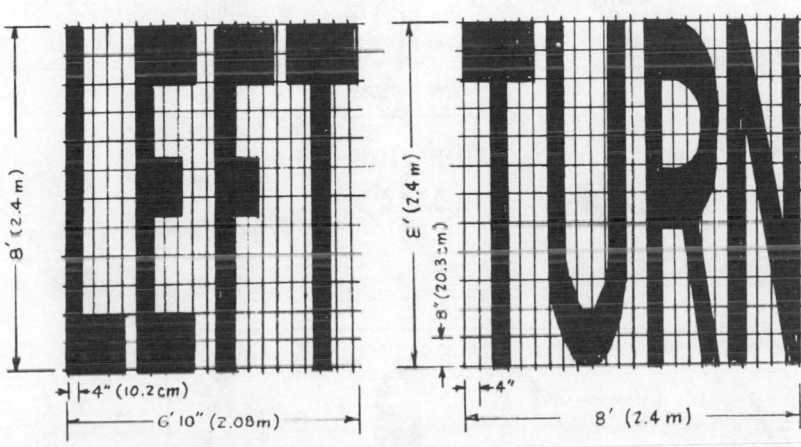

*Elongated letters for street marking*

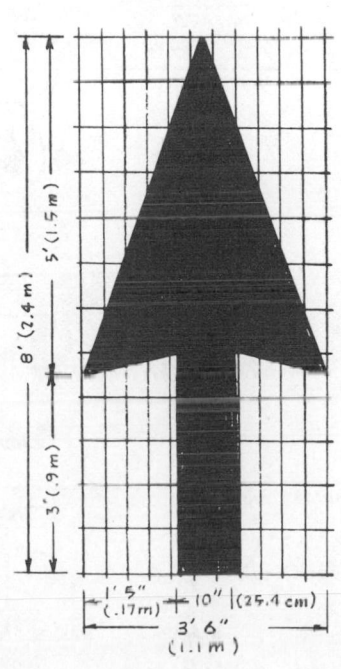

## Visual Background Interference

It is easy to forget that the visual background within which a sign or signal must be observed may interfere with successful location of the sign and/or with easy reading of the sign or recognition that a signal light is illuminated or not. Two major background interference factors should be considered:

1. Glare from the sun or from very bright artificial lights
2. General visual clutter from objects, other signs, crowds of people, or vehicles

When such interference is anticipated, it is important to provide a sufficient border area around the significant elements of the sign to ensure that the observer can focus on and read the symbols or words and/or recognize that a signal light is on.

For symbolic or alphanumeric signs, a good rule of thumb for the border area is to provide approximately the same amount of clear bor-

 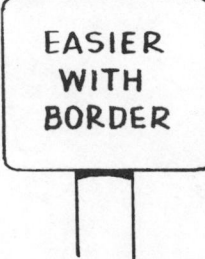 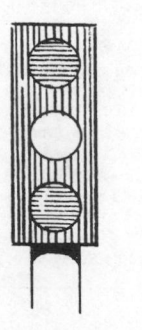 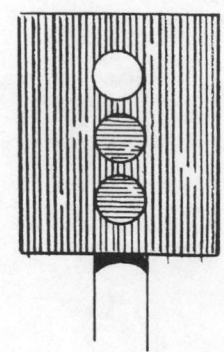

*Black, background shield*

der area as is required for the symbology (including the clear space between the lines and the lettering).

For outdoor traffic-light signals, when glare from the sun is a common condition for signal viewing (i.e., when the signal light has to be located so that a driver looks into the setting sun on a daily basis), a shield should be provided around the lights. Such shields should

be painted matte black to accentuate the contrast between the illuminated signals and the adjacent shield. As a rule of thumb, the shield border around the lights should be about two to three times the diameter of the lights.

BRIGHT — HEADLIGHTS | DIM | PARKING LIGHTS | FOG LIGHTS | PASSING LIGHTS | EMERGENCY VEH. (ROTATING)

   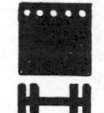 

INTERIOR LIGHT | INSTR. PANEL ILLUMINATION *or* | FRONT RUNNING LTS. (TRACTOR) | REAR RUNNING LTS. (TRAILER) | SIDE RUNNING LTS. (TRAILER)

  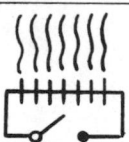   

EMERGENCY HAZARD (FLASHING) | AIR CONDITIONING | HEATER (ON-OFF) | (ADJUST) | DEFROSTER | VENTILATING FAN

ENGINE OIL | ENG OIL PRESSURE | AMMETER / GENERATOR *or* | | TEMPERATURE | WATER TEMP

FUEL | CHOKE | TIME (HRS.) | RADIO (ON-OFF-VOL.) | (STA-SELECT) | HOOD RELEASE

WIPER | WASHER | BOTH | HORN | CIGARETTE LIGHTER | ASH TRAY

WINDSHIELD

## PICTORIAL SYMBOLS FOR VEHICLE CONTROLS

Examples of automotive pictorial symbols are given on page 514.[3]

### Federal Standards

Certain controls in automobiles are required to have international pictorial symbols, as indicated by the accompanying table, which is taken from *Control Location, Identification and Illumination—Passenger Cars, Multipurpose Passenger Vehicles, Trucks and Buses,* Motor Vehicle Safety Standard no. 101. Note, however, that from a good human engineering standpoint, all labels, whether they consist of words or symbols, should always be illuminated for night use. Other important principles include the following:

If a control is rotatable, the symbol or word should be placed on the panel next to the control rather than on the control so that the label will always appear upright.

A word label should accompany any pictorial symbol in order to ensure symbol intelligibility.

TABLE 1—Control Identification and Illumination

| COLUMN 1 | COLUMN 2 | COLUMN 3 | COLUMN 4 | COLUMN 5 |
|---|---|---|---|---|
| Motor Vehicle Equipment Control | Word or Abbreviation | Permissible Symbol | Alternate Permissible Symbol | Illumination |
| Engine Start | Engine Start[1] | None | None | |
| Engine Stop | Engine Stop[1] | None | None | Yes[1] |
| Manual Choke | Choke | None | None | |
| Hand Throttle | Throttle | None | None | |
| Automatic Vehicle Speed Control | | None | None | Yes |
| Headlamps and Taillamps | Lights[2] | | [2,4] | |
| Vehicular Hazard Warning Signal | Hazard | | [4] | Yes |
| Clearance Lamps | Clearance Lamps[3] or CL LPS | | [3,4] | Yes |
| Identification Lamps | Identification Lamps or ID LPS | None | None | Yes |
| Windshield Wiping System | Wiper or Wipe | | | Yes |
| Windshield Washing System | Washer or Wash | | | Yes |
| Windshield Defrosting and Defogging System | Defrost or Def | None | None | Yes |
| Heating and Air Conditioning System | | None | None | Yes |

[1] Use when engine control is separate from the key locking system.
[2] Use also when clearance, identification lamps and/or side marker lamps are controlled with the headlamp switch.
[3] Use also when clearance lamps, identification lamps and/or side marker lamps are controlled with one switch other than the headlamp switch.
[4] Framed areas may be filled.

[3]For a more complete description, see Henry Dreyfuss, *Symbol Sourcebook: An Authoritative Guide to International Graphic Symbols,* McGraw-Hill Book Company, New York, 1972.

## PICTORIAL SYMBOLS FOR AGRICULTURAL EQUIPMENT

The various pictorial symbols shown in the accompanying illustrations were developed by the Tractor Technical Committee of the Society of Automotive Engineers (a more complete description and explanation may be found in SAE J389a).

The purpose of these symbols is to provide a more universal type of communication that overcomes language barriers when a particular piece of equipment may be used in countries where operators speak different languages.

The color combinations and the sizes of these symbols require adjustment to each particular application, (e.g., viewing distance, ambient lighting conditions, or the necessity to indicate the urgency of action by the use of color, such as red for warning or danger and yellow for caution).

*Note:* From a human engineering point of view, it is desirable also to include word labels with each of the pictorials; although the pictorial may be generally meaningful, a word label provides additional assurance that the user will understand the meaning upon first seeing it. Obviously, the words must be in the language of the particular user group.

By law, all farm vehicles using public roads must display the triangular symbol shown in the sketch below on the rear of the machinery.

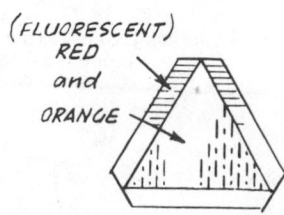

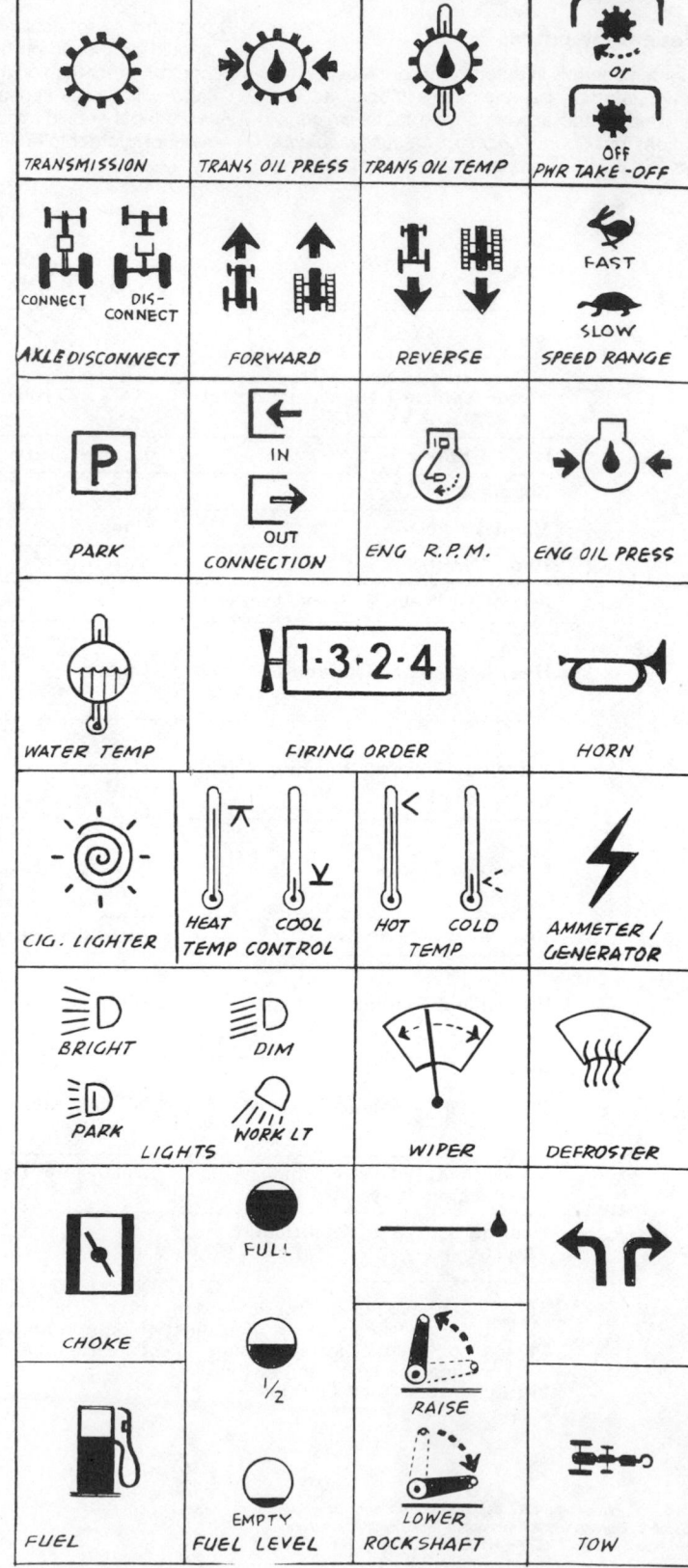

## OTHER SYMBOLIC AND PICTORIAL GRAPHICS

Properly designed and/or selected symbolic or pictorial graphics can be used to reduce an observer's "reading time," but they must be chosen carefully, either because they are already standardized or on the basis of experimental testing.

### Abstract Symbols[4]

Abstract symbols should not be used for identification purposes where the user is not familiar with the symbols. Although the various engineering professions have developed standardized symbology to aid them in the preparation and rapid reading of schematics, such symbology is unfamiliar to the average user of control panels or products.

Quasi-abstract analog symbology can be used to a limited degree for control panels and product labeling. Examples are shown in the accompanying illustrations.

Man-made landmarks are best represented by recognizable silhouettes, as illustrated by the accompanying sketches, which show representations of a school, a church, and a racetrack. If size does not permit, fall back on the triangle with an appropriate call-out to identify the landmark.

Other common man-made as well as natural features are frequently presented in graphic form, as shown in the accompanying sketches.

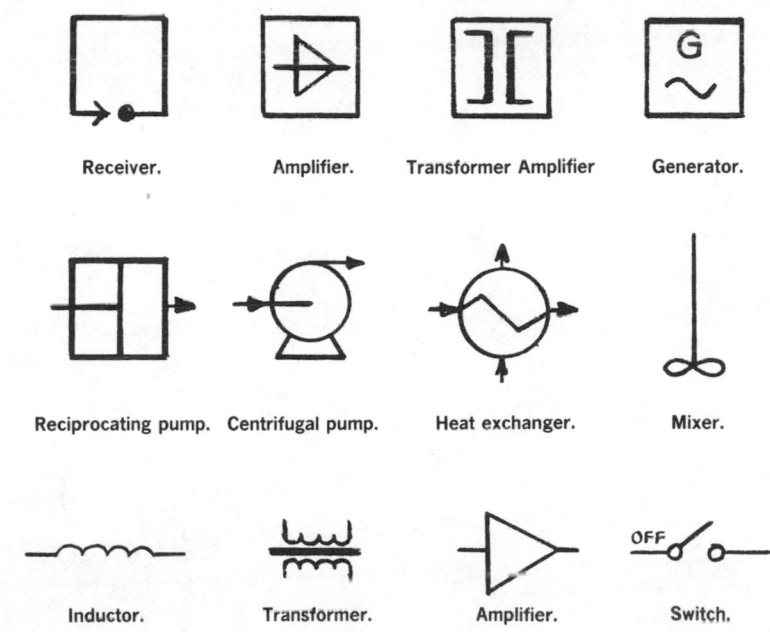

Receiver.   Amplifier.   Transformer Amplifier   Generator.

Reciprocating pump.   Centrifugal pump.   Heat exchanger.   Mixer.

Inductor.   Transformer.   Amplifier.   Switch.

Cathode ray tube.

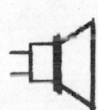

Loudspeaker.

Male connector.

Female connector.

[4]For standardized symbology recommended for various engineering disciplines, refer to the American Standards Association, 1430 Broadway, New York, N.Y. 10018.

## Map Graphics

Although the science of cartography provides considerable guidance in the development and use of map graphics, which should be used as a basic reference, certain key factors should be kept in mind in creating special geographic displays that are typically wall-mounted and viewed by a group of observers. These factors are represented in the accompanying illustrations.

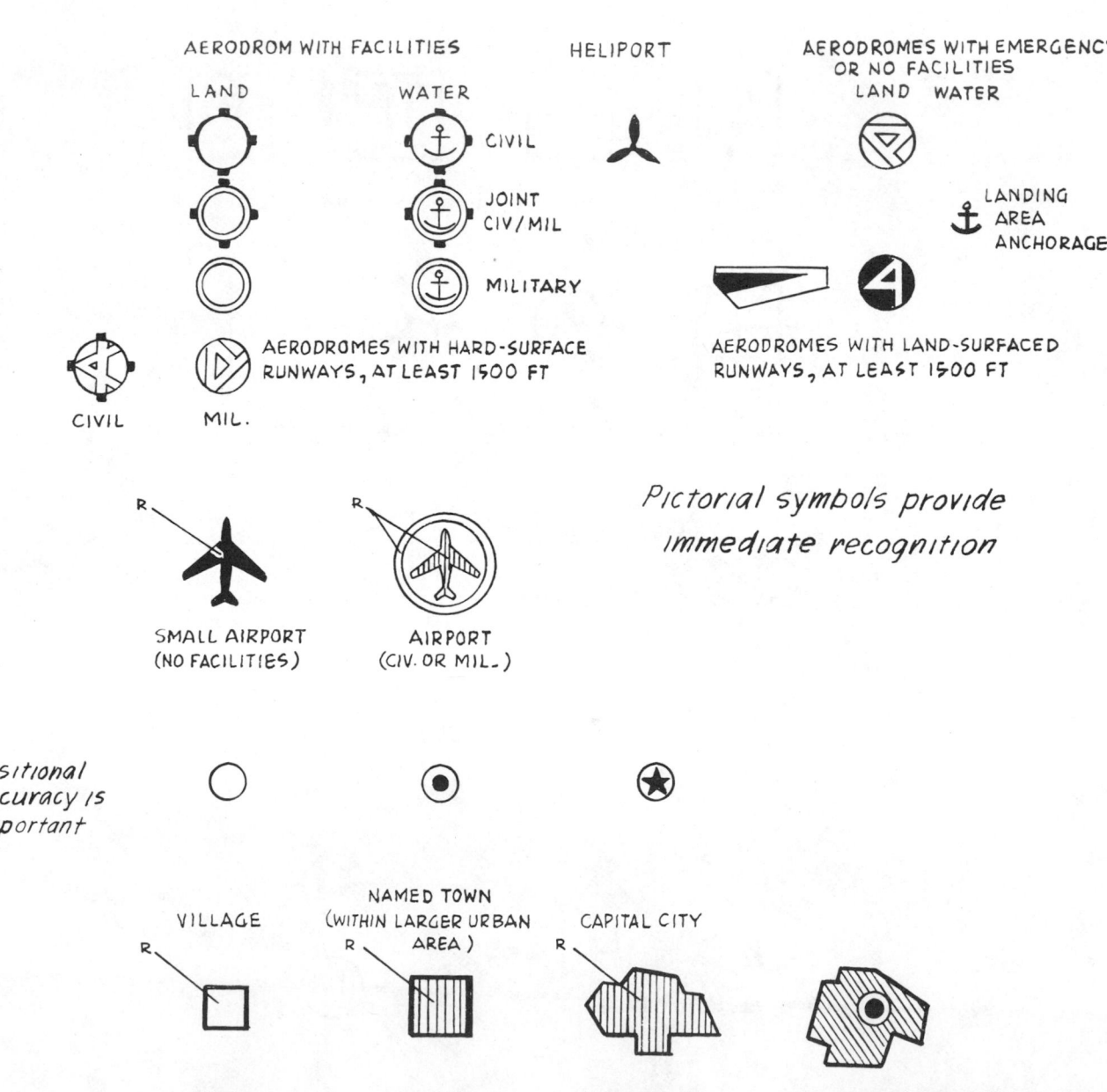

## Man-Made Features

A combination of geometric and pictorial symbols used together.

Pictorial symbols provide immediate recognition.

Positional accuracy is important.

Significance is indicated by size, but is further amplified by shape.

To define a point precisely, use dots or round circles. Then create additional significance via the triangle or square. Finally, indicate the approximate plan shape of a known area, such as a city.

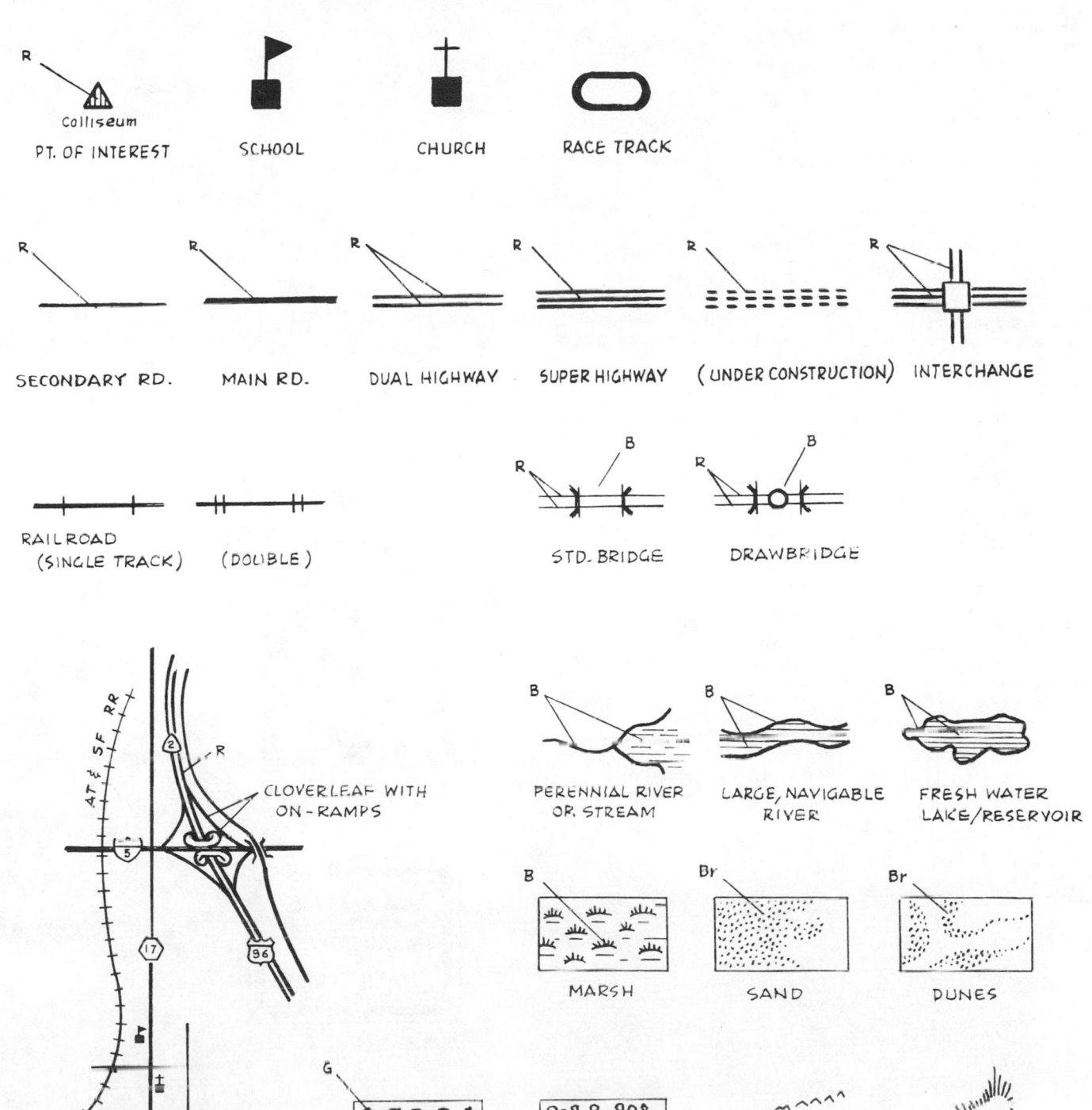

R
Colliseum
PT. OF INTEREST     SCHOOL     CHURCH     RACE TRACK

SECONDARY RD.     MAIN RD.     DUAL HIGHWAY     SUPER HIGHWAY     (UNDER CONSTRUCTION)     INTERCHANGE

RAILROAD
(SINGLE TRACK)     (DOUBLE)     STD. BRIDGE     DRAWBRIDGE

CLOVERLEAF WITH
ON-RAMPS

PERENNIAL RIVER
OR STREAM     LARGE, NAVIGABLE
RIVER     FRESH WATER
LAKE/RESERVOIR

MARSH     SAND     DUNES

ORCHARD     WOODS     MT. RANGE     SINGLE MT.

## SAFETY-RELATED SIGNS

A primary objective for safety-related signs is to make them conspicuous. This requires the use of the following special features:

Unusual border patterns such as the black and yellow stripes shown in the accompanying illustration.

Large lettering for the critical words "caution," "warning," "danger," etc.

A bold red border (the color red is most often associated with danger).

Self-luminous paint for the sign background, which tends to make the sign appear to be lighted.

Standardized symbols (see the examples presented later in this section).

The position of safety-related signs is also critical in terms of making sure that they will be seen. Avoid placing such signs at right angles to the expected normal viewing direction; i.e., the intended observer should be exposed to a safety sign as far in advance of arriving at the actual hazard site as possible.

The sign should always be located where it will not become inadvertently covered if someone places an object in front of the hazard site.

If the hazard is especially critical and it is important to warn a person not to pass close by, it is often desirable to add a flashing red light that can be seen well in advance.

Be careful about accepting so-called warning-sign size standards; manufacturers of such signs often stock a given warning sign in several sizes, but it may be that none of the signs are large enough for the sighting or reading distance involved.

Safety-related signs should be made from materials and/or painted with paints that are extremely resistant to wear and deterioration. Since signs often are not frequently replaced, the critical aspect of conspicuousness may be lost after a time, and thus the sign will fail to draw the attention of passing observers.

## Examples of Caution and Danger Signs

The Office of Safety and Health Administration (OSHA) has attempted to standardize the general format and design of caution and danger signs for use in industrial establishments so that workers will recognize them quickly and understand what they mean. Manufacturers of safety equipment (including signing) generally can supply these signs, and thus designers should not have to create special signs of their own.

As shown in the accompanying sketches, the upper panel of the caution sign should consist of yellow letters on a black background; the lower section should have a yellow background with black letters.

Certain special pictorial graphics have been standardized, as shown in the accompanying illustration of a biological hazard sign. It is particularly important not to create an alternative pictorial graphic to replace these standardized symbols because this would tend to confuse the observer.

In the case of a danger or warning sign, as illustrated by the accompanying sketches, the top panel should consist of the word "danger" in white letters on a red oval background. The lower section should consist of black letters on a white background.

Signs are usually available in various sizes. Select the size that is compatible with the greatest required viewing distance.

Additional examples of various types of safety signing are shown in the accompanying illustrations.

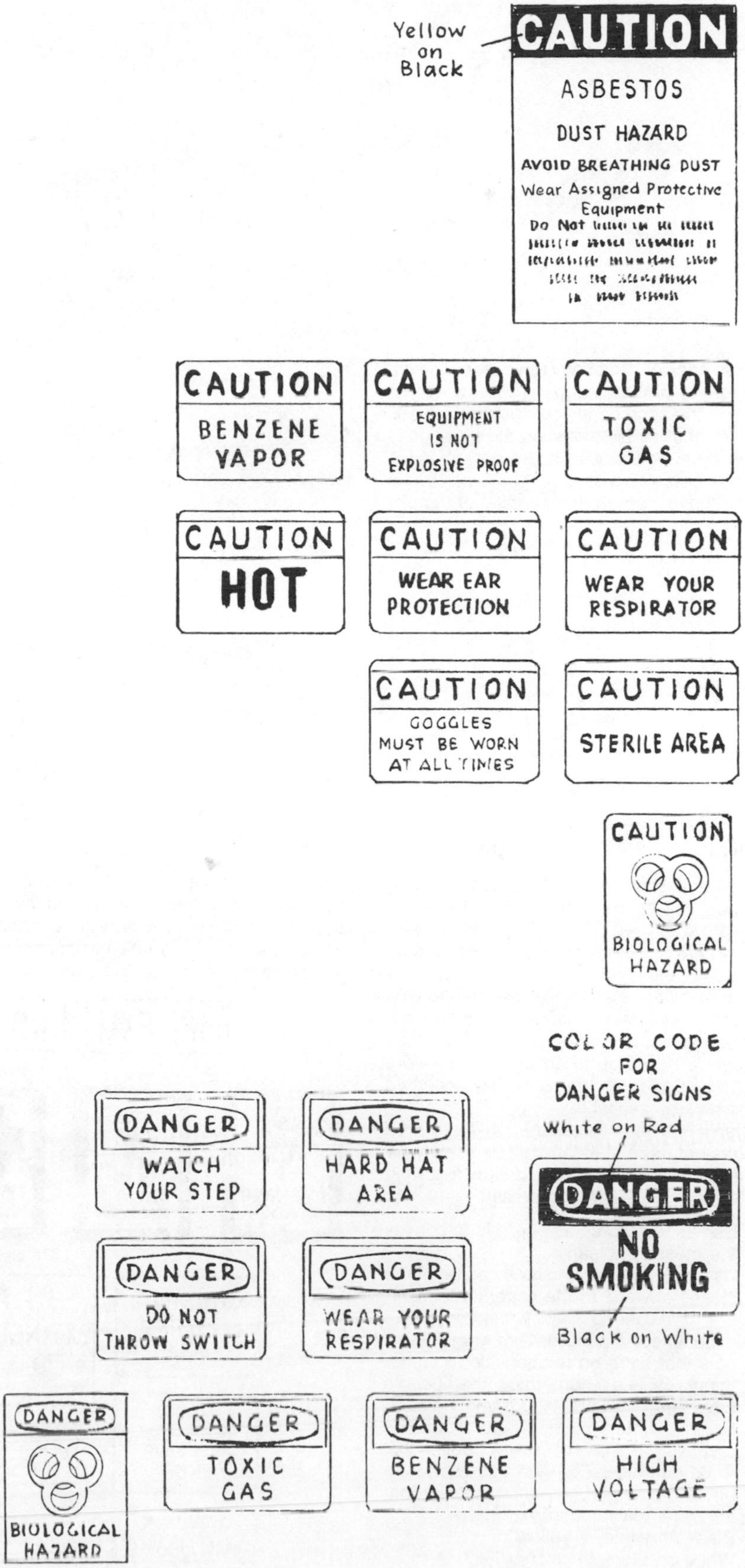

Yellow on Black

**CAUTION**
ASBESTOS
DUST HAZARD
AVOID BREATHING DUST
Wear Assigned Protective Equipment
Do Not ...

**CAUTION** BENZENE VAPOR

**CAUTION** EQUIPMENT IS NOT EXPLOSIVE PROOF

**CAUTION** TOXIC GAS

**CAUTION** HOT

**CAUTION** WEAR EAR PROTECTION

**CAUTION** WEAR YOUR RESPIRATOR

**CAUTION** GOGGLES MUST BE WORN AT ALL TIMES

**CAUTION** STERILE AREA

**CAUTION** BIOLOGICAL HAZARD

**DANGER** WATCH YOUR STEP

**DANGER** HARD HAT AREA

**DANGER** DO NOT THROW SWITCH

**DANGER** WEAR YOUR RESPIRATOR

COLOR CODE FOR DANGER SIGNS
White on Red

**DANGER** NO SMOKING

Black on White

**DANGER** BIOLOGICAL HAZARD

**DANGER** TOXIC GAS

**DANGER** BENZENE VAPOR

**DANGER** HIGH VOLTAGE

Radioactive warning signs generally should contain the standardized pictorial symbol shown in these examples.

Reddish-Purple

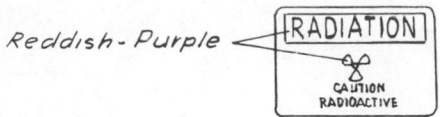
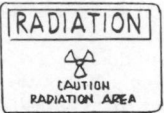

Other special pictorials have been developed for various hazardous environments, as shown in the accompanying illustration. Note, however, that all pictorials should be accompanied by a word set to make sure that the observer gets the message, since he or she may not have learned the meaning of the pictorial.

## Safety-Related Advisory Signs

The examples shown in the accompanying sketches represent typical advisory signs which are safety-related but which do not necessarily fall into the warning or caution categories.

Such signs are also available from the typical safety equipment manufacturer and are designed to meet OSHA standards. Do not try to create a new type of advisory sign before checking to see whether a similar accepted sign is already available.

When it has been determined that a desired sign has not been standardized by some governmental agency, then consider the following in developing a new advisory sign:

1. Use as few words as possible to convey the message clearly.
2. Use a pictorial whenever it can be made clearly obvious to the expected observers, considering their background and experience. Do not invent vague symbols that have no relationship to something real, i.e., something that people have seen and will recognize.
3. Use maximum figure-ground contrasts and avoid the use of "pretty" colors just to be different. The best approach is to utilize the same color combinations that are used for standardized signs:
   Black on white or yellow
   White on black or red

When in doubt about colors to use on a facility that is already color-coded, such as a fire box or circuit box, provide a rectangular white background on which words or symbols are painted in black.

## USE OF COLORS FOR SIGNING AND MARKING SAFETY HAZARDS[6]

*Red* should be used to identify:
Fire protection equipment, fire apparatus, and fire alarm boxes
Fire blanket boxes
Fire buckets or pails
Fire exit signs
Fire extinguishers
Industrial fire hydrants
Fire pumps
Fire sirens
Post indicator valves for sprinkler systems
Cans containing flammable liquids
Stop bars on hazardous machines

*Orange* should be used to identify:
Dangerous parts of machinery
Energized equipment that could cut, crush, shock, or otherwise injure people
The inside of a cabinet door to indicate when the door is open, when guards around moving parts are removed, or an electrical circuit box is open, exposing high voltage potential

*Yellow* should be used to identify:
Possible physical hazards such as being struck, falling, stumbling, tripping, or being caught between moving and stationary devices. Yellow and black striping around the hazard is recommended

*Green* should be used to identify:
Safety equipment
First-aid equipment

*Blue* should be used to identify:
Situations in which equipment should not be started or moved
Situations in which equipment is being repaired or worked on
Electrical circuit boxes

*Purple* should be used to identify:
Radiation hazards including x-rays, alpha rays, beta rays, gamma rays, neutrons, protons, deuterons, and mesons. Yellow should be used in combination with purple for markers such as tags, labels, signs, and floor marking

*Black and white* should be used as basic colors for designating traffic and housekeeping markings. Black and white striping and/or checkerboard borders and solid areas may be used for added conspicuousness.

## Recommended Munsell Designations for Color Coding Visual Displays

### Color-Code Preferences for Four Different Color-Code Schemes (Munsell Color Designations)

| Eight-Color Code | | Seven-Color Code | | Six-Color Code | | Five-Color Code | |
|---|---|---|---|---|---|---|---|
| n | p | n | p | n | p | n | p |
| 1R | 999 | 5R | 1008 | 1R | 999 | 1R | 999 |
| 9R | 892 | 3YR | 890 | 3YR | 890 | 7YR | 884 |
| 1Y | 946 | 5Y | 1128 | 9Y | 1131 | 7GY | 960 |
| 7GY | 960 | 1G | 1103 | 5G | 1101 | 1B | 1093 |
| 9G | 1099 | 7BG | 1095 | 5B | 1087 | 5P | 1007 |
| 5B | 1087 | 7PB | 1133 | 9P | 1005 | | |
| 1P | 1135 | 3RP | 1003 | | | | |
| 3RP | 1003 | | | | | | |

*Note:* n = book notation in Munsell Color System. p = Munsell Production Number. R = red. Y = yellow. G = green. B = blue. P = purple. RP = reddish purple, YR = yellowish green, etc.
*Source:* Adapted from Conover and Craft, *The Use of Color in Coding Displays,* WADC, WPAFB, WADC-TR-55-471, October 1958.

## EQUIPMENT COLOR-CODING GUIDELINES

### Equipment Color Codes

| Color | Coding |
|---|---|
| **Red:** | |
| Meaning | Fire protection equipment and apparatus |
| Typical use | Fire alarm boxes; fire exit signs; fire extinguishers, buckets, and pails; fire hose locations; fire blanket boxes; fire hydrants, pumps, and sirens; fire sprinkler piping |
| Meaning | Danger |
| Typical use | Danger signs; safety cans or other portable containers of flammable liquids having a flash point at or below 80° F (26.9° C) (excluding shipping containers); red lights on temporary obstructions or construction |
| Meaning | Stop |
| Typical use | Stop button used for emergency stopping of machinery; emergency stop bars on hazardous machinery |
| **Orange:** | |
| Meaning | Dangerous parts of machines or energized equipment which may cut, crush, shock, or otherwise injure; purpose is to emphasize danger when enclosure doors or guards are open and a safety hazard exists |
| Typical use | Safety starting buttons; exposed edges of pulleys, gears, cutting devices, rollers, power jaws, etc. |
| **Yellow:** | |
| Meaning | Physical hazards, such as stumbling, tripping, falling, or striking against an object |
| Typical use | Handrails, guardrails, or top and bottom treads of stairways where caution is needed; lower pulley blocks and cranes; pillars, posts, and columns which might be struck; materials-handling equipment, such as industrial tractors, trucks, trailers, forklifts, conveyors, or gantry cranes; piping systems containing dangerous materials |
| Meaning | Caution |
| Typical use | Caution signs |
| **Green:** | |
| Meaning | Safety |
| Typical use | Safety bulletin boards |
| Meaning | Location of first-aid equipment |
| Typical use | Gas masks; first-aid kits; stretchers; safety deluge showers; safety signs |
| **Blue:** | |
| Meaning | Caution against starting, using, or moving equipment under repair or in use |
| Typical use | Scaffolding and ladders; electrical controls; valves |
| **Purple:** | |
| Meaning | Radiation hazards (used in combination with yellow for tags, labels, signs, and floor markers) |
| Typical use | Containers of radioactive materials; disposal cans for contaminated materials; signal lights to indicate when radiation-producing machines are in operation |
| **Black and/or white:** | |
| Meaning | Traffic and housekeeping markings |
| Typical use | Directional signs; dead ends of aisles or passageways |

[6]For additional information, refer to *Safety Color Code for Marking Physical Hazards,* ANSI Z53.1.

### Electrical Conductor Color Coding

| Instructions | Cable Coding Standards* | | |
| | No. of Conductors | Basic Color | Tracer |
|---|---|---|---|
| 1. Enter table at number of conductors desired to be coded. | 1 | Black | None |
| | 2 | White | None |
| | 3 | Red | None |
| 2. Colors appearing to the right of the entry number are the appropriate combination for that particular number of conductors. | 4 | Green | None |
| | 5 | Orange | None |
| | 6 | Blue | None |
| | 7 | White | Black |
| | 8 | Red | Black |
| 3. Example: If a cable consists of 12 conductors, the twelfth color combination would be black and white, the eighth color combination would be red and black, and so on. | 9 | Green | Black |
| | 10 | Orange | Black |
| | 11 | Blue | Black |
| | 12 | Black | White |
| | 13 | Red | White |
| | 14 | Green | White |
| | 15 | Blue | White |
| | 16 | Black | Red |
| | 17 | White | Red |
| | 18 | Orange | Red |
| | 19 | Blue | Red |
| | 20 | Red | Green |
| | 21 | Orange | Green |

*Note:* Only saturated colors should be used. Special care should be taken to select hues which are not easily confused (i.e., red and orange) and which are bright enough to be seen under typically poor lighting conditions (blue may be confused with black if the blue is too dark).

*If the cable has concentrically laid conductors, the first combination or color applies to the center conductor; if the cable contains various sizes of conductors, the first color applies to the largest conductor, and the sequence continues in order of conductor size.

*Source: Human Factors Engineering Design for Army Material,* MIL-HDBK-759, 1975.

### Hydraulic Conductor Color Coding

| Function | Color | Function Definition |
|---|---|---|
| Intensified pressure | Black | Pressure in excess of supply pressure induced by a booster or intensifier |
| Supply pressure | Red | Pressure of power-actuating fluid |
| Charging pressure | Intermittent red | Pump inlet pressure higher than atmospheric pressure |
| Reduced pressure | Intermittent red | Auxiliary pressure lower than supply pressure |
| Metered flow | Yellow | Fluid at a controlled flowrate (other than pump delivery) |
| Exhaust | Blue | Return of power-actuating fluid to reservoir |
| Intake | Green | Subatmospheric pressure, usually on intake side of pump |
| Drain | Green | Return of leakage of control-actuating fluid to reservoir |
| Inactive | Blank | Fluid within the circuit but not serving a functional purpose during phase being represented |

*Source: Human Factors Engineering Design for Army Material,* MIL-HDBK-759, 1975.

## Gas Cylinder Color Coding

| | Area to Be Coded | | | |
|---|---|---|---|---|
| Type of Gas | A (Top) | B (First Band) | C (Second Band) | Remainder of Cylinder |
| Fuel: | | | | |
| Petroleum, liquefied | Yellow | Orange | Yellow | Yellow |
| Petroleum, nonliquefied | Yellow | White | Yellow | Yellow |
| Hydrogen | Yellow | Black | Yellow | Yellow |
| Manufactured gas | Brown | Yellow | Yellow | Yellow |
| Acetylene | Yellow | Yellow | Yellow | Yellow |
| Refrigerant: | | | | |
| Ammonia | Brown | Yellow | Orange | Orange |
| Freon | Orange | Orange | Orange | Orange |
| Methyl chloride | Yellow | Brown | Orange | Orange |
| Oxidizing gas: | | | | |
| Oxygen | Green | Green | Green | Green |
| Oxygen, aviator's | Green | White | Green | Green |
| Air, oil-pumped | Black | Green | Green | Black |
| Air, water-pumped | Black | Green | Black | Black |
| Helium, oxygen | White | Green | Black | Black |
| Oxygen, carbon dioxide | Gray | Green | Black | Black |
| Inert gas: | | | | |
| Carbon dioxide | Gray | Gray | Gray | Gray |
| Helium, oil-pumped | Gray | Orange | Gray | Gray |
| Helium, oil-free | Gray | Orange | Orange | Gray |
| Nitrogen, oil-pumped | Gray | Black | Gray | Gray |
| Nitrogen, water-pumped | Gray | Black | Black | Gray |
| Fire extinguisher: | | | | |
| Carbon dioxide | Red | Red | Red | Red |
| Methyl bromide | Red | Brown | Red | Red |

## Gas Cylinder Color Code Meanings

| Color | Fed. Std. 595* | Meaning |
|---|---|---|
| Orange-yellow | 13538 | Flammable materials; materials commonly known to be flammable |
| Brown | 10080 | Toxic and poisonous materials; extremely hazardous to personnel |
| Light blue | 15102 | Anesthetics and harmful materials |
| Green | 14110 | Oxidizing materials; materials which readily furnish oxygen for combustion |
| Aircraft gray | 16473 | Physically dangerous because of state of temperature, pressure, etc. |
| Insignia red | 11136 | Fire-extinguishing materials |
| Jet | 17038 | No significant meaning |
| Insignia white | 17875 | For general use where specified |
| Orange | 12246 | No significant meaning |
| Middlestone | 30266 | For use on selected groups for segregation purposes |

*Federal Standard 595, *Colors*.

**Gas Cylinder Color Coding
Requirements**

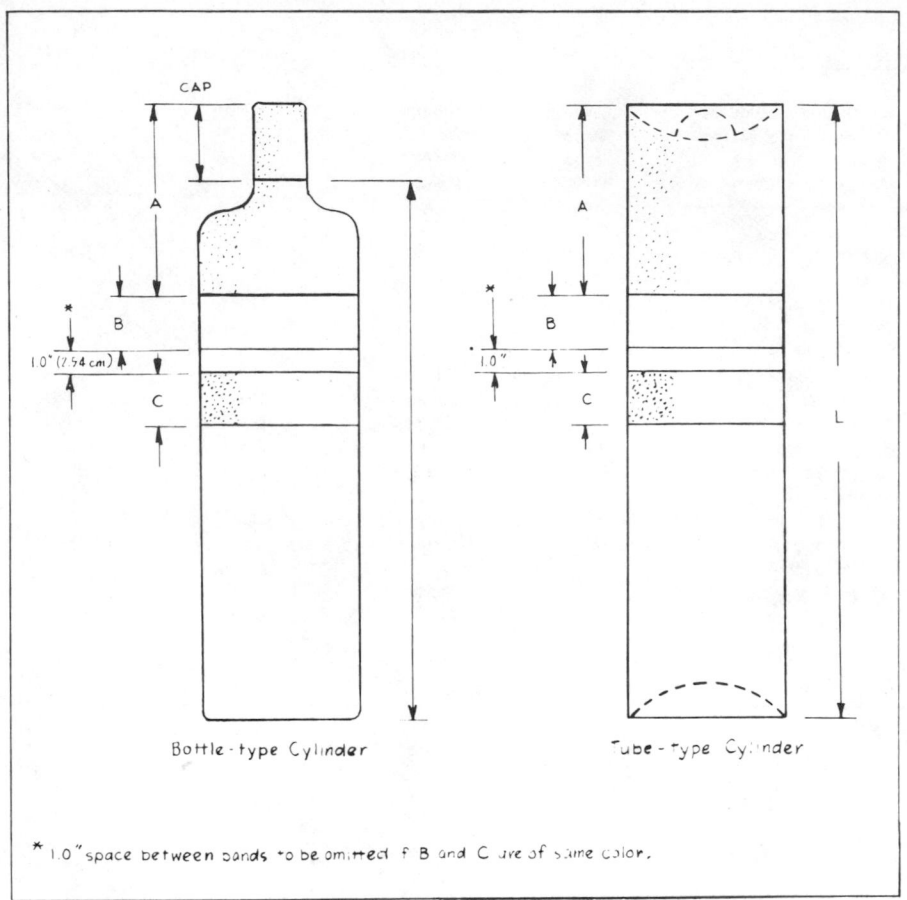

* 1.0" space between bands to be omitted if B and C are of same color.

**Dimensions of Color Coded Areas (in)**

| Cylinder Length L | Cap and Shoulder A | Color Bands B and C |
|---|---|---|
| 30.0 (76.2 cm) and under | Bottle-type: one-fourth of L-cap | 2.0 (5.08 cm) |
|  | Tube-type: one-fourth of L |  |
| Over 30.0 | (independent of cylinder length) | 3.0 (7.62 cm) |

## Pneumatic Conductor Color Coding

Color code standards for marking *Fluid Lines* and *Valves* for military equipment are identified as follows (after Air Force, Ballistic Missile Division):

*Fuel:* red
*Rocket Oxidizer:* green, gray
*Rocket Fuel:* red, gray
*Water Injection:* red, gray-red
*Inerting:* orange, green
*Lubrication:* yellow
*Hydraulic:* blue, yellow
*Pneumatic:* orange, blue
*Instrument Air:* orange, gray
*Coolant:* blue
*Breathing Oxygen:* green
*Air Conditioning:* brown, gray
*Monopropellant:* yellow, orange
*Fire Protection:* brown
*De-Icing:* gray
*Rocket Catalyst:* yellow, green
*Compressed Gas:* orange
*Electrical Conduit:* brown, orange
*All Other:* white

Dangerous products being shipped are to be labeled in the following manner, as required by the Interstate Commerce Commission:

*Red* letters on *White* background: poisons, explosives, poisonous gases, and tear gas
*Black* letters on *Green* background: compressed gases
*Black* letters on *Red* background: inflammable liquids and fireworks
*Black* letters on *Yellow* background: inflammable solids and oxidizing materials
*Black* letters on *White* background: acids

Following are some of the military specifications that should be referred to with respect to the color and marking of military equipment:

MIL-C-3702: *Cable, Power, Electrical, Ignition, High Tension*
MIL-M-8555: *Missiles, Guided: Design and Construction*
MIL-M-11745: *Marking and Labeling of U.S. Army Marine Craft*
MIL-M-13231: *Marking of Electronic Items*

MIL-M-19590: *Marking of Commodities and Containers of Radioactive Materials*

Refer to the following military standards:

MIL-STD-101: *Color Code for Pipelines and Compressed Gas Cylinders*
MIL-STD-129: *Marking for Shipment and Storage*
MIL-STD-161: *Identification of Bulk Petroleum Products and Missile Fuels*
MIL-STD-172: *Color Code for Containers of Liquid Propellants*
MIL-STD-195: *Marking of Connections for Electric Assemblies*
MIL-STD-450: *Signs for Contaminated or Dangerous Areas*
MIL-STD-709: *Ammunition Color Coding*

*Note:* The reader is advised to check current specifications, standards, and local ordinances and regulations as they may apply, since such regulatory materials continue to undergo modification and updating.

## QUANTITATIVE READOUT DIGITAL DISPLAYS

The digital readout generally provides the most rapid and accurate method for presenting purely quantitative information. In selecting or designing such displays, consider the following:

1. The numeral characters should be as legible as possible in terms of character style, height/width ratio, stroke width, and figure contrast.

2. Mechanical counter readouts (drum type) should increase in value by an upward movement of the drum. The face of each drum (and its numerals) should not be buried deeply in the counter window. Individual drums that separate numerals are not desirable because they tend to obscure the numbers and make the total set of numbers harder to read. Wide separators between drums are similarly undesirable. However, such separators are useful for imprinting decimal points and/or commas when the numerical value contains many characters, e.g., 1,000,370.

3. Numbers should always read left to right; vertical arrays are undesirable.

4. Whenever possible, select display systems in which you can use optimized characters (including the drum type, folding character types, etc.). However, when it is necessary to use other types because of a requirement for more rapid response, dot and line matrices are permissible. The accompanying sketches illustrate line or bar matrices that can be used to create effective numeric characters. Although sloping matrix characters have been used extensively, these are less desirable. The slope should never exceed about 11°. Electrically pulsed counter rates should not exceed 50 per second.

5. Although LED (light-emitting diode) displays are effective for use in well-controlled ambient light conditions and/or for nighttime applications, LCDs (liquid crystal displays) are more effective for most general applications because they are easier to see under all ambient light conditions.

6. Dot matrix characters are generally more readable than other matrix-type displays if there is a sufficient number of dots to create the critical features of the harder-to-read characters, i.e., 2, 3, 4, 5, 6, 8, and 9. The 9 × 7 matrix shown in the accompanying illustration is the best format. Note that, although only full steps are required for the vertical dimension, half steps are required horizontally in order to provide proper diagonal strokes for numbers like the 4 that is illustrated.[6]

7. Electronically generated digital characters should be as free of brightness variations, jitter, and flicker as possible.

8. Although various colors are acceptable when they have special meaning, the preferred colors are black on white and white on black for general use (and/or a bluish color if the display medium is a CRT). In the latter case, avoid long-persistence phosphors, which may produce ghost images.

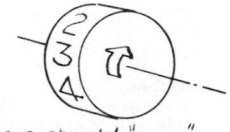

Numbers should "snap" into position - should not follow one another faster than 2/sec. to be read.

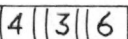

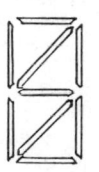

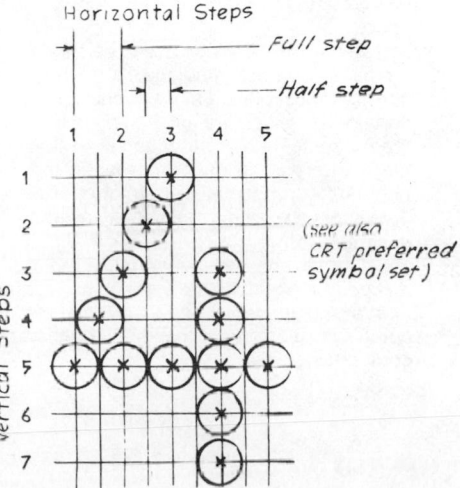

Horizontal Steps — Full step — Half step

(see also CRT preferred symbol set)

Vertical Steps

[6] A number of recent studies have indicated that, although general recommendations for the design of alphanumeric characters suggest that properly designed solid or continuous-stroke characters provide more error-free, rapid readout, matrix-type numerical characters are just as good, if not better, in some cases. Examination of such studies indicates that these conclusions are influenced by the fact that the displays studied typically do not control all the critical variables to draw a general conclusion that would contradict the previous study results. From a practical point of view, therefore, almost any type of digital character generation technique can be made to produce an adequately legible readout if the design of the characters is such that a sharp, evenly lighted character is produced that contains the critical pattern features that make it discernible from another character that tends to look like it.

## DIALS AND GAUGES

Dials and gauges are used primarily to provide quantitative information. However, as opposed to a digital display (which provides extremely precise quantitative information), the dial or gauge also provides some additional information in the form of advance warning, rate of change, and/or opportunity to make "cross-dial" extrapolation; this is due to the fact that the pointer position and motion act as an additional qualitative cue as to what is happening.

First, and perhaps foremost, whenever possible avoid the use of dials and/or gauges in which double pointers or scales are used. A single scale and pointer format is best because it provides the least reading error.

As shown in the accompanying sketches, both round (or half-round) and linear or rectangular formats are possible. It has not been shown that either format results in any difference in readout effectiveness.

However, the accompanying illustrations comparing the round and rectangular gauge formats provide some interesting insights into the possible value of the rectangular format over the round format:

When several gauges are placed side by side, it is obvious that the relative positions of the pointers are more readily assessed.

The direction of pointer increase and/or the value level is always clear in the rectangular format, whereas in the round dial, it is more difficult to tell by the position of the pointer just where it is, depending on which quadrant of the dial the pointer is in; i.e., when the pointer is on the left or top half of the dial, the direction of pointer movement for increase and decrease is "natural," but when the pointer is in the right or bottom half of the dial, the increase and decrease direction of motion is reversed.

In the case of an automobile speedometer, for example, the horizontal scale can be placed higher on the instrument panel (that is, the entire scale can be placed higher); this makes it easier for the driver to glance at the instrument without taking his or her eyes too far from the road.

Often, several rectangular gauges can be placed closer together than round dials. This, of course, depends on the particular instrument package shape behind the panel.

When several round dials have to be clustered for ease in check readings (i.e., when an operator must glance at them quickly to ascertain that things are functioning properly), the instruments can be oriented so that all the pointers are aligned alike when the instruments are indicating normal operation. This makes a single pointer misalignment easy to see.

Vertical arrays should be oriented so that normal pointer positions are vertical; horizontal arrays should be oriented so that normal pointer positions are horizontal (see the accompanying sketches).

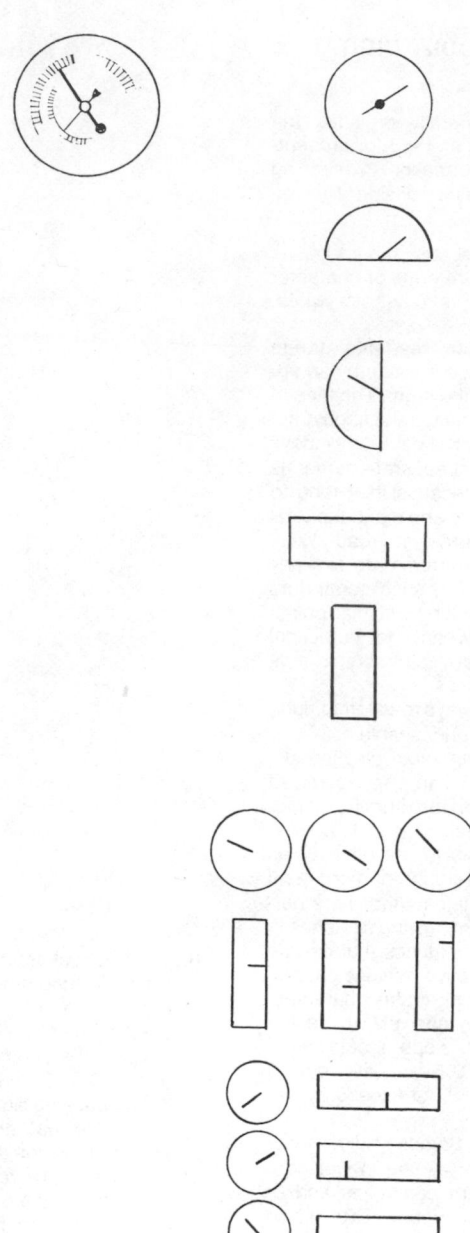

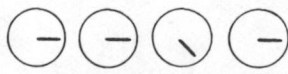

## Dial Size

The size of a dial or rectangular gauge normally should be determined by the number of scale graduation marks required (which is a matter of the inherent precision required). Instrument manufacturers have tended to standardize their instrument package sizes for the sake of production convenience and therefore have sometimes put more markings on the instrument face than can be accurately read. The spacing of scale graduation marks must be great enough so that the observer can discriminate between one mark and another, and also see the relation between the marks and the pointer, without taking an inordinate amount of time peering at the instrument.

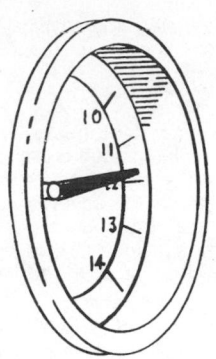

In addition, the size of the dial depends on how deeply inset the face is below the bezel. When the manufacturer's case design includes a deeply inset dial face, the bezel tends to limit the viewing angle, with the result that (depending on the viewing angle) some of the numbers and/or scale marks may not be visible. Some designers have created poor formats to overcome this problem, such as placing the numbers "inside" the scale to make sure that they are not obscured. As will be noted later, this format is counter to good pointer-scale relationships and results in a pointer that partially covers up the scale numbers.

## Dial Size in Relation to Required Number of Scale Marks

Too many scale marks, too close together, make dial reading difficult and error-prone. In fact, it is impractical to have a dial that is so small that the inner annulus of the scale is less than 1 in (2.5 cm) in diameter. Use the accompanying table as a guide in determining the size of dial necessary to accommodate the required number of scale marks. For example, if the viewing distance is approximately 12 ft (3.7m) and if 50 scale graduation marks are required, the inner annulus of the scale must be at least 5 in (13 cm) in diameter.

**Minimum Diameter of Inner Annulus at Various Viewing Distances***

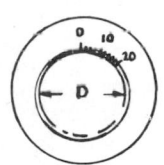

| No. of Scale Marks | 20 in | 3 ft | 6 ft | 12 ft | 20 ft |
|---|---|---|---|---|---|
| 50 | | 1.3 in | 2.6 in | 5.0 in | 9.0 in |
| 100 | 1.4 in | 2.6 in | 5.0 in | 10.0 in | 17.0 in |
| 150 | 2.0 in | 3.9 in | 8.0 in | 15.0 in | 26.0 in |
| 200 | 2.9 in | 5.0 in | 1.0 in | 21.0 in | 34.0 in |
| 250 | 3.5 in | 6.4 in | 13.0 in | 26.0 in | 43.0 in |
| 300 | 4.0 in | 7.7 in | 15.0 in | 31.0 in | 51.0 in |
| 350 | 5.0 in | 9.0 in | 18.0 in | 36.0 in | 60.0 in |

A common design error occurs when the designer believes that a specified level of accuracy must be provided and therefore crowds the scale marking in order to get a certain number of marks within the constraints of the dial-face diameter, as shown in the accompanying illustration. The result is that the observer cannot differentiate between the marks and not only makes errors in reading but also spends more time reading; thus the basic objective is lost.

The designer should either reconsider the need for the level of accuracy thought to be required (i.e., the instrument itself may not provide the level of accuracy it was assumed to) or use fewer marks, since the observer can probably interpolate the pointer position between fewer, properly spaced marks more accurately and reliably than he or she would be able to do with the poorly marked dial.

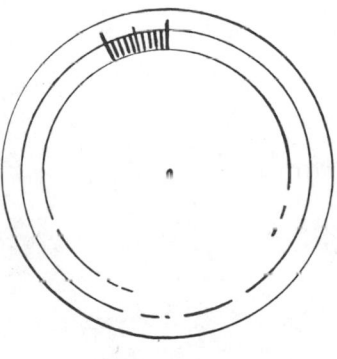

*POOR*

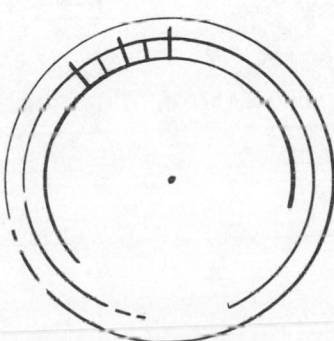

*PREFERRED*

The following are some typical scale-marking design errors to be avoided:

1. Dots instead of lines
2. Thick marks
3. Marks joined by a heavy base line
4. Long marks spaced close together
5. Irregularly spaced marks about an arc, which cause the observer to shift his or her spatial reference
6. Uneven spacing, which leaves the observer in doubt as to the value of each mark
7. Significant spacing variation between the ends of the marks, which, because of the tight radius of the dial, makes it hard for the eye to follow along the scale-pointer axis

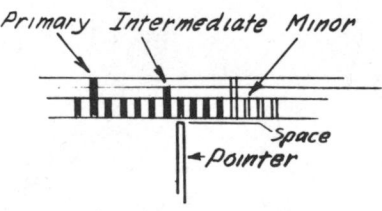

## Pointer-Scale Relationships

For ease of reading, the scale-pointer relationship should be as shown in the accompanying sketch. The point tip should be the same width as the scale mark, and it should fall just short of the mark.

Placing the pointer just inside the scale annulus ensures that it will never be confused with a mark. However, the plane of the scale and the pointer also should be the same; otherwise, the observer may misread the scale-point association because he or she was looking at the instrument from an angle (parallax). The accompanying sketches illustrate the proper arrangement for pointer and scale planes.

Finally, the numbers associated with the scale should be outside the scale ring so that the pointer will not obscure the number.

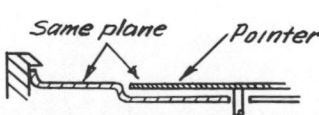

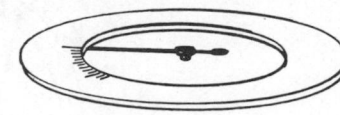

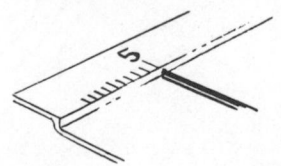

## Instrument Scale-Marking Guidelines

Experiments by military psychologists produced the scale-marking specifications shown below; scales were read with fewer errors and faster response when these dimensional characteristics were present. The 28-in (71-cm) viewing distance is typical of the aircraft cockpit situation. For other viewing distances, the proportions should be maintained, even though values are increased or decreased; i.e., multiply each dimension times the viewing distance, or

$$\text{Dimension at 28 in} \times \frac{x \text{ in}}{28}$$

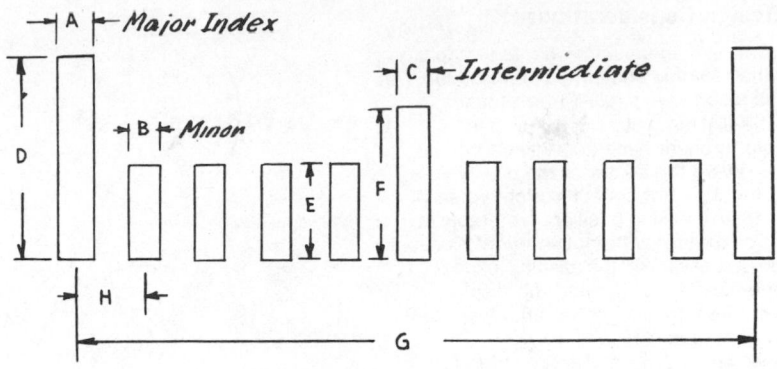

### Scale Mark Dimensions for 28-in (71-cm) Viewing Distance

| Dimension | Black on White | White on Black |
|---|---|---|
| A width | 0.035 in (0.069mm) | 0.125 in (0.032 mm) |
| B width | 0.025 in (0.054mm) | 0.125 in (0.032 mm) |
| C width | 0.030 in (0.076mm) | 0.125 in (0.032 mm) |
| D length | 0.22 in (0.56mm) | 0.22 in (0.56 mm) |
| E length | 0.10 in (0.25mm) | 0.10 in (0.25mm) |
| F length | 0.16 in (0.41mm) | 0.16 in (0.41 mm) |

*Note:* Although the major, intermediate, and minor index mark widths are shown to vary for black on white markings, it is suggested that all mark widths be equal for most applications so that a pointer tip can be used that is the same width as all the marks. Use the minor index width to define the pointer tip.

## Direction of Pointer Movement

Scale numbering should reflect the expected direction of pointer movement; i.e., pointers should move to the right, up, or clockwise to indicate a value increase.

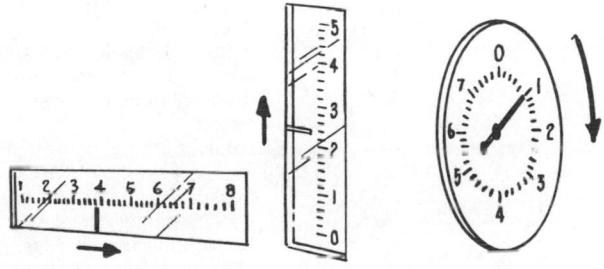

## Scale Interval

For most people, scale intervals of 1 through 10 (or multiples thereof) are easiest to interpret without error. The next easiest are intervals of 2, 4, 6, 8, and 10. However, intervals such as 3, 6, and 9 (or 4, 8, and 12) can cause confusion.

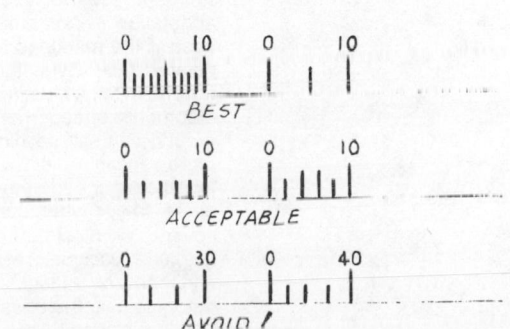

## Other Design Considerations

Make sure that numerals are oriented upright rather than slanted and turned on their side. The exception to this rule is the situation in which the dial may rotate, in which case it is desirable to orient each numeral so that its base is toward the center of the dial, which makes the numeral appear upright when it rotates to an indexing position, preferably at the top of the dial. Alternative orientations may be necessary if the display exposes only one-half of the dial and the display is oriented to emphasize an up-and-down direction of motion.

If a scale covers only part of a circumference, attempt to balance the number positions so that they present a symmetrical appearance.

Consider the use of patterned or colored range strips when the primary readout is to determine when a system is performing within specified ranges, i.e., when it is less important to read specific values.

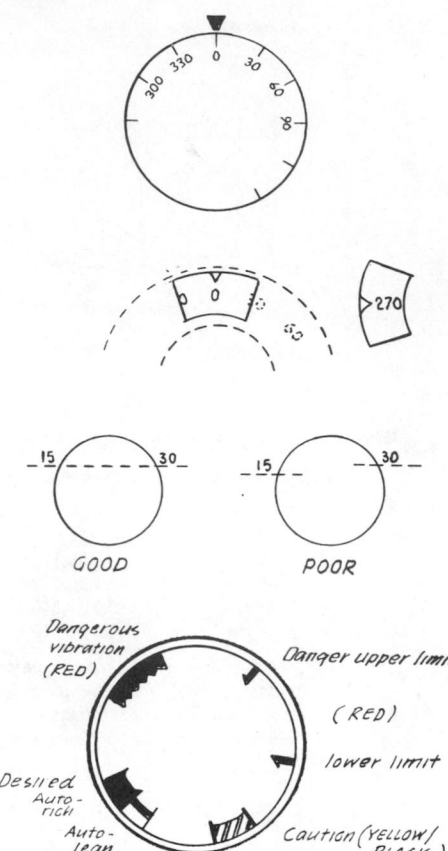

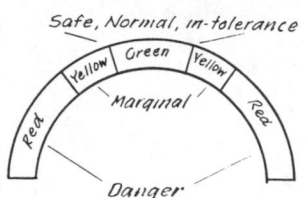

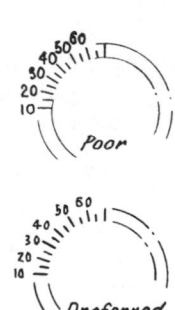

Although it is always desirable to optimize the size and shape of numerals used on dial displays, circumstances may dictate a smaller instrument face. It is better to use smaller numerals and thus avoid the clutter and illegibility that result when characters of a so-called optimum size are crowded together.

Whenever the dial scale is of finite length, be sure to leave a sufficient break between the beginning and end of the scale so that observers do not mistakenly assume that the scale continues and/or that they should continue reading around the scale, but with some special extrapolation.

When several dial or rectangular, scaled instruments or gauges are mounted within a single work station or on the same instrument panel, make sure that the scale breakdown is alike on all the instruments; i.e., do not mix two different scale breakdowns, as shown in the accompaning illustration.

Instrument faces for use when night or dark adaptation is not critical are more easily seen if the markings are black on a light-colored face. Even though the instrument may have to be illuminated at night (e.g., an automobile speedometer), the lighter background can still be dimly lit and provide a more readable display. In the case of the automobile speedometer, the display is also easier to see when sunlight reflects on the cover. For blackout conditions, however (airplane cockpits, etc.), light characters and marks on a black background are mandatory to help the observer maintain proper dark adaptation.

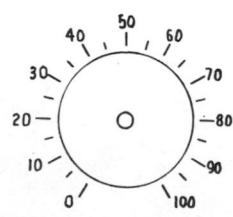

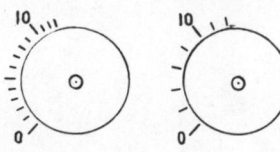

## Design of Pointers for Dial and Gauge Instruments

For maximum reading accuracy, a pointer should be of equal width throughout so that the parallel sides visually project to match the parallel-sided scale marks to which the pointer must be associated.

The width of the pointer should be exactly the same width as the scale marks. That is why, in a previous discussion of scale marking, it was noted that all scale marks should be the same width.

In many applications, however, the thin pointer is not conspicuous enough, especially when the user has to depend on a brief, quick glance to see how a particular system is performing. Thus, it is necessary to increase the width of the pointer, i.e., to make it bolder and more obvious during a quick scanning look. In such cases, the pointer should be designed as shown in the accompanying sketch. Note, however, that although the main stem of the pointer is enlarged, it is still tapered at the tip so that the width of the tip still matches the scale-mark width.

Generally speaking, if a pointer has a "tail" on it, the observer can judge slight movements of the pointer better than if the pointer is tailless. The tail should not be more than one-third the length of the pointing segment, however; otherwise, there may be confusion as to which end is the pointing end.

Above all, avoid artistic pointer designs.

Pointer illumination is often difficult. A translucent dial face provides a good background for silhouetting a black pointer. Another useful technique, shown in the accompanying sketch, consists of a plastic pointer ring inside a scale ring. Both can be either back-or edge-lighted with considerable success.

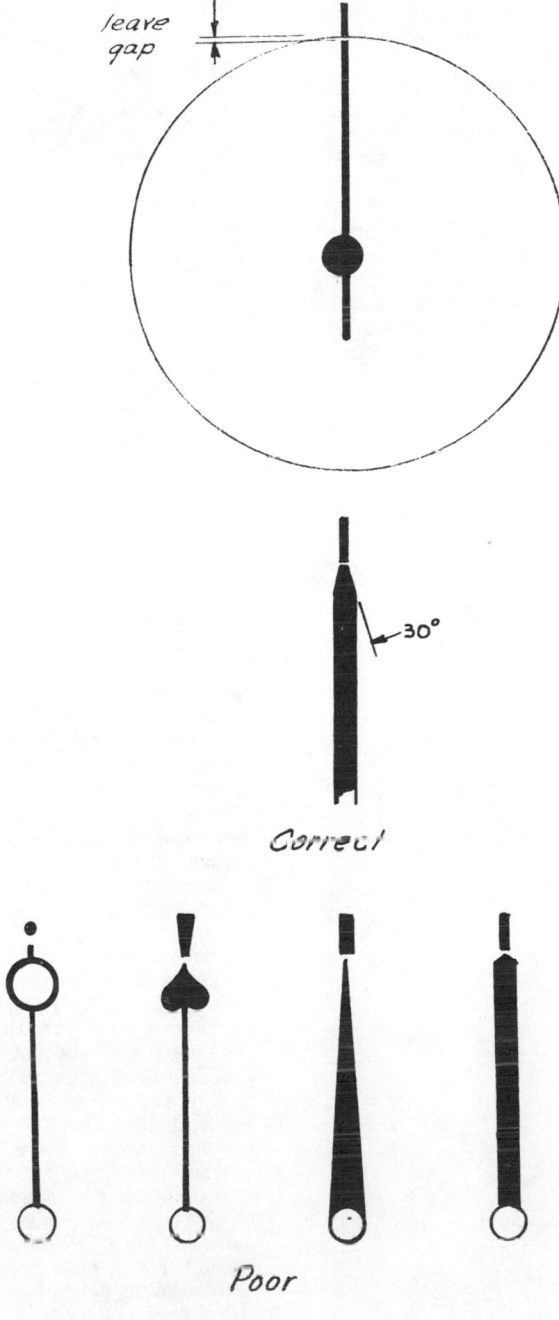

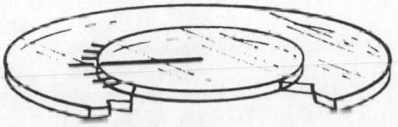

## Multiple Pointer Display Considerations

First, avoid multipointer instruments if at all possible. When absolutely necessary, consider the following:

A problem of pointer visibility and discrimination can occur when two or more pointers are necessary. (Three pointers are to be avoided if at all possible.) Several typical design errors are shown in the accompanying sketches. The two pointers are the same color, the shorter pointer tip is too far from the line of marking, and the tails of both pointers are the same length.

The accompanying illustration shows several ways to improve the visibility and discriminability of the dual pointer system. The color contrast and the variation in the length of pointer tails help clarify pointer identification. The length of the shorter pointer has been increased so that the tip of the pointing end is not so far from the line of marking, and a point has replaced the blunt tip, illustrated in the first example. Other ideas for aiding discrimination between pointers are shown in the accompanying illustrations.

When two pointers are required on the same instrument dial, the upper one is generally too far from the surface on which its corresponding scale marking is located. One way to solve this problem, as illustrated by the accompanying sketch, is to bend the upper point in order to bring the pointer tip closer to the dial surface. Another method is to place the scale marks on a raised plane so that the marks are about midway between the two pointers.

When the particular instrument requires use of a "moving scale" which must be referenced against an indexing mark or pointer, it is usually necessary to make the index marker or pointer more conspicuous than an ordinary pointer because of the lack of length and pointer or index motion. The accompanying sketches illustrate acceptable design practice. Special attention is called to the double scaled instrument shown in the accompanying illustration. Here, making the central pointer pictorial has made it more conspicuous and identifiable; i.e., it represents the ship on which the operator is working. In other words, the operator knows immediately that the ship symbol represents the ship's heading, while the diamond pointer represents the "true" heading.

When a moving-tape type of gauge is used in order to provide an expanded scale, the indexing pointer can be designed either as a pointer positioned at the side or as an indexing or "lubber" line across the tape. In some cases two pointers may be required: one to indicate where the observer's position is now, and the other to tell the observer which way to go.

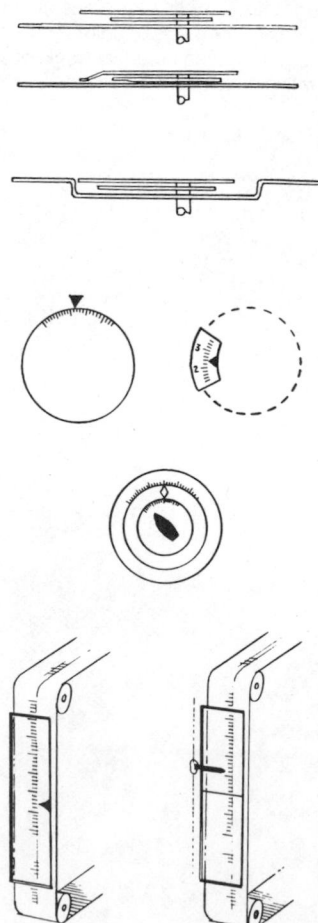

Many other pointer concepts and designs are possible, as the accompanying illustrations demonstrate.

In the case of the inset dial with its own separate pointer, care must be taken not to make the smaller display so small that it cannot be read. One should begin with the smaller one and make it legible; it then will dictate how large the primary display must be.

Various semipictorial pointer combinations are possible, as the "flap indicator" instrument illustrates. The familiar flap shape provides both an easy identification device and a quick indication of where the flap position is at any given moment.

Various navigation instruments require an entirely different approach to pointer design, as illustrated by the accompanying sketch. The cross pointers move independently to indicate the positional relationship between where the pilot is and where he or she should steer to obtain the proper flight path. The double scale system provides accurate guidance information.

## Example of Wristwatch Dial Design

Although the design shown in the accompanying sketch shows some artistic license, it also illustrates how one designer managed to apply a number of the principles discussed above. This watch, incidentally, can be read under the most severe sunlight as well as nighttime lighting conditions. Note these features: coplanar number marks and hands, a primary minute hand that does not cover the blocks, a second hand that is adjacent to the minute and second marks, extremely good contrast, and excellent differentiation between the hands due to size and shape.

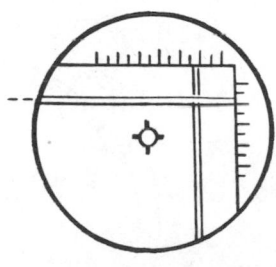

Flight Path cross-pointer

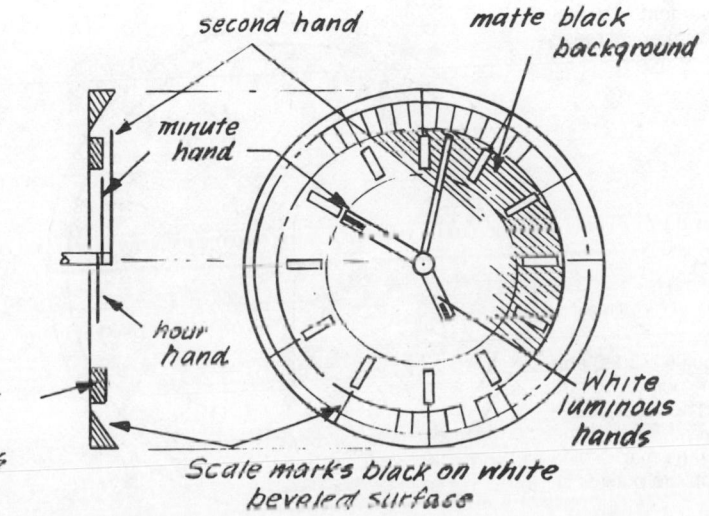

second hand

matte black background

minute hand

hour hand

Number blocks luminous white

White luminous hands

Scale marks black on white beveled surface

**Example of a Poorly Designed Aircraft Instrument versus the Correct Format**

The accompanying illustration of a doppler navigation display shows a typical layout created by an instrument manufacturer.

*Both scale breakdowns should have been similar.*

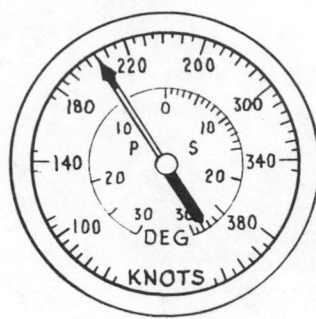

*Numerals should have been positioned outside of the scale marks so that the respective pointer will not cover marks or numerals.*

The two major human engineering errors in this format are typically the result of the designer's attempt to create symmetry (i.e., positional balance of the numbers for the outside scale). The accompanying sketch illustrates how the instrument face should have been laid out. Although symmetry is desirable in certain cases, it is irrelevant here. It is more important to maintain a similar scaling factor between the two scales. Equally important is the principle of keeping pointer-scale and number relationships separated so that the pointers do not obscure any scale marks or numbers, or at least those which must be visually associated. Note that, in the second sketch, the smaller, or inner, scale does not contain the lesser scale marks because they are too close together. Pilots can estimate these as accurately as they can actually "read" the marking-pointer relationship, and they can do it faster. "P" and "S" refer to port and starboard degree readings.

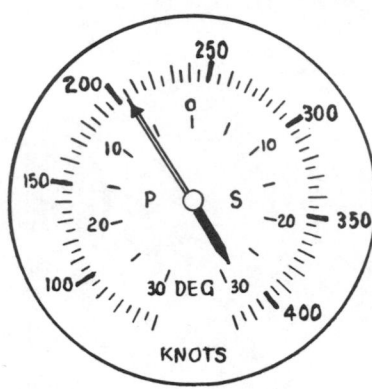

The typical Mach Number display format shown in the accompanying illustration ignores several important human engineering principles. The scale break is at the top of the display rather than at the bottom; i.e., the zero position should be at the bottom so that the initial direction of pointer movement (e.g., upward) is naturally associated with an increase in speed.

*Correct layout*

In addition, the numerals, scale marks, and pointer are arranged so that the pointer covers up the marks and numerals. The inner scale-mark annulus should be "even" so that the pointer tip falls short of any of the marks. The numerals should be located on the outside of the scale markings.

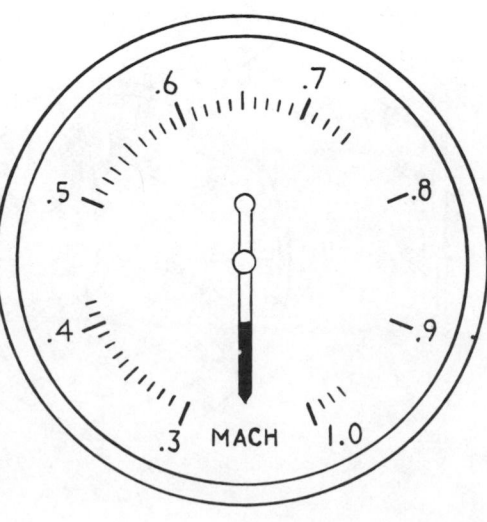

## Example of a Thermostat Display and Control

The accompanying sketch shows a popular thermostat control unit that combines the display of temperature, the displayed setting, and the controller all in a seemingly compact package. In an informal survey, over 50 percent of the persons asked reported the wrong temperature reading. They had read the scale at the top, which is the scale for setting the control. Unfortunately, the values of setting did not coincide with the temperature readout value.

This illustrates several poor human engineering features. In addition to the obvious problem cited above, the index marks are gold on a white background, which makes them extremely difficult to see through the plastic (knob) cover.

The pointers cover the index marks; thus at certain angles, it is difficult to tell exactly where the pointer is pointing.

The accompanying sketch illustrates a preferred display approach and also the application of a number of the good human engineering principles discussed earlier in this section.

The scales are vertically oriented, which is more typical of people's expectation of colder or warmer. The pointer does not cover the scale on the control side, and the adjacency of this scale, the pointer, and the control knob ensures proper association. To further ensure that pointers are not confused, an actual thermometer is provided, although this could just as well be instrumented by means of another pointer.

## Examples of Functional-Group Display Optimization

The accompanying sketches illustrate the optimization of automobile driving instrument displays:

The accompanying sketch illustrates several salient features of conventional instruments and back-lit indicators. The advisory indicators to the left advise the driver of conditions that should be checked. The combined advisory lights and the fuel, oil, temperature, and ampere indicators advise the driver when a condition is not optimum. The speedometer instrument has an evenly spaced scale. The high-beam advisory light is close to the nominal line of sight as the driver watches the road ahead. Although round dials also could be used in the same relative positions, the rectangular scalar instruments take up less space, thus allowing the entire display grouping to be raised to a position near the driver's nominal line of sight.

Using more exotic solid-state display techniques, considerably more information can be displayed in essentially the same or less space as illustrated in the accompanying sketch. Everything on the display can be displayed by purely electrical means, including pointer position. In this case, only speed, fuel quantity, time, and mileage would appear on a continuous basis. Other information would appear only as required, thus reducing the operator's visual load and the number of distractions from his or her external viewing task.

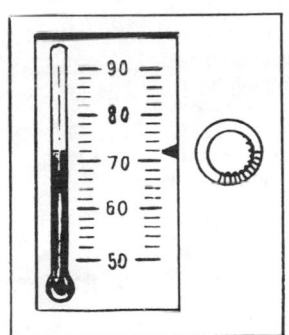

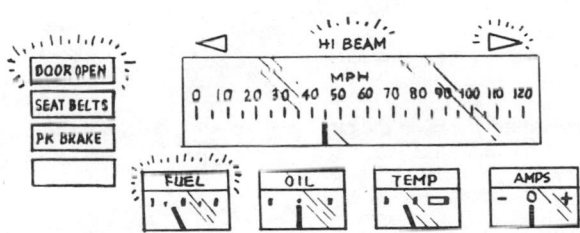

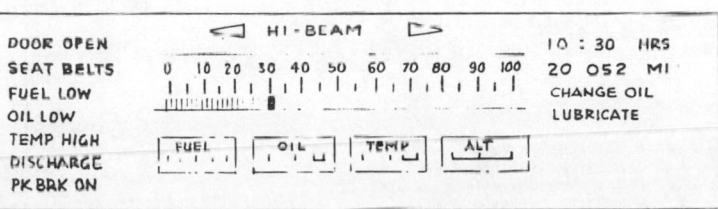

## ORIENTING AND POSITIONING DISPLAYS

Orienting and positioning oneself or one's equipment requires a special kind of display. Although a quantitative reference often is needed, it is important to display some pictorially represented reference for ease of interpretation.

The two displays shown in the accompanying sketches[7] illustrate how pictorial as well as quantitative information can be combined. The first sketch shows a realistic representation of a submarine in a dive. The second sketch shows a simpler yet somewhat pictorial representation.

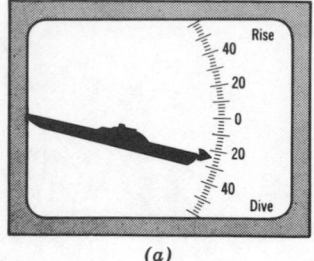
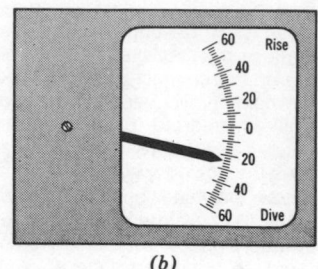

*(a)*                    *(b)*

Aircraft orientation is a typical system requiring special pictorial representation of the current attitude of the aircraft. As shown in the accompanying sketches,[8] two alternatives are potentially available. In the first sketch, a picture-window view of how the earth's horizon tilts relative to the aircraft is provided. The second sketch shows a different perspective in that a model represents the banking of the aircraft. The two lower sketches are more typical of how an actual instrument might be designed in order to provide quantitative references. The tilting horizon is often referred to as an "inside-out" display, and the tilting model as an "outside-in" display.

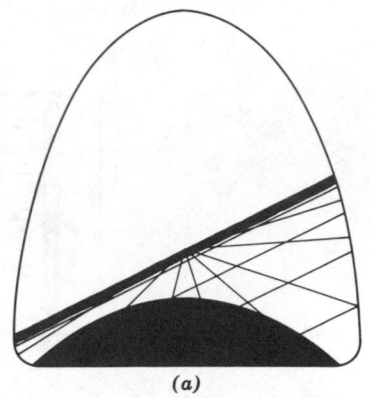
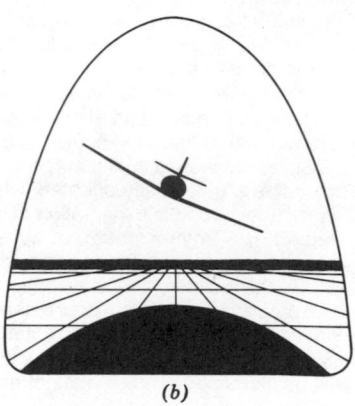

*(a)*                    *(b)*

If the pilot or operator cannot see outside the aircraft or vehicle at all, the outside-in display format is better. However, if the operator may switch from the instruments to an actual outside view, it is best to use the inside-out format.

Horizon-ball instruments can be particularly confusing unless care is exercised in the graphics presentation. In the accompanying illustration, note the contrast between the sky and earth portions of the ball and the suggested approach for making the simulated aircraft symbol readable both during the day and at night (under cockpit illumination).

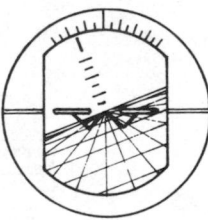

INSIDE-OUT            OUTSIDE-IN

Because the sky-earth contrast is emphasized, the pilot is never confused if he or she should happen to reference the instrument while the aircraft is inverted.

As an objective, the designer should make the orientation instruments look as "realistic" as practical, while at the same time keeping the realism detail as simple as possible so that this type of detail does not make it difficult for the pilot to detect or recognize scale or reference indices.

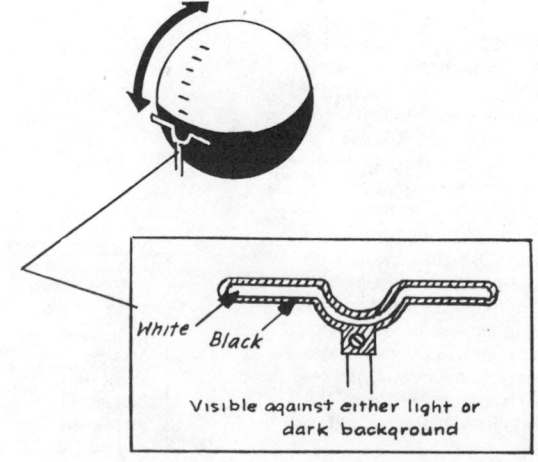

White    Black

Visible against either light or dark background

[7]Illustrations after C. R. Kelley, *Manual and Automatic Control,* John Wiley & Sons, Inc., New York, 1968, p. 93.

[8]Illustrations after C. R. Kelley, *Manual and Automatic Control,* John Wiley & Sons, Inc., New York, 1968, p. 117.

An aircraft or similar vehicle also needs a display of its geographic position and directional attitude. A typical plan position indicator (PPI) would provide a picture of the aircraft on a simulated map, as illustrated by the accompanying sketches.[9] Several possibilities exist, of course. The first sketch shows a north-oriented map with a moving and turning aircraft symbol; the second sketch shows a north-oriented moving map with a turning but not-moving aircraft symbol; the third sketch shows a turning map with a moving but not-turning symbol; and the fourth sketch shows a turning map with a stationary, not-moving aircraft symbol.

Although each format has good and poor features (the symbol is moving in the wrong direction, the map is upside down, etc.), the one shown in the fourth sketch is most like VFR experience (and most like the pilot's expectancy).

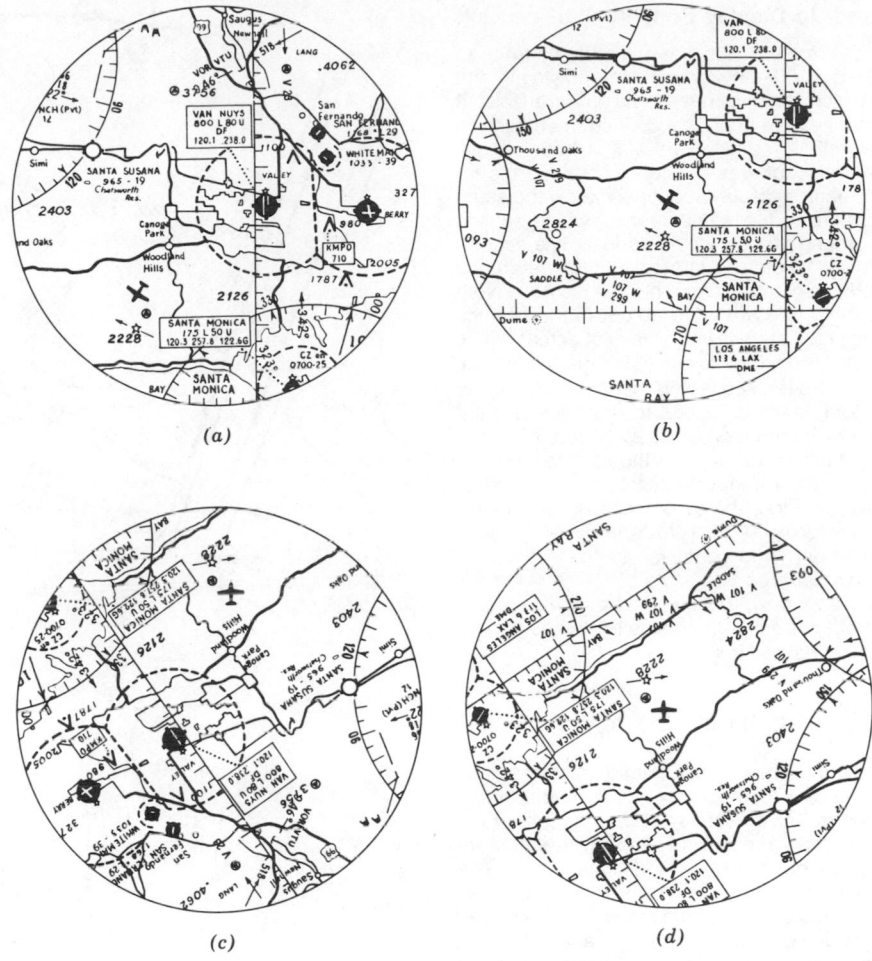

*(a)*      *(b)*

*(c)*      *(d)*

Integration of both types of positioning and orienting displays is desirable from the user's point of view. The accompanying sketch illustrates a typical format used by military aircraft. Note the simplification of the PPI from a detailed map to a CRT-type of display, in which a command directional marker is provided for the pilot to "fly to,"

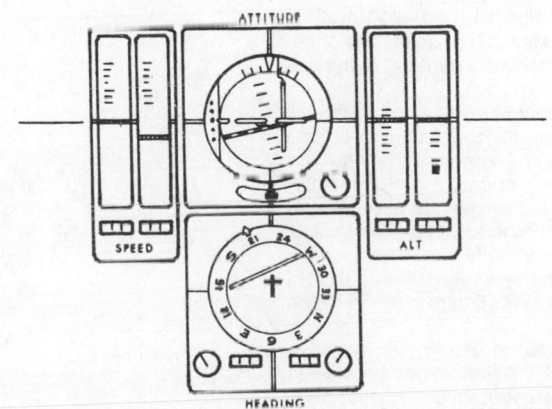

[9]Illustrations after C. R. Kelley, *Manual and Automatic Control*, John Wiley & Sons, Inc., New York, 1968, p. 97.

## Head-Up Display Concept

Because pilots may intermittently fly from an IFR to a VFR environment when flying in and out of clouds, they need to be prepared to shift instantly from their attitude instrumentation to the actual "outside" view. The head-up display concept was created to provide the instrumented display at approximately the same position on the windscreen at which outside reference objects would appear. The accompanying sketch illustrates attitude reference marks projected onto the windscreen. Note that they are of the inside-out format, coinciding precisely with what the pilot actually sees when it is possible to see out of the windshield.

It has been suggested that a similar display system would be useful for automobiles, but for key information such as speed, fuel, and malfunction warnings. Although the concept has merit, it is not practical for a number of reasons. First, the automobile driver faces a myriad of lights at night, and lighted indices would only add confusion to the driver's already cluttered view. In addition, if rain were on the window, there would be a serious dazzle effect from both outside lights and the additional lighted indices.

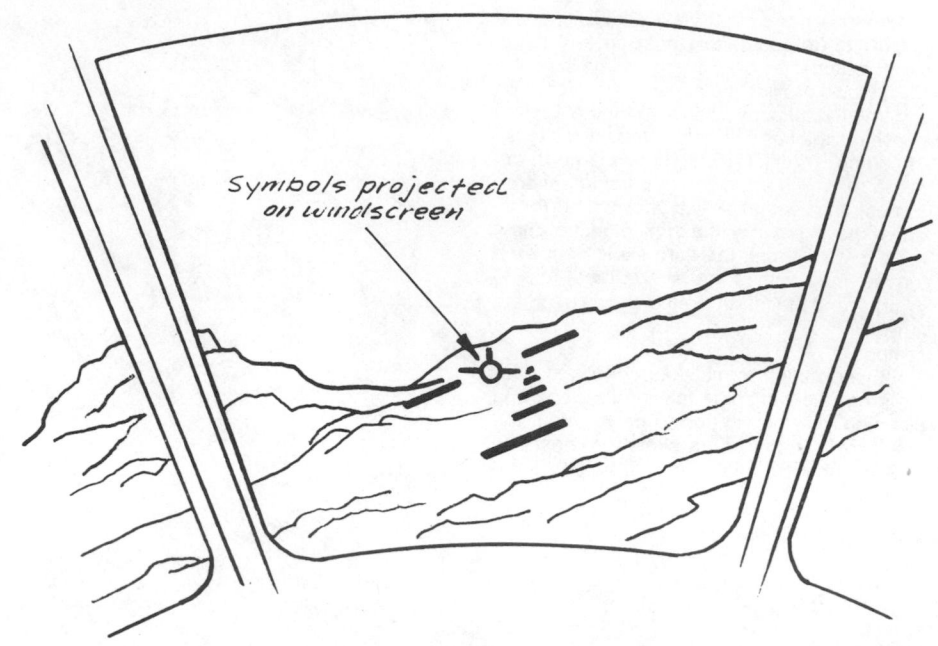

Symbols projected on windscreen

## LUMINOUS IMAGE DISPLAYS

Electronically generated image displays such as the CRT (cathode-ray tube) offer maximum flexibility for presenting a wide variety of images and have the advantage of almost instantaneous image modification (as evidenced by the television systems widely used in homes, business, and industry). The following guidelines are offered to assist the designer in making specific choices regarding types of tubes, size, and display mode.

### General

CRT-type displays should be viewed from a position perpendicular to the screen; thus the display should be packaged and/or positioned so that the intended observer normally assumes the proper viewing position.

The overall size of the CRT screen should be determined on the basis of the smallest significant detail size that must be visually resolved by the eye at its expected viewing distance.

The screen should be protected from direct high ambient light because it tends to reflect and obscure images and/or excites the phosphor, reducing the inherent contrast of the display.

CRT displays should be packaged so that all adjustment controls for the tube are on the front of the package, i.e., so that they are accessible to both an operator and a maintenance technican. Purely service controls can be placed under a hinged cover to minimize tampering by inexperienced operators.

A transparent safety screen should always be provided over the CRT to prevent radiation or implosion injury.

Except for critical detection (i.e., radar or sonar), CRT displays should not be viewed under complete blackout ambient light conditions.

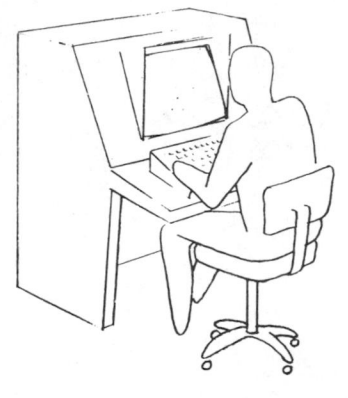

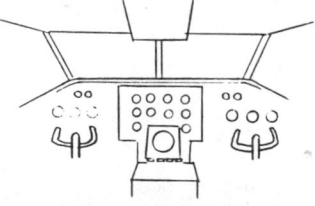

## Desirable CRT Performance Characteristics

**TUBE SIZE** The diameter of the tube should be a minimum of 7 in (18 cm) for plan position target search operations, 10 to 14 in (25 to 36 cm) for central-area target detection efficiency, and 12 to 19 in (30 to 48 cm) for typical console operator viewing distances (all signal types). Larger tube sizes are acceptable for TV-type viewing and status-board displays that may be viewed by several operators. Sizes larger than 48 in (122 cm) are not recommended because of the loss in brightness and contrast as the viewers are forced to stand further away from the screen.

**TARGET SYMBOL SIZE** The minimum size of target pips should be about 12 min of arc measured at the eye, assuming fairly ideal viewing conditions. The preferred size is about 20 min of arc. Alphanumeric symbols should be at least 12 min of arc even with optimum viewing conditions, and about 25 min of arc when conditions are not good (poor viewing angle, high ambient light and glare, etc.).

**MINIMUM TARGET SEPARATION** To obtain effective target separation for detection purposes, the minimum separation must be at least 0.1 min of arc.

**DISPLAY RESOLUTION** The maximum number of lines possible within the state of the art is always desirable for complex motion displays. Typical TV scan lines and symbolic character height is 10 for adequate character legibility. For digitally generated images that may include handwritten characters, the scan lines and symbolic character height should be about 125 lines per inch.

**TV DISPLAY ASPECT RATIO** The standard width/height ratio is 4:3, but ratios of 5:7 or 2:3 provide the greatest legibility.

**ACCEPTABLE BANDWIDTH** The acceptable bandwidth is 4.0 to 10 MHz.

**GRAY LEVEL** There should be a minimum of five levels for TV; a single level (black and white) is acceptable for most digitally generated images.

**DISPLAY BRIGHTNESS** A line brightness of 50 fL ($\pm$ 40) is required under normal ambient light levels; there should be lower adjustable brightness for lower ambient light conditions.

**BRIGHTNESS CONTRAST** The contrast ratio should be as near 90 percent as is practicable.

**PHOSPHOR PERSISTENCE** The absolute minimum for target detection tasks is 0.1 s (2 to 3 s is much preferred). As a rule of thumb, images should not persist beyond the time necessary for the eye to detect the presence of a target; prolonged persistence only confuses the image. P-4 and P-7 phosphors are suggested for alphanumeric and/or discrete image displays that change frequently.

**FLICKER** Display pulse rates should be compatible with critical flicker frequency response of the eye (CFF); i.e., the particular phosphor-driver combination should not generate pulses in the 30- to 55-Hz range.

**GEOMETRIC DISTORTION** Displacement of any image element should not exceed 2 percent of the image height.

**COLOR MISREGISTRATION** For additive color systems, the maximum acceptable misregistration is $\pm$ 05 percent.

**PPI DISPLAY SWEEP RATE** Any rate between 1 and 70 rpm is acceptable, but target detectability is enhanced at the slower rates.

**ALPHANUMERIC CHARACTER WIDTH/HEIGHT RATIO** A ratio of 2:3 to 3:5 is best for maximum legibility and fastest recognition.

**ALPHANUMERIC CHARACTER STROKE WIDTH** Ratios of 2:6 to 1:10 are recommended.

**ALPHANUMERIC CHARACTER STYLE** The closer the character style is to the character legibility guidelines noted elsewhere for printed matter, the more readable the character is. However, when rapid readout is not an absolute requirement, modified, matrix-type characters are acceptable. Although sloping (italic) characters are frequently used and are acceptable for single values (i.e., when there is at least one character-height separation between character lines), vertical characters are preferred.

**DISPLAY FORMATS** Display formats should be compatible with the manner in which the display image is generated; i.e., a plan position indicator sweeps an arc, and therefore the shape of the display should provide a round or circular face, and a rectangular face should be used for horizontal or vertically swept display systems, used typically for TV motion pictures and/or alphanumeric status or message boards. For image previewing, a scrolling technique is suggested so that an operator (e.g., a bank proof operator) can key in information from one image while viewing the next.

**VIEWING DISTANCE** The optimum for the typical console situation is 18 to 20 in (45.7 to 50.8 cm), e.g., a 12- to 19-in (30.5- to 48.3-cm) screen. A maximum of about 20 ft (6.0 m) should be planned as the limit, depending on the size of the display.

**VIEWING ANGLE** This should not be less than 30° from the perpendicular axis. For seated operator console operation, the operator's sight line should not exceed about 30° vertically or horizontally, i.e., about 15° on either side of the operator's normal center-of-the-display viewing angle.

**ADJUSTABILITY PROVISIONS** As a minimum, capability should be provided for the operator to adjust CRT brightness and contrast. In addition, however, it is also desirable to provide capability for operator tilt adjustment of the CRT or CRT cover face so that the operator can remove disturbing reflections and glare from the display.

**POSITIVE VS NEGATIVE IMAGE** A positive image (i.e., a dark image on a light background) is preferred when the display is used under normal or bright ambient lighting conditions. A negative image (i.e., a light image against a dark background) is preferred under dimout or blackout ambient lighting conditions, since it is desirable to reduce the amount of light that could affect the observer's dark adaptation level.

**COLOR** For normal ambient lighting conditions, use a phosphor color that is near white in color (bluish white, blue, and green are also acceptable, however). For dimout or blackout ambient lighting conditions, use an orange or reddish phosphor to minimize reduction in the observer's dark adaptation level.

When a filter cover is used over the display tube, use a neutral density filter to minimize distortion of the color from the CRT.

Under dimout ambient lighting conditions where reflections may be a serious problem, use a circularized Polaroid filter or a cross-polarized filter-light system to prevent reflections and/or tube phosphor excitement.

By placing the light filter 90° to the CRT filter, it is possible to provide illumination for panel controls without reducing the brightness contrast of the CRT signal.

Multicolored CRT displays are suggested when it is desirable to present typical TV movies, target display encoding for easier differentiation (e.g., enemy versus friendly targets), or other graphic information when it is helpful to differentiate between scales, curves, callouts, etc. Red, green, blue, and yellow are the most reliably differentiated colors.

## General Checklist for CRT Displays[10]

Whenever possible, mount the scope face perpendicular to the operator's normal line of sight. If the operator is standing, his or her line of sight is about 5° downward; seated, it is 15 to 20° downward. Ideally, the operator's line of sight should be perpendicular to the tube center.

Recommended viewing distance is 16 to 20 in (40.6 to 50.8 cm).

Tube size should be proportional to the required resolution. Scopes as small as 2 to 5 in (5.1 to 12.7 cm) in diameter may be used for special cases, e.g., infrequent calibration or tuning.

Scopes with a 5- to 7-in (12.7 to 17.8-cm) diameter may be used when they adequately display the necessary information and when no plotting is required.

For plotting, the scope diameter should be 10 in (25.4 cm) or larger.

The shape of the CRT should be compatible with the information format, i.e., round for plan position displays (PPI) and rectangular for A-scan presentations, document images, etc.

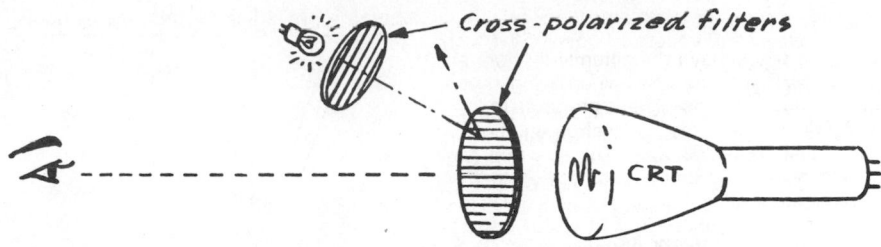

Cross-polarized lighting system for CRTs.

[10]*Human Factors Engineering Design for Army Materiel,* MIL-HDBK-769, 1975.

For a CRT displaying synthetic video, as opposed to raw or unprocessed video, the system should be capable of presenting a symbol of the required size (at least 20 min of arc) for any return resolvable by the system. For synthetic video, in which signals are clipped for digital presentation to the discrete points resulting from digital conversion, the number of points per inch is

$$\frac{\text{Number of points per tube diameter}}{\text{Tube diameter in inches}}$$

For a 14-in (36-cm) screen at 1024 points per tube diameter,

$$\frac{1024}{14} = 73 \text{ points per inch}$$

Symbols should be drawn as point-to-point representations. Symmetrical symbols require an even number of points. For the screen described above (with a viewing distance of 16 in, 41 cm), a 20-min visual angle would subtend 0.093 in (0.236 cm), or 6.77 points. A symmetrical square would have sides extending over 8 CRT points, or ⅛ in (0.28 cm). The area in space which is masked by a symbol equals

$$\frac{\text{Number of points on symbol diameter}}{\text{Number of points on screen diameter}} \times$$

range of coverage

If the screen described above covers 100 km (62 mi) the area masked by the symbol is

$$\frac{8}{1024} \times 100 \text{ km} = 0.782 \text{ km}$$

If the symbol is accompanied by identifying alphanumeric information, each character covers an area representing an area 0.782 × 0.782 km² (0.289 mi²) in the real world. If the scale size is changed to magnify the area covered by the CRT, the area masked by the symbol decreases. For 4× scale (25-km, 16-mi coverage), the area masked by a single symbol is

$$\frac{0.750}{4} \times 0.188 \text{ km}^2$$

Control devices used in conjunction with CRTs (to permit an operator to request information or direct action against a target) should provide a centering accuracy of approximately ±4 CRT points. All such devices (e.g., joy sticks, tracking balls, "stiff sticks," and light pens) require a switch action to indicate that the operator's action is directed to the area currently designated.

A light pen (or pointing device) recognizes only bright line structures specifically called by the computer program and falling within a field of approximately one symbol diameter. To avoid problems, the light-pennable character should not be blink-coded, two individually pennable characters must be separated by at least one symbol, and the cord connecting the pen to the equipment should be attached at a point where it will cause minimal interference with the display viewing and will not accidentally catch on console features.

## Summary of Coding Method for Symbols

| Code | Number of Steps in Code | Evaluation | |
|---|---|---|---|
| Alphanumerics | Unlimited | Excellent | High information-handling rate; unlimited number of coding steps |
| Geometrics | 20 or more | Excellent | Certain shapes easily recognized; many coding steps |
| Color | 4 | Excellent | Difficulty in techniques of reproducing for CRT; objects easily and quickly identified |
| Blink | 2 | Poor | Distracting and fatiguing; interacts poorly with other codes; best for attention getting; few steps in code |
| Brightness | 2 | Poor | Limited number of steps; fatiguing; detrimental to decoding performance |
| Line lengths | 4 | Fair | Limited number of steps; clutters displays |
| Angular orientation | 12 | Fair | 95 percent of estimates correct within 15 percent |
| Inclination | 24 or more | Fair | Many coding steps, especially with combinations |
| Visual number (dots) | 6 | Fair | Few steps |
| Combinations | Unlimited | Good | Avoid overloading symbols with too much information; complex combination can degrade decoding speed and accuracy |

Point resolution is conventionally a fixed percentage of display size. On a 12-in (30-cm) CRT, resolution is about 85 points per inch (1023 × 1023 per display surface). A 4-ft (1.2-m) display of the same matrix would have a resolution of about 21 points per inch. This has little effect on alphanumeric data but puts a definite limitation on graphic displays. On such a 4-ft (1.2-m) screen, lines or points must be separated by 0.05 in (0.13 cm). In both CRT and large-screen displays, line thickness is also proportional to display area.

A 12-in (30-cm) CRT is contained within a visual angle of 41° for a 16-in (41-cm) viewing distance. An operator is able to scan 8.6 in (22 cm) (+15°) comfortably without head movement. The same viewing angle is available at 64 in (163 cm) on a 4-ft (1.2-m) screen. If the larger screen is viewed at a greater distance, the total angular view is decreased; however, element sizes must be increased. With a 16-ft (4.9-m) viewing distance, a 4-ft (1.2-m) screen can be viewed comfortably by a small working group; however, for detail resolution comparable to that of a 12-in (30-cm) CRT, the larger screen can represent only 4 in (10.2 cm) of the CRT display. A choice must be made between the area which can be shown and the amount of detail that can be presented.

Ambient illuminance should not contribute more than 25 percent of screen brightness through diffuse reflection and/or phosphor excitation. The ambient illuminance in the CRT area should have appropriate intensity and color with respect to other visual tasks, e.g., setting controls, reading instruments, inspecting maps, and performing various maintenance and housekeeping tasks, but it should not interfere with the visibility of signals on the CRT display.

The luminance range of surfaces immediately adjacent to scopes should be between 10 and 100 percent of the screen background luminance. With the exception of emergency indicators, light sources in the immediately surrounding area should not be brighter than scope signals.

Viewing hoods or glare-reduction devices should be provided when very faint signals must be detected, and/or a suitable light control filter system should be used to maintain maximum ratios of target signal to CRT background luminance.

Electrically or optically generated displays should conform to MIL-STD-884. Radar CRTs display information more effectively when a symbolic shape code is used in association with, or in place of, raw video pips. The following encoding principles are suggested: The number of discrete symbols should be held to a minimum. Symbol meanings should be standardized.

Symbols should be recognizable without reference to some comparison standard.

If combination codes are used, the respective meanings should be decodable separately without confusion. Two codes maximum combination.

Symbol shapes should reflect a natural relationship to the event or type of target the symbol is supposed to represent.

Codes should be easy to learn and retain.

Symbols should differ widely in shape; avoid variations of a single form, e.g., a circle and an ellipse.

Redundant cues (i.e., those which associate a symbol with a particular size) increase coding efficiency.

External modifiers surrounding a symbol should be avoided.

Reserve blinking coding for emergency situations. There should be no more than two blink rates.

Two brightnesses are the maximum for brightness coding.

Four line lengths can be identified with minimum error.

No more than three blip sizes should be used.

| Video Signal | Vertical Resolution | | Horizontal Resolution | |
|---|---|---|---|---|
| | Center (Lines) | Corner (Lines) | Center (Lines) | Corner (Lines) |
| Monochrome | 400 | 400 | 800 | 700 |
| Red, green, or blue | 400 | 400 | 800 | 700 |

## Typical TV Equipment Standards[11]

Television signal characteristics should be based on a 525-line scanning standard, interlaced 2:1, with 60 fields and 30 frames per second. A video bandwidth of 4.5 MHz should be used for color picture signal transmission. A video bandwidth of 10 MHz should be used for monochrome picture signals.

## Color Picture Monitors

**SCAN SIZE** The normal scan should provide a display in which all four corners of the raster are visible. Width and height controls should have sufficient range to vary the raster size from $-10$ to $+20$ percent while maintaining specified linearity.

**RESOLUTION** The limiting horizontal and vertical resolution of the monitor should be as specified in the above table for both composite and noncomposite operation.

**ASPECT RATIO** The width/height ratio of the raster should be 4:3.

## Monochrome Large-Screen TV Projector

**SCAN SIZE** Width and height controls should have sufficient range to vary raster size from $-10$ to $+20$ percent of the nominal dimensions.

**ASPECT RATIO** The width/height aspect ratio of the normal picture should be 4:3.

**RESOLUTION** The limiting horizontal resolution with the specified brightness level should be at least 800 lines at the picture center and 700 lines at the corners. Vertical resolution should be at least 400 lines, attained when a composite or noncomposite monochrome video signal is applied to the projector input.

**BRIGHTNESS** The projector should provide brightness levels as shown in the table below. Variation in any area of the screen plane should not exceed $\pm30$ percent of the center of the plane.

**GRAY-SCALE REPRODUCTION** Nine shades of gray and the white background should be distinguishable.

**GEOMETRIC DISTORTION** The combined effects of all distortion should not displace any point on the projected display from its correct position more than 1.0 percent of the picture height.

**TRAPEZOIDAL DISTORTION** The projector should be capable of correcting "keystone" or trapezoidal distortion resulting from vertical tilt of the screen, i.e., screen tilt from the perpendicular to the optical axis of the projector within $\pm15°$.

**INTERLACE** Displacement of any scanning line from a center position between lines of the alternate field should not exceed 10 percent of the distance between the lines of the alternate field.

**WHITE BALANCE** The TV monitor should be capable of producing, from a monochrome input signal, a white that corresponds to CIE illuminant "C" ($x = 0.310$, $y = 0.316$).

## TV Quality Variations

At resolutions of 8, 10, and 12 lines, the quality of TV equipment has little significant effect on the accuracy or speed with which standard alphanumerics can be read. At 6 lines, readability is good with high-quality TV (i.e., a minimum of 945 lines).

For group viewing of TV, a minimum vertical resolution of 15 lines/character height is recommended when small visual angles are involved. At 15 lines/character height to total display height is $\frac{1}{32}$, and 16 rows of characters can be put on the screen (i.e., as long as the screen visual angle is within 8 min of arc).

For a symbol resolution of 15 lines, the

recommended maximum viewing distances for various TV monitor sizes are as follows:

27-in (69-cm) monitor—18 ft (5.5 m)
24-in (61-cm) monitor—15 ft (4.6 m)
21-in (53-cm) monitor—13 ft (3.9 m)
17-in (43-cm) monitor—11 ft (3.4 m)

For pictorial TV viewing, the recommended maximum viewing distances are as follows:

21- to 30-in (53- to 76-cm) monitor—20 to 30 ft (6.1 to 9.2 m)
19- to 23-in (48- to 58-cm) monitor—10 to 20 ft (3.1 to 6.1 m)
17- to 19-in (43- to 48-cm) monitor—6 to 10 ft (1.8 to 3.1 m)
15- to 17-in (38- to 43-cm) monitor—30 in to 6 ft (76 cm to 1.8 m)
9-in (23-cm) monitor—18 to 30 in (46 to 76 cm)

## Effect of TV Surround Brightness

The mean value of surround brightness preferred by viewers of broadcast TV is shown in the accompanying graph (plotted for three surround areas at each of five values of peak screen luminance).

Ambient illumination glare should be minimized; i.e., sources should not be located within 60° of the observer's central visual field.

The optimum brightness distribution on the surface of the TV display is approximately 17 fL measured from the central axis and not less than 13 fL measured at the largest angle of view off center. Assuming an ambient light level of 1 fc on the display, this permits viewing in about 10:1 contrast for symbols with regard to the background.

Brightness ratios required for comfortable viewing of large-screen displays are determined by locating two values: (1) the minimum ratio required for adequate viewing and (2) the maximum measure of brightness without annoying aftereffects. The maximum brightness for group displays should not be more than 35 fL. Higher brightness may produce afterimages if the display is viewed for extended periods. An increase in brightness over the 15- to 35-fL maximum contributes no significant improvement in visual acuity.

## Monochrome Television Project

| Image Size | Screen Brightness (fL) |
|---|---|
| 6 × 8 ft | 62 |
| 9 × 12 ft | 28 |
| 12 × 16 ft | 16 |
| 15 × 20 ft | 10 |
| 24 × 32 ft | 4 |

*Note:* 1.0 ft = 0.3 m.

[11]Defense Communications Agency, *Television Technical Characteristics*, vol. I, *Picture Generation and Display Equipment*, AD 805-174, November 1965.

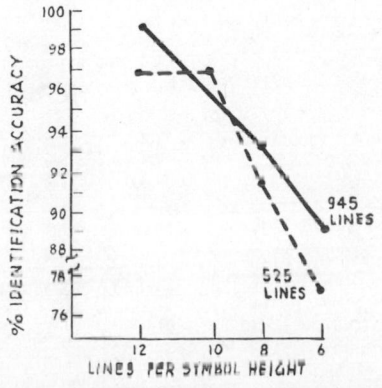

Accuracy of symbol identification for good-quality versus low-cost TV.

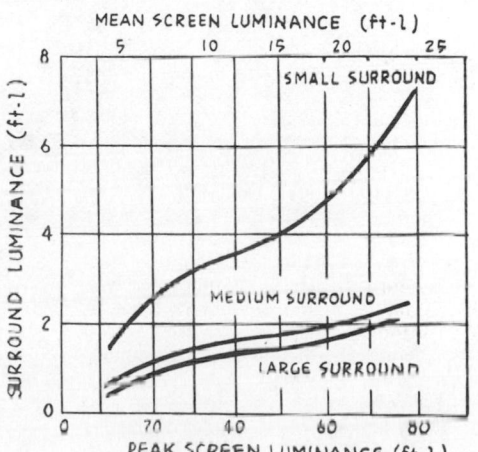

## CRT Symbol Characteristics

**DOT MOSAIC**  The coarsest acceptable mosaic is 5 × 7, as shown in the accompanying illustration. Even these may be marginal compared with larger matrices. Only 35 decoded lines are required. If only numbers and a limited number of symbols are required, such as horizontal or vertical lines or crosses, the number of dots may be further reduced to 35.

**STROKE MOSAIC**  Bar, stroke, or segment patterned symbols can be created as shown in the accompanying illustration. Such matrices are suitable only for numerical characters, not letters. Readability is not as good as with dot matrices if speed and accuracy are required. The seven-segment format is the minimum configuration that is acceptable. Additional segments can be added to improve the legibility of characters. The character format can be slanted, but legibility suffers if the slope is more than 5–7°.

Dot matrices are more readable than segmented types because of the capability to produce curved portions of characters.

**SYMBOL SPACING**  At low brightness levels (1 fL), spacing should be 25 percent of character dimension; at high brightness levels, spacing should approach 200 percent for maximum readability of a single character. For typical applications (e.g., clear text messages or grouped numbers), use 50 percent of average character width for between-character spacing within a word or number group and 75 to 100 percent spacing between words or groups. Spacing between lines of characters should be at least 50 percent of character height.

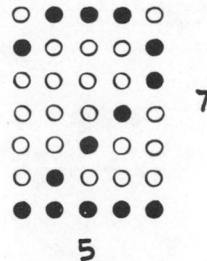

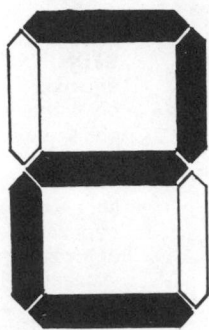

Dot matrices are more readable than segmented types because of the capability to produce apparent curved portions of characters.

A 5 × 7 matrix of dot characters is minimally acceptable; a 7 × 9 matrix, as shown in the illustration on page 545, is preferred.

ABCDEFGHIJKLM
NOPQRSTUVWXYZ
1234567890

| Minimum character height | 3.1–4.2 mm |
|---|---|
| Maximum character height for 5 × 7 dot matrix | 4.5 mm |
| Width/height ratio | 3:4–4:5 |
| Stroke/height ratio | 1:8–1:6 |
| Minimum number of raster lines | 10 lines |

**Suggested CRT Character Optimized
Set Using a 7 × 9 Matrix**

The suggested character formation shown in
the accompanying illustration contains most of
the alphanumerics and supporting symbols required for data processing. They are designed
to minimize recognition errors, or at least possible confusion between one character and
another. Although lowercase letters have
been designed, they are not recommended
because it is difficult to create lowercase features that are not easily confused, especially
if the observer has to read information rapidly.

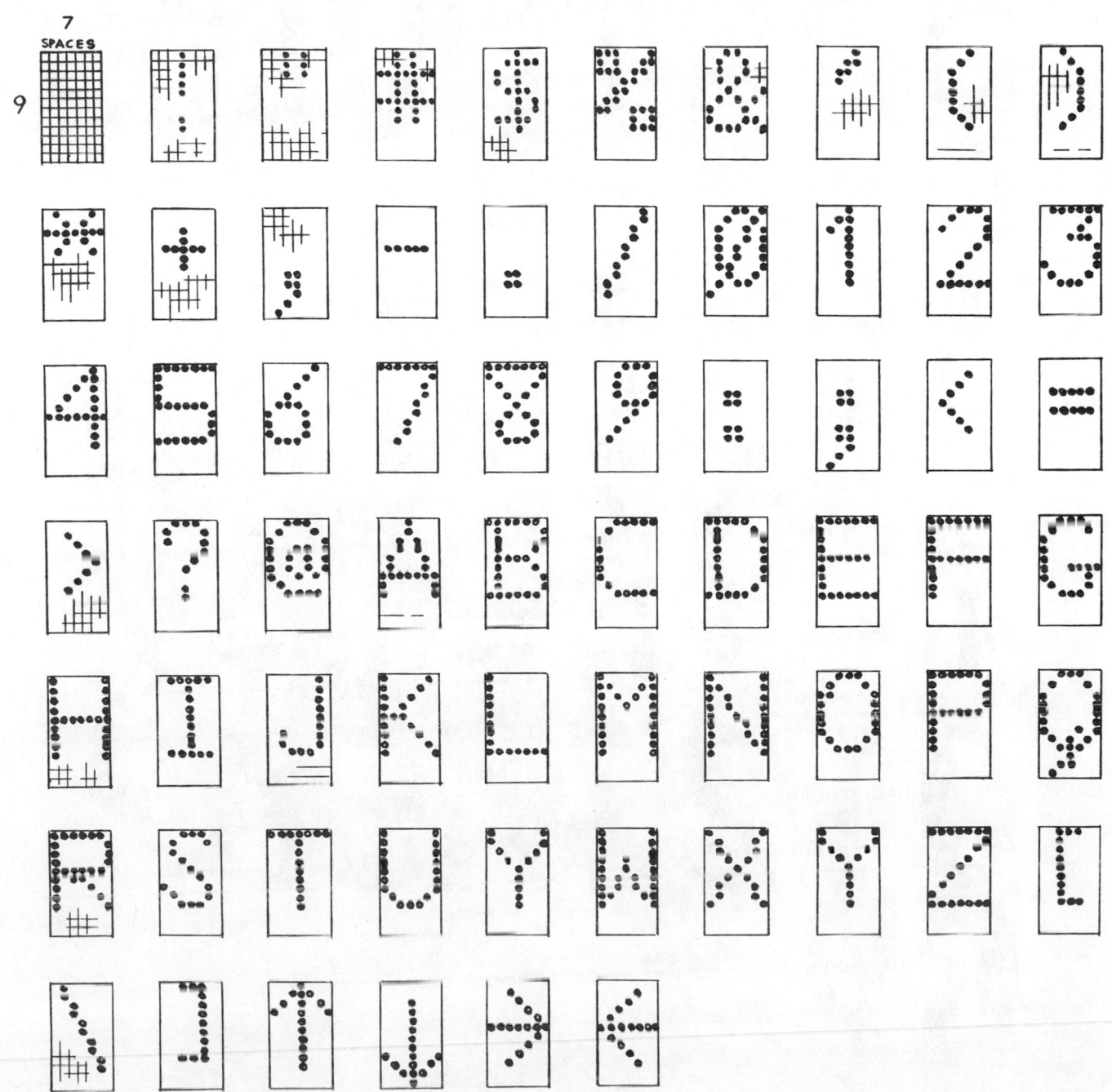

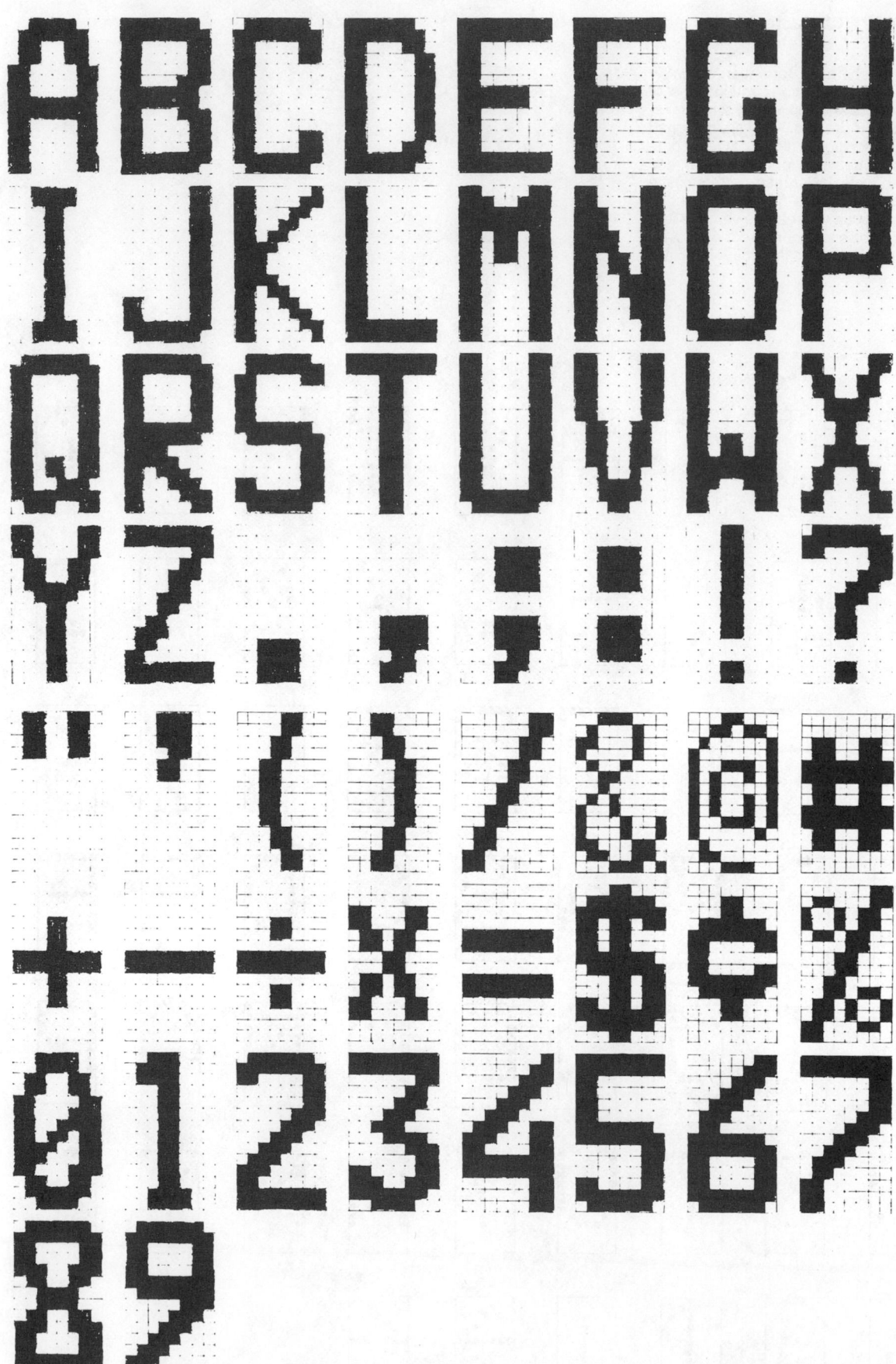

**Example of an 8 × 14 matrix character set.**

**General CRT Phosphor Characteristics and Applications**

| Identification | CIE Coordinates | | Persistence | Fluoresence | Phosphoresence | Decay Time (m/s) | Typical Applications |
|---|---|---|---|---|---|---|---|
| | x | y | | | | | |
| P-1 | 0.218 | 0.712 | M | YG | YG | 24 | Radar and test equipment oscilloscopes |
| P-2 | 0.279 | 0.534 | M | YG | YG | 35–100 | Oscilloscopes |
| P-3 | 0.523 | 0.469 | M | YG | YO | | |
| P-4 | 0.270 | 0.300 | M/S | W | W | 25 | Monochrome TV |
| P-5 | 0.169 | 0.132 | M/S | B | B | 25 | Photography |
| P-6 | 0.338 | 0.374 | S | | W | | |
| P-7 | 0.357 | 0.537 | L | Y | YG | | Radar, sonar, and oscilloscopes |
| | | | M/S | B | B | | |
| | 0.151 | 0.032 | M/S | W | YG | 40–60 | |
| P-11 | 0.139 | 0.148 | L | B | B | 25 80 | Photography |
| P-12 | 0.605 | 0.394 | M/S | O | O | | Radar |
| P-14 | 0.504 | 0.443 | M | YO | O | | Displays where repetition rate is 2 to 4 s after excitation is removed |
| | 0.150 | 0.093 | M/S | B | O | | |
| | | | M/S | PB | YO | | |
| P-15 | 0.246 | 0.439 | VS | G | G | | TV pickup of photographs by flying spot scanning |
| P-16 | 0.175 | 0.003 | VS | BP | BP | 0.12 | TV pickup of photographs by flying spot scanning |
| P-17 | 0.302 | 0.390 | L | Y | BP | | Military displays |
| | | | VS | G | BP | | |
| | | | S | YW, BW | Y | | |
| P-18 | 0.333 | 0.347 | M | W | W | | Low-frame-rate TV |
| P-19 | 0.572 | 0.422 | L | O | O | | Radar |
| P-20 | 0.444 | 0.536 | M/S | YG | YG | 0.01–2 | High-visibility displays |
| P-21 | 0.539 | 0.373 | M | RO | RO | | Radar |
| P-22 | 0.155 | 0.060 | M | B | B | 25 | Color TV |
| | 0.285 | 0.600 | M | YG | YG | 60 | |
| | 0.675 | 0.325 | M | OR | OR | 0.9 | |
| P-23 | 0.375 | 0.390 | | | | | Interchangeable with P-4 |
| P-24 | 0.245 | 0.441 | S | G | G | | Flying spot scanner tubes |
| P-25 | 0.557 | 0.430 | M | O | O | | Long-persistence displays up to 10 s |
| P-26 | 0.582 | 0.416 | VL | O | O | | Radar |
| P-27 | 0.674 | 0.326 | M | RO | RO | 27 | Color TV monitors |
| P-28 | 0.370 | 0.540 | L | YG | YG | 0.5 s | Radar and sonar |
| P-29 | (P₂ + | P₂₅) | M | W | YW | | Aircraft indicators and radar |
| P-31 | 0.193 | 0.420 | M/S | G | G | 4 | Oscilloscopes |
| P-32 | | | L | PB | YG | | Radar |
| P-33 | 0.559 | 0.440 | VL | O | O | | Radar |
| P-34 | 0.235 | 0.364 | VL | BG | YG | 40 s | Oscilloscopes and radar |
| P-35 | 0.286 | 0.420 | MS | G | B | 1 | Oscilloscopes |
| P-36 | 0.400 | 0.543 | VS | YG | YG | 0.25 | Flying spot scanning tubes |
| P-37 | 0.143 | 0.208 | VS | B | B | 0.155 | Flying spot scanning tubes |
| P-38 | 0.561 | 0.437 | VL | O | O | 1040 | Integrating phosphor for low-repetition-rate displays and radar |
| P-39 | 0.223 | 0.698 | L | YG | YG | 150 | Integrating phosphor for low-repetition-rate displays and radar |
| P-40 | 0.276 | 0.3117 | M | W | YG | | Integrating phosphor for low-repetition-rate displays and radar |

B = blue, G = green, O = orange, P = purple, W = white, Y = yellow.
VL = very long, 1 s or over.
L = long, 100 ms to 1.0 s.
M = medium, 1 ms to 100 ms.
M/S = medium short, 10 μs to 1.0 ms.
S = short, 1 μs to 10 μs.
VS = very short, less than 1 μs.

*Special Notes:*
P-1—High efficiency, resolution, and resistance to burn; lacks low-level or short persistence.
P-2—Decrease in decay with increase in beam current.
P-4—Sulfide version.
P-7—High efficiency and resistance to burn; amber filter required to obtain long persistence.
P-19—Slow refresh rate for flickerless display; low light output; low burn resistance.
P-22—Sulfide blue and green, vanadate for red.
P-25—Desired low-level persistence; high resistance to burn, low light output.
P-26—Slow refresh rate for flickerless display; low light output and burn resistance.

*(continued)*

P-31—Curve has blue peak at 450 nm; high efficiency, resolution, and resistance to burn; lacks low-level or short persistence.

P-33—Decay decreases with beam current decreases; burns rapidly when used with stationary or slow-moving beam.

P-34—IR stimulatable; Y-phosphor.

P-35—Resists burning compared with P-11.

P-39—Similar to P-1 but with longer decay.

P-40—Similar to P-7 but with longer decay.

P-37—Similar to P-11 but with shorter decay.

P-36—Similar to P-20 but with shorter decay.

*Source:* After H. R. Luxenberg and R. L. Kuehn, *Display System Engineering,* McGraw-Hill Book Company, New York, 1968.

## CRT Flicker Thresholds of the Average Observer

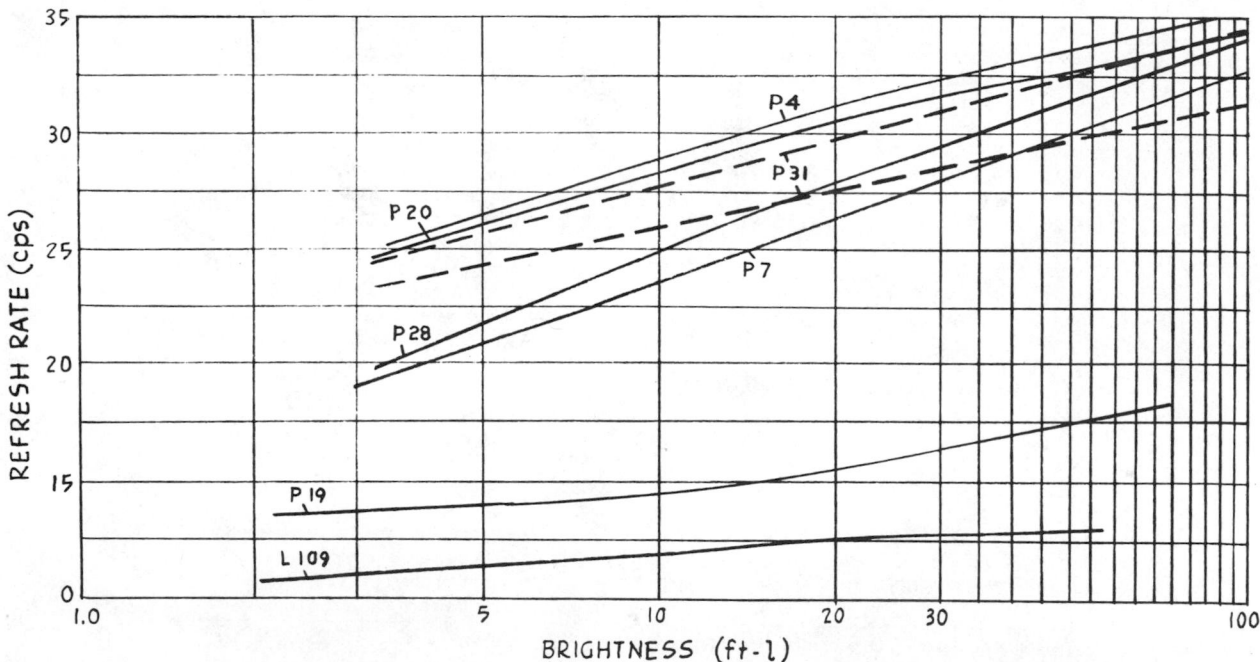

For character displays the pulse rate should be greater than about 30 to 40 Hz so that the characters will not appear to blink. Some flicker may be noticed with average display brightness values unless the repetition rate is maintained at least at 50 Hz.

Flicker will not be noticed in TV at 60 fields per second unless the display brightness exceeds about 180 fL; 50 fields per second is usually acceptable if the display brightness is reduced to 30 fL.

## CRT Display Format (Radar and Sonar)

Electronic displays used for detecting, identifying, and tracking target information usually are presented in either rectangular or polar-coordinate form, i.e., an A-scan or PPI (plan position indicator) format. For obvious reasons, the A-scan display face should generally appear as a rectangularly shaped display, and the PPI should appear circular.

A variety of formats are available, as shown in the accompanying illustrations.

Regardless of the type of scope presentation, certain features, such as the following, should be designed according to good human engineering design principles.

**DISPLAY SIZE** For a single operator involved in a target search task, use a relatively small display (7 to 14 in, 18 to 36 cm) to reduce the area for visual scanning. For tracking targets, use a tube size of 10 to 17 in (25 to 43 cm). If more than one person must work at the CRT station, consider diameters of 24 to 30 in (61 to 76 cm).

**DISPLAY CURSORS** Cursors help the operator relate target position to actual range and bearing values. They should always be accompanied by indexing scales laid over or around the display face and/or electrically connected to digital panel readouts. Electronic cursors are preferred because they do not produce the parallax problems that mechanical cursors often create.

**GRID OVERLAYS** Target position can be interpreted much faster if a grid overlay is supplied. The more accurate the reading requirement, however, the more elaborate the grid structure must become. Grids are more confusing than useful on displays that are less than about 14 in (36 cm) in diameter. Even here it may be desirable to create a grid that will not obscure small target traces.

For example, the minimum spacing between range rings on a polar display should be on the order of 1° (36 min visual angle subtended at the eye), or about 0.50 in (1.27 cm) at an average 18-in (46-cm) viewing distance. A mixture of solid and dashed grid lines or range rings helps reduce the potential for obscuring targets and also aids in identifying major range and bearing elements (numbered)

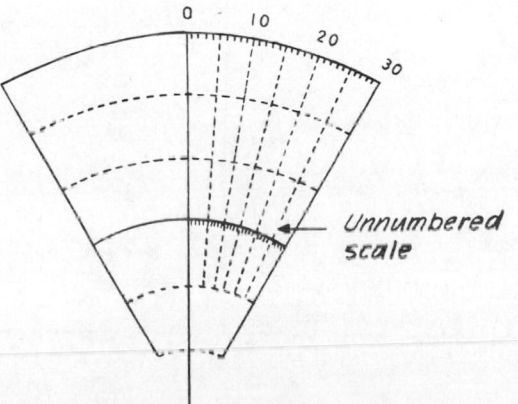

*Unnumbered scale*

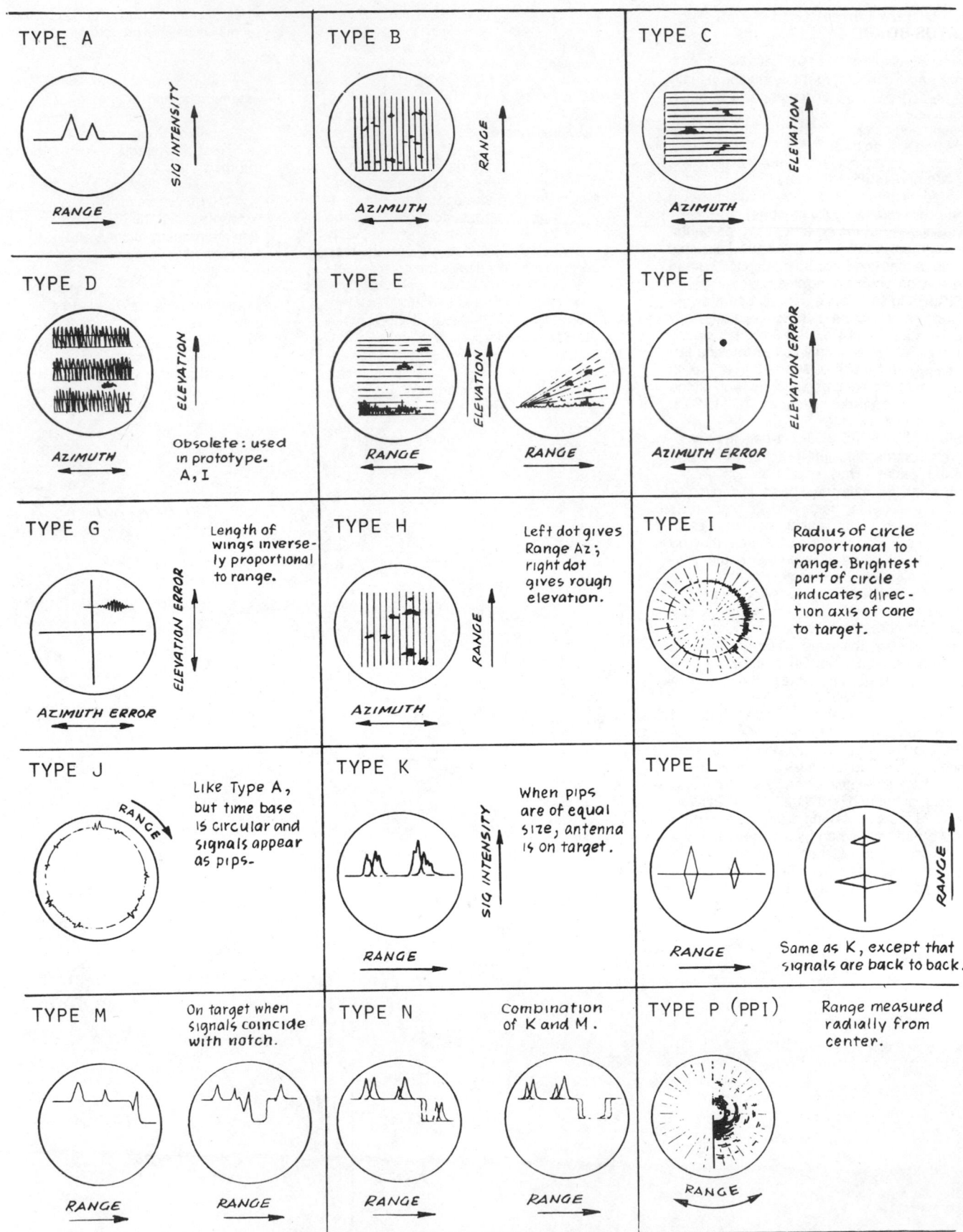

**TYPE A** — RANGE / SIG INTENSITY

**TYPE B** — AZIMUTH / RANGE

**TYPE C** — AZIMUTH / ELEVATION

**TYPE D** — AZIMUTH / ELEVATION — Obsolete: used in prototype. A, I

**TYPE E** — RANGE / ELEVATION — RANGE / ELEVATION

**TYPE F** — AZIMUTH ERROR / ELEVATION ERROR

**TYPE G** — AZIMUTH ERROR / ELEVATION ERROR — Length of wings inversely proportional to range.

**TYPE H** — AZIMUTH / RANGE — Left dot gives Range Az; right dot gives rough elevation.

**TYPE I** — Radius of circle proportional to range. Brightest part of circle indicates direction axis of cone to target.

**TYPE J** — RANGE — Like Type A, but time base is circular and signals appear as pips.

**TYPE K** — RANGE / SIG INTENSITY — When pips are of equal size, antenna is on target.

**TYPE L** — RANGE / RANGE — Same as K, except that signals are back to back.

**TYPE M** — RANGE / RANGE — On target when signals coincide with notch.

**TYPE N** — RANGE / RANGE — Combination of K and M.

**TYPE P (PPI)** — RANGE — Range measured radially from center.

Scope types.

## LARGE-SCREEN, MAP, AND STATUS-BOARD DISPLAYS

A wide variety of techniques, formats, and arrangements have been used to provide general information display for a large number of people to use on a concurrent basis (as illustrated by the accompanying sketch). One should consider the following guidelines in developing such display arrangements:

1. Avoid too much detail on the large map or status board if the amount of information creates crowding of alphanumerics, groups of data, or complexity because of the inability to provide the necessary spacing to clarify symbols, call-outs, or other related information. Leave the detail to individual operator displays.

2. Avoid using too many different colors on maps, in target symbols, in alphanumeric headings, etc.; it becomes difficult to keep the color codes in mind.

3. If moving targets are displayed on a large map, keep the number at a minimum and display only those which move fairly slowly.

4. If the display area must have low ambient illumination or blackout in order for individual operator displays to be used effectively, use white or luminescent markings against a dark background for maps and status boards to help operators maintain dark adaptation.

5. If color-coded target information is to be used on the large map display, use only a neutral color, such as gray, for the map background; this allows the color targets to have maximum effect (contrast).

6. When projectors are used to project information on a large map display, make sure the projectors are positioned so that they are not readily visible to the operating personnel, i.e., so that the projector lights do not create glare problems. Also consider the projector position in terms of minimizing the possibility that individuals will walk across and interfere with the projected image.

7. Determine and provide the proper alphanumeric and/or symbol sizes on the large-screen displays in terms of the maximum viewing distances at which each set of characters or symbols must be read.

8. When a large display has to be created using two or more projectors, design the system so that the points where images join are properly aligned and so that one or more of the projections do not flicker or jitter.

9. Consider various coding techniques to help operators easily discriminate between new and old data, friendly and enemy targets, and moving and fixed targets; such techniques as flashing symbols and brighter versus darker symbols can be used.

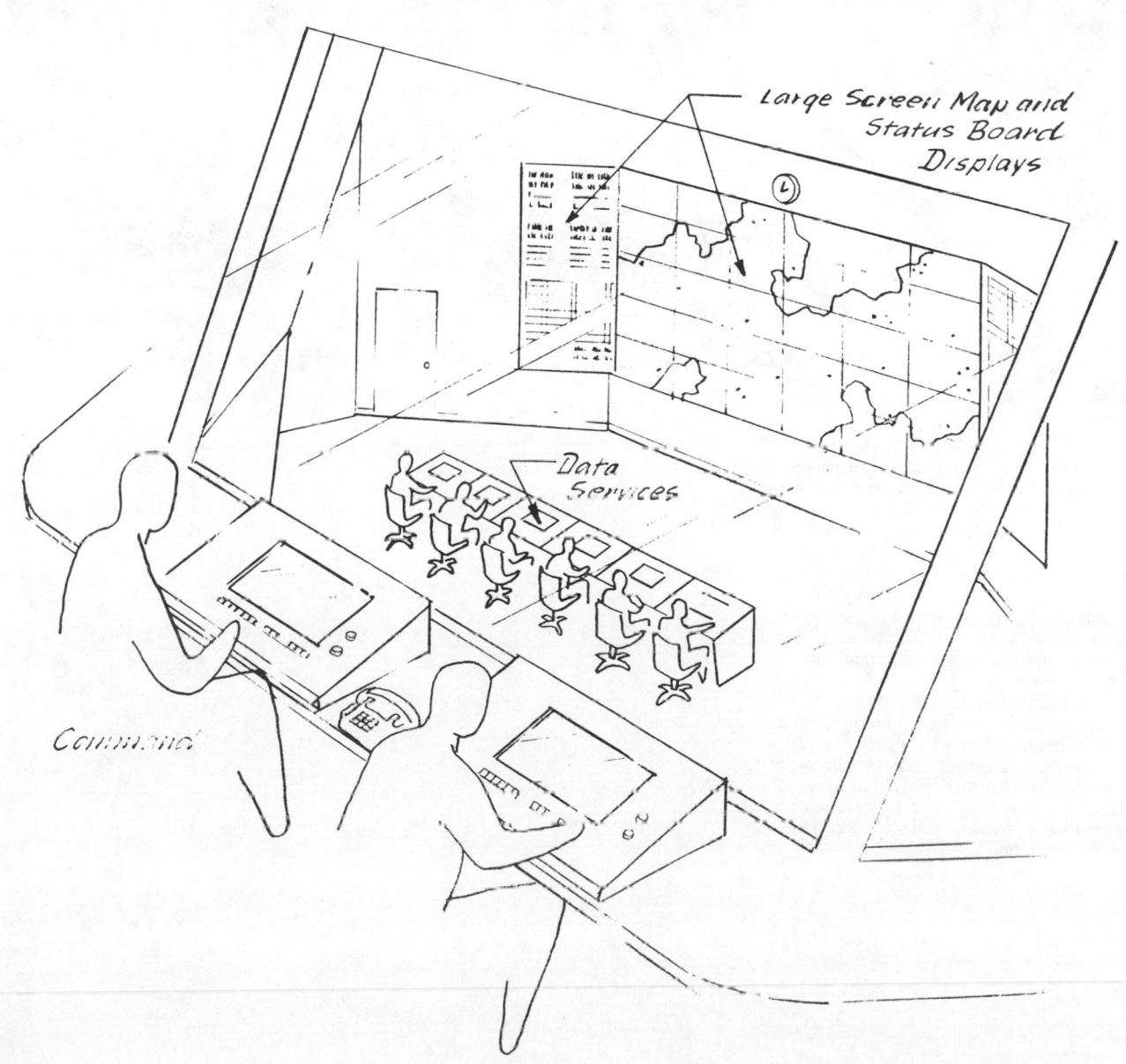

**Lighted Status Boards**[12]

A variety of light and word status-board systems are currently available for use in installations where fairly high ambient lighting levels make it difficult to use other types of displays.

The dot pattern provides a means for creating extremely legible alphanumerics that can be observed from rather widely dispersed viewing angles. Such systems can be controlled from a central operator station and/or via automatic computer to provide periodic update of the information.

Typical applications include the air terminal schedule board shown in the accompanying illustration, command and control status and pictorial reference displays, sports scoreboards, and many others.

Such systems can be constructed for a wide range of size requirements, from the individual operator control station to the largest advertising sign.

[12]Illustration by permission of Ferranti-Packard Limited, Mississauga, Ontario, Canada.

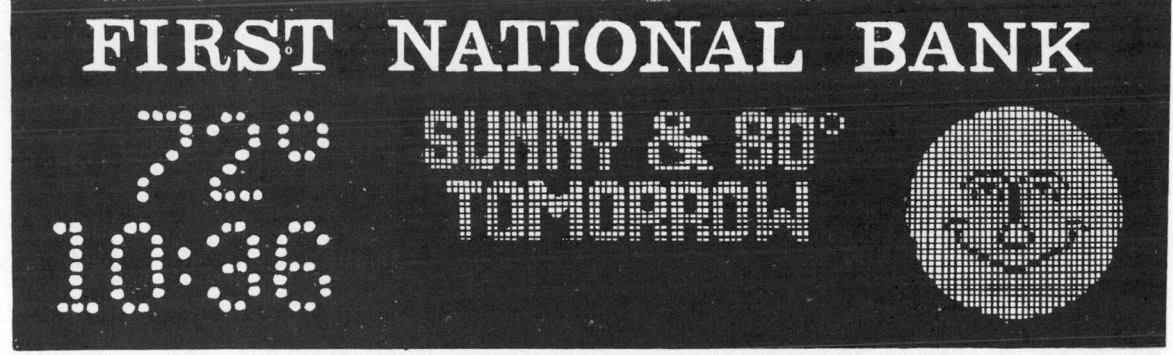

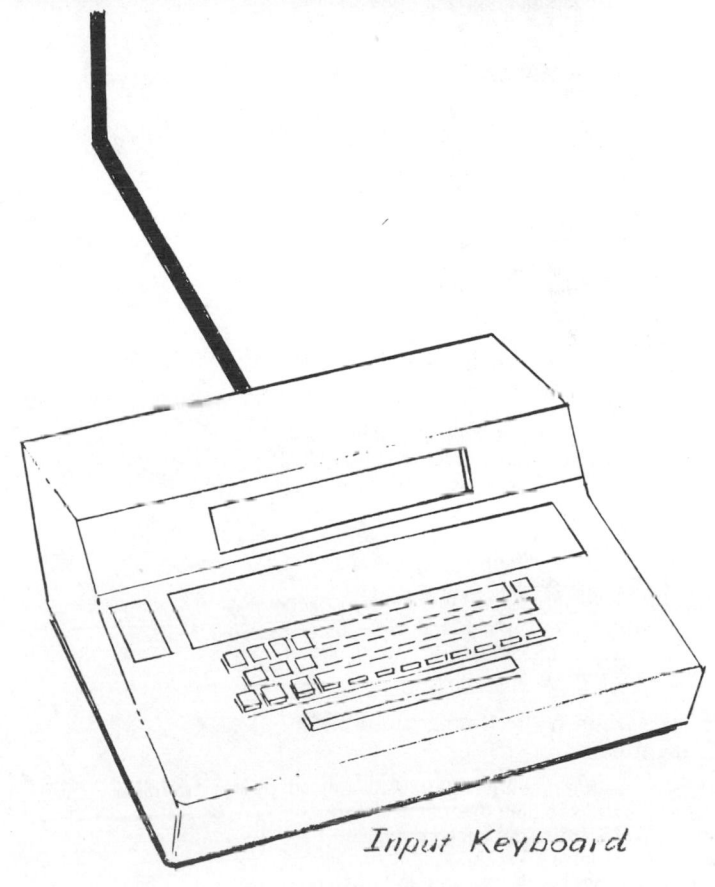

### Variable-Matrix Animated Signs

Current technology provides several techniques for creating animated graphics, some with sufficient refinement to faithfully reproduce actual photographs in large scale. As the accompanying sketches illustrate, such sign techniques are useful for extremely critical visuals, such as the highway warning display, and for purely advertising displays as well.

Control systems provide either repetitive or programmed presentations or manual input control as implied by the console keyboard unit shown in the accompanying illustration.

### OPTICALLY PROJECTED DISPLAYS

The symbol size should be approximately 10 to 15 min of arc resolved at the observer's eye.

The aspect ratio should be approximately 1.33:1.48.

The symbol stroke should be about one-sixth to one-tenth the character height.

The viewing distance should be approximately four times the image width.

The viewing angle should be between 20 and 30° from the centerline of the display.

Image luminance should be at least 10 fL.

The direction of light and dark contrast is not important for legibility.

The ambient light impinging on the screen should not exceed about 0.02 fL.

The contrast ratio should be about 500:1, measured with no film in the projector.

Matte screens are generally recommended because screen brightnesses are essentially equal at all viewing angles. However, they are only about 85 to 90 percent efficient. When light output from the projector is limited, a lenticular screen should be used (for approximately 1.5 to 2.0 gain).

*Input Keyboard*

## Projectors

Most inexperienced people who must periodically set up and operate still or motion picture projectors recognize that such devices are generally poorly engineered from the human factors point of view. It is hoped that the following general recommendations will generate improved machine designs in the future:

Provide an easily understandable system for setting up the equipment, i.e., for erecting supports for movie reels; attaching the reels; threading the film (preferrably automatic); inserting slides (right side up); attaching the slide case, tray, or a carousel; adjusting the projection angle, focus, and brightness; etc.

Provide an easily understandable set of controls, properly labeled so that they can be seen in a dimly lit room.

Provide sufficient lengths of power cord, audio cord, and remote control cord, color-coded so that they are not easily confused.

Use low-noise motors and fan systems and/or insulate so that the projector noise will not interfere with audio communication.

Design the package so that it does not weigh over 25 lb (11.3 kg) and is small enough for one-person transport.

Design for ease of lamp replacement and regular servicing.

## Projector Brightness

To determine the required projector light output, take the projection surface area (2 ft²) and divide this by the screen gain. Multiply this by the recommended screen brightness (footlamberts) shown below:

1. For motion pictures:
   - 5 fL     Minimum (marginal for some observers)
   - 10 fL     Satisfactory
   - 15 fL     Excellent
   - 20 fL     Maximum (flicker threshold for some observers)
2. For slides:
   - 1 fL     Absolute minimum (difficult to distinguish color from black and white)
   - 1.5 fL     Minimum for gross images
   - 5 fL     Minimum for detailed slide information
   - 10 fL     Satisfactory
   - 20 fL     Excellent
3. For projected TV:
   - 2 fL     Gross black-and-white images
   - 10 fL     Maximum (to avoid flicker thresholds)

## Black versus White-Background Slide Presentation

When slides are presented in a relatively dark room for extended periods of time, it is recommended that the slide backgrounds be dark with white graphics. This allows the observer's eyes to become dark-adapted and remain so until the presentation is completed. If, on the other hand, only one or two slides are presented at a time, with intermittent raising of the ambient room illumination, and/or if slides are presented in a medium to fairly high ambient illumination environment, it is best to present black graphics on a light background.

## Colored Backgrounds

Colored backgrounds provide useful relief and stimulation in slide presentations; in addition, colors can be used to present information. Colored background should, however, be of a light shade so that black graphics are maximally legible. More saturated color should be reserved only for graphics themselves (i.e., without overlayed graphics or alphanumerics).

## Slide Preparation

The proportion of lantern slides should be approximately 7:10. The ratio of alphanumeric character height on the screen to the most distant observer should be about 1:300, i.e., 1.0-in (2.5-cm) characters for a 25-ft (7.6-m) viewing distance.

As a general rule, no character should be less than about 0.45 in (1.1 cm) high, with a stroke width of approximately 0.006 in (0.015 cm).

In preparing a chart for projection, enlarge the copy about three times so that characters are about 0.125 in (0.318 cm) high.

## Making Slides for Opaque Projection

Alphanumeric characters typed in all capital letters will be legible when converted to a 2- by 2-in (5.1- ×5.1-cm) slide when the material fits within a 3.0- ×4.5-in (7.6- ×11.4-cm) area. Illustrations should fit the horizontal format of the screen; i.e., the width should be about 1½ times the height.

As a general rule, limit line copy to about 10 lines, with five to seven words per line. The space between rows of characters should be at least one-half the height of the characters, and preferably a full character separation.

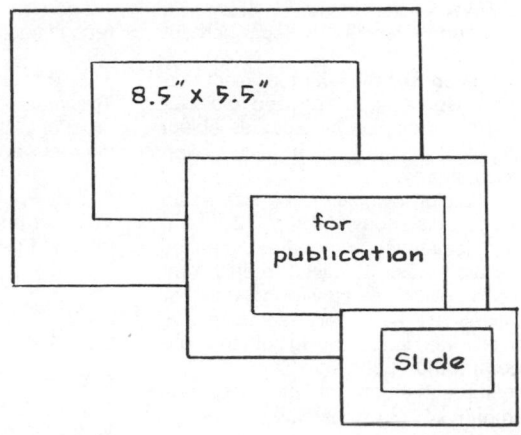

$$\begin{bmatrix} 1.0'' = 2.54 \text{ cm} \\ 1.0' = 0.30 \text{ m} \end{bmatrix}$$

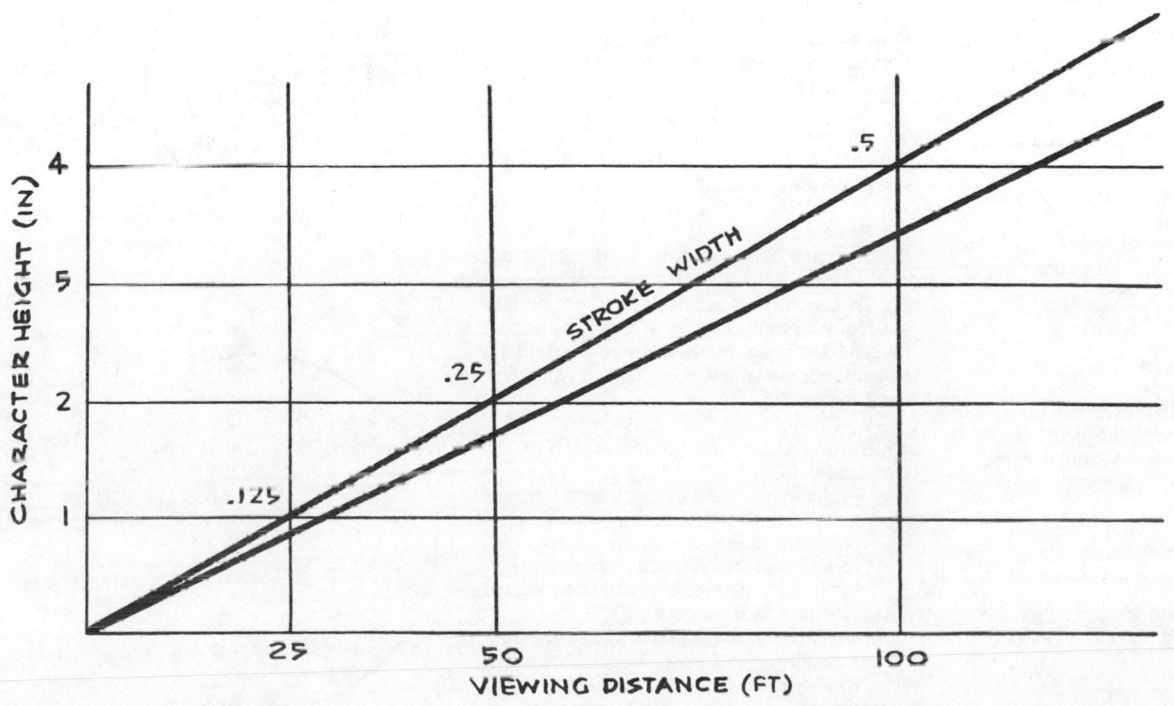

# HAND-HELD OPTICAL AIDS

Optical aids are used primarily to offset reduction in visibility due to extreme viewing distance. As a general rule, viewing with two eyes is preferred over single-eye viewing because it preserves a certain amount of depth perception.

Optical aids are most useful for object recognition and inspection, as opposed to detection. The latter is difficult because as object size is magnified, so too is the relative motion of the optical image.

For identification of objects the size of an airplane up to a distance of 3000 yd (2742 m), a $2\times$ magnification is appropriate. Increases in magnification make it difficult to shift from naked-eye to optically aided viewing. If higher magnifications are necessary, some means should be provided for steadying both the optical device and the observer.

Visibility through an optical device depends on the amount of light transmitted. Thus the more separate glass surfaces there are in an optical system, the greater the loss of light. All such surfaces should be antireflection-coated.

### General Magnification Criteria

| Distance | Magnification |
|---|---|
| 1000 yd— (914 m) | 1–1.5$\times$ |
| 3000 yd— (2742 m) | 2–3$\times$ |
| 5000 yd— (4570 m) | 3–4$\times$ |
| Above 5000 yd* | 6–8$\times$ |

*Should be tripod-mounted.

When an optical device is to be used for night viewing, provide reticle illumination with variable intensity control from 0 to about 0.02 fL.

Binocular devices should have means for adjusting the interocular distance independently from individual eye viewing distance.

## Direct-View Optical Instrument Design[13]

### 1. User Visual Considerations
Visual spectrum: 400 nm ($400\times10^{-9}$ m) through 700 nm
Visual acuity: 60 s of arc for the emmetropic eye
Maximum tolerable astigmatism: ¼ diopter
Vernier acuity: 10 s of arc
Stereoscopic acuity: 12 s of arc
Extraocular limits for onset of blurred vision and/or possible fusion loss:
  *a.* Abduction: 15 min
  *b.* Adduction: 30 min
  *c.* Supraduction: 15 min
Stimulation-intensity range:
  *a.* Smallest detectable (threshold) using rod vision: $10^{-6}$ mL; cone vision: 1 mL
  *b.* Largest tolerable cone vision: $10^4$ mL

### 2. Optical Parameters
Magnification: between 1 and 20 power
  *a.* Rifle, pistol sights: 4 power
  *b.* Monoculars or binoculars: 8 power

*Note:* If more than one magnification is required, use discrete magnifications rather than a "zoom" technique, unless accuracy is not important.

### 3. Field of View
The field of view should be compatible with the intended use, such as search and detection, recognition, identification, or aiming.

### 4. Entrance Pupil
The entrance pupil should be equal to the product of the magnification and the exit pupil diameter and is therefore defined by these.

### 5. Exit Pupil
The exit pupil should be consistent with the intended use:

  *a.* Daylight use: not less than 3.0 mm
  *b.* Twilight and darker use: not less than 7.0 mm

### 6. Eye Relief
A long eye relief (25 mm) is desirable on vehicular-mounted sights to protect the user from gun recoil, observing on the move, etc.; 15 mm permits the user to wear glasses.

### 7. Eyepiece Adjustment
Fixed-focus eyepieces set between 0.50 and 1.00 diopter may be used for 4 power or less. Eyepiece dioptric adjustments and a scale ($+$ 3.0 diopters, although 4.0 diopters is more desirable—in 0.5-diopter increments) should be provided on all instruments over 4 power.

### 8. Axial Resolution
Axial resolution should be equal to, or better than, 60 s of arc divided by the magnification in order to create an "eye-limited" instrument.

### 9. Aberrations
Aberrations should be according to standard optical practice.

### 10. Luminous Transmission
Luminous transmissions should be greater than 50 percent. The elements should be antireflection-coated, except for the focal plane components.

### 11. Parallax
The reticle should be focused to the target range of primary interest.

### 12. Reticles
Reticle lines should subtend 2.0 min at the eye. (Lines are preferred to dots.) A ring should be used, as opposed to a spot and/or a small crosshair. Reticle illumination should be provided for instruments to be used during twilight or lower ambient conditions (with dimming capability).

### 13. Binocular and Biocular Considerations
  *a.* Provide interpupillary adjustment scaled from 58 to 73 mm (1-mm intervals). Magnification differences between two barrels should not exceed 2 percent, and luminous transmission difference should not exceed 5 percent.
  *b.* Collimation, i.e., alignment of the binoculars, should not exceed the following at the eye: 15.0 to 40.0 min/arc divergence, or 25.0 min/arc convergence.

### 14. Filters
Consider the use of neutral density filters to reduce glare from the sun or other high-level light sources.

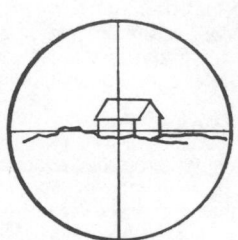

*Vertical / Horizontal Reference.*

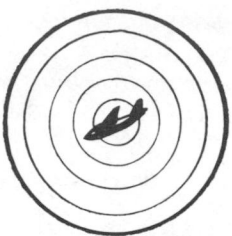

*Two-dimensional reference without requiring sight to be rotated.*

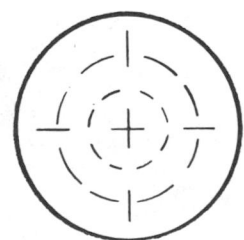

*Broken lines = less target interference.*

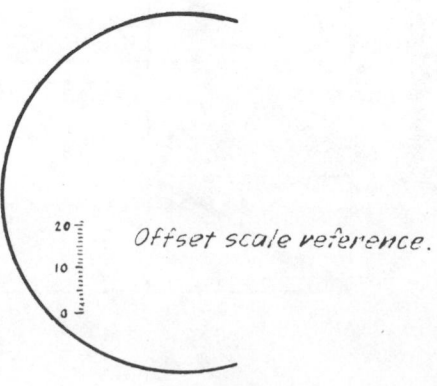

*Offset scale reference.*

[13]*Human Factors Engineering Design for Army Materiel,* MIL-HDBK-759, March 1975.

A — SUPERCILIARY ARCH REQUIREMENT ————————————— .6875″ (1.7463 cm)

B — NASAL BONE REQUIREMENT ————————————————— .8750″ (2.2225 cm)

C — GREATER ALAR CARTILAGE REQUIREMENT ———————— 1.25″ (3.175 cm)

D — SEPTAL CARTILAGE REQUIREMENT —————————————— 1.75″ (4.8387 cm)

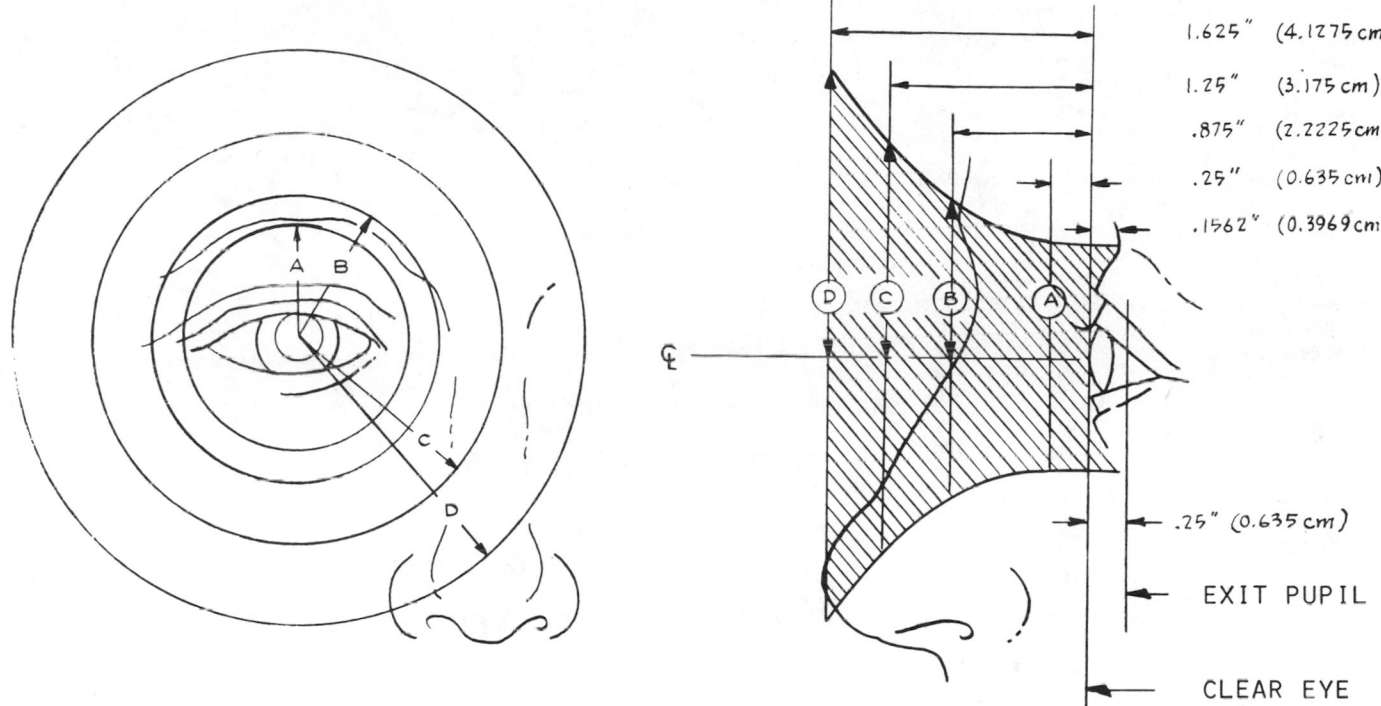

1.625″ (4.1275 cm)
1.25″ (3.175 cm)
.875″ (2.2225 cm)
.25″ (0.635 cm)
.1562″ (0.3969 cm)

.25″ (0.635 cm)

EXIT PUPIL

CLEAR EYE

**Anatomical limits on axially symmetrical ocular metal parts.**

## 15. Eyecups

Use appropriate eyecups to cushion the user's head from the device and to keep stray light from entering the eyepiece and/or the observer's eye.

## Reticle Design Suggestions

Use the minimum number of reticle lines or circles necessary to provide adequate reference; too many lines interfere with the view of the target.

Provide a stable frame of reference by means of vertical and horizontal reticle lines. Avoid too many intervening angular lines, except when the viewing task does not require vertical and horizontal reference; concentric circles allow the observer to judge the size of an object that may be changing direction, e.g., an aircraft target.

Use broken reticle lines when too many lines tend to obscure the target.

If a scale reference is required, consider placing the scale to one edge of the reticle plate, thus minimizing the potential interference of the scale with the primary viewing area.

Optical aids that may be used where there is the possibility of extremely bright flashes (e.g., an atomic flash) should be equipped with an automatic filter system, since the hazard of flash blindness is increased by the optical system.

### Telescopic Sight Reticle Visibility

A collimator sight permits observation of a reticle with the eye focused at infinity, or with the same eye accommodation as required to view a target, and does not require that the eye be precisely positioned. Light from an illuminated reticle passes through a collimating lens placed one focal length from the reticle, rendering the light rays leaving the reticle parallel. Collimated light from the reticle combines with light from the target scene by use of a coated glass plate (beamsplitter), permitting light both to pass straight through from one face and to reflect light striking the opposite face. By using dichroic filters for beam splitters, color discrimination can be made between the target and the reticle; i.e., dichroic filters reflect what they do not transmit, and the ratio of reflectance to transmittance can be made to vary with wavelength.

Experiments by the U.S. Army Human Engineering Laboratory[14] indicate that a red background (beam splitter) with a blue-green reticle is a preferred combination for all typical field viewing conditions, with a blue-green background and a red-orange reticle as an effective alternative. Experiments showed that a red view of the world with a light-blue reticle pattern of reasonable brightness that remains just visible against bright backgrounds, including the sky, provides the best operator discrimination performance.

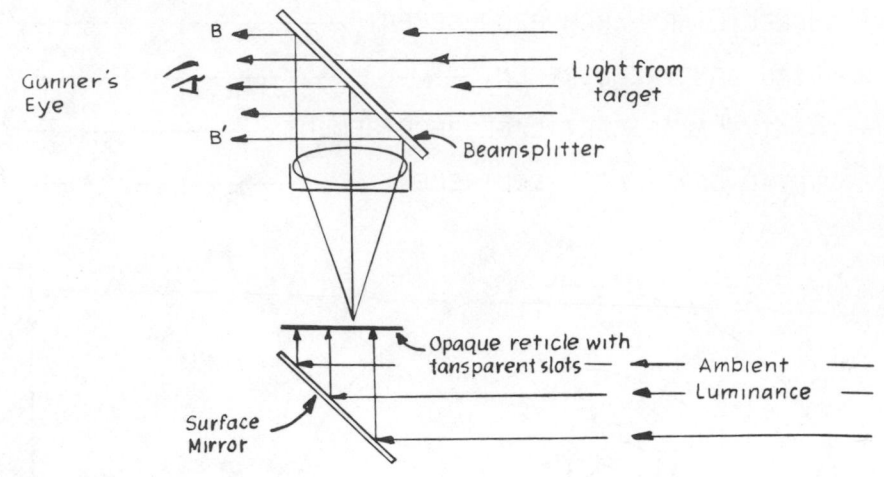

TRANSMITTED MODE

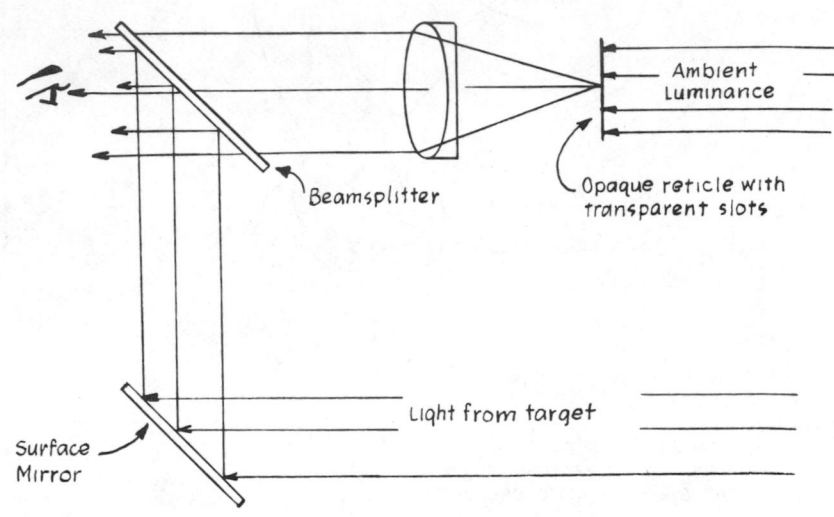

REFLECTED MODE

Schematic of a reflex-collimating sight

[14]N. W. Doss and R. R. Kramer, *Experiment for the Selection of Reflex-collimating Sight Components*, Technical Note 4-76, U.S. Army Human Engineering Laboratory, Aberdeen Proving Ground, Md., August 1976.

## GUIDELINES FOR SELECTING A TONAL VERSUS A SPEECH MODE

### 1. Use a Tonal Signal:

When immediate action is required on the part of the listeners, i.e., when vocal explanations or directions are not necessary for the listeners to know what the signal means and what they should do

When a specific point in time (that has no absolute value) is to be indicated, e.g., when the sound of a gong tells the listeners that something has happened or is about to happen, that they should be prepared for a message, etc.

When a spoken message would compromise the security of a situation, i.e., when a coded tonal signal would be unrecognizable to persons not privy to the code

When noise conditions are unfavorable for receiving spoken messages

When speech channels are overloaded

When a spoken message could annoy listeners for whom it is not intended or when the spoken message could mask other messages

When the intended listeners are familiar with the tonal signal implication or the tonal signal code

When it is desired to use the simplest audio signal

### 2. Use a Spoken Message:

When more message flexibility is needed than a tonal signal can convey

When it is necessary to identify the source of the information

When listeners have not had training in a special tonal signal code

When there is a need for rapid two-way exchanges of information

When the intended information deals with a future time and when preparation is required, e.g., during a countdown preparatory to initiating some operation

When use of a tonal signal countdown could result in a miscount

When operational stress surrounding the intended listeners could cause them to forget the meaning of a tonal signal code

## AUDITORY DISPLAY SELECTION GUIDE

| Display Configuration | Use Criteria | Human Engineering Considerations |
|---|---|---|
| Steady continuous tone<br>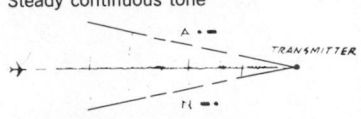 | Use in both ears for tracking, i.e., both ears equally loud for "on target," shifting left or right when "off target." | Select the frequency which is least subject to other nonverbal auditory signals or typical vehicle system noises. |
| Prerecorded spoken message<br>*Now hear this!*<br>*"this is your captain speaking"*<br>*calling Doctor Brown . . . come to*<br>*emergency!* | Use when communication flexibility is necessary.<br>Use when it is necessary to identify the source of a message.<br>Use when a stressful situation might cause the listener to forget the meaning of a coded signal.<br>Use when a simple coded signal cannot adequately give directions or instructions to the listener.<br>Use when ambient masking noise characteristics obviate the use of simple tonal signals.<br>Use when other complex tonal signal possibilities have already been exhausted, i.e., have been assigned and cannot be duplicated. | Utilize the most concise message possible that still retains maximum intelligibility and minimum ambiguity. |
| Intermittent tone | Use for "start and stop" timing.<br>Use for continuous information where rate of change of input is low.<br><br>Use for irregularly occurring signals, i.e., alarms.<br><br>Use as indicator of speed, i.e., a change in rate of signal occurrence. (Faster signal repetition indicates an increase in vehicle speed.)<br>Use for continuous-wave message transmission, i.e., Morse code. | Avoid a "train" of signals which might be confused with similar electrical system noise. |
| Warble and undulating tones<br>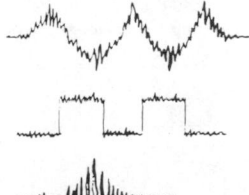 | Use when masking noise characteristics are unknown or cover a broad frequency spectrum in random pattern.<br><br>Use pitch differences to represent "up and down" relationships.<br><br>Use for locating a null position such as in orienting a radio beam. | Avoid a signal that sounds like any characteristic navigation signal or radio signal such as might occur when two carriers present a "beat frequency" effect. Where distraction is a critical factor, abrupt onset of the audio signal should be avoided, as well as apparent movement of the signal in space. |
| Bell, buzzer, horn, siren, and whistle | Use bell for fire alarm.<br>Use horn for emergency warning, equipment malfunction, pressure leak, hatch open, etc.<br>Use whistle (typical of "Bosun's whistle") for "All Hands Alert" (followed by further communication).<br>Use buzzer for individual operator alert to receive further communication.<br>Use Claxon for "Report to Duty Stations." | The signal should be "distinctive" and unlikely to be obscured by other noises.<br>Where crew members' attention may be highly concentrated on a task, use signals with relatively high alerting capacity.<br>Signals should be at least 10 dB above the ambient noise level. They should be repeated so as not to be missed.<br>Caution signals should be provided with reset and volume controls.<br>Consider a multistage warning or alarm system, i.e., an initial tone signal followed by the critical signal, followed by speech where appropriate, to indicate the specific nature of the hazard.<br>Alarm signals should focus attention on the first 0.5 s, with all essential information within the first 2.5 s. Use "sudden" onset.<br>Concentrate signal energy between 250 and 2500 Hz, with signal readily identifiable on the basis of components below 2000 Hz. |
| Microphones<br> | Provide microphone installation in all pressure helmets.<br>Provide microphones for each operator position for shirt-sleeve operation. Microphones should be head-mounted for all situations where the operator will have both hands occupied. | Provide maximum convenience and minimum discomfort. |

**AUDITORY DISPLAY SELECTION GUIDE** *(continued)*

| Display Configuration | Use Criteria | Human Engineering Considerations |
|---|---|---|

Telephone handsets

Use at operator stations where the operator does not perform communications as a primary activity and can pick up the handset with essentially no interference with his or her regular duties.

Use in situations where it is desirable to reduce the fatigue caused by wearing a headset.

Provide earphones in all pressure helmets.

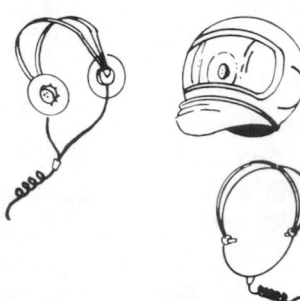

Earphones

Provide two portable head-mounted earphones for each operator position for shirt-sleeve operation.

The comfort and convenience of the operator as well as audio quality must be considered so that the operator will use the equipment as designed.

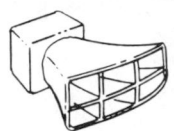

Loudspeakers

Use loudspeakers only when it is desirable to communicate with more than one person at the same time and when it is anticipated that one or more of the parties concerned will not be near a phone or will be wearing headsets or a space helmet (i.e., a public address system).

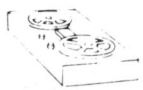

Audio recorders and playbacks

Use wherever it is desirable to preserve an audio communication for:

1. A record of normal operating messages
2. A record of emergency message transmission for evaluation of a mission failure or catastrophe

Use for presenting a prerecorded message (verbal alarm, procedural instruction, etc.).

Use as an on-board test, evaluation, or research analysis aid where the operator's activity precludes making a manual record and/or to capture critical messages that may be used to analyze events prior to an accident (e.g., an on-board pilot–air controller message recorder).

All units should be packaged for convenience of use and servicing.

Units should be packaged and located for maximum protection from destruction during the accident.

## TYPE OF SYSTEM IN RELATION TO APPLICATION

Although the choice of system may be obvious in most cases, there are specific operational reasons for selecting one type of communications system over another. The following guidelines are intended to help the planner or designer decide which type of system to use for a particular operational environment:

Public address system: Use for nonspecific emergency announcements for all diversely located listeners. Do not use for routine messages intended for one individual if the announcement may interfere with localized communication.

Intercom system: Use for small group-specific, two-way communication and/or for individual communication between people who are physically isolated from one another, where

telephone dialing would create a considerable waste of time (e.g., for communications between a shop supervisor and personnel at factory work stations). Do not use where conversational privacy is necessary.

Telephone: Use for situations requiring individual conversational privacy.

Radio: Use for situations where there are long distances between the conversants and/or when one or more of the conversants are moving from one location to another, (e.g., a security guard on foot or a police officer in a patrol car).

Closed-circuit TV: Use where face-to-face visual reference and/or reference to the object under discussion adds considerably to communication reliability.

*Type of Communicating Interface*

Handset (telephone): Use when at least one hand can always be free to hold the handset.

Headset (earphone or earphones and microphone): Use when both hands are often occupied.

Wireless microphone: Use when a fixed microphone or hand-held microphone with cable would restrict the user's freedom to move about.

## GENERAL PRINCIPLES FOR AUDITORY DISPLAY DESIGN

**SITUATION COMPATIBILITY** The effectiveness of any auditory display depends on proper consideration of the sound environment within which the display must operate; i.e., a spoken message is easily obscured by other spoken messages, a particular tonal signal is easily masked by other similar tonal signals, and any auditory signal or message may be masked by frequent, extremely loud bursts of noise such as would occur during a battlefield bombardment or the noise of a jet aircraft afterburner.

**COMMON SIGNAL CONVENTIONS** Certain types of signals (sirens, bells, claxons, etc.) are recognized and associated with certain activities, such as those of fire fighters and the police. Such characteristic signals should not be used for other purposes when the situation is such that the more common convention is in use. Other natural relationships, such as high frequencies being associated with "up" and low frequencies being associated with "down," should be taken into account when considering the use of pitch as an auditory code.

**STANDARDIZATION** Once a particular auditory signal code is established for a given op-

erating situation, the same signal should not be designated for some other display.

**LISTENER SENSITIVITY** The signal frequency of auditory displays should be compatible with the midrange of the ear's response curve for both pitch and loudness; i.e., avoid the use of signals at the extreme ends of the sensitivity curves, where response reliability is more easily masked.

**DUAL MODE** When the noise environment is unknown or suspected of being difficult to penetrate, consider one or both of the following: (1) using a shifting frequency signal that passes through the entire noise spectrum and (2) combining the auditory signal with a visual signal.

**QUANTIFIABLE SIGNALS** When the tonal signal information is to be quantifiable, provide

a reference tone, e.g., a base-line loudness or pitch against which the primary signal can be compared.

**PRIVACY** If a signal must occur in an area in which only certain personnel should be privy to its purpose and others are not to be unduly annoyed, select a simple bell tone that can be recognized among ambient speech sounds without being loud.

**SELECT LISTENERS** Consider a simple repetition code signal to indicate who is to respond, e.g., to answer a phone or to respond to a previously established set of task events.

**ANNOYANCE** Avoid the use of extremely loud signals that may startle listeners, and add to the overall noise levels, or interfere with local speech activity. When the ambient noise level gets too high, use earphones.

## AUDITORY DISPLAY FOR WARNING PURPOSES: GENERAL OBJECTIVES FOR USE AND DESIGN

Auditory warning displays are particularly good because they do not require the intended listener to be facing the signal source, and, with the exception of impossibly loud noise environments, they can be designed to cope with most acoustic interferences. In addition, because the audio channel of the operator is typically last to be affected by such stresses as dynamic oscillation of the listener, vibration, and high altitude, audio signals can be used under these stress conditions more successfully than visual warning devices.

| Operational Objective | Guidelines |
|---|---|
| Audibility in high-noise: level environment | The signal energy should be concentrated within a narrow frequency band that is lowest in the background noise spectrum. A signal-to-noise ratio of 5:1 should be maintained if at all possible. The audio signal power at the listener's ear should be approximately 300 mW. |
| | If the ambient noise is such that the warning signal would have to reach intensity levels that could be painful or injurious, provide the listener with a headset or a combination headset within an acoustic helmet. |
| | If the nature of the ambient noise is such that it is loud across most of the audible frequency range, consider the use of a repeating signal (with shutoff control) and provide a noise operated gain control to maintain a compatible signal-to-noise ratio as the background noise changes. |
| | Use a noise-canceling microphone so that background noise will not be picked up along with the warning signal or the speaker's voice. Consider delaying the signal to one of the listener's ears (by 0.01 s) or reversing the phase of one earphone of the headset. |
| Fast operator response | Make the signal brief and concentrate the signal energy at the beginning. If a canned voice signal is used, provide a short alerting signal (0.5 s) just before the speech message to prepare the listener. |
| Operator distraction | Provide the operator with a temporary "silencing control" so that he or she can prevent the warning signal from interfering with a particularly critical ongoing task. The shutoff control must, however, have an automatic reset feature so that the warning system is not permanently disabled. Do not provide the operator with a volume control to quiet an interfering warning signal; the operator will generally turn it down to the point where he or she may never be alerted to an emergency. |
| Interference of warning signal with other important signals | Deliver the warning signal to one earphone via a headset and the other signals to the other earphone. Alternate the warning and general signal input positions every 2–3 s. Avoid frequency bands normally used by common communication or navigation signals. |
| Reduction of signal confusion | Avoid selecting warning signals whose particular characteristics are easily confused with common signals such as navigation signals, Morse code, sonar returns, radio or electronic countermeasure signals, and typical electrical interference. |
| Transmission mode | Where practicable, send warning signals via a different system from those being used for speech communications. |

## NONVERBAL AND ALARM SIGNALS

### 1. Two-Phase Signals
Attention should be focused on the first 0.5 s of the alerting signal and then upon the first 2.0 s of the identifying or action signal. Alerting signals should precede action signals by a time interval of 2.0 to 8.0 s for isolated signals and of 0.3 to 2.0 s for action signals occurring in sequence. When the alerting signal consists of a "glissando," the audible signal should spend at least 0.1 s in each octave band from lowest to highest.

Where masking is a critical factor, warning signals should be concentrated in frequency bands that are unused or little used by primary communication (command, navigation, advisory, etc.). Major concentration of energy should be between 250 and 2500 Hz/s and the signal should be readily identifiable on the basis of sound components below 2000 Hz. Where possible, avoid the following:

a. Trains of impulses that resemble electrical interference
b. Simple glissandi which might be confused with sounds made by carriers, i.e., beat frequency and oscillator effects
c. Periodic impulses similar to radar signals
d. Random noise similar to that generated by such equipment as an air conditioner
e. Musical sounds
f. Noise similar to static or sporadic radio signals
g. Modulated or interrupted tones resembling navigation signals or coded radio transmissions

Where warning signals delivered to headsets might mask other essential auditory signals (e.g., navigation and communication signals), separate channels may be provided to direct the warning signal to one ear and the background signal to the other ear. Presentation of warning signals may also be dichotic and alternating with the dichotic presentation of communication and navigation signals.

Where earphones cover both ears of the operator during normal system operation, the warning signal should be directed to the operator's headset as well as externally.

Audio warning signals should be at least 10 dB above the loudest expected ambient noise level. Such signals should not exceed a maximum intensity of 110 dB in areas (near the sound source) which must occasionally be occupied by unprotected personnel for as long as 10 min duration. The signal energy should be concentrated in the frequency band in which background noise is lowest.

### 2. Single-Phase Signals
All essential information should be contained in the first 0.5 s of the signal. Considerations regarding masking and signal strength are the same as for the two-phase signal.

### 3. False Alarms
Audio warning systems should be designed in such a way that any possibility of false alarms is minimized.

### 4. System Independence
The circuitry of audio warning systems should be independent of the system in which they are to provide warning information so as to preclude warning system failure in the event of primary system failure.

### 5. Operator Control
All audio warning systems or devices should be provided with a test button, switch, or other means of verifying the audio aspects of the signal at any time. Whether audio warning signals are designed to terminate automatically, by manual control, or both, an automatic "reset function" should be provided. The reset function should be controlled by a sensing mechanism which should recycle the signal system to a specified condition as a function of time or the state of the signaling system. The signal duration should be at least 0.5 s and should continue until an appropriate response is made by the operator. The completion of the proper corrective action should automatically terminate the warning signal. A shutoff switch should be provided that is controllable by the operator, the sensing mechanism, or both, depending on careful consideration of the operational situation and personnel safety factors. (Caution signals should be provided with manual reset and volume controls.) Shutoff switches should not prevent the presentation of other warning signals and should reset automatically after an interval of time.

### 6. Steady, Continuous Tonal Signals
Signal frequency should be confined between 400 and 1500 Hz.

The signal should exceed its masked threshold by at least 15 dB. Do not exceed a sound power level of 135 dB for intermittent operation (120 dB for extended periods). Set the signal level 60 dB or more above the absolute threshold.

For listening in noise, select a signal midway between the masked threshold and about 110 dB.

Provide automatic volume control when the dynamic range is 20 to 30 dB or less.

### 7. Intermittent Tonal Signals
Signal frequency should be confined between 400 and 1500 Hz.

The signal should exceed its masked threshold by at least 15 dB. Do not exceed a sound power level of 135 dB.

Set the signal level 60 dB or more above the absolute threshold. Select a signal midway between the masked threshold (noise) and about 110 dB.

Provide automatic volume control when the dynamic range is 20 to 30 dB or less.

### 8. Warble or Undulating Tonal Signals
The signal frequency should be confined between 500 and 1000 Hz. Use a pitch rise and fall rate of about 1 to 3 Hz. Where an alerting signal consists of a glissando, the audible signal should remain at least 0.1 s in each octave band from the lowest to the highest frequency.

Where the operator's task is to detect frequency change, set the signal at least 30 dB above the absolute threshold.

## VARIOUS SIGNAL DEVICES AND CHARACTERISTICS

Buzzers: Usually low-intensity ranges, low frequency (150 to 400 Hz). Generally suitable for relatively quiet situations where the distinctive sound commands attention but does not cause "alarm."

Bells: Normally higher intensity and frequency than buzzers. Suitable for situations where the ambient noise is higher (although they can also be used at lower intensity in fairly quiet environments such as a bank or department store, where they provide communication to operating personnel without disturbing customers).

Horns: Usually high intensity (90 to 100 dB) and middle to lower frequency. However, horns can be used with higher intensity and higher frequency up to 3000 to 4000 Hz. Because they generally are of longer duration, they are not suitable for quiet environments, but rather for high-noise-level situations where urgent emergency warning is desired.

Chimes: Similar to quiet bells, typically in the frequency range of 500 to 1000 Hz. Suitable primarily for situations where the resident personnel require signaling without disturbing customers or visitors. Excellent for hospitals and similar operating situations.

Sirens: Can be made to penetrate almost any noise environment. The combination of high intensity and constantly changing frequency is not easily masked by spurious or continuous ambient noise conditions. Primarily for emergency signaling by police, fire fighters, etc.

**Characteristic Intensity Ranges and Predominant Frequencies for Specific Signaling Devices**

| Signal | Average Intensity At 10 ft | Average Intensity At 3 ft | Predominant Audible Frequency |
|---|---|---|---|
| *For Large Areas (High Intensity)* | | | |
| 4-in bell | 65–77 | 75–83 | 1000 |
| 6-in bell | 74–83 | 84–94 | 600 |
| 10-in bell | 85–90 | 95–100 | 300 |
| Horn | 90–100 | 100–110 | 5000 |
| Siren | 100–110 | 110–121 | 7000 |
| *For Small Areas (Low Intensity)* | | | |
| Heavy-duty buzzer | 50–60 | 70 | 200 |
| Light-duty buzzer | 60–70 | 70–80 | 400–1000 |
| 1-in bell | 60 | 70 | 1100 |
| 2-in bell | 62 | 72 | 1000 |
| 3-in bell | 63 | 73 | 650 |
| Chime | 69 | 78 | 500–1000 |

## VERBAL (SPEECH) SIGNALS

### 1. Word Selection

In selecting words for use in auditory warning signals, priority should be given to aptness, conciseness, and intelligibility (in that order). When a set of standard words are chosen, i.e., a phonetic alphabet (Able = A, Baker = B, Charlie = C, etc.), select words of two or more syllables rather than one-syllable words and test to make sure that no two words in the set sound similar.

### 2. Talker Characteristics

Voice communication systems should be designed to make maximum use of features that will preserve and enhance the intelligibility of speech produced by the "average talker" with a generalized "American accent" (for the United States), average intensity level (approximately 70 to 75 dB sound pressure level [SPL]), and average speech rate. When a message is to be conveyed in another language, select a talker whose basic language training and experience are in the particular language; e.g., extreme care should be taken in selecting an English-speaking translator to provide messages in German, Spanish, or French to make sure that his or her pronunciations will not be misconstrued by the listeners because of improper emphasis, colloquial discrepancies, etc.

### 3. Presentation

Verbal signals that occur frequently (e.g., status and caution) should be presented in a fairly formal manner, as opposed to an impersonal one. Although emphasis is important in circumstances where a sense of urgency must be conveyed, care should be taken to train the talker not to convey emotional stress or anxiety.

Less frequent and more urgent signals may be less formal and delivered in a more personal manner to instill confidence.

### 4. Signal Characteristics

Speech signals should fall within the range of 200 to 6100 Hz.

The audio signal power should be approximately 300 mW at the listener's ear.

The signal-to-noise ratio should be at least 5:1.

Provide distortion-free signals and minimize masking effects that may occur as a result of nonuniform frequency response or a lack of uniform amplification.

Maintain intelligibility nearly equal to that available from a given system at sea level by using a pressure-sensitive device to adjust the gain of the amplifier (as pressure changes, e.g., at higher altitudes).

Provide microphones and earphones that have been designed to give uniform frequency response at all altitudes.

Verbal warning signals should be "processed" only when necessary to increase or preserve intelligibility, i.e., by increasing the strength of consonant sounds relative to vowel strength. Where speech signals must be relatively intense because of high ambient noise, use "peak clipping" to protect the listener against auditory overload.

Where speech communication signals are presented dichotically via earphones, one of the following methods should be used for the recorded warning signals:

a. Dichotic presentation (individual input to each ear)

b. Distinctive, nonmasking, nonspeech component to identify the speech as being part of the warning signal, i.e., superimposed on a speech component

c. A very distinctive voice for recording warning signals, e.g., a female voice rather than the typical male voice.

Where confusion might exist, avoid scrambled speech effects that might be confused with "monkey chatter" from adjacent channels.

### 5. Verbal Warning Devices

a. Initial alerting signal (nonspeech) to attract the listener's attention and to designate the general problem

b. A brief, standardized speech signal (message) which identifies the specific condition and suggests the appropriate action that should follow

Verbal alarms for critical functions should be at least 20 dB above the speech interference level (SIL) at the operating position of the intended listener. SIL describes the effectiveness of noise in masking speech. It is the average (in decibels) of the sound levels of masking noise in the 600- to 1200-, 1200- to 2400-, and 2400- to 4800-octave bands.

For most applications where "killer" warning messages are required, use a mature male voice; although female voices often provide a unique variation to attract attention, there is some evidence that certain listener groups, such as pilots, sometimes lack confidence in the training and experience of the female operator.

The warning message should always be presented in a calm manner, with the intent of eliciting a rational reaction from the listener.

## TELEPHONE HANDSET PACKAGING AND DESIGN

### Packaging

The key factors in packaging the telephone handset are the following:

1. The earphone and mouthpiece should be accommodated to the user's head so that regardless of the size of the head or its unique shape, the user can hold the earphone close to the ear, and the mouthpiece close to the lips (without breaking contact with the ear). Ideally, the axis of the speech projection should coincide with the receiver axis, as shown in the accompanying sketch. Because of differences in head size, this is not possible without an adjustable handle. An acceptable compromise is shown in the accompanying specification.

2. The handle should fit the hand comfortably. The accompanying sketch indicates the maximum acceptable dimensions. Avoid square cross-sectional shapes and edges.

3. When a "talk switch" is provided on the handset, place it as shown in the accompanying sketch so that the handset can be used with either hand.

4. Use coiled or retractable cords with handsets to minimize cord tangles. Handset cords that are used in fixed operator stations are easily knocked on the floor. Unless there is a need for a longer cord, shorten the cord so that it prevents the handset from ever striking the floor and causing damage to the handset.

5. The total weight of the handset should not exceed 10 to 11 oz (0.28 to 0.31 kg).

6. Where normal communicating requirements necessitate the frequent use of more than one handset, locate the various handsets for convenience according to use priority. Consider the following potential user problems:

   a. Cord interference; i.e., the cord of the phone in use should not drag across another phone and knock it off its hook or base.

   b. Nominally locate the phone for left-hand use so that the user can use his or her right hand for writing (although this inconveniences left-handed operators, they are in the minority). Alternatively, provide a flexible configuration that allows users to arrange their phones for their own convenience.

   c. Phones used in vehicles should be provided with "securing cradles" and/or other devices that will prevent the movement of the vehicle from unseating the handset.

7. Handset bases should be designed so that they are not easily displaced, especially if the base has a dialing device. Consider the use of rubber feet, suction cups, and/or other special holding devices.

8. In-handle channel and line selection concepts may be helpful for some applications as long as the following do not occur:

   a. The dial or push buttons cause the handle to be awkward to hold onto.

   b. The dial or push buttons are too small for accurate finger manipulation, or the push buttons are placed too close together.

   c. The handset becomes too heavy.

9. Except in special cases, any color should be acceptable for the handset package. A given color can be used to identify a special channel, i.e., a "hot line."

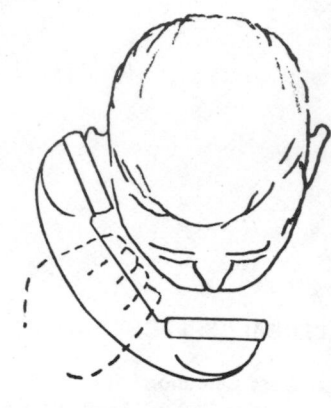

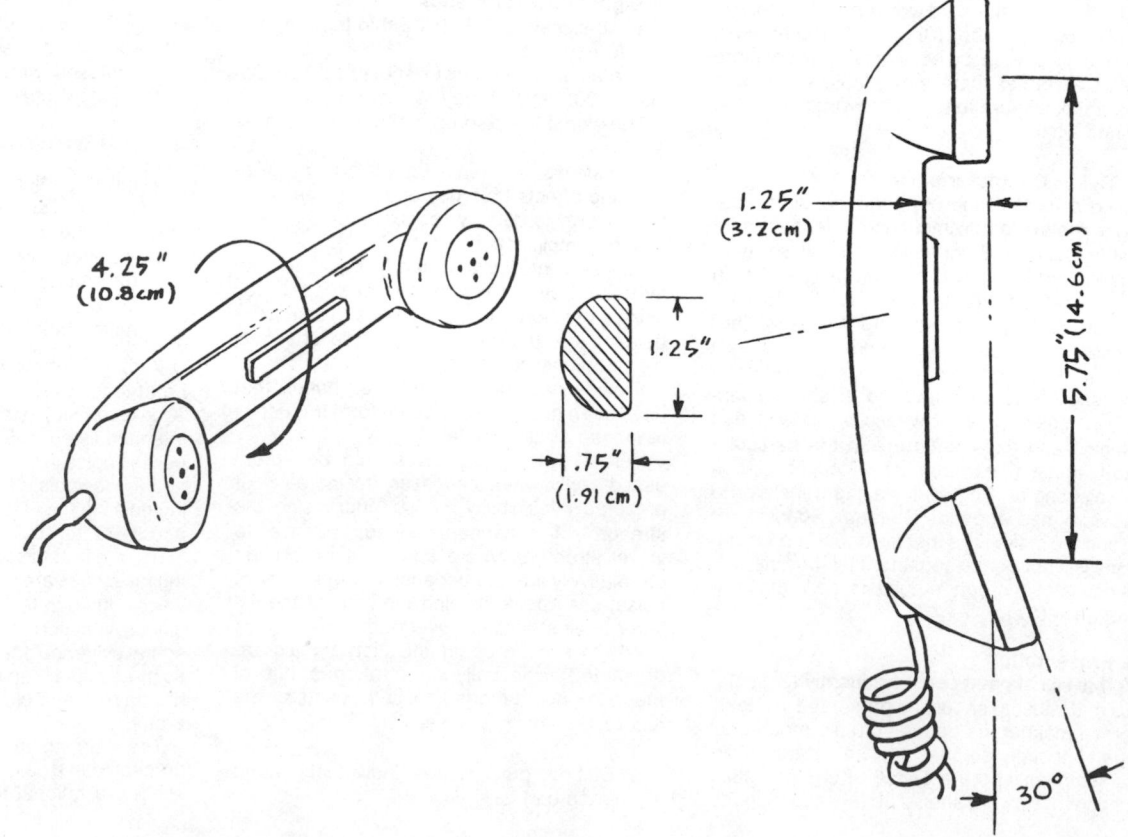

**Special Considerations**

A specially designed handset that will slide easily beneath the soldier's helmet

A console-mounted phone to keep the desk clear (or because the console has no horizontal surface) and a phone that is inset to keep it from being accidentally knocked from its cradle

A conference phone supplement so that several individuals can participate in the same telephone conversation

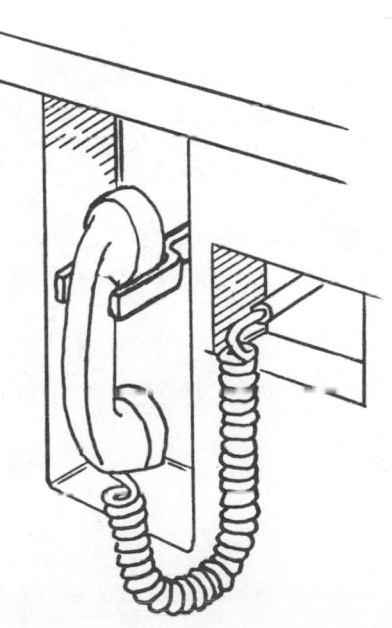

*Conference speaker*

## WALL-MOUNTED PUBLIC AND PRIVATE TELEPHONES

The height of the wall-mounted telephone should be determined by the convenience of the critical use elements, i.e., the handset, the dial or push-button pad, a fixed mouthpiece, a coin slot, etc.

For adult users, the phone package typically should be located between waist and eye height. Generally this would place the mid-point of the phone box about 5 ft (1.5 m) above the ground or floor.

For public places frequented by wheelchair users, at least one phone should be placed at a lower level so that it can be operated easily by the person sitting in a wheelchair. The highest element to be used by the person in a wheelchair (dial, coin slot, etc.) should not be more than about 4 ft (1.2 m) above ground or floor level.

From a human engineering point of view, dials are less preferred than push-button selector types, either for the typical desk-mounted phone or for other types, including wall-mounted phones, handset-mounted phones, etc.

Use a push-button phone rather than dial phone to minimize errors and difficulties for the orthopedically handicapped.

Provide coin insert slots that permit the coin to be inserted vertically rather than laid into a flat, disk opening. This allows the user to maintain a more secure grasp of the coin.

Mount the phone so that there is sufficient space for a wheelchair user to approach parallel to the front of the phone box.

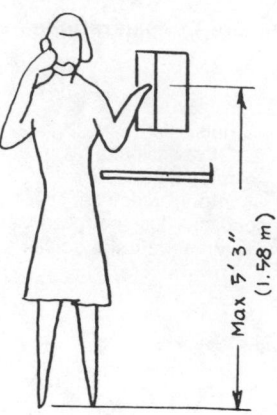

Max 5' 3" (1.58 m)

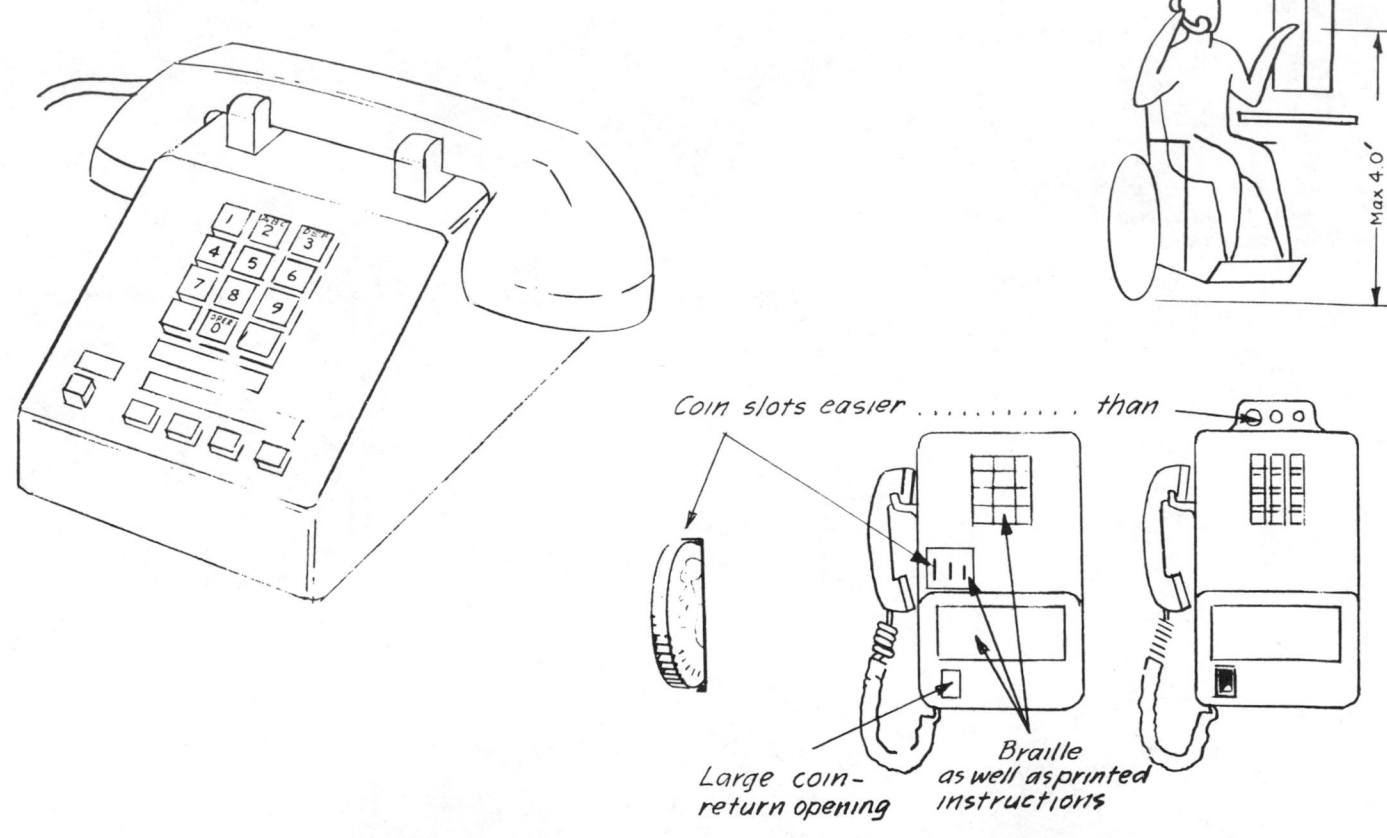

Max 4.0' (1.2 m)

Coin slots easier .......... than

Large coin-return opening

Braille as well as printed instructions

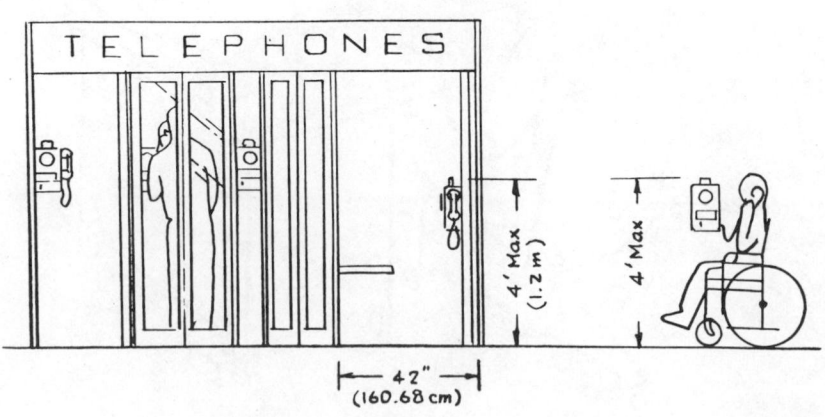

TELEPHONES

4' Max (1.2 m)

4' Max

42" (160.68 cm)

## Operating Characteristics

The microphone (mouthpiece) should be:

1. Highly sensitive to acoustic speech signals.
2. Of sufficiently high quality to provide faithful transduction of the speech signal into an electrical signal.
3. Capable of rejecting other acoustic signals and noise.
4. Designed for "close talking." Do not use large pressure-gradient velocity-type microphones; instead, use noise-canceling types in the low-frequency sound range (100 dB overall SPL). Avoid "Ribbon Mikes" for close talking.
5. Designed so that the talker has to talk into the axis of the microphone, so that the talker's lips naturally fall close to the microphone, and so that the operator's hand never covers the microphone holes inadvertently.
6. Designed to have a smooth frequency response between 200 and 6100 Hz.
7. Designed to provide a dynamic range (when working with the selected amplifier) great enough to admit a minimum of a 50-dB variation in signal input. A close-talking microphone should not overload with signals as high as 130 dB.
8. Equipped with a noise shield for applications where the ambient noise level is extremely high (above 100 dB). The shield volume should be at least 250 cm³, but should not be so large as to be unwieldy for the operator to handle and use. The shield should fit tightly against the operator's face with normal hand pressure or head-strap tension; i.e., the operator should not have to press the shield with considerable force in order to get it to fit. Provide exhalation holes in the shield, but place these as far from the actual microphone as practicable. Sound-absorbing material should be enclosed in acoustically transparent, waterproof material. The shield (if worn) should not impede the talker's mouth or jaw motion, thus causing speech distortion.

those of the remainder of the system.

2. Provide a dynamic range, without appreciable distortion (at least 40 dB), capable of power-handling capacity for peak amplifier output.
3. Provide a combination earphone and socket (or cushion) for which the earphone sensitivity and the earcup attenuation together will provide an adequate signal-to-noise ratio of 5:1.
4. Provide an earphone cushion which is comfortable for long-duration use and which is not prone to developing fungus or collecting moisture.
5. Connect the earphones to operate out of phase.
6. Delay the signal reaching one ear about 500 μs.
7. Use binaural rather than monaural headsets if listeners will be in an intense noise environment.

8. Provide a sidetone (delay of 0.05 s) to make the talker increase his or her voice level of effort.
9. Provide sufficient electrical power to drive the peak sound pressure level to 131 dB when using two earphones.
10. Provide a gain control with a dynamic range sufficient to make the signal at least 15 dB more intense than the ambient noise.
11. Provide a reduction in frequency range below 500 Hz and above 4000 Hz if it results in an increase in the average power of the audio signal.
12. Provide a uniform frequency response of receiver and headset between 300 and 4000 Hz to avoid unpredictable distortions.
13. Provide a pressure-operated gain control switch to compensate for the effects of altitude in aircraft cabins.

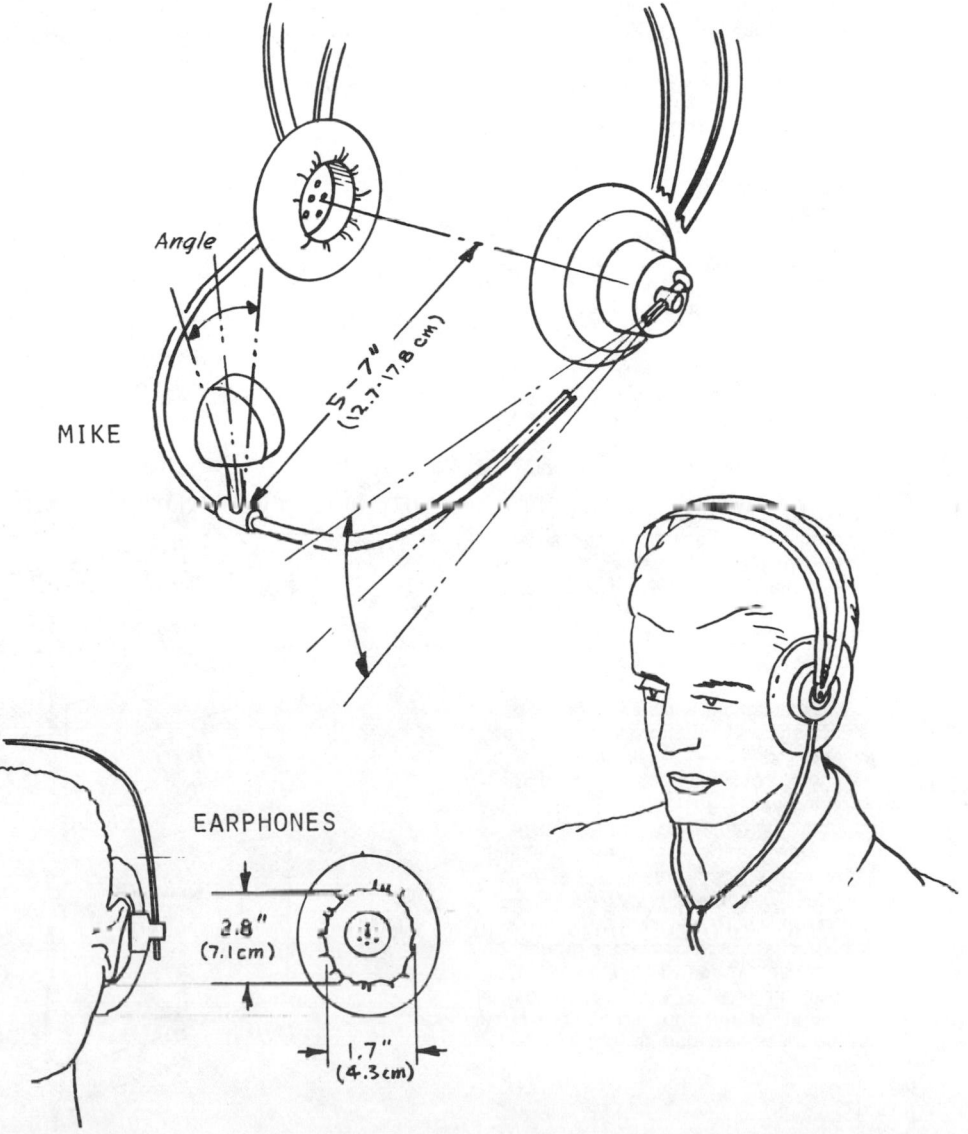

## PACKAGING AND DESIGN OF HEAD-MOUNTED MICROPHONES AND EARPHONES

### Packaging

Provide complete adjustability of headset components.

There should be lightweight, minimum tension to hold the earphones snugly against the head

Provide a soft cushion which is large enough so that it does not press on the user's ear but which seals around the ear to minimize external noise interference.

### Operating Characteristics

1 Provide smooth frequency response characteristics which are as broad as

## LOUDSPEAKERS

Loudspeakers should be subject to the same frequency response characteristics as other elements of the communication system, except that loudspeakers for use in multispeaker installations (i.e., where several speech channels are to be monitored simultaneously) should respond uniformly over the range of 100 to 8000 Hz.

### Use Considerations

1. Where an individual loudspeaker is provided a single operator, locate the speaker as close as practicable to the operator (within 3 ft, 0.9 m) so that the operator can reduce the signal of his or her particular speaker and thus refrain from adding to the overall noise level of an operating area.
2. Locate speakers and microphones so as to minimize speaker feedback into the microphone.
3. Where several channels are to be monitored simultaneously by means of loudspeakers, separate the speakers by at least 10° with respect to the principal listener so that he or she can localize the separate speaker sounds. If additional channel differentiation becomes necessary (i.e., if the 10° does not provide clear definition), utilize low-pass filtering ($f_c$ = 1800 Hz) to signals fed to loudspeakers on one side of the operator, high-pass filtering ($f_c$ = 2000 Hz) to signals fed to loudspeakers on the opposite side of the operator, and unfiltered signals to loudspeakers in front of, or behind, the operator.
4. Position speakers approximately at the head (or ear) height of listeners, except in auditoriums, where the speaker should be at podium height or higher, and in multilevel work areas, where it may be important to recognize that a sound emanates from a position below the operator.

## AUDIO RECORDER AND PLAYBACK PACKAGING AND DESIGN

### Packaging

Portable recorders should obviously be small and light in weight. This is particularly important for the typical "pocket recorders" that business people and engineers find so useful. Although no studies have defined the size, shape, and weight limits for such devices, the general specifications shown in the accompanying illustration are suggested as a design objective for packaging a hand-held recorder that:

1. Fits a shirt pocket
2. Lies easily in the hand
3. Does not weigh so much that it is uncomfortable in the pocket or hand
4. Is simple to operate because of the manner in which the controls are laid out and labeled

*Note:* The microphone should be positioned where it can be seen easily and positioned close to the talker's lips, and a pilot light should indicate when the machine is recording. A separate microphone jack should be located on the edge opposite to the one shown in the accompanying sketch.

Obviously, recording equipment is available in all sizes and shapes; variations are dictated by many things, including the size and shape of the internal and external components, usage parameters, and (in the case of home entertainment equipment) aesthetic objectives.

The two examples shown in the sketches at the left of page 569 (not representative of any particular brand of recorder) illustrate some of the packaging considerations for typical popular recording units.

The small "carry-around," cassette type of recorder shown in the accompanying sketch illustrates some good organizational objectives. For example, the operating controls are placed on top, where they can easily be seen when the package is placed on the ground, on a table, or on a car seat. The operating push buttons are arranged in a sequence that is easy to understand and remember. They could also be color-coded as follows:

STOP—red
PLAY—blue
RECORD—yellow

Note that the EJECT button is *not* located with the other operating controls.

The typical table-model recorder shown in the sketch on page 569 illustrates another important packaging principle: The controls are on a beveled front panel so that the operator can see them when sitting in a chair and when standing up to change tapes.

Although the above ideas may seem simple, inspection of a great many current recorder packages will provide evidence that designers often fail to think about the ease of operator interface.

### Field Portability

Field telephones, radios, and other electronic devices that have to be carried on a continuing basis should be packaged so that they can be operated easily while the operator is standing, sitting, or "on the run." They obviously should not be heavy; they should weigh no more than 15 lb (6.8 kg) and should be lighter if the soldier also has to carry a backpack. The shape should be compatible with the person's anatomical characteristics; that is, the package should not be wider than the operator's

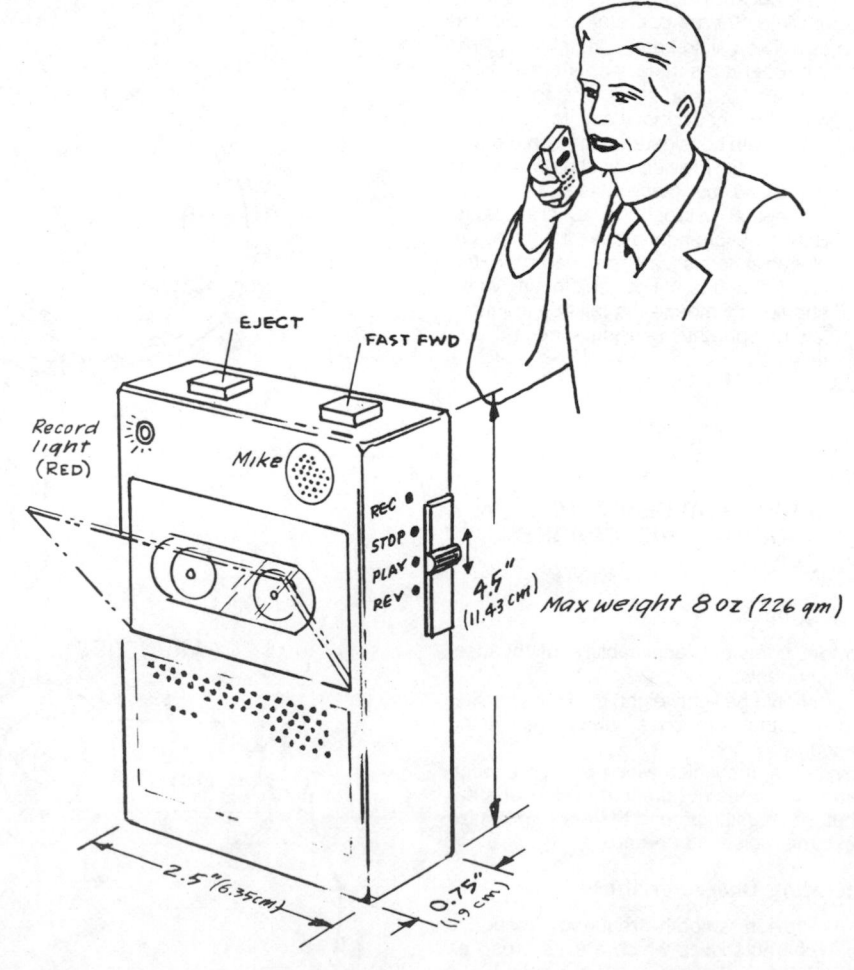

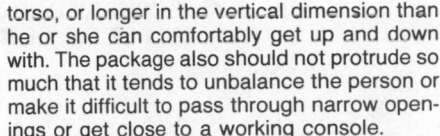

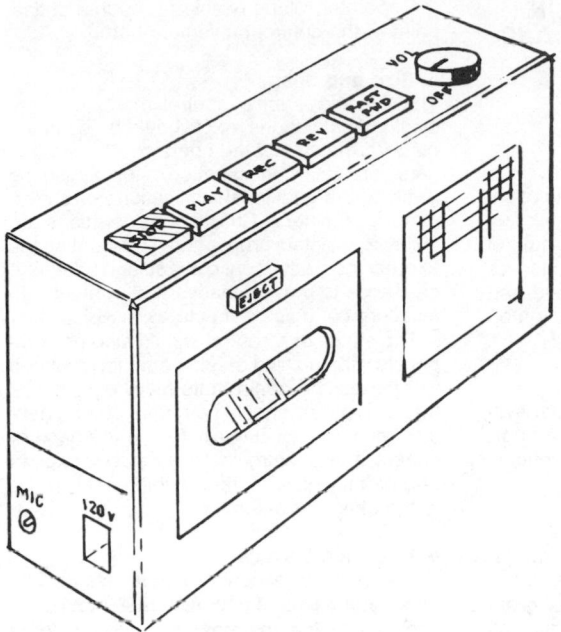

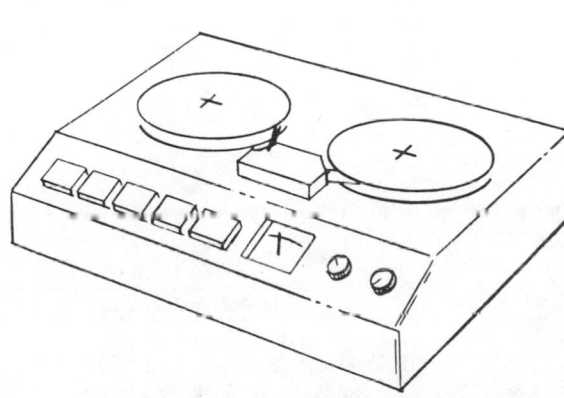

torso, or longer in the vertical dimension than he or she can comfortably get up and down with. The package also should not protrude so much that it tends to unbalance the person or make it difficult to pass through narrow openings or get close to a working console.

As illustrated in the accompanying sketches, the controls should be placed where the operator can see and reach them without having to search blindly or twist or reach behind his or her back to adjust them.

The package should be easy to don and doff. Do not use harness designs which require considerable fastening or unfastening or awkward adjustments to make the package fit into the right place on the body.

If for some reason the total weight and/or the size and shape of the equipment make it impossible to create a comfortable and convenient single package, divide it into two or more components. However, make sure that the controls that must be operated while the person is moving are in front.

Helmet-mounted, battery-operated radios have been created for unique operational conditions such as for use by aircraft carrier flight-deck personnel. Care should be taken to minimize the weight and bulk of such units because they can make the wearer "top-heavy," especially when wind currents are high. Controls for the radio should be attached to the helmet if at all possible so that in the event the helmet must suddenly be removed, the operator does not become entangled in wires between the helmet and a separate control box.

## Operating Characteristics

Audio recorders should be subject to the same frequency response restrictions as microphones, headsets, and loudspeakers (see the requirements in the sections on these topics). In addition, recorders should provide the following:

1. Negligible speed variation through the use of a stable, constant-speed drive system.
2. A convenient method for removing and replacing cassettes or tape spools. It is particularly important to provide graphics and labels on the equipment to "pictorialize" an especially confusing procedure for inserting the cassette right side up or the tape in the proper position.
3. Adequate audio monitoring and/or visual monitoring and control of intensity levels, battery condition, etc.
4. Where appropriate, a simple method for reversing the playback system and for locating a particular tape segment.
5. A method for manipulating the level of the auditory sidetone (while recording) to keep the speaker's signal level within a required range for satisfactory recording.
6. An automatic shutoff control to reduce tape drive wear.

## AMPLIFIER, TRANSMITTER, AND RECEIVER CHARACTERISTICS

Bandwidth: Provide an essentially "flat" audio frequency response between 200 and 6100 Hz.

Gain: Provide sufficient dynamic range and gain to handle the range of instantaneous pressures of the primary input objectives (speech, music, etc.) and to develop the necessary signal level at the headset or loudspeaker terminals.

Noise: Amplifier background noise should be less than that introduced by the microphone.

Control: Provide automatic (nonlinear) gain control (AGC) if noisy environments are anticipated. Talker sidetone should be taken from the system ahead of the AGC. Use an attack time of 0.1 and a release time of about 10 s in "limiter" amplifiers.

Frequency pre-emphasis: Speech system input devices should employ frequency pre-emphasis wherever practicable. This should consist of a positive-slope frequency characteristic no greater than 9 dB per octave, over the range 150 to 4800 Hz. Where transmission equipment utilizes pre-emphasis and where peak clipping is not used, reception equipment should employ frequency de-emphasis of characteristics that are complementary to those of the pre-emphasis feature. De-emphasis should be a negative-slope frequency response not greater than 9 dB per octave over the frequency range 150 to 4800 Hz.

System limitations: Where system engineering necessities require speech-transmission bandwidths narrower than 150 to 4800 Hz, the frequency range should be limited to the band 200 to 3000 Hz.

Clipping: Where speech signals are to be transmitted over channels showing less than 15 dB peak-speech-to-RMS-noise ratios, peak clipping of 12 to 20 dB should be employed at system input and should be preceded by frequency pre-emphasis as specified above, except that peak clipping should not be employed in single sideband voice-transmission systems.

## IMPORTANT CONSIDERATIONS IN CONTROL SELECTION, DESIGN, AND USE

### 1. Type of Control
The control should be chosen as though it were an extension of the operator's limb; i.e., it should be operable in terms of the natural motions of the arm, wrist, finger, leg, ankle, or foot, and it should not require awkward and unnatural positioning, extension, or motion on the part of the operator.

### 2. Feedback
The control interface and basic controller system should provide feedback so that the operator knows at all times what his or her input is accomplishing.

### 3. Resistance
There should be sufficient resistance to operator inputs to dampen spurious inputs, but not so much that the operator has to put great force into the control, that his or her muscles are quickly fatigued or that the operator has difficulty maintaining the nominal operating position.

### 4. Position of the Control
Controls should be placed where they do not require the operator to assume awkward body positions or make frequent long reaching movements. The position should reflect consideration of the excursion requirements of the control system so that there is no chance that the operator will be unable to reach a critical point in the control movement path.

### 5. Size and Shape
The size and shape of control interfaces (handles, knobs, buttons, etc.) should be compatible with the size of the operator's hands, fingers, or feet. Consideration should also be given to the additional implications of shoes, gloves, or mittens. Critical considerations are ability to maintain proper grip, sufficient space to preclude inadvertent contact, and adequate clearance to prevent inadvertent contact with adjacent controls or structures.

The shape of a control should also be compatible with the kind of grip or motion required to operate the control interface; e.g., knobs that are rotated should be round, oblong handles must not accidentally turn in the hand or fingers, and a control that is supposed to point at a certain scale mark or numeral should be in the shape of a pointer.

### 6. Interface Surface
The surface of a control handle should depend on the type of operation required; i.e., in some cases the operator may need a smooth surface in order to change positions on the handle or knob without having his or her fingers stick to the surface, while in other cases the operator may need serrations or knurling in order to apply a firm grip for maximum force.

### 7. One-Hand versus Two-Hand Operation
Two hands often provide more precision or force. However, two-handed controls should not be used if one hand is required to operate another control simultaneously with the other.

## HAND-OPERATED VERSUS FOOT-OPERATED CONTROLS

| Hand-Operated Controls | Foot-Operated Controls |
|---|---|
| Very precise inputs possible. | Limited precision capability. |
| Limited force capability. | Great force capability. |
| Because hand controls are usually located closer to the eyes, controls are easier to find and observe during the operation. | Foot controls usually are located far from the eyes and therefore are difficult to find and generally cannot be observed easily during operation. |
| Hand controls usually take up less space. | Foot controls usually require more space. |
| Wide range of motions available, e.g., push, pull, rotate, press, and squeeze. | Only a few motions are available, e.g., leg push and ankle-toe press. |
| Two hands can perform entirely different operations with relative ease. | Two feet generally can perform only similar operations (although one can press while the other is being withdrawn). |
| Can usually be operated as easily from a standing as from a sitting position. | Operator can use only one foot at a time while standing, and precision while standing is poor. |
| Operator can perform rapid, repetitive motions. | Repetitive inputs are difficult and very slow. |
| Operator can utilize different fingers to make extremely rapid and complex, successive inputs; also, complex combinations of fingers can be used at the same time. | Use of toes is impractical; thus the operation is limited to successive, alternate, or coincident two-foot operations. |

*Note:* The above are generalizations for the average person, i.e., obviously, there are exceptional people who may develop unusual pedal skills.

Although many types of special control systems can be used for unique cases such as musical instruments, one should normally consider manual or pedal controls that are typically expected by the average user.

## CONTROL SELECTION GUIDE

| Control Configuration | Use Criteria | Human Engineering Considerations |
|---|---|---|
| Hand-operated push buttons: Square push button 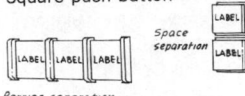 | Use for noncritical operations which start an action, select a channel, or turn on a piece of equipment. Use for "event stack" combination lighted push buttons. | Locate within easy reach of the operator, considering the constraints of pressurized garments, acceleration restraint systems, etc. |
| Round (guarded) push button  | Use for critical operations (such as for a "panic" button) to stop a "runaway" equipment operation, fire and emergency escape sequence, etc. | Locate so that inadvertent operation is impossible (i.e., so that the switch cannot be actuated by bumping it accidentally). |
| Round (extended) push button  | Use in conjunction with a primary hand controller for changing control mode, initiating communication, etc. Use for multibutton keyboard arrays and circuit-breaker panels. | Locate within easy reach of the operator's thumb, considering the constraints of a pressurized glove. Locate so that the switch will not be activated accidentally during normal manipulation of the parent controller. |
| Bar-type push button 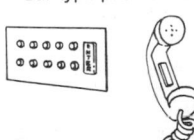 | Use for either finger or thumb operation of a communication handset. Use as an "entry" tab for inserting preset numbers into a computer from a keyboard. | Consider location within keyset configuration for optimum operational sequence. Consider position on the handset for convenient use by fingers or thumb for both left- and right-hand operation. |
| Toggle switch 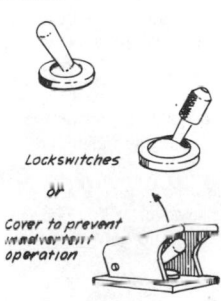 | Use a standard two-position toggle switch for START/STOP, ON/OFF, etc. Use a "momentary" (spring-return) toggle switch for check-reading an instrument or circuit or use a spring-return toggle switch for a "slewing" operation. Use a three-position toggle switch for combining alternatives of a single function such as AUTOMATIC, MANUAL, or OFF. Note: Do not use four-position switches unless absolutely necessary because of space limitations. | Consider particularly the hazards associated with accidental contact with the toggle switch, and also the difficulties of operating in a pressurized glove. |
| Rocker Switch  | Use in lieu of a toggle switch when the protruding toggle handle would be a considerable hazard, i.e., when a crew member could be injured by bumping or striking the handle or handle cover accidentally. Use an illuminated rocker switch in lieu of a toggle switch where illumination conditions make it impossible to see the toggle position, assuming this requirement to be critical. Use in lieu of a toggle switch when panel space is inadequate. | Special precautions should be taken in placement of rocker switches to obviate accidental operation. Minimize intermixing of toggle switches and rocker switches in the same panel area; i.e., use one or the other. |
| Rotary selector control (mechanical detents)  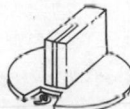 | Use (where it is desirable to select two or more—but no more than 24—channels, circuits, etc., with a single control) to save operator time and panel space which would be required by separate control switches. Note: Do not use for continuous frequency selection, audio intensity control, panel illumination control, etc. | Make the "pointing" end of the switch knob obviously different from the nonpointing end. Provide mechanical detents so that the operator can "feel" and "hear" when the control is properly positioned. The knob should be shaped like a bar or pointer to differentiate this type of control from continuous-function controls, which have round knobs. Provide an index mark on both the top and the end of the pointer or provide a cutout in the knob skirt that exposes the numerical setting. |

## CONTROL SELECTION GUIDE

| Control Configuration | Use Criteria | Human Engineering Considerations |
|---|---|---|
| Rotary adjustment control (continuous motion)  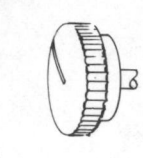 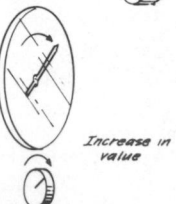 | Use where smooth, continuous adjustment of audio intensity, light intensity, random radio frequency selection, etc., is required. *Note:* Do not use on detent-type switches.<br><br>Use to provide continuous control of a related display pointer. | Use a round knob, which allows ease of finger positioning. Use suitable knob surfaces which provide necessary finger security but which will not interfere with manipulation dexterity. Use a larger-diameter knob when "fine" adjustment is required. Maintain proper direction-of-motion relationships. Place the control where the operator's hand will not cover up the display when he or she is adjusting the display. Use the proper CD-ratio to ensure ease and accuracy of display adjustment. |
| Rotary valve control 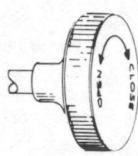 | Use to control flow of materials, fuel, air, liquid oxygen, etc. | Use a round handle configuration suitable for gripping by the hand as a whole, but with sufficient finger clearance for pressurized glove operation. Whenever possible, locate valve handles far enough from other control knobs (which are similar in shape) to avoid direction-of-motion confusion (i.e., accepted usage for valves is for a clockwise turn to provide a decrease in flow, whereas the same motion in other types of controls provides an increase in functional value). In addition to placement separation, emphasize clear, unambiguous marking and other appropriate code techniques to minimize the chance of direction-of-motion of operator error. |
| Thumb-wheel controls (continuous smooth motion) | Use for limited-range, continuous adjustment functions (panel light dimmer, audio volume, etc.). *Note:* Use only as a second choice to a normally mounted finger knob in the event some advantages may accrue in terms of saving panel space. | Coarse serrations should be provided on the edge of the knob. Sufficient exposure of the knob rim must be provided for convenient manipulation with pressurized gloves, but guarded against accidental activation. |
| Thumb-wheel controls (discrete stepping motion) 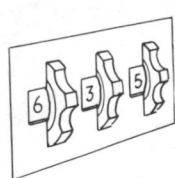 | Use in conjunction with detented counter-type selector control and display combinations. | The edge of the thumb wheel should provide protrusions, or "teats," which coincide with the position of the numbers on the appropriate counter drum. These devices should be placed and/or guarded so that they cannot be accidentally displaced, or bumped, causing injury to personnel or damage to garments. |
| Push-pull controls:<br>Stirrup handle  | Use for an ON/OFF control which (because of mechanical reasons) requires more force than can be applied by a finger-operated control. | Locate controls so that they cannot be accidentally struck by personnel and cause injury or garment damage or interfere with normal or emergency crew movement. Make the stirrup opening large enough to accommodate four fingers (including gloves when worn). |
| T-shaped handle  | Use in lieu of a stirrup handle when space is not available for whole-hand operation. | Make handle extensions on each side of the central shaft long enough to accommodate the index and middle fingers on one side and the third and fourth fingers on the other. |
| Ball handle  | Use in place of the above when the orientation of a stirrup or T-shaped handle would cause serious inconvenience to the operator and possibly delay his or her total reaction time in the event of an emergency. | Select a ball diameter that will comfortably accept a "fist grip" if the force requirement is high and the thumb and at least two fingers if the force requirement is low. |

## CONTROL SELECTION GUIDE (continued)

| Control Configuration | Use Criteria | Human Engineering Considerations |
| --- | --- | --- |
| Levers (console type) | Use for main power setting to provide forward vehicle thrust or machine acceleration. | The handle length should be long enough to accommodate a four-finger grip. |
| | Use for vehicle attitude control. | All lever-type controls must be designed and located so that the full control motion (excursion) is within the limits of reach of the restrained operator wearing a pressurized garment. |
| | Use for discrete (bang-bang type) power application (e.g., translation). | |
| Levers (extended arm operation) | Use where medium to heavy manual force application is required. | |
| Hand wheel | Use for opening and closing pressure-tight hatches where extremely heavy force application may be required, for high-torque valves, etc.* | The diameter of the wheel and the diameter of the wheel rim must be compatible with operator dimensions, considering the constraints of pressurized garments. |
| | | The operator must be restrained in a weightless environment. |
| | Use for low-torque, multirevolution cranking operations where many turns are made in rapid fashion.* | Select an optimum diameter for the expected force, rate, and operator position variables involved. |
| Steering column levers† | Use for automobiles, trucks, buses, etc. Use when it is desirable to bring critical control functions within "fingertip" reach of the driver (i.e., for easy reach from the normal hands-on-steering-wheel position).<br>Typical functions:<br>  Automatic gear selection<br>  Turn signal operation | Consider typical driver control placement and motion expectancy, i.e., certain functions for right-hand operation and others for left-hand operation.<br>Avoid too many different functions or motions on a single lever—especially if inadvertent "cross-coupling" might occur. Use only one lever per hand.<br>Use rotation action sparingly. If used, provide knurling to indicate to the driver that this motion is possible. |

## CONTROL SELECTION GUIDE *(continued)*

| Control Configuration | Use Criteria | Human Engineering Considerations |
|---|---|---|
| Slide-type panel control† 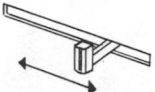 | Typical functions:<br>    Environmental control | |
| 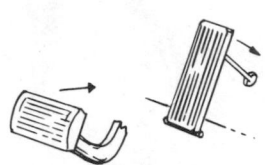 | Use foot-operated controls to free the driver's hands and allow simultaneous functional operations.<br>Typical functions:<br>    Foot-operated dimmer switch<br>    Accelerator pedal<br>    Service brake pedal<br>    Clutch pedal | Optimize pedal and switch locations to simplify the driver's problem of "finding" a particular control without looking at it and to help ensure that the driver will not inadvertently operate a control that is adjacent to the one he or she wants to operate.<br>Place controls within the driver's leg reach capabilities, considering the required control excursion. Mount the control so that its major motion axis is compatible with the natural motion axis of the operator's limbs. |
| Aircraft rudder pedals† 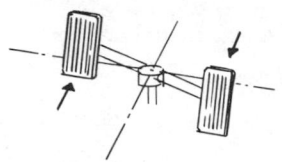 | Use for directional control of aircraft. | Provide an optimum force regime, i.e., enough resistance to minimize inadvertent input, but not so much as to cause strain or fatigue. |
| Hip-operated switch bar† 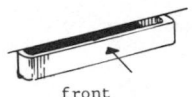<br>front | Use for machine operation where both hands are normally busy at the same time (commercial laundry press, industrial stamping press, etc.) either to start or to stop an operation. | Locate the control appropriately for the user's hip height. Minimize resistance and/or make the contact surface large in order to avoid bruising the user's hip. |
| 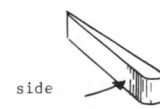<br>side | Use for transportation system turnstile so that passengers can pass through while holding bundles. | |
| Foot bar 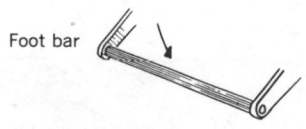 | Use as above for machine operation, but only when it is expected that the operator can afford to stand safely on one foot while the other foot operates the foot bar. (As an alternative, a switch mat can also be used to open doors.) | Determine convenient location on the basis of the nominal working position of the operator. Optimize switch position in relation to door opening speed to accommodate typical user walking speed. |

*Note:* This table was initially created for application to manned space vehicle system design. It has been expanded here to include additional systems and product applications.

  \*These suggestions originally were made with reference to space systems, in which operators normally would be "weightless."

  †These are additions to the original space systems guidelines to provide suggestions for typical terrestrial systems, e.g., highway vehicles, aircraft, and industrial machines.

## CONTROL MOTION EXPECTANCY (DIRECTION STEREOTYPES)

Some control motions seem more natural to people as a result of either innate characteristics (musculoskeletal configuration, handedness, etc.) or habit patterns that have been learned because most products are designed in a certain way.

Perhaps the most reliable motion stereotype is that associated with a steering wheel; practically everyone (both experienced and novice drivers) expects to turn the wheel clockwise to cause the vehicle to turn right.

Lever motion stereotypes vary depending on the location and position of the lever.

When the lever is positioned as shown in the accompanying sketch, a forward position usually implies to the operator that whatever he or she is controlling will increase in speed. Pulling the lever back is expected to reduce speed and stop the machine or vehicle.

When the lever is placed alongside the operator, he or she normally expects to pull up on the lever to "brake" the machine or vehicle.

An exception to the "forward-to-increase" and "back-to-decrease" expectancy is the gear-shifting lever (see the accompanying illustration).

Other lever-type controls generally relate to the movement of some subcomponent (as opposed to control of vehicle motions, noted above). The direction of motion of the lever is expected to reflect the direction in which the operator wants the subcomponent to move (as illustrated in the accompanying sketch).

Special vehicle controls, such as those in an airplane, have been standardized because of the way in which the controlling elements (e.g., the rudder, elevators, and ailerons) are placed on the vehicle. A typical lever control (joy stick) is expected to operate the vehicle or change vehicular direction when it is moved as shown in the accompanying sketch; i.e., the pilot moves the stick forward to cause the nose of the aircraft to pitch down and pulls back on the stick to cause the nose to pitch up. Right-left roll is accomplished by very "natural" right-left stick movements.

Another typical aircraft control raises and lowers the landing gear. The pilot expects the control motion to coincide with the desired position of the landing gear. In order to ensure this expected relationship, the control should be positioned as shown in the accompanying illustration.

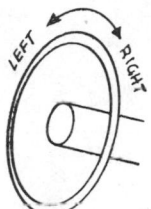

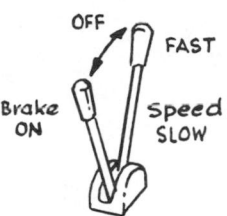

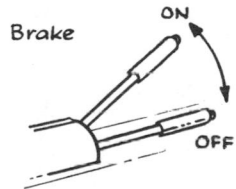

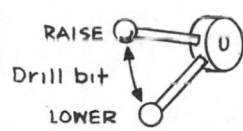

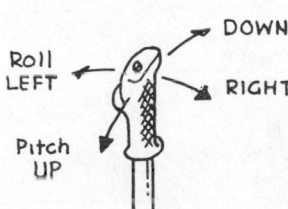

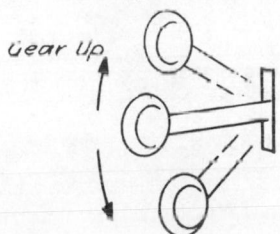

Aircraft throttle levers should be positioned as shown in the accompanying sketch in order to provide the expected changes in aircraft speed, i.e., forward for faster and aft for slower.

However, when engines are used for braking, a secondary "aft" motion of the controller is required. This motion must be thought of as braking, although there is an actual increase in engine power as the control is pulled farther aft. To preclude inadvertent use of the reverse thrust, the control is usually designed so that the pilot has to "pull up" before pulling the control aft.

Certain high-performance aircraft have adjustable wing positions; i.e., the wings are normally placed in the forward position for landing (slower speeds) and in an aft position for cruising (high speeds). This creates a direction-of-motion conflict in control operation. Although the novice operator might expect the direction of motion of both the controls and the wings to be alike, experienced pilots generally expect most of the lever-type controls in the cockpit to move forward when they want to increase the speed of the aircraft. Thus, in this special case, the controller should move forward to reflect increasing aircraft speed, even though the actual control action sweeps the wings aft.

Tiller bars are in actuality "levers." Typically associated with directional control of marine-craft, they cause the craft's rudder to move right or left and thus change the direction of boat movement.

As illustrated in the accompanying sketches, the tiller control could be positioned in at least two different ways.

The first configuration provides the more natural control motions for novice boat handlers because they use the tiller as a "pointing device."

On the other hand, experienced boat handlers are more used to the second tiller configuration, and use of the first configuration could very well confuse the experienced sailor.

Whenever practical, a steering-wheel type of controller is recommended because it provides a natural controller–vehicle movement relationship for both the novice and the experienced boat handler.

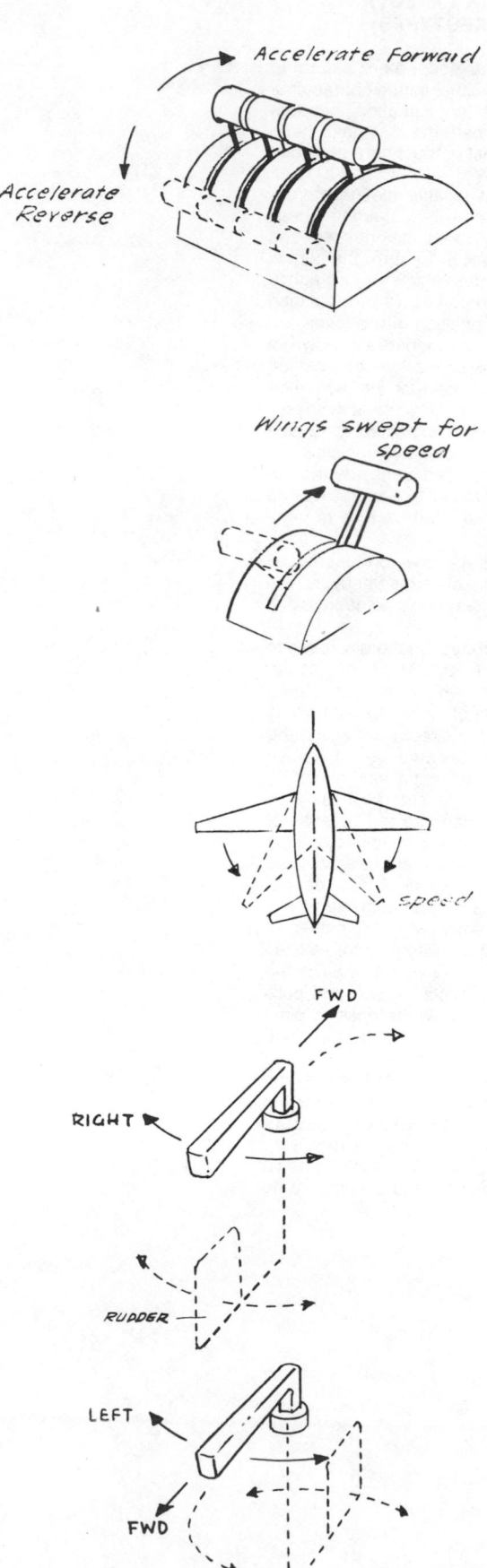

The well-established rudder bar stereotype shown in the accompanying illustration applies to foot-operated directional control of aircraft and other, similar vehicular control configurations (e.g., sleds). Although there is an obvious discrepancy between controller motions and controlling element movement (e.g., the rudder), the natural expectancy of moving the right foot to effect a turn to the right and the left foot to effect a turn to the left is more significant (primarily because the operator seldom has knowledge of the actual control system characteristics).

If pedals replace the bar, toe pressure should control vehicle direction, and heel pressure should control braking.

A common type of panel-mounted control is the pull-type knob illustrated in the accompanying sketch. Although the typical control movement direction seems to be opposite to that described earlier for other controls (i.e., one might expect to push the control forward for ON, GO, or OPERATE conditions), long-standing convention has created an opposite expectancy; i.e., the control is pulled out for an operational mode and is pushed in to cause the operation to cease.

Recently, slide-type controls have been introduced for a variety of control operations (including control of vehicular functions, such as the environmental system, and control of radios, e.g., station selection and volume control). When slide controllers are used (not necessarily recommended as the preferred type of controller), the direction of control movement should be as shown in the accompanying illustrations.

ON/OFF function switches, regardless of their particular configuration (toggles, rockers, etc.), should operate as shown in the accompanying illustrations; i.e., movement of the control upward or to the right or depression of the top or right portion of the rocker switch should cause an ON condition.

Rotary-motion switches (including both the detented, discrete-positioning and the continuously variable potentiometer types) should move clockwise for an increase in "value." However, this is true only when the control knob is operated only by the right hand and/or is located on a panel facing the operator.

When such a control is positioned for left-hand use only, the direction of motion should be as shown in the accompanying illustration because the operator shifts from a clockwise stereotyped expectancy to a spatial expectancy relationship; i.e., the operator rotates the control surface forward for an increase in functional value.

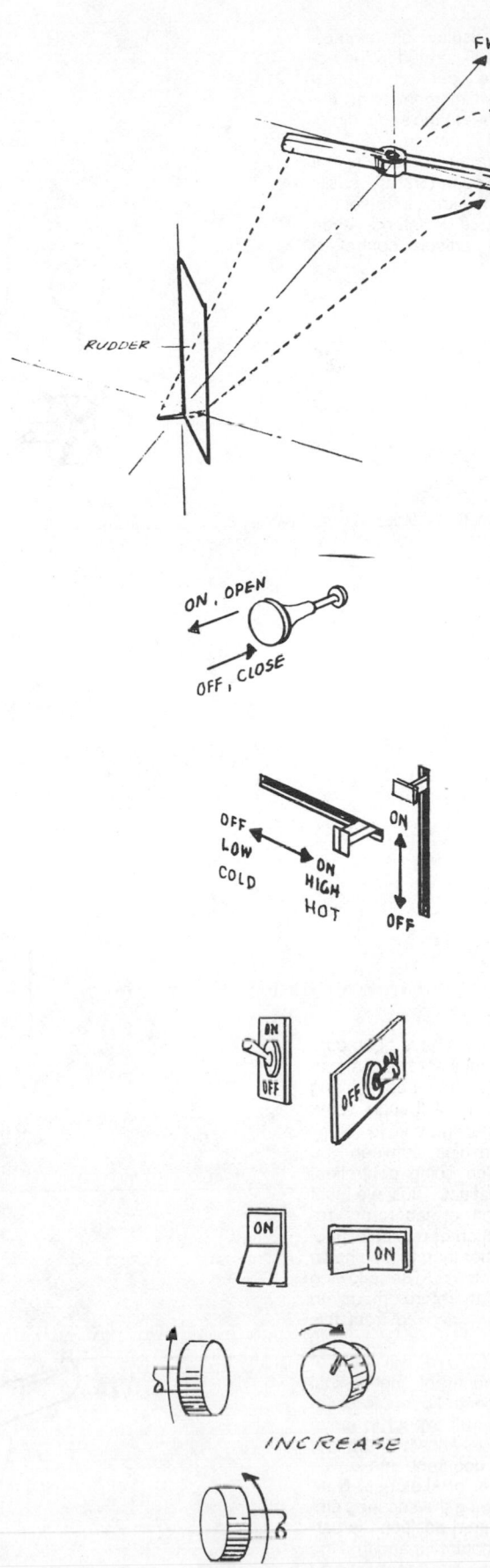

Control and visual display direction-of-motion expectancies are illustrated in the accompanying sketch. The operator's expectancy is for the displayed element (e.g., the instrument pointer) to move in the same direction in which the control is moved.

Use of left-hand-operated controls is not recommended because operators are easily confused and often turn the knob in the opposite direction from what they intend—even after they have used the particular control arrangement.

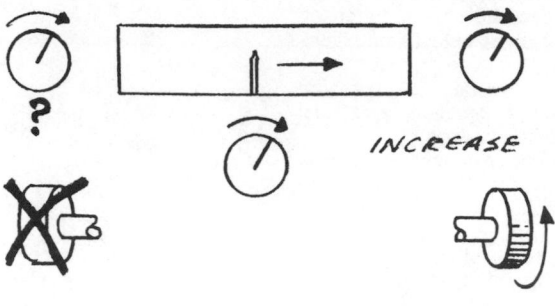

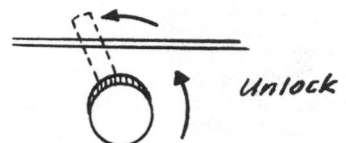

Some control relationships have been confused by the fact that various hardware manufacturers created devices without considering the operator expectancies. Although some people have sufficient background to "see" the mechanical relationships between the controller and the device being controlled, others rely entirely on instinct. Thus, we must depend on extrapolation of laboratory research dealing with direction-of-motion expectations using controls other than those shown in the accompanying sketches. The following recommendations are therefore made in terms of "expert opinion" derived from this research.

Assume that the typical user will have at least a reasonable amount of mechanical background for judging control operation relationships; i.e., he or she will expect to move the rotary latching pawl as shown in the first sketch, and to move the doorknob and key as shown in the second set of sketches. Note that the principle of "moving the securing element away from its secured position" is carried out for each type of controller and that the direction of movement varies depending on which hand is used to operate the control.

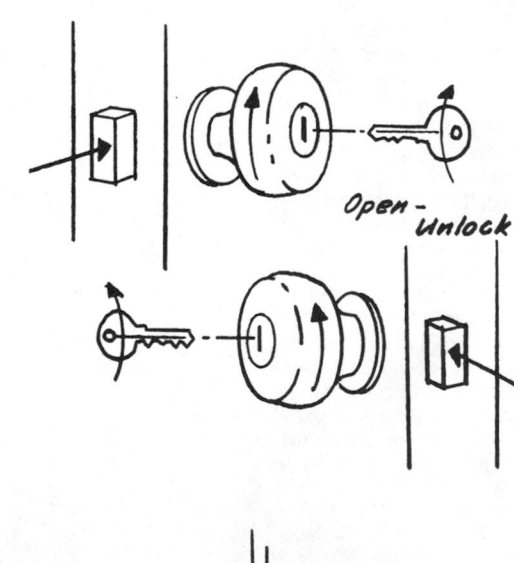

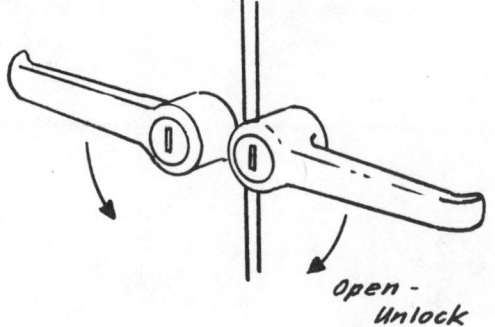

Control position–direction-of-motion relations may vary depending on the location of the control with respect to the operator.

Although, as a general rule, one can translate control motions from one operating plane to another merely by assuming that the operator "faces" the control panel, this does not carry through to all conditions (as shown in the accompanying illustration). Typically, controls alongside an operator are not "faced" until the control position moves aft of the operator's chest. Once aft of this point, the operator typically turns to face the control.

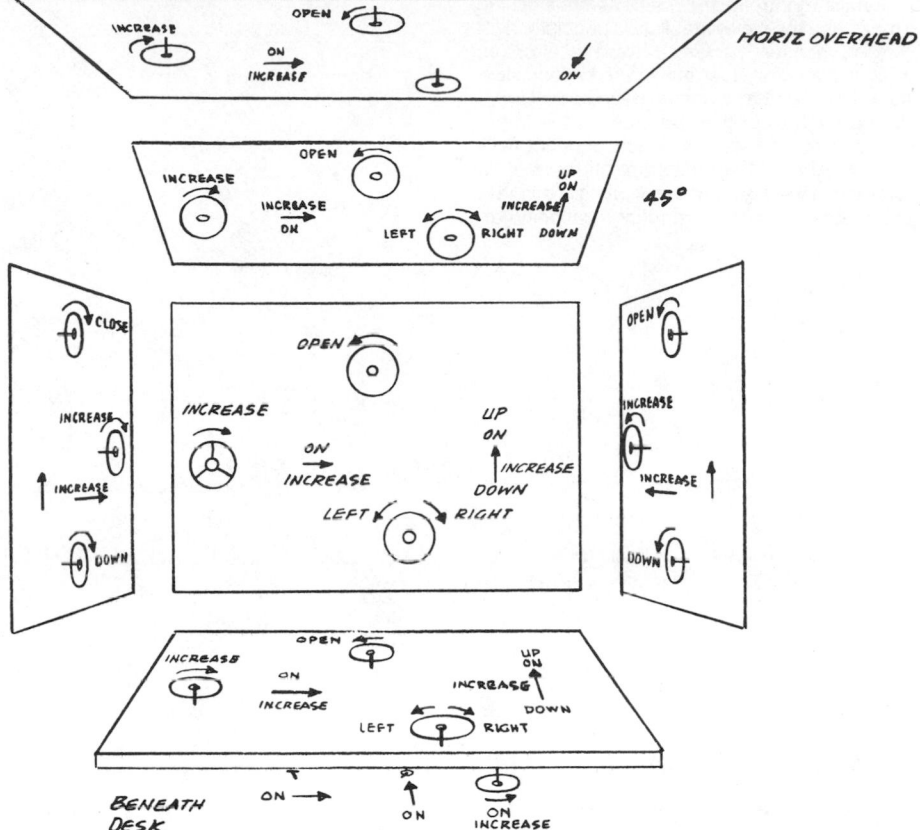

The illustration of the antenna extension is separated from the first illustration because it represents a unique direction-of-motion principle; i.e., a clockwise movement of the wheel or crank is always expected to increase the antenna length or height, regardless of which plane the control is mounted on.

Antenna
extension

Key-pad key arrangements imply direction-of-motion relationships even though the actual key movement is in only one direction.

Data entry keyboard arrangements were somewhat standardized without reference to known human-oriented key arrangement research. Thus we have a situation in which industry convention creates a powerful constraint against applying what might be considered a more efficient organization of key-pad push buttons. The accompanying sketch illustrates current data entry key-pad convention. Basically, it represents a principle of reading from the bottom and to the right, i.e., increasing values from left to right and from bottom to top.

Communications keyboards, on the other hand, have typically been laid out more in con-formance to the results of the aforementioned key arrangement research; i.e., they read from the top to the bottom and from left to right.

Although it is not recommended that the data entry key-pad arrangement be changed, the following discrepancies should be noted, as they may have implications for some newer keyboards:

1. Although one reads systematically the numbers from bottom to top, the zero and final entry is at the bottom; i.e., normally the zero is a final digit entry prior to entry.
2. It is normal to read from the top down; thus the data entry arrangement precipitates extra initial search.

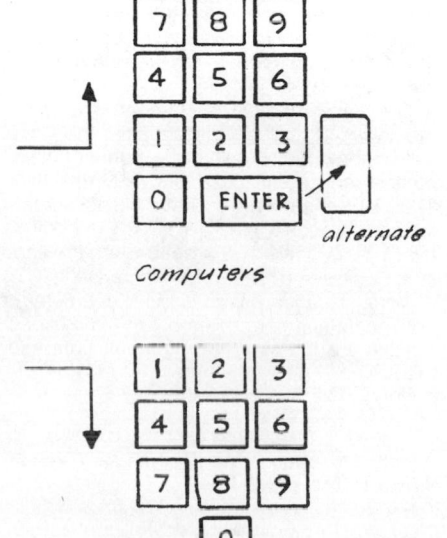

Rotary controls for discrete positioning should use the well-established principle of "a moving pointer versus a fixed scale," as shown in the first illustration. The second illustration shows that when values are imprinted on the knob skirt, they progress in the opposite direction to that in which one normally expects to read them (i.e., the principle of "a fixed index versus a moving scale"). The latter configuration leads to frequent positioning errors.

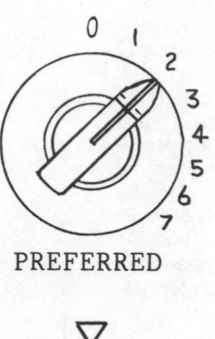

PREFERRED

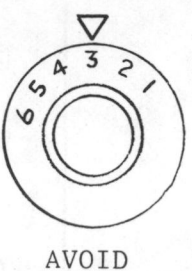

AVOID

## CONTROL AND DISPLAY DIRECTION-OF-MOTION RELATIONSHIPS

The most reliable expectancy occurs when the right-hand arrangements illustrated by the two upper sketches are used.

A weaker expectancy occurs when the arrangements for left-hand operation shown in the lower sketches are used.

These expectancies occur because of the apparent mechanical relationships that the operator observes; i.e., the operator, purposely or not, observes the directional movement of the control perimeter and the scale pointer as moving together. In the upper vertical display, the pointer must be adjacent to the side-mounted control. In the lower vertical scale, the pointer is moved to the other side so that it, too, is adjacent to the left side-mounted control.

The lower vertical scale-control arrangements are less desirable because the visual expectancy conflicts with the normal "blind" control movement expectancy noted earlier. Thus, if by chance the operator did not actually look at the scale (i.e., adjusted the controls blind), he or she would probably turn the knob in a clockwise direction for "increase."

*Note:* Because of the circular geometry of both scale and control knob, left-hand expectancies are the same as right-hand arrangement.

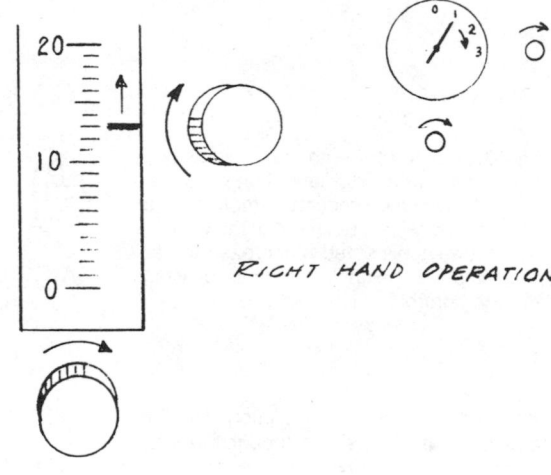

RIGHT HAND OPERATION

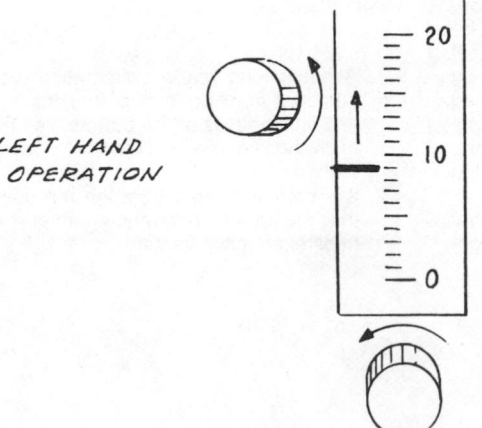

LEFT HAND OPERATION

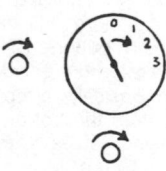

Valve operation is a special case of predetermined convention creating expectancies opposite to those established by human behavioral research. The single valve direction-of-motion stereotype convention is mechanically related; i.e., the valve is turned to "unscrew" a valve seat and thus allow liquid, gas, or other substances to "flow." In other words, to increase the functional factor, the valve handle is turned counterclockwise—as opposed to the behaviorally oriented principle of a clockwise motion to effect an increase in functional value or factor.

This stereotype is so well established now that any attempt to reverse the standard should be avoided. This applies to single valves only, as will be explained below.

When dual valve handles are involved, i.e., when one is controlled with the right hand and the other is controlled with the left hand, the above convention ceases to remain constant. The average user becomes somewhat confused and tends to revert to another direction-of-motion referencing philosophy. Although this stereotype is not as well established as others, it involves the following interpretive factors:

1. Motion toward the user to obtain flow increase
2. Mirror-image operation, i.e., both hands performing similar motions

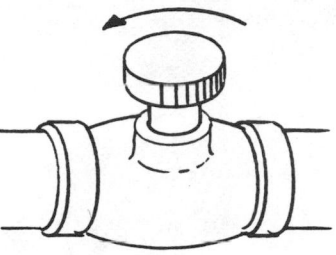

INCREASE

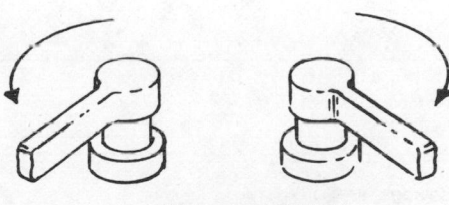

INCREASE

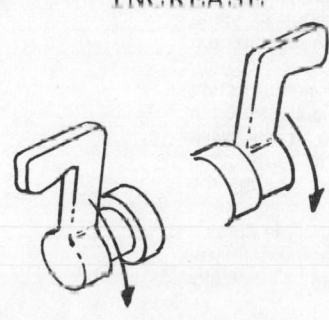

Automotive, foot-operated control conventions are well established, and even though they appear to violate some of the previously noted direction-of-motion stereotypes derived from laboratory research, these conventions should not be reversed.

The accompanying sketches show the typical foot-operated controls for automobiles, trucks, and buses. As indicated, the motion required to accomplish each function is forward—regardless of the vehicle's motion relationship.

The automotive gear-shifting functional motion position convention is also well established and is in fact standardized by U.S. governmental regulations.

It should be noted, however, that if it were not for such previous standardization, a more operator-compatible configuration could be created that probably would reduce the confusion that makes the training of new drivers more difficult. A so-called optimized configuration is illustrated in the accompanying sketches. These are presented primarily as ideas that could possibly be transferred to some as-yet-unproposed control system.

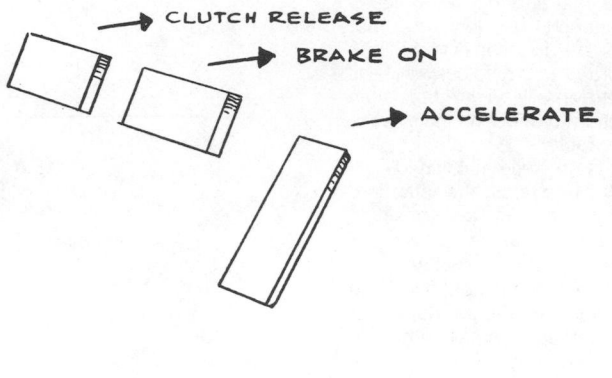

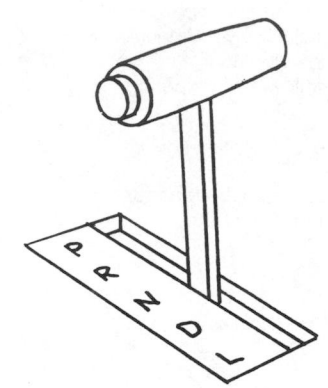

Automatic transmission

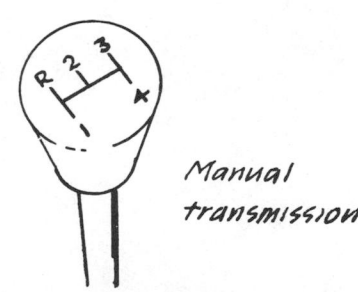

Manual transmission

Connector-fastener devices are often required to secure cables, hoses, and other, similar elements to each other or to some fixed element. Thus these connector-fasteners are in fact controls.

For most applications, the well-established direction-of-motion stereotype is illustrated in the accompanying sketches; i.e., a clockwise motion secures and tightens the connection, and a counterclockwise motion loosens the connection.

*Note:* Exceptions to this rule include nuts on wheels, where the stereotype must be reversed to prevent the movement of the wheel from loosening the nut.

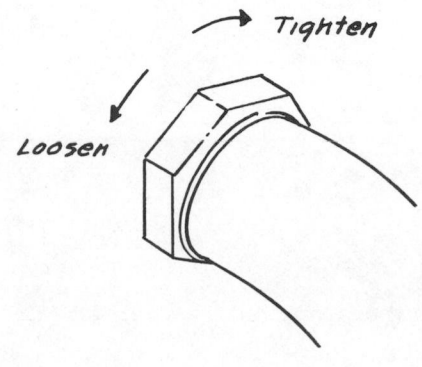

## CONTROL OPERATION EXPECTATIONS VERSUS CONTROLLER APPEARANCE

One should not use a controller configuration that is confusing to the potential operator. The following typical expectations should be observed:

1. A push button should extend out from the panel and thus imply that it is to be pressed or pushed in. Some push button and indicator displays are made to look alike; i.e., they both are flush with the panel, making it difficult to tell which one is a control and which is merely an indicator.

2. A pointer-shaped knob implies that the device is a positioning control that can be positioned in discrete steps.

3. A circular-shaped knob implies that the knob turns continuously. It also makes it easier for the operator to reposition his or her fingers for multiturn manipulation.

4. The waferlike knob indicates that it is to be pulled out. If it also is to be turned, it should have serrations around the edges, implying rotational possibilities. (Do not use serrations if the knob is not intended to be turned.)

5. The rocker-shaped control implies that either end of the switch cap can be pushed. However, the angular cues may not be apparent if the control is positioned at a poor visual angle. Make sure that the operational cues are apparent in the use position before selecting this type of control.

6. The toggle-switch handle provides a readily apparent cue that the handle can be alternately positioned from one side to the other—especially when several of the switches are arranged in a horizontal or vertical pattern.

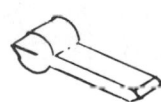

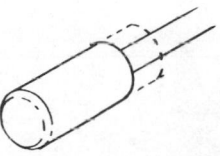

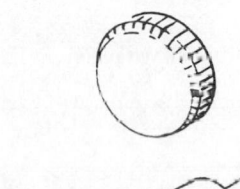

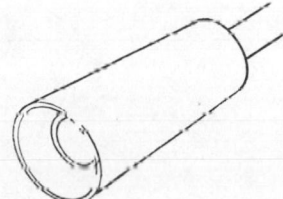

### Deliberate Confusion

Sometimes deliberate confusion is created by the appearance of a control device, as illustrated by the sketches at the right.

Lack of an obvious "pointing" feature makes it possible for the operator to turn a selector knob to the wrong position. Symmetrical bar handles on rotary knobs should be avoided.

Confusion with other panel-mounted features often leads the operator to believe the control is something other than a controller; i.e., it may look like a lighting fixture or a panel fastener.

An otherwise effective pointer-selector knob may be so distorted that when the operator manipulates the control in the dark or without actually looking at it, he or she tends to point the handle at the desired setting because the handle feels like a pointer.

A sleeve handle that is designed to slide inward is seldom recognized as performing that type of movement.

An extremely thin, waferlike rotating knob is often mistaken for a push button.

Push buttons deeply inset in the end of a control handle are usually not observed when the operator views the handle from the side. Thus the operator fails to find the button switch.

## CONTROL HANDLE AND KNOB SIZE

Both the size and shape of a control handle or knob can affect the ease with which an operator can manipulate the control. The size and shape of the handle must be compatible with the size of the hand. The following general considerations are important when selecting and/or designing control handles and knobs:

When considerable force has to be applied to a lever or steering wheel, the diameter of the handle must be large enough so that the user's hand and finger surface contact is maximized—but not so large that a firm grip cannot be maintained.

If a precise knob rotation is required, the knob should be large enough so that all the user's fingers and the thumb can be placed on the knob rim and/or surround the knob edges if considerable torque is also required.

If limited panel space requires the use of very small knobs, make sure that there is sufficient knob depth so that the operator can have as much surface contact as possible; for knobs with very small diameters, add serrations and knurling to improve the operator's grip.

Always analyze the relative position that the operator's hand will be in when he or she manipulates the knob or handle; this may suggest a different type of knob or handle.

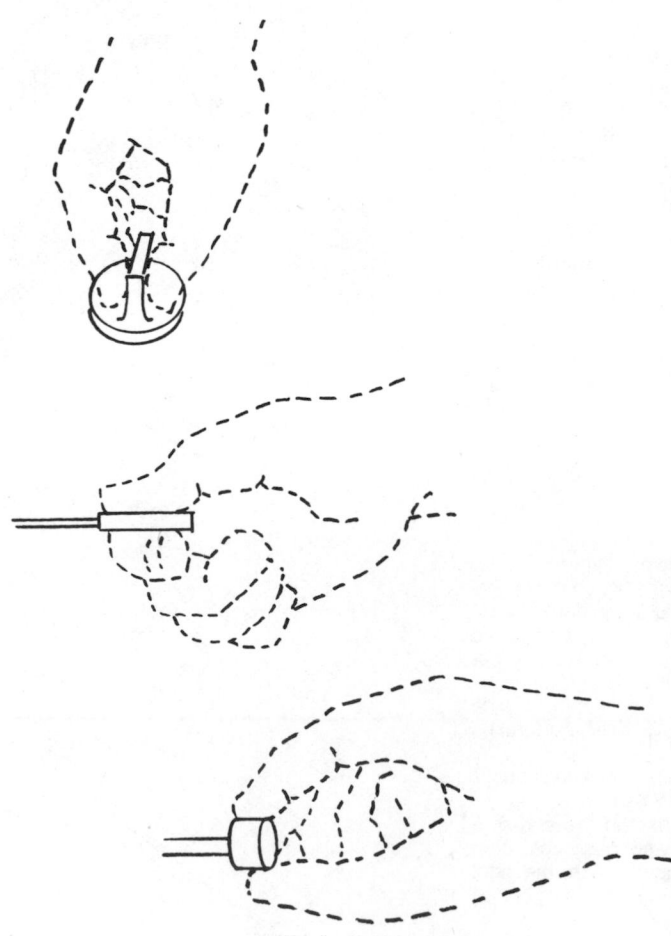

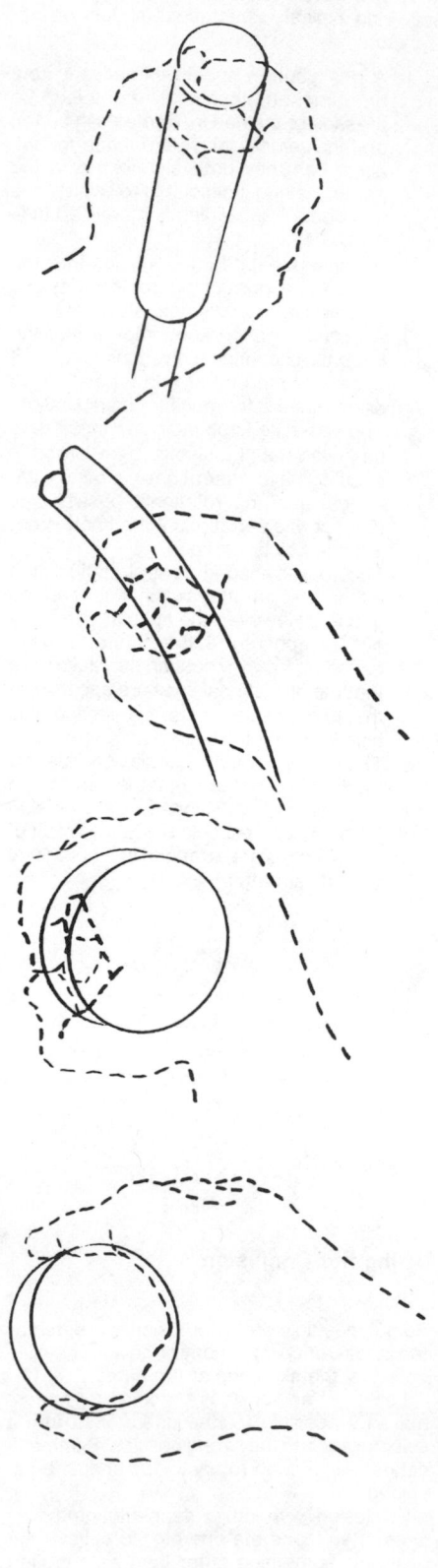

## CONTROL HANDLE SHAPE

The shape of a control handle should be determined on the basis of how the operator is expected to grip the handle (depending on the relative position of the handle control placement, position, and expected control motions). Some typical considerations are illustrated by the accompanying sketches:

A pistol-grip handle should taper toward the bottom, as illustrated, because of the foreshortening of the operator's fingers.

For maximum grip strength and security, use a round or oval handle of equal dimensions (cross section). The typical practice of providing finger convolutions on the handle is not desirable because few people need to position their fingers in exactly the same place. To improve grip, add knurling or serrations to the handle's sides.

A round grip provides greater flexibility for the operator if his or her orientation with respect to the handle must be changed.

When finger- or thumb-operated push buttons are required on the handle, position them where the intended finger or thumb "naturally" falls.

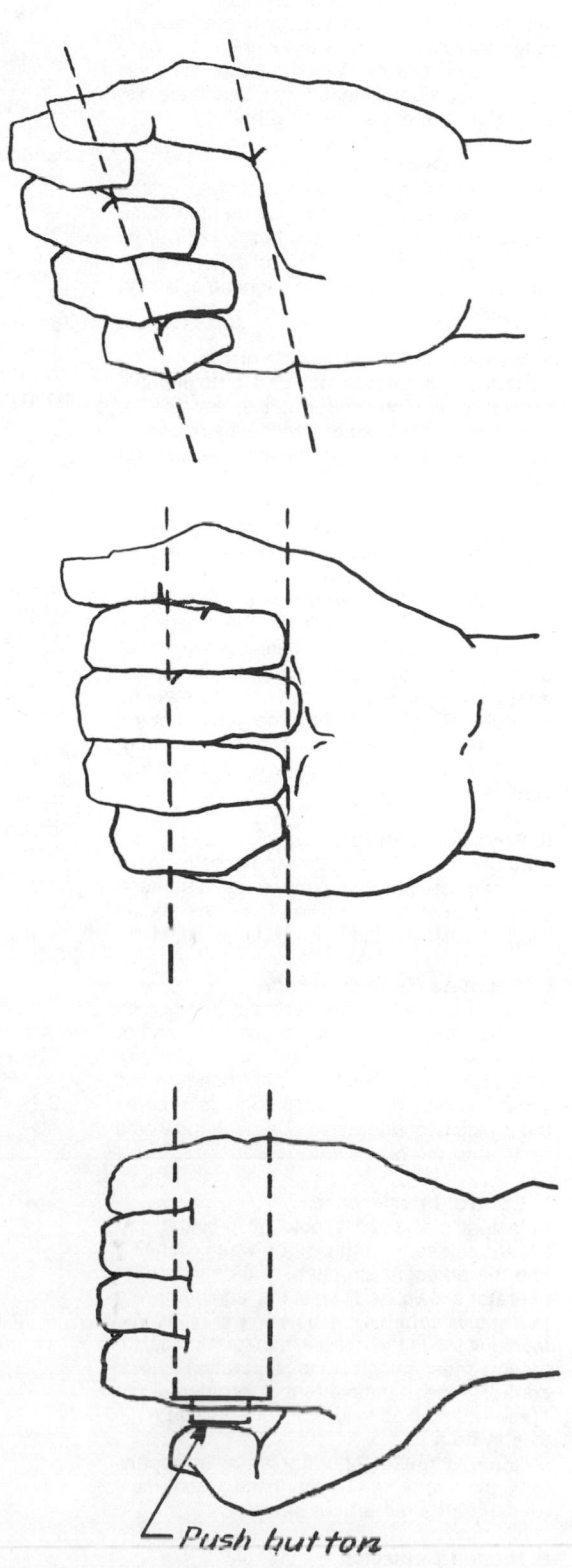

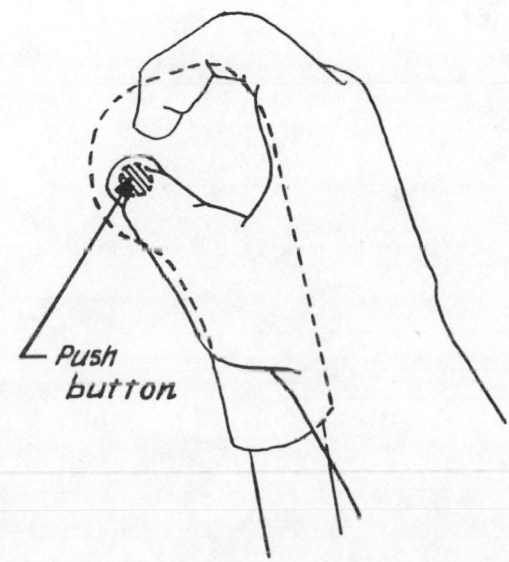

Push button

Push button

## FUNCTIONAL PRINCIPLES FOR DECIDING ON THE LOCATION AND ARRANGEMENT OF CONTROLS

### 1. Functionally Related Controls

Consider those controls which relate to a common system to be functionally related and arrange them so that they are in the same general area on a panel. When a panel display is affected by a particular control, locate the display and control near each other.

### 2. Sequence of Use

When several controls are normally operated in a certain sequence, arrange them in some systematic order, i.e., from left to right, from top to bottom, etc. This is particularly important when "missing" a step might be critical to an operating sequence.

### 3. Most Frequently Used Controls

To reduce operator fatigue and/or to improve control operation performance, locate the most frequently used or continuous-operation controls where they are most convenient for the operator.

### 4. Emergency Controls

Any control that requires rapid response time in an emergency should be located where the operator does not have to search for or reach the control. For example, if the operator's hands are typically on a steering control most of the time, emergency controls should be within a few inches' reach from the steering control, except when the emergency control could be foot-operated, in which case the principle applies to any normally continuous foot control operation.

### 5. Precision Controls

Although limited precision is potentially possible for foot-operated control, as a general principle precise controlling functions should be performed by the operator's hands.

### 6. Alternate Hand Operation

Consider the potential emergency in which an operator may not be able to use one hand or the other, e.g., when a battle wound has incapacitated a pilot's right or left hand. In this situation, location of critical controls such as the joy stick for operation by either hand would be vital to the pilot's safe return.

### 7. Control Interference

Sufficient clearance should be provided between adjacent controls, between controls and the adjacent structure, and between the operator's own body and the equipment so that critical controls can be easily grasped and manipulated in the normal manner. Special attention must be given to separating critical controls whose inadvertent or accidental operation could lead to loss of control or damage to a system.

Controls that the operator normally reaches for without looking at them (blind reach) must be separated more than others.

### 8. Hidden Controls

All controls normally operated by a pilot, driver, or other operator should be visible from the nominal operating position; i.e., avoid hiding operational controls behind the seat, underneath work surfaces, or behind unlabeled cover panels.

### 9. Work-Load Distribution

As a general principle, equalize the work load of the control operator; i.e., distribute the control activity between his or her two hands and two feet (recognizing the limitations of limbs and preferred hand, etc.). The primary objective should be to avoid overloading the operator's right hand just because he or she is right-handed and is more agile and dexterous when using this hand.

Also, avoid assigning control operations to two limbs when a simultaneous motion geometry or motion pattern may cause motor response disruption, e.g., "patting one's head while rubbing one's stomach." Typical interference patterns occur when one hand is required to move a lever forward, while the other hand has to move a similar lever aft; when one hand operates a crank in the clockwise direction, while the other hand has to rotate a potentiometer also in the clockwise direction; or when an operator has to perform intricate hand manipulations and foot manipulations at the same time (a feat requiring considerable skill and practice).

*Caution:* The reader will no doubt recognize that there are possible conflicts among the various principles enumerated above. The temptation is to ignore the principles because of these apparent conflicts. Rather than ignoring them, think of them as having varying priorities; i.e., at times one principle has a higher priority than others, depending upon the particular design problem and operating situation. Thus, the designer should examine all the principles and place them in proper perspective according to his or her system's operating situation. Above all, do not ignore the principles, for they are individually valid objectives.

## CONTROL LOCATION AND REACHING REQUIREMENTS

Avoid the temptation to place controls where operators have to reach for them, just because they are "within reach." Several typical errors are illustrated by the accompanying sketches:

Distant reaching causes operators to shift from their normal operating position; it often requires that they "look for the control" and take their eyes and attention away from the roadway or displays they should be looking at; it takes extra time, during which they may miss an important cue from an instrument or the outside environment; and it sometimes creates considerable physical inconvenience.

Do not place one control behind another so that operators have to take the time to reach around the intervening control.

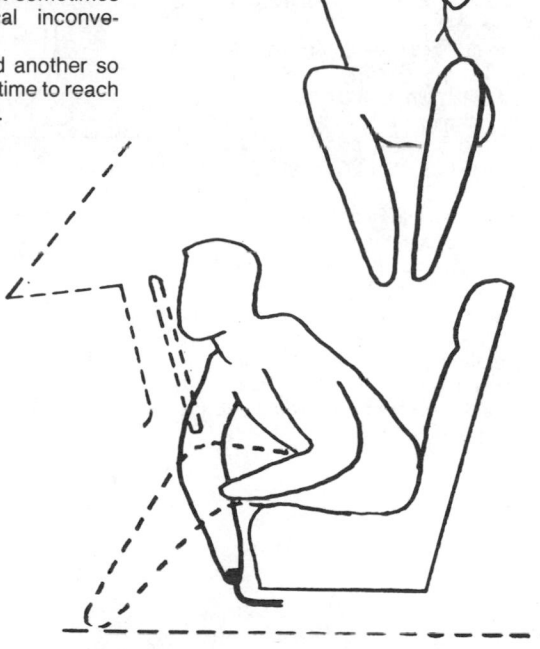

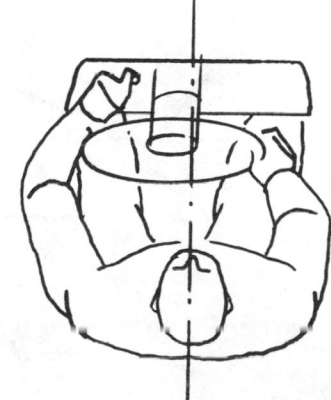

Panel controls should be within a few inches' reach of the normal hand position while driving.

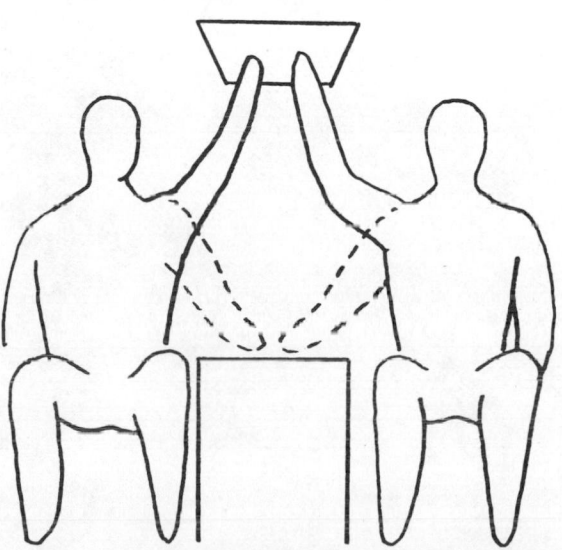

Operators often need to be in their normal operating position to make a proper seat adjustment. Making them lean over does not help.

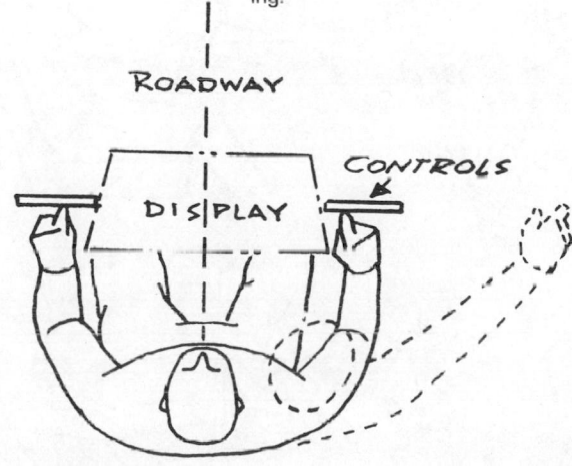

Shared controls should be equally convenient to both operators.

## STEERING WHEELS

Wheels are traditionally provided for steering automobiles, trucks, and buses. For maximum comfort the wheel should be tilted about 60°, and the wheel diameter should be 14 to 15 in (35.6 to 38 cm).

When power steering is not provided for large, heavy vehicles, more force may be required to turn the wheel. Thus, the steering wheel should be oriented horizontally, and the wheel diameter should be increased to 18 to 20 in (45.7 to 50.8 cm).

When power steering is provided for larger vehicles, the steering wheel should be tilted to a more comfortable position for long-duration driving, and the wheel diameter should be reduced so that the driver can reach more easily across the wheel.

For typical passenger cars, vans, and smaller trucks, the steering-wheel angle and size should approximate the optimum configuration noted above; i.e., the wheel tilt should be between 45 and 60°, and wheel diameter should be between about 14 and 16 in (35.6 and 40.6 cm).

For special vehicles that are low to the ground (e.g., sports cars), the wheel tilt should be between 60 and 90°, and the diameter should be 14 to 15 in (35.6 to 38 cm).

*Note:* The spoke arrangements illustrated for each of the vehicle classes should be appropriate for the type of steering-wheel manipulation; i.e., the number, positioning, and size of spokes and wheel rims should vary to provide gripping ease and force.

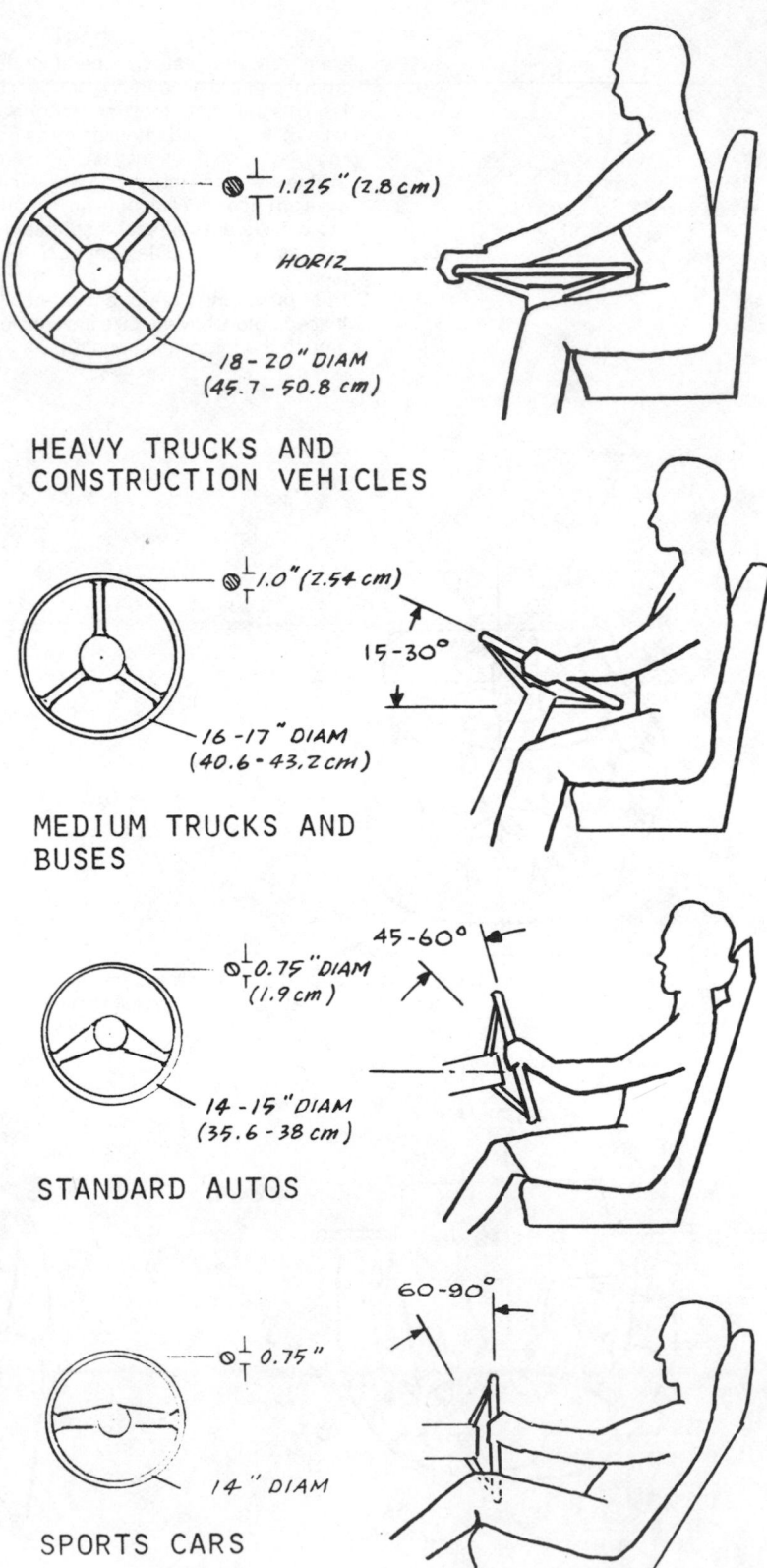

1.125" (2.8 cm)
HORIZ
18 - 20" DIAM (45.7 - 50.8 cm)

HEAVY TRUCKS AND CONSTRUCTION VEHICLES

1.0" (2.54 cm)
15 - 30°
16 - 17" DIAM (40.6 - 43.2 cm)

MEDIUM TRUCKS AND BUSES

45 - 60°
0.75" DIAM (1.9 cm)
14 - 15" DIAM (35.6 - 38 cm)

STANDARD AUTOS

60 - 90°
0.75"
14" DIAM

SPORTS CARS

Steering-wheel rims should be round or oval. Avoid special decorative convolutions or ridges that interfere with gripping.

However, provide gripping aids on the underneath side of the rim to reduce the possibility that the driver's grip will be disturbed by a sudden wheel motion or slipping when he or she attempts to apply force during a turning maneuver.

Wheel spokes should be oval and of the recommended dimensions so that the driver can be assured of a firm, comfortable grip if he or she elects to use the spokes for steering.

Note that the oval dimensions for high-force wheels for large vehicles are larger than those recommended for smaller vehicles and vehicles that are provided with power steering assist.

When practicable, limit steering-wheel displacement to about 120° for the primary turning range so that the driver is able to keep both hands on the wheel during the maneuver (without repetitive hand replacement).

Lock-to-lock limits for maximum turn should be between 1.5 to 1 (steering-wheel versus vehicle-wheel turns) and 3.0 to 1.

Although very strong drivers can exert more force than indicated above, they may not apply it in time to avoid an accident. If power steering is provided to reduce turning force requirements, be sure to provide sufficient emergency, mechanical advantage to keep force requirements below 50 lb (23 kg).

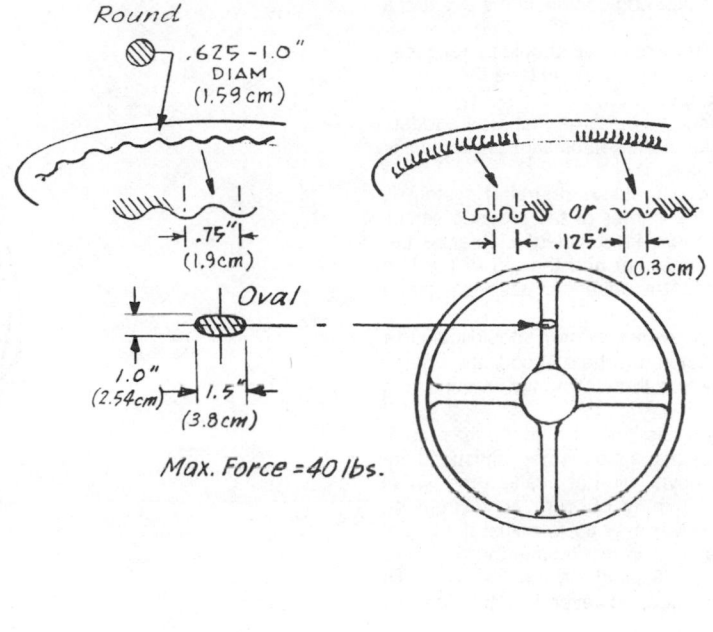

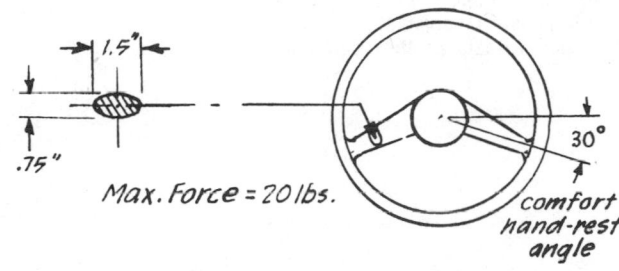

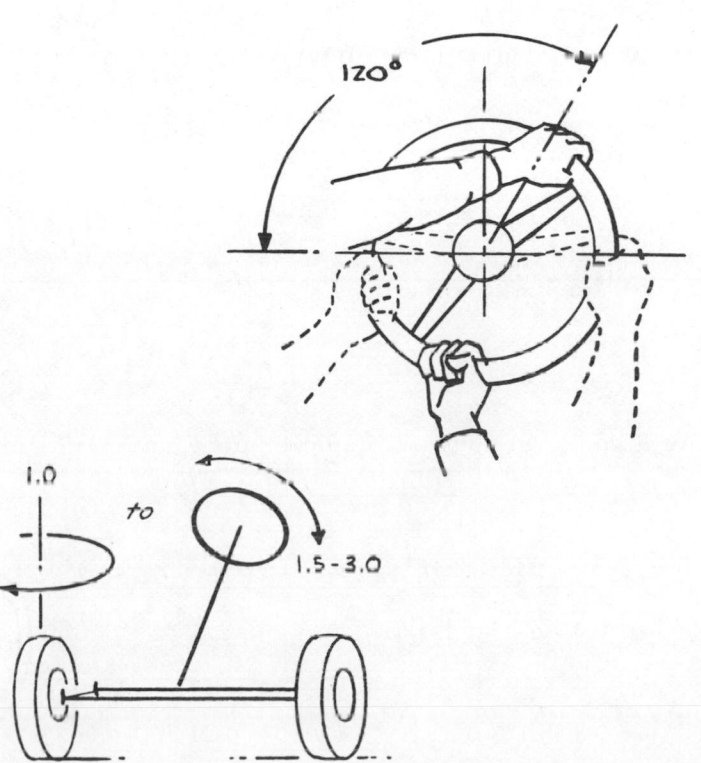

When steering wheels are tilted, they should be designed so that the driver cannot be impaled on the steering column in the event of a crash.

The steering-wheel hub should be recessed with respect to the wheel rim (see the accompanying sketch).

The hub should be large, flat, and padded; i.e., avoid decorative hubs with protruding or pointed profiles.

If an air-bag (passive restraint) system is installed in the center of the steering wheel, make sure there is sufficient clearance between the wheel rim and the rim of the bag housing so that the driver can easily grasp and grip the wheel.

The steering-wheel system should include a collapsible column to help absorb the energy of a driver who is thrown into the wheel during a crash.

Tiltable wheels are not generally recommended; they make it extremely difficult to ensure that a driver will not position the wheel where the rim may obscure an instrument, and/or the driver may try to adjust the wheel tilt while the car is in motion and thus interfere with steering efficiency. When the wheel tilt option *is* provided, however, limit the tilt to no more than 15°.

Do not use oddly shaped wheels; drivers cannot perform multiturn manipulations as efficiently with such wheels as they can with round wheels.

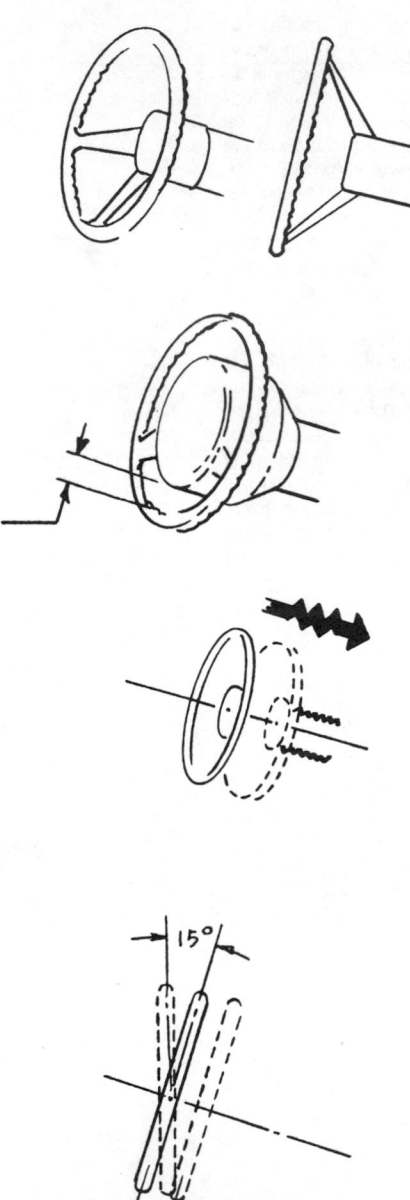

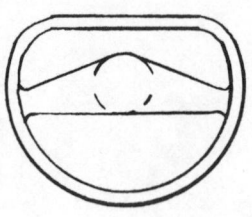

## AIRCRAFT CONTROLLERS

### Wheel-Type Joy Stick

The wheel-type joy stick is generally used for most aircraft other than military fighters where lack of lateral space within the cockpit dictates a single lever-type joy stick.

The so-called wheel generally is configured as a half wheel in order that the pilot can see panel instruments beyond the wheel.

Key factors in designing and positioning the wheel control are the following:

Fore-aft excursion should be within comfortable reach of the pilot; the pilot should not have to lean forward for pitch-down movements or rear back for pitch-up movements, and his or her normal position for flight control should be comfortable (i.e., the upper arm should hang approximately vertically).

Wheel handles should be a comfortable distance apart and tilted so that the pilot can apply maximum aft pull force when required.

Clearance must be provided between the wheel and the pilot's knees and between the vertical control shaft and the pilot's seat, and there must be clearance aft of the pilot's elbow when pulling the wheel aft.

When possible, design the system so that full rotation is no more than 40°.

An arc pattern is acceptable, but a straight-line pull is easier for the pilot.

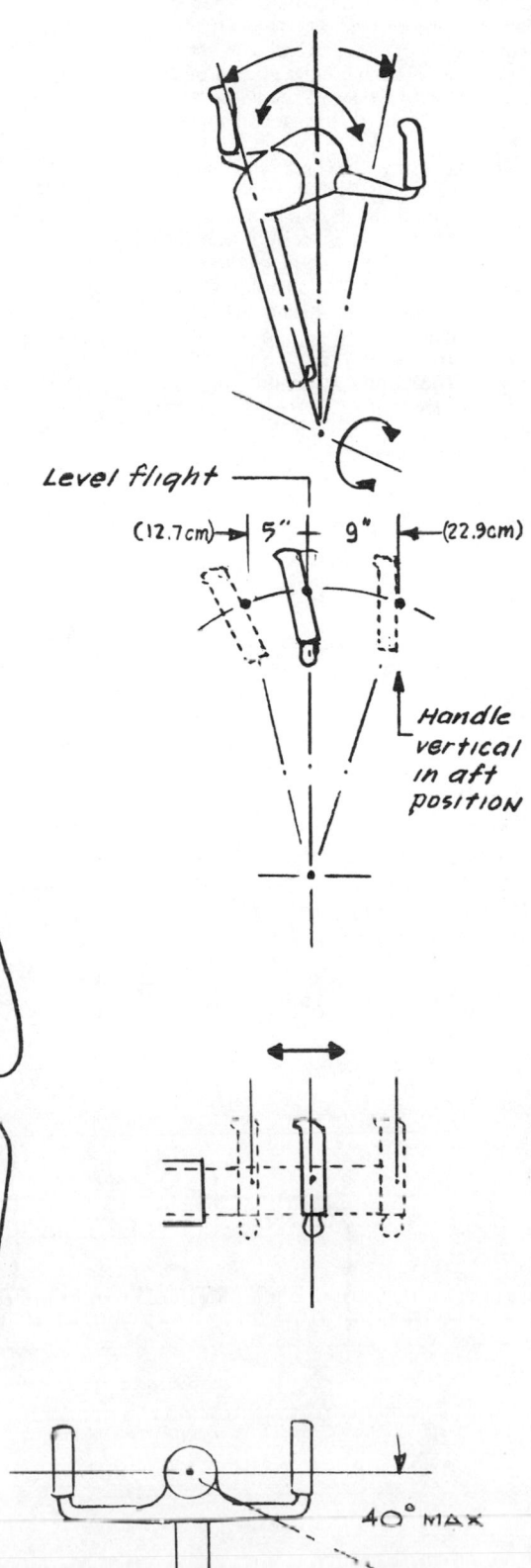

Level flight

(12.7cm) → 5"    9" ← (22.9cm)

Handle vertical in aft position

Knee

Elbow

Seat

40° MAX

A variety of wheel configurations are possible and equally acceptable, as long as certain dimensional criteria are maintained. The accompanying illustrations provide guidelines:

For small aircraft, where a wheel of minimum size is desirable, be sure that the handles are at least 12 in (30.5 cm) apart. Handgrips should be round or oval. A 1.125-in (2.858-cm) diameter is optimum. (Oval dimensions are illustrated.) The handgrip should be at least 6 in (15 cm) long and should be tilted inward, as shown in the accompanying sketch.

A very effective wheel configuration is shown in the accompanying illustration. Not only is it very comfortable, but it also provides useful feedback in terms of control position. Note, however, that the lateral handle spacing must be at least 15 in (38.1 cm) to prevent the handles from striking the pilot's knees during a turn.

Standards suggested for U.S. Air Force equipment are shown in the accompanying illustration. The handgrips are 7 in (17.78 cm) long and are slightly curved. Pods are provided for thumb switches.

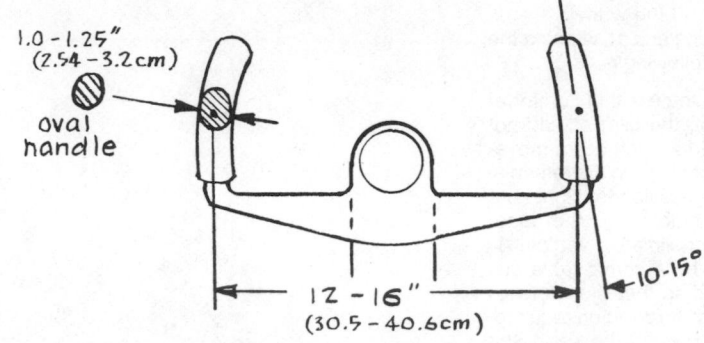

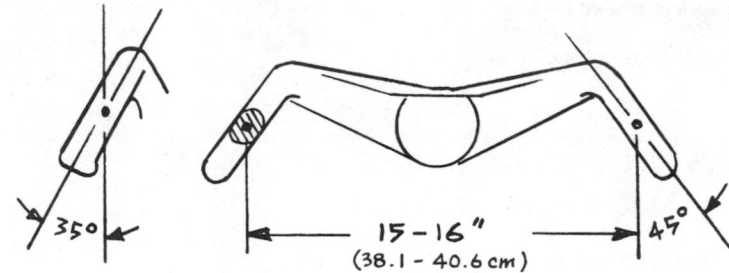

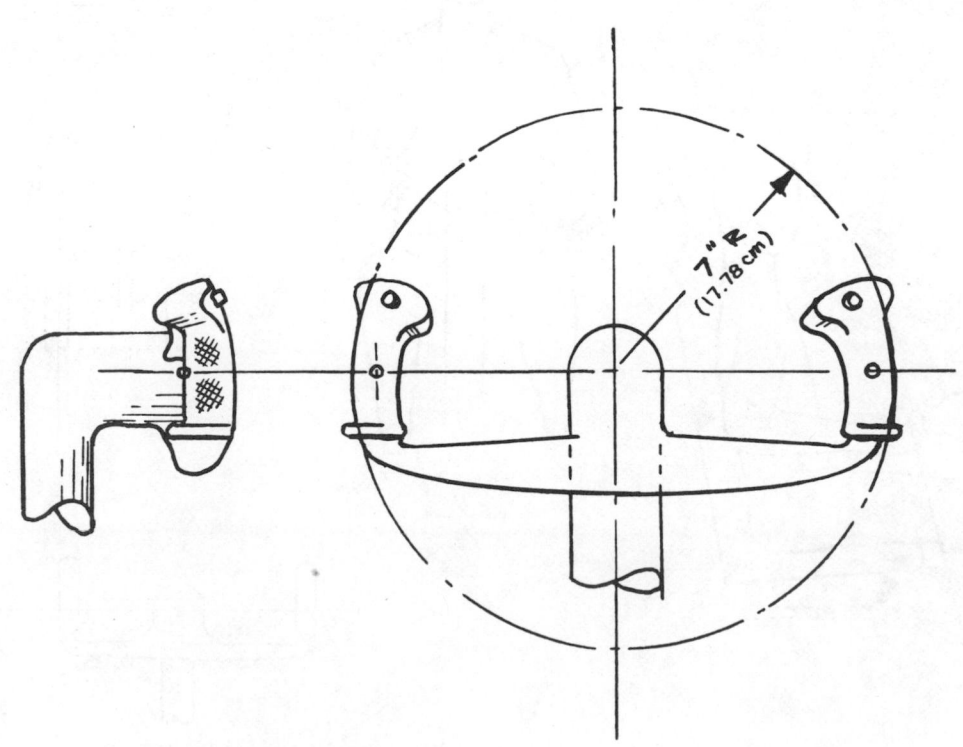

## Center-Post-Type Joy Stick

The simple post-type joy stick is generally used for very small aircraft where the fuselage width does not permit use of a wheel. Although many of the clearance problems are similar to those described for the wheel-type controller, one special problem must be addressed, i.e., lateral clearance between the pilot's legs for moving the joy stick from side to side.

The guidelines shown with the accompanying illustrations may help the designer develop a satisfactory joy-stick arrangement in the cockpit.

Note especially the height of the joy-stick handle with respect to the seat. Although the handle can be located higher without affecting controllability, it can easily interfere with the pilot's view of the panel instruments.

The control system should not require the operator to employ more than about 20 lb (9 kg) of force in order to move the joy stick to maximum excursions in any direction.

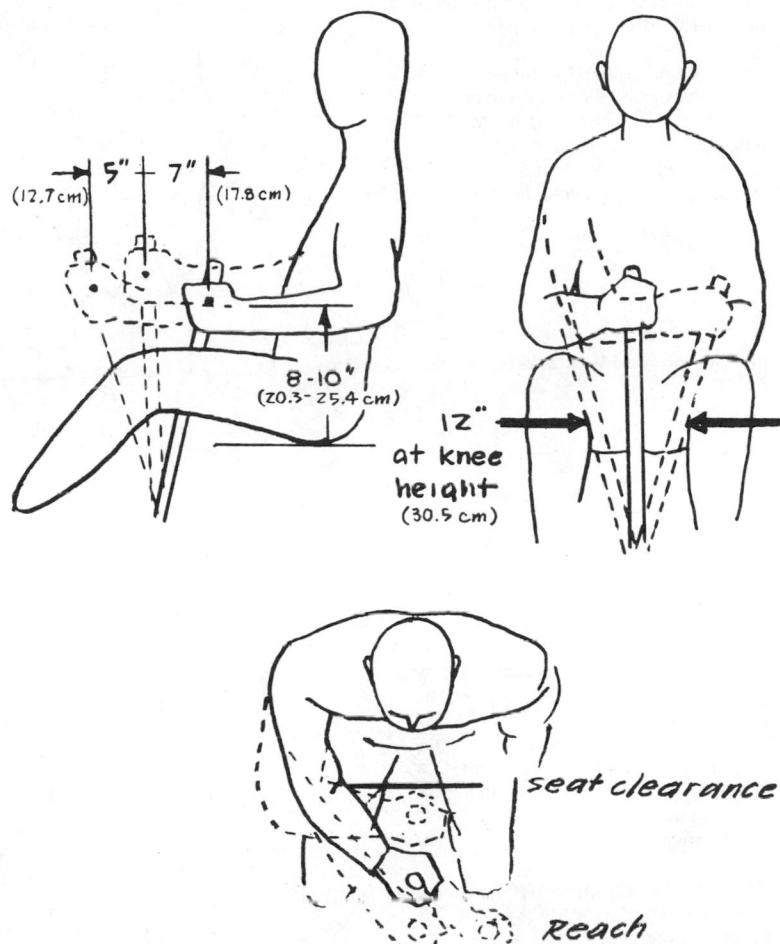

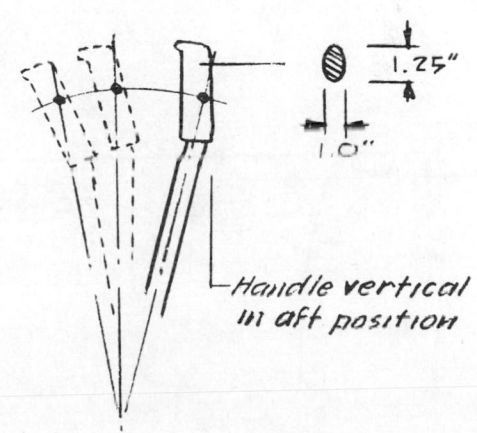

The joy-stick handle shown in the accompanying illustration is typical of pistol-grip types often designed for use in military aircraft, i.e., where additional switches and/or controls are added to the handle so that they are convenient during critical flight modes. Note especially the dimensional suggestions for the gun or bomb-release trigger.

Also note the suggested knurling on each side of the grip to aid the pilot in maintaining a secure grip on the handle.

Switches should be placed far enough apart to minimize inadvertent switch actuation.

A "heel rest" is recommended at the bottom of the handle so that the pilot can rest his or her hand and reduce the necessity to grip the handle tightly during relaxed flight conditions.

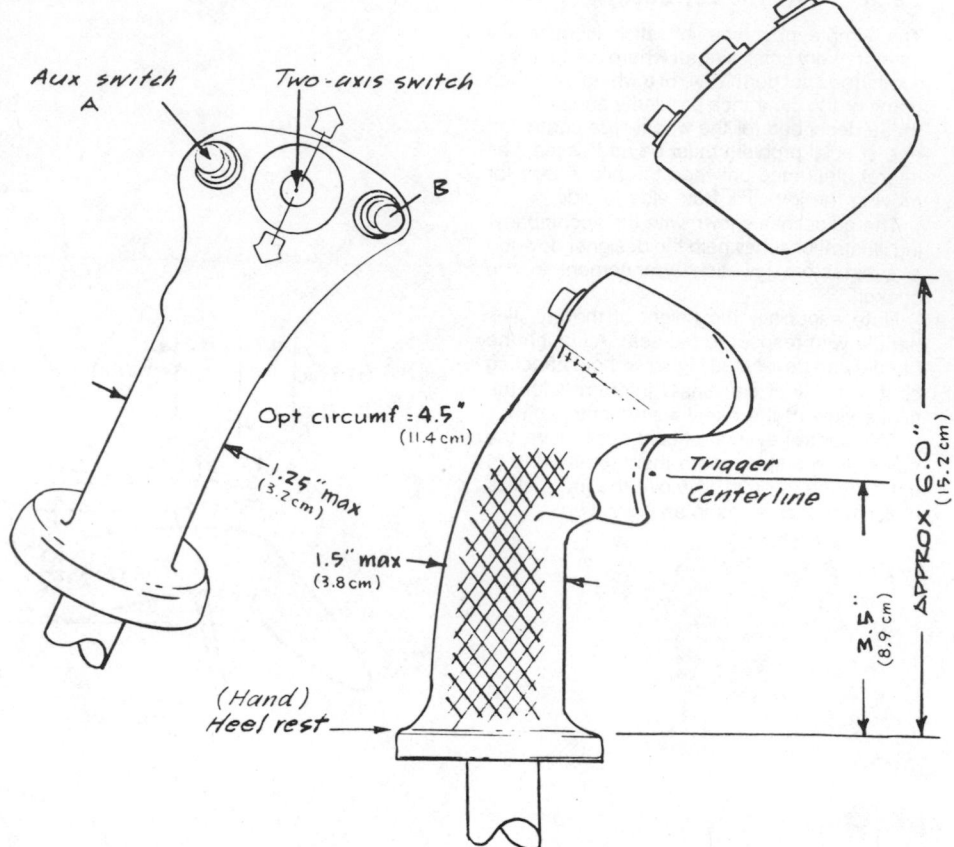

## Six-Function Joy Stick

Six functions are probably the maximum number that should be considered for a single controller; more functions make it difficult for pilots to keep from actuating the wrong one, either because they forget which switch is which or because of inadvertent motion inputs (e.g., cross coupling).

The accompanying illustration shows one possible six-function configuration.

*Note:* The specific geometry and dimensions are suggested guidelines and should not be construed as absolute.

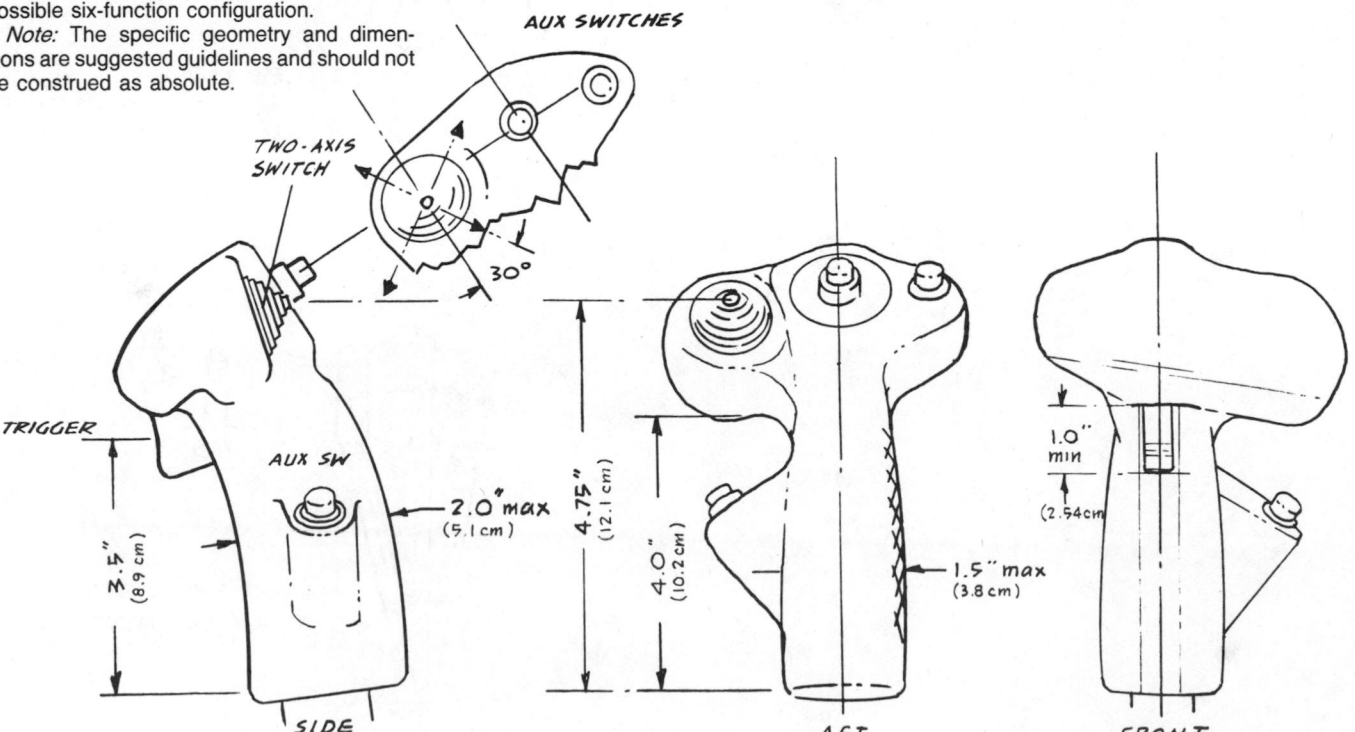

## Side-Stick Controller

A side-stick type of joy stick may be required when sophisticated display and control packages must be placed between the pilot's knees (thus eliminating the possibility of the normal joy stick). Obviously, this greatly reduces the range of controller movement. Instead of full arm motions, pilots must rely almost entirely on wrist motion, which also reduces the amount of force they can apply. A typical electric stick is shown in the accompanying sketch. Fore-aft movement should be limited to plus or minus 30° and side-to-side motion to plus or minus 45°. Control move-

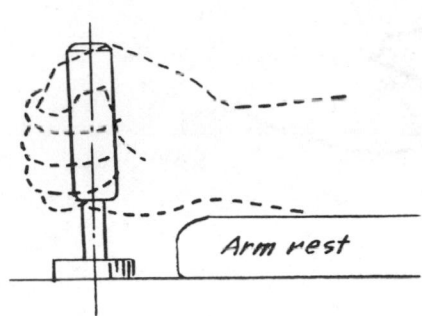

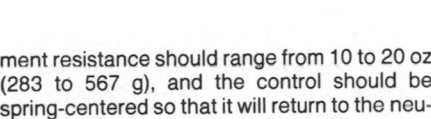

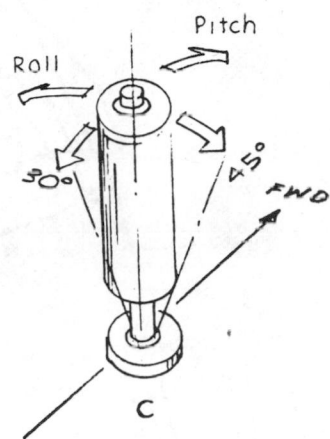

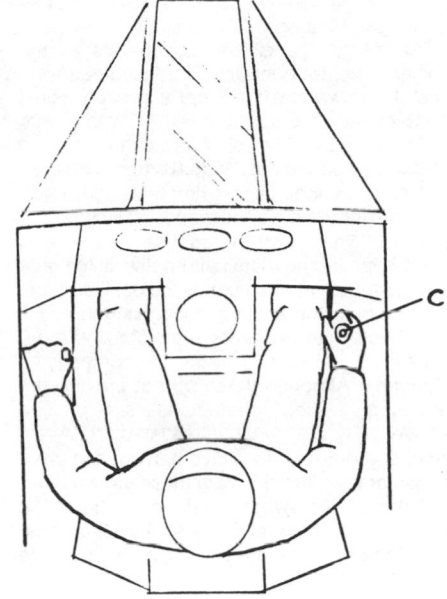

ment resistance should range from 10 to 20 oz (283 to 567 g), and the control should be spring-centered so that it will return to the neutral position.

Theoretically, the most efficient side controller would be one that does not require the pilot to make any arm motion; i.e., the controller should be designed so that the pilot's arm can remain firmly on an armrest, with only natural wrist motions required to manipulate the controller. Experimental devices such as that shown in the sketch at the right have been created to simulate the natural motion geometry of the human wrist.

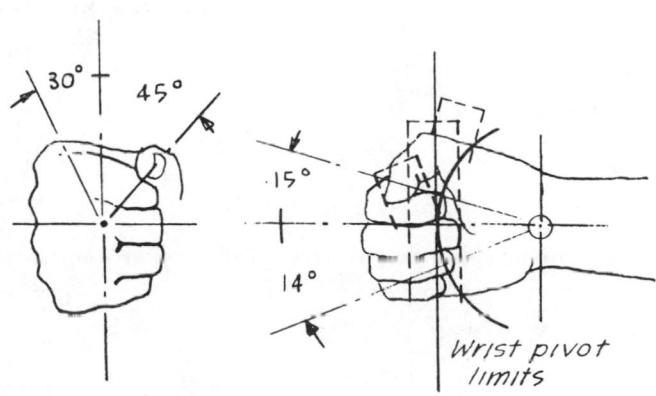

The shape and actual size of handles for side-stick controllers should be approximately the same as those suggested for the normal joy stick (center type), as shown below.

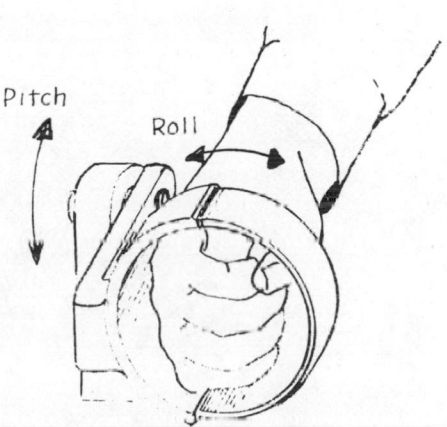

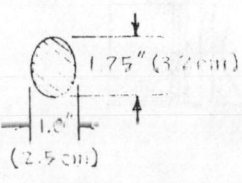

## PENCIL JOY STICK

The pencil joy stick gets its name from the fact that it is generally held like a pencil or pen. It is useful for console situations in which an operator manipulates cursors or other electronic graphics on a CRT display.

The more "pencil-like" the device is, the easier it is to operate. The accompanying sketch shows that the ideal approach would be to provide the operator with a "wand" with which the basic controller arm can be moved through a cone about 2 in (5.0 cm) in diameter.

A cup positioned approximately at desk surface height receives the wand, making it possible to "write" with the controller.

Perhaps the next best alternative is to place the control pivot point below the desk surface, as shown in the accompanying sketch.

Various manufacturers offer other alternatives, as shown in the accompanying sketches. Although these do not provide the optimum "writing" characteristics discussed above, they are generally satisfactory. Avoid selecting devices that have handle diameters considerably different from those shown here.

Although the switch-handle configuration is acceptable, the smaller joy stick with a separate, left-hand-operated switch on the console is preferred.

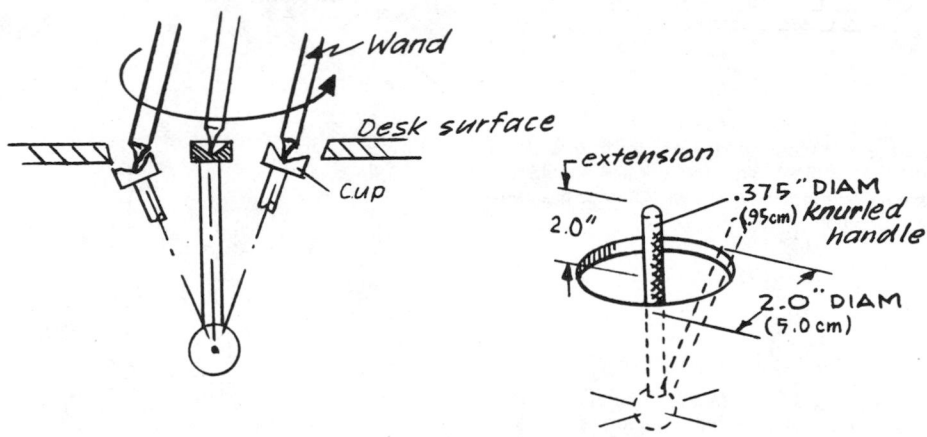

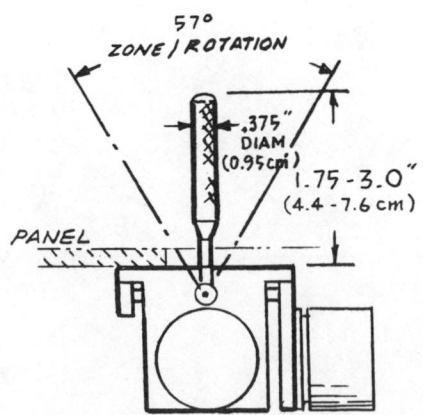

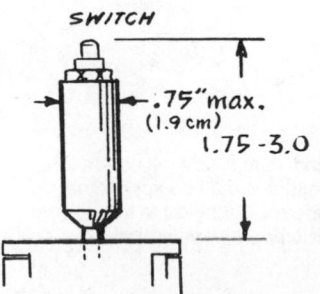

Pencil joysticks should be self-centering

Least preferred – awkward, easy to disturb stick position while operating switch – place switch for operation with other hand.

## JOY-BALL CONTROLLER

Although slightly less desirable than the pencil joy stick from the point of view of operator accuracy, the joy-ball controller is sometimes preferred in order to eliminate the interference problem occasionally associated with the protruding joy stick.

A separate switch should be provided to disengage the joy ball when it is not being used.

Joy balls are not self-centering and therefore have the disadvantage that one can easily leave the controller in a position in which the display hook or cursor is completely out of view. When joy balls are used, it is desirable to design the control system so that the displayed elements will always remain in view on the CRT.

The accompanying sketch illustrates available joy-ball hardware.

### "Mouse" Controller

The "mouse" positioning controller is a fairly recent development designed to replace the other types of coordinate position designation and cursor positioning. When experimentally compared with other devices, the mouse was found to be easier to use, faster, and more accurate.[15] As shown in the accompanying sketch, X/Y coordinate wheels generate positional signals, making it possible to use the device on any flat surface, as opposed to placing a controller on crowded console surfaces or having to cover portions of a display, as in the case of light pens and the like.

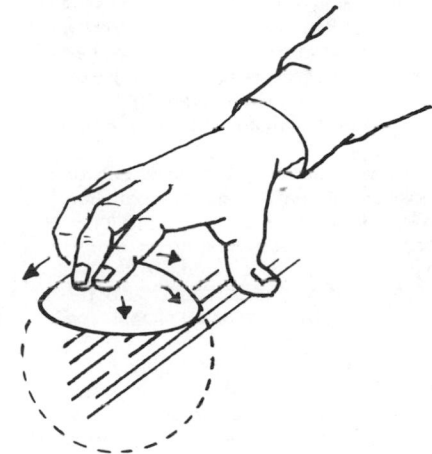

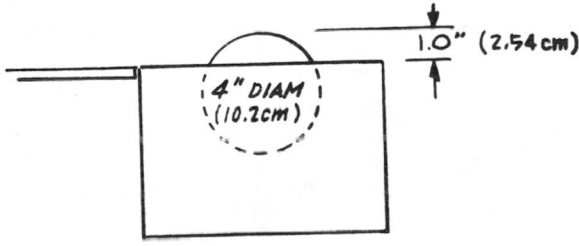

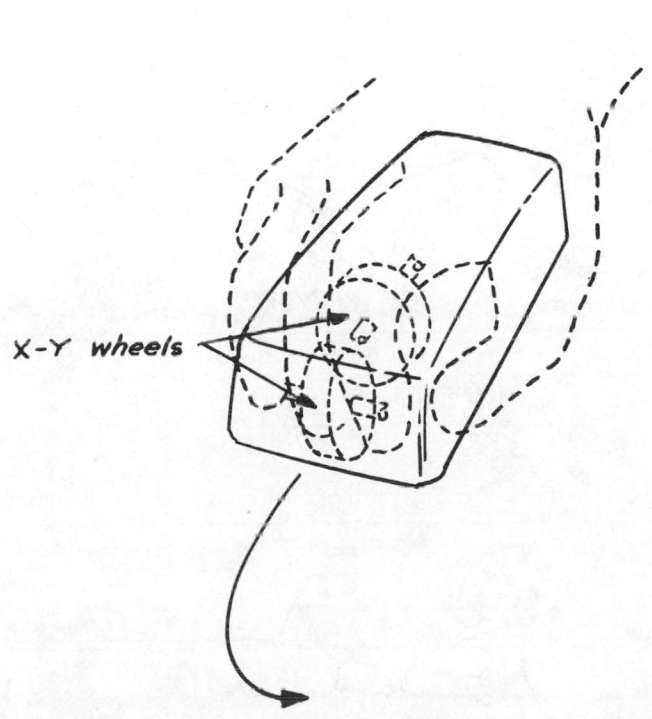

X-Y wheels

[15]S. K. Card, et al., "Evaluation of Mouse, Rate-controlled Isometric Joystick, Step Keys, and Text Keys for Text Selection on a CRT," *Ergon,* vol. 21, p. 8, 1978.

## GEAR-SHIFTING CONTROLS (AUTOMOTIVE)

Although there are inherent conflicts with recognized human engineering principles regarding the direction of movement of lever controls, automotive gear-shifting patterns have been in use sufficiently long that people have come to expect them to articulate in relatively standard patterns. Mechanical and automatic gear systems are illustrated in the accompanying sketches.

The several dimensional guidelines shown in these sketches are not to be taken as hard-and-fast requirements; rather, they are guidelines for designing control handles that will be acceptable to most drivers, regardless of their hand size.

There should be enough spacing between gear positions so that the operator can "feel" the change in position, i.e., so that he or she can locate the new position from an adjacent one in the same direction. In addition, the lever should also be designed so that the lever pawl is spring-loaded, which gives additional "feel" to the positioning action.

A separate action should be required of the operator in order to select the reverse position (to preclude inadvertently placing the lever into reverse gear when the vehicle is moving forward). A push button is recommended for the console-mounted system, and a "pull toward the driver" for the column-mounted system.

Incidentally, the column-mounted automatic lever position is preferred over the console configuration because in this position the position displays are easy to see from the driver's normal "eyes-on-the-road" sight line.

The force required to move a gear-shifting lever should not exceed about 10 lb (4.5 kg), nor should it be less than 2 lb (0.9 kg) to preclude inadvertent actuation.

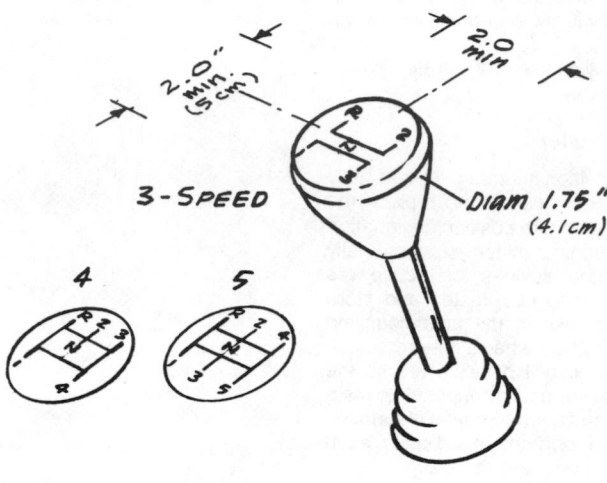

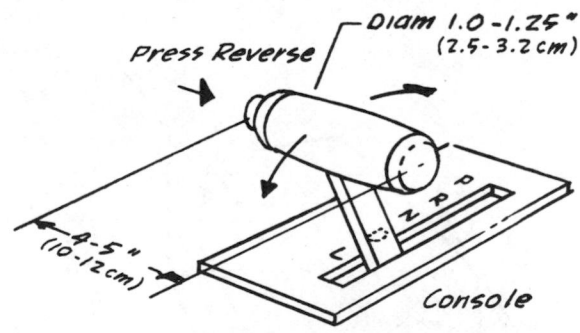

AUTOMATIC

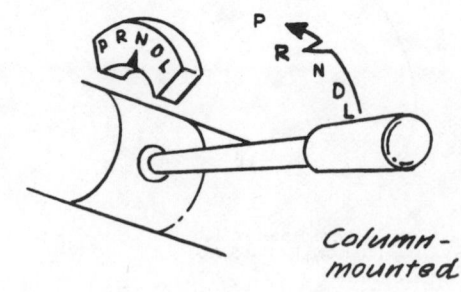

Column-mounted

## Other Gear-Shifting Lever Handle Configurations

The key factors to consider in the design of any gear-shifting lever handle interface are the following:

Make the motion compatible with the user's natural expectations; i.e., the user should push forward or lift up on the lever to cause an increase in function, value, or forward motion and should pull back for the opposite effects.

To prevent inadvertent movement into a particular gear, provide some mechanical deterrent that requires the operator to perform a special, secondary action such as moving the lever in and out of a slotted channel or pressing an auxiliary lever.

Provide a handle design that is compatible with the user's hand shape and size and with the strength of his or her fingers, such as for squeezing (see the accompanying sketch).

Avoid certain types of release concepts, such as those illustrated in the accompanying sketches.

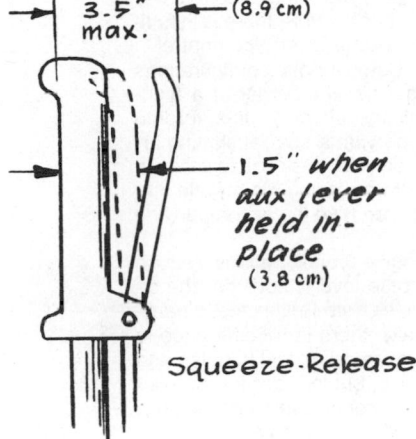

3.5" max. — (8.9 cm)

1.5" when aux lever held in-place (3.8 cm)

Squeeze-Release

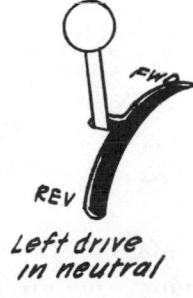

FWD

REV

Left drive in neutral

Spring-load against out board side of slot.

Right drive slow forward

Slotted Neutral Position for Dual-Drive System

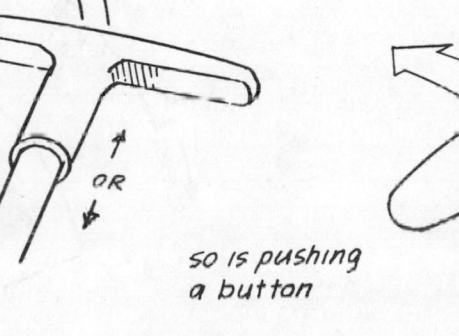

Pulling on lever at same time it is being translated is very awkward!

OR

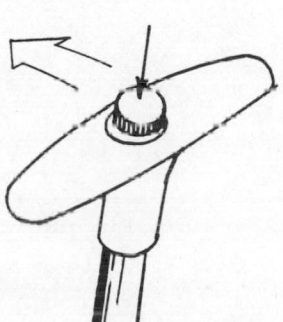

so is pushing a button

AVOID

## AIRCRAFT THROTTLE CONTROLS

Aircraft throttle control concepts are perhaps among the best established of all control items. In general, these concepts follow good human engineering direction-of-motion principles and thus have extremely good expectancy response characteristics.

The only exception might be the configuration, shown in the accompanying sketch, in which the single pull-type knob is sometimes used for smaller aircraft where there is insufficient space to provide a lever-type control.

When special power-boost configurations are required (e.g., the afterburner in a high-performance military aircraft), the throttle quadrant should provide a special slotting arrangement that requires the pilot to make a special motion before placing the throttle into the after burner mode (see the accompanying sketch).

When dual engine configurations require two separate throttle levers, consider the dimensional guidelines noted in the accompanying sketch. Actually, there is no difference in terms of use between cylindrical handles and ball-shaped handles, but in order for the pilot to articulate the individual levers comfortably, the dimensions shown should be used.

When thrust reversing is provided, it is necessary to include a safety feature that will prevent inadvertent engine reversing. One typical method is shown in the accompanying sketch; before the pilot can pull the levers back into the reverse regime, they must be lifted.

*Note:* When multiple cylindrical handles are used on throttle levers, it is recommended that the handles have slight depressions that the pilot can "feel" to help in differentiating one handle from another.

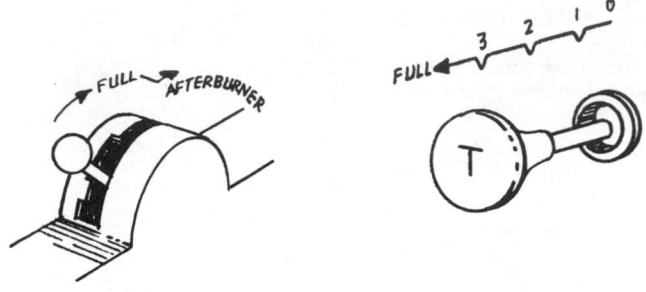

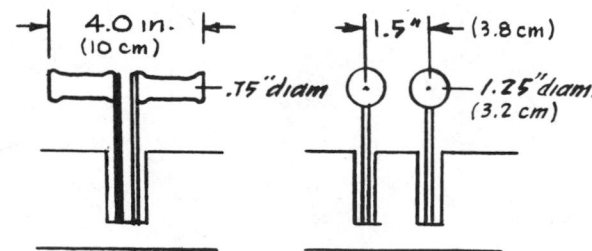

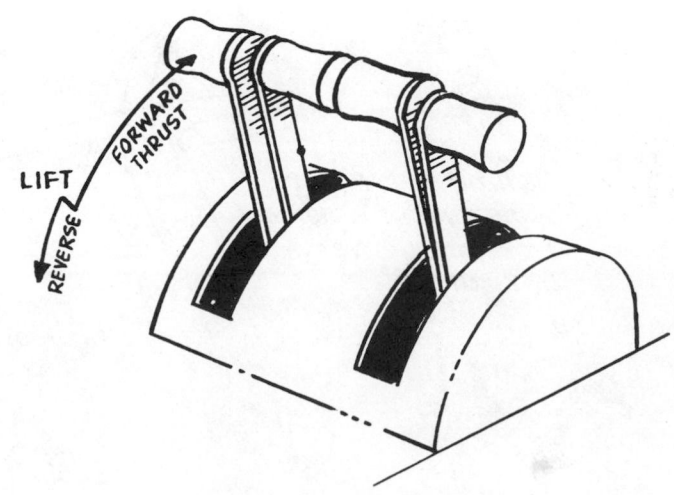

## AUTOMOTIVE ACCELERATOR PEDALS

The experienced driver has an uncanny ability to find and operate a variety of accelerator pedal configurations, as long as they are generally located slightly to the right and forward of the seat. However, if the placement and angle of the pedal depart appreciably from the guidelines shown in the accompanying sketches, the driver may soon become fatigued because of the necessity to twist the body or extend the leg and ankle to fit a poorly configured accelerator pedal arrangement. Key considerations are the following:

The driver should always be able to rest his or her heel on the floor or on a heel stirrup while holding or depressing the pedal.

The pedal should be big enough so that both the small- and the large-footed driver can press the pedal with the ball of the foot.

The "release angle" (nonactive) of the pedal must not be too steep; otherwise, the driver's ankle tires easily during the time he or she releases pressure on the pedal.

The force required to depress the pedal should not exceed about 20 lb (9.1 kg) and should not be less than 10 lb (4.5 kg) in order to support the foot weight.

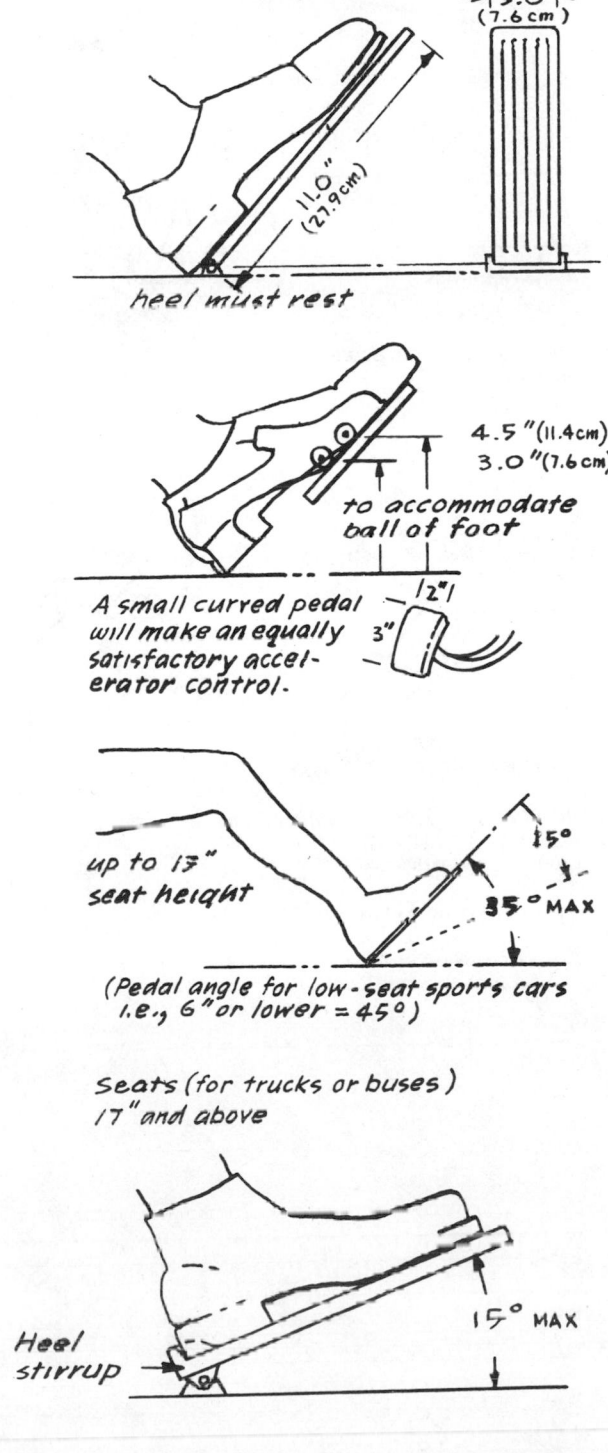

Suggested Pedal Dimensions

3.0" (7.6 cm)

11.0" (27.9 cm)

heel must rest

4.5" (11.4 cm)
3.0" (7.6 cm)
to accommodate ball of foot

A small curved pedal will make an equally satisfactory accelerator control.

2"

3"

up to 17" seat height

15°
35° MAX

(Pedal angle for low-seat sports cars i.e., 6" or lower = 45°)

Seats (for trucks or buses) 17" and above

Heel stirrup

15° MAX

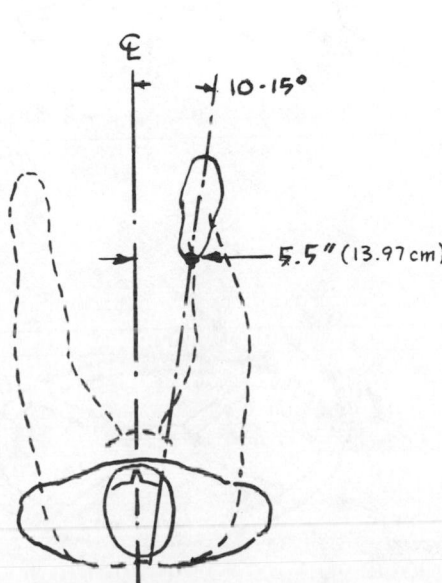

10-15°

5.5" (13.97 cm)

### HIGH-FORCE MANUAL CONTROLS[16]

When there is no alternative and an operator is required to apply considerable force to a controller, consider the following guidelines:

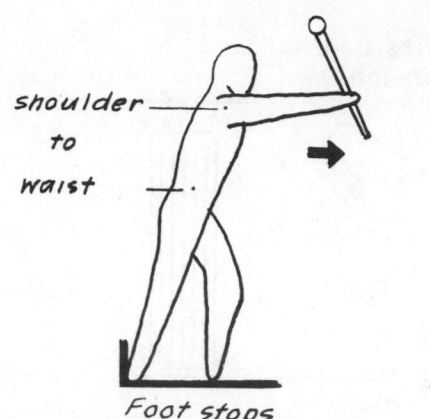
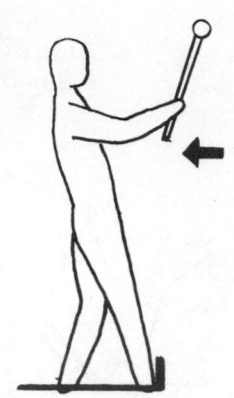

Position the levers between the operator's waist and shoulder levels.

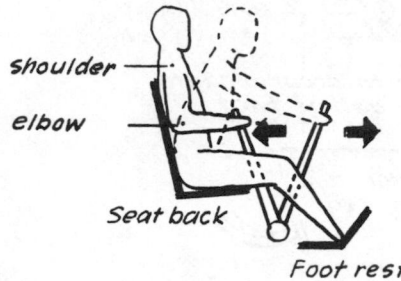
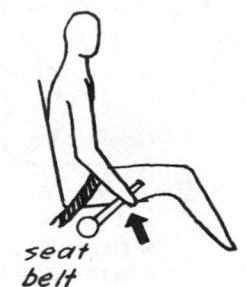
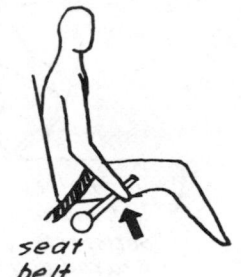

Provide something the operator can pull or pull against, or secure the operator so that his or her body is not displaced.

Increase the lever length or the wheel diameter, but not to the point where the driver cannot reach the lever or wheel comfortably.

Provide the proper size of handle to maximize gripping capability.

Allow the operator to use two hands.

Consider the use of a foot-operated control.

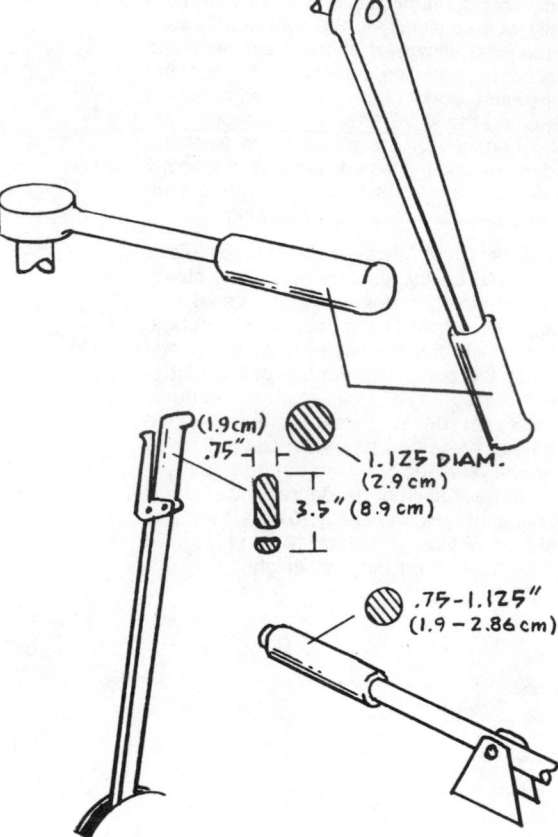
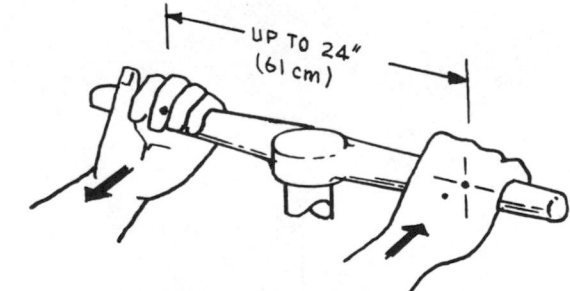
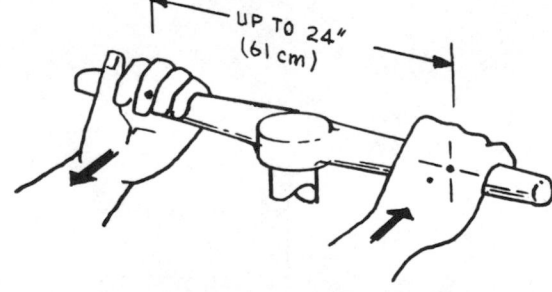
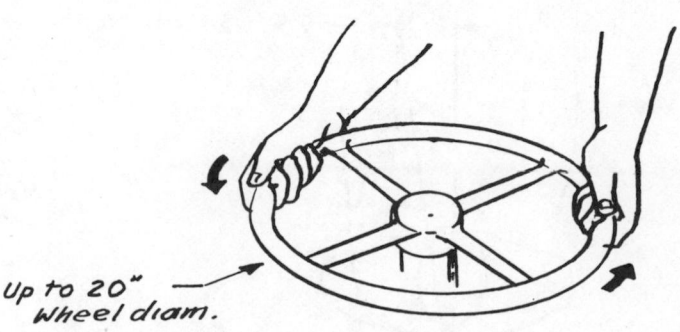

[16] See the section on human strength capabilities in Chap. 4.

## Maximum Force Applications

When operators are required to apply the absolute maximum force of which they are capable, use a foot-operated controller. In order for such operators to obtain their maximum capability, they should be seated so that they can push against a solid backrest, as illustrated in the accompanying sketch. The point at which the ball of the operator's foot contacts the pedal should be approximately at the same height as the operator's seat reference point, with the seat backrest at the angle shown.

It is necessary to provide a seat that is adjustable fore and aft, however, so that operators with different leg lengths can be in a position that will allow them to obtain the optimum upper and lower leg angles shown.

Remember, however, that people cannot apply maximum force for sustained periods, nor can they apply maximum force repeatedly over a long period of time.

## Vehicular Pedal Relationships

A typical array of clutch, service brake, and accelerator pedals should provide an approximately common pedal level when the pedals are in the nonactuated position. Since the brake and clutch pedals generally involve some leg motion, they should be suspended from above; the accelerator pedal involving only ankle movement must pivot from below. The centerlines of the brake and clutch pedals should be at least 8 to 10 in (20 to 25 cm) apart and should be placed symmetrically in front of the driver. If the vehicle has an automatic transmission (no clutch), place the brake pedal on the driver's centerline; the pedal should be approximately 6 to 8 in (15 to 20 cm) wide so that it can be operated by either foot.

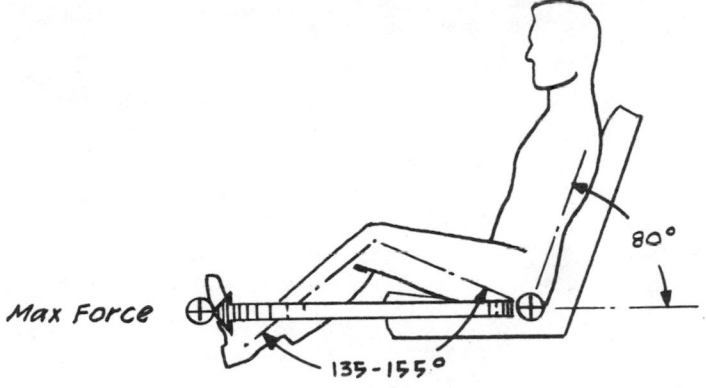

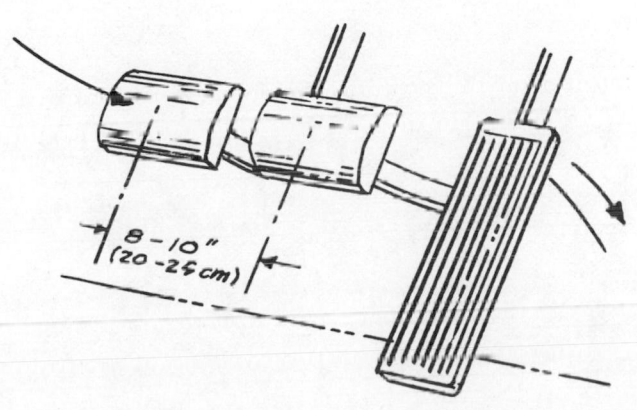

## FOOT-OPERATED CONTROLS

### Brake Pedals

Since many brake pedal controls require some amount of force, the geometric relationship between the operator's leg and foot and the position and angle of the pedal is an important consideration; i.e., depending on the height of the seat, not only must foot pedals be placed within reach, but they also must operate in a direction that is compatible with the force application vector.

A—A high seat position wherein the operator presses down more than he or she presses forward. Brake forces should not exceed about 20 lb (9.0 kg). The braking stroke should articulate downward and forward in a straight line, as shown.

B—A midposition seat height wherein the operator presses about equally forward and down. Brake forces should not exceed about 40 lb (18.0 kg). The stroke should follow a reversed curvilinear pattern, as shown.

C—A low seat height wherein the operator can obtain maximum force because he or she can take maximum advantage of the seat back. However, the maximum braking requirement should not exceed about 140 lb (63 kg) for vehicles operated by males and females and 180 lb (82 kg) for vehicles operated only by males. The stroke should follow a reversed curvilinear pattern, as shown.

*Note:* In all cases, the optimum angular relationship between the lower leg and the pedal surface plane should be approximately 90°. The accompanying illustration shows the general relationships that must be considered with respect to force vector and hip, knee, and ankle pivot points.

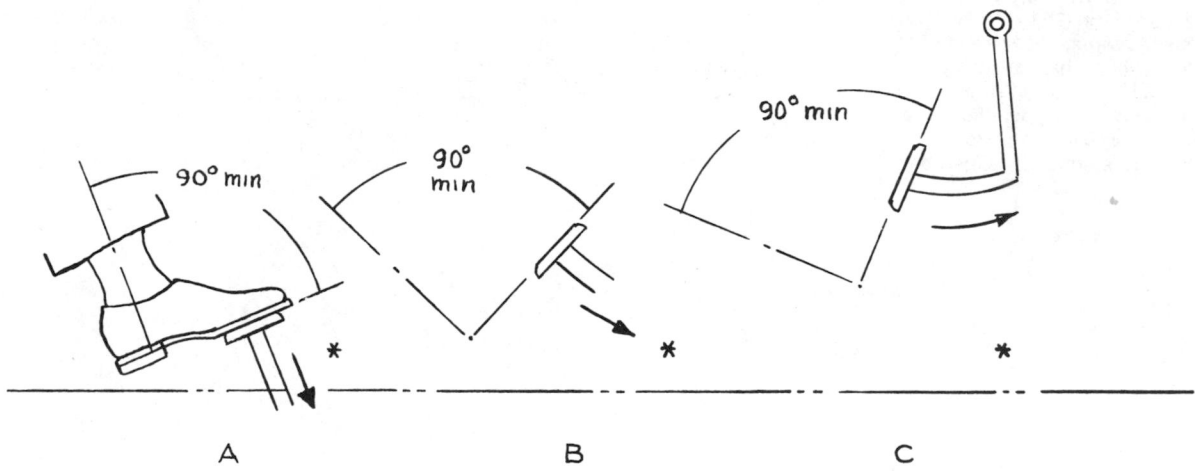

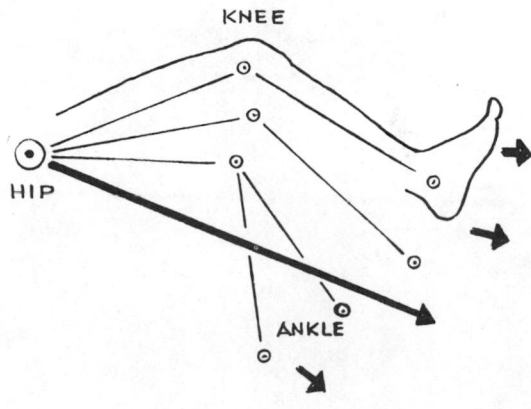

## Pedal Considerations for Vehicles with High Driver Positions

For some applications, such as the driver station of buses or heavy cargo tractors, the driver's position is typically quite high, and the plane of foot pedals is at a low angle (20° or less). In such cases it is desirable to select a pedal which includes a heel stirrup, as shown in the accompanying sketch, because operators tend to "miss" placing the heel at the pivot point of the pedal and therefore have to apply too much force because the foot is too far back on the pedal.

Heel stirrup designs are also appropriate when, for some reason, one cannot place the fulcrum or heel reference on the floor of the vehicle, i.e., when the foot is suspended in midair.

If two stirrup-type pedals are used together (as for aircraft rudder pedals), the spacing between the pedals should be at least 3.0 in (7.6 cm) but not more than 12.0 in (30.5 cm). The preferred and most comfortable spacing is about 6.0 in (15 cm). Lateral separation between the pedals can vary acceptably between 4 and 8 in (10 and 20 cm). The pedals should rotate freely on spindles and should be made of rubber or provided with some other nonskid feature.

### Bicycle Pedals

| Adult Bicycle | Youth Bicycle | Toddler Tricycle |
|---|---|---|
| $r = 7.0$ in (17.6 cm) | $h = 6.0$ in (15.2 cm) | $r = 4.0$ in (10.2 cm) |
| $w = 4.0$ in (10.2 cm) | $w = 3.5$ in (8.9 cm) | $w = 2.5$ in (6.4 cm) |
| $d = 3.0$ in (7.6 cm) | $d = 2.5$ in (6.4 cm) | $d = 2.0$ in (5.0 cm) |

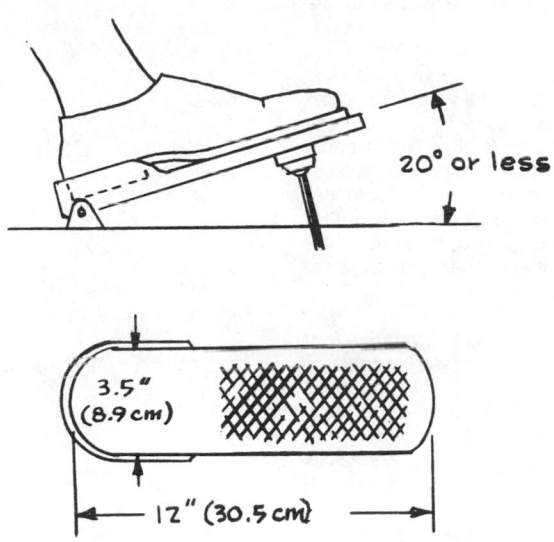

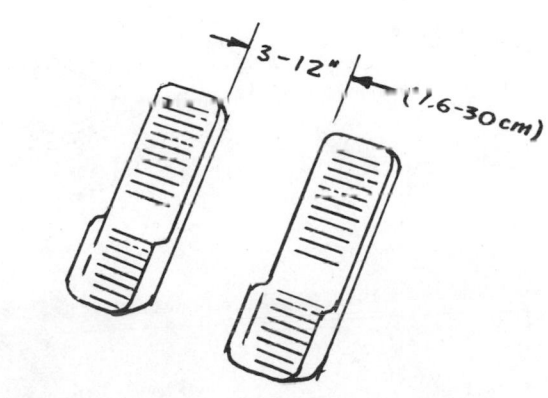

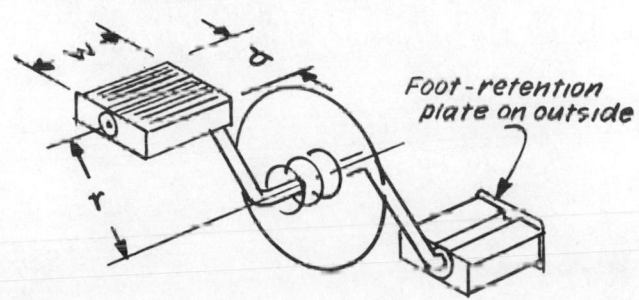

## Foot Switches (Standing Operation)

Generally speaking, foot-operated controls are not recommended because it is difficult for standing operators to operate a pedal without losing their balance. Proper location requires that the foot pedal be placed so that a minimum of movement is required to reach it. By the same token, the choice of location should take into account possible inadvertent actuation.

Portable foot pedals are suggested for many operations since they allow individual operators to place the pedal where it is most convenient for the particular task. There are no current standards regarding switch pedal size, either the mechanically operated type or the pressure-sensitive type (see the accompanying sketches). However, avoid pedals that are too small, since they are difficult to locate without looking for them. The best angle for the mechanically operated type is about 10° or less.

If it appears advantageous to operate a foot switch alternately with both feet because of some other requirement to change operator position, either a bar-type foot or a knee switch configuration should be considered. This type of alternate foot or leg access is particularly important for applications where a safety cutoff feature is necessary. The significant advantages are that an operator does not have to look for the switch and does not have to be very accurate. The accompanying illustration notes the important dimensional considerations. The pedal-bar height should not exceed about 6.0 in (15 cm), and it should not extend into the normal footprint zone.

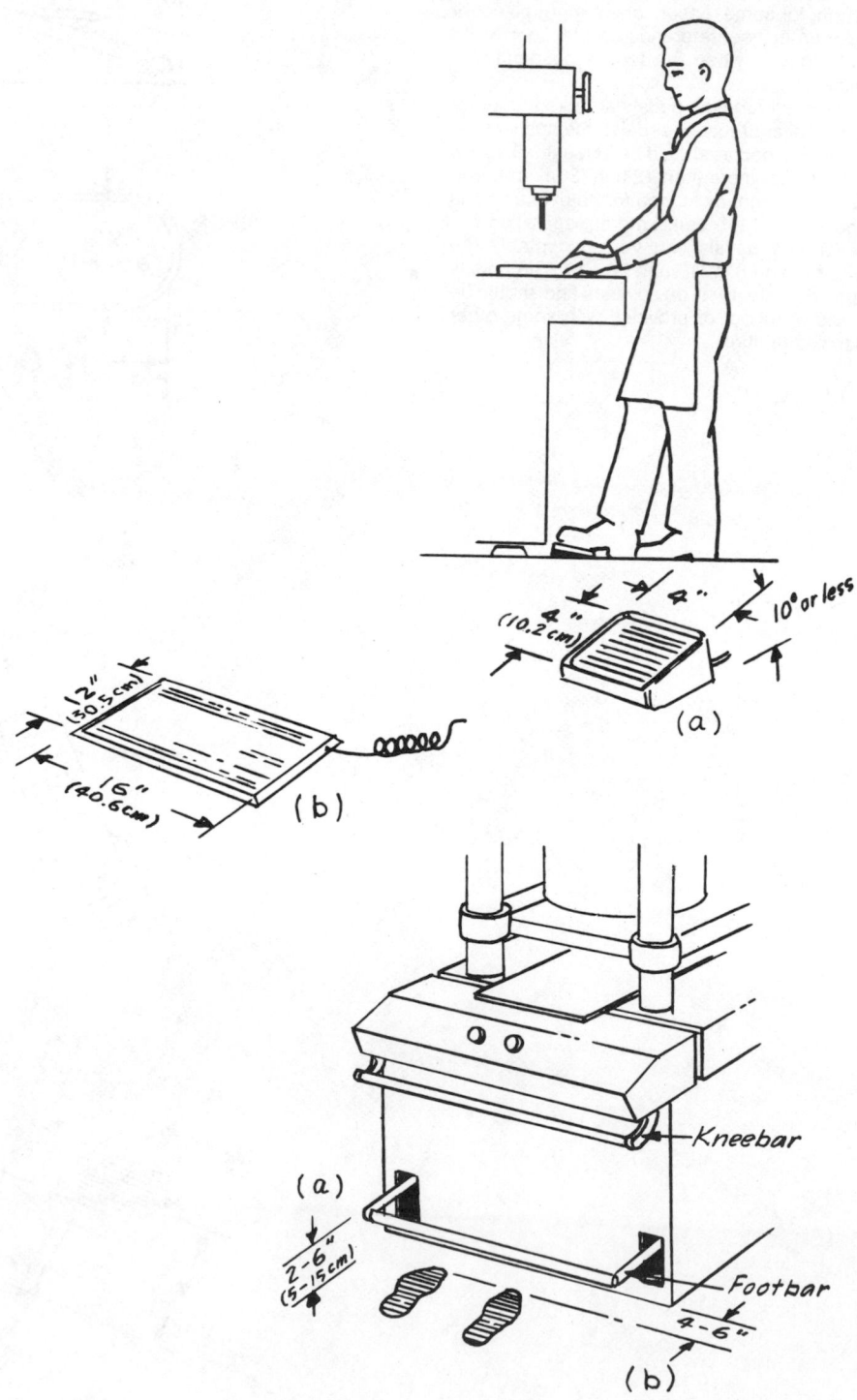

## Knee-Operated Switch (Seated Operator)

A knee-operated control is useful to free an operator's hands for other tasks. The primary objective is to position the knee bar so that operators of various sizes can use it easily and also operate it regardless of variations in their seating positions.

The knee bar should be made large enough to account for these variations. The accompanying illustration shows recommended distances forward of the nominal seat reference (seat-pan–backrest juncture) and above the floor for the center of the knee bar. Usually the knee bar should be about 6.0 in (15.2 cm) tall by 8.0 in (20.3 cm) long in order to accommodate both small female and large male operators.

*Note:* All corners facing the operator should be radiused, and the leading edge should curve outward slightly to minimize the possibility that operators will injure their knees as they slide into place (see the accompanying sketch).

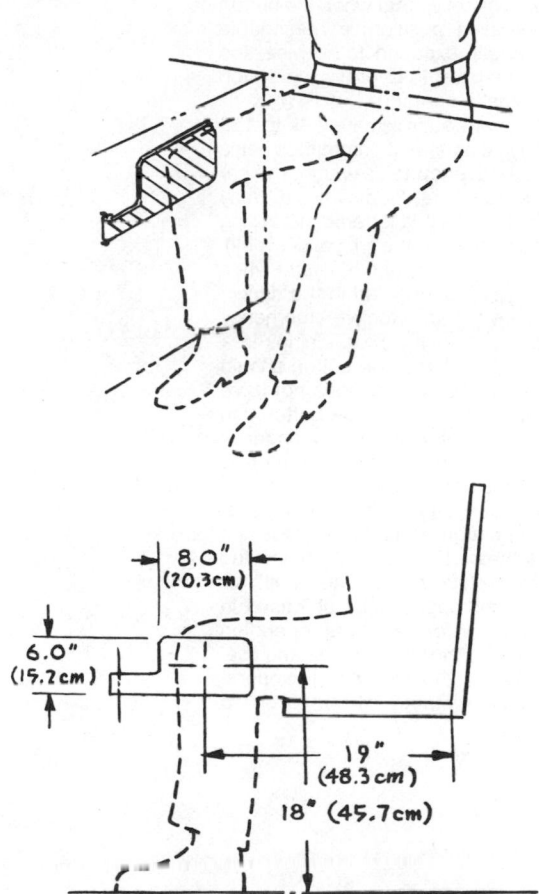

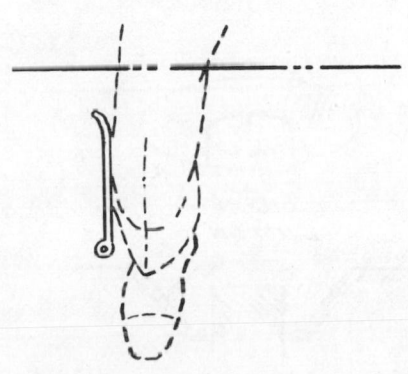

## PUSH BUTTONS

A variety of push-button sizes and shapes are acceptable as long as certain conditions are maintained. The smallest diameter should not be less than about ¼ in (0.64 cm) for bare-hand operation or ½ in (1.3 cm) when gloves are worn.

Normally, the button should extend far enough above a panel so that when the button is in the depressed position, a reasonable amount of it is still exposed (otherwise, the operator may not be able to press the switch far enough to make contact).

When buttons are approximately ¾ to 1 in (1.9 to 2.5 cm) in diameter, it sometimes helps to provide a concave top to the button (finger centering). An additional "location aid" may be provided in the form of a tapered, recessing bezel (such as is found in the typical doorbell).

Illuminating the push button is useful when the button needs to be located in the dark.

The push button is an effective emergency control device for shutting off machines in a hurry. In such applications, the button should be larger so that the operator does not have to be so accurate, i.e., so that the button can be operated either with the end of a finger or with the heel of the hand. As the button is made larger, consider a concave surface.

When it becomes important to preclude inadvertent push-button actuation, recess the button so that the button is actually below the adjacent panel surface plane. The "well" for such configurations needs to be at least 1 in (2.5 cm) in diameter for bare-hand operation and at least 2 in (5.1 cm) for gloved-hand use. Tapering the sides of the well helps the operator find the button without having to be so accurate.

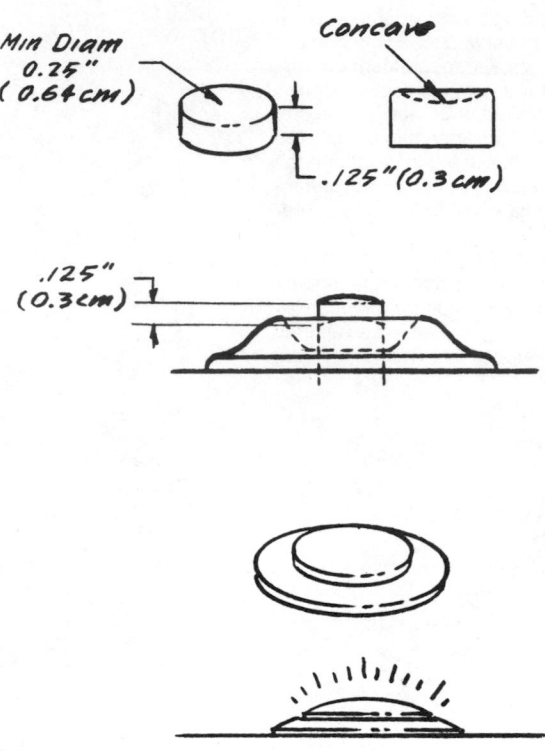

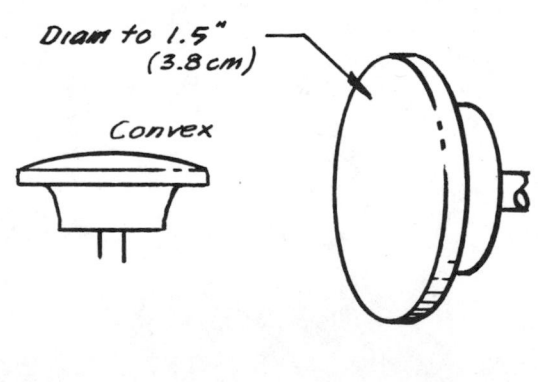

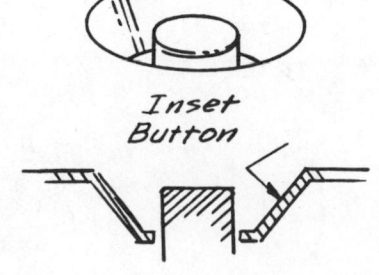

## Grouping Push Buttons

As indicated by these guidelines, adequate button separation is required to avoid inadvertent operation of adjacent switches.

### Minimum Button Separation

| Arrangement | A | B | C | D | E | F |
|---|---|---|---|---|---|---|
| Vertical plane: | | | | | | |
| No gloves | 0.75 in (1.9 cm) | 0.625 in (1.6 cm) | 0.1875 in (0.476 cm) | 0.25 in (0.64 cm) | 1.25 in (3.17 cm) | 0.75 in (1.9 cm) |
| With gloves | 1.75 in (9.45 cm) | 1.625 in (4.12 cm) | 0.375 in (0.95 cm) | 1.25 in (3.17 cm) | 2.25 in (5.72 cm) | 1.75 in (4.45 cm) |
| Horizontal plane: | | | | | | |
| No gloves | 1.0 in (2.54 cm) | 0.875 in (2.22 cm) | 0.4375 in (1.11 cm) | 0.50 in (1.27 cm) | 1.50 in (3.81 cm) | 1.0 in (2.54 cm) |
| With gloves | 2.0 in (5.1 cm) | 1.875 in (4.76 cm) | 0.375 in (0.95 cm) | 1.25 in (3.17 cm) | 2.25 in (5.72 cm) | 2.0 in (5.1 cm) |
| Under severe vibration or oscillation | 3.0 in (7.6 cm) | | | 3.0 in (7.6 cm) | | 3.0 in (7.6 cm) |
| *For blind selection* | 6.0 in (15 cm) apart in front of operator; 12 in (31 cm) apart when buttons are located in the peripheral areas. | | | | | |

*Note:* The above guidelines apply to any shape of button.

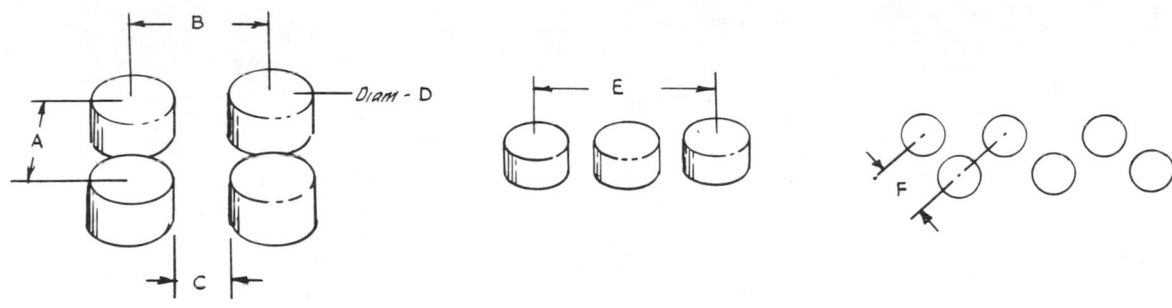

## Push-Button Shape

From a biomechanical point of view, it makes no difference whether one selects a round, square, or rectangular push button. However, the square or rectangular button provides more useful area for labeling, and obviously a rectangular button laid on its side provides more label space than one that is standing on end.

Shape can also be used as a cue for determining the functional significance of a particular push button. This is especially important when advisory indicators are used in combination with push buttons. A round push button is seldom mistaken for an advisory indicator.

Off-the-shelf push-button configurations such as those shown in the accompany illustration are popular for many applications. The separator bar between buttons is also helpful in preventing the operator from accidentally pressing an adjacent button. These devices are generally available for either horizontal or vertical button arrays. The dimensions shown are representative.

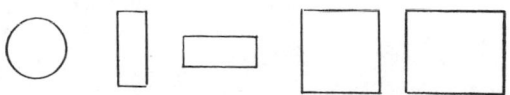

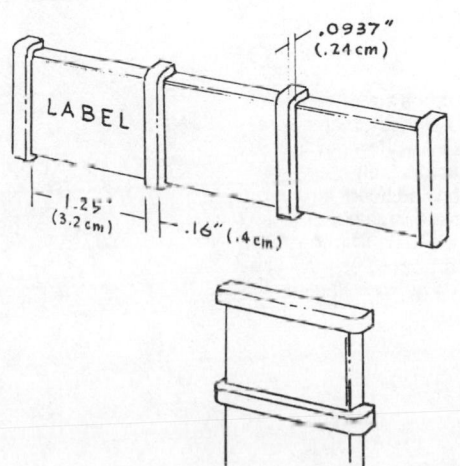

When inadvertent button operation could be serious, other, more accentuated guard systems should be considered. The one illustrated is effective because the height of the guard prevents the operator's finger from "slopping over" into the adjacent button area, and it also allows the labels to be seen with minimum interference.

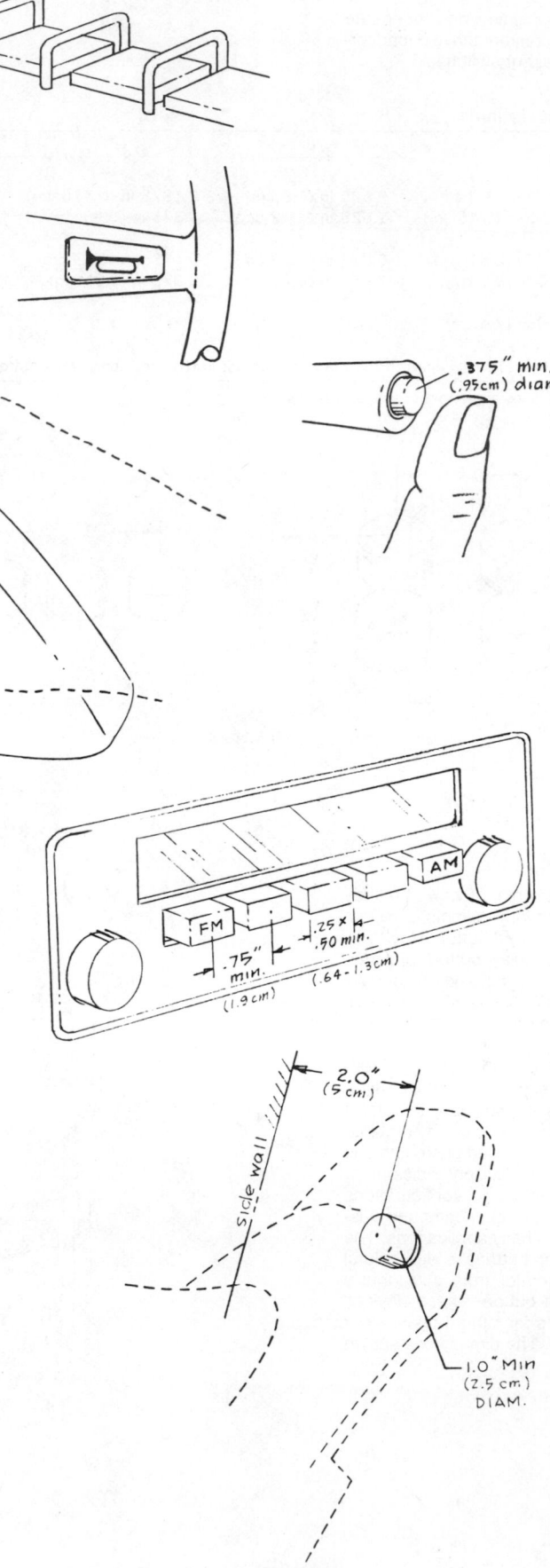

*Guards*

Push buttons can, of course, take on the shape of a particular surround without a reduction in their effectiveness. An example might be the "horn button" that is placed on a steering-wheel spoke; this button can be slightly tapered to conform more aesthetically with the shape of the spoke.

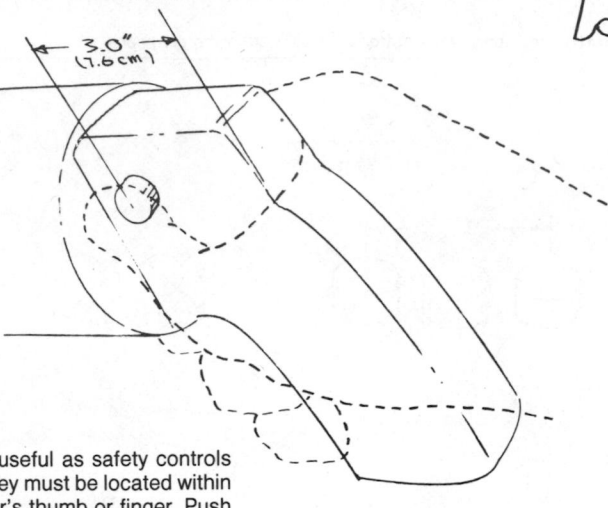

Push buttons are useful as safety controls for hand tools, but they must be located within reach of the operator's thumb or finger. Push buttons should be at least 0.25 in (0.64 cm) in diameter and should extend about 0.125 in (0.318 cm)

A push button located in the end of a handle should be visible from the operator's normal eye reference position.

Push buttons used for special equipment in vehicles (e.g., a radio) should not be so small that the driver must look at the switch to select the right button. Also remember that the driver may be wearing gloves, which requires more separation between the switches.

Foot-operated push buttons are often useful to alleviate the work load placed on an operator's hands. More than one button per foot is not recommended, however. Make sure that there is space between any adjacent structure and the button so that there is plenty of room for the foot. However, too much space may not be desirable since an operator may use the adjacent structure as a guide in finding the button.

Push-button forces should be within the range of 10 to 20 oz (283 to 567 g) (although they can be as high as 40 oz, 1134 g) in order to reduce the possibility of inadvertent actuation.

Push-button action should be positive; i.e., there should be elastic resistance aided by slight sliding friction, starting slowly, building rapidly, and with a final sudden drop to indicate activation. An audible snap action is helpful under high-noise–level conditions, although the snap action must not be too heavy if the button is to be operated frequently over extended periods of time.

## Contact Switches

Although not a true push button, the contemporary "contact switch" is often used to replace the mechanical push button. Since this type of switch provides no three-dimensional cue for the operator, it is important to provide extremely clear graphic delineation of each switch boundary.

It must also be made clear, by means of appropriate labels, which are switch functions and which are purely visual readouts, since both the switches and the readouts are flush with the panel surface.

*Note:* Contact switches should not be used where continuous, long-duration operations are expected, since it has been noted that operators tend to strike the push-button surfaces harder because of a lack of tactile or auditory feedback. Some complain of sore fingers after a long operating session.

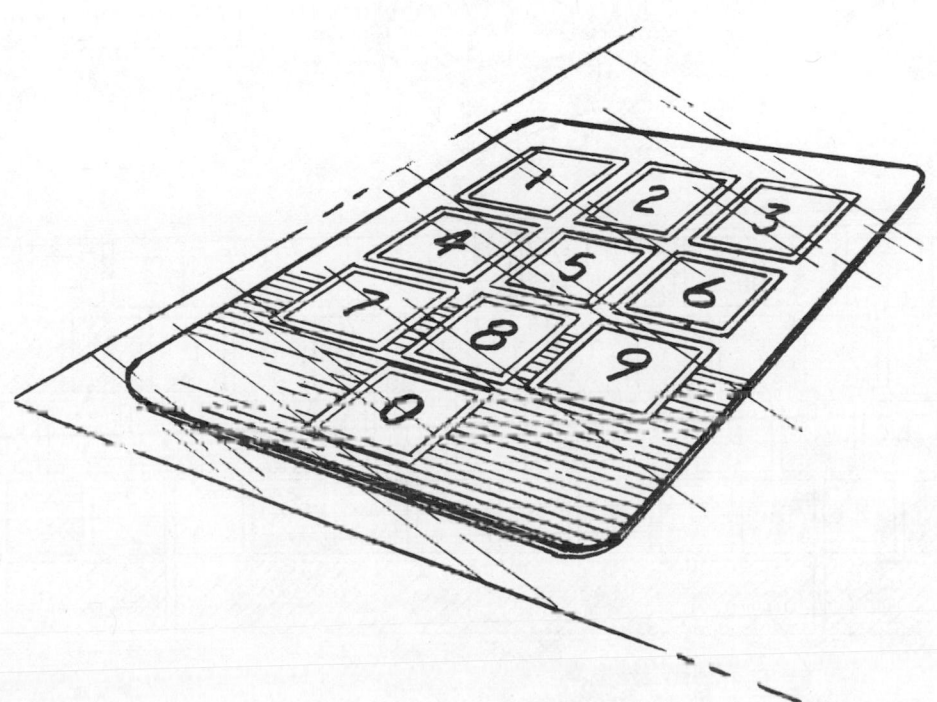

## TYPEWRITER KEYBOARDS

Often referred to as the "QWERTY keyboard," the so-called standard arrangement of certain keys should be maintained in order not to confuse experienced typists. The guidelines provided here should be followed. Spare keys can be utilized for special functional requirements.

### Typewriter Keyboard Operating Characteristics

Typewriter keyboards should be designed so that they can accept key strokes at a rate of about 20 per second, with short burst rates of up to 50 key strokes per second.

The force required to actuate the typical typewriter key should not exceed about 5 oz (142 g). However, to reduce inadvertent actuation, there should be at least 2 oz (57 g) of resistance provided. An adjustable force control is highly desirable, especially for mechanical machines.

Provide an acoustic shroud around the portion of the typewriter housing the principal noise-producing elements of the machine.

Provide a visible and intelligible centering display to aid the typist in identifying where the key will strike the sheet of paper, i.e., permanent indexing marks that will not be covered by the typewriter ribbon or other elements of the machine.

The ON/OFF control should be placed where it can be readily found, even by the inexperienced operator. Provide some type of indication to let the operator know when the machine is on and design the system so that, in the event the operator forgets to turn off the machine, it will automatically shut itself off after a given amount of nonoperating time.

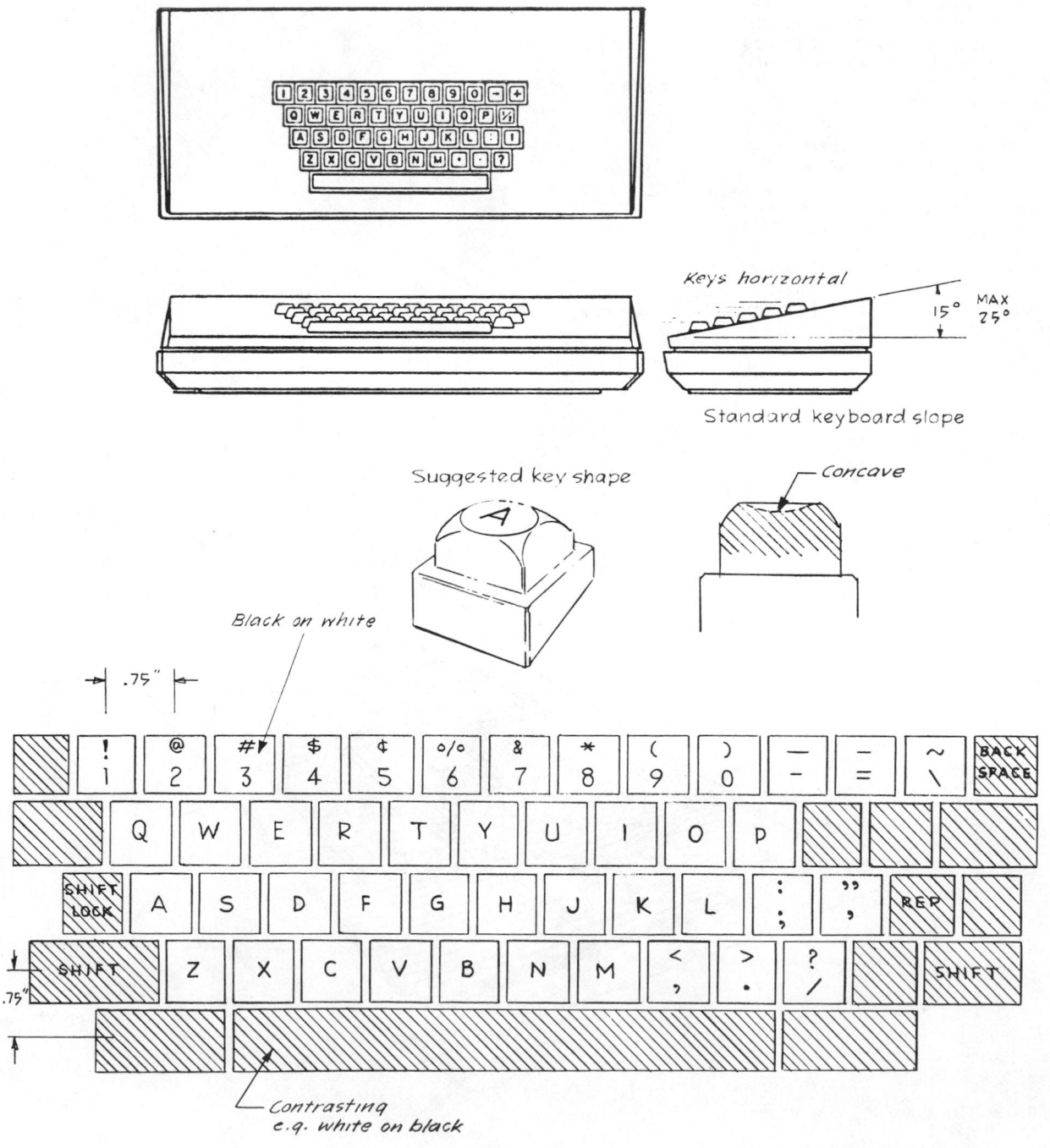

Keys horizontal    15° MAX 25°

Standard keyboard slope

Suggested key shape    Concave

Black on white    .75"

.75"

Contrasting e.g. white on black

## CALCULATOR KEYBOARDS

Only the numerical keys on calculator keyboards have been somewhat standardized. Additional standardization would help the operator, at least for certain common functions.

The accompanying sketch illustrates suggestions for standardizing the common keying elements. The numerical keys are arranged in the order found on most calculators. The "clear" key is placed in the upper left-hand corner because this is the natural point to begin an operation; i.e., the operator first needs to make sure that the machine is clear for the first entry. The primary functions are arranged from bottom to top on the left of the numerical keys. Note that these are of a contrasting color. The "=" key is placed at the top of this series because the operator normally proceeds from this final operation to the display. Other functions should be separated from these more frequently used keys, and they should be in a contrasting color. Note that the "clear" and "=" keys are color-coded. Although the suggested colors are somewhat arbitrary, they do have significance; i.e., green suggests that the machine is ready, and yellow suggests caution before striking this final entry.

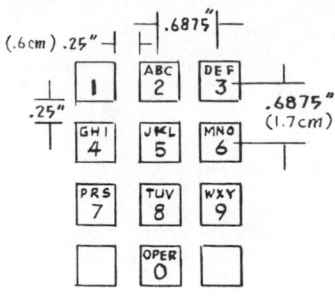

## TELEPHONE KEYBOARDS

The arrangement and dimensions shown in the accompanying illustration have been standardized for use with U.S. telephone sets.

The buttons are concave to aid the user in centering his or her finger on the button. The buttons are $^{11}/_{16}$ in (1.7 cm) square.

Although there are no hard-and-fast rules governing these dimensions, the arrangement of numbers and letters was determined experimentally.

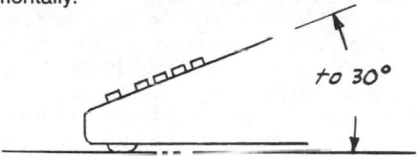

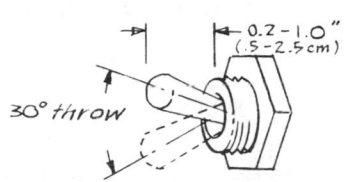

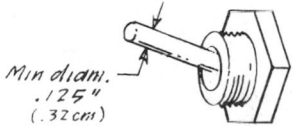

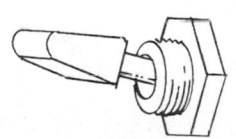

## TOGGLE SWITCHES

Toggle switches come in a variety of shapes and sizes, most of which are satisfactory. One should, however, choose a toggle switch in which the "throw" or displacement is at least 30° so that the operator can tell at a glance which position the switch is in.

A toggle-switch handle should be at least 0.5 in (1.3 cm) long, and preferably no longer than 1.0 in (2.5 cm).

For smaller, shorter switch handles, the resistance of the switch should not exceed about 8 oz (227 g); for larger switches, it should not be more than 15 oz (425 g).[17]

The sketches at the right illustrate typical handle shapes that are generally available. The first is the most typical. The second is often representative of very small toggle switches. Such a shape is not desirable when the handle is long, however, because the user can be injured if he or she bumps against the handle. The third handle shown provides added tactile identification when a switch is used in the dark. The fourth handle is especially good, not only because it makes it easier to identify the switch position, but also because the special handle can usually be obtained in various colors, making it possible to color-code certain switches.

Switch separation is especially important when several switches are arranged side by side or vertically in columns. Note the different spacing requirements for vertical and horizontal arrangements.

Although three-position toggle-switch hardware is available, it is not recommended. It is not possible to maintain the 30° "throw" when more than two positions are provided. It is suggested that, when three or more switch positions are required, some other type of switch be selected.

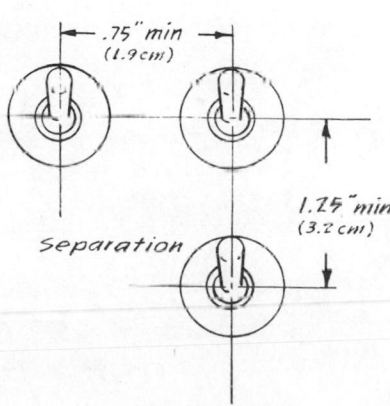

[17]Toggle switches should snap into place with an audible click.

## ROCKER SWITCHES

Rocker switches are a good substitute for the more common toggle switch to provide good physical indication of switch position. Like toggle switches, rocker switches, to be effective with regard to visual identification of position, should be designed so that the slope of the handle from the nominal plane is about 30°, and the nominal plane should be at least 0.125 in (0.32 cm) above the plane of the panel on which the switch is mounted. ·

Rocker switches may be broad and thus provide space for labeling, or they may be made narrower, as illustrated. In either case, however, the switches should be mounted so that the switch centerlines are no closer than 0.75 in (1.9 cm) when the switches are arranged side by side.

Switch resistance should be between about 8 and 12 oz (227 and 340 g) so that the switch is easy to operate and yet is not easily disturbed inadvertently.

Rocker switches typically can be obtained in different colors and thus can be color-coded for ease in differentiation. In fact, the two faces of a single switch usually can be obtained in two different colors, which makes it even easier to tell which switches are activated and which are not.

Rocker switches (like toggle switches) should be designed to snap into position with an audible click.

The rocker switch is preferred over the toggle switch when switches are laid out on a horizontal panel (see the accompanying sketch) because the toggle switch tends to protrude in this arrangement and is easy to become snagged on if an operator has to reach across it.

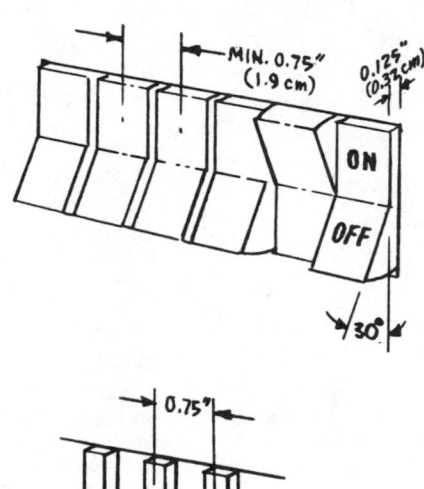

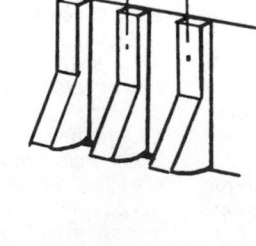

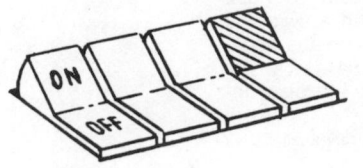

## ROUND CONTROL KNOBS

Round knobs should be used only for applications in which the rotation is continuous and smooth and/or when multirevolutions are required.

The smallest useful knob should have minimum dimensions of 0.25 in (0.6 cm) in diameter × 0.75 in (1.9 cm) in depth, and the knob should be either serrated or knurled. A maximum resistance of 2 to 4 oz (57 to 113 g) should be the most force expected by an operator. Such configurations should be used only sparingly, such as for a clock setting or for gross adjustment of certain noncritical displays.

The preferred minimum size of the knob for convenient operation of volume or illumination potentiometer controls is 0.50 × 0.50 in (1.3 × 1.3 cm), with appropriate serrations to ensure comfortable and effective gripping. The upper limit to ensure ability to make fine adjustments is approximately 4 oz (113 g) of resistance. The edge of the knob can be tapered slightly, although this adds nothing to the effectiveness of the knob. Avoid severe tapering, however, for the operator's fingers may tend to slip from the knob.

For most normal panel layouts, select a knob whose dimensions are approximately as shown in the accompanying sketch. These are easily accommodated even by an operator who is wearing gloves. Resistance can vary from 4 to 10 oz (113 to 283 g) as the diameter of the knob is increased.

Be particularly careful in accepting some of the decorative knobs produced by some manufacturers, for they sometimes possess features that make the knob difficult to handle. An acceptable alternative to the previous shape is shown in the accompanying sketch.

*Note:* Choose serration patterns that provide good gripping characteristics, i.e., sharper peaks or points rather than "humps."

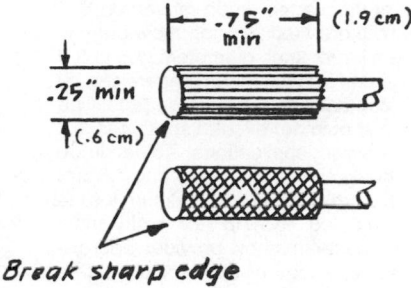

*Break sharp edge*

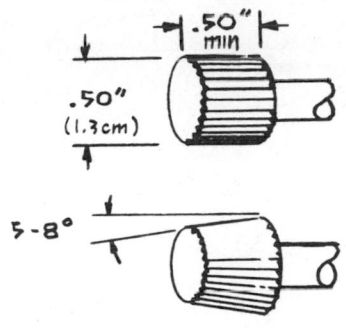

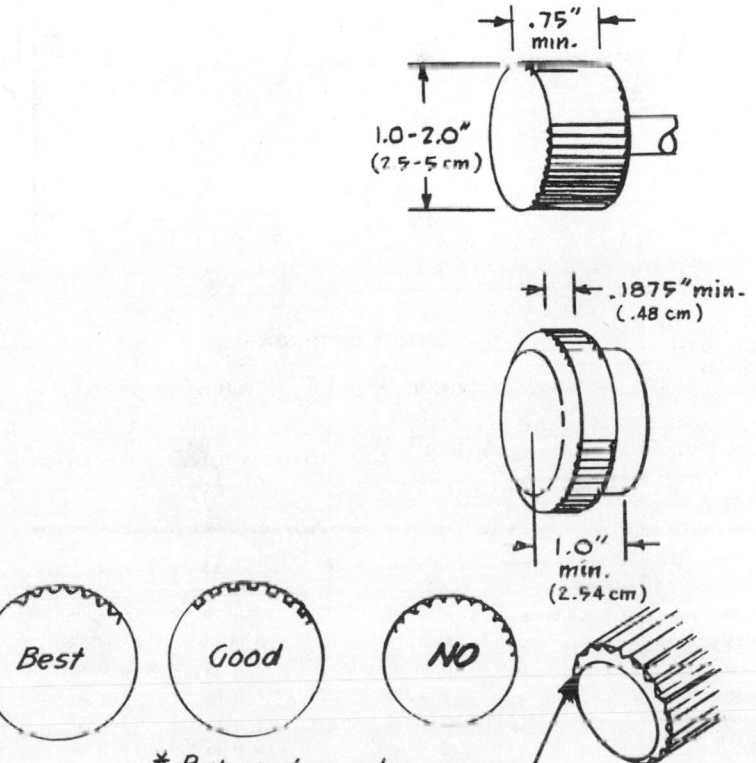

*But - radius edge*

Control knob or switch or potentiometer operating characteristics should include viscous damping as opposed to static resistance. A certain amount of resistance is desirable (approximately 2.0 in/oz in order to minimize the possibility of inadvertent knob movement).

Generally, more precise knob movement is possible with larger knob diameters (1.5 to 3.0 in, 3.8 to 7.6 cm) because more fingers can be brought in contact with the knob surface. The knob torque should not exceed about 4 in/oz.

For high-torque applications, select knob diameters between 2 and 3 in (5.1 and 7.6 cm) or 3.25 in (8.2 cm) maximum. Maximum torque should not exceed 15 ft/lb (for husky male operators). The table below provides approximate values for knobs of different diameters and depths. Instead of serrations, use knurling on the edges of high-torque knobs for maximum gripping. The edges should be slightly beveled so as not to injure the operator's hand.

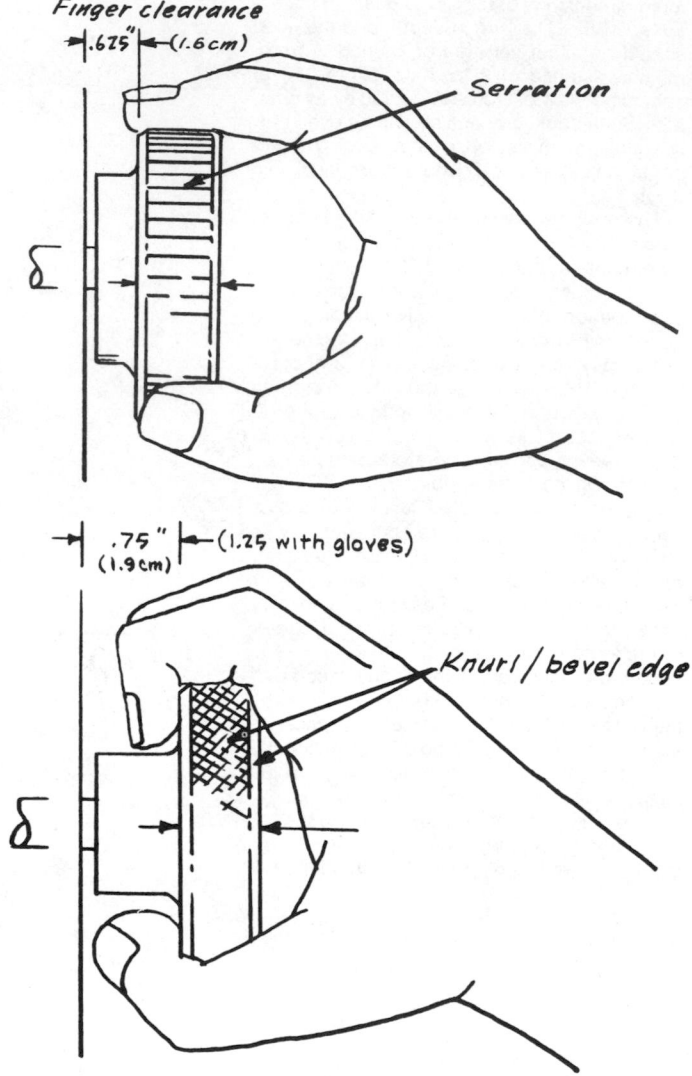

| KNOB DIAMETER (INCHES) | KNOB DEPTH (INCHES) | | | |
|---|---|---|---|---|
| | FINGER GRIPPED | | FULL-HAND GRIPPED | |
| | 0.50 | 1.0 | 0.50 | 1.0 |
| 0.50 | 5 in. lb | 6 in. lb | 11 in. lb | 16 in. lb |
| 0.75 | 6 in. lb | 8 in. lb | 20 in. lb | 29 in. lb |
| 1.00 | 8 in. lb | 10 in. lb | 5 ft lb | 6 ft lb |
| 1.50 | 13 in. lb | 15 in. lb | 7 ft lb | 10 ft lb |
| 2.00 | 20 in. lb | 24 in. lb | 11 ft lb | 13 ft lb |
| 3.00 | 6 ft lb | 6 ft lb | 14 ft lb | 16 ft lb |

## Control and Display Movement

The amount of control movement required to effect a display movement or change is referred to as the "control/display ratio" or "CD ratio." Although it should be optimized for each situation whenever possible, some general guidelines are shown in the accompanying illustration. A complete rotation of a control knob should cause about 1 to 2 in (2.5 to 5.1 cm) of movement of the displayed element, e.g., the pointer on the scalar instrument.

In the case of an auditory signal used to find a station, about 30 to 60° of rotation of the knob should be used to "null out" the station position with minimum necessity for unnecessary searching.

A dual-mode sensitivity capability may be required for some kinds of control-display relationships. For example, if a pencil joy stick is being used to control an electronic hook on a CRT display, a single CD ratio would be unsatisfactory since the operator typically needs to "get into the vicinity of a target" as rapidly as possible, after which he or she requires a more sensitive CD ratio in order to finally pinpoint the target location. As a general rule, the gross CD ratio should be approximately 1:1; i.e., a full sweep of the joy stick should translate the hook all the way across the screen. However, for the fine adjustment around the target site, the control lever should move approximately 1½ to 2 times as far (relatively speaking) as the hook moves on the scope.

**STACKED (OR GANGED) KNOBS** Up to three knobs may be ganged on the same shaft center to save space. This should not be done, however, unless space is really at a premium. When this concept seems necessary, consider the following in order to effect the best operation and the least operator error:

1. The largest knob should be used for operation 1, as shown in the accompanying illustration, so that the operator works outward and thus is less apt to disturb a previous setting.
2. Gross tuning should be assigned to the smaller knob, and fine tuning to the larger one.
3. If the two knobs work sometimes together and sometimes independently, design them so that the operator depresses the top knob to engage the bottom one, with spring loading to return the upper knob to its independent status.

The "ideal dimensioning" for ganged controls has been determined experimentally, as illustrated. The accompanying sketch shows unintentional contact conditions to look out for.

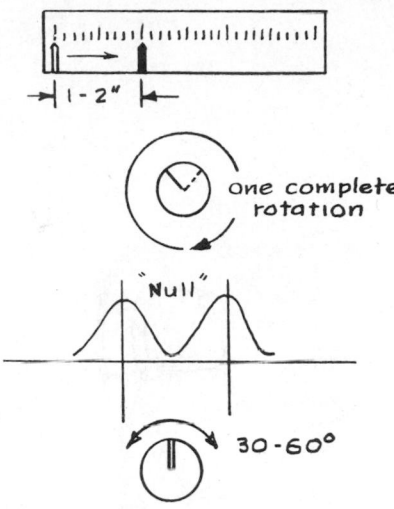

one complete rotation

"Null"

30-60°

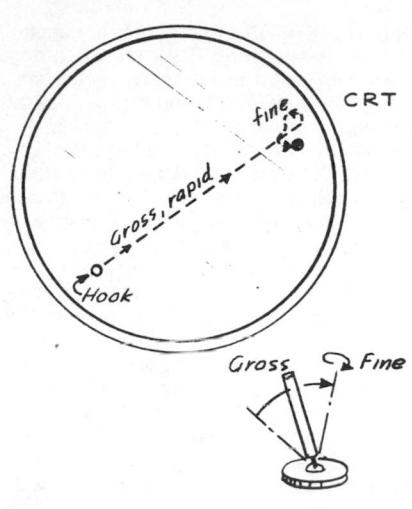

fine  CRT

Gross, rapid

Hook

Gross   Fine

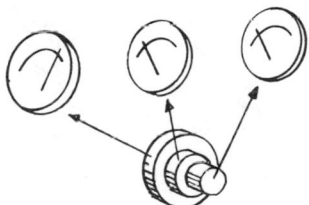

Largest knob associated with left hand dial, smallest with right hand dial.

gross adjustment

fine adjust

MIN DIAM 1.125"(3.2 cm)

Coarse vs fine adjustment designation

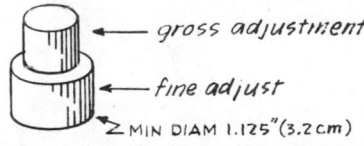

press to engage lower knob

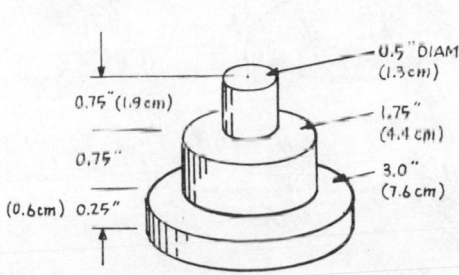

0.5" DIAM (1.3 cm)

0.75"(1.9 cm)

1.75" (4.4 cm)

0.75"

3.0" (7.6 cm)

(0.6 cm) 0.25"

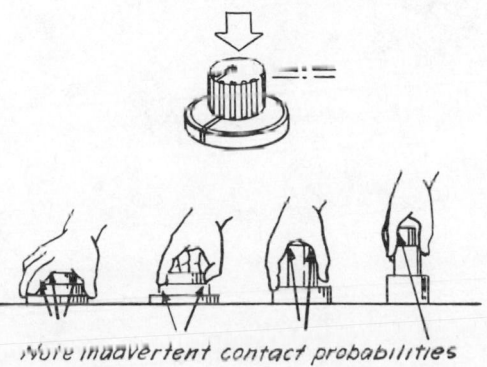

Note inadvertent contact probabilities

**SKIRTED KNOBS** Knob skirts are useful for at least two reasons: They provide a convenient place for a more visible pointer and/or for numbers or labels, and they help prevent the operator's fingers from scrubbing the panel surface. A properly designed skirt can prevent the operator's fingers from covering up the pointer or numbers. The design shown in the accompanying sketch provides relatively optimum relationships to accomplish the above objectives.

**COMBINATION KNOBS AND CRANKS** When it is desired to provide for rapid multiturning for initial gross setting, followed by a need for more accurate setting, the combination knob and crank design is useful. Two versions are shown. The folding crank handle is recommended for applications where it is probable that an operator might inadvertently bump the control or catch his or her clothing on the extended handle. Suggested dimensions for such control knobs are as follows:

Minimum radius—0.625 in (1.6 cm)
Maximum resistance—2 lb (0.9 kg)

Cranks should not be mounted directly in front of an operator, particularly on the operator's centerline, but rather to one side or the other. If considerable torque is required, place the crank control as shown. If, however, very rapid cranking is required, place the crank either in front but to one side of the operator, or preferably, on an angle to one side of the operator, as illustrated.

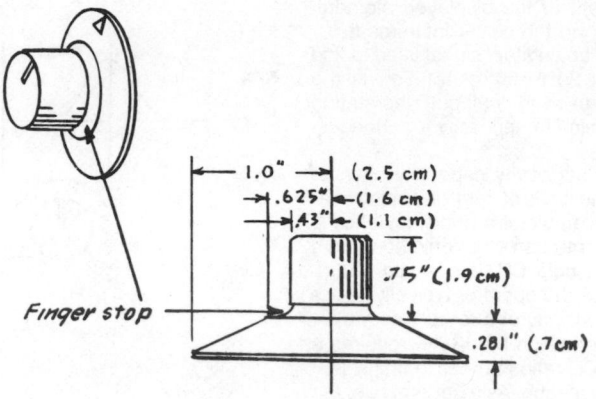

Finger stop

1.0" (2.5 cm)
.625" (1.6 cm)
.43" (1.1 cm)
.75" (1.9 cm)
.281" (.7 cm)

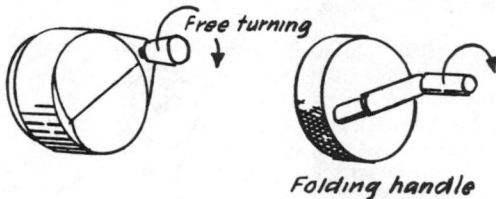

Free turning

Folding handle

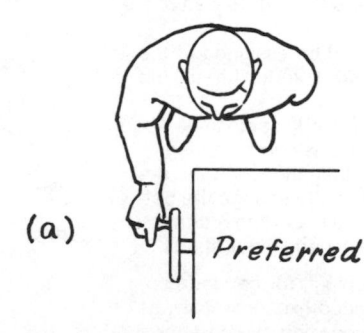

(a) Preferred

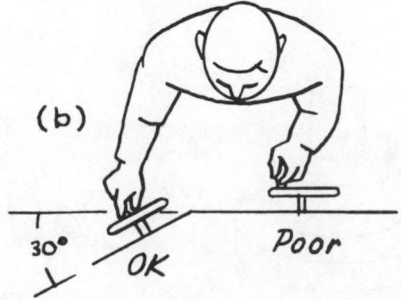

(b)

30°   OK        Poor

## WHEEL AND CRANK CONTROLS

Wheel and crank controls frequently are use-ful for machine applications as well as for elec-tronic console operations. The guidelines shown in the accompanying illustration and table provide approximate design goals when any of the separate parameters may already be fixed.

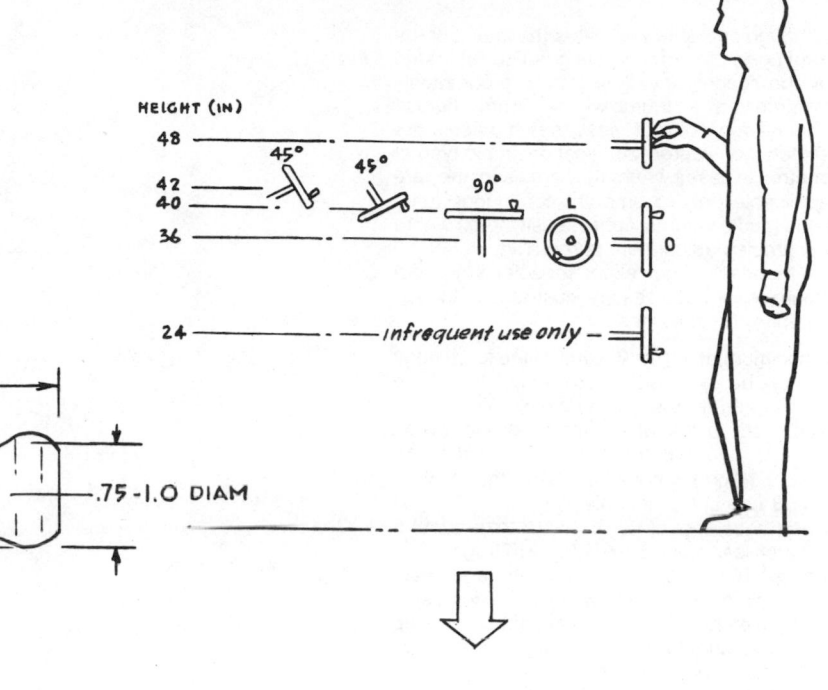

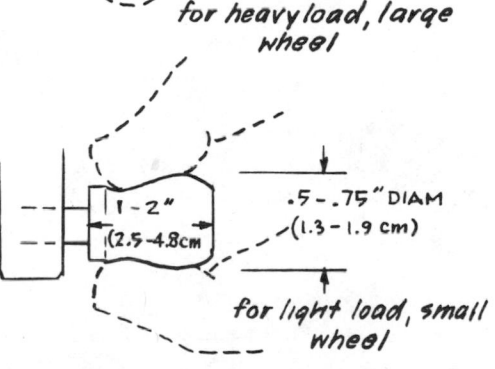

4.0" MIN (10.2 cm)

.75-1.0 DIAM

for heavy load, large wheel

1-2" (2.5-4.8 cm)

.5-.75" DIAM (1.3-1.9 cm)

for light load, small wheel

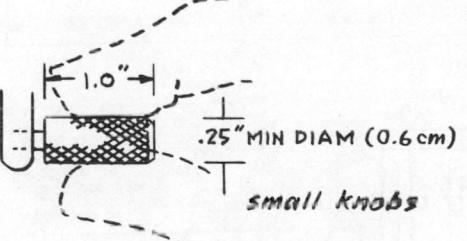

1.0"

.25" MIN DIAM (0.6 cm)

small knobs

Note: Handles must rotate freely.

### Optimum Force Limits versus Control Position

| Height, in | Position, deg | Type | Size Handwheel W, Diameter in, Crank C, Radius, in At Torque of | | |
|---|---|---|---|---|---|
| | | | 0 in-lb | 0 in-lb | 90 in-lb |
| 24 | 0 | W | 3–6 | 10 | 16 |
| 36 | 0 | W | 3–8 | 10–16 | 16 |
| | L | W | 3–6 | 10 | 10 |
| | 0 | C | 1½–4½ | 4½–7½ | 4½–7½ |
| 39 | 90 | W | 3–10 | 10–16 | 16 |
| | 90 | C | 2½–4½ | 4½–7½ | 4½–7½ |
| 40 | 45 | W | 3–6 | 6–16 | 10–16 |
| | −45 | C | 2½–7½ | 4½–7½ | 4½–7½ |
| 42 | 45 | W | 3–6 | 10 | 10–16 |
| | 45 | C | 2½–4½ | 2½–4½ | 4½ |
| 48 | 0 | W | 3–6 | 8–16 | 10–16 |
| | 0 | C | 2½–4½ | 4½ | 4½–7½ |

*Note:* C = crank; W = wheel. 1 in = 2.54 cm; 1 lb = 0.4536 kg.

## THUMB WHEELS

Although operable with either the thumb or the forefinger, the control configuration illustrated by the accompanying sketches is commonly referred to as a "thumb wheel." From a human engineering point of view, thumb wheels are seldom considered the most desirable type of control. In cases where they are used, they are applicable only to smooth, continuous functions (audio volume, light intensity, etc.) where nonprecise adjustment is required.

The principal guidelines for selecting or designing these devices are illustrated in the accompanying sketches:

A minimum of 1.0 in (2.5 cm) of the knob edge must be exposed, and/or about 90° of the circumference of the wheel.

The width of the wheel can be as narrow as 0.125 in (0.318 cm), but a minimum of 0.15 in (0.38 cm) is preferred. More than 0.50 in (1.3 cm) is undesirable.

Resistance should not be more than about 6 oz or less than 2 oz (170 to 579 g).

Acceptable multiple arrangements and spacing minimums are shown in the accompanying sketches. Note that horizontally arrayed wheels should not be arranged one above the other or one in front of another.

Thumb wheels should not be used if the operator has to wear gloves.

Wheel motions upward or to the right should provide "increasing function."

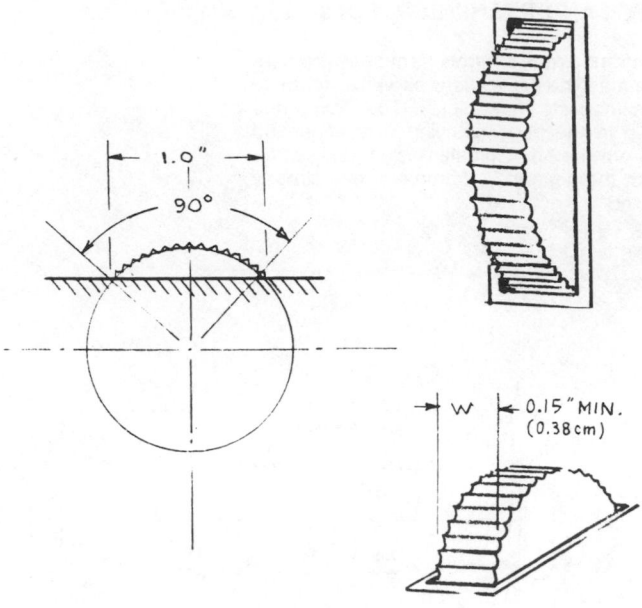

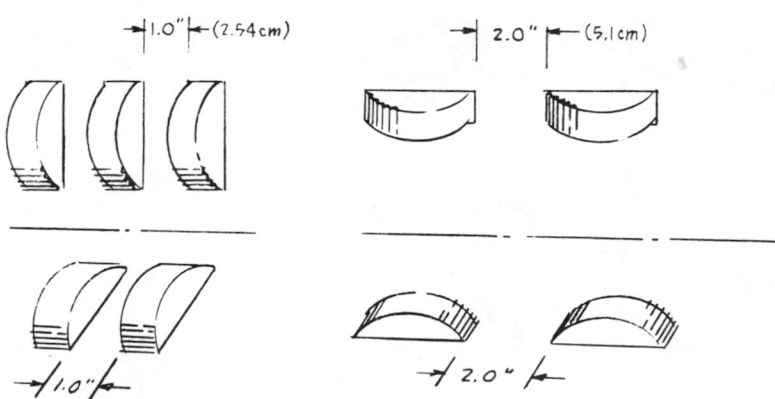

## Discrete-Positioning Thumb Wheels

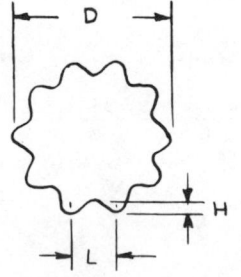

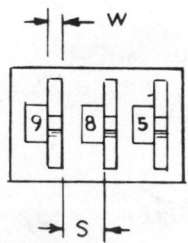

| | D (Diameter) | L (Trough Distance) | W (Width) | D (Depth) | S (Separation) | Resistance |
|---|---|---|---|---|---|---|
| Minimum | 1.5 in (3.8 cm) | 0.45 in (1.14 cm) | 0.1 in (2.54 cm) | ⅛ in (0.3 cm) | 0.4 in (1.0 cm) | 1 in-lb (0.12 m·N) |
| Maximum | 2.5 in (6.4 cm) | | | ½ in (1.3 cm) | | 3 in-lb (0.34 m·N) |

## POINTER KNOBS

In contrast to the smooth, round knob recommended for continuous, multiturn controls, a "pointer-shaped" knob should be used for controls when the operator may wish to switch from one discrete position to another. The switch should have mechanical detents so that the operator can "feel" and perhaps hear when the knob is properly seated or positioned.

A prime objective of the pointer knob is to provide (1) a sufficiently discriminable pointing shape so that the operator knows which end of the knob is the pointing end and (2) adequate depth so that the operator can get a firm enough grip to turn the knob.

Various knob shapes are generally acceptable as long as certain details are not compromised. For example, the accompanying sketches show typical "bar-shaped" controls that provide adequate pointing characteristics. The $d$ dimension should always be at least 0.50 in (1.3 cm); the $l$ dimension should be at least 0.50 in (1.3 cm); and the $w$ dimension should not be less than 0.25 in (0.6 cm).

The third knob shape is generally a better configuration than the first two because of the sloping pointer end, which makes it easier for the operator's eye to follow the pointer directly to the point at which it becomes adjacent to an index mark on the panel.

The fourth pointer shape provides additional length to assist the operator when more force is required to position the knob.

The minimum torque requirement for pointer knobs should be about 10 in/oz (283 g); the maximum should be around 45 in/oz (1275 g). The switches should have elastic resistance which builds up and then decreases as the switch "seats."

When panel space dictates use of a very small switch knob, slight variations are possible, as shown in the accompanying illustration. The left-hand knob is not especially desirable, but it can be used under extreme conditions. The others are preferred because they are more comfortable to use.

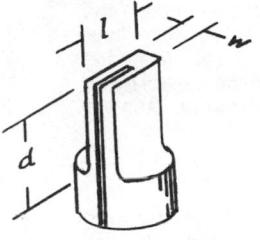

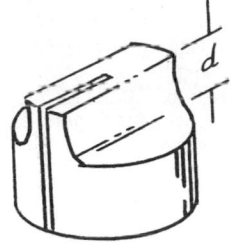

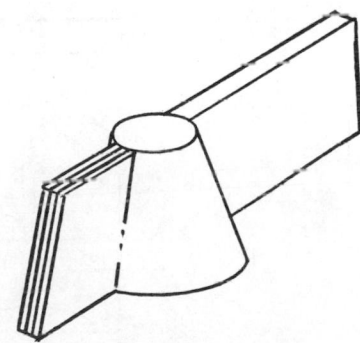

When it is known that a pointer knob will be used at night or under darkened ambient lighting conditions, the knob pointer should be illuminated. An effective method is to use a transparent plastic knob, cover it with a coat of transluminescent white paint or material, and then place an opaque coating over that. The pointer should then be inscribed through the outer opaque coating so that the white inner coating shows through and so that either edge lighting or back lighting from the control panel can shine through the pointer areas (see the accompanying sketch).

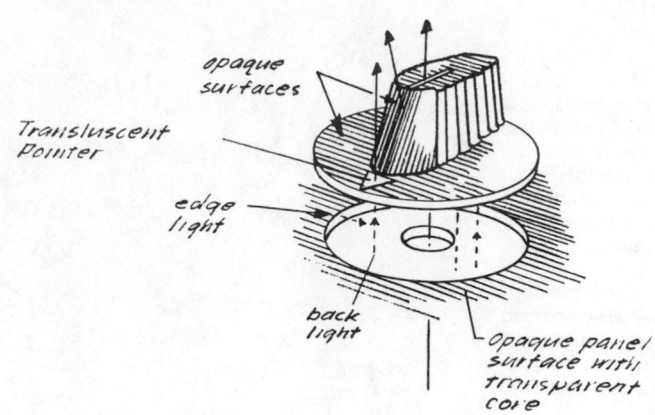

Lighted pointer knob.

### Idealized Pointer Knob

The idealized pointer-knob shape shown in the accompanying sketch illustrates several desirable features to look for when selecting a knob:

A 45° pointer slope ensures a better view of the pointer tip from most viewing positions.

The long, tapering handle provides good leverage for rotating the knob, without having to cover the pointer tip with the fingers.

All sharp corners are eliminated.

The skirt provides a finger rest so that the fingers do not scrub the panel.

Not only is the configuration attractive, but also it will never be confused with round knobs, nor will the operator be in doubt as to which is the pointing end of the knob, whether he or she is looking at the knob or not.

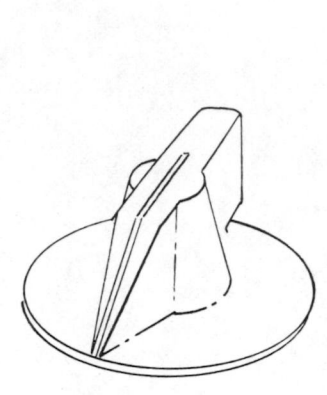

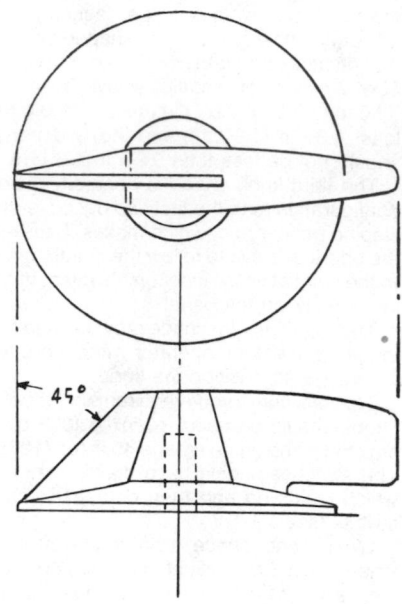

### Example of Lighted Pointer Knob Design

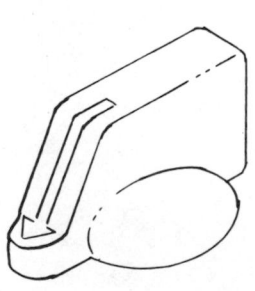

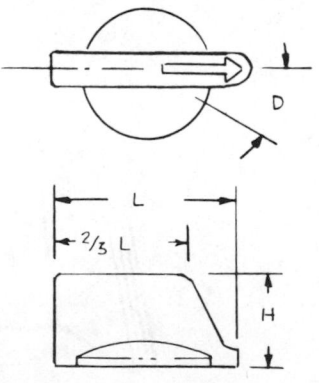

| Operating Method | Pointer, in | | | D (Displacement in Degrees) | | |
|---|---|---|---|---|---|---|
| | L (Length) | W (Width) | H (Depth) | Visual Positioning | Nonvisual Positioning | Resistance, oz* |
| Minimum | 1.00 | 0.25 | 0.50 | 15 | 30 | 12 |
| Maximum | 3.00 | 1.00 | 3.00 | 40 | 40 | 48 |
| | | | | (90 when large separations are necessary) | | |

*Notes:* Military Standard MS 25166, *Knob, Pointer, Illuminated,* meets, to a certain degree, the above requirements. 1 in = 2.54 cm; 1 oz = 28 g.

*Use elastic resistance which builds up and then decreases as each detent is approached so that the control will snap into position and cannot easily stop between adjacent positions.

## OTHER FINGER-OPERATED CONTROLS

Although a variety of control shapes are suitable for operation by the fingers, it is important to select controls that are large enough so that the operator can manipulate them conveniently and with a minimum of discomfort. The guidelines shown in the accompanying sketches provide useful information for selecting appropriate control devices.

Finger-operated levers should have resistances of 30 oz (850 g) or less. The action should include definite positional detents so that the operator knows the control is properly positioned.

The handle-length minimum for resistances of 30 oz (850 g) or less should be about 1.0 in (2.5 cm). When the resistance exceeds 30 oz (850 g), the handle length must be increased. The maximum resistance should not exceed about 3 lb (1.4 kg), in which case the handle length needs to be increased to at least 1.5 in (3.8 cm).

The slide switch resistance should not exceed about 10 oz (283 g).

A single "push-push" type of wall switch is preferred over the more typical two-button configuration because it removes the necessity for the user to decide which position the switch is in. The larger switch plate makes it possible to operate the switch with an arm or elbow when the hands are occupied.

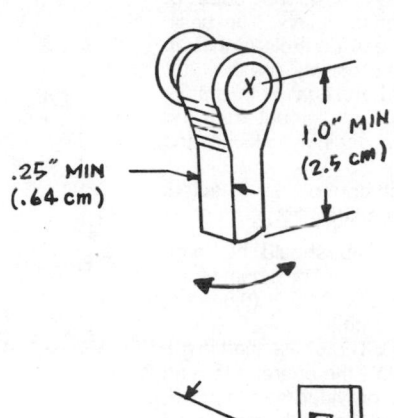

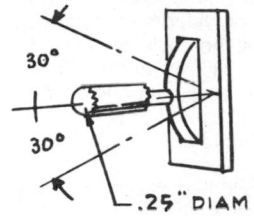

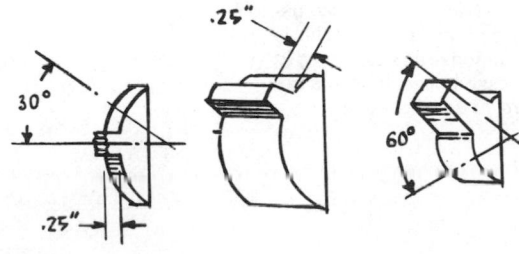

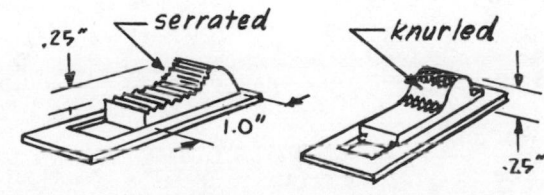

### Fingertip-Operated Slide Controls

Although these controls can be used either for continuous-motion or for discrete, detented switching functions, they are not generally recommended for either one. However, they can be used for fairly coarse operating functions when operating error may be of little consequence and/or when the time taken by an operator to get the control positioned does not introduce other response delays. The basic problem with using these controls is that, in order to operate them precisely, the operator must steady his or her hand against the panel, using a long, total-arm movement from the shoulder, which is exceedingly difficult and very inaccurate.

Nevertheless, if such controls are selected, consider the following guidelines:

The control knob or tab should be large enough to get hold of, e.g., minimums for $a$ = 0.50 in (1.3 cm); $b$ = 0.25 in (0.64 cm); and $c$ = 0.50 in (1.3 cm).

The handle should be offset on its shaft in the direction that will allow the operator to read whatever labeling is provided on the panel; i.e., the labeling will normally be placed above the horizontal slide or to the left or right of the vertical slide, depending on which hand is used to operate the slide.

The slide resistance should not exceed about 2 lb (0.9 kg).

The handle should be oriented as shown in the accompanying sketches, and there should normally be a pointer inscribed in the handle edges.

When multiple slide arrangements are made, sufficient separation must be maintained so that labels can be seen and easily associated with the slide they apply to. (This is one of the most common errors observed in the design of slide assemblies. It occurs because of the desire to save space.)

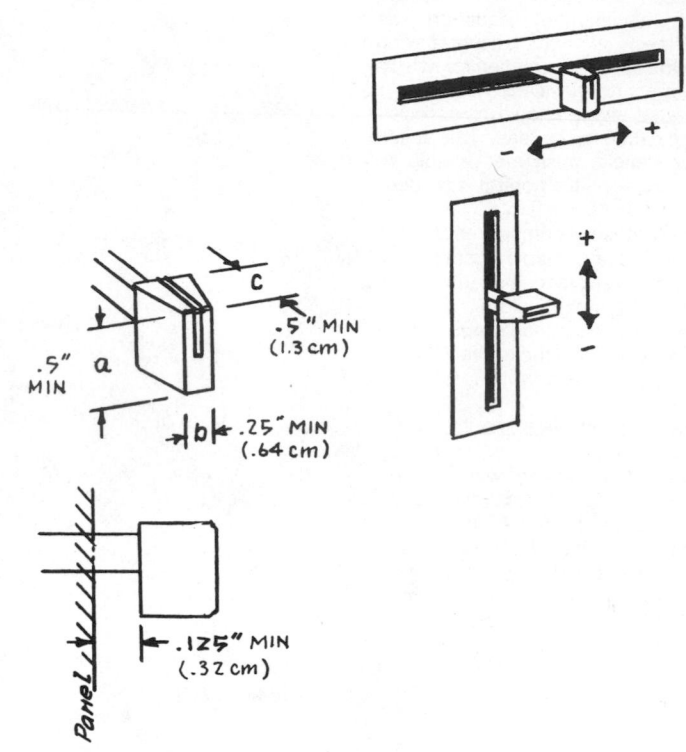

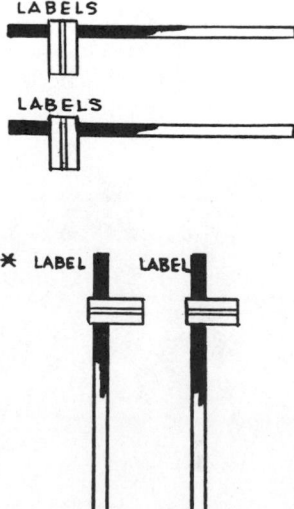

\* Label on left when viewed from the left; right viewed from right.

## Vehicular Column-Mounted Controls

Variously referred to as "stalks," "wands," and "fingertip controls," column-mounted controls are used primarily to bring certain critical control functions within reach of the driver's fingertips as his or her hands hold the steering wheel. In recent years more and more control functions have been located here, and the potential for operator error has increased. Consider the following guidelines:

Do not employ more than one lever on each side of the column.

Do not employ more than two lever motions per lever, plus use of an "end-mounted" push button, except when it can be demonstrated that another motion (i.e., rotation of the lever handle) can be accomplished without inadvertent input of one of the other lever motions.

Mount the levers and handles so that they are approximately within fingertip reach (e.g., 3.0 in, 7.6 cm, from the steering-wheel rim), but not closer, so that the lever handle cannot be bumped as the driver manipulates the steering wheel. When labels are required on the lever handle, these must be visible from the driver's nominal eye reference position; i.e., the driver should not have to shift his or her body to see around the steering wheel.

The minimum force resistance should be about 6 oz (170 g); the maximum should be 20 oz (567 g).

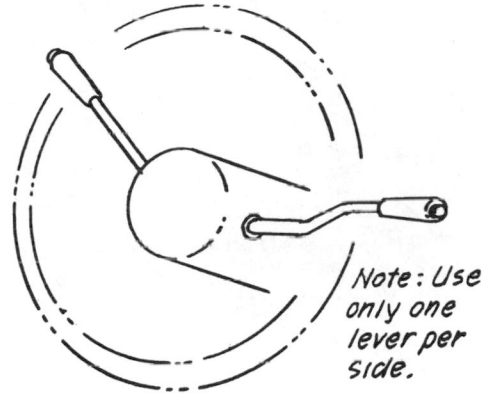

Note: Use only one lever per side.

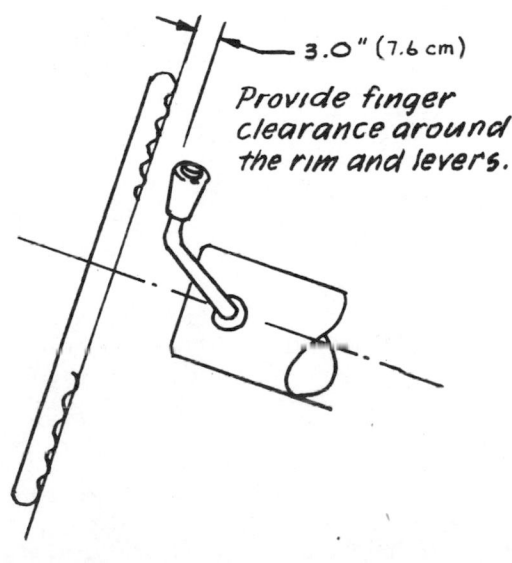

3.0" (7.6 cm)

Provide finger clearance around the rim and levers.

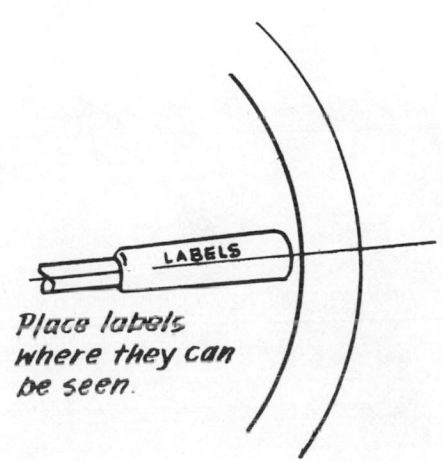

Place labels where they can be seen.

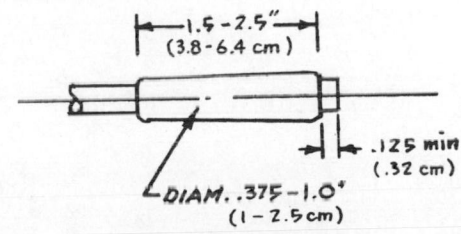

1.5-2.5" (3.8-6.4 cm)

.125 min (.32 cm)

DIAM. .375-1.0" (1-2.5 cm)

for adequate gripping.

# DRIVING CONTROLS FOR THE ORTHOPEDICALLY HANDICAPPED

**Guide for Determining Special Control Requirements for Handicapped Drivers**

| Handicap | Complete hand control | Hand control extension | Complete foot control | Spinner knob | Steering cuff | Right-side directional | Left-side gear shift | Left-foot gas pedal | Parking brake extension | Hand clutch | Hand brake | Hand dimmer switch | Automatic transmission | Power steering | Power brakes | Power windows | Power six-way seat adjustment | Full-view mirror |
|---|---|---|---|---|---|---|---|---|---|---|---|---|---|---|---|---|---|---|
| Left leg | | | | | | | | | x | x | | | x | x | | | | |
| Right leg | | | | | | | | x | | | | | x | x | x | | | |
| Both legs | x | | | x | | | | | x | | | | x | x | x | x | x | |
| Right arm | | | | x | | | x | | | | | | x | x | | | | |
| Left arm | | | | x | | | | | | | | | x | x | | | | |
| Both arms | | | x | | | | | | | | | | x | x | x | x | | |
| Left leg and left arm | | | | x | | | x | | | | | | x | x | x | | | |
| Right leg and right arm | | | | x | | | x | x | | | | | x | x | x | | | |
| Left arm and right leg | | | | x | | | | x | | | | | x | x | x | | | |
| Right arm and left leg | | | | x | | | x | | x | | | | x | x | x | | | |
| Quadriplegic (all four limbs) | x | | | x | x | | | | x | | | | x | x | x | x | x | |
| Balance high paraplegic | x | | | x | | | | | x | | | | x | x | x | x | x | |
| | | | | | | | | | | | | | | | | | | |
| Neck rotation restriction | | | | | | | | | | | | | | | | | | x |
| (lack of height) | x | | | | | | | | | | | | x | x | x | x | | |

*Source:* M. Less, et al., *Hand Controls and Assistive Devices for the Physically Disabled Driver,* Human Resources Center, Albertson, N.Y., 1977.

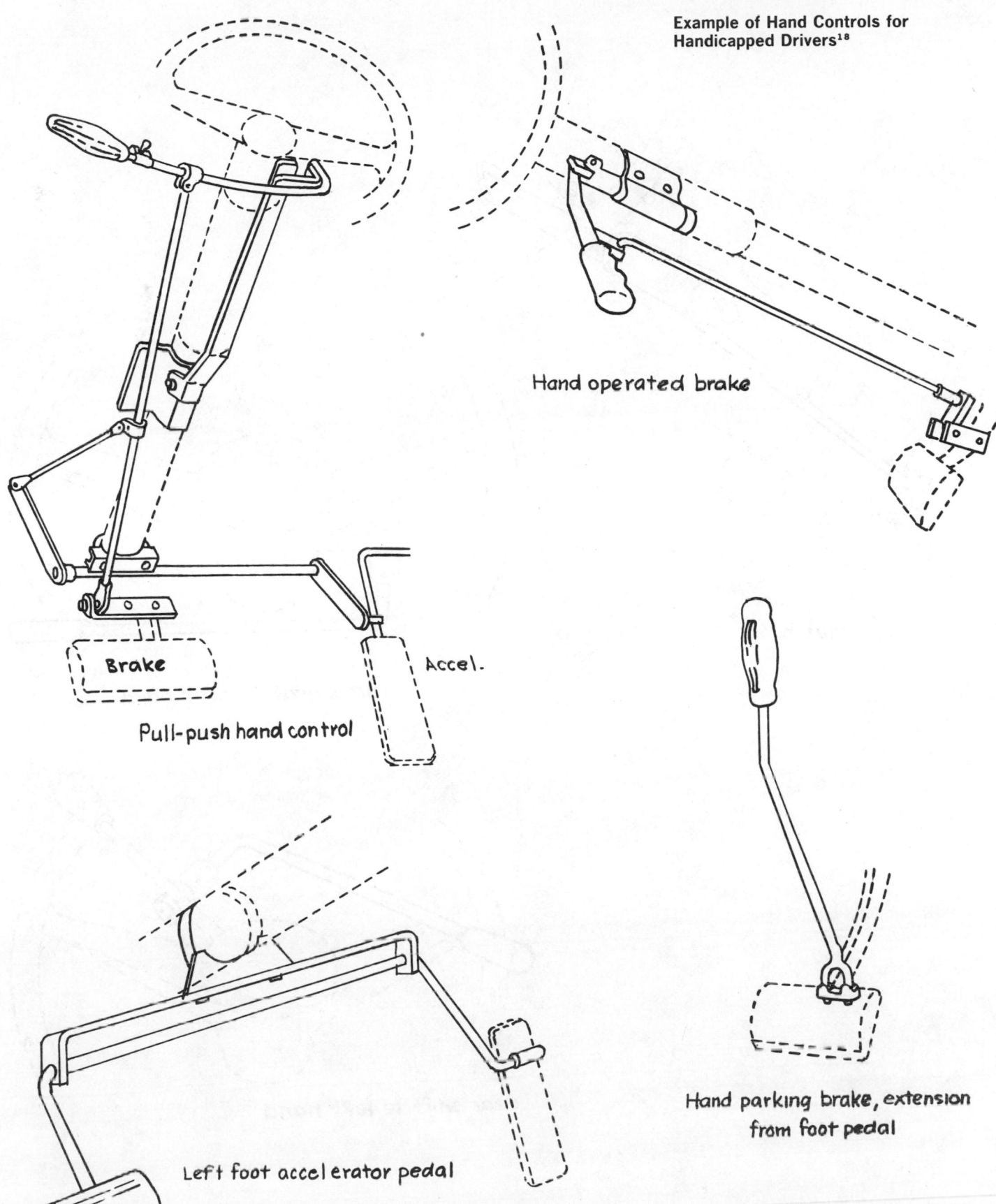

**Example of Hand Controls for Handicapped Drivers**[18]

Hand operated brake

Pull-push hand control

Brake

Accel.

Left foot accelerator pedal

Hand parking brake, extension from foot pedal

[18]M. Less, et al., *Hand Controls and Assistive Devices for the Physically Disabled Driver*, Human Resources Center, Albertson, N.Y., 1977.

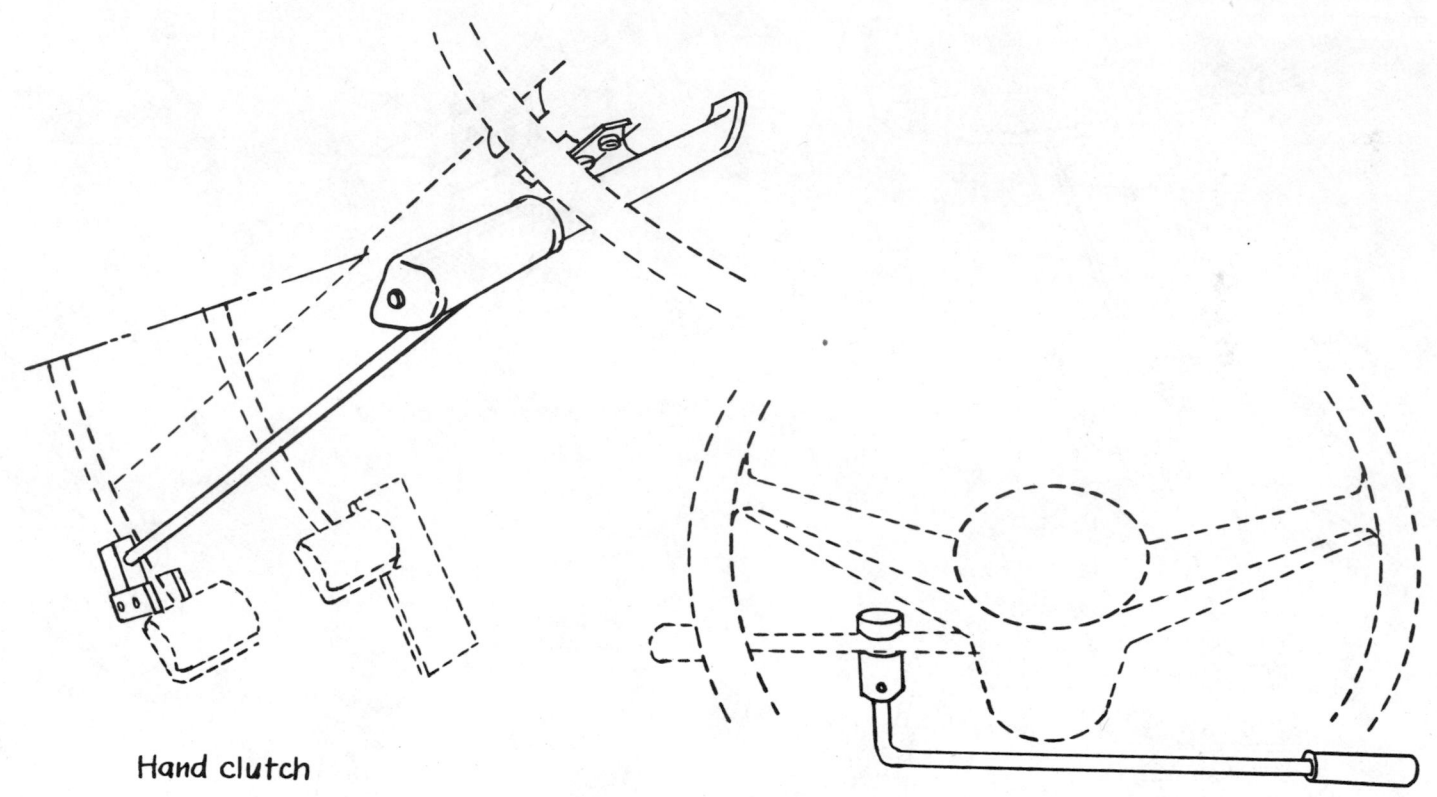

Hand clutch

Turn signal to right hand

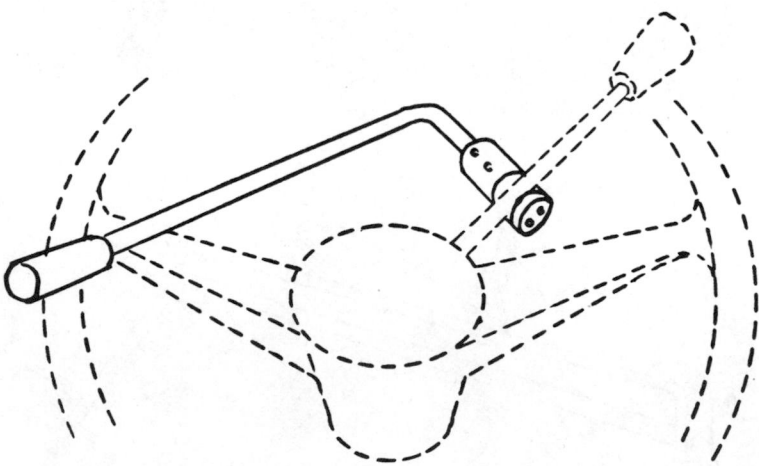

Gear shift to left hand

Steering aids generally require a custom fit for each particular handicap; i.e., although some individuals may be able to use the simple spinner knob, others cannot grip such a device and therefore require a "stirrup" device, as shown in the accompanying sketch.

*Note:* All spinner-type aids should have some means for adjusting the position of the spinner on the steering-wheel rim (although for 95 percent of the user population, a four-o'clock position is usually best).

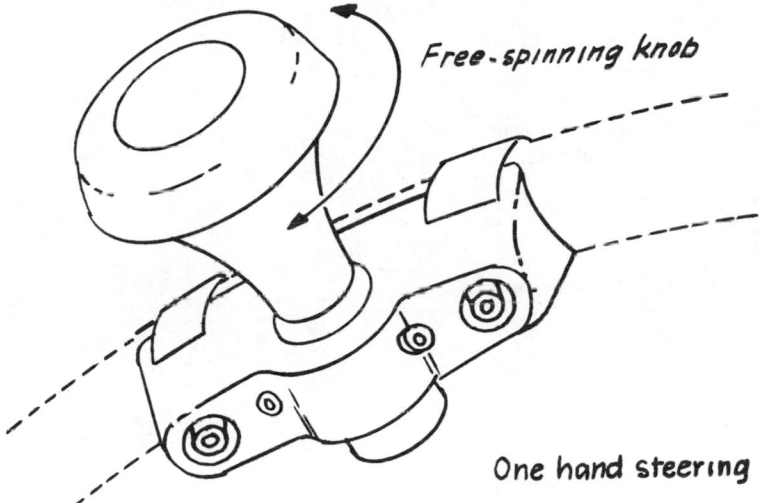

Free-spinning knob

One hand steering

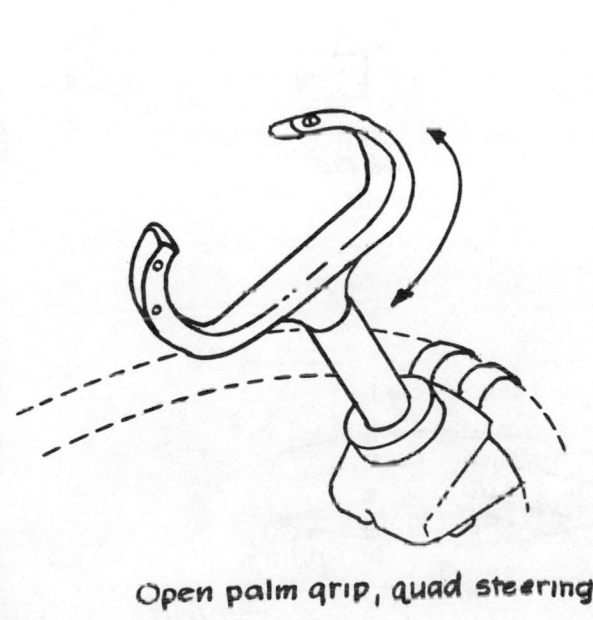

Open palm grip, quad steering

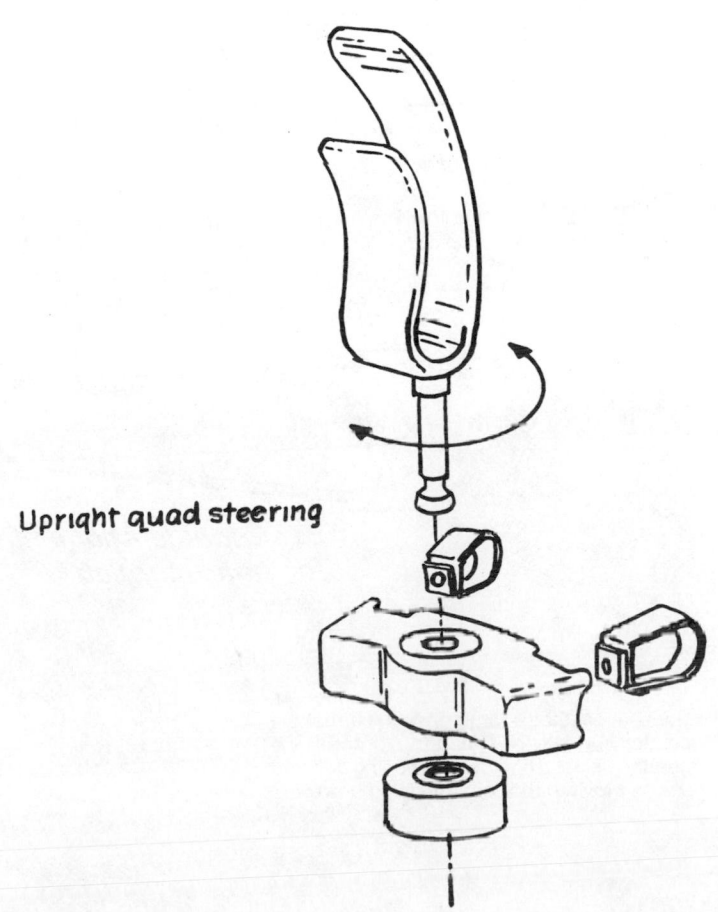

Upright quad steering

## VALVE HANDLES

Since valve handles generally are associated with the possibility of considerable force application on the part of the user, the handle should be designed so that the operator can place his or her hand in a comfortable position and grip the handle with a minimum possibility of pain or injury.

For higher force applications, the round handle with knurled edges is preferred. The edges should be radiused, as shown in the accompanying sketch, and the dimensions shown should be followed in selecting an appropriate handle.

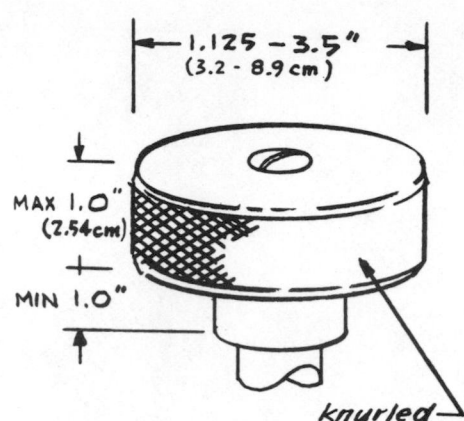

For lower force applications, consider the handle shapes shown in the accompanying illustrations.

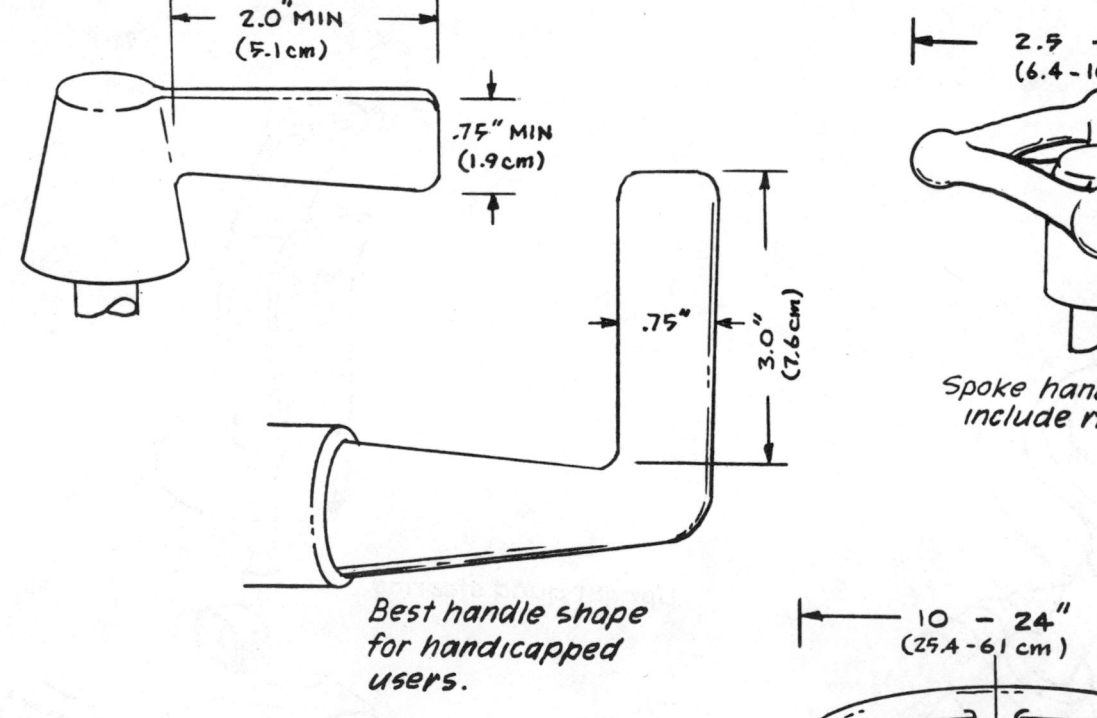

Best handle shape for handicapped users.

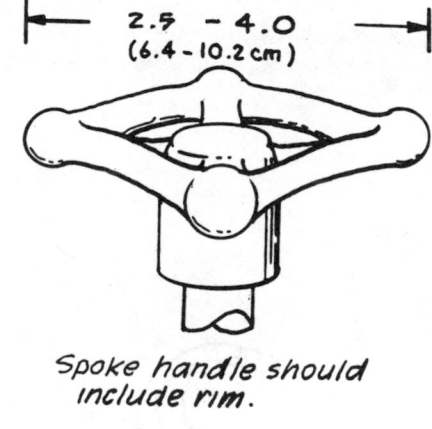

Spoke handle should include rim.

For very high force valve requirements, consider a wheel that is large enough so that the operator can use both hands. To assist the operator, select a rim that has ribbing or serrations to provide additional gripping power.

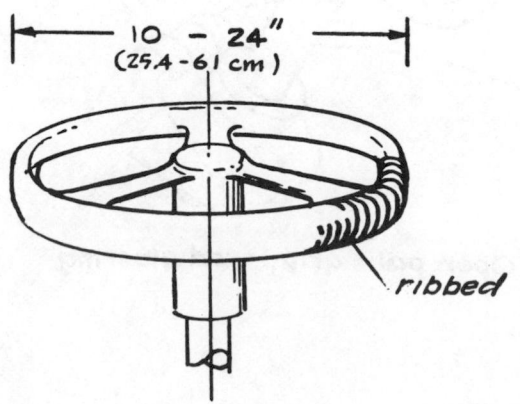

## PULL-TYPE CONTROLS

The size and shape of a pull-type control handle should be compatible with the manner in which the hand can most easily grasp and manipulate the handle, i.e., with the fingers or the whole hand.

Even for very light forces, the handle should be large enough so that the user is not forced to struggle to get hold of, and maintain proper grip on, the device. A simple "fluted" handle, as shown in the accompanying sketch, provides the best gripping characteristics. Such handles are suitable for forces between 1.0 and 4.0 lb (0.45 and 1.8 kg).

The T-shaped handle is effective for light to medium forces. Although the T-shaped handle can be operated with two fingers (i.e., one on each side of the shaft), it should not be used if the force requirement exceeds about 15 lb (6.8 kg).

For higher forces, use the dimensional guidelines shown in the accompanying illustration; i.e., the handle should be large enough so that the user can get two fingers on each side of the shaft.

A stirrup-shaped handle is sometimes useful to help keep objects such as belts or straps or the operator's clothing from becoming caught behind the handle (which may occur with the T-shaped handle). In the case of this handle, since the ends are closed, it is critical to provide enough space inside the stirrup so that the operator's hand and fingers are not crushed when force is applied.

*Note:* Both the T-shaped handle and the stirrup-shaped handle should be either round or oval; i.e., avoid rectangular shapes with sharp edges, which may injure the user's hand.

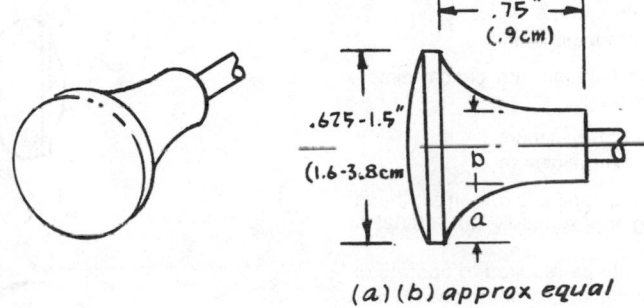

(a)(b) approx equal

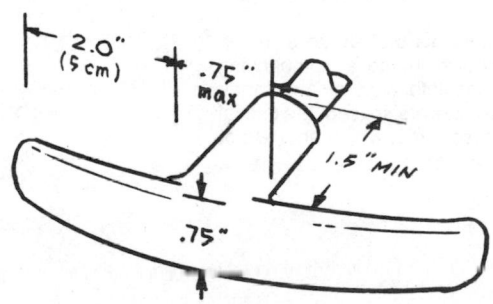

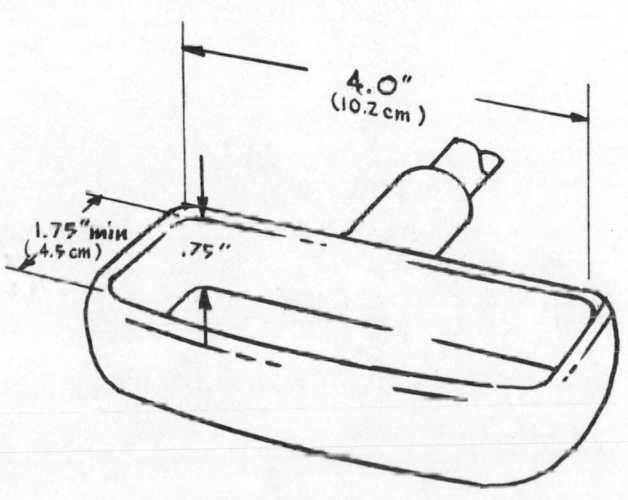

## DOOR HANDLES

Although round knobs are used extensively for architectural purposes in the United States, L-shaped handles are used more frequently in other countries. The L-shaped handle is preferred because it is more easily used by handicapped persons and children, who have difficulty gripping and applying force to the round knob. If round knobs are used, their dimensions should be as follows:

Diameter—2.0 in (5.1 cm)
$d$—1.125 in (2.86 cm)
$c$—1.0 in (2.5 cm) minimum

For L-shaped handles, the critical dimensions are as follows:

$1$ = 3.0 in (7.6 cm) minimum
$c$ = 1.0 in (2.5 cm) minimum

An oval-shaped handle is preferred, the dimensions being approximately $1.0 \times 0.625$ in ($2.5 \times 1.6$ cm).

The maximum force required to operate either type of handle should not exceed 6 ft/lb (88 N/m).

Emergency bar controls requiring two-hand operation, such as those illustrated in the accompanying sketches, are difficult for children and many handicapped persons to use. To be acceptable, the unlatching force should not exceed 10 lb (45 N). The height of the bar above the floor should be about 36 in (0.9 m). The bar diameter should be about 1.0 in (2.5 cm).

Door latching controls that utilize a thumb latch lever, as shown in the accompanying sketch, are also difficult for children and handicapped persons to operate. If used, the critical factor is thumb lever force, which should not exceed 1.0 lb (0.45 kg).

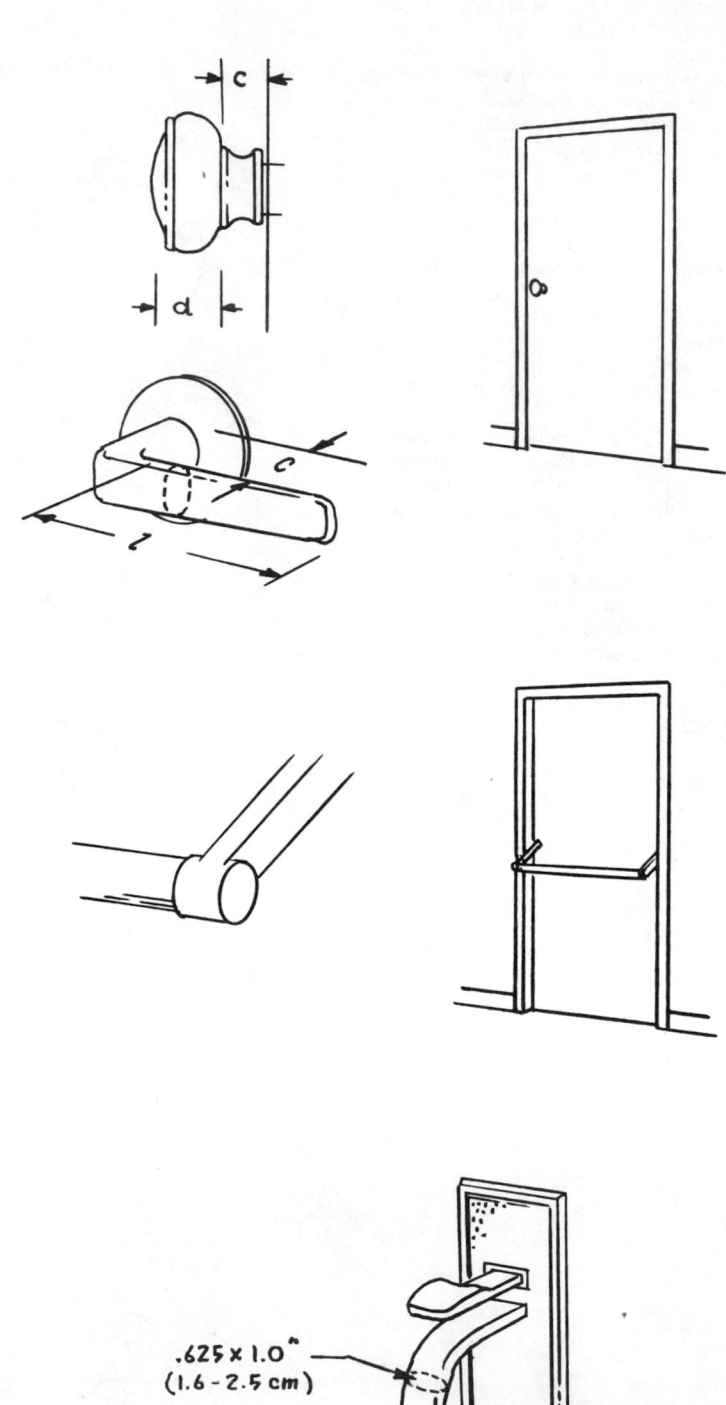

.625 × 1.0″
(1.6 – 2.5 cm)

Handles and latching hardware used for sliding doors are often poorly designed because the manufacturer was not concerned about the user's manipulatory capabilities. The major problem invariably relates to a latching or locking device which is too small for the operator to obtain sufficient gripping area. As a result, many people, especially children and the elderly, cannot apply enough force to operate the latch.

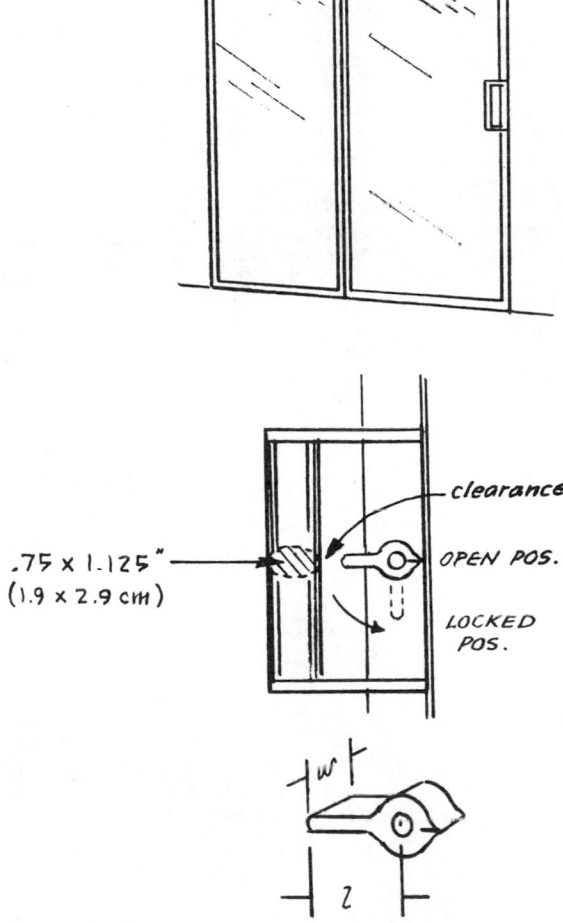

The latch handle should provide a surface sufficiently large for adequate gripping. The following dimensions are considered minimum:

$w$ = 0.625 in (1.6 cm)
$l$ = 1.25 in (3.18 cm)

Although not particularly critical, the door handle should be oval for maximum comfort, and the preferred dimensions should be approximately 0.75 × 1.125 in (1.9 × 2.9 cm).

*Special Note:* With regard to the selection of door handles for architectural purposes, it should be noted that opening and closing a door may often be an emergency matter; thus designers should be particularly careful not to allow a desire to select aesthetically pleasing but difficult-to-operate designs to overshadow the basic functional objective. Many acceptable handle designs can be found which are also pleasing. Above all, *avoid knobs or handles that have sharp edges or decorative patterns.*

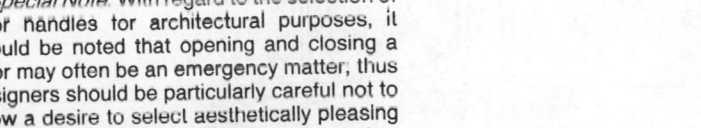

## Vehicle Door Handles

The shape of the exterior vehicular door handle shown in the accompanying sketch is preferred, although other shapes may be acceptable in certain circumstances.

The handle shown in the accompanying illustration is a closed loop in which the latch is operated by pushing the button with the thumb. The desirable feature of this type of handle is that the operator is not confronted with a movement of the handle when the latch releases the door. Note that the handle for the right-hand door is oriented differently from that for the left-hand door. This forces users to position themselves properly to keep the door from knocking them over when it is pulled open.

Although more awkward to use, the lift-type handle shown in the accompanying sketch is reasonably acceptable.

Although this handle is often found on some vehicles, it is extremely hard to use because it has to be grasped from an awkward angle, and the pulling motion is even more awkward. When it is used, the handles on each side must be mounted so that the operator is not forced to stand in front of the door. The handle shown is for the right side of the car and is intended to be operated by the right hand, as the operator stands aft of the door.

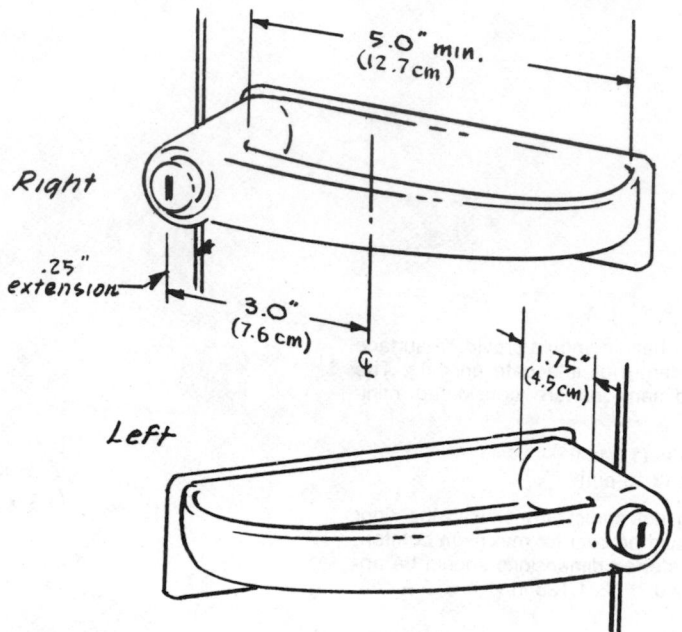

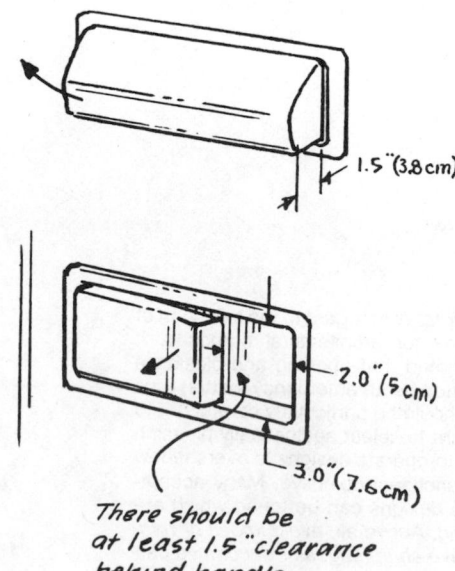

There should be at least 1.5" clearance behind handle.

Interior vehicular door handles should be placed where they are easily located by the user (sometimes in an emergency), but they should not become a hazard in the event the occupant is inadvertently thrown against the handle.

A preferred handle is illustrated in the accompanying sketch. Note that it is inset and that the open end is toward the front.

The important hand and finger clearance dimensions are shown for an optimum handle. In addition, however, there must be sufficient space behind the handle (i.e., inside the housing) so that the occupant can slip his or her fingers behind it. This requires a minimum of 1.0 in (2.5 cm) of clearance.

The "loop handle," which is a commonly used interior vehicular handle, is shown in the accompanying sketch. Although this is an acceptable configuration, many manufacturers tend to make the handle and space around it too small. As a minimum, the user should be able to get at least two fingers either inside the loop or past the end of the handle. The dimensions shown are recommended.

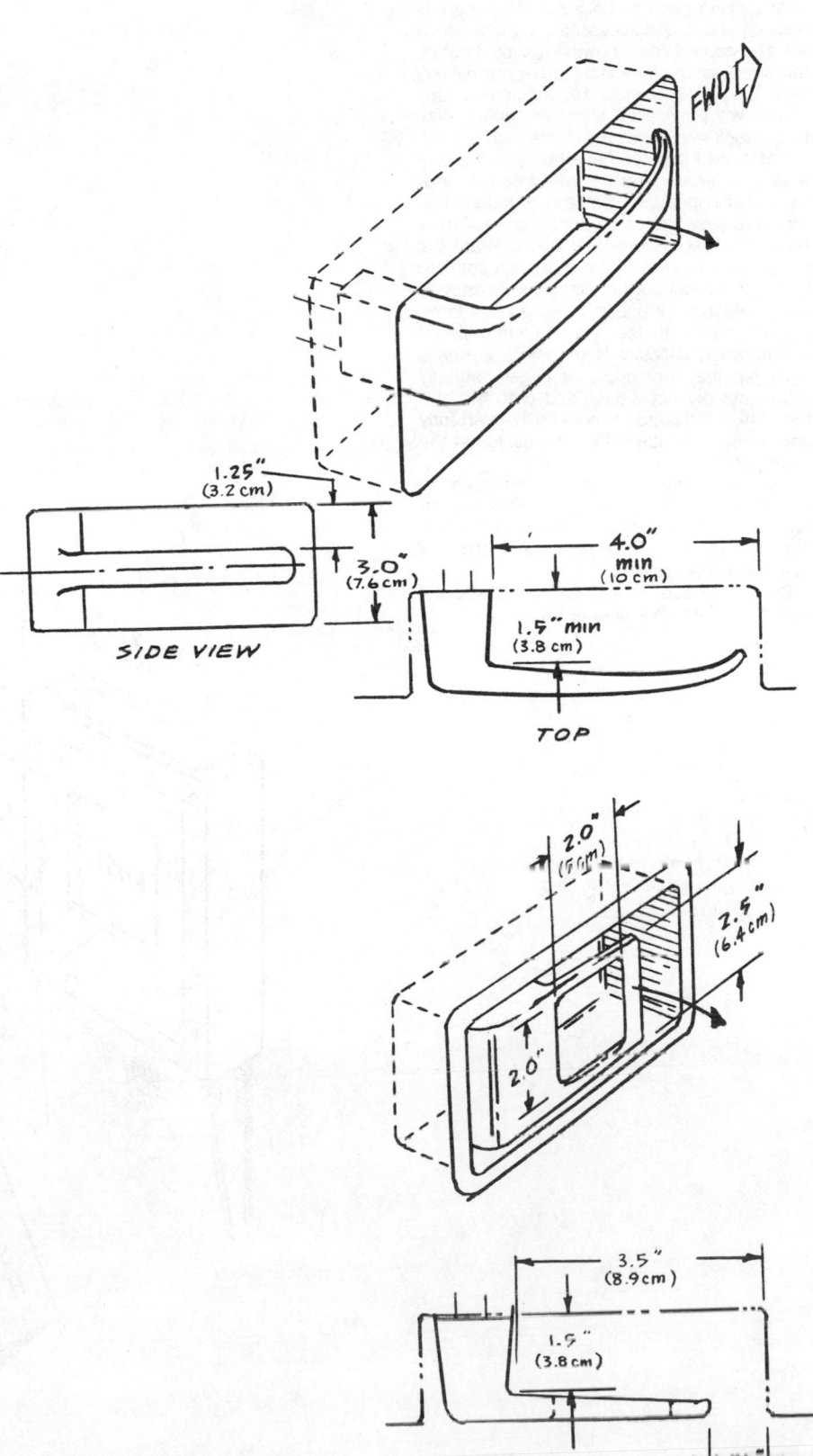

FWD

1.25" (3.2 cm)

3.0" (7.6 cm)

SIDE VIEW

4.0" min (10 cm)

1.5" min (3.8 cm)

TOP

2.0" (5 cm)

2.5" (6.4 cm)

2.0"

3.5" (8.9 cm)

1.5" (3.8 cm)

1.25" (3.2 cm)

TOP

## VEHICLE LOCKS AND SAFETY-BELT BUCKLES

The plunger-type control shown in the accompanying sketch is preferred over other types of door locking controls because the control is typically visible and accessible on the window-sill. The control knob, however, should not be too small since some locks are more difficult to lift. The dimensions and approximate shape shown will provide an effective control even for the lock with a fairly high force requirement.

Safety-belt buckles have become more or less standardized for all vehicles; i.e., they usually are operated by means of a push button to release the catch. It has been shown, however, that buckles that approximate the dimensions illustrated are more acceptable. First, the overall buckle dimension is easy to grasp. Second, the button release is large enough to be actuated by the thumb or finger with minimum difficulty. Note that the button is inset (so that the thumb or finger naturally slides into the depression and onto the button). Third, the button cannot be inadvertently depressed and thus unlock the buckle at the wrong time.

The push-button buckle configuration shown in the accompanying sketch is also effective. In fact, it is more convenient to operate and is easier to see from the normal user's eye reference position.

Button pressure should not exceed about 4 lb (1.8 kg) for either type of buckle.

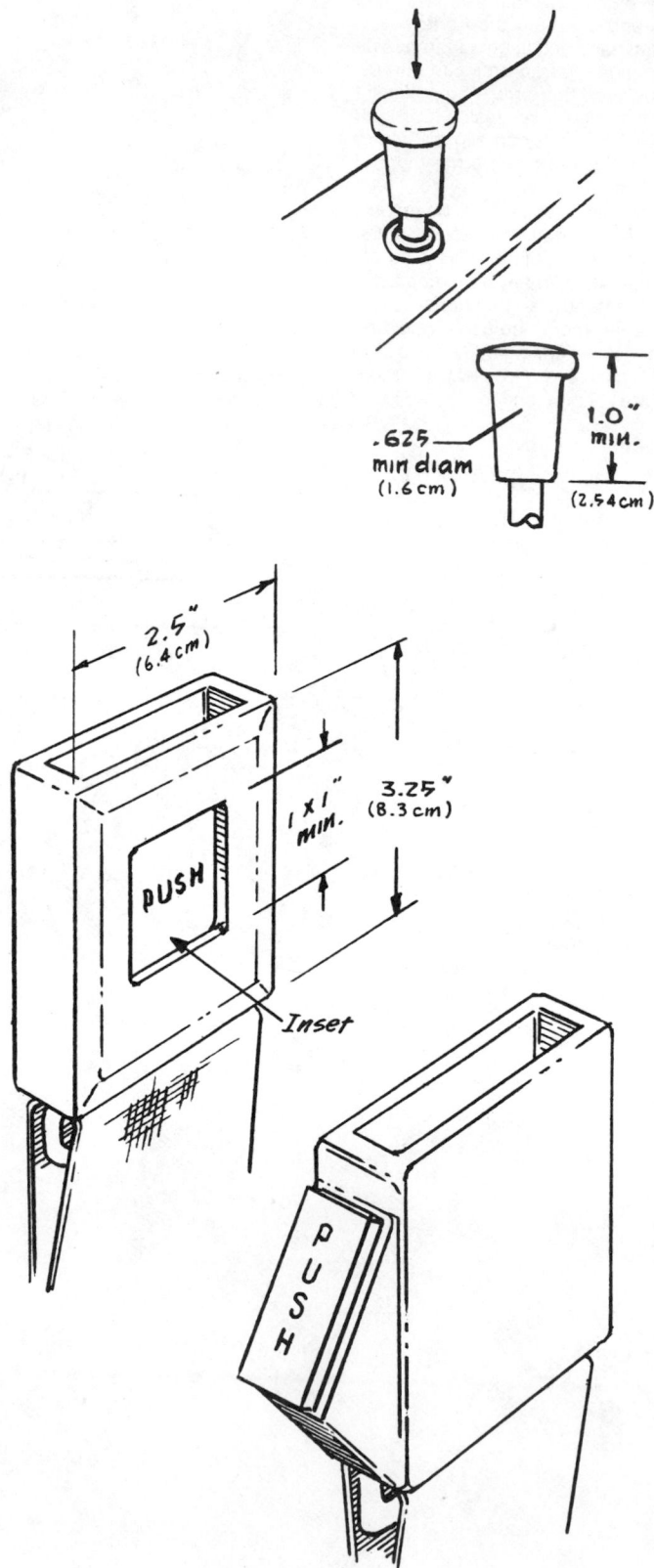

## COMBINATION CABINET HANDLE AND LATCHING MECHANISMS

Although typically thought of only as handles, the combination handle and latching mechanisms illustrated in the accompanying sketches are actually controls.

Equipment racks and consoles generally have a removable chassis and panels that require a combination handle and latching mechanism. Since the handle is also used to carry the chassis around after it is removed, it should be fixed. To provide the best combination of handle and latch release, the handle concept shown in the accompanying sketch is suggested; i.e., the handle is fixed, but it contains a push-button release in a position that is convenient for thumb operation. For ease of carrying, the handle cross section should be oval, and the dimensions shown should be used to make it easy to reach the release button while the hand is placed firmly on the handle.

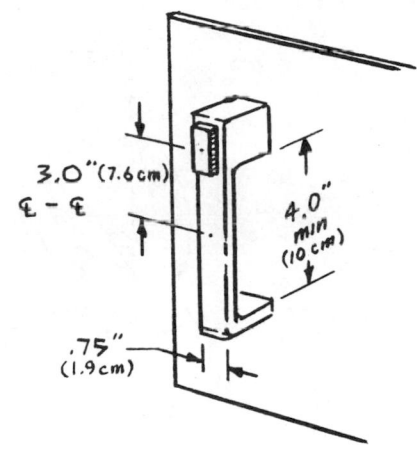

Filing cabinets also require a combination handle and latch release. Since the cabinet drawer is not generally carried around, a single handle is sufficient. A convenient combination of handle and latch release tab is illustrated in the accompanying sketch.

Note the dimensions suggested for the tab release handle. A common fault of most similar hardware is that the tab is too small and shallow for easy operation. The handle should be oval in shape, and/or the edges of the rectangular handle should be radiused so that the user does not injure his or her fingers while pulling out the drawer.

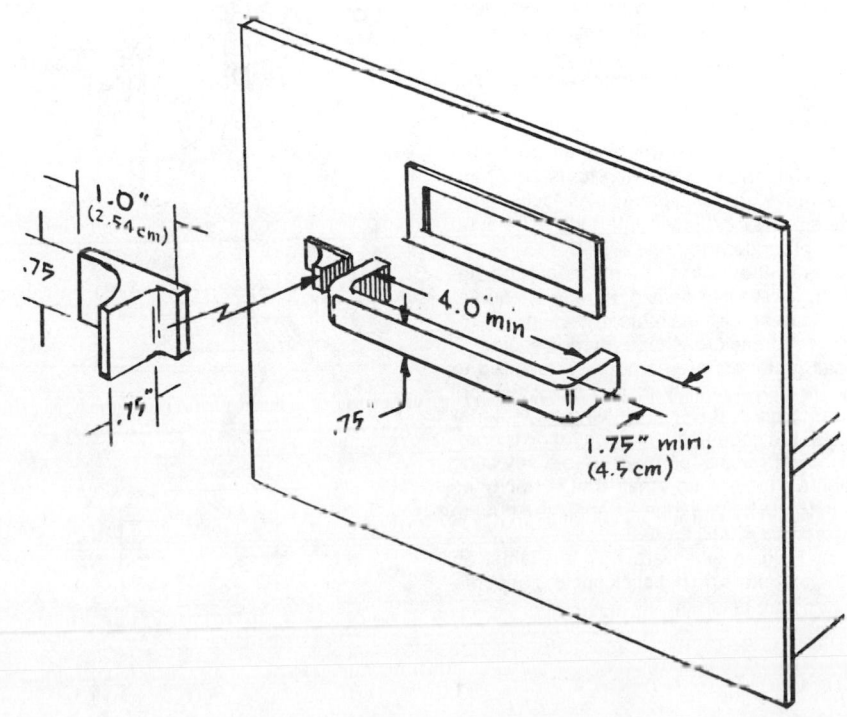

## SUITCASE AND BRIEFCASE LATCHES

These latches are in actuality controls, even though they are not usually thought of in that way.

The primary problem with most of these devices is that the manufacturer has not provided adequate finger clearance to "lift" the primary locking tab. The dimensional guidelines illustrated in the accompanying sketches will provide a more acceptable latching device.

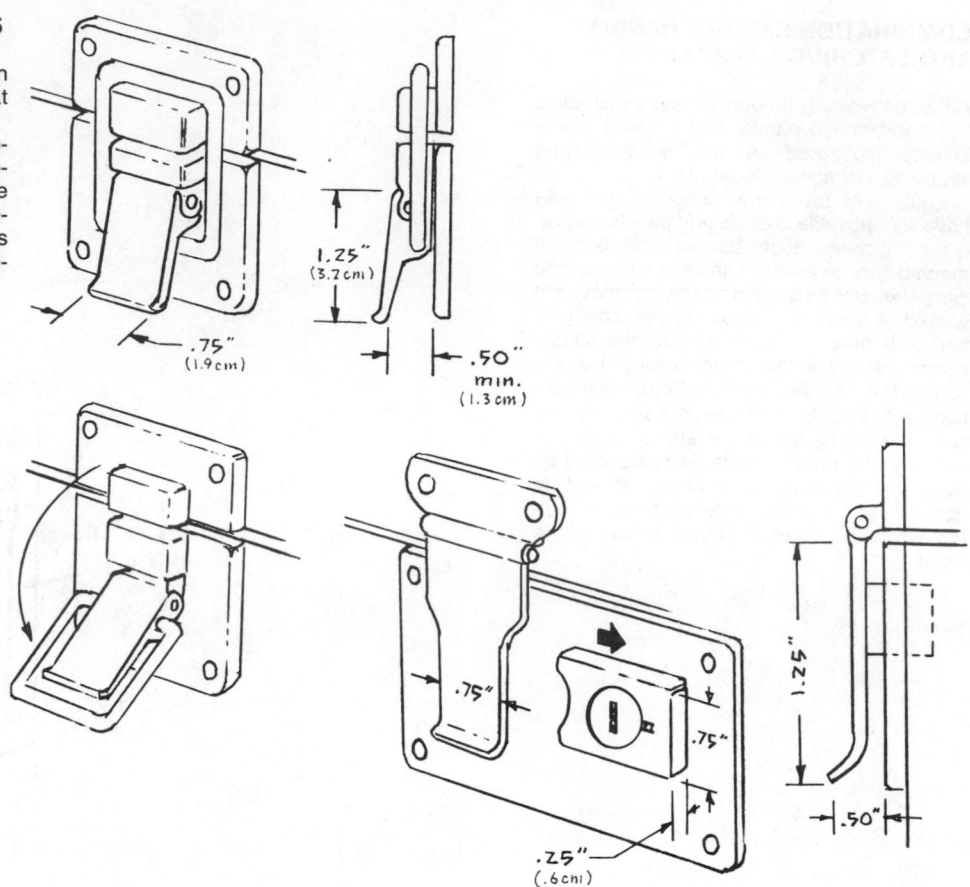

## OTHER LATCHING CONTROLS

Folding handle configurations are often desirable in order to provide unobstructed stacking of equipment or cases. The design shown in the accompanying sketch is typical. The critical dimensions required to make such handles accessible are provided. A folding handle should, however, be designed with a strong spring detent to prevent it from inadvertently folding while the user is trying to use the handle to move the equipment. Although such handles are not ordinarily intended to be used for this purpose, they often are.

Although the ½-in (1.3-cm) clearance behind the folded handle is generally adequate, more clearance is desirable if it will not compromise the basic package interior space.

Small fasteners are sometimes required to eliminate the necessity to find a special tool such as a screwdriver. The wing-tip fastener shown in the accompanying illustration is commonly used for this purpose. Avoid selecting fasteners that are so small that it becomes difficult to get hold of them or apply the necessary force to rotate them.

*Note:* Folding types are also available, so that the fasteners do not stick out and become a hazard to passersby.

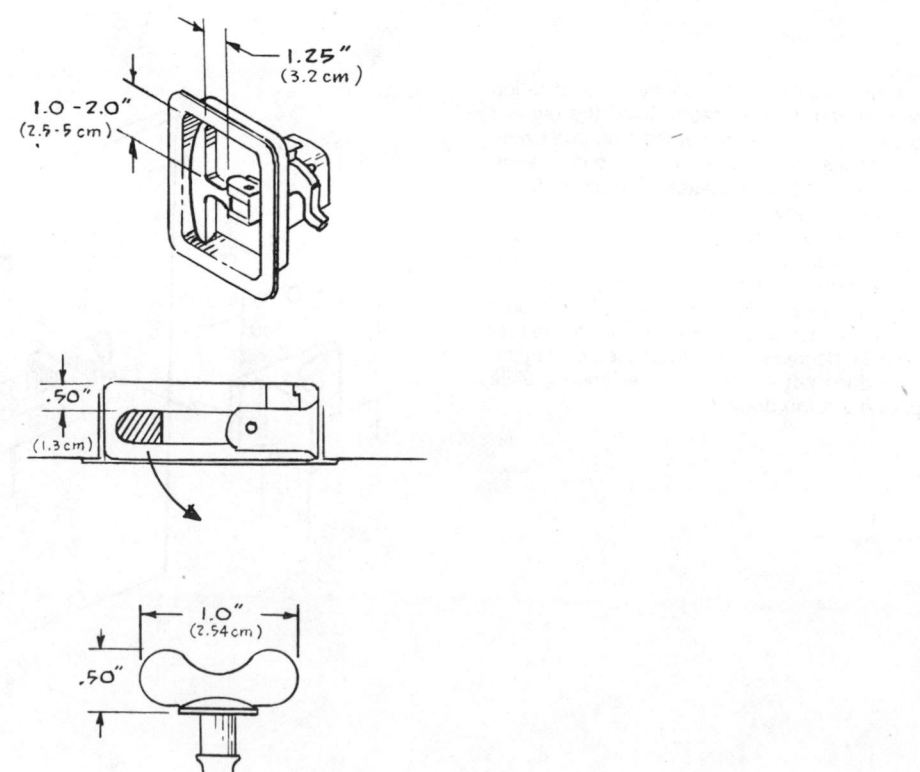

## SPECIAL INTEGRATED HANDLES AND CONTROLS

### Slide Switches

Key considerations in the design and/or selection of slide switches for applications such as a flashlight are sufficient contact surface area, vertical extension above the operating surface, and serration and surface shaping to ensure that the operator can apply proper force to move the switch.

The concave switch shape shown in the accompanying sketch should be at least 0.75 in (1.9 cm) long by 0.25 in (0.63 cm) wide. The switch travel should be at least 0.25 in (0.63 cm), but no more than 0.625 in (1.588 cm), to provide tactile feedback without requiring too long a throw to make electrical contact.

The alternate tab and slide shown in the accompanying illustration should extend about 0.375 in (0.953 cm) above the working surface, but should be at least 0.375 in (0.953 cm) wide to provide adequate thumb-contact surface. Neither switch configuration should require more than about 10 in/oz (283.5 g) force to operate.

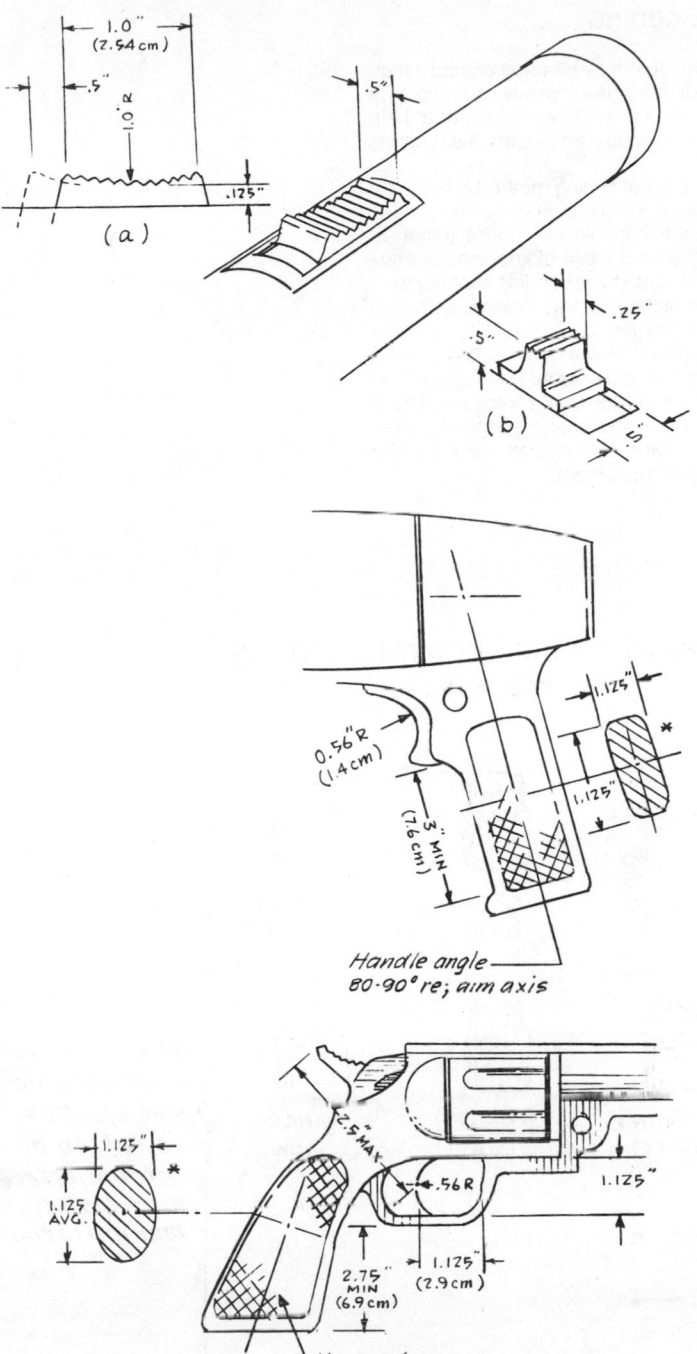

### Trigger Controls

Triggers used on power tools and hand-held weapons should be shaped and should provide clearance dimensions approximately as shown in the accompanying illustrations. Other handle and handle-trigger dimensional relationships are also important to effective operation. In selecting or designing trigger installations, take special precaution against potential finger entrapment and snagging of gloves.

## CONTROL CODING

Control knobs that may be used on more than one operator console or piece of equipment should be coded in some way in order to help the operator recognize and locate the controls more easily.

The most natural coding method is to locate the common controls in the same general place on each console or control panel. In addition, consider the use of knob shape and/or color as an additional coding technique.

Shape (or tactile) coding requires selecting shapes that are not confused with one another, regardless of the size of the control knob. In addition, the shaped knob should not turn, causing it to feel as if it were a different shape. The color and shape codes shown in the illustration below have been developed for use on military equipment.

## U.S. NAVY RADAR CONTROL CODES
### (Ref: MIL-STD-91528)

TUNING (ORANGE)    GAIN (RED)    INTENSITY (BLUE)

DIMMING (WHITE)    FOCUS (VIOLET)    RANGE (YELLOW)

*Note: Only the knob caps are shape/color-coded, so as not to interfere with knob manipulation.*

MARKER (GREEN)

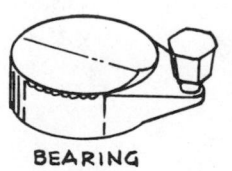

BEARING (GRAY)

## U.S. ARMY VEHICLE CONTROL CODES

LIGHTS    IGNITION    FUEL    POWER TRAIN    SPECIAL PURPOSE EQUIP.

ACCESSORY EQUIP.

## U.S. AIR FORCE AIRCRAFT CONTROL CODES

SUPER-CHARGER    MIXTURE    LANDING GEAR    THROTTLE

POWER    CARB. AIR    FLAPS    R.P.M.

FIRE EXTINGUISHER

Shape code standards for panel and vehicle controls.

The coded shapes shown in the accompanying sketches were developed for use on agricultural machinery by Henry Dreyfuss. They are used on a number of contemporary farm machines today.

It should be noted that, although shape coding has its place, in some situations it has little value. For example, the shapes shown here were developed without consideration of the fact that the typical machinery operator probably wears heavy gloves and thus cannot "feel" and identify the shapes.

It is often better to rely on the natural identification coding that an operator typically uses, e.g., location, and make sure that the location of the controls on new machinery is similar to that of the controls on previously used machines.

TRANSMISSION

THROTTLE

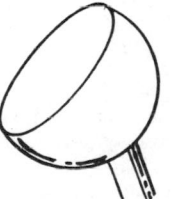

PLATFORM OR
REEL LIFT

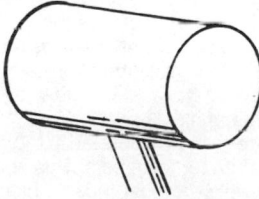

GROUND SPEED

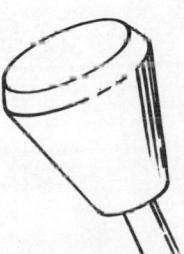

SEAT ADJ.

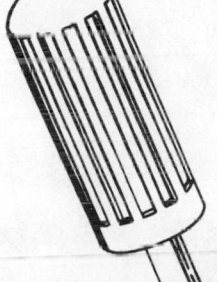

SEPARATOR RACHET

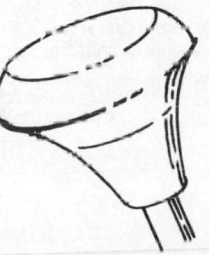

AUGER

## CONTROL LABELING

As a general guideline, all controls should be labeled. Exceptions, of course, are control devices that everyone recognizes and knows how to operate and controls that are located where the operator typically cannot see them. Examples of these exceptions include the following:

Automobile steering wheels and aircraft joy sticks
Automobile foot pedals and aircraft rudder pedals
Door and equipment cover latches
Switch mats, etc.

Rotating controls should not be labeled on the control knob or handle because the label becomes difficult to read when the knob is rotated.

When labels associated with controls are placed on the panel adjacent to the control, position the label so that the control knob will not obscure the label from the expected position from which the observer looks at the control. That is, when the eye reference is below the control, place the label below the knob; when the eye reference is above the control, place the label above the control; and when the eye reference is to one side of the control, place the label on the observer's side.

Otherwise, try to be consistent in positioning control labels; i.e., minimize the alternating positioning of labels above and below on the same operating panel.

When a control may be operated in the dark, provide illumination for the labels. This applies to all control labels, not just those which the designer may have decided were important. The reason for this is that operators make their own decisions relative to importance, and an unlabeled control may be inadvertently operated and/or confuse an operator because he or she assumes that the illumination is malfunctioning.

Refer to the section on visual displays for other recommendations regarding labeling.

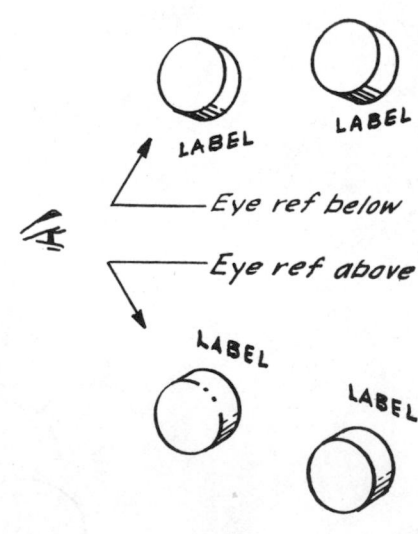

# CONTROL DYNAMICS

## 1. Feedback

All control systems should be designed to provide maximum information to the operator concerning the effects of the operator's input. This may consist of intrinsic feedback, extrinsic feedback, or both. Intrinsic feedback is that which is generated by the controller device, e.g., by "feel." Extrinsic feedback is that which emanates either from observation of an external visual condition or from a visual display especially provided to indicate the effect of controller inputs.

## 2. Control Input Mode

Continuous controlling or tracking is accomplished by either a compensatory or a pursuit mode. Compensatory tracking consists of adjusting a controller to maintain a vehicle, reticle, or pointer coincident with some visual reference. Pursuit tracking consists of "following" some randomly deviating reference. Compensatory tracking modes are more effective for acceleration control or when feedback inputs are infrequent. Otherwise, pursuit tracking modes are more efficient.

## 3. Controller Characteristics

Elastic resistance is desirable in a controller because it provides useful feedback from the controller movement or displacement; i.e., displacement can be quickly related to the changes in resistance. Design the controller system so that it provides distinct resistance gradients at critical control positions.

Viscous damping should be provided to aid in executing smooth control input at a prescribed rate.

Inertial resistance in the control system is desirable to enhance smooth control input and minimize the possibility of accidental or inadvertent activation. It should not be excessive, however, or it will become difficult to change adjustment direction. A certain amount of inertial resistance can, however, compensate for frictional drag that may be difficult to eliminate from the system.

Static friction should be minimized (e.g., resistance to control movement initiation). Not only does it provide no feedback, but also it often causes an excessive input force to overcome, which the operator may have difficulty correcting once he or she has observed the "overshoot."

In order for an operator to judge resistance, the input force required should be at least in the 5- to 10-lb (2.3- to 4.5-kg) range or above. Obviously, the resistance for manual controllers should be less than that for pedal controllers. Resistances above 30 to 40 lb (13.6 to 18 kg) should be avoided, however, in order to prevent fatigue.

Continuous tracking controllers should be designed to automatically return to a neutral reference position.

Sliding (coulomb) friction should be minimized in controller systems since it provides no meaningful input to the operator and may contribute to fatigue.

## 4. Control/Display Ratio

A low control/display (CD) ratio occurs when one unit of movement of the controller equals one unit of movement of the controlled element (e.g., the instrument pointer). Low CD ratios are desirable when it is important to obtain a gross system change with minimum controller movements. A high CD ratio is one in which the controller adjustment unit is greater than the unit motion of the system. High CD ratios are desirable for "fine" adjustment. Although specific control systems may require either a low or a high CD ratio because the nature of the control task is fairly constant, other systems may require a varying CD ratio program, such as for steering a vehicle or in using a joy-stick controller to manipulate an electronic "hook" to capture a target on a CRT display.

## 5. Detrimental Characteristics

Dead space, lag, backlash, and excessive resistance should be avoided in controller systems. Dead space occurs when movement of a controller does not immediately result in movement of the vehicle or visual display element. Dead space should be eliminated as much as possible, especially for sensitive or high CD ratio systems. Lag is a constant delay that may be inherent in the system design. With high CD ratios (1:6 to 1:3), lag is more serious for compensatory tracking. However, when CD ratios are low, lag may improve performance. Backlash is the tendency of a system to reverse when the control input ceases. If backlash cannot be minimized, the system display gain should be as low as possible.

## 6. System Aiding

Aided acceleration or rate tracking (i.e., allowing a single control movement to affect two variables, such as position and rate or acceleration) should be considered in order to shift the burden of differentiation, integration, and algebraic addition from the operator. Quickening (or display augmentation) should be considered for vehicle control systems where the "apparent" response of the system to control adjustments is delayed (for example, in ships or submarines). The quickened display shows the operator what control action should be taken. Predictor display concepts should be considered when the future state of the system is not directly predictable, i.e., when the control system computes and displays the predicted position of the vehicle at some selected future time. This permits operators to assess the predicted results of any current input they may desire to make, thus allowing them to readjust the input in the event the first input would have an undesired result. The predictor display provides a surrogate for operator prediction (e.g., when a driver looks ahead and predicts that his or her steering input is too much or too little to properly negotiate a curve at the vehicle's current speed).[19]

---

[19]Because of the typically unique variations among more complex continuous-control systems, generalizations regarding specification of CD ratios and system dynamics are inadvisable. Therefore, the reader should consult someone who is experienced in complex control systems design and/or refer to the following excellent discussions of the general subject of control system dynamics: C. R. Kelley, *Manual and Automatic Control: A Theory of Manual Control and Its Application to Manual and to Automatic Systems*, John Wiley & Sons, Inc., New York, 1968; and E. C. Poulton, *Tracking Skill and Manual Control*, Academic Press, Inc., New York, 1974.

## GUIDELINES FOR THE DESIGN AND SELECTION OF FASTENERS

Use the minimum number of fasteners compatible with requirements for securing components. In designing equipment packages and covers, use tongue and slot or similar techniques to help reduce the number of fasteners required.

Use "captive" fasteners to avoid loss of all or part of the fastening device.

It should be readily apparent when a fastener has been "released."

Work space should be provided around a fastener for fingers, tools, or "wrenching space."

Standardized fasteners should be used wherever possible to minimize stocking requirements and the number of different tools required.

Use manually operated fasteners where suitable to avoid obtaining or carrying tools.

Use fasteners that are operable by commonly used rather than special tools.

Use the same type and size of fastener consistently for a given application.

Screws, bolts, or nuts with different threads also should have clearly different physical sizes to minimize their being interchanged.

Consider how a worn or stripped fastener can be removed before selecting it; avoid stud fasteners that are an integral part of the machine or housing. Use screws rather than rivets.

Select safety fasteners for applications where vibration could result in loosening and perhaps loss of the fastener during operation.

Fasteners that are operable by hand should also be rugged enough to withstand operation by some type of tool.

Spring-load catches so that they lock on contact, rather than requiring a separate locking step.

Use "long latch" catches to minimize the possibility of inadvertent latch release.

When a latch has a handle, locate the latch release near the handle so that it can be operated with one hand.

Screw heads should have deep slots to reduce tool slippage and damage to the screw head.

Place the screw head so that offset screwdrivers are not required.

When bolts and nuts are used, make sure there is clearance to both ends.

Keep bolts as short as possible so that they will not snag personnel or equipment. Do not locate bolts where inadvertent replacement by a longer bolt might interfere with the movement of an adjacent mechanism or

puncture adjacent cover or "skin."

Coarse threads are preferable to fine threads for low torques because they reduce the possibility of cross threading.

Avoid the use of left-hand threads unless the system requirements demand them; then identify both bolts and nuts clearly by a suitable marking, shape, or color.

Cotter keys or pins should fit snugly, but they should not have to be "driven out."

Use safety wire only when self-locking fasteners cannot withstand the expected vibration or stress. Attach safety wire so that it is easy to remove and replace.

Use retainer rings which hold with a positive snap action. Avoid rings which become difficult to remove when they are worn.

Provide retainer chains to capture fasteners which have to be completely removed but which could easily become lost.

Avoid the use of fastener systems in which a captive spring could suddenly drive the fastener into the face of the technician once the fastener is released or in which the spring itself pops out and becomes lost. Similarly, avoid systems in which a washer easily becomes detached and lost without the technician's knowing about it. Avoid springs that are so strong that a technician has to use a special tool to depress the spring before engaging the basic fastener.

**FASTENER, TYPES AND DESIGN
CONSIDERATIONS**[20]

| Type | Description |
|---|---|
| | Adjustable pawl fastener.<br>As knob is tightened, the pawl moves along its shaft to pull back against the frame. 90° rotation locks, unlocks fastener. |
| | "Dzus"-type fastener with screwdriver slot.<br>Three-piece one-quarter–turn fastener. Spring protects against vibration. 90° rotation locks, unlocks fastener. |
| | Wing head. "Dzus" type.<br>90° rotation locks, unlocks fastener. |
| | Captive fastener with knurled, slotted head.<br>Retaining washer holds the threaded screw captive. |
| | Draw-hook latch.<br>Two-piece, spring latch, base unit and striker. When engagement loop is hooked over striker, depressing lever closes unit against force of springs. Lever is raised to unhook. |
| | Trigger-action latch.<br>One-piece, bolt latch. Depressing trigger releases bolt, which swings 90° under spring action and opens latch. To close, move bolt back into position. |
| | Snapslide latch.<br>One-piece snapslide. Latch is opened by pulling lever back with finger to engage release lever. |
| | Hook latch.<br>Hook engages knob on striker plate. Handle is pulled up locking in place. To release, reverse procedure. |

[20]*Human Factors Engineering Design for Army Materiel, MIL-HDBK-769, 1975.*

**Types of Fasteners**

| Type | Description | Maintainability Considerations | Approximate Operating Time |
|------|-------------|-------------------------------|------------|
| * Knurled | Adjustable pawl fastener. As knob is tightened the pawl moves along its shaft to pull back against the frame. 90° rotation locks, unlocks fastener. | 1. No tools required. | 01 min |
| | "Dzus" type fastener with screwdriver slot. Three-piece, one-quarter–turn fastener. Spring protects against vibration 90° rotation locks, unlocks fastener. | 1. Tools may be required. 2. Should not be used for front panel fasteners or in structural applications. Preferred type for lightweight panels other than front panels. | 05 min |
| 1.0" .5" | Wing head. "Dzus" type. 90° rotation locks, unlocks fastener. | 1. No tools required 2. Should not be used for front panel fasteners or in structural applications. Preferred type for lightweight panels other than front panels. | 04 min |
| .5" Knurled .5" extension | Captive fastener with knurled, slotted head. The threaded screw is made captive by a retaining washer. | 1. Tools may be required. 2. Operating time depends on number of turns required. | 04 min |

*When gloves may be worn, increase all dimensions by 0.125 in (0.318 cm).
Each of the fasteners shown in the accompanying table is evaluated from the point of view of ease of use in the context of equipment accessibility for maintenance.

**Types of Latches**

| Type | Description | Maintainability Considerations | Approximate Operating Time |
|---|---|---|---|
| | Draw hook latch. Two-piece, spring latch, base unit and striker. Engagement loop is hooked over striker and lever is depressed, closing unit against force of springs. Lever is raised to unhook. | 1. Relatively slow action. 2. Requires considerable force to disengage loops. | 03 min |
| | Trigger action latch. One-piece, bolt latch. Latch is opened by depressing a trigger to release bolt which swings 90° under spring action. To close move bolt back into position. | 1. Extremely fast action type latch. 2. Strong spring action might cause personal injury. | 01 min |
| | Snapslide latch. One piece snapslide. Latch is opened by pulling lever back with finger to engage release lever. | 1. Fast action. | 02 min |
| | Hook latch. Hook engages knob on striker plate. Handle is pulled up locking in place. To release reverse procedure. | 1. Relatively slow action. 2. Takes up room on equipment. | 03 min |

**Types of Fastener Head Styles**

| Type | Description | Maintainability Considerations |
|------|-------------|-------------------------------|
| 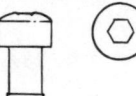 | Fillister head, slotted. Smaller in diameter than round head, but has deeper slot. (FF·S·92 Type 1, Style 4S)* | 1. Requires common screwdriver. 2. Deep slot not easily stripped. |
| 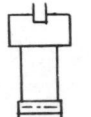 | Fillister head, hexagon socket. Same as above except with "Allen" type socket. | 1. Required "Allen" wrench not always available. 2. "Allen" wrench usually takes longer to use than common screwdriver. |
| 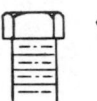 | Knurled head, slotted. This is an exaggerated fillister with a knurled head and a screwdriver slot. | 1. Can be used with fingers or common screwdriver. 2. Preferred type for retaining front panels or removable cover plates. |
|  | Hexagon head. Standard head for machine bolts and screws. (FF-S-92, Type II. Style 10P)* | 1. Requires use of wrench or "spintite." |
|  | Hexagon head, slotted. Same as above except with added screwdriver slot. (FF-S-92. Type I. Style 10S)* | 1. Wrench, "spintite," or common screwdriver can be used to remove or install. |
|  | 82° flathead, cross-recessed. (FF-S-92. Type III. Style 2C)* | 1. Cross-recessed screwdriver not always available. |
|  | 82° flathead, slotted. (FF-S-92. Type I. Style 2S)* | 1. Not suitable on thin panel. |
| 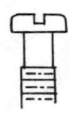 | 92° oval head, slotted. Similar to flathead but with rounded head. (FF-S-91. Type I. Style 6S)* | 1. Allows deeper slot than flathead. 2. MIL-E-16400 specifies this type with cup washer for rack-mounted panels. |
| | Pan head, slotted. Large diameter with high outer edges for maximum driving power. (FF-S-91, Type I, Style 9S)* | 1. Standard type for panels other than rack-mounted panels and front panels. |

*Styles recommended by Federal Specification FF-S-92 for use wherever possible, in order to keep inventories to a minimum.

$d$ = minimum slot depth = 0.0625 in (0.1588 cm).

**Types of Captive Screws**

| Type | Description | Design Considerations |
|------|-------------|-----------------------|
| Undercut | Captive screw with undercut. Panel is tapped to screw size. | 1. Easily installed and removed. |
| Spring | Captive screw, spring loaded. | 1. Easily installed.<br>2. Fast action. |
| | Captive screw which is forced through rubber or plastic grommet. | 1. Frequent use can cause excessive wear on rubber or plastic grommet resulting in a loose fit. |
| Hook | Captive screw with hook which falls into undercut of screw. | 1. Hook can be easily jarred loose. |
| Upset | Captive screw with upset thread. | 1. No removal features. |
| | Captive screw with nut staked in place after assembly. | 1. Time consuming to put in place and remove. |
| Cotter pin | Captive screw with cotter pin forced in place after assembly. | 1. Cotter pin can work loose if subjected to excessive vibration. |

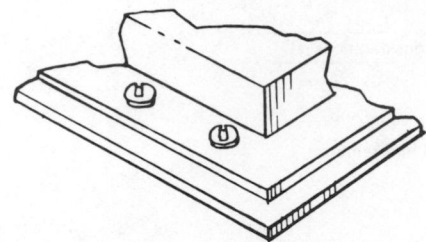

**Chassis Mounted on Horizontal Shelf (Secured by Screw Fasteners through Flange)**

| Securing Method | Maintainability Considerations |
|---|---|

| | |
|---|---|
| | **Description:** Captive screw.<br>**Advantages:** No loss of screws and washers. All work performed from one side.<br>**Disadvantage:** Space required for captive device.<br>**Tools Required:** Screwdriver.<br>**Operating Time:** Approximately 6 min per fastener. |
| | **Description:** Screw into tapped note with flat washer and lock washer.<br>**Advantage:** All work performed from one side.<br>**Disadvantages:** Time required to position washers. Possible loss of screws and washer.<br>**Tools Required:** Screwdriver.<br>**Operating Time:** Approximately 8 min per fastener. |
| | **Description:** Screw through clearance holes with flat washer, lock washer, and nut.<br>**Advantages:** No significant advantages.<br>**Disadvantages:** Possible loss of screws, washers, and nuts. Requires access to both sides of shelf. Two-handed operation. Time required to position washers and nuts.<br>**Tools Required:** Screwdriver and wrench or "spintite."<br>**Operating Time:** Approximately 8 min per fastener. |
| | **Description:** Screw through clearance holes with lock nut.<br>**Advantage:** No washers required.<br>**Disadvantages:** Lock nut difficult to turn. Requires access to both sides of shelf. Two-handed operation. Possible loss of screws and nuts. Time required to position nut.<br>**Tools Required:** Screwdriver and wrench or "spintite."<br>**Operating Time:** Approximately 14 min per fastener. |

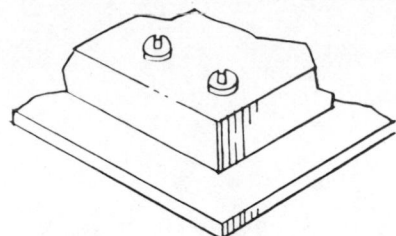

## Chassis Mounted on Horizontal Shelf (Secured by Screw Fasteners through Chassis)

| Securing Method | Maintainability Considerations |
|---|---|
| | Description: Captive screw. |
| | Advantage: No loss of screws or washers. All work performed from one side. No separate handling washers. One-handed operations. |
| | Disadvantage: Alignment of screw can be difficult. Space required for captive device. |
| | Tools Required: Screwdriver. |
| | Operating Time: Approximately 6 min per fastener. |
| | Description: Stud through chassis with flat washer, lock washer, and nut. |
| | Advantage: No screw alignment problem. Studs act as locating pins. |
| | Disadvantages: Possible loss of nuts and washers. Chassis must be lifted over studs. |
| | Tools Required: Wrench or "spintite." |
| | Operating Time: Approximately 7 min per fastener. |
| | Description: Screw into stand-off with flat washer and lock washer. |
| | Advantage: Very little screw alignment problem. |
| | Disadvantages: Possible loss of screws and washers. Chassis must be lifted over stand-off. |
| | Tools Required: Screwdriver. |
| | Operating Time: Approximately 8 min per fastener. |

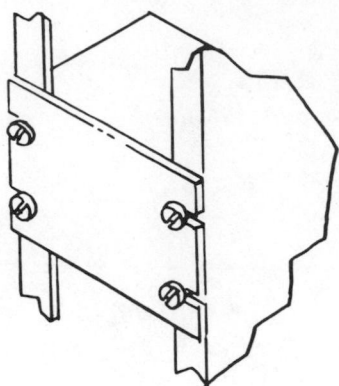

**Chassis Mounted on Vertical Rack (Secured by Screw Fasteners into Frame)**

| Securing Method | Maintainability Considerations |
| --- | --- |

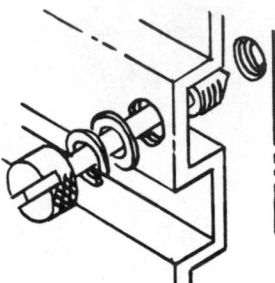

Description: Captive screw. Slotted head-type preferred for front panels.

Advantages: No loss of screws and washers. All work performed from one side. No separate handling of washers. One-handed operation.

Disadvantage: Space required for captive device.

Tools Required: Screwdriver. None if thumb screw is used.

Operating Time: Approximately 6 min per fastener.

1:0 X .75" MIN

Description: Thumb screw with lock washer and flat washer.

Advantage: No tools required. One-sided operation.

Disadvantage: Possible loss of screws and washers.

Tools Required: None.

Operating Time: Approximately 8 min per fastener.

Description: Screw into tapped hole with flat washer and lock washer. Preferred for panels other than front panels.

Advantage: All work performed from one side.

Disadvantage: Time required to position washers. Possible loss of screws and washers.

Tools Required: Screwdriver.

Operating Time: Approximately 8 min per fastener.

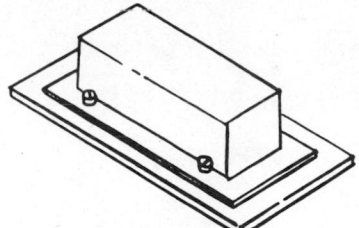
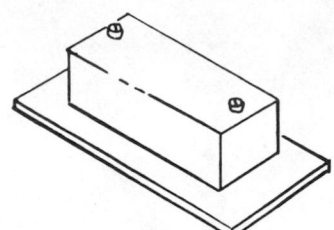

## Chassis Mounted on Horizontal Shelf (Secured by Quick-Acting Fasteners)

| Securing Method | Maintainability Considerations |
|---|---|

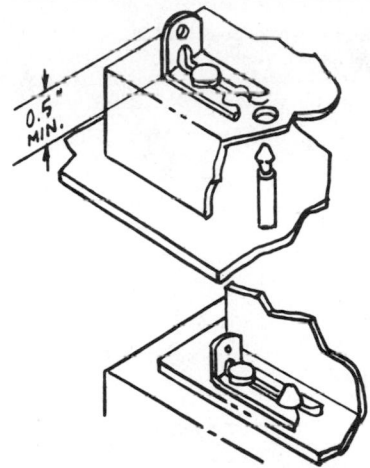

Description: Snap-slide latch.
Advantages: Fast action. No tools required.
Disadvantages: Flange-mounted method may be difficult to operate when side clearance is limited.
Tools Required: None.
Operating Time: Approximately 0.02 min per fastener.

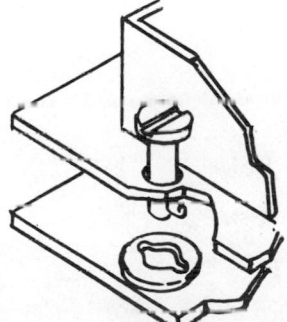

Description: "Dzus"-type fastener.
Advantages: Operating time approximately one-tenth that of captive screw-type fastener.
Disadvantage: Requires use of screwdriver.
Tools Required: Screwdriver.
Operating Time: Approximately 0.05 min per fastener.

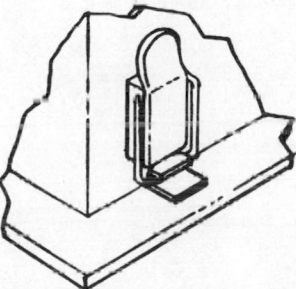

Description: Spring or drawhook latch.
Advantage: Requires no tools.
Disadvantages: Requires space to operate. Closing may require more time due to difficulty in engaging drawhook. Strong spring action might cause personnel injury.
Tools Required: None.
Operating Time: Approximately 0.03 min per fastener.

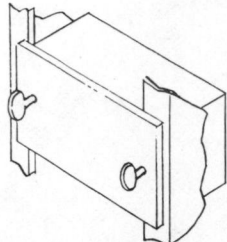

**Chassis Mounted on Vertical Rack (Front Panel Secured by Quick-Acting Fasteners)**

| Securing Method | Maintainability Considerations |
| --- | --- |

| | |
| --- | --- |
| Description: | Push-button latch. |
| Advantages: | Integral handle provides easy removal of chassis. Fast release. Closes with snap action. |
| Disadvantages: | Requires outside space for handle. Does not pull panel against gasket as well as other types. |
| Tools Required: | None. |
| Operating Time: | Approximately 0.03 min per fastener. |

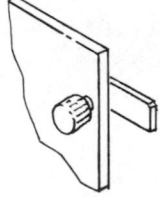

| | |
| --- | --- |
| Description: | Pawl latch. Operates with 90° turn. |
| Advantage: | Requires minimum inside space. |
| Disadvantage: | Difficult to operate when much pull-up is required. |
| Tools Required: | None. |
| Operating Time: | Approximately 0.05 min per fastener. |

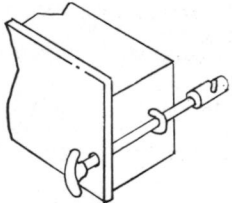

| | |
| --- | --- |
| Description: | Cam-action. Operates with 90–180° turn of handle. |
| Advantage: | Fairly fast release operation. |
| Disadvantage: | Initial engagement of cam may be difficult. |
| Tools Required: | None. |
| Operating Time: | Approximately 0.05 min per fastener. |

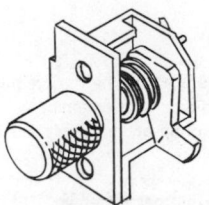

| | |
| --- | --- |
| Description: | Adjustable pawl latch. Operating handle pulls pawl toward panel. |
| Advantage: | No significant advantage over the other types. |
| Disadvantages: | Requires more time than other types. Excessive inside space. |
| Tools Required: | None. |
| Operating Time: | Approximately 0.1 min per fastener. |

When frequent removal of a chassis is necessary, the time required for dismounting a hinged cover may be significant. For example, dismounting a chassis using two standard hinges may require the removal of from four to six screws, taking up to 5 min or more. Use of separable hinges can reduce this time to a fraction of a minute.

Any separable hinge must, however, include features to prevent accidental separation under shock or vibration conditions. The hinges shown in the accompanying illustration should not be used to bear the weight of a heavy chassis when the equipment is closed; i.e., guide pins or screw fasteners should be provided to support a heavy chassis.

## SEPARABLE HINGES

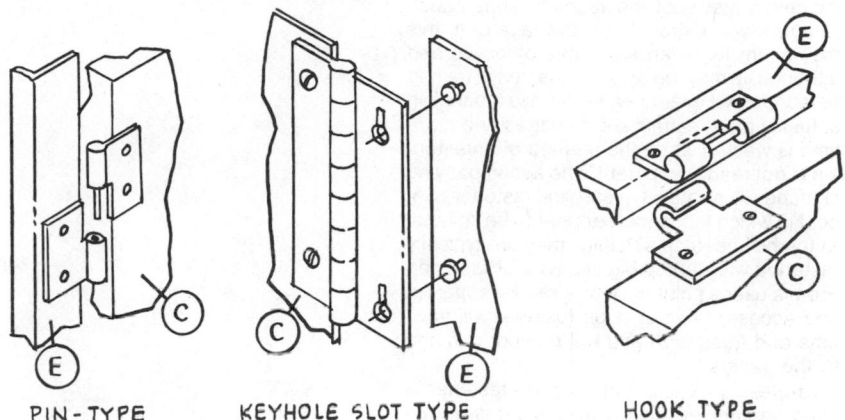

PIN-TYPE     KEYHOLE SLOT TYPE     HOOK TYPE

E = EQUIPMENT
C = CHASSIS

## CHASSIS AND DRAWER SLIDE FASTENERS

Slide-mounted equipment chassis systems (and file-drawer slide systems) using ball or roller bearing slides are often improperly designed in terms of the locking and fastening features. Such slide systems should, of course, provide an automatic locking system to prevent a drawer or chassis from being accidentally pulled all the way out of a cabinet. Care should be taken, however, to design the slide-locking system so that it *is possible* to override the lock and fully remove the chassis or drawer. More often than not, this is an extremely complex and confusing operation because the basic procedure is not evident in the hardware itself. Obvious features (i.e., latches or buttons) should be provided on the slide-locking system, and/or clear instructions should appear directly on the hardware. Remember that people seldom remember where they have put an instruction booklet.

More flexible slide-rotational systems may appear even more complex to the user, and thus careful consideration must be given to the release latching concepts and to the visibility and recognition factors associated with locking and unlocking both the horizontal and rotational elements of the slide system mechanisms.

Consider the following guidelines:

The chassis should be easily and quickly detachable for removal from a cabinet without the need for hand tools.

Detaching the chassis from the slide should not require placing the hands in a position that might lead to injury if the chassis suddenly moves.

The slides should automatically lock in the open position and/or in a rotated position securely enough to prevent dislodging the chassis.

The slides should be easily removable from the cabinet for replacement.

Special tools should not be required to remove or repair slides.

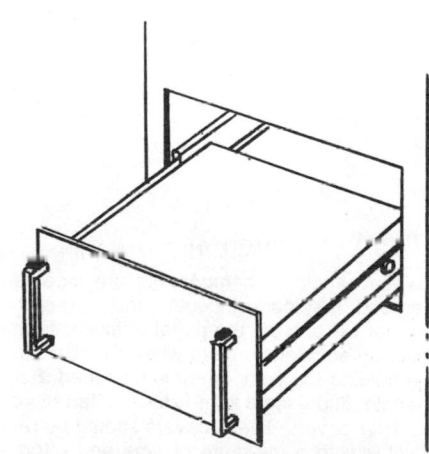

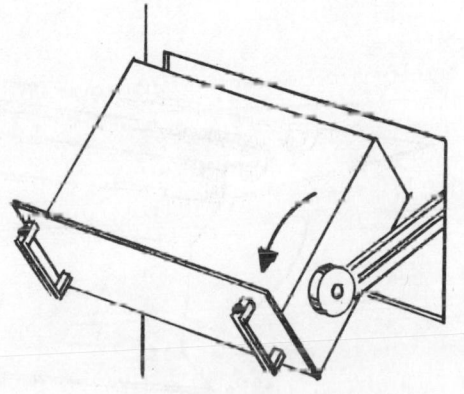

## BLIND FASTENERS

Blind fasteners are, of course, commonly used where the designer assumes that the fastening device may spoil the aesthetic appearance of a product. Although, on the face of it, this may seem to be an admirable objective and although it may be argued that only trained personnel will ever need to remove a panel, in actuality more panels are damaged and more time is wasted when the method of unfastening is not readily apparent. The accompanying sketches illustrate a typical blind fastener concept in which trim panels appear to be "glued" to the bulkhead. In actuality, they are typically fastened with the spring-clip type of fastener. From a user's point of view, a readily apparent and accessible screw-type fastener will save time and frustration and will prevent damage to the panels.

Another desirable type of blind fastener is used on cabinet doors where there is no dynamic environment involved, in which case a magnetic catch is preferred over the spring-clip type of fastener.

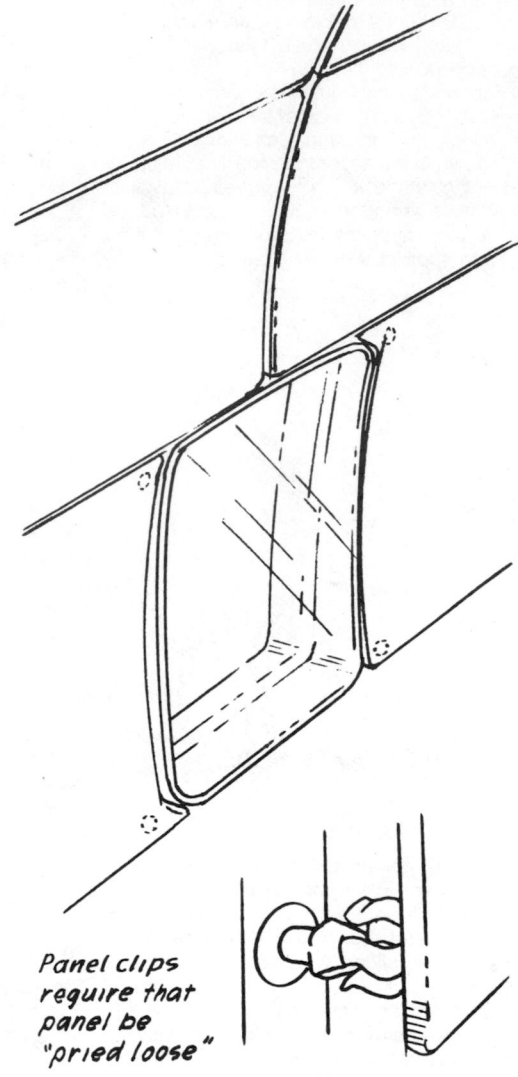

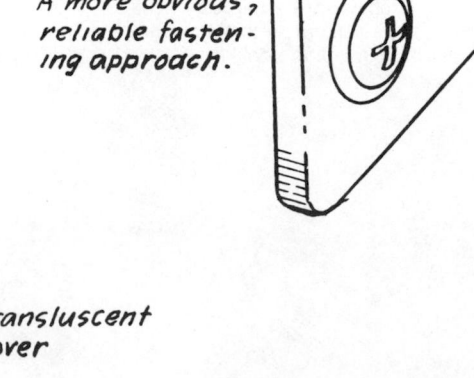

Panel clips require that panel be "pried loose"

A more "obvious", reliable fastening approach.

When panels or components are located overhead, fastener concepts should be chosen that will prevent the panel or fixture from falling on someone's head when the last device holding the component is loosened. For example, fluorescent light fixtures often have dust filter covers. These covers should be removable with a minimum of time and effort, and they should be held on one side so that the filter cover can be hinged down and held while lamps are replaced.

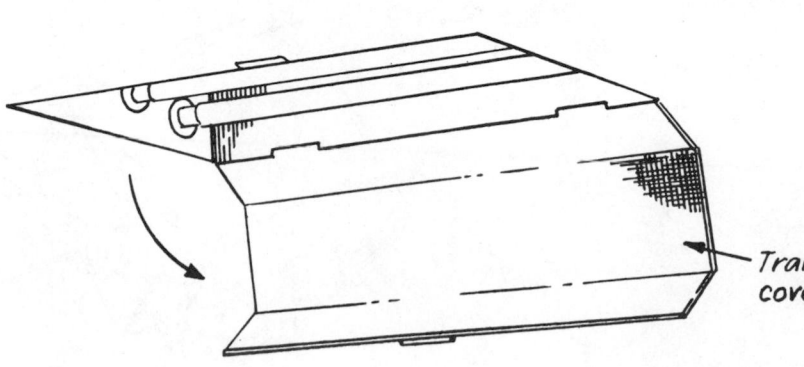

Transluscent cover

## GUIDELINES FOR THE DESIGN AND SELECTION OF TOOLS

### Hand-Held Tools

The following are key factors to consider:

Mechanical factors
Anthropometric factors
Anatomical considerations
Physical characteristics of the tool
Ergonometric considerations
The work environment

**MECHANICAL FACTORS** A hand tool is used to extend and reinforce the strength and effectiveness of the operator's limb or limbs while he or she is performing a specific task. Forces generated within the human musculoskeletal system are transmitted to the workpiece through the tool. Reactive forces are generated from the workpiece to the human structure via the tool. Stresses on the human result from excessive forces and poor posture, which frequently are due to poor tool design. Worse yet, when operators are applying a large force on a tool and it slips or breaks, they often find themselves in an unstable position, or they may be injured from the recoil of force.

**ANTHROPOMETRIC FACTORS** The dimensional characteristics of human operators are critical to the effectiveness with which they can grasp, hold, manipulate, or guide a hand tool. Particularly important to effective tool use is the design of tool handles, not only so that they can be gripped properly, but also so that the tool can be properly guided during a given application of force.

**ANATOMICAL CONSIDERATIONS** The tool design must reflect adequate consideration of the fact that the operator's wrist and arm have specific rotational characteristics and that whenever tool action requires thrust, rotation, or both, limb rotational axes are therefore important to proper tool manipulation. Basic to the tool handling is the posture that must be taken in order to permit the best mechanical advantage.

**PHYSICAL CHARACTERISTICS OF THE TOOL** The basic tool characteristics to be considered include the following:

The tool weight should balance close to the point of support.

The tool material should not conduct electricity.

Tool handles should have a fairly high coefficient of friction to minimize slipping.

Nonporous, hard materials should be used to preclude the retention of toxic substances or the collection of sharp materials, which may become embedded in grip areas.

The overall weight should be compatible both with the need to reduce the potential for fatigue when lifting or holding the tool and with the need to minimize the amount of pressure the operator is required to exert in order to make the tool perform as intended.

**ERGONOMIC CONSIDERATIONS** The matter of human energy expenditure and how this energy is best exerted should underlie the conceptualization of each new tool design. Such factors as human mobility, dexterity, equilibrium, and posture, as well as consideration of potential ballistic movement advantage, all should receive the full attention of the tool designer.

**THE WORK ENVIRONMENT** Not all work environments may lend themselves to a single tool-operator interface concept; i.e., the tool may have to be used under less than optimum conditions, such as overhead, near the floor or ground, with inadequate lighting, or on an unstable platform. Consideration should be given to the potential difficulties imposed by radical temperature, ventilation, and humidity conditions and/or by the requirement to wear heavy clothing and gloves.

Power hand tools subject an operator to additional stresses that must be considered. Specific considerations relate to the following:

**ROTATING COMPONENTS** Proper guards against inadvertent contact with moving blades or gears must be provided.

**ELECTRICAL SHOCK** Appropriate grounding and/or contact isolation between the tool and the operator must be provided.

**HEAT** Heat generated either by motors or by the interaction between the tool blade and the workpiece poses a dual hazard for operators, i.e., they may receive actual burns, or their escape reaction when they touch a hot surface may cause them to drop the tool or enter into a series of dangerous motions (falling from a ladder, backing into an obstacle, dropping the tool against the body, etc.).

**FLYING PARTICLES OR SPARKS** Power tools frequently generate flying particles or sparks, thus creating hazards not only for the operator but also for persons nearby. Consideration should be given early in the tool conceptualization to providing appropriate protection from such hazards.

**POWER CONTROL** Unintentional power activation and the lack of an expedient power cutoff are both critical to the safety of power tool operators.

**HAZARDS TO THE NONOPERATING HAND** Many injuries caused by power tools result from the fact that the nonoperating hand is accidentally placed where it should not be; e.g., the operator may be using his or her nonoperating hand to hold material, an electric cord, or some other object out of the way. In conceptualizing a hand-held power tool, one should carefully examine the various procedural sequences wherein an operator may place the nonoperating hand in an exposed position. Such an examination may suggest that the tool be specifically designed so that the operator has to place the nonoperating hand on the tool in order to ensure that the hand is kept out of the way.

Other types of special tools create unique problems. Typical of these are soldering tools, which can cause burns if the operator touches a tip that is still hot or lays the tool on flammable materials, for example. Welding operations and the tools used to perform them create another hazard in terms of the "blinding" effects of the extremely bright torch and also the attendant possibility of burns.

It should be noted that the standard approach has been to accept many of the above hazards as part of the job and to place the burden of protection on users, i.e., to recommend wearing eye protectors, using special electrical grounding devices, wearing gloves, etc. Obviously, in many cases these are the only methods available to reduce the hazard potential. However, the designer should, in each new tool design, review such hazards and attempt to remove them whenever possible in the design itself. When this cannot be accomplished, the designer should assume the responsibility for providing appropriate warning labels on the tool and/or properly worded warning instructional materials to accompany the tool. After all, the designer should know better than anyone else what hazards a new tool presents

## GENERAL PRINCIPLES FOR TOOL DESIGN[21]

Make the tool easy to carry close to the body with one hand.

Make the tool easy to set down or hang up.

When the tool includes a "weight-holding function," separate this from force, guidance, and control functions.

Provide specific handholds (otherwise, the operator may take hold of the tool in a hazardous manner).

Make the tool compact and light so that it is easy to carry, handle, and store.

Align the tool's center of gravity with the grasping hand so that the operator will not have to overcome rotational movement or torque of the tool.

Make tool grasping surfaces resilient and slip-resistant.

Round off all edges and corners of the tool package and avoid adding sharp, protruding elements.

Make flywheels solid and without spokes; round the edges.

Provide nonaccess guards around all moving parts.

Use low-voltage electrical power and double insulation.

Use safety-secured couplings with high-pressure, compressed air or fluid power tools to prevent inadvertent disconnection and line whipping.

Provide flanges on the tool handle to prevent the hand from slipping forward as force is applied and/or on the aft part of the handle to keep a heavy tool from slipping out of the hand when it is being carried.

Orient the tool handle so that the operator's wrist can remain in the most natural position while he or she is applying force or guidance inputs.

When an auxiliary control must be manipulated while the operator is holding the tool, place the control where the thumb or fingers of the holding hand can reach and operate it without disturbing the holding position or place the control where it can be operated by the other hand.

If sleeve-type handle covers are provided, make sure they do not inadvertently slip, rotate, or come off.

Color-code certain tool components, e.g., safety switches.

Insulate contact surfaces to preclude electrical shock or high-temperature interface with the operator's hand or body.

Use a battery-powered design where practical to avoid entanglements with an electric cord.

Although it may not be possible to control noise produced by the tool-work interface, the designer should attempt to minimize internal noise that may be produced by bearings and other internal components.

[21]Leo Greenberg and Don B. Chaffin, *Workers and Their Tools: A Guide to the Ergonomic Design of Hand Tools and Small Presses,* Pendell Publishing Co., via an educational gift from AMP Incorporated of Harrisburg, Pa.

## TOOL DESIGN CHECKLIST

Work position
Work material holding
Tool grasp interface
Visibility
Tool operation and guidance
Prevention of debris and sparks
Tool balance and weight
Tool holder
Electric cord position and control
Guards for moving parts
Tool storage
Tool adjustment
Vibration and heat attenuation
Emergency safety switch
Prevention of parts loosening
Prevention of electrical shorts
Tool maintenance and repair

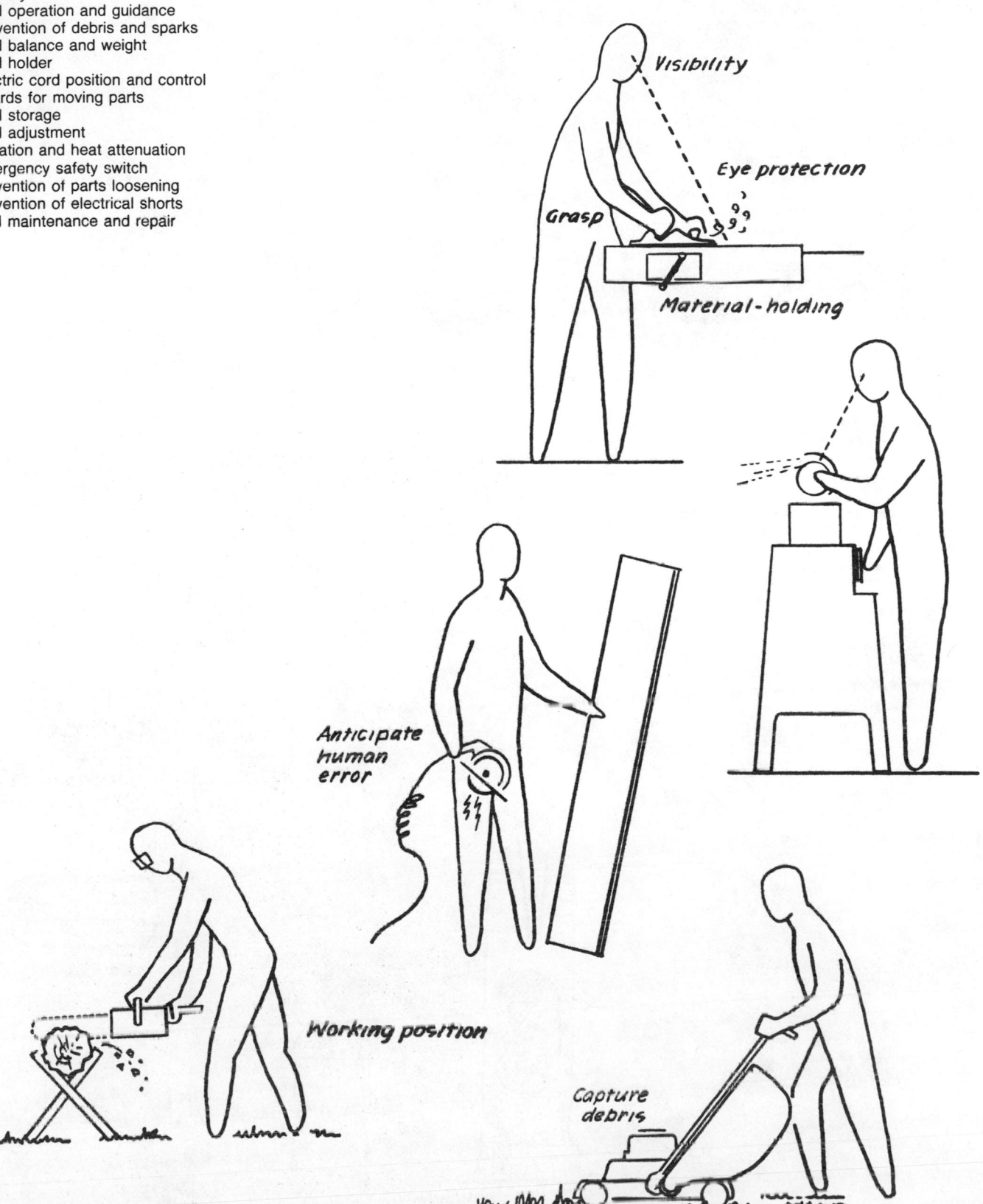

In considering various tools, it is almost always important to examine the relationship between the handle and the method or manner in which the tool will be used; i.e., the designer should consider whether it will be used merely to hold the tool or whether it must be used to rotate, push, pull, or guide the tool, for example. Equally important are the problems associated with gripping and applying pressure. Some important handle considerations are illustrated by the accompanying sketches:

Not only must screwdrivers and chisels be gripped tightly, but also they usually require that pressure be exerted toward the work. Generally, it is desirable to place the handle-tool longitudinal axis in line with the operator's arm in order to minimize fatigue and reduce the chance of "skewing" the tool and causing it to slip off the work.

Screwdrivers have to be rotated, and thus they should have round handles. Slight grooves or knurling helps keep the hand from slipping. Screwdrivers should have handle diameters that are compatible with the force expected to be applied. Except for very small, finger-operated screwdrivers, handles should be about 1.0 to 1.5 in (2.5 to 3.8 cm) in diameter.

Other tools should have semiflat sides. In the case of a hammer, the flat side aids in guiding the stroke. In the case of a chisel, the flat side helps control the lateral position of the blade.

Handles for pliers and similar two-handled tools should have flat sides, but with radiused edges to decrease the possibility of injuring the hands and fingers.

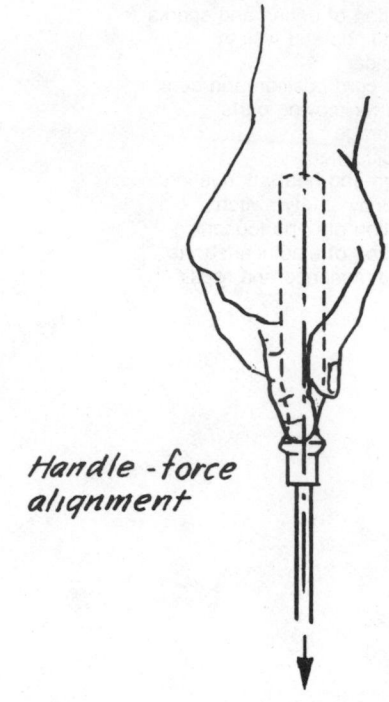

*Handle-force alignment*

SCREWDRIVER

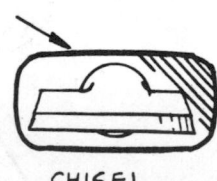

CHISEL

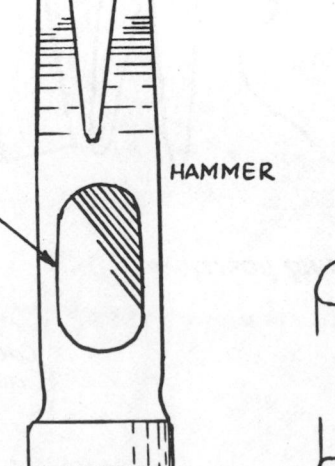

HAMMER

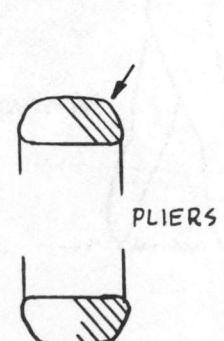

PLIERS

### Tool Design Guidelines: Examples

Tools which are held in the hand for a considerable time or which require continuous gripping and possible pressure applications should be designed not only so that they fit the hand and fingers but also so that stresses are not placed on the wrist and forearm.

A typical surgical instrument shown in the accompanying sketches tended to roll in the surgeon's fingers and produced considerable fatigue in the fingers and thumb. A redesign (shown in the accompanying illustrations) by Professor Tichauer of New York University eliminated most of the problems by changing the tool geometry so that the main action of the thumb opposing the second, third, and fourth fingers and the stabilizing action of the second, fourth, and fifth fingers were made easier. In addition, the redesigned instrument lessened sharp wrist angle.

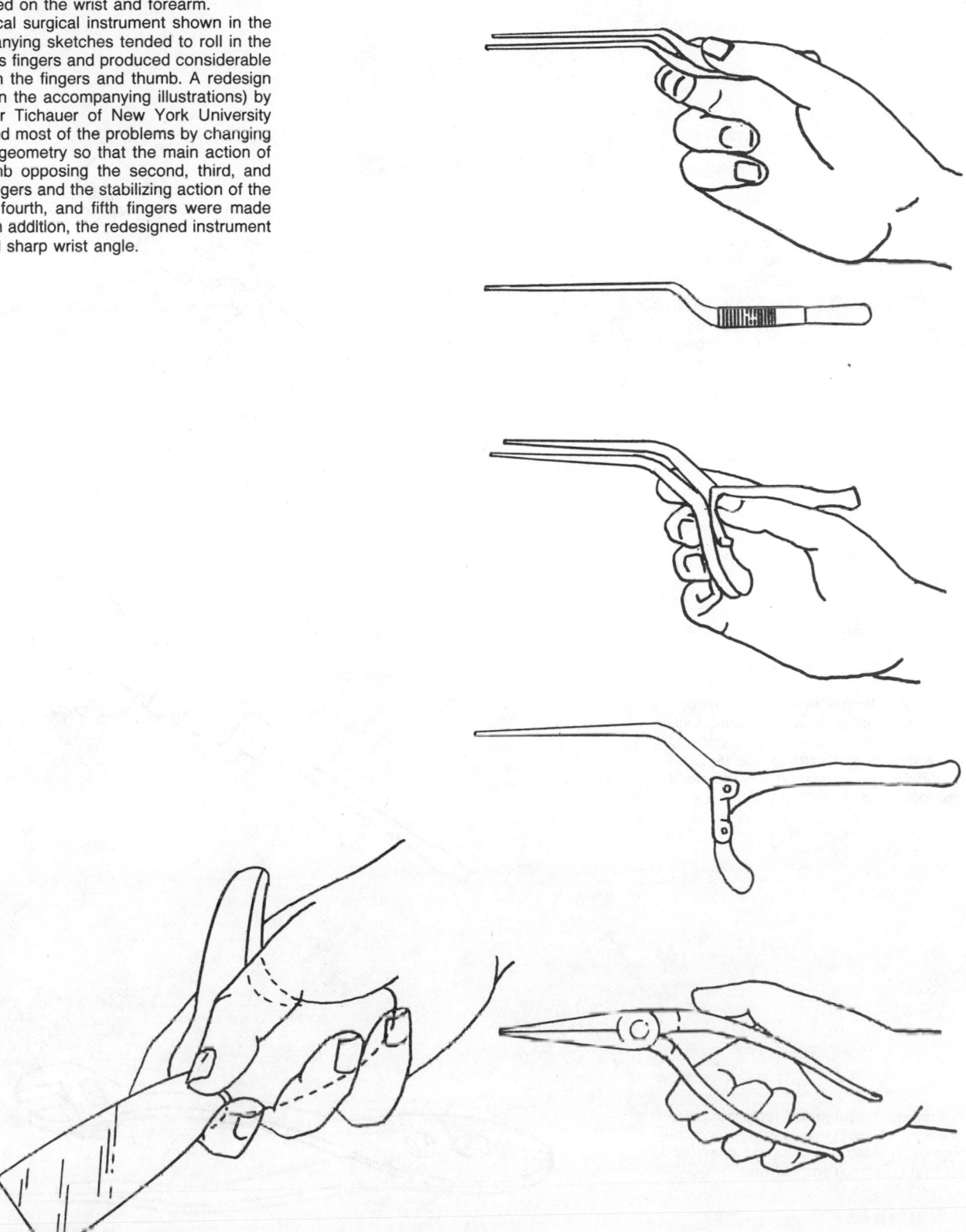

## Some Simple Errors to Avoid

Avoid glossy tool surfaces whenever critical labels or scales must be read. The tool surface in these areas should be a light color with black markings so that the visual contrast is maximized.

A typical caliper tool can be anodized so that the background surface provides good contrast for black scale marks and numbers. Consult the section on displays for scale-marking suggestions.

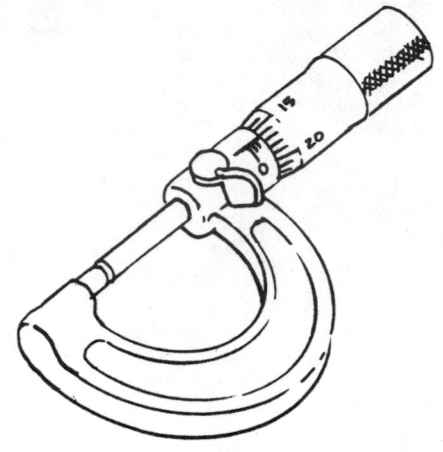

Avoid some of the picturesque but useless handle shapes used by many tool manufacturers. The handle shown in the accompanying sketch not only is awkward to use but also creates considerable stress on the operator's hand and often may incur bruising.

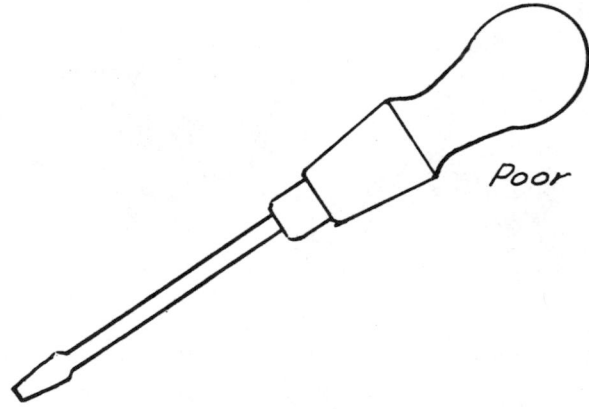

*Poor*

Do not create "all-purpose" tools that probably do not do anything very well. The accompanying sketch illustrates a typical combination tool that is difficult to hold onto.

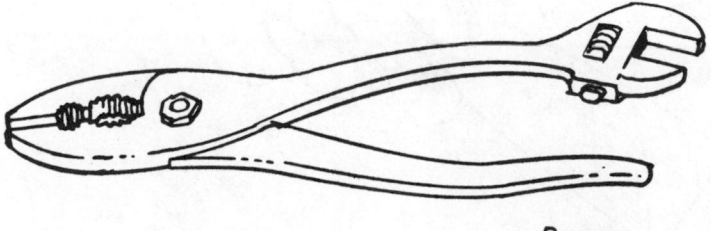

*Poor*

## SPECIFIC TOOL DESIGN RECOMMENDATIONS

**HAMMERS** Handles for hammers should be designed to aid the user in obtaining a secure grip. As a general rule, the size of the handle should fit the user's hand; i.e., the size has little relation to the overall hammer size (except in terms of structural strength).

The suggestions concerning handle size and shape illustrated by the accompanying sketch are appropriate for most hammers. A slight depression not only aids gripping but also tends to keep the hand away from the end of the handle, which avoids the tendency to allow the hammer to slip out of the hand. Note the flat sides, which help keep the hammer from rotating in the hand.

**KEYHOLE SAWS** Note in the accompanying sketch that the handle is placed at a slight angle (approximately 10 to 15°).

**HANDSAWS** The handgrip should be approximately 45° to the blade axis (not the back side of the blade).

Note that the handles of both types of saws have flat sides to help the user keep the vertical orientation of the blade under control.

The stirrup handle must have a sufficient opening so that the user's fingers are not pinched. Note that, although there is a slight curvature on the outside of the rear part of the handle, the inside is straight.

A properly placed thumb stall will help the user control the blade angle and will relieve stress due to the application of force to the sawing operation.

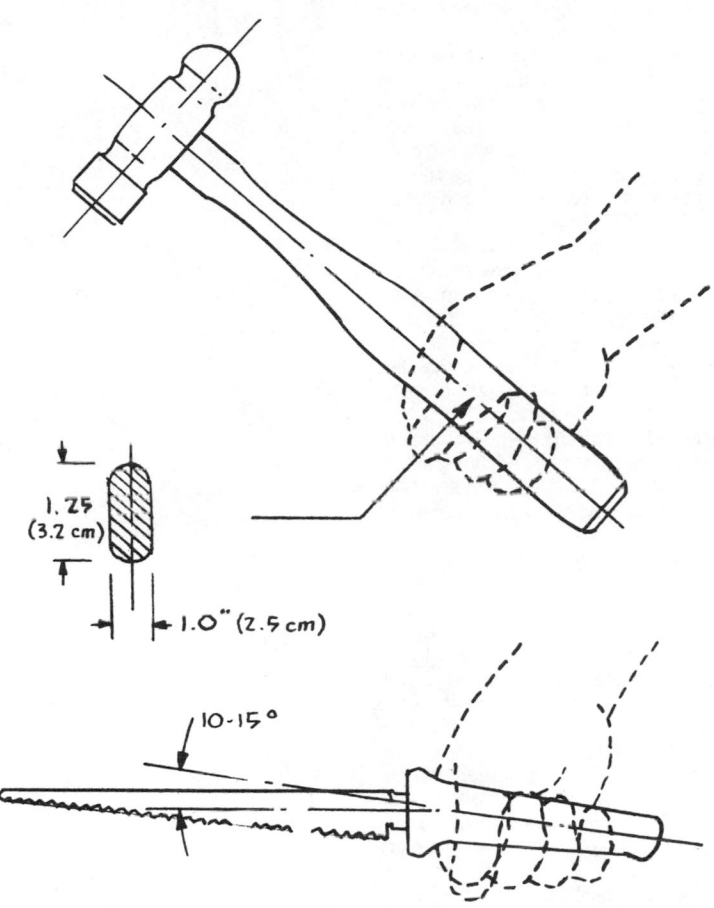

1.25 (3.2 cm)

1.0" (2.5 cm)

10-15°

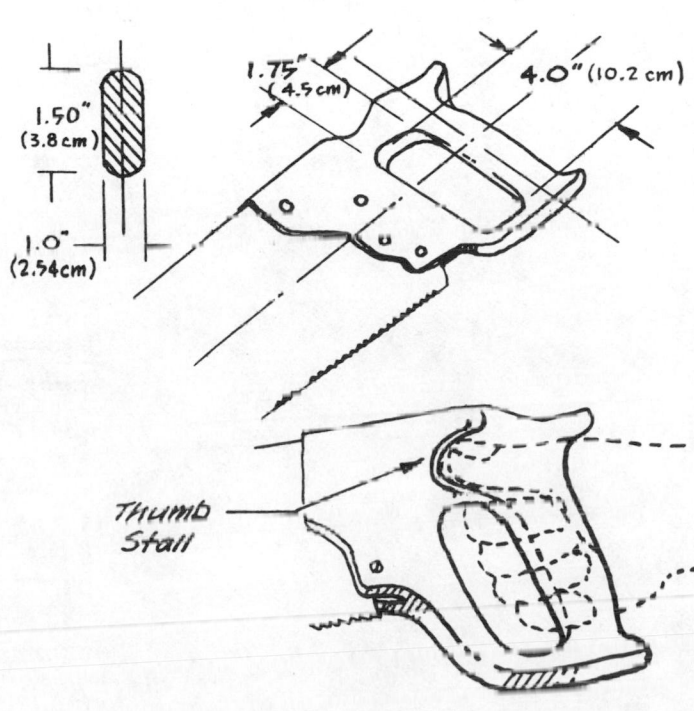

1.50" (3.8 cm)

1.0" (2.54 cm)

1.75" (4.5 cm)

4.0" (10.2 cm)

Thumb Stall

**SCREWDRIVERS** A long cylindrical handle is more appropriate for the screwdriver than some of the odd shapes that manufacturers seem to want to use. Slight "fluting" helps increase the gripping force. The flutes should not have sharp edges, however. The end of the handle should be rounded. A finger stop with an intervening thumb and finger depression is recommended. The handle length shown in the accompanying sketch is the minimum for medium-sized to large screwdrivers; diameters should be kept to between 1.125 and 1.50 in (3.2 and 3.81 cm).

Smaller screwdrivers are, of course, useful for small jobs, and the dimensions should be appropriate for the typical finger manipulation requirements. Diameters less than about 0.50 in (1.3 cm) should not be used, however, and the handle must be long enough so that the operator can get at least two fingers and the thumb in contact with the handle.

**PISTOL-GRIP TOOL HANDLES** The accompanying sketch illustrates general dimensional requirements to make the pistol grip compatible with the typical range of adult hands. As indicated, the critical dimensions relate to finger clearance for a trigger finger (if a trigger loop is provided), the reach distance between the centers of the trigger and the main grip point, and the necessary grip surface for the third, fourth, and little fingers. Note also that a finger stop is suggested at the bottom of the handle and that the back of the handle at the point where it contacts the main tool body is appropriately radiused.

Note the absence of fluted finger depressions on the forward edge of the handle. This common practice is not recommended because people's desired finger positions vary and the fluting forces the hands into undesirable positions if the flutes do not fit.

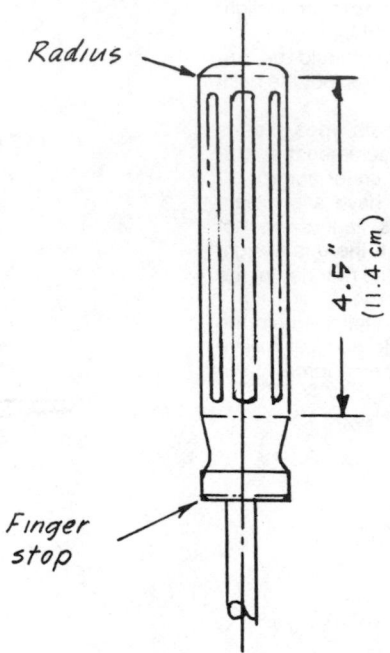

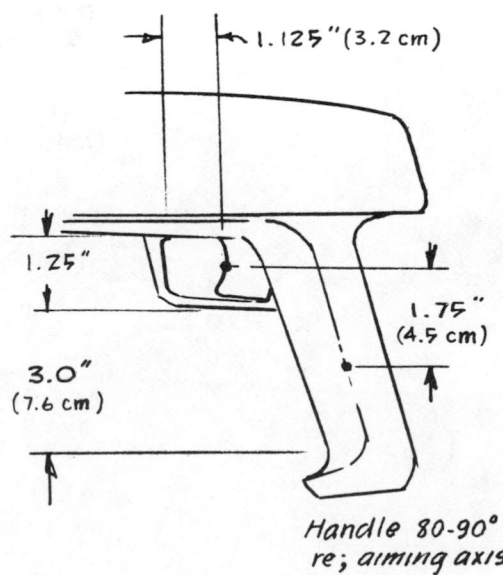

## Handle Angles for Typical Power Tools

The accompanying sketches illustrate suggested handle angles to make typical power tools easier to use in normal working positions.

The hand plane requires a combined downward and forward thrust; thus the 45° handle is generally best.

The power-saw handle should be approximately the same as the plane handle, although the weight of the saw eliminates the need for a downward thrust. The angle, however, is generally the most comfortable for guiding the saw.

Tools of this type should also be designed so that, from his or her typical eye reference position, the operator can easily see whatever guidance reference is provided (usually a slot in the baseplate).

A reverse angle is desirable for tools such as the sabre saw; otherwise, operators' hands and wrists are often forced into a cramped position, and/or operators may assume an awkward body position in order to see what they are doing.

*Note:* The insides of all the above handles should be straight; handle cross sections should generally be oval, but with flat sides.

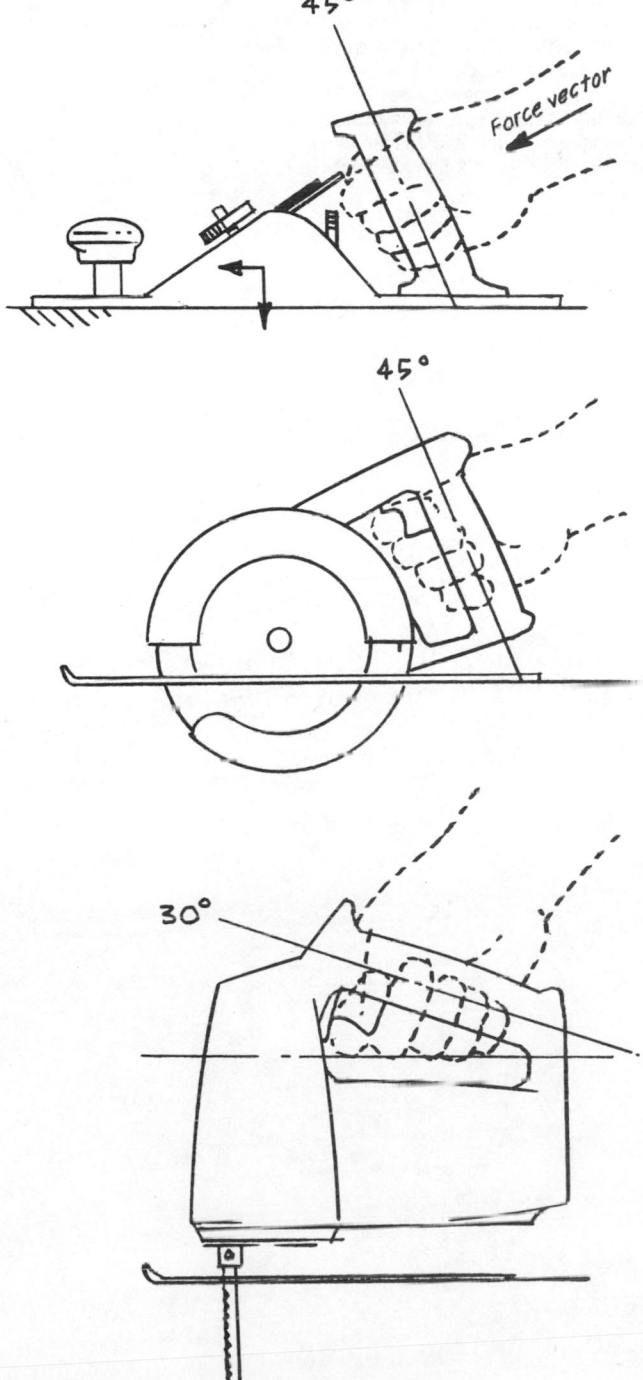

Hand-held drills, more than some other power tools, are often used in a great variety of positions. Therefore, the handle should be generally perpendicular to the drill axis, or fairly near to it. As indicated in the accompanying sketches, it is suggested that the smaller tool have a slight handle tilt, whereas the larger tool should have perpendicular handles.

Few tool manufacturers provide a visual alignment reference; without such a reference, the operator must try to guess at what angle he or she is holding the tool. Aiming marks should be placed on the tool housing to assist the operator in aligning the direction of the drill.

The accompanying sketch of a small soldering tool is self-explanatory in that it shows a recommended departure from the average, straight handle-tip configuration. Note, however, the tapered handle (since the shorter fingers are at the bottom of the handle).

The small hand-held paint sprayer is often used in many positions, much like the small hand drill. Thus the slight handle tilt is all that should be used—if any.

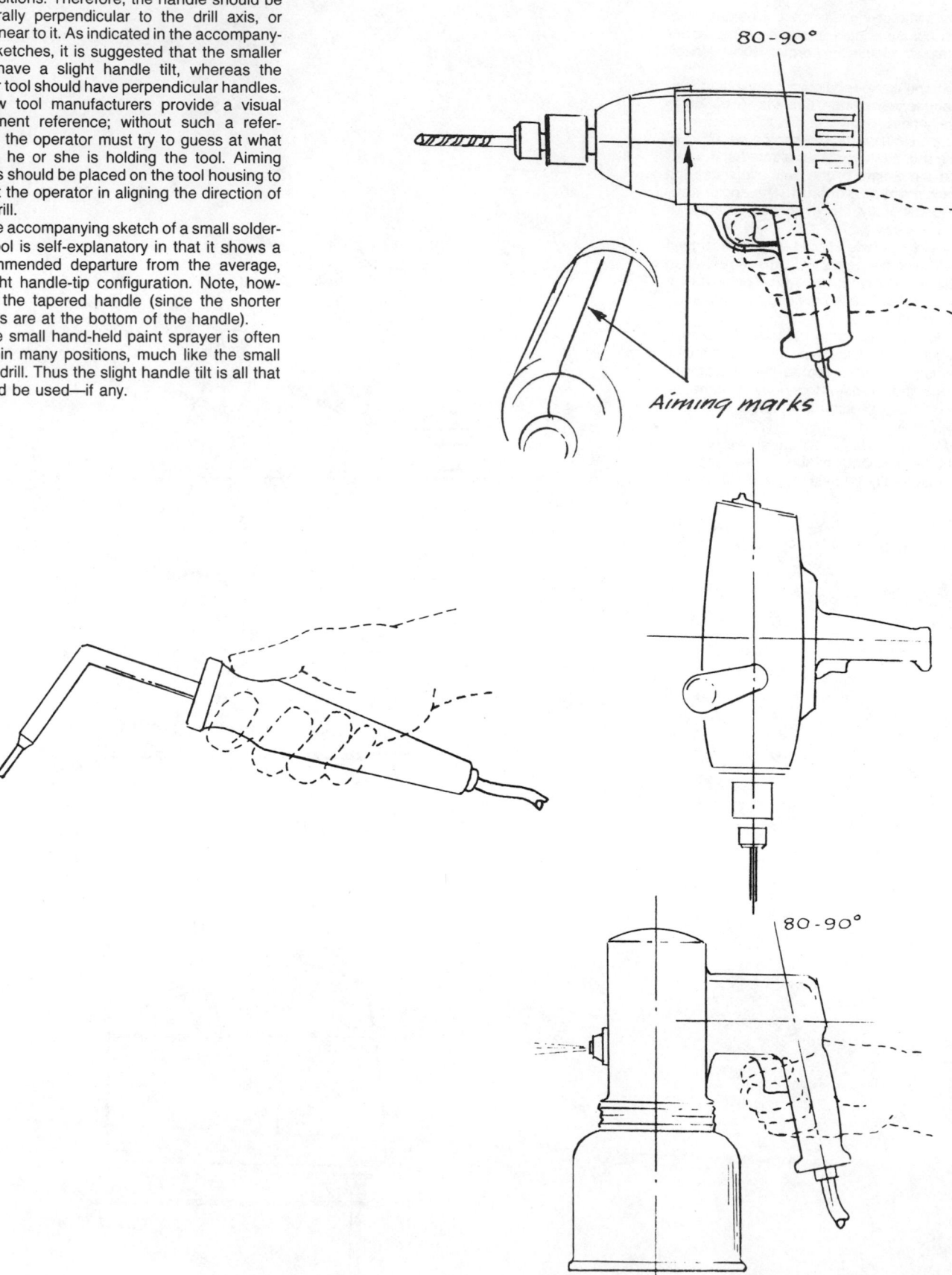

Larger, heavier tools such as the chain saw shown in the accompanying sketch, generally require two handles to help the operator handle the weight and keep the tool properly balanced. The best handle orientation for most such devices is similar to that illustrated. Note that the main guiding and operating handle is essentially in alignment with the blade and main housing. The auxiliary handle is generally used to help support the tool. However, it should be a "looping" configuration so that the operator can grasp it at more than one point as he or she changes saw position and direction. It is suggested that the loop continue across both sides so that the tool can be used by either left- or right-handed operators.

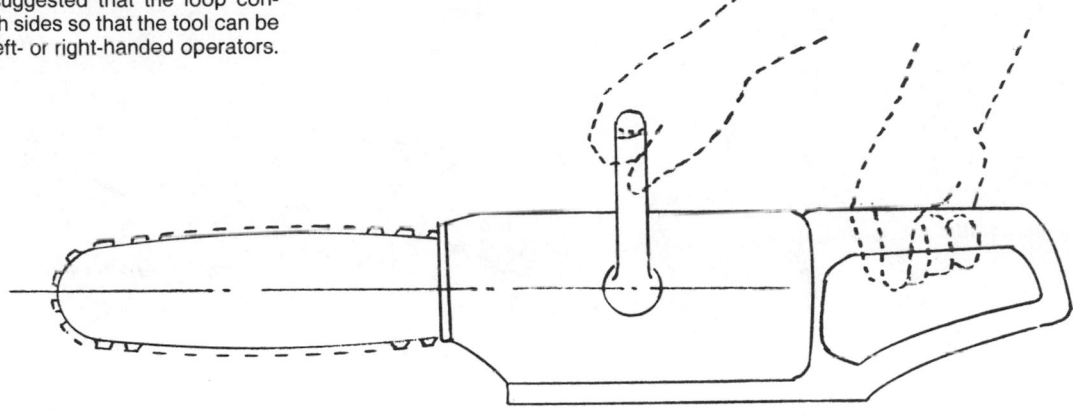

Smaller, household power tools also require special consideration with regard to the size and shape of handles and related controls. The electric knife is a case in point. The hand and wrist in the first sketch are in a more comfortable position than the ones in the second sketch (which shows an actual product configuration).

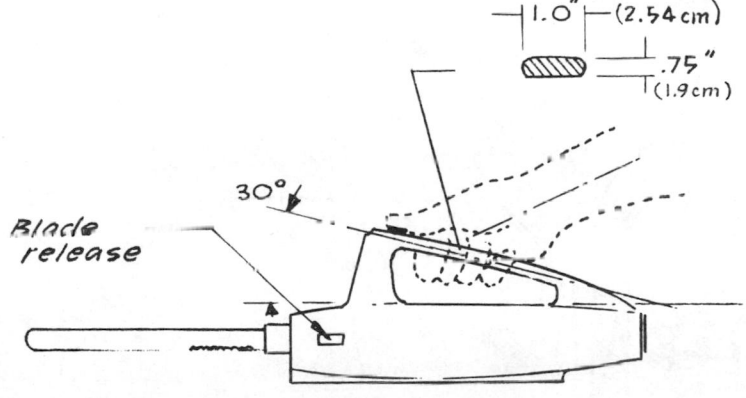

The tool shown in the first sketch was also designed so that the operator could not operate the blade-release switch on the side. This was done so that the operator would not grab the blades with the left hand, with the cutting edges in the palm of the hand.

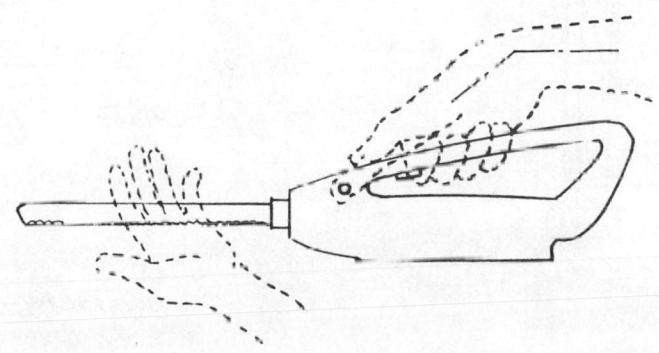

Although pincer-type tools are made in a variety of sizes and shapes, they have common features that often cause the user problems. Most of these have to do with the size and shape of the handles.

The accompanying sketch illustrates several suggestions concerning handles. Most handles should be long enough to support the entire set of fingers and handgrip, and the maximum handle spread should not be so great that persons with very small hands are unable to open the jaws of the tool fully.

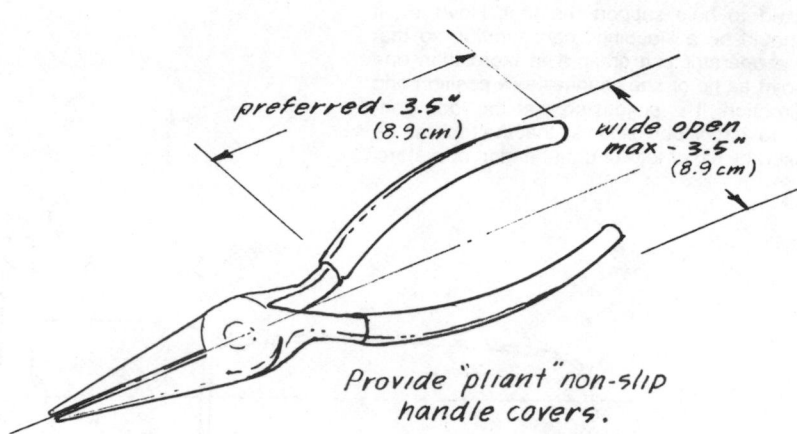

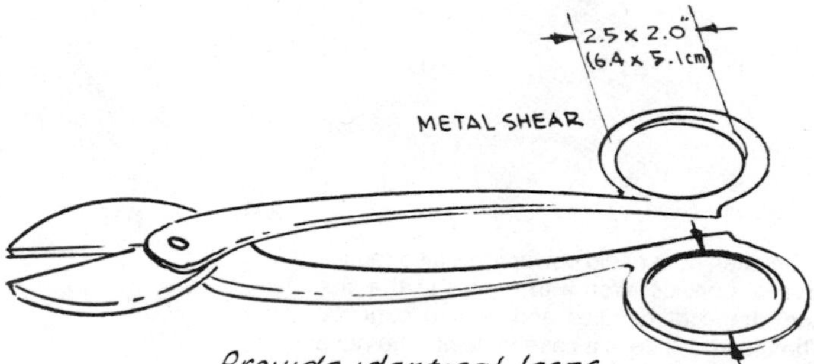

preferred - 3.5"
(8.9 cm)

wide open
max - 3.5"
(8.9 cm)

Provide "pliant" non-slip
handle covers.

Scissors and other cutting tools generally need "loop" handles in order to help the operator maintain the optimum finger position for tool force and guidance applications.

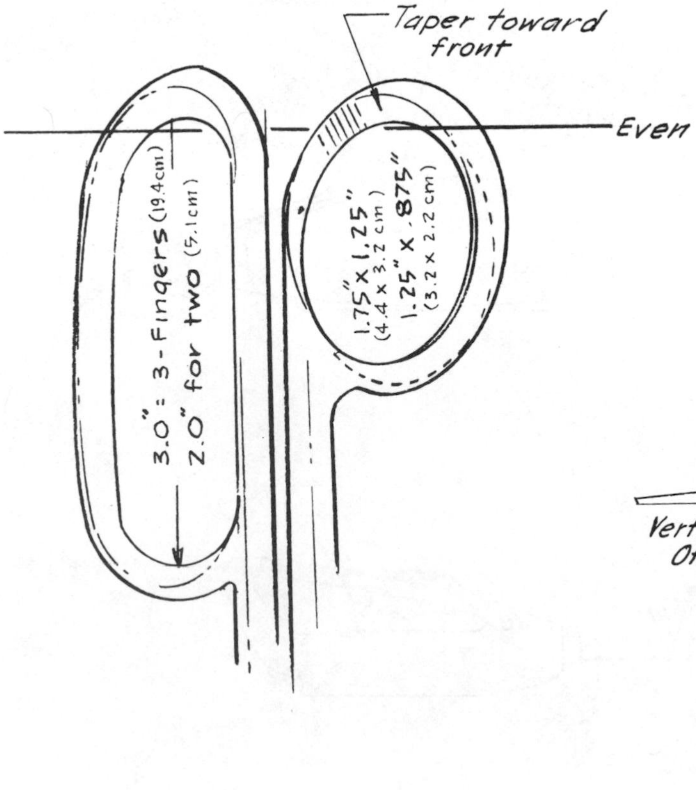

2.5 x 2.0"
(6.4 x 5.1cm)

METAL SHEAR

Provide identical loops
for either right or left
hand use.

Taper toward
front

Even

3.0" = 3-Fingers (19.4cm)
2.0" for two (5.1cm)

1.75" x 1.25"
(4.4 x 3.2 cm)
1.25" x .875"
(3.2 x 2.2 cm)

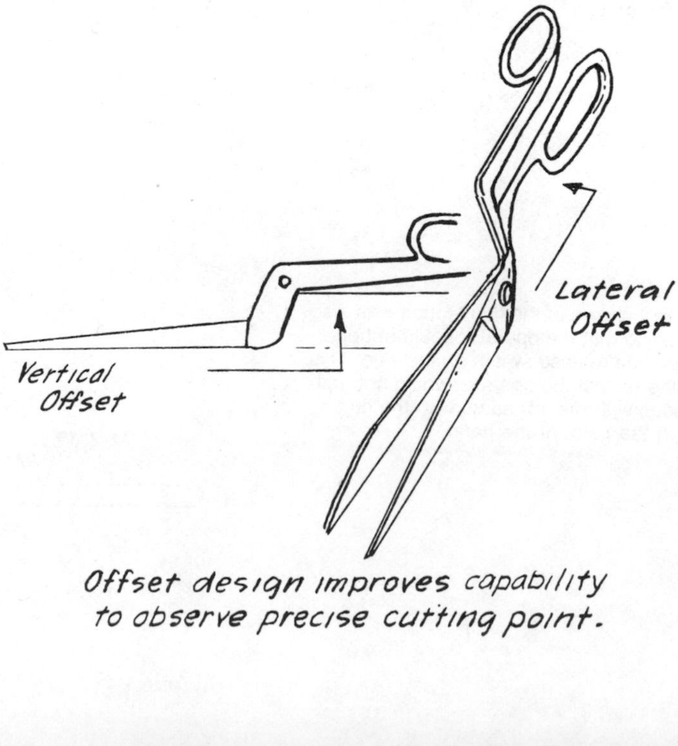

Lateral
Offset

Vertical
Offset

Offset design improves capability
to observe precise cutting point.

## TOOL DESIGN TO MINIMIZE OPERATOR INJURY

Typical hazards include the following:

Laceration due to exposed blades, points, edges, rotating wheels or gears, rough edges, burrs, etc.

Bruises and contusions due to coming in contact with accessible pinch points and bumping against sharp corners or edges or protruding structures or fasteners

Abrasions due to exposure to rapidly moving machine components, brushing against a rough surface, or slipping of the hand from a rough handgrip

Skeletal fractures or breaks due to inadvertent entrapment between two articulating machine components, gears, or belts or due to unforeseen component reaction against the operator's limb

Eye penetration due to a lack of eye protection from flying particles

Burns due to access to hot surfaces

Electrocution due to lack of insulation from inherent electrical current and/or possible failures

"White fingers" (Raynaud's syndrome) due to excessive, continued exposure to tool and hand vibration

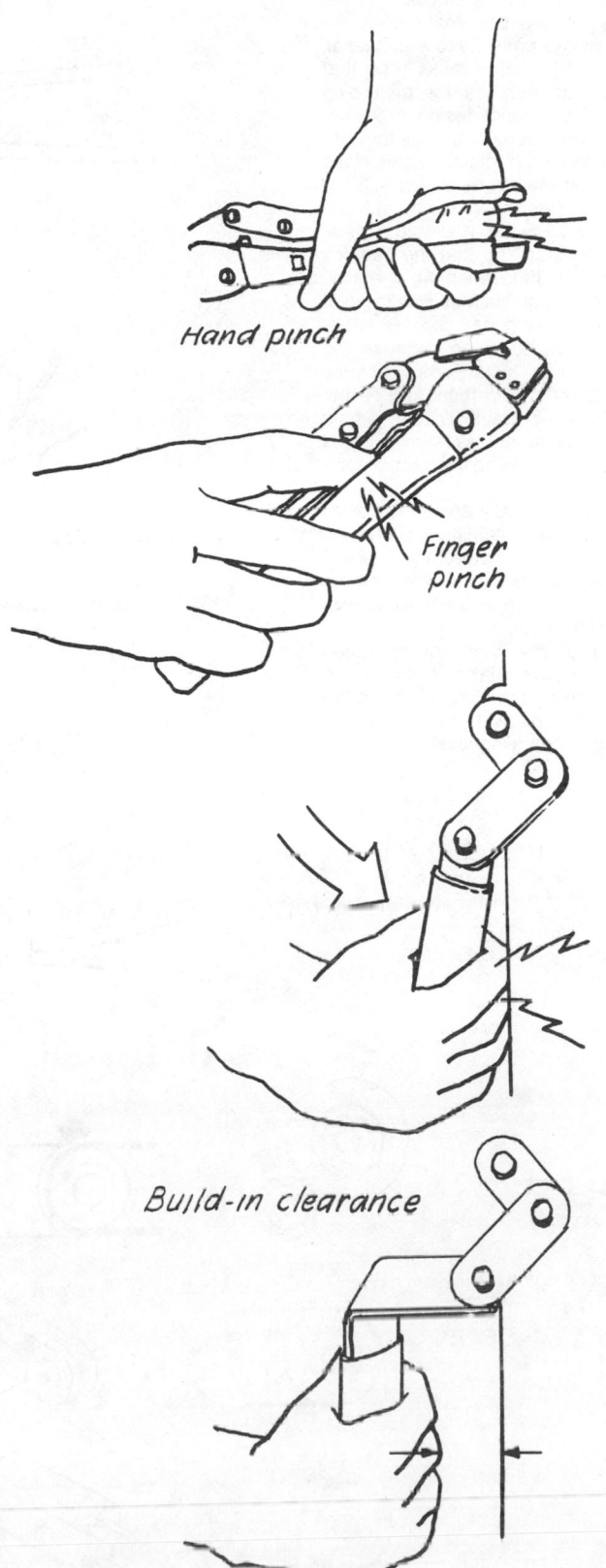

Hand pinch

Finger pinch

Build-in clearance

The accompanying sketches illustrate a number of typical problem situations that, it is hoped, will provide the reader with ideas regarding an even wider array of tools.

Whether the tool being planned is a very simple hand-actuated device or a complex powered device, a complete analysis should be made of the potential personal modes. Although it is not always possible to eliminate all the injury potential (i.e., one must hope that users will follow instructions for their own safety), many fairly simple design modifications can change an otherwise unsafe tool into a relatively safe one. Be particularly conscious of the hazards that the average user will not see or expect. Handles that close completely catch the operator unaware. A simple expedient is to create a "stop" so that the handles do not fully close and thus pinch the operator.

Similarly, a larger tool, such as illustrated in the accompanying sketches, can be especially dangerous, not so much because the operator cannot see the hazards, but because people tend to hurry and to focus not on the hazards but on the work activity. The simple change shown in the second sketch illustrates one possible way of removing the hazard completely.

One should be especially conscious of all potential pinch points in machine tools having automatically moving components. The accompanying sketches illustrate most of the potential pinch points often associated with automatic machine tools.

The guarding of the point of operation should be such that the relation between the size of the opening and its distance from the hazard point conforms with the guidelines shown in the accompanying illustration.

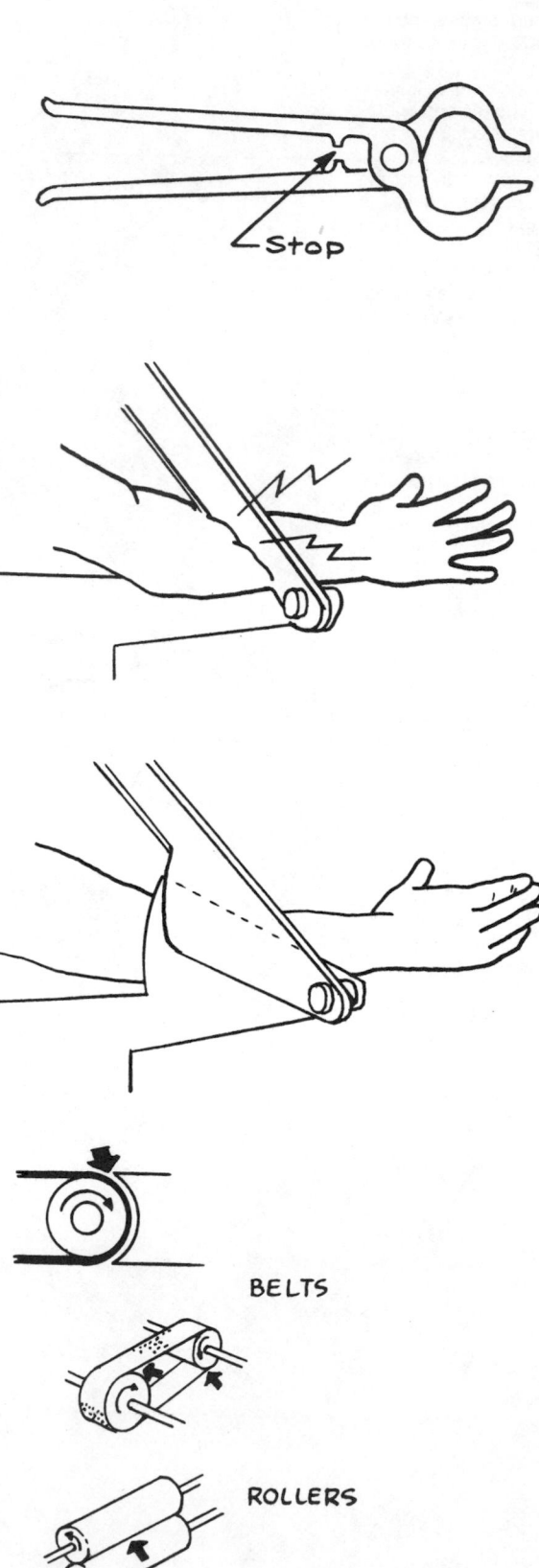

**Guarding Limitations for Ram Opening**

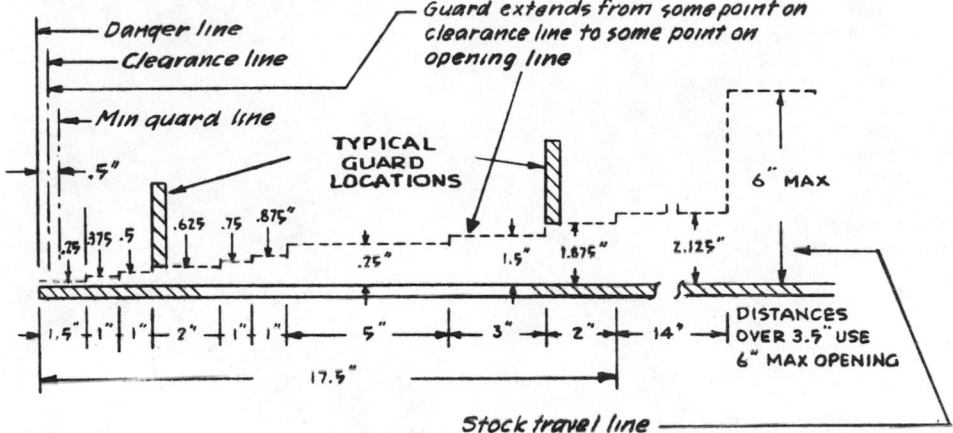

Relationship between allowable opening in barrier guard and its distance from the point of operation. (*Source:* OSHA Standards Section 1910.217.)

| Distance of Opening from Point of Operation Hazard, in | Maximum Width of Opening, in |
|---|---|
| ½ to 1½ | ¼ |
| 1½ to 2½ | ⅜ |
| 2½ to 3½ | ½ |
| 3½ to 5½ | ⅝ |
| 5½ to 6½ | ¾ |
| 6½ to 7½ | ⅞ |
| 7½ to 12½ | 1¼ |
| 12½ to 15½ | 1½ |
| 15½ to 17½ | 1⅞ |
| 17½ to 31½ | 2⅛ |

**Guarding Limitations for Punch Press Ram Stroke**

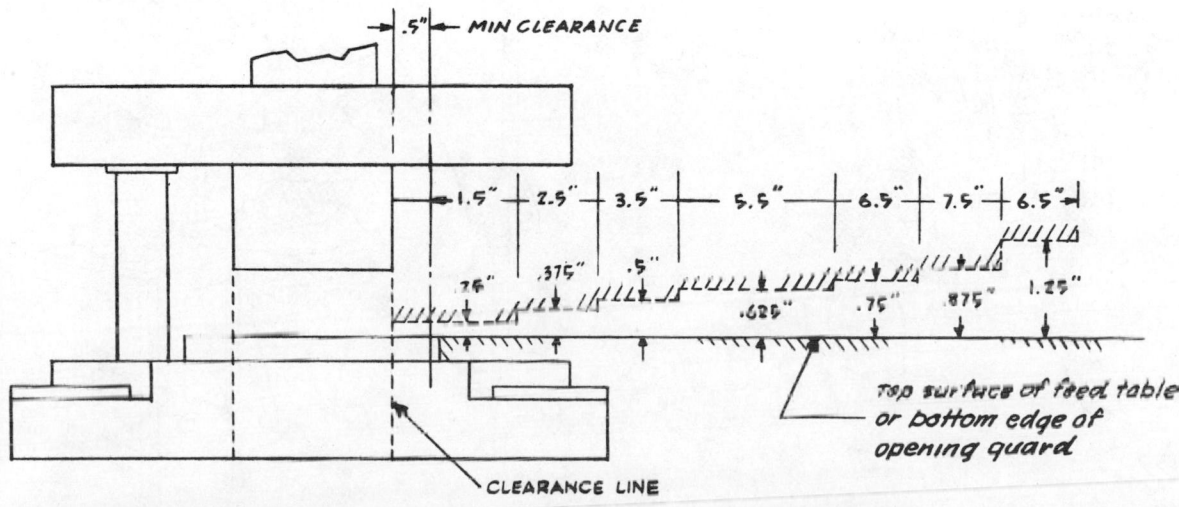

Guarding punch press by limitation of ram stroke. When enclosure of die is necessary, size of openings should not exceed that shown for various distances of fingers from die. (*Source:* "Safe Openings for Some Point of Operation Guards," Technical Guide No. 2, American Mutual Insurance Alliance, May 1966, p. 10.)

Do not fall into the trap of allowing hazards to exist because eliminating them would get in the way of the operation or make it difficult to service or maintain the tool. It takes only a little ingenuity to devise hazard barriers that are still compatible with proper operation and maintenance. The accompanying sketch illustrates a good solution to the problems of guarding a typical rotating component, providing an opportunity to inspect the component during operation, and making it accessible for part changes and/or other servicing.

Sometimes the severity of the problem is such that it cannot be solved merely by using machine guards. Other safety schemes such as those shown in the accompanying sketches may be required. Too often such schemes have to be added because the machine tool was not properly designed in the first place.

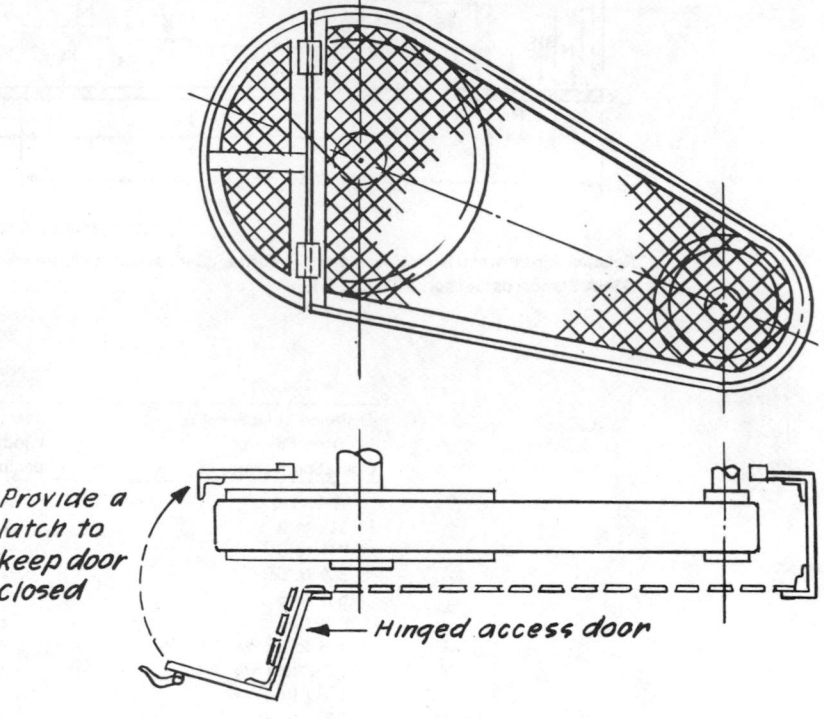

Provide a latch to keep door closed

← Hinged access door

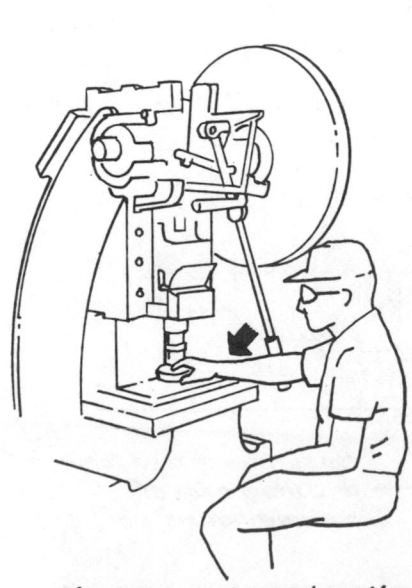

Machine automatically pushes hand away at critical moment,

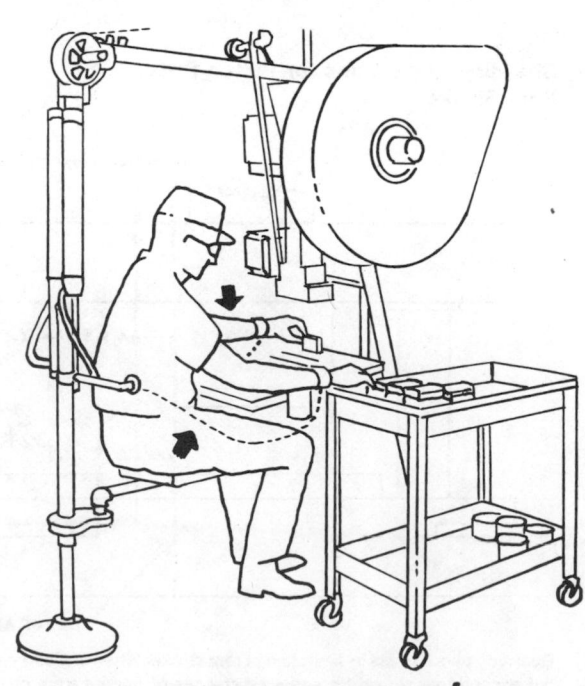

or    pulls hands away !

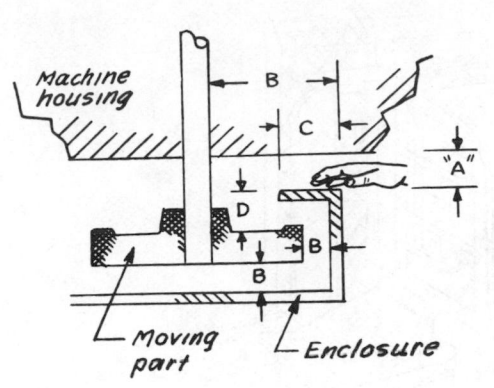

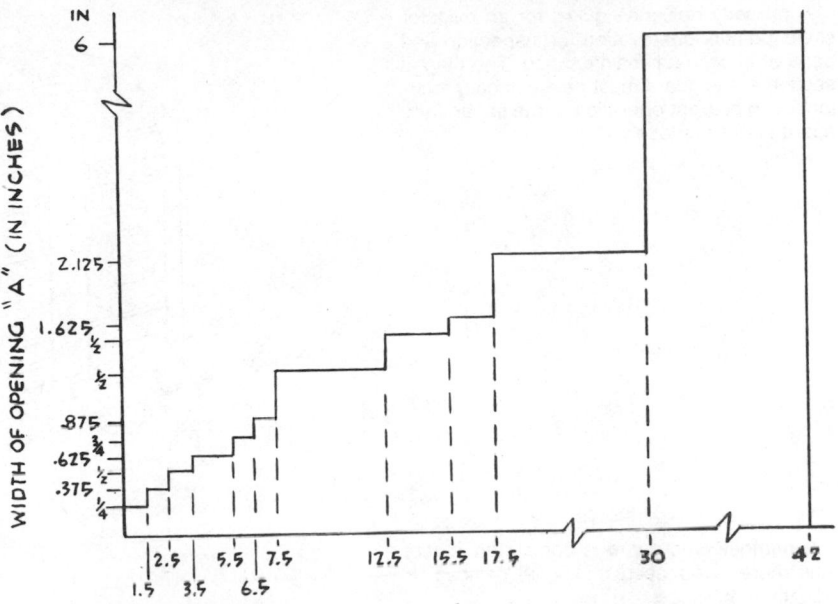

DISTANCE TO MOVING PART ( B, or C+D, whichever is greater - inches )

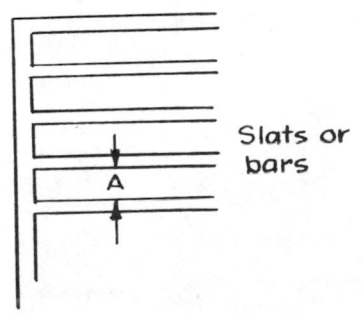

Slats or bars

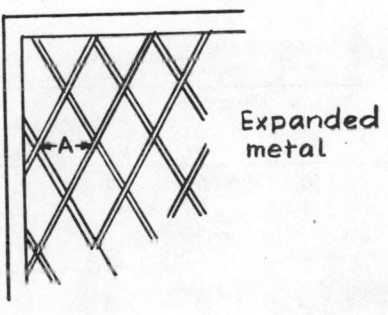

Expanded metal

## Standard Materials and Dimensions for Machinery Guards

| | Material | Clearance from Moving Part at All Points, in | Size of Filler Materials | | Minimum Height of Guard from Floor or Platform Level, ft and in |
|---|---|---|---|---|---|
| | | | Largest Mesh or Opening Allowable, B, in | Minimum Gauge (U.S. Standard) or Thickness | |
| | Woven wire | Under 2 | 3/8 | No. 16-3/8 in | 7 0 |
| | | 2-4 | 1/2 | No. 16-1/2 | 7 0 |
| | | 4-15 | 2 | No. 12-2 | 7 0 |
| | Expanded metal | Under 4 | 1/2 | No. 18-1/3 in | 7 0 |
| | | 4-15 | 2 | No. 13-2 | 7 0 |
| | Perforated metal | Under 4 | 1/3 | No. 20-1/3 in | 7 0 |
| | | 4-15 | 2 | No. 14-2 | 7 0 |
| | Sheet metal | Under 4 | | No. 22 | 7 0 |
| | | 4-15 | | No. 22 | 7 0 |
| | Wood or metal strips crossed | Under 4 | 3/8 | | |
| | | 4-15 | 2 | | |
| | Wood or metal strips not crossed | Under 4 | 1/2 the width | 3/4-in wood or No. 16 metal | 7 0 |
| | | 4-15 | One width | | |
| | Plywood plastic, or equivalent | Under 4 | | 1/4 in. | |
| | | 4-15 | | 1/4 in | 7 0 |
| | standard railing | Minimum of 15 Maximum of 20 | | | 6 |

*Source:* ANSI Standard B15.

A properly designed guard for an alligator shear permits observation for inspection and ease of access for maintenance. The hinged section of the guard must be electrically interlocked to prevent operation of the shear if the guard is not in place.

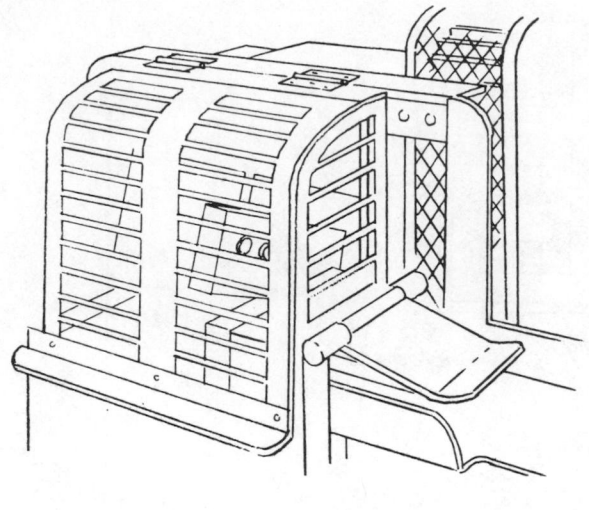

A photoelectric guard is completely static; thus there is no need for signal devices or adjustments.

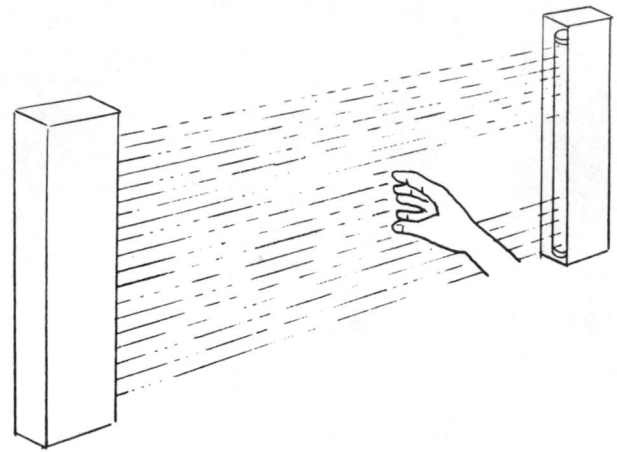

**VIBRATING TOOLS** When selecting or designing reciprocating or vibrating tools, the frequency and amplitude spectrum of the vibrations transmitted onto the operator should be evaluated. The risk of somatic resonance response is highest when the amplitude exceeds 100 $\mu$m within the frequency band 3 to 125 Hz.

When tools are activated by a rotating mechanism, the maximum torque transmitted on the axis of rotation of the forearm should be below 12 in-lb (1.4 N·m). A two-handed power tool should be designed with an angle of 120° between the gripping axes of both hands, and the handles should be isolated to minimize as much vibratory transmission as possible.

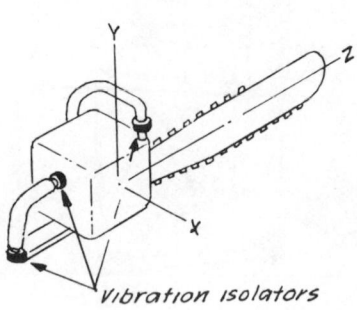

Vibration isolators

## Human Resistance to Electrical Current

| Body Area | Resistance, ohms |
|---|---|
| Dry skin | 100,000–600,000 |
| Wet skin | 1,000 |
| Internal body, hand to foot | 400–600 |
| Ear to ear | About 100 |

| Effect | Direct | | 60 Hz | | 10,000 Hz | |
|---|---|---|---|---|---|---|
| | Men | Women | Men | Women | Men | Women |
| Slight sensation on hand | 1 | 0.6 | 0.4 | 0.3 | 7 | 5 |
| Perception threshold | 5.2 | 3.5 | 1.1 | 0.7 | 12 | 8 |
| Shock—not painful, muscular control not lost | 9 | 6 | 1.8 | 1.2 | 17 | 11 |
| Shock—painful, muscular control not lost | 62 | 41 | 9 | 6 | 55 | 37 |
| Shock—painful, let-go threshold | 76 | 51 | 16 | 10.5 | 75 | 50 |
| Shock—painful and severe, muscular contractions, breathing difficult | 90 | 60 | 23 | 15 | 94 | 63 |
| Shock—possible ventricular fibrillation effect from 3-5 shocks | 500 | 500 | 100 | 100 | | |
| Short shocks lasting $t$ s | | | 165/$t$ | 165/$t$ | | |
| High-voltage surges | 50* | 50* | 13.6* | 13.6* | | |

*Energy in watt-seconds, or joules.

## GENERAL GUIDELINES FOR FURNITURE AND LARGE APPLIANCES

In designing or selecting furnishings, the following general considerations should be kept in mind:

User fit: Careful consideration should be given to the dimensions and anatomic characteristics of users to make sure that the furniture fits them, supports them properly, and adjusts to their activities, i.e., rest, work, and recreation.

User efficiency: Consideration should be given to what users have to do to and with the furniture in terms of arrangement of elements, articulation of components, operation of special controls, accessibility within components, and visual impact of finishes.

Interactive characteristics: Consideration should be given to how one piece of furniture interacts with another positionally and in terms of orientation and interference.

Safety: Consideration must always be given to potential safety hazards, including sharp contact points, fragility of the structure, balance and stability, and flammability.

Housekeeping: Someone has to clean around, move, store, and rearrange furniture at some time. The design should aid, not hinder, such operations.

Good furniture design can be created for any styling objective, but good human engineering must come first.

The accompanying illustrations have been made purposely simple to establish the idea that the basic furniture configurational requirements generally have little to do with aesthetic flourishes. The designer, using his or her imagination, can add the styling flourishes around any basic structural concept.

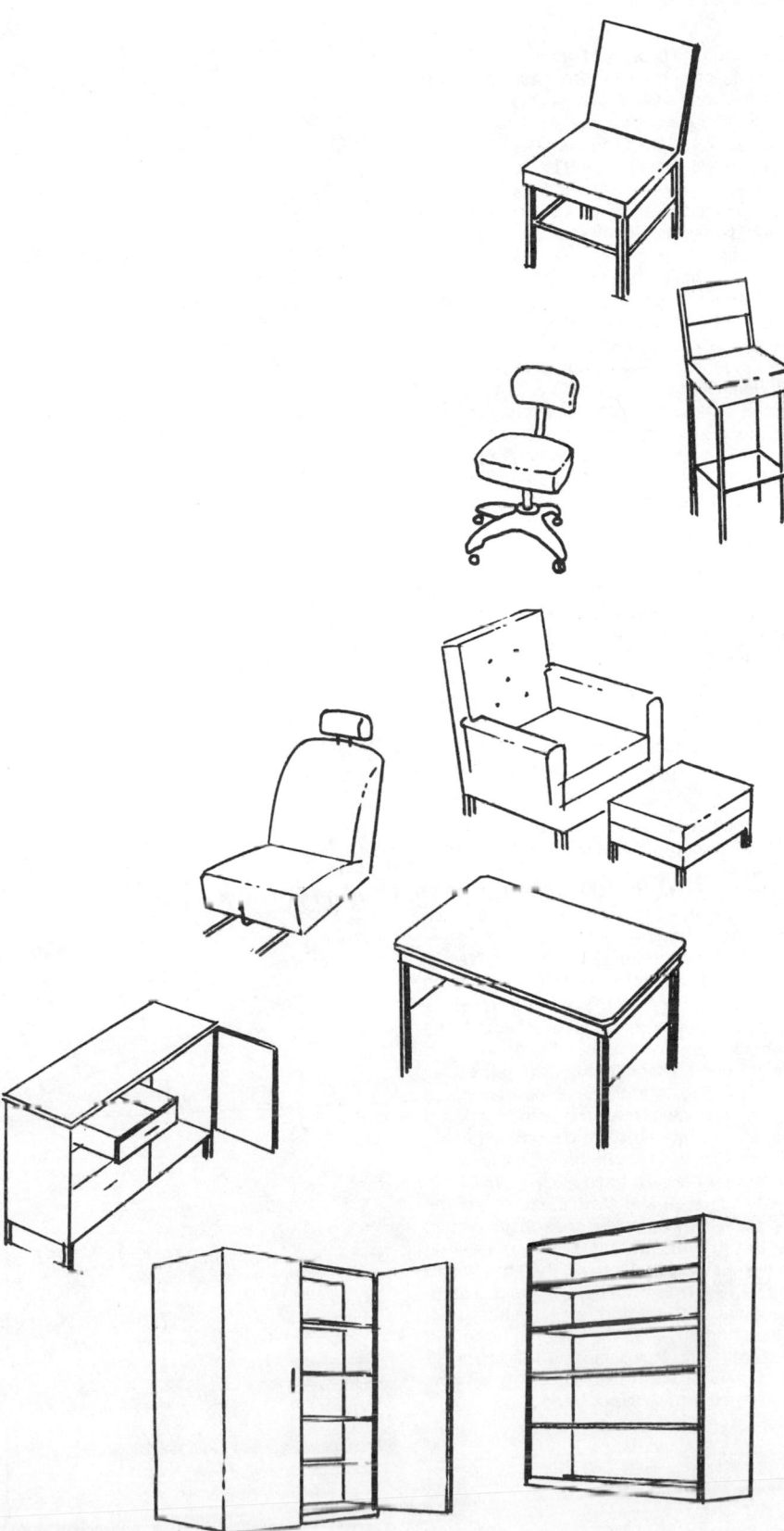

## DESIGN CHARACTERISTICS OF A GOOD SEAT

Whether designing a chair, a davenport, or a pilot or vehicle driver seat, one should start with the following critical features as a base line:

Good posture: To provide good sitting posture, i.e., one in which the back and neck muscles are least strained, the seat pan should tilt back approximately 5° to shift the upper torso weight and thus cause the torso to rest fully against the seat back. The angle between the seat pan and the seat backrest must be approximately 105° to keep the torso against the backrest and yet not force the occupant to lean his or her head forward in order to balance it properly.

Surface support: The seat should be approximately 19 in (48.3 cm) wide to keep large occupants from "lapping over." The seat-pan length should be about 17 in (43.2 cm). A longer seat strikes the back of the short person's legs; a shorter one provides too little support for the longer-legged person's thighs. The seat back should be about 20 in (50.8 cm) high so that even the tallest person's shoulder blades are supported.

Seat-pan height: The best compromise seat-pan height for a full range of male and female adults for normal sitting (i.e., chairs) is about 17 in (43.2 cm) at the leading edge of the seat pan. If the seat pan is higher, the short female occupant's legs will dangle. Even the 17 in (43.2 cm) assumes that she will be wearing shoes with a 1.5- to 2.0-in (3.8- to 5.1-cm) heel. Although a slightly lower seat pan would be desirable for very short-legged people, the lower height would make it very difficult for tall people to get up out of the seat.

For other types of seats, all the above guidelines, except for the seat height, should be used.

Stools can be higher, but a footrest should always be provided to maintain the basic foot-to-seat-height relationship.

Vehicle seats should generally be lower, at least for the driver, because the driver needs to extend his or her legs to operate the pedals. The optimum height for an automobile driver's seat should vary from about 12 to 13 in (30 to 33 cm) above the nominal floor (heel resting position for accelerator pedal operation). In order to accommodate the range of leg reaches from that of the 5th-percentile female driver to that of the 95th-percentile male driver, the seat should have a minimum of 8 in (20 cm) of fore-aft adjustment; i.e., when the seat is all the way back, the seat height should be 12 in (30 cm) above the heel point, and when it is all the way forward, the seat height should be 13 in (33 cm). This seat-height variation, coupled with the fore-aft adjustment, allows the short driver to see over the hood and still reach the foot pedals, and the lower aft position allows the taller driver to sit low enough to clear his or her head.

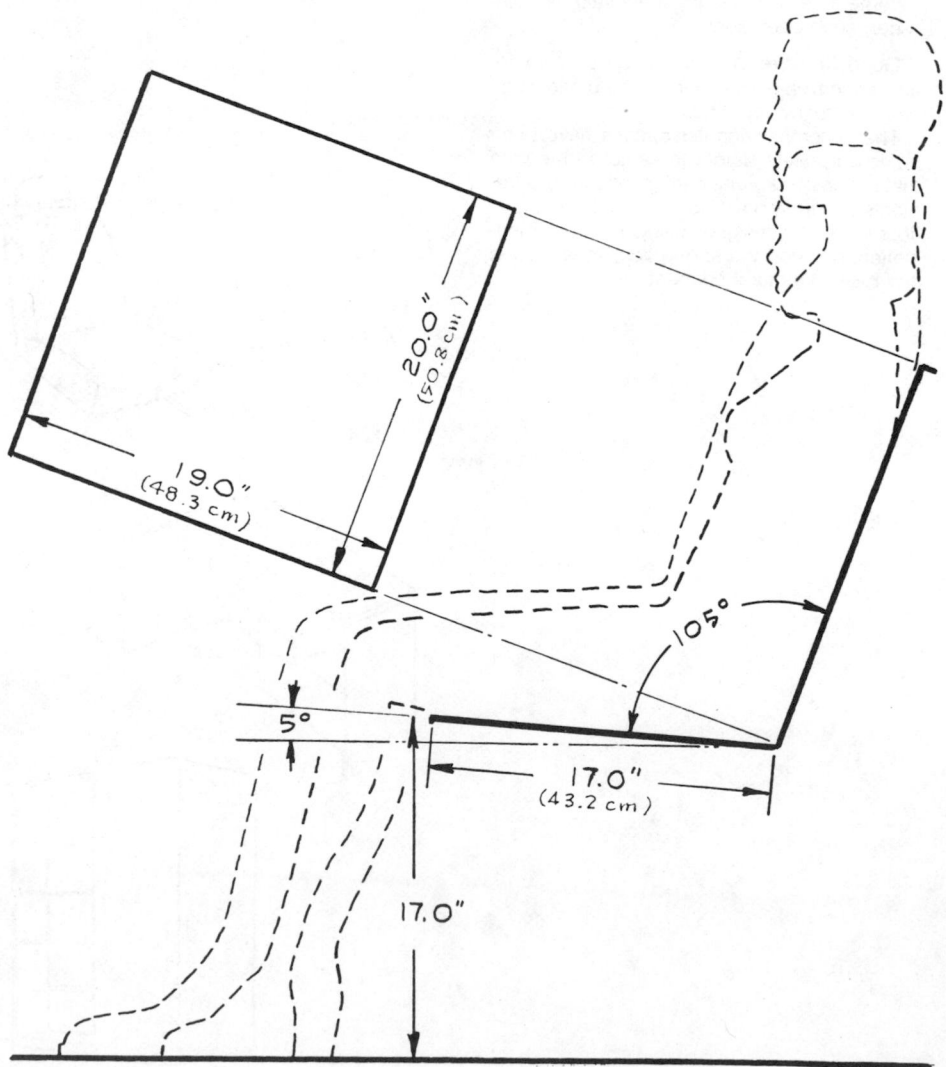

## Seat Design: Cautions and Suggestions

Except in unique cases (such as the astronaut couch designed to help absorb launch and reentry *g* forces), avoid the use of seat-pan or backrest contouring. Sling-type seats are especially bad. Contouring might work if every seat were designed specifically for each occupant, but this is not practical. No two people need contours in the same place, and a contour in the wrong place is much worse than none at all. Fortunately, most people are appropriately padded so that they can accommodate well to a flat seat pan or seat backrest.

The so-called bucket seat should be avoided, especially for the seat pan. The concave seat pan causes the occupant's thighs to roll inward; as a result, although the occupant may think the seat is fine at first, he or she will soon discover that it causes discomfort and, in the case of a vehicle operator's seat, that it may even make it difficult to move the legs laterally.

On the other hand, a concave backrest may help reduce the effects of sideway in automobile driving. The concave feature should be applied only to the lateral axis of the seat back; the backrest should be flat or straight up and down.

Deep cushions and/or softly sprung cushions should be avoided for both the seat pan and the backrest. As the accompanying sketch illustrates, the upper part of the backrest pushes the occupant's torso forward, his or her own weight compresses the lower back and seat cushions to reduce the desired included angle, and the forward cushion cuts off circulation in the legs.

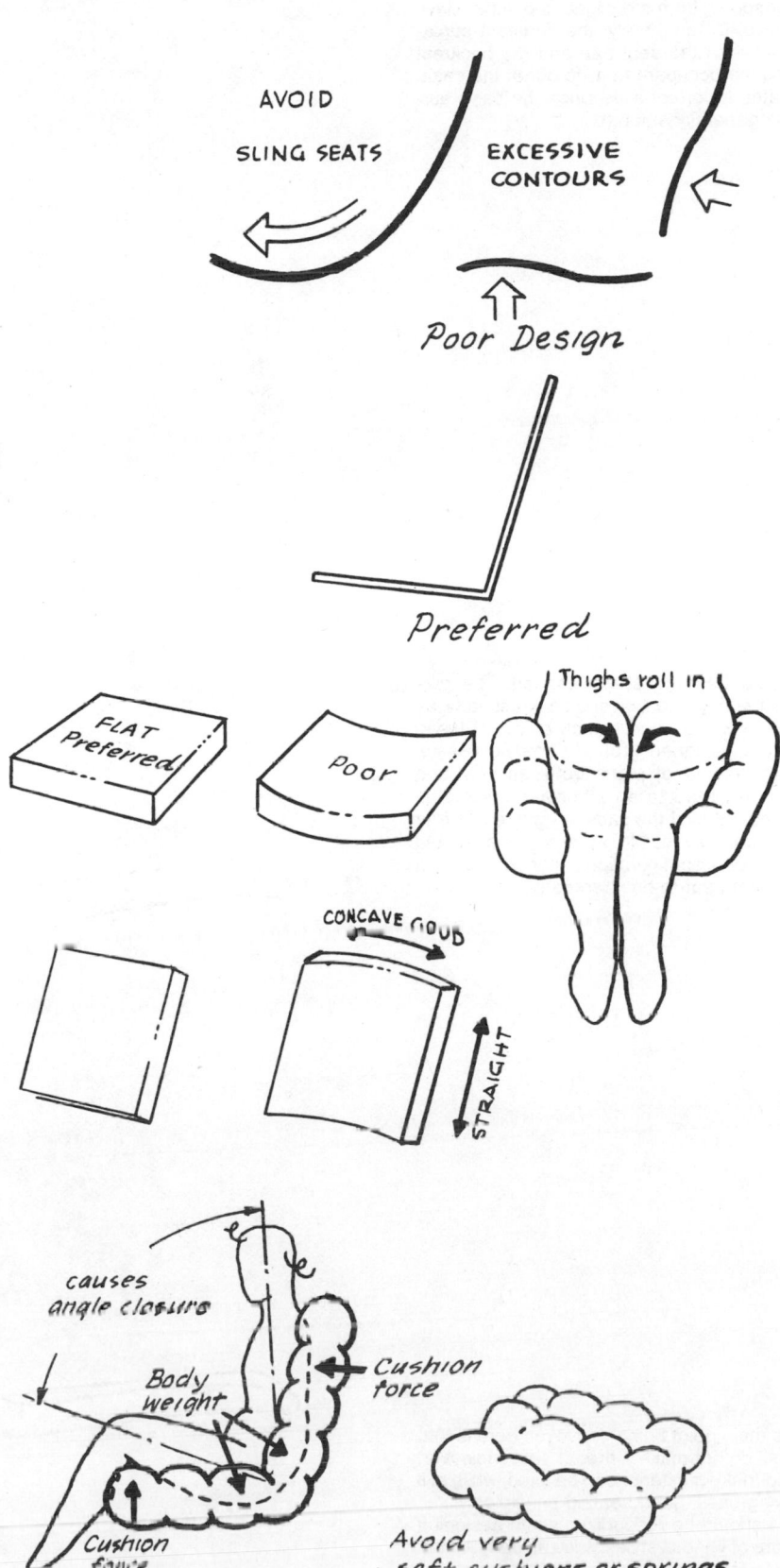

### A Typical Production Design Error

The first sketch below illustrates a common mistake in the design of stack chairs. The seat is cast of one-piece fiber glass or other plastic. The shape is, from a designer's point of view, attractive. Unfortunately, the constant curvature between the seat pan and the backrest causes the occupant to slide out of the chair. It creates a perfect slide since the basic surface is generally smooth.

Frequently such chairs can also be purchased with added seat and backrest pads, as shown in the second sketch below. If these pads are designed properly, the necessary, distinct change of plane between seat and back is returned to an otherwise deplorable seat, especially if the pads are made of fabric rather than plastic. In addition, one should look for pad configurations that also ensure the proper seat-to-back angle of 105°.

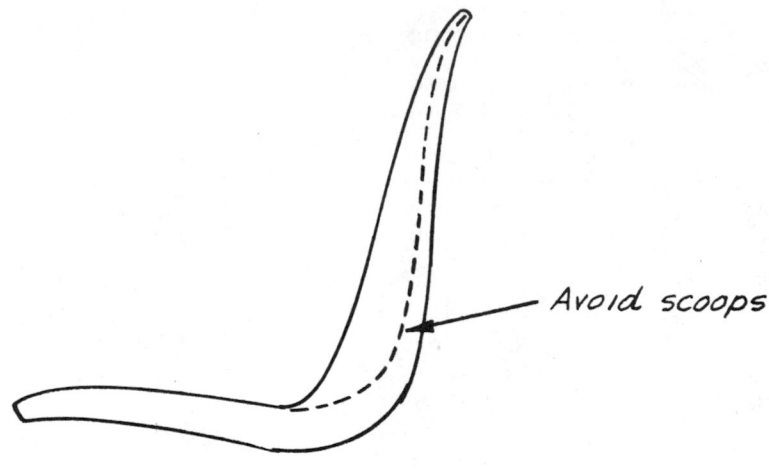

*Avoid scoops*

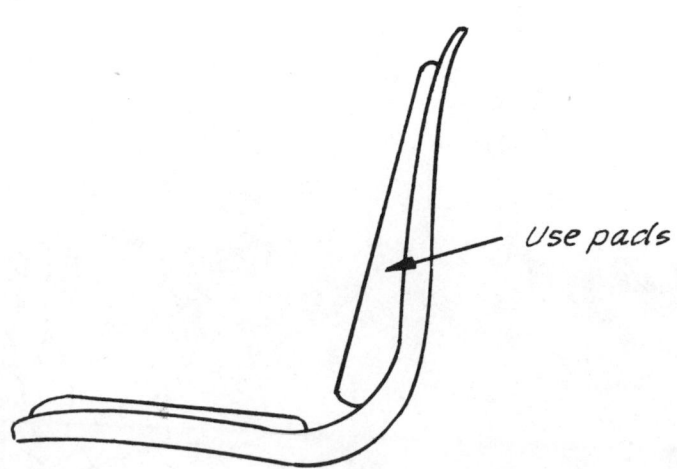

*Use pads*

The main point being conveyed here is that "looks" do not make a proper seat. However, a good-looking exterior can be used, while still following basic good seating principles.

This should be recognized when designing furniture of various styles, such as early American, Mediterranean, or French provincial.

## Designed-in Hazards

Designers often unknowingly create hazards for the consumer, especially those associated with chairs and seats. A few of the typical designed-in hazards are noted below and illustrated by the accompanying sketches:

The seat structure should not have exposed sharp protrusions, corners, or edges with which someone can come in contact.

The assembly hardware should not project so that someone can snag his or her clothing or hands on brackets, protruding bolts, or other elements.

Folding seats should not have obvious pinch points where a person could inadvertently place his or her hand while folding or unfolding the seat.

Swivel chairs should have bases that will support the typical user who insists on leaning back too far. Four legs or supports should be used, rather than three. Use caster wheels that will not skid on slick floors.

So that people will not trip over the chair base, avoid chair base designs that stick out beyond the general perimeter of the chair.

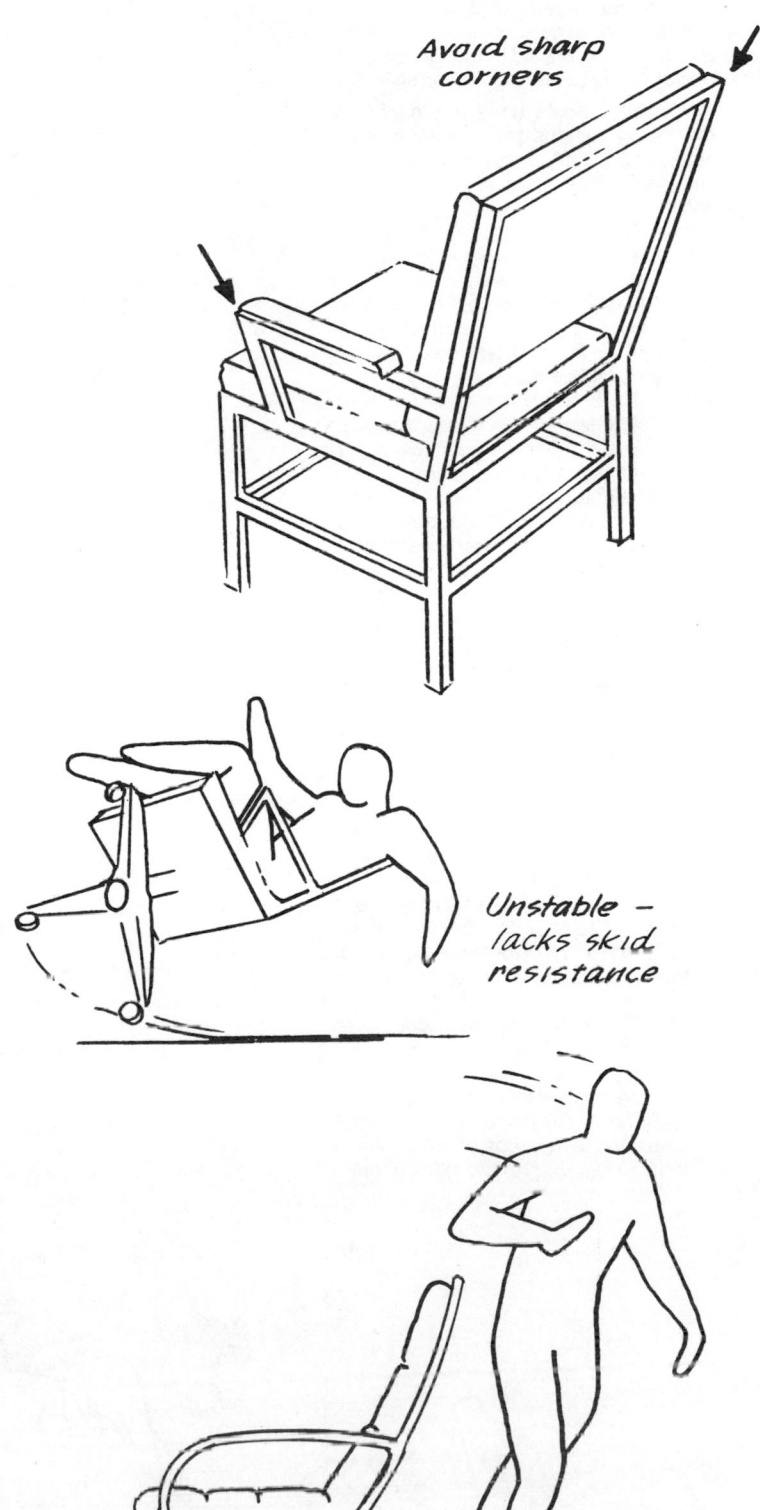

Avoid sharp corners

Unstable – lacks skid resistance

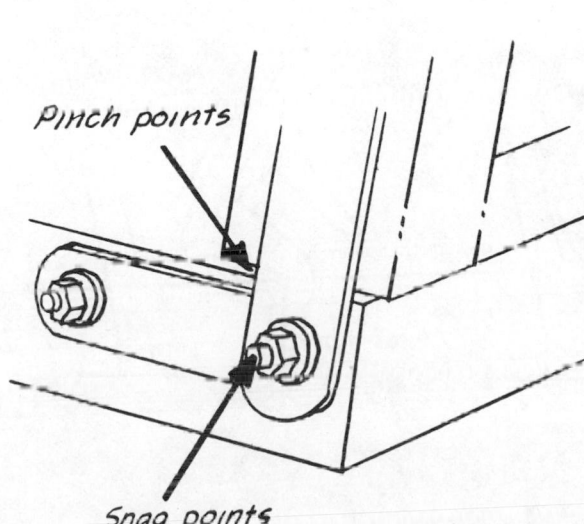

Pinch points

Snag points

Tripping hazards

## HOUSEHOLD SEATING

Avoid the tendency to believe that chairs, divans, and other seats used around the home, in hotels, in office waiting rooms, etc., are no more than objects of decor. This attitude has caused many a backache. The following suggestions will aid in the design and/or selection of more appropriate furnishings:

1. All seating should have approximately the basic dimensional characteristics noted previously in terms of seat size and angles for maintenance of proper posture.

2. Chairs and/or stools used in conjunction with eating should be selected at the same time that tables are chosen so that the seat-height–table-height relationship is proper. A 30-in (76.2-cm) table requires an 18-in (45.7-cm) seat; a 29-in (73.7-cm) table requires a 17-in (43.2-cm) seat. Snack bars and counters should be designed with a particular seat height in mind. Since production chairs or counter stools come in several standardized heights, select the desired height of chair first and design the table to fit.

3. Sofas, divans, and other overstuffed seats almost invariably suffer from a lack of concern for good seating principles. Either look for the following or design to these guidelines:

   a. Observe the previously defined seat-length criteria—most divan seats are too long.

   b. Observe the previously defined seat-height criteria—most divans are too low. However, some easy chairs are too high.

   c. Select firm cushions—most production furniture is too soft.

   d. Provide kick room at the front of the seat or divan; people have to place their feet as near their center of gravity as they can in order to get out of the typically low overstuffed chair or divan.

   e. If the chair is a reclining type, make sure that the base of the backrest coincides with the surface of the seat; i.e., make sure that as the backrest folds back, the hinge point does not cause the lower edge of the backrest to raise higher than the seat cushion.

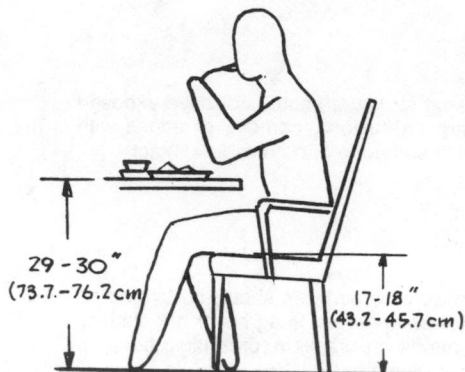

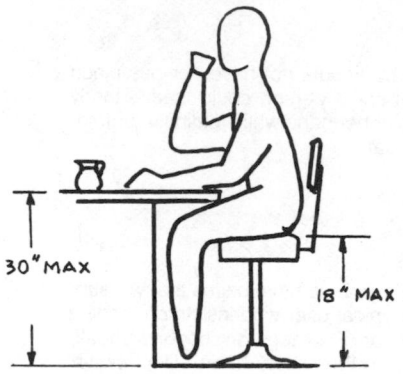

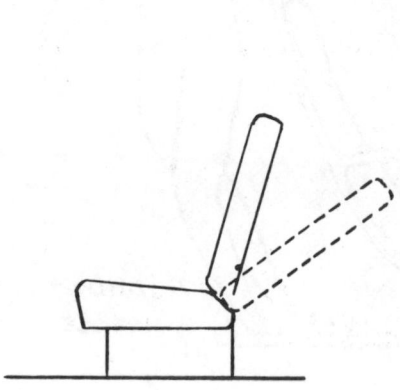

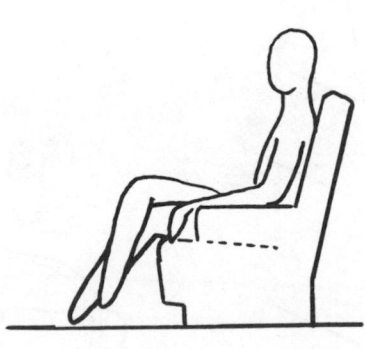

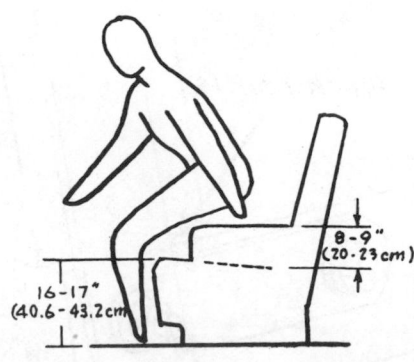

## OFFICE SEATING

Although considerably more attention has been paid to the improvement of certain kinds of office seating (especially secretarial chairs) in recent years, there is still a tendency to glamorize office furniture to the detriment of its basic purpose, i.e., to improve the comfort and productivity of the office worker.

**EXECUTIVE CHAIRS** Most executive chairs have seat pans that are too long—the previously noted dimensional and angular guidelines apply here as well as to any other seat.

Since the executive chair generally becomes the personal chair of the person to whom it is assigned, provide height adjustment. Also, provide casters for added mobility and a spring-adjustable backrest so that the user can either maintain an alert working position or lean back for more relaxed conversation.

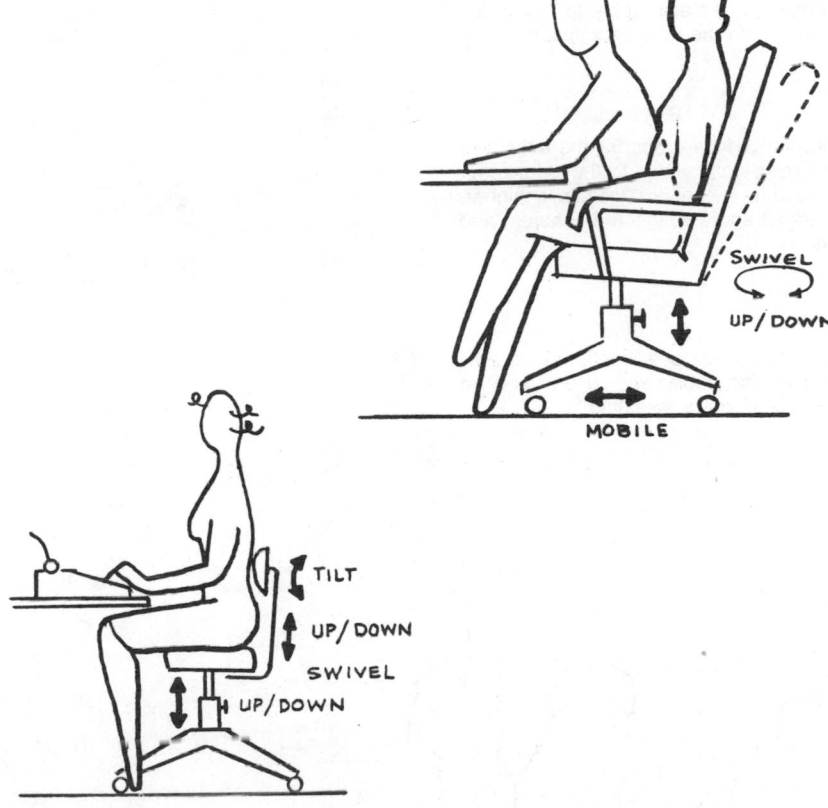

**SECRETARIAL CHAIRS** Provide personal height adjustment for both the seat and the backrest (a most important feature for the secretary). The backrest should also tilt so that it can conform to the individual worker's lumbar support needs.

**WORKBENCH STOOLS** Provide essentially the same adjustments for personalization that were suggested for the secretarial chair, except do not provide castors; i.e., this seat must "sit still." Make sure that the vertical adjustment systems remain secure if the chair is lifted by the backrest.

*Note:* Some experimental stools have been suggested which have tilting seats (i.e., for a proposed "half-sit, half-lean" position). There is no evidence that such a device improves a worker's productivity.

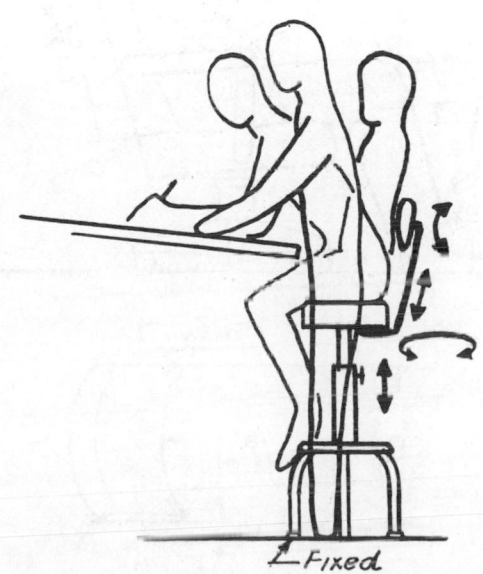

## Guidelines for Typical Office Chairs

The secretarial chair should have dimensions and seat and backrest angles similar to those shown in the accompanying illustration. The height adjustment should be the secure plunger type that is spring-loaded so that the adjustment is easy to make.

The basic seat and backrest dimensions for the secretarial chair also apply to the draftsman's chair, with the noted exceptions.

Although the dimensions for the executive chair are somewhat similar to those for other chairs, notable exceptions include a higher, solid backrest and a suggested headrest and armrests.

*Note:* Although the seat back may be covered with plastic, fabric should be used on seat pans to minimize sweating.

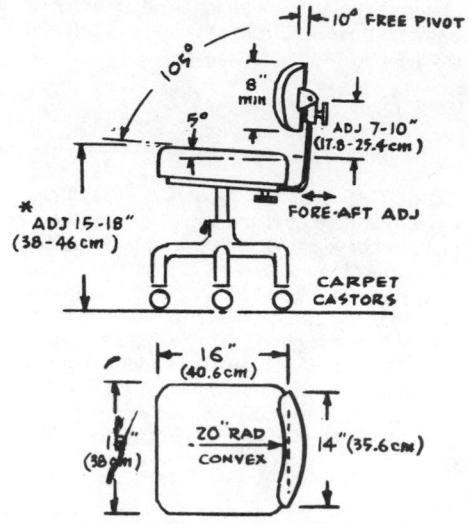

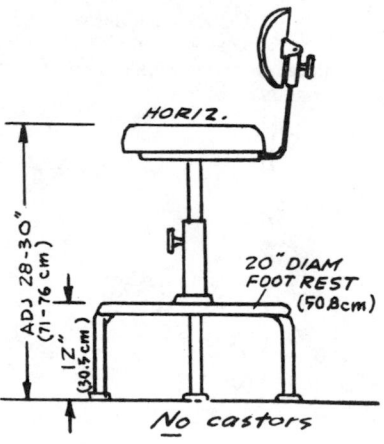

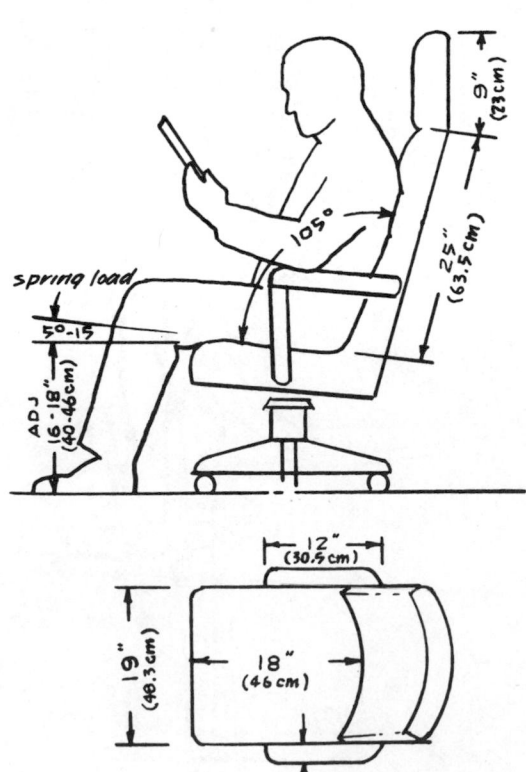

( ✱ 1.0" = 2.54 cm )

## SPECIAL WORK SEATS AND CHAIRS

When special working situations suggest the need for special seating, consider the possibility of designing the work seat so that it can be adjusted more specifically to each individual worker's size and to the workplace (task situation).

Regardless of the several possibilities for making a seat adjustable, certain basic dimensions should be retained. These include the ones shown in the accompanying illustration, i.e., a proper seat-pan size and a minimum back support area.

In addition, consider the following possibilities in terms of adjustable seating:

Either a mobile, caster-borne chair or a swivel seat if the chair has to be secured to a deck or other structure. Swivel seats should be designed so that the seat can be secured in specific positions for situations such as aboard ship. An adjustment lock (spring-loaded) every 45° or so is usually suitable.

Fore-aft adjustment in the seat pan may also be helpful for a seat that is secured to a deck (so that the worker can move closer to the work).

Vertical adjustment should be provided so that each worker can seek the best working height for his or her own height.

Although the nominal seat-pan–backrest angle should be about 105°, provide a swivel backrest that will allow the cushion to conform to the occupant's own back contour.

Armrests should also be considered if the seat is used for long durations. Armrests should adjust up and down so that the occupant can easily slip in and out of the seat. Armrests (see the accompanying sketch) can also tilt inward to serve as a semirestraint system if the seat is used on a moving platform such as a ship or tractor. The addition of a "wraparound" section to the backrest also helps secure the occupant against lateral forces.[22]

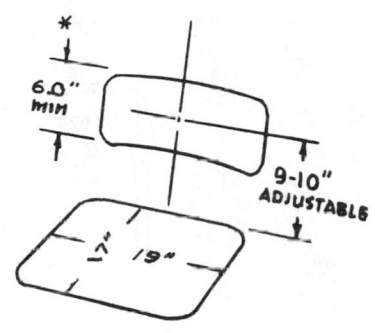

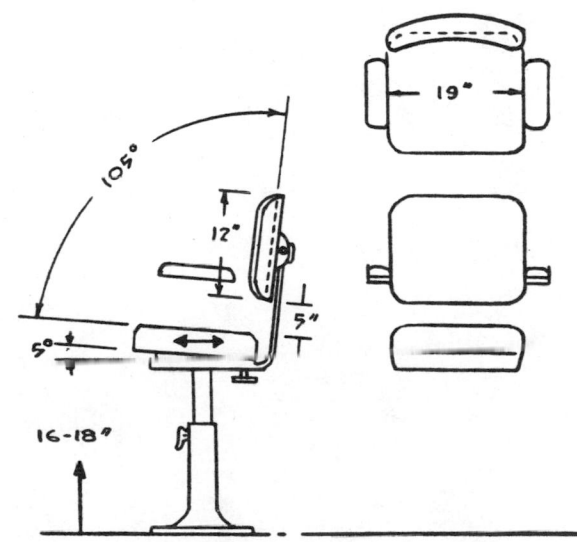

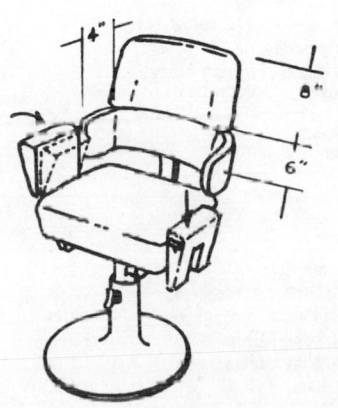

[22]Adjustment handles should always be designed and located so that the operator can use them easily, but the controls should not get in the way and possibly cause personal injury.

## DESK-CHAIRS FOR SCHOOLROOM APPLICATIONS

Using the table below and the reference numbers in the accompanying sketch, one can determine the general recommended dimensions for various aspects of a typical schoolroom desk-chair.

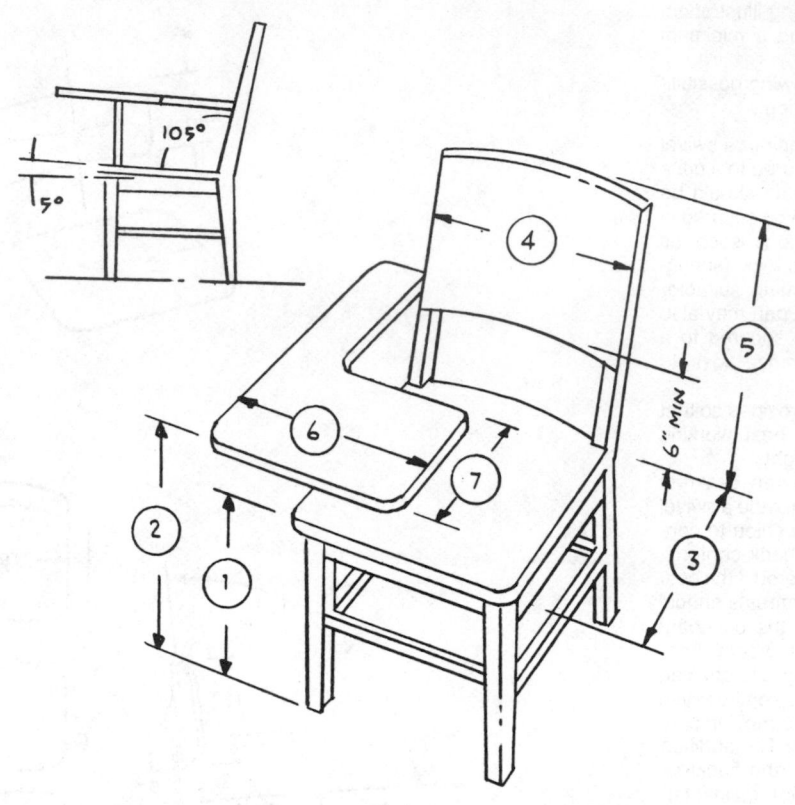

Although these dimensions provide a fairly good compromise solution, it is highly recommended that adjustable chairs be used for at least the lower grades, since children vary greatly in size at these ages.

| | Dimensions | | | | | | |
| | 1 | 2 | 3 | 4 | 5 | 6 | 7 |
|---|---|---|---|---|---|---|---|
| Kindergarten | 11 in | 18 in | 14 in | 9 in | 10 in | 11 in | 7 in |
| Grades 1 through 3 | 13 in | 19 in | 14 in | 10 in | 11 in | 11 in | 10 in |
| Grades 4 through 6 | 13 in | 22 in | 14 in | 15 in | 13 in | 12 in | 10 in |
| Grades 7 through 9 | 15 in | 25 in | 15 in | 17 in | 15 in | 12 in | 12 in |
| Grades 10 through 12 | 17 in | 29 in | 17 in | 19 in | 20 in | 16 in | 12 in |
| Adults | 17 in | 29 in | 17 in | 19 in | 20 in | 16 in | 12 in |

## AUDITORIUM SEATING

Although more austere seating dimensions and interseat spacing are typically specified for auditorium seating, the following suggestions are considered to be more adequate in terms of providing patrons with the proper convenience and comfort.

Seats should be reasonably wide (20 in, 51 cm), backrests should be high (24 in, 61 cm), and the spacing between seats should be about 6 in (15 cm) to minimize inter-elbow interference. Armrests should be about 7 in (18 cm) above the seat-pan surface level to prevent the occupant's shoulders from becoming hunched. Armrests should be wide (a minimum of 4 in, 10 cm).

Maximum comfort is realized when the seat-pan surface is angled upward about 5° but is flat (no contours). A flat backrest is best, although a concave shape (radius of 20 in, 50 cm) is acceptable.

Seats should be padded, but not to the extent that occupants sink into soft, billowy cushions, which become uncomfortable after a lengthy occupancy.

Seats should be staggered as shown in the accompanying illustration so that patrons have a good view between the heads of persons sitting in front of them. To calculate a given row spacing $(x)$ and seat spacing $(y)$ for unrestricted view of the stage $(a)$ from any seat, proportional to the distance $(d)$, apply the following formula: $a = kd$ (where $k$ is a constant and $y = t/x$, $t$ being the thickness of one head); i.e., if $x = 900$, $y = 500$, and $t = 200$, then $k = 0.33$. Thus 9 m (30 ft) from the front of the stage, $a = 0.33 \times 9 = 3$ m (9.8 ft); i.e., 3 m (9.8 ft) of the width of the stage can be seen without interruption, which is one-third of the average 9-m (30-ft) procenium opening.[23]

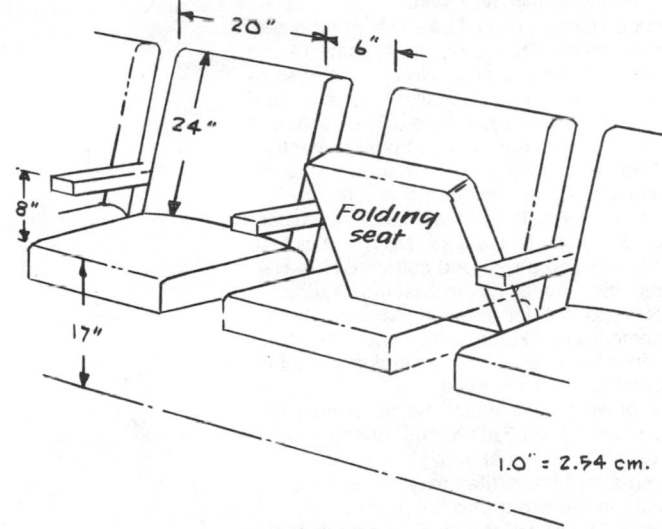

1.0″ = 2.54 cm.

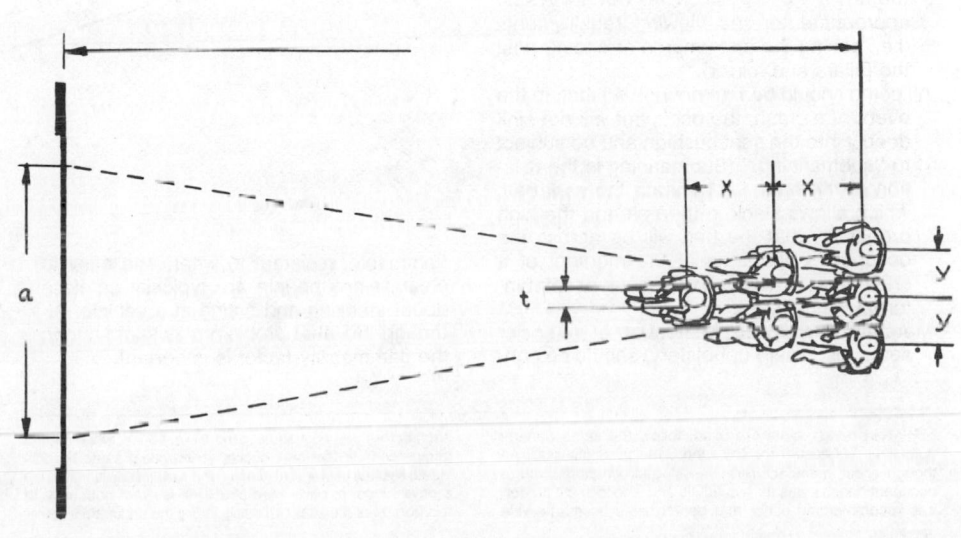

[23]R. Ham, *Theatre Planning,* University of Toronto Press, Toronto.

## AUTOMOTIVE SEATING

Although vehicle passengers could be provided options that are not recommended for the vehicle driver, there is considerable question as to whether this is advisable. For example, the frequently provided fully tiltable seat back may be desirable to allow a front-seat passenger to lie back to sleep or rest, but it obviously places that passenger in an unsafe position in the event of a crash. Seat belts and/or passive air bags will not be effective if such a passenger is not in the right position.

The basic criteria noted earlier apply to the automobile seat. In addition, several other features should be considered:

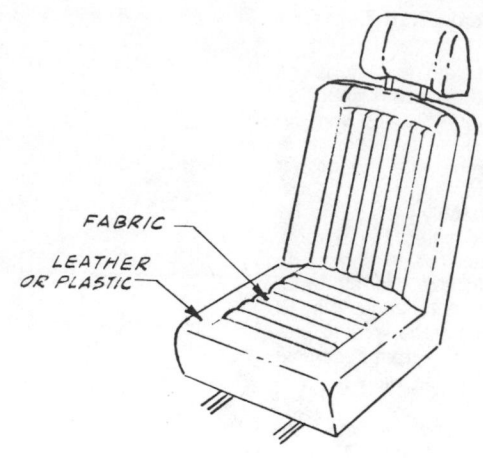

The major portion of the seat should be covered with fabric so that the seat "breathes." Leather or plastic may be used on the edges to make it easier to slide in and out of the seat and to minimize wear.

A ribbed pattern should be sewn into the seat pan and backrest covers as illustrated in the accompanying sketch. Note that the seat-back pattern runs vertically. This helps control sidesway resistance. The seat-pan cover pattern runs in the opposite direction. This helps keep the occupant from sliding forward during a sudden deceleration.

A headrest should be provided for all passengers to reduce the possibility of whiplash injury during a rear-end collision. However, the driver needs to see past the headrest in order to look for traffic in the rear. The accompanying sketch shows one method for improving rear vision. Headrests should be provided on back seats too.

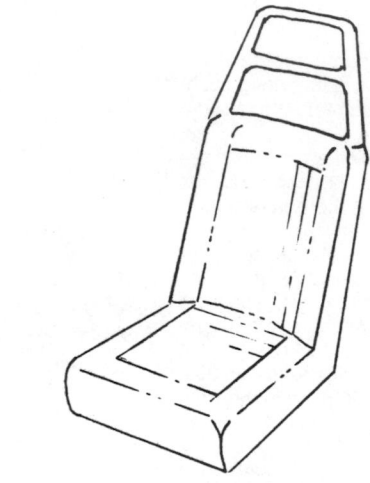

The driver's seat should be adjustable fore and aft at least 8 in (20 cm). Although other adjustments are appreciated by many who can afford them, they may place the occupant in the wrong position for a safety belt or air bag to be effective. The back of the driver's seat should never be designed to lay back more than a few degrees.

The driver's seat should be constructed so that the occupant will not sink into the cushion more than about 1 in (2.5 cm) (unless the seat is equipped with a height adjustment control). Otherwise, it is very possible that the driver's eye position will not always be appropriate for the viewing requirements (i.e., seeing the instruments or seeing past the pillars and mirror).

All seats should be firm enough so that, in the event of a crash, the occupant will not sink deeply into the seat cushion and be subject to "submarining." (Submarining is the rotation of the buttocks beneath the seat belt, which allows slack in the belt and the high probability that the belt will be across the occupant's abdomen at the moment of a crash impact and cause serious or fatal internal injuries).[24]

Materials used in the fabrication of vehicular seats (especially upholstery) should be non-flammable, resistant to wear, and easy to clean, since people are typically careless about smoking and eating in a vehicle. Although the latter problem may seem minor, the flammability factor is important.

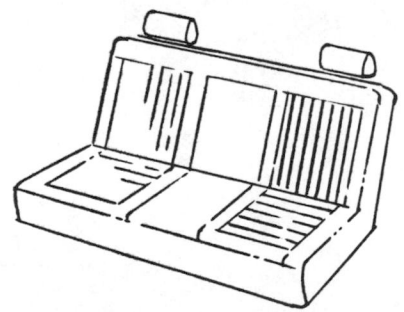

---

[24]When bench seats are used, follow the same general guidelines suggested for any other type of single seat. Although most manufacturers design their so-called three-occupant bench seats for two adults and a child in the center, it is recommended either that bench seats be made wide enough for three adults (using the 19-in, 48-cm, width criteria per person) or that only bucket or individual seats be provided. Avoid a bench seat in which the center position sits atop a drive tunnel or other hard structure. Sudden bottoming of this portion of the seat can easily injure the occupant's spine.

## Example of the Application of Human Engineering Principles to the Design of an Automobile Seat

The accompanying sketches illustrate the following points:

1. A tubular steel frame provides the basic dimensional and seat-pan–backrest geometry.

2. A thin, concave steel plate is welded to the frame to preserve the desired backrest form, i.e., one that reduces the effects of sidesway.

3. Rubberized upholsterer's strips are laced tightly across the seat-pan frame, and a 1-in (2.5-cm) sheet of high-density fire-resistant padding is laid over both the seat pan and the backrest. This is covered with upholsterer's muslin to hold the pads in place and prevent scrubbing and sloughing of the padding.

4. A transparent plastic sacklike cover is drawn tightly over the headrest frame and secured to the frame. This provides a somewhat resilient, fixed headrest that fits all occupants and still provides "see-through" capability so that the driver can look to the rear.

5. A combination fabric and Naugahyde cover is used over the seat pan and seat back. Fabric is used in the center sections and is "ribbed" as shown in order to allow the seat to breathe and to minimize lateral sway and forward slippage of the occupant. The borders of the cover are smooth to allow the occupant to slip in and out of the seat easily and to provide good wear capabilities.

6. The seat adjustment control is placed in the midfront position to be accessible to either hand. A simple upward pull deactivates a spring-return latch so that the seat can be moved fore and aft. Note that this construction is unlike typical contemporary construction. A 220-lb (100-kg) occupant would not sink into the seat pan more than 0.5 in (1.3 cm), except if the car hit a bump—in which case the lacing would absorb the shock without causing the occupant to "bottom out." By not allowing heavy occupants to sink into a soft cushion, as opposed to allowing lightweight occupants to sink less, the designer was able to hold eye references and head clearance tolerances to a fine degree. Many current soft-cushioned seats allow the eye reference to vary as much as 2.0 in (5.1 cm) as a result of the varying weights of different occupants.

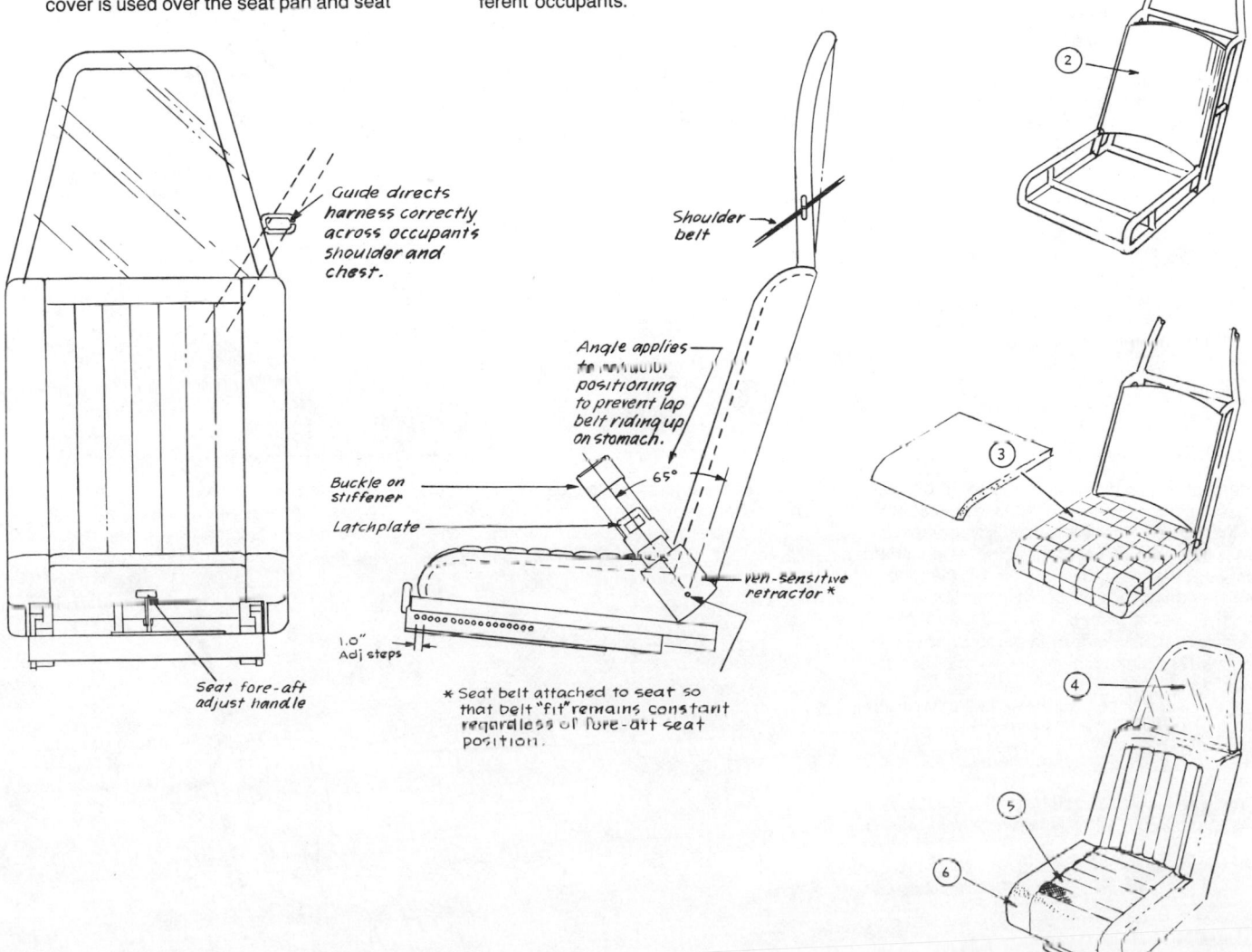

Guide directs harness correctly across occupant's shoulder and chest.

Seat fore-aft adjust handle

Shoulder belt

Angle applies to individual positioning to prevent lap belt riding up on stomach.

Buckle on stiffener

Latchplate

65°

1.0" Adj steps

* Seat belt attached to seat so that belt "fit" remains constant regardless of fore-aft seat position.

veri-sensitive retractor *

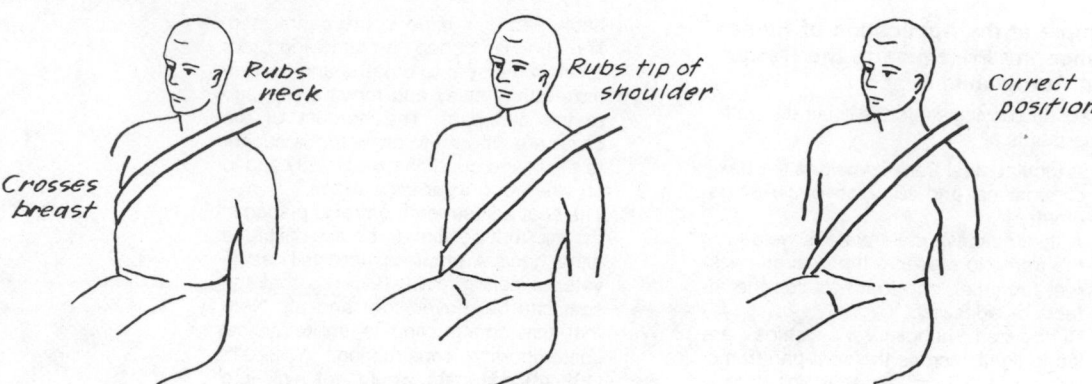

## Safety Belts[25]

Not only must safety belts be designed to restrain a vehicle occupant properly during a crash, but they also should be comfortable to wear; otherwise (when they have a choice), people will try to avoid wearing them.

The accompanying sketches illustrate the most common complaints about seat belts (the chief complaints are associated with the torso belt). In the first sketch, the belt rides against the wearer's neck. In the second sketch, the belt rides atop the wearer's shoulder tip. The third sketch illustrates where the belt should fall, i.e., across the chest, midway between the breasts.

The accompanying sketch illustrates the optimized belt-crossing pattern for a torso belt to ensure that it is comfortable for a range of adult wearers from the 5th-percentile female to the 95th-percentile male (United States population).

These specifications are related to a standard 50th-percentile anthropomorphic dummy, seated properly in the mid-adjustment position of an adjustable front automobile seat. The cross-hatched envelope represents a compliance envelope used to determine whether a particular seat belt meets the comfort criteria.

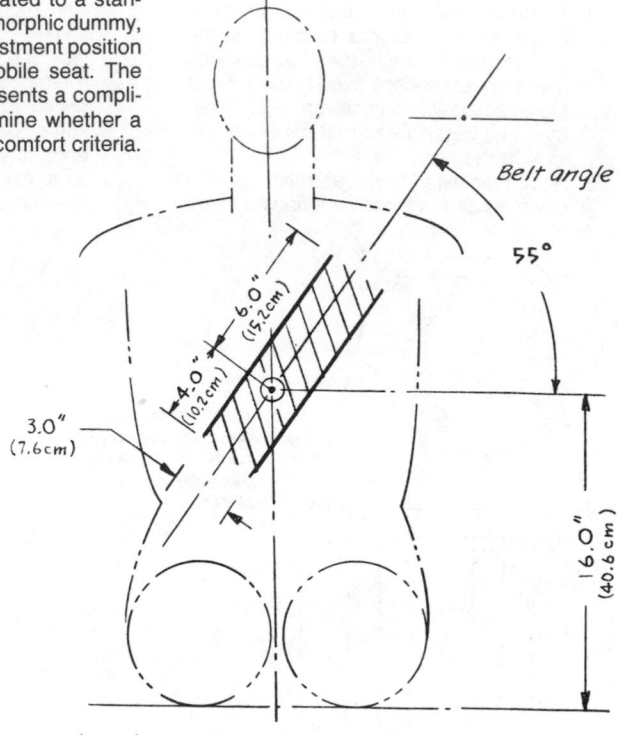

## Child Restraint Systems

Although a variety of methods could be used to restrain children in vehicles, the accompanying sketch illustrates what is perhaps the best approach. Rearward-facing safety seats provide the best distribution of the child's weight during a frontal crash. In addition, when such a seat is placed in the front seat of the vehicle, the child faces the person who is driving, which makes it easier to monitor the child's activities.

Key issues to be addressed in the design of such a seat are the following:

Use an X-type lap belt or harness for maximum security.

Provide a single release buckle with a latching system that a child generally cannot tamper with.

Provide a suitable method for securing the seat to the vehicle so that it will not shift with lateral movements of the car or rebound from either rear or frontal crash forces.

Provide different seat sizes for children in different age groupings, e.g., six months to one year, two to three years, and four to six years. Generally by the seventh year, the child can be secured by a standard active lap belt.

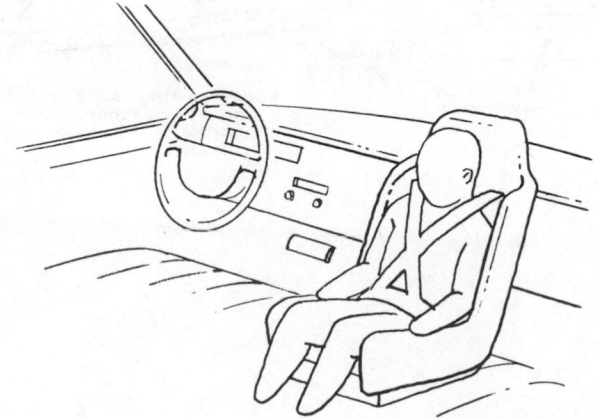

Special attention should be directed to the problem of rear passengers. Because the rear passengers must be squeezed in between the wheel wells in most automobile body styles, contemporary seat designs are particularly poor. It should be remembered that rear passengers are exactly the same as front passengers, and therefore the same principles for good seating practices noted earlier should be followed. Primarily, the seat dimensions should be adequate for either three or two passengers—not two and one-half! The seat shape should be similar to that shown in the first sketch; i.e., a straight backrest without uncomfortable bolstering at the outer edges should be provided (see the second sketch). Headrests should be provided for at least the two outside passengers.

It should be noted that rear passengers also need headroom. Many designs seem to reflect a lack of interest in adequate headroom and body room for rear passengers, perhaps because of the faulty notion that the rear seat is not often occupied.

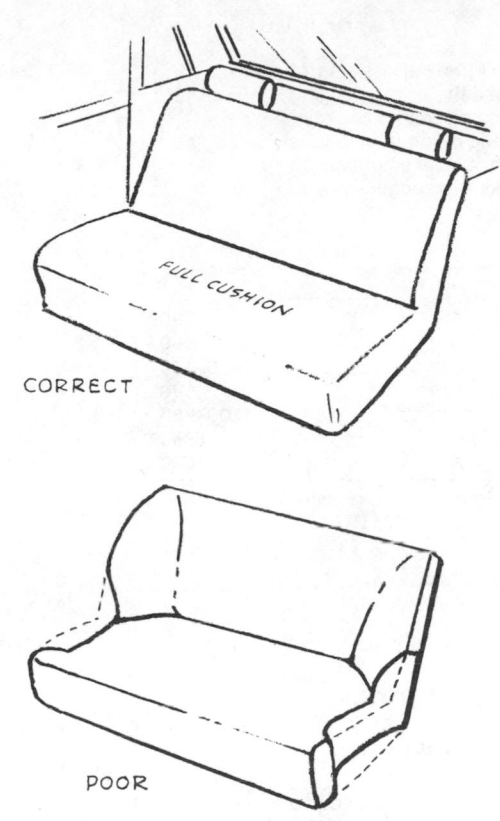

## COMMERCIAL TRANSPORT PASSENGER SEATING

In spite of the fact that people probably will continue to travel even though seats are uncomfortable, public reaction would improve if aircraft, train, and bus passenger seats were designed to "fit." Instead of designing the seats after the fuselage size has made it necessary to crowd people, basic seating criteria (along with other human factors) should dictate how big the fuselage should be.

The basic dimensional and seat angle guidelines noted earlier apply to passenger seats as well as to other types of seating. Several other factors should be considered, some of which are illustrated by the accompanying sketch. Key considerations should include the following:

A 5° seat-pan pitch with a 105° seat-to-backrest angle for normal upright seat position.
Continuous seat recline adjustment to 45°.
A backrest height of 24 in (61 cm) before the bottom of the headrest begins.
A minimum space between armrests of 19 in (48 cm) (more is desirable for long-haul applications).
The seat-pan angle should automatically adjust to a horizontal position as the seat backrest reclines so that the occupant can stretch out his or her legs.
Side-by-side passengers (when provided with intermediate armrests) should be given either separate armrests 2 in (5 cm) wide or a large 4- to 5-in (10- to 13-cm) common armrest between them. When a minimum seating width has to be observed, design the intermediate armrests so that they can be folded back out of the way.
If tray tables are integrated into the back of the seats, design them so that the tray table will remain level even though the seat to which it is attached is reclined.
Passenger seat controls, flight attendant call buttons, and light and communications controls should be located where passengers can see and find them easily without getting out of their seats.
Seat cushions should not be overly soft. They should be flat and should be made of, and covered with, nonflammable materials that "breathe."
No sharp projections, corners, or edges should be exposed that would create a hazard either during the normal movement of people within the vehicle or during an emergency escape.
As a minimum, provide lap belts to secure occupants against rough air or crash conditions. This applies also to intercontinental buses. Although lap belts are also recommended for rail vehicles and local buses, it would perhaps be unrealistic to expect passengers to use them.

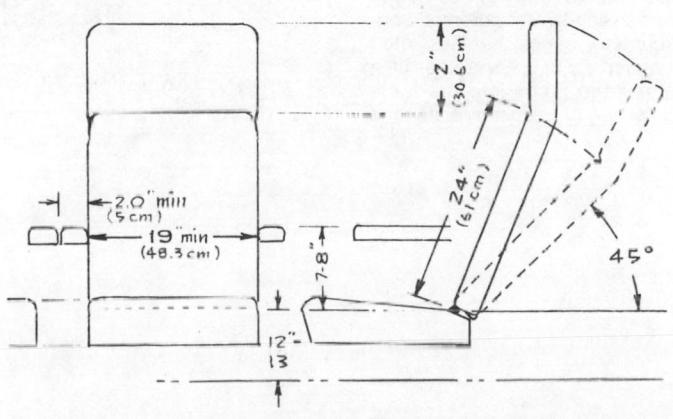

## Bus Seating

Levis and others studied several aspects of bus design and developed the accompanying recommendations regarding seating. The principal issue leading to these recommendations was ease of access, with special emphasis on the problems of the older passenger.

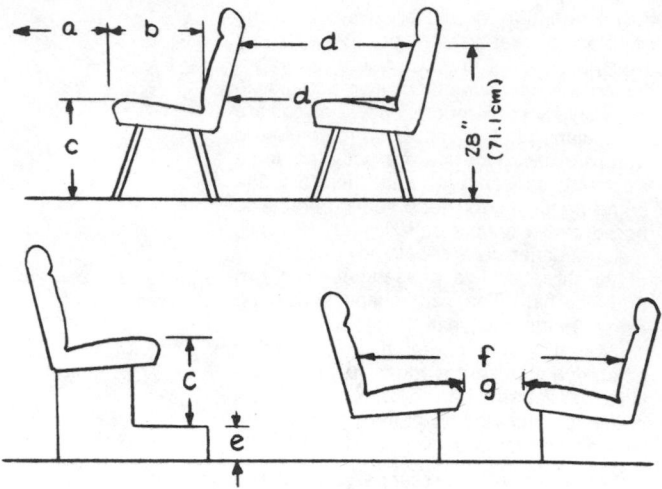

| Dimension | Preferred, mm | Acceptable, mm |
|---|---|---|
| a | 310 | 280 |
| b | 400 | 380 |
| c | 432 | 400–460 |
| d | 720 | 680 |
| e | 200 | 100–250 |
| f | 1460 | 1360 |
| g | 700 | 600 |

*Source:* J. A. Levis, "The Seated Bus Passenger—A Review," *Applied Ergonomics,* vol. 9, no. 3, pp. 143–150, 1978.

## GUIDELINES FOR THE SELECTION AND DESIGN OF TABLES, DESKS, COUNTERS, AND WORK BENCHES

The key issues that should be addressed to ensure compatibility between the user and the table, desk, counter, or workbench are the following:

Proper working height
Adequate leg and foot clearance
Adequate work area

The accompanying illustrations provide nominal dimensional guidelines for tables that are used for writing, reading, or eating; for desks that are used for writing, storage, and conferences; and for special secretarial desks, including support for a typewriter.

The work surface height shown represents the maximum, and the knee clearances represent minimums. The dimensions given for the tops of tables and desks obviously are somewhat arbitrary. However, they are suggested for the following good reasons:

The width must be sufficient so that the worker is not continually knocking things off the surface.

Depths should be sufficient so that the user's legs and feet do not extend beyond the opposite side and inadvertently trip someone who might be passing by. All potential contact points (corners, edges, handles, etc.) should be rounded so that if the user or a passerby happens to bump into the furniture, he or she will not receive a serious injury.

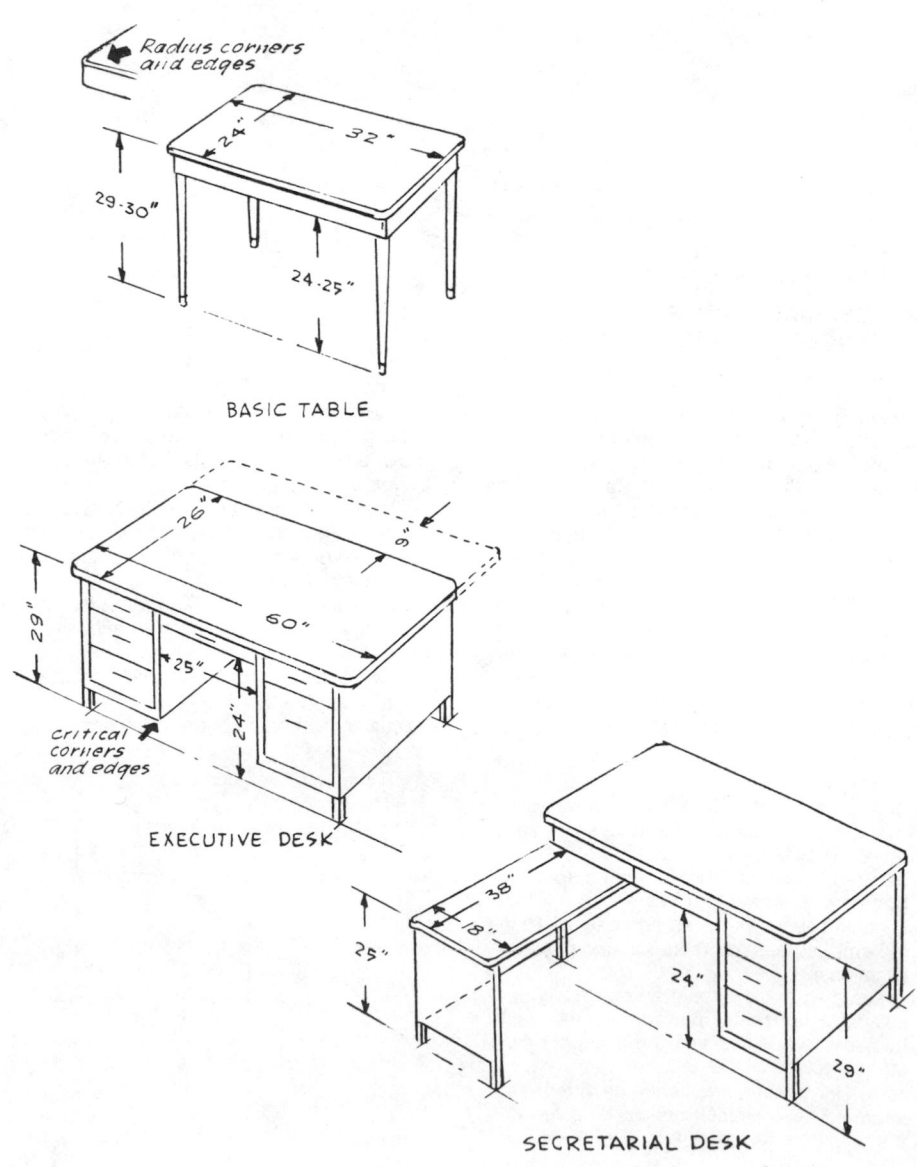

BASIC TABLE

EXECUTIVE DESK

SECRETARIAL DESK

Drafting tables generally should be as adjustable as possible in order to satisfy a variety of work requirements. Consider the possibility of the user's choosing between a sitting and a standing position, working on a flat versus a vertical surface, or being able to work from several sides of the table. When adjustment hardware is designed, make sure that it provides the necessary ease and the necessary security so that the user's valuable time is not wasted and so that the table position cannot shift at a critical moment (e.g., because the individual leaned a little too heavily on the edge of the tabletop).

If a single fixed (all-around) drafting table is desired, consider the general dimensional guidelines shown in the accompanying illustration. This set of specifications will provide a good compromise to suit a wide range of drawing activities and sizes of workers. Note the footrest, which implies that the table serves for both a sitting (with stool) and a standing work position. A 28- to 29-in (71- to 74-cm) stool height works well with the specified table dimensions.

A typical service counter at which people stand across from one another should be between 40 and 41 in (102 and 104 cm).

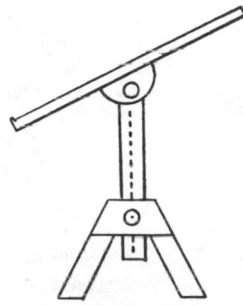

FOR DRAFTING

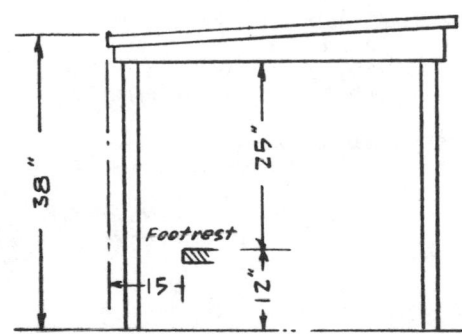

38"   25"   Footrest   21"   15"

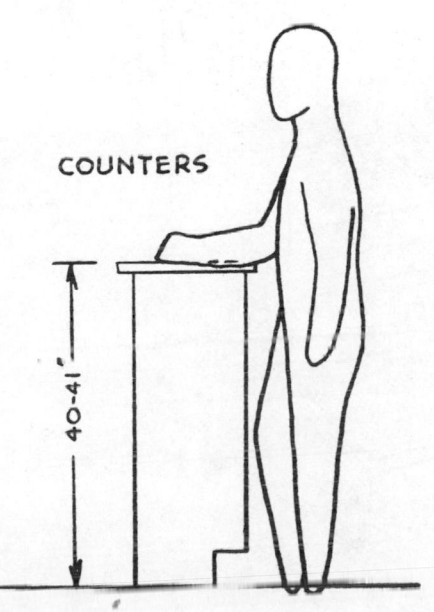

COUNTERS

40-41"

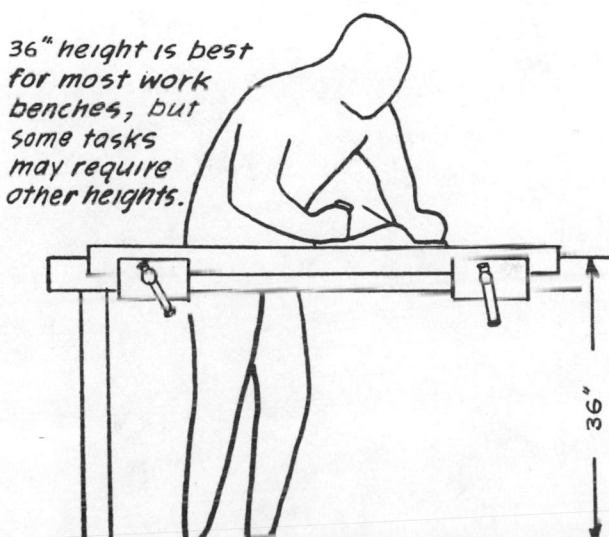

WORK BENCHES

36" height is best for most work benches, but some tasks may require other heights.

36"

The typical household counter top should be approximately 36 in (91 cm) high in order for it to be reasonably comfortable for both the male and female members of the family. Studies have shown, however, that people prefer a working height that specifically fits them, which suggests that designers should consider ways to create adjustable counter heights. This would be especially helpful to alleviate the problems that children obviously have in using the bathroom sink.

Counter tops should be about 24 in (61 cm) deep to provide a good compromise between maximum useful work space and the ability to reach easily to the back of the counter. Counter edges around sinks should contain a lip to catch liquids. The edges should be radiused, however, so that anyone bumping against the edge is not injured (avoid the popular sharp-edged Formica top).

Provide kick space below the front of the supporting cabinet.

There are fewer restrictions with regard to many other types of tables that are used in the typical household or lobby, such as end tables and coffee tables. Some considerations that might be observed in designing or selecting such tables are the following:

End tables should not be higher than the armrests of an adjacent chair.

Coffee tables should not be lower than about 12 in (30 cm) or higher than about 18 in (46 cm).

Credenzas or magazine tables that are usually interacted with from a standing position should be about 36 in (91 cm) high.

Very low tables must definitely not have sharp corners or edges. This is especially true of glass-topped tables.

The legs of single-pedestal occasional tables should not extend beyond the perimeter of the top and thus create tripping hazards.

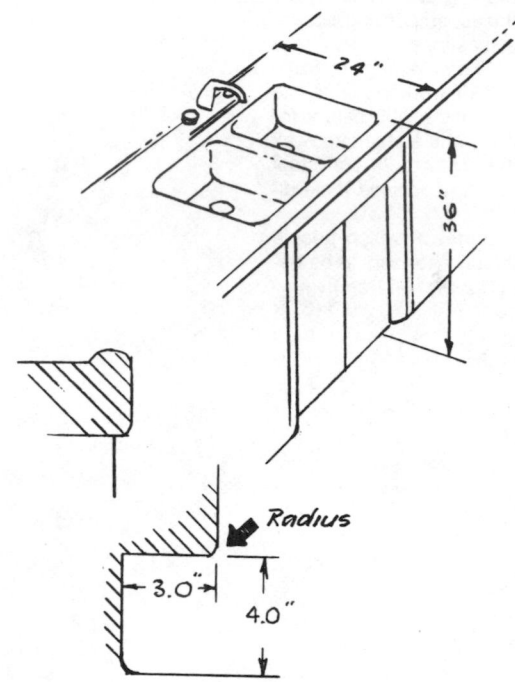

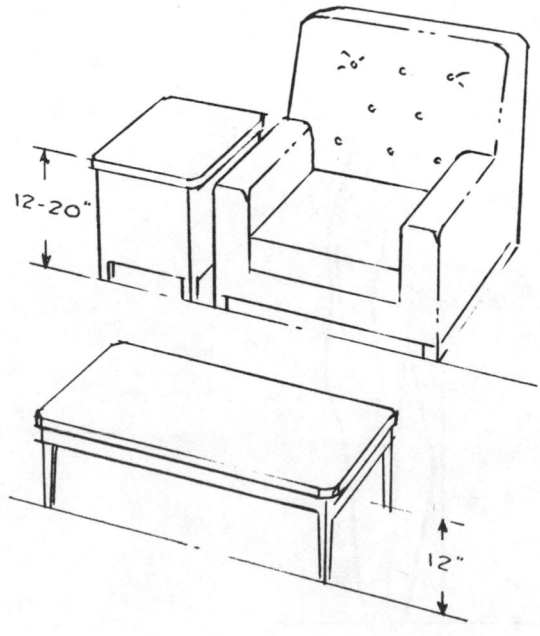

### The Critical Table Leg

Table leg concept, design, and positioning often cause the consumer as many problems as any other aspect of the table, or more. A few of the most common errors are discussed below:

The legs on dining tables are obviously not in view when diners attempt to take their places at the table. Placing the legs where diners are least likely to sit is an important design objective. A second objective is to avoid creating fancy styling features on the leg that protrude precisely at knee height.

Although single-pedestal tables are probably the least annoying to the user (see the accompanying sketch), special styling such as that illustrated creates an extremely annoying problem for the diner. However, it may be more annoying for the table's owner when he or she discovers that the legs have been damaged (because they provide an ideal place for people to rest their feet).

Small occasional tables such as the one shown in the accompanying sketch are easily tipped because the legs often have been extended beyond the perimeter of the top. The more legs there are, the less likely this is to happen, and the more stable the table is (in spite of the belief that it is easier to level a three-legged table).

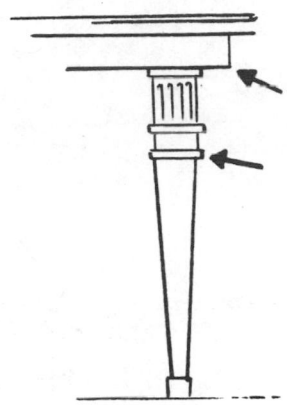

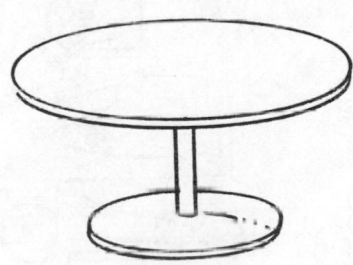

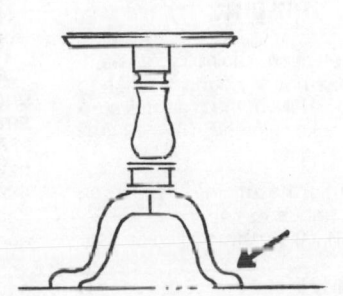

### Horizontal Dimensions of Tables

People must have room for their feet and elbows as they sit around a table. The accompanying sketches provide important guidelines for designing or selecting the proper size of table for the particular application.

The guidelines shown are for adults. The dimensions can be reduced for special tables to be used by children. For children between the ages of ten and thirteen, the dimensions can be reduced by about 15 percent; for younger children, they can be reduced by about 20 percent.

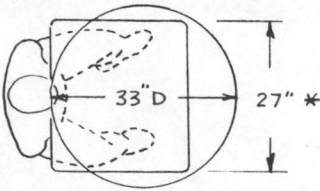

*Single diner*

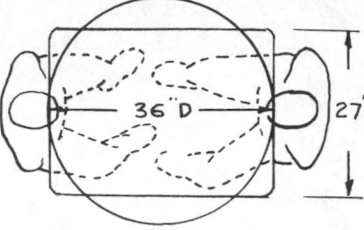

*"Facing" couple*

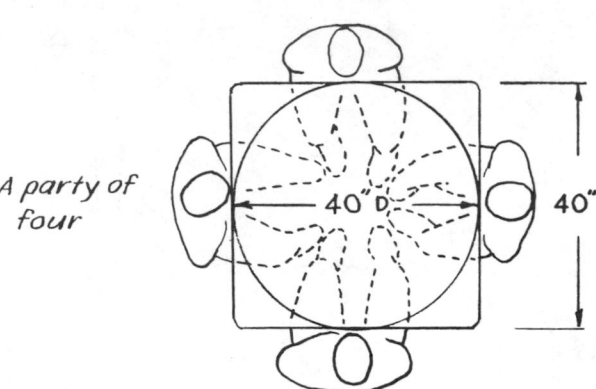

*A party of four*

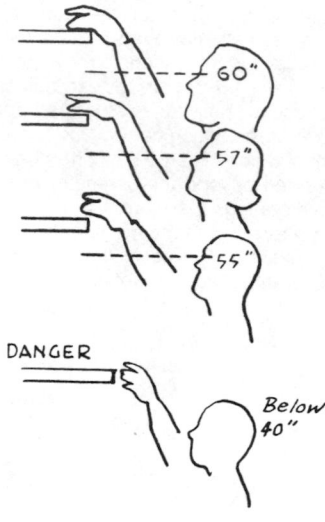

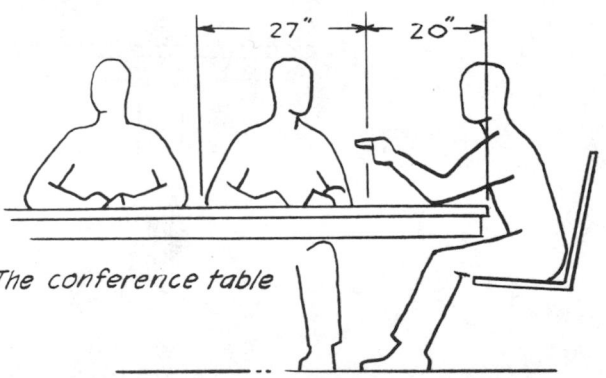

*The conference table*

\* 1.0" = 2.54 cm.

### CABINETS AND SHELVING

In designing or selecting cabinetry or shelf-type storage units, a primary consideration to keep in mind is the nature of the potential storage and retrieval tasks involved. This requires attention to the following:

The size and weight of the articles to be stored
Whether the user needs to see the article
How high and how far in the particular user can reach
Strength and mobility limitations of the user at various positions, e.g., stretching, bending, stooping, and kneeling

As indicated by the illustrations at the right, one must keep in mind the size differences among the possible users.

Avoid the temptation to assume that shelf height is the only consideration. The sketch at the right illustrates the importance of shelf depth as well as shelf height. Although it may appear wasteful not to take advantage of the total available depth of a cabinet, this is not necessarily an advantage to the user.

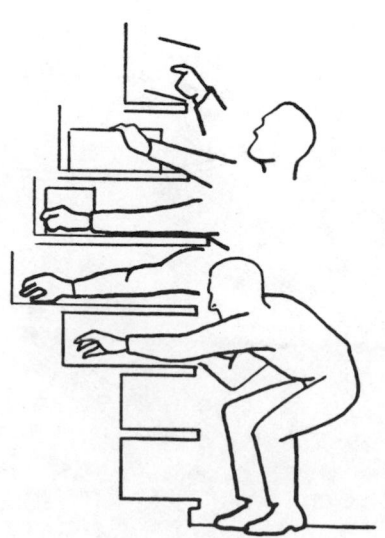

## Typical Shelving Parameters

A—For small, light articles such as books that are stored on edge

B—For larger but lightweight articles that are easy to grasp

C—For light to medium-weight articles that are easy to grasp with two hands

D—For medium-weight articles that require the use of two hands

E—For medium-weight to heavy articles that are not too large and are easily managed with two hands

F—For fairly heavy, large articles that require both hands and an optimum weight-carrying position

G—For medium-weight articles that are small enough to be lifted to the appropriate weight-carrying position

H—For light to medium-weight articles that are easy to grasp and lift

I—Recommended height to be beyond the reach of toddlers

*When user is confined, keep stored materials within reach.*

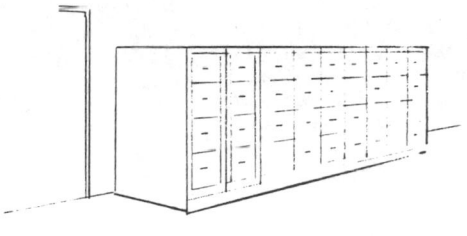

*Wasted wall and floor space.*

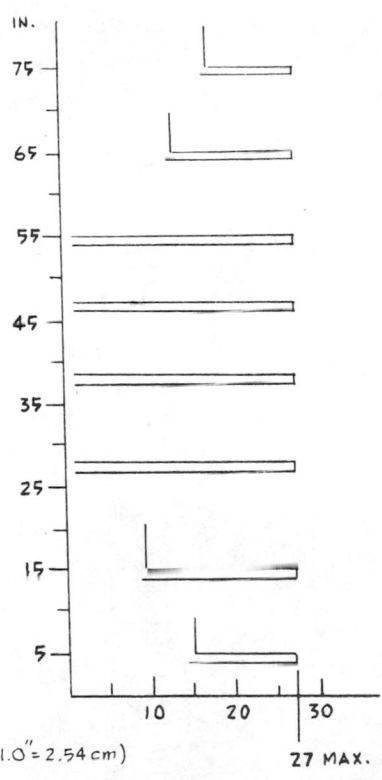

(1.0″ = 2.54 cm)     27 MAX.

*Better approach!*

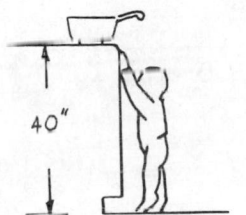

40″

## Storage Cabinets

General philosophies regarding storage cabinet selection and design are gradually changing to provide more efficiency for the user as well as more effective use of space.

The need to rearrange one's immediate work space from time to time suggests mobile cabinetry such as shown in the illustration above. In addition to allowing the cabinet to be brought to a better position, this type of filing arrangement positions the stored materials so that the user does not have to reach or strain to inspect them.

Typical filing cabinets generally are relatively deep and narrow. Not only do they take up valuable floor area, but they also are easily upset when the user inadvertently opens the upper drawers without thinking about the imbalance they are creating.

It is generally more desirable to take advantage of available wall space without losing floor area, as is illustrated in the sketch above. Special cabinets are available in this format and should be considered whenever floor space is at a premium.

Key operating issues to address are the following:

Reach
Materials visibility
Flexibility for rearrangement
Cabinet stability
Drawer control
Handle and latch operability
Compartment identification
In-cabinet materials separation and control
Cabinet mobility

### Specialized Furniture[26]

The accompanying sketches illustrate general dimensional guidelines used by architects for the development and/or procurement of display shelving for libraries and supermarkets.

The key factors in selecting and/or designing such furnishings are (1) the relationships between what the user is doing and for how long he or she may do it and (2) the transfer of books or other items between the shelf and the service container, by the person who services the shelf.

It is recommended that furnishings such as these be designed for ease of vertical shelf adjustment and also in terms of precluding the development of deep shelves, which make the user's task of retrieving objects more difficult.

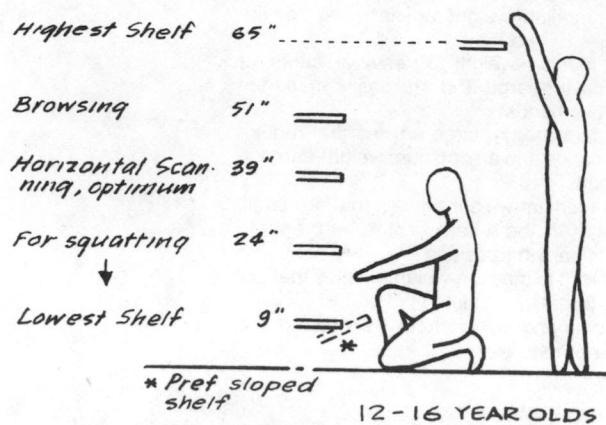

LIBRARY SHELVES

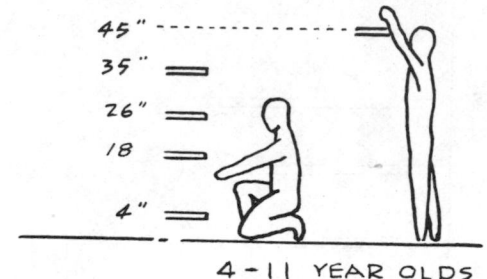

MARKET SHELVES

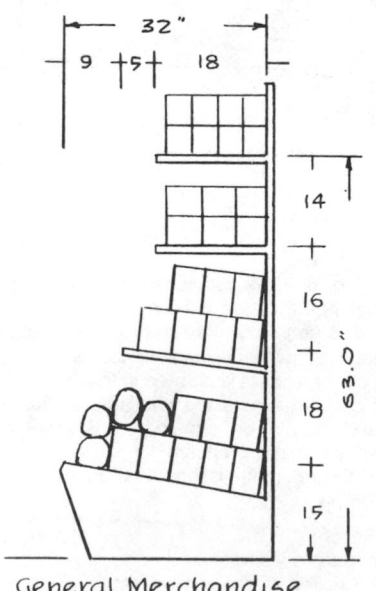

General Merchandise

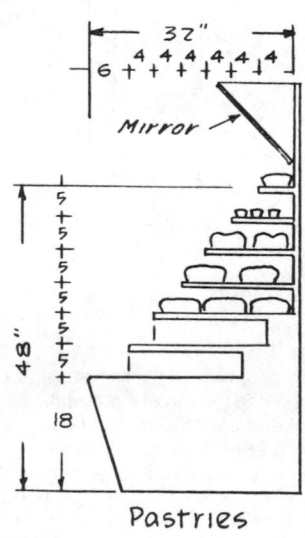

Pastries

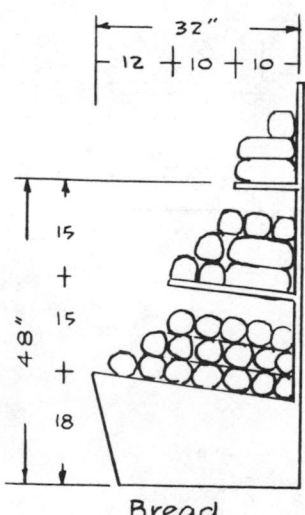

Bread

[26]J. De Chiara and J. H. Callender, *Time-Saver Standards for Building Types,* McGraw-Hill Book Company, New York, 1973, pp. 276, 616.

## BED DESIGN

The following are standard bed sizes:

Single      6 ft 6 in × 3 ft (1.9 × 0.91 m)
Twin       6 ft 6 in × 3 ft 3 in (1.9 × 0.99 m)
Three-Quarter   6 ft 6 in × 4 ft (1.9 × 1.22 m)
Double      6 ft 6 in × 4 ft 6 in (1.9 × 1.37 m)

Recent sales records, however, show that most people now purchase queen- or king-size beds. There is good reason for wanting a larger bed. The standard length of 6 ft 6 in (1.9 m) does not provide the space needed at the ends of a bed; e.g., many people sleep with their arms above their heads, and most people who are of average height or taller do not like to have covers cramp their feet. Beds (or, more appropriately, mattresses) should be at least 80 in (203.2 cm) long. The width of a double bed is more a matter of choice. Two people who do not like to touch each other need a bed that is at least 72 in (182.9 cm) wide.

The top of the mattress of an adult's bed should be approximately 18 in (45.7 cm) high above the floor. If the bed is lower, it is difficult to get out of and to make up. If it is higher, it is difficult to get into.

There should be at least 30 in (76.2 cm) of vertical clearance between bunk beds. A ladder should be provided to aid in getting in and out of the upper bed.

Children's beds should be designed so that the side rails (when up) are too high for the youngster to crawl over (at least 29 in, 73.7 cm, above the mattress). One side rail should be adjustable so that the child can be easily retrieved from the bed. Vertical open rail sides are suggested so that an adult can see into the bed to make sure that everything is all right, but the rails must be spaced close together so that the youngster cannot get his or her head through the openings (a maximum of 3 in, 7.6 cm). All structure and hardware must be "childproof" and nontoxic.

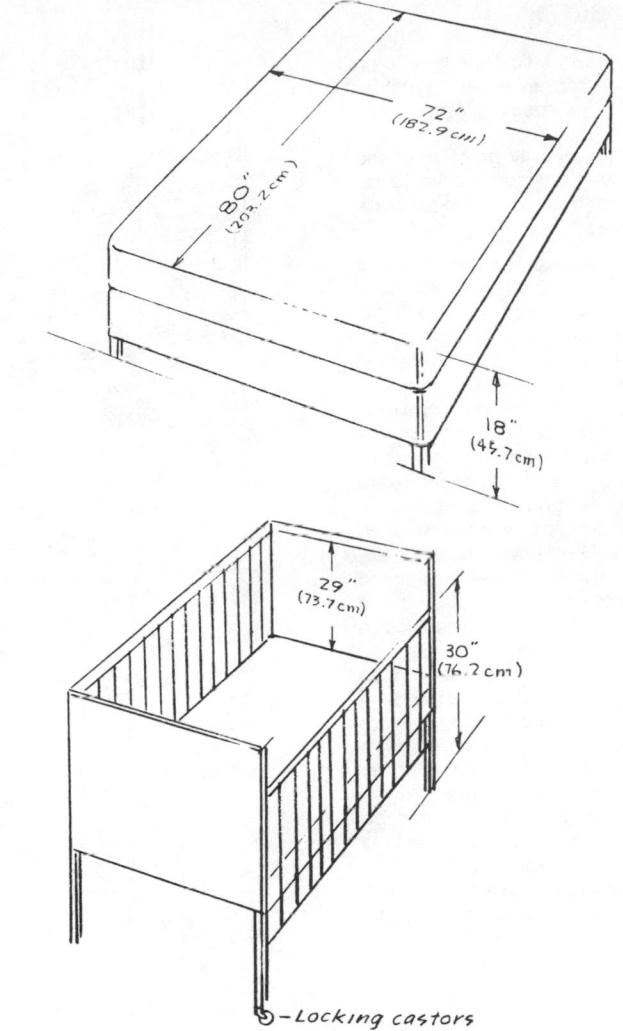

⊗ — Locking castors

Headroom is a significant consideration in the design of bunk beds. Head clearance (C) should actually be sufficient so that a person can sit down on the bed to don or doff shoes and sit up in the bed (at least partially) to read. For youth beds, this head clearance should be at least 26 in (66 cm); for adult bunk beds, it should be at least 30 in (76 cm).

In a recent study of shipboard berthing conditions, it was found that head clearance was essentially absent; vertical clearances as little as 18 in (46 cm) were found.

In the design of bunk beds, it is also critical that the support for the upper bed be sufficiently rigid to prevent it from sagging under the weight of a heavy upper occupant and encroaching on the head clearance of the lower occupant.

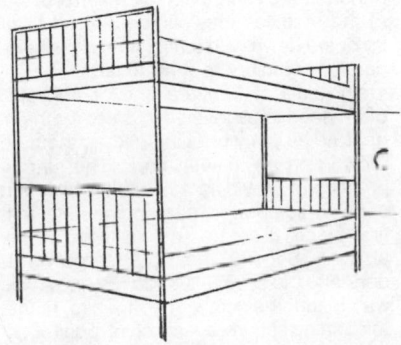

## FURNITURE MOBILITY

The ability to move and adjust furniture is an important human factor since most furniture owners do not have the special aids available to moving specialists.

In considering the mobility problem of the user and the eventual design required to improve mobility, keep in mind the typical problems itemized below:

To move furniture or appliances, one must first be able to get hold of the article. Although it is perhaps undesirable to put handholds where they will detract from the appearance of the furniture, appropriately concealed handles or handholds should be provided.

A heavy piece of furniture should be provided with casters so that it can be moved away from a wall without damage to the wall or adjacent furniture. Select caster hardware that is especially designed to roll easily on carpet. On furniture items with an attached mirror (see the accompanying sketch), design the mirror so that it is easily removed and replaced—preferably without the need for tools. Mirrors should be provided with a height adjustment.

Heavy appliances that ordinarily will be positioned in close quarters (i.e., beside or in between built-in cabinetry) should be provided with a built-in slide-roller system so that the item can be rolled straight in or out (see the accompanying sketch of a refrigerator). Although hardware add-ons can be purchased by the homeowner, such devices seldom work well. Besides, it is the manufacturer's responsibility to design and build a complete appliance, not one the buyer has to add to.

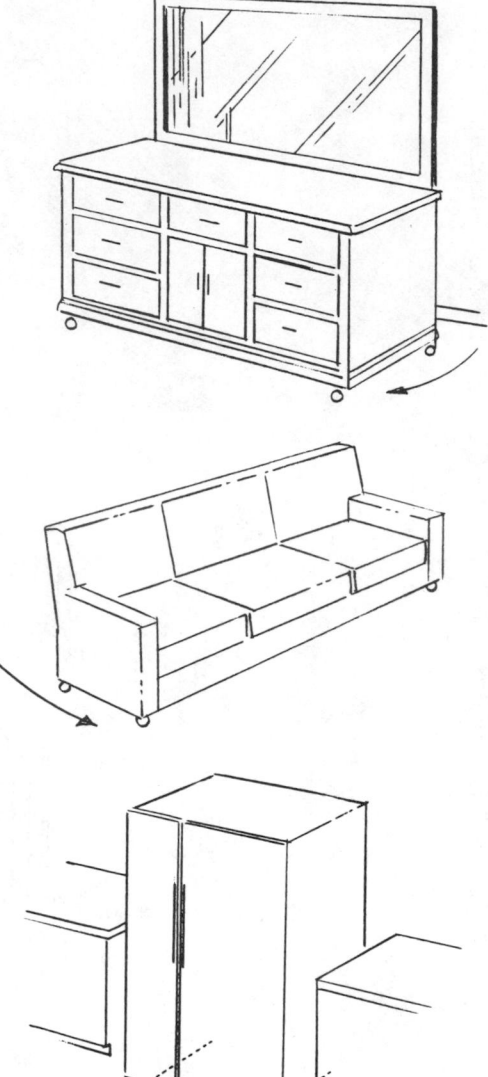

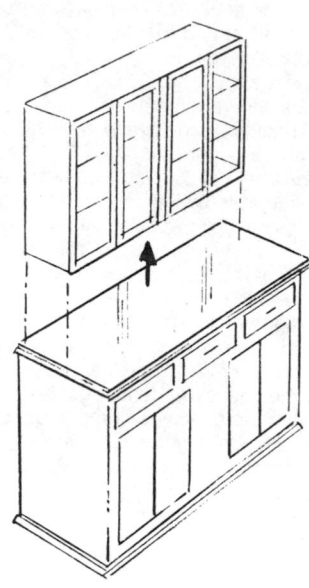

All furniture should be designed with standard access dimensions in mind since architectural doorways are well standardized and are not likely to change. The clearance on each side of a piece of furniture when it is being moved through a doorway is critical to those handling the furniture. The owner also worries about movers scratching furniture when it has to be passed through a doorway.

If the overall dimensions of a particular piece of furniture are going to be too large for it to pass in one piece through the typical door, up a staircase, or down a hallway, design the furniture to be easily disassembled so that it can be taken in and out in several pieces. Although most manufacturers are very careful to size or modularize their various pieces for ease of transport, in very few cases do they also remember to provide handholds. Even though modularized, many of the modules may still be difficult to pick up or set down with a minimum possibility of damage or injury to the mover.

Certain items of furniture and/or appliances often need to be leveled in order for them to operate properly or even to keep them from rocking on an uneven floor. The standard practice of providing a leveling screw is not considered good human engineering. In the first place, it creates an awkward working position. Second, most people do not understand the system and do not know which way to turn the screw or bolt. Third, to make the setting secure, a wrench is required, and it is impossible to arrive at a proper adjustment without much trial and error. A leveling system should be designed that can be operated by the foot.

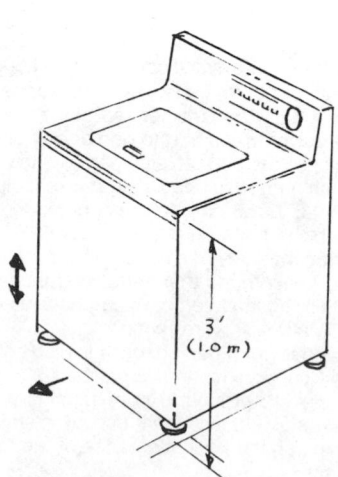

3'
(1.0 m)

## SPECIAL SAFETY CONSIDERATIONS IN THE DESIGN AND SELECTION OF FURNITURE AND APPLIANCES

Avoid sharp corners and edges.

Avoid cabinet hardware (i.e., handles) with sharp corners or edges.

Provide drawer stops so that the user cannot inadvertently pull the drawer completely out of the cabinet.

Provide latches to preclude drawers from drifting open.

Place the hinge point of adjacent cabinet doors (or structures) so that, when opening a door, the user will not pinch his or her fingers between the handle and the adjacent structure or door handle.

Provide hand and finger clearance at the base of cabinets or appliances that may be picked up and transported from one place to another.

Provide locks on cabinets that are used to store drugs, medicine, or other products that should be kept from children.

Avoid the use of glass shelves. If they are used, make sure that the glass is thick enough to support the expected loads and/or provide enough support to prevent the glass from breaking.

Eliminate all sharp edges and corners on glass shelves.

If slide-out shelves are provided, make sure that there are mechanical stops to prevent the shelf from being inadvertently pulled all the way out. Provide an edge around a slide-out shelf so that stored objects will not slide off the shelf when it is being manipulated.

Provide framing around all glass doors. Provide friction hinges or other devices to prevent glass doors from swinging shut and breaking.

Avoid glass-top tables without suitable edge frames. If edge frames are not used, make sure that all corners and edges are rounded. A slight color tint is recommended so that everyone can see the glass.

Avoid placing glass doors in the lower portions of any cabinet unless the glass is shatter-proof.

Provide adequate insulation within the surfaces or structure of all appliances that produce heat.

Place controls for heating appliances where children cannot get to them and/or place controls beneath a cover that can be secured from access by children.

Make sure that all handles used on heating equipment are insulated.

Provide proper insulation to preclude shock from any electrical appliance.

Design all electrical appliances so that electrical shorts are not probable under expected use conditions. Provide fuses within the equipment to prevent external electrical hazards (i.e., lightning) from entering the equipment.

Design electrical appliances so that potentially dangerous electrical exposures are inaccessible for the service or repair person.

Provide a visual indication that an appliance is on or off. This includes flush stove-top burners that do not appear to be on.

Provide heavy-duty electrical lead-ins and plugs with sufficient surface area to encourage the use of the plug rather than the wire to disconnect the lead from a wall outlet.

Provide implosion-proof covers over TV display tubes.

Select and/or design freestanding tables, cabinets, stools, etc., to minimize the probability of their tipping over.

Design folding furnishings to preclude pinched fingers as the elements are being retracted or deployed. Make sure that there is an easily recognized catching "feel" or sound to let the user know when lids, legs, etc., are securely fastened.

Use only nonflammable upholstery materials on furniture.

Locate the controls for such appliances as refrigerators, washing machines, and ironers where children are not able to reach them and/or place such controls under well-latched covers.

Design portable packages so that they are not heavy enough to cause strain from carrying, so that the package can be carried easily without injury to the person's fingers, and so that the package content is properly balanced throughout the typical retrieval-transport-deposit cycle. Provide suitably sized and positioned carrying handles for such internal components as equipment chassis.

Provide automatic electrical disconnects for servicing so that when a cabinet drawer is unlatched, all electrical circuitry is shut down (until the technician is ready to perform further testing, in which case provide a separate bypass switch within the cabinet or chassis).

**Safety Suggestions for Common
Furniture and Appliance Problem
Areas**

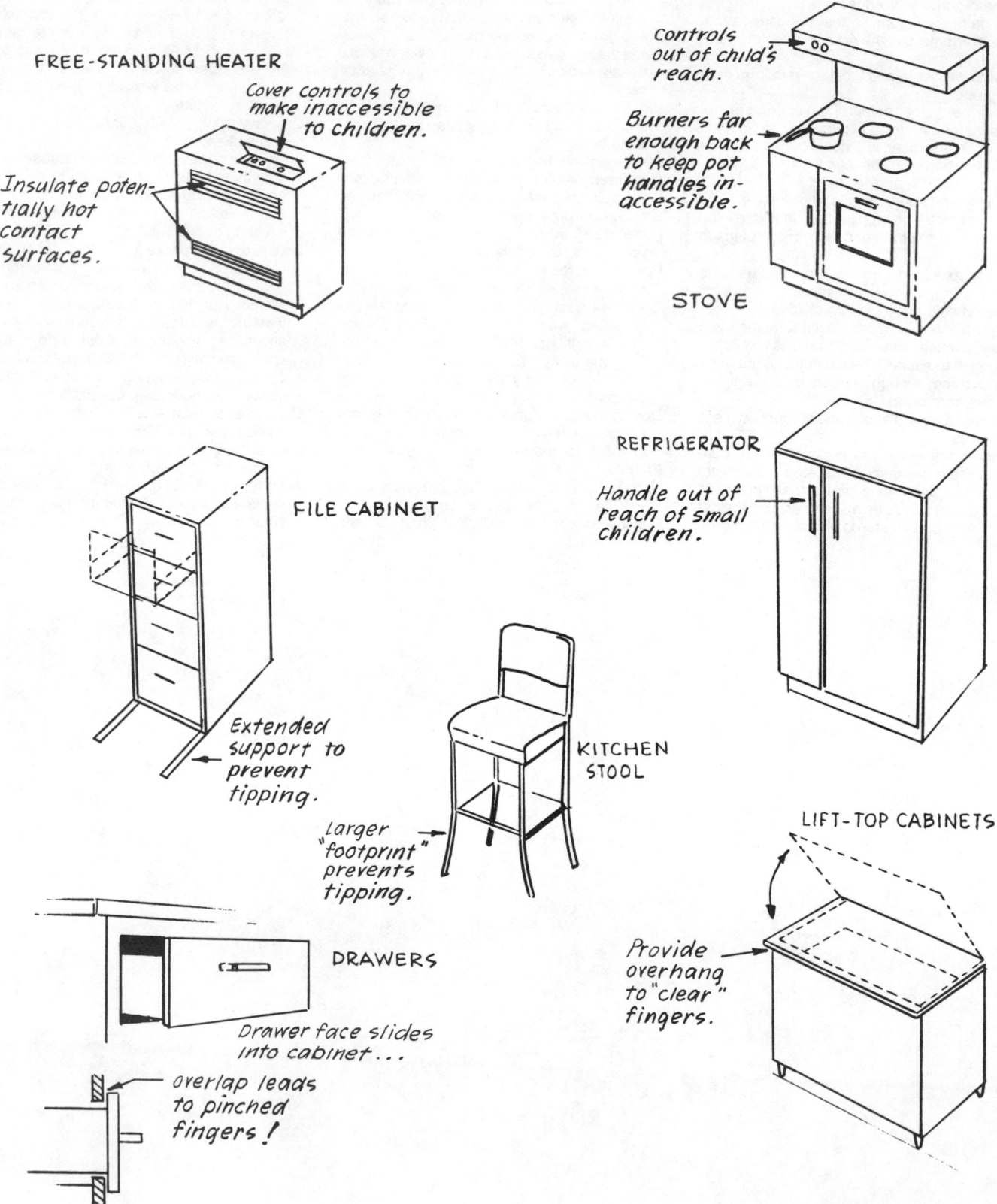

FREE-STANDING HEATER

Cover controls to make inaccessible to children.

Insulate potentially hot contact surfaces.

Controls out of child's reach.

Burners far enough back to keep pot handles inaccessible.

STOVE

FILE CABINET

Extended support to prevent tipping.

REFRIGERATOR

Handle out of reach of small children.

KITCHEN STOOL

Larger "footprint" prevents tipping.

LIFT-TOP CABINETS

Provide overhang to "clear" fingers.

DRAWERS

Drawer face slides into cabinet...

overlap leads to pinched fingers!

# 4

# Human
# Factors
# Data

# INTRODUCTION

It would obviously take a small library to provide all the human factors data required for a full understanding of how the human functions. In previous chapters, an attempt has been made to include design-related interpretations of the way the human interacts with various design situations and to recommend solutions to problems where research and/or experience has provided guidelines.

It would be impossible to anticipate all the possible design problems with which the designer may be confronted. The purpose of this chapter is to acquaint the reader with the basic human factors data that designers commonly need to assist them in seeking their own solutions. The chapter does not provide a complete exposure to basic human factors data; rather, it presents carefully selected information that experience has shown to be related to the majority of design questions. It is hoped that this brief compilation will aid the reader when he or she is unable to find a particular design guideline within an earlier chapter.

## CHAPTER ORGANIZATION

This chapter first defines the basic character-istics of the human body, including its various subsystem responses in a physical sense. This is followed by a series of sections dealing with the more significant human-system interactions, e.g., motor response, work output response, response to the thermal environment, visual response, and auditory response.

## ADDITIONAL REFERENCES

Since this chapter provides only a very limited coverage of human factors data, the following list of additional references has been included. Readers are encouraged to refer to these sources to expand their personal understanding of the various human factors that may be extremely important to their design work.

DAMON, ALBERT, et al.: *The Human Body in Equipment Design,* Harvard University Press, Cambridge, Mass., 1966.

DEGREENE, KENYON B.: *Systems Psychology,* McGraw-Hill Book Company, New York, 1970.

McCORMICK, ERNEST J.: *Human Factors in Engineering Design,* 4th ed., McGraw-Hill Book Company, New York, 1975.

MURRELL, K. F. H.: *Human Performance in Industry,* Reinhold Publishing Corporation, New York, 1965.

OLISHIFSKI, JULIAN B., and FRANK E. McELROY (eds.): *Fundamentals of Industrial Hygiene,* National Safety Council, Chicago, 1971.

PARKER, JAMES F., and VITA R. WEST (eds.): *Bioastronautics Data Book,* 2d ed., National Aeronautics and Space Administration, 1973.

ROEBUCK, JOHN A., et al.: *Engineering Anthropometry Methods,* John Wiley & Sons, Inc., New York, 1975.

STEEN, EDWIN B., and ASHLEY MONTAGU: *Anatomy and Physiology,* vols. I and II, Barnes & Noble, Inc., New York, 1961.

VAN COTT, HAROLD P., and ROBERT G. KINKADE (eds.): *Human Engineering Guide to Equipment Design,* Tam's Books, Inc., Los Angeles, 1972.

WOODSON, WESLEY E., and DONALD W. CONOVER: *Human Engineering Guide for Equipment Designers,* 2d ed., University of California Press, Berkeley, 1964.

# GENERAL BODY CHARACTERISTICS

### 1. Temperature

As indicated in the accompanying sketch, normal body temperatures range from a low skin temperature of about 80°F (26.6°C) to a high internal temperature of about 99.6°F (37.5°C). At about 79.5°F (26.4°C), the heart fails.

### 2. Weight

Adult males normally weigh from about 110 to 215 lb (50 to 97 kg); adult females weigh from about 92 to 175 lb (42 to 79 kg).

The adult brain weight varies from about 38.8 to 60 oz (1100 to 1700 g) for males and from about 37 to 55 oz (198 to 1560 g) for females.

The adult male heart weighs about 10 oz (280 g); the adult female heart weighs about 8 oz (226 g).

Water makes up about 54 to 60 percent of total body weight; muscles, 36 to 42 percent; fat, 18 to 20 percent; and bone, about 18 percent.

Other elementary composition percentages are as follows:

Carbon       50 percent
Oxygen       20 percent
Hydrogen     10 percent
Nitrogen     8.5 percent
Calcium      4 percent
Phosphorus   2.5 percent
Potassium    1 percent
Sulfur       0.8 percent
Sodium       0.4 percent
Chlorine     0.4 percent
Magnesium    0.1 percent
Iron         0.01 percent
Manganese    0.001 percent
Iodine       0.00005 percent
Trace elements—

### 3. Quantity of Blood

Males—1.5 gal (567 cm³); females—0.875 gal (330 cm³)

### 4. Surface Area of the Skin

Males—2.2 m²; females—1.9 m²

### 5. Lung Capacity

Males—4.5 to 9.5 qt; females—3.3 to 5.7 qt

### 6. Air Intake per Breath

Resting: males—0.79 qt; females—0.36 qt
Light work: males—1.77 qt; females—0.91 qt
Heavy work: males—2.15 qt; females—0.93 qt

### 7. Number of Breaths per Minute at Rest

Males—14 to 18; females—20 to 22

The deepest intake (vital capacity) for males is about 5.18 qt; for females it is 3.17 qt.

### 8. Heartbeats per Minute

Sleeping—55 to 60 beats per minute
Awake—70 beats per minute

### 9. Blood Pressure

120/80 mm at age 20

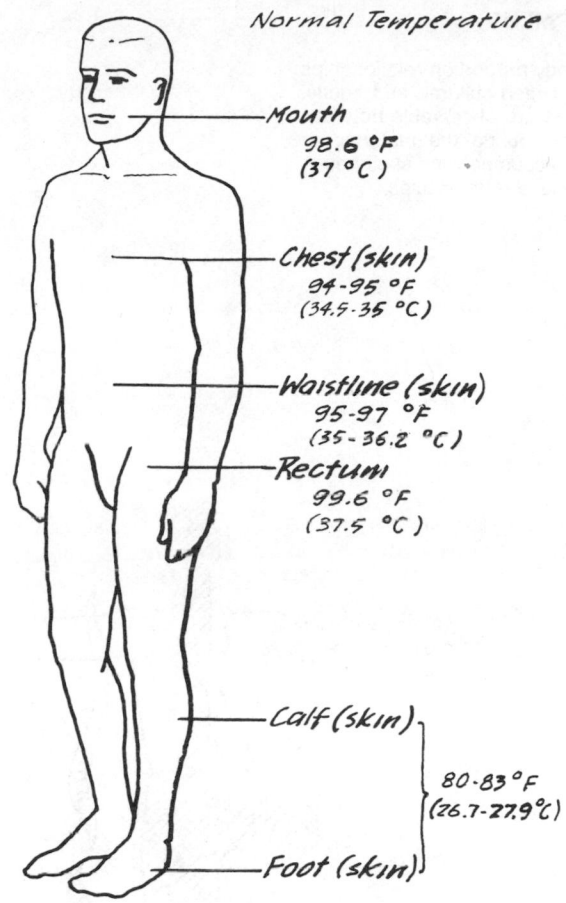

Normal Temperature

Mouth
98.6 °F
(37 °C )

Chest (skin)
94-95 °F
(34.5-35 °C)

Waistline (skin)
95-97 °F
(35-36.2 °C)

Rectum
99.6 °F
(37.5 °C )

Calf (skin)

80-83 °F
(26.7-27.9°C)

Foot (skin)

## BODY DEVELOPMENT

Certain general body proportion relationships
are observable between children and adults.
Similar relationships are observable between
males and females during the maturing or
growth cycle. The accompanying sketches illustrate some of these relationships.

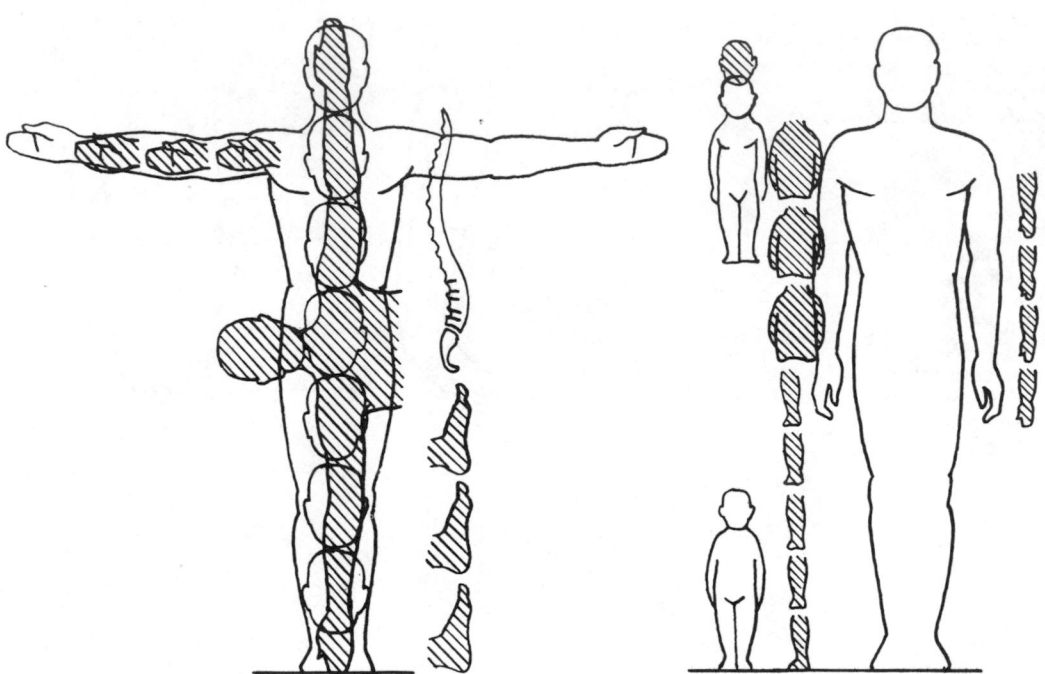

Proportional relationships between body components and between child and adult size.

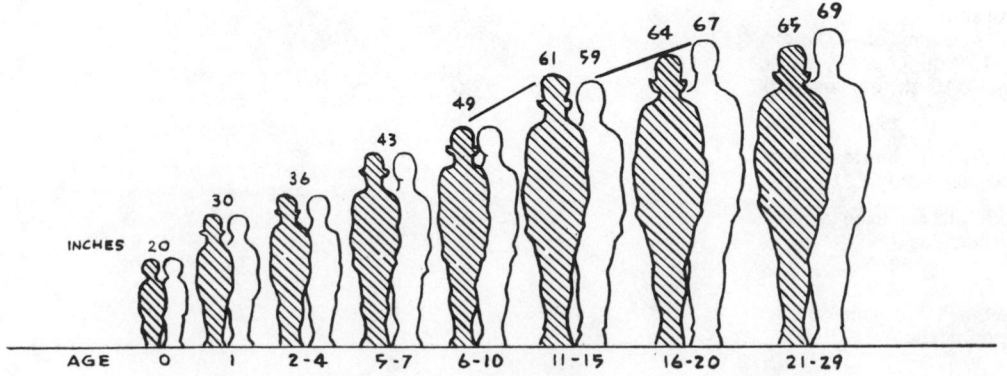

Male-female growth pattern. (Note the reversal that occurs during the years 11 through 20.)

## BODY BUILD

Sometimes referred to as "somatotype," one's dominant body build can be described as either "endomorphy," "mesomorphy," or "ectomorphy." The accompanying sketches illustrate the general differences between these body types.

Endomorphs tend to have a soft, round form with loose and flabby tissue, small bones, and a spherical head. The body has a relatively low density and is physically weak. Mean stature is around 66 in (168 cm), and mean weight is 179 lb (81 kg).

Mesomorphs tend to have a more massive, solid form with a squarish head, large bones, and heavy muscles. They are often muscle-bound and somewhat physically awkward. Mean stature is about 68 in (173 cm), and mean weight is 141 lb (64 kg).

Ectomorphs tend to have a slender body and limbs, a small head, and a small face with delicate features. They generally are more agile. Mean stature is around 69 in (175 cm),

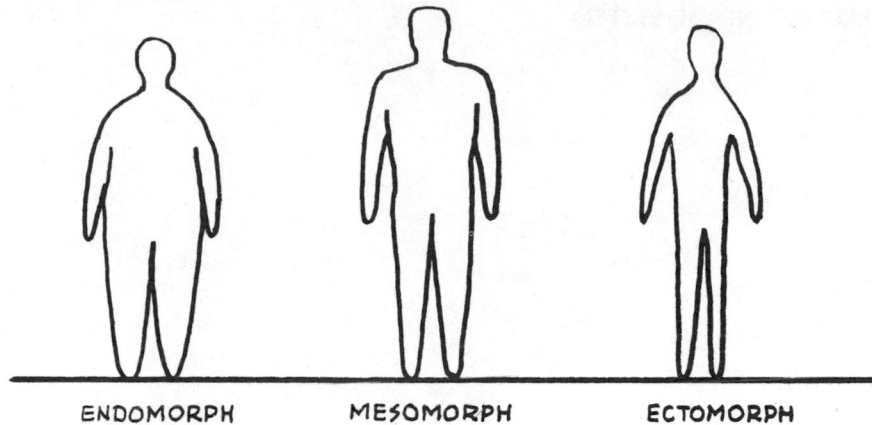

ENDOMORPH　　　MESOMORPH　　　ECTOMORPH

and mean weight is 141 lb (64 kg).

The principal value in recognizing differences in body build is that accommodating the human body requires careful consideration of more than just size. That is, in addition to differences in need for clearance, one must sometimes account for variations in mobility and agility.

## BODY SURFACE AREA

### Body Surface Area in Relation to Stature and Weight[1]
**Area, m²**

| Body Weight, kg | Height, cm | | | | | | | | | | | | | | | | | | | | | | | | |
|---|---|---|---|---|---|---|---|---|---|---|---|---|---|---|---|---|---|---|---|---|---|---|---|---|---|
| | 20 | 30 | 40 | 50 | 60 | 70 | 80 | 90 | 100 | 110 | 120 | 130 | 140 | 150 | 160 | 170 | 180 | 190 | 200 | 210 | 220 | 230 | 240 | 250 | 260 |
| 5 | .18 | .20 | .23 | .26 | .29 | .33 | .37 | .42 | .48 | .55 | .62 | | | | | | | | | | | | | | |
| 10 | | .35 | .36 | .38 | .41 | .44 | .48 | .52 | .57 | .64 | .69 | .76 | | | | | | | | | | | | | |
| 15 | | | | .54 | .57 | .60 | .63 | .67 | .72 | .77 | .83 | .89 | | | | | | | | | | | | | |
| 20 | | | | | | .68 | .72 | .76 | .80 | .85 | .91 | .97 | 1.03 | | | | | | | | | | | | |
| 25 | | | | | | | .80 | .84 | .88 | .93 | .98 | 1.03 | 1.09 | 1.15 | | | | | | | | | | | |
| 30 | | | | | | | | .92 | .96 | 1.01 | 1.05 | 1.10 | 1.16 | 1.22 | 1.28 | | | | | | | | | | |
| 35 | | | | | | | | | 1.04 | 1.08 | 1.12 | 1.17 | 1.23 | 1.29 | 1.35 | 1.42 | | | | | | | | | |
| 40 | | | | | | | | | | 1.11 | 1.15 | 1.20 | 1.25 | 1.30 | 1.36 | 1.42 | 1.48 | 1.55 | | | | | | | |
| 45 | | | | | | | | | | | 1.23 | 1.27 | 1.32 | 1.37 | 1.43 | 1.48 | 1.54 | 1.61 | | | | | | | |
| 50 | | | | | | | | | | | 1.30 | 1.34 | 1.39 | 1.44 | 1.49 | 1.54 | 1.60 | 1.67 | 1.74 | | | | | | |
| 55 | | | | | | | | | | | 1.37 | 1.42 | 1.46 | 1.50 | 1.55 | 1.61 | 1.67 | 1.73 | 1.80 | | | | | | |
| 60 | | | | | | | | | | | 1.44 | 1.48 | 1.52 | 1.57 | 1.62 | 1.67 | 1.73 | 1.79 | 1.85 | 1.92 | | | | | |
| 65 | | | | | | | | | | | | 1.54 | 1.58 | 1.63 | 1.68 | 1.73 | 1.79 | 1.85 | 1.91 | 1.97 | | | | | |
| 70 | | | | | | | | | | | | 1.61 | 1.65 | 1.70 | 1.75 | 1.80 | 1.85 | 1.91 | 1.96 | 2.02 | 2.08 | | | | |
| 75 | | | | | | | | | | | | 1.68 | 1.72 | 1.76 | 1.81 | 1.86 | 1.91 | 1.96 | 2.02 | 2.07 | 2.13 | | | | |
| 80 | | | | | | | | | | | | 1.74 | 1.78 | 1.82 | 1.86 | 1.91 | 1.96 | 2.02 | 2.07 | 2.13 | 2.18 | 2.25 | | | |
| 85 | | | | | | | | | | | | 1.81 | 1.84 | 1.88 | 1.92 | 1.97 | 2.02 | 2.07 | 2.13 | 2.18 | 2.24 | 2.31 | | | |
| 90 | | | | | | | | | | | | 1.87 | 1.90 | 1.94 | 1.98 | 2.03 | 2.08 | 2.13 | 2.18 | 2.24 | 2.30 | 2.36 | | | |
| 95 | | | | | | | | | | | | | 1.97 | 2.01 | 2.05 | 2.09 | 2.14 | 2.18 | 2.24 | 2.30 | 2.36 | 2.42 | 2.48 | | |
| 100 | | | | | | | | | | | | | 2.03 | 2.07 | 2.12 | 2.16 | 2.20 | 2.24 | 2.30 | 2.35 | 2.41 | 2.47 | 2.54 | | |
| 105 | | | | | | | | | | | | | 2.10 | 2.14 | 2.18 | 2.22 | 2.26 | 2.31 | 2.35 | 2.41 | 2.47 | 2.53 | 2.60 | | |
| 110 | | | | | | | | | | | | | 2.17 | 2.21 | 2.24 | 2.28 | 2.32 | 2.36 | 2.41 | 2.47 | 2.53 | 2.58 | 2.65 | 2.73 | |
| 115 | | | | | | | | | | | | | 2.23 | 2.27 | 2.30 | 2.33 | 2.38 | 2.42 | 2.47 | 2.53 | 2.58 | 2.64 | 2.71 | 2.78 | |
| 120 | | | | | | | | | | | | | | 2.33 | 2.36 | 2.39 | 2.43 | 2.48 | 2.53 | 2.58 | 2.63 | 2.70 | 2.77 | 2.84 | 2.93 |
| 125 | | | | | | | | | | | | | | 2.39 | 2.42 | 2.45 | 2.49 | 2.53 | 2.58 | 2.63 | 2.69 | 2.76 | 2.83 | 2.90 | 2.97 |
| 130 | | | | | | | | | | | | | | 2.44 | 2.47 | 2.51 | 2.54 | 2.59 | 2.63 | 2.68 | 2.75 | 2.82 | 2.88 | 2.95 | 3.02 |
| 135 | | | | | | | | | | | | | | 2.50 | 2.53 | 2.56 | 2.60 | 2.64 | 2.69 | 2.74 | 2.81 | 2.87 | 2.93 | 3.00 | 3.08 |
| 140 | | | | | | | | | | | | | | 2.55 | 2.58 | 2.62 | 2.66 | 2.70 | 2.74 | 2.80 | 2.87 | 2.93 | 2.98 | 3.06 | |
| 145 | | | | | | | | | | | | | | 2.61 | 2.63 | 2.67 | 2.71 | 2.75 | 2.80 | 2.86 | 2.92 | 2.98 | 3.04 | | |
| 150 | | | | | | | | | | | | | | 2.66 | 2.69 | 2.73 | 2.77 | 2.81 | 2.86 | 2.92 | 2.97 | 3.03 | 3.09 | | |
| 155 | | | | | | | | | | | | | | 2.72 | 2.74 | 2.78 | 2.83 | 2.87 | 2.92 | 2.97 | 3.03 | 3.08 | | | |
| 160 | | | | | | | | | | | | | | 2.77 | 2.80 | 2.83 | 2.88 | 2.92 | 2.97 | 3.02 | 3.08 | | | | |
| 165 | | | | | | | | | | | | | | | 2.86 | 2.89 | 2.93 | 2.97 | 3.02 | 3.07 | | | | | |
| 170 | | | | | | | | | | | | | | | 2.91 | 2.94 | 2.98 | 3.03 | 3.07 | | | | | | |
| 175 | | | | | | | | | | | | | | | 2.96 | 2.99 | 3.03 | 3.08 | | | | | | | |
| 180 | | | | | | | | | | | | | | | 3.01 | 3.04 | 3.08 | | | | | | | | |
| 185 | | | | | | | | | | | | | | | 3.06 | 3.09 | | | | | | | | | |

[1] Albert Damon et al., *The Human Body in Equipment Design*, Harvard University Press, Cambridge, Mass., 1966, p. 155. Area in square meters.

SKELETAL LINKAGE SYSTEM

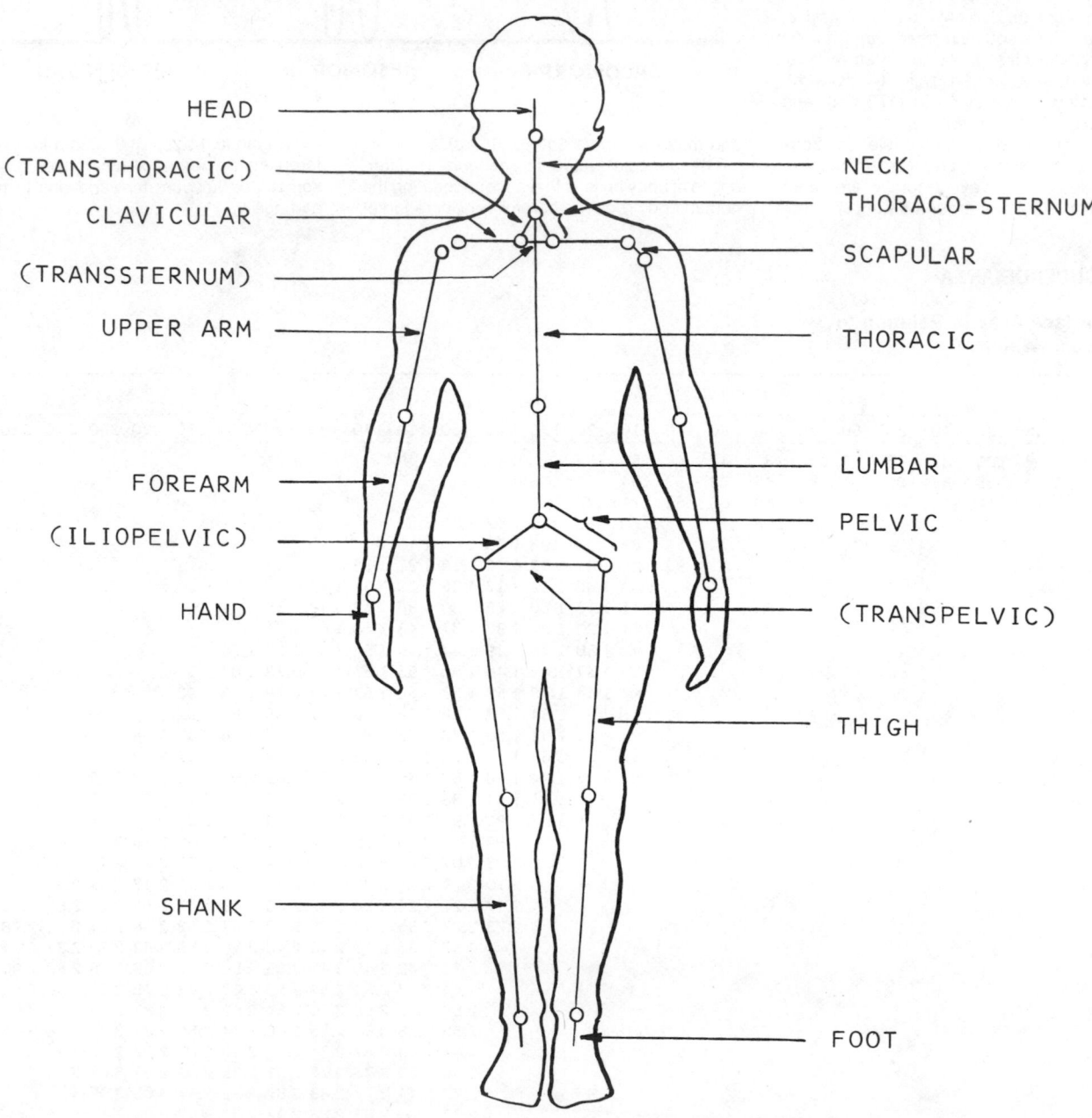

HEAD

(TRANSTHORACIC)

CLAVICULAR

(TRANSSTERNUM)

UPPER ARM

FOREARM

(ILIOPELVIC)

HAND

SHANK

NECK

THORACO-STERNUM

SCAPULAR

THORACIC

LUMBAR

PELVIC

(TRANSPELVIC)

THIGH

FOOT

## SKELETAL NOMENCLATURE AND CHARACTERISTICS

The skeleton is the basic support structure for the body, and thus it is an important consideration when designing equipment that interfaces with the body.

The dry, fat-free adult male skeleton weighs approximately 8 lb (3.6 kg); the adult female skeleton weighs approximately 6 lb (2.7 kg).

Selected skeletal nomenclature is shown in the accompanying sketches. These are the terms that have been found to be most useful when working with special scientific references dealing with typical body functions and equipment design.

In addition, certain terms are commonly used in connection with particular design problems; e.g., the lumbar vertebra is a key reference when working with seating design.

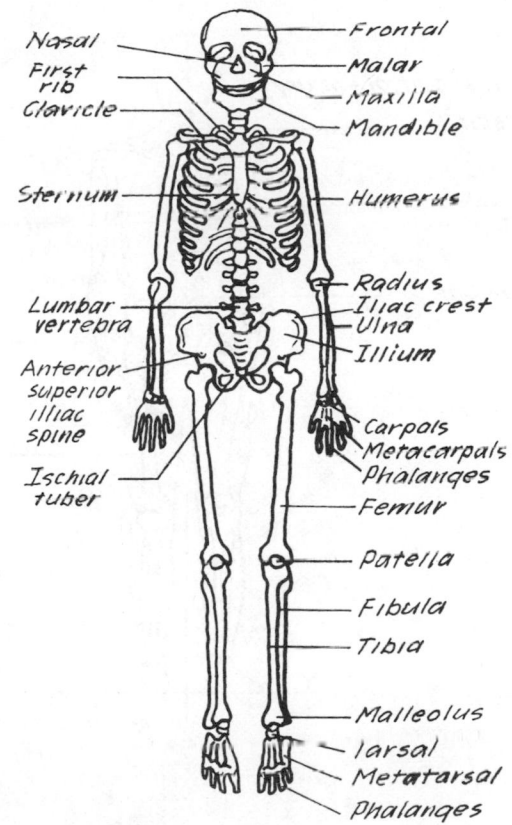

Of the total weight of the skeleton, the skull makes up about 20 percent; the ribs, 18 percent; the arms, 18 percent; and the legs, about 45 percent.

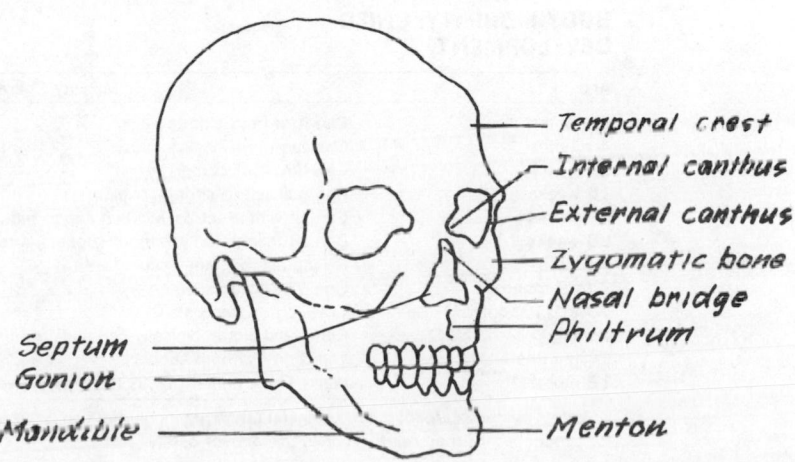

## BODY MOBILITY: HINGE POINTS AND CENTERS OF GRAVITY

Mobility and posture control depend on the skeletal hinge-point locations and the weight and centers of gravity of various body components. The accompanying table and illustration present key hinge-point and c/g locations and the approximate weight of key body segments.

|  | Weight, lb (kg) | Percent of Total Weight |
|---|---|---|
| Head | 10.7 (4.86) | 6.9 |
| Trunk and neck | 70.7 (32.1) | 46.1 |
| Upper arms | 10.1 (4.59) | 6.6 |
| Lower arms | 6.4 (2.9) | 4.2 |
| Hands | 2.6 (1.18) | 1.7 |
| Upper legs | 33.0 (14.98) | 21.5 |
| Lower legs | 14.7 (6.67) | 9.6 |
| Feet | 5.2 (2.36) | 3.4 |

*Note:* Example total—153.4     100.0

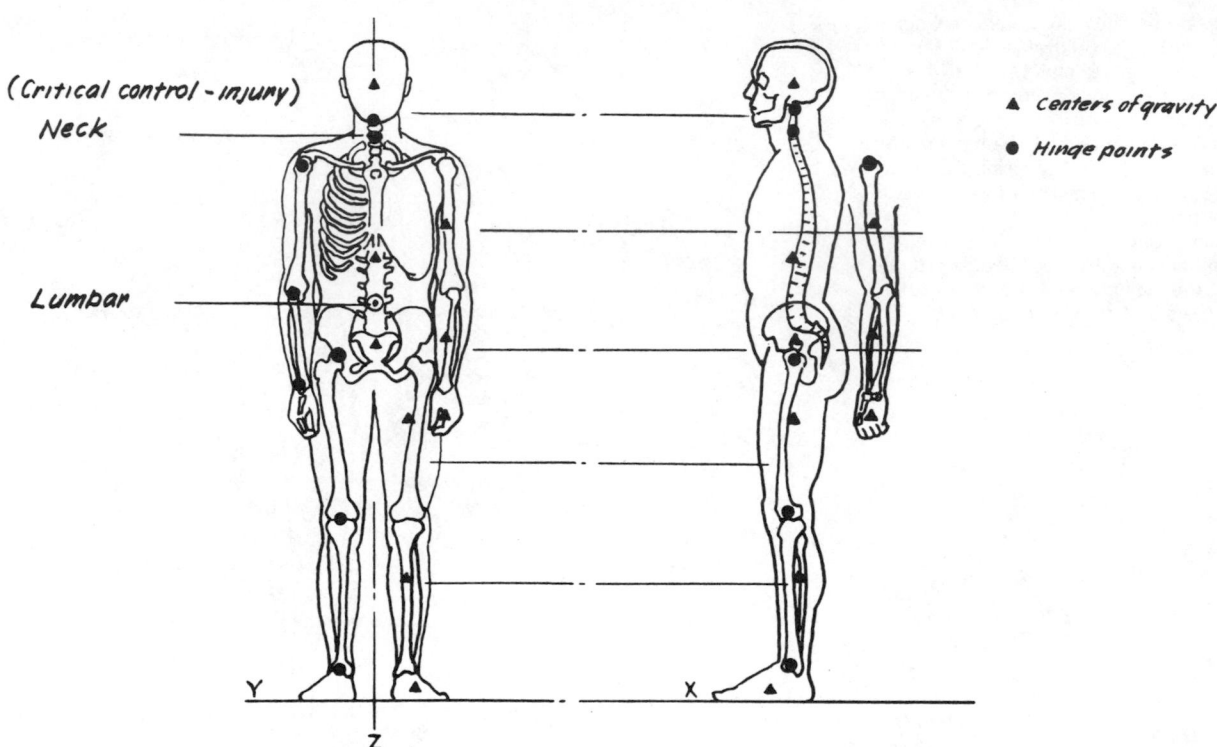

(Critical control - injury)
Neck

Lumbar

▲ Centers of gravity

● Hinge points

## BODY MOBILITY: CHILD DEVELOPMENT

| Age | Mobility Characteristic |
|---|---|
| 2 months | Can turn from side to back. |
| 4 months | Can turn from back to side. |
| 6 months | Can turn over completely. |
| 16 weeks | Can pull self to sitting position. |
| 20 weeks | Can sit with erect back, when supported. |
| 28 weeks | Can sit momentarily without support, after being placed in a sitting position. |
| 6 months | Rolling and hitching occur. |
| 7 to 9 months | Creeping occurs. |
| 10 to 11 months | Crawls; can stand alone. |
| 1 year | Can stand alone; typically can shift from merely standing to walking. |
| 14 months | Walks well without support. |
| 18 months | Walks like an adult (i.e., is not stiff-legged). |

*Note:* The above are, of course, general approximations that can be used to establish operational conditions for products that may relate to child use and/or safety.

**TYPICAL LIMITS OF BODY MOVEMENTS[2]**

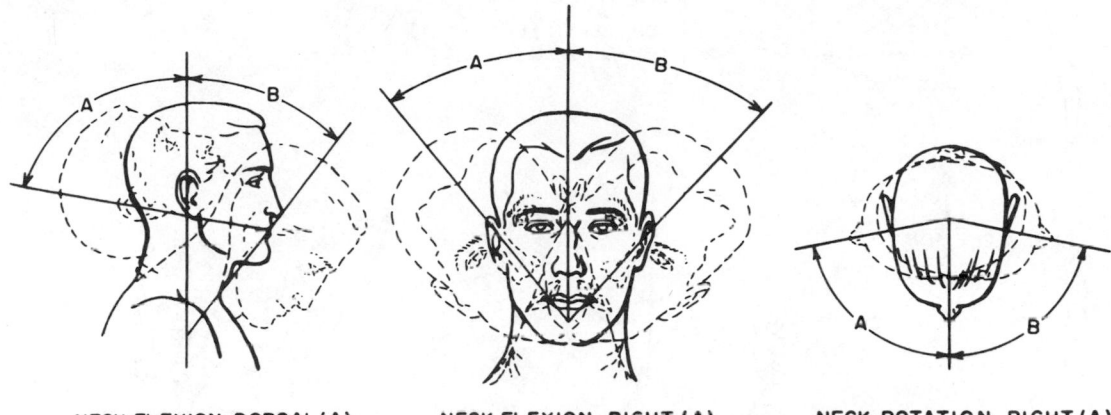

NECK FLEXION, DORSAL (A), VENTRAL (B)

NECK FLEXION, RIGHT (A), LEFT (B)

NECK ROTATION, RIGHT (A), LEFT (B)

Other body movement characteristics are illustrated by the accompanying sketches and tables. Although joint mobility decreases only slightly with age for healthy individuals between the ages of 20 and 60, the incidence of arthritis beyond age 45 suggests that designers give serious consideration to this limiting factor.

**Range of Movement at the Neck Joint***

|  | Average | SD |
|---|---|---|
| Ventral flexion | 60° | 12 |
| Dorsal flexion | 61° | 27 |
| Right-left flexion | 41° | 7 |
| Right-left rotation | 79° | 14 |

*Male civilians.

ANTHROPOMETRY

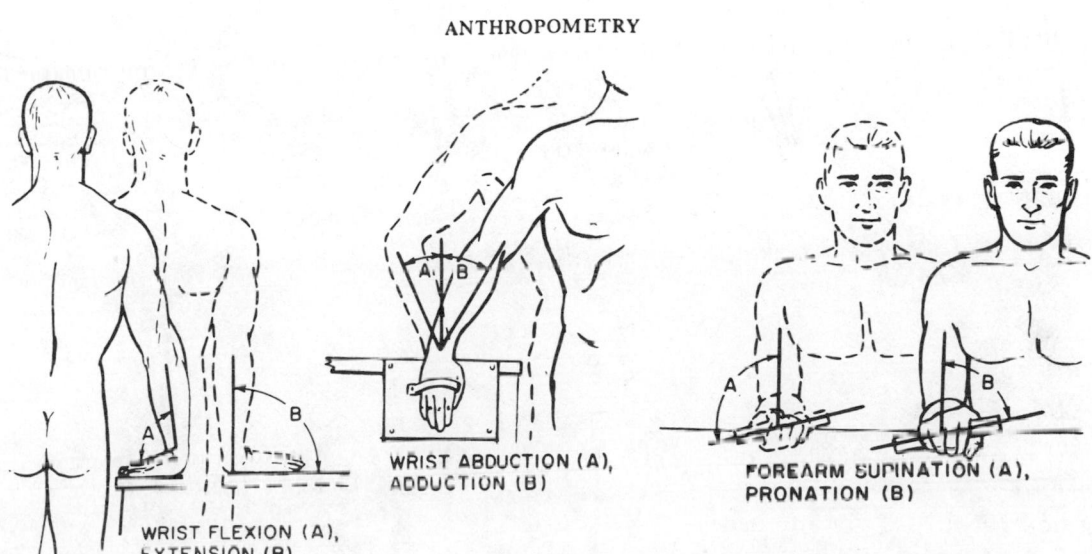

WRIST FLEXION (A), EXTENSION (B)

WRIST ABDUCTION (A), ADDUCTION (B)

FOREARM SUPINATION (A), PRONATION (B)

[2]C. T. Morgan et al. (eds.), *Human Engineering Guide to Equipment Design*, McGraw-Hill Book Company, New York, 1963, pp. 553–555.

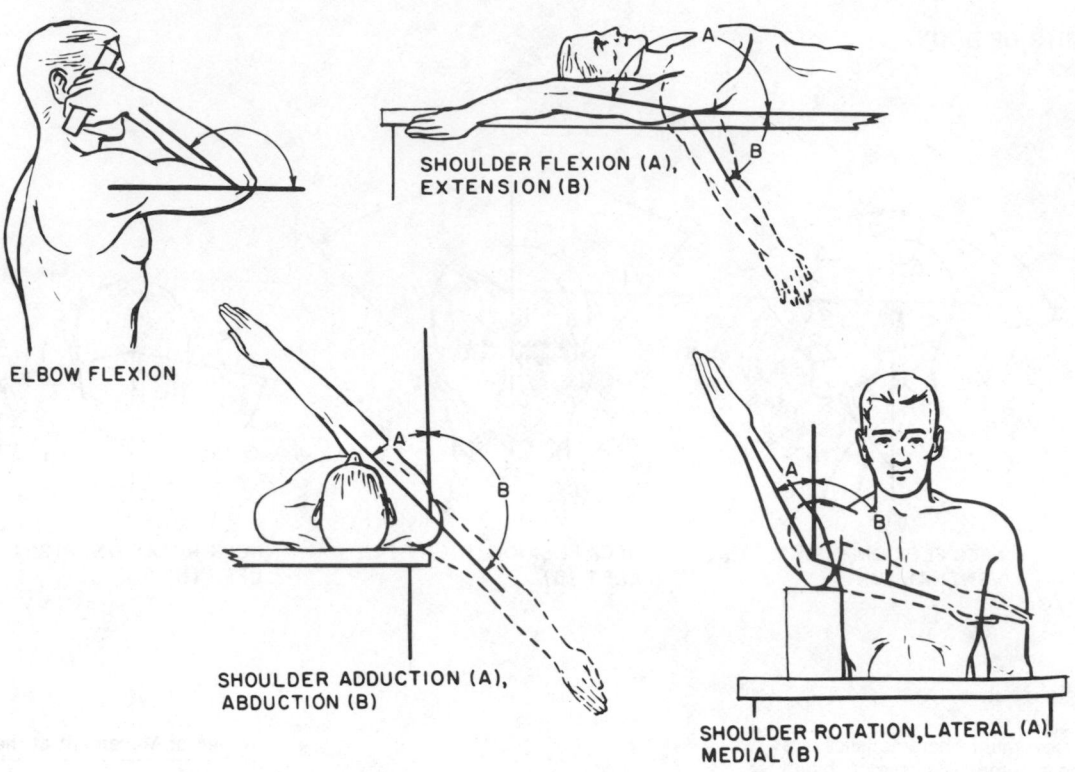

ELBOW FLEXION

SHOULDER FLEXION (A),
EXTENSION (B)

SHOULDER ADDUCTION (A),
ABDUCTION (B)

SHOULDER ROTATION, LATERAL (A),
MEDIAL (B)

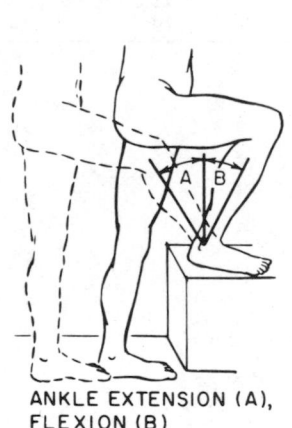

ANKLE EXTENSION (A),
FLEXION (B)

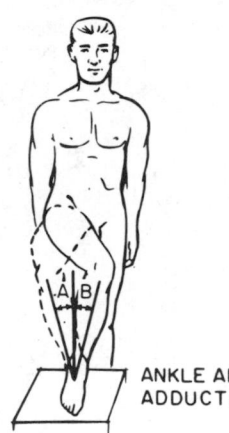

ANKLE ABDUCTION (A),
ADDUCTION (B)

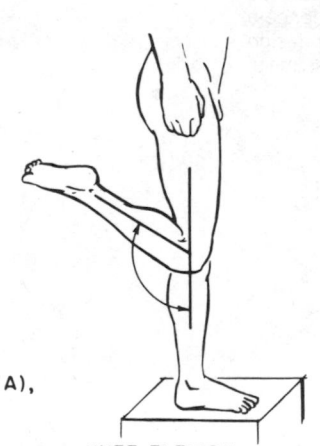

KNEE FLEXION,
STANDING

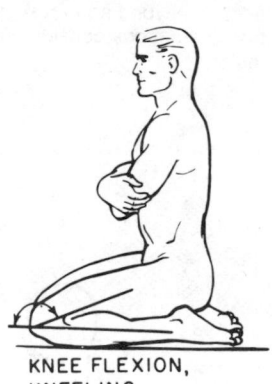

KNEE FLEXION,
KNEELING

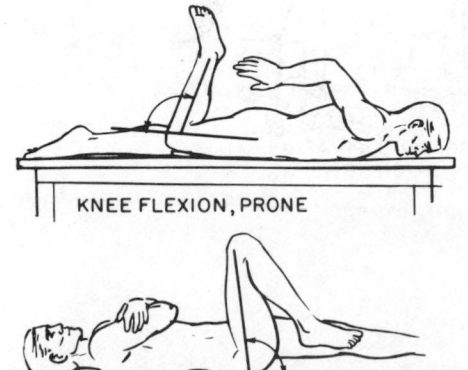

KNEE FLEXION, PRONE

HIP FLEXION

KNEE ROTATION,
MEDIAL (A),
LATERAL (B)

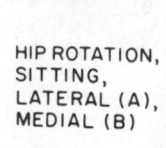

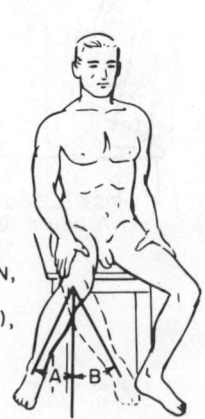

HIP ROTATION,
SITTING,
LATERAL (A),
MEDIAL (B)

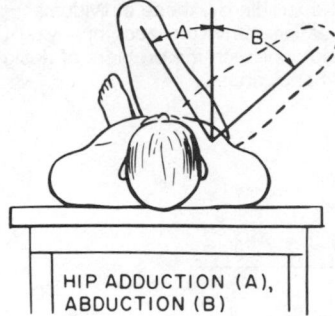

HIP ADDUCTION (A),
ABDUCTION (B)

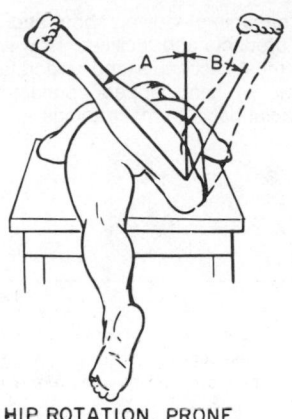

HIP ROTATION, PRONE,
MEDIAL (A),
LATERAL (B)

### Range of Movement at the Joints of the Hand and Arm of Male Air Force Personnel

| Movement [2] | Range (deg) | |
|---|---|---|
| | Avg. | S.D. |
| Wrist flexion | 90 | 12 |
| Wrist extension | 99 | 13 |
| Wrist adduction | 27 | 9 |
| Wrist abduction | 47 | 7 |
| Forearm supination | 113 | 22 |
| Forearm pronation | 77 | 24 |
| Elbow flexion | 142 | 10 |
| Shoulder flexion | 188 | 12 |
| Shoulder extension | 61 | 14 |
| Shoulder adduction | 48 | 9 |
| Shoulder abduction | 134 | 17 |

*Note:* Females generally exceed males with respect to the range of joint movements, except in the case of the knee joint (see the accompanying table). Ordinarily, fat men and women have less movement capability than slender ones. People with muscular body builds have less than those with average builds; however, because of the effect of special exercise or training, one should not assume that these generalizations always apply.

### Range of Movement at the Joints of the Foot and Leg of Male Air Force Personnel

| Movement | Range (deg) | |
|---|---|---|
| | Avg. | S.D. |
| Ankle flexion | 35 | 7 |
| Ankle extension | 38 | 12 |
| Ankle adduction | 24 | 9 |
| Ankle abduction | 23 | 7 |
| Knee flexion | | |
| Standing | 113 | 13 |
| Kneeling | 159 | 9 |
| Prone | 125 | 10 |
| Knee rotation | | |
| Medial | 35 | 12 |
| Lateral | 43 | 12 |
| Hip flexion | 113 | 13 |
| Hip adduction | 31 | 12 |
| Hip abduction | 53 | 12 |
| Hip rotation (sitting) | | |
| Medial | 31 | 9 |
| Lateral | 30 | 9 |
| Hip rotation (prone) | | |
| Medial | 39 | 10 |
| Lateral | 34 | 10 |

### Average Increase in Range of Joint Movement of Women over Men

| Movement | Difference (deg) |
|---|---|
| Wrist flexion and extension | 14 |
| Wrist adduction and abduction | 11 |
| Elbow flexion and extension | 8 |
| Shoulder abduction (rearward) | 2 |
| Ankle flexion and extension | 4 |
| Knee flexion and extension | 0 |
| Hip flexion | 3 |

## ORTHOPEDIC HANDICAPS RELATIVE TO MOBILITY

In spite of either congenital or permanent or temporary physical handicaps, many people are still capable of moving about and utilizing various products and facilities. However, designers must recognize and understand the limitations of handicapped consumers and take special pains to make things accessible to, and operable by, these individuals.

The definitions in the accompanying table may help cope with the problem of designing for the handicapped.

| Term | Definition |
|---|---|
| Abduction | The movement of an extremity (arm or leg) away from the middle of the body. |
| Adduction | The movement of an extremity (arm or leg) toward the middle of the body. |
| Amputee | A person who has suffered the loss of a limb. |
| Arthritis | A chronic disease of the joints that may limit limb positioning direction or extent and/or the use of the hand and fingers for gripping. |
| Ataxia | Loss of coordinate movement-position sense as a result of malfunction of the lower part of the brain. |
| Athetosis | Repetitive, involuntary movements due to brain damage (impossible to control even with treatment). |
| Bilateral | On both sides. |
| Cerebral palsy | A complex handicap with variations resulting from brain damage and manifested in poor muscle control and/or paralysis. |
| Congenital | Existing at birth. |
| Cord lesion | A disease of, or injury to, the spinal cord. |
| Diplegia | Paralysis affecting similar parts on both sides of the body. |
| Dysautonomia | A genetic disease affecting the autonomic nervous system. |
| Dystonia musculorum deformans | A rare chronic disease marked by involuntary, irregular contortions of the muscles of the trunk and extremities. Symptoms appear chiefly during walking, at which time the contortions twist the body forward and sideways in a grotesque fashion. |
| Encephalitis | A group of infectious diseases of the central nervous system, characterized by inflammation of brain tissues. |
| Epilepsy | An irregular electrical discharge of the brain, causing a temporary lapse of consciousness (known as a "seizure"). |
| Erb's palsy | Paralysis of a group of muscles of the shoulder and upper arm. The arm hangs limp, the hand rotates inward, and normal movements are lost. |
| Friedrich's ataxia | Progressive paralysis of the lower limbs. Symptoms depend upon the area or areas of the nervous system involved. |
| Hemiparesis | Partial paralysis of one side of the body. |
| Hemiplegia | Complete paralysis of one side of the body. |
| Little's disease | The general term for all forms of cerebral spastic diplegia. |
| Multiple sclerosis | Deterioration of the brain or spinal cord; involves weakness in coordination, strong jerking movements of the arms and legs, euphoria, and scanning speech. |
| Muscular dystrophy | A progressive disease of the muscular tissue, resulting in diminished strength. |
| Paraparesis | Partial paralysis of the legs, such as might be caused by polio. |
| Paraplegia | Total paralysis of the legs due to spinal cord or brain injury. |
| Prosthesis | A biomechanical device fitted to the human body to replace a part that is missing, e.g., an artificial limb. |
| Quadriparesis | Complete paralysis of the body below the arms and some dysfunction in the arms and hands short of paralysis. |
| Rheumatoid arthritis | A systemic disorder resulting in inflammatory changes, atrophy, and rarefaction of the bones. Leads to crippling deformities. |
| Spastic paraplegia (infantile) | Spastic paralysis occurring in early childhood as a result of injuries during birth. |
| Vertebrae cervical | The vertebrae in the neck area of the backbone. (C-6 is the sixth vertebra from the head.) |
| Vertebrae, thoracic | The 12 vertebrae in the chest area of the backbone. (T-11 is the eighteenth vertebra from the head.) |
| Vertebra, lumbar | The fifth vertebra in the abdominal area of the backbone. (L-5 is the twenty-fourth vertebra from the head.) |
| Vertebra, sacral | The "tailbone," which is a combination of the sacrum and coccyx fused together |

*Note:* The commonly used terms "hemiplegia," "paraplegia," and "quadriplegia" are often misused. "Hemiparesis," "paraparesis," or "quadriparesis" is the correct term if some function remains; i.e., "plegia" denotes that no function remains.

## WHOLE-BODY WEIGHT (NUDE)[3]

Whole-body weights for various populations are shown in the accompanying table. Values for only the 5th, 50th and 95th percentiles are provided, since these are the most significant for the majority of design requirements. For additional population percentiles, refer to the references provided in the section on body dimensions.

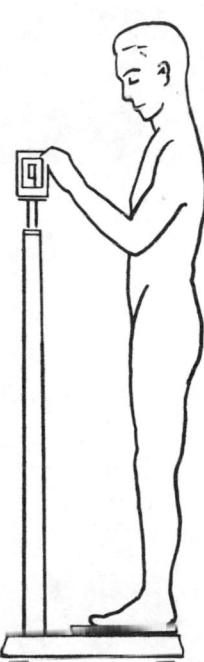

| | Percentile | | |
|---|---|---|---|
| | 5th | 50th | 95th |
| Adults: | | | |
| Males | 124 lb | 168 lb | 224 lb |
| Females | 104 lb | 139 lb | 208 lb |
| Boys: | | | |
| Age 17 | 120 lb | 149 lb | 195 lb |
| Age 14 | 84 lb | 112 lb | 153 lb |
| Age 12 | 67 lb | 84 lb | 121 lb |
| Age 6 | 37 lb | 44 lb | 59 lb |
| Age 2 | 23 lb | 27 lb | 31 lb |
| Girls: | | | |
| Age 17 | 92 lb | 120 lb | 162 lb |
| Age 14 | 81 lb | 108 lb | 146 lb |
| Age 12 | 62 lb | 88 lb | 120 lb |
| Age 6 | 35 lb | 44 lb | 54 lb |
| Age 2 | 22 lb | 24 lb | 31 lb |
| Adults age 70 and over: | | | |
| Males | 105 lb | 144 lb | 196 lb |
| Females | 93 lb | 135 lb | 191 lb |
| Truck and bus drivers: | | | |
| Male | 138 lb | 175 lb | 227 lb |
| Females | 100 lb | 128 lb | 181 lb |
| Airline pilots (Male) | 134 lb | 168 lb | 201 lb |
| Flight attendants (Female) | 102 lb | 117 lb | 133 lb |
| Law enforcement officers: | | | |
| Males | 146 lb | 181 lb | 230 lb |
| Females | — | — | — |

**Average Male Adult Weights for Other Populations**

| Nationality | Weight, lb |
|---|---|
| Chinese | 120 |
| English | 143 |
| Filipino | 118 |
| French | 149 |
| Greek | 143 |
| Hawaiian | 172 |
| Indian | 123 |
| Japanese | 115 |
| Korean | 124 |
| American Indians: | |
| Hopi | 135 |
| Navajo | 139 |
| Zuni | 125 |

Note: 1.0 lb = 0.45 kg.

[3]These data have been compiled from the sources listed at the end of this section.

## PHYSICAL DIMENSIONS

The science of measuring the human body is *anthropometry*. Anthropometrists have been measuring a wide variety of people in many ways for many years. More than 350 different measurements have been taken at one time or another. Although measuring techniques have become more or less standardized, there is no assurance that each anthropometrist measured his or her particular sample of subjects, or each particular dimension, in the same manner. Thus one has to assume a certain caution in comparing data from various populations.

A great share of anthropometric data has been generated for the purpose of scientific comparison of populations, and not necessarily for the purpose of providing design-related information. This is evident from the fact that subjects are measured in stiff and unnatural positions, and in the nude. Also, since anthropometric surveys are extremely expensive, they are not repeated periodically to update information as people and populations change. Because of these and other unique conditions, it is important that the user of anthropometric data recognize the limitations of such data and treat the information merely as a point of departure. That is, anthropometric data are most useful as an aid in selecting a sample of "live" test subjects, who then should be used to test any proposed design.

Unfortunately, not all surveys include all the dimensions one might need for a given design problem. The most complete surveys have almost always been conducted on military personnel, and most of these have dealt with males.

In spite of these shortcomings, designers should make a point of referring to anthropometric data whenever a design involves a "fit" problem. Also, care should be taken to refer to data that are representative of the pertinent user population, since a particular survey may not have examined the dimension in question and thus may provide no data on it. To assist the designer, the following information on body dimensions has been organized according to the population groups in which product designers are most commonly interested. It will be noted that there are some obvious gaps in the tables where data for a given dimension and population are unavailable. Also, it should be pointed out that, in some cases, a substitute value has been entered when (in the author's opinion) a similar population seemed to justify using a particular dimensional value for the population in question. Although such liberty is often frowned upon by conservative anthropometrists, it is excused here on the basis that *none of these data should be used directly for design purposes, but rather as starting points to establish requirements for selecting a dynamic (live) test subject sample.* There is no substitute for a good, full-scale mockup evaluation to validate "fit" of the human body.

## BASIC U.S. ADULT CONSUMER ANTHROPOMETRIC REFERENCE[4]

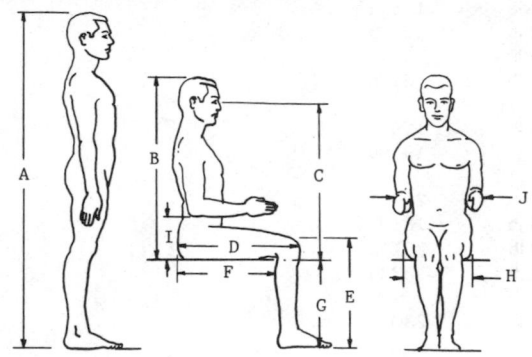

```
A - Standing Height
B - Sitting Height
C - Seated Eye Height
D - Upper Leg Length
E - Knee Height
F - Seat Length
G - Seat Height
H - Seat Width
I - Elbow Rest Height
J - Elbow Room
```

Basic Percentile Dimensions
of U. S. Adult Population
(18 to 79 years)

| %  | A | | B | | C | | D | | E | | F | | G | | H | | I | | J | | Weight | |
|----|------|------|------|------|------|------|------|------|------|------|------|------|------|------|------|------|------|------|------|------|------|------|
| 5  | 63.6 | 59.0 | 33.2 | 30.9 | 28.7 | 27.4 | 21.3 | 20.4 | 19.3 | 17.9 | 17.3 | 17.0 | 15.5 | 14.0 | 12.2 | 12.3 | 7.4 | 7.1 | 13.7 | 12.3 | 126 | 104 |
| 10 | 64.5 | 59.8 | 33.8 | 31.4 | 29.3 | 27.8 | 21.8 | 20.9 | 20.0 | 18.2 | 17.9 | 17.3 | 16.0 | 14.2 | 12.5 | 12.7 | 8.0 | 7.6 | 14.3 | 12.9 | 134 | 111 |
| 20 | 66.0 | 61.1 | 34.4 | 32.2 | 30.0 | 28.4 | 22.3 | 21.3 | 20.4 | 18.6 | 18.4 | 17.9 | 16.4 | 14.7 | 13.1 | 13.3 | 8.5 | 8.2 | 15.0 | 13.5 | 144 | 118 |
| 30 | 66.8 | 61.8 | 34.9 | 32.6 | 30.5 | 28.7 | 22.7 | 21.7 | 20.7 | 19.1 | 18.8 | 18.2 | 16.7 | 15.1 | 13.4 | 13.6 | 8.9 | 8.5 | 15.5 | 14.1 | 152 | 125 |
| 40 | 67.6 | 62.4 | 35.3 | 33.1 | 30.9 | 29.0 | 23.0 | 22.1 | 21.1 | 19.3 | 19.2 | 18.6 | 17.0 | 15.4 | 13.7 | 14.0 | 9.2 | 8.9 | 16.0 | 14.6 | 159 | 131 |
| 50 | 68.3 | 62.9 | 35.7 | 33.4 | 31.3 | 29.3 | 23.3 | 22.4 | 21.4 | 19.6 | 19.5 | 18.9 | 17.3 | 15.7 | 14.0 | 14.3 | 9.5 | 9.2 | 16.5 | 15.1 | 166 | 137 |
| 60 | 68.8 | 63.7 | 36.0 | 33.8 | 31.7 | 29.6 | 23.6 | 22.6 | 21.7 | 19.8 | 19.8 | 19.2 | 17.6 | 16.0 | 14.3 | 14.7 | 9.8 | 9.5 | 17.0 | 15.6 | 173 | 144 |
| 70 | 69.7 | 64.4 | 36.5 | 34.2 | 32.0 | 29.8 | 23.9 | 22.9 | 22.0 | 20.1 | 20.1 | 19.5 | 17.8 | 16.3 | 14.6 | 15.1 | 10.2 | 9.7 | 17.5 | 16.3 | 181 | 152 |
| 80 | 70.6 | 65.1 | 36.9 | 34.6 | 32.5 | 30.2 | 24.4 | 23.4 | 22.4 | 20.5 | 20.5 | 19.9 | 18.2 | 16.6 | 14.9 | 15.6 | 10.6 | 10.1 | 18.1 | 17.1 | 190 | 164 |
| 90 | 71.8 | 66.4 | 37.6 | 35.2 | 33.0 | 30.7 | 24.8 | 24.0 | 22.9 | 21.0 | 21.0 | 20.6 | 18.8 | 17.0 | 15.5 | 16.4 | 11.0 | 10.7 | 19.0 | 18.3 | 205 | 182 |
| 95 | 72.8 | 67.1 | 38.0 | 35.7 | 33.5 | 31.0 | 25.2 | 24.6 | 23.4 | 21.5 | 21.6 | 21.0 | 19.3 | 17.5 | 15.9 | 17.1 | 11.6 | 11.0 | 19.9 | 19.3 | 217 | 199 |

[4]H. W. Stoudt et al., *Weight, Height and Selected Body Dimensions of Adults, United States 1960–1962,* Public Health Service Publication no. 1000, ser. 11, no. 8.

## STANDING HEIGHT (STATURE)[5]

Standing height is the primary indicator by which one selects general test subjects to evaluate various designs when there is no specific aspect of the design pertaining to an individual body component. The dimension is pertinent for adjusting head clearances.

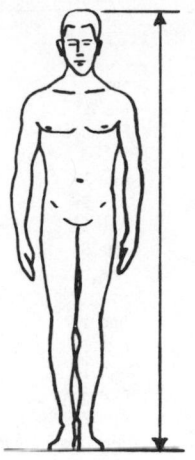

|  | Percentile | | |
|---|---|---|---|
|  | 5th | 50th | 95th |
| Adults: | | | |
| Males | 63.6 in | 68.3 in | 72.8 in |
| Females | 59.0 in | 62.9 in | 67.1 in |
| Boys: | | | |
| Age 17 | 65.1 in | 69.4 in | 72.6 in |
| Age 14 | 58.0 in | 63.2 in | 69.1 in |
| Age 12 | 54.4 in | 58.0 in | 63.5 in |
| Age 6 | 41.8 in | 45.1 in | 49.1 in |
| Age 2 | 31.7 in | 33.2 in | 35.8 in |
| Girls: | | | |
| Age 17 | 60.0 in | 64.1 in | 67.6 in |
| Age 14 | 57.8 in | 62.3 in | 66.5 in |
| Age 12 | 54.1 in | 58.8 in | 63.0 in |
| Age 6 | 41.3 in | 45.0 in | 47.9 in |
| Age 2 | 30.4 in | 33.0 in | 35.4 in |
| Adults age 70 and over: | | | |
| Males | 61.3 in | 66.2 in | 69.5 in |
| Females | 55.3 in | 61.8 in | 64.9 in |
| Truck and bus drivers: | | | |
| Males | 65.1 in | 69.8 in | 74.3 in |
| Females | 58.9 in | 63.5 in | 68.0 in |
| Airline pilots (Male) | 66.0 in | 70.0 in | 73.9 in |
| Flight attendants (Female) | 62.5 in | 65.4 in | 68.8 in |
| Law enforcement officers: | | | |
| Males | 66.6 in | 70.0 in | 74.0 in |
| Females | — | — | — |

## EYE HEIGHT (STANDING)

This dimension is pertinent to the location of visual displays and/or the sizing of visual obstructions, where a small person may have to see over the obstruction or over someone who is taller. As a general "rule of thumb," the eye height of males is about 5.2 in (13 cm) less than their standing height; for females it is about 4.8 in (12.2 cm) less. Similarly, when people stand normally (i.e., with some slump), their eye height lowers by about 1.6 in (4.1 cm) for males and by about 1.2 in (3.0 cm) for females.[6]

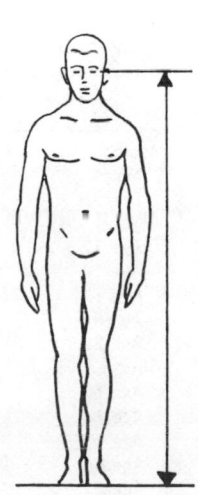

|  | Percentile | | |
|---|---|---|---|
|  | 5th | 50th | 95th |
| Adults: | | | |
| Males | 60.8 in | 64.7 in | 68.6 in |
| Females | 57.3 in | 60.3 in | 65.3 in |
| Boys: | | | |
| Age 17 | — | — | — |
| Age 14 | — | — | — |
| Age 12 | — | — | — |
| Age 6 | — | — | — |
| Age 2 | — | — | — |
| Girls: | | | |
| Age 17 | — | — | — |
| Age 14 | — | — | — |
| Age 12 | — | — | — |
| Age 6 | — | — | — |
| Age 2 | — | — | — |
| Adults age 70 and over: | | | |
| Males | 58.3 in | 61.1 in | 64.9 in |
| Females | 50.8 in | 55.8 in | 60.8 in |
| Truck and bus drivers: | | | |
| Males | 59.6 in | 63.4 in | 67.5 in |
| Females | — | | — |
| Airline pilots (Male) | 61.0 in | 65.0 in | 68.9 in |
| Flight attendants (Female) | 58.0 in | 60.9 in | 64.3 in |
| Law enforcement officers: | | | |
| Males | 61.6 in | 65.0 in | 69.0 in |
| Females | — | — | — |

[5]The data in this table and the following tables have been compiled from the several references cited at the end of this section.
[6]Rules of thumb apply only to adults.

## OVERHEAD REACH (STANDING)

This dimension is pertinent to locating controls that are overhead. It should be used in conjunction with stature because, although a short person must be able to reach a control, it should not be so low that it becomes an obstruction for the taller person. Use this dimension to select subjects for evaluating the accessibility of objects on high shelves.

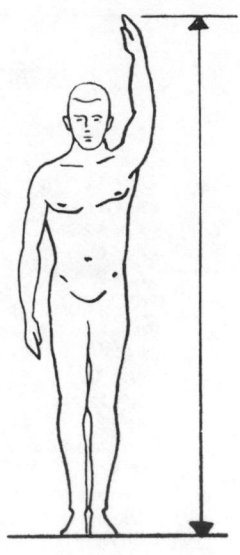

| | Percentile | | |
|---|---|---|---|
| | 5th | 50th | 95th |
| Adults: | | | |
| Males* | 82.0 in | 88.0 in | 94.0 in |
| Females* | 73.0 in | 79.0 in | 86.0 in |
| Boys: | | | |
| Age 17 | 77.7 in | 84.1 in | 88.6 in |
| Age 14 | 70.5 in | 77.3 in | 83.1 in |
| Age 12 | 64.7 in | 70.7 in | 78.9 in |
| Age 6 | 50.3 in | 53.3 in | 58.2 in |
| Age 2 | 38.2 in | 41.7 in | 48.1 in |
| Girls: | | | |
| Age 17 | 72.6 in | 77.7 in | 81.9 in |
| Age 14 | 66.1 in | 75.2 in | 83.1 in |
| Age 12 | 65.0 in | 72.1 in | 77.4 in |
| Age 6 | 48.1 in | 53.0 in | 56.8 in |
| Age 2 | 38.7 in | 41.6 in | 45.6 in |
| Adults age 70 and over: | | | |
| Males | — | — | — |
| Females | 69.5 in | 75.5 in | 82.5 in |
| Truck and bus drivers: | | | |
| Males* | 81.0 in | 88.1 in | 93.7 in |
| Females | — | — | — |
| Airline pilots* (Male) | 85.0 in | 91.5 in | 95.1 in |
| Flight attendants* (Female) | 65.8 in | 68.9 in | 72.3 in |
| Law enforcement officers: | | | |
| Males* | 85.0 in | 89.7 in | 95.2 in |
| Females | — | — | — |

*Estimated.

## FORWARD REACH (STANDING)

This dimension is pertinent to the selection of test subjects who will be evaluating reach conditions within the design situation. This dimension should be used in conjunction with leg-reach dimensions when the design situation calls for a seated operator to operate both hand and foot controls.

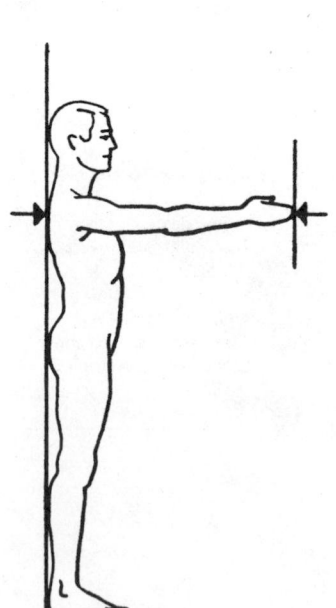

| | Percentile | | |
|---|---|---|---|
| | 5th | 50th | 95th |
| Adults: | | | |
| Males | 31.9 in | 34.6 in | 37.3 in |
| Females | 29.7 in | 31.8 in | 34.1 in |
| Boys: | | | |
| Age 17 | 25.7 in | 28.7 in | 32.0 in |
| Age 14 | 23.8 in | 26.9 in | 29.3 in |
| Age 12 | 22.2 in | 24.2 in | 26.8 in |
| Age 6 | 17.6 in | 19.1 in | 20.9 in |
| Age 2 | 14.0 in | 15.8 in | 18.1 in |
| Girls: | | | |
| Age 17 | 23.9 in | 26.7 in | 28.8 in |
| Age 14 | 22.8 in | 25.5 in | 28.4 in |
| Age 12 | 21.8 in | 24.3 in | 26.4 in |
| Age 6 | 16.6 in | 18.7 in | 20.6 in |
| Age 2 | 13.7 in | 15.1 in | 16.8 in |
| Adults age 70 and over: | | | |
| Males | — | 33.5 in | — |
| Females | — | — | — |
| Truck and bus drivers: | | | |
| Males | 33.0 in | 35.8 in | 38.4 in |
| Females | — | — | — |
| Airline pilots (Male) | 32.9 in | 34.3 in | 37.0 in |
| Flight attendants (Female) | 29.0 in | 31.0 in | 33.3 in |
| Law enforcement officers: | | | |
| Males | 32.3 in | 34.8 in | 37.7 in |
| Females | — | — | — |

## MAXIMUM BODY WIDTH

As indicated by the lack of data below, this measurement is seldom taken during typical anthropometric surveys. However, it is pertinent to lateral clearance requirements in design work. It would be useful, for instance, in selecting test subjects to evaluate lateral corridor clearance.

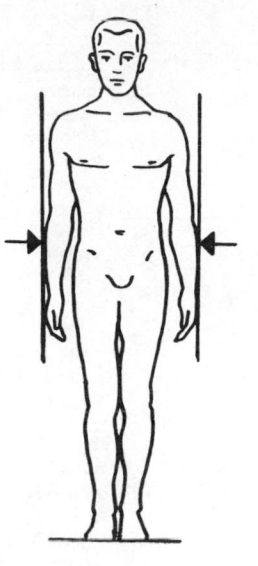

|  | Percentile | | |
|---|---|---|---|
|  | 5th | 50th | 95th |
| Adults: | | | |
| Males* | 18.8 in | 20.9 in | 22.8 in |
| Females | | | |
| Boys: | | | |
| Age 17 | | | |
| Age 14 | | | |
| Age 12 | | | |
| Age 6 | | | |
| Age 2 | | | |
| Girls: | | | |
| Age 17 | | | |
| Age 14 | | | |
| Age 12 | | | |
| Age 6 | | | |
| Age 2 | | | |
| Adults age 70 and over: | | | |
| Males | | | |
| Females | | | |
| Truck and bus drivers: | | | |
| Males | | | |
| Females | | | |
| Airline pilots | | | |
| Flight attendants | | | |
| Law enforcement officers: | | | |
| Males | | | |
| Females | | | |

*Combined Air Force personnel and college students.

## CROTCH HEIGHT

This dimension is a secondary indicator of leg length and is pertinent to design situations in which a person is required to step over an obstacle.

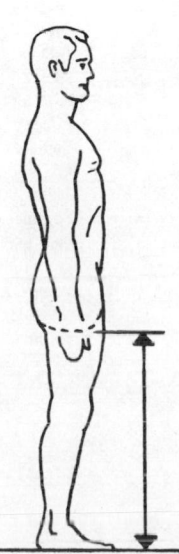

|  | Percentile | | |
|---|---|---|---|
|  | 5th | 50th | 95th |
| Adults: | | | |
| Males | 30.0 in | 33.0 in | 36.1 in |
| Females | 26.8 in | 29.3 in | 32.0 in |
| Boys: | | | |
| Age 17 | — | — | — |
| Age 14 | — | — | — |
| Age 12 | 25.1 in | 27.8 in | 30.7 in |
| Age 6 | 17.9 in | 19.9 in | 21.7 in |
| Age 2 | 11.5 in | 13.4 in | 17.3 in |
| Girls: | | | |
| Age 17 | — | — | — |
| Age 14 | — | — | — |
| Age 12 | 24.9 in | 28.0 in | 30.6 in |
| Age 6 | 17.8 in | 20.2 in | 21.8 in |
| Age 2 | 11.5 in | 13.4 in | 17.3 in |
| Adults age 70 and over: | | | |
| Males | | | |
| Females | | | |
| Truck and bus drivers: | | | |
| Males | | | |
| Females | | | |
| Airline pilots | | | |
| Flight attendants | | | |
| Law enforcement officers: | | | |
| Males | | | |
| Females | | | |

## SITTING HEIGHT

This dimension is pertinent to the establishment of proper overhead clearances for seated persons. It is particularly important in the design of driver work stations.

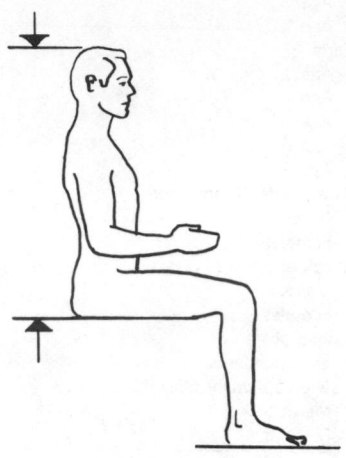

| | Percentile | | |
|---|---|---|---|
| | 5th | 50th | 95th |
| Adults: | | | |
| Males | 33.2 in | 35.7 in | 38.0 in |
| Females | 30.9 in | 33.4 in | 35.7 in |
| Boys: | | | |
| Age 17 | 33.7 in | 35.9 in | 37.8 in |
| Age 14 | 29.4 in | 32.3 in | 35.6 in |
| Age 12 | 28.2 in | 29.9 in | 32.5 in |
| Age 6 | 23.0 in | 25.0 in | 26.8 in |
| Age 2 | 19.0 in | 20.6 in | 21.9 in |
| Girls: | | | |
| Age 17 | 31.8 in | 33.7 in | 35.9 in |
| Age 14 | 29.9 in | 32.4 in | 34.8 in |
| Age 12 | 28.0 in | 30.4 in | 32.9 in |
| Age 6 | 22.8 in | 24.9 in | 26.5 in |
| Age 2 | 18.9 in | 19.8 in | 21.5 in |
| Adults age 70 and over: | | | |
| Males | 31.8 in | 34.3 in | 36.7 in |
| Females | 28.1 in | 32.1 in | 34.8 in |
| Truck and bus drivers: | | | |
| Males | 34.3 in | 36.3 in | 38.2 in |
| Females | — | — | — |
| Airline pilots (Male) | — | — | — |
| Flight attendants (Female) | 32.4 in | 34.3 in | 36.1 in |
| Law enforcement officers: | | | |
| Males | 34.1 in | 36.3 in | 38.5 in |
| Females | — | — | — |

## EYE HEIGHT (SITTING)

This dimension is pertinent to the design of work stations in which visual displays and/or outside viewing requires accommodation of a range of operator sizes. As a rule of thumb the eye height of males is about 5.2 inch (13 cm) less than their sitting height; that of females is about 4.8 in (12.2 cm) less. Similarly, when people sit normally (with some slump), their eye height lowers by about 1.2 in (3.0 cm) for males and by about 1.2 in (3.0 cm) for females.

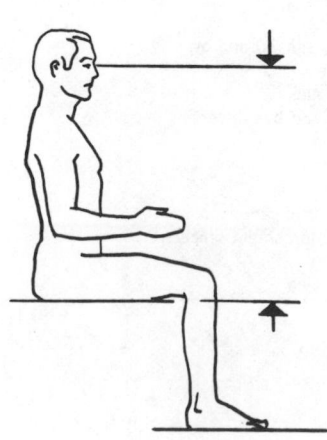

| | Percentile | | |
|---|---|---|---|
| | 5th | 50th | 95th |
| Adults: | | | |
| Males | 28.7 in | 31.3 in | 33.5 in |
| Females | 27.4 in | 29.3 in | 31.0 in |
| Boys: | | | |
| Age 17 | 29.7 in | 31.6 in | 33.2 in |
| Age 14 | 25.2 in | 28.2 in | 30.7 in |
| Age 12 | 23.6 in | 25.4 in | 27.0 in |
| Age 6 | 18.3 in | 20.5 in | 22.2 in |
| Age 2 | 15.9 in | 17.6 in | 18.8 in |
| Girls: | | | |
| Age 17 | 26.9 in | 29.0 in | 31.1 in |
| Age 14 | 25.1 in | 28.0 in | 30.4 in |
| Age 12 | 23.6 in | 25.6 in | 27.6 in |
| Age 6 | 17.8 in | 20.6 in | 22.4 in |
| Age 2 | 15.2 in | 16.7 in | 18.1 in |
| Adults age 70 and over: | | | |
| Males | — | 26.8 in | — |
| Females | — | — | — |
| Truck and bus drivers: | | | |
| Males | 27.7 in | 29.6 in | 31.6 in |
| Females | — | — | — |
| Airline pilots (Male) | 29.4 in | 31.5 in | 33.5 in |
| Flight attendants (Female) | 28.1 in | 29.9 in | 31.7 in |
| Law enforcement officers: | | | |
| Males | 29.1 in | 31.3 in | 33.5 in |
| Females | — | — | — |

## BUTTOCK-TO-POPLITEAL LENGTH

This dimension is pertinent to seat length. Although it is desirable to provide adequate support for the larger person, it is the shorter person who will have the most problems if this dimension is ignored in the test subject sample.

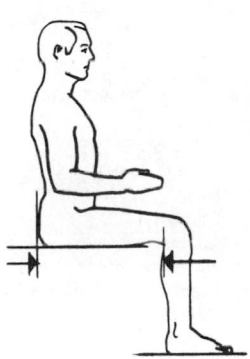

| | Percentile | | |
|---|---|---|---|
| | 5th | 50th | 95th |
| Adults: | | | |
|   Males | 17.3 in | 19.5 in | 21.6 in |
|   Females | 17.0 in | 18.9 in | 21.0 in |
| Boys: | | | |
|   Age 17 | | | |
|   Age 14 | | | |
|   Age 12 | | | |
|   Age 6 | | | |
|   Age 2 | | | |
| Girls: | | | |
|   Age 17 | | | |
|   Age 14 | | | |
|   Age 12 | | | |
|   Age 6 | | | |
|   Age 2 | | | |
| Adults age 70 and over: | | | |
|   Males | 17.0 in | 18.9 in | 21.2 in |
|   Females | 17.0 in | 18.7 in | 20.0 in |
| Truck and bus drivers: | | | |
|   Males | 20.2 in | 22.2 in | 24.0 in |
|   Females | 18.4 in | 20.5 in | 22.3 in |
| Airline pilots | | | |
| Flight attendants (Female) | 17.4 in | 19.0 in | 20.6 in |
| Law enforcement officers: | | | |
|   Males | — | — | — |
|   Females | — | — | — |

## MIDSHOULDER HEIGHT (SITTING)

This dimension is pertinent to the location of controls (i.e., most controls should not be located above the shoulder) and to the location of the anchor point for seat belts (i.e., the belt should not depart aft of the shoulder horizontally or at a negative angle; otherwise, it will create discomfort and tend to push the occupant downward during a frontal-impact crash).

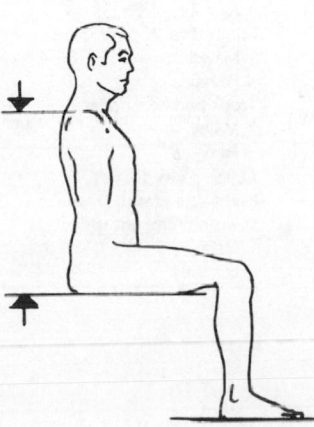

| | Percentile | | |
|---|---|---|---|
| | 5th | 50th | 95th |
| Adults: | | | |
|   Males | 21.0 in | — | 25.0 in |
|   Females | 18.0 in | — | 25.0 in |
| Boys: | | | |
|   Age 17 | — | — | — |
|   Age 14 | 18.1 in | 20.1 in | 21.2 in |
|   Age 12 | 17.8 in | 19.4 in | 21.5 in |
|   Age 6 | 13.7 in | 15.4 in | 17.0 in |
|   Age 2 | 11.1 in | 12.3 in | 13.3 in |
| Girls: | | | |
|   Age 17 | — | — | — |
|   Age 14 | 19.1 in | 21.2 in | 22.6 in |
|   Age 12 | 17.6 in | 19.5 in | 21.0 in |
|   Age 6 | 13.6 in | 15.1 in | 16.5 in |
|   Age 2 | 11.1 in | 12.3 in | 14.0 in |
| Adults age 70 and over: | | | |
|   Males | 20.9 in | 24.0 in | 27.2 in |
|   Females | — | — | — |
| Truck and bus drivers: | | | |
|   Males | 21.2 in | 24.3 in | 27.5 in |
|   Females | — | — | — |
| Airline pilots | — | — | — |
| Flight attendants (Female) | 20.1 in | 22.9 in | 25.7 in |
| Law enforcement officers: | | | |
|   Males | — | — | — |
|   Females | — | — | — |

## BUTTOCK-TO-KNEE LENGTH

This dimension is pertinent to establishing knee clearance for the seated operator. It should be used in conjunction with knee-height and thigh-clearance dimensions.

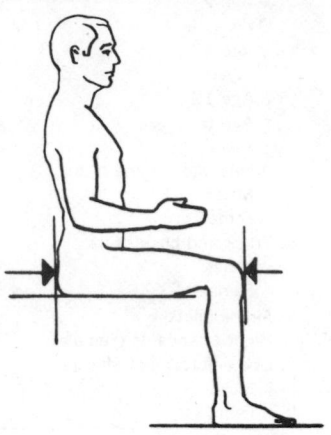

| | Percentile | | |
|---|---|---|---|
| | 5th | 50th | 95th |
| **Adults:** | | | |
| Males | 21.3 in | 23.3 in | 25.2 in |
| Females | 20.4 in | 22.4 in | 24.6 in |
| **Boys:** | | | |
| Age 17 | 21.2 in | 23.1 in | 25.0 in |
| Age 14 | 19.2 in | 21.4 in | 23.5 in |
| Age 12 | 17.7 in | 19.5 in | 21.5 in |
| Age 6 | 12.9 in | 14.1 in | 15.6 in |
| Age 2 | 8.5 in | 9.3 in | 10.5 in |
| **Girls:** | | | |
| Age 17 | 20.0 in | 21.7 in | 23.5 in |
| Age 14 | 19.4 in | 21.2 in | 23.2 in |
| Age 12 | 17.9 in | 20.1 in | 22.0 in |
| Age 6 | 12.5 in | 14.2 in | 15.4 in |
| Age 2 | 7.4 in | 9.4 in | 10.6 in |
| **Adults age 70 and over:** | | | |
| Males | 21.0 in | 22.6 in | 24.4 in |
| Females | 19.9 in | 22.2 in | 23.9 in |
| **Truck and bus drivers:** | | | |
| Males | 22.7 in | 24.6 in | 26.8 in |
| Females | 20.6 in | 22.9 in | 24.9 in |
| Airline pilots (Male) | 22.0 in | 23.6 in | 25.6 in |
| Flight attendants (Female) | 21.2 in | 22.6 in | 24.2 in |
| **Law enforcement officers:** | | | |
| Males | 22.6 in | 24.2 in | 26.1 in |
| Females | — | — | — |

## POPLITEAL HEIGHT (SITTING)

This dimension is pertinent to the establishment of appropriate seat heights. It is also pertinent to the selection of test subjects who will be used to evaluate the relationships between a vehicle seat and foot controls.

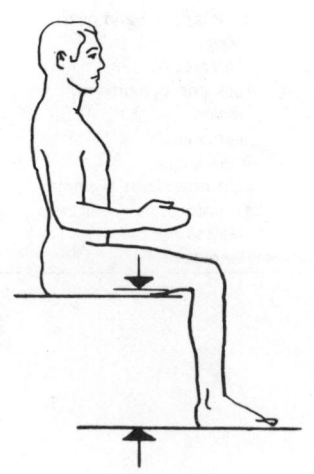

| | Percentile | | |
|---|---|---|---|
| | 5th | 50th | 95th |
| **Adults:** | | | |
| Males | 15.5 in | 17.3 in | 19.3 in |
| Females | 14.0 in | 15.7 in | 17.5 in |
| **Boys:** | | | |
| Age 17 | — | — | — |
| Age 14 | — | — | — |
| Age 12 | 13.2 in | 14.6 in | 16.1 in |
| Age 6 | 10.4 in | 11.5 in | 12.6 in |
| Age 2 | — | — | — |
| **Girls:** | | | |
| Age 17 | — | — | — |
| Age 14 | — | — | — |
| Age 12 | 13.0 in | 14.7 in | 16.3 in |
| Age 6 | 10.2 in | 11.3 in | 12.5 in |
| Age 2 | — | — | — |
| **Adults age 70 and over:** | | | |
| Males | 15.2 in | 16.6 in | 17.9 in |
| Females | 13.5 in | 15.6 in | 17.2 in |
| **Truck and bus drivers:** | | | |
| Males | 15.7 in | 17.5 in | 19.7 in |
| Females | — | — | — |
| Airline pilots (Male) | 15.7 in | 17.0 in | 18.2 in |
| Flight attendants (Female) | 15.9 in | 17.1 in | 18.5 in |
| **Law enforcement officers:** | | | |
| Males | — | — | — |
| Females | — | — | — |

## KNEE HEIGHT (SITTING)

This dimension is pertinent to the establishment of knee clearance. It should be used in conjunction with buttock-to-knee length and thigh-clearance dimensions.

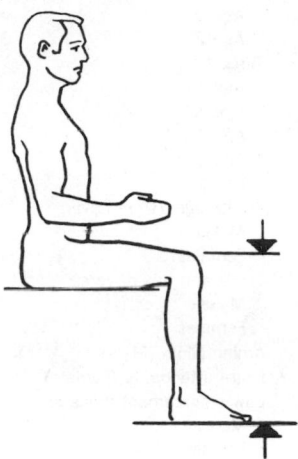

| | Percentile | | |
| --- | --- | --- | --- |
| | 5th | 50th | 95th |
| Adults: | | | |
| Males | 19.3 in | 21.4 in | 23.4 in |
| Females | 17.9 in | 19.6 in | 21.5 in |
| Boys: | | | |
| Age 17 | 19.4 in | 21.8 in | 23.4 in |
| Age 14 | 18.5 in | 20.3 in | 22.6 in |
| Age 12 | 17.2 in | 18.6 in | 20.7 in |
| Age 6 | 12.4 in | 13.9 in | 15.3 in |
| Age 2 | 8.1 in | 9.4 in | 10.2 in |
| Girls: | | | |
| Age 17 | 17.8 in | 19.5 in | 21.2 in |
| Age 14 | 17.7 in | 19.5 in | 21.1 in |
| Age 12 | 17.0 in | 18.6 in | 20.2 in |
| Age 6 | 12.4 in | 13.5 in | 14.8 in |
| Age 2 | 8.0 in | 9.4 in | 10.0 in |
| Adults age 70 and over: | | | |
| Males | 19.0 in | 20.7 in | 22.7 in |
| Females | 17.3 in | 19.4 in | 20.9 in |
| Truck and bus drivers: | | | |
| Males | 20.1 in | 21.7 in | 23.5 in |
| Females | — | — | — |
| Airline pilots (Male) | 20.1 in | 21.7 in | 23.3 in |
| Flight attendants (Female) | 19.1 in | 20.4 in | 21.9 in |
| Law enforcement officers: | | | |
| Males | 20.4 in | 22.0 in | 23.7 in |
| Females | — | — | — |

## THIGH CLEARANCE (SITTING)

This dimension is pertinent to the establishment of the proper clearance between a seat and the lower edge of a desk and to the establishment of the space between the driver's seat and a vehicle steering wheel.

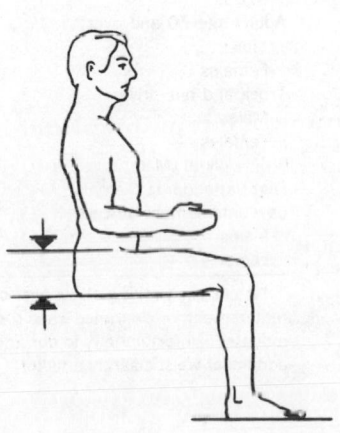

| | Percentile | | |
| --- | --- | --- | --- |
| | 5th | 50th | 95th |
| Adults: | | | |
| Males | 4.3 in | 5.7 in | 6.9 in |
| Females | 4.1 in | 5.4 in | 6.9 in |
| Boys: | | | |
| Age 17 | 4.8 in | 6.1 in | 6.9 in |
| Age 14 | 4.3 in | 5.5 in | 6.4 in |
| Age 12 | 4.2 in | 5.0 in | 5.8 in |
| Age 6 | 3.1 in | 3.6 in | 4.3 in |
| Age 2 | 2.5 in | 3.3 in | 3.9 in |
| Girls: | | | |
| Age 17 | 4.6 in | 5.4 in | 6.8 in |
| Age 14 | 4.6 in | 5.4 in | 6.1 in |
| Age 12 | 4.2 in | 4.9 in | 6.1 in |
| Age 6 | 3.0 in | 3.7 in | 4.5 in |
| Age 2 | 2.1 in | 3.1 in | 3.7 in |
| Adults age 70 and over: | | | |
| Males | 4.1 in | 5.2 in | 6.6 in |
| Females | 4.0 in | 5.2 in | 6.5 in |
| Truck and bus drivers: | | | |
| Males | 5.0 in | 5.9 in | 6.9 in |
| Females | — | — | — |
| Airline pilots (Male) | 4.8 in | 5.6 in | 6.5 in |
| Flight attendants (Female) | — | — | — |
| Law enforcement officers: | | | |
| Males | — | — | — |
| Females | — | — | — |

## ELBOW-TO-FINGERTIP LENGTH

This dimension is pertinent to special problems such as evaluating whether a control may have been placed too close to the operator, with the result that (because of the seat) the operator cannot get his or her arm back far enough to use the control properly.

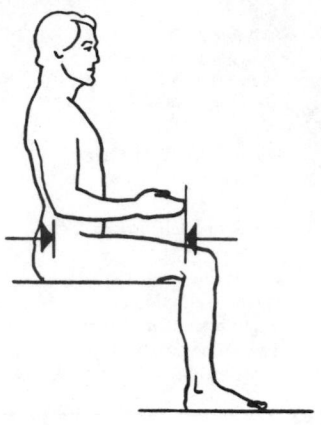

|  | Percentile | | |
|  | 5th | 50th | 95th |
| --- | --- | --- | --- |
| Adults: | | | |
|   Males | — | — | — |
|   Females | — | — | — |
| Boys: | | | |
|   Age 17 | 17.5 in | 18.7 in | 20.0 in |
|   Age 14 | 15.6 in | 17.3 in | 19.1 in |
|   Age 12 | 14.4 in | 15.6 in | 17.3 in |
|   Age 6 | 10.9 in | 12.0 in | 13.2 in |
|   Age 2 | 7.6 in | 8.9 in | 9.9 in |
| Girls: | | | |
|   Age 17 | 15.5 in | 16.8 in | 18.0 in |
|   Age 14 | 15.2 in | 16.5 in | 17.9 in |
|   Age 12 | 14.0 in | 15.7 in | 17.3 in |
|   Age 6 | 10.7 in | 11.7 in | 12.8 in |
|   Age 2 | 7.7 in | 8.8 in | 9.4 in |
| Adults age 70 and over: | | | |
|   Males | 17.2 in | 18.3 in | 19.5 in |
|   Females | — | — | — |
| Truck and bus drivers: | | | |
|   Males | 17.4 in | 18.8 in | 20.4 in |
|   Females | 15.1 in | 16.7 in | 18.2 in |
| Airline pilots (Male) | 17.6 in | 18.9 in | 20.2 in |
| Flight attendants (Female) | 16.0 in | 17.2 in | 18.3 in |
| Law enforcement officers: | | | |
|   Males | — | — | — |
|   Females | — | — | — |

## WAIST DEPTH

This dimension is pertinent to the establishment of clearance requirements between the backrest of a chair or seat and the leading edge of a worktable, desk, or steering wheel.

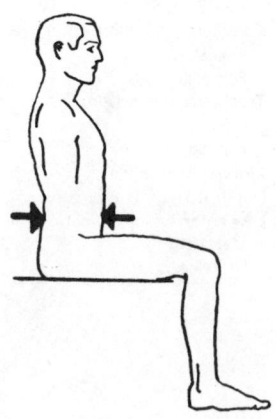

|  | Percentile | | |
|  | 5th | 50th | 95th |
| --- | --- | --- | --- |
| Adults: | | | |
|   Males | 7.1 in | 9.7 in | 12.3 in |
|   Females | 5.8 in | 6.6 in | 7.9 in* |
| Boys: | | | |
|   Age 17 | | | |
|   Age 14 | | | |
|   Age 12 | | | |
|   Age 6 | | | |
|   Age 2 | | | |
| Girls: | | | |
|   Age 17 | | | |
|   Age 14 | | | |
|   Age 12 | | | |
|   Age 6 | | | |
|   Age 2 | | | |
| Adults age 70 and over: | | | |
|   Males | | | |
|   Females | | | |
| Truck and bus drivers: | | | |
|   Males | | | |
|   Females | | | |
| Airline pilots (Male) | 7.6 in | 9.9 in | 10.4 in |
| Flight attendants (Female) | 5.2 in | 5.8 in | 6.5 in |
| Law enforcement officers: | | | |
|   Males | | | |
|   Females | | | |

*In certain working situations, one should consider the additional requirement for clearance when the pregnant woman must be accommodated. Unfortunately, to our knowledge, there are no data on this additional waist-clearance factor.

## ELBOW REST HEIGHT

Not only is this dimension pertinent to the establishment of armrest heights, but it also provides a basis for establishing the level of a writing surface and/or the approximate position of the middle row of a keyboard, the location of a joy-stick handle or control wheel, etc.

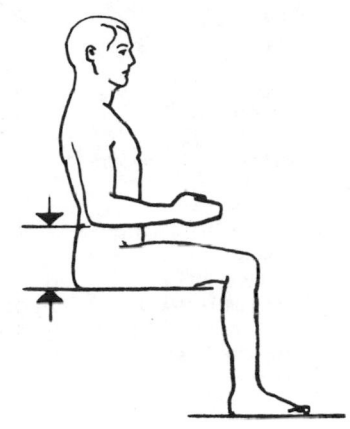

|  | Percentile | | |
| --- | --- | --- | --- |
|  | 5th | 50th | 95th |
| Adults: | | | |
|   Males | 7.4 in | 9.5 in | 11.6 in |
|   Females | 7.1 in | 9.2 in | 11.0 in |
| Boys: | | | |
|   Age 17 | | | |
|   Age 14 | | | |
|   Age 12 | | | |
|   Age 6 | | | |
|   Age 2 | | | |
| Girls: | | | |
|   Age 17 | | | |
|   Age 14 | | | |
|   Age 12 | | | |
|   Age 6 | | | |
|   Age 2 | | | |
| Adults age 70 and over: | | | |
|   Males | 6.5 in | 8.6 in | 10.6 in |
|   Females | 6.4 in | 8.4 in | 10.0 in |
| Truck and bus drivers | | | |
|   Males | — | — | — |
|   Females | — | — | — |
| Airline pilots (Male) | 7.4 in | 9.1 in | 10.8 in |
| Flight attendants (Female) | 7.7 in | 9.6 in | 11.0 in |
| Law enforcement officers: | | | |
|   Males | — | — | — |
|   Females | — | — | — |

## BUTTOCK-TO-HEEL LENGTH

This dimension is pertinent to the establishment of the distance of foot-operated controls from the seat reference point.

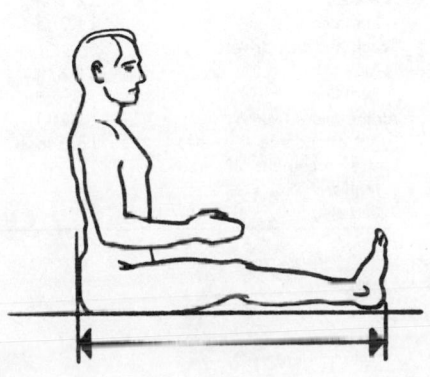

|  | Percentile | | |
| --- | --- | --- | --- |
|  | 5th | 50th | 95th |
| Adults: | | | |
|   Males | 39.0 in | 42.0 in | 46.0 in |
|   Females | 34.0 in | 37.8 in | 41.2 in |
| Boys: | | | |
|   Age 17 | — | — | — |
|   Age 14 | 31.7 in | 34.1 in | 39.1 in |
|   Age 12 | 30.4 in | 33.7 in | 37.3 in |
|   Age 6 | 22.5 in | 24.7 in | 26.4 in |
|   Age 2 | 15.1 in | 16.5 in | 18.3 in |
| Girls: | | | |
|   Age 17 | — | — | — |
|   Age 14 | 33.1 in | 36.8 in | 39.0 in |
|   Age 12 | 30.2 in | 33.7 in | 37.1 in |
|   Age 6 | 22.4 in | 24.6 in | 27.2 in |
|   Age 2 | 14.1 in | 17.0 in | 18.0 in |
| Adults age 70 and over: | | | |
|   Males | — | — | — |
|   Females | — | — | — |
| Truck and bus drivers | | | |
|   Males | — | — | — |
|   Females | — | — | — |
| Airline pilots (Male) | 38.3 in | 42.3 in | 46.3 in |
| Flight attendants (Female) | — | — | — |
| Law enforcement officers: | | | |
|   Males | — | — | — |
|   Females | — | — | — |

## SHOULDER BREADTH

This dimension is pertinent to the establishment of lateral clearance between persons who may be required to sit side by side and to the establishment of the lateral clearance requirement for a worker who may have to squeeze into a tight space to work on an item of equipment.

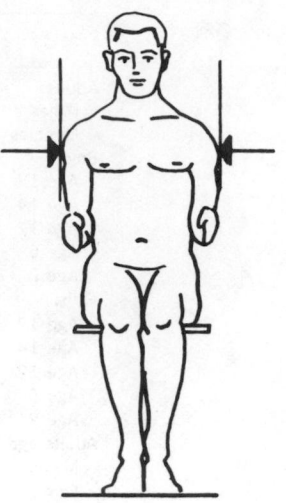

| | | Percentile | |
|---|---|---|---|
| | 5th | 50th | 95th |
| Adults: | | | |
|   Males | 16.4 in | 17.9 in | 19.6 in |
|   Females | 14.4 in | 15.7 in | 17.6 in |
| Boys: | | | |
|   Age 17 | 15.5 in | 17.4 in | 19.2 in |
|   Age 14 | 13.7 in | 15.3 in | 17.4 in |
|   Age 12 | 12.3 in | 13.8 in | 15.6 in |
|   Age 6 | 9.3 in | 11.0 in | 12.5 in |
|   Age 2 | 8.0 in | 8.9 in | 9.8 in |
| Girls: | | | |
|   Age 17 | 14.2 in | 15.6 in | 17.3 in |
|   Age 14 | 13.4 in | 14.8 in | 16.7 in |
|   Age 12 | 12.2 in | 13.8 in | 15.8 in |
|   Age 6 | 10.0 in | 10.9 in | 11.9 in |
|   Age 2 | 7.6 in | 9.0 in | 9.7 in |
| Adults Age 70 and over: | | | |
|   Males | 15.6 in | 17.0 in | 18.5 in |
|   Females | — | — | — |
| Truck and bus drivers: | | | |
|   Males | 16.9 in | 18.3 in | 19.9 in |
|   Females | — | — | — |
| Airline pilots (Male) | 16.5 in | 17.9 in | 19.4 in |
| Flight attendants (Female) | 14.9 in | 16.0 in | 17.0 in |
| Law enforcement officers: | | | |
|   Males | 17.6 in | 19.4 in | 21.4 in |
|   Females | — | — | — |

## HIP BREADTH (SITTING)

This dimension is pertinent to the establishment of seat widths, particularly the lateral clearance between armrests.

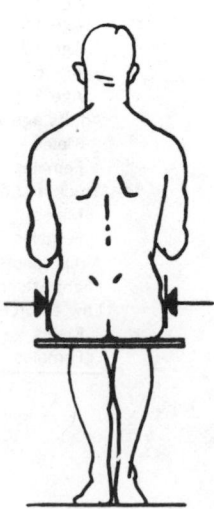

| | | Percentile | |
|---|---|---|---|
| | 5th | 50th | 95th |
| Adults: | | | |
|   Males | 12.2 in | 14.0 in | 15.9 in |
|   Females | 12.3 in | 14.3 in | 17.1 in |
| Boys: | | | |
|   Age 17 | 11.9 in | 13.1 in | 14.8 in |
|   Age 14 | 10.4 in | 11.8 in | 13.6 in |
|   Age 12 | 9.2 in | 10.7 in | 12.8 in |
|   Age 6 | 7.5 in | 8.3 in | 9.4 in |
|   Age 2 | 6.9 in | 7.5 in | 8.4 in |
| Girls: | | | |
|   Age 17 | 12.0 in | 13.5 in | 15.7 in |
|   Age 14 | 11.2 in | 12.7 in | 14.4 in |
|   Age 12 | 9.6 in | 11.3 in | 13.5 in |
|   Age 6 | 7.4 in | 8.4 in | 9.3 in |
|   Age 2 | 6.6 in | 7.4 in | 8.4 in |
| Adults age 70 and over: | | | |
|   Males | 12.1 in | 13.6 in | 15.5 in |
|   Females | 11.7 in | 14.0 in | 16.8 in |
| Truck and bus drivers: | | | |
|   Males | 13.2 in | 14.5 in | 16.3 in |
|   Females | — | — | — |
| Airline pilots (Male) | 12.1 in | 13.2 in | 14.4 in |
| Flight attendants (Female) | 13.3 in | 14.5 in | 15.6 in |
| Law enforcement officers: | | | |
|   Males | — | — | — |
|   Females | — | — | — |

## FOREARM-TO-FOREARM BREADTH

This dimension is pertinent to the establishment of seat and armrest separation. It also should be considered along with shoulder breadth to establish lateral body clearance for side-by-side seating.

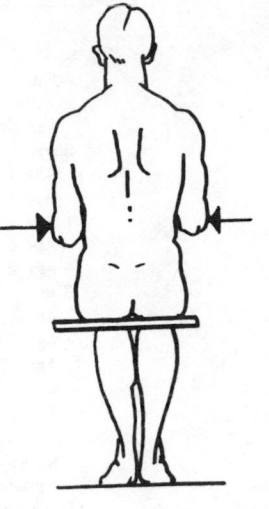

| | Percentile | | |
| --- | --- | --- | --- |
| | 5th | 50th | 95th |
| Adults: | | | |
|   Males | 13.7 in | 16.5 in | 19.9 in |
|   Females | 12.3 in | 15.1 in | 19.3 in |
| Boys: | | | |
|   Age 17 | | | |
|   Age 14 | | | |
|   Age 12 | | | |
|   Age 6 | | | |
|   Age 2 | | | |
| Girls: | | | |
|   Age 17 | | | |
|   Age 14 | | | |
|   Age 12 | | | |
|   Age 6 | | | |
|   Age 2 | | | |
| Adults age 70 and over: | | | |
|   Males | 14.0 in | 16.4 in | 18.7 in |
|   Females | 13.1 in | 15.7 in | 19.1 in |
| Truck and bus drivers: | | | |
|   Males | 16.6 in | 19.4 in | 23.1 in |
|   Females | 14.1 in | 16.5 in | 20.2 in |
| Airline pilots (Male) | 15.2 in | 17.2 in | 19.8 in |
| Flight attendants (Female) | 11.6 in | 13.0 in | 14.6 in |
| Law enforcement officers: | | | |
|   Males | — | — | — |
|   Females | — | — | — |

## INTERPUPILLARY BREADTH

This dimension is pertinent to the design of eyeglasses, binoculars, and other optical aids.

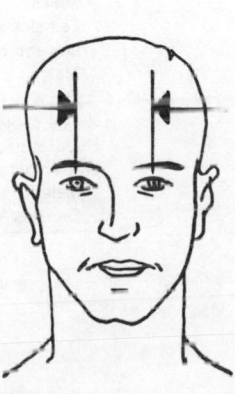

| | Percentile | | |
| --- | --- | --- | --- |
| | 5th | 50th | 95th |
| Adults: | | | |
|   Males | 2.27 in | — | 2.74 in |
|   Females | 1.41 in | — | 2.94 in |
| Boys: | | | |
|   Age 17 | | | |
|   Age 14 | | | |
|   Age 12 | | | |
|   Age 6 | | | |
|   Age 2 | | | |
| Girls: | | | |
|   Age 17 | | | |
|   Age 14 | | | |
|   Age 12 | | | |
|   Age 6 | | | |
|   Age 2 | | | |
| Adults age 70 and over: | | | |
|   Males | | | |
|   Females | | | |
| Truck and bus drivers: | | | |
|   Males | 2.28 in | 2.44 in | 2.80 in |
|   Females | — | — | — |
| Airline pilots (Male) | 2.27 in | 2.49 in | 2.74 in |
| Flight attendants (Female) | — | — | — |
| Law enforcement officers: | | | |
|   Males | — | — | — |
|   Females | — | — | — |

## HEAD BREADTH

This dimension is pertinent to the design of helmets and other head-mounted gear.

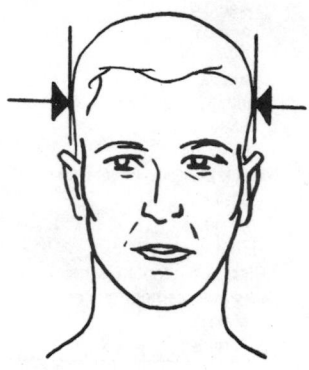

| | Percentile | | |
|---|---|---|---|
| | 5th | 50th | 95th |
| Adults: | | | |
|   Males | 5.6 in | 6.0 in | 6.4 in |
|   Females | 5.4 in | 5.7 in | 6.1 in |
| Boys: | | | |
|   Age 17 | 5.7 in | 6.0 in | 6.4 in |
|   Age 14 | 5.5 in | 5.9 in | 6.2 in |
|   Age 12 | 5.4 in | 5.7 in | 6.1 in |
|   Age 6 | 5.2 in | 5.5 in | 5.9 in |
|   Age 2 | 4.9 in | 5.2 in | 5.5 in |
| Girls: | | | |
|   Age 17 | 5.5 in | 5.8 in | 6.1 in |
|   Age 14 | 5.5 in | 5.7 in | 6.1 in |
|   Age 12 | 5.4 in | 5.7 in | 5.9 in |
|   Age 6 | 5.1 in | 5.4 in | 5.7 in |
|   Age 2 | 4.7 in | 5.0 in | 5.4 in |
| Adults age 70 and over: | | | |
|   Males | 5.8 in | 6.1 in | 6.4 in |
|   Females | — | — | — |
| Truck and bus drivers: | | | |
|   Males | 5.7 in | 6.0 in | 6.3 in |
|   Females | — | — | — |
| Airline pilots (Male) | 5.7 in | 6.1 in | 6.4 in |
| Flight attendants (Female) | 5.4 in | 5.7 in | 6.1 in |
| Law enforcement officers: | | | |
|   Males | 5.7 in | 6.1 in | 6.5 in |
|   Females | — | — | — |

## HEAD LENGTH

This dimension is pertinent to the design of helmets and other head-mounted gear. It should be used in conjunction with head breadth.

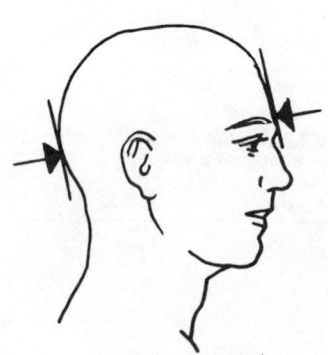

| | Percentile | | |
|---|---|---|---|
| | 5th | 50th | 95th |
| Adults: | | | |
|   Males | 7.3 in | 7.7 in | 8.2 in |
|   Females | 7.0 in | 7.4 in | 7.7 in |
| Boys: | | | |
|   Age 17 | 7.2 in | 7.7 in | 8.0 in |
|   Age 14 | 7.2 in | 7.6 in | 8.1 in |
|   Age 12 | 6.9 in | 7.5 in | 7.7 in |
|   Age 6 | 6.6 in | 7.2 in | 7.6 in |
|   Age 2 | 6.5 in | 6.9 in | 7.3 in |
| Girls: | | | |
|   Age 17 | 6.9 in | 7.4 in | 7.8 in |
|   Age 14 | 7.0 in | 7.4 in | 7.9 in |
|   Age 12 | 6.8 in | 7.3 in | 7.8 in |
|   Age 6 | 6.6 in | 7.1 in | 7.4 in |
|   Age 2 | 6.1 in | 6.7 in | 6.9 in |
| Adults age 70 and over: | | | |
|   Males | 7.3 in | 7.7 in | 8.1 in |
|   Females | — | — | — |
| Truck and bus drivers: | | | |
|   Males | 7.2 in | 7.6 in | 8.1 in |
|   Females | — | — | — |
| Airline pilots (Male) | 7.2 in | 7.8 in | 8.6 in |
| Flight attendants (Female) | 6.9 in | 7.3 in | 7.7 in |
| Law enforcement officers: | | | |
|   Males | 7.3 in | 7.8 in | 8.2 in |
|   Females | — | — | — |

## HEAD LENGTH (MAXIMUM)

This dimension is useful in the design of special headgear that requires a full face mask. These values include an arbitrary 1-in (2.5-cm) additional clearance.

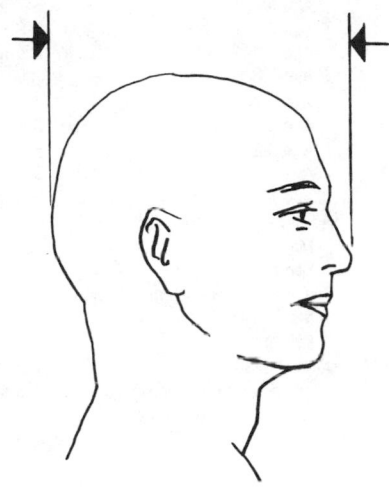

| | Percentile | | |
|---|---|---|---|
| | 5th | 50th | 95th |
| Adults: | | | |
|   Males | | | |
|   Females | | | |
| Boys: | | | |
|   Age 17 | | | |
|   Age 14 | | | |
|   Age 12 | | | |
|   Age 6 | | | |
|   Age 2 | | | |
| Girls: | | | |
|   Age 17 | | | |
|   Age 14 | | | |
|   Age 12 | | | |
|   Age 6 | | | |
|   Age 2 | | | |
| Adults age 70 and over: | | | |
|   Males | | | |
|   Females | | | |
| Truck and bus drivers: | | | |
|   Males | 8.1 in | 8.7 in | 9.2 in |
|   Females | | | |
| Airline pilots | | | |
| Flight attendants | | | |
| Law enforcement officers: | | | |
|   Males | | | |
|   Females | | | |

## HEAD HEIGHT (MAXIMUM)

This dimension is pertinent to the design of fully enclosed headgear or helmets. As suggested by the lack of data for various populations, the values indicated below were derived from a very small sample of male and female subjects selected at random and provide some idea of the larger head height characteristics.

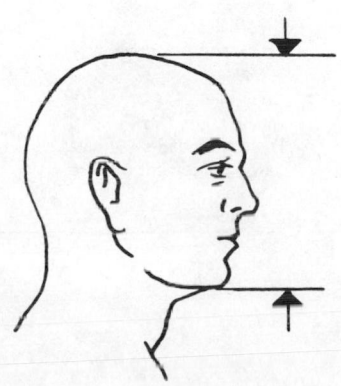

| | Percentile | | |
|---|---|---|---|
| | 5th | 50th | 95th |
| Adults: | | | |
|   Males | — | — | 10.2 in |
|   Females | — | — | 9.9 in |
| Boys: | | | |
|   Age 17 | 7.8 in | 8.6 in | 9.1 in |
|   Age 14 | 7.5 in | 8.1 in | 8.9 in |
|   Age 12 | 7.5 in | 7.8 in | 8.5 in |
|   Age 6 | 6.6 in | 7.3 in | 8.0 in |
|   Age 2 | 6.3 in | 6.9 in | 7.5 in |
| Girls: | | | |
|   Age 17 | 7.5 in | 8.1 in | 8.7 in |
|   Age 14 | 7.3 in | 8.0 in | 8.5 in |
|   Age 12 | 7.0 in | 7.8 in | 8.5 in |
|   Age 6 | 6.3 in | 7.2 in | 7.6 in |
|   Age 2 | 6.1 in | 6.6 in | 7.2 in |
| Adults age 70 and over: | | | |
|   Males | | | |
|   Females | | | |
| Truck and bus drivers: | | | |
|   Males | | | |
|   Females | | | |
| Airline pilots | | | |
| Flight attendants | | | |
| Law enforcement officers: | | | |
|   Males | | | |
|   Females | | | |

## CHIN-TO-EYE HEIGHT

This dimension is pertinent to the design of headgear in which eye protective elements are integrated into the helmet. Although these values represent a very small subject sample, they provide some idea of the larger chin-to-eye heights. They should be considered in conjunction with head height.

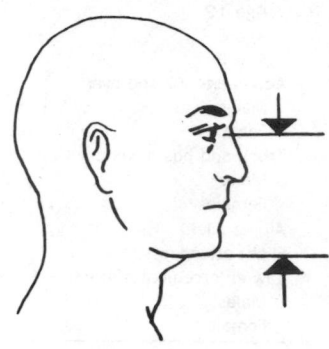

|  | Percentile | | |
|---|---|---|---|
|  | 5th | 50th | 95th |
| **Adults:** | | | |
| Males | 4.3 in | — | 5.0 in |
| Females | 3.8 in | — | 4.25 in |
| **Boys:** | | | |
| Age 17 | 4.1 in | 4.4 in | 4.8 in |
| Age 14 | 3.8 in | 4.3 in | 4.7 in |
| Age 12 | 3.7 in | 4.0 in | 4.3 in |
| Age 6 | 3.2 in | 3.5 in | 3.8 in |
| Age 2 | 2.9 in | 3.2 in | 3.6 in |
| **Girls:** | | | |
| Age 17 | 3.7 in | 4.2 in | 4.5 in |
| Age 14 | 3.9 in | 4.1 in | 4.5 in |
| Age 12 | 3.6 in | 4.0 in | 4.4 in |
| Age 6 | 3.1 in | 3.5 in | 3.8 in |
| Age 2 | 2.8 in | 3.1 in | 3.4 in |
| **Adults age 70 and over:** | | | |
| Males | | | |
| Females | | | |
| **Truck and bus drivers:** | | | |
| Males | | | |
| Females | | | |
| Airline pilots (Male) | 4.3 in | — | 5.1 in |
| Flight attendants (Female) | 3.8 in | — | 4.6 in |
| **Law enforcement officers:** | | | |
| Males | | | |
| Females | | | |

*Note:* By subtracting the above dimensions from maximum head height, it can be observed that the top of a person's head is approximately 5 in (12.7 cm) above his or her eye height; this gives some idea of how much clearance should be provided above an eye reference for a seated operator. A good rule of thumb is to provide a minimum of 7 in (18 cm) of head clearance for bareheaded operators.

## HAND LENGTH

This dimension is pertinent to the design of gloves, mittens, and other devices for protecting the hands. This dimension is also useful in selecting test subjects for evaluating hand-held or hand-manipulated devices such as handrails and gun and joy-stick grips.

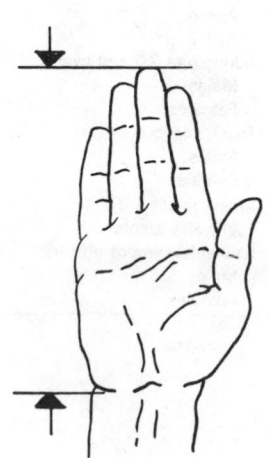

|  | Percentile | | |
|---|---|---|---|
|  | 5th | 50th | 95th |
| **Adults:** | | | |
| Males | 7.0 in | 7.6 in | 8.2 in |
| Females | 6.4 in | 6.9 in | 7.4 in |
| **Boys:** | | | |
| Age 17 | 6.8 in | 7.4 in | 7.9 in |
| Age 14 | 6.3 in | 7.0 in | 7.6 in |
| Age 12 | 5.7 in | 6.3 in | 7.0 in |
| Age 6 | 4.6 in | 5.0 in | 5.7 in |
| Age 2 | 3.6 in | 3.9 in | 4.2 in |
| **Girls:** | | | |
| Age 17 | 6.1 in | 6.7 in | 7.2 in |
| Age 14 | 6.1 in | 6.7 in | 7.2 in |
| Age 12 | 5.6 in | 6.4 in | 7.0 in |
| Age 6 | 4.5 in | 4.9 in | 5.4 in |
| Age 2 | 3.3 in | 3.8 in | 4.3 in |
| **Adults age 70 and over:** | | | |
| Males | 7.0 in | 7.4 in | 8.0 in |
| Females | — | — | — |
| **Truck and bus drivers:** | | | |
| Males | 7.1 in | 7.6 in | 8.0 in |
| Females | — | — | — |
| Airline pilots (Male) | 6.9 in | 7.5 in | 8.0 in |
| Flight attendants (Female) | 6.3 in | 6.8 in | 7.3 in |
| **Law enforcement officers:** | | | |
| Males | 7.0 in | 7.6 in | 8.2 in |
| Females | — | — | — |

## HAND BREADTH AT THUMB

This dimension is pertinent to the design of hand protective gear and is also useful in defining lateral hand clearance.

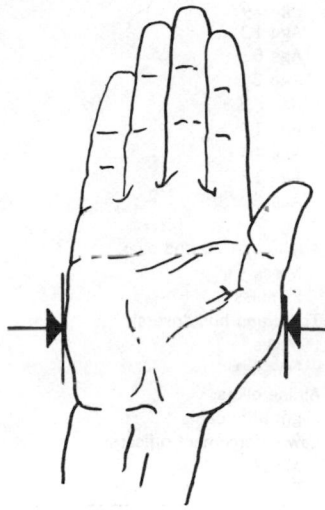

| | Percentile | | |
|---|---|---|---|
| | 5th | 50th | 95th |
| Adults: | | | |
|   Males | | | |
|   Females | | | |
| Boys: | | | |
|   Age 17 | | | |
|   Age 14 | | | |
|   Age 12 | | | |
|   Age 6 | | | |
|   Age 2 | | | |
| Girls: | | | |
|   Age 17 | | | |
|   Age 14 | | | |
|   Age 12 | | | |
|   Age 6 | | | |
|   Age 2 | | | |
| Adults age 70 and over: | | | |
|   Males | | | |
|   Females | | | |
| Truck and bus drivers | | | |
|   Males | | | |
|   Females | | | |
| Airline pilots (Male) | 3.7 in | 4.1 in | 4.4 in |
| Flight attendants (Female) | 3.2 in | 3.6 in | 4.0 in |
| Law enforcement officers: | | | |
|   Males | | | |
|   Females | | | |

## HAND BREADTH AT METACARPAL

This dimension is pertinent to the design of hand protective gear.

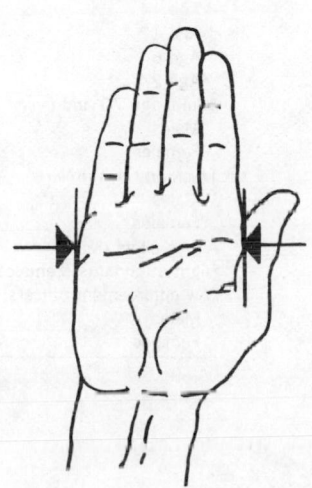

| | Percentile | | |
|---|---|---|---|
| | 5th | 50th | 95th |
| Adults: | | | |
|   Males | 3.1 in | 3.4 in | 3.8 in |
|   Females | 2.7 in | 3.0 in | 3.4 in |
| Boys: | | | |
|   Age 17 | 3.2 in | 3.5 in | 3.7 in |
|   Age 14 | 2.8 in | 3.2 in | 3.6 in |
|   Age 12 | 2.6 in | 2.9 in | 3.3 in |
|   Age 6 | 2.1 in | 2.4 in | 2.6 in |
|   Age 2 | 1.7 in | 1.9 in | 2.2 in |
| Girls: | | | |
|   Age 17 | 2.7 in | 2.9 in | 3.3 in |
|   Age 14 | 2.7 in | 2.9 in | 3.2 in |
|   Age 12 | 2.5 in | 2.8 in | 3.1 in |
|   Age 6 | 2.0 in | 2.3 in | 2.5 in |
|   Age 2 | 1.6 in | 1.8 in | 2.1 in |
| Adults age 70 and over: | | | |
|   Males | 3.1 in | 3.3 in | 3.6 in |
|   Females | — | — | — |
| Truck and bus drivers: | | | |
|   Males | 3.2 in | 3.5 in | 3.8 in |
|   Females | — | | — |
| Airline pilots (Male) | 3.2 in | 3.5 in | 3.8 in |
| Flight attendants (Female) | 2.7 in | 2.9 in | 3.1 in |
| Law enforcement officers: | | | |
|   Males | 3.3 in | 3.5 in | 3.8 in |
|   Females | — | — | — |

## HAND THICKNESS AT METACARPAL

This dimension is pertinent to the design of hand protective gear and also to the establishment of hand clearance.

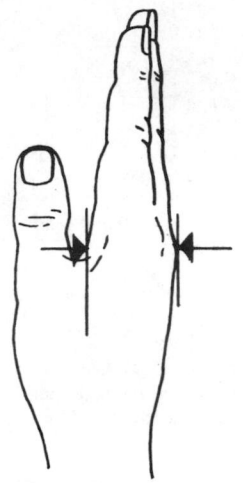

|  | Percentile | | |
|---|---|---|---|
|  | 5th | 50th | 95th |
| Adults: | | | |
| Males | 1.1 in | 1.2 in | 1.3 in |
| Females | 0.8 in | 1.0 in | 1.1 in |
| Boys: | | | |
| Age 17 | | | |
| Age 14 | | | |
| Age 12 | | | |
| Age 6 | | | |
| Age 2 | | | |
| Girls: | | | |
| Age 17 | | | |
| Age 14 | | | |
| Age 12 | | | |
| Age 6 | | | |
| Age 2 | | | |
| Adults age 70 and over: | | | |
| Males | | | |
| Females | | | |
| Truck and bus drivers: | | | |
| Males | | | |
| Females | | | |
| Airline pilots | | | |
| Flight attendants | | | |
| Law enforcement officers: | | | |
| Males | | | |
| Females | | | |

## FOOT LENGTH[7]

This dimension is pertinent to the design of footgear and also to the establishment of clearances for the foot (see the discussion of clothing effects).

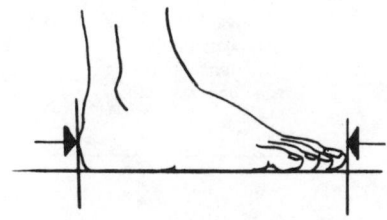

|  | Percentile | | |
|---|---|---|---|
|  | 5th | 50th | 95th |
| Adults: | | | |
| Males | 9.7 in | 10.4 in | 11.2 in |
| Females | 8.7 in | 9.4 in | 10.2 in |
| Boys: | | | |
| Age 17 | 9.5 in | 10.6 in | 11.1 in |
| Age 14 | 9.0 in | 10.0 in | 11.0 in |
| Age 12 | 8.4 in | 9.2 in | 10.1 in |
| Age 6 | 6.5 in | 7.0 in | 8.0 in |
| Age 2 | 4.5 in | 5.1 in | 5.7 in |
| Girls: | | | |
| Age 17 | 8.6 in | 9.2 in | 10.1 in |
| Age 14 | 8.5 in | 9.2 in | 10.2 in |
| Age 12 | 8.3 in | 9.1 in | 9.8 in |
| Age 6 | 6.3 in | 7.0 in | 7.6 in |
| Age 2 | 4.2 in | 5.0 in | 5.5 in |
| Adults age 70 and over: | | | |
| Males | 9.7 in | 10.2 in | 10.9 in |
| Females | — | — | — |
| Truck and bus drivers: | | | |
| Males | 9.6 in | 10.4 in | 11.3 in |
| Females | — | — | — |
| Airline pilots (Male) | 9.8 in | 10.5 in | 11.3 in |
| Flight attendants (Female) | 8.7 in | 9.4 in | 10.1 in |
| Law enforcement officers: | | | |
| Males | — | — | — |
| Females | — | — | — |

[7]Measured with subjects standing.

## BALL OF FOOT WIDTH[a]

This dimension is pertinent to the design of footgear and also to the establishment of foot clearance.

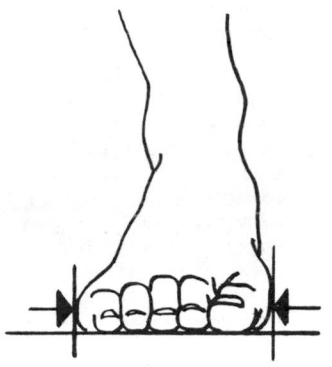

|  | Percentile | | |
|---|---|---|---|
|  | 5th | 50th | 95th |
| Adults: | | | |
|   Males | 3.5 in | 3.9 in | 4.3 in |
|   Females | 3.2 in | 3.6 in | 4.0 in |
| Boys: | | | |
|   Age 17 | 3.7 in | 4.1 in | 4.4 in |
|   Age 14 | 3.4 in | 3.9 in | 4.3 in |
|   Age 12 | 3.1 in | 3.5 in | 4.0 in |
|   Age 6 | 2.5 in | 2.8 in | 3.1 in |
|   Age 2 | 1.8 in | 2.1 in | 2.4 in |
| Girls: | | | |
|   Age 17 | 3.2 in | 3.5 in | 3.9 in |
|   Age 14 | 3.1 in | 3.5 in | 3.9 in |
|   Age 12 | 3.0 in | 3.4 in | 3.8 in |
|   Age 6 | 2.4 in | 2.7 in | 3.0 in |
|   Age 2 | 1.6 in | 1.9 in | 2.2 in |
| Adults age 70 and over: | | | |
|   Males | 3.6 in | 3.9 in | 4.3 in |
|   Females | — | — | — |
| Truck and bus drivers: | | | |
|   Males | 3.7 in | 4.0 in | 4.3 in |
|   Females | — | — | — |
| Airline pilots (Male) | 3.5 in | 3.8 in | 4.1 in |
| Flight attendants (Female) | 3.2 in | 3.4 in | 3.8 in |
| Law enforcement officers: | | | |
|   Males | | | |
|   Females | | | |

## SHOULDER CIRCUMFERENCE

This dimension is pertinent to the design of garments. It also is useful in establishing the general description of a particular test subject who may be used for dynamic testing of various physical mockups.

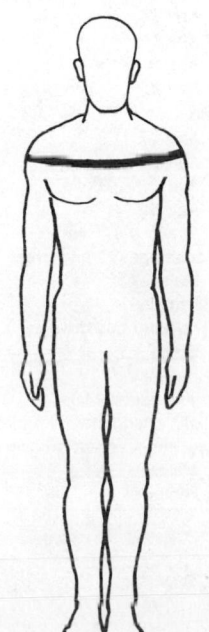

|  | Percentile | | |
|---|---|---|---|
|  | 5th | 50th | 95th |
| Adults: | | | |
|   Males | | | |
|   Females | | | |
| Boys: | | | |
|   Age 17 | | | |
|   Age 14 | | | |
|   Age 12 | | | |
|   Age 6 | | | |
|   Age 2 | | | |
| Girls: | | | |
|   Age 17 | | | |
|   Age 14 | | | |
|   Age 12 | | | |
|   Age 6 | | | |
|   Age 2 | | | |
| Adults age 70 and over: | | | |
|   Males | | | |
|   Females | | | |
| Truck and bus drivers: | | | |
|   Males | | | |
|   Females | | | |
| Airline pilots (Male) | 41.9 in | 45.3 in | 47.7 in |
| Flight attendants (Female) | 35.6 in | 37.6 in | 39.7 in |
| Law enforcement officers: | | | |
|   Males | | | |
|   Females | | | |

[a]Measured with subjects standing.

## CHEST CIRCUMFERENCE

This dimension can be used in conjunction with shoulder circumference for garment design and the definition of the body characteristics of test subjects.

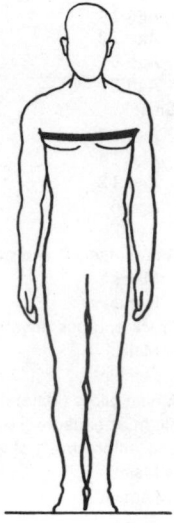

| | Percentile | | |
|---|---|---|---|
| | 5th | 50th | 95th |
| **Adults:** | | | |
| Males | 34.3 in | 39.0 in | 44.9 in |
| Females | 30.4 in | 34.3 in | 40.5 in |
| **Boys:** | | | |
| Age 17 | 32.5 in | 36.2 in | 40.8 in |
| Age 14 | 27.5 in | 31.3 in | 36.1 in |
| Age 12 | 25.6 in | 28.1 in | 32.1 in |
| Age 6 | 20.7 in | 22.6 in | 25.3 in |
| Age 2 | 18.0 in | 19.0 in | 20.6 in |
| **Girls:** | | | |
| Age 17 | 29.7 in | 32.3 in | 36.3 in |
| Age 14 | 27.6 in | 30.8 in | 34.6 in |
| Age 12 | 25.2 in | 28.7 in | 33.4 in |
| Age 6 | 20.3 in | 22.3 in | 24.3 in |
| Age 2 | 17.3 in | 18.4 in | 20.5 in |
| **Adults age 70 and over:** | | | |
| Males | 33.0 in | 38.1 in | 44.1 in |
| Females | 30.1 in | 35.0 in | 39.9 in |
| **Truck and bus drivers:** | | | |
| Males | — | — | — |
| Females | — | — | — |
| Airline pilots (Male) | 36.3 in | 38.8 in | 41.3 in |
| Flight attendants (Female) | 31.2 in | 33.7 in | 36.4 in |
| **Law enforcement officers:** | | | |
| Males | 35.4 in | 40.1 in | 45.6 in |
| Females | — | — | — |

## WAIST CIRCUMFERENCE

This dimension can be used in conjunction with shoulder and chest circumference for garment design and the definition of the body characteristics of test subjects.

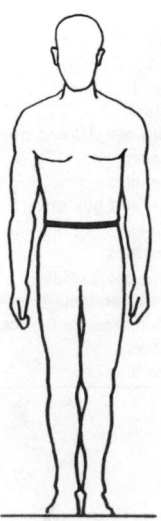

| | Percentile | | |
|---|---|---|---|
| | 5th | 50th | 95th |
| **Adults:** | | | |
| Males | 28.4 in | 34.8 in | 42.9 in |
| Females | 24.1 in | 29.2 in | 39.1 in |
| **Boys:** | | | |
| Age 17 | 26.5 in | 29.6 in | 35.9 in |
| Age 14 | 23.7 in | 26.8 in | 32.7 in |
| Age 12 | 22.1 in | 24.8 in | 29.7 in |
| Age 6 | 18.0 in | 20.3 in | 23.0 in |
| Age 2 | 15.5 in | 17.3 in | 19.3 in |
| **Girls:** | | | |
| Age 17 | 24.9 in | 27.8 in | 34.4 in |
| Age 14 | 23.1 in | 26.8 in | 31.9 in |
| Age 12 | 21.8 in | 25.1 in | 32.0 in |
| Age 6 | 18.1 in | 20.2 in | 23.1 in |
| Age 2 | 14.8 in | 17.4 in | 19.4 in |
| **Adults age 70 and over:** | | | |
| Males | 28.3 in | 35.8 in | 41.3 in |
| Females | 25.9 in | 33.2 in | 39.9 in |
| **Truck and bus drivers:** | | | |
| Males | — | — | — |
| Females | — | — | — |
| Airline pilots (Male) | 29.0 in | 32.0 in | 35.0 in |
| Flight attendants (Female) | 22.8 in | 24.5 in | 26.4 in |
| **Law enforcement officers:** | | | |
| Males | 30.1 in | 35.3 in | 42.3 in |
| Females | — | — | — |

## HIP (BUTTOCK) CIRCUMFERENCE (STANDING)

This dimension can be used in conjunction with hip circumference considerations for garment design and the definition of the body characteristics of test subjects.

| | Percentile | | |
|---|---|---|---|
| | 5th | 50th | 95th |
| **Adults:** | | | |
| Males | — | — | — |
| Females | — | — | — |
| **Boys:** | | | |
| Age 17 | 32.8 in | 35.7 in | 40.7 in |
| Age 14 | 28.3 in | 32.4 in | 37.4 in |
| Age 12 | 26.0 in | 29.2 in | 34.5 in |
| Age 6 | 20.5 in | 22.6 in | 26.0 in |
| Age 2 | 18.6 in | 20.5 in | 22.7 in |
| **Girls:** | | | |
| Age 17 | 32.1 in | 36.1 in | 41.6 in |
| Age 14 | 29.9 in | 34.1 in | 38.8 in |
| Age 12 | 26.1 in | 30.7 in | 35.9 in |
| Age 6 | 20.5 in | 23.1 in | 25.5 in |
| Age 2 | 18.1 in | 20.2 in | 22.1 in |
| **Adults age 70 and over:** | | | |
| Males | — | — | — |
| Females | — | — | — |
| **Truck and bus drivers:** | | | |
| Males | — | — | — |
| Females | — | — | — |
| Airline pilots (Male) | 35.5 in | 37.8 in | 40.1 in |
| Flight attendants (Female) | 33.4 in | 35.4 in | 37.8 in |
| **Law enforcement officers:** | | | |
| Males | — | — | — |
| Females | — | — | — |

## HIP (BUTTOCK) CIRCUMFERENCE (SITTING)

This dimension is pertinent to the design of seats and seatbelts. When it is used in conjunction with hip breadth, subjects can be selected to represent a larger user population for evaluating seat widths and lap belt lengths.

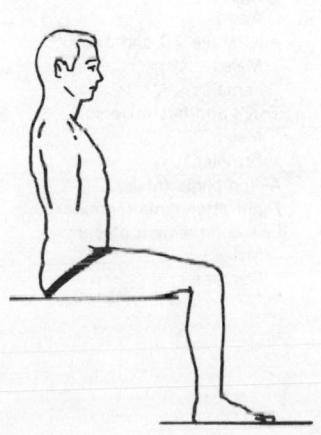

| | Percentile | | |
|---|---|---|---|
| | 5th | 50th | 95th |
| **Adults:** | | | |
| Male | | | |
| Female | | | |
| **Boys:** | | | |
| Age 17 | | | |
| Age 14 | | | |
| Age 12 | | | |
| Age 6 | | | |
| Age 2 | | | |
| **Girls:** | | | |
| Age 17 | | | |
| Age 14 | | | |
| Age 12 | | | |
| Age 6 | | | |
| Age 2 | | | |
| **Adults age 70 and over:** | | | |
| Males | | | |
| Females | | | |
| **Truck and bus drivers:** | | | |
| Males | | | |
| Females | | | |
| Airline pilots (Male) | 38.9 in | 41.7 in | 44.5 in |
| Flight attendants (Female) | 37.0 in | 39.2 in | 41.4 in |
| **Law enforcement officers:** | | | |
| Males | | | |
| Females | | | |

## HEAD CIRCUMFERENCE

This dimension is pertinent to the design of headgear.

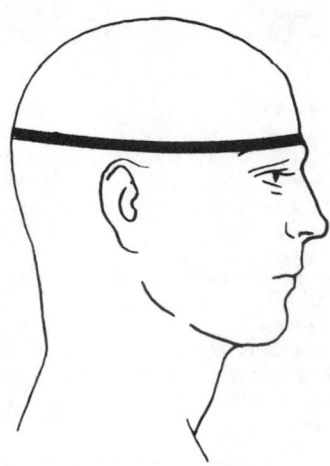

| | Percentile | | |
|---|---|---|---|
| | 5th | 50th | 95th |
| Adults: | | | |
|   Males | — | — | — |
|   Females | — | — | — |
| Boys: | | | |
|   Age 17 | 21.1 in | 22.3 in | 23.3 in |
|   Age 14 | 20.4 in | 21.5 in | 22.8 in |
|   Age 12 | 20.1 in | 21.1 in | 22.3 in |
|   Age 6 | 19.2 in | 20.4 in | 21.2 in |
|   Age 2 | 18.5 in | 19.3 in | 20.4 in |
| Girls: | | | |
|   Age 17 | 20.4 in | 21.5 in | 22.6 in |
|   Age 14 | 20.3 in | 21.2 in | 22.3 in |
|   Age 12 | 19.8 in | 20.9 in | 21.9 in |
|   Age 6 | 19.1 in | 19.8 in | 20.8 in |
|   Age 2 | 18.1 in | 18.8 in | 19.7 in |
| Adults age 70 and over: | | | |
|   Males | — | — | — |
|   Females | — | — | — |
| Truck and bus drivers: | | | |
|   Males | — | — | — |
|   Females | — | — | — |
| Airline pilots (Male) | 21.9 in | 22.5 in | 23.1 in |
| Flight attendants (Female) | 20.8 in | 21.5 in | 22.2 in |
| Law enforcement officers: | | | |
|   Males | 21.6 in | 22.6 in | 23.6 in |
|   Females | — | — | — |

## NECK CIRCUMFERENCE

This dimension may be pertinent to the design of special garments and/or devices worn about the user's neck.

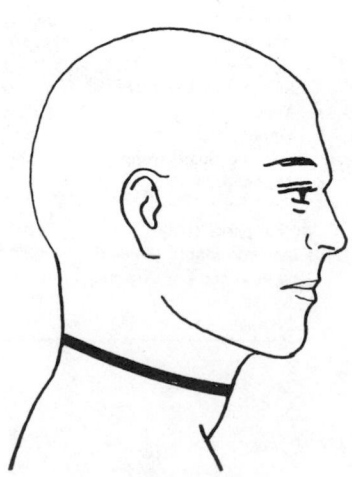

| | Percentile | | |
|---|---|---|---|
| | 5th | 50th | 95th |
| Adults: | | | |
|   Males | — | — | — |
|   Females | — | — | — |
| Boys: | | | |
|   Age 17 | 12.4 in | 14.2 in | 15.0 in |
|   Age 14 | 11.5 in | 12.7 in | 13.8 in |
|   Age 12 | 10.7 in | 11.6 in | 12.6 in |
|   Age 6 | 9.2 in | 10.2 in | 11.0 in |
|   Age 2 | 8.6 in | 9.6 in | 10.1 in |
| Girls: | | | |
|   Age 17 | 11.3 in | 12.2 in | 13.3 in |
|   Age 14 | 10.9 in | 11.7 in | 13.1 in |
|   Age 12 | 10.4 in | 11.4 in | 12.8 in |
|   Age 6 | 9.1 in | 10.0 in | 10.9 in |
|   Age 2 | 8.2 in | 9.2 in | 9.7 in |
| Adults age 70 and over: | | | |
|   Males | — | — | — |
|   Females | — | — | — |
| Truck and bus drivers: | | | |
|   Males | — | — | — |
|   Females | — | — | — |
| Airline pilots (Male) | 14.3 in | 15.0 in | 15.7 in |
| Flight attendants (Female) | 11.1 in | 11.8 in | 12.6 in |
| Law enforcement officers: | | | |
|   Males | — | — | — |
|   Females | — | — | — |

## STATURE STATISTICS FOR OTHER POPULATIONS

With the exception of anthropometric surveys conducted by military organizations, most other population surveys include only a few dimensional statistics. The most common (and one which provides a point of departure for comparing populations) is stature. The following compilation of statures for different populations presented in the accompanying table is useful when a particular product design is intended for one or more of the specific populations noted.

## RACIAL AND ETHNIC ORIGINS OF THE U.S. POPULATION

**Racial and Ethnic Origins of U.S. Population (from Census Bureau Data, April 1970)**

| Group | Number, in Thousands | Percent |
|---|---|---|
| White | 177,784 | 87.5 |
|   Spanish-speaking | 10,115 | 4.9 |
| Black | 22,580 | 11.1 |
| Other | 2,883 | 1.4 |
|   Indian | 793 | 0.4 |
|   Japanese | 591 | 0.3 |
|   Chinese | 435 | 0.2 |
|   Filipino | 343 | 0.2 |
|   Other | 720 | 0.4 |

| Group or Nationality | Percentile, cm | | |
|---|---|---|---|
| | 5th | 50th | 95th |
| Australian Army males | 164.0 | 173.0 | 184.0 |
| Bulgarian civilian males | 163.6 | 170.2 | 176.8 |
| Bulgarian women | 151.6 | 157.8 | 164.0 |
| Canadian civilian males | 153.9 | 159.8 | 173.5 |
| Chinese civilian males | — | 166.4 | — |
| English, RAF male pilots | 166.1 | 177.0 | 185.9 |
| English, male Navy recruits | 164.7 | 174.5 | 184.2 |
| Filipino civilian males | — | 163.1 | — |
| French civilian males | — | 172.5 | — |
| French civil pilots | 168.1 | 173.1 | 178.0 |
| German civilian males | 167.1 | 176.8 | 187.5 |
| Greek Army males | 160.9 | 170.5 | 180.3 |
| Hawaiian civilian males | — | 171.2 | — |
| Indian civilian males | — | 167.9 | — |
| Iranian Army males | 157.5 | 166.7 | 176.6 |
| Israeli civilian males | — | 168.7 | — |
| Italian Army males | 160.2 | 170.7 | 180.7 |
| Japanese Air Force, male pilots | 159.4 | 166.7 | 175.0 |
| Korean Army males | 154.5 | 164.0 | 172.7 |
| Korean Air Force, male pilots | 159.6 | 165.6 | 173.7 |
| Latin American Army males | 157.0 | 165.9 | 177.0 |
| Norwegian (unknown males) | 171.5 | 177.5 | 183.5 |
| Swedish civilian males | — | 174.0 | — |
| Swedish civilian females | 164.1 | — | — |
| Thai Army males | 155.0 | 163.5 | 172.0 |
| Turkish Army males | 160.6 | 169.0 | 179.2 |
| Vietnamese Army males | 151.6 | 160.4 | 169.6 |
| U.S. military: | | | |
|   Army ground forces, male | 163.8 | 174.4 | 185.6 |
|   Army ground forces, female | 152.6 | 162.8 | 174.1 |
|   Army aviators, male | 164.2 | 174.6 | 185.0 |
|   Air Force, male pilots | 167.2 | 177.3 | 187.7 |
|   Air Force, ground, female | 156.8 | 162.7 | 168.6 |
|   Marine Corps, male | 168.2 | 174.5 | 180.8 |
|   Navy, enlisted, male | 171.7 | 177.6 | 183.5 |
|   Navy, pilots, male | 168.2 | 177.5 | 187.7 |
|   Navy, divers, male | 170.0 | 176.2 | 182.2 |
| U.S. FAA civil pilots, male | 170.4 | 176.7 | 183.2 |

## ANTHROPOMETRIC REFERENCES

The following references were used to compile the previous data on weight and body dimensions. Note that some liberties have been taken; i.e., when a certain dimension was not available for a population, a substitution was made from the most likely (similar) population. This was done, however, only when it was considered likely that the borrowed data would be quite similar to that for the population designated.

*Anthropometric Source Book,* NASA Reference Publication 1024, vols. I, II, and III, 1978 (prepared for the National Aeronautics and Space Administration, Scientific and Technical Information Office, by Webb Associates, Yellow Springs, Ohio).

DAMON, ALBERT, et al.: *The Human Body in Equipment Design,* Harvard University Press, Cambridge, Mass., 1966.

GARRETT, JOHN W., and KENNETH W. KENNEDY: *A Collation of Anthropometry,* AMRL-TR-68-1, vols. I and II, Aerospace Medical Research Laboratory, Aerospace Medical Division, Air Force Systems Command, Wright-Patterson Air Force Base, Ohio, 1971.

HAMILL, P. V. V., et al.: *Body Weight, Stature, and Sitting Height: White and Negro Youths 12–17 Years,* U.S. Department of Health, Education, and Welfare, Public Health Service ser. 11, no. 126, 1973.

MALINA, R. M., et al.: *Body Dimensions and Proportions: White and Negro Children 6–11 Years,* U.S. Department of Health, Education, and Welfare, Public Health Service ser. 11, no. 143, 1973.

MALINA, R. M., et al.: *Selected Body Measurements of Children 6–11 Years,* U.S. Department of Health, Education, and Welfare, Public Health Service ser. 11, no. 123, 1973.

MARTIN, J. I., et al.: *Anthropometry of Law Enforcement Officers,* Technical Document 442, U.S. Naval Electronics Laboratory Center, San Diego, 1975.

SNYDER, RICHARD G, et al.: *Physical Characteristics of Children as Related to Death and Injury for Consumer Product Safety Design,* UM-HSRI-BI-75-5, Highway Safety Research Institute, Ann Arbor, Mich., 1975.

WHITE, ROBERT M.: "Anthropometric Measurements on Selected Populations of the World," in Alphonse Chapanis (ed.), *Ethnic Variables in Human Factors Engineering,* The Johns Hopkins Press, Baltimore, 1975.

The reader is referred to the following for selected military population anthropometrics:

CHURCHILL, EDMUND, et al.: *Anthropometry of U.S. Army Aviators—1970,* U.S. Army Natick Laboratories Technical Report 72-52-CE, Natick, Mass., 1971.

CHURCHILL, EDMUND, et al.: *Anthropometry of Women of the U.S. Army—1977,* Report no. 2, Natick-TR-77-024, U.S. Army Natick Research and Development Command, Natick, Mass.

CLAUSER, CHARLES E., et al.: *Anthropometry of Air Force Women,* AMRL-TR-70-5, Aerospace Medical Research Laboratory, Aerospace Medical Division, Air Force Systems Command, Wright-Patterson Air Force Base, Ohio, 1972.

GIFFORD, E. C.: *Compilation of Anthropometric Measures on U.S. Navy Pilots,* NAMC-AGEL-437, Naval Air Material Command, Air Crew Equipment Laboratory, Philadelphia, 1960.

WHITE, ROBERT M., and EDMUND CHURCHILL: *The Body Size of Soldiers: U.S. Army Anthropometry 1966,* U.S. Army Natick Laboratories, Natick, Mass., 1971.

Special acknowledgment is made of the invaluable assistance of Mr. Robert M. White in reviewing and contributing to the foregoing special anthropometric tables. Mr. White, formerly with the U.S. Army Laboratory in Natick, Massachusetts, has specialized for many years in the collection of anthropometric data for various populations and is recognized as an authority in this area.

## CORRELATION AMONG BODY DIMENSIONS[9]

Some body dimensions correlate rather well with others, making it possible to select test subjects for special design evaluation activities.

However, although ordinarily one would expect tall persons to have long legs, arms, and torsos, and short persons to have short legs, arms, and torsos, to ensure that a particular design does not cause problems for special members of the expected user population, it is important not to make too many assumptions about dimensional correlations. The accompanying table presents those correlations that indicate safe assumptions and those that do not.

In addition, consider the following suggestions:

For reach problems select a variety of subjects who variously have short stature, short legs, and short arms.

For clearance problems select a variety of subjects who variously are tall, have long legs and/or sitting heights, and have wide shoulders and hip breadths.

If the weight of a subject is pertinent (in addition to the other dimensional characteristics noted above), consider, for example, one or two heavy but small subjects, for they may sink into a seat and spoil your plans for stabilizing a range of eye references or head clearances.

| Dimension | Correlation with Stature | Correlation with Chest Circumference |
|---|---|---|
| Weight | | 0.820 |
| **Torso** | | |
| Sitting height | 0.722 | |
| Trunk height | 0.587 | |
| Cervical height | 0.948 | |
| Waist circumference | | 0.745 |
| Hip circumference | | 0.744 |
| Bideltoid | | 0.706 |
| Cross-back width | | 0.486 |
| Hip breadth | | 0.649 |
| Shoulder circumference | | 0.806 |
| Neck circumference | | 0.627 |
| **Arm** | | |
| Inside arm length | 0.718 | |
| Sleeve length | 0.624 | |
| Inseam | 0.819 | |
| Forearm-hand length | 0.640 | |
| Shoulder-elbow length | 0.660 | |
| Elbow breadth | | 0.677 |
| Wrist circumference | | 0.556 |
| **Leg** | | |
| Lower leg | 0.688 | |
| Patella height | 0.795 | |
| Buttock-knee height | 0.751 | |
| Outseam | 0.886 | |
| Total crotch length | | 0.467 |
| Crotch-thigh circumference | | 0.731 |

## EFFECTS OF CLOTHING ON HUMAN BODY DIMENSIONS[10]

As noted previously, most human body dimensions were obtained using nude subjects. For practical applications, therefore, one must consider the effect that clothing, shoes, gloves, hats, or other items of apparel may have on specific base-line dimensions. The accompanying table provides rough approximations of how much clothing adds to, or subtracts from, given body dimensions. However, when a particular design problem indicates the possibility that some particular clothing item may influence clearance, reach, etc., one should select an appropriate subject sample and have subjects put on the specific clothing items pertinent to the expected user environment. Remember that clothing not only increases clearance dimensions but also often creates considerable restrictions to mobility and reach capability.

Key questions with regard to what type of clothing should be considered include whether an operation will be performed in shirt sleeves, at low temperatures, under reduced pressure, or in other special environments.

**Effect of Various Types of Clothing on Human Body Dimensions**

| Dimension | Street Clothes Men | Street Clothes Women | Winter Clothes Men | Winter Clothes Women | Heavy Flight Clothing | Pressure Suits* Unpressurized | Pressure Suits* Pressurized |
|---|---|---|---|---|---|---|---|
| Weight | 5 lb | 3½ lb | 10 lb | 7 lb | 12–15 lb | 21 lb | 21 lb |
| Stature | 1 in | ½–3¾ in | 1 in | ½–3¾ in | 3 in | 3½ in | 2½ in |
| Vertical reach | 1 | ½–3¾ | 1 | ½–3¾ | 1 | (−2½) | (−16½) |
| Eye height, standing | 1 | ½–3¾ | 1 | ½–3¾ | 1 | (−3½) | 2½ |
| Crotch to floor | 1 | ½–3¾ | 1 | ½–3¾ | 1 | (−1) | (−1) |
| Foot length | 1¼ | ½ | 1½ | ½–¾ | 1 | 1 | 1 |
| Foot width | ½–¾ | ¼–(−½)† | ½–1 | ¼–½ | ¾ | ¾ | ¾ |
| Head length | — | — | —‡ | | 4½ | 4½ | 4½ |
| Head width | — | — | —‡ | | 4½ | 4½ | 4½ |
| Hand length | — | — | ¾ | ½ | ½ | ½ | ¼ |
| Hand width | — | — | ½ | ¼ | ½ | ½ | 1 |
| Hand thickness | — | — | ½ | ¼ | ½ | ¾ | 1¼ |
| Fist circumference | — | — | 1 | ¾ | 1 | 1¼ | 3 |
| Shoulder width | ½ | ¼ | 2–3 | 1 | 1½ | 1 | ½ |
| Hip width | ½ | ¼ | 2–3 | 1 | 1½ | 1 | 2¾ |
| Elbow-to-elbow width | ¾ | ¼ | 2–3½ | 1–1½ | 1 | 6 | 9 |
| Thigh clearance | ½ | ¼ | 1 | ¾ | 2 | 1¾ | 2 |
| Forearm-to-fist length | ½ | ¼ | ¾ | ½ | 1 | 1½ | 5½ |

*Certain pressurized garments are designed for seated position and therefore shorten certain dimensions. The helmet, however, tends to rise under pressure.

†Women's dress shoes confine and shrink foot width.

‡An Army steel helmet is approximately 12 by 10¼ in.

[9]W. E. Woodson and D. W. Conover, *Human Engineering Guide for Equipment Designers,* University of California Press, Berkeley, 1964.

[10]W. E. Woodson and D. W. Conover, *Human Engineering Guide for Equipment Designers,* University of California Press, Berkeley, 1964.

**FUNCTIONAL DIMENSIONS**[11]

**Grasping Reach for Air Force Men to a Horizontal Plane at the Seat Reference Level (SRL) (Seat-Pan Angle—6°; Backrest—13°)**

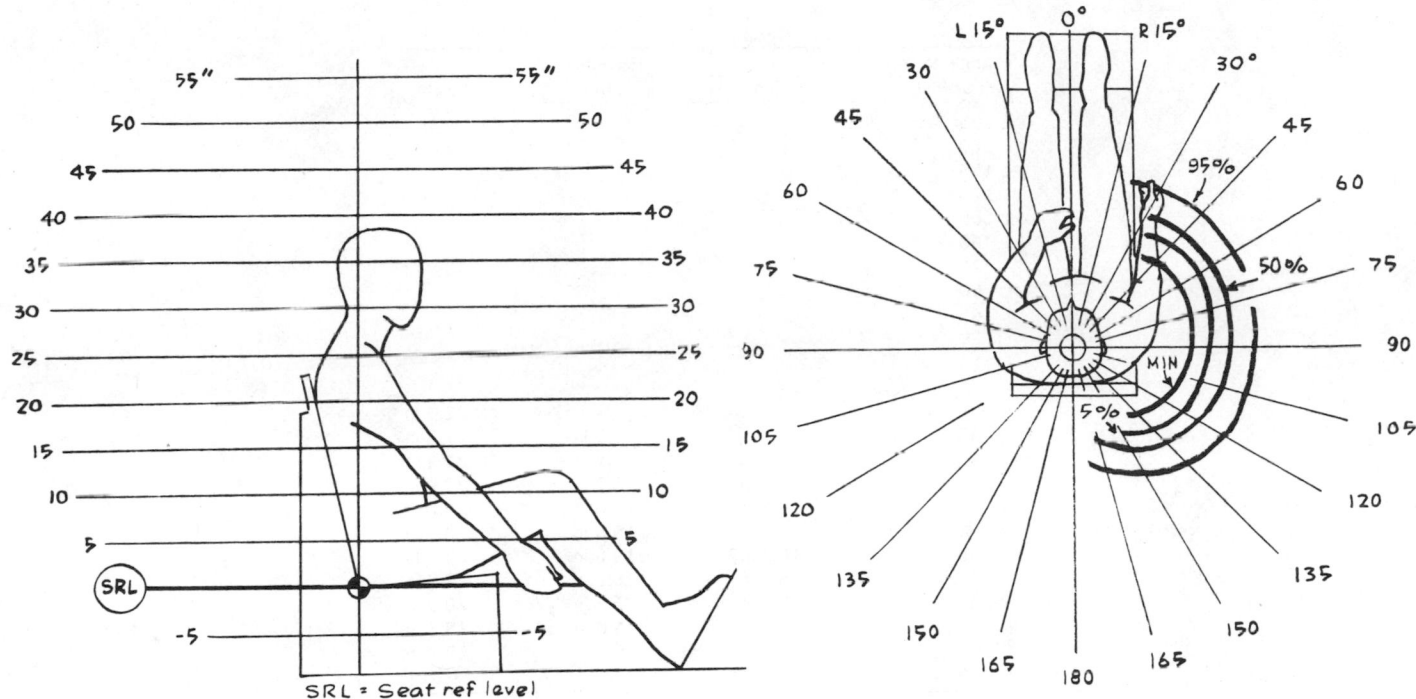

SRL = Seat ref level

| Angle to Left or Right | Minimum | Percentile | | |
|---|---|---|---|---|
| | | 5th | 50th | 95th |
| L 165 | | | | |
| L 150 | | | | |
| L 135 | | | | |
| L 120 | | | | |
| L 105 | | | | |
| L 90 | | | | |
| L 75 | | | | |
| L 60 | | | | |
| L 45 | | | | |
| L 30 | | | | |
| L 15 | | | | |
| 0 | | | | |
| R 15 | | | | |
| R 30 | | 17.50 in | 20.75 in | 25.00 in |
| R 45 | 16.25 in | 19.50 in | 21.75 in | 26.00 in |
| R 60 | 17.50 in | 20.50 in | 22.25 in | 26.25 in |
| R 75 | 17.25 in | 20.00 in | 22.25 in | 26.00 in |
| R 90 | 17.00 in | 19.50 in | 22.25 in | 25.50 in |
| R 105 | 16.25 in | 18.75 in | 22.00 in | 25.25 in |
| R 120 | 15.00 in | 18.25 in | 20.75 in | 24.50 in |
| R 135 | 13.00 in | 16.50 in | 19.00 in | 23.50 in |
| R 150 | | 14.00 in | 16.50 in | 20.25 in |
| R 165 | | | 13.00 in | 17.00 in |
| 180 | | | | |

[11]*Anthropometric Source Book*, NASA Reference Publication 1024, 1978.

**Grasping Reach for Air Force Women
to a Horizontal Plane at the SRL**

| Angle to Left or Right | Percentile | | |
|---|---|---|---|
| | 5th | 50th | 95th |
| L 165 | | | |
| L 150 | | | |
| L 135 | | | |
| L 120 | | | |
| L 105 | | | |
| L 90 | | | |
| L 75 | | | |
| L 60 | | | |
| L 45 | | | |
| L 30 | | | |
| L 15 | | | |
| 0 | | | |
| R 15 | | | 22.0 in |
| R 30 | | 16.2 in | 21.7 in |
| R 45 | 14.0 in | 17.5 in | 22.2 in |
| R 60 | 15.2 in | 18.7 in | 23.0 in |
| R 75 | 16.2 in | 19.0 in | 23.7 in |
| R 90 | 16.7 in | 19.5 in | 23.7 in |
| R 105 | 16.0 in | 19.0 in | 23.0 in |
| R 120 | 15.2 in | 18.2 in | 22.0 in |
| R 135 | 13.0 in | 16.5 in | 20.5 in |
| R 150 | | 13.0 in | 18.7 in |
| R 165 | | | 15.7 in |
| R 180 | | | |

**Grasping Reach for Air Force Men to
a Horizontal Plane 5 in (13 cm)
above the SRL**

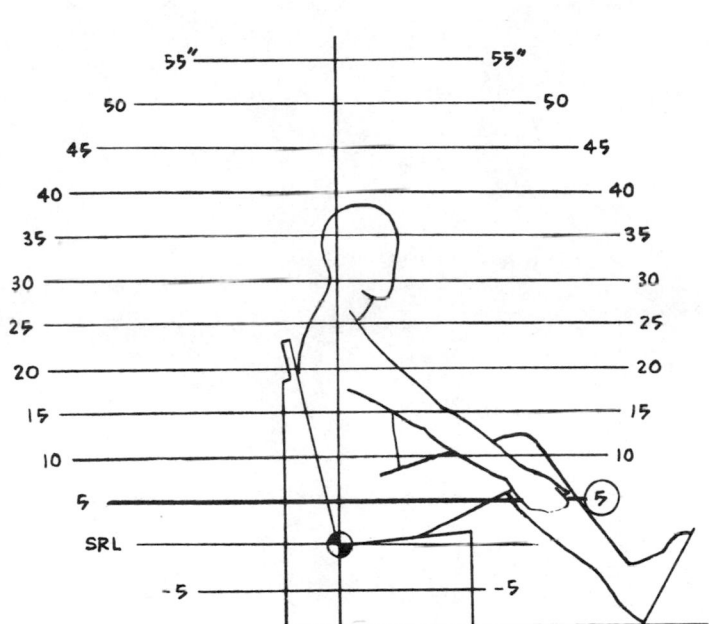

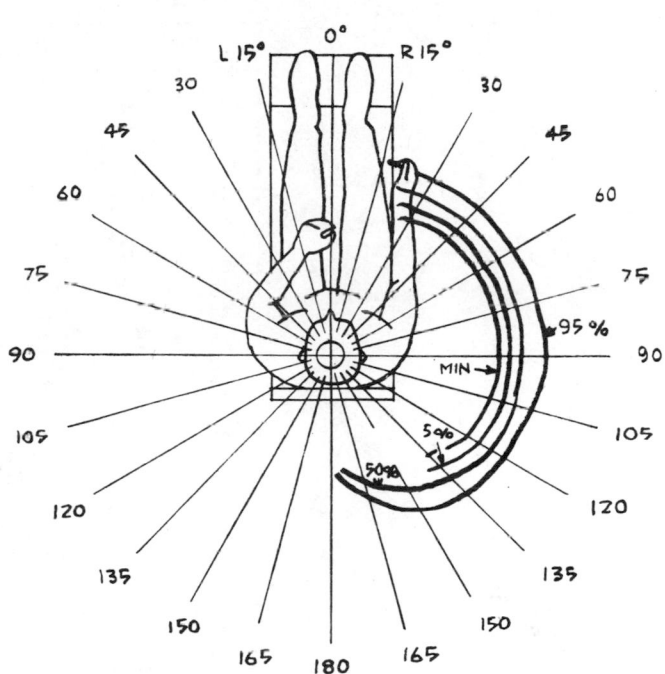

| Angle to | | Percentile | | |
| Left or Right | Minimum | 5th | 50th | 95th |
|---|---|---|---|---|
| L 165 | | | | |
| L 150 | | | | |
| L 135 | | | | |
| L 120 | | | | |
| L 105 | | | | |
| L 90 | | | | |
| L 75 | | | | |
| L 60 | | | | |
| L 45 | | | | |
| L 30 | | | | |
| L 15 | | | | |
| 0 | | | | |
| R 15 | | | | |
| R 30 | 22.00 in | 23.75 in | 26.00 in | 29.50 in |
| R 45 | 23.50 in | 25.25 in | 27.25 in | 30.00 in |
| R 60 | 23.75 in | 25.75 in | 27.75 in | 30.00 in |
| R 75 | 24.00 in | 25.75 in | 27.50 in | 30.25 in |
| R 90 | 24.00 in | 25.75 in | 27.50 in | 30.75 in |
| R 105 | 23.75 in | 25.25 in | 27.00 in | 30.00 in |
| R 120 | 23.00 in | 24.50 in | 26.50 in | 29.00 in |
| R 135 | 21.50 in | 22.75 in | 25.00 in | 28.00 in |
| R 150 | | | 22.25 in | 25.75 in |
| R 165 | | | 19.25 in | 21.25 in |
| 180 | | | | |

**Grasping Reach for Air Force Women
to a Horizontal Plane 6 in (15 cm)
above the SRL[12]**

| Angle to | Percentile | | |
|---|---|---|---|
| Left or Right | 5th | 50th | 95th |
| L 165 | | | |
| L 150 | | | |
| L 135 | | | |
| L 120 | | | |
| L 105 | | | 10.5 in |
| L  90 | | | 11.5 in |
| L  75 | | | 14.5 in |
| L  60 | | | 16.0 in |
| L  45 | | | 18.0 in |
| L  30 | | | 20.0 in |
| L  15 | | | — |
| 0 | | | |
| R  15 | 20.0 in | 22.5 in | 26.5 in |
| R  30 | 21.0 in | 23.0 in | 27.5 in |
| R  45 | 21.5 in | 23.7 in | 28.0 in |
| R  60 | 23.2 in | 25.0 in | 28.0 in |
| R  75 | 23.7 in | 25.0 in | 28.5 in |
| R  90 | 23.7 in | 25.2 in | 28.5 in |
| R 105 | 23.2 in | 25.0 in | 27.7 in |
| R 120 | 22.0 in | 24.0 in | 26.2 in |
| R 135 | 20.7 in | 23.0 in | 25.5 in |
| R 150 | — | 20.0 in | 24.0 in |
| R 165 | — | 16.2 in | 21.0 in |
| R 180 | | | |

[12]Note that although the reach charts for males and females have been paired throughout this section, only a single illustration of the male has been provided for each pair. However, it is important also to note that the reference levels above the SRL for the data on females are not quite the same as those for the data on males. For example, the data on males corresponding to the accompanying table were taken at 5 in (13 cm) above the SRL, while the above data were taken at 6 in (15 cm) above the SRL. This slight variation in reference level occurs throughout the remaining pairs of data.

**Grasping Reach for Air Force Men to
a Horizontal Plane 10 in (25 cm)
above the SRL**

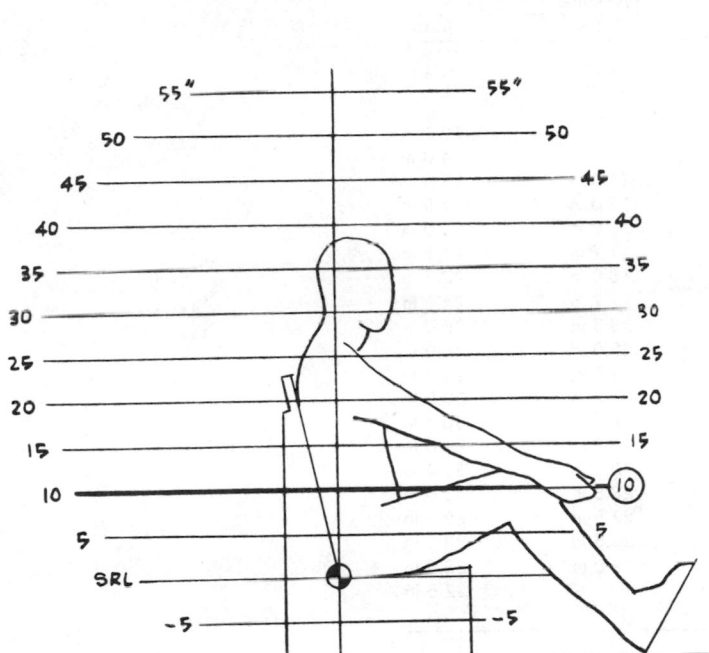

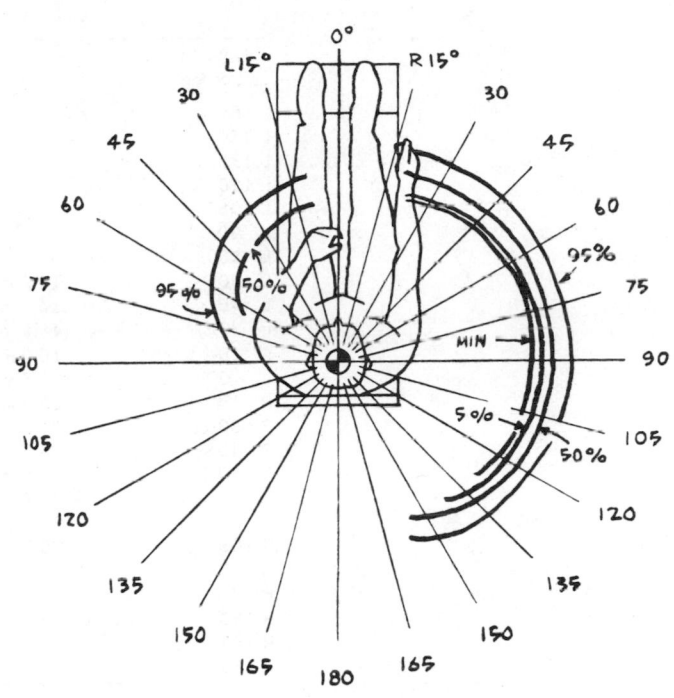

| Angle to Left or Right | Minimum | Percentile | | |
|---|---|---|---|---|
| | | 5th | 50th | 95th |
| L 165 | | | | |
| L 150 | | | | |
| L 135 | | | | |
| L 120 | | | | |
| L 105 | | | | |
| L 90 | | | | 13.50 in |
| L 75 | | | | 17.25 in |
| L 60 | | | 16.50 in | 21.00 in |
| L 45 | | | 19.50 in | 23.25 in |
| L 30 | | | 21.00 in | 24.75 in |
| L 15 | | | 22.00 in | 26.25 in |
| 0 | | | | |
| R 15 | | | | |
| R 30 | 26.25 in | 27.00 in | 29.25 in | 33.00 in |
| R 45 | 27.25 in | 28.25 in | 30.50 in | 33.75 in |
| R 60 | 28.00 in | 29.00 in | 30.75 in | 33.50 in |
| R 75 | 28.25 in | 29.25 in | 30.75 in | 33.50 in |
| R 90 | 28.25 in | 29.25 in | 31.00 in | 33.50 in |
| R 105 | 27.75 in | 28.75 in | 30.50 in | 32.75 in |
| R 120 | 26.75 in | 27.75 in | 29.75 in | 31.50 in |
| R 135 | — | 26.25 in | 28.25 in | 30.75 in |
| R 150 | — | — | 25.25 in | 28.75 in |
| R 165 | | | | |
| 180 | | | | |

**Grasping Reach for Air Force Women
to a Horizontal Plane 12 in (31 cm)
above the SRL**

| Angle to | Percentile | | |
| Left or Right | 5th | 50th | 95th |
|---|---|---|---|
| L 165 | | | |
| L 150 | | | |
| L 135 | | | |
| L 120 | | | 12.7 in |
| L 105 | | | 14.0 in |
| L 90 | | 11.0 in | 15.5 in |
| L 75 | | 13.0 in | 17.5 in |
| L 60 | 12.2 in | 15.0 in | 20.0 in |
| L 45 | 14.5 in | 17.7 in | 21.5 in |
| L 30 | 16.5 in | 10.0 in | 22.7 in |
| L 15 | 19.0 in | 21.7 in | 24.5 in |
| 0 | 21.5 in | 23.5 in | 26.0 in |
| R 15 | 23.0 in | 25.0 in | 28.0 in |
| R 30 | 24.0 in | 26.0 in | 29.2 in |
| R 45 | 25.5 in | 27.2 in | 30.0 in |
| R 60 | 26.5 in | 28.2 in | 30.7 in |
| R 75 | 26.7 in | 28.2 in | 31.0 in |
| R 90 | 27.2 in | 28.5 in | 31.0 in |
| R 105 | 26.5 in | 28.5 in | 31.0 in |
| R 120 | — | 27.5 in | 29.5 in |
| R 135 | — | 25.5 in | 28.2 in |
| R 150 | — | 19.0 in | 25.0 in |
| R 165 | | | 22.5 in |
| R 180 | | | |

**Grasping Reach for Air Force Men to
a Horizontal Plane 15 in (38 cm)
above the SRL**

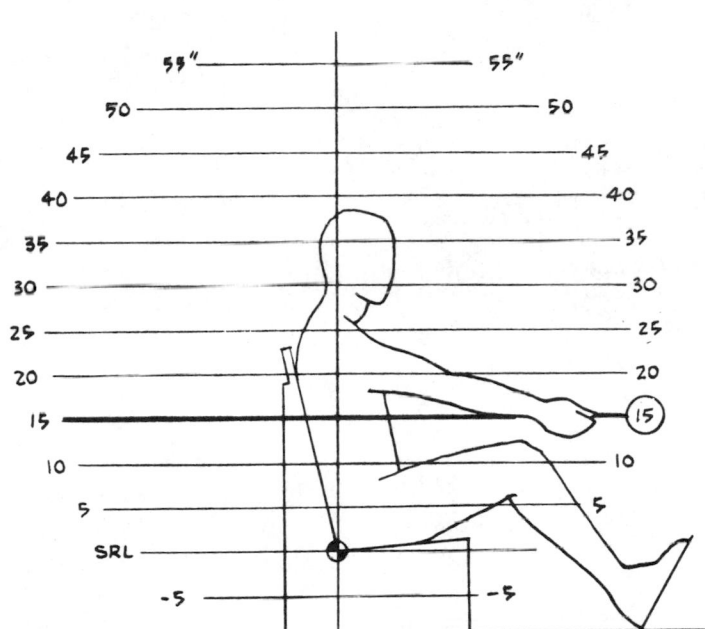

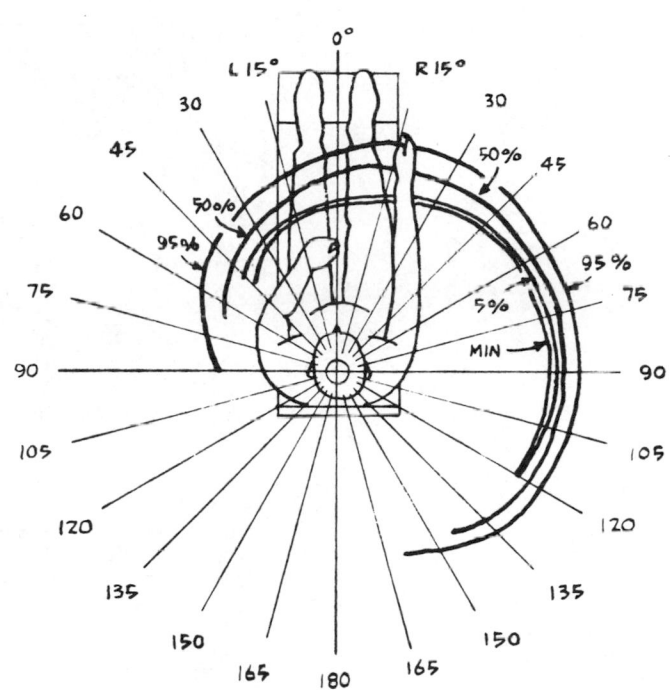

| Angle to Left or Right | Minimum | Percentile | | |
|---|---|---|---|---|
| | | 5th | 50th | 95th |
| L 165 | | | | |
| L 150 | | | | |
| L 135 | | | | |
| L 120 | | | | |
| L 105 | | | | |
| L 90 | | | | 17.50 in |
| L 75 | | | | 20.00 in |
| L 60 | | | 19.25 in | 23.00 in |
| L 45 | | 19.00 in | 21.50 in | 25.75 in |
| L 30 | 21.00 in | 21.75 in | 24.00 in | 27.25 in |
| L 15 | 22.50 in | 23.25 in | 26.00 in | 28.75 in |
| 0 | 24.25 in | 24.75 in | 28.75 in | 31.00 in |
| R 15 | 26.00 in | 26.60 in | 30.50 in | 34.00 in |
| R 30 | 28.25 in | 28.50 in | 31.50 in | 35.00 in |
| R 45 | 29.50 in | 30.00 in | 32.75 in | 35.50 in |
| R 60 | 30.00 in | 31.00 in | 32.50 in | 34.75 in |
| R 75 | 30.00 in | 31.50 in | 32.50 in | 34.75 in |
| R 90 | 30.25 in | 31.00 in | 32.50 in | 34.75 in |
| R 105 | 30.00 in | 30.75 in | 32.25 in | 34.50 in |
| R 120 | 29.00 in | 29.50 in | 32.00 in | 33.75 in |
| R 135 | — | — | 30.00 in | 32.50 in |
| R 150 | — | — | — | 29.50 in |
| R 165 | | | | |
| 180 | | | | |

**Grasping Reach for Air Force Women
to a Horizontal Plane 18 in (46 cm)
above the SRL**

| Angle to Left or Right | Percentile | | |
|---|---|---|---|
| | 5th | 50th | 95th |
| L 165 | | | |
| L 150 | | | |
| L 135 | | | |
| L 120 | | | 14.0 in |
| L 105 | | 11.0 in | 15.5 in |
| L  90 | 10.5 in | 13.0 in | 17.2 in |
| L  75 | 11.7 in | 15.0 in | 19.7 in |
| L  60 | 14.0 in | 17.7 in | 21.0 in |
| L  45 | 16.7 in | 19.7 in | 23.0 in |
| L  30 | 18.7 in | 21.5 in | 24.2 in |
| L  15 | 20.0 in | 23.0 in | 26.0 in |
| 0 | 22.5 in | 24.7 in | 27.5 in |
| R  15 | 24.2 in | 26.2 in | 29.5 in |
| R  30 | 25.5 in | 27.5 in | 30.2 in |
| R  45 | 26.7 in | 28.7 in | 31.0 in |
| R  60 | 27.7 in | 29.5 in | 32.0 in |
| R  75 | 27.7 in | 29.7 in | 32.0 in |
| R  90 | 28.0 in | 30.0 in | 31.7 in |
| R 105 | 27.5 in | 30.2 in | 32.2 in |
| R 120 | — | 28.7 in | 31.0 in |
| R 135 | — | — | 28.2 in |
| R 150 | — | — | 15.0 in |
| R 165 | | | |
| 180 | | | |

**Grasping Reach for Air Force Men to
a Horizontal Plane 20 in (51 cm)
above the SRL**

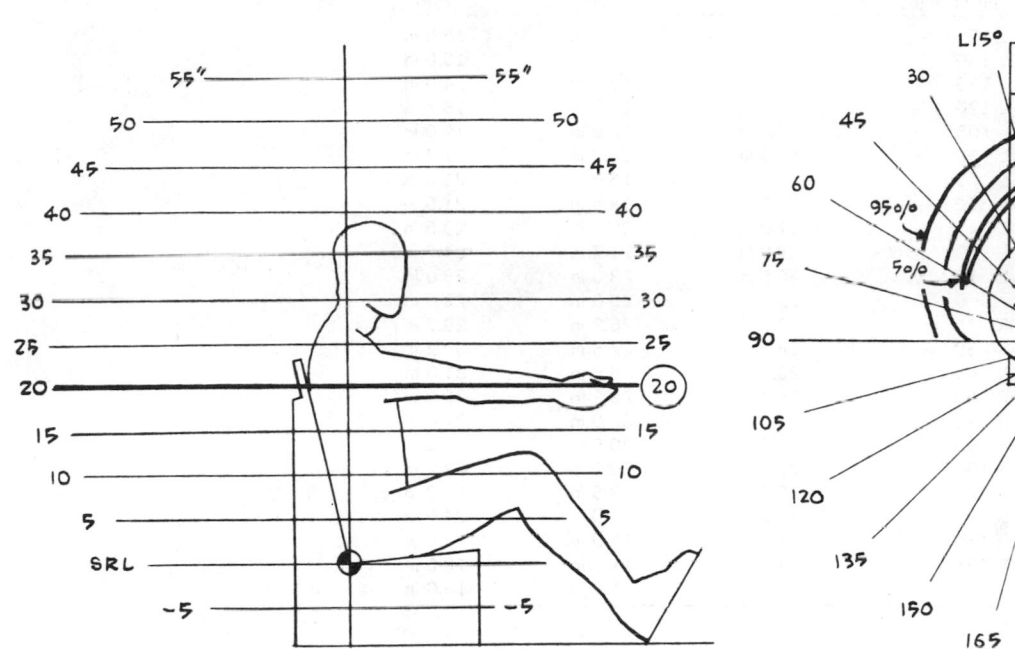

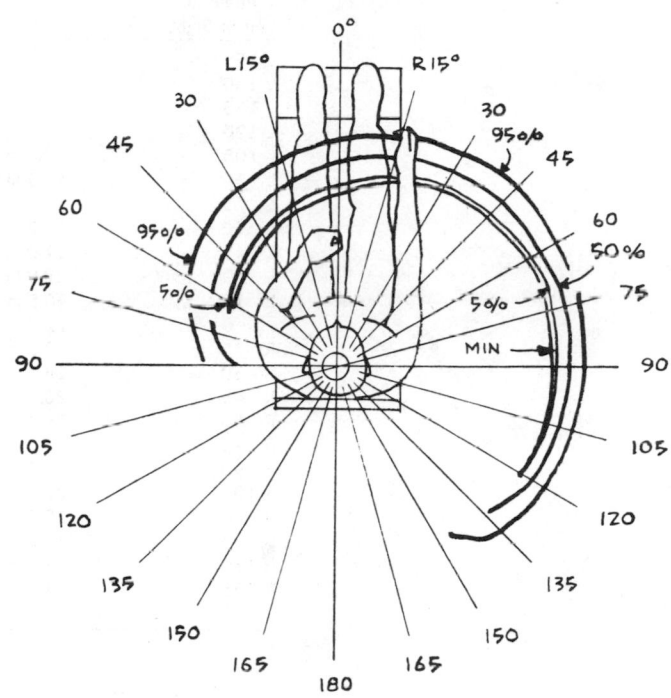

| Angle to | | Percentile | | |
| Left or Right | Minimum | 5th | 50th | 95th |
|---|---|---|---|---|
| L 165 | | | | |
| L 150 | | | | |
| L 135 | | | | |
| L 120 | | | | |
| L 105 | | | | |
| L 90 | | | 14.00 in | 18.75 in |
| L 75 | | | 18.00 in | 21.50 in |
| L 60 | 17.00 in | 17.50 in | 20.50 in | 24.50 in |
| L 45 | 18.25 in | 19.50 in | 22.75 in | 26.75 in |
| L 30 | 20.25 in | 21.50 in | 24.75 in | 28.25 in |
| L 15 | 22.50 in | 23.50 in | 26.75 in | 29.75 in |
| 0 | 25.00 in | 25.50 in | 28.75 in | 31.75 in |
| R 15 | 27.25 in | 28.00 in | 30.50 in | 34.00 in |
| R 30 | 29.00 in | 30.00 in | 32.00 in | 35.75 in |
| R 45 | 30.50 in | 31.00 in | 33.50 in | 36.25 in |
| R 60 | 31.50 in | 32.00 in | 33.75 in | 36.25 in |
| R 75 | 31.50 in | 32.25 in | 34.00 in | 36.50 in |
| R 90 | 31.75 in | 32.25 in | 34.00 in | 36.00 in |
| R 105 | 31.50 in | 31.75 in | 33.50 in | 35.75 in |
| R 120 | — | 30.50 in | 33.00 in | 35.50 in |
| R 135 | — | — | — | 34.50 in |
| R 150 | | | | |
| R 165 | | | | |
| 180 | | | | |

**Grasping Reach for Air Force Women
to a Horizontal Plane 24 in (61 cm)
above the SRL**

| Angle to Left or Right | Percentile | | |
|---|---|---|---|
| | 5th | 50th | 95th |
| L 165 | | 9.0 in | 15.0 in |
| L 150 | | 9.0 in | 16.0 in |
| L 135 | | 10.7 in | 14.0 in |
| L 120 | | 10.0 in | 16.7 in |
| L 105 | 8.0 in | 12.2 in | 19.0 in |
| L 90 | 10.0 in | 14.7 in | 17.7 in |
| L 75 | 11.5 in | 16.0 in | 21.0 in |
| L 60 | 14.2 in | 18.5 in | 21.5 in |
| L 45 | 17.0 in | 20.0 in | 23.5 in |
| L 30 | 19.0 in | 21.7 in | 24.7 in |
| L 15 | 20.5 in | 23.0 in | 26.0 in |
| 0 | 22.0 in | 25.0 in | 28.0 in |
| R 15 | 23.5 in | 26.2 in | 29.5 in |
| R 30 | 25.0 in | 27.5 in | 30.2 in |
| R 45 | 26.2 in | 28.5 in | 31.0 in |
| R 60 | 26.7 in | 29.2 in | 32.0 in |
| R 75 | 27.0 in | 30.0 in | 32.0 in |
| R 90 | 27.5 in | 30.5 in | 32.0 in |
| R 105 | 27.2 in | 30.2 in | 32.2 in |
| R 120 | 13.0 in | 28.5 in | 31.0 in |
| R 135 | 11.0 in | 14.0 in | 27.0 in |
| R 150 | 9.0 in | 12.0 in | 22.0 in |
| R 165 | 8.2 in | 11.1 in | 18.0 in |
| 180 | — | 11.0 in | 16.0 in |

**Grasping Reach for Air Force Men to
a Horizontal Plane 25 in (64 cm)
above the SRL**

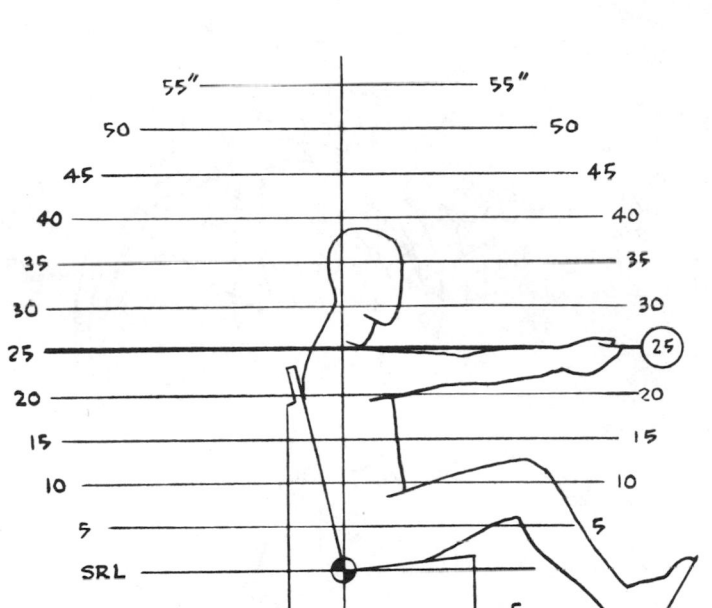

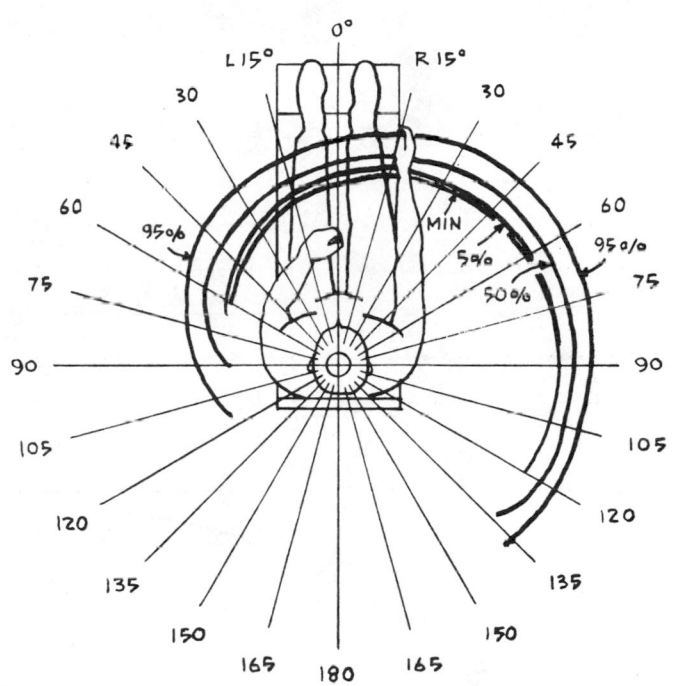

| Angle to | | Percentile | | |
|---|---|---|---|---|
| Left or Right | Minimum | 5th | 50th | 95th |
| L 165 | | | | |
| L 150 | | | | |
| L 135 | | | | |
| L 120 | | | | |
| L 105 | | | | 17.75 in |
| L 90 | | | 15.75 in | 20.25 in |
| L 75 | | | 19.25 in | 22.25 in |
| L 60 | 17.75 in | 18.25 in | 21.50 in | 24.75 in |
| L 45 | 19.25 in | 20.00 in | 23.25 in | 27.25 in |
| L 30 | 21.50 in | 22.50 in | 25.00 in | 28.50 in |
| L 15 | 23.25 in | 24.00 in | 27.00 in | 29.75 in |
| 0 | 25.00 in | 26.25 in | 28.50 in | 31.50 in |
| R 15 | 27.25 in | 28.25 in | 30.25 in | 32.60 in |
| R 30 | 29.25 in | 30.25 in | 32.50 in | 35.25 in |
| R 45 | 30.50 in | 31.00 in | 33.50 in | 35.75 in |
| R 60 | 31.00 in | 31.50 in | 33.75 in | 37.00 in |
| R 75 | 31.50 in | 32.00 in | 33.50 in | 36.50 in |
| R 90 | 31.75 in | 32.25 in | 33.75 in | 36.25 in |
| R 105 | 31.25 in | 31.50 in | 33.50 in | 36.00 in |
| R 120 | — | 30.50 in | 33.25 in | 35.50 in |
| R 135 | | | | 35.00 in |
| R 150 | | | | |
| R 165 | | | | |
| 180 | | | | |

**Grasping Reach for Air Force Men to
a Horizontal Plane 30 in (76 cm)
above the SRL**

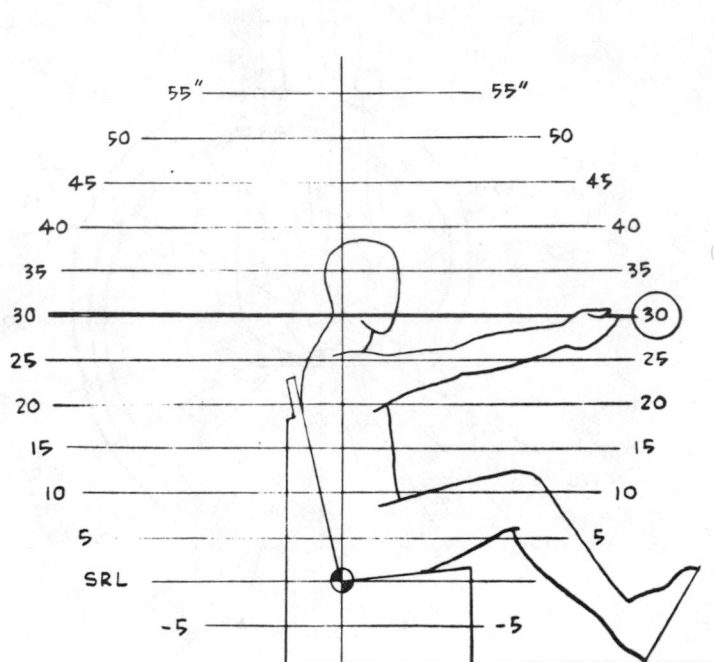

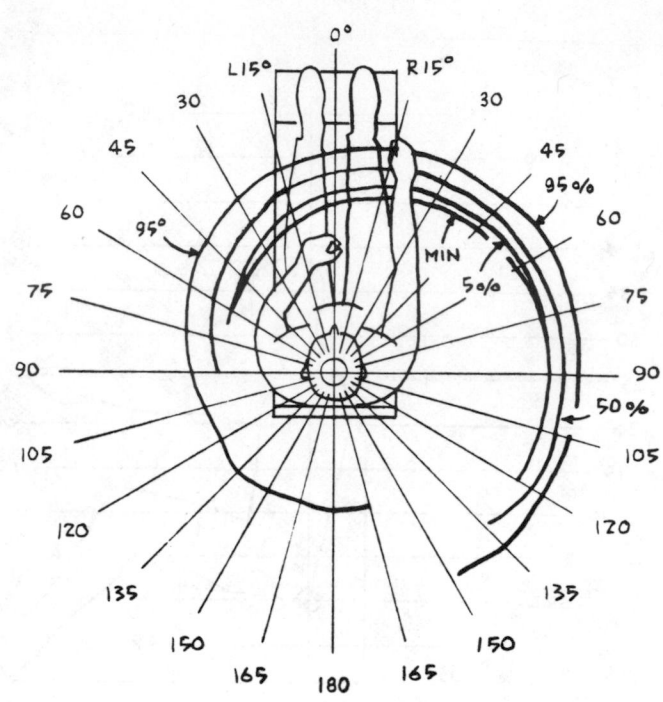

| Angle to Left or Right | Minimum | Percentile | | |
|---|---|---|---|---|
| | | 5th | 50th | 95th |
| L 165 | | | | 18.75 in |
| L 150 | | | | 19.25 in |
| L 135 | | | | 20.00 in |
| L 120 | | | | 18.75 in |
| L 105 | | | | 19.00 in |
| L 90 | | | 16.75 in | 20.75 in |
| L 75 | | | 18.75 in | 22.50 in |
| L 60 | 17.00 in | 17.25 in | 20.75 in | 24.50 in |
| L 45 | 18.25 in | 19.00 in | 22.50 in | 26.50 in |
| L 30 | 19.75 in | 21.50 in | 24.50 in | 28.25 in |
| L 15 | 22.00 in | 23.75 in | 26.75 in | 29.50 in |
| 0 | 23.75 in | 25.50 in | 28.50 in | 31.00 in |
| R 15 | 26.00 in | 27.25 in | 29.75 in | 33.00 in |
| R 30 | 27.75 in | 29.00 in | 31.50 in | 34.25 in |
| R 45 | 28.75 in | 30.25 in | 32.25 in | 34.75 in |
| R 60 | 30.00 in | 31.00 in | 32.75 in | 35.75 in |
| R 75 | 30.75 in | 31.25 in | 33.00 in | 35.50 in |
| R 90 | 31.00 in | 31.25 in | 33.25 in | 35.75 in |
| R 105 | 30.75 in | 31.00 in | 33.00 in | 35.25 in |
| R 120 | — | 30.25 in | 32.50 in | 34.75 in |
| R 135 | | | | 34.50 in |
| R 150 | | | | — |
| R 165 | | | | 19.50 |
| 180 | | | | 20.25 |

**Grasping Reach for Air Force Women
to a Horizontal Plane 30 in (76 cm)
above the SRL**

| Angle to Left or Right | Percentile | | |
|---|---|---|---|
| | 5th | 50th | 95th |
| L 165 | 7.2 in | 12.5 in | 19.2 in |
| L 150 | 6.2 in | 12.0 in | 16.5 in |
| L 135 | 6.7 in | 8.7 in | 15.2 in |
| L 120 | 7.0 in | 10.7 in | 17.0 in |
| L 105 | 6.5 in | 12.0 in | 18.0 in |
| L 90 | 8.7 in | 13.0 in | 17.2 in |
| L 75 | 10.0 in | 15.5 in | 20.0 in |
| L 60 | 13.0 in | 17.5 in | 21.0 in |
| L 45 | 15.0 in | 19.0 in | 22.0 in |
| L 30 | 17.0 in | 20.5 in | 24.2 in |
| L 15 | 18.2 in | 22.0 in | 25.2 in |
| 0 | 20.0 in | 23.0 in | 27.0 in |
| R 15 | 21.5 in | 24.5 in | 28.2 in |
| R 30 | 22.5 in | 25.7 in | 29.0 in |
| R 45 | 23.2 in | 27.5 in | 29.7 in |
| R 60 | 24.5 in | 27.7 in | 30.5 in |
| R 75 | 25.2 in | 28.5 in | 30.2 in |
| R 90 | 25.7 in | 28.7 in | 31.0 in |
| R 105 | 26.0 in | 29.0 in | 31.0 in |
| R 120 | 16.2 in | 26.2 in | 29.5 in |
| R 135 | 12.7 in | 19.5 in | 27.5 in |
| R 150 | 11.0 in | 16.2 in | 23.5 in |
| R 165 | 10.5 in | 15.5 in | 22.0 in |
| 180 | 9.5 in | 15.0 in | 20.0 in |

**Grasping Reach for Air Force Men to
a Horizontal Plane 35 in (89 cm)
above the SRL**

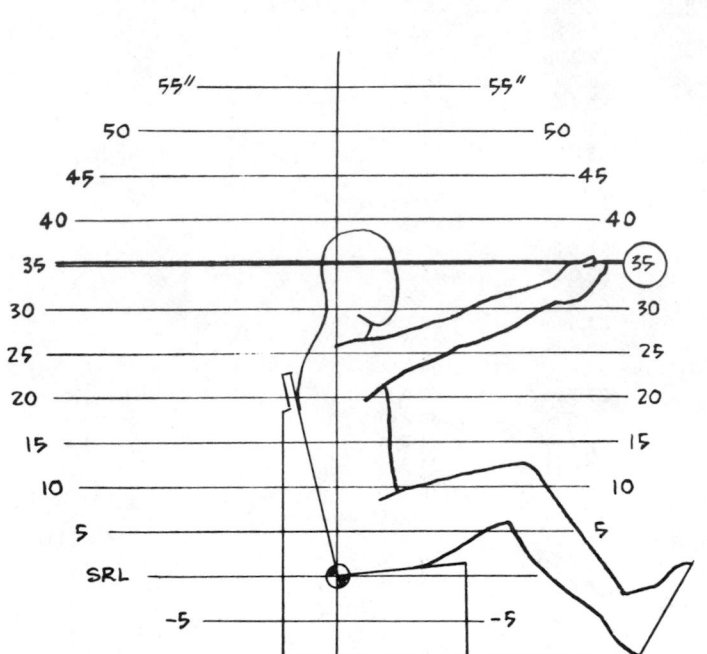

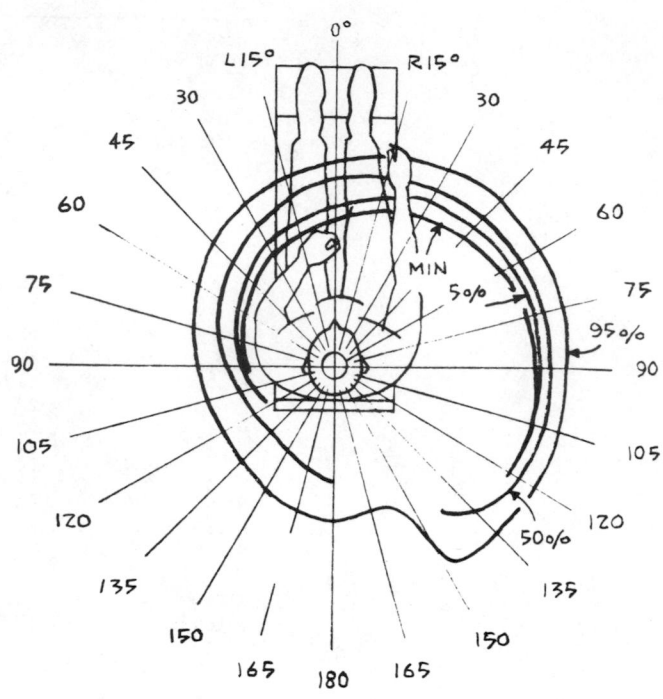

| Angle to | | Percentile | | |
| Left or Right | Minimum | 5th | 50th | 95th |
|---|---|---|---|---|
| L 165 | | | 14.75 in | 21.00 in |
| L 150 | | | 13.75 in | 20.00 in |
| L 135 | | | 13.25 in | 19.00 in |
| L 120 | | 10.75 in | 13.25 in | 18.75 in |
| L 105 | | 12.25 in | 14.00 in | 18.75 in |
| L 90 | 12.75 in | 13.75 in | 15.50 in | 20.00 in |
| L 75 | 14.25 in | 15.00 in | 17.25 in | 21.00 in |
| L 60 | 15.25 in | 16.00 in | 18.75 in | 21.50 in |
| L 45 | 16.25 in | 17.25 in | 20.50 in | 24.75 in |
| L 30 | 18.00 in | 19.25 in | 22.50 in | 26.25 in |
| L 15 | 19.25 in | 21.00 in | 24.75 in | 27.00 in |
| 0 | 20.75 in | 22.25 in | 26.50 in | 28.50 in |
| R 15 | 22.75 in | 24.75 in | 27.75 in | 31.00 in |
| R 30 | 24.50 in | 26.75 in | 29.25 in | 32.75 in |
| R 45 | 26.75 in | 28.25 in | 30.50 in | 33.75 in |
| R 60 | 28.00 in | 29.00 in | 31.00 in | 33.75 in |
| R 75 | 28.75 in | 29.50 in | 31.25 in | 34.00 in |
| R 90 | 29.00 in | 29.75 in | 31.25 in | 33.50 in |
| R 105 | 29.00 in | 29.75 in | 31.50 in | 33.50 in |
| R 120 | 28.50 in | 29.00 in | 31.00 in | 33.50 in |
| R 135 | | | 28.50 in | 33.50 in |
| R 150 | | | | 31.50 in |
| R 165 | | | | 21.75 in |
| 180 | | | 16.50 in | 22.25 in |

**Grasping Reach for Air Force Women
to a Horizontal Plane 36 in (91 cm)
above the SRL**

| Angle to | Percentile | | |
| Left or Right | 5th | 50th | 95th |
| --- | --- | --- | --- |
| L 165 | 9.0 in | 13.0 in | 19.5 in |
| L 150 | 8.0 in | 11.5 in | 17.7 in |
| L 135 | 7.2 in | 10.2 in | 16.0 in |
| L 120 | 7.2 in | 10.0 in | 15.5 in |
| L 105 | 7.2 in | 10.5 in | 15.2 in |
| L  90 | 7.7 in | 11.5 in | 16.0 in |
| L  75 | 8.2 in | 13.0 in | 17.2 in |
| L  60 | 10.0 in | 14.2 in | 18.0 in |
| L  45 | 11.5 in | 15.5 in | 19.5 in |
| L  30 | 13.2 in | 17.2 in | 21.5 in |
| L  15 | 14.2 in | 19.0 in | 22.7 in |
|   0 | 16.2 in | 20.5 in | 24.0 in |
| R  15 | 17.5 in | 21.5 in | 14.7 in |
| R  30 | 18.5 in | 22.5 in | 26.0 in |
| R  45 | 19.2 in | 24.0 in | 27.0 in |
| R  60 | 20.7 in | 25.0 in | 27.7 in |
| R  75 | 21.0 in | 25.5 in | 28.0 in |
| R  90 | 22.2 in | 26.2 in | 28.7 in |
| R 105 | 21.2 in | 26.2 in | 28.7 in |
| R 120 | 18.2 in | 25.0 in | 27.7 in |
| R 135 | 12.5 in | 19.0 in | 25.7 in |
| R 150 | 10.0 in | 17.2 in | 23.5 in |
| R 165 | 10.2 in | 16.0 in | 22.0 in |
|  180 | 9.5 in | 15.2 in | 21.2 in |

**Grasping Reach for Air Force Men to
a Horizontal Plane 40 in (102 cm)
above the SRL**

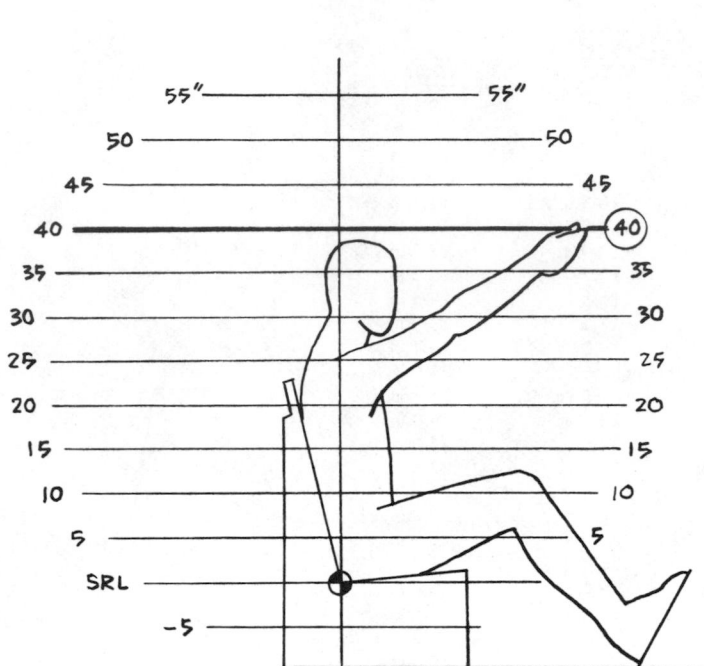
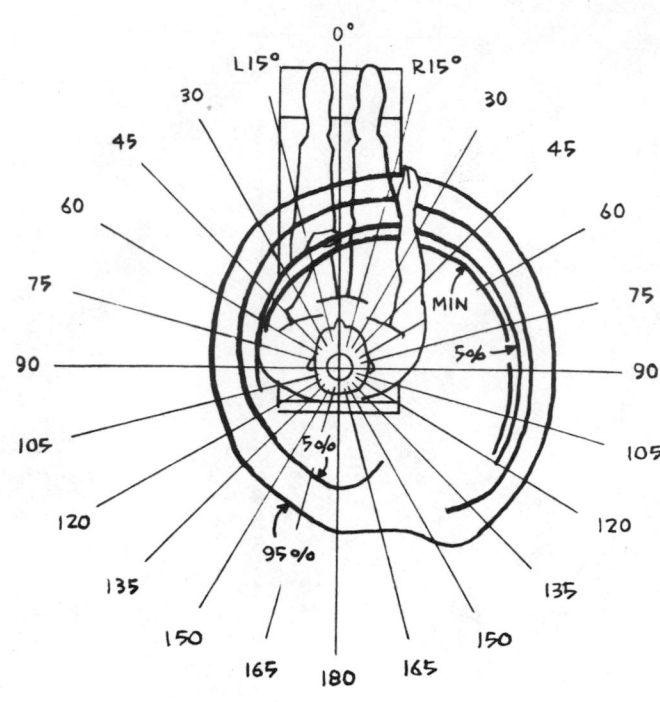

| Angle to | | Percentile | | |
| Left or Right | Minimum | 5th | 50th | 95th |
|---|---|---|---|---|
| L 165 | | | 15.50 in | 21.50 in |
| L 150 | | | 14.75 in | 20.00 in |
| L 135 | | | 14.00 in | 19.25 in |
| L 120 | | 11.25 in | 13.25 in | 18.50 in |
| L 105 | | 11.75 in | 13.25 in | 18.25 in |
| L  90 | 12.00 in | 12.25 in | 13.75 in | 18.25 in |
| L  75 | 12.25 in | 12.50 in | 15.00 in | 18.75 in |
| L  60 | 12.50 in | 13.25 in | 16.25 in | 20.00 in |
| L  45 | 13.00 in | 14.00 in | 17.75 in | 21.50 in |
| L  30 | 13.75 in | 15.50 in | 19.50 in | 23.50 in |
| L  15 | 15.25 in | 17.00 in | 21.25 in | 24.50 in |
|   0 | 17.00 in | 19.00 in | 23.00 in | 25.75 in |
| R  15 | 18.75 in | 21.00 in | 24.50 in | 28.50 in |
| R  30 | 21.00 in | 22.75 in | 26.25 in | 30.50 in |
| R  45 | 23.25 in | 24.75 in | 27.75 in | 31.50 in |
| R  60 | 24.25 in | 25.50 in | 28.00 in | 31.25 in |
| R  75 | 25.00 in | 26.00 in | 28.00 in | 31.50 in |
| R  90 | 25.00 in | 26.25 in | 28.25 in | 31.50 in |
| R 105 | 25.75 in | 26.75 in | 28.50 in | 31.75 in |
| R 120 | | 26.25 in | 28.75 in | 31.50 in |
| R 135 | | | 27.00 in | 31.00 in |
| R 150 | | | | 29.25 in |
| R 165 | | | 16.75 in | 23.75 in |
|   180 | | | 17.75 in | 23.50 in |

**Grasping Reach for Air Force Women
to a Horizontal Plane 42 in (107 cm)
above the SRL**

| Angle to Left or Right | Percentile | | |
|---|---|---|---|
| | 5th | 50th | 95th |
| L 165 | 5.0 in | 10.2 in | 17.0 in |
| L 150 | 4.2 in | 9.0 in | 15.0 in |
| L 135 | 3.7 in | 8.5 in | 13.7 in |
| L 120 | 3.5 in | 8.0 in | 13.0 in |
| L 105 | 3.2 in | 8.0 in | 12.5 in |
| L 90 | 3.5 in | 8.0 in | 13.0 in |
| L 75 | 3.7 in | 8.7 in | 14.5 in |
| L 60 | 4.0 in | 9.5 in | 16.2 in |
| L 45 | 4.7 in | 10.5 in | 16.0 in |
| L 30 | 5.5 in | 11.5 in | 17.0 in |
| L 15 | 6.5 in | 12.5 in | 17.7 in |
| 0 | 7.5 in | 14.0 in | 18.5 in |
| R 15 | 9.0 in | 16.0 in | 19.0 in |
| R 30 | 10.0 in | 17.0 in | 20.5 in |
| R 45 | 11.2 in | 17.5 in | 22.0 in |
| R 60 | 12.0 in | 19.0 in | 22.5 in |
| R 75 | 13.0 in | 20.0 in | 23.5 in |
| R 90 | 14.0 in | 20.0 in | 24.0 in |
| R 105 | 14.0 in | 20.5 in | 24.0 in |
| R 120 | 12.0 in | 18.5 in | 23.5 in |
| R 135 | 9.2 in | 15.5 in | 21.2 in |
| R 150 | 7.5 in | 14.0 in | 19.7 in |
| R 165 | 6.5 in | 12.2 in | 19.0 in |
| 180 | 5.5 in | 11.0 in | 18.7 in |

**Grasping Reach for Air Force Men to
a Horizontal Plane 45 in (114 cm)
above the SRL**

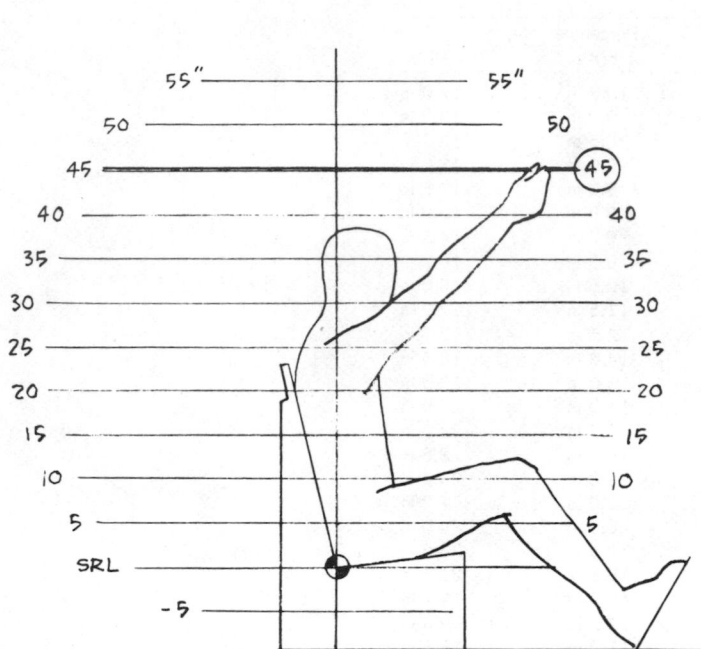

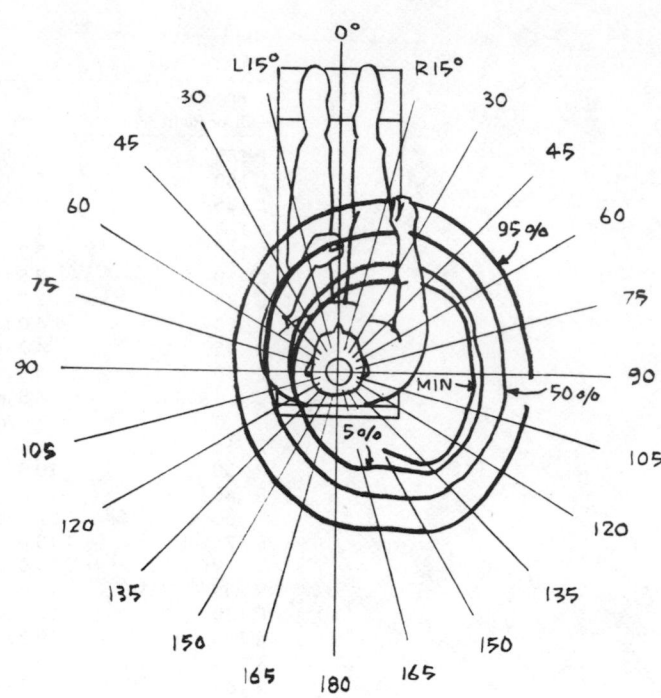

| Angle to | | Percentile | | |
| Left or Right | Minimum | 5th | 50th | 95th |
|---|---|---|---|---|
| L 165 | | 10.50 in | 14.00 in | 20.00 in |
| L 150 | 8.5 in | 8.75 in | 12.25 in | 18.25 in |
| L 135 | 7.50 in | 7.75 in | 11.00 in | 16.75 in |
| L 120 | 7.00 in | 7.50 in | 10.50 in | 15.50 in |
| L 105 | 6.75 in | 7.25 in | 10.25 in | 15.00 in |
| L 90 | 6.75 in | 7.25 in | 10.50 in | 15.00 in |
| L 75 | 6.75 in | 7.50 in | 11.00 in | 15.25 in |
| L 60 | 7.00 in | 7.75 in | 12.00 in | 16.25 in |
| L 45 | 7.50 in | 8.50 in | 13.50 in | 18.25 in |
| L 30 | 8.50 in | 9.50 in | 15.00 in | 19.75 in |
| L 15 | 10.00 in | 11.00 in | 16.50 in | 21.25 in |
| 0 | 11.25 in | 12.75 in | 18.25 in | 22.75 in |
| R 15 | 13.00 in | 15.50 in | 20.00 in | 24.75 in |
| R 30 | 14.75 in | 17.50 in | 22.00 in | 26.25 in |
| R 45 | 17.25 in | 19.00 in | 23.50 in | 27.00 in |
| R 60 | 19.25 in | 20.50 in | 24.00 in | 27.25 in |
| R 75 | 19.50 in | 20.50 in | 24.00 in | 27.50 in |
| R 90 | 19.75 in | 21.00 in | 24.25 in | 27.75 in |
| R 105 | 20.25 in | 21.50 in | 24.50 in | 28.00 in |
| R 120 | 19.75 in | 21.25 in | 24.50 in | 27.75 in |
| R 135 | 18.75 in | 20.00 in | 23.25 in | 27.75 in |
| R 150 | | 15.50 in | 20.75 in | 26.00 in |
| R 165 | | 14.75 in | 18.00 in | 22.75 in |
| 180 | | 12.75 in | 16.50 in | 21.50 in |

## Functional Reach Limits for the Typical Automobile Driver (5th-Percentile Female)[13]

On the basis of automobile seating geometry (see the accompanying sketch, which shows a typical family vehicle), maximum reach limits of a representative (U.S. driver population) 5th-percentile female driver were defined (see the accompanying table). Reach limits are described with reference to the standard H point used by the automobile industry. Limits are shown for both the left and right hands. Because of the constraints of the door panel on the left, values are included only for three distances from the steering-wheel centerline. The two central value sets represent reach around a typical steering wheel with a 15-in (38-cm) diameter. The first value represents reach limits when the subject is not constrained by a shoulder belt; i.e., she can lean forward and/or to the side. The second value represents limits when a tight shoulder belt is worn (i.e., the belt does not allow the driver to lean forward or to the side).

Reach limits for other conditions, e.g., different seating configurations, obviously would be different.[14]

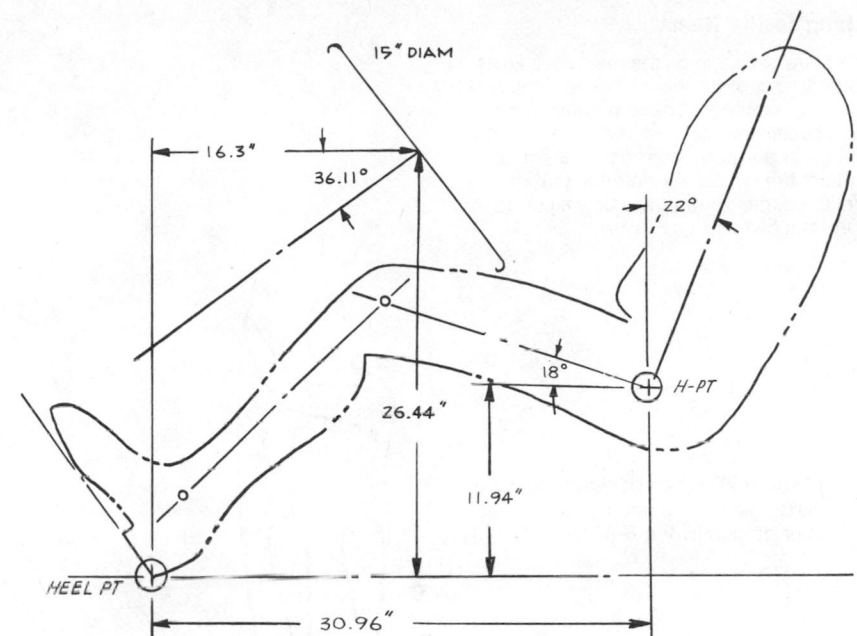

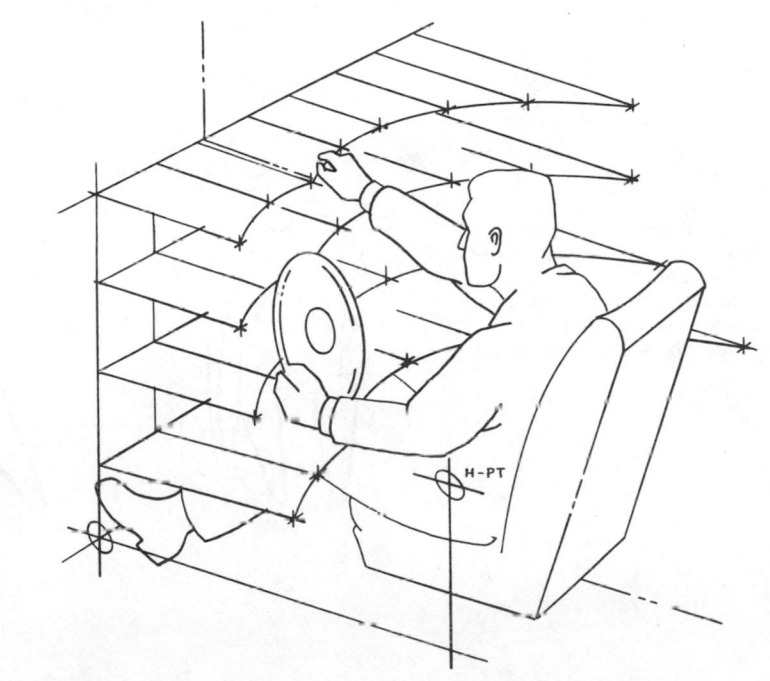

[13]W. E. Woodson et al.—Driver Eye Position and Control Reach Anthropometrics, MFI Report 71-117, 1971 (DOT Contract FH-11-7619).

[14]For a complete description of reach limits for other seating configurations, see "Driver Hand Control Reach," SAE-J287, in *SAE Handbook*, part 2, 1977 or later.

| *a** | Left Hand | | | | Right Hand | | | | |
|---|---|---|---|---|---|---|---|---|---|
| | 16 In | 8 In | 4 In | C/L | C/L | 4 In | 8 In | 16 In | 24 In |
| 35 in | 21.8 | 25.1 | 26.0 | 25.5 | 24.4 | 25.4 | 25.1 | 22.4 | 16.1 |
| | 15.6 | 19.2 | 20.0 | 19.0 | 21.2 | 21.3 | 20.7 | 16.0 | 0 |
| 20 in | 28.6 | 30.1 | 31.2 | 29.6 | 29.3 | 30.2 | 30.5 | 29.2 | 25.2 |
| | 23.8 | 25.9 | 25.7 | 24.9 | 26.2 | 26.6 | 26.4 | 23.1 | 20.0 |
| 15 in | 29.8 | 31.7 | 31.1 | 29.8 | 27.9 | 30.1 | 30.6 | 30.2 | 26.7 |
| | 24.4 | 26.2 | 25.0 | 23.5 | 23.0 | 25.9 | 26.7 | 25.0 | 20.1 |
| 10 in | 29.5 | 31.8 | 31.3 | | | 29.8 | 30.4 | 30.5 | 26.0 |
| | 24.0 | 26.1 | 25.2 | | | 26.0 | 26.3 | 25.1 | 19.9 |
| 5 in | 29.7 | 30.5 | | | | 29.6 | 28.6 | 25.2 | |
| | 22.8 | 24.7 | | | | 25.0 | 23.5 | 18.9 | |
| −3 in | 26.4 | 28.1 | | | | 27.8 | 27.5 | 23.6 | |
| | 16.7 | 20.0 | | | | 21.8 | 0 | 0 | |

*a = reach level above or below the H point.

## Reaching to the Rear[15]

Few data are available on the rear reach limits of an adult, except for those shown in the accompanying sketches. These data are based on the specific problem of reaching for the latch plate of an active seat-belt system, and they reflect the typical interference that an individual encounters from the side wall or door of an automobile.

The first set of sketches illustrate reaching with the inside hand for a latch plate located on the upper portion of the B pillar.

SRP

The second set of sketches illustrate reaching for a latch plate located near the floor, beside the seat or aft of the seat. Here reaching is by the outside arm and hand. This assumes that there is at least 3 in (7.6 cm) of clearance between the seat and the door and/or structure.

The last set of sketches illustrate reaching over the seat back with the outside arm and hand.

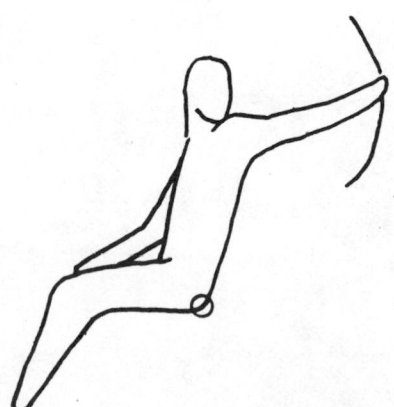

[15]W. E. Woodson et al., *Development of Specifications for Passive Belt Systems*, MFI Report 78-109, 1978 (DOT Contract DOT-HS-7-01617).

The accompanying chart combines the reach limits of a 5th-percentile woman for the inside arm versus the combined outside arm reaches.

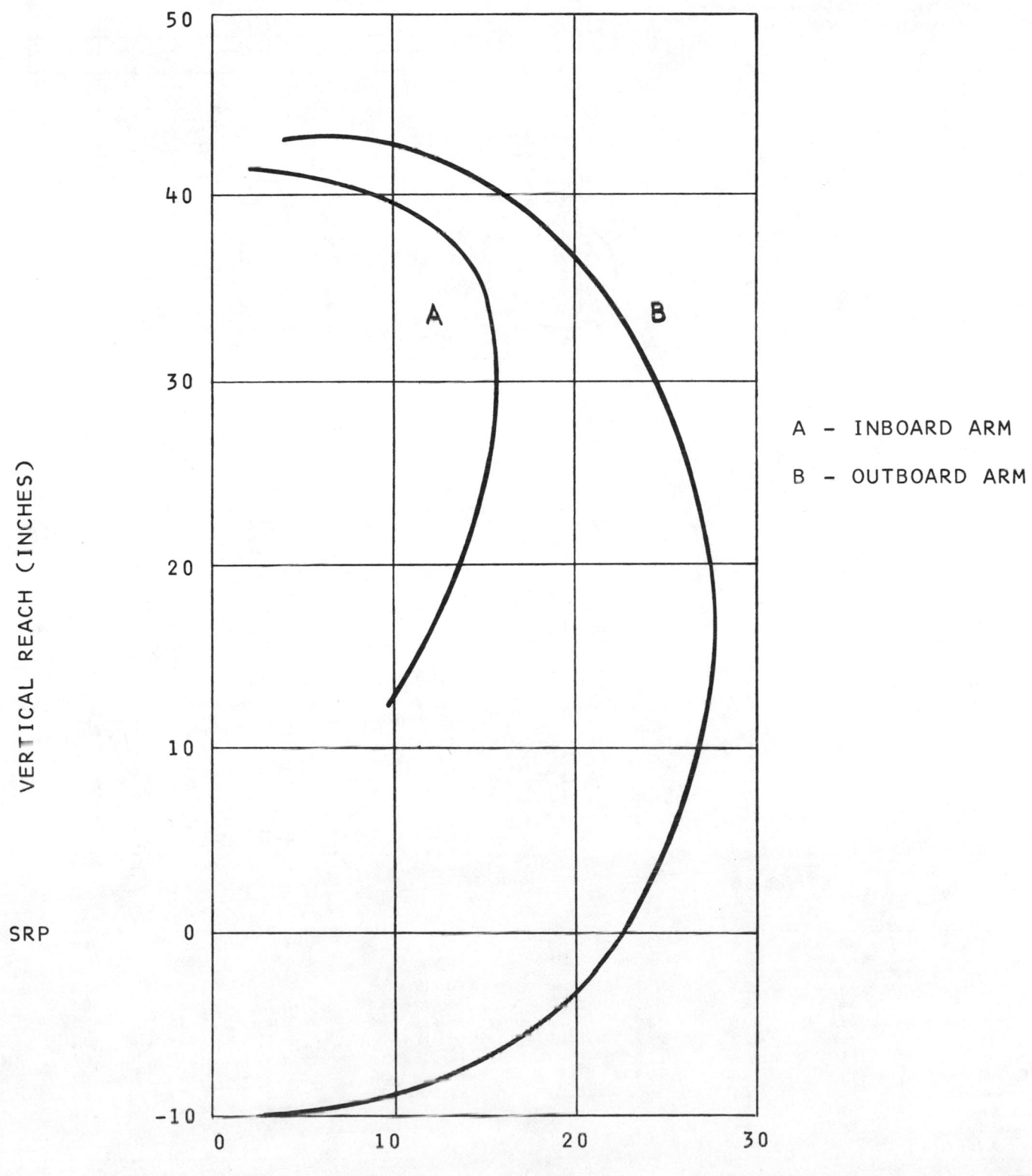

A — INBOARD ARM

B — OUTBOARD ARM

VERTICAL REACH (INCHES)

SRP

AFT REACH (INCHES)

### Functional Reach from a Wheelchair[16]

The approximate reach-limit values shown in the accompanying graphs were derived on the basis of a sample of 91 male and 36 female subjects confined to a wheelchair. Note the differences between the maximum and the comfortable reach limits, a subjective but important consideration in design.

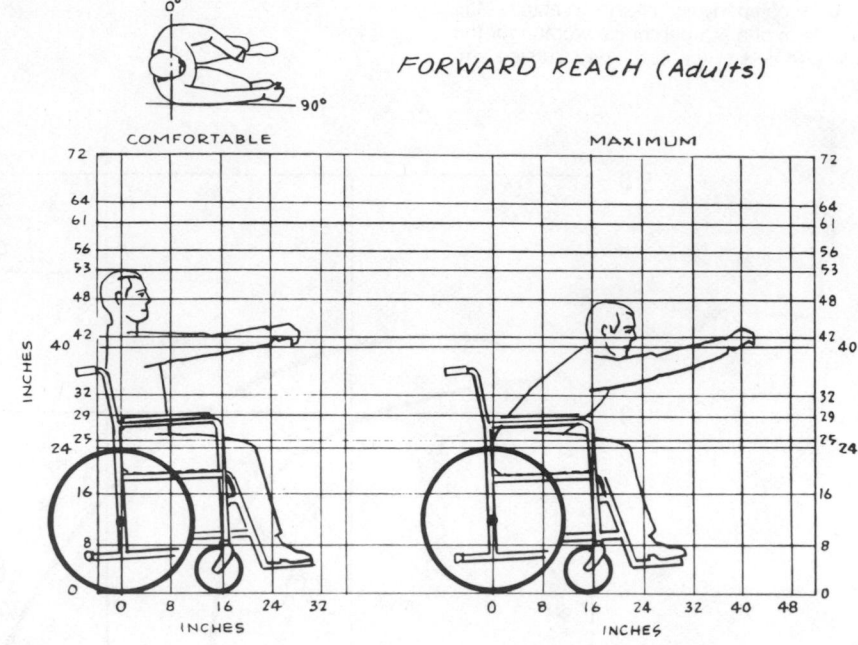

FORWARD REACH (Adults)

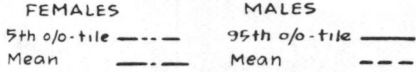

| FEMALES | MALES |
|---|---|
| 5th o/o-tile ---- | 95th o/o-tile ——— |
| Mean ——·— | Mean ——— |

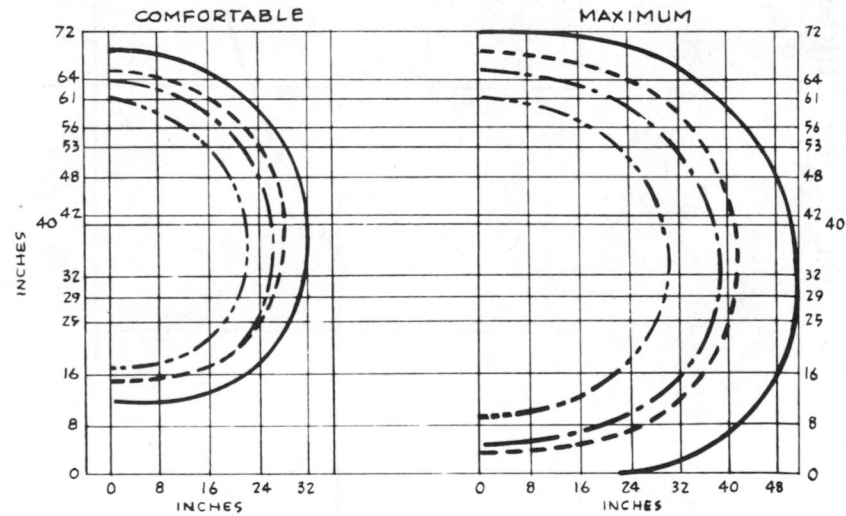

| Male | 5th Percentile | Mean | 95th Percentile | SD |
|---|---|---|---|---|
| Floor to vertex | 49.1 | 52.4 | 55.7 | 2.0 |
| Floor to eye | 44.7 | 48.1 | 51.5 | 2.1 |
| Floor to shoulder | 38.4 | 40.8 | 43.2 | 1.5 |
| Floor to elbow | 24.9 | 27.3 | 29.7 | 1.5 |
| Floor to thigh | 23.3 | 25.4 | 27.5 | 1.3 |
| Shoulder width | 14.3 | 16.8 | 19.3 | 1.5 |
| Hip width | 12.2 | 14.1 | 16.0 | 1.2 |
| Toe projection | 3.7 | 5.1 | 6.5 | 0.9 |
| Lower leg length | 15.7 | 17.3 | 18.9 | 1.0 |
| Age in years | | 34.7 | | |

| Female | 5th Percentile | Mean | 95th Percentile | SD |
|---|---|---|---|---|
| Floor to vertex | 46.9 | 50.4 | 53.8 | 2.1 |
| Floor to eye | 42.7 | 46.4 | 50.1 | 2.2 |
| Floor to shoulder | 35.4 | 39.3 | 42.2 | 1.8 |
| Floor to elbow | 23.2 | 26.7 | 30.2 | 2.1 |
| Floor to thigh | 22.8 | 24.7 | 26.6 | 1.2 |
| Shoulder width | 13.5 | 15.1 | 16.7 | 1.0 |
| Hip width | 12.6 | 14.3 | 16.6 | 1.8 |
| Toe projection | 2.6 | 4.1 | 5.6 | 0.9 |
| Lower leg length | 14.4 | 16.2 | 18.0 | 1.1 |
| Age | | 32.1 | | |

[16]W. F. Floyd et al., "A Study of the Space Requirements of Wheelchair Users," *Paraplegia*, vol. 4, no. 1, pp. 24–37, May 1966.

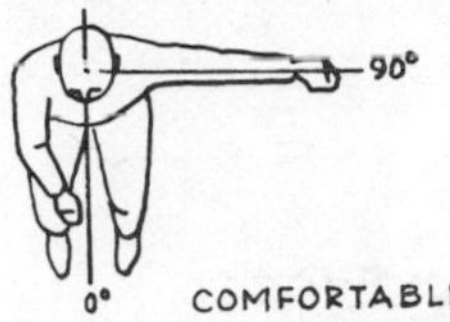

## LATERAL REACH (Adults)

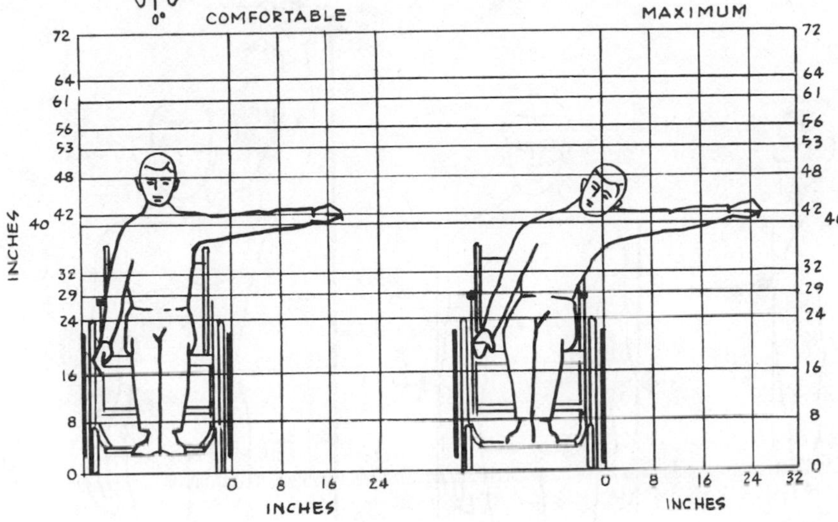

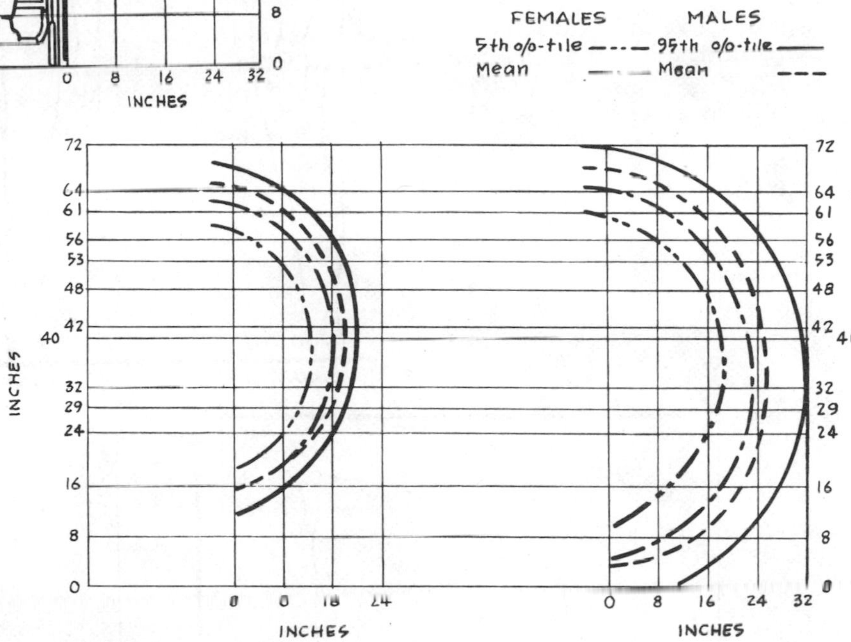

COMFORTABLE

MAXIMUM

INCHES

INCHES

INCHES

| FEMALES | MALES |
|---|---|
| 5th o/o-tile ---- | 95th o/o-tile ——— |
| Mean —·—·— | Mean ---- |

INCHES

INCHES

## DIAGONAL REACH (Adults)

| FEMALES | MALES |
|---|---|
| 5th o/o-tile ----- | 95th o/o-tile ——— |
| Mean —·—·— | Mean ---- |

COMFORTABLE

MAXIMUM

INCHES

INCHES

INCHES

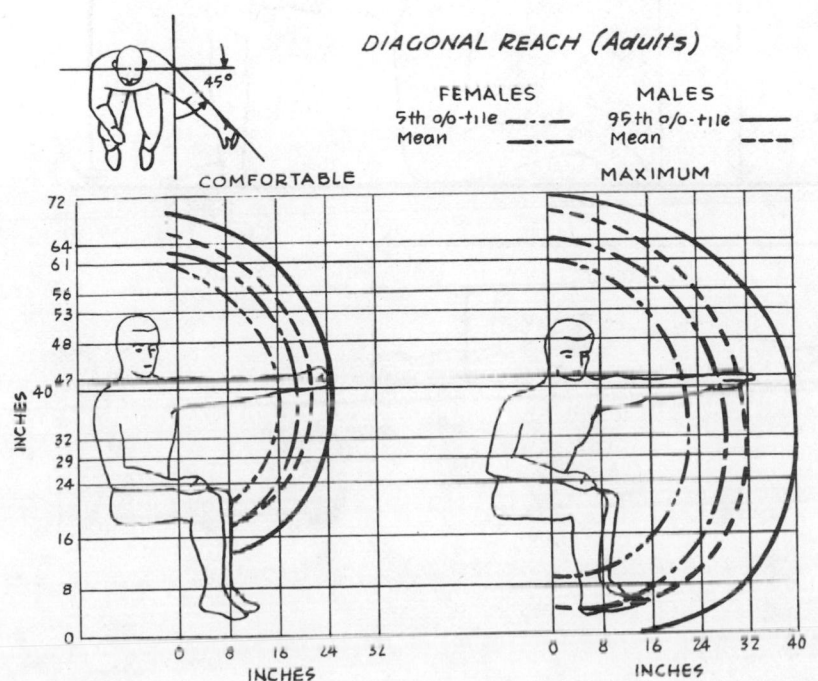

## ANTHROPOMETRIC DATA RELATIVE TO TYPICAL WORKING POSITIONS[17]

The accompanying illustrations and tables provide useful information about work-space dimensioning. However, keep in mind that these data were derived from measurements of male Air Force personnel, and although they represent a somewhat functional dimensional guide, each work space must be given specific attention in terms of its own unique requirements.

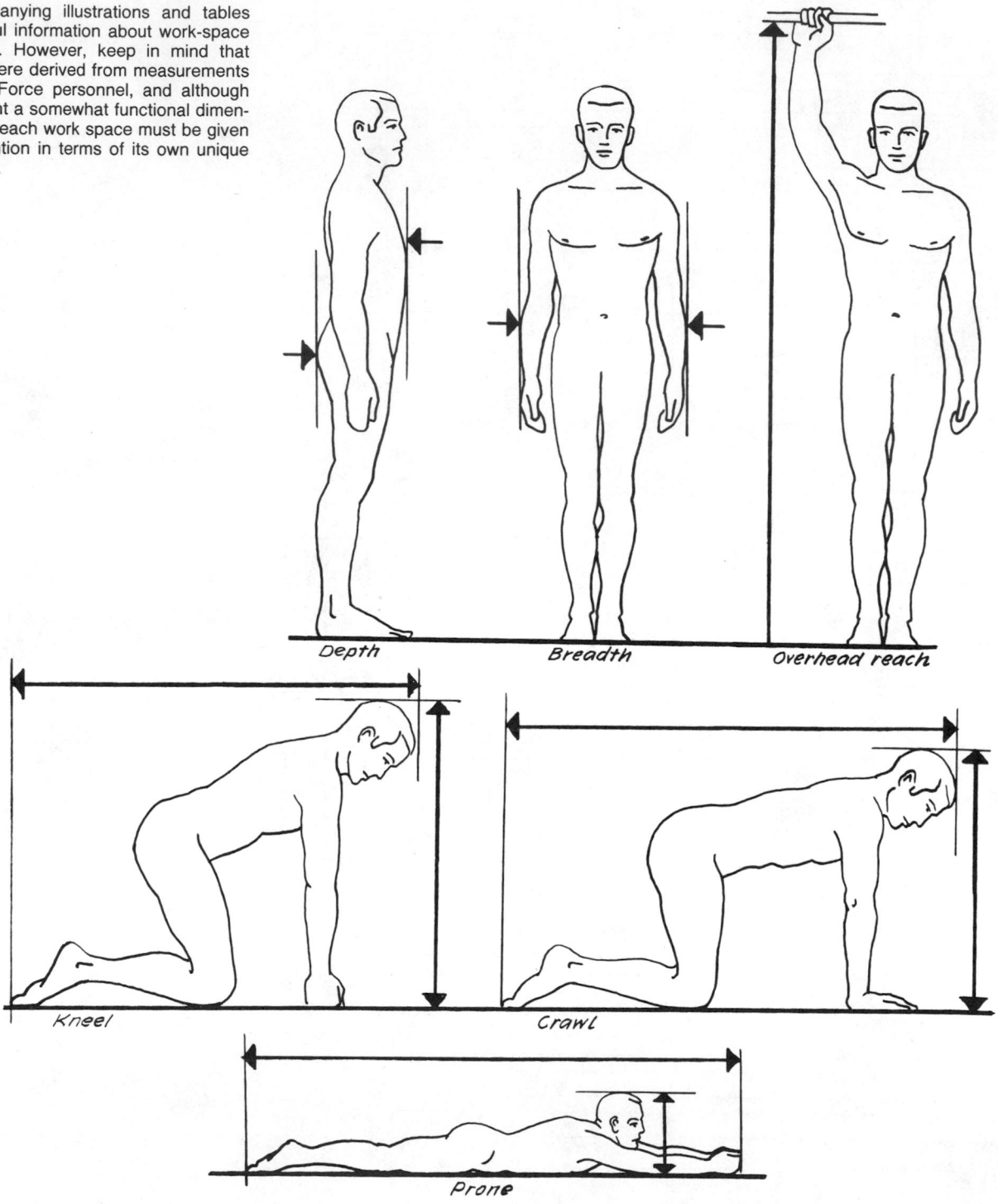

Depth     Breadth     Overhead reach

Kneel     Crawl

Prone

[17]H. T. E. Hertzberg et al., *The Anthropometry of Working Positions*, WADC Technical Report 54-520, Wright-Patterson Air Force Base, Ohio; and K. W. Kennedy and B. E. Filler, *Aperture Sizes and Depths of Reach for One- and Two-Handed Tasks*, AMRL-TR-66-27, Aerospace Medical Research Laboratory, Wright-Patterson Air Force Base, Ohio, 1966.

| Measurement | Number of Subjects | 5th Percentile | Mean | 95th Percentile | SD |
|---|---|---|---|---|---|
| Maximum body depth | 118 | 10.1 | 11.5 | 13.0 | 0.88 |
| Maximum body breadth | 40 | 18.8 | 20.9 | 22.8 | 1.19 |
| Overhead reach | 40 | 76.8 | 82.5 | 88.5 | 3.33 |
| Kneeling height | 40 | 29.7 | 32.0 | 34.5 | 1.57 |
| Kneeling length | 40 | 37.6 | 43.0 | 48.1 | 3.26 |
| Crawling height | 40 | 26.2 | 28.4 | 30.5 | 1.30 |
| Crawling length | 40 | 49.3 | 53.2 | 58.2 | 2.61 |
| Prone height | 40 | 12.3 | 14.5 | 16.4 | 1.28 |
| Prone length | 40 | 84.7 | 90.1 | 95.8 | 3.41 |

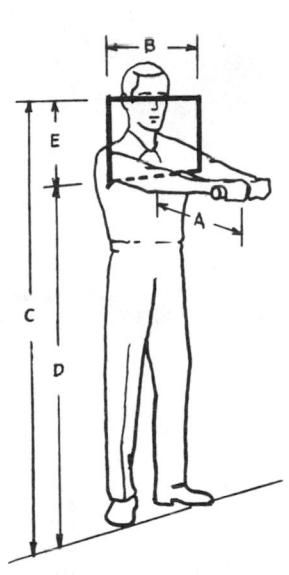

### Standing, Forward Reach (Both Arms)

| | Percentiles | | | | |
|---|---|---|---|---|---|
| | 5th | 25th | 50th | 75th | 95th |
| A. Depth of reach<br>Range: 17.50 to 25.25<br>SD: 1.50 | 19.25 in | 21.00 in | 22.25 in | 22.75 in | 24.50 in |
| B. Breadth of aperture<br>Range: 15.00 to 20.25<br>Mean: 17.69<br>SD: 1.19 | 15.50 in | 17.00 in | 17.75 in | 18.50 in | 19.50 in |
| C. Floor to top of aperture<br>Range: 58.75 to 70.50<br>SD: 2.34 | 61.00 in | 63.50 in | 65.25 in | 66.50 in | 69.00 in |
| D. Floor to bottom of aperture<br>Range: 51.25 to 61.75<br>Mean: 56.09<br>SD: 2.05 | 52.25 in | 54.75 in | 56.00 in | 57.25 in | 59.00 in |
| E. Vertical dimension of aperture | (1) | (1) | (1) | (1) | (1) |

(1) 16.75 in.

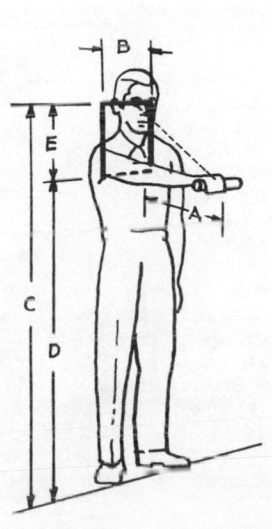

### Standing, Forward Reach (Preferred Arm)

| | Percentiles | | | | |
|---|---|---|---|---|---|
| | 5th | 25th | 50th | 75th | 95th |
| A. Depth of reach<br>Range: 19.50 to 27.50<br>Mean: 23.61<br>SD: 1.82 | 20.25 in | 22.25 in | 23.75 in | 25.00 in | 26.75 in |
| B. Breadth of aperture: 12.00 | | | | | |
| C. Floor to top of aperture<br>Range: 58.25 to 70.50<br>Mean: 64.88<br>SD: 2.36 | 61.00 in | 63.25 in | 65.00 in | 66.25 in | 69.00 in |
| D. Floor to bottom of aperture<br>Range: 51.25<br>Mean: 56.09<br>SD: 2.05 | 52.25 in | 54.75 in | 56.00 in | 57.25 in | 59.00 in |
| E. Vertical dimension of aperture | (1) | (1) | (1) | (1) | (1) |

(1) 16.75 in

### Standing, Lateral Reach (Preferred Arm)

| | Percentiles | | | | |
|---|---|---|---|---|---|
| | 5th | 25th | 50th | 75th | 95th |
| A. Depth of reach<br>Range: 21.75 to 28.63<br>Mean: 24.65<br>SD: 1.51 | 22.00 in | 23.50 in | 24.75 in | 25.75 in | 26.75 in |
| B. Breadth of aperture 10.00 | | | | | |
| C. Floor to top of aperture<br>Range: 58.25 to 70.00<br>Mean: 64.70<br>SD: 2.32 | 60.75 in | 63.25 in | 64.25 in | 66.00 in | 68.75 in |
| D. Floor to bottom of aperture<br>Range: 51.25 to 61.75<br>Mean: 56.09<br>SD: 2.05 | 52.25 in | 54.75 in | 56.00 in | 57.25 in | 59.00 in |
| E. Vertical dimension of aperture | (1) | (1) | (1) | (1) | (1) |

(1) 16.5 in.

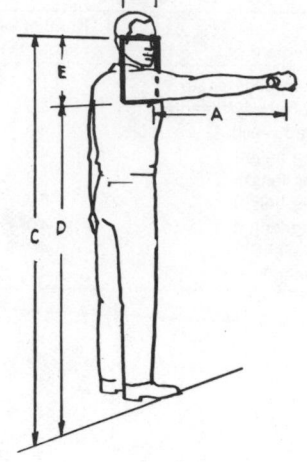

### Seated, Forward Reach (Both Arms)

| | Percentiles | | | | |
|---|---|---|---|---|---|
| | 5th | 25th | 50th | 75th | 95th |
| A. Depth of reach<br>Range: 14.00 to 23.50<br>Mean: 18.26<br>SD: 2.15 | 15.00 in | 16.50 in | 17.75 in | 19.50 in | 22.25 in |
| B. Breadth of aperture<br>Range: 13.50 to 18.75<br>Mean: 16.12<br>SD: 1.25 | 13.75 in | 15.25 in | 16.00 in | 17.00 in | 18.25 in |
| C. Floor to top of aperture.<br>Range: 39.25 to 51.00<br>Mean: 43.25<br>SD: 2.05 | 19.75 in | 41.75 in | 43.00 in | 44.25 in | 46.50 in |
| D. Floor to bottom of aperture†<br>Range: 32.50 to 41.75<br>Mean: 36.59<br>SD: 1.59 | 34.25 in | 35.50 in | 36.50 in | 37.50 in | 39.00 in |
| E. Vertical dimension of aperture | (1) | (1) | (1) | (1) | (1) |

(1) 12.25 in.

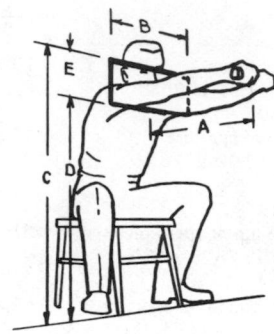

### Cross-Legged Seated, Forward Reach (Both Arms)

| | Percentiles | | | | |
|---|---|---|---|---|---|
| | 5th | 25th | 50th | 75th | 95th |
| A. Depth of reach<br>Range: 13.50 to 22.25<br>Mean: 17.08<br>SD: 1.91 | 13.75 in | 15.75 in | 16.75 in | 18.25 in | 20.00 in |
| B. Breadth of aperture<br>Range: 13.50 to 18.50<br>Mean: 15.89<br>SD: 1.54 | 13.75 in | 14.75 in | 16.00 in | 16.75 in | 17.75 in |
| C. Floor to top of aperture<br>Range: 22.25 to 30.50<br>Mean: 25.30<br>SD: 1.54 | 22.75 in | 24.25 in | 25.25 in | 26.25 in | 28.00 in |
| D. Floor to bottom of aperture<br>Range: 17.00 to 23.25<br>Mean: 19.23<br>SD: 1.19 | 17.00 in | 18.50 in | 19.25 in | 20.00 in | 21.25 in |
| E. Vertical dimension of aperture | (1) | (1) | (1) | (1) | (1) |

(1) 11.0 in.

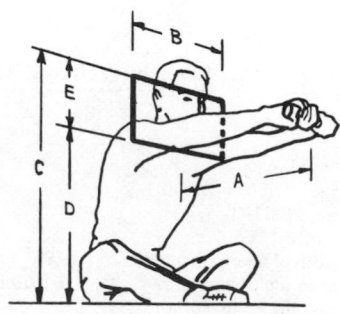

**Selected Working-Position Data for
U.S. Army Men and Women (Wearing
Typical Army Clothing)[18]**

Overhead reach height—standing with heels 23 cm apart and toes 15 cm
from wall; arms extended overhead with fists touching and against wall; 1st
phalanges horizontal. Measured from floor to highest point on 1st phalanges.

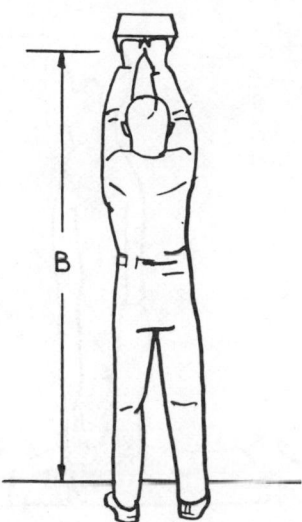

Overhead reach breadth—standing with heels 23 cm apart and toes 15 cm
from wall; arms extended overhead with fists touching and against wall; 1st
phalanges horizontal. Measured horizontally across arms or shoulders,
whichever is wider.

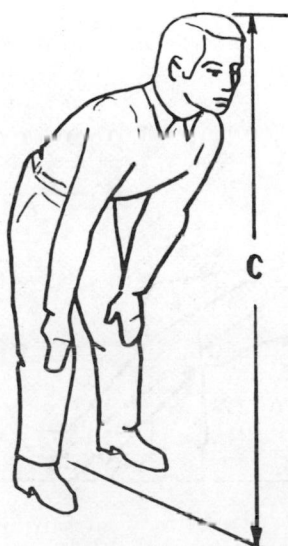

Bent torso height—standing with feet 30 cm apart; bending over and placing
palms of the hands on kneecaps; elbows and knees locked; looking forward;
head tilted as far back as possible. Measured from floor to top of head.

Bent torso breadth—standing with feet 30 cm apart; bending over and placing the palms of the hands on kneecaps; elbows and knees locked; looking
forward; head tilted as far back as possible. Measured as maximum horizontal distance across shoulders.

[18]The illustrations on this page and on pp. 766 and 767 refer
to position specifications and values in the table on p. 768.

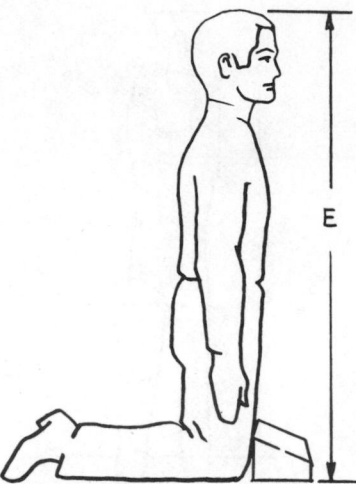

Kneeling height—kneeling with toes extended and lightly touching rear wall; torso erect with arms hanging loosely at sides. Measured from floor to top of head.

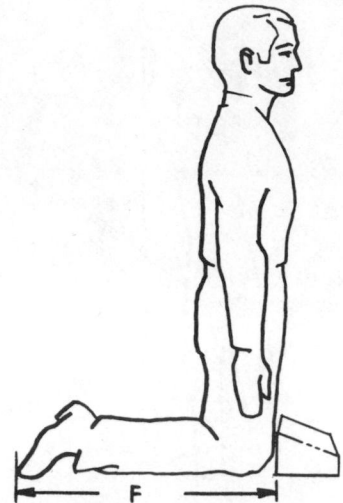

Kneeling leg length—kneeling with toes extended and lightly touching rear wall; torso erect with arms hanging loosely at sides. Measured from wall to anterior portion of both knees.

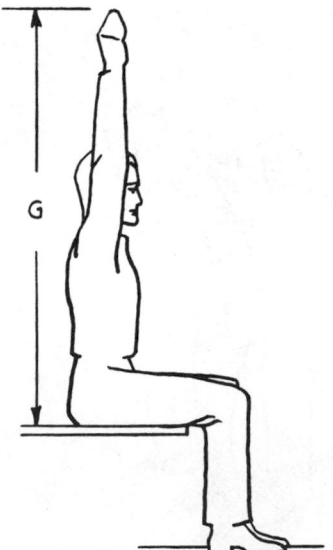

Overhead reach, sitting—sitting erect; right side against wall; right arm extended upward with palm flat against wall and fingers extended. Measured from sitting surface to tip of middle finger.

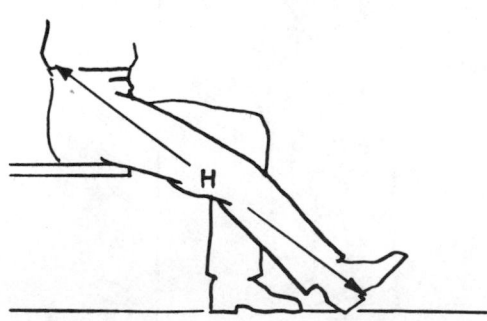

Functional leg length—sitting erect on edge of chair; right leg extended forward with knee straightened. Measured from heel along axis of leg to posterior waist.

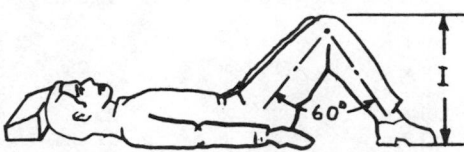

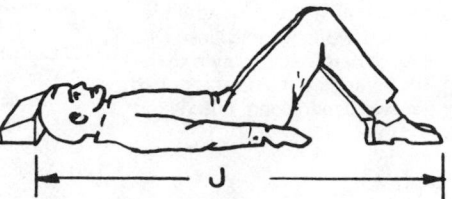

Bent knee height, supine—lying supine; knees raised until angle between upper and lower legs approximates 60°; toes lightly touching wall. Measured from floor to highest point on knees.

Horizontal length, knees bent—lying supine; knees raised until angle between upper and lower legs approximates 60°; toes lightly touching wall. Measured from wall to top of head.

Functional reach, extended—standing erect; looking straight ahead; right shoulder extended as far forward as possible while back of left shoulder firmly against wall; arm horizontal. Measured from wall to tip of index finger.

## SELECTED FUNCTIONAL DIMENSIONS[19]

Few definitive studies have been conducted for the purpose of establishing the functional dimensions of the human as he or she pursues day-to-day living and working activities. One recent study of shipboard habitability requirements (using 5th-percentile males) produced the following general dimensional criteria:

**Anthropometric Data for U.S. Male and Female Personnel: Common Working Positions, Percentile Values**

| | 5th Percentile | | 95th Percentile | |
|---|---|---|---|---|
| | Men | Women | Men | Women |
| A. Overhead reach | 78.9 in | 73.0 in | 90.8 in | 84.7 in |
| B. Overhead reach, breadth | 13.9 in | 12.4 in | 16.5 in | 14.9 in |
| C. Bent torso height | 49.4 in | 44.4 in | 59.0 in | 54.6 in |
| D. Bent torso breadth | 16.1 in | 14.5 in | 19.0 in | 17.1 in |
| E. Kneeling height | 48.0 in | 45.1 in | 53.9 in | 51.3 in |
| F. Kneeling leg length | 25.2 in | 23.3 in | 29.7 in | 27.8 in |
| G. Overhead reach, sitting | 50.3 in | 46.2 in | 57.9 in | 54.9 in |
| H. Functional leg length | 43.5 in | 39.2 in | 50.3 in | 46.7 in |
| I. Bent knee height, supine | 17.6 in | 16.3 in | 21.1 in | 19.5 in |
| J. Horizontal length, knee bent | 59.4 in | 55.2 in | 68.1 in | 64.5 in |
| K. Functional reach | 33.2 in | 28.9 in | 39.8 in | 36.5 in |

*Source:* MIL-STD-1472B, *Notice—2,* May 10, 1978.

Active reach and grasp across a typical workbench, drafting table, or similar work surface is limited to about 36 in (91 cm). Although added reach is possible for a lower work surface, loss of balance is quite likely.

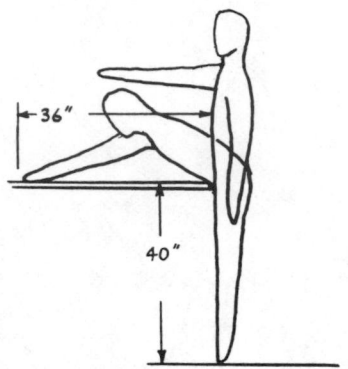

Extended reach and grasp from a seated position is limited to about 42 in (107 cm) without considerable effort. Refer to the reach envelope references in the previous description of driver reach studies.

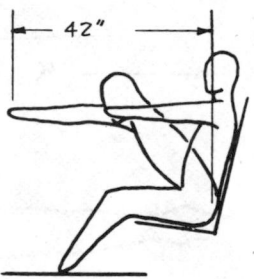

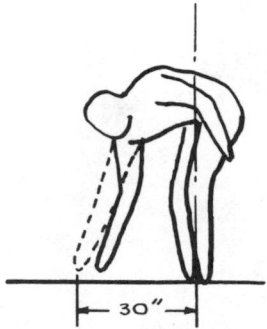
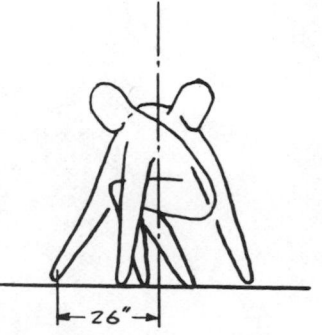

Reaching from a stooping or squatting position is limited to the values shown in the accompanying sketches.

Reaching from a fixed ladder is at best a precarious activity. One should not expect an individual to reach to the side more than about 48 in (122 cm).

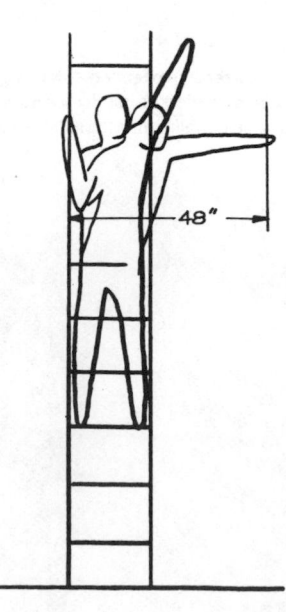

[19]W. E. Woodson, et al., *Habitability in Shipboard Living Spaces,* vol. II, *Supporting Data: General,* prepared for the U.S. Naval Ships Engineering Center, Hyattsville, Md., 1972.

## Functional Clearance Requirements

Berthing space should be large enough to permit stretching the legs, raising the arms overhead, and turning easily without rolling out of bed. The dimensions shown are suggested.

The clearance between bunks should be sufficient to allow the individual to sit up. There should be enough clearance in front of bunks so that the individual can make up the bed without bumping into the opposite wall.

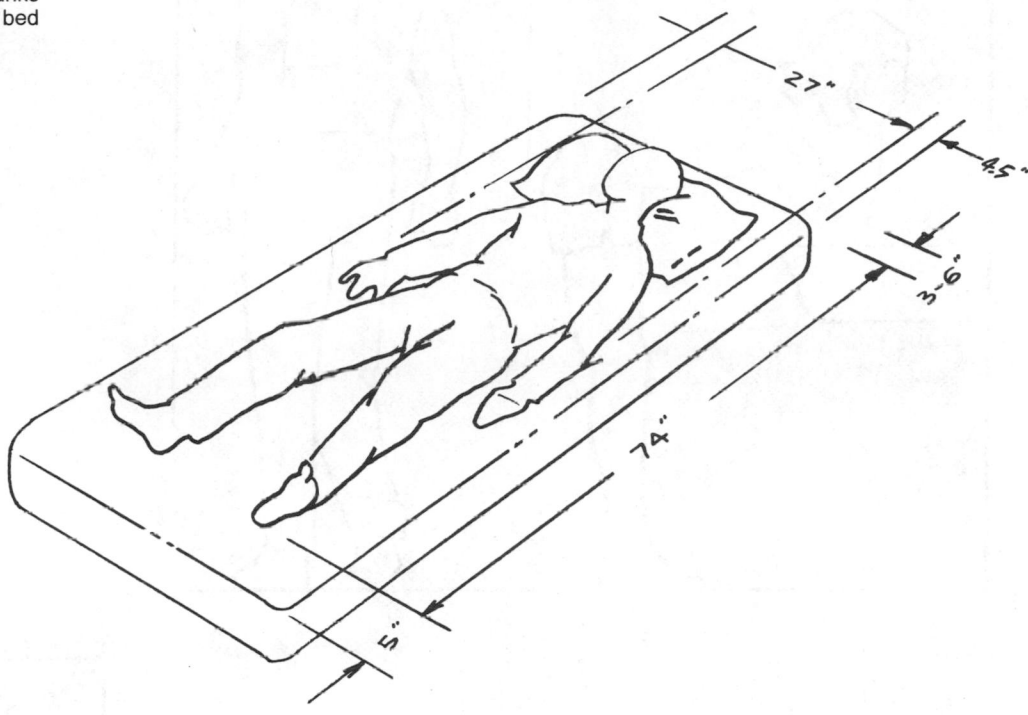

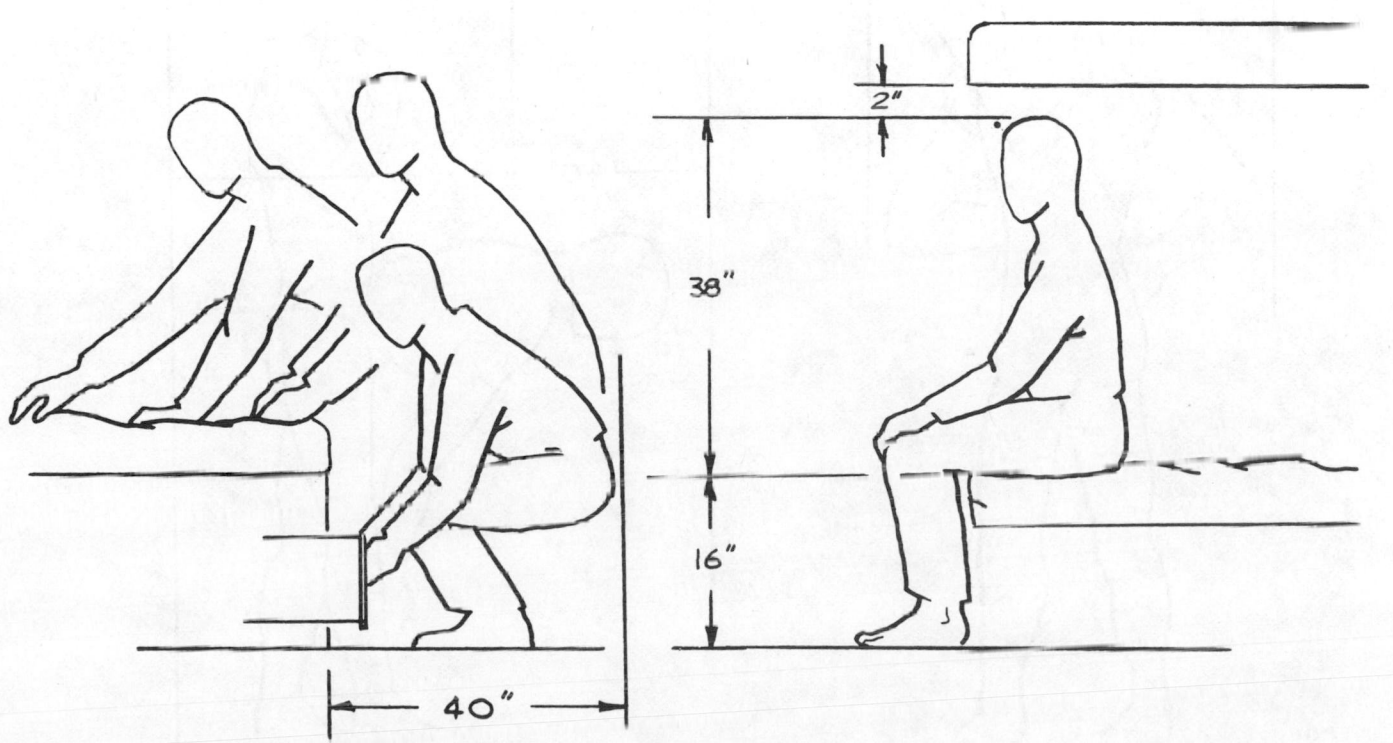

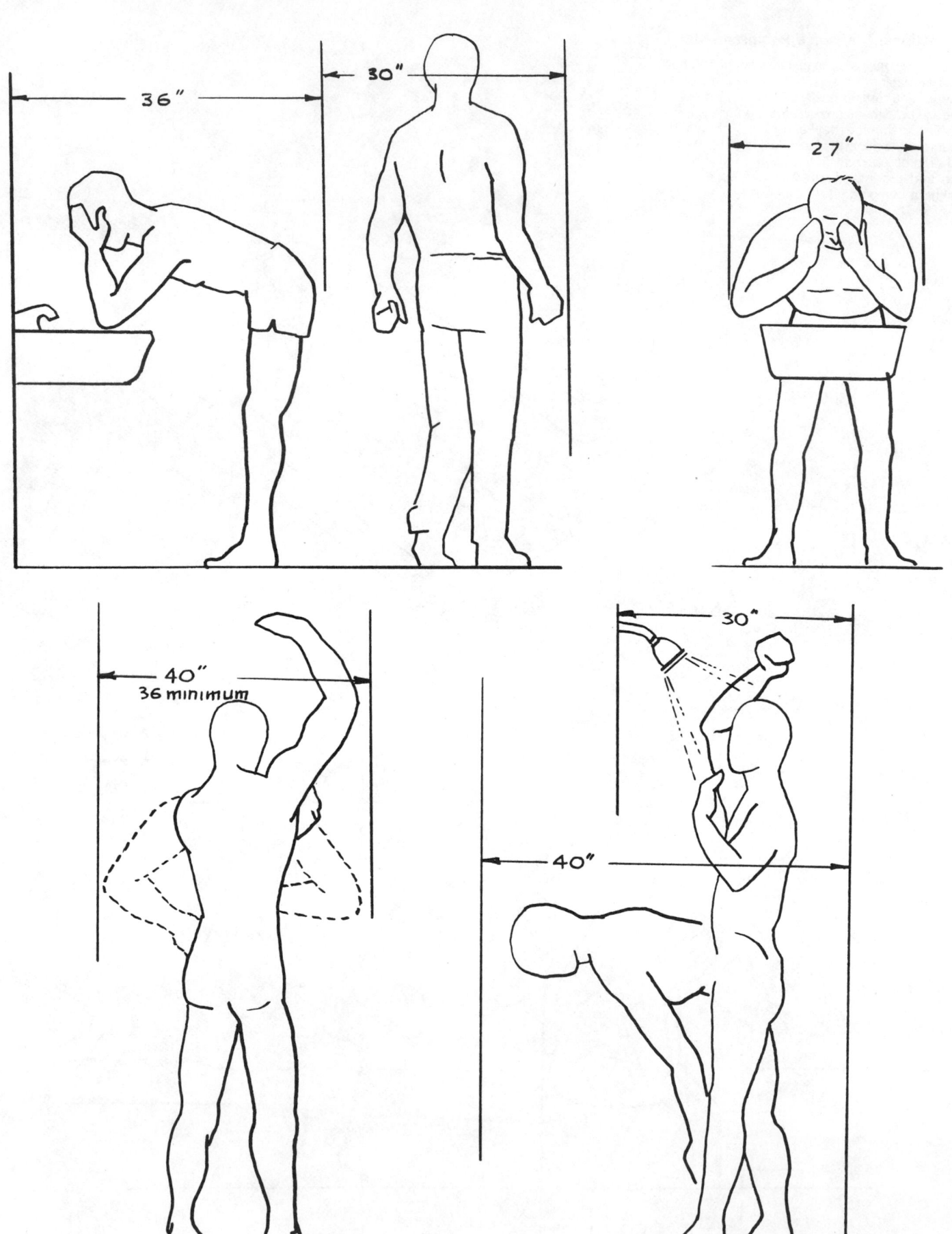

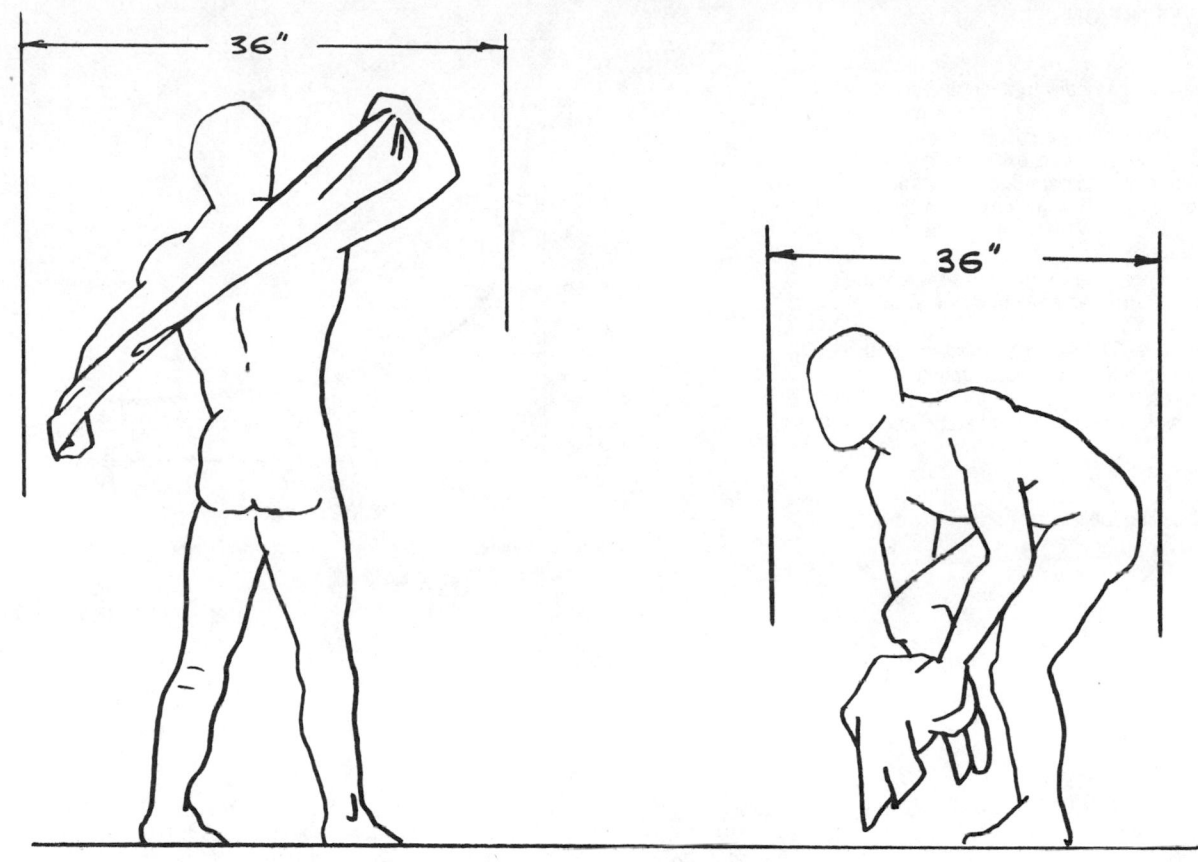

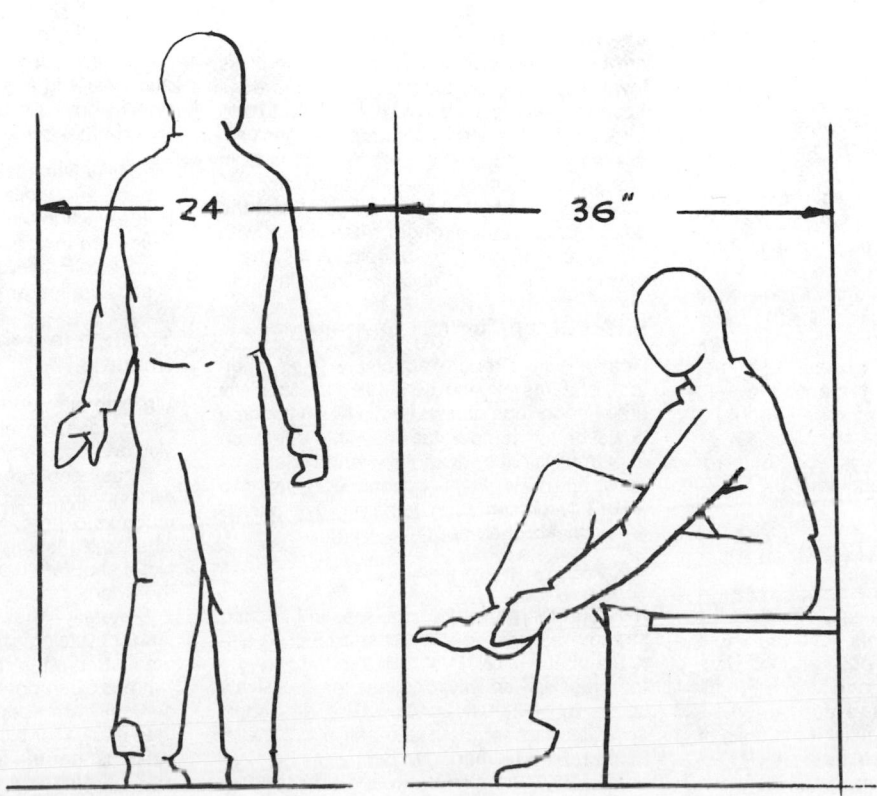

## BASIC DEFINITIONS

*Strength:* The maximum force that muscles can exert isometrically in a single voluntary effort.

*Isometric strength (static):* The maximum force that muscles can exert when muscle length remains constant during contraction.

*Isotonic strength:* The maximum force that muscles can exert when muscle tension is kept constant.

*Concentric force:* The force exerted when the muscle is shortened against an external resistance.

*Eccentric force:* The force exerted when the muscle lengthens passively against an external force.

*Effort:* Physiological strain, both static and dynamic.

*Work:* Dynamic effort (i.e., force times displacement).

*Endurance:* The ability to continue work or exert force.

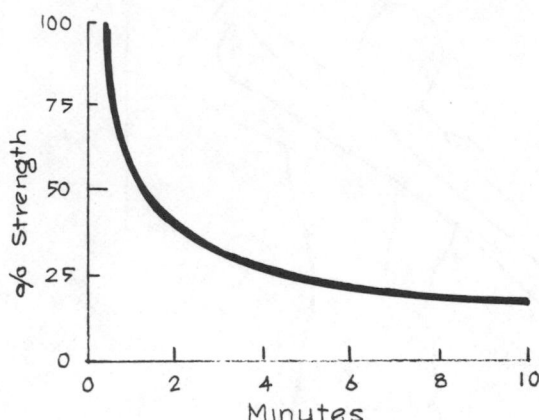

Typical endurance time in relation to force requirements.

## GENERALITIES RELATIVE TO MUSCLE FORCE

Muscle force is a function of the following.

**MUSCLE TENSION**  Muscle tension is maximum when the length of the muscle is greatest and momentarily there is no change in length. Muscle force decreases as the muscle shortens and as its rate of shortening increases.

**MECHANICAL ADVANTAGE**  Power is applied at the point of muscle attachment (i.e., the long bones are the lever arms, and the joints are the fulcrums). Thus, for example, when muscle force of extension is applied at the elbow, power is greatest when the elbow is flexed. However, optimum mechanical advantage occurs at the midpoint of full elbow travel.

Optimum mechanical advantage more than compensates for the shortened muscle, therefore providing maximum strength at the midpoint. Human muscles in maximum contraction can exert considerable force (as much as 1000 lb, 453 kg), but such forces cannot be fully utilized directly because all muscles work at some mechanical disadvantage, thus reducing output but increasing rate of movement.

## Age and Sex in Relation to Strength

In general, adult females are only about two-thirds as strong as adult males.

In terms of age, individuals have maximum strength between the ages of 30 and 40. Usually there is a rapid development in strength between the ages of 13 and 19, and this development slows somewhat between the ages of 20 and 25. This is followed by a slower increase in strength to the maximum between

25 and 30 years of age. People begin to lose about 10 percent of their strength by age 40, 15 percent by age 50, 20 percent by age 60, and at least 25 percent by age 65.

## Effect of Body Build

As a rule, people with larger body builds have more strength, although less powerfully built individuals may require less oxygen for a given task requiring strength. Slender persons often are best at performing rapidly fatiguing tasks involving strenuous exercise. Physique does not necessarily correlate with the ability to perform moderate exercise. Normal persons usually show a 30 to 50 percent increase in strength after about 12 weeks of training.

Strength is affected by health, diet, and the use of drugs, and strength often varies with diurnal conditions (e.g., people usually have maximum strength at about midmorning).

## Effect of the Thermal Environment

Heat affects strength adversely (e.g., when temperatures exceed 85°F (29°C), especially under conditions of high humidity). In general, however, low temperature has little effect except in terms of body mobility and finger dexterity. When individuals become acclimated to a hot environment, they generally gain back a great deal of their normal strength.

## Effect of Altitude

Although there usually is no loss in handgrip strength as one goes from sea level to altitudes of about 12,000 ft (3660 m), endurance for muscular activity declines progressively above about 8000 ft (2440 m). Strength endurance declines to about one-half by 20,000 ft (6100 m). Partial handgrip paralysis begins at around 24,000 ft (7320 m), and an individual

reaches the point of unconsciousness at about 26,000 ft (7930 m). Although some acclimatization occurs, this requires several weeks or months and therefore is of little practical consequence. Lack of oxygen is a major factor in the decrease of muscle efficiency.

## Effect of Acceleration

Although accelerations up to 5 *g*'s do not affect strength, endurance is affected. Arm movements are effective up to about 6 *g*'s, and wrist and finger movements are effective up to about 12 *g*'s. Practical considerations include the following:

Forces acting against the direction of acceleration are decreased.

Forces acting with the direction of acceleration are increased.

Forces acting perpendicular to the direction of acceleration are least affected.

## Strength in Relation to Emotional Condition

Strength may increase under stress (i.e., fear, panic, rage, or even excitement). However, skill and accuracy generally are degraded.

It has been demonstrated that, with hypnosis, pull with forearm flexion may increase as much as 26 percent. Increases may also occur when the maximum effort is preceded by a pistol shot or when the subject shouts during the effort.

Generally speaking, psychological rather than physiological factors determine maximum strength in the "real world."

It has been noted that white-collar workers generally are about 10 to 20 percent weaker than manual or blue-collar workers. The implication is that the latter are used to a rougher and more strength-demanding environment.

## Effect of Body and Limb Position on Strength Capacity

### 1. Body Position

When individuals are not restricted in terms of body position and are provided with appropriate supporting and/or anchoring facilities, they generally will assume a position from which they can apply their maximum force capability. However, this does not necessarily mean that this is always the best position for maintaining lesser force applications for extended periods of time.

Since there is usually a reciprocal response during force applications (e.g., lifting, pushing, and pulling), it is important to provide appropriate support and anchoring conditions, such as a flat, level floor or deck; a solid, stable seat and seat backrest; or a footrest.

### 2. Limb Position

Both limb position and direction of force application are important variables in determining the amount of force an individual can apply. Handgrip forces generally are greater if the gripping task is close to the individual's body than if it is at arm's length. Arm strength is greater if the individual can push against a backrest or footrest. Maximum leg force occurs when the individual's knee is slightly bent (in a seated position with the leg just "short" of maximum extension and with the ball of the foot at approximately the same level as the individual's buttocks). Maximum arm force occurs when the force can be applied approximately at shoulder level. For the seated individual, pull force is greatest when the object is positioned at nearly maximum arm length; push force is greatest when the object is positioned at about half the full arm extension.

Lifting capabilities depend on the size, shape, and gripping characteristics of the package being lifted and on the distribution of weight within the package. For example, a package that is too large to allow the individual to wrap his or her arms around it, to grip it securely, or to offset poorly distributed weight (the package's c/g is too far from the individual's own c/g) cannot be lifted or carried without the probability that the individual will drop it, will lose his or her balance, and/or will suffer strain and possibly some semipermanent or permanent injury.

Selected strength capabilities for various lifting and force-application situations are provided on the following pages. Since some of the data pertain only to adult males, a good rule of thumb for applying the guidelines to females is that females generally are about one-third weaker than males.

## HAND DYNAMOMETER STRENGTH

Although there is some disagreement on this point, hand dynamometer strength is generally considered a fairly good indicator of an individual's overall ability to apply force and thus is useful for selecting specific subject groups to test designs involving human force applications. The accompanying table provides general dynamometer strength capabilities for various populations. Values are for the right hand only.

| Subject Population | Age Range or Mean | Maximum Squeeze | | | |
| --- | --- | --- | --- | --- | --- |
| | | Range | 5th Percentile | Mean | 95th Percentile | SD |
| U.S. Air Force, male personnel | 18–25 | 90–203 lb | 105 lb | 134 lb | 164 lb | 18 |
| U.S. Navy, female personnel | 22 | 46–119 lb | 58 lb | 73 lb | 87 lb | 8.8 |
| U.S. civilian males | 20–30 | | 74 lb | 108 lb | 142 lb | 21 |
| U.S. industrial workers (female) | 32 | 46–119 lb | 57 lb | 74 lb | 91 lb | 10.3 |
| U.S. truck and bus drivers (male) | 37 | 80–190 lb | 91 lb | 121 lb | 151 lb | 18.1 |
| U.S. elderly males | 81 | | 35 lb | 64 lb | 92 lb | 17.3 |

*Source:* Albert Damon et al., *The Human Body in Equipment Design*, Harvard University Press, Cambridge, Mass., 1966, pp. 219–220.

### Grip Strength for Children Ages 3 to 10 (KP)

| Age | 10th Percentile | Median | 90th Percentile | Maximum |
| --- | --- | --- | --- | --- |
| 3 | 3.2 | 4.3 | 6.8 | 8.0 |
| 5 | 5.1 | 7.2 | 10.1 | 12.1 |
| 7 | 6.2 | 11.0 | 14.8 | 17.9 |
| 10 | 12.9 | 16.7 | 22.0 | 26.2 |

*Source:* Clyde L. Owings, *Strength Characteristics of U.S. Children for Product Safety Design*, 011903-F, University of Michigan, Ann Arbor, for the Product Safety Commission, 1975.

## GENERAL HUMAN STRENGTH-LIMIT EXPECTATIONS

Since people do not always use the best method for applying force, one should not plan on using maximum capability values for design. The nominal values shown in the accompanying illustrations are suggested for various common force-application activities.

Although people can lift more than the values shown in the accompanying illustration, one should not expect more from an operator (i.e., these are 5th-percentile values when the individual tries to lift using only back-muscle power).

The lower illustrations represent maximum force expectations for adult males. For females, one should cut the values approximately in half.

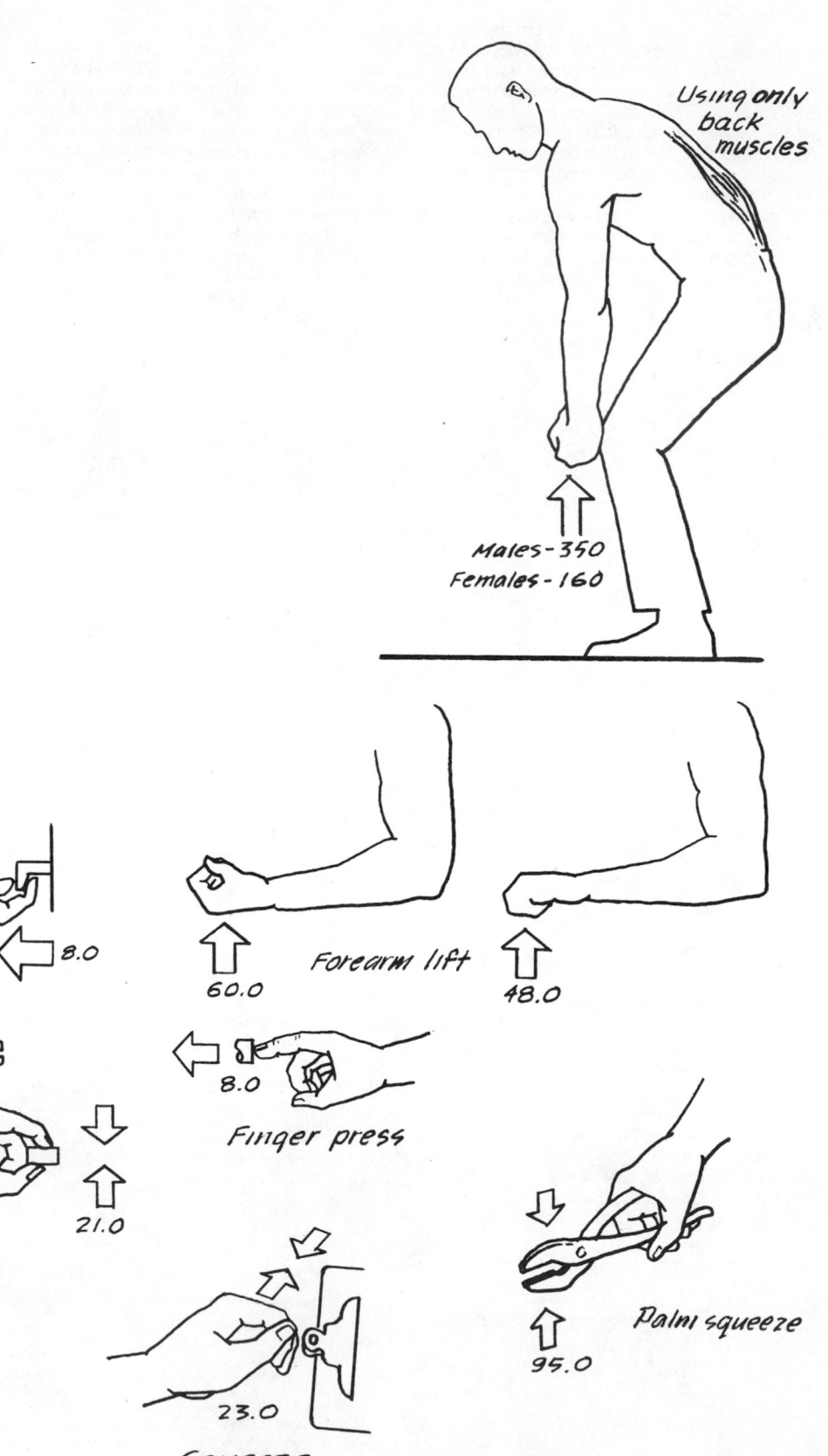

Using only back muscles

Males - 350
Females - 160

Finger pull    8.0

Forearm lift    60.0    48.0

Pinch    21.0    21.0

Finger press    8.0

Squeeze    23.0

Palm squeeze    95.0

## COMMON CONTROL OPERATIONAL FORCE LIMITS

The accompanying sketches and force values represent approximate upper limits for situations involving typical civilian user populations (male and female adults).

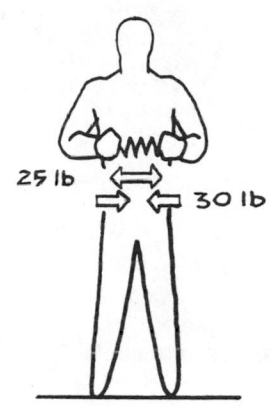

### 1. Two-Hand Scissor-Action Force

The suggested force limit is based on the assumption that the two hands are about 6 to 8 in (15 to 20 cm) apart. If the action takes place above or below normal elbow level, force capabilities diminish rapidly.

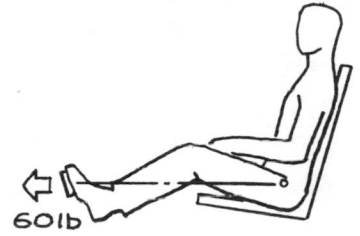

### 2. Pedal Force from a Seated Position

In the typical driving scenario, a driver often is surprised and must apply brake force quickly. Thus the suggested upper force limit accounts not only for the weaker members of the user population but also for the psychological disadvantages of being faced with an emergency. When power assist can be provided, this force can and should be considerably reduced.

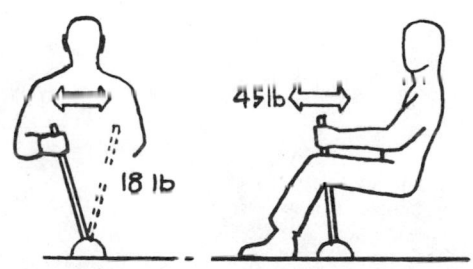

### 3. Lever or Joy-Stick Forces from a Seated Position

Note that fore-aft force capacities are greater than lateral capacities because of the support advantages provided by a seat and toe board or rudder pedals.

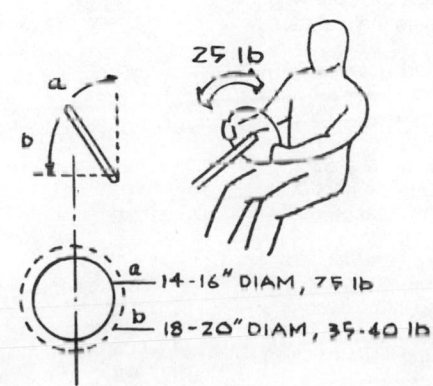

### 4. Vehicle Steering Forces

For vehicles operated by both males and females, the upper force-limit objective should be 25 lb (11.3 kg) or less in order to account for an emergency response (e.g., failure of the power steering). For heavier vehicles, a higher force may be acceptable, but the wheel angle and diameter should be increased as indicated in the accompanying sketch.

## FORCE LIMITS FROM A STANDING POSITION

General upper force-limit approximations for typical control operations are illustrated in the accompanying sketches. These assume that the point of force application has been optimized (i.e., between waist and shoulder height above the floor).

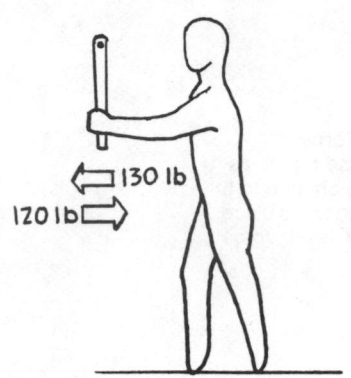

### 1. Force Limits for Lever-Type Controls

Although slightly higher values could be expected if both hands were used, the values are representative of the upper design limit, assuming that the operator has no special aids to help in retaining his or her primary body position (i.e., something to push against).

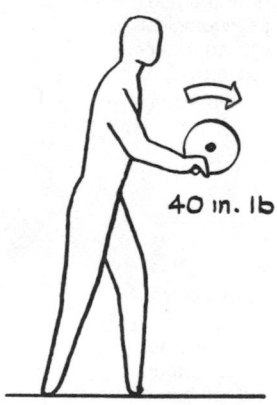

### 2. Force Limits for Hand Cranking

The best hand-cranking positions are illustrated in the accompanying sketches. The value shown assumes a wheel diameter between about 4.5 and 8.0 in (11.4 and 20 cm). Cranking efficiency and capability to apply maximum forces diminish rapidly when wheel diameters are either smaller or greater than noted.

Note that force limits should be lower when the cranking plane is parallel to the operator's own frontal plane.

If hand cranks are located overhead, the upper force-limit values should be reduced by 40 to 50 percent.

*Note:* Physical activity incurs some physical debt as a consequence of the physiological work load and/or the work environment. If the work rate is low enough, a muscle group can function almost indefinitely. However, at high rates muscles soon fatigue and, in some cases, may cease to function effectively altogether. The values illustrated in the accompanying sketches apply to reasonably slow and infrequent manipulation of the particular control.

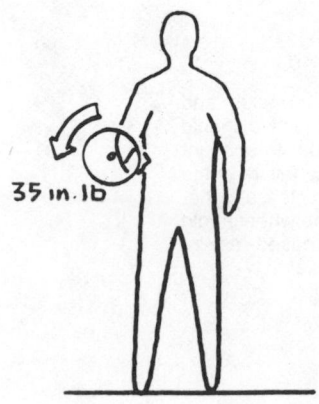

## STRENGTH IN RELATION TO CONTROL OPERATION

### Hand Controls[20]

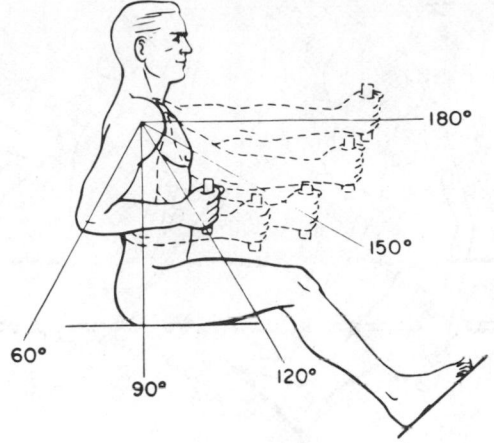

### Maximum Force Exerted in the Sitting Position on a Vertical Handgrip at Various Elbow Angles by the Right Arm of Male College Students

| Direction of force | Elbow angle (deg) | Percentiles (lb) | | | S.D. |
|---|---|---|---|---|---|
| | | 5th | 50th | 95th | |
| Push | 60 | 34 | 92 | 150 | 38 |
| | 90 | 36 | 86 | 154 | 33 |
| | 120 | 36 | 103 | 172 | 43 |
| | 150 | 42 | 123 | 194 | 45 |
| | 180 | 50 | 138 | 210 | 49 |
| Pull | 60 | 24 | 63 | 74 | 23 |
| | 90 | 37 | 88 | 135 | 30 |
| | 120 | 42 | 104 | 154 | 31 |
| | 150 | 56 | 122 | 190 | 26 |
| | 180 | 52 | 120 | 171 | 37 |
| Left | 60 | 20 | 52 | 87 | 19 |
| | 90 | 18 | 50 | 97 | 23 |
| | 120 | 22 | 53 | 100 | 26 |
| | 150 | 20 | 54 | 104 | 25 |
| | 180 | 20 | 50 | 104 | 26 |
| Right | 60 | 17 | 42 | 82 | 20 |
| | 90 | 16 | 37 | 68 | 18 |
| | 120 | 15 | 34 | 62 | 17 |
| | 150 | 15 | 33 | 64 | 18 |
| | 180 | 14 | 34 | 62 | 24 |
| Up | 60 | 20 | 49 | 82 | 18 |
| | 90 | 20 | 56 | 106 | 22 |
| | 120 | 24 | 60 | 124 | 24 |
| | 150 | 18 | 56 | 118 | 28 |
| | 180 | 14 | 43 | 88 | 22 |
| Down | 60 | 20 | 51 | 89 | 21 |
| | 90 | 26 | 53 | 88 | 20 |
| | 120 | 26 | 58 | 98 | 23 |
| | 150 | 20 | 47 | 80 | 18 |
| | 180 | 17 | 41 | 82 | 18 |

### Maximum Force Exerted in the Sitting Position on a Vertical Handgrip at Various Elbow Angles by the Left Arm of Male College Students

| Direction of force | Elbow angle (deg) | Percentiles (lb) | | | S.D. |
|---|---|---|---|---|---|
| | | 5th | 50th | 95th | |
| Push | 60 | 22 | 79 | 164 | 31 |
| | 90 | 22 | 83 | 172 | 35 |
| | 120 | 26 | 99 | 180 | 42 |
| | 150 | 30 | 111 | 192 | 48 |
| | 180 | 42 | 126 | 196 | 47 |
| Pull | 60 | 26 | 64 | 110 | 23 |
| | 90 | 32 | 80 | 122 | 28 |
| | 120 | 34 | 94 | 152 | 34 |
| | 150 | 41 | 112 | 188 | 37 |
| | 180 | 50 | 116 | 172 | 37 |
| Left | 60 | 12 | 32 | 62 | 17 |
| | 90 | 10 | 33 | 72 | 19 |
| | 120 | 10 | 30 | 68 | 18 |
| | 150 | 8 | 29 | 66 | 20 |
| | 180 | 8 | 30 | 64 | 20 |
| Right | 60 | 17 | 50 | 83 | 21 |
| | 90 | 16 | 48 | 87 | 22 |
| | 120 | 20 | 45 | 89 | 21 |
| | 150 | 15 | 47 | 113 | 27 |
| | 180 | 13 | 43 | 92 | 22 |
| Up | 60 | 15 | 44 | 82 | 18 |
| | 90 | 17 | 52 | 100 | 22 |
| | 120 | 17 | 54 | 102 | 25 |
| | 150 | 15 | 52 | 110 | 27 |
| | 180 | 9 | 41 | 83 | 23 |
| Down | 60 | 18 | 46 | 76 | 18 |
| | 90 | 21 | 49 | 92 | 20 |
| | 120 | 21 | 51 | 102 | 23 |
| | 150 | 18 | 41 | 74 | 16 |
| | 180 | 13 | 35 | 72 | 15 |

[20] C. T. Morgan et al. (eds.), *Human Engineering Guide to Equipment Design*, McGraw-Hill Book Company, New York, 1963, pp. 560, 562–567.

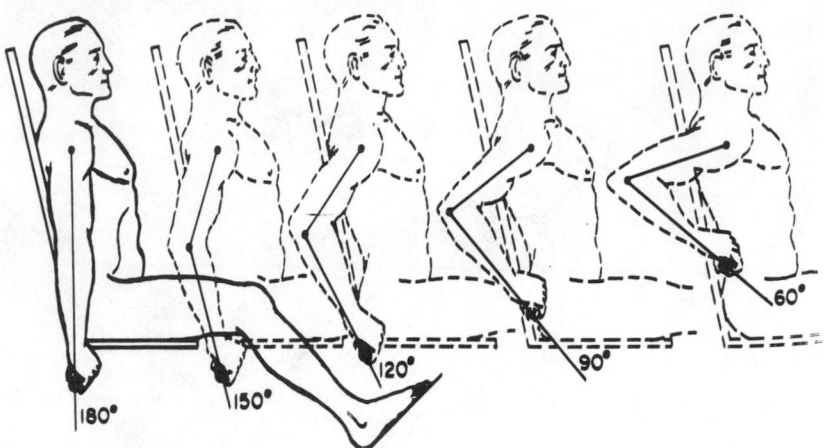

| Maximum Force Exerted in the Sitting Position with the Wrist Pronated on a Horizontal Handgrip at Various Elbow Angles by the Right Arm of Male College Students | | | | | | Maximum Force Exerted in the Sitting Position with the Wrist Pronated on a Horizontal Handgrip at Various Elbow Angles by the Left Arm of Male College Students | | | | |
|---|---|---|---|---|---|---|---|---|---|---|---|
| Direction of force | Elbow angle (deg) | Percentiles (lb) | | | S.D. | Direction of force | Elbow angle (deg) | Percentiles (lb) | | | S.D. |
| | | 5th | 50th | 95th | | | | 5th | 50th | 95th | |
| Push | 60 | 40 | 94 | 156 | 36 | Push | 60 | 33 | 86 | 138 | 35 |
| | 90 | 25 | 65 | 100 | 24 | | 90 | 27 | 60 | 93 | 28 |
| | 120 | 23 | 46 | 70 | 15 | | 120 | 17 | 43 | 71 | 17 |
| | 150 | 18 | 40 | 66 | 18 | | 150 | 15 | 37 | 69 | 18 |
| | 180 | 17 | 32 | 59 | 12 | | 180 | 12 | 32 | 59 | 13 |
| Pull | 60 | 13 | 37 | 50 | 16 | Pull | 60 | 20 | 39 | 64 | 18 |
| | 90 | 14 | 32 | 54 | 13 | | 90 | 17 | 37 | 65 | 18 |
| | 120 | 13 | 26 | 43 | 10 | | 120 | 12 | 30 | 56 | 14 |
| | 150 | 12 | 29 | 48 | 10 | | 150 | 15 | 32 | 52 | 13 |
| | 180 | 11 | 28 | 48 | 12 | | 180 | 16 | 34 | 61 | 15 |
| Right | 60 | 19 | 41 | 72 | 19 | Right | 60 | 20 | 42 | 66 | 15 |
| | 90 | 12 | 31 | 64 | 15 | | 90 | 17 | 38 | 60 | 12 |
| | 120 | 9 | 26 | 53 | 13 | | 120 | 17 | 34 | 53 | 8 |
| | 150 | 9 | 21 | 39 | 11 | | 150 | 17 | 31 | 54 | 11 |
| | 180 | 10 | 19 | 34 | 7 | | 180 | 15 | 28 | 41 | 8 |
| Left | 60 | 16 | 48 | 73 | 18 | Left | 60 | 18 | 36 | 51 | 15 |
| | 90 | 16 | 39 | 59 | 15 | | 90 | 11 | 27 | 54 | 11 |
| | 120 | 15 | 34 | 47 | 11 | | 120 | 10 | 22 | 39 | 10 |
| | 150 | 18 | 32 | 45 | 7 | | 150 | 9 | 23 | 53 | 16 |
| | 180 | 16 | 31 | 57 | 13 | | 180 | 10 | 20 | 49 | 13 |
| Up | 60 | 23 | 49 | 79 | 20 | Up | 60 | 22 | 57 | 100 | 22 |
| | 90 | 28 | 69 | 112 | 29 | | 90 | 37 | 77 | 123 | 24 |
| | 120 | 41 | 91 | 138 | 30 | | 120 | 45 | 91 | 145 | 30 |
| | 150 | 43 | 99 | 165 | 38 | | 150 | 58 | 100 | 159 | 32 |
| | 180 | 35 | 95 | 156 | 35 | | 180 | 47 | 101 | 171 | 11 |
| Down | 60 | 23 | 81 | 158 | 35 | Down | 60 | 18 | 74 | 139 | 35 |
| | 90 | 22 | 83 | 142 | 35 | | 90 | 23 | 75 | 136 | 34 |
| | 120 | 37 | 92 | 161 | 35 | | 120 | 29 | 75 | 148 | 40 |
| | 150 | 40 | 90 | 154 | 34 | | 150 | 39 | 79 | 136 | 29 |
| | 180 | 41 | 87 | 143 | 31 | | 180 | 34 | 76 | 138 | 31 |

### Maximum Force Exerted in the Sitting Position with the Wrist Supinated on a Horizontal Handgrip at Various Elbow Angles by the Right Arm of Male College Students

| Direction of force | Elbow angle (deg) | Percentiles (lb) | | | S.D. |
|---|---|---|---|---|---|
| | | 5th | 50th | 95th | |
| Push | 60 | 34 | 96 | 172 | 39 |
| | 90 | 25 | 65 | 117 | 24 |
| | 120 | 20 | 43 | 71 | 17 |
| | 150 | 17 | 36 | 59 | 14 |
| | 180 | 12 | 32 | 58 | 15 |
| Pull | 60 | 16 | 51 | 93 | 25 |
| | 90 | 13 | 43 | 74 | 19 |
| | 120 | 11 | 40 | 63 | 17 |
| | 150 | 11 | 37 | 66 | 17 |
| | 180 | 15 | 39 | 73 | 19 |
| Right | 60 | 18 | 44 | 73 | 19 |
| | 90 | 18 | 39 | 72 | 24 |
| | 120 | 17 | 34 | 64 | 15 |
| | 150 | 15 | 32 | 60 | 14 |
| | 180 | 14 | 29 | 48 | 12 |
| Left | 60 | 13 | 36 | 70 | 17 |
| | 90 | 13 | 31 | 48 | 12 |
| | 120 | 12 | 30 | 46 | 11 |
| | 150 | 12 | 31 | 52 | 14 |
| | 180 | 10 | 28 | 44 | 10 |
| Up | 60 | 17 | 45 | 78 | 22 |
| | 90 | 21 | 63 | 107 | 27 |
| | 120 | 41 | 88 | 143 | 33 |
| | 150 | 37 | 103 | 161 | 40 |
| | 180 | 51 | 113 | 165 | 34 |
| Down | 60 | 20 | 59 | 132 | 35 |
| | 90 | 17 | 80 | 143 | 37 |
| | 120 | 29 | 92 | 148 | 13 |
| | 150 | 37 | 93 | 150 | 35 |
| | 180 | 44 | 87 | 135 | 32 |

### Maximum Force Exerted in the Sitting Position with the Wrist Supinated on a Horizontal Handgrip at Various Elbow Angles by the Left Arm of Male College Students

| Direction of force | Elbow angle (deg) | Percentiles (lb) | | | S.D. |
|---|---|---|---|---|---|
| | | 5th | 50th | 95th | |
| Push | 60 | 35 | 89 | 176 | 42 |
| | 90 | 25 | 59 | 104 | 27 |
| | 120 | 15 | 40 | 80 | 18 |
| | 150 | 13 | 38 | 69 | 30 |
| | 180 | 14 | 30 | 47 | 10 |
| Pull | 60 | 23 | 54 | 87 | 23 |
| | 90 | 13 | 42 | 68 | 21 |
| | 120 | 14 | 40 | 66 | 18 |
| | 150 | 16 | 40 | 62 | 15 |
| | 180 | 17 | 40 | 70 | 18 |
| Right | 60 | 16 | 38 | 64 | 12 |
| | 90 | 12 | 32 | 46 | 12 |
| | 120 | 14 | 31 | 55 | 13 |
| | 150 | 12 | 32 | 62 | 15 |
| | 180 | 12 | 29 | 43 | 9 |
| Left | 60 | 17 | 42 | 81 | 20 |
| | 90 | 16 | 33 | 52 | 12 |
| | 120 | 14 | 28 | 45 | 8 |
| | 150 | 12 | 26 | 43 | 10 |
| | 180 | 8 | 27 | 44 | 10 |
| Up | 60 | 20 | 49 | 89 | 22 |
| | 90 | 24 | 75 | 131 | 29 |
| | 120 | 38 | 94 | 152 | 33 |
| | 150 | 44 | 104 | 164 | 36 |
| | 180 | 45 | 111 | 173 | 40 |
| Down | 60 | 20 | 58 | 138 | 41 |
| | 90 | 23 | 80 | 160 | 43 |
| | 120 | 35 | 84 | 136 | 33 |
| | 150 | 43 | 84 | 136 | 29 |
| | 180 | 36 | 78 | 124 | 28 |

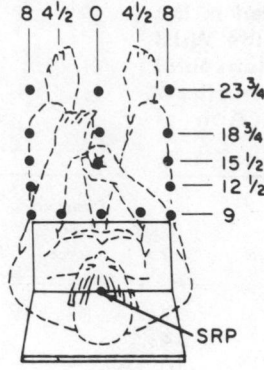

### Maximum Force Exerted in Pushing on an Aircraft Control Stick by the Right Arm of Male Air Force Personnel in the Sitting Position [1]

| Control distance from SRP (in.) | Control distance from mid-plane of body (in.) | Percentiles (lb) | | |
|---|---|---|---|---|
| | | 5th | 50th | 95th |
| 9 | 0 | 26 | 46 | 67 |
| | 4½ (left) | 18 | 33 | 54 |
| | 8 (left) | 12 | 29 | 44 |
| | 4½ (right) | 34 | 58 | 82 |
| | 8 (right) | 37 | 65 | 95 |
| 12½ | 8 (left) | 18 | 36 | 68 |
| | 8 (right) | 43 | 74 | 102 |
| 15½ | 0 | 43 | 86 | 160 |
| | 8 (left) | 23 | 60 | 118 |
| | 8 (right) | 53 | 100 | 164 |
| 18¾ | 0 | 64 | 124 | 177 |
| | 8 (left) | 36 | 72 | 114 |
| | 8 (right) | 70 | 125 | 198 |
| 23¾ | 0 | 54 | 106 | 141 |
| | 8 (left) | 29 | 64 | 104 |
| | 8 (right) | 56 | 100 | 147 |

[1] Stick is grasped 13½ in. above the SRP.

### Maximum Force Exerted in Pulling on an Aircraft Control Stick by the Right Arm of Male Air Force Personnel in the Sitting Position

| Control distance from SRP (in.) | Control distance from mid-plane of body (in.) | Percentiles (lb) | | |
|---|---|---|---|---|
| | | 5th | 50th | 95th |
| 9 | 0 | 34 | 57 | 86 |
| | 4½ (left) | 28 | 45 | 66 |
| | 8 (left) | 26 | 40 | 67 |
| | 4½ (right) | 39 | 62 | 88 |
| | 8 (right) | 39 | 58 | 86 |
| 12½ | 8 (left) | 33 | 53 | 77 |
| | 8 (right) | 49 | 80 | 108 |
| 15½ | 0 | 54 | 83 | 113 |
| | 8 (left) | 39 | 64 | 98 |
| | 8 (right) | 55 | 89 | 119 |
| 18¾ | 0 | 56 | 86 | 127 |
| | 8 (left) | 45 | 74 | 108 |
| | 8 (right) | 58 | 99 | 126 |
| 23¾ | 0 | 62 | 102 | 138 |
| | 8 (left) | 51 | 90 | 129 |
| | 8 (right) | 58 | 103 | 133 |

**Maximum Force Exerted to the Left on an Aircraft Control Stick by the Right Arm of Male Air Force Personnel in the Sitting Position**

| Control distance from SRP (in.) | Control distance from mid-plane of body (in.) | Percentiles (lb) | | |
|---|---|---|---|---|
| | | 5th | 50th | 95th |
| 9 | 0 | 30 | 47 | 66 |
| | 4½ (left) | 31 | 49 | 67 |
| | 8 (left) | 24 | 44 | 65 |
| | 4½ (right) | 26 | 46 | 78 |
| | 8 (right) | 26 | 44 | 72 |
| 12½ | 8 (left) | 23 | 44 | 70 |
| | 8 (right) | 22 | 39 | 59 |
| 15½ | 0 | 24 | 38 | 52 |
| | 8 (left) | 20 | 35 | 58 |
| | 8 (right) | 24 | 40 | 70 |
| 18¾ | 0 | 8 | 32 | 53 |
| | 8 (left) | 16 | 30 | 56 |
| | 8 (right) | 22 | 39 | 70 |
| 23¾ | 0 | 14 | 29 | 46 |
| | 8 (left) | 11 | 21 | 49 |
| | 8 (right) | 20 | 37 | 66 |

**Maximum Force Exerted to the Right on an Aircraft Control Stick by the Right Arm of Male Air Force Personnel in the Sitting Position**

| Control distance from SRP (in.) | Control distance from mid-plane of body (in.) | Percentiles (lb) | | |
|---|---|---|---|---|
| | | 5th | 50th | 95th |
| 9 | 0 | 23 | 38 | 49 |
| | 4½ (left) | 31 | 48 | 64 |
| | 8 (left) | 34 | 55 | 74 |
| | 4½ (right) | 15 | 27 | 51 |
| | 8 (right) | 12 | 22 | 43 |
| 12½ | 8 (left) | 31 | 48 | 70 |
| | 8 (right) | 16 | 24 | 46 |
| 15½ | 0 | 20 | 28 | 39 |
| | 8 (left) | 25 | 43 | 63 |
| | 8 (right) | 13 | 22 | 49 |
| 18¾ | 0 | 15 | 25 | 35 |
| | 8 (left) | 22 | 36 | 61 |
| | 8 (right) | 14 | 24 | 50 |
| 23¾ | 0 | 13 | 20 | 30 |
| | 8 (left) | 19 | 31 | 48 |
| | 8 (right) | 12 | 22 | 51 |

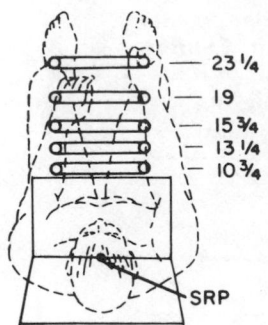

23¼
19
15¾
13¼
10¾

SRP

### Maximum Force Exerted in Pushing on an Aircraft Control Wheel by the Right Arm of Male Air Force Personnel in the Sitting Position [1]

| Control distance from SRP (in.) | Control position (deg) | Percentiles (lb) | | |
|---|---|---|---|---|
| | | 5th | 50th | 95th |
| 10¾ | 0 | 52 | 86 | 135 |
| | 45 (left) | 48 | 84 | 149 |
| | 90 (left) | 32 | 67 | 125 |
| | 45 (right) | 40 | 67 | 128 |
| | 90 (right) | 19 | 52 | 112 |
| 13¼ | 90 (left) | 32 | 54 | 93 |
| | 90 (right) | 25 | 51 | 83 |
| 15¾ | 0 | 61 | 90 | 155 |
| | 90 (left) | 32 | 59 | 139 |
| | 90 (right) | 32 | 53 | 102 |
| 19 | 0 | 64 | 121 | 235 |
| | 90 (left) | 37 | 88 | 171 |
| | 90 (right) | 33 | 67 | 140 |
| 23¼ | 0 | 105 | 171 | 242 |
| | 90 (left) | 82 | 131 | 211 |
| | 90 (right) | 49 | 117 | 197 |

[1] Wheel grips are 18 in. above the SRP and 15 in. apart.

### Maximum Force Exerted in Pulling on an Aircraft Control Wheel by the Right Arm of Male Air Force Personnel in the Sitting Position

| Control distance from SRP (in.) | Control position (deg) | Percentiles (lb) | | |
|---|---|---|---|---|
| | | 5th | 50th | 95th |
| 10¾ | 0 | 44 | 66 | 102 |
| | 45 (left) | 40 | 67 | 111 |
| | 90 (left) | 23 | 55 | 109 |
| | 45 (right) | 39 | 67 | 97 |
| | 90 (right) | 18 | 43 | 87 |
| 13¼ | 90 (left) | 33 | 67 | 120 |
| | 90 (right) | 31 | 60 | 102 |
| 15¾ | 0 | 66 | 94 | 145 |
| | 90 (left) | 42 | 71 | 144 |
| | 90 (right) | 49 | 80 | 130 |
| 19 | 0 | 73 | 106 | 169 |
| | 90 (left) | 60 | 88 | 127 |
| | 90 (right) | 61 | 94 | 149 |
| 23¼ | 0 | 77 | 125 | 182 |
| | 90 (left) | 73 | 117 | 162 |
| | 90 (right) | 74 | 110 | 186 |

### Maximum Force Exerted to the Left on an Aircraft Control Wheel by the Right Arm of Male Air Force Personnel in the Sitting Position

| Control distance from SRP (in.) | Control position (deg) | Percentiles (lb) | | |
|---|---|---|---|---|
| | | 5th | 50th | 95th |
| 10¾ | 0 | 26 | 46 | 88 |
| | 45 (left) | 21 | 54 | 123 |
| | 90 (left) | 23 | 47 | 91 |
| | 45 (right) | 31 | 54 | 120 |
| | 90 (right) | 21 | 42 | 104 |
| 13¼ | 90 (left) | 26 | 44 | 86 |
| | 90 (right) | 25 | 45 | 99 |
| 15¾ | 0 | 27 | 46 | 112 |
| | 90 (left) | 27 | 43 | 82 |
| | 90 (right) | 29 | 50 | 86 |
| 19 | 0 | 25 | 44 | 95 |
| | 90 (left) | 22 | 43 | 76 |
| | 90 (right) | 33 | 52 | 104 |
| 23¼ | 0 | 20 | 39 | 86 |
| | 90 (left) | 21 | 38 | 73 |
| | 90 (right) | 26 | 55 | 109 |

### Maximum Force Exerted to the Right on an Aircraft Control Wheel by the Right Arm of Male Air Force Personnel in the Sitting Position

| Control distance from SRP (in.) | Control position (deg) | Percentiles (lb) | | |
|---|---|---|---|---|
| | | 5th | 50th | 95th |
| 10¾ | 0 | 20 | 48 | 96 |
| | 45 (left) | 24 | 69 | 121 |
| | 90 (left) | 27 | 59 | 101 |
| | 45 (right) | 24 | 51 | 118 |
| | 90 (right) | 15 | 54 | 112 |
| 13¼ | 90 (left) | 21 | 52 | 98 |
| | 90 (right) | 19 | 51 | 111 |
| 15¾ | 0 | 27 | 59 | 97 |
| | 90 (left) | 19 | 53 | 96 |
| | 90 (right) | 20 | 46 | 91 |
| 19 | 0 | 30 | 63 | 104 |
| | 90 (left) | 27 | 46 | 94 |
| | 90 (right) | 22 | 41 | 87 |
| 23¼ | 0 | 35 | 60 | 98 |
| | 90 (left) | 26 | 42 | 82 |
| | 90 (right) | 22 | 40 | 68 |

## Rotational Force Applications

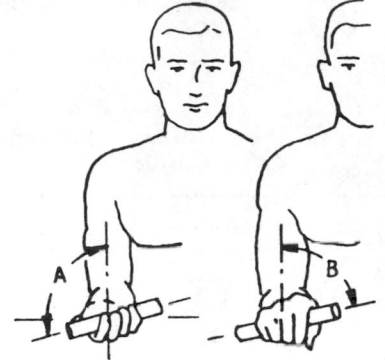

### Maximum Force on a Vertical Handgrip by Wrist-Forearm Supination (A) and Pronation (B)*

| Hand | Movement | Percentile | | | SD |
| | | 5th | 50th | 95th | |
|------|----------|-----|------|------|-----|
| Right | A | 35 lb | 64 lb | 93 lb | 18 |
| | B | 29 lb | 71 lb | 119 lb | 28 |
| Left | A | 30 lb | 62 lb | 88 lb | 16 |
| | B | 31 lb | 71 lb | 132 lb | 31 |

*Male college students.

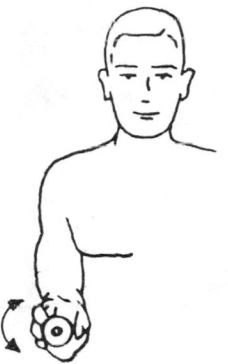

### Maximum Force on Various Sizes of Round Knobs with Knurled Edges*

| Knob Diameter, in | Mean Torque, in-oz | SD |
|-------------------|--------------------|-----|
| 0.25 | 19.6 | 5.4 |
| 0.50 | 45.9 | 13.5 |
| 0.75 | 93.1 | 33.1 |
| 1.00 | 116.0 | 35.5 |
| 1.25 | 132.9 | 37.7 |
| 1.50 | 146.8 | 37.5 |
| 2.00 | 210.2 | 48.9 |
| 3.00 | 477.7 | 136.6 |
| 4.00 | 698.0 | 173.9 |
| 5.00 | 973.4 | 262.8 |

*Male Air Force personnel.

## Handwheels (Partial Turn)

Considerably higher forces are possible for two-handed operation of handwheels with large diameters. For short turns using a vertical or horizontally mounted wheel with a diameter of, say, 20 or more in (51 cm), a strong male adult can exert as much as 120 in/lb. The wheel column and/or rim should be mounted approximately at the operator's elbow level. The accompanying table provides guidelines for establishing handwheel force requirements.

## Steering-Wheel Forces (16-in-, 41-cm-, Diameter Wheel)[21]

Although numerous studies of steering-wheel force have been performed in static laboratory settings, the true, expected force response of drivers while driving is considerably different because emergency conditions requiring a sudden application of force—say, to overcome a failure of the power steering—introduce a psychological factor. The steering-wheel force data shown in the accompanying

[21]B. F. Pierce et al., Human Force Considerations in the Failure of Power Assisted Devices, MFI Report 73-105, 1970 (DOT-HS-230-2-396).

### Optimum Size of Handwheels for Partial Turns (Handwheel Facing Standing Operator)

| Torque, in-lb | Wheel Diameter, in (38 to 48 in above the Floor) | Wheel Diameter, in (below 38 in or above 48 in) |
|---------------|----------------------------------------------------|---------------------------------------------------|
| 20–40 | 6.0 | 10.0 |
| 40–60 | 10.0 | 16.0 |
| 60–90 | 10.0 | 16.0 |
| Above 90 | 16.0 | 16.0 |

Source: K. F. H. Murrell, Human Performance in Industry, Reinhold Publishing Corporation, New York, 1965, p. 242.

table represent typical responses of randomly selected male and female drivers in the U.S. civilian population.

Steering-wheel turning rates attained by the same subjects were as shown in the accompanying table.

| Trial | N | Mean | 5th %-tile |
|-------|-----|--------|------------|
| 1 | 182 | 41.6 lb | 14.4 lb |
| 2 | 182 | 36.2 lb | 17.1 lb |
| 3 | 182 | 38.3 lb | 19.5 lb |

| Trial | N | Mean | 5th %-tile |
|-------|-----|--------|------------|
| 1 | 182 | 4.3 lb | 0.5 lb |
| 2 | 182 | 6.6 lb | 1.8 lb |
| 3 | 182 | 7.3 lb | 2.2 lb |

## Pedal Force-Limit Expectations[22]

The maximum amount of force that an individual can apply to foot pedals depends on several factors, including the angular relationships between the seat back and the seat pan, the distances between the seat and the pedal, and the angle and size of the pedal. Perhaps the most important consideration, however, is whether the operator is expected to operate the pedal using only ankle flexion. The accompanying tables provide a number of force-limit values for various seating geometries and force-application conditions. Data are for young male populations.

In actual practice, however, consider the following guidelines:

For typical automobile braking, limit force requirements to approximately 60 lb (27 kg).

For typical accelerator pedal operation, limit forces to 20 lb (9.1 kg) (minimum 10 lb, 4.5 kg).

For aircraft rudder pedals, limit forces to 20 lb (9.1 kg) for ankle operation and to 150 lb (68 kg) for full-leg operation.

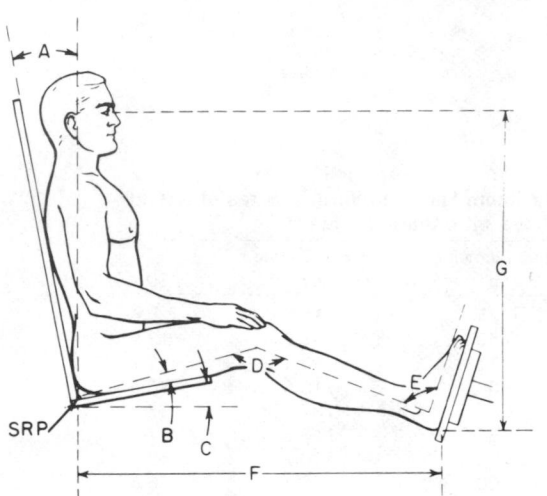

Full-leg, maximum force mode.

### Maximum Force that Can Be Exerted in Extension of the Leg at the Hip and Knee by British Male Civilians in the Sitting Position

| Test conditions | | | | Avg. force (lb) |
|---|---|---|---|---|
| A | B | C | D | |
| 0 | 0 | 0 | 90 | 63 |
| 0 | 0 | 0 | 113 | 89 |
| 0 | 0 | 0 | 135 | 156 |
| 0 | 5 | 0 | 164 | 559 |
| 0 | 6 | 0 | 94 | 73 |
| 0 | 8 | 0 | 93 | 87 |
| 0 | 10 | 0 | 80 | 77 |
| 0 | 10 | 0 | 90 | 59 |
| 0 | 10 | 0 | 135 | 270 |
| 0 | 10 | 0 | 165 | 346 |
| 0 | 15 | 0 | 149 | 227 |
| 0 | 15 | 0 | 160 | 845 |
| 0 | 15 | 0 | 169 | 530 |
| 0 | 16 | 0 | 129 | 319 |
| 0 | 17 | 0 | 117 | 212 |
| 0 | 17 | 0 | 151 | 684 |
| 0 | 33 | 0 | 106 | 184 |

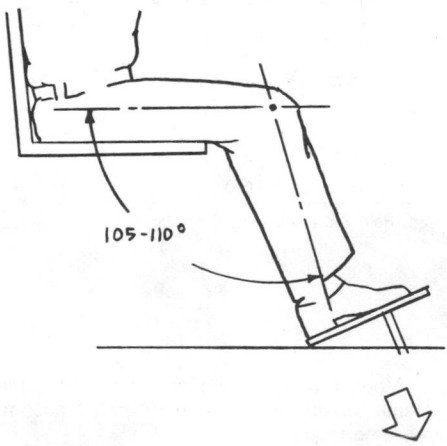

Ankle flexion mode.

### Maximum Force that Can Be Exerted in Extension of the Ankle by Male Air Force Personnel Under Various Test Conditions [1]

| Test conditions [2] | | | | Percentiles (lb) | | | |
|---|---|---|---|---|---|---|---|
| A | E | F | G | 5th * | 50th | 95th * | S.D. |
| 13 | 10 | 35¼ | 37 | 14 | 120 | 227 | 64.8 |
| 13 | 10 | 38½ | 37 | 54 | 149 | 244 | 57.7 |
| 13 | 30 | 35¼ | 37 | 25 | 137 | 249 | 68.3 |
| 13 | 30 | 38½ | 37 | 64 | 184 | 303 | 72.7 |
| 13 | 50 | 35¼ | 37 | 24 | 88 | 152 | 38.9 |
| 13 | 50 | 38½ | 37 | 48 | 133 | 218 | 51.7 |
| 13 | 10 | 35¼ | 39 | 15 | 111 | 207 | 58.6 |
| 13 | 10 | 38½ | 39 | 37 | 135 | 232 | 59.4 |
| 13 | 30 | 35¼ | 39 | 26 | 137 | 249 | 67.8 |
| 13 | 30 | 38½ | 39 | 60 | 178 | 297 | 72.0 |
| 13 | 50 | 35¼ | 39 | 22 | 91 | 161 | 42.4 |
| 13 | 50 | 38½ | 39 | 50 | 132 | 213 | 49.4 |
| 13 | 10 | 35¼ | 41 | 18 | 118 | 218 | 60.8 |
| 13 | 10 | 38½ | 41 | 35 | 122 | 209 | 53.1 |
| 13 | 30 | 35¼ | 41 | 32 | 143 | 254 | 67.5 |
| 13 | 30 | 38½ | 41 | 50 | 175 | 300 | 76.1 |
| 13 | 50 | 35¼ | 41 | 23 | 98 | 174 | 46.0 |
| 13 | 50 | 38½ | 41 | 50 | 138 | 226 | 53.5 |

[22]C. T. Morgan et al. (eds)., *Human Engineering Guide to Equipment Design,* McGraw-Hill Book Company, New York, 1963, pp. 569, 570.

## LIFTING AND CARRYING

Variables to be considered are the following

### 1. Task Variables

Location of the object to be lifted (accessibility)
Size and shape of the object
Height from and/or to which the object is lifted
Weight and weight distribution of the object
Relation between the object's c/g and points of grasp
Whether the object is lifted by handles
Working-position characteristics (e.g., awkward rather than normal)
Manipulatory accuracy requirements
Frequency and duration of lifting and carrying movements

### 2. Human Variables

Age and sex of the person or persons doing the lifting
Body dimensions (stature, arm length, etc.)
Physical fitness
Experience and training

### 3. Environmental Variables

Thermal (temperature, humidity, and ventilation)
Dynamic (platform motion and vibration)

The most important determinant of lifting force is the distance of the feet from the grasping axis (lifting force is greatest if the lift weight is in the same vertical plane as the body, and it decreases rapidly as the weight moves away from the body). The best height for lifting is approximately fingertip height, with the hands hanging freely at the sides. Above this height, lift capacity decreases very rapidly; below this height, it decreases more slowly. With the load near the floor, capacity is only about 75 percent of best height capacity.

The maximum individual loads carried in either hand by means of handgrips should be about 60 lb (27 kg) for short distances and 35 lb (14 kg) for longer distances. Bulky articles should not weigh more than about 30 lb (13 kg). In general, a package is considered heavy when it reaches about 35 percent of a person's own body weight. Less strain occurs if equal weights are carried in a balanced manner, i.e., half in one hand and half in the other. Carrying one heavy weight a given distance incurs less physiological cost than carrying two weights, each weighing half as much, the same distance in two trips.

Handle size and shape, as well as position, make a considerable difference in the individual's acceptance of maximum weight.

### General Guidelines with Respect to Lifting

Assuming that a package is appropriately designed so that it can be picked up, carried, and deposited without damage to the package or injury to the individual, the general weight values shown in the accompanying tables can be used as guidelines in the design of hand-carried packages.

The accompanying sketches are meant to emphasize significant considerations in developing an appropriate package size, shape, and weight configuration. First, one must be able to get hold of the package; thus the designer needs to consider the position the person has to get into and how his hands and fingers can be accommodated to obtain a firm grip. Second, the package should be designed so that the individual can assume a comfortable and safe carrying position (minimum back or arm strain, minimum pressure on the hands or fingers, ability to "see ahead," etc.). Third, the position where the package is to be deposited is important, i.e., if the final resting level is high, the package weight cannot be as great.

### Some Typical Industrial Standards

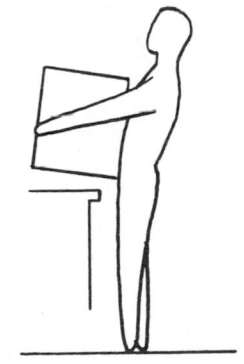

#### Reasonable Weight Limits for Occasional Lifting

| Age, Years | Male | Female |
|---|---|---|
| 14–16 | 33 lb | 22 lb |
| 16–18 | 42 lb | 26 lb |
| 18–20 | 51 lb | 31 lb |
| 20–35 | 55 lb | 33 lb |
| 35–50 | 46 lb | 29 lb |
| Over 50 | 35 lb | 22 lb |

*Source:* From the Swiss Accident Insurance Institute.

#### Maximum Acceptable Weight of Lift for Adult Males

| Height of Lift | Percent of Population | | | | |
|---|---|---|---|---|---|
| | 90 | 75 | 50 | 25 | 10 |
| Floor level to knuckle height | 52 lb | 59 lb | 66 lb | 73 lb | 80 lb |
| Knuckle height to shoulder height | 51 lb | 56 lb | 62 lb | 68 lb | 73 lb |
| Shoulder height to arm reach | 48 lb | 53 lb | 60 lb | 67 lb | 72 lb |

*Source:* From Liberty Mutual Insurance Company.

## Lifting Packages without Handles (Young Adult Males Typical of Industrial and Military Populations)[23]

The shape of the package has an appreciable effect on a person's ability to pick it up, lift it, and transport it, as shown by the accompanying sketches. It should also be noted that whether the package is cylindrical or rectangular affects how closely or snugly it can be grasped and/or how effectively it can be manipulated; i.e., at certain points in the handling sequence a rectangular shape provides advantages, and at other points a cylindrical shape provides advantages.

The accompany sketches illustrate a continuing lifting and carrying sequence for a person with a 5th- to a 95th-percentile physique.

PACKAGE SIZE:   12×12×12"         12×12×18"         6×8×36"

| Lifting and Carrying Height above the Floor | A | B | C |
|---|---|---|---|
| 36 in | 75 lb | 95 lb | 110 lb |
| 48 in | 55 lb | 75 lb | 95 lb |
| 60 in | 40 lb | 50 lb | 50 lb |

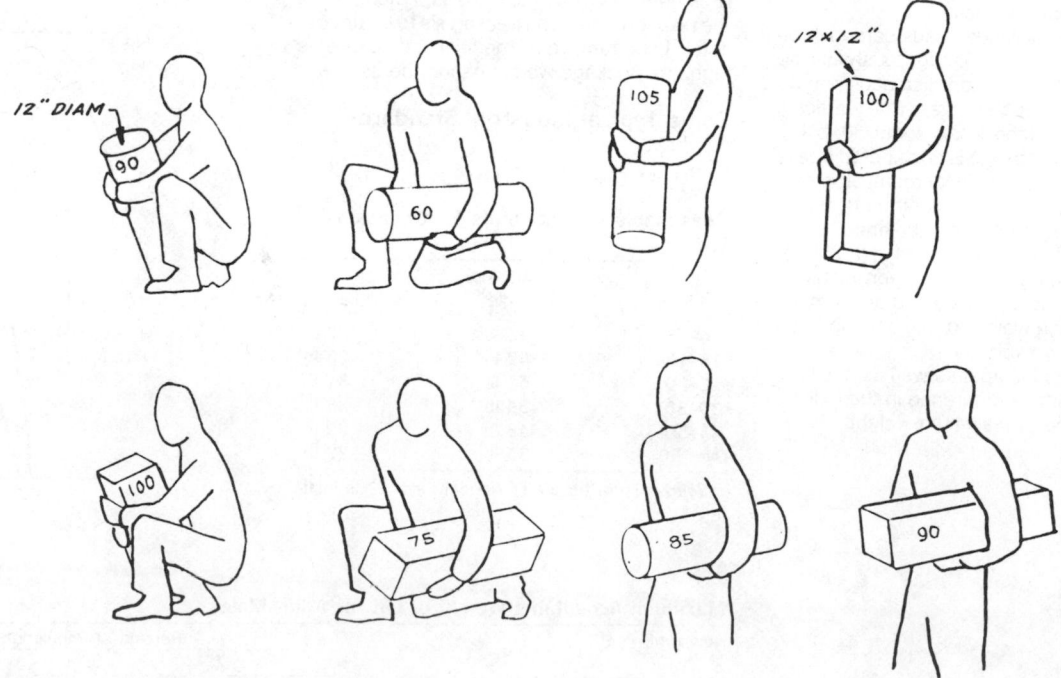

[23]W. E. Woodson and D. W. Conover, *Human Engineering Guide for Equipment Designers,* University of California Press, Berkeley, 1964.

## Weight in Relation to Carrying Mode

The ease with which various packages can be carried depends not only on the weight but also on the type and placement of the handles. The accompanying sketches illustrate common situations and suggested upper weight limits for typical package-carrying configurations:

When the package (e.g., a suitcase) has smooth sides, it can be carried against the person's hip or leg without too much stress. The weight should not exceed more than about 45 lb (20 kg) for males and 35 to 40 lb (14 to 18 kg) for females.

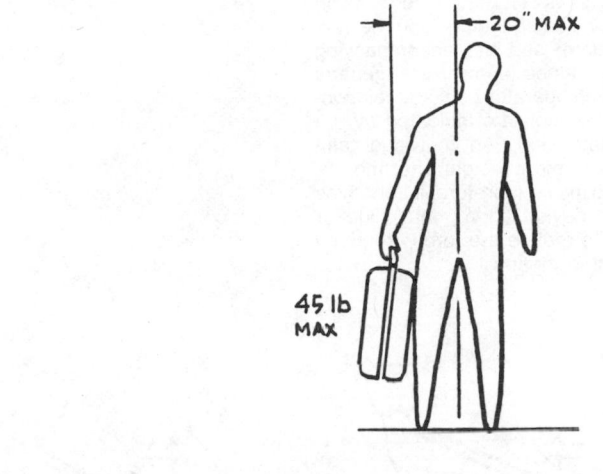

When the package surfaces are not smooth (as is the case with electronic equipment chassis, etc.), limit weight to about 35 lb (14 kg).

Although the typical male can carry an equipment package weighing up to about 60 lb (27 kg) for a short distance, this much weight should involve two persons carrying the package by means of appropriately located dual handles. Two men should not be required to carry 100 lb (45 kg) very far or more than 200 lb (90 kg) more than a short distance.

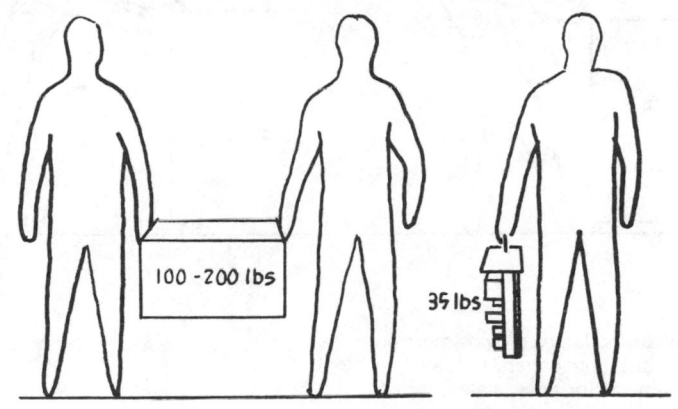

The removal and replacement of equipment units from a rack at different heights require consideration of the suggested maximum package weights in relation to working height, as indicated in the accompanying sketch.

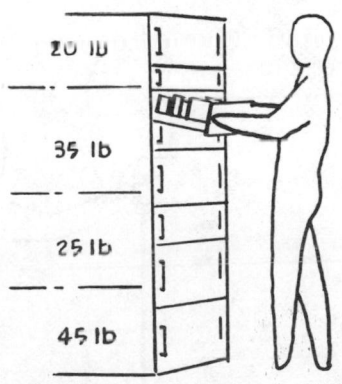

Backpack weight-carrying guidelines are indicated in the accompanying sketch.

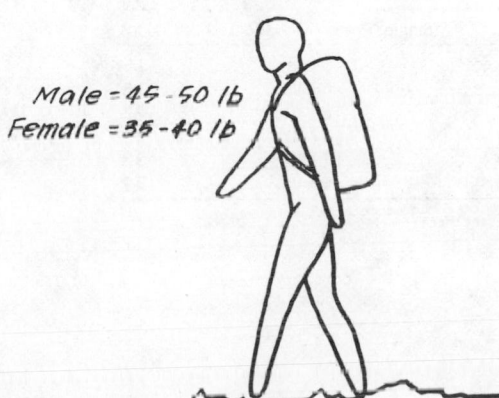

## CAPACITY TO MOVE LARGE OBJECTS

The capacity of an individual to move large objects, such as items of furniture, depends on several factors, including the position the individual can assume (see the accompanying sketches). The values shown are general guidelines for each operator-package relationship (forces for women are indicated by the values in parentheses). An individual can move considerably more weight after the object has begun to move; therefore, it behooves the designer to devise appropriate slide or caster systems to reduce the forces required to put the object in motion.

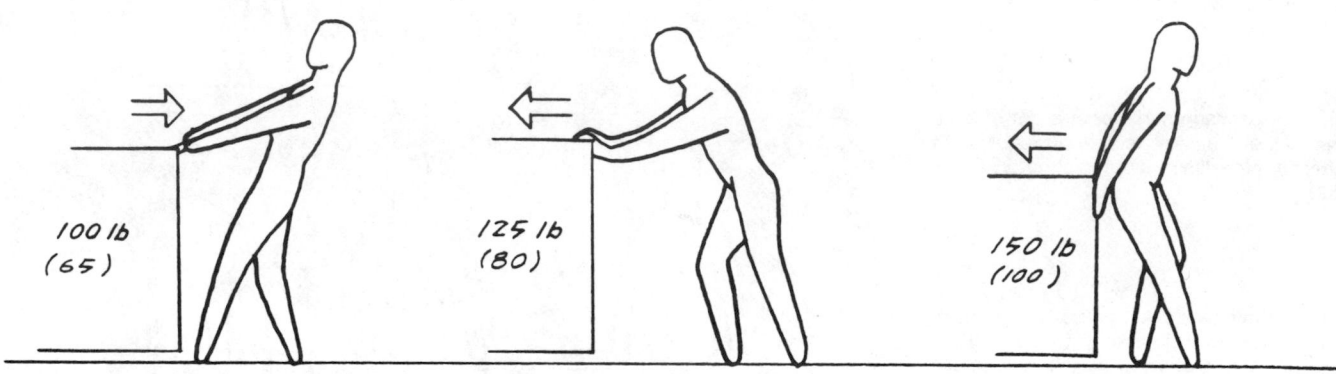

100 lb (65)     125 lb (80)     150 lb (100)

The accompanying tables provide more precise maximal static force capabilities, as determined by Kroemer[24] for adult males, for various application modes. Note that these are for conditions that include structural "push-off" support.

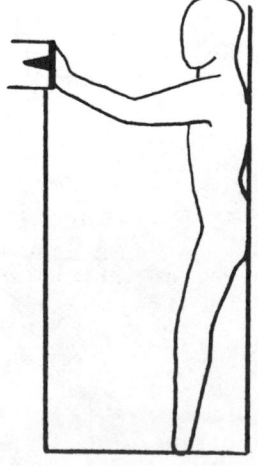

| Distance | Force | |
|---|---|---|
| | Means | SD |
| | **Both Hands** | |
| 50 | 131 lb | 32 |
| 60 | 150 lb | 36 |
| 70 | 221 lb | 61 |
| 80 | 289 lb | 90 |
| 90 | 220 lb | 68 |
| 100 | 145 lb | 57 |
| | **Single, Preferred Hand** | |
| 50 | 59 lb | 15 |
| 60 | 67 lb | 16 |
| 70 | 81 lb | 22 |
| 80 | 117 lb | 32 |
| 90 | 111 lb | 38 |
| 100 | 96 lb | 39 |

[24] K. H. E. Kroemer, *Push Forces Exerted in Sixty-Five Common Working Positions*, AMRL-TR-68-143, Aerospace Medical Research Laboratory, Wright-Patterson Air Force Base, Ohio, 1968.

| Distance | Force | |
|---|---|---|
| | Means | SD |
| a. 80 | 149 lb | 40 |
| 100 | 174 lb | 48 |
| 120 | 175 lb | 37 |
| b. 80 | 161 lb | 36 |
| 100 | 164 lb | 52 |
| 120 | 184 lb | 31 |
| c. 80 | 141 lb | 33 |
| 100 | 152 lb | 44 |
| 120 | 194 lb | 2 |

*Note:* $a$ = 50 in; $b$ = 70 in; c = 90 in.; *a, b,* and *c* represent the height of the force plate.

| Distance | Force | |
|---|---|---|
| | Means | SD |
| a. 70 | 171 lb | 38 |
| 80 | 192 lb | 40 |
| 90 | 178 lb | 32 |
| b. 60 | 130 lb | 25 |
| 70 | 157 lb | 28 |
| 80 | 163 lb | 32 |
| c. 60 | 117 lb | 29 |
| 70 | 139 lb | 29 |
| 80 | 143 lb | 30 |

*Note:* $a$ = 60 in; $b$ = 70 in; $c$ = 80 in.

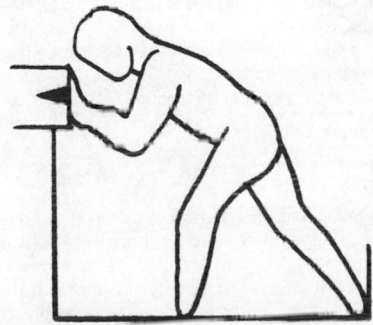

| Distance | Force | |
|---|---|---|
| | Means | SD |
| a. 70 | 140 lb | 33 |
| 80 | 155 lb | 35 |
| 90 | 132 lb | 30 |
| b. 70 | 123 lb | 28 |
| 80 | 122 lb | 28 |
| 90 | 120 lb | 18 |
| c. 70 | 97 lb | 21 |
| 80 | 101 lb | 21 |
| 90 | 109 lb | 18 |

*Note:* $a$ = 70 in; $b$ = 80 in; $c$ = 90 in.

## BODY EQUILIBRIUM (BALANCE AND BODILY ORIENTATION)

Maintenance of body equilibrium results from sensory inputs from visual, auditory, semicircular canal, skin, and kinesthetic and visceral sources.

Body verticality can be fairly accurately judged when the inputs from the various sources represent an accurate picture of the relationships between the body and the earth's gravity. However, if for some reason information from one or more of the sources becomes distorted (e.g., abnormal dynamic conditions affect the semicircular canals, and high *g* forces from a rapid vehicular movement cause high skin pressures, movement of the internal organs, or even confusing visual patterns or auditory sound delays), one's assessment of his or her body position may reflect considerable variation from the true condition of the body.

### Perception of Body Verticality

Both visual and postural verticality are determined largely by the joint action of visual and gravitational forces. Visually, it is important that external visual references be available and that they be compatible with the earth reference. Gravitation *(g)* force exerts a considerable influence on our perception of verticality (which is in the direction of action of the resultant *g* force). The average threshold for sensing body tilt in any direction is 2 to 3°, although under certain conditions tilts of as

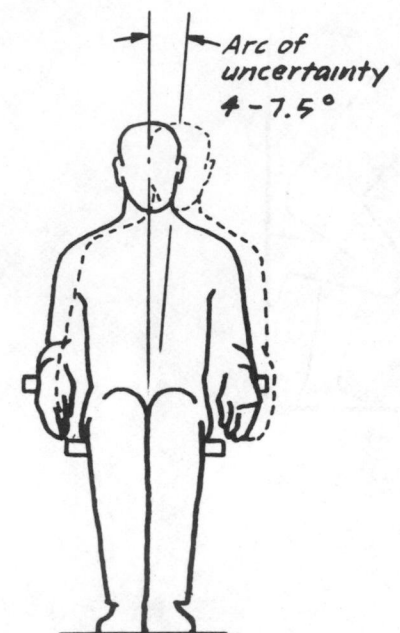

*Arc of uncertainty 4 – 7.5°*

much as 15° may not be detected (e.g., an aircraft banking or pitching at night or under foggy or cloudy conditions).

Blindfolded subjects generally can reorient themselves to a vertical position within about 4 to 5° (the arc of uncertainty). When subjects sit on soft seats, this increases to about 7.5°. The error is usually in the direction of the initial tilt away from verticality. One tends to adapt to a slight tilt, which explains in part why some pilot trainees continue to "fly with one wing down."

## BODY SWAY

The body is ordinarily moving to some degree; i.e., it is not rigid, but is swaying in one direction or another. Fore-aft sway is about 50 to 75 percent greater than side-to-side sway when the body is in a standing position. Sway is independent of body height or weight. It increases if the body is held tense. Sway obviously increases when there is no visual input. There is no relationship between static equilibrium and the ability to maintain dynamic equilibrium, i.e., to remain upright under unstable platform conditions.

Sway can be induced by certain noncompatible visual cues, e.g., illusions of motion (nystagmus). An interrelationship between vision and labyrinthine (inner ear) cues occurs with body movements (rotation). Nystagmus is a reflex eye movement interrupted by saccadic eye movements in the opposite direction to rotational movement. Upon rotation of the body at a constant, maintained rate, the following occurs:

1. Compensatory eye movement after about a 50- to 80-ms delay from the start of rotation
2. Opposite eye movement after the preceding body motion has decelerated to zero (called "post-rotational nystagmus")
3. Eye movements in the original direction once post-rotational nystagmus or eye movement has decelerated (called "inverse nystagmus")

This sequence may take as long as 10 s after the start of rotation.

[25]W. E. Woodson and D. W. Conover, *Human Engineering Guide for Equipment Designers,* University of California Press, Berkeley, 1964.

## ILLUSIONS ASSOCIATED WITH BODY MOVEMENT[25]

| Condition of Observer | Condition of Visual Stimulus Object(s) | Nature of Illusion |
|---|---|---|
| Starts motion | Stationary | Observer sees visual stimulus moving in opposite direction, feels stationary. |
| Stationary | Starts motion | Observer sees object stationary, feels self moving in direction opposite to object's motion. |
| Accelerating rotation to 15 rpm, dark room with only target illuminated | Slow revolution around observer's head, in same direction as his acceleration | Object loses motion, is displaced in direction opposite to its direction of movement.* |
| Above rotation continued at 15 rpm | As above | Observer loses own sensation of motion, object appears to move in opposite direction with greater velocity than before.* |
| Strong rotation suddenly stops | Fixed | At first, object and subject both seem to rotate. Then observers may feel fixed while visual stimuli rotate about them in a direction opposite to that of their previous motion. This illusion is called the *oculo-gyral illusion.* After all apparent motion stops, subject still feels vaguely unpleasant sensations.* |

*When experiments are conducted in a lighted room these illusions do not occur.

Some illusions related to body movement are identified in the accompanying table, including several that are not necessarily associated with nystagmus. It should be noted, however, that these are not universally observed by all people or described in the same manner by those who do observe them.

## VISUAL AND GRAVITATIONAL CONFLICT

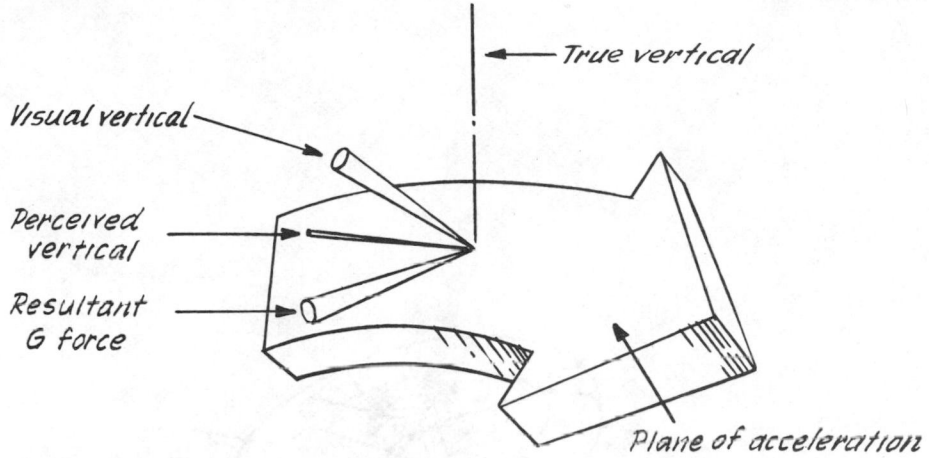

Both visual and postural orientation are influenced by the visual framework. Both are likewise influenced by resultant $g$ forces acting on the body. There is some controversy over which cues, visual or gravitational, are predominant as determinants of our orientation, but it may safely be said that our judgments, in cases of conflict between the two cues, represent a compromise. Such a conflict is present, for example, in an airplane flying in conditions of zero visibility, having an angular acceleration giving a resultant $g$ force on the body, say 45° from the vertical. The cockpit, say 90° from the vertical when this force is acting, represents the whole visual field. Under this conflict of visual and gravitational cues (90 vs 45°), one feels the vertical to be somewhere between them, with the bulk of the evidence indicating that the judgment is closer to the gravitational than to the visual forces. The condition illustrated, in which a pilot's orientation with respect to the earth is disturbed, is called "aviator's vertigo." Visual indicators which show the plane's true attitude to the pilot, and proper training in the use of these instruments, help suppress any actions on his part that might be taken on the basis of erroneous orientation.

### Sensitivity to Movement

LINEAR ACCELERATION

    VERTICAL, 4–12 CM/SEC$^2$

    HORIZONTAL, 12–20 CM/SEC$^2$

ANGULAR ACCELERATION     →

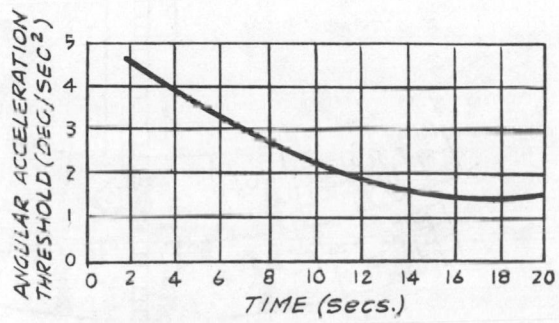

Under normal flying conditions the motion threshold is 102° per second per second.

## CORIOLIS FORCES

Although most people are aware of the discomfort sometimes associated with riding on a rotating platform such as a merry-go-round, they are not aware of what would happen to them if they could not see the stationary earth reference outside the rotating platform. This particular phenomenon poses an important question for the space station designer who may be considering "spinning" the station to create artificial gravity conditions for astronauts on board. Although the idea of reproducing normal earth gravitational effects will aid astronauts in several ways (both physiologically and operationally), the coriolis forces reacting on the station's occupants will introduce some strange conditions which must be overcome by careful arrangement of on-board facilities. For example, the coriolis force will affect astronauts differently, depending on the direction in which they are moving and also on how far from the hub of the station they are located. The accompanying sketch shows at least three different conditions for a station rotating counterclockwise. Experiments have indicated that although some people can adapt to this unique condition, others cannot. Even assuming that only astronauts who can adapt will be aboard, the on-board facilities and equipment should be designed and arranged to minimize rather than add to the astronauts' difficulties in coping with coriolis forces.

Man's Direction of movement →

Direction of Coriolis force ⬇C

## SIMPLE VERSUS COMPLEX REACTION TIME

In practical terms, one should think more in terms of total response time than in terms of simple reaction time. The term "reaction time" generally is considered at two levels. The first is called "simple reaction time," which is the shortest time between the moment a sensory receptor is stimulated and the time some body element reacts, as when a speck of dust strikes an eyelash, and the eyelid immediately closes to protect the eye. Practically speaking, however, scientists measuring simple reaction time typically have a test subject respond to a stimulus by pressing, say, a key—thus allowing the scientist to measure and compare the reaction time of various sensory channels (vision, audition, etc.).

The second level is called "complex reaction time," in which the scientist attempts to include human information-processing time; e.g., the subject may be asked to recognize one stimulus from among several and to respond by selecting one of several response modes.

Since complex reactions are more akin to real-world situations, the designer is more concerned with complex reaction time or, more realistically, total response time. In a sense, response time is a function of several factors, including the following:

1. The sensory channel through which the stimulus is initiated
2. The signal or stimulus characteristics
3. The complexity of the signal
4. The signal rate
5. Whether anticipatory provisions are present
6. The response mode, e.g., the body member used

## SIMPLE REACTION-TIME COMPARISONS OF SENSORY INPUT CHANNELS[26]

## EFFECTS OF SIGNAL CHARACTERISTICS ON REACTION TIME

Signal intensity: The greater the intensity of the signal, the faster will be the reaction time.

Signal anticipation: When the signal is anticipated, reaction time is typically shorter.

Practice: Reaction time tends to be reduced with practice.

Pacing: If operators can set their own pace, they can often react faster to known signals.

Signal quality: Operators generally can react faster to a high-pitched sound, a brighter light, a larger visual target, a longer-duration signal, and a signal emanating from a particular location (i.e., a visual signal that strikes the center rather than the periphery of the eye). There is no difference in simple reaction time between a steady and a flashing light signal. However, when one intermittent light signal must be distinguished on the basis of flash rate, reaction time is related to the length of the flash.

Likelihood of signal appearance: The least likely signals will have the longest reaction times.

Signal format: When signals are arranged sequentially or are meaningfully grouped, reaction time is typically shortened; i.e., reaction time is a function of the number of signals that occur in the next sequential step or is proportional to the number of signal groups. When two successive signals occur within 0.1 s, they are generally treated as a single signal. When the interval between signals is 0.5 s or greater, the operator is capable of responding to each signal separately.

Overload: Although an operator can adjust to excessive signal rates by relying on memory for short bursts, total response failure may occur when rates are too high for too long.

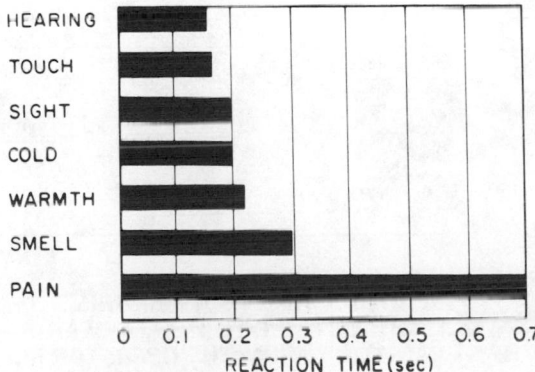

[26]C. T. Morgan et al. (eds.), *Human Engineering Guide to Equipment Design*, McGraw-Hill Book Company, New York, 1963, p. 229.

Using the typical key response technique, scientists have determined the reaction times for various sensory input channels shown in the accompanying graph.

Signals should not occur at rates faster than about two per second unless some means are provided for anticipating the signal. Avoid alerting periods shorter than 0.1 s.

## COMPLEX REACTION TIME

The following estimates have been made regarding the several components of a basic, complex reaction time, i.e., speed of perception to action:

Stimulus detection and neural transit time
0.1 s
Brain recognition time     0.4 s
Decision-making time     up to 4.0 s
Motor response time (minimum)     6.0 s

## AGE AND SEX DIFFERENCES

The reaction time of adult males is generally faster than that of adult females, and, as one might expect, older people have a longer reaction time. Typical comparisons are shown in the accompanying table.

| Age | Males Light, s | Males Sound, s | Female Light, s | Female Sound, s |
|---|---|---|---|---|
| 20 | 0.24 | 0.23 | 0.32 | 0.31 |
| 30 | 0.22 | 0.19 | 0.26 | 0.20 |
| 40 | 0.26 | 0.24 | 0.34 | 0.30 |
| 50 | 0.27 | 0.25 | 0.36 | 0.30 |
| 60 | 0.38 | 0.37 | 0.44 | 0.42 |

## SOME COMMON RESPONSE TIMES

| Simple Movement Response | Time, s |
|---|---|
| Moving eyes (focusing on new spot) | 0.48 |
| Walking (one pace) | 0.60 |
| Sitting down in a chair | 1.32 |
| Standing up from a sitting position | 1.62 |

Other generalizations of interest include the following:

It generally takes about 20 percent longer to respond with the feet than with the hands. The preferred hand is usually about 3 percent faster than the nonpreferred one.

Everyone has a refractory period of about 2 to 3 s, regardless of a stimulus demand rate, which means that a second stimulus arriving within 0.5 s will be treated together with the first (e.g., two successive signals arriving 0.1 s apart will be treated as a single signal).

Simple reaction times usually can be reduced by as much as 40 percent by providing an alerting signal (approximately 2 to 8 s in advance for isolated response tasks and 2 to 3 s in advance for sequential tasks).

## EFFECT OF NUMBER OF RESPONSE CHOICES

As one might expect, when the number of response choices increases, the reaction time is lengthened. The accompanying table illustrates this point.

| Number of Choices | Mean Reaction Time, s |
|---|---|
| 1 | 0.20 |
| 2 | 0.35 |
| 3 | 0.40 |
| 4 | 0.45 |
| 5 | 0.50 |
| 6 | 0.55 |
| 7 | 0.60 |
| 8 | 0.60 |
| 9 | 0.65 |
| 10 | 0.65 |

## SELECTED MANUAL RESPONSE-TIME EXAMPLES

Although many studies have been carried out for the purpose of establishing response times for a variety of manual tasks, space permits citing only a few examples to demonstrate the variability that must naturally accompany the specifics of a given task.

### Reach Distance and Terminal Movement Condition

| Reach Distance, in | Terminal Conditions, s Always the Same | Terminal Conditions, s Varied |
|---|---|---|
| 10 | 0.03 | 0.041 |
| 20 | 0.045 | 0.066 |
| 30 | 0.062 | 0.093 |

## EXPECTED VERSUS UNEXPECTED STIMULUS RESPONSE TIMES

Simple reaction-time reference     0.20 s
Forewarned     0.51 s
Surprised     0.61 s

## MOVEMENT RESPONSE END-POINT CONTROL

When the operator has to control the final movement end point, his or her total response time is greater than when the end point is automatically controlled (e.g., the operator moves a stylus or control to a mechanical "stop").

| Distance, in | Mechanical Stop, s | Operator-Controlled |
|---|---|---|
| 6 | 0.28 | 0.47 |
| 12 | 0.36 | 0.57 |
| 20 | 0.46 | 0.67 |
| 30 | 0.58 | 0.79 |

## END-POINT PRECISION

The more precise or accurate the operator has to be in making a control movement, the longer his or her response time will be.

| Precision, in | Time, s |
|---|---|
| 0.50 | 0.11 |
| 0.25 | 0.27 |
| 0.125 | 0.48 |
| 0.0625 | 0.67 |

## EFFECT OF ADDED MANUAL FORCE ON RESPONSE TIME

| Force Required, lb | Time, s |
|---|---|
| 2 | 0.02 |
| 4 | 0.04 |
| 6 | 0.06 |
| 8 | 0.08 |
| 10 | 0.10 |
| 50 | 0.33 |

## REACTION TIME TO A WARNING LIGHT RELATIVE TO ITS ANGULAR VARIATION FROM THE OPERATOR'S NOMINAL LINE OF SIGHT (WHILE TRACKING)

| Angular Variation, | Time, s |
|---|---|
| 20 | 0.40 |
| 40 | 0.40 |
| 80 | 0.80 |
| 100 | 1.30 |

## EFFECTS OF DISTRACTION

| Type of Distraction | Without Distraction | With Distraction |
|---|---|---|
| Continuous | 0.218 | 0.220 |
| Continuous with periodic interruptions | 0.202 | 0.206 |
| Random/intermittent | 0.204 | 0.220 |

## EFFECT OF REACH DISTANCE ON TOGGLE-SWITCH OPERATION

The response times shown in the accompanying table are those of an operator who, in addition to making the simple movement to the switch location in response to a warning light, had to decide which hand to take off a joy stick and then proceed to reach for and manipulate a toggle switch.

| Distance, in | Time, s |
|---|---|
| 6 | 0.76 |
| 9 | 0.66 |
| 15 | 0.84 |
| 18 | 0.77 |

Note the randomness of the responses, which is due, of course, to the decision-making step in the total response scenario.

## EFFECT OF TOGGLE-SWITCH SEPARATION OR SPACING

| Distance Apart, in | Time, s |
|---|---|
| 1.0 | 0.28 |
| 2.0 | 0.24 |
| 4.0 | 0.35 |
| 8.0 | 0.40 |

Note that when switches are too close, response time is slightly longer; when they are too far apart, response time is significantly longer.

## COMPARISON OF RESPONSE TIMES FOR VARIOUS DIGIT-INPUT DEVICES[27]

10-key pad (calculator type)    12.0 s
10 × 10 matrix keyboard    13.0 s
10 separate levers    17.0 s
10 separate rotary knobs    18.0 s

## VISUAL SCANNING RESPONSE TIME

Two examples of visual scanning response time are shown below. The first involves the common task of pilots when they are required to transfer their line of sight from looking at instruments inside the cockpit to scanning outside the cockpit to look for other aircraft:

1. Vision shifts from outside to inside 1.5 s
2. Vision shifts from inside to outside 2.39 s

The difference is explained by the greater problem of accommodating from the near-vision instrument-watching task to the far-vision blank-sky, target search task.

A further example of flying response deals with the difference between making a recovery flight maneuver using instruments and making the same recovery using a VFR (visual flight reference):

1. Instrument recovery    11.0 s
2. VFR recovery    9.2 s

The difference here is explained by the naturalness of a VFR, making it unnecessary for the pilot to perform extra data processing from abstract instrument graphics.

A second scanning example relates to the positioning of panel displays in the typical automobile. When instruments are clustered high, near the normal roadway line of sight, response to their indications is faster than when the instruments are farther down and are spread out on the instrument panel:

Close to line of sight    1.9 s
Normal panel position    2.2 s
Bottom of panel, spread out    2.9 s

### Other Factors

Various other factors, of course, can obviously affect operator response time. These include physical and psychological stress, constraints within the workplace, and constrictions caused by clothing. Often, too, response times are shorter when the task is paced; i.e., operators can usually operate faster when they are paced than when they pace themselves. Response times vary less under paced conditions.

Finally, one must always consider the response of the mechanical system, since in many cases the operator's response is conditioned by feedback received from the machine.

## SYSTEM RESPONSE-TIME CONSIDERATIONS

Whereas the discussion so far has dealt with the general question of how human reaction time affects a system, the following comments deal with how system response time affects the human—or how machine delays are perceived by the human operator.

Gallaway[28] notes that "silences longer than about 4 s are perceived as unacceptable breaks" in both human (conversation) and man-machine operations (especially in interactive human-computer systems). He defines the following general response-time considerations:

- As a delay (response time) is extended beyond the user's concept of the amount of work being done by a system, frustration increases and affects the quantity and quality of the user's performance.
- Response-time delays are more than a mere inconvenience to the user; i.e., rather than a straight-line decrease in efficiency as delay increases, there are sudden drops in mental efficiency.
- Performing most tasks requires an operator to "hold" a body of information in mind, and the longer this information has to be held, the greater is the chance of forgetting (after 2.0 s, most individuals become aware of waiting, and short-term memory failure begins to occur).
- Response times that exceed 2.0 s should follow only a task that an operator perceives as being completed or one that is structured to appear as a completion point.
- Longer response times are more acceptable to an operator following completion of an important or long task than following routine, short tasks.
- Although a response signal can communicate more than one message concurrently, the response time that should be sought is that demanded by the task in the group which demands the fastest response time.
- When a response to an operator input exceeds about 5.0 s, it is desirable to indicate to the operator that some process is being performed by the machine, e.g., "transaction being processed."
- Systems with a high degree of variation in response times for a given task are more frustrating to operators. In principle, the range of acceptable variation in delay is that range within which the human cannot detect differences, i.e.,:

  1. Response time of 0 to 2 s: variance should be about +5 percent.
  2. Response time of 2 to 4 s: variance should be about +10 percent.
  3. Response time of 4 to 30 s: variance should be about +15 percent.

- In general, delays greater than 15 s rule out conversational interactions. When delays exceed this amount, the operator should be freed from physical or mental captivity (waiting on the machine) and allowed to turn to another activity until some display requests that he or she return to the operation.
- Response times greater than 4 to 15 s generally are too long for conversation requiring short-term memory and probably inhibit problem-solving activity and/or create frustration for a data-entry task.
- Response times greater than 2 to 4 s may be inhibiting for tasks requiring a high level of concentration. A 2-s wait at a terminal when the operator is absorbed and emotionally committed to task completion will seem an interminable wait for most operators.
- Response times less than 2 s are desirable when the operator has to remember information through several responses.
- Response to a signal (e.g., feedback to a pressing key) should be set at less than 100 ms.

---

[27] The task was to enter 10 different digits.

[28] G. R. Gallaway, *Response Times to User Activities in Interactive Man-Machine Computer Systems*, Report no. 277-0014601, NCR Corp., Dayton, Ohio, 1978.

### Other References for Human Reaction Time

BAILEY, G. B., and R. PRESGRAVE: *Basic Motion Timestudy,* McGraw-Hill Book Company, New York, 1958.

BROADBENT, D. E.: *Successive Responses to Simultaneous Stimuli,* Report FPRC-934, Flying Personnel Research Committee, RAF Institute of Aviation Medicine, Farnborough, England, 1955.

BROWN, J. S., et al.: *Discrete Movements toward and away from the Body in the Horizontal Plane,* Report no. 6, Project no. 2, U.S. Naval Training Devices Center, Port Washington, N.Y., 1948.

BROWN, J. W., and A. T. SLATER-HAMMEL: "Discrete Movements in the Horizontal Plane as a Function of Their Length and Direction," *Journal of Experimental Psychology,* vol. 39, no. 84, 1949.

DAMON, ALBERT, et al.: *The Human Body in Equipment Design,* Harvard University Press, Cambridge, Mass., 1966.

DEUPREE, R. H., and J. R. SIMON: "Reaction Time and Movement Time as a Function of Age, Stimulus Duration and Task Difficulty," *Ergonomics,* vol. 6, no. 403, 1963.

FORBES, G.: "The Effect of Certain Variables on Visual and Auditory Reaction Times," *Journal of Experimental Psychology,* vol. 35, no. 153, 1945.

GIBBS, C. B., et al.: *Reaction Times to the Flashing Light Signals of Cars: The Effects of Varying the Frequency and Duration of the Flash,* Report APU-245, Applied Psychology Research Unit, Medical Research Council, Cambridge, England, 1955.

HYMAN, R.: "Stimulus Information as a Determinant of Reaction Time," *Journal of Experimental Psychology,* vol. 45, no. 188, 1953.

MACKWORTH, J. F., and N. H. MACKWORTH: "The Overlapping of Signals for Decisions," *American Journal of Psychology,* vol. 69, no. 26, 1956.

MILTON, J. L., et al.: *Pilot Reaction Time: The Time Required to Comprehend and React to Contact and Instrument Recovery Problems,* Report TSEAA-694-13a, Aerospace Medical Research Laboratory, Wright Air Development Center, Wright-Patterson Air Force Base, Ohio, 1947.

MORGAN, C. T., et al. (eds.): *Human Engineering Guide to Equipment Design,* McGraw-Hill Book Company, New York, 1963.

PROVINS, K. A.: *A Study of Some Factors Affecting Speed of Cranking,* Report no. 53/755, Royal Naval Personnel Research Committee, Medical Research Council, London, 1953.

REID, L. S., and J. G. HOLLAND: *The Influence of Stimulus Similarity and Stimulus Rate: Third of a Series of Reports on Experimental Analysis of Complex Task Performance,* WADC-TR-54-146, Aerospace Medical Research Laboratory, Wright-Patterson Air Force Base, Ohio, 1954.

## VOLUNTARY CONTROLS

Voluntary control of human body components, as distinguished from involuntary movement, is achieved by means of the human brain (cortex). Voluntary movements are generally classified as "tension movement" or "ballistic movement." In the first case, there is a slow, intense movement created by contraction of antagonistic muscles operating against one another with unequal tension. In the second case, movements are free and generally more rapid, since the simultaneous operation of antagonistic muscles is at a minimum.

Voluntary movements are part of a perceptual-motor process involving coordination between one or more sensors (visual, auditory, tactile, and/or proprioceptive), the cortex, and the musculoskeletal (motor response) system. The perceptual-motor system is a complex feedback system involving at least three major loops (see the accompanying diagram).[29] As this diagram illustrates, the human control system involves such concepts as oscillation, inertia, damping, reverberation, and lag, and its characteristics at any given time depend on many factors, including mental set, practice, instructions, motivation, the physical characteristics of the workplace, and the environment (temperature, light, sound, vibration, etc.).

In general, the human functions as a nonlinear system but becomes more linear as the task becomes more predictable.

The human system often introduces its own "noise," which may affect the ultimate response. Motor coordination cannot be generalized from task to task, although steadiness in any given individual remains constant.

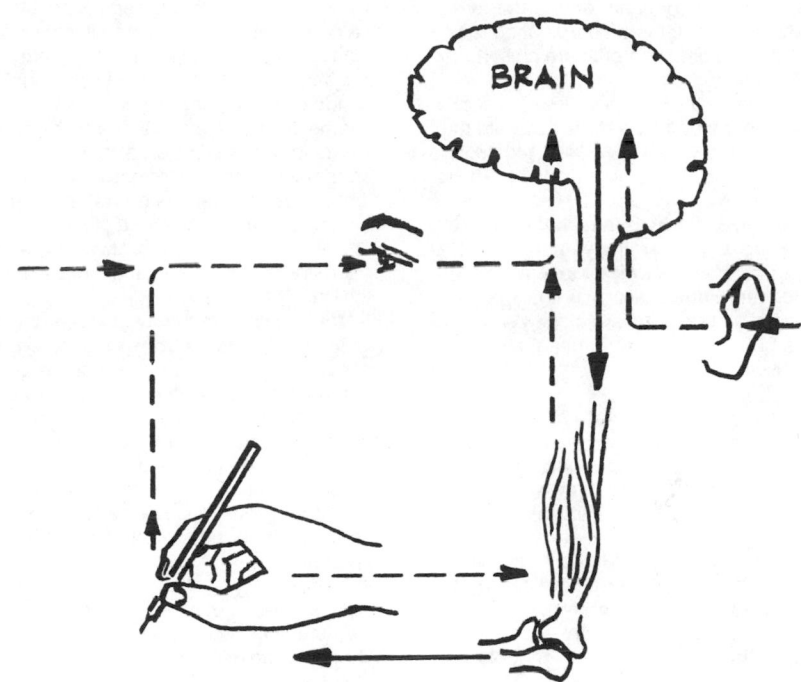

Although the perceptual-motor response may be driven primarily by specific sensory inputs (visual or auditory), continuous feedback is received by other sources (tactual, kinesthetic, proprioceptive).

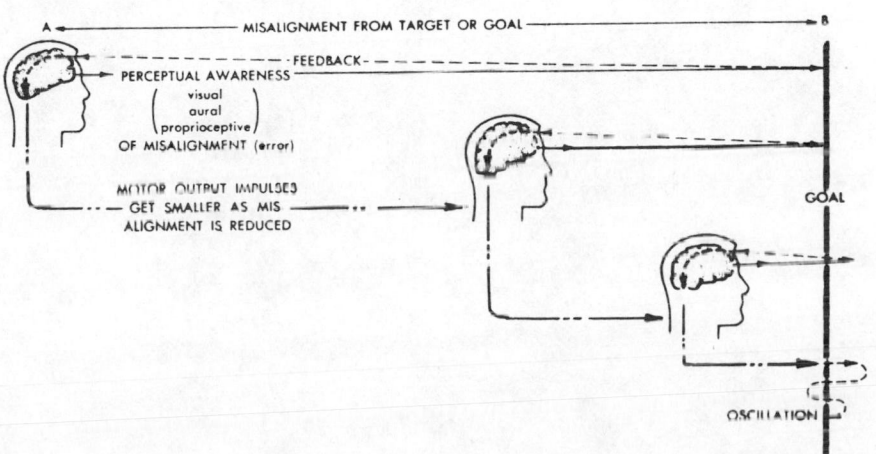

[29]W. E. Woodson and D. W. Conover, *Human Engineering Guide for Equipment Designers*, University of California Press, Berkeley, 1964.

## GENERALIZATIONS

Eye-hand coordination is generally more accurate and reliable than ear-hand coordination.

Sensorimotor control between any of the sensory channels and the hands is generally more accurate and reliable than that between the sensory channel and the feet.

Hand and arm movement coordination is better when these movements are close to the body and symmetrical.

Arm movements that progress forward and/or away from the body are more accurate than arm movements that are directed toward the body. The same is true of leg movements.

Right-handed individuals are more proficient at making clockwise movements than at making counterclockwise movements; left-handed individuals are just the opposite. However, all people make clockwise movements better with the right hand, and counterclockwise movements better with the left hand.

Generally, a person can rotate his or her hand and wrist more precisely in one direction than another, i.e., the right hand clockwise, and the left hand counterclockwise.

Multiple arm and/or leg translatory movements are more efficient when they are similar, i.e., moving the left hand to the left or forward while at the same time moving the right hand to the right and forward.

A person can apply force more accurately to two simultaneously operated controls (i.e., one for each hand) when the controls are located symmetrically with respect to the body and when the directions of movement are similar.

When separate and different kinds of controls are operated simultaneously, there is a high probability that the operation of one or another control will suffer in terms of operator input efficiency.

Combined movements often result in "cross talk" e.g., trying to push and precisely rotate a control at the same time almost invariably introduces inaccuracy in one of the motions.

Excessive disparity between two manual operations (e.g., patting one's head while at the same time rubbing the stomach) often results in a complete breakdown of the sensorimotor response of one of the activities.

Continuous feedback is desirable in order for most control movements to remain optimized, i.e., to maintain an accurate direction, force application, and/or rate of movement.

On the basis of proprioceptive feedback, individuals judge extent of movement more accurately than movement force applied, and force more accurately than duration of movement.

Although visual control is more important while an individual is learning a new perceptual-motor task, as performance becomes habitual, proprioceptive feedback (or "feel") may become the more important feedback resource.

When less than 0.5 s per movement is required, blind movements are as accurate as when using visual positioning.

When the rate of movement is constant, accuracy diminishes as the interval between movements increases.

It takes about 0.04 s to develop maximum tension in the muscles of the human arm.

People tend to compensate for changes in the rate of sensory inputs versus movement rates; i.e., they move a control at a higher rate when the control display ratio is less sensitive.

Friction that is proportional to the rate of movement (viscous friction), in contrast to coulomb friction, is often advantageous in many control systems. Similarly, inertia in the form of a heavy control moved directly by the operator may also improve performance.

The control ratio (i.e., the ratio of control movement to indicator movement) generally has more effect on the time required to make settings within a given degree of tolerance than inertia, friction, or size of the control. Primary adjustment times decrease with increasing distance moved by a controlled marker (pointer or "hook") for each revolution of a control. Secondary adjustment times, however, increase with increases of marker movement per control movement or revolution. Overall time is minimum for gear ratios (e.g., pointer movement in inches per control revolution is between about 1 and 2). Although this ratio is optimum when fine tolerances are required, it may be advantageous to use higher ratios (6:1) in order to arrive at the general, desired setting. For joy-stick controls, the optimum condition for tolerances of 0.10 in (0.25 cm) is about 2½ units of stick movement for one unit of indicator movement.

## HAND-POSITIONING ACCURACY OF THE BLIND[30]

Although people become extremely accurate at positioning their hands with practice (as in the case of a concert pianist), without benefit of sight the accuracy with which one can point or position the arm or hand is reduced. The accompanying sketch illustrates this point. Note that an individual's blind positioning response is most accurate when he or she

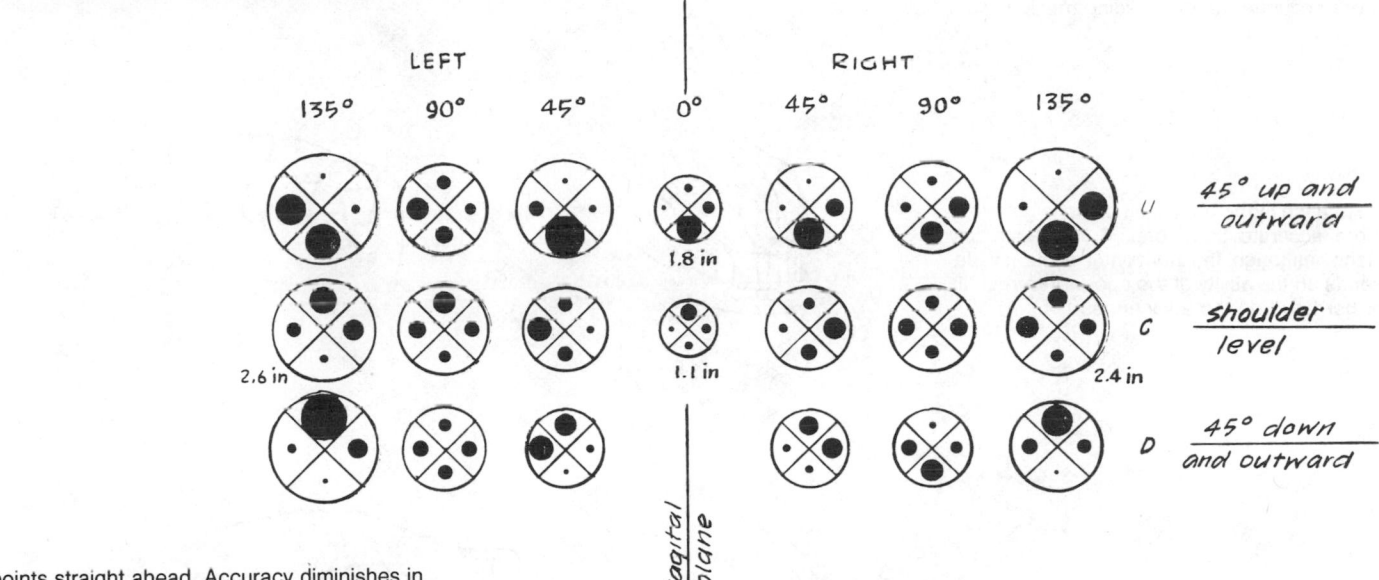

points straight ahead. Accuracy diminishes in all other directions. Several values are shown to give an approximate idea of how accuracy in peripheral points may vary from the most accurate point directly in front of the operator. The dark circles within each larger circle denote accuracy within general areas; i.e., the smaller circles denote greater accuracy.

The study from which these data were derived required subjects to sit in a level seat. When an individual is sitting or standing on a tilting platform, inaccuracies increase rapidly and nonuniformly.

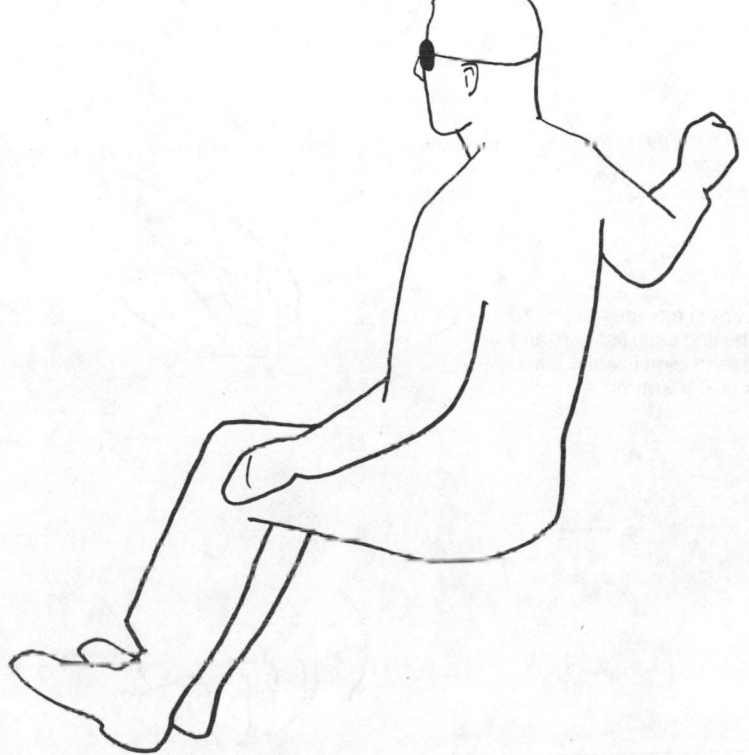

[30]P. M. Fitts, and C. Crannell, *Location Discrimination: Accuracy of Reaching Movements to Twenty-Four Different Areas*, U. S. Air Force Air Materiel Command Technical Report no. 5833, 1950.

## FINGER DEXTERITY AND CONTROL OPERATION ACCURACY

Although certain people develop considerable dexterity with practice, the average person is able to perform certain types of control manipulations more accurately than others, as the accompanying sketches illustrate.

Rotational manipulation is more accurate than sliding manipulation or movement of thumb or finger wheels, although the latter is more accurate than the sliding manipulation.

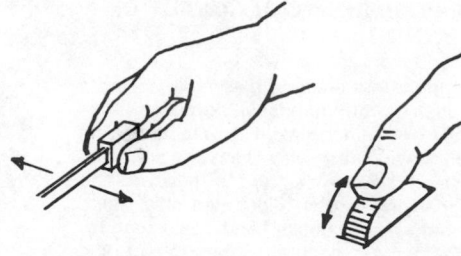

Rotation in the horizontal plane is generally more accurate than rotation in the vertical plane, although the horizontal accuracy depends on the ability of the operator to rest his or her hand on the adjacent surface.

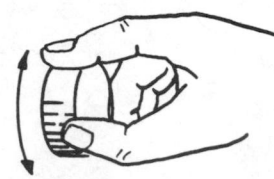

A push button is located and pressed more accurately when it is positioned as shown in the first sketch than when it is positioned as shown in the second sketch.

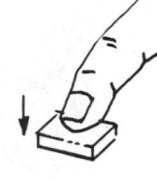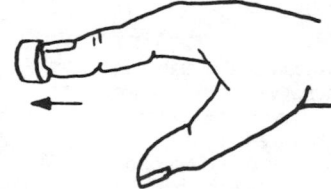

A pencil-sized joy stick is manipulated more precisely than one requiring a full fist grip, and the accuracy is increased significantly if the operator can rest his or her arm on a nearby horizontal surface.

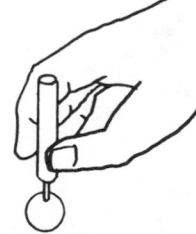

An L-shaped handle is more accurately positioned than a round knob, such as a doorknob.

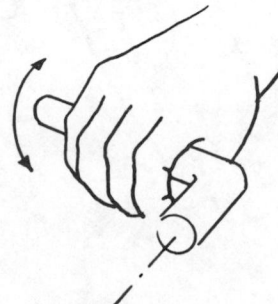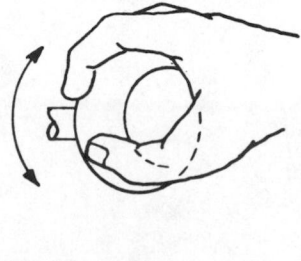

## VISUAL POSITIONING ACCURACY

Visual positioning accuracy is a function of motor patterns such as direction and distance of movement. The accompanying chart illustrates the three important components: (A) primary movement time, (B) average reaction time (250 ms), and (C) secondary movement time. Primary movement time increases with distance, but less than in direct proportion to distance. Secondary movement time (e.g., small corrective adjustments) is relatively constant for distance in excess of about 10 cm.

In general and within reason, the relative time required for visual positioning movements is about the same for short and long distances because a higher velocity is typically used as distance increases.

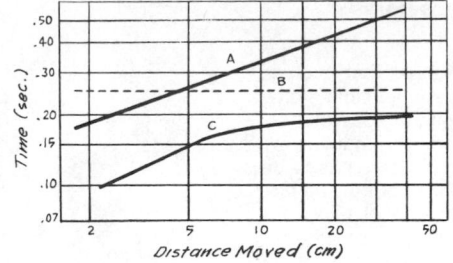

## CRANKING MOVEMENTS

Maximum turn or cranking rate is obtained when the cranking radius is between 3 and 5 cm. However, maximum turn rate decreases with increasing load. For example, the curve shown in the accompanying graph is for zero load where the maximum turn rate is about 275 r/min. At 5000 g/cm, the turn rate becomes about 240 r/min.

In general, friction aids the discrimination of small load changes. However, inertia usually provides smoother cranking performance.

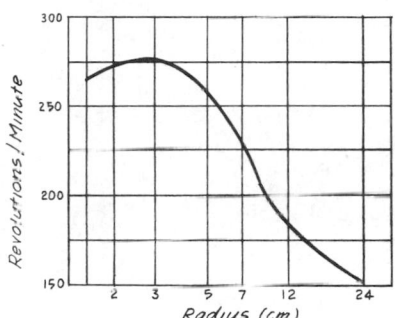

## CONTINUOUS-ADJUSTMENT CONTROL RESPONSE

Common continuous-adjustment controlling tasks include steering and controlling speed and altitude. The controlled system may be a highway vehicle, an aircraft, a spacecraft, a ship, or a submarine, or the movement and positioning of an external object may be involved, as when a crane operator acquires, transports, or deposits a package that is attached to the crane cable system.

Control effectiveness depends on at least the following factors:

The ability of the operator to *anticipate* what is going to happen when he or she provides input to the system

The ability of the operator to predict what will happen when he or she makes specific system inputs

Feedback on a timely basis about what is happening as the operator makes control inputs

How much differentiation, integration, and/or algebraic addition the control and display task requires of the operator

How well the specific control and display devices provide compatible relationships between the operator's sensory, perceptual, and motor and physical characteristics

Control effectiveness is degraded by the following factors:

Long delays between inputs, e.g., perceived changes in incoming information, results of operator inputs on system condition, or direct feedback from controller manipulation

Too much noise in the system, e.g., extraneous signals, dynamic disturbances, or mechanical artifacts such as "dead space," "stiction," and force irregularities

Incompatibilities between control and display direction and rate of motion

Controller force requirements that are either too high or too low

Incompatibility of the position, direction, and range of movement of the controller with the operator's musculoskeletal system and operating position

A requirement that the wrong body element be used, e.g., the hand versus the foot, the left hand versus the right hand, or the whole limb versus the hand and fingers

The following general statements about human efficiency with respect to continuous-adjustment control activities should be considered.

Humans are more efficient:

When they can make larger motions, since their own proprioceptive feedback mechanisms impart information about what they are doing.

When the movements they make are in the *same* direction in which the object, system, or displayed element moves.

When the rate of change of their control movement is similar to that of the controlled object or displayed element.

When sufficient information is supplied (naturally or artificially) to allow them to "predict" what is going to happen if they maintain their present control input and/or modify it to some extent. To predict, they must also know the general limits of their control system's response range.

When the control forces are not too high and are approximately equal throughout the controller movement range; i.e., high initiating forces (stiction) require the operator to suddenly compensate, once the controller is put into motion.

When there is some friction in the control system to minimize the effect of external dynamic disturbances and/or inequalities in their own musculoskeletal systems (asymmetry, tremor, fatigue, etc.). However, variable force within certain systems may be desirable to provide the operator with cues relative to the position or condition of the system, e.g., increasingly higher force as a brake pedal approaches "lock-up," toward the limits of a steering-wheel excursion to minimize overcontrolling, or force cues to indicate position of a transmission shift lever.

When they are properly positioned and secured; i.e., a seated operator with appropriate handrests or armrests is less influenced by dynamic disturbance and by problems of maintaining body equilibrium.

In terms of small, accurate inputs, when they primarily rely on hand and finger movement to manipulate a controller.

When they are not required to manipulate too many separate controls in an integrated manner, i.e., when they do not have to perform sequential operation of several hand and/or foot controls while at the same time carrying out a primary continuous-adjustment task (several switching inputs on a joy stick or steering wheel, etc.).

When they do not have to hold a control device (and their arms or legs in suspension) for long periods of time or hold the control in a fixed position for extended periods.

When they are provided appropriate system aids to relieve them of complex mental information processing to differentiate, integrate, extrapolate, or perform algebraic additions during the control tasks.

When the controller-display setup allows them to see what they are doing, e.g., when their hands do not cover up displayed information, and when vehicle elements do not obscure the external view.

## Pursuit versus Compensatory Tracking

When performing a tracking task that involves matching a continuously variable input signal with an output of the system, the operator requires information indicating the input to be matched and the output with which it is to be matched. When an operator's control and display system provides separate indications for input and for output, this is referred to as a "pursuit task"; i.e., the operator makes inputs into the control and then to a display element to "follow" a target (system input). On the other hand, when input and output signals are presented to the operator in terms of a difference between the system and the operator's control input, the task is called "compensatory" or "error-reducing."

Pursuit tracking is generally a more "natural" task. In addition, the operator is better able to keep track of input changes and to anticipate them, and the operator can also keep better track of the position of the controlled element. Pursuit tracking is the preferred mode for simple systems, particularly where input frequency is relatively high.

Compensatory tracking is generally less natural (and thus requires training). However, when the dynamic characteristics of the system are complex, compensatory tracking modes minimize the probability that operators will confuse input with the results of their own control actions. Another important advantage is that compensatory tracking information can be displayed in less space because in error-reducing formats the total output range does not have to be displayed.

In general, a pursuit display is best when high-frequency inputs are present, and a compensatory display is best when low-frequency inputs are present. A pursuit display allows the operator to anticipate. It is best for higher-order control systems; the higher the frequency of desired output, the more superior a pursuit mode is to a compensatory mode. With rate controls, a pursuit display is superior as long as the desired output and input are time-varying. With rate controls, a pursuit display is superior except when the cutoff frequency is about 0.1 Hz or less. When the display must be small (because of lack of space) and when the change in desired output is very slow, a compensatory display mode is superior. A compensatory display is best when an "aided system" supplies derivative terms where positional errors are corrected (i.e., the operator performs better when making corrective responses proportional to only the error).

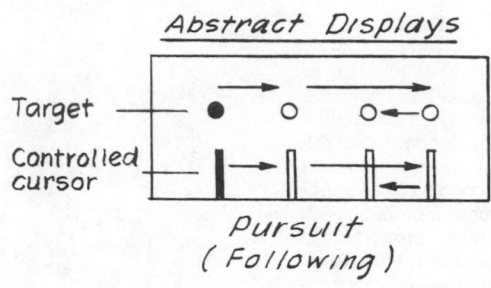

Abstract Displays

Target — Controlled cursor

Pursuit (Following)

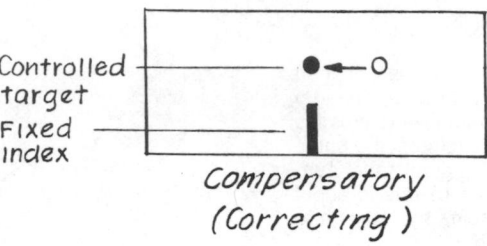

Controlled target — Fixed index

Compensatory (Correcting)

A pursuit display should be used when:

1. The course contains high frequencies.
2. It is a zero-order control system.
3. The operator must know actual output, not just error.

A compensatory display should be used when:

1. The system is aided or quickened.
2. The display must be kept small, but output range is large, and/or the precision requirements are high.

Either display can be used when:

1. The course is simple.
2. The display needs to be kept small, but the output range is large, and/or the precision requirements are high.[31]

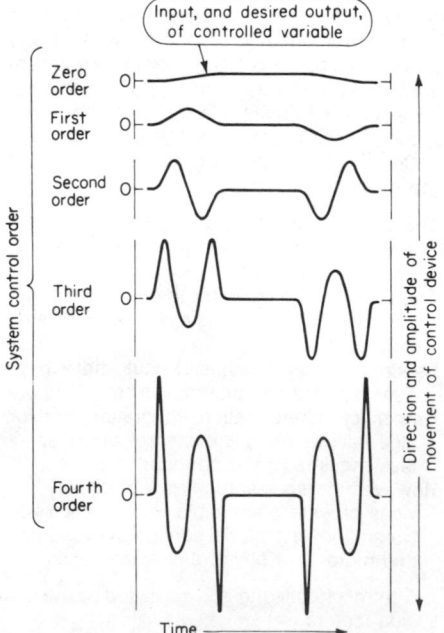

Control order illustrated by changes in a controlled variable and in its first four derivatives. Each line represents the changes over time that would have to be made with control systems of various control orders to make the controlled variable correspond to the input.

[31]For a more complete discussion of human tracking capabilities and limitations and for detailed recommendations regarding the design of closed-loop systems, see E. J. McCormick, *Human Factors in Engineering and Design*, 4th ed., McGraw-Hill Book Company, New York, 1976, p. 220.

## WORK EFFICIENCY

If one thinks of humans as machines that produce work output, they are much like factories, i.e., they take in raw materials in the form of food, water, and air, and from these they produce materials for maintaining their bodies, power to operate the various body systems, and energy to perform work. The efficiency with which work is performed is a ratio of output to input. The following equation describes the process:

Work efficiency =

$$\frac{\text{work done (e.g., foot-pounds)}}{\text{work done} + \text{heat produced}}$$

It is estimated that human efficiency generally does not exceed about 20 to 25 percent. According to Shilling,[32] a typical 154-lb (70-kg) adult male has the fuel and energy requirements for typical activities shown in the accompanying table. The approximate amount of energy required for one day of typical activities are estimated to be as shown in the accompanying table.

## COMPARISON OF ENERGY EXPENDITURES BY MALES AND FEMALES FOR SELECTED WORK TASKS

### Energy Expenditure in Relation to Body Position and Activity

### Specific Task Output Costs

| Activity | Calories per Hour |
|---|---|
| Sleeping | 65 |
| Resting (awake but lying still) | 77 |
| Standing at attention | 115 |
| Typewriting rapidly | 140 |
| Sweeping the floor (38 strokes per minute) | 169 |
| Walking slowly (2.6 mi/h) | 200 |
| Performing carpentry, sheetmetal work, or painting | 240 |
| Walking moderately fast | 300 |
| Sawing wood | 480 |
| Swimming | 500 |
| Running (5.3 mi/h) | 570 |
| Walking up steps | 1100 |

| | Calories |
|---|---|
| 8 h of sleep | 520 |
| 3 h of light work | 510 |
| 8 h of medium to heavy work | 1920 |
| 5 h of sitting at rest | 500 |
| Day's total | 3450 |

| Occupation | Type of Work | Kilocalories per Day Men | Kilocalories per Day Women |
|---|---|---|---|
| Bookkeeper | Sedentary, light manual | 2400 | 2000 |
| Secretary | Sedentary, light manual | 2700 | 2250 |
| Bus driver | Sedentary, medium manual | 3000 | 2500 |
| Letter carrier | Walking, climbing stairs, and carrying mailbag | 3000 | 2500 |
| Machine operator | Sedentary, heavy manual | 3300 | 2750 |
| Stonemason | Heavy manual | 3600 | 3000 |
| Laborer | Heavy manual, lifting and carrying | 3900 | 3250 |
| Carpenter | Manual, continuous (hammering and sawing) | 3900 | 3250 |
| Ballet dancer | Vigorous, continuous whole-body movements | 3900 | 3250 |
| Coal miner | Standing, extremely heavy body and limb movements | 4200 | — |
| Lumberjack | Standing, extremely heavy body and limb movements | 4500 | — |
| Athlete | Marathon running, swimming, and football | 4800 | 4250 |

| Posture and Mobility Involvement | Kilocalories per Minute | Kilocalories per Hour |
|---|---|---|
| Sitting | 0.3 | 20.0 |
| Kneeling | 0.5 | 30.0 |
| Crouching | 0.5 | 30.0 |
| Standing | 0.6 | 35.0 |
| Bending over while standing | 0.8 | 50.0 |
| Walking (medium rate) | 1.7–3.5 | 100–200 |
| Climbing (10° grade minus load) | 0.75-m climbing height | 0–400 |
| Work involving one arm | 0.7–2.2 | 40–120 |
| Work involving two arms | 1.5–3.0 | 80–160 |

*Note:* The above values represent general estimates and should not be used except to establish approximations for evaluating the consequence of proposed design or task configurations.

*Source:* G. Lehmann, *Praktische Arbeitsphysiologie*, Georg Thieme Verlag, Stuttgart, 1953.

| Activity | Working Conditions | Kilocalories per Minute |
|---|---|---|
| Walking | Without load, level surface (3km/h) | 1.7 |
| | (4km/h) | 2.1 |
| Walking | With load carried on back (10 kg/4km/h) | 3.6 |
| | (30 kg/4km/h) | 5.3 |
| | (50 kg/4km/h) | 8.1 |
| Climbing | 16°/11.5 m/min (no load) | 8.3 |
| | (20-kg load) | 10.5 |
| | (50-kg load) | 16.0 |
| Cycling | 16 km/h | 5.2 |
| Housekeeping | Cooking | 1.0–2.0 |
| | Light cleaning, ironing, | 2.0–3.0 |
| | scrubbing floor | 4.0–5.0 |

[32] C. W. Shilling, *The Human Machine*, U.S. Naval Institute, Annapolis, Md., 1955.

## WORK, HEAT, AND OXYGEN COSTS

The human machine oxidizes its fuel in order to obtain energy. The body oxidizes fuel at temperatures of about 37°C and converts chemical energy directly into work.

Human energy exchange is expressed primarily in terms of heat. When work is performed, 10 to 30 percent of the energy expenditure is applied to the external load, and the rest comes out as heat.

Oxygen costs (e.g., heat production) for typical activities are listed in the accompanying tables.[33]

**Everyday Activities**

| Activity | Oxygen Consumption, l/min | Equivalent Heat Production, kcal/min |
|---|---|---|
| Asleep | | |
|   Sleeping, men over 40 | 0.22 | 1.1 |
|   Sleeping, men aged 30–40 | 0.24 | 1.2 |
|   Sleeping, men aged 20–30 | 0.24 | 1.2 |
|   Sleeping, men aged 15–20 | 0.25 | 1.3 |
| Resting | | |
|   Lying fully relaxed | 0.24 | 1.2 |
|   Lying moderately relaxed | 0.26 | 1.3 |
|   Lying awake, after meals | 0.28 | 1.4 |
|   Sitting at rest | 0.34 | 1.7 |
| Very light activity—seated | | |
|   Writing | 0.36 | 1.8 |
|   Riding in automobile | 0.40 | 2.0 |
|   Typing | 0.46 | 2.3 |
|   Polishing | 0.48 | 2.4 |
| Very light activity—standing | | |
|   Relaxed | 0.36 | 1.8 |
|   Drafting | 0.38 | 1.9 |
|   Taking lecture notes | 0.40 | 2.0 |
|   Peeling potatoes | 0.42 | 2.1 |
| Light activity—seated | | |
|   Playing musical instruments | 0.58 | 2.9 |
|   Repairing boots and shoes | 0.60 | 3.0 |
|   At lecture | 0.60 | 3.0 |
|   Assembling weapons | 0.72 | 3.6 |
| Light activity—standing | | |
|   Entering ledgers | 0.52 | 2.6 |
|   Washing clothes | 0.74 | 3.7 |
|   Ironing | 0.88 | 4.4 |
|   Scrubbing | 0.94 | 4.7 |
| Light activity—moving | | |
|   Slow movement about room | 0.50 | 2.5 |
|   Vehicle repairs | 0.68 | 3.4 |
|   Slow walking | 0.76 | 3.8 |
|   Washing | 0.84 | 4.2 |
| Moderate activity—lying | | |
|   Creeping, crawling, prone resting maneuvers | 1.14 | 5.7 |
|   Crawling | 1.22 | 6.1 |
|   Swimming breaststroke at 1 mi/h | 1.36 | 6.8 |
|   Swimming crawl at 1 mi/h | 1.40 | 7.0 |
| Moderate activity—sitting | | |
|   Rowing for pleasure | 1.00 | 5.0 |
|   Cycling at 8–11 mi/h | 1.14 | 5.7 |
|   Cycling rapidly | 1.38 | 6.9 |
|   Trotting on horseback | 1.42 | 7.1 |
| Moderate activity—standing | | |
|   Gardening | 1.16 | 5.8 |
|   Chopping wood | 1.24 | 6.2 |
|   Baseball pitching | 1.30 | 6.5 |
|   Shoveling sand | 1.36 | 6.8 |
| Moderate activity—moving | | |
|   Golf | 1.08 | 5.4 |
|   Table tennis | 1.16 | 5.8 |
|   Tennis | 1.26 | 6.3 |
|   Army drill | 1.42 | 7.1 |
| Heavy activity—lying | | |
|   Leg exercises, average | 1.50 | 7.5 |
|   Swimming breaststroke at 1.6 mi/h | 1.64 | 8.2 |
|   Swimming backstroke at 1.0 mi/h | 1.66 | 8.3 |
|   Lying on back, head raising | 1.76 | 8.8 |
| Heavy activity—sitting | | |
|   Cycling rapidly, own pace | 1.66 | 8.3 |
|   Cycling at 10 mi/h, heavy bicycle | 1.78 | 8.9 |
|   Cycling in race (100 mi in 4 h 22 min) | 1.96 | 9.8 |
|   Trotting on horseback | 1.96 | 9.8 |
| Heavy activity—standing | | |
|   Chopping wood | 1.50 | 7.5 |
|   Shoveling sand | 1.54 | 7.7 |
|   Sawing wood by hand | 1.60 | 8.0 |
|   Digging | 1.78 | 8.9 |
| Heavy activity—moving | | |

[33]P. Webb, "Work, Heat, and Oxygen Cost," in J. F. Parker and V. R. West (eds.), *Bioastronautics Data Book* (2d ed., National Aeronautics and Space Administration), 1973.

**Everyday Activities** *(continued)*

| Activity | Oxygen Consumption, l/min | Equivalent Heat Production, kcal/min |
|---|---|---|
| Skating at 9 mi/h | 1.56 | 7.8 |
| Playing soccer | 1.66 | 8.3 |
| Skiing at 3 mi/h on level | 1.80 | 9.0 |
| Climbing stairs at 116 steps/min | 1.96 | 9.8 |
| Very heavy activity—sitting | | |
| Cycling at 13.2 mi/h | 2.00 | 10.0 |
| Rowing with two oars at 3.5 mi/h | 2.20 | 11.0 |
| Galloping on horseback | 2.28 | 11.4 |
| Sculling (97 strokes/min) | 2.52 | 12.6 |
| Very heavy activity—moving | | |
| Fencing | 2.10 | 10.5 |
| Playing squash | 2.10 | 10.5 |
| Playing basketball | 2.28 | 11.4 |
| Climbing stairs | 2.40 | 12.0 |
| Extreme activity | | |
| Wrestling | 2.60 | 13.0 |
| Marching at double | 2.66 | 13.3 |
| Endurance marching | 2.96 | 14.8 |
| Harvard Step Test | 3.22 | 16.1 |

**Specialized Activities**

| Activities | Oxygen Consumption, l/min | Equivalent Heat Production, kcal/min |
|---|---|---|
| Engineering tasks | | |
| Medium assembly work | 0.58 | 2.9 |
| Welding | 0.60 | 3.0 |
| Sheet metal work | 0.62 | 3.1 |
| Machining | 0.66 | 3.3 |
| Punching | 0.70 | 3.5 |
| Machine fitting | 0.90 | 4.5 |
| Heavy assembly work—noncontinuous | 1.02 | 5.1 |
| Driving vehicles and piloting aircraft | | |
| Driving a car in light traffic | 0.26 | 1.3 |
| Night flying—DC-3 | 0.32 | 1.6 |
| Piloting DC-3 in level flight | 0.34 | 1.7 |
| Piloting helicopters | 0.36 | 1.8 |
| Instrument landing—DC-4 | 0.50 | 2.5 |
| Piloting light aircraft in rough air | 0.54 | 2.7 |
| Taxiing DC-3 | 0.58 | 2.9 |
| Piloting bomber aircraft in combat | 0.58 | 2.9 |
| Driving car in heavy traffic | 0.64 | 3.2 |
| Driving truck | 0.66 | 3.3 |
| Driving motorcycle | 0.70 | 3.5 |
| Moving over rough terrain on foot | | |
| Flat, firm road    2.5 mi/h | 0.56–0.98 | 2.8–4.9 |
| Grass path    2.5 mi/h | 0.64–1.02 | 3.2–5.1 |
| Stubble field    2.5 mi/h | 0.80–1.22 | 4.0–6.1 |
| Deeply plowed field    2.0 mi/h | 0.98–1.38 | 4.9–6.9 |
| Steep 45° slope    1.5 mi/h | 0.98–1.38 | 4.9–6.9 |
| Plowed field    3.3 mi/h | 1.56 | 7.8 |
| Soft snow, with 44-lb load    2.5 mi/h | 4.2 | 21.0 |
| Load carrying | | |
| Walking on level with 58-lb load, trained men | | |
| 2.1 mi/h | 0.38 | 1.9 |
| 2.7 mi/h | 0.58 | 2.9 |
| 3.4 mi/h | 0.92 | 4.6 |
| 4.1 mi/h | 1.66 | 8.3 |
| Walking on level with 67-lb load, trained men | | |
| 2.1 mi/h | 0.46 | 2.3 |
| 2.7 mi/h | 0.58 | 2.9 |
| 3.4 mi/h | 1.02 | 5.1 |
| 4.1 mi/h | 1.68 | 8.4 |

*(continued)*

### Specialized Activities *(continued)*

| Activities | Oxygen Consumption, l/min | Equivalent Heat Production, kcal/min |
|---|---|---|
| Walking on level with 75-lb load, trained men | | |
| 2.1 mi/h | 0.50 | 2.5 |
| 2.7 mi/h | 0.68 | 3.4 |
| 3.4 mi/h | 1.04 | 5.2 |
| 4.1 mi/h | 1.72 | 8.6 |
| Walking up 36% grade with 43-lb load, sedentary men | | |
| 0.5 mi/h | 1.34 | 6.7 |
| 1.0 mi/h | 2.46 | 12.3 |
| 1.5 mi/h | 3.20 | 16.0 |
| Swimming on surface | | |
| Breaststroke | | |
| 1 mi/h | 1.40 | 7.0 |
| 2 mi/h | 5.80 | 29.0 |
| 3 mi/h | 19.40 | 97.0 |
| Crawl | | |
| 1 mi/h | 1.80 | 9.0 |
| 2 mi/h | 3.60 | 18.0 |
| 3 mi/h | 9.60 | 48.0 |
| Butterfly | | |
| 1 mi/h | 2.40 | 12.0 |
| 2 mi/h | 5.80 | 29.0 |
| 3 mi/h | 15.00 | 75.0 |
| Walking under water | | |
| Walking in tank    minimal rate | 0.58 | 2.9 |
| Walking on muddy bottom    minimal rate | 1.10 | 5.5 |
| Walking in tank    maximal rate | 1.44 | 7.2 |
| Walking on muddy bottom    maximal rate | 1.68 | 8.4 |
| Movement in snow | | |
| Skiing in loose snow    2.6 mi/h | 1.62 | 8.1 |
| Sled pulling—hard snow    2.2 mi/h | 1.72 | 8.6 |
| Snowshoeing (bearpaw)    2.5 mi/h | 1.74 | 8.7 |
| Skiing on level    3.0 mi/h | 1.80 | 9.0 |
| Sled pulling—low drag, medium snow    2.0 mi/h | 1.94 | 9.7 |
| Snowshoeing—trail type    2.5 mi/h | 2.06 | 10.3 |
| Walking, 12–18 in snow, breakable crust    2.5 mi/h | 2.54 | 12.7 |
| Skiing on loose snow    5.2 mi/h | 2.92 | 14.6 |
| Snowshoeing—trail type    3.5 mi/h | 2.96 | 14.8 |
| Skiing on loose snow    8.1 mi/h | 4.12 | 20.6 |
| Measured work at different altitudes | | |
| Bicycle ergometer | | |
| Workload    Altitude | | |
| 430 kg·m/min    720 mm Hg | 1.02 | 5.1 |
| 430 kg·m/min    620 mm Hg | 0.98 | 4.9 |
| 430 kg·m/min    520 mm Hg | 1.08 | 5.4 |
| Mountain climbing | | |
| 880–1037 kg·m/min    610 mm Hg | 1.84–2.20 | 9.2–11.0 |
| 566–786 kg·m/min    425 mm Hg | 1.54–1.90 | 7.7–9.5 |
| 393–580 kg·m/min    370 mm Hg | 1.28–2.10 | 6.4–10.5 |

## WORK AND REST

Approximately 4 kcal/min has been established as a desirable upper limit of the average energy cost of work. This requires that there be some rest in order for the body to compensate for the excess above this limit. A formula for estimating the amount of rest required for a given amount of work is

$$R = \text{rest} = \frac{T(K - S)}{K - 1.5}$$

where $R$ = rest (in minutes), $T$ = total working time, $K$ = the average kilocalories per minute of work, and $S$ = kilocalories per minute as the adopted standard. The value 1.5 approximates the resting level in kilocalories per minute.

If an $S$ value of 4 kcal/min is used and we want to calculate $R$ for a 1-h period (e.g., $T$ = 60 min), the formula becomes

$$R = \frac{60(K - 4)}{K - 1.5}$$

Then one can project a curve such as that shown in the accompanying graph. Of course, individual differences in energy considered as an acceptable standard make this formula purely an approximation.

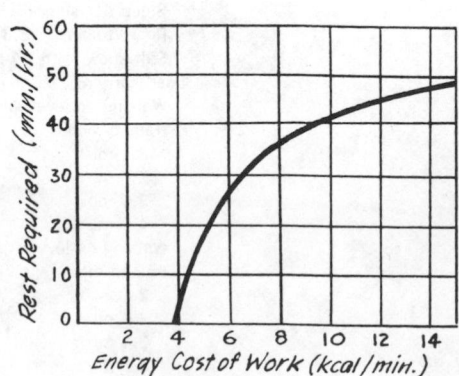

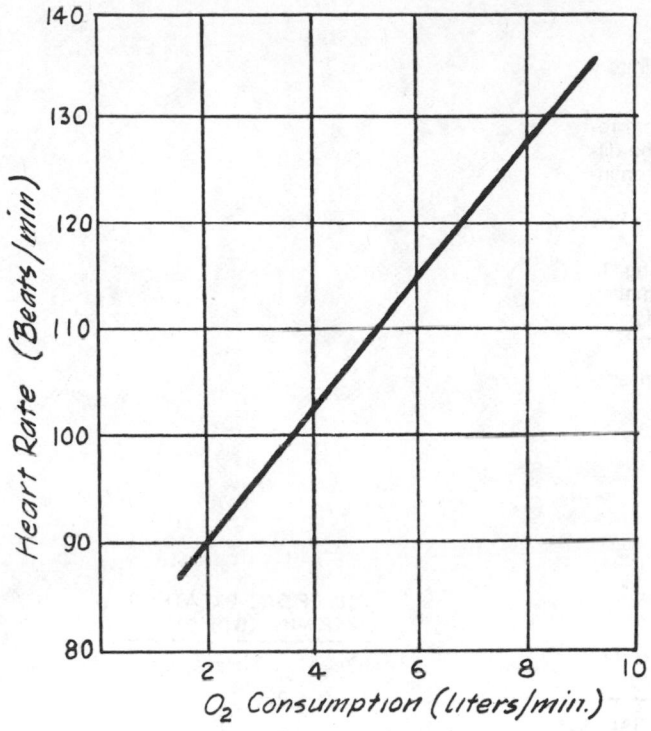

Heart rate versus oxygen consumption.

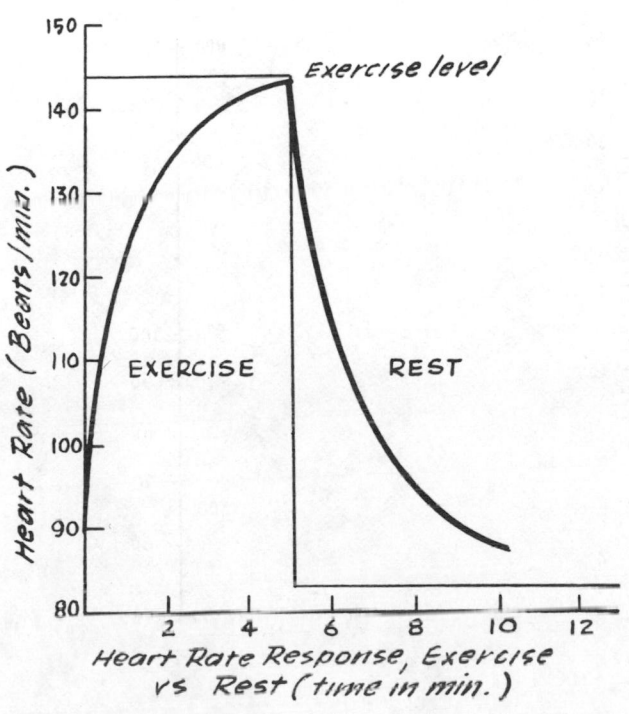

Heart rate: exercise versus rest.

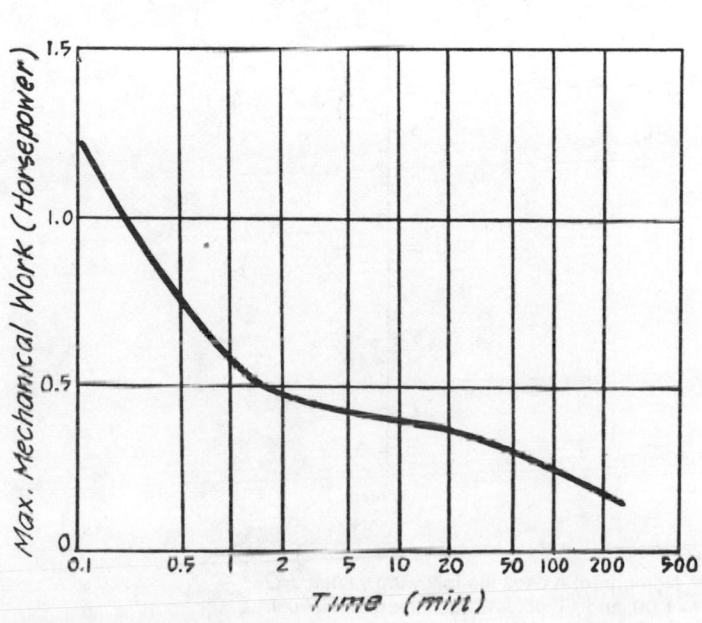

Work output versus work duration.

## CALCULATING OXYGEN COSTS

To determine heat output from respiratory data:

1. Calculate oxygen consumption from respiratory ventilation volume and the difference in $O_2$ concentration between inspired and expired air.
2. Correct volume to 0°C, 760 mmHg, dry (STPD).
3. Select heat output corresponding to each unit volume of $O_2$ by approximating the subject's diet or on the basis of his or her measured respiratory quotient.

The accompanying nomograms help simplify the calculations.

A.

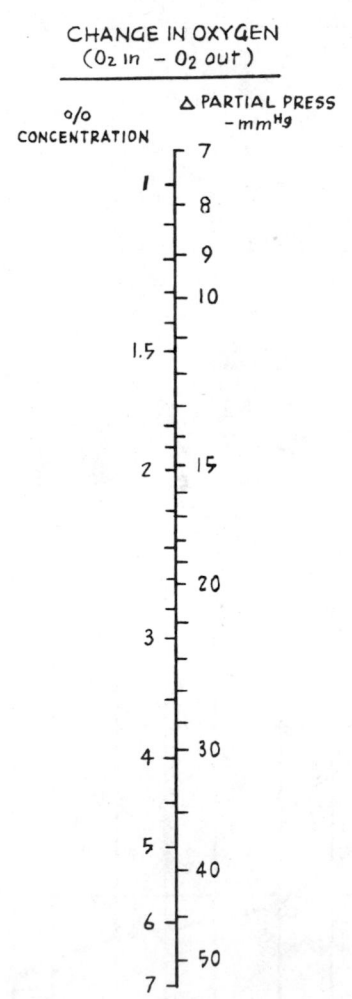

CHANGE IN OXYGEN
($O_2$ in – $O_2$ out)

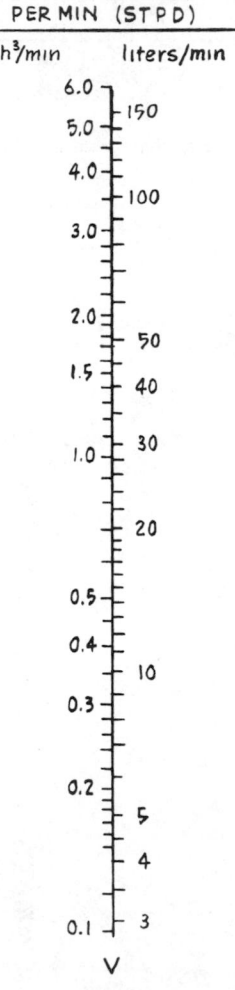

VOL. OF GAS BREATHED
PER MIN (STPD)

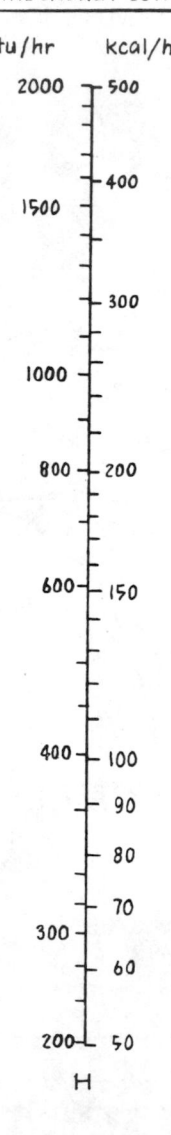

TOTAL ENERGY OUTPUT

Nomogram A uses the following values: *RQ* = 1.00, and 1 L of $O_2$ is equivalent to 5.0 kcal, permitting direct calculation of heat output *(H)* in Btu per hour from oxygen uptake *(U)* and ventilation rate *(V)*. Alternatively, *U* can be calculated from *H* and *V*, or *V* can be calculated from *U* and *H*.

B.

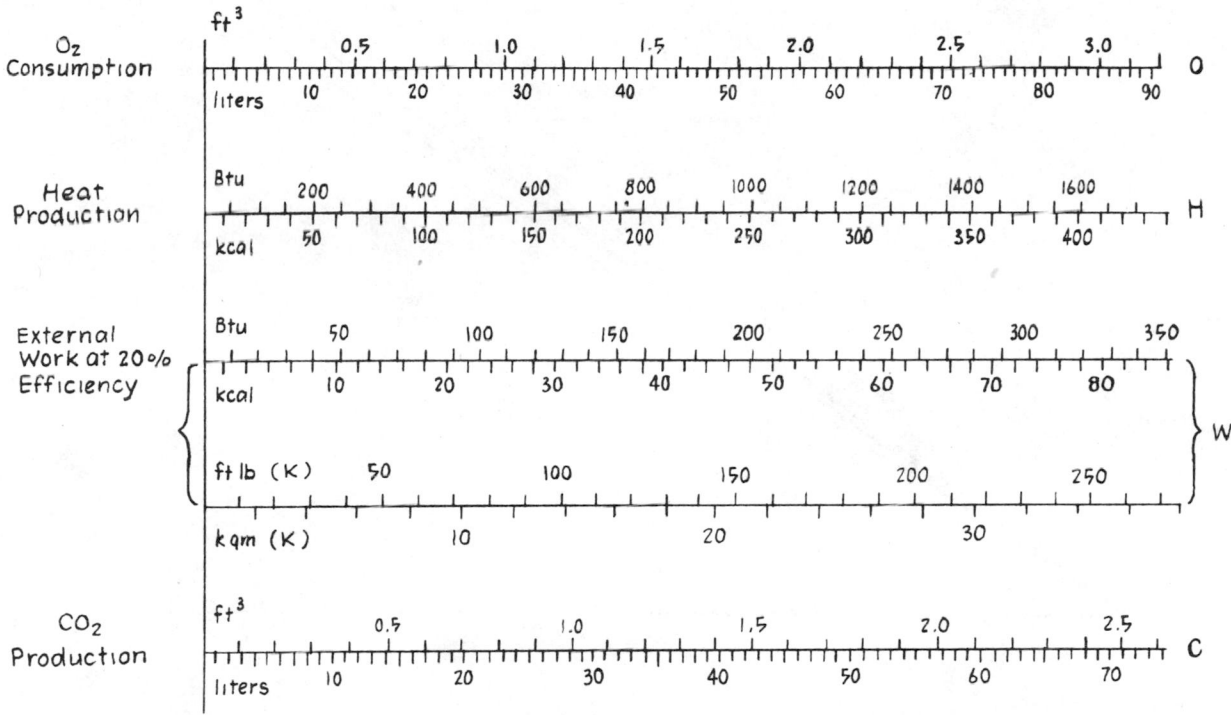

Nomogram B uses the following values: *RZ* = 0.82, and 1 L of $O_2$ is equivalent to 4.825 kcal. This nomogram allows one to interrelate, by drawing straight vertical lines, the values for oxygen consumption *(O)*, heat output *(H)*, external work output *(W)*, and carbon dioxide production *(C)* at typical conversion rates. *H* may be as much as 3 percent lower or 5 percent higher than the quoted value at any specific oxygen consumption, depending on the *RQ*, which equals 0.7 for a pure fat diet and 1.00 for a pure carbohydrate diet. The values given in the third and fourth lines should be modified if the efficiency changes. Typical ranges are 5 to 25 percent, with an average of 20 percent; thus the listed work output may increase by three-fourths if the task is one that can be performed at high efficiency (e.g., bicycling). Conversely, the true value may be reduced by three-fourths if the function is inefficiently performed (e.g., high-speed walking),

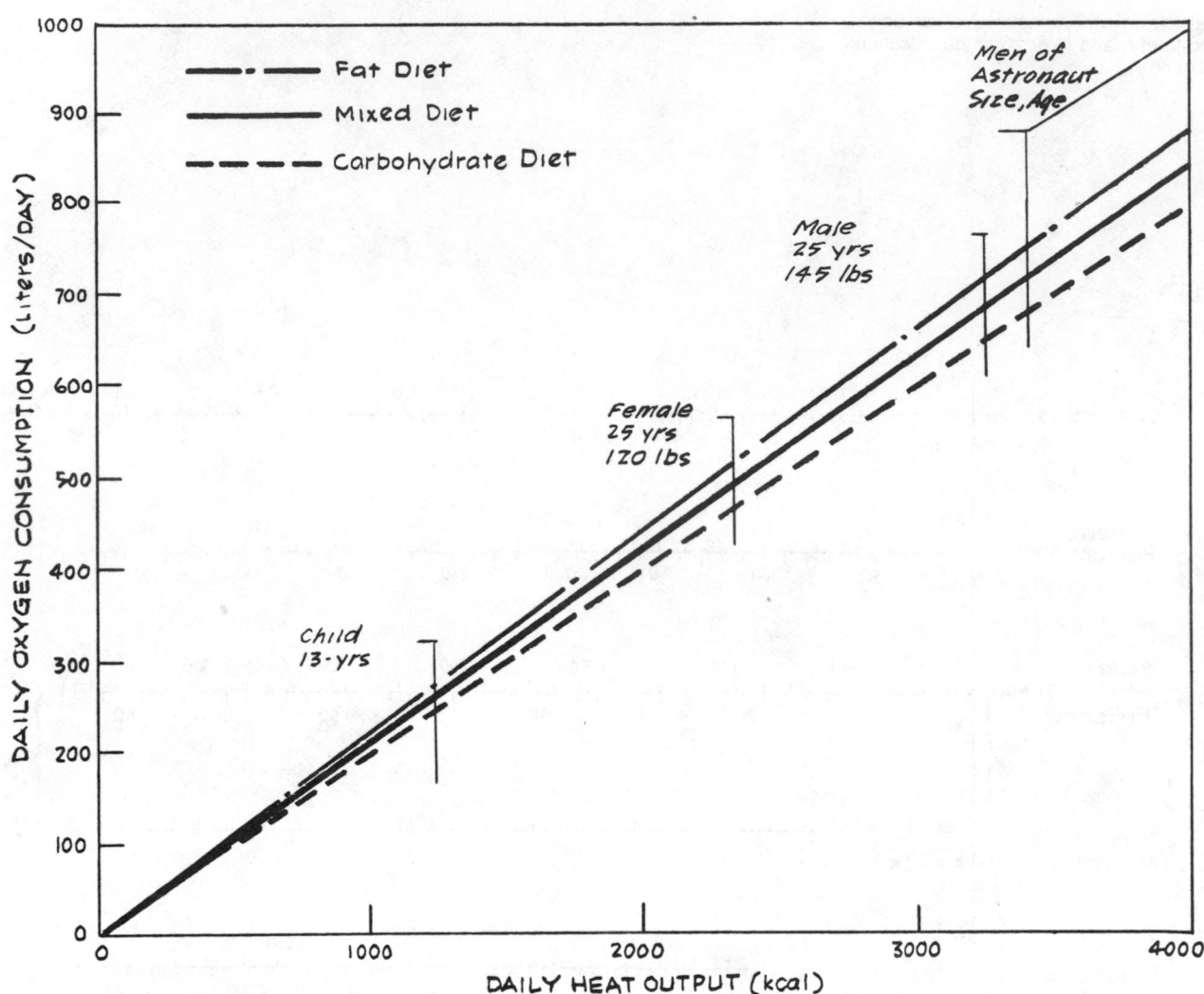

Estimate of total daily O₂ cost and heat produced for normally active people, as affected by body size. (J. F. Parker and V. R. West [Eds.], *Bioastronautics Data Book,* 2d ed., National Aeronautics and Space Administration, 1973, p. 852.)

## THERMAL RESPONSE OF THE BODY

Normal body temperatures and cooling requirements for various regions of the body are given in the accompanying table. At 62°F (16.7°C) body cooling begins. Above 78°F (25.5°C) the body gains heat. The body adapts to changes as great as 20 percent within the comfort range through evaporative cooling when it is hot and through exercise when it is cold.

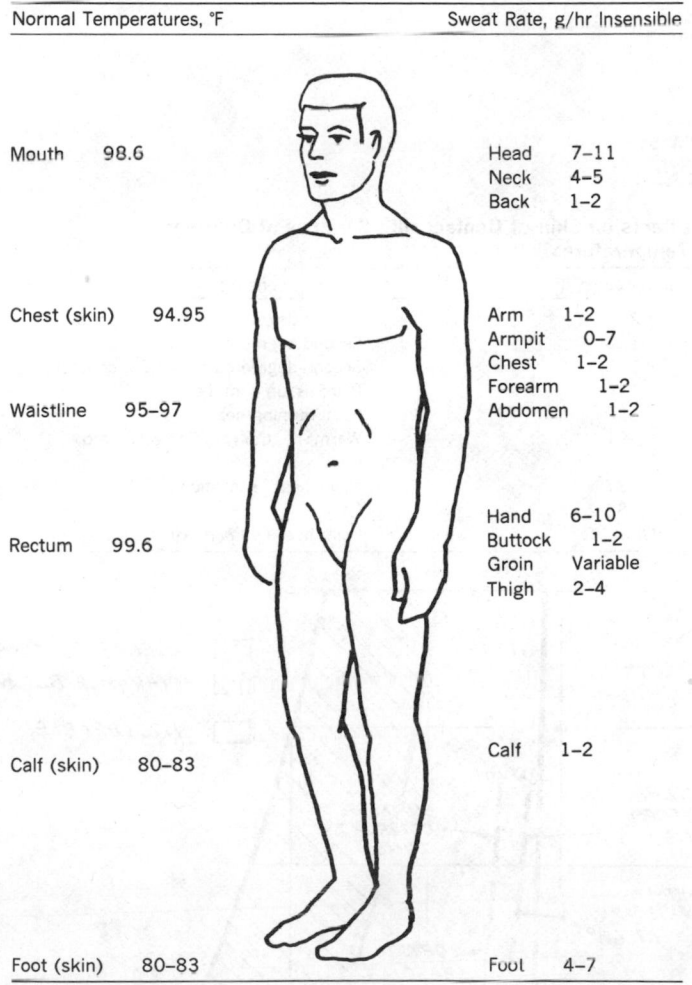

| Normal Temperatures, °F | | Sweat Rate, g/hr Insensible | |
| --- | --- | --- | --- |
| Mouth | 98.6 | Head | 7–11 |
| | | Neck | 4–5 |
| | | Back | 1–2 |
| Chest (skin) | 94.95 | Arm | 1–2 |
| | | Armpit | 0–7 |
| | | Chest | 1–2 |
| | | Forearm | 1–2 |
| Waistline | 95–97 | Abdomen | 1–2 |
| Rectum | 99.6 | Hand | 6–10 |
| | | Buttock | 1–2 |
| | | Groin | Variable |
| | | Thigh | 2–4 |
| Calf (skin) | 80–83 | Calf | 1–2 |
| Foot (skin) | 80–83 | Foot | 4–7 |

| Region | Preferred Temperature, °F | Heat Loss, Btu/h | Area, ft² | Skin Conductance, Btu/ft²·°F |
| --- | --- | --- | --- | --- |
| Head | 94.4 | 15.9 | 2.15 | 1.61 |
| Chest | 94.4 | 32.6 | 1.83 | 3.87 |
| Abdomen | 94.4 | 17.9 | 1.29 | 3.02 |
| Back | 94.4 | 49.3 | 2.43 | 4.31 |
| Buttocks | 94.4 | 33.0 | 1.94 | 3.70 |
| Thighs | 91.4 | 47.7 | 3.55 | 1.76 |
| Calves | 87.5 | 58.0 | 2.15 | 2.06 |
| Feet | 83.6 | 39.7 | 1.29 | 1.96 |
| Arms | 91.4 | 33.4 | 1.07 | 4.10 |
| Hands | 83.5 | 63.5 | 0.75 | 6.45 |

## Response in Relation to Skin Temperatures

### General Responses

| Temperature of Skin, °F | Response |
|---|---|
| 98 | Very hot |
| 96 | Unpleasantly warm |
| 94 | Slightly warm |
| 93 | Comfortable |
| 91 | Comfortably cool |
| 88 | Slightly too cool |
| 86 | Unpleasantly cool |
| 84 | Very cold |

### Temperature and Its Effect on Comfort of the Extremities

| | Skin Temperature, °F | |
|---|---|---|
| | Hands | Feet |
| Minimum acceptable | 68 | 73 |
| Tolerable | 68–59 | 73–64 |
| Intolerable | 59–50 | 64–55 |
| Numbness sets in | <50 | <55 |

### Effects on Skin of Contact with Surfaces of Different Temperatures

| Temperature, °F | Sensation or Effect |
|---|---|
| 212 | Second-degree burn on 15-s contact |
| 180 | Second-degree burn on 30-s contact |
| 160 | Second-degree burn on 60-s contact |
| 140 | Pain; tissue damage |
| 120 | Pain; burning heat |
| 91 | Warm; neutral (physiological zero) |
| 54 | Cool |
| 37 | "Cool heat" sensation |
| 32 | Pain |
| Below 32 | Pain; tissue damage (freezing) |

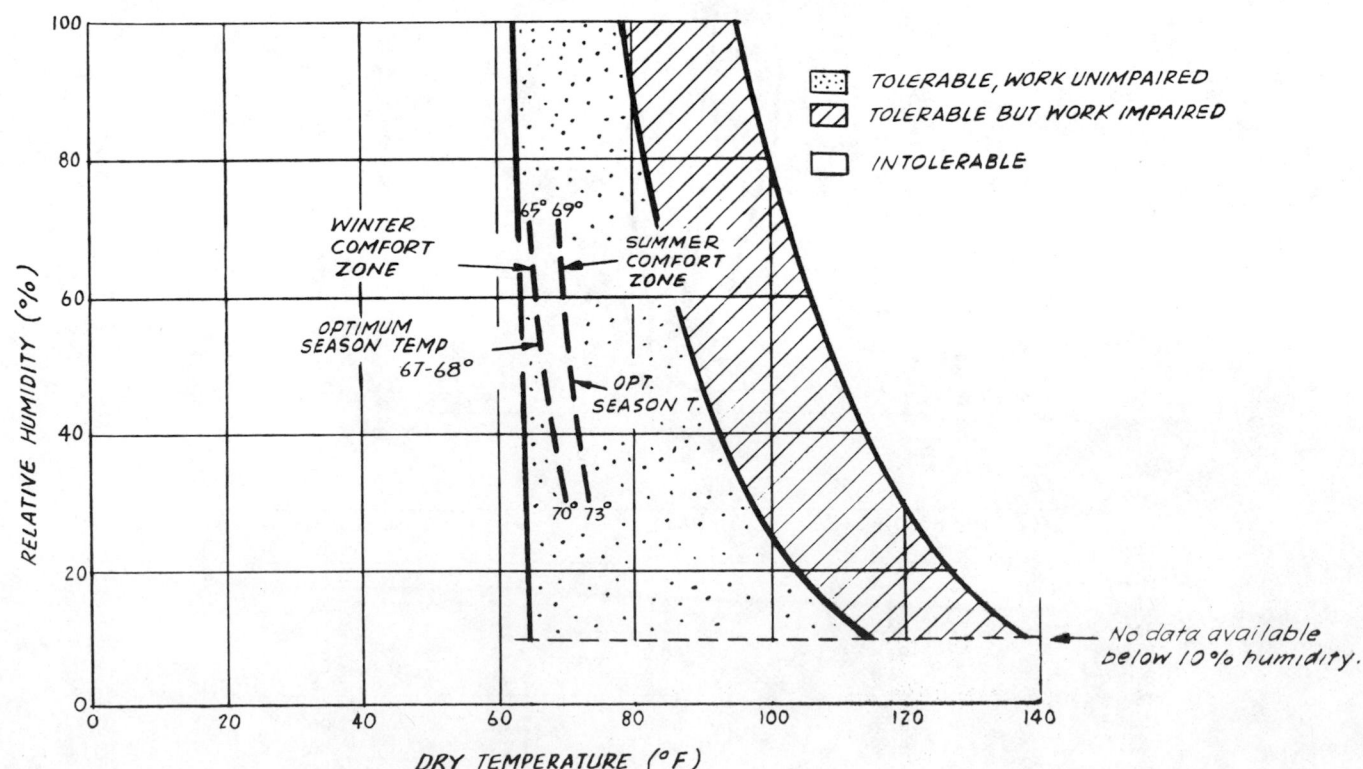

Tolerable temperature-humidity requirements (with conventional clothing).

## TOLERANCE TO HIGH TEMPERATURES[34]

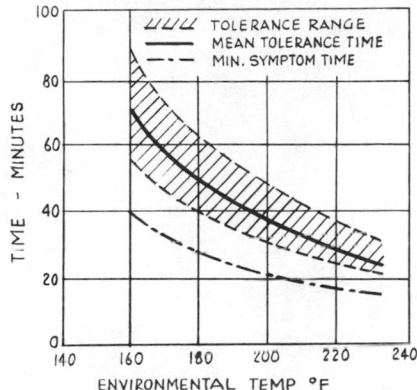

Mean and range of tolerance time and minimum symptom time for high temperature.

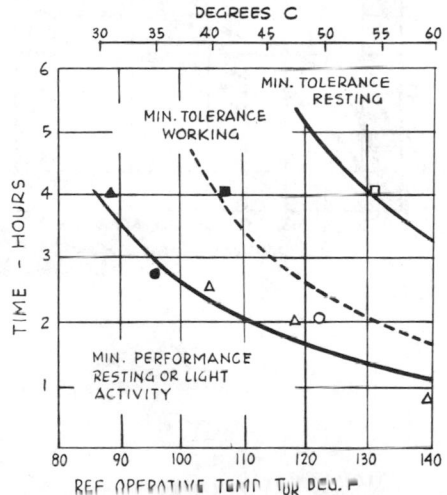

$T_{OR}$ = Operative temp at 0.79 in. Hg vapor pressure

● Heavy pursuitmeter test
▲ Mixed test battery
■ Working males
△ Wireless Teleg. test
○ Visual vigilance test
□ Resting males

Performance and tolerance limits in the quasi-compensable zone for lightly dressed men.

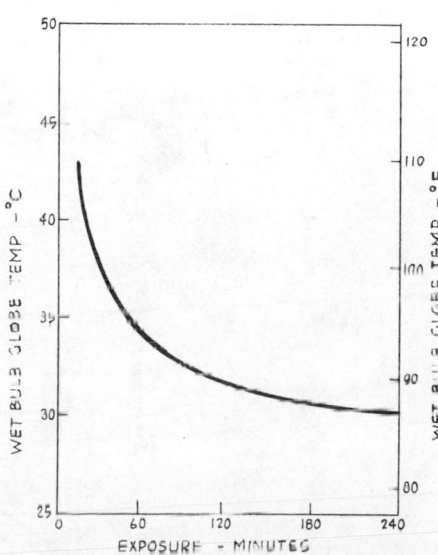

Upper limits of exposure for unimpaired mental performance.

[34]*Criteria for a Recommended Standard-Occupational Exposure to Hot Environments*, National Institute for Occupational Safety and Health, USGPO-HSM72-10269, 1972.

## TOLERANCE TO LOW TEMPERATURES

The accompanying graph shows tolerance times for endurance to cold temperatures. Subjects were wearing various types of clothing, were seated, and were performing light work (air velocity, approximately 200 ft/min, 61 m/min; barometric pressure, 1 atm).

A—1-clo (light coveralls)
B—2-clo (woolen underwear, coveralls, and jacket)
C—3-clo (intermediate-weight flight clothing)
D—4-clo (heavy flight clothing)

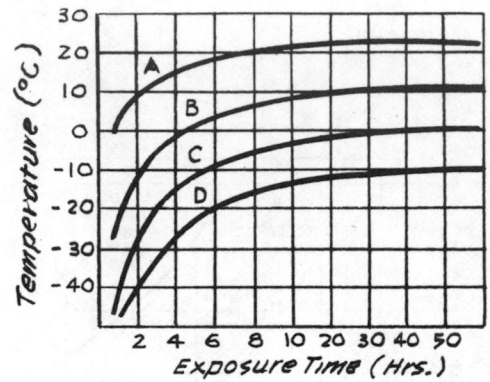

### Effects of Cold on Selected Task Performances

A—tactile sensitivity, bare hand
B—simple visual reaction time
C—manual skill

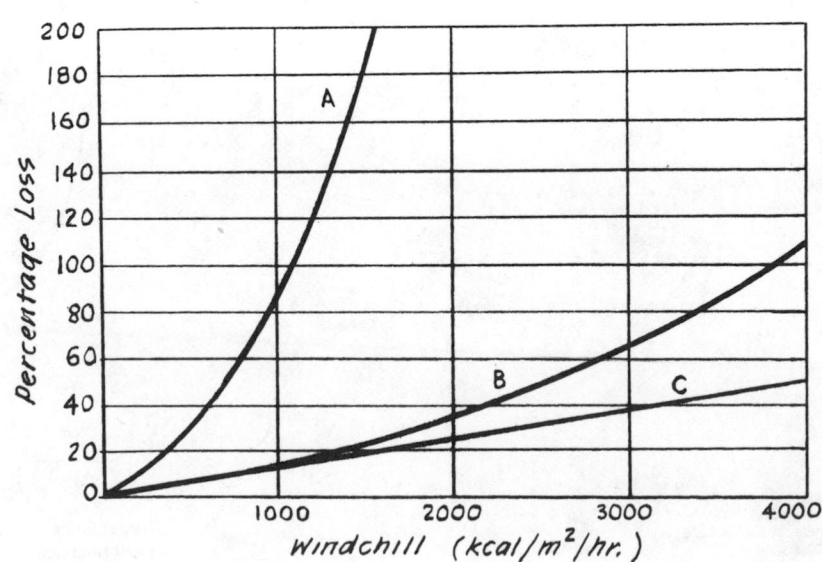

**Wind-Chill Nomograph[35]**

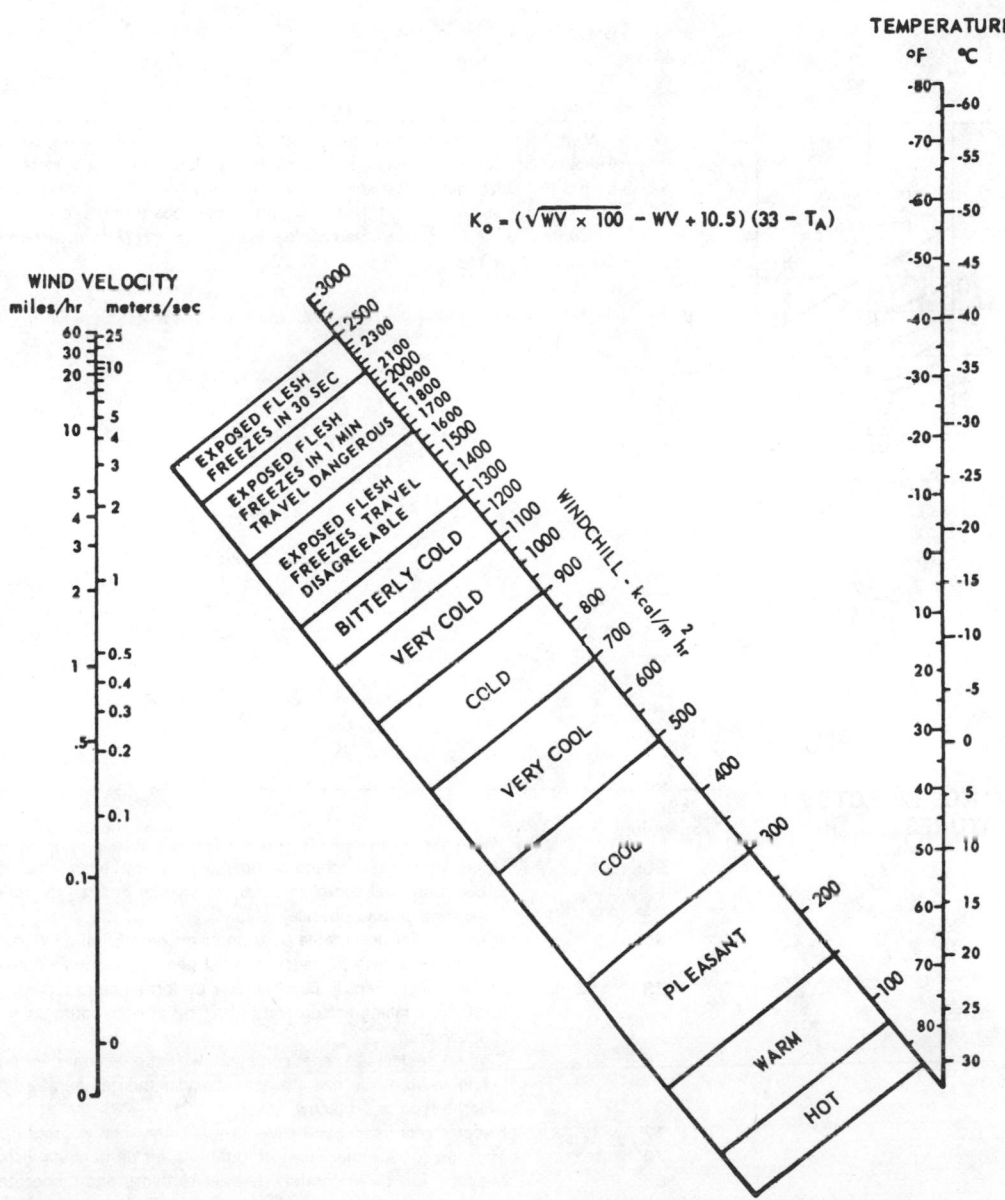

$$K_o = (\sqrt{WV \times 100} - WV + 10.5)(33 - T_A)$$

NOTE:

    The windchill index does not account for physiological adaptations or adjustments and should not be used in a rigorous manner. It is based on field measurements during World War II of the rate of cooling of a container of water.

[35]J, F, Parker and V. R. West (eds.), *Bioastronautics Data Book,* 2d ed., National Aeronautics and Space Administration, 1973.

## TEMPERATURE AND VENTILATION REQUIREMENTS

Ventilation requirements in relation to net airspace and the removal of body odor are shown in the accompanying graph. The graph illustrates how the intensity of body odors in a given area depends on the rate of flow of odor-free air.

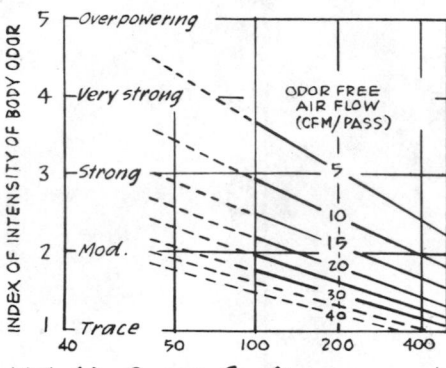

Net Air Space (cu ft per person)

**General Minimum Fresh-Air Requirements for Typical Ventilation Needs**

| Type of Occupancy | Airspace per Person, ft³ | Ventilation per Person, ft³/min |
|---|---|---|
| Sedentary adults | 100 | 25 |
| | 200 | 16 |
| | 300 | 12 |
| | 500 | 7 |
| Adult laborers | 200 | 23 |
| Grade school children | 100 | 29 |
| | 200 | 21 |
| | 300 | 17 |
| | 500 | 11 |

*Note:* These are minimum requirements during the normal heating season, with or without recirculation and without air conditioning. Norms for rooms at sea level range from 5 to 50 ft³/min per person, e.g., 7.5 for theatres; 15 for restaurants, hospital rooms, and apartments; and 30 for private offices, boardrooms, etc.

*Source:* R. A. McFarland, *Human Factors in Air Transport Design,* McGraw-Hill Book Company, New York, 1946, pp. 119, 121.

## HUMAN PERFORMANCE EFFECTS AT VARIOUS TEMPERATURES

| Effective Temperature °F* | Performance Effects |
|---|---|
| 90 | Upper limit for continued occupancy over any reasonable period of time. |
| 80–90 | Expect universal complaints, serious mental and psychomotor performance decrement, and physical fatigue. |
| 80 | Maximum for acceptable performance even of limited work; work output reduced as much as 40–50 percent; most people experience nasal dryness. |
| 78 | Regular decrement in psychomotor performance expected; individuals experience difficulty falling asleep and remaining asleep; optimum for bathing or showering. |
| 75 | Clothed subjects experience physical fatigue, become lethargic and sleepy, and feel warm; unclothed subjects consider this temperature optimum without some type of protective cover. |
| 72 | Preferred for year-round sedentary activity while wearing light clothing. |
| 70 | Midpoint for summer comfort; optimum for demanding visual-motor tasks. |
| 68 | Midpoint for winter comfort (heavier clothing) and moderate activity, but slight deterioration in kinesthetic response; people begin to feel cool indoors while performing sedentary activities. |
| 66 | Midpoint for winter comfort (very heavy clothing), while performing heavy work or vigorous physical exercise. |
| 64 | Lower limit for acceptable motor coordination; shivering occurs if individual is not extremely active. |
| 60 | Hand and finger dexterity deteriorates, limb stiffness begins to occur, and shivering is positive. |
| 55 | Hand dexterity is reduced by 50 percent, strength is materially less, and there is considerable (probably uncontrolled) shivering. |
| 50 | Extreme stiffness; strength applications accompanied by some pain; lower limit for unprotected exposure for more than a few minutes. |

*These temperature effects are based on relatively still air and normal humidity (40 to 60 percent). Higher temperatures are acceptable if airflow is increased and humidity is lowered (a shift of from 1 to 4°); lower temperatures are less acceptable if airflow increases (a shift upward of 1 to 2°).

## VISUAL SYSTEM CHARACTERISTICS

The primary organ through which visual responses are made is, of course, the eye. The accompanying sketches show the key elements of the visual receptor and liken them to the camera and the photographic process.

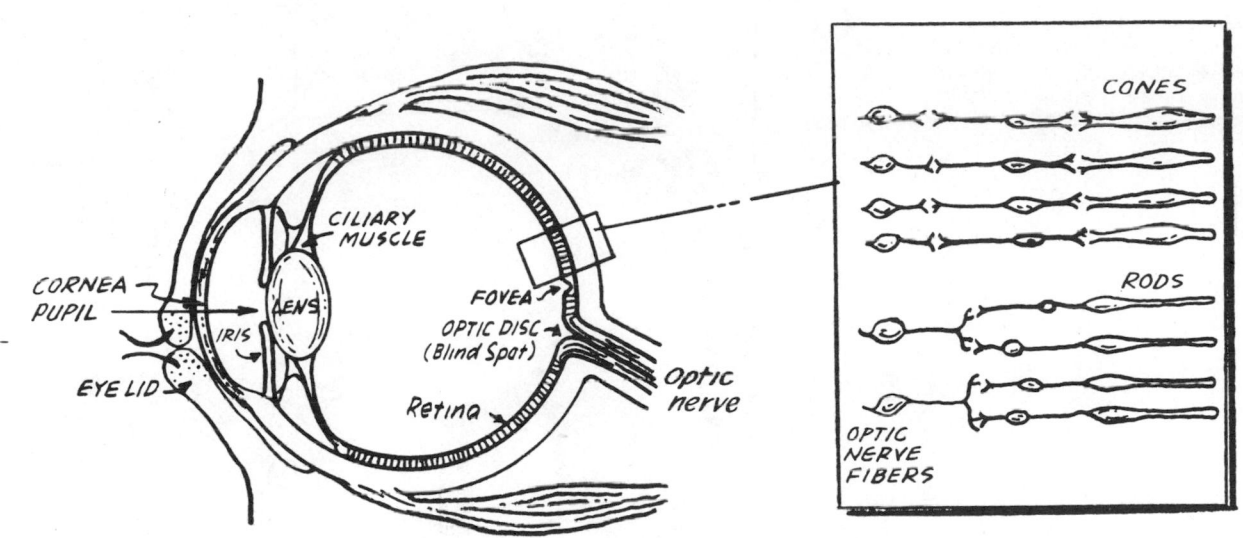

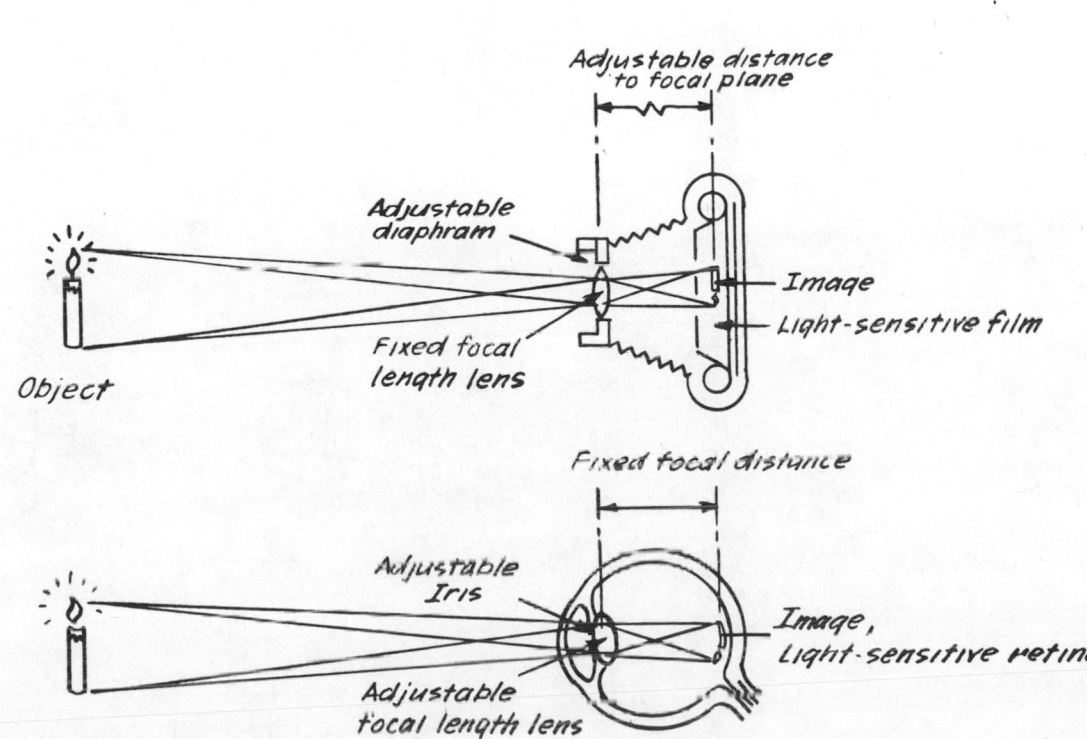

### Image and Retinal System

The image of a moving object moves across the retina when the eyes are held still, providing information of the movement through sequential receptor stimulation along the path.

### Eye and Head Movement System

When the eye follows a moving object, the image remains more or less stationary upon the retina, although the person still sees movement.

The image and retinal system and the eye and head movement system sometimes disagree, leading to visual illusions that are often confusing and misleading.

### Size Constancy

The image of an object on the retina halves in size with each doubling of viewing distance of the object. However, it does not appear to shrink so much because the brain compensates for the shrinkage of the image with distance.

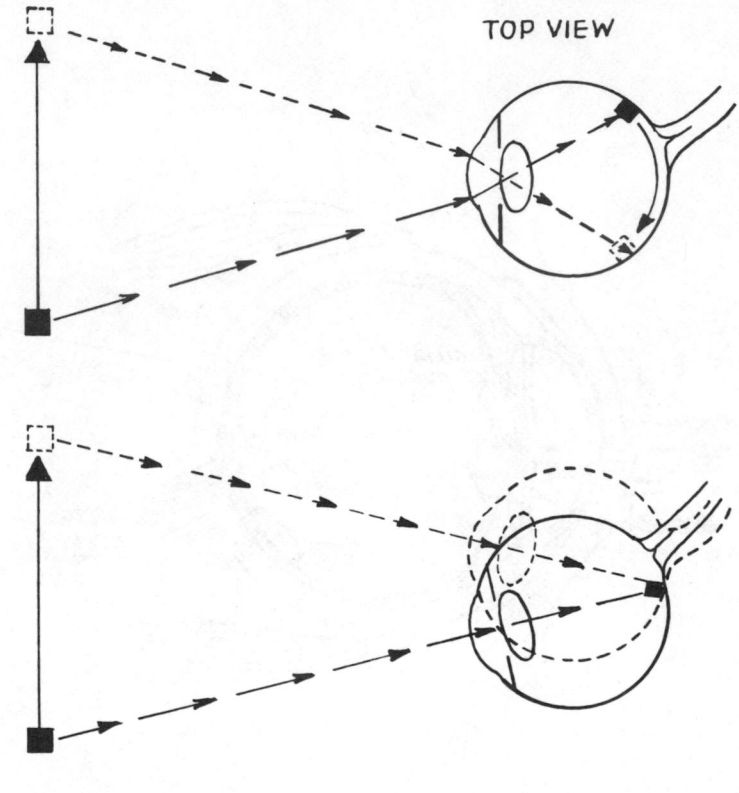

TOP VIEW

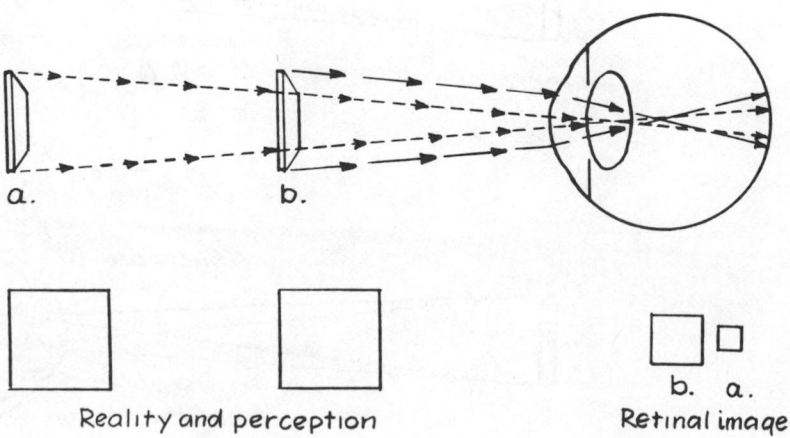

a.          b.

Reality and perception

b.    a.

Retinal images

## ACCOMMODATION AND CONVERGENCE

The size of a retinal image is calculated by using the equation $AB/ab = An/an$, when the size of the object and its viewing distance are known.

### Amplitude of Accommodation

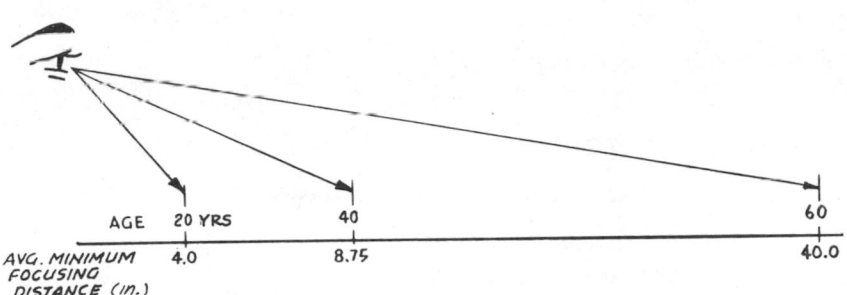

Although a child can focus on an object that is as close as 2.4 in (6.1 cm), an adult can focus no closer than 6 in (15.2 cm). Hardening of the lens with age reduces the focusing capacity of older persons' eyes.

Accommodation versus age.

### Convergence

Convergence acts like a range finder, providing depth perception.

#### Diameter of the Pupil in Millimeters

| Age | In Daylight | At Night | Difference |
|-----|-------------|----------|------------|
| 20 | 4.7 | 8.0 | 3.3 |
| 30 | 4.3 | 7.0 | 2.7 |
| 40 | 3.9 | 6.0 | 2.1 |
| 50 | 3.5 | 5.0 | 1.5 |
| 60 | 3.1 | 4.1 | 1.0 |
| 70 | 2.7 | 3.2 | 0.5 |
| 80 | 2.3 | 2.5 | 0.2 |

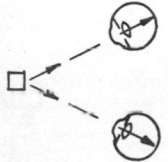

### Accommodation in Relation to Age

The nearest points at which an individual can see objects distinctly when using maximum accommodation are shown in the accompanying table; i.e., the object must be as close as, or closer than, that corresponding to the specified accommodative power (in diopters).

| Age | Minimum for Age | Normal for Age |
|-----|-----------------|----------------|
| 17 | 8.8 | 11.8 |
| 20 | 8.1 | 11.1 |
| 25 | 6.9 | 9.9 |
| 30 | 5.7 | 8.7 |
| 40 | 2.8 | 5.8 |
| 45 | 0.6 | 3.6 |

Source: Air Force Manual 160-1.

### EXTERNAL AND INTERNAL CUES TO DEPTH AND DISTANCE

| | External Cues |
|---|---|
| Linear perspective | Apparent convergence of parallel lines and related effects |
| Apparent size | A strong cue to distance of objects of known size and texture |
| Motion parallax | Relative angular motion as either head or objects move |
| Interposition | Nearer objects eclipse more distant ones |
| Aerial perspective | Contrast and color loss due to aerosols; useless in free space |
| Shading | A cue to three-dimensional form of objects (not to distance) |
| Apparent intensity | A cue only to distance of effective "point sources" |
| | Internal Cues |
| Accommodation | Relatively unimportant |
| Convergence | Useful limit is about 20 m |
| Binocular disparity | Most important intrinsic cue to depth and distance |

Note: All cues except for the last two can be utilized by a single eye, and by extension, in uniocular optical devices.

## RADIANT ENERGY AND THE EYE

Light is visually evaluated radiant energy. As shown in the accompanying illustration, the visible portion of the electromagnetic radiant energy spectrum is extremely narrow—about 380 to 770 nm (1 nm = 1 mu = $10^{-9}$ m).

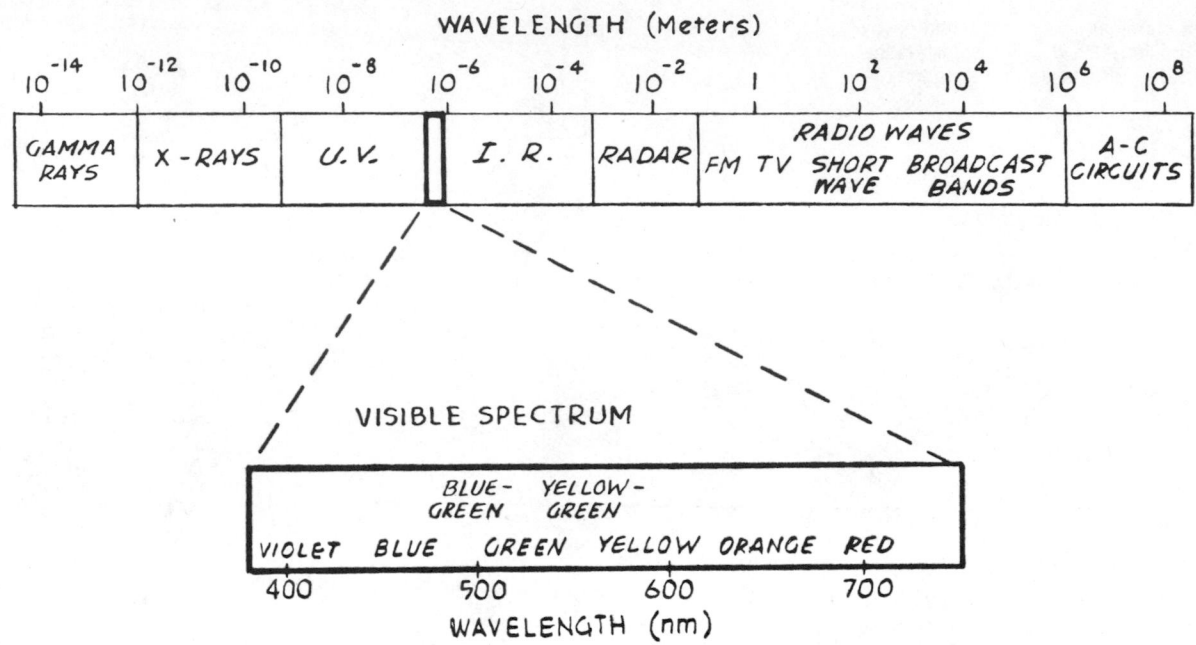

## CHARACTERISTICS OF VISION

| Parameter | Vision |
|---|---|
| Sufficient stimulus | Light-radiated electromagnetic energy in the visible spectrum |
| Spectral range | Wavelengths from 400 to 700/$\mu$ (violet to red) |
| Spectral resolution | 120 to 160 steps in wavelength (hue) varying from 1 to 20/$\mu$ |
| Dynamic range | ~90 dB (useful range) for rods = 0.00001 to 0.004 mL; cones = 0.004 mL to 10,000 mL |
| Amplitude resolution $\Delta I/I$ | Contrast = $\Delta I/I$ = 0.015 |
| Response rate for successive stimuli | ~0.1 s |
| Reaction time for simple muscular movement | ~0.22 s |
| Best operating range | 500 to 600/$\mu$ (green-yellow) 10 to 200 fc |
| Indications for use | 1. Spatial orientation required. 2. Spatial scanning or search required. 3. Simultaneous comparisons. 4. Multidimensional material presented. 5. High ambient noise levels. |

## Intensity Relationships

It takes only one-billionth of a lambert to excite the eye (absolute threshold).

The minimum ratio at which light can be seen is called the "contrast threshold." A light of very low intensity can be seen against a dark background. However, the light must be much brighter to be seen against a bright background, as shown in the accompanying table.

| Log of Illuminance Just Visible at the Eye, fc | Log of Background Brightness, mL |
|---|---|
| Photopic vision: | |
| − 5 | 4 |
| − 6.25 | 2 |
| − 7.60 | 0 |
| Mesopic vision: | |
| − 8.3 | − 2 |
| − 8.8 | − 4 |
| Scotopic vision: | |
| − 9.6 | − 6 |
| − 9.8 | Total darkness |

Visibility also depends on light target area (visual angle), as shown in the accompanying table.

| Just visible brightness, log fL | − 2.2 | − 2.7 | −3.25 | −4.5 | −4.75 | −5.6 | −5.8 |
|---|---|---|---|---|---|---|---|
| Diameter of target, min/arc | 0.3 | 0.6 | 1.0 | 6.0 | 10.0 | 60.0 | 100.0 |

The threshold visibility of a flashing light depends on the total energy in the flash, which is the intensity of the flash multiplied by the duration. (For flashes longer than 0.1 s, visibility depends on intensity alone.)

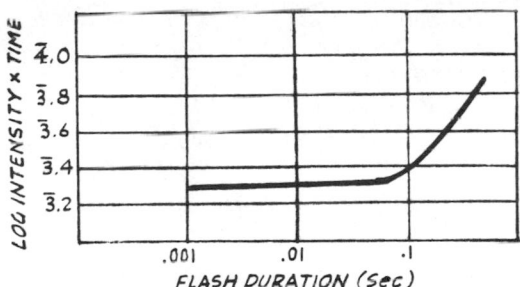

## PHOTOPIC AND SCOTOPIC RESPONSE FUNCTIONS

Photopic (foveal) vision occurs under high illumination conditions, where the luminosity function is mediated by the cones. Scotopic (peripheral) vision occurs under low illumination conditions, where the luminosity function is mediated by the rods. A third type of vision (mesopic) occurs when both cones and rods are activated, i.e., under low or intermediate illumination conditions, such as at dusk or dawn.

In the fovea, where visual acuity is at a maximum, vision is mediated entirely by cones. Although cones are distributed throughout the peripheral retina, cone density is considerably less than in the foveal region, and therefore acuity is reduced in the periphery. At the limits of peripheral vision, cone vision may be completely absent, and only the rods mediate. As a result, the eye detects only light. Therefore, a faint light will disappear if one looks directly at it.

It should be noted that while each of the curves is shown at the same luminosity scale in the accompanying graph, this does not mean that the eye has equal peak luminous responses in the photopic and scotopic modes.

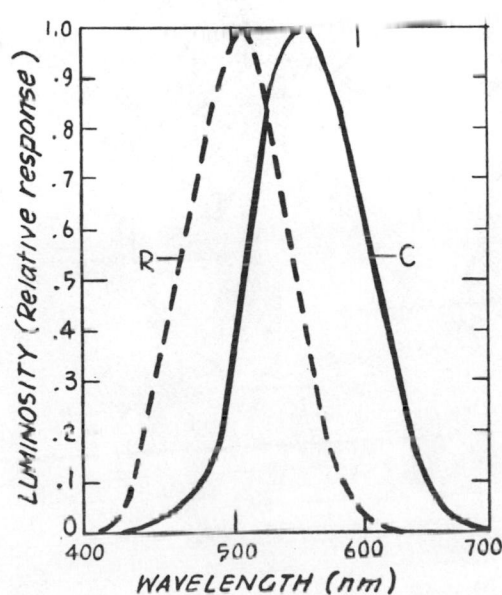

Standard luminosity functions (R—Rod, or photopic vision; C—Cone, or scotopic vision).

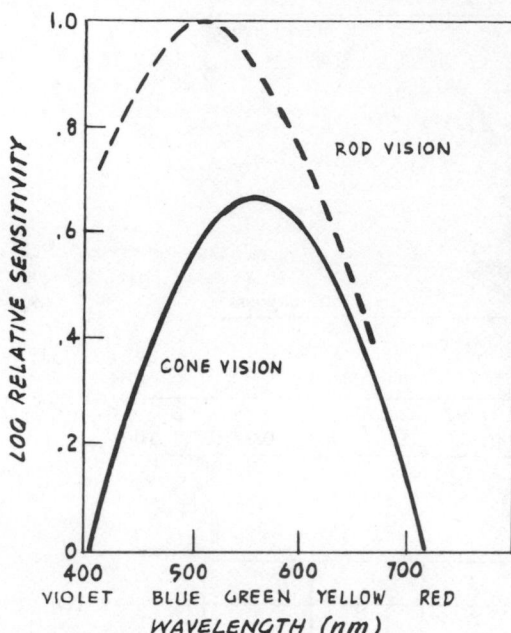

Spectral response of retinal rods and cones, with relative
amount of radiant energy for absolute threshold
vision as a function of the wavelength of light.

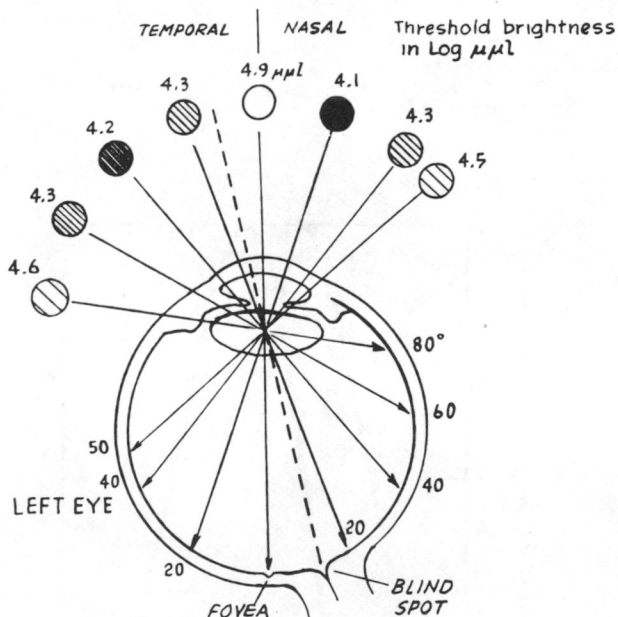

Sensitivity of different parts of the retina at night.

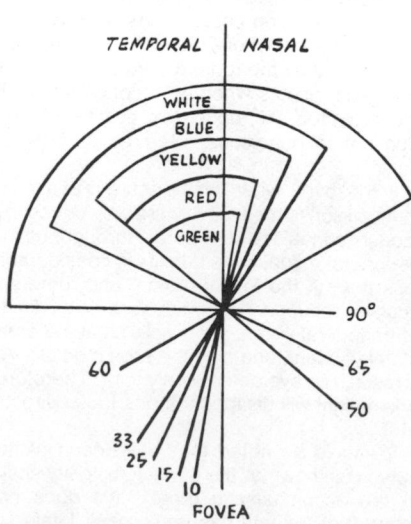

Sensitivity of different parts of the retina in
daylight.

## VISUAL ACUITY[36]

The term "visual acuity" refers to the resolving power of the eye or its ability to detect fine detail.

There is a marked relationship between visual acuity and the distribution of retinal rods and cones, as illustrated in the first graph.

Since there are more cones in the central area, acuity (photopic vision) is better in this area.

Photopic vision, in which rods are most used, is better farther from the foveal area of the retina.

The relationship between field brightness and minimum perceptible brightness difference (another measure of visual acuity) is shown in the second graph.

Although there is no appreciable change in the sensitivity of the cones (B) used for daylight conditions, there is a considerable change in the sensitivity of the rods (A), which are used for nighttime conditions.

At the lowest intensity of light, the eye can see a line whose thickness subtends a visual angle of 10 min. At very high intensities, the eye can see a line subtending less than 1 s of visual angle. These values assume maximum contrast between the lines and the background.

The relationship between contrast and image size is illustrated by a standard "parallel bar" target under a brightness level of 30 mL. As shown in the accompanying table, as contrast is increased, minimum size and spacing between parallel bars can be decreased without rendering the separation invisible. However, with decreasing contrast, target size and separation must be increased in order to maintain threshold acuity.

| Contrast, % | 45 | 8 | 5 | 3 | 2.8 |
|---|---|---|---|---|---|
| Visual angle min | 1 | 2 | 4 | 10 | 16 |

Visual acuity also varies under different spectral illuminants. When the background brightness level is 0.075 fc, the relationships shown in the accompanying table hold true.

| Visual acuity, % | 52 | 70 | 75 | 68 | 63 |
|---|---|---|---|---|---|
| Wavelength (Mμ) | 485 | 520 | 590 | 625 | 665 |

When higher illumination levels are available, however, the relationship between illuminant color and acuity for black-and-white targets is negligible. Luminance contrast, color contrast, illumination level, and exposure time are more important to acuity than color is.

Moving targets need to be larger as the rate of movement increases in order to provide the necessary acuity to distinguish target detail (see the accompanying table).

| Visual acuity, % | 100 | 56 | 30 | 19 | 10 |
|---|---|---|---|---|---|
| Velocity, °/s | 0 | 50 | 100 | 150 | 200 |

Acuity deteriorates with age, as shown in the accompanying graph. The designer must remember that older persons' eyes require more light and larger detail.

[36]J. W. Wulfeck et al., *Vision in Military Aviation*, WADC Technical Report 58-399, 1958.

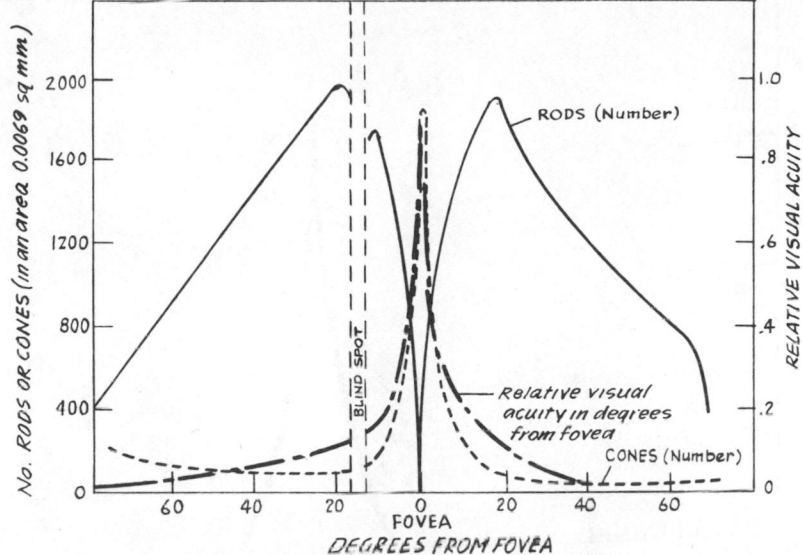

The relationship of visual acuity to the distribution of rods and cones.

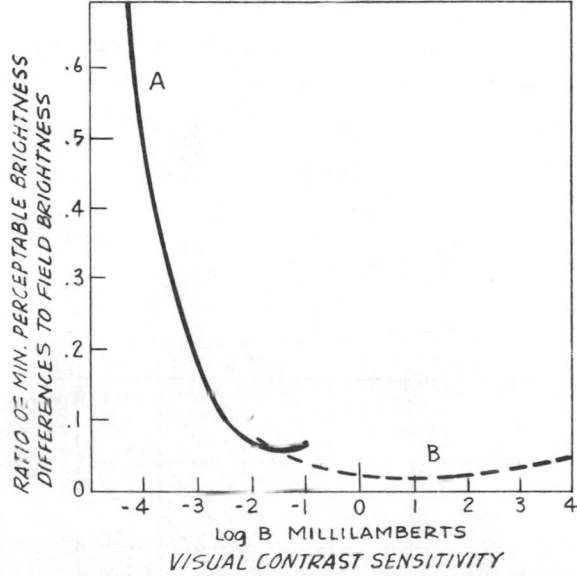

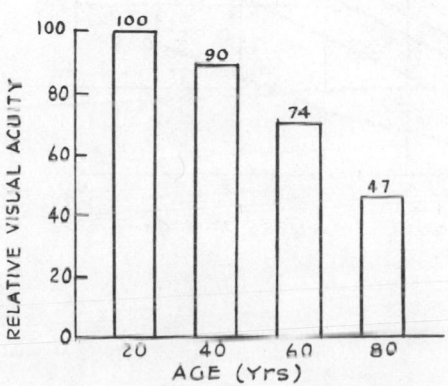

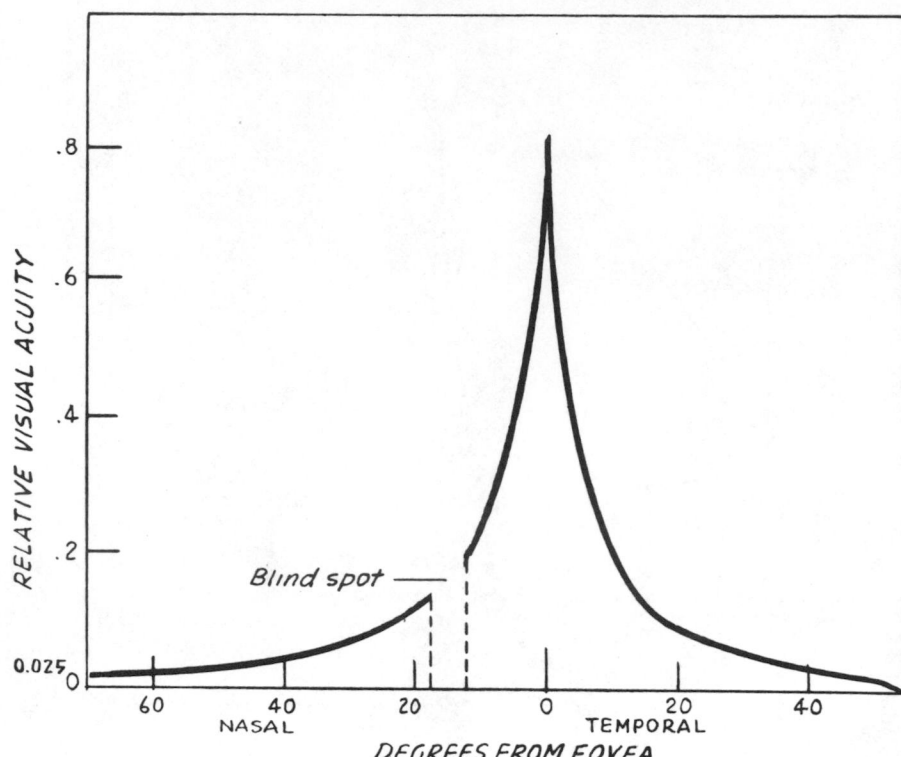

Daylight visual acuity for different parts of the eye.

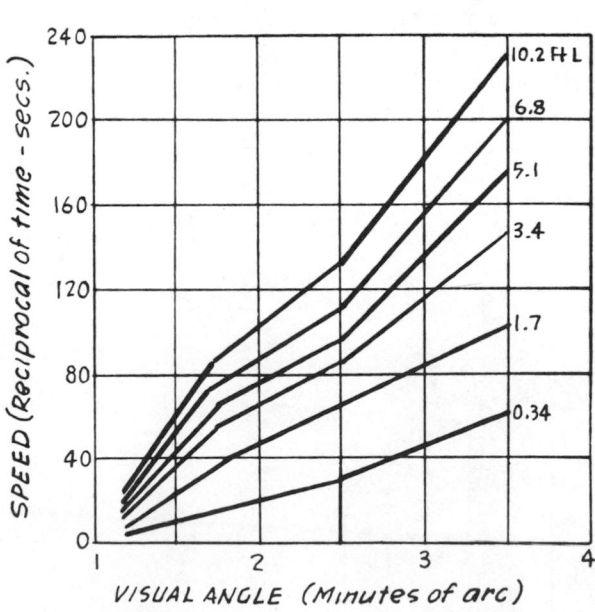

Acuity versus time.

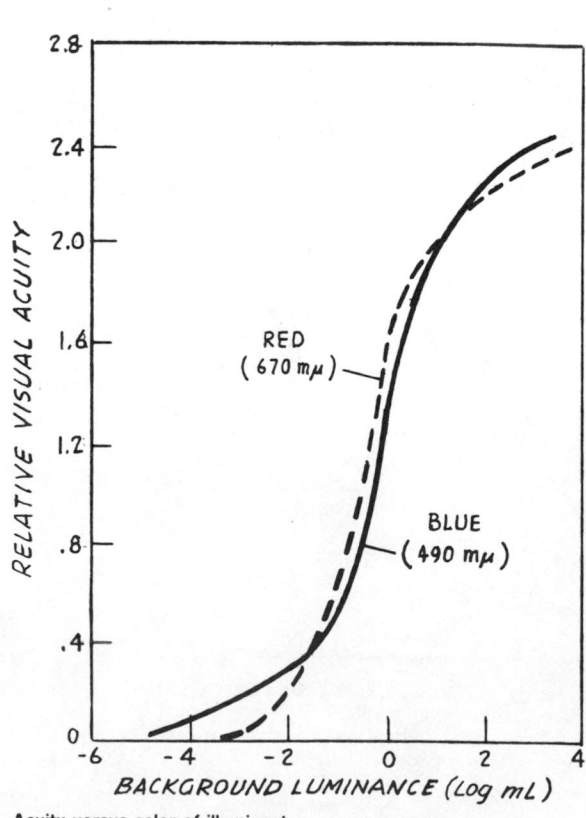

Acuity versus color of illuminant.

## RELATIONSHIP BETWEEN CONTRAST, VISUAL ANGLE, AND BRIGHTNESS IN DETERMINING WHETHER WE SEE OR NOT

The accompanying three-dimensional chart illustrates the relationship between three critical variables that determine whether a person sees or does not see an object. When the combination of variables intersects behind the curve, the eye cannot discriminate, whereas any intersecting point in front of the curve provides good seeing conditions.

Good seeing is also affected by the brightness relationships between all elements within the visual field. As the accompanying table shows, it is desirable to maintain different brightness relationships between the primary visual task and immediate and distant visual phenomena.

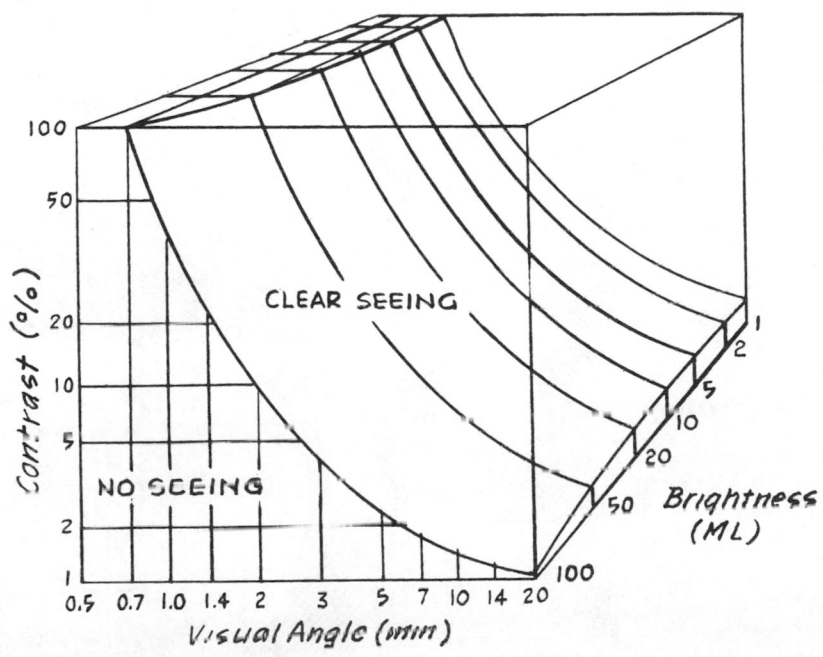

### Recommended Brightness Ratios within the Visual Field

| Areas | Recommended Maximum Luminance Ratio | |
|---|---|---|
| | Office | Industrial |
| Task and adjacent surroundings | 3:1 | |
| Task and adjacent darker surroundings | | 3:1 |
| Task and adjacent lighter surroundings | | 1:3 |
| Task and more remote darker surfaces | 5:1 | 10:1 |
| Task and more remote lighter surfaces | 1:5 | 1:10 |
| Luminaires (or windows, etc.) and surfaces adjacent to them | | 20:1 |
| Anywhere within normal field of view | | 40:1 |

## VISUAL FIELD

### Monocular Field

The general field that can be seen by the single eye is illustrated in the accompanying chart, which shows how the brow and the nose limit the field above and toward the nose. This chart also shows the extent to which some colors are more perceptible than others as a result of the distribution of retinal sensors; i.e., one can perceive blue farther out on the periphery than red. The outer limit denotes that only white light is recognizable.

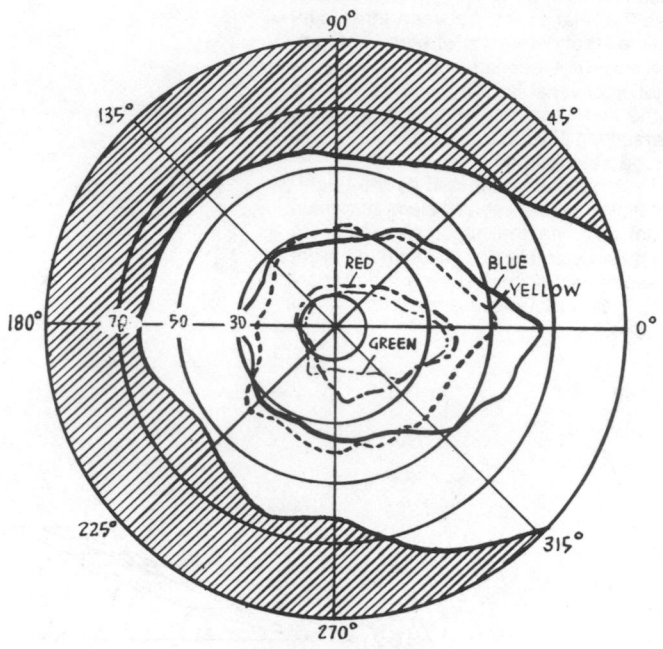

### Binocular Field

Overlapping of the two binocular fields is illustrated in the accompanying chart and sketch. Note that the 15° cone about the major visual axis is always recommended as the limit for locating warning lights that must be seen while watching a specific display.

The accompanying sketch illustrates the general limiting variable for peripheral vision.

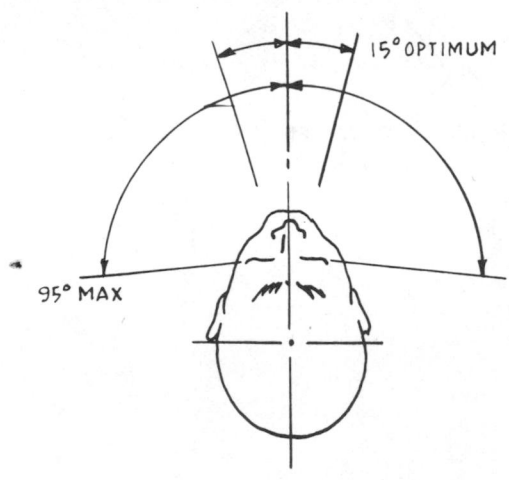

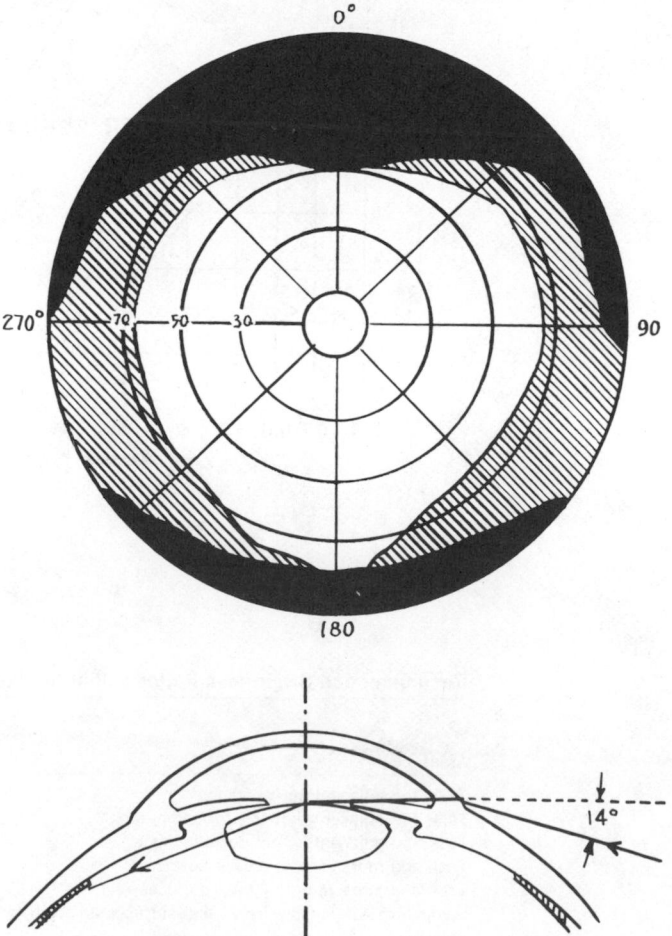

### The Limits of the Visual Field under Various Kinds of Restraint

| Movement Permitted | Type of Field and Factors Limiting Field | Horizontal Limits | | Vertical Limits | |
|---|---|---|---|---|---|
| | | Temporal Ambinocular Field (Each Side) | Nasal Binocular Field (Each Side) | Field Angle Up | Field Angle Down |
| Moderate movements of head and eyes assumed as: | Range of fixation | 60° | | 45° | |
| Eyes: 15° right or left 15° up or down | Eye deviation (assumed) | 15° | 15° | 15° | 15° |
| | Peripheral field from point of fixation | 95° | (45°) | 46° | 67° |
| Head: 45° right or left 30° up or down | Net peripheral field from central fixation | 110° | 60° | 61° | 82° |
| | Head rotation (assumed) | 45° | 45° | 30° | 30° |
| | Total peripheral field (from central body line) | 155° | 105° | 91° | 112° |
| Head fixed Eyes fixed (central position with respect to head) | Field of peripheral vision (central fixation) | 95° | 60° | 46° | 67° |
| Head fixed Eye maximum deviation | Limits of eye deviation (= range of fixation) | 74° | 55° | 48° | 66° |
| | Peripheral field (from point of fixation) | 91° | Approx(5°) | 18° | 16° |
| | Total peripheral field (from central head line) | 165° | 60° | 66° | 82° |
| Head maximum movement Eyes fixed (central with respect to head) | Limits of head motion (= range of fixation) | 60° 72° | 72° | 45° 80° | 90° |
| | Peripheral field (from point of fixation) | 95° | 60° | 46° | 67° |
| | Total peripheral field (from central body line) | 167° | 132° | 126° | 157° |
| Maximum movement of head and eyes | Limits of head motion | 72° | 72° | 80° | 90° |
| | Maximum eye deviation | 74° | 55° | 48° | 66° |
| | Range of fixation (from central body line) | 146° | 127° | 128° | 156° |
| | Peripheral field (from point of fixation) | 91° | Approx (5°) | 18° | 16° |
| | Total peripheral field (from central body line) | 237° | 132° | 146° | 172° |

J. W. Wulfeck, et al., *Vision in Military Aviation,* WADC Technical Report 58-399, 1958.

## MONOCULAR DAYLIGHT ACUITY RELATIVE TO CENTRAL ACUITY[37]

The relative acuity for one-eye daytime vision in various parts of the visual field is shown in the accompanying chart. Note that the best vision is within the small ring at the center of the field. Under the most favorable conditions, ½ min of arc may be discriminated. The figures on the larger rings illustrate how many times larger an object must be to be seen clearly.

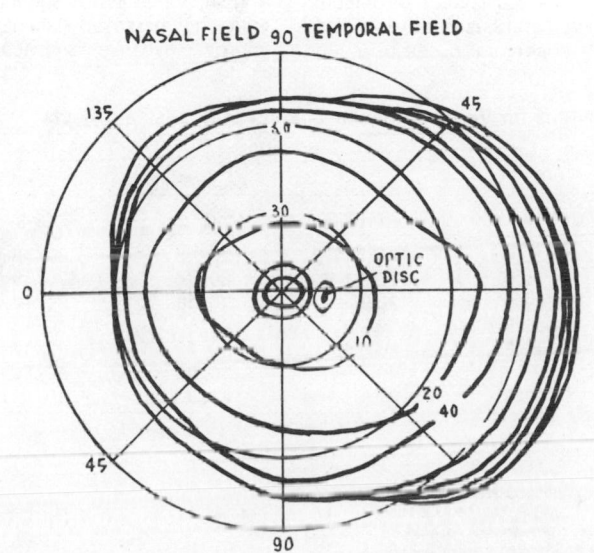

NASAL FIELD 90 TEMPORAL FIELD

[37]After W. S. Duke-Elder, *Test Book of Ophthalmology,* The C. V. Mosby Company, St. Louis, 1934.

## Loss of Peripheral Vision with Increases in Observer Motion[38]

As an observer (e.g., vehicle operator) moves forward within a visual field, his or her peripheral field sensitivity is reduced. It has been estimated that an individual can lose up to 40 percent of side vision when concentrating on a central visual task, as one typically does when driving. Objects at the side of the road tend to become blurred as speed increases and as the operator fixates farther down the road, as illustrated in the accompanying graph.

It has been determined that 75°/s is the angular speed at which visual acuity begins to deteriorate. This is called "dynamic acuity." The accompanying table gives angular speeds of objects along the path of travel for various speeds at various distances from the line of travel. Objects as close as 10 ft (3.1 m) are easily seen at walking and running speeds of about 8 mi/h (12.9 km/h), but at speeds of 18 mi/h (29 km/h), an individual begins to lose clear vision of nearby objects, e.g., objects 10 ft (3.1 m) away or closer.

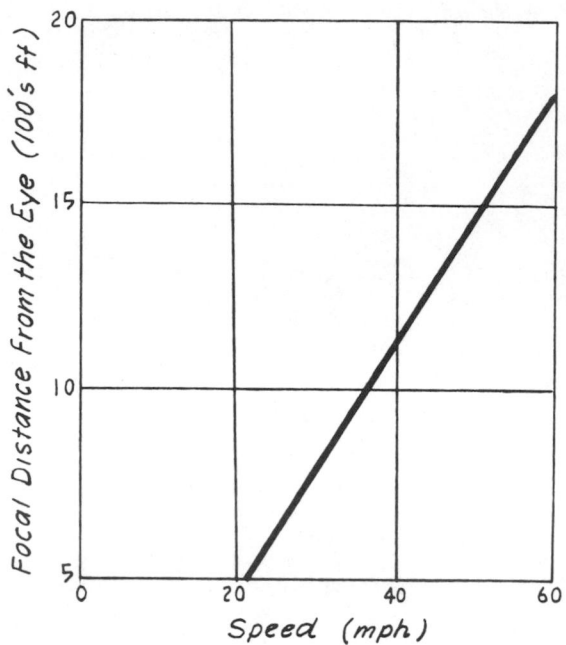

### Angular Speeds of Objects along Path of Travel for Various Speeds and Distances

| Miles per Hour | Feet per Second | Degrees and Seconds | | | | | | | | | |
|---|---|---|---|---|---|---|---|---|---|---|---|
| 100 | 146 | 164° 24′ | 149° 22′ | 135° 20′ | 122° 34′ | 111° 24′ | 101° 10′ | 92° 24′ | 84° 46′ | 78° 6′ | 72° 16′ |
| 80 | 118 | 160° 46′ | 142° 34′ | 126° 6′ | 111° 4′ | 99° 26′ | 89° 2′ | 80° 16′ | 72° 50′ | 66° 30′ | 61° 6′ |
| 70 | 102 | 157° 10′ | 136° 24′ | 119° 6′ | 103° 8′ | 91° 8′ | 80° 44′ | 72° 10′ | 65° 2′ | 59° 6′ | 54° 4′ |
| 60 | 88 | 154° 24′ | 131° 8′ | 111° 26′ | 95° 28′ | 82° 42′ | 72° 32′ | 64° 18′ | 57° 38′ | 52° 8′ | 47° 30′ |
| 50 | 74 | 149° 6′ | 123° 14′ | 101° 16′ | 85° 34′ | 72° 62′ | 63° 20′ | 55° 44′ | 49° 40′ | 44° 42′ | 40° 38′ |
| 40 | 58 | 141° 18′ | 110° 50′ | 88° 4′ | 71° 14′ | 60° 14′ | 51° 36′ | 45° 2′ | 39° 52′ | 35° 44′ | 32° 22′ |
| 30 | 44 | 131° 8′ | 95° 28′ | 72° 32′ | 57° 38′ | 47° 30′ | 40° 18′ | 34° 54′ | 30° 46′ | 27° 30′ | 24° 50′ |
| 20 | 29 | 110° 50′ | 71° 44′ | 51° 36′ | 39° 12′ | 32° 22′ | 27° 12′ | 23° 24′ | 20° 34′ | 18° 18′ | 16° 32′ |
| 15 | 22 | 95° 28′ | 57° 38′ | 40° 18′ | 30° 46′ | 24° 50′ | 20° 48′ | 17° 52′ | 15° 40′ | 14° | 12° 38′ |
| 10 | 15 | 73° 6′ | 41° 8′ | 28° 6′ | 21° 16′ | 17° 4′ | 14° 16′ | 12° 34′ | 10° 44′ | 9° 32′ | 8° 36′ |
| 8 | 12 | 61° 56′ | 33° 24′ | 22° 38′ | 17° 4′ | 13° 42′ | 11° 26′ | 9° 48′ | 8° 36′ | 7° 38′ | 6° 54′ |
| 5 | 7 | 38° 36′ | 19° 12′ | 13° 20′ | 10° | 8° 2′ | 6° 42′ | 5° 50′ | 5° 2′ | 4° 28′ | 4° 2′ |
| 2 | 3 | 17° 4′ | 8° 36′ | 5° 4′ | 4° 18′ | 3° 26′ | 2° 52′ | 2° 28′ | 2° 10′ | 1° 56′ | 1° 44′ |
| | | 10 | 20 | 30 | 40 | 50 | 60 | 70 | 80 | 90 | 100 |

The accompanying table[39] shows acuity ranges for various angular speeds of objects at the side of the road and illustrates variations among groups of observers. It can be seen that visual acuity falls off even at relatively low angular velocities. It should be noted, however, that increased illumination considerably improves dynamic visual acuity. And when vehicle windows are larger, acuity is increased because the eye has a better opportunity to pick up and track the object.

### Visual Acuity Ranges for Various Angular Speeds of Roadside Objects

| Distance to Object, ft | 60 mi/h | | | | 30 mi/h | | | |
|---|---|---|---|---|---|---|---|---|
| | Angular Velocity, °/s | Visual Acuity | | | Angular Velocity, °/s | Visual Acuity | | |
| | | Gr 1 | Gr 2 | Gr 3 | | Gr 1 | Gr 2 | Gr 3 |
| 1000 | 5.0 | 20/51 | 20/48 | 20/38 | 2.5 | 20/51 | 20/48 | 20/38 |
| 500 | 10.0 | 20/51 | 20/48 | 20/39 | 5.0 | 20/51 | 20/48 | 20/38 |
| 100 | 47.5 | 20/64 | 20/54 | 20/42 | 24.75 | 20/53 | 20/48 | 20/29 |
| 80 | 57.5 | 20/73 | 20/50 | 20/45 | 30.75 | 20/54 | 20/49 | 20/39 |
| 60 | 72.5 | 20/96 | 20/70 | 20/52 | 40.25 | 20/59 | 20/51 | 20/41 |
| 40 | 95.5 | 20/154 | 20/99 | 20/70 | 57.5 | 20/73 | 20/59 | 20/45 |
| 20 | 131.0 | 20/317 | 20/317 | 20/121 | 95.5 | 20/154 | 20/99 | 20/70 |

[38] P. L. Connolly, *Visual Considerations of Man, the Vehicle and the Highway*, part II, SAE Paper SP-279, 1966.
[39] P. L. Connolly, *Visual Considerations of Man, the Vehicle, and the Highway*, part II, SAE Paper SP-279, 1966.

## DARK ADAPTATION

The transition from high- to low-level light requires adaptation time. Adaptation from low to higher light levels is approximately instantaneous, although it can be annoying or even slightly painful.

Although the time required to become dark-adapted varies both among subjects and within an individual eye, the general form of the dark adaptation curve remains somewhat characteristic. Note the transition point at which adaptation is fairly rapid, after which it levels off and then proceeds farther, but at a slower pace.

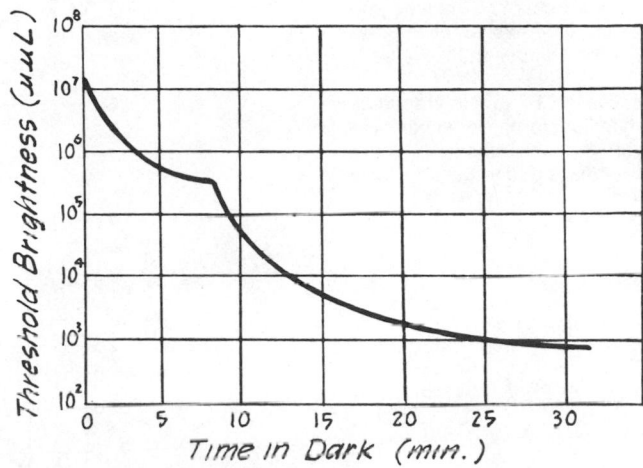

Course of dark adaptation showing intensity just barely discernible as function of time spent in dark.

Dark adaptation patterns vary depending on the color of the light, as illustrated in the accompanying graph.

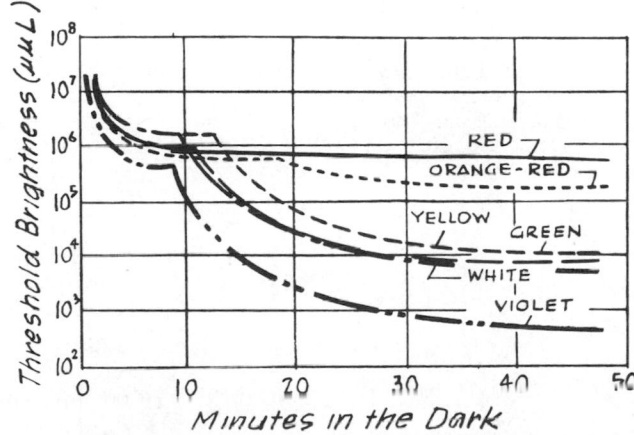

Dark adaptation measured with light of different wavelengths. Although lights were all equated in brightness initially, they are no longer equally bright even at cone threshold. The differences are further exagerated during rod dark adaptation.

The accompanying graph shows dark adaptation as a function of the region of the retina being stimulated.

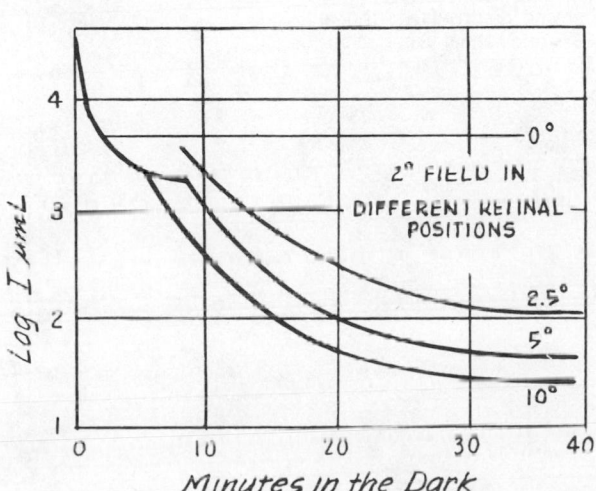
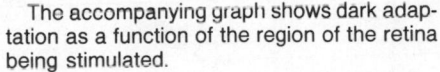

## CRITICAL FUSION FREQUENCY (CFF)

CFF is the point at which successive light flashes blend into a continuous light. A flickering light that is on 50 percent of the time and off 50 percent of the time flashes at the rate of 10 Hz; it appears brighter than a steady light.

The relation between CFF and retinal illumination $(l)$ with white light for three retinal locations is shown in the accompanying chart. Crossing of the curves is due to the shift from rod to cone vision.

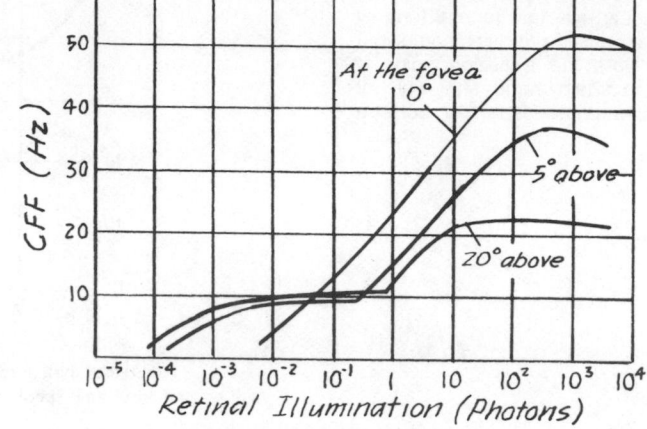

The relation between CFF and various target-area sizes is shown in the accompanying graph. The small areas (2 to 3°) stimulate only the cones, whereas for larger areas (6 and 19°) a double function yields a higher CFF.

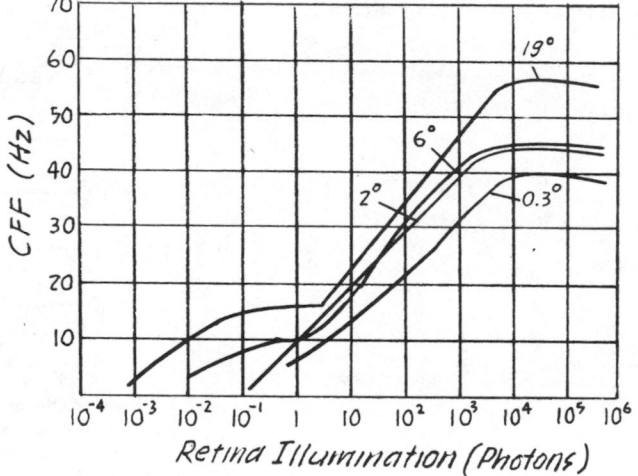

The accompanying graph shows the relation between CFF and retinal illumination for different spectral regions.

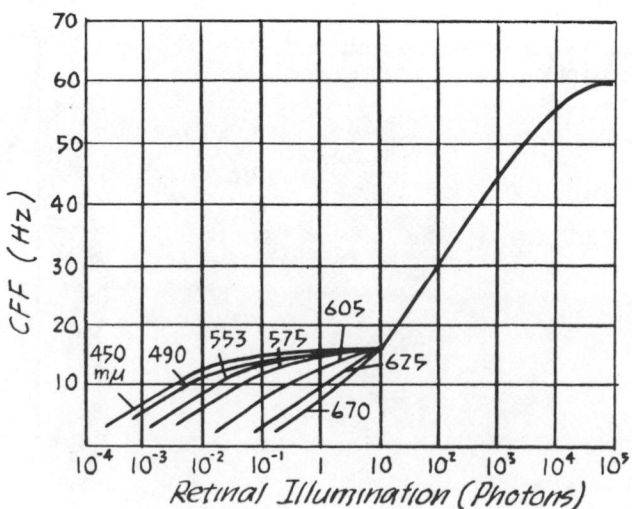

## GLARE

### Effect of Glare on Performance

| Function | Effects |
| --- | --- |
| Visibility | Glare of 5 fc located 5° above the line of vision reduces visibility about as much as replacing a 100-fc lighting system with a 1-fc system. An object that can be seen at 20 ft with 5-fc glare at 5° could be seen at 45 ft with no glare. Glare becomes less significant as the brightness of an object is raised, and is greater for a large object seen in low contrast to its background than for a small, high contrast object, such as a printed letter. |
| Size threshold | Glare greatly increases size threshold, particularly for objects of low contrast. The effect of glare diminishes as contrast and illumination of surrounds are increased, and thus is more easily controlled at high levels of illumination. Threshold size for an object of 5% contrast under 1 fL with a 5-fc glare 5° above line of vision = 30.0 min visual angle; for an object at 100 fL with no glare = 0.63 min. |
| Contrast threshold | Glare greatly increases the contrast threshold. The degree of increase depends in complex fashion upon the brightness of task background, the glare illumination produced by each glare source at the eye on a plane perpendicular to the line of sight to the task, and the angle of separation between the glare source and the line of sight. |
| Muscular tension | Glare significantly increases muscular tension—in some cases up to 30%. Although visual tasks may be carried out under severe handicaps with no loss in immediate accuracy, a decrease in efficiency is apparent with a considerable increase in nervous muscular tension. Subjects not compensating for glare effects may show a considerable drop in performance but no increase in muscular tension. |
| Vision through eyeglasses | The number of persons wearing glasses is sufficiently large and the glare effect sufficiently severe so that reflections from eyeglasses should be considered in placing light sources. When the light source is behind the head, no glare from eyeglasses will exist if the light source is 30° or more above the line of vision, 40° or more below the line of vision, or at an angle of 15° or more with the axis of symmetry of the head. |

## COLOR PERCEPTION

We can only generalize with respect to what each individual perceives in the case of a particular color. The accompanying chromaticity diagram has been created to identify particular wavelength characteristics that normally lead to an individual's identification of colors. The points A, B, C, and E represent standardized light sources under which colors are designated. The dominant wavelength and purity of a color are determined by reference to a particular illuminant. This diagram is based on the concept that the eye actually has three separate mechanisms of color response and that any wavelength stimulates all these mechanisms, not just one, in varying degrees.

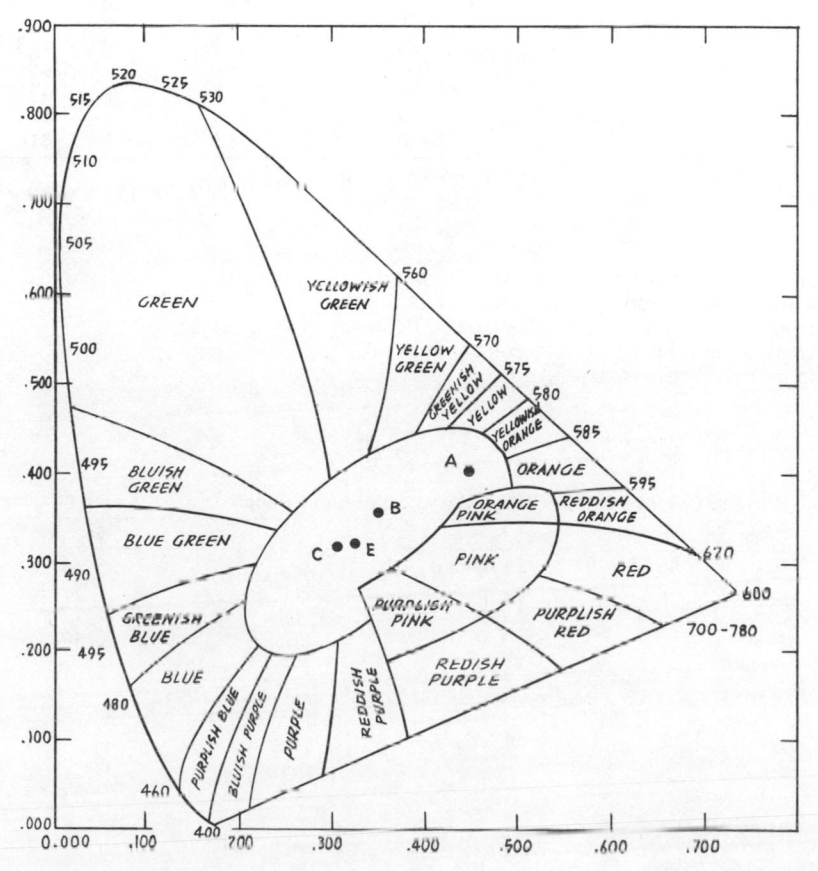

## Color Identification Capability

People use the term "color" to identify only one aspect of color, i.e., hue. As the accompanying color model illustrates, color consists of three attributes—hue, brightness, and saturation. A fully saturated *red* hue can be modified considerably and still be perceived as red by color-normal observers; i.e., when one moves toward the center of the model, the so-called red color becomes less saturated (less intense), and when one moves toward the upper end of the model, the color becomes pale. In addition, as one moves around the circumference of the model, a given color tends to blend with its neighboring hue; e.g., red moving toward yellow first appears orange-red, then orange, then reddish yellow, and finally yellow.

Although some observers can discriminate 150 or more hues under proper viewing conditions, the average person can accurately and reliably verbalize only about eight or nine hues. The accompanying table and nomogram[40] indicate probable color discrimination limits.

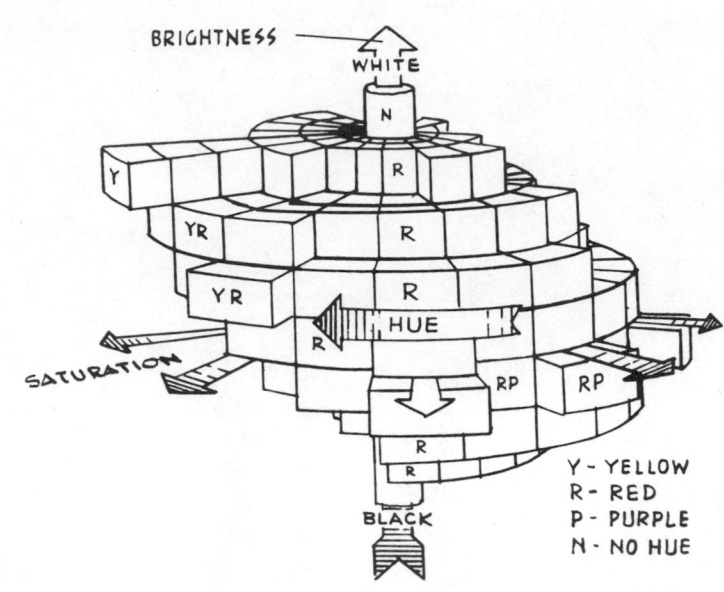

Y – YELLOW
R – RED
P – PURPLE
N – NO HUE

### Nine Equally Discriminable Surface Colors

| Hues | Code Number | Munsell Book Number | Excitation Purity | Dominant Wavelength |
|------|-------------|---------------------|-------------------|---------------------|
| 1 | 1.5 | 3R | 37.2 | 629 |
| 2 | 3 | 9R | 65.8 | 596 |
| 3 | 5.5 | 9YR | 81.8 | 582 |
| 4 | 8.5 | 1GY | 76.0 | 571 |
| 5 | 11.5 | 3G | 27.5 | 538 |
| 6 | 15 | 7BG | 35.0 | 491 |
| 7 | 18 | 9B | 56.5 | 481 |
| 8 | 20.5 | 9PB | 52.7 | 460 |
| 9 | 24 | 3RP | 36.5 | 510 |

Since many people have some level of color deficiency, it is desirable to select colors that both color-normal and color-deficient individuals can safely identify. The accompanying table defines such colors.

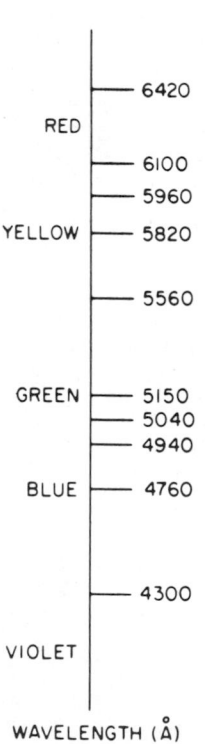

### Surface Colors for Color-Normal and Color-blind People

| Color | Spec. No.* | Color | Spec. No.* |
|-------|-----------|-------|-----------|
| Red | 1110 | Gray | 1625 |
| Orange | 1210 | Buff | 1745 |
| Yellow | 1310 | White | 1755 |
| Blue | 10B 7/6 | Black | 1770 |
| Purple | 2715 | | |

*From Fed. Spec. II-C-595 except for blue, which is from Munsell (1942).

[40]C. T. Morgan et al. (eds.), *Human Engineering Guide to Equipment Design*, McGraw-Hill Book Company, New York, 1963, p. 83.

## What People Do Perceive versus What They Can Perceive

Surface color recognition depends on three distinct factors: the color of the light source, the color of the reflecting surface or surfaces, and the state of the observer's visual system. Even though the object—say, a red box—may be a single basic color, this color can be modified both by the color of the light falling directly on it and/or by the color of the light reflected by nearby surfaces. And, of course, the color of the top surface (which receives the full impact of the light source) will not appear the same as the color of the side of the box that is in shadow or the side that is receiving bounce light from another surface (which may be of a different color). Fortunately, there is usually enough color constancy so that these variations are still accounted for, and thus the observer compensates and calls the box "red."

The paler a color is, the more easily it is influenced by the color of the light source and by the color of the light being reflected by nearby surfaces. In addition, colors are greatly influenced by the level of illumination and by the inherent reflectivity characteristics of the surface being viewed. For example, in a dim light, colors tend to take on a grayish and less vital appearance. A highly polished surface may reflect high-intensity illumination so effectively that the basic hue is indistinguishable.

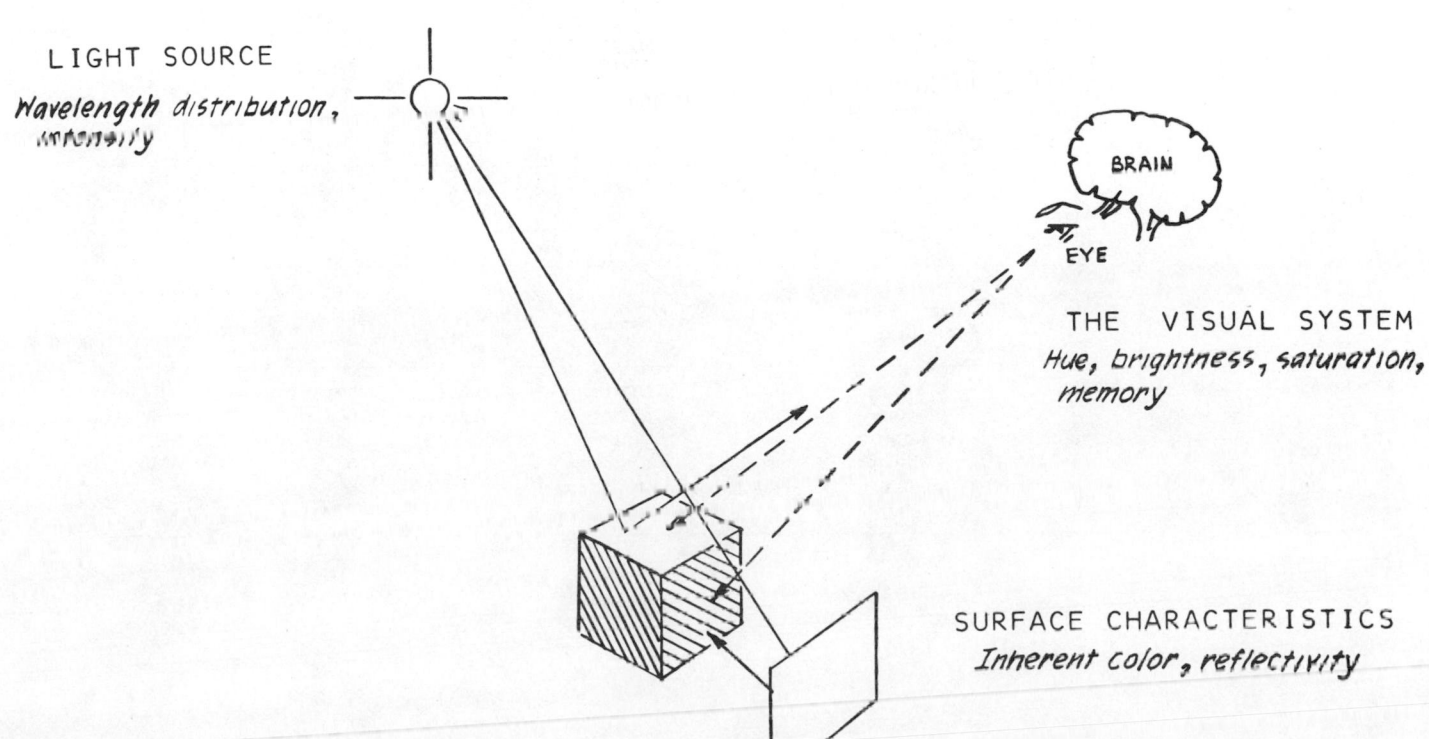

LIGHT SOURCE
Wavelength distribution, intensity

BRAIN

EYE

THE VISUAL SYSTEM
Hue, brightness, saturation, memory

SURFACE CHARACTERISTICS
Inherent color, reflectivity

## Effect of Monochromatic Light on Colored Objects

If the wavelength distribution of the light source is modified so that certain wavelengths are missing or considerably reduced (red light, yellow light, green light, etc.) the inherent color characteristics of a surface reflect a different color impression. Thus, as indicated in the accompanying table, a yellow object will appear to be red-orange when viewed under a red light.

## Effect of Standard Luminaires[41]

The accompanying tables give a general description of the effects that common fluorescent and mercury or filament luminaires have on colored surfaces. Both the lighting level and the color of the light affect the appearance of colored surfaces. Filament lamps and warm fluorescent lamps, which are deficient in blue, emphasize the redness of a surface color and thus accent warm hues.

### Effect of Colored Light on Colored Objects

| Object Color | Red Light | Blue Light | Green Light | Yellow Light |
|---|---|---|---|---|
| White | Light pink | Very light blue | Very light green | Very light yellow |
| Black | Reddish black | Blue black | Greenish black | Orange black |
| Red | Brilliant red | Dark bluish red | Yellowish red | Bright red |
| Light blue | Reddish blue | Bright blue | Greenish blue | Light reddish blue |
| Dark blue | Dark reddish purple | Brilliant blue | Dark greenish blue | Light reddish purple |
| Green | Olive green | Green blue | Brilliant green | Yellow green |
| Yellow | Red orange | Light reddish brown | Light greenish yellow | Brilliant light orange |
| Brown | Brown red | Bluish brown | Dark olive brown | Brownish orange |

### Color Effects of White Fluorescent Lamps

| | Cool White | Deluxe Cool White | Warm White | Deluxe Warm White | Daylight | White | Soft White—Natural |
|---|---|---|---|---|---|---|---|
| Lamp appearance; effect on neutral surfaces | White | White | Yellowish white | Yellowish white | Bluish white | Pale yellowish white | Pinkish white |
| Effect on "atmosphere" | Neutral to moderately cool | Neutral to moderately cool | Warm | Warm | Very cool | Moderately warm | Warm, pinkish |
| Colors strengthened | Orange, yellow, blue | All nearly equal | Orange, yellow | Red, orange, yellow, green | Green, blue | Orange, yellow | Red, orange |
| Colors grayed | Red | None appreciably | Red, green, blue | Blue | Red, orange | Red, green, blue | Green, blue |
| Remarks | Blends with natural daylight | Best over-all color rendition; simulates natural daylight | Blends with incandescent light | Excellent color rendition; simulates incandescent light | Usually replaceable with CW | Usually replaceable with CW or WW | Usually replaceable with CWX or WWX |

[41]A. O. Halse, *The Use of Color in Interiors*, McGraw-Hill Book Company, New York, 1968, pp. 41, 42.

Color Effects of Mercury and Filament Lamps

| | Mercury | White Mercury | Color-Improved Mercury | Deluxe White Mercury | Filament |
|---|---|---|---|---|---|
| Lamp appearance; effect on neutral surfaces | Greenish blue white | Greenish white | Yellowish white | White | Yellowish white |
| Effect on "atmosphere" | Very cool, greenish | Moderately cool, greenish | Warm, yellowish | Moderately cool | Warm |
| Colors strengthened | Yellow, green, blue | Yellow, green, blue | Yellow, green | Orange, yellow, blue | Red, orange, yellow |
| Colors grayed | Red, orange | Red, orange | Blue | Green | Blue |
| Remarks | Poor overall color rendering | | Color rendering often acceptable, but not equal to any white fluorescent | Color rendering good; compares favorably with CWX fluorescent | Excellent color rendering |

## Recommended Lamp Colors for Various Tasks and Locations

Art studios, art classes, etc. ..................................................CWX
Apparel shops—sports or daytime wear ....................................CWX
   Evening wear ...........................................................WWX
Barber and beauty shops ....................................................WWX
Bowling alleys ...............................................................CW
Color inspection and grading ................................................CWX
Classrooms ..................................................................CW
Delicatessens ...............................................................CWX
Drugstores ..................................................................CWX
Florists .....................................................................CWX
General office areas ........................................................CW
General work areas—no critical color work ..................................CW
Hardware stores .............................................................CWX
Homes and apartments .......................................................WWX
Hotel, motel guest rooms ....................................................WWX
"Intimate" areas—lounges, restaurants, areas of residential character, etc. ...WWX
Jewelry stores ..............................................................CWX
Private offices .......................................................CWX or WWX
Shoe stores .................................................................WWX
Supermarkets, groceries, meat markets .......................................CWX
Variety stores ..............................................................CWX

## Classification of Color Vision

| Type of Color Vision | Incidence (Men),* % | Number of Color Receptors† | Spectral Colors Confused | Colors Probably Seen |
|---|---|---|---|---|
| Trichromatic: | | | | |
| Normal | | 3 | Some reduction in wavelength discrimination | Reds look darker than normal |
| Deuteranomalous | 4.69 | 3 | Some reduction in wavelength discrimination | Reds look darker than normal |
| Protanomalous | 0.89 | 3 | Some reduction in wavelength discrimination | Reds look darker than normal |
| Tritanomalous | Not known | | | |
| Dichromatic: | | | | |
| Deuteranope | 1.42 | 2 | Red, yellow, and green | Yellow, white, and blue |
| Protanope | 1.21 | 2 | Red, and yellow, and green | Green (?), white, and blue |
| Tritanope | 0.008 | 2 | Blue and green | Reds look darker than normal |
| Monochromatic | 0.001 | 1 | All | White and black |

\* Average of several surveys.

† Assuming trichromatic color vision theory.

## Average Reaction Times for Each Class of Color Vision

| Subjects | Reaction Time, s |
|---|---|
| Normals | 0.497 |
| Deuteranomals | 0.788 |
| Protanomals | 0.708 |
| Deuteranopes | 0.855 |
| Protanopes | 0.984 |
| All abnormals | 0.945 |

## Psychological Response to Color

Although color researchers have not always been able to quantify the precise effects of various colors or light levels on humans, their research and experience seem to indicate that certain colors and light conditions often elicit typical repeatable reactions. Illumination level and quality do contribute to impressions of compartment size, to an atmosphere of warmth or coolness, or generate moods or feelings of feeling repose, pleasantness, excitement, boredom, or depression. Color can have a considerable bearing on whether food looks appetizing and on the appearance of one's skin.

The following four basic psychological response characteristics are perhaps most important to the development of appropriate color and lighting schemes:

1. Certain colors make a space appear larger than it actually is, while others cause a space to "close in" on the observer.
2. Certain colors cause a space to seem warm, while others make it seem cool.
3. Some colors appear to have a definite effect on the mood of the observer; i.e., some colors may be stimulating, while others are quieting.
4. Some colors seem to clash with each other and therefore produce a feeling of irritation to observers who are especially sensitive to color incompatibilities.

## Color in Relation to Size of Space

Average observers perceive dark and/or saturated colors as advancing toward them, and light and/or desaturated colors as receding away from them. If we wish to make a compartment appear less crowded, we should use light and unsaturated colors. Any structure or piece of furniture that is dark in color will make a space appear crowded. The level of illumination contributes to this appearance of being crowded in that a dimly lit space generally appears smaller than one that has a higher, more evenly distributed level of light. Therefore, with few exceptions, good illumination and lighter colors are desirable.

## Color in Relation to Temperature

So-called cool colors (e.g., blues and greens) make a space seem cooler, while so-called warm colors (such as reds, browns, and oranges) make a space seem warmer. Although little quantitative evidence of any direct correlation between cool or warm colors and actual physiological temperature variation exists, cool colors usually are preferred for facilities located in hot climates, and warmer colors are preferred for facilities located in cold climates. However, since many colors considered cool or warm also have other mood-generating properties, it is important not to overemphasize these temperature-related characteristics. Thus a warm color may be more important in terms of its ability to generate a feeling of comfort or relaxation, especially if a compartment is well air-conditioned and therefore does not need "cooling off" by means of color.

## Color in Relation to Mood

Most individuals tend to associate various "artificial" colors with their personal experiences with nature's own color schemes. The accompanying table provides a fairly complete list of known color-mood associations as well as other psychological responses in terms of food, odors, etc. Perhaps the most important consideration in selecting a color scheme is avoidance of traditional institutional gray and white. A variety of colors helps enliven the environment and provide relief from the boredom of neutral colors.

## Typical Colors and Their Effect on Humans

| Description | Response |
|---|---|
| Pale pink | Slightly warm feeling, pleasing when associated with odors. |
| Pink | Warm, soft, pleasing when associated with odors and taste. |
| Reddish-pink | Warm feeling, slightly stimulating, pleasing association with taste. |
| Scarlet | Very warm feeling, stimulating and exciting. |
| RED | Warm feeling, stimulating, generally exciting, but calm to the extent of indicating a protective quality, pleasing to the appetite, associated with danger. |
| Reddish-orange | Warm, stimulating, exciting, cheerful, pleasing association with taste. |
| ORANGE | Warm, stimulating, exciting, cheerful, pleasing association with taste. |

Any of the above reflect light that is flattering to skin and complexion, thus making people appear healthy and normal (as opposed to cool colors which create an appearance of being pale). Scarlet through ORANGE colors advance slightly toward the observer.

| | |
|---|---|
| Yellowish-orange | Warm, somewhat exciting, cheerful, express a feeling of comfort, associated with pleasing taste. |

*(continued)*

**Typical Colors and Their Effect on Humans** *(continued)*

| Description | Response |
| --- | --- |
| YELLOW | Warm, somewhat exciting, cheerful, comfortable, associated with pleasing taste. |
| Pale yellow | Warm, cheerful, associated with pleasing odor and taste, softness and comfort. |
| Greenish-yellow | May be associated with feeling of slight warmth or coolness depending on other colors used; not associated with good taste. |
| Yellowish-green | May be associated with feeling of slight warmth or coolness depending on other colors used; calming, somewhat neutral, may or may not be associated with good taste. |
| GREEN | Generally cool, slightly cheerful, comfortable, calming, associated with pleasing refreshing odor. |
| Bluish-green | Cool, calming, associated with good taste. |
| BLUE | Cool, comfort, protective, calming, although may be slightly depressing if other colors are dark, associated with bad taste. |

GREEN through BLUE will advance toward observer if dark shades are used.

| Description | Response |
| --- | --- |
| Pale blue | Cool, soft, calming, tends to neutralize if other colors are pale, reflected light makes skin appear pale. |
| Lavender | Slightly cool, calming, soft, associated with pleasing odor, but with bad taste. |
| Violet | Slightly warm, calming, associated with bad taste. |
| Royal blue | Rich, substantial, may be slightly depressing if used with other dark colors, associated with bad taste. |
| Purple | Rich, protective, calming, may be depressing, associated with pleasing odor, but bad taste. |

Violet through Purple will advance toward observer, creates a feeling of heaviness.

| Description | Response |
| --- | --- |
| Hot pink (a yellowish cast) | Very warm, stimulating, exciting, cheerful. |
| Rose (a bluish cast in pink range) | Neutral relative to warmth, comfortable, calm, associated with pleasing odor. |
| Fluorescent orange | Warm, stimulating, exciting, cheerful, extremely conspicuous. |
| Fluorescent red | Warm, stimulating, exciting, cheerful, extremely conspicuous. |
| Fluorescent yellow | Warm, stimulating, slightly irritating but cheerful. |
| Chartreuse | Slightly warm, cheerful, associated with bad taste. |
| Olive | Warm, comfortable, slightly depressing, associated with bad taste. |
| Cream | Slightly warm, comfortable, calming, clean, reflected light enhances skin tone. |
| Buff | Warm, comfortable, calm, soft; good blend with other colors. |
| Tan | Warm, very comfortable; good blend with other colors. |
| Reddish-brown | Warm, comfortable, cheerful, slightly stimulating. |
| Brown | Warm, comfortable, rich, substantial, protective, may be slightly depressing. |

Fluorescent Orange and Red, Olive, Reddish-brown and Browns advance toward observer.

| Description | Response |
| --- | --- |
| WHITE | Neutral, sterile, clean, fresh, stark, crisp; may appear hard or soft depending on lighting color; may appear harsh, glaring. |
| Off-white | Neutral, clean, fresh. |
| Light gray | Neutral, clean, fresh, calming, soft, comfortable; critical that the tint is compatible with other colors (e.g., bluish, pinkish, yellowish, etc.) |
| GRAY | Neutral, comfortable, calming, slightly hard; critical that the tint is compatible with other colors used (e.g., bluish, pinkish, yellowish, etc.) |
| Dark gray | Neutral, comfortable, may be depressing; critical that the tint is compatible with other colors used (e.g., bluish, pinkish, yellowish, etc.); substantial, heavy, advances toward the observer. |
| Flat black | Solid, heavy, comfortable, generally neutral, advances toward the observer, gives the impression of being dirty, vagueness, recedes. |
| Deep black | Protective, depressing, heavy, substantial, advances toward observer in small amounts, but may recede in large amounts (e.g., painting a ceiling black). |
| Gold/brass | Rich, comfortable, warm (tint important for compatibility with other colors); slightly advancing. |
| Silver/aluminum | Neutral, cold, hard, clean, stiff and uncomfortable, sterile, lifeless, recedes. |
| Light wood grain | Warm, comfortable, quiet. |
| Dark wood grain | Warm, comfortable, quiet, protective, slightly depressing. |
| Light leather grain | Warm, comfortable, cheerful, soft. |
| Dark leather grain | Warm, comfortable, protective, soft, slightly depressing, advancing. |

*Source: U.S. Navy Shipboard Color Coordination Guidance Manual*, NAVSEA 0929-002-7010 (prepared by Man Factors, Inc., 1975).

## PSYCHOLOGICAL RESPONSE TO AMOUNT AND COLOR OF LIGHT

A certain amount of confusion has been generated as a result of conflicting specifications for illumination for good seeing versus desired psychological response. The designer of interiors, in particular, must accommodate consumers in terms of both aspects; i.e., good seeing is essential to ensure adequate visual task performance, but it is also necessary to create a lighting environment that produces a desired psychological response. For example, although a high level of illumination may not be necessary in order for the occupants of a space to perform a visual task, the illumination level may have to be high so that they will "feel" that there is adequate light.

An important aspect of the lighting of interior spaces is the use of interior surfaces that provide adequate reflectance, as indicated in the accompanying table.

**Suggested Reflectance Ranges for Interiors**

| Interior Surface or Element | Percentage Reflectance Desired |
|---|---|
| Ceilings | 60–90%. Surface should be white, or a very pale tint that is compatible with the hue of the interior wall. A minimum of 80% is required for effective visual performance of indirect lighting methods. |
| Walls | 50–85%. Higher reflectance is required above standing waist height when the wall consists of two different reflectance values (i.e., a lower "dado" is used to reduce the expected soiling of lower wall surfaces). |
| Windows or glass walls | Fabric or other drapery provisions should maintain an average reflectance of 15–45% (or more). |
| Furniture | 30–40%. It is important to use higher reflectance on the tops of desks or tables to avoid extreme brightness contrast between the top and reading materials. Matte finishes are required to prevent glare. |
| Floors | 15–35%. Middle to high values are preferred in order to ensure visual detection of possible obstructions on the floor that might cause tripping. |

Although little quantitative evidence exists regarding the psychological effect on general alertness of light levels, the accompanying table indicates suggested requirements.

| Alertness-Level Requirement | Average Light Level, fL |
|---|---|
| Maximum mental alertness required for highly complex mental task performance | 50 and above |
| Medium mental alertness required for routine manual tasks, leisure reading, and/or stimulating social activity | 40 |
| Minimum mental alertness required for nondemanding social intercourse and/or perceptual-motor performance (dining, dressing, personal hygiene, etc.) | 30 |
| Rest, mental alertness for minimal interaction with other people (e.g., as in a bar or private dining environment) | 15 |
| Sleep | Below 3 |

The color of the light source also affects an individual's interpretation of brightness or intensity (see the accompanying graph). If the level is too high (A), surface colors seem to be faded and unnatural; if the level is too low, the colors appear dim and cold (C). Illuminants falling within the area (B) appear normal. The graph suggests that warmer light is more acceptable to most people when the brightness level is low.

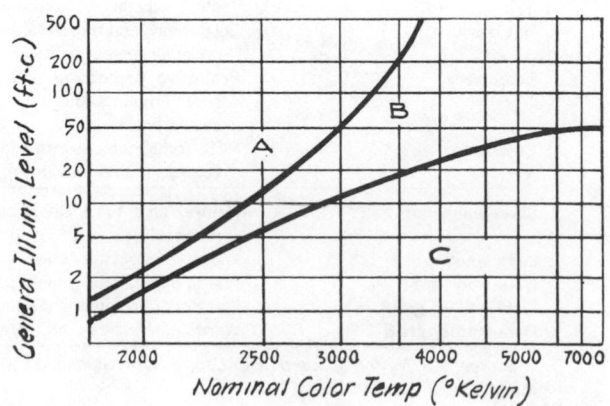

**AUDITORY SYSTEM
CHARACTERISTICS**[42]

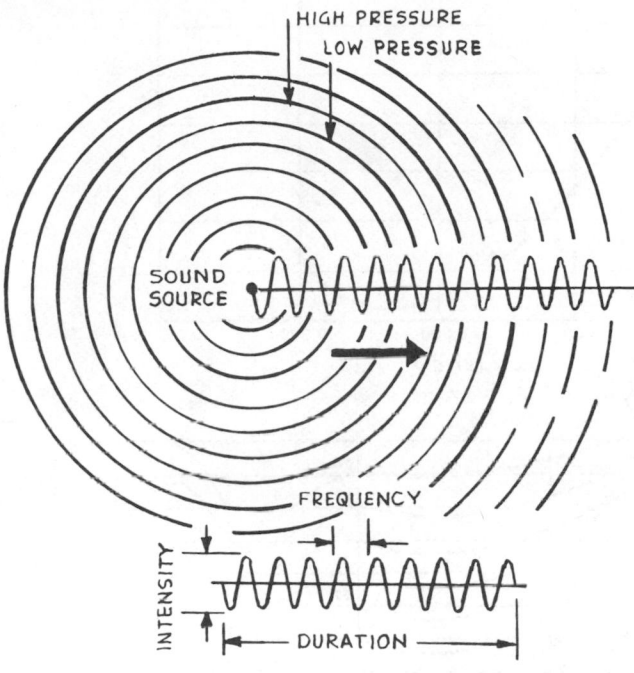

PURE TONE

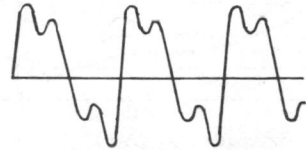

PERIODIC SOUND

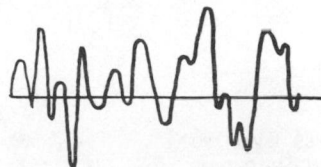

NON-PERIODIC SOUND

WHITE NOISE

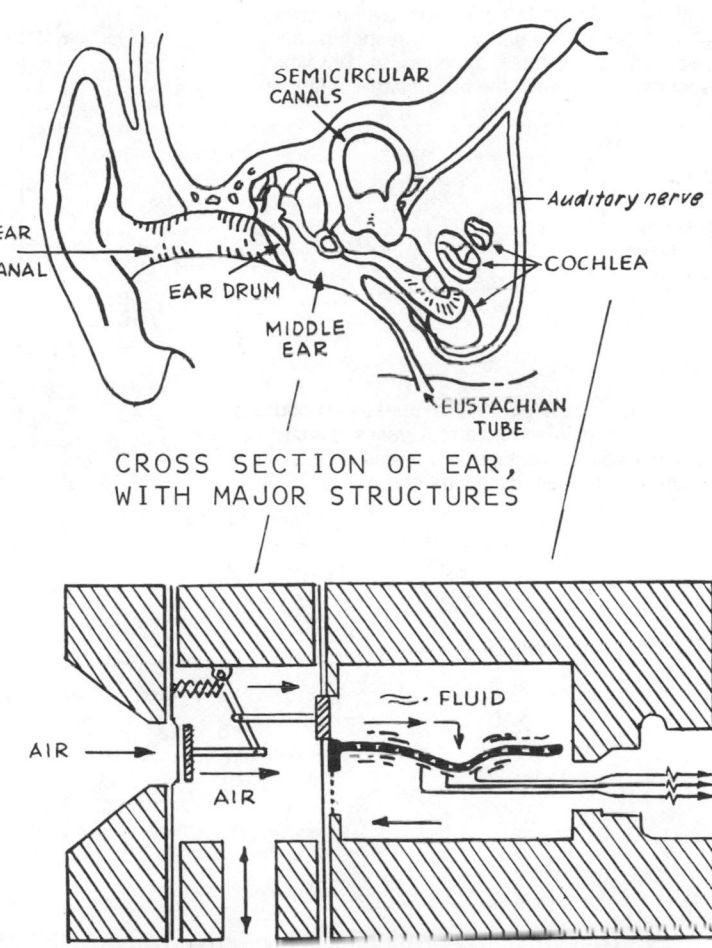

CROSS SECTION OF EAR,
WITH MAJOR STRUCTURES

Functional diagram of ear. Sound waves impinge on outer ear and cause ear drum to vibrate. Vibrations are conducted via lever action of middle-ear bones to oval window, which actuates fluid-filled inner ear. Portions of nerve endings in cochlea are selectively excited, and their outputs terminate in brain.

[42]After W. E. Woodson and D. W. Conover, *Human Engineering Guide for Equipment Designers,* University of California Press, Berkeley, 1964.

## AUDITION PARAMETERS

The accompanying graph shows thresholds for hearing (B) and feeling (A). Generally, hearing is considered a subjective phenomenon, while sound is considered an objective phenomenon. The ear typically responds in a relatively predictable manner to physical sounds, as shown in the accompanying table.

| Sound | Hearing |
|---|---|
| Amplitude | |
| Pressure | Loudness |
| Intensity | |
| Frequency | Pitch |
| Energy distribution | Quality |

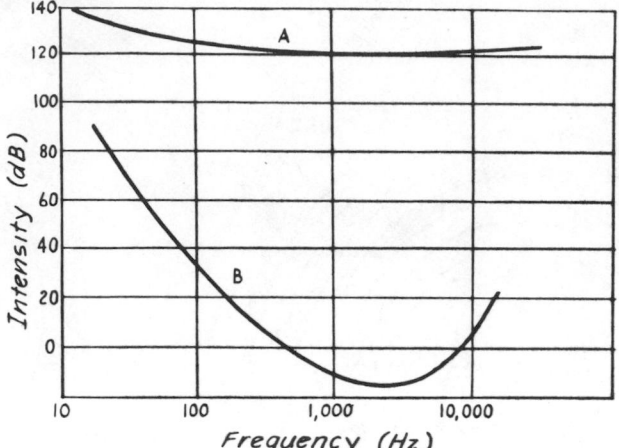

Human audibility curve.

Loudness is the physical response to sound pressure and intensity, and it varies directly with the sound pressure and intensity, as indicated in the accompanying graph.

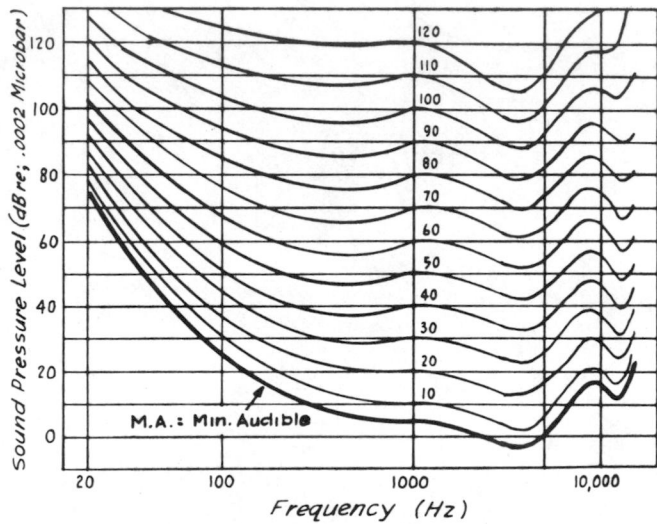

Equal loudness contours.

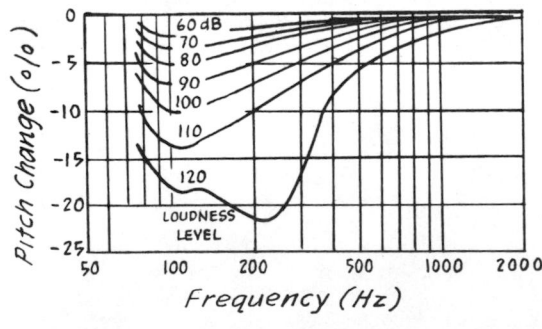

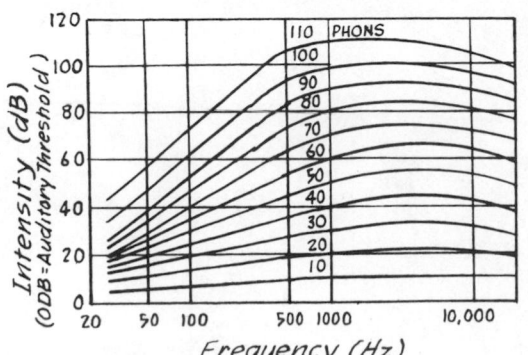

Contours of constant loudness level are shown in the accompanying graph. The curves illustrate the amount by which the pitch of a pure tone is shifted when the loudness level is changed from 40 dB to the level of contour. For example, a 100-Hz tone will seem 10 percent lower in pitch when the loudness level shifts from 40 to 100 dB, but one of 500 Hz will shift only 2 percent lower.

Loudness-level contours showing all pure tones on each curve sound equally loud. Observations with tones that differ considerably from the reference tone, however, are more difficult to make, and the error for the combined result is on the order of 1 to 2 dB.

Loudness as a function of intensity and frequency is illustrated in the upper diagram. Subjective loudness in "sones" is represented vertically above the intensity-frequency plane. The heavier curves progressing from front to rear are equal-loudness contours for pure tones.

Pitch as a function of frequency is illustrated in the lower diagram. The upper curve shows that subjective pitch (in "mels") increases less and less rapidly as the stimulus frequency increases in a linear fashion. The lower curve shows that subjective pitch increases as stimulus frequency is increased logarithmically. Our tempered musical scale is a logarithmic scale. The pitch of a 1000-Hz tone, 40 dB above threshold, is defined as 1000 mels.

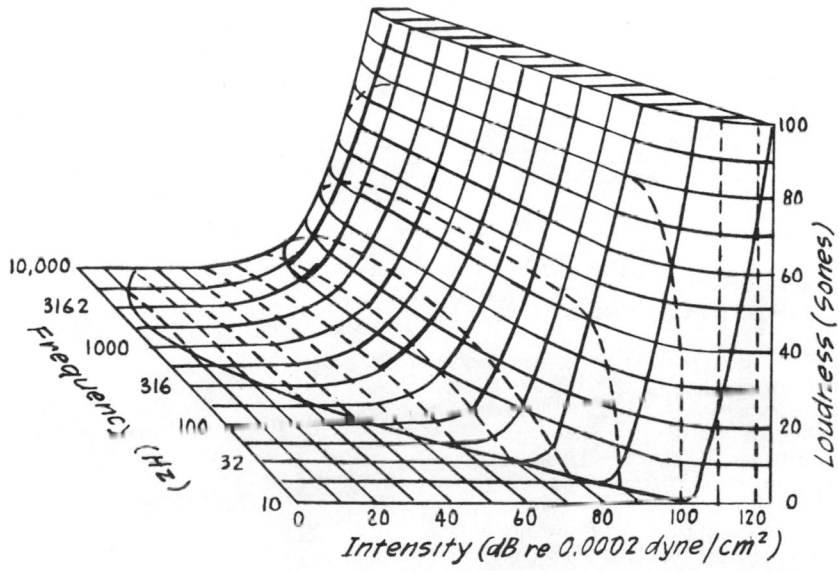

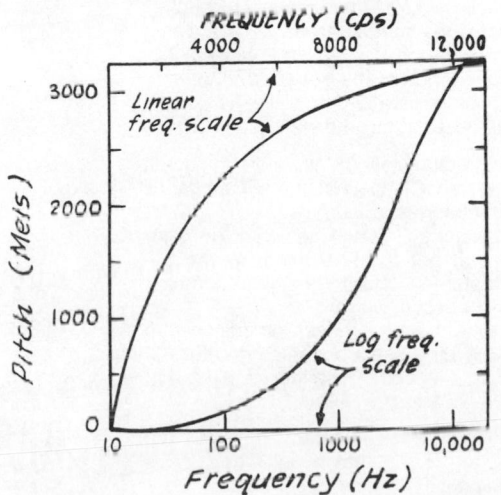

**EFFECT OF FREQUENCY ON DIFFERENTIAL THRESHOLD** $\Delta f/f$ as a function of frequency for five different sensation levels is shown in the accompanying graph. Between 500 and 8000 Hz, $\Delta f/f$ remains relatively constant. Between 62 and 500 Hz, $\Delta f$ is approximately constant. At 45 Hz, $\Delta f/f$ versus $f$ shows an inflection point and the tendency to bend downward, with the tendency being more marked at higher sensation levels (because of subjective harmonics).

**EFFECT OF INTENSITY OF TEST TONES** $\Delta f/f$ as a function of sensation level for seven different frequencies is shown in the accompanying graph. Sensitivity increases as sensation increases, up to 40 dB (0= threshold of hearing at that frequency). Note that the effect is more pronounced for lower tones.

When the loudness level of a 1024-Hz tone (as well as its duration) is varied, accuracy of discrimination is reduced by the decrease in loudness; the effect is most pronounced for short durations.

**SENSITIVITY TO DIFFERENCES IN LOUDNESS: EFFECT OF FREQUENCY** Sensitivity to differences in loudness of pure tones as a function of frequency for seven different intensity levels is shown in the accompanying graph (energy of the wave is the intensity measure). On the basis of these data, it is apparent that the greatest number of loudnesses that are distinguishable at any frequency is 370 at 1300 Hz.

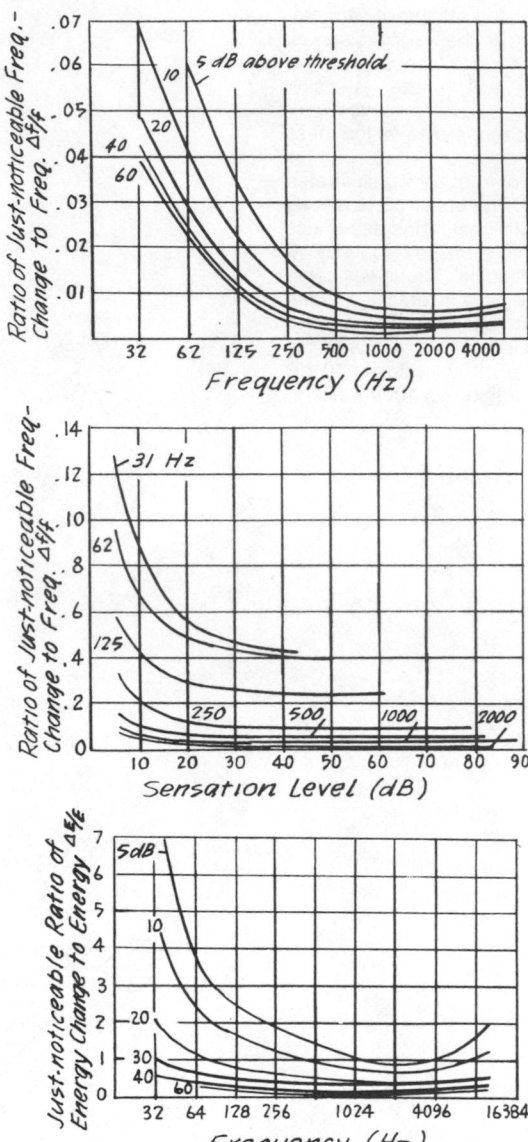

## RESPONSE TO COMBINED TONES

When two tones are combined, even though they are of the same frequency, their loudness is the summation of both tones. However, if the phase of one tone is altered, the same pitch is heard, but with less loudness. If two tones are identical in terms of phase, they cancel each other out, and neither is heard.

Beats will occur (i.e., loudness builds and fades) when two tones that are slightly different in frequency are presented at the same time. Beats occur at a frequency equal to the difference between the two tones. The pitches of the two tones merge and will be heard as a single tone, the pitch of which is halfway between the pitches of the separate tones.

When the separation of tones is varied, beats will occur in three distinct stages:

At close frequencies (6 beats per second) beats will be very distinct and will appear as smooth changes in loudness.

As the frequency interval between tones increases (8 beats per second), beats become faster, and thus the resultant tone appears to throb or pulsate.

The two tones become more prominent as the frequency differential becomes greater (20 beats per second), resulting in a "buzz" rather than a throb.

The accompanying diagram illustrates the several phenomena that occur when tones are combined.

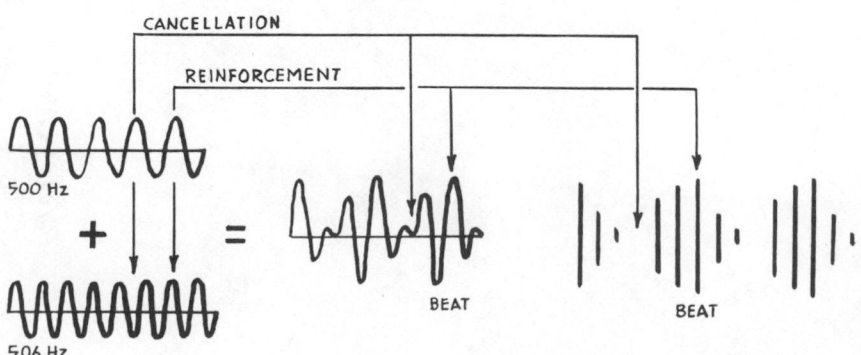

The effect of different intensity interactions upon the experiencing of beats is shown in the first graph. At the upper and lower extremes, only one tone is heard (a strong tone masking out a weaker one). Between 20 and 60 dB, the absolute intensity is unimportant. Beats are most apparent when the two generators are approximately equal in intensity.

## THRESHOLDS FOR SHORT TONES WITH DIFFERENT RATES OF REPETITION

Threshold signal-to-noise ratio for masked thresholds as a function of repetition rate of the sound is shown in the second graph. Differences between curves for different durations are almost exactly equivalent to the differences in total energy produced (difference is independent of repetition rate). Note that the shape of curves is similar regardless of frequency and that threshold decreases continuously as repetition rate increases. Discontinuity between the 2- and 5-per-second rates is more marked, indicating that the ear does little integrating of energy beyond 200-ms duration.

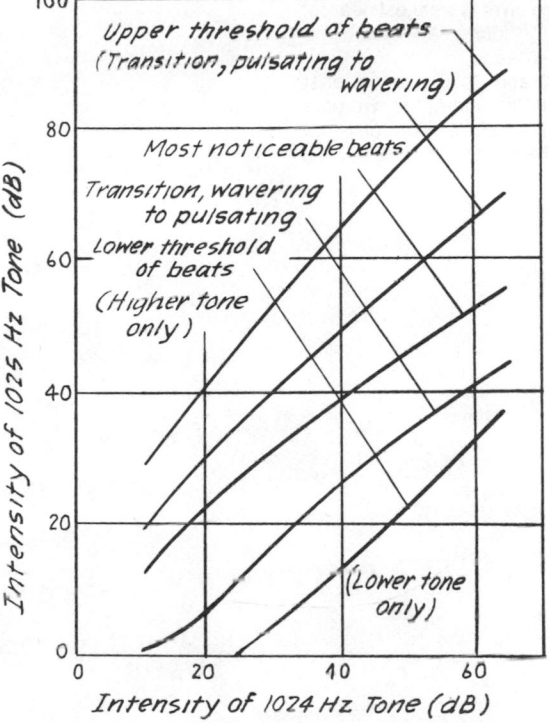

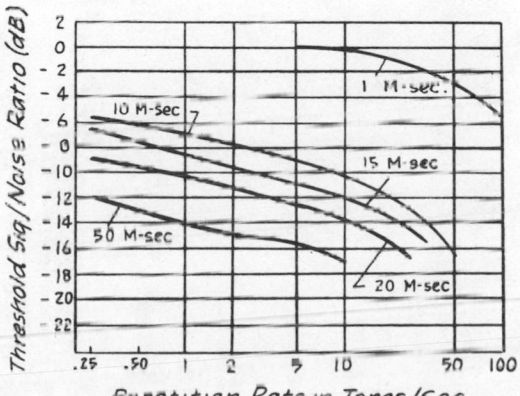

## LOCALIZATION OF SOUND

Listeners can determine the direction from which a sound emanates accurately when the sound source is to their right or left or directly above or below them but less accurately when the sound source is in front of them, to the rear, or at intermediate levels above or below their heads.

It is often disconcerting to listeners if they can see the lips of an orator or the motions of an orchestra but the sound emanates (via loudspeakers) from another direction.

**DEPENDENCE OF LOCALIZATION UPON FREQUENCY** The average error of localization in degrees at various frequencies is illustrated in the graph at the bottom of this page. Accuracy is worst between 2000 and 4000 Hz; otherwise, localization is relatively constant. Note that the average error is smallest nearest the median plane and increases toward the side of the listener's head.

For stimuli of 200 and 2000 Hz, the least amount of intensity difference at the two ears required to produce a shift in localization is 0.9 dB at 200 Hz and 0.2 dB at 2000 Hz.

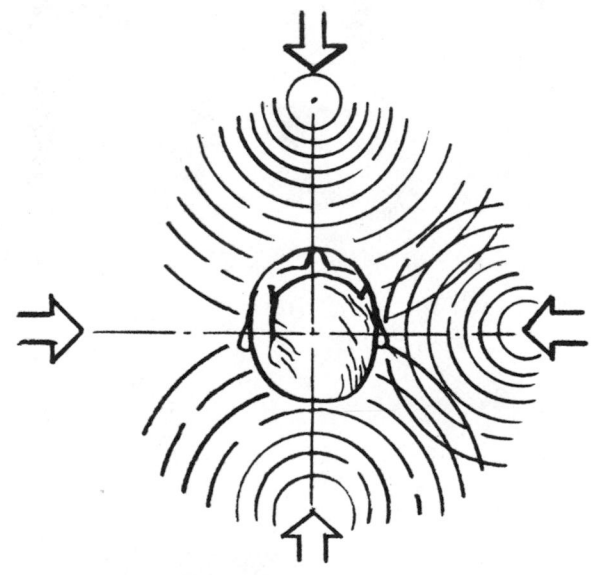

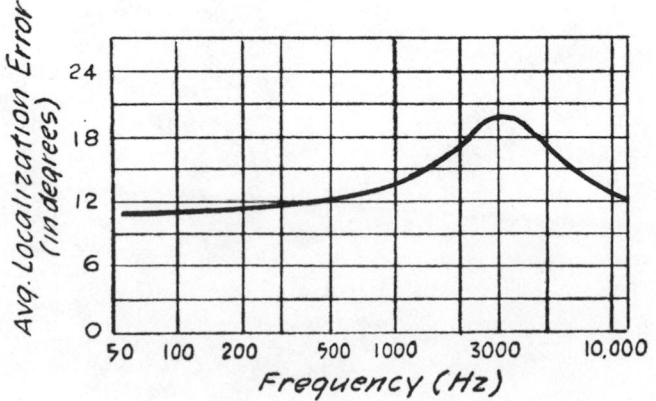

## MASKING

When one sound masks another, intensity and frequency relationships determine the extent of the masking.

The accompanying graph shows the relation between masking produced by a white noise and the effective level of the noise; i.e., the effective level is the amount of noise power in a narrow frequency band (the critical band) centered about the frequency of the masked threshold. The graph indicates that the function is essentially independent of the frequency of the masked sinusoid.

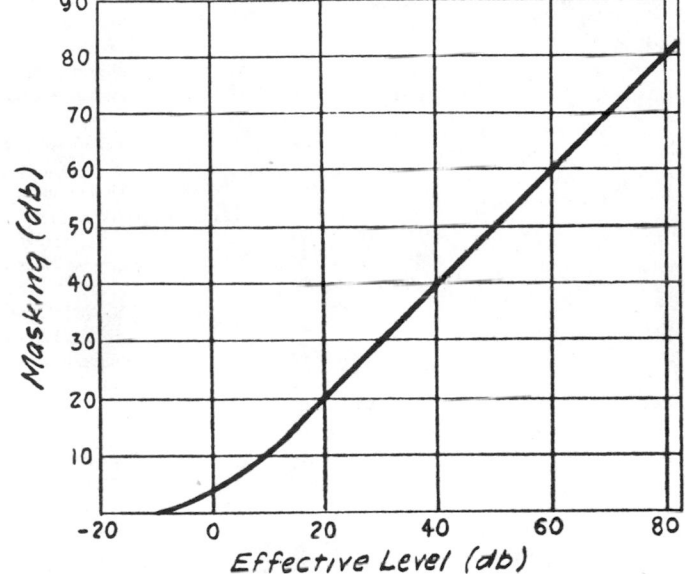

A pure tone more effectively masks a tone higher than itself than one that is lower. Masking is greatest when the masking frequency is close to the "wanted" signal frequency.

Masking increases with increases in the intensity of the masking tone. The accompanying graph illustrates masking at frequencies from 200 to 1200 Hz produced by a 1200 Hz tone.

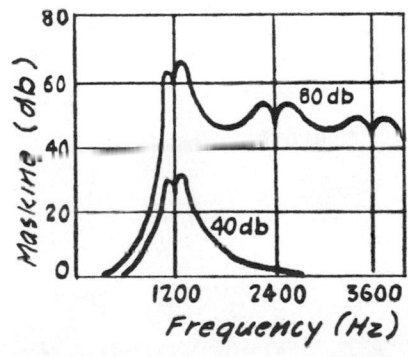

**LOUDNESS THRESHOLD OF A MASKED TONE** Contours for masked threshold (i.e., the sound pressure necessary to make a pure tone just audible in the presence of an interfering sound) at various noise levels are shown in the accompanying graph. Note that thresholds generally parallel one another, with approximately 20-dB intervals between masking noise levels.

Pitch discrimination becomes worse as the signal-to-noise ratio changes from infinity (no noise) to zero. Pitch discrimination is, however, better at some frequencies than others; e.g., 500 Hz is superior to 800 Hz.

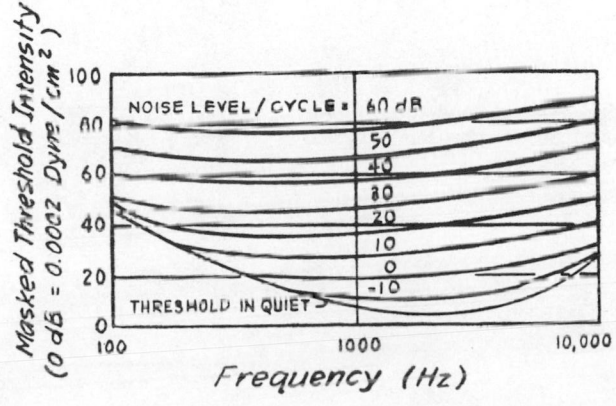

## EFFECTS OF NOISE ON HUMAN PERFORMANCE

The term "noise" refers to "unwanted" sounds. One should consider not only the annoyance aspects of unwanted sounds but also the deleterious effects that noise may have on individual task performance. The accompanying table presents a general analysis of how various noise levels affect humans and their ability to perform typical tasks.

| Noise Level, dB | Effects |
|---|---|
| 100 | Serious reduction in alertness. Attention lapses occur, although attention duration is usually not affected. Temporary hearing loss occurs if no protection is provided in the region 600–1200 Hz. Most people will consider this level unacceptable, and 8 hr is the maximum duration they will accept. |
| 95 | Considered to be the upper acceptance level for occupied areas where people expect the environment to be noisy. Temporary hearing loss often occurs in the range of 300–1200 Hz. Speech will be extremely difficult, and people will be required to shout, even though they may be talking directly into a listener's ear. |
| 90 | At least half of the people in any given group will judge the environment as being too noisy, even though they expected a noisy environment. Some temporary hearing loss in the range of 300–1200 Hz occurs. Skill errors and mental decrements will be frequent. The annoyance factor is high, and certain physiological changes often occur (e.g., the pupils dilate, the blood pressure increases, and the stroke volume of the heart may decrease). Listening to a radio is impossible without good earphones. The maximum duration that most people will accept is 8 hr. |
| 85 | The upper acceptance level (noise expected) in the range of 150–1200 Hz. Some hearing loss occurs in the range of 300–1200 Hz. This is considered the upper comfort level, although some cognitive performance decrement can be expected, especially where decision making is necessary. |
| 80 | Conversation is difficult (i.e., people have to converse in a loud voice less than 1 ft apart). It is difficult to think clearly after about 1 hr. There may be some stomach contraction and an increase in metabolic rate. Strong complaints can be expected from those exposed to this level in confined spaces, and 8 hr is the maximum duration acceptable within the frequency range 1200–4800 Hz. |
| 75 | Too noisy for adequate telephone conversation. A raised voice is required for conversants 2 ft apart. Most people will still judge the environment as being too noisy. |
| 70 | The upper level for normal conversation, even when conversants are close together (at a distance of 6 ft people will have to shout). Although persons such as industrial workers and shipboard personnel who are used to working in a noisy environment will accept this noise level, unprotected telephone conversation will be difficult (upper phone level is 68 dB). |
| 65 | The acceptance level when people expect a generally noisy environment. Intermittent personal conversation is acceptable. About half of the people in a given population will experience difficulty sleeping. |
| 60 | The upper limit for spaces used for dining, social conversation, and sedentary recreational activities. Most people will rate the environment as "good" for general daytime living conditions. |
| 55 | The upper acceptance level for spaces where quiet is expected (150–2400 Hz). People will have to raise their voices slightly to converse over distances greater than 8 ft. This level of noise will awaken about half of a given population about half of the time. It is still annoying to people who are especially sensitive to noise. |
| 50 | Acceptable to most people where quiet is expected. About 25 percent will be awakened or delayed in falling asleep. Normal conversation is possible at distances up to 8 ft. |
| 40 | Very acceptable to all. The recommended upper level for quiet living spaces, although a few people may still have sleep problems. |
| 30 | Necessary for specialized listening tasks (e.g., threshold signal detection). |
| Below 30 | Introduces additional problems; i.e., low-level intermittent sounds become disturbing. Some people have difficulty getting used to the extreme quiet, and a few may become psychologically disturbed. |

*Note:* The above represents an amalgamation of many studies and contains a general interpretation of a wide variety of subject samples and testing conditions.

When noise levels exceed 100 dB, potentially serious consequences occur as shown in the accompanying table.

| Noise Level, dB | Spectrum | Duration | Effects |
|---|---|---|---|
| 105 | Jet engine | 2 min | Reduced visual acuity, stereoscopic acuity, and near-point accommodation and permanent hearing loss when exposure continues over a long period (months) |
| 110 | Machinery noise | 8 hr | Chronic fatigue and digestive disorders |
| 120 | Broadband | 1 hr | Loss of equilibrium |
| 150 | 1–100 Hz | 2 min | Reduced visual acuity, chest-wall vibration, changes in respiratory rhythm, and a "gagging" sensation* |

*Subjects were wearing so-called protective aids to prevent hearing loss.

## GENERAL NOISE EXPOSURE LIMITS

### Deafness Avoidance

| Octave Band, Frequency in Hz | Maximum Permissible Sound Pressure Level, dB | |
| --- | --- | --- |
| | Occasional (1 hr or Less) | Continuous (Period of Months) |
| 38–75 | 125 | 115 |
| 75–150 | 120 | 110 |
| 150–300 | 120 | 110 |
| 300–600 | 120 | 105 |
| 600–1200 | 115 | 100 |
| 1200–2400 | 110 | 95 |
| 2400–4800 | 105 | 90 |
| 4800–9600 | 110 | 95 |

| Exposure Time | With Ears Unprotected | With Earplugs | With Earplugs and Earmuffs |
| --- | --- | --- | --- |
| 8 hr | 100 | 112 | 120 |
| 1 hr | 108 | 120 | 128 |
| 5 min | 120 | 132 | 140 |
| 30 s | 130 | 142 | 150 |

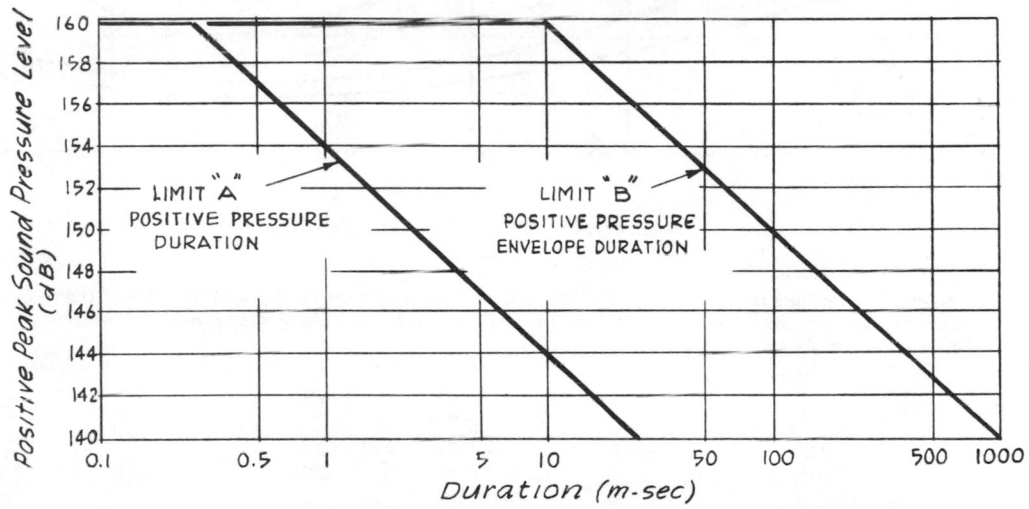

Duration (milliseconds) Impulse noise limits. Positive peak sound pressure level versus positive pressure duration should be below limit A; positive peak sound pressure level versus positive pressure envelope duration should be below level B (HEL Std. 1-63R).

## Office of Safety and Health Administration (OSHA) Noise Standards

### Occupational Safety and Health Act Permissible Daily Noise Exposure

| Duration, hr | Sound Level, dBA |
| --- | --- |
| 8 | 90 |
| 6 | 92 |
| 4 | 95 |
| 3 | 97 |
| 2 | 100 |
| 1.5 | 102 |
| 1 | 105 |
| 0.5 | 110 |
| 0.25 | 115 |

Note: When exposure is intermittent at different levels, the fraction $C_1/T_1 + C_2/T_2 \ldots C_n/T_n$ should not exceed unity to meet the exposure limit. $C_n$ = total exposure time at the specified level; $T_n$ = total exposure time permitted at the specified level.

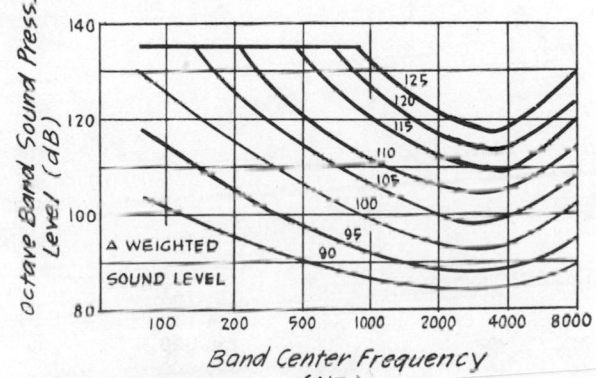

Contours for determining equivalent A-weighted sound level. This graph is used in interpreting octave-band sound levels according to OSHA provisions.

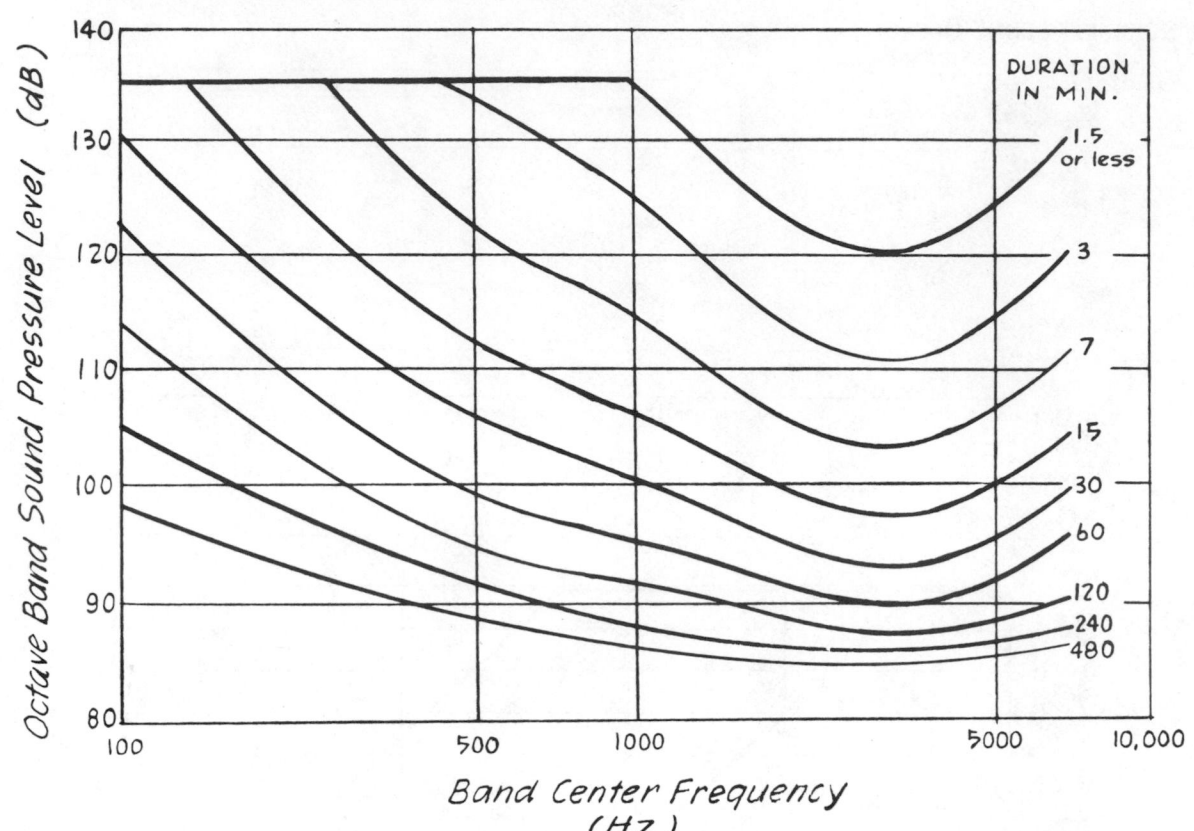

Damage risk contours for one exposure per day (left ordinate) and/or one-third octave or narrower (right ordinate) bands of noise. (From Committee on Hearing, Bioacoustics and Biomechanics, National Academy of Science, 1964.)

**Speech Interference Limits**

| Distance, ft | Acoustic Absorption of Area, Sabines* | Speech Level, dB | | | |
|---|---|---|---|---|---|
| | | Shout | Very Loud | Raised Voice | Normal |
| 0.5 | Any value | 90 | 84 | 78 | 72 |
| 1 | Any value | 84 | 78 | 72 | 66 |
| 2 | Any value | 78 | 72 | 66 | 60 |
| 4 | Below 100 | 78 | 72 | 66 | 60 |
| | Above 100 | 72 | 66 | 60 | 54 |
| 8 or more | Below 100 | 78 | 72 | 66 | 60 |
| | 100–400 | 72 | 66 | 60 | 54 |
| | 400–1600 | 66 | 60 | 54 | 48 |

*The sabine is a unit of absorption equal to the absorption of 1 ft² of surface that is totally sound absorbent. As more absorption is present, the less interference can be tolerated because speech energy is partially absorbed. The above table is for the male voice. For the female voice, decrease values 5 db.

**Comfort Limits**

| Octave Band, Frequency in Hz | Comfort Criteria, dB | |
|---|---|---|
| | Normally Noisier Environment | Normally Quiet Environment |
| 38–75 | 100 | 80 |
| 75–150 | 95 | 70 |
| 150–300 | 90 | 60 |
| 300–600 | 85 | 55 |
| 600–1200 | 75 | 50 |
| 1200–2400 | 65 | 50 |
| 2400–4800 | 60 | 50 |
| 4800–9600 | 55 | 45 |

**Speech Intensity Characteristics Relative to Establishment of Speech Interference Noise Limits**

| Description | Typical Intensity of Source, dB | Distance | Message Characteristic |
|---|---|---|---|
| Soft whisper | | 3–6 in | Secret communication |
| Audible whisper | 44–69 | 8–20 in | Confidential or personal |
| Normal voice | 50–75 | 20–60 in | Personal or small group |
| Loud voice | 56–81 | 5–8 ft | Nonpersonal, intermediate group |
| Very loud voice | 62–87 | To 20 ft | Group address |
| Shouting | 68–93 | Upper limit | Hailing or emergency communication |

**Subjective Effects of Changes in Sound Level**

| Sound-Level Change, dB | Change in Apparent Loudness |
|---|---|
| 3 | Barely perceptible |
| 5 | Clearly noticeable |
| 10 | Either twice as loud or twice as soft |
| 20 | Much louder or softer |

**Upper Noise-Level Limits Recommended for Military Facilities***

| Sound Level, dBA | Type of Activity | Communication Equivalent | Office Application |
|---|---|---|---|
| 108 | Maximum design limit for AMC equipment (hearing protection required) | No direct communication | Not recommended |
| 100 | Armored vehicles (hearing protection required) | Electrically aided communication satisfactory with attenuating helmet or headset; limited shouted communication possible with difficulty | Not recommended |
| 90 | Materiel which is beyond the state of the art of meeting 85 dBA (hearing protection required) | Shouted communication possible at short distances (1–2 ft) | Not recommended |
| 85 | Acceptable level for unprotected hearing for 8-hr exposures | Shouted communication possible at several feet (3–4 ft); telephone use difficult | Not recommended |
| 75 | Maintenance shops, garages, and keypunch areas | Occasional telephone use and occasional direct communication at up to 5 ft is acceptable | Not recommended |
| 65 | Operation centers, mobile command and communication centers, computer rooms, word processing centers, kitchens, and laundries | Frequent telephone use and frequent direct communication at up to 5 ft is acceptable | Business machine offices |
| 55 | Drafting rooms, laboratories, and conferences with two or three people | No difficulty with telephone use and occasional direct communication at up to 15 ft | Shop offices and general secretarial areas |
| 45 | Libraries, conference rooms, command and control centers, theatres, and sleeping areas | No difficulty with direct communication | General offices |
| 35 | Recording studios and large conference rooms | Areas requiring unusually extreme quiet | Executive offices |

*These recommended levels are in consonance with the requirements of MIL-STD-1472 and MIL-STD-1474.

**Steady-State Noise-Level Limits for Person-to-Person Communications (U.S. Army Standards)**

| Octave Band Limits, Hz | | Center Frequency | | Noise Level, dB | |
|---|---|---|---|---|---|
| A | B | A | B | A | B |
| 37.5– 75 | 44– 87 | 53 | 63 | 79 | 77 |
| 75 – 150 | 87– 175 | 106 | 125 | 73 | 72 |
| 150 – 300 | 175– 350 | 212 | 250 | 68 | 67 |
| 300 – 600 | 350– 700 | 425 | 500 | 64 | 63 |
| 600 –1200 | 700– 1400 | 850 | 1000 | 62 | 61 |
| 1200 –2400 | 1400– 2800 | 1700 | 2000 | 60 | 59 |
| 2400 –4800 | 2800– 5600 | 3400 | 4000 | 58 | 58 |
| 4800 –9600 | 5600–11200 | 6800 | 8000 | 57 | 57 |

*Note:* A—non-electrically aided (commercial frequencies ASA Z24.10-1953); B—non-electrically aided (preferred frequencies ASA S1.6-1960).

*Source:* Human Engineering Laboratory HEL Standard S-1-63B.

### Commercial Work-Space Noise Criteria[43]

Noise criteria curves (NC) are widely used as work-space design criteria (see the accompanying graph). The accompanying table identifies typical work spaces with the appropriate NC curves that should be applied.

NC curves are referenced to preferred octave bands (lower abscissa) and to commercial bands (upper abscissa). When using the commercial frequency, the NC number is also the speech interference level (SIL) for that particular spectrum.

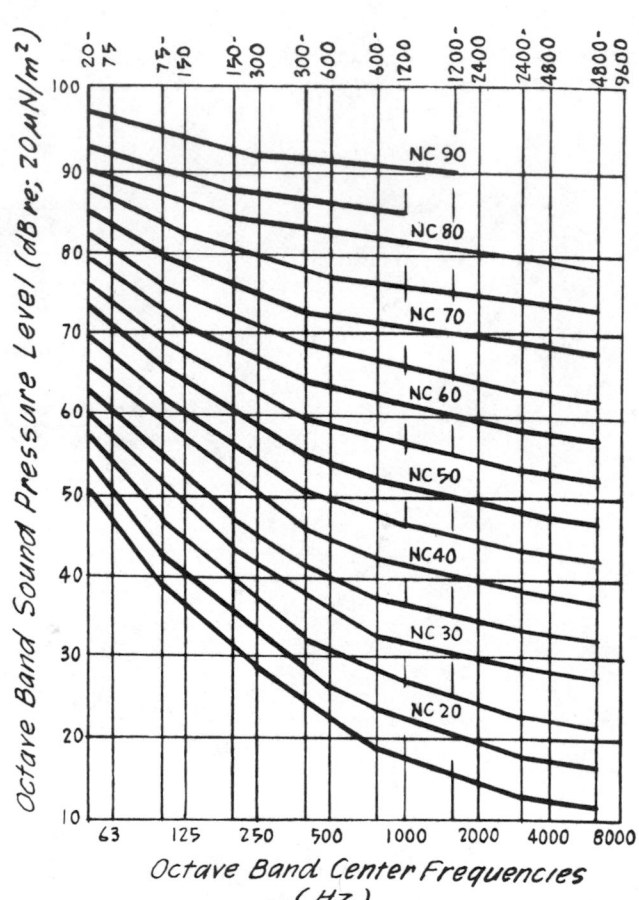

**Recommended NC Curves for Various Work Spaces**

| NC Curve | Type of Work Space | Communication Equivalent | Office Application |
|---|---|---|---|
| 90 | | Noise-attenuating headset required | Not recommended |
| 80 | | Communication very difficult; telephone use unsatisfactory | Not recommended |
| 70–80 | | Raised voice range 1–2 ft; shouting range 3–6 ft; telephone use very difficult | Not recommended |
| 60–70 | | Raised voice range 1–2 ft; telephone use difficult | Not recommended |
| 55–60 | | Very noisy, not suited for office; telephone use difficult | Not recommended |

*(continued)*

[43]D. C. Hodge, and G. R. Garinther, in J. F. Parker and V. R. West (eds.), *Bioastronautics Data Book*, 2d ed., National Aeronautics and Space Administration, 1973, pp. 693–750.

**Recommended NC Curves for Various Work Spaces** *(continued)*

| NC Curve | Type of Work Space | Communication Equivalent | Office Application |
|---|---|---|---|
| 55 | Spacecraft during nonpowered flight | | |
| 50–55 | | Unsatisfactory for conferences of over 3 people; telephone use slightly difficult; normal voice at 2 ft, raised voice at 3 ft | Areas with typists and accounting machines |
| 40–50 | Restaurants, sports coliseums | Conferences at 4- to 5-ft table; telephone use slightly difficult; normal voice at 3–6 ft; raised voice at 6–12 ft | Large drafting rooms |
| 35–40 | | Conferences at 6- to 8-ft table; telephone use satisfactory; normal voice at 6–12 ft | Medium-sized offices |
| 30–35 | Libraries, hospitals, motion picture theatres, home sleeping areas, assembly halls | Quiet office; conferences at 15-ft table; normal voice at 10–30 ft | Private or semiprivate offices; reception rooms; conference rooms for up to 20 people |
| 25–30 | Courtrooms, churches, home sleeping areas, assembly halls, hotels and apartments, TV studios, music rooms, schoolrooms | Very quiet offices; large conferences | Executive offices; conference rooms for 50 people |
| 20–25 | Legitimate theatre, concert halls, broadcasting studios | | |

## REVERBERATION

Reverberating sounds tend to mix with direct sounds, thus affecting the clarity with which the sounds are perceived by the listener.

Reverberation time in relation to speech intelligibility is illustrated in the accompanying graph.

Reverberation time in relation to room size for both speech and music sounds is illustrated in the accompanying graph.

Suggested reverberation objectives for specific listening requirements are noted in the accompanying table.

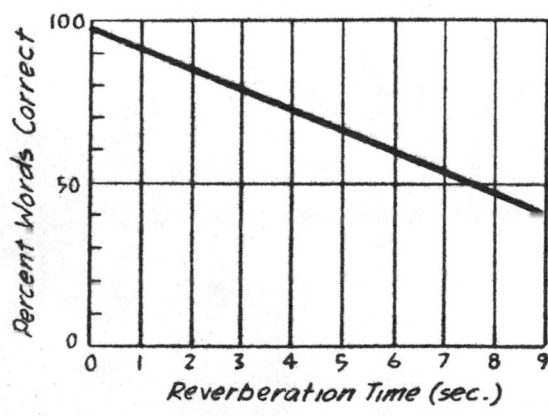

| Listening Requirement | Reverberation Times, s |
|---|---|
| Optimum for speech (too dead for music) | Below 1.0 |
| Good for speech, fair for music | 1.0–1.5 |
| Fair for speech, good for music | 1.5–2.0 |
| Poor for speech, good for liturgical music | 2.0 + |

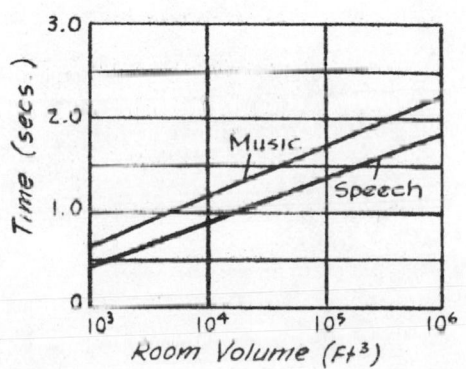

## ULTRASONIC, LOW-FREQUENCY, AND INFRASONIC SOUND

### Ultrasonic Noise Limits

Sound pressure level (SPL) should not exceed 75 dB in one-third-octave bands centered at 8 to 16 kHz or at 20 to 31.5 kHz in order to preclude unpleasant subjective reactions and prevent TTS.

### Low-Frequency and Infrasonic Noise Limits

To prevent injury in the low-frequency and infrasonic noise range (1 to 100 Hz), the limits noted in the accompanying table should not be exceeded.

| Frequency, Hz | SPL, dB | Duration, min/day | |
|---|---|---|---|
| 1–7 | 150 | 4 | Should use earplugs to reduce unpleasant sensations |
| 8–11 | 145 | 4 | Should use earplugs to reduce unpleasant sensations |
| 12–20 | 140 | 4 | Should use earplugs to reduce unpleasant sensations |
| 21–100 | 135 | 20 | Without earplugs |
| 21–100 | 150 | 20 | With earplugs |

*Note:* Refers to pure tones or octave bands with center frequencies as indicated. "Duration" refers to one exposure per day with a 24-hr gap between successive exposures.

### Selected References: Noise Control Standards

*Acoustics: Assessment of Occupational Noise Exposure for Hearing Conservation Purposes,* Recommendation R 1999, International Standards Organization, Geneva, Switzerland, 1971.
*Criteria for a Recommended Standard—Oc-cupational Exposure to Noise,* Report HSM 73-11—1, National Institute for Occupational Safety and Health, Cincinnati, 1972.
*Hazardous Noise Exposure,* Air Force Regulation 161-35, Department of the Air Force, 1973.
*Hearing Conservation Program,* BUMEDINST 6260.6B, U.S. Navy, Bureau of Medicine and Surgery, 1970.
*Information on Levels of Environmental Noise Requisite to Protect Public Health and Welfare with Adequate Margin of Safety,* Report 550/9-74-004, Environmental Protection Agency, Office of Noise Abatement and Control, 1974.
*Noise and Conservation of Hearing,* TB MED 251, The Surgeon General's Office, 1972.
*Noise Limits for Army Material,* MIL-STD-1474A(MI), U.S. Department of Defense, 1975.

## RESPONSE TO TACTILE PRESSURE

Stimulus thresholds for pressure on significant regions of the body are as shown in the accompanying table.

| Region | Pressure, g/mm² |
|---|---|
| Tip of the finger | 3 |
| Back of the finger | 5 |
| Front of the forearm | 8 |
| Back of the hand | 12 |
| Abdomen | 26 |
| Back of the forearm | 33 |
| Thick parts of the sole of the foot | 250 |

Relative accuracy of judging location and spatial separation and number of pressure stimuli on the volar surface of the forearm (2-s interval between successive stimulations) are as shown in the accompanying table.

| Axis | Point Separation, mm | |
|---|---|---|
| | Successive Points | Simultaneous Points |
| Longitudinal | 9.16–9.64 | 12.67–15.73 |
| Transverse | 9.40–9.55 | 14.60–15.02 |

*Note:* Weight of probe = 20 g.

### Amount of Pressure Relative to Accuracy of Judging Location

| Region | Weight | | | | |
|---|---|---|---|---|---|
| | 12 g | 27 g | 40 g | 57.5 g | 67.5 g |
| Back of hand | | | | | |
| Mean | 7.21 | 6.93 | 6.86 | 6.69 | 6.76 |
| SD | 4.7 | 4.6 | 4.4 | 4.3 | 4.4 |
| Volar surface of forearm: | | | | | |
| Mean | 7.19 | 7.13 | 6.14 | 6.75 | 6.59 |
| SD | 4.5 | 4.5 | 3.7 | 4.1 | 3.8 |

*Note:* Tactile adaptation time tends to occur and is longer for heavier pressures. People generally improve with practice.

## COMFORT CONSIDERATIONS

A different kind of tactile response, and one that is often ignored, involves the pressure of seat-belt webbing on the user's body. The most critical complaints concern too much pressure on the occupant's shoulder or chest. As is illustrated in the accompanying sketch, a typical torso belt crosses the chest at a diagonal, and although it should approximately bisect the chest near the sternum, it often is poorly laid out; as a result, the belt rides across the individual's breast (which is especially annoying to females). A recent study indicates that the pressure of a torso belt on the occupant's chest or shoulder should not exceed about 0.9 lb (4 N); otherwise, at least 50 percent of typical automobile passengers will complain. (Note that this is not the force coming from a retractor, but the pressure, or force, measured at the point where the webbing presses against the occupant's chest or shoulder.)

# LINEAR ACCELERATION[44]

## Forward Acceleration Effects ($+G_x$)

2–3 $G_x$: Increased weight and abdominal pressure; progressive slight difficulty in focusing and slight spatial disorientation, each subsiding with experience; 2 $G_x$ tolerable at least to 24 hr, 4 $G_x$ up to at least 60 min.

3–6 $G_x$: Progressive tightness in chest (6 $G_x$, 5 min), chest pain, loss of peripheral vision, difficulty in breathing and speaking, blurring of vision, effort required to maintain focus.

6–9 $G_x$: Increased chest pain and pressure; breathing difficult, with shallow respiration from position of nearly full inspiration; further reduction in peripheral vision, increased blurring, occasional tunneling, great concentration to maintain focus; occasional lacrimation; body, legs, and arms cannot be lifted at 8 $G_x$; head cannot be lifted at 9 $G_x$.

9–12 $G_x$: Breathing difficulty severe; increased chest pain; marked fatigue; loss of peripheral vision, diminution of central acuity, lacrimation.

15 $G_x$: Extreme difficulty in breathing and speaking; severe vise-like chest pain; loss of tactile sensation; recurrent complete loss of vision.

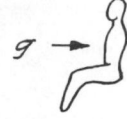

## Backward Acceleration Effects ($-G_x$)

Similar to those of $+G_x$ acceleration with modifications produced by reversal of force vector. Chest pressure reversed, hence breathing easier; pain and discomfort from outward pressure toward restraint harness manifest at about $-8$ $G_x$; with forward head tilt cerebral hemodynamic effects manifest akin to $-G_z$; distortion of vision at $-6$ to $-8$ $G_x$; feeling of insecurity from pressure against restraint.

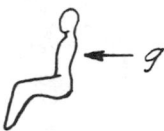

## Lateral Acceleration Effects ($\pm G_y$) (Little Information Available)

3 $G_y$: Discomfort after 10 s; pressure on restraint system, feeling of supporting entire weight on clavicle; inertial movement of hips and legs, yawing and rotation of head toward shoulder; petechiae and bruising; engorgement of dependent elbow with pain.

5 $G_y$ (14.5 s): External hemorrhage, severe postrun headache.

## Positive Acceleration Effects ($+G_z$)

1 $G_z$: Equivalent to the erect or seated terrestrial posture.

2 $G_z$: Increase in weight, increased pressure on buttocks, drooping of face, and soft body tissues.

2-½ $G_z$: Difficult to raise oneself.

3–4 $G_z$: Impossible to raise oneself, difficult to raise arms and legs, movement at right angles impossible; progressive dimming of vision after 3 to 4 s, progressing to tunneling of vision.

4-½ to 6 $G_z$: Diminution of vision, progressive to blackout after about 5 s; hearing and

then consciousness lost if exposure continued; mild to severe convulsions in about 50 percent of subjects during or following unconsciousness, frequently with bizarre dreams; occasionally paresthesias, confused states and, rarely, gustatory sensations; no incontinence; pain not common, but tension and congestion of lower limbs with cramps and tingling; inspiration difficult; loss of orientation for time and space up to 15 s postacceleration.

## Negative Acceleration Effects ($-G_z$)

$-1$ $G_z$: Unpleasant but tolerable facial suffusion and congestion.

$-2$ to $-3$ $G_z$: Severe facial congestion, throbbing headache; progressive blurring, graying, or occasionally reddening of vision after 5 s; congestion disappears slowly, may leave petechial hemorrhages, edematous eyelids.

$-5$ $G_z$: Five seconds, limit of tolerance, rarely reached by most subjects.

## Body Movement versus Acceleration

$+2$ $G_z$: Walking and movement along a ladder against acceleration very difficult.

$+3$ $G_z$: Walking, crawling, and movement along a ladder against acceleration impossible; unaided escape from vehicle impossible; parachute donning time increased from 17 to 75 s.

$+4$ $G_z$: Movement at right angles to acceleration vector impossible.

$+5$ $G_x$: Difficult to hold feet forward on rudder pedals.

$+6$ to $+7$ $G_z$: Extremely difficult to reach face curtain ejection seat firing mechanisms (Christie, 1961).

$+8$ $G_x$: Arms, legs, and body cannot be lifted.

$+9$ $G_x$: Unsupported head cannot be lifted although use of counterweighted headgear permits motion up to $+12$ $G_x$.

$+25$ $G_x$: Hand and wrist movement still possible (Collins et al., 1958).

[44]T. M. Fraser, in J. F. Parker and V. R. West (eds.), *Bioastronautics Data Book*, 2d ed., National Aeronautics and Space Administration, 1973, pp. 149–219.

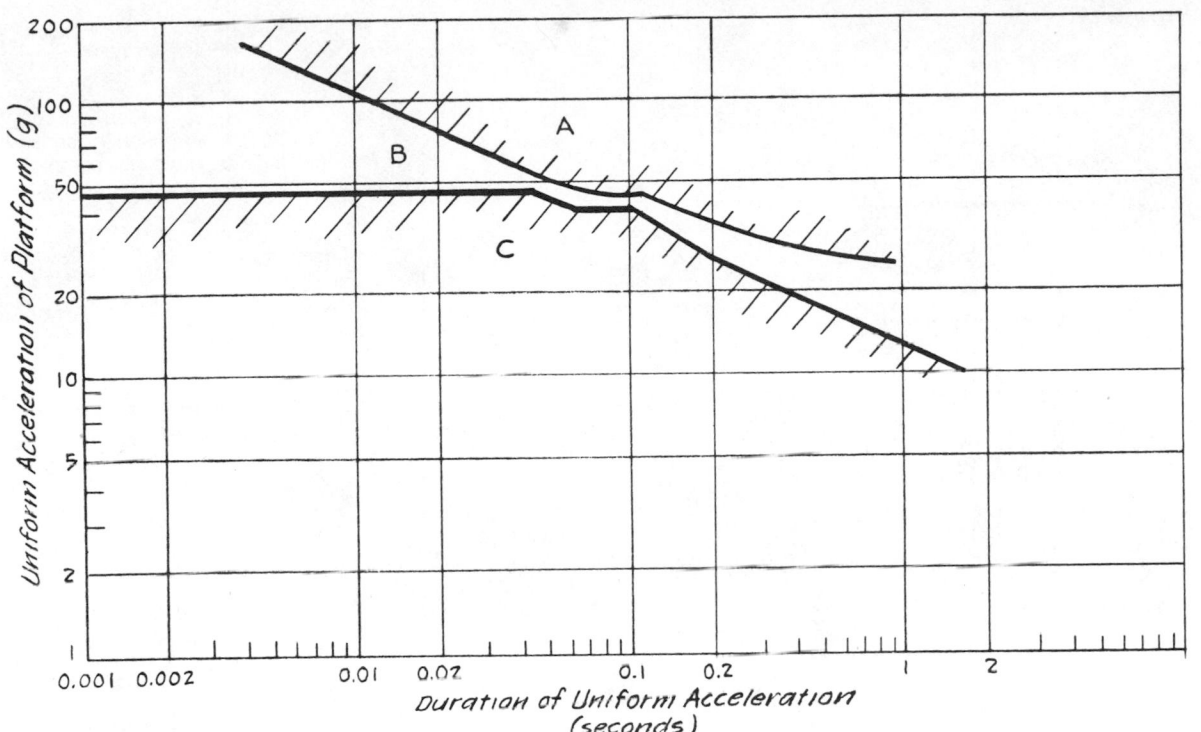

Maximum tolerance limits to transverse acceleration. A—severe injury; B—moderate injury; C—voluntary exposure (uninjured or debilitated). (After A. Damon et al., *The Human Body in Equipment Design,* Harvard University Press, Cambridge, Mass., 1966, p. 267.)

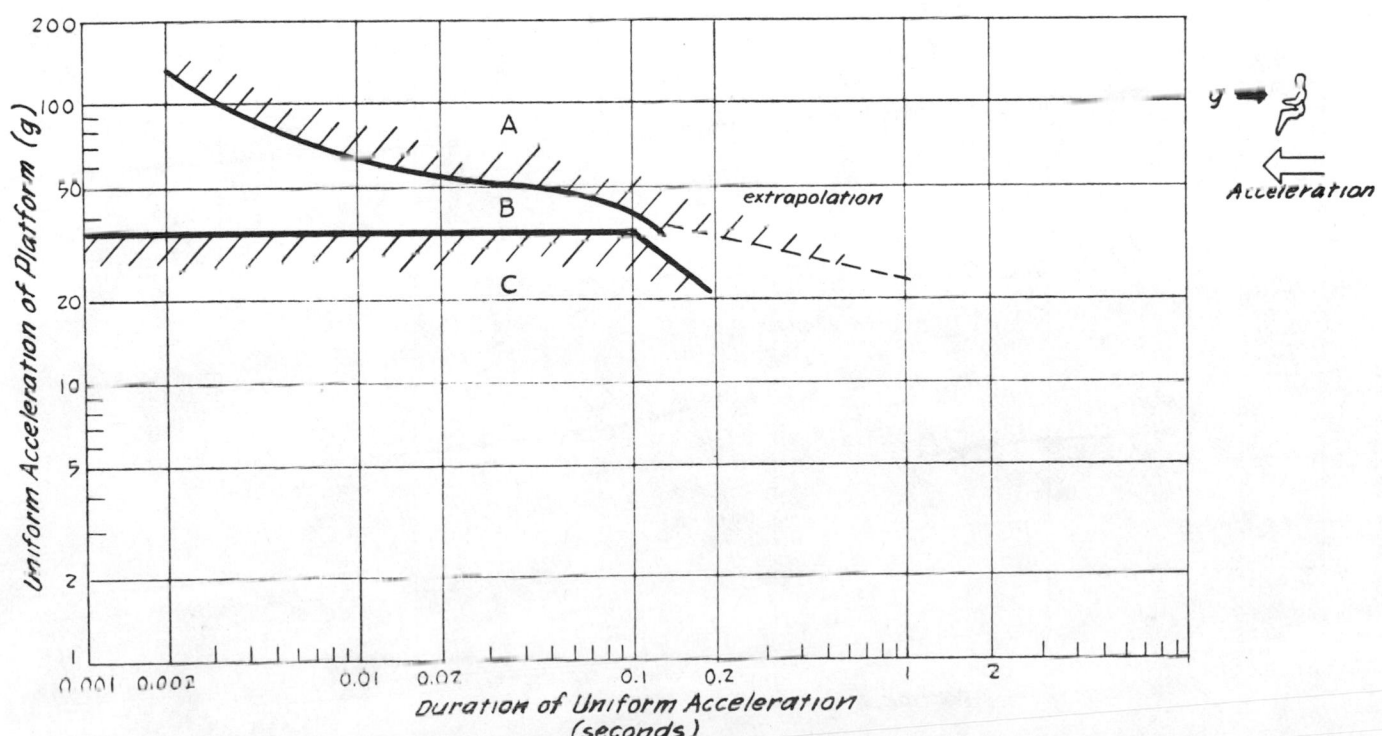

Maximum tolerance limits to transverse acceleration. A—severe injury; B—moderate injury; C—voluntary exposure (uninjured or debilitated). (After A. Damon et al., *The Human Body in Equipment Design,* Harvard University Press, Cambridge, Mass., 1966, p. 268.)

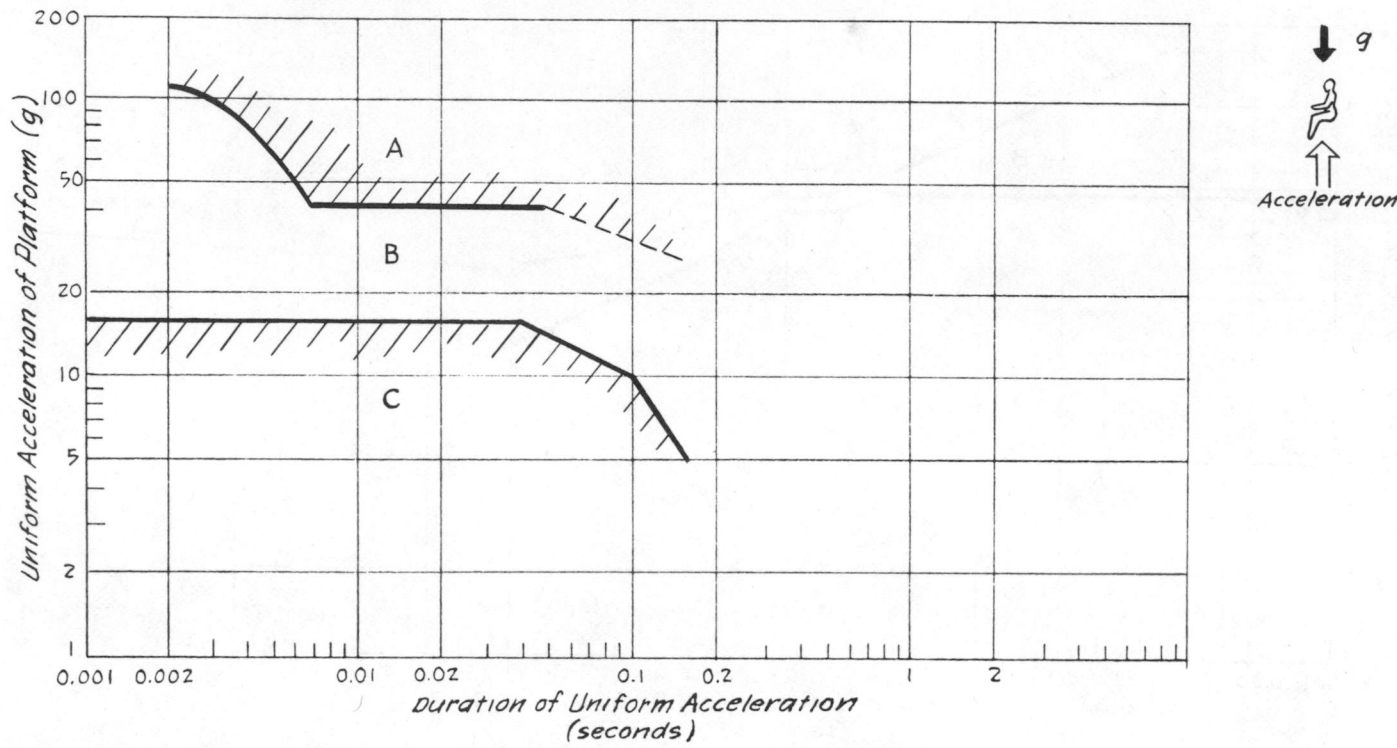

Maximum tolerance limits to positive acceleration. A—severe injury; B—moderate injury; C—voluntary exposure (uninjured or debilitated). (After A. Damon et al., *The Human Body in Equipment Design,* Harvard University Press, Cambridge, Mass., 1966, p. 269.)

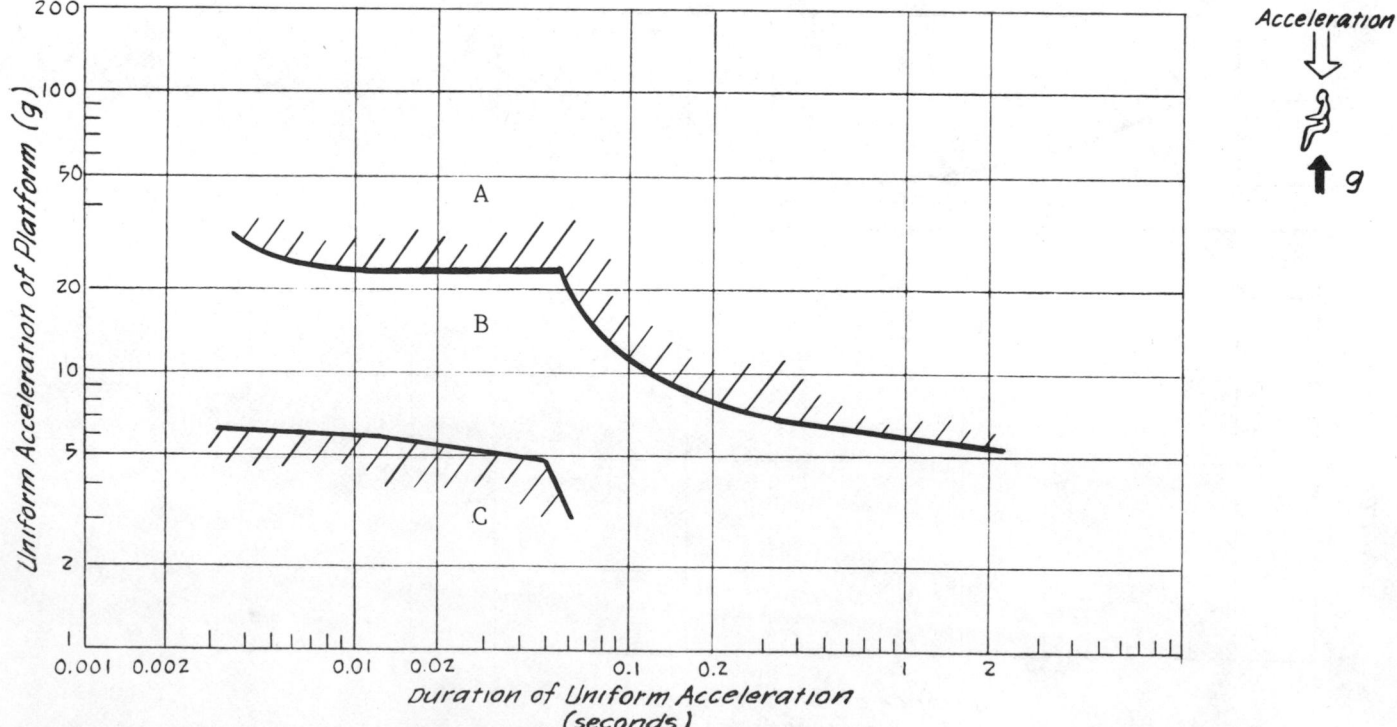

Maximum tolerance limits to negative acceleration. A—severe injury; B—moderate injury; C—voluntary exposure (uninjured or debilitated). (After A. Damon et al., *The Human Body in Equipment Design,* Harvard University Press, Cambridge, Mass., 1966, p. 270.)

## ROTARY ACCELERATION[45]

Most people can tolerate rotation rates up to 6 r/min in any axis or combination of axes, but they cannot tolerate rates in the region of 12 to 30 r/min and rapidly become disoriented and nauseated above 6 r/min unless they have been prepared by gradual rate increases.

However, rotation rates of 60 r/min for up to 3 to 4 s around the y axis (pitch) and around the z axis (spin) not only are tolerated but also elicit responses of pleasantness. The rate becomes intolerable again at around 80 r/min in the pitch mode and at about 90 r/min in the spin mode. There are, of course, exceptional people, such as professional skaters, who may spin for as long as 12 s at about 420 r/min.

Long-duration exposures to the pitch mode have been endured for as long as 60 min at 6 r/min by selected subjects.

The center of rotation within the body determines the nature of the resulting effects. It has been estimated that unconsciousness from circulatory effects probably occurs in humans after 3 to 10 s at 160 r/min when the center of rotation is about the heart and at 180 r/min when the center of rotation is about the iliac crest.

## Tolerance to Tumbling[46]

Theoretical limits for human tolerance to tumbling, based on animal work and limited human experience. The two curves at left describe safe and unsafe regions and damages likely to be incurred in simple tumbling about two axes of the body. The right-hand graph estimates tumbling tolerances for the specific case of tumbling combined with deceleration from 35 to 15 g.

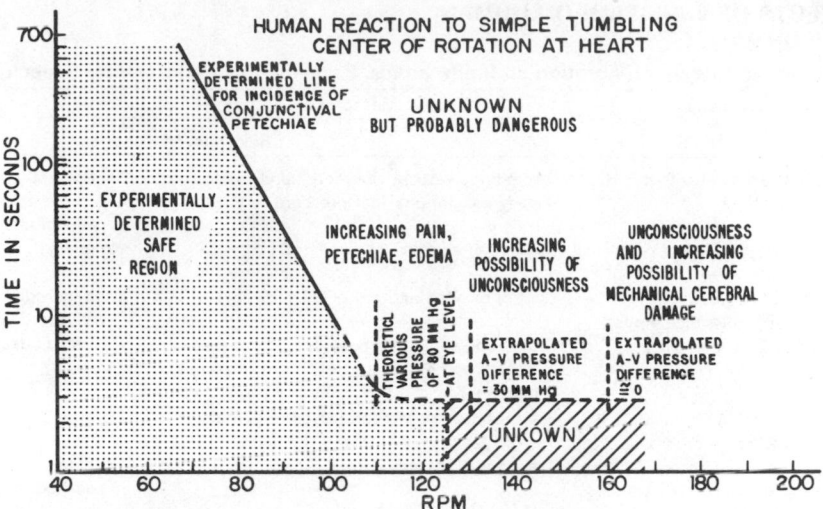

Human reaction to simple tumbling center of rotation at heart.

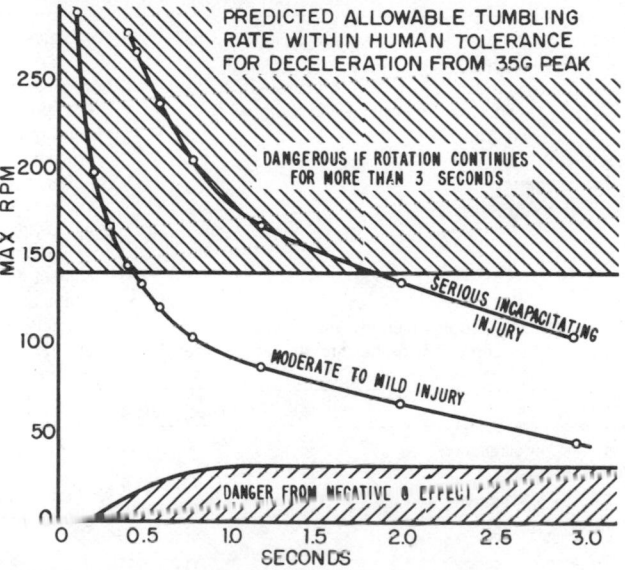

Time required to decelerate to 15g from 35g initial peak.

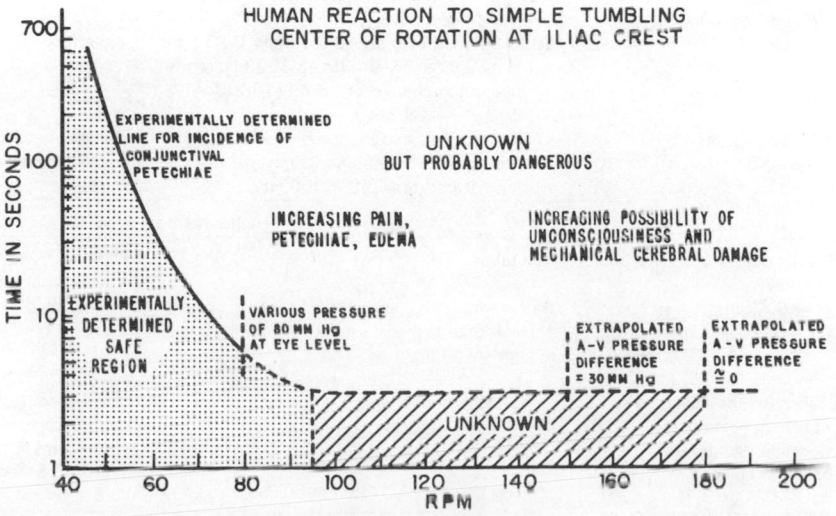

Human reaction to simple tumbling center of rotation at iliac crest.

[45]T. M. Fraser, in J. F. Parker and V. R. West (eds.), *Bioastronautics Data Book*, 2d ed., National Aeronautics and Space Administration, 1973, pp. 191–219.

[46]*Human Factors Engineering*, AFSC DII 1-3, ser. 1-0, General Air Force Systems Command, Wright-Patterson Air Force Base, Ohio, 1977.

## EFFECTS OF VIBRATION ON HUMAN PERFORMANCE[47]

### Summary of Effects of Vibration on Biodynamics, Psychomotor Performance, Speech, Hearing, and Higher Mental Processes

| Vibration Conditions | Measures | Effect | Source |
|---|---|---|---|
| *Biodynamic Mechanisms* | | | |
| $\pm 0.15$–$0.35$ $g_z$ at 0.9–6.5 Hz, low amplitude | Whole body vertical vibration, hand tremor, body equilibrium, foot pressure | Foot pressure constancy impaired at 3.5 to 6.5 Hz, error increase with intensity; no residual effects | Schmitz, Simons, and Boettcher, 1960 |
| $\pm ng_x \pm ng_y$ for ½ hr | Body sway equilibrium | No effects | Hornick, Boettcher, and Simons 1961 |
| $\pm g_z$ 2–20 Hz (intensities = ⅓ short-term tolerance limits) | Control of pitch and roll of a chair | Wide individual differences; decrement between 3 and 12 Hz, worst at 6 Hz | Coermann, Magid, and Lange, 1962 |
| $\pm g_z$ at 0, 2, 5, and 8 Hz | Orientation (orienting body position to face targets at 15, 30, and 60° from reference plane) | Only small decrement in accuracy; mean error < 0.5° | Ayoub, c. 1969 |
| $\pm 0.03 - \pm 0.41$ $g_z$ at 0, 3, 5, and 8 Hz | Leg muscular power (on bicycle ergometer) | No effects | Harrison, 1969 |
| Various peak-to-peak accelerations at 1 Hz with 3 Hz, and 2 Hz with 6 Hz | Arm-hand steadiness | Positional errors significantly related to rms and frequency of vibration; 90% of error was periodic; 1 Hz with 3-Hz combination produced larger error; small (0.5–1 $g$) differences in acceleration had no effect | Clarke et al., 1965 |
| *Psychomotor Performance* | | | |
| +0.25 $g_z$ at 2.4–9.5 Hz | Time to pick up markers and place in small circular areas | Completion time worst at 3.4 and 4.8 Hz | Guignard and Irving, 1960 |
| +0.5 rms $g_z$ at 2–30 Hz (13-Hz peak power) | Digital decimal input with push button, toggle switch, rotary switch, and thumbwheel controls | Accuracy unaffected; insert times increased by 4%; push buttons and toggle switches were most rapidly used, with the former preferred; thumbwheels were most accurate | Dean et al., 1967 |
| 0, 0.2, 0.4, 0.6, and 0.8 rms $g_z$ for 5 min | Same | No effects for 0.2 and 0.4 rms $g_z$; significant increase in insert time for 0.6 and 0.8 rms $g_z$; speed: push buttons > *rotary switches > thumbwheels; error rate: push buttons highest and thumbwheels lowest for high intensity vibrations | Dean, Farrell, and Hitt, 1967 |
| $\pm g_x$ and $\pm g_y$ at 0.33 and 0.80 Hz at amplitude of $\pm 6.3$ and $\pm 7.0$ in | Nut and bolt assembly and disassembly; placement of probe through various sized holes | No effects at 0.33 Hz; time required increased by 30% at 0.80 Hz with no increase in accuracy | Seeman and Williams, 1966 |
| *Speech Intelligibility* | | | |
| $\pm g_z$ at 10, 20, 30, 40, and 50 Hz | Intelligibility | Most effect at 10 and 20 Hz | Nixon, 1962 |
| 0.5 $g_z$ sinusoidal at 6 Hz; 0.75 $g_z$ at 4 and 8 Hz; 1.0 $g_z$ at 2–20 Hz | Intelligibility and quality | No effect on intelligibility at 65 dB; "quality" poorer than control condition | Nixon and Sommer, 1963 |
| *Audition* | | | |
| 5-Hz sinusoidal, 5-Hz random amplitude, 4- to 12-Hz random frequency | Frequency (pitch) change (1200 for 1600 Hz) at 86-dB tones of 0.25-s duration every second—detection | No effect | Weisz, Goddard, and Allen 1965 |
| $\pm g_z$ | 1200 Hz at 86 dB presented every 0.25 s for 1 s against a 74-dB, 30- to 3000-Hz white noise; pitch change at 86 dB (1600 for 1200 Hz)—detection | No effect | Holland, 1966 |
| +1 $g_z \pm 0.7$ $g_z$ at 15 Hz (amplitude 0.036 in) for 30 min | TTS determined as function of vibration and noise versus noise alone (acoustical frequencies from 250–6000 Hz) | Extremely small vibration effect at low tone frequencies only | Guignard and Coles, 1965 |
| *Higher Mental Processes* | | | |
| $\pm 0.15 - 0.35$ $g_z$ at 2.5 and 3.5 Hz | Mental addition | No effect | Schmitz, Simons, and Boettcher, 1960 |
| $+g_z$ at 5, 7, and 11 Hz | Pattern matching and discrimination | No effect | Buckhout, 1964 |
| 0.40 rms $g_z$ random vibration | Navigational tasks in simulated low-altitude, high-speed flight | No effect | Schohan, Rawson, and Soliday, 1965; Soliday and Schohan, 1965 |
| No vibration; no noise, no vibration; noise only; vibration plus noise; postvibration +4.0 $g_z$ at 70 Hz | Continuous counting at a given rate | Decrement, especially during 5–7 min of exposure; residual effects noted; 70% of decrement attributed to vibration (30% to noise) Ss over 36 showed greater decrement | Ioseliani, 1967 |

*Symbol > here indicates faster than.

[47]R. J. Hornick, in J. F. Parker and V. R. West (eds.), *Bioastronautics Data Book*, 2d ed., National Aeronautics and Space Administration, 1973, pp. 297–348.

## +1 $G_z \pm ng_z$ Vibration Effects on Tracking Ability, Seated Subjects

| Vibration Conditions | Tracking Task | Effect | Source |
|---|---|---|---|
| 2.5 Hz, $\pm0.18$ and 0.35 $g$; 3.5 Hz, $\pm0.15$ and 0.30 $g$, 1.5-hr duration | Subjects and wheel on shaker; static display; single axis compensatory tracking task in horizontal axis | Error a function of frequency and intensity; error greater at 3.5 Hz; incomplete recovery after exposure | Schmitz, Simons, and Boettcher, 1960 |
| 0.9–5.5 Hz at $\pm0.15$, 0.25, and 0.35 $g$, 10-min duration | Same | Error elicited, but no effect of frequency or intensity (possibly due to short duration); incomplete recovery after exposure | Schmitz, Simons, and Boettcher, 1960 |
| $\pm0.20$ $g$ at 5 Hz, 1-hr duration | Side-arm controller, single axis (horizontal) compensatory tracking | 40% decrement in task; incomplete recovery in 15-min postvibration period | Ayoub, c. 1969 |
| 1–27 Hz at four subjective reaction levels—from "perceptible" to "alarming" | Aircraft wheel and column; two axis compensatory; horizontal tracking with wheel, vertical with column | Error related to subjective intensity level; greatest error at 10–18 Hz; greater error in horizontal axis in vibration and nonvibration conditions, but vibration has greater relative effect on vertical control | Chaney and Parks, 1964 |
| 2, 4, 6, 8, 11, and 15 Hz with single amplitudes of 0.06 and 0.13 in | Two axis compensatory; static base display; control stick | Slight improvement for 0.06 in; over 40% error increase at 0.13 in; greatest error at 8 Hz | Catterson, Hoover, and Ashe, 1962 |
| 5 Hz, $\pm0.26$–0.36 $g$; 7 Hz, $\pm0.29$–0.41 $g$; 11 Hz, $\pm0.55$–0–0.77 $g$; 5-min duration | Two axis compensatory; side-mounted stick controller, second-order dynamics | Greatest error produced in vertical tracking axis; 5 and 11 Hz have greatest error in vertical axis; 5 Hz produces greatest error in horizontal; vertical error about 40% greater in vibration compared to static condition; amplitude levels equivalent to 25–35% of the 1-min human tolerance level degrades tracking | Buckhout, 1964 |
| 5 Hz, $\pm0.10$–0.26 $g$, 20-min duration | Two axis compensatory; side-mounted controller | Horizontal tracking not affected; vertical tracking significantly impaired at and above $\pm0.21$ $g$, about 40% greater than in a nonvibration condition | Harris, Chiles, and Touchstone, 1964 |
| 5 Hz, $\pm0.10$–0.26 $g$; 7 Hz, $\pm0.15$–0.30 $g$; 11 Hz, $\pm0.25$–0.62 $g$; 20-min duration | Two axis compensatory; rate control, side-stick | Horizontal tracking not affected; vertical axis tracking, first impaired at $\pm0.20$ $g$ for 5 Hz, $\pm0.25$ $g$ for 7 Hz, and $\pm0.37$ $g$ for 11 Hz, these levels being about 20% of the 1-min human tolerance limits; decrements in tracking were up to 40% greater than nonvibration error | Harris and Shoenberger, 1966 |

## Effects of Long Duration Random $g_z$ Vibration

| Vibration Conditions | Tracking Task | Effect | Source |
|---|---|---|---|
| Aircraft F-111 dynamics; input 0.017–0.291 rms $g$ but closed-loop control inputs resulted in 0.028–0.291 rms $g$; 1.5-hr duration | Center-stick and side stick controllers, two axis terrain following and heading changes | Horizontal control not affected; higher rms $g$ inputs result in greater pitch control error; no trend for pitch control error to increase as a function of time; pitch error and altitude error for side-stick is about 50% of that with the center-stick, with resulting rms $g$ reduced by 20% with the sidestick; control error increases as function of task difficulty (displayed terrain) | Soliday and Schohan, 1964 |
| 1–5 Hz, shaped PSD | Two axis, with several controller configurations and dynamics | Best performance with side-stick controller with arm support | Torle, 1965 |
| Aircraft, fighter dynamics, input 0.33 rms $g$, but closed-loop control inputs resulted in 0.40 rms $g$; 1-hr duration | Center-stick, simulated terrain following; three task difficulty levels—simulated flat, rolling, and mountainous terrain | Primary error relationship with tracking task difficulty, greatest error with roughest terrain; altitude error does not increase as function of time | Soliday and Schohan, 1965 |
| Closed-loop rms $g$ levels varied between 0.05 and 0.40 rms $g$ as function of simulated gust speed, and pilot inputs; 3-hr duration | Similar to preceding | No altitude control error trend as function of time; increase in simulated "crashes" at end of runs suggesting fatigue | Schohan, Rawson, and Soliday, 1965 |
| 1–12 Hz, peak energy at 1 and 7 Hz, with 0.10, 0.15, and 0.20 rms $g$; 4-hr duration | Side-stick controller for two axis terrain following and avoidance over three simulated terrain types—flat, rolling, mountainous; open-loop platform dynamics | No trend for error to increase as function of time for two easiest terrains; error increase as a function of time for only the mountainous terrain after 2.5 hr of exposure; error not related to rms $g$ intensity; error residual in postvibration period; greatest error for most difficult displayed terrain | Hornick and Lefritz, 1966 |

*(continued)*

**Effects of Long Duration Random $g_z$ Vibration** *(continued)*

| Vibration Conditions | Tracking Task | Effect | Source |
|---|---|---|---|
| 1–6 Hz, 0.12 and 0.16 rms $g$; two PSD shapes, one with power peak at 2 Hz, one at 5 Hz; 6-hr duration | Two axis, open-loop platform dynamics; primarily side-stick controller, with occasional use of center-stick | No trend for error to increase as function of time; error not a function of rms $g$ intensity; absolute error greater in vertical axis; greater error in spectrum with 5-Hz power peak | Holland, 1967 |
| CH-46A helicopter spectrum, variable rms $g$ and PSD shapes; seven 40-min tests during 6-hr period | Side-stick controller, two axis | No error with this very simple tracking task | Dean, McGlothlen, and Monroe, 1964 |
| UH-1B helicopter spectrum, and frequency sweeps, 4 Hz to over 100 Hz | Variety of controls and servo systems | Greatest tracking error near 5 Hz; position control more accurate in 5- to 25-Hz range; force control more accurate above 25 Hz | Rosenberg and Segal, 1966 |

## RESPONSE TO VEHICULAR VIBRATION

Oborne[48] has suggested that subjects' responses to vibration provides a reasonable prediction of the acceptability of various vibration levels for transportation systems. The first table suggests response categories based on a wide variety of research. The second table categorizes the responses of subjects in Oborne's latest studies.

| Author(s) | Description | Average Intensity Levels, m/s², rms below approx 20 Hz |
|---|---|---|
| Oborne, 1977 | Just comfortable | 0.90–1.30 |
| *Theoretical Curves* | | |
| Lippert, 1946 | Slightly disagreeable (9–16 Hz) | 0.15–1.59 |
| Goldman, 1948 | Threshold of discomfort (4–12 Hz) | 0.24–1.04 |
| Janeway, 1948 | Recommended limit (6–20 Hz) | 0.25 |
| Soliman, 1968 | Lower level of threshold annoyance (4–22 Hz) | 0.42 |
| ISO, 1974 | ½-hr reduced comfort level (4–8 Hz) | 0.41 |
| *Field Studies* | | |
| Zand, 1932 | Threshold of unpleasantness | 0.86 |
| Best, 1945 | Noticeable or objectionable | 0.59 |
| Getline, 1955 | Satisfactory or rough | 0.67 |
| Jacklin and Liddell, 1933 | Disturbing | 0.69 |

## Comfort Response to Hovercraft Vibration

| Label | Definition | RMS Acceleration, m/s², 0–4 Hz | 8–16 Hz |
|---|---|---|---|
| Very comfortable | Willing to put up with this level for a long journey | Below 0.48 | Below 0.15 |
| Comfortable | Willing to put up with the present level for about 1½ hr | 0.48–0.95 | 0.15–0.85 |
| Just comfortable | Not be willing to put up with this level for more than about ½ hr | 0.95–1.30 | 0.85–1.30 |
| Uncomfortable | Would only tolerate this level again for a short journey | 1.30–1.80 | 1.30–1.95 |
| Very uncomfortable | Would not consider traveling in any form of transport with so high a level | Above 1.80 | Above 1.80 |

[48]D. J. Oborne, "Vibration and Passenger Comfort: Can Data from Subjects Be Used to Predict Passenger Comfort?" *Applied Ergonomics*, vol. 9, no. 3, pp. 155–161, 1978.

## VIBRATION TOLERANCE LIMITS[49]

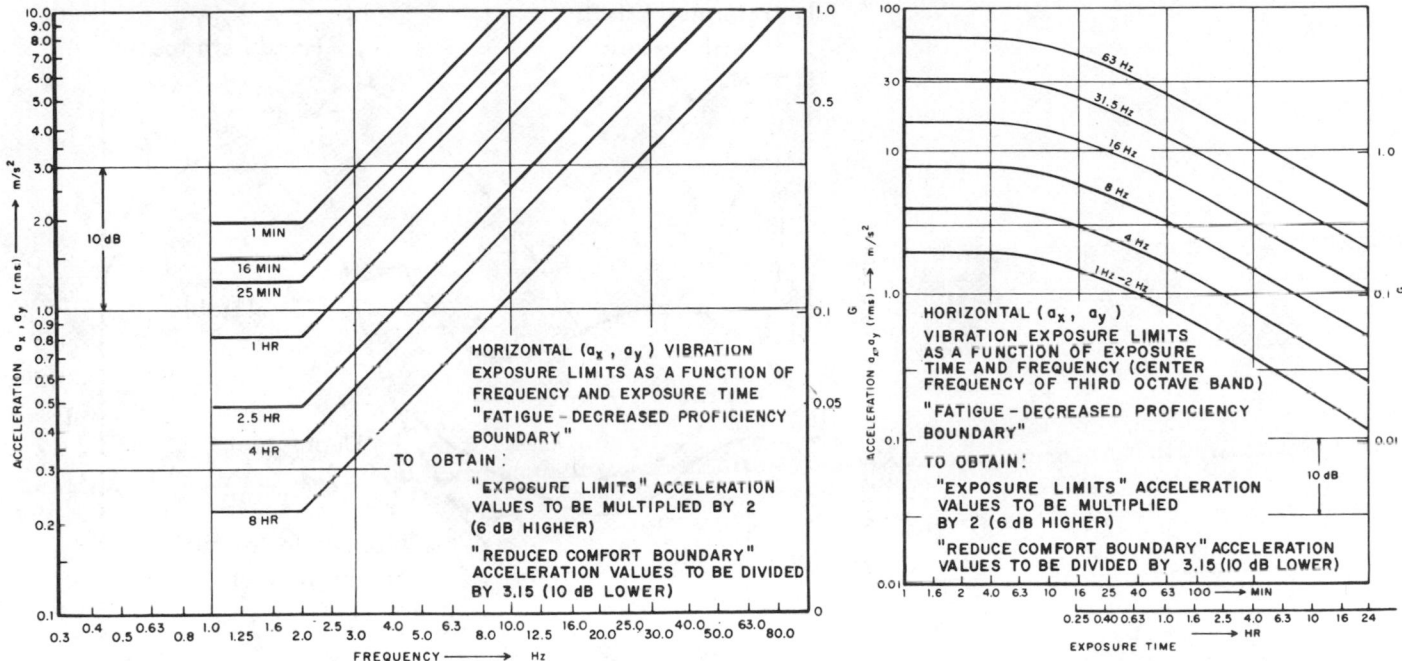

HORIZONTAL ($a_x$, $a_y$) VIBRATION EXPOSURE LIMITS AS A FUNCTION OF FREQUENCY AND EXPOSURE TIME "FATIGUE – DECREASED PROFICIENCY BOUNDARY"

TO OBTAIN:

"EXPOSURE LIMITS" ACCELERATION VALUES TO BE MULTIPLIED BY 2 (6 dB HIGHER)

"REDUCED COMFORT BOUNDARY" ACCELERATION VALUES TO BE DIVIDED BY 3.15 (10 dB LOWER)

HORIZONTAL ($a_x$, $a_y$) VIBRATION EXPOSURE LIMITS AS A FUNCTION OF EXPOSURE TIME AND FREQUENCY (CENTER FREQUENCY OF THIRD OCTAVE BAND)

"FATIGUE – DECREASED PROFICIENCY BOUNDARY"

TO OBTAIN:

"EXPOSURE LIMITS" ACCELERATION VALUES TO BE MULTIPLIED BY 2 (6 dB HIGHER)

"REDUCED COMFORT BOUNDARY" ACCELERATION VALUES TO BE DIVIDED BY 3.15 (10 dB LOWER)

## Vertical Vibration Limits

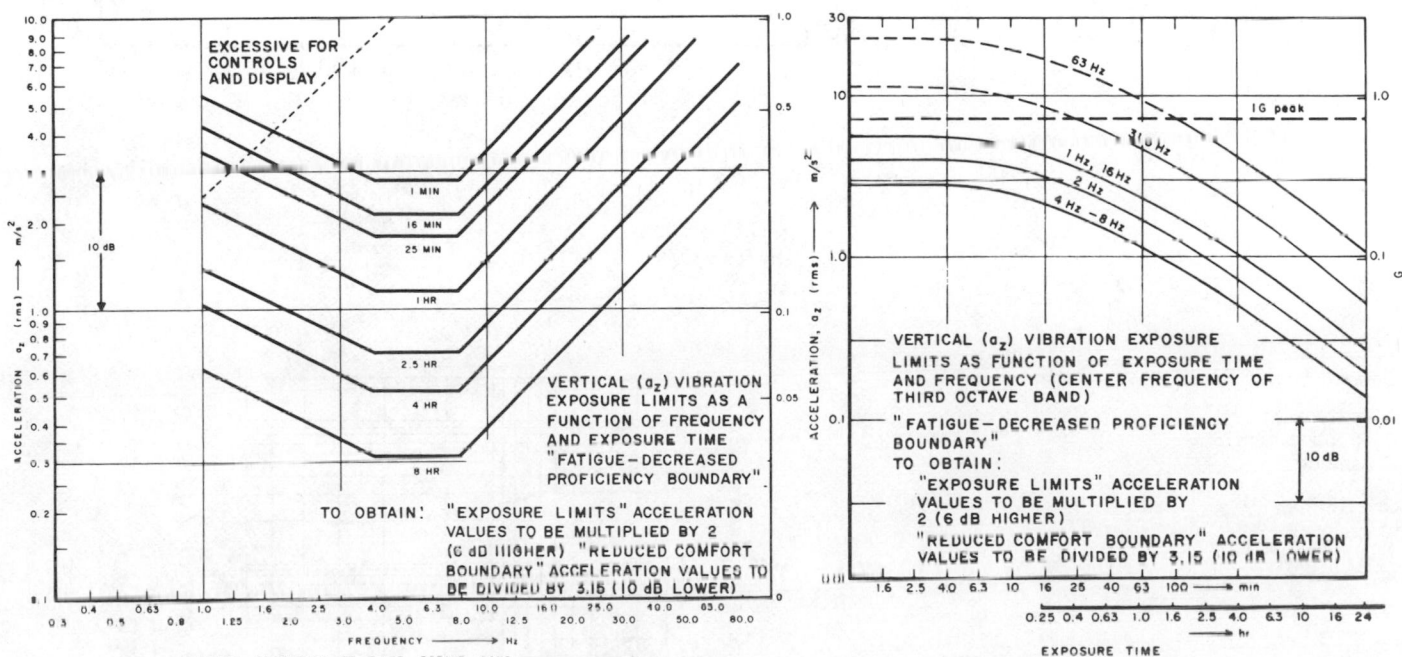

VERTICAL ($a_z$) VIBRATION EXPOSURE LIMITS AS A FUNCTION OF FREQUENCY AND EXPOSURE TIME "FATIGUE – DECREASED PROFICIENCY BOUNDARY".

TO OBTAIN: "EXPOSURE LIMITS" ACCELERATION VALUES TO BE MULTIPLIED BY 2 (6 dB HIGHER) "REDUCED COMFORT BOUNDARY" ACCELERATION VALUES TO BE DIVIDED BY 3.15 (10 dB LOWER)

VERTICAL ($a_z$) VIBRATION EXPOSURE LIMITS AS FUNCTION OF EXPOSURE TIME AND FREQUENCY (CENTER FREQUENCY OF THIRD OCTAVE BAND)

"FATIGUE – DECREASED PROFICIENCY BOUNDARY"
TO OBTAIN:
"EXPOSURE LIMITS" ACCELERATION VALUES TO BE MULTIPLIED BY 2 (6 dB HIGHER)
"REDUCED COMFORT BOUNDARY" ACCELERATION VALUES TO BE DIVIDED BY 3.15 (10 dB LOWER)

[49]*Design Handbook, Human Factors Engineering,* AFSC DH 1-3, ser. 1-0, Department of the Air Force, Headquarters Aeronautical Systems Division, Wright-Patterson Air Force Base, Ohio.

**COMPOSITE LIMIT REFERENCE FOR**
**VARIOUS OPERATING CONDITIONS[50]**

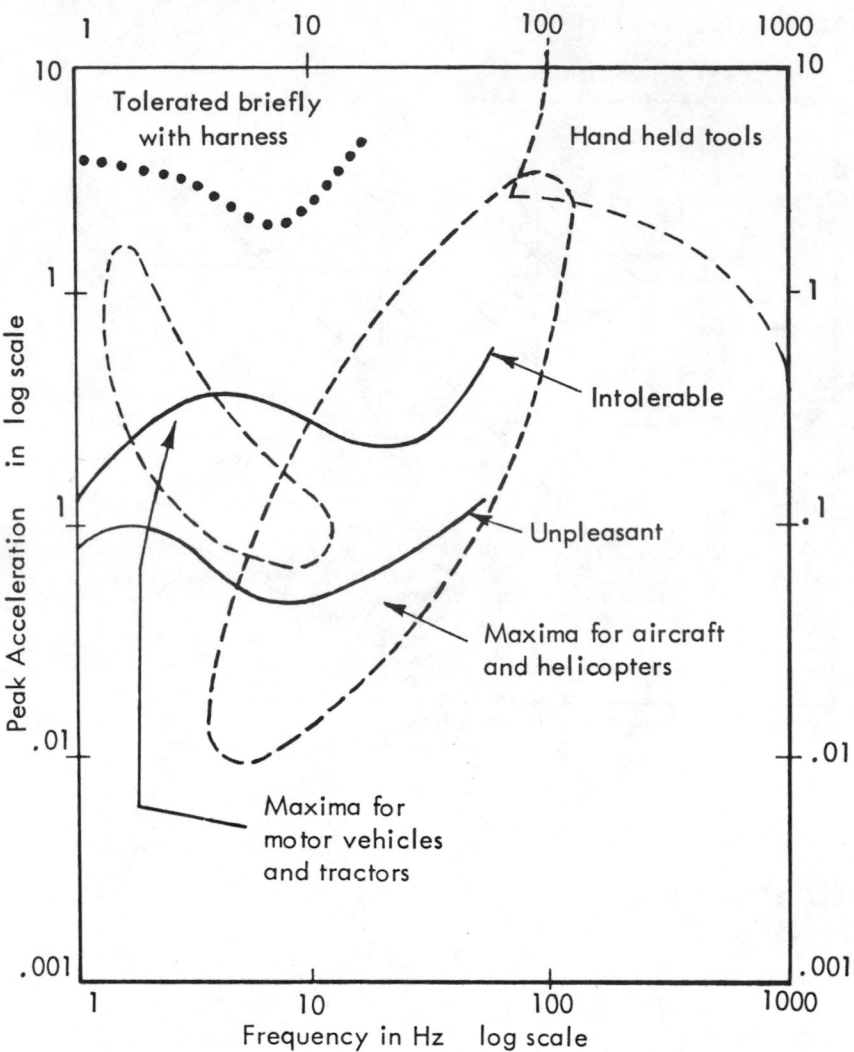

*General Human Reactions to Vibration*

A—severe
B—troublesome
C—smooth

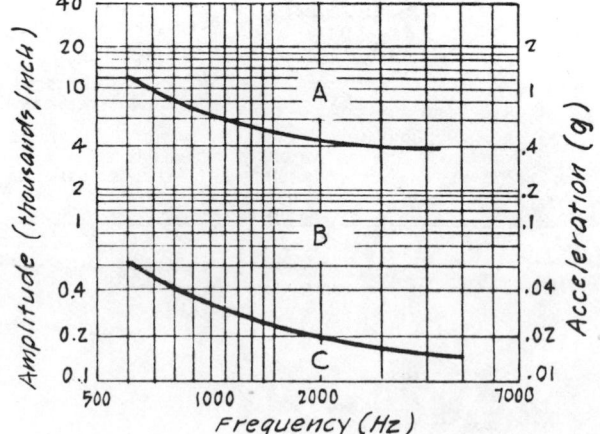

[50]*Human Factors for Designers of Naval Equipment,* Medical Research Council, Royal Naval Personnel Research Committee, 1970, chap. 1.

## OXYGEN[51]

The effects of excess oxygen and the lack of oxygen are shown in the accompanying table.

| Partial Pressure of Oxygen, mm of Hg | Percent of Oxygen in Dry Air at Sea Level Pressure | Effect |
|---|---|---|
| | | Oxygen Excess |
| 456 | 60 | Onset of oxygen poisoning after some hours. |
| 167 | 22 | Limit set in RN to control fire hazard in charcoal filters in nuclear submarines. |
| | | Normal |
| 160 | 21 | Normal atmospheric level |
| | | Oxygen Lack |
| 137 | 18 | Accepted limit of alertness. Loss of night vision. Earliest sign—dilation of the pupils. |
| 114 | 15 | Performance seriously impaired. Hallucinations, excitation, apathy. |
| 100 | 13 | Coordination impaired. Emotional upset. |
| 84 | 11 | Paralysis, loss of memory. Irreversible unconsciousness. |
| 46 | 6 | Death before symptoms apparent. |

*Note:* The effect of falling oxygen is insidious, because it dulls the brain and prevents realization of danger.

The accompanying graph shows that a lack of oxygen results from reduced atmospheric pressure. At altitudes above 2000 m, the fall in atmospheric pressure requires compensation in the inspired air. By the same token, at depths below sea level, similar compensation is required to offset the effects of excess oxygen.

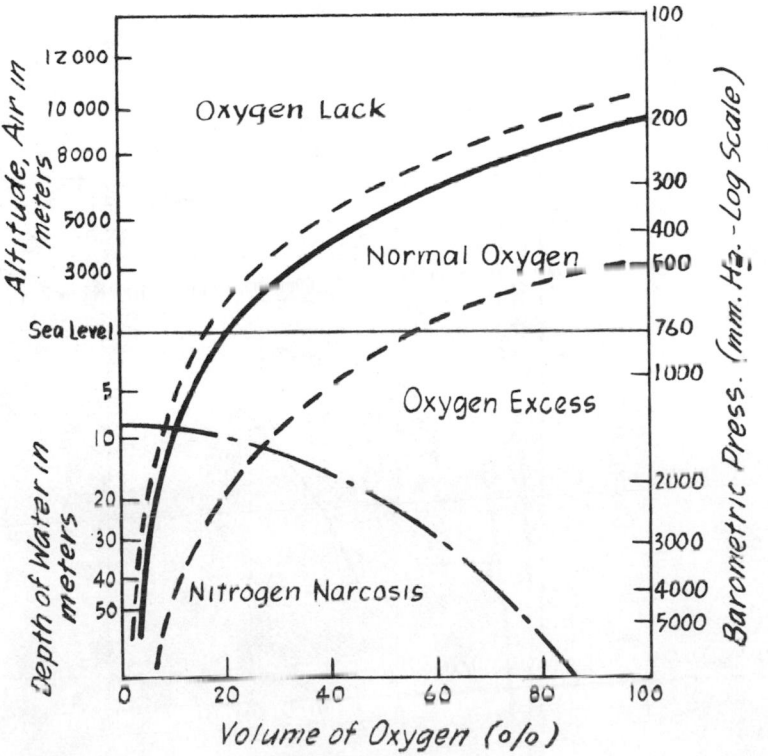

Oxygen concentration, nitrogen narcosis, and atmospheric pressure.

[51]*Human Factors for Designers of Naval Equipment,* Medical Research Council, Royal Naval Personnel Research Committee, 1970, chap. 1.

**Major Effects of Altitude without Supplemental Oxygen or Pressurization, Approximate Time of Consciousness, and Recommended Altitude Limits for Various Situations**

| Altitude, ft | Major Effects | Appoximate Time of Consciousness, s | Recommended Limits |
|---|---|---|---|
| 5,000 | Loss of central night vision initiated | — | |
| 8,000 | | — | Begin supplemental oxygen for routine flights |
| 10,000 | Loss of peripheral night vision, complex coordination, orientation, hearing, memory, and intellectual skills initiated | — | Maximum without supplemental oxygen in routine situation |
| 15,000 | Loss of tracking skills and reaction (decision) time initiated | — | |
| 16,000 | Loss of emotional control initiated | — | |
| 17,000 | Loss of simple coordination and vision initiated | — | |
| 18,000 | | — | Maximum without supplemental oxygen in emergency |
| 20,000 | | — | Begin cabin pressurization for routine flights |
| 23,000 | Decompression sickness initiated | — | |
| 25,000 | | 116 | |
| 28,000 | | 69 | Maximum without cabin pressurization |
| 30,000 | | 54 | Begin positive-pressure breathing for routine flights |
| 35,000 | | 32 | Maximum for demand oxygen in routine situation |
| 40,000 | | 23 | |
| 42,000 | | — | Maximum for positive-pressure breathing in routine situation Maximum for cabin pressurization |
| 43,000 | | — | Maximum for demand oxygen in emergency |
| 45,000 | | — | Maximum for positive-pressure breathing in emergency Begin use of pressure suit |

*Source:* Adapted from U.S.A.F. Manual 80-1.

## Oxygen Safety Limits

Chronic oxygen poisoning can occur at or above 460 mm of mercury. Short-term memory deteriorates after about 10 min of breathing pure oxygen.[52]

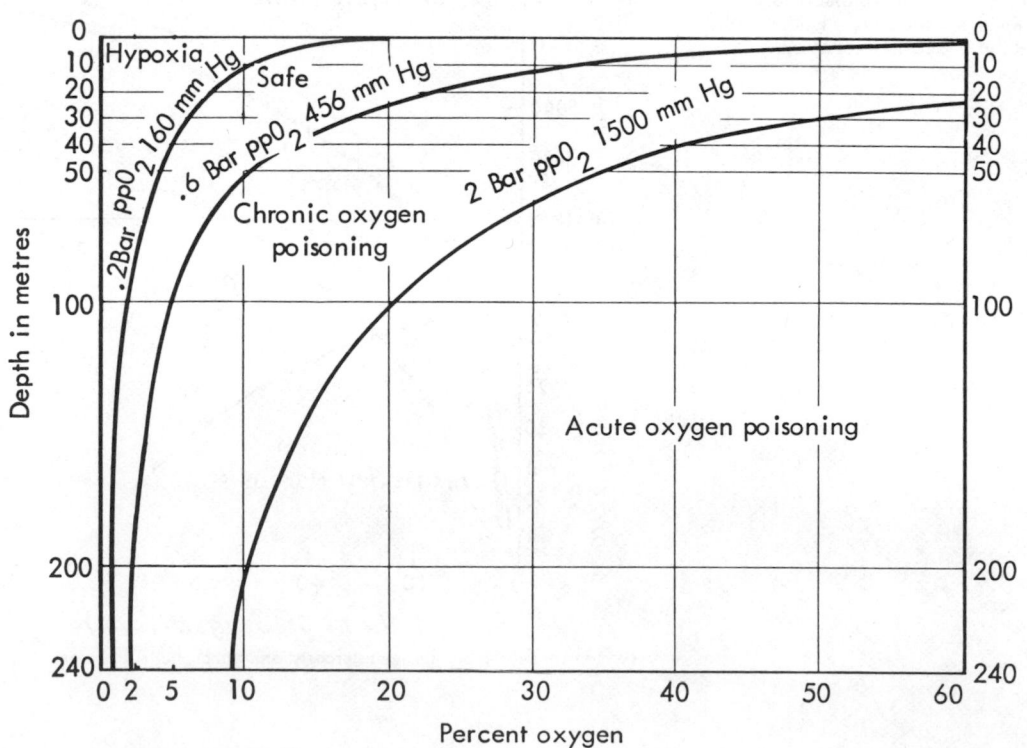

[52]*Human Factors for Designers of Naval Equipment,* Medical Research Council, Royal Naval Personnel Research Committee, 1970, chap. 1.

## CARBON DIOXIDE

The air we breathe normally contains about 0.03 percent of carbon dioxide. The concentration of carbon dioxide should be maintained below 0.5 percent. The necessary ventilation rate to maintain this level is illustrated in the accompanying table.

| | Oxygen Consumption per Person at Sea Level | Ventilation Rate per Person | | | |
|---|---|---|---|---|---|
| | | Sea Level | 5000 ft | 10,000 ft | 15,000 ft |
| At rest | 0.008 ft³/min | 1.2 ft³/min | 1.4 ft³/min | 1.7 ft³/min | 2.1 ft³/min |
| Moderate activity | 0.028 ft³/min | 3.9 ft³/min | 4.7 ft³/min | 6.7 ft³/min | 6.9 ft³/min |
| Vigorous activity | 0.056 ft³/min | 8.7 ft³/min | 9.7 ft³/min | 11.7 ft³/min | 14.5 ft³/min |

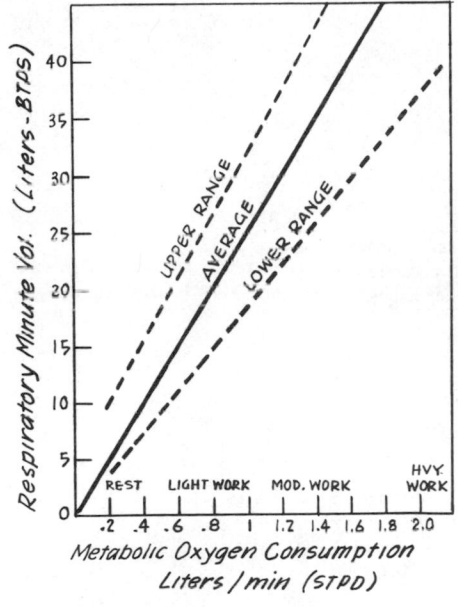

### Maximum Limits for Carbon Dioxide Exposure

| Maximum Limit, % | Situation | Remarks |
|---|---|---|
| 0.5 | General limit for 8-hr industrial exposure | Threshold limit value (TLV) laid down by the American Conference of Government Industrial Hygienists (ACGIH). |
| 1 | 90-day limit for continuous exposure in nuclear submarines (to be reduced to 0.5%) | Future design limit for removal machinery is 0.5%. |
| 3 | Limit for short-term exposure in conventional submarines | Brain functions deteriorate above this level. Heavy breathing. |

### Effects of Continuous Exposure to Carbon Dioxide over a Number of Weeks

| Level, % | Effect |
|---|---|
| 0.5–1 | Probably no significant physiological, psychological, or adaptive changes occur. |
| 1–3 | Gradual changes in the composition of the blood and body fluids, which may lead to irreversible changes in the tissues on prolonged exposure. |
| 3 and above | Deterioration in performance, alteration of basic body functions, and irreversible changes in the tissues |

## CARBON MONOXIDE

The effects of carbon monoxide poisoning are outlined in the accompanying table. Carbon monoxide is particularly dangerous because it is odorless and colorless.

Carboxyhemoglobin levels of over 5 percent have been known to produce intellectual deterioration, and excessive exposure over long periods often leads to arterial and heart disease.

$$\frac{\text{CO in vol/million}}{6} = \text{carboxyhemoglobin \%}$$

### Effects of Carbon Monoxide Exposure

| Atmosphere, Volumes per million | Carbon Monoxide in Blood, % Carboxyhemoglobin | Effects |
|---|---|---|
| 0–60 | 0–10 | None subjectively noticeable, but initial visual and psychomotor impairment is revealed in objective tests. |
| 60–120 | 10–20 | Tightness across forehead, slight headache, flushed complexion. |
| 120–180 | 20–30 | Headache with throbbing in temples, breathlessness from any exertion. |
| 180–240 | 30–40 | Severe headache, weakness, dizziness, dimness of vision, nausea, and vomiting with possibility of collapse. |
| 240–300 | 40–50 | All preceding symptoms with increased pulse rate and respiration and greater possibility of collapse. |
| 300–360 | 50–60 | Loss of consciousness, with increased or irregular respiration, rapid pulse, and possibility of coma with convulsions. |
| 360–480 | 60–00 | Coma, convulsions, depressed heart action, respiratory failure, and possibility of death. |

## OZONE

Under certain conditions three atoms of oxygen combine to form one atom of ozone, which is poisonous in both short- and long-term exposure. An individual who is exposed for about 10 min to 1 volume per million of ozone is likely to feel irritation in the nose and throat, to cough, and to experience visual degradation; 100 volumes per million may injure body tissues, and 1000 volumes per million produces fatigue, depression, and pain in the head and body. Eventual unconsciousness and death may result.

## OTHER TOXIC GASES

The accompanying table[53] shows permissible limits for other toxic gases, excess exposure to which may endanger human health, both temporarily and permanently. The American Conference of Government Industrial Hygienists (ACGIH) has established threshold limit values (TLV) on the basis of an 8-hr working-day exposure. Maximum permissible concentrations (MPC) for 90-day exposures have been established by the British Royal Navy for use in nuclear submarine environments.

**Maximum Permissible Concentrations of Toxic Gases**

| Substance | Formula | Threshold Limit Value for an 8-hr Working Day,* TLV | Maximum Permissible Concentration for a 90-day Exposure,† MPC90 |
|---|---|---|---|
| Ammonia | $NH_3$ | 25 vpm (to be reduced) | 25 vpm |
| Arsine | $AsH_3$ | 0.05 vpm | 0.01 vpm |
| Carbon dioxide | $CO_2$ | 0.5% | 1% (to be reduced to 0.5%) |
| Carbon monoxide | CO | 50 vpm | 25 vpm |
| Hydrocarbons (not methane) | | Not listed as a group | 40 mgm/m³ (8 mgm/m³ aromatic) |
| Hydrogen chloride | HCl | 5 vpm | 0.1 vpm |
| Hydrogen fluoride | HF | 3 vpm | 0.1 vpm |
| Mercury | Hg | Alkyl (skin effect) 0.01 mgm/m³ Non-alkyl 0.05 mgm/m³ | 0.05 mgm/m³ |
| Chlorine | Cl | 1 vpm | 0.1 vpm |
| Methane | $CH_4$ | Explosive | 1.2% (risk of explosion) |
| Methyl alcohol (including formaldehyde and formic acid) | $CH_3OH$ | 200 vpm | 10 vpm |
| Nitrogen dioxide | $NO_2$ | 5 vpm | 0.5 vpm |
| Ozone | $O_3$ | 0.1 vpm | 0.02 vpm |
| Phosgene | $COCl_2$ | 0.1 vpm | 0.05 vpm |
| Stibine | $SbH_3$ | 0.1 vpm | 0.05 vpm |
| Sulphur dioxide | $SO_2$ | 5 vpm | 1 vpm |
| Benzene | $C_6H_6$ | 25 vpm | 1 vpm |
| Toluene | $C_7H_8$ | 100 vpm | 4 vpm |
| Xylene | $C_9H_{12}$ | 100 vpm | 4 vpm |

*Note:* Vpm—volumes of gas or vapor per million of air. Mg/m³—milligrams of solid material, particles, or dust per cubic meter of air.

\*Standard of the American Conference of Government Industrial Hygienists.

†MODN Book of Reference 1326, 1973.

[53]After *Human Factors for Designers of Naval Equipment,* Medical Research Council, Royal Naval Personnel Research Committee, 1970, chap. 1.

## RADIATION AND IONIZATION[54]

The accompanying chart shows wavelengths of electromagnetic radiation and how they are used, indicating the region in which damage to living tissue can occur.

Nuclear radiation may damage the tissues of the body. An acute dose of 500 rad of penetrating ionizing radiation is likely to kill about half of an exposed population within 30 days. However, if the dose is distributed slowly over 10 years, there is only an increased risk that an exposed person will develop cancer. An individual who receives an acute dose of 200 rad of penetrating ionizing radiation equally distributed over his or her body is likely to feel sick for a day or two before recovering. With doses of over 5000 rad, the brain is directly affected; the person may collapse within minutes and die within hours.

Ultraviolet light, after repeated exposure, may produce localized cancer of the skin. Laser beams, microwaves, and very intense radar waves may produce temporary tissue damage.

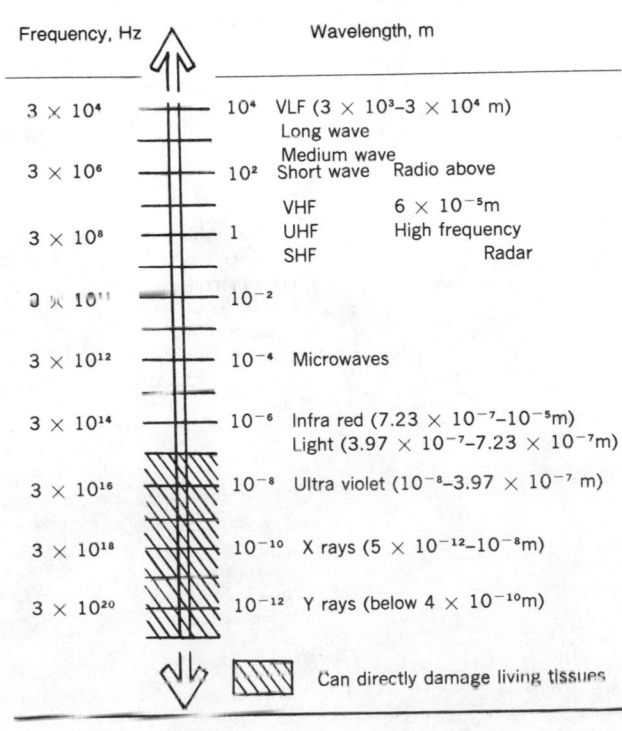

[54]*Human Factors for Designers of Naval Equipment,* Medical Research Council, Royal Naval Personnel Research Committee, 1970, chap. 1.

## U.S. BASIC RADIATION PROTECTION STANDARDS

| Exposed Parts | Permissible Accumulated Dose |
|---|---|
| Whole body | |
| Head and trunk | 1.25 rem in any calendar quarter |
| Active blood-forming organs | |
| Gonads | 5 ($N$-18) during a lifetime |
| Lens of the eye | ($N$-present age) |
| Skin of whole body | 7.5 rem in any calendar quarter |
| Thyroid | 30 rem in any calendar year |
| Hands and forearms | 18.75 rem in any calendar quarter |
| Feet and ankles | 75 rem in any calendar year |

The accompanying table shows the effects of acute, whole-body, external radiation exposure and the probable therapy required.[55] (A rem is the dosage of an ionizing radiation that will have the same biological effect as 1 part R of x-ray or gamma-ray dosage.)

| 0.25 rems | 25–100 rems | 100–200 rems | 200–300 rems | 300–600 rems | 600 rems or more |
|---|---|---|---|---|---|
| *Immediate Effects* | | | | | |
| No detectable clinical effects | Slight transient reductions in lymphocytes and neutrophils. | Nausea and fatigue with possible vomiting above 125 rems.* | Nausea and vomiting on first day. | Nausea, vomiting, and diarrhea in first few hours. | Nausea, vomiting, and diarrhea in first few hours. |
| | Disabling sickness not common; exposed individuals should be able to proceed with usual duties. | Reduction in lymphocytes and neutrophils with delayed recovery. | Latent period up to 2 weeks or perhaps longer. | Latent period with no definite symptoms, perhaps as long as 1 week. | Short latent period with no definite symptoms in some cases during first week. |
| *Delayed Effects* | | | | | |
| Delayed effects may occur | Delayed effects possible, but serious effects on average individual very improbable. | Delayed effects may shorten life expectancy in the order of 1%. | Following latent period, the following symptoms appear but are not severe: loss of appetite, and general malaise, sore throat, pallor, petechiae, diarrhea, moderate emaciation. | Epilation, loss of appetite, general malaise, and fever during second week, followed by hemorrhage, purpura, petechiae, inflammation of mouth and throat, diarrhea, and emaciation in the third week. | Diarrhea, hemorrhage, purpura, inflammation of mouth and throat, fever toward end of first week. Rapid emaciation and death as early as the second week, with eventual death of up to 100% of exposed individuals. |
| | | | Recovery likely in about 3 months unless complicated by poor previous health or superimposed injuries or infections. | Some deaths in 2 to 6 weeks. Possible eventual death to 50% of the exposed individuals for about 500 rems. | |
| *Therapy* | | | | | |
| | Reassurance is probably the only therapy needed. | Patients should be kept under hematological surveillance. | Antibiotics should be administered as indicated. | Antibiotics and blood transfusions should be administered as indicated. | Bone marrow transplantations may be tried. Electrolyte balance should be maintained. |

*Occurs on the first day or so following irradiation, followed by a "latent period" up to 2 weeks or more, during which the patient has no disabling illness and can proceed with his regular occupation. The usual symptoms, such as loss of appetite and malaise, may reappear, but if they do, they are mild.

[55] *Design Handbook, Human Factors Engineering,* AFSC DH 1-3, Ser. 1-0, Department of the Air Force, Headquarters Aeronautical Systems Division, Wright-Patterson Air Force Base, Ohio.

## EFFECT OF ELECTRICAL CURRENT ON THE HUMAN BODY

Body resistance to electrical current varies over a wide range, making it difficult to define dangerous potentials exactly. Hand-to-hand resistance varies from 1000 to 4000 $\Omega$ even with good contact. Hand-to-foot resistance is slightly less. With dry skin, resistances as high as 250,000 $\Omega$ are possible. Under ordinary working conditions when the hands are sweaty, a resistance of 5000 $\Omega$ is commonly expected. Safety recommendations suggest 1000 $\Omega$ as the highest resistance value. The accompanying table identifies some of the potential effects that can be expected for various current levels.

| Current, mA | Effects |
|---|---|
| Less than 1 | Usually not felt, i.e., no sensation. |
| 1–2 | There is a sensation of shock, but not pain. The individual can let go because muscular control is not lost. |
| 3–15 | Painful shock occurs, but the individual can still let go. |
| 15–20 | Painful shock occurs, and the individual may not be able to let go because control of the immediately adjacent muscles is affected. |
| 20–50 | Very painful shock occurs, plus severe muscular contractions. Breathing typically becomes difficult. |
| 50–100 | Ventricular fibrillation (a heart condition that results in instant death) is very likely. |
| 100–200 | Same as above, except that the results are certain. |
| 200 and over | Severe burns occur as well as muscular contractions so severe that the chest muscles stop the heart during the duration of the shock. |

## SKIN REACTION TO CHEMICAL SUBSTANCES

The accompanying table presents a partial list of chemical substances and their reaction on the skin.

| Agent | Reaction |
|---|---|
| Acids: | |
| Acetic | Dermatitis and ulcers |
| Carbolic | Irritation and erosion, eczema, and anesthesia |
| Chromic | Ulcers (chrome holes on the skin), inflammation, and perforation of the nasal septum |
| Hydrochloric | Irritation and ulceration |
| Hydrofluoric | Severe burning, erosion ulcers, and blisters |
| Lactic | Ulcers (if strong solution) |
| Nitric | Severe burns and ulcers |
| Oxalic | Local caustic action on the skin |
| Sulfuric | Corrosive action on the skin and severe inflammation of the mucous membranes |
| Alkalis: | |
| Calcium cyanamide | Irritation and ulceration |
| Calcium oxide | Dermatitis, burns, and ulcers |
| Potassium hydroxide | Severe burning, persistent ulcers, and loss of fingernails |
| Sodium hydroxide | |
| Sodium silicate | Thickening of the skin and ulcers on the fingers |
| Sodium or potassium cyanide | Blisters and ulcers |
| Salts: | |
| Antimony and its compounds | Irritation and eczematous eruptions |
| Arsenic | Skin darkening, perforation of the nasal septum, eczema around the mouth and nose, and possible loss of nails or hair |
| Barium | Eczema and cyanosis of skin |
| Bromine | Brownish stains and skin erruptions |
| Chromium (hexavalent compounds) | Chrome holes on the skin, perforation of the nasal septum, and eczematous eruptions |
| Mercury compounds | Corrosion and irritation and mercurial eczema |
| Sodium | Burns and ulcers |
| Zinc chloride | Ulcers of the skin and nasal septum |
| Solvents: | |
| Acetone | Dry (defatted) skin |
| Benzene | Dry (defatted) skin |
| Carbon disulfide | Dry (defatted) skin |
| Chlorinated phenols | Severe eruptions |
| Petroleum distillates | Acne and epithelioma |
| Trichlorethylene | Dry, cracked skin |
| Turpentine | Red, blistered skin and eczema |
| Dyes: | |
| Chlorinated compounds | Blisterlike eruptions |
| Dinitrochlorobenzene | Blisterlike eruptions |
| Nitro and nitroso compounds | Red skin and eczematous eruptions |
| Phenyl hydrazine | Blisterlike skin eruptions |
| Insecticides: | |
| Chlorophenols | Red skin, and blisters |
| Creosote | Pustular eczema, warts, and epithelioma |
| Fluorides | Severe burns and dermatitis |
| Pyrethrum | Red skin, blisters, and pimples |
| Rotenone | Red skin and blisters |
| Resins: | |
| Coal tar, pitch, and asphalt | Acute dermatitis, acne, inflammation, epitheliomatous cancer, eczema, and ulcers |
| Synthetics, e.g., phenol-formaldehyde | Extremely red and itchy skin |
| Synthetic waxes, e.g., chloronaphthalenes and chlorodiphenyls | Dermatitis and acne |

*Source:* Julian B. Olishifski and Frank E. McElroy (eds.), *Fundamentals of Industrial Hygiene*, National Safety Council, Chicago, 1971.

## HUMAN INFORMATION-PROCESSING CAPABILITIES

Humans can transmit only about 5 to 10 bits of information per second (b/s). They can transmit about 2 b/s when the stimuli they receive are fairly well structured, although this can often be doubled by adding appropriate coding or anchoring of the input. The relationship between bits per item and the bits-per-second limitation depends on what is referred to as "attention switch time" (e.g., 0.1 to 0.2 s). Thus an operator can accept no more than two or three items of data per second.

As the speed of information-processing demand increases, the number of errors typically increases. Thus the overall information transfer rate tends to remain constant at about $10^{-2}$ b/s. Through careful input display design, however, this rate can be increased to $10^{-3}$ or even $10^{-6}$ under good conditions. The following error classification scheme, which was suggested by Kidd,[56] helps characterize types of information-processing errors and why they occur:

1. Failure to detect a signal: input overload or underload and/or actual interference
2. Misidentification: insufficient cues
3. Improper weighting of informational factors and/or selection of input factors: poor or inadequate conceptualizations or evaluation of action choices
4. Action failure: A wrong action at the right time or a right action at the wrong time

## INFORMATION STORAGE CAPACITY

Information storage is of two distinct types: long-term and short-term. Short-term memory storage capacity is generally limited to about 30 lb or eight individual items. The human generally organizes stored information in terms of sensory modality (visual, auditory, etc.). The most significant storage problem occurs because of the potential interference between old ("held") information and new items that present themselves during the holding period. This accounts for the frequent "reversal errors" in information processing. As a general principle, human memory is more effectively utilized as a means of orienting and sequencing information than as a depository for isolated data or symbolic items.

## INPUT CAPACITY

The total sensory input capacity of the human system is about $10^9$ b/s (as compared with output capacity of 10 b/s). The five basic input categories for the visual channel are:

Relative position
Shapes
Brightness
Color
Movement

For the auditory channel they are:

Pitch
Loudness
Rhythm
Timbre (the quality that allows one to distinguish different voices, instruments, or special auditory displays)

## DECISION MAKING

Three basic kinds of information must be available for an operator to make good decisions:

Information regarding policies and objectives
Information regarding possible alternatives and consequences
Information about the current state of the system

It is important to recognize that, in making decisions, people pursue two different courses of action:

1. Evaluating likely outcomes
2. Keeping options open as long as possible

## STRESS SUSCEPTIBILITY

Human information processing is subject to a variety of stresses that may affect the efficiency with which information is received, processed, and acted upon. Two basic factors are important to consider:

1. The state of arousal of the human system (alertness)
2. Potential skill deterioration due to disorganization, cumulative disruption, and/or fatigue

The effects of, and reactions to, skill breakdown have been categorized as shown in the table at the top of the page.

## INFORMATION DISPLAY IN RELATION TO DECISION MAKING[57]

Although research has shown that, on the basis of information acquisition only, added display complexity generally degrades operator performance, decision-making performance is not always similarly affected:

### Skill Breakdown

| Effect | Reaction |
| --- | --- |
| Failure of selective attention fails Perceptual disorganization | Narrowing of attention Reduction in the size of the data sample and actions upon it and reduction in filtering efficiency |
| Cumulative disruption | Temporary halts (complete stop-pages) with start-overs |

The relationship between "fact density" and decision adequacy may be curvilinear; i.e., where there is a decreasing slope under low-density conditions and/or an increasing slope under high-density conditions, actual decision-making performance may improve. For example, although complexity of information acquisition degrades overall performance when fact density is low, decision-making performance may actually increase with increases in fact density.

When information is compressed, an inverted "U-shaped" relationship occurs, with moderate levels of compression producing the best decision-making performance; i.e., the critical variable is not merely symbol count, etc., but also the subjective weighting of the particular facts being compressed.

At levels of low fact density, high compression, and high display clutter, simple coding (e.g., color) may be superior to more elaborate encoding combinations. However, at other levels of fact compression and clutter, a double coding is usually more effective (e.g., color plus size).

Subject motivation (incentive) works only at low informational levels. At high levels of motivation, degraded performance is often mediated by scattered attention.

Increases in perceptual clutter sometimes increase performance rather than degrading it, as one might expect, because we are used to working in a particularly cluttered type of information-transfer environment. The more random an irrelevancy, the more performance is facilitated because of our need to have irrelevancies for a "figure-ground" decision-making background.

It is important, however, to recognize the importance of input validity to decision making, since humans quickly recognize and react to the futility of trying to decide on information that destroys or degrades the quality of their decisions.

[56]J. S. Kidd, "Some Sources of Load and Constraints on Operator Performance in a Simulated Radar Air Traffic Control Task," USAF WADD TR-60-612, 1961.

[57]After W. T. Singleton, "The Ergonomics of Information Presentation," *Applied Ergonomics*, vol. 2, no. 4, pp. 213–220, 1971.

**AMOUNT OF INFORMATION IN
ABSOLUTE JUDGMENT FOR VARIOUS
STIMULUS DIMENSIONS**

| Stimulus Dimension | Number of Levels Which Can Be Discriminated | Bits of Information Transmitted* |
|---|---|---|
| Audition (single dimension) | | |
| Pure tones | 5 | 2.3 |
| Loudness | 5 | 2.3 |
| Audition (combination of six variables including frequency and intensity, on-time fraction, duration, spatial location, and rate/interruption | 150 | 7.2 |
| Vision (single dimension), e.g., pointer and linear scale | 9 | 3.1 |
| Visual size | 7 | 2.8 |
| Hue | 9 | 3.1 |
| Brightness | 5 | 2.3 |
| Vision (combinations): | | |
| Size, brightness, and hue | 17 | 4.1 |
| Hue and saturation | 15 | 3.9 |
| Position of dot in square | 24 | 4.6 |
| Odor (single dimension) | 4 | 2.0 |
| Odor (combination, e.g., kind, intensity, and number) | 16 | 4.0 |
| Taste: | | |
| Saltiness | 4 | 1.9 |
| Sweetness | 3 | 1.7 |

*Lack of correspondence of bit values is due to rounding to the nearest whole number.

"Although we cannot neatly separate different sequential phases of information processing (i.e., sensation, perception, decision, response, etc.), it should be noted that the bottleneck probably is the brain rather than the sensory mechanisms."[58] It has been estimated that information reduction from reception through intermediate processes to permanent storage is as shown in the accompanying table.

| Process | Maximum Flow of Information, b/s |
|---|---|
| Sensory reception | 1,000,000,000 |
| Nerve connections | 3,000,000 |
| Consciousness | 16 |
| Permanent storage | 0.7 |

[58]E.J. McCormick, *Human Factors Engineering*, 3d ed., McGraw-Hill Book Company, New York, 1970, p. 92.

## VISUAL AND AUDITORY CODING METHODS[59]

### Summary of Certain Visual and Auditory Coding Methods

(Numbers refer to number of levels which can be discriminated on an absolute basis under optimum conditions.)

| | |
|---|---|
| Alphanumeric | Single numerals, 10; single letters, 26; combinations, unlimited. Good; especially useful for identification; uses little space if there is good contrast. Certain items easily confused with each other. |
| Color | Hues, 9; hue, saturation, and brightness combinations, 15–24. Particularly good for searching and counting tasks; poorer for identification tasks; trained observers can use many codes (up to 24). Affected by some lights; problem with color-defective individuals. |
| Geometric shapes | 15 or more. Generally useful coding system, particularly in symbolic representation; good for CRTs. Shapes used together need to be discriminable; some sets of shapes more difficult to discriminate than others. |
| Visual angle | 24. Generally satisfactory for special purposes such as indicating direction, angle, or position on round instruments like clocks, CRTs, etc. |
| Size of forms (such as squares) | 5. Takes considerable space. Use only when specifically appropriate; preferably use less than 5. |
| Visual number | 6. Use only when specifically appropriate, such as to represent numbers of items. Takes considerable space; may be confused with other symbols. |
| Brightness of lights | 3–4. Use only when specifically appropriate. Preferably limit to two levels; weaker signals may be masked. |
| Flash rate of lights | 4. Limited applicability. Preferably limit to two levels; combination of individual flashes and controlled time intervals may have special application, such as lighthouse signals and naval communications. |
| Sound frequency | 5. For untrained listeners, use less than 5 levels; space widely apart, but avoid multiples and low and high frequencies; intensity should be 30 dB above threshold. Frequency changes easier to detect than single frequencies; combinations usually require training except for clearly distinguishable sounds such as bells, buzzers, and sirens. |
| Sound intensity | 4. Preferably use less than 4. Intensity changes easier to detect than single intensities; for pure tones restrict to 1000–4000 Hz, but preferably use wide band. |
| Sound duration | Use clear-cut differences, preferably 2 or 3. |
| Sound direction | Difference in intensity to two ears should be distinct; particularly useful for directional information (i.e., right versus left). |

## KEEPING TRACK OF SEQUENTIAL EVENTS[60]

At slow event-presentation rates (one per second or less), individual events falling into two categories can be discriminated one at a time and mentally tallied with considerable accuracy.

As either the number of categories or the trial length increases, the number of errors per trial increases, generally very rapidly. For a given trial length, one should present the operator with fewer categories with multiple occurrences rather than large numbers of categories with fewer occurrences.

At fast presentation rates, the ON/OFF ratio has little effect on performance as long as the ON time is sufficient for accurate stimulus perception. Intermediate-rate–short-stimulus ON time tends to optimize performance. Slow-rate–short-stimulus ON time degrades performance.

It is generally desirable to "pace" operator performance by means of cues as to the expected occurrence of a next event.

In general, the slower the rate of presentation, the better the performance, a truism that nevertheless is important to remember.

Irrelevant information along with relevant information does not necessarily degrade performance, provided that it is introduced in a manner that allows operators to pace themselves.

Irregular rates, on the other hand, disrupt performance because the operator cannot anticipate the occurrence of the next stimulus.

Stimuli with a "natural" or built-in order result in better performance. Whenever possible, sort out information categories prior to displaying information to the operator and align visual information elements spatially so that each category appears in its own unique location each time it is presented.

Subtractions are more difficult than additions in setting up informational content, and they should be kept to a minimum in any given sequence.

Information of greater value generally is "remembered" best, although typically at the expense of the remaining information.

The number of multiple categories displayed simultaneously can be increased to three with no significant degradation when the information rate is held constant.

## SUMMARY GUIDELINES RELATIVE TO INFORMATION-PROCESSING OPTIMIZATION

Multidimensional coding typically results in more information transmission than single-dimension coding.

Information-processing capacity is adversely affected by both load and speed, i.e., the number of sources should be limited.

When time sharing of sensory inputs occurs, signals should be separated temporally (preferably by 0.5 s or more).

When operators can control the input rate (are self-pacing) and have some method of identifying the more important input (if they have a choice), auditory signals are generally more durable than visual signals.

Simultaneous presentation of the same information via two sensory modalities usually increases the probability of reception.

Signals presented on one channel (such as audition) can serve as cues to facilitate the use of another channel.

Time sharing of visual tasks adversely affects primarily those tasks with greater uncertainty and those which depend on short-term memory; thus speed stress does not affect all tasks equally.

[59]E. J. McCormick, *Human Factors Engineering*, 3d ed., McGraw-Hill Book Company, New York, 1970, p. 113.

[60]After R. A. Monty, "Keeping Track of Sequential Events: Implications for the Design of Displays," *Ergonomics*, vol. 16, no. 4, 1973.

## INFORMATION TRANSFER FACILITATION

Certain information transfer facilitators are at the disposal of the designer and should be utilized whenever possible to maximize the human's processing response. The accompanying table reviews some of the key facilitators that can be used.

| Facilitator | Remarks |
|---|---|
| Familiar patterns | Present information to operators in a form that is already familiar to them. |
| Visualization | Whenever practical, utilize the natural tendency of most individuals to try to visualize, even if the presentation mode is via channels other than the visual one. |
| Context | Provide as complete a picture as possible in a logically organized manner. |
| Normal relationships | Present informational components and component-observer relationships that are natural (alphanumeric characters right side up, high-low values oriented according to specific use format, etc.). |
| Minimal extrapolation | Avoid requiring the observer to extrapolate; i.e., present quantities, values, and patterns in their intended form. |
| Reference bases | Provide a continuous reference (a scale value, a map, a comparison background, etc.). |
| Timing | Time inputs to avoid overload and delay; present them in a logical sequence and provide immediate feedback to the operator's responses or queries. |
| Noise | Control input competition, interference, and distractions. |
| Conspicuousness | Emphasize informational cues via adequate intensity, contrast, and special encoding to ensure maximum arousal and attention. |
| Expectation | Anticipate the operator's mental set and prepare the operator to receive the information. |
| Meaningfulness | Clarify the value, necessity, and urgency of the information transfer. |

*Note:* For additional enlightenment with regard to human information processing, see D. E. Rumelhart, *Introduction to Human Information Processing*, John Wiley & Sons, Inc., New York, 1977.

## IMPORTANCE OF BEHAVIORAL EXPECTANCY

Designers should recognize the importance of anticipating what people will do with the equipment or facilities they are about to design. Otherwise, it is very likely that the consumer will misuse the product or facility, possibly suffer injuries that lead to litigation, and/or select a competitive product when next making a purchase.

The following section presents a number of different types of expectancy data, some of which are based on research studies, and others on the experience of designers and human factors specialists.

## BEHAVIORAL CHANGE FROM CHILDHOOD TO ADULTHOOD

The accompanying table lists the general changes that the human goes through from infancy to adulthood. Although there are obvious individual variations, the generalizations may help the designer understand a particular age group's behavior with respect to the design under consideration.

| Age | Behavioral and Motor Development Description |
|---|---|
| 1 month | Lifts head when held to another's shoulder. Makes crawling motions when prone. Lifts head unsteadily when prone. Turns head laterally when prone. |
| 2 months | Holds head erect for a short time when held to another's shoulder. Lifts head when held horizontally on back. Lifts chest a short distance when prone. Makes vertical arm thrusts in random play when on back. |
| 3 months | Holds head erect and steady when held to another's shoulder. Turns from back to side. Pushes or elevates self by using arms when in a prone position. |
| 4 months | Holds head steady when carried or when swayed. Tries to sit up when on back. Sits with resistant body pressure when supported by pillows. Hands frequently open. Thumb opposition appears in grasping. |
| 5 months | Rolls from back to stomach. Sits with slight prop. Picks up cube from table on contact. |
| 6 months | Sits momentarily without support if placed in a favorable leaning position. Grasps with simultaneous flexion of fingers. Retains transient hold of two cubes, one in each hand. Can turn from stomach to back, and back again to stomach. Rolling and hitching (in leg region) occur. |
| 7 months | Tends to unilateral reaching and manipulation. Rotates wrist freely in manipulation. Scoops or rakes hand to secure a pellet. Picks cube deftly and directly from table. Begins crawling. Leans toward object of curiosity, reaches for it, and then handles, pulls, sucks, shakes, and rattles it (which often results in damage to the object and harm to the baby but which is valuable as a means of learning). |
| 8 months | Sits momentarily without support. Raises self to sitting position. Picks up pellet with partial finger prehension. Thumb opposition appears in grasping objects. |

*(continued)*

| Age | Behavioral and Motor Development Description   *(continued)* |
|---|---|
| 9 months | Sits alone. Use thumb opposition in seizing cube. Makes a locomotive reaction in prone position. |
| 10 months | Pulls self to standing position. Crawls for toys. |
| 12 months | Stands with support. Creeps or hitches along. Walks with help. Shows a preference for one hand on reaching. Scribbles imitatively with crayon. Has acquired a fairly mature pattern of reaching. Can use a cup with both hands; later, with practice, can use a cup with one hand. Can pull off socks, shoes, caps, and mittens. Can hold pencil or crayon. Interested in self-feeding. |
| 15 months | Stands alone. Walks alone. Can grasp spoon and insert in dish. Uses toys more. Throws and picks up objects and throws them again; puts them in and takes them out of receptacles. Often destructive in play with toys because of poor muscle coordination. |
| 18 months | Climbs stairs or chair. Throws ball into box. Scribbles spontaneously and vigorously. Tries to put on clothes; wants to bathe self, brush hair, and use a toothbrush. Pulls toys, carries and hugs a doll or stuffed animal, imitates adult activities (such as reading), and actively gets into everything. |
| 21 months | Walks attended on street. Walks backward. Differentiates between stroking scribbles and circular scribbles. Play is solitary. Feels, pats, and pounds toys. Strings wooden beads and pulls them in and takes them out of holes in tops of boxes. Transports blocks in wagons rather than building with them. Imitates activities of persons in the environment. |
| 2 years | Runs. Piles a tower of six blocks with good coordination. Imitates vertical and horizontal strokes. Plays simple catch and toss with ball. Can operate a kiddie car around a chair. Can remove all garments unless buttoned in back. Can open boxes, unscrew lids from bottles and jars, turn leaves of books, insert pegs in a pegboard, string beads, and use scissors. Tends to use one hand more than the other. Makes deliberate swimming movements and tends to remain in prone position in water. |
| 2½ years | Goes up and down stairs alone. Piles seven or eight blocks with coordination. Tries to stand on one foot. Copies vertical line. |
| 3 years | Draws circle from copy. Creases a piece of paper neatly. Aligns a card to an edge. Can feed self with a fork and spread butter with a knife. Cannot yet use a knife to cut. Much given to dawdling. Most vulnerable to accidents (boys more so than girls). Most accidents occur in home, Thursday through Saturday. |
| 4 years | Draws from copy. Traces diamond path. Hand preference established between the ages of 4 and 6. Can ascend and descend ladders. Can learn tricycling, jumping rope, balancing on a rail or top of wall, roller skating, ice skating (double runners), and dancing. Perception of short distances is similar to that of an adult. Ability to perceive differences in form increases gradually from 2 to 6 years. |
| 5 years | Draws triangle from copy. Draws prism from copy. Between 5 and 6 becomes proficient in throwing and catching balls. Between 5 and 6 can make recognizable letters, though writing is slow, laborious, and poor at this age. Can select medium-sized objects from a group of small, medium-sized, and large objects. Concept of roundedness is well established. Is able to estimate weights of different objects with fair accuracy; can tell the difference between 3- and 15-g weights when they are the same size. |
| 6–12 years | Can handle all garment fasteners in any way. Should be able to bathe self, dress, tie shoes, and comb hair. Can use tools to make simple objects like boats and wagons. Nursery school and kindergarten children can use scissors in following the outlines of pictures, model with clay, make cookies, sew, paint, crayon, paste, copy simple geometric figures, and help with simple household tasks such as carrying glasses and pitchers of milk without spilling. Can roller-skate, ice-skate (single runners), swim, dive, ride a bicycle, and keep time to music by skipping or dancing. Girls surpass boys in skills involving finer muscles (painting, sewing, weaving, and hammering); boys excel in skills involving grosser muscles (throwing a basketball, kicking a soccer ball, and doing a standing and running broad jump). Handedness is established; with guidance (and willingness on the child's part), handedness can be changed *before* age 6, but not easily after that age. Concepts of space become defined. From using weights and rulers, comes to learn the meaning of ounces, pounds, inches, feet, yards, and even miles. School arithmetic helps formulate more definite ideas of space and distance. |
| 12–14 years | Time of rapid physical change, leading to confusion, feelings of |

*(continued)*

| Age | Behavioral and Motor Development Description |
|---|---|
| | motor development comparable to that of peers is extremely important; otherwise, is unable to take part in games and sports, which in turn has an adverse effect on social adjustments. Tests to measure agility, control, strength, and static balance show the greatest increase for boys after 14 years of age. For girls, improvement comes up to age 14 and then lags, mainly because of change of interests. Speed of voluntary movement increases continually from beginning of early adolescence to the end of that period. A 13-year-old has six-sevenths as much speed as a 17-year-old. Among boys, physical strength doubles between the ages of 12 and 16. Among girls, the greatest increase in strength comes near the time of the menarche. At all ages after puberty boys are stronger than girls, and this superiority increases with age. Girls usually attain their maximum strength at about age 17, and boys at about age 21 or 22. |
| 17–21 years | A transitional period during which the adjustments to a mature status and to mature levels of behavior (begun during early adolescence) normally are gradually completed. The increase in physical strength that accompanies growth of the muscular system motivates the older adolescent to make use of his or her newly acquired strength. Complicated skills can be acquired during this period. |

*Source: Design Reference Guide* (prepared by Man Factors, Inc., for the Consumer Product Safety Commission, Contract no. CPSC-C-75-010s, 1976).

## ATTITUDES OF VARIOUS AGE GROUPS TOWARD SAFETY

Since safety should be of the utmost concern to the designer, it is important to recognize the general nature of human behavior and attitudes as they relate to the general question of safety.

| Age Group | Attitudinal Description |
|---|---|
| Infants and young children | Have a great curiosity about everything; anything new or unusual will motivate them to explore, unless the newness is so pronounced that it frightens them. |
| School-age children | Pressures to adjust to peer groups, to others at home and at school, to extracurricular activities, and to community activities tend to make them more cautious. However, the importance of conforming and of being accepted by the peer group tempts them to take chances that result in more serious accidents than those suffered by younger children. |
| Teen-agers | Because of the physiological changes that are occurring, tend toward restlessness, social antagonism, resistance to authority, erratic behavior, instability, and often daring behavior. |
| Young adults | Because of a gain in control over their larger bodies, plus the increase in physical strength that accompanies growth of the muscular system, are motivated to make use of their newly acquired strength, take more chances, and attempt more daring and demanding feats. |
| Persons in the early and middle years of adulthood | Adjusting to new life patterns and playing new roles such as spouse, parent, and breadwinner make them more cautious, more oriented toward efficiency and cost effectiveness, and willing to take only well-calculated risks. |
| Older adults, retirement age | Because of decreasing strength, lowered efficiency of sensory organs, reduced reaction speed, and increased physical handicaps, are less willing to take risks and generally exercise greater caution. |
| The elderly | For a variety of reasons (including physical infirmities and general perceptual narrowing), move more slowly and cautiously when engaging in all activities, their caution sometimes creating dangers when they drive, travel on foot, and use products and facilities. Most significantly, older people tend to lack patience, to be unaware of obvious hazards, and to expect others to "look out" for them. |

*Source: Design Reference Guide* (prepared by Man Factors, Inc., for the Consumer Product Safety Commission, Contract no. CPSC-C-75-0102, 1976).

**GENERAL BEHAVIORAL
EXPECTANCIES DATA TABLE FOR
DESIGNERS**

| Behavior Pattern | Safety Implications | Design Considerations |
|---|---|---|
| Built-in habits and/or natural behaviors and associations: | Some appearances, operations, and procedures seem more natural than others to people; hence their first impulse is to interpret, operate, or use a device according to these expectations. | If the product is designed so that it either is or appears to be contrary to natural expectations, the user frequently makes mistakes, some of which could lead to hazardous situations. |
| 1. People tend to assume that an object, device, or package is small enough to get hold of and also light enough to pick up. | If an object actually is very heavy, users may suffer severe strain or lose their balance and fall. | Either the user should be warned or the design should include some type of fastening to prevent the object from being lifted. |
| 2. People assume that a vehicle is designed to "fit" their physical characteristics and thus try to adapt to it, whether they can do so effectively or not. | Lack of proper fit often causes users to assume a precarious position or attempt a reach effort that not only makes their situation insecure but also forces them to assume a fatiguing position or apply uncontrolled forces and manipulatory patterns. | The vehicle should be adjustable when it cannot be made to fit all users by a single configuration. |
| 3. People have learned (and therefore expect) that water faucets and liquid valve handles rotate counterclockwise to increase the flow of a liquid, gas, or steam. | If such valves are designed contrary to expectations, when such valves have to be turned off rapidly (in an emergency), individuals frequently will turn them in the wrong direction first and then make a correction— which could be too late. | Valve handles should be marked and mounted according to standard practice. No liquid valve should be incorporated into a design that does not operate according to this standard. |
| 4. Electrically powered control switches are expected to move upward, to the right, or clockwise to effect electrical continuity, i.e. to turn power on.* | If such switches are designed contrary to expectations, people may turn an electrical device on when they think they have turned it off, thus exposing themselves or others to a functioning system which they assume is off. | Design or select electrical switches and control handles that are compatible with the expected directions of motion. |
| 5. People expect that when they move a vehicle control to the right, or clockwise, the vehicle will move in that direction. | Incompatibilities with user expectations result in confusion and often in loss of control. | In the design of vehicles, vehicle control motions should be similar. If the operator must turn around to face a control, the vehicle-control relationships become confusing. Avoid such arrangements if at all possible. |
| 6. People expect that when they actuate a knob or switch that causes an instrument pointer and/or some part of a machine to move, the directions of motion of both the control and the pointer (or part) will be the same. | Incompatibilities between the motion of the instrument pointer (or part) and that of the control element lead to erroneous dial settings and movements of subelements, often causing delays to correct the error and sometimes considerable danger to the user. | Knob and switch motion directions should be consistent with motion compatibility rules, and no control relationship should be used that does not produce a consistent natural response.† |
| 7. People have become conditioned to certain color meanings (e.g., red for "danger," "fire," and "hot"; green for "OK," "go," and "acceptable"; yellow or amber for "caution," "look out for," and "yield"; and blue (in the case of electrical devices) for "cold" or "cool." | Misuse of colors may cause observers to ignore a caution or danger signal or to assume the presence of one that does not actually exist. Either condition creates confusion and delay. | Observe current color-coding standards (which may differ for various product categories). |
| 8. People's attention is drawn to bright and vivid colors, bright lights, loud noises, flashing lights, and repeated and undulating sounds. | People may be distracted by an inadequately selected display and fail to attend to an important task detail, or an important task may not be given attention when needed because the device, signal, or sound was not designed to be conspicuous. When their | Use stimuli of adequate intensity when attention needs stimulation; do *not* use high-intensity stimulation when distractions should be minimized. |

*(continued)*

| Behavior Pattern | Safety Implications | Design Considerations |
|---|---|---|
| | attention is drawn to a bright light, individuals may disturb their own visual adaptation level to the point where they cannot see necessary visual detail (as when they face oncoming automobile headlights while driving). | |
| 9. Dark, dull colors appear to advance toward the observer; lighter, brighter colors appear to recede. | People may react instinctively (i.e., "duck") to miss a dark object above their heads, even though there is ample clearance. | Observe normal brightness and color impression conventions in selecting colors for interior surfaces. |
| 10. People expect to face auditory signals in order to corroborate and/or reinforce reception. | Placing an auditory warning signal behind an individual could cause the person to displace his or her eye reference as a result of turning the head or body. If this occurs at the wrong moment, it could cause a vehicle operator to miss seeing something in front of the vehicle. | Locate auditory warning signals so that they appear to come from in front of the operator and also locate them close to head height. |
| 11. People assume a relationship between objects on the basis of their spatial proximity (i.e., they assume that things are related somehow when they are located or grouped together). | If a label or control appears to be associated with a display or another control, the individual's inclination is to use that control to effect a modification of the nearby display, whereas it may be the wrong control. | Careful positioning and/or arrangement of controls, displays, labels, or instructions is required to ensure the proper associations. |
| 12. People generally regard products as being safe; that is, they:<br>a. Assume that a system is *ready* to go and hence do not check.<br>b. Grab a handrail assuming that it is strong enough to support them.<br>c. Assume that a system is not on and hence proceed to touch or manipulate without caution.<br>d. Assume that no one else has turned the system on and therefore do not check to make sure. | Although the assumption of "safe" should not be made, the fact that it is tends to preclude users from *thinking* safety, thus reducing the likelihood that they will anticipate possible hazards and make the necessary check. | Try to design the product so that it cannot be used improperly. If that is impractical, provide guards or other means to cause the user to think about the hazards. Finally, if neither of the above is practical, take steps to warn the user by means of labels or instructions, appropriately coded to imply caution or warning. |
| 13. When about to lose their balance or fall, people instinctively reach for and grab the nearest thing. | Sometimes the things most convenient for grasping are not strong enough to support the individual, or they may be hot or slippery or cause an electrical shock. | Configure the design to provide appropriate emergency supports, or else make sure that hazardous components that might be used for support are not within reach. |
| 14. People instinctively use their hands first to test or explore. | Many injuries occur because users touch an object with their hands or fingers, even when it was intended that the task be performed by means of an intermediate device of some kind. | Either make the product suitable for handling or else make it quite clear that its use requires a device supplied to obviate the necessity for using one's hands. |
| 15. People expect a seat height to be at a level they are used to; a stair angle or tread pattern similar to one with which they are familiar and equally spaced in terms of riser height and tread depth; and a handhold or railing of a size and shape that can be gripped securely. | Individuals tend to adjust their posture, weight, mass, and gait to what they expect. When their expectations are not fulfilled, they are apt to lose their balance, take a wrong step, or trip. | Adhere to recognized standards and guidelines so that furnishings and architectural features are consistent with learned use patterns and are compatible with human physical dimensions. |

*(continued)*

| Behavior Pattern | Safety Implications | Design Considerations |
|---|---|---|
| **Lack of knowledge and/or experience:**<br>1. Across the broad consumer population there is a lack of knowledge about, and experience with, the technological features of many products, e.g.:<br>a. Structural integrity<br>b. Mechanical relationships<br>c. Electricity<br>d. Thermal characteristics<br>e. Gaseous and combustible materials and conditions leading to explosion or fire<br>f. Implications regarding center of gravity and balance<br>g. Acceleration, velocity, and inertial factors | Since consumers generally are not prepared either to anticipate or to analyze conditions or the possible results of incompatibilities among physical elements and phenomena, they inadvertently initiate events that lead to improper and hazardous interactions. And because many consumers do not know what makes structures collapse or tip over, what causes materials to ignite or explode, or what inertial forces will do either to others or to themselves, they do not approach the product prepared to deal with these hazards. | Analyze the potentially hazardous interactions between the naive user and the various product elements to make sure that every means possible of precluding misuse has been designed into the product, that misuse has been warned against, and/or that the potential severity of the hazard has been minimized. |
| 2. Most people know very little about their own physical, sensorimotor, or cognitive characteristics or limitations. Thus they do not recognize:<br>a. How they might cause muscle strain, sprains, or skeletal fracture<br>b. How easily they can be thrown off balance<br>c. How vertigo is induced<br>d. How momentum and centrifugal forces affect their balance or ability to reach or apply force<br>e. What is required for them to see effectively and reliably, hear effectively, and feel (touch) accurately<br>f. The importance of adequate sensory feedback in order to control their actions properly or how various internal and external stresses can modify their interpretation of the feedback phenomena | Product users often do not recognize that they are not getting adequate inputs through their sensory channels or that they are reacting to illusory phenomena, upon which they are basing their decisions, rather than to the true informational input. | Learn more about how the human body works, and about its basic characteristics and limitations and then design in such a way as to minimize overstressing, creating illusory stimuli, or creating sensory contradictions and uncertainties. |
| 3. The average consumer is not familiar with much of the technical terminology used for identification | Typical users do not understand the terminology, or they may misinterpret it and thus proceed on an erroneous basis to misuse the product. | Avoid highly technical terms (i.e., use standard nomenclature and abbreviations) and/or test new terms to make sure that they are clear to nontechnically oriented individuals. |
| **Inattention:**<br>1 Inattention is a phenomenon exhibited by everyone at one time or another, either as a natural trait or because of external and internal stresses, fatigue, boredom, or lack of motivation. | Lapse of attention often results in failure to observe a hazardous condition, read a label or instruction correctly, understand an instruction fully, see a warning light, or watch where one's feet, hands, head, or other body and limb components are going and whether they are clearing obstructions. | Anticipate the possibility of inattention in conceptualizing product design; i.e., think of the design in *use* terms. Where the possibility of inattention may be critical to the proper operation and safety of the product-user system, steps must be taken to preclude misuse and/or to provide means for attracting the user's attention. |
| **Distraction:**<br>1. Many people are easily distracted, either by certain aspects of the product's features or by the procedures for its use. | Distraction from the primary product use task may cause users to miss seeing an important element or event, to fail to observe the path of a part of their bodies or the | Analyze the product design and potential use procedures to eliminate distractions and/or to provide a stimulus to maintain user procedural continuity and reliability, or else design to |

*(continued)*

| Behavior Pattern | Safety Implications | Design Considerations |
|---|---|---|
| | bodily position they have assumed, or to break their train of thought, causing them to omit a step or fail to remember what they have completed. | minimize the possible effects of procedural errors. The environment within which the product will be used must be known, and affective influences must be guarded against. |
| Hurrying: <br> 1. Most people tend to hurry at one time or another. | When individuals hurry, they tend not to observe or to react carefully; they may not have time to think out or arrive at decisions based on a full analysis of a given situation. Thus, oversights may lead to inadvertent omissions, inappropriate responses, or even inaccuracies in sensorimotor performance. | Recognize the probability that a user will be in a hurry and design the product so that the use procedures help pace the user's response and/or so that the possibility of a hazardous consequence of an error or omission is minimized. |
| Deliberate risks and shortcuts: <br> 1. Although risk taking and frequent shortcutting of proper procedures are more typical of certain users (e.g., young people), almost everyone takes some risk or shortcut at one time or another. | When a calculated risk cannot be based on a full evaluation of the potential failure modes and consequences, a product use failure probably will occur. Usually the user has neither the knowledge nor the time to make such an evaluation; hence the probability of failure is high. | Anticipate the possible risk-taking and shortcutting behaviors and design the product in such a way as to preclude shortcutting or make the consequences of omissions less serious. |
| 2. Many people refuse to look at and read signs or other visual warnings. | Failure to see or read labels, instructions, or signs leaves the individual uninformed and therefore more prone to error. | Make critical signs and signals conspicuous, legible, visible, and understandable and place them where the user is expected to be looking. |
| Complacency, neglect, and overconfidence: <br> 1. Because of overconfidence, individuals often proceed without thinking, observing, checking, or reading instructions or labels. <br> 2. Complacency tends to dull people's sensitivity and alertness. <br> 3. Some people fail to exercise caution because of a neglectful and disorganized approach to life. | When people assume that they know how something should be operated, they often make faulty judgments that lead to errors. In some product use situations, alertness is vital to safe performance. | Anticipate possible lack of interest or overconfidence on the part of the user and design in such a way as to alert and stimulate the user to possible critical aspects of product use and operation. The design should be made as simple as possible, and potential hazards should be conspicuously identified and brought to the user's attention. The design should preclude the operational steps' being performed out of sequence. |
| Forgetting: <br> 1. Although everyone occasionally forgets things, some people by their very nature seem to forget more regularly than others or to forget because of stress or fatigue. | Forgetting a critical step, for whatever reason, can have disastrous consequences. People also can forget about something they ordinarily know is dangerous or potentially injurious. | Anticipate the possibility of forgetfulness on the part of the user and design out critical features where such procedural omissions could lead to injury. Where necessary and possible, the design should provide some technique to monitor whether the user has forgotten a critical step. |
| Confusion: <br> 1. People are easily confused by things that are unfamiliar. | Lack of familiar features not only lengthens the time required for a user to respond (perhaps making the response too late for safety) but also may freeze the individual into inaction, resulting in an unsafe condition. | Avoid designing products which are so entirely new that nothing about them (shapes, colors, nomenclature, etc.) is familiar to the user. |
| 2. People are easily confused by complexity (in numbers, sizes, shapes, arrangements, concomitant operations, etc.). | Complexity places considerable demand on human perceptual-motor and cognitive processes, and under these circumstances any outside stress could lead to a collapse of the human system response. | Simplify all operator involvements; i.e., complex decisions should be made by the designer of the product. |
| 3. Most people are confused by | Not only will proper identification | Do not assume that people can |

*(continued)*

| Behavior Pattern | Safety Implications | Design Considerations |
|---|---|---|
| lack of identification of operating interface elements and by the inability to see or find a display or control or to locate the spot where the hand should be placed in order to pick up the device. | improve understanding, but it will also reduce the time necessary to figure out what to do and how to do it. Too much time and/or faulty deductions can lead to misuse or other human errors affecting operator and equipment safety. | recognize product features and understand them without labels or instructions (often labels or instructions are not included because of aesthetic design objectives or cost economies). |
| 4. Many people are unable to perceive mechanical relationships, geometric relationships, or pictorial representations and to interpret these correctly. | If the designer assumes that all people have his or her skill in interpreting relationships and graphics and if the designer uses these to communicate with the typical consumer, the result will be confusion and misinterpretation by the user, resulting in possible misuse. | Avoid using special graphics or depending on relationships that are familiar only to technical people (i.e., design for the nontechnical user). |
| Lack of manual dexterity, skill, or practice:<br>1. Many people either lack the innate capacity for dextrous manipulation or control of their bodies and limbs or else have not developed (through practice) the manual skills needed for many tasks involving the fingers, hands, arms, legs, or feet. | When the use of a product (and the ultimate safety of the operation) is highly dependent on manual dexterity or skill and practice, there is a great opportunity for error at one time or another—error that could lead to an unsafe manipulation or operation. | Try to design products in such a way that they do not require great dexterity, precision, or speed or highly sensitive response to various feedbacks. Designers should understand the innate characteristics of the human sensorimotor servosystem and design so that lags in the human system are taken into account. |
| 2. Some manipulatory and other manual skills are especially degraded when the specific human-product relationship is not optimum (e.g., the position of the operator in relation to the task, the extent and direction of movement, and the rate and rate of change of movement). | Increased physical and mental strain reduces the ability to coordinate body, limb, hand, and finger movements and to make precise direction, rate, and force inputs, and it reduces attention and perceptual awareness of errors. | Human-product design relationships should be "natural," convenient, minimally demanding, and not subject to potential misuse. |
| 3. Some manipulatory skills are especially degraded by gloves and clothing. | Clothing may snag and preclude completion of movement; bulk may cause inadvertent contact. | Provide *ample,* not just minimum, clearance between components. |
| Spurious autonomic responses:<br>1. Fear, anxiety, and uncertainty often lead to irregular, uncontrolled motor responses.<br>2. Panic often leads to unexpected responses (e.g., withdrawal, inaction, and violent and uncontrolled input to a device or control).<br>3. Pain leads to automatic recoil in most cases, although the opposite may be true of small children. | When responses cannot be predicted under these unusual conditions, many undesirable reactions may occur that have adverse consequences in terms of safety, e.g., a driver may freeze to the wheel or accelerator or suddenly exert forces or movements opposite to those which should be exerted, or a person may jump back after receiving an electrical shock, a burn, a sharp cut, or a blow to the body, twisting his or her body suddenly in an effort to recover from a shift in body mass, etc. | Be careful about deciding how users may react to very sudden disturbances in normal situations; these usually do not allow time for thinking, since they are the result of completely involuntary and unpredictable reactions. The design should preclude the possibility or probability of such situations and provide "wiggle room" for the possible reactions so that they do not set off a chain of hazardous events. |
| Special behaviors:<br>1. Children:<br>a. Put things into their mouths. | Choking and strangulation. | Make products too large to reach the child's throat. |
| b. Put their hands and fingers in holes and other openings (including electric outlets). | Entrapment, burns, cuts, shock, electrocution. | Make openings too small and provide covers that children cannot remove. |
| c. Try to squeeze into or through openings. | Entrapment. | Make opening too small for children to fit through. |
| d. Climb up on anything they can (e.g., boxes, furniture, and ladders). | Falling. | Eliminate climbing aids, i.e., handles or openings that can be used as a foothold or handhold |

*(continued)*

| Behavior Pattern | Safety Implications | Design Considerations |
|---|---|---|
| e. Crawl into boxes, sacks, closets, pipes, etc. | Exposure to hazards within. | Provide secure closures and locks that cannot be operated by children. |
| f. Pull on cords, clothes, and tablecloths. | Pulling objects on top of child. | Use cordless systems and place electrical outlets where cords will not be within reach. |
| g. Pick things up and shake them or pound them on other people or other things. | Poking an object into the eyes and breaking windows. | Use soft material that is too heavy to pick up. |
| h. Pound on all vertical surfaces with their hands (including mirrors and glass doors). | Cuts. | Use shatterproof glass or plastic. |
| i. Try to open containers, doors, and drawers and to push all such things closed. | Exposure to hazardous substances or medicine; mashing, pinching, or cutting fingers; and dumping contents of a drawer on self. | Use lids that children cannot open, drawer stops, and large, obvious handles. |
| j. Try to push chairs, stools, and low tables. | Pushing into glass and tipping furniture over on self. | Add sufficient weight and provide proper support and balance. |
| k. Try to reach for things on the tops of counters, stoves, and pieces of furniture. | Burns, cuts from falling glass objects, and bruises from falling objects. | Increase the height beyond the reach of children and place controls and burners out of their reach. |
| l. Try to take everything apart, including toys, books, and all kinds of devices. | Swallowing small parts. | Make as indestructible as possible and difficult to disassemble. |
| m. Try to turn handles on appliances, doors, and other equipment. | Being exposed to hazards outdoors, turning on electrical power or gas, and starting machines, the operation of which is hazardous to the child. | Place handles out of the reach of children and design them so that children cannot operate them (i.e., design handles that require great force, complex coordination, etc.). |
| n. Try to ride, operate, or drive any kind of vehicle they can climb upon or get into. | A variety of injuries if machine moves. | Provide locks to keep children out. |
| o. Push all buttons and pull all levers. | Starting an operation that could lead to hazards if uncontrolled. | Provide locks to keep children out, cover push-button panel, and require special secondary movement before lever will move. |
| p. Put things into any other things that have holes in them. | Equipment malfunction that leads to a variety of hazards. | Cover openings. |
| q. Like to push things over (to see them fall). | Cuts and bruises. | Design so that weight and/or balance precludes tipping. |
| r. Like to poke their fingers and other objects into other people's (especially other children's) mouths, ears, and eyes. | Various physical injuries. | No known precaution except to make all objects less pointed. |
| 2. The aged and handicapped: | | |
| a. Shuffle when they walk and therefore trip over things. | Falling and striking objects. | Provide smooth walkway surfaces and mark steps with a highly contrasting color. |
| b. Hold objects with a weak grip and often unsteadily and therefore tend to drop them. | Cuts while trying to retrieve a broken object. | Use large handles, shatterproof materials. |
| c. Do not see well or observe carefully and hence bump into, trip and fall over, or fall from irregular surfaces or stairs. | Falling, striking objects, and tumbling on stairs. | Provide good illumination and high contrast at changes of grade. |
| d. Do not exert a normal effort to control their posture and hence tend to lose their balance easily. | Falling and bumping into objects, receiving injuries on contact. | Make all walking surfaces level, keep all related objects free from sharp corners or edges, and provide strategically located handholds. |
| e. Do not hear well or listen carefully and hence often are unaware of impending danger indicated by sounds. | Entering a hazardous zone unaware of impending danger, e.g., an approaching vehicle. | Provide visual warning and use an audio warning signal that uniquely penetrates consciousness. |
| f. Cannot apply as much force as younger or unimpaired persons and hence approach and try to | Misapplication of force, leading to operational hazard and potential injury. | Reduce operating force, optimize control to preclude misuse, and encourage proper use by |

(continued)

| Behavior Pattern | Safety Implications | Design Considerations |
|---|---|---|
| manipulate things in an unconventional manner. | | making it the only way in which a product can be used. |
| g. Have a limited perceptual awareness and are less apt to think problems through; hence they observe less and attend less to what they are doing. | Failure to notice a potential hazard, leading to a variety of consequences. | Maximize perceptual cues (i.e., more contrast and more conspicuous features) and minimize complexity. |

*These expectancies are peculiar to Americans. Europeans expect the reverse, i.e., to turn a switch down or to the left in order to turn power on.

†See W. E. Woodson and D. W. Conover, *Human Engineering Guide for Equipment Designers*, University of California Press, Berkeley, 1964.

## TYPES OF SPACE

Besides needing enough space in order to move about and perform various tasks, people react to space in a variety of ways. Several researchers have defined the space surrounding the individual in terms of the limits within which people categorically respond (see the accompanying sketch). *Intimate space* is that area in which a person tends not to allow anyone to intrude unless intimate relationships are expected. *Personal space* is that area within which a person allows only selected friends or fellow workers with whom personal discussion is mandatory. *Social space* is that area within which the individual expects to make purely social contacts on a temporary basis. And, finally, *public space* is that area within which the individual does not expect to have direct contact with others. Obviously, the more intimate the spatial relationship becomes, the more people resist intrusion by others. Personal space factors are important in establishing the privacy requirements for architectural design.

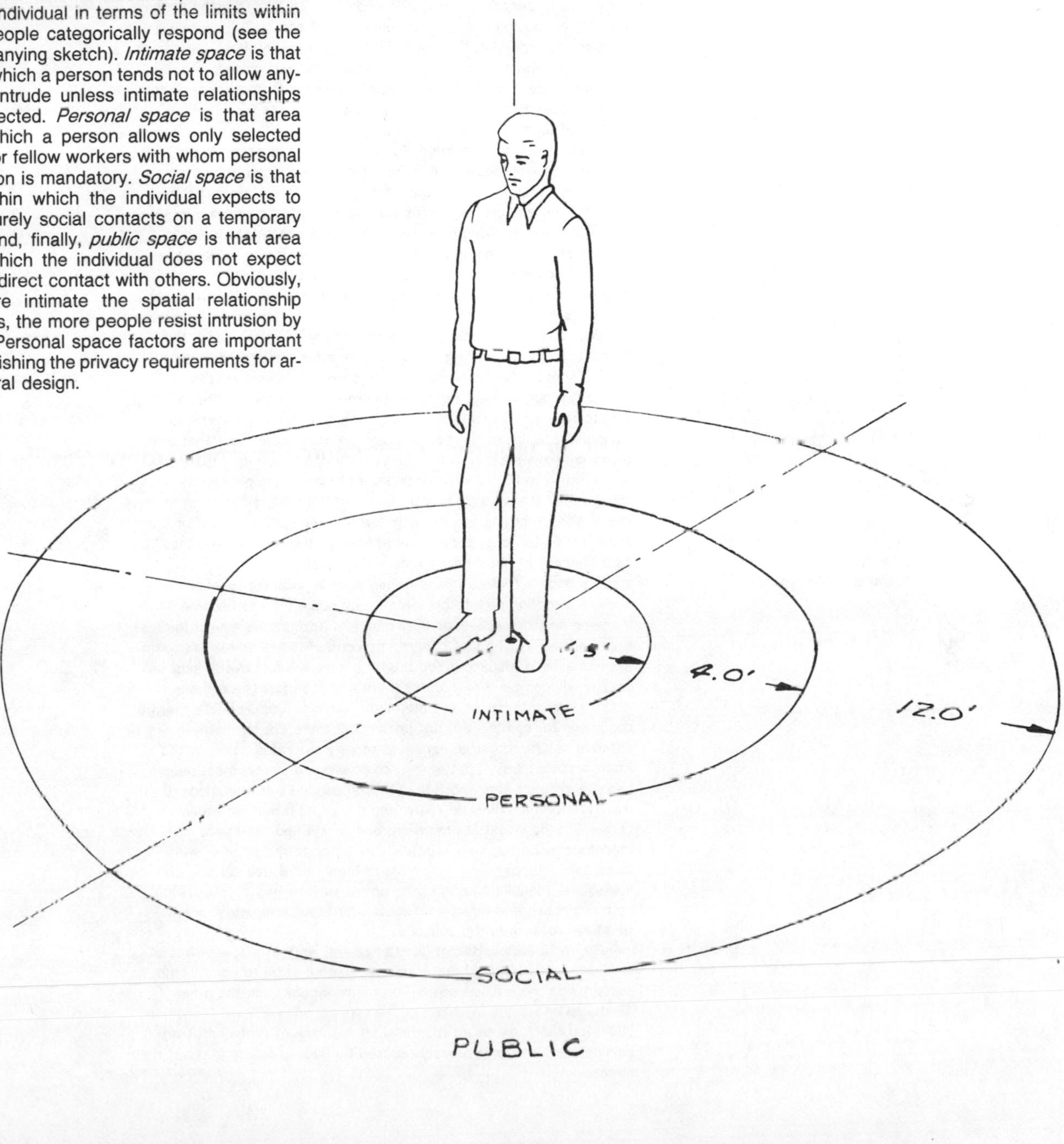

INTIMATE · PERSONAL · SOCIAL · PUBLIC · 1.5′ · 4.0′ · 12.0′

## TYPICAL SUBJECTIVE RESPONSES TO SELECTED SPATIAL FEATURES

Although few research data have been generated with regard to how people respond to specific spatial factors (at least in terms of being able to prescribe precise, quantitative guidelines), it is important for the designer to reflect on potentially negative reactions that often result when a given space is not made compatible with what the user expects in terms of the size, shape, organization, color, and illumination of a particular space. The considerations listed in the accompanying table are suggested as a checklist for the designer.

| Space Characteristic | Probable Response |
|---|---|
| Size (generally volume) | If the space is too small for the number of people, furnishings, equipment, or other objects that occupy it, people will consider it to be crowded. Although they may accept a crowded condition on a temporary basis, they will object to living or working in such a space for extended periods of time. If the space is too large for the people, furnishings, equipment, or other objects that occupy it, people will consider it "unfriendly," inconvenient, and/or overly demanding in terms of communicating, travel distance, maintenance, etc. Although they may accept the "barnlike" atmosphere for temporary periods, they will object to living or working in such a space for extended periods of time. |
| Shape (generally proportion) | If the space is out of proportion (too narrow, wide, long, high, etc.) for the intended use, people will consider it awkward and often distracting or oppressive. Although they may accept proportional distortion on a short-term basis (i.e., as they pass through briefly), they will object to living or working in such a space for extended periods of time. If the space contains such distortions as all curved surfaces, acute wall junctures, and too many projections or surface changes, people will consider it confusing and difficult to maneuver in and/or furnish. Although they may accept such distortions (or even consider them interesting) on a temporary or one-time basis, they will object to living or working in such a space for extended periods of time. It should also be noted that blind people depend on the constant proportions of right-angle corners to aid them in negotiating a space; such individuals are easily confused by curved surfaces, walls that are not at right angles, and periodic projections that imply they may have reached a turning point. When a ceiling is extremely high relative to the lateral dimension of a space, people feel as though they are working in a pit and that the walls are closing in on them. When a ceiling is extremely low and the space in front of the observer is very long, people feel as though the room is "endless" or as if they will hit their heads unless they duck. |
| Color and illumination | If a space is dark (unless this is required for a particular operation, such as a motion picture presentation), people tend to become lethargic and less active, or they may feel anxious. As a rule, the less bright a room is, the less cheerful it seems. A small space will seem even smaller. If a space is too bright, people will feel overly exposed, or they will complain of glare or thermal discomfort (even though actual glare in terms of accepted light levels or inappropriate thermal conditions for comfort are not present). If there are too many different colors, too large expanses of very saturated color, or too many and too "busy" patterns of color within a space, most people become irritated after more than a brief exposure to the space. If there is too little color, no visual pattern, or no other decorative "break" in the visual environment, people will find the space monotonous, boring, and eventually irritating to the point of wanting to escape. Although isolated points of highly reflective surface provide interest, all-metallic and highly reflective surface treatments create both subjective and directly objective interference for most people who have to work in the space. |
| Windows | Generally, most people do not like to live and work in a space that is devoid of windows. First and foremost, people seem to need visual contact with the outside world. Too many windows, on the other hand, can cause the following possible negative reactions: too much glare, too much exposure (fishbowl effect), lack of protection from outside elements, true anxiety (caused by floor-to-ceiling glass at high elevations). *(continued)* |

| Space Characteristic | Probable Response |
|---|---|
| Space organization | The internal components within a space and the traffic corridors and entrance and exit locations will seem either well organized or badly organized. The furnishings, partitions, decorative objects, etc., will appear as being either organized or disorganized, depending on the observer's ability to comprehend what things are and where they are with respect to his or her vantage point. Key behavioral response issues are: apparent capability to find one's way to specific locations, apparent ease for interacting and communicating with others with whom the individual must associate, apparent privacy provisions necessary to perform individual tasks. Although these are sometimes conflicting needs, the people who use a space will perform on the basis of how well each of these factors has been executed for *them,* not for the designer or the boss. The organization of internal space components obviously interacts with all the other space characteristics; i.e., the individual perceives and reacts to the combined effects of size, shape, color and illumination, windows, and organization simultaneously. A significant behavioral response will be an individual's interpretation of whether sufficient options are available for local modification of his or her own portion of the space. Even though people may never require a modification, they react to their own space in terms of permanently established restrictions that eventually elicit the feeling that the space is too small, the wrong shape, too dark, or isolated from the rest of the world, for example. |
| Furnishings | As a general rule, people are sensitive to improperly proportioned furniture, i.e., furniture that is too large, too small, or the wrong shape for the space in which it is placed. Although the designer normally tries to select furnishings that are properly proportioned for the space he or she has created, this may ultimately restrict the efficiency of the individual (e.g., a desk or storage cabinet may be too small). Thus, although the general visual proportions of furniture in relation to space must be taken into account to avoid negative observational responses, shortchanging the individual in terms of specific furniture and use requirements soon stimulates an even stronger negative response. |

## Unique Responses to the Individual's Location within a Space

Observation and study of what people do when they occupy spaces provide several important thoughts with respect to where a designer expects to place people. For example:

People in auditoriums will not sit next to a side wall, especially if the wall is very high.

People generally dislike sitting facing a wall that is too close, unless they can look out a window.

People generally prefer to sit in a position from which they can observe the entrance to the room and also observe other people, but they do not like to sit so that they are directly observed by others (or think they are).

People dislike having to face other people (in a waiting room, airplane, train, etc.) as they come out of a rest room.

People coming into an auditorium from the rear will tend to take the back seats, not those farther toward the front. Similarly, they will take the seat nearest an aisle, rather than one toward the center of a pew or row of seats.

In a conference situation, most people will take a seat at the corner of the conference table, not an end position or one halfway along the side of the table.

People generally will sit facing a light-colored rather than a dark or highly (saturated) colored or patterned wall.

When there is a row of chairs or stools, most people will seek out one that is not next to one that is occupied.

## PSYCHOLOGICAL MOOD

Many attempts have been made to define basic human needs in relationship to designing an environment that is satisfying to the eventual customer, user, or occupant of living and working spaces. The accompanying model illustrates one such attempt to formalize human needs.

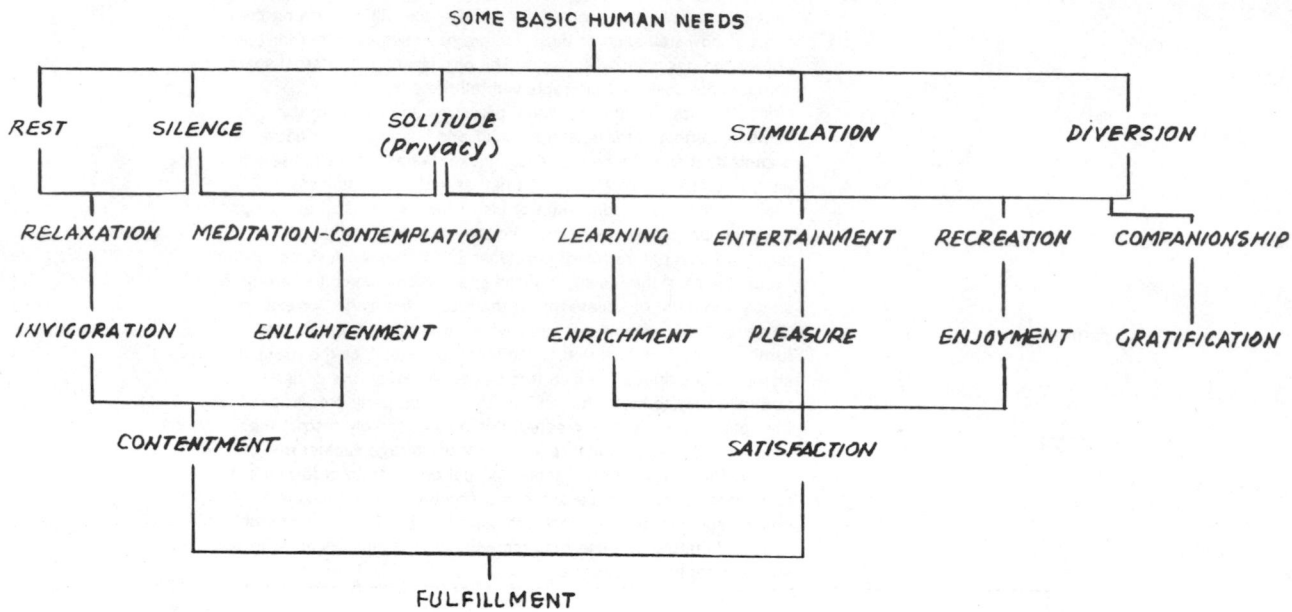

SOME BASIC HUMAN NEEDS

REST    SILENCE    SOLITUDE (Privacy)    STIMULATION    DIVERSION

RELAXATION    MEDITATION-CONTEMPLATION    LEARNING    ENTERTAINMENT    RECREATION    COMPANIONSHIP

INVIGORATION    ENLIGHTENMENT    ENRICHMENT    PLEASURE    ENJOYMENT    GRATIFICATION

CONTENTMENT    SATISFACTION

FULFILLMENT

Complete fulfillment probably is beyond the expectation of most people. However, certain negative factors may seriously affect moods, attitudes, and eventual response to the environment. Some of these factors are the following:

Boredom
Apathy
Anger
Fatigue
Depression
Distress
Frustration
Anxiety
Pain

Although any or all of these may not be controllable through design, they can be influenced by it, i.e., by the nature of the task, by the environment within which the task is performed, and by how much stress or lack of stimulation the task involves.

## EXPECTANCIES WITH RESPECT TO PRODUCT APPEARANCE

Although there are apparent relationships between various aspects of a product's appearance and the attitudes individuals may express toward it (especially in terms of what people decide to buy), there is little quantitative information to assist the designer in knowing what to do to make a product more acceptable. Key issues regarding product appearance include the following:

Size, shape, and proportion
Color
Texture
Apparent functionality
Compatibility with contemporary concepts (and/or with the style of a particular period, e.g., early American)

It becomes obvious that each of these characteristics hinges upon the particular visual environment within which the product will operate. For example, the size of furnishings needs to be compatible with the size of the space within which the furniture will be placed; the color of the product should be compatible not only with its general surroundings but also with the colors of the other products within these surroundings; the texture of the product should fit with the other textures used in the operating environment; the product must appear to have a purpose within the given environment; and, finally, the product should not appear outdated, unless it is obvious that it is meant to look like an antique, for example.

There is some validity to the theory that an attractive product will be treated with more respect than one that looks "cheap" or poorly designed. However, such concepts are difficult to quantify and write specifications for. On the other hand, it is important to recognize that most people soon neglect products that are hard to maintain in terms of good appearance.

A balance is required between making a product conspicuous or gaudy (to attract the user) and making it so dull and functional that the user tends to neglect it or tire of it after a short period of time.

Almost any product can be made to look attractive *after* its basic functionality has been established. That is, although attractiveness is important at the point of initial purchase, the consumer soon becomes disenchanted if the product is not functional and easy to use.

A product should be attractive from all expected viewing angles; i.e., there should be apparent continuity and completeness from each point of view.

A product should have the appearance of being substantial and sturdy; i.e., it should not appear too fragile and easily damaged.

## HUMAN FATIGUE

The problem of fatigue is very important to the consumer, and therefore it should be of concern to the designer; i.e., at least a design should be such that it does not result in activities that are unnecessarily fatiguing in terms of operation or maintenance.

A design may create two different types of fatigue: physiological fatigue, in which the operator's muscles are overstressed, and psychological fatigue (or mental fatigue), which may be caused by design-induced stress, i.e., complexity, high accuracy demands, or environmental implications, such as noise.

## MUSCLE FATIGUE

Muscle fatigue occurs because of certain biochemical reactions within the muscle. Nerve impulses acting on a muscle fiber initiate a series of chemical reactions which result from, and contribute to, muscle contractions. Adenosine triphosphate breaks down under the influence of enzymes to create adnosine diphosphate, which in turn releases energy to enable the muscle to perform work. The triphosphate has to be regenerated before another contraction can occur. Energy required for this to happen is supplied by a breakdown of glycogen to lactic acid. Lactic acid is a poisonous by-product that must be removed by oxidation to carbon dioxide and water. The oxygen thus removes the by-products of the energy-producing reaction, which may continue some time after the muscular activity has ceased. The energy for muscular activity thus comes from a reaction which does not depend primarily on the presence of oxygen and which allows work to be done even when the immediate supply of oxygen is insufficient. This often permits the body to make a sudden extreme effort that might otherwise be impossible if the energy always had to be obtained directly from the oxidation of some substance within the muscle fiber.

Oxygen may come either from that stored in red muscle fiber or from the blood supply. As long as the supply is adequate to preclude lactic acid buildup, work is classified as "aerobic." If, however, the rate of work exhausts the reserve of oxygen, the work is said to be "partly anaerobic," and the muscle builds up an oxygen debt; i.e., there is an accumulation of lactic acid in the muscle and bloodstream, causing pain or muscular fatigue.

## PSYCHOLOGICAL FATIGUE

The exact nature of psychological or mental fatigue is less clearly understood since it is extremely difficult to isolate and quantify causes and effects. Mental fatigue also confounds the researcher because opposite conditions may produce essentially the same mental or psychological aberration; i.e., one can become just as fatigued by boredom as by overwork. Unfortunately, we know of many instances in which individuals have been able to summon some reserve during a crisis, and thus we tend to believe that, with the proper stimulation, most people can continue mental activity indefinitely. Considerable evidence to the contrary, however, shows us that when an individual works too near mental capacity for long periods, almost any emergency that suddenly occurs may push the individual beyond his or her capacity to cope, the result often being a complete collapse or disorientation. Finally, the question of the threshold of psychological fatigue is further confounded by the fact that an individual's threshold may be stressed to within a few degrees of tolerance by preoperating conditions (prior activities), with the result that he or she has no tolerance to cope with an overly demanding mental task or situation.

### Design Implications Relative to Minimizing Potential Muscle Fatigue

Avoid the following:

Designs that require operators to apply near-maximum force capacities over many cycles and for long periods of time

Designs that require continuous, rapid, repetitive muscle contractions for long periods, e.g., pounding, tapping, cranking, or push-pull cycling

Designs that force operators to "hold" some device in a fixed position for long periods without intermittent rest periods

Designs that require operators to maintain an upright posture for long periods without adequate body support (as in the case of a seat)

Designs that require operators to make very long reaches, frequently and for extended periods of time

Designs that require operators to stand or sit in an awkward position and to hold their arms above their heads for a long period of time

Designs that require operators to work in a "bent-over" or squatting position or in a position on their stomachs or backs, with the accompanying stress of holding the head and arms in a strained position

Designs that require operators to bend over and straighten up frequently and over a long period of time

Workplace layouts that require many steps, repeated again and again over a long period of time

Workplace layouts that require operators to sit "askew" (in a twisted position) in order to watch a display and at the same time operate some control (especially a foot control)

Workplace layouts that require operators to hold one foot above a foot control (between pedal depressions) for long periods of time

Workplace layouts that require operators to continuously move their heads from side to side or up and down

Workplace layouts that require operators to step up and down frequently for long periods of time

Tool designs that require operators to hold and push a tool against a work surface or component to maintain contact pressure

Tool designs that require operators to hold a very heavy tool in a precise position for a long period of time

Tool or other equipment designs that require operators to maintain a very tight grip to keep the tool in place (especially if the grip must be maintained for a long period of time)

## POSTURE IN RELATION TO FATIGUE

Incorrect posture produces both physical and mental fatigue. As the lower sketches illustrate, the least fatigue occurs when the body can be kept in balance, since that is when there is the least demand on the muscular system to keep the body upright. The musculoskeletal system is most nearly balanced in the standing position. Even in this position, the various sets of muscles require fairly constant flexing to minimize fatigue; i.e., if the body were maintained in a "stiff," erect position, as when a person is at attention, the muscles would be in a state of continuous tension, and thus considerable fatigue would result. On the other hand, a generally erect and balanced position in which the muscles are constantly flexing a slight amount uses up the least amount of energy and produces the minimum amount of fatigue.

When an individual sits, the same principle of muscle balance and minor flexing is desirable. However, in order to create this balance, the body needs to be positioned so that the head mass is easily erected in a generally vertical axis above the torso and buttock masses.

As the accompanying sketches illustrate, the best and least fatiguing posture is one in which it is easy to position the head approximately in a vertical column above the torso and buttock masses (in a seat). If a seat, for example, is sloped too far back, as shown in the second sketch, the individual will naturally pitch his or her head forward to try to reestablish the columnar effect. If the seat is at 90°, as shown in the third sketch, the individual naturally pitches his or her head back to reestablish the columnar effect. Either of the latter conditions produces considerable strain on the neck and back muscles; this is why it is important to provide seats that have the proper seat-pan and seat-back angles.

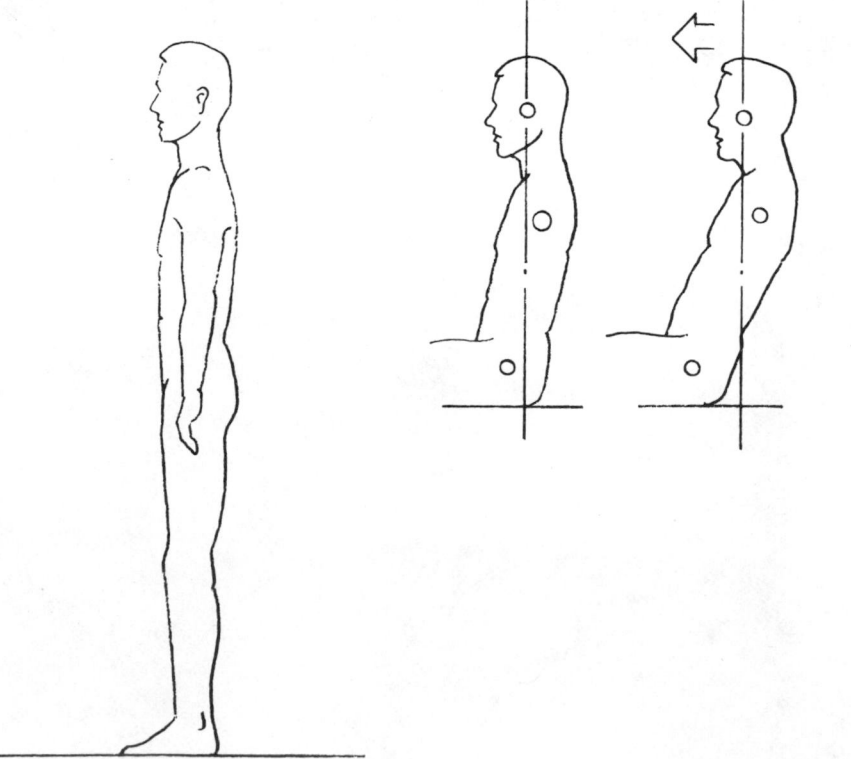

**Common Fatigue-Producing Postures**

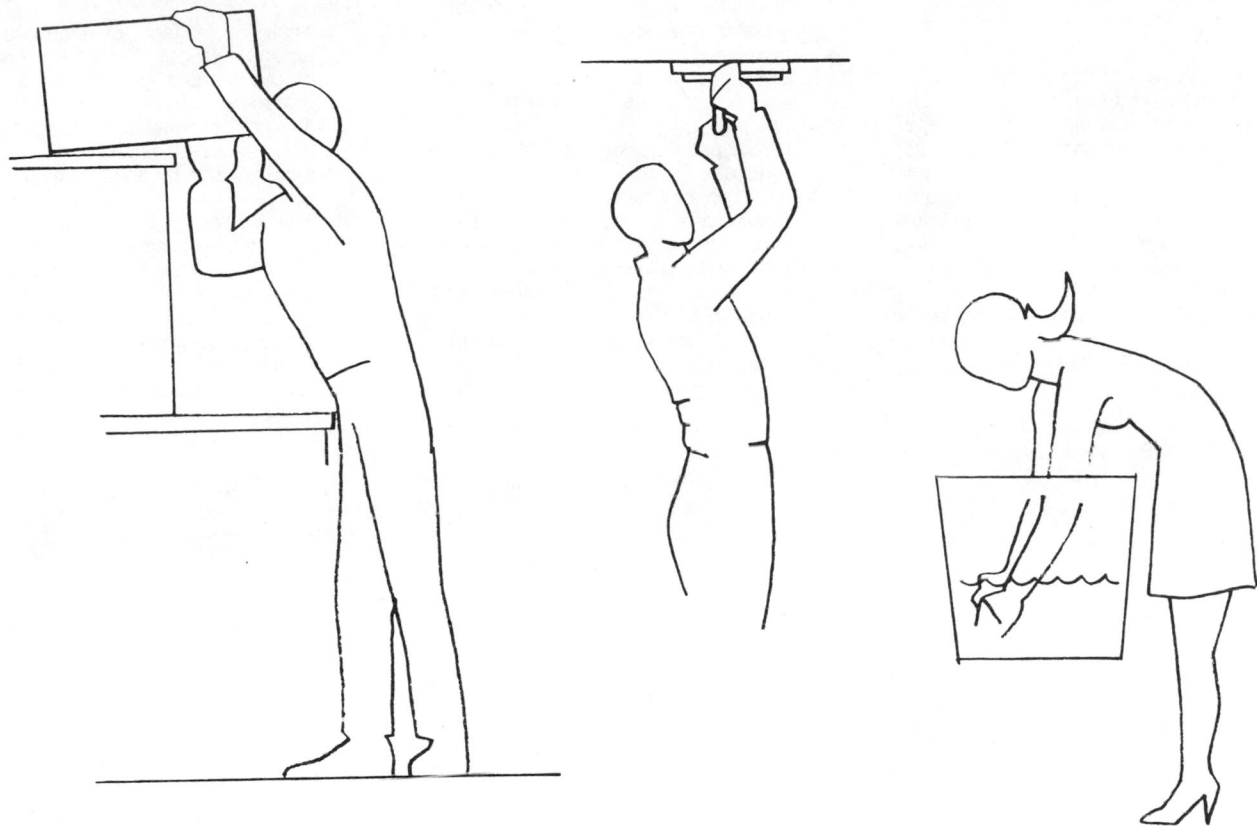

Repetitive lifting to an awkward position. An extended work task located above the worker's head. Bending over for extended periods.

Lack of attention during work-station design planning often results in the worker's being subjected to severe strain, especially if his or her task requires assumption of these poor postures for a considerable length of time. Neural and skeletal disorders often can be traced to the workplaces illustrated.

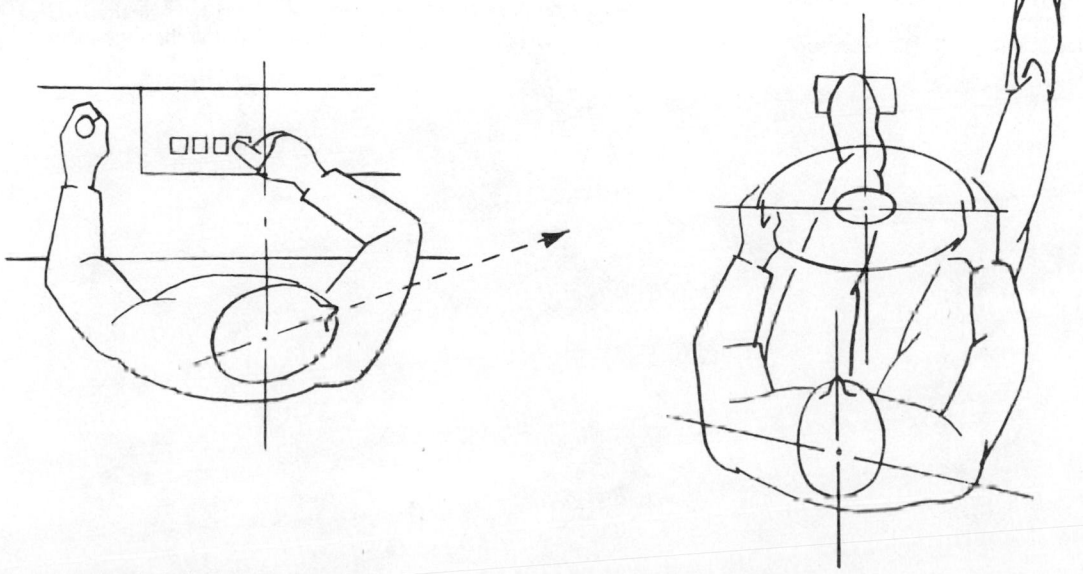

Sitting for long periods in a skewed position.

## EQUIPMENT DESIGN AND OPERATING FEATURES KNOWN TO CONTRIBUTE TO MENTAL FATIGUE

1. Too many separate visual displays have to be monitored simultaneously.
2. Visual display formats require extrapolation rather than providing directly usable information.
3. Visual display detail is considerably greater than required; e.g., there are more scale marks than are warranted by the inherent accuracy of the instrumentation.
4. The legibility of visual display details is borderline, requiring unnecessarily close scrutiny in order to detect, recognize, and interpret what is being displayed.
5. Visual displays vibrate because they are not properly shock-mounted.
6. Visual displays are not adequately illuminated, or there are uncontrolled glare sources within the critical viewing envelope.
7. Simultaneous audio communications and/or excessive background noise is present.
8. Overly precise control adjustments are required. ,
9. Poor control dynamics, in terms of force-displacement, control-display, direction-of-motion, and/or movement ratio incompatibilities, are present.
10. There are long delays in informational feedback, i.e., long periods between signals or changes in equipment status.
11. There is a lack of timely indication of whether the equipment is functioning properly.
12. Continuous manual monitoring or control tasks are required that could just as well be automatic, with periodic operator alerting.
13. There is a lack of standardization among various similarly operated pieces of equipment, thus requiring operators to shift their point of reference.
14. The control panel layout is poorly organized, making it necessary for the operator to search for appropriate panel elements.
15. The workplace environment is inadequately controlled in terms of:
    a. Lighting, temperature, humidity, ventilation, noise, vibration, acceleration, pressure, etc.
    b. Support furnishings (standing platform, seating, writing surface, reference storage, restraint system, etc.).
    c. Space, e.g., clearance.

## WORK OVERLOAD VERSUS BOREDOM

Mental stress and/or fatigue may be due either to an overload condition or to sheer boredom. Generally it is rather obvious when a design creates an overload condition, i.e., when the operator is required to do too many things at the same time. However, boredom is more difficult to cope with in terms of what can be done to a particular design. It has been suggested that one of the primary methods for dealing with the problem of potential boredom is to make sure that the human is given tasks that are more suitable for his or her unique capabilities, namely, planning and decision making. Alternatively, the machine or equipment should be given those tasks which are more difficult for the human and/or which are monotonous and unchallenging. The following general guidelines are suggested:

*To prevent work overload:*

Sequence tasks rather than creating overlaps.
Make individual tasks short.
Minimize task precision requirements.

*To prevent boredom:*

Provide task variety.
Distribute tasks equally throughout the work period.
Assign the operator only significant tasks and make it clearly evident that the operator rather than the machine is in control.

# 5

# Human
# Engineering
# Methods

## INTRODUCTION

Whenever practicable, human engineering specialists should be used to help identify and solve human engineering problems during product or system development. However, this is not always possible, and thus the purpose of this chapter is to provide a number of guidelines to assist designers in doing their own human engineering.

The materials presented in this chapter follow somewhat closely the human engineering program practices of the U.S. military services as these have been spelled out in human engineering specifications and standards. They deal, for instance, with requirements to perform certain analyses during the early conceptualization of systems, with human factors contributions to, and responsibility for, monitoring the human engineering aspects of hardware design; and with the test and evaluation of the design at various stages during and following the design and development process.

Although these guidelines were originally developed for use in military systems development, they have been found to be equally applicable to any hardware development program, large or small. Because the methods are based on a logical and systematic process of (1) establishing the proper role of the human in the system, (2) designing the human-machine interfaces to fit the human's capabilities and limitations, (3) evaluating and testing to see that the design does fit, and (4) properly training the human to finally close the loop and thus assure reliable, total human-machine performance effectiveness, they should apply to all products that are used or operated by humans.

## A GENERAL PHILOSOPHY

There is a tendency to believe that human factors engineering occurs naturally either as part of the typical engineering process or as the result of independent scientific endeavors. However, the truth is that a wedding of these is required wherein the results of scientific research are integrated into the design on a timely basis. To accomplish this successfully, the engineer or designer must be alert to the need for human factors inputs soon enough to incorporate good human engineering principles, and the human factors scientist must be sufficiently cognizant of the designer's technical problems to perform human factors research in a manner that is applicable to design. So-called human engineers are those individuals who have gained the necessary experience and knowledge to perceive design needs and interpret scientific information in a way that allows these needs to be met. In other words, the human engineering specialist is part engineer-designer and part human factors scientist.

The methods and techniques discussed in this chapter have been developed over a period of many years and have been designed specifically to solve human engineering problems. The methods to be covered include:

System requirements definition
Functional requirements definition
Human-machine function allocation
End-item design requirements definition
Personnel requirements definition
Human engineering application
System test and evaluation

The techniques covered include:

Function-flow block diagraming
Information-flow analysis
Operational sequence diagraming
Task analysis
Time-line analysis
Link analysis
Mockups and simulation
Human-machine performance testing

For help in more complex human factors experimenting, the reader is urged to get in touch with a qualified behavioral, physiological, or medical scientist.[1]

## HUMAN FACTORS ENGINEERING EMPHASIS

The following quotes[2] from Air Force Regulation 800-15 are provided to indicate the importance that military agencies attach to human engineering aspects of materiel development:

> This regulation establishes the policies and responsibilities for incorporating the human engineering, biomedical, manning, test and evaluation, and training aspects of human factors into the mainstream engineering and program management effort of all acquisition programs and conceptual studies. It applies to each Air Force organization engaged in, or supporting, acquisition programs as defined in AFR 800-2.

Specific requirements include the following:

1. *Human Factors Engineering (HFE)* Human performance is an integral part of total system or equipment performance whenever human interface is required.
   a. HFE is concerned with those engineering and management tasks required to provide for effective human performance (both operations and maintenance) in a system. Accordingly, HFE must receive management consideration (in terms of both program priorities and development approach) equivalent to that given all other components of the system.

   b. HFE is a part of the mainstream engineering effort throughout the system life cycle. It is that component of systems engineering which seeks to optimize the system by integrating the human performance necessary to operate, maintain, support, and control the system in its intended operational environment (see AFR 800-3).
2. *Policy on the Management of Human Factors Engineering*
   HFE must be an integral part of R&D planning, conceptual study efforts, exploratory, advanced, and engineering development projects, equipment procurements, modifications, and system acquisition programs where the intended end product has human performance as an integral part.

   The system or program manager will take special precautions to coordinate HFE elements with integrated logistic support (ILS) elements (see AFR 800-8).

ELEMENTS OF HUMAN FACTORS ENGINEERING

In each of the following elements the functions to be performed represent activities to be managed and controlled. The consideration of HFE requirements must begin with the inception of the system or equipment life cycle. The scope of this effort depends on the nature and the type of system or equipment program, and must be tailored to meet specific program objectives. These elements are:

a. *The Human Engineering Element*— Human engineering is the application of knowledge about human capabilities and their limitations to the system or equipment design, to achieve desired system performance requirements through the most effective use of man's performance capability.
b. *The Biomedical Element*—The Biomedical element includes every area that requires provisions for the promotion of health and safety—and for the protection, sustenance, escape, survival, and recovery of personnel employed within the total system environment. This support will be provided for operations, maintenance and support personnel under both normal and emergency conditions. It will include health protection from conditions resulting from system functions for any personnel who are not included in the total system complex (see AFR 161-2), but who will be affected by the system.
c. *Manpower and Personnel Requirements Element*—This element includes the development of manpower and personnel requirements to insure that enough trained people are available to operate, maintain, control, and support the system or equipment. Data developed in managing this element serves as the basis for manpower and personnel planning and programming decisions. These HFE requirements must be considered early enough in the acquisition life cycle to permit the training and assignment of trained people to manpower authorizations at the initial operational date of each system.
d. *Training Element*—This HFE element includes all training provided, conducted, or managed by the using command, ATC or the contractor. It incorporates, as a minimum, the trained

personnel requirements, training plan, training equipment development, training, training support data, and training facilities. All components of this element will be defined, designed, procured, or conducted based on (and justified by) a task analysis of system requirements. This task analysis will be performed in accordance with the ISD process when feasible and cost effective (see AFM 50-2). The subelements of this element are:
   (1) System Trained Personnel Requirements
   (2) Training Plan (TP)
   (3) Training Equipment Development (TED)
   (4) Training Facilities
   (5) Training Support Data (TSD)
e. *The Human Factors Test and Evaluation Element*—This element is part of the system test effort and will be conducted as directed in AFR 80-14. It is concerned with determining whether Air Force personnel, with system training, can in fact operate, maintain and support the system in its intended operational environment. HFE testing generally assesses these factors:
   (1) Whether human engineering requirements and criteria have been incorporated into the system design and are adequate.
   (2) If biomedical and safety criteria have been met.
   (3) If the system provides for efficient human performance in its intended operational environment. Additionally, HFE testing can consider the following factors:
      (a) If personnel planning information (PPI) is appropriate, complete, and adequate.
      (b) If job performance aids are effective and adequate.
      (c) If training and training equipment requirements have been met.
f. *Annex to System Test Plan*—The coordination of HFE inputs for developmental test and evaluation is the responsibility of the HFE Team assigned to the program office.
   (1) This input may be prepared as a separate annex to the system test plan, so that the HFE test and evaluation will be consistent with the total system objectives.
   (2) The test plan for operational testing of the system or equipment must identify the specific responsibilities for carrying out each HFE part of the test, and must provide sufficient access to the system and equipment to make valid HFE testing possible.

Each of the U.S. military services imposes similar requirements on the hardware system developer relative to human engineering, and each service has a series of instructions, notices, etc., that identify these requirements. Authority for such requirements is established by Department of Defense Directive DOD DIR 5000.1, *Acquisition of Major Defense Systems*. Designers who are about to engage in hardware development for the military should acquaint themselves with the most recent issue of DID DIR 5000.1 and with other documents of the particular service for which they intend to design.

All the services utilize two principal human engineering references: MIL-H-46855, *Human*

---

[1]An excellent introduction to methods for conducting human factors experiments is presented in Alphonse Chapanis, *Research Techniques in Human Engineering*, The Johns Hopkins Press, Baltimore, 1959.
[2]U.S. Department of the Air Force, AF Regulation 800-15.

*Engineering Requirements for Military Systems, Equipment and Facilities,* and MIL-STD-1472, *Human Engineering Design Criteria for Military Systems, Equipment and Facilities.*

Although, obviously, not all products are designed for the military, nor do all products have to meet the rigid requirements of the services, the basic objectives defined by these documents are useful guides for the development of any product or facility. This is particularly true since no similar requirements have been defined by domestic, commercial, or private agencies. That is, the need for human engineering in all product design is similar even though there are no formal requirements to which one can address one's intended design program.

Whenever feasible, it is advisable to enlist the services of a knowledgable human engineering specialist to assist in resolving human factors problems, especially when the project is intended for a U.S. military service. This is also recommended for other programs, although it is often more cost-effective to utilize a consultant, especially if the product development is of insufficient magnitude or complexity to warrant hiring a full-time human engineering staff.

## ORGANIZATION/ACTIVITY INTEGRATION

| GROUPS | ACTIVITIES | END PRODUCTS |
|---|---|---|
| — SYSTEM ENGINEERING | ○MISSION ANALYSIS | ○OPERATIONAL REQUIREMENTS |
| — ENGINEERING DESIGN | ●FUNCTION ANALYSIS | ○FUNCTIONAL REQUIREMENTS |
| — INTEGRATED LOGISTICS | ●DESIGN TRADE-OFFS | ●FUNCTION ALLOCATIONS |
| — SAFETY ENGINEERING | ●DESIGN REQUIREMENTS SPECIFICATION | ●DESIGN PERFORMANCE REQUIREMENTS & TEST PLAN |
| — TEST ENGINEERING | ●PRELIMINARY DESIGN | ●PERSONNEL REQUIREMENTS |
| — HUMAN ENGINEERING | ●DETAILED DESIGN | ○DESIGN SPECIFICATIONS |
| — RELIABILITY/QUALITY CONTROL | ●INTEGRATED LOGISTICS PLANNING | ○DESIGN DRAWINGS |
| — TRAINING | ●MANNING/TRAINING REQUIREMENTS | ○LOGISTICS PLAN |
| — PUBLICATIONS | ○SUPPORT SYSTEMS/PRODUCTION PLANNING | ○PUBLICATIONS PLAN |
| | ●TEST/EVALUATION PLANNING | ●TRAINING PLAN |
| | | ○FACILITIES PLAN |
| | | ○PRODUCTION PLAN |
| | | ○PUBLICATIONS PLAN |

HARDWARE
SOFTWARE
FIRMWARE
FACILITIES
TOOLS, AIDS, PUBLICATIONS
TRAINED PERSONNEL
SPARES

Note: Human Engineering should interact with all groups as indicated in the left column, and their activities should include involvement in each of the items shown in the second column (open circles indicate secondary responsibilities, darkened circles indicate a major contribution), with significant responsibility for end products in the third column (darkened circles indicate major contributions).

## ESTIMATING THE HUMAN ENGINEERING EFFORT REQUIRED FOR SYSTEM DEVELOPMENT

The checklist below may be used as a reference for preparing the human engineering program for most system development programs. A minimum of 2 percent of the total program cost is required for effective human engineering participation.

**Typical, Expected Human Engineering Services in System Development**

| Service Category | Deliverable End Product | Labor, in Days, by Subsystem | | | | | | | | Total Labor, in Days |
|---|---|---|---|---|---|---|---|---|---|---|
| | | A | B | C | D | E | F | G | H | |
| A. Program Planning | HF Prog Plan | | | | | | | | | |
| B. Requirements Determination | | | | | | | | | | |
|    1. Documentation review | Report | | | | | | | | | |
|    2. Field studies | Report | | | | | | | | | |
| C. Concept Development | | | | | | | | | | |
|    1. Systems | Report | | | | | | | | | |
|    2. Equipment | Report | | | | | | | | | |
|    3. Facilities | Report | | | | | | | | | |
| D. Analysis | | | | | | | | | | |
|    1. Information flow | Info. fl. chart | | | | | | | | | |
|    2. Time line | T. L. analysis | | | | | | | | | |
|    3. Performance prediction | Report | | | | | | | | | |
|    4. Gross task | Gross T. A. | | | | | | | | | |
|    5. Fine task | Fine T. A. | | | | | | | | | |
|    6. Operation sequence and interface | OSD 1-4 | | | | | | | | | |
|    7. Human-machine | HM fl. chart | | | | | | | | | |
|    11. Hazards | Report | | | | | | | | | |
|    12. Habitability | Report | | | | | | | | | |
|    13. Trade studies | Report | | | | | | | | | |
|    14. Allocation studies | Alloc. Table | | | | | | | | | |
|    15. Information requirements | Info. Req. Table | | | | | | | | | |
|    16. Cont./disp./com. | C/D/C Table | | | | | | | | | |
| E. Design and Development | | | | | | | | | | |
|    1. Performance specs | Spec. | | | | | | | | | |
|    2. Hardware specs | Spec. | | | | | | | | | |
|    3. Software specs | Spec. | | | | | | | | | |
|    5. Mock-up preparation | Mock-up (1–4) | | | | | | | | | |
|    6. Dynamic simulation | Report | | | | | | | | | |
|    7. Consultation | | | | | | | | | | |
|    8. Design guidance | Design guide | | | | | | | | | |
|    9. Design review | | | | | | | | | | |
|    10. Cont. disp. hardware select | | | | | | | | | | |
|    11. Panel layout | Drawings | | | | | | | | | |
|    12. Work space arrangement | Drawings | | | | | | | | | |
|    13. Equipment procedures | Procedures man'l | | | | | | | | | |
|    14. Detail design | Dwgs. & doc. | | | | | | | | | |
|    15. Training programs | Training pkg. | | | | | | | | | |
|    16. Operator manuals | Manual | | | | | | | | | |
|    17. Maintenance manuals | Manual | | | | | | | | | |
|   Design Verification | | | | | | | | | | |
|    1. Checklist preparation | Checklist | | | | | | | | | |
|    2. Appraisal by checklist | Report | | | | | | | | | |
|    3. Test planning | Test plans | | | | | | | | | |
|    4. Test and evaluation | Report | | | | | | | | | |
|    5. Test support | | | | | | | | | | |
|    6. Follow-up studies | Report | | | | | | | | | |
|   Applied Research | Report | | | | | | | | | |
|    TOTALS | | | | | | | | | | |

## GENERAL GUIDELINES FOR DEFINING THE HUMAN FACTORS IMPLICATIONS OF VARIOUS PRODUCT DESIGN PROGRAMS AND ENSURING PROPER HUMAN FACTORS ENGINEERING INPUT TO THE FINAL PRODUCT DESIGN

### Point of Entry

One does not always have the luxury or benefit of entering every design program at the very beginning; thus the methods and techniques for human engineering must be tailored for each design program. The important considerations are discussed below.

### 1. Preconceptual Stage

The ideal point of entry for any program is before the customer has made any specific hardware designations. In other words, the design engineer has an opportunity to work with the customer to define the basic operational requirement. In such cases, the design engineer should consider a "total systems analysis program" to systematically define:

a. The mission and operational requirements
b. The functions that are required to accomplish each mission event
c. The performance requirements for each function
d. The allocation of functions to hardware, software, or human elements

### 2. Conceptual Stage

The next most opportune point of entry for the design engineer is at the "system design concept" stage, where the designer, although given general directions with respect to certain hardware elements, is allowed to make trade-offs relative to how to put various hardware elements together into a (preliminary) base-line system. In such cases, the design engineer not only should consider the analytic steps outlined above but also should include additional analyses to define:

a. The best design approach for accomplishing each hardware functional assignment, i.e., subsystems trade-off studies
b. Preliminary operator, maintainer, and user task descriptions
c. A preliminary manning and training requirements definition
d. Preliminary information-flow and operational sequence diagrams

### 3. Predesign Stage

The next best point of entry for the design engineer is at the "preliminary design" stage, where the designer, although constrained by preconditions and by the concepts of the customer, still has an opportunity to perform a number of additional design trade-off studies to provide a rationale for possible modification and improvement of the initial concepts. During this stage, the designer not only should review (and possibly modify) the previously noted analyses but also should consider the following:

a. Human-machine mockup studies

b. Human-machine simulation studies
c. Fundamental (but applied) human factors research studies
d. Time-line and link analyses
e. Refined task analysis and task description

### 4. Detailed Design Stage

In many cases, the engineering designer may not be given the opportunity to perform the preceding preparatory activities, in which case he or she essentially is told what to design and/or what system components to select and adapt into a system or product. Still, at this stage, the designer should consider a limited analysis program that includes all the above analytic steps, but at some lesser degree of refinement. As a minimum, the designer should consider the following:

a. Creating a statement of the system or product purpose
b. Developing a function-flow schematic
c. Creating information-flow and operational sequence diagrams for the critical operating and maintaining operations
d. Performing link analyses for all key operator, maintainer, and user human-equipment interfaces
e. Identifying critical skill requirement specifications that imply the need for operator, maintainer, and user indoctrination and/or training
f. Performing detailed hazards and safety analyses.
g. Creating and evaluating critical human-machine mockups of key operator, maintainer, and user interfaces (control stations, equipment handling, facilities layouts, etc.)

## Design Process Documentation

It has become a way of life in modern design to document how and why one has arrived at a specific design concept and configuration. This is necessary because of the number of approval steps required before the design can be released to production, and it may also be an important resource later if questions concerning product liability arise. In any case, it makes good sense to systematically examine and document each step of the design process simply to satisfy one's own desire to create the most cost-effective product. The following design steps are typical of those which should be fully documented.

### 1. Customer Requirement

The general purpose of the proposed product or system should be clearly defined at the very beginning, before any serious design work is begun. Even though one may have some preconceived design ideas, it is important to "back up" and make sure that the basic purpose for which the product is to be used is clearly understood, both by the customer and by the manufacturer or designer.

### 2. Mission Definition

The proposed system should be examined in terms of mission or use objectives, wherein "performance" criteria are established. These will eventually be the measure of how successful the design is.

### 3. Constraints Environment

On the basis of the specific mission or use objectives, one must identify and clarify all the constraints within which the system must operate and/or within which the design must be adjusted, e.g., cost.

### 4. A Use Scenario

A dynamic, operating scenario should be created as early as possible to illustrate how the mission objectives are going to be accomplished when the system is put into operation. This is particularly desirable for the purpose of communicating the operational concept to non-engineering-oriented clients or to members of management who do not have the time to read complicated written descriptions.

### 5. Operational Requirements

Detailed performance specifications should be developed that establish quantitative requirements for the sensitivity, accuracy, speed, tolerance, reliability, delivery schedule, and service and maintenance of the system.

### 6. Function Definition

All the functions and subfunctions required to accomplish the proposed mission should be identified and described. These should include both development and operation or mission functions, since the development may require as much planning as the operation.

### 7. Function Allocation

Trade-off analyses should be performed to determine the most cost-effective assignment of basic mission functions, i.e., to hardware, software, firmware, and/or human operators or maintainers. More human factors expertise is required at this part of product or system conceptualization than at any other.

### 8. Detailed Functional Requirements Definition

At this point, hardware functions and human functions are examined both independently and collectively to define the specific requirements that end items must accomplish. It is at this point that human engineering principles should be introduced in order to make sure that the design concepts that may be proposed are going to be compatible with human capabilities and limitations.

### 9. System, Equipment, and Facilities Design Concept Definition

Preliminary design trade-offs should be made, and preliminary base-line hardware drawings should be prepared, in order to provide a "first look" at the total human-machine system concept. It is important that human factors be given high priority during these trade off studies. The base-line system provides the first significant hardware review point, at which major decisions will be made concerning whether to proceed with the concept, rework it, or drop it entirely because of technical or economic difficulties.

### 10. Detailed Design

At this point, development tasks are divided into hardware and personnel development activities; i.e., the engineers and designers pursue the hardware and software side of the development, and human factors specialists pursue the personnel requirements side. However, close interaction is still vital to the successful accomplishment of both activities since the hardware subsystem and the personnel eventually must join to demonstrate a successful match between human and machine to ultimately produce effective system performance.

Human engineering activities during detailed design should include the following:

a. Human engineering design input: Human engineering principles are applied at each stage of design and include critical review of initial design concepts, approval of drawings before release, and evaluation of mockups. Some human factors research may be required when previously established human engineering practices and criteria are insufficient to make appropriate design decisions.

b. Task analysis: As design concepts materialize, operator, maintainer, and user tasks should be described to reflect how, where, and with what operators and maintainers will be required to perform salient operating and maintaining tasks in order to exercise control of the hardware system.

c. Hazards analysis: Although gross analyses should have been made during concept formulation, it is not until design details begin to emerge that one can identify the critical probabilities of certain hazards due to equipment failures, operator failures, or unique interactions between the hardware system and the environment within which it will operate.

d. Time-line analysis: Although the early mission scenario should provide some general description of time relationships, it is not until design details become available that one can estimate the operating and maintaining timing characteristics of system operation. At this point, it becomes evident whether the planned use of the operator or maintainer is feasible, i.e., whether the tasks imposed on the human are within his or her capability to respond or whether the operator or maintainer is apt to be overloaded or underloaded.

e. Link analysis: The link analysis allows one to examine the organization of human tasks in terms of logical grouping and sequencing of activities. It should be done at both the general workplace and the control interface level.

f. Manning requirements analysis: Gross manning requirements are examined during the early planning and conceptual stage of system development. However, it is not until operator and maintainer tasks have been defined in more detail that one has sufficient information to clearly define what kinds of people and how many people will be required to man the system once it is put into operation.

g. Training requirements analysis: Plans for training operators, maintainers, and general users derive from the task descriptions noted above. Training objectives and a training plan should be developed as quickly as possible following detailed design because early development of key training equipment or facilities may be involved.

h. Training aids, equipment, and facilities design: As soon as training requirements are defined, training support items can be defined, including classroom aids, part task trainers, training simulators, special training facilities, and supporting documentation and other materials. As noted above, because it is necessary to provide certain long-lead-time items at the same time that prime hardware is delivered, one may have to perform an initial training requirements analysis prior to having all the design information that normally is generated by the task analyses and hardware human engineering. These items include such things as complex simulators and special training facilities.

i. Technical publications: Certain special publications must be prepared in time to be delivered along with the product and/or slightly before the beginning of training. These include operations and maintenance service manuals (which may become teaching aids). Human engineering attention should be given to the conceptualization and final preparation of these documents so that they will be made maximally helpful to the user. In military system development, it is required that these manuals be field-tested to ensure their use effectiveness.

j. Human test and evaluation: The military requires an early and continuing human engineering test and evaluation program during system development. It considers that human engineering evaluations at each stage of analysis and design are actually tests of the adequacy of the human engineering effort. The final test, however, is one in which all elements of the total system (including hardware, software, firmware, documentation aids, and trained personnel) are tested together as the ultimate measure of total system effectiveness. It is here that the design, procedures, and training are tested to demonstrate whether all development objectives have been properly met.

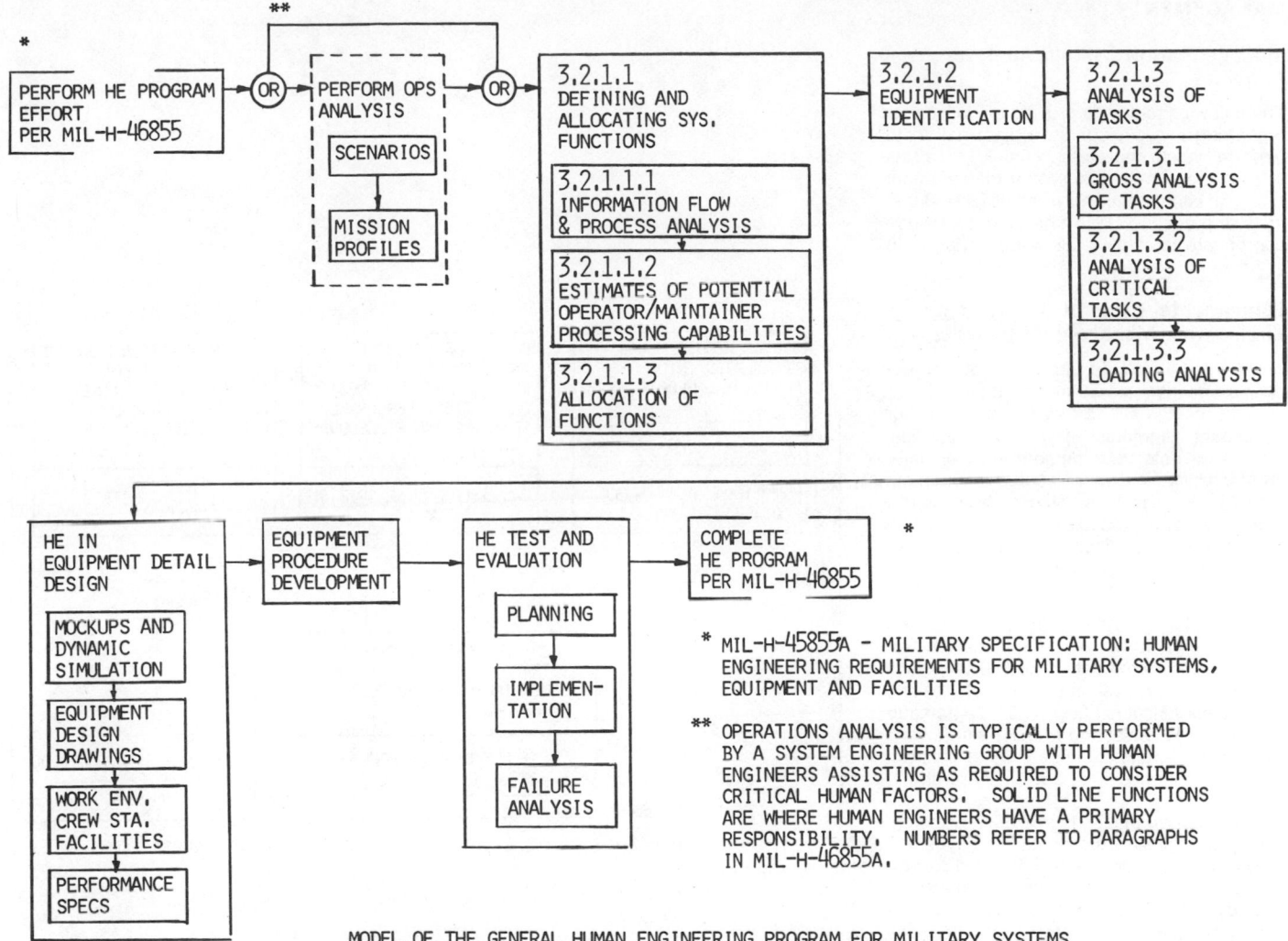

*

PERFORM HE PROGRAM EFFORT PER MIL-H-46855

OR

** PERFORM OPS ANALYSIS

SCENARIOS

MISSION PROFILES

OR

3.2.1.1 DEFINING AND ALLOCATING SYS. FUNCTIONS

3.2.1.1.1 INFORMATION FLOW & PROCESS ANALYSIS

3.2.1.1.2 ESTIMATES OF POTENTIAL OPERATOR/MAINTAINER PROCESSING CAPABILITIES

3.2.1.1.3 ALLOCATION OF FUNCTIONS

3.2.1.2 EQUIPMENT IDENTIFICATION

3.2.1.3 ANALYSIS OF TASKS

3.2.1.3.1 GROSS ANALYSIS OF TASKS

3.2.1.3.2 ANALYSIS OF CRITICAL TASKS

3.2.1.3.3 LOADING ANALYSIS

HE IN EQUIPMENT DETAIL DESIGN

MOCKUPS AND DYNAMIC SIMULATION

EQUIPMENT DESIGN DRAWINGS

WORK ENV. CREW STA. FACILITIES

PERFORMANCE SPECS

EQUIPMENT PROCEDURE DEVELOPMENT

HE TEST AND EVALUATION

PLANNING

IMPLEMEN-TATION

FAILURE ANALYSIS

COMPLETE HE PROGRAM PER MIL-H-46855

*

* MIL-H-45855A – MILITARY SPECIFICATION: HUMAN ENGINEERING REQUIREMENTS FOR MILITARY SYSTEMS, EQUIPMENT AND FACILITIES

** OPERATIONS ANALYSIS IS TYPICALLY PERFORMED BY A SYSTEM ENGINEERING GROUP WITH HUMAN ENGINEERS ASSISTING AS REQUIRED TO CONSIDER CRITICAL HUMAN FACTORS. SOLID LINE FUNCTIONS ARE WHERE HUMAN ENGINEERS HAVE A PRIMARY RESPONSIBILITY. NUMBERS REFER TO PARAGRAPHS IN MIL-H-46855A.

MODEL OF THE GENERAL HUMAN ENGINEERING PROGRAM FOR MILITARY SYSTEMS

## MANAGEMENT

### Management of System Analysis Activities

The following materials provide guidelines for organizing any development project in order to perform system analyses in a timely and effective manner. The key to proper management is to plan each of the analysis activities so that they all proceed concurrently, but to include appropriate iterations to allow each analysis to influence the others.

### Fundamental Cycle of the System Engineering Management Process

The accompanying diagram[3] illustrates a process for defining a system on a total basis so that the design will reflect requirements for equipment, computer programs, facilities, procedural data, and personnel in an integrated fashion. It provides source requirement data for the development of specifications, test plans, and procedures and the backup data required to define, contract, design, develop, produce, install, check out, and test the system.

Step 1: Identify system requirements and translate these into basic functional requirements, i.e., statements of operation. These should be portrayed in the form of top- and first-level functional flow block diagrams to portray sequential and parallel interactions of functions. It is not necessary to proclaim a solution at this time, but simply to understand its use.

Step 2: Analyze functions and associated criteria and translate these into design requirements in sufficient technical detail to provide criteria for *(a)* designing equipment and/or computer programs and defining facility equipment and intersystem interfaces and *(b)* determining requirements for personnel, training, training equipment, and procedural data. These requirements should be recorded on requirements allocation sheets and time-line sheets.

Step 3: Concurrently with step 2, conduct system design engineering studies to *(a)* determine selection of alternative functions and function sequences; *(b)* determine design, personnel, training, and procedural data re-

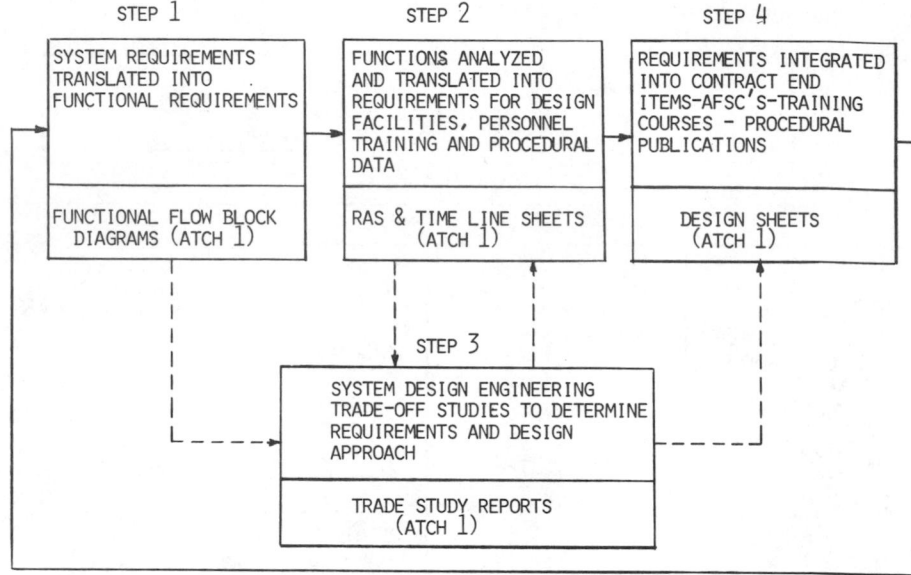

FUNDAMENTAL CYCLE OF THE SYSTEM ENGINEERING MANAGEMENT PROCESS

quirements imposed by the functions; *(c)* determine the best way to satisfy design requirements; and *(d)* select the best design approach for integrating design requirements into contractual end items (CEI) of equipment, computer software programs, firmware, and/or operator/maintainer/user roles.

Step 4: Prepare CEIs in terms of specific performance, design, and test requirements.

*Note:* The above steps are iterative in the sense that, as each step proceeds, analytic information develops that suggests modifications to previous steps, which in turn require successive modifications of later steps. The key to success in this process is thorough documentation so that all members of a design team can remain apprised of all results of analysis, decisions, and current conceptual status.

[3]After Air Force Systems Command, *Systems Engineering Management Procedures,* AFSCM 375-5, 1966.

## System Engineering Documentation and Application Criteria[4]

| Documentation | Description | Basic Purpose | Application |
|---|---|---|---|
| Functional flow block diagram | Identifies and sequences the system and system element functions that must be accomplished in order to achieve system project objectives. | 1. Facilitates development of system requirements in view of the basic operations that must be accomplished to achieve system objectives.<br>2. Develops the basis for establishing intersystem functional interfaces, as well as identifying system relationships. | All system programs and projects for which the definition or acquisition phase is applicable or directed. Also selected projects in the advanced development and exploratory development elements. |
| Requirement allocation sheet (RAS) | Defines the requirements and constraints pertaining to each of the flow diagram functions and apportions these requirements to equipment, facilities, personnel, and procedural data. | 1. Facilities development of system element requirements on a functional, system basis rather than purely a hardware basis.<br>2. Facilitates correlation of hardware, computer programs facility personnel, and procedural data requirements to the functions these system elements are accomplishing.<br>3. Identifies trade-off studies required to determine more detailed system definition. | |
| Trade study report | Documents the trade-offs and back-up rationale pertaining to the functional diagram and requirements developed on the RAS, design sheet, schematic, time-line sheets, and other system engineering, documentation. | Facilitates (1) systematic consideration of all possible solutions in view of defined system constraints and (2) selection of the best solution. | |
| Time-line sheet | Presents system functions against a time base in their required sequence of accomplishment. | 1. Facilitates cognizance of time element and specific sequence parallel relationships between functions in the development of system element.<br>2. Used to evaluate design effectiveness in terms of reaction time, performance time, maintenance down-time, equipment and personnel utilization time. | |
| Schematic block diagram | Schematically identifies and represents hardware computer programs and facility subsystem and item component functional interfaces and interrelationships. | 1. To facilitate development of hardware and computer programs facility items in view of constraining interfaces.<br>2. Used to allocate requirements developed on the RAS contract end item. | |
| Design sheet | Identifies hardware and computer program and facility end item performance design, test requirements. Becomes section 3 and 4 of part detail specification. (reference AFSCM 378-1) | Defines hardware computer program and facilities performance, design, and test criteria on an end item basis. | |
| Facility interface sheet | Identifies functional and physical interfaces between equipment and facilities on an end item basis. | Supplements the RAS to collect and further define specific equipment interfaces with the facility on an end item rather than functional basis. | Any system programs and projects for which a definition or acquisition phase is applicable or directed and that will result in equipment that has complex interfaces with facilities. |
| End item maintenance sheet (manual) | Summarizes maintenance requirements on a specific end item, subassembly, and component basis. | Facilitates systematic and complete development of maintenance requirements for each system, end item, and subassemblies component. | Any system programs or projects for which a definition or acquisition phase is applicable or directed and that results in relatively complex, nonstandard hardware, facilities, or computer program development which will require major logistic support. |
| Maintenance sheets (automated) | Summarize maintenance requirements on a specific end item, subassembly, and component basis. Provides data for configuration management, computer program and detail maintenance data elements. May be modified for manual use. | 1. Facilitates systematic and complete development of maintenance requirements for each system end item, and subassemblies components.<br>2. Facilitates sorting and combining selected elements of logistics data in support of logistics activities. | Any system programs or projects for which a definition or acquisition phase is applicable or directed and that results in relatively complex nonstandard hardware, facilities or computer program development which will require major logistic support. |

[4]Air Force Systems Command, *Systems Engineering Management Procedures*, AFSCM 375-5, 1966.

*(continued)*

**System Engineering Documentation
and Application Criteria** *(continued)*

| Documentation | Description | Basic Purpose | Application |
|---|---|---|---|
| Maintenance loading sheet | Correlates maintenance functions and RAS (including frequency of occurrence, time for accomplishment, etc.) to personnel, MGE and spares. | 1. Facilitates determination of the quantity of MGE, personnel, and spares required to maintain the system.<br>2. Provides an input to system effectiveness studies in terms of utilization factors. | All system programs and projects for which definition or acquisition phase is applicable or directed and that will result in or require any one of the following:<br>1. High launch rates and involving large number of flight vehicles or end items.<br>2. Large numbers of different end items of a relatively complex and nonstandard nature.<br>3. Large numbers of end items developed in widely dispersed areas. |
| MGE utilization sheet | Identifies MGE quantities by specific use location. | 1. Facilitates identification of total MGE quantity requirements.<br>2. Provides input to maintenance loading sheet. | |
| Personnel utilization sheet | Identifies maintenance personnel effort by specific maintenance location. | 1. Facilitates identification of total maintenance personnel requirements.<br>2. Provides input to maintenance loading sheet. | |
| Calibration requirements summary | Summarizes equipment calibration requirements at each echelon of calibration. | Provides a convenient summary to define system calibration and measuring standard requirements. | All system programs and projects for which an acquisition phase is applicable or directed that will be turned over to an Air Force using command or that have large number of end items requiring periodic calibration. |
| Equipment provisioning figure | Defines MGE end item ordering data. | Provides MGE provisioning data to AFLC on an end item basis. | All system programs and projects for which an acquisition phase is applicable or directed and that will result in MGE development which will be procured by AFLC. |

**Identifying Key Elements for a System
Development Program**

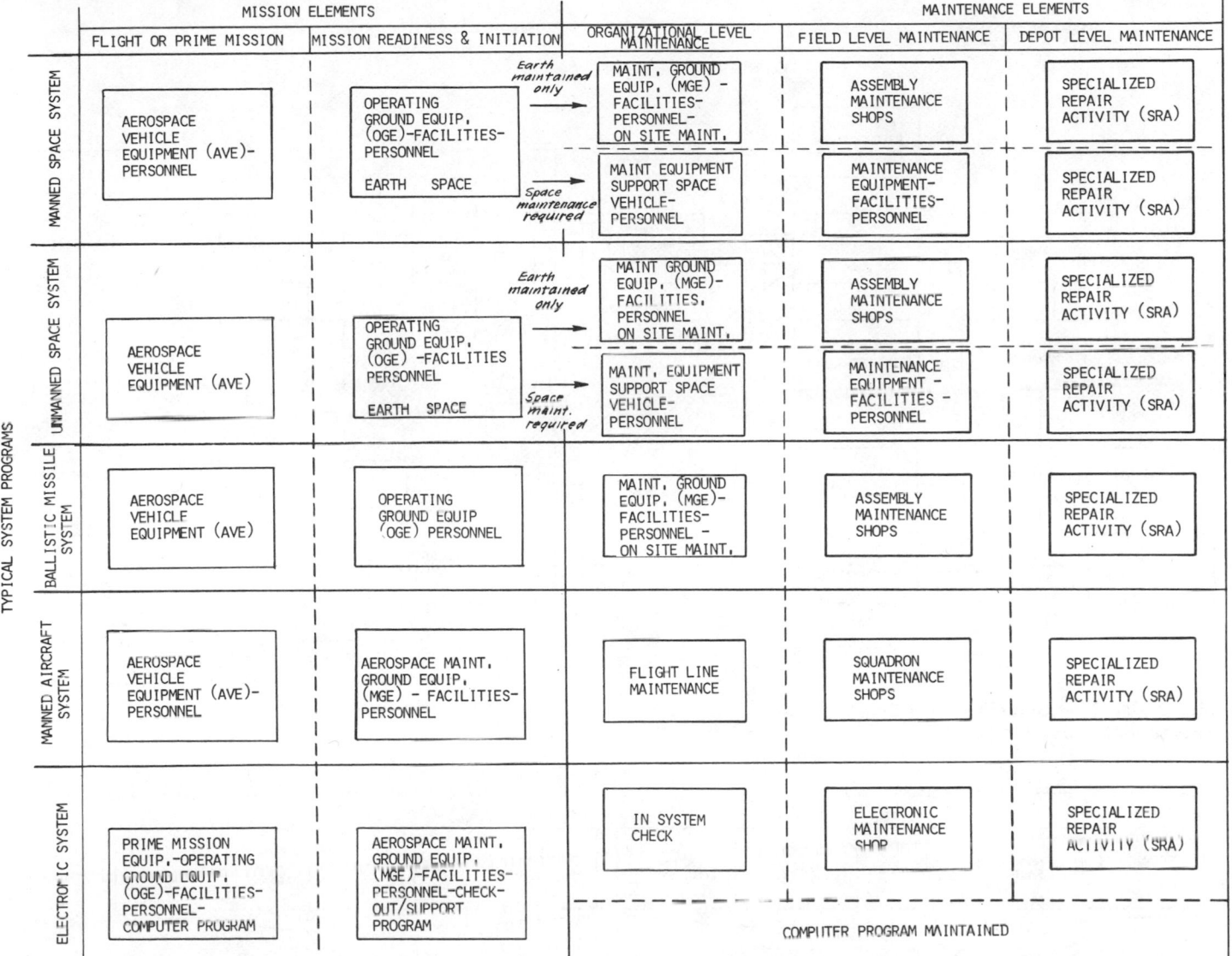

| | FLIGHT OR PRIME MISSION | MISSION READINESS & INITIATION | ORGANIZATIONAL LEVEL MAINTENANCE | FIELD LEVEL MAINTENANCE | DEPOT LEVEL MAINTENANCE |
|---|---|---|---|---|---|
| **MANNED SPACE SYSTEM** | AEROSPACE VEHICLE EQUIPMENT (AVE) -PERSONNEL | OPERATING GROUND EQUIP. (OGE)-FACILITIES-PERSONNEL  EARTH   SPACE | *Earth maintained only* → MAINT. GROUND EQUIP. (MGE) -FACILITIES-PERSONNEL-ON SITE MAINT.  *Space maintenance required* → MAINT EQUIPMENT SUPPORT SPACE VEHICLE-PERSONNEL | ASSEMBLY MAINTENANCE SHOPS  MAINTENANCE EQUIPMENT-FACILITIES-PERSONNEL | SPECIALIZED REPAIR ACTIVITY (SRA)  SPECIALIZED REPAIR ACTIVITY (SRA) |
| **UNMANNED SPACE SYSTEM** | AEROSPACE VEHICLE EQUIPMENT (AVE) | OPERATING GROUND EQUIP. (OGE) -FACILITIES PERSONNEL  EARTH   SPACE | *Earth maintained only* → MAINT GROUND EQUIP. (MGE)-FACILITIES. PERSONNEL ON SITE MAINT.  *Space maint. required* → MAINT. EQUIPMENT SUPPORT SPACE VEHICLE-PERSONNEL | ASSEMBLY MAINTENANCE SHOPS  MAINTENANCE EQUIPMENT - FACILITIES - PERSONNEL | SPECIALIZED REPAIR ACTIVITY (SRA)  SPECIALIZED REPAIR ACTIVITY (SRA) |
| **BALLISTIC MISSILE SYSTEM** | AEROSPACE VEHICLE EQUIPMENT (AVE) | OPERATING GROUND EQUIP (OGE) PERSONNEL | MAINT. GROUND EQUIP. (MGE)-FACILITIES-PERSONNEL -ON SITE MAINT. | ASSEMBLY MAINTENANCE SHOPS | SPECIALIZED REPAIR ACTIVITY (SRA) |
| **MANNED AIRCRAFT SYSTEM** | AEROSPACE VEHICLE EQUIPMENT (AVE)-PERSONNEL | AEROSPACE MAINT. GROUND EQUIP. (MGE) - FACILITIES-PERSONNEL | FLIGHT LINE MAINTENANCE | SQUADRON MAINTENANCE SHOPS | SPECIALIZED REPAIR ACTIVITY (SRA) |
| **ELECTRONIC SYSTEM** | PRIME MISSION EQUIP.-OPERATING GROUND EQUIP. (OGE)-FACILITIES-PERSONNEL-COMPUTER PROGRAM | AEROSPACE MAINT. GROUND EQUIP. (MGE)-FACILITIES-PERSONNEL-CHECK-OUT/SUPPORT PROGRAM | IN SYSTEM CHECK | ELECTRONIC MAINTENANCE SHOP | SPECIALIZED REPAIR ACTIVITY (SRA) |
| | | | COMPUTER PROGRAM MAINTAINED | | |

*(Column group headers: MISSION ELEMENTS (FLIGHT OR PRIME MISSION, MISSION READINESS & INITIATION); MAINTENANCE ELEMENTS (ORGANIZATIONAL LEVEL MAINTENANCE, FIELD LEVEL MAINTENANCE, DEPOT LEVEL MAINTENANCE). Row group label: TYPICAL SYSTEM PROGRAMS.)*

*Note:* Although the accompanying examples are typical of military and space programs, similar elemental breakdowns can be developed for almost any system development program, including architectural, highway, communications, and product systems. The principal implication is that all systems should be approached from a total operational requirements point of view; i.e., the human factor permeates the development process wherever an operator, maintainer, or user must provide some input or perceive some output.

## Human-Machine System Model

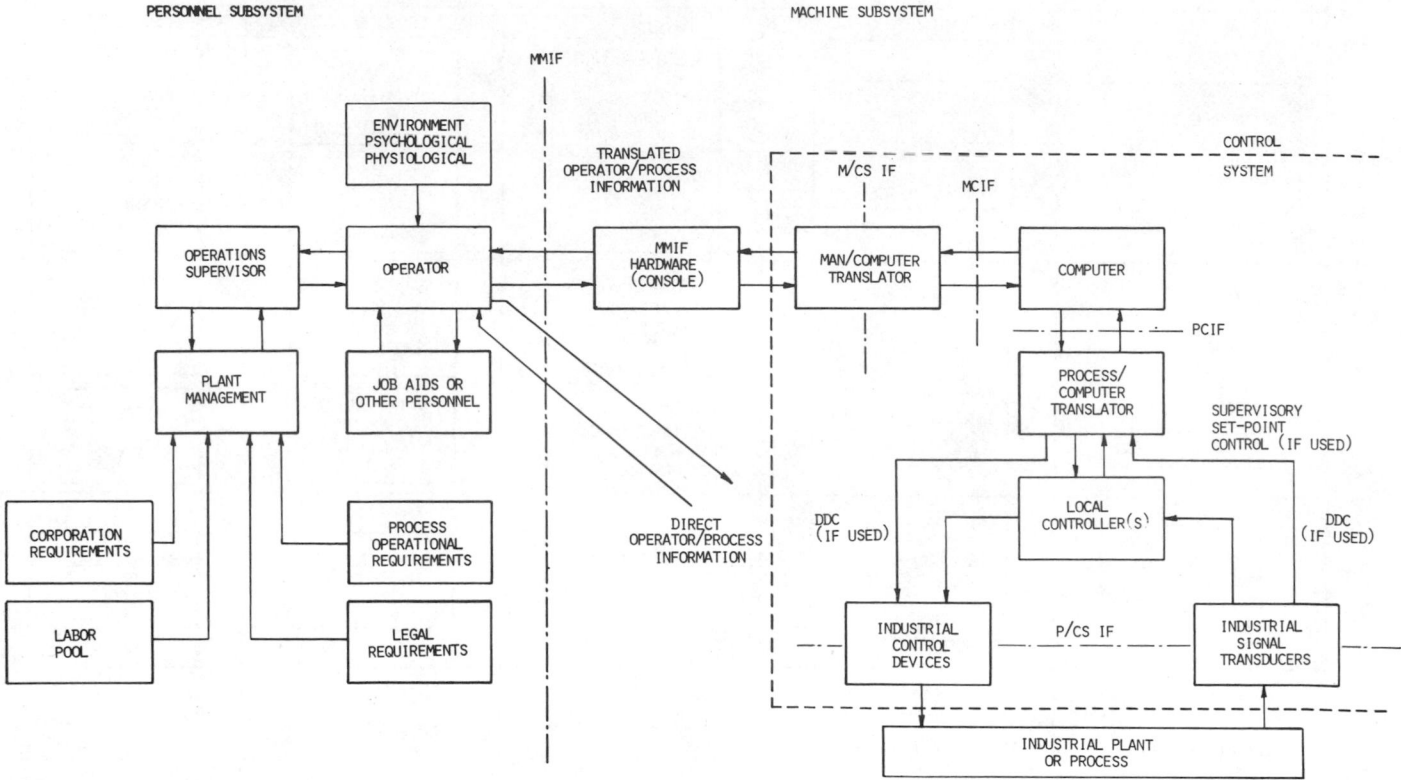

THE MAN-MACHINE SYSTEM MODEL

## Human-Machine Guideline Flow Chart
## (Personnel Subsystem)

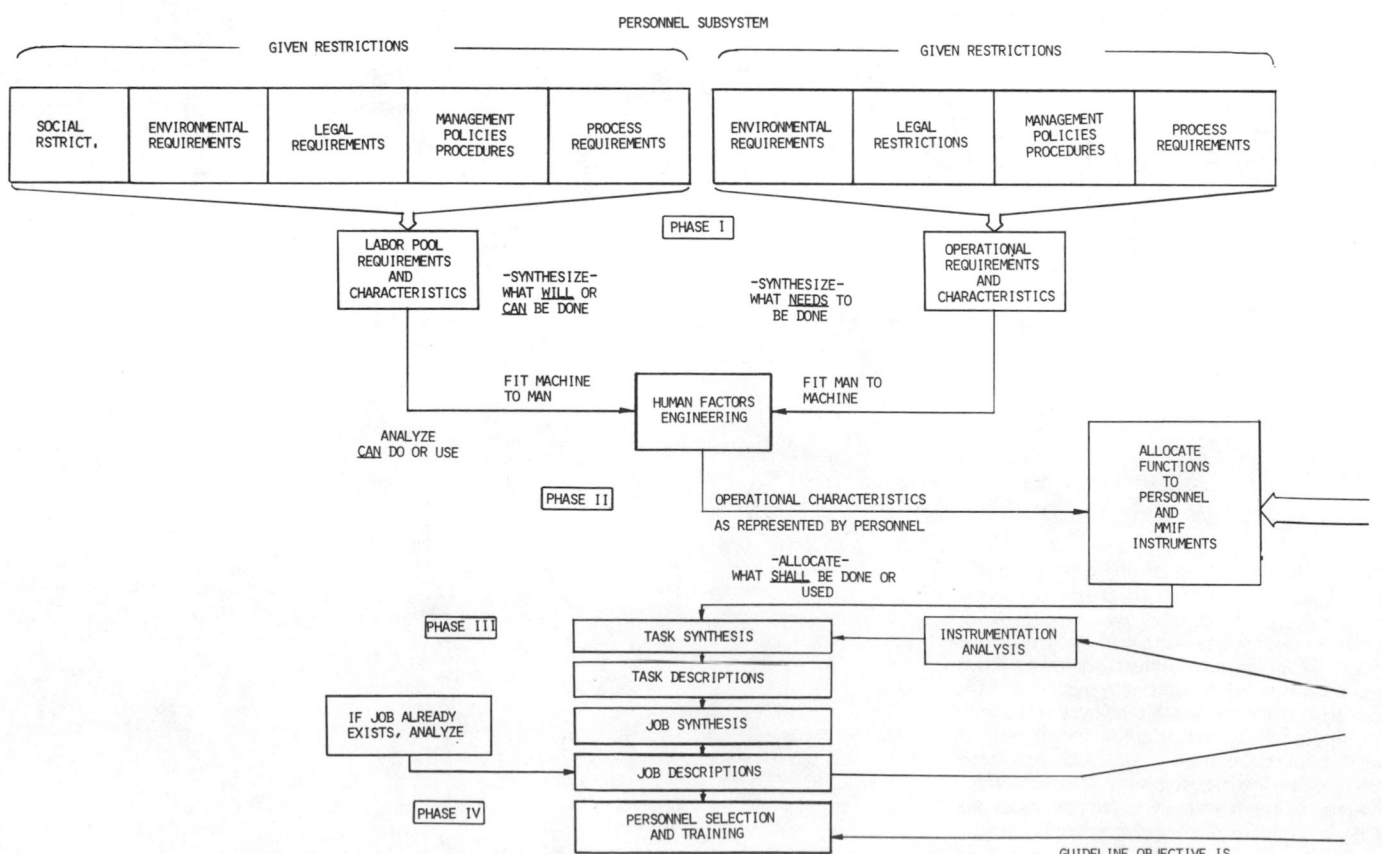

## Human-Machine Guideline Flow Chart
## (Process Subsystem)

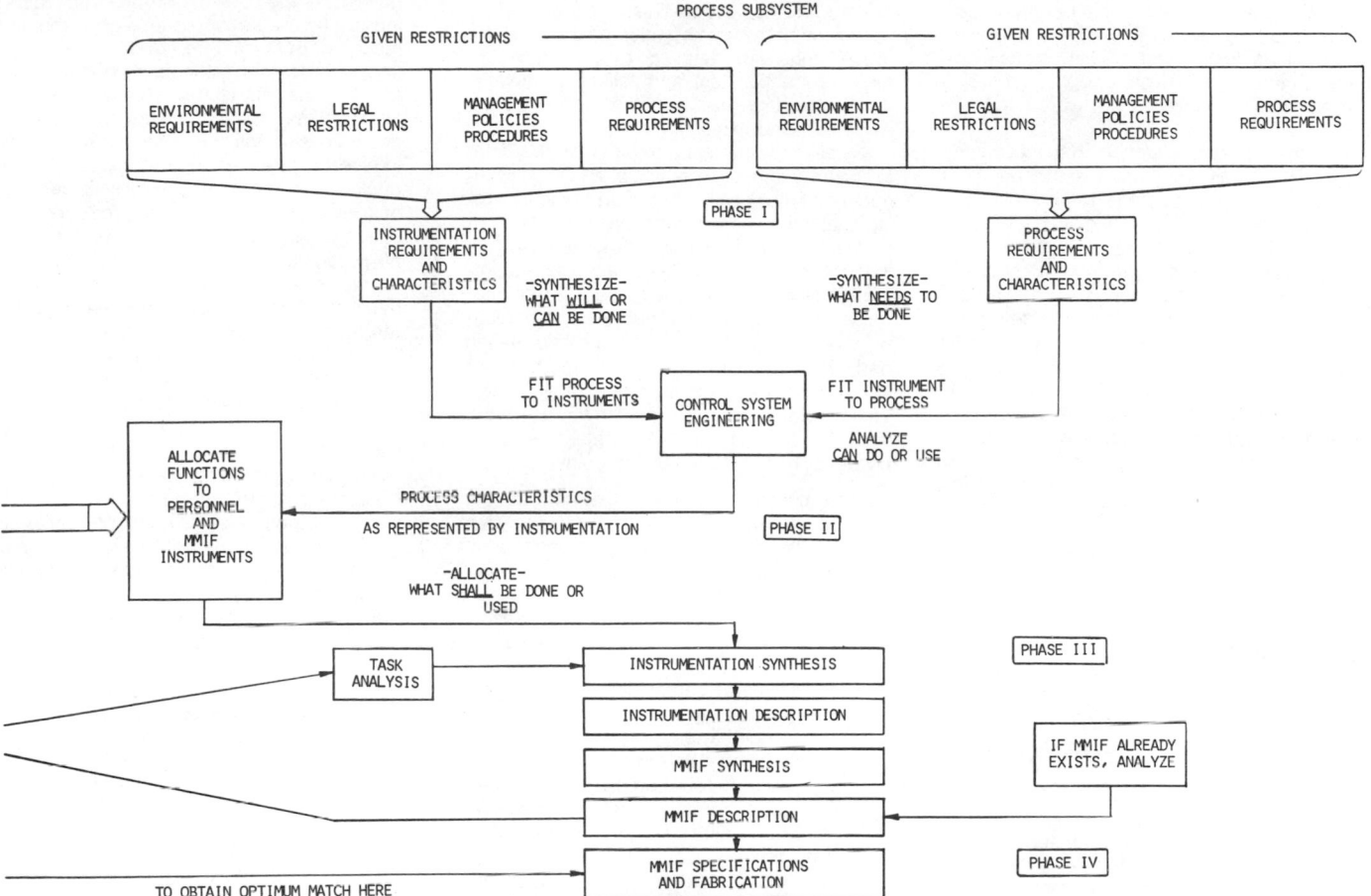

## Steps in System Engineering Analysis

Although the following discussion relates primarily to how the military regards system engineering analysis and, more specifically, to how human factors are included in these analyses, the general steps and analysis objectives are equally applicable (although to varying degrees) to any new product or system design development.

In general, system engineering analyses include the following.

### 1. Mission Requirements Analysis
This involves defining the purpose, objectives, environment, and constraints that any proposed system design is to address. The human factor is an important aspect of this initial set of definitions, especially in terms of the environment in which the system is expected to operate and in terms of certain constraints that the human in the system may place on any proposed design. The environment will relate to physiological limitations of the human; personnel availability and capability may relate to how and where the human can be used.

### 2. Functional Requirements Analysis
Defining the specific functions that must be performed in order to complete the proposed mission successfully is obviously a critical step

that must be taken before one can decide what to design. Functional requirements are "actions" that have to be accomplished. Before one decides whether actions are accomplished by machines or humans, one should make sure that the functional requirements are clearly defined as actions—not as preconceived notions or ideas about equipment or people.

An important methodology or technique is to develop a graphic model of the function hierarchy in the form of block diagrams; i.e., functional block diagraming is a key part of functional requirements analysis. Functional block diagraming is typically done at several levels, starting with the "top level," where a very gross picture of major functions is shown, each of which will eventually be broken down at several lower levels until a specific critical end-item requirement will emerge, e.g., a piece of equipment, a component, a part, a training aid, a tool, or, most important, a trained operator or maintainer.

### 3. Function Allocation
As indicated above, to this point, function definitions have been pure action requirements, devoid of decisions about whether hardware or people should perform the functions. The next step, and one in which the human factor becomes extremely important, is to decide which functions can be done more effectively

by machines and which functions can be done more effectively by people. It is critical at this point to have human factors specialists participate in these function-allocating decisions.

### 4. Detailed Functional Requirements Definition
Although a gross description has already been made for each function defined in step 2, now that allocations have been made (at least preliminarily) as to which functions go to the hardware designer and which go to the human or personnel subsystem designer, each of these specialists proceeds to define his or her respective functional responsibilities in sufficient detail to establish a basis for preliminary design development. Remember that the engineer and the human factors specialist (although each looks at his or her assigned functions from a unique point of view and technical objective) must coordinate with each other in order to ensure an impedance match between machine and human requirements.

### 5. Preliminary Design Specifications Analysis
As the more detailed requirements are completed for each hardware and/or human function, these are now analyzed in terms of a possible design solution. Engineers are concerned with hardware and facilities designs, and human factors engineers are concerned

primarily with the "design of the personnel subsystem." However, this is the point at which the human factors engineer must also be concerned with what the engineer perceives and how he or she executes hardware and facilities designs, for these designs eventually dictate exactly what people in the system have to do to use the hardware and facilities.

It is in this latter situation that the human factors engineer must provide the engineer with human engineering design principles and criteria so that the design engineer will come up with designs that are compatible with human operator and maintainer capabilities and limitations. During engineering preliminary design, various studies are made to determine which design approach is best, both from an operational and a cost point of view. The human factors engineer should work closely with the engineers and designers during these studies so that design trade-offs will include consideration of operator and maintainer capabilities and limitations. The optimization of designs must reflect both good engineering and good human engineering.

During predesign studies, it may be necessary to perform both engineering and human engineering tests to verify certain assumptions or demonstrate that one design is better than another. It is at this point that mockups are extremely valuable, allowing the designer and the human factors specialist to work together to determine the best compromises between engineering and human engineering objectives.

For large, complex systems, preliminary design typically involves a number of different engineering specialists. These may include, for example, air frame, control systems, avionics, electrical, computer, and other hardware and software specialists or groups. In addition, the typical company usually will have support engineering groups such as integrated logistics, reliability, quality control, factory production, and publications. Each of these groups, along with the primary engineering groups, has primary responsibilities for both the prime system hardware and other end items such as a maintenance and supply logistics plan, a special production facility, development plans for test and other facilities, and finally the human engineering program plan that typically is required by military procurement contracts.

Kept purposely separate from the above is the preliminary design of the engineering test function or functions. It is important to address this independently because of the degree to which human factors should be part of this activity. Although engineering tests are a recognized part of any development program, human engineering testing often is neglected. A final system's performance effectiveness can be measured only in terms of the combined human-machine output. It therefore is critical that the human engineering test requirements be included in the overall system test planning.

A final output of preliminary design, then, is a base-line proposal of what the system is to be like in terms of hardware, software, facilities, support elements, and, most significantly, the operator and maintainer element. Typically, a detailed production plan and operational support plan should be proposed so that

the customer is able to assess not only the basic design concept but also the general costs and schedules required to acquire the system and operate it. The end product is a specification that can be used as the basis for proceeding with detailed design, development, production, test, and delivery of the system. Acquisition or production contracts are written around this specification.

It is perhaps important to pause at this point to illustrate some of the typical analytic procedures that may be used for each of the foregoing analyses. The following examples or samples are meant to show various approaches that different analysts have taken to accomplish the analyses just described. Every analyst tends to feel more at home with a particular style or format for developing information and documenting it for general review by his or her peers, which explains the slight differences among the examples. However, for the sake of clarifying analytic objectives, one should consider the following points:

a. Some sort of numerical system should be created so that each of the analyses can be identified and traced back to initial mission objectives and/or basic functional requirements. Nothing is more embarrassing than to come up with a design for which one cannot find a basic requirement!

b. In addition to verbal descriptions of requirements and/or analytic conclusions, one should attempt to create graphic models that show the flow from one analysis or requirement to another. Block diagraming is a favorite technique, as evidenced by the numerous examples that follow.

c. Functions should be identified by a verb rather than by a noun that describes an equipment or human component.

d. End items should be identified by a noun since these eventually must be applied to a drawing.

e. System schematic diagrams should illustrate both normal and alternative pathways, and all loops must be closed.

f. Drawings, study documentation, test documentation, and other end products should include the identification number of the functions to which they pertain.

## How to Decide How Much Analysis and Time Should Be Spent on the Analysis of the Human Factor in Product Development

Obviously, the same amount of analysis is not required for all product designs, either because of the inherent need or because of the amount of time and economic resources that can be made available. The following suggestions may be helpful as a guide in determining the extent to which human engineering analysis should be performed.

### 1. Need
Need should always be the first consideration. In the case of some products, lack of proper preparatory analysis may lead to costly consequences in terms of redesign, loss of sales potential, and the even more serious possibility of litigation in the event of personal injury

suits. Some guidelines are as follows:

If a proved product design is merely being updated or refined and if its in-use track record indicates reliable operation, good human-machine performance, and no apparent safety problems, extensive human engineering analysis is not usually cost-effective, unless the proposed modifications could possibly introduce hardware- or user-induced failures. In the latter case, human engineering analysis generally is required only on those aspects of the product modification which are new and suspect in terms of potential user misunderstanding, performance failures due to previously learned habit patterns, and/or unique hazards that the modification may introduce.

If the product is brand new but fairly simple in terms of concept and operation, the human engineering analysis requirement is probably minimal. Here it may be necessary to develop only a simple descriptive scenario of how the product will be used. This must be done in order to force the designer to review all the basic operations and maintenance interfaces in order to determine what human engineering design principles should be applied during the design process.

If the product is generally new and relatively complex, a more complete human engineering analysis may be required in order (a) to ensure that the basic concept is compatible with user capabilities and limitations and (b) to derive a more complete set of human engineering design requirements for each subelement that has to be designed. As the system becomes more complex (particularly a new one), there is generally a greater need to perform these human engineering analyses for products that must define manning and training requirements.

Either modified or new products or systems that pose special hazards for users (operators and maintainers) usually require more human engineering analysis—especially analyses of hazards and of failure modes and effects. Typical of these are systems involving excessive energy-producing features; extremes of vibration, noise, or temperature; radiation; toxic contamination; and/or unusual extraterrestrial space conditions. Such systems require the most thorough and complete human engineering analysis because their basic abnormality requires that special precautions be taken concerning the manner in which humans participate in the system and the way they must be protected, not only from the system and the environment but also from their own performance failures.

### 2. Time Allowance
Time often is a constraining factor in terms of the kind and amount of human engineering analysis effort that can be accomplished before actual product design is commenced. On the other hand, time constraints are too often used as an excuse for not performing the necessary analysis; e.g., arbitrary decisions are made to forgo human engineering analysis in order to meet design schedules (which often are arbitrary and should not be allowed to preclude proper preparation for design). The following are suggested guidelines with respect

to providing human engineering analysis time:

A design modification or even a new design that has sufficient operating performance precedence to indicate that there are no significant reasons to doubt that users can operate and maintain the new product as efficiently and safely as the previous or similar one generally does not require extensive preliminary human engineering analyses and therefore much time for analysis. The human engineering concern here is chiefly one of monitoring detailed hardware selection and design on the basis of recognized human engineering design practices, principles, and criteria.

In the case of a design that, although somewhat new, is required as quickly as practicable to meet an urgent operating need, the time allotted to human engineering analysis may be limited; i.e., although one would prefer to take more time, conditions exist that make it mandatory to produce the new or modified product as quickly as possible. Typical examples might include a special weapon needed for use in urgent battle conditions, a device to control traffic at a particularly dangerous intersection, and a safety device to prevent misoperation of an agricultural machine. In such cases, the urgency is established by a higher authority on the basis of the high probability that delaying production will add to already mounting problems within the operational situation. Not only is there insufficient time to perform the desired human engineering analyses, but also the chances are great that the product will already be designed and in use before any reasonably efficient analysis could be completed. Typically this kind of urgency should dictate eliminating human engineering analysis only when the item being designed is for specific, limited application and very limited production runs.

When there is even a shred of doubt about a new design with respect to whether the human's role is compatible with his or her inherent capabilities or about the possible hazards that could exist for the user, time should not be used as an excuse to eliminate desirable human engineering analyses. Rather, the approach should be to examine the product design problem realistically and recommend that sufficient time be provided to do the analyses. This is particularly important for new designs which push the hardware state of the art, which are particularly complex in nature (involving many hardware elements and many human participants), or which appear to require personnel skills that are borderline relative to the capabilities of the expected user population. Generally, if enough time has been provided to do typical systems engineering analyses, there is also enough time to include appropriate human engineering analyses.

## 3. Level of Effort
The level of human engineering analysis effort (e.g., the number of actual worker-hours) is or should be a function of need versus time—not just cost. At one extreme one should avoid "make-work" analyses costing many worker-hours; at the other extreme one should not expect to derive much benefit from an analysis which was limited to the efforts of too few analysts or which was performed too quickly. Some of the factors to consider in making decisions about the level of effort for human engineering analyses are as follows:

If the design involves a military customer, the level of human engineering analysis effort will invariably be specified by the customer. If this is not clearly spelled out in a request for proposal (RFP), it is best to provide a complete estimate based on the requirements of MIL-H-46855, *Human Engineering Requirements for Military Systems, Equipment and Facilities.* Ordinarily, in reviewing the proposal, the military customer will define how much analysis is required. If the customer does not do so, steps must be taken to clarify this matter before signing the contract.

For designs that do not involve a military or other governmental customer, one should estimate the requirements for a human engineering analysis effort on the basis of the following:

a. If the operating concept is not radically different from that of previous designs demonstrated to be effectively and safely operated by the expected user, the human engineering analysis can probably be limited to a brief identification of operating sequences and/or communications- or information-flow diagrams to help identify significant user-hardware interface requirements.

b. If only one or two new elements are being added to an otherwise older and proved design, analyses should be done only on the new additions. That is, it should not be necessary to perform full-blown, in-depth function allocation, information-flow, or operational sequence analyses on the entire system.

c. On the other hand, if the new design includes untried techniques or procedural concepts, advanced hardware, or applications in unusual environments, the human engineering analysis should be as complete as necessary to ensure that the concept is compatible with the human user's characteristics, capabilities, and limitations. It is wise in this case to make sure that there are enough *trained* human engineering analysts to complete a thorough analysis of all design elements before these are committed to the drawing board.

A good rule of thumb for estimating the amount of effort that should be devoted to human engineering analysis is that such analysis should be roughly 50 percent of the total engineering analysis. This is required because the human engineering analyst generally has to be involved not only in his or her own analysis efforts but also in those of the engineering analysts.

## 4. Cost
Actual cost predictions for any kind of analysis activity cannot, of course, be made without addressing the specific system or product or without knowing specific manpower and salary rates. However, certain cost factors can be noted that are perhaps useful in approaching the task of costing an analytic program. Some of these are as follows:

Costs will vary depending on the types of analysts required. Human engineering analysis may require a variety of analysts, depending on the type of system under consideration. To start with, the analyses we are dealing with require not only a special kind of person who has had experience and is familiar with the objectives, parameters, and subject matter of human-machine system or product interface design, but also persons with special backgrounds in some instances. For example, in a manned space system, many of the analyses will involve the expertise of an aerospace surgeon or physiologist. Complex information systems may require the expertise of a senior-level psychologist. As a minimum, a qualified human factors engineering analyst is usually a senior-level scientist. It can be seen, then, that the salary level of the people usually required will not be low. This is not to say, however, that lower-salaried individuals cannot learn to be qualified human engineering analysts. In many cases, junior analysts can be used when only minimal analysis is required.

Whether manual or computer analysis techniques will be used also is a determinant in the cost of human engineering analysis. This, of course, depends on the in-house computer facilities and on the desirability of writing programs for the analyses, a question that relates to the extent of the analysis requirement and the desire to minimize the time needed to perform the analyses.

Cost is determined by the size and complexity of the system, i.e., the number of subsystems within the system that involve interactions between equipment elements and people elements. The more elements there are, the greater the number of analyses and thus the greater the costs. As a rule of thumb, one can consider analysis costs as follows:

a. A simple product (a hand tool, a household appliance, an item of furniture, etc.) should require no more than a few hours to analyze the basic functional requirements clearly and determine the interface characteristics and specific design requirements.

b. A more complex product (an equipment console, a test instrument, an electromechanical tool, a fairly simple machine tool, etc.) should require no more than a few days to perform the necessary analyses to establish the above requirements clearly.

c. A multielement system that involves a single operator (e.g., a driver-vehicle system) may require anywhere from a month to several months to perform analyses sufficient to clarify all the human-machine interactions and establish conceptual objectives and design goals.

d. A large, complex, multielement system involving many systems and subsystems and large numbers of operators and support and maintenance personnel (an aerospace system, a ship system, a command and control system, etc.) may require as long as a year or more to complete all the required human engineering analyses. This type of system typically will involve human engineering analyses at different stages, first in the customer's house and later in the contractor's facility. A major reason why this analysis effort is extended is that a great deal of the analyst's time necessarily must be spent in seeking out the necessary information to commence and complete the analyses. And, because of the number of subsystems (each requiring separate analyses), coordination and integration of the various analyses require time over and above that needed for the simple act of constructing block diagrams or flow diagrams or constructing system model graphics and writing descriptive material relative to these graphics. Finally, in almost all complex system analyses, many iterations are required before the final analyses reflect all the intervening trade-offs and decisions that result from system refinement. It is at this level that trained human engineering analysts should always be used.

### Military Requirements[5]

Analysis should include application of human engineering techniques as follows.

#### 1. Defining and Allocating System Functions

The functions that must be performed by the system in achieving its objective should be analyzed. Human engineering principles and criteria should be applied to specify human-equipment performance requirements for system operation, maintenance, and control functions and to allocate system functions to (a) automatic operation and maintenance, (b) manual operation and maintenance, or (c) some combination of these.

#### 2. Information-Flow and Information-Processing Analysis

Analyses should be performed to determine the basic information flow and processing required to accomplish the system objective and should include decisions and operations without reference to any specific machine implementation or level of human involvement.

#### 3. Estimates of Potential Operator and Maintainer Processing Capabilities

Plausible human roles (e.g., those of operator, maintainer, programmer, decision maker, communicator, and monitor) in the system should be identified. Estimates of processing capability in terms of load, accuracy, rate, and time delay should be prepared for each potential operator and maintainer information-processing function. These estimates should be used initially to determine allocation of functions and should later be refined at appropriate times for use in the definition of operator and maintainer information requirements and control, display, and communication requirements. In addition, estimates should be made of the likely effects on these capabilities of

[5]*Human Engineering Requirements for Military Systems, Equipment and Facilities,* *MIL-H-46588A.

implementation or nonimplementation of human engineering design recommendations. Results from studies in accordance with "studies" requirement may be used as supportive inputs for these estimates.

#### 4. Allocation of Functions

From projected operator and maintainer performance data, cost data, and known constraints, the contractor should conduct analyses and trade-off studies to determine which system functions should be machine-implemented and which should be reserved for the human operator and maintainer.

#### 5. Equipment Identification

Human engineering principles and criteria should be applied along with all other design requirements to identify and select the equipment to be operated, maintained, or controlled by humans. The selected design configuration should reflect human engineering inputs, expressed in quantified or "best-estimate" quantified terms, to satisfy the functional and technical design requirements and to ensure that the equipment will meet the applicable criteria contained in MIL-STD-1472, as well as other human engineering criteria specified by the contract.

#### 6. Analysis of Tasks

The analyses shall provide one of the bases for making design decisions, e.g., determining, to the extent practicable, before hardware fabrication, whether system performance requirements can be met by combinations of anticipated equipment and personnel and assuring that human performance requirements do not exceed human capabilities. These analyses should also be used as basic information for developing preliminary manning levels, equipment procedures, and skill, training, and communications requirements. Those gross tasks identified during human engineering analysis which are related to end items of equipment to be operated or maintained by humans and which require critical human performance, reflect possible unsafe practices, or are subject to promising improvements in operating efficiency should be further analyzed, with the approval of the procuring activity.

#### 7. Analysis of Critical Tasks

Further analysis of critical tasks should identify (a) the information required by the human, including cues for task initiation; (b) the information available to the human; (c) the evaluation process; (d) the decision reached after evaluation; (e) the action taken; (f) the body movements required by the action taken; (g) the work-space envelope for the human required by the action taken; (h) the work space available to the human; (i) the location and condition of the work environment; (j) the frequency and tolerances of the action; (k) the time base; (l) feedback informing the human of the adequacy of his or her actions; (m) the tools and equipment required; (n) the number of personnel required and their specialty and experience; (o) the job aids or reference required; (p) the special hazards involved; (q) the operator interaction where more than one crew member is involved; (r) the operational limits of human performance; and (s) the operational

limits of the machine (state of the art). The analysis should be performed for all affected missions and phases, including degraded modes of operation.

#### 8. Loading Analysis

Individual and crew workload analysis should be performed and compared with performance criteria.

*Note:* Critical tasks are those which, if not accomplished in accordance with system requirements, will most likely have adverse effects on cost, system reliability, efficiency, effectiveness, or safety.

a. Jeopardized performance of an authorized mission
b. Degradation of the circular error probability (CEP) to an unacceptable level
c. Delay of a mission beyond acceptable time limits
d. Improper operation resulting in a system "no-go," inadvertent weapons firing, or failure to achieve operational readiness alert
e. Exceeding of predicted times for maintenance personnel and maintenance ground equipment to complete maintenance tasks
f. Degradation of system equipment below reliability requirements
g. Damaging of system equipment resulting either in a return to a maintenance facility for major repair or in unacceptable costs, spare requirements, or system downtime
h. A serious compromise of weapon system security
i. Injury to personnel

### Checklist for Reviewing System Engineering Analyses

#### 1. Functional Analysis Development

a. Do mission descriptions reflect an accurate interpretation of mission and threat analyses and tactical and nontactical objectives as provided by the personnel data analysis, and are they related to personnel planning data requirements?
b. Are operations and maintenance concepts delineated in sufficient detail to permit application in subsequent task analysis and personnel requirements determination?
c. Do operations and maintenance concepts reflect a correct and complete understanding of personnel capabilities and quantitative limitations?
d. Have functional descriptions and operations system diagrams or other diagramatic presentations of functional relations been updated to reflect progress in system design?
e. Has the installation schedule been modified to reflect any changes?
f. Have detailed equipment descriptions been prepared that provide an accurate and complete description of system equipment as required for personnel planning data purposes? Are they comprehensible to readers who are not engineers?
g. Are equipment descriptions accompanied by illustrative flow diagrams and/or pictures?

*h.* Are any items so unclear or undefined that further study and delineation are required?

*i.* Does a lack of threat analyses or installation schedule data suggest new developments preparation of this portion of the personnel planning data?

### 2. Human-Machine Assignment

*a.* Have the human-machine assignments accomplished during contract definition been updated?

*b.* Have operations systems diagrams or other diagrams been updated? Do the diagrams depict all assignments?

*c.* Have marginal assignments and problem areas identified during contract definition been resolved?

*d.* Have additional data sources and techniques employed in the resolutions been clearly identified?

*e.* Have maintenance assignments been amplified to reflect progress in equipment definition?

*f.* Do assignments reflect an understanding of personnel capabilities?

*g.* Do problem areas still exist that need further study? Should these problems be included in the new developments concurrent studies program?

### 3. External Load Definition

*a.* Have external load data prepared during contract definition been updated to reflect any additional contractor analyses?

*b.* Have all input sources been clearly identified?

*c.* Have levels of activity been clearly related to input sources?

*d.* Have input sources and levels of activity been clearly related to equipment positions?

*e.* Do any problem areas or data gaps exist that indicate inclusion in the new developments concurrent study program?

### 4. Task Analysis Data

*a.* Have all identified system functions involving humans been included in the analysis?

*b.* Do task descriptions clearly delineate human inputs and outputs and methods of presenting feedback concerning response adequacy?

*c.* Are all task inputs clearly identified by data source?

*d.* Have task sequences been clearly delineated? Do these sequences reflect the results of the external load analyses in that points of peak load are identified in relation to task sequence?

*e.* Have all tasks requiring rapid, difficult, and/or perceptual-motor combinations of action been fully amplified as required in the personnel planning data specification?

*f.* Have operations systems diagrams or other diagramatic depictions of task sequences been prepared in appropriate areas? Are these complete, and do they reflect the external load analysis data?

*g.* Have task sequences been organized to reflect different system mission and tactical and nontactical objectives as outlined in the functional analysis?

*h.* Have estimates of task time and task frequency been provided? Has the basis for such estimates been clearly identified?

*i.* Have performance standards been provided as outlined in the personnel planning data specifications?

*j.* Are there any areas of task description or analysis that are sufficiently incomplete or inadequately justified so as to require further special study?

### 5. Position Structure Data

*a.* Has the guidance contained in the personnel planning data specification been reflected in developing the position structure?

*b.* Have operations systems diagrams or other diagramatic depictions been prepared to support and illustrate the position structure? Are they in sufficient detail?

*c.* Do the task groupings reflect the results of external load analyses?

*d.* Does the position structure reflect the requirements for interaction with other systems within the activity as outlined in the new developments system interface study?

*e.* Have position descriptions been prepared in accordance with specifications?

### 6. Knowledge and Skill Requirements

*a.* Have the knowledge and skill requirements developed during contract definition been updated to reflect the additional detail available concerning system function, operator and maintenance tasks, and position structure?

*b.* Do the requirements accurately reflect the guidance contained in reference documents?

*c.* Are the requirements documented in the format and level of detail required by the customer?

*d.* Are the knowledge and skill requirements supported by identified task and position statements?

## Preparing a Human Engineering Plan

Military system development typically calls for submission of a formal human engineering plan as part of a contractor's proposal. Although the content and level of detail may vary, the plan should consider inclusion of the following types of information:

1. Introduction: This is a general statement of the purpose and scope of the proposed effort and includes references to specific regulatory documents to which the plan relates. Key among these documents are MIL-H-46855 (the basic human engineering specification) and MIL-STD-1472 (the basic human engineering standard).

2. Initial guidance meeting: Usually it is required that, upon award of a contract, an initial proposal review meeting be conducted during which the military agency's and the contractor's human engineering personnel will get together to hash out any parts of the proposed human engineering plan that may not be quite to the agency's liking.

3. Personnel and organization: The agency wants to know specifically who in the contractor's organization will work on the program and where these people are in the contractor's organization. It is expected that the human engineering function will be in a position to have an appropriate impact on the entire development program and that it will not be relegated to some obscure level that will be ineffectual.

4. System and task analysis: Human engineering is expected to play an important role in the entire basic system, subsystem, and design analysis effort to ensure that the human factor is given appropriate emphasis at each stage of development.

5. Human engineering design assist: It is important to indicate where and when human engineering will be applied, by whom, and how this effort is to be documented.

6. Mockups: The military expects human engineering to play a significant role in deciding not only what mockups will be constructed but also how they will be used. The military expects mockups to be used as design tools, not merely sales gimmicks.

7. Human engineering tests: The agency wants to know what kinds of specific human engineering tests will be conducted and how and when they will be conducted.

8. Human engineering design verification plan: A separate design verification is required to identify the points at which specific human factors requirements are evaluated. These include all the above as well as special demonstration tests.

9. Documentation and reporting: The military agency will expect to have complete documentation of key phases of the human engineering effort, including monthly reports, special study reports, and all key design reviews. Appropriate engineering drawings are to be attached to all design reviews. If human engineering standards have been compromised for any reason, this is to be reported, including reasons for the compromise.

10. Schedule: A detailed schedule for accomplishing the human engineering effort is to be provided. It should be keyed to the contractor's overall milestone schedule, and the schedule should indicate the level of effort throughout the program.

## MISSION DESCRIPTION

### Establishing System Mission Requirements

The following example of a mission analysis provides insight into the types of information one should consider in defining the mission of the proposed system, the performance expectations, constraints, and basic insights regarding probable functions.[6]

### General Mission—Strategic Bombing of Targets within 1200-Mi Range

| | System Description |
|---|---|
| General: | Midrange ballistic missile |
| Range: | Maximum 1200 mi, minimum 400 mi |
| Payload: | 10,000 lbs |
| Propulsion: | Solid |
| Guidance: | Inertial; not susceptible to ECM jamming |
| Launch capability: | All missiles off ground within 5 min following strike order |
| Areas of deployment: | All climatological and geographic conditions north of 45° N. latitude |
| Mobility: | Temporary site, 15-min lead time to initiation of redeployment |
| Missile: | Roadable, transportable on trailer-erector vehicle |
| Nature of site: | Soft with provision of mobile living quarters for squadron personnel; surveyed bench mark required at each site location |
| Logistics: | Supply from closest air base, maximum separation 200 mi; helicopter requirement for personnel and components |

| Mission | Mission Requirements |
|---|---|
| 1. Strategic bombardment of military, industrial, and urban targets with high-yield weapons | a. System performance such as to minimize enemy capability for:<br>  (1) Detection of missile<br>  (2) Adequate early warning and dispersal<br>  (3) Interception of missile or ECM<br>  (4) Retaliation<br>  (5) Strategic support of military<br>b. Satisfactory guidance accuracy<br>c. Effective control of warhead burst altitude<br>d. Short reaction time-strike order to launch<br>e. Mobility of weapon system<br>f. Low initial dollar cost per operational missile (and associated equipment)<br>g. Low operations and maintenance costs<br>h. Moderate manpower requirements |

| Mission Requirements | Performance Requirements |
|---|---|
| Missile flight performance | a. Engine ignition upon firing signal<br>b. Acceleration and velocity per program<br>c. Attitude control within tolerance limits<br>d. Nose cone separation per program (following engine burnout) |
| Guidance accuracy | a. Circular error probability (.50) radius of 3 mi |
| Warhead burst altitude control | a. Detonation at preselected altitude $\pm$ 5000 ft |
| Reaction time | a. All operational missiles launched within 15 min following strike order<br>  (1) Possible requirement for automated launch subsystem<br>  (2) Continuous monitoring of missile readiness required, implying need for fairly extensive display subsystem<br>  (3) Possible requirement for highly skilled maintenance personnel<br>b. Missiles continuously on alert (combat standby)<br>c. Downtime for maintenance not in excess of 10% |
| Mobility of system | a. Site capable of initiating redeployment within 15 min<br>b. Total time required for site setup (to launch time) not in excess of 30 min |

---

[6]D. Meister, et al. *The Impact of Manpower Requirements and Personnel Resources Data on System Design*, AMRL-TR-68-44, Air Force Systems Command Wright-Patterson Air Force Base, Ohio.

| System Constraints | Description |
| --- | --- |
| Dollars | Limited; total available for R&D—$100 million |
| Schedule time | R&D complete by December 1960; production by 1964 |
| Physical resources | Unlimited for purposes of this system |
| Availability of manpower | Extremely limited in re AF 5 to 7 level personnel with missile and electronic AFSCs |

| Environmental Programs | Description |
| --- | --- |
| Climate and weather | Arctic, continental, and marine climates as found in Europe north of 45° N. latitude. Because of alert requirements, system must be capable of all-weather 24-h operation. |
| Wind | System must be capable of launch in winds up to 45 mi/h. |
| Temperature and humidity | See "climate and weather." |
| Geography | Installation may be located on any terrain accessible by roads of less than 7% grades. Level area of finite dimensions required for site. Tundra may provide a problem. Trees may provide some cover with respect to aerial reconnaissance. |
| Atmospheric composition and contaminants | Not relevant to mobile open site and solid propellant operations. |
| Lighting and audition | Lighting and auditory noise anticipated as design problems in trailer interiors. Communication system requires lines from living quarters to operations trailer. Also need alternative means of communications between wings, squadrons, and sites. |
| Safety hazards | Accidental detonation of warhead, premature ignition of missile power plant, toppling of erect missile, falls from work platform, electric shock. Requirement for "buddy" system may increase manpower requirements. |

## Mission Segments

| Segment Start Time, min | Segment | Segment Demarcation |
| --- | --- | --- |
| Variable by distance, time starts at site | Transport system equipment to launching site | Deployment order |
| | Assemble missile and prepare pad | Reach site |
| | Erect missile | Missile moved to pad |
| | Activate and check out missile guidance and control subsystems | Missile in launch attitude |
| | Insert mission and target data | Arm and fuse test OK |
| | Maintain alert status* | Warhead burst altitude set |
| | Arm warhead | Arm order—redeploy order |
| | Launch missile | Fusing check complete strike order |
| | Prepare for redeployment† | Missile launched |
| | | Trailers ready to move |

*This segment is not required if missile is to be launched immediately.
†This segment is not required if missile is launched.

# SYSTEM ANALYSIS
## Mission Description

**System Functions**

| Mission Segment | System Function |
| --- | --- |
| Transport system equipment to launching site | 1. Supply automotive energy to transport:<br>   *a.* Missile<br>   *b.* Warhead and nose cone<br>   *c.* Guidance modules and spares<br>   *d.* Transporter erector<br>   *e.* Operations trailer<br>   *f.* Electrical power generator trailer<br>   *g.* Crew quarters, mess, administrative, and security trailers<br>2. Control above prime movers as necessary for trip from air base to site. Communications equipment and procedures required for communicating among vehicles during move. Will status of missile and warhead have to be monitored during move? If so, how will movement of vehicles affect monitoring displays? What must and can be monitored during move? |
| Assemble missile and prepare pad | 1. Transport W/H, nose cone, and guidance modules to missile.<br>2. Mate components. Mating equipment required. Implications for weights to be lifted by personnel.<br>3. Attach components and physical links. Stress importance of identifying components for fast assembly.<br>4. Lay portable pad support for missile. |
| Erect missile | 1. Control angular movement of missile from horizontal to vertical position.<br>2. Secure missile on pad. |
| Activate and check out missile guidance and control subsystems | 1. Activate gyros and other components.<br>2. Activate test instruments.<br>3. Check yaw accuracy.<br>4. Check roll accuracy.<br>5. Check pitch accuracy.<br>6. Identify, remove, and replace malfunctioning units.<br>7. Communicate possible no-go situation to launch trailer.<br>8. Test warhead arming and fusing components. |
| Insert mission and target data | 1. Orient missile in azimuth.<br>2. Insert trajectory tape.<br>3. Set warhead burst altitude (bombardment mission). |
| Maintain alert status* | 1. Periodically monitor all loops for in-tolerance functioning.<br>2. Provide warning when malfunction exists.<br>3. Isolate out-of-tolerance condition to specific module. Because of requirement for fast reaction, malfunction diagnosis will have to be highly automated.<br>4. Remove and replace module.<br>5. Report existing and anticipated missile out-of-commission time. |
| Arm warhead upon receipt of strike order | 1. Arm warhead. Safety precautions.<br>2. Recheck fusing system. |
| Launch missile upon receipt of strike order | 1. Make final subsystem checks (probably all automatic).<br>2. Activate firing circuit.<br>3. Clear launch area.<br>4. Ignite engine. |
| Prepare for site redeployment† upon receipt of order | 1. Remove missile from pad to TV.<br>2. Disassemble missile, nose cone, and guidance.<br>3. Load nose cone, guidance, and spares in helicopters.<br>4. Disassemble and stow launch pad.<br>5. Ready all trailers for movement. |

*This function is not required if missile is to be launched immediately.
†This function is not required if missile is launched.

## Example of a Nonmilitary Mission Description

To demonstrate the importance of a thorough mission description for domestic as well as military systems, the following abbreviated mission description for a proposed motor home is presented.

### 1. General Mission Description

A motor home is desired that will serve as a "home on wheels" for a family of four. It will be designed to provide a self-sustained family environment for up to 2 weeks, with the capability of traveling on normal, improved roadways and urban streets in a safe and reliable manner. The design should be compatible with the typical adult male's driving and maintenance skills, except for specialized services for overhaul.

### 2. Performance Requirements

a. Normal cruising range shall be 500 mi (805 km) without refueling.

b. It shall be capable of operating on improved roads and grades up to and including 20°.

c. It shall be able to cruise on level grade up to and including 60 mi/h (97 km/h).

d. It must be able to turn within a 30-ft (9.2-m) radius.

e. It shall be able to be stopped from a speed of 60 mi/h (97 km/h) in 100 ft (30.5 m), on normally dry and sand-free roadways.

f. It shall provide the capability to store, expend, prepare, and service a family of two adults and two children in terms of food, water, and waste management; on-board clothing and baggage; and eating, sleeping, and recreational facilities.

g. Self-contained electrical power shall be provided for up to 24 hr.

h. Maximum on-the-road safety shall be provided in terms of vehicle systems performance, structural integrity, passenger restraint, driver performance efficiency, cargo restraint, and nonflammability.

### 3. Constraints

a. The purchase cost shall not exceed $35,000.

b. The power plant shall operate on typically available fuels.

c. The overall external dimensions of the vehicle shall not exceed the following: width—12 ft (3.7 m); height—13 ft (4.0 m); length—35 ft (10.7 m).

d. Dry weight shall not exceed 10,000 lb (4530 kg).

e. The power plant shall operate efficiently at altitudes up to 12,000 ft (0600 m) and under thermal conditions from −30 to +120° C.

f. Internal comfort conditions must be provided when external conditions are as follows: noise—100 dB; temperature—from −30 to +120° C; humidity—95 percent.

g. The vehicle shall remain upright under winds up to and including 50 mi/h (81 km/h) without tie-downs.

h. It shall be possible to change a tire when the vehicle is isolated from typical service facilities.

### 4. Special Safety Requirements

a. The vehicle shall meet all current federal motor vehicle safety standards applicable to motor homes.

b. A three-point, active safety-belt system shall be provided for at least four occupants—specifically, for the driver and the right-front passenger and for at least two persons in designated "riding positions" in the rear.

c. All interior surfaces and structure shall provide on-the-road crash protection.

d. Nonflammable materials shall be used throughout the interior.

e. Appropriate, centrally located power and utility cutoff controls shall be provided to preclude possible hazards from leaving certain utilities, such as the stove, on when they are not being used.

f. A minimum of two exits shall be provided —one at one end and one side, and the other at the other end and side.

### 5. Special Security Requirements

a. All exit doors shall contain dead-bolt safety locks operable both from inside and from outside the vehicle.

b. All windows shall be openable from the inside, but shall have suitable vandal-resistant locking devices operable only from the inside.

### 6. Other Requirements

a. All utility elements shall be designed for temporary external hookup to water and electrical power. All electrical equipment shall utilize 120 V with suitable conversion to dc for on-the-road functioning.

b. External convenience outlets shall be provided for portable external floodlighting plug-in.

c. Internal furnishings shall include but not be limited to:

(1) Kitchen furnishings (e.g., range, oven, refrigerator, sink, utensil storage, and temporary garbage storage).

(2) Dining furnishings (e.g., table for four with seating; storage for dishes, flatware, and crystal; and pantry for dry foodstuffs).

(3) Sleeping furnishings (e.g., beds for two adults and two children and clothing storage).

(4) Bath furnishings (e.g., chemical toilet, washbasin, shower, and linen storage).

(5) Recreational furnishings (e.g., AM-FM radio, portable TV, and book rack).

*Note:* All storage areas and devices shall be designed to retain stored items during travel.

(6) Interior illumination shall provide appropriate lighting levels and seeing quality for each activity area noted above.

d. Driver-station controls and displays shall include as a minimum the following:

(1) Primary driving controls (e.g., steering, acceleration, braking, and ignition), windshield wiper and washer, windshield defogger and defroster, air conditioning (to be separate from cabin air conditioning), starter, warning horn, headlight dimmer, cruise control, rearview mirrors (interior and exterior), and exterior lights (e.g., headlights and running lights).

(2) Communications (e.g., CB and AM-FM radio, plus three-track stereo radio and intercom).

### 7. Typical Mission

Parking:
    Urban (driveway, curbside, filling station)
    Unimproved (campsite, desert, snow, etc.)
Under way:
    Urban streets
    Freeway
    Tunnel
    Bridge
    Country road
Temporary living:
    Complete 2-week living cycle under various environments (hot, cold, rain, snow, and wind)

## FUNCTIONAL REQUIREMENTS DEFINITION

### Functional Block Diagraming

Block diagraming is perhaps the most familiar method of indicating basic system organization and function. Often, however, what purports to be functional block diagraming is really equipment block diagraming, as is readily evident in the appearance of blocks labeled "display console," "data entry panel," "tape recorder," drum storage," etc. Functional blocks should be concerned with what is *done* rather than with the specific realization of the means to do it. Functional block diagrams should not be allowed to evolve into equipment block diagrams prematurely; i.e., the decision concerning which functions should be performed by a piece of equipment rather than a human operator should not be made until the full scope of functional requirements has been clearly defined.

For example, "detection" is a functional term, whereas "detector's console" is not and already assumes an allocation of a function to human and machine. Since allocation of functions should always follow development of the initial system concept, it is essential to avoid equipment representation and its implication that the function already has been allocated, to either a human *or* a machine. A premature human-machine allocation may overlook the possibility that a human can perform a given task with greater cost effectiveness than a machine. By the same token, to assign a human a function that is really beyond his or her reliable limits just to save the cost of another machine is equally foolish. However, as the various trade-offs are considered, the original block diagrams may be refined for each of the alternatives under consideration. The accompanying examples of block diagrams include some tentative assignments of hardware, software, and personnel. This is proper —but only after some initial analysis of human and machine capabilities has been performed.

The first illustration shows elements of a hierarchical block structure. Some of the essential features of block diagraming are as follows:

1. An expanding series of diagrams gives successively more detailed information on each functional block. This detail may be carried to as many levels as are required to drive out some end item (operator, machine, component, etc.).
2. Functions are numbered in a manner that preserves the continuity of function and logical breakout from function origin.
3. The top-level diagram should show the system development process itself as well as the operational functioning of the system to be developed. In other words, the system being developed is *within* a larger system—which is the system for accomplishing the development.

4. Branching can and should be shown as indicated in the top-level diagram. Once the particular system in the example becomes operational, it is either in combat service or in a maintenance state, both of which require additional breakout with finer-level diagrams. The introduction of branching in functional block diagraming provides greater flexibility and facilitates the transition to information flow charting (which comes later).
5. The diagram should be organized so that one can easily find the input and follow through the function blocks to the resulting output. This should be evident at each level. In fact, if one finds that at a

particular level it is impossible to follow from left to right, from input to output, an error in functional designation has been made (a useful method of cross-checking).

In the second illustration a more detailed (though not complete) block diagram of function block number 5.0 from the first diagram is shown. Note that this diagram—as may sometimes be desired—is of mixed levels (second and third). Note also that, except for the associated systems, there is not yet any implication of human-machine function allocation.

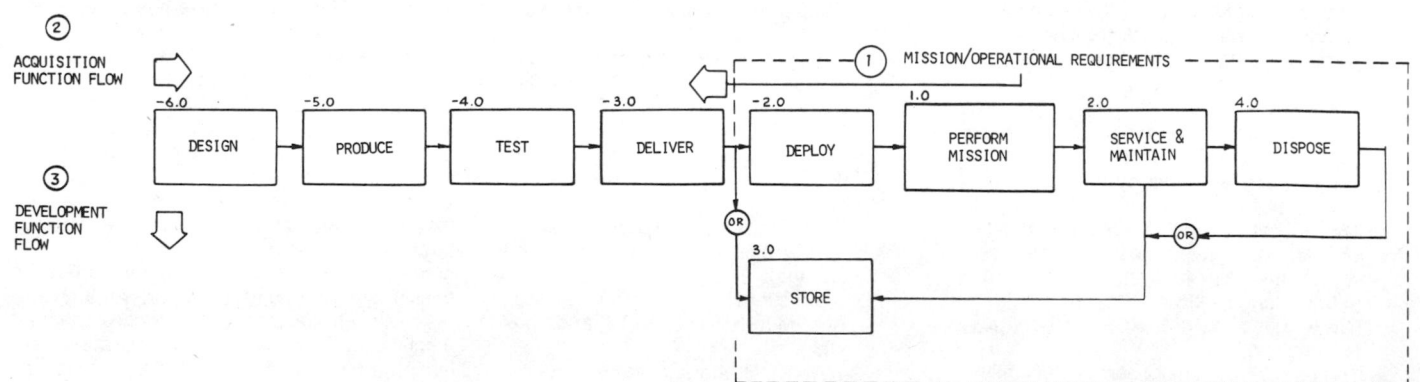

**General model for functional flow block diagraming.**

**Functional and Subfunctional Block
Diagraming**

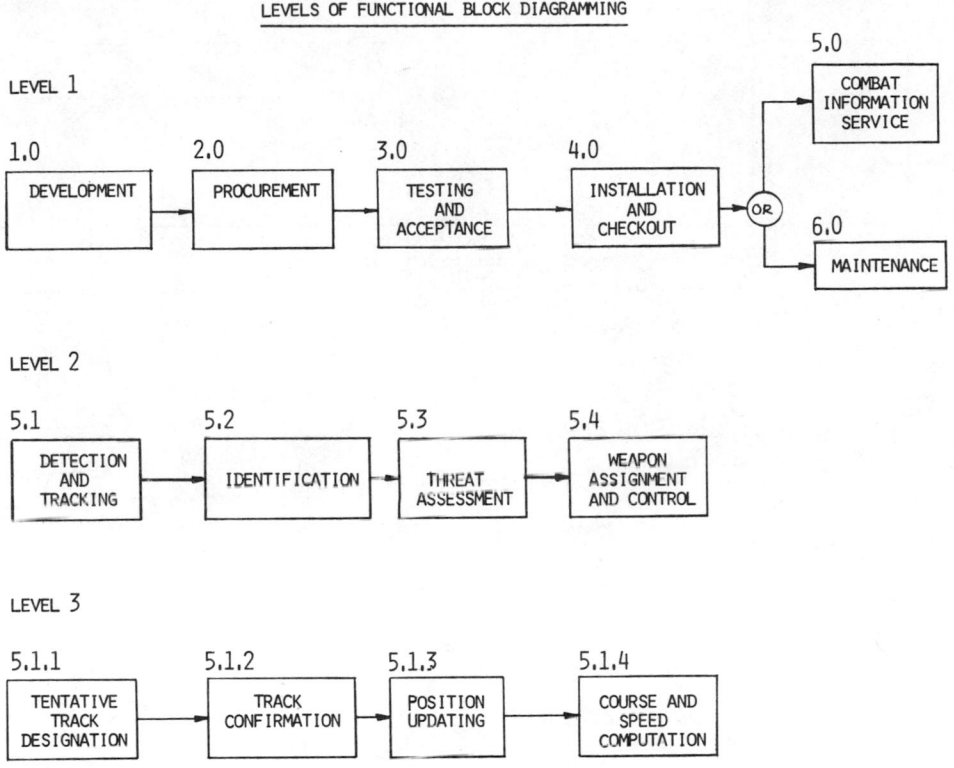

LEVELS OF FUNCTIONAL BLOCK DIAGRAMMING

**Example: Top-Level Functional Block
Diagram for an Aeronautical System**

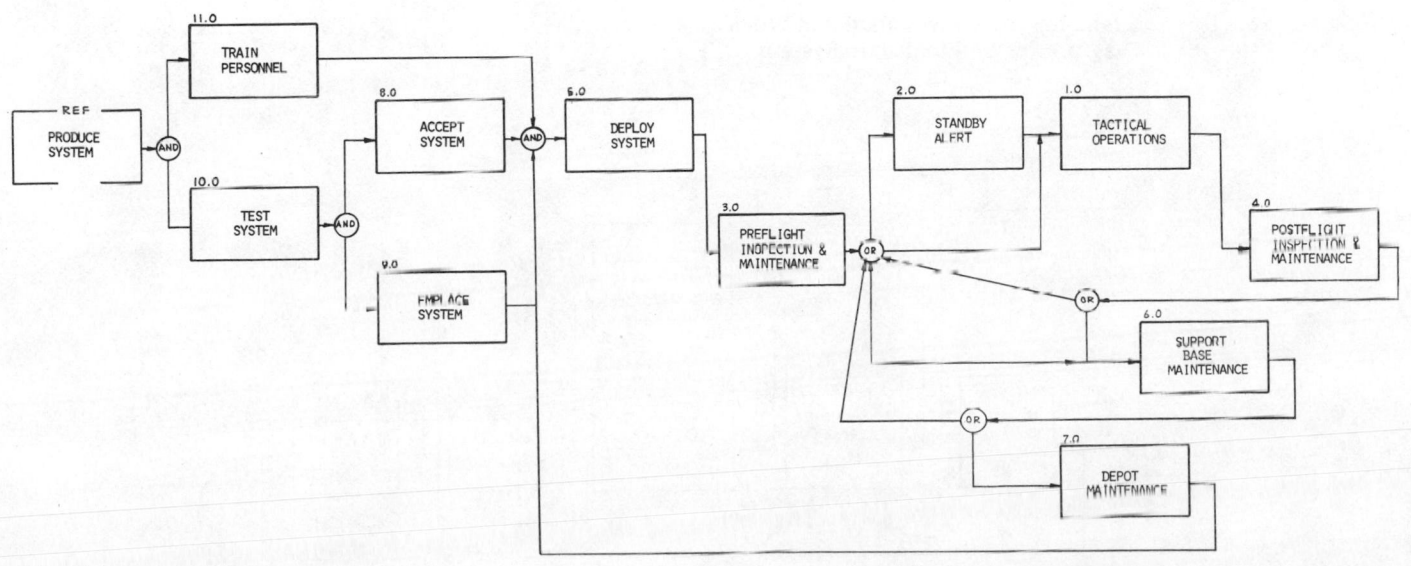

**Example: First-Level Functional Block
Diagram for an Aeronautical System**

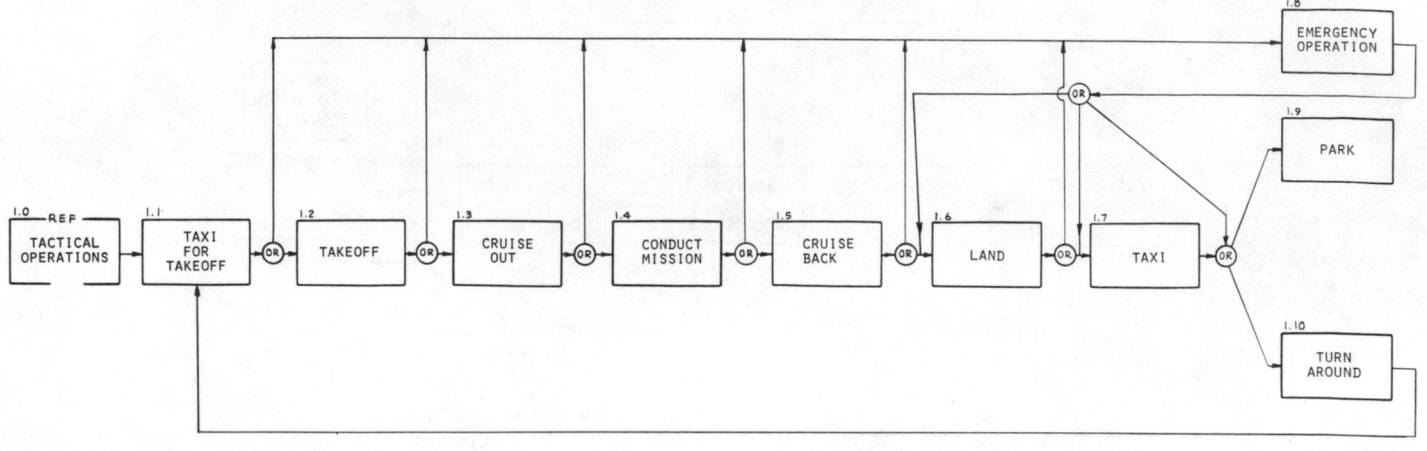

**Example: Top-Level Functional Block
Diagram for an Electronic System**

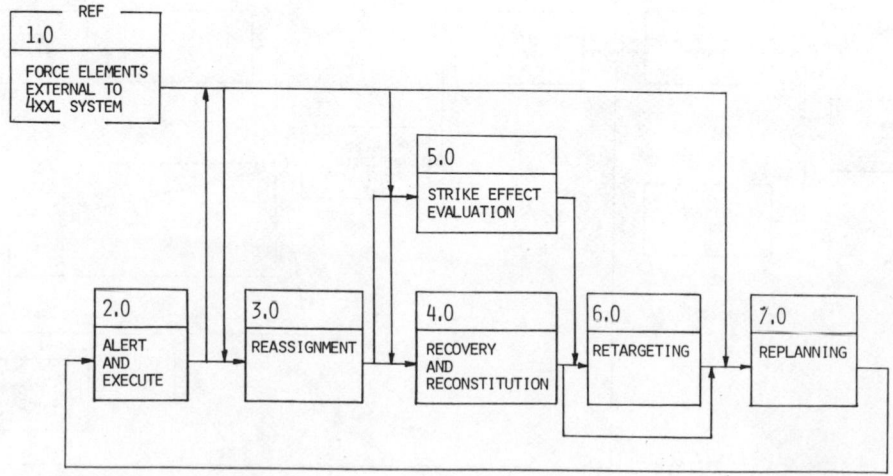

**Example: First-Level Functional Block Diagram for an Electronic System**

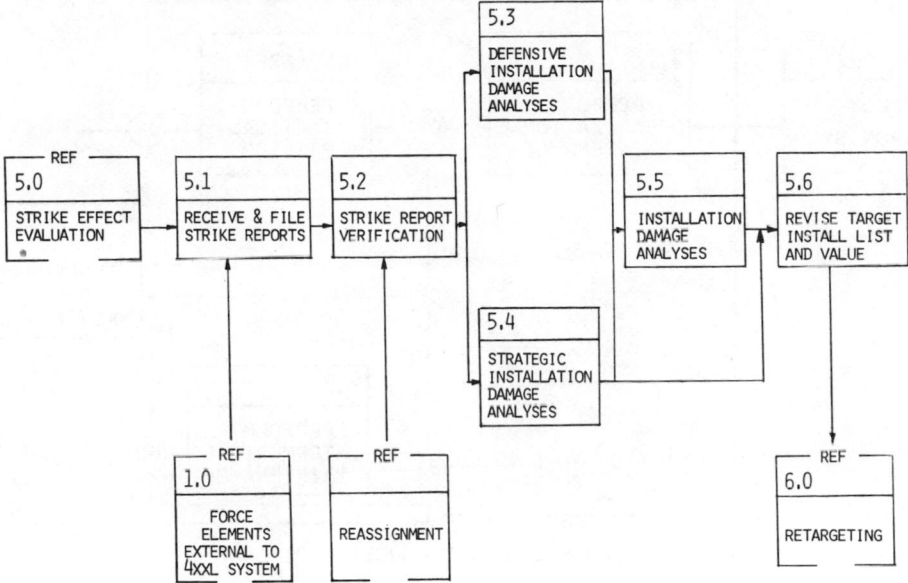

**Example: Second-Level Functional Block Diagram for an Electronic System**

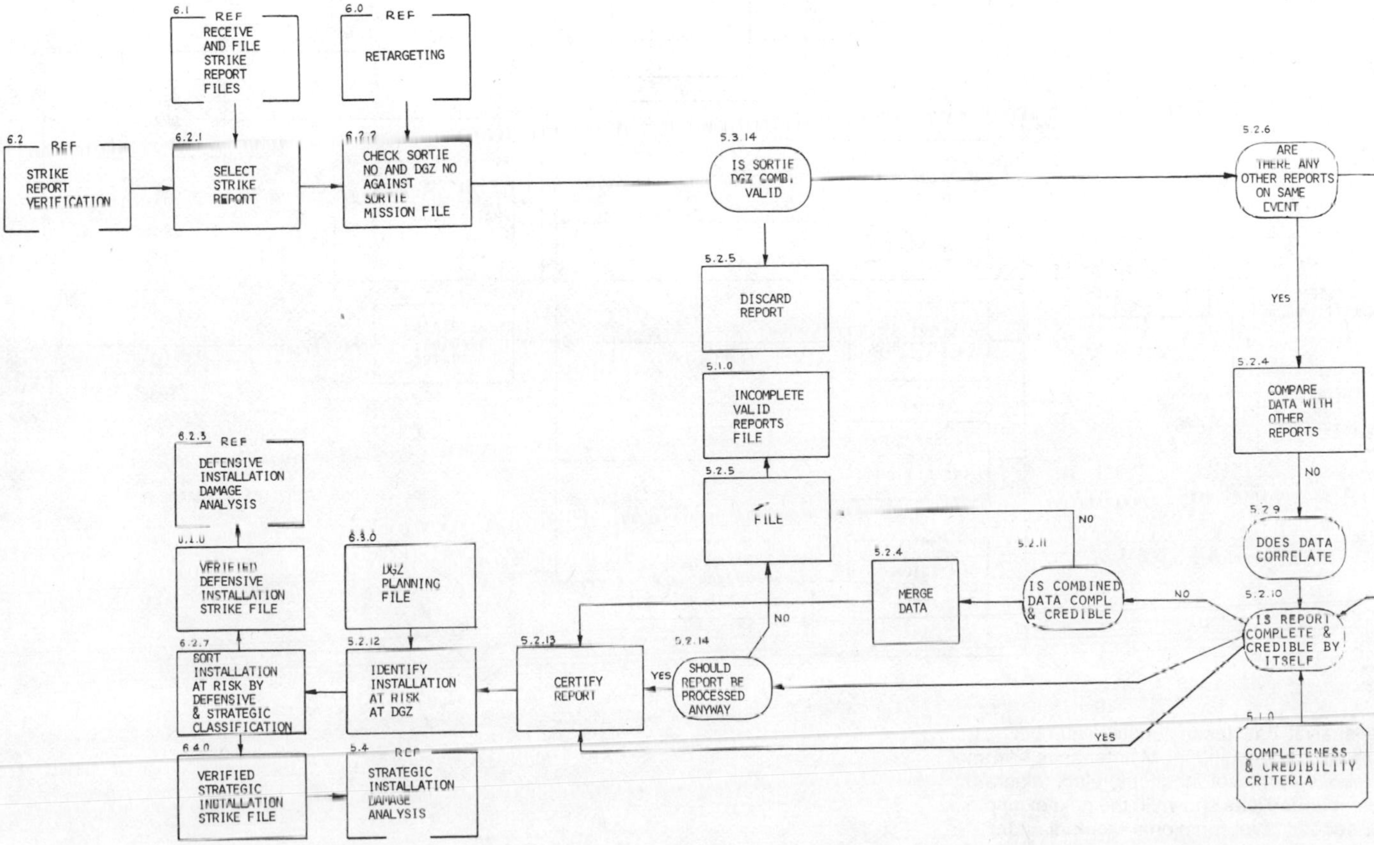

**Example: Expanding a Maintenance
Function Series**

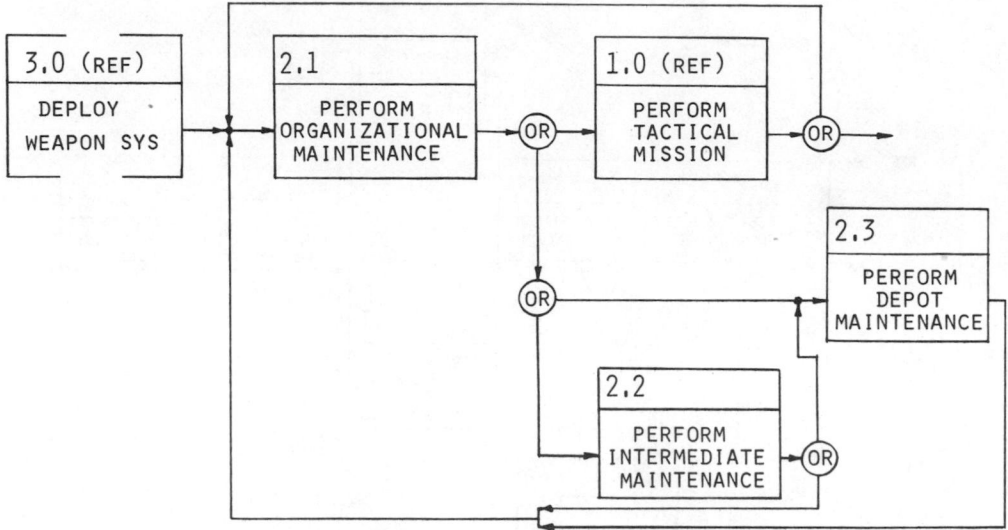

FIRST LEVEL MAINTENANCE FUNCTION DIAGRAM

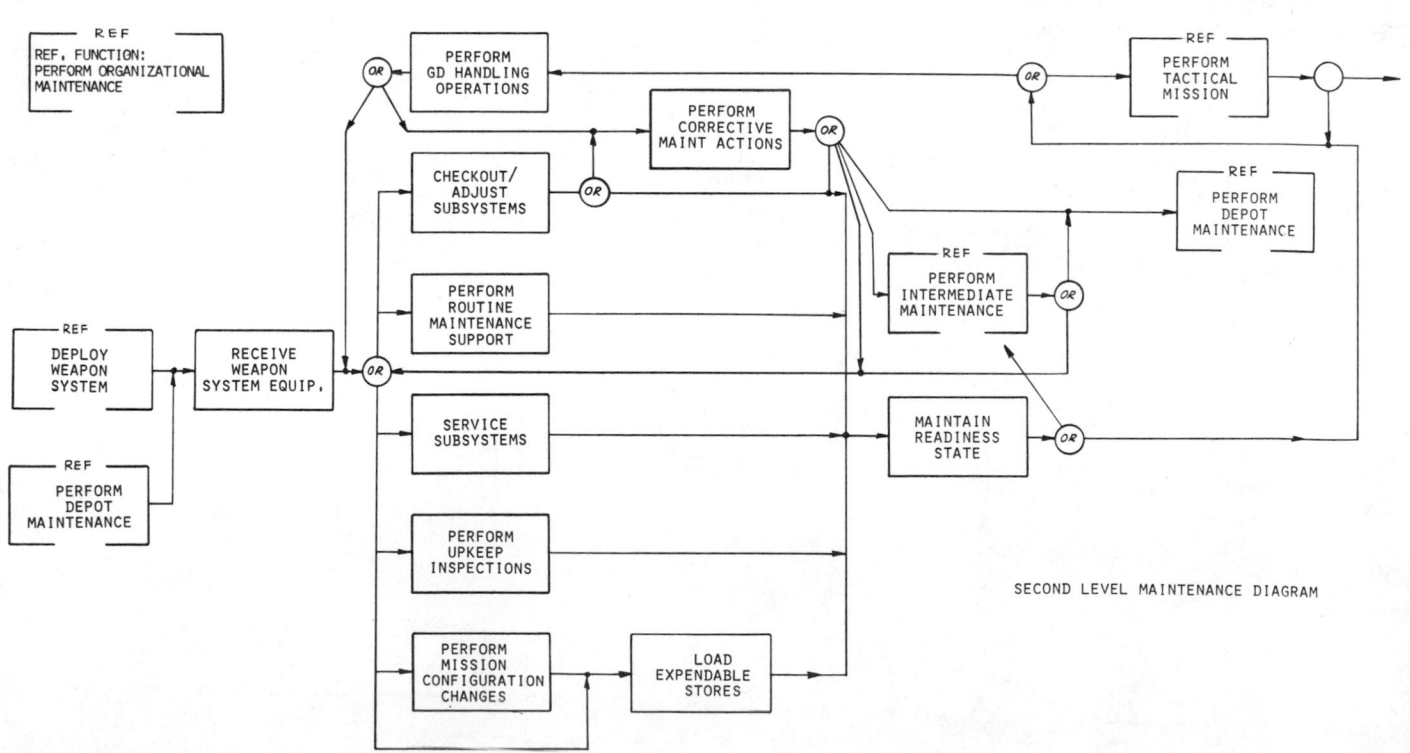

SECOND LEVEL MAINTENANCE DIAGRAM

The analyst has taken function number 2.1,
"Perform Organizational Maintenance," from
the above first-level functional block diagram
and expanded it as shown in the next example
of a second-level functional block diagram.

**Example: Second-Level Functional Block Diagram for Facilities to Support a Nuclear Missile System**

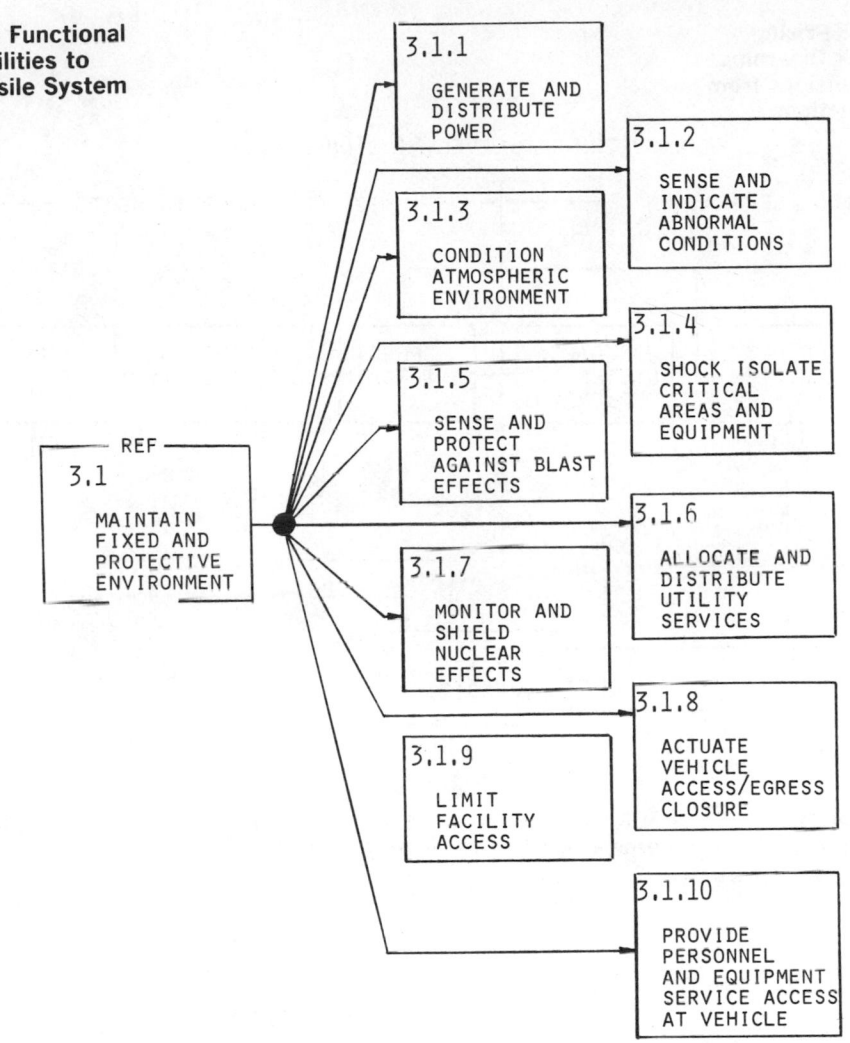

**Example: Third Level Functional Block Diagram for Support Facilities for a Missile System**

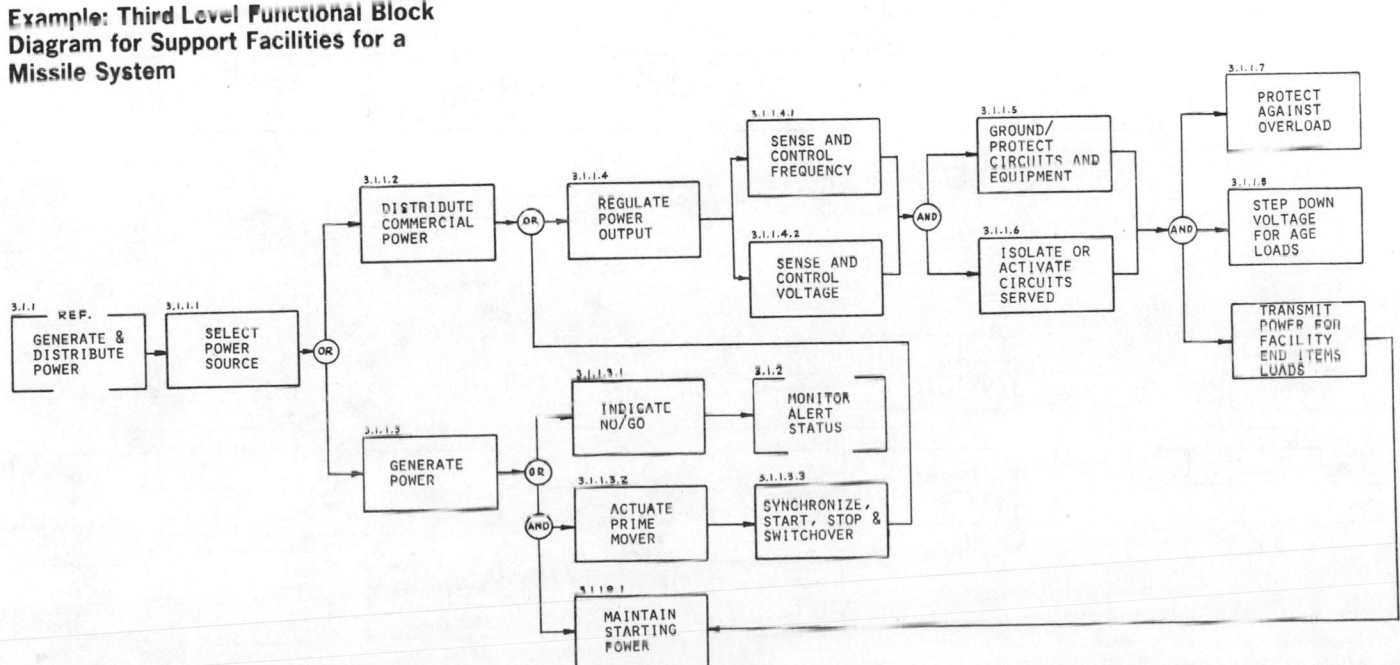

**Example: Hand-Held Missile-Firing
Weapon—Functional Block Diagrams
Illustrating Sub-Level Extensions from
a Series of First-Level Functions**

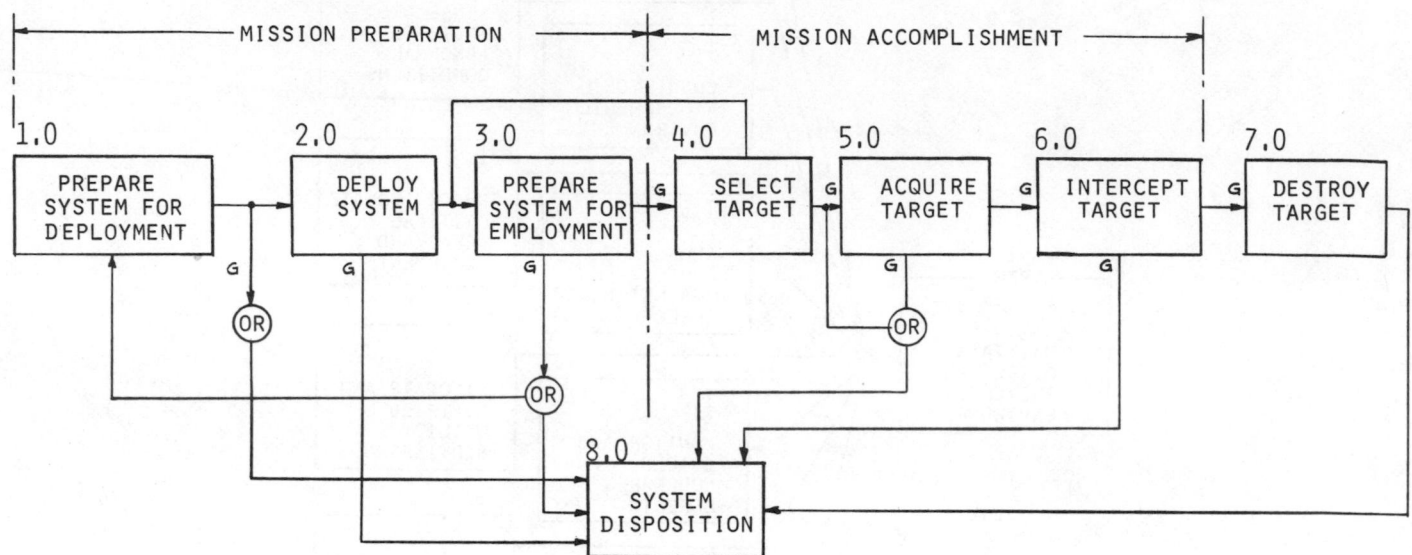

XXXX SYSTEM TOP LEVEL MISSION FUNCTIONS

**Example: Use of Functional Block
Diagraming to Define Maintenance
Tasks as a Prelude to Task Analysis**

Note: Maintenance functions have already
been assigned to the human in the system.

XXX DISK DRIVE
FUNCTIONAL ANALYSIS

1.0 INSTALL

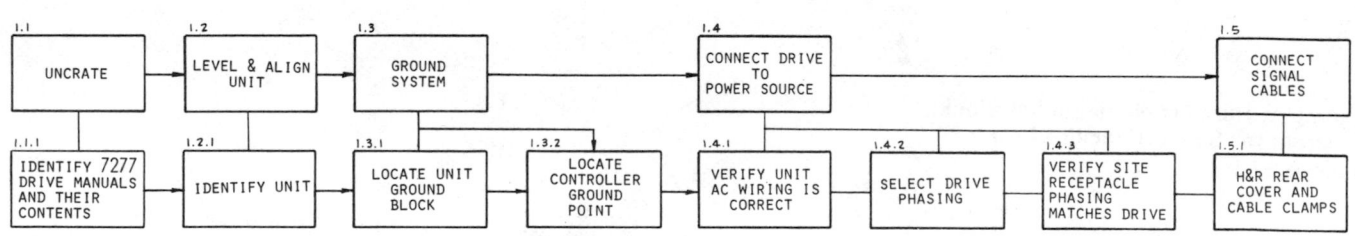

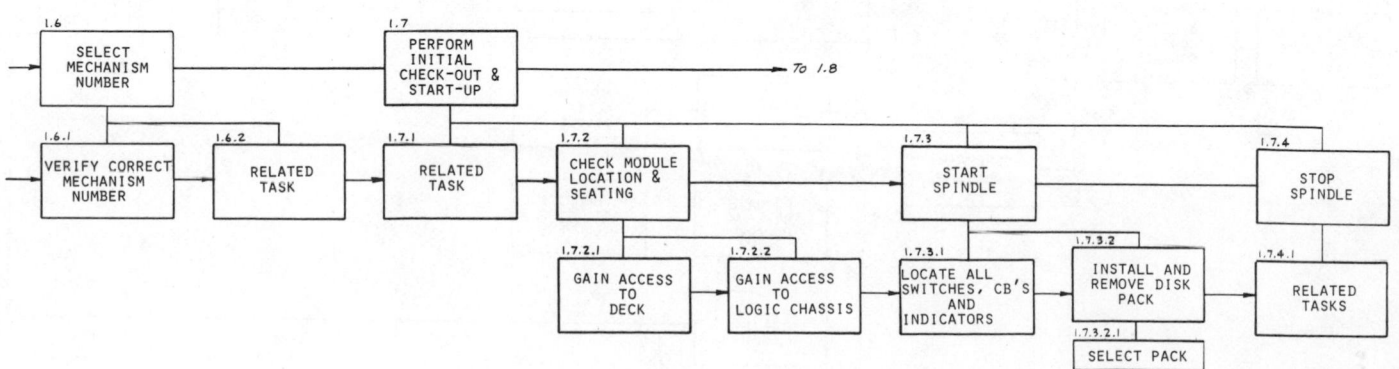

XXX  DISK DRIVE
FUNCTIONAL ANALYSIS

2.0   PREVENTIVE MAINTENANCE

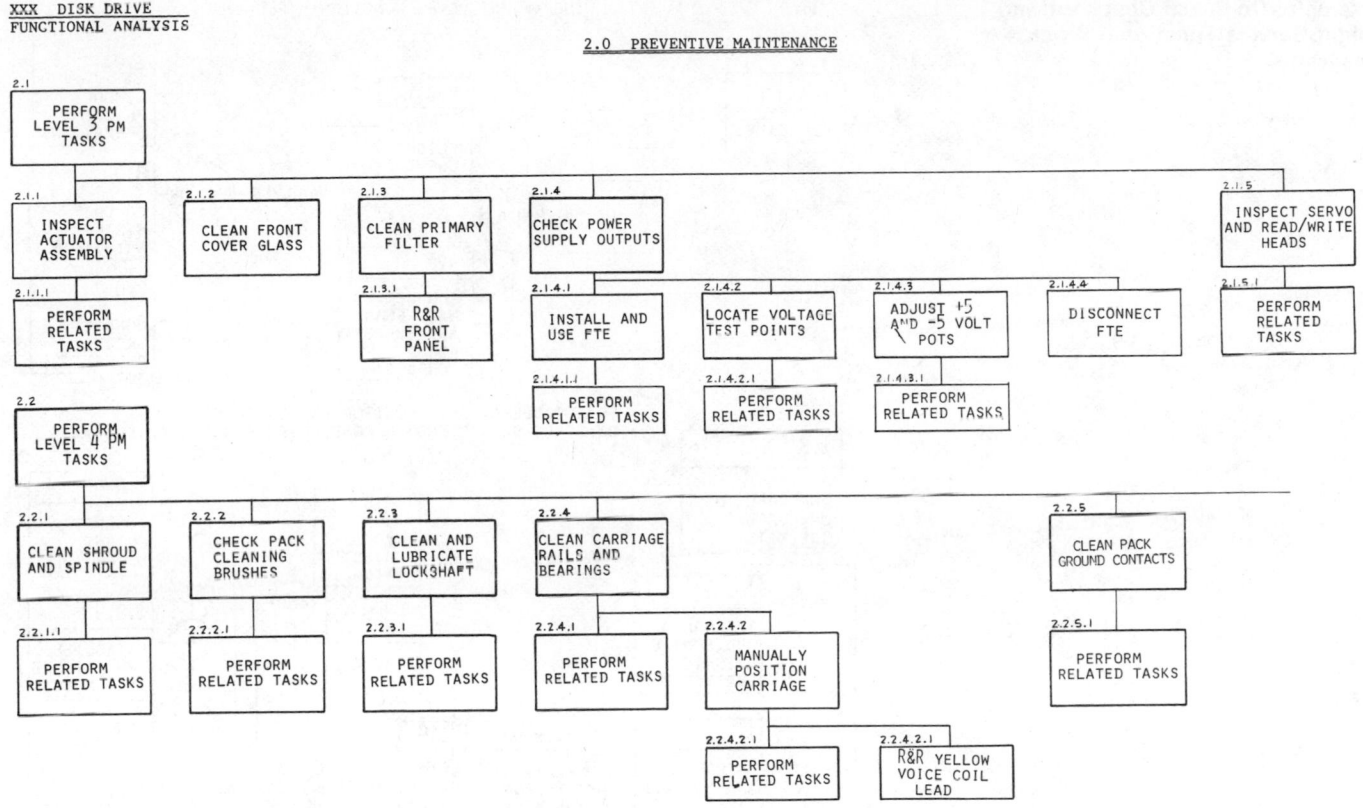

**Example: Training Function Block Diagram**

XXX DISK DRIVE
FUNCTIONAL ANALYSIS

3.0  CORRECTIVE MAINTENANCE

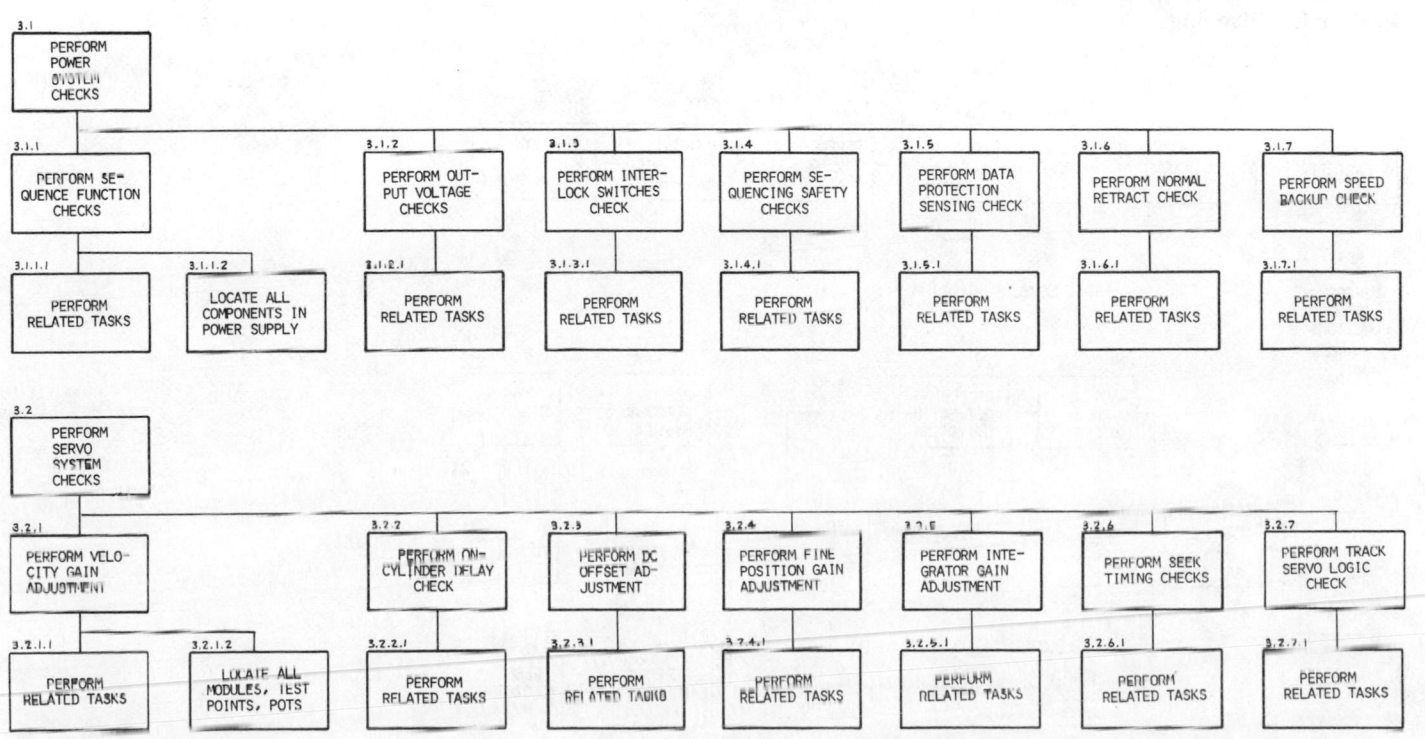

**Example: On-Board Checkout and
Flight Service Functional Block
Diagram**

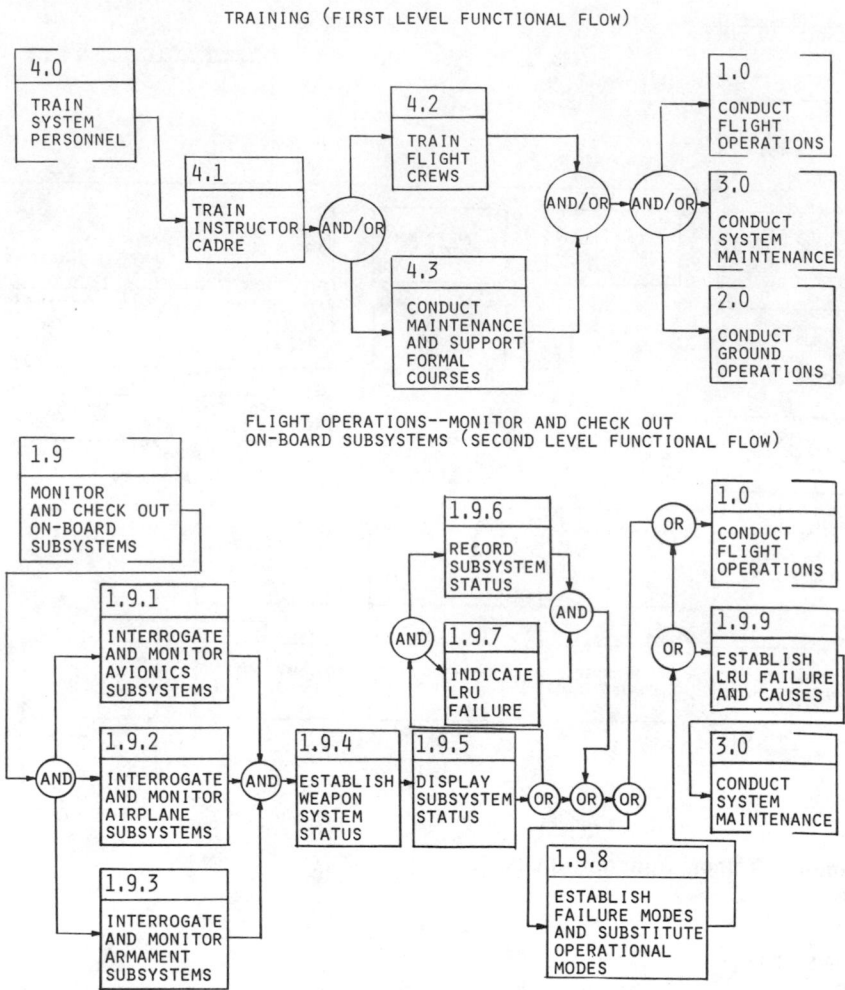

TRAINING (FIRST LEVEL FUNCTIONAL FLOW)

FLIGHT OPERATIONS--MONITOR AND CHECK OUT
ON-BOARD SUBSYSTEMS (SECOND LEVEL FUNCTIONAL FLOW)

**Example: Flow Diagraming Applied to
Residential Planning**

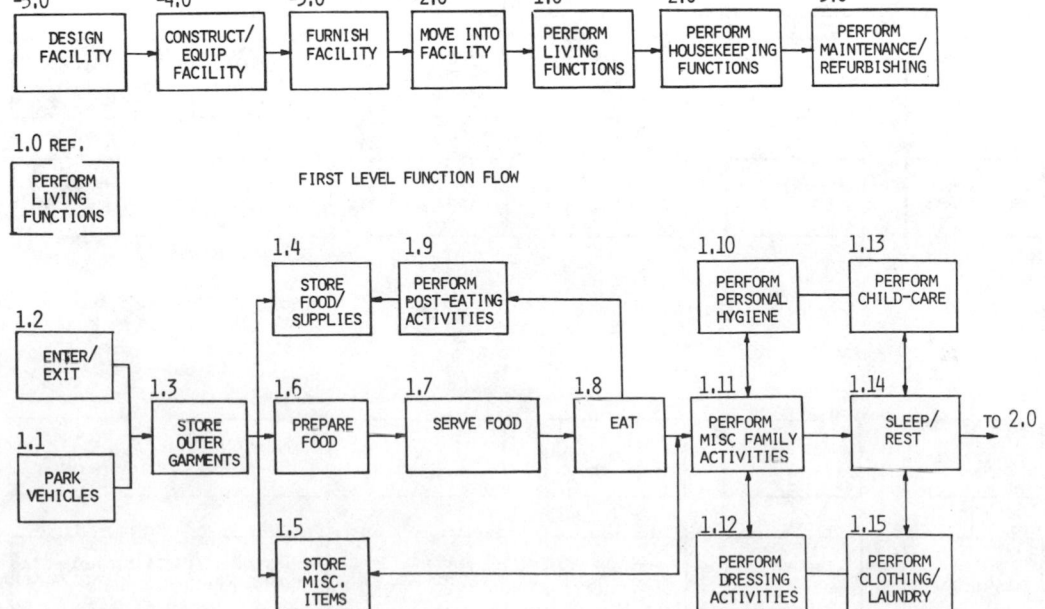

TOP LEVEL FUNCTION FLOW

FIRST LEVEL FUNCTION FLOW

## Analysis of Functions to Further Define the Requirements and Possible Constraints Associated with Accomplishing Each Function

Although functional block diagraming provides an immediate picture of all the probable functions that may be required to accomplish mission objectives, it is usually necessary to examine each of the proposed functions in terms of specifics regarding sublevel requirements for each function and in terms of the possible constraints that may affect the way in which each function can be most effectively accomplished.

Generally, if the system is complex, the many functions that must be performed cross more than one engineering discipline (including human factors). Therefore, one should consider the most knowledgeable individual for each function and assign this person the responsibility of expanding on the functional requirements. Typically, the assigned human engineer or engineers should work with each of the specialists who are to expand on the functional requirements.

Functional requirements analysis should be thoroughly documented as illustrated in the accompanying example. Unfortunately, this brief example fails to identify the particular function number. Therefore, when an analyst is given a set of functions for expanding and documenting requirements and constraints, each function that he or she presents should have an identifying number as well as the title.

Note in this example that the analyst has keyed the set of functions to specific mission objectives; this important feature of the documentation makes it easier for others to relate analyses to mission requirements that may have been specified in initial new development documentation or a request for proposal.

### Example: Analysis Documentation of Detailed Functional Requirements

| Mission Objectives | Functional Requirements | Constraints |
|---|---|---|
| Illuminate ground targets from an aircraft. | 1. Acquire and identify the target.<br>2. Position the aircraft for illumination of the target.<br>3. Illuminate the target. | 1. Illumination system must utilize a laser with specific characteristics.<br>2. Illumination must be accomplished from a maximum slant range of 20,000 ft.<br>3. Operation must be possible under both day and night conditions.<br>4. Control of illuminating beam must have an accuracy of 0.37 mil.<br>5. Control from a low-performance aircraft (0–300 knots). |

## FUNCTION ALLOCATION

### Human-Machine Function Allocation Analysis

Theoretically, decisions regarding how to accomplish each function established by the functional block diagraming and functional requirements analysis should not be made until one is sure that all the necessary functions have been specified and agreed upon. This is to prevent some preconceived idea of how to accomplish a function without serious consideration of several possibilities, particularly whether the function is more appropriate to human or to machine implementation. There are, of course, exceptions, i.e., when past experience has demonstrated that people cannot do certain things and therefore some hardware device *must be used*.

The human machine function allocation step is extremely important because it allows knowledgeable individuals to contemplate the pros and cons of using a human rather than a machine to accomplish each of the proposed functions. Obviously, one should strive to select the alternative that provides the most cost-effective solution.

Human factors specialists should enter into this phase of the analysis since they are generally more knowledgeable with respect to what humans can or cannot do well, risks that may be too great to expose the human to, severe problems with regard to selecting the right individuals to perform the function, problems associated with training the individual to do an adequate job, and so on. The following guidelines and analytic examples are provided to aid the reader in addressing the function allocation analysis task.

### General Considerations for Human versus Machine Function Allocation

ENVIRONMENTAL CONSTRAINTS The human's physiological tolerance to certain operating environments is limited; therefore, one has to make an early decision whether to incur the necessary costs and complexity to protect and support the human under severe environmental conditions (extreme atmospheric pressure, acceleration, temperature, noise, vibration, or radiation and/or potential emergency situations produced by explosive blasts, fire, atmospheric or chemical contamination, etc.).

SENSORY ISOLATION To perform useful tasks within a control environment, humans must be able to receive sensory inputs (information) at levels commensurate with their inherent sensory channel threshold capabilities; e.g., humans can see only so far and hear a signal that is only so soft, and these direct perceptions are easily degraded by various interfering environments. In fact, sensory inputs may be distorted, causing humans to make perceptual errors, i.e., to misinterpret what they see, hear, or feel.

SPEED AND ACCURACY Human response cannot compete with the capacity of a machine in terms of speed and accuracy; thus, functional allocations to humans must be made on the basis of their capacity.

OVERLOAD Humans are fairly limited compared with machines in terms of how much information they can absorb and handle at one time, how many things they can monitor or control at one time, and how effectively they can maintain cognizance of a situation for extended periods of time or under severe physiological and psychological stress conditions.

PHYSICAL STRENGTH Humans are extremely limited compared with machines in terms of how much force they can apply, and for how long.

STORAGE CAPACITY Humans' capacity to store large amounts of information over the long term is extremely great, but their ability to retrieve information quickly is sometimes extremely limited and unreliable; a machine, however, can store almost any amount of data and recall it almost immediately. On the other hand, the machine's capacity to store and retrieve is entirely limited to what is designed into it.

HUMAN-MACHINE PERFORMANCE SURVEILLANCE Humans (compared with machines) are relatively poor "self-monitors" and are easily influenced by emotional factors and by environmental and operational distortions.

**INTERPRETATION OF, AND RESPONSE TO, UNEXPECTED EVENTS** Humans possess the unique capacity to constantly reevaluate a situation, change their approach, and invent new ideas on the basis of unexpected events and operating conditions. They often can continue with an alternative or less-than-perfect procedure, whereas a machine may quit completely. As noted earlier, a machine does only what it was designed to do; i.e., its capability is limited by the designer's capacity to anticipate all events and conditions of operation.

**FATIGUE** Humans' capacity and functional capabilities are subject to short- and long-term fatigue effects, whereas machines *can* be designed to be almost fatigue-resistant.

**LEARNING** Humans generally require some finite learning period to perform a new function. A machine begins its operation immediately and theoretically requires neither initial training nor proficiency refreshment.

**COST** As long as humans are used properly (within their basic physiological and psychological limits), they often are the least expensive component of a system. Although one obviously must account for the costs of supporting them (housing, pay, medical expenses, etc.), in many cases this has to be done anyway. One must be careful not to try to duplicate human capabilities completely, for such duplication by machine may end up costing much more. A thorough cost comparison is the only sound method for deciding whether to use the human or the machine for given functions. In the case of military systems, the customer must provide complete support for the human. On the other hand, domestic systems and products typically do not require such complete support for the human.

## Checklist for Making Decisions Relative to Human-Machine Coupling

In order for humans to complement the capability of a system, they must be coupled with machines in a way that will allow them to utilize their capabilities to the maximum. Thus the following should be considered.

**GENERAL CONCEPTUAL PRINCIPLES FOR HUMAN-MACHINE SYSTEM COUPLING**

1. Select the sensorimotor link which makes the best use of human capacity, sensitivity, and reliability. Avoid coupling via a particular link merely on the basis of tradition or because it may appear that a particular hardware implementation is less expensive, easier to design, or already available.
2. Choose a coupling approach that maximizes total system effectiveness; do not choose an approach on the basis of whether it is easy or hard to automate a function.
3. Couple humans with machines in such a way that they are not compelled to work at peak limits all or most of the time.
4. Couple humans with machines in such a way that they can recognize or feel that their contribution is meaningful and important. Avoid giving humans machine-serving responsibilities.
5. Couple humans with machines in such a way that information flow and informa-

tion processing are natural; this minimizes learning time and the probability of confusion or errors.

6. Select coupling methods that do not require extremely precise manipulations; continuous, repetitive movements; frequent, laborious, and lengthy calculations where accuracy is critical; or physical contributions that demand reaching one's upper strength limits.
7. Couple humans with machines as though they might at some time have to assume control (even though the nominal mode may be automatic).
8. Use hardware to aid the human; do not use the human to complement a predetermined hardware concept.

**GENERAL PHYSIOLOGICAL CONSIDERATIONS IN HUMAN-MACHINE ALLOCATION ANALYSIS[7]**

Actual physiological hazards to operator health obviously must be considered while determining function allocation to human or machine. However, one must also examine the other boundaries in terms of how (although they may not lead directly to personal injury) conditions might stress operators sufficiently to induce them to make errors and in terms of how these errors might in turn lead to potential system loss and/or eventual injury to operators and others.

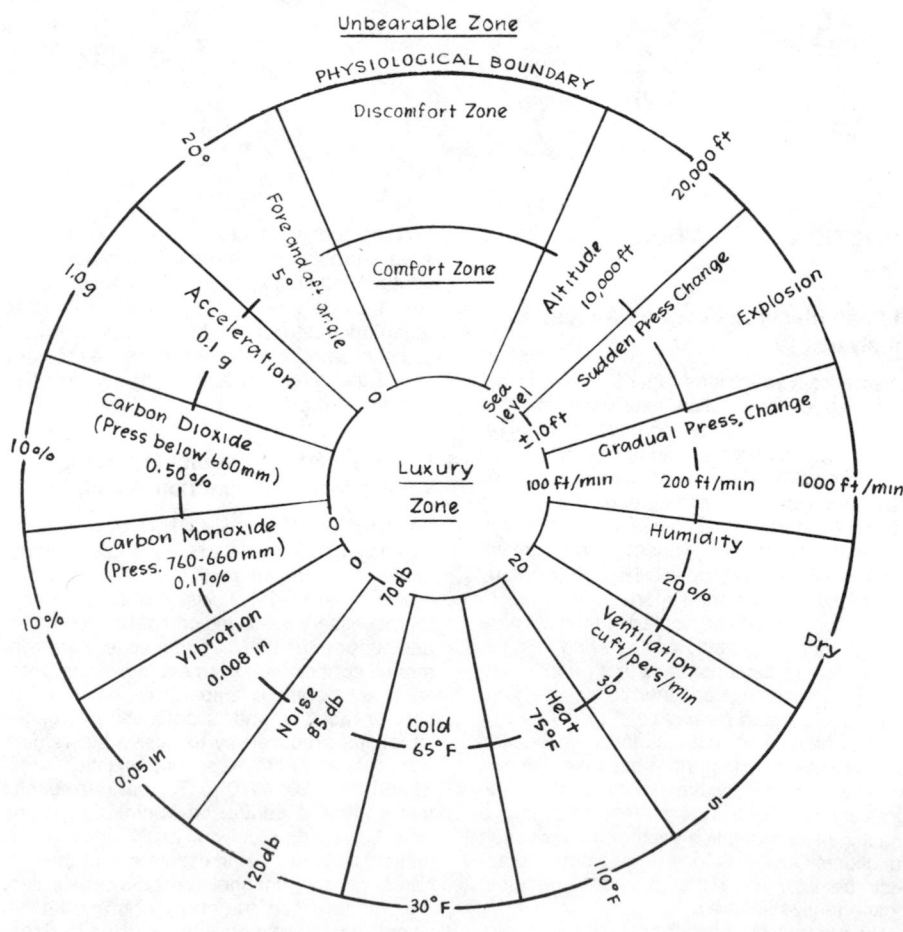

[7]R. A. McFarland, *Human Factors in Air Transportation: Occupational Health and Safety*, McGraw-Hill Book Company, New York, 1952, p. 705.

It is suggested that the designer prepare and have at hand a list of potential injury modes (cuts, bruises, fractures, amputations, burns, internal ruptures, eye penetration, asphyxiation, etc.). This should be used as a checklist to evaluate the proposed human-machine allocation decisions.

## Example: Function Allocation Analysis and Screening Worksheet

| Hypothetical Tracking Functions | Inherent Operator Capabilities | | | | | Inherent Equipment Capabilities | | | TOTAL | SCORE | PROPOSED ALLOCATION | | | |
| | Detecting Signals in the Presence of High-Noise Environment (X5) | Recognizing Objects under Varying Conditions of Perception (X4) | Handling Unexpected Occurrences or Low-Probability Events (X4) | Reasoning Inductively (X1) | Profiting from Experience (X2) | Responding Quickly to Signals (X3) | Performing Precise Routine Repetitive Operations (X2) | Computing and Handling Large Amounts of Stored Information Quickly and Accurately (X4) | Operator | Machine | Operator | Bcth | Equipment | Software |
|---|---|---|---|---|---|---|---|---|---|---|---|---|---|---|
| 1. Determine if target tracks in system | 25 | 8 | 12 | 3 | 6 | 9 | 8 | 4 | 81 | 41 | X | | | |
| 2. Actuate sequence | 5 | 4 | 4 | 1 | 2 | 3 | 2 | 4 | 20 | 24 | | X | | |
| 3. Put next target in track list under close control | 5 | 4 | 4 | 1 | 2 | 9 | 10 | 4 | 21 | 43 | | | X | X |
| 4. Advance hook on display to track coordinates | 5 | 4 | 4 | 1 | 4 | 9 | 10 | 4 | 21 | 43 | | | X | X |
| 5. Determine if target video present | 20 | 8 | 8 | 3 | 4 | 9 | 8 | 4 | 70 | 39 | X | | | |
| 6. Determine if hook lines up with present target position, etc. | 20 | 8 | 8 | 3 | 6 | | 8 | 4 | 73 | 40 | X | | | |

**Example: Alternative Man-Machine Function Allocation Analysis**

**Sample Function Allocation Chart**

| Function Number | Function Title | Function Requirements | Environmental Considerations | Operational Considerations | Potential Human Role | Desirable Equipment Role | Human Function Description | Equipment Function Description |
|---|---|---|---|---|---|---|---|---|
| 1.0 | Prepare system for deployment. | Verify system operational readiness, and prepare system for shipment. | Temperature, humidity. Ventilation. Illumination. Noise. | Time urgency | Inspect equipment visually. Use test equipment. Test operating condition of all system components. Remove and replace components in shipping containers. Load shipping on vehicles. | Measure subsystem component functioning. Provide lift and load of system components onto transport vehicles. | | |
| 1.3 | Verify system operational readiness. | Verify weapon. Test set and trainers' operation readiness. | (Same as above) | (Same as above) | Visual inspection and test of weapon using test set visual inspection and test of test set and trainers using portable test equipment: Replacement of lost accessories such as covers, seals, etc. | Measure subsystem/ component functioning. | | |
| 1.3.1 | Verify weapon operational readiness. | Remove weapon from container. Prepare weapon for testing. Prepare test set. Perform Post test. Shut down and Remove weapon. Return weapon to container. | (Same as 1.0) | (Same as 1.0) | Remove weapon from container. Prepare weapon for test. Prepare test set. Perform test. Perform post-test shutdown and remove weapon. Return weapon to container. | Test set = provide go-no/go indications of weapon functional readiness. | | |
| 1.3.1.4 | Test impulse generator. | Connect weapon impulse-generator circuit to test set impulse generator circuit. Actuate weapon impulse-generator circuit. Measure by weapon impulse generator. | (Same as 1.0) | (Same as 1.0) | Operate a switch to connect weapon/test set impulse circuits. Operate weapon warm-up lever. Read weapon impulse generator output on test set indicator. | Switch to connect weapon/test. Set impulse circuits. (See 4.2.1.2) Visual indication of impulse generator output. | Actuate impulse generator circuit switch interface device on test set. Actuate weapon warm-up lever. Observe impulse-generator indicator on test set. | Provide switch knob. (See 4.2.1.2.1) Provide impulse generator indicator on test set. |

**Checking and Guidance Control**

| Alternative 1: Function Description | Alternative 2: Function Description | Alternative 3: Function Description | Alternative 4: Function Description |
|---|---|---|---|
| Technician at missile with portable test equipment initiates inputs and checks and records outputs. If equipment is out of tolerance, technician diagnoses fault and repairs. | Technician at missile connects cabling on orders of operator at console. Operator initiates input and checks and records output. If equipment is out of tolerance, technician traces fault using operator-communicated outputs to diagnose. Technician then repairs. | Technician at missile connects cabling. Operator starts console sequence. Operator monitors console for red-light indication. If red light flashes, technician and operator use machine information to diagnose fault. Machine records all outputs. | Sequence is part of automatic total missile checkout sequence. Operator receives audible no-go signal, and red light indicates malfunctioning unit. Operator communicates to technician which unit to replace. Machine records all outputs. |

**Typical Factors to Be Considered in Trading off Alternative Ways of Checking Guidance and Control Loops**

| Alternative 1: Function Description | Alternative 2: Function Description | Alternative 3: Function Description | Alternative 4: Function Description |
|---|---|---|---|
| Geography:<br>  Mobility requirement precludes permanent missile cover. Canvas shelter only to be allowed.<br>Weather:<br>  All-weather arctic operational requirement limits likelihood of technician's remaining long under canvas shelter only.<br>Wind:<br>  45-mi/h wind could endanger person on missile. Test equipment could easily be lost. | Geography:<br>  Mobility requirement precludes permanent missile cover. Canvas shelter only to be allowed.<br>Weather:<br>  All-weather arctic operational requirement limits likelihood of technician's remaining long under canvas shelter only.<br>Wind:<br>  45-mi/h wind could endanger person on missile. | Geography:<br>  Mobility requirement precludes permanent missile cover. Canvas shelter only to be allowed.<br>Weather:<br>  Operator could check from van technician. Would be required outside only for short periods in case of malfunction.<br>Wind:<br>  Technician would be exposed to wind for short periods only. Would not need too much freedom of movement, and hence could be lashed if necessary in high wind. | Geography:<br>  Mobility requirement may seriously impair the space needed for the equipment needed to automate.<br>Weather:<br>  Technican would be outside only for time required to remove and replace diagnosed unit.<br>Wind:<br>  Minimum exposure time for technician; very little mobility needed for technician. |
| Type of output:<br>  Information display<br>  Portable voltmeter<br>  Pointer indicator<br>Cost:<br>  Low fabrication<br>  Low instrumentation<br>Time:<br>  Short to achieve design<br>  Long to operate<br>Skill level:<br>  Fairly high<br>  (must test and repair)<br>Manning:<br>  Possible for one person to test and maintain | Type of output:<br>  Information display<br>  Panel voltmeter<br>  Digital<br>Cost:<br>  Medium fabrication<br>  Medium fabrication<br>Time:<br>  Longer to achieve design<br>  Shorter to operate<br>Skill level:<br>  Operator—fairly low<br>  Maintenance—fairly high (must diagnose)<br>Manning:<br>  Two people needed<br>  Communication needed | Type of output:<br>  Information display<br>  Panel light<br>Cost:<br>  High instrumentation<br>  High fabrication<br>Time:<br>  Long to achieve design<br>  Short to operate<br>Skill level:<br>  Operator—medium<br>  Maintenance—low (remove and replace)<br>Manning:<br>  Two people needed<br>  Communication needed | Type of output:<br>  Information display<br>  Panel light and signal<br>  Malfunctioning unit displayed<br>Cost:<br>  Very high instrumentation<br>  Very high fabrication<br>Time:<br>  Very long to achieve design<br>  Very short to operate<br>Skill level:<br>  Operator—low<br>  Maintenance—low (remove and replace); machine diagnoses fault)<br>Manning:<br>  Two people needed<br>  Both can assume other duties at same time—needed only when equipment malfunctions |
| Accuracy:<br>  Possibility of misreading meter, especially if meter must be read to close limits | Accuracy:<br>  Likelihood of misreading small numbers | Accuracy:<br>  Dependent on equipment reliability | Accuracy:<br>  Dependent on equipment reliability; error probability on part of operator practically zero |

| Start Time (from Initial Function) | Function | Human-Machine Combination | Start Function Cue | Start Action Required | Remarks |
|---|---|---|---|---|---|
| $T - 15$ min (duration 2 min) | Activate gyros and other components | Human-initiated | Completion of secure missile | Manually operated switch | Action well within human capabilities. No unreasonable time pressure. Environmental program likely to demand that function be performed indoors (therefore remotely). |
| $T = 15$ min (duration 2 min) | Activate test instruments | Human-initiated | Completion of secure missile | Manually operated switch | Action well within human capabilities. No unreasonable time pressure. Environmental program likely to demand that function be performed indoors (therefore remotely). |
| $T = 17$ min (duration 2 min) | Check yaw accuracy | Machine check<br>Human monitor | Circuit warmup complete (either machine-indicated or automatic sequence) | Either manually operated switch or automatic program | 2-min time duration for operation. Precludes use of human from time stress standpoint. Automatic operation should include no-go indication to human. |
| $T = 19$ min (duration 2 min) | Check roll accuracy | Machine check<br>Human monitor | Yaw accuracy checked (automatically sequenced) | Automatic program | 2-min time duration for operation. Precludes use of human from time stress standpoint. Automatic operation should include no-go indication to human. |

| Functional Requirement | Constraints | Design Approaches or Alternatives | Recommendations and Rationale | Comments and/or Associated Tasks |
|---|---|---|---|---|
| | | | | |

## INFORMATION FLOW CHARTING

This technique is often employed to illustrate the flow of information in terms of operations and decisions required to accomplish the functions identified in a system function block diagram. Like block diagraming, information flow charting may be used at various levels of detail. The initial information flow charts should be concerned with gross functions without regard to whether functions are performed by humans or by machines. Information flow charts prepared subsequent to tentative human-machine function allocation will reflect this allocation in the decisions, operations, and branching that are represented.

In that it records the sequence of operations and decisions that must be performed in order to satisfy a definite system function, the information flow chart is similar to the flow chart used by computer programmers. Both charts are based on binary choice decisions and intervening operations. That most decisions can be reduced to a binary situation is evidenced by the vast array of problems that can be computerized using simple binary logic. There are two important reasons for using binary decision logic in all information flow charting:

1. It expedites communication through the use of simple yet universally applicable conventions.
2. It provides for easy translation of information flow charts into logic flow charts for computerized sections of the system.

Like block diagraming, information flow charting can be used at various levels of specificity. A decision at a general level may split into several decisions at a more detailed level. For example:

*General Level*

Any targets need identification processing?

*Specific Level*

Any newly entered targets need ID processing?
Any target tracks need confirmation of tentative ID?

Any confirmed IDs need rechecking?

Each of these more detailed decisions may have associated with it one or more detailed operations. Similarly, an *operation* at a general level may break down into more detailed decisions and operations. In the accompanying example, human functions are represented by a single symbol, and machine functions are represented by two concentric symbols:

*General Level*

Call up track.

*Specific Level*

Enter track digits.
Press TN call-up button.
Load track data in buffers.
Display track data.
TN readout correct?
Proceed with operation.

It is not necessary that the flow chart be prepared to a uniform level of specificity. For many situations it may be entirely appropriate to treat certain parts of a process in only general terms and focus in on other, more critical aspects by going into greater detail. The analyst must keep his or her purpose in mind and peg the level of detail accordingly. Usually those parts of a flow chart presented only in general terms initially are broken down into greater detail as the development progresses. The flow chart on page 929 illustrates gross-level detection and tracking functions. Note that at this level the chart is applicable to virtually any detection and tracking system; the decisions and operations are common to all such systems. Even here, however, the power of the flow chart is apparent because it makes one begin to think of implementation alternatives, such as:

1. By what means can any given signal set be compared with known targets in the system?
2. How can probable targets be marked so that their appearance can be recognized readily?

The information flow chart on page 930 shows tracking function at a finer level of detail. Here, each machine decision and operation is represented by two concentric symbols to differentiate them from human functions. Note that the format for these charts utilizes a narrow column at the left-hand side of the page for the chart proper, consisting only of symbols and connecting lines, and a wide column for text statements keyed line by line with respective flow-chart symbols. Note also that flow paths are always complete—every path either recirculates or eventually terminates in a valid exit—with no ends left dangling. This fact is extremely important and is what makes the information flow chart such a powerful tool. The technique imposes a discipline upon analysts, requiring them to consider alternatives that otherwise might be overlooked. The result is thoroughness and logical closure which could never be attained by conventional block diagraming techniques or narrative descriptions.

## Rules for Data-Flow Diagraming Analysis[8]

1. Show components in the same relative position as in equipment to the extent that this is practical.
2. Include only information that is needed to perform specified operational control, service, and/or maintenance.
3. Present only enough information to identify components; they are neither pictured nor schematized.
4. Emphasize chiefly the nature of the signal flow between components, particularly of points where power or energy checks are made.
5. Be concerned only with the information- and energy-flow characteristics of the interface actions, not with the elec-

[8]Based on original material from *Data Flow: The General Problem and a Cognitive Model*, MRL-TDR-62-42, Behavioral Sciences Laboratory, 6570th Aerospace Medical Research Laboratories, Air Force Systems Command, Wright-Patterson Air Force Base, Ohio, 1962.

tromechanical characteristics of components.

6. Strive for similarity of appearance (e.g., graphics), whether data flow among major components, assemblies, or subassemblies is being depicted.

7. Base the analysis on a particular configuration of control settings for the prime equipment and/or test equipment or on induced special conditions such as broken feedback loops or servos set to a particular position (these control settings or special conditions are indicated on the diagram). Note that all loops should close on each type or level of diagram.

8. Show electrical characteristics of the signal to be monitored or checked at each operating or maintenance test point; complex time and shape characteristics of the signal are keyed to tables at the edge of the diagram or to separate, referenced pages.

9. Omit information about the nature of the signal flow within units which are replaceable or can be thrown away, at the level of maintenance for which the diagram was prepared.

10. Every functional system, subsystem, piece of equipment, and component should be interrelating numbering systems so that each end item is traceable to its original functional requirement.

**Example: Gross-Level Flow Chart for
Detection and Tracking**

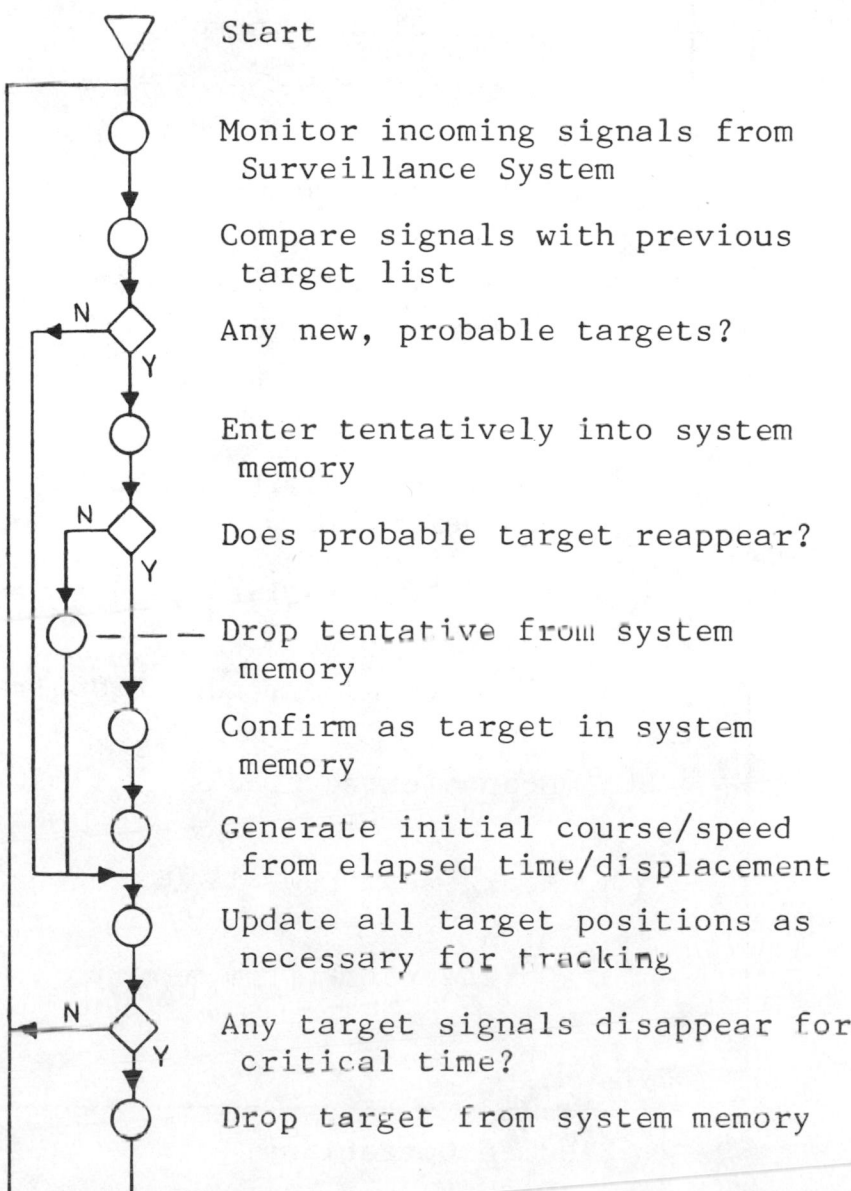

Start

Monitor incoming signals from Surveillance System

Compare signals with previous target list

Any new, probable targets?

Enter tentatively into system memory

Does probable target reappear?

Drop tentative from system memory

Confirm as target in system memory

Generate initial course/speed from elapsed time/displacement

Update all target positions as necessary for tracking

Any target signals disappear for critical time?

Drop target from system memory

Note that no human-machine function allocation has been assumed at this level.

**PORTION OF A FLOW CHART OF A HYPO-
THETICAL TRACKING FUNCTION**

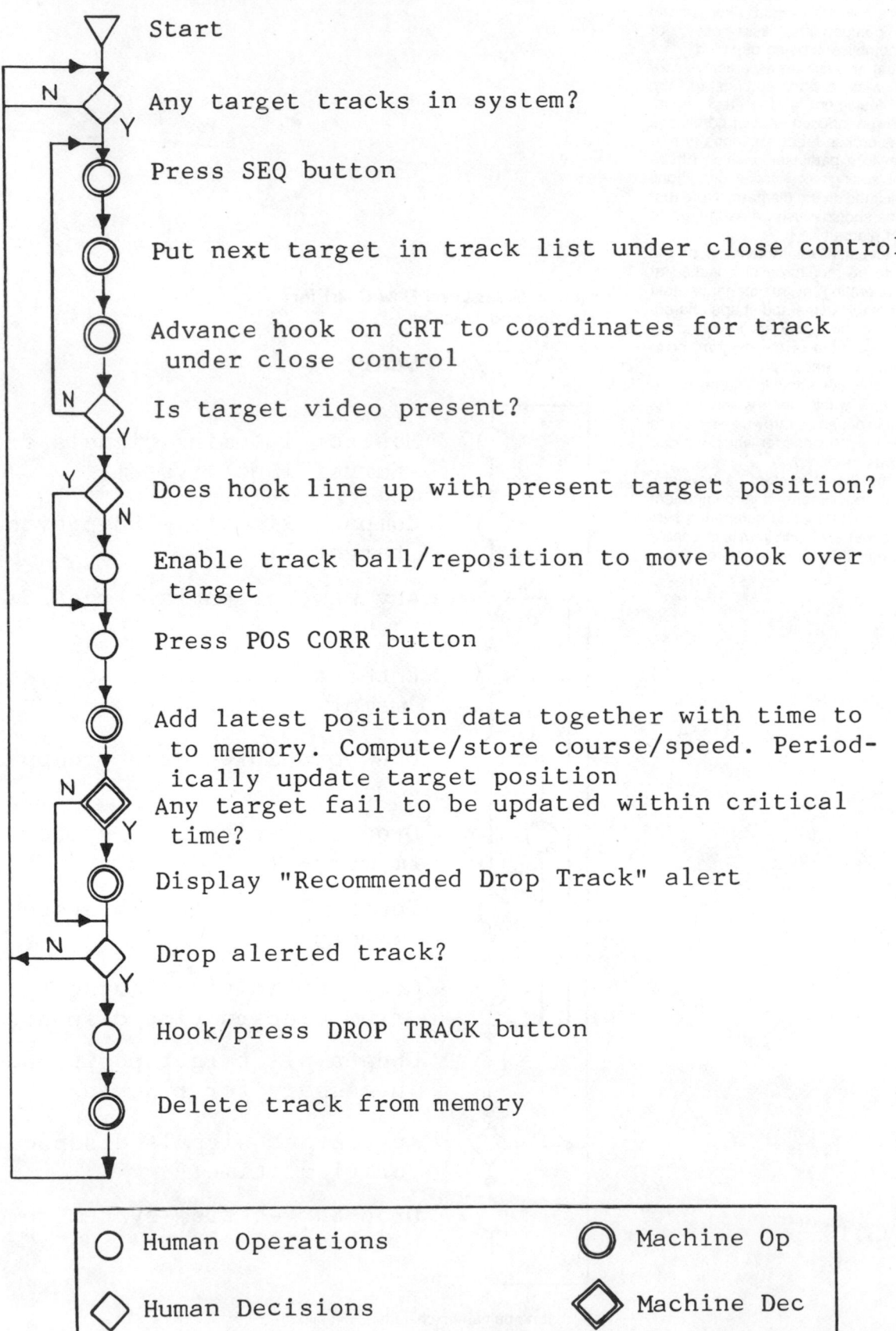

Start

Any target tracks in system?

Press SEQ button

Put next target in track list under close control

Advance hook on CRT to coordinates for track
under close control

Is target video present?

Does hook line up with present target position?

Enable track ball/reposition to move hook over
target

Press POS CORR button

Add latest position data together with time to
to memory. Compute/store course/speed. Periodically update target position

Any target fail to be updated within critical
time?

Display "Recommended Drop Track" alert

Drop alerted track?

Hook/press DROP TRACK button

Delete track from memory

○ Human Operations          ◎ Machine Op

◇ Human Decisions          ◈ Machine Dec

EXAMPLE FROM MIL-H-46855

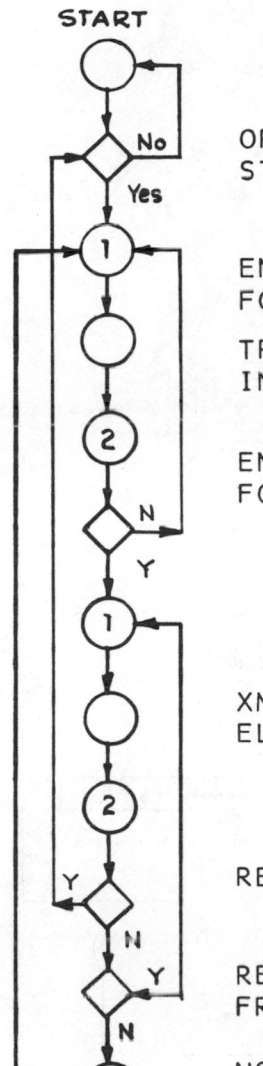

ORIGINATE MSG TO
STATION 2 NOW?

ENABLE STA. 1 TERMINAL
FOR 1 ⟶ 2; DISABLE FOR 2 ⟶ 1

TRANSMIT 1 ⟶ 2 CALL SIGNAL
IN ELECT. FORM

ENABLE STA. 1 TERMINAL
FOR 2 ⟶ 1; DISABLE FOR 1 ⟶ 2

XMIT 1 ⟶ 2 MSG IN
ELECTRICAL FORM

RECEIVE OK FROM STA. 2?

RECEIVE REPEAT REQUEST
FROM STA. 2?

NO ACKNOWLEDGEMENT
FROM STA. 2

NOTES

1. SYMBOLS:   ◇ DECISION        ◯ OPERATION

2.        ALL LOOPS MUST CLOSE

3.        AT THE FUNCTIONAL LEVEL OF USAGE, DECISIONS
          AND OPERATIONS ARE PHRASED IN TERMS OF <u>SYSTEM</u>
          FUNCTIONS RATHER THAN OPERATOR TASKS AND CIRCUIT
          BEHAVIOR

SAMPLE FLOW CHART.  STATION 1 (MAN-MACHINE)
FUNCTIONS IN TWO-STATION INTERCOM

Sample flow chart. Station 1 (human-machine) functions in two-station intercom.

## Information-Flow Analysis: Other Applications

The technique of analyzing information requirements by graphic means is applicable to situations other than the design of complex operational systems. As illustrated in the diagram[9] below, information flow is critical to the organization of a production system. As the diagram shows, the following features are delineated:

1. Information input sources: Groups such as engineering design, blueprint checking, and reproduction, each supplying information to production
2. Information channels: Various checkpoints through which information passes and where it is probably modified
3. Information receivers: Various production departments which make use of the information

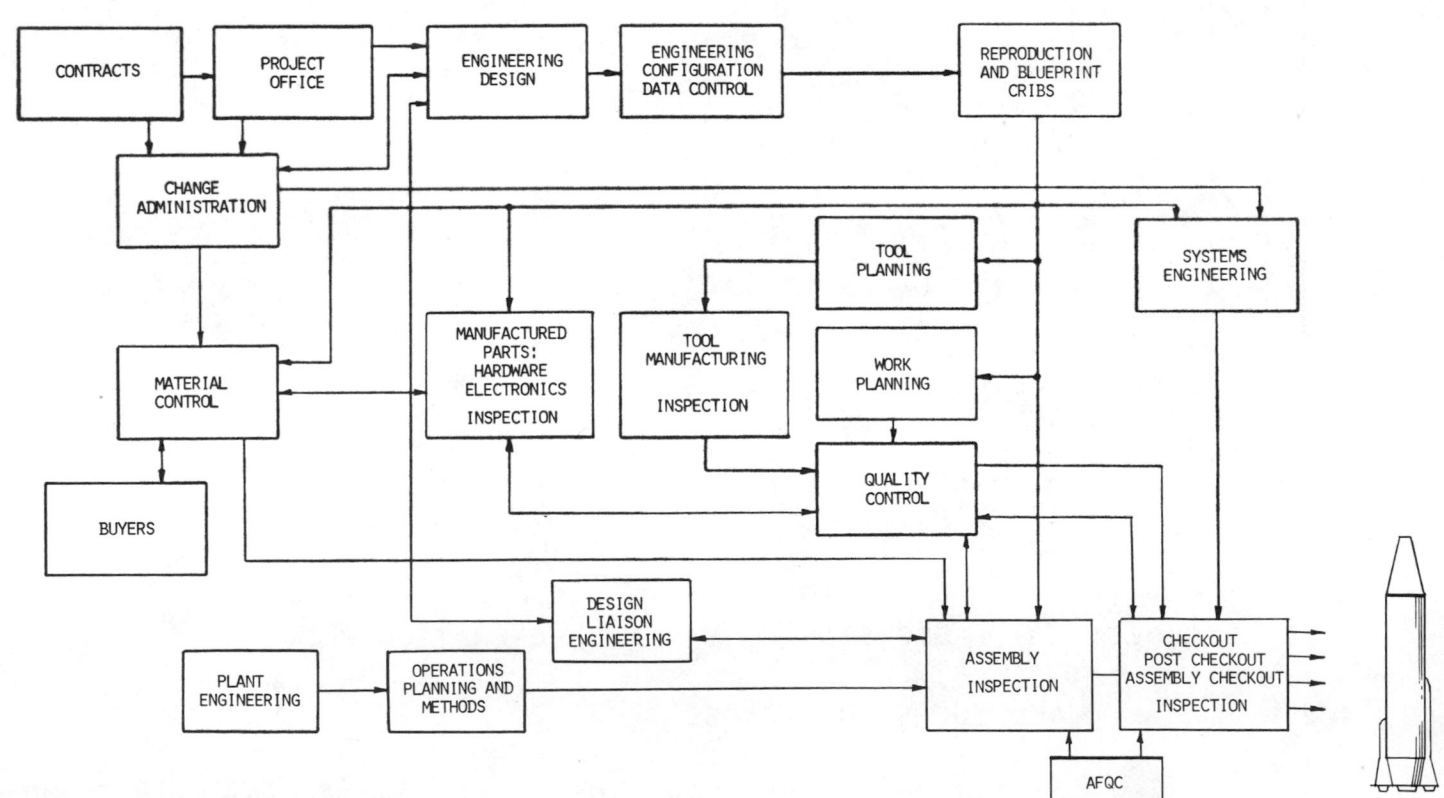

[9]From D. Meister and G. F. Rabideau, *Human Factors Evaluation in System Development*, John Wiley & Sons, Inc., New York, 1965, pp. 284–285.

**General Human-Machine Models**

**INTERNAL INFORMATION-FLOW MODEL OF
THE HUMAN OPERATOR**

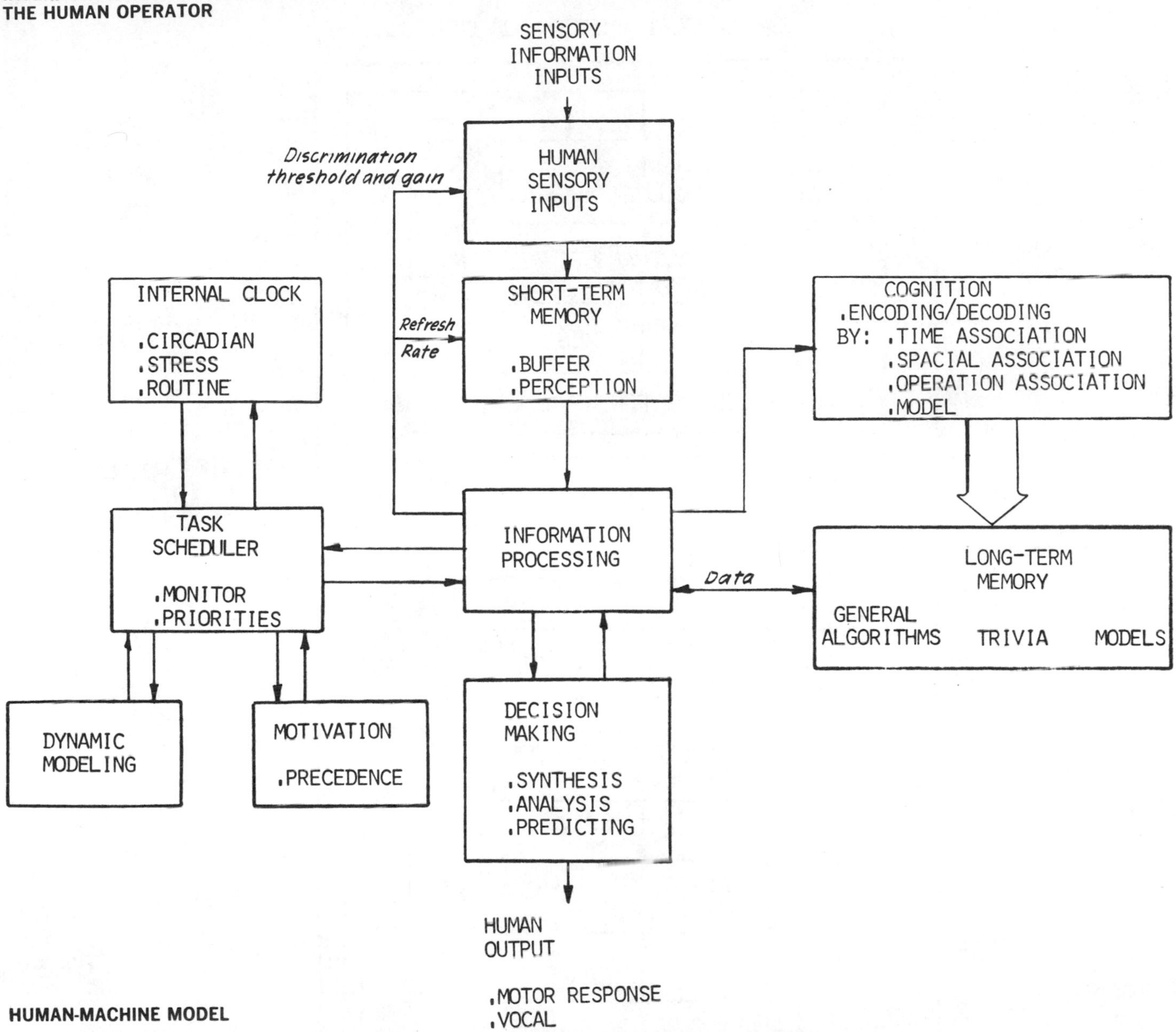

SENSORY
INFORMATION
INPUTS

HUMAN
SENSORY
INPUTS

*Discrimination
threshold and gain*

SHORT-TERM
MEMORY

.BUFFER
.PERCEPTION

*Refresh
Rate*

INTERNAL CLOCK

.CIRCADIAN
.STRESS
.ROUTINE

COGNITION
.ENCODING/DECODING
BY: .TIME ASSOCIATION
.SPACIAL ASSOCIATION
.OPERATION ASSOCIATION
.MODEL

TASK
SCHEDULER

.MONITOR
.PRIORITIES

INFORMATION
PROCESSING

*Data*

LONG-TERM
MEMORY

GENERAL
ALGORITHMS    TRIVIA    MODELS

DYNAMIC
MODELING

MOTIVATION

.PRECEDENCE

DECISION
MAKING

.SYNTHESIS
.ANALYSIS
.PREDICTING

HUMAN
OUTPUT

.MOTOR RESPONSE
.VOCAL

**HUMAN-MACHINE MODEL**

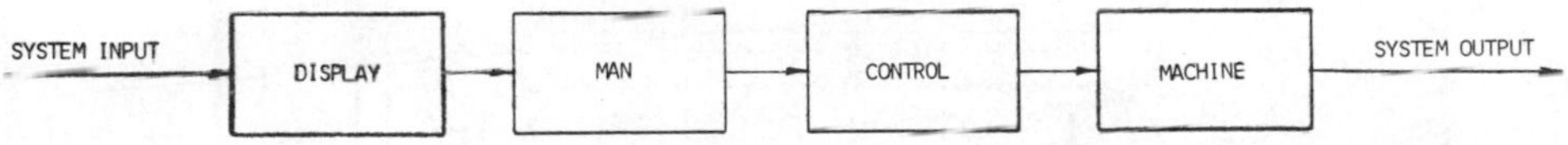

SYSTEM INPUT → DISPLAY → MAN → CONTROL → MACHINE → SYSTEM OUTPUT

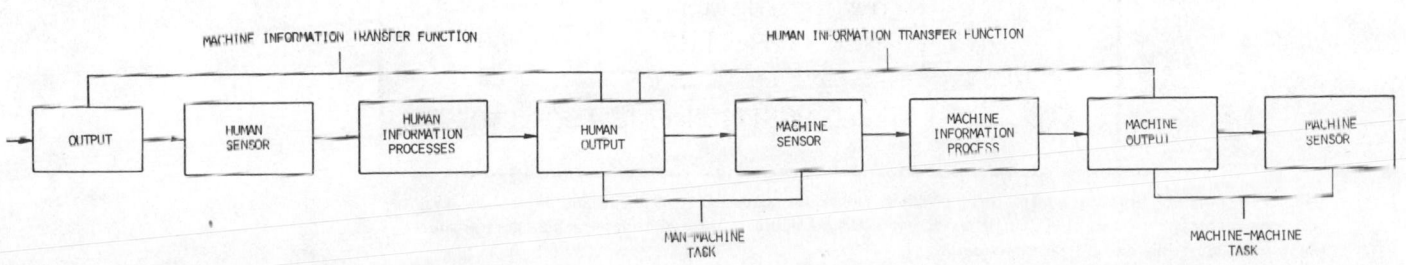

MACHINE INFORMATION TRANSFER FUNCTION          HUMAN INFORMATION TRANSFER FUNCTION

OUTPUT | HUMAN SENSOR | HUMAN INFORMATION PROCESSES | HUMAN OUTPUT | MACHINE SENSOR | MACHINE INFORMATION PROCESS | MACHINE OUTPUT | MACHINE SENSOR

MAN-MACHINE
TASK

MACHINE-MACHINE
TASK

**FLOW CHARTING FOR DRIVER MODEL
DEVELOPMENT**

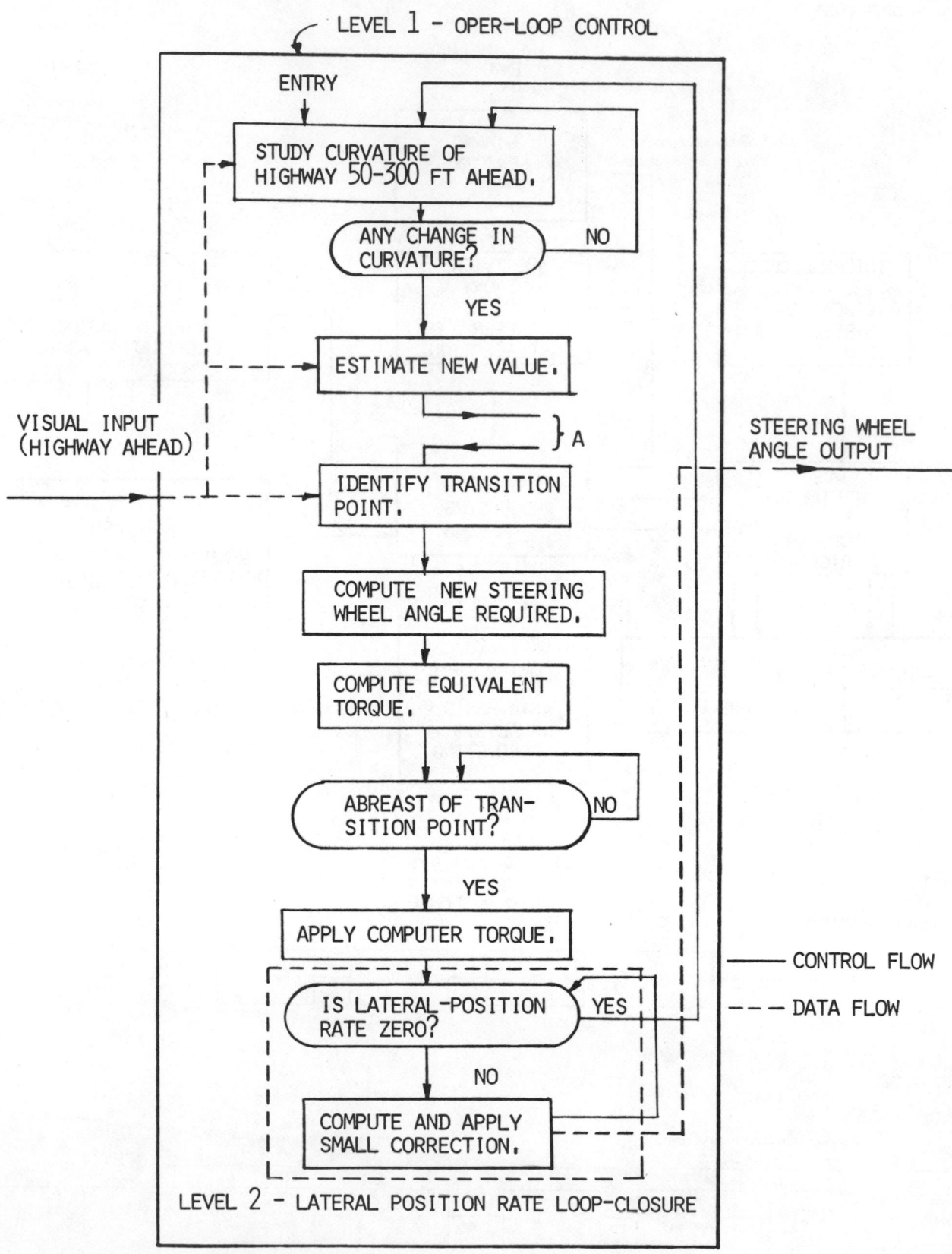

Level 1: Open-loop curvature control using preview. The driver scans the road ahead and detects points at which the highway curvature changes. He or she estimates the new curvature while approaching the transition point, translating this into a new wheel setting.

LEVEL 2

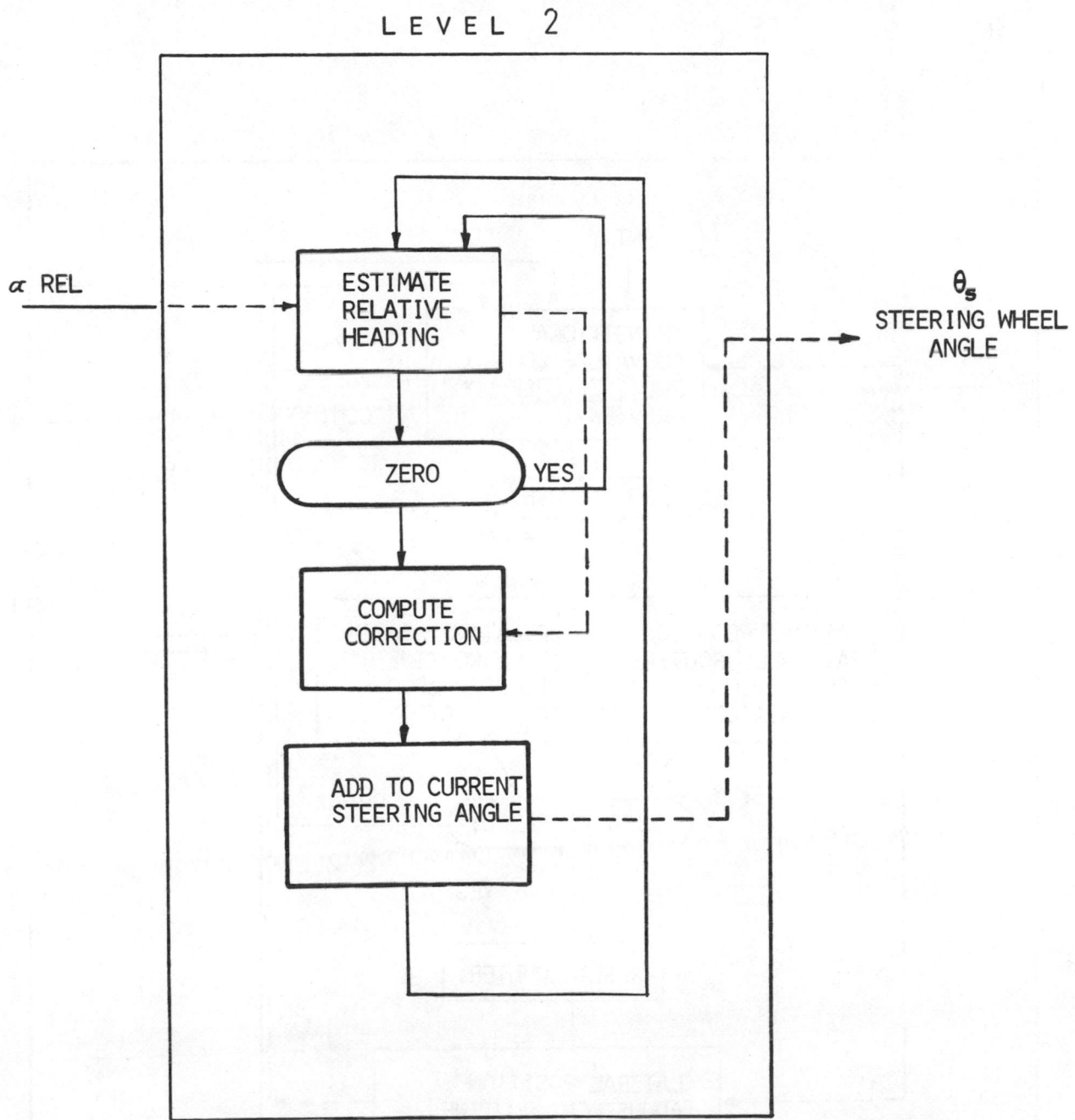

Level 2: Closed-loop control of relative heading and/or relative path angle and/or lateral position rate.

**OVERALL FLOW CHART FOR DRIVER INFOR-
MATION PROCESSING IN VEHICLE STEER-
ING**

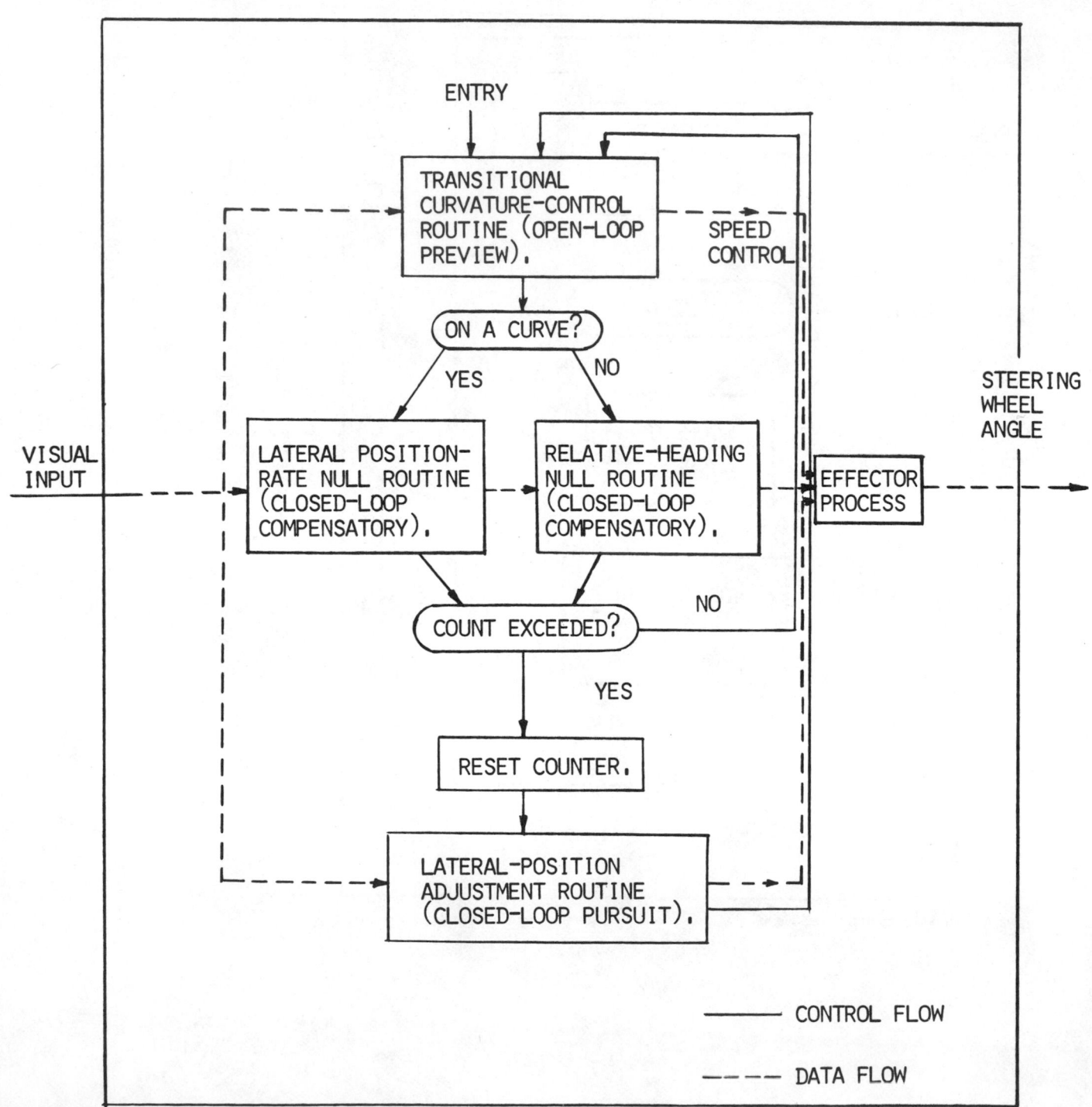

Level 3: Closed-loop control of lateral position. Since the precision of estimation of relative heading, relative path angle, or lateral position rate is limited to about 3 bits per observation and thus will have a definite threshold, the second-level control policy fails to precisely null out relative-heading errors. A third-level policy is required for direct estimation of lateral position and its closed-loop adjustment. This necessarily creates a second or higher-order control system.

**FLOW CHART FOR SPEED CONTROL AD-
JUSTMENT ROUTINE TO MEET MAXIMUM
LATERAL ACCELERATION CONSTRAINT**[10]

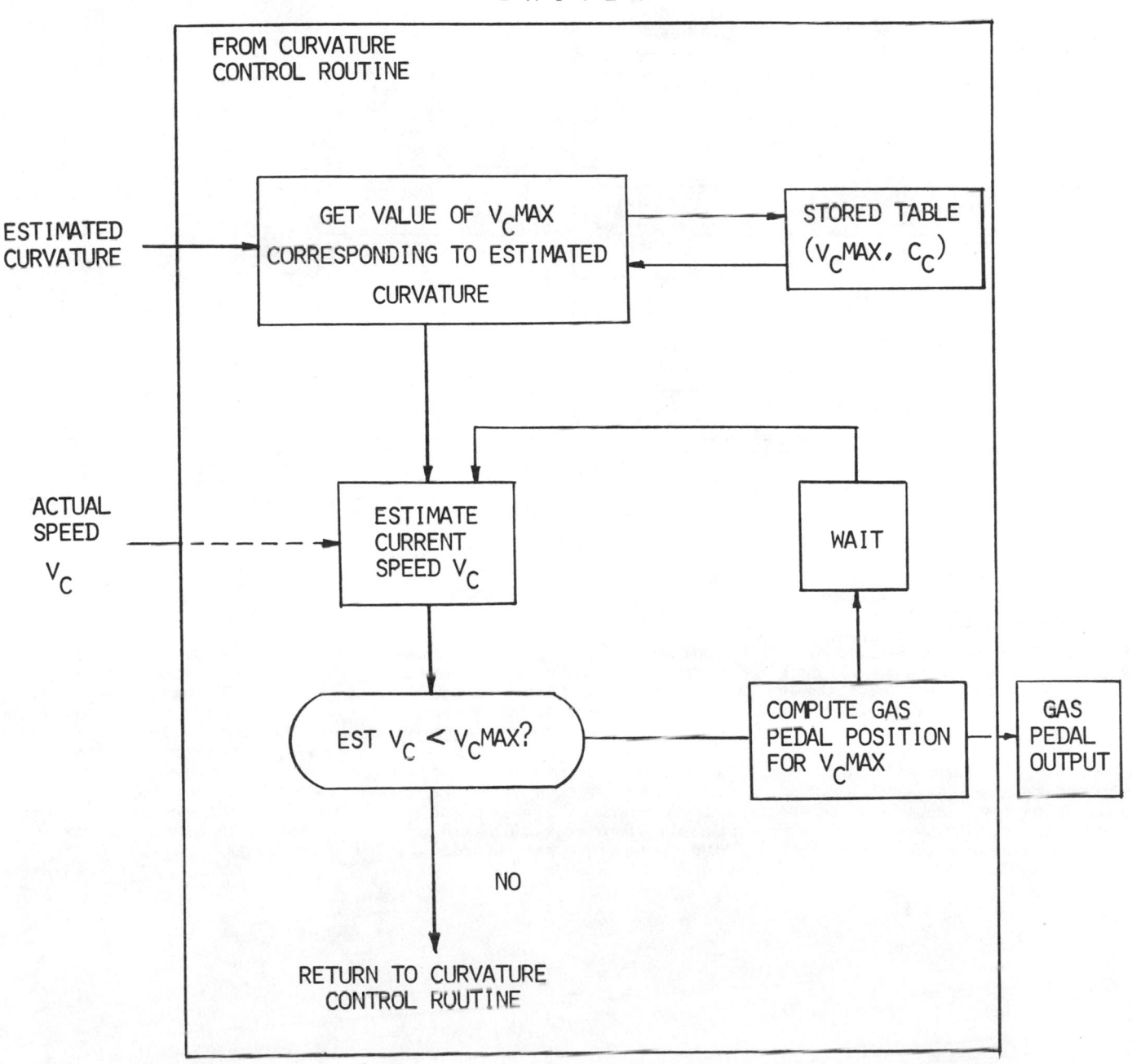

D R I V E R

FROM CURVATURE
CONTROL ROUTINE

ESTIMATED
CURVATURE

GET VALUE OF $V_C$MAX
CORRESPONDING TO ESTIMATED
CURVATURE

STORED TABLE
($V_C$MAX, $C_C$)

ACTUAL
SPEED
$V_C$

ESTIMATE
CURRENT
SPEED $V_C$

WAIT

EST $V_C$ < $V_C$MAX?

COMPUTE GAS
PEDAL POSITION
FOR $V_C$MAX

GAS
PEDAL
OUTPUT

NO

RETURN TO CURVATURE
CONTROL ROUTINE

[10]E. R. F. W. Crossman and H. Szostak, "Man-Machine
Models for Car Steering," *Proceedings of the Fourth Annual
NASA-University Conference on Manual Control,* NASA SP-
192, University of Michigan, Ann Arbor, Mar. 21-23, 1968.

## HUMAN-MACHINE SYSTEM MODELING

### Modeling the Car-Driver System[11]

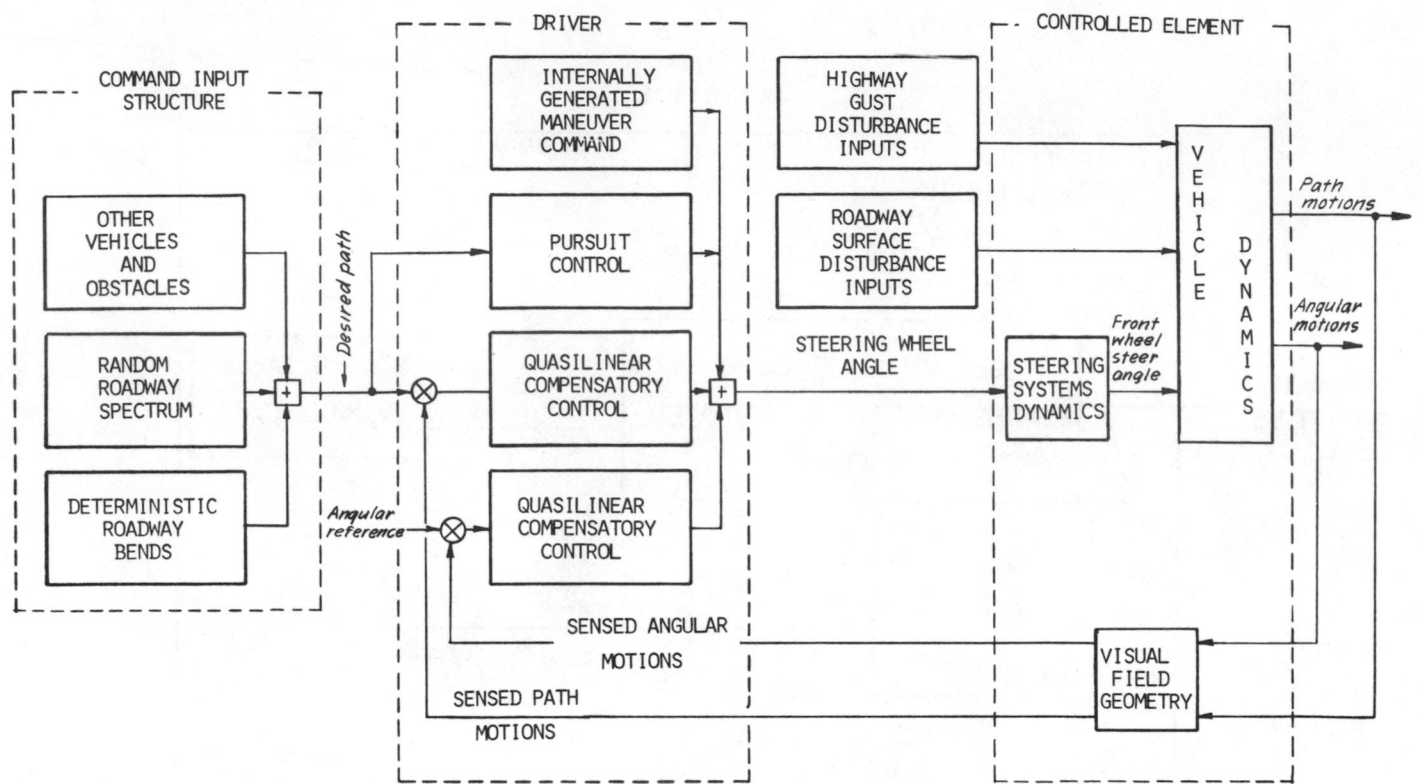

**Topology of the closed-loop control structure.**

Three general levels of control structure are illustrated in the accompanying model; these correspond to three levels of driver behavior and consequent system performance. The levels are as follows:

Precognitive: This involves executing a learned maneuver in an open-loop manner (command input structure). Command comes from within the driver after being triggered by some pattern or stimulus in the visual and/or proprioceptive field (e.g., turning into a driveway, overtaking, or passing).

Pursuit: This involves taking advantage of a knowledge of the system input to structure driver "feed forward" to improve performance (pursuit control block). Pursuit behavior is the combined open-loop–closed-loop characteristic. The open-loop feed-forward element provides the driver output that causes the vehicle output to (nearly) duplicate command input, while the closed-loop portion of the system acts as a vernier control to reduce any residual errors.

Compensatory: This implies an operation on a perceived error between the actual vehicle motion and the desired motion or input quantity (quasi-linear compensatory control). The compensatory level differs from the pursuit level in that only errors are the basis for control, and command inputs are not used to structure feed forward to the driver's output.

[11]D. H. Weir, and D. T. McRuer, "Models for Steering Control of Motor Vehicles," *Proceedings of the Fourth Annual NASA-University Conference on Manual Control,* NASA SP-192, University of Michigan, Ann Arbor, Mar. 21–23, 1968.

## Modeling Operator Visual and Motion Cues[12]

General hypotheses regarding utilization of motion cues include the following:

Motion cues cannot be used and will be ignored when tracking a random-appearing command input with a compensatory display.

With any input display combination except that shown in the accompanying model, motion cues may be utilized.

Addition of motion cues will greatly enhance performance when considerable lead equalization and/or effective-time-delay reduction in the visual feedback would be beneficial.

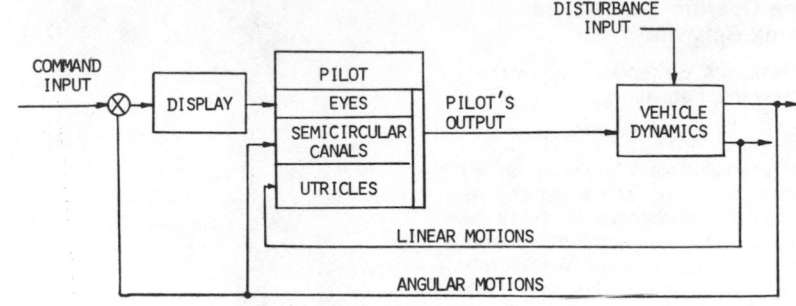

BASIC STRUCTURE OF MULTIMODALITY PILOT MODEL.

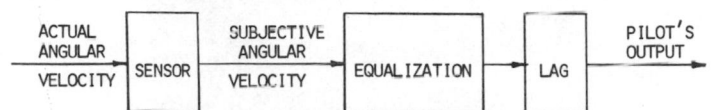

ELEMENTS OF THE SEMICIRCULAR CANAL PATH.

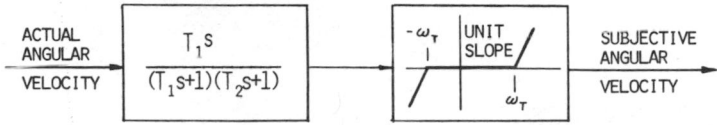

SEMICIRCULAR CANAL DYNAMICS.

With motion cues, the gain in the visual path may be considerably increased over what it would be for fixed-base operation, the lead may be reduced, and the effective time delay may be increased.

The motion feedbacks will dominate at high frequencies, the visual feedback will dominate at low frequencies, and they will be of comparable magnitude in the region of 2 to 4 rd/s.

The utricular path will not be used unless the linear acceleration feedback is more favorable than the semicircular canal feedback.

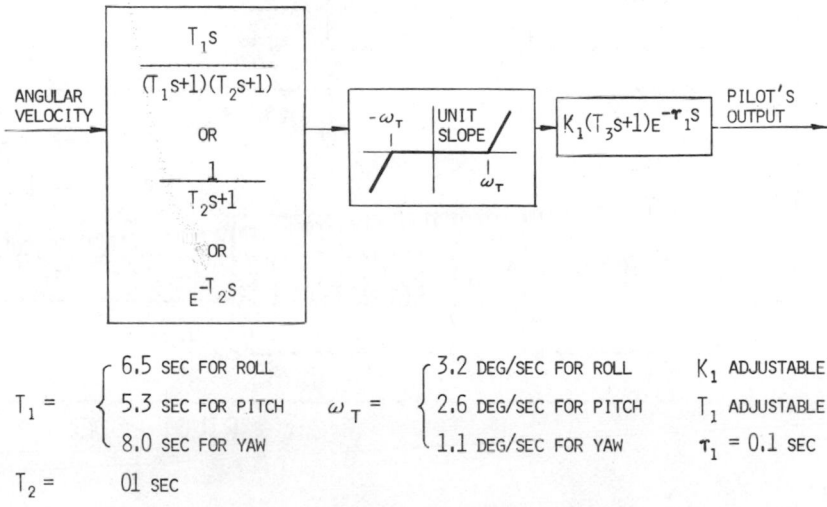

$$T_1 = \begin{cases} 6.5 \text{ SEC FOR ROLL} \\ 5.3 \text{ SEC FOR PITCH} \\ 8.0 \text{ SEC FOR YAW} \end{cases} \qquad \omega_T = \begin{cases} 3.2 \text{ DEG/SEC FOR ROLL} \\ 2.6 \text{ DEG/SEC FOR PITCH} \\ 1.1 \text{ DEG/SEC FOR YAW} \end{cases}$$

$K_1$ ADJUSTABLE
$T_3$ ADJUSTABLE
$\tau_1 = 0.1$ SEC

$$T_2 = 01 \text{ SEC}$$

MODEL FOR SEMICIRCULAR CANAL PATH.

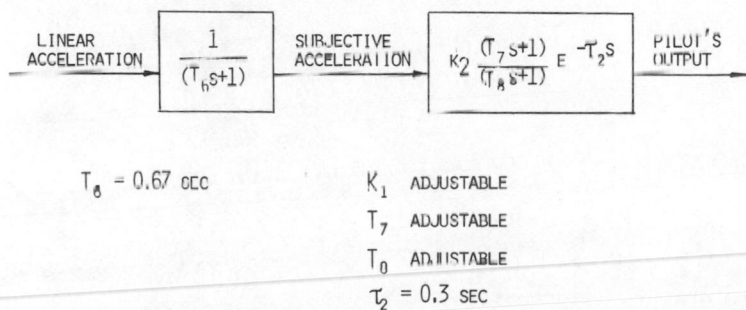

$T_6 = 0.67$ SEC

$K_2$ ADJUSTABLE
$T_7$ ADJUSTABLE
$T_8$ ADJUSTABLE
$\tau_2 = 0.3$ SEC

MODEL FOR UTRICULAR PATH.

[12]R. L. Stapleford, "Multimodality Pilot Model for Visual and Motion Cues," *Proceedings of the Fourth Annual NASA-University Conference on Manual Control* NASA SP-192, University of Michigan, Ann Arbor, Mar. 21–23, 1968.

## Modeling Operator Instrument Monitoring Behavior[13]

Vehicle dynamics are represented by the linear, time-invariant equations

$$\mathbf{x}(t) = A\,\mathbf{x}(t) + B\,\mathbf{u}(t) + \mathbf{w}(t)$$

where $\mathbf{x}$ is vehicle state vector, $\mathbf{u}$ is pilot's control or input vector, and $\mathbf{w}$ (which represents external disturbances such as wind gusts) is a zero-mean, gaussian white-noise vector with covariance matrix $\mathbf{W} \cdot y_d(t)$ represents the displayed variable, i.e., linear combinations of vehicle states. Thus the equation

$$y_d(t) = C\,\mathbf{x}(t)$$

allows for the possibility that not all the vehicles may be explicitly displayed.

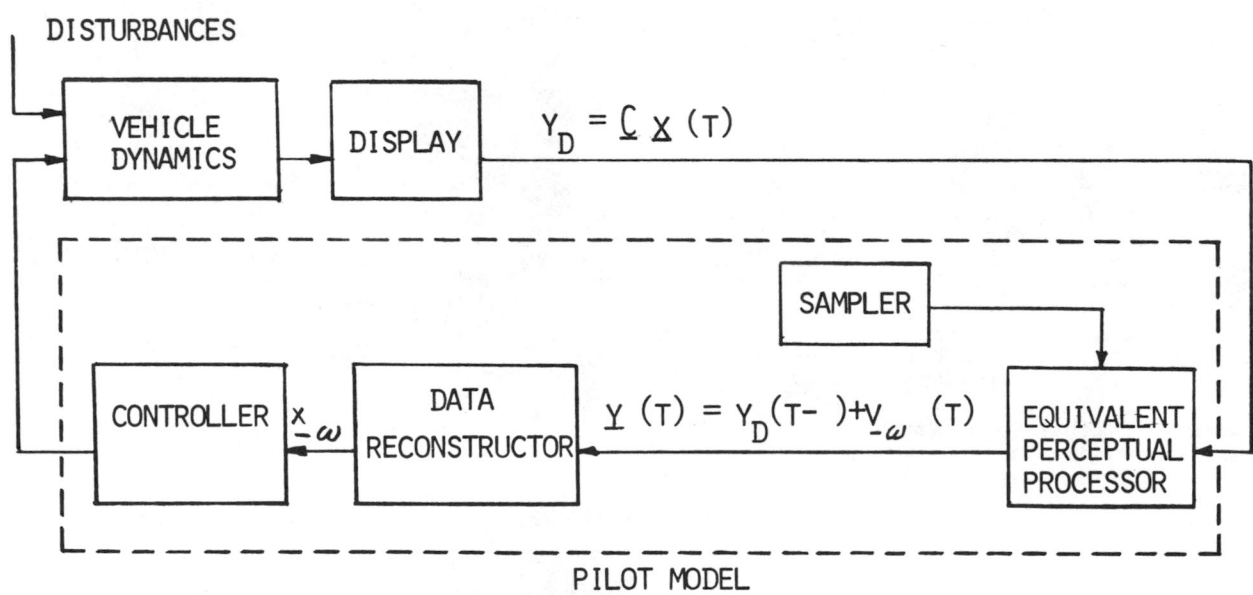

PILOT-VEHICLE-DISPLAY SYSTEM MODEL

[13]Sheldon Baron and David L. Kleinman, "The Human as an Optimal Controller and Information Processor," *Proceedings of the Fourth Annual NASA-University Conference on Manual Control*, NASA SP-192, University of Michigan, Ann Arbor, Mar. 21–23, 1968.

**Modeling Manual Instrument Landing Control and Display Systems[14]**

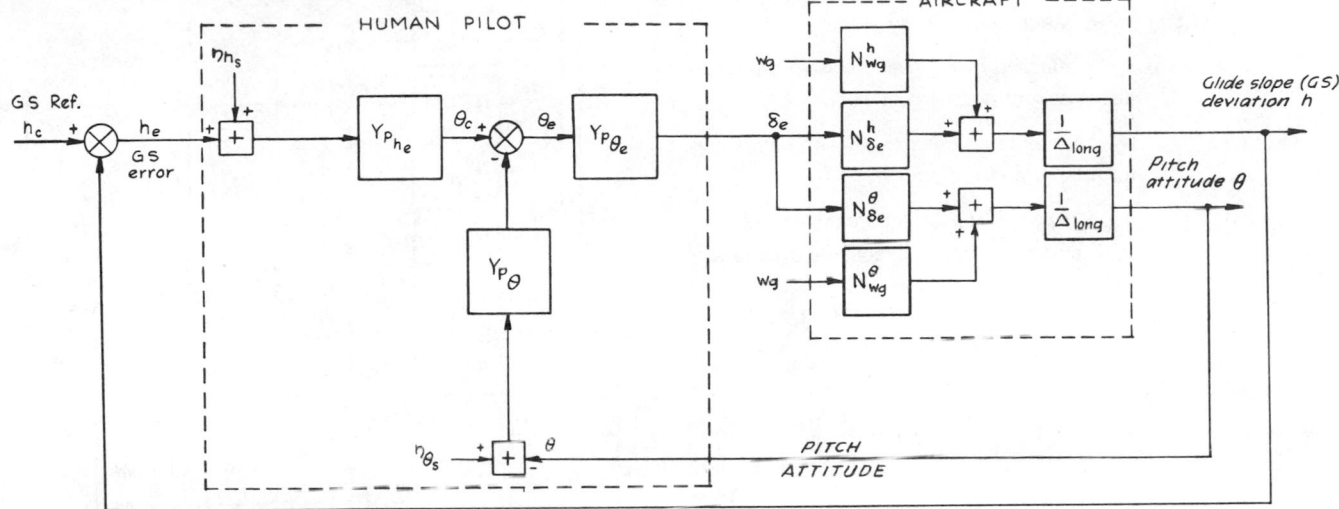

$Y_{p\theta_e} = e^{\tau e_0 s} = e^{-0.3s}$

$Y_{p\theta} = K_\theta \left(s + \dfrac{1}{T_{L\theta}}\right) e^{-\tau\theta a}$

$\qquad = -0.63(s + 0.9)e^{-0.4s}$

$Y_{ph_e} = Kh_e \left(s + \dfrac{1}{T_{Lh}}\right) e$

$\qquad = -.0013(s + 1.1)e^{-.4s}$

$\eta h_s$, $\eta\theta_s$ are sampling remnant

$\delta_e$ = elevator control

$w_g$ = normal gust velocity

$N_\alpha^\beta$ are air frame transfer function numerators for $\alpha$ input and $\beta$ outpout

$\Delta_{long}$ is characteristic airframe transfer function denominator for longitudinal dynamics

Height control system topology for ILS manual control with pitch attitude and glide slope deviation displays.

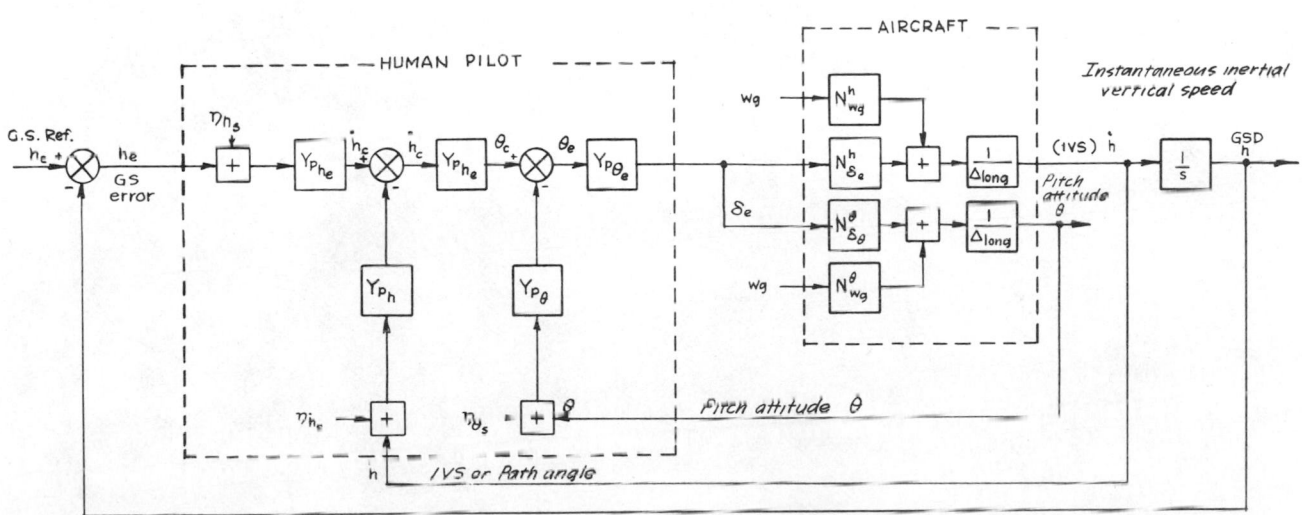

Height control system topology for ILS manual control with instantaneous inertial vertical speed or path angle display added.

[14]W. F. Clement et al., "Application of Systems Analysis Theory for Manual Control Displays to Aircraft Instrument Landing," *Proceedings of the Fourth Annual NASA-University Conference on Manual Control,* NASA SP-192, University of Michigan, Ann Arbor, 1968.

$Y_{p\phi_e} = {}^{re}0^s = e^{-0.3s}$    $\eta y_s$, $\eta\psi$, $\eta\phi_s$ are sampling remnant

$y_{p\phi} = K_\phi(s + 1/T_{L\phi})$    $\delta_a$ = aileron control

$\quad = 1.33(s + 1.5)$    $v_g$ = lateral gust velocity

$Y_{p\psi a} = K_{\psi a} \pm 1.0$    $p_g$ = effective rolling gust velocity from spanwise distribution of normal gust velocity

$Y_{p\psi} = K_\psi \,_\psi\, 1.9$    $N\beta\alpha$ are airframe transfer function numerators for $\alpha$ input and $\beta$ output

$Y_{py_e} = K_{y_e} = 0.0018$    $\Delta_{lat}$ = characteristic airframe transfer function denominator for lateral-directional dynamics

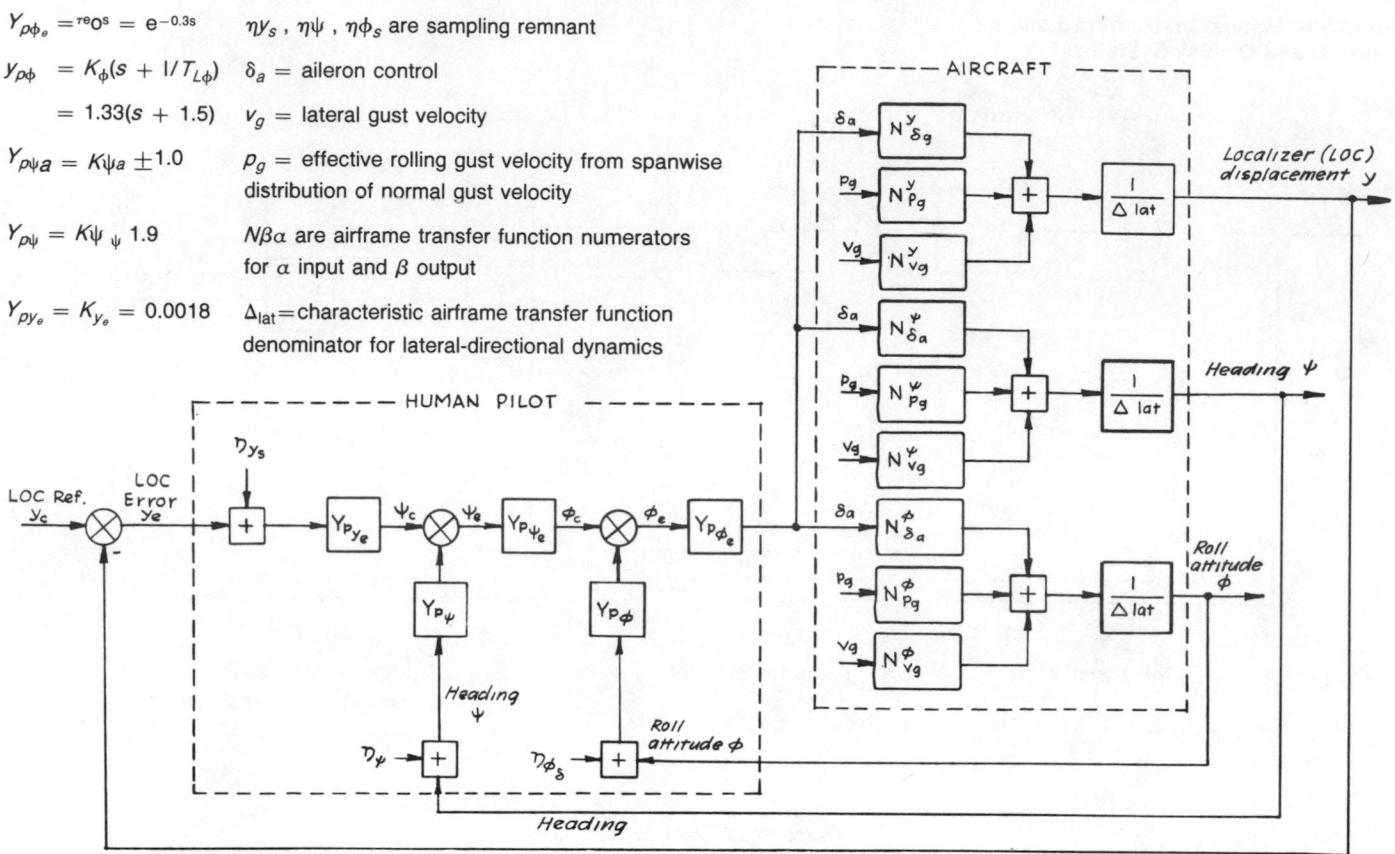

**Track (lineup) control system topology for ILS manual control with roll attitude, heading, and localizer deviation displays.**

**OPFRATIONAL SEQUENCE DIAGRAMING**

Example: Operational Sequence
Diagram (OSD)[15]

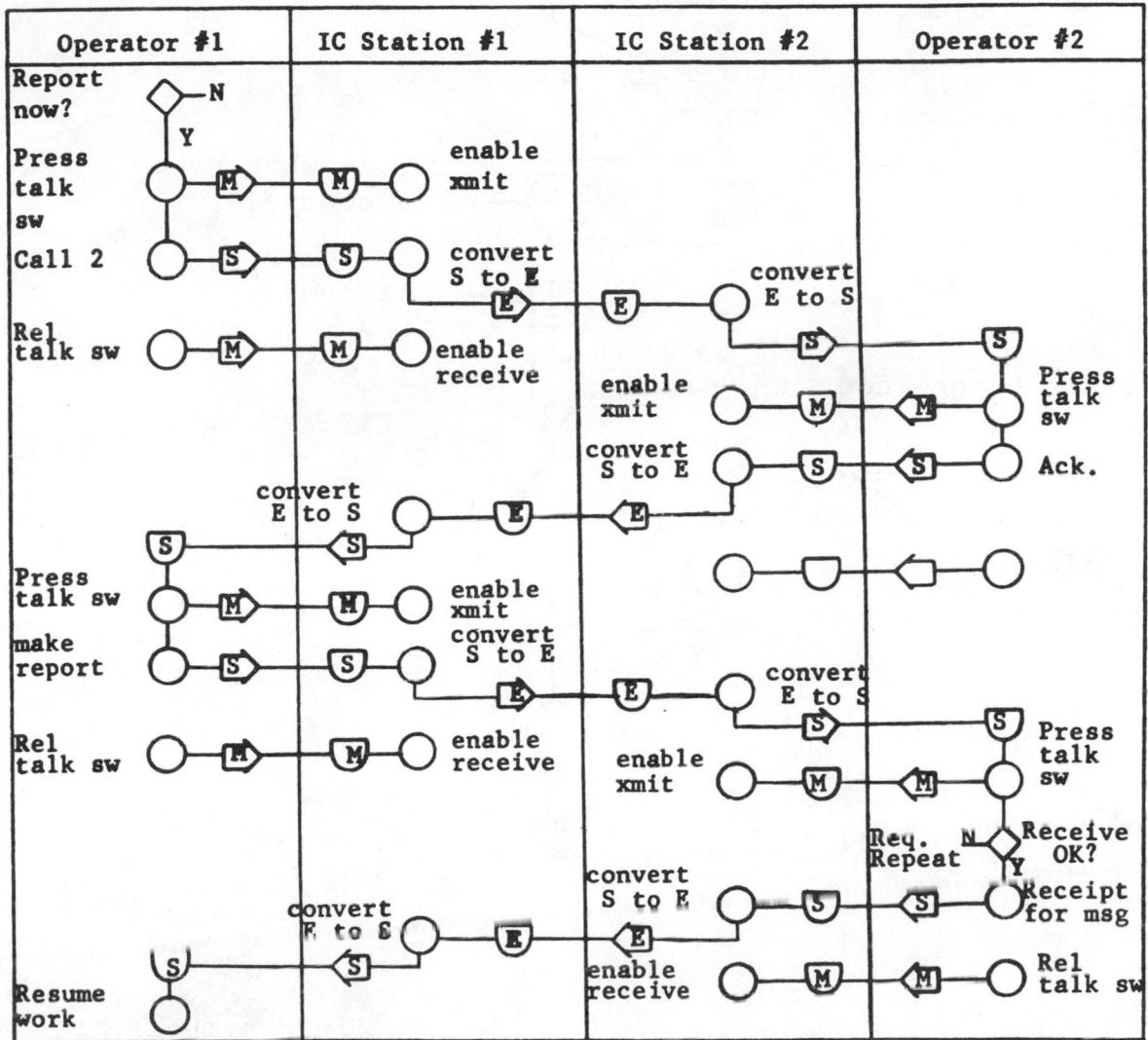

Notes on Operational Sequence Diagram

Symbols

◇ Decision

○ Operation

▷ Transmission

▽ Receipt

◗ Delay

□ Inspect, Monitor

▽ Store

Links

M  mechanical or manual

E  electrical

V  visual

S  sound

etc.

Stations or subsystems are shown by columns
Sequential time progresses down the page

Two-station intercom, with station no. 1 acting as originator.

[15]From MIL-H-46855.

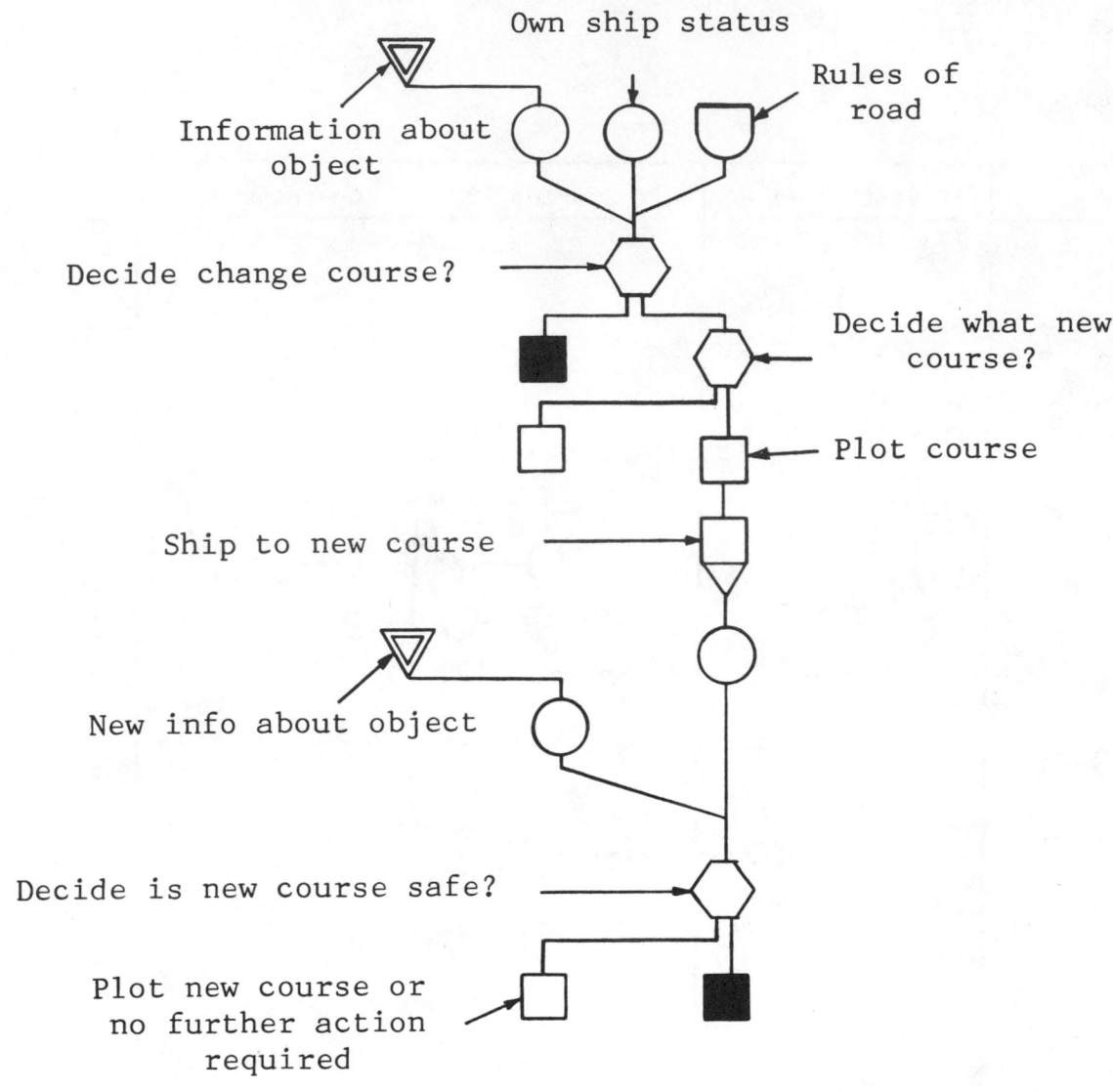

Own ship status

Rules of road

Information about object

Decide change course?

Decide what new course?

Plot course

Ship to new course

New info about object

Decide is new course safe?

Plot new course or no further action required

⬡  Processing or decision

▢  Action; e.g., control operation

▽  Transmitted information

○  Received information; e.g., Indicator display

⌴  Previously stored information; e.g., Knowledge

Single lined symbols represent manual operations

▣  Double lined symbols are automated operations

■  Solid symbols indicate inaction or no information

◪  Partial information or incorrect operations due to noise or error sources in the system

Y  "AND" logic (Lines join before entering element or separate after leaving)

⊻  "OR" logic (Separate lines into or out of an element)

In equations, capital letters stand for elements. A bar over a capital letter indicates the element receives only part of the input required to take place. The subscript 0 represents a null state; 1 represents an only active state. The letters a,b,c as subscripts indicate the multiple outputs of an element. A dot indicates an "and" relationship; a + represents "or."

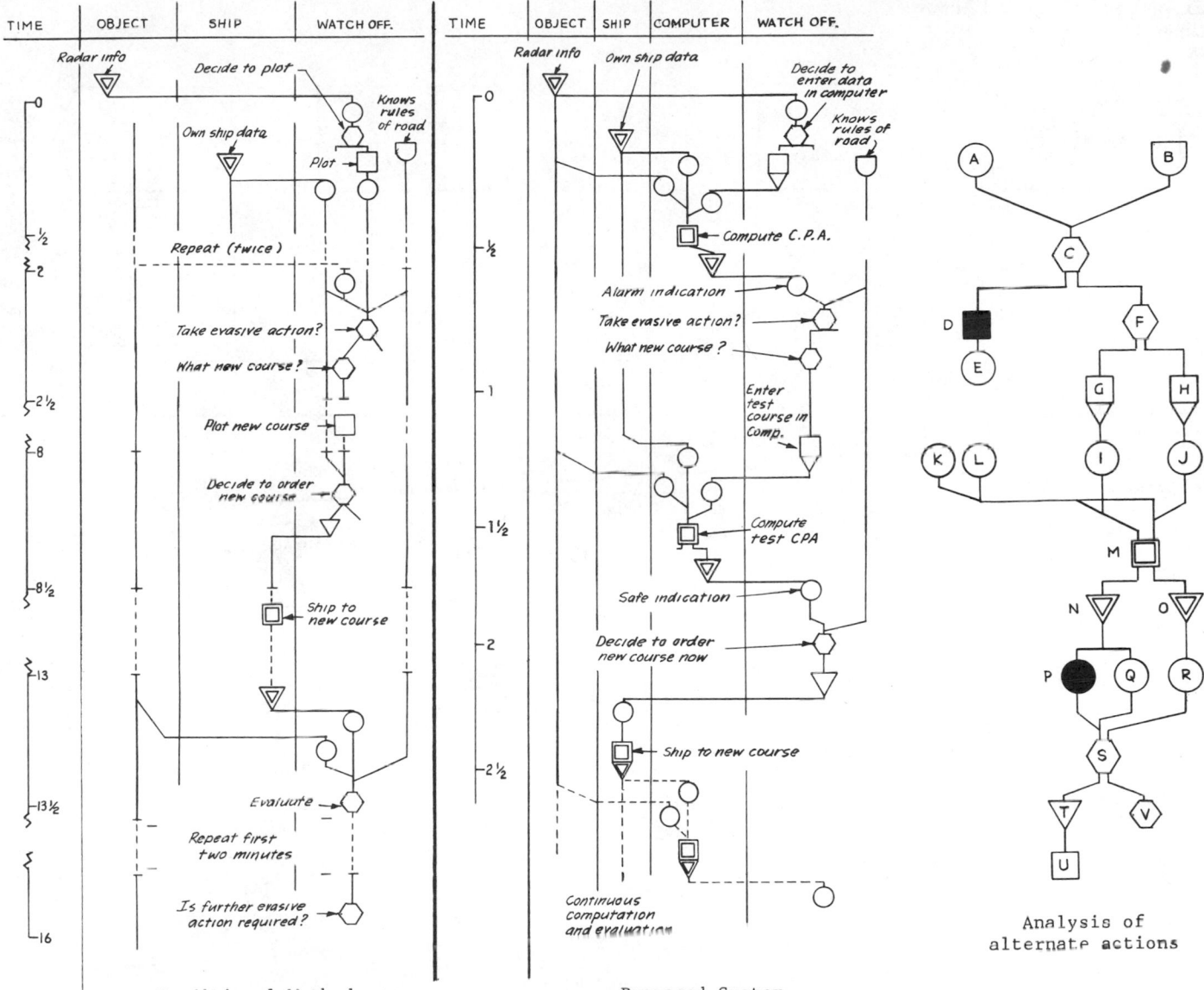

Traditional Method

Proposed System

Analysis of
alternate actions

$$\Lambda_1 \quad D_1 \to C_1 \to [\,(D_1 \cdot F_0) \to E_1\,] + (D_0\, F_1) \tag{1}$$

$$F_1 \to [\,(G_1 \to I_1) \cdot H_0\,] + [\,G_0 \cdot (H_1 \cdot J_1)\,] \cdot \overline{M} \tag{2}$$

$$K_1 \cdot L_1 \cdot (I_1 + J_1) \to M_1 \tag{3}$$

$$(H_1 \to J_1) \cdot K_1 \cdot L_1 \to M_d \to (N_1 \cdot O_0) \tag{4}$$

$$\to (P_0\, Q_1) \cdot S_d \to T_1 \to U_1$$

$$(G_1 \to I_1) \cdot K_1 \to L_1 \to M_b \to (N_0 \cdot O_1) \to R_1 \to S_b \to V_1 \tag{5}$$

Logic analysis of personnel and equipment operations. (From *Human Factors Design Standards for the Fleet Ballistic Missile Weapon System*, NAVWEPS OD 18412A, Washington, 1963.)

**Example: Decision and Action Diagram**

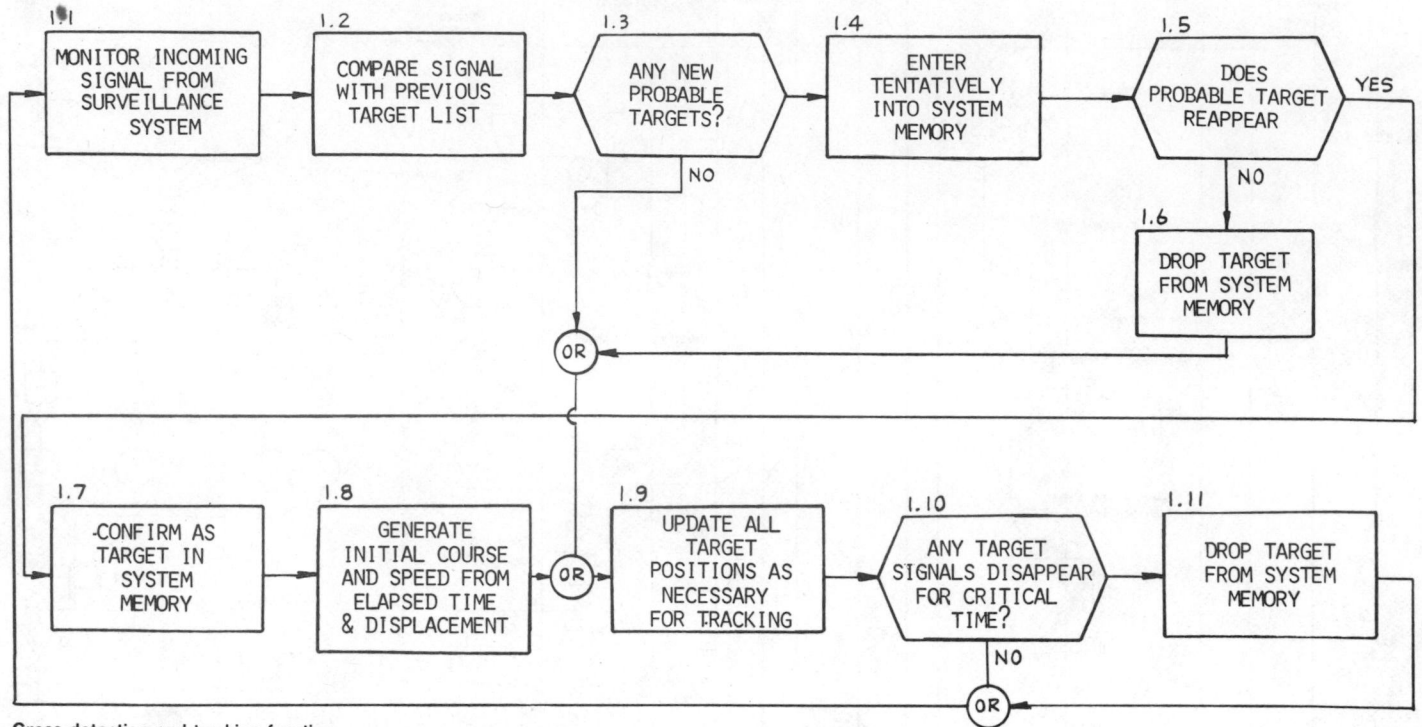

Gross detection and tracking function.

**Example: Operator Decision/Action, Operational Sequence Analysis**

Scenario: *Command Data Upload*
Operator: *Ground Controller/System Analyst*

| Decision and Action Flow | Cue, Decision, and Action | Notes | Time |
|---|---|---|---|
| 1. ▽ | Ground controller:<br>  Transmission in progress via GCS 1.<br>  Alarm bell sounds; fault message appears on CRT screen indicating fault in GCS 1 transmitter. Advises recovery system is attempting to resolve problem. Message from recovery system indicates GCS 2 is used by analysis center to monitor SV-2. Operator is given option of preempting GCS 2 or bypassing scheduled command upload. | | |
| 2. ◇ Y<br>see note<br>3. ◯ | Preempt? | "No" decision requires schedule readjustment. | |
| 4. ▽ Y | Authorizes recovery system to preempt; notifies ground system analyst of fault in GCS 1; continues normal operation.<br>Ground system analyst: | Operational sequence switches to analyst. | |
| 5. ◇<br>6. ◯<br>7. ▽ | Phone call from ground controller advises of fault in GCS 1 transmitter.<br>Calls up GCS 1 status display to determine degree of fault isolation accomplished.<br>Types in command.<br>Receives display showing last. . . . | | |

*Note:*
Cue ▽
Decision ◇
Action ◯

### Example: Semipictorial Operating-Flow Diagraming Technique

In this example, the analyst attempts to illustrate the flow of transactions within a banking system. The diagram below pictorializes a current banking system to provide a base line for comparison with a proposed system (second diagram). Using the diagram on page 948 as a point of departure, the analyst divides operating sequences into functional cycles and proceeds to depict the flow of information and activity functions commencing with the initial cue to begin daily operations within a branch bank facility. In this analysis it should be noted that proposed new equipment elements have already been established. However, the purpose of the analysis was to clarify where, when, and how these would function within the desired flow of information, decision, and action protocol of a current banking institution. The diagram on page 949 shows the symbology for operator and equipment functions.

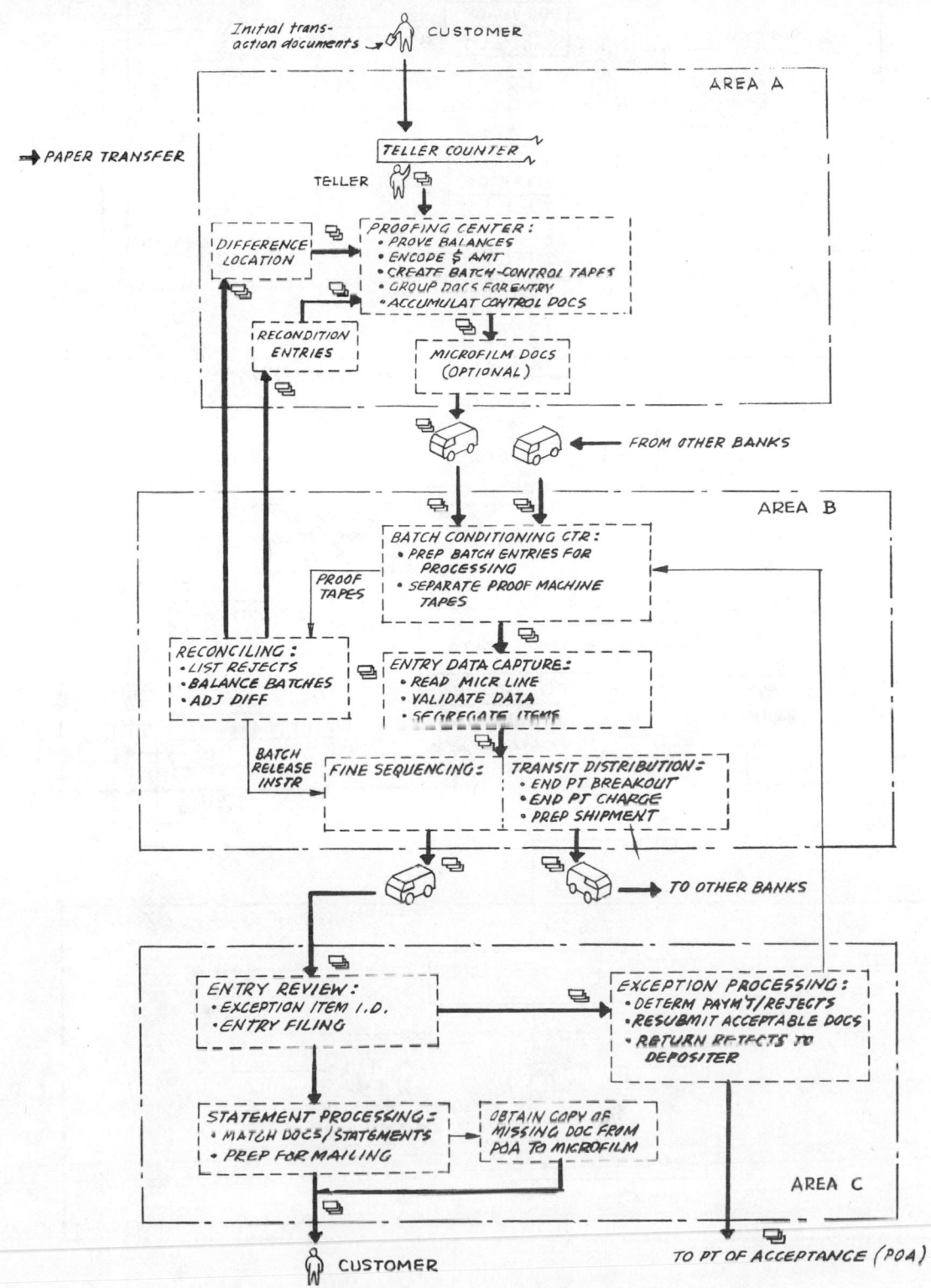

ORIGINAL BANKING OPERATION

**Example: Semipictorial OSD Technique**

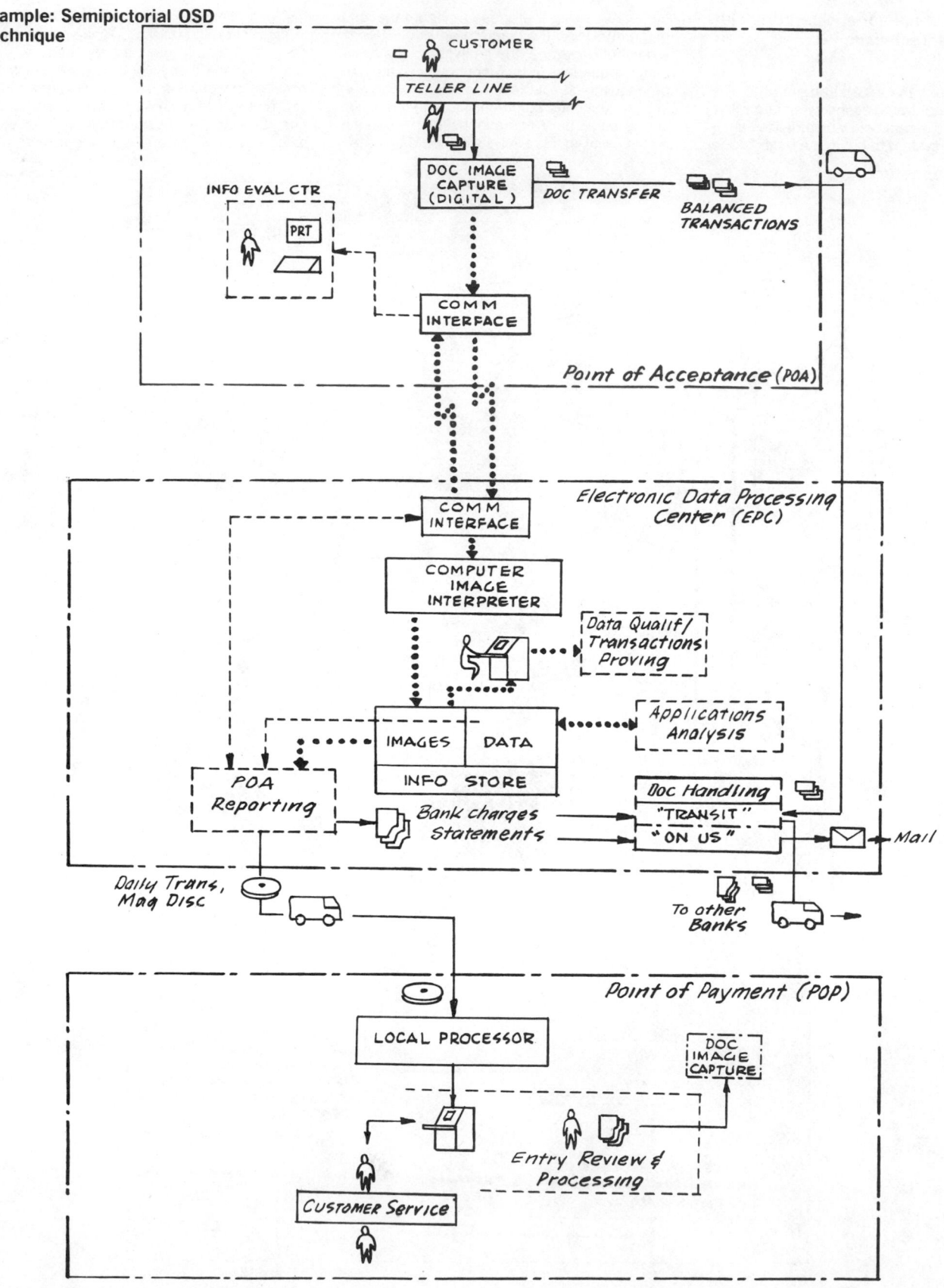

Proposed New System Using Automated Data Capture & Routing

SEQUENCE CYCLE    *START OF DAY CYCLE*

| FUNCTION/OPERATION | EPC | BRANCH |
|---|---|---|
| | SUPERVISOR | POA BANK OFFICER |

1. Cue to begin day's processing
2. Initialize EPC - verify power "on"
   . Turn power "on"
3. Initiate "start-of-day program"
4. Enter start-of-day info (date, time, EPC, security info, etc.)
5. Initiate subsystem/load sequence
6. Initiate subsystem/reply status
   (Manually) start/error correct, CR perform special sequence
7. Establish inter-EPC communications
8. Review, purge, re-order file storage system
9. Verify initialization complete
   . Repeat all or part of initialization
10. Activate POA equipment
11. Verify POA equipment activated
    . Telephone POA to activate equipment
12. Establish communications with POA
    . Direct alternate communication activation
13. Verify comm established/ILU activated

CONT.

LEGEND: Operator/Equipment Functions

□  Equipment (Automatic Process)

○  Operator Action (Non Equip)

▽  Decision Point

▣ (boxed ▽)  Automated Decision

⊖ (circled ▽)  Operator Decision

◙ (boxed ○)  Operator/Equip Interaction (visual, keying, activation, adjusting, load/unload)

## TASK ANALYSIS

As the term implies, task analysis involves the examination of what a particular design requires of an operator, maintainer, or product user who must perform a monitoring, control, or maintenance task. The sooner one can examine and define the task implications of a proposed task, the sooner he or she will know whether the design is compatible with the user's capabilities and limitations.

Task analyses and task descriptions serve two main purposes, among others. First, as noted above, the task analysis tells us not only whether the design is feasible but also whether it is optimum from the user's standpoint. Second, the task analysis provides important information for the personnel subsystem specialist who must plan and devise personnel acquisition and training programs for the system.

Preliminary task analyses should be performed as soon as possible during or directly following the initial conceptual phases of a development program; thereafter, they should be regularly refined and updated until the task description becomes a procedural guide.

The accompanying materials are meant to provide guidance in task analysis through the several examples of descriptive materials and documenting formats. It will be observed that various analysts have taken different approaches to documenting their analysis and have used different formats for describing tasks. Obviously, there are no set rules, and therefore these examples merely illustrate the types of information that various analysts felt were necessary to provide useful documentation—for assisting in design review and/or for developing manning and training requirements.

**Example: Preliminary Task Analysis**

```
Function 4.1.2:
Load xx projectile into        PRELIMINARY OPERATOR FUNCTIONAL
x-1 breech block               REQUIREMENTS WORKSHEET
```

```
SYSTEM/OP CONSTRAINTS:  Loading rate = 2/Minute; Sea State 4

TASK DESCRIPTION:  Monitor ammunition shaft delivery signal, open door,
    position projectile carriage, transfer projectile to carriage, move
    loaded carriage to loading position, open gun breech, transfer pro-
    jectile to breech, check position, adjust fuse, close breech, notify
    fire control officer "ready".

PHYSIOLOGICAL CONSIDERATIONS:  Maximum lift capability for one man is
    60 pounds; higher weights require special sling or method for direct
    roll-transfer of projectile.

PSYCHOLOGICAL CONSIDERATIONS:  Visual codes for identification and spec-
    ial calibration markings must be legible as well as visible under
    battle (red) lighting environment.

SAFETY CONSIDERATIONS:  Automatic door closure on ammunition delivery
    system and possible effects of ship roll require special indoctri-
    nation and drill to prevent accidents (e.g., closure of door on
    personnel) or loss of control of projectile.
```

**Example: Task Analysis
Documentation Format Showing One
Analyst's Approach for Defining Task
Characteristics at Both a Gross and a
Refined Level**

## Gross Task Analysis

| Task | Display | | Decisions | Subtasks Listed | Characteristic Errors or Malfunctions |
| | Description | Critical Values | | | |
|---|---|---|---|---|---|
| A. Set up before first run | Position of brake | Not in 9 o'clock position | Release brake—move to 9 o'clock position | 1 | |
| | Recall of last link | Copy light since last inking (est.) | Measure ink—add if necessary | 18, 19, 20, 21, 22 | |

## Analysis of Subtasks

| Subtask or Task | Display-Control Descriptions | Control Action | Indication of Response Adequacy | Objective Criterion of Response Adequacy | Characteristic Errors or Malfunctions |
|---|---|---|---|---|---|
| A. Release brake brake | Brake | Turn clockwise (up) | Brake stop in 9 o'clock position | | |
| B. Attach the stencil | Wheel | Turn | Stencil head clamp available | | |
| | Stencil head clamp | Lift left end | Stencil head clamp loosens | | |

| Function | Human Functional Element | Task | Subtask |
|---|---|---|---|
| Activate gyros and other components | Make power ON input | 1. Turn on ac and dc missile guidance power | a. Check to assure that all circuit breakers are "in" |
| | | | b. Position ac missile power guidance switch at ON |
| | | | c. Position dc missile guidance power switch at ON |
| Activate test instruments | Make power ON input | 2. Turn on missile preflight analyzer power | a. Check to assure that test patchboard is inserted and locked |
| | | | b. Position analyzer master power switch at ON |
| | | | c. Insert test tape in tape reader and lock in place |
| | Make analyzer self-test initiating input | 3. Start analyzer self-test test sequence | a. Position test switch at SELF-TEST |
| | Monitor self-test readout | | b. Check test result window for indication of analyzer self-test result |
| | Provide corrective inputs and monitoring when self-test yields a "no-go" | Alternative procedure: isolate and replace malfunctioning analyzer module | a. Check code displayed in malfunctioning module window against modular code table |
| | | | b. Remove module from analyzer and take to storage |
| | | | c. Obtain replacement module from storage; take to analyzer and insert it |
| | | | d. Repeat task 3 |

## Suggested Information Requirements for Expanded Task Analyses

| Item | Description |
| --- | --- |
| Operator and maintainer position | Position-type title or specialty rating that it is believed the task should be assigned. |
| Job operation | Job operation performance description, i.e., operation performed in support of a mission function, usually performed at a single location as a unit of work, in the sense that it has a well-defined beginning and ending. |
| Duty | Larger units of work under a job operation assigned to an individual in execution of a position. Duties are made up of operationally similar tasks within a given operator or maintainer position. An example might be "carries out overall subsystem test on XX radar." |
| Task title | A short title that indicates what an individual does (in functional terms). |
| Task index number | Task-identifying number, usually corresponding with OSD numbering system. |
| Data sources | Major sources of data useful for later verification (equipment, location, performance criteria, etc.). |
| Work-area location | Where task is accomplished. |
| Task description | Includes:<br>Equipment used<br>Displays used<br>Controls used<br>Actions (detailed, sequential description of each task step)<br>Support equipment and aids used<br>Feedback re individual's actions |
| Type of task | Qualities or characteristics of task (fixed procedure, variable procedure, motor skill, system analysis, circuit analysis, display interpretation, etc.). |
| Frequency of performance | Once, twice, etc.; hourly; per shift, daily, weekly, monthly. |
| Performance time | Estimated hours and minutes job or task will take; cite maximum permissible time, etc. |
| Criticality | Effect of failure to perform on the success or failure of job or mission; potential for personal injury. |
| Newness | Extent to which the task is new to the using agency. |
| Other positions | Other tasks whose performance depends on, or interacts with, the performance of other personnel position types (indicate functional interrelations). |
| Safety factors | Any known or suspected hazards, sources, or preventive requirements. |
| Tools and equipment | Tools, test equipment, protective gear, reference manuals, etc. |
| Skills and knowledge | Types and level of skill or knowledge pertinent to the selection and training of individuals for the job or tasks. |
| Physical characteristics | Special physical requirements necessary for effective task performance (e.g., body size, strength, dexterity, coordination, physiological tolerance to specific and anticipated stresses). |

**Example: Analysis to Break Out Tasks
and Subtasks for Human Functions**

### Behavior Classification for Task Analysis

| Processes | Activities | Specific Behaviors |
|---|---|---|
| Perceptual processes | Searching for and receiving information | Detect<br>Inspect<br>Observe<br>Read<br>Receive<br>Scan<br>Survey |
| | Identifying objects, actions, and events | Discriminate<br>Identify<br>Locate |
| Mediational processes | Information processing | Categorize<br>Calculate<br>Encode<br>Compute<br>Interpolate<br>Itemize<br>Tabulate<br>Translate |
| | Problem solving and decision making | Analyze<br>Calculate<br>Choose<br>Compare<br>Compute<br>Estimate<br>Predict<br>Plan |
| Communication processes | | Advise<br>Answer<br>Communicate<br>Direct<br>Indicate<br>Inform<br>Instruct<br>Request<br>Transmit |
| Motor processes | Simple, discrete tasks | Activate<br>Close<br>Connect<br>Disconnect<br>Join<br>Move<br>Press<br>Set<br>Raise<br>Lower<br>Hold |
| | Complex, continuous tasks | Adjust<br>Align<br>Regulate<br>Synchronize<br>Track<br>Transport |

*Source:* Adapted from C. Berliner, et al., "Behaviors, Measures and Instruments for Performance Evaluation in Simulated Environments," paper presented at a Symposium and Workshop on Quantification of Human Performance, Albuquerque, N.M., 1964.

**Example: Format for Analyzing Tasks
and Developing Task Descriptions**

Position: Pilot
Duty: Vehicle checkout and start

| Task | Time, min | Control | Elements Activity | Indication | Remarks (Precautions, etc.) |
|------|-----------|---------|-------------------|------------|------------------------------|
| 1.1.1 Engine start | 0.1 | | Observe | Ground crew signal | Procedure performed per checklist |
| | 0.05 | APU Battery SW | Press/monitor | On | Decision to use APU |
| | 0.05 | APU Fuel SW | Press/monitor | On | |
| | 0.05 | APU Start | Press | | |
| | 0.5 | | Monitor | Start | |
| | 0.05 | Engine fuel Master SW | Press | On | |
| | 0.05 | AUX Fuel pumps | Press | On | |
| | 0.05 | | Observe | Fuel pressure | Variable time estimate |
| | 0.05 | Engine start | Press | | |
| | 0.05 | | | Engine parameters | |

**Example: Action and Information
Requirements Analysis Form**

| | Approach Requirements Analysis | | | Trade-off Information and DATA Integration | |
|---|---|---|---|---|---|
| Approach-land Functional Requirements | Action Requirements | Information Requirements | Related Information Requirements, Sources, and Problems | Related Accident Features | Related Survey Commentary |
| 1.0 Initiate pre-approach procedures | 1.0.1 Review approach information | 1.0.1.1 Approach orientation 1.0.1.2 Approach constraints   Approach   Requirements   Obstacles   Hazards   Weather   Minima | Approach plate data   Obstacle locations   Course and path data   Terrain characteristics   Hazards   Minimum decision altitudes Position data | Data misinterpreted not used effectively. Hazards misappraised. Navigation positioning errors. | Cannot remember all details. Study time is limited while setting up approach. Improve to emphasize basic data-critical data holder, e.g., go-around hdg. and altitude. Need clearer picture of position situation. |
| | 1.0.2 Coordinate approach with control | 1.0.2.1 Communication Path designation Unique limitations and   constraints Environmental   conditions Barometric pressure | Coordination and confirmation of approach clearance Altimeter setting | Clearances and procedures are misunderstood; not followed; in error. Altimeter misset; misread. Confusion of inches mercury versus millibars sea level versus field elevation reference | Need procedures for better coordination between airplane and traffic control to improve understanding of situation and control intent. Improve altimetry presentation method. Standardize setting references —field elevation preferred for landing. Redundant setting checks. Radio chatter and changes. |

**Requirements Allocation Sheet**

| Requirements Allocation Sheet | Functional Diagram Title and Number | | Equipment Identification | | Personnel and Training Equipment Requirements | | | | |
|---|---|---|---|---|---|---|---|---|---|
| | Nomenclature and Number of CEI | | | Detail Spec, CEI, Master Control Number | | | | | |
| Function Name and Number | Design Requirements | Facility Requirements | Nomenclature | | Tasks | Time Required | Performance Required | Training and Training Equip. Required | Procedural Data Requirements |
| 3.3.1.3 Electrical continuity check | Line voltage- 0–150 Ω | None | Line tester | 1.3 | Attach clips. Turn on-off and SW and knob. Adjust Monitor | Estimated 120's | Read to ± 5 Ω | Actual tester | *Note:* Instructions in maintenance management |
| 3.3.1.4 Hydraulic line press check | Line Pressure- 350 lb/in² | None | Hydraulic press tester | 2.4 | Remove protection cap. Attach quick-connect line connectors. Open line valve. Read Meter | Estimated: 325's | Read to ± 2 lb/in² | Hydraulic pressure tester | See instructions in checkout manual *Note:* Be sure line valve is in closed position before making line connection. |

## Example: Task Analysis Worksheet

| 1. Job Operation: Prepare Swimmer Delivery Vehicle (SDV) for Launch | 2. Task Title: Replenish Breathing Gas Supply | 3. Location of Task: On Dock or in Water | 4. Source Reference: SDV Specification | 5. Operational Site: World Wide |
|---|---|---|---|---|

EQUIPMENT DATA | PERSONNEL DATA

| 6. Equipment Required | CSE Required | 7. Source Ref. | 8. Number of personnel | 9. Recommended Position Title | 10. Applicable Rating |
|---|---|---|---|---|---|
| Breathing gas supply pump & containers, press gauge | Breathing gas supply containers | none | 2 | Vehicle Maintenance Man | BM |

TASK PERFORMANCE DATA

| 11. Time Required Hours 0 Minutes 25 | 12. Frequency Day Week Month unknown | 13. Probable Error Minimal on land moderate in water | 14. Speed ☐1. Not critical ☑2. Moderately critical ☐3. Highly critical | 15. Positioning/Handling Equip, Special care ☐1. Little ☐3. Considerable ☑2. Moderate |
|---|---|---|---|---|

**16. Manipulating Controls**
- ☐ None
- ☑ Hand Valves
- ☐ Toggle Switches
- ☐ Selector Switches
- ☐ Pushbuttons
- ☐ Cont. Audio Feedback
- ☐ Cont. Visual Feedback

**17. Source of Special Dangers**
- ☐ None
- ☐ Mechanical
- ☐ Electrical
- ☑ Explosive
- ☐ Temperature
- ☐ Volatile Fuels
- ☐ Toxic Substance
- ☐ Other

TRAINING DATA

| 18. Nature of Procedure ☐ Fixed ☑ Motor Skill ☐ Circuit Analysis ☑ Variable ☐ System Analysis | 19. Technical Manual Function ☐ Primary ☑ Secondary | 20. Training Difficulty ☐ Easy ☐ Hard ☑ Moderate | 21. Training Equipment ☐ Complex ☑ Simple ☐ None |
|---|---|---|---|

| 22. Technical Manual Title: Life Support System, Maintenance & Replenishment of SDV-TR-013 | 23. Training Course Title: Life Support System, Repair and Replenishment of SDV |
|---|---|

| 24. Task Performance Date ☐ Estimated ☑ Tried in mockup ☐ Tried, hardware | 25. Equipment Development Status ☐ Design Plan ☐ Preliminary Design ☐ Mockup ☑ Prototype ☐ Production Design ☐ Production Equipment | 26. Narrative Task Information; Remove skin cover plate over life support quick-connect fitting. Attach gas supply nozzle to boat, open supply valve, and monitor flow rate. Shut off valve when pressure reaches 2800 psi. Disconnect supply and check for leaks at nozzle. Replace skin cover. |
|---|---|---|

## Example: Providing Inputs to Design from Task Analysis

**TASK CRITICALITY** This is not an independent output; rather, it is deduced from the nature of the task and is presented to the engineer with the task description in the form of a note to the task description.

There are three major steps in the derivation of task criticality:

1. *Identify the potential errors which can be made in performance of the task.* This is largely a matter of considering the elements of the task and the perceptual, motor, and decision-making demands imposed on the operator. Thus, in the case of a simple task which involves *(a)* reading a pressure gauge regulating the internal pressure of a rocket and *(b)* stopping a pump at a specified pressure, errors may manifest themselves in two ways:
   *a.* Failing to stop at the prescribed point
   *b.* Stopping the pump before the prescribed point
2. *Identify the effect of each potential error on system operation.* In the example above, failing to stop the pump at the prescribed point may result in overpressurization and bursting of the rocket being pressurized. Stopping before the prescribed point will result in underpressurization, which will cause certain sensitive instruments requiring a pressurized atmosphere to function erratically.
3. *Estimate the relative criticality of the potential errors.* Criticality may be scaled in terms of categories such as loss of personnel, destruction of the system, mission failure or abort, mission degradation, and mission delay. In these terms, overpressurization may be more critical than underpressurization, since it may result in explosion of the rocket and destruction of the rocket and launch pad as

well as loss of life, while underpressurization is less critical, since the mission may (though not necessarily will) be degraded.

Pointing out a task as being critical to the engineer "flags" that task as one requiring special consideration in design. Among the solutions which are possible (certainly the list is not exhaustive) are the following:

1. Replacing the human with an automatic means of accomplishing the function if the desired level of correct performance cannot be achieved in any other way
2. Providing means to reduce the probability of error, e.g., assigning a special feedback device to warn the operator when the task is being performed incorrectly
3. Assigning the task to only highly skilled personnel

Task criticality is highly related to specification of task difficulty and error likelihood. Since the engineer thinks in terms of physical effects on the system, it is preferable to flag the task as being critical without indicating that it also has a high difficulty or error-likelihood index. The provision of quantitative indices of a highly precise nature, such as probability of operator error to four figures (e.g., .0013), is not advised, since the engineer cannot interpret the quantitative values in design-relevant terms. A gross categorization of task difficulty, such as the following three-part scale, is as much precision as the engineer can handle in design terms:

1. Simple, routine
2. Somewhat difficult
3. Very difficult

Moreover, it is necessary to apply the above scale only to those few very critical tasks which merit design attention, not to all tasks.

**TASK DURATION** Task duration should be considered in two ways:

1. As a system requirement, i.e., the time within which the task must be performed

in order to accomplish a given system function
2. As an anticipated human performance capability, i.e., the time within which the operator can actually perform the task

Item 1 is a criterion against which item 2 can be evaluated as satisfying or failing to satisfy system time requirements.

As a system requirement, a task may have to be accomplished in so short a time either that the operator cannot physically perform it or that the probability of the operator's making an error will be substantially increased because of the time loading. In either case, special attention must be drawn to such a task. If the system time requirement is inflexible, it may be necessary to automate the function involved (to eliminate the operator) or else to redesign the manner in which the task can be performed or the equipment is to be operated or maintained.

Task duration is, of course, not critical unless the system's required response time is also critical to the successful accomplishment of the mission. Hence, it is necessary to analyze the mission segment in terms of its time demands before examining any individual task duration. Information required in order to perform task duration analysis will include:

1. System performance time requirements
2. A description of the tasks to be performed by personnel in each mission segment
3. The estimated time required to perform the task

Of these informational requirements, the last is the most difficult to secure because it requires data on the performance time capability of personnel, such as the time required to hook up an umbilical connection. That information can be secured from previous comparable systems in which similar or identical tasks have been timed or from the body of general human performance data in the literature. Neither of these sources is readily available.

## Example: Task Duration Analysis for SM-X "Assemble Missile" Segment

| Function or Task | Output | | Duration | | Input | | Remarks |
|---|---|---|---|---|---|---|---|
| | Not Earlier Than | Not Later Than | | | Earliest | Latest | |
| A. Transport W/H, etc. | | | | | | | ① Output cannot be made earlier than time missile clears track bed. |
|   2. Position crane | — | T + 1 | 1 min | | — | T + 0 | |
|   5. Rotate nose cone, forward | ① | T + 3 | 1 | | T + 2② | T + 2.5 | |
|   6. Position truck ahead of transporter erector | T + 3 | T + 4 | 1 | | T + 2.5 | T + 3 | ② Time of initiation depends upon rate at which a second technician hoists nose cone. |
| B. Mate missile and nose cone | | | | | | | ③ Output can start as soon as components are mated. |
|   1. Use crane and truck to mate components | — | T + 6 | 2 | | T + 3.5 | T + 4.5 | |
| C. Attach nose cone to missile | | | | | | | ④ This task is also performed by a second technician, who exerts a pacing effect. |
|   1. Install bolts to attach components | ③ | T + 10 | 4④ | | T + 5.5 | T + 7 | |

*Source:* D. Meister, et al., *The Impact of Manpower Requirements and Personnel Resources Data on System Design,* AMRL-TR-68-44, Air Force Systems Command, Wright-Patterson Air Force Base, Ohio.

## TASK DIFFICULTY (ERROR LIKELIHOOD)[16]

The human factors specialist is especially concerned about task difficulty because this, in turn, may lead to a higher error probability, with its attendant effects on mission accomplishment. Task difficulty arises because system requirements are incompatible with and overload the skill capability of the individuals assigned to perform the task.

Task difficulty is not the same as error likelihood. A difficult task need not automatically have a higher error probability if personnel of higher skill are available to compensate for the increased task difficulty. The significance of task difficulty is intensified when the task is also critical to the accomplishment of the mission. Such difficult and critical tasks automatically demand redesign because their attendant error probability cannot be accepted. As in the case of maximum task durations which the operator's performance cannot meet, it may be necessary to automate the performance of the task, relax the accuracy requirement (thus implicitly accepting a higher error probability), or redesign the task to simplify it.

The determination of task difficulty must be made by analyzing the individual task in terms of the inputs which initiate the task (e.g., a verbal message) and the outputs which accomplish the task (e.g., a switch action). The human factors specialist will look for the following characteristics, which may (not necessarily will) indicate an excessively difficult task:

1. The input which initiates the task requires excessively precise visual discriminations or fine motor responses.
2. The operator's response to the initiating inputs must be performed so quickly that he or she has problems keeping up with the initiating inputs.
3. The accuracy demanded of the operator in responding to the initiating inputs is excessive (e.g., heading error must be within 0.5°).
4. The task must be coordinated extremely precisely with other tasks performed by other personnel.
5. The environment in which the task must be performed tends to degrade task performance (e.g., high levels of noise or acceleration).
6. Information from multiple sources (e.g., several displays on a control panel) must be integrated by the operator in order to make a decision.

7. There is less than the desirable amount of information available on the basis of which a decision must be made or an action taken.
8. The task is composed of many subtask elements, the correct performance of which is necessary to task performance, but the amount of feedback provided (knowledge of the correctness or incorrectness of subtask accomplishment) is inadequate.
9. Short-term memory requirements for task performance are excessive (e.g., memory for long sequences or target coordinates).

The design solutions available for reducing task difficulty include the following:

1. Providing additional training or selecting more highly skilled personnel
2. Simplifying the task by such means as combining information sources, providing additional feedback, subdividing the task among several operators, or changing the manner in which the task must be performed
3. Reducing system requirements by accepting a higher error probability, longer response time, etc.

## Example: Task Analysis versus Training Requirements Definition

JOB: _____    TASK ANALYSIS/TRAINING OBJECTIVES & CRITERIA WORKSHEET    Page No. 1    By: _____    Date: _____

| TASK NUMBER | TASK, SUB-TASK, STEP, ELEMENT DESCRIPTION | INSTRUCTIONAL OBJECTIVES/CRITERIA | | | CRITERION ITEM |
| | | PERFORMANCE | CONDITIONS | CRITERION | |
|---|---|---|---|---|---|
| 0.000 | Describe duties, tasks and task elements in the order of their occurrence and in abbreviated form. Arrange generally to show input (display, information received), mediation (diagnosis, decision), and output (action/feedback). | Write specific statement of intent, i.e., what learner is expected to be able to do and how you will know when he is able to do it (i.e., terminal behavior). | Specify important conditions under which performance is to occur, tools he will be given, job aids, prescribed procedures; also anything learner will be denied. | State standard of acceptable performance. | Write test question that matches objective in performance and conditions. |
| | Note any unique constraints that would affect task, such as: environmental factors (noise, illumination, temperature), stress (time limits, etc.), physical (location considerations that may add to task difficulties, etc.). | | | | |
| | Indicate whether order of task occurrence is critical or non-critical. | | | | |
| | Make sure task descriptions are stated in behavioral terms! | | | | |
| | - - - - - - - - - - - - - - - - - - - - - | - - EXAMPLE - - - | - - - - - - - | - - - - - - | - - - - - - |
| | Corrective Maintenance: Radar Bright Display Equipment | | | | |
| 3.0 | Determine if isolation amplifier is performing its design function. | | | | |
| 3.1 | Compare input and output signals of amplifier to ensure no objectionable noise or gain distortion introduced. | Be able to connect test equipment to power source and test points, read voltage values and for alignment | Given a Sperry synchroscope. With reference manual but without assistance or supervision | In accordance with procedures prescribed by TM 7-864 and Instruction Manual 8-60. Perform ks in current | Given a Sperry synchroscope, connect it to power supply, check voltages in accordance |

[16] D. Meister et al., *The Impact of Manpower Requirements and Personnel Resources Data on System Design*, AMRL-TR-68-44, Air Force Systems Command, Wright-Patterson Air Force Base, Ohio.

**Example: Personnel Requirements
Documentation**

<u>PERSONNEL REQUIREMENTS INFORMATION</u>

Job Category:_____Duty:_____

Operation:_____Task:_____

Date:_____

<u>Task Analysis Information</u>

Task Performance Frequency:_____Performance Time:_____

Task Requirements:
  Perceptual Level:_____
  Decision Level:_____
  Stress Conditions:_____
  Motor Skills:_____
  Physical Requirements:
        Height:_____Strength:_____Weight:_____Other_____

Recommended Changes in Task Structure and Human Requirements:_____
_____

Personnel Used in Task Verification:_____

Company Procedures Knowledge Task Requirements:_____

Recommended Personnel Selection Requirements:

  General Aptitudes:_____
  General Training/Education:_____

Recommended Pre-selection Testing:

  Performance:_____
  Intelligence:_____
  Interests:_____
  Personality:_____
  Special:_____
  Scoreable Checklist Performance:_____

In-Plant Training Recommendations:

  Requires No Training:_____
  Requires Familiarization:_____
  Requires Special Training:_____

Action Taken by Personnel Department:

Date Report Forwarded:_____Date Report Returned:_____

**Example: Quantitative/Qualitative
Personnel Requirements Inventory
Documentation**

Position Title: Liquid Fuel Systems Maintenance

Specialist/Technician                    Air Force Specialty Code (AFSC) 56850B/70B

| Duties and Tasks | Performance Reliability | Skill Level | Skill Type and Proficiency Availability | Training Requirements |
|---|---|---|---|---|
| 1.0 Connect propellant transporter(s) to fixed-base propellant transfer system<br>Possible Human Errors<br>Cross-connection of hoses<br>Misalignment of coupling | .9975 | 5 | 1. Position Summary<br>The LFMS/T is responsible, under the direction of the Fuel Specialist/Supervisor, for the conduct of propellant loading and unloading operations at the launch site, including emergency or damage control activities if required. He decontaminates the surface area, the launch duct or the sump, or the rocket as required. He directs emplacement and removal of mobile PTS equipment and seals fixed portions of the PTS. LFMS/T's assigned to maintenance teams perform unscheduled maintenance on PTS fixed equipment at the launch complex. Any LFMS/T's performing PTS procedures at the launch complex should know both the emergency and normal loading and unloading procedures.<br>2. Environment<br>a. Launch Site<br>(1) The LFMS/T controls traffic into and from the launch complex security area. He directs positioning of propellant transporters and the interconnecting of PTS elements. | The training should be designed to train selected Air Force personnel to:<br>1. Operate control panels to determine malfunctions in the propellant transfer system.<br>2. Perform interconnections of the electrical and mechanical mobile equipment.<br>3. Operate and maintain the propellant transfer system components and the various items of GOF/GSE including the propellant transporters, propellant holding tanks, nitrogen storage tanks, and propellant transfer control unit.<br>4. Perform emergency unloading procedures for the PTS.<br>5. Inspect storage tanks, disconnects, valves, piping, pressure regulators, filters, pressure gages, and temperature gages for leakage, corrosion, damage, and wear; also solenoids, relays, switches, chassis of the transfer control panel, sensors, amplifiers, motors, terminals, interconnecting cabling and instrumentation for proper operation. Instruction should include removal and replacement procedures for the above listed items. |
| 2.0 Prepare to transfer propellant from transporter(s) to ready-storage vessels (RSV's)<br>Valves in system left open<br>System not completely bled<br>RSV vents not opened | .9960 | 5 | | |
| 3.0 Transfer propellant from transporter(s) Ready Storage Vessels (RSV's)<br>Leaks in system due to improper connection or maintenance<br>Failure to close valves when transporter is empty | .9978 | 5 | | |

**Example: Maintenance Training
Requirements Analysis**

| Maintenance Level | Maintenance Task Category | Support Equipment (AGERD) | Skills and Knowledges Required to Do the Job | Category of Skills or Knowledges to Be Taught or Strengthened | Equipment and/or Training Recommendations |
|---|---|---|---|---|---|
| Organizational Unscheduled Sub-system Troubleshooting (Continued) | | | | f. Oral Verbalization: Communications procedures between maintenance personnel during task performance. | |
| Intermediate Unscheduled LRU Mechanical and Electrical Repair | 4. LRU (Mechanical and Electrical) Repair: involves disassembly in varying degrees, visual inspections, running of tests and measurements (e.g., clearance tolerances, pressure loss and leakage tests, torque tightening checks). Certain minor servicing functions such as lubrication or dehydrator recharging is accomplished. Repair consists mainly of parts | a. Regulated air supply.<br>b. Power supply.<br>c. Valve and flange caps.<br>d. Pressure and height gauges.<br>e. Depth and thickness micrometers.<br>f. Flowmeter.<br>g. Replacement parts.<br>h. Standard electronic AGE.<br>i. Cleaning materials.<br>j. Anti-Q valve test | a. Knowledge of appearance, location, nomenclature and fitting sequences of LRU parts.<br>b. Knowledge of part functioning principles and relationship to LRU and fitting principles.<br>c. Be able to use precision mechanical tools, and other associated AGE.<br>d. Knowledge of visual appearance of acceptable part condition.<br>e. Know how to perform each step of procedures. | a. Nonverbal Identification: Identification of LRU parts configuration.<br>b. Nonverbal Detection: Spatial judgments in parts fitting tasks.<br>c. Learning facts: Knowledge of part appearance, location nomenclature, fitting sequence, visual cues of acceptable condition, gauge scale readings, test setup.<br>d. Learning Principles: Parts fitting, parts and LRU | 1. Transparencies (animated and static): For presentation of facts and principles on LRU and precision measuring tools.<br>2. Self-Teaching Programmed Texts: Using appropriate illustrations to produce identification skills of parts and nomenclature knowledge to aid assembly and disassembly tasks. Also used to teach the |

*(continued)*

**Example: Maintenance Training Requirements Analysis**

| Maintenance Level | Maintenance Task Category | Support Equipment (AGERD) | Skills and Knowledges Required to Do the Job | Category of Skills or Knowledges to Be Taught or Strengthened | Equipment and/or Training Recommendations |
|---|---|---|---|---|---|
| Organizational Unscheduled Sub-system Troubleshooting (Continued) | | | | f. Oral Verbalization: Communications procedures between maintenance personnel during task performance. | |
| | replacement. Some external surface damage repair. Assembly follows with running of functional "quality assurance" tests. | set (0500). | f. Be able to make minor repair decisions regarding extent of repair required from test results. <br> g. Be able to make minor maintenance decisions. | functioning, and relationships. <br> e. Learning Procedures: Testing, inspection, and test setup procedures. Disassembly and assembly procedures. <br> f. Using Principles: Evaluating test results. Apply principles in performance of tasks. <br> g. Positioning Movement: Using precision tools in making measurements, fitting parts together. | concept and use of scales and interpolating readings of scales on precision instruments. <br> 3. Real Precision Tools: For active student use to permit perceptual and motor skill practice and learning on appropriate LRU's and setups (for task specific tools only). <br> 4. Bench Setups with Real LRU's and Support AGE: The nature of mechanical work requires the development of perceptual and motor skills, spatial. |

## Acceptable Task Verbs

*Activate:* To initiate system operation by movement of a control.

*Actuate:* To move a control switch (push button, rotary selector or round knob, toggle, lever, etc.).

*Adjust:* To reset a control as a result of new or nonprogrammed information. (Refers usually to nondetented control hardware.)

*Advise:* To notify others verbally.

*Align:* To bring control and display elements into coincidence or to match physical hardware for the purpose of assembly or checking of assembly.

*Assemble:* To join major assemblies and subassemblies. (Includes all attach actions plus mating of component interfaces.)

*Calculate:* To compute numerically (either mentally or physically).

*Calibrate:* To check accuracy, deviation, or variation by comparison with a standard.

*Carry:* To transport, except without mechanical assistance.

*Check, test:* To carry out those activities associated with verifying operational readiness.

*Clean:* To remove dirt, grease, or other contaminants.

*Connect:* To couple matched connection hardware (electrical, hydraulic, or pneumatic lines or nonflow mechanical or structural supporting interfaces).

*Decide:* To select a course of action on the basis of previous analysis.

*Depress:* To actuate a push-button control.

*Disassemble:* To disjoin major assemblies and subassemblies. (Includes all detach actions.)

*Disconnect:* To decouple matched connection hardware (see *connect*).

*Drain:* To empty liquid contents from tanks, reservoirs, lines, etc.

*Evaluate:* To analyze information for the purpose of establishing its validity and significance.

*Fasten:* To secure common fastener hardware (nuts and bolts, latches, Zeus fasteners, screws, etc.).

*Fill:* To introduce fuel, oil, hydraulic fluid, etc., into tanks or reservoirs.

*Flush:* To introduce fluids into tanks, reservoirs, or lines for the purpose of decontamination.

*Insert:* To put into (e.g., a bolt into a hole).

*Inspect:* To make a programmed, critical visual examination of equipment for obvious defects.

*Install:* To carry out the initial activity of attaching components or subcomponents.

*Interpret:* To analyze information for the purpose of establishing its meaning.

*Listen:* To carry out an auditory activity for the purpose of extracting specific information.

*Measure:* To make a qualitative and/or quantitative determination of the operating condition of a component or subcomponent.

*Monitor:* To carry out those activities associated with maintaining cognizance of a system's status (e.g., observing panel status lights).

*Note:* To carry out a perceptual activity (visual, auditory, etc.) for the purpose of extracting specific information from display hardware and/or other operational phenomena.

*Observe:* To carry out a visual activity for the purpose of maintaining an awareness of a general condition.

*Operate:* To carry out those activities associated with the *use* of consoles and test equipment.

*Place:* To move a toggle switch or lever arm in a specific direction.

*Position:* To physically orient an object for the purpose of accomplishing a specified task.

*Read:* To carry out a visual activity for the purpose of extracting specific information from display hardware.

*Record:* To fill out checklists, log sheets, or other required forms.

*Remove:* To detach components or subcomponents, usually in connection with maintenance activities; to move from the original position (e.g., raising an engine cover prior to start).

*Repair:* To carry out those activities associated with restoring damaged, worn-out, or malfunctioning equipment to an operable condition.

*Replace:* To attach components or subcomponents, usually in connection with maintenance activities.

*Secure:* The reverse of *operate*.

*Service:* To perform the scheduled cleanup, lubrication, and replenishment necessary to prepare equipment for use.

*Set:* To place a rotary control pointer in a predetermined position.

*Transfer:* To carry or transport something (with the aid of equipment) from one location to another.

*Transport:* To move an assembly, unit, component, etc., from one place to another with the assistance of a mechanical aid; includes steering or guiding the mechanical aid.

*Troubleshoot:* To isolate a fault within a system, subsystem, or component.

*Unfasten:* To release common fastener hardware (see *fasten*).

*Verify:* To confirm that an expected condition exists.

*Withdraw:* To retract from (e.g., the reverse of insert).

## DESIGN STUDIES

### Human Engineering Design Analysis

Once initial hardware functions are assigned and basic performance requirements are established for each specified hardware end item, the engineer and/or designer commences to explore possible design alternatives. The accompanying sample outline would be used as one proceeds through the design process. Note that this example indicates that the designer must look not only at the design itself but also at the probable test requirements that will be needed to evaluate the eventual performance of the hardware.

| DESIGN SHEET | NOMENCLATURE | CEI NO OR CRITICAL COMPONENT CODE IDENTIFICATION |
|---|---|---|
| | | DETAIL SPEC NO_____ |
| | (A) | (B) |

<u>REQUIREMENTS FOR DESIGN AND TEST</u>

```
3.0        Requirements
3.1        Performance
  3.1.1        Functional Characteristics
    3.1.1.1    Primary Performance Characteristics
    3.1.1.2    Secondary Performance Characteristics
  3.1.2        Operability
    3.1.2.1    Reliability
    3.1.2.2    Maintainability
    3.1.2.3    Useful Life
    3.1.2.4    Environment
    3.1.2.5    Transportability
    3.1.2.6    Human Performance
    3.1.2.7    Safety
3.2        CEI Definition
  3.2.1        Interface Requirements
    3.2.1.1    Schematic Arrangement
    3.2.1.2    Detailed Interface Definition
  3.2.2        Component Identification
    3.2.2.1    Gov't Furnished Property List
    3.2.2.2    Engineering Critical Components List
    3.2.2.3    Logistics Critical Components List
3.3        Design Construction
  3.3.1        General Design Features
  3.3.2        Selection of Specifications and Standards
  3.3.3        Materials, Parts and Processes
  3.3.4        Standard and Commercial Parts
  3.3.5        Moisture and Fungus Resistance
  3.3.6        Corrosion of Metal Parts
  3.3.7        Interchangeability and Replaceability
  3.3.8        Workmanship
  3.3.9        Electromagnetic Interference
  3.3.10       Identification and Marking
  3.3.11       Storage

4.0        Quality Assurance Provisions
4.1        Category I Test
  4.1.1        Engineering Test and Evaluation
  4.1.2        Preliminary Qualification Tests
  4.1.3        Formal Qualification Tests
             etc.
```

**Example: Design Trade-Off Study, Operator Implication Analysis**

| Design Alternatives | Operator Outputs | Evaluation |
|---|---|---|
| 1. Portable, hand-held illuminator with open sight | Target search activity | Direct vision is the most efficient method. Both pilot and operator will participate. When one detects or identifies the target, he or she will inform the other of the target location. Feasible under daylight conditions. |
| | Target detection and identification | Visual angle must be at least 5 mils for detection of target. Target detection at maximum range of 20,000 ft will be limited to targets of 200 ft or greater in diameter. Smaller targets will require a reduction in range. This activity is feasible with the stated limitations. |
| | Aircraft control | Aircraft control is a factor in that the pilot must be aware of the target location in order to properly fly the plane. A comparison of systems must therefore take this activity into account. Given that the pilot is aware of the target location, the activity is feasible. |
| | Etc. | Etc. |

**Example: Human Engineering Design Trade-Off Analysis**

```
              HUMAN ENGINEERING DESIGN TRADEOFF

                                            DESIGN APPROACHES
   DESIGN CRITERIA                            1   2   3

   DESIGN FOR OPERABILITY

    Number of Operations
    Complexity or Difficulty of Operations
    Duration of Operations/Sequence of Operations
    Relation of Operations to Crew Capability
    Training Requirements
    Degree to Which Operations Can Be Integrated With
      Other Mission Activities
    Control/Display Relationships
    Probability of Error Occurrence
    Probability of Error Detection
    Information Availability
    Information Accessibility
    Controllability
    Visibility - Lighting Requirements
    Number of Operators Required

   DESIGN FOR MAINTAINABILITY

    Ease of Malfunction Detection
    Ease of Malfunction Isolation - Diagnosis
    Preventive - Scheduled Maintenance Requirements
    Equipment Accessibility
    Ease of Removal/Replacement
    Ease of Repair
    Support Requirements - Tools, Test Equipment, etc.
    Requirements for Job Aids or Manuals
    Availability of Redundant Systems and/or Spaces
    Number/Duration of Operations for Checkout,
      Alignment, Installation, Calibration
    Requirements for Special Handling
    Number of Technicians Required
    Special Tools/Equipment Requirements

   DESIGN FOR SAFETY

    Safety Hazards
    Safety Provisions
```

**Example: Developing a Quantitative Trade-Off Matrix**

The technique of making weighted estimates for evaluating available design alternatives is illustrated by the accompanying chart. Each requirement is assigned a weighting factor. The alternative with the highest score (right-hand column) is the recommended approach.

| Approach | C/D Relationship | Display Viewing | Access | Information | Response Time | Impact on Workload | Impact on Training | Totals |
|---|---|---|---|---|---|---|---|---|
| Percent of Weight | 25 | 15 | 15 | 5 | 20 | 10 | 10 | 100 |
| 1a | Weight x Rating Factor | | | | | | | |
| 1b | | | | | | | | |
| 2 | | | | | | | | |
| etc. | | | | | | | | |
| etc. | | | | | | | | |
| etc. | | | | | | | | |

Rating Factors:
| | | |
|---|---|---|
| Excellent | 5 | Minimal |
| Very Good | 4 | Low |
| Good | 3 | Medium |
| Fair | 2 | Significant |
| Poor | 1 | Highly Significant |
| Very Poor | 0 | |

**Example: Design Alternatives Trade-Off Study Analysis Documentation**

| Design Alternatives | Functions | Operator Outputs | Operator Inputs | Controls-Displays Needed |
|---|---|---|---|---|
| 1. Portable hand-held illuminator with open sight | 1. Acquire and identify target | Target search activity<br>Detection and identification | Direct visual inputs from target<br>Direct visual inputs from target | None<br>None |
| | 2. Position aircraft | Operation of aircraft controls by pilot | Direct visual inputs from target | Aircraft controls and displays |
| | 3. Illuminate target | Acquire target in open sight; initiate tracking | Direct visual inputs from target | Open sighting device for alignment of laser beam with target |
| | | Actuate illuminator control | Tactical inputs from control | Illuminator on-off control |
| | | Track target | Direct visual inputs from target | Open sighting device |
| 2. Second alternative | Etc. | Etc. | Etc. | Etc. |

## Design Approach Trade-Off Studies Documentation

| NOMENCLATURE<br>Trade-Off Release Energy to Unfasten Stage 1 | | COMPARISON MATRIX OF DESIGN APPROACHES | | SELECTION |
|---|---|---|---|---|
| *FUNCTIONAL & TECHNICAL DESIGN<br>REQUIREMENTS | | 1 | 2 | |
| Operability<br><br>Insure response of actuator within 200 milliseconds of valve release | Stored pneumatic energey shall be released by a valve to activate Stage 1 unfastening mechanisism. Valve flow parameters must insure positive and rapid energy release to the unfastening mechanism. All connectors shall comply with military standards for threads and fittings (Reference function 1.2.1.14 and RAS Doc. 2-00002, 14 Feb, 1964) | SOLONOID VALVE<br><br>A solonoid valve operated by a 28 VDC signal from guidance and control will release energy to actuating mechanism.<br><br>*DISCUSSION<br><br>Pro<br>1. Solenoid valve is an off the shelf component with standard line connectors.<br><br>Con<br>1. Solenoids are heavy.<br><br>2. Voltage spikes would be caused in DC line.<br><br>3. A heavy load would be imposed on diodes in the guidance/control autopilot programmer switch. | EXPLOSIVE VALVE<br><br>An explosive valve operated by a 28 VDC signal from guidance and control will release energy to actuating mechanism.<br><br>*DISCUSSION<br><br>Pro<br>1. Explosive valve is an off the shelf component with standard line connectors.<br><br>2. Low power is required to fire explosive squibb.<br><br>3. No valve leakage prior to actuation.<br><br>4. Instant response.<br><br>5. Light weight.<br><br>6. No voltage spikes would be caused in DC line.<br><br>Con<br>1. Valve cannot be functionally checked, but has good reliability record. | Performance 2, 1<br><br>Reliability 2, 1<br><br>Procurability 2, 1<br><br>---<br><br>SELECTION<br><br>Solution 2<br><br>Note: See section 4 of trade study report for reasons for selection. |

* For example purposes only representative requirements and partial discussion are listed.

## Example: Format for Documenting Design Approach Trade-Off Analyses

The last example of design trade-off documentation was selected not only to illustrate an effective format for documenting the conclusion from the analysis but also to point out a typical flaw; i.e., there is no indication that the designer considered any of the human factors that might be pertinent to the trade-off decision.

It must be emphasized that to integrate human and machine properly, human factors should always be included in the basic trade-off variables. Avoid the temptation to believe that human factors can be solved after all the mechanical, structural, electrical, or engineering decisions have been made. Painting the box the right color will not correct inherent design constraints that may preclude the operator from performing his or her task satisfactorily.

For help in making proper human factors trade-offs during design work, the reader is referred to other parts of this handbook and/or to the government human engineering standard MIL-STD-1472B if the system under consideration is being developed for a military agency.

As a general rule, when considering human factors during design trade-off study, the following major issues should be addressed and generally given priority in the order shown:

1. The hazard and safety aspects of alternative designs
2. The human performance reliability probability of each design
3. The human energy cost of each design
4. The training implication of each design
5. The manpower cost (in terms of skill demand)
6. The probability that operators or maintainers will accept alternative designs.

It may be appropriate to test alternative design concepts by mocking them up and/or by simulating certain features in order to confirm or deny the acceptability of a particular design. Refer to the section on mockups later in this chapter.

## Example: Human Factors Design Requirements

Once a concept has been adopted, it is desirable to establish human engineering design requirements to assure implementation commensurate with the capabilities of personnel expected to man the system.

### FUNCTIONAL REQUIREMENTS

1. The sighting device must provide a variable magnification factor capability from 2 to 9 power. A field of view of 2 to 7° is considered sufficient for the tracking requirements of the system.
2. The sighting device must be common to the task of sighting landmarks and targets and for aiming the illuminator before and during the illumination task.

### OPERATOR DESIGN REQUIREMENTS

1. The sighting device must have a coated lens to reduce glare.
2. The sighting device must be stabilized for long ranges of sighting where higher magnification powers are used.
3. Integral with the sighting device must be an internal display of a compass rose, graduated in milliradians (mil). This scale must provide accuracy of reading to within $\pm 1°$. It shall not be necessary for the operator to move his or her head to read the compass display while viewing through the telescope.
4. The image seen through the sighting device shall be stabilized common to the beam stabilization used for the illuminator such that the operator shall see the exact duplicate of that at which the laser beam is pointed.
5. The sighting device shall have a variable focus control, and the magnification power must be readily selectable while operating.
6. A captive eyepiece cover and one for the objective lens shall be provided for protection against inclement weather, sand, etc.

## TIME-LINE ANALYSIS

Time-line analysis is a technique used to help derive human performance requirements by showing (diagramatically) the functional and temporal relationships between tasks as well as the task loadings for any combination of tasks.

The time-line analysis is displayed by means of a chart or series of charts. Although numerous formats have been developed and used by various analysts, the most useful type is shown in the illustration below. In this format it can be seen that tasks are grouped by operator and displayed as a two-dimensional graph that indicates the estimated amount of the operator's time which is occupied at various intervals along a common time scale.

With this type of graphic the analyst can easily spot those intervals during a mission in which a given operator may be overloaded. It is obvious that an operator cannot accomplish two tasks simultaneously if both occupy a single perceptual-motor channel or decision-making response 100 percent of the time. The time-line chart exposes such conditions if it is properly developed. When such conditions are spotted, it is apparent that one of two things must be done—either a task will have to be given to another operator, or the operator must be provided some type of machine assist (e.g., an operation becomes automatic rather than manual).

The task load estimates come from several sources. For example, the task may be the same as, or similar to, a task required in another system which is in actual operation. Task time information from previous systems is generally the most reliable since it has been verified in practice. When such information is not documented, the next best source of information is from operators who perform or have performed similar tasks. It is desirable to get estimates from several operators since there is frequent variation in their estimates. The human engineer generally has to probe the task question with the operator in fairly good detail to provide the operator with a basis upon which to make an estimate. It is important, for instance, to clarify the fact that in some cases two tasks could occupy almost 100 percent of the operator's time if one task involved a different perceptual-motor channel than the other. For example, an operator can usually monitor an aural channel almost full time and still monitor a visual display almost full time. When experienced operators are not available, the human engineer along with knowledgeable equipment engineers may have to make an "expert guess" about the task. The human engineer will have to break the task down into its simplest elements and extrapolate from what he or she knows, based upon human performance studies, about division of attention.

The time line may be made up of a single, continuous chart from the beginning to the end of a mission, or there may be several charts, each of which expands a particularly critical segment of the mission. The time scale should be commensurate with task complexity; i.e., 5-min intervals may be all that is necessary for simple tasks, while 5-s intervals may be required for more complex tasks. Whatever interval is used, however, should be common for the total group of tasks and operators when they interact.

## Example: Time-Line Analyses (Gross Levels)

### SYSTEM OPERATION SCENARIO

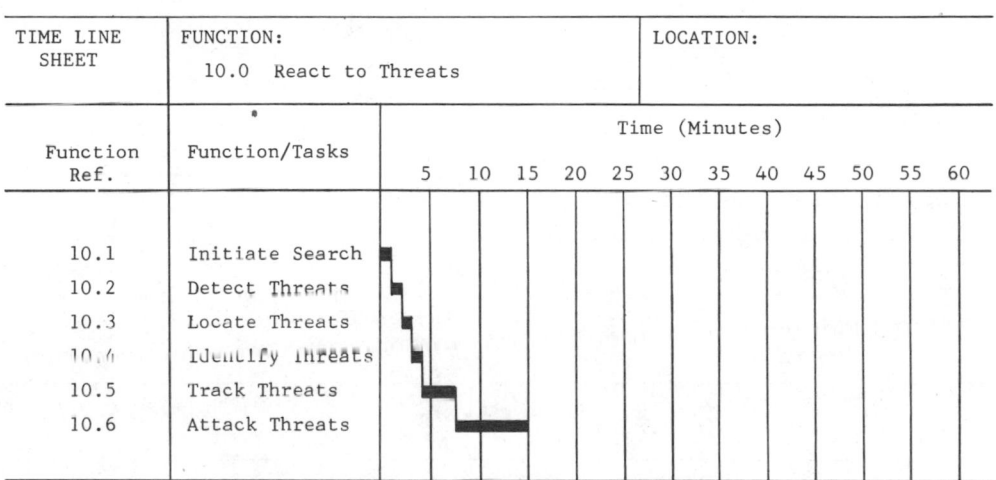

### OPERATOR TASK SCENARIO

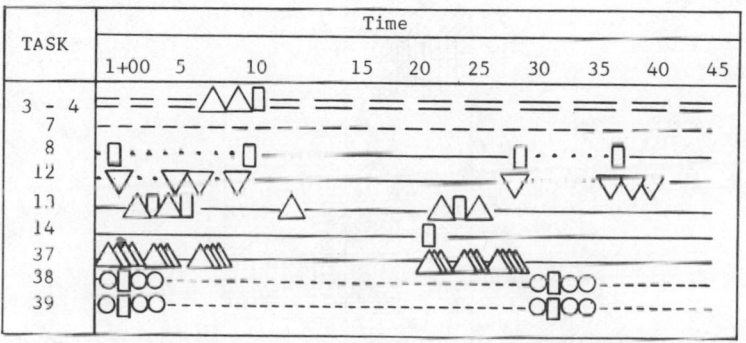

LEGEND:

△ Manipulation of Control or Mechanical Feature

▽ Manual Recording of Data: Write, Plot

○ Visual Checking, Observing, Discrete Indications

▯ Decision Making, Advising

**Example: Time Line (Top Level)**

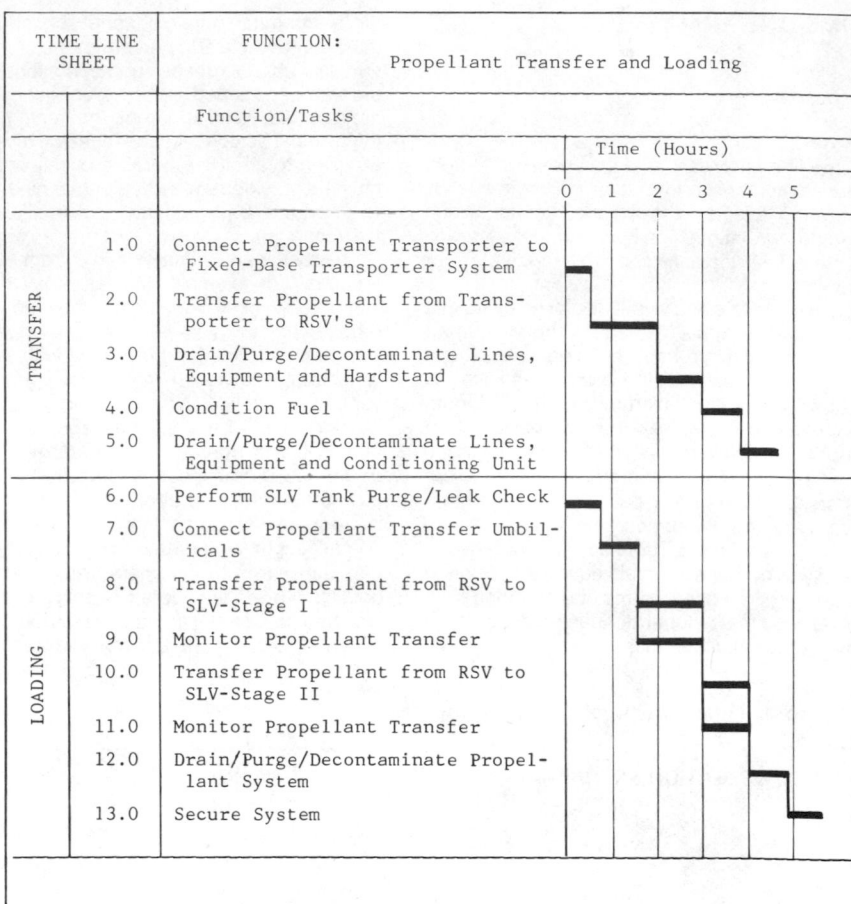

**Example: Time Line (Detailed Level)**

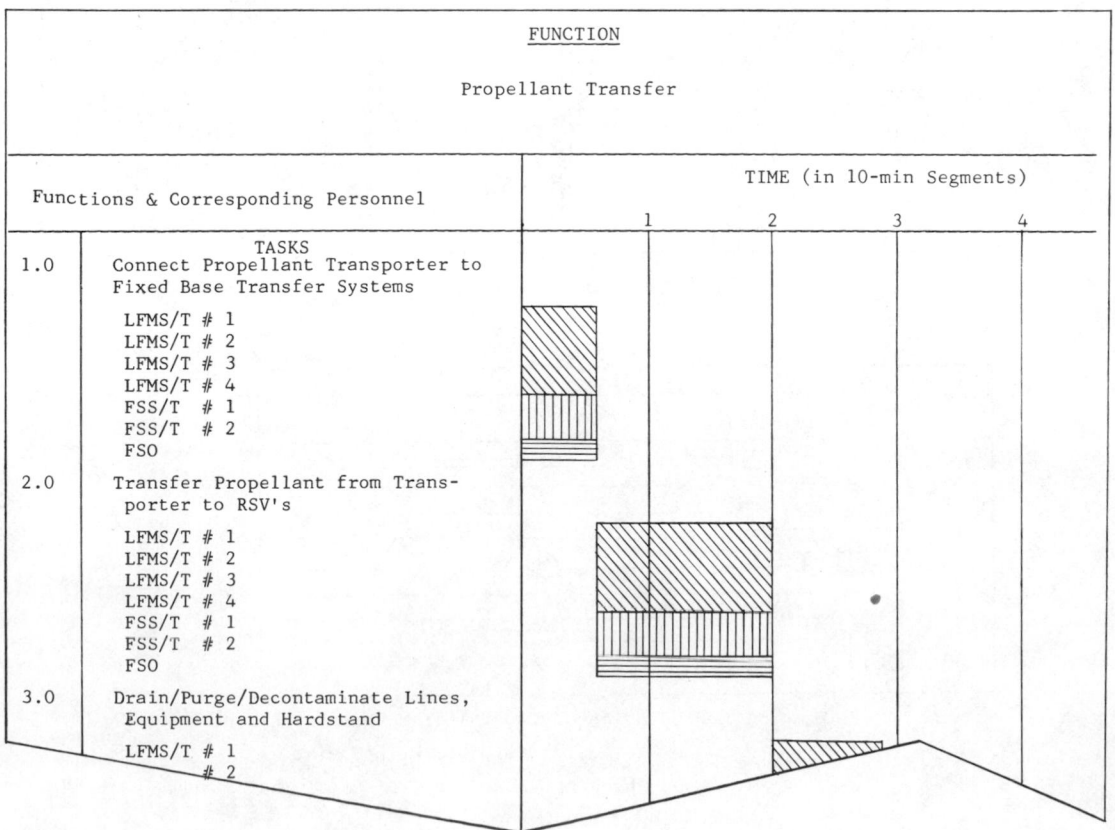

**Example: Time-Line–Workload Analysis Profile**

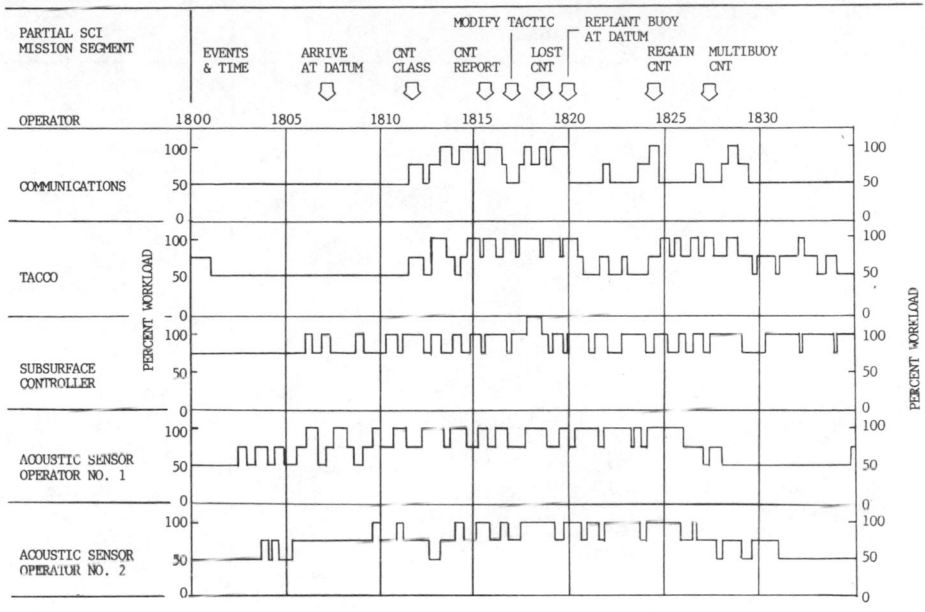

Example:  Time-Line/Workload Analysis Profile

## LINK ANALYSIS

### Link-Analysis Procedure

Link analysis is a technique for developing a best arrangement of components on a control panel, functional elements within a work station, elements and traffic flow within a work area, or humans and machines in a system. This technique is used only after decisions have been made regarding what items are to be on the panel, what equipment will be used in the system, how many people will be used, etc. Thus, link analysis is used to place the system components, be they items on a panel or people and machines in a workroom, in the "best arrangement" on the basis of criteria (such as minimum distance between workers or fewest movements between workers and equipment) important for the system under evaluation.

The term "link" as used here refers to any connection between a human and a machine or between one person and another. If one person must talk or physically contact another (i.e., hand that person a message), this is represented by a link. If a person must see a display or reach a control on a machine, he or she has a link to that machine. Ordinarily, any links between machines can be neglected unless the link possesses some quality which might cause the system to operate inefficiently (i.e., excessive length of a hardwire connection between two machines might result in high line loss in power). In these cases, links between machines are also included in the link analysis.

A typical example of a link analysis for a system involving four operators and four pieces of equipment is shown below.

**STEP 1**   Draw a circle for every person in the system and label it with a code number:

①

Draw a square for every item of equipment and label it with a code letter:

Ⓐ

**STEP 2**   Determine the type of link between each of the equipments and operators, equipments and equipments, and operators and operators. The different link types should be coded, such as:

——————— control links
——————— visual links
· · · · · talk links

The three codes are fairly standard for link analysis use. Other codes required can be made up by the individual doing the analysis.

**STEP 3**   Establish the "link value" for each link. The link value is based on two factors: (1) the importance of the link in accomplishing the mission assigned to the system under evaluation and (2) the number of times (frequency) the link is used in completing the mission. A number, between 1 and 3, is assigned for both frequency and importance, with the higher number representing maximum importance and frequency. As an example, if a particular control movement on the panel was very critical in fulfilling the equipment use (3) and it was frequently activated during the equipment use (3), the link value for that particular control link would be 9 (3 × 3). If, on the other hand, a link was infrequently used (1) but was critical when it was needed (3), the link value for that particular link would be 3 (1 × 3). Thus, the link value

is found by multiplying the frequency rating by the importance rating.

Selection of the frequency and importance ratings are usually based on past experience, either of the individual doing the analysis or someone familiar with similar systems already in use. In addition, the engineer responsible for the equipment design should be of assistance.

**STEP 4**   Prepare an analysis chart (see illustration on page 968) of the link values established for the system under evaluation. For each operator, show all the links (with each link value) associated with that operator. Do the same for each piece of equipment. For each operator and equipment item, add up the total of the link values and write this number to the right side. This provides an idea of the priority of equipment use and the operators most active in the system operation.

**STEP 5**   Prepare a schematic diagram (or series of diagrams) of possible arrangements. It is preferable to make this schematic to scale by cutting components out of paper and laying them on a scaled drawing of the space available (panel face, work-space floor plan, etc.). Starting with the operator or machine with the highest total link value, place the remaining components around it, moving them as necessary to minimize link crossings and shorten links, especially those with high link values. If conflicts occur between links, it may be necessary to reassess the original link values. The evaluation and rearrangement continue until the "best fit" solution is obtained. (See the final layout of the sample system on the bottom of page 968.) It should be emphasized that additional changes may be required in system layout once full-scale mockups or early hardware make actual system layout and evaluation possible.

OPERATORS                    EQUIPMENTS

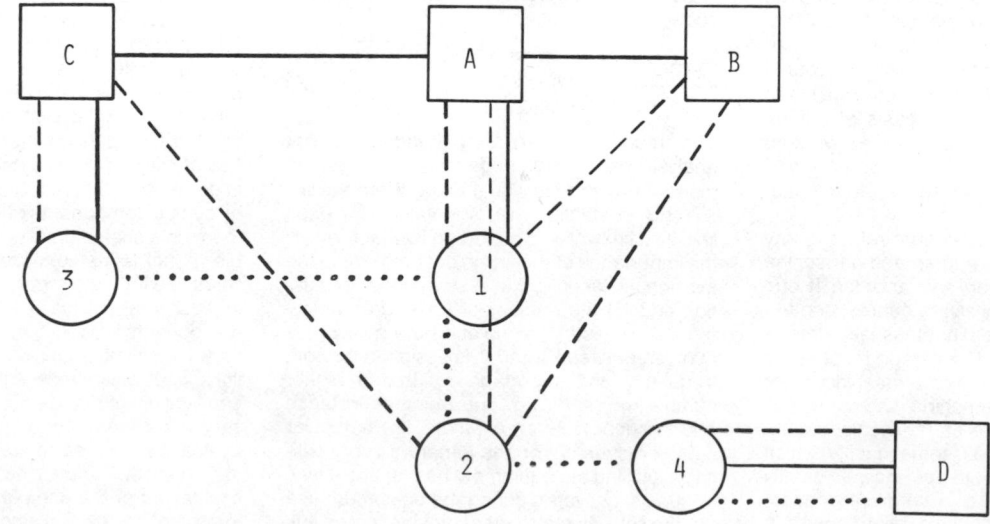

Sample Analysis Chart for Link Values

Schematic of Final System Layout

**Example: Use of Link-Analysis Technique to Evaluate Proposed Physical Layout of Components and People**

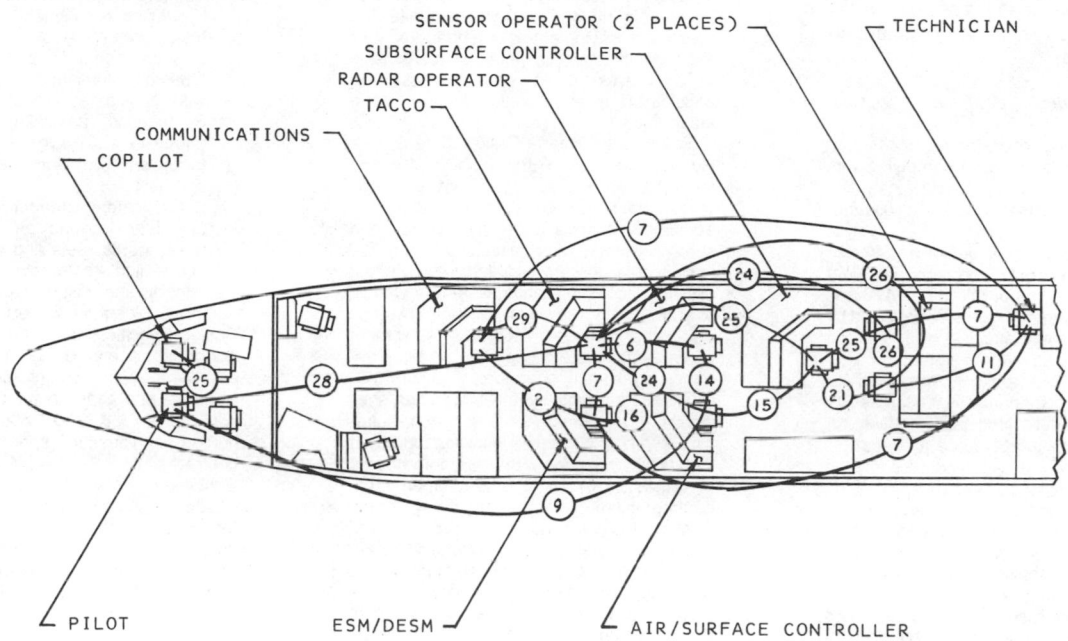

NUMBER IN CIRCLE PRESENTS RELATIVE CRITICALITY X FREQUENCY
OF COMMUNICATIONS FOR A GIVEN FUNCTION/TIME PERIOD

## Link Analysis Applied to Control Panels[17]

Link analysis may be conducted on preliminary drawings of a proposed control panel layout by imagining the operational steps an operator would go through (from the point of turning on power to shutting down the panel or system). An effective approach is to take a pencil and draw lines representing the path the operator's hand would take and/or the points where the operator's eye must observe a visual display. Each step is numbered in order. As the first sketch illustrates, one can quickly observe some unnecessary steps and repetitious pathways. Although one can make an alternative drawing based on the original link analysis, it is suggested that a more efficient method is to "cut out" the panel components and lay these out on a basic panel drawing (i.e., one that has only the edges shown). In this way, several alternatives can be tried before a second link analysis is performed to see whether the new layout provides an improvement.

During the link analysis one should also remain sensitive to the quality (understandability) of proposed panel component labels; i.e., as one proceeds through the link steps, it should become obvious that similar label nomenclature sometimes is confusing or that abbreviations are not appropriate (because one finds oneself wondering what an abbreviation means).

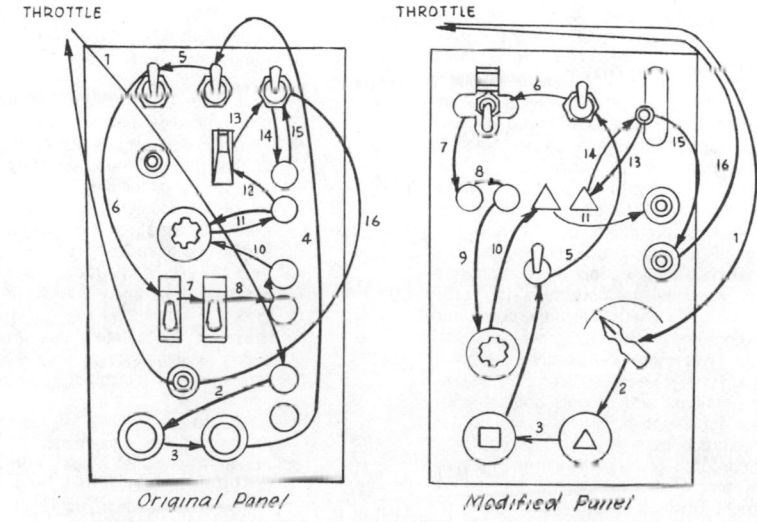

Original Panel     Modified Panel

[17]H. Older, "A Procedure for Representing Results of Human Engineering Studies," paper presented at the Naval Research Human Engineering Conference, October 1953.

## HUMAN ENGINEERING IN DESIGN

The following are excerpts from MIL-H-46855, the human engineering specification used by the U.S. military services. These excerpts pertain specifically to the human engineering effort during design.

*Human Engineering in Equipment Detail Design*
During detail design of equipment, the human engineering inputs, made in complying with the analysis requirements of previous paragraphs (re: system analysis), as well as other appropriate human engineering inputs, shall be converted into detail equipment design features. Design of the equipment shall meet the applicable criteria of MIL-STD-1472 and other human engineering criteria specified by the contract. Human engineering provisions in the equipment shall be evaluated for adequacy during design reviews. Personnel assigned human engineering responsibilities by the contractor shall participate in design reviews and engineering change proposal reviews of equipment end items to be operated or maintained by man. Human engineering requirements during equipment detail design are specified in the following paragraphs.

*Studies, Experiments and Laboratory Tests* —The contractor shall conduct experiments, laboratory tests (including dynamic simulation), and studies required to resolve human engineering and life support problems specific to the system. Human engineering and life support problem areas shall be brought to the attention of the procuring activity, and shall include the estimated effect on the system if the problem is not studied and resolved. These experiments, laboratory tests and studies shall be accomplished in a timely manner, i.e., such that the results may be incorporated in equipment design. The performance of any major study effort shall require approval by the procuring activity.

Mockups and Models—At the earliest practical point in the development program and well before fabrication of system prototypes, full-scale three-dimensional mockups of equipment involving critical human performance (such as an aircrew compartment, maintenance work shelter, or a command control console) shall be constructed. The proposed Human Engineering Program Plan shall specify mockups requiring procuring activity approval and modification to reflect changes. The workmanship shall be no more elaborate than is essential to determine the adequacy of size, shape, arrangement, and panel content of the equipment for use by man. The most inexpensive materials practical shall be used for fabrication. These mockups and models shall provide a basis for resolving access, workspace and related human engineering problems, and incorporating these solutions into system design. In those design areas where equipment involves critical human performance and where human performance measurements are necessary, functional mockups shall be provided, subject to prior approval by the procuring activity. The mockups shall be available for inspection as determined by the procuring activity. Upon approval by the procuring activity, scale models may be substituted for mockups. Disposition of mockups and models, after they have served the purposes of the contract, shall be as directed by the procuring activity.

Dynamic Simulation—Dynamic simulation techniques shall be utilized as a human engineering design tool when necessary for the detail design of equipment requiring critical human performance. Consideration shall be given to use of various models for the human operator, as well as man-in-the-loop simulation. While the simulation equipment is intended for use as a design tool, its potential relationship to, or use as, training equipment shall be considered in any plan for dynamic simulation.

Equipment Detail Design Drawings— Human engineering principles and criteria shall be applied to equipment drawings during detail design to assure that the equipment can be efficiently, reliably and safely operated and maintained. The following drawings are included: panel layout drawings, communication system drawings, overall layout drawings, control drawings and other drawings depicting equipment important to system operation and maintenance by human operators. The approval of these drawings by the contractor shall signify that human engineering requirements are incorporated thereon and that the design complies with applicable criteria of MIL-STD-1472 and other human engineering criteria specified by the contract.

Work Environment, Crew Stations and Facilities Design—Human engineering principles and criteria shall be applied to detail design of work environments, crew stations and facilities to be used by man in the system. The approval of drawings, specifications and other documentation of work environment, crew stations and facilities by the contractor shall signify that human engineering requirements are incorporated thereon and that the design complies with applicable criteria of MIL-STD-1472 and other human engineering criteria specified by the contract. Design of work environment, crew stations and facilities which affect human performance, under normal, unusual and emergency conditions, shall consider at least the following where applicable:

a. Atmospheric conditions, such as composition, volume, pressure and control for decompression, temperature, humidity and air flow.
b. Weather and climate aspects, such as hail, snow, mud, arctic, desert and tropic conditions.
c. Range of accelerative forces, positive and negative, including linear, angular and radial.
d. Acoustic noise (steady state and impulse), vibration, and impact forces.
e. Provision for human performance during weightlessness.
f. Provision for minimizing disorientation.
g. Adequate space for man, his movement, and his equipment.
h. Adequate physical, visual, and auditory links between men and men, and men and their equipment, including eye position in relation to display surfaces, control and external visual areas.
i. Safe and efficient walkways, stairways, platforms and inclines.
j. Provisions for minimizing psycho/physiological stresses.
k. Provisions to minimize physical or emotional fatigue, or fatigue due to work-rest cycles.
l. Effects of clothing and personal equipment, such as full and partial pressure suits, fuel handler suits, body armor, polar clothing, and temperature regulated clothing.
m. Equipment handling provisions, including remote handling provisions and tools when material and environment require them.
n. Protection from chemical, biological, toxicological, radiological, electrical and electromagnetic hazards.
o. Optimum illumination commensurate with anticipated visual tasks.
p. Sustenance and storage requirements (i.e., oxygen, water and food), and provision for refuse management.
q. Crew safety protective restraints (shoulder, lap and leg restraint systems, inertia reels and similar items) in relation to mission phase and control and display utilization.

Human Engineering in Performance and Design Specifications—The provisions of performance and design specifications, prepared by the contractor, shall conform to applicable human engineering criteria of MIL-STD-1472 and other human engineering specified by the contract.

*Equipment Procedure Development*— Based upon the human performance functions and tasks identified by human engineering analyses, the contractor shall apply human engineering principles and criteria to the development of procedures for operating, maintaining or otherwise using the system equipment. This effort shall be accomplished to assure that the human functions and tasks identified through human engineering analysis are organized and sequenced for efficiency, safety and reliability and to assure that the results of this effort shall be reflected in the development of training and technical publications. The approval of these publications by the contractor shall signify that the human engineering requirements are incorporated therein.

*Human Engineering in Test and Evaluation*—The contractor shall establish and conduct a test and evaluation program to: (1) assure fulfillment of applicable requirements herein; (2) demonstrate conformance of system, equipment and facility design to human engineering design criteria; (3) confirm compliance with performance requirements where man is a performance determinant; (4) secure quantitative measures of system performance which are a function of man-machine interaction; and (5) determine whether undesirable design or procedural features have been introduced. (The fact that these functions may occur at various stages in system or equipment development shall not preclude final human engineering verification of the complete system. Both operator and maintenance tasks shall be performed as described in approved test plans during the final system test.)

Planning—Human engineering testing shall be incorporated into the test and evaluation program and shall be integrated into engineering design tests, contractor demonstrations, R & D acceptance tests and other major development tests. Compliance with human engineering requirements shall be tested as early as possible. Human engineering findings from early testing shall be used in planning and conducting later tests.

Implementation—The human engineering test and evaluation program, contained in approved test plans, shall be implemented by the contractor. Test documentation (e.g., checklists, data sheets, questionnaires, schedules, operating procedures, test procedures) shall be available at the test site. Human engineering portions of all tests shall include, where applicable, the following:

a. A simulation (or actual conduct where possible) of mission or work cycle.

b. Tests in which human participation is critical with respect to speed, accuracy, reliability or cost.

c. A representative sample of noncritical scheduled and unscheduled maintenance tasks.

d. Proposed job aids.

e. Utilization of personnel who are representative of the range of the intended military user population in terms of skills, size, and strength and wearing suitable military garments and equipment which are appropriate to the tasks, and approved by the procuring activity.

f. Collection of task performance data.

g. Identification of discrepancies between required and obtained task performance.

h. Criteria for the acceptable performance of the test.

Failure Analysis—All failures occurring during, or as a result of test and evaluation shall be subjected to a human engineering review to differentiate between failures due to equipment alone, man-equipment incompatibilities and those due to human error. The procuring activity shall be notified of design deficiencies which contribute to human error.

Cognizance and Coordination—The human engineering program shall be coordinated with maintainability, system safety, reliability, personnel, training and other related programs, and shall be integrated into the total system program. The human engineering portion of any analysis, design or test and evaluation program shall be conducted under the direct cognizance of personnel assigned human engineering responsibility by the contractor.

## HUMAN ENGINEERING SURVEILLANCE DURING THE DESIGN PROCESS

It does little good to express a resolve to think about the human factor during the design process or to pass out human engineering design handbooks. Experience has shown that the designer has too many other factors to think about and thus will probably forget about the human element until decisions are made and constraints are designed into the system, equipment, or product. Thus, whether a human factors specialist is assigned as a "watchdog" or whether some member of the design staff is given the human engineering responsibilities, constant surveillance is required. The following suggestions are offered to help designers plan a reasonably effective program for monitoring the human engineering aspects of design:

1. Identify specific design efforts which are expected to be accomplished during the program or project and which

appear to have potential human operator, maintainer, or user interface features (they will contain displays, controls, handles, or fasteners; they must be manually lifted, moved, or carried; they probably will produce noise, vibration, heat, radiation, or toxic fumes; etc.).

2. Identify the designer who will be responsible for each design and determine the schedule for that design to begin.

3. Make initial contact with each designer and review the human factors concerns with each one. Discuss human engineering objectives and concepts. Set up a schedule for repeat contacts at regular intervals to review the progress of the design and offer to consult any time the designer has a question about human factors. Above all, be available and point out that you are there to help, not criticize.

4. Review all top-level drawings and from these identify lower-level drawings that have pertinent human engineering features. Review these quickly so that the drawing review cycle is not delayed. Prepare a written evaluation with recommendations. Then discuss the recommendations directly with the designer before passing the drawing on to the next person. Document the results of the discussion, indicating whether your recommendations are accepted or rejected. If rejected, document the reason.[18]

5. Participate in or observe pertinent engineering mockup evaluations and laboratory tests.

6. Participate in vendor product evaluations and assist in component selection.

7. Participate in final drawing-release review, insisting on sign-off responsibility before drawings can be released to manufacturing.

8. Participate in customer design reviews.

9. Monitor factory production and assembly of pertinent parts of the system. Last-minute "glitches" often appear in the assembly shop, especially with respect to maintenance-type tasks (inaccessibility of fasteners, connection reversal features, etc.).

10. Participate in prototype testing. Although test engineers are skilled at detecting, analyzing, and reporting hardware failures, they generally do not recognize design-induced human failures that can be caught during this type of testing.

## SUGGESTED STEPS TO ENSURE PROPER HUMAN ENGINEERING IN PRODUCT AND SYSTEM DESIGN

1. Review the mission, purpose, and use scenario for the proposed product or

[18]Documentation of human factors during design is becoming more and more urgent as legal aspects of design induced injuries place the burden of safety on the designer as well as the manufacturer.

system until you fully understand everything possible about the basis for the product or system and the conditions under which it is to be used.

2. Where practical, engage the services of a qualified human engineering specialist to assist and advise throughout the development of the proposed design and its final production. When this is impractical, consider the following steps.

3. Acquire and/or develop a "human engineering design checklist" to be used throughout the design development and production cycle. Although general checklist examples can be found (some of which are provided in this chapter), take time to create a checklist that is tailored to the nature of your proposed design. This checklist typically requires modification as the design development progresses since, as one comes to know more about the design features, additional user-hardware interface features generate additional human engineering questions that must be addressed and monitored throughout the design cycle.

4. Acquire appropriate human engineering design guides and references. If a qualified human factors engineering specialist is available, he or she probably will already have these. If not, obtain copies of at least the following:

a. MIL-STD-1472B, *Human Engineering Design Criteria for Military Systems, Equipment and Facilities* (for products or systems oriented toward the military).

b. Van Cott and Kincaide, *Human Engineering Guide to Equipment Design,* Government Printing Office, Washington.

c. J. F. Parker and V. R. West (eds.), 2d ed., *Bioastronautics Data Book,* National Aeronautics and Space Administration, Washington, 1973.

5. Review initial design concept ideas (either your own or those of subordinates who will be responsible for each portion of the design) to make sure that all potential user interface aspects of the proposed *design concept* have been properly identified and considered in each idea. The purpose of this step is to avoid establishing constraints to good human engineering practice.

6. Using the above human engineering checklist, monitor each design activity as it progresses from preliminary through detailed design steps, making sure that the human factors are kept constantly in mind and that good human engineering practice, principles, and criteria are being considered during each step of the design process. Keep a running record of whether human engineering principles and criteria are being applied and, if compromises are being made, of how and why these are being introduced in the design. This record may be extremely important, for example, if the military customer requires evidence that you have not ignored human

factors and/or justification for abrogating some human engineering principle or if a customer brings suit against you, and you need evidence to prove that you did the best you could to prevent misuse of the product, i.e., that you took all practicable means to minimize the probability of misuse or potential hazard to the customer.

7. Use mockups to "test" the efficacy of all user-hardware interface designs, using "real people" as subjects. Examine and evaluate the mockup-operator interfaces in terms of human performance efficiency (time, error, inconvenience, comfort, inadvertent hazard potentials, etc.). Record these observations and seek appropriate design modifications. Modify the mockup and reevaluate.

8. Perform experiments when necessary to establish design criteria where previously cited reference guides do not provide adequate information for design decisions. This may require development of special, dynamic, real-time simulations of procedural and environmental conditions.

9. Whenever possible, fabricate a hardware prototype and evaluate this under real-life conditions (using typical expected user subjects as well as expert operators). Obtain quantitative performance measures of the total user-hardware operation to prove that the combined human-machine operation is satisfactory. Make sure that all deficiencies are fed back into the design cycle and that appropriate design modifications are made. Retest if necessary.

10. Critically review final production drawings to make sure they are correct before being released for final production fabrication and assembly.

11. Identify critical areas where lack of proper production quality control might result in poor user-hardware interface results because of "sloppy" manufacturing or assembly procedures. Take steps to establish necessary procedural control and inspection to preclude mistakes in the factory.

12. Perform production (field) tests of production hardware before approving it for final delivery to the customer. These tests should include "man-in-the-loop" exercising, not merely visual inspection and/or test of the hardware and software components alone.

### Typical Methods for Human Engineering Evaluation of System and Component Design

| Method | Purpose | Typical Use |
|---|---|---|
| Worker and consumer interrogation | To define general product user needs | When a system or product is to be redesigned or new ones are to be developed, it is desirable to ask people who are using a similar product or system how they feel about it in terms of adequacy to do the job, ease with which the job can be done, and/or special problems that occur with the present hardware, facility, or tools. |
| Human-machine operational observation | To define general, dynamic, and environmental factors associated with product use | Although interrogation of the user may include demonstration, it is advantageous to observe an operation covertly over an extended period of time in order to have a more objective assessment of problems the operator may have. |
| Personal operation experience | To provide the designer with a personal "feel" for the problems a worker may have identified or may have observed from a distance | When practicable, it is recommended that the designer actually operate or use a product similar to the one he or she intends to redesign and/or create in order to experience at firsthand the problems or needs which a worker has pointed out or which the designer may have observed from a distance. |
| Time and motion study | To measure task performance against a time base and thus identify critical product or task conditions | Especially useful for systems in which worker output is a significant factor, i.e., in which increase in rate of production is a primary objective of the system. The method provides quantitative information that helps establish priorities for design and procedural improvement, either through redesign and/or through new design. |
| Preliminary design review | To evaluate the general, overall probability that a design will meet original operational objectives | Preliminary design drawings provide the first general picture of what a system or product will look like and how it will probably work. It is especially important at this stage to examine the product-user relationships to make sure that these are compatible with operator, maintainer, and user characteristics and limitations, prior to spending too much time with detailed design efforts. |
| Mockup evaluation | To evaluate both general and specific three-dimensional relationships between various elements of a proposed system and/or specific relationships between a specific part of the product or system and its user | Minature-scale mockups provide an overall view of a system that could not be viewed effectively in full scale. Full-scale mockups provide one-to-one evaluation of direct product-operator physical relationships and quasi-operating conditions. |

*(continued)*

**Typical Methods for Human Engineering Evaluation of System and Component Design** *(continued)*

| Method | Purpose | Typical Use |
|---|---|---|
| Simulation | To investigate and verify dynamic aspects of a proposed design prior to final commitment. To provide an opportunity to explore alternatives both during predesign and during detailed design | A proposed design concept may be examined and/or tested at several levels prior to committing it to preparation of preliminary or detailed design. It is important to include the human user during such simulations since the ultimate effectiveness of the product includes the typical user input, both intentional and unintentional. Simulation should be in "real time" and should include representative environmental, dynamic, and personnel elements, depending upon the stage of design at which the simulation effort is being performed. Evaluative methods should include quantitative measures of system or product-operator performance. |
| Detailed design drawing review | To verify human-machine and product-user compliance with good human engineering practice | Methodical analysis of detailed design drawings prior to their release to production is vital to prevent typical oversights of human-machine interface problems. This is critical because of the probability that changes may have been made since original human engineering inputs were made at the predesign stage. |
| Prototype testing | To verify the operating and maintenance effectiveness of the system or product under actual operating conditions | Whenever practicable, a prototype of the product or system should be tested prior to committing it to production. Such tests should include use of operator, maintainer, and user personnel who are representative of the final user population. Although initial tests may be made by special test personnel in order to ensure safety of operation, the ultimate question is whether typical users can and will operate the system or use the product as planned. Quantitative measurement of human-machine performance should be accomplished whenever the complexity, integrity, or safety of the system is critical. |
| Field evaluation | To provide a continuing assessment of system or product effectiveness under typical use and environmental conditions, in order that service requests can be effectively met, design modification requirements can be defined, and/or new concepts can be anticipated | Systematic follow-up using formal data and information retrieval methods provides a means for a manufacturer to maintain adequate customer relations and to develop insight useful for expanding the product's capability. To provide future human engineering guidelines, useful human factors data should be collected at the same time that pure hardware data are being collected. |
| Human factors experiments | To support current and/or future design requirements | Special human factors experiments may be required at various times, either during a product development and/or independently. Cognizance of applicable human factors research information is required at the beginning of any development program to assist in defining the human role in system operation. When research data do not exist, special studies may be required to define human capabilities with respect to a specific, proposed application. Whether to perform such research depends on the criticality of the question and/or the time available to perform the research. Such research should have as broad an application as possible, as long as it provides the necessary answers to the immediate problem; i.e., when the answer is too specific, the cost is increased over the long run, since the research may have to be repeated for the next program. |

## PRODUCT EVALUATION CHECKLIST

The following is a checklist for evaluating whether a product is compatible with user behavioral expectancies:

1. Is it obvious what the product is and what it is to be used for?
2. Is it obvious how the product is to be used?
3. Is the product simple to use, i.e., is it simple to prepare for use, to begin to use or operate, to continue to operate, to stop using or operating, and to place in a stored or nonoperational condition?
4. Are there hidden hazards to the operator or user, i.e., hazards that cannot be readily observed when looking at the product?
5. Will use of the product create hazards for others not involved in its use?
6. Could the product be operated and misused by someone who should not be using it, such as a child?
7. Could the product be used for some purpose for which it was not intended?
8. Should the product be used only under certain conditions, and is this immediately obvious to the potential user?
9. Are potential product failures or conditions under which the product should not be operated or used identified for the user in time to avoid misuse?
10. Would use by someone who is fatigued, under the influence of drugs or alcohol, or otherwise incapacitated be hazardous to the user or others—and if so, does the design provide appropriate use-prevention features?
11. Are there potential hazards associated with the product when it is not being used, when it is stored, or when it is otherwise unattended—and if so, does the design provide appropriate use-prevention features?
12. Is it critical that the product be serviced or maintained in a specific manner—and if so, are the servicing requirements made clear to the user?
13. Is the product prone to damage or failure—and if so, does it contain a fail-safe mechanism?
14. Is the product easy and convenient to service and maintain without special training—and if not, is this clear to the user?
15. Can the product be used when it is not functioning properly without creating a hazard to the user?
16. Is the product, in any way, outside the normal user's physical, mental, or normal behavioral capacities (e.g., in terms of size, weight, or operation)?

## CHECKLISTS

### Preliminary Human Engineering Checklist for Initial Hardware Selection and Design Analysis

The checklist below provides a starting point for gross evaluation, but it should be expanded for detailed design evaluation. Evaluation designations are as follows: S—satisfactory; C—compromise but acceptable; and U—unsatisfactory. Various other designations are also possible to include evaluative indications of temporary status.

A. EQUIPMENT OPERATION
  1. CONSOLE SHAPE/SIZE
     a. Desk height, area
     b. Control reach
     c. Display view
     d. Body, limb clearance

  2. PANEL LOCATION
     a. Frequency of use
     b. Sequence of use
     c. Emergency response
     d. Multi-operator use

  3. PANEL LAYOUT
     a. Functional grouping
     b. Sequential organization
     c. Identification
     d. Spacing for clearance

  4. DISPLAYS
     a. Functional compatibility for intended purposes
     b. Intelligibility of information content
     c. Control interaction
     d. Legibility; figures, pointers, scales
     e. Visibility; illumination, parallax
     f. Location
     g. Identification

  5. CONTROLS
     a. Functional compatibility for intended purpose
     b. Location, motion, excursion and force
     c. Display interaction
     d. Spacing, clearance, size
     e. Identification

B. ASSEMBLY - SERVICES - MAINTENANCE
  1. INSTALLATION, SERVICE, & MAINT ACCESSIBILITY
     a. Location, size of openings
     b. Covers, fastening/removal
     c. Identification

  2. EQUIPMENT HANDLING/TRANSPORT
     a. Size/shape/weight/balance
     b. Handling clearance
     c. Handling aids
     d. Instructions/labels/warnings

  3. CHASSIS LAYOUT, PACKAGING
     a. Ease of handling.
     b. Access to components for test, (component) replacement
     c. Identification
     d. Hazard/Damage protection

  4. CABLES/LINES/CONNECTIONS
     a. Ease and security of assembly-disassembly
     b. Connection error
     c. Identification
     d. Access, test, trouble-shooting replacement

C. SYSTEM SAFETY
  1. PERSONNEL HAZARDS
     a. Shock
     b. Burns: direct, chemical
     c. Hearing damage
     d. Tripping/falling
     e. Pinching
     f. Cutting
     g. Bumping

  2. EQUIPMENT DAMAGE
     a. Electrical overload, short, ground
     b. Mechanical overload, strip, bend, rupture, break
     c. Explosion/fire

D. GENERAL
  1. LABELS/MARKING
     a. Intelligibility
     b. Legibility
     c. Location, spacing
     d. Permanence

  2. EQUIPMENT FINISH
     a. Color
     b. Texture
     c. Reflectivity

  3. STORAGE
     a. Location
     b. Volume
     c. Material accessibility, security

  4. WORK AREA ILLUMINATION
     a. Light level; range, control
     b. Distribution, contrast
     c. Color

(Evaluation columns: S C U)

**Example: Human Engineering Design Checklist**

Although each human engineering checklist should be tailored to fit each particular program, the accompanying example may provide a useful point of departure for the individual who is attempting to construct a checklist for the first time. This example is provided courtesy of the Human Factors Group at the Electric Boat Division of General Dynamics.

---

## HUMAN FACTORS CHECKLIST

## FOR

## DESIGN ENGINEERS

## GENERAL DYNAMICS

*Quincy Division*

---

TABLE OF CONTENTS

USES

Use this checklist to perform a static evaluation to ensure that the constraints and recommendations of human engineering are in the design of components and total system development.

This checklist will assist engineering personnel in designing equipment in accordance with human capabilities and limitations. Used early in a design program, it will ensure that human engineering principles will be incorporated, at relatively low cost, in ultimate ship design. Used later in the program, the checklist serves as a final check that human engineering principles are in the design documentation.

The following procedure is recommended when evaluating components and systems with the checklist:

Review the table of contents to find the sections to be used for the check.

Compare the remaining checklist items with the design characteristics and check minimum acceptable, marginal acceptable, or reject, opposite each item.

Itemize the number of reject items and explain them in a memorandum to human factors personnel. Discrepancies not resolved in a conference between design engineers and human factors personnel will be submitted to the customer through Contracts for clarification.

**HUMAN FACTORS CHECKLIST**

PASS | MARGINAL | REJECT

1. *SAFETY*
   a. Conspicuous placards are mounted adjacent to high voltage, extremely cold, very hot, etc, equipment.
   b. A hazard-alerting device is provided to warn personnel of impending or existing hazards; e.g., fire, presence of combustible or asphyxiating gas, radiation, etc.
   c. A guard is provided on all moving parts of machinery and transmission equipment, including pulleys, belts, gears, and blades, etc., in which personnel may become injured or entangled.
   d. Self-locking or other foolproof devices are incorporated on elevating stands and work platforms to prevent accidental or inadvertent collapse.
   e. Some form of anchor or outrigger is employed on stands with high centers of gravity.
   f. Where applicable, the center of gravity of equipment is distinctly marked.
   g. Handrails are provided on platforms, stairs, and around floor openings, or wherever personnel may fall from an elevation.
   h. Safety bars or chains are attached across stair or step openings on a platform to prevent falling.
   i. Automatic shut-off devices are provided on fuel service equipment to prevent overflow and spillage.
   j. Portable hand-operated fire extinguishers are provided in areas where fire hazards exist or may be created.
   k. Emergency doors and exists are constructed so that they are readily accessible, unobstructed, and quick-opening; i.e., can be opened by a single motion of hand or foot.

1

l. Eye baths, showers, and other first-aid equipment are readily available in areas where toxic materials are handled.
m. Provision is made for neutralization or flushing of harmful materials spilled on equipment or personnel.
n. Areas of operation or maintenance where special protective clothing, tools, or equipment are necessary (such as insulated shoes, nonsparking tools, gloves, or suits) are specifically identified, and appropriate action taken to ensure the availability of such items concurrent with complete development of the hardware.
o. "No Step" markings are incorporated where applicable.
p. Weight capacity is indicated on stands, hoists, lifts, jacks, and similar weight-bearing equipment to prevent overloading.
q. Jacking and hoisting points are conspicuously and unambiguously identified.
r. Wiring is routed through plugs and connectors so that removal of a plug or connector does not expose hot leads.
s. Pipe lines carrying liquid, gas, steam, etc., are clearly and unambiguously labeled or coded as to contents, pressure, heat or cold, and any other hazardous properties.
t. Skid-proof flooring and stair or step treads are provided.
u. Clearance for fingers is provided in the design of telescopic steps or ladders.
v. Provisions should be made to prevent personnel from coming into contact with voltages in excess of 70 volts rms. Do not locate adjustment screws or other commonly worked-on parts near unprotected high voltages or hot parts.
w. Provide interlocks where potentials exceed 70 volts rms, with a means of bypassing for servicing with a proper warning indicator.
x. Provide voltage dividers with test points for measurement of voltages in excess of 100 volts.

y. Ground all external metal parts, control shafts, and bushings. Antenna or transmission line terminals should be at ground potential except with regard to the energy to be radiated.
z. Provide rotating antenna assemblies with local power safety switch at the antenna.
aa. Provide protection to personnel from imploding cathode ray tubes.
bb. Doors or hinged covers should have rounded corners and should have positive-acting hold-open devices. The length of projecting and over-hanging edges should be held to a minimum, and all edges and corners should be rounded.
cc. Design locking mechanisms for doors and drawers to prevent injury to the operator when the lock is released.
2. *AMBIENT ENVIRONMENT*
2.1 *Noise*
   a. No one is exposed to noise levels that exceed 150 db, no matter how much the noise level in the ear canal has been reduced.
   b. The amount of noise produced has been held at acceptable levels by careful acoustical design of new equipment, modification or redesign of present equipment, muffling of exhausts, etc.
   c. The amount of noise transmitted has been held at acceptable levels by increasing the distance between the work area and the source of noise, constructing barriers between the work area and the source of noise, sound-treating the work area to reduce vibration, placing equipment on vibration mounts, or other means.
   d. Where it is not possible to reduce noise levels through equipment design, personnel are provided with protective devices while working in high-intensity noise areas.
2.2 *Air-Conditioning*
   a. Air-conditioning is provided if the effective temperature exceeds 90 degrees Fahrenheit.

PASS MARGINAL REJECT

2.3 *Illumination*

a. Sharp gradients in illumination ratios (10 to 1 or greater) are avoided and gradients less than 8:1 surrounding displays should be avoided.

b. Local, direct lighting is provided for equipment which is not properly lighted by the general, diffuse lighting.

c. The intensity level and corresponding types of illumination for various visual tasks are in accord with Table IV, MIL-STD-803.

3. *WORK SPACE LAYOUT*

3.1 *Anthropometric Requirements*

a. The location, size, etc. of equipment is such that the equipment will be easily operated and maintained by at least the 5th to 95th percentile group of the Navy population.

b. Demands placed upon personnel in the performance of their duties lie within the following limits (it should be noted that these dimensions must be altered to allow for encumbrances).

Minimum overhead height for standing position:
73 inches

Maximum allowable overhead reach:
76 inches

Minimum height required for crawling:
31 inches

Maximum allowable depth of reach:
23 inches

Minimum dimension for passing body width:
20 inches

Maximum dimension for passing body thickness:
13 inches

c. Sufficient room to accommodate the hand is provided in the grasping of all handles.

d. Handles on cabinets and consoles are recessed, when practical, to eliminate projections on the cabinet surfaces.

e. All cabinets, consoles, and work surfaces requiring that an operator stand or sit in close proximity to their front surface contain a kick space four inches deep by four inches high at the base.

4

e. Arm rests are provided at all consoles. These rests are a part of the console or a part of the operator's chair.

f. Console arm supports provide at least 8 usable inches (preferably 12 inches) of resting surface projecting horizontally across the front of the console.

g. Arm rests integral with the back of the operator's chair are a minimum of two inches wide by 10 inches long.

h. If the operator must record data, a writing surface 12 inches in depth is provided.

i. Knee and foot room beneath the panel surfaces is provided, with minimum dimensions as follows: 26 inches high by 20 inches wide by 18 inches deep.

j. Desk tops, writing tables, and other work surfaces provided for seated operation are 30 inches above the floor.

k. Most important and frequently used displays near line of normal operator vision.

4. *CONSOLE AND PANEL DESIGN*

4.1 *Displays*

4.1.1 *PRINCIPLES*

a. The information displayed to each person who operates a piece of system equipment is limited to that information which is necessary to the specific actions or decisions requested of him.

b. Information is presented to the operator in directly usable form; that is, requirements for decoding, transpositioning, interpolation, etc., are minimized.

c. Displays are designed so that the failure of the display circuitry is immediately and readily apparent to the operator.

6

PASS MARGINAL REJECT

3.2 *Standing Operations*

a. Visual displays on vertical panels are mounted in an area no higher than 70 inches and no lower than 40 inches above the standing surface.

b. Precise reading indicators are placed in an area no higher than 64 inches and no lower than 48 inches above the standing surface.

c. Controls are mounted on vertical panels in an area no higher than 70 inches and no lower than 30 inches above the standing surface.

d. Precision controls, or those operated frequently, are located 40 to 55 inches from the standing surface.

e. Work benches and other work surfaces provided for standing operations are 32–38 inches above the floor.

f. Convenient work surfaces to support job instruction manuals, worksheets, etc., are provided where necessary for standing operators of control-display panels.

3.3 *Seated Operations*

a. Where continuous monitoring or control is required of a seated operator, controls and displays are mounted on a sloped console surface.

b. For normal seated operations, the slope of the control-display panel surface begins 30 inches from the floor, with the over-all console height not exceeding 48 inches, thereby allowing the operator's direct line of sight to extend beyond the console.

c. If the operator's direct line of sight is not required to extend beyond the console, the over-all console height does not exceed (but may extend) 65 inches from the floor.

d. If the over-all console height is between 48 and 65 inches, the upper panel surface is inclined from the vertical toward the operator.

5

d. Failure of display circuitry does not cause a failure in the equipment associated with the display.

e. Crucial visual checks identified by attention-getting devices (e.g., visual or aural signals).

f. Information presentation is by means of most suitable type of display (e.g., pictorial, numerical, etc.) for task being performed.

g. Probability of confusion among instruments is minimal.

h. Instruments used only for maintenance made inconspicuous to operators not performing maintenance.

i. Precision of information is not greater or less than required to meet job task requirements.

4.1.2 *LABELING*

a. Trade names and other irrelevant information deleted.

b. Easy to read under expected conditions of illumination.

c. Labels do not obscure other needed information.

d. Labels in capital, extended copy in lower case letters (11 point type; line of 14–28 picas).

e. Each control and display is identified as to function.

f. Labels appear either on or immediately adjacent to (preferably above) the controls and displays to be identified.

g. Labels are located to preclude association of a label with the wrong control or display.

h. The location of labels in relation to controls and displays is consistent on all system equipment.

i. Labels are brief.

j. Labels clearly indicate the function being displayed or controlled.

7

PASS  MARGINAL  REJECT

k. Highly similar names are avoided.

l. Abbreviations, where required, are common or meaningful, and conform with MIL-STD-12 and ANA Bulletin 261.

m. Abstract symbols (squares, Greek alphabet, etc.) are not used as labels. Common, meaningful symbols such as the percent sign, plus sign, etc., are acceptable.

n. Lettering on panels is black, color No. 37038, Federal Standard 595.

4.1.3 *CODING*

a. Applicable coding techniques have been selected from the following methods.
    a. Color
    b. Size
    c. Location
    d. Shape

b. Optimum use is made of coding techniques for identification of functionally related displays.

c. Optimum use is made of coding techniques for indication of the relationship between displays (e.g. all emergency switches painted red). NOTE: Items d through g refer to color coding of indicator lights.

d. Red is used to alert an operator that the system or any portion of the system is inoperative. Also used for fire protection, danger, and stop warnings.

e. Amber is used to caution an operator that a condition exists which is marginal but that the system can still operate.

f. Green is used to indicate that a unit or component is within tolerance, or a condition is satisfactory or safe.

PASS  MARGINAL  REJECT

g. White is used to indicate those system conditions that are not intended to provide a right or wrong implication (such as indications of alternative functions), or are indicative of transitory conditions where such indication does not imply success of operations.

h. Lights used to denote emergency conditions (personnel or equipment disaster), and only those indicating such conditions, are coded in flashing red.

i. The flash rate for flashing warning lights is between three and five flashes per second, with "on" time being approximately equal of "off" time.

j. Colors for coding controls are chosen from the following list.
| Red | 11105 | Amber | 13538 |
| Green | 14187 | White | 17875 |
| Black | | Gray | |

k. If it is considered imperative to relate a control to its corresponding display by means by color coding, the display and control are the same color.

l. If shape coding is selected as a coding scheme or combined with other codes for purposes of differentiating between controls, the application of the code is uniform throughout the system.

m. Control shapes, if used for coding purposes, are identifiable visually.

n. If blind operation of controls is a requirement, the shapes selected for coding are tactually discriminable.

o. Blue is used for covered electrical outlets and fuse boxes.

p. Flammable materials should be yellow.

q. Toxic and poisonous materials should be brown.

r. Glare shields, control columns, control wheels, control stick grips and plastic plate-edge lighted panels: black-colored.

8

9

s. Emergency exists and exit releases: orange yellow.

t. Instrument panels: medium gray.

4.1.4 *SCALES, DIALS, COUNTERS*

a. Numbers and letters are large enough for accurate reading at normal distance.

b. Reflected light does not create illusion warning is "ON" or obscure reading.

c. Where given, operating conditions always fall within a certain range on the scale; these areas are made readily identifiable by means of coding. Coding may be used to convey such information as desirable operating range, dangerous operating level, etc.

d. Reliance on color coding must be modified when other than white illumination is used.

e. Where scale color coding is used, the meaning of the colors conforms to paragraph 6.1.2 of MIL-STD-803.

f. Wherever possible, scales start at zero.

g. Scale graduations progress by one or five or 10 units, or decimal multiples thereof.

h. The increase in numerical progression reads clockwise, from left to right, or from the bottom up.

i. The number of minor or intermediate marks is not greater than nine.

j. Maximum contrast is used between scale face and markings.

k. On stationary scales, all numbers are upright.

l. On moving scales, all numbers are upright at the reading position.

m. The magnitude of the reading increases with a clockwise movement of the pointer.

n. Clockwise movement of a scale pointer results from clockwise movement of an associated rotary control, or movement forward, upward, or to the right of an associated lever or switch.

o. Where positive and negative values around a zero value are displayed, the zero is located at the 12 o'clock position (preferred) or the 9 o'clock position.

p. There is an obvious scale break between the two ends of a scale (not less than 1-½ divisions), except on multi-revolution instruments such as the clock.

q. For ease of monitoring a group of circular scaletype indicators, if a stable value of given operating conditions is present, the displays are arranged:
  1) in rows, with pointers normally aligned horizontally; i.e., the 9 o'clock position reflecting normal operating conditions; or
  2) in columns, with pointers normally aligned vertically; i.e., the 12 o'clock position reflecting normal operating conditions.

r. Where space is limited, numerals are placed inside of the graduation marks to avoid constriction of the scale.

s. Where space is not limited, the numbers are placed outside of the marks to avoid having the numbers covered by the pointer.

t. The pointer is located to the right of vertical scales, and at the bottom of horizontal scales.

u. For linear scales, numerals are placed on the side of the graduation marks away from the pointer to avoid having the numbers covered by the pointer.

v. The pointer or lubber line is at the 12 o'clock position for right-left directional information, or at the 9 o'clock position for up-down information.

w. If a scale is used for setting a given number, e.g., tuning a receiver to a desired wave-length, the unused portion of the dial face is covered, where advisable.

10

11

| | PASS | MARGINAL | REJECT |
|---|---|---|---|

x.  If the unused portion of a scale is covered, the open window is large enough to permit at least one numbered graduation to appear at each side of any setting.

y.  If a scale is used in tracking, or as a heading indicator, the entire dial face is exposed.

z.  Counters are used, where feasible, for presenting large ranges of quantitative data where continuous trend indication is not necessary and where quick, precise reading is required.

aa.  Numbers change by snap action rather than by continuous movement.

bb.  Counters are mounted as close to the panel surface as possible to maximize viewing angle and minimize parallax and shadows.

cc.  The height-to-width ratio of numerals for counter displays is 1:1, rather than 5:3 as recommended for dials and scales.

dd.  Numerals do not follow each other faster than about two per second if the observer is expected to read the numbers consecutively.

ee.  Counters used to indicate sequencing of equipment are designed to reset automatically upon completion of the sequence; manual provision for resetting also is provided.

ff.  Rotation of a counter reset knob is clockwise to increase the counter indication.

gg.  Printers are used when a permanent record of quantitative data is desired.

hh.  Plotters are used where a permanent record of continuous graphic data is desired.

4.1.5  *CATHODE-RAY TUBES*

a.  Wherever possible, the scope face is perpendicular to the operator's normal line of sight.

| | PASS | MARGINAL | REJECT |
|---|---|---|---|

b.  The ambient illumination in the CRT area is sufficiently high for other visual functions (reading instruments), but does not interfere with the visibility of the signals on the CRT display.

c.  Ambient room illumination will be minimized.

d.  Scopes are adequately shielded from the light when room illumination is sufficiently high for other visual tasks.

e.  The brightness of surfaces immediately adjacent to the scope falls in a range from screen brightness to 10 percent of screen brightness.

f.  Surfaces immediately adjacent to the scope have a dull matte finish.

g.  Shape conforms to configuration of presentation; PPI, round; A-scan, rectangular; etc.

h.  Grid markers available for interpolation.

i.  Adequate spacing used between range rings on polar displays (no less than ½U spacing).

j.  Suggested viewing distance 16U for polar grid observation.

4.1.6  *INDICATOR AND LEGEND LIGHTS*

a.  Legend lights are used primarily to display qualitative information that requires an immediate reaction by the operator or that his attention be called to an important system status.

b.  Legend lights required to denote personnel or equipment disaster, caution or impending danger, and master summation go/no-go conditions are discriminately larger than all other legend-type displays.

c.  Legend lights required to denote the conditions indicated in 3. above are brighter than other legend displays.

d.  Legend light lettering is visible and legible whether or not the display is energized (applies to single-legend displays only).

| | PASS | MARGINAL | REJECT |
|---|---|---|---|

e.  Spacing between adjacent edges of round, simple indicator lights is sufficient for unambiguous labeling and convenient bulb removal.

f.  Simple indicator lights used to denote emergency conditions are 1 inch in diameter and present a red flashing light.

g.  Simple indicator lights used to denote cautionary conditions are 1 inch in diameter and present a steady amber light.

h.  Simple indicator lights used to denote master summation conditions are 1 inch in diameter and present a steady red or green light.

i.  Word warning lights are used where possible to specify an emergency condition.

j.  All displays necessary to support an operator activity, or sequence of activities, are grouped together; e.g., subsystem grouping.

k.  Displays are arranged in relation to one another to reflect the sequence of use or the functional relations of the components they represent, in that order of preference.

l.  Distinct, functional areas set apart for purposes of ready identification are outlined by black lines approximately ⅛ inch wide.

m.  Master caution, master warning, and summation lights used to indicate the condition of the entire subsystem are set apart from the lights which show the status of the subsystem components.

n.  Indicators displaying critical functions are located within 30 degrees of the normal line of sight.

o.  Transilluminated indicators are several times as bright as the background against which they appear, but not so bright as to dazzle the operator.

p.  When transilluminated indicators are used under varied ambient illumination, a dimming control is provided.

| | PASS | MARGINAL | REJECT |
|---|---|---|---|

q.  If panels are to be used outdoors, provision is made to prevent reflected sunlight from making the indicators appear illuminated.

r.  Where many lights are located on a control panel, a master light test control is incorporated.

s.  Indicators which are not tested by a master light test control are designed for "press-to-test" bulb testing.

t.  Provision is made for bulb removal from the front of the display panel without the use of tools, or other equally rapid and convenient means.

u.  The filter caps and legend plates of indicator lights are physically coded or captive so as to preclude the possibility of a mismatch.

v.  Pushbuttons are used where a control or an array of controls is needed for momentary contact or for activating a locking circuit in a high frequency-of-use situation.

w.  Button surfaces are concave to fit the finger, or provide a high degree of frictional resistance to prevent slipping.

x.  Buttons provide "snap feel" or an audible click to indicate that the control has been activated.

y.  A channel or cover guard is provided when prevention of accidental activation is imperative.

z.  The minimum diameter for fingertip operation is at least ½ inch.

aa.  The minimum diameter for emergency controls which can be activated by thumb or heel of the hand is at least ¾ inch.

bb.  Button displacement is between ⅛ inch and ¾ inches.

cc.  Button resistance is between 10 ounces and 40 ounces.

| | PASS | MARGINAL | REJECT |
|---|---|---|---|

dd. Foot pushbuttons are used only in those cases where an operator is likely to have both hands occupied at the time the pushbutton requires activation.

ee. Under normal conditions, foot pushbuttons are designed for the toe operation (by the ball of the foot) rather than heel operation.

ff. Foot pushbuttons provide "snap feel" or an audible click to indicate they have been actuated.

gg. The button diameter is at least ½ inch.

hh. Displacement for normal operation is at least ½ inch.

ii. Displacement for actuation by heavy boots is at least 1 inch.

jj. Displacement for controls operated by ankle flexion only does not exceed 2-½ inches.

kk. Displacement for controls operated by leg movement does not exceed 4 inches.

ll. Button resistance when foot will not rest on control is at least 4 pounds.

mm. Button resistance will range from 10 to 20 pounds.

nn. Toggle switches are used for those control functions which require two discrete positions when space limitations are severe.

oo. In general, toggle switches are vertically oriented, with the up position representing ON, and the down position representing OFF.

pp. Where inadvertent operation is considered serious, those toggle switches susceptible to accidental activation are guarded (usually with a channel guard).

qq. Where prevention of accidental activation is of primary importance, a cover guard is used.

16

| | PASS | MARGINAL | REJECT |
|---|---|---|---|

rr. Control tip diameter is between ⅛ inch and 1 inch.

ss. Lever arm length is between ¾ inch and 2 inches.

tt. Minimum displacement between adjacent control positions is at least 30 degrees.

uu. Maximum displacement between adjacent control positions does not exceed 120 degrees.

vv. Switch resistance is between 10 ounces and 40 ounces.

**4.1.7 ROTARY SELECTOR SWITCH**

a. Rotary selector switches are used for discrete functions where three or more positions are required.

b. The number of positions incorporated into any one rotary switch does not exceed 24.

c. The pointer knob has a tapered tip, unless shape coding has been applied to the knobs.

d. Shape coding is utilized where a group of rotary switches, used for widely different functions, is placed on one panel and switch confusion is likely.

e. The position of the pointer knob relative to the scale minimizes parallax.

f. A moving pointer on a fixed scale is used, unless complete justification exists for a moving scale with a fixed index.

g. Maximum pointer width does not exceed 1 inch.

h. Displacement (between adjacent detents) for visual positioning is at least 15 degrees.

i. Displacement for non-visual positioning is at least 30 degrees.

j. Resistance is between 12 ounces and 48 ounces.

17

| | PASS | MARGINAL | REJECT |
|---|---|---|---|

**4.1.8 LEVERS**

a. Levers are used where large amounts of force or displacement are involved.

b. Levers are used where multi-dimensional movement of the control is required.

c. Levers placed close to each other and not readily discriminable from each other are coded.

d. All levers are labeled as to function and direction of motion.

e. If fine or continuous adjustments must be made with the lever, support is provided for the elbow, forearm, or wrist.

f. The diameter of spherical handles grasped by the fingers is between ¾ and 3 inches.

g. The diameter of spherical handles grasped by the hand is between 1-½ inches and 3 inches.

h. Maximum displacement does not exceed 14 inches for fore-aft movement, or 38 inches for lateral movement.

i. Minimum resistance for aircraft joysticks is at least 5 pounds.

j. For one-hand operation along the fore-aft axis, maximum resistance of the lever handle does not exceed the following limits:
 1) Lever 10 inches forward from seat reference point: 30 pounds.
 2) Lever 16–24 inches forward from seat reference point: 40 pounds.

k. For operation along the lateral axis, maximum resistance of the lever handle does not exceed the following limits.
 1) One-hand operation (lever 10–19 inches forward of seat reference point): 20 pounds.

18

| | PASS | MARGINAL | REJECT |
|---|---|---|---|

 2) Two-hand operation (lever 10–19 inches forward of seat reference point): 30 pounds.

l. Knobs are used where precise, accurate adjustments of a continuous variable are required, and where little force is necessary.

m. For knobs grasped by the fingertips:
 1) Minimum depth is at least ¾ inch.
 2) Minimum diameter is at least 1 inch.
 3) Maximum diameter does not exceed 4 inches.

n. For knobs grasped with thumb and finger encircled:
 1) Minimum diameter is at least 1 inch
 2) Maximum diameter does not exceed 3 inches.

o. For knobs grasped with the palm:
 1) Minimum diameter is at least 1-½ inches.
 2) Maximum diameter does not exceed 3 inches.

p. Knob displacement is determined by the desired control/display ration.

q. Maximum resistance for fingertip operation of small (1-inch diameter) knobs does not exceed 4-½ ounces.

r. Maximum resistance for fingertip operation of larger knobs (greater than 1-inch diameter) does not exceed 6 ounces.

**4.1.9 CRANKS**

a. Cranks are used for tasks requiring many rotations of the control.

b. Minimum crank radius is at least 1 inch.

c. Maximum radius with minimum load and high rate (up to 275 rpm) does not exceed 4-½ inches.

19

**PASS / MARGINAL / REJECT**

d. The grip handle is designed so that it turns freely around its shaft.

### 4.1.10 HANDWHEEL

a. Handwheels, designed for two-hand operation, are used where the breakout or rotational forces are too large to be overcome with a one-hand control.

b. Minimum handwheel diameter is at least 7 inches.

### 4.1.11 PEDALS

a. Pedals are used where a large amount of displacement and force is required and foot activation is desired.

b. Pedals return to their null position when force is removed.

c. Pedal size is at least 1-½ inch by 3-½ inches.

d. Pedal displacement falls within the following limits.
   1) Minimum, for normal operation: 1 inch
   2) Minimum, with heavy boots: 1 inch
   3) Maximum, by ankle flexion only: 2-½ inches
   4) Maximum, by leg movement: 7 inches

e. Pedal resistance falls within the following limits.
   1) Minimum, foot not resting on control: 4 pounds
   2) Minimum, where foot may rest on control: 10 pounds
   3) Maximum, with ankle flexion only: 20 pounds
   4) Maximum, with leg movement: 80 pounds

## 4.2 Control/Display Relationships

### 4.2.1 ARRANGEMENTS

a. All controls having sequential relations, or having to do with a particular function or operation, or which are operated together, are grouped together, along with the associated displays.

**PASS / MARGINAL / REJECT**

b. When a control is associated with a transilluminated indicator, the indicator is located so as to be immediately and unambiguously associated with the control.

c. When a control is associated with a transilluminated indicator, the indicator is located above the control.

d. If a control knob is adjacent to the instrument it controls, it is located so that the control or the hand normally used for setting does not obscure the indicator.

e. Lights are used to display equipment response and not merely switch position, unless the switch position cannot be made apparent by proper design and labeling of the control.

f. Warning lights are integral with, or adjacent to, the levers, switches, or other control devices by which the operator is to take action.

### 4.2.2 PRECAUTIONS

a. The control is located or oriented so that the operator is not likely to hit it or move it accidentally in the normal sequence of control movements.

b. Physical barriers are placed around the control, or it is recessed or shielded.

c. The control is covered or guarded.

d. Interlocks are provided so that extra movement or the prior operation of a related or locking control is required.

e. Resistance is built into the control so that definite or sustained effort is required to actuate it.

## 5. MAINTAINABILITY

### 5.1 Unitization

a. Unless structurally or functionally not feasible, all equipment is designed so that rapid and easy removal and replacement of malfunctioning units can be accomplished by one operator.

**PASS / MARGINAL / REJECT**

b. Where possible, units serving the same function in different applications are designed to be interchangeable.

c. The number of inputs and outputs from each unit is kept to a minimum by grouping of functions so that a minimum of crisscrossing of signals between units is required.

d. Where consistent with system design requirements, functions are so utilized that it is possible to check and adjust each unit separately.

### 5.2 Handling

a. Where possible, units are small and light enough for one man to handle and carry; i.e., weight of removable units is held below 45 pounds.

b. Units in excess of 45 pounds have provision for two-man lift where the lifting height is not in excess of five feet and where the total weight is not in excess of 90 pounds.

c. Units weighing over 90 pounds have provision for mechanical or power lift.

d. All units weighing 45 pounds or more are prominently labeled with their weight.

e. All units designed to be removed and replaced are provided with handles, or other suitable provision made for grasping, handling, and carrying.

f. Wherever possible, handles or grasp areas are located over the center of gravity so that the unit does not swing or tilt when lifted.

g. Handles which are to be used by the ungloved hand are at least 5-½ inches in length and 2 inches in depth (inside measurements), and ½-inch in diameter for units under 25 pounds, or ¾-inch in diameter for units over 25 lbs.

h. Where feasible, rests or stands are provided and incorporate provisions for test equipment, tools, and manuals.

**PASS / MARGINAL / REJECT**

i. Irregular, fragile, or awkward extensions, such as cables, wave guides, hoses, etc., are easily removable before a unit is handled.

j. Where feasible, guides, tracks, and stops are provided to facilitate handling and to prevent damage to units and components.

### 5.3 Covers and Cases

a. The proper orientation of a unit within its case is obvious, either through design of the case or by means of appropriate labels.

b. Where possible, cases are designed to lift off units, rather than units lifted out of cases.

c. The method of opening a cover is obvious, either from the construction of the cover itself or from an instruction plate permanently attached to the outside of the cover.

d. It is obvious when a cover is in place but not secured.

e. Sharp edges and corners on cases and covers are avoided.

f. Where space permits, hinged covers are used to reduce the number of fasteners required.

|  | PASS | MARGINAL | REJECT |
|---|---|---|---|

**5.4**　*Mounting of Units*
a.　Wherever possible, identical screw and bolt heads are used, thereby enabling various panels and components to be removed with one type of tool.
b.　Units which are frequently pulled out of their installed position for checking are mounted on roll-out racks, slides, or hinges.
c.　Guide pins or their equivalent are provided on units for alignment during mounting.
d.　Limit stops which can be conveniently over-ridden are provided on roll-out racks and drawers.
e.　Interchangeable units are coded (color, labels, etc.) to indicate the correct unit and its orientation for replacement.

**5.5**　*Location of Components*
a.　Parts are mounted in an orderly array on a "two-dimensional" surface and not "stacked" on one another; e.g., the lower layer not supporting the upper layer of units.
b.　Large parts which are difficult to remove are mounted so that they do not prevent access to other parts.
c.　There is sufficient space to use test equipment and other required tools without difficulty or hazard.
d.　Structural members of the units and chassis do not prevent access to components.
e.　All throwaway assemblies or parts are accessible without removal of other components.
f.　Delicate components are located or guarded so that they will not be damaged while the unit is being handled or worked on.
g.　If screwdriver adjustments must be made blind, mechanical guides are provided, or the screws mounted so that the screwdriver will not fall out of line.
h.　Sensitive adjustments are located or guarded so that they will not be disturbed.

24

i.　Internal controls are not located close to dangerous voltages.
j.　Units are laid out so that a minimum of place-to-place movement is required of the operator during checkout.
h.　Bills of material or maintenance nomenclature are not placed on the front of the panel.
l.　Where possible, bills of material or maintenance nomenclature are placed behind the control panel, in order to facilitate troubleshooting and maintenance.

**5.6**　*General Accessibility*
a.　All access covers which are not completely removable are self-supporting in the open position.
b.　If instructions applying to a covered unit are lettered on a hinged door, the lettering is properly oriented for reading when the door is open.
c.　Sliding, rotating, or hinged units to which rear access is required are free to open or rotate their full distance and remain in the open position without being supported by hand.
d.　Check points, adjustment points, cables and connectors, and labels are accessible, and where possible, face the maintenance man.
e.　Bulkheads, brackets, other units, etc., do not interfere with removal or opening of covers of units within which work must be done.
f.　Unless a unit is completely self-checking, provision is made for checking the operation of the unit in the operating condition without the use of special rigs and harnesses.
g.　An opening with no cover is used unless this is likely to degrade system performance.
h.　A sliding or hinged cap is used if dirt, moisture, or other foreign materials are a problem.
i.　A quick-opening cover plate is used if a cap will not meet stress requirements.

25

j.　Removal of any replaceable unit requires opening or removal of a minimum number of covers or panels (preferably one).
k.　Wherever possible, units are located so that no other equipment must be removed to gain access.
l.　Where it is necessary to place one unit behind another, the unit requiring most frequent access is most accessible to the user.
m.　When equipment is of a highly critical nature and its maintenance requires highly specialized skills, access to units maintained by one operator does not require removal of equipment maintenance by a second operator.
n.　Units are located and mounted so that access to them may be achieved without danger to personnel from electrical charge, heat, sharp edges and points, moving parts, chemical contamination, or other sources.

**5.7**　*Size of Apertures*
a.　Openings and work spaces provided for adjusting and handling units are large enough to permit the required activity.
b.　Where possible, openings and work spaces permit adequate view of the components being manipulated.
c.　Inserting empty hand-held flat: 2-¼ by 4-¼ inches.
d.　Smallest square hole through which empty hand can be inserted: 3-½ by 3-½ inches.
e.　Inserting a miniature vacuum tube, held with the thumb and first two fingers, up to the center knuckle of the middle finger: 2-½ by 2-½ inches.
f.　Inserting a large vacuum tube (CRC 1625) with base diameter 1-⅜ inches, height 4-½ inches, excluding pins and grid cap: 4 by 4 inches.
g.　Inserting a box or electronic assembly, grasped by handles on front, into an aperture: ½-inch clearance on each side of the box.
h.　Reaching through aperture with both hands to a depth of six to 25 inches: width—¾ depth of reach; height—5 inches, 6 with gloves.

26

i.　Reaching in full arm's length (to shoulders) straight ahead with both arms: width—19-½ inches; height—5 inches, 6 with gloves.

**5.8**　*Conductors*
a.　Cables are routed so that they cannot be walked on or used for hand holds.
b.　Cables are routed so that they are accessible for inspection and repair.
c.　Cables are routed so that they need not be bent or twisted sharply or repeatedly during repair.
d.　Cables containing individually insulated conductors with a common sheath are coded as specified in ARDCM 80-5.

**5.9**　*Connectors*
a.　Plugs which require no more than one turn, or other quick-disconnect plugs, are used whenever feasible.
b.　Connectors are located far enough apart so that they can be grasped firmly for connection and disconnection.
c.　The rear of plug connectors is accessible for test and service, except where potting, sealing, or other considerations preclude this.
d.　Plugs or receptacles are provided with alignment pins, or other alignment devices.
e.　Alignment pins project beyond electrical pins in plugs or receptacle.
f.　Where a reasonable possibility exists for unintentional interchange of connectors, plugs are designed so that it is impossible to insert the wrong plug in a receptacle.
g.　Plugs or receptacles are arranged so that the alignment pins are oriented in the same direction throughout the system where possible.
h.　Connecting plugs and receptacles are identified by color, shape, or other acceptable means.

27

i. Plugs and receptacles have painted stripes, arrows, or other indications to show the position of alignment pins for proper insertion.

j. The system is designed so that all "hot" contacts are socket contacts.

5.10 *Fasteners*

a. Maximum use is made of tongue-and-slot catches to minimize the number of fasteners required.

b. The number and diversity of fasteners are the minimum commensurate with requirements for stress, bonding, etc.

c. Where feasible, the same size and type of fasteners are used for all covers and cases.

d. Captive fasteners are used where possible.

e. Screws with different threads are of different sizes.

f. If compatible with stress and load considerations, fasteners for mounting assemblies, subassemblies etc., fasten or unfasten with a maximum of one complete turn.

g. Bolts requiring high torque are provided with external grip heads.

h. Captive bolts and nuts are used in situations where the dropping of these small items into the equipment will cause damage or create a difficult removal problem.

5.11 *Test Points*

a. Where a unit is not completely self-checking in its operation condition, appropriate test points are provided which provide a measure of unit output.

b. Only such primary test points as are necessary to determine that a unit is malfunctioning are provided.

c. Primary test points are located and coded to be readily distinguishable from secondary test points.

d. Where feasible, primary test points are grouped in a line or matrix reflecting the sequence of tests to be made.

e. Primary test points used in adjusting the unit are located close to the controls and displays used in the adjustment.

f. Where feasible and not in conflict with other requirements, a secondary test point is supplied at the input and output of each part or throwaway component.

g. Sufficient test points are provided so that it is not necessary to remove subassemblies from assemblies in troubleshooting.

h. Each test point is marked to be readily identifiable.

5.12 *Lubrication*

a. Units containing mechanical components requiring lubrication is specified by a label at or near the lube port.

b. Where lubrication is required, the type lubricant to be used and the frequency of lubrication is specified by a label at or near the lube port.

5.13 *Unit Labeling*

a. Wherever possible, the outside covering of manufactured parts such as resistors, condensers, and tubes, is stamped or coded with relevant information concerning electrical characteristics of the part.

b. Where space permits, terminals are labeled with the same code symbol as the wire attached to it.

c. If surface labels must be used, silk-screened or stamped labels are used in preference to stenciled labels or decals.

d. Labels are not hidden by units or parts.

5.14 *Equipment Color*

a. The console interior of operational and maintenance equipment is Gray. Color 26622, Fed. Std. 595, where maintenance and troubleshooting are required within the console.

b. Where maintenance and troubleshooting within the console will not be required, an economical internal protective finish is used.

28

29

---

## Sample Human Engineering Checklist (for Vehicles)

**GENERAL ACCOMMODATIONS** How appropriate are the general passenger, driver, and package carrying and storage accommodations?

| | Satisfactory | Unsatisfactory |
|---|---|---|
| Ingress and egress clearance for both normal and emergency use and for the handicapped and the elderly | | |
| Interior space dimensions (passenger width, head, and leg clearance) | | |
| Temperature and ventilation | | |
| Ride characteristics (sway, oscillation, and vibration) | | |
| Interior and exterior package storage space (amount, location, and accessibility) | | |
| Crash protection (restraint systems, noninjurious interior, handholds, and shatterproof glazing) | | |
| Noise minimization | | |
| Adequate seating (size, shape, seat-back angles, adjustability, properly secured) | | |

**DRIVER WORK STATION** How appropriate are the layout of the driver's work station and the provisions for proper vehicle control?

| | Satisfactory | Unsatisfactory |
|---|---|---|
| Geometric relationship between seat, controls, and external viewing (to accommodate 5th-percentile female through 95th-percentile male drivers) | | |
| Control and display location and arrangement and detail interface designs | | |
| Exterior viewing capability (front, side, and rear, i.e., clear angle of view, including normal and adverse weather conditions) | | |
| Control system adequacy (steering force and rate, brake force, clutch force and excursion, accelerator force and excursion, and transmission lever pattern and force characteristics) | | |

**GENERAL SAFETY**

| | Satisfactory | Unsatisfactory |
|---|---|---|
| Cab integrity (front, rear, and side crash; roll-over) | | |
| Nonflammable interior materials | | |
| Flame barriers between passenger compartment engine and rear fuel compartment or tank | | |
| Fail-safe systems (power windows, seats, steering, and brakes) | | |
| Non-brake-lockup and skid control | | |
| Blowout-proof tires | | |
| Energy-absorption systems (front, rear, and side impact) | | |
| Noninjurious pedestrian-vehicle interfaces (sheet metal, handles, and trim) | | |
| Rupture-proof fuel tank | | |
| Steering-wheel lock | | |
| Rearview mirrors | | |

Exterior signal lights (stop, turn, running, and backup) _____ _____

Audio backup signal (commercial, construction, and agricultural vehicles; buses and trucks) _____ _____

Maintenance interfaces (fans, belts, open gears, electrical connections, jacking, fuel service, and sharp corners and edges on sheet metal) _____ _____

## Sample Human Engineering Checklist (for Architectural Facilities)

**SITE PLANNING**   How appropriate is the layout of structures and related roads and walkways?

|  | Satisfactory | Unsatisfactory |
|---|---|---|
| Main visitor reception point obvious from the expected normal approach | _____ | _____ |
| Main visitor and vehicle entry obvious from the expected approach street | _____ | _____ |
| Facility address conspicuous and readable from typical "first-view" point or points | _____ | _____ |
| Visitor parking convenient to visitor entry from the street, pedestrian approach and from visitor parking | _____ | _____ |
| Convenience of employer and employee entrances | _____ | _____ |
| Convenience and separation from visitors of services and delivery entrances | _____ | _____ |
| Interbuilding pedestrian convenience | _____ | _____ |
| Separation of noise-producing or hazard-associated facilities from other occupied buildings | _____ | _____ |
| Access for the handicapped (from parking area to and from buildings) | _____ | _____ |
| Signing adequacy (number, location, and size) | _____ | _____ |

**BUILDING DESIGN**   How appropriate are the general size, shape, interior organization, and orientation of the basic building structures?

|  | Satisfactory | Unsatisfactory |
|---|---|---|
| Entrances and exits (number, location, size, door manipulability, identification, illumination, and accessibility to the handicapped) | _____ | _____ |
| Exterior and interior entrance and exit thresholds (size, shape, door clearance, protection from adverse weather, accessibility to the handicapped, illumination, steps, ramps, and guardrails) | _____ | _____ |
| Working- and living-space size, shape, access, illumination, ventilation, heating and cooling, acoustics, and service outlets | _____ | _____ |
| Intraspace organization for ease of traffic flow, control of specifically restricted areas, ease of communication, privacy, and flexibility for alternative use | _____ | _____ |
| Convenience of common-use spaces (rest rooms, storage, central utilities, etc.) | _____ | _____ |
| Intralevel traffic convenience (normal, emergency, and for service personnel and the handicapped) | _____ | _____ |
| Interior signing effectiveness (conspicuous, visible from strategic vantage points, braille for the blind, and illumination) | _____ | _____ |

**STAIR, RAMP, ESCALATOR, AND ELEVATOR DESIGN**

|  | Satisfactory | Unsatisfactory |
|---|---|---|
| Optimized stair configuration (riser and tread dimensions and incline angle, width, number of steps in relation to rest platforms, approach and departure platform dimensions, handrails, illumination, and nonslip tread) | _____ | _____ |
| Optimized ramp configuration (ramp angle for the handicapped, width, ramp length in relation to rest platform, handrails, approach and departure platform dimensions, nonslip tread, and illumination) | _____ | _____ |
| Escalator characteristics (stair riser and tread dimensions, lead-in dimensions, handrails, rate of movement, and visual contrast and illumination) | _____ | _____ |
| Elevator characteristics (interior size, entry dimensions, safety doors, interior handrails, operator control location and layout, including braille identification, ventilation, illumination, and rate of movement) | _____ | _____ |

**DOORS, WINDOWS, HARDWARE, AND FIXTURES**

|  | Satisfactory | Unsatisfactory |
|---|---|---|
| Door and doorway characteristics (clearances, door-swing interference, and type and location of handles and locks in terms of use by the handicapped) | _____ | _____ |
| Window characteristics (ventilation and drafts; ease of opening, closing, and locking; safety relative to inadvertent contact; glare control; and ease of cleaning and repair) | _____ | _____ |
| Plumbing fixture characteristics (accessibility, including to handicapped users; ease of manipulation; absence of injury-producing features in the event of inadvertent contact; accessibility for repair; and appropriate positioning of fixtures that require perimeter clearance, e.g., sink, toilet, bidet, and urinal) | _____ | _____ |
| External fixture location, i.e., hose bibs, electrical outlets, and luminaires | _____ | _____ |
| Interior lighting fixtures (type, number, location, and switch access) | _____ | _____ |
| Environmental system (heating and ventilating distribution, outlet and return locations, automatic temperature sensor control location, and temperature control and display configuration) | _____ | _____ |

**FLOORS**

|  | Satisfactory | Unsatisfactory |
|---|---|---|
| Level, nonslip, well-defined grade changes | _____ | _____ |
| Adequate illumination, nonglare surfaces, and nonillusory patterns | _____ | _____ |
| Ramps at level changes for the handicapped | _____ | _____ |

**GENERAL SAFETY**

|  | Satisfactory | Unsatisfactory |
|---|---|---|
| Nonflammable materials | _____ | _____ |
| Adequate number, type, location, and marking of emergency exit routes and exits | _____ | _____ |
| Grounded electrical systems and fixtures | _____ | _____ |
| Adequate normal and emergency illumination | _____ | _____ |
| No sharp projections or edges | _____ | _____ |
| View ports in doors that swing in both directions; no revolving doors | _____ | _____ |
| Safety glass or plastic in floor-length windows or doors | _____ | _____ |

## HAZARDS ANALYSIS

**General Procedural Model for Safety Analysis (for both Predesign and Postdesign Analyses)[19]**

**Example: Safety Analysis Model[20]**

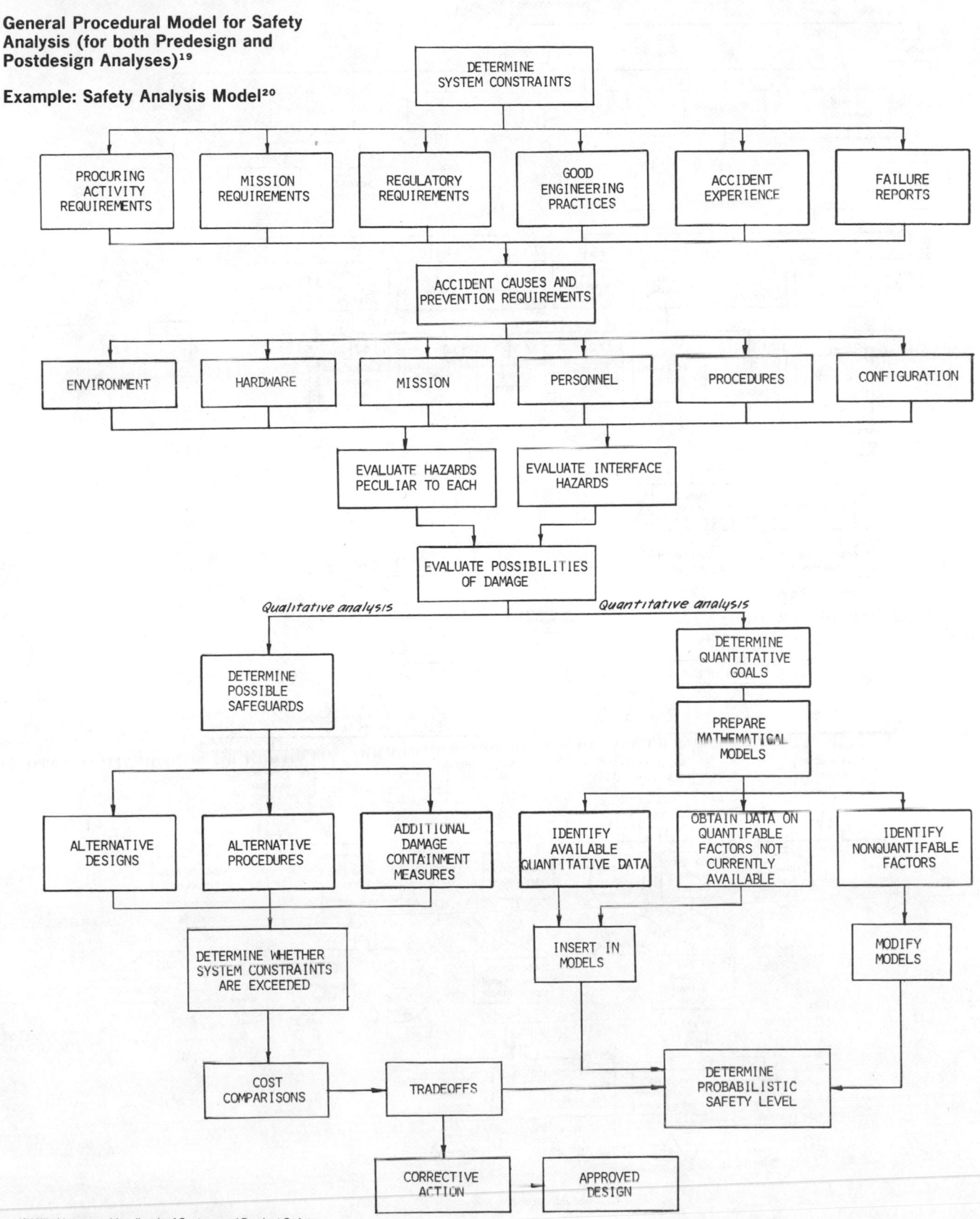

[19]Willie Hammer, *Handbook of System and Product Safety*, Prentice-Hall, Inc. Englewood Cliffs, N.J., 1972, p. 87.

[20]Willie Hammer, *Handbook of System and Product Safety*, Prentice-Hall, Inc., Englewood Cliffs, N.J., 1972, p. 99.

## Example: Safety Analysis Model[20]

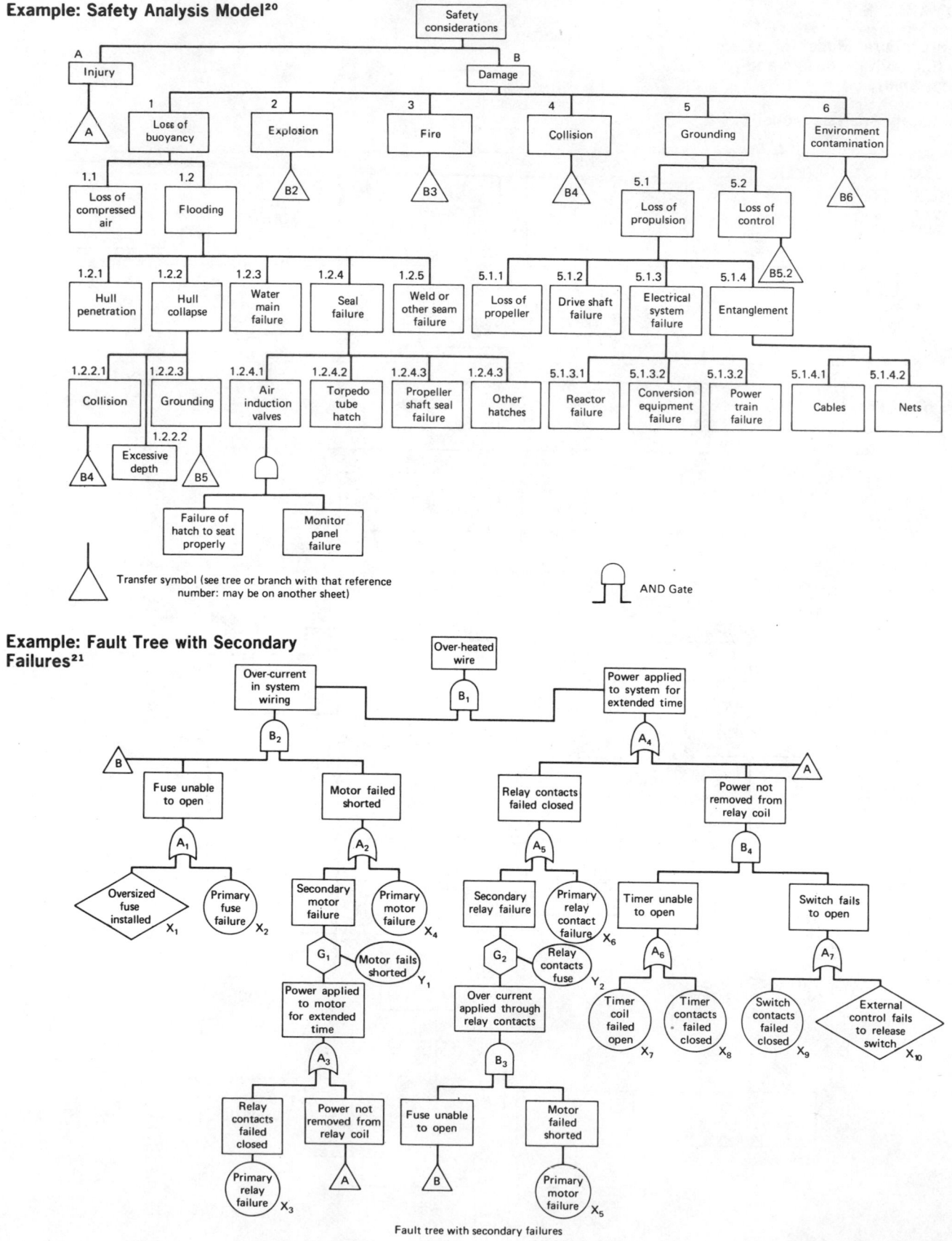

Transfer symbol (see tree or branch with that reference number; may be on another sheet)

AND Gate

## Example: Fault Tree with Secondary Failures[21]

Fault tree with secondary failures

[21]Willie Hammer, *Handbook of System and Product Safety*, Prentice-Hall, Inc., Englewood Cliffs, N.J., 1972, p. 243.

## Failure Modes and Effects during Product Manufacture

Although many product development project managers are aware of the value of an analysis of failure modes and effects for the operational aspects of a design, they fail to consider the importance of failures that may occur during the manufacture of the product. The following is a list of some of the failure modes and effects that should be examined during manufacture:

Leads severed, broken, or pulled from components during assembly

Connector pins bent, broken, improperly soldered, or actually missing

Insulation cut, torn, or abraded, exposing conductors

Glass components cracked, ruining hermetic seals

Tubing bent, reducing flow of fluids or air

Insulation melted by soldering or welding

Poor welds that break or separate under stress

Diodes, transistors, or other delicate semiconductors damaged by currents generated in improperly grounded welding or soldering equipment

Improper clamping of wire bundles or cables

Improper heat treating of metal parts

Overstressing of parts during an assembly

Improper engagement of threaded components during assembly

Use of improper bolt or screw lengths

Unidirectional components installed backward

Failure to install lock washers, safety wires, or similar devices to prevent loosening or loss of connectors

Failure to install critical components such as heat shields, bonds, or grounds

Inadequate clearances between parts, causing them to rub or bind

Inadvertent gouging, scoring, or denting of metal surfaces

Failure to install seals, gaskets or "O" rings properly

Inadvertent contamination of materials or components

Installation of cables or lines wherein the wrong connections are made

Failure to close and seal covers properly, thereby allowing moisture to seep in or lubricants or other substances to leak out

Damage to components during testing, allowing transient electrical surges to exceed critical limits

Misalignment of components

Damage of material surfaces during a polishing, painting, or finishing process

Tools, screws, or other elements accidentally left inside a critical unit as it is closed and sealed

Failure to add appropriate lubricant

*Note:* Any of the above could be prevented or at least minimized by proper design.

## Potential Injury during the Manufacturing Process

Equally important to the overall success of product manufacture is an analysis of the potential failures that could occur during the manufacturing process, leading to personal injury in the fabrication and assembly plant. The following are some of the more critical considerations:

Eye injuries due to sparks, flying particles, fumes, sprays, or extremely high intensity light (from a welder's tool)

Injuries from cutting tools, sharp materials, etc.

Injuries due to components falling on a worker

Injuries sustained when a worker is drawn into moving parts of fabricating and assembly machinery

Falls on slick floors

Skin damage from toxic materials and substances being handled during the manufacturing process

Respiratory damage due to breathing toxic fumes

Fractures, bruises, cuts, abrasions, or more serious injuries sustained when a worker is caught between heavy parts being assembled

Injuries sustained when a worker is struck by in-factory transport systems, such as cranes, forklifts, and elevators

Injuries due to explosions and/or fire in the plant

## Example: Format for Failure Modes and Effects Analysis[22]

### Failure Modes and Effects Analysis

| 1. Subsystem | | 2. Dwg No. | | | 3. Prepared by | | 4. Date |
|---|---|---|---|---|---|---|---|
| Item | Failure Modes | Cause of Failure | Possible Effects | Probability of Occurrence | Criticality | | Possible Action to Reduce Failure Rate or Effects |
| Motor case | Rupture | a. Poor workmanship<br>b. Defective materials<br>c. Damage during transportation<br>d. Damage during handling<br>e. Overpressurization | Destruction of missile | 0.0006 | Critical | | Close control of manufacturing processes to ensure that workmanship meets prescribed standards. Rigid quality control of basic materials to eliminate defectives. Inspection and pressure testing of completed cases. Provision of suitable packaging to protect motor during transportation. |
| Propellant grain | a. Cracking<br>b. Voids<br>c. Bond separation | a. Abnormal stresses from cure<br>b. Excessively low temperatures<br>c. Aging effects | Excessive burning rate, overpressurization, motor case rupture during otherwise normal operation | 0.0001 | Critical | | Carefully controlled production. Storage and operation only within prescribed temperature limits. Suitable formulation to resist effects of aging. |
| Liner | a. Separation from motor case<br>b. Separation from motor grain or insulation | a. Inadequate cleaning of motor case after fabrication<br>b. Use of unsuitable bonding material<br>c. Failure to control bonding process properly | Excessive burning rate Overpressurization Case rupture during operation | 0.0001 | Critical | | Strict observance of proper cleaning procedures. Strict inspection after cleaning of motor case to ensure that all contaminants have been removed. |

[22]Willie Hammer, *Handbook of System and Product Safety*, Prentice-Hall, Inc., Englewood Cliffs, N.J., 1972, p. 151

## ERROR ANALYSIS

### Example: Format for Human Error Analysis during Human Engineering Design Review

#### Error Analysis

Control—Power Toggle

| Type of Error | Effect | Probability of Occurrence | Probability of Detection | Recommendations |
|---|---|---|---|---|
| Inadvertent actuation | Unintentional activation or deactivation | Great—due to location and proximity to other controls | Moderate—feedback from displays | Relocate to periphery of console—add guards |
| Substitution | Erroneous sequence of operations | Great—switch is located adjacent to other toggles | Moderate—feedback from displays | Relocate to periphery—consider use of pushbutton |
| Adjustment | Selection of STANDBY instead of ON or OFF | Little—labeling is adequate | Moderate—feedback from displays | None |

### Task and Procedure Hazards Analysis[23]

#### Procedures Analysis

| Task | Danger | Effect | Cause | Corrective or Preventive Measures |
|---|---|---|---|---|
| Charge nitrogen pressure vessel | 1. A loose hose may whip. | Personnel could be injured or equipment damaged. | Hose failure; connection failure; failure to tighten connection adequately. | Tie down, chain, or sandbag hose at close intervals. Personnel wear hard hats and face shields. Establish torque values for tightening connections. Warning and caution notes in procedures. |
| | 2. Vessel bursts. | Fragmentation. Fragments may injure personnel or damage nearby equipment. | Inadequate strength. | Use high safety factor design. Provide warning against overpressurizing system. Do not expose pressure vessel to heat. Incorporate relief and safety valves. Test vessel to ensure that it will carry required pressure. |
| | 3. High-velocity gas escapes. | Gas may blow solid particles into eyes or against skin. Loss of gas may cause system to become inoperative due to lack of pressure. | Leak; hose failure; loosening fitting on pressurized system; crack. | Procedures to provide warnings to depressurize system before attempting to disassemble connectors. Personnel to wear face shields. |

A thorough analysis of the procedures required to accomplish each operator and maintainer task is desirable in order to isolate critical interfaces in which (1) the design should be reconsidered, (2) special safety procedures and training requirements should be defined, and/or (3) the steps the manufacturer has taken to provide maximum safety for the customer should be documented (in the event that a legal action is taken by an injured customer in the future). The accompanying sample format demonstrates the approach of one analyst.

For some complex systems, it is desirable to extend this analysis by estimating operator and maintainer reliability for each of the more critical tasks wherein human failures have a significant impact on system performance or possible human injury. An example of such an estimate is illustrated in the table below.

### Example: Procedural Analysis for Identifying Potential Human Error

| System Function | Type of Equipment | Task | Potential Error | Suggestions for Equipment Design | Suggestions for Procedure |
|---|---|---|---|---|---|
| Prepare nose cone for inspection | Crane, sling, hand tools | Remove nose cone afterbody | Removal of bolts in improper sequence | Number bolt holes to indicate proper sequence | Insert caution in tech manual |
| Check out trajectory control test set | Functional test set | Perform simulated nose cone test | Misinterpretation of indicator lights | Legend lenses should be consistent and unambiguous | Spell out meaning of abbreviations in checklist |
| Inspect transport trailer for damage | Trailer, checklist | Inspect as indicated by checklist | Personal injury | None | Secure brakes before crawling under trailer |
| Complete nose cone high-pressure test | Test equipment, hand tools | Disconnect nose cone from test equipment | Exposure to high pressure | None | Ensure that pressure is vented before disconnecting hoses |

[23]Willie Hammer, *Handbook of System and Product Safety*, Prentice-Hall, Inc., Englewood Cliffs, N.J., 1972, pp. 197, 199.

## Example: Predicting Human Reliability

### Means and Standard Deviations of Ratings and Reliability Estimates for the Task Elements

| Task Element | Rating Mean | Rating SD | Reliability Estimate | Task Element | Rating Mean | Rating SD | Reliability Estimate |
|---|---|---|---|---|---|---|---|
| Read technical instructions | 8.3 | 2.2 | 0.9918 | Install nuts, plugs, and bolts | 4.6 | 1.7 | 0.9979 |
| Read time (Brush Recorder) | 8.2 | 2.1 | 0.9921 | Install union | 4.5 | 1.8 | 0.9979 |
| Read electrical or flow meter | 7.0 | 2.8 | 0.9945 | Lubricate "O" ring | 4.5 | 2.5 | 0.9979 |
| Inspect for loose bolts and clamps | 6.4 | 1.9 | 0.9955 | Remove initiator simulator | 4.1 | 1.9 | 0.9983 |
| Position multiple position electrical switch | 6.3 | 2.4 | 0.9957 | Install protective cover (friction fit) | 4.1 | 2.2 | 0.9983 |
| Mark position of component | 6.2 | 2.1 | 0.9958 | Read time (watch) | 4.1 | 2.1 | 0.9983 |
| Install lockwire | 6.0 | 2.3 | 0.9961 | Verify switch position | 4.1 | 1.9 | 0.9983 |
| Inspect for bellows distortion | 6.0 | 2.7 | 0.9961 | Inspect for lock wire | 4.1 | 2.1 | 0.9983 |
| Install Marman clamp | 6.0 | 1.8 | 0.9961 | Close hand valves | 4.0 | 2.6 | 0.9983 |
| Install gasket | 6.0 | 2.1 | 0.9962 | Install drain tube | 4.0 | 2.1 | 0.9983 |
| Inspect for rust and corrosion | 5.9 | 2.1 | 0.9963 | Install torque wrench adapter | 3.9 | 1.7 | 0.9984 |
| Install "O" ring | 5.7 | 2.2 | 0.9965 | Open hand valves | 3.8 | 2.6 | 0.9985 |
| Record reading | 5.7 | 2.3 | 0.9966 | Position two-position electrical switch | 3.8 | 1.5 | 0.9985 |
| Inspect for dents, cracks, and scratches | 5.6 | 2.4 | 0.9967 | Spray leak detector | 3.7 | 2.0 | 0.9986 |
| Read pressure gauge | 5.4 | 2.2 | 0.9969 | Verify component removed or installed | 3.5 | 2.4 | 0.9988 |
| Inspect for frayed shielding | 5.4 | 2.3 | 0.9969 | Remove nuts, plugs, and bolts | 3.5 | 1.7 | 0.9988 |
| Inspect for QC seals | 5.3 | 2.6 | 0.9970 | Install pressure cap | 3.4 | 1.6 | 0.9988 |
| Tighten nuts, bolts, and plugs | 5.3 | 2.6 | 0.9970 | Remove protective closure (friction fit) | 3.2 | 1.6 | 0.9990 |
| Apply gasket cement | 5.3 | 2.3 | 0.9971 | Remove torque wrench adapter | 3.0 | 1.6 | 0.9991 |
| Connect electrical cable (threaded) | 5.2 | 2.2 | 0.9972 | Remove reducing adapter | 3.0 | 1.7 | 0.9991 |
| Inspect for air bubbles (leak check) | 5.0 | 2.2 | 0.9974 | Remove Marman clamp | 3.0 | 1.7 | 0.9991 |
| Install reducing adapter | 4.9 | 1.6 | 0.9975 | Remove pressure cap | 2.8 | 1.8 | 0.9992 |
| Install initiator simulator | 4.9 | 2.5 | 0.9975 | Loosen nuts, bolts, and plugs | 2.8 | 1.3 | 0.9992 |
| Connect flexible hose | 4.9 | 2.4 | 0.9975 | Remove union | 2.7 | 1.4 | 0.9993 |
| Position "zero in" knob | 4.8 | 1.6 | 0.9976 | Remove lockwire | 2.7 | 1.5 | 0.9993 |
| Lubricate bolt or plug | 4.7 | 1.6 | 0.9979 | Remove drain tube | 2.6 | 1.4 | 0.9993 |
| Position hand valves | 4.6 | 1.6 | 0.9979 | Verify light illuminated or extinguished | 2.2 | 1.6 | |

**Human Error Potential as a Basis for Design Analysis[24]**

**Error Prevention through Good Design**

| Causes of Primary Errors | Preventive Measures (Taken by Designer or Methods Engineer) |
| --- | --- |
| 1. Improvising procedures that are lacking in the field | 1. Provide adequate instructions. |
| 2. Following prescribed but incorrect procedures | 2. Ensure that procedures are correct. |
| 3. Failure to follow prescribed procedures | 3. Ensure that procedures are not too lengthy, too fast, or too slow for good performance, and are not hazardous or awkward. |
| 4. Lack of adequate planning for error or unusual conditions | 4. Provide backout or emergency procedures in instructions. |
| 5. Lack of understanding of procedures. | 5. Ensure that instructions are easy to understand. |
| 6. Lack of awareness of hazards. | 6. Provide warnings, cautions, or explanations in instructions. |
| 7. Untimely activation of equipment. | 7. Provide interlocks or timer lockouts. Provide warning or caution notes against activating equipment unless disconnected or disengaged from load, or other damaging conditions. |
| 8. Errors of judgment, especially during periods of stress. | 8. Minimize requirements for making hurried judgments, especially at critical times, through programmed contingency measures. |
| 9. Critical components installed incorrectly. | 9. Provide designs permitting such components to be installed only in the proper ways. Use asymetric configurations on mechanical equipment or electrical connectors; use female or male threads or different-sized connections on critical valves, filters, or other components in which direction of flow is important. |
| 10. Exceeding prescribed limitations on load, speed, or other parameter. | 10. Provide governors and other parameter limiters. Provide warnings on: exceeding limitations, inadequate strength of stressed parts, and use of excessive mechanical leverage. |
| 11. Lack of suitable tools or equipment. | 11. Ensure that need for special tools or equipment is minimized; develop and provide those that are necessary; stress their need in instructions. |
| 12. Interference with normal habits. | 12. Ensure that recognition and activation patterns are in accordance with usual practices and expectancies. |
| 13. Lack of data on which to make correct or timely decisions. | 13. Ensure that response time is adequate for corrective action; if not, provide automatic corrective devices. |
| 14. Hampered activities because of interference between personnel. | 14. Ensure that space is adequate to perform required activities simultaneously. |
| 15. Inability to concentrate because of unsafe conditions or equipment. | 15. Ensure that personnel must not work close to unguarded moving parts, hot surfaces, sharp edges, or other dangers. |
| 16. Error or delay in use of controls. | 16. Avoid proximity, interference, awkward location, or similarity of critical controls. Locate control close to readout. Locate readouts above control so hand or arm making adjustment does not block out readout instrument. Ensure that controls are labeled prominently for easy understanding. |
| 17. Error or delay in reading instruments. | 17. Ensure that instruments are labeled and designed for easy understanding; do not require reader to turn head or move body; and that visibility problems due to glare or lack of light, legibility, viewing angle, contrast, or reflections are avoided. Provide direct readings of specific parameters so operator does not have to interpret. |
| 18. Inadvertent activation of controls. | 18. For critical functions provide controls that cannot be activated inadvertently: use torque types instead of push buttons. Provide guards over critical switches. |
| 19. Controls activated in wrong order. | 19. Place functional controls in sequence in which they are to be used. Provide interlocks where sequences are critical. |
| 20. Control settings by operator not precise enough. | 20. Provide controls that permit making settings or adjustments without need for extremely fine movements. Use click type controls. |

(continued)

[24]Willie Hammer, *Handbook of System and Product Safety*, Prentice-Hall, Inc., Englewood Cliffs, N.J., 1972, p. 72.

**Error Prevention through Good Design** *(continued)*

| Causes of Primary Errors | Preventive Measures (Taken by Designer or Methods Engineer) |
|---|---|
| 21. Controls broken by excessive force. | 21. Ensure that controls are adequate to withstand maximum stress an operator could apply. Provide warning and caution notes for those devices that could be overstressed. |
| 22. Failure to take action at proper time because of faulty instruments. | 22. Provide procedures to calibrate instruments periodically, or provide the means to ensure during operation that they are working correctly. |
| 23. Confusion in reading critical instruments because of instrument clutter. | 23. Make critical instruments most prominent or locate in easiest to read area. |
| 24. Failure to note critical indication. | 24. Provide suitable auditory or visual warning device that will attract operator's attention to problem. |
| 25. Involuntary reaction or inability to perform properly because of pain due to burns, electrical shock, puncture wound, or impact. | 25. Insulate or guard against hot surfaces, "live" electrical conductors, sharp objects, and hard surfaces. |
| 26. Fatigue. | 26. Avoid placing on operator severe and tiring physical and mental requirements such as loads, concentration times, vibration, personal stress, and awkward positions. |
| 27. Vibration and noise cause irritation and inability to read meters and settings or to operate controls. | 27. Provide vibration isolators or noise elimination devices. |
| 28. Irritation and loss of effectiveness due to high temperature and humidity. | 28. Provide environmental control. Prevent entrance or generation of heat or moisture from external sources or from internal equipment or processes. |
| 29. Loss of effectiveness due to lack of oxygen, or to presence of toxic gas, airborne particulate matter, or odors. | 29. Prevent generation or entrance of contaminants into the occupied space. Provide suitable life support equipment. Avoid presence near occupied areas of lines or equipment containing hazardous gases or liquids. |
| 30. Degradation of capabilities due to extremely low temperatures. | 30. Ensure that design provides for adequate heating or insulation, protective shelter, equipment, or clothing. |
| 31. Fixation or hypnosis. | 36. Avoid procedures or designs that require visual concentrations for long periods of time. Avoid humming equipment. Provide alternate reference points. Provide procedures to relieve monotony. |
| 32. Disorientation or vertigo. | 32. Provide adequate reference points or means to maintain orientation. |
| 33. Slipping and falling. | 33. Incorporate friction surfaces or devices, guard rails, access hole covers on floor openings, or protective harness in designs. |
| 34. Inattention. | 34. Avoid long intervals between procedural steps. Provide female voice on audio devices to attract attention. Provide bright, colorful, and pleasant work areas. |

## General Checklist for Safety Achievement via Appropriate Analysis and Design[25]

### Special Safety Considerations

Because of the special nature of some products, it is sometimes necessary to consider safety from more than one point of view. These problems are best explained through examples:

A hand-held weapon will always have certain inherent hazards because of the fact that the device is designed purposely to fire a projectile. Experience has shown that such a weapon should always have some type of safety mechanism to help prevent unplanned firings. The position and specific articulation of the safety feature should be such that the device is convenient to use after the weapon is placed in the firing position; this is necessary in order to minimize the possibility of discharging the weapon while the user is carrying it or is mounting it to the firing position. However, the position and articulation of the safety device should not be such that it can be activated by inadvertent brushing of the device as the weapon is being carried or lifted to firing position or at any time the weapon is being serviced. In addition, since weapons often may be found by children, the safety device should be designed so that a child *cannot* operate and thus remove the safety feature.

A power tool or implement should always have a safety feature to prevent inadvertent operation, and the safety device must be convenient to operate once the tool is in the operating position. Designing and locating the safety device require careful consideration of *all* possible points in the handling cycle in order to make sure that careless handling will not also deactivate the device. In addition, there should be an automatic deactivation of power when the tool is set down temporarily so that no one else could inadvertently pick up the tool and accidentally start the motor, thinking that it had been turned off.

Child-proof bottle caps are generally required by law on all medicine bottles. Remember, however, that some medicines must be quickly accessible to certain persons (e.g., an elderly heart patient), and it is therefore vital that the individual needing to open the bottle can do so easily and rapidly. Confusing safety caps or ones that require considerable finger strength may make the medicine inaccessible to the person who needs it as quickly as possible.

Safety belts obviously serve a basic safety need. However, in addition to providing security for the passenger during a crash, they must also be easy to remove in an emergency—either by the individual wearer or by a rescuer, in the event the wearer may be unconscious. One must consider both the location and the method of operation of the release device in terms of how difficult it might be to find and operate under abnormal conditions, such as in the dark, when the vehicle is not upright, and when it is not possible to open a door.

Standardization plays an important part in safety in a variety of situations and should be kept in mind during any hazards analysis. The following considerations are important:

1. Critical safety features (door handles, safety latches, light switches, etc.) should always be illuminated so that they can be located and operated at night.
2. Standard safety devices (light switches, door handles, restraint system release handles, fire extinguishers, first-aid boxes, etc.) should be located where people expect to find them.
3. Safety signs and signals should be located where people normally will be looking, and they must not have the potential of being hidden by an object placed in front of them or by individuals standing in front of them at a critical moment.
4. Safety devices should operate in a standardized manner; i.e., one device should not have a switch or handle that moves in one direction, while another, similar device has a handle that moves in the opposite direction. The split second required to correct an action may result in injury.
5. A safety device should never be used that may malfunction because of loss of electrical power. A safety system should provide maximum security throughout any anticipated operating cycle; i.e., avoid a system which may provide an initial safety condition but which, because of the nature of the device, expends its safety capability in "one shot" and thus is unavailable for continuing hazardous conditions. The automobile air bag is a case in point. It is effective only on the first impact, deflating shortly thereafter. Thus, if the crash scenario continues, the passenger no longer has any safety protection.

Toys present a particularly difficult safety problem in that they are subject to breakage by the child and thus present hazards after they are broken as well as before.

In order to resolve some of the special problems exemplified by the above, one should systematically analyze both the normal use and potential abnormal or misuse modes of the particular product. In doing this, keep in mind the following critical human characteristics:

1. The users' level of intelligence (both intended and unintended users)
2. The users' experience with this particular device
3. The various environmental conditions under which the device may be used
4. The physical characteristics of the individual user population (both intended and unintended users), i.e., size, strength, mobility, dexterity, visual and auditory capacity, reaction time, and possible handicaps

To correct safety deficiencies, consider the following:

1. Remove the hazard altogether if possible.
2. If that is not possible, try to minimize the potential effects of the hazard.
3. Provide a barrier between the hazard and the user.
4. Warn the user of the potential hazard.

Finally, do not simply assume that your solution to a safety problem works. Test it.

### Safety Measures

| Accident Prevention | Damage Minimization and Control |
|---|---|
| 1. Hazard elimination | 1. Isolation |
| 2. Hazard level limitation | a. Distance |
|   a. Intrinsic safety | b. Energy absorption |
|   b. Limit-level sensing control | c. Deflection |
|   c. Continuous monitor and automatic control | d. Containment |
| 3. Lockouts, lockins, and interlocks |   Hazard |
|   a. Isolation |   Operation |
|   b. Lockouts and lockins |   Personnel |
|   c. Interlocks |   Material |
| 4. Fail safe designs |   Critical equipment |
|   a. Fail passive | 2. Personal protective equipment |
|   b. Fail active |   a. Programmed dangerous operation |
|   c. Fail operational |   b. Investigations and corrections |
| 5. Failure minimization |   c. Emergencies |
|   a. Monitoring | 3. Minor loss acceptance |
|   b. Warning | 4. Escape and survival |
|   c. Safety factors and margins |   a. Point of no-return warning |
|   d. Failure rate reduction |   b. Crashworthiness designs |
|     Derating |   c. Escape and survival equipment |
|     Timed replacements |   d. Escape and survival procedures |
|     Screening | 5. Rescue |
|     Redundancy |   a. Procedures |
| 6. Backout and recovery |   b. Equipment |
|   a. Normal sequence restoration | |
|   b. Aborting entire operation | |
|   c. Inactivating only malfunctioning equipment | |
|     Automatic | |
|     Manual | |

[25]Willie Hammer, *Handbook of System and Product Safety*, Prentice-Hall, Inc., Englewood Cliffs, N.J., 1972, p. 253.

# MOCKUPS

## Use of Mockups for Human Engineering Test and Evaluation

Mockups of products are widely used both as a design tool and for marketing purposes in industry. Unfortunately, management perceives the value of the latter use more readily than the value of the former. This section attempts to champion the use of mockups for purposes of design evaluation, especially for validating design features that involve user interfaces.

## Types of Mockups

Two distinct types of mockups, each used for different purposes, are of interest. The first is the miniature-scale mockup, which is valuable in examining three-dimensional relationships between several fairly large features that would be difficult to view all at once if they were demonstrated in full scale. The second is the full-scale mockup, which is valuable in examining the immediate three-dimensional relationships between the system being mocked up and an operator or maintainer, in terms of space, reach, visibility, and convenience.

## Levels of Mockup Complexity and Detail

Although many people think of a mockup as a fairly substantial structure simulating the three-dimensional attributes of a work station or space layout, there are several levels of mockup implementation. Each has its place, depending upon the needs of the designer and perhaps the time and money available to create the mockup. The following descriptions and use suggestions provide a guide to help designers decide upon which level of mockup implementation they need to accomplish their objectives.

### 1. Paper Mockup

The paper mockup is the least expensive and time-consuming to prepare. It is constructed by cutting out scaled, two-dimensional elements and/or a drawing and placing these on a plan view of a work area or on a wall representing the vertical surface of some cabinet that will eventually hold the proposed control panel.

For a miniature-scale workplace or facilities arrangement problem, one can create an excellent first-step evaluation of various alternatives; i.e., one can try out several possible arrangements of consoles, equipment racks, desks, or other furnishings to determine the arrangement that provides for the fewest operator steps, the best clearances for traffic flow, the best viewing angles, and so on. Such mockups can be made out of paper or cardboard (the latter tends to lay flat more readily and is less affected by drafts, which often blow paper cutouts off the table). Typically, one starts out with a substantial baseboard marked off in square feet (to whatever scale is chosen by the designer). Then the equipment items are cut out of either paper or cardboard (to scale) and are labeled. The cutouts are then manipulated into various alternative positions until the best arrangement is obtained, considering all the critical factors of traffic flow, intercommunication between operator

stations, access for maintenance, and a general feeling of spaciousness or lack of clutter. This type of mockup is very effective for architectural planning, such as for determining room size or equipment arrangement or even for revamping office space or work space in an existing building or office.

The paper mockup is also suitable for initial examination of full-scale relationships. For example, one can place a full-scale drawing of a proposed control panel on a wall in order to determine at what height the panel should be placed or what the best orientation of the panel would be from a nominal operator position (either standing or sitting). Such a mockup can also be used for initial evaluation of the arrangement of controls and displays; i.e., one can simulate going through the proposed sequence of panel operation and quickly discover that items are arranged in the drawing in such a way that too many hand crossovers occur, that things used most often are too far out of reach, that one tends to cover a display while operating a control, or that the proposed labels and the proposed arrangement of panel elements are confusing, making the panel elements hard to locate quickly. Since no construction is required, it is easy and inexpensive to try several alternatives until the best one is found. The paper wall mockup is also useful in establishing the size of labeling required for viewing from alternate positions, such as from across the room. Finally, this inexpensive approach provides the first opportunity to evaluate space and reach problems using test subjects at the extremes of body size to make sure that both large and small operators are accommodated equally well.

It should be emphasized that the paper mockup is not to be regarded as a plaything that is beneath the dignity of the practicing professional. Experience has demonstrated many times that more costly mockups have not provided any more useful information than a paper mockup would have provided, and considerable time is often wasted on more exotic mockups, not to mention the cost. Finally, it should also be noted that a paper mockup will provide important criteria for developing more sophisticated mockups later on during the design program.

### 2. Soft Three-Dimensional Mockup

A soft mockup is one that is made of wood and cardboard or of special inexpensive laminates such as Foam Core. The term "soft" relates primarily to the fact that the materials are easy to cut and assemble, as opposed to metal. The general approach consists of creating a simple wood understructure, to which sheets of cardboard or Foam Core are attached to provide the "skin," e.g., a console, panel rack, or workplace enclosure. To these skins are attached panel drawings or simulated controls and displays that may be cut out of stiff cardboard upon which pictorials of instruments or controls and/or function labels are mounted (to give an added three-dimensional appearance not unlike that of the real hardware elements).

The principal value of the full-scale three-dimensional mockup is its ability to represent spatial interrelationships between an operator and the controls and displays he or she must monitor and manipulate, arm reach and viewing angle parameters, and clearances for feet,

legs, and body (including seats). For early stages of design analysis, it usually is not necessary to enclose a console completely, for example. Only the front of the console is important. The back of the console need not be enclosed, and thus it is not necessary to spend the time and money required to mock up several alternative console shapes.

Three-dimensional miniature-scale mockups can be made of wood, preferably some lightweight wood that is easy to fabricate (e.g., balsam). Mockup elements (consoles, panel racks, desks, chairs, machine tools, etc.) are more easily manipulated if small magnets are embedded into the base of each piece and if the baseboard representing a floor space or area is made of a suitable board upon which a metal plate is mounted. In this way, the equipment and furnishing models are easily moved about and yet stay in place.

A second-level soft mockup may be desirable once the general pattern of a console has been established. That is, in place of the paper or cardboard display and control simulations, actual instruments can be mounted on the control panels. These displays and controls can in turn be made active to various levels, depending on the purpose of the simulation. That is, instruments can be wired to demonstrate illumination, or they can be instrumented to the extent of tying into a computer program, which allows one to simulate and evaluate alternative display and control system parameter concepts.

A third level of sophistication for the soft mockup consists in mounting the mockup on dynamic platforms that simulate, for example, the motions of an aircraft, space vehicle, or ship. Generally, one would not go to this level of time and cost unless there was some solid indication that the dynamic aspects of the operator-control situation were especially critical to the ultimate functioning of the human-machine system. It becomes obvious that this level of mockup begins to approach the capability of becoming a training simulator. Therefore, one can equate the time and cost in terms of both a design tool and an eventual training tool that may be required by the customer.

Although soft mockups are recognized as a useful tool for examining three-dimensional operator-station problems, it should be noted that they are also extremely useful for examining the problems of maintenance technicians. Once again, both miniature-scale and full-scale mockups have their place, for the reasons noted above. For example, the miniature-scale model provides an excellent method for examining proposed layouts for production and for service and maintenance facilities. Using the same approach discussed above, scale models of prime vehicles, support vehicles, structural features, etc., all can be easily viewed on the small-scale simulation. An added benefit of the scale model is that, when different groups of people are involved (such as the contractor and the customer, who may be many miles away), photographs can be taken of several alternative layouts that have been tried, and these can be mailed back and forth to increase the level of communication between the interested parties. In fact, the photographic technique provides an excellent method for preparing for a design review briefing. For example, the designer can take pictures of each alternative arrangement that he or she has tried and evaluated. By making these into slides, it is possible within a few minutes to run through and explain the rationale for making a final arrangement recommendation. This saves a lot of questions from those who would like to know whether the designer has explored all possibilities.

Finally, in the case of mockups that will not be involved in any kind of dynamic evaluations (i.e., mockups which are static, which will not have to support heavy components, or which will not be sat upon or leaned against by persons involved in the evaluations), it is important to use materials that are easily and quickly modifiable by the designer so that it is not necessary to wait for a mockup specialist to make changes. On the other hand, some mockups will have to have sufficient structural integrity to support heavy components, persons who may climb upon them, or accelerative forces from moving platform dynamics. These generally are all-wood structures and in certain areas are covered with a plywood skin.

### 3. Hard Mockup

Generally speaking, the so-called hard mockup is used to define the detailed assembly aspects of a final production design. For example, the hard mockup will be made of metal and will include the precise details of the inner structure and outer skin, which the designer, along with the production engineer, will explore and for which they will define types of attachment hardware for internal components, external components, trim, etc. A good example is the fuselage mockup that is used to define the routing and attachment of cables, hydraulic lines, pneumatic lines, control cables, etc. Although the designer generally prescribes where these components go and in general how they will be attached and connected, it is not certain what obstacles there may be within the basic fuselage structure. Thus the hard mockup provides a precisely scaled breadboard, so to speak, on which the designer can determine how best to route and fasten components. Once these decisions are made and final routings and fastenings are actually mounted in the mockup, the mockup becomes a reference for production-line personnel. Typically these hard mockups are retained for the life of a production run and sometimes beyond (to be used for later modification analyses).

The hard mockup also provides one other important contribution. In deciding how to assemble large, complex elements of a hardware system, it becomes important to work out a plan for assembly that minimizes conflicts among assembly operations. For example, in very tight spaces, various assembly operations cannot all work in the space simultaneously, and the mockup allows one to determine the most efficient order in which to assemble components.

From a purely human engineering point of view, the hard mockup provides an excellent opportunity to evaluate specific human-machine interface problems. These include maintenance tasks (accessibility to components and parts that have to be inspected, removed and replaced, calibrated, or adjusted in the field).

### 4. Other Mockups

**STYLING BUCKS AND MODELS**    The vehicular industry has long used and depended on the clay model to "shape" or "style" the exterior of a new-model car, truck, or bus. Over a basic frame, modeling clay is laid and shaped to represent the proposed exterior lines of the new model (and alternatives). Such models are made in both miniature- and full-scale versions. A special clay material allows the model to be painted so that the full scope of appearance elements can be evaluated for eye appeal.

For smaller products, either the structure and clay approach can be used, or package appearance design can be modeled from other more substantial materials such as fiberglass or thermo-setting plastics. This type of model provides sufficient structural integrity to allow more specific user testing; i.e., a model of a hand tool, hair dryer, or electric knife can be picked up and manipulated to test manipulability and ease of handling.

**PORTABILITY MODELS**    Many small products require handling by the user, operator, or maintenance technician; i.e., they must be picked up, positioned, operated, donned or doffed, transported, and set down again. Before final package decisions are made, such models provide an important step in validating concepts before the product is finally produced. Such models should simulate the dimensional and dynamic aspects of the product so that one can determine whether the expected range of users can, in fact, manipulate the package with ease and surety. Often, until the package configuration is tested in full-scale model form, one cannot be sure whether the weight and balance characteristics of the proposed package will stress users or cause them to drop the package or lose their balance.

**ACCESSIBILITY MOCKUPS**    Although many military projects require key demonstrations of accessibility for maintenance at certain points during system development, these tend to pertain to major access points in an aircraft fuselage, a tank shell, or other prime hardware unit. These represent only the minimal areas where maintenance access mockups should be considered. For example, any small piece of equipment such as a radio, radar, or sonar unit or a piece of test equipment is a subject for verifying maintenance accessibility by means of a mockup. One should consider early mockups before packaging constraints preclude optimizing access; i.e., by means of an early mockup of various package configuration alternatives, one might be encouraged to change the shape or orientation of the package in order to provide a more convenient opening position when the unit is finally installed within close-fitting quarters. A second level of mockup could examine the positioning order of components so that a technician will not have to remove several interfering components to get at the one that is most often expected to need inspection, removal, and replacement. These mockups are especially useful for evaluating various types of slides, fasteners, latches, and handles, as well as for various approaches to the articulation (for accessibility) and/or the removal and replacement of chassis and printed circuit boards.

### How to Determine What Type and Level of Mockup to Use

There are no hard-and-fast rules for deciding what type of mockup to use or to what level of sophistication one should go in developing the mockup and/or operational simulation. The following general guidelines may be helpful.

**AVAILABLE TIME**    Although one should make every effort to find the time to "be sure" (i.e., by mocking up and verifying a complex, three-dimensional concept before committing it to final design), the time available before a design has to be completed and the product delivered may make it impossible to develop and evaluate the mockup. There have been instances where an extensive mockup and design simulation program was completed months after a system had been delivered to the customer. In order to avoid such errors, it is vital that both the customer and the contractor recognize and prepare reasonable schedules so that, if a mockup or simulator is vital, its capabilities can be used in time to influence hardware design.

**CRITICALITY OF THE HUMAN-MACHINE INTERACTIONS**    The following are considered critical interactions:

1. The operator and maintainer working space is limited because of external confining restraints (e.g., a small cockpit, close maintenance quarters, low overhead, or limited crawl space). Such space limitations are particularly critical if the worker or operator may have to wear heavy protective garments. The space should be evaluated with large test subjects wearing the most constraining type of garment.
2. Escape envelope constraints are especially critical for obvious safety reasons. Mockups should be used to evaluate not only the envelope shape and size but also the position of the exit with respect to potentially difficult conditions under which the escape envelope must be traversed and the position of the final exit with respect to opening and departure (e.g., an exit on the top of an aircraft fuselage may be too high for a small person to get to, especially if high $g$ forces are working against the escapee). Escape system and procedure mockups are vital for determining whether escape-time criteria can be met by a particular design configuration.
3. Visibility constraints and/or adequacy should be verified using three-dimensional mockups. Although general anthropometric criteria provide ball-park estimates of how to arrange operators and equipment, windows, glare shields, etc., the final verification requires *in situ* examination by selected ranges of subjects of both large and small stature. Sight lines to both internal displays and external viewing requirements need verification before a final operator-station configuration is accepted.
4. Reach constraints should be verified for certain operator and maintainer workplaces, especially if the operator or maintenance technician has to be restrained

by a safety belt or limited access space. The question includes the reach limits not only of small individuals (possibly constrained by special garments) but also of large individuals, who may not have sufficient clearance to retract an arm or leg.

5. Lighting parameters are often critical where it is very important to provide low but even illumination of all visual displays, labels, and controls or where it is possible that reflections and glare sources may interfere with proper display or external, out-the-window viewing. Such mockups should be evaluated under all the expected ambient lighting conditions. It is wishful thinking to believe that one can identify and adjust a design for all possible glare or reflection problems.

6. Ingress and egress mockups are an important consideration, not only for the questions of emergency escape noted above, but also for normal entry and exit. This is especially important in the case of vehicles for which older and disabled individuals are expected to be among the user population.

7. Control and display arrangement for ease of location, identification, and speed of use should be mocked up and evaluated in real time by a number of typical user subjects. This is especially important when the control operation is complex and when mistakes may be costly to the final operation. Although a cursory link analysis of a preliminary layout drawing provides the first indication of whether a panel layout appears to be logical and convenient, it may not tell the full story; i.e., when the panel is associated with other displays and controls and possibly a constraining seat, the otherwise adequate panel arrangement may prove to be less effective than it appeared in the drawing analysis.

8. Component-rack interactions may present difficult mobility problems for the maintenance technician, and therefore they should be examined in the three-dimensional mockup setup. Typical problems that show up in the mockup evaluation include the following: the technician cannot bend over far enough to lift a chassis out of the lower portion of a rack; there is not enough clearance to open an access door fully; an access opening is so small that the technician's hand or arm blocks the view of the object he or she is trying to reach or manipulate; and there is an intervening hazard when the technician reaches across a moving belt or gear to adjust or remove some component.

## UTILITY OF A PARTICULAR MOCKUP LEVEL TO PROVIDE NECESSARY ANSWERS

There are those who swear by mockups, just as there are those who think that mockups are a waste of time and money. Mockups are often made more elaborate than necessary, usually because an individual manager or engineer likes to have the most lifelike representation possible to show a customer. Although this is impressive and may have value from a marketing point of view, it may be a waste in terms of its design verification value. On the other hand, making the mockup too minimal may prevent one from finding answers to the questions outlined above. The designer must decide what kind of mockup will provide the needed answers; any expansion beyond this requirement is probably a waste of effort and money.

This problem is particularly difficult to resolve in the case of mockups that tend to be real-life simulations. That is, how far to go with simulation is always a moot question. In general, one can never completely simulate the real world under any circumstances, short of creating the actual article and testing it in a real-world environment. Basically, that is the role of a prototype article, e.g., a prototype aircraft, a prototype automobile, or a prototype piece of test equipment. Short of this, mockups and simulations of partial task elements are often more cost-effective.

**MOCKUP ADJUSTABILITY** To be maximally effective as a design evaluation tool, a mockup should be made as adjustable as possible. As noted above, there often is a tendency to feel that a mockup can be made to reflect all the "right" features the first time and that the only purpose is to prove that one's design analysis is correct. Almost without fail, one finds that changes are required after the evaluation. However, unless the mockup is constructed in a fashion that allows for easy modification, it may be necessary to fabricate an entirely new one. Before constructing any mockup, therefore, look ahead and identify all the features that you may suspect could require adjustment during or after the first evaluations. This will allow you to make these changes and reevaluate them with a minimum of time and cost. It will also allow you to demonstrate quickly to customers why you have chosen the final configuration you have—merely by showing them the several versions you tested.

## Miniature-Scale Mockups and Models

Miniature-scale mockups and models are especially useful for evaluating the multicomponent layout and arrangement of large elements such as clusters of buildings and large structures, weapons and troops in a battle configuration, multiple aircraft and support equipment on a flight deck, and equipment, machines, and furnishings within a building, office, laboratory, or factory. Depending on the overall number of elements, one should select the scale to fit the evaluation purposes. Models need not be of fine detail, but they should be fairly precisely scaled so that traffic flow and work-space clearance can be realistically assessed. It is suggested that 1/6-scale models be used for interior room arrangements if it becomes desirable to evaluate any detail on model equipment or furnishings (position of displays, lines of sight, illumination possibilities, etc.).

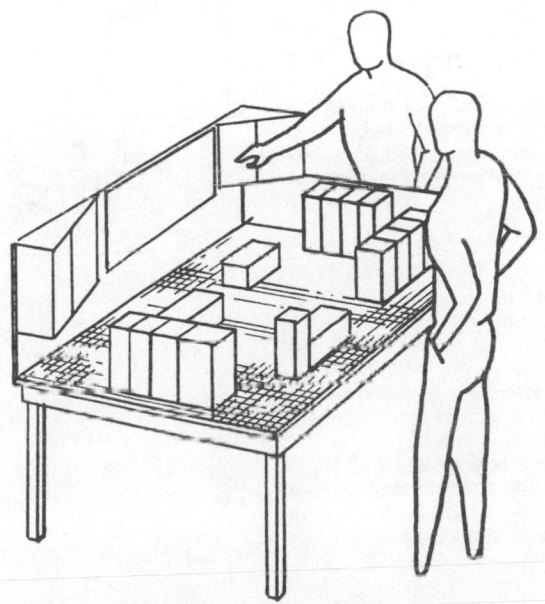

*to provide vantage point to observe total arrangement proposals*

As long as the mockup representation involves a flat floor, it is suggested that a metal plate inscribed in a 1-ft (0.3-m) grid be used as the primary reference base. This makes it easy to assess clearances quickly.

Model equipment, furniture, and machinery should be magnetized so that, although these elements stay put and are not easily disarranged or knocked over, they are still easily moved from one place to another.

Use miniature-scale models to:

Evaluate traffic flow
Establish visual sight lines
Route utilities
Evaluate workplace convenience and maneuvering space
Evaluate production-line flow
Position common-access elements
Make a preliminary lighting and color analysis
Determine the best escape routes
Determine access for maintenance and service
Decide on preliminary decor

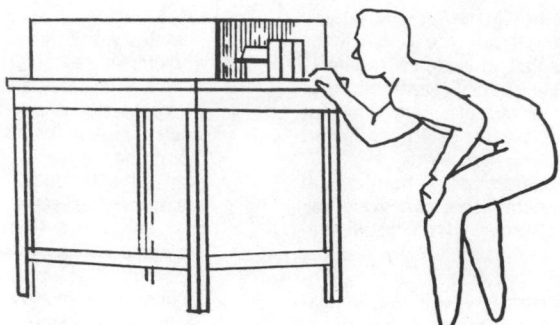

*for preliminary examination as actual operator would see it*

## Use of Drawings and Paste-Ups

Early evaluation of proposed control panel positioning and layouts can be accomplished inexpensively by pinning a drawing on a wall or, better yet, by pinning or attaching self-adhering panel element patterns on a cork or felt board. If a metal wall is available, similar results can be obtained using two-sided tape or flexible magnetized tape on the back of panel elements. It is recommended that independent panel elements be placed over the panel drawing because this permits moving the elements around on the board.

As illustrated by the accompanying sketches, both standing- and sitting-operator relationships can be evaluated. The following key issues can be resolved by these techniques:

Identifying reach problems
Analyzing operational sequence and functional grouping
Determining proper spacing of components and positioning of proposed labels
Establishing the best height for visual displays and determining whether a hand or arm obscures a display

When it appears desirable to change the plane of a particular panel, an inexpensive independent structure can be devised upon which the drawings and/or cutouts can be pasted (as illustrated in the accompanying sketch). A wooden structure can be covered by cardboard or Foam Core to provide sufficient panel rigidity. A light metal panel can be easily attached to the structure and/or Foam Core surfaces, making it possible to utilize the magnet technique for attaching panel components independently.

**Adjustable, Erector Set Mockups for Operator Control and Display Positioning Definition**

As opposed to establishing control and display positioning using general reach criteria, it is highly desirable to establish optimum positions by placing an operator in a three-dimensional framework, unobstructed by preconceived consoles, racks, bulkheads, etc. As the accompanying illustrations show, an open framework provides an ideal method for adapting controls or displays to the operator. Control and display elements should be mounted on fully adjustable supports so that the control or display panel can be moved around until the best position is found in terms of line of sight or reach.

Controls should be mounted in such a way that the control device can be moved through its intended excursion, especially joy sticks, steering wheels, foot pedals, etc.

It is also possible to attach key visibility elements (window posts or frames, rearview mirrors, etc.) to evaluate fields of view.

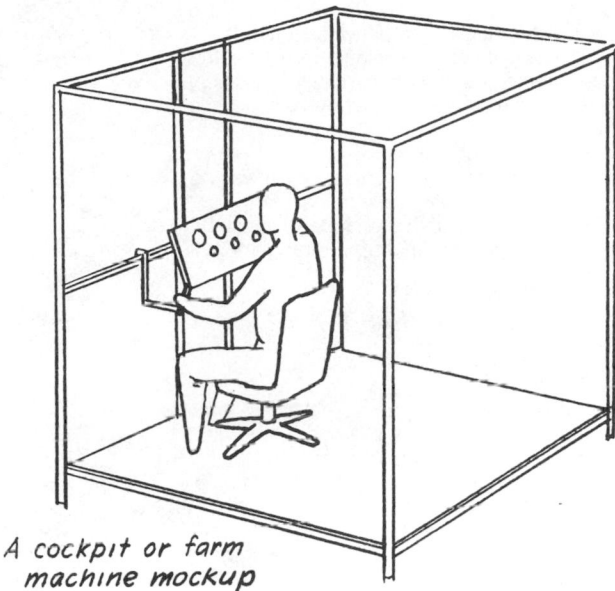

*A cockpit or farm machine mockup*

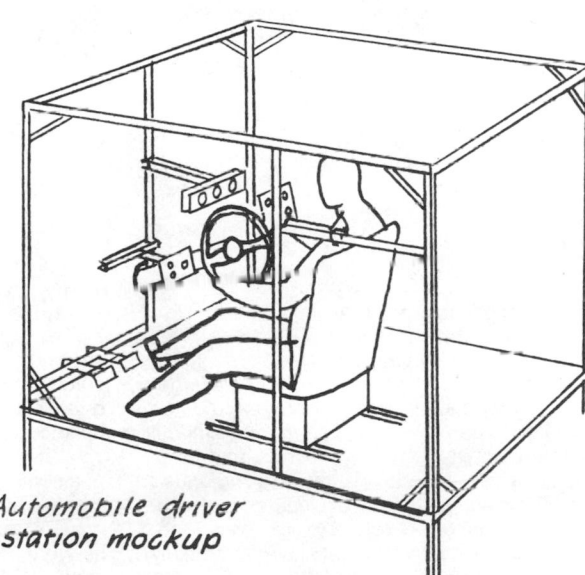

*Automobile driver station mockup*

As the accompanying sketches illustrate, service and maintenance questions can also be evaluated with the Erector Set mockup approach.

*Service equipment rack mockup*

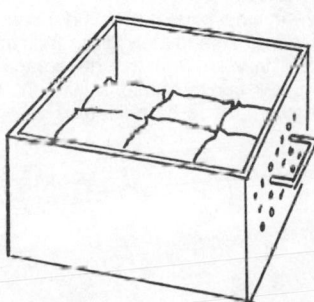

*Portable package/handle-locating mockup (note sandbag weights)*

## Simple Graphic Art Techniques for Mockup Fabrication

One should seek the simplest techniques commensurate with demonstrating the significant human-machine interface characteristics that are to be evaluated. Two of these are illustrated by the accompanying sketches.

Simulated three-dimensional visual control components made from readily available materials, if executed properly, give a realistic impression of the components that an operator would expect to see on a control panel. The clear plastic over the dial face allows one to anticipate annoying reflections.

The use of movable components makes it possible to try more than one arrangement.

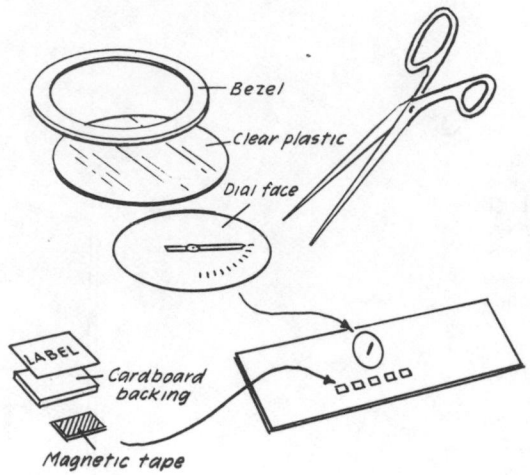

An inexpensive yet realistic-looking console can be constructed of a wood frame and smooth skin; the latter can be painted (using a water-base paint) to reflect the desired color scheme and demonstrate the desired contrast between the sheet metal and the control panels (assuming they are of a different color or shade).

Panels should be taped both inside and outside to provide structural integrity. When properly taped, the mockup will support considerable weight (e.g., actual metal panels and small instruments).

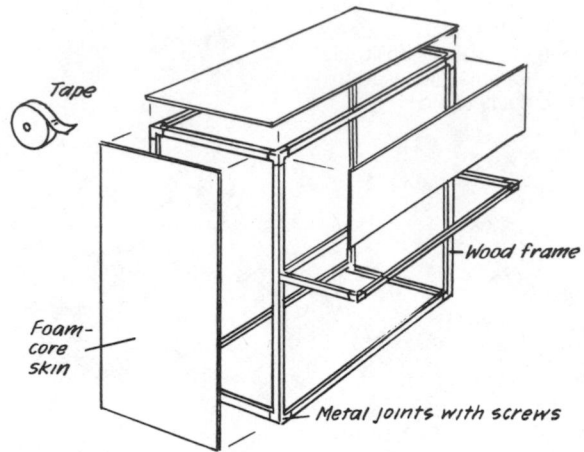

## Experimenting with Alternative Panel Layouts

More often than not, the designer has to utilize off-the-shelf hardware for most control panel designs. It is important, therefore, to recognize that many of the components may be larger "behind the panel" than they appear to be on the front side of the panel. In making cutouts of standard instruments, switches, and other components, it is wise to draw in (using dotted lines) the behind-the-panel clearance requirements so that one is not apt to arrange components closer together than they can actually be placed when it comes time to mount the actual instruments.

In experimenting with alternative panel layouts, one should make a checklist of the principal human engineering features that should be kept in mind, i.e., functional organization, sequence of use, frequency of use, and primacy or importance. In addition, however, one can also examine a number of factors that are not in the typical human engineering guides. For example, although there are criteria for spacing to prevent inadvertent activation and criteria for size of label letters, etc., spacing often is a matter of general appearance, e.g., balance, symmetry, and absolute clarity. One can "see" these characteristics only when experimenting with several different arrangements and spacings. For example, it may become obvious that, because the panel has to be very small, crowding obviously creates confusion. By making slight alterations and/or adding separator lines around certain related functions, one can alleviate the confusion. One can also tell rather quickly when the size variation among labels is not sufficient to provide an immediately clear indication of function levels. Even though criteria have been established by human engineers, these sometimes need to be adjusted.

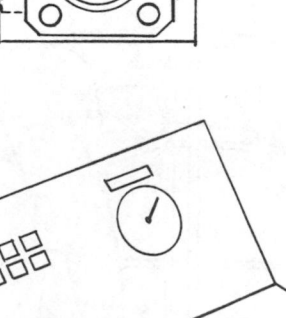

### Human-Machine System and Environment Simulation as a Design Tool

Any physical mockup is in a sense a simulation of the human-machine environment. As the system becomes more complex, however, it is often desirable to enlist the aid of more exotic apparatus, coupled with computer-aided scenario generation. As the accompanying sketch illustrates, one may require simulation (and control) of both external and internal environmental parameters, since all these influence how well the human-machine system functions in various modes. Not shown is the possibility of including dynamic properties such as oscillation and movement of the subject oper-

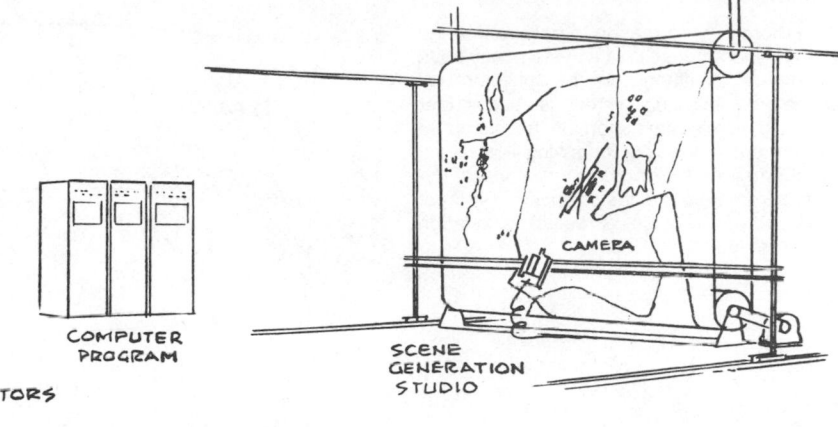

ator. The extent of realism depends on several things, including how important each variable may be to the eventual performance of the operator, how well it *can* be simulated (i.e., a poor simulation is worse than none), how much time is available to develop the rather complex simulation system, and, of course, whether the cost is justified. If the cost exceeds the cost of the ultimate hardware system, it usually is not justified. Avoid the temptation to create an exotic simulation just because it is a design challenge.

### Mobility Evaluation Mockups

Although there seems to be an abundance of information regarding architectural interfaces wherein the user climbs something, moves about, holds onto something, etc., it is always advisable to evaluate these in each specific application by means of a mockup. This is particularly true whenever the user is encumbered with things that must be carried or by

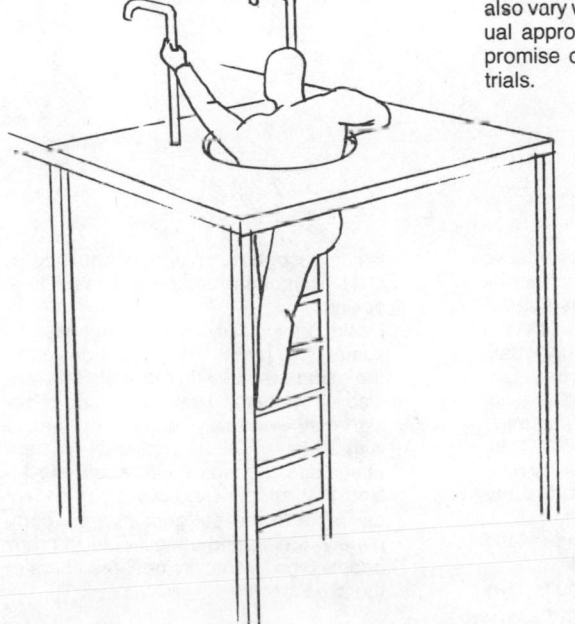

special restrictive clothing that must be worn. In the accompanying sketch, for example, the critical considerations are the exact relationship between the ladder below the opening and between the opening and the hand assist rails above the opening. A difference of just a few inches may be very important in preventing users from missing a step or losing their balance. The positions of these elements may also vary with the direction in which the individual approaches the opening. The best compromise can be determined only by physical trials.

### Visual Display Element Evaluation

Typically, one should rely on the criteria provided elsewhere in this handbook to determine the type of characters to use for labels. However, display nomenclature may be difficult to establish without evaluating the aspect of understandability. A simple tachistoscopic slide presentation is suggested, wherein one can test several alternative nomenclatures, abbreviations, or pictorial symbols to determine which ones lead to the fewest errors in interpretation. As illustrated, several subjects can be used simultaneously to evaluate the alternative display nomenclature, symbols, or even other visual parameters, such as size and color.

### Dynamic Simulators

When there is good reason to believe that the dynamic characteristics of a system may have a considerable influence on the behavioral response of the eventual system users, consider using a dynamic simulation of the expected environment. In the accompanying illustration, a total integration of both external visual dynamic parameters and the physical movement of the operator provides a realistic simulation of the real world.

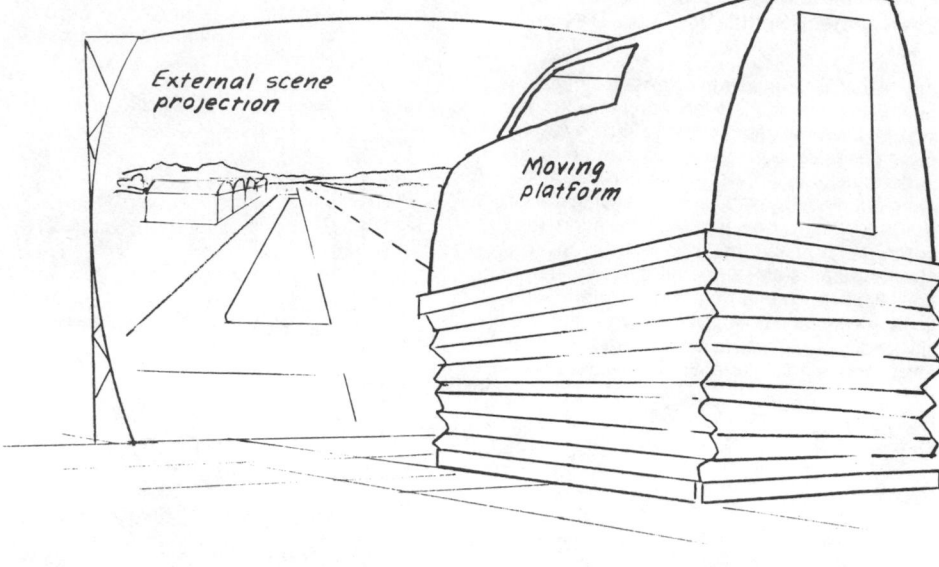

When more critical physical conditions such as positive, negative, or reduced gravity forces may affect the efficiency of an operator, consider the use of special simulators such as the centrifuge, acceleration and deceleration sled, or water tank (for weightless simulation), as illustrated in the accompanying sketch.

Other types of simulation to consider might include the pressure chamber or heat chamber.

*Caution:* Whenever dynamic simulation may involve conditions that explore stress levels that could be hazardous, enlist the aid of specialists to help design and monitor the test.

### Choosing Subjects for Mockup Evaluations

Although there may be good reasons (i.e., time and funding constraints) for limiting mockup evaluation to the designer and his or her immediate associates, the results may be far from satisfactory. There are several reasons for this:

1. Designers may be biased toward a design they have just created; i.e., because it is their own "brainchild," they really do not want to see any problems that the design may create for the user.

2. Designers are not always representative of the user population in terms of either physical or psychological characteristics; i.e., a designer may be extremely large or small, very strong or very weak, or particularly skilled in terms of equipment manipulation and understanding, or the designer may tend to "read into" the operation what the design is intended to demand because he or she already knows what is expected of the user.

3. A male designer may not understand how a woman will react to a design.

4. Designers, because they are adults, may fail to recognize or understand how a child or younger person will react to a design.

5. Designers typically assume that all consumers will react to a product design in the same way they do (after all, they are also consumers). However, since a design may eventually be used by people with different ethnic, cultural, or language backgrounds, things about the design that appear perfectly clear and/or desirable to the designer may be completely foreign and undesirable to many people who eventually become users of the product.

## Guidelines for Selecting Mockup Evaluators

The following guidelines are provided to aid the designer in selecting subjects to use as evaluators.

**POPULATION** Select evaluators directly from the expected user population, such as military personnel (as opposed to civilians), adults (as opposed to children), or members of a particular professional or vocational group.

**BODY SIZE** Select evaluators from specific or general populations on the basis of recognized anthropometric surveys. For example, several professions (Air Force, Army, and Navy personnel; astronauts; bus drivers; etc.) may be limited by anthropometric selection criteria; i.e., they will be only so tall or so heavy because a particular agency has established selection criteria that preclude taking applicants outside these dimensional ranges. On the other hand, not only do civilian populations include persons with a wider range of body sizes, but also the range is expanded because of inclusions of both males and females and both young and old people.

**STRENGTH** When the mockup evaluation involves lifting and carrying the product and/or manipulating controls that may have fairly heavy force requirements, the evaluator population should represent mainly the weaker members of the expected user population.

**INTELLIGENCE, EXPERIENCE, AND TRAINING** In certain cases, it would be wasteful of human capability to try to make a product acceptable to everyone; i.e., some people will never be able to cope with very complex equipment operations, even with extensive training. When the operation of the proposed product normally would be restricted to specialists already known to have the requisite knowledge and skills, there is no need to compromise the proficiency of the human-machine system (i.e., to make it unduly simple). In such cases, evaluators should be chosen who are already knowledgeable about the proposed design and therefore can evaluate the mockup or simulation from a skilled operator's point of view.

On the other hand, if the product may be operated by unskilled individuals (and therefore may create a critical training requirement), evaluators should be chosen from an unskilled group in order to determine what there may be about the design that is confusing or difficult to manipulate. This provides an opportunity to determine what modifications may be necessary to make the design more "trainable" and less apt to create special personnel selection requirements for the eventual customer.

**PRODUCT ACCEPTABILITY** In evaluating a proposed design (via a mockup), one should consider a highly varied sample of evaluators, i.e., the sample should include people of both sexes and of all sizes, ages, and backgrounds in order to allow for the random inclusion of personal attitudes and opinions, physical difficulties, etc. Although such subject samples should be generally random in terms of all these varying characteristics, it is often necessary to make sure that specific human characteristics are represented. This is important whenever the designer suspects that certain characteristics of the design may create special problems for certain members of the user population. For example, in evaluating seat belts, it is known that large and small people have certain kinds of problems. Although one wishes to obtain a general evaluation of how people may expect to use a system or how they may react to certain features of convenience, it is important to make sure that a sufficient number of the large and small evaluators evaluate those features of any belt system that cause belt systems not to fit properly. Thus, although the sample should generally represent the random attitudes of a broad sample of users, one should add a few subjects who are representative of the extreme anthropometric characteristics of the general sample.

**TECHNOLOGICAL VARIABILITY** There is a tendency to believe that just because we are a technologically oriented society, all product users understand and can cope with modern machines. Studies have shown that this is not true. In some societies, people have not had the opportunity to grow up with modern technology; therefore, not only are they unable to cope with the demands of modern machines, but they also may completely reject machine-age products. This presents a difficult problem for designers, for seldom do they have an opportunity to study the population in question, nor are there typically any study data to help them determine what a particular population will do or expect. The principal recourse under these circumstances is to search for subjects who may be recently from the area in question, using these evaluators as a screening group to determine whether there will be major difficulties as a result of certain characteristics of the design that have been mocked up for evaluation.

## Guidelines for Determining Number of Mockup Evaluators

First of all, the designer may be confronted by the problem of convincing management that it is important to evaluate a mockup in more than a cursory sense. For example, many management people believe that *they are the final judge* in determining whether a design is satisfactory. Although it is probably true that some manager will have the final say, it is dangerous to assume that such a person has the knowledge to make judgments regarding effective human-machine performance. Convincing such a manager to rely on a more scientific approach to product evaluation, one that reliably reflects the specific user's response to the design, requires considerable diplomacy on the part of the designer. Consider the following in deciding whom to use as evaluators and how many to use.

**EVALUATING REACH AND CLEARANCE** Although a fairly large, random sample of subjects is required when one wishes to establish some statistical distribution for reach and clearance, in most cases it is satisfactory to evaluate these parameters in a proposed design (via a mockup) using a minimum of about five large subjects to evaluate clearance and about five small subjects to evaluate reach. One should select these groups on the basis of the particular population for which the product is being designed, such as a military or civilian group. For clearance, select five subjects whose stature is at approximately the 95th percentile and whose weight is also at about the 95th percentile. For reach, select five subjects whose stature is at about the 5th percentile (for a mixed user population, use 5th-percentile females). For all practical purposes, selecting five subjects at each end of the stature range will provide sufficient variability in the other characteristics to provide an adequate mix of characteristics that also relate to clearance and reach, e.g., variations in hip breadth, buttock-to-knee length, knee height (sitting), head height (sitting), eye height (sitting), head height, shoulder height, shoulder breadth, abdominal depth, and lengths of upper and lower arm segments. In addition, one usually can depend on a reasonable variation among 5 or 10 subjects in terms of overall body and limb mobility.

**EVALUATING WEIGHT CHARACTERISTICS OF A PORTABLE PACKAGE** The minimum number of subjects required to evaluate the weight characteristics of a package mockup is about 10. One should select relatively small subjects since not only is the size of the subject a general indicator of overall strength, but also the ability to lift is a function of the length of a person's arms, the size of his or her hands, and the general girth of the individual's abdomen, chest, and shoulders. One or two of the subjects should be fairly fat, since the fat individual may have difficulty holding the package close to his or her chest and stomach.

**EVALUATING CONTROL MANIPULABILITY** Since a number of significant variables influence how well an individual may be able to manipulate a control device, a minimum of 20 subjects should be used. In general, the subject sample should be selected at random, since one hopes to sample a variety of individual differences, including anthropometric differences and differences in mobility, dexterity, steadiness, strength, sensorimotor proficiencies, and psychological expectations. Depending on the expected user population, both age and sex should be represented about equally throughout the sample; i.e., the subject sample should consist of half males and half females, and there should be a wide distribution of ages.

**TEAM PROCEDURE: MOCKUP AND SIMULATION EVALUATIONS** When more complex mockups are used to simulate team function and procedural response, a minimum of three complete teams should be used; i.e., as opposed to having an evaluation made by a single team of subjects, repeat the evaluation with at least two additional teams of subjects. This is desirable because one is never sure whether the first team is either unusually good or unusually poor, nor can one be sure, if the performances of the two teams vary significantly, which one's performance is truly representative of the system's quality in terms of design effectiveness. The response of the third team provides a means for breaking a tie, in addition to indicating which response performance is probably more representative of the design quality.

**COMFORT AND CONVENIENCE EVALUATION** An evaluation of the comfort and convenience aspects of a product or system mockup is perhaps the most difficult type of evaluation to perform, because in spite of all attempts to focus the evaluator's attention on significant design areas, each individual makes a different internal interpretation of what constitutes comfort or convenience. Therefore, it is recommended that a minimum of 50 subjects be used for this type of mockup evaluation. It is further suggested that, in performing this kind of evaluation, a special evaluative approach be taken, i.e., one in which subjects are forced to focus on, and respond to, specific, predetermined aspects of the design, one at a time—or at least short groups of aspects that can be separated in terms of operational steps. An example would be the case of evaluating seat belts in an automobile. Here, one might segregate the several typical use sequences, i.e., entering the vehicle and donning the belt, adjusting the seat and closing the door, simulating operating the vehicle (looking out the windows, reaching for the controls, looking at the displays, etc.), and finally doffing the belt and getting out of the car. Upon completing each of these sequences, the subject would be asked to rate various features of the belt system as they pertained only to that particular operating sequence.

Also important in this area of comfort and convenience are dynamic and environmental factors. Although one can obtain a general, first approximation of consumer response in the static mockup, it is probable that a person's responses would be modified after actually experiencing the seat belt under more realistic conditions, i.e., after riding for awhile over various road conditions.

### Establishing the Basis for Test Subject Selection

#### PRODUCT FORM AND FIT PROBLEMS
When the problem involves physical relationships between an operator and some hardware element (seat, console, cockpit, furniture, etc.), test subjects should be selected to represent the critical dimensional limits of expected operators. First, it is important to have several subjects who represent the largest people who may have problems of clearance and several subjects who represent the smallest people who may have problems of reach. Second, there should be a scattering of subject sizes in between the largest and the smallest in order to provide assurance that some in-between dimensional incompatibility is not overlooked. In a sense, we have a stratified sample of sizes because we know that certain physical accommodation problems are more critical at the extremes.

It is standard practice in the military services to attempt to fit 90 percent of the population, i.e., from the 5th- to the 95th-percentile population of a particular military sample. This means, however, that, in the event the product is to be used by both male and female operators, the test sample should range from the 5th-percentile female through the 95th-percentile male (as opposed to ranging from the 5th- to the 95th-percentile male when only male operators are expected).

#### OTHER PRODUCT FACTORS
When the questions to be answered do not relate to the physical size of the expected user population, a more random sample of test subjects should be used. For example, when we are investigating sensorimotor response relative to some proposed display and control system, body size has less to do with the subject's re-

sponse, and therefore we need a random sample of subject skills, experience, motivation, and attitudes in order to assess the average response of an expected user population.

If the problem has something to do with human strength, it may be desirable to select more subjects who are weaker, since these are the users who may have more difficulty with problems of control force, lifting, and so on. It should be noted that one cannot determine how much strength a subject can apply merely on the basis of his or her size, although it can generally be assumed that females are weaker than males.

If the product may be used by older people, it is important to include a number of elderly persons, both male and female, to reflect the compatibilities or incompatibilities of the proposed product design with typical age-related infirmities, e.g., poor vision, reduced mobility and strength, and slower response time.

Similarly, if handicapped users are expected, a variety of handicapped test subjects should be included to reflect the appropriate range of infirmities and difficulties they may have, such as being confined to a wheelchair.

Ethnic differences, such as differences in language or technological background, may be an important aspect of human factors testing, and the appropriate selection of test subjects who represent unique design interface barriers must be considered. For example, a test to determine how intelligible certain pictorial symbols are to people of various nationalities might be appropriate.

#### GENERAL PRODUCT ACCEPTANCE
For most consumer acceptance tests or surveys, it is desirable to take a completely random

sampling approach since the normal sales objective is to reach as broad a population of purchasers as possible. In such cases, it is up to the customer to establish the limits to which he or she expects to promote the product, i.e., to the world, to a specific country, or to a specific part of a country. It should be observed that this type of sampling generally relates to consumer opinions and attitudes, which change frequently. Therefore, not only should sampling be approached on an area-by-area basis, but also a successive series of surveys should be carried out to allow for opinion shifts. A good example is provided by the automobile industry. Some people have to get used to a new model of an automobile; i.e., the new body style may be sufficiently radical to "turn them off." However, after 6 months to a year, they are accustomed to the new style, and it becomes the reference by which they measure satisfaction with other new models.

### Developing a Composite Anthropometric Population

When anthropometric survey data on several specific populations are available and when it is desirable to know the distribution of these dimensions as a group rather than as separate populations, a composite population can be synthesized. An example is shown in the accompanying illustration, wherein a certain troop transport is to carry 30 percent cadet pilots, 45 percent gunners, and 25 percent nurses, whose respective heights are given in the graph. The computation is shown in the table on page 1003, with the final computation plotted (dashed line) on the initial chart.[26]

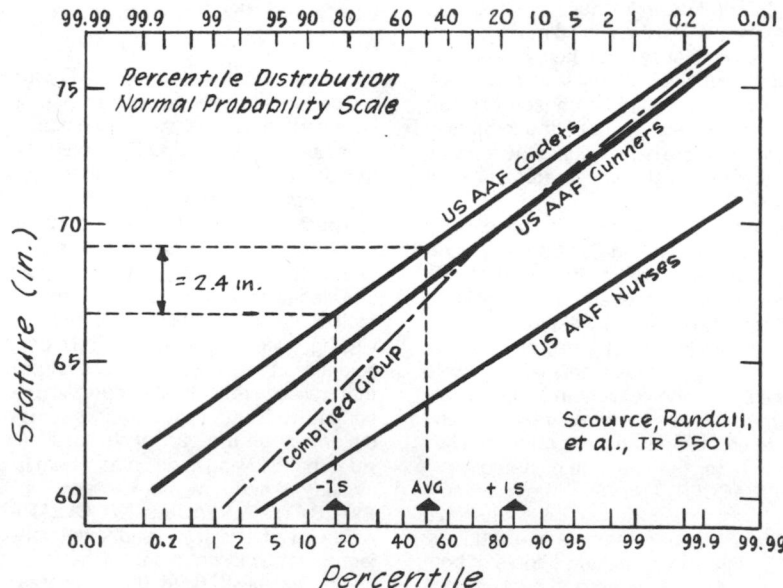

Illustration of the use of normal probability paper for developing composite populations (Roebuck, 1975).

[26]After J. A. Roebuck et al., *Engineering Anthropometry Methods,* John Wiley & Sons, Inc., New York, 1975, pp. 158, 159.

### A Method of Forming a Composite Population Dimension Distribution from Three Populations with Different Means and Standard Deviations (Roebuck, 1957)*

| | | Cadets | | Gunners | | Nurses | | Combined |
|---|---|---|---|---|---|---|---|---|
| | Stature, in | Percentile | Percent in New Population | Percentile | Percent in New Population | Percentile | Percent in New Population | Population Percentile |
| Column No. Operation | (1) | (2) | (3) Col. (2) × .30 | (4) | (5) Col. (4) × .45 | (6) | (7) Col. (6) × .25 | (8) (3) + (5) + (7) |
| | 61 | 0.03 | 0.009 | 0.3 | 0.135 | 12.5 | 3.13 | 3.27 |
| | 64 | 1.4 | 0.42 | 6.0 | 2.7 | 59.0 | 14.75 | 17.87 |
| | 67 | 18.0 | 5.4 | 36.0 | 16.2 | 94.6 | 23.7 | 45.3 |
| | 70 | 64.0 | 19.2 | 80.0 | 36.0 | 99.9 | 24.98 | 80.18 |
| | 73 | 94.6 | 28.4 | 98.0 | 44.1 | 100 | 25 | 97.5 |

*Percentiles determined from accompanying figure = 67.4 in; 95% — 72.2 in; 5% — 60.4 in.

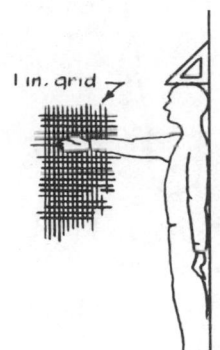

## Measuring and Weighing Test Subjects

In order to verify that one has representative test subjects (i.e., to represent established 5th and 95th percentiles), one should not depend entirely on the subjects' reports of their height or weight. In addition, in many instances the designer needs one or two other measurements that average subjects will not know about themselves. The rules for measuring are quite simple as long as a few rules are followed.

### 1. Stature
To measure stature, have the subject remove his or her shoes and stand with the head, shoulder blades, buttocks, and heels flush against a wall. Using a draftsman's triangle (see the accompanying illustration), measure the subject's height from the floor.

### 2. Arm Reach
Ask the subject, while he or she is still standing against the wall, to raise and point the arm straight forward with the palm of the hand facing inward and with the fingers extended. Measure the distance from the wall to the tip of the longest finger, usually the middle finger.

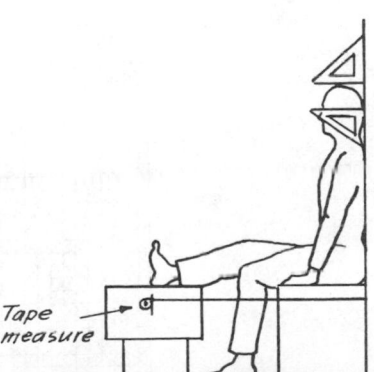

### 3. Leg Reach
Have the subject sit on a flat box with the head, shoulder blades, and buttocks against a wall. An adjustable seat and footstool should be used so that the subject's feet will be flat on the floor. Have the subject raise the right leg and lay it on the footstool. Older people may need help. Measure the distance from the wall to the subject's heel. A small triangle will help define this distance. Do not be disturbed by the fact that some people cannot fully straighten out their legs. However, have the subject straighten the leg as much as possible.

### 4. Weight
A standard health scale should be used. Subjects should be as lightly clothed as practical (see the accompanying illustration). This will require a nurse or another woman when the subject is a female.

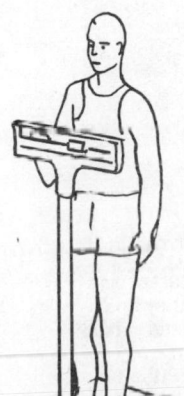

### 5. Strength
Although not absolutely valid, hand dynamometer strength is a relatively good indicator of an individual's overall strength capability. Hand dynamometers are not easy to find unless one is purchased for permanent retention. Such instruments often can be borrowed from a college or university that has an anthropology department. Usually one can also be found at a local hospital. Although instructions could be presented here, each dynamometer may be different, and thus one should ask for instructions from those from whom the particular instrument is borrowed.

The above measurements are usually sufficient for most design work where one is trying to make sure that test subjects are representative of a particular population's extremes. If, however, other measurements are required for some special design purpose, it is recommended that the designer seek help from a trained anthropologist.

## SPECIAL TOOLS

### Two-Dimensional Drafting Templates

Two-dimensional templates representing the physical dimensions of the human body are useful for initial estimates of operator-equipment interface "fit." Although some templates are available (see the accompanying illustration), one generally has to fabricate his or her own set of manikins to whatever scale is desired. To accomplish the typical range of drawing applications satisfactorily, one should develop and use both male and female templates in both the 95th- and the 5th-percentile ranges. The 5th percentile is needed to evaluate reach problems, and the 95th to evaluate clearance problems. Both will be required to evaluate eye-level reference requirements, since both tall and short people must be accommodated.

Two-dimensional drafting templates designed for automotive use can be ordered from the Society of Automotive Engineers, 400 Commonwealth Drive, Warrendale, PA 15096.

Two-dimensional templates representing U.S. Air Force personnel can be ordered from the 6570th Aerospace Medical Research Laboratory, Wright-Patterson Air Force Base, Ohio 45433, Attention: Mr. Kenneth W. Kennedy.

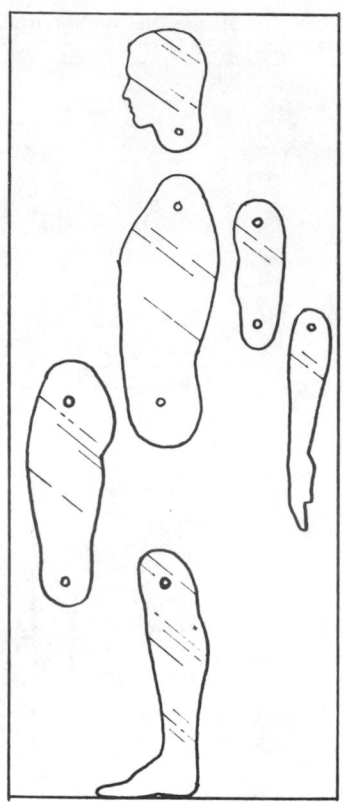

Unassembled

The accompanying illustrations show how an articulated manikin is cut out and joined together to provide an articulating template for drawing around. It is suggested that manikin parts be made of (at least) $\frac{1}{16}$-in (0.16-cm) transparent plastic (preferably fluorescent orange) so that one can see through the template while drawing.

Following are specifications for three of the most often-used templates, i.e., a 5th-percentile male, 95th-percentile adult male, and a 5th-percentile adult female. Note that these templates are representative of nude dimensions, and thus clothing allowances should be made during design.

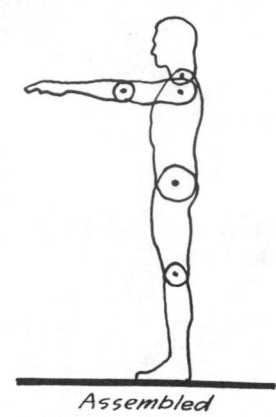

Assembled

### 5th-Percentile Civilian Male Manikin

Manikins can be made out of transparent plastic. Parts should be riveted so that they articulate freely. The thickness of the plastic should be sufficient to provide adequate rigidity for ease of tracing around the manikin (the optimum thickness will be a function of the selected scale and overall size of the finished manikin).

**95th-Percentile Civilian Male Manikin**

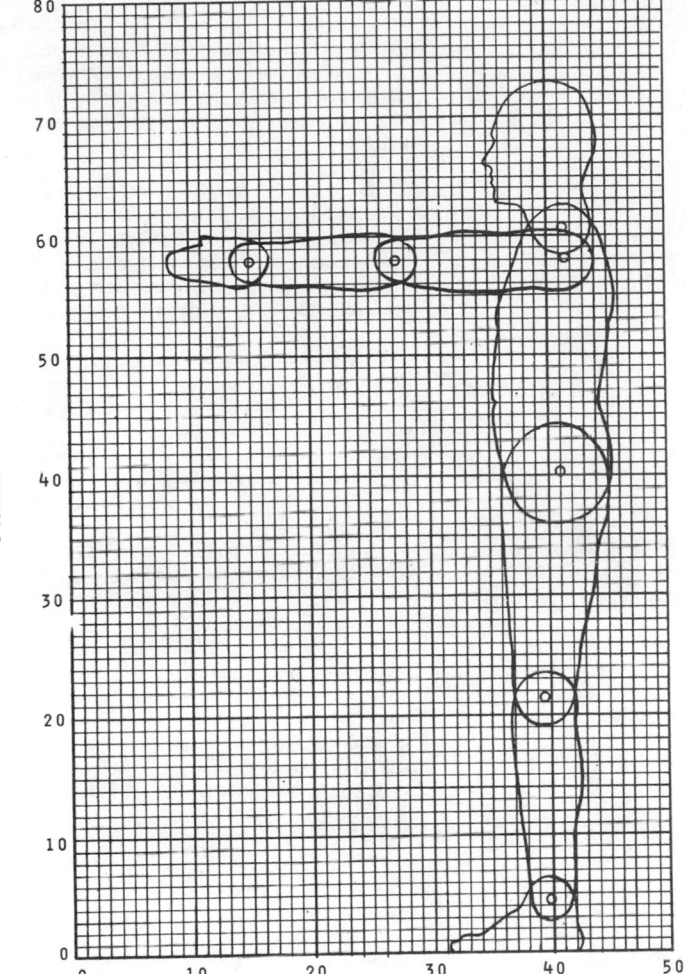

**5th-Percentile Civilian Female Manikin**

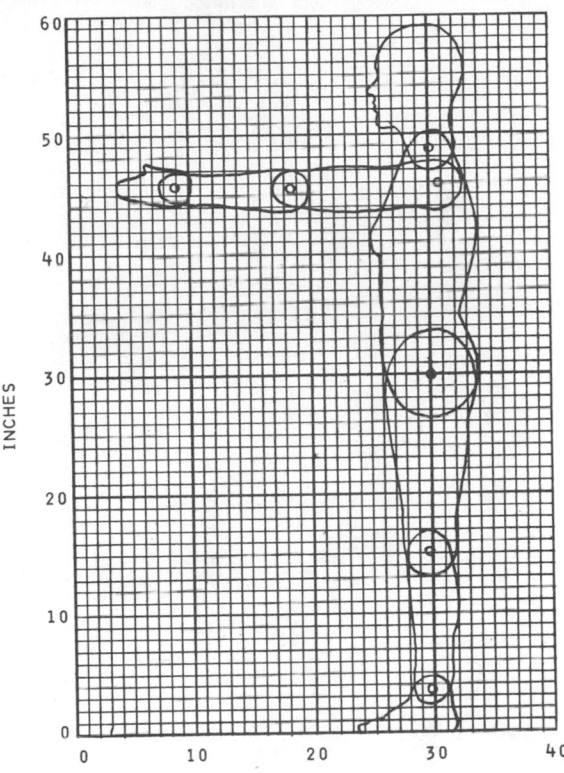

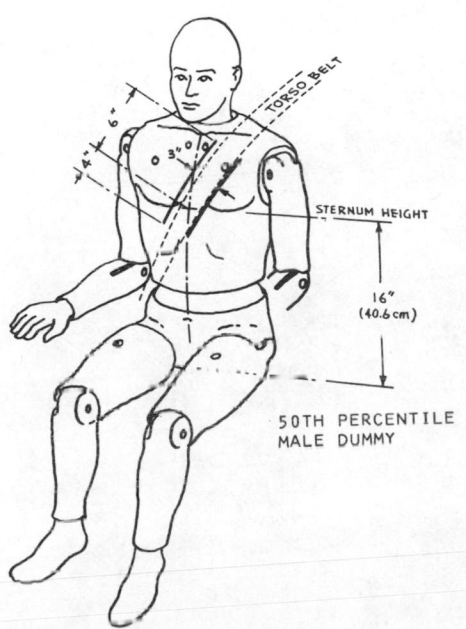

50TH PERCENTILE
MALE DUMMY

STERNUM HEIGHT

16"
(40.6 cm)

### The Articulating Anthropomorphic Dummy

The principal purpose of the articulating anthropormorphic dummy has been for sled or crash testing. Since the dummy is designed to articulate at key points (i.e., head and neck, shoulders, elbow, wrist, waist, hip, knee, and ankle), it responds to crash loads *somewhat* similarly to the way the actual human body does. Dummies can be instrumented (using force sensors) to measure dynamic forces at various points on the dummy. Although the articulating anthropomorphic dummy is extremely useful, one must not assume that it actually responds identically to the way a human body would. It is considerably less flexible.

Recently, the dummy has been used as a bench mark for establishing other design crite-

ria. For example, the dummy can be marked to help evaluate the crossing pattern of a proposed seat belt; i.e., parallel lines spaced 3 in (7.6 cm) apart and at an angle of about 55° will define the envelope in which a torso belt should cross the dummy's chest. This makes it possible to test whether the proposed belt will fit the general range of seat belt users (from a 5th-percentile female adult to a 95th-percentile male adult).

Although similar dummies have been designed to represent the 5th-percentile female American driver and the 95th-percentile male American driver, the 50th-percentile dummy is the only one recognized at the present time for crash testing purposes by the U.S. government and the automobile industry.

## Commonly Needed Tools for Human Engineering Testing and Design Evaluation

Although there will be times when more elaborate testing facilities and measuring equipment may be needed during particular human engineering design programs, the ones listed in the accompanying table are most commonly required.

| Type of Tool | Typical Uses |
|---|---|
| Anthropometric tools:<br>  Flexible tape<br>  Anthropometer (designers can fabricate their own less expensively)<br>  Adjustable-height chair<br>  Adjustable-height footstool<br>  Health scale | To select test subjects for evaluating a mockup. The important measurements most often taken include standing height, sitting height, arm length, leg length, hip breadth (sitting), and weight. |
| Force-measuring tools:<br>  Spring scales (0–10, 0–40, and 0–100 lb; 0–4.5, 0–18, and 0–45 kg)<br>  Torque meter (0–10 oz. 28 g and 0–40 oz, 1130 g)<br>  Strain gauges (0–10 lb, 0–4.5 kg) | To test forces required to operate various manual and pedal controls, forces required to open doors, weights of packages that must be manually moved or transported, or forces that a seat belt imposes on an occupant's body. |
| Sound-level measurement equipment:<br>  Sound-level meter<br>  Half-octave band filter | To measure sound levels of noisy environments. Equipment should provide means for measuring both an overall sound level and levels within critical frequency bands. |
| Light-level measurement equipment:<br>  Light meter | To measure light levels falling on surfaces within the visual field and the light output of specific illuminated displays (down to the specific display markings). |
| Temperature, humidity, and airflow measurement equipment:<br>  Wet- and dry-bulb temperature-measuring units<br>  Volometer | To measure ambient room conditions, surface temperatures, airflow from ventilators, etc. |
| Oxygen-consumption measurement equipment | To measure the energy demand of the task on the subject. |
| Heart-rate and pulse-rate measurement equipment | To measure task stress imposed on the subject. |
| Galvanic-skin-response measurement equipment | To measure task, environment, and psychological stress imposed on the subject. |
| Time and motion measurement equipment:<br>  Stopwatch<br>  Automatic timer<br>  Color motion picture camera<br>  Variable-speed (slow-motion) projector | To obtain and measure task activity information about subject for later analysis. |
| Audio tape recorder(s):<br>  Pocket type<br>  Voice-actuated type | To record verbal conversations, make verbal notations, etc., during a test. |
| Gas chromatographic equipment | To record and identify gaseous conditions in a closed ecological system. |
| Slide projector<br>  (with statistoscopic shutter) | To present timed-exposure targets or other information to test legibility and interpretability of labels, nomenclature, type style, and figure-ground contrast ratio. |
| Strip recorders | To provide hard copy data for later analysis. |

## Home-Made Anthropometer

The accompanying sketch illustrates how a home-made anthropometer should be dimensioned and fabricated. Its application is illustrated in the lower sketch.

It is suggested that the anthropometer be constructed out of approximately ⅛-in (0.32-cm) aluminum and that it be hinged as shown at the top by means of a wing nut so that the two portions of the device can be tightly secured once the dimension is set.

The two probes at the lower end of the device should be slightly rounded on the ends so that one will not accidentally poke the subject with the sharp corners.

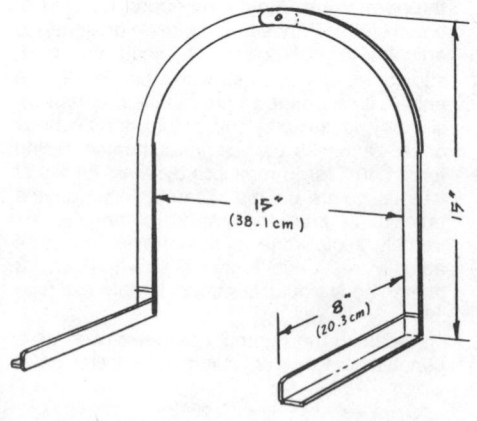

## TESTING

### General Considerations for Conducting Human Engineering Tests

First, it is highly recommended that an experienced human factors specialist be employed to assist in the design and conduct of human engineering tests whenever this is practicable. It is particularly important if the proposed test deals with several highly interactive variables that require complex experimental designs in order to isolate and establish the significance of certain variables relative to human performance. It is equally important to acquire help when a particular test may involve possible hazards to the test subjects.

However, designers can conduct certain types of tests on their own as long as they keep in mind certain problems associated with testing that involves human subjects. The following should be considered:

1. A sufficient number of test subjects and trials should be used to ensure that the test results do not represent some sporadic incident not representative of the true performance of the subjects.
2. Sufficient procedural control should be provided so that one can safely infer that, "when this adjustment or input occurs, with these conditions, this is the typical response or human output." Generally one must keep certain features constant while a single variable is adjusted so that, in the end, it is possible to correlate specific conditions with specific human responses.
3. In most cases, test subjects should not be privy to the real purpose or objectives of the test; they should be informed only of the procedures they are to follow. This helps avoid "expectancy biases," wherein subjects may attempt to give the experimenter the response they think the experimenter wants, rather than performing according to their individual or unique response characteristics.
4. The test procedure should cause entry and sequential aspects of the test to be randomly presented; i.e., it should not be possible for the subject to "learn" a sequence or anticipate a step because of start cues. Attempt to remove any possibility that "order effects" show up in the final subject response.
5. When a particular hardware configuration is being evaluated in terms of whether subjects perform better on one version than another, make sure that alternative configurations are equally well represented (in terms of quality of realism, workmanship, and to visual attractiveness).
6. Experimenters should remain as nonconspicuous as possible during evaluations in order to minimize inadvertent cues they might provide the test subjects during a test sequence.
7. Attempt to set up the test so that the trials are long enough to identify the difference between "learning" and informed response. However, trials should not be so long that fatigue produces artifacts in the results that have little to do with the basic evaluation objectives. Note, however, that fatigue effects may be a significant test objective, in which case one must design the test so as to differentiate between design-induced fatigue and fatigue that results merely from the fact that the test subjects are generally tired or bored with the activity.
8. Whenever possible, design all tests in such a way that quantitative (as well as qualitative) results can be documented. Avoid general opinion-oriented evaluations unless these are quantified in terms of forced choice ratings or rankings by test subjects. Incidentally, this is desirable even for general VIP evaluations of a proposed design. Although some visitors may object to being used as test subjects, it is possible to formalize their review in a manner which is not objectionable to them but which is organized sufficiently to allow one to tabulate and compare opinions.
9. Select the right subjects; e.g., avoid using engineers to represent a typical nonengineering user population.
10. Obtain important demographic information about each subject so that, in the event it becomes necessary to evaluate test results in terms of subject differences, you do not have to try to find a particular subject again. In fact, attempt to screen out subjects who for some reason may not be able to complete a test series or who may not be available for a second test.
11. When a test involves visual acuity (visual display, illumination level, etc.), either select subjects on the basis of their having normal acuity (by giving the subjects an eye test), or specifically seek out and include subjects who have the typical visual anomalies you expect among the eventual user population. For example, if the user population will include older people, who normally require higher illumination levels and larger visual detail to see properly, be sure to include subjects who are older and may have to wear corrective lenses.
12. When the design to be tested may be used by persons with various handicaps, select a broad range of handicapped subjects because, although a given subject may have a particular type of handicap, each such individual tends to vary in terms of how he or she copes with various mobility, manipulative, visual, or auditory tasks.
13. Carefully plan any human engineering task in terms of the procedures to be used and practice conducting the test prior to conducting the main test. Remember that, once a test has begun, it should be continued to completion; i.e., avoid having to stop in the middle of a test sequence because you have forgotten something or have lost your place in the test sequence or because the data-recording system has failed. Nothing aggravates test subjects more than feeling that the experimenter is not sure of what he or she is doing.
14. If a particular design evaluation involves special considerations, such as time of day (i.e., both day and night ambient illumination conditions) or special clothing (i.e., light summer clothing and/or heavy winter clothing), be sure to include these conditions in the test.
15. Make sure that the mockup and/or test apparatus is fully checked out and will not fail during the test. Be especially careful to see that there are no unforeseen hazards either for the subjects or for the experimenters. If there appear to be certain stress conditions associated with the test, make sure that none of the subjects who are selected have tendencies or conditions that could be compromised during the stressing phases of the test (a potential for epileptic seizure, heart failure, etc.).

### Field Testing and Evaluation

Nothing provides a more conclusive demonstration of adequate design than the actual field test of a prototype model. Typically this is done in the case of airplanes, highway vehicles, seacraft, agricultural machines, etc. Too often, however, such tests rely on the opinions of so-called expert test personnel, e.g., test pilots or test drivers. Although there is often a good reason to utilize these experts (for reasons of safety and because they are perhaps more skilled and perceptive in their analysis of features that may not be quite right), the real proof of design acceptability comes when the intended user (who lacks professional skill) can also operate the system.

Although the main objective of most field tests is to prove that the system will do what it is supposed to do, too often one may not obtain the type of information from such tests to suggest modifications that will make the system easier to operate by the average user. In addition, the field test should provide other information that is important to the eventual user; e.g., previously described tasks can be verified, training objectives can be confirmed, and training aids can be evaluated.

A field test should be designed with the following objectives in mind:

It will demonstrate the reliability of the hardware under all mission and environmental conditions.

It will demonstrate that the training program provided all the necessary skills for the operator or maintainer to cope with the operational conditions under which the system is to be used.

It will pinpoint any hardware, software, operator interface, or procedural discrepancies in such a way that one knows what to do to correct deficiencies in any of these areas (e.g., hardware, software, human interface, procedures, and training).

It will allow one to verify whether estimates of manning level are correct.

It will allow one to evaluate whether estimates of time factors were correct (service turnaround, loading, boarding, emergency escape, etc.)

It will help determine any safety hazards that may not have been anticipated.

It will provide an initial impression of consumer acceptability of the product.

## Methods and Techniques for Human Engineering Field Testing

**OPERATOR PERFORMANCE MEASUREMENT** A quantitative measure of operator and maintainer performance in terms of time and error should be provided; i.e., in addition to the probable effort to elicit personal evaluations by the test subjects and/or expert test operators, actual measurement of the performance of these participants should always be generated wherever practicable.

**COMPLETE TEST OPERATIONAL SCENARIO** The field test should be designed and implemented so that one systematically exercises the human-machine system through all phases of the operating scenario. This should include both normal and emergency conditions. The scenario should also be complete in terms of all the human-system interactions: visual, auditory, mobility, dexterity, communications, decision making, and control.

**REAL-WORLD ENVIRONMENT** The test action should occur on site, i.e., on real roadways and streets, in rough water, during the day and at night, and under different environmental conditions (rain, snow, heat, cold, high wind, fog, etc.).

**PERFORMANCE RECORDING** In addition to the actual instrumentation of operator interface responses noted above, consider the use of other covert techniques such as voice recorders, motion picture cameras, and human physiological sensors (pulse, heart rate, galvanic skin resistance, etc.).

**OPERATOR OPINION** Provide a means for documenting the opinions of test pilots, drivers, and/or typical test subjects, not only in terms of posttest debriefing, but also in terms of on-line evaluation during the testing (e.g., a voice recorder). Debriefing questionnaires should be designed to elicit specific, design-related comments, as opposed to general verbal descriptions of what the subject felt about the equipment.

**TEST OBSERVERS** Where appropriate, consider the use of trained independent observers to take notes while the test is under way. The independent observer may be important because the actual test subject is sometimes too busy to observe critical events as they are happening.

Trained observers are also effective in performing general population activities. For example, strategically located observers can observe and document how many drivers are wearing seat belts, how pedestrians behave with respect to crossing streets or using paths on a campus, or how individuals utilize a particular piece of equipment, such as a ticket vendor, a coin-operated laundry machine, an elevator, or an escalator.

Equip observers with tabulating devices where appropriate (stopwatch, hand counter, sketch pad, questionnaire and documentation record pad, etc.).

**TEST SUBJECT INDOCTRINATION** All field test subjects should be properly indoctrinated before they participate in any field testing. Depending on the complexity of the test, try to develop a brief statement of the purpose and objectives of the test and a general overview of what will be expected of the subject prior to, during, and following the test. This should be written down so that all subjects receive the same instructions. In some cases, it may also be desirable to provide written instructions during the test; i.e., so that each subject receives exactly the same instructions, each step is prepared and presented on a slide or by means of a recording or is read to the subject by the experimenter.

**TEST CONDUCT** Although it is generally best to leave the subjects alone as much as possible in order to reduce the possible influence of an experimenter's method of instruction, voice inflection, or body language, it is sometimes desirable to have the subjects verbalize about what they are doing while the test is in progress. This should be part of the indoctrination, with possible reminders by the experimenter if the subjects seem to forget.

When several different experimenters are to be used, allow them to practice until they demonstrate that they perform consistently and in a similar fashion.

In certain kinds of field testing, it will be necessary to assign the responsibility for safety to a given individual; i.e., although each experimenter is responsible for the safety of his or her particular subject during the test, general overall safety monitoring should be provided. This is especially important when several equipment or vehicle components are engaged in an interactive operation.

Safety equipment should be provided to deal with emergencies in certain types of field tests, such as fire trucks, rescue equipment and personnel, and medical personnel.

When extended-duration field tests are being conducted, it may be necessary to provide housing, food, or other support for both the test personnel and the test subjects.

**IN-THE-FIELD TEST DATA ANALYSIS** Depending on the nature, location, and duration of a particular field test, it may be important to provide facilities and equipment for analyzing data on the spot. Such data may be important to the test conductor in terms of deciding whether to make modifications to the test schedule or procedures. Alternatively, it may be possible to transfer data via various telecommunication methods to a base site or to a laboratory where large computer facilities are available. Such remote analysis is often required when the basic system is operating in the air or in space.

**HAZARDOUS-ENVIRONMENT TESTS**
When a test involves the use of test personnel or subjects in a hazardous environment, it is extremely important to fully instrument both the system and the people involved so that the test conductors can monitor in real time the exact status of the hardware and the individuals involved, such as an astronaut aboard a space vehicle or a pilot performing unusually hazardous maneuvers in an aircraft. Fail-safe communication links must be provided so that information links are not broken as a result of physical conditions or environmental anomalies.

Wherever practicable, rescue personnel must be made available for quick emergency response. In addition, system experts should be at hand to advise the test personnel aboard the system how to correct problems, how to eject or abandon the system, etc. All possible emergency events that can be anticipated should be analyzed, and procedures should be practiced before the test is begun. Above all, provide the test personnel as much on-board capability as possible for taking care of their own emergency; they may not have enough time to communicate with, and receive instructions from, the base test conductor.

## The "Cooper Rating System"

The use of "expert operators" to evaluate the apparent controllability of a complex system can provide an effective method, and possibly the only method, for determining whether a system is acceptable. This requires not only a stable of expert operators but also a method for quantifying the opinions of these experts. The Cooper Rating System was developed to allow experienced pilots to evaluate the quality of a new aircraft in terms of its controllability. It is recommended that some method similar to that outlined in the table on page 1009 be devised for any system in which the quality characteristics may be critical to the safety and/or general consumer acceptability of the product.

Of special importance, however, is the recognition by the "expert" evaluators that they may have skills that are not particularly representative of those of the general consumer population. Therefore, in evaluating a system, the expert must view the system from the consumer's point of view, i.e., put himself or herself in the position (physically, psychologically, and in terms of experience) of the consumer. The designer should identify the probable differences between the intended user and the expert and make sure that the expert is indoctrinated.

## Cooper Pilot Rating System*

| Operating Conditions | Adjective Rating | Numerical Rating | Description | Primary Mission Accomplished | Can Be Landed |
|---|---|---|---|---|---|
| Normal operation | Satisfactory | 1 | Excellent, includes optimum. | Yes | Yes |
| | | 2 | Good, pleasant to fly. | Yes | Yes |
| | | 3 | Satisfactory, but with some mildly unpleasant characteristics. | Yes | Yes |
| Emergency operation | Unsatisfactory | 4 | Acceptable, but with unpleasant characteristics. | Yes | Yes |
| | | 5 | Unacceptable for normal operation. | Doubtful | Yes |
| | | 6 | Acceptable for emergency condition only. | Doubtful | Yes |
| No operation | Unacceptable | 7 | Unacceptable even for emergency condition. | No | Doubtful |
| | | 8 | Unacceptable—dangerous. | No | No |
| | | 9 | Unacceptable—uncontrollable. | No | No |
| | Catastrophic | 10 | Motions possibly violent enough to prevent pilot escape. | No | No |

*E.W. Vinje and D.P. Miller, Interpretation of Pilot Opinion by Application of Multiloop Models to a VTOL Flight Simulator Task, United Aircraft Corporation, Third Annual NASA University Conference on Manual Control, NASA SP-144, March 1967.

## Using a Random Sample of Consumers to Help Identify User/Design Problems

By exposing a large number of typical consumers to currently available systems or equipment, one can generally determine what problems they will have in operating the equipment. The test should be designed so that each subject "uses" the product in systematic steps and then rates the problems he or she has on a scale from "No Problem" to "Serious Problem." The accompanying sample test asks subjects to rate different aspects of automobile seat belts in several different vehicles.

**EXAMPLE TEST PROCEDURE** The following are examples of subject data, procedures and data collection questionnaire used to conduct a consumer evaluation of various automotive seat belt systems. Although it is uniquely designed for a specific type of test, the general implications are applicable to a variety of tests in which an experimenter conducts a consumer subject test of actual prototype hardware. It is based on complete subject involvement in use of the hardware.

```
PASSIVE RESTRAINT SYSTEM TEST


       Subject Demographic Data Record


Subject No. _____
Male:____Female:____Height:_____Weight:_____I.D. No:_____

Age Category:  Young____, Middle____, Elderly____

Vehicle Driven:
   a. Primary;  Make_____Model_____Year_____
   b. Secondary:Make_____Model_____Year_____

Primary Car Belt System: Lap Belt Only              _____
                         Separate Lap/Shoulder Belts _____
                         Combined Lap/Shoulder Belt   _____

Primary Seat Belt Usage: Most of the Time            _____
                         Part of the Time            _____
                         Never Use                   _____

If Not Used, Why?        Inconvenient                _____
                         Confusing                   _____
                         Uncomfortable               _____
                         Other                       _____
```

Detailed Test Procedures for:
Passive Belt Systems Evaluation

1.0  INFORMATION HANDOUT

Each subject will be given the following handout which he will read prior to commencing the test sequence:

"Because passive restraint systems will be installed in all cars after the year 1982 models, the U.S. Department of Transportation would like to make sure that such systems will be safe, comfortable, and convenient.

"A passive belt system is one that you, as a vehicle occupant don't have to put on. It is designed so that as you close the car door it goes around you automatically, securing you safely in the seat. And when you open the door the belt is removed from around you automatically so that you can get out of the car easily.

"We have selected several different passive belt systems which we would like you to help us evaluate. We are interested in knowing how comfortable and convenient each of these belt systems seems to you.

"So as you get in and out of each of the cars today be alert to any confusion, inconvenience, or discomfort you experience with each system. We are not evaluating the design of the vehicles! Only the design and operation of the belt system.

"Immediately after each part of the test we will ask you some questions that will allow us to record your observations and reactions while the experience is still fresh in your mind."

2.0  TEST CONDUCTOR'S STATEMENT

The following statement will be read verbatim to each test subject by the Test Conductor:

This is an overview of what you will be asked to do.

• We are going to have you evaluate 5 different passive belt systems.

• You will be asked to get into each car in turn, close the door, adjust the seat to your preferred driving position, then imitate several driving tasks such as: adjust and look into rearview mirrors; look over your shoulder as though you were checking traffic before making a turn; turn the steering wheel; and operate several of the driver controls—after which you will get out of the car.

At certain points in this procedure you will be asked specific questions about the effect the belt system had on your ability to perform various tasks and on your general feelings of comfort (or discomfort).

• We will treat the cars and belt systems in pairs testing each one separately, then comparing the second car/system with the previous one.

• Each question should be answered in terms of rating a possible problem. If you feel there was no problem, I will mark a zero for that question. If you experienced a minor problem I will mark the question with a "1", and so on. The evaluation ratings are as follows:

No problem (for you)   = 0
Minor problem          = 1
Moderate problem       = 2
Serious problem        = 3

Your participation in this test will help us decide what is needed to make all passive belt systems more acceptable from the standpoint of comfort and convenience.

Do you have any questions before we begin?

[Note to Examiner:  Distinguish between hang-ups on parts of car (door-lock, etc.) vs. hang-ups caused by belt.]

Subject No._____

Test Conductor_____

First or Second Test     Car A B C D E F G

<u>Detailed Test Procedures for Passive Belt System Evaluation</u>

| Test Conductor Instructions | Questions for Subjects | No Problem | | | Serious Problem |
|---|---|---|---|---|---|
| <u>Read to S</u>: We want your opinion about this passive belt system in terms of how it may (or may not) interfere with your ability to get into this car and get ready to drive. After you are in the car I will ask you a series of questions about how easy it was to open the door, get into the seat, and close the door after you. I also will ask whether the belts got in the way, rubbed your chest or neck or seemed too loose or too tight. | <u>Test Sequence No. 1 - Entering Car</u> | | | | |
| Now please open the car door, get into the driver's seat and close the door. Re-arrange or straighten the seat belt if you feel it is necessary. | 1.  Did the seat belt make it hard for you to open the car door all the way? | 0 | 1 | 2 | 3 |
| | 2.  Did the belt tend to re-close the door, making it difficult for you to get into the car? | 0 | 1 | 2 | 3 |
| Ask this question only when S is entering this car for the first time. | 3.  When you first saw the belt were you in doubt as to how you were supposed to get into the car? | 0 | 1 | 2 | 3 |
| | 4.  Did you have any difficulty getting past the belt and into the seat? | 0 | 1 | 2 | 3 |

<u>Detailed Test Procedures for Passive Belt System Evaluation</u>

| Test Conductor Instructions | Questions for Subjects | No Problem | | | Serious Problem |
|---|---|---|---|---|---|
| | <u>Test Sequence No. 1 (continued)</u> | | | | |
| | 5.  Did the belt interfere with your closing the door? | 0 | 1 | 2 | 3 |
| Omit with 2-point system. | 6.  Did the belt scrub across your legs as you closed the door? | 0 | 1 | 2 | 3 |
| | 7.  Did the belt scrub across your shoulder or chest as you closed the door? | 0 | 1 | 2 | 3 |
| | 8.  Did the belt annoy your neck after the car door was closed? | 0 | 1 | 2 | 3 |
| | 9.  Was the belt too tight across your chest or shoulder? | 0 | 1 | 2 | 3 |
| If a motorized belt system, ask question No. 10. | 10. Did the belt graze, or pass uncomfortably close to, your head or face? | 0 | 1 | 2 | 3 |
| Test conductor answers this question. Omit with 2-point system. | 11. Was the lap portion of the belt lying across S's lap (where it <u>should</u> be for safety) or was it pulled up, onto his/her stomach (where it might cause injury?) | 0 | 1 | 2 | 3 |
| | 12. Did the belt seem too loose after it was around you? | 0 | 1 | 2 | 3 |

Detailed Test Procedures for Passive Belt System Evaluation

| Test Conductor Instructions | Questions for Subjects | No Problem | | Serious Problem | |
|---|---|---|---|---|---|
| | Test Sequence No. 1 (continued) | | | | |
| | 13. Did the belt tend to entrap your arms? | 0 | 1 | 2 | 3 |
| | 14. Did the movement of the belt disarrange your clothing? | 0 | 1 | 2 | 3 |
| Ask questions No. 15 and 16 only if S is evaluating a belt system that includes a manual or automatic convenience hook or puller/rotating arm system. Read to S. Please exit the vehicle.<br><br>Explain use of convenience hook.<br><br>Read to S before entering vehicle: Some passive restraint systems release from the convenience hook automatically. The belt positions itself around the driver without any driver involvement. Would you please enter the vehicle? Then touch the ignition key. | 15. Did the rate at which the belt enveloped you startle you? (Explain the term "envelop" if necessary) | 0 | 1 | 2 | 3 |
| | 16. Did you have any problem keeping your arms from being entrapped as you were being enveloped by the belt? | 0 | 1 | 2 | 3 |
| | 17. While entering did you have any difficulty getting past the belt and into the seat? | 0 | 1 | 2 | 3 |
| | 18. If you had a convenience hook would you use it regularly for extra comfort while entering and exiting? | Yes | | No | |
| | Comments_____<br>_____<br>_____<br>_____<br>_____<br>_____<br>_____ | | | | |

Detailed Test Procedures for Passive Belt System Evaluation

| Test Conductor Instructions | Questions for Subjects | No Problem | | Serious Problem | |
|---|---|---|---|---|---|
| Read to S: You probably will have to adjust your seat in order to be able to reach and see the driving controls and instruments comfortably. I am going to ask you some questions (after you have adjusted the seat) to find out whether or not this seat belt interferes with seat adjusting and how the belt feels during and after seat adjustment.<br><br>Now please adjust the seat to the position you would prefer for comfortable driving. Do not adjust the backrest, however. | Test Sequence No. 2 - Adjusting Seat<br><br>1. Did the belt restrict your ability to adjust the seat? | 0 | 1 | 2 | 3 |
| | 2. After you had adjusted the seat, | | | | |
| | a. Did the belt lie against your neck? | 0 | 1 | 2 | 3 |
| | b. Was the belt too tight across your chest? | 0 | 1 | 2 | 3 |
| | c. Was the belt too loose? | 0 | 1 | 2 | 3 |
| Omit with 2-point system. | d. Had the belt slipped from your lap, up and onto your stomach? | 0 | 1 | 2 | 3 |
| Experimenter answer: | (Was it necessary for S to readjust the belt after he finished adjusting the seat?) | 0 | 1 | 2 | 3 |

## Human Performance Measurement during Design Testing and Evaluation

As a general rule, human performance is measured in terms of how many errors a subject makes and/or how long it takes the subject to perform specific tasks. However, human subjects can also help judge the adequacy of a design either by rating the design against some set of standard criteria or by rating several designs against one another. The table below provides a brief description of several techniques for performance and/or design acceptance adequacy.

| Method | Technique | Statistic | Application |
|---|---|---|---|
| Adjustment (average error) | The subject adjusts a stimulus until it is subjectively equal to or is in some relationship to a criterion. | Average of settings (average error of settings measures precision). | Absolute threshold; equality; equal intervals; equal ratios. |
| Minimal change (limits) | The experimenter varies the stimulus upward and/or downward. The subject signals its apparent relation to a criterion. | Average value of stimulus at transition point of subject's judgment. | All thresholds; equality. |
| Pair comparison | Stimuli are presented in pairs. Each stimulus is paired with each other stimulus. The subject indicates which of each pair is greater with respect to a given attribute. | Proportion of judgment calling one stimulus greater than another. Proportions sometimes are translated into scale values via assumption of normal distribution of judgments. | Order; equal intervals (under distribution assumption). |
| Constant stimuli | Several comparison stimuli are paired at random with a fixed standard. The subject indicates whether each comparison is greater or less than the standard. (This is a special case of the pair-comparison technique.) | Size of difference threshold equals stimulus distance between 50 and 75 percentage points on psychometric function. | All thresholds; equality; equal intervals; equal ratios. |
| Quantal | Various fixed increments are added to a standard, with no time interval between. Each increment is added several times in succession. The subject indicates the apparent presence or absence of the increment. | Size of sensory quantum equals distance between intercepts of rectilinear psychometric function. | Differential thresholds. |
| Order of merit | A group of stimuli, presented simultaneously, are set up in apparent rank order by subject. | Average or median rank assigned by subjects. | Order. |
| Rating scale | Each of a set of stimuli is given an absolute rating in terms of some attribute. The rating may be numerical or descriptive. | Average or median rating assigned by subjects. | Order; equal intervals; stimulus rating. |

## Organizing and Documenting Field Test Activities

The importance of documenting test results of hardware is well known, and such tests are recognized as absolutely necessary. Documenting how well people perform is equally important. Although this might appear to be an unnecessary admonition, too often it is believed that performance problems should merely be noted in a general way and that the actual performance of the people operating the equipment does not need to be measured. However, if a thorough program of human engineering has been pursued to this point, certain human performance standards will have been established. Therefore, it makes common sense to test the equipment user as well as the equipment.

As noted earlier, a final human-machine system test should examine not only the effectiveness of the hardware design but also the procedures, the supporting documentation (e.g., technical manuals), and the training.

Special test observation and recording forms should be devised to prompt field test observers and to provide them a place to record their observations. The accompanying sample format was created for the field testing of an aeronautical system. This test was oriented primarily toward field service and maintenance operations. Although only two of the test sheets are shown, they give an idea of what the field test observers were told to look for. Each type of test should have its own forms, tailored to fit the particular application and the features that will probably be observed.

HUMAN ENGINEERING FIELD EVALUATION FORM

TASK TIME RECORD:    TASK: _____    EVALUATOR_____

Preparation _____                                       DATE_____
Performance _____    _____
Clean Up   _____    _____
Total      _____    _____
                      _____

| TASK DIFFICULTY | | APPARENT PROBLEM |
|---|---|---|
| 1. Didn't know how to start. | ☐ | |
| 2. Didn't know what to do next. | ☐ | |
| 3. Missed a procedural step. | ☐ | |
| 4. Got steps out of order. | ☐ | |
| 5. Repeated step unnecessarily. | ☐ | |
| 6. Initiated action too late; too soon. | ☐ | |
| 7. Difficulty coordinating activity of two or more persons. | ☐ | |

RECOMMENDED ACTIONS:

_____

                                          PROCEDURAL PROBLEMS

HUMAN ENGINEERING FIELD EVALUATION FORM

TASK TIME RECORD:    TASK:_____    EVALUATOR_____

Preparation _____                                       DATE_____
Performance _____    _____
Clean Up   _____    _____
Total      _____    _____
                      _____

| TASK DIFFICULTY | | APPARENT PROBLEM |
|---|---|---|
| 1. Difficulty locating proper display or control. | ☐ | |
| 2. Confused meaning of label. | ☐ | |
| 3. Couldn't read display detail. | ☐ | |
| 4. Mis-read display. | ☐ | |
| 5. Missed warning light. | ☐ | |
| 6. Difficulty positioning control. | ☐ | |
| 7. Set control incorrectly. | ☐ | |
| 8. Disturbed adjacent control. | ☐ | |
| 9. Didn't recognize color code. | ☐ | |

RECOMMENDED ACTION:

_____

                                          DISPLAY-CONTROL PROBLEMS

## BIBLIOGRAPHIC

### A Selected Bibliography

As one might expect, the literature relating to human factors is huge since it covers not only research in the various directly related disciplines of anthropology, biology, medicine, physiology, and psychology but also research in the indirectly related disciplines of engineering and physics, which must deal with the human element from a purely physical point of view. The following bibliography in no way purports to be either comprehensive or definitive, but rather representative of some of the important references that have been found to be useful over a wide range of typical problem areas in human factors design.

The bibliography has been divided into several categories to make it easier for the reader to find certain types of information quickly:

The sources listed under Basic Human Engineering References have been generated by human factors specialists for their own use in carrying out human engineering design efforts.

The sources listed under Basic Health and Safety References, although not necessarily generated by human factors specialists, are still important and are often referred to by such specialists.

The sources listed under Basic Architectural References have been generated by architects but provide data that are useful to the human factors specialist.

The sources listed under Basic Highway Engineering References deal with highway design problems, which were discussed in earlier chapters.

The sources listed under Other Useful Human Factors References include a general cross section of items that are especially important in supplementing the material presented in the previous chapters.

### BASIC HUMAN ENGINEERING REFERENCES

Altman, James W., Angeline C. Marchese, and Barbara W. Marchiando: *Guide to Design of Mechanical Equipment for Maintainability*, Behavioral Sciences Laboratory, ASD Technical Report 61-381, August 1961.

American Society for Training and Development: *Training and Development Handbook: A Guide to Human Resource Development*, McGraw-Hill Book Company, New York, 1976.

*Anthropometric Source Book*, vol. I, *Anthropometry for Designers*, NASA Reference Publication 1024, National Aeronautics and Space Administration, 1978.

*Anthropometric Source Book*, vol. II, *A Handbook of Anthropometric Data*, NASA Reference Publication 1024, National Aeronautics and Space Administration, 1978.

*Anthropometric Source Book*, vol. III, *Annotated Bibliography of Anthropometry*, NASA Reference Publication 1024, National Aeronautics and Space Administration, 1978.

Cornog, D. Y., and F. C. Rose: *Legibility of Alphanumeric Characters and Other Symbols*, vol. II, *A Reference Handbook*, National Bureau of Standards, 1967.

*Design Handbook, Series 1-0, General: System Safety*, 2d ed., Air Force Systems Command, AFSC DH 1-6, 1969.

*Design Handbook, Series 1-0, General: Environmental Engineering*, 2d ed., Air Force Systems Command, AFSC DH 1-5, 1971.

*Design Handbook, Series 1-0, General: Human Factors Engineering*, 2d ed., Air Force Systems Command, AFSC DH 1-3, 1972.

*Design Handbook, Series 1-0, General: Maintainability (for Ground Electronic Systems)*, 2d ed., Air Force Systems Command, AFSC DH 1-9, 1973.

*Design Handbook, Series 2-0, Aeronautical Systems: Life Support*, Air Force Systems Command, AFSC DH 2-8, 1971.

*Design Handbook, Series 2-0, Aeronautical Systems: Crew Stations and Passenger Accommodations*, 2d ed., Air Force Systems Command, AFSC DH 2-2, 1972.

*Design Handbook, Series 3-0, Space and Missile Systems: Ground Equipment and Facilities*, Air Force Systems Command, AFSC DH 3-3, 1969.

*Engineering Design Handbook*, AMCP 706-134, Headquarters, U.S. Army Material Command, October 1972.

Farrell, Richard J., and John M. Booth: *Design Handbook for Imagery Interpretation Equipment*, Report no. D180-19063-1, Boeing Aerospace Company, December 1975.

Geer, Charles W.: *Analyst's Guide for the Analysis Sections of MIL-H-46855*, Report no. D180-19476-1, Boeing Aerospace Company, June 30, 1976.

Greenberg, Leo, and Don B. Chaffin: *Workers and Their Tools: A Guide to the Ergonomic Design of Hand Tools and Small Presses*, AMP Incorporated, no date.

*Human Engineering Design Criteria for Military Systems, Equipment and Facilities*, MIL-STD-1472B.

*Human Engineering Requirements for Military Systems, Equipment and Facilities*, MIL-H-46855

*Human Factors for Designers of Naval Equipment*, vol. I, Medical Research Council, 1970.

*Human Factors for Designers of Naval Equipment*, vol. II, Medical Research Council, 1971.

Meister, David, and Dennis J. Sullivan: *Guide to Human Engineering Design for Visual Displays*, The Bunker-Ramo Corporation, Contract no. N00014-68-C-0278, Aug. 30, 1969.

*Military Standardization Handbook: Human Factors Engineering Design for Army Material*, MIL-HDBK-759, Mar. 12, 1975.

Parker, James F., and Vita R. West (eds.): *Bioastronautics Data Book*, 2d ed., National Aeronautics and Space Administration, Washington, 1973.

Robbins, D. H.: *Guidebook on Anthropomorphic Test Dummy Usage*, Michigan University, Highway Safety Research Institute, Ann Arbor, Mar. 31, 1977.

Woodson, W. E., and D. W. Conover: *Human Engineering Guide for Equipment Designers*, 2d ed., University of California Press, Berkeley, 1964.

Wulfeck, Joseph W., et al.: *Vision in Military*, WADC Technical Report 58-399, ASTIA Document no. AD 207780, Wright Air Development Center, November 1958.

*Journals and Periodicals Relevant to Human Engineering*

Aerospace Medicine
Applied Ergonomics
Ergonomics
Human Factors

### BASIC HEALTH AND SAFETY REFERENCES

*Accident Prevention Manual for Industrial Operations*, 7th ed., National Safety Council, Chicago, 1974.

Hammer, Willie: *Handbook of System and Product Safety*, Prentice-Hall, Inc., Englewood Cliffs, N.J., 1972.

*Handbook of Occupational Safety and Health*, National Safety Council, Chicago.

Handley, William: *Industrial Safety Handbook*, McGraw-Hill Book Company, New York, 1970.

*Heating and Cooling for Man in Industry*, American Industrial Hygiene Association, Akron, Ohio, 1970, 1975.

*Industrial Noise Manual*, American Industrial Hygiene Association, Akron, Ohio, 1966.

*Industrial Ventilation: A Manual of Recommended Practice*, American Conference of Governmental Industrial Hygienists, Committee on Industrial Ventilation, Lansing, Mich., 1974.

*OSHA Standards Checklists*, parts 1910 and 1926, National Safety Council, Chicago.

*Safety Guide for Health Care Institutions*, National Safety Council and American Hospital Association, 1972.

*Threshold Limit Values for Chemical Substances and Physical Agents in the Workroom Environment*, American Conference of Governmental Industrial Hygienists, Cincinnati. (Issued annually.)

U.S. Department of Health, Education, and Welfare: *The Industrial Environment—Its Evaluation and Control*, Publication no. 614, 1965.

U.S. Department of Housing and Urban Development: *A Design Guide for Home Safety*, Contract no. H-1113, January 1972.

U.S. Department of Labor: *Occupation Health Hazards: Their Evaluation and Control*, Bulletin 198, 1968.

U.S. Department of Transportation, National Highway Traffic Safety Administration: *Federal Motor Vehicle Safety Standards and Regulations*.

U.S. Public Health Service: *Industrial Noise—A Guide to Its Evaluation and Control*, Publication no. 1572, 1967.

### BASIC ARCHITECTURAL REFERENCES

*Access to the Environment*, vol. 1, U.S. Department of Housing and Urban Development, Office of Policy Development and Research, no date.

*Access to the Environment*, vol. 2, U.S. Department of Housing and Urban Development, Office of Policy Development and Research, no date.

*Access to the Environment*, vol. 3, U.S. Department of Housing and Urban Development, Office of Policy Development and Research, no date.

Allen, Rex Whitaker, and Ilona Von Karolyi: *Hospital Planning Handbook*, John Wiley & Sons, Inc., New York, 1976.

Atkin, William Wilson, and Joan Adler: *Interiors Book of Restaurants*, Whitney Library of

Design, Whitney Publications, Inc., New York, 1960.

BLANKENSHIP, EDWARD G.: *The Airport,* Frederick A. Praeger, Inc., New York, 1974.

CALLENDER, JOHN HANCOCK (ed.): *Time-Saver Standards for Architectural Design Data,* McGraw-Hill Book Company, New York, 1974.

DE CHIARA, JOSEPH, and JOHN HANCOCK CALLENDER: *Time-Saver Standards for Building Types,* McGraw-Hill Book Company, New York, 1973.

END, HENRY: *Interiors Book of Hotels and Motor Hotels,* Whitney Library of Design, Whitney Publications, Inc., New York, 1963.

FRIED, C., and R. S. GIBSON: *Handbook of Color Notation Systems,* Technical Memorandum 10-61, ULSL Army Ordnance Human Engineering Laboratories, Aberdeen Proving Ground, Md., 1961.

GALVIN, PATRICK J.: *Kitchen Planning Guide for Builders and Architects,* Structures Publishing Company, 1972.

GOLDSMITH, SELWYN: *Designing for the Disabled,* 2d ed., McGraw-Hill Book Company, New York, 1968.

HOEL, LESTER A., and ERVIN S. ROSZNER: *Transit Station Planning and Design: State of the Art,* Carnegie-Mellon University, April 1976.

HOPF, PETER S.: *Designer's Guide to OSHA,* McGraw-Hill Book Company, New York, 1975.

*IES Lighting Handbook,* 5th ed., Illuminating Engineering Society, New York, 1972.

KIRA, ALEXANDER: *The Bathroom,* The Viking Press, Inc., New York, 1976.

KONCELIK, JOSEPH A.: *Designing the Open Nursing Home,* Dowden, Hutchinson & Ross, Inc., 1976.

LAWSON, FRED: *Designing Commercial Food Service Facilities,* The Whitney Library of Design, Whitney Publications, Inc., New York, 1973.

LAWSON, FRED: *Hotels, Motels and Condominiums: Design, Planning and Maintenance,* The Architectural Press Ltd., 1976.

LION, EDGAR: *Shopping Centers: Planning, Development, and Administration,* John Wiley & Sons, Inc., New York, 1976.

LYTLE, R. J.: *Farm Builder's Handbook,* 2d ed., Structures Publishing Company, 1973.

NEUFERT, ERNST: *Architects' Data,* Archon Books, The Shoestring Press, Inc., Hamden, Conn., 1970.

PAUL, SAMUEL: *Apartments: Their Design and Development,* Reinhold Book Corporation, New York, 1967.

PETERS, PAULHANS, and FRIEDEMANN WILD: *Design and Planning Factories,* Van Nostrand Reinhold Company, 1972.

PILE, JOHN: *Interiors 3rd Book of Offices,* Whitney Library of Design, Whitney Publications, Inc., New York, 1976.

PUSHKAREV, BORIS: *Urban Space for Pedestrians,* The M.I.T. Press, Cambridge, Mass., 1975.

RAMSEY, CHARLES G., and HAROLD R. SLEEPER: *Architectural Graphic Standards,* John Wiley & Sons, Inc., New York, 1970.

TAYLOR, ANNE P., and GEORGE VLASTOS: *School Zone: Learning Environments for Children,* Van Nostrand Reinhold Company, 1975.

TREGENZA, PETER: *The Design of Interior Circulation: People and Buildings,* Van Nostrand Reinhold Company, 1976.

## BASIC HIGHWAY ENGINEERING REFERENCES

BAKER, ROBERT F. (ed.): *Handbook of Highway Engineering,* Van Nostrand Reinhold Company, 1975.

CAPELLE, DONALD G., et al. (eds.): *An Introduction to Highway Transportation Engineering,* Institute of Traffic Engineers, 1968.

U.S. Department of Commerce, Bureau of Public Roads, *Manual on Uniform Traffic Control Devices for Streets and Highways,* June 1961.

POST, THEODORE J., et al.: *A Users' Guide to Positive Guidance,* U.S. Department of Transportation, Contract no. DOT-HS-11-8864, June 1977. *Symbol Signs,* The American Institute of Graphic Arts, November 1974.

## OTHER USEFUL HUMAN FACTORS REFERENCES

APPLE, JAMES M.: *Material Handling Systems Design,* The Ronald Press Company, New York, 1972.

ARCHITECTURAL RECORDS: *Behavioral Architecture: Toward an Accountable Design Process,* McGraw-Hill Book Company, New York, 1977.

BAKKER, CORNELIS B., and MARIANNE K. BAKKER-RABDAU: *No Trespassing! Explorations in Human Territoriality,* Chandler & Sharp Publishers, Inc., 1973.

BARNES, RALPH M.: *Motion and Time Study: Design and Measurement of Work,* 6th ed., John Wiley & Sons, Inc., New York, 1968.

BENSON, O. O., and H. STRUGHOLD (eds.): *Physics and Medicine of the Atmosphere and Space,* John Wiley & Sons, Inc., New York, 1960.

BIRREN, FABER: *Color Psychology and Color Therapy,* University Books, Inc., New Hyde Park, N.Y., 1961.

BIRREN, FABER: *Light, Color and Environment,* Van Nostrand Reinhold Company, 1969.

BIRREN, FABER: *Color for Interiors Historical and Modern,* Whitney Library of Design, Whitney Publications, Inc., New York, no date.

BORKO, H.: *Computer Applications in the Behavioral Sciences,* Prentice-Hall, Inc., Englewood Cliffs, N.J., 1962.

BRUNING, JAMES L., and B. L. KINTZ: *Computational Handbook of Statistics,* Scott, Foresman and Company, Glenview, Ill., 1968.

CARROOL, ROBERT F., et al.: *Guidelines for the Design of Man-Machine Interfaces for Process Control,* Purdue University, School of Engineering, Lafayette, Ind., 1976.

CARY, JANE RANDOLPH: *How to Create Interiors for the Disabled,* Pantheon Books, a division of Random House, Inc., New York, 1978.

CHAPANIS, ALPHONSE: *Research Techniques in Human Engineering,* The Johns Hopkins Press, Baltimore, 1959.

CHAPANIS, ALPHONSE: *Man-Machine Engineering,* Wadsworth Publishing Company, Inc., Belmont, Calif., 1965.

CHAPANIS, ALPHONSE (ed.): *Ethnic Variables in Human Factors Engineering,* The Johns Hopkins Press, Baltimore, 1975.

COHEN, JACOB, and PATRICIA COHEN: *Applied Multiple Regression/Correlation Analysis for the Behavioral Sciences,* John Wiley & Sons, Inc., New York, 1975.

CROSSAN, RICHARD M., and HAROLD W. NANCE: *Master Standard Data: The Economic Approach to Work Measurement,* McGraw-Hill Book Company, New York, 1962.

CUNNINGHAM, C. E., and WILBERT COX: *Applied Maintainability Engineering,* John Wiley & Sons, Inc., New York, 1972.

DAMON, ALBERT, HOWARD W. STOUDT, and ROSS A. MCFARLAND: *The Human Body in Equipment Design,* Harvard University Press, Cambridge, Mass., 1966.

DE CHIARA, JOSEPH, and LEE KOPPELMAN: *Planning Design Criteria,* Van Nostrand Reinhold Company, 1969.

DE CHIARA, JOSEPH, and LEE E. KOPPELMAN: *Site Planning Standards,* McGraw-Hill Book Company, New York, 1978.

DEGREENE, KENYON B.: *Systems Psychology,* McGraw-Hill Book Company, New York, 1970.

DOELLE, LESLIE L.: *Environmental Acoustics,* McGraw-Hill Book Company, New York, 1972.

DREYFUSS, H.: *The Measure of Man: Human Factors in Design,* Whitney Library of Design, Whitney Publications, Inc., New York, 1960.

DREYFUSS, HENRY: *Designing for People,* Paragraphic Books, New York, 1967.

DREYFUSS, HENRY: *Symbol Sourcebook: An Authoritative Guide to International Graphic Symbols,* McGraw-Hill Book Company, New York, 1972.

DRURY, C. G., and J. G. FOX (eds.): *Human Reliability in Quality Control,* Taylor & Francis Ltd., 1975.

ECKMAN, D. P.: *Systems: Research and Design,* John Wiley & Sons, Inc., New York, 1961.

EDWARDS, ALLEN L.: *Statistical Methods for the Behavioral Sciences,* Rinehart & Company, Inc., New York, 1954.

FANGER, P. O.: *Thermal Comfort: Analysis and Applications in Environmental Engineering,* McGraw-Hill Book Company, New York, 1973.

FITTS, PAUL M., and MICHAEL I. POSNER: *Human Performance,* Wadsworth Publishing Company, Inc., Belmont, Calif., 1968.

FLYNN, JOHN E., and ARTHUR W. SEGIL: *Architectural Interior Systems,* Van Nostrand Reinhold Company, 1970.

FOGEL, L. J.: *Biotechnology: Concepts and Applications,* Prentice-Hall, Inc., Englewood Cliffs, N.J., 1963.

FOOTT, SYDNEY: *Handicapped at Home,* Quick Fox, New York, 1977.

FREUND, JOHN E., PAUL E. LIVERMORE, and IRWIN MILLER: *Manual of Experimental Statistics,* Prentice-Hall, Inc., Englewood Cliffs, N.J., 1960.

GAEL, SIDNEY, and LAWRENCE E. REED: *Personnel Equipment Data: Concept and Content,* Aerospace Medical Research Laboratories, ASD Technical Report 61-739, December 1961.

GAGNE, ROBERT M. (ed.): *Psychological Principles in System Development,* Holt, Reinhart and Winston, Inc., New York, 1962.

GERATHERWOHL, S. J.: *Principles of Bioastronautics,* Prentice-Hall, Inc., Englewood Cliffs, N.J., 1963.

GERRITSEN, FRANS: *Theory and Practice of Color: A Color Theory Based on Laws of Perception,* Van Nostrand Reinhold Company, 1974.

GOSHEN, CHARLES E. (ed.): *Psychiatric Archi-*

*tecture,* The American Psychiatric Association, 1961.

GRANDJEAN, E.: *Fitting the Task to the Man: An Ergonomic Approach,* Taylor & Francis Ltd., 1969.

GRANDJEAN, ETIENNE: *Ergonomics of the Home,* Taylor & Francis Ltd., 1973.

GREEN, ISAAC, et al.: *Housing for the Elderly: The Development and Design Process,* Van Nostrand Reinhold Company, 1975.

HALSE, ALBERT O.: *The Use of Colors in Interiors,* McGraw-Hill Book Company, New York, 1968.

HALL, EDWARD: *The Hidden Dimension,* Doubleday & Company, Inc., Garden City, N.Y., 1966.

HAMILL, PETER V. V., FRANCIS E. JOHNSTON, and WILLIAM GRAMS: *Height and Weight of Children,* U.S. Department of Health, Education, and Welfare, 1970.

HAMILL, PETER V. V., FRANCIS E. JOHNSTON, and STANLEY LEMESHOW: *Body Weight, Stature, and Sitting Height: White and Negro Youths 12–17 Years,* U.S. Department of Health, Education, and Welfare, 1973.

HAMILL, PETER V. V., FRANCIS E. JOHNSTON, and STANLEY LEMESHOW: *Height and Weight of Youths 12–17 Years,* U.S. Department of Health, Education, and Welfare, 1973.

HAMILL, PETER V. V., FRANCIS E. JOHNSTON, and STANLEY LEMESHOW: *Height and Weight of Children: Socioeconomic Status,* U.S. Department of Health, Education, and Welfare, 1977.

HARRIGAN, JOHN E., and JANET R. HARRIGAN: *Human Factors Program for Architects, Interior Designers and Clients,* Blake Printing and Publishing, San Luis Obispo, Calif., 1976.

HARRIS, D. H., and F. B. CHANEY: *Human Factors in Quality Assurance,* John Wiley & Sons, Inc., New York, 1969.

HEALY, RICHARD J.: *Design for Security,* John Wiley & Sons, Inc., New York, 1968.

HOWELL, WILLIAM CARL, and IRWIN L. GOLDSTEIN: *Engineering Psychology: Current Perspectives in Research,* Appleton Century Crofts, New York, 1971.

KELLEY, CHARLES R.: *Manual and Automatic Control,* John Wiley & Sons, Inc., New York, 1968.

KELLY, KENNETH L., and DEANE B. JUDD: *Color: Universal Language and Dictionary of Names,* National Bureau of Standards, December 1976.

KULLER, RIKARD (ed.): *Architectural Psychology,* Dowden, Hutchinson, & Ross, Inc., 1973.

LANG, JON, et al.: *Designing for Human Behavior: Architecture and the Behavioral Sciences,* Dowden, Hutchinson, & Ross, Inc., 1974.

LAURIE, GINI: *Housing and Home Services for the Disabled: Guidelines and Experiences in Independent Living,* Harper & Row, Publishers, Incorporated, New York, 1977.

LEE, DOUGLAS H. D.: *Physiological Objectives in Hot Weather Housing,* U.S. Department of Housing and Urban Development, June 1969.

LOSEE, J. E., et al.: *Methods for Computing Manpower Requirements for Weapon Systems under Development,* Technical Report no. 61-361, Aerospace Systems Division, Wright-Patterson Air Force Base, Ohio.

LUXENBERG, H. R., and RUDOLPH L. KUEHN (eds.): *Display Systems Engineering,* McGraw-Hill Book Company, New York, 1968.

McCORMICK, ERNEST J.: *Human Factors in Engineering and Design,* 4th ed., McGraw-Hill Book Company, New York, 1976.

McFARLAND, ROSS A.: *Human Factors in Air Transport Design,* McGraw-Hill Book Company, New York, 1946.

McFARLAND, ROSS A.: *Human Factors in Air Transportation: Occupational Health and Safety,* McGraw-Hill Book Company, New York, 1953.

McGRATH, JAMES J., ALBERT HARABEDIAN, and DONALD N. BUCKNER: *Review and Critique of the Literature on Vigilance Performance,* Technical Report 206-1, Human Factors Research, Inc., December, 1959.

MALINA, ROBERT M., PETER V. V. HAMILL, and STANLEY LEMESHOW: *Selected Body Measurements of Children 6–11 Years,* U.S. Department of Health, Education, and Welfare, 1973.

MALINA, ROBERT M., PETER V. V. HAMILL, AND STANLEY LEMESHOW: *Body Dimensions and Proportions, White and Negro Children 6–11 Years,* U.S. Department of Health, Education, and Welfare, 1974.

MEISTER, DAVID: *Human Factors: Theory and Practice,* John Wiley & Sons, Inc., New York, 1971.

MEISTER, DAVID: *Behavioral Foundations of System Development,* John Wiley & Sons, Inc., New York, 1976.

MEISTER, DAVID, and GERALD F. RABIDEAU: *Human Factors Evaluation in System Development,* John Wiley & Sons, Inc., New York, 1965.

MEISTER, D., D. J. SULLIVAN, and W. G. ASKREN: *The Impact of Manpower Requirements and Personnel Resources Data on System Design,* Aerospace Medical Research Laboratories, AMRL-TR-68-44, no date.

MICHELSON, WILLIAM (ed.): *Behavioral Research Methods in Environmental Design,* Dowden, Hutchinson, & Ross, Inc., 1975.

MOOS, RUDOLF H.: *The Human Context: Environmental Determinants of Behavior,* John Wiley & Sons, Inc., New York, 1976.

MURRELL, K. F. H.: *Human Performance in Industry,* Reinhold Publishing Corporation, New York, 1965.

OLISHIFSKI, JULIAN B., and FRANK E. McELROY (eds.): *Fundamentals of Industrial Hygiene,* National Safety Council, 1971.

PANERO, JULIUS: *Anatomy of Interior Designers,* 3d ed., Whitney Library of Design, Whitney Publications, Inc., New York, 1962.

PARSONS, H. M.: *Man-Machine System Experiments,* The Johns Hopkins Press, Baltimore, 1971.

PERIN, CONSTANCE: *With Man in Mind: An Interdisciplinary Prospectus for Environmental Design,* The M.I.T. Press, Cambridge, Mass., 1970.

PETERS, GEORGE A.: *Product Liability and Safety,* Coiner Publications, Ltd., 1971.

PIERCE, J. R.: *Symbols, Signals and Noise: The Nature and Process of Communication,* Harper & Brothers, New York, 1961.

PIERMAN, BRIAN C.: "Color in the Health Care Environment," *Proceedings of a Special Workshop Held at the National Bureau of Standards,* Gaithersburg, Md., Nov. 16,

1976, NSB SP-516, September 1978.

POULTON, E. C.: *Environment and Human Efficiency.* Charles C. Thomas, Publisher, Springfield, Ill., 1970.

POULTON, E. C.: *Tracking Skill and Manual Control,* Academic Press, Inc., New York, 1974.

PROPST, ROBERT: *The Office: A Facility Based on Change,* Herman Miller, Inc., 1968.

ROEBUCK, J. A., K. H. E. KROEMER, and W. G. THOMSON: *Engineering Anthropometry Methods,* John Wiley & Sons, Inc., New York, 1975.

ROTH, E. M. (ed.): *Compendium of Human Responses to the Aerospace Environment,* vol. I, secs. 1–6; vol. II, secs. 7–9; vol. III, *A Descriptive Model for Determining Optimal Human Performance in Systems,* NASA CR-1205(I), Clearinghouse for Federal Scientific and Technical Information, Springfield, Va.

RUMELHART, DAVID E.: *Introduction to Human Information Processing,* John Wiley & Sons, Inc., New York, 1977.

SELL, R. C., et al.: *Human Factors in Work, Design and Production,* Taylor & Francis Ltd., 1977.

SELL, S. B., and C. A. BERRY: *Human Factors in Jet and Space Travel,* The Ronald Press Company, New York, 1961.

SINGLETON, W. T., R. S. EASTERBY, and D. C. WHITFIELD (eds.): *The Human Operator in Complex Systems,* Taylor & Francis Ltd., 1967.

SINGLETON, W. T., J. G. FOX, and D. WHITFIELD (eds.): *Measurement of Man at Work,* Taylor and Francis Ltd., 1971.

SITTIG, MARSHALL: *Hazardous and Toxic Effects of Industrial Chemicals,* Noyes Data Corp., Park Ridge, N.J., 1979.

SOMMER, ROBERT: *Personal Space: The Behavioral Basis of Design,* Prentice-Hall, Inc., Englewood Cliffs, N.J., 1969.

SOMMER, ROBERT: *Tight Spaces: Hard Architecture and How to Humanize It,* Prentice-Hall, Inc., Englewood Cliffs, N.J., 1974.

SPECTOR, W. S. (ed.): *Handbook of Biological Data,* WADC Technical Report 65-273, ASTIA AD 110501, 1956.

STEVENS, S. S. (ed.): *Handbook of Experimental Psychology,* John Wiley & Sons, Inc., New York, 1951.

STOUDT, HOWARD W., ALBERT DAMON, ROSS A. McFARLAND, and JEAN ROBERTS: *Skinfolds, Body Girths, Biacromial Diameter, and Selected Anthropometric Indices of Adults,* U.S. Department of Health, Education, and Welfare, 1973.

TEICHNER, W. H., and D. OLSON: *Predicting Human Performance in Space Environments,* NASA CR-1370, Clearinghouse for Scientific and Technical Information, Springfield, Va.

TICHAUER, E. R.: *The Biomechanical Basis of Ergonomics: Anatomy Applied to the Design of Work Stations,* John Wiley & Sons, Inc., New York, 1978.

TIFFIN, J., and E. J. McCORMICK: *Industrial Psychology,* Prentice-Hall, Inc., Englewood Cliffs, N.J., 1965.

VAN COTT, HAROLD P., and ROBERT G. KINKADE (eds.): *Human Engineering Guide to Equipment Design,* McGraw-Hill Book Company, New York, 1972.

WARD, COLIN (ed.): *Vandalism,* Van Nostrand

Reinhold Company, 1973.

WHEELER, VIRGINIA HART: *Planning Kitchens for Handicapped Homemakers,* The Institute of Physical Medicine and Rehabilitation, New York University Medical Center, New York, no date.

WOODWORTH, ROBERT S., and HAROLD SCHLOS-BERG: *Experimental Psychology,* Holt, Rein-hart and Winston, Inc., New York, 1965.

YERGES, LYLE F.: *Sound, Noise, and Vibration Control,* Van Nostrand Reinhold Company, 1969.

## INFORMATION SOURCES

### Organizations from Which Technical Expertise in Human Factors May Be Obtained

Aerospace Medical Research Laboratories
Aerospace Medical Division
Air Force Systems Command
Wright-Patterson Air Force Base, Ohio 45433

Electronic Systems Division
Scientific and Information Division
Air Force Systems Command (ESTI)
L. G. Hanscom Field
Bedford, Mass. 01731

Federal Aviation Administration
Civil Aeromedical Research Institute
Psychology Branch
Aero Center, P.O. Box 1082
Oklahoma City, Okla.

Federal Aviation Administration
Human Factors Branch
Research Division
National Aviation Facilities Experimental Cen-
ter
Atlantic City, N.J.

National Academy of Sciences
Highway Research Board
2101 Constitution Ave., N.W.
Washington, D.C. 20418

National Aeronautics and Space Administra-
tion
Biomedical Division, Office of Life Science
Office of Space Science
National Aeronautics and Space Administra-
tion
Washington, D.C. 20546

National Bureau of Standards
Human Factors, Center for Consumer Product
Technology
Washington, D.C. 20234

National Safety Council
425 North Michigan Ave.
Chicago, Ill. 60611

USAF Human Resources Laboratory
USAF Headquarters
Brooks Air Force Base, Tex. 78235

USAF School of Aerospace Medicine
USAF—SAM/VNE
Brooks Air Force Base, Tex. 78235

U.S. Army Armor
Human Research Laboratory Unit
Ft. Knox, Ky. 40121

U.S. Army Human Engineering Laboratories
Aberdeen Proving Ground, Md. 21005

U.S. Army Missile Command
Human Factors Research Unit
Redstone Arsenal, Ala. 35808

U.S. Army Research Institute of Environmental Medicine
Natick, Mass. 01762

U.S. Department of Transportation
National Highway Traffic Safety Administration
400 Seventh St., S.W.
Washington, D.C. 20591

U.S. Naval Medical Research Laboratory
Bureau of Medicine and Surgery
Human Engineering Branch
Submarine Base
New London, Conn.

U.S. Navy Electronics Center
Human Factors Division
San Diego, Calif. 92100

U.S. Navy Office of Naval Research
Human Engineering Branch (Code 455)
Washington, D.C. 20360

U.S. Postal Service Laboratory
Human Factors Branch
11711 Parklane Dr.
Rockville, Md. 20852

### Biblographic Services

Clearinghouse for Federal Scientific and
Technical Information
Springfield, Va. 22151

Defense Documentation Center
Cameron Station
Alexandria, Va. 22314

### Professional and Technical Organizations from Which Human Factors Technical Information May Be Available

American Medical Association
535 North Dearborne St.
Chicago, Ill. 60610

American Society of Safety Engineers
5 North Wabash Ave.
Chicago, Ill. 60602

Association of American Railroads
Division of System Studies
Research and Test Department
1920 L St., N.W.
Washington, D.C. 20036

Flight Safety Foundation, Inc.
468 Park Ave. S.
New York, N.Y. 10016

The Human Factors Society, Inc.
1124 Montana Ave., Suite B
Box 1369
Santa Monica, Calif.

Illuminating Engineering Society
345 East 47th St.
New York, N.Y. 10017

National Academy of Sciences
Transportation Research Board
2101 Constitution Ave., N.W.
Washington, D.C. 20418

### Other U.S. Government Agencies from Which Human Factors Information May Be Available

Federal Aviation Agency
Office of System Engineering Management
Hq. 800 Independance Ave., S.W., AEM-20
Washington, D.C. 20591

Federal Highway Administration
Human Factors Branch
Washington, D.C. 20590

Federal Railroad Administration
Human Factors Office, Railway Safety Research
2100 Second St., S.W.
Washington, D.C. 20590

U.S. Department of Commerce
National Bureau of Standards
Office of Codes and Safety Standards
Connecticut Ave. and Van Ness St., N.W.
Washington, D.C. 20234

U.S. Department of Health, Education, and
Welfare
Public Health Service
Washington, D.C. 20201

U.S. Department of Health, Education, and
Welfare
Public Health Service
Bureau of Occupational Safety and Health,
Occupational Injury and Disease Control
1014 Broadway
Cincinnati, Ohio 45202

U.S. Department of the Interior
Bureau of Mines
18th and C Sts., N.W.
Washington, D.C. 20006

U.S. Department of Labor
Fourteenth St. and Constitution Ave., N.W.
Washington, D.C. 20210

### Sources for Obtaining Regulations, Specifications, and Standards

American National Standards Institute, Inc.
1430 Broadway
New York, N.Y. 10018.
(Request complete catalog.)

National Institute of Occupational Safety and
Health
U.S. Department of Health, Education, and
Welfare
5600 Fishers La.
Rockville, Md. 20857
(Request publications catalog.)

Society of Automotive Engineers
2 Pennsylvania Plaza
New York, N.Y. 10001

U.S. Air Force regulations and manuals: Director of Administrative Services, USAF Headquarters, The Pentagon, Washington, D.C.

U.S. Air Force specifications, standards, ex-
hibits, bulletins, and manuals:
Aeronautical systems: Director of Adminis-
trative Services, Headquarters, Aeronauti-
cal Systems Division, Air Force Systems
Command, Wright-Patterson Air Force
Base, Ohio
Command and control systems: Director of
Administrative Services, Headquarters,
Electronic Systems Division, Air Force Sys-
tems Command, L. G. Hanscom Field, Bed-
ford, Mass.
Missile systems: Director of Administrative
Services, Headquarters, Ballistic Systems
Division, Air Force Systems Command, Air
Force Unit Post Office, Los Angeles, Calif.

U.S. Army regulations published by the Department of the Army: The Adjutant General, Headquarters, Department of the
Army, The Pentagon, Washington, D.C.

U.S. Army specifications, standards, and technical instructions:
Ordnance systems: U.S. Army Ordnance, Human Engineering Laboratories, Aberdeen Proving Ground, Md.
Signal systems: U.S. Army Signal Corps Laboratories, Ft. Monmouth, N.J., Attn: Adjutant.
Missile systems: Redstone Arsenal, Human Factors Unit, Huntsville, Ala.

U.S. Navy specifications, regulations, standards, manuals, and instructions:
U.S. Department of the Navy, Bureau of Naval Personnel, Washington, D.C.
U.S. Department of the Navy, Bureau of Ships, Washington, D.C.
U.S. Department of the Navy, Bureau of Naval Weapons, Washington, D.C.
U.S. Department of the Navy, Bureau of Aeronautics, Washington, D.C.

U.S. Consumer Product Safety Commission
Washington, D.C. 20207
Request publications relating to consumer product safety and the Consumer Product Safety Act.)

U.S. Department of Defense
Defense Supply Agency
Washington, D.C.
(Request *Consolidated Index of DoD Specifications and Standards.*)

U.S. Department of Labor
Occupational Safety and Health Administration
Washington, D.C. 20210

## Foreign Institutions from Which Human Factors Information May be Obtained[27]

### ENGLAND

**Institute for Consumer Ergonomics**
University of Technology
Loughborough, Leicestershire LE11 3TU
Loughborough, England

**Ergonomics Information Analysis Centre**
Department of Engineering Production
University of Birmingham
Birmingham, England

### FRANCE

**Conservatoire des Arts et Metiers**
Department Physiologie du Travail et Ergonomie
41 Gay-Lussac
Paris 5, France

### GERMANY

**Federal Institute for Occupational Safety and Accident Research**
435 Martener Street
Dortmund-70, F.R. Germany

### JAPAN

**Department of Systems Science**
Tokyo Institute of Technology
2-12-1, Ohakayama, Megro
Tokyo 152, Japan

### NETHERLANDS

**Institute for Perception TNO**
5 Kampweg
Soesterberg, Netherlands

### SWEDEN

**Department of Psychology**
University of Uppsala
Trädgårdsgatan 20
Uppsala, Sweden

## METRICS

### Dimension, Quantity, Amount, and Rate

| | |
|---|---|
| acre | = 43,560 square feet |
| | = 4,840 square yards |
| | = 4,047 square meters |
| | = $1.562 \times 10^{-3}$ square miles |
| ampere-hour | = $3.600 \times 10^{3}$ coulombs |
| | = $3.731 \times 10^{-2}$ faradays |
| angstrom unit (A) | = $3.937 \times 10^{-9}$ |
| | = $1 \times 10^{-4}$ microns (mu) |
| | = $1 \times 10^{-8}$ centimeters |
| astronomical unit (AU) | = $1.495 \times 10^{8}$ kilometers |
| atmosphere | = 14.7 pounds per square inch |
| | = 76.0 centimeters of mercury |
| | = 29.92 inches of mercury |
| | = $3.39 \times 10^{1}$ feet of water |
| | = 1.033 kilograms per square centimeter |
| | = $1.033 \times 10^{4}$ kilograms per square meter |
| | = 1.058 tons per square foot |
| bar | = $9.869 \times 10^{-1}$ atmospheres |
| | = $1 \times 10^{6}$ dynes per square centimeter |
| | = $1.020 \times 10^{4}$ kilograms per square meter |
| | = $2.089 \times 10^{3}$ pounds per square foot |
| | = $1.45 \times 10^{1}$ pounds per square inch |
| British thermal unit | = $1.0409 \times 10^{1}$ liter-atmosphere |
| | = $1.055 \times 10^{10}$ ergs |
| | = $7.781 \times 10^{2}$ foot-pounds |
| | = $2.520 \times 10^{2}$ gram-calories |
| | = $3.927 \times 10^{-4}$ horsepower-hours |
| | = $1.055 \times 10^{3}$ joules |
| | = $1.0758 \times 10^{2}$ kilogram-meters |
| | = $2.928 \times 10^{-4}$ kilowatthours |
| British thermal unit per hour | = $2.162 \times 10^{-1}$ foot-pounds per second |
| | = $7.0 \times 10^{-2}$ gram-calories per second |
| | = $3.929 \times 10^{-4}$ horsepower |
| | = $2.931 \times 10^{-1}$ watts |
| British thermal unit per minute | = $1.296 \times 10^{1}$ foot-pounds per second |
| | = $2.356 \times 10^{-2}$ horsepower |
| | = $1.757 \times 10^{1}$ watts |
| British thermal unit per square foot per minute | = $1.22 \times 10^{-1}$ watts per square inch |
| candle per square centimeter | = 3.146 lamberts |
| candle per square inch | = $4.870 \times 10^{-1}$ lamberts |
| centigrade (degrees) | = $°C \times 9/5 + 32$ Fahrenheit (degrees) |
| | = $°C + 273.18$ Kelvin (degrees) |
| centimeter | = $3.281 \times 10^{-2}$ feet |
| | = $3.937 \times 10^{-1}$ inches |
| | = $1 \times 10^{-5}$ kilometers |
| | = $6.214 \times 10^{-6}$ miles |
| | = $3.937 \times 10^{2}$ mils |
| | = $1.094 \times 10^{-2}$ yards |
| | = $1 \times 10^{4}$ microns |
| | = $1 \times 10^{8}$ angstrom units |
| centimeter-dyne | = $1.020 \times 10^{-3}$ centimeter-grams |
| | = $1.020 \times 10^{-8}$ meter-kilograms |
| | = $7.375 \times 10^{-8}$ pound-feet |
| centimeter-gram | = $9.807 \times 10^{2}$ centimeter dynes |
| | = $1 \times 10^{-5}$ meter-kilograms |
| | = $7.233 \times 10^{-5}$ pound-feet |
| centimeter of mercury | = $4.461 \times 10^{-1}$ foot of water |
| | = $2.785 \times 10^{1}$ pounds per square foot |
| | = $1.934 \times 10^{-1}$ pounds per square inch |

*(continued)*

[27]The term "ergonomics" is generally used in lieu of "human factors."

## Dimension, Quantity, Amount, and Rate *(continued)*

| | |
|---|---|
| centimeter per second | $= 1.969$ feet per minute |
| | $= 3.281 \times 10^{-2}$ feet per second |
| | $= 3.6 \times 10^{-2}$ kilometers per hour |
| | $= 1.943 \times 10^{-2}$ knots |
| | $= 6.0 \times 10^{-1}$ meters per minute |
| | $= 2.237 \times 10^{-2}$ miles per hour |
| | $= 3.728 \times 10^{-4}$ miles per minute |
| centimeter per second per second | $= 3.281 \times 10^{-2}$ feet per second per second |
| | $= 3.6 \times 10^{-2}$ kilometers per hour per second |
| | $= 2.237 \times 10^{-2}$ miles per hour per second |
| circumference | $= 6.283$ radians |
| coulomb | $= 1.036 \times 10^{-5}$ faradays |
| cubic centimeter | $= 3.531 \times 10^{-5}$ cubic feet |
| | $= 6.102 \times 10^{-2}$ cubic inches |
| | $= 1.308 \times 10^{-6}$ cubic yards |
| | $= 2.642 \times 10^{-4}$ gallons (U.S. liquid) |
| | $= 1.057 \times 10^{-3}$ quarts (U.S. liquid) |
| | $= 2.113 \times 10^{-3}$ pints (U.S. liquid) |
| | $= 1 \times 10^{-6}$ cubic meters |
| | $= 1 \times 10^{-3}$ liters |
| cubic foot | $= 2.832 \times 10^{4}$ cubic centimeters |
| | $= 1.728 \times 10^{3}$ cubic inches |
| | $= 2.832 \times 10^{-2}$ cubic meters |
| | $= 3.704 \times 10^{-2}$ cubic yards |
| | $= 7.48052$ gallons (U.S. liquid) |
| | $= 2.832 \times 10^{1}$ liters |
| | $= 5.984 \times 10^{1}$ pints (U.S. liquid) |
| | $= 2.992 \times 10^{1}$ quarts (U.S. liquid) |
| cubic foot per minute | $= 4.72 \times 10^{2}$ cubic centimeters per second |
| | $= 1.247 \times 10^{1}$ gallons per second |
| | $= 4.720 \times 10^{-1}$ liters per second |
| | $= 6.243 \times 10^{1}$ pounds water per minute |
| | $= 6.46317 \times 10^{-1}$ million gallons per day |
| | $= 4.48831 \times 10^{2}$ gallons per minute |
| cubic inch | $= 1.639 \times 10^{1}$ cubic centimeters |
| | $= 5.787 \times 10^{-4}$ cubic feet |
| | $= 1.639 \times 10^{-5}$ cubic meters |
| | $= 2.143 \times 10^{-5}$ cubic yards |
| | $= 4.329 \times 10^{-3}$ gallons (U.S. liquid) |
| | $= 1.639 \times 10^{-2}$ liters |
| | $= 3.463 \times 10^{-2}$ pints (U.S. liquid) |
| | $= 1.732 \times 10^{-2}$ quarts (U.S. liquid) |
| cubic meter | $= 1 \times 10^{6}$ cubic centimeters |
| | $= 3.531 \times 10^{1}$ cubic feet |
| | $= 6.1023 \times 10^{4}$ cubic inches |
| | $= 1.308$ cubic yards |
| | $= 2.642 \times 10^{2}$ gallons (U.S. liquid) |
| | $= 1 \times 10^{3}$ liters |
| | $= 2.113 \times 10^{3}$ pints (U.S. liquid) |
| | $= 1.057 \times 10^{3}$ quarts (U.S. liquid) |
| cubic yard | $= 7.646 \times 10^{5}$ cubic centimeters |
| | $= 2.7 \times 10^{1}$ cubic feet |
| | $= 4.6656 \times 10^{4}$ cubic inches |
| | $= 7.646 \times 10^{-1}$ cubic meters |
| | $= 2.02 \times 10^{2}$ gallons (U.S. liquid) |
| | $= 7.646 \times 10^{2}$ liters |
| | $= 1.6159 \times 10^{3}$ pints (U.S. liquid) |
| | $= 8.079 \times 10^{2}$ quarts (U.S. liquid) |
| cubic yard per minute | $= 4.5 \times 10^{-1}$ cubic feet per second |
| | $= 3.367$ gallons per second |
| | $= 1.274 \times 10^{1}$ liters per second |
| day | $= 8.64 \times 10^{4}$ seconds |
| | $= 1.44 \times 10^{3}$ minutes |
| degree (angle) | $= 1.745 \times 10^{-2}$ radians |
| | $= 3.6 \times 10^{3}$ seconds (angle) |
| degree per second | $= 1.745 \times 10^{-2}$ radians per second |
| | $= 1.667 \times 10^{-1}$ revolutions per minute |
| | $= 2.778 \times 10^{-3}$ revolutions per second |

*(continued)*

### Dimension, Quantity, Amount, and Rate *(continued)*

| | |
|---|---|
| dram (apothecaries' or troy weight) | $= 1.3714 \times 10^{-1}$ ounces (avoirdupois)<br>$= 1.25 \times 10^{-1}$ ounces (troy) |
| dram (U.S. fluid or apothecaries' weight) | $= 3.6967$ cubic centimeters |
| dram | $= 1.7718$ grams<br>$= 2.7344 \times 10^{1}$ grains<br>$= 6.25 \times 10^{-2}$ ounces |
| dyne | $= 1.020 \times 10^{-3}$ grams<br>$= 1 \times 10^{-7}$ joules per centimeter<br>$= 1 \times 10^{-5}$ joules per meter (newtons)<br>$= 1.020 \times 10^{-6}$ kilograms<br>$= 7.233 \times 10^{-5}$ poundals<br>$= 2.248 \times 10^{-6}$ pounds |
| dyne per square centimeter | $= 1 \times 10^{-6}$ bars<br>$= 1 \times 10^{-2}$ ergs per square millimeter<br>$= 9.869 \times 10^{-7}$ atmospheres<br>$= 2.953$ inches of mercury (at 0 degrees Celsius)<br>$= 4.015 \times 10^{-4}$ inches of water (at 4 degrees Celsius) |
| ell | $= 1.143 \times 10^{2}$ centimeters<br>$= 4.5 \times 10^{1}$ inches |
| em, pica | $= 1.67 \times 10^{-1}$ inches<br>$= 4.233 \times 10^{-1}$ centimeters |
| erg | $= 9.486 \times 10^{-11}$ British thermal units<br>$= 1.0$ dyne-centimeter<br>$= 7.376 \times 10^{-8}$ foot-pounds<br>$= 2.389 \times 10^{-8}$ gram-calories<br>$= 1.020 \times 10^{-3}$ gram-centimeters<br>$= 3.725 \times 10^{-14}$ horsepower-hours<br>$= 1.0 \times 10^{-7}$ joules<br>$= 2.389 \times 10^{-11}$ kilogram-calories<br>$= 1.020 \times 10^{-8}$ kilogram-meters<br>$= 2.773 \times 10^{-14}$ kilowatthours<br>$= 2.773 \times 10^{-11}$ watthours |
| erg per second | $= 5.668 \times 10^{-9}$ British thermal units per minute<br>$= 1.0$ dyne-centimeter per second<br>$= 4.426 \times 10^{-6}$ foot-pounds per minute<br>$= 7.3756 \times 10^{-8}$ foot-pounds per second<br>$= 1.341 \times 10^{-10}$ horsepower<br>$= 1.433 \times 10^{-9}$ kilogram-calories per minute<br>$= 1.0 \times 10^{-10}$ kilowatts |
| faraday | $= 2.68 \times 10^{1}$ ampere-hours<br>$= 9.649 \times 10^{4}$ coulombs |
| faraday per second | $= 9.65 \times 10^{4}$ amperes (absolute) |
| fathom | $= 6.0$ feet<br>$= 1.8288$ meters |
| foot | $= 3.048 \times 10^{1}$ centimeters<br>$= 3.948 \times 10^{-4}$ kilometers<br>$= 3.048 \times 10^{-1}$ meters<br>$= 1.645 \times 10^{-4}$ nautical miles<br>$= 1.894 \times 10^{-4}$ statute miles<br>$= 1.2 \times 10^{4}$ mils |
| footcandle | $= 1.0764 \times 10^{1}$ lumens per square meter (lux) |
| foot of water | $= 2.95 \times 10^{-2}$ atmospheres<br>$= 8.826 \times 10^{-1}$ inches of mercury<br>$= 3.048 \times 10^{-2}$ kilograms per square centimeter<br>$= 3.048 \times 10^{2}$ kilograms per square meter<br>$= 6.243 \times 10^{1}$ pounds per square foot<br>$= 4.335 \times 10^{-1}$ pounds per square inch |
| foot per minute | $= 5.080 \times 10^{-1}$ centimeters per second<br>$= 1.667 \times 10^{-2}$ feet per second<br>$= 1.829 \times 10^{-2}$ kilometers per hour<br>$= 3.048 \times 10^{-1}$ meters per minute<br>$= 1.136 \times 10^{-2}$ miles per hour |
| foot per second | $= 3.048 \times 10^{1}$ centimeters per second<br>$= 1.097$ centimeters per hour<br>$= 5.921 \times 10^{-1}$ knots<br>$= 1.829 \times 10^{1}$ meters per minute |

*(continued)*

### Dimension, Quantity, Amount, and Rate *(continued)*

| | |
|---|---|
| | $= 6.818 \times 10^{-1}$ miles per hour |
| | $= 1.136 \times 10^{-2}$ miles per minute |
| feet per second per second | $= 3.048 \times 10^{1}$ centimeters per second per second |
| | $= 1.097$ kilometers per hour per second |
| | $= 3.048 \times 10^{-1}$ meters per second per second |
| | $= 6.818 \times 10^{-1}$ miles per hour per second |
| foot per 100 feet | $= 1.0$ per cent grade |
| foot-pound | $= 1.286 \times 10^{-3}$ British thermal units |
| | $= 1.356 \times 10^{7}$ ergs |
| | $= 3.241 \times 10^{-1}$ gram-calories |
| | $= 5.050 \times 10^{-7}$ horsepower-hours |
| | $= 1.356$ joules |
| | $= 3.241 \times 10^{-4}$ kilograms-calories |
| | $= 1.383 \times 10^{-1}$ kilogram-meters |
| | $= 3.766 \times 10^{-7}$ kilowatthours |
| foot-pounds per minute | $= 1.286 \times 10^{-3}$ British thermal units per minute |
| | $= 1.667 \times 10^{-2}$ foot-pounds per second |
| | $= 3.030 \times 10^{-5}$ horsepower |
| | $= 3.241 \times 10^{-4}$ kilogram-calories per minute |
| | $= 2.260 \times 10^{-5}$ kilowatts |
| foot-pound per second | $= 4.6263$ British thermal units per hour |
| | $= 7.717 \times 10^{-2}$ British thermal units per minute |
| | $= 1.818 \times 10^{-3}$ horsepower |
| | $= 1.945 \times 10^{-2}$ kilogram-calories per minute |
| | $= 1.356 \times 10^{-3}$ kilowatts |
| gallon | $= 3.785 \times 10^{3}$ cubic centimeters |
| | $= 1.337 \times 10^{-1}$ cubic feet |
| | $= 2.31 \times 10^{2}$ cubic inches |
| | $= 3.785 \times 10^{-3}$ cubic meters |
| | $= 4.951 \times 10^{-3}$ cubic yards |
| | $= 3.785$ liters |
| gallon (liquid, imperial) | $= 1.20095$ gallons (U.S. liquid) |
| gallon (U.S.) | $= 8.3267 \times 10^{-1}$ gallons (imperial) |
| gallon of water | $= 8.337$ pounds of water |
| gallon per minute | $= 2.228 \times 10^{-3}$ cubic feet per second |
| | $= 6.308 \times 10^{-2}$ liters per second |
| | $= 8.0208$ cubic feet per hour |
| grain | $= 3.657 \times 10^{-2}$ drams (avoirdupois) |
| grain per imperial gallon | $= 1.4286 \times 10^{1}$ parts per million |
| grain per U.S. gallon | $= 1.7118 \times 10^{1}$ parts per million |
| | $= 1.4286 \times 10^{2}$ pounds per million |
| gram | $= 9.807 \times 10^{2}$ dynes |
| | $= 3.527 \times 10^{-2}$ ounces (avoirdupois) |
| | $= 3.215 \times 10^{-2}$ ounces (troy) |
| | $= 7.093 \times 10^{-2}$ poundals |
| | $= 2.205 \times 10^{-3}$ pounds |
| gram-calorie | $= 3.9683 \times 10^{-3}$ British thermal units |
| | $= 4.184 \times 10^{7}$ ergs |
| | $= 3.086$ foot-pounds |
| | $= 1.5596 \times 10^{-6}$ horsepower-hours |
| | $= 1.162 \times 10^{-6}$ kilowatthours |
| | $= 1.162 \times 10^{-3}$ watthours |
| gram-calorie per second | $= 1.4286 \times 10^{1}$ British thermal units per hour |
| gram-centimeter | $= 2.343 \times 10^{-8}$ kilogram-calories |
| gram per centimeter | $= 5.6 \times 10^{-3}$ pounds per inch |
| gram per cubic centimeter | $= 6.243 \times 10^{1}$ pounds per cubic foot |
| | $= 3.613 \times 10^{-2}$ pounds per cubic inch |
| horsepower | $= 4.244 \times 10^{1}$ British thermal units per minute |
| | $= 3.3 \times 10^{4}$ foot-pounds per minute |
| | $= 5.50 \times 10^{2}$ foot-pounds per second |
| | $= 1.068 \times 10^{1}$ kilogram-calories per minute |
| | $= 7.457 \times 10^{-1}$ kilowatts |
| | $= 7.457 \times 10^{2}$ watts |
| horsepower (metric) | $= 9.863 \times 10^{-1}$ horsepower |
| horsepower | $= 1.014$ horsepower (metric)     *(continued)* |

### Dimension, Quantity, Amount, and Rate *(continued)*

| | |
|---|---|
| horsepower-hour | $= 2.547 \times 10^3$ British thermal units |
| | $= 2.6845 \times 10^{13}$ ergs |
| | $= 1.98 \times 10^6$ foot-pounds |
| | $= 6.4119 \times 10^5$ gram-calories |
| | $= 2.684 \times 10^6$ joules |
| | $= 6.417 \times 10^2$ kilogram-calories |
| | $= 2.737 \times 10^5$ kilogram-meters |
| inch | $= 2.540$ centimeters |
| | $= 1.578 \times 10^{-5}$ miles |
| | $= 2.54 \times 10^1$ millimeters |
| | $= 1 \times 10^3$ mils |
| | $= 2.778 \times 10^{-2}$ yards |
| | $= 2.54 \times 10^8$ angstrom units |
| inch of mercury | $= 3.342 \times 10^{-2}$ atmospheres |
| | $= 1.133$ feet of water |
| | $= 3.453 \times 10^{-2}$ kilograms per square centimeter |
| | $= 3.453 \times 10^2$ kilograms per square meter |
| | $= 7.073 \times 10^1$ pounds per square foot |
| | $= 4.912 \times 10^{-1}$ pounds per square inch |
| inch of water (at 4 degrees Celsius) | $= 2.458 \times 10^{-3}$ atmospheres |
| | $= 7.355 \times 10^{-2}$ inches of mercury |
| | $= 2.54 \times 10^{-3}$ kilograms per square centimeter |
| | $= 5.781 \times 10^{-1}$ ounces per square inch |
| | $= 5.204$ pounds per square foot |
| | $= 3.613 \times 10^{-2}$ pounds per square inch |
| joule | $= 9.486 \times 10^{-4}$ British thermal units |
| | $= 1 \times 10^7$ ergs |
| | $= 7.736 \times 10^{-1}$ foot-pounds |
| | $= 2.389 \times 10^{-4}$ kilogram-calories |
| | $= 1.020 \times 10^{-1}$ kilogram-meters |
| | $= 2.778 \times 10^{-4}$ watthours |
| joule per centimeter | $= 1.020 \times 10^4$ grams |
| | $= 1 \times 10^7$ dynes |
| | $= 1.10^2$ joules per meter |
| | $= 7.233 \times 10^2$ poundals |
| | $= 2.248 \times 10^1$ pounds |
| kilogram | $= 9.80665 \times 10^5$ dynes |
| | $= 7.093 \times 10^1$ poundals |
| | $= 2.2046$ pounds |
| | $= 3.5274 \times 10^1$ ounces (avoirdupois) |
| | $= 9.842 \times 10^{-4}$ tons (long) |
| | $= 1.102 \times 10^{-3}$ tons (short) |
| kilogram-calorie | $= 3.968$ British thermal units |
| | $= 3.086 \times 10^3$ foot-pounds |
| | $= 1.558 \times 10^{-3}$ horsepower-hours |
| | $= 4.183 \times 10^3$ joules |
| | $= 1.163 \times 10^{-3}$ kilowatt-hours |
| kilogram-calorie per minute | $= 5.143 \times 10^1$ foot-pounds per second |
| | $= 9.351 \times 10^{-2}$ horsepower |
| | $= 6.972 \times 10^{-2}$ kilowatts |
| kilogram-meter | $= 9.296 \times 10^{-3}$ British thermal units |
| | $= 9.807 \times 10^7$ ergs |
| | $= 7.233$ foot-pounds |
| | $= 9.807$ joules |
| | $= 2.723 \times 10^{-6}$ kilowatt-hours |
| kilogram per cubic meter | $= 6.243 \times 10^{-2}$ pounds per cubic foot |
| | $= 3.613 \times 10^{-5}$ pounds per cubic inch |
| kilogram per meter | $= 6.72 \times 10^{-1}$ pounds per foot |
| kilogram per square centimeter | $= 9.80665 \times 10^5$ dynes per square centimeter |
| | $= 9.678 \times 10^{-1}$ atmospheres |
| | $= 3.281 \times 10^1$ feet of water |
| | $= 2.896 \times 10^1$ inches of mercury |
| | $= 2.048 \times 10^3$ pounds per square foot |
| | $= 1.422 \times 10^1$ pounds per square inch |
| kilogram per square meter | $= 9.679 \times 10^{-5}$ atmospheres |
| | $= 9.807 \times 10^{-5}$ bars |
| | $= 3.281 \times 10^{-3}$ feet of water |
| | $= 2.896 \times 10^{-3}$ inches of mercury |

*(continued)*

### Dimension, Quantity, Amount, and Rate   *(continued)*

|  |  |
|---|---|
|  | $= 2.048 \times 10^{-1}$ pounds per square foot |
|  | $= 1.422 \times 10^{-3}$ pounds per square inch |
| kilometer | $= 3.281 \times 10^{3}$ feet |
|  | $= 3.937 \times 10^{4}$ inches |
|  | $= 6.214 \times 10^{-1}$ statute miles |
|  | $= 5.396 \times 10^{-1}$ nautical miles |
|  | $= 1.0936 \times 10^{3}$ yards |
| kilometer per hour | $= 2.778 \times 10^{1}$ centimeters per second |
|  | $= 5.468 \times 10^{1}$ feet per minute |
|  | $= 9.113 \times 10^{-1}$ feet per second |
|  | $= 5.396 \times 10^{-1}$ knots |
|  | $= 1.667 \times 10^{1}$ meters per minute |
|  | $= 6.214 \times 10^{-1}$ miles per hour |
| kilometer per hour per second | $= 2.778 \times 10^{1}$ centimeters per second per second |
|  | $= 9.113 \times 10^{-1}$ feet per second per second |
|  | $= 6.214 \times 10^{-1}$ miles per hour per second |
| kilowatt | $= 5.692 \times 10^{1}$ British thermal units per minute |
|  | $= 4.426 \times 10^{4}$ foot-pounds per minute |
|  | $= 7.376 \times 10^{2}$ foot-pounds per second |
|  | $= 1.341$ horsepower |
|  | $= 1.434 \times 10^{1}$ kilogram-calories per minute |
| kilowatthour | $= 3.413 \times 10^{3}$ British thermal units |
|  | $= 3.6 \times 10^{13}$ ergs |
|  | $= 2.655 \times 10^{6}$ foot-pounds |
|  | $= 8.5985 \times 10^{5}$ gram-calories |
|  | $= 1.341$ horsepower-hours |
|  | $= 8.605 \times 10^{2}$ kilogram-calories |
| knot | $= 6.080 \times 10^{3}$ feet per hour |
|  | $= 1.8532$ kilometers per hour |
|  | $= 1.0$ nautical miles per hour |
|  | $= 1.151$ statute miles per hour |
|  | $= 2.027 \times 10^{3}$ yards per hour |
|  | $= 1.689$ feet per second |
|  | $= 5.148 \times 10^{1}$ centimeters per second |
| lambert | $= 3.183 \times 10^{-1}$ candles per square centimeter |
|  | $= 2.054$ candles per square inch |
| light year | $= 5.9 \times 10^{12}$ miles |
|  | $= 9.46091 \times 10^{12}$ kilometers |
| liter | $= 1 \times 10^{3}$ cubic centimeters |
|  | $= 3.531 \times 10^{-2}$ cubic feet |
|  | $= 6.102 \times 10^{1}$ cubic inches |
|  | $= 1.308 \times 10^{-3}$ cubic yards |
|  | $= 2.642 \times 10^{-1}$ gallons (U.S. liquid) |
|  | $= 2.113$ pints (U.S. liquid) |
|  | $= 1.057$ quarts (U.S. liquid) |
| liter per minute | $= 5.886 \times 10^{-4}$ cubic feet per second |
| lumen | $= 7.958 \times 10^{-2}$ spherical candle power |
| lumen per square foot | $= 1.0$ footcandles |
|  | $= 1.076 \times 10^{1}$ lumens per square meter |
| lux | $= 9.29 \times 10^{-2}$ footcandles |
| meter | $= 1 \times 10^{10}$ angstrom units |
|  | $= 5.4681 \times 10^{-1}$ fathoms |
|  | $= 3.281$ feet |
|  | $= 3.937 \times 10^{1}$ inches |
|  | $= 5.396 \times 10^{-4}$ nautical miles |
|  | $= 6.214 \times 10^{-4}$ statute miles |
|  | $= 1.094$ yards |
| meter per minute | $= 1.667$ centimeters per second |
|  | $= 3.281$ feet per minute |
|  | $= 5.468 \times 10^{-2}$ feet per second |
|  | $= 6.0 \times 10^{-2}$ kilometers per hour |
|  | $= 3.238 \times 10^{-2}$ knots |
|  | $= 3.728 \times 10^{-2}$ miles per hour |
| meter per second | $= 1.968 \times 10^{2}$ feet per minute |
|  | $= 3.281$ feet per second |
|  | $= 6.0 \times 10^{-2}$ kilometers per minute |
|  | $= 2.237$ miles per hour |
|  | $= 3.728 \times 10^{-2}$ miles per minute |

*(continued)*

## Dimension, Quantity, Amount, and Rate   *(continued)*

| | |
|---|---|
| meter per second per second | = 3.281 feet per second per second |
| | = 3.6 kilometers per hour per second |
| | = 2.237 miles per hour per second |
| mil | = $2.54 \times 10^{-3}$ centimeters |
| | = $8.333 \times 10^{-5}$ feet |
| | = $1.0 \times 10^{-3}$ inches |
| | = $2.54 \times 10^{-8}$ kilometers |
| | = $2.778 \times 10^{-5}$ yards |
| mile (nautical) | = $6.076 \times 10^{3}$ feet |
| | = 1.853 kilometers |
| | = $1.853 \times 10^{3}$ meters |
| | = 1.1516 statute miles |
| | = $2.0254 \times 10^{3}$ yards |
| mile (statute) | = $5.280 \times 10^{3}$ feet |
| | = $6.336 \times 10^{4}$ inches |
| | = 1.609 kilometers |
| | = $8.684 \times 10^{-1}$ nautical miles |
| | = $1.760 \times 10^{3}$ yards |
| | = $1.69 \times 10^{-13}$ light years |
| mile per hour | = $4.470 \times 10^{1}$ centimeters per second |
| | = $8.8 \times 10^{1}$ feet per minute |
| | = 1.467 feet per second |
| | = 1.6093 kilometers per hour |
| | = $2.862 \times 10^{-2}$ kilometers per minute |
| | = $8.864 \times 10^{-1}$ knots |
| | = $2.682 \times 10^{1}$ meters per minute |
| | = $1.667 \times 10^{-2}$ miles per minute |
| mile per hour per second | = $4.47 \times 10^{1}$ centimeters per second per second |
| | = 1.467 feet per second per second |
| | = 1.6093 kilometers per hour per second |
| | = $4.47 \times 10^{-1}$ meters per second per second |
| mile per minute | = $2.682 \times 10^{3}$ centimeters per second |
| | = $8.8 \times 10^{1}$ feet per second |
| | = 1.6093 kilometers per minute |
| | = $8.684 \times 10^{-1}$ knots per minute |
| millimeter | = $3.281 \times 10^{-3}$ feet |
| | = $3.937 \times 10^{-2}$ inches |
| | = $6.214 \times 10^{-7}$ miles |
| | = $3.937 \times 10^{1}$ mils |
| | = $1.094 \times 10^{-3}$ yards |
| minute (angle) | = $1.667 \times 10^{-2}$ degrees |
| | = $2.909 \times 10^{-4}$ radians |
| minute (time) | = $9.9206 \times 10^{-5}$ weeks |
| | = $6.944 \times 10^{-4}$ days |
| | = $1.667 \times 10^{-2}$ hours |
| newton | = $1.0 \times 10^{5}$ dynes |
| ohm (international) | = 1.0005 ohms (absolute) |
| ounce | = $4.375 \times 10^{2}$ grains |
| | = $2.8349 \times 10^{1}$ grams |
| | = $6.25 \times 10^{-2}$ pounds |
| ounce (fluid) | = 1.805 cubic inches |
| | = $2.957 \times 10^{-2}$ liters |
| ounce (troy) | = 1.097 ounces (avoirdupois) |
| ounce per square inch | = $4.309 \times 10^{3}$ dynes per square centimeter |
| | = $6.25 \times 10^{-2}$ pounds per square inch |
| parsec | = $1.9 \times 10^{13}$ miles |
| | = $3.084 \times 10^{13}$ kilometers |
| part per million | = $5.84 \times 10^{-2}$ grains per U.S. gallon |
| | = $7.016 \times 10^{-2}$ grains per imperial gallon |
| | = 8.345 pounds per million gallons |
| pint (liquid) | = $4.732 \times 10^{2}$ cubic centimeters |
| | = $1.671 \times 10^{-2}$ cubic feet |
| | = $2.887 \times 10^{1}$ cubic inches |
| | = $4.732 \times 10^{-4}$ cubic meters |
| | = $6.189 \times 10^{-4}$ cubic yards |
| | = $1.25 \times 10^{-1}$ gallons |
| | = $4.732 \times 10^{-1}$ liters |

*(continued)*

### Dimension, Quantity, Amount, and Rate   *(continued)*

| | |
|---|---|
| Planck's constant | $= 6.6256 \times 10^{-27}$ erg-seconds |
| pound | $= 2.56 \times 10^2$ drams |
| | $= 4.448 \times 10^5$ dynes |
| | $= 7.0 \times 10^3$ grains |
| | $= 4.5359 \times 10^2$ grams |
| | $= 4.536 \times 10^{-1}$ kilograms |
| | $= 1.6 \times 10^1$ ounces |
| | $= 3.217 \times 10^1$ poundals |
| | $= 5.0 \times 10^{-4}$ short tons |
| pound (avoirdupois) | $= 1.4583 \times 10^1$ ounces (troy) |
| pound (troy) | $= 1.3166 \times 10^1$ ounces (avoirdupois) |
| poundal | $= 1.3826 \times 10^4$ dynes |
| | $= 1.41 \times 10^1$ grams |
| | $= 1.383 \times 10^{-3}$ joules per centimeter |
| | $= 1.383 \times 10^{-1}$ joules per meter (newtons) |
| | $= 1.41 \times 10^{-2}$ kilograms |
| | $= 3.108 \times 10^{-2}$ pounds |
| pound-foot | $= 1.356 \times 10^7$ centimeter-dynes |
| | $= 1.3825 \times 10^4$ centimeter-grams |
| | $= 1.383 \times 10^{-1}$ meter-kilograms |
| pound of water | $= 1.602 \times 10^{-2}$ cubic feet |
| | $= 2.768 \times 10^1$ cubic inches |
| | $= 1.198 \times 10^{-1}$ gallons |
| pound of water per minute | $= 2.670 \times 10^{-4}$ cubic feet per second |
| pound per cubic foot | $= 1.602 \times 10^{-2}$ grams per cubic centimeter |
| | $= 1.602 \times 10^1$ kilograms per cubic meter |
| | $= 5.787 \times 10^{-4}$ pounds per cubic inch |
| pound per cubic inch | $= 2.768 \times 10^1$ grams per cubic centimeter |
| | $= 1.728 \times 10^3$ pounds per cubic foot |
| pound per foot | $= 1.488$ kilograms per meter |
| pound per inch | $= 1.786 \times 10^2$ grams per centimeter |
| pound per square foot | $= 4.725 \times 10^{-4}$ atmospheres |
| | $= 1.602 \times 10^{-2}$ feet of water |
| | $= 1.414 \times 10^{-2}$ inches of mercury |
| | $= 4.882$ kilograms per square meter |
| | $= 6.944 \times 10^{-3}$ pounds per square inch |
| pound per square inch | $= 6.804 \times 10^{-2}$ atmospheres |
| | $= 2.307$ feet of water |
| | $= 2.036$ inches of mercury |
| | $= 7.031 \times 10^2$ kilograms per square meter |
| | $= 1.44 \times 10^2$ pounds per square foot |
| | $= 7.2 \times 10^{-2}$ short tons per square foot |
| | $= 7.03 \times 10^{-2}$ kilograms per square meter |
| quadrant (angle) | $= 9.0 \times 10^1$ degrees |
| | $= 5.4 \times 10^3$ minutes |
| | $= 1.571$ radians |
| | $= 3.24 \times 10^5$ seconds |
| quart (dry) | $= 6.72 \times 10^1$ cubic inches |
| quart (liquid) | $= 9.464 \times 10^2$ cubic centimeters |
| | $= 3.342 \times 10^{-2}$ cubic feet |
| | $= 5.775 \times 10^1$ cubic inches |
| | $= 9.464 \times 10^{-4}$ cubic meters |
| | $= 1.238 \times 10^{-3}$ cubic yards |
| | $= 2.5 \times 10^{-1}$ gallons |
| | $= 9.463 \times 10^{-1}$ liters |
| radian | $= 5.729 \times 10^1$ degrees |
| | $= 3.438 \times 10^3$ minutes |
| | $= 6.366 \times 10^{-1}$ quadrants |
| | $= 2.063 \times 10^5$ seconds |
| radian per second | $= 5.729 \times 10^1$ degrees per second |
| | $= 9.549$ revolutions per minute |
| | $= 1.592 \times 10^{-1}$ revolutions per second |
| radian per second per second | $= 5.7296 \times 10^2$ revolutions per minute per minute |
| | $= 9.549$ revolutions per minute per second |
| | $= 1.5492 \times 10^{-1}$ revolutions per second per second     *(continued)* |

### Dimension, Quantity, Amount, and Rate  *(continued)*

| | |
|---|---|
| ream | = 500 sheets |
| revolution per minute | = 6.0 degrees per second |
| | = $1.047 \times 10^{-1}$ radians per second |
| | = $1.667 \times 10^{-2}$ revolutions per second |
| revolution per minute per minute | = $1.745 \times 10^{-3}$ radians per second per second |
| | = $1.667 \times 10^{-2}$ revolutions per minute per second |
| | = $2.778 \times 10^{-4}$ revolutions per second per second |
| revolution per second | = $3.6 \times 10^{2}$ degrees per second |
| | = 6.283 radians per second |
| | = 60 revolutions per minute |
| revolution per second per second | = 6.283 radians per second per second |
| | = $3.6 \times 10^{3}$ revolutions per minute per minute |
| | = $6.0 \times 10^{1}$ revolutions per minute per second |
| second (angle) | = $2.778 \times 10^{-4}$ degrees |
| | = $1.667 \times 10^{-2}$ minutes |
| | = $4.848 \times 10^{-6}$ radians |
| slug | = $1.459 \times 10^{1}$ kilograms |
| | = $3.217 \times 10^{1}$ pounds |
| sphere (solid angle) | = $1.257 \times 10^{1}$ steradians |
| square centimeter | = $1.973 \times 10^{5}$ circular mils |
| | = $1.076 \times 10^{-3}$ square feet |
| | = $1.550 \times 10^{-1}$ square inches |
| | = $1.0 \times 10^{-4}$ square meters |
| | = $3.861 \times 10^{-11}$ square miles |
| | = $1.196 \times 10^{-4}$ square yards |
| square foot | = $2.296 \times 10^{-5}$ acres |
| | = $9.29 \times 10^{2}$ square centimeter |
| | = $1.44 \times 10^{2}$ square inches |
| | = $9.29 \times 10^{-2}$ square meters |
| | = $3.587 \times 10^{-8}$ square miles |
| | = $1.111 \times 10^{-1}$ square yards |
| square inch | = $1.273 \times 10^{6}$ circular mils |
| | = 6.452 square centimeters |
| | = $6.944 \times 10^{-3}$ square feet |
| | = $6.452 \times 10^{2}$ square millimeters |
| | = $7.716 \times 10^{-4}$ square yards |
| | = $1.0 \times 10^{6}$ square mils |
| square kilometer | = $1.076 \times 10^{7}$ square feet |
| | = $1.550 \times 10^{9}$ square inches |
| | = $1.0 \times 10^{6}$ square meters |
| | = $3.861 \times 10^{-1}$ square miles |
| | = $1.196 \times 10^{6}$ square yards |
| square meter | = $1.076 \times 10^{1}$ square feet |
| | = $1.55 \times 10^{3}$ square inches |
| | = $3.861 \times 10^{-7}$ square miles |
| | = 1.196 square yards |
| square mile | = $6.40 \times 10^{2}$ acres |
| | = $2.788 \times 10^{7}$ square feet |
| | = 2.590 square kilometers |
| | = $3.098 \times 10^{6}$ square yards |
| square millimeter | = $1.076 \times 10^{-5}$ square feet |
| | = $1.55 \times 10^{-3}$ square inches |
| square yard | = $2.066 \times 10^{-4}$ acres |
| | = 9.0 square feet |
| | = $1.296 \times 10^{3}$ square inches |
| | = $8.361 \times 10^{-1}$ square meters |
| | = $3.228 \times 10^{-7}$ square miles |
| steradian | = $7.968 \times 10^{-2}$ spheres |
| | = $1.592 \times 10^{-1}$ hemispheres |
| | = $6.366 \times 10^{-1}$ spherical right angles |
| | = $3.283 \times 10^{3}$ square degrees |
| temperature (degrees Celsius) + 273 | = 1.0 absolute temperature (degrees Kelvin) |
| temperature (degrees Celsius) + 17.78 | = 1.8 temperature (degrees Fahrenheit) |
| temperature (degrees Fahrenheit) + 460 | = 1.0 absolute temperature (degrees Rankin) |

*(continued)*

### Dimension, Quantity, Amount, and Rate  *(continued)*

| | |
|---|---|
| temperature (degrees Fahrenheit) $-$ 32 | = 5/9 temperature (degrees Celsius) |
| ton (long) | = $2.24 \times 10^3$ pounds |
| ton (metric) | = $2.205 \times 10^3$ pounds |
| ton (short) | = $2.0 \times 10^3$ pounds |
| | = $9.0718 \times 10^2$ kilograms |
| | = $3.2 \times 10^4$ ounces |
| | = $8.9287 \times 10^{-1}$ tons (long) |
| | = $9.078 \times 10^{-1}$ tons (metric) |
| ton (short) per square foot | = $9.765 \times 10^3$ kilograms per square meter |
| | = $1.389 \times 10^1$ pounds per square inch |
| watt | = 3.4129 British thermal units per hour |
| | = $5.688 \times 10^{-2}$ British thermal units per minute |
| | = $1.0 \times 10^7$ ergs per second |
| | = $4.427 \times 10^1$ foot-pounds per minute |
| | = $7.378 \times 10^{-1}$ foot-pounds per second |
| | = $1.341 \times 10^{-3}$ horsepower |
| | = $1.433 \times 10^{-2}$ kilogram-calories per minute |
| | = $1.0 \times 10^{-3}$ kilowatts |
| watthour | = 3.413 British thermal units |
| | = $3.6 \times 10^{10}$ ergs |
| | = $2.656 \times 10^3$ foot-pounds |
| | = $8.605 \times 10^2$ gram-calories |
| | = $1.341 \times 10^{-3}$ horsepower-hours |
| | = $8.605 \times 10^{-1}$ kilogram-calories |
| | = $1.0 \times 10^{-3}$ kilowatt-hours |
| watt (international) | = 1.000165 watts (absolute) |
| week | = $1.68 \times 10^2$ hours |
| | = $1.008 \times 10^4$ minutes |
| | = $6.048 \times 10^5$ seconds |
| yard | = $9.144 \times 10^1$ centimeters |
| | = $9.144 \times 10^{-1}$ meters |
| | = $4.934 \times 10^{-4}$ nautical miles |
| | = $5.682 \times 10^{-4}$ statute miles |
| year | = $3.65256 \times 10^2$ days (mean solar) |
| | = $8.7661 \times 10^3$ hours (mean solar) |

## Conversion Table for Barometric Pressure Units[29]

| | Atmospheres | Newtons per Square Meter | Bars | Millibars | Kilograms per Square Centimeter | Grams per Square Centimeter (Centimeter H₂O) | Millimeters of Hg | Inches of Hg | Pounds per Square Inch |
|---|---|---|---|---|---|---|---|---|---|
| 1 Atmosphere = | 1 | $1.013 \times 10^5$ | 1.013 | 1013 | 1.033 | 1033 | 760 | 29.92 | 14.70 |
| 1 Newton/m² (N/m²) = | $0.9869 \times 10^{-5}$ | 1 | $10^{-5}$ | 0.01 | $1.02 \times 10^{-5}$ | 0.0102 | 0.0075 | $0.2953 \times 10^{-3}$ | $0.1451 \times 10^{-3}$ |
| 1 bar = | 0.9869 | $10^5$ | 1 | 1000 | 1.02 | 1020 | 750.1 | 29.53 | 14.51 |
| 1 millibar (mb) = | $0.9869 \times 10^{-3}$ | 100 | 0.001 | 1 | 0.00102 | 1.02 | 0.7501 | 0.02953 | 14.51 |
| 1 kg/cm² = 1 gm/cm² | 0.9681 | $0.9807 \times 10^5$ | 0.9807 | 980.7 | 1 | 1000 | 735 | 28.94 | 14.22 |
| (1 cm H₂O) = | 968.1 | 98.07 | $0.9807 \times 10^{-3}$ | 0.9807 | 0.001 | 1 | 0.735 | 0.02894 | 0.01422 |
| 1 mm Hg = | 0.001316 | 133.3 | 0.001333 | 1.333 | 0.00136 | 1.36 | 1 | 0.03937 | 0.01934 |
| 1 in Hg = | 0.0334 | 3386 | 0.03386 | 33.86 | 0.03453 | 34.53 | 25.4 | 1 | 0.4910 |
| 1 lb/in² = | 0.06804 | 6895 | 0.06895 | 68.95 | 0.0703 | 70.3 | 51.70 | 2.035 | 1 |

[28] J. F. Parker and V. R. West (eds.), *Bioastronautics Data Book*, 2d ed., National Aeronautics and Space Administration, Washington, 1973, p. 3.

Equivalent Pressures, Altitudes, and
Depths[28]

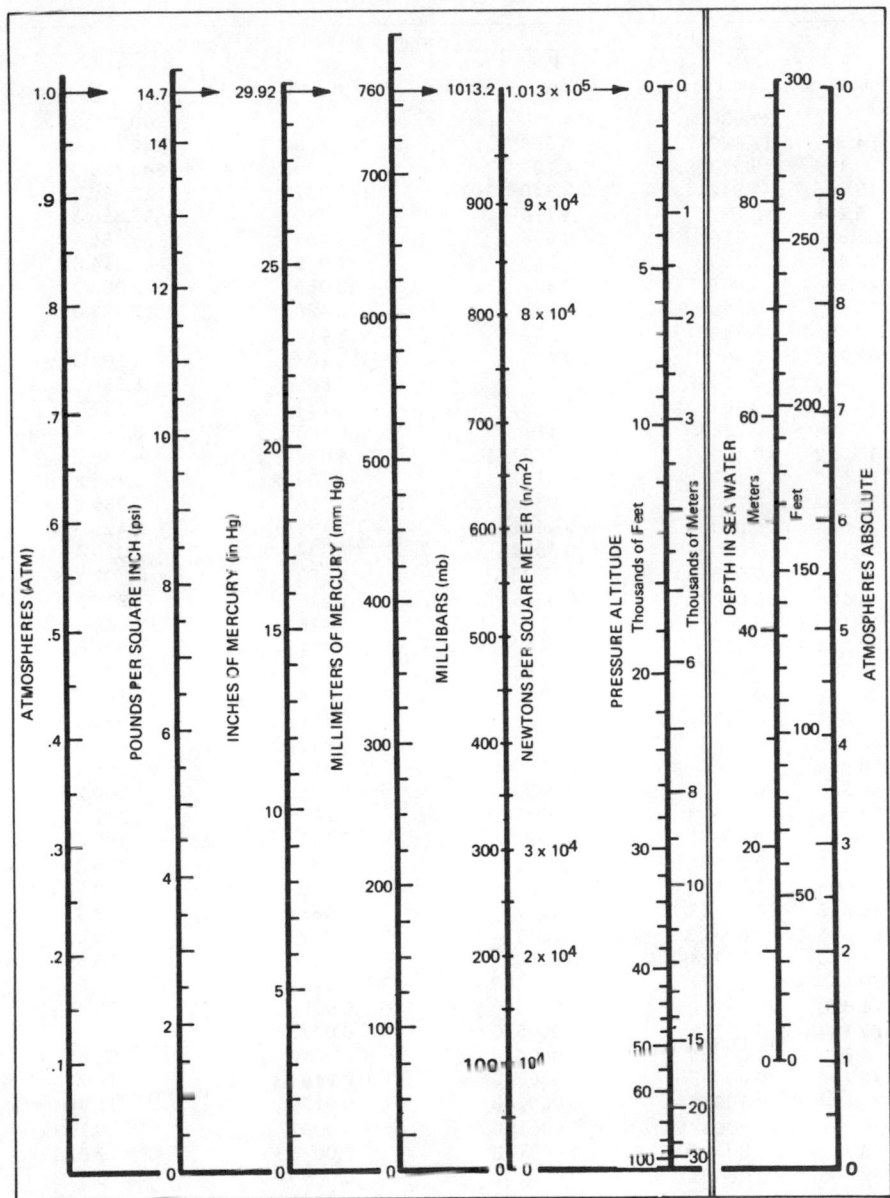

[28]J. F. Parker and V. R. West (eds.), *Bioastronautics Data Book*, 2d ed., National Aeronautics and Space Administration, Washington, 1973, p. 4.

## High-Altitude–Space-Pressure Environment

| Altitude | | | Pressure Equivalents | | | | Mean Free Path, ft |
|---|---|---|---|---|---|---|---|
| Feet | Miles | Kilometers | Inches of Hg | Pounds per Square Inch Absolute | Millimeters (Torr) of Hg | Microns | |
| 0 | 0 | 0 | 29.930 | 14.700 | 760.222 | 760,222.0 | $2.176 \times 10^{-7}$ |
| 15,000 | 2.841 | 4.572 | 17.420 | 8.556 | 442.468 | 442,468.0 | 3.457 |
| 30,000 | 5.682 | 9.144 | 9.572 | 4.701 | 243.129 | 243,129.0 | 5.807 |
| 45,000 | 8.523 | 13.716 | 4.842 | 2.378 | 122.987 | 122,987.0 | $1.119 \times 10^{-6}$ |
| 60,000 | 11.364 | 18.288 | 2.277 | 1.118 | 57.836 | 57,835.8 | 2.293 |
| 70,000 | 13.258 | 21.336 | 1.396 | 0.686 | 35.458 | 35,458.4 | 3.716 |
| 75,000 | 14.205 | 22.860 | 1.099 | 0.540 | 27.915 | 27,914.6 | 4.740 |
| 80,000 | 15.152 | 24.384 | 0.869 | 0.427 | 22.065 | 22,065.0 | 6.035 |
| 85,000 | 16.099 | 25.908 | 0.689 | 0.338 | 17.493 | 17,493.0 | 7.670 |
| 90,000 | 17.046 | 27.432 | 0.548 | 0.269 | 13.914 | 13,914.1 | 9.732 |
| 95,000 | 17.992 | 28.956 | 0.437 | 0.215 | 11.102 | 11,102.3 | $1.233 \times 10^{-5}$ |
| 100,000 | 18.939 | 30.480 | 0.350 | 0.172 | 8.885 | 8,884.9 | 1.559 |
| 105,000 | 19.886 | 32.004 | 0.281 | 0.138 | 7.132 | 7,132.3 | 1.968 |
| 110,000 | 20.833 | 33.528 | 0.226 | 0.111 | 5.743 | 5,742.9 | 2.504 |
| 115,000 | 21.780 | 35.052 | 0.183 | 0.0897 | 4.638 | 4,638.0 | 3.175 |
| 120,000 | 22.727 | 36.576 | 0.148 | 0.0727 | 3.759 | 3,759.2 | 4.009 |
| 125,000 | 23.674 | 38.100 | 0.120 | 0.0591 | 3.056 | 3,055.6 | 5.042 |
| 130,000 | 24.621 | 39.624 | 0.0982 | 0.0482 | 2.493 | 2,493.0 | 6.315 |
| 135,000 | 25.568 | 41.148 | 0.0803 | 0.0394 | 2.039 | 2,039.4 | 7.879 |
| 140,000 | 26.515 | 42.672 | 0.0659 | 0.0323 | 1.673 | 1,672.8 | 9.793 |
| 145,000 | 27.462 | 44.196 | 0.0542 | 0.0266 | 1.376 | 1,376.2 | $1.213 \times 10^{-4}$ |
| 150,000 | 28.409 | 45.720 | 0.0447 | 0.0220 | 1.135 | 1,135.4 | 1.497 |
| 155,000 | 29.356 | 47.244 | 0.0370 | 0.0182 | 0.939 | 939.0 | 1.841 |
| 160,000 | 30.303 | 48.768 | 0.0306 | 0.0150 | 0.778 | 778.0 | 2.227 |
| 165,000 | 31.250 | 50.292 | 0.0254 | 0.0125 | 0.644 | 644.4 | 2.692 |
| 170,000 | 32.197 | 51.816 | 0.0210 | 0.0103 | 0.534 | 533.9 | 3.253 |
| 175,000 | 33.144 | 53.340 | 0.0174 | 0.00855 | 0.442 | 442.0 | 3.906 |
| 180,000 | 34.091 | 54.864 | 0.0144 | 0.00706 | 0.365 | 365.0 | 4.678 |
| 185,000 | 35.038 | 56.388 | 0.0118 | 0.00582 | 0.301 | 300.7 | 5.612 |
| 190,000 | 35.985 | 57.912 | 0.00973 | 0.00478 | 0.247 | 247.2 | 6.748 |
| 200,000 | 37.879 | 60.960 | 0.00653 | 0.00321 | 0.166 | 166.0 | 9.814 |
| 205,000 | 38.825 | 62.484 | 0.00532 | 0.00261 | 0.135 | 135.2 | $1.180 \times 10^{-3}$ |
| 210,000 | 39.773 | 64.008 | 0.00431 | 0.00212 | 0.110 | 109.6 | 1.417 |
| 215,000 | 40.720 | 65.532 | 0.00348 | 0.00171 | 0.0884 | 88.37 | 1.709 |
| 220,000 | 41.667 | 67.056 | 0.00279 | 0.00137 | 0.0709 | 70.89 | 2.071 |
| 225,000 | 42.614 | 68.580 | 0.00224 | 0.00109 | 0.0566 | 56.59 | 2.522 |
| 230,000 | 43.561 | 70.104 | 0.00177 | 0.000868 | 0.0449 | 44.91 | 3.088 |
| 235,000 | 44.508 | 71.628 | 0.00140 | 0.000685 | 0.0354 | 35.43 | 3.802 |
| 240,000 | 45.455 | 73.153 | 0.00109 | 0.000573 | 0.0278 | 27.76 | 4.707 |
| 245,000 | 46.402 | 74.676 | 0.000851 | 0.000418 | 0.0216 | 21.62 | 5.864 |
| 250,000 | 47.349 | 76.200 | 0.000658 | 0.000323 | 0.0167 | 16.72 | 7.353 |
| 255,000 | 48.296 | 77.724 | 0.000505 | 0.000248 | 0.0128 | 12.83 | 9.284 |
| 260,000 | 49.242 | 79.248 | 0.000385 | 0.000189 | 0.00977 | 9.769 | $0.0118 \times 10^{0}$ |
| 265,000 | 50.189 | 80.772 | 0.000290 | 0.000143 | 0.00738 | 7.376 | 0.01538 |
| 270,000 | 51.136 | 82.296 | 0.000217 | 0.000107 | 0.00552 | 5.522 | 0.02036 |
| 275,000 | 52.083 | 83.820 | 0.000162 | 0.0000795 | 0.00412 | 4.117 | 0.02697 |
| 280,000 | 53.030 | 85.344 | 0.000121 | 0.0000593 | 0.00307 | 3.068 | 0.03570 |
| 285,000 | 53.977 | 86.868 | 0.0000901 | 0.0000443 | 0.00229 | 2.289 | 0.04727 |
| 290,000 | 54.924 | 88.392 | 0.0000672 | 0.0000330 | 0.00171 | 1.707 | 0.06257 |
| 295,000 | 55.871 | 89.916 | 0.0000502 | 0.0000246 | 0.00127 | 1.274 | 0.08281 |
| 300,000 | 56.818 | 91.440 | 0.0000375 | 0.0000184 | 0.000952 | 0.952 | 0.1118 |
| 350,000 | 66.288 | 106.680 | $0.335 \times 10^{-5}$ | $0.165 \times 10^{-5}$ | 0.0000852 | 0.0852 | 1.629 |
| 400,000 | 75.758 | 121.920 | $0.631 \times 10^{-6}$ | $0.310 \times 10^{-6}$ | 0.0000160 | 0.0160 | 13.81 |
| 450,000 | 85.227 | 137.160 | 0.248 | 0.122 | $0.631 \times 10^{-5}$ | 0.00631 | 60.32 |
| 500,000 | 94.697 | 152.400 | 0.138 | $0.678 \times 10^{-7}$ | 0.351 | 0.00351 | 151.3 |
| 550,000 | 104.167 | 167.640 | $0.879 \times 10^{-7}$ | 0.432 | 0.223 | 0.00223 | 278.7 |
| 600,000 | 113.636 | 182.880 | 0.592 | 0.291 | 0.150 | 0.00150 | 447.0 |
| 650,000 | 123.106 | 198.120 | 0.411 | 0.202 | 0.105 | 0.00105 | 675.9 |
| 700,000 | 132.576 | 213.360 | 0.292 | 0.144 | $0.742 \times 10^{-6}$ | 0.000742 | 986.1 |
| 750,000 | 142.046 | 228.600 | 0.212 | 0.104 | 0.537 | 0.000537 | 1,408.0 |
| 800,000 | 151.515 | 243.840 | 0.156 | $0.764 \times 10^{-8}$ | 0.395 | 0.000395 | 1,956.0 |
| 850,000 | 160.985 | 259.080 | 0.116 | 0.569 | 0.294 | 0.000294 | 2,676.0 |
| 900,000 | 170.455 | 274.320 | $0.874 \times 10^{-8}$ | 0.429 | 0.222 | 0.000222 | 3,611.0 |
| 950,000 | 179.924 | 289.560 | 0.666 | 0.324 | 0.169 | 0.000169 | 4,811.0 |
| 1,000,000 | 189.394 | 304.800 | 0.513 | 0.252 | 0.130 | 0.000130 | 6,325.0 |
| 1,100,000 | 208.333 | 335.280 | 0.612 | 0.153 | $0.792 \times 10^{-7}$ | 0.0000792 | 10,550.0 |
| 1,200,000 | 227.273 | 365.760 | 0.195 | $0.959 \times 10^{-9}$ | 0.496 | 0.0000496 | 17,040.0 |
| 1,300,000 | 246.212 | 396.240 | 0.126 | 0.616 | 0.319 | 0.0000319 | 26,760.0 |

*(continued)*

**High-Altitude–Space-Pressure Environment** *(continued)*

| Altitude | | | Pressure Equivalents | | | | |
|---|---|---|---|---|---|---|---|
| Feet | Miles | Kilometers | Inches of Hg | Pounds per Square Inch Absolute | Millimeters (Torr) of Hg | Microns | Mean Free Path, ft |
| 1,400,000 | 265.152 | 426.720 | $0.825\times10^{-9}$ | 0.405 | 0.210 | 0.0000210 | 40,740.0 |
| 1,500,000 | 284.091 | 457.200 | 0.552 | 0.271 | 0.140 | 0.0000140 | 60,930.0 |
| 1,600,000 | 303.030 | 487.680 | 0.376 | 0.185 | $0.956\times10^{-8}$ | $0.956\times10^{-5}$ | 89,830.0 |
| 1,700,000 | 321.970 | 518.160 | 0.260 | 0.128 | 0.661 | 0.661 | 130,000.0 |
| 1,800,000 | 340.909 | 548.640 | 0.182 | $0.893\times10^{-10}$ | 0.462 | 0.462 | 185,800.0 |
| 1,900,000 | 359.849 | 579.120 | 0.129 | 0.631 | 0.326 | 0.326 | 263,900.0 |
| 2,000,000 | 378.788 | 609.600 | $0.917\times10^{-10}$ | 0.450 | 0.233 | 0.233 | 371,100.0 |
| 2,100,000 | 397.727 | 640.080 | 0.659 | 0.324 | 0.167 | 0.167 | 516,200.0 |
| 2,200,000 | 416.667 | 670.560 | 0.478 | 0.235 | 0.121 | 0.121 | 713,300.0 |
| 2,300,000 | 435.606 | 701.040 | 0.348 | 0.171 | $0.884\times10^{-9}$ | $0.884\times10^{-8}$ | 978,400.0 |

*Source:* 0 to 300,000 ft, U.S. Standard Atmosphere, 1966. CONDITION: 30° N, July day, Geometric Altitude.
*Source:* Over 300,000 ft, U.S. Standard Atmosphere, 1962.

# REFERENCES
## Metrics

## Units of Luminance Conversion

| | Nit | Stilb | Bougie Nectomètre Carré | Apostilb | Milli-apostilb | Micro-apostilb | Lambert | Milli-lambert | Micro-lambert | Foot-lambert | Candle per Square Foot | Candle per Square Inch |
|---|---|---|---|---|---|---|---|---|---|---|---|---|
| 1 Nit (nt) = 1 Candela = m² | 1 | $10^{-4}$ | $10^4$ | 3.14 | $3.14\times10^3$ | $3.14\times10^6$ | $3.14\times10^{-4}$ | $3.14\times10^{-1}$ | $3.14\times10^2$ | $2.919\times10^{-1}$ | $9.29\times10^{-2}$ | $6.452\times10^{-4}$ |
| 1 Stilb (sb) = 1 Candela = cm² | $10^4$ | 1 | $10^8$ | $3.14\times10^4$ | $3.14\times10^7$ | $3.14\times10^{10}$ | 3.14 | $3.14\times10^3$ | $3.14\times10^6$ | $2.919\times10^3$ | $9.29\times10^2$ | 6.452 |
| 1 Bougie Hectomètre = 1 Candela | $10^{-4}$ | $10^{-8}$ | 1 | $3.14\times10^{-4}$ | $3.14\times10^{-1}$ | $3.14\times10^2$ | $3.14\times10^{-8}$ | $3.14\times10^{-5}$ | $3.14\times10^{-2}$ | $2.919\times10^{-5}$ | $9.29\times10^{-6}$ | $6.452\times10^{-8}$ |
| 1 Apostilb (asb) = 1 Candela = $\pi \times$ m² | $3.183\times10^{-1}$ | $3.183\times10^{-5}$ | $3.183\times10^3$ | 1 | $10^3$ | $10^6$ | $10^{-4}$ | $10^{-1}$ | $10^2$ | $9.29\times10^{-2}$ | $2.957\times10^{-2}$ | $2.054\times10^{-4}$ |
| 1 Milli-apostilb (masb) = 1 Candela = $\pi \times 1000 \times$ m² | $3.183\times10^{-4}$ | $3.183\times10^{-8}$ | 3.183 | $10^{-3}$ | 1 | $10^3$ | $10^{-7}$ | $10^{-4}$ | $10^{-1}$ | $9.29\times10^{-5}$ | $2.957\times10^{-5}$ | $2.054\times10^{-7}$ |
| 1 Micro-apostilb (μasb) = 1 Candela = $\pi \times 10^6 \times$ m² | $3.183\times10^{-7}$ | $3.183\times10^{-11}$ | $3.183\times10^{-3}$ | $10^{-6}$ | $10^{-3}$ | 1 | $10^{-10}$ | $10^{-7}$ | $10^{-4}$ | $9.29\times10^{-8}$ | $2.957\times10^{-8}$ | $2.054\times10^{-10}$ |
| 1 Lambert (L) = 1 Candela = $\pi \times$ cm² | $3.183\times10^3$ | $3.183\times10^{-1}$ | $3.183\times10^7$ | $10^4$ | $10^7$ | $10^{10}$ | 1 | $10^3$ | $10^6$ | $9.29\times10^2$ | $2.957\times10^2$ | 2.054 |
| 1 Milli-lambert (mL) = 1 Candela = $\pi \times 10^3 \times$ cm² | 3.183 | $3.183\times10^{-4}$ | $3.183\times10^4$ | 10 | $10^4$ | $10^7$ | $10^{-3}$ | 1 | $10^3$ | $9.29\times10^{-1}$ | $2.957\times10^{-1}$ | $2.054\times10^{-3}$ |
| 1 Micro-lambert (μL) = 1 Candela = $\pi \times 10^6 \times$ cm² | $3.183\times10^{-3}$ | $3.183\times10^{-7}$ | $3.183\times10$ | $10^{-2}$ | 10 | $10^4$ | $10^{-6}$ | $10^{-3}$ | 1 | $9.29\times10^{-4}$ | $2.957\times10^{-4}$ | $2.054\times10^{-6}$ |
| 1 Foot-lambert (ftL) = 1 Candela = $\pi \times$ ft² | 3.426 | $3.426\times10^{-4}$ | $3.426\times10^4$ | 10.764 | $1.0764\times10^4$ | $1.0764\times10^7$ | $1.0764\times10^{-3}$ | 1.0764 | $1.0764\times10^3$ | 1 | 0.3183 | $2.14\times10^{-3}$ |
| 1 Candle per Square foot = 1 Candela = ft² | $1.0764\times10$ | $1.0764\times10^{-3}$ | $1.0764\times10^5$ | $3.382\times10$ | $3.382\times10^4$ | $3.382\times10^7$ | $3.382\times10^{-3}$ | 3.382 | $3.382\times10^3$ | 3.14 | 1 | $6.944\times10^{-3}$ |
| 1 Candle per Square inch = 1 Candela = inch² | $1.55\times10^3$ | $1.55\times10^{-1}$ | $1.55\times10^7$ | $4.869\times10^3$ | $4.869\times10^6$ | $4.869\times10^9$ | $4.869\times10^{-1}$ | $4.869\times10^2$ | $4.869\times10^5$ | $4.524\times10^2$ | $1.44\times10^2$ | 1 |

**Illumination Units**

| NUMBER OF → MULTIPLIED BY "F" EQUALS NUMBER OF ↓ | LUMEN/mm² | LUMEN/cm² = PHOT | LUMEN/in² | LUMEN/ft² = FOOT-CANDLE | MILLIPHOT | LUMEN/m² = METER-CANDLE = LUX | LUMEN/hm² = HECTOMETER-CANDLE | LUMEN/km² = KILOMETER-CANDLE | LUMEN/mi² = MILE-CANDLE | LUMEN/naut. mi² = SEA-MILE-CANDLE |
|---|---|---|---|---|---|---|---|---|---|---|
| LUMEN/mm² | 1 | $10^{-2}$ | $1.550 \times 10^{-3}$ | $1.076 \times 10^{-5}$ | $10^{-5}$ | $10^{-6}$ | $10^{-10}$ | $10^{-12}$ | $3.861 \times 10^{-13}$ | $2.912 \times 10^{-13}$ |
| LUMEN/cm² = PHOT | $10^{2}$ | 1 | .1550 | $1.076 \times 10^{-3}$ | $10^{-3}$ | $10^{-4}$ | $10^{-8}$ | $10^{-10}$ | $3.861 \times 10^{-11}$ | $2.912 \times 10^{-11}$ |
| LUMEN/in² | $6.452 \times 10^{2}$ | 6.452 | 1 | $6.944 \times 10^{-3}$ | $6.452 \times 10^{-3}$ | $6.452 \times 10^{-4}$ | $6.452 \times 10^{-8}$ | $6.452 \times 10^{-10}$ | $2.491 \times 10^{-10}$ | $1.878 \times 10^{-10}$ |
| LUMEN/ft² = FOOT-CANDLE | $9.290 \times 10^{4}$ | $9.290 \times 10^{2}$ | 144 | 1 | .9290 | $9.290 \times 10^{-2}$ | $9.290 \times 10^{-6}$ | $9.290 \times 10^{-8}$ | $3.587 \times 10^{-8}$ | $2.705 \times 10^{-8}$ |
| MILLIPHOT | $10^{5}$ | $10^{3}$ | $1.550 \times 10^{2}$ | 1.076 | 1 | $10^{-1}$ | $10^{-5}$ | $10^{-7}$ | $3.861 \times 10^{-8}$ | $2.912 \times 10^{-8}$ |
| LUMEN/m² = METER-CANDLE = LUX | $10^{6}$ | $10^{4}$ | $1.550 \times 10^{3}$ | 10.76 | 10 | 1 | $10^{-4}$ | $10^{-6}$ | $3.861 \times 10^{-7}$ | $2.912 \times 10^{-7}$ |
| LUMEN/hm² HECTOMETER-CANDLE | $10^{10}$ | $10^{8}$ | $1.550 \times 10^{7}$ | $1.076 \times 10^{5}$ | $10^{5}$ | $10^{4}$ | 1 | $10^{-2}$ | $3.861 \times 10^{-3}$ | $2.912 \times 10^{-3}$ |
| LUMEN/km² = KILOMETER-CANDLE | $10^{12}$ | $10^{10}$ | $1.550 \times 10^{9}$ | $1.076 \times 10^{7}$ | $10^{7}$ | $10^{6}$ | $10^{2}$ | 1 | .3861 | .2912 |
| LUMEN/mi² = MILE-CANDLE | $2.590 \times 10^{12}$ | $2.590 \times 10^{10}$ | $4.014 \times 10^{9}$ | $2.788 \times 10^{7}$ | $2.590 \times 10^{7}$ | $2.590 \times 10^{6}$ | $2.590 \times 10^{2}$ | 2.590 | 1 | .7541 |
| LUMEN/naut. mi² = SEA-MILE-CANDLE | $3.435 \times 10^{12}$ | $3.435 \times 10^{10}$ | $5.324 \times 10^{9}$ | $3.697 \times 10^{7}$ | $3.435 \times 10^{7}$ | $3.435 \times 10^{6}$ | $3.435 \times 10^{2}$ | 3.435 | 1.326 | 1 |

"A" (columns) — "B" (rows)

ONE UNIT OF "A" IS SAME ILLUMINATION
AS "F" UNITS OF "B"

**Tristimulus Values of the
Equal-Energy Spectrum in the CIE
System (2°)**

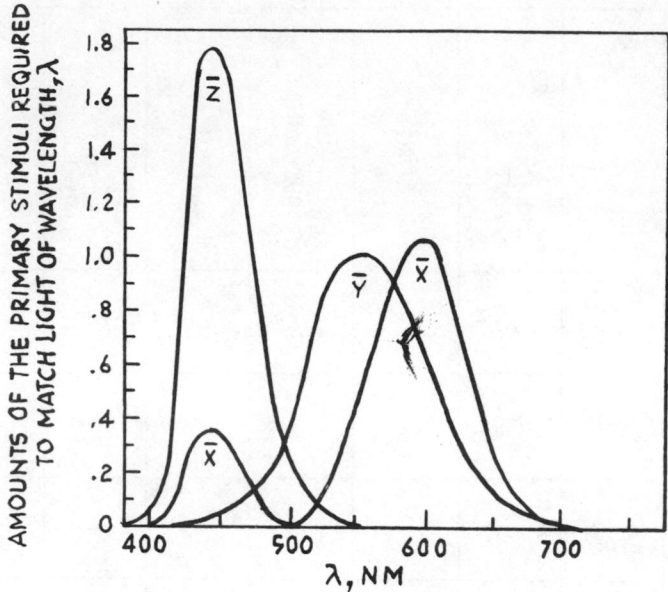

CIE tristimulus values x̄, ȳ, and z̄ of unit energy at indicated wavelengths.

## Tristimulus Values

| Wavelength λ, nm | $\bar{x}\lambda$ | $\bar{y}\lambda$ | $\bar{z}\lambda$ | Wavelength λ, nm | $\bar{x}\lambda$ | $\bar{y}\lambda$ | $\bar{z}\lambda$ |
|---|---|---|---|---|---|---|---|
| 380 | 0.0014 | 0.0000 | 0.0065 | 580 | 0.9163 | 0.8700 | 0.0017 |
| 385 | 0.0022 | 0.0001 | 0.0105 | 585 | 0.9786 | 0.8163 | 0.0014 |
| 390 | 0.0042 | 0.0001 | 0.0201 | 590 | 1.0263 | 0.7570 | 0.0011 |
| 395 | 0.0076 | 0.0002 | 0.0362 | 595 | 1.0567 | 0.6949 | 0.0010 |
| 400 | 0.0143 | 0.0004 | 0.0679 | 600 | 1.0622 | 0.6310 | 0.0008 |
| 405 | 0.0232 | 0.0006 | 0.1102 | 605 | 1.0456 | 0.5668 | 0.0006 |
| 410 | 0.0435 | 0.0012 | 0.2074 | 610 | 1.0026 | 0.5030 | 0.0003 |
| 415 | 0.0776 | 0.0022 | 0.3713 | 615 | 0.9384 | 0.4412 | 0.0002 |
| 420 | 0.1344 | 0.0040 | 0.6456 | 620 | 0.8544 | 0.3810 | 0.0002 |
| 425 | 0.2148 | 0.0073 | 1.0391 | 625 | 0.7514 | 0.3210 | 0.0001 |
| 430 | 0.2839 | 0.0116 | 1.3856 | 630 | 0.6424 | 0.2650 | 0.0000 |
| 435 | 0.3285 | 0.0168 | 1.6230 | 635 | 0.5419 | 0.2170 | 0.0000 |
| 440 | 0.3483 | 0.0230 | 1.7471 | 640 | 0.4479 | 0.1750 | 0.0000 |
| 445 | 0.3481 | 0.0298 | 1.7826 | 645 | 0.3608 | 0.1382 | 0.0000 |
| 450 | 0.3362 | 0.0380 | 1.7721 | 650 | 0.2835 | 0.1070 | 0.0000 |
| 455 | 0.3187 | 0.0480 | 1.7441 | 655 | 0.2187 | 0.0816 | 0.0000 |
| 460 | 0.2908 | 0.0600 | 1.6692 | 660 | 0.1649 | 0.0610 | 0.0000 |
| 465 | 0.2511 | 0.0739 | 1.5281 | 665 | 0.1212 | 0.0446 | 0.0000 |
| 470 | 0.1954 | 0.0910 | 1.2876 | 670 | 0.0874 | 0.0320 | 0.0000 |
| 475 | 0.1421 | 0.1126 | 1.0419 | 675 | 0.0636 | 0.0232 | 0.0000 |
| 480 | 0.0956 | 0.1390 | 0.8130 | 680 | 0.0468 | 0.0170 | 0.0000 |
| 485 | 0.0580 | 0.1693 | 0.6162 | 685 | 0.0329 | 0.0119 | 0.0000 |
| 490 | 0.0320 | 0.2080 | 0.4652 | 690 | 0.0227 | 0.0082 | 0.0000 |
| 495 | 0.0147 | 0.2586 | 0.3533 | 695 | 0.0158 | 0.0057 | 0.0000 |
| 500 | 0.0049 | 0.3230 | 0.2720 | 700 | 0.0114 | 0.0041 | 0.0000 |
| 505 | 0.0024 | 0.4073 | 0.2123 | 705 | 0.0081 | 0.0029 | 0.0000 |
| 510 | 0.0093 | 0.5030 | 0.1582 | 710 | 0.0058 | 0.0021 | 0.0000 |
| 515 | 0.0291 | 0.6082 | 0.1117 | 715 | 0.0041 | 0.0015 | 0.0000 |
| 520 | 0.0633 | 0.7100 | 0.0782 | 720 | 0.0029 | 0.0010 | 0.0000 |
| 525 | 0.1096 | 0.7932 | 0.0573 | 725 | 0.0020 | 0.0007 | 0.0000 |
| 530 | 0.1655 | 0.8620 | 0.0422 | 730 | 0.0014 | 0.0005 | 0.0000 |
| 535 | 0.2257 | 0.9149 | 0.0298 | 735 | 0.0010 | 0.0004 | 0.0000 |
| 540 | 0.2904 | 0.9540 | 0.0203 | 740 | 0.0007 | 0.0003 | 0.0000 |
| 545 | 0.3597 | 0.9803 | 0.0134 | 745 | 0.0005 | 0.0002 | 0.0000 |
| 550 | 0.4334 | 0.9950 | 0.0087 | 750 | 0.0003 | 0.0001 | 0.0000 |
| 555 | 0.5121 | 1.0002 | 0.0057 | 755 | 0.0002 | 0.0001 | 0.0000 |
| 560 | 0.5945 | 0.9950 | 0.0039 | 760 | 0.0002 | 0.0001 | 0.0000 |
| 565 | 0.6784 | 0.9786 | 0.0027 | 765 | 0.0001 | 0.0000 | 0.0000 |
| 570 | 0.7621 | 0.9520 | 0.0021 | 770 | 0.0001 | 0.0000 | 0.0000 |
| 575 | 0.8425 | 0.9154 | 0.0018 | 775 | 0.0000 | 0.0000 | 0.0000 |
| 580 | 0.9163 | 0.8700 | 0.0017 | 780 | 0.0000 | 0.0000 | 0.0000 |
| | | | | Totals | 21.3713 | 21.3714 | 21.3715 |

## Useful Physical Constants

| | |
|---|---|
| Acceleration of gravity (g) | = 32.17 ft/s$^2$ |
| | = 980.6 cm/s$^2$ |
| Velocity of sound in dry air at 0°C and 1 atm | = 33,136 cm/s |
| | = 1,089 ft/s |
| Heat of fusion of water | = 79.7 cal/g |
| | = 144 Btu/lb |
| Heat of vaporization of water at 1.0 atm | = 540 cal/g |
| | = 970 Btu/lb |
| Specific heat of air | = $C_p$ = 0.238 cal/g (°C) |
| Density of air at 0°C and 760 mm | = 0.991293 g/cm$^3$ |
| Velocity of light (c) | = 2.997902 $\times$ 10$^{10}$ cm/s |
| Avogadro's number (N) | = 6.061 $\times$ 10$^{23}$ molecules per gram-mole |
| Pi | = 3.14159265 |
| Naperian-logarithm base | = 2.71828183 |
| Radiation absorbtion dose (rad) | = 1.0 $\times$ 10$^2$ erg/g |
| Roentgen | = 8.3 $\times$ 10$^{-1}$ rad |

## Greek Alphabet

| | | | | | | | | | | | |
|---|---|---|---|---|---|---|---|---|---|---|---|
| Alpha | A | $\alpha$ | Iota | I | $\iota$ | Rho | P | $\rho$ |
| Beta | B | $\beta$ | Kappa | K | $\kappa$ | Sigma | $\Sigma$ | $\varsigma$ |
| Gamma | $\Gamma$ | $\gamma$ | Lambda | $\Lambda$ | $\lambda$ | Tau | P | $\tau$ |
| Delta | $\Delta$ | $\delta$ | Mu | M | $\mu$ | Upsilon | Y | $\upsilon$ |
| Epsilon | E | $\epsilon$ | Nu | N | $\nu$ | Phi | $\Phi$ | $\phi$ |
| Zeta | Z | $\zeta$ | Xi | $\Xi$ | $\xi$ | Chi | X | $\chi$ |
| Eta | H | $\eta$ | Omicron | O | $o$ | Psi | $\Psi$ | $\psi$ |
| Theta | $\Theta$ | $\theta$ | Pi | $\Pi$ | $\pi$ | Omega | $\Omega$ | $\omega$ |

# Index

## About the Author

**Wesley E. Woodson** is President, Man Factors, Inc., a human factors research and consulting firm in San Diego, California. Experienced in both military and domestic system and product design, he has been an active designer of man-machine system concepts covering aerospace, man-in-space, architectural, highway and transportation, surface and subsurface, interiors, electro-mechanical, electronic, and business machines, and computer systems. He is the author of *Human Engineering Guide for Equipment Designers* in addition to over 200 technical reports and articles dealing with a wide variety of human factors research and application. In 1974 Mr. Woodson was the recipient of the Jack Kraft Award for significant efforts to extend application of human factors principles and methods to new areas of endeavor.